About Island Press

Island Press is the only nonprofit organization in the United States whose principal purpose is the publication of books on environmental issues and natural resource management. We provide solutions-oriented information to professionals, public officials, business and community leaders, and concerned citizens who are shaping responses to environmental problems.

In 2005, Island Press celebrates its twenty-first anniversary as the leading provider of timely and practical books that take a multidisciplinary approach to critical environmental concerns. Our growing list of titles reflects our commitment to bringing the best of an expanding body of literature to the environmental community throughout North America and the world.

Support for Island Press is provided by the Agua Fund, The Geraldine R. Dodge Foundation, Doris Duke Charitable Foundation, Ford Foundation, The George Gund Foundation, The William and Flora Hewlett Foundation, Kendeda Sustainability Fund of the Tides Foundation, The Henry Luce Foundation, The John D. and Catherine T. MacArthur Foundation, The Andrew W. Mellon Foundation, The Curtis and Edith Munson Foundation, The New-Land Foundation, The New York Community Trust, Oak Foundation, The Overbrook Foundation, The David and Lucile Packard Foundation, The Winslow Foundation, and other generous donors.

The opinions expressed in this book are those of the authors and do not necessarily reflect the views of these foundations.

Ecosystems and Human Well-being:
Current State and Trends, Volume 1

Ecosystems and Human Well-being: Current State and Trends, Volume 1

Edited by:

Rashid Hassan
University of Pretoria
South Africa

Robert Scholes
Council for Science and Industrial Research
South Africa

Neville Ash
UNEP World Conservation
Monitoring Centre
United Kingdom

Findings of the Condition and Trends Working Group
of the Millennium Ecosystem Assessment

Washington • Covelo • London

The Millennium Ecosystem Assessment Series

Ecosystems and Human Well-being: A Framework for Assessment
Ecosystems and Human Well-being: Current State and Trends, Volume 1
Ecosystems and Human Well-being: Scenarios, Volume 2
Ecosystems and Human Well-being: Policy Responses, Volume 3
Ecosystems and Human Well-being: Multiscale Assessments, Volume 4
Our Human Planet: Summary for Decision-makers

Synthesis Reports (available at MAweb.org)

Ecosystems and Human Well-being: Synthesis
Ecosystems and Human Well-being: Biodiversity Synthesis
Ecosystems and Human Well-being: Desertification Synthesis
Ecosystems and Human Well-being: Human Health Synthesis
Ecosystems and Human Well-being: Wetlands and Water Synthesis
Ecosystems and Human Well-being: Opportunities and Challenges for Business and Industry

No copyright claim is made in the work by: N.V. Aladin, Rob Alkemade, Vyacheslav Aparin, Andrew Balmford, Andrew J. Beattie, Victor Brovkin, Elena Bykova, John Dixon, Nikolay Gorelkin, Terry Griswold, Ward Hagemeijer, Jack Ives, Jacques Lemoalle, Christian Leveque, Hassane Mahamat, Anthony David McGuire, Eduardo Mestre Rodriguez, Mwelecele-Malecela-Lazaro, Oladele Osibanjo, Joachim Otte, Reidar Persson, Igor Plotnikov, Alison Power, Juan Pulhin, Inbal Reshef, Ulf Riebesell, Alan Rodgers, Agnes Rola, Raisa Toryannikova, employees of the Australian government (C. Max Finlayson), employees of the Canadian government (Randy G. Milton, Ian D. Thompson), employees of WHO (Robert Bos), employees of the U.K. government (Richard Betts, John Chilton), and employees of the U.S. government (Jill Baron, Kenneth R. Hinga, William Perrin, Joshua Rosenthal, Keith Wiebe). The views expressed in this report are those of the authors and do not necessarily reflect the position of the organizations they are employees of.

ISLAND PRESS is a trademark of The Center for Resource Economics.

Library of Congress Cataloging-in-Publication data.

Ecosystems and human well-being : current state and trends : findings of
the Condition and Trends Working Group / edited by Rashid Hassan, Robert
Scholes, Neville Ash.
     p. cm.—(The millennium ecosystem assessment series ; v. 1)
    Includes bibliographical references and index.
    ISBN 1-55963-227-5 (cloth : alk. paper)—ISBN 1-55963-228-3 (pbk. : alk. paper)
    1. Human ecology. 2. Ecosystem management. 3. Biotic communities.
4. Biological diversity. 5. Ecological assessment (Biology) I. Hassan,
Rashid M. II. Scholes, Robert. III. Ash, Neville. IV. Millennium Ecosystem
Assessment (Program). Condition and Trends Working Group. V. Series.
GF50.E264 2005
333.95—dc22

                                    2005017196

British Cataloguing-in-Publication data available.

Printed on recycled, acid-free paper ♾

Book design by Maggie Powell
Typesetting by Coghill Composition, Inc.

Manufactured in the United States of America
10 9 8 7 6 5 4 3 2 1

Millennium Ecosystem Assessment: Objectives, Focus, and Approach

The Millennium Ecosystem Assessment was carried out between 2001 and 2005 to assess the consequences of ecosystem change for human well-being and to establish the scientific basis for actions needed to enhance the conservation and sustainable use of ecosystems and their contributions to human well-being. The MA responds to government requests for information received through four international conventions—the Convention on Biological Diversity, the United Nations Convention to Combat Desertification, the Ramsar Convention on Wetlands, and the Convention on Migratory Species—and is designed to also meet needs of other stakeholders, including the business community, the health sector, nongovernmental organizations, and indigenous peoples. The sub-global assessments also aimed to meet the needs of users in the regions where they were undertaken.

The assessment focuses on the linkages between ecosystems and human well-being and, in particular, on "ecosystem services." An ecosystem is a dynamic complex of plant, animal, and microorganism communities and the nonliving environment interacting as a functional unit. The MA deals with the full range of ecosystems—from those relatively undisturbed, such as natural forests, to landscapes with mixed patterns of human use and to ecosystems intensively managed and modified by humans, such as agricultural land and urban areas. Ecosystem services are the benefits people obtain from ecosystems. These include *provisioning services* such as food, water, timber, and fiber; *regulating services* that affect climate, floods, disease, wastes, and water quality; *cultural services* that provide recreational, aesthetic, and spiritual benefits; and *supporting services* such as soil formation, photosynthesis, and nutrient cycling. The human species, while buffered against environmental changes by culture and technology, is fundamentally dependent on the flow of ecosystem services.

The MA examines how changes in ecosystem services influence human well-being. Human well-being is assumed to have multiple constituents, including the *basic material for a good life*, such as secure and adequate livelihoods, enough food at all times, shelter, clothing, and access to goods; *health,* including feeling well and having a healthy physical environment, such as clean air and access to clean water; *good social relations,* including social cohesion, mutual respect, and the ability to help others and provide for children; *security,* including secure access to natural and other resources, personal safety, and security from natural and human-made disasters; and *freedom of choice and action,* including the opportunity to achieve what an individual values doing and being. Freedom of choice and action is influenced by other constituents of well-being (as well as by other factors, notably education) and is also a precondition for achieving other components of well-being, particularly with respect to equity and fairness.

The conceptual framework for the MA posits that people are integral parts of ecosystems and that a dynamic interaction exists between them and other parts of ecosystems, with the changing human condition driving, both directly and indirectly, changes in ecosystems and thereby causing changes in human well-being. At the same time, social, economic, and cultural factors unrelated to ecosystems alter the human condition, and many natural forces influence ecosystems. Although the MA emphasizes the linkages between ecosystems and human well-being, it recognizes that the actions people take that influence ecosystems result not just from concern about human well-being but also from considerations of the intrinsic value of species and ecosystems. Intrinsic value is the value of something in and for itself, irrespective of its utility for someone else.

The Millennium Ecosystem Assessment synthesizes information from the scientific literature and relevant peer-reviewed datasets and models. It incorporates knowledge held by the private sector, practitioners, local communities, and indigenous peoples. The MA did not aim to generate new primary knowledge but instead sought to add value to existing information by collating, evaluating, summarizing, interpreting, and communicating it in a useful form. Assessments like this one apply the judgment of experts to existing knowledge to provide scientifically credible answers to policy-relevant questions. The focus on policy-relevant questions and the explicit use of expert judgment distinguish this type of assessment from a scientific review.

Five overarching questions, along with more detailed lists of user needs developed through discussions with stakeholders or provided by governments through international conventions, guided the issues that were assessed:

- What are the current condition and trends of ecosystems, ecosystem services, and human well-being?

- What are plausible future changes in ecosystems and their ecosystem services and the consequent changes in human well-being?

- What can be done to enhance well-being and conserve ecosystems? What are the strengths and weaknesses of response options that can be considered to realize or avoid specific futures?

- What are the key uncertainties that hinder effective decision-making concerning ecosystems?

- What tools and methodologies developed and used in the MA can strengthen capacity to assess ecosystems, the services they provide, their impacts on human well-being, and the strengths and weaknesses of response options?

The MA was conducted as a multiscale assessment, with interlinked assessments undertaken at local, watershed, national, regional, and global scales. A global ecosystem assessment cannot easily meet all the needs of decision-makers at national and sub-national scales because the management of any

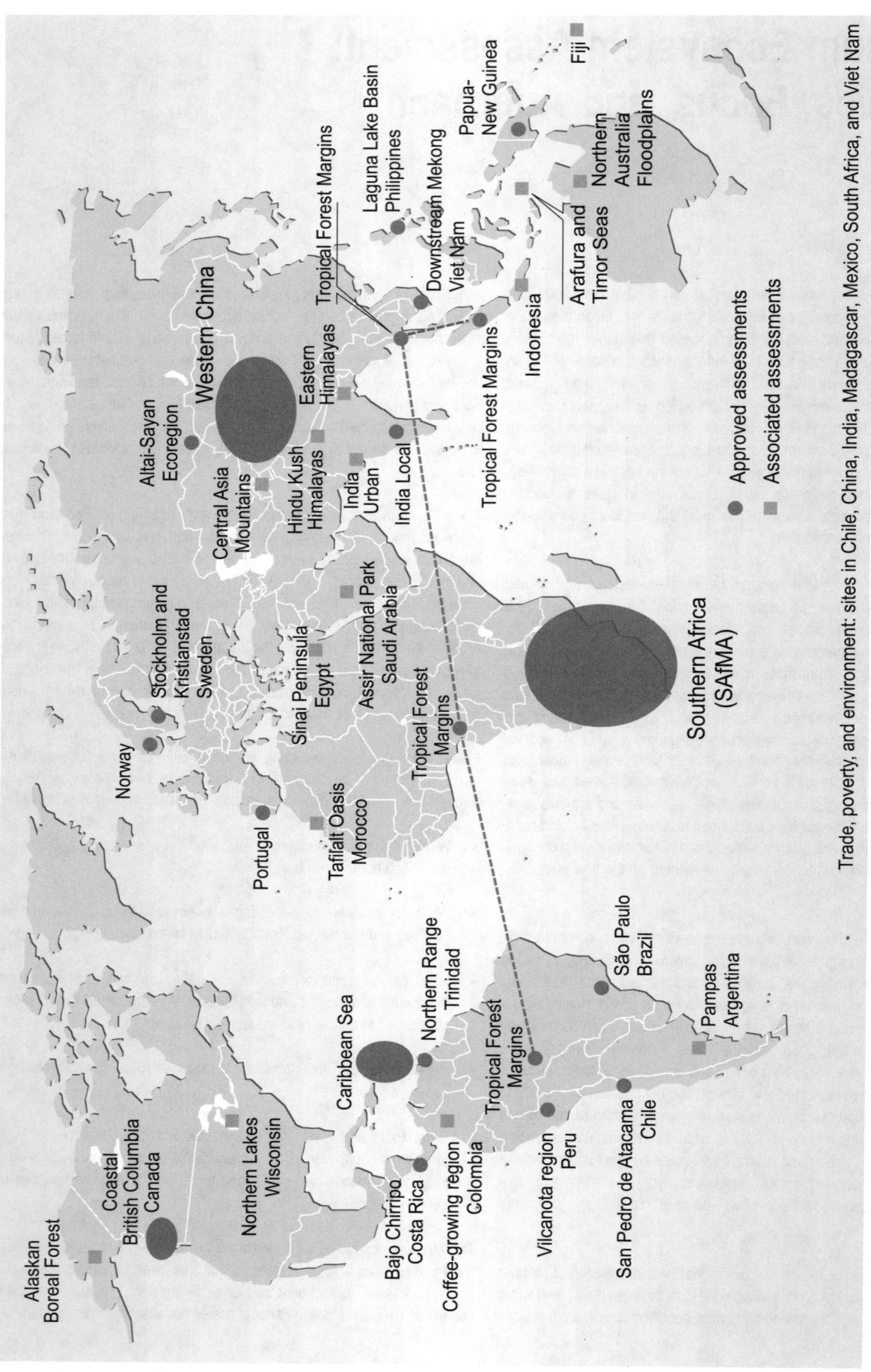

Alaskan
Boreal Forest

Coastal
British Columbia
Canada

Northern Lakes
Wisconsin

Caribbean Sea

Northern Range
Trinidad

Bajo Chirripo
Costa Rica

Coffee-growing region
Colombia

Tropical Forest
Margins

São Paulo
Brazil

Vilcanota region
Peru

Pampas
Argentina

San Pedro de Atacama
Chile

Portugal

Norway

Stockholm and
Kristianstad
Sweden

Tafilalt Oasis
Morocco

Sinai Peninsula
Egypt

Assir National Park
Saudi Arabia

Tropical Forest
Margins

Southern Africa
(SAfMA)

Altai-Sayan
Ecoregion

Western China

Central Asia
Mountains

Eastern
Himalayas

Hindu Kush
Himalayas

India
Urban

India Local

Tropical Forest Margins

Indonesia

Arafura and
Timor Seas

Tropical Forest Margins

Laguna Lake Basin
Philippines

Downstream Mekong
Viet Nam

Papua-
New Guinea

Northern
Australia
Floodplains

Fiji

● Approved assessments

■ Associated assessments

Eighteen assessments were approved as components of the MA. Any institution or country was able to undertake an assessment as part of the MA if it agreed to use the MA conceptual framework, to centrally involve the intended users as stakeholders and partners, and to meet a set of procedural requirements related to peer review, metadata, transparency, and intellectual property rights. The MA assessments were largely self-funded, although planning grants and some core grants were provided to support some assessments. The MA also drew on information from 16 other sub-global assessments affiliated with the MA that met a subset of these criteria or were at earlier stages in development.

Trade, poverty, and environment: sites in Chile, China, India, Madagascar, Mexico, South Africa, and Viet Nam

ECOSYSTEM TYPES   ECOSYSTEM SERVICES

SUB-GLOBAL ASSESSMENT	COASTAL	CULTIVATED	DRYLAND	FOREST	INLAND WATER	ISLAND	MARINE	MOUNTAIN	POLAR	URBAN	FOOD	WATER	FUEL and ENERGY	BIODIVERSITY-RELATED	CARBON SEQUESTRATION	FIBER and TIMBER	RUNOFF REGULATION	CULTURAL, SPIRITUAL, AMENITY	OTHERS
Altai-Sayan Ecoregion			●	●	●			●			●	●	●	●		●		●	
San Pedro de Atacama, Chile		●	●								●	●		●				●	●
Caribbean Sea	●					●	●							●					
Coastal British Columbia, Canada	●			●	●			●			●			●		●	●	●	
Bajo Chirripo, Costa Rica		●		●	●						●	●		●		●	●	●	●
Tropical Forest Margins				●	●						●			●		●	●	●	●
India Local Villages		●			●			●			●	●	●			●		●	●
Glomma Basin, Norway		●		●	●						●	●	●			●		●	●
Papua New Guinea	●					●		●			●	●		●		●	●	●	●
Vilcanota, Peru		●	●				●				●			●			●	●	●
Laguna Lake Basin, Philippines	●	●			●			●		●	●	●	●	●	●	●	●	●	●
Portugal	●		●	●	●	●				●	●	●		●	●	●		●	●
São Paulo Green Belt, Brazil	●									●	●	●	●	●	●	●	●	●	●
Southern Africa	●		●	●	●					●	●	●		●		●	●	●	●
Stockholm and Kristianstad, Sweden	●			●	●						●	●		●	●	●		●	●
Northern Range, Trinidad				●			●				●	●			●	●	●	●	●
Downstream Mekong Wetlands, Viet Nam	●	●		●	●			●			●	●	●	●	●	●		●	●
Western China		●	●	●	●						●	●		●	●	●	●	●	●
Alaskan Boreal Forest				●	●						●	●		●					●
Arafura and Timor Seas	●						●	●			●	●	●		●	●		●	●
Argentine Pampas		●						●			●	●						●	●
Central Asia Mountains								●			●	●		●					●
Colombia coffee-growing regions		●		●				●			●	●						●	●
Eastern Himalayas			●	●							●			●					●
Sinai Peninsula, Egypt					●	●		●			●	●		●				●	●
Fiji	●										●	●						●	●
Hindu Kush-Himalayas	●										●	●					●	●	●
Indonesia			●	●	●			●		●	●	●	●	●	●	●	●	●	●
India Urban Resource				●							●	●						●	●
Tafilalt Oasis, Morocco		●	●	●							●	●						●	●
Northern Australia Floodplains					●			●			●	●		●			●	●	●
Assir National Park, Saudi Arabia		●		●	●						●	●				●	●	●	●
Northern Highlands Lake District, Wisconsin																			

particular ecosystem must be tailored to the particular characteristics of that ecosystem and to the demands placed on it. However, an assessment focused only on a particular ecosystem or particular nation is insufficient because some processes are global and because local goods, services, matter, and energy are often transferred across regions. Each of the component assessments was guided by the MA conceptual framework and benefited from the presence of assessments undertaken at larger and smaller scales. The sub-global assessments were not intended to serve as representative samples of all ecosystems; rather, they were to meet the needs of decision-makers at the scales at which they were undertaken. The sub-global assessments involved in the MA process are shown in the Figure and the ecosystems and ecosystem services examined in these assessments are shown in the Table.

The work of the MA was conducted through four working groups, each of which prepared a report of its findings. At the global scale, the Condition and Trends Working Group assessed the state of knowledge on ecosystems, drivers of ecosystem change, ecosystem services, and associated human well-being around the year 2000. The assessment aimed to be comprehensive with regard to ecosystem services, but its coverage is not exhaustive. The Scenarios Working Group considered the possible evolution of ecosystem services during the twenty-first century by developing four global scenarios exploring plausible future changes in drivers, ecosystems, ecosystem services, and human well-being. The Responses Working Group examined the strengths and weaknesses of various response options that have been used to manage ecosystem services and identified promising opportunities for improving human well-being while conserving ecosystems. The report of the Sub-global Assessments Working Group contains lessons learned from the MA sub-global assessments. The first product of the MA—*Ecosystems and Human Well-being: A Framework for Assessment,* published in 2003—outlined the focus, conceptual basis, and methods used in the MA. The executive summary of this publication appears as Chapter 1 of this volume.

Approximately 1,360 experts from 95 countries were involved as authors of the assessment reports, as participants in the sub-global assessments, or as members of the Board of Review Editors. The latter group, which involved 80 experts, oversaw the scientific review of the MA reports by governments and experts and ensured that all review comments were appropriately addressed by the authors. All MA findings underwent two rounds of expert and governmental review. Review comments were received from approximately 850 individuals (of which roughly 250 were submitted by authors of other chapters in the MA), although in a number of cases (particularly in the case of governments and MA-affiliated scientific organizations), people submitted collated comments that had been prepared by a number of reviewers in their governments or institutions.

The MA was guided by a Board that included representatives of five international conventions, five U.N. agencies, international scientific organizations, governments, and leaders from the private sector, nongovernmental organizations, and indigenous groups. A 15-member Assessment Panel of leading social and natural scientists oversaw the technical work of the assessment, supported by a secretariat with offices in Europe, North America, South America, Asia, and Africa and coordinated by the United Nations Environment Programme.

The MA is intended to be used:

- to identify priorities for action;

- as a benchmark for future assessments;

- as a framework and source of tools for assessment, planning, and management;

- to gain foresight concerning the consequences of decisions affecting ecosystems;

- to identify response options to achieve human development and sustainability goals;

- to help build individual and institutional capacity to undertake integrated ecosystem assessments and act on the findings; and

- to guide future research.

Because of the broad scope of the MA and the complexity of the interactions between social and natural systems, it proved to be difficult to provide definitive information for some of the issues addressed in the MA. Relatively few ecosystem services have been the focus of research and monitoring and, as a consequence, research findings and data are often inadequate for a detailed global assessment. Moreover, the data and information that are available are generally related to either the characteristics of the ecological system or the characteristics of the social system, not to the all-important interactions between these systems. Finally, the scientific and assessment tools and models available to undertake a cross-scale integrated assessment and to project future changes in ecosystem services are only now being developed. Despite these challenges, the MA was able to provide considerable information relevant to most of the focal questions. And by identifying gaps in data and information that prevent policy-relevant questions from being answered, the assessment can help to guide research and monitoring that may allow those questions to be answered in future assessments.

Contents

Foreword

The Millennium Ecosystem Assessment was called for by United Nations Secretary-General Kofi Annan in 2000 in his report to the UN General Assembly, *We the Peoples: The Role of the United Nations in the 21st Century.* Governments subsequently supported the establishment of the assessment through decisions taken by three international conventions, and the MA was initiated in 2001. The MA was conducted under the auspices of the United Nations, with the secretariat coordinated by the United Nations Environment Programme, and it was governed by a multistakeholder board that included representatives of international institutions, governments, business, NGOs, and indigenous peoples. The objective of the MA was to assess the consequences of ecosystem change for human well-being and to establish the scientific basis for actions needed to enhance the conservation and sustainable use of ecosystems and their contributions to human well-being.

This volume has been produced by the MA Condition and Trends Working Group and assesses the state of knowledge on ecosystems and their services, the drivers of ecosystem change, and the consequences of ecosystem change for human well-being. The material in this report has undergone two extensive rounds of peer review by experts and governments, overseen by an independent Board of Review Editors.

This is one of four volumes (*Current State and Trends, Scenarios, Policy Responses,* and *Multiscale Assessments*) that present the technical findings of the Assessment. Six synthesis reports have also been published: one for a general audience and others focused on issues of biodiversity, wetlands and water, desertification, health, and business and ecosystems. These synthesis reports were prepared for decision-makers in these different sectors, and they synthesize and integrate findings from across all of the Working Groups for ease of use by those audiences.

This report and the other three technical volumes provide a unique foundation of knowledge concerning human dependence on ecosystems as we enter the twenty-first century. Never before has such a holistic assessment been conducted that addresses multiple environmental changes, multiple drivers, and multiple linkages to human well-being. Collectively, these reports reveal both the extraordinary success that humanity has achieved in shaping ecosystems to meet the needs of growing populations and econo-

mies and the growing costs associated with many of these changes. They show us that these costs could grow substantially in the future, but also that there are actions within reach that could dramatically enhance both human well-being and the conservation of ecosystems.

A more exhaustive set of acknowledgments appears later in this volume but we want to express our gratitude to the members of the MA Board, Board Alternates, Exploratory Steering Committee, Assessment Panel, Coordinating Lead Authors, Lead Authors, Contributing Authors, Board of Review Editors, and Expert Reviewers for their extraordinary contributions to this process. (The list of reviewers is available at www.MAweb.org.) We also would like to thank the MA Secretariat and in particular the staff of the Condition and Trends Working Group Technical Support Unit for their dedication in coordinating the production of this volume, as well as the World Conservation Monitoring Centre, which housed this TSU.

We would particularly like to thank the Co-chairs of the Condition and Trends Working Group, Dr. Rashid Hassan and Dr. Robert Scholes, and the TSU Coordinator, Neville Ash, for their skillful leadership of this Working Group and their contributions to the overall assessment.

Dr. Robert T. Watson
MA Board Co-chair
Chief Scientist, The World Bank

Dr. A.H. Zakri
MA Board Co-chair
Director, Institute for Advanced Studies
United Nations University

Preface

The *Current State and Trends* assessment presents the findings of the Condition and Trends Working Group of the Millennium Ecosystem Assessment. This volume documents the current condition and recent trends of the world's ecosystems, the services they provide, and associated human well-being around the year 2000. Its primary goal is to provide decision-makers, ecosystem managers, and other potential users with objective information and analyses of historical trends and dynamics of the interaction between ecosystem change and human well-being. This assessment establishes a baseline for the current condition of ecosystems at the turn of the millennium. It also assesses how changes in ecosystems have affected the underlying capacity of ecosystems to continue to provide these services in the near future, providing a link to the Scenarios Working Group's report. Finally, it considers recent trends in ecosystem conditions that have been the result of historical responses to ecosystem service problems, providing a link to the Responses Working Group's report.

Although centered on the year 2000, the temporal scope of this assessment includes the "relevant past" to the "foreseeable future." In practice, this means analyzing trends during the latter decades of the twentieth century and extrapolating them forward for a decade or two into the twenty-first century. At the point where the projections become too uncertain to be sustained, the Scenarios Working Group takes over the exploration of alternate futures.

The Condition and Trends assessment aims to synthesize and add to information already available from other sources, whether in the primary scientific literature or already in assessment form. In many instances this information is not reproduced in this volume but is built upon to report additional findings here. So this volume does not, for example, provide an assessment of the science of climate change per se, as that is reported in the findings of the Intergovernmental Panel on Climate Change, but the findings of the IPCC are used here as a basis to present information on the consequences of climate change for ecosystem services.

A summary of the process leading to this document is provided in Figure A.

The document has three main parts plus a synthesis chapter and supporting material. (See Figure B.) After the introductory material in Part I, the findings from the technical assessments are presented in two orthogonal ways: Part II deals with individual categories of ecosystem services, viewed across all the ecosystem types from which they are derived, while Part III analyses the various systems from which bundles of services are derived. Such organization allows the chapters to be read as standalone documents and assists readers with thematic interests. In Part IV, the synthesis chapter pulls out the key threads of findings from the earlier parts to construct an integrated narrative of the key issues relating ecosystem change (through changes in ecosystem services) to impacts on human well-being.

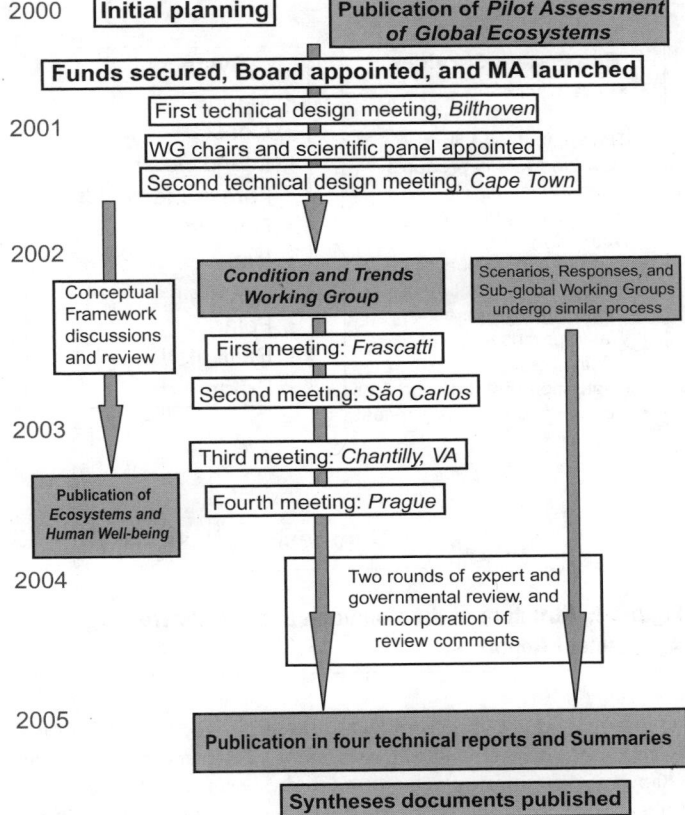

Figure A. Schedule of the Condition and Trends Working Group Assessment

Appendices provide an extensive glossary of terms, abbreviations, and acronyms; information on authors; and color graphics.

Part I: General Concepts and Analytical Approaches

The first part of this report introduces the overarching conceptual, methodological, and crosscutting themes of the MA integrated approach, and for this reason it precedes the technical assessment parts. Following the executive summary of the MA conceptual framework volume (*Ecosystems and Human Well-being: A Framework for Assessment*), which is **Chapter 1,** the analytical approaches to a global assessment of ecosystems and ecosystem services are outlined in **Chapter 2. Chapter 3** provides a summary assessment of the most important changes in key indirect and direct drivers of ecosystem change over the last part of the

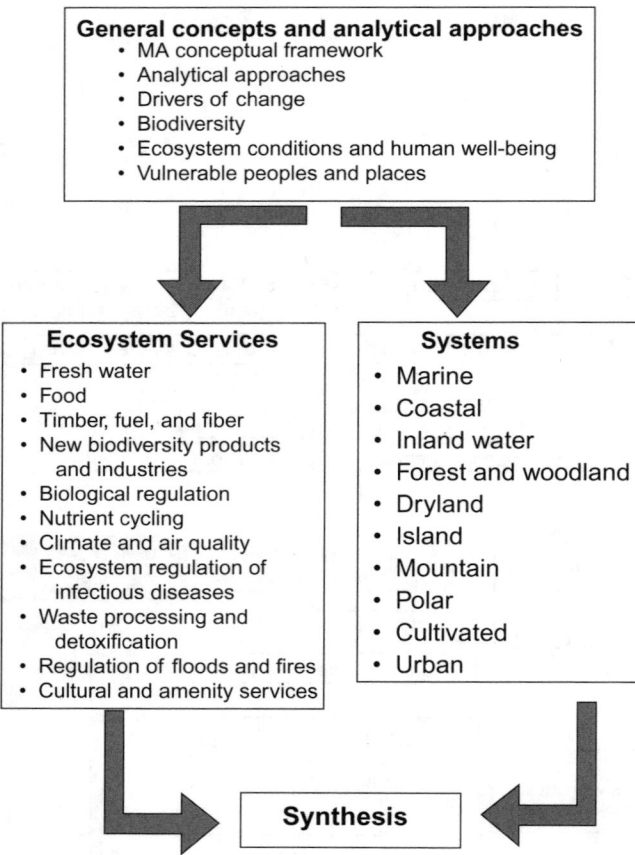

General concepts and analytical approaches
- MA conceptual framework
- Analytical approaches
- Drivers of change
- Biodiversity
- Ecosystem conditions and human well-being
- Vulnerable peoples and places

Ecosystem Services
- Fresh water
- Food
- Timber, fuel, and fiber
- New biodiversity products and industries
- Biological regulation
- Nutrient cycling
- Climate and air quality
- Ecosystem regulation of infectious diseases
- Waste processing and detoxification
- Regulation of floods and fires
- Cultural and amenity services

Systems
- Marine
- Coastal
- Inland water
- Forest and woodland
- Dryland
- Island
- Mountain
- Polar
- Cultivated
- Urban

Synthesis

Figure B. Structure of the Condition and Trends Working Group Assessment Report

twentieth century, and considers some of the key interactions between these drivers (the full assessment of drivers, of which this chapter is a summary, can be found in the *Scenarios* volume, Chapter 7). The remaining chapters in Part I—on biodiversity (**Chapter 4**), human well-being (**Chapter 5**), and vulnerability (**Chapter 6**)—introduce issues at a global scale but also contain a synthesis of material drawn from chapters in Parts II and III.

Each of these introductory overarching chapters aims to deal with the general issues related to its topic, leaving the specifics embedded in later chapters. This is intended to enhance readability and to help reduce redundancy across the volume. For example, **Chapter 2** seeks to give an overview of the types of analytical approaches and methods used in the assessment, but not provide a recipe for conducting specific assessments, and **Chapter 3** aims to provide the background to the various drivers that would otherwise need to be discussed in multiple subsequent chapters.

Biodiversity provides composition, structure, and function to ecosystems. The amount and diversity of life is an underlying necessity for the provision of all ecosystem services, and for this reason **Chapter 4** is included in the introductory section rather than as a chapter in the part on ecosystem services. It outlines the key global trends in biodiversity, our state of knowledge on biodiversity in terms of abundance and distribution, and the role of biodiversity in the functioning of ecosystems. Later chapters consider more fully the role of biodiversity in the provision of ecosystem services.

The consequences of ecosystem change for human well-being are the core subject of the MA. **Chapter 5** presents our state of

knowledge on the links between ecosystems and human well-being and outlines the broad patterns in well-being around the world.

Neither the distribution of ecosystem services nor the change in these services is evenly distributed across places and societies. Certain ecosystems, locations, and people are more at risk from changes in the supply of services than others. **Chapter 6**, on vulnerable peoples and places, identifies these locations and groups and examines why they are particularly vulnerable to changes in ecosystems and ecosystem services.

Part II: An Assessment of Ecosystem Services

The Condition and Trends assessment sets out to be comprehensive in its treatment of ecosystem services but not exhaustive. The list of "benefits that people derive from ecosystems" grows continuously with further investigation. The 11 groups of services covered by this assessment deal with issues that are of vital importance almost everywhere in the world and represent, in the opinion of the Working Group, the main services that are most important for human well-being and are most affected by changes in ecosystem conditions. The MA only considers ecosystem services that have a nexus with life on Earth (biodiversity). For example, while gemstones and tidal energy can both provide benefits to people, and both are found within ecosystems, they are not addressed in this report since their generation does not depend on the presence of living organisms. The ecosystem services assessed and the chapter titles in this part are:

Provisioning services:
- Fresh Water
- Food
- Timber, Fiber, and Fuel
- New Products and Industries from Biodiversity

Regulating and supporting services:
- Biological Regulation of Ecosystem Services
- Nutrient Cycling
- Climate and Air Quality
- Human Health: Ecosystem Regulation of Infectious Diseases
- Waste Processing and Detoxification
- Regulation of Natural Hazards: Floods and Fires

Cultural services:
- Cultural and Amenity Services

Each of the chapters in this section in fact deals with a cluster of several related ecosystem services. For instance, the chapter on food covers the provision of numerous cereal crops, vegetables and fruits, beverages, livestock, fish, and other edible products; the chapter on nutrient cycling addresses the benefits derived from a range of nutrient cycles, but with a focus on nitrogen; and the chapter on cultural and amenity services covers a range of such services, including recreation, aesthetic, and spiritual services. The length of the treatment afforded to each service reflects several factors: our assessment of its relative importance to human well-being; the scope and complexity of the topic; the degree to which it has been treated in other assessments (thus reducing the need for a comprehensive treatment here); and the amount of information that is available to be assessed.

Part II considers services from each of the four MA categories: provisioning, regulating, cultural, and supporting services. Each service chapter has been developed to cover the same types of information. First the service is defined. Then, for each service, the spatial distribution of supply and demand is quantified, along with recent trends. The direct and indirect drivers of change in the service are analyzed. And finally the consequences of the changes in the service for human well-being are examined and quantified to the degree possible.

Examples are given of the responses by decision-makers at various levels (from the individual to the international) to issues relating to change in service supply. Both successful and unsuccessful interventions are described, as supportive material for the *Policy Responses* volume.

Part III: An Assessment of Systems from which Ecosystem Services Are Derived

The Condition and Trends Working Group uses the term "systems" in describing these chapters rather than the term "ecosystems." This is for several reasons. First, the "systems" used are essentially reporting units, defined for pragmatic reasons. They represent easily recognizable broad categories of landscape or seascape, with their included human systems, and typically represent units or themes of management or intervention interest. Ecosystems, on the other hand, are theoretically defined by the interactions of their components.

The 10 selected systems assessed here cover much larger areas than most ecosystems in the strict sense and include areas of system type that are far apart (even isolated) and that thus interact only weakly. In fact, there may be stronger local interactions with embedded fragments of ecosystems of a different type rather than within the nominal type of the system. The "cultivated system," for instance, considers a landscape where crop farming is a primary activity but that probably includes, as an integral part of that system, patches of rangeland, forest, water, and human settlements.

Second, while it is recognized that humans are always part of ecosystems, the definitions of the systems used in this report take special note of the main patterns of human use. The systems are defined around the main bundles of services they typically supply and the nature of the impacts that human use has on those services.

Information within the systems chapters is frequently presented by subsystems where appropriate. For example, the forest chapter deals separately with tropical, temperate, and boreal forests because they deliver different services; likewise, the coastal chapter deals explicitly with various coastal subsystems, such as mangroves, corals, and seagrasses.

The 10 system categories and the chapter titles in this part are:
- Marine Fisheries Systems
- Coastal Systems
- Inland Water Systems
- Forest and Woodland Systems
- Dryland Systems
- Island Systems
- Mountain Systems
- Polar Systems
- Cultivated Systems
- Urban Systems

Definitions for these system categories can be found in Box 1.3 in Chapter 1. These system categories are not mutually exclusive, and some overlap spatially. For instance, mountain systems contain areas of forest systems, dryland systems, inland water systems, cultivated systems, and urban systems, while coastal systems include components of all of the above, including mountain systems. Due to this overlap, simple summations of services across systems for global totals should be avoided (an exercise that the MA has avoided in general): some may be double-counted, while others may be underrepresented. Notwithstanding these caveats, the systems have been defined to cover most of the Earth's surface and not to overlap unnecessarily. In many instances the boundaries between systems are

diffuse, but not arbitrary. For instance, the coastal system blends seamlessly into the marine system on the one hand and the land systems on the other. The 50-meter depth distinction between coastal and marine separates the systems strongly influenced by actions on the land from those overwhelmingly influenced by fishing. There is significant variation in the area of coverage of each system.

The system definitions are also not exhaustive, and no attempt has been made to cover every part of the global surface. Although ~99% of global surface area has been covered in this assessment, there are just over 5 million square kilometers of terrestrial land surface not included spatially within any of the MA system boundaries. These areas are generally found within grassland, savanna, and forest biomes, and they contain a mix of land cover classes—generally grasslands, degraded forests, and marginal agricultural lands—that are not picked up within the mapping definitions for the system boundaries. However, while these excluded areas may not appear in the various statistics produced along system boundaries, the issues occurring in these areas relating to ecosystem services are well covered in the various services chapters, which do not exclude areas of provision outside MA system boundaries.

The main motivation for dealing with "systems" as well as "services" is that the former perspective allows us to examine interactions between the services delivered from a single location. These interactions can take the form of trade-offs (that is, where promoting one service reduces the supply of another service), win-win situations (where a single management package enhances the supply of several services), or synergies, where the simultaneous use of services raises or depresses both more than if they were independently used.

The chapters in Part III all present information in a broadly similar manner: system description, including a map and descriptive statistics for the system and its subsystems; quantification of the services it delivers and their contribution to well-being; recent trends in the condition of the system and its capacity to provide services; processes leading to changes in the system; the choices and resultant trade-offs between systems and between services within the system; and the contributions of the system to human well-being.

Part IV: Synthesis

Chapter 28 does not intend to be a summary. That task is left to the summaries or Main Messages of each chapter and to the Summary at the start of this volume. Instead, the synthesis chapter constructs an integrated narrative, tracing the principal causes of ecosystem change, the consequences for ecosystems and ecosystem services, and the resultant main impacts on human well-being. The chapter considers the key intellectual issues arising from the Condition and Trends assessment and presents an assessment of our underlying knowledge on the consequences of ecosystem change for people.

Supporting material for many of the chapters, and further details of the Millennium Ecosystem Assessment, including of the various sub-global assessments, plus a full list of reviewers, can be found at the MA Web site at www.MAweb.org.

Rashid Hassan
University of Pretoria, South Africa

Robert Scholes
Council for Science and Industrial Research, South Africa

Neville J. Ash
UNEP-World Conservation Monitoring Centre

Acknowledgments

First and foremost, we would like to thank the MA Condition and Trends Working Group for their hard work, and for all the stimulating discussions we had over the course of the project. Special thanks are also due to the MA Secretariat staff who worked tirelessly on this project:

Walter V. Reid—Director

Administration
Nicole Khi—Program Coordinator
Chan Wai Leng—Program Coordinator
Belinda Lim—Administrative Officer
Tasha Merican—Program Coordinator

Sub-Global
Marcus J. Lee—Technical Support Unit Coordinator and MA Deputy Director
Ciara Raudsepp-Hearne—TSU Coordinator

Condition and Trends
Neville J. Ash—TSU Coordinator
Dalène du Plessis—Program Assistant
Mampiti Matete—TSU Coordinator

Scenarios
Elena Bennett—TSU Coordinator
Veronique Plocq-Fichelet—Program Administrator
Monika B. Zurek—TSU Coordinator

Responses
Pushpam Kumar—TSU Coordinator
Meenakshi Rathore—Program Coordinator
Henk Simons—TSU Coordinator

Engagement and Outreach
Christine Jalleh—Communications Officer
Nicolas Lucas—Engagement and Outreach Director
Valerie Thompson—Associate

Other Staff
John Ehrmann—Lead Facilitator
Keisha-Maria Garcia—Research Assistant
Lori Han—Publications Manager
Sara Suriani—Conference Manager
Jillian Thonell—Data Coordinator

Interns
Emily Cooper, Elizabeth Wilson, Lina Cimarrusti

We would like to acknowledge the contributions of all of the authors of this book, and the support provided by their institutions that enabled their participation. We would like to thank the host organizations of the MA Technical Support Units—WorldFish Center (Malaysia); UNEP-World Conservation Monitoring Centre (United Kingdom); Institute of Economic Growth (India); National Institute of Public Health and the Environment (Netherlands); University of Pretoria (South Africa), U.N. Food and Agriculture Organization; World Resources Institute, Meridian Institute, and Center for Limnology of the University of Wisconsin-Madison (all in the United States); Scientific Committee on Problems of the Environment (France); and International Maize and Wheat Improvement Center (Mexico)—for the support they provided to the process.

We thank several individuals who played particularly critical roles: Linda Starke, Nigel Varty, and Lynn Newton for editing the report; Hyacinth Billings and Caroline Taylor for providing invaluable advice on the publication process; Maggie Powell for preparing the page design and all the Figures and Tables; Elizabeth Wilson for helping to proof the Figures and Tables; Carol Inskipp and Gill Bunting for checking chapter citations and references; and Ian May, Corinna Ravilious, and Simon Blythe for the preparation of numerous graphics and GIS-derived statistics. And we thank the other MA volunteers, the administrative staff of the host organizations, and colleagues in other organizations who were instrumental in facilitating the process: Mariana Sanchez Abregu, Isabelle Alegre, Adlai Amor, Emmanuelle Bournay, Herbert Caudill, Habiba Gitay, Helen Gray, Sherry Heileman, Norbert Henninger, Toshi Honda, Francisco Ingouville, Timothy Johnson, Humphrey Kagunda, Brygida Kubiak, Nicolas Lapham, Liz Levitt, Elaine Marshall, Christian Marx, Stephanie Moore, John Mukoza, Arivudai Nambi, Laurie Neville, Adrian Newton, Carolina Katz Reid, Liana Reilly, Philippe Rekacewicz, Carol Rosen, Anne Schram, Jeanne Sedgwick, Tang Siang Nee, Darrell Taylor, Tutti Tischler, Dan Tunstall, Woody Turner, Mark Valentine, Gillian Warltier, Elsie Vélez Whited, Kaveh Zahedi, and Mark Zimsky.

For technical assistance with figures and references in Chapter 13, we thank Natalia Ungelenk and Silvana Schott, and for their work in developing, applying, and constructing tables from the Gridded Rural-Urban Mapping Project, which was used not only in Chapter 27 but in several others as well, we would like to thank Francesca Pozzi, Greg Booma, Adam Storeygard, Bridget Anderson, Greg Yetman, and Lisa Lukang. Kai Lee, Terry McGee, and Priscilla Connolly deserve special mention for their review of Chapter 27, as does Maria Furhacker for her review of Chapter 15.

We thank the members of the MA Board and its chairs, Robert Watson and A.H. Zakri, the members of the MA Assessment Panel and its chairs, Angela Cropper and Harold Mooney, and the members of the MA Review Board and its chairs, José Sarukhán and Anne Whyte, for their guidance and support for this Working Group. We also thank the current and previous Board Alternates: Ivar Baste, Jeroen Bordewijk, David Cooper, Carlos Corvalan, Nick Davidson, Lyle Glowka, Guo Risheng, Ju Hongbo, Ju Jin,

Kagumaho (Bob) Kakuyo, Melinda Kimble, Kanta Kumari, Stephen Lonergan, Charles Ian McNeill, Joseph Kalemani Mulongoy, Ndegwa Ndiang'ui, and Mohamed Maged Younes. We thank the past members of the MA Board whose contributions were instrumental in shaping the MA focus and process, including Philbert Brown, Gisbert Glaser, He Changchui, Richard Helmer, Yolanda Kakabadse, Yoriko Kawaguchi, Ann Kern, Roberto Lenton, Corinne Lepage, Hubert Markl, Arnulf Müller-Helbrecht, Seema Paul, Susan Pineda Mercado, Jan Plesnik, Peter Raven, Cristián Samper, Ola Smith, Dennis Tirpak, Alvaro Umaña, and Meryl Williams. We wish to also thank the members of the Exploratory Steering Committee that designed the MA project in 1999–2000. This group included a number of the current and past Board members, as well as Edward Ayensu, Daniel Claasen, Mark Collins, Andrew Dearing, Louise Fresco, Madhav Gadgil, Habiba Gitay, Zuzana Guziova, Calestous Juma, John Krebs, Jane Lubchenco, Jeffrey McNeely, Ndegwa Ndiang'ui, Janos Pasztor, Prabhu L. Pingali, Per Pinstrup-Andersen, and José Sarukhán. We thank Ian Noble and Mingsarn Kaosa-ard for their contributions as members of the Assessment Panel during 2002.

We would particularly like to acknowledge the input of the hundreds of individuals, institutions, and governments (see list at www.MAweb.org) who reviewed drafts of the MA technical and synthesis reports. We also thank the thousands of researchers whose work is synthesized in this report. And we would like to acknowledge the support and guidance provided by the secretariats and the scientific and technical bodies of the Convention on Biological Diversity, the Ramsar Convention on Wetlands, the Convention to Combat Desertification, and the Convention on Migratory Species, which have helped to define the focus of the MA and of this report.

We also want to acknowledge the support of a large number of nongovernmental organizations and networks around the world that have assisted in outreach efforts: Alexandria University, Argentine Business Council for Sustainable Development, Asociación Ixacavaa (Costa Rica), Arab Media Forum for Environment and Development, Brazilian Business Council on Sustainable Development, Charles University (Czech Republic), Cambridge Conservation Forum, Chinese Academy of Sciences, European Environmental Agency, European Union of Science Journalists' Associations, EIS-Africa (Burkina Faso), Forest Institute of the State of São Paulo, Foro Ecológico (Peru), Fridtjof Nansen Institute (Norway), Fundación Natura (Ecuador), Global Development Learning Network, Indonesian Biodiversity Foundation, Institute for Biodiversity Conservation and Research–Academy of Sciences of Bolivia, International Alliance of Indigenous Peoples of the Tropical Forests, IUCN office in Uzbekistan, IUCN Regional Offices for West Africa and South America, Permanent Inter-States Committee for Drought Control in the Sahel, Peruvian Society of Environmental Law, Probioandes (Peru), Professional Council of Environmental Analysts of Argentina, Regional Center AGRHYMET (Niger), Regional Environmental Centre for Central Asia, Resources and Research for Sustainable Development (Chile), Royal Society (United Kingdom), Stockholm University, Suez Canal University, Terra Nuova (Nicaragua), The Nature Conservancy (United States), United Nations University, University of Chile, University of the Philippines, World Assembly of Youth, World Business Council for Sustainable Development, WWF-Brazil, WWF-Italy, and WWF-US.

We are extremely grateful to the donors that provided major financial support for the MA and the MA Sub-global Assessments: Global Environment Facility; United Nations Foundation; David and Lucile Packard Foundation; World Bank; Consultative Group on International Agricultural Research; United Nations Environment Programme; Government of China; Ministry of Foreign Affairs of the Government of Norway; Kingdom of Saudi Arabia; and the Swedish International Biodiversity Programme. We also thank other organizations that provided financial support: Asia Pacific Network for Global Change Research; Association of Caribbean States; British High Commission, Trinidad & Tobago; Caixa Geral de Depósitos, Portugal; Canadian International Development Agency; Christensen Fund; Cropper Foundation, Environmental Management Authority of Trinidad and Tobago; Ford Foundation; Government of India; International Council for Science; International Development Research Centre; Island Resources Foundation; Japan Ministry of Environment; Laguna Lake Development Authority; Philippine Department of Environment and Natural Resources; Rockefeller Foundation; U.N. Educational, Scientific and Cultural Organization; UNEP Division of Early Warning and Assessment; United Kingdom Department for Environment, Food and Rural Affairs; United States National Aeronautic and Space Administration; and Universidade de Coimbra, Portugal. Generous in-kind support has been provided by many other institutions (a full list is available at www.MAweb.org). The work to establish and design the MA was supported by grants from The Avina Group, The David and Lucile Packard Foundation, Global Environment Facility, Directorate for Nature Management of Norway, Swedish International Development Cooperation Authority, Summit Foundation, UNDP, UNEP, United Nations Foundation, United States Agency for International Development, Wallace Global Fund, and World Bank.

Reader's Guide

The four technical reports present the findings of each of the MA Working Groups: Condition and Trends, Scenarios, Responses, and Sub-global Assessments. A separate volume, *Our Human Planet*, presents the summaries of all four reports in order to offer a concise account of the technical reports for decision-makers. In addition, six synthesis reports were prepared for ease of use by specific audiences: Synthesis (general audience), CBD (biodiversity), UNCCD (desertification), Ramsar Convention (wetlands), business and industry, and the health sector. Each MA sub-global assessment will also produce additional reports to meet the needs of its own audiences.

All printed materials of the assessment, along with core data and a list of reviewers, are available at www.MAweb.org. In this volume, Appendix A contains color maps and figures. Appendix B lists all the authors who contributed to this volume. Appendix C lists the acronyms and abbreviations used in this report and Appendix D is a glossary of terminology used in the technical reports. Throughout this report, dollar signs indicate U.S. dollars and ton means tonne (metric ton). Bracketed references within the Summary are to chapters within this volume.

In this report, the following words have been used where appropriate to indicate judgmental estimates of certainty, based on the collective judgment of the authors, using the observational evidence, modeling results, and theory that they have examined: very certain (98% or greater probability), high certainty (85–98% probability), medium certainty (65%–58% probability), low certainty (52–65% probability), and very uncertain (50–52% probability). In other instances, a qualitative scale to gauge the level of scientific understanding is used: well established, established but incomplete, competing explanations, and speculative. Each time these terms are used they appear in italics.

Ecosystems and Human Well-being: Current State and Trends, Volume 1

Summary: Ecosystems and Their Services around the Year 2000

Core Writing Team: Robert Scholes, Rashid Hassan, Neville J. Ash
Extended Writing Team: Condition and Trends Working Group

CONTENTS

Human Well-being and Life on Earth

- Human well-being depends, among other things, on the continued supply of services obtained from ecosystems.
- Human actions during the last 50 years have altered ecosystems to an extent and degree unprecedented in human history. The consequences for human well-being have been mixed. Health and wealth have, on average, improved, but the benefits are unevenly distributed and further improvement may be limited by an insufficient supply of key ecosystem services.
- Biological diversity is a necessary condition for the delivery of all ecosystem services. In most cases, greater biodiversity is associated with a larger or more dependable supply of ecosystem services. Diversity of genes and populations is currently declining in most places in the world, along with the area of near-natural ecosystems.

Inescapable Link between Ecosystem Condition and Human Well-being

All people depend on the services supplied by ecosystems, either directly or indirectly. Services are delivered both by "near-natural" ecosystems, such as rangelands, oceans, and forests, and by highly managed ecosystems such as cultivated or urban landscapes.

Human well-being, by several measures and on average across and within many societies, has improved substantially over the past two centuries and continues to do so. The human population in general is becoming better nourished. People live longer, and incomes have risen. Political institutions have become more participatory. In part these gains in well-being have been made possible by exploiting certain ecosystem services (the provisioning services, such as timber, grazing, and crop production), sometimes to the detriment of the ecosystem and its underlying capacity to continue to provide these and other services. Some gains have been made possible by the unsustainable use of other resources. For example, the increases in food production have been partly enabled by drawing on the finite supply of fossil fuels, an ecosystem service laid down millions of years ago.

The gains in human well-being are not distributed evenly among individuals or social groups, nor among the countries they live in or the ecosystems of the world. The gap between the advantaged and the disadvantaged is increasing. For example, a child born in sub-Saharan Africa is 20 times more likely to die before age five than a child born in an industrial country, and this ratio is higher than it was a decade ago. People living in urban areas, near coasts, and in systems with high ecosystem productivity in general have above-average well-being. **People living in drylands and mountainous areas, both characterized by lower ecosystem productivity, tend to have below-average, and more variable, well-being.**

Populations are growing faster in ecosystems characterized by low well-being and low ecosystem productivity than in high well-being, high productivity areas. Figure C1, which uses GDP as a proxy for human well-being, illustrates this situation. Trends are similar for other measures of human well-being, such as infant mortality rate. [5, 6, 16, 22]

Many human and ecological systems are under multiple severe and mutually reinforcing stresses. The causes include the direct and indirect impacts of extraction of services themselves, as well as the unintended side effects of other human activities. Certain linked ecological-human systems, by virtue of their structure or location, are more sensitive to stress than others. Examples include freshwater, coastal, mountain, island, and dryland systems.

Some groups of people are disproportionately likely to experience loss of well-being associated with declining levels of ecosystem services. The billion people poorest people in the world mostly live in rural areas where they are directly dependent on croplands, rangelands, rivers, seas, and forests for their livelihoods. For them especially, mismanagement of ecosystems threatens survival. Among better-off and urban populations, ecosystem changes affect well-being in less direct ways, but they remain important. They are partly buffered by technology and the ability to substitute some resources with others, but they also remain ultimately dependent on ecosystems for the basic necessities of life. Impacts are experienced differentially as a function of adaptive capacity, which can be manifested at the individual, household, community, national, or regional level. The groups ultimately responsible for the loss or decline of ecosystem services are often not the ones that bear the immediate impacts of their decline.

A large and growing number of people are at high risk of adverse ecosystem changes. The world is experiencing a worsening trend of human suffering and economic losses from natural disasters. Over the past four decades, for example, the number of weather-related disasters affecting at least a million people has increased fourfold, while economic losses have increased tenfold. The greatest loss of life has been concentrated in developing countries. Ecosystem transformation has played a significant, but not exclusive, role in increasing the vulnerability of people to such disasters. Examples are the increased susceptibility of coastal populations to tropical storms when mangrove forests are cleared and the increase in downstream flooding that followed land use changes in the upper Yangtze River. [16]

Special Role of Biodiversity in Supplying Ecosystem Services

In some cases, biodiversity can be treated as an ecosystem service in its own right, such as when it is the basis of nature-based tourism or the regulation of diseases. In other respects, it is a necessary condition underpinning the long-term provision of other services, such as food and clean fresh water. **Variation among genes, populations, and species and the variety of structure, function, and composition of ecosystems are necessary to maintain an acceptable and resilient level of ecosystem services in the long term.** [1]

For ecosystem functions such as productivity and nutrient cycling, the level, constancy of the service over time, and resilience to shocks all decline over the long term if biodiversity declines (*established but incomplete*). In general, there is no sudden biodiversity threshold below which ecosystem services fail. Quantifying the relationship between biodiversity and levels of ecosystem function has only been achieved in a few experimental situations and remains an area of active research. The amount and type of biodiversity required varies from service to service. **Regulatory services generally need higher levels of biodiversity than provisioning services do.** [11]

Changes in species composition can alter ecosystem processes even if the number of species present remains unchanged or increases. Thus, conserving the composition of communities rather than simply maximizing species numbers is more likely to maintain higher levels of ecosystem services. Reduction of the number of species, especially if the species lost are locally rare, may have a hardly detectable effect on ecosystem services in the short term. However, there is evidence from terres-

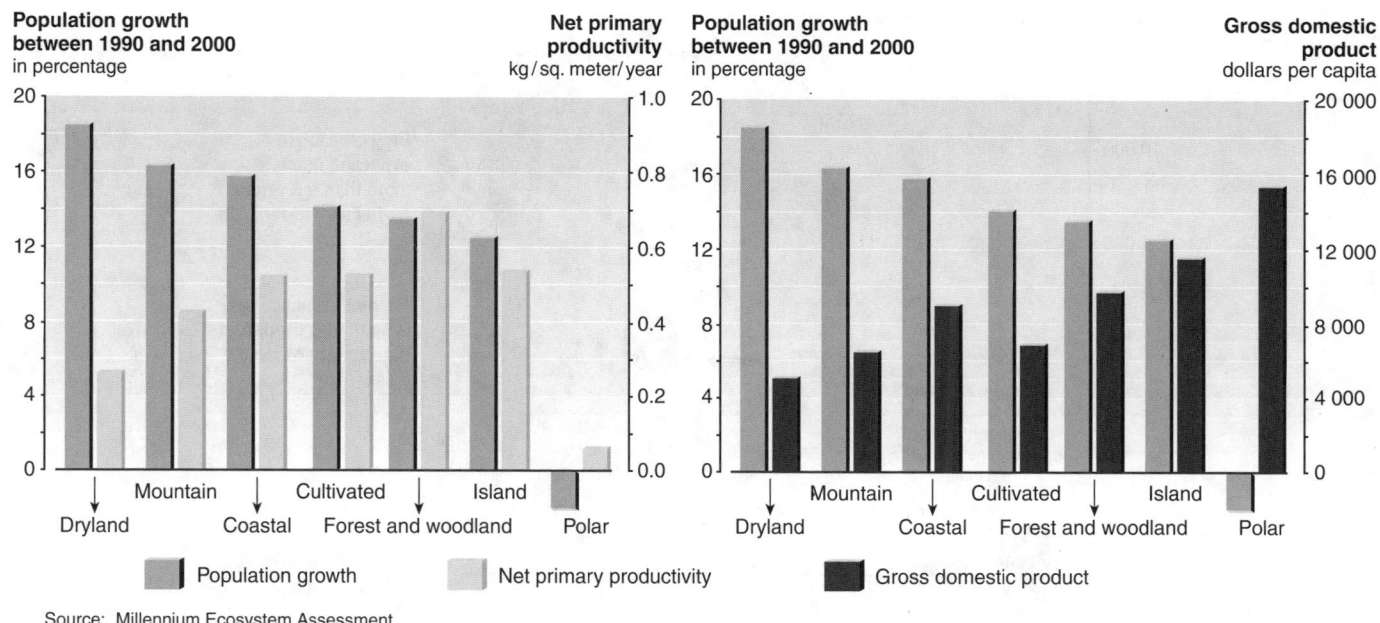

Figure C1. Population Growth Rates in 1990–2000, Per Capita GDP, and Ecosystem Productivity in 2000 in MA Ecological Systems

trial and aquatic systems that a rich regional species pool is needed to maintain ecosystem stability in the face of a changing environment in the long term. [11]

The integrity of the interactions between species is critical for the long-term preservation of human food production on land and in the sea. For example, pollination is an essential link in the production of food and fiber. Plant-eating insects and pathogens control the populations of many potentially harmful organisms. The services provided by coral reefs, such as habitat and nurseries for fish, sediment stabilization, nutrient cycling, and carbon fixing in nutrient-poor environments, can only be maintained if the interaction between corals and their obligate symbiotic algae is preserved. [11]

The preservation of genetic variation among crop species and their wild relatives and spatial heterogeneity in agricultural landscapes are considered necessary for the long-term viability of agriculture. Genetic variability is the raw material on which plant breeding for increased production and greater resilience depends. In general practice, agriculture undermines biodiversity and the regulating and supporting ecosystem services it provides in two ways: through transforming ecosystems by converting them to cultivated lands (extensification) and through the unintended negative impacts of increased levels of agricultural inputs, such as fertilizers, biocides, irrigation, and mechanical tillage (intensification). Agroforestry systems, crop rotations, intercropping, and conservation tillage are some of the agricultural techniques that maintain yields and protect crops and animals from pests without heavy investment in chemical inputs. [11]

A large proportion of the world's terrestrial species are concentrated in a small fraction of the land area, mostly in the tropics, and especially in forests and on mountains. Marine species are similarly concentrated, with the limited area of coral reefs, for example, having exceptionally high biodiversity. Most terrestrial species have small geographical ranges, and the ranges are often clustered, leading to diagnosable "hotspots" of both richness and endemism. These are frequently,

but not exclusively, concentrated in isolated or topographically variable regions such as islands, peninsulas, and mountains. The African and American tropics have the highest recorded species numbers in both absolute terms and per unit of area. Endemism is also highest there and, as a consequence of its isolation, in Australasia. Locations of species richness hotspots broadly correspond with centers of evolutionary diversity. Available evidence suggests that across the major taxa, tropical humid forests are especially important for both overall diversity and their unique evolutionary history. [4]

Among plants and vertebrates, the great majority of species are declining in distribution, abundance, or both, while a small number are expanding. Studies of African mammals, birds in cultivated landscapes, and commercially important fish all show the majority of species to be declining in range or number. Exceptions can be attributed to management interventions such as protection in reserves and elimination of threats such as overexploitation, or they are species that thrive in human-dominated landscapes. In some regions there may be an increase in local biodiversity as a result of species introductions, the long-term consequences of which are hard to foresee. [4]

The observed rates of species extinction in modern times are 100 to 1,000 times higher than the average rates for comparable groups estimated from the fossil record (*medium certainty*). (See Figure C2.) The losses have occurred in all taxa, regions, and ecosystems but are particularly high in some—for instance, among primates, in the tropics, and in freshwater habitats. Of the approximately 1,000 recorded historical extinctions, most have been on islands. Currently and in the future, the most threatened species are found on the mainland, particularly in locations of habitat change and degradation. **The current rate of biodiversity loss, in aggregate and at a global scale, gives no indication of slowing, although there have been local successes in some groups of species. The momentum of the underlying drivers of biodiversity loss, and the consequences of this loss, will extend many millennia into the future.** [4]

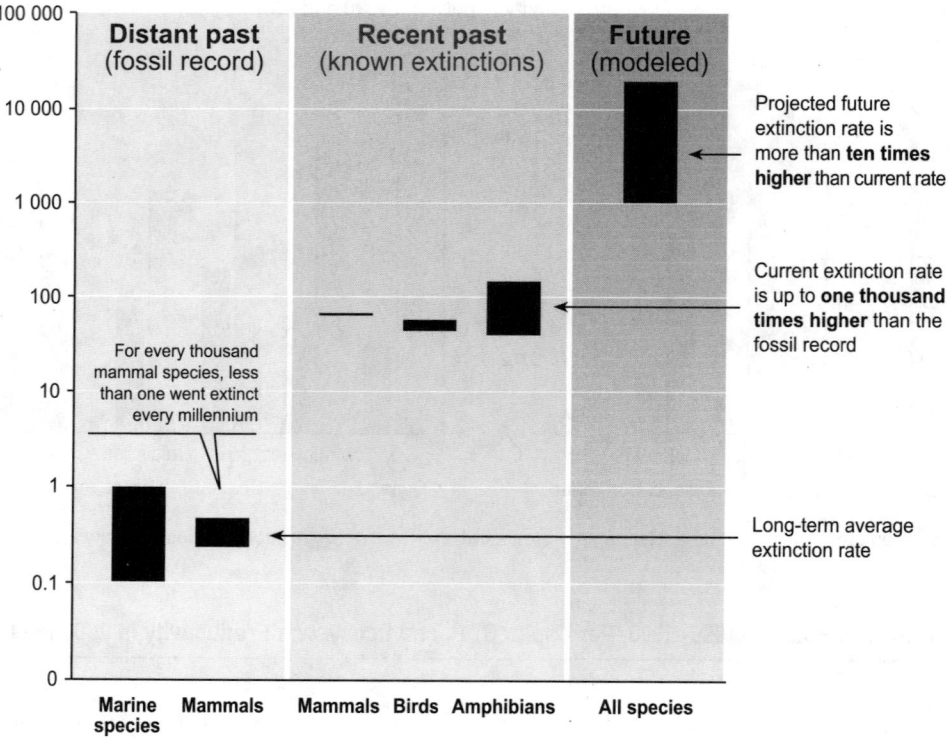

Figure C2. Species Extinction Rates Determined from the Fossil Record, from Observation, and from Estimation of Projected Rates

Less than a tenth of known plant and vertebrate species have been assessed in terms of their vulnerability to extinction ("conservation status"). Birds have the lowest proportion (12%) threatened with global extinction (defined as a high certainty of loss from throughout its range) in the near-to-medium term (*high certainty*). Among mammal species, 23% are threatened with extinction (*high certainty*). Of the amphibia for which sufficient information is available to make an assessment, 32% are threatened (*medium certainty*). For cycads (an ancient group of plants), 52% of the species are threatened, as are 25% of conifer species (*high certainty*). [4]

The taxonomic groups with the highest proportion of threatened species tend to be those that rely on freshwater habitats. Extinction rates, based on the frequency of threatened species, are broadly similar across terrestrial biomes (broad ecosystem types). Most terrestrial extinctions during the coming century are predicted to occur in tropical forests, because of their high species richness. [4]

Factors Causing Changes in Ecosystems

Increasing Demand for Ecosystem Services

Increasing consumption per person, multiplied by a growing human population, are the root causes of the increasing demand for ecosystem services. The global human population continues to rise, but at a progressively slower rate. The population increased from 3 billion to 6 billion between 1960 and 2000 and is likely to peak at 8.2–9.7 billion around the middle of the twenty-first century. Migration to cities and population growth within cities continue to be major demographic trends. The world's urban population increased from about 200 million to 2.9 billion over the past century, and the

number of cities with populations in excess of 1 million increased from 17 to 388. (See Figure C3.) [3]

Overall demand for food, fiber, and water continue to rise. Improvements in human well-being, enabled by economic growth, almost invariably lead to an increase in the per capita demand for provisioning ecosystem services such as food, fiber, and water and in the consumption of energy and minerals and the production of waste. **In general, the increase in demand for provisioning services is satisfied at the expense of supporting, regulating, and cultural ecosystem services.** Efficiency gains permitted by new technology reduce per capita consumption levels below what they would have been without technological and behavioral adaptation, but they have tended not to keep pace with growth in demand for provisioning services. [3]

Increasing Pollution and Waste

Ecosystem problems associated with contaminants and wastes are in general growing. Some wastes are produced in nearly direct proportion to population size (such as sewage). Others, such as domestic trash and home-use chemicals, reflect the affluence of society. Where there is significant economic development, waste loadings tend to increase faster than population growth. In some cases the per capita waste production subsequently decreases, but seldom to the pre-growth level. The generation of industrial wastes does not necessarily increase with population or development state, and it may often be reduced by adopting alternate manufacturing processes. The neglect of waste management leads to impairment of human health and well-being, economic losses, aesthetic value losses, and damages to biodiversity and ecosystem function. [3, 15]

The oversupply of nutrients (eutrophication) is an increasingly widespread cause of undesirable ecosystem

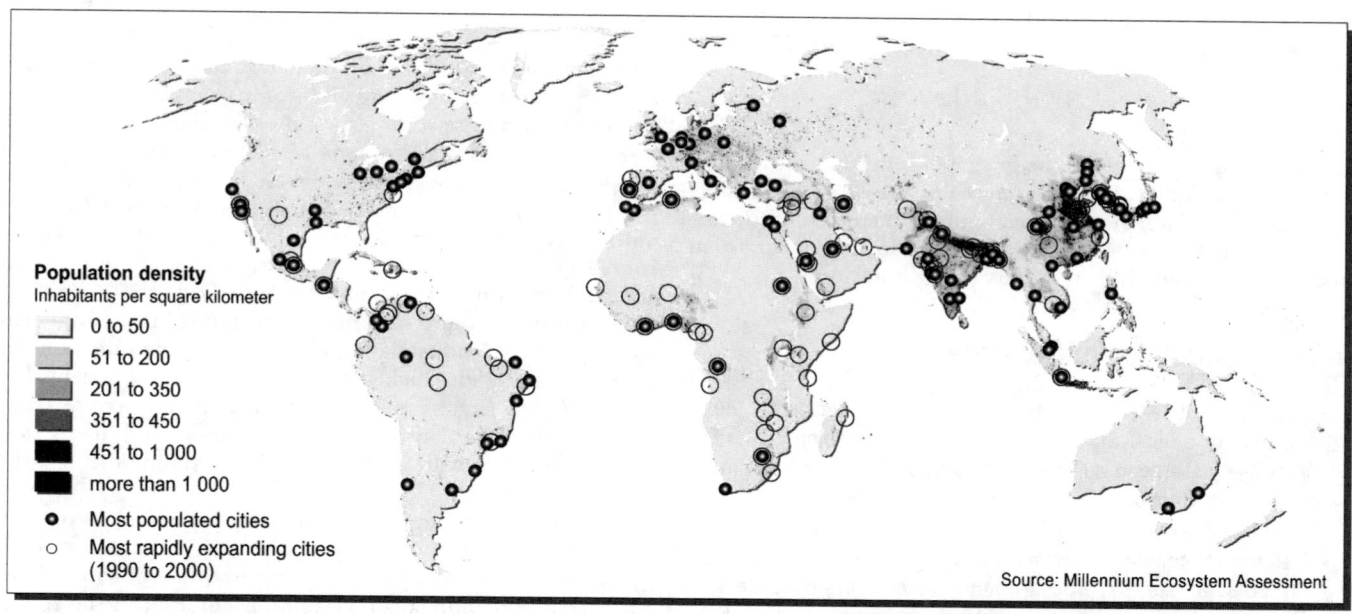

Figure C3. Human Population Density in 1995 and the Most Populated and Rapidly Changing Cities in 1990-2000

change, particularly in rivers, lakes, and coastal systems. Nutrient additions on the land, including synthetic fertilizers, animal manures, the enhancement of N-fixation by planted legumes, and the deposition of airborne pollutants, have resulted in approximately a doubling of the natural inputs for reactive nitrogen in terrestrial ecosystems and an almost fivefold increase in phosphorus accumulation. The reduction of biodiversity at the species and landscape levels has permitted nutrients to leak from the soil into rivers, the oceans, and the atmosphere. Emissions to the atmosphere are a significant driver of regional air pollution and the buildup of the greenhouse gas nitrous oxide (and, to a small extent, methane). [3, 12, 19, 20]

Global Trade

The increasing volume of goods and services that are traded internationally, the distance that they are moved, the mobility of people, and the connectivity of local and global economies have all increased the spatial separation between cause and effect in ecosystem change. **Without appropriate regulation, global trade can be a key driver of overharvesting of resources such as high-value timber and marine resources.** Trade pressures and opportunities underlie patterns of land use change in many parts of the world. The movement of people and goods is an important vector in the spread of diseases and non-native invasive organisms. [3]

Changing Climate

The effects of climate change on ecosystems are becoming apparent, especially in polar regions, where on average temperatures are now warmer than at any time in the last 400 years and the Antarctic peninsular is one of the most rapidly warming regions on the planet; in mountains, where there has been widespread glacier retreat and loss of snowpack; and in coastal systems, where coral reefs in particular have been affected by sea temperature warming and increased carbon dioxide concentrations. Although many of the potential effects of climate change on ecosystem service provision to date have not been clearly distinguishable from short-term variations, **climate change over the next century**

is projected to affect, directly and indirectly, all aspects of ecosystem service provision. [3, 13, 19, 24, 25]

Overexploitation of Natural Resources

If a renewable natural resource is used at a faster rate than it is replenished, the result is a decline in the stock and eventually a decrease in the quantity of the resource that is available for human use. Overfishing, overgrazing, and overlogging are widespread examples of overexploitation. In the process of fishing, logging, mining, cultivation, and many other human activities, unintended collateral damage is done to ecosystems, affecting the supply of both the target resource and other services as well. When the net supply of ecosystem services is so damaged that it fails to recover spontaneously within a reasonable period after the level of the action causing the damage is reduced, the ecosystem is degraded. **Significant areas of forest, cultivated land, dryland rangelands, and coastal and marine systems are now degraded, and the degraded area continues to expand.** [4]

Changing Land Use and Land Cover

Current rates of land cover change are greatest for tropical moist forests and for temperate, tropical, and flooded grasslands, with >14% of each of these lost between 1950 and 1990. Temperate broadleaf forests, Mediterranean forests, and grasslands had already lost more than 70% of their original extent by 1950. The rates of loss in these forest types have now slowed, and in some cases the forest area has expanded. Deforestation and forest degradation are currently focused in the tropics. Data on changes in boreal forests are especially limited. [4, 21]

Habitat loss is the fastest-growing threat to species and populations on land and will continue to be the dominant factor for the next few decades. Fishing is the dominant factor reducing populations and fragmenting the habitats of marine species and is predicted to lead to local extinctions, especially among large, long-lived, slow-growing species and endemic species. [4]

Habitat fragmentation (the reduction of natural cover into smaller and more disconnected patches) compounds the effects of habitat loss. The disruptive effects of fragmentation extend hundreds of meters inwards from the edges of the patches, making small patches highly vulnerable to loss of species and functions. [11]

Invasion by Alien Species

In a wide range of terrestrial, marine, and freshwater ecosystems, accidental or voluntary introduction of non-native species by humans has altered local biological community interactions, triggering dramatic and often unexpected changes in ecosystem processes and causing large monetary and cultural losses. [3, 4, 23]

Trends in Ecosystem Services

- The supply of certain ecosystem services has increased at the expense of others. Significant gains in the provision of food and fiber have been achieved through habitat conversion, increased abstraction and degradation of inland waters, and reduced biodiversity.
- Fish cannot continue to be harvested from wild populations at the present rate. Deep-ocean and coastal fish stocks have changed substantially in most parts of the world and the harvests have begun to decline and will continue to do so.
- The supply of fresh water to people is already inadequate to meet human and ecosystem needs in large areas of the world, and the gap between supply and demand will continue to widen if current patterns of water use will continue.
- Declining trends in the capacity of ecosystems to render pollutants harmless, keep nutrient levels in balance, give protection from natural disasters, and control the outbreaks of pests, diseases, and invasive organisms are apparent in many places.

The main trends in key ecosystem services over the last 50 years are summarized in Table C1. Individual ecosystem services are discussed below in further detail.

Provisioning Services

Food

Major inequalities exist in access to food despite the more than doubling of global production over the past 40 years. An estimated 852 million people were undernourished in 2000–02, up 37 million from 1997–99. [8] There are important differences in the regional trends: the number of undernourished people in China is declining, while the number in Africa continues to increase. Of the global undernourished, 1% live in industrial countries, 4% live in countries in transition, and the remaining 95% are found in developing countries.

Figure C4 demonstrates that the economic value of food production is also not evenly distributed around the world, both because of the uneven distribution of natural factors such as climate and nutrient supply and because the prices obtained for food products vary according to demand and wealth. The impacts of activities associated with food production on other ecosystem services are unevenly distributed as well.

New cultivars of wheat, maize, and rice, coupled with increased inputs of fertilizers, irrigation, and an expansion of the cultivated area, were the main factors underlying the 250% increase in total cereal production since 1960. **The rate of increase of cereal production has slowed over the last decade,** for reasons that are uncertain but that include a long-term decline in the real price of cereals, a saturation in the per capita cereal consumption in many countries, a temporary decline in the use of cereals as livestock feed in the 1970s and 1980s, the declining quality of land in agricultural production, and diminishing returns to efforts aimed at improving yields of maize, wheat, and rice.

Adequate nutrition requires a diverse diet, containing sufficient micronutrients and protein as well as calories. The world's poorest people continue to rely on starchy staples, which leads to protein, vitamin, and mineral deficiencies. **Demand for high-value, protein-rich products such as livestock and fish has increased** with rising incomes in East and Southeast Asia (7% annual growth in livestock production over past 30 years). **The accelerating demand for animal protein is increasingly met by intensive ("industrial" or "landless") production systems,** especially for chicken and pigs. While these systems have contributed to large increases in production, they create serious waste problems and put increased pressure on cultivated systems and fisheries to provide feed inputs (and are thus not truly "landless").

The dietary changes that accompany increasing income can improve health; however, overconsumption, leading to obesity and heart disease, is also a growing health problem (65% of Americans and more than 17 million children in developing countries are overweight). Calorie intake is only 20% higher per capita in industrial countries than in developing countries on average, but protein intake is 50% higher and fat intake is almost twice as high.

Harvest pressure has exceeded maximum sustainable levels of exploitation in one quarter of all wild fisheries and is likely to exceed this limit in most other wild fisheries in the near future. In every ocean in the world, one or more important targeted stocks have been classified as collapsed, overfished, or fished to their maximum sustainable levels, and at least one quarter of important commercial fish stocks are overharvested (*high certainty*). Although fish consumption has doubled in developing countries in the last three decades, the per capita annual consumption has declined by 200 grams since 1985, to 9.2 kilograms per person (excluding China). Fish products are heavily traded, and approximately 50% of fish exports are from developing countries. Exports from developing countries and the Southern Hemisphere presently offset much of the demand shortfall in European, North American, and East Asian markets.

The growth in demand for fish protein is being met in part by aquaculture, which now accounts for 22% of total fish production and 40% of fish consumed as food. **Marine aquaculture has not to date relieved pressure on wild fisheries, because the food provided to captive fish is partly based on wild-harvested fish products.**

Government policies are significant drivers of food production and consumption patterns, both locally and globally. Investments in rural roads, irrigation, credit systems, and agricultural research and extension serve to stimulate food production. Improved access to input and export markets boosts productivity. Opportunities to gain access to international markets are conditioned by international trade and food safety regulations and by a variety of tariff and non-tariff barriers. Selective production and export subsidies stimulate overproduction of many food crops. This translates into relatively cheap food exports that benefit international consumers at the expense of domestic taxpayers and that undermine the welfare of food producers in poorer countries.

Wild terrestrial foods are locally important in many developing countries, often bridging the hunger gap created by stresses such as droughts and floods and social unrest. Wild foods are important sources of diversity in some diets, in

Table C1. Trends in the Human Use of Ecosystem Services and Enhancement or Degradation of the Service around the Year 2000

Service	Sub-category	Human Use[a]	Enhanced or Degraded[b]	Notes	MA Chapter
Provisioning Services					
Food	Crops	↑	↑	Food provision has grown faster than overall population growth. Primary source of growth from increase in production per unit area but also significant expansion in cropland. Still persistent areas of low productivity and more rapid area expansion, e.g., sub-Saharan Africa and parts of Latin America.	C8.2
	Livestock	↑	↑	Significant increase in area devoted to livestock in some regions, but major source of growth has been more intensive, confined production of chicken, pigs, and cattle.	C8.2
	Capture Fisheries	↓	↓	Marine fish harvest increased until the late 1980s and has been declining since then. Currently, one quarter of marine fish stocks are overexploited or significantly depleted. Freshwater capture fisheries have also declined. Human use of capture fisheries has declined because of the reduced supply, not because of reduced demand.	C18 C8.2.2 C19
	Aquaculture	↑	↑	Aquaculture has become a globally significant source of food in the last 50 years and, in 2000, contributed 27% of total fish production. Use of fish feed for carnivorous aquaculture species places an additional burden on capture fisheries.	C8 Table 8.4
	Wild plant and animal food products	NA	↓	Provision of these food sources is generally declining as natural habitats worldwide are under increasing pressure and as wild populations are exploited for food, particularly by the poor, at unsustainable levels.	C8.3.1
Fiber	Timber	↑	+/−	Global timber production has increased by 60% in the last four decades. Plantations provide an increasing volume of harvested roundwood, amounting to 35% of the global harvest in 2000. Roughly 40% of forest area has been lost during the industrial era, and forests continue to be lost in many regions (thus the service is degraded in those regions), although forest is now recovering in some temperate countries and thus this service has been enhanced (from this lower baseline) in these regions in recent decades.	C9.ES C21.1
	Cotton, hemp, silk	+/−	+/−	Cotton and silk production have doubled and tripled respectively in the last four decades. Production of other agricultural fibers has declined.	C9.ES
	Wood fuel	+/−	↓	Global consumption of fuelwood appears to have peaked in the 1990s and is now believed to be slowly declining buts remains the dominant source of domestic fuel in some regions.	C9.ES
Genetic resources		↑	↓	Traditional crop breeding has relied on a relatively narrow range of germplasm for the major crop species, although molecular genetics and biotechnology provide new tools to quantify and expand genetic diversity in these crops. Use of genetic resources also is growing in connection with new industries base on biotechnology. Genetic resources have been lost through the loss of traditional cultivars of crop species (due in part to the adoption of modern farming practices and varieties) and through species extinctions.	C26.2.1

(continues)

Table C1. continued

Service	Sub-category	Human Use[a]	Enhanced or Degraded[b]	Notes	MA Chapter
Biochemicals, natural medicines, and pharmaceuticals		↑	↓	Demand for biochemicals and new pharmaceuticals is growing, but new synthetic technologies compete with natural products to meet the demand. For many other natural products (cosmetics, personal care, bioremediation, biomonitoring, ecological restoration), use is growing. Species extinction and overharvesting of medicinal plants is diminishing the availability of these resources.	C10
Fresh water		↑	↓	Human modification to ecosystems (e.g., reservoir creation) has stabilized a substantial fraction of continental river flow, making more fresh water available to people but in dry regions reducing river flows through open water evaporation and support to irrigation that also loses substantial quantities of water. Watershed management and vegetation changes have also had an impact on seasonal river flows. From 5% to possible 25% of global freshwater use exceeds long-term accessible supplied and require supplied either through engineered water transfers of overdraft of groundwater supplies. Between 15% and 35% of irrigation withdrawals exceed supply rates. Fresh water flowing in rivers also provides a service in the form of energy that is exploited through hydropower. The construction of dams has not changed the amount of energy, but it has made the energy more available to people. The installed hydroelectric capacity doubled between 1960 and 2000. Pollution and biodiversity loss are defining features of modern inland water systems in many populated parts of the world.	C7
Regulating Services					
Air quality regulation		↑	↓	The ability of the atmosphere to cleanse itself of pollutants has declined slightly since preindustrial times but likely not by more than 10%. Then net contribution of ecosystems to this change is not known. Ecosystems are also a sink for tropospheric ozone, ammonia, NO_x, SO_2, particulates, and CH_4, but changes in these sinks were not assessed.	C13.ES
Climate regulation	Global	↑	↑	Terrestrial ecosystems were on average a net source of CO_2 during the nineteenth and early twentieth century and became a net sink sometime around the middle of the last century. The biophysical effect of historical land cover changes (1750 to present) is net cooling on a global scale due to increased albedo, partially offsetting the warming effect of associated carbon emissions from land cover change over much of that period.	C13.ES
	Regional and local	↑	↓	Changes in land cover have affected regional and local climates both positively and negatively, but there is a preponderance of negative impacts. For example, tropical deforestation and desertification have tended to reduce local rainfall.	C13.3 C11.3
Water regulation		↑	+/−	The effect of ecosystem change on the timing and magnitude of runoff, flooding, and aquifer recharge depends on the ecosystem involved and on the specific modifications made to the ecosystem.	C7.4.4

Erosion regulation			↑	↓	Land use and crop/soil management practices have exacerbated soil degradation and erosion, although appropriate soil conservation practices that reduce erosion, such as minimum tillage, are increasingly being adopted by farmers in North America and Latin America.	C26
Water purification and waste treatment			↑	↓	Globally, water quality is declining, although in most industrial countries pathogen and organic pollution of surface waters has decreased over the last 20 years. Nitrate concentration has grown rapidly in the last 30 years. The capacity of ecosystems to purify such wastes in limited, as evidenced by widespread reports of inland waterway pollution. Loss of wetlands has further decreased the ability of ecosystems to filter and decompose wastes.	C7.2.5 C19
Disease regulation			↑	+/−	Ecosystem modifications associated with development have often increased the local incidence of infectious diseases, although major changes in habitats can both increase or decrease the risk of particular infectious diseases.	C14
Pest regulation			↑	↓	In many agricultural areas, pest control provided by natural enemies has been replaced by the use of pesticides. Such pesticide use has itself degraded the capacity of agroecosystems to provide pest control. In other systems, pest control provided by natural enemies is being used and enhanced through integrated pest management. Crops containing pest-resistant genes can also reduce the need for application of toxic synthetic pesticides.	C11.3
Pollination			↑	↓[c]	There is *established but incomplete* evidence of a global decline in the abundance of pollinators. Pollinator declines have been reported in at least one region or country on every continent except for Antarctica, which has no pollinators. Declines in abundance of pollinators have rarely resulted in complete failure to produce seed or fruit, but more frequently resulted in fewer seeds or in fruit of reduced viability or quantity. Losses in populations of specialized pollinators have directly affected the reproductive ability of some rare plants.	C11 Box 11.2
Natural hazard regulation			↑	↓	People are increasingly occupying regions and localities that are exposed to extreme events, thereby exacerbating human vulnerability to natural hazards. This trend, along with the decline in the capacity of ecosystems to buffer from extreme events, has led to continuing high loss of life globally and rapidly rising economic losses from natural disasters.	C16 C19
Cultural Services						
Cultural diversity			NA	NA		
Spiritual and religious values			↑	↑	There has been a decline in the numbers of sacred groves and other such protected areas. The loss of particular ecosystem attributes (sacred species or sacred forests), combined with social and economic changes, can sometimes weaken the spiritual benefits people obtain from ecosystems. On the other hand, under some circumstances (e.g., where ecosystem attributes are causing significant threats to people), the loss of some attributes may enhance spiritual appreciation for what remains.	C17.2.3

(continues)

Table C1. continued

Service	Sub-category	Human Use[a]	Enhanced or Degraded[b]	Notes	MA Chapter
Knowledge systems		NA	NA		
Educational values		NA	NA		
Inspiration		NA	NA		
Aesthetic values		↑	↓	The demand for aesthetically pleasing natural landscapes has increased in accordance with increased urbanization. There has been a decline in quantity and quality of areas to meet this demand. A reduction in the availability of and access to natural areas for urban residents may have important detrimental effects on public health and economies.	C17.2.5
Social relations		NA	NA		
Sense of place		NA	NA		
Cultural heritage values		NA	NA		
Recreation and ecotourism		↑	+/−	The demand for recreational use of landscapes is increasing, and areas are increasingly being managed to cater for this use, to reflect changing cultural values and perceptions. However, many naturally occurring features of the landscape (e.g., coral reefs) have been degraded as resources for recreation.	C17.2.6 C19
Supporting Services					
Soil formation		†	†		
Photosynthesis		†	†		
Primary production		†	†	Several global MA systems, including drylands, forest, and cultivated systems, show a trend of NPP increase for the period 1981 to 2000. However, high seasonal and inter-annual variations associated with climate variability occur within this trend on the global scale	C22.2.1
Nutrient cycling		†	†	There have been large-scale changes in nutrient cycles in recent decades, mainly due to additional inputs from fertilizers, livestock waste, human wastes, and biomass burning. Inland water and coastal systems have been increasingly affected by eutrophication due to transfer of nutrients from terrestrial to aquatic systems as biological buffers that limit these transfers have been significantly impaired.	C12
Water cycling		†	†	Humans have made major changes to water cycles through structural changes to rivers, extraction of water from rivers, and, more recently, climate change.	C7

[a] For provisioning services, human use increases if the human consumption of the service increases (e.g., greater food consumption); for regulating and cultural services, human use increases if the number of people affected by the service increases. The time frame is in general the past 50 years, although if the trend has changed within that time frame, the indicator shows the most recent trend.

[b] For provisioning services, we define enhancement to mean increased production of the service through changes in area over which the service is provided (e.g., spread of agriculture) or increased production per unit area. We judge the production to be degraded if the current use exceeds sustainable levels. For regulating and supporting services, enhancement refers to a change in the service that leads to greater benefits for people (e.g., the service of disease regulation could be improved by eradication of a vector known to transmit a disease to people). Degradation of a regulating and supporting service means a reduction in the benefits obtained from the service, either through a change in the service (e.g., mangrove loss reducing the storm protection benefits of an ecosystem) or through human pressures on the service exceeding its limits (e.g., excessive pollution exceeding the capability of ecosystems to maintain water quality). For cultural services, degradation refers to a change in the ecosystem features that decreases the cultural (recreational, aesthetic, spiritual, etc.) benefits provided by the ecosystem. The time frame is in general the past 50 years, although if the trend has changed within that time frame the indicator shows the most recent trend.

[c] *Low to medium certainty.* All other trends are *medium to high certainty.*

Legend

↑ = Increasing (for human use column) or enhanced (for enhanced or degraded column)

↓ = Decreasing (for human use column) or degraded (for enhanced or degraded column)

+/− = Mixed (trend increases and decreases over past 50 years or some components/regions increase while others decrease)

NA = Not assessed within the MA. In some cases, the service was not addressed at all in the MA (such as ornamental resources), while in other cases the service was included but the information and data available did not allow an assessment of the pattern of human use of the service or the status of the service.

† = The categories of "human use" and "enhanced or degraded" do not apply for supporting services since, by definition, these services are not directly used by people. (Their costs or benefits would be double-counted if the indirect effects were included). Changes in supporting services influence the supply of provisioning, cultural, or regulating services that are then used by people and may be enhanced or degraded.

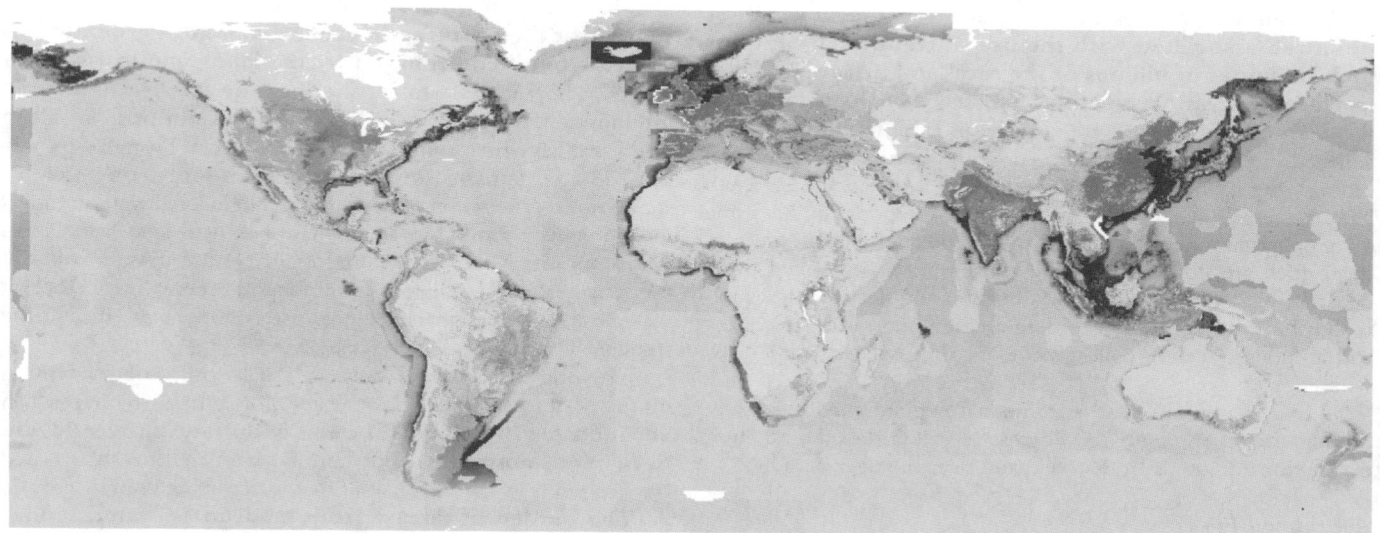

Terrestrial Production
(dollars/sq. km./yr)
300,000

0

Marine Production
(dollars/sq. km./yr)
200,000

0

Figure C4. Spatial Distribution of Value of Food Production for Crops, Livestock, and Fisheries, 2000. This Figure was constructed by multiplying the harvest derived from all regions of the world by the average product price obtained in that region. (Data for Iceland were only available aggregated to the rectangular area shown.) A color version of this map appears in Appendix A (see Figure 8.2).

that they are highly nutritious and are often not labor-intensive to collect or prepare. Although they have significant economic value, in most cases wild foods are excluded from economic analysis of natural resource systems as well as official statistics, so the full extent of their importance is poorly quantified.

Wood for Timber and Pulp

The absolute harvest of timber is projected, with *medium certainty*, to increase in the future, albeit at a slower rate than over the past four decades. [9] The high growth in timber harvests since 1960 (60% and 300% for sawlogs and pulpwood respectively) has slowed in recent years. Total forest biomass in temperate and boreal regions increased over this period but decreased in mid-latitude and tropical forests. Demand for hardwoods is a factor in tropical deforestation, but is typically not the main driver. Conversion to agricultural land, a trend often underlain by policy decisions, is overall the major cause. **A third of timber is harvested from plantations rather than naturally regenerating forests, and this fraction is projected to grow.** Plantations currently constitute 5% of the global forest area. In general, plantations provide a less diverse set of ecosystem services than natural forests do.

Most trade in forest products is within-country, with only about 25% of global timber production entering international trade. However, international trade in forest products has increased three times faster in value than in harvested volume. The global value of timber harvested in 2000 was around $400 billion, about one quarter of which entered in world trade, representing some 3% of total merchandise traded. Much of this trade is among industrial countries: the United States, Germany, Japan, the United Kingdom, and Italy were the destination of more than half of the imports in 2000, while Canada, the United States, Sweden, Finland, and Germany account for more than half of the exports.

The global forestry sector annually provides subsistence and wage employment of 60 million work years, with 80% in the developing world. There is a trend in increasing employment in sub-tropical and tropical regions and declining employment in temperate and boreal regions.

Biomass Energy

Wood and charcoal remain the primary source of energy for heating and cooking for 2.6 billion people. [9] Global consumption appears to have peaked in the 1990s and is now believed to be slowly declining as a result of switching to alternate fuels and, to a lesser degree, more-efficient biomass energy technologies. Accurate data on fuelwood production and consumption are difficult to collect, since much is produced and consumed locally by households. The global aggregate value of fuelwood production per capita has declined in recent years, easing concerns about a widespread wood energy crisis, although local and regional shortages persist.

Serious human health damages are caused by indoor pollution associated with the use of traditional biomass fuels in homes of billions of the rural and urban poor that lack adequate smoke venting. In 2000, 1.6 million deaths and the equivalent of 39 million person-years of ill health (disability-adjusted life years) were attributed to the burning of traditional biomass fuels, with women and children most affected. Health hazards increase where wood shortages lead to poor families using dung or agricultural residues for heating and cooking. Where adequate fuels are not available, the consumption of cooked foods declines, with adverse effects on nutrition and health. Local fuelwood shortages contribute to deforestation and result in lengthy and arduous travel to collect wood in rural villages, largely by women.

While examples of full commercial exploitation of modern biomass-based energy technologies are still fairly modest, their production and use is likely to expand over the next decades.

Agricultural Fibers

Global cotton production has doubled and silk production has tripled since 1961, with major shifts in the production regions. [9] The total land area devoted to cotton production has stayed virtually constant; area expansion in India and the United States was offset by large declines in Pakistan and the former Soviet Union. These shifts have impacts on land available for food crops and on water resources, since much of the cotton crop is irrigated. Silk production shifted from Japan to China. Production of wool, flax, hemp, jute, and sisal has declined.

Fresh Water

Water scarcity has become globally significant over the last four decades and is an accelerating condition for roughly 1–2 billion people worldwide, leading to problems with food production, human health, and economic development. Rates of increase in a key water scarcity measure (water use relative to accessible supply) from 1960 to the present averaged nearly 20% per decade globally, with values of 15% to more than 30% per decade for individual continents. Although a slowing in the global rate of increase in use is projected between 2000 and 2010, to 10% per decade, the relative use ratio for some regions is likely to remain high, with the Middle East and North Africa at 14% per decade, Latin America at 16%, and sub-Saharan Africa at 20%. [7]

Contemporary water withdrawal is approximately 10% of global continental runoff, although this amounts to between 40% and 50% of the continental runoff to which the majority of the global population has access during the year.

Population growth and economic development have driven per capita levels of water availability down from 11,300 to about 5,600 cubic meters per person per year between 1960 and 2000. Global per capita water availability is projected (based on a 10% per decade rate of growth of water use, which is slower than the

past decades) to drop below 5,000 cubic meters per person per year by 2010 (*high certainty*).

Terrestrial ecosystems are the major global source of accessible, renewable fresh water. Forest and mountain ecosystems are associated with the largest amounts of fresh water—57% and 28% of the total runoff, respectively. These systems each provide renewable water supplies to at least 4 billion people, or two thirds of the global population. Cultivated and urban systems generate only 16% and 0.2%, respectively, of global runoff, but due to their close proximity to humans they serve from 4.5–5 billion people. Such proximity is associated with nutrient and industrial water pollution.

More than 800 million people currently live in locations so dry that there is no appreciable recharge of groundwater or year-round contribution by the landscape to runoff in rivers. They are able to survive there by drawing on "fossil" groundwater, by having access to piped water, or by living along rivers that have their source of water elsewhere. From 5% to possibly 25% of global freshwater use exceeds long-term accessible supplies and is now met either through engineered water transfers or overdraft of groundwater supplies (*medium certainty*). In North Africa and the Middle East, nonsustainable use (use in excess of the long-term accessible renewable supply) represents 43% of all water use, and the current rate of use is 40% above that of the sustainable supply (*medium certainty*).

Growing competition for water is sharpening policy attention on the need to allocate and use water more efficiently. **Irrigation accounts for 70% of global water withdrawals (over 90% in developing countries),** but chronic inefficiencies in irrigated systems result in less than half of that water being used by crops.

The burden of disease from inadequate water, sanitation, and hygiene totals 1.7 million deaths and the loss of up to 54 million healthy life years per year. Some 1.1 billion people lack access to improved water supply and more than 2.6 billion lack access to improved sanitation. It is *well established* that investments in clean drinking water and sanitation show a close correspondence with improvement in human health and economic productivity. Half of the urban population in Africa, Asia, and Latin America and the Caribbean suffer from one or more diseases associated with inadequate water and sanitation.

The management of fresh water through dams, levees, canals, and other infrastructure has had predominantly negative impacts on the biodiversity of inland waters and coastal ecosystems, including fragmentation and destruction of habitat, loss of species, and reduction of sediments destined for the coastal zone. The 45,000 existing large dams (more than 15 meters high) generate both positive and negative effects on human well-being. Positive effects include flow stabilization for irrigation, flood control, and hydroelectricity. Negative effects include health issues associated with stagnant water and the loss of services derived from land that has become inundated. A significant economic consequence of soil erosion is the reduction of the useful life of dams lower in the drainage basin due to siltation.

Genetic Resources

The exploration of biodiversity for new products and industries has yielded major benefits for humanity and has the potential for even larger future benefits. [10] The diversity of living things, at the level of the gene, is the fundamental resource for such "bioprospecting." While species-rich environments such as the tropics are in the long term expected to supply the majority of pharmaceutical products derived from ecosystems, bioprospecting to date has yielded valuable products from a wide variety of environments, including temperate forests and grass-lands, arid and semiarid lands, freshwater ecosystems, mountain and polar regions, and cold and warm oceans.

The continued improvements of agricultural yields through plant breeding and the adaptation of crops to new and changing environments, such as increased temperatures, droughts, and emerging pests and diseases, requires the conservation of genetic diversity in the wild relatives of domestic species and in productive agricultural landscapes themselves.

Regulating Services

The Regulation of Infectious Diseases

Ecosystem changes have played a significant role in the emergence or resurgence of several infectious diseases of humans. [14] The most important drivers are logging, dam building, road building, expansion of agriculture (especially irrigated agriculture), urban sprawl, and pollution of coastal zones. There is evidence that ecosystems that maintain a higher diversity of species reduce the risks of infectious diseases in humans living within them; the pattern of Lyme disease in North America is one example. **Natural systems with preserved structure and characteristics are not receptive to the introduction of invasive human and animal pathogens brought by human migration and settlement.** This is indicated for cholera, kala-azar, and schistosomiasis (*medium certainty*).

Increased human contact with ecosystems containing foci of infections raises the risk of human infections. Examples occur where urban systems are in close contact with forest systems (associated with malaria and yellow fever) and where cultivated lands are opened in forest systems (hemorrhagic fevers or hantavirus). Major changes in habitats can both increase or decrease the risk of a particular infectious disease, depending on the type of land use, the characteristics of the cycle of disease, and the characteristics of the human populations. Although disease emergence and re-emergence due to ecosystem alteration can occur anywhere, people in the tropics are more likely to be affected in the future due to their greater exposure to reservoirs of potential disease and their greater vulnerability due to poverty and poorer health infrastructure.

Regulation of Climate, Atmospheric Composition, and Air Quality

Ecosystems are both strongly affected by and exert a strong influence on climate and air quality. [13] **Ecosystem management has significantly modified current greenhouse gas concentrations.** Changes in land use and land cover, especially deforestation and agricultural practices such as paddy rice cultivation and fertilizer use, but also rangeland degradation and dryland agriculture, made a contribution of 15–25% to the radiative forcing of global climate change from 1750 to present.

Ecosystems are currently a net sink for CO_2 and tropospheric ozone, while they remain a net source of methane and nitrous oxide. About 20% of CO_2 emissions in the 1990s originated from changes in land use and land management, primarily deforestation.

Terrestrial ecosystems were on average a net source of CO_2 during the nineteenth and early twentieth centuries; they became a net sink sometime around the middle of the last century (*high certainty*) and were a sink for about a third of total emissions in the 1990s (energy plus land use). The sink may be explained partially by afforestation, reforestation, and forest management in North America, Europe, China, and other regions and partially by the fertilizing effects of nitrogen deposition and increasing atmo-

spheric CO_2. The net impact of ocean biology changes on global CO_2 fluxes is unknown.

The potential of terrestrial ecosystem management to alter future greenhouse gas concentrations is significant through, for instance, afforestation, reduced deforestation, and conservation agriculture. However, the potential reductions in greenhouse gases remain much smaller than the projected fossil fuel emissions over the next century (*high certainty*). The management of ecosystems for climate mitigation can yield other benefits as well, such as biodiversity conservation.

Ecosystems also modify climate through alteration of the physical properties of Earth's surface. For instance, deforestation in snowy regions leads to regional cooling of land surface during the snow season due to increase in surface albedo and to warming during summer due to reduction in transpiration (water recycled by plants to atmosphere). Positive feedbacks involving sea surface temperature and sea ice propagate this cooling to the global scale. The net physically mediated effect of conversion of high-latitude forests to more open landscapes is to cool the atmosphere (*medium certainty*). Observations and models indicate, with *medium certainty*, that **large-scale tropical and sub-tropical deforestation and desertification decrease the precipitation in the affected regions.** The mechanism involves reduction in within-region moisture recycling and an increase in surface albedo. [14]

Tropospheric ozone is both a greenhouse gas and an important pollutant. It is both produced and destroyed by chemical reactions in the atmosphere, and about a third of the additional tropospheric ozone produced as a result of human activities is destroyed by surface absorption in ecosystems. The capacity of the atmosphere to convert pollutants harmful to humans and other life forms into less harmful chemicals is largely controlled by the availability of hydroxyl radicals. The global concentration of these is believed to have declined by about 10% over the past centuries.

Detoxification of Wastes

Depending on the properties of the contaminant and its location in the environment, wastes can be rendered harmless by natural processes at relatively fast or extremely slow rates. The more slowly a contaminant is detoxified, the greater the possibility that harmful levels of the contaminant will occur. Some wastes, such as nutrients and organic matter, are normal components of natural ecosystem processes, but the anthropogenic loading rates are often so much higher than the natural throughput that they significantly modify the ecosystem and impair its ability to provide a range of services, such as recreation and appropriate-quality fresh water and air. **The costs of reversing damages to waste-degraded ecosystems are typically large, and the time scale for remediation is long. In some cases, rehabilitation is effectively impossible.** [15]

Protection from Floods

The impact of extreme weather events is increasing in many regions around the world. [7, 16, 19] For example, flood damage recorded in Europe in 2002 was higher than in any previous year. Increasing human vulnerability, rather than increasing physical magnitude or frequency of the events themselves, is overall the primary factor underlying the rising impact. People are increasingly occupying regions and locations that are exposed to flooding—settling on coasts and floodplains, for instance—thus exacerbating their vulnerability to extreme events. **Ecosystem changes have in some cases increased the severity of floods,** however, for example as a result of deforestation in upland areas and the loss of mangroves. Local case studies have shown that appropriate management of ecosystems contributes to reduction of vulnerability to extreme events.

Cultural Services

Human societies have developed in close interaction with the natural environment, which has shaped their cultural identity, their value systems, and indeed their economic well-being. Human cultures, knowledge systems, religions, heritage values, social interactions, and the linked amenity services (such as aesthetic enjoyment, recreation, artistic and spiritual fulfillment, and intellectual development) have always been influenced and shaped by the nature of the ecosystem and ecosystem conditions in which culture is based. **Rapid loss of culturally valued ecosystems and landscapes has led to social disruptions and societal marginalization in many parts of the world.** [17]

The world is losing languages and cultures. At present, the greatest losses are occurring in situations where languages are not officially recognized or populations are marginalized by rapid industrialization, globalization, low literacy, or considerable ecosystem degradation. Especially threatened are the languages of 350 million indigenous peoples, representing over 5,000 linguistic groups in 70 countries, which contain most of humankind's traditional knowledge. Much of the traditional knowledge that existed in Europe (such as knowledge on medicinal plants) has also gradually eroded due to rapid industrialization in the last century. [17]

The complex relationships that exist between ecological and cultural systems can best be understood through both "formal knowledge" and "traditional knowledge." Traditional knowledge is a key element of sustainable development, particularly in relation to plant medicine and agriculture, and the understanding of tangible benefits derived from traditional ecological knowledge such as medicinal plants and local species of food is relatively well developed. However, understanding of the linkages between ecological processes and social processes and their intangible benefits (such as spiritual and religious values), as well as the influence on sustainable natural resource management at the landscape level, remains relatively weak. [17]

Many cultural and amenity services are not only of direct and indirect importance to human well-being, they also represent a considerable economic resource. (For example, nature- and culture-based tourism employs approximately 60 million people and generates approximately 3% of global GDP.) Due to changing cultural values and perceptions, there is an increasing tendency to manage landscapes for high amenity values (such as recreational use) at the expense of traditional landscapes with high cultural and spiritual values. [17]

Supporting Services

There are numerous examples of both overabundance and insufficiency of nutrient supply. Crop yields and nutritional value in parts of Africa, Latin America, and Asia are strongly limited by poor soils, which have become even more depleted by farming with low levels of nutrient replenishment. On the other hand, overfertilization is a major contributor to environmental pollution through excess nutrients in many areas of commercial farming in both industrial and developing countries.

The capacity of terrestrial ecosystems to absorb and retain the nutrients supplied to them either as fertilizers or from the deposition of airborne nitrogen and sulfur has been undermined by the radical simplification of ecosystems into large-scale, low-diversity agricultural landscapes. Ex-

cess nutrients leak into the groundwater, rivers, and lakes and are transported to the coast. Treated and untreated sewage released from urban areas adds to the load. The consequence of the excessive and imbalanced nutrient load in aquatic ecosystems is an explosion of growth of certain plants (particularly algae) and a loss of many other forms of life, a syndrome known as eutrophication. The decomposing residues of the plants (often compounded by organic pollutants) deplete the water of oxygen, creating anaerobic "dead zones" devoid of life forms that depend on oxygen. Such dead zones have been discovered in many lakes and estuaries and off the mouths of several large rivers, and they are expanding.

How Are Key Ecological Systems Doing?

The systems where multiple problems are occurring at the same time, seriously affecting the well-being of hundreds of millions of people, are:

- wetlands, including rivers, lakes, and salt and saltwater marshes, where water abstraction, habitat loss and fragmentation, and pollution by nutrients, sediments, salts, and toxins have significantly impaired ecosystem function and biodiversity in most major drainage basins;
- the arid parts of the world, where a large, growing, and poor population often coincides with water scarcity, cultivation on marginal lands, overgrazing, and overharvesting of trees;
- particular coastal systems, notably coral reefs, estuaries, mangroves, and urbanized coasts, where habitat loss and fragmentation, overharvesting, pollution, and climate change are the key issues; and
- tropical forests, where unsustainable harvesting and clearing for agriculture threatens biodiversity and the global climate.

The majority of ecosystems have been greatly modified by humans. Within 9 of the 14 broad terrestrial ecosystem types (biomes), one fifth to one half of the area has been transformed to croplands, mostly over the past two centuries. Tropical dry forests are the most affected by cultivation, with almost half of the biome's native habitats replaced with cultivated lands. Temperate grasslands, temperate broadleaf forests, and Mediterranean forests have each experienced more than 35% conversion. Only the biomes unsuited to crop plants (deserts, boreal forests, and tundra) are relatively intact. (See Table C2.) [4]

Freshwater Systems: Wetlands, Rivers, and Lakes

It is *established but incomplete* that **inland water ecosystems are in worse condition overall than any other broad ecosystem type,** and it is *speculated* that about half of all freshwater wetlands have been lost since 1900 (excluding lakes, rivers, and reservoirs). The degradation and loss of inland water habitats and species is driven by water abstraction, infrastructure development (dams, dikes, levees, diversions, and so on), land conversion in the catchment, overharvesting and exploitation, introduction of exotic species, eutrophication and pollution, and global climate change. [20]

Clearing or drainage for agricultural development is the principal cause of wetland loss worldwide. It is estimated that by 1985, 56–65% of available wetland had been drained for intensive agriculture in Europe and North America, 27% in Asia, 6% in South America, and 2% in Africa. **The construction of dams and other structures along rivers has resulted in fragmentation of almost 40% of the large river systems in the world.** This

Table C2. Comparative Table of Systems as Reported by the Millennium Ecosystem Assessment. Note that these are linked human and ecological systems and often are spatially overlapping. They can therefore be compared but they should not be added up. Figure C1 presents data on human well-being by system type graphically.

System and Subsystem	Area[a] (million sq. km.)	Share of Terrestrial Surface of Earth (percent)	Population Density (people per square km.) Urban	Population Density (people per square km.) Rural	Growth rate (percent 1990–2000)	GDP per Capita (dollars)	Infant Mortality Rate[b] (deaths pers 1,000 live births)	Mean NPP (kg. carbon per sq. meter per year)	Share of Systems Covered by PAs[c] (percent)	Share of Area Transformed[d] (percent)
Marine	349.3	68.6[e]	–	–	–	–	–	0.15	0.3	-
Coastal	**17.2**	**4.1**	**1,105**	**70**	**15.9**	**8,960**	**41.5**	**–**	**7**	**–**
Terrestrial	6.0	4.1	1,105	70	15.9	8,960	41.5	0.52	4	11
Marine	11.2	2.2[e]	–	–	–	–	–	0.14	9	–
Inland water[f]	**10.3**	**7.0**	**817**	**26**	**17**	**7,300**	**57.6**	**0.36**	**12**	**11**
Forest/woodlands	**41.9**	**28.4**	**472**	**18**	**13.5**	**9,580**	**57.7**	**0.68**	**10**	**42**
Tropical/sub-tropical	23.3	15.8	565	14	17	6,854	58.3	0.95	11	34
Temperate	6.2	4.2	320	7	4.4	17,109	12.5	0.45	16	67
Boreal	12.4	8.4	114	0.1	−3.7	13,142	16.5	0.29	4	25
Dryland	**59.9**	**40.6**	**750**	**20**	**18.5**	**4,930**	**66.6**	**0.26**	**7**	**18**
Hyperarid	9.6	6.5	1,061	1	26.2	5,930	41.3	0.01	11	1
Arid	15.3	10.4	568	3	28.1	4,680	74.2	0.12	6	5
Semiarid	22.3	15.3	643	10	20.6	5,580	72.4	0.34	6	25
Dry subhumid	12.7	8.6	711	25	13.6	4,270	60.7	0.49	7	35
Island	**7.1**	**4.8**	**1,020**	**37**	**12.3**	**11,570**	**30.4**	**0.54**	**17**	**17**
Island states	4.7	3.2	918	14	12.5	11,148	30.6	0.45	18	21
Mountain	**35.8**	**24.3**	**63**	**3**	**16.3**	**6,470**	**57.9**	**0.42**	**14**	**12**
300–1,000m	13.0	8.8	58	3	127	7,815	48.2	0.47	11	13
1,000–2,500m	11.3	7.7	69	3	20.0	5,080	67.0	0.45	14	13
2,500–4,500m	9.6	6.5	90	2	24.2	4,144	65.0	0.28	18	6
> 4,500m	1.8	1.2	104	0	25.3	3,663	39.4	0.06	22	0.3
Polar	**23.0**	**15.6**	**161**[g]	**0.06**[g]	**−6.5**	**15,401**	**12.8**	**0.06**	**42**[g]	**0.3**[g]
Cultivated	**35.3**	**23.9**	**786**	**70**	**14.1**	**6,810**	**54.3**	**0.52**	**6**	**47**
Pasture	0.1	0.1	419	10	28.8	15,790	32.8	0.64	4	11
Cropland	8.3	5.7	1,014	118	15.6	4,430	55.3	0.49	4	62
Mixed (crop and other)	26.9	18.2	575	22	11.8	11,060	46.5	0.6	6	43
Urban	**3.6**	**2.4**	**681**	**–**	**12.7**	**12,057**	**36.5**	**0.47**	**0**	**100**
GLOBAL	**510**	**–**	**681**	**13**	**16.7**	**7,309**	**57.4**	**–**	**4**	**38**

[a] Area estimates based on GLC2000 dataset for the year 2000 except for cultivated systems where area is based on GLCCD v2 dataset for the years 1992–93 (C26 Box 1).

[b] Deaths of children less than one year old per 1,000 live births.

[c] Includes only natural protected areas in IUCN categories I to VI.

[d] For all systems except forest/woodland, area transformed is calculated from land depicted as cultivated or urban areas by GLC2000 land cover dataset. The area transformed for forest/woodland systems is calculated as the percentage change in area between potential vegetation (forest biomes of the WWF ecoregions) and current forest/woodland areas in GLC2000. Note: 22% of the forest/woodland system falls outside forest biomes and is therefore not included in this analysis.

[e] Percent of total surface of Earth.

[f] Population density, growth rate, GDP per capita, and growth rate for the inland water system have been calculated with an area buffer of 10 kilometers.

[g] Excluding Antarctica.

is particularly the case in river systems with parts of their basins in arid and semiarid areas. [20]

The water requirements of aquatic ecosystems are in competition with human water demands. Changes in flow regime, transport of sediments and chemical pollutants, modification of habitat, and disruption of the migration routes of aquatic biota are some of the major consequences of this competition. Through consumptive use and interbasin transfers, **several of the world's largest rivers no longer run all the way to the sea for all or part of the year** (such as the Nile, the Yellow, and the Colorado). [7]

The declining condition of inland waters is putting the services derived from these ecosystems at risk. The increase in pollution to waterways, combined with the degradation of wetlands, has reduced the capacity of inland waters to filter and assimilate waste. Water quality degradation is most severe in areas where water is scarce—arid, semiarid, and dry subhumid regions. Toxic substances and chemicals novel to the ecosystem are reaching waterways in increasing amounts with highly uncertain long-term effects on ecosystems and humans. [20]

Estimates are that between 1.5 billion and 3 billion people depend on groundwater supplies for drinking. Groundwater is the source of water for 40% of industrial use and 20% of irrigation globally. In arid countries this dependency is even greater; for example, Saudi Arabia supplies nearly 100% of its irrigation requirement through groundwater. Overuse and contamination of groundwater aquifers are known to be widespread and growing problems in many parts of the world, although many pollution and contamination problems that affect groundwater supplies have been more difficult to detect and have only recently been discovered. [7]

Inland waters have high aesthetic, artistic, educational, cultural, and spiritual values in virtually all cultures and are a focus of growing demand for recreation and tourism. [20]

Dryland Systems: Deserts, Semiarid, and Dry Subhumid Rangelands

Drylands cover 41% of Earth's land surface and are inhabited by more than 2 billion people, about one third of the human population. **Semiarid drylands are the most vulnerable to loss of ecosystem services** (*medium certainty*), because they have a relatively high population in relation to the productive capacity of the system. [22].

Desertification is the process of degradation in drylands, where degradation is defined as a persistent net loss of capacity to yield provisioning, regulating, and supporting ecosystem services. **Worldwide, about 10–20% of drylands are judged to be degraded** (*medium certainty*). The main causes of dryland degradation are grazing with domestic livestock and cutting of trees at rates exceeding the regrowth capacity of the ecosystem, inappropriate cultivation practices that lead to erosion and salinization of the soil, and climate change, which is affecting rates of evapotranspiration and precipitation.

Where the limits to sustainable cultivation and pastoralism have been reached, the promotion of alternative livelihoods such as production of crafts, tourism-related activities, and even aquaculture (such as aquatic organisms of high market value, cultured in often abundant drylands' low-quality water, within evaporation-proof containers) can take some pressure off dryland ecosystems and their services. [22]

Wetlands in drylands, such as oases, rivers, and marshes, are disproportionately important in terms of the biodiversity that they support and the ecosystem services they provide. [20, 22]

It is *well established* that desertification has adverse impacts in non-dryland areas, often many thousands of kilometers away. For example, dust storms resulting from reduced vegetative cover lead to air quality problems, both locally and far away. Drought and loss of land productivity are dominant factors that cause people to migrate from drylands to better-serviced areas. [22]

Forests, Including Woodlands and Tree Plantations

The global area of naturally regenerating forest has declined throughout human history and has halved over the past three centuries. **Forests have effectively disappeared in 25 countries, and more 90% of the former forest cover has been lost in a further 29 countries.** [21]

Following severe deforestation in past centuries, forest cover and biomass in North America, Europe, and North Asia are currently increasing due to the expansion of forest plantations and regeneration of natural forests. From 1990 to 2000, the global area of temperate forest increased by almost 3 million hectares per year, of which approximately 1.2 million hectares were was planted forest. The main location of deforestation is now in the tropics, where it has occurred at an average rate exceeding 12 million hectares per year over the past two decades. (See Figure C5.) **Taken as a whole, the world's forests are not managed in a sustainable way, and there is a total net decrease in global forest area, estimated at 9.4 million hectares per year.** In absolute terms, the rate and extent of woodland loss exceeds that of forests.

The decline in forest condition is caused, among other factors, by the low political power of human communities in forest areas in many countries; deforestation due to competitive land use and poor management; slow change of traditional, wood-oriented forest management paradigms; the lack of forest management on landscape-ecosystem basis; acceleration of natural and human-induced disturbance regimes during the last decade (possibly linked to climate change); and illegal harvest in many developing countries and countries with economies in transition, often linked to corruption. [21]

In addition to the 3.3 billion cubic meters of wood delivered by forests annually, numerous non-wood forest products are important in the lives of hundreds of millions people. Several studies show that the combined economic value of "nonmarket" (social and ecological) services often exceeds the economic value of direct use of the timber, but the nonmarket values are usually not considered in the determination of forest use. Wooded landscapes are home to about 1.2 billion people, and **350 million of the world's people, mostly the poor, depend substantially for their subsistence and survival on local forests.** Forests and woodlands constitute the natural environment and almost sole source of livelihood for 60 million indigenous people and are important in the cultural, spiritual, and recreational life of communities worldwide. [21]

Terrestrial ecosystems, and wooded lands in particular, are taking up about a fifth of the global anthropogenic emissions of carbon dioxide, and they will continue to play a significant role in limiting global climate change over the first decades of this century. Tree biomass constitutes about of 80% of terrestrial biomass, and **forests and woodlands contain about half of the world's terrestrial organic carbon stocks.** Forests and woodlands provide habitat for half or more of the world's known terrestrial plant and animal species, particularly in the tropics. [21]

Marine and Coastal Systems

All the oceans of the world, no matter how remote, are now affected by human activities. Ecosystem degradation associated with fishing activities is the most widespread and dominant impact, with pollution as an additional factor on coastal shelves, and habitat loss a factor in populated coastal areas. [18, 19]

Global fish landings peaked in the late 1980s and are now declining (*medium certainty*). There is little likelihood of this declining trend reversing under current practices. Fishing pressure is so strong in some marine systems that the biomass of targeted species, especially larger fishes as well as those caught incidentally, has been reduced by 10 times or more relative to levels prior to the onset of industrial fishing. **In addition to declining landings, the average trophic level of global landings is declining** (in other words, the high-value top-predator fish are being replaced in catches by smaller, less preferred species), and the mean size of caught fish is diminishing in many species, including yellowfin and bigeye tuna. [18]

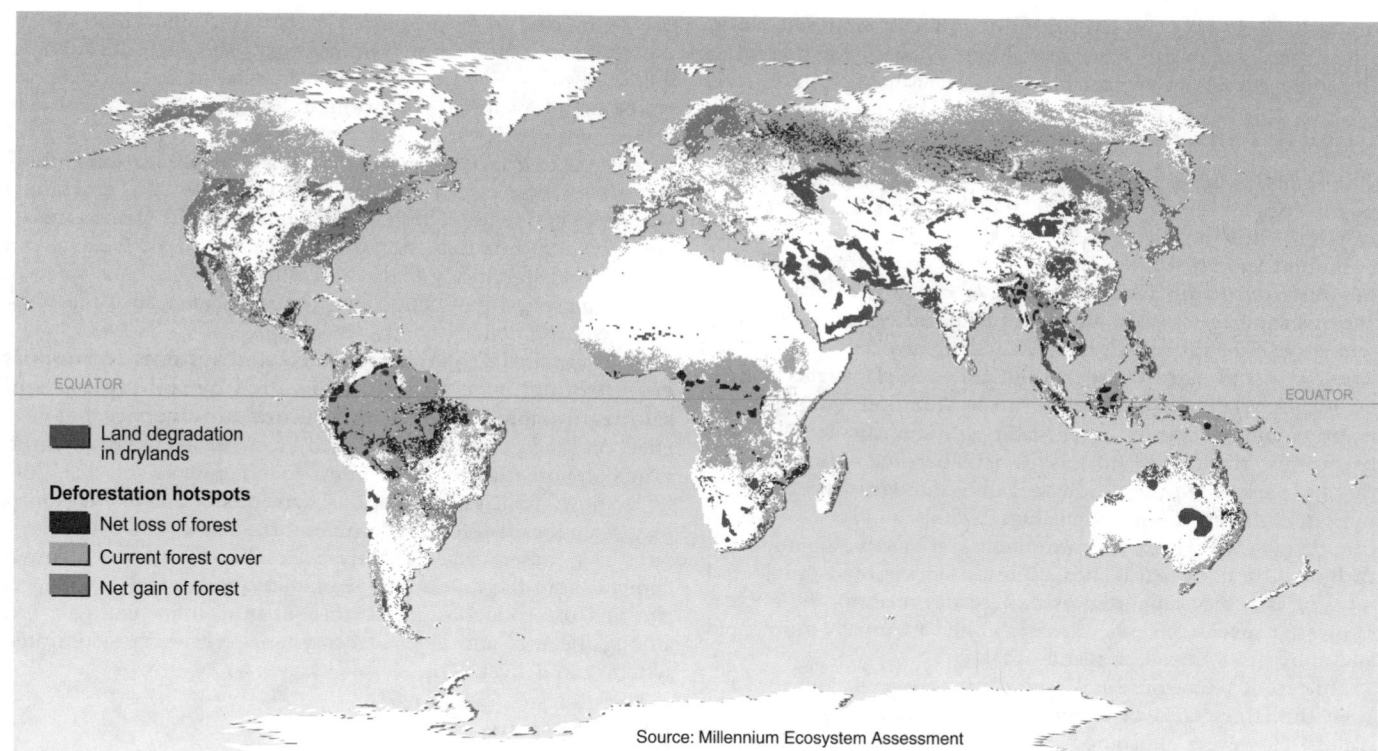

Source: Millennium Ecosystem Assessment

Figure C5. Locations Reported by Various Studies as Undergoing High Rates of Land Cover Change in the Past Few Decades. In the case of forest cover change, the studies refer to the period 1980–2000 and are based on national statistics, remote sensing, and to a limited degree expert opinion. In the case of land cover change resulting from degradation in drylands (desertification), the period is unspecified but inferred to be within the last half-century, and the major study was entirely based on expert opinion, with associated high levels of uncertainty. Change in cultivated area is not shown. Note that areas showing little current change are often locations that have already undergone major change.

Industrial fleets are fishing further offshore and deeper to meet the global demand for fish. Until a few decades ago, depth and distance from coasts protected much of the deep-ocean fauna from the effect of fishing. Massive investments in the development of fishing capacity has led to fleets that now operate in all parts of the world's oceans, including polar areas, at great depths, and in low-productivity tropical zones. These trawl catches are extracted from easily depleted accumulations of long-lived species. The biomass of large pelagic fishes exploited by long liners, purse seiners, and drift netters have also plummeted. **Some fisheries that collapsed in recent decades show no signs of recovering**, such as Newfoundland cod stocks in the northwest Atlantic and orange roughy in New Zealand. [18]

Oil spills, depletions of marine mammals and seabirds, and ocean dumping also contribute to degradation in marine systems, especially at local and regional scales. Although major oils spills are infrequent, their impacts are severe when they do occur. Overfishing and pollution affect marine mammals and seabirds through declining food availability. An estimated 313,000 containers of low-intermediate radioactive waste dumped in the Atlantic and Pacific Oceans since 1970 pose a significant threat to deep-sea ecosystems should the containers leak. [18]

Coastal ecosystems are among the most productive yet highly threatened systems in the world. Approximately 35% of mangroves for which data are available and 20% of coral reefs are estimated to have been destroyed, and a further 20% of corals degraded globally since 1960. Degradation is also a severe problem, both from pressures originating within the coastal zones and from the negative impacts of upstream land uses. Upstream freshwater diversion has meant a 30% decrease worldwide in water and sediment delivery to estuaries, which are key nursery areas and fishing grounds. [19] Knowledge of cold-water corals is limited, and new large reefs are still being discovered. Cold-water coral reefs are estimated to have high species diversity, the biggest threat to which comes from fishery trawling activities.

The main indirect drivers of coastal ecosystem change are related to development activities on the land, particularly in areas adjacent to the coast. Approximately 17% of the world lives within the boundaries of the MA coastal system (up to an elevation of 50 meters above sea level and no further than 100 kilometers from a coast), and approximately 40% live in the full area within 100 kilometers of a coast. The absolute number is increasing through a combination of in-migration, high reproduction rates, and tourism. Physical demand on coastal space is increasing through urban sprawl, resort and port development, and aquaculture, the impacts of which extend beyond the direct footprints due to pollution, sedimentation, and changes in coastal dynamics. Destructive fishing practices, overharvesting, climate change, and associated sea level rise are also important threats to coastal habitats, including forests, wetlands, and coral reefs.

Nearly half of the coastal population has no access to improved sanitation and thus faces increasing risks of disease as well as decreasing ecosystem services as a result of pollution by human wastes. Harmful algal blooms and other pathogens affecting the health of both humans and marine organisms are on the rise. [19] Nitrogen loading to the coastal zone has doubled worldwide and has driven coral reef community shifts. Alien species invasions have also altered coastal ecosystems and threaten both marine species and human well-being. [18]

Island Systems

The ability of island systems to meet the rising demands of local populations for services has declined considerably, such that some islands are now unable to meet such demands without importing significant services from elsewhere. **Biodiversity loss and habitat destruction on islands can have more immediate and serious repercussions than on continental systems,** as a consequence of the relatively restricted genetic diversity, small population sizes and narrow distribution ranges of plants and animals on islands. Many studies show that specialization, coupled with isolation and endemism, make island ecosystems especially sensitive to disturbances. Island species have become extinct at rates that have exceeded those observed on continents, and the most important driver of wild population declines and species extinction on islands has been the introduction of invasive alien species. Although the idea that islands are more susceptible to biological invasion is poorly supported by current information, the impacts of invasive species once they are established are usually more rapid and more pronounced on islands. [23]

In recent years tourism, especially nature-based tourism, has been the largest area of economic diversification for inhabited islands. However, unplanned and unregulated development has resulted in ecosystem degradation, including pollution, and loss of coral reefs, which is undermining the very resource on which the tourism sector is based. Alternative, more environmentally and culturally sensitive forms of tourism ("ecotourism") have developed in some areas. [23]

Cultivated Systems: Croplands, Planted Pastures, and Agroforestry

Cultivated lands are ecosystems highly transformed and managed by humans for the purpose of providing food and fiber, often at the expense of other ecosystem services. **More land was converted to cropland in the 30 years after 1950 than in the 150 years between 1700 and 1850, and one quarter of Earth's terrestrial surface is now occupied by cultivated systems.** (See Figure C6.) Within this area, one fifth is irrigated. [26]

As the demand for food, feed, and fiber has increased, farmers have responded both by expanding the area under cultivation (extensification) and by raising yields per unit land and per unit time (intensification). **Over the past 40 years, in global aggregate, intensification has been the primary source of increased output,** and in many regions (including in the European Union, North America, Australia, and recently China) the extent of land under cultivation has stabilized or even contracted. However, countries with low productivity and high population pressure—conditions that apply in much of sub-Saharan Africa—continue to rely mainly on expansion of cultivated areas for increasing food productivity. In Asia (outside of China), almost no high-productivity land remains available for the expansion of agriculture. Area expansion usually brings more marginal land (steeper slopes, poorer soils, and harsher climates) into production, often with unwelcome social and environmental consequences. Urban expansion is a growing cause of displacement of cultivation, but the area involved remains small in global terms. [26]

Increases in the yields of crop production systems due to increased use of inputs over the past 40 years have reduced the pressure to convert other ecosystems into cropland. Twenty million square kilometers of natural ecosystem have

been protected from conversion to farmland since 1950 due to more intensive production. On the other hand, intensification has increased pressure on inland water ecosystems due to increased water withdrawals for irrigation and to nutrient and pesticide leakage from cultivated lands, with negative consequences for freshwater and coastal systems, such as eutrophication. Intensification also generally reduces biodiversity within agricultural landscapes and requires higher energy inputs in the form of mechanization and the production of chemical fertilizers. Especially in systems that are already highly intensified, the marginal value of further increased production must be weighed against the additional environmental impacts. [26]

The intrinsic capacity of cultivated systems to support crop production is being undermined by soil erosion and salinization and by loss of agricultural biodiversity, but their effect on food production is masked by increasing use of fertilizer, water, and other agricultural inputs. (See Figure C7.) [8, 22, 26]

National policies, international agreements, and market forces play a significant role in determining the fate of ecosystems services as a consequence of cultivation. They all influence farmer choices about the scale and type of cultivation as well as the level and mix of production inputs that, in turn, influence trade-offs among the mix and level of ecosystem services that cultivated systems can deliver. [26]

Urban Systems

Urban areas currently cover less than 3% of the total land area of Earth, but they contain an increasing fraction of the world's population. Currently about half of the world's people live in urban areas. The urban requirements for ecosystem services are high, but it could be just as stressful if the same number of people, with similar consumption and production patterns, were dispersed over the rural landscape. In general, **the well-being of urban dwellers is higher than that of their rural neighbors,** as measured by wealth, health, and education indicators. Urban centers facilitate human access to and management of certain ecosystem services through, for example, the scale economies of piped water systems. [27]

Nevertheless, urban developments pose significant challenges with respect to ecosystem services and human well-being. The problems include inadequate and inequitable access to ecosystem services within urban areas, degradation of ecosystems adjoining urban areas, and pressures on distant ecosystems resulting from production, consumption, and trade originating in urban areas. [27]

In affluent countries, the negative impacts of urban settlements on ecosystem services and human well-being have been delayed and passed on to future generations or displaced onto locations away from the urban area. While urban developments in other parts of the world have been quite different, this trend and its political implication remain significant. [27]

Interrelated problems involving local water, sanitation, waste, and pests contribute a large share of the urban burden of disease, especially in low-income settlements. The consumption and production activities driving long-term, global ecosystem change are concentrated in urban centers, especially upper-income settlements. [27]

Urbanization is not inherently bad for ecosystems: ecosystems in and around urban areas can provide a high level of biodiversity, food production, water services, comfort, amenities, cultural values, and so on if well managed. When the loss of ecosystem services due to urban activities is systematically addressed,

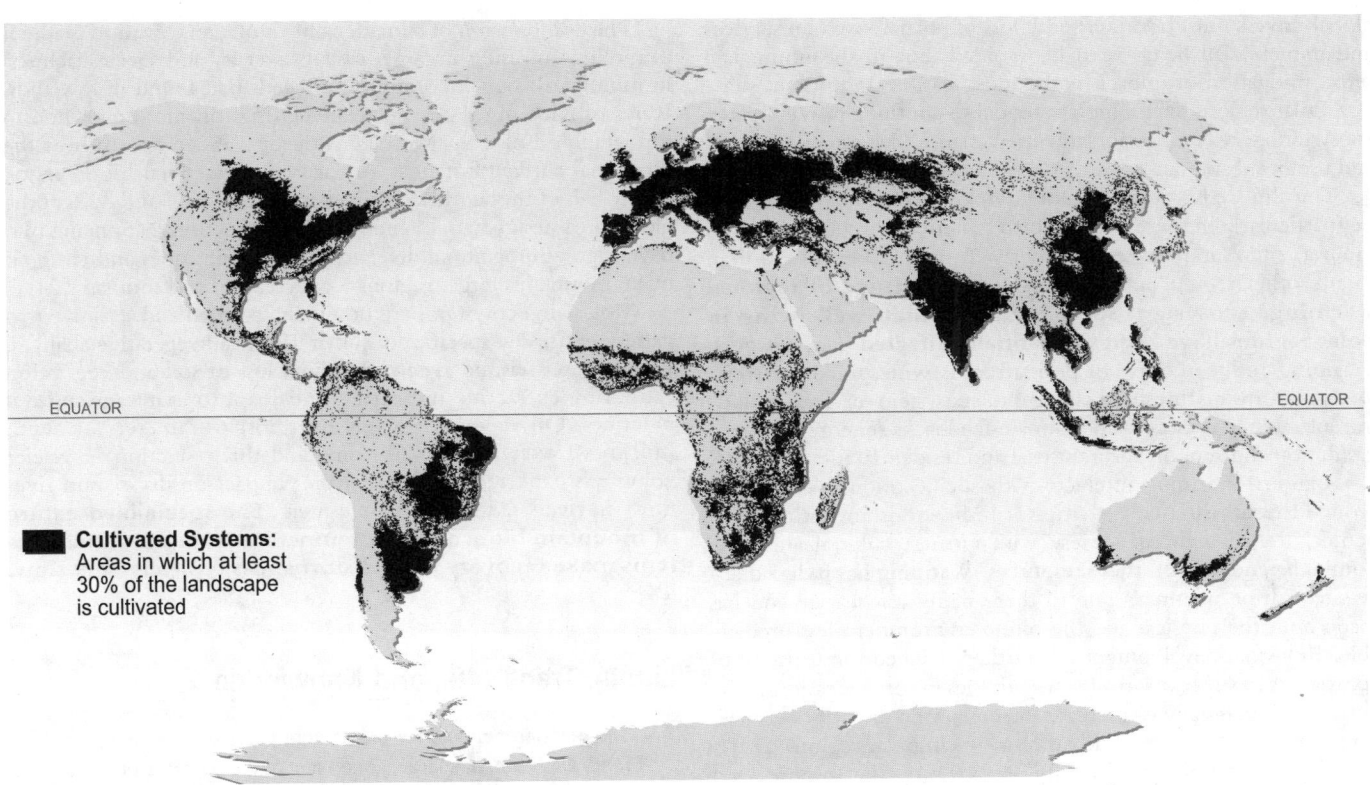

EQUATOR

EQUATOR

Cultivated Systems:
Areas in which at least
30% of the landscape
is cultivated

Figure C6. Extent of Cultivated Systems in 2000. Cultivated systems (defined in the MA to be areas in which at least 30% of the landscape comes under cultivation in any particular year) cover 24% of the terrestrial surface.

Indices = 100 in 1990

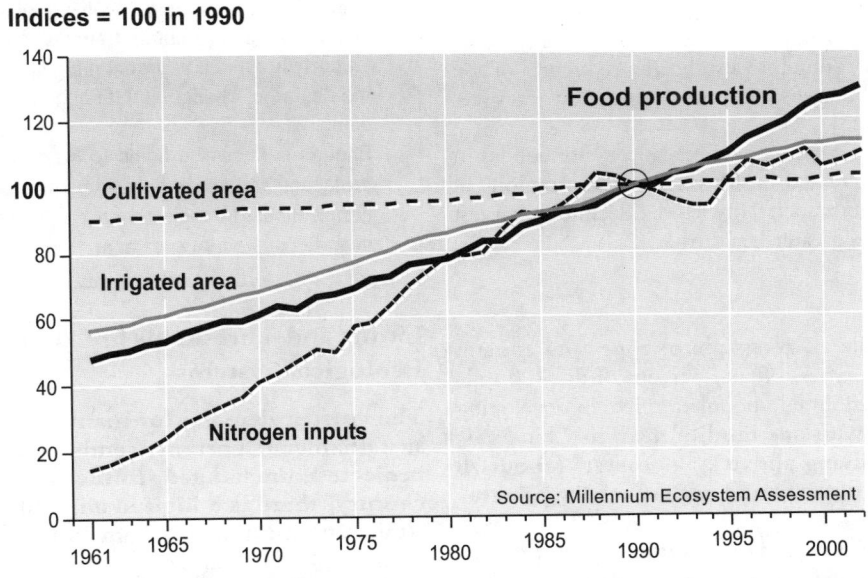

Figure C7. Trends in the Factors Related to Global Food Production since 1961. There have been significant trade-offs in cultivated systems between food production and water availability for other uses (due to irrigation), as well as water quality (due to increased nutrient loading). The role of improved crop varieties has also been extremely significant since 1961.

these losses can be greatly reduced. With a few exceptions, however, there is little evidence of cities taking significant steps to reduce their global ecosystem burdens. A city may be sustained by ecosystem services derived from an area up to 100 times larger than the city itself. [27]

Polar Systems

Direct, locally caused impacts of human activities on polar regions have been modest, and the most significant causes of change mainly originate outside the polar region. These global drivers must be addressed if loss of polar ecosystem services and human well-being is to be avoided, but in the immediate term, mitigation of impacts is the most feasible and urgent strategy. Polar regions have a high potential to continue providing key ecosystem services, particularly in wetlands, where biodiversity and use of subsistence resources are concentrated. [25]

The climate has warmed more quickly in portions of the Arctic (particularly in the western North American Arctic and central Siberia) and Antarctic (especially the Antarctic peninsula) than in any other region on Earth. **As a consequence of regional warming, ecosystem services and human well-being in polar regions have been substantially affected** (*high certainty*). Warming-induced thaw of permafrost is widespread in Arctic wetlands, causing threshold changes in ecosystem services, including subsistence resources and climate feedbacks (energy and trace gas fluxes) and support for industrial and residential infrastructure.

Regional warming interacts with socioeconomic change to reduce the subsistence activities of indigenous and other rural people, the segment of society with greatest cultural and economic dependence on these resources. Warming has reduced access to marine mammals due to there being less sea ice and has made both the physical and the biotic environment less predictable. Industrial development has further reduced the capacity of ecosystems to support subsistence activities in some locations. The net effect is generally to increase the economic disparity between rural subsistence users and urban residents in polar regions. [25]

Changes in polar biodiversity are affecting the resources on which Arctic people depend for their livelihoods. Important changes include increased shrub dominance in Arctic wetlands, which contributes to summer warming trends and alters forage available to caribou; changes in insect abundance that alter food availability to wetland birds and energy budgets of reindeer and caribou; increased abundance of snow geese that are degrading Arctic wetlands; and overgrazing by domestic reindeer in parts of Fennoscandia and Russia. There has also been a reduction of top predators in Antarctic food webs, altering marine food resources in the Southern Ocean. [25]

Increases in persistent organic pollutants and radionuclides in subsistence foods have increased health risks in some regions of the Arctic, but diet changes associated with the decline in harvest of these foods are usually a greater health risk. [25]

Mountain Systems

Mountain systems straddle all geographical zones and contain many different ecosystem types. Ninety percent of the 720 million people in the global mountain population live in developing and transition countries, with one third of them in China. **Almost all of the people living above 2,500 meters (about 70 million people) live in poverty and are especially vulnerable to food insecurity.**

Human well-being in lowland areas often depends on resources originating in mountain areas, such as timber, hydroelectricity, and water. Indeed, river basins from mountain systems supply nearly half of the human population with water, including in some regions far from the mountains themselves, and **loss of ecosystem functions in mountains increases environmental risks in both mountains and adjacent lowland areas.** However, there is rarely a systematic reinvestment of benefits derived from mountain systems in the conservation of upland resources. Mountains often represent political borders, narrow key transport corridors, or refuges for minorities and political opposition, and as such they are often focal areas of armed conflicts. [24]

The compression of climatic zones along an elevation gradient in mountains results in large habitat diversity and species richness in mountains, which commonly exceeds that found in lowlands. Rates of endemism are also relatively high in mountains due to topographic isolation. Mountains occupy about one fifth of the terrestrial surface but host a quarter of terrestrial biodiversity, nearly half of the world's biodiversity "hotspots," and 32% of the global area designated for biodiversity protection. Mountains also have high ethnocultural diversity. Scenic landscapes and clean air make mountains target regions for recreation and tourism. [24]

Mountain ecosystems are unusually exposed and sensitive to a variety of stresses, specifically climate-induced vegetative changes, volcanic and seismic events, flooding, loss of soil and vegetation caused by extractive industries, and inappropriate agricultural practices. On average, glaciers have lost 6–7 meters of depth (thickness) over the last 20 years, and this reduction in glacier volume is expected to have a strong impact on dry-season river flows in rivers fed largely by ice melt. **The specialized nature of mountain biota and low temperatures in mountain systems make recovery from disturbances typically very slow.** [24]

Limits, Trade-offs, and Knowledge

- **The growing demand for provisioning services, such as water, food, and fiber, has largely been met at the expense of supporting, regulating, and cultural ecosystem services.**
- **For some provisioning services, notably fresh water and wild-harvested fish, demand exceeds the available supply in large and expanding parts of the world.**
- **Some ongoing, large-scale human-induced ecosystem changes, such as those involving loss of biodiversity, climate change, excessive nutrient supply, and desertification, are effectively irreversible. Urgent mitigation action is needed to limit the degree of change and its negative impacts on human well-being.**
- **Enough is known to begin to make wiser decisions regarding protection and use of ecosystem services. Making this information available to decision-makers is the purpose of the Millennium Ecosystem Assessment.**

Limits and Thresholds in Coupled Human-Ecological Systems

The current demand for many ecosystems services is unsustainable. If current trends in ecosystem services are projected, unchanged, to the middle of the twenty-first century, there is a high likelihood that widespread constraints on human well-being will result. This highlights the need for globally coordinated adaptive responses, a topic further explored in the MA *Scenarios* and *Policy Responses* volumes.

Some limits to the degree of acceptable ecosystem change represent the level of tolerance by society, reflecting the trade-offs that people are willing (or forced) to make between different aspects of well-being. They are "soft limits," since they are socially determined and thus move as social circumstances change. Many such limits are currently under international negotiation, indicating that some key ecosystem services are approaching levels of concern. Examples are the amounts of fresh water allocated to different countries in shared basins, regional air quality norms, and the acceptable level of global climate change.

Other limits are a property of the ecological system itself and can be considered "hard limits." Two types of hard limit are of concern. The first is nonlinearity, which represents a point beyond which the loss of ecosystem services accelerates, sometimes abruptly. An example is the nitrogen saturation of watersheds: once the absorptive capacity of the ecosystem is exceeded, there is a sudden increase in the amount of nitrogen leaking into the aquatic environment. The second type is a true system threshold that, if crossed, leads to a new regime from which return is difficult, expensive, or even impossible. An example is the minimum habitat area required to sustain a viable population of a given species. If the area falls below this, eventual extinction is inevitable. We have fallen below this limit for many thousands of species (*medium certainty*).

Abrupt and possibly irreversible change may not be widely apparent until it is too late to do much about it. The dynamics of both ecological and human systems have intrinsic inertia—the tendency to continue changing even when the forces causing the change are relieved. The complexity of coupled human–ecological systems, together with our state of partial knowledge, make it hard to predict precisely at what point such thresholds lie. The overexploitation of wild fisheries is an example of a threshold that has already been crossed in many regions. [6, 13, 18, 25]

Thresholds of abrupt and effectively irreversible change are known to exist in the climate-ocean-land system (*high certainty*), although their location is only known with *low to medium certainty*. For example, it is *well established* that a decrease in the vegetation cover in the Sahara several thousand years ago was linked to a decrease in rainfall, promoting further loss of cover, leading to the current dry Sahara. It is *speculated* that a similar mechanism may have been involved in the abrupt decrease in rainfall in the Sahel in the mid-1970s. There are potential thresholds associated with climate feedbacks on the global carbon cycle, but large uncertainties remain regarding the strength of the feedback processes involved (such as the extent of warming-induced increases in soil respiration, the risk of large-scale dieback of tropical forests, and the effects of CO_2, nitrogen, and dust fertilization on carbon uptake by terrestrial and marine ecosystems). [12, 13, 22, 25]

Current human-induced greenhouse gas emissions to the atmosphere are greater than the capacity of global ecosystems to absorb them (*high certainty*). The oceans and terrestrial ecosystems are currently absorbing only about half of the carbon emissions resulting from fossil fuel combustion. As a result, the atmospheric concentration of CO_2 is rising, along with other greenhouse gases, leading to climate change. Although land use management can have a significant impact on CO_2 concentrations in the short term, future trends in atmospheric CO_2 are likely to depend more on fossil fuel emissions than on ecosystem change. [13]

Nitrogen additions to the environment are approaching critical limits in many regions. The increasing extent of oxygen-poor "dead zones" in freshwater or coastal ecosystems that have received elevated inputs of nutrients—nitrogen and phosphorus, in particular—over long periods of time is a symptom of the degree to which the nutrient retention capacity of terrestrial and freshwater systems has been overloaded. [12]

The capacity of Earth as a whole to render other waste products of human activities relatively harmless is unknown. It is *well established* that at high loading rates of wastes such as persistent organic pollutants, heavy metals, and radionuclides, the local ecosystem capacity can be overwhelmed, allowing waste accumulation to the detriment of human well-being and

the loss of ecosystem biodiversity [15]. A potential nonlinear response, currently the subject of intensive scientific research, is the atmospheric capacity to cleanse itself of air pollution (in particular, hydrocarbons and reactive nitrogen compounds). This capacity depends on chemical reactions involving the hydroxyl radical, the atmospheric concentration of which has declined by about 10% (*medium certainty*) since preindustrial times. [13]

Understanding the Trade-offs Associated with Our Actions

The growth in human well-being over the last several decades has come in large part through increases in provisioning services, usually at the expense of other services. In particular:

* The substantial increase in the production of food and fiber has expanded the area of cultivated systems (including plantation forests) at the expense of semi-natural ecosystems such as forests, rangelands, and wetlands. It has largely been achieved as a result of large inputs of nutrients, water, energy, and pesticides, with deleterious consequences for other ecosystems and the global climate.
* Clearing and transformation of previously forested land for agricultural and timber production, especially in tropical and sub-tropical forests, has reduced the land's capacity to regulate flows of water, store carbon, and support biological diversity and the livelihoods of forest-dwelling people.
* Harvesting of fish and other resources from coastal and marine systems (which are simultaneously under pressure from elevated flows of nutrients, sediments, and pollutants from the land) has impaired these systems' capacity to continue to deliver food in the future.

This assessment has shown that although we have many of the conceptual and analytical tools to illustrate the existence of trade-offs, the detailed information required to quantify adequately even the main trade-offs in economic terms is generally either lacking or inaccessible. An example of a tool useful for trade-off analysis is the valuation of ecosystem services, but such valuations have only been done for a few services and in a few places. The MA has also shown that failure to fully comprehend the trade-offs associated with particular actions has, in many instances, resulted either in a net decrease in human well-being or in an increase that is substantially less than it could have been. Examples of this include the loss of non-wood products and watershed services from overlogged forests, the loss of timber and the declines in offshore fisheries and storm protection from conversion of mangroves to aquaculture, and the loss of wetland products from conversion to intensive agriculture. (See Figure C8.) The continued tendency to make decisions on a sectoral basis prevents trade-offs from being fully considered.

Several independently derived international goals and commitments are interconnected via the ecosystems they affect. Thoughtful and informed consideration of trade-offs and synergies would be best achieved by coordinated implementation. An example of the importance of ecosystem service trade-offs in the pursuit of human well-being is provided by the Millennium Development Goals. In meeting the goal of reducing hunger, for instance, progress toward the goal of environmental sustainability could be compromised, and vice versa. A narrowly sectoral approach often simply displaces problems to other sectors. Ecosystem approaches, as adopted by the Convention on Biological Diversity, the Ramsar Convention on Wetlands, the Food and Agriculture Organization, and others, show promise for improving the future condition of services and human

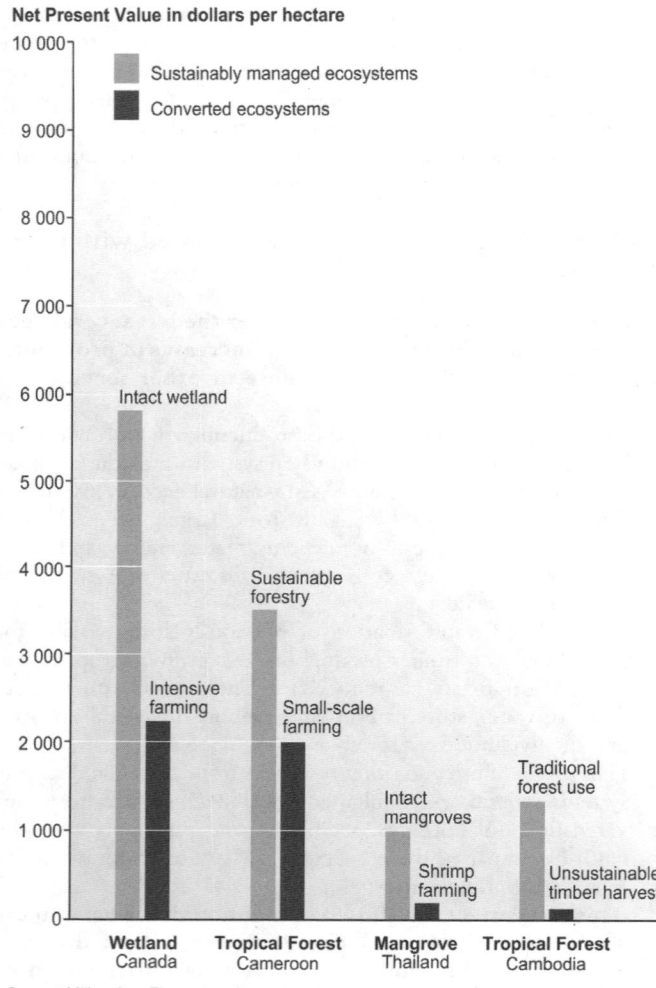

Net Present Value in dollars per hectare

Source: Millennium Ecosystem Assessment

Figure C8. Economic Benefits under Alternate Management Practices. In each case, the net benefits from the more sustainably managed ecosystem are greater than those from the converted ecosystem even though the private (market) benefits would be greater from the converted ecosystem. (Where ranges of values are given in the original source, lower estimates are plotted here.)

well-being as a whole, specifically by balancing the objectives of economic development and ecosystem integrity. In managing ecosystems, a balance needs to be found between provisioning services on the one hand and supporting, regulating, cultural and amenity services on the other hand. [7, 28]

Knowledge and Uncertainty

The experience of this assessment has been that it is hard to demonstrate, quantitatively and unequivocally, the widely accepted and intuitive link between ecosystem changes and changes in human well-being. There are several reasons for this. First, the impacts of ecosystem change on well-being are often subtle, which is not to say unimportant; impacts need not be blatant to be significant. Second, human well-being is affected by many factors in addition to the effects of ecosystem services. Health outcomes, for example, are the combined result of ecosystem condition, access to health care, economic status, and myriad other factors. Unequivocally linking ecosystem changes to changes

in well-being, and vice versa, is especially difficult when the data are patchy in both cases, as they usually are. Analyses linking well-being and ecosystem condition are most easily carried out at a local scale, where the linkages can be most clearly identified, but information on ecosystems and human well-being is often only available in highly aggregated form, for instance at the national level. Spatially explicit data with sub-national resolution would greatly facilitate future assessments. [2]

The availability and accuracy of data sources and methods for this assessment were greatest for provisioning services, such as crop yield and timber production. Direct data on regulating, supporting, and cultural services such as nutrient cycling, climate regulation, or aesthetic value are difficult to obtain, making it necessary to use proxies, modeled results, or extrapolations from case studies. Data on biodiversity have strong biases toward the species level, large organisms, temperate systems, and species used directly by people. [2, 4, 28]

Knowledge for quantifying ecosystem responses to stress is equally uneven. Methods to estimate crop yield responses to fertilizer application, for example, are well developed, but methods to quantify relationships between ecosystem services and human well-being, such as the effects of altered levels of biodiversity on the incidence of diseases in humans, are at an earlier stage of development. Thousands of novel chemicals, including long-lived synthetic pharmaceuticals, are currently entering the biosphere, but there are few systematic studies to understand their impact on ecosystems and human well-being. [2, 28]

Observation systems relating to ecosystem services are generally inadequate to support informed decision-making. Some previously more-extensive observation systems have declined in recent decades. For example, substantial deterioration of hydrographic networks is occurring throughout the world. The same is true for standard water quality monitoring and the recording of biological indicators. [7]

Both "traditional" and "formal" knowledge systems have considerable value for achieving the conservation and sustainable use of ecosystems. The loss of traditional knowledge has significantly weakened the linkages between ecosystems and cultural diversity and cultural identity. This loss has also had a direct negative effect on biodiversity and the degradation of ecosystems, for instance by exceeding traditionally established norms for resource use. This knowledge is largely oral. As significant is the loss of languages, which are the vehicle by which cultures are communicated and reproduced. [17]

A Call for Action

Despite the gaps in knowledge, **enough is known to indicate the need for urgent collective action, building on existing activities, to mitigate the further loss of ecosystem services.** It is *well established* that inadequate access to ecosystem services currently is an important factor in the low well-being of a large fraction of the global population and is likely to constrain improvements in well-being in the future.

Urgency is indicated because in situations where the probability of effectively irreversible, negative impacts is high, where the human and natural systems involved have high inertia, and where knowledge of the consequences is incomplete, early action to reduce the rate of change is more rational than waiting until conditions become globally intolerable and potentially irreversible. **Collective action is required** because uncoordinated individual action is necessary but insufficient to mitigate the many issues that have large-scale underlying causes, mechanisms, or consequences. Coordinated action at all levels of social organization—from local to global—is called for if the many islands of local failure are not to coalesce into expanding regions of degradation and if problems with global reach are to be managed. Coordinated action is also required to enable islands of local success to be expanded and propagated in distant locations.

The history of human civilization has many examples of social upheaval associated with ecosystem service failure at the local or regional scale. There are many current examples where the demands on ecosystems are exceeding the limits of the system to supply ecosystem services. Global-scale examples are given in this report, and local and regional examples are found in the *Multiscale Assessments* volume. **Two things are different now compared with any other time in history: human impacts are now ubiquitous and of greater intensity than at any time in the past, and in most cases we can no longer plead ignorance of the consequences.** Whereas in the past, natural disasters, pollution, or resource depletion led to local hardships, realignment of power, and the regional migration of people to better-serviced areas, in the present era the impacts are global in reach. Displacement of the problem to other places and future generations, or starting afresh in a new place, are no longer viable options.

A turning point in the growth of the human population on Earth is likely by mid-century. As the *Scenarios* and *Policy Responses* volumes show, the opportunity and technical means exist to provide food, water, shelter, a less-hazardous environment, and a better life to the existing population, and even to the additional 3 billion people likely to inhabit Earth by the middle of the twenty-first century, but we are currently failing to achieve this. We are also undermining our capacity to do so in the future by failing to take actions that will reduce the risk of adverse changes in Earth's ecological systems that will be difficult and costly to reverse.

Reducing the pressure on critical systems and services will be neither easy nor cost-free, but it is certain that net human well-being is better served by maintaining ecosystems in a condition that is capable of providing adequate levels of essential services than by trying to restore such functions at some future time.

Chapter 1

MA Conceptual Framework

BOXES

FIGURES

This chapter provides the summary of Millennium Ecosystem Assessment, *Ecosystems and Human Well-being: A Framework for Assessment* (Island Press, 2003), pp. 1–25, which was prepared by an extended conceptual framework writing team of 51 authors and 10 contributing authors.

Main Messages

Human well-being and progress toward sustainable development are vitally dependent upon improving the management of Earth's ecosystems to ensure their conservation and sustainable use. But while demands for ecosystem services such as food and clean water are growing, human actions are at the same time diminishing the capability of many ecosystems to meet these demands.

Sound policy and management interventions can often reverse ecosystem degradation and enhance the contributions of ecosystems to human well-being, but knowing when and how to intervene requires substantial understanding of both the ecological and the social systems involved. Better information cannot guarantee improved decisions, but it is a prerequisite for sound decision-making.

The Millennium Ecosystem Assessment was established to help provide the knowledge base for improved decisions and to build capacity for analyzing and supplying this information.

This chapter presents the conceptual and methodological approach that the MA used to assess options that can enhance the contribution of ecosystems to human well-being. This same approach should provide a suitable basis for governments, the private sector, and civil society to factor considerations of ecosystems and ecosystem services into their own planning and actions.

1.1 Introduction

Humanity has always depended on the services provided by the biosphere and its ecosystems. Further, the biosphere is itself the product of life on Earth. The composition of the atmosphere and soil, the cycling of elements through air and waterways, and many other ecological assets are all the result of living processes—and all are maintained and replenished by living ecosystems. The human species, while buffered against environmental immediacies by culture and technology, is ultimately fully dependent on the flow of ecosystem services.

In his April 2000 Millennium Report to the United Nations General Assembly, in recognition of the growing burden that degraded ecosystems are placing on human well-being and economic development and the opportunity that better managed ecosystems provide for meeting the goals of poverty eradication and sustainable development, United Nations Secretary-General Kofi Annan stated that:

> *It is impossible to devise effective environmental policy unless it is based on sound scientific information. While major advances in data collection have been made in many areas, large gaps in our knowledge remain. In particular, there has never been a comprehensive global assessment of the world's major ecosystems. The planned Millennium Ecosystem Assessment, a major international collaborative effort to map the health of our planet, is a response to this need.*

The Millennium Ecosystem Assessment was established with the involvement of governments, the private sector, nongovernmental organizations, and scientists to provide an integrated assessment of the consequences of ecosystem change for human well-being and to analyze options available to enhance the conservation of ecosystems and their contributions to meeting human needs. The Convention on Biological Diversity, the Convention to Combat Desertification, the Convention on Migratory Species, and the Ramsar Convention on Wetlands plan to use the

findings of the MA, which will also help meet the needs of others in government, the private sector, and civil society. The MA should help to achieve the United Nations Millennium Development Goals and to carry out the Plan of Implementation of the 2002 World Summit on Sustainable Development. It has mobilized hundreds of scientists from countries around the world to provide information and clarify science concerning issues of greatest relevance to decision-makers. The MA has identified areas of broad scientific agreement and also pointed to areas of continuing scientific debate.

The assessment framework developed for the MA offers decision-makers a mechanism to:

- *Identify options that can better achieve core human development and sustainability goals. All countries and communities are grappling with the challenge of meeting growing demands for food, clean water, health, and employment.* And decision-makers in the private and public sectors must also balance economic growth and social development with the need for environmental conservation. All of these concerns are linked directly or indirectly to the world's ecosystems. The MA process, at all scales, was designed to bring the best science to bear on the needs of decision-makers concerning these links between ecosystems, human development, and sustainability.

- *Better understand the trade-offs involved—across sectors and stakeholders—in decisions concerning the environment.* Ecosystem-related problems have historically been approached issue by issue, but rarely by pursuing multisectoral objectives. This approach has not withstood the test of time. Progress toward one objective such as increasing food production has often been at the cost of progress toward other objectives such as conserving biological diversity or improving water quality. The MA framework complements sectoral assessments with information on the full impact of potential policy choices across sectors and stakeholders.

- *Align response options with the level of governance where they can be most effective.* Effective management of ecosystems will require actions at all scales, from the local to the global. Human actions now directly or inadvertently affect virtually all of the world's ecosystems; actions required for the management of ecosystems refer to the steps that humans can take to modify their direct or indirect influences on ecosystems. The management and policy options available and the concerns of stakeholders differ greatly across these scales. The priority areas for biodiversity conservation in a country as defined based on "global" value, for example, would be very different from those as defined based on the value to local communities. The multiscale assessment framework developed for the MA provides a new approach for analyzing policy options at all scales—from local communities to international conventions.

1.2 What Is the Problem?

Ecosystem services are the benefits people obtain from ecosystems, which the MA describes as provisioning, regulating, supporting, and cultural services. (See Box 1.1.) Ecosystem services include products such as food, fuel, and fiber; regulating services such as climate regulation and disease control; and nonmaterial benefits such as spiritual or aesthetic benefits. Changes in these services affect human well-being in many ways. (See Figure 1.1.)

The demand for ecosystem services is now so great that trade-offs among services have become the rule. A country can increase food supply by converting a forest to agriculture, for example, but in so doing it decreases the supply of services that may be of equal or greater importance, such as clean water, timber, ecotourism destinations, or flood regulation and drought control. There are many indications that human demands on ecosystems will grow still greater in the coming decades. Current estimates of 3 billion more people and a quadrupling of the world economy by 2050 imply a formidable increase in demand for and consumption of biological and physical resources, as well as escalating impacts on ecosystems and the services they provide.

The problem posed by the growing demand for ecosystem services is compounded by increasingly serious degradation in the capability of ecosystems to provide these services. World fisheries are now declining due to overfishing, for instance, and a significant amount of agricultural land has been degraded in the past half-century by erosion, salinization, compaction, nutrient depletion, pollution, and urbanization. Other human-induced impacts on ecosystems include alteration of the nitrogen, phosphorous, sulfur, and carbon cycles, causing acid rain, algal blooms, and fish kills in rivers and coastal waters, along with contributions to climate change. In many parts of the world, this degradation of ecosystem services is exacerbated by the associated loss of the knowledge and understanding held by local communities—knowledge that sometimes could help to ensure the sustainable use of the ecosystem.

This combination of ever-growing demands being placed on increasingly degraded ecosystems seriously diminishes the prospects for sustainable development. Human well-being is affected not just by gaps between ecosystem service supply and demand but also by the increased vulnerability of individuals, communities, and nations. Productive ecosystems, with their array of services, provide people and communities with resources and options they can use as insurance in the face of natural catastrophes or social upheaval. While well-managed ecosystems reduce risks and vulnerability, poorly managed systems can exacerbate them by increasing risks of flood, drought, crop failure, or disease.

Ecosystem degradation tends to harm rural populations more directly than urban populations and has its most direct and severe impact on poor people. The wealthy control access to a greater share of ecosystem services, consume those services at a higher per capita rate, and are buffered from changes in their availability (often at a substantial cost) through their ability to purchase scarce ecosystem services or substitutes. For example, even though a number of marine fisheries have been depleted in the past century, the supply of fish to wealthy consumers has not been disrupted since fishing fleets have been able to shift to previously underexploited stocks. In contrast, poor people often lack access to alternate services and are highly vulnerable to ecosystem changes that result in famine, drought, or floods. They frequently live in locations particularly sensitive to environmental threats, and they lack financial and institutional buffers against these dangers. Degradation of coastal fishery resources, for instance, results in a decline in protein consumed by the local community since fishers may not have access to alternate sources of fish and community members may not have enough income to purchase fish. Degradation affects their very survival.

Changes in ecosystems affect not just humans but countless other species as well. The management objectives that people set for ecosystems and the actions that they take are influenced not just by the consequences of ecosystem changes for humans but also by the importance people place on considerations of the intrinsic value of species and ecosystems. Intrinsic value is the value of something in and for itself, irrespective of its utility for someone else. For example, villages in India protect "spirit sanctuaries" in relatively natural states, even though a strict cost-benefit calculation might favor their conversion to agriculture. Similarly, many countries have passed laws protecting endangered species based on the view that these species have a right to exist, even if their protection results in net economic costs. Sound ecosystem management thus involves steps to address the utilitarian links of people to ecosystems as well as processes that allow considerations of the intrinsic value of ecosystems to be factored into decision-making.

The degradation of ecosystem services has many causes, including excessive demand for ecosystem services stemming from economic growth, demographic changes, and individual choices. Market mechanisms do not always ensure the conservation of ecosystem services either because markets do not exist for services such as cultural or regulatory services or, where they do exist, because policies and institutions do not enable people living within the ecosystem to benefit from services it may provide to others who are far away. For example, institutions are now only beginning to be developed to enable those benefiting from carbon sequestration to provide local managers with an economic incentive to leave a forest uncut, while strong economic incentives often exist for managers to harvest the forest. Also, even if a market exists for an ecosystem service, the results obtained through the market may be socially or ecologically undesirable. Properly managed, the creation of ecotourism opportunities in a country can create strong economic incentives for the maintenance of the cultural services provided by ecosystems, but poorly managed ecotourism activities can degrade the very resource on which they depend. Finally, markets are often unable to address important intra- and intergenerational equity issues associated with managing ecosystems for this and future generations, given that some changes in ecosystem services are irreversible.

The world has witnessed in recent decades not just dramatic changes to ecosystems but equally profound changes to social systems that shape both the pressures on ecosystems and the opportunities to respond. The relative influence of individual nation-states has diminished with the growth of power and influence of a far more complex array of institutions, including regional

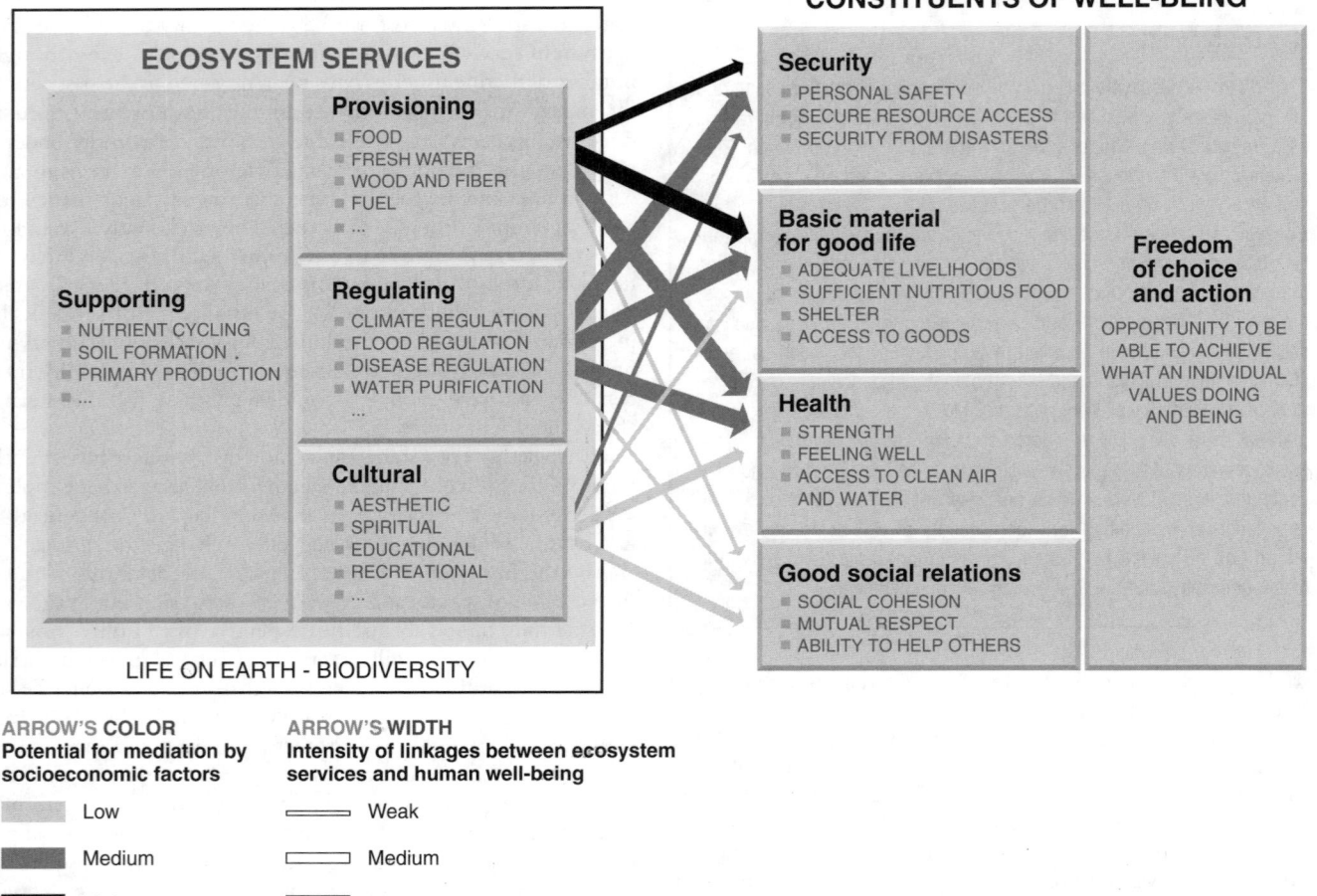

Figure 1.1. Linkages between Ecosystem Services and Human Well-being. This Figure depicts the strength of linkages between categories of ecosystem services and components of human well-being that are commonly encountered and includes indications of the extent to which it is possible for socioeconomic factors to mediate the linkage. (For example, if it is possible to purchase a substitute for a degraded ecosystem service, then there is a high potential for mediation.) The strength of the linkages and the potential for mediation differ in different ecosystems and regions. In addition to the influence of ecosystem services on human well-being depicted here, other factors—including other environmental factors as well as economic, social, technological, and cultural factors—influence human well-being, and ecosystems are in turn affected by changes in human well-being. (Millennium Ecosystem Assessment)

governments, multinational companies, the United Nations, and civil society organizations. Stakeholders have become more involved in decision-making. Given the multiple actors whose decisions now strongly influence ecosystems, the challenge of providing information to decision-makers has grown. At the same time, the new institutional landscape may provide an unprecedented opportunity for information concerning ecosystems to make a major difference. Improvements in ecosystem management to enhance human well-being will require new institutional and policy arrangements and changes in rights and access to resources that may be more possible today under these conditions of rapid social change than they have ever been before.

Like the benefits of increased education or improved governance, the protection, restoration, and enhancement of ecosystem services tends to have multiple and synergistic benefits. Already, many governments are beginning to recognize the need for more effective management of these basic life-support systems. Examples of significant progress toward sustainable management of biological resources can also be found in civil society, in indigenous and local communities, and in the private sector.

1.3 Conceptual Framework

The conceptual framework for the MA places human well-being as the central focus for assessment, while recognizing that biodiversity and ecosystems also have intrinsic value and that people take decisions concerning ecosystems based on considerations of well-being as well as intrinsic value. (See Box 1.2.) The MA conceptual framework assumes that a dynamic interaction exists between people and other parts of ecosystems, with the changing human condition serving to both directly and indirectly drive change in ecosystems and with changes in ecosystems causing changes in human well-being. At the same time, many other factors independent of the environment change the human condition, and many natural forces are influencing ecosystems.

The MA focuses particular attention on the linkages between ecosystem services and human well-being. The assessment deals with the full range of ecosystems—from those relatively undisturbed, such as natural forests, to landscapes with mixed patterns of human use and ecosystems intensively managed and modified by humans, such as agricultural land and urban areas.

A full assessment of the interactions between people and ecosystems requires a multiscale approach because it better reflects the multiscale nature of decision-making, allows the examination of driving forces that may be exogenous to particular regions, and provides a means of examining the differential impact of ecosystem changes and policy responses on different regions and groups within regions.

This section explains in greater detail the characteristics of each of the components of the MA conceptual framework, moving clockwise from the lower left corner of the Figure in Box 1.2.

1.3.1 Ecosystems and Their Services

An ecosystem is a dynamic complex of plant, animal, and microorganism communities and the nonliving environment interacting as a functional unit. Humans are an integral part of ecosystems. Ecosystems provide a variety of benefits to people, including provisioning, regulating, cultural, and supporting services. Provisioning services are the products people obtain from ecosystems, such as food, fuel, fiber, fresh water, and genetic resources. Regulating services are the benefits people obtain from the regulation of ecosystem processes, including air quality maintenance, climate regulation, erosion control, regulation of human diseases, and water purification. Cultural services are the nonmaterial benefits people obtain from ecosystems through spiritual enrichment, cognitive development, reflection, recreation, and aesthetic experiences. Supporting services are those that are necessary for the production of all other ecosystem services, such as primary production, production of oxygen, and soil formation.

Biodiversity and ecosystems are closely related concepts. Biodiversity is the variability among living organisms from all sources, including terrestrial, marine, and other aquatic ecosystems and the ecological complexes of which they are part. It includes diversity within and between species and diversity of ecosystems. Diversity is a structural feature of ecosystems, and the variability among ecosystems is an element of biodiversity. Products of biodiversity include many of the services produced by ecosystems (such as food and genetic resources), and changes in biodiversity can influence all the other services they provide. In addition to the important role of biodiversity in providing ecosystem services, the diversity of living species has intrinsic value independent of any human concern.

The concept of an ecosystem provides a valuable framework for analyzing and acting on the linkages between people and the environment. For that reason, the "ecosystem approach" has been endorsed by the Convention on Biological Diversity, and the MA conceptual framework is entirely consistent with this approach. The CBD states that the ecosystem approach is a strategy for the integrated management of land, water, and living resources that promotes conservation and sustainable use in an equitable way. This approach recognizes that humans, with their cultural diversity, are an integral component of many ecosystems.

In order to implement the ecosystem approach, decision-makers need to understand the multiple effects on an ecosystem of any management or policy change. By way of analogy, decision-makers would not make a decision about financial policy in a country without examining the condition of the economic system, since information on the economy of a single sector such as manufacturing would be insufficient. The same need to examine the consequences of changes for multiple sectors applies to ecosystems. For instance, subsidies for fertilizer use may increase food production, but sound decisions also require information on whether the potential reduction in the harvests of downstream fisheries as a result of water quality degradation from the fertilizer runoff might outweigh those benefits.

For the purpose of analysis and assessment, a pragmatic view of ecosystem boundaries must be adopted, depending on the questions being asked. A well-defined ecosystem has strong interactions among its components and weak interactions across its boundaries. A useful choice of ecosystem boundary is one where a number of discontinuities coincide, such as in the distribution of organisms, soil types, drainage basins, and depth in a waterbody. At a larger scale, regional and even globally distributed ecosystems can be evaluated based on a commonality of basic structural units. The global assessment being undertaken by the MA reports on marine, coastal, inland water, forest, dryland, island, mountain, polar, cultivated, and urban regions. These regions are not ecosystems themselves, but each contains a number of ecosystems. (See Box 1.3.)

People seek multiple services from ecosystems and thus perceive the condition of given ecosystems in relation to their ability to provide the services desired. Various methods can be used to assess the ability of ecosystems to deliver particular services. With those answers in hand, stakeholders have the information they need to decide on a mix of services best meeting their needs. The MA considers criteria and methods to provide an integrated view of the condition of ecosystems. The condition of each category of ecosystem services is evaluated in somewhat different ways, although in general a full assessment of any service requires considerations of stocks, flows, and resilience of the service.

1.3.2 Human Well-being and Poverty Reduction

Human well-being has multiple constituents, including the basic material for a good life, freedom of choice and action, health, good social relations, and security. Poverty is also multidimensional and has been defined as the pronounced deprivation of well-being. How well-being, ill-being, or poverty are experienced and expressed depends on context and situation, reflecting local physical, social, and personal factors such as geography, environment, age, gender, and culture. In all contexts, however, ecosystems are essential for human well-being through their provisioning, regulating, cultural, and supporting services.

Human intervention in ecosystems can amplify the benefits to human society. However, evidence in recent decades of escalating human impacts on ecological systems worldwide raises concerns about the spatial and temporal consequences of ecosystem changes detrimental to human well-being. Ecosystem changes affect human well-being in the following ways:

- *Security* is affected both by changes in provisioning services, which affect supplies of food and other goods and the likelihood of conflict over declining resources, and by changes in regulating services, which could influence the frequency and magnitude of floods, droughts, landslides, or other catastrophes. It can also be affected by changes in cultural services as, for example, when the loss of important ceremonial or spiritual attributes of ecosystems contributes to the weakening of social relations in a community. These changes in turn affect material well-being, health, freedom and choice, security, and good social relations.

- ***Access to basic material for a good life*** is strongly linked to both provisioning services such as food and fiber production and regulating services, including water purification.

- ***Health*** is strongly linked to both provisioning services such as food production and regulating services, including those that influence the distribution of disease-transmitting insects and of irritants and pathogens in water and air. Health can also be

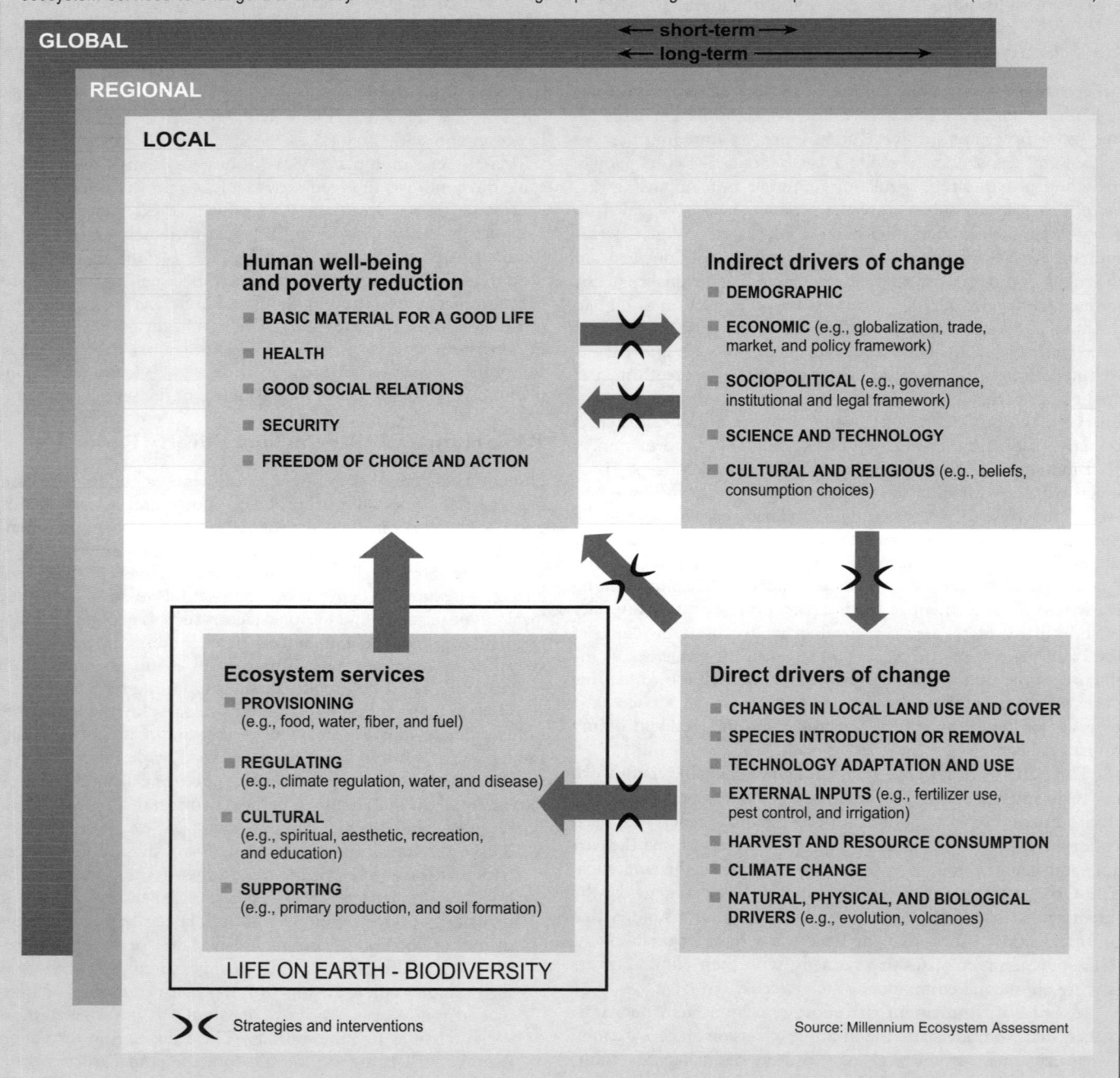

BOX 1.2
Millennium Ecosystem Assessment Conceptual Framework

Changes in factors that indirectly affect ecosystems, such as population, technology, and lifestyle (upper right corner of figure), can lead to changes in factors directly affecting ecosystems, such as the catch of fisheries or the application of fertilizers to increase food production (lower right corner). The resulting changes in the ecosystem (lower left corner) cause the ecosystem services to change and thereby affect human well-being.

These interactions can take place at more than one scale and can cross scales. For example, a global market may lead to regional loss of forest cover, which increases flood magnitude along a local stretch of a river. Similarly, the interactions can take place across different time scales. Actions can be taken either to respond to negative changes or to enhance positive changes at almost all points in this framework (black cross bars).

Source: Millennium Ecosystem Assessment

linked to cultural services through recreational and spiritual benefits.

- *Social relations* are affected by changes to cultural services, which affect the quality of human experience.
- *Freedom of choice and action* is largely predicated on the existence of the other components of well-being and are thus

influenced by changes in provisioning, regulating, or cultural services from ecosystems.

Human well-being can be enhanced through sustainable human interactions with ecosystems supported by necessary instruments, institutions, organizations, and technology. Creation of these through participation and transparency may contribute to

BOX 1.3

Reporting Categories Used in the Millennium Ecosystem Assessment

The MA used 10 categories of systems to report its global findings. (See Table.) These categories are not ecosystems themselves; each contains a number of ecosystems. The MA reporting categories are not mutually exclusive: their areas can and do overlap. Ecosystems within each category share a suite of biological, climatic, and social factors that tend to differ across categories. Because these reporting categories overlap, any place on Earth may fall into more than one category. Thus, for example, a wetland ecosystem in a coastal region may be examined both in the MA analysis of "coastal systems" as well as in its analysis of "inland water systems."

Millennium Ecosystem Assessment Reporting Categories

Category	Central Concept	Boundary Limits for Mapping
Marine	Ocean, with fishing typically a major driver of change	Marine areas where the sea is deeper than 50 meters.
Coastal	Interface between ocean and land, extending seawards to about the middle of the continental shelf and inland to include all areas strongly influenced by the proximity to the ocean	Area between 50 meters below mean sea level and 50 meters above the high tide level or extending landward to a distance 100 kilometers from shore. Includes coral reefs, intertidal zones, estuaries, coastal aquaculture, and seagrass communities.
Inland water	Permanent water bodies inland from the coastal zone, and areas whose ecology and use are dominated by the permanent, seasonal, or intermittent occurrence of flooded conditions	Rivers, lakes, floodplains, reservoirs, and wetlands; includes inland saline systems. Note that the Ramsar Convention considers "wetlands" to include both inland water and coastal categories.
Forest	Lands dominated by trees; often used for timber, fuelwood, and non-timber forest products	A canopy cover of at least 40% by woody plants taller than 5 meters. The existence of many other definitions is acknowledged, and other limits (such as crown cover greater than 10%, as used by the Food and Agriculture Organization of the United Nations) are also reported. Includes temporarily cut-over forests and plantations; excludes orchards and agroforests where the main products are food crops.
Dryland	Lands where plant production is limited by water availability; the dominant uses are large mammal herbivory, including livestock grazing, and cultivation	Drylands as defined by the Convention to Combat Desertification, namely lands where annual precipitation is less than two thirds of potential evaporation, from dry subhumid areas (ratio ranges 0.50–0.65), through semiarid, arid, and hyper-arid (ratio <0.05), but excluding polar areas; drylands include cultivated lands, scrublands, shrublands, grasslands, semi-deserts, and true deserts.
Island	Lands isolated by surrounding water, with a high proportion of coast to hinterland	Islands of at least 1.5 hectares included in the ESRI ArcWorld Country Boundary dataset.
Mountain	Steep and high lands	As defined by Mountain Watch using criteria based on elevation alone, and at lower elevation, on a combination of elevation, slope, and local elevation range. Specifically, elevation >2,500 meters, elevation 1,500–2,500 meters and slope >2 degrees, elevation 1,000–1,500 meters and slope >5 degrees or local elevation range (7 kilometers radius) >300 meters, elevation 300–1,000 meters and local elevation range (7 kilometers radius) >300 meters, isolated inner basins and plateaus less than 25 square kilometers extent that are surrounded by mountains.
Polar	High-latitude systems frozen for most of the year	Includes ice caps, areas underlain by permafrost, tundra, polar deserts, and polar coastal areas. Excludes high-altitude cold systems in low latitudes.
Cultivated	Lands dominated by domesticated plant species, used for and substantially changed by crop, agroforestry, or aquaculture production	Areas in which at least 30% of the landscape comes under cultivation in any particular year. Includes orchards, agroforestry, and integrated agriculture-aquaculture systems.
Urban	Built environments with a high human density	Known human settlements with a population of 5,000 or more, with boundaries delineated by observing persistent night-time lights or by inferring areal extent in the cases where such observations are absent.

freedoms and choice as well as to increased economic, social, and ecological security. By ecological security, we mean the minimum level of ecological stock needed to ensure a sustainable flow of ecosystem services.

Yet the benefits conferred by institutions and technology are neither automatic nor equally shared. In particular, such opportunities are more readily grasped by richer than poorer countries and people; some institutions and technologies mask or exacerbate environmental problems; responsible governance, while essential, is not easily achieved; participation in decision-making, an essential element of responsible governance, is expensive in time and resources to maintain. Unequal access to ecosystem services has often elevated the well-being of small segments of the population at the expense of others.

Sometimes the consequences of the depletion and degradation of ecosystem services can be mitigated by the substitution of knowledge and of manufactured or human capital. For example, the addition of fertilizer in agricultural systems has been able to offset declining soil fertility in many regions of the world where people have sufficient economic resources to purchase these inputs, and water treatment facilities can sometimes substitute for the role of watersheds and wetlands in water purification. But ecosystems are complex and dynamic systems and there are limits to substitution possibilities, especially with regulating, cultural, and supporting services. No substitution is possible for the extinction of culturally important species such as tigers or whales, for instance, and substitutions may be economically impractical for the loss of services such as erosion control or climate regulation. Moreover, the scope for substitutions varies by social, economic, and cultural conditions. For some people, especially the poorest, substitutes and choices are very limited. For those who are better off, substitution may be possible through trade, investment, and technology.

Because of the inertia in both ecological and human systems, the consequences of ecosystem changes made today may not be felt for decades. Thus, sustaining ecosystem services, and thereby human well-being, requires a full understanding and wise management of the relationships between human activities, ecosystem change, and well-being over the short, medium, and long term. Excessive current use of ecosystem services compromises their future availability. This can be prevented by ensuring that the use is sustainable.

Achieving sustainable use requires effective and efficient institutions that can provide the mechanisms through which concepts of freedom, justice, fairness, basic capabilities, and equity govern the access to and use of ecosystem services. Such institutions may also need to mediate conflicts between individual and social interests that arise.

The best way to manage ecosystems to enhance human well-being will differ if the focus is on meeting needs of the poor and weak or the rich and powerful. For both groups, ensuring the long-term supply of ecosystem services is essential. But for the poor, an equally critical need is to provide more equitable and secure access to ecosystem services.

1.3.3 Drivers of Change

Understanding the factors that cause changes in ecosystems and ecosystem services is essential to designing interventions that capture positive impacts and minimize negative ones. In the MA, a "driver" is any factor that changes an aspect of an ecosystem. A direct driver unequivocally influences ecosystem processes and can therefore be identified and measured to differing degrees of accuracy. An indirect driver operates more diffusely, often by al-

tering one or more direct drivers, and its influence is established by understanding its effect on a direct driver. Both indirect and direct drivers often operate synergistically. Changes in land cover, for example, can increase the likelihood of introduction of alien invasive species. Similarly, technological advances can increase rates of economic growth.

The MA explicitly recognizes the role of decision-makers who affect ecosystems, ecosystem services, and human well-being. Decisions are made at three organizational levels, although the distinction between those levels is often diffuse and difficult to define:

- by individuals and small groups at the local level (such as a field or forest stand) who directly alter some part of the ecosystem;
- by public and private decision-makers at the municipal, provincial, and national levels; and
- by public and private decision-makers at the international level, such as through international conventions and multilateral agreements.

The decision-making process is complex and multidimensional. We refer to a driver that can be influenced by a decision-maker as an endogenous driver and one over which the decision-maker does not have control as an exogenous driver. The amount of fertilizer applied on a farm is an endogenous driver from the standpoint of the farmer, for example, while the price of the fertilizer is an exogenous driver, since the farmer's decisions have little direct influence on price. The specific temporal, spatial, and organizational scale dependencies of endogenous and exogenous drivers and the specific linkages and interactions among drivers are assessed in the MA.

Whether a driver is exogenous or endogenous to a decision-maker is dependent upon the spatial and temporal scale. For example, a local decision-maker can directly influence the choice of technology, changes in land use, and external inputs (such as fertilizers or irrigation), but has little control over prices and markets, property rights, technology development, or the local climate. In contrast, a national or regional decision-maker has more control over many factors, such as macroeconomic policy, technology development, property rights, trade barriers, prices, and markets. But on the short time scale, that individual has little control over the climate or global population. On the longer time scale, drivers that are exogenous to a decision-maker in the short run, such as population, become endogenous since the decision-maker can influence them through, for instance, education, the advancement of women, and migration policies.

The indirect drivers of change are primarily:

- demographic (such as population size, age and gender structure, and spatial distribution);
- economic (such as national and per capita income, macroeconomic policies, international trade, and capital flows);
- sociopolitical (such as democratization, the roles of women, of civil society, and of the private sector, and international dispute mechanisms);
- scientific and technological (such as rates of investments in research and development and the rates of adoption of new technologies, including biotechnologies and information technologies); and
- cultural and religious (such as choices individuals make about what and how much to consume and what they value).

The interaction of several of these drivers, in turn, affects levels of resource consumption and differences in consumption both within and between countries. Clearly these drivers are changing—population and the world economy are growing, for instance, there are major advances in information technology and biotechnology, and the world is becoming more interconnected. Changes in these drivers are projected to increase the demand for and consumption of food, fiber, clean water, and energy, which will in turn affect the direct drivers. The direct drivers are primarily physical, chemical, and biological—such as land cover change, climate change, air and water pollution, irrigation, use of fertilizers, harvesting, and the introduction of alien invasive species. Change is apparent here too: the climate is changing, species ranges are shifting, alien species are spreading, and land degradation continues.

An important point is that any decision can have consequences external to the decision framework. These consequences are called externalities because they are not part of the decision-making calculus. Externalities can have positive or negative effects. For example, a decision to subsidize fertilizers to increase crop production might result in substantial degradation of water quality from the added nutrients and degradation of downstream fisheries. But it is also possible to have positive externalities. A beekeeper might be motivated by the profits to be made from selling honey, for instance, but neighboring orchards could produce more apples because of enhanced pollination arising from the presence of the bees.

Multiple interacting drivers cause changes in ecosystem services. There are functional interdependencies between and among the indirect and direct drivers of change, and, in turn, changes in ecological services lead to feedbacks on the drivers of changes in ecological services. Synergetic driver combinations are common. The many processes of globalization lead to new forms of interactions between drivers of changes in ecosystem services.

1.3.4 Cross-scale Interactions and Assessment

An effective assessment of ecosystems and human well-being cannot be conducted at a single temporal or spatial scale. Thus the MA conceptual framework includes both of these dimensions. Ecosystem changes that may have little impact on human well-being over days or weeks (soil erosion, for instance) may have pronounced impacts over years or decades (declining agricultural productivity). Similarly, changes at a local scale may have little impact on some services at that scale (as in the local impact of forest loss on water availability) but major impacts at large scales (forest loss in a river basin changing the timing and magnitude of downstream flooding).

Ecosystem processes and services are typically most strongly expressed, are most easily observed, or have their dominant controls or consequences at particular spatial and temporal scales. They often exhibit a characteristic scale—the typical extent or duration over which processes have their impact. Spatial and temporal scales are often closely related. For instance, food production is a localized service of an ecosystem and changes on a weekly basis, water regulation is regional and changes on a monthly or seasonal basis, and climate regulation may take place at a global scale over decades.

Assessments need to be conducted at spatial and temporal scales appropriate to the process or phenomenon being examined. Those done over large areas generally use data at coarse resolutions, which may not detect fine-resolution processes. Even if data are collected at a fine level of detail, the process of averaging in order to present findings at the larger scale causes local patterns or anomalies to disappear. This is particularly problematic for processes exhibiting thresholds and nonlinearities. For example, even though a number of fish stocks exploited in a particular area might have collapsed due to overfishing, average catches across all stocks (including healthier stocks) would not reveal the extent of

the problem. Assessors, if they are aware of such thresholds and have access to high-resolution data, can incorporate such information even in a large-scale assessment. Yet an assessment done at smaller spatial scales can help identify important dynamics of the system that might otherwise be overlooked. Likewise, phenomena and processes that occur at much larger scales, although expressed locally, may go unnoticed in purely local-scale assessments. Increased carbon dioxide concentrations or decreased stratospheric ozone concentrations have local effects, for instance, but it would be difficult to trace the causality of the effects without an examination of the overall global process.

Time scale is also very important in conducting assessments. Humans tend not to think beyond one or two generations. If an assessment covers a shorter time period than the characteristic temporal scale, it may not adequately capture variability associated with long-term cycles, such as glaciation. Slow changes are often harder to measure, as is the case with the impact of climate change on the geographic distribution of species or populations. Moreover, both ecological and human systems have substantial inertia, and the impact of changes occurring today may not be seen for years or decades. For example, some fisheries' catches may increase for several years even after they have reached unsustainable levels because of the large number of juvenile fish produced before that level was reached.

Social, political, and economic processes also have characteristic scales, which may vary widely in duration and extent. Those of ecological and sociopolitical processes often do not match. Many environmental problems originate from this mismatch between the scale at which the ecological process occurs, the scale at which decisions are made, and the scale of institutions for decision-making. A purely local-scale assessment, for instance, may discover that the most effective societal response requires action that can occur only at a national scale (such as the removal of a subsidy or the establishment of a regulation). Moreover, it may lack the relevance and credibility necessary to stimulate and inform national or regional changes. On the other hand, a purely global assessment may lack both the relevance and the credibility necessary to lead to changes in ecosystem management at the local scale where action is needed. Outcomes at a given scale are often heavily influenced by interactions of ecological, socioeconomic, and political factors emanating from other scales. Thus focusing solely on a single scale is likely to miss interactions with other scales that are critically important in understanding ecosystem determinants and their implications for human well-being.

The choice of the spatial or temporal scale for an assessment is politically laden, since it may intentionally or unintentionally privilege certain groups. The selection of assessment scale with its associated level of detail implicitly favors particular systems of knowledge, types of information, and modes of expression over others. For example, non-codified information or knowledge systems of minority populations are often missed when assessments are undertaken at larger spatial scales or higher levels of aggregation. Reflecting on the political consequences of scale and boundary choices is an important prerequisite to exploring what multi- and cross-scale analysis in the MA might contribute to decision-making and public policy processes at various scales.

1.4 Values Associated with Ecosystems

Current decision-making processes often ignore or underestimate the value of ecosystem services. Decision-making concerning ecosystems and their services can be particularly challenging because different disciplines, philosophical views, and schools of thought assess the value of ecosystems differently. One paradigm of value, known as the utilitarian (anthropocentric) concept, is based on the principle of humans' preference satisfaction (welfare). In this case, ecosystems and the services they provide have value to human societies because people derive utility from their use, either directly or indirectly (use values). Within this utilitarian concept of value, people also give value to ecosystem services that they are not currently using (non-use values). Non-use values, usually known as existence values, involve the case where humans ascribe value to knowing that a resource exists even if they never use that resource directly. These often involve the deeply held historical, national, ethical, religious, and spiritual values people ascribe to ecosystems—the values that the MA recognizes as cultural services of ecosystems.

A different, non-utilitarian value paradigm holds that something can have intrinsic value—that is, it can be of value in and for itself—irrespective of its utility for someone else. From the perspective of many ethical, religious, and cultural points of view, ecosystems may have intrinsic value, independent of their contribution to human well-being.

The utilitarian and non-utilitarian value paradigms overlap and interact in many ways, but they use different metrics, with no common denominator, and cannot usually be aggregated, although both paradigms of value are used in decision-making processes.

Under the utilitarian approach, a wide range of methodologies has been developed to attempt to quantify the benefits of different ecosystem services. These methods are particularly well developed for provisioning services, but recent work has also improved the ability to value regulating and other services. The choice of valuation technique in any given instance is dictated by the characteristics of the case and by data availability. (See Box 1.4.)

Non-utilitarian value proceeds from a variety of ethical, cultural, religious, and philosophical bases. These differ in the specific entities that are deemed to have intrinsic value and in the interpretation of what having intrinsic value means. Intrinsic value may complement or counterbalance considerations of utilitarian value. For example, if the aggregate utility of the services provided by an ecosystem (as measured by its utilitarian value) outweighs the value of converting it to another use, its intrinsic value may then be complementary and provide an additional impetus for conserving the ecosystem. If, however, economic valuation indicates that the value of converting the ecosystem outweighs the aggregate value of its services, its ascribed intrinsic value may be deemed great enough to warrant a social decision to conserve it anyway. Such decisions are essentially political, not economic. In contemporary democracies these decisions are made by parliaments or legislatures or by regulatory agencies mandated to do so by law. The sanctions for violating laws recognizing an entity's intrinsic value may be regarded as a measure of the degree of intrinsic value ascribed to them. The decisions taken by businesses, local communities, and individuals also can involve considerations of both utilitarian and non-utilitarian values.

The mere act of quantifying the value of ecosystem services cannot by itself change the incentives affecting their use or misuse. Several changes in current practice may be required to take better account of these values. The MA assesses the use of information on ecosystem service values in decision-making. The goal is to improve decision-making processes and tools and to provide feedback regarding the kinds of information that can have the most influence.

1.5 Assessment Tools

The information base exists in any country to undertake an assessment within the framework of the MA. That said, although new

---

BOX 1.4

Valuation of Ecosystem Services

Valuation can be used in many ways: to assess the total contribution that ecosystems make to human well-being, to understand the incentives that individual decision-makers face in managing ecosystems in different ways, and to evaluate the consequences of alternative courses of action. The MA uses valuation primarily in the latter sense: as a tool that enhances the ability of decision-makers to evaluate trade-offs between alternative ecosystem management regimes and courses of social actions that alter the use of ecosystems and the multiple services they provide. This usually requires assessing the change in the mix (the value) of services provided by an ecosystem resulting from a given change in its management.

Most of the work involved in estimating the change in the value of the flow of benefits provided by an ecosystem involves estimating the change in the physical flow of benefits (quantifying biophysical relations) and tracing through and quantifying a chain of causality between changes in ecosystem condition and human welfare. A common problem in valuation is that information is only available on some of the links in the chain and often in incompatible units. The MA can make a major contribution by making various disciplines better aware of what is needed to ensure that their work can be combined with that of others to allow a full assessment of the consequences of altering ecosystem state and function.

The ecosystem values in this sense are only one of the bases on which decisions on ecosystem management are and should be made. Many other factors, including notions of intrinsic value and other objectives that society might have (such as equity among different groups or generations), will also feed into the decision framework. Even when decisions are made on other bases, however, estimates of changes in utilitarian value provide invaluable information.

---

data sets (for example, from remote sensing) providing globally consistent information make a global assessment like the MA more rigorous, there are still many challenges that must be dealt with in using these data at global or local scales. Among these challenges are biases in the geographic and temporal coverage of the data and in the types of data collected. Data availability for industrial countries is greater than that for developing ones, and data for certain resources such as crop production are more readily available than data for fisheries, fuelwood, or biodiversity. The MA makes extensive use of both biophysical and socioeconomic indicators, which combine data into policy-relevant measures that provide the basis for assessment and decision-making.

Models can be used to illuminate interactions among systems and drivers, as well as to make up for data deficiencies—for instance, by providing estimates where observations are lacking. The MA makes use of environmental system models that can be used, for example, to measure the consequences of land cover change for river flow or the consequences of climate change for the distribution of species. It also uses human system models that can examine, for instance, the impact of changes in ecosystems on production, consumption, and investment decisions by households or that allow the economy-wide impacts of a change in production in a particular sector like agriculture to be evaluated. Finally, integrated models, combining both the environmental and human systems linkages, can increasingly be used at both global and sub-global scales.

The MA incorporates both formal scientific information and traditional or local knowledge. Traditional societies have nurtured and refined systems of knowledge of direct value to those societies but also of considerable value to assessments undertaken at regional and global scales. This information often is unknown to science and can be an expression of other relationships between society and nature in general and of sustainable ways of managing natural resources in particular. To be credible and useful to decision-makers, all sources of information, whether scientific, traditional, or practitioner knowledge, must be critically assessed and validated as part of the assessment process through procedures relevant to the form of knowledge.

Since policies for dealing with the deterioration of ecosystem services are concerned with the future consequences of current actions, the development of scenarios of medium- to long-term changes in ecosystems, services, and drivers can be particularly helpful for decision-makers. Scenarios are typically developed through the joint involvement of decision-makers and scientific experts, and they represent a promising mechanism for linking scientific information to decision-making processes. They do not attempt to predict the future but instead are designed to indicate what science can and cannot say about the future consequences of alternative plausible choices that might be taken in the coming years.

The MA uses scenarios to summarize and communicate the diverse trajectories that the world's ecosystems may take in future decades. Scenarios are plausible alternative futures, each an example of what might happen under particular assumptions. They can be used as a systematic method for thinking creatively about complex, uncertain futures. In this way, they help us understand the upcoming choices that need to be made and highlight developments in the present. The MA developed scenarios that connect possible changes in drivers (which may be unpredictable or uncontrollable) with human demands for ecosystem services. The scenarios link these demands, in turn, to the futures of the services themselves and the aspects of human welfare that depend on them. The scenario building exercise breaks new ground in several areas:

- development of scenarios for global futures linked explicitly to ecosystem services and the human consequences of ecosystem change,
- consideration of trade-offs among individual ecosystem services within the "bundle" of benefits that any particular ecosystem potentially provides to society,
- assessment of modeling capabilities for linking socioeconomic drivers and ecosystem services, and
- consideration of ambiguous futures as well as quantifiable uncertainties.

The credibility of assessments is closely linked to how they address what is not known in addition to what is known. The consistent treatment of uncertainty is therefore essential for the clarity and utility of assessment reports. As part of any assessment process, it is crucial to estimate the uncertainty of findings even if a detailed quantitative appraisal of uncertainty is unavailable.

1.6 Strategies and Interventions

The MA assesses the use and effectiveness of a wide range of options for responding to the need to sustainably use, conserve, and restore ecosystems and the services they provide. These options include incorporating the value of ecosystems in decisions, channeling diffuse ecosystem benefits to decision-makers with focused local interests, creating markets and property rights, educating and dispersing knowledge, and investing to improve ecosystems and the services they provide. As seen in Box 1.2 on

the MA conceptual framework, different types of response options can affect the relationships of indirect to direct drivers, the influence of direct drivers on ecosystems, the human demand for ecosystem services, or the impact of changes in human well-being on indirect drivers. An effective strategy for managing ecosystems will involve a mix of interventions at all points in this conceptual framework.

Mechanisms for accomplishing these interventions include laws, regulations, and enforcement schemes; partnerships and collaborations; the sharing of information and knowledge; and public and private action. The choice of options to be considered will be greatly influenced by both the temporal and the physical scale influenced by decisions, the uncertainty of outcomes, cultural context, and the implications for equity and trade-offs. Institutions at different levels have different response options available to them, and special care is required to ensure policy coherence.

Decision-making processes are value-based and combine political and technical elements to varying degrees. Where technical input can play a role, a range of tools is available to help decision-makers choose among strategies and interventions, including cost-benefit analysis, game theory, and policy exercises. The selection of analytical tools should be determined by the context of the decision, key characteristics of the decision problem, and the criteria considered to be important by the decision-makers. Information from these analytical frameworks is always combined with the intuition, experience, and interests of the decision-maker in shaping the final decisions.

Risk assessment, including ecological risk assessment, is an established discipline and has a significant potential for informing the decision process. Finding thresholds and identifying the potential for irreversible change are important for the decision-making process. Similarly, environmental impact assessments designed to evaluate the impact of particular projects and strategic environmental assessments designed to evaluate the impact of policies both represent important mechanisms for incorporating the findings of an ecosystem assessment into decision-making processes.

Changes also may be required in decision-making processes themselves. Experience to date suggests that a number of mechanisms can improve the process of making decisions about ecosystem services. Broadly accepted norms for decision-making process include the following characteristics. Did the process:

- bring the best available information to bear?
- function transparently, use locally grounded knowledge, and involve all those with an interest in a decision?
- pay special attention to equity and to the most vulnerable populations?
- use decision analytical frameworks that take account of the strengths and limits of individual, group, and organizational information processing and action?
- consider whether an intervention or its outcome is irreversible and incorporate procedures to evaluate the outcomes of actions and learn from them?
- ensure that those making the decisions are accountable?
- strive for efficiency in choosing among interventions?
- take account of thresholds, irreversibility, and cumulative, cross-scale, and marginal effects and of local, regional, and global costs, risk, and benefits?

The policy or management changes made to address problems and opportunities related to ecosystems and their services, whether at local scales or national or international scales, need to be adaptive and flexible in order to benefit from past experience, to hedge against risk, and to consider uncertainty. The understanding of ecosystem dynamics will always be limited, socioeconomic systems will continue to change, and outside determinants can never be fully anticipated. Decision-makers should consider whether a course of action is reversible and should incorporate, whenever possible, procedures to evaluate the outcomes of actions and learn from them. Debate about exactly how to do this continues in discussions of adaptive management, social learning, safe minimum standards, and the precautionary principle. But the core message of all approaches is the same: acknowledge the limits of human understanding, give special consideration to irreversible changes, and evaluate the impacts of decisions as they unfold.

Chapter 2

Analytical Approaches for Assessing Ecosystem Condition and Human Well-being

Coordinating Lead Authors: Ruth DeFries, Stefano Pagiola
Lead Authors: W.L. Adamowicz, H. Resit Akçakaya, Agustin Arcenas, Suresh Babu, Deborah Balk, Ulisses Confalonieri, Wolfgang Cramer, Fander Falconí, Steffen Fritz, Rhys Green, Edgar Gutiérrez-Espeleta, Kirk Hamilton, Racine Kane, John Latham, Emily Matthews, Taylor Ricketts, Tian Xiang Yue
Contributing Authors: Neville Ash, Jillian Thönell
Review Editors: Gerardo Ceballos, Sandra Lavorel, Gordon Orians, Stephen Pacala, Jatna Supriatna, Michael Young

Main Messages

Many tools are available to assess ecosystem condition and support policy decisions that involve trade-offs among ecosystem services. Clearing forested land, for example, affects multiple ecosystem services (such as food production, biodiversity, carbon sequestration, and watershed protection), each of which affects human well-being (such as increased income from crops, reduced tourism value of biodiversity, and damage from downstream flooding). Assessing these trade-offs in the decision-making process requires scientifically based analysis to quantify the responses to different management alternatives. Scientific advances over the past few decades, particularly in computer modeling, remote sensing, and environmental economics, make it possible to assess these linkages.

The availability and accuracy of data sources and methods for this assessment are unevenly distributed for different ecosystem services and geographic regions. Data on provisioning services, such as crop yield and timber production, are usually available. On the other hand, data on regulating, supporting, and cultural services such as nutrient cycling, climate regulation, or aesthetic value are seldom available, making it necessary to use indicators, model results, or extrapolations from case studies as proxies. Systematic data collection for carefully selected indicators reflecting trends in ecosystem condition and their services would provide an improved basis for future assessments. Methods for quantifying ecosystem responses are also uneven. Methods to estimate crop yield responses to fertilizer application, for example, are well developed. But methods to quantify relationships between ecosystem services and human well-being, such as the effects of deteriorating biodiversity on human disease, are at an earlier stage of development.

Ecosystems respond to management changes on a range of time and space scales, and careful definition of the scales included in analyses is critical. Soil nutrient depletion, for example, occurs over decades and would not be captured in an analysis based on a shorter time period. Some of the impact of deforestation is felt in reduced water quality far downstream; an analysis that only considers the forest area itself would miss this impact. Ideally, analysis at varying scales would be carried out to assess trade-offs properly. In particular, it is essential to consider nonlinear responses of ecosystems to perturbations in analysis of trade-offs, such as loss of resilience to climate variability below a threshold number of plant species.

Ecosystem condition is only one of many factors that affect human well-being, making it challenging to assess linkages between them. Health outcomes, for example, are the combined result of ecosystem condition, access to health care, economic status, and myriad other factors. Interpretations of trends in indicators of well-being must appropriately account for the full range of factors involved. The impacts of ecosystem change on well-being are often subtle, which is not to say unimportant; impacts need not be drastic to be significant. A small increase in food prices resulting from lower yields will affect many people, even if none starve as a result. Tracing these impacts is often difficult, particularly in aggregate analyses where the signal of the effect of ecosystem change is often hidden by multiple confounding factors. Analyses linking well-being and ecosystem condition are most easily carried out at a local scale, where the linkages can be most clearly identified.

Ultimately, decisions about trade-offs in ecosystem services require balancing societal objectives, including utilitarian and non-utilitarian objectives, short- and long-term objectives, and local- and global-scale objectives. The analytical approach for this report aims to quantify, to the degree possible, the most important trade-offs within different ecosystems and among ecosystem services as input to weigh societal objectives based on comprehensive analysis of the full suite of ecosystem services.

2.1 Introduction

This report systematically assesses the current state of and recent trends in the world's ecosystems and their services and the significance of these changes for human livelihoods, health, and well-being. The individual chapters draw on a wide variety of data sources and analytical methods from both the natural and social sciences. This chapter provides an overview of many of these data and methods, their basis in the scientific literature, and the limitations and possibilities for application to the assessment of ecosystem condition, trends, and implications for human well-being. (See Figure 2.1.)

The Millennium Ecosystem Assessment's approach is premised on the notion that management decisions generally involve trade-offs among ecosystem services and that quantitative and scientifically based assessment of the trade-offs is a necessary ingredient for sound decision-making. For example, decisions to clear land for agriculture involve trade-offs between food production and protection of biological resources; decisions to extract timber involve trade-offs between income from timber sales and watershed protection; and decisions to designate marine protected areas involve trade-offs between preserving fish stocks and the availability of fish or jobs for local populations. Accounting for these trade-offs involves quantifying the effects of the management decision on ecosystem services and human-well being in comparable units over varying spatial and temporal scales.

The next section of this chapter discusses data and methods for assessing conditions and trends in ecosystems and their services. Individual chapters of this report apply these methods to identify the implications of changes in ecosystem condition (such as forest conversion to cropland) for ecosystem services (such as flood protection). Rigorous analyses of these linkages are a key prerequisite to quantifying the effects on human well-being (such as damage from downstream flooding).

The third section discusses data and methods for quantifying the effects of changes in ecosystem services on human well-being, including human health, economic costs and benefits, and poverty and other measures of well-being, and on the intrinsic value of ecosystems. These methods provide a framework for assessing management decisions or policies that alter ecosystems, based on comprehensive information about the repercussions for human well-being from intentional or unintentional alteration of ecosystem services.

The final section of this chapter discusses approaches for assessing trade-offs from management decisions. These approaches aim to quantify, in comparable units, the repercussions of a decision for the full range of ecosystem services. The approaches must also account for the varying spatial and temporal scale over which management decisions alter ecosystem services. Decisions to clear forests, for example, provide immediate economic benefits for local interests but contribute to an increase of greenhouse gases in the atmosphere, with longer-term implications at the global scale.

While this chapter provides a general overview of the available methods and data sources and their applicability to the assessment, individual chapters provide detailed descriptions of data sources used in reference to a particular ecosystem or service. Core data sets used by all chapters to ensure consistency and comparability among the different ecosystems are described in Appendix 2.1.

The data sources and methods used in this report were generally not developed explicitly for this assessment. Yet the combination of approaches—including computer modeling, natural resource and biodiversity inventories, remote sensing and geographic information systems, traditional knowledge, case studies,

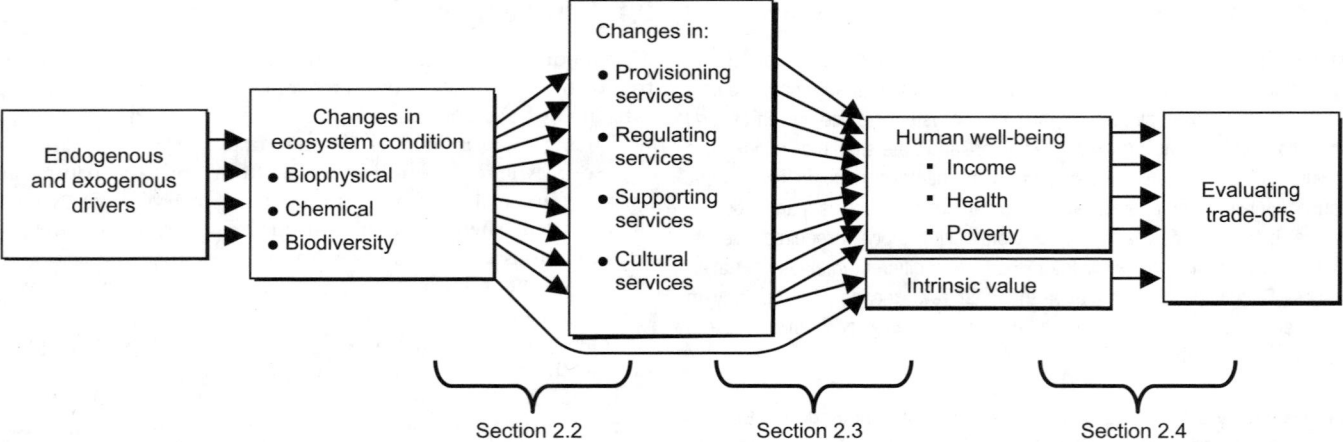

Figure 2.1. Linking Ecosystem Condition to Well-being Requires Assessing Ecosystem Condition and Its Effect on Services, the Impact on Human Well-being and Other Forms of Value, and Trade-offs among Objectives

indicators of ecosystem conditions and human well-being, and economic valuation techniques—provides a strong scientific foundation for the assessment. Systematic data collection for carefully selected indicators reflecting trends in ecosystem condition and their services would provide a basis for future assessments.

2.2 Assessing Ecosystem Condition and Trends

The foundation for analysis is basic information about each ecosystem service (Chapters 7–17) and spatially defined ecosystem (Chapters 18–27). Deriving conclusions about the important trends in ecosystem condition and trade-offs among ecosystem services requires the following basic information:

- What are the current spatial extent and condition of ecosystems?
- What are the quality, quantity, and spatial distributions of services provided by the systems?
- Who lives in the ecosystem and what ecosystem services do they use?
- What are the trends in ecosystem condition and their services in the recent (decades) and more distant past (centuries)?
- How does ecosystem condition, and in turn ecosystem services, respond to the drivers of change for each system?

The availability of data and applicability of methods to derive this basic information (see Table 2.1) vary from ecosystem to ecosystem, service to service, and even region to region within an ecosystem type. For example, the U.N. Food and Agriculture Organization reports data on agricultural products, timber, and fisheries at the national level (e.g., FAO 2000a). Although data reliability is sometimes questionable due to known problems such as definitions that vary between data-submitting countries, data on provisioning ecosystem services with value as commodities are generally available. On the other hand, data on the spatial distribution, quantity, and quality of regulating, supporting, and cultural services such as nutrient cycling, climate regulation, or aesthetic value have generally not been collected, and it is necessary to use indicators, modeled results, or extrapolations from case studies as proxy data. Within a given ecosystem service or geographic system, resource inventories and census data are generally more readily available and reliable in industrial than developing countries.

The following sections provide overviews of each of these data sources and analytical approaches used throughout the report.

2.2.1 Remote Sensing and Geographic Information Systems

The availability of data to monitor ecosystems on a global scale is the underpinning for the MA. Advances in remote sensing technologies over the past few decades now enable repeated observations of Earth's surface. The potential to apply these data for assessing trends in ecosystem condition is only beginning to be realized. Moreover, advances in analytical tools such as geographic information systems allow data on the physical, biological, and socioeconomic characteristics of ecosystems to be assembled and interpreted in a spatial framework, making it feasible to establish linkages between drivers of change and trends in ecosystem services.

2.2.1.1 Remote Sensing

Ground-based surveys for mapping vegetation and other biophysical characteristics can be carried out over limited areas, but it would be an enormous undertaking to carry out globally comprehensive ground-based surveys over the entire surface of Earth. Remote sensing—broadly defined as the science of obtaining information about an object without being in direct physical contact (Colwell 1983)—is the primary data source for mapping the extent and condition of ecosystems over large areas. Moreover, remote sensing provides measurements that are consistent over the entire area being observed and are not subject to varying data collection methods in different locations, unlike ground-based measurements. Repeated observations using the same remote sensing instrument also provide measurements that are consistent through time as well as through space.

Most remote sensing data useful to assess ecosystem conditions and trends are obtained from sensors on satellites. (See Table 2.2.) Satellite data are generally digital and consequently amenable to computer-based analysis for classifying land cover types and assessing trends. There are several types of digital remotely sensed data (Jensen 2000). Optical remote sensing provides digital images of the amount of electromagnetic energy reflected or emitted from Earth's surface at various wavelengths. Active remote sensing of long-wavelengths microwaves (radar), short-wavelength laser light (lidar), or sound waves (sonar) measures the amount of backscatter from electromagnetic energy emitted from the sensor itself.

The spatial resolution (area of ground observed in a picture element or pixel), temporal resolution (how often the sensor re-

Table 2.1. Data Sources and Analytical Approaches for Assessing Ecosystem Condition and Trends

Type of Information Required	Data Source or Analytical Method						
	Remote Sensing and GIS	Natural Resource and Biodiversity Inventories	Socioeconomic Data	Ecosystem Models	Indicators of Ecosystem Condition	Indigenous and Traditional Knowledge	Case Studies of Ecosystem Response to Drivers
Current spatial extent and condition of ecosystem	X	X			X		
Quality, quantity, and spatial distributions of services provided by system		X		X			
Human populations residing in and deriving livelihoods from system			X			X	X
Trends in ecosystem conditions and services	X	X		X	X	X	X
Response of ecosystem condition and services to drivers				X	X	X	X

cords imagery from a particular area), spectral resolution (number of specific wavelength intervals in the electromagnetic spectrum to which the sensor is sensitive), and radiometric resolution (precision in the detected signal) determine the utility of the data for a specific application. For example, data with very high spatial resolution can be used to map habitats over local areas, but low temporal resolution limits the ability to map changes over time.

A key element in the interpretation of remote sensing data is calibration and validation with in situ data. Ground-based data aids the interpretation of satellite data by identifying locations of specific features in the land surface. These locations can then be pinpointed on the satellite image to obtain the spectral signatures of different features. Ground-based data are also critical to test the accuracy and reliability of the interpretation of satellite data. Linking ground-based with satellite data poses logistical challenges if the locations required are inaccessible. Moreover, the land surface is often heterogeneous so that a single pixel observed by the satellite contains multiple vegetation types. The ground observations then need to be scaled to the spatial resolution of the sensor. Despite these challenges, ground-based data for calibration and validation are central to the effective use of satellite data for ecosystem assessment.

Analyses of satellite data are a major contribution to assessments of ecosystem conditions and trends, especially over large areas where it is not feasible to perform ground surveys. Technological challenges such as sensor drift and sensor degradation over time, lack of data continuity, and persistent cloud cover, particularly in humid tropics, are challenges to routine application of satellite data to monitor ecosystem condition. Ground observations and local expertise are critical to accurate interpretation of satellite data.

Satellite data contribute to several types of information needs for assessments of ecosystem condition, including land cover and land cover change mapping, habitat mapping for biodiversity, wetland mapping, land degradation assessments, and measurements of land surface attributes as input to ecosystem models.

2.2.1.1.1 Mapping of land cover and land cover change

Over the last few decades, satellite data have increasingly been used to map land cover at national, regional, continental, and global scales. During the 1980s, pioneering research was conducted to map vegetation at continental scales, primarily with data acquired by the U.S. National Oceanographic and Atmospheric Administration's meteorological satellite, the Advanced Very High Resolution Radiometer. Multitemporal data describing seasonal variations in photosynthetic activity were used to map vegetation types in Africa (Tucker 1985) and South America (Townshend 1987). In the 1990s, AVHRR data were used to map land cover globally at increasingly higher spatial resolution, with the first global land cover classification at 1x1 degree resolution (approximately 110x110 kilometers) (DeFries and Townshend 1994), followed by 8x8 kilometer resolution (DeFries 1998) and finally 1x1 kilometer resolution (Loveland and Belward 1997; Hansen 2000).

Global satellite data also have enabled mapping of fractional tree cover to further characterize the distributions of forests over Earth's surface (DeFries 2000). At pantropical scales, AVHRR data have been used to map the distribution of humid forests (Malingreau 1995; Mayaux 1998), and radar data provide useful information for mapping land cover types where frequent cloud cover presents difficulties for optical data (DeGrandi 2000; Saatchi 2000; Mayaux et al. 2002). A suite of recently launched sensors, including MODIS, SPOT Vegetation, and GLI, provide globally comprehensive data to map vegetation types with greater accuracy due to improved spectral, spatial, and radiometric resolutions of these sensors (Friedl 2002). The GLC2000 land cover map derived from SPOT Vegetation data provides the basis for the MA's geographic designation of ecosystems (Bartholome and Belward 2004; Fritz et al. 2004). (See Appendix 2.1.)

One of the most significant contributions to be gained from satellite data is the identification and monitoring of land cover change, an important driver of changes in ecosystem services.

Table 2.2. Satellite Sensors for Monitoring Land Cover, Land Surface Properties, and Land and Marine Productivity

Platform	Sensor	Spatial Resolution at Nadir	Date of Observations
Coarse Resolution Satellite Sensors (> 1 km)			
NOAA–TIROS (National Oceanic and Atmospheric Administration–Television and Infrared Observation Satellite)	AVHRR (Advanced Very High Resolution Radiometer)	1.1km (local area coverage) 8km (global area coverage)	1978–present
SPOT (Systéme Probatoire pour la Observation de la Terre)	VEGETATION	1.15km	1998–present
ADEOS-II (Advanced Earth Observing Satellite)	POLDER (Polarization and Directionality of the Earth's Reflectances)	7km x 6km	2002–present
SeaStar	SeaWIFS (Sea viewing Wide Field of View)	1km (local coverage); 4km (global coverage)	1997–present
Moderate Resolution Satellite Sensors (250 m–1 km)			
ADEOS-II (Advanced Earth Observing Satellite)	GLI (Global Imager)	250m–1km	2002–present
EOS AM and PM (Earth Observing System)	MODIS (Moderate Resolution Spectroradiometer)	250–1,000m	1999–present
EOS AM and PM (Earth Observing System)	MISR (Multi-angle Imaging Spectroradiometer)	275m	1999–present
Envisat	MERIS (Medium Resolution Imaging Spectroradiometer)	350–1,200m	2002–present
Envisat	ASAR (Advanced Synthetic Aperature Radar)	150–1,000m	2002–present
High Resolution Satellite Sensors (20 m–250 m)[a]			
SPOT (Systéme Probatoire pour la Observation de la Terre)	HRV (High Resolution Visible Imaging System)	20m; 10m (panchromatic)	1986–present
ERS (European Remote Sensing Satellite)	SAR (Synthetic Aperture Radar)	30m	1995–present
Radarsat		10–100m	1995–present
Landsat (Land Satellite)	MSS (Multispectral Scanner)	83m	1972–97
Landsat (Land Satellite)	TM (Thematic Mapper)	30m (120m thermal-infrared band)	1984–present
Landsat (Land Satellite)	ETM+ (Enhanced Thematic Mapper)	30m	1999–present
EOS AM and PM (Earth Observing System)	ASTER (Advanced Spaceborne Thermal Emission and Reflection Radiometer)	15–90m	1999–present
IRS (Indian Remote Sensing)	LISS 3 (Linear Imaging Self-scanner)	23m; 5.8m (panchromatic)	1995–present
Very High Resolution Satellite Sensors (< 20 m) [a]			
JERS (Japanese Earth Resources Satellite)	SAR (Synthetic Aperature Radar)	18m	1992–98
JERS (Japanese Earth Resources Satellite)	OPS	18mx24m	1992–98
IKONOS		1m panchromatic; 4m multispectral	1999–present
QuickBird		0.61m panchromatic; 2.44m multispectral	2001–present
SPOT–5	HRG–HRS	10m; 2.5m (panchromatic)	2002–present

Note: The list is not intended to be comprehensive.

[a] Data were not acquired continuously within the time period.

Data acquired by Landsat and SPOT HRV have been the primary sources for identifying land cover change in particular locations. Incomplete spatial coverage, infrequent temporal coverage, and large data volumes have precluded global analysis of land cover change. With the launch of Landsat 7 in April 1999, data are obtained every 16 days for most parts of Earth, yielding more comprehensive coverage than previous Landsat sensors. Time series of Landsat and SPOT imagery have been applied to identify and measure deforestation and regrowth mainly in the humid tropics (Skole and Tucker 1993; FAO 2000a; Achard 2002). Deforestation is the most measured process of land cover change at the regional scale, although major uncertainties exist about absolute area and rates of change (Lepers et al. 2005).

Data continuity is a key requirement for effectively identifying land cover change. With the exception of the coarse resolution AVHRR Global Area Coverage observations over the past 20 years, continuous global coverage has not been possible. DeFries et al. (2002) and Hansen and DeFries (2004) have applied the AVHRR time series to identify changes in forest cover over the last two decades, illustrating the feasibility of using satellite data to detect these changes on a routine basis. Continuity of observations in the future is an essential component for monitoring land cover change and identifying locations with rapid change. For long-term data sets that cover time periods longer than the lifetime of a single sensor, cross calibration for a period of overlap is necessary. Moreover, classification schemes used to interpret the satellite data need to be clearly defined and flexible enough to allow comparisons over time.

2.2.1.1.2 Applications for biodiversity

There are two approaches for applying remote sensing to biodiversity assessments: direct observations of organisms and communities and indirect observations of environmental proxies of biodiversity (Turner et al. 2003). Direct observations of individual organisms, species assemblages, or ecological communities are possible only with hyperspatial, very high resolution (~1m) data. Such data can be applied to identify large organisms over small areas. Airborne observations have been used for censuses of large mammal abundances spanning several decades, for example in Kenya (Broten and Said 1995).

Indirect remote sensing of biodiversity relies on environmental parameters as proxies, such as discrete habitats (for example, woodland, wetland, grassland, or seabed grasses) or primary productivity. This approach has been employed in the US GAP analysis program (Scott and Csuti 1997). Another important indirect use of remote sensing is the detection of habitat loss and fragmentation to estimate the implications for biodiversity based on species-area relationships or other model approaches. (See Chapter 4.)

2.2.1.1.3 Wetland mapping

A wide range of remotely sensed data has been used to map wetland distribution and condition (Darras et al. 1998; Finlayson et al. 1999; Phinn et al. 1999). The utility of such data is a function of spatial and spectral resolutions, and careful choices need to be made when choosing such data (Lowry and Finlayson in press). The NOAA AVHRR, for example, observes at a relatively coarse nominal spatial resolution of 1.1 kilometer and allows only the broad distribution of wetlands to be mapped. More detailed observations of the extent of wetlands can be obtained using finer resolution Landsat TM (30 meters) and SPOT HRV (20 meters) data. As with all optical sensors, the data are frequently affected by atmospheric conditions, especially in tropical coastal areas where humidity is high and the presence of water beneath the vegetation canopy cannot be observed.

Remotely sensed data from newer spaceborne hyperspectral sensors, Synthetic Aperture Radar, and laser altimeters provide more comprehensive data on wetlands. Although useful for providing present-day baselines, however, the historical archive is limited, in contrast to the optical Landsat, AVHRR, and SPOT sensors, which date back to 1972, 1981, and 1986 respectively.

Aerial photographs have been acquired in many years for over half a century at fine spatial resolutions and when cloud cover is minimal. Photographs are available in a range of formats, including panchromatic black and white, near-infrared black and white, true color, and color infrared. Stereo pairs of photographs can be used to assess the vertical structure of vegetation and detect, for example, changes in the extent and height of mangroves (Lucas et al. 2002).

The European Space Agency's project Treaty Enforcement Services using Earth Observation has assessed the use of remote sensing for wetland inventory, assessment, and monitoring using combinations of sensors in support of wetland management. The approach has been extended through the GlobWetland project and its Global Wetland Information Service project to provide remotely sensed products for over 50 wetlands across 21 countries in Africa, Europe, and North and Central America. The project is designed to support on-the-ground implementation of the Ramsar Convention on Wetlands.

2.2.1.1.4 Assessing land degradation in drylands

Interpretation of remotely sensed data to identify land degradation in drylands is difficult because of large variations in vegetation productivity from year-to-year variations in climate. This variability makes it problematic to distinguish trends in land productivity attributable to human factors such as overgrazing, soil salinization, or burning from variations in productivity due to inter-annual climate variability or cyclical drought events (Reynolds and Smith 2002). Land degradation is defined by the Convention to Combat Desertification as "reduction or loss, in arid, semi-arid and dry sub-humid areas, of the biological or economic productivity of rainfed cropland, irrigated cropland, or ranges, pastures, forests, and woodlands resulting from land uses or from a process or combination of processes, including processes arising from human activities and habitation patterns." Quantifying changes in productivity involves an established baseline of land productivity against which changes can be assessed. Such a baseline is often not available. Furthermore, the inherent variability in year-to-year and even decade-to-decade fluctuations complicates the definition of a baseline.

One approach to assess land productivity is through rain-use efficiency, which quantifies net primary production (in units of biomass per unit time per unit area) normalized to the rainfall for that time period (Prince et al. 1990). Rain-use efficiency makes it possible to assess spatial and temporal differences in land productivity without the confounding factor of climate variability. Several models are available to estimate net primary production, as described later, with some using remotely sensed vegetation indices such as the Normalized Difference Vegetation Index (ratio of red to infrared reflectance indicating vegetative activity) as input data for the models. Studies have examined patterns in NDVI, rain-use efficiency, climate, and land use practices to investigate possible trends in land productivity and causal factors (e.g., Prince et al. 1990; Tucker et al. 1991; Nicholson et al. 1998).

The European Space Agency's TESEO project has examined the utility of remote sensing for mapping and monitoring desertification and land degradation in support of the Convention to Combat Desertification (TESEO 2003). Geostationary satellites such as Meteosat operationally provide basic climatological data, which are necessary to estimate rain-use efficiency and distinguish climatic from land use drivers of land degradation. Operational meteorological satellites, most notably the Advanced Very High Resolution Radiometer, have provided the longest continuous record for NDVI from the 1980s to the present. More recently launched sensors such as VEGETATION on-board SPOT and MODIS on-board the Earth Observation System have been designed specifically to monitor vegetation. Satellite data also identify locations of fire events and burn scars to provide information on changes in dryland condition related to changes in fire regime (Giglio et al. 1999). Applications of microwave sensors such as ERS are emerging as possible approaches to map and monitor primary production. Microwave sensors are sensitive to the amount of living aboveground vegetation and moisture content of the upper soil profile and are appropriate for identifying changes in semiarid and arid conditions.

Advancements in the application of remote sensing for mapping and monitoring land degradation involve not just technical issues but institutional issues as well (TESEO 2003). National capacities to use information and technology transfer currently limit the possible applications.

2.2.1.1.5 *Measurements of land surface and marine attributes as input to ecosystem models*

Satellite data, applied in conjunction with ecosystem models, provide spatially comprehensive estimates of parameters such as evapotranspiration, primary productivity, fraction of solar radiation absorbed by photosynthetic activity, leaf area index, percentage of solar radiation reflected by the surface (albedo) (Myneni 1992; Sellers 1996), ocean chlorophyll (Doney et al. 2003), and species distributions (Raxworthy et al. 2003). These parameters are related to several ecosystem services. For example, a decrease in evapotranspiration from the conversion of part of a forest to an urban system alters the ability of the forest system to regulate climate. A change in primary production relates to the food available for humans and other species. The satellite-derived parameters provide an important means for linking changes in ecosystem condition with implications for their services—for example, linking changes in climate regulation with changes in land and marine surface properties. (See Chapter 13.)

2.2.1.2 *Geographic Information Systems*

To organize and analyze remote sensing and other types of information in a spatial framework, many chapters in this report rely on geographic information systems. A GIS is a computer system consisting of computer hardware and software for entering, storing, retrieving, transforming, measuring, combining, subsetting, and displaying spatial data that have been digitized and registered to a common coordinate system (Heywood 1998; Johnston 1998). GIS allows disparate data sources to be analyzed spatially. For example, human population density can be overlain with data on net primary productivity or species endemism to identify locations within ecosystems where human demand for ecosystem services may be correlated with changes in ecosystem condition. Locations of roads can be entered into a GIS along with areas of deforestation to examine possible relationships between the two variables. The combination of remote sensing, GIS, and Global Positioning Systems for field validation is powerful for assessing trends in ecosystem condition (Hoffer 1994; ICSU 2002a).

GIS can be used in conjunction with remote sensing to identify land cover change. A common approach is to compare recent and historical high-resolution satellite images (such as Landsat Thematic Mapper). For example, Figure 2.2 illustrates the changes in forest cover between 1992 and 2001 in Mato Grosso, Brazil. Achard et al. (2002) have used this approach in 100 sample sites located in the humid tropical forests to estimate tropical deforestation.

GIS has also been applied in wilderness mapping, also known as "mapping human impact." These exercises estimate human influence through geographic proxies such as human population density, settlements, roads, land use, and other human-made features. All factors are integrated within the GIS and either summed up with equal weights (Sanderson 2002) or weighted according to perceptions of impact (Carver 2002). This exercise has been carried out at regional scales (for example Lesslie and Maslen 1995; Aplet 2000; Fritz 2001) as well as on a global scale (for example, UNEP 2001; Sanderson 2002). Sanderson et al. (2002) used the approach to estimate the 10% wildest areas in each biome of the world. The U.N. Environment Programme's Global Biodiversity (GLOBIO) project uses a similar methodology and examines human influence in relation to indicators of biodiversity (UNEP 2001).

A further application of GIS and remote sensing is to test hypotheses and responses of ecosystem services to future scenarios (Cleland 1994; Wadsworth and Treweek 1999). For example, GIS is used in the MA's sub-global assessment of Southern Africa to predict the degree of fuelwood shortages for the different districts of Northern Sofala Province, Mozambique, in 2030. This is done by using the GIS database showing available fuelwood per district in 1995 and projecting availability in 2030, assuming that the current trend of forest degradation of 0.05 hectares per person per year will continue. This allows identification of districts where fuelwood would be most affected.

GIS is also applicable for assessing relationships between health outcomes and environmental conditions (see Chapter 14) and for mapping risks of vulnerable populations to environmental stressors (see Chapter 6). The spatial displays aim to delineate the places, human groups, and ecosystems that have the highest risk associated with them. Examples include the "red data" maps depicting critical environmental situations (Mather and Sdasyuk 1991), maps of "environmentally endangered areas" (National Geographic Society 1989), and locations under risk from infrastructure expansion (Laurance et al. 2001), biodiversity loss (Myers et al. 2000), natural hazards, impacts from armed conflicts (Gleditsch et al. 2002), and rapid land cover change (Lepers et al. 2005). The analytical and display capabilities can draw attention to priority areas that require further analysis or urgent attention. Interactive Internet mapping is a promising approach for risk mapping but is currently in its infancy.

2.2.2 Inventories of Ecosystem Components

Inventories provide data on various ecosystem components relevant to this assessment. The most common and thorough types of inventories relate to the amount and distribution of provisioning services such as timber and agricultural products. Species inventories also provide information useful for assessing biodiversity, and demographic data provide essential information on human populations living within the systems.

2.2.2.1 *Natural Resource Inventories*

Many countries routinely conduct inventories of their natural resources. These generally assess the locations and amounts of

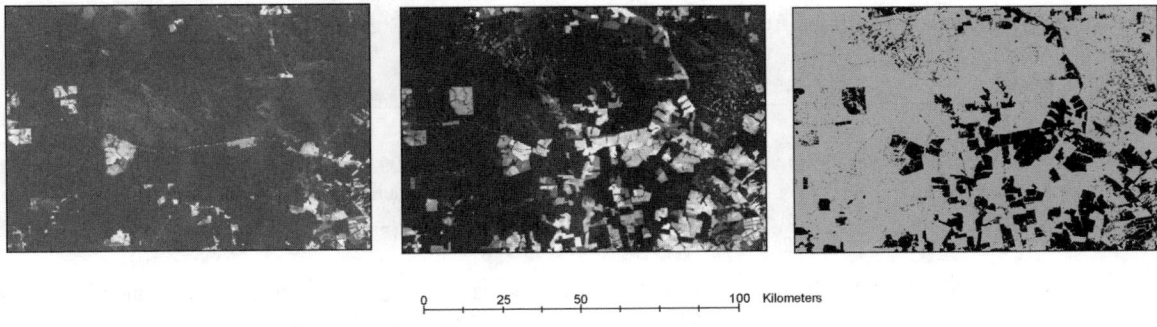

Figure 2.2. Subset of Landsat ETM+ Scenes for an Area in the State of Mato Grosso, Brazil Acquired August 6, 1992 (left) and July 30, 2001 (middle). Light to dark shades represent radiance in band 3 (.63–.62). The difference between the dates indicates deforestation in black (right). The area includes approximately 5534'25"W, 1154'20"S (bottom right corner).

economically important ecosystem services such as timber, agricultural products, and fisheries. FAO periodically publishes compilations of the national-level statistics in forest resources, agricultural production, fisheries production, and water resources. (See Table 2.3.) These statistics are widely used throughout this report. They are in many cases the only source of globally comprehensive data on these ecosystem services. Meta-analyses of local natural resource inventories also provide information on ecosystem condition and trends (Gardner et al. 2003), although they are not spatially comprehensive.

Although the assessment of ecosystem conditions and trends relies heavily on data from resource inventories, there are a number of limitations. First, questions remain about varying methods and definitions used by different countries for data collection (Matthews 2001). For example, several studies based on analysis of satellite data indicate that the FAO Forest Resource Assessment overestimates the rate of deforestation in some countries (Steininger 2001; Achard 2002; DeFries 2002). For fisheries, there are no globally consistent inventories of fisheries and fishery resources. Efforts to develop them are only starting, with the implementation of the FAO Strategy for Improving Information on Status and Trends of Capture Fisheries, which was adopted in 2003 in response to concerns about the reliability of fishery data (FAO 2000b).

Second, resource inventories are often aggregated to the national level or by sub-national administrative units. This level of aggregation does not match the ecosystem boundaries used as the reporting unit for the MA. Third, data quality is highly uneven, with greater reliability in industrial than developing countries. In many countries, deforestation "data" are actually projections based on models rather than empirical observations (Kaimowitz and Angelsen 1998). Fourth, statistics on the production of an ecosystem service do not necessarily provide information about the capacity of the ecosystem to continue to provide the service. For example, fisheries catches can increase for years through "mining" of the stocks even though the underlying biological capability of producing fish is declining, eventually resulting in a collapse. Finally, inventories for noncommodity ecosystem services, particularly the regulating, supporting, and cultural services, have not been systematically carried out.

2.2.2.2 Biodiversity Inventories

Inventories of the biodiversity of ecosystems are far less extensive than those of individual natural resources with value as commodities. Only a small fraction of biodiversity is currently monitored and assessed. This is probably because there are few perceived economic incentives to inventory biodiversity per se and because

biodiversity is a complex phenomenon that is difficult to quantify and measure. (See Chapter 4.) Nonetheless, biodiversity inventories can provide a general sense of the relative biodiversity importance (such as richness, endemism) of ecosystems; they can illuminate the impacts of different human activities and management policies on biodiversity; and, when targeted at service-providing taxa or functional groups (pollinators, for instance), they can link changes in biodiversity within these groups directly to changes in the service provided.

Biodiversity inventories are conducted at a range of spatial scales, which are chosen to best address the issue or question at hand. Most, however, can be usefully grouped into three distinct categories: global inventories, regional inventories, and local inventories. Because biodiversity is complex, inventories typically focus on one aspect of biodiversity at a time, such as species richness or habitat diversity. A few examples of inventories at each of these scales illustrate their relative strengths, limitations, and utilities for the MA.

At the global scale, only a handful of biodiversity inventories exist. These typically provide species lists for relatively well-known taxa, based on relatively large spatial units. For example, the World Conservation Monitoring Centre (1992) compiled species inventories of mammals, birds, and swallowtail butterflies for all nations in the world. The World Wild Fund for Nature is conducting an inventory of all vertebrates and plants in each of the world's 867 terrestrial ecoregions (defined by WWF as relatively large units of land or water containing a distinct assemblage of natural communities and species, with boundaries that approximate the original extent of natural communities prior to major land use change).

These inventories are useful for documenting overall patterns of biodiversity on Earth, in order to indicate global priorities for biodiversity conservation or areas of high-expected threat (Sisk et al. 1994; Ceballos and Brown 1995; Dinerstein 1995). Their utility for focused analyses is limited, however, by the coarse units on which they are based and their restriction to mostly vertebrate taxa (which are not often the most important for the provision of ecosystem services).

In addition, the World Conservation Union–IUCN has been producing Red Data Books and Red Lists of Threatened Species since the 1960s. Currently, the IUCN Red List is updated annually (see www.redlist.org). The criteria for listing are transparent and quantitative. The IUCN Red List is global in coverage and is the most comprehensive list of threatened species, with almost all known bird, mammal, and amphibian species evaluated; there are plans for complete coverage of reptiles in the next few years. Data on fish species include FISHBASE (Frose and Pauly 2000), Ceph-

Table 2.3. Examples of Resource Inventories Applicable to Assessing Ecosystem Condition and Trends

Type	Source	Description
Forest Resources		
Forest area and change	FAO, *Global Forest Resources Assessment*	Published every 10 years (1980, 1990, 2000). Provides national and global estimates of total forest area and net changes during the preceding decade, as well as information on plantations, forest ownership, management, and environmental parameters such as forest fires and biomass volumes.
Forest products	FAO, *State of the World's Forests*	Published every two years. Provide summary tables of national and regional production statistics for major categories of industrial roundwood, pulp, and paper.
	ITTO, *Annual Review and Assessment of the World Timber Situation*	Published annually. Tabular databases on volume and value of production, consumption, and trade among ITTO producer and consumer countries. Time series for five years prior to publication.
Wood energy	IEA, *Energy Statistics and Balances of OECD and Non-OECD Countries* (four reports)	Published every two years. IEA data since 1994–95 have covered combustible renewables and waste in national energy balances, including disaggregated data for production and consumption of wood, charcoal, black liquor, and other biomass. Data provided at national and various regional aggregate levels.
Agricultural Resources		
Agricultural land, products, and yields	*FAOSTAT-Agriculture* (data available on-line)	Time series data since 1961 on extent of agricultural land use by country and region, production of primary and processed crops, live animals, primary and processed animal products, imports and exports, food balance sheets, agricultural inputs, and nutritional yield of many agricultural products.
Specific products	Member organizations of the Consultative Group on International Agricultural Research	Issue-specific datasets on crops, animals, animal products, agricultural inputs, and genetic resources. Variety of spatial and temporal scales.
Fish Resources		
Fish stocks	FAO, *Review of the State of World Fishery Resources: Marine Fisheries*	Tabular information on the state of exploitation, total production, and nominal catches by selected species groups for major world fisheries.
Marine and inland fisheries	FAO, *FISHSTAT* (data available on-line at www.fao.org/fi/statist/statist.asp)	Databases on fishery production from marine capture and aquaculture, fish commodity production, and trade. Global, regional, and national data. Time series range from 20 to 50 years.
	FAO, *The State of World Fisheries and Aquaculture*	Published every two years. Data on five-year trends in fisheries production, utilization, and trade for the world and for geographic and economic regions. National data for major fishing countries. Also provides extensive analysis of fishery issues.
	FAO, *Yearbook of Fishery Statistics*	Updated annually. Includes aquaculture production and capture production by country, fishing area, principal producers, and principal species. Also trade data in fishery products.
	FAO, *Fisheries Global Information System,* at www.fao.org/fi/figis	Information on aquatic species, marine fisheries, fisheries issues, and, under development in collaboration with regional fishery bodies, the state of marine resources and inventories of fisheries and fishery resources.
	International Center for Living Aquatic Management, *FishBase 2000*	Database on more than 27,000 fish species and references. Many datasets incomplete.
Freshwater/Inland Water Resources		
Water resources	FAO, *AQUASTAT*	Global data on water resources and irrigation by country and region. Information on average precipitation, total internal water resources, renewable groundwater and surface water, total renewable water resources, and total exploitable water resources.
	State Hydrological Institute (Russia) and UNESCO, *World Water Resources and Their Use,* 1999	Global database on surface water resources and sectoral use. Includes water use forecasts to 2025.

BASE (Wood et al. 2000), ReefBase (Oliver et al.), and the Census of Marine Life (O'Dor 2004). Freshwater fish species are also being evaluated on a region basis for inclusion in the IUCN Red Lists.

Inventories at regional or continental scales are generally of higher overall quality and are more common than global data. Many of these data sets are based on grids of varying resolution. Examples include data on vertebrates in sub-Saharan Africa (grid size 1 degree or approximately 110 square kilometers) (Balmford et al. 2001), birds in the Americas (grid size 611,000 square kilometers) (Blackburn and Gaston 1996), several taxa of plants and animals in Britain (grid size 10 square kilometers) (Prendergast et al. 1993), and terrestrial vertebrates and butterflies in Australia (grid size 1 degree) (Luck et al. 2004). These grid-based inventories, as well as others based on political boundaries (countries, states) are based on arbitrary units that rarely reflect ecosystem boundaries. As a result, their utility is limited in assessing the biodiversity of a particular ecosystem. Some regional-scale inventories are based on ecological units, including a study on vertebrates, butterflies, tiger beetles, and plants for 116 WWF ecoregions in North America (Ricketts et al. 1999).

All these regional inventories can be used to understand patterns of biodiversity and endangerment (e.g., Ceballos and Brown 1995) and to link these patterns to threats and drivers operating at regional scales (e.g., Balmford et al. 2001; Ricketts in press). As is often the case, these data sets are most complete and dependable in the industrial world, although data are improving in many developing regions.

Because many ecosystem services (such as pollination and water purification) are provided locally, local-scale biodiversity inventories are often the most directly valuable for assessing those services. There are thousands of local inventories in the literature, comparing biodiversity between ecosystem types, among land use intensities, and along various environmental gradients. This literature has not been systematically compiled, and it is not possible to list all the studies here.

We illustrate the types of available data here with biodiversity studies in agricultural landscapes dominated by coffee cultivation. Local inventories in these landscapes have quantified the decline in both bird (Greenberg et al. 1997) and arthropod (Perfecto et al. 1997) diversity with increasing intensification of coffee production. Other studies have shown a decline in moth (Ricketts et al. 2001) and bird (Luck and Daily 2003) diversity with increasing distance from remnant patches of forest. Most relevant to ecosystem services supporting coffee production, the diversity and abundance of coffee-visiting bees declines with increasing distance from forest (Ricketts in press) and with increasing intensification (Klein et al. 2002).

Local inventories offer data that can directly inform land use policies and illuminate trade-offs among ecosystem services for decision-makers. Unfortunately, they are often time- and resource-intensive. In addition, the results are only relevant to the specific taxon and location under study, so general lessons are often difficult to glean. However, the collective results of many such studies can lead to useful general guidelines and principles.

Another method of compiling results from many biodiversity inventories is to examine the collections of museums and herbaria (Ponder et al. 2001). These house enormous amounts of information, accumulated sometimes over centuries of study. Furthermore, museums are beginning to use information technologies and the Internet to pool their information into aggregate databases, such that records from any museum can be searched (e.g., Edwards et al. 2000). These aggregate databases are an invaluable resource for studying the distribution of biodiversity. Museum and herbaria records, however, often contain a variety of spatial, temporal, and taxonomic biases and gaps due to the ad hoc and varying interests of collecting scientists (Ponder et al. 2001). These biases must be carefully considered when using museum data to assess biodiversity status and trends.

Ideally, data for characterizing biodiversity in the individual systems and its response to changes in ecosystem condition would be collected routinely according to an appropriate sampling strategy that meets the needs of the specified measures. Most often this is not the case, however, and data assimilated for other purposes are used, such as routine or sporadic surveys and observations made by naturalists. Generally such observations relate only to the most obvious and common species, especially birds and sometimes mammals, butterflies, and so on.

2.2.2.3 Demographic and Socioeconomic Data on Human Populations

Because the MA considers human populations as integral components of ecosystems, data on the populations living within the systems are one of the foundations for this analysis. Demographic and socioeconomic data provide information on the distributions of human populations within ecosystems, a prerequisite to analyzing the dependence of human well-being on ecosystem services.

Most information on the distribution and characteristics of human population is collected through population censuses and surveys. Nearly all countries of the world conduct periodic censuses (see www.census.gov/ipc/www/cendates/cenall.pdf); most countries conduct them once per decade. Census data are collected and reported by administrative or political units, such as counties, provinces, or states. These administrative boundaries generally do not correspond to the geographic boundaries of ecosystems.

To address this mismatch, the most recent version of the Gridded Population of the World (version 3) (CIESIN et al. 2004; CIESIN and CIAT 2004) contains population estimates for over 350,000 administrative units converted to a grid of latitude-longitude quadrilateral cells at a nominal spatial resolution of 5 square kilometers at the equator (Deichmann et al. 2001). The accuracy depends on the quality and year of the input census data and the resolution of the administrative units. Other data sets show how population is distributed relative to urban areas, roads, and other likely population centers, such as LandScan, which uses many types of ancillary data, including land cover, roads, nighttime lights, elevation and slope, to reallocate populations within administrative areas to more specific locations (Dobson 2000).

There are large data gaps on poverty distribution and access to ecosystem services such as fresh water (UNDP 2003). Some census data include resource use such as fuelwood and water source (Government of India 2001), but inventories on the use of ecosystems services are not generally available to establish trends. Increasingly, however, censuses and large-scale surveys are beginning to include questions on resource use. The World Bank's Living Standards Measurement Survey, for example, is introducing modules on resource use (Grosh and Glewwe 1995). As most nationally representative socioeconomic and demographic surveys are not georeferenced beyond administrative units, they must be used with care when making inferences at the moderate and high resolutions often used in ecological data analysis.

By combining census information about human settlements with geographic information, such as city night-time lights from satellite data, a new global database indicates urban areas from rural ones (CIESIN et al. 2004). These can be applied to distinguish urban and rural land areas in different ecosystems and to infer implications for resource use. (See Chapter 27.)

2.2.3 Numerical Simulation Models

Numerical models are mathematical expressions of processes operating in the real world. The ecological and human interactions within and among ecosystems are complex, and they involve physical, biological, and socioeconomic processes occurring over a range of temporal and spatial scales. Models are designed as simplified representations to examine assumptions and responses to driving forces.

Models span a wide range in complexity with regard to processes and spatial and temporal scales. Simple correlative models use statistical associations established where data are adequate in order to predict responses where data are lacking. For example, the CLIMEX model (Sutherst 1995) predicts the performance of an insect species in a given location and year in response to climate change based on previously established correlations from comparable locations and previous years. Dynamic, process-based models, on the other hand, are sets of mathematical expressions describing the interactions among components of a system at a specified time step. For example, the CENTURY model simulates fluxes of carbon, water, and nitrogen among plant and soil pools within a grassland ecosystem (Parton 1988). An emerging class of models, such as IBIS (Foley 1996) and LPJ (Sitch et al. 2003), incorporate dynamic processes but also simulate the dynamics of interacting species or plant functional types. Such models have been applied at the site, regional, and global scales to investigate ecosystem responses to climate change scenarios and increasing atmospheric carbon dioxide concentrations (e.g., Cramer et al. 2004).

Table 2.4 lists categories of models useful for the assessment of ecosystem condition and services. These models address various aspects of ecosystem condition. For example, hydrologic models can be used to investigate the effects of land cover changes on flood protection, population models can assess the effects of habitat loss on biodiversity, and integrated assessment models can synthesize this information for assessing effects of policy alternatives on ecosystem condition. Assessments rely on models to:

- **Fill data gaps.** As noted, data to assess trends in ecosystem condition and their services are often inadequate, particularly for regulating, supporting, and cultural services. Models are used to address these deficiencies. For example, Chapter 13 uses results from four ecosystem models (McGuire 2001) to estimate the impacts of changes in land use, climate, and atmospheric composition on carbon dioxide emissions from ecosystems.

- **Quantify responses of ecosystem services to management decisions.** One of the major tasks for the MA is to assess how changes in ecosystem condition alter services. Does removal of forest cover within a watershed alter flood protection? Does conversion to cropland alter climate regulation? Models can be used to simulate changes in the ecosystem condition (such as land cover) and estimate the response (in stream flow, for instance). A hydrologic model (e.g., Liang 1996) can quantify the change in stream flow in response to removal of forest cover. A land surface model linked to a climate model (e.g., Sellers 1986) can quantify the change in water and energy fluxes to the atmosphere from a specified change in land cover and the resulting effect on surface temperature. To the extent that models are adequate representations of reality, they provide an important tool for quantifying

the effects of alternative management decisions on ecosystem services.

- **Predict long-term ecological consequences of altered ecosystem condition.** Many human activities affect ecosystem condition only after a time lag. As a consequence, some effects of ecosystem management are not observed for many years. In such cases, models can be used to predict long-term ecological consequences. For example, the effect of timber harvest on the persistence of threatened species such as the spotted owl can be assessed using habitat-based metapopulation models (Akçakaya and Raphael 1998).

The reliability of long-term model predictions depends on the level of understanding of the system, the amount and quality of available data, the time horizon, and the incorporation of uncertainty. Predictions about simpler systems (such as single-species dynamics) are more reliable than those about complex systems (such as community composition and dynamics), because of the higher level of understanding ecologists have for simpler systems. The amount and quality of the data determine the uncertainty in input parameters, which in turn affect the reliability of the output. Longer-term predictions are less reliable because these uncertainties are compounded over time. Even uncertain predictions can be useful, however, if the level of uncertainty can be objectively quantified. Complex models can also identify shifts in ecosystem regime, such as the sudden loss of submerged vegetation in shallow lakes subject to eutrophication (Scheffer et al. 2001), and nonlinear responses to drivers.

- **Test sensitivities of ecosystem condition to individual drivers or future scenarios.** Observed changes in ecosystem condition result from the combined responses to multiple drivers. Changes in soil fertility in a rangeland, for example, reflect the combined response to grazing pressure, climate variations, and changes in plant species. Direct observations of soil fertility do not enable understanding of which driver is causing the response or how the drivers interact. A series of model simulations, changing one or more drivers for each model run, facilitates understanding of the response of soil fertility to each of the drivers. To the extent that models represent processes realistically, model simulations can identify nonlinear and threshold responses of ecosystems to multiple drivers. For example, neither overfishing nor pollution alone may lead to precipitous declines in fish stocks, but the combined response could have unanticipated effects on fish stocks.

- **Assess future viability of species.** Quantitative methods and models for assessing the chances of persistence of species in the future are collectively called population viability analysis. Models used in PVAs range from unstructured single-population models to metapopulation models with explicit spatial structure based on the distribution of suitable habitat (Boyce 1992; Burgman 1993). PVA provides a rigorous methodology that can use different types of data, incorporate uncertainties and natural variabilities, and make predictions that are relevant to conservation goals. PVA is most useful when its level of detail is consistent with the available data and when it focuses on relative (comparative) rather than absolute results and on risks of decline rather than extinction (Akçakaya and Sjögren-Gulve 2000). An important advantage of PVA is its rigor. In a comprehensive validation study, Brook et al. (2000) found the risk of population decline predicted by PVA closely matched observed outcomes, there was no significant bias, and population size projections did not differ significantly from reality. Further, the predictions of five PVA software packages they tested were highly concordant. PVA results can also be

Table 2.4. Examples of Numerical Models for Assessing Condition and Trends in Ecosystems and Their Services

Type of Model	Description	Examples of Models
Climate and land-atmosphere models	Land surface models of exchanges of water, energy, and momentum between land surface and atmosphere.	Sellers et al. 1986; Liang et al. 1996
Watershed and hydrologic models	Large basin models of hydrologic processes and biogeochemical exchanges in watersheds.	Fekete et al. 2002; Green et al. in press; Seitzinger and Kroeze 1998
Population and metapopulation models	Models of dynamics of single populations predicting future abundance and trends, risk of decline or extinction, and chance of growth. They can be scalar, structured (e.g., age-, stage-, and/or sex-based), or individual-based and incorporate variability, density dependence, and genetics. Metapopulation models focus on the dynamics of and interactions among multiple populations, incorporating spatial structure and dispersal and internal dynamics of each population. Their spatial structure can be based on the distribution and suitability of habitat, and they can be used to assess species extinction risks and recovery chances.	Akçakaya 2002; Lacy 1993
Community or food-web models	Models focusing on the interactions among different trophic levels (producers, herbivores, carnivores) or different species (e.g., predator-prey models).	Park 1998; USDA 1999
Ecosystem process models	Models that include both biotic and abiotic components and that represent physical, chemical, and biological processes in coastal, freshwater, marine, or terrestrial systems. They can predict, for example, vegetation dynamics, including temporal changes in forest species and age structure.	Pastorok et al. 2002
Global terrestrial ecosystem models	Models of biogeochemical cycling of carbon, nitrogen, and other elements between the atmosphere and biosphere at the global scale, including vegetation dynamics, productivity, and response to climate variability.	Field et al. 1995; Foley et al. 1996; McGuire et al. 2001; Sitch et al. 2003
Multi-agent models	Agents are represented by rules for behavior based on interactions with other actors or physical processes.	Moss et al. 2001
Integrated assessment models	Models that assemble, summarize, and interpret information to communicate to decision-makers.	Alcamo et al. 1994

tested for single models by comparing predicted values with those observed or measured in the field (McCarthy 2001).

- **Understand the dynamics of social environmental interactions.** Individually based methods such as multiagent modeling are increasingly used to understand social and environmental interactions. Multiagent behavioral systems seek to model social-environment interactions as dynamic processes (see Moss et al. 2001). Human actors are represented as software agents with rules for their own behavior, interactions with other social agents, and responses to the environment. Physical processes (such as soil erosion) and institutions or organizations (such as an environmental regulator) may also be represented as agents. A multiagent system could represent multiple scales of vulnerability and produce indicators of multiple dimensions of vulnerability for different populations. Multiagent behavioral systems have an intuitive appeal in participatory integrated assessment. Stakeholders may identify with particular agents and be able to validate a model in qualitative ways that is difficult to do for econometric or complex dynamic simulation models. However, such systems require significant computational resources (proportional to the number of agents), and a paucity of data for validation of individual behavior is a constraint.

Models are useful tools for ecosystem assessments if the selection of models, input data, and validation are considered carefully for particular applications. A model developed with data from one location is not directly applicable to other locations. Moreover, data to calibrate and validate models are often difficult to obtain. The appropriateness of a model for an assessment task also depends as much on the capacity of the model variables to capture the values and interests of the decision-making and stakeholding communities as on the accuracy of the underlying scientific data.

2.2.4 Indicators of Ecosystem Condition and Services

An indicator is a scientific construct that uses quantitative data to measure ecosystem condition and services, drivers of changes, and human well-being. Properly constituted, an indicator can convey relevant information to policymakers. In this assessment, indicators serve many purposes, for example:

- as easily measured quantities to serve as surrogates for more difficult to measure characteristics of ecosystem condition—for example, the presence of fecal coliform in a stream is relatively easy to measure and serves a surrogate for poor sanitation in the watershed, which is more difficult to measure.

- as a means to incorporate several measured quantities into a single attribute as an indicator of overall condition—for example, the widely used Index of Biotic Integrity is an indicator of aquatic ecosystem condition (Karr et al. 1986). The IBI is an additive index combining measures of abundances of different taxa. The individual measures can be weighted according to the importance of each taxa for aquatic health.

- as a means to communicate effectively with policy-makers regarding trends in ecosystem conditions and services—for example, information on trends in disease incidence reflects trends in disease control as a "regulating" ecosystem service. The former can be readily communicated to a policymaker.

- as a means to measure the effectiveness of policy implementation.

Identifying and quantifying the appropriate indicators is one of the most important aspects of the chapters in this report because it is simply not possible to measure and report all aspects of ecosystems and their relation to human well-being. It is also important to identify appropriate indicators to establish a baseline against which future ecosystem assessments can be compared.

Indicators are designed to communicate information quickly and easily to policy-makers. Economic indicators, such as GDP, are highly influential and well understood by decision-makers. Measures of poverty, life expectancy, and infant mortality directly convey information about human well-being. Some environmental indicators, such as global mean temperature and atmospheric carbon dioxide concentrations, are becoming widely accepted as measures of anthropogenic effects on global climate. Measures of ecosystem condition are far less developed, although some biophysical measures such as spatial extent of an ecosystem and agricultural output are relatively easy to quantify. There are at this time no widely accepted indicators to measure trends in supporting, regulating, or cultural ecosystem services, much less indicators that measure the effect of changes in these services on human well-being. Effective indicators meet a number of criteria (NRC 2000). (See Box 2.1.)

The U.S. National Research Council (NRC 2000) identifies three categories of ecological indicators. First, the extent and status of ecosystems (such as land cover and land use) indicate the coverage of ecosystems and their ecological attributes. Second, ecological capital, further divided into biotic raw material (such as total species diversity) and abiotic raw materials (such as soil nutrients), indicates the amount of resources available for providing services. Finally, indicators of ecological functioning (such as lake trophic status) measure the performance of ecosystems.

Table 2.5 provides examples of three major types of indicators used in this report. (Indicators of human well-being and their utility for measuring how well-being responds to changes in ecosystem services are described later in this chapter.)

- **Indicators of direct drivers of change.** No single indicator represents the totality of the various drivers. Some direct drivers of change (see MA 2003 and Chapter 3) have relatively straightforward indicators, such as fertilizer usage, water consumption, irrigation, and harvests. Indicators for other drivers, including invasion by non-native species, climate change, land cover conversion, and landscape fragmentation, are not as well developed, and data to measure them are not as readily available. Measures such as the per capita "ecological footprint," defined as the area of arable land and aquatic ecosystems required to produce the resources used and assimilate wastes produced per person (Rees 1992), attempt to quantify the demand on ecosystem services into a single indicator. (See Chapter 27.)

- **Indicators of ecosystem condition.** Indicators of biophysical condition of ecosystems do not directly reflect the cause and effect of the drivers but nevertheless can contribute to policy formulation by directing attention to changes of importance. To determine causal relationships, models of interactions among variables must be used. As an analogy with human health, an increase in body temperature indicates infection that warrants further examination. As an example in the biophysical realm, declining trends in fish stocks can trigger investigations of possible causal mechanisms and policy alternatives. Indicators of ecosystem condition include many dimensions, ranging from the extent of the ecosystem to demographic characteristics of human populations to amounts of chemical contaminants (The H. John Heinz III Center for Science, Economics, and the Environment 2002).

- **Indicators of ecosystem services.** Indicators for the provisioning services discussed in Chapters 7–17 generally relate to commodity outputs from the system (such as crop yields or fish) and are readily communicable to policy-makers. Indicators related to the underlying biological capability of the system to maintain the production through supporting and regulating services are a greater challenge. For example, indicators measuring the capability of a system to regulate climate, such as evapotranspiration or albedo, are not as readily interpretable for a policy-maker.

Indicators are essential, but they need to be used with caution (Bossel 1999). Over-reliance on indicators can mask important changes in ecosystem condition. Second, while it is important that indicators are based on measurable quantities, the selection of indicators can be biased toward attributes that are easily quantifiable rather than truly reflective of ecosystem condition. Third, comparing indicators and indices from different temporal and spatial scales is challenging because units of measurement are often inconsistent. Adding up and combining factors has to be done very carefully and it is crucial that the method for combining individual indicators is well understood.

Indicators of biodiversity are particularly important for this assessment. Indicators of the amount and variability of species within a defined area can take many forms. The most common measures are species richness—the number of species—and species diversity, which is the number of species weighted by their relative abundance, biomass, or other characteristic, as in Shannon-Weiner or other similar indices (Rosenzweig 1995).

These two simple measures do not capture many aspects of biodiversity, however. They do not differentiate between native and invasive or introduced species, do not differentiate among species in terms of sensitivity or resilience to change, and do not focus on species that fulfill significant roles in the ecosystem (such as pollinators and decomposers). Moreover, the result depends on the definition of the area and may be scale-dependent. The measures also may not always reflect biodiversity trends accurately. For example, ecosystem degradation by human activities may temporarily increase species richness in the limited area of the impact. Thus refinements of these simple measures provide more insights into the amount of biodiversity. (See Box 2.2.)

Aggregate indicators of trends in species populations such as the Index of Biotic Integrity for aquatic systems (Karr and Dudley 1981) and the Living Planet Index (Loh 2002) use existing data sets to identify overall trends in species abundance and, by implication, the condition of the ecosystems in which they occur. The

---

BOX 2.1

Criteria for Effective Ecological Indicators (NRC 2000)

- Does the indicator provide information about changes in important processes?
- Is the indicator sensitive enough to detect important changes but not so sensitive that signals are masked by natural variability?
- Can the indicator detect changes at the appropriate temporal and spatial scale without being overwhelmed by variability?
- Is the indicator based on well-understood and generally accepted conceptual models of the system to which it is applied?
- Are reliable data available to assess trends and is data collection a relatively straightforward process?
- Are monitoring systems in place for the underlying data needed to calculate the indicator?
- Can policymakers easily understand the indicator?

Table 2.5. Examples of Indicators to Assess Ecosystem Condition and Trends

Characteristic Described by Indicator	Example of Indicator	Category of Indicator	Availability of Data for Indicator	Units
Direct drivers of change				
Land cover conversion	area undergoing urbanization	ecological state	high	hectares
Invasive species	native vs. non-native species	ecological capital	medium	percent of plant species
Climate change	annual rainfall	ecological state	high	millimeters per year
Irrigation	water usage	ecological functioning	high	cubic meters per year
Ecosystem condition				
Condition of vegetation	landscape fragmentation	ecological state	medium	mean patch size
Condition of soil	soil nutrients	ecological capital	medium	nutrient concentration
	soil salinization	ecological state	low	salt concentration
Condition of biodiversity	species richness	ecological capital	low	number of species/unit area
	threatened species	ecological functioning	medium	percent of species at risk
	visibility of indicator species	ecological functioning	low-medium	probability of extinction
Condition of fresh water	presence of contaminants	ecological state	high	concentration of pollutants index of biotic integrity
Ecosystem service				
Production service	food production	ecological functioning	high	yield (kilograms per hectare per year)
Capacity to mitigate floods	change in stream flow per unit precipitation	ecological capital	low	discharge (cubic meters per second)
Capacity for cultural services	spiritual value	ecological capital	low	?
Capacity to provide biological products	biological products of potential value	ecological capital	low	number of products or economic value

Note: See section 2.3.4 for indicators of human well-being.

BOX 2.2

Indicators of Biodiversity

The following is a sample of the types of indicators that can be used to monitor status and trends in biodiversity. The list is not exhaustive, and specific choice of indicators will depend on particular scale and goals of the monitoring program.

- **Threatened species:** the number of species that are in decline or otherwise classified as under threat of local or global extinction.
- **Indicator species:** species that can be shown to represent the status or diversity of other species in the same ecosystem. Indicator species have been explored as proxies for everything from whole ecosystem restoration (e.g., Carignan and Villard 2002) to overall species richness (e.g., MacNally and Fleishman 2002). The phrase "indicator species" is also used broadly to include several of the other categories listed here.
- **Umbrella species:** species whose conservation is expected to confer protection of other species in the same ecosystem (for example, species with large area requirements). If these species persist, it is assumed that others persist as well (Roberge and Angelstam 2004).
- **Taxonomic diversity:** the number of species weighted by their evolutionary distinctiveness (Mace et al. 2003). This indicator is increased with both high species richness and high levels of taxonomic diversity among species. Care is needed that the indicator of taxonomic diversity represents lineage in evolutionary history.
- **Endemism:** the number of species found only in the specific area (e.g., Ricketts in press). Note that this is a scale-dependent measure: as the area assessed increases, higher levels of endemism will result.
- **Ecological role:** species with particular ecological roles, such as pollinators and top predators (e.g., Kremen et al. 2002).
- **Sensitive or sentinel species:** trends in species that react to changes in the environment before other species, especially changes due to human activities (e.g., de Freitas Rebelo et al. 2003). Similar to the famous "canary in the coal mine," monitoring these sensitive species is thought to provide early warning of ecosystem disruption.
- **Aggregate indicators:** indices that combine information about trends in multiple species, such as the Living Planet Index, which aggregates trends in species abundances in forest, fresh water, and marine species (Loh 2002), and the Index of Biotic Integrity, which combines measures of abundances of different taxa in aquatic systems (Karr and Dudley 1981).

Living Planet Index is an aggregation of three separate indices, each the average of trends in species abundances in forest, freshwater, and marine biomes. It can be applied at national, regional, and global levels. The effectiveness of such an aggregate indicator depends on availability and access to data sets on a representative number of species, which is particularly problematic in many developing countries.

The number of species threatened with extinction is an important indicator of biodiversity trends. Using this indicator requires that a number of conditions to be met, however. First, the criteria used to categorize species into threat classes must be objective and transparent and have a scientific basis. Second, the changes in the status of species must reflect genuine changes in the conservation status of the species (rather than changes in knowledge or taxonomy, for example). Third, the pool of species evaluated in two different time periods must be comparable (if more threatened species are evaluated first, the proportion of threatened species may show a spurious decline).

The IUCN Red List of Threatened Species mentioned earlier meets these conditions. The criteria used in assigning species to threat categories (IUCN 2001) is quantitative and transparent yet allows for flexibility and can incorporate data uncertainties (Akçakaya 2000). The IUCN Red List database also records whether or not a species has been evaluated for the first time. For species evaluated previously, the assessment includes reasons for any change in status, such as genuine change in the status of the species, new or better information available, incorrect information used previously, taxonomic change affecting the species, and previous incorrect application of the Red List criteria. Finally, the complete coverage of some taxonomic groups helps make evaluations comparable, although the fact that new species are being evaluated for other groups must be considered when calculating measures such as the proportion of threatened species in those groups.

2.2.5 Indigenous, Traditional, and Local Knowledge

Traditional knowledge broadly represents information from a variety of sources including indigenous peoples, local residents, and traditions. The term indigenous knowledge is also widely used referring to the knowledge held by ethnic minorities from the approximately 300 million indigenous people worldwide (Emery 2000). The International Council for Science defines TK as "a cumulative body of knowledge, know-how, practices and representation maintained and developed by peoples with extended histories of interaction with the natural environment. These sophisticated sets of understandings, interpretations and meanings are part and parcel of a cultural complex that encompasses language, naming and classification systems, resource use practices, ritual, spirituality and worldview" (ICSU 2002b).

TK and IK are receiving increased interest as valuable sources of information (Martello 2001) about ecosystem condition, sustainable resource management (Johannes 1998; Berkes 1999; 2002), soil classification (Sandor and Furbee 1996), land use investigations (Zurayk et al. 2001), and the protection of biodiversity (Gadgil et al. 1993). Traditional ecological knowledge is a subset of TK that deals specifically with environmental issues.

Pharmaceutical companies, agribusiness, and environmental biologists have all found TEK to be a rich source of information (Cox 2000; Kimmerer 2000). TEK provides empirical insight into crop domestication, breeding, and management. It is particularly important in the field of conservation biology for developing conservation strategies appropriate to local conditions. TEK is also

useful for assessing trends in ecosystem condition (Mauro and Hardinson 2000) and for restoration design (Kimmerer 2000), as it tends to have qualitative information of a single local record over a long time period.

Oral histories can play an important role in the field of vulnerability assessment, as they are especially effective at gathering information on local vulnerabilities over past decades. Qualitative information derived from oral histories can be further developed as storylines for further trends and can lead into role playing simulations of new vulnerabilities or adaptations (Downing et al. 2001).

However, TK has for a long time not been treated equally to knowledge derived from formal science. Although Article 27 of the Universal Declaration of Human Rights of 1948 protects Intellectual Property, the intellectual property rights of indigenous people have often been violated (Cox 2000). The Convention on Biological Diversity of 1992 for the first time established international protocols on the protection and sharing of national biological resources and specifically addressed issues of traditional knowledge. In particular, the parties to the convention agree to respect and preserve TK and to promote wide applications and equitable sharing of its benefits (Antweiler 1998; Cox 2000; Singhal 2000).

The integration of TEK with formal science can provide a number of benefits, particularly in sustainable resource management (Johannes 1998; Berkes 2002). However, integrating TEK with formal science is sometimes problematic (Antweiler 1998; Fabricus et al. 2004). Johnson (1992) cites the following as reasons why integrating TEK is difficult:

- Traditional environmental knowledge is disappearing and there are few resources to document it before it is lost.
- Translating concepts and ideas from cultures based on TEK (mainly oral-based knowledge systems) into the concepts and ideas of formal science is difficult.
- Appropriate methods to document and integrate TEK are lacking, and natural scientists often criticize the lack of rigor of the traditional anthropological methods for interviewing and participant observation.
- Integrating TEK and formal science is linked to political power, and TEK is often seen as subordinate.

Moreover, existing practices of TEK, such as forest management, are not necessarily sustainable (Antweiler 1998).

It has been repeatedly pointed out that if TEK is integrated it needs to be understood within its historical, socioeconomic, political, environmental, and cultural location (Berkes 2002). This implies that the ratio of local to scientific knowledge will vary depending on the case and situation (Antweiler 1998). The limitations and shortcomings of integrating TEK and formal science must be addressed, and the methods chosen to collect this knowledge should take the location-specific environments in which they operate into account (Singhal 2000). Integration can also be hindered by different representations of cross-scale interactions, nonlinear feedbacks, and uncertainty in TEK and formal science (Gunderson and Holling 2002). Due to this high degree of uncertainty, it is essential to validate and compare both formal and informal knowledge (Fabricus et al. 2004).

There have been general concerns about scaling up TEK to broader spatial scales, as this traditional knowledge is seldom relevant outside the local context (Forsyth 1999; Lovell et al. 2002). Moreover, analysts warn of a downplaying of environmental problems when TEK is overemphasized. On the other hand, researchers have also warned that efforts to integrate or bridge different knowledge systems will lead inevitably to the compartmentalization and distillation of traditional knowledge into a form that is understandable and usable by scientists and resource managers alone (Nadasdy 1999).

Despite these limitations, TEK—if interpreted carefully and assessed appropriately—can provide important data on ecosystem conditions and trends. The most promising methods of data collection are participatory approaches, in particular Participatory Rural Appraisal (Catley 1996). PRA is an alternative to unstructured visits to communities, which may be biased toward more accessible areas, and to costly, time-consuming questionnaire surveys (Chambers 1994). PRA was developed during the early 1990s from Rapid Rural Appraisal, a cost-effective and rapid way of gathering information. RRA was criticized as being too "quick and dirty" and not sufficiently involved with local people. PRA tries to overcome the criticisms of RRA by allowing recipients more control of problem definition and solution design and by carrying out research over a longer period (Zarafshani 2002; Scoones 1995). Activities such as interviewing, transects, mapping, measuring, analysis, and planning are done jointly with local people (Cornwall and Pratt 2003).

Participatory methods have their limitations: First, they only produce certain types of information, which can be brief and superficial. Second, the information collected may reflect peoples' own priorities and interests. Third, there might be an unequal power relation among participants and between participants and researchers (Cooke and Kothari 2001). Glenn (2003) warns that a rush to obtain traditional knowledge can be biased toward pre-existing stereotypes and attention to vocal individuals who do not necessarily reflect consensus.

The MA sub-global assessments used a wide range of participatory research techniques to collect and integrate TEK and local knowledge into the assessment process. In addition to PRA (Pereira 2004), techniques such as focus group workshops (Borrini-Feyerabend 1997), semi-structured interviews with key informants (Pretty 1995), forum theater, free hand and GIS mapping, pie charts, trend lines, timelines, ranking, Venn diagrams, problem trees, pyramids, role playing, and seasonal calendars were used (Borrini-Feyerabend 1997; Jordan and Shrestha 1998; Motteux 2001).

2.2.6 Case Studies of Ecosystem Responses to Drivers

Case studies provide in-depth analyses of responses of ecosystem conditions and services to drivers in particular locations. For example, the study of the Yaqui Valley in Mexico illustrates the response of birds, marine mammals, and fisheries to upland runoff generated by increasing fertilizer use in the heavily irrigated valley (Turner II et al. 2003). Evidence generated from a sufficient number of case studies allows general principles to emerge about ecosystem responses to drivers. Case studies, which can analyze relationships in more detail than would be possible with nationally aggregated statistics or coarse resolution data, also illustrate the range of ecosystem responses to drivers in different locations or under different biophysical conditions.

Few studies have been undertaken to synthesize information from case studies. One such effort analyzed 152 sub-national case studies investigating the response of tropical deforestation to economic, institutional, technological, cultural, and demographic drivers (Geist and Lambin 2001, 2002). The analysis revealed complex relationships between drivers and deforestation in different regions of the tropics, indicating challenges for generic and widely applicable land-use policies to control deforestation. The MA does not carry out such extensive meta-analyses, but rather uses their results where available as well as results from individual case studies from the scientific literature.

Drawing conclusions from case studies must be done with caution. First, individual studies do not generally use standard protocols for data collection and analysis, so comparisons across case studies are difficult. Second, researchers make decisions about where to carry out a case study on an individual basis, so biases might be introduced from inadequate representation from different locations. Third, unless a sufficient number of case studies are available it is not prudent to draw general conclusions and extrapolate results from one location to another. In spite of these limitations, case studies can illustrate possible linkages between ecosystem response and drivers and can fill gaps generated by lack of more comprehensive data when necessary.

2.3 Assessing the Value of Ecosystem Services for Human Well-being

This section addresses the data and methods for assessing the linkages between ecosystem services and human well-being.

2.3.1 Linking Ecosystem Condition and Trends to Well-being

Ecosystem condition is only one of many factors that affect human well-being, making it challenging to assess linkages between them. Health outcomes, for example, are the combined result of ecosystem condition, access to health care, economic status, and myriad other factors. Interpretations of trends in indicators of well-being must appropriately account for the full range of factors involved.

The impacts of ecosystem change on well-being are often subtle, which is not to say unimportant; impacts need not be drastic to be significant. A small increase in food prices resulting from lower yields as a result of land degradation will affect the well-being of many people, even if none starve as a result.

Two basic approaches can be used to trace the linkages between ecosystem condition and trends and human well-being. The first attempts to correlate trends in ecosystem condition to changes in human well-being directly, while the second attempts to trace the impact to the groups affected through biophysical and socioeconomic processes. For example, the impact of water contamination on the incidence of human disease could be estimated by correlating measures of contaminants in water supplies with measures of the incidence of gastrointestinal illnesses in the general population, controlling for other factors that might affect the relationship. Alternatively, the impact could be estimated by using a dose-response function that relates the incidence of illness to the concentration of contaminants to estimate the increase in the probability of illness, then combining that with estimates of the population served by the contaminated water to arrive at a predicted total number of illnesses.

Both approaches face considerable problems. Efforts to correlate ecosystem condition with human well-being directly are difficult because of the presence of multiple confounding factors. Thus the incidence of respiratory illness depends not only on the concentration of airborne contaminants but also on predisposition to illness through factors such as nutritional status or the prevalence of smoking, exposure factors such as the proportion of time spent outdoors, and so on. Analyses linking well-being and ecosystem condition are most easily carried out at a local scale, where the linkages can be most clearly identified.

2.3.2 Measuring Well-being

Human well-being has several key components: the basic material needs for a good life, freedom and choice, health, good social relations, and personal security. Well-being exists on a continuum with poverty, which has been defined as "pronounced deprivation in well-being." One of the key objectives of the MA is to identify the direct and indirect pathways by which ecosystem change can affect human well-being, whether positively or negatively.

Well-being is multidimensional, and so very hard to measure. All available measures have problems, both conceptual (are they measuring the right thing, in the right way?) and practical (how do we actually implement them?). Moreover, most available measures are extremely difficult to relate to ecosystem services.

Economic valuation offers a way both to value a wide range of individual impacts (some quite accurately and reliably, others less so) and, potentially but controversially, to assess well-being as a whole by expressing the disparate components of well-being in a single unit (typically a monetary unit). It has the advantage that impacts denominated in monetary units are readily intelligible and comparable to other benefits or to the costs of intervention. It can also be used to provide information to examine distributional, equity, and intergenerational aspects. Economic valuation techniques are described in the next section.

Health indicators address a key subset of impacts of ecosystem services on well-being. They are an important complement to economic valuation because they concern impacts that are very difficult and controversial to value. Some health indicators address specific types of health impacts; others attempt to aggregate a number of health impacts. Likewise, poverty indicators measure a dimension of well-being that is often of particular interest. These, too, are described later in the chapter.

Numerous other well-being indicators (such as the Human Development Index) have been developed in an effort to capture the multidimensionality of well-being into a single number, with varying degrees of success. Although these indicators are arguably better measures of well-being, they tend not to be very useful for assessing the impact of ecosystems, as many of the dimensions they add (literacy, for instance) tend not to be sensitive to ecosystem condition. These aggregate indicators and the limitations they face are described near the end of this chapter.

2.3.3 Economic Valuation

One of the main reasons we worry about the loss of ecosystems is that they provide valuable services—services that may be lost or diminished as ecosystems degrade. The question then immediately arises: how valuable are these services? Or, put another way, how much worse off would we be if we had less of these services? We need to be able to answer these questions to inform the choices we make in how to manage ecosystems.

Economic valuation attempts to answer these questions. It is based on the fact that human beings derive benefit (or "utility") from the use of ecosystem services either directly or indirectly, whether currently or in the future, and that they are willing to "trade" or exchange something for maintaining these services. As utility cannot be measured directly, economic valuation techniques are based on observation of market and nonmarket exchange processes. Economic valuation usually attempts to measure all services in monetary terms, in order to provide a common metric in which to express the benefits of the diverse variety of services provided by ecosystems. This explicitly does not mean that only services that generate monetary benefits are taken into consideration in the valuation process. On the contrary, the es-

sence of most work on valuation of environmental and natural resources has been to find ways to measure benefits that do not enter markets and so have no directly observable monetary benefits. The concept of Total Economic Value is a framework widely used to disaggregate the utilitarian value of ecosystems into components (Pearce 1993). (See Box 2.3.)

Valuation can be used in many different ways (Pagiola et al. 2004). The MA uses valuation primarily to evaluate trade-offs between alternative ecosystem management regimes that alter the use of ecosystems and the multiple services they provide. This approach focuses on assessing the value of changes in ecosystem services resulting from management decisions or other human ac-

---

BOX 2.3

Total Economic Value

The concept of total economic value is widely used by economists (Pearce and Warford 1993). This framework typically disaggregates the utilitarian value of ecosystems into direct and indirect use values and non-use values:

- **Direct use values** are derived from ecosystem services that are used directly by humans. They include the value of *consumptive uses,* such as harvesting of food products, timber for fuel or construction, medicinal products, and hunting of animals for consumption, and of *non-consumptive uses,* such as the enjoyment of recreational and cultural amenities like wildlife and bird watching, water sports, and spiritual and social utilities that do not require harvesting of products. Direct use values correspond broadly to the MA notion of provisioning and cultural services. They are typically enjoyed by people located in the ecosystem itself.
- **Indirect use values** are derived from ecosystem services that provide benefits outside the ecosystem itself. Examples include the natural water filtration function of wetlands, which often benefits people far downstream; the storm protection function of coastal mangrove forests, which benefits coastal properties and infrastructure; and carbon sequestration, which benefits the entire global community by abating climate change. This category of benefits corresponds broadly to the MA notion of regulating and supporting services.
- **Option values** are derived from preserving the option to use in the future services that may not be used at present, either by oneself (*option value*) or by others or heirs (*bequest value*). Provisioning, regulating, and cultural services may all form part of option value to the extent that they are not used now but may be used in the future.
- **Non-use values** refer to the value people may have for knowing that a resource exists even if they never use that resource directly. This kind of value is usually known as *existence value* (or, sometimes, *passive use value*). This is one area of partial overlap with non-utilitarian sources of value (see the section on intrinsic value).

The TEV framework does not have any direct analog to the MA notion of supporting services of ecosystems. Rather, these services are valued indirectly, through their role in enabling the ecosystem to provide provisioning and enriching services.

Valuation is usually relatively simple in the case of direct use value, and then increasingly difficult as one moves on to indirect use value, option value, and non-use value.

tions. This type of valuation is most likely to be directly policy-relevant.

Economic valuation has also been used to derive the total value of ecosystem services at a given time (e.g., Costanza et al. 1997) and to simulate the value of ecosystem services in an integrated Earth system model (Boumans et al. 2002). Efforts to estimate the total value of the services being provided by ecosystems at any one time, if conducted properly, can provide useful information on their contribution to economic activity and to well-being. Their usefulness for policy is limited, however, as it is rare for all ecosystem services to be completely lost (and even then, this would usually only happen over time). This chapter, therefore, focuses on methods useful for assessing changes in ecosystem services. (For further discussion of the difference between these approaches, see Bockstael et al. 2000 and Pagiola et al. 2004.)

2.3.3.1 Valuation Methods

Many methods for measuring the utilitarian values of ecosystem services are found in the resource and environmental economics literature (Mäler and Wyzga 1976; Freeman 1979; Hufschmidt et al. 1983; Mitchell and Carson 1989; Pearce and Markandya 1989; Braden and Kolstad 1991; Hanemann 1992; Freeman 1993; Pearce 1993; Dixon et al. 1994; Johansson 1994; Pearce and Moran 1994; Barbier et al. 1995; Willis and Corkindale 1995; Seroa da Motta 1998; Garrod and Willis 1999; Seroa da Motta 2001; Pearce et al. 2002; Turner et al. 2002; Pagiola et al. in review). Table 2.6 summarizes the main economic valuation techniques.

Some techniques are based on actual observed behavior data, including some methods that deduce values indirectly from behavior in surrogate markets, which are hypothesized to have a direct relationship with the ecosystem service of interest. Other techniques are based on hypothetical rather than actual behavior data, where people's responses to questions describing hypothetical markets or situations are used to infer value. These are generally known as "stated preference" techniques, in contrast to those based on behavior, which are known as "revealed preference" techniques. Some techniques are broadly applicable, some are applicable to specific issues, and some are tailored to particular data sources. As in the case of private market goods, a common feature of all methods of economic valuation of ecosystem services is that they are founded in the theoretical axioms and principles of welfare economics. These measures of change in well-being are reflected in people's willingness to pay or willingness to accept compensation for changes in their level of use of a particular service or bundle of services (Hanemann 1991; Shogren and Hayes 1997). These approaches have been used extensively in recent years, in a wide range of policy-relevant contexts.

A number of factors and conditions determine the choice of specific measurement methods. For instance, when the ecosystem service in question is privately owned and traded in the market, its users have the opportunity to reveal their preferences for that service compared with other substitutes or complementary commodities through their actual market choices, given relative prices and other economic factors. For this group of ecosystem services a demand curve can be derived from observed market behavior, and this allows changes in well-being to be estimated. However, many ecosystem services are not privately owned and not traded, and hence their demand curves cannot be directly observed and measured. Alternative methods have been used to derive values for such ecosystem services.

Valuation is a two-step process. First, the services being valued have to be identified. This includes understanding the nature of the services and their magnitude, and how they would change if the ecosystem changed; knowing who makes use of the services, in what way and for what purpose, and what alternatives they have; and establishing what trade-offs might exist between different kinds of services an ecosystem might provide. The bulk of the work involved in valuation actually concerns quantifying the biophysical relationships. In many cases, this requires tracing through and quantifying a chain of causality. (See Figure 2.3 for an example.) Valuation in the narrow sense only enters in the second step in the process, in which the value of the impacts is estimated in monetary terms.

2.3.3.1.1 Changes in productivity

The most widely used technique, thanks to its broad applicability and its flexibility in using a variety of data sources, is known as the change in productivity technique. It consists of tracing through chains of causality (such as those illustrated in Figure 2.3) so that the impact of changes in the condition of an ecosystem can be related to various measures of human well-being. Such impacts are often reflected in goods or services that contribute directly to human well-being (such as production of crops or of clean water), and as such are often relatively easily valued. The valuation step itself depends on the type of impact but is often straightforward:

- The net value in reductions in irrigated crop production resulting from reduced water availability is easy to estimate, for example, as crops are often sold. (Even so, it is a very common error to use the reduction in the gross value of crop production rather than the net value. Using gross value omits the costs of production and so overestimates the impact.)

- Where the impact is on a good or service that is not marketed or where observed prices are unreliable indicators of value, the valuation can become more complex. The impact of hydrological changes on use of water for human consumption, for example, once again begins by tracing through chains of causality to estimate the changes in the quantity and quality of water available to consumers. This is itself often difficult. The prices typically charged to consumers for this water, moreover, are not reliable measures of the value of the water to consumers, as they are set administratively, with no regard for supply and demand (indeed, in most cases water fees do not even cover the cost of delivering the water to consumers, let alone the value of the water itself). The value of an additional unit of water can be estimated in various ways, such as the cost of alternative sources of supply (cost-based measures are described later) or asking consumers directly how much they would be willing to pay for it (contingent valuation, described later). Note that it is very important to use the value of an additional unit of water, since some amount of water is, of course, vital for survival. Thus an additional unit of water will be very valuable when water is scarce, but much less so when water is plentiful. In this case, as in many others, averages can be misleading.

- When the impact is on water quality rather than quantity, the impact on well-being might be reflected in increased morbidity or even mortality. Again, the process begins by tracing through chains of causality, for example by using dose-response functions that tie concentrations of pollutants to human health. Valuing the impact on health itself can then be done in a number of ways (see cost of illness and human capital, in the next section).

- In some cases, the impact is on relatively intangible aspects of well-being, such as aesthetic benefits or existence value.

Table 2.6. Main Economic Valuation Techniques (Adapted from Pagiola et al. forthcoming)

Methodology	Approach	Applications	Data Requirements	Limitations
Revealed preference methods				
Change in productivity	trace impact of change in environmental services on produced goods	any impact that affects produced goods	change in service; impact on production; net value of produced goods	data on change in service and consequent impact on production often lacking
Cost of illness, human capital	trace impact of change in environmental services on morbidity and mortality	any impact that affects health (e.g., air or water pollution)	change in service; impact on health (dose-response functions); cost of illness or value of life	dose-response functions linking environmental conditions to health often lacking; underestimates, as it omits preferences for health; value of life cannot be estimated
Replacement cost (and variants, such as relocation cost)	use cost of replacing the lost good or service	any loss of goods or services	extent of loss of goods or services; cost of replacing them	tends to overestimate actual value
Travel cost method	derive demand curve from data on actual travel costs	recreation	survey to collect monetary and time costs of travel to destination; distance traveled	limited to recreational benefits; hard to use when trips are to multiple destinations
Hedonic prices	extract effect of environmental factors on price of goods that include those factors	air quality, scenic beauty, cultural benefits	prices and characteristics of goods	requires vast quantities of data; very sensitive to specification
Stated preference methods				
Contingent valuation (CV)	ask respondents directly their willingness to pay for a specified service	any service	survey that presents scenario and elicits willingness to pay for specified service	many potential sources of bias in responses; guidelines exist for reliable application
Choice modeling	ask respondents to choose their preferred option from a set of alternatives with particular attributes	any service	survey of respondents	similar to CV; analysis of the data generated is complex
Other methods				
Benefits transfer	use results obtained in one context in a different context	any for which suitable comparison studies are available	valuation exercises at another, similar site	can be widly inaccurate, as many factors vary even when contexts seem "similar"

Particular efforts have been made in recent years to develop techniques to value such impacts, including hedonic price, travel cost, and contingent valuation methods.

2.3.3.1.2 Cost of illness and human capital

The economic costs of an increase in morbidity due to increased pollution levels can be estimated using information on various costs associated with the increase: any loss of earnings resulting from illness; medical costs such as for doctors, hospital visits or stays, and medication; and other related out-of-pocket expenses. The estimates obtained in this manner are interpreted as lower-bound estimates of the presumed costs or benefits of actions that result in changes in the level of morbidity, since this method disregards the affected individuals' preference for health versus illness and restrictions on non-work activities. Also, the method assumes that individuals treat health as exogenous and does not recognize that individuals may undertake defensive actions (such as using special air or water filtration systems to reduce exposure to pollution) and incur costs to reduce health risks.

When this approach is extended to estimate the costs associated with pollution-related mortality (death), it is referred to as the human-capital approach. It is similar to the change-in-productivity approach in that it is based on a damage function relating pollution to productivity, except that in this case the loss in productivity is that of human beings, measured in terms of expected lifetime earnings. Because it reduces the value of life to the present value of an individual's future income stream, the human-capital approach is extremely controversial when applied to mortality. Many economists prefer, therefore, not to use this approach and to simply measure the changes in the number of deaths (without monetary values) or measures such as disability-adjusted life years (described later).

2.3.3.1.3 Cost-based approaches

The cost of replacing the services provided by the environmental resource can provide an order of magnitude estimate of the value of that resource. For example, if ecosystem change reduces the water filtration services, the cost of treating water to make it meet the required quality standards could be used. The major underlying assumptions of these approaches are that the nature and extent of physical damage expected is predictable (there is an accurate damage function available) and that the costs to replace or restore

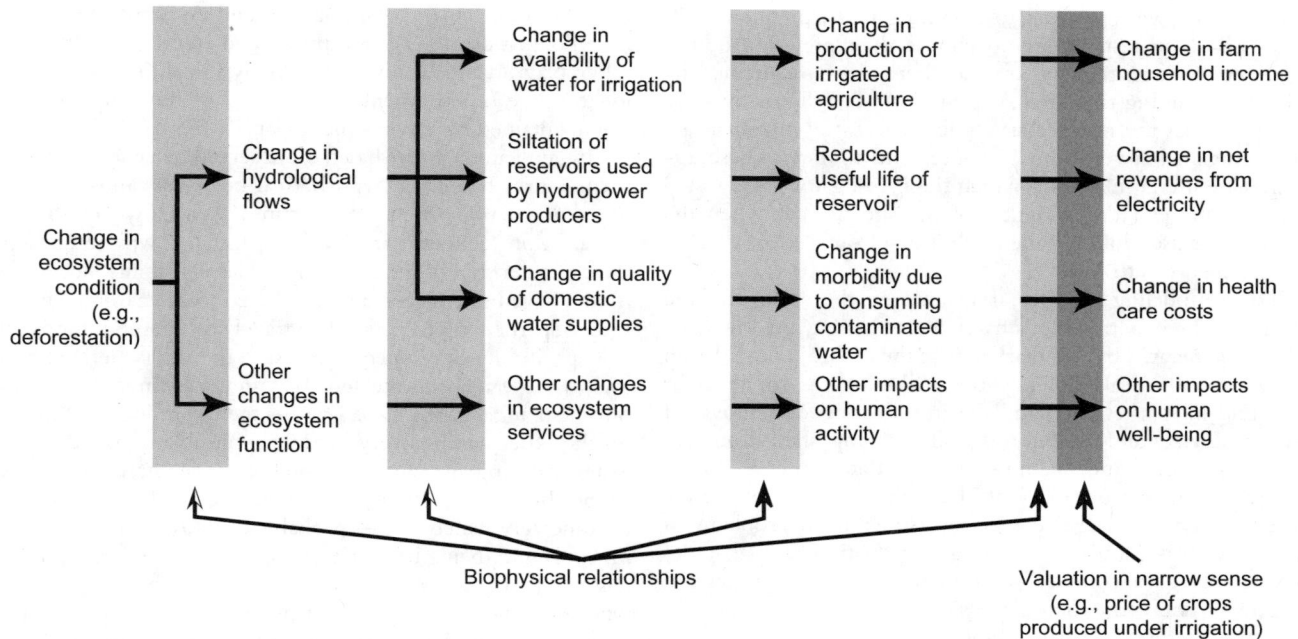

Figure 2.3. Valuing the Impact of Ecosystem Change (Adapted from Pagiola et al. forthcoming)

damaged assets can be estimated with a reasonable degree of accuracy. It is further assumed that the replacement or restoration costs do not to exceed the economic value of the service. This assumption may not be valid in all cases. It simply may cost more to replace or restore a service than it was worth in the first place—for example, because there are few users or because their use of the service was in low-value activities.

As there are often multiple ways that replacement costs could be estimated (for example, the value of lost water filtration services could be estimated based on the cost of restoring the wetland that had provided the service, the cost of treating water to meet quality standards, or the cost of obtaining suitable water from another source), the cheapest option should be considered as the replacement cost estimate. Because of these problems, cost-based approaches are generally thought to provide an upper-bound estimate of value.

2.3.3.1.4 Hedonic analysis

The prices paid for goods or services that have environmental attributes differ depending on those attributes. Thus, a house in a clean environment will sell for more than an otherwise identical house in a polluted neighborhood. Hedonic price analysis compares the prices of similar goods to extract the implicit value that buyers place on the environmental attributes. This method assumes that markets work reasonably well, and it would not be applicable where markets are distorted by policy or market failures. Moreover, this method requires a very large number of observations, so its applicability is limited.

2.3.3.1.5 Travel cost

The travel cost method is an example of a technique that attempts to deduce value from observed behavior in a surrogate market. It uses information on visitors' total expenditure to visit a site to derive their demand curve for the site's services. The technique assumes that changes in total travel costs are equivalent to changes in admission fees. From this demand curve, the total benefit visitors obtain can be calculated. (It is important to note that the

value of the site is not given by the total travel cost; this information is only used to derive the demand curve.) This method was designed for and has been used extensively to value the benefits of recreation, but it has limited utility in other settings.

2.3.3.1.6 Contingent valuation

Contingent valuation is an example of a stated preference technique. It is carried out by asking consumers directly about their willingness-to-pay to obtain an environmental service. A detailed description of the service involved is provided, along with details about how it will be provided. The actual valuation can be obtained in a number of ways, such as asking respondents to name a figure, having them choose from a number of options, or asking them whether they would pay a specific amount (in which case, follow-up questions with higher or lower amounts are often used).

CV can, in principle, be used to value any environmental benefit simply by phrasing the question appropriately. Moreover, since it is not limited to deducing preferences from available data, it can be targeted quite accurately to ask about the specific changes in benefits that the change in ecosystem condition would cause. Because of the need to describe in detail the good being valued, interviews in CV surveys are often quite time-consuming. It is also very important that the questionnaire be extensively pre-tested to avoid various sources of bias.

CV methods have been the subject of severe criticism by some analysts. A "blue-ribbon" panel was organized by the U.S. Department of Interior following controversy over the use of CV to value damages from the 1989 Exxon Valdez oil spill. The report of this panel (NOAA 1993) concluded that CV can provide useful and reliable information when used carefully, and it provided guidance on doing so. This report is generally regarded as authoritative on appropriate use of the technique.

2.3.3.1.7 Choice modeling

Choice modeling (also referred to as contingent choice, choice experiments, conjoint analysis, or attribute-based stated choice method) is a newer approach to obtaining stated preferences. It consists of asking respondents to choose their preferred option from a set of alternatives where the alternatives are defined by attributes (including the price or payment). The alternatives are designed so that the respondent choice reveals the marginal rate of substitution between the attributes and money. These approaches are useful in cases in which the investigator is interested in the valuation of the attributes of the situation or when the decision lends itself to respondents choosing from a set of alternatives described by attributes.

Choice modeling has several advantages: the control of the stimuli is in the experimenter's hand, as opposed to the low level of control generated by real market data; the control of the design yields greater statistical efficiency; the attribute range can be wider than found in market data; and the introduction or removal of products, services and attributes is easily accomplished (Louviere et al. 2000; Holmes and Adamowicz 2003; Bateman et al. 2004). The disadvantages associated with the technique are that the responses are hypothetical and therefore suffer from problems of hypothetical bias (similar to contingent valuation) and that the choices can be quite complex when there are many attributes and alternatives. The econometric analysis of the data generated by choice modeling is also fairly complex.

2.3.3.1.8 Benefits transfer

A final category of approach is known as benefits transfer. This is not a methodology per se but rather refers to the use of estimates obtained (by whatever method) in one context to estimate values in a different context. For example, an estimate of the benefit obtained by tourists viewing wildlife in one park might be used to estimate the benefit obtained from viewing wildlife in a different park. Alternatively, the relationship used to estimate the benefits in one case might be applied in another, in conjunction with some data from the site of interest ("benefit function transfer"). For example, a relationship that estimates tourist benefits in one park, based in part on their attributes such as income or national origin, could be used in another park, but with data on income and national origin of that park's visitors.

Benefits transfer has been the subject of considerable controversy in the economics literature, as it has often been used inappropriately. A consensus seems to be emerging that benefit transfer can provide valid and reliable estimates under certain conditions. These conditions include the requirement that the commodity or service being valued be very similar at the site where the estimates were made and the site where they are applied and that the populations affected have very similar characteristics. Of course, the original estimates being transferred must themselves be reliable in order for any attempt at transfer to be meaningful.

2.3.3.1.9 Summary of valuation methods

Each of these approaches has seen extensive use in recent years, and considerable literature exists on their application. These techniques can and have been applied to a very wide range of issues (McCracken and Abaza 2001), including the benefits of ecosystems such as forests (Bishop 1999; Kumari 1995; Pearce et al. 2002; Merlo and Croitoru in press), wetlands (Barbier et al. 1997; Heimlich et al. 1998), and watersheds (Aylward 2004; Kaiser and Roumasset 2002). Other studies have focused on the value of particular ecosystems services such as water (Young and Haveman 1985), non-timber forest benefits (Lampietti and Dixon 1995; Bishop 1998), recreation (Bockstael et al. 1991; Mantua at al. 2001; Herriges and Kling 1999), landscape (Garrod and Willis 1992; Powe et al. 1995), biodiversity for medicinal or industrial

uses (Simpson et al. 1994; Barbier and Aylward 1996), natural crop pollination (Ricketts in press), and cultural benefits (Pagiola 1996; Navrud and Ready 2002). Many valuation studies are cataloged in the Environmental Valuation Reference Inventory Web site maintained by Environment Canada (EVRI 2004).

In general, measures based on observed behavior are preferred to measures based on hypothetical behavior, and more direct measures are preferred to indirect measures. However, the choice of valuation technique in any given instance will be dictated by the characteristics of the case and by data availability. Several techniques have been specifically developed to cater to the characteristics of particular problems. The travel cost method, for example, was specifically developed to measure the utility derived by visitors to sites such as protected areas and is of limited applicability outside that particular case. The change in productivity approach, on the other hand, is very broadly applicable to a wide range of issues. Contingent valuation is potentially applicable to any issue, simply by phrasing the questions appropriately and as such has become very widely used—probably excessively so, as it is easy to misapply and, being based on hypothetical behavior, is inherently less reliable than measures based on observed behavior. For some types of value, however, stated preference methods may be the only alternative. Thus, existence value can only be measured by stated preference techniques.

In some cases, the value of a given benefit can be estimated in several ways. For example, the value of water purification might be estimated by the avoided health impacts (an application of change in productivity), by the avoided costs of treating water (an application of replacement costs), or by asking consumers for their willingness to pay for clean water (an application of contingent valuation). In such cases, it is appropriate to take the lowest figure as the estimate of the value of the benefit. It would make little sense to consider water purification to be worth 100 based (for example) on willingness to pay if treating the water to achieve the same result would only cost 10.

2.3.3.2 Putting Economic Valuation into Practice

Whatever valuation method is used, framing the question to be answered appropriately is critical. In most policy-relevant cases, the concern is over changes in the level and mix of services provided by an ecosystem. At any given time, an ecosystem provides a specific "flow" of services, depending on the type of ecosystem, its condition (the "stock" of the resource), how it is managed, and its socioeconomic context. A change in management (whether negative, such as deforestation, or positive, such as an improvement in logging practices) will change the condition of the ecosystem and hence the flow of benefits it is capable of generating. It is rare for all ecosystem services to be lost entirely; a forested watershed that is logged and converted to agriculture, for example, still provides a mix of provisioning, regulating, supporting, and cultural services, even though both the mix and the magnitude of specific services will have changed.

The typical question being asked, then, is whether the total value of the mix of services provided by an ecosystem managed in one way is greater or smaller than the total value of the mix provided by that ecosystem if it were managed in another way. Consequently, an assessment of this change in the value is typically most relevant to decision-makers. Where the change does involve the complete elimination of ecosystem services, such as the conversion of an ecosystem through urban expansion or road-building, then the change in value would equal the total economic value of the services provided by the ecosystem. Measurements of this total value can also be useful to policy-makers as an economic indicator, just as measures of gross domestic product or genuine savings provide policy-relevant information on the state of the economy.

Assessing the change in value of the ecosystem services caused by a management change can be achieved either by explicitly estimating the change in value or by separately estimating the value of ecosystem services under the current and the alternative management regime and then comparing them. If the loss of a given service is irreversible, then the loss of the option of using that service in the future ("option value") should also be included. An important caveat here is that the appropriate comparison is between the ecosystem with and without the management change; this is not the same as a comparison of the ecosystem before and after the management change, as many other factors will typically also have changed.

The actual change in the value of the benefits can be expressed either as a change in the value of the annual flow of benefits, if these flows are relatively constant, or as a change in the value of all future flows. The latter is equivalent to the change in the capital value of the ecosystem and is particularly useful when future flows are likely to vary substantially over time. (It is important to bear in mind that the capital value of the ecosystem is not separate and additional to the value of the flows of benefits it generates; rather, the two are intimately linked in that the capital value is the value of all future flows of benefits.)

Estimating the change in the value of the flow of benefits provided by an ecosystem begins by estimating the change in the physical flow of benefits. This is illustrated in Figure 2.3 for a hypothetical case of deforestation that affects the water services provided by a forest ecosystem. As noted earlier, the bulk of the work involves quantifying the biophysical relationships. Thus, valuing the change in production of irrigated agriculture resulting from deforestation requires estimating the impact of deforestation on hydrological flows, determining how changes in water flows affect the availability of water to irrigation, and then estimating how changes in water availability affect agricultural production. Only at the end of this chain does valuation in the strict sense occur—in putting a value on the change in agricultural production, which in this instance is likely to be quite simple, as it is based on observed prices of crops and agricultural inputs. The change in value resulting from deforestation then requires summing across all the impacts.

Clearly, tracing through these chains requires close collaboration between experts in different disciplines—in the deforestation example, between foresters, hydrologists, water engineers, and agronomists as well as economists. It is a common problem in valuation that information is only available on some links in the chain and often in incompatible units. An increased awareness by the various disciplines involved of what is needed to ensure that their work can be combined with that of others would facilitate more thorough analysis of such issues, including valuation.

In bringing the various strands of the analysis together, there are many possible pitfalls to be wary of. Inevitably, some ecosystem benefits will prove impossible to estimate using any of the available techniques, either because of lack of data or because of the difficulty of extracting the desired information from them. To this extent, estimates of value will be underestimates. Conversely, there is an opposite danger that benefits (even if accurately measured) might be double-counted.

As needed, the analysis can be carried out either from the perspective of society as a whole or from the perspective of individual groups within society. When the analysis is undertaken from the societal perspective, it should include all costs and benefits associated with ecosystem management decisions, which

should be valued at their opportunity cost to society (sometimes known as "shadow prices"). In contrast, focusing on a particular group usually requires focusing on a subset of the benefits provided by an ecosystem, as that group may receive some benefits but not others; groups located within an ecosystem, for example, typically benefit most from provisioning services but little from regulating services, whereas downstream users receive few benefits from provisioning services but many benefits from regulating services. It also requires using estimates of value specific to that group (the value of additional water, for example, will be different depending on whether it is used for human consumption or irrigation). The analysis can thus allow for distributional impacts and equity considerations to be taken into account, as well as overall impacts on well-being at the societal level.

This type of disaggregation is also very useful for understanding the incentives that particular groups face in making their ecosystem management decisions. Many ecosystems are mismanaged, from a social perspective, precisely because most groups that make decisions about ecosystem management perceive only a subset of the benefits it provides (Pagiola and Dixon 2001). Understanding how the benefits and costs of ecosystem management are distributed across different groups can also help design mechanisms to align their incentives with those of society (Pagiola and Platais in press).

Assessing the impact of ecosystem change almost always requires comparing costs and benefits at different times. In economic analysis, this is achieved by discounting future costs and benefits so that all are expressed in today's monetary units (Portney and Weyant 1999). Because discounting makes future benefits appear smaller, this practice has been controversial, and some have called for use of lower (perhaps even zero) discount rate when assessing environmental issues. Discount rates, however, reflect preferences for current as opposed to future consumption. Whatever discount rate is chosen, it should be applied in all evaluations involving choices between outcomes occurring at different times.

Similarly, estimating the impact of changes in management on future flows of benefits allows for intergenerational considerations to be taken into account. Here, too, the bulk of the work involved concerns predicting the change in future physical flows; the actual valuation in the narrow sense forms only a small part of the work. Predicting the value that future generations will place on a given service is obviously difficult. Technical, cultural, or other changes could result in the value currently placed on a service either increasing or decreasing. Often, the best that can be done is to simply assume that current values will remain unchanged. If trends suggest that a particular change in values will occur, that can be easily included in the analysis. Such predictions are notoriously unreliable, however.

2.3.4 Indicators of Specific Dimensions of Well-being

Well-being cannot be measured solely in terms of income, nor can non-income aspects of well-being always be expressed in monetary terms. This section reviews several indicators that seek to capture specific aspects of well-being which economic valuation often captures imperfectly, if at all, including health, poverty, and vulnerability.

2.3.4.1 Health Indicators

Biological responses involved in human disease phenomena are among the most important set of parameters for assessing environmental quality, and measures in support of environmental protection are often justified on the basis of their impact on human health (Moghissi 1994).

Health indicators have been used extensively to monitor the health of populations and are usually defined in terms of health outcomes of interest. The majority of health indicators so far developed, however, have no direct reference to the environment; examples include simple measures of life expectancy or cause-specific mortality rates, where no attempt has been made to estimate any portion of these health outcomes attributable to the environment. An Environmental Health Indicator can be seen as a measure that summarizes, in easily understandable and relevant terms, some aspect of the relationship between the environment and health that is amenable to action (Corvalan 1996). It is a summarized measure both of health outcomes and hazard exposures, which represents an underlying causal relationship between an environmental exposure and a health consequence (Pastides 1995). As with all indicators, appropriate EHIs vary according to the problem and the context.

EHIs can be constructed by linking aggregate data (linkage-based), by identifying environmental indicators with a health linkage (exposure-based), or by identifying health indicators with an environmental linkage (outcome-based). There are special complexities in the identification of EHIs since the incidence of many environmentally related diseases cannot be easily traced back to specific environmental exposures (Kjellström 1995). The Driving forces-Pressure-State-Exposure-Effect-Action framework, which has been proposed by the World Health Organization, is a widely accepted conceptual framework to guide the development of EHIs. The Driving Forces component refers to the factors that motivate and push the environmental processes involved (population growth, technological and economic development, policy intervention, and so on). The drivers result in the generation of pressures, normally expressed through human occupation or exploitation of the environment, and may be generated by all sectors of economic activity. In response to these pressures, the state of the environment is often modified, producing hazards. Exposure refers to the intersection between people and the hazards in the environment. These exposures lead to a wide range of health effects, ranging from well-being through morbidity or mortality (Briggs 1999).

EHIs are needed to monitor both trends in the state of the environment and trends in health resulting from exposures to environmental risk factors. They are useful also to compare areas or countries in terms of their environmental health status, to assess the effects of policies and other interventions on environmental health, and to investigate potential links between environment and health (Briggs 1999). EHIs use a variety of units, but many are expressed in disability-adjusted life years: the sum of life years lost due to premature mortality and years lived with disability, adjusted for severity (Murray 1994, 1997).

Usable EHIs depend heavily on the existence of known and definable links between environment and health. Difficulties in establishing these relationships (due, for example, to the complexity of confounding effects and the problems of acquiring reliable exposure data) inhibit the practical use of many potential indicators and make it difficult to establish core indicator sets (Corvalan 2000). Thus, the presence of environmental changes does not translate automatically into health outcomes, and the incidence of many environmentally related diseases cannot be easily traced back to specific environmental exposures. Many broader environmental issues, such as deforestation, loss of biodiversity, soil degradation, and climate change have a much less direct link to health. Although the effects of ecosystem disturbance on human health may be relatively direct, they may also occur at the end of long,

complex causal webs, dependent on many intermediate events. When these effects are subtle and indirect, often entailing complex interactions with social conditions, their measurement through indicators is often difficult.

WHO, by assigning weight factors in the form of estimated environmental fraction of reported DALYs for relevant diseases, has estimated that 23% of the global burden of disease is related to environmental factors (WHO 1997).

Sets of specific EHIs have been proposed to monitor both environmental quality and population health levels on a national basis, encompassing different types of hazards (chemical, physical, and biological) and modifications in several ecosystems, such as forests, agroecosystems, and urban ecosystems (Confalonieri 2001). In addition, indicators have recently been proposed to monitor the interactions between human health effects and the quality of specific ecosystems, including oceans (Dewailly 2002), freshwater ecosystems (Morris 2002), and urban systems (Hancock 2002). Table 2.7 shows simple examples of how changes in ecosystem services generate hazards to human health and how these can be measured by EHIs.

Health impact assessment provides a framework and a systematic procedure to estimate the health impact of a proposed intervention or policy action on the health of defined population groups. HIA produces hypothetical health trade-offs of adopting different courses of action (Scott-Samuel et al. 2001). These estimates may be converted in monetary values, to facilitate comparisons with non-health impacts. Applying an HIA typically involves a prospective assessment of a program or intervention before implementation, although it may be carried out concurrently or retrospectively. The HIA gathers opinions and concerns regarding the proposed policy, uses knowledge of health determinants regarding the expected impacts of the proposed policy or intervention, and describes the expected health impacts using both quantitative and qualitative methods, as appropriate.

2.3.4.2 Poverty and Equity

Possibly the most closely watched impacts of ecosystem changes are those that pertain to poverty. Although poverty has historically been defined in strictly economic terms, in recent years a broader understanding of poverty has increasingly been used, in which poverty is understood as encompassing not only deprivation of materially based well-being but also a broader deprivation of opportunities (World Bank 2001). The MA conceptual framework recognizes five linked components of poverty: the necessary material for a good life, health, good social relations, security, and freedom and choice.

Despite the broader understanding of poverty, most poverty indicators still pertain to monetary measures of well-being. Income has been most widely used as a poverty indicator. In recent years, however, many analysts have argued that consumption is a better measure, as it is more closely related to well-being and reflects capacity to meet basic needs through income and access to credit. It also avoids the problem of income flows being erratic at certain times of the year—especially in poor agrarian economies—which can cause reporting errors. Income-based poverty indicators are easier to compare with other variables such as wages. They are also more widely collected, in contrast to consumption data that are seldom collected, thereby limiting the possibility of undertaking comparative analyses.

Monetary-based indicators have the further limitation that they cannot reflect individuals' feeling of well-being and their access to basic services. A household's ability to address risks and threats (and hence, its feeling of well-being) can change dramati-

Table 2.7. Examples of Ecosystem Disruption and Environmental Health Indicators

Ecosystem	Service	Change	Hazard	Human Health Outcome	Indicators
Coastal	waste processing	organic overload	microbes	diarrhea; cholera	incidence
Urban	air quality regulation	air pollution	CO; NO_x; SO_2	asthma	morbidity; body burden of metals
Freshwater	water filtration	depletion	poor hygiene	diarrhea	childhood mortality
Tropical forest	regulation of water and nutrient cycles	deforestation	infections	malaria; arbovirus infections	incidence
Agroecosystem	food production	pesticides	toxic exposure	reproduction problems	fertility rates
Freshwater/marine	provision of fish	overharvesting	depletion of fish resource	reduced consumption of fish protein	protein deficiency

cally even as income and consumption remain stable. Factoring in the effect of vulnerability, analysts estimate that monetary-based indicators can understate poverty and inequality by around 25% (World Bank 2001). In response, efforts have been made to develop non-monetary-based poverty indicators such as outcomes relating to health, nutrition, or education, as well as composite indices of wealth (Wodon and Gacitúa-Marió 2001). These alternative poverty indicators, however, face methodological and data collection issues that make comparisons between countries difficult.

Poverty measures are defined relative to a poverty line (the cutoff separating the poor from the non-poor). Many types of poverty measures exist, but the most commonly used are the headcount index (a measure of poverty incidence, which computes the number of people or share of the population below the poverty line), the poverty gap (a measure of the depth of poverty, which describes how far below the poverty line people are), and the squared poverty gap (a measure of poverty severity, which combines both poverty gap and inequality among the poor). A related set of measures is used to measure inequality, including the Gini coefficient (a measure between 0 and 1, with 0 representing perfect equality and 1 perfect inequality) and the Atkinson index (which incorporates the strength of societal preference for equality).

Most countries determine their own poverty line, making international comparisons of poverty data conceptually and practically difficult. Poverty lines in rich countries are characterized by a higher purchasing power than in poorer nations, making comparisons subject to possible inaccurate interpretation (World Bank 2003). In response, an international poverty line was established in order to measure poverty across countries. The dollar-a-day poverty line (this has been updated to $1.08 a day, in 1993 prices) was chosen. It is converted to local currency units using purchasing power parity exchange rates. However, the non-uniform derivation of the PPP changes the relative value of expenditures between countries and may affect poverty comparisons. The World Bank, for example, uses the PPP-based international poverty line to arrive at comparable aggregate poverty estimates across countries, but it relies mostly on national poverty lines in its poverty analysis.

Reliable and consistent poverty analyses require uniform and high-quality data that are in many cases—especially in developing countries—not available. The Living Standards Measurement Study program was established to develop methods to monitor progress in improving standards of living, in identifying the impacts of policy reforms on well-being, and in establishing a common language by which research proponents and policy-makers could communicate (Grosh and Glewwe 1995). LSMS surveys are used to gather data on a gamut of household activities, many of which are used as poverty indicators. Well-being is measured by consumption; hence in most LSMS research on poverty, measurement of consumption is heavily emphasized in the surveys. With the strong interest in addressing poverty issues in the context of sustainable development, there are current efforts to expand the scope of the LSMS surveys to include variables pertaining to natural resource and environmental management. Exploratory efforts are being undertaken to possibly include a module on environmental health in the LSMS research.

The link between poverty and ecosystem services is established by monitoring changes in ecosystem services and observing how they change poverty measures. The issues of whether the poor are agents or victims of environmental degradation (or both) and of possible trade-offs between ecosystem condition and the well-being of the poor are both burning topics among scholars and policy-makers (Reardon and Vosti 1997; World Bank 2002). Recent work has documented that the poor tend to rely heavily on goods and services provided by the environment and thus are particularly vulnerable to their degradation (Cavendish 1999; Vedeld et al. 2004).

2.3.4.3 Other Indicators

A great number of other indicators can be used to assess various dimensions of human well-being. For example, several indicators help measure progress toward achieving the Millennium Development Goals in addition to the poverty and health indicators just described (World Bank 2002). Adult literacy rates measure educational attainment, and indicators such as net enrollment ratios in primary education or the proportion of students starting grade 1 who reach grade 5 can measure progress toward the goal of universal primary education (MDG 2). The ratio of girls to boys at various levels of education, the ratio of literate females to males, the share of women in nonagricultural employment, and the share of seats in parliament held by women can be used to measure progress toward the goal of promoting gender equality (MDG 3). And maternal mortality ratios and the proportion of births attended by skilled personnel can be used to measure progress toward improving maternal health (MDG 5). These and many other indicators can provide valuable insights, but they are often difficult to relate to ecosystem condition as they are also affected by many other factors. (Note that risk and vulnerability indicators are discussed in Chapter 6.)

2.3.5 Aggregate Indicators of Human Well-being

Several indicators are in use as aggregate indicators of human well-being. The most commonly used, of course, is the gross domestic product, which is a measure of economic activity. This indicator has long been known to be imperfect, even for the narrow purpose of measuring economic activity, let alone as a measure of overall well-being. The limitations of GDP as an indicator have led to substantial efforts to improve it and to develop alternatives.

The linkage between human well-being and national accounting is not particularly straightforward, since GDP, for example, includes both consumption of produced goods—yielding direct benefits for well-being—and investment in physical capital—yielding future benefits for well-being. Moreover, many factors, including the enjoyment of environmental amenities, are not captured in the value of consumption recorded in the national accounts.

Recent results in the theory of environmental accounting make the linkage between asset accounting and well-being explicit. Hamilton and Clemens (1999) show that there is a direct link between the change in the value of all assets (including produced and natural assets) and the present value of social well-being: declining asset values, measured at current shadow prices, imply future declines in social well-being. Dasgupta and Mäler (2000) and Asheim and Weitzman (2001) have extended these results. The World Bank has been publishing estimates of adjusted net saving for roughly 150 countries since 1999 (World Bank 2003). Relying on internationally available data sets, these estimates adjust traditional measures of saving to reflect investments in human capital; depreciation of produced capital; depletion of minerals, energy, and forests; and damages from emissions of carbon dioxide.

Efforts to develop alternative indicators of well-being include composite indices that capture the multidimensionality of well-being. Early attempts to develop composite indices include the Weighted Index of Social Progress (Estes 1984, 1988) and the Physical Quality of Life Index (Morris 1979). More recently, the Human Development Index (UNDP 1998, 2003), which combines measures of life expectancy, literacy, education enrollment, and GDP per capita, has been widely used. The Human Poverty Index is similar, but with different variables for industrial and developing countries, while the Gender-related Development Index adjusts for disparities in achievement for men and women (UNDP 2003). None of these indicators include environmental variables explicitly. One indicator that does is the Calvert-Henderson Quality of Life Indicator, which includes measures of environmental, social, and economic conditions (Flynn 2000; Henderson 2000).

Composite indicators suffer from the arbitrariness of the weighting of their different components, however. Some authors prefer to simply list the components individually, without attempting to aggregate them into a single measure. Thus the World Bank provides a wide selection of indicators in its annual *World Development Indicators* publication (World Bank 2004), and UNDP provide a variety of indicators in addition to the aggregated HDI in the annual *Human Development Report* (UNDP 2003). Many of these indicators have substantial limitations from the perspective of the MA, as they are extremely difficult to relate to environmental conditions.

2.3.6 Intrinsic Value

Economic valuation attempts to measure the utilitarian benefits provided by ecosystems. In addition, many people ascribe ecological, sociocultural, or intrinsic values to the existence of ecosystems and species and, sometimes, to inanimate objects such as "sacred" mountains.

Some natural scientists have articulated a theory of value of ecosystems in reference to the causal relationships between parts of a system—for example, the value of a particular tree species to control erosion or the value of one species to the survival of another species or an entire ecosystem (Farber et al. 2002). At a global scale, different ecosystems and their species play different roles in the maintenance of essential life-support processes (such as energy conversion, biogeochemical cycling, and evolution). The magnitude of this ecological value is expressed through indicators such as species diversity, rarity, ecosystem integrity (health), and resilience. The concept of ecological value is captured largely in the "supporting" aspect of the MA's definition of ecosystem services.

What might be called sociocultural value derives from the value people place on elements in their environment based on different worldviews or conceptions of nature and society that are ethical, religious, cultural, and philosophical. A particular mountain, forest, or watershed may, for example, have been the site of an important event in their past, the home or shrine of a deity, the place of a moment of moral transformation, or the embodiment of national ideals. These values are expressed through, for example, designation of sacred species or places, development of social rules concerning ecosystem use (for instance, "taboos"), and inspirational experiences.

For many people, sociocultural identity is in part constituted by the ecosystems in which they live and on which they depend—these help determine not only how they live, but also who they are. To some extent, this kind of value is captured in the concept of cultural ecosystem services and can be valued using economic valuation techniques. To the extent, however, that ecosystems are tied up with the very identity of a community, the sociocultural value of ecosystems transcends utilitarian preference satisfaction. These values might be elicited by using, for example, techniques of participatory assessment (Campell and Luckert 2002).

The notion that ecosystems have intrinsic value is based on a variety of points of view. Intrinsic value is a basic and general concept that is founded on many and diverse cultural and religious worldviews. Among these are indigenous North and South American, African, and Australian cultural worldviews, as well as the major religious traditions of Europe, the Middle East, and Asia. In the Judeo-Christian-Islamic tradition of religions, human beings are attributed intrinsic value on the basis of having been created in the image of God. Some commentators have argued that plant and animal species, having also been created by God and declared to be "good," also have intrinsic value on the same basis (Barr 1972; Zaidi 1981; Ehrenfeld and Bently 1985).

In some American Indian cultural worldviews, animals, plants, and other aspects of nature are conceived as relatives, born of one universal Mother Earth and Father Sky (Hughes 1983). The essential oneness of all being, *Brahman*, which lies at the core of all natural things, is basic to Hindu religious belief (Deutch 1970). Closely related to this idea is the moral imperative of *ahimsa*, non-injury, extended to all living beings. The concept of *ahimsa* is also central to the Jain environmental ethic (Chapple 1986).

In democratic societies, the modern social domain for the ascription of intrinsic value is the parliament or legislature (Sagoff 1998). In other societies a sovereign power ascribes intrinsic value, although this may less accurately reflect the actual values of citizens than do parliamentary or legislative acts and regulations. The metric for assessing intrinsic value is the severity of the social and legal consequences for harming what society has deemed to be intrinsically valuable.

2.4 Assessing Trade-offs in Ecosystem Services

The challenge to decision-making is to make effective use of new information and tools in this changing context in order to improve the decisions that intend to enhance human well-being and provide for a sustainable flow of ecosystem services. Perhaps the most important traditional challenge in decision-making about ecosystems is the complex trade-off faced when making decisions that will negatively affect or otherwise alter ecosystems. Increasing the flow of one service from a system, such as provision of timber, may decrease the flow from others, such as carbon sequestration or the provision of habitat. In addition, benefits, costs, and risk are not allocated equally to everyone, so any intervention will change the distribution of human well-being—another trade-off. Improved provision of appropriate information can help in assessing the trade-offs among ecosystem services resulting from policy decisions.

Understanding the impact of ecosystem management decisions would be simplest if all impacts were expressed in common units. If information on the impact of ecosystem change is presented solely as a list of consequences in physical terms—so much less provision of clean water, perhaps, and so much more production of crops—then the classic problem of comparing apples and oranges applies.

The purpose of economic valuation is to make the disparate services provided by ecosystems comparable to each other by measuring their relative contribution to human well-being. As utility cannot be measured directly, economic valuation usually attempts to measure all services in monetary terms. This is purely a matter of convenience, in that it uses units that are widely recognized, saves the effort of having to convert values already expressed in monetary terms into some other unit, and facilitates comparison with other activities that also contribute to well-being, such as spending on education or health. In particular, it puts the impacts of ecosystem change into units that are readily understood by decision-makers and the lay public. When all im-

pacts of ecosystem change are expressed in these terms, then they can readily be introduced into frameworks such as cost-benefit analysis in order to assess policy alternatives.

Other metrics are occasionally proposed. Some analysts, for example, have advocated the use of energy units (Odum and Odum 1981; Hall et al. 1986), arguing that as all goods and services are ultimately derived from natural resources by expending energy, energy is the real source of material wealth. These approaches can provide valuable insights into particular issues. For purposes such as the MA, however, these approaches have several disadvantages—in particular, they have no direct link to human well-being, and they require a considerable effort to convert a wide variety of impacts into common units.

Efforts to place everything into common units will necessarily remain incomplete, however, sometimes because of lack of data and sometimes because value arises not from utilitarian benefits but from intrinsic value or from another source of value. Societies have many objectives, only some of them purely utilitarian. Furthermore, the value of an ecosystem service varies, depending on whether a critical threshold for ecosystem condition or human well-being is crossed (Farber et al. 2002). In other words, placing everything into common units is sometimes impossible and frequently undesirable. It is important to stress, however, that even incomplete efforts to express impacts in common units can be helpful by reducing the number of different dimensions that need to be taken into considerations.

Graphical depictions of the trade-offs in ecosystem services associated with alternative policy options can provide useful input to decision-makers. "Spider diagrams" such as that in Figure 2.4 can depict the amount of ecosystem services associated with different management alternatives. For example, Figure 2.4 depicts hypothetical trade-offs among five ecosystem services associated with an expansion of cropland in a forested area: food production, carbon sequestration, species richness, soil nutrients, and base streamflow. Comparison of the ecosystem services available before forest conversion to cropland with the services after forest conversion allows a decision-maker to account for the full suite of ecosystem services affected by the conversion. The approach requires quantifiable and measurable indicators for each of the services depicted. The quantities depicted can be an absolute measure (such as tons of carbon stored) relative to a previous quantity, to a relevant average quantity (for the area, for instance, or for the biome), or to an ideal "sustainable" amount.

The degree to which the diagram effectively communicates trade-offs in ecosystem services depends on the explicit definition of the values on the axes and the ability to quantify them. A series of diagrams for varying time since forest clearing and for varying

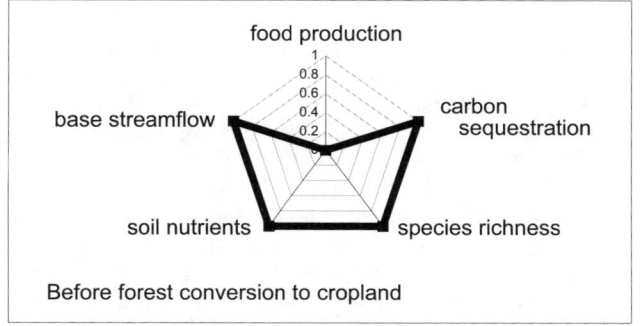

Before forest conversion to cropland

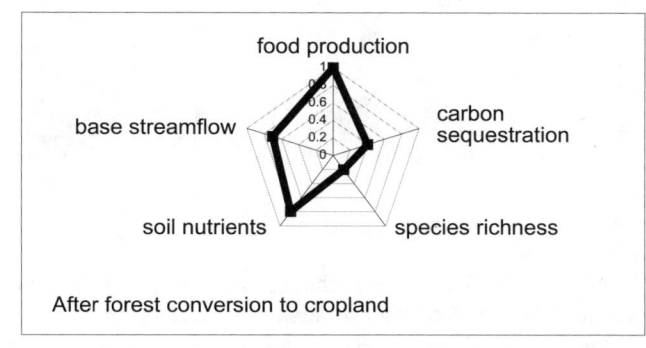

After forest conversion to cropland

Figure 2.4. Hypothetical Trade-offs in a Policy Decision to Expand Cropland in a Forested Area. Indicators range from 0 to 1 for low to high value of service. The values of the indicators vary according to the spatial and temporal scales of interest.

spatial scales of interest could be used to inform decision-makers about the effects on ecosystem services for the varying scales of analysis. When a large number of management alternatives are to be compared, they can be portrayed either in a series of spider diagrams or across all management alternatives, as in Figure 2.5 (Heal et al. 2001a).

Depictions of ecosystem services associated with predefined management alternatives, as in Figures 2.4 and 2.5, are simple and readily communicable to decision-makers but are often unable to account for non-linearities and thresholds in responses of ecosystem services to management decisions. When such phenomena are present, figures such as Figure 2.6 can help assess choices. For example, application of nitrogen fertilizer involves a trade-off between increasing crop yields and decreasing coastal fisheries if nitrate leaching leads to hypoxia in downstream coastal locations, as it has in the Mississippi Delta (Donner and Kucharik 2003). Balancing an objective of maximum crop yields with minimum damage to coastal fisheries requires knowledge of the response curves of each service to nitrogen fertilizer application. In this example, fertilizer application beyond point "A" results in negligible increase in crop yield but substantial nitrate leaching. A decision to apply fertilizer greater than point "A" trades small increases in crop yield for large increases in nitrate leaching. A decision to apply fertilizer less than point "A" trades small decreases in nitrate leaching for forgone large increases in crop yield. To the extent that the shape of the response curves can be quantified, management alternatives can account for these types of nonlinear responses to determine the most desirable alternative.

Portraying interactions among multiple ecosystem services graphically quickly becomes complex and unwieldy. Heal et al. (2001a) suggest constructing "production possibility frontiers" to model combinations in the amounts of ecosystem services possible to achieve a management objective. For example, possible combinations of ecosystem services such as carbon storage and timber

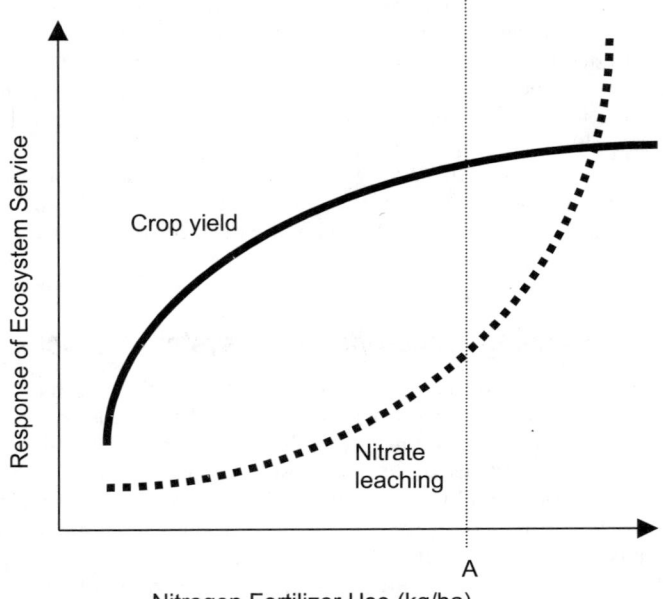

Figure 2.6. Example of Nonlinear Responses of Two Ecosystem Services (Crop Yields and Coastal Fisheries) to Application of Nitrogen Fertilizer

production can be modeled to achieve varying levels of water purification. The optimal mix of these services can then be selected, depending on the management objectives.

Multicriteria analysis provides another formal framework to help assess choices in the presence of multiple, perhaps contradictory, objectives (Falconí 2003). In a multicriteria analysis, a matrix is constructed showing how each of the alternatives under consideration ranks relative to the other alternatives, according to each criterion. This impact matrix, which may include quantitative, qualitative, or both types of information, allows the best alternative to the decision or analysis problem to be found (Munda 1995; Martínez-Alier et al. 1998). A vast number of multicriteria methods have been developed and applied for different policy purposes in different contexts (Munda 1995). The main advantage of such models is that they make it possible to consider a large number of data, relations, and objectives that are generally present in a specific real-world decision problem, so that the decision problem at hand can be studied in a multidimensional fashion. When different conflicting evaluations are taken into consideration, however, a multicriteria problem is mathematically ill defined. The application of the different methods can lead to different solutions. In some cases, solutions that satisfy multiple objectives may not be possible.

Consideration of the trade-offs involves clear definitions about the spatial and temporal scales of interest. How are future impacts on ecosystem services included in the analysis? Over what time frame should these impacts be considered? Does the alteration in ecosystem services affect human well-being distant in space from the ecosystem change (such as through downstream effects or atmospheric transport)? How are impacts that cross administrative or ecosystem boundaries incorporated in the analysis? Assessments need to be conducted within a scale domain appropriate to the processes or phenomena being examined. Cost-benefit analysis has often fallen short in the past in part because the spatial and temporal boundaries it used did not encompass all the impacts of the proposed interventions (Dixon et al. 1994). This same weakness applies to all assessment methodologies: they will only

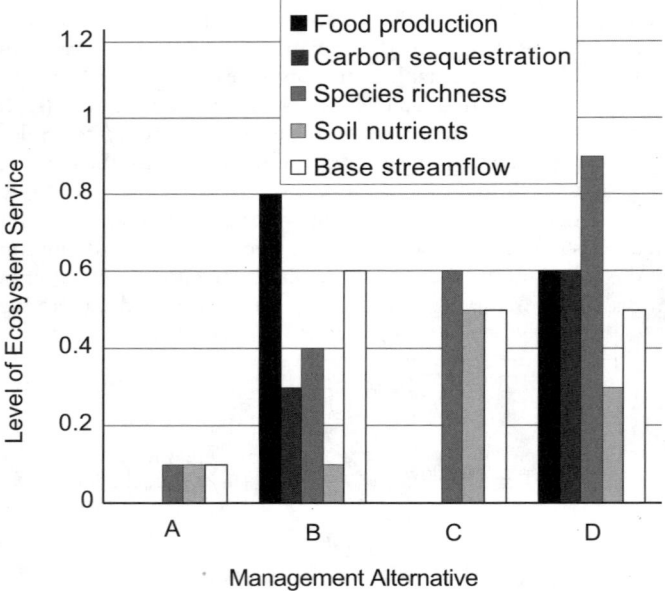

Figure 2.5. Portrayal of Hypothetical Trade-offs in Ecosystem Services Associated with Management Alternatives for Expanding Cropland in a Forested Area. Indicators range from 0 to 1 for low to high value of service. See text for management alternatives. (Adapted from Heal et al. 2001b).

be meaningful if the spatial and temporal scales of the analysis have been carefully defined. Too narrow a definition of either could result in a misperception of the problems. For example, if soil nutrients decline over time under agricultural use, the perceived impact on that dimension depends crucially on the time period chosen for the indicators.

Appendix 2.1. Core Data Sets Used by the MA to Assess Ecosystem Condition and Trends

The Millennium Ecosystem Assessment has involved the development and distribution of a range of data sets and indicators. Although the overall MA products primarily consist of syntheses of findings from existing literature, the data and indicators developed or presented within the MA play important roles both in presenting information on the links between ecosystems and human well-being and in establishing year 2000 "baseline" conditions for reference in future global and sub-global assessments.

For many central themes of the MA, there are multiple available data sets on which elements of the assessment could be based and from which different conclusions could be drawn. For example, there is a range of land cover data sets available based on information from different satellite sensors and interpretation techniques, from which different statistics on land cover could be generated. To ensure consistency of analysis and comparability of results across the chapters and working groups of the MA, a small number of MA "core data sets" were selected. (See Appendix Table 2.1.) Although chapter teams also made use of alternative data sets, applicable findings are in each case also presented based on an analysis with the various core data sets, and the strengths and weaknesses of these data sets are assessed for the particular application in the chapters.

A description of the choice of MA systems, the main reporting unit for the Condition and Trends Working Group, can be found in the Preface. Appendix Table 2.2 presents the updated system boundary definitions, adding detail to the brief system descriptions given in Box 1–3 of Chapter 1.

Data management procedures were developed for the use of data sets in the MA. A Web-based data catalogue recorded metadata for all data sets used in the MA. Data Archives were established at CIESEN, the World Data Center For Biodiversity and Ecology, and UNEP–WCMC for all data in categories 4–6 of Appendix Table 2.3, as well as for some data in category 2 if they were used for a significant portion of analysis in a particular chapter. MA archived data are freely accessible to any user, and all archived data sets are accompanied by metadata in the ISO metadata standard (ISO 19115: Geographic Information).

Appendix Table 2.1. Summary of MA Core Datasets

Core Dataset	Brief Description	Lead Agencies
Global land cover	Global Land Cover 2000 dataset; a global product of land cover in year 2000, based on SPOT Vegetation satellite data	EU JRC, with regional networks
Human population density	an updated Gridded Population of the World dataset, referenced to year 2000, and including a rural/urban split, including a point database of human settlements >5,000 people, an urban mask (polygons), and a complete urban-rural gridded surface	CIESIN, with World Bank and IFPRI
Protected areas	the 13th UN List of Protected Areas, from which a "snapshot" of the extent of Protected Areas in the year 2000 has been generated, as a baseline dataset for the MA	UNEP-WCMC, with WCPA
Subnational agricultural statistics	sub-national time series and single year crop production data including area, production, and yield, available for the globe	IFPRI, with wider consortium
Climate	0.5-degree dataset of monthly surface climate extending from 1901 to 2000 over global land areas, excluding Antarctica 10-minute mean monthly surface climate grids for the 1961–90 period covering a similar area	University of East Anglia CRU and University of Oxford, UK
Human well-being indicators	sub-national infant mortality, malnutrition, and GDP data; global data, although malnutrition index only available for the developing world	CIESIN
Areas of rapid land cover change	a synthesis of the knowledge of areas affected by rapid land cover change during the last 20 years for various change classes, including deforestation, cropland and pasture expansion, soil degradation and desertification, urban expansion, and exceptional fire events	IGBP/IHDP, LUCC, GOFC/GOLD
Global MA reporting "units"	datasets delineating MA system boundaries (see Appendix Table 2.2), biomes and biogeographical realms, and socioeconomic regional reporting units	various

Appendix Table 2.2. MA System Boundary Definitions

MA System	Description
Coastal	The area between the interpolated 50 m bathymetry and 50 m elevation contours from the ETOPO2 dataset. The 50 m inland contour is constrained to a maximum distance of 100 km.
Cultivated	Agricultural classes from version 2 of the Global Land Cover Characteristics Dataset. Cropland, pasture, and mosaic (or mixed) agriculture and other land use classes are included.
Dryland	A subset of the aridity zone map published in the World Atlas of Desertification. Aridity zones are derived from an Aridity Index calculated as the ratio of precipitation to potential evapotranspiration. The zones hyper-arid, arid, semiarid, and dry subhumid are included in the dryland system.
Forest and woodland	Derived from the Global Land Cover 2000 Dataset. Extracted classes are broadleaved, needle-leaved, mixed tree cover, regularly flooded (such as mangroves) and burnt tree cover, and a mosaic tree cover/other natural vegetation class (classes 1 to 10 of the global classification).
Inland water	Includes major rivers, wetlands, lakes, and reservoirs as compiled in the Global Lakes and Wetlands Database–Level 3.
Island	Oceanic and coastal islands as defined by ESRI's ArcWorld Country Boundaries dataset. Approximately 11,925 islands are represented and include those listed as members of the Alliance of Small Island States and the Small Island Developing States Network.
Marine	The marine system boundary is defined from the interpolated 50 m bathymetry (from the ETOPO2 dataset) seaward. Longhurst's biome classification provides subsystem categorizations.
Mountain	Derived from UNEP-WCMC's mountain dataset, using criteria of altitude, slope, and local elevation range. Altitudinal life zones form subsystem reporting units.
Polar	Arctic and sub-arctic vegetation types define the northern hemisphere portion of the polar system. Vegetation types are delineated from a combination of global and regional land cover maps from remote imagery. Antarctica forms the southern portion of the polar system.
Urban	Derived from the Global Land Cover 2000 Dataset artificial surfaces class (class 22 in the global legend).

Appendix Table 2.3. Data Handling Procedures in the MA

Data Application in the MA	Data Handling Procedures
1. Peer-reviewed or validated datasets cited in MA reports	full citation in MA report
2. Peer-reviewed or validated datasets used in MA analysis (e.g., to calculate area, quantity), map, or table but unmodified	full citation in MA report included in MA Data Catalog may be included in datasets available for online access as part of MA outreach
3. Non-peer-reviewed datasets cited in MA reports	dataset critically assessed; quality and validity of the dataset reviewed by chapter team before incorporating results from the source into an MA Report following materials sent to the Working Group Technical Support Unit: title of dataset; location (URL if available); institution responsible for maintaining the data; information on the availability of the data to other researchers; contact details for one or two people who can be contacted for further information about the source
4. Non-peer-reviewed datasets used in MA analysis, map, or table but unmodified	procedures in category 3 followed included in MA Data Catalog included in MA Data Archive if possible (particularly if a key dataset for the analysis) may be included in datasets available for online access as part of MA outreach
5. Data modified in an MA analysis or new datasets produced through existing peer-reviewed data; considered an "MA Dataset"	dataset critically assessed; quality and validity of the dataset reviewed by chapter team before incorporating results from the source into an MA Report. MA Metadata Standards followed included in MA Data Catalog and MA Data Archive made freely available to other users
6. MA Core Datasets	MA Metadata Standards followed included in MA Data Catalog and Data Archive made freely available to other users
7. MA Heritage Datasets — datasets representing a valuable "baseline" condition for year 2000 (e.g., NDVI data)	MA Metadata Standards followed included in MA Data Catalog and MA Data Archive made freely available to other users

References

Achard, F., Eva, H., Stibig, H. J., Mayaux, P., Gallego, J. and Richards, T., 2002: Determination of deforestation rates of the world's humid tropical forests. *Science,* **297,** 999–1002.

Akçakaya, H.R., 2002: RAMAS GIS: Linking landscape data with population viability analysis. Version 4. *Applied Biomathematics.,* Setauket, New York.

Akçakaya, H.R. and M.G. Raphael, 1998: Assessing human impact despite uncertainty: viability of the northern spotted owl metapopulation in the northwestern USA. *Biodiversity and Conservation,* **7,** 875–894.

Akçakaya, H.R. and P. Sjogren-Gulve, 2000: Population viability analysis in conservation planning: an overview. *Ecological Bulletins,* **48,** 9–21.

Akçakaya, H.R., Ferson, S., Burgman, M. A., Keith, D. A., Mace, G. M., Todd, C. R., 2000: Making consistent IUCN classifications under uncertainty. *Conservation Biology,* **14,** 1001–1013.

Antweiler, C., 1998: Local knowledge and local knowing: An anthropological analysis of contested 'cultural products' in the context of development. *Anthropos,* **93(406),** 469–494.

Aplet, G., Thomson, J. and Wilbert, M., 2000: *Indicators of wildness. Using attributes of the land to assess the context of wildness.* Proc. RMRS-P-15, USDA Forest Service, Rocky Mountain Research Station, Ogden, UT.

Asheim, G.B. and M.L. Weitzman, 2001: Does NNP growth indicate welfare improvement? *Economics Letters,* **73,** 233–39.

Aylward, B., 2004: Land Use, Hydrological function and economic valuation. In *Forests, Water and People in the Humid Tropics,* M. Bonnell and L.A. Bruijnzeel (eds.), Cambridge University Press, Cambridge.

Balmford, A., J.L. Moore, T. Brooks, N. Burgess, L.A. Hansen, P. Williams, and C. Rahbek, 2001: Conservation conflicts across Africa. *Science,* **291,** 2616–2619.

Barbier, E.B., M. Acreman, and D. Knowler, 1997: *Economic Valuation of Wetlands,* IUCN, Cambridge.

Barbier, E.B. and B.A. Aylward, 1996: Capturing the Pharmaceutical Value of Biodiversity in a Developing Country. *Environmental and Resource Economics,* **8,** 157–181.

Barbier, E.B., G. Brown, S. Dalmazzone, C. Folke, M. Gadgil, et al. 1995: The economic value of biodiversity. In: *Chap 12 in UNEP: Global Biodiversity Assessment.,* Cambridge University Press, Cambridge, UK, 823–914.

Barr, J., 1972: Man and nature: The ecological controversy and the Old Testament. *Bulletin of the John Rylands Library,* **55,** 9–32.

Bartholome, E. M. and Belward A. S., 2004, GLC2000; a new approach to global land cover mapping from Earth Observation data, *International Journal of Remote Sensing* (in press)

Bateman, I., R. Carson, B. Day, M. Hanemann, N. Hanley, et al. 2004: *Environmental Valuation with Stated Preference Methods: A Manual.* Edward Elgar.

Berkes, F., 1999: *Sacred Ecology: Traditional Ecological Knowledge and Resource Management.* Taylor and Francis, Philadelphia and London, UK.

Berkes, F., 2002: Cross-scale institutional linkages: Perspectives from the bottom up. In: *The Drama of the Commons,* E. Ostrom, T. Dietz, N. Dolak, P.C. Stern, S. Stonich, and E.U. Weber (eds.), National Academy Press, Washington, DC, 293–322.

Bishop, J.T., 1998: *The Economics of Non Timber Forest Benefits: An Overview.* Environmental Economics Programme Paper No. GK 98–01, IIED, London.

Bishop, J.T., 1999: *Valuing Forests: A Review of Methods and Applications in Developing Countries.,* IIED, London.

Blackburn, T.M. and K.J. Gaston, 1996: Spatial patterns in the species richness of birds in the New World. *Ecography,* **19.**

Bockstael, N.E., K.E. McConnell, and I.E. Strand, 1991: Recreation. In: *Measuring the Demand for Environmental Quality,* J.B. Braden and C.D. Kolstad (eds.), Contributions to Economic Analysis No. 198, Amsterdam, North Holland.

Bockstael, N.E., A. M. Freeman, III, R. J. Kopp, P. R. Portney, and V. K. Smith, 2000: On measuring economic values for nature. *Environ. Sci. Technol.,* **34** (8), 1384–1389.

Borrini-Feyerabend, G., 1997: *Beyond Fences: Seeking Social Sustainability in Conservation.,* International Union for the Conservation of Nature, Kasparek-Verlag, Gland, Switzerland.

Bossel, H., 1999: *Indicators for Sustainable Development: Theory, Method, Application.,* International Institute for Sustainable Development, Winnipeg, Canada, 124pp.

Boumans, R., R. Costanza, J. Farley, M.A. Wilson, R. Portela, J. Rotmans, F. Villa, and M. Grasso, 2002: Modeling the dynamics of the integrated earth system and the value of global ecosystem services using the GUMBO model. *Ecological Economics,* **41,** 529–560.

Boyce, M.S., 1992: Population viability analysis. *Annual Review of Ecology and Systematics,* **23,** 481–506.

Braden, J.B. and C.D. Kolstad (eds.), 1991: *Measuring the Demand for Environmental Quality. Contributions to Economic Analysis No. 198,* North-Holland, Amsterdam.

Briggs, D., 1999: *Environmental Health Indicators: Framework and Methodologies.,* WHO/SDE/OEH/99.10, Geneva, 117 pp.

Brook, B.W., O'Grady, J. J., Chapman, A. P., Burgman, M. A., Akçakaya, H. R., and Frankham, R., 2000: Predictive accuracy of population viability analysis in conservation biology. *Nature,* **404,** 385–387.

Broten, M.D., and M. Said, 1995: Population trends in and around Kenya's Masai Mara Reserve. In: *Serengeti II, Dynamics, Management, and Conservation of an Ecosystem,* A.R.E. Sinclair and P. Arcese (Eds.), University of Chicago Press, Chicago, IL.

Burgman, M.A., Ferson, S. and Akçakaya, 1993: *Risk Assessment in Conservation Biology.* Chapman and Hall, London, UK, 314 pp.

Campell, B. and M. Luckert (eds.), 2002: *Uncovering the Hidden Harvest: Valuation Methods for Woodland and Forest Resources.* Earthscan, London.

Carignan, V. and M.-A. Villard, 2002: Selecting indicator species to monitor ecological integrity: A review. *Environmental Monitoring and Assessment,* **78(1),** 45–61.

Carver, S., Evans, A. and Fritz, S., 2002: Wilderness attribute mapping in the United Kingdom. *International Journal of Wilderness,* **8(1),** 24–29.

Catley, A.P., and Aden, A., 1996: Use of participatory rural appraisal (PRA) tools for investigating tick ecology and tick-borne disease in Somaliland. *Tropical Animal Health and Production,* **28(1).**

Cavendish, W., 1999: *Empirical Relationships in the Poverty-Environment Relationship of African Rural Households.* Working Paper No. WPSS 99–21, Centre for the Study of African Economies, Oxford University, Oxford.

Ceballos, G. and J.H. Brown, 1995: Global patterns of mammalian diversity, endemism, and endangerment. *Conservation Biology,* **9,** 559–568.

Chambers, R., 1994: Participatory Rural Appraisal (PRA): Analysis of Experience. *World Development,* **22(9),** 1253–1268.

Chapple, C.K., 1996: Non-injury to animals: Jaina and Buddhist perspectives. In: *Animal Sacrifices: Religious Perspectives on the Use of Animals in Science,* T. Regan (ed.), Temple University Press, Philadelphia, PA.

CIESIN, IFPRI, and CIAT, 2004: Global Rural-Urban mapping Project (GRUMP): Urban Extents (alpha version). Center for International Earth Science Network (CIESIN), Columbia University; International Food Policy Research Institute (IFPRI), Washington, DC; Centro Internacional de Agricultura Tropical (CIAT), Palisades, NY. Available at Available at http://beta.sedac.ciesin.columbia.edu/gpw.

CIESIN and CIAT, 2004: Gridded Population of the World (GPW), Version 3 beta. Center for International Earth Science Network (CIESIN), Columbia University, and Centro Internacional de Agricultura Tropical (CIAT), Palisades, NY. Available at Available at http://beta.sedac.ciesin.columbia.edu/gpw.

Cleland, D.T., Crow, T. R., Hart, J. B., and Padley, E. A., 1994: Resource Management Perspective: Remote Sensing and GIS Support for Defining, Mapping, and Managing Forest Ecosystems. In: *Remote Sensing and GIS in Ecosystem Management,* V.A. Sample (ed.), 243–264.

Colwell, R.N., 1983: *Manual of Remote Sensing, 2nd Edition.,* American Society of Photogrammetry and Remote Sensing, Falls Church, VA.

Confalonieri, U.E.C., 2001: Environmental Change and Health in Brazil: Review of the Present Situation and Proposal for Indicators for Monitoring these Effects. In: *Human Dimensions of Global Environmental Change. Brazilian Perspectives,* D.J. IN: Hogan, & Tolmasquin, M. T. (ed.), Brasileira De Ciencias, R. Janeiro, 43–77.

Cooke, B. and U. Kothari (eds.), 2001: *Participation and the New Tyranny?* Zed Books, London.

Cornwall, A. and G. Pratt (eds.), 2003: *Pathways to Participation: Reflections on PRA.* ITDG Publishing, UK.

Corvalan, C., Briggs, S., and Kjellström, T., 1996: *Development of Environmental Health Indicators.,* UNEP, FAO and WHO, Geneva, 19–53 pp.

Corvalan, C., Briggs, S., and Nielhuis, G. (ed.), 2000: *Decision-Making in Environmental Health. From Evidence to Action.* Taylor & Francis, London and New York, 278 pp.

Costanza, R., R. d'Arge, R. de Groot, S. Farber, M. Grasso, et al. 1997: The value of the world's ecosystem services and natural capital. *Nature,* **387(253–260).**

Cox, P.M., 2000: Will tribal knowledge survive the millennium? *Science,* **287(5450),** 44–45.

Cramer, W., A. Bondeau, S. Schaphoff, W. Lucht, B. Smith, and S. Sitch, 2004: Tropical forests and the global carbon cycle: Impacts of atmospheric carbon dioxide, climate change and rate of deforestation. *Philosophical Transactions of the Royal Society Series B,* **359,** 331–343.

Darras, S., M. Michou, and C. Sarrat, 1998: *IGBP-DIS Wetland Data Initiative: A First Step Towards Identifying a Global Delineation of Wetlands,* IGBP-DIS Office, Toulouse, France.

Dasgupta, P. and K.-G. Mäler, 2000: National net product, wealth, and social well-being. *Environment and Development Economics,* **5(Parts 1 & 2),** 69–93.

de Freitas Rebelo, M., M.C.R. do Amaral, and W.C. Pfeiffer, 2003: High Zn and Cd accumulation in the oyster Crassostrea rhizophorae, and its relevance as a sentinel species. *Marine Pollution Bulletin,* **46(10),** 1354–1358.

DeFries, R., Hansen, M., Townshend, J. R. G., and Sohlberg, R., 1998: Global land cover classifications at 8km spatial resolution: The use of training data derived from Landsat Imagery in decision tree classifiers. *International Journal of Remote Sensing,* **19(16),** 3141–3168.

DeFries, R., Hansen, M., Townshend, J., Janetos, A. and Loveland, T., 2000: A new global data set of percent tree cover derived from remote sensing. *Global Change Biology,* **6,** 247–254.

DeFries, R., Houghton, R. A., Hansen, M., Field, C., Skole, D. L. and Townshend, J., 2002: Carbon emissions from tropical deforestation and regrowth based on satellite observations for the 1980s and 90s. *Proceedings of the National Academies of Sciences,* **99(22),** 14256–14261.

DeFries, R.S. and J.R.G. Townshend, 1994: NDVI-derived land cover classification at global scales. *International Journal of Remote Sensing,* **15(17),** 3567–3586.

DeGrandi, F., Mayaux, P., Malingreau, J.-P., Rosenqvist, A., Saatchi, S. and Simard, M., 2000: New perspectives on global ecosystems from wide area radar mosaics: Flooded forest mapping in the tropics. *International Journal of Remote Sensing,* **20,** 1235–1250.

Deichmann, U., D. Balk, and G. Yetman, 2001: Transforming Population Data for Interdisciplinary Usages: From census to grid. NASA Socioeconomic Data and Application Center (SEDAC). Available at http://sedac.ciesin.columbia.edu/plue/gpw/GPWdocumentation.pdf.

Deutch, E., 1970: Vedanta and ecology. In: *Indian Philosophical Annual,* T.M.P. Mahadevan (ed.), University of Madras, India.

Dewailly, E., et. al., 2002: Indicators of Ocean and Human Health. *CAN. J. PUBL. HEALTH,* 93(suppl. 1), 534–538.

Dinerstein, M., Graham, D. J., Webster, A. L. et. al., 1995: *Conservation Assessment of the Terrestrial Ecoregions of Latin America and the Caribbean,* World Bank and World Wildlife Fund, Washington, D.C.

Dixon, J.A., L.F. Scura, R.A. Carpenter, and P.B. Sherman, 1994: *Economic Analysis of Environmental Impacts.* Earthscan, London.

Dobson, J.E., Bright, P.R., Coleman, R. C., Durfee and Worley, B. A., 2000: Landscan: A global population database for estimating populations at risk. *Photogrammetric Engineering and Remote Sensing,* **66(7),** 849–857.

Doney, S.C., D.M. Glover, S.J. McCue, and M. Fuentes, 2003: Mesoscale variability of Sea-viewing Wide Field-of-View Sensor (SeaWIFS) satellite ocean color: Global patterns and spatial scales. *Journal of Geophysical Research,* **108(C2),** 10.1029/2001JC000843.

Donner, S.D. and C.J. Kucharik, 2003: Evaluating the impacts of land management and climate variability on crop production and nitrate export across the Upper Mississippi Basin. *Global Biogeochemical Cycles,* **17(3),** doi:10.129/2001GB001808.

Downing, T. E., R. Butterfield, S. Cohen, S. Huq, R. Moss, A. Rahman, Y. Sokona, and L. Stephen, 2001: *Climate Change Vulnerability: Linking Impacts and Adaptation,* University of Oxford, Oxford.

Edwards, J.L., M.A. Lane, and E.S. Nielsen, 2000: Interoperability of biodiversity databases: Biodiversity information on every desktop. *Science (Washington D C),* **289(5488),** 2312–2314.

Ehrenfeld, D. and P.J. Bently, 1985: Judaism and the practice of stewardship. *Judaism,* **34,** 301–311.

Emery, A., 2000: *Integrating Indigenous Knowledge in Project Planning and Implementation.* The World Bank. The Canadian International Development Agency. Washington, D.C.

Estes, R., 1984: *The Social Progress of Nations.* Praeger Publishers, New York.

Estes, R., 1988: *Trends in World Social Development: The Social Progress of Nations, 1970–1987.* Praeger, New York.

EVRI, 2004: Environment Valuation Reference Inventory. Environment Canada. Available at www.evri.ca.

Fabricius, C., R. Scholes, and G. Cundill, 2004: Mobilising knowledge for ecosystem assessments. *Paper developed for a conference on Bridging Scales and Epistemologies,* Alexandria, Egypt.

Falconí, F., 2003: *Economía y desarrollo sostenible: Matrimonio feliz o divorcio anunciado.,* FLASCO, Quito, Ecuador.

FAO Food and Agriculture Organization of the United Nations, 2000a: *Global Forest Resource Assessment 2000,* Rome, 511 pp.

FAO, 2000b: *State of World Fisheries and Aquaculture.* Rome.

Farber, S.C., R. Costanza, and M.A. Wilson, 2002: Economic and ecological concepts for valuing ecosystem services. *Ecological Economics,* **41,** 375–392.

Field, C.B., Randerson, J. T. and Malmstrom, C. M., 1995: Global net primary production: Combining ecology and remote sensing. *Remote Sensing of Environment,* **51,** 74–88.

Finlayson, C.M., N.C. Davidson, A.G. Spiers, and N.J. Stevenson, 1999: Global wetland inventory—status and priorities. *Marine and Freshwater Research,* **50,** 717–727.

Flynn, P., 2000: Research Methodology. In: *IN Calvert-Henderson Quality of Life Indicators,* J. Henderson, Lickerman, J. and Flynn, P. (ed.), Maryland: Calvert Group, USA.

Foley, J., Prentice, I. C., Ramankutty, S., Levis, D., Pollard, D., Sitch, S. and Haxeltine, A., 1996: An integrated biosphere model of land surface processes, terrestrial carbon balance, and vegetation dynamics. *Global Biogeochemical Cycles,* **10,** 603–629.

Forsyth, T., 1999: Science, myth and knowledge: Testing Himalayan environmental degradation in Thailand. *Geoforum,* **27,** 375–392.

Freeman, A.M., 1979: *The Benefits of Environmental Improvements, Theory and Proactive.* Johns Hopkins University Press, Baltimore, MD.

Freeman, A.M., 1993: *The Measurement of Environmental and Resource Values: Theory and Methods.* Resources for the Future, Washington, D.C.

Friedl, M.A., McIver, D. K., Hodges, J. C. F., Zhang, X. Y., Muchoney, et al. 2002: Global land cover mapping from MODIS: algorithms and early results. *Remote Sensing of Environment,* **83(1–2),** 287–302.

Fritz, S., E. Bartholemé, A. Belward, A. Hartley, H.-J. Stibig, et al. 2004: *Harmonisation, Mosaicing and Production of the Global Land Cover 2000 Database,* EUR 20849/EN.

Fritz, S., See, L., and Carver, S., 2001: A fuzzy modeling approach to wild land mapping in Scotland. In: *Innovations in GIS 7,* D. Martin and P. Atkinson (eds.), Taylor and Francis, London.

Frose, R. and D. Pauly, 2000: *Fishbase 2000, Concepts, Design, and Data Sources,* ICLARM, Los Baños, Philippines, distributed with 4 CD ROMs.

Gadgil, M., F. Berkes, and C. Folke, 1993: Indigenous knowledge for biodiversity conservation. *Ambio,* **22,** 151–156.

Gardner, T.A., I.M. Cote, J.A. Gill, A. Grant, and A.R. Watkinson, 2003: Long-term region-wide declines in Caribbean corals. *Science,* **301,** 958–960.

Garrod, G. and K. Willis, 1992: The Environmental economic impact of woodland: A two-stage hedonic price model of the amenity value of forestry in Britain. *Applied Economics,* **24,** 715–728.

Garrod, G.D. and K.G. Willis, 1999: *Economic Evaluation of the Environment.* Edward Elgar, Cheltenham.

Geist, H.J. and E.F. Lambin, 2001: *What Drives Tropical Deforestation? A Meta-analysis of Proximate and Underlying Causes of Deforestation Based on Subnational Case Study Evidence,* LUCC Report Series No. 4, Louvain-la-Neuve, Belgium.

Geist, H.J. and E.F. Lambin, 2002: Proximate causes and underlying forces of tropical deforestation. *BioScience,* **52(2),** 143–150.

Giglio, L., J.D. Kendall, and C.O. Justice, 1999: Evaluation of global fire detection algorithms using simulated AVHRR infrared data. *International Journal of Remote Sensing,* **20(10),** 1947–1986.

Gleditsch, N.P., M. Wallenstein, M. Erikson, M. Sollenberg, and H. Strand, 2002: Armed conflict 1946–2000: A new dataset. *Journal of Peace Research,* **39(5),** 615–637.

Glenn, R., 2003: Appendix H: Traditional Knowledge. In: *Cumulative Environmental Effects of Oil and Gas Activities on Alaska's North Slope,* N.R. Council (ed.), The National Academies Press, Washington, D.C, pp. 232–233.

Government of India, 2001: *Census of India 2001,* Office of the Registrar General, New Delhi.

Green, P., C.J. Vörösmarty, M. Meybeck, J. Galloway, and B.J. Peterson, in press: Pre-industrial and contemporary fluxes of nitrogen through rivers: A global assessment based on typology. *Biogeochemistry.*

Greenberg, R., P. Bichier, A.C. Angon, and R. Reitsma, 1997: Bird populations in shade and sun coffee plantations in central Guatemala. *Conservation Biology,* **11,** 448–459.

Grosh, M. and P. Glewwe, 1995: *A Guide to Living Standards Surveys and Their Data Sets.* LSMS Working Paper No. 120, World Bank, Washington, D.C.

Gunderson, L. and C.S. Holling, 2002: *Panarchy: Understanding transformations in human and natural systems.* Island Press, Washington, D.C.

Hall, C., C. Cleveland, and R. Kaufmann, 1986: *Energy and Resource Quality.* Wiley Interscience, New York.

Hamilton, K. and M. Clemens, 1999: Genuine savings rates in developing countries. *World Bank Economic Review,* **13(2),** 333–356.

Hancock, T., 2002: Indicators of Environmental Health in the Urban Setting. *Canadian Journal of Public Health,* **93(suppl. 1),** S45–S51.

Hanemann, W.M., 1991: Willingness to pay and willingness to accept: How much can they differ? *American Economic Review,* **81(3),** 635–647.

Hanemann, W.M., 1992: Preface. In: *Pricing the European Environment,* S. Navrud (ed.), Scandinavian University Press, Oslo.

Hansen, M. and R. DeFries, 2004: Detecting long term forest change using continuous fields of tree cover maps from 8km AVHRR data for the years 1982–1999. *Ecosystems.* **7(7),** 695–716.

Hansen, M., DeFries, R., Townshend, J. R. G., and Sohlberg, R., 2000: Global land cover classification at 1km spatial resolution using a classification tree approach. *International Journal of Remote Sensing,* **21(6),** 1331–1364.

Heal, G., G. Daily, P.R. Ehrlich, J. Salzman, C. Boggs, et al. 2001a: Protecting natural capital through ecosystem service districts. *Stanford Environmental Law Journal,* **20(2),** 333–364.

Heal, G., G.C. Daily, P.R. Ehrlich, J. Salzman, C. Boggs, et al. 2001b: Protecting natural capital through ecosystem service districts. *Stanford Environmental Law Journal,* **20(2),** 333–364.

Heimlich, R.E., K.D. Weibe, R. Claasen, D. Gadsy, and R.M. House, 1998: *Wetlands and Agriculture: Private Interests and Public Benefits.* Agricultural Economic Report No. 765.10, ERS, USDA, Washington, D.C.

Henderson, H.J., Lickerman, J., and Flynn, P (ed.), 2000: *Calvert-Henderson Quality of Life Indicators.* Maryland: Calvert Group.

Herriges, J.A. and C.L. Kling (eds.), 1999: *Valuing Recreation and the Environment: Revealed Preference Methods in Theory and Practice.* Edward Elgar, Northampton.

Heywood, I., Cornelius, S. and Carver, S., 1998: *An Introduction to Geographical Information Systems.* Addison Wesley Longman, New York.

Hoffer, R.M., 1994: Challenges in Developing and Applying Remote Sensing to Ecosystem Management. In: *Remote Sensing and GIS in Ecosystem Management,* V.A. Sample (ed.), 25–40. Island Press, Washington, D.C.

Holmes, T. and W. Adamowicz, 2003: Attribute Based Methods. In: *A Primer on Nonmarket Valuation,* P.A. Champ, K.J. Boyle, and T.C. Brown (eds.), Kluwer.

Hufschmidt, M.M., D.E. James, A.D. Meister, B.T. Bower, and J.A. Dixon, 1983: *Environment, Natural Systems, and Development: An Economic Valuation Guide.* Johns Hopkins University Press, Baltimore, MD.

Hughes, J.D., 1983: *American Indian Ecology.* Texas Western Press, El Paso, TX.

ICSU (International Council for Science), 2002a: *Series on Science for Sustainable Development, No. 8: Making Science for Sustainable Development More Policy Relevant: New Tools for Analysis,* Paris, France, 28 pp.

ICSU, 2002b: *Science, Traditional Knowledge and Sustainable Development.* ICSU Series on Science for Sustainable Development No. 4, International Council for Science, Paris, 24 pp.

IUCN (World Conservation Union), 2001: *IUCN Red List Categories: Version 3.1,* Species Survival Commission, Gland, Switzerland and Cambridge, UK.

Jensen, J.R., 2000: *Remote Sensing of the Environment: An Earth Resource Perspective.* Prentice Hall, Upper Saddle River, New Jersey.

Johannes, R.E. (ed.), 1998: *Traditional Ecological Knowledge: A Collection of Essays.* IUCN, Gland, Switzerland.

Johansson, P.O., 1994: *The Economic Theory and Measurement of Environmental Benefits.* Cambridge University Press, Cambridge, UK.

Johnson, M. (ed.), 1992: *Lore: Capturing Traditional Environmental Knowledge.* Denne Cultural Institute, International Development Research Centre, Ottawa, Canada.

Johnston, C.A., 1998: *Geographical Information Systems in Ecology.* Blackwell Science Ltd, London.

Jordan, G.H. and B. Shrestha, 1998: *Integrating geomatics and participatory techniques for community forest management: Case studies from the Yarsha Khola watershed, Dolakha District,* ICIMOD, Kathmandu, Nepal.

Kaimowitz, D. and A. Angelsen, 1998: *Economic Models of Tropical Deforestation: A Review,* CIFOR, Bogor, Indonesia.

Kaiser, B. and J. Roumasset, 2002: Valuing indirect ecosystem services: The case of tropical watersheds. *Environment and Development Economics,* **7,** 701–714.

Karr, J.R. and D.R. Dudley, 1981: Ecological perspective on water quality goals. *Environmental Management,* **5,** 55–68.

Karr, R.J., K.D. Fausch, P.L. Angermeier, P.R. Yant, and I.J. Schlosser, 1986: *Assessment of biological integrity in running waters: A method and its rationale,* Illinois Natural History Survey Special Publication No. 5, Champaign, IL.

Kimmerer, R.W., 2000: Native knowledge for native ecosystems. *Journal of Forestry,* **98(8),** 4–9.

Kjellström, T.a.C., 1995: Framework for the Development for Environmental Health Indicators. *World Health Statistical Quarterly,* **48,** 144–154.

Klein, A.M., I. Steffan-Dewenter, D. Buchori, and T. Tscharntke, 2002: Effects of land-use intensity in tropical agroforestry systems on coffee flower-visiting. *Conservation Biology,* **16,** 1003–1014.

Kremen, C., N.M. Williams, and R.W. Thorp, 2002: Crop pollination from native bees at risk from agricultural intensification. *Proceedings of the National Academy of Sciences—US,* **99,** 16812–16816.

Kumari, K. 1995: An Environmental and economic assessment of forest management options: A case study of Malaysia." Environment Department Working Paper No.26, World Bank, Washington, DC.

Lacy, R.C., 1993: VORTEX: A computer simulation model for population viability analysis. *Wildlife Research,* **20,** 45–65.

Lampietti, J. and J.A. Dixon, 1995: *To See the Forest for the Trees: A Guide to Non-Timber Forest Benefits.* Environment Department Paper No. 13, World Bank, Washington, D.C.

Laurance, W.F., M.A. Cochrane, S. Bergen, P.M. Fearnside, P. Delamonica, et al. 2001: The Future of the Brazilian Amazon. *Science,* **291(5503),** 438–439.

Lepers, E., E.F. Lambin, A.C. Janetos, R. DeFries, F. Achard, N. Ramankutty, and R.J. Scholes, 2005: A synthesis of rapid land-cover change information for the 1981–2000 period. *BioScience,* **55 (2),** 19–26.

Lesslie, R. and M. Maslen, 1995: *National Wilderness Inventory Handbook of Procedures, Content and Usage.* 2nd ed. ed. Australian Government Publishing Service, Canberra, Australia.

Liang, X., Lettenmaier, D. P. and Wood, E. F., 1996: One-dimensional statistical dynamic representation of sub-grid spatial variability of precipitation in the two-layer variable infiltration capacity model. *Journal of Geophysical Research,* **101((D16) 21),** 403–422.

Loh, J., 2002: *Living Planet Report 2002,* World Wildlife Fund International, Gland, Switzerland.

Louviere, J., D. Henscher, and J. Swait, 2000: *Stated Choice Methods—Analysis and Application.* Cambridge University Press, Cambridge, UK.

Loveland, T.R. and A.S. Belward, 1997: The IGBP-DIS global 1km land cover data set, DISCover: first results. *International Journal of Remote Sensing,* **18(15),** 3289–3295.

Lovell, C., A. Madondo, and P. Moriarty, 2002: The question of scale in integrated natural resource management. *Conservation Ecology,* **5(2),** 25.

Lowry, X. and C.M. Finlayson, in press: *A Review of Spatial Datasets for Wetland Inventory in Northern Australia,* Department of the Environment and Heritage, Supervising Scientist, Australian Government, Canberra, Australia.

Lucas, R.M., J.C. Ellison, A. Mitchell, B. Donnelly, M. Finlayson, and A.K. Milne, 2002: Use of stereo aerial photography for assessing changes in the extent and height of mangroves on tropical Australia. *Wetlands Ecology and Management,* **10(2),** 159–173.

Luck, G. and G. Daily, 2003: Tropical countryside bird assemblages: richness, composition, and foraging differ by landscape context. *Ecological Applications,* **13(1),** 235–247.

Luck, G.W., T.H. Ricketts, G.C. Daily, and M. Imhoff, 2004: Spatial conflict between people and biodiversity. *Proceedings of the National Academy of Sciences,* **101,** 5732–5736.

Mace, G.M., J.L. Gittleman, and A. Purvis, 2003: Preserving the tree of life. *Science (Washington D C),* **300(5626),** 1707–1709.

MacNally, R. and E. Fleishman, 2002: Using "indicator" species to model species richness: Model development and predictions. *Ecological Applications,* **12(1),** 79–92.

Mäler, K.-G. and R.E. Wyzga, 1976: *Economic Measurement of Environmental Damage.* OECD, Paris.

Malingreau, J.P., F. Achard, G. D'Souza, H. J. Stibig, J. D'Souza, C. Estreguil, and H. Eva, 1995: AVHRR for global tropical forest monitoring: The lessons of the TREES project. *Remote Sensing Reviews,* **12,** 29–40.

Mantua, U., M. Merlo, W. Sekot, and B. Welcker, 2001: *Recreational and Environmental Markets for Forest Enterprises: A New Approach Towards Marketability of Public Goods,* CABI Publishing, Wallingford.

Martello, M., 2001: A paradox of virtue?: "Other" knowledges and environment-development politics. *Global Environmental Politics,* **1,** 114–141.

Martínez-Alier, J., G. Munda, and J. O'Neill, 1998: Weak comparability of values as a foundation of ecological economics. *Ecological Economics,* **26(3),** 277–286.

Mather, J. and G. Sdasyuk, 1991: *Global Change: Geographic Approaches.* University of Arizona Press, Tucson, Arizona.

Matthews, E., 2001: *Understanding the FRA 2000: Forest Briefing No. 1.,* World Resources Institute, Washington, D.C., 12 pp.

Mauro, F. and P.D. Hardinson, 2000: Traditional knowledge of indigenous and local communities. *Ecological Applications,* **105(5),** 1263–1269.

Mayaux, P., G.F. DeGrandi, Y. Rauste, M. Simard, and S. Saatchi, 2002: Regional scale vegetation maps derived from the combined L-band GRFM and C-band CAMP Wide Area Radar Mosaics of Central Africa. *International Journal of Remote Sensing,* **23(7),** 1261–1282.

Mayaux, P., Achard, F. and Malingreau, J. P., 1998: Global tropical forest area measurements derived from coarse resolution satellite imagery: A comparison with other approaches. *Environmental Conservation,* **25(1),** 37–52.

McCarthy, M.A., Possingham, H. P., Day, J. R. and Tyre, A. J., 2001: Testing the accuracy of population viability analysis. *Conservation Biology,* **15,** 1030–1038.

McCracken, J.R. and H. Abaza, 2001: *Environmental Valuation: A Worldwide Compendium of Case Studies.* Earthscan, London.

McGuire, A.D., S. Sitch, J.S. Clein, R. Dargaville, G. Esser, et al. 2001: Carbon balance of the terrestrial biosphere in the twentieth century: Analysis of CO_2, climate and land use effects with four process-based ecosystem models. *Global Biogeochemical Cycles,* **15,** 183–206.

Merlo, M. and L. Croitoru (Eds.), in press: *Valuing Mediterranean Forests: Towards Total Economic Value,* CABI Publishing, Wallingford.

Millennium Ecosystem Assessment, 2003: *Ecosystems and Human Well-being: A Framework for Assessment.* Island Press, Washington, DC.

Mitchell, R.C. and R. Carson, 1989: *Using Surveys to Value Public Goods: The Contingent Valuation Method.* Resources for the Future, Washington, DC.

Moghissi, A.A., 1994: Life Expectancy as a Measure of Effectiveness of Environmental Protection. *Environment International,* **20,** 691–692.

Morris, M.D., 1979: *Measuring the Condition of the World's Poor: The Physical Quality of Life index.* Pergamon Press, New York.

Morris, R.D.a.C., 2002: Environmental Health Surveillance: Indicators for freshwater ecosystems. *Canadian Journal of Public Health,* **93(suppl. 1),** 539–544.

Moss, S., C. Pahl-Wostl, and T.E. Downing, 2001: Agent-based integrated assessment modeling: The example of climate change. *Integrated Assessment,* **2(1),** 17–30.

Motteux, N., 2001: *The development and coordination of catchment fora through the empowerment of rural communities.* WRC Research Reports 1014/1/01, Water Research Commission, South Africa.

Munda, G., 1995: *Multicriteria Evaluation in a Fuzzy Environment.* Physica-Verlag, Heidelberg.

Murray, C.J.L., 1994: Quantifying the burden of disease: The technical basis of disability—adjusted life.years. *BULL. WHO.,* **72,** 429–455.

Murray, C.J.L., 1997: Global mortality, disability, and the contribution of risk factors: Global burden of disease study. *Lancet,* **349,** 1436–1442.

Myers, N., R.A. Mittermeier, C.G. Mittermeier, G.A.B. daFonesca, and J. Kent, 2000: Biodiversity hotspots for conservation priorities. *Nature,* **403,** 853–857.

Myneni, R.B., G. Asrar, D. Tanre, and B. J. Choudhury, 1992: Remote sensing of solar radiation absorbed and reflected by vegetated land surfaces. *IEEE Transactions on Geoscience and Remote Sensing,* 302–314.

Nadasdy, P., 1999: The politics of TEK: Power and the "integration" of knowledge. *Arctic Anthropology,* **36,** 1–18.

National Geographic Society, 1989: *Endangered Earth.* National Geographic Society, Washington, DC.

Navrud, S. and R.C. Ready (eds.), 2002: *Valuing Cultural Heritage: Applying Environmental Valuation Techniques to Historic Buildings, Monuments and Artifacts.* Edward Elgar, Cheltenham, UK.

Nicholson, S.E., C.J. Tucker, and M.B. Ba, 1998: Desertification, drought, and surface vegetation: An example from the West African Sahel. *Bulletin of the American Meteorological Society,* **79,** 815–829.

NOAA (National Oceanic and Atmospheric Administration), 1993: Report of the NOAA Panel on Contingent Valuation. *Federal Register,* **58(10, Friday January 15),** 4602–4614.

NRC (National Research Council), 2000: *Ecological Indicators for the Nation.* National Academy Press, Washington, D. C.

O'Dor, R., 2004: A census of marine life. *BioScience,* **54(2),** 92–93.

Odum, H.T. and E.C. Odum, 1981: *Energy Basis for Man and Nature.* McGraw Hill, New York.

Oliver, J., M. Noordeloos, Y. Yusuf, M. Tan, N. Nayan, C. Foo, and F. Shahriyah: ReefBase: A Global Information System on Coral Reefs [Online]. Cited May 22 2004. Available at http://www.reefbase.org.

Pagiola, S., 1996: *Economic Analysis of Investments in Cultural Heritage: Insights from Environmental Economics.* World Bank, Washington, DC.

Pagiola, S. and J.A. Dixon, 2001: Local Costs, Global Benefits. In: *Valuation of Biodiversity Benefits: Selected Studies,* OECD (ed.), OECD, Paris.

Pagiola, S. and G. Platais, in press: *Payments for Environmental Services: From Theory to Practice.* World Bank, Washington, DC.

Pagiola, S., K. von Ritter, and J.T. Bishop, 2004: *Assessing the Economic Value of Ecosystem Conservation.* Environment Department Working Paper No.101. World Bank, Washington, D.C.

Pagiola, S., G. Acharya, and J.A. Dixon, in review: *Economic Analysis of Environmental Impacts.* Earthscan, London.

Park, R.A., 1998: *AQUATOX for Winfdoes: A modular toxic effects model for aquatic ecosystems.,* U. S. Environmental Protection Agency, Washington, D. C., 3–13 pp.

Parton, W.J., Stewart, J. W. B. and Cole, C. V., 1988: Dynamics of C, N, P and S in grassland soils: a model. *Biogeochemistry,* **5,** 109–131.

Pastides, H., 1995: An Epidemiological Perspective on Environmental Health Indicators. *HEALTH STAT. Q.,* **48,** 139–143.

Pastorok, R.S., Bartell, S. M., Ferson, S. and Ginzburg (Eds.), 2002: *Ecological modelling in Risk Assessment: Chemical Effects on Populations, Ecosystems, and Landscapes.* Lewis Publishers, Boca Raton, Florida.

Pearce, D.W., 1993: *Economic Values and the Natural World.* Earthscan, London, 144 pp.

Pearce, D.W. and A. Markandya, 1989: *The Benefits of Environmental Policy: Monetary Valuation.* OECD, Paris.

Pearce, D.W. and D. Moran, 1994: *The Economic Value of Biodiversity.* Earthscan, London, 192 pp.

Pearce, D.W., D. Moran, and D. Biller, 2002: *Handbook of Biodiversity Valuation: A Guide for Policy Makers.* OECD, Paris.

Pereira, H., 2004: *Ecosystem Services and Human Well-Being: A Participatory Study in a Mountain Community in Northern Portugal,* Subglobal Assessment Report, Millennium Ecosystem Assessment.

Perfecto, I., J. N. Vandermeer, P. Hanson, and V. Cartin, 1997: Arthropod biodiversity loss and the transformation of a tropical agro-ecosystem. *Biodiversity and Conservation,* **6,** 935–945.

Phinn, S., L. Hess, and C.M. Finlayson, 1999: An Assessment of the Usefulness of Remote Sensing for Wetland Monitoring and Inventory in Australia. In: *Techniques for Enhanced Wetland Inventory, Assessment and Monitoring,* C.M. Finlayson and A.G. Spiers (eds.), Supervising Scientist Report 147, Canberra, Australia, 44–82.

Ponder, W.F., G.A. Carter, P. Flemons, and R.R. Chapman, 2001: Evaluation of museum collection data for use in biodiversity assessment. *Conservation Biology,* **15(3),** 648–657.

Portney, P.R. and J.P. Weyant, 1999: *Discounting and Intergenerational Equity.* Resources for the Future, Washington, D.C.

Powe, N.A., G.D. Garrod, and K.G. Willis, 1995, Valuation of urban amenities using an hedonic price model. *Journal of Property Research,* **12,** 137–147.

Prendergast, J.R., R.M. Quinn, J.H. Lawton, B.C. Eversham, and D.W. Gibbons, 1993: Rare species, the coincidence of diversity hotspots and conservation strategies. *Nature,* **365,** 335–337.

Pretty, J., 1995: *Regenerating agriculture: Policies and practice for sustainability and self reliance.* Earthscan Publications Ltd., London, 320 pp. pp.

Prince, S.D., E. Brown DeColstoun, and L.L. Kravitz, 1990: Evidence from rain-use efficiencies does not indicate extensive Sahelian desertification. *Global Change Biology,* **4,** 359–374.

Raxworthy, C.J., E. Martinez-Meyer, N. Horning, R.A. Nussbaum, G.E. Schneider, M.A. Ortega-Huerta, and A.T. Peterson, 2003: Predicting distributions of known and unknown reptile species in Madagascar. *Nature,* **426,** 837–841.

Reardon, T. and S.A. Vosti, 1997: Poverty-Environment Links in Rural Areas of Developing Countries. In: *Sustainability, Growth, and Poverty Alleviation: A Policy and Agroecological Perspective,* S.A. Vosti and T. Reardon (eds.), Johns Hopkins University Press for IFPRI, Baltimore.

Rees, W., 1992: Ecological footprints and appropriated carrying capacity: What urban economics leaves out. *Environment and Urbanization,* **4(2),** 121–130.

Reynolds, J.R. and M.S. Smith (eds.), 2002: *Global Desertification: Do Humans Cause Deserts?* Vol. DWR 88Dahlem Workshop Report, Berlin, 438 pp. pp.

Ricketts, T.H., in press: Do tropical forest fragments enhance pollinator activity in nearby coffee crops? *Conservation Biology.*

Ricketts, T.H., E. Dinerstein, D.M. Olson, and C. Louckes, 1999: Who's where in North America: Patterns of species richness and the utility of indicator taxa for conservation. *Bioscience,* **49,** 369–381.

Ricketts, T.H., G.C. Daily, P.R. Ehrlich, and J.P. Fay, 2001: Countryside biogeography of moths in a fragmented landscape: Biodiversity in native and agricultural habitats. *Conservation Biology,* **15,** 378–388.

Roberge, J.-M. and P. Angelstam, 2004: Usefulness of the umbrella species concept as a conservation tool. *Conservation Biology,* **18(1),** 76–85.

Rosenzweig, M.L., 1995: *Species diversity in space and time.* Cambridge University Press, Cambridge, 436 pp.

Saatchi, S., Nelson, B., Podest, E. and Holt, J., 2000: Mapping land cover types in the Amazon basin using 1km JERS-1 mosaic. *International Journal of Remote Sensing,* **21,** 1201–1234.

Sagoff, M., 1998: Aggregation and deliberation in valuing environmental public goods: A look beyond contingent valuation. *Ecological Economics,* **24,** 213–230.

Sanderson, E.W., Jaiteh, M. Levy, M. A., Redford, K. H., Wannebo, A. V. and Woolmer, G., 2002: The human footprint and the last of the wild. *BioScience,* **52(10),** 891–904.

Sandor, J.A. and L. Furbee, 1996: Indigenous knowledge and classifications of soils in the Andes of southern Peru. *Soil Science Society of America,* **60,** 1502–1512.

Scheffer, M., S.R. Carpenter, J. Foley, Prentice, I. C., Ramankutty, S., Levis, D., Pollard, D., Sitch, S. and Haxeltine, A., C. Folke, and B. Walker, 2001: Catastrophic shifts in ecosystems. *Nature,* **413,** 591–596.

Scoones, I., 1995: PRA and anthropology: Challenges and dilemmas. *PLA Notes,* **24,** 17–20.

Scott, J.M. and B. Csuti, 1997: Gap analysis for biodiversity survey and maintenance. In: *Biodiversity II: Understanding and Protecting our Biological Resources,* M.L. Reaka-Kudla, D.E. Wilson, and E.O. Wilson (eds.), Joseph Henry Press, Washington, D.C., 321–340.

Scott-Samuel, A., M. Birley, and K. Ardern, 2001: *The Merseyside Guidelines for Health Impact Assessment.,* Department of Public Health Liverpool, Liverpool, UK.

Sellers, P.J., Los, S. O., Tucker, C. J., Justice, C. O., Dazlich, D., Collatz, C. J. and Randall, D. A., 1996: A revised land surface parameterization (SiB2) for atmospheric GCMs. Part II: The generation of global fields of terrestrial biophysical parameters from satellite data. *Journal of Climate,* **9,** 706–737.

Sellers, P.J., Mintz, Y., Sud, Y. C. and Dalmer, A., 1986: A simple biosphere model (SiB) for use with general circulation models. *Journal of Atmospheric Science,* **43(6),** 505–531.

Seroa da Motta, R., 1998: *Manual para Valoração Econômica de Recursos Ambientais.* MMA, Brasília.

Seroa da Motta, R. (ed.), 2001: *Environmental Issues and Policy Making in Developing Countries.* Edgar Elgar Publishing, London.

Shogren, J. and J. Hayes, 1997: Resolving differences in willingness to pay and willingness to accept: A reply. *American Economic Review,* **87,** 241–244.

Simpson, D.R., R.A. Sedjo, and J.W. Reid, 1994: *Valuing Biodiversity for Use in Pharmaceutical Research,* Resources for the Future, Washington, DC.

Singhal, R., 2000: A model for integrating indigenous and scientific forest management potentials and limitations for adaptive learning. In: *Forestry, Forest Users and Research: New Ways of Learning,* A. Lawrence (ed.), ETFRN (European Tropical Forest Research Network) Publications Series 1, Wagingen, The Netherlands.

Sisk, T., A.E. Launer, K.R. Switky, and P.R. Ehrlich, 1994: Identifying extinction threats: Global analyses of the distribution of biodiversity and the expansion of the human enterprise. *BioScience,* **44,** 592–604.

Sitch, S., B. Smith, I.C. Prentice, A. Arneth, A. Bondeau, et al. 2003: Evaluation of ecosystem dynamics, plant geography and terrestrial carbon cycling in the LPJ dynamic global vegetation model. *Global Change Biology,* **9,** 161–185.

Skole, D. and C. Tucker, 1993: Tropical deforestation and habitat fragmentation in the Amazon: satellite data from 1978 to 1988. *Science,* **260,** 1905–1910.

Steininger, M.K., Tucker, C. J., Townshend, J. R. G., Killeen, T. J., Desch, A., Bell, V. and Ersts, P., 2001: Tropical deforestation in the Bolivian Amazon. *Environmental Conservation,* **28(2),** 127–134.

Sutherst, R.W., Maywald, G. F. and Skarratt, D. B., 1995: Predicting insect distributions in a changed climate. In: *Insects in a Changing Environment,* R. Harrington and N.E. Stork (eds.), Academic Press, London, 59–91.

TESEO (Treaty Enforcement Services using Earth Observation), 2003: *Treaty Enforcement Services using Earth Observation (TESEO): Desertification.* University of Valencia, EOS.D2C, Chinese Academy of Forestry, European Space Agency.

The H. John Heinz III Center for Science, Economics, and the Environment, 2002: *The State of the Nation's Ecosystems: Measuring the Lands, Waters, and Living Resources of the United States.* Cambridge University Press, Cambridge, U.K.

Townshend, J.R.G., Justice, C. O. and Kalb, V. T., 1987: Characterization and classification of South American land cover types using satellite data. *International Journal of Remote Sensing,* **8,** 1189–1207.

Tucker, C.J., H.E. Dregne, and W.W. Newcomb, 1991: Expansion and contraction of the Saharan Desert from 1980 to 1990. *Science,* **253,** 299–301.

Tucker, C.J., Townshend, J. R. G. and Goff, T. E., 1985: African land-cover classification using satellite data. *Science,* **227,** 369–375.

Turner II, B.L., P.A. Matson, J. McCarthy, R.W. Corell, L. Christensen, et al. 2003: Illustrating the coupled human-environment system for vulnerability analysis: Three case studies. *Proceedings of the National Academies of Sciences,* **100(14),** 8080–8085.

Turner, K., J. Paavloa, P. Cooper, S. Farber, V. Jessamy, and S. Georgiou, 2002: *Valuing Nature: Lessons Learned and Future Research Directions.* CSERGE Paper No. EDM-2002–05, CSERGE, London.

Turner, W., S. Spector, N. Gardiner, M. Fladeland, E. Sterling, and M. Steininger, 2003: Remote sensing for biodiversity science and conservation. *Trends in Ecology and Evolution,* **18(6),** 306–314.

UNDP (United Nations Development Programme), 1998: *Human Development Report 1998.* New York, NY.

UNDP, 2003: *Human Development Report 2003: Millennium Development Goals: A Compact Among Nations to End Human Poverty.,* United Nations Development Programme, Published by Oxford University Press, New York.

UNEP (United Nations Environment Programme), 2001: *GLOBIO. Global Methodology for Mapping Human Impacts on the Biosphere.* Environment Information and Assessment Technical Report UNEP/DEWA/TR.01–3, UNEP, Nairobi (Kenya).

USDA (U.S. Department of Agriculture), 1999: Forest Vegetation Simulator website. USDA Forest Service, Forest Management Service Center, Fort Collins, CO. Available at http://www.fs.fed.us/fmsc/fvs.

Vedeld, P., A. Angelsen, A. Sjaastad, and G. Kobugabe Berg, 2004: *Counting on the Environment: Forest Incomes and the Rural Poor.* Environment Department Paper No.98. World Bank, Washington, D.C.

Wadsworth, R. and J. Treweek, 1999: *Geographical Information Systems for Ecology.* Addison Wesley Longman Limited, Essex, UK.

WCMC (World Conservation Monitoring Centre), 1992: *Global Biodiversity: Status of the Earth's Living Resources.* Cambridge, UK.

WHO (World Health Organization), 1997: *Health and Environmental in Sustainable Development: Five Years after the Earth Summit,* Geneva.

Willis, K.G. and J.T. Corkindale (eds.), 1995: *Environmental Valuation: New Perspectives.* CAB International, Wallingford.

Wodon, Q. and E. Gacitúa-Marió (eds.), 2001: *Measurement and Meaning: Combining Quantitative and Qualitative Methods for the Analysis of Poverty and Social Exclusion in Latin America.* World Bank, Washington, D.C.

Wood, J.B., C.L. Day, and R.K. O'Dor, 2000: CephBase: testing ideas for cephalopod and other species-level databases. *Oceanography,* **13,** 14–20.

World Bank, 2001: *World Development Report 2000/2001: Attacking Poverty.* Oxford University Press, Oxford, 335 pp.

World Bank, 2002: *World Development Indicators 2002.* World Bank, Washington, DC, 432 pp.

World Bank, 2002: *Linking Poverty Reduction and Environmental Management: Policy Challengers and Opportunities.,* Department for International Development, European Commission, United Nations Development Programme, and World Bank, Washington, D.C.

World Bank, 2003: *World Development Indicators 2003,* World Bank, Washington, D.C.

World Bank, 2004: *World Development Indicators 2004,* World Bank, Washington, D.C.

Young, R.A. and R.H. Haveman, 1985: Economics of water resources: A survey. In: *Handbook of Natural Resource and Energy Economics Vol. II,* A.V. Kneese and J.L. Sweeney (eds.), North Holland, Amsterdam.

Zaidi, I.H., 1981: On the ethics of man's interaction with the environment: An Islamic Approach. *Environmental Ethics,* **3(1),** 35–47.

Zarafshani, K., 2002: Some reflections on the PRA approach as a participatory inquiry for sustainable rural development: An Iranian perspective. Paper presented at the *Proceedings of the 18th Annual Conference.* AIAEE (Association for International Agricultural and Extension Education), Durban, South Africa.

Zurayk, R., F. el-Awar, S. Hamadeh, S. Talhouk, C. Sayegh, A.-G. Chehab, and K. al Shab, 2001: Using indigenous knowledge in land use investigations: A participatory study in a semi arid mountainous region of Lebanon. *Agriculture, Ecosystems, and Environment,* **86,** 247–262.

Chapter 3
Drivers of Ecosystem Change: Summary Chapter

Author: Gerald C. Nelson

Main Messages

This chapter provides a summary of the assessment of global drivers of ecosystem change that appears as Chapter 7 of the MA *Scenarios* volume. A driver is any natural or human-induced factor that directly or indirectly causes a change in an ecosystem. A direct driver unequivocally influences ecosystem processes. An indirect driver operates more diffusely, by altering one or more direct drivers.

The MA categories of indirect drivers of change are demographic, economic, sociopolitical, scientific and technological, and cultural and religious. Important direct drivers include climate change, nutrient pollution, land conversion leading to habitat change, overexploitation, and invasive species and diseases.

Changes in ecosystem services are almost always caused by multiple, interacting drivers that work over time and over level of organization and that happen intermittently. Changes in ecosystem services can feed back to alter drivers.

3.1 Introduction

This chapter provides a summary of the assessment of global drivers of ecosystem change that appears as Chapter 7 of the MA *Scenarios* volume (with references included). The chapter examines the two right boxes in the MA conceptual framework: indirect and direct drivers. (See Chapter 1 for the diagram and description of the conceptual framework.) It is important to recognize that neither this discussion nor the chapter in the *Scenarios* volume covers the remaining two boxes for the framework—the mechanisms by which drivers interact with specific ecosystems to alter their condition and ability to deliver services. Those topics are left to the individual condition and service chapters in this volume and to later chapters in the *Scenarios* report.

The MA definition of a driver is any natural or human-induced factor that directly or indirectly causes a change in an ecosystem. A direct driver unequivocally influences ecosystem processes. An indirect driver operates more diffusely, by altering one or more direct drivers. The MA categories of indirect drivers of change are demographic, economic, sociopolitical, scientific and technological, and cultural and religious. Important direct drivers include climate change, plant nutrient use, land conversion leading to habitat change, and invasive species and diseases.

3.2 Changes in Key Indirect Drivers

3.2.1. Demographic

Global population doubled in the past 40 years and increased by 2 billion people in the last 25 years, reaching 6 billion in 2000. Developing countries have accounted for most population growth in the past quarter-century, but there is now an unprecedented diversity of demographic patterns across regions and countries. Some high-income countries such as the United States are still experiencing high rates of population growth, while some developing countries such as China, Thailand, North Korea, and South Korea have very low rates.

Urban areas now contain about half the world's population yet cover less than 3% of the terrestrial surface. Regional rates of urbanization vary widely. High-income countries typically have populations that are 70–80% urban. Some developing-country regions, such as parts of Asia, are still largely rural, while Latin America, at 75% urban, is indistinguishable from high-income countries in this regard.

World population will likely peak before the end of the twenty-first century at less than 10 billion. The global population growth rate peaked at 2.1% a year in the late 1960s and fell to 1.35% a year by 2000, when global population reached 6 billion. Population growth over the next several decades is expected to be concentrated in the poorest, urban communities in sub-Saharan Africa, South Asia, and the Middle East. Populations in all parts of the world are expected to experience substantial aging during the next century. While industrial countries will have the oldest populations, the rate of aging could be extremely fast in some developing countries.

3.2.2 Economic

Global economic activity increased nearly sevenfold between 1950 and 2000. Despite the population growth just described, average income per person almost doubled during this period. However, dramatic regional variations in per capita income growth existed. As per capita income grows, the structure of consumption changes, with wide-ranging potential for effects on ecosystem condition and services. With rising per capita incomes, the share of additional income spent on food declines and the consumption of industrial goods and services rises. The composition of people's diets changes, with less consumption of starchy staples (rice, wheat, potatoes) and more of fat, meat, fish, fruits, and vegetables.

Energy and materials intensity (the energy use per unit of economic output) tend to decline with rising levels of GDP per capita. In other words, energy and material productivity—the inverse of energy intensity—improve in line with overall macroeconomic productivity. However, growth in productivity has historically been outpaced by economic output growth. Hence, materials and energy use have risen in absolute terms over time.

Domestic policy distortions such as taxes and subsidies can have serious economic and environmental consequences, both in the country where they are implemented and elsewhere. Subsidies to conventional energy are estimated to have been $250–300 billion a year in the mid-1990s. The 2001–03 average subsidies paid to the agricultural sectors of OECD countries were over $300 billion annually. OECD protectionism and subsidies cost developing countries more than $20 billion a year in lost agricultural income.

Policies that distort international trade flows also have negative economic consequences. Nations with lower trade barriers, more open economies, and transparent government processes tend to have higher per capita income growth rates. International trade is an important source of economic gains, as it enables comparative advantage to be exploited and it accelerates the diffusion of more-efficient technologies and practices.

3.2.3 Sociopolitical

Sociopolitical drivers encompass the forces influencing decision-making and include the quantity of public participation in decision-making, the makeup of participants in public decision-making, the mechanisms of dispute resolution, the role of the state relative to the private sector, and levels of education and knowledge. Over the past 50 years, there have been significant changes in sociopolitical drivers. There is a declining trend in centralized authoritarian governments and a rise in elected democracies. The role of women is changing in many countries, average levels of formal education are increasing, and there has been a rise in civil society (such as increased involvement of NGOs and grassroots organizations in decision-making processes). The trend toward democratic institutions has helped empower local communities, women, and resource-poor households. There has been an increase in multilateral environmental agreements. The importance of the state relative to the private sector—as a supplier of goods and services, as a source of employment, and as a source of innovation—is declining.

3.2.4 Cultural and Religious

To understand culture as a driver of ecosystem change, it is most useful to think of culture as the values, beliefs, and norms that a group of people share. In this sense, culture conditions individuals' perceptions of the world, influences what they consider important, and suggests courses of action that are appropriate and inappropriate. Broad comparisons of whole cultures have not proved useful because they ignore vast variations in values, beliefs, and norms within cultures. Nevertheless, cultural differences clearly have important impacts on direct drivers. Cultural factors, for example, can influence consumption behavior (what and how much people consume) and may be a particularly important driver of environmental change.

3.2.5 Science and Technology

The development and diffusion of scientific knowledge and technologies that exploit that knowledge have profound implications for ecological systems and human well-being. The twentieth century saw tremendous advances in the understanding of how the world works physically, chemically, biologically, and socially and in the applications of that knowledge to human endeavors. Productivity improvements from application of science and technology are estimated to have accounted for more than one third of total GDP growth in the United States from 1929 to the early 1980s, and for between one third and two thirds of GDP growth in OECD countries over the period 1947 to 1973.

The impact of science and technology on ecosystem services is most evident in the case of food production. Much of the increase in agricultural output over the past 40 years has come from an increase in yields per hectare rather than an expansion of area under cultivation. For instance, wheat yields rose 208%, rice yields rose 109%, and maize yields rose 157% in the past 40 years in developing countries. At the same time, unintended effects of technological advances can lead to the degradation of ecosystem services. For example, eutrophication of freshwater systems and hypoxia in coastal marine ecosystems result from excess application of inorganic fertilizers. Advances in fishing technologies have contributed significantly to the depletion of marine fish stocks.

3.3 Changes in Key Direct Drivers

For terrestrial ecosystems, the most important direct drivers of change in ecosystem services in the past 50 years, in the aggregate, have been land cover change (in particular, conversion to cropland) and the application of new technologies, which have contributed significantly to the increased supply of services such as food, timber, and fiber.

Deforestation and forest degradation affect 8.5% of the world's remaining forests, nearly half of which are in South America. Deforestation and forest degradation have been more extensive in the tropics over the past few decades than in the rest of the world, although data on boreal forests are especially limited, and the extent of the loss in this region is less well known. Approximately 10% of the drylands and hyper-arid zones of the world are considered degraded, with the majority of these areas in Asia. Cropped areas currently cover approximately 30% of Earth's surface.

For marine ecosystems and their services, the most important direct driver of change in the past 50 years, in the aggregate, has been fishing. Improved marine fishing technology has made it possible to extract considerable fish biomass from the marine system. In fact, humankind has probably reached the maximum (and in some places has exceeded) levels of fish biomass removal before significant ecosystem changes are induced. In the Gulf of Thailand, for example, higher-trophic animals are no longer present and the system is dominated by lower-trophic species with a high biomass turnover. Research in West Africa and the North Atlantic indicates similar changes. FAO estimates that about half of the wild marine fish stocks for which information is available are fully exploited and offer no scope for increased catches.

For freshwater ecosystems and their services, depending on the region, the most important direct drivers of change in the past 50 years include modification of water regimes, invasive species, and pollution, particularly high levels of nutrient loading. The introduction of non-native invasive species is one of the major causes of species extinction in freshwater systems. For example, marine and estuarine waters in North America are heavily invaded, mainly by crustaceans and mollusks in a pattern corresponding to trade routes.

Over the past four decades, excessive nutrient loading has emerged as one of the most important direct drivers of ecosystem change in terrestrial, freshwater, and marine ecosystems. Synthetic production of nitrogen fertilizer has been the key driver for the remarkable increase in food production that has occurred during the past 50 years. Nitrogen application has increased fivefold since 1960, but as much as 50% of the nitrogen fertilizer applied is lost to the environment, depending on how well the application is managed. Since excessive nutrient loading is largely the result of applying more nutrients than crops can use, it harms both farm incomes and the environment.

Excessive nitrogen loading can cause algal blooms, decreased drinking water, eutrophication of freshwater ecosystems (a process whereby excessive plant growth depletes oxygen in the water), hypoxia in coastal marine ecosystems (substantial depletion of oxygen resulting in die-offs of fish and other aquatic animals), nitrous oxide emissions contributing to global climate change, and air pollution by nitrogen oxides in urban areas. Improvements in nitrogen use efficiency require more investment in technologies that achieve greater congruence between crop nitrogen demand and nitrogen supply from all sources and that do not reduce farmer income.

Phosphorus application has increased threefold since 1960, with a steady increase until 1990 followed by leveling off at a level approximately equal to 1980's applications. These changes are mirrored by phosphorus accumulation in soils, which maintains high levels of phosphorus runoff that can cause eutrophication of freshwaters and coastal waters.

Many ecosystem services are reduced when inland waters and coastal ecosystems become eutrophic. Water from lakes that experience algal blooms is more expensive to purify for drinking or other industrial uses. Eutrophication can reduce or eliminate fish populations. Possibly the most striking change is the loss of many of the cultural services provided by lakes. Foul odors of rotting algae, slime-covered lakes, and toxic chemicals produced by some blue-green algae during blooms keep people from swimming, boating, and otherwise enjoying the aesthetic value of lakes.

Climate change in the past century has already had a measurable impact on ecosystems. Earth's climate system has changed since the preindustrial era, in part due to human activities, and is projected to continue to change throughout the twenty-first century. During the last 100 years, the global mean surface temperature has increased by about 0.6 Celsius, precipitation patterns have changed spatially and temporally, and global average sea level rose between 0.1 and 0.2 meters. The global mean surface temperature is projected to increase from 1990 to 2100 by 1.4–5.8 Celsius, accompanied by more heat waves. Precipitation patterns are projected to change, with most arid and semiarid areas becoming drier and with an increase in heavy precipitation events, leading to an increased incidence in floods and drought. The MA scenarios project a sea level rise of 9–88 centimeters.

Observed changes in climate, especially warmer regional temperatures, have already affected biological systems in many parts of the world. There have been changes in species distributions, population sizes, and the timing of reproduction or migration events, as well as an increase in the frequency of pest and disease outbreaks, especially in forested systems.

Biological invasions are a global phenomenon, affecting ecosystems in most biomes. Human-driven movement of organisms, deliberate or accidental, has caused a massive alteration of species ranges, overwhelming the changes that occurred after the retreat of the last Ice Age. In some ecosystems, invasions by alien organisms and diseases have resulted in extinction of native species or a huge loss in ecosystem services. In the United States, for example, invasions of non-native plants, animals, and microbes are thought to be responsible for 42% of the decline of native species now listed as endangered or threatened.

The threats that biological invasions pose to biodiversity and to ecosystem-level processes translate directly into economic consequences such as losses in crops, fisheries, forestry, and grazing capacity. Mismanagement of semiarid grasslands, for instance, combined with climatic changes has caused woody plant invasion by native bushes and loss of grazing lands in North and South America. However, introductions of alien species can also be beneficial in terms of human population; most food is produced from introduced plants and animals.

3.4 Driver Interactions

Changes in ecosystem services are almost always caused by multiple, interacting drivers that work over time (such as population and income growth interacting with technological advances that lead to climate change) and over level of organization (such as local zoning laws versus international environmental treaties) and that happen intermittently (such as droughts, wars, and economic crises).

Changes in ecosystem services can feed back to alter drivers. For example, changes in ecosystems create new opportunities for and constraints on land use, induce institutional changes in response to perceived and anticipated resource degradation, and give rise to social effects such as changes in income inequality.

No single conceptual framework captures the broad range of case study evidence. This chapter ends with a few selected examples of driver interactions and ecosystem consequences.

In all regions of the humid tropics, deforestation is primarily the result of a combination of commercial wood extraction, permanent cultivation, livestock development, and the extension of overland transport infrastructure. However, many regional variations on this general pattern are found. Deforestation driven by swidden agriculture is more widespread in upland and foothill zones of Southeast Asia than in other regions. Road construction by the state followed by colonizing migrant settlers, who in turn practice slash-and-burn agriculture, is most frequent in lowland areas of Latin America, especially in the Amazon Basin. Pasture creation for cattle ranching is causing deforestation almost exclusively in the humid lowland regions of mainland South America. Expansion of smallholder agriculture and fuelwood extraction for domestic uses are important causes of deforestation in Africa. These regional differences mostly come from varying mixes of economic, institutional, technological, cultural, and demographic factors underlying the direct causes of deforestation.

Agricultural intensification is usually defined as substantial use of purchased inputs, especially fertilizer, in combination with new plant varieties that respond well to the increased inputs. Globally, intensification has been a major contributor to the doubling of food production over the last 40 years. Drivers of intensification include new crop and fertilizer production technologies, development of markets and transportation infrastructure, and changes in credit and price policies.

Urbanization provides another illustration of interactions. Though only about 2% of Earth's land surface is covered by built-up area, the effect of urban systems on ecosystems extends well beyond urban boundaries. Three processes of urban change appear to be of relevance for ecosystem change: the growth of urban population (urbanization), the growth of built-up area (urban growth), and the spreading of urban functions into the urban hinterland connected with a decrease in the urban-rural gradient in population density, land prices, and so on (urban sprawl).

Chapter 4

Biodiversity

Coordinating Lead Authors: Georgina Mace, Hillary Masundire, Jonathan Baillie
Lead Authors: Taylor Ricketts, Thomas Brooks, Michael Hoffmann, Simon Stuart, Andrew Balmford, Andy Purvis, Belinda Reyers, Jinliang Wang, Carmen Revenga, Elizabeth Kennedy, Shahid Naeem, Rob Alkemade, Tom Allnutt, Mohamed Bakarr, William Bond, Janice Chanson, Neil Cox, Gustavo Fonseca, Craig Hilton-Taylor, Colby Loucks, Ana Rodrigues, Wes Sechrest, Alison Stattersfield, Berndt Janse van Rensburg, Christina Whiteman
Contributing Authors: Robin Abell, Zoe Cokeliss, John Lamoreux, Henrique Miguel Pereira, Jillian Thönell, Paul Williams
Review Editors: Gerardo Ceballos, Sandra Lavorel, Gordon Orians, Steve Pacala

---

*This appears in Appendix A at the end of this volume.

Main Messages

Biodiversity—the diversity of genes, populations, species, communities, and ecosystems—underlies all ecosystem processes. Ecological processes interacting with the atmosphere, geosphere, and hydrosphere determine the environment on which organisms, including people, depend. Direct benefits such as food crops, clean water, clean air, and aesthetic pleasures all depend on biodiversity, as does the persistence, stability, and productivity of natural systems.

For many ecosystem services, local population extinctions are more significant than global extinctions—human communities depend for their well-being on populations of species that are accessible to them. The most appropriate measures and indicators of biodiversity depend on the value or service being assessed and involve a consideration of the components of biodiversity that are involved (from genes, individuals, populations, species, and communities to ecosystems) and the service that is being delivered.

Knowledge of biodiversity is uneven, with strong biases toward the species level, large animals, temperate systems, and components of biodiversity used by people. This results in gaps in knowledge, especially regarding the status of tropical systems, marine and freshwater biota, plants, invertebrates, microorganisms, and subterranean biota.

Most estimates of the total number of species on Earth lie between 5 million and 30 million. Of this total, roughly 2 million species have been formally described; the remainder are unknown or unnamed. The overall total could be higher than 30 million if poorly known groups such as deep-sea organisms, fungi, and microorganisms including parasites have more species than currently estimated.

Most macroscopic organisms have small, often clustered, geographical ranges, leading to diagnosable centers of both diversity and endemism, which are frequently concentrated in isolated or topographically variable regions (islands, mountains, peninsulas). A large proportion of the world's terrestrial biodiversity at the species level is concentrated in a small area of the world, mostly in the tropics. The Neotropics and Afrotropics have the highest species richness. Endemism is also high in these regions and, as a consequence of its isolation, in Australasia. Even among the larger and more mobile species such as the terrestrial vertebrates, more than one third of all species have ranges less than 1,000 square kilometers. In contrast, local and regional diversity of microorganisms appears to be more similar to large-scale and global diversity, indicating greater dispersal, larger range sizes, and lower levels of regional species clustering.

Across a range of measures, tropical forests are outstanding in their levels of biodiversity at and above the species level. Regions of high species richness broadly correspond with centers of evolutionary diversity, and available evidence suggests that across major taxa, tropical moist forests are especially important for both overall variability and unique evolutionary history. Species richness, family richness, and species endemism are all highest for this biome, even after accounting for area and productivity.

Over the past few hundred years humans may have increased the species extinction rate by as much as three orders of magnitude. This estimate is uncertain because the extent of extinctions in undescribed taxa is unknown, because the status of many described species is poorly known, because it is difficult to document the final disappearance of very rare species, and because there are extinction lags between the impact of a threatening process and the resulting extinction. However, the most definite information, based on recorded extinctions of known species over the past 100 years, indicates extinction rates are around 100 times greater than rates characteristic of species in the fossil record. Other less direct estimates, some of which refer to extinctions hundreds of years into the future, estimate extinction rates 1,000 to 10,000 times higher than rates recorded among fossil lineages.

Between 12% and 52% of species within well-studied higher taxa are threatened with extinction, according to the *IUCN Red List*. Less than 10% of named species have been assessed in terms of their conservation status. Of those that have, birds have the lowest percentage of threatened species at 12%. The patterns of threat are broadly similar for mammals and conifers, which have 23% and 25% of species threatened, respectively. The situation with amphibians looks similar, with 32% threatened, but information is more limited. so this may be an underestimate. Cycads have a much higher proportion of threatened species, with 52% globally threatened. In regional assessments, taxonomic groups with the highest proportion of threatened species tended to be those that rely on freshwater habitats. Threatened species show continuing declines in conservation status, and species threat rates tend to be highest in the realms with highest species richness.

The main causes of species extinction are changing from a historical trend of introductions and overexploitation affecting island species to present-day habitat loss and degradation affecting continental species. While the vast majority of recorded extinctions since 1500 have occurred on oceanic islands, continental extinctions are now as common as island extinctions. Approximately 50% of extinctions over the past 20 years occurred on continents. This trend is consistent with the observation that most terrestrial species threatened with extinction are continental. Despite the growing importance of habitat loss and degradation, species introductions and overexploitation also remain significant threats to biodiversity on continents and islands.

Climate change, which contributes to habitat change, is becoming the dominant driver, particularly in vulnerable habitats. Under climate change, endemic montane, island, and peninsula species are especially vulnerable, and coastal habitats such as mangroves, coral reefs, and coastal wetlands are especially at risk from resulting sea level rises. Both recent empirical evidence and predictive modeling studies suggest that climate change will increase population losses. In some regions there may be an increase in local biodiversity—usually as a result of species introductions, the long-term consequences of which are hard to foresee.

Among a range of higher taxa, the majority of species are currently in decline. Studies of amphibians globally, African mammals, birds in intensively managed agricultural lands, British butterflies, Caribbean corals, waterbirds, and fishery species show the majority of species to be declining in range or number. Those species that are increasing have benefited from management interventions such as protection in reserves or elimination of threats such as overexploitation or are species that tend to thrive in human-dominated landscapes.

The majority of biomes have been greatly modified by humans. Between 20% and 50% of 9 of the 14 biomes have been transformed to croplands. Tropical dry forests are the most reduced by cultivation, with almost half of the biome's native habitats replaced with cultivated lands. Three other biomes—temperate grasslands, temperate broadleaf forests, and Mediterranean forests—have experienced 35% or more conversion. Biomes least reduced by cultivation include deserts, boreal forests, and tundra. While cultivated lands provide many provisioning services, such as grains, fruits, and meat, habitat conversion to agriculture typically leads to reductions in native biodiversity.

Homogenization, the process whereby species assemblages become increasingly dominated by a small number of widespread, human-adapted

species, represents further losses in biodiversity that are often missed when only considering changes in absolute numbers of species. The many species that are declining as a result of human activities tend to be replaced by a much smaller number of expanding species that thrive in human-altered environments.

We lack comprehensive global-scale measures to assess whether the internationally agreed target of significantly reducing the rate of loss of biodiversity by 2010 will be met. However, our understanding of the dynamics of drivers, and particularly of lag times from changes in drivers to eventual impacts on biodiversity, suggest it is most unlikely to be achievable. The 2010 target, as agreed at WSSD in 2002 and adopted by the parties to the Convention on Biological Diversity, is an important goal for biodiversity management. It is probably too late to reverse the near-term trends in biodiversity loss given the lag times in ecosystem responses. Until critical drivers are mitigated, most declines seem likely to continue at the same or increased rates, although there is evidence that biodiversity loss is slowing or even recovering for some habitats (such as temperate woodlands) and species (temperate birds, for example).

4.1 Introduction

Biodiversity is fundamental to ecosystem functioning. Extrinsic or abiotic factors, such as climate and geophysical conditions, help to determine the boundaries of ecosystems (Colwell and Lees 2000; Gaston 2000). But within these boundaries, intrinsic or biotic factors such as the abundance, distribution, dynamics, and functional variation among biodiversity components of ecosystems regulate the magnitude and variability of ecosystem processes, such as production or decomposition. (See Chapter 11.) Together, these extrinsic and intrinsic factors determine the specific properties of an ecosystem, such as its stability, its fertility, or its susceptibility to invasion. They also determine the type of ecosystem found, such as drylands, forest or woodland, or inland waters.

The benefits that humans derive from ecosystems are known as ecosystem services (see Chapter 1) and include breathable air, fertile soils, and productive forests and fisheries, as well as many cultural benefits such as recreational hunting or inspirational values. Such ecosystem services are obtained only if ecosystems include the biodiversity that guarantees the functional processes necessary to deliver them.

This chapter focuses on the fundamental aspects of biodiversity that underpin all ecosystem processes and that are valued in their own right. Biodiversity relevant to particular services is documented in the Chapters 7 to 17 of this volume, while biodiversity as one element in the management of particular ecosystems for the delivery of services is discussed in Chapters 18 to 27. This chapter describes what is known about biodiversity globally, the nature of biodiversity variation and its measurement, the main drivers of change, and the observed trends in distribution, variation, and abundance of biodiversity.

4.1.1 Biodiversity and Its Assessment

Biodiversity is the diversity among living organisms in terrestrial, marine, and other aquatic ecosystems and the ecological complexes of which they are part. It includes diversity within and between species and the diversity of ecosystems. In addition to the important role of biodiversity in providing ecosystem services, it also has intrinsic value, independent of any human concern.

In addition to its intrinsic value, the roles of biodiversity in the provision of ecosystem services can be summarized under the following headings:

- *supporting roles* include the underpinning of ecosystems through structural, compositional, and functional diversity;
- *regulatory roles* through the influence of biodiversity on the production, stability, and resilience of ecosystems;
- *cultural roles* from the nonmaterial benefits people derive from the aesthetic, spiritual, and recreational elements of biodiversity; and
- *provisioning roles* from the direct and indirect supply of food, fresh water, fiber, and so on.

All these roles are strongly interrelated, and it is rarely possible to separate them in practice. Yet defining roles is an essential step in assessing biodiversity: any measures should be relevant to the role being examined and to the purpose of the assessment (The Royal Society 2003). For example, a biologist wishing to assess the changing status of biodiversity in a wetland before and after land use changes in the watershed might turn to the most widely available information—trends in bird population sizes. People interested in birds would regard this as important, but if the observer were concerned about overall species richness, the bird data could be insufficient or even misleading. Due to their unusual dispersal ability, birds might be relatively well buffered from the effects of habitat change. The consequences of the land use change on less vagile species, such as plants, invertebrates, or below-ground biota could be very different. Similarly, if the effect on ecosystem services were of most interest, then other species and measures other than population size will be more informative. If provisioning services were under examination, then the assessment would be better focused on the abundance and distribution of the ecosystem components essential for food or fiber production. Thus, given the complexity of biodiversity, the most readily available measures rarely reflect the real attribute of interest for any particular role (The Royal Society 2003).

Biodiversity is commonly measured at the levels of genes, species or ecosystems. At each of these, measures may represent one or many of the following:

- *Variety,* reflecting the number of different types. For example, this could refer to different species or genes, such as how many bird species live in a particular place or how many varieties of a genetic crop strain are in production.
- *Quantity and quality,* reflecting how much there is of any one type. Variation on its own will only rarely meet people's needs. For example, for many provisioning services (food, fresh water, fiber) the quantity or the quality matter more than the presence of a particular genetic variety, species, or ecosystem.
- *Distribution,* reflecting where that attribute of biodiversity is located. For example, having all the world's pollinators present but only in a single location will not meet the needs of the plants that depend on them. Many ecosystem services are location-specific. For instance, human and natural communities need to be close to wetlands to benefit from their regulatory roles.

In practice, the relevant measure and attribute depends on the role being assessed. For example, many benefits of biodiversity depend on the functional and structural variability in species, whereas most provisioning services and many regulatory services depend more on the quantity and distribution of populations and ecosystems. Long-term sustainability of many services depends on the maintenance of genetic variability. Ultimately, maintaining variability in any biodiversity component provides options for the future, even if not all variants have an obvious role to play. Thus, variability plays a special role, which probably explains why it is generally emphasized in discussions of biodiversity value.

Table 4.1 summarizes the importance of quantity versus variability among different biodiversity components in relation to ecosystem services. Broadly speaking, and according to our present level of understanding, variability is more significant at the genetic and species levels, whereas quantity and distribution are more significant at the population and ecosystem levels. For most ecosystem services, local loss of biodiversity (population reduction or local extinction) is most significant; but for future option values and for certain services such as genetic variability and bioprospecting, global loss is the primary consideration.

4.1.2 The Diversity and Evolution of Life

Living organisms were originally divided into two kingdoms: animal and vegetable (the Animalia and the Plantae), but more recently it has become clear that this simple division does not reflect the true diversity of life. The five Kingdom scheme that followed divided all living organisms into Monera (bacteria), Protista (single-celled organisms), Fungi, Plants, and Animals. In terms of either numerical diversity or phylogenetic diversity (measuring the degree of independent evolutionary history), however, it is now clear that this too misrepresents the diversity of life.

Most organisms are very small (microscopic), and DNA and RNA studies reveal that the living world is more appropriately divided into three groups: the Bacteria, the Archaea (a group once included with the bacteria but now shown to be as different from them as they both are from the rest), and the rest—the Eukaryotae. Bacteria and Archaea have no well-defined nucleus and are referred to as Prokaryotae (or prokaryotes). The Eukaryotae (or eukaryotes) have a well-defined nucleus and comprise the animals, plants, fungi, and protists. A fourth group of biological entities, the viruses, are not organisms in the same sense that eukaryotes, archaeans, and bacteria are, and so they are not included. However, they are of considerable biological importance.

Life arose on Earth 3.5–4.5 billion years ago, and for probably the first 1–2 billion years there were only prokaryotes. The first definitive fossils of eukaryotes are found about 2 billion years ago, but they started to proliferate quite rapidly and the multicellular

eukaryotes appear about 1.5 billion years ago. The first animals appeared much later, around 700 million years ago for many soft-bodied marine invertebrates, such as the sponges, jellyfish, soft corals, and worms. By about 500 million years ago an abundant fossil record includes marine invertebrates with exoskeletons, vertebrates, and plants. All phyla existing today appear shortly after. Today's diverse assemblage of mammals, birds, and flowering plants appeared within the past 70 million years, but it is not until about 7 million years ago that humans in their most primitive form appeared, and not until 100,000–200,000 years ago that modern humans appeared.

Evolutionary biologists believe that all existing life is derived from a single, common ancestral form. The fact that millions of species live on Earth today is a consequence of processes leading to speciation. Speciation involves the splitting of a single species lineage. It occurs in three different ways: allopatric, parapatric, and sympatric. Allopatric speciation is speciation by geographic isolation and requires the imposition of a barrier that prevents individuals in the two lineages from interbreeding with one another. For most animals, geographical isolation has been the most important barrier, and the larger and more vagile the animal, the wider the barrier must be. As a result, allopatric speciation in most animals can take place only in large geographic areas where substantial barriers, such as wide water gaps or isolated mountains exist. In parapatric speciation there is no complete geographic isolation, but lineages diverge across environmental gradients. Sympatric speciation is speciation without geographic isolation. Plants, for example, commonly speciate via a duplication of their chromosomes, a process that can be accomplished in a single generation. The different process and conditions required for speciation results in a great variation in the rate of speciation. However, in general the process is slow, usually taking millions of years.

The short and clustered branches on the molecular tree of life (see Figure 4.1) illustrate the relatively close and recent relationships among the organisms with which we are most familiar and that dominate most biodiversity assessments (Plants, Animals, Fungi). However, the microorganisms that dominate the branches of the evolutionary tree are extremely important in any assessment of biodiversity. These groups include most of the forms that are the main providers of most regulating and supporting services and that are key to many provisioning services (Nee 2004).

4.1.3 Practical Issues for Ecosystem Assessment

The term ecosystem can be applied to any functioning unit with biotic and abiotic elements, ranging from tiny pockets of life to the entire planet. Hence there are some practical issues to address in determining units for analysis and assessment. The Millennium Ecosystem Assessment uses ecosystems as a unit for assessment based on the definition adopted by the Convention on Biological Diversity: "a dynamic complex of plant, animal and microorganism communities and their nonliving environment interacting as a functional unit" (UN 1992). As such, ecosystems do not have clearly definable boundaries, and any classification, no matter how many categories it has, can become somewhat arbitrary. A practical approach to this problem is to build up a series of map overlays of significant factors, mapping the location of discontinuities, such as in the distribution of organisms, the biophysical environment (soil types, drainage basins, depth in a water body), and spatial interactions (home ranges, migration patterns, fluxes of matter). A useful ecosystem boundary for analysis is then the place where a number of these discontinuities coincide.

Based on this general methodology, different systems for classifying terrestrial ecosystem classifications have been developed.

Table 4.1. Measures of Biodiversity at Different Levels. The measures reflect different service benefits. In practice, some kinds of measures are more significant than others. The bold text reflects the most significant measures for ecosystem services.

Level	Importance of Variability	Importance of Quantity and Distribution
Genes	**adaptive variability for production and resilience to environmental change, pathogens, etc.**	local resistance and resilience
Populations	different populations retain local adaptation	**local provisioning and regulating services, food, fresh water**
Species	**the ultimate reservoir of adaptive variability, representing option values**	**community and ecosystem interactions are enabled through the co-occurrence of species**
Ecosystems	different ecosystems deliver a diversity of roles	**the quantity and quality of service delivery depends on distribution and location**

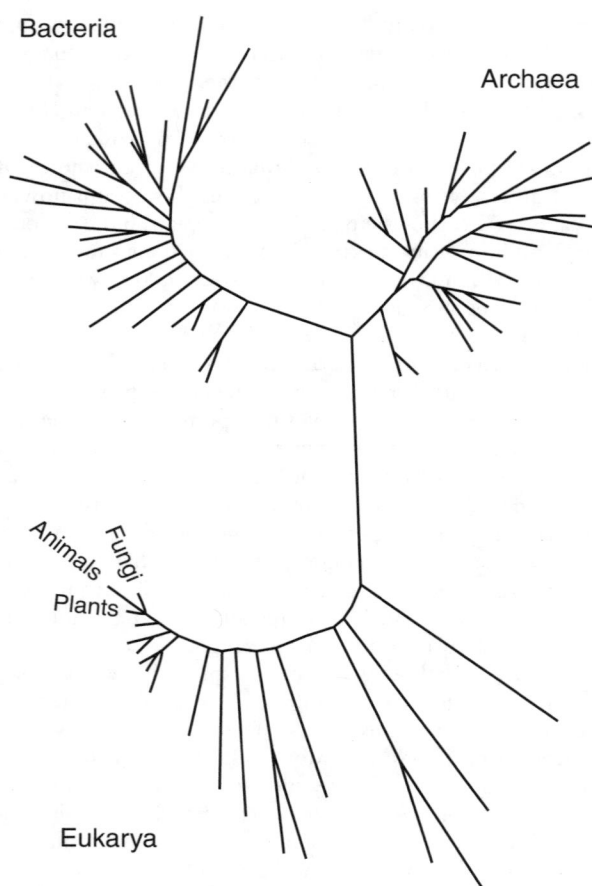

Figure 4.1. Tree of Life: Biodiversity through a Molecular Lens.
This scheme is based on ssRNA gene-sequence data and shows
the relationships of organisms in the three main domains of life—
Bacteria, Archaea, and Eukarya (creatures with cells like our own).
Visible organisms are found among the plants, animals, and fungi.
Not only are these groups just twigs on the tree of life, many of their
members are invisible as well. (Nee 2004)

(See Table 4.2.) Generally, ecosystems can be characterized by
either community structure and functioning or species composition or by a combination of the two. Spatially, ecosystem maps
have been derived through various techniques, such as modeling
(using climatic parameters for example), mapping (from remotely
sensed images or delineation of species extents), or a combination
of both.

Different classifications serve different purposes and may yield
different results. For example, the result of an analysis between
five broad global biomes and six global terrestrial ecosystem classifications is shown in Figure 4.2. The ecosystem classifications
were chosen to capture a range of the varying techniques that
have been used to map ecosystem boundaries. The five broad
biomes include desert (both hot and cold deserts), forest and
woodland, grassland (includes grassland, savanna, steppe, and
shrub), mixed, and tundra. The mixed class comprises the mixed
mountain classes of FAO, the mixed mountain and island systems
of Udvardy (1975), and the Mediterranean forests, woodland, and
scrub class of WWF. It is difficult to divide mixed classes accurately between the remaining broad biome classes, so they were
classified as a separate class.

There is reasonable agreement in area between some of the
biomes and less agreement among others. The biomes that are

reasonably consistent across ecosystem maps are forest and woodland, desert, and tundra. Delineation of grasslands is less consistent, and the reported grassland area differs across ecosystem maps
by as much as 30%. Forest and woodland, the most predominant
biome, is represented at between 42% and 53% of the terrestrial
land surface (approximately 55 million to 73 million square kilometers). These results illustrate the implications of different
choices of global ecosystem classifications for assessment, particularly as relates to the grassland biome.

Table 4.2 illustrates the methods used to define the ecosystem
boundaries, the purpose for which they were classified, and the
scale at which they were mapped. These are variables that should
be considered in order to determine the appropriateness of a classification for a particular assessment.

In this chapter and elsewhere in this assessment, the WWF
terrestrial biomes, built up from the classification of terrestrial
ecoregions, were chosen to assess magnitude, distribution, condition, and trend of terrestrial biodiversity. (See Figure 4.3 in Appendix A.) Currently there is no equivalent classification for
marine ecosystems. A separate set of freshwater biomes, used to
classify freshwater ecoregions, is in preparation by WWF and The
Nature Conservancy.

4.2 Current Status of Biodiversity

This section presents information on the global status of biodiversity, measured at the scale of biogeographic realms, biomes, species, populations, and genes. Under each heading, the significance
of that level is introduced, followed by information on what is
known about its current condition

4.2.1 Biogeographic Realms

Biogeographic realms are large spatial regions within which ecosystems share a broadly similar biota. Eight terrestrial biogeographic realms are typically recognized, corresponding roughly to
continents (for example, the Afrotropical realm). Terrestrial biogeographic realms reflect freshwater biodiversity patterns reasonably well, but marine biogeographic realms are poorly defined.

4.2.1.1 Definition and Measurement

Similar ecosystems (tropical moist forests, for instance) share processes and major vegetation types worldwide, but their species
composition varies markedly among the world's eight biogeographic realms (Olson et al. 2001). For example, the major tree
species in tropical moist forests in Southeast Asia differ from those
dominating tropical moist forests in South America. There is substantial variation in the extent of change and degradation to biodiversity among the biogeographic realms, and they face different
combinations of drivers of change. In addition, the options for
mitigating or managing drivers vary among realms. Although
realms map roughly onto continents, they differ from continents
in important ways as a result of biogeographic history.

4.2.1.2 Current Status of Biogeographical Realms

Biogeographic realms vary widely in size. The largest is the Palearctic, followed by the Afrotropical and Nearctic realms; the
smallest is Oceania. (See Table 4.3.) These area estimates are based
on terrestrial area only, although the realm boundaries can be applied to inland water ecosystems with slight modifications of the
boundaries to ensure that they do not cut across freshwater ecoregions or biomes (habitat types). Among terrestrial realms, net
primary productivity (Imhoff et al. 2004) and biomass (Olson et
al. 1980) values are highest in the Neotropics, followed closely by

Table 4.2. Description of Six Common Global Ecosystem Classifications

Ecosystem Classification	Description	Use	Spatial Resolution
Bailey Ecoregions (Bailey and Hogg 1986)	Bailey and Hogg developed a hierarchical classification including domains, divisions, and provinces that incorporates bioclimatic elements (rainfall and temperature)—based largely on the Koppen-Trewartha climatic system, altitude, and landscape features (soil type and drainage). Macroclimate defines the highest classification level and increasing numbers of variables are used to describe more detailed regional classifications.	Intended to demarcate ecologically similar areas to predict the impact of management and global change (Bisby 1995).	1: 30,000,000 scale (Bailey 1989)
FRA Global Ecological Zones (FAO 2001)	FAO's classification is based on the Koppen-Trewartha climate system and combined with natural vegetation characteristics that are obtained from regional ecological or potential vegetation maps.	Developed for the "Global Forest Resources Assessment 2000" as a way to aggregate information on forest resources.	useful at 1: 40,000,000 scale (FAO 2001)
Holdridge Life Zones (Holdridge 1967)	Holdridge's life zones are derived using three climatic indicators: biotemperature (based on the growing season length and temperature); mean annual precipitation; and a potential evapotranspiration ratio, linking biotemperature with annual precipitation to define humidity provinces. R. Leemans, then at IIASA, prepared the digital spatial data.	Initially derived to incorporate into models of global climate change.	0.5° geographic latitude/ longitude
Ramankutty Global Potential Vegetation (Ramankutty and Foley 1999)	Derived from a combination of satellite data and the Haxeltine and Prentice potential natural vegetation data. In places that are not dominated by humans, satellite-derived land cover (mainly the DISCover dataset) is used as a measure of potential vegetation. In places dominated by anthropogenic land cover, the Haxeltine and Prentice data set was used to fill in the gaps.	Initially derived to facilitate the analysis of cultivation land use practices and global natural or "potential" vegetation. Potential vegetation is regarded as the vegetation most likely to currently exist without the impact of human activities.	5 minute geographic latitude/ longitude
Udvardy's Biogeographical Realms and Provinces (Udvardy 1975)	This system combines physiognomic and biogeographical approaches. The physical structure of the dominant vegetation in combination with distinctive flora and fauna compositions defines the boundaries.	The classification has been used for biogeographical and conservation purposes. IUCN, for example, has used this map as a basis for assessing the representativness of global projected areas.	usable at 1: 30,000,000 scale
WWF Terrestrial Ecoregions of the World (Olson et al. 2001)	WWF ecoregions have been delineated through the combination of existing global ecoregion maps, global and regional maps of the distribution of selected groups of plants and animals, and vegetation types and through consultation with regional experts. Ecoregions identify relatively large units of land containing a distinct assemblage of natural communities and species, with boundaries that approximate the original extent of natural communities prior to major land use change.	A tool to identify areas of outstanding biodiversity and representative communities for the conservation of biodiversity.	variable, based on global or regional source (1:1 million to 1:7.5 million); useful at scales of 1:1 million or higher

the Afrotropical and Indo–Malayan realms. The least productive is the Antarctic realm.

Land cover composition also varies widely between realms. Because realms are defined biogeographically, and not by dominant habitat type, each realm typically contains a mix of land cover types as mapped by GLC2000 (USGS-EDC 2003). (See Figure 4.4 in Appendix A.) Some biogeographic realms, however, are dominated by a single land cover type. For example, more than 40% of the Australasian realm consists of herbaceous cover and more than 40% of the Neotropics consist of broadleaf forests. In each biogeographic realm, significant areas have been converted from native habitats to agriculture and urban land uses. All realms have experienced at least 10% habitat conversion, and the Indo–Malayan realm has by far the largest percentage of agricultural and urban lands (54%).

Partly in response to this land conversion, nations in all biogeographic realms have designated formal protected areas to conserve native ecosystems. Protection (IUCN classes I–IV) (WCMC 2003) of terrestrial biogeographic realms ranges between 4.0 and 9.5%. The realms with the greatest proportion of protected land area are Oceania (9.5%) and the Nearctic (7.8%). The Indo-Malayan (4.8%) and Palearctic (4.0%) realms contain the lowest proportion of protected land area. The Palearctic is the largest, and although only 4.0% is protected, it contains the largest total protected land area. The vast majority of protected areas have been designed to protect terrestrial ecosystems and biodiversity features, which has led to relative under-protection of inland water and marine biodiversity. (See Chapters 18, 19, and 20.)

The extent of inland water systems is greatest in the Nearctic and Palearctic realms (for example, lakes and peatlands). The

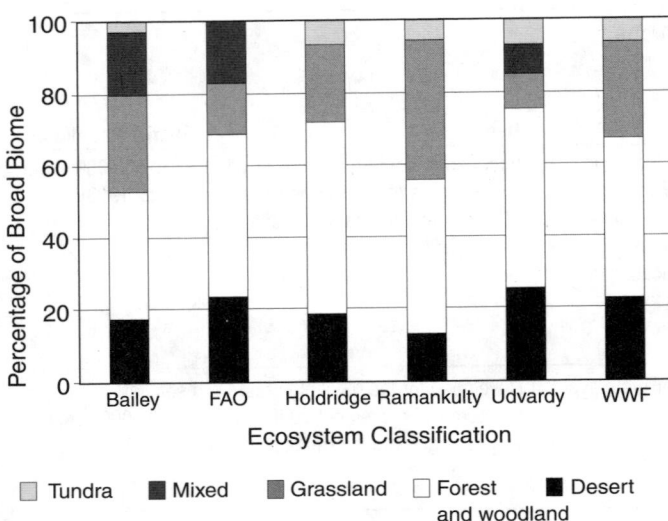

Figure 4.2. Area of Broad Biomes as Estimated by Six Ecosystem Classifications

Nearctic realm has by far the largest proportion of the world's lakes (Revenga and Kura 2003). In terms of water volume, however, the Neotropic and the Indo-Malayan realms contribute the most discharge into the oceans. Australasia contributes the least, with only 2% of the world's freshwater discharge (Fekete et al. 1999). The extent and distribution of inland water ecosystems has not been exhaustively documented at the global or regional scale. And while the biogeographic and ecological classification of inland water ecosystems is less well developed than for terrestrial

ecosystems, more than 50 classifications are in use (see, e.g., Asian wetland classification system: Finlayson 2002; Darwall and Revenga in prep).

Each biogeographic realm contains a range of major habitat types or biomes. The Indo-Malayan, Oceanic, and Neotropical realms are dominated by tropical forest and grassland biomes, while the polar realms (Palearctic, Nearctic) contain higher proportions of tundra and boreal forest. The Afrotropics are dominated by tropical grasslands. Although dominated by different biomes, most realms contain similar biome richness. All but Oceania include 9–11 of the 14 terrestrial biomes. Oceania is composed mostly of low, tropical islands and is dominated by tropical forest and tropical grassland biomes.

In part due to differences in biome richness and composition, biogeographic realms differ markedly in species and family richness, at least for the four vertebrate classes for which data exist. Figure 4.5 shows species richness among realms based on presence or absence records of terrestrial vertebrates (birds, mammals, and reptiles) in each of the 825 WWF terrestrial ecoregions (WWF 2004). This is supplemented by an analysis of extent of occurrence polygon data for amphibians and threatened birds (Baillie et al. 2004; BirdLife 2004b). The Neotropics are by far the most species-rich realm, both overall for terrestrial vertebrates and for each of the four taxa. (See Figure 4.5a.) Other realms containing high proportions of tropical forests (such as Indo-Malayan) also show high species richness in terrestrial vertebrates. With the exception of Antarctica, Oceania is the least species-rich realm due to its small overall land area and the relatively species-poor faunas typical of islands.

Biodiversity at the level of families is more similar among biogeographic realms (see Figure 4.5b) except for Oceania and Antarctica. These patterns differ somewhat among some inland water

Table 4.3. Magnitude and Biodiversity of the World's Eight Terrestrial Biogeographic Realms. Realms are mapped in Figure 4.3.

	Size, Productivity, and Protection				Richness				Endemism				Family Richness				Family Endemism			
Biogeographic Realm	Area (x10⁵km²)	Mean NPP (10¹⁰gC/yr/cell)[a]	Biomass (kgC/m²)	Percent Protected (IUCN I-IV)	Amphibians	Birds	Mammals	Reptiles	Amphibians	Birds	Mammals	Reptiles	Amphibians	Birds	Mammals	Reptiles	Amphibians	Birds	Mammals	Reptiles
AA	92.5	25.7	3.9	5.1	545	1,669	688	1,305	515	1,330	614	1,209	6	93	35	20	3	20	18	3
AN	32.8	0.0	0.0	0.9	0	36	0	0	0	4	0	0	0	15	0	0	0	0	0	0
AT	217.3	40.7	4.3	6.5	930	2,228	1,161	1,703	913	1,746	1,049	1,579	15	94	52	22	8	11	14	3
IM	85.2	43.1	5.7	4.8	882	2,000	940	1,396	722	758	544	1,094	11	100	43	26	3	1	2	3
NA	204.2	14.2	4.5	7.6	298	696	481	470	235	58	245	175	11	67	30	27	8	0	2	0
NT	193.8	64.5	6.2	5.1	2,732	3,808	1,282	2,561	2,660	3,217	1,061	2,258	12	93	49	39	7	24	23	7
OC	0.5	24.3	3.7	9.5	3	272	15	50	3	157	10	26	1	38	4	9	0	1	0	0
PA	527.4	10.5	2.9	4.0	395	1,528	903	774	255	188	472	438	13	97	44	21	5	0	0	0

[a] Grid cells were 0.25° cells, roughly 28x28km at the equator.

Key

AA	Australasian	AT	Afrotropical	NA	Nearctic	OC	Oceanic
AN	Antarctic	IM	Indo-Malayan	NT	Neotropical	PA	Palearctic

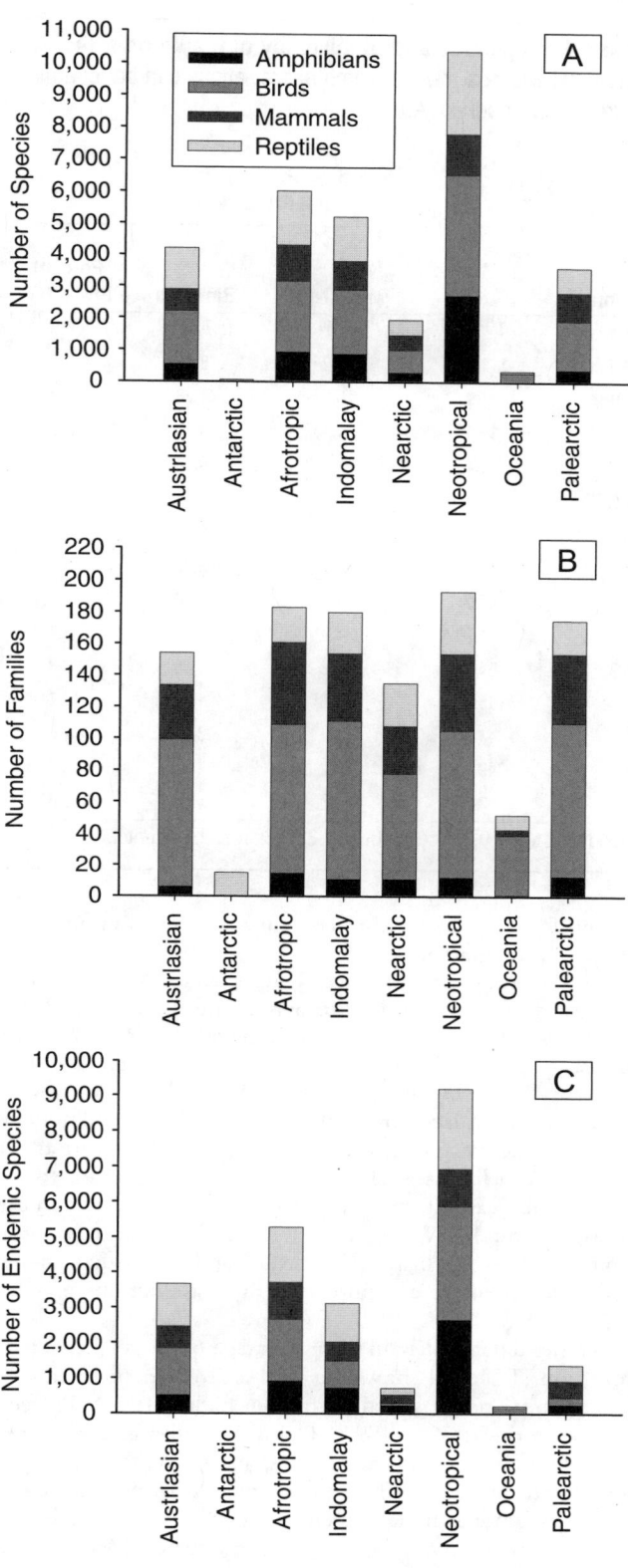

Figure 4.5. Diversity Comparisons for Eight Terrestrial Biogeographic Realms. The comparisons shown are for species richness (A), family richness (B), and endemism (C).

groups. The Neotropics have more than twice as many freshwater fish families as the Nearctic and Palearctic, and the Afrotropic and Indo-Malayan realms are only slightly behind the Neotropics (Berra 2001).

The number of species restricted to single realms (realm endemics) closely mirrors species richness patterns, at least for the four vertebrate classes assessed here. (See Figure 4.5c.) The Neotropics contain not only the greatest number of terrestrial vertebrate species but also the greatest number that occur only there. In all realms, however, the percentage of endemic species compared with total species richness is substantial (34–88%). Oceans, deserts, and other barriers to dispersal have resulted in vertebrate terrestrial faunas that are largely unique to each continent. We do not know how this pattern compares to patterns of realm endemism in nonvertebrates.

4.2.2 Biomes

4.2.2.1 *Definition and Measurement*

Biomes represent broad habitat and vegetation types and span across biogeographic realms (for example, the tundra biome is found in both Palearctic and Nearctic realms). Biomes are useful units for assessing global biodiversity and ecosystem services because they stratify the globe into ecologically meaningful and contrasting classes.

Throughout this chapter, and elsewhere in the MA, the 14 biomes of the WWF terrestrial biome classification are used, based on WWF terrestrial ecoregions (Olson et al. 2001). The nested structure of this classification, with finer-scale ecoregions nested into both biomes and biogeographic realms, allows assessments to be scaled up or down depending on the objectives. Furthermore, several datasets are already available and others continue to be associated with the WWF classification (such as vertebrate and plant species distribution data, threatened species, area-based estimates of net primary productivity, and land cover). The biome-level boundaries have very good resolution and accuracy, as they are based on the finer-scale ecoregions and are of an appropriate scale and number for global reporting.

These boundaries are based on the original or potential extent of these ecosystems or biomes, and do not take human-induced land cover changes into account. The extent of the ecosystems or biomes before the extensive changes brought about with the rise of the human population and industrialization in the modern era will probably never be known. We refer to this earlier, less altered state as "original," while recognizing that climatic and environmental changes have always caused change and movements in Earth's ecosystems. Therefore the global classifications can only be an approximation of the original boundaries of these ecosystems. The difference between original and current extent can be significant and forms an important component of the assessment of biodiversity loss.

There is no comparable global classification of freshwater biomes, but WWF and The Nature Conservancy are developing a major new biome classification for fresh water, to be completed in 2005. Terrestrial biomes tell us little by themselves about the size or type of freshwater habitat, which in turn has an enormous influence on the kind and number of species occurring there. For instance, a major river system can be adjacent to a very small basin, and both may fall within the same terrestrial biome, but they can contain vastly different assemblages of aquatic species. Freshwater biomes in the forthcoming classifications will be based largely on a combination of system size and type (such as large rivers versus small lakes), connectivity to coastal zones (such as total connectivity for islands), and overarching climatic conditions (such as temperate versus tropical or dry versus moist).

Like freshwater biomes, marine biome classification is less developed than that for terrestrial systems. The dynamic nature and the relative lack of natural boundaries in oceanic ecosystems make biogeographic divisions problematic, and there is no standard classification scheme. Nonetheless, several classifications of the marine realm exist, some based on biogeography (such as Briggs 1974), others on oceanographic and hydrological properties, and still others on ecological features, such as using the distribution of species assemblages in relation to seasonal characteristics of local and regional water masses (Ford 1999). Longhurst (1995) classified the world's oceans into four ecological domains and 56 biogeochemical provinces, largely on the basis of estimates of primary production rates and their changes over time. (See chapter 18.) Hayden et al. (1984) subdivided Dietrich's (1963) 12 marine realms into oceanic realms and coastal regions on the basis of physical and chemical properties including salinity, temperature, and seasonal movement of water and air masses.

Two marine classification systems have been used more widely. First, Bailey (also based on Dietrich 1963, 1998) includes oceanic ecoregions in his global classifications, mapping 14 marine divisions spread between the three domains. Continental shelves (less than 200 meters water depth) are distinguished; other divisions are delineated on the ocean surface based on four main factors: latitude and major wind systems (determining thermal zones) and precipitation and evaporation (determining salinity).

Second, Sherman and Alexander's (1986) system of large marine ecosystems delineates 62 regions of ocean encompassing near-coastal areas from river estuaries to the seaward boundary of continental shelves and the seaward margins of coastal current systems. They are relatively large regions (greater than 200,000 square kilometers), characterized by distinct bathymetry, hydrography, biological productivity, and trophically dependent populations. This approach aims to facilitate regional ecosystem research, monitoring, and management of marine resources and focuses on the products of marine ecosystems (such as the fish harvest). In general, no marine biome classification scheme has successfully covered the wide range of oceanic depths and addressed the lack of regional uniformity, thus complicating a global assessment of marine biodiversity.

4.2.2.2 Current Status of Major Terrestrial Biomes

The world's 14 terrestrial biomes vary in total area by two orders of magnitude, from nearly 35 million square kilometers (deserts and dry shrublands) to 350,000 square kilometers (mangroves). (See Table 4.4.)

Biomes also vary widely in per-area measures of plant biomass (Olson et al. 1980) and net primary productivity (Imhoff et al. 2004). Net primary productivity is the net amount of carbon fixed by plants through photosynthesis (after subtracting respiration) and represents the primary energy source for the world's ecosystems (Vitousek et al. 1986). Tropical moist forests show high levels of both standing biomass and annual productivity, while other biomes, such as temperate coniferous forests and boreal forests, have high biomass despite low annual (and more seasonal) productivity.

Each biome mapped in Figure 4.3, while typically dominated by the expected vegetation cover, actually comprises a complex mosaic of different land cover types as mapped by GLC2000. (See Figure 4.6 in Appendix A.) This heterogeneity is due in part to fine-scale mixture of ecosystems within these broadly defined biomes. For example, boreal forests are composed primarily of co-

Table 4.4. Magnitude and Biodiversity of the World's 14 Terrestrial Biomes. Key to biome abbreviations can be found in Figure 4.3 in Appendix A.

Biome	Size, Productivity, and Protection			
	Area	Mean NPP	Biomass	Percent Protected
	(x10⁵km²)	(10¹⁰gC/yr/cell)[a]	(kgC/m²)	(IUCN I-IV)
TMF	231.6	74.2	8.41	5.5
TDF	31.9	45.2	4.28	4.9
TCF	16.3	44.4	5.69	2.5
TeBF	135.4	28.3	4.48	3.8
TeCF	42.2	26.1	8.72	8.9
BF	118.5	11.0	6.19	6.3
TG	216.3	40.7	3.92	5.5
TeG	146.9	17.6	2.18	1.9
FG	11.2	34.4	3.10	8.7
MG	54.5	15.8	2.08	3.8
T	115.6	3.8	1.03	13.7
MF	44.9	21.1	3.30	2.8
D	349.1	6.2	1.18	3.7
M	3.5	41.4	4.64	8.6

[a] Grid cells were 0.25° cells, roughly 28x28km at the equator.

niferous forest land cover but contain a substantial proportion of shrublands and grasslands.

Another cause of land cover heterogeneity within biomes is conversion of native habitats to agriculture, pastures, and other human land uses. Indeed, in over half the biomes, 20–50% of land area has been converted to human use. Tropical dry forests are the most affected by cultivation, with almost half of the biome's native habitats replaced by cultivated lands. Three additional biomes—temperate grasslands, temperate broadleaf forests, and Mediterranean forests—have experienced 35% or more conversion. Biomes least affected by cultivation include deserts, boreal forests, and tundra. While cultivated lands provide many provisioning services (such as grains, fruits, and meat), habitat conversion to intensive agriculture leads to reductions in native biodiversity.

Biomes differ widely in the percentage of the total area under protection. Table 4.4 shows the total area under protection, including only lands classified in the four highest IUCN Protected Area categories (IUCN 1994). Flooded grasslands, tundra, temperate coniferous forests, mangroves, and boreal forests have the highest percentage area under protection—perhaps because these biomes are among the least useful for competing land uses, such as agriculture. Conversely, temperate grasslands, Mediterranean forests, and tropical coniferous forests are the least protected biomes.

To compare species richness among biomes, a similar methodology used to determine species richness at the level of realms has been applied. Tropical biomes have the highest levels of overall species richness, as well as the highest richness for each of the four taxa analyzed. (See Figure 4.7.) This is true of tropical moist forest, but also, perhaps surprisingly, of tropical grasslands and savannas and tropical dry forests, the second and fourth richest biomes

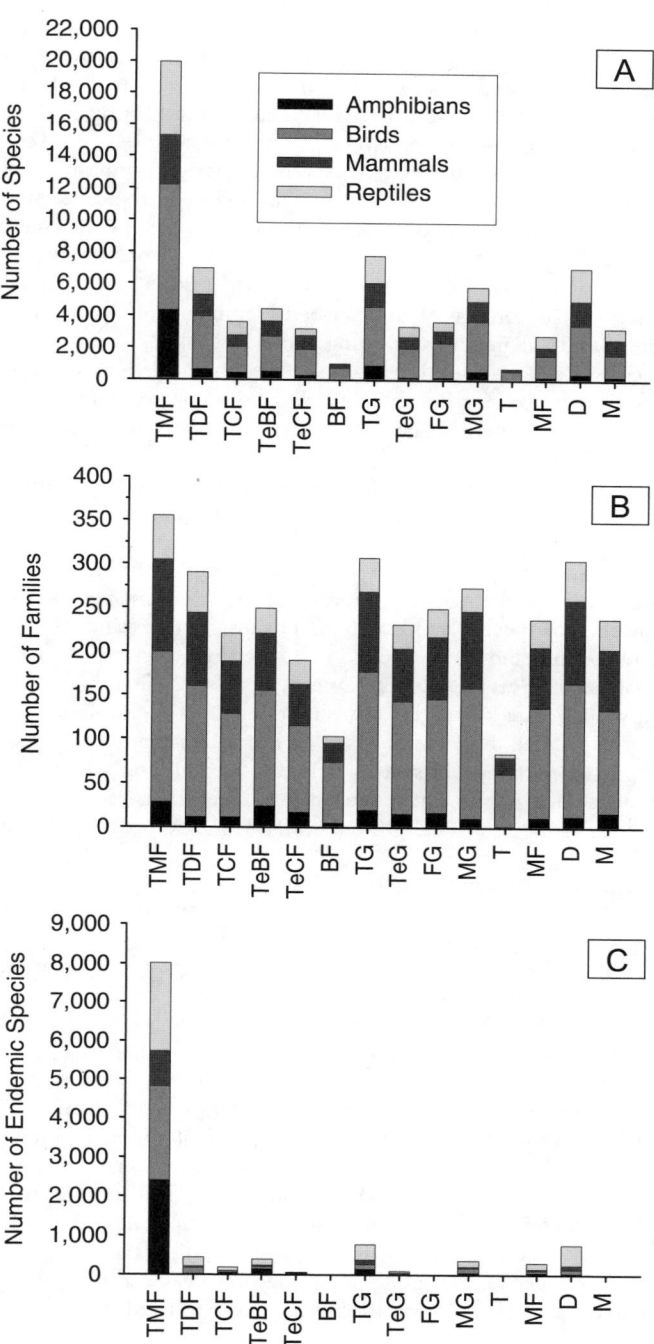

Figure 4.7. Diversity Comparisons for 14 Terrestrial Biomes. The comparisons shown are for species richness (A), family richness (B), and endemism (C). Biome codes as in Figure 4.3 (in Appendix A).

The number of biome-endemic species—that is, species found in a certain biome and nowhere else—varies widely among biomes. Tropical moist forests contain by far the highest number of endemic species, an order of magnitude more than any other biome. This pattern again may be the result of high speciation rates in this biome, as well as relatively smaller range sizes in lower latitudes (Rosenzweig 1995; Gaston 2000).

The relative richness of the world's biomes, however, may be influenced by their relative sizes as well. Biomes vary enormously in area, as noted earlier, and species richness is well known to increase with the area sampled (Rosenzweig 1995). Therefore, although both tropical moist forests and tropical grasslands contain high total richness, this may be due in part to the fact that they represent two of the largest biomes. Figure 4.8 plots species richness against area for the 14 biomes. In fact, the two are not statistically related (p > 0.75).

4.2.3 Species

The classification of living organisms into manageable groups greatly facilitates their study. The hierarchical system of classification used today is largely based on evolutionary relationships. The major categories, from the most inclusive to the smallest groups Kingdom-Phylum-Class-Order-Family-Genus-Species. It is at the level of species that living organisms are most widely known, both by common and scientific names.

4.2.3.1 *Definition and Measurement*

Although natural historians have been classifying living organisms into species since at least classical times, there is still no consensus on how this is best done (Hey 2001). Since the middle of the twentieth century, the dominant idea of how to define the term "species" has been the biological species concept (Mayr 1963), which defines species as groups of interbreeding natural populations whose members are unable to successfully reproduce with members of other such groups. Gene flow within a species leads to cohesion, whereas the lack of gene flow between different species means they are independent evolutionary lineages. Species therefore have natural and objective boundaries under this view, and so are natural units for biodiversity assessment.

Another hierarchy to which species belong is the evolutionary "family tree," or phylogeny, that links them all. In some well-studied groups (such as angiosperms (APG 1998) and birds (Sibley and Monroe 1990)), current taxonomic classification largely (and increasingly) reflects evolutionary relationships, such that species in a given taxon are all thought to share a more recent common

overall. Deserts and Mediterranean grasslands are also relatively rich biomes for terrestrial vertebrate species.

Tropical moist forests also contain the greatest diversity of higher taxa and therefore represent the greatest store of Earth's evolutionary history. The five biomes richest in terrestrial vertebrate species are also the five richest in families, although differences among biomes are not as pronounced. Tropical moist forests, therefore, contain many more species per family on average, suggesting that this biome has experienced higher rates of species diversification within families.

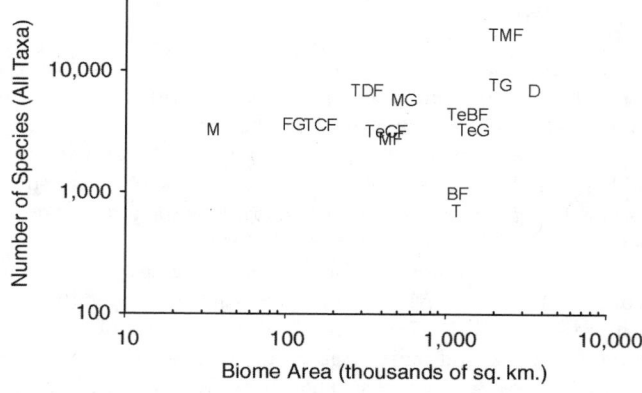

Figure 4.8. Species Richness of 14 Terrestrial Biomes in Relation to Biome Area. Biome codes as in Figure 4.3 (in Appendix A).

ancestor with each other than with species in other taxa. Higher taxonomic groupings then represent increasing levels of independent evolutionary history. In less well known groups, by contrast, classifications may not (and may not even attempt to) reflect phylogeny.

Regardless of how phylogenetic groups are recognized and named, decisions about the taxonomic rank (genus, family, and so on) of the various groups are arbitrary (Avise and Johns 1999). Many genera of insects, for instance, originated earlier than most avian families. Unlike biological species, higher taxonomic categories and lower taxonomic categories, like subspecies or races, have no natural boundaries.

Therefore species have advantages over other levels in the classificatory hierarchy and are useful units for biodiversity assessment. Some problems with using species as a unit for biodiversity assessment remain—both theoretical and practical; they can often be overcome or ameliorated with care, but they should never be overlooked (Isaac et al. 2004; Mace 2004). (See Box 4.1.)

4.2.3.2 How Many Species Are There?

Estimates of the total number of eukaryotic species vary greatly, most commonly falling between 5 million and 30 million (May 1992). The uncertainty stems from the fact that most taxonomic work is concentrated away from the most species-rich taxa (Gaston and May 1992) and regions (Gaston 1994a). In addition, the intensity of taxonomic work is actually declining (Godfray 2002). The discussion here is restricted to eukaryotic species. In the prokaryotes, different methods for recognizing and naming species, as well as severe problems with incomplete knowledge, make assessments and comparisons of species richness unreliable (Ward 2002; Curtis et al. 2002; Nee 2003).

Many methods of estimating total species numbers are based in some way on numbers of known, named species. Uncertainties around these estimates themselves pull in opposing directions. On the one hand, the lack of comprehensive systematic databases results in underestimates of known species numbers (Sugden and Pennisi 2000). On the other hand, the extent of synonymy between named taxa results in overestimates (May and Nee 1995). Several ongoing initiatives, such as Species 2000, the Integrated Taxonomic Information System, and the Global Biodiversity Information Facility, aim to eliminate these problems by providing up-to-date, electronic catalogues of known species (Bisby et al. 2002).

In total, summing across taxa suggests that the number of known species on the planet lies at around 1.75 million (Heywood and Watson 1995; Groombridge and Jenkins 2002). (See Figure 4.9.) It has, however been shown that some of these figures are underestimates; for example, mollusks are now believed to number 100,000 known species (Peeters and Van Goethem 2003). Further, current rates of species description average 15,000 species per year (Stork 1993), less than 1% of the known total, and hence at least another 135,000 species are likely to have been described over the decade since 1995, bringing the total of known species toward 2 million (Peeters et al. 2003).

A range of techniques exist for estimating the total species richness of the planet (May 1988). These can be grouped into two main classes (Stork 1997)—methods based on ratios of known to unknown species and those based on the extrapolation of samples (see Table 4.5)—with more speculative techniques based on scaling rules between species and body size (May 1990a), specialist opinion (Gaston 1991), and community pattern (Godfray et al. 1999).

Methods based on ratios between known and unknown species have a long history but were first brought to high profile by Raven (1983). Specifically, he extrapolated the known 2:1 ratio of tropical to temperate vertebrate species to the existing 2 million known species—most of which are temperate insects—to estimate that there should be two as-yet-undescribed tropical insects for each temperate species, for a total of 3–5 million species. Stork and Gaston (1990) used similar logic (based on the percentage of British insects that are butterflies) to estimate the total numbers of insects at 4.9–6.6 million. Hodkinson and Casson (1991) extrapolated the percentage of undescribed Hemiptera in samples from Sulawesi to all insects, suggesting a total of 1.84–2.57 million species, while Hodkinson (1992) generalized this argument to suggest the number of species could be estimated at approximately 5 million, based on percentages of undescribed species in studies from the tropics.

The development of the second method—extrapolation of samples—is much more recent and was first developed by Erwin (1982). In studies of beetle species inhabiting tropical trees on Panama, he recorded high levels of both richness and local endemism. Extrapolating these figures globally, he estimated the total number of species at 30 million. His assumptions and methods have been tested and refined (Stork 1988; Hammond 1994; Ødegaard 2000; Sørensen 2003; Novotny et al 2002), and this method now suggests a lower global species richness of 4–6 million.

In general, there continues to be much debate in the literature regarding estimates of species richness, even among well-studied groups such as the extant seed plants. Lower estimates for seed plants range from 223,000 (Scotland and Wortley 2003) to 270,000 and 320,000 (May 1992; Prance et al. 2000), while higher estimates range up to 422,000 (Govaerts 2001; Bramwell 2002), although the higher figure is somewhat controversial (Thorne 2002; Scotland and Wortley 2003).

Several other particularly poorly known groups of organisms present additional problems for the estimation of global species richness (May 1995). Based on extrapolations of box-core samples from the seafloor, Grassle and Maciolek (1992) suggested a total of 10 million marine macrofaunal species; this may be rather high, but clearly enormous deep-sea species richness remains undiscovered. Likewise, the known global total of 72,000 fungi is certainly a large underestimate; based on the ratio of fungi to plants in Britain, Hawksworth (1991) estimated the global number to be closer to 1.5 million. Maybe most important, parasitic richness remains largely unknown: if the possibility that there is at least one host-specific parasite for all metazoan or vascular plant species is borne out (Toft 1986), the number of estimated species could double.

4.2.3.3 Variation in Species Richness in Time and Space

While the number of species on the planet is hard to estimate, its variability across space and time is much harder. Nearly all patterns of species richness are known with greater confidence for terrestrial than for either marine or freshwater systems. Species are unevenly distributed over Earth's surface (Rosenzweig 1995) and across phylogenetic space: species' ages and histories vary widely (May 1990b). Considerable data have recently been compiled that allow the identification of numerous patterns of variation, but these remain restricted to tiny subsets of all species, and so their general applicability remains unknown. Nevertheless, for lack of any truly comprehensive datasets, these data form the basis for the rest of this section.

For many purposes, species are not all equal—in particular those species with long independent evolutionary histories and

BOX 4.1

Species in Theory and Practice

Species concepts based on gene flow and its limits, such as the biological species concept, are not applicable to asexual taxa. They are also inadequate for "pansexual" taxa, such as some bacteria, where gene flow can be common between even very dissimilar types. However serious these concerns are in theory, they rarely matter for biodiversity assessment because the data collected on such groups are usually insufficient for the problems to emerge.

These and other issues have, however, led to a proliferation of species concepts: there are dozens in current use (Claridge et al. 1997; Mayden 1997), though most share the feature that species are independent evolutionary lineages. Most of the concepts—whether based on gene flow, ecological separation, or morphological distinctiveness—tend to give similar answers in most cases, for two reasons. First, most species have a considerable history of independent evolution—maybe millions of years—and have evolved morphological, ecological, and reproductive characters that set them apart from other species. Second, most populations within species share common ancestors with other populations in the very recent past, so they are barely differentiated at all. Borderline cases, where different criteria disagree, are relatively rare (Turner 1999).

Application of the phylogenetic species concept, however, may lead to the recognition of very many more species than when other concepts are used. A phylogenetic species is "the smallest group of organisms that is diagnosably distinct from other such clusters and within which there is a parental pattern of ancestry and descent" (Cracraft 1983); any diagnosable difference, however small, is deemed a sufficient basis for describing a new species. Taxonomic revisions that apply this concept to a taxon for the first time typically roughly double the number of species recognized (Agapow et al. 2004).

Most theoretical species concepts, like the biological one, are not very operational: they define the sort of entity a species should be but do not provide a method for delimiting them (Mayden 1997). In practice, simpler, perhaps informal decision rules are typically used to determine how many species to describe (Quicke 1993), with these rules differing among major taxa (Claridge et al. 1997). Even within a group, taxonomists lie on a continuum from "lumpers" (who recognize few species, which will consequently tend to be widespread) to "splitters" (who recognize many species, which often have restricted distributions), with obvious consequences for biodiversity assessment (Hull 1997).

The recognition that a full catalogue of the world's species is hundreds of years away, at current rates of description, has prompted initiatives to simplify the jobs of describing and defining animal and plant species (Godfray et al. 1999; Hebert et al. 2003; Tautz et al. 2003) and calls for a program to sequence DNA from all the world's biota (Wilson 2003). These initiatives are controversial and are currently only at the trial stage.

Species are the major taxonomic unit for counting biodiversity: species lists are important for both monitoring and broad-scale priority setting (Mace et al. 2003). However, species may differ in the weighting they receive, to reflect differences in their perceived biodiversity value. In addition to species of recognized economic importance, four other categories of species that might receive more weight are keystones (whose loss from a system would lead to large-scale changes in it), indicators (whose

sensitive requirements mean that their abundance reflects overall system health), flagships (charismatic species whose plight attracts publicity), and umbrellas (flagships whose conservation in situ would automatically help conserve many other species) (Meffe and Carroll 1994). More weight might also be assigned to species that are at risk of extinction, or rare, or have restricted distributions (e.g., Myers et al. 2000).

There is no consensus about exactly how any of these weights should be determined nor their relative importance. Phylogenetic information can also be considered, by weighting species or locations according to the amount of unique evolutionary history they embody (Vane-Wright et al. 1991; Faith 1992).

These ways of augmenting information in species lists may be of little use when species lists are very incomplete (Mace et al. 2003), which they can be for even well-known taxa. Then, any comparisons between regions, systems, or taxa that do not control for variation in sampling effort run the risk of serious error. The picture is even cloudier when sampling effort differences are compounded with differences in species concept. Counts of higher taxa (such as genera or families) might be more robust than species counts to sampling differences among regions, and so they may be pragmatic choices despite the loss of precision incurred (Balmford et al. 1996). Some very broad-scale comparisons among groups (bacteria versus mammals, for example) are practically meaningless because the differences in taxonomic practice are so great (Minelli 1993). Comparisons over time are hampered by the taxonomic instability that results from discovery of new species and changes in species concepts and by changing information about previously known species (Mace et al. 2003).

Because of these considerations, the interpretation of biodiversity measures based on species numbers is not always straightforward. Such measures are most likely to be useful when the taxonomy of the group is apparently almost complete (that is, few species remain to be discovered), when the sampling and taxonomic effort has been equal among the units being compared, or when sampling and effort have at least been measured in a way permitting correction for sampling biases. In addition, it is clearly important that taxonomic practice, including the choice of species concept, be reasonably consistent.

These requirements mean that species-based approaches are much more useful when applied to unusually well known taxa or well-known parts of the world (such as birds and mammals or Northern temperate regions) rather than to other taxonomic groups or less well documented systems (such as nematodes or freshwater and marine systems). The wealth of data available for the best-known groups permits very useful comparisons to be made between places in, for example, how many species there are, how many are threatened with extinction, or how many are threatened by overexploitation. However, patterns seen in a single group may be specific to that group (Prendergast et al. 1993).

Different lineages have different ecological requirements and biogeographical histories, so they naturally may have different patterns of diversity and trends: consequently, no single taxon is sure to be a good surrogate for biodiversity as a whole. If comparisons are intended to reflect overall biodiversity, they should therefore be replicated using multiple taxa wherever possible.

few surviving relatives contain irreplaceable genetic diversity. Measures of phylogenetic diversity reflect this and can sometimes be approximated by higher taxon diversity.

Global species richness maps exist for mammals (terrestrial species only) (see Figure 4.10 in Appendix A), amphibians (see Figure 4.11 in Appendix A), scleractinian corals (Veron 2000), the 239 bumblebee species of the genus *Bombus* (Williams 1998), marine finfish species across FAO region and freshwater finfish by continent (Froese and Pauly 2003) (see Figure 4.12 in Appendix A), plants (see Figure 4.13 in Appendix A) (Barthlott et al. 1999),

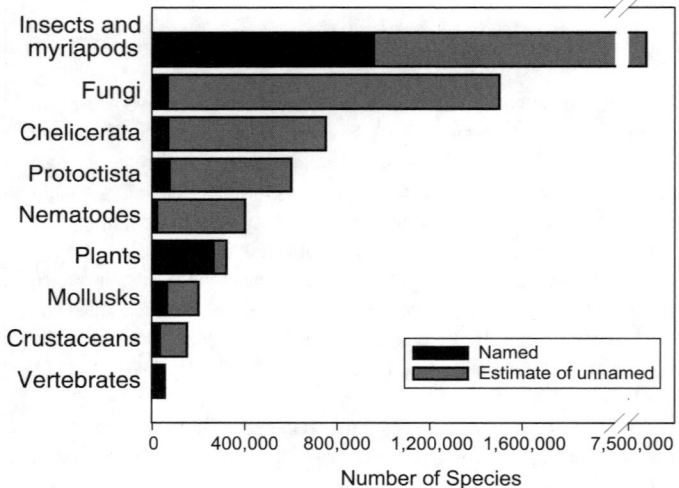

Figure 4.9. Estimates of Proportions and Numbers of Named Species and Total Numbers in Groups of Eukaryote Species (following Groombridge and Jenkins 2002)

Table 4.5. Estimates of Number of Species Worldwide

Estimate	Reference	Method
30 million	Erwin 1982	extrapolation from samples
3–5 million	Raven 1983	ratios known:unknown species
10–80 million	Stork 1988	extrapolation from samples
4.9–6.6 million	Stork and Gaston 1990	ratios known:unknown species
1.84–2.57 million	Hodkinson and Casson 1991	ratios known:unknown species
5 million	Hodkinson 1992	ratios known:unknown species
4–6 million	Novotny et al. 2002	extrapolation from samples

and freshwater fish by river basin (multimedia.wri.org/water sheds_2003/gm2.html). The lack of distributional data for invertebrates generally (in particular, for aquatic species) is clearly a major limitation on inference from these data; some regional data sets exist, but these are so heavily skewed toward north temperate regions as to have limited value in a global assessment. Another limitation of these data is their static nature: they reflect current extent of occurrence, not historical range, which can often be very different (Channell and Lomolino 2000), and they fail to reflect temporal variation within species' ranges—for example, for migratory species (Gómez de Silva Garza 1996). Further limitations come from wholesale sampling artifacts: for instance, the Congo Basin and New Guinea are particularly poorly sampled for all taxa, likely leading to an underrepresentation of species richness in these areas.

The most obvious pattern emerging from these data is that for most taxa the tropics hold much higher species richness than do the temperate, boreal, and polar regions. Figure 4.14 demonstrates this by plotting the number of species in each 5-degree latitudinal band for all terrestrial mammals, threatened birds (as global bird data are not yet available), and amphibians. As expected from the species-area relationship (Rosenzweig 1995), some of this pattern is explained by variation in landmass across

latitudinal bands. However, species richness is much higher in the tropics than would be expected based on area alone, peaking around the equator for all taxa (rather than in northern high latitudes, as would be predicted based on area alone).

The other pattern apparent from Figures 4.10–4.13 is the broadly similar distribution of diversity between taxa. Thus, for example, species richness per grid cell is tightly correlated between mammals and amphibians. Differences seem likely to be driven by particular biological traits. Birds, for example, have the ability to disperse over water more than most of the taxa mapped here, and so occur in larger numbers on islands, while ectothermic reptiles flourish in desert regions generally impoverished in other taxa. Other differences are less easily explained, such as the high richness of mammal species in East Africa and of amphibians in the Atlantic forest. In general, these differences will increase with increasing evolutionary distance (and hence often corresponding ecological differences) between taxa (Reid 1998): less correlation is expected between mammal and coral distributions, for instance, than between mammal and bird distributions.

Macroecological patterns of freshwater and marine species richness are less well understood. Diversity of pelagic predators seems to peak at intermediate latitudes (20–30° N and S), where tropical and temperate species ranges overlap (Worm et al. 2003). Several studies have documented a latitudinal gradient in the shallow-water benthos, with decreasing richness toward the poles, but data on nematodes suggest that no latitudinal trend exists (see Snelgrove 1999, and references therein). A recent global assessment of local stream insect richness found peaks in generic richness near 30–40° N latitude, though the study compared individual stream surveys rather than summing values across all latitudinal bands (Vinson and Hawkins 2003).

4.2.3.4 Geographic Centers of Endemism and Evolutionary Distinctiveness

Interacting with geographic variation in species richness is variation among species in range size. Most species have small range sizes (Gaston 1996), although there is variation within this general pattern. Among the vertebrates, the more mobile species, such as birds, tend to have large ranges, while those of less mobile species, such as amphibians, generally have much smaller ranges. (See Figure 4.15.) Nevertheless, the shape of frequency distributions of species' range sizes appears to be similar across all taxa examined to date (with the median range size consistently an order of magnitude smaller than the mean), probably because shared processes are shaping these distributions (Gaston 1998). The small range size of most species has important consequences for the conservation of biological diversity, given the widespread inverse correlation between species' range size and extinction risk (Purvis et al. 2000b).

Not only do most species have small ranges, but these narrowly distributed species tend to co-occur in "centers of endemism" (Anderson 1994). Such centers have traditionally been identified through the overlap of restricted-range species, found using threshold approaches that consider only species with distributions smaller than a given percentile or area (Hall and Moreau 1962). Among vertebrates, almost all such centers of endemism lie in isolated or topographically varied regions. This is true for both geographical isolates, such as mountains and peninsulas, and real land isolates—islands (Baillie et al. 2004). Maybe as a consequence of this, they also tend to be near the coast.

The degree to which this pattern is found for other taxa, and in particular in the aquatic realm, is unclear, but evidence from analysis of scleractinian corals and selected fish, mollusks, and lob-

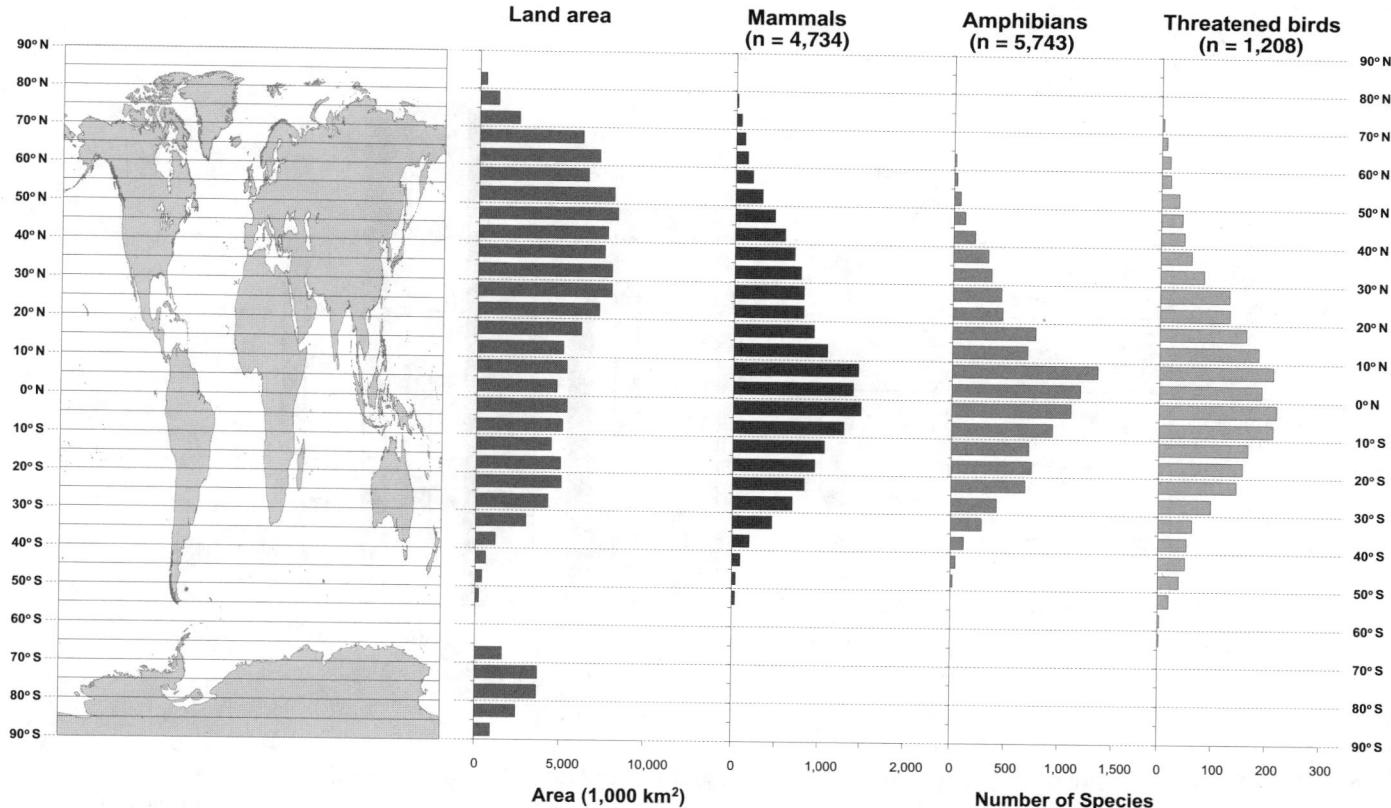

Figure 4.14. Variation in Species Richness across 5-degree Latitudinal Bands for All Mammal (Terrestrial Only), Amphibian, and Threatened Bird Species, Shown in Relation to Total Land Area per Latitudinal Band

sters suggests that coral reef centers of endemism also tend to be isolated, either by distance or by current flow (Roberts et al. 2002).

Centers of endemism are also concentrated in the tropics. Centers of endemism across birds, mammals, and amphibians tend to overlap (Baillie et al. 2004), and a broadly similar pattern is expected for plant endemism as well (WWF and IUCN 1994, 1995, 1997; Myers et al. 2000), although Mediterranean regions are more important as centers of endemism for plants than for vertebrates.

The range area and endemism patterns characteristic of the vertebrates (as well as of the plants, possibly) do not appear to represent the situation for invertebrates or microorganisms. Despite the fact that the data are extremely sparse and species have rarely been comprehensively identified locally, let alone mapped, various lines of evidence suggest that patterns of spatial turnover for these groups may be very different. While it is known that local endemism can be very high for some invertebrates in certain areas, this measure—calculated as the ratio of local to regional richness—varies widely. In Amazonia, for example, these ratios varied from about 80% for some moth species (indicating low endemism) to less than a few percent for earthworms (indicating very high endemism and spatial turnover) (Lavelle and Lapied 2003).

Species richness in soils is important for many ecosystem processes, but this habitat has been relatively poorly studied compared with aboveground systems (Fitter 2005). Microbial diversity is known to be high, though quantification at both local and global scales is limited by the technical issues of standardizing methods for defining microbial species. Richness of larger soil organism

varies: some groups appear to be locally very diverse relative to global or regional diversity. This seems to be especially the case for smaller organisms and those with high dispersal abilities (through wind and water, for instance). Currently poorly understood, species richness in soils may be best explained through a better understanding of the temporal and spatial variability of the physical properties of soil as a habitat (Fitter 2005).

More generally, it has been suggested that the extent of local endemism correlates negatively with the dispersal capabilities of the taxon. Interpreting this pattern more broadly, and using extensive inventories of free-living protists and other microbial eukaryotes in a freshwater pond and a shallow marine bay, Finlay and Fenchel (2004) suggested that most organisms smaller than 1 millimeter occur worldwide wherever their required habitats are found. This can result from almost unrestricted dispersal driven by huge population sizes and very small body size, with the consequently low probability of local extinction. Organisms larger than 10 millimeters are much less abundant and rarely cosmopolitan. In Finlay and Fenchel's data, the 1–10 millimeter size range accommodates a transition from a more-or-less cosmopolitan to a regionally restricted distribution.

More detailed studies can reveal different spatial richness patterns within taxa and in different major biomes. For example, in one study of Neotropical mammals, dryland habitats were shown to be more diverse in endemic mammalian species than were the tropical forests (Mares 1992). Marine biota reveal a similar overall decline in diversity with increasing latitude to that observed in terrestrial realms, but the strength and slope of the gradient are subject to regional, habitat, and organismal features (Hillebrand 2004). Detailed studies of the species richness of fish and inverte-

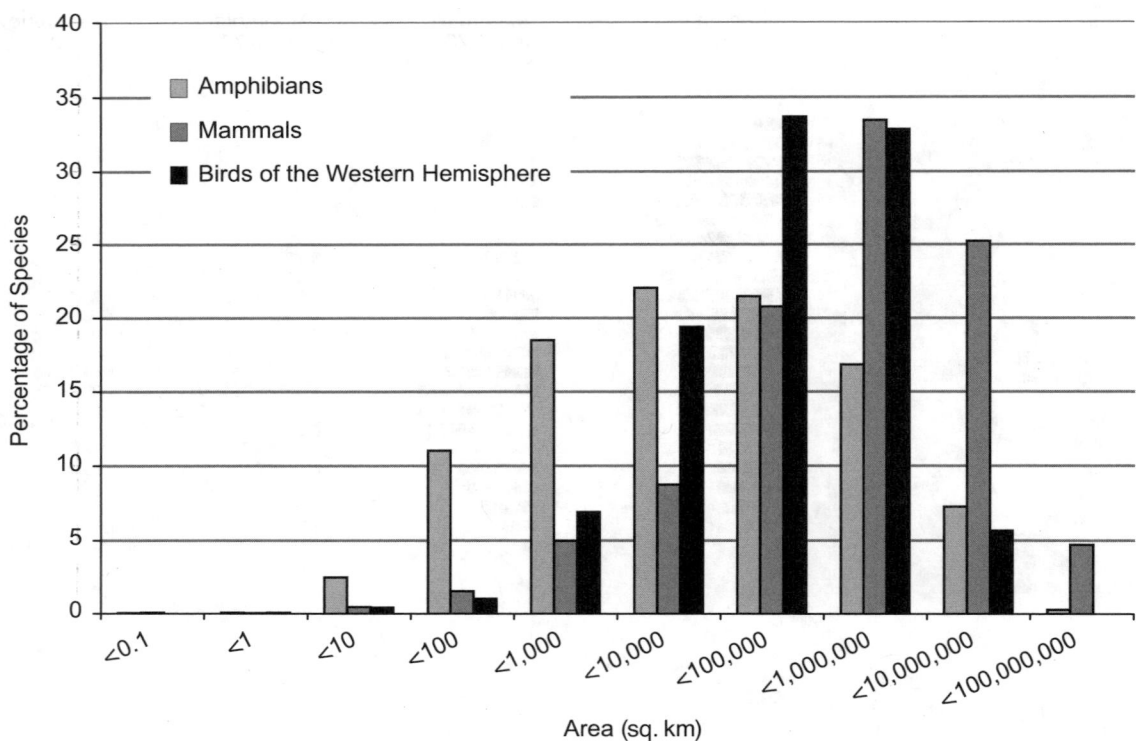

Figure 4.15. Frequency Distribution of Log10 Transformed Range Sizes for Mammal, Bird, and Amphibian Species. Mammal, terrestrial species only: n = 4,734, mean = 1.7×10^6 sq. km.; median = 2.5×10^5 sq. km.; bird, species endemic to the Western Hemisphere only: n = 3,980, mean = 2.1×10^6 sq. km., median = 4.0×10^5 sq. km.; amphibians: n = 4,141, mean = 3.2×10^5 sq. km., median = 1.3×10^4 sq. km. Data-deficient species are excluded. The log10 transformation makes the distribution look slightly left-skewed, but in fact the untransformed distribution is strongly right-skewed—that is, most species have very small range sizes.

brates in the Atlantic showed no clear trends but seemed to be related to sea-surface temperature or nitrate concentrations (Macpherson 2002).

In addition to the variability of species richness across geographic space, species richness varies over time. There is enormous variation between species in terms of their evolutionary age or the time since divergence from their closest relative (Faith 1992). Comprehensive phylogenetic data allowing evolutionary relationships to be drawn across entire species groups remain sparse. However, it is possible to use taxonomic relationships to approximate evolutionary relationships (Vane-Wright et al. 1991) in order to measure evolutionary distinctiveness among species. As with species richness, the few data that exist for terrestrial taxa indicate that tropical rainforests are regions with the greatest number of taxa with lengthy independent evolutionary history—for example, for plant families (Williams et al. 1994) and primates and carnivores (Sechrest et al. 2002). The applicability of this variation in higher taxon diversity in aquatic systems remains largely untested, however, and the massive phylum diversity in the sea (32 of 33 phyla occur in the sea, compared with just 12 on land) suggests some important differences here (May 1994).

Based on the notion that conserving global biodiversity requires preserving these spatial and temporal patterns, one recent analysis investigated the extent to which species diversity is covered by the current network of protected areas (Rodrigues et al. 2004). The analyses were based on distribution maps of 11,633 species of terrestrial vertebrates and found that at least 12% of all species analyzed do not occur in protected areas. This rises to 20% of threatened species, the loss of which would result in the disappearance of at least 38 threatened genera. (See Table 4.6.)

Most species not found in protected areas are concentrated in tropical regions, and within these in centers of endemism: mainly islands and tropical mountain areas (Rodrigues et al. 2004).

Equivalent analyses are not possible yet for hyper-rich taxa such as plants or insects. However, the results for vertebrates indicate that taxa with higher levels of endemism (smaller range sizes) are proportionally less covered by protected areas; if so, the number of plant and insect species not found in any protected areas may be higher than for terrestrial vertebrates (Rodrigues et al. 2004). Freshwater species are also likely to be poorly covered, as most currently existing protected areas were not created focusing on freshwater habitats; even when species-rich freshwater systems occur in protected areas, they are not necessarily protected. The coverage of marine richness is surely tiny, with only about 0.5% of the world's oceans covered by protected areas (Chape et al. 2003).

4.2.4 Populations

4.2.4.1 Definition and Measurement

The term population is used in many different fields and at many scales, resulting in a number of different definitions (Wells and Richmond 1995). Most definitions identify a population as a geographical entity within a species that is distinguished either ecologically or genetically (Hughes et al. 1997). The genetically based definition ("Mendelian population") is a reproductive community of sexual and cross-fertilizing individuals that share a common gene pool (Dobzhansky 1950). This is measured by assessing gene flow and genetic variation. The demographically based definition identifies populations based on groups of individuals

Table 4.6. Numbers of Gap Species and Genera of Mammals, Birds, Amphibians, and Freshwater Turtles and Tortoises in the Current Global Protected Area Network. (Adapted from Rodrigues et al. 2004) Values in parentheses are percentage of all taxa/threatened taxa analyzed within a given group. A threatened genus is one in which all species are threatened. For birds, it was only possible to evaluate gaps for threatened species and genera. Data for mammals (terrestrial species only) and threatened birds are as in Figure 4.25, and for amphibians, as in chapter 20; data for turtles based on the EMYSystem World Turtle Database 2003 (Iverson et al. 2003).

Group	Median Species Range Size (square kilometers)	Number of Species	Numbers of Gap Species	Number of Genera	Numbers of Gap Genera
Threatened and non-threatened					
Mammals	247,341	4,735	258 (5.5%)	1,091	14 (1.3%)
Birds	n.a.	9,917	n.a.	2,085	n.a.
Turtles	309,172	273	21 (7.7%)	84	2 (2.4%)
Amphibians	7,944	5,454	913 (16.7%)	445	9 (2.0%)
Threatened					
Mammals	22,902	1,063	149 (14.0%)	194	14 (7.2%)
Birds	4,015	1,171	232 (19.8%)	128	15 (11.7%)
Turtles	167,611	119	12 (10.1%)	21	0 (0%)
Amphibians	896	1,543	411 (26.6%)	65	9 (13.9%)
All taxa analyzed	38,229	11,633	1,424 (12.2%)	n.a.	n.a.

that are sufficiently isolated to have independent population dynamics (Luck et al. 2003).

For some purposes, it is useful to categorize groups of organisms that may not correspond to a Mendelian or demographic population. For example, a group of bees in a field might be a population worthy of study. A population can also be defined as a unit that is important for conservation (conservation unit), such as evolutionary significant units, (Moritz 1994; Crandall et al. 2000) or for management (management units), such as fish stocks. Populations may also be defined in relation to the services that they provide. Thus, a service-providing unit would be that section of a population that is essential for providing a specific ecosystem service (Luck et al. 2003).

The definition of population used in the MA is more general and could lead to a number of different interpretations of a population's boundary. It is "a group of individuals of the same species, occupying a defined area and usually isolated to some degree from other similar groups." Specification of the way in which the term population is being used is clearly important, given the great diversity of uses of the term.

4.2.4.2 Current Status of Population-Level Biodiversity

Populations are an important aspect of biodiversity as they are widely understandable units and are the ones most often monitored, exploited, and managed by people. Change in the status of populations provides insight into the status of genetic diversity, as the extinction of a population may represent the loss of unique genetic material. Populations are also the level at which we can best observe the relationship between biodiversity and ecosystem functioning. Most of the services provided by ecosystems require a large number of local populations (Hughes et al. 1997). For example, erosion control requires a number of different local plant populations. The loss of these local populations may have profound effects on erosion but limited impact on the overall status of the species involved. Thus it is important to focus on the condition of local populations if we are concerned with the maintenance of ecosystem processes and the provision of ecosystem services.

There are a number of ways that the condition of populations can be measured: the total number of populations in a given area, the total number of individuals within each population, the geographic distribution of populations, and the genetic diversity within a population or across populations (Luck et al. 2003). The most common measures are assessments of the distribution and abundance.

Populations are dynamic and are continually changing due to variation in births and deaths, immigration, and emigration. At any one time a species will likely have some populations that are increasing while others are decreasing, and it may be going extinct. A species can have many different structures, ranging from one continuous population of individuals, to disjunct populations of individuals with some exchange of individuals among them (known as a metapopulation) (Wells and Richmond 1995) and to disjunct populations that are completely isolated. Although there is great variation in abundance and distribution, the majority of species have small distributions (see Hughes et al. 1997) and therefore small populations. Small numbers of individuals or limited distributions result in such populations being more susceptible to extinction due to stochastic events (Gilpin and Soulé 1986; Lande 1993) such as a hurricane or fire, random demographic effects (Richter-Dyn and Goel 1972; Goodman 1987), the potential negative effects of limited genetic variability (Soulé 1980); or simply because a threat process such as habitat loss, exploitation, or introduced species is more likely to drive to extinction a species that is restricted in distribution or composed of few individuals.

Given the magnitude of populations, it is little surprise that there are few comprehensive global datasets. One example is the global inventory of population estimates and trends for waterbirds maintained since 1996 by Wetlands International. The most recent (third) edition (Wetlands International 2002) listed 2,271 biogeographic populations of 868 species of waterbirds. Other organizations, such as IUCN–the World Conservation Union, BirdLife International, NatureServe, UNEP World Conservation Monitoring Centre, FAO, and the European Nature Information System, collect data on species distributions and in some cases populations.

But the quality of the population data remains poor, and where data do exist the species tend to be either commercially important (such as fish stocks), charismatic (such as tigers and elephants), or threatened with extinction. There is also a significant regional bias, with the least data available in regions such as the tropics, where population numbers are likely the highest. Another useful source of data for trends on populations is the Global Population Dynamics Database (NERC 1999), with 5,000 separate time series available, ranging from annual counts of mammals or birds at individual sampling sites to weekly counts of zooplankton and other marine fauna.

Despite these limitations, population-level information is extremely useful for a range of applications for assessments of biodiversity and ecosystem services.

4.2.5 Genes and Genomes

Genes are sequences of nucleotides in a particular segment (locus) of a DNA molecule. Slightly different sequences (alleles) at a locus may result in protein variants differing in amino acid sequence, which may have different biochemical properties and thus cause phenotypic differences in morphology, physiology or the behavior of individuals. The allele that causes sickle-cell anemia in humans, for example, is the result of a single nucleotide substitution (adenine replaced by guanine) in the second position of the sixth codon of the beta-globin gene.

The complete genetic material of a species constitutes its genome. Eukaryotic genomes are organized into discrete longitudinal bodies in the nucleus, called chromosomes. The number, size, and shape of chromosomes within species are usually constant, but often differ between species. The human genome has 46 chromosomes and about 3.2 billion nucleotides, containing about 30,000 to 40,000 genes.

Biodiversity at the within-species level is usually measured by genetic diversity, which refers to the variety of alleles and allele combinations (genotypes) present in a species. Genetic diversity is reflected in the differences among individuals for many characters, from DNA sequences and proteins to behavioral and morphological traits such as eye, skin, and hair color in humans. This diversity allows populations to evolve by means of changing relative frequency of different alleles to cope with environmental changes, including new diseases, pests, parasites, competitors and predators, pollution, and global change. Naturally outbreeding species with large populations usually possess large stores of genetic diversity that confer differences among individuals in their responses to any environmental change.

Numerous species have been observed to evolve in response to environmental change as a result of genetic diversity. For example, industrial melanism has evolved in about 200 species of moths in areas subject to industrial pollution (Majerus 1998), and resistance to insecticides, herbicides, antibiotics, and other biocontrol agents has evolved in numerous "pest" species (McKenzie 1996).

The plentiful genetic diversity in many plant and animal species has been exploited extensively by humans through artificial selection to generate numerous breeds specialized in providing various service products such as meat, milk, eggs, fiber, guidance, hunting, companion, and aesthetics. (See also Chapter 10 for a discussion of genetic bioprospecting.) In contrast, species lacking genetic diversity usually have difficulty adapting to environmental changes and face increased risk of extinction because any environmental change that harms one individual is likely to harm other individuals to a similar extent. It has been demonstrated that inbred populations lacking genetic diversity have lower fitness and are less adaptable to new environmental challenges than the outbred populations they are derived from (Reed et al. 2003).

Genetic diversity is also important in maintaining the reproductive and survival ability (reproductive fitness) of individuals in outbreeding species even in a stable environment. In naturally outbreeding species, loss of genetic diversity usually leads to the homogeneity within individuals and thus reduced reproductive fitness (inbreeding depression) and increased risk of extinction. The U.S. endangered Florida panther, a subspecies restricted to a small relic population of approximately 60–70 individuals in southern Florida, has very low levels of genetic diversity revealed by different genetic markers. As a result, Florida panthers suffer from inbreeding depression evidenced by an extraordinarily high frequency of morphological abnormalities ("cow lick" patterns in their fur and kinked tails), cardiac defects, undescended testis, and poor semen quality (Roelke et al. 1993).

Inbreeding depression, interacting with environmental and demographic stochasticity, is believed to contribute to the extinction of populations (Saccheri et al. 1998). In many inbred species and populations, the effects of inbreeding cease to be a problem, probably because most mutations deleterious under inbreeding become selectively removed, and the populations that survive are those that no longer possess such alleles. However, usually numerous populations become extinct and only a very small fraction survive this inbreeding and selection process (Frankham 1995).

A variety of methods can be used to measure genetic diversity. (See Box 4.2.)

Generally, plenty of genetic variation can be found within an outbreeding species at various organization levels, within individuals, between individuals within a population, and between populations. From a functional point of view, genetic variation can be classified as neutral and adaptive. The rich neutral genetic diversity is (arguably) revealed by using various molecular markers. In a typical large outbreeding species, about 80% of microsatellite loci are polymorphic, which have on average 5~10 distinctive alleles and heterozygosities of 0.6~0.8 (Frankham et al. 2002). The adaptive variation is also abundant within various species, although more difficult to identify and quantify than neutral variation. A study on the plant of white clover (*Trifolium repens*), a stoloniferous perennial species, provides a good example (Dirzo and Raven 2003). Individual plants taken from a population growing in a 1-hectare field in North Wales were screened for those genes associated with different characters of known adaptive importance. Among 50 clones selected from the field, all but a few differed in the combinations of genes affecting their fitness in nature.

The current magnitude and distribution of genetic diversity within a species depends on the effects and interactions of several evolutionary forces (such as mutation, selection, migration, and genetic drift) over the long evolutionary history of the species. Mutations are sudden changes in the nucleotide sequence of genes or the organization of genes in a genome and are the ultimate source of new genetic variation. Migration is the exchange of genes between populations. It changes the distribution of genetic variation directly and its magnitude indirectly when interacting with other evolutionary forces. Selection is the nonrandom transmission of alleles or allele combinations between generations, depending on their adaptive values in a given environment. It acts to either maintain or deplete genetic variation, depending on the way it operates. Genetic drift refers to the random changes in allele frequency over time due to sampling (reproduction and survival) in a genetically small population. It usually reduces genetic variation.

BOX 4.2
Measuring Genetic Diversity

Like biodiversity at other levels, genetic diversity within a species can be measured in many different ways. A simple measurement is the proportion of polymorphic loci among all loci sampled. A locus is regarded as polymorphic if two or more alleles coexist in the population and the most frequent allele has a frequency smaller than 99%. The proportions of polymorphic loci for proteins revealed by electrophoresis are about 30% in mammalian species.

Genetic diversity is measured more appropriately by allelic diversity (the average number of alleles per locus) and gene diversity (average heterozygosity across loci). These measures are not suitable for DNA sequences, however, because the extent of genetic variation at the DNA level is generally quite extensive. When long DNA sequences are considered, each sequence in the sample may be different from the other sequences. In such cases, these measures cannot discriminate among different loci or populations and are therefore no longer informative about genetic diversity. More appropriate measures of genetic diversity for DNA sequences are the average number of nucleotide differences between two homologous sequences randomly chosen from a population and the number of segregating nucleotide sites in a sample of sequences.

In practice, the genetic diversity of a population is assessed by sampling a number of individuals, genotyping them at some marker loci, and calculating one or more of the diversity measurements. Various markers, including enzymes and other proteins, microsatellites (simple sequence repeats or short tandom repeats), RAPD (random amplified polymorphic DNA), AFLP (amplified fragment length polymorphism), RFLP (restriction fragment length polymorphism), SNP (single nucleotide polymorphism), and DNA sequences, can be assayed to assess the genetic diversity of a population.

However, caution should be exercised in measuring and comparing the genetic diversity between different populations. First, any measurement of genetic diversity suffers from sampling errors. To obtain a reliable estimate, a large number of individuals should be sampled and genotyped at a large number of marker loci. Second, different measures of genetic diversity cannot be compared directly. Gene diversity, for example, is de-termined by not only allelic diversity but also allele frequencies. Third, different kinds of markers usually show different degrees of diversity in a population. Genetic diversity detected by microsatellites is typically much greater than that by proteins. In large outbreeding species, the number of alleles and heterozygosity per polymorphic locus is typically 5–10 and 0.6–0.8, respectively, for microsatellites but around 2 and 0.3, respectively, for proteins. When comparing the genetic diversity among populations, the same diversity measurement should be calculated for the same set of markers assayed from large samples.

A species is usually not homogenous genetically, and the genetic diversity within it can be partitioned at different hierarchic levels, between populations, between individuals within a population, and within individuals (for nonhaploids). Usually Wright's F statistics (Wright 1969) are used to describe the hierarchical genetic structure of a species. When the observed and expected heterozygosity averaged across populations of a species are denoted by H_I and H_S, respectively, and the expected heterozygosity for the entire species is denoted by H_T, the F statistics can be expressed as $F_{IS} = 1 - (H_I / H_S)$, $F_{ST} = 1 - (H_S / H_T)$, and $F_{IT} = 1 - (H_I / H_T)$ for a diallelic locus (Nei 1987). The heterozygosity (H_I, H_S, and H_T) and thus F statistics can be determined from various genetic markers. F_{IS} indicates the reduction of within-individual diversity relative to within-population diversity, and is determined mainly by the mating system (such as selfing, random mating) of the species. F_{ST} measures the between-population diversity as a proportion of the total diversity of the entire species. It is determined by the balance between the homogenizing force of migration among populations and the opposing force of local drift within populations. Habitat fragmentation may lead to excessive inbreeding within and differentiation between populations in the short term, and to extinction or speciation in the long term. F_{IT} indicates the reduction of within-individual diversity relative to the total diversity of the species, and is determined by both the mating system and the subdivision (isolation) of the species. The relationship of the three measures is $(1 - F_{IT}) = (1 - F_{IS})(1 - F_{ST})$.

Despite the well-established theory concerning the genetic structure of populations, empirical data are mostly limited to a relatively restricted set of species and situations, most commonly related to agriculture. Even less common are continuing assessments over time and space that would allow inferences about the large-scale and long-term trends in genetic diversity.

The genetic diversity harbored within a population or species varies greatly among loci, depending on the mutation and selection forces acting on them. Proteins, for example, generally have much less genetic variation due to their functional (selective) constraints and low mutation rate than molecular markers (such as microsatellites). For protein variation as assessed by electrophoresis, only about 28% loci are polymorphic and 7% loci are heterozygous in an average individual, both being much smaller than those for microsatellites (Frankham et al. 2002).

Genetic diversity is reduced at loci subject to directional selection and increased at loci under balancing selection, compared with that of neutral loci. For example, the major histocompatibility complex loci are involved in fighting disease, combating cancer, and controlling transplant acceptance or rejection and are thus believed to be under balancing selection. The MHC contains over 100 loci falling into three main groups, termed class I, II, and III. In vertebrates, MHC loci exhibit the highest polymorphism of all known functional loci. The human MHC (called HLA), for example, have 67, 149, and 35 alleles at the class I HLA-A, HLA-B, and HLA-C loci and 69, 29, and 179 at the class II DPB, DQB, and DRB loci, respectively (Hedrick and Kim 2000).

The amount of diversity also depends on the effective size (*Ne*) of a population, defined as the size of the idealized Wright-Fisher population (a diploid monoecious species with random mating including self-fertilization, with constant size and discrete generations and with an equal probability of reproduction and survival among individuals) that would give rise to the variance of change in gene frequency or the rate of inbreeding observed in the actual population under consideration. In populations with small to intermediate values of *Ne*, most loci are effectively neutral and their genetic diversity is predominantly determined by genetic drift and is lost at a rate of $1/2$ *Ne* per generation. Therefore large populations tend to have higher genetic diversity than small populations.

Reductions in the size of large populations will have major consequences for their diversity, even if the reduction is only for a short period. Hence, populations fluctuating in sizes tend to have less diversity than might be expected from their average size. Most endangered species and populations are found to possess lower genetic diversity than related, nonendangered species with large population sizes. Of 38 endangered mammals, birds, fish, insects, and plants, 32 had lower genetic diversity than related nonendangered species (Frankham 1995). A survey of allozyme genetic diversity in major taxa showed that the average heterozygosity within species is lower in vertebrates (6.4%) than in invertebrates (11.2%) or plants (23%), possibly due to the usually smaller population sizes in the former (Ward et al. 1992).

Local adaptation shapes the distribution of genetic diversity at selected loci among populations and geographic regions. A good example is the human sickle cell anemia allele, whose distribution (in Africa, the Mediterranean, and Asia) coincides with that of malaria. More variation is found between populations for loci that confer adaptations to local conditions. The distribution of diversity also depends on population structure and mating system. Species capable of long-range migration (such as flying birds and insects) tend to have small geographic intraspecific variation.

4.3 Anthropogenic Drivers

In the past, major changes to the world's biota appear to have been driven largely by processes extrinsic to life itself, such as climate change, tectonic movements leading to continental interchange, and even extra-terrestrial events in the case of the late Tertiary changes. (See Chapter 3.) While these processes remain important, current changes in biodiversity result primarily from processes intrinsic to life on Earth, and almost exclusively from human activities—rapid climate change, land use change, exploitation, pollution, pathogens, the introduction of alien species, and so on. These processes are known as anthropogenic direct drivers.

Having provided an overview of the current status of global diversity in the preceding sections, the current processes leading to change are considered here. Although the interactions are complex and often synergistic, it is important to distinguish among the main causes of biodiversity loss in order to identify, propose, and implement effective response strategies. The most important direct impacts on biodiversity are habitat destruction (Bawa and Dayanandan 1997; Laurance et al. 2001; Tilman et al. 2001), the introduction of alien species (Everett 2000; Levine 2000), overexploitation (Pauly et al. 2002; Hutchings and Reynolds 2004), disease (Daszak et al. 2001), pollution (Baillie et al. 2004), and climate change (Parmesan et al. 1999; McLaughlin et al. 2002; Walther et al. 2002; Thomas et al. 2004a, 2004b).

In order to provide information on existing linkages between anthropogenic drivers of change in species richness patterns and the rate and nature of such changes, indices for such linkages based on the most prominent anthropogenic drivers have been calculated based on expert knowledge. (See Figure 4.16 in Appendix A.) Although subjective, these indices are the best information currently available. Their aim is not to provide exact information on the existing trends between anthropogenic drivers and biodiversity patterns but rather to provide a general overview of such trends.

The figure indicates that habitat change is presently the most pervasive anthropogenic driver, with habitat fragmentation, introduced alien species, and exploitation being the next most common drivers. Threats such as disease, pollution, and climate change are identified as having slightly less impact, but it should be noted that these estimates are based on a projection until 2010. Threats such as disease (Baillie et al. 2004) and climate change will likely play a much greater role in the near future (Thomas et al. 2004a, 2004b). Where trend estimates have been made, all the main direct drivers are expected to increase in intensity. The various drivers have also been rated by the extent to which the process is believed to be reversible. Climate change and the introduction of

invasive alien species are highlighted as the two drivers that are most difficult to reverse. Certainty of these estimates is highest for the most common drivers and interactions at the species, population, and biome level and lowest at the genetic level.

4.3.1 Habitat Change, Loss, and Degradation

The land use requirements of a large and growing human population have led to very high levels of conversion of natural habitat. Loss of habitat area through clearing or degradation is currently the primary cause of range declines in species and populations. When areas of high human activity and significant human land transformation (Easterling et al. 2001; Harcourt et al. 2001) are spatially congruent with areas of high species richness or endemism (Balmford and Long 1994; Fjeldså and Rahbek 1998; Freitag et al. 1998; Ceballos and Ehrlich 2002), the negative implications for biodiversity are greatly exacerbated.

Agricultural land is expanding in about 70% of countries, declining in 25%, and roughly static in 5% (FAO 2003). Forest cover alone is estimated to have been reduced by approximately 40% in historical times (FAO 1997). This decline continues, with about 14.6 million hectares of forest destroyed each year in the 1990s, resulting in a 4.2% loss of natural forest during this time period (FAO 2000b). Other habitats types have experienced even greater change in historical times, such as tropical, sub-tropical, and temperate grasslands, savannas, and shrublands as well as flooded grasslands. (Habitat change is described further later in this chapter.)

A major issue in habitat and land use change is habitat fragmentation, which has severe consequences for many species. Fragmentation is caused by natural disturbance (fires or wind, for instance) or by human-driven land use change and habitat loss, such as the clearing of natural vegetation for agriculture or road construction, which leads previously continuous habitats to become divided. Larger remnants, and remnants that are close to other remnants, are less affected by fragmentation. Small fragments of habitat can only support small species populations, which therefore tend to be vulnerable to extinction. Moreover, small fragments of habitat may have altered interior habitat. Habitat along the edge of a fragment has a different climate and favors different species to the interior. Small fragments are therefore unfavorable for species that require interior habitat. Fragmentation affects all biomes, including, in particular, forests. Globally, over half of the temperate broadleaf and mixed forest biome and nearly one quarter of the tropical rain forest biome have been fragmented or removed by humans, as opposed to only 4% of the boreal forest. Overall, Europe has faced the most human-caused fragmentation and South America has the least (Wade et al. 2003).

Species that disappear most quickly from fragmented terrestrial landscapes often have large area requirements and are primary-habitat specialists that avoid the modified habitats (Tilman et al. 1994; Laurance et al. 2001). Some species are also particularly vulnerable to so-called edge effects, where the area of land at the edge of the habitat patch is altered and less suitable for the species (Woodroffe and Ginsberg 1998). Species that are specialized to particular habitats and those with poor dispersal ability suffer from fragmentation more than generalist species with good dispersal ability. Species with naturally unstable populations may also be intrinsically vulnerable to fragmentation, presumably because their fluctuating populations are likely to fall below some critical threshold. Likewise, organisms with low rates of population growth may be less likely to recover from population declines and suffer a greater loss of genetic diversity (via genetic drift and inbreeding) during population bottlenecks.

River fragmentation, which is the interruption of a river's natural flow by dams, inter-basin transfers, or water withdrawal, is an indicator of the degree that rivers have been modified by humans (Ward and Stanford 1989). An analysis of river fragmentation and flow regulation (Revenga et al. 2000) assessing 227 large river systems around the world, with the exception of South Asia and Australia, shows that 60% of the world's large rivers are highly or moderately fragmented. Waterfalls, rapids, riparian vegetation, and wetlands are some of the habitats that disappear when rivers are regulated or impounded (Dynesius 1994).

Fragmentation has also affected 90% of the water volume in these rivers. All river systems with parts of their basins in arid areas or that have internal drainage systems are highly fragmented. The only remaining large free-flowing rivers in the world are found in the tundra regions of North America and Russia and in smaller coastal basins in Africa and Latin America. (See Chapter 20.)

Even though dam construction has greatly slowed in many industrial countries (and some countries, such as the United States, are even decommissioning a few dams), the demand and untapped potential for dams is still high in many parts of the world, particularly in Asia, Latin America, and Turkey. As of 2003, around 1,500 dams over 60 meters are planned or under construction around the world (WWF and WRI 2004). The basins with the largest number of dams planned or under construction include the Yangtze River in China with 46 dams, La Plata basin in Argentina with 27 dams, and the Tigris and Euphrates basin with 26 (WWF and WRI 2004).

While many species disappear or decline in fragmented habitats, others can increase dramatically. Species that favor habitat edges or disturbed habitats, that readily tolerate the surrounding matrix, or whose predators or competitors have declined often become more abundant after fragmentation (Laurance et al. 2001). For instance, common species that adapt well to standing water habitats often replace stream-adapted species in river systems with many dams. In addition, the matrix commonly supports abundant populations of exotic weeds or generalist animals that can invade habitat fragments.

4.3.2 Invasive Alien Species

Humans have been responsible for introducing animals and plants to new areas for thousands of years (Milberg and Tyrberg 1993). With improvements in transportation and the globalization of trade, however, the introduction of non-native species to new habitats or ecosystems has greatly increased (e.g., Gaston et al. 2003). Most introductions fail (Mack et al. 2000), but when they are successful and become established as invasive alien species—defined as those species introduced outside their normal area of distribution whose establishment and spread modify ecosystems, habitats, or species, with or without economic or environmental harm—they can have a major impact on native biodiversity. Invasive alien species may threaten native species as direct predators or competitors, as vectors of disease, or by modifying the habitat (for example, the impact of herbivores on plant communities) or altering native species dynamics.

The causes of introductions are many. Some are intentional (a species released for hunting or introduced as a biological control, for example), but more commonly they are unintentional (introduced with traded goods such as lumber, for instance, or in the ballast water of ships or through the pet trade). Although species that have recently extended their native range or have experienced major changes in species dynamics within their native range are not considered as alien invasive species, the negative impact on other aspects of biodiversity can be just as serious.

Homogenization is partially a consequence of invasive alien species, along with the extirpation of native endemic species and habitat alterations (Rahel 2002). For example, European settlers introduced fish into North America for sport and for food, and fish faunas across the continental United States have become more similar through time. On average, pairs of states have 15.4 more species in common now than before European settlement of North America. The 89 pairs of states that formerly had no species in common now share an average of 25.2 species. Introductions have played a larger role than extirpations of local endemic species in homogenizing fish faunas (Rahel 2000). At the same time, North American fish species (such as the rainbow trout) have become established in Europe, leading to further homogenization of fish faunas between Europe and North America.

Invasive alien species have been a major cause of extinction, especially on islands and in freshwater habitat. In the latter, the introduction of alien species is the second leading cause of species extinction (Hill et al. 1997; Harrison and Stiassny 1999), and on islands it compares with habitat destruction as the lead cause of extinction over the past 20 years (Baillie et al. 2004). Islands such as Guam (Fritts and Rodda 1998; Wiles et al. 2003), Hawaii (Atkinson et al. 2000), New Zealand (Atkinson and Cameron 1993), and the Mascarenes (Cheke 1987) provide clear examples of the devastating influence invasive alien species continue to have on native biodiversity. Awareness about the importance of stemming the tide of invasive alien species is increasing, but effective implementation of preventative measures is lacking (Simberloff 2000). The rate of introductions continues to be extremely high; for example, in New Zealand plant introductions alone have occurred at a rate of 11 species per year since European settlement in 1840 (Atkinson and Cameron 1993).

The water hyacinth (*Eichhorina crassipes)* and the European zebra mussel (*Dreissena polymorpha*) are just two examples of the many alien species that have significantly altered the ecosystems in which they have successfully invaded, with major implications for native biodiversity as well as economic ramifications.

The water hyacinth has had negative effects on fisheries, hydroelectric production, agriculture, human health, and economies across the tropics. A native to the Amazon basin, it has invaded more than 50 countries on five continents, sometimes taking over entire river and lake systems (Barrett 1989). It was introduced both intentionally and unintentionally, specifically to help purify water from waste treatment facilities and for use as an ornamental aquarium plant. Lake Victoria, which borders Uganda, Tanzania, and Kenya, is the most dramatic example of the havoc water hyacinth can wreak on an ecosystem. First sighted in 1989, water hyacinth now covers 90% of Lake Victoria's shoreline. This thick mat of water hyacinth competes with the native plants, fish, and frogs for oxygen, often causing asphyxiation and massive die-offs (see www.state.gov/g/oes/ocns/inv/cs/2299.htm) and costing local economies millions of dollars (McNeely 1996).

In 1988 the European zebra mussel was transported to Lake St. Clair (in the United States and Canada) in the ballast water of a transatlantic freighter. Within 10 years the mussel had spread to all five neighboring great lakes (USGS 2000). The mussels form massive colonies and tend to clog underwater structures, such as intake pipes for power plants and other industrial infrastructure. Their efficiency at filtering the water and removing alga and microorganisms has greatly increased clarity and also resulted in reduced food availability for larval fish and many invertebrates, which could cause a shift in fish species composition (Griffiths 1993). The mussel has also greatly reduced the population of native mussels (Masteller and Schloesser 1991). The economic costs of these alien mussels for U.S. and Canadian water users has been estimated by the U.S. Fish and Wildlife Services at about $5 billion over the next 10 years (USGS 2000).

4.3.3 Introduced Pathogens

As with alien species, the process of globalization, with increased international travel and commerce, has greatly facilitated the spread of pathogens. This process has been further assisted by an increase in the conditions under which pathogens thrive, such as very high population densities in domestic plants or animals, or species living in suboptimal conditions due to rapid environmental change. As these processes intensify, newly emerging diseases may become an even greater threat to species (Daszak et al. 2001). When diseases become established in a population, initial declines may be followed by chronic population depression, which in turn increases the population's vulnerability to extinction. In some cases pathogens can cause catastrophic depopulation of the naïve host species and even extinction (Daszak et al. 2000).

Parallels between human and wildlife emerging infectious diseases extend to early human colonization of the globe and the dissemination of exotic pathogens. For instance, the impacts within Africa of rinderpest were severe. Transmitted by a highly pathogenic morbillivirus, enzootic to Asia, the disease was introduced into Africa in 1889. It wiped out more than 90% of the Kenya's buffalo population and had secondary effects on predator populations and local extinctions of the tsetse fly (Daszak et al. 2000). It also had serious consequences for the human population, leading to famine and subsequently the spread of tsetse. (See Chapter 14 for more on human infectious disease agents.)

Over the last decade, a number of pathogens introduced directly or indirectly by human activities have caused large-scale declines in several wildlife species (Dobson and Foufopoulos 2001). One example is the 20% decline of the lion population in the Serengeti, Tanzania (Roelke-Parker et al. 1996). The epidemic was caused by the canine distemper virus, transmitted to the wild carnivores from domestic dogs introduced by the local communities surrounding the park. African wild dogs are also believed to have been affected by this virus. Their local extinction from the Serengeti in 1991 was concurrent with epizootic canine distemper in sympatric domestic dogs (Roelke-Parker et al. 1996). More surprisingly, canine distemper has also spread from the terrestrial to aquatic habitats. A canine distemper virus infection has caused mortality in seals on a number of occasions in the former Soviet Republics (Stone 2000).

Infectious disease is currently a serious problem in aquaculture, not only to the fish being farmed but to wild populations as well. When infected farmed fish escape from aquaculture facilities, they can transmit these diseases and parasites to wild stocks, creating further pressure on them. For instance, infectious salmon anemia, a deadly disease affecting Atlantic salmon, poses a serious threat to the salmon farming industry. It was first detected in Norwegian salmon farms in 1984, from which it is believed to have spread to other areas, being detected in Canadian salmon (1996), in Scotland (1999), and in U.S. farms (2001) (Doubleday 2001; Goldburg et al. 2001). Norwegian field studies observed that wild salmon often become heavily infected with sea lice (parasites that eat salmon flesh) while migrating through coastal waters, with the highest infection levels occurring in salmon-farming areas (Goldburg et al. 2001).

Introduced diseases have been implicated in the local extinction of a number of species and the global (species) extinction of seven amphibians, three birds, and one plant over the past 20 years (Baillie et al. 2004). However, the first proven example of extinction by infection occurred when a microsporidian parasite killed the captive remnant population of the Polynesian tree snail, *Partula turgida* (Daszak et al. 2000). The actual number of amphibians that have gone extinct due to disease is almost certainly much higher than seven species, as 122 species have been identified as "possibly extinct" (not formally "extinct" until extensive surveys to establish their disappearance have been completed), with 113 having disappeared since 1980. The explanation for this rapid decline is not well understood, but disease and climate change are the most commonly cited reasons (Stuart et al. 2004). In 1998, a previously unknown chytrid fungus named *Batrachochytrium dendrobatidis* was discovered and is believed to be a major cause of amphibian decline (Berger et al. 1998; Longcore et al. 1999).

4.3.4 Overexploitation

People have exploited wildlife throughout history, and even in ancient times the extinction of some species was caused through unsustainable harvesting levels. However, exploitation pressures have increased with the growing human population. Although sustainable exploitation of many species is theoretically achievable, many factors conspire to make it hard to achieve in practice, and overexploitation remains a serious threat to many species and populations. Among the most commonly overexploited species or groups of species are marine fish and invertebrates (FAO 2000a, see section 5.5.1.5), trees, animals hunted for bushmeat, and plants and animals harvested for the medicinal and pet trade (IIED and Traffic 2002; TRAFFIC 2002).

Most industrial fisheries are either fully or overexploited (FAO 2000a), as documented later in this chapter. An increasing number of studies are highlighting the inherent vulnerability of marine species to overexploitation (Hoenig and Gruber 1990; Griffiths 1993; Huntsman et al. 1999; Reynolds et al. 2001; Dulvy et al. 2003). Particularly susceptible species tend to be both valuable and relatively easy to catch as well as having relatively "slow" life history strategies (Reynolds et al. 2002). Thus species such as large groupers, croakers, sharks, and skates are particularly vulnerable (Baillie et al. 2004). Although the response of species and ecosystems to severe depletions is extremely complex (Jackson et al. 2001; Hutchings and Reynolds 2004), there is increasing evidence that many marine populations do not recover from severe depletion, even when fishing has stopped (Hutchings 2000; Baillie et al. 2004; Hutchings and Reynolds 2004). (See Chapter 18 for more on exploitation of marine fisheries.)

Many of the current concerns with overexploitation of bushmeat—wild meat taken from the forests by local people for income or subsistence—are similar to those of fisheries, where sustainable levels of exploitation remain poorly understood and where the offtake is difficult to effectively manage. Although the true extent of exploitations is poorly known, it is clear that rates of offtake are extremely high in the tropical forest throughout the world (Anstey 1991; Robinson and Redford 1991; Bennett et al. 2000; FitzGibbon et al. 2000). Unsustainable levels of hunting are believed to be of great concern for a large number of target species, many of which are extremely high profile, such as gorillas, chimpanzees, and elephants. The loss of species or populations due to exploitation will not only have ecological implications, it will greatly affect the food security and livelihoods of the communities that depend on these resources.

The trade in wild plants and animals and their derivatives is poorly documented but is estimated at nearly $160 billion (IIED and Traffic 2002). It ranges from live animals for the food and pet trade (such as parrots, tropical fish, and turtles) to ornamental plants and timber (such as rattan, orchids, and mahogany). An array of wildlife products and derivatives, such as food, exotic leather goods, musical instruments, and even medicines, can be found in markets around the world.

Because the trade in wild animals and plants crosses borders between countries, the effort to regulate it requires international cooperation to safeguard certain species from overexploitation. The Convention on International Trade in Endangered Species of Wild Fauna and Flora is an international governmental agreement aimed at ensuring that international commercial trade in species of wild animals and plants does not threaten their survival. Today CITES provides varying degrees of protection to more than 30,000 species of animals and plants, whether they are traded as live specimens, fur coats, or dried herbs. CITES only applies to international trade, leaving most of the national trade in wild species poorly regulated and monitored in many countries.

In freshwater systems, trade in wild plants and animals is seriously threatening some species. Three quarters of Asia's freshwater turtles, for instance, are listed as threatened, many due to increase in trade. For example, on average there are over 30 tons per year of all imported turtle shells into Taiwan alone. The total trade may add up to several times this amount (TRAFFIC 2002).

4.3.5 Climate Change

The detectable impact of human actions on the rate and direction of global environmental change is already being felt on global biodiversity. Modern climate change may have been a contributing factor in the extinction of at least one species, the golden toad (*Bufo periglenes*) (Pounds et al. 1999), and present evidence suggests strong and persistent effects of such change on both plants and animals, evidenced by substantial changes to the phenology and distribution of many taxa (Parmesan and Yohe 2003; Root et al. 2003). For example, there have been substantial advances in the dates of bird nesting, budburst, and migrant arrivals across the Holarctic, and in the same region both birds and butterflies have shown considerable northward range expansions (Parmesan et al. 1999; Walther et al. 2002). Climate change is not likely to affect all species similarly. Certain species or communities will be more prone to extinction than others due to the direct or underlying effects of such change, and risk of extinction will increase especially for those species that are already vulnerable. Vulnerable species often have one or more of the following features: limited climatic ranges, restricted habitat requirements, reduced mobility, or isolated or small populations.

Best estimates suggest that present climate change trends will continue (Watson 2002) and that these changes will have substantial impacts on biodiversity, with some scenarios indicating that as many as 30% of species will be lost as a consequence of such change (Thomas et al. 2004a). Although past climate variation may not have caused many extinctions (Huntley and Webb 1989; Roy et al. 1996), modern change is likely to have a considerably greater effect owing to interactions between rapid climate change and substantial anthropogenic habitat destruction and alteration (Hill et al. 1999; Sala et al. 2000; Warren et al. 2001; Walther et al. 2002). See Chapter 3 of this volume and Chapter 7 of the *Scenarios* volume for more information on climate change and other drivers.

4.3.6 Changing Threat Processes over Time

An examination of bird extinctions over the past 500 years identifies introduced species as the main cause of bird extinction, followed by exploitation and then habitat loss (Baillie et al. 2004). However, dominant drivers attributed to currently threatened birds highlight habitat loss as the greatest threat, followed by exploitation and, last, introduced species (Baillie et al. 2004; Bird-

Life 2004b). This shift in dominant drivers of bird extinction can be explained by the rapid increase in habitat destruction over the last century. This, combined with other threat processes, has resulted in a greater number of mainland bird species becoming threatened with extinction (see BirdLife 2004b).

Just as habitat change has replaced introduced species as the dominant cause of extinction for birds, the dominant driver could easily change again in the near future. For example, climate change could become the dominant cause of extinction (Thomas et al. 2004a). However, as the main drivers of extinction continue to intensify, it will be increasingly difficult to disentangle the main cause of extinction, as the interactions between them will become increasingly complex.

4.4 Recent Trends in Biodiversity

The beginning of this chapter presented an overview of the status of different components of biodiversity. This section presents information about rates and patterns of change in each of these components. Because of the lack of data, genetic diversity is omitted from consideration here. Although genetic diversity is lost from declining and fragmented populations at rates that can be estimated and measured, hardly any data exist to estimate this or its impact in most places and species. As more complete information is available regarding genetic diversity of cultivated species, a further description of agricultural genetic diversity can be found in Chapter 26.

Even for better-studied taxa and for the data-rich parts of the world, monitoring schemes that allow for a quantification of biodiversity trends have been operating for a few decades at most. The initial ecological conditions at the time such schemes were implemented are used as baselines against which subsequent changes are assessed. However, in most cases these are not "pristine" conditions, and in fact may correspond to ecosystems that have already suffered significant change in their biodiversity levels. The "shifting baseline syndrome" was first described for fisheries science (Pauly 1995; Sheppard 1995; Jackson 2001; Jackson et al. 2001), with the observation that every new generation of scientists accepts as a baseline the stock size and species composition that occurred at the beginning of their careers, using this to evaluate changes and propose management recommendations. The implication of the shifting baseline syndrome not only for fisheries but also for conservation science in general is that as biodiversity erodes, so do our targets for its conservation (Balmford 1999).

Our ignorance on the characteristics of pristine ecosystems often makes it difficult to understand whether observed short-term changes in biodiversity correspond to true trends or to noise created by natural fluctuations. This reinforces the need for long-term monitoring programs, as well as making the best use of existing historical evidence, even if only as anecdotal records (Pauly 1995). Some of the more important datasets collected on trends in the amount of biodiversity are presented here, although it is very difficult to extrapolate from any of these to infer a trend in the amount of species-level biodiversity, either globally or regionally.

4.4.1 Populations

Species are generally composed of a number of populations. Therefore, assessing all populations within a species is the same thing as a species-level assessment. In some cases, a species comprises only one population. Thus there is natural overlap when assessing trends in populations and species. The distinction between the two is further blurred by the fact that many studies monitor the status of all populations that make up a species distribution, as well as taxa where only a subset of populations are represented. Here we focus on large-scale analyses that provide insight into trends in either the distribution or abundance of populations, and in many cases the examples contain species-level assessments. We first discuss population trends on a global scale and then highlight trends in specific taxonomic groups. Where possible, we focus on long-term studies, as for these there is greater certainty that observed trends are not the result of short-term fluctuations (Ranta et al. 1997).

Little is known about the rate of loss of populations on a global scale. Hughes et al. (1997) present an extremely rough estimate of the global loss of populations by first estimating the total number of populations in the world (their intermediate estimate is about 3 billion populations). They then estimate a rate of habitat loss in the tropics of 0.08%, and conclude that roughly 16 million populations are being lost per year in tropical forests alone.

The best available estimate of global trends in populations is WWF's Living Planet Index. Time-series population data has been collected from a number of sources over the past 30 years. The LPI is calculated by averaging three ecosystem-based population indices, including 555 terrestrial species, 323 freshwater species, and 267 marine species (Loh 2002; Loh and Wackermagel 2004). Between 1970 and 2000, the LPI dropped by approximately 40%. During this time there were declines of approximately 30% in the terrestrial species population index, 30% in the marine species population index, and 50% in the freshwater species population index. The dependence of the index on relatively long-term datasets available in the published literature results in a strong taxonomic and regional bias. It also means many small, remote, and often threatened populations being overlooked. Such populations are difficult to monitor, and thus measures of their abundance are rarely consistently reported (Gaston 1994). However, it does clearly demonstrate that for well-known taxa and regions, the trends are consistently downward.

4.4.1.1 Birds

Although birds are one of the best-studied groups, we lack data on population trends for the majority of species. However, important studies of specific regions or groups of birds provide insight into overall trends. Here the findings are presented from a few examples of the large-scale bird population studies, including a global study of waterbird populations, a large-scale study of bird populations in Europe, and a study of range decline in Central and South America.

Waterbirds—bird species that are ecologically dependent on wetlands and other freshwater habitats—and particularly migratory waterbirds are probably one of the best-studied groups of animals on Earth (Rose and Scott 1997). Global-level information on the status and trends of waterbirds by biogeographic population is compiled and regularly updated by Wetlands International through its International Waterbird Census and published as *Waterbird Population Estimates* (Wetlands International, 2002). More detailed information is also available for waterbird species in North America, compiled by the U.S. Geological Service, and for the Western Palaearctic and Southwest Asia, prepared by Wetlands International (e.g., Delany et al. 1999). For African-Eurasian waterbird populations, comprehensive analyses have been compiled for ducks, geese, and swans (Anatidae) (e.g., Scott and Rose 1996) and waders (Charadrii) (Stroud et al. 2004). Although distributional data are available for other regions, comprehensive in-

formation on status and trends of waterbirds is generally lacking (Revenga and Kura 2003).

Despite the variations in availability of information, trend data show that in every region the proportion of populations of waterbirds in decline exceeds those that are increasing. At the global level, 41 % of known populations are decreasing, 36 % are stable, and 19 % are increasing (Wetlands International 2002). (See Table 4.7.) Asia and Oceania are the regions of highest concern for the conservation of waterbirds. In Africa and the Neotropics, more than twice as many known populations are decreasing than increasing. In Europe and North America, waterbird population numbers seem to be more equally distributed among the three categories (stable, increasing, and decreasing). It is important to note, however, that these data are more readily available for smaller populations, which are more likely to be in decline.

Trends in bird populations more generally have been best documented in Europe, North America, and Australia. In Europe, trend data are available from the Pan-European Common Bird Monitoring Scheme, currently implemented in 18 countries (Gregory et al. 2003). The data show trends in common in widespread farmland and woodland birds since 1980. (See Figure 4.17.) On average, populations of woodland birds in Europe have remained stable over the last 20 years, although their numbers have fluctuated in response to winter conditions (trend 1980–2002 = −2%). Populations of common and widespread farmland birds, in contrast, have declined sharply, especially in the 1980s, and the downward trend continues at a slower rate (trend 1980–2002 = −29%). This rapid decrease is believed to reflect a severe deterioration in the quality of farmland habitats in Europe, affecting both birds and other elements of biodiversity.

In Central and South America, where population-level data on bird species are scarce, BirdLife International has devised a different approach to measuring the decline in species richness. In Figure 4.18 (in Appendix A), a density map depicts the areas where threatened bird species used to occur but now no longer do so (mapped at a resolution of ¼ degree grid cell) (BirdLife 2004b). This measures a decline in occupancy (measured as a decline in extent of occurrence), a variable typically correlated to abundance (Brown 1984; He and Gaston 2000). In the Neotropics, some 230 globally threatened birds—approximately 50% of threatened species that occur in the region—have become extinct across significant parts of their range. (This high proportion is not surprising, as many threatened species are classified as so based on declining trends in their ranges/populations (IUCN 2001)). On average, approximately 30% of their total ranges has been lost, varying from tiny areas of less than 100 square kilometers (approximately 40 species) to considerable areas of greater than 20,000 square kilometers (approximately 70 species).

This analysis is based on a review of areas or sites where species were recorded historically but not recently, or where habitat loss or other threats seem certain to have resulted in their disappearance. In some areas, up to 20 species have disappeared—the highest recorded density of local extirpations of globally threatened bird species in the world. Losses of range are inevitably associated with a reduction in the total numbers of individuals and hence, an increasing risk of extinction.

4.4.1.2 Mammals

Global estimates of changes in populations exist for many mammals. Ceballos and Ehrlich (2002) used a dataset consisting of all ranges of terrestrial mammals of Australia and subsets of ranges for terrestrial mammals of Africa, South East Asia, Europe, and North and South America, consisting of roughly 4% of about 4,650

Table 4.7. Waterbird Population Trends (Revenga and Kura 2003, based on Wetlands International 2002)

Geographic Region	Population Trend				Number of Populations		
	Stable	Increasing	Decreasing	Extinct	With Known Trend	Lacking Trend	Total Number
Africa	141	62	172	18[a]	384	227	611
Europe	83	81	100	0	257	89	346
Asia	65	44	164	6	279	418	697
Oceania	51	11	42	28	138	241	379
Neotropics	100	39	88	6	234	306	540
North America	88	62	68	2	220	124	344
Global total[b]	**404**	**216**	**461**	**60**	**1,138**	**1,133**	**2,271**

[a] Most extinctions in Africa have been on small islands.

[b] Global totals do not equal the sum of the column because a population is often distributed in more than one Ramsar region.

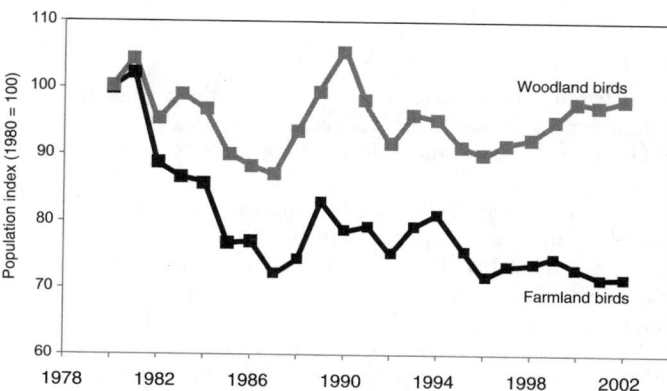

Figure 4.17. Trends in Common Farmland and Woodland Birds in Europe since 1980 (data courtesy of the Pan-European Common Bird Monitoring Scheme)

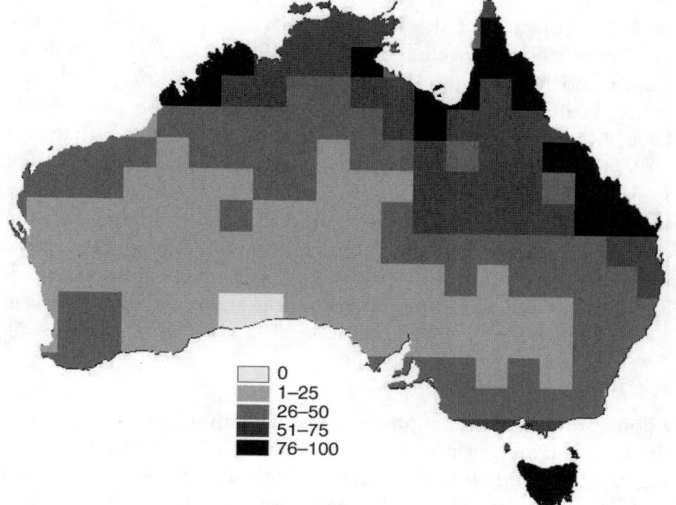

Figure 4.19. Percentage of Mammals That Have Disappeared from Each 2 Degree by 2 Degree Quadrant in Australia during Historic Times (Ceballos and Ehrlich 2002)

known mammal species to compare historic and present ranges. In their sample, declining mammal species had collectively lost 50% of their continental populations as judged by loss in area. In Australia, where the data were most comprehensive, the proportion of declining species was 22%. The greatest population declines occurred in northeastern Australia and Tazmania. (See Figure 4.19.) If this were representative of all regions, it would suggest a greater than 10% loss of mammal populations. However, this may not be indicative of other areas, as Australian mammals have been among the most prone to extinction (Hilton-Taylor 2000).

There are a few important datasets on population trends in large mammals. For example, the IUCN Species Survival Commission has monitored trends in rhinoceros populations in Africa and Asia for over 20 years (Khan 1989; Cumming et al. 1990; Foose and van Strien 1997; Emslie and Brooks 1999). This dataset reveals highly divergent trends between the different rhinoceros species. Two species, the southern white rhinoceros (*Ceratotherium simum simum*) and the Indian rhinoceros (*Rhinoceros unicornis*), have experienced long-term increases for the last century under very strict conservation and management regimes, whereas the black rhinoceros (*Diceros bicornis*), northern white rhinoceros (*C. simum cottoni*), and the Sumatran rhinoceros (*Dicerorhinus sumatrensis*) have suffered from catastrophic declines, mainly due to illegal hunting. In the case of the black rhinoceros, intensive conservation measures have stabilized the situation since the early 1990s. For the northern white rhinoceros and the Javan rhinoc-

eros (*R. sondaicus*), the trends are uncertain (both have perilously small populations).

Whale populations are monitored by the Scientific Committee of the International Whaling Commission. Their data indicate significantly increasing population trends for four whale stocks involving three species: gray whale *Eschrichtius robustus* eastern north Pacific; bowhead whale *Balaena mysticetus* Bering-Chukchi-Beaufort Seas stock; humpback whale *Megaptera novaeangliae* western north Atlantic; and humpback whale *M. novaeangliae* Southern Hemisphere south of 60° S in summer. These increasing trends reflect population recoveries following a period of very heavy harvesting pressure; at present, the datasets are not available to compare current whale population levels with historical estimates (although recalculation of the whaling commission's catch data might make this possible in the future). So although these data indicate some recovery in certain whale populations, it is in the context of major overall declines since the onset of commercial whaling. Recent analyses based on genetic markers indicate that these declines may have been even more dramatic than previously thought (Roman and Palumbi 2003).

While there are few strictly freshwater mammal species, some are considered freshwater system–dependent or semi-aquatic mammals, given that they spend a considerable amount of time in fresh water. Unfortunately, population trend data on most of these species are lacking, but information does exist for some well-studied species (such as the pygmy hippopotamus and some otter populations in Europe) (Revenga and Kura 2003). A group for which there is more information on population trends, given their precarious conservation status, is the freshwater cetaceans or river dolphins. There are five species of river dolphins and one species of freshwater porpoise living in large rivers in Asia and South America. Populations of river cetaceans have declined rapidly in recent years, driven by habitat loss and degradation (Reeves et al. 2000).

While trends in populations of single species or small taxonomic groups provide useful information, multispecies datasets are more useful for identifying general overall trends. In Figure 4.20, trend data from various IUCN/SSC sources have been assembled into trend categories (reflecting changes over the last 20 years) in order to provide an overall picture of population or abundance trends among 101 large mammal species in Africa (data provided courtesy of the IUCN/SSC, with particular reference to Oliver 1993; Nowell and Jackson 1996; Oates 1996; Sillero-Zubiri and Macdonald 1997; Woodroffe et al. 1997; Mills and Hofer 1998; Barnes et al. 1999; East 1999; Emslie and Brooks 1999; Moehlman 2002). From this figure, it can be seen that over 60% of the species are clearly decreasing, and another almost 20% are in the "stable or decreasing" category. Only 4% of the species are clearly increasing. The Figure also shows that a larger fraction of the species with smaller populations is declining than those with larger populations. This overall heavily negative trend is likely to be indicative of a deteriorating environmental situation over much of the African continent.

4.4.1.3 Amphibians

Populations of many amphibians are declining in several parts of the globe. Different possible causes have been suggested, including habitat change (mainly affecting small-scale freshwater habitats such as ponds and streams), disease, climate change, acid precipitation, habitat loss, and increased UV-B irradiation. Houlahan et al. (2000) used data from 936 populations to examine global

trends in amphibian populations. The studies that were analyzed ranged from 2 to 31 years in duration. Their findings suggest a relatively rapid decline from the late 1950s peaking in the 1960s, followed by a reduced decline to the present. Alford et al. (2001) later reanalyze the same data and suggested that the global decline may have begun in the 1990s.

Regardless of the exact timeframe, it is commonly accepted that amphibian populations have recently declined on a global scale. This is supported by the recent IUCN-SSC/CI-CABS/NatureServe Global Amphibian Assessment (Stuart et al. 2004). Out of the 4,048 amphibian species (70.9%) for which trends have been recorded, 61.0% (2,468 species) are estimated to be declining, 38.3% (1,552 species) are stable, and 0.69% (28 species) are increasing (Baillie et al. 2004). The report estimated that there are presently 435 more amphibians listed in the IUCN higher categories of threat then would have been in 1980 and that between 9 and 122 species have gone extinct during this time period (Stuart et al. 2004).

4.4.1.4 Reptiles

Global trends in reptiles have not been synthesized to the same extent as they have for amphibians. The fact that IUCN has only assessed the conservation status of 6% of the 8,163 described reptiles indicates how little is known about their global status and trends (Baillie et al. 2004). Reptiles share many of the same environments and are susceptible to many of the same threats as amphibians, and it has therefore been suggested that they may be experiencing similar or greater declines (Gibbins et al. 2000), but this remains to be rigorously tested.

Turtles and tortoises are among the best-studied reptiles. Within this group, large declines have been identified in the marine turtles, with six of the seven species listed as threatened by IUCN (Baillie et al. 2004). Overall rates of decline are unknown for turtles and tortoises, but reports on the trade of Southeast Asian freshwater turtles indicate that many of these species are rapidly declining. TRAFFIC Southeast Asia estimates trade volumes at a minimum of 13,000 tons of live turtles in 1999 (TRAFFIC Southeast Asia 2001, see section 5.4.5) and that this trade is increasing. IUCN is now conducting a Global Reptile Assessment that will soon help clarify the status and trends of this group.

4.4.1.5 Fish

Little is known about the majority of fish populations, but the global decline of commercially important fish stocks or populations is relatively well documented (e.g., Jackson et al. 2001; Myers and Worm 2003; Hutchings and Reynolds 2004).

Data on trends of some 600 fish populations covering more than 100 species can be found at fish.dal.ca/ to myers/data.html, usually in terms of trends in spawning stock biomass. Summarized data on the overall status of fish stocks, based on catch statistics in their SOFIA report, are available from FAO in *The State of the World's Fisheries and Aquaculture* reports produced every two years (see FAO 2000a).

The data available to FAO at the end of 1999 identified 590 "stock" items. For 441 (75%) of these, there is some information on the state of the stocks and, although not all of this is recent, it is the best that is available. The stock items are classified as underexploited (U), moderately exploited (M), fully exploited (F), overexploited (O), depleted (D), or recovering (R), depending on how far they are, in terms of biomass and fishing pressure, from the levels corresponding to full exploitation. Full exploitation is taken as being loosely equivalent to maximum sustainable yield (equivalent to being harvested at the biological limit). The

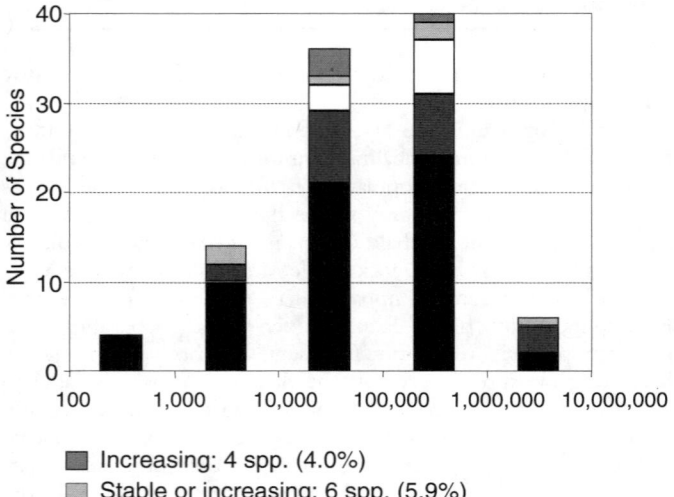

Increasing: 4 spp. (4.0%)
Stable or increasing: 6 spp. (5.9%)
Stable: 9 spp. (8.9%)
Stable or decreasing: 20 spp. (19.8%)
Decreasing: 61 spp. (60.4%)

Figure 4.20. Trends in 101 African Large Mammal Species

overall state of the fish stocks being monitored according to these classifications is shown in Figure 4.21.

Figure 4.21 indicates that 28% (R + D + O) of the fish stocks being assessed have declined to levels below which a maximum sustainable yield can be taken and that a further 47% (F) require stringent management (which may or may not already be in place) to prevent decline into a similar situation. In total, 75% of these stocks (R + D + O + F) need management to prevent further declines or to bring about recovery in spawning stock biomass. Conversely, 72% (F + M + U) of the stocks are still capable of producing a maximum sustainable yield. These data have also been broken down regionally and are available in FAO (2002).

The State of the World's Fisheries and Aquaculture report (FAO 2000a) identifies trends since 1974 in each stock classification, as a percentage of the total number of fish stocks being assessed by FAO. The percentage of underexploited stocks (U + M) has declined steadily, while the proportion of stocks exploited beyond maximum sustainable yield levels (O + D + R) has increased steadily over this time period. If these data are representative of fisheries as a whole, they indicate an overall declining trend in spawning stock biomass for commercially important fish species over the last 30 years.

The FAO data (FAO 2000a) demonstrate that there is significant increase in the exploitation of deep-sea fish stocks, such as populations of orange roughy (*Hoplostethus atlanticus*), alfonsinos (*Berycidae*), and dories (*Zeidae*). Many of these species have slow growth rates, and it is not yet clear that the methods established to fish them sustainably will be successful.

Little is known about the status of most shark populations (Castro et al. 1999). Baum et al. (2003) used the largest shark dataset covering the north Atlantic to assess declines of coastal and oceanic shark populations. Shark declines are believed to be occurring as a result of increased bycatch from pelagic long-line fisheries and direct exploitation for shark fins. Baum et al. (2003) found that all recorded shark species within the study area, with the exception of makos, have experienced a decline of more than 50% in the past 8–15 years. Sharks grow and reproduce slowly, so even if exploitation were stopped, their recovery would be slow.

The use of catch statistics to assess freshwater stocks, which is common practice with marine species, is difficult because much of the inland catch is underreported by a factor of three or four, according to FAO (FAO 1999; FAO 2000a). Nevertheless, FAO's last major assessment of inland fisheries (FAO 1999) reported that most inland capture fisheries that rely on natural reproduction of the stocks are overfished or are being fished at their biological limit.

Some large lakes have been systematically studied because of their importance as a fishery resource. The North American Great Lakes are a case in point. Annual fish stock assessments are conducted for commercially important salmonoid species, such as lake trout and Pacific salmon, and for their prey species (such as alewife, rainbow smelt, bloater, sculpin, and lake herring) (USGS Great Lakes Science Center 2003). The prey population assessments for the five lakes show that with the exception of Lake Superior, whose status is mixed but improving, populations of prey species in the other four lakes are all decreasing (USGS Great Lakes Science Center 2003). With respect to predator species in the lakes, many native species, such as lake trout and sturgeon, are found in vastly reduced numbers and have been replaced by introduced species (Environment Canada and U.S. EPA 2003).

Other regularly assessed lakes include Lake Victoria and Lake Tanganyika in Africa. These also show a decline in native fisheries and replacement with exotic species. The most widely known and frequently cited is the disappearance of over 300 haplochromine cichlids in Lake Victoria and the decline or disappearance of most of the riverine fauna in the east and northeastern forests of Madagascar (Stiassny 1996). There is also documented evidence of the threatened fish fauna of crater lakes in western Cameroon and the South African fish fauna, which has 63 % of its species endangered, threatened, or "of special concern" (Moyle 1992; Lévêque 1997).

The few examples of riverine fish assessments show that many inland fisheries of traditional importance have also declined precipitously. The European eel fishery, for example, has steadily declined over the last 30 years (Kura et al. 2004). By the mid-1980s, the number of new glass eels (eel juveniles) entering European rivers had declined by 90%. Recent figures show that this has now dropped to 1% of former levels (Dekker 2003).

Other fish stocks for which there is longer-term catch and status information include Pacific and Atlantic salmon in North America, fisheries of the Rhine and Danube Rivers in Europe, and fisheries of the Pearl River in China. All of these have declined to just a small fraction of their former levels due to overexploitation, river alteration, and habitat loss, putting some of these species at serious risk of extinction (Balcalbaca-Dobrovici 1989; Lelek 1989; Liao et al. 1989; WDFW and ODFW 1999).

Finally, even fisheries that until recently were reasonably well managed, such as the caviar-producing sturgeons in the Caspian Sea, and fisheries from relatively intact rivers such as the Mekong in Southeast Asia are rapidly declining (Kura et al. 2004.). For example, while almost all 25 species of sturgeon in the world have been affected to some degree by habitat loss, fragmentation of rivers by dams, pollution, and overexploitation, much of the recent decline in the catch of caviar-producing sturgeon is a direct result of overfishing and illegal trade (De Meulenaer and Raymakers 1996; WWF 2002). Major sturgeon populations have already declined by up to 70% (WWF 2002).

4.4.1.6 Corals

A meta-analysis of trends in Caribbean corals reveals that there has been a significant decline over the past three decades and although the decline has slowed, the trend persists. The average hard coral cover on reefs has been reduced by 80%, from around 50% to 10% cover, in three decades (Gardner et al. 2003). This significant trend supports the notion that coral reefs are globally threatened (Hodgson and Liebeler 2002; Hughes et al. 2003). (See Chapter 19.)

4.4.1.7 Invertebrates

Invertebrates represent the greatest proportion of eukaryotic biodiversity, but we know virtually nothing about their distributions,

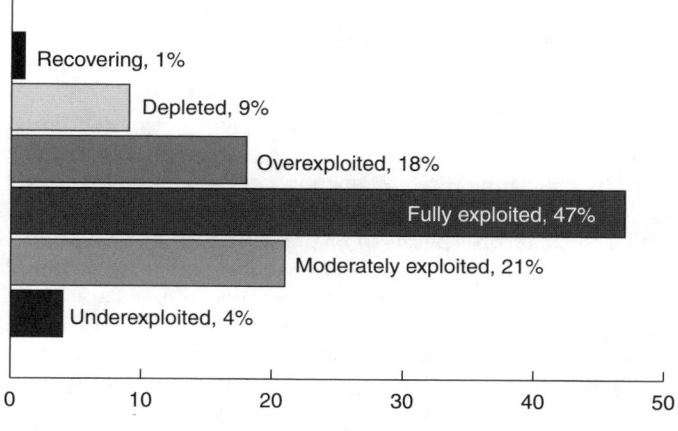

Figure 4.21. The State of Fish Stocks, 1999 (FAO 2000)

populations, or associated trends. However, especially among the insects a few well-studied groups such as butterflies, moths, and dragonflies (Hilton-Taylor 2000) may provide some insight. One example—a regional study of butterflies in the Netherlands (van Swaay 1990)—identifies decline as the most common trend over the past century. Of the 63 species assessed, 29 (46%) decreased or became extinct, 17 (27%) experienced little change, and 7 (11%) appeared to have increased their range. The remaining 10 species (16%) tended to fluctuate in range. Butterflies have also been relatively well monitored in the UK. Although many common and widespread species are believed to have increased in range and abundance (Pollard et al. 1995), the overall trend appears to be one of decline. A recent study examining population and regional extinctions indicates that British butterfly distributions have decreased by 71% over the past 20 years (Thomas et al. 2004b). This was found to be much higher than both birds and plants.

Although trends of well-studied insects indicate that this group shows similar trends of decline to other taxonomic groups, some studies indicate that insects in specific habitat types may be relatively resistant (Karg and Ryszkowski 1996). Understanding the general trends associated with insects is extremely important as it will provide much greater insight into global trends in biodiversity.

Two other groups of invertebrates that have been studied in more detail are freshwater mollusks and crustaceans. There are many lists on freshwater mollusks at national and regional levels, a number of which are available on the Internet (e.g. species .enviroweb.org/omull.html and www.worldwideconchology .com/DatabaseWindow.html). These databases, however, are not standardized or comparable; therefore an assessment of the current status and trends of freshwater mollusks at the global level is difficult. Existing lists are also biased toward terrestrial and marine groups.

The United States is one of the few countries in which the conservation status of freshwater mollusks and crustaceans has been widely assessed. Half of the known U.S. crayfish species and two thirds of U.S. freshwater mollusks are at risk of extinction (Master et al. 1998), with severe declines in their populations in recent years. Furthermore, at least 1 in 10 of the freshwater mollusks is likely to have already gone extinct (Master et al. 1998). The alarming rate of extinction of freshwater mollusks in eastern North America is even more pronounced. According to the U.S. Federal Register, less than 25% of the present freshwater bivalves appear to have stable populations. The status of gastropods is much less known. Of 42 species of extinct gastropods in the United States, 38 were reported from the Mobile Bay Basin in Southern North America (Bogan 1997).

4.4.1.8 Plants

Information on global trends in the status of plants is lacking, but overall population declines are likely given the high rates of habitat modification and deforestation described earlier, along with other threats, such as overexploitation, alien invasive species, pollution, and climate change. In addition, 12 of the 27 documented global extinctions over the past 20 years have been plants (Baillie et al. 2004).

The sixth meeting of the Conference of the Parties of the Convention on Biological Diversity adopted the Global Strategy for Plant Conservation. This strategy highlights monitoring the status and trends of global plant diversity as one of the objectives (UNEP 2002b), which it is hoped will lead to greater insight into global trends.

Cycads are one of the few groups where the conservation status of all species has been assessed and trend data exist. Population trends are available for 260 species of cycads (Cycadopsida, 288 species in total). Of these, 79.6% (207 species) are declining, 20.4% (53 species) are stable, and none are considered to be increasing (Baillie et al. 2004).

Another important dataset is available for trends in wood volume and biomass. FAO's *Forest Resources Assessment 2000* indicates opposing trends between the tropics and nontropics in terms of both volume and above-ground woody biomass over the period 1990–2000. There has been a decreasing trend in the tropics, compared with an increasing trend in the nontropics. These data should be interpreted with caution, due in part to problems of data compatibility between countries (see FAO 2001b and Chapter 21 for more details). Note also that no distinction is made here between undisturbed forest, secondary forests, and plantations.

4.4.1.9 Conclusion on Population

Measuring change in populations is important for understanding the link between biodiversity and ecosystem function, as significant changes in populations can have important implications for the function of ecosystems long before any species actually goes extinct (e.g., Jackson et al. 2001; Springer et al. 2003).

The data presented in this section represent a brief assessment of the types of data that are available on the trends in populations. Although the datasets described are not easily comparable with each other and are certainly not collectively representative of biodiversity as a whole, a few basic conclusions can be drawn.

Both declining and increasing trends can be documented from available studies; in most cases, declining trends appear to outweigh increasing trends, often by a considerable margin, and some increasing trends can be related to very specific situations (for example, population recovery following periods of intensive harvesting or successful reintroduction programs). Overall, the emerging evidence suggests that for macroorganisms, especially those with small areas of distribution, most populations are declining as a result of human activities and are being replaced by individuals from a much smaller number of expanding species that thrive in human-altered environments. The result will be a more homogenized biosphere with lower diversity at regional and global scales (McKinney and Lockwood 1999).

4.4.2 Species

4.4.2.1 Current Extinction Rates

The evolution of new species and the extinction of others is a natural process. Species present today represent only 2–4% of all species that have ever lived (May et al. 1995). Over geological time there has been a net excess of speciation over extinction that has resulted in the diversity of life experienced today. However, the high number of recent extinctions suggests that the world might now be facing a rapid net loss of biodiversity. This can be tested by comparing recent extinction rates to average extinction rates over geological time.

The fossil record appears to be punctuated by five major mass extinctions (Jablonski 1986), the most recent of which occurred 65 million years ago. However, the majority of extinctions have been spread relatively evenly over geological time (Raup 1986), enabling estimates of the average length of species' lifetimes through the fossil record. Studies of the marine fossil record indicate that individual species persisted for periods ranging from 1 million to 10 million years (May et al. 1995). These data probably underestimate background extinction rates, because they are necessarily largely derived from taxa that are abundant and widespread in the fossil record.

Using a conservative estimate of 5 million as the total number of species on the planet, we would therefore expect anywhere between five extinctions per year to roughly one extinction every two years (for all 5 million species on the planet). As noted earlier, recent extinctions have been best studied for birds, mammals, and amphibians, and in these groups over the past 100 years roughly 100 species have become extinct. This is in itself similar to background extinction rates, but these groups represent only 1% of described species.

Assuming for the moment that the susceptibility to extinction of birds, mammals, and amphibians is similar to species as a whole, then 100 times this number of species (10,000 species) were lost over the past 100 years. But this assumption of equivalent extinction risk is very uncertain, and given the additional uncertainty over the total number of species on the planet, it is preferable to convert these data into a relative extinction rate, measured as the number of extinctions per million species per year (Pimm et al. 1995). A background extinction rate of 0.1–1 E/MSY then corresponds to the average marine fossil species lifetimes. Mammalian background extinction rates are also believed to be within these limits, falling within a range of 0.21 E/MSY (strictly for lineages rather than species, but provides a conservative estimate (Alroy 1998; Regan et al. 2001) and 0.46 E/MSY (Foote 1997).

Measuring recent extinction rates is difficult, not only because our knowledge of biodiversity is limited, but also because even for the best studied taxa there is a time lag between the decline toward extinction and the actual loss of species. In the case of extinctions caused by habitat loss, in particular, it may take thousands of years before a restricted remnant population is finally driven to extinction (Diamond 1972).

With this in mind, it is possible to use recent documented extinctions to make a very conservative estimate of current extinction rates, though this is limited because only a few taxonomic groups have been reasonably well analyzed for extinctions. There are approximately 21,000 described species of birds, mammals, and amphibians. The roughly 100 documented extinctions for these groups during the past century yields an E/MSY of 48, which is 48 to 476 times greater than the background extinction rate of 0.1 to 1. If "possibly extinct" species are included in this analysis, the total number of extinctions and possible extinctions over the past 100 years for these groups is 215 species, which results in an E/MSY that is 102 to 1,024 higher than background rates. Broken down by taxonomic group, mammals have the highest E/MSY (64) followed by birds (at 45) and finally amphibians (40). If possibly extinct species are considered, however, then amphibians have the highest E/MSY at 167 followed by mammals with 68 and finally birds with 59. (It should be noted that mammals have not been completely assessed for possibly extinct species (see Baillie et al. 2004).)

A broad range of techniques have been used to estimate contemporary extinction rates, including estimates based on both direct drivers (such as habitat destruction) and indirect drivers (such as human energy consumption) of extinction (Myers 1979; Myers 1988; Reid 1992; Smith et al. 1993; Ehrlich 1994; Mace and Kunin 1994; Pimm and Brooks 1999; Regan et al. 2001; Baillie et al. 2004; also see MA *Scenarios,* Chapter 10). Many of these studies give rise to estimates of E/MSY that are 1,000 to 10,000 higher than background rates (Pimm and Brooks 1999), generally higher than the conservative estimate for birds, mammals, and amphibians based on documented extinctions. (See Figure 4.22 in Appendix A.) Estimates based on documented extinctions are likely to be underestimates because the *IUCN Red List* is very

conservative in recording species as actually extinct and because many extinctions have probably been missed due to limited survey effort for most taxonomic groups.

The trend in species extinction rates can be deduced by putting together extinction rates characteristic of well-recorded lineages in the fossil record, recorded extinctions from recent times, and estimated future extinction rates based on the approaches just described. All these estimates are uncertain because the extent of extinctions of undescribed species is unknown, because the status of many described species is poorly known, because it is difficult to document the final disappearance of very rare species, and because there are extinction lags between the impact of a threatening process and the resulting extinction (which particularly affects some modeling techniques). However, the most definite information, based on recorded extinctions of known species over the past 100 years, indicates extinction rates are around 100 times greater than rates characteristic of comparable species in the fossil record. Other less direct estimates, some of which refer to extinctions hundreds of years into the future, estimate extinction rates 1000 to 10,000 times higher than rates recorded among fossil lineages.

Current anthropogenically caused extinction is not solely a characteristic of contemporary societies. Since the initial revelations that humanity greatly inflated extinction rates with stone-age technology (Martin and Wright 1967; Martin and Klein 1984), large quantities of new data have demonstrated significant extinction episodes occurred with, for example, the arrival of people in Australia 46,000 years ago (Roberts et al. 2002), in the Americas 12,000 years ago (Alroy 2001), in Madagascar (Goodman and Patterson 1997) and the Pacific 2,000 years ago (Steadman 1995), and elsewhere (MacPhee 1999).

4.4.2.2 Current Levels of Threat to Species

At a global level, nearly 850 species have been recorded as becoming extinct or at least extinct in the wild since 1500 (Baillie et al. 2004). Species extinctions represent the final point in a series of population extinctions; in fact, distinct populations may be being lost at a rate much faster than species overall, with serious negative consequences for local ecosystem function (Hughes et al. 1997).

The most extensive global dataset on trends in species richness is the *IUCN Red List of Threatened Species* (see www.redlist.org and Baillie et al. 2004). The *IUCN Red List* is formalized through the application of categories and criteria (IUCN 2001) that are based on assessments of extinction risk (Mace and Lande 1991). These criteria are now broadly used in many parts of the world and have been adapted for use at multiple scales.

The *2004 IUCN Red List of Threatened Species* is based on assessments of 38,047 species. Of these, 7,266 animal species and 8,321 plant species (15,547 species in total) have been placed in one of the IUCN Categories of Threat (vulnerable, endangered, or critically endangered). However, the *IUCN Red List* needs to be interpreted with caution, because for most taxonomic groups the assessments are very incomplete and heavily biased toward the inclusion of the most threatened species. As of 2004, assessments of almost every species have been completed for three animal groups (mammals, birds, and amphibians) and two plant groups (conifers and cycads). The number of species in each IUCN Red List Category for all five of these groups is given in Table 4.8. Reptiles have not yet been completely assessed.

In all five of these groups, the proportions of species in categories of high extinction risk are much greater than would be expected if species were becoming extinct at rates typically observed over geological time. The levels of threat are lowest among

Table 4.8. Number of Species in IUCN Red List Categories for Comprehensively Assessed Taxonomic Groups (Baillie et al. 2004)

Class	EX	EW	Subtotal	CR	EN	VU	Subtotal	LR/cd	NT	DD	LC	Total
Animals												
Mammals	73	4	77	162	352	587	1,101	64	587	380	2,644	4,853
Birds	129	4	133	179	345	689	1,213	0	773	78	7,720	9,917
Amphibians	34	1	35	427	761	668	1,856	0	359	1,290	2,203	5,743
Plants												
Conifers	0	0	0	17	43	93	153	26	53	59	327	618
Cycads	0	2	2	47	39	65	151	0	67	18	50	288

See IUCN 2001 for more details on the definitions of the Red List categories.

Key

EX extinct	EN endangered	LR/cd lower risk/	NT near threatened
EW extinct in the wild	VU vulnerable	conservation dependent	DD data deficient
CR critically endangered			LC least concern

birds, where 12% of species are threatened (vulnerable + endangered + critically endangered). There has been a trend of increasing threat between 1988 and 2004, as measured by the movement of species into more threatened Red List Categories (BirdLife 2004b). The relatively low level of threat in birds is possibly related to their tendency to be highly mobile, resulting in their generally wide geographic distributions.

The pattern of distribution of threat categories among species is broadly similar for mammals and conifers, with 23% (1,101) and 25% (153) respectively of the species being globally threatened. Based on the evidence from comprehensive regional assessments (e.g., Stein et al. 2000), it is more than possible that future studies will show this very high level of threat to be typical of the current global situation among most groups of terrestrial species.

The situation with amphibians is broadly similar: 32% (1,856) globally threatened. However, the true level of threat among amphibians is probably masked by the fact that 23% of the species are classified as data-deficient (compared with 8% for mammals and 10% for conifers). The overall conservation situation of amphibians will probably eventually prove to be much worse than the mammal and conifer situations and might be typical of the higher levels of threat associated with freshwater (or freshwater-dependent) species (Master et al. 2000). Amphibian extinction risk has been retrospectively analyzed back to the early 1980s, and shows a similar rate of decline to that of birds (BirdLife 2004b), but with a greater number of the more seriously threatened species declining (Baillie et al. 2004).

The cycad situation is much worse, with 52% (151) of species globally threatened. This is possibly reflective of the relict nature of these ancient species, with most species now surviving only in very small populations.

Species are not all equal: some represent much more evolutionary history than others (Vane-Wright et al. 1991). If extinctions were randomly distributed across the tree of life, surprisingly little evolutionary history would be lost (Nee and May 1997). However, extinctions are far from phylogenetically random: there is strong taxonomic selectivity in the current extinction crisis, with the result that the loss of evolutionary history is much more than that expected were species to be lost randomly with respect to their taxonomic affiliation (Purvis et al. 2000a).

There is a clear trend for higher levels of threat among the larger species. Of the mammals, for example, 38% (81) of the Artiodactyla (antelopes, cattle, sheep, and so on), 82% (14) of the Perissodactyla (horses, rhinos, and tapirs), 39% (114) of the Primates, 100% (2) of the Proboscidea (elephants), and 100% (5) of Sirenia (dugongs and manatees) are globally threatened (Baillie et al. 2004). Among the birds, high levels of threat are particularly apparent among orders such as Apterygiformes (kiwis) with 100% (4) threatened, Sphenisciformes (penguins) 57% (10), Pelecaniformes (cormorants, pelicans, and so on) 26% (17), Procellariiformes (albatrosses and petrels) 47% (62), Ciconiiformes (storks, ibises, and spoonbills) 21% (28), Galliformes (pheasants, partridges, quails, and so on) 27% (78), Gruiformes (cranes, bustards, rails, and so on) 33% (76), Columbiformes (doves and pigeons) 22% (75), and Psittaciformes (parrots) 29% (109) (Baillie et al. 2004). These orders include species that are flightless, ground-dwelling, particularly vulnerable to alien predators, and edible or economically valuable. The most noteworthy result from the threat analysis of amphibians is the particularly large proportion of globally threatened salamanders—46% (234) of the total number of threatened amphibians. Salamanders are often long-lived, slow-breeding species, with limited ability to disperse over significant distances.

The *IUCN Red List* does not yet include comprehensive datasets for taxonomic groups confined to freshwater ecosystems. Nor have there been any complete assessments of any invertebrate groups. Some important regional datasets are becoming available, however, for example for North America, compiled by NatureServe. A summary and analysis of these data for the United States are presented in Stein et al. (2000). NatureServe uses a different system for categorizing levels of threat, and their categories are not strictly comparable with those of IUCN. Nevertheless, for the purposes of this assessment their system does broadly indicate levels of extinction risk and is therefore useful in determining trends.

Based on an assessment of 20,439 species, NatureServe determined that one third of the U.S. flora and fauna appears to be of conservation concern. NatureServe has comprehensively assessed the status of every U.S. species in 13 taxonomic groups, and the percentage of each of these species that is at risk is shown in Figure 4.23. The most noteworthy finding of this study is that the species groups relying on freshwater habitats—mussels, crayfishes, stoneflies, fishes, and amphibians—exhibit the highest levels of risk. Sixty-nine percent of freshwater mussels are at risk. Dragon-

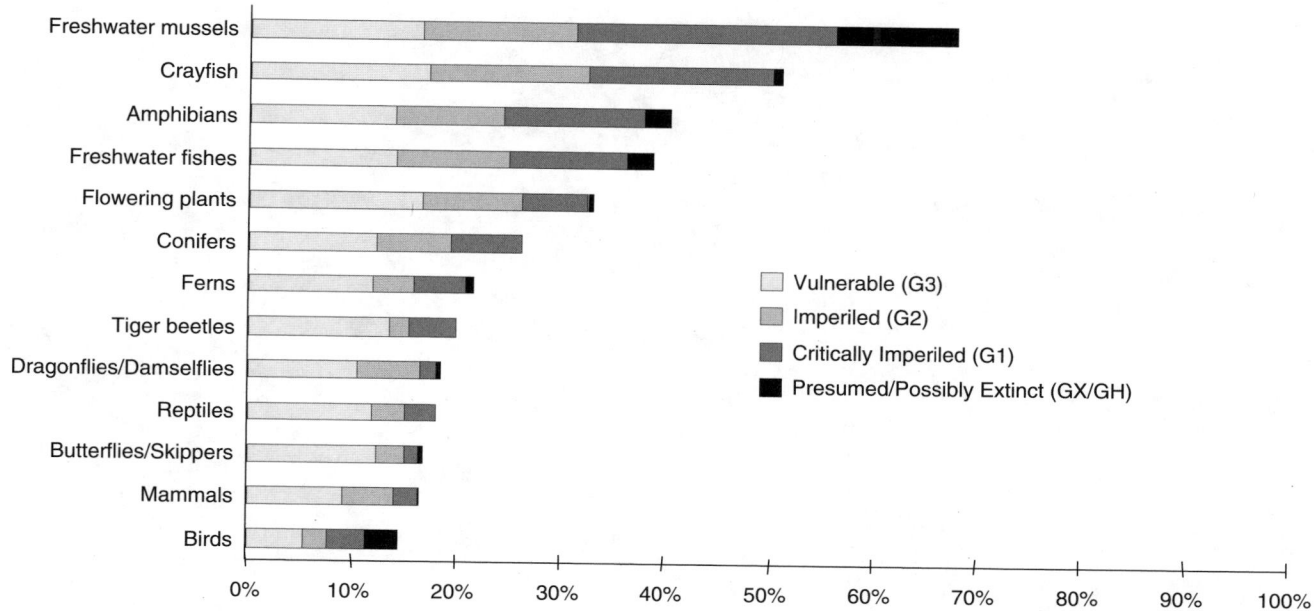

Figure 4.23. Percentage of Species at Risk in the United States, by Plant and Animal Groups (Stein et al. 2000, reproduced with permission of NatureServe)

flies and damselflies seem to be an exception to this pattern since, despite being freshwater-dependent, their threat level is relatively low. The threat levels are also very high in the United States for flowering plants (33%).

No comprehensive assessment has yet been carried out on the threat levels for any marine group. However, work is progressing fast to assess the status of all the chondrichthyan fishes (sharks, rays, and chimaeras). To date, the IUCN/SSC Shark Specialist Group has assessed one third (373 species) of the world's chondrichthyans (out of a total of approximately 1,100 species), and 17.7% are listed as threatened (critically endangered, endangered, or vulnerable), 18.8% near-threatened, 37.5% data-deficient, and 25.7% of least concern (Baillie et al. 2004). However, it is not at all clear that sharks and rays are good indicators of overall biodiversity trends in marine ecosystems. In view of the life history strategies for these species (slow-breeding, long-lived), it is likely that they are more threatened than some other marine groups.

4.4.2.3 *Traits Associated with Threat and Extinction*

The patterns of threat and extinction are not randomly distributed among species (Bennett and Owens 1997; Gaston and Blackburn 1997b; Owens and Bennett 2000). Ecological traits demonstrated to be associated with high extinction risk (even after controlling for phylogeny) include high trophic level, low population density, slow life history or low fecundity, and small geographical range size (Bennett and Owens 1997; Purvis et al. 2000b). For primates and carnivores, these traits together explain nearly 50% of the total between-species variation in extinction risk, and much of the remaining variation can be accounted for by external anthropogenic factors that affect species irrespective of their biology (Purvis et al. 2000b).

However, different taxa are threatened by different mechanisms, which interact with different biological traits to affect extinction risk. For bird species, extinction risk incurred through persecution and introduced predators is associated with large body size and long generation time but is not associated with degree of specialization, whereas extinction risk incurred through habitat loss is associated with habitat specialization and small body size

but not with generation time Owens and Bennett (2000). For Australian marsupials, the risk of extinction has been found to be better predicted by geographical range overlap with sheep (Fisher et al. 2003).

Extinction risk is not independent of phylogeny, presumably because many of the biological traits associated with higher extinction risk tend to co-occur among related species. Among birds, for example, families that contain significantly more threatened species than average are the parrots (Psittacidae), pheasants and allies (Phasianidae), albatrosses and allies (Procellariidae), rails (Rallidae), cranes (Gruidae), cracids (Cracidae), megapodes (Megapodidae), and pigeons (Columbidae) (Bennet and Owens 1997). There is also a positive relationship between the proportion of species in a taxon that are considered to be threatened and the evolutionary age of that taxon, both for the global avifauna and the avifauna of the New World (Gaston and Blackburn 1997b).

The majority of recorded species extinctions since 1500 have occurred on islands. A total of 72% of recorded extinctions in five animal groups (mammals, birds, amphibians, reptiles, and mollusks) were of island species (Baillie et al. 2004). Island flora and fauna were especially vulnerable to the human-assisted introduction of predators, competitors, and diseases, whereas species on continents were not so ecologically naive. However, predictions of future extinctions stem from the ongoing loss of continental, tropical forests; hence 452 of a total of 1,111 threatened bird species are continental (Manne et al. 1999). A shift from island to mainland extinctions is consistent with a recent examination of extinctions over the past 20 years, where island and mainland extinctions were roughly equal (Baillie et al. 2004).

4.4.2.4 *Geographical Patterns of Threat and Extinction*

The geography of threat and extinction is far from even, with the majority of threatened species concentrated in tropical and warm temperate endemic-rich "hotspots" (Myers et al. 2000). Figure 4.24 shows the locations of known mammal, bird, and amphibian extinctions since 1500. The different patterns between these three groups are striking. Mammal extinctions are concentrated in the Caribbean and Australia. In both cases, these are thought to be

Figure 4.24. Locations of Extinct and Extinct in the Wild Mammal, Bird, and Amphibian Species since 1500 (Baillie et al. 2004)

second waves of human-induced extinction, following the over-exploitation of the Pleistocene (MacPhee 1999); in any case, the current mammalian fauna in these regions is but a modest sample of the native fauna prior to human arrival, particularly in terms of medium- and large-sized mammals (Woods and Sergile 2001; Brook and Bowman 2004). The remainder of the recorded mammalian extinctions are widely scattered, most being on oceanic islands.

Avian extinctions are overwhelmingly concentrated on oceanic islands, especially on Hawaii and New Zealand (Steadman 1995), with very few elsewhere. With few exceptions, oceanic island avifaunas have lost most of their endemic species over the last 1,000 years.

The highest number of recorded amphibian extinctions is on Sri Lanka. However, the current wave of amphibian extinction, which appears to be accelerating, is concentrated in montane areas from Honduras south to northern Peru, in the Caribbean islands, in eastern Australia, and perhaps in the Atlantic Forest of southern Brazil.

Maps of species richness of threatened mammals and birds are presented in Figure 4.25 (in Appendix A). (For a species richness map of threatened amphibians, see Chapter 20.) The maps show interesting similarities and differences. They all show concentrations of threatened species in hotspots (Myers et al. 2000), in particular in the Andes, southern Brazil, West Africa, Cameroon, the Albertine Rift of Central Africa, the Eastern Arc Mountains of Tanzania, eastern Madagascar, Sri Lanka, the Western Ghats of India, the eastern Himalayas, central China, mainland Southeast Asia, and Borneo.

The mammal map is noteworthy in that there is at least one threatened mammal species in most parts of the world. In addition to the geographic regions just listed, important concentrations of threatened mammals also occur in the eastern Amazon basin,

southern Europe, Kenya, Sumatra, Java, the Philippines, and New Guinea. Interestingly, MesoAmerica, Australia, and the Caribbean islands appear to have relatively low numbers of threatened mammals. However, it should be noted that patterns of threat will appear low in areas where the vulnerable species have already gone extinct (which may be the case in the Caribbean and Australia) and that threatened mammals with extremely small distributions will not be easily viewed on the map (such as many restricted-range montane species in MesoAmerica).

The bird map differs in that the importance of oceanic islands is emphasized. Other areas that are of great importance for threatened birds but not listed earlier include the Caribbean islands, the Cerrado woodlands of Brazil, the highlands of South Africa, the plains of northern India and Pakistan, Sumatra, the Philippines, the steppes of central Asia, eastern Russia, Japan, southeastern China, and New Zealand. As with mammals, MesoAmerica and Australia are relatively unimportant for threatened birds. But so are the Amazon basin, Europe, Java, and New Guinea.

Amphibians generally have much more restricted ranges than birds and mammals (see Chapter 20), and threatened amphibian species therefore occupy a much smaller global area, a very different picture to mammals. In the small areas where they are concentrated, however, threatened amphibians occur more densely than either mammals or birds (up to 44 species per half-degree grid square, compared with 24 for both mammals and birds) (Baillie et al. 2004). The majority of the world's known threatened amphibians occur from Mexico south to northern Peru and on the Caribbean islands. Most of the other important concentrations of globally threatened amphibians mirror the patterns of threat for mammals and birds, although eastern Australia and the southwestern Cape region of South Africa are also centers of threatened amphibians. The paucity of data from certain parts of the world probably results in serious underestimation of the concentrations of threatened amphibians, especially in the Albertine Rift, Eastern Himalayas, much of mainland Southeast Asia, Sumatra, Sulawesi, the Philippines, and Peru.

Lack of comprehensive geographic and threat assessment for other species groups precludes the presentation of maps for other taxa. Given the similarity between patterns of threatened species for mammals, birds, and amphibians, many other taxonomic groups such as reptiles, fish, invertebrates, and plants may demonstrate broadly similar patterns. However, there are also likely to be many differences. For example, distribution patterns of threatened reptiles (in particular, lizards) are likely to highlight the importance of many arid ecosystems. It is already known that some distribution patterns of threatened plants do not match those of most animal groups, the most notable examples being the Cape Floral Region and Succulent Karoo of South Africa and the deserts of the southwestern United States and northern Mexico. Patterns of threat in marine ecosystems will of course be completely different, and data on these patterns are still largely unavailable.

One potentially useful device for understanding variation in threat intensity across areas is the concept of extinction filters, whereby prior exposure to a threat selectively removes those organisms that are most vulnerable to it, leaving behind a community that is more resilient to similar threats in the future (Balmford 1996). This idea can explain temporal and spatial variation in species' vulnerability to repeated natural changes in the past (such as glaciation events). It may also shed light on the contemporary and future impact of anthropogenic threats. For example, the impact of introduced rats on island-nesting seabirds appears less marked on islands with native rats or land crabs, which have selected for resilience to predators (Atkinson 1985). In a similar fashion, corals may be less likely to bleach in response to rising sea temperatures in areas where they have been repeatedly exposed to temperature stresses in the past (Brown et al. 2000; Podesta and Glynn 2001; West and Salm 2003).

One consequence of the global patterns of extinction and invasion is biotic homogenization. This is the process whereby species assemblages become increasingly dominated by a small number of widespread, human-adapted species. It represents further losses in biodiversity that are often missed when only considering local changes in absolute numbers of species. The many species that are declining as a result of human activities tend to be replaced by a much smaller number of expanding species that thrive in human-altered environments. The outcome is a more homogenized biota with lower diversity at regional and global scales. One effect is that in some regions where diversity has been low because of isolation, the biotic diversity may actually increase—a result of invasions of non-native forms (for example, some continental areas such as the Netherlands as well as oceanic islands). Recent data also indicate that the many losers and few winners tend to be nonrandomly distributed among higher taxa and ecological groups, enhancing homogenization.

4.4.2.5 Conclusion on Species

The rate of species extinction is several orders of magnitude higher than the natural or background rate, even in birds, where the level of threat is the lowest among the assessed taxa. And the great majority of threatened species continue to decline. The geography of declines and extinctions is very uneven and concentrated in particular areas, especially in the humid tropics. Past geographic extinction patterns vary markedly between mammals, birds, and amphibians, but future patterns (as indicated by patterns of currently threatened species) are likely to be more closely correlated. The limited data that exist suggest that biodiversity is more severely threatened in freshwater ecosystems than in terrestrial ecosystems. Studies suggest that ancient taxonomic lineages are particularly prone to extinction (Gaston and Blackburn 1997b; Purvis et al. 2000a). Biodiversity trends in marine ecosystems are yet to emerge, although from the limited data available, it appears that the general trends are not fundamentally different from those in terrestrial ecosystems.

4.4.3 Biomes

Rates of loss of natural land cover for the world's biomes can be measured using a unique dataset on land use change, the HYDE dataset (Klein Goldewijk 2001). This dataset uses information on historical population patterns and agriculture statistics to estimate habitat conversion between 1950 and 1990, based on maps of 0.5-degree resolution. These data indicate that by 1950 all but two biomes (boreal forests and tundra) had lost substantial natural land cover to croplands and pasture. (See Figure 4.26.) Mediterranean forests and temperate grassland biomes had experienced the most extensive conversion, with roughly only 30% of native vegetation cover remaining in 1950.

Loss of native habitat cover has continued, with most biomes experiencing substantial additional percentages of native land cover between 1950 and 1990. The tropical dry broadleaf forests biome has lost the highest percentage of additional habitat (16.1%); only tundra has lost very little if anything to agricultural conversion in those 40 years.

The percentage of remaining habitat in 1950 is highly correlated with rates of additional loss since then. This result indicates that, in general, patterns of human conversion among biomes have remained similar over at least the last century. For example, boreal forests had lost very little native habitat cover through until 1950 and have lost only a small additional percentage since then. In contrast, the temperate grasslands biome had lost nearly 70% of its native cover by 1950 and has lost an additional 15.4% since then. Two biomes appear to be exceptions to this pattern: Mediterranean forests and temperate broadleaf forests. Both of these biomes had lost the majority of their native habitats by 1950 but since then have lost less than 2.5% further habitat. These biomes contain many of the world's most established cities and most extensive surrounding agricultural development (Europe, the

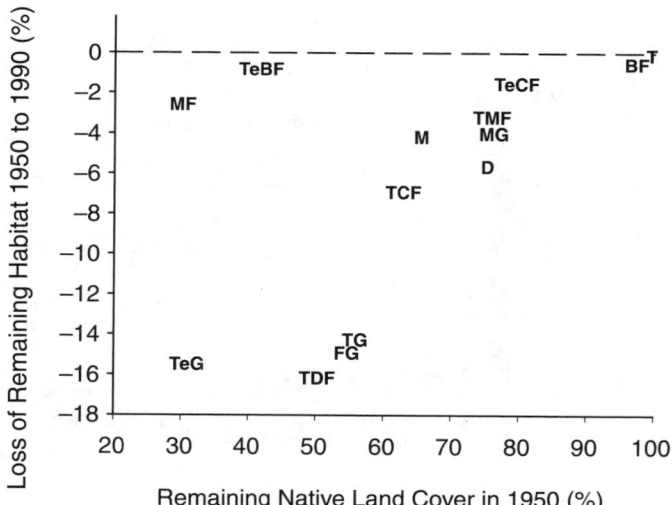

Figure 4.26. Relationship between Native Habitat Loss by 1950 and Additional Losses between 1950 and 1990. Biome codes as in Figure 4.3 (in Appendix A).

United States, the Mediterranean basin, and China). It is possible that in these biomes the most suitable land for agriculture had already been converted by 1950.

In addition to the total amount of habitat loss, the spatial configuration of loss can strongly affect biodiversity. Habitat fragmentation typically accompanies land use change, leaving a complex landscape mosaic of native and human-dominated habitat types. Quantitative data on habitat fragmentation are difficult to compile on the scale of biomes or realms, but habitat fragmentation typically endangers species by isolating populations in small patches of remaining habitat, rendering them more susceptible to genetic and demographic risks as well as natural disasters (Laurance et al. 1997; Boulinier et al. 2001).

Changes in the biodiversity contained within the world's biomes are generally assessed in terms of the species they contain. Changes in species, however, are difficult to measure. Species abundances can fluctuate widely in nature, making it difficult at times to detect a true decline in abundance. And, as described earlier, species extinctions are difficult to count, as the vast majority of species on Earth have yet to be described, extinctions are still relatively rare among known species, and establishing an extinction with confidence is difficult.

Given these difficulties, a reasonable indicator of current and likely future change in biodiversity within a biome is the number of species facing significant extinction risk. The threatened species identified by IUCN are used in this analysis. As such, the analysis is limited to terrestrial vertebrates, which represent less than 1% of the total species on Earth and may not fully represent patterns in other taxa.

Biomes differ markedly in the number of threatened species they contain (see Figure 4.27), with tropical moist forests housing by far the largest number. The percentage of total species that are endangered, however, is more similar among biomes with temperate coniferous forests approaching a similar percentage as tropical moist forests. Comparing these two patterns of threat suggests that higher absolute losses of species in tropical moist forests may be expected, with more similar rates of extinction in other biomes.

4.4.4 Biogeographic Realms

Like biomes, biogeographic realms differ markedly in the amounts of habitat conversion to agriculture before and since

1950. (Klein Goldewijk 2001). (See Figure 4.28.) By 1950, for example, the Indo-Malayan realm had already lost almost half its natural habitat cover. In all realms, at least a quarter of the area had been converted to other land uses by 1950. (These findings exclude Oceania and Antarctica due to lack of data.)

In the 40 years from 1950 to 1990, habitat conversion continued in nearly all biogeographic realms. More than 10% of the land area of the temperate northern realms of the Nearctic and Palearctic as well as the Neotropical realm has been converted to cultivation. Although these realms are currently extensively cultivated and urbanized, the amount of land under cultivation and pasture seems to have stabilized in the Nearctic, with only small increases in the Palearctic in the last 40 years. Within the tropics, rates of conversion to agriculture range from very high in the Indo-

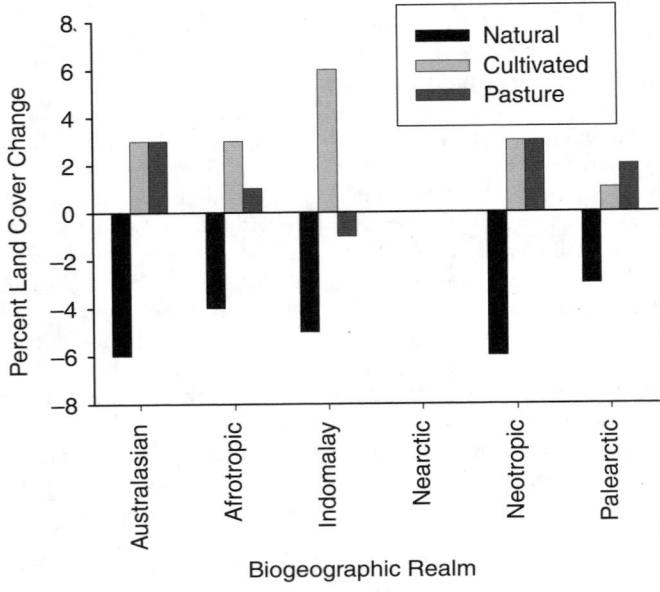

Figure 4.28. Percentage Change (1950–90) in Land Area of Biogeographic Realms Remaining in Natural Condition or under Cultivation and Pasture. Two biogeographic realms are omitted due to lack of data: Oceania and Antarctica.

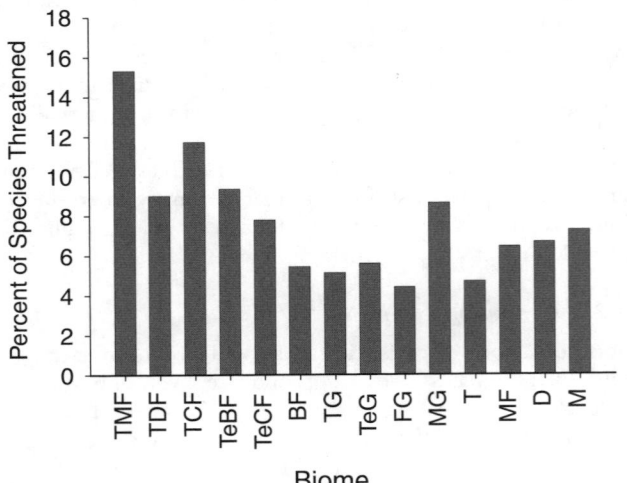

Figure 4.27. Patterns of Species Threat among the World's 14 Terrestrial Biomes. The figures show the raw numbers of threatened species (i.e., ranked as Critically Endangered, Endangered, or Vulnerable by the IUCN) and the percentage of each biome's species that are threatened. Reptiles have not been completely assessed. Biome codes as in Figure 4.3 (in Appendix A).

Malayan realm to moderate in the Neotropics and the Afrotropics, although land cover change has not yet stabilized and shows large increases, especially in cropland area since the 1950s. Australasia has relatively low levels of cultivation and urbanization, but these have also increased in the last 40 years.

As with biomes, the number of threatened vertebrate species differs widely among biogeographic realms (Baillie et al. 2004). (See Figure 4.29.) The largest numbers are found in tropical realms (Neotropic, Indo-Malay, and Afrotropic), while the Nearctic, Oceania, and Antarctica realms hold the least. The percentage of total species that are endangered, however, shows a very different pattern. Most strikingly, over 25% of species in Oceania are threatened, more than twice the percentage of any other realm. The high rates of species threat in Oceania are likely due to well-known factors that endanger island faunas, including high rates of endemism, severe range restriction, and vulnerability to introduced predators and competitors (Manne et al. 1999). Although the Neotropics contain many threatened species, the extraordinary richness of this realm results in a lower percentage of threatened species than Oceania. Therefore, based on species threat levels, we can expect a larger absolute change in biodiversity (measured as expected species extinctions) in the tropical continents, but the highest rates of extinction on tropical islands.

4.5 Improving Our Knowledge of Biodiversity Status and Trends

Biodiversity is a complex concept and so, therefore, is its measurement. Ideally, for any particular assessment, measures of biodiversity would reflect those aspects particularly relevant to the context (Royal Society 2003). For an assessment of ecosystem services for example, biodiversity assessment should be based on measures that are relevant to the provision of services and to human well-being. Unfortunately, the information currently available on global biodiversity is limited, and the data presented in this chapter are therefore rather general. As our understanding of the role of biodiversity improves, so does the potential for better and more relevant measures to be developed. This section considers the need for better indicators of biodiversity status and the specific context of indicators to measure progress against the

2010 biodiversity target, and some clear gaps that will have to be filled if we are to make progress in understanding trends in biodiversity and their consequences are highlighted.

4.5.1 Indicators of Global Biodiversity Status

Documenting trends in biodiversity and the actions and activities that affect it requires suitable indicators. Indicators in this sense are a scientific construct that uses quantitative data to measure biodiversity, ecosystem condition and services, or drivers of change. A useful indicator will provide information about changes in important processes, be sensitive enough to detect important changes but not so sensitive that signals are masked by natural variability, detect changes at the appropriate temporal and spatial scale without being overwhelmed by variability, be based on well-understood and generally accepted conceptual models of the system to which it is applied, be based on reliable data to assess trends and have a relatively straightforward data collection process, have monitoring systems in place for the underlying data needed to calculate the indicator, and be easily understood by policy-makers (NRC 2000 and see also Chapter 2).

Unfortunately, as noted earlier, most existing biological measures, especially those reflecting species richness or various aspects of species diversity, do not reflect many important aspects of biodiversity, especially those that are significant for the delivery of ecosystem services. In addition, few measures have been repeated to allow for a fair assessment of trends over time. Care also needs to be taken in the interpretation of these measures and their use as indicators. For example, these simple measures of species richness may not differentiate between native and invasive or introduced species, differentiate among species in terms of sensitivity or resilience to change, or focus on species that fulfill significant roles in the ecosystem (such as pollinators or decomposers). Moreover, many measures depend on the definition of the area and may be scale-dependent, and they may not always reflect biodiversity trends accurately.

Aggregate indicators of trends in species populations such as the Index of Biotic Integrity for aquatic systems (Karr and Dudley 1981) and the Living Planet Index (Loh and Wackermagel 2004; Loh et al. 2005) use published data on trends in populations of a variety of wild species to identify overall trends in species abundance and, by implication, the condition of the ecosystems in

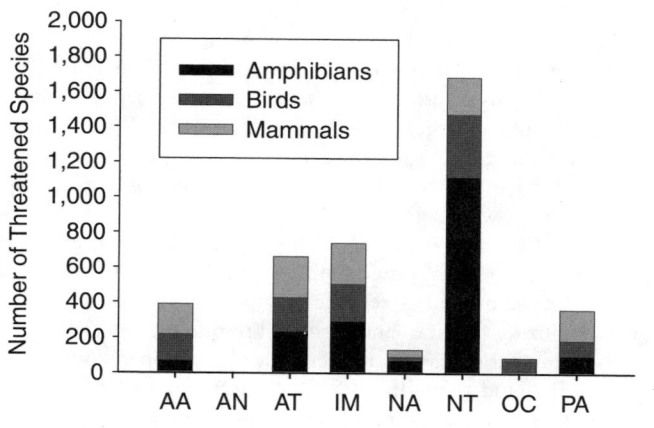
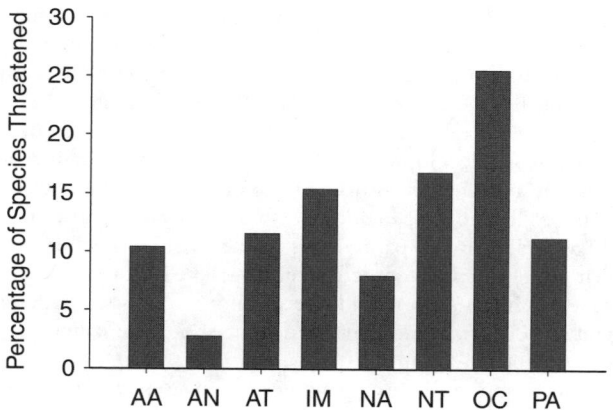

Figure 4.29. Patterns of Species Threat among the World's Eight Terrestrial Biogeographic Realms. The figures show the raw numbers of threatened species (i.e., ranked as Critically Endangered, Endangered, or Vulnerable by the IUCN) and the percentage of each realm's species that are threatened. Reptiles have not been completely assessed. Realm codes as in Table 4.3.

which they occur. The LPI can be applied at national, regional, and global levels, as described earlier. Although it is based on a large number of population trends, various sampling biases affect the index, though with care these biases can be addressed (Loh et al. 2005). Being based on local population abundances, the LPI may be an appropriate biodiversity indicator for ecosystem services, especially with careful sampling of the populations included in its calculation. A complementary index is the Red List Index derived from the *IUCN Red List* of threatened species (Butchart et al. 2004). Red List Indices illustrate the relative rate at which a particular set of species changes in overall threat status (that is, projected relative extinction-risk), based on population and range size and trends as quantified by Red List categories. RLIs can be calculated for any representative set of species that has been fully assessed at least twice. The RLI for the world's birds shows that that their overall threat status has deteriorated steadily during 1988–2004 in all biogeographic realms and ecosystems. A preliminary RLI for amphibians for 1980–2004 shows similar rates of decline (Butchart et al 2005). Both these indexes (LPI and RLI) synthesize much detailed information into a few compelling data points and are being used for assessing progress against the Convention on Biological Diversity's 2010 target.

Currently there has been much less attention paid to the development of indicators for aspects of biodiversity other than species and populations. One recent attempt to collate and synthesize all up-to-date estimates of global trends in population size or habitat extent could find global estimates of habitat change (spanning at least five years since 1992) for only four major biomes (tropical forest, temperate and boreal forest, seagrass, and mangroves) (Balmford et al. 2003a). Neither has there been a focused process to measure the intensity and trends in the key drivers of biodiversity change or the implementation and effectiveness of response options. (See Chapter 5 of the *Policy Responses* volume.) It is clear that a broader set of biodiversity indicators is required, with indicators that are aligned against valued aspects of biodiversity. The adoption of the 2010 biodiversity target makes this task still more urgent (Balmford et al. 2005; Green et al. 2005).

4.5.2 The CBD 2010 Biodiversity Target

In April 2002, at the Sixth Conference of the Parties of the Convention on Biological Diversity, 123 Ministers described a desire to halt biodiversity loss and further committed themselves to actions to "achieve, by 2010, a significant reduction of the current rate of biodiversity loss at the global, regional and national levels as a contribution to poverty alleviation and to the benefit of all life on earth" (Decision VI/26; CBD Strategic Plan). Carrying this message forward, the world's leaders, at the World Summit on Sustainable Development, set a target for "a significant reduction in the current rate of loss of biological diversity" by 2010. This target has now been adopted formally by the parties to the CBD as well as by all participants in the WSSD. The same or a similar target is being adopted at regional levels. For example, the European Union Council adopted a more ambitious target and agreed in 2001 "that biodiversity decline should be halted . . . by 2010" (European Council 2001).

Apart from the data gathering needed to assess progress against the target, its formulation poses some technical challenges, especially for its implementation at a global level. First, the measures of biodiversity to be used as indicators need to be available over a sufficient time period, need to have been measured or estimated consistently, and need to be relevant to the goals. At global level, consistent and repeated measures of biodiversity are rather few, although there are better options at national and regional scales.

Second, because the target is "a significant reduction in the current rate of loss of biological diversity," it requires that the rate of loss has declined, not that there is no more biodiversity loss, or even recovery. In this sense, especially given the time scale, this makes achieving the target more realistic. To demonstrate any change in rate, however, at least three estimates of the measure need to be available for a period of time prior to 2010. Two measures at different points in time can only give information about absolute change, not changes in the rate.

Third, consideration needs to be given to timelines against which progress will be measured. In highly degraded systems, the target may be achieved simply because the system is so reduced that further loss has to be at a slower rate. The choice of baseline against which changes are measured will affect the decision about whether or not the target has been met. A slight increase from a recent low biodiversity score might more readily be perceived as a reduction in rate than the same increase compared with a historically longer and greater decline.

4.5.2.1 The Development of Indicators for the 2010 Target

Efforts to develop indicators for the 2010 target have progressed on a number of fronts, most notably through the work of the CBD. The WSSD was followed by the Open-Ended Intersessional Meeting on the Multi-Year Programme of Work, which recommended that CBD COP7 "establish specific targets and timeframes on progress toward the 2010 target" and "develop a framework for evaluation and progress, including indicators" (UNEP 2003a, 2003b, 2003c).

At the Ninth Meeting of the Subsidiary Body on Scientific, Technical and Technological Advice, it was recommended that development and adoption of indicators of biodiversity loss might be accomplished through a pilot phase between the Seventh and Eighth Conference of the Parties to test a limited set of indicators for their suitability and feasibility, to be implemented by national institutes of the Parties and international organizations with relevant data and expertise (Recommendation IX/13). The recommendations from SBSTTA 9 were carried forward at COP 7 in February 2004 via Decision VII/30, in which the Parties agreed to the following seven focal areas for indicator development:

- reducing the rate of loss of the components of biodiversity, including: (i) biomes, habitats and ecosystems; (ii) species and populations; and (iii) genetic diversity;
- promoting sustainable use of biodiversity;
- addressing the major threats to biodiversity, including those arising from invasive alien species, climate change, pollution, and habitat change;
- maintaining ecosystem integrity, and the provision of goods and services provided by biodiversity in ecosystems, in support of human well-being;
- protecting traditional knowledge, innovations and practices;
- ensuring the fair and equitable sharing of benefits arising out of the use of genetic resources; and
- mobilizing financial and technical resources, especially for developing countries, in particular least developed countries and small island developing States among them, and countries with economies in transition, for implementing the Convention and the Strategic Plan.

It was further agreed that goals and sub-targets would be established, and indicators identified, for each of the focal areas. At the time of writing, eight indicators had been identified for immediate testing and a further 13 were under development. (See Table 4.9.)

Table 4.9. Focal Areas, Indicators for Immediate Testing, and Indicators for Future Development, Agreed to by SBSTTA 10, February 2005

Focal Area	Indicator for Immediate Testing	Possible Indicators for Development by SBSTTA or Working Groups
Status and trends of the components of biological diversity	trends in extent of selected biomes, ecosystems, and habitats	
	trends in abundance and distribution of selected species	
	coverage of protected areas	
	change in status of threatened species (Red List indicator under development)	
	trends in genetic diversity of domesticated animals, cultivated plants, and fish species of major socioeconomic importance	
Sustainable use	area of forest, agricultural, and aquaculture ecosystems under sustainable management	proportion of products derived from sustainable sources
Threats to biodiversity	nitrogen deposition	
	numbers and cost of alien invasions	
Ecosystem integrity and ecosystem goods and services	marine trophic index	application to freshwater and possibly other ecosystems
	water quality in aquatic ecosystems	incidence of human-induced ecosystem failure
	connectivity/fragmentation of ecosystems	health and well-being of people living in biodiversity-based, resource-dependent communities
		biodiversity used in food and medicine
Status of traditional knowledge, innovations, and practices	status and trends of linguistic diversity and numbers of speakers of indigenous languages	further indicators to be identified by a Working Group
Status of access and benefit-sharing		indicator to be identified by a Working Group
Status of resource transfers	official development assistance provided in support of the Convention (OECD-DAC-Statistics Committee)	indicator for technology transfer

The first measures to move from the indicators for development to indicators for immediate testing seem likely to be the Red List Index (Butchart et al. 2004) and possibly also new measures of the extent and quality of key habitats. Coral reef extent may be assessed using established methods (Gardner et al. 2003), and there are likely to be other potential habitats whose extent can be assessed, using remote sensing techniques (such as mangroves).

4.5.2.2 Prospects for Meeting the 2010 Target

Despite the fact that at the time of writing there is no agreement on a complete set of indicators to be used for the 2010 target, various lines of evidence indicate that it is unlikely to be met. First, as evident from information in the preceding section, trends are still downwards for most species and populations, and the rate of decline is generally not slowing. The same is true also for data presented in aggregate indices such as the Living Planet Index (Loh and Wackermagel 2004), the Red List Index (Baillie et al. 2004), and the Pan-European Common Bird Monitoring Scheme (Gregory et al. 2003).

In the case of both the simple and aggregate measures, there are a few exceptions of species and ecosystems where declines are slowing or have been reversed. For example, the reduction in decline rates for some temperate woodland bird species and the recovery of some large mammals in Africa are testament to the potential success of effective management. These cases, however, generally result from management interventions that have been in place for many years and in some cases decades and they are in the minority.

For the species and habitats that showed continuing decline in 2004, prospects for meeting the 2010 target will depend on sources of inertia and the time lag between a management intervention and the response. Natural sources of inertia correspond to the time scales inherent to natural systems; for example, all external factors being equal, population numbers grow or decline at a rate corresponding to the average turnover time or generation time. Even though meeting the 2010 target does not require recovery, many natural populations have generation times that limit the long-lasting improvements that can be realistically expected between now and 2010. On top of this is anthropogenic inertia resulting from the time scales inherent in human institutions for decision-making and implementation (MA 2003). For most systems these two sources of inertia will lead to delays of years, and more often decades, in slowing and reversing a declining biodiversity trend. This analysis assumes that the drivers of change could indeed be halted or reversed in the near term, although there is currently little evidence that any of the direct or indirect drivers are slowing or that any are well controlled at large to global scale.

The delay between a driver affecting a system and its consequences for biodiversity change can be highly variable. In the case of species extinctions this process has been well studied, and habitat loss appears to be one direct driver for which lag times will be long . In studies of African tropical forest bird species, the time from habitat fragmentation to species extinction has been estimated to have a half-life of approximately 50 years for fragments of roughly 1,000 hectares (Brooks et al. 1999). In Amazonian forest fragments of less than 100 hectares, half of the bird species were lost in less than 15 years, whereas fragments larger than 100 hectares lost species over time scales of a few decades to perhaps a century (Ferraz et al. 2003).

On the one hand, these time lags mean that estimates of current extinction rates may be underestimates of the ultimate legacy of habitat loss. For example, for African primate populations it is estimated that over 30% of species that will ultimately be lost as a result of historical deforestation still exist in local populations (Cowlishaw 1999). On the other hand, the time lags offer opportunities for interventions to be put in place to slow or reverse the trends, so long as in this case the period to habitat recovery is shorter than the time to extinction.

4.5.3 Key Gaps in Knowledge and Data

Certain gaps in knowledge and data relating to biodiversity are almost certain to prove critical over coming years, and efforts are urgently needed to gather this information, particularly if biodiversity indicators are to become more reliable and informative.

- Data are sparse for certain key taxa—especially invertebrates, plants, fungi, and significant groups of microorganisms, including those in the soil. These groups are especially important for ecosystem services, yet global syntheses and trend information on even significant subsets are entirely missing. It seems likely that both extinction rates and local diversity and endemism may be lower among microorganisms than in the well-studied groups, suggesting that intense monitoring may not be so important. However this remains to be validated. Taxonomy as a discipline underpins much of this work yet is in decline worldwide.
- Conservation assessment has proceeded at increased intensity over recent decades. However, knowledge of biodiversity trends falls far behind knowledge of status. Too often assessments are undertaken using new methods, new measures, or new places. Trends, which are critical to current questions, rely on a time series of comparable measures.
- Local and regional datasets are generally of higher quality and cover longer time periods than global data. A better understanding of the relationship of local to global processes and the development of techniques to allow local dynamics to inform large-scale assessments would allow rapid progress to be made in large to global-scale assessments.
- There are far fewer studies at the genetic level than for populations, species, and ecosystems, yet these are significant components of biodiversity for assessing present and future adaptability to changing environments.
- Marine and freshwater areas are less well known than terrestrial areas. Among terrestrial habitats, biodiversity trends in biomes such as drylands and grasslands are less well known than trends in forests.
- The impacts of biodiversity change on ecosystems services are still poorly understood. Even where knowledge is better, there are almost no studies documenting the trends over time.

Alongside new data, approaches to long-term, large-scale continuous monitoring of biodiversity and attitudes to data sharing will need to be developed, as well as the infra-structure and technical and human resources that such an effort will require.

4.6 A Summary of Biodiversity Trends

The evidence presented in this chapter supports three broad conclusions about recent and impending changes in the amount and variability of biodiversity: there have been and will continue to be substantial changes that are largely negative and largely driven by people; these changes are varied—taxonomically, spatially, and temporally; and the changes are complex, in several respects.

First, changes are substantial and predominantly negative. Although there are very real limitations in the extent and quality of our knowledge of the changing state of nature, we already have overwhelming evidence that humans have caused the loss of a great deal of biodiversity over the past 50,000 years and that rates of loss have accelerated sharply over the past century. Current rates of species extinction are at least two orders of magnitude above background rates and are expected to rise to at least three orders above background rates.

Among extant species, 20% of all species in those groups that have been comprehensively assessed (mammals, birds, amphibians, conifers, and cycads) are believed to be threatened with extinction in the near future. For birds (the only taxon for which enough data are available), this proportion has increased since 1988 (BirdLife 2004a). Even among species not threatened with extinction, the past 20–40 years have seen substantial declines in population size or the extent of range in most groups monitored. These include European and North American farmland birds, large African mammals, nearly 700 vertebrate populations worldwide (Loh 2002), British birds, waders worldwide (IWSG et al. 2002), British butterflies and plants (Thomas et al. 2004a), amphibians worldwide (Houlahan et al. 2000; Alford and Pechmann 2001; Stuart et al. 2004), and most commercially exploited fish. These declines in populations are broadly mirrored by declines in the extent and condition of natural habitats (Jenkins et al. 2003).

Second, changes are varied. Rates of biodiversity decline, although very largely negative, vary widely on at least three dimensions. Taxonomically, certain groups appear more vulnerable to change than others: thus amphibians, and freshwater organisms in general, exhibit higher levels of threat and steeper rates of population decline than do better-known groups such as birds or mammals (Houlahan et al. 2000; Alford and Pechmann 2001; Loh 2002). Within groups, phylogenetically distinct, ancient, and species-poor lineages seem consistently to be faring disproportionately badly. Some generalist species are expanding their ranges, either naturally or as invasive aliens, whereas many ecological specialists are in decline.

Spatially, most species losses to date have been concentrated on islands. Disproportionately high rates of contemporary habitat conversion in endemic-rich areas of the tropics, where areas of dense human settlement and high species richness tend to coincide, mean that impending extinctions are particularly concentrated in tropical island and montane systems. In temperate regions, in contrast, substantial historical reductions in habitat extent have led to relatively few global extinctions (due in part to species having larger ranges at higher latitudes). Currently, populations and habitats are expanding in some temperate regions, such as temperate forests (Jenkins et al. 2003). Freshwater and marine patterns are less well documented.

Temporally, two patterns stand out. The first is that the scale of loss is in general increasing (although it is important to note that, both on land and at sea, preindustrial human-caused losses were also very substantial (Jackson 1994; Jackson et al. 2001)). The second pattern is that the anthropogenic drivers of loss are also changing; for example, invasive species and overexploitation were the predominant causes of bird extinctions in historic times, while habitat conversion, especially to agriculture, is the most significant driver currently facing threatened species (Baillie et al. 2004; BirdLife 2004a), with climate change predicted to emerge as another major threat in the near future (Thomas et al. 2004a).

Third, changes are complex. Besides variety, the overriding feature of biodiversity is its complexity. Patterns of biodiversity loss are in turn correspondingly complex, in several respects. Species, populations, and ecosystems differ not just in their exposure but also in their vulnerability to anthropogenic drivers of change. In addition, complex interactions within communities mean that changes in the abundance of one species will often have broad-ranging effects through a system. (See also Chapter 12.) One well-documented example is the recent switch by Aleutian island killer whales to hunting sea otters instead of pinnipeds (likely triggered by fishing-related declines in pinnipeds); this has greatly reduced sea otter numbers, allowing the population and grazing pressure of sea urchins to increase, in turn leading to a dramatic decline in kelp density (Estes et al. 1998). In Australia, the deliberate introduction of African grasses (such as gamba grass, *Andropogon gayanus*) to native woody savannas has also increased the intensity of frequent, very intense fires due to the highly flammable nature of the introduced grasses (Rossiter et al. 2003); as elsewhere (D'Antonio and Vitousek 1992), changes in the fire regime in turn reduced native tree and shrub cover, thereby accelerating the invasion of fire-tolerant aliens and resulting in a wholesale ecosystem shift from woody vegetation to open grassland.

Another aspect of the complexity is that community dynamics mean threats themselves rarely operate in isolation (Myers 1995). The impact of climate change, for example, is predicted to be far more marked where habitat transformation and fragmentation blocks the movement of species in response to shifting climate (Thomas et al. 2004a), a hypothesis recently supported by data on U.K. butterflies (Warren et al. 2001). Similarly, there is now growing evidence of synergistic effects of increased UV-B exposure, acidification, and pathogens on declining amphibian populations (Kiesecker and Blaustein 1995; Long et al. 1995) and of synergistic effects between logging, forest fragmentation, and fire in tropical forests (Cochrane et al. 1999; Cochrane 2003).

It is also becoming clear that often ecosystems respond not linearly to external changes but in a stepwise manner (Myers 1995). Thus cumulative biotic or abiotic pressures that at first appear to have little effect may lead to quite sudden and unpredictable changes once thresholds are crossed (Scheffer et al. 2001). Moreover, such thresholds may become lower as anthropogenic impacts simplify systems and reduce their intrinsic resilience to change. (See also Chapter 12.) One well-studied example is the sudden switch in 1983 from coral to algal domination of Jamaican reef systems. This followed several centuries of overfishing of herbivores, which left the control of algal cover almost entirely dependent on a single species of sea urchin, whose populations collapsed when exposed to a species-specific pathogen (Hughes 1994; Jackson 1997). As a result, Jamaica's reefs shifted (apparently irreversibly) to a new low diversity, algal-dominated state with very limited capacity to support fisheries (McManus et al. 2000). Given their potential importance, much more work is needed on whether threshold effects such as this are typical, how reversible they are, and where thresholds lie.

Extrapolation from current trends suggests that both the amount and variability of nature will continue to decline over much of Earth (UNEP 2002a; Jenkins et al. 2003). The exception is likely to be in some industrial countries, where forest cover may continue to increase and, with it, the population sizes of many forest-dependent species. In contrast, clearance of natural habitats, reductions of populations, and the associated loss of populations and indeed species look set to persist and even accelerate across much of the tropics and across many if not most aquatic systems. Particularly vulnerable areas include cloud forests, coral reefs, mangroves (threatened by the synergistic effects of climate change and habitat clearance), all but the very largest blocks of tropical forest, and most freshwater habitats. Particularly vulnerable taxa include large marine species, large-bodied tropical vertebrates, and many freshwater groups (Jenkins et al. 2003).

References

Agapow, P.-M., O.R.P. Bininda-Emonds, K.A. Crandall, J.L. Gittleman, G.M. Mace, J.C. Marshall, and A. Purvis, 2004: The impact of species concept on biodiversity studies. *Quarterly Review of Biology, 79*, 161–179.

Alford, R.A. P.M. Dixon and J.H.K. Pechmann, 2001: Global Amphibian population declines. *Nature, 412*, 499–500.

Alroy, J., 1998: Equilibrial diversity dynamics in North American mammals. In: *Biodiversity dynamics, turnover of populations, taxa, and communities*, M.L. McKinney and J.A. Drake (eds.), Columbia University Press, New York, 232–287.

Alroy, J., 2001: A multi-species overkill simulation of the end-Pleistocene megafaunal mass extinction. *Science, 292*, 1893–1986.

Anderson, S., 1994: Area and endemism. *Quarterly Review of Biology, 69*, 451–471.

Anstey, S., 1991: *Wildlife Utilisation in Liberia (World Wide Fund for Nature and FDA Wildlife Survey)*. World Wide Fund for Nature and FDA, Gland, Switzerland.

APG (Angiosperm Phylogeny Group), 1998: An ordinal classification for the families of flowering plants. *Annals of the Missouri Botanic Garden, 85*, 531–553.

Atkinson, A.I.E., 1985: The spread of commensal species of *Rattus* to oceanic islands and their effects on island avifaunas. In: *The Conservation of Island Birds. Case Studies for the Management of Threatened Bird Species*, P.J. Moors (ed.), ICBP, Cambridge.

Atkinson, C.T., R.J. Dusek, K.L. Woods, and W.M. Iko, 2000: Pathogenicity of avian malaria in experimentally-infected Hawaii Amakihi. *Journal of Wildlife Diseases, 36(2)*, 197–204.

Atkinson, I.A.E. and E.K. Cameron, 1993: Human Influence On the Terrestrial Biota and Biotic Communities of New-Zealand. *Trends in Ecology & Evolution, 8(12)*, 447–451.

Avise, J.C. and G.C. Johns, 1999: Proposal for a standardized temporal scheme of biological classification for extant species. *Proceedings of the National Academy of Science USA, 96*, 7358–7363.

Bailey, R.G., 1989: Explanatory supplement to Ecoregions map of the continents. *Environmental Conservation, 16*, 307–309.

Bailey, R.G., 1998: *Ecoregions: the ecosystem geography of oceans and continents*. Springer-Verlag, New York.

Bailey, R.G. and H.C. Hogg, 1986: A world ecoregions map for resource partitioning. *Environmental Conservation, 13*, 195–202.

Baillie, J.E.M., C. Hilton-Taylor, and S.N. Stuart, 2004: *2004 IUCN Red List of Threatened Species. A Global Species Assessment*. IUCN, Gland, Switzerland.

Balcalbaca-Dobrovici, N., 1989: The Danube River and it's Fisheries. In: *Proceedings of the International Large River Symposium*, D.P. Dodge (ed.). 106, 455–468.

Balmford, A., 1996: Extinction filters and current resilience: the significance of past selection pressures for conservation biology. *Trends in Ecology & Evolution, 11(5)*, 193–196.

Balmford, A., 1999: (Less and less) great expectations. *Oryx, 33*, 87–88.

Balmford, A. and A. Long, 1994: Avian endemism and forest loss. *Nature, 372*, 623–624.

Balmford, A., M.J.B. Green, and M.G. Murray, 1996: Using higher-taxon richness as a surrogate for species-richness: I. Regional tests. *Proceedings of the Royal Society of London, series B Biological Sciences, 263*, 1267–1274.

Balmford, A., R.E. Green, and M. Jenkins, 2003a: Measuring the changing state of nature. *Trends in Ecology & Evolution, 18(7)*, 326–330.

Balmford, A., K.J. Gaston, S. Blyth, A. James, and V. Kapos, 2003b: Global variation in terrestrial conservation costs, conservation benefits, and unmet conservation needs. *Proceedings of the National Academy of Sciences of the United States of America*, **100(3)**, 1046–1050.

Balmford, A., P. Crane, A. Dobson, R.E. Green, and G.M. Mace, 2005: The 2010 challenge: data availability, information needs, and extraterrestrial insights. *Philosophical Transactions of the Royal Society of London B.* 360:221–228.

Barnes, R.F.W., G.C. Craig, H.T. Dublin, G. Overton, W. Simmons, and C.R. Thouless, 1999: African Elephant Database 1998. IUCN, Gland, Switzerland.

Barrett, S.C., 1989: Water Weed Invasions. *Scientific American,* **October,** 90–97.

Barthlott, W., N. Biedinger, G. Braun, F. Feig, G. Kier, and J. Mutke, 1999: Terminological and methodological aspects of the mapping and analysis of global biodiversity. *Acta Botanica Fennica,* **162,** 103–110.

Baum, J.K., R.A. Myers, D.G. Kehler, B. Worm, S.J. Harley, and P.A. Doherty, 2003: Collapse and conservation of shark populations in the Northwest Atlantic. *Science,* **299(5605),** 389–392.

Bawa, K.S. and S. Dayanandan, 1997: Socioeconomic factors and tropical deforestation. *Nature,* **386,** 562–563.

Bennett, E.L., A. Nyaoi, and J. Sompud, 2000: Saving Borneo's Bacon: The Sustainability of Hunting in Sarawak and Sabah. In: *Hunting for Sustainability,* J.G. Robinson and E.L. Bennett (eds.), Columbia Press, New York, 305–324.

Bennett, P.M. and I.P.F. Owens, 1997: Variation in extinction risk among birds: Chance or evolutionary predisposition? *Proceedings of the Royal Society of London Series B-Biological Sciences,* **264(401–408).**

Berger, L., R. Speare, P. Daszak, D.E. Green, A.A. Cunningham, et al. 1998: Chytridiomycosis causes amphibian mortality associated with population declines in the rainforests of Australia and Central America. *Proceedings of the National Academy of Sciences of the United States of America,* **95,** 9031–9036.

Berra, T.M., 2001: *Freshwater fish distribution.* Academic Press, San Diego.

BirdLife, 2004a: *Threatened Birds of the World 2004.* Birdlife International, Cambridge, UK.

BirdLife, 2004b: *State of the World's Birds 2004—Indicators for our changing world.* BirdLife International, Cambridge, U.K.

Bisby, F.A., 1995: Characterization of biodiversity. In: *Global Biodiversity Assessment,* V.H. Heywood (ed.), Cambridge University Press, Cambridge.

Bisby, F.A., J. Shimura, M. Ruggiero, J. Edwards, and C. Haeuser, 2002: Taxonomy, at the click of a mouse. *Nature,* **418,** 367.

Bogan, A.E., 1997: The Silent Extinction. *American Paleontologist,* **5(1),** 2–4.

Boulinier, T., J.D. Nichols, J.E. Hines, J.R. Sauer, C.H. Flather, and K.H. Pollock, 2001: Forest fragmentation and bird community dynamics: Inference at regional scales. *Ecology,* **82(4),** 1159–1169.

Bramwell, D., 2002: How many plant species are there? *Plant Talk,* **28,** 32–33.

Briggs, J., 1974: *Marine Zoogeography.* McGraw-Hill, New York.

Brook, B.W. and D. Bowman, 2004: The uncertain blitzkrieg of Pleistocene megafauna. *Journal of Biogeography,* **31,** 517–523.

Brooks, T.M., S.L. Pimm, and J.O. Oyugi, 1999: Time lag between deforestation and bird extinction in tropical forest fragments. *Conservation Biology,* **13,** 1140–1150.

Brown, B.E., R.P. Dunne, M.S. Goodson, and A.E. Douglas, 2000: Bleaching patterns in coral reefs. *Nature,* **404,** 142–143.

Brown, J.H., 1984: On the relationship between abundance and distribution of species. *American Naturalist,* **124,** 255–279.

Butchart, S.H.M., A.J. Stattersfield, L.A. Bennun, S.M. Shutes, H.R. Ackakaya, et al. 2004: Measuring global trends in the status of biodiversity: Red List Indices for Birds. *PLoS. Biology,* **2(12),** e383.

Butchart, S.H.M., A.J. Stattersfield, J. Baillie, L.A. Bennun, S.N. Stuart, et al. 2005, Using Red List Indices to measure progress towards the 2010 target and beyond. *Philosophical Transactions of the Royal Society. Series B.,* **360,** 255–268.

Castro, J.I., C.M. Woodley, and R.L. Brudek, 1999: *A preliminary evaluation of the status of shark species (FAO Fisheries Technical Paper no. 380).* FAO (Fisheries), Rome.

Ceballos, G. and P.R. Ehrlich, 2002: Mammal population losses and the extinction crisis. *Science,* **296,** 904–907.

Channell, R. and M.V. Lomolino, 2000: Dynamic biogeography and the conservation of endangered species. *Nature,* **403,** 84–86.

Chape, S., S. Blyth, L. Fish, P. Fox, and M. Spalding, 2003: 2003 United Nations List of Protected Areas. I.a. UNEP (ed.)Gland, Switzerland.

Cheke, A.S., 1987: An ecological history of the Mascarene Islands, with particular reference to extinctions and introductions of land vertebrates. In: *Studies of Mascarine Island birds,* A.W. Diamond (ed.), Cambridge University Press, Cambridge, 5–89.

Claridge, M.F., H.A. Dawah, and M.R. Wilson, 1997: *Species: The Units of Biodiversity.* Chapman & Hall, London.

Cochrane, M.A., 2003: Fire science for rainforests. *Nature,* **421(6926),** 913–919.

Cochrane, M.A., A. Alencar, M.D. Schulze, C.M. Souza, D.C. Nepstad, P. Lefebvre, and E.A. Davidson, 1999: Positive feedbacks in the fire dynamic of closed canopy tropical forests. *Science,* **284(5421),** 1832–1835.

Colwell, R.K. and D.C. Lees, 2000: The mid-domain effect: geometric constraints on the geography of species richness. *Trends in Ecology and Evolution,* **15,** 70–76.

Cowlishaw, G., 1999: Predicting the Pattern of Decline of African Primate Diversity: an Extinction Debt from Historical Deforestation. *Conservation Biology,* **13,** 1183–1193.

Cracraft, J., 1983: Species concepts and speciation analysis. *Current Ornithology,* **1,** 159–187.

Crandall, K.A., O.R.P. Bininda-Emonds, G.M. Mace, and R.K. Wayne, 2000: Considering evolutionary processes in conservation biology. *Trends in Ecology & Evolution,* **15(7),** 290–295.

Crooks, K.R. and M.E. Soule, 1999: Mesopredator release and avifaunal extinctions in a fragmented system. *Nature,* **400(6744),** 563–566.

Cumming, D.H.M., R.F. du Toit, and S.N. Stuart, 1990: *African Elephants and Rhinos. Status Survey and Conservation Action Plan.,* IUCN, Gland, Switzerland.

Curtis, T.P., W.T. Sloan, and J.W. Scannell, 2002: Estimating prokaryotic diversity and its limits. *Proceedings of the National Academy of Sciences of the United States of America,* **99(16),** 10494–10499.

D'Antonio, C.M. and P.M. Vitousek, 1992: Biological invasions by exotic grasses, the grass-fire cycle, and global change. *Annual Review of Ecology and Systemantics,* **23,** 63–87.

Darwall, W. and C. Revenga ((in prep)): Inland Water Ecosystems. In The State of the Worlds Protected Areas. IUCN, Gland, Switzerland and Cambridge.

Daszak, P., A.A. Cunningham, and A.D. Hyatt, 2000: Emerging Diseases of Wildlife—Threats to Biodiversity and Human Health. *Science,* **287**(443–449).

Daszak, P., A.A. Cunningham, and A.D. Hyatt, 2001: Anthropogenic environmental change and the emergence of infectious diseases in wildlife. *Acta Tropica,* **78,** 103–116.

De Meulenaer, T. and C. Raymakers, 1996: *Sturgeons of the Caspian Sea and the International Trade in Caviar.,* TRAFFIC, TRAFFIC Europe., Brussels, Belgium.

Dekker, W., 2003: *Eel stocks dangerously close to collapse.,* ICES (International Council for the Exploration of the Sea), Copenhagen, Denmark.

Delany, S.N., C. Reyes, E. Hubert, S. Pihl, E. Rees, L. Haanstra, and A. van Strien, 1999.: *Results from the International Waterbird Census in the Western Palearctic and Southwest Asia, 1995 and 1996.* 54, Wetlands International Publication No. 54., Wageningen, The Netherlands:.

Diamond, J.M., 1972: Biogeographic kinetics: estimation of relaxation times for avifaunas of Southwest Pacific Islands. *Proceedings of the National Academy of Sciences, USA,* **69,** 3199–3203.

Dietrich, G., 1963: *General Oceanography: An Introduction.* John Wiley, New York.

Dirzo, R. and P.H. Raven, 2003: Global state biodiversity and loss. *Annual Review of Environmental Resources,* **28,** 137–167.

Dobson, A. and J. Foufopoulos, 2001: Emerging infectious pathogens of wildlife. *Philosophical Transactions of the Royal Society of London B,* **356,** 1001.

Dobzhansky, T., 1950: Mendelian populations and their evolution. *American Naturalist,* **84,** 401–418.

Doubleday, W.G., 2001: Is Atlantic Salmon Aquaculture a Threat to Wild Stocks in Altantic Canada? *ISUMA: Canadian Journal of Policy Research 2,* **2(1).**

Dulvy, N.K., Y. Sadovy, and J.D. Reynolds, 2003: Extinction vulnerability in marine populations. *Fish and Fisheries,* **4,** 25–64.

Dynesius, M.a.C.N., 1994: Fragmentation and Flow Regulation of River Systems in the Northern Third of the World. *Science,* **266,** 753–762.

East, R., 1999: African Antelope Database 1998. IUCN, Gland, Switzerland.

Easterling, W.E., J.R. Brandle, C.J. Hays, Q. Guo, and D.S. Guertin, 2001: Simulating the impact of human land use change on forest composition in the Great Plains agroecosystems with the Seedscape model. *Ecological Modelling,* **140,** 163–176.

Ehrlich, P.R., 1994: Energy use and biodiversity loss. *Philosophical Transactions of the Royal Society of London B: Biological Sciences,* **344,** 99–104.

Emslie, R. and M. Brooks, 1999: *African Rhino. Stauts Survey and Conservation Action Plan.* IUCN, Gland, Switzerland.

Environment Canada and U.S. EPA (Environmental Protection Agency), 2003: Great Lakes Atlas. Living Resources. Environment Canada and U.S. EPA. Ottawa, Canada.

Erwin, T.L., 1982: Tropical forests: their richness in Coleoptera and other Arthropod species. *The Coleopterist's Bulletin,* **36,** 74–75.

Estes, J.A., M.T. Tinker, T.M. Williams, and D.F. Doak, 1998: Killer Whale Predatoin on Sea Otters Linking Oceanic and Nearshore Ecosystems. *Science,* **282,** 473–476.

European Council, 2001: *Presidency Conclusions.* SN/200/1/01 REV1, Goteburg Council.

Everett, R.A., 2000: Patterns and pathways of biological invasions. *Trends in Ecology & Evolution,* **15,** 177–178.

Faith, D.P., 1992: Conservation evaluation and phylogenetic diversity. *Biological Conservation,* **61,** 1–10.

FAO (Food and Agriculture Organization of the United Nations), 1997: *The State of the World's Forests 1997.* Rome, Italy.

FAO, 1999: *Review of the State of World Fishery Resources: Inland Fisheries. FAO Inland Water Resources and Aquaculture Service, Fishery Resources Division.* FAO Fisheries Circular No. 942, Rome.

FAO, 2000a: *The State of the World's Fisheries and Aquaculture 2000.* FAO, Rome.

FAO, 2000b: *Global Forest Resource Assessment 2000.* Food and Agriculture Organization, Rome.

FAO, 2001a: *Global Forest Resources Assessment 2000 Main Report.* FAO Forest Paper 140, Food and Agriculture Organization of the United Nations, Rome, 479 pp.

FAO, 2001b: *Forest Resources Assessment 2000.* FAO Forestry Paper 140, FAO, Rome.

FAO, 2002: *State of the World's Fisheries and Aquaculture Report 2002.* FAO.

FAO, 2003: *State of the World's Forest 2003.* FAO, Rome.

Fekete, B., C.J. Vörösmarty, and W. Grabs, 1999: *Global, Composite Runoff Fields Based on Observed River Discharge and Simulated Water Balance.* World Meteorological Organization Global Runoff Data Center Report No. 22., Koblenz, Germany.

Ferraz, G., G.J. Russell, P.C. Stouffer, R.O. Bierregaard, S.L. Pimm, and T.E. Lovejoy, 2003: Rates of species loss from Amazonian forest fragments. *Procedings of the National Academy of Sciences of the USA,* **100(24),** 14069–14073.

Finlay, B.F. and T. Fenchel, 2004: Cosmopolitan Metapopulations of Free-Living Microbial Eukaryotes. *Protist,* **155,** 237–244.

Finlayson , C. M., G. W. Begg, J. Howes, J. Davies, K. Tagi, and L. Lowry, 2002. Asian Wetland Inventory, Wetlands International, Netherlands.

Fisher, D.O., S.P. Blomberg, and I.P.F. Owens, 2003: Extrinsic versus intrinsic factors in the decline and extinction of Australian marsupials. *Proceedings of the Royal Society of London Series B-Biological Sciences,* **270,** 1801–1808.

Fitter, A.H., 2005. Darkness visible: reflections on underground ecology. *Journal of Ecology,* **93,** 231–243.

FitzGibbon, C.D., H. Mogaka, and J. Fanshawe, 2000: Threatened Mammals, Subsistence Harvesting, and High Human Population Densities: A recipe for disaster? In: *Hunting for Sustainability,* J.G. Robinson and E.L. Bennett (eds.), Columbia Press, New York, 154–167.

Fjeldså, J. and C. Rahbek, 1998: Continent-wide conservation priorities and diversification processes. In: *Conservation in a changing world,* G.M. Mace, A. Balmford, and J.R. Ginsberg (eds.), Cambridge University Press, Cambridge, UK, 139–160.

Foose, T.J. and N. van Strien, 1997: *Asian Rhinos. Status Survey and Conservation Action Plan (2nd Edition).* IUCN, Gland, Switzerland.

Foote, M., 1997: Estimating taxonomic durations and preservation probability. *Palaeobiology,* **23,** 278–300.

Ford, R.G., 1999: Defining Marine Ecoregions of the Pacific Continental United States. In: *Terrestrial ecoregions of North America: a conservation assessment.,* T.e.a. Ricketts (ed.), Island Press, Washington DC.

Frankham, R., 1995: Conservation genetics. *Annual Review Genetics,* **29,** 305–327.

Frankham, R., B.J. D., and D.A. Briscoe, 2002: *Introduction to Conservation Genetics.* Cambridge University Press, Cambridge, UK.

Freitag, S., A.O. Nicholls, and A.S. van Jaarsveld, 1998: Dealing with established reserve networks and incomplete distribution data sets in conservation planning. *South African Journal of Science,* **94,** 79–86.

Fritts, T.H. and G.H. Rodda, 1998: The role of introduced species in the degradation of island ecosystems: A case history of Guam. *Annual Review of Ecology and Systematics,* **29,** 113–140.

Froese, R. and D.E. Pauly, 2003: FishBase.World Wide Web electronic publication. [Internet] Cited 12 November 2003. Available at www.fishbase.org.

Gardner, T.A., I.M. Cote, J.A. Gill, A. Grant, and A.R. Watkinson, 2003: Long-term region-wide declines in Caribbean corals. *Science,* **301,** 958–960.

Gaston, K.J., 1991: The magnitude of global insect species richness. *Conservation Biology,* **5,** 283–296.

Gaston, K.J., 1994: *Rarity.* Chapman and Hall, London.

Gaston, K.J., 1994a: Spatial patterns of species distribution: how is our knowledge of the global insect fauna growing? *Biological Conservation,* **67,** 37–40.

Gaston, K.J., 1996: Species-range-size distributions: patterns, mechanisms and implications. *Trends in Ecology and Evolution,* **11,** 197–201.

Gaston, K.J., 1998: Species-range size distributions: products of speciation, extinction and transformation. *Philosophical Transactions of the Royal Society of London B,* **353,** 219–230.

Gaston, K.J., 2000: Global patterns in biodiversity. *Nature,* **405,** 220–227.

Gaston, K.J. and R.M. May, 1992: Taxonomy of taxonomists. *Nature,* **356,** 281–282.

Gaston, K.J. and T.M. Blackburn, 1997b: Evolutionary age and risk of extinction in the global avifauna. *Evolutionary Ecology,* **11,** 557–565.

Gaston, K. J., Jones, A. G., Hänel, C. & Chown, S. L. 2003 Rates of species introduction to a remote oceanic island. *Proceedings of the Royal Society of London Series B-Biological Sciences* **270,** 1091–1098.

Gibbins, J.W., D.E. Scott, T.J. Ryan, K.A. Buhlmann, T.D. Tuberville, et al. 2000: The global decline of reptiles, déjà vu amphibians. *BioScience,* **50,** 653–666.

Gilpin, M.E. and M.E. Soulé, 1986: Minimum viable populations: processes of species extinctions. In: *Conservation Biology—the Science of Scarcity and Diversity,* M.E. Soule (ed.), Sinauer Associates, Michigan, 19–34.

Godfray, H.C.J., 2002: Challenges for taxonomy: the discipline will have to reinvent itself if it is to flourish. *Nature,* **417,** 17–19.

Godfray, H.C.J., O.T. Lewis, and J. Memmott, 1999: Studying insect diversity in the tropics. *Philosophical Transactions of the Royal Society of London B,* **354,** 1811–1824.

Goldburg, R.J., M.S. Elliott, and R.L. Naylor, 2001: *Marine Aquaculture in the United States: Environmental Impacts and Policy Options.* Pew Oceans Commission. Arlington, Virgina.

Gómez de Silva Garza, H., 1996: The conservation importance of semiendemic species. *Conservation Biology,* **10,** 674–675.

Goodman, D., 1987: The demography of chance extinction. In: *Viable populations for conservation,* M.E. Soule (ed.), Cambridge University Press, Cambridge, 11–34.

Goodman, S.M. and B.D. Patterson (eds.), 1997: *Natural change and Human Impact in Madagascar.* Smithsonian Institution, Chicago.

Govaerts, R., 2001: How many species of seed plants are there? *Taxon,* **50,** 1085–1090.

Grassle, J.F. and N.J. Maciolek, 1992: Deep-sea species richness: regional and local diversity estimates from quantitative bottom samples. *American Naturalist,* **139,** 313–341.

Green, R.E., A. Balmford, P.R. Crane, G.M. Mace, J.R. Reynolds, and R.K. Turner, 2005: Responding to the Johannesburg Challenge: A Framework for Improved Monitoring of Biodiversity. *Conservation Biology* **19,** 56–65.

Gregory, R.D., P. Vorisek, A.J. van Strien, M. Eaton, and S.R. Wotton, 2003: *From bird monitoring to policy-relevant indicators. A report to the European Topic Centre on Nature Protection and Biodiversity.*

Griffiths, R.W., 1993: The changing environment of Lake St. Clair. Paper presented at the *Third International Zebra Mussel Conference,* Toronto, Canada.

Groombridge, B. and M.D. Jenkins, 2002: *World Atlas of Biodiversity.* University of California Press, Berkeley.

Hall, B.L. and R.E. Moreau, 1962: A study of the rare birds of Africa. *Bulletin of the British Museum (Natural History) Zoology,* **8,** 313–378.

Hammond, P.M., 1992: Species inventory. In: *Global Biodiversity,* B. Groombridge (ed.), Chapman & Hall, London, UK, 17–39.

Harcourt, A.H., S.A. Parks, and R. Woodroffe, 2001: Human density as an influence on species/area relationships: double jeopardy for small African reserves? *Biodiversity and Conservation,* **10,** 1011–1026.

Harrison, I.J. and M.J. Stiassny, 1999: The Quiet Crisis: A Preliminary Listing of the Freshwater Fishes of the World that Are Extinct or "Missing in Action." In: *Extinctions in Near Time,* MacPhee (ed.), New York: Kluwer Academic/Plenum Publishers, New York, 271–331.

Hawksworth, D.L., 1991: The fungal dimension of biodiversity: magnitude, significance and conservation. *Mycological Research,* **95,** 641–655.

Hayden, B., G. Ray, and R. Dolan, 1984: Classification of coastal and marine environments. *Environmental Management,* **11,** 199–207.

He, F. and K.J. Gaston, 2000: Estimating species abundance from occurrence. *American Naturalist,* **156,** 553–559.

Hebert, P.D.N., A. Cywinska, S.L. Ball, and J.R. De Waard, 2003: Biological identification through DNA barcodes. *Proceedings of the Royal Society of London, series B Biological Sciences,* **270,** 313–321.

Hedrick, P.W. and T.J. Kim, 2000: Genetics of complex polymorphisms: parasites and maintenance of MHXC variation. In: *Evolutionary Genetics from Mol-*

ecules to Morphology, R.S.S.C.K. Krimbas (ed.), Cambridge University Press, Cambridge, UK, 204–234.

Hey, J., 2001: The mind of the species problem. *Trends in Ecology and Evolution,* **16,** 326–329.

Heywood, V.H. and R.T. Watson (eds.), 1995: *Global Biodiversity Assessment.* United Nations Environment Programme., Cambridge, U.K.

Hill, G., J. Waage, and G. Phiri, 1997: The Water Hyacinth Problem in Tropical Africa. In: *Proceedings of the International Water Hyacinth Consortium,* e. E.S. Delfosse and N.R. Spencer (ed.) Washington, DC:.

Hill, J.K., C.D. Thomas, and B. Huntley, 1999: Climate and habitat availability determine 20th century changes in a butterfly's range margin. *Proceedings of the Royal Society of London B,* **266,** 1197–1206.

Hillebrand, H., 2004: Strength, slope and variability of marine latitudinal gradients. *Marine Ecology-Progress Series,* **273,** 251–267.

Hilton-Taylor, C.C., 2000: *2000 IUCN Red List of Threatened Species.* IUCN, Gland, Switzerland and Cambridge, UK, 61 pp.

Hodgson, G. and J. Liebeler, 2002: *The Global Coral Reef Crisis: Trends and Solutions.* Reef Check Foundation, Los Angeles.

Hodkinson, I.D., 1992: Global insect diversity revisited. *Journal of Tropical Ecology,* **8,** 505–508.

Hodkinson, I.D. and D. Casson, 1991: A lesser predilection for bugs: Hemiptera (Insecta) diversity in tropical rain forests. *Biological Journal of the Linnean Society,* **43,** 101–109.

Hoenig, J.M. and S.H. Gruber, 1990: Life-history patterns in the elasmobranches: implications for fisheries management. In: *Elasmobranchs as Living Resources: Advances in the Biology, Ecology, Systematics, and the Status of Fisheries. Technical Report NMFS 90,* H.L. Pratt Jr., S.H. Gruber, and T. Taniuchi (eds.), US Department of Commerce National Oceanic and Atmospheric Administration (NOAA), 1–16.

Houlahan, J.E., C.S. Findlay, B.R. Schmidt, A.H. Meyer, and S.L. Kuzmin, 2000: Quantitative evidence for global amphibian population declines. *Nature,* **404,** 752–755.

Hughes, J.B., G.C. Daily, and P.R. Ehrlich, 1997: Population diversity: its extent and extinction. *Science,* **278,** 689–692.

Hughes, T.P., 1994: Catastrophes, phase shifts, and large-scale degradation of a Caribbean coral reef. *Science,* **265,** 1547–1551.

Hughes, T.P., A.H. Baird, D.R. Bellwood, M. Card, S.R. Connolly, et al. 2003: Climate Change, Human Impacts, and the Resilience of Coral Reefs. *Science,* **301,** 929–933.

Hull, D.L., 1997: The ideal species concept -and why we can't get it. In: *Species: The Units of Biodiversity,* M.F. Claridge, H.A. Dawah, and M.R. Wilson (eds.), Chapman and Hall, London, 357–380.

Huntley, B. and T. Webb, 1989: Migration: species' response to climatic variations caused by changes in the earth's orbit. *Journal of Biogeography,* **16,** 5–19.

Huntsman, G.R., J. Potts, R.W. Mays, and D. Vaughan, 1999: Groupers (Serranidae, Epinephelinae): endangered apex predators of reef communities. In: *Life in the Slow Lane: ecology and Conservation of Long-lived Marine Animals. Symposium 23,* J.A. Musick (ed.), American Fisheries Society, Bethesda, MD, USA, 217–231.

Hutchings, J.A., 2000: Collapse and recovery of marine fishes. *Nature,* **406,** 882–885.

Hutchings, J.A. and J.D. Reynolds, 2004: Marine fish stock population collapses: consequences for recovery and extinction risk. *BioScience,* **54,** 297–309.

IIED (International Institute for Environment and Development) **and Traffic,** 2002: *Making a Killing or Making a Living: Wildlife trade, trade controls and rural livelihoods.* London, UK.

Imhoff, M.L., L. Bounoua, T.H. Ricketts, C. Loucks, A.H. Harris, and W.T. Lawrence, 2004: Human Consumption of Net Primary Production. *Nature.* **429,** 870–3.

Imhoff, M.L., L. Bounoua, T.H. Ricketts, C. Loucks, R. Harriss, and W.T. Lawrence, in review: The geography of consumption: Human appropriation of NPP.

Isaac, N.J.B., J. Mallett, and G.M. Mace, 2004: Taxonomic inflation: its influence on macroecology and conservation. *Trends in Ecology and Evolution,* **19(9),** 464–469.

IUCN (World Conservation Union), 2001: *IUCN Red List Categories and Criteria. Version 3.1.* IUCN, Gland, Switzerland.

IUCN, 1994: *Guidelines for Protected Area Management Categories,* IUCN Commission on National Parks/World Conservation Monitoring Centre, The World Conservation Union, Gland, Switzerland

Iverson, J.B., A.R. Kiester, L.E. Hughes, and A.J. Kimerling, 2003: The EMY System World Turtle Database 2003. Oregon State University. Available at http://emys.geo.orst.edu.

IWSG (International Wader Study Group) D.A. Stroud, N.C. Davidson, R. West, D.A. Scott, L. Haanstra, O. Thorup, B. Ganter, and S. Delany (compilers), 2002: *Status of migratory wader populations in Africa and Western Eurasia in the 1990s.* 15, International Wader Studies 15.

Jablonski, D., 1986: Background and mass extinctions: the alternation of macroevolutionary regimes. *Science,* **231,** 129–133.

Jackson, J.B.C., 1994: Constancy and change of life in the sea. *Philosophical Transactions of the Royal Society Series B,* **344,** 55–60.

Jackson, J.B.C., 1997: Reefs since Columbus. *Coral Reefs,* **16,** Suppl.: S23—S32.

Jackson, J.B.C., 2001: What was natural in the coastal oceans? *Proceedings of the National Academy of Sciences, USA,* **98,** 5411–5418.

Jackson, J.B.C., M.X. Kirby, W.H. Berger, K.A. Bjorndal, L.W. Botsford, et al., 2001: Historical Overfishing and the Recent Collapse of Coastal Ecosystems. *Science,* **293,** 629–637.

Jenkins, R.K.B., L.D. Brady, M. Bisoa, J. Rabearivony, and R.A. Griffiths, 2003: Forest disturbance and river proximity influence chameleon abundance in Madagascar. *Biological Conservation,* **109(3),** 407–415.

Karg, J. and L. Ryszkowski, 1996: Animals in arable land. In: *Dynamics of an agricultural landscape,* L. Ryszkowski, French N. and Kedziora A (ed.), Panstwowe Wydawnictwo Rolnicze i Lesne, Poznan, 138–172.

Karr, J.R. and D.R. Dudley, 1981: Index of Biotic Integrity for Aquatic Systems. *Environmental Management.,* **5,** 44–68.

Khan, M.b.M.K., 1989: *Asian Rhinos. An Action Plan for their Conservation.* IUCN, Gland, Switzerland.

Kier, G., J. Mutke, W. Kueper, H. Kreft, and W. Barthlott, 2002: *Richness of vascular plant species of the terrestrial ecoregions of the world.* Unpublished report, University of Bonn Botanical Institute, Bonn, 2002.

Kiesecker, J.M. and A.R. Blaustein, 1995: Synergism between Uv-B Radiation and a Pathogen Magnifies Amphibian Embryo Mortality in Nature. *Proceedings of the National Academy of Sciences of the United States of America,* **92(24),** 11049–11052.

Klein Goldewijk, K., 2001: Estimating global land use change over the past 300 years: the HYDE database. *Global Biogeochemical Cycles,* **15,** 417–434.

Koh, L.P., R.R. Dunn, N.S. Sodhi, R.K. Colwell, H.C. Proctor, and V.S. Smith, 2004: Species coextinctions and the biodiversity crisis. *Science,* **305(5690),** 1632–1634.

Kura, Y., C. Revenga, E. Hoshino, and G. Mock, 2004: *Fishing for Answers: Making Sense of the Global Fish Crisis,* World Resources Institute, Washington, DC.

Lande, R., 1993: Risks of Population Extinction from Demographic and Environmental Stochasticity and Random Catastrophes. *American Naturalist,* **142 (- 6),** 911–927.

Laurance, W.F., M.A. Cochrane, S. Bergen, P.M. Fearnside, P. Delamonica, C. Barber, S. D'Angelo, and T. Fernandes, 2001: The future of the Brazilian Amazon. *Science,* **291,** 438–439.

Laurance, W.F., R.O. Bierregaard Jr., C. Gascon, R.K. Didham, A.P. Smith, et al. 1997: Tropical forest fragmentation: synthesis of a diverse and dynamic discipline. In: *Tropical Forest Fragments,* W.F. Laurance and R.O. Bierregaard (eds.), University of Chicago Press, 502–514.

Lavelle, P. and E. Lapied, 2003: Endangered earthworms of Amazonia: an homage to Gilberto Righi. *Pedobiologia,* **47(5–6),** 419–427.

Lelek, A., 1989: The Rhine River and Some of it's Tributaries Under Human Impact in the Last Two Centuries. In: *Proceedings of the International Large River Symposium.,* D.P. Dodge (ed.). 106, 469–487.

Lévêque, C., 1997.: *Biodiversity dynamics and conservation: the freshwater fish of tropical Africa.* Cambridge University Press., Cambridge, UK.

Levine, J.M., 2000: Species-diversity and biological invasions: relating local-process to community pattern. *Science,* **288,** 852–854.

Liao, G.Z., K.X. Lu, and X.Z. Xiao, 1989: Fisheries Resources of the Pearl River and their Exploitation. In: *Proceedings of the International Large River Symposium,* D.P. Dodge (ed.). 106, 561–568.

Loh, J., 2002: *Living Planet Report 2002.* WWF-World Wild Fund for Nature, 2002, 35 pp.

Loh, J. and M. Wackermagel, 2004: *Living Planet Report 2004.* WWF International, Gland, Switzerland, 40 pp.

Loh, J., R.E. Green, T. Ricketts, J. Lamoreux, M. Jenkins, V. Kapos, and J. Randers, 2005: The Living Planet Index: using species population time series to track trends in biodiversity. *Philosophical Transactions of the Royal Society of London B.* **360,** 289–295.

Long, L.E., L.S. Saylor, and M.E. Soulé, 1995: A pH/UV B synergism in amphibians. *Conservation Biology,* **9,** 1301–1303.

Longcore, J.E., A.P. Pessier, and D.K. Nichols, 1999: Batrachochytrium dendrobatidis gen. and sp. nov., a chytrid pathogenic to amphibians. *Mycologia*, **91**, 219–227.

Longhurst, A., 1995: Seasonal cycles of pelagic production and consumption. *Progress in Oceanography*, **36(2)**, 77–167.

Lovejoy, T.E., R.O. Bierregard, A.B. Rylands, J.R. Malcolm, C.E. Quintela, et al. 1986: Edge and other effects of isolation on Amazon forest fragments. In: *Conservation Biology: The science of scarcity and diversity.*, M.E. Soule (ed.) Sinauer, Sunderland, Massachusettes.

Luck, W.G., G.C. Daily, and P.R. Ehrlich, 2003: Population diversity and eco-system services. *Trends in Ecology and Evolution*, **18**, 331–336.

Mace, G.M., 2004: The role of taxonomy in species conservation. *Philosophical Transactions of the Royal Society of London Series B-Biological Sciences*, **359(1444)**, 711–719.

Mace, G.M. and R. Lande, 1991: Assessing extinction threats: toward a re-evaluation of IUCN threatened species categories. *Conservation Biology*, **5**, 148–157.

Mace, G.M. and W. Kunin, 1994: Classifying threatened species: means and ends. *Philosophical Transactions of the Royal Society of London B*, **344(1307)**, 91–97.

Mace, G.M., J.L. Gittleman, and A. Purvis, 2003: Preserving the Tree of Life. *Science*, **300**, 1707–1709.

Mack, R.N., D. Simberloff, W.M. Lonsdale, H. Evans, M. Clout, and F.A. Bazzaz, 2000: Biotic invasions: Causes, epidemiology, global consequences, and control. *Ecological Applications*, **10(3)**, 689–710.

MacPhee, R.D.E., 1999: *Extinctions in Near Time: Causes, Contexts, and Consequences*. Kluwer Academic/Plenum Publishers, New York, USA.

Macpherson, E., 2002: Large-scale species-richness gradients in the Atlantic Ocean. *Proceedings of the Royal Society of London Series B-Biological Sciences*, **269(1501)**, 1715–1720.

Majerus, M.E.N., 1998: *Melanism: Evolution in Action*. Oxford University Press, Oxford, UK.

Manne, L.L., T.M. Brooks, and S.L. Pimm, 1999: Relative risk of extinction of passerine birds on continents and islands. *Nature*, **399**, 258–261.

Mares, M.A., 1992: Neotropical Mammals and the Myth of Amazonian Biodiversity. *Science*, **255(5047)**, 976–979.

Martin, P.S. and H.E. Wright, 1967: *Pleistocene Extinctions: the Search for a Cause*. Yale University Press, Newhaven, USA.

Martin, P.S. and R.G. Klein, 1984: *Quarternary Extinctions: a Prehistoric Revolution*. University of Arizona Press, Tuscon, USA.

Masteller, E.C. and D.W. Schloesser, 1991: Infestation and impact of zebra mussels on the native unionid population at Presque Isle State Park, Erie, Pennsylvania. Paper presented at the *Second Annual Zebra Mussel Research Conference*, Rochester, NY.

Master, L.L., S.R. Flack, and B.A. Stein, 1998: *Rivers of Life: Critical Watersheds for Protecting Freshwater Biodiversity*. Arlington, Virginia.

Master, L.L., B.A. Stein, L.S. Kutner, and G.A. Hammerson, 2000: Vanishing assets. Conservation status of US species. In: *Precious Heritage. The Status of Biodiversity in the United States,*, B.A. Stein, L.S. Kutner, and J.S. Adams (eds.), Oxford University Press, Oxford, New York, 93–118.

May, R.M., 1988: How many species are there on earth? *Science*, **241**, 1441–1449.

May, R.M., 1990a: How many species? *Philosophical Transactions of the Royal Society of London B*, **330**, 293–304.

May, R.M., 1990b: Taxonomy as destiny. *Nature*, **347**, 129–130.

May, R.M., 1992: How many species inhabit the earth? *Scientific American*, **267(4)**, 42–48.

May, R.M., 1994: Biological diversity: differences between land and sea. *Philosophical Transactions of the Royal Society of London B*, **343**, 105–111.

May, R.M., 1995: Conceptual aspects of the quantification of the extent of biological diversity. In: *Biodiversity: Measurement and Estimation*, D.L. Hawksworth (ed.), Chapman & Hall, London, UK, 13–20.

May, R.M. and S. Nee, 1995: The species alias problem. *Nature*, **378**, 447–448.

May, R.M., R.H. Lawton, and N.E. Stork, 1995: Assessing extinction rates. In: *Extinction Rates*, J.H. Lawton and R.M. May (eds.), Oxford University Press, Oxford, UK, 1–24.

Mayden, R.L., 1997: A hierarchy of species concepts: the denouement in the sage of the species problem. In: *Species: The Units of Biodiversity*, M.F. Claridge, H.A. Dawah, and M.R. Wilson (eds.), Chapman and Hall, London, 381–424.

Mayr, E., 1963: *Animal Species and Evolution*. Belknap Press, Cambridge MA.

McKenzie, J.A., 1996: *Ecological and Evolutionary Aspects of Insecticide Resistance*. Academic Press and R.G. Lands Co., Austin, TX.

McKinney, M.L. and J.L. Lockwood, 1999: Biotic homogenization: a few winners replacing many losers in the next mass extinction. *Trends in Ecology & Evolution*, **14**, 450–453.

McLaughlin, J.F., J.J. Hellmann, C.L. Boggs, and P.R. Ehrlich, 2002: Climate change hastens population extinctions. *Proceedings of the National Academic of Science, USA*, **99**, 6070–6074.

McManus, J.W., L.A.B. Menez, K.N. Kesner-Reyes, S.G. Vergara, and M.C. Ablan, 2000: Coral reef fishing and coral-algal phase shifts: implications for global reef status. *Ices Journal of Marine Science*, **57(3)**, 572–578.

McNeely, J.A., 1996: Human dimensions of invasive alien species: How global perspectives are relevant to China. In: *Conserving China's Biodiversity (II)*, J.S. Peter, S. Wang, and Y. Xie (eds.), China Environmental Science Press, Beijing, 169–181.

Meffe, G.K. and C.R. Carroll, 1994: *Principles of conservation biology*. Sinauer, Sunderland, MA.

Milberg, P. and T. Tyrberg, 1993: Naive Birds and Noble Savages a Review of Man Caused Prehistoric Extinctions of Island Birds. *Ecography*, **16(3)**, 229–250.

Millennium Ecosystem Assessment, 2003: *Ecosytems and human well-being: a framework for assessment*. Island Press, Washington DC.

Mills, G. and H. Hofer, 1998: *Hyaenas: Status Survey and Conservation Action Plan*. IUCN, Gland, Switzerland.

Minelli, A., 1993: *Biological Systematics: The State of the Art*. Chapman & Hall, London.

Moehlman, P.D., 2002: *Equids: Zebras, Asses and Horses. Status Survey and Conservation Action Plan.*, IUCN, Gland, Switzerland.

Moritz, C., 1994: Defining 'evolutionarily significant units' for conservation. *Trends in Ecology and Evolution*, **9**, 373–375.

Moyle, P.B.a.R.A.L., 1992: Loss of biodiversity in aquatic ecosystems: evidence from fish faunas. In: *Conservation biology: the theory and practice of nature conservation, preservation, and management.*, S.K. P.L. Fiedler and Jain (ed.), Chapman and Hall., New York, NY:, 127–169.

Myers, N., 1979: *The Sinking Ark: a New Look at the Problem of Disappearing Species*. Pergamon Press, London, UK.

Myers, N., 1988: Threatened biotas: "hotspots" in tropical forests. *Environmentalist*, **8**, 187.

Myers, N., 1995: Environmental Unknowns. *Science*, **269(5222)**, 358–360.

Myers, N., R.A. Mittermeier, C.G. Mittermeier, G.A.B. da Fonseca, and J. Kent, 2000: Biodiversity hotspots for conservation priorities. *Nature*, **403**, 853–858.

Myers, R.A. and B. Worm, 2003: Rapid worldwide depletion of predatory fish communities. *Nature*, **423**, 280–283.

Nee, S., 2003: Unveiling prokaryotic diversity. *Trends in Ecology & Evolution*, **18(2)**, 62–63.

Nee, S., 2004: More than meets the eye—Earth's real biodiversity is invisible, whether we like it or not. *Nature*, **429(6994)**, 804–805.

Nee, S. and R.M. May, 1997: Extinction and the loss of evolutionary history. *Science*, **278**, 692–694.

Nei, M., 1987: *Molecular Evolutionary Genetics*. Columbia University Press, New York.

NERC (Natural Environment Research Council), 1999: The Global Population Dynamics Database. NERC Centre for Population Biology, Imperial College. Available at http://www.sw.ic.ac.uk/cpb/cpb/gpdd.html.

Novotny, V., Y. Basset, S.E. Miller, G.D. Weiblen, B. Bremer, L. Cizek, and P. Drozd, 2002: Low host specificity of herbivorous insects in a tropical forest. *Nature*, **416**, 841–844.

Nowell, K. and P. Jackson, 1996: *Wild Cats. Status Survey and Conservation Action Plan*. IUCN, Gland, Switzerland.

NRC (National Research Council), 2000: *Ecological Indicators for the Nation*. National Academy Press, Washington, D. C.

Oates, J.F., 1996: *African Primates: Status Survey and Conservation Action Plan (Revised edition)*. IUCN, Gland, Switzerland.

Ødegaard, F., O.H. Diserud, S. Engen, and K. Aagaard, 2000: The magnitude of local host specificity for phytophagous insects and its implications for estimates of global species richness. *Conservation Biology*, **14**, 1182–1186.

Oliver, W.L.R., 1993: *Pigs, Peccaries and Hippos. Status Survey and Conservation Action Plan*. IUCN, Gland, Switzerland.

Olson, D.M., E. Dinerstein, E.D. Wikramanayake, N.D. Burgess, G.V.N. Powell, et al. 2001: Terrestrial ecoregions of the world: a new map of life on earth. *BioScience*, **51**, 933–938.

Olson, J.S., J.A. Watts, and L.J. Allison, 1980: Major World Ecosystem Complexes Ranked by Carbon in Live Vegetation: A Database. Available at http://cdiac.esd.ornl.gov/ndps/ndp017.html.

Owens, I.P.F. and P.M. Bennett, 2000: Ecological basis of extinction risk in birds: Habitat loss versus human persecution and introduced predators. *Proceedings of the National Academy of Sciences of the United States of America,* **97,** 12144–12148.

Parmesan, C. and G. Yohe, 2003: A globally coherent fingerprint of climate change impacts across natural systems. *Nature,* **421,** 37–42.

Parmesan, C., N. Ryrholm, C. Steganescu, J.K. Hill, C.D. Thomas, et al. 1999: Poleward shifts in geographical ranges of butterfly species associated with regional warming. *Nature,* **399,** 579–583.

Pauly, D., 1995: Anecdotes and the shifting baseline syndrome of fisheries. *Trends in Ecology & Evolution,* **10,** 430.

Pauly, D., V. Christensen, S. Guenette, T.J. Pitcher, U.R. Sumaila, C.J. Walters, R. Watson, and D. Zeller, 2002: Towards sustainability in world fisheries. *Nature,* **418,** 689–695.

Peeters, M. and J.L. Van Goethem, 2003: Zoological diversity. In: *Biodiversity in Belgium,* M. Peeters, A. Franklin, and J.L. Van Goethem (eds.), Royal Belgian Institute of Natural Sciences, Brussels, 93–216.

Peeters, M., A. Franklin, and J.L. Van Goethem (eds.), 2003: *Biodiversity in Belgium.* Royal Belgian Institute of Natural Sciences, Brussels, Belgium, 416 pp.

Pimm, S.L. and T.M. Brooks, 1999: The sixth extinction: how large, how soon, and where? In: *Nature and Human Society: the Quest for a Sustainable World,* P.H. Raven (ed.), National Academy Press, Washington, D.C., USA, 46–62.

Pimm, S.L., G.J. Russell, J.L. Gittleman, and T.M. Brooks, 1995: The future of biodiversity. *Science,* **269,** 347–350.

Podesta, G.P. and P.G. Glynn, 2001: The 1997–1998 El Nino event in Panama and Galapagos: an update of thermal stress indices relative to coral bleaching. *Bulletin of Marine Sciences,* **69,** 43–59.

Pollard, E., D. Moss, and T.J. Yates, 1995: Population trends of common British butterflies at monitored sites. *Journal of Applied Ecology,* **32,** 9–16.

Pounds, J.A., M.P.L. Fogden, and J.H. Campbell, 1999: Biological response to climate change on a tropical mountain. *Nature,* **398,** 611–615.

Prance, G.T., H. Beentje, J. Dransfield, and R. Johns, 2000: The tropical flora remains undercollected. *Annals of the Missouri Botanical Garden,* **87(1),** 67–71.

Prendergast, J.R., R.M. Quinn, J.H. Lawton, B.C. Eversham, and D.W. Gibbons, 1993: Rare species, the coincidence of diversity hotspots and conservation strategies. *Nature,* **356,** 335–337.

Purvis, A., P.-M. Agapow, J.L. Gittleman, and G.M. Mace, 2000a: Nonrandom Extinction and the Loss of Evolutionary History. *Science,* **288,** 328–330.

Purvis, A., J.L. Gittleman, G.C. Cowlishaw, and G.M. Mace, 2000b: Predicting extinction risk in declining species. *Proceedings of the Royal Society of London, series B Biological Sciences,* **267,** 1947–1952.

Quicke, D.L.J., 1993: *Principles and techniques of contemporary taxonomy.* Chapman & Hall, London.

Rahel, F.J., 2000: Homogenization of fish faunas across the United States of America. *Science,* **288,** 854–856.

Rahel, F.J., 2002: Homogenisation of Freshwater Faunas. *Annual Review of Ecology and Systematics,* **33,** 291.

Ranta, E., V. Kaitala, and P. Lundberg, 1997: The spatial dimension in population fluctuations. *Science,* **278(5343),** 1621–1623.

Raup, D.M., 1986: Biological extinction in Earth history. *Science,* **231,** 1528–1533.

Raven, P.H., 1983: The challenge of tropical biology. *Bulletin of the Entomological Society of America,* **29(1),** 4–12.

Reed, D.H., E.H. Lowe, D.A. Briscoe, and R. Frankham, 2003: Fitness and adaptation in a novel environment: Effect of inbreeding, prior environment, and lineage. *EVOLUTION,* **57,** 1822–1828.

Reeves, R.R., B.D. Smith, and T. Kasuya, 2000: *Biology and Conservation of Freshwater Cetaceans in Asia. IUCN Species Survival Commission Paper No. 23,* viii. IUCN-The World Conservation Union, Gland, Cambridge, 152 pp.

Regan, H.M., R. Lupia, A.N. Drinnan, and M.A. Burgman, 2001: The currency and tempo of extinction. *The American Naturalist,* **157(1).**

Reid, W.V., 1992: How many species will there be? In: *Tropical Deforestation and Species Extinction,* T.C. Whitmore and J.A. Sayer (eds.), Chapman & Hall, London, UK, 55–73.

Reid, W.V., 1998: Biodiversity hotspots. *Trends in Ecology and Evolution,* **13,** 275–280.

Revenga, C. and Y. Kura, 2003: *Status and Trends of Biodiversity in Inland Water Ecosystems.* Secretariat of Convention on Biodiversity, Montreal.

Revenga, C., J. Brunner, N. Henninger, K. Kassem, and R. Payne, 2000: *Pilot Analysis of Global Ecosystems: Freshwater Systems,* World Resources Institute, Washington DC.

Reynolds, J.D., S. Jennings, and N.K. Dulvy, 2001: Life histories of fishes and population responses to exploitation. In: *Conservation of Exploited Species,* J.D. Reynolds, G.M. Mace, K.H. Redford, and J.G. Robinson (eds.), Cambridge University Press, Cambridge, UK., 147–168.

Reynolds, J.D., N.K. Dulvy, and C.R. Roberts, 2002: Exploitation and other threats to fish conservation. In: *Handbook of Fish Biology and Fisheries, Volume 2: Fisheries,* P.J.B. Hart and J.D. Reynolds (eds.). 2, Blackwell Publishing, Oxford, UK, 319–341.

Richter-Dyn, N. and N.S. Goel, 1972: On the extinction of a colonising species. *Theoretical Population Biology,* **3,** 406–433.

Roberts, C.M., C.J. McClean, J.E.N. Veron, J.P. Hawkins, G.R. Allen, et al. 2002: Marine biodiversity hotspots and conservation priorities for tropical reefs. *Science,* **295,** 1280–1284.

Robinson, J.G. and K. Redford, 1991: *Neotropical Wildlife Use and Conservation.* Chicago University Press, Chicago.

Rodrigues, A.S.L., S.J. Andelman, M.I. Bakarr, L. Boitani, T.M. Brooks, et al. 2004: Effectiveness of the global protected area network in representing species diversity. *Nature,* **428(6983),** 640–643.

Roelke, M.E., J. Martenson, and S.J. O'Brien, 1993: The consequences of demographic reduction and genetic depletion in the endangered Florida panther. *Curr. Biol.,* **3,** 340–350.

Roelke-Parker, M.E., L. Munson, C. Packer, R. Kock, S. Cleaveland, et al. 1996: A canine distemper virus epidemic in Serengeti lions (Panthera leo). *Nature,* **379,** 441–445.

Roman, J. and S.R. Palumbi, 2003: Whales before Whaling in the North Atlantic. *Science,* **301,** 508–510.

Root, T.L., J.T. Price, K.R. Hall, S.H. Schneider, C. Rosenzweig, and J.A. Pounds, 2003: Fingerprints of global warming on wild animals and plants. *Nature,* **421,** 57–60.

Rose, P.M. and D.A. Scott, 1997: *Waterfowl Population Estimates—Second Edition.,* Wetlands International Publication No. 44. : Wetlands International., Wageningen, The Netherlands.

Rosenzweig, M.L., 1995: *Species Diversity in Space and Time.* Cambridge University Press, Cambridge, UK.

Rossiter, N.A., S.A. Setterfield, M.M. Douglas, and L.B. Hutley, 2003: Testing the grass-fire cycle: alien grass invasion in the tropical savannas of northern Australia. *Diversity and Distributions,* **9(3),** 169–176.

Roy, K., J.W. Valentine, D. Jablonski, and S.M. Kidwell, 1996: Scales of climatic variability and time averaging in Pleistocene biotas: implications for ecology and evolution. *Trends in Ecology & Evolution,* **11,** 458–462.

Saccheri, I., M. Kuussaari, M. Kankare, P. Vikman, W. Fortelius, and I. Hanki., 1998: Inbreeding and extinction in a butterfly metapopulation. *Nature,* **392,** 491–494.

Scheffer, M., S.R. Carpenter, J.A. Foley, C. Folke, and B.H. Walker, 2001: Catastrophic shifts in ecosystems. *Science,* **413,** 591–596.

Scotland, R.W. and A.H. Wortley, 2003: How many species of seed plants are there? *Taxon(52),* 101–104.

Scott, D.A. and P.M. Rose, 1996: *Atlas of Anatidae populations in Africa and western Eurasia.* 41, Wetlands International Publication No. 41, Wageningen, The Netherlands.

Sechrest, W., T.M. Brooks, G.A.B. Fonseca, W.R. Konstant, R.A. Mittermeier, A. Purvis, A.B. Rylands, and J.L. Gittleman, 2002: Hotspots and the conservation of evolutionary history. *Proceedings of the National Academy of Sciences of the United States of America,* **99,** 2067–2071.

Sheppard, C., 1995: The Shifting Baseline Syndrome. *Marine Pollution Bulletin,* **30,** 766–767.

Sherman, K. and L. Alexander (eds.), 1986: *Variability and management of Large Marine Ecosystems. AAAS Selected Symposium 99,* Westview Press, Inc., Boulder, Colorado.

Sibley, C.G. and B.L. Monroe, 1990: *Distribution and Taxonomy of Birds of the World.* Yale University Press, New Haven.

Sillero-Zubiri, C. and D. Macdonald, 1997: *The Ethiopian Wolf. Status Survey and Conservation Action Plan.* IUCN, Gland, Switzerland.

Simberloff, D., 2000: Extinction-proneness of island species—Causes and management implications. *Raffles Bulletin of Zoology,* **48(1),** 1–9.

Smith, F.D.M., R.M. May, R. Pellew, T.H. Johnson, and K.R. Walter, 1993: Estimating extinction rates. *Nature,* **364,** 494–496.

Snelgrove, P.V.R., 1999: Getting to the Bottom of Marine Biodiversity: Sedimentary Habitats—Ocean bottoms are the most widespread habitat on Earth and support high biodiversity and ecosystem services. *BioScience,* **49,** 129–138.

Sørensen, L.L., 2003: Stratification of the spider fauna in a Tanzanian forest. In: *Arthropods of Tropical Forests,* Y. Basset, V. Novotny, S.E. Miller, and R.L. Kitching (eds.), Cambridge University Press, Cambridge, UK, 92–101.

Soulé, M.E., 1980: Thresholds for survival: maintaining fitness and evolutionary potential. In: *Conservation biology: an evolutionary-ecological perspective*, M.E. Soule and B.A. Wilcox (eds.), Sinauer Associates, Sunderland, Mass, 151–169.

Soulé, M.E., 1988: Reconstructed dynamics of rapid extinctions in chaparral requiring birds in urban habitat islands. *Conservation Biology*, **2**, 75–92.

Springer, A.M., J.A. Estes, G.B. van Vliet, T.M. Williams, D.F. Doak, E.M. Danner, K.A. Forney, and B. Pfister, 2003: Sequential megafaunal collapse in the North Pacific Ocean: An ongoing legacy of industrial whaling? *Proceedings of the National Academic of Science, USA*, **100**, 12223–12228.

Steadman, D.W., 1995: Prehistoric extinctions of Pacific island birds: biodiversity meets zooarcheology. *Science*, **267**, 1123–1131.

Stein, B.A., L.S. Kutner, and J.S. Adams, 2000: *Precious Heritage: the status of biodiversity in the United States*. Oxford University Press, New York.

Stiassny, M.L.J., 1996: An overview of freshwater biodiversity: with some lessons from African fishes.. *Fisheries*, **21**, 7–13.

Stone, R., 2000: Virology: Canine Virus Blamed in Caspian Seal Deaths. *Science*, **289(5487)**, 2017–2018.

Stork, N.E., 1988: Insect diversity: facts, fiction and speculation. *Biological Journal of the Linnean Society*, **35**, 321–337.

Stork, N.E., 1993: How many species are there? *Biodiversity and Conservation*, **2**, 215–232.

Stork, N.E., 1997: Measuring global biodiversity and its decline. In: *Biodiversity II*, M.L. Reaka-Kudla, D.E. Wilson, and E.O. Wilson (eds.), National Academy of Sciences, Washington DC, USA, 41–68.

Stork, N.E. and K.J. Gaston, 1990: Counting new species one by one. *New Scientist*, **1729**, 43–47.

Stroud, D.A., N.C. Davidson, R. West, D.A. Scott, L. Hanstra, O. Thorup, B. Ganter, and S.c.o.b.o.t.I.W.S.G. Delany, 2004: Status of migratory wader populations in Africa and Western Eurasia in the 1990s. *International Wader Studies*, **15**, 1–259.

Stuart, S.N., J.S. Chanson, N.A. Cox, B.E. Young, A.S.L. Rodrigues, D.L. Fischman, and R.W. Waller, 2004: Status and trends of amphibian declines and extinctions worldwide. Cited 14th October 2004. Available at www .sciencexpress.org.

Sugden, A. and E. Pennisi, 2000: Diversity digitized. *Science*, **289**, 2305.

Tautz, D., P. Arctander, A. Minelli, R.H. Thomas, and A.P. Vogler, 2003: A plea for DNA taxonomy. *Trends in Ecology and Evolution*, **18**, 70–74.

The Royal Society, 2003: Measuring biodiversity for conservation. London, UK, The Royal Society.

Thomas, C.D., A. Cameron, R.A. Green, M. Bakkenes, L.J. Beaumont, et al. 2004a: Extinction risk from climate change. *Nature*, **427**, 145–148.

Thomas, J.A., M.G. Telfer, D.B. Roy, C.D. Preston, J.J.D. Greenwood, et al., 2004b: Comparative losses of British butterflies, birds, and plants and the global extinction crisis. *Science*, **303(5665)**, 1879–1881.

Thorne, R.F., 2002: How many species of seed plants are there? *Taxon*, **51**, 511–512.

Tilman, D., R.M. May, C.L. Lehman, and M.A. Nowak, 1994: Habitat destruction and the extinction debt. *Nature*, **371**, 65–66.

Tilman, D., J. Fargione, B. Wollf, C. D'Antonio, A. Dobson, R. Howarth, D. Schindler, W.H. Schlesinger, D. Simberloff, and D. Swackhamer, 2001: Forecasting agriculturally driven global environmental change, *Science* **(292)**, 281–284.

Toft, C.A.P., 1986: Coexistence in organisms with parasitic lifestyles. In: *Community Ecology*, J.M. Diamond and T.J. Case (eds.), Harper and Row, New York, USA, 445–463.

TRAFFIC Southeast Asia, 2001: *An Overview of the Trade in live South-east Asian Freshwater Turtles, An Information Paper for the 17th Meeting of the CITES Animals Committee*. German CITES Scientific Authority (Federal Agency for Nature Conservation), Federal Ministry for the Environment, Nature Conservation and Nuclear Safety, Hanoi, Viet Nam, 30 July to 3 August 2001.

TRAFFIC, 2002: Traffic international. Available at www.traffic.org.

Turner, G.F., 1999: What is a fish species? *Rev. fish biol. fisher.*, **9**, 281–297.

Udvardy, M.D.F., 1975: *A classification of the biogeographical provinces of the world*. Occasional Paper No. 18, IUCN, Morges, Switzerland.

UN (United Nations), 1992: *Rio Declaration on Environment and Development*. New York.

UNEP (United Nations Environment Programme), 2002a: *United Nations, 2002: Report on the sixth meeting of the Conference of the Parties to the Convention on Biological Diversity (UNEP/CBD/COP/6/20/Part 2) Strategic Plan Decision VI/26*.

UNEP, 2002b: *Global strategy for plant conservation*. UNEP/CBD/COP/6/20, UNEP.

UNEP, 2003a: *United Nations Environment Programme, 2003: Integration of outcome-orientated targets into the programmes of work for the Convention taking into account the 2010 biodiversity target, the Global Strategy for Plant Conservation, and relevant targets set by the World Summit on Sustainable Development (UNEP/CBD/SBSTTA/9/14). Convention for Biological Diversity*.

UNEP, 2003b: *United Nations Environment Programme, 2003: 2010 The global biodiversity challenge (UNEP/CBD Meeting Report). Convention on Biological Diversity.*, UNEP.

UNEP, 2003c: *United Nations Environment Programme, 2003: Report of the Subsidiary Body on Scientific, Technical and Technological Advice on work of its ninth meeting (UNEP/CBD/7/4). Convention on Biological Diversity*.

USGS (U.S. Geological Survey), 2000: Zebra mussels cause economic and ecological problems in Great Lakes (GLSC Fact Sheet). Available at www.glsc .usgs.gov.

USGS, Great Lakes Center, 2003: *Great Lakes: Long Term Fish Stock Assessment and Monitoring.*, Ann Arbor, Michigan.

USGS-EDC, 2003: Global land cover characteristics data base. Available at http://edcdaac.usgs.gov/glcc/globdoc2_0.html#files.

van Swaay, C.A.M., 1990: An assessment of the changes in butterfly abundance in the Netherlands during the 20th Century. *Biological Conservation*, **52**, 287–302.

Vane-Wright, R.I., C.J. Humphries, and P.H. Williams, 1991: What to protect?—systematics and the agony of choice. *Biological Conservation*, **55**, 235–254.

Veron, J.E.N., 2000: *Corals of the World*. Australian Institute of Marine Science, Townsville, Australia.

Vinson, M.R. and C.P. Hawkins, 2003: Broad-scale geographical patterns in local stream inset genera richness. *Ecogeography*, **26(6)**, 751–767.

Vitousek, P.M., P.R. Ehrlich, A.H. Ehrlich, and P.A. Matson, 1986: Human appropriation of the products of photosynthesis. *Bioscience*, **36**, 368–373.

Wade, T. G., K. H. Riitters, J. D. Wickham, and K. B. Jones, 2003. Distribution and causes of global forest fragmentation. *Conservation Ecology*, 7: 7. [online] URL: http://www.consecol.org/vol7/iss2/art7

Walther, G.-R., E. Post, P. Convey, A. Menzel, C. Parmesan, T.J.C. Beebee, J.-M. Fromentin, O. Hoegh-Guldberg, and F. Bairlein, 2002: Ecological responses to recent climate change. *Nature*, **416**, 389–395.

Ward, B.B., 2002: How many species of prokaryotes are there? *Procedings of the National Academy of Sciences of the USA*, **99**, 10234–10236.

Ward, J.V. and J. A. Stanford, 1989: Riverine Ecosystems: The Influence of Man on Catchment Dynamics and Fish Ecology. In: *Proceedings of the International Large River Symposium*, D.P. Dodge (ed.). 106 Ottawa, Canada.

Ward, R.D., D.O.F. Skibinski, and M. Woodwark, 1992: Protein heterozygosity, protein structure, and taxonomic differentiation. *Evolutionary Biology*, **26**, 73–159.

Warren, M.S., J.K. Hill, J.A. Thomas, J. Asher, R. Fox, et al. 2001: Rapid responses of British butterflies to opposing forces of climate and habitat change. *Nature*, **414**, 65–69.

Watson, R.T., 2002: *Climate Change 2001: Synthesis Report*. Cambridge University Press, Cambridge.

WCMC (World Conservation Monitoring Centre), 2003: World database on Protected Areas., IUCN-WCPA/UNEP-WCMC.

WDFW and ODFW (Washington Department of Fish and Wildlife and Oregon Department of Fish and Wildlife), 1999: *Status Report: Columbia River Fish Runs and Fisheries*, Joint Columbia River Management Staff, Vancouver, 1938–1998 pp.

Wells, J.V. and M.E. Richmond, 1995: Populations, metapopulations, and species populations: What are they and who should care? *Wildlife Society Bulletin*, **23**, 458–462.

West, J.M. and R.V. Salm, 2003: Resistance and resilience to coral bleaching: implications for coral reef conservation and management. *Conservation Biology*, **17**, 956–967.

Wetlands International, 2002: *Waterbird Population Estimates*. Third Edition ed. *Wetlands International Global Series No. 12*, Wageningen, The Netherlands, 226 pp.

Wiles, G.J., J. Bart, R.E. Beck, and C.F. Aguon, 2003: Impacts of the brown tree snake: Patterns of decline and species persistence in Guam's avifauna. *Conservation Biology*, **17(5)**, 1350–1360.

Williams, P.H., 1998: Key-sites for conservation area-selection methods for biodiversity. In: *Conservation in a changing world*, G.M. Mace, A. Balmford, and J.R. Ginsberg (eds.), Cambridge University Press, Cambridge, UK, 211–250.

Williams, P.H., C.J. Humphries, and K.J. Gaston, 1994: Centres of seed-plant diversity: the family way. *Proceedings of the Royal Society of London B,* **256,** 67–70.

Wilson, E.O., 2003: The encyclopedia of life. *Trends in Ecology & Evolution,* **18,** 77–80.

Woodroffe, R. and J.R. Ginsberg, 1998: Edge effects and the extinction of populations inside protected areas. *Science,* **280(5372),** 2126–2128.

Woodroffe, R., J. Ginsburg, and D. Macdonald, 1997: *The African Wild Dog. Status Survey and Conservation Action Plan.,* IUCN, Gland, Switzerland.

Woods, C.A. and F.E. Sergile (eds.), 2001: *Biogeography of the West Indies— Patterns and Perspectives.* CRC Press, Boca Raton, Florida.

Worm, B., H.K. Lotze, and R.A. Myers, 2003: Predator diversity hotspots in the blue ocean. *Proceedings of the National Academic of Science, USA,* **100,** 9884–9888.

Wright, S., 1969: *Evolution and the Genetics of Populations.* Vol. 2, *The Theory of Gene Frequencies,* University of Chicago Press, Chicago, IL.

WWF (World Wide Fund for Nature), 2002: WWF Factsheet: Sturgeon. Factsheet prepared for the 12th Meeting of the Conference of the Parties to CITES Santiago, 3–15th November 2002.

WWF, 2004: Wildfinder: an ecoregion-species database. [Database] World Wildlife Fund-US. Available at http://www.worldwildlife.org/wildfinder.

WWF and IUCN, 1994: *Centres of Plant Diversity, Volume 1: Europe, Africa, South West Asia and the Middle East.* Vol. 1, IUCN Publications Unit for WWF and IUCN., Cambridge.

WWF and IUCN, 1995: *Centres for Plant Diversity, Volume 2: Asia, Australasia and the Pacific.* Vol. 2, IUCN Publications Unit for WWF & IUCN, Cambridge.

WWF and IUCN, 1997: *Centres for Plant Diversity, Volume 3: The Americas.* Vol. 3, Publications Unit for WWF & IUCN, Cambridge.

WWF and WRI (World Resources Institute), 2004: *Rivers at Risk: Dams and the future of freshwater ecosystems.* WWF International, London, UK., 47 pp.

Chapter 5

Ecosystem Conditions and Human Well-being

Coordinating Lead Authors: Marc Levy, Suresh Babu, Kirk Hamilton
Lead Authors: Valerie Rhoe, Alessandro Catenazzi, Ma Chen, Walter V. Reid, Debdatta Sengupta, Cai Ximing
Contributing Authors: Andrew Balmford, William Bond
Review Editors: David Rapport, Linxiu Zhang

---

*This appears in Appendix A at the end of this volume.

Main Messages

Over historical time frames, human well-being has on aggregate improved by several orders of magnitude. Incomes have increased, populations have grown, life expectancies have risen, and political institutions have become more participatory. In the global aggregate, human well-being continues to expand, although there are variations across geographical regions.

Well-being is not distributed evenly among individuals, countries, or social groups. Inequality is high, and gaps between the well-off and the disadvantaged are increasing. A child born in sub-Saharan Africa is 20 times more likely to die before age five than a child born in an OECD country, and this ratio is higher than it was a decade ago.

The ecosystems as classified by the Millennium Ecosystem Assessment vary in the degree to which they harbor high values of human well-being. For example, cultivated systems and coastal ecosystems tend to be characterized by high human well-being, while drylands are characterized by low human well-being.

The degree to which well-being varies across ecosystems is not the same everywhere. Variance in human well-being is highest in Asia and sub-Saharan Africa and lowest in the OECD. For example, per capita incomes in Asia are 40% higher in the coastal zones than they are in the drylands, while in OECD countries there is no significant difference between incomes in the two systems.

Populations are increasing in ecosystems characterized currently by low well-being. Whereas historically populations have tended to shift from low-productivity ecosystems to high-productivity ecosystems or to urban areas, today there are signs that the relative concentration of people in less productive ecosystems is going up. The concentration of poor in less-favored lands of Asia and sub-Saharan Africa is an example.

These trends signify that there are a large number of people at risk of adverse ecosystem changes. Approximately 1.1 billion people survive on less than $1 per day of income, most of them in rural areas where they are highly dependent on agriculture, grazing, and hunting for subsistence. For these people, degradation and declining productivity of ecosystems threatens their survival.

Ecosystem changes affect human well-being in a variety of different ways. Often the impacts of ecosystem changes are shifted from the groups responsible for them onto other groups. Sometimes ecosystem change is embedded in distributional conflicts over resources, with one group improving at another group's expense. Often impacts are experienced differentially as a function of levels of coping capacity—such differences can manifest themselves at the individual, household, regional, or national level.

Among less poor populations, ecosystem changes affect well-being in more subtle but not necessarily less important ways. Declines in incomes, loss of culturally important natural resources, and increases in threats to health can be expected to accompany declining ecosystem health.

5.1 Introduction

This chapter provides an overview of the primary patterns and trends in human well-being and summarizes what is known broadly about the connections between human well-being and ecosystems. It does not substitute for the more detailed findings found in the other chapters of this volume, but rather provides an overarching empirical foundation for assessing human well-being and ecosystem change side by side.

5.2 Dimensions of Human Well-being

This section reviews how the different dimensions of human well-being are measured, the extent of our knowledge about them, and the primary measurement gaps. (See also Chapter 2 for further description of the measurement of human well-being.) The benefits of qualitative data and new approaches for assessing human well-being are acknowledged, but the use of this information is limited in this global assessment because of difficulties comparing data across countries and aggregating information within units of ecosystems. Five dimensions of human well-being are recognized in the MA: basic material for a good life, freedom and choice, health, good social relations, and security (MA 2003).

5.2.1 Basic Material for a Good Life

The basic materials for a good life include adequate income, household assets, food, water, and shelter. Considerable effort goes into measuring and monitoring these dimensions of well-being. Systems of national accounts generate regular estimates of gross national income and GNP; these estimates are made roughly comparable through calculation of purchasing power parity, estimates that correct for price differences across countries. Food production is measured by central governments and by international organizations such as the Food and Agriculture Organization. FAO also measures national per capita dietary energy supply for most countries. (See Chapter 8 for further information.) Micronutrient information on vitamin A, iodine, and iron is collected at the community level but is not as widely available as DES. National-level water estimates are available, though these tend to be less precise than income and food measures. Of all these measures, shelter is probably the most poorly measured. The distribution of the quality of substandard housing is not known with any confidence.

Increasingly, measures of these basic material dimensions of well-being are being carried out through detailed household surveys (Deaton 1997). Living standard measurement surveys began collecting standardized measures of well-being at the household level in 1985. Approximately 100 developing countries have carried out such surveys. Other large-scale household survey efforts include demographic and health surveys and national-level multiple indicator cluster surveys. Surveys such as these provide information about nutrition, housing type, household assets, and access to water, health care, and sanitation.

Although great effort goes into these measurement efforts, they do not provide a complete enough picture to support a full understanding of the distribution of well-being and its relationship to ecosystem services. Comparable measures of water and sanitation access, for example, are scarce, because terminologies, methodologies, and measurement priorities differ from place to place (Millennium Project 2004). Even the measurements that receive the greatest level of effort, the national income measures, are inadequate for many purposes. They imperfectly capture economic activity outside the formal sector and the value from subsistence activities, for example, which is often highly important among populations vulnerable to decline in ecosystem services.

Especially problematic is the concept of poverty. Although poverty is the focus of many local, national, and international policies, and although many countries measure it, there is little congruence of measurement efforts and, as a result, little ability to portray clear patterns and trends (Deaton 2003; Reddy and Pogge

2002). Some efforts to measure poverty do so strictly in terms of income. For example, the World Bank relies heavily on household survey data to estimate the number of people living on less than $1 per day, and this measure is a prominent component of the Millennium Development Goal targets.

However, measures of income poverty have been criticized because of problems of comparability and relevance. Comparability is difficult because of the challenges involved in comparing prices, which is quite important at such low thresholds of income. Relevance is a problem where human well-being is often heavily dependent on non-income factors such as household assets (stoves, bicycles, toilets, housing materials, and so on) and output from subsistence activities (cultivation, hunting, and fishing). In spite of such measurement and comparability difficulties, an attempt is made here to provide an overview of trends in selected human well-being indicators and ecosystems.

5.2.2 Freedom and Choice

Freedom is defined as the range of options a person has in deciding on and realizing the kind of life to lead. At a broad scale, only a few of the many specific phenomena that are relevant to this dimension of well-being are measured at all, and many of those that are measured are problematic.

The degree to which political institutions are participatory is not measured by any intergovernmental agency, chiefly because of disputes over what are appropriate measures. The most widely used measure in the scientific literature is the polity database, which provides annual measures of democratic institutions for 160 countries (Jaggers and Gurr 1995). The polity data, and similar efforts such as the Heritage Foundation's index of civil liberties, provide comparable, clear measures of national political institutions that emphasize electoral institutions and constraints on chief executives. There are no comparable data that measure citizen participation in decision-making at regional or local levels, although this dimension of freedom and choice has been well connected to ecosystem management and human well-being (Ostrom 1990; Ostrom et al. 2002).

Education is a clear aspect of well-being that enhances life prospects. Comparable international measures are poor. A frequently used measure is literacy, which is a component of the Human Development Index. Literacy is hard to measure accurately and comparably (Bruns et al. 2003). Moreover, it represents only a small aspect of education. Some countries collect comparable data on the percent of the population (often broken down by gender) enrolled in school, but this too is incomplete. More relevant for life prospects is the school completion rate, but this is not well measured.

5.2.3 Health

Human health is measured in a variety of ways, and knowledge about broad trends and patterns concerning health is good. Life expectancy, infant mortality, and child mortality are measured fairly intensively. Most central governments collect these vital statistics and publish them; international organizations such as the World Health Organization and World Bank collate and harmonize these measures. As a result, high-quality country-level time series on these measures are available. Some scholars have been able to construct time series going back several centuries by relying on a range of vital statistics collections (Maddison 2001).

Knowledge of health aspects of human well-being is more limited when it comes to more precise dimensions such as subnational patterns or specific disease prevalences. There are no consistent monitoring or measurement efforts that measure health outcomes at a sub-national level. Infant mortality and child mortality are measured through a set of coordinated household survey efforts, including the DHS and MICS, and these can be used to generate estimates for sub-national regions, on the order of about 10 regions per country. Such surveys are not carried out in every country, however, and no more than about once per decade. As a result, it is difficult to portray the distribution of human health at a level of resolution more precise than national boundaries.

When it comes to measuring health outcomes in terms of specific diseases, monitoring and surveillance are also less complete than for vital statistics. Few disease incidence statistics are collected across a significant number of countries in a comparable enough fashion to permit robust tracking of patterns. The World Health Organization collects disease-specific data by country, but reports primarily at the level of six world regions because of limitations in comparability (WHO 2004).

5.2.4 Good Social Relations

Humans enjoy a state of good social relations when they are able to realize aesthetic and recreational values, express cultural and spiritual values, develop institutional linkages that create social capital, show mutual respect, have good gender and family relations, and have the ability to help others and provide for their children (MA 2003; Dasgupta and Serageldin 1999). This aspect of human well-being is not well measured, largely because it is more difficult to observe directly. Partly as a result of recent scholarship identifying the importance of social relations and social capital in explaining a range of important public policy outcomes (e.g., Putnam et al. 1993; OECD 2001), interest in measuring this dimension has increased considerably in the past decade. Although comparable quantitative measurements remain very primitive, there are case studies that illustrate the sensitivity of ecosystem changes on good social relations. (See Box 20.12 in Chapter 20, on North America's Great Lakes and Invasive Species.) Some research has noted that high levels of economic development are sometimes associated with poor social relations (Jungeilges and Kirchgässner 2002). Problems such as suicide and divorce are observed consequences of such dynamics.

5.2.5 Security

Humans can be said to live in a state of security when they do not suffer abrupt threats to their well-being. Chapter 6 indicates that people within the geographical region of a threat are differently susceptible to its negative effects. Those who are poor, sick, or malnourished generally have fewer assets and coping strategies and are more likely to be more severely affected.

Some of the most salient threats are organized violence, economic crises, and natural disasters. Comparable measures of organized violence are available for international warfare and civil war, but generally not for banditry and other forms of crime. One prominent collection of data on war is the Uppsala conflict database, which attempts to document all political conflicts resulting in 25 or more deaths for the period 1946–2003 (Gleditsch et al. 2002). It provides measures of the frequency of war as well as estimates of the magnitude of war, and rough information on geographical extent. Political violence is not evenly distributed across the world; it is especially concentrated in the poor countries of the world, though not all poor countries experience violence and not all violence takes place among the poor. Figure 5.1, showing the distribution of political conflict, maps the combined incidence of low-intensity (25 deaths or fewer), middle-intensity, and high-intensity (1,000 deaths or more) conflicts for the period 1975–2003. If a region had all three types of conflicts during each of

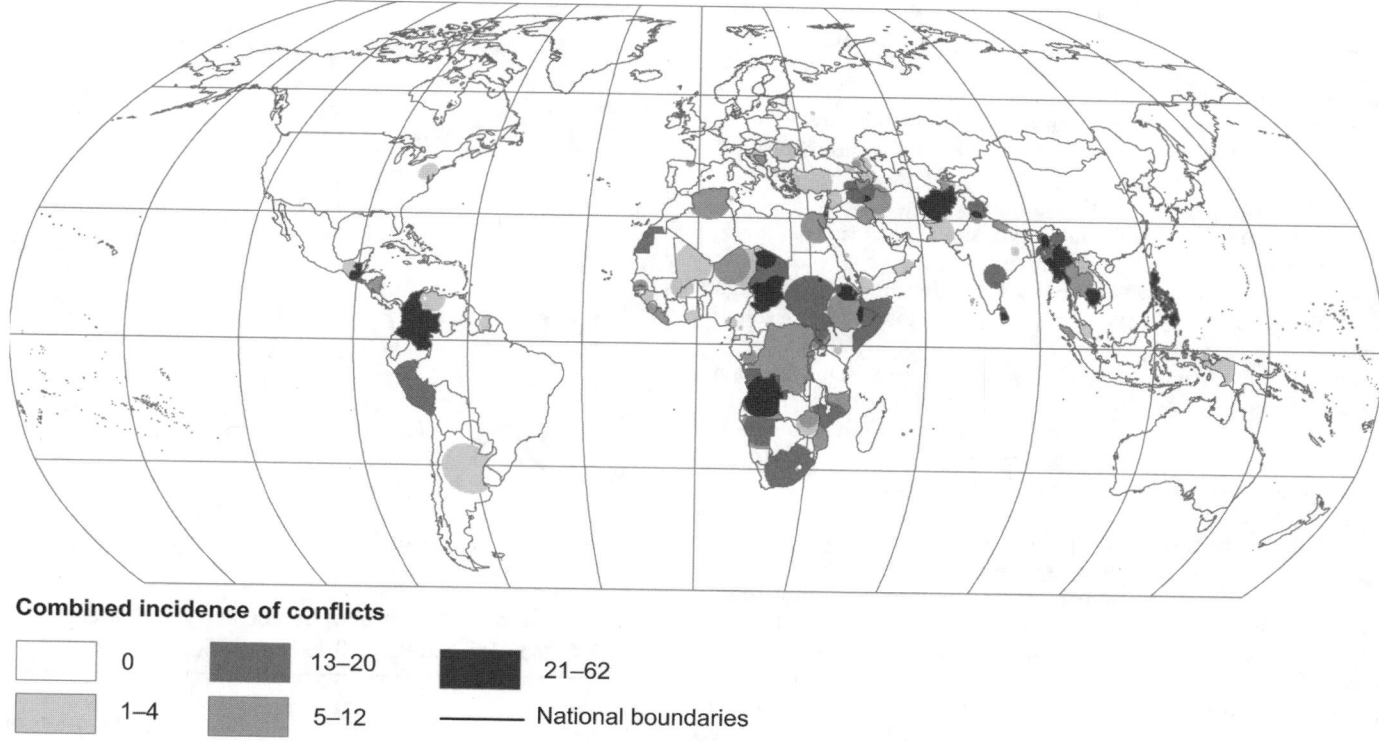

Combined incidence of conflicts

□	0	▨	13–20	■	21–62
▨	1–4	▨	5–12	——	National boundaries

Figure 5.1. Distribution of Internal Political Conflict, 1975–2003 (Robinson Projection)

these 29 years, it would have a value of 87; the observed range is 0–62.

Economic crises—hyperinflation, depressions, exchange-rate shocks, and so on—are well measured by international financial institutions such as the International Monetary Fund, the World Bank, and various regional and national financial authorities. Yet the direct impact of such shocks on ecosystems have not been studied or documented adequately.

Natural disasters are not measured well, though various international organizations and research centers are seeking to improve measurement (Guha-Sapir and Below 2002). The most glaring deficiency in efforts to measure natural disasters is in the area of human impacts. Although some insurance companies undertake considerable efforts to quantify insured economic losses due to natural disasters, many of the grossest effects on human well-being are not insured economic losses, but rather loss of life and shelter in poor communities. (Further information on natural disasters can be found in Chapters 6 and 16.)

5.2.6 Aggregations

A wide variety of efforts to aggregate the multiple dimensions of human well-being have been attempted, the most prominent of which is the Human Development Index. The HDI is endorsed by the intergovernmental community through the U.N. General Assembly and widely used in policy assessments and the scholarly literature. A large number of countries now calculate their own HDIs, typically reporting at sub-national levels. (Other aggregations enjoy less support and are used less widely; see Chapter 2.) The HDI aggregates measures of economic well-being (per capita income), health (life expectancy), and education (literacy and enrollment). It does not take into account cultural or social aspects of well-being, and it considers security dimensions only insofar as they are reflected in economic and health outcomes. It is not

meant to be an all-encompassing measure of well-being but rather a useful indicator of development consistent with the development-as-freedom approach pioneered by Sen (1999).

It is acknowledged that human well-being is not equally distributed among different social groups, including among men and women. In response, the U.N. Development Programme in 1995 began calculating the Gender-related Development Index (Prescott-Allen 2001).

Beyond measuring current well-being, there is the important question of sustaining well-being in the future. Asset accounting (an outgrowth of "green" national accounting) provides the necessary framework for assessing sustainability. As Hamilton and Clemens (1999) and Dasgupta and Mäler (2000) show, the change in real wealth—genuine savings—is equal to the change in the discounted future flows of well-being measured in dollars. The World Bank publishes figures on "adjusted net saving" that account for depreciation of assets; investment in human capital; depletion of minerals, energy and forests; and selected pollution damages (World Bank 2004). While ecosystem services are not directly valued in the published figures, the framework is robust enough to incorporate the economic value of the degradation of ecosystem "assets" where these values have been estimated.

5.3 Patterns and Trends in the Distribution of Human Well-being

5.3.1 Global Trends

In the aggregate, human well-being has improved dramatically since the advent of agriculture first made possible the accumulation of wealth. Incomes have risen, life expectancy has gone up, food supplies have risen, culture has become enriched, and political institutions have become more participatory. Two exceptions to this generalization have been the trends in warfare and hunger. Battle deaths (both combatant and civilian) peaked in the middle of the twentieth century, as a consequence of the intensity of the two world wars. Since 1945 they have declined. The second exception is the number of hungry people, which is now increasing. Although the size of world population is not a direct measure of well-being, it constitutes a fundamental background measure and is therefore included in this summary.

Of these trends, population growth shows clear signs of leveling off. (See Chapter 3.) Per capita incomes, life expectancy, and democratization do not yet show signs of leveling off, although they have increased historically at different rates. The absolute number of hungry people began to rise in 1995/97 (FAO 2003), as described in Chapter 8. Warfare patterns are not stable enough to identify clear trends, although the past 50 years have been comparatively peaceful in historical terms. Cultural trends are not susceptible to simple generalizations. Some observers have argued that the global reach of a relatively homogenous mass media threatens local cultural institutions, while others have argued that knowledge of cultural traditions has been able to spread globally; placing values on recent changes such as these is difficult.

Many dimensions of human well-being that can be measured on a large scale, then, have increased considerably over the past 10 centuries and in the aggregate shows signs of continued expansion. Figure 5.2 shows estimates of human life expectancy and per capita income over the past 2,000 years, demonstrating the enormous improvements in basic material aspects of well-being over this long time frame. Figure 5.3 shows trends in the level of democracy and warfare since 1800 in 25-year increments; signs of progress are also visible in these more social dimensions of well-being. As Modelski and Perry (2002) demonstrate, the percentage of the world's population living under democratic institutions has increased steadily for several centuries and crossed the 50% mark in the 1990s. War deaths are lower today than they were in the first half of the twentieth century, but not low in longer historical comparison.

5.3.2 Distributional Patterns

Human well-being is not evenly distributed across individuals, social groups, or nations. Inequality across national boundaries is

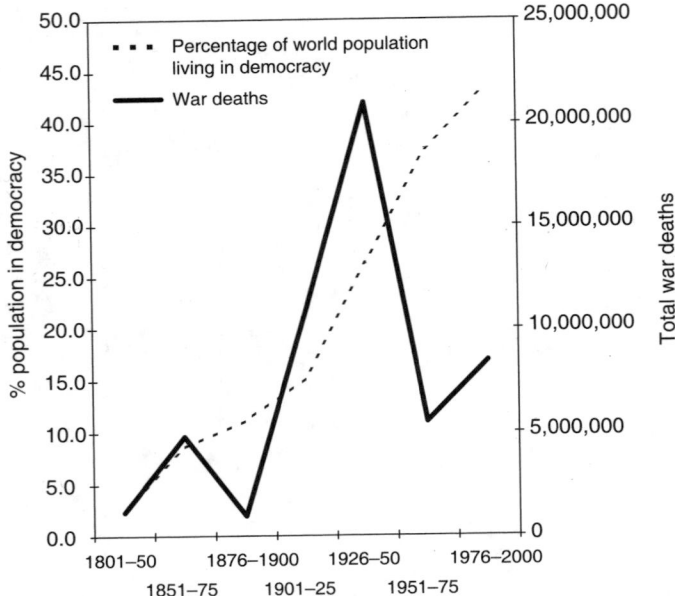

Figure 5.3. Trends in the Level of Democracy and Warfare since 1800 (Modelski and Perry III 1991 and 2002; Sarkees 2000)

high by historical standards. Prior to the industrial revolution in the nineteenth century, national differences in economic output per capita were relatively low; as countries and regions accelerated at different rates of industrialization, they generated dramatic differences in economic growth rates (Maddison 2001; Baudrillard 1998). Cross-national income inequality has increased over the past decade (World Bank 2003). National inequality is apparent in other measures of well-being as well. A child born in sub-Saharan Africa is 20 times more likely to die before age five than a child born in an OECD country, and this ratio is higher than it was a decade ago.

At the individual level, the gap between the world's poorest and the world's richest individuals has increased over the past two decades, though whether this signifies a general increase in inequality among the world's population is the subject of debate (UNDP 2003).

Recent increases in human well-being have been most pronounced in East Asia, where income-poverty levels have been reduced by approximately half since 1990. However, there have been systematic decreases in human well-being in many countries over the last decade. Although differences in growth rates are common, absolute declines in well-being on this scale have been rare. During the 1980s only four countries experienced declines in their rankings in the Human Development Index; during the 1990s, 21 countries registered declines, 14 of which were in sub-Saharan Africa (UNDP 2003). Hundreds of millions of people are living in countries with economic growth rates too low to permit significant poverty reductions (UN Millennium Project 2004).

The overall global pattern of human well-being, therefore, is one in which aggregate levels are continuing to increase at historical rates, although a large number of individuals appear to be stuck at very low levels of well-being.

5.3.3 Spatial Patterns

Human well-being is not evenly distributed with respect to the world's ecosystems. At a global level, there are a limited number of measures of human well-being available through which to assess patterns across ecosystem boundaries. Population totals and

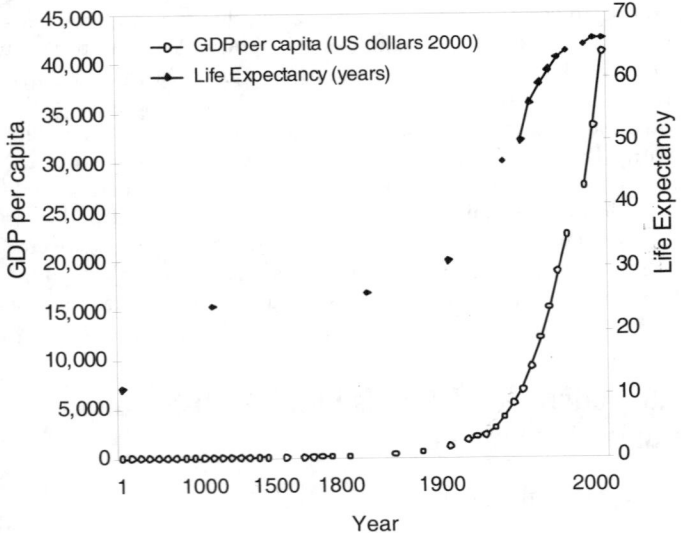

Figure 5.2. Trends for Life Expectancy and Per Capita Income, Past 2,000 Years (Maddison 2001; World Bank 2004; Kremer 1993; Haub 1995; United Nations 1953)

densities, infant mortality rates, GDP, and GDP per capita can be calculated using spatial data derived from sub-national sources.

The primary indicators for the MA systems and subsystems are reported in Tables 5.1 to 5.4 and shown graphically in Figure 5.4. (These aggregations do not take into account the urban system; a separate set of aggregations is reported in Chapter 27.) Of these indicators, infant mortality is the most spatially representative measure of human well-being because it is available for most countries at a sub-national level. The global distribution of infant mortality rates is provided in Figure 5.5 (in Appendix A).

At the broadest level of generalization, it is clear that infant mortality rates are highest within drylands and that most of the world's population and GDP is located within cultivated systems. One way to assess well-being across ecosystems more precisely is to compare the fraction of the world's population found within each ecosystem to the fraction of the world's GDP within each ecosystem. If well-being were distributed randomly with respect to ecosystem boundaries, these two numbers would be approximately equal in each ecosystem. Where the ratio of population fraction to GDP fraction is much higher than one, people are less

Table 5.1. Human Well-being Indicators, by MA System

System	GDP	Infant Mortality Rate	Area	Population	Population Density
	(billion dollars)	(deaths per thousand live births)	(million sq. km.)	(billion)	(people per sq. km.)
Coastal	9,148	41.5	6.0	1.0	169.7
Cultivated	27,941	54.3	35.3	4.1	116.2
Drylands	10,395	66.6	59.9	2.1	35.2
Forest	11,406	57.7	41.9	1.2	28.4
Inland Water	10,215	57.6	29.1	1.4	48.1
Island	7,029	30.4	7.1	0.6	85.5
Mountain	7,890	57.9	31.9	1.2	38.2
Polar (Arctic)	96	12.8	8.1	0.0	0.7

Table 5.2. Population Growth within MA Systems, 1990–2000

System	Change in Population	Net Change in Population	Change in Population per Square Kilometer
	(million)	(percent)	
Cultivated	505.7	14.1	14.3
Dryland	329.6	18.5	5.5
Inland Water	203.5	17.0	7.0
Mountain	171.0	16.3	5.4
Forest	142.1	13.5	3.4
Coastal	140.3	15.9	23.3
Island	67.0	12.3	9.5
Polar	−117.9	−6.5	0.0

Table 5.3. Distribution of Dryland Population Growth, 1990–2000

Region	Increase in Dryland Population	Share of World Population Increase
	(million)	(percent)
Asia	180	54.5
Former Soviet Union	5	1.4
Latin America	21	6.4
Northern Africa	65	19.6
OECD	10	3.2
Sub-Saharan Africa	49	14.9
World	**330**	**100.0**

well off in relative terms; where it is much lower than one, people are better off.

As seen in Figure 5.6, this ratio varies considerably across the MA systems, and this variation is correlated with differences in measured infant mortality rates. The drylands emerge as clearly an ecosystem characterized by low levels of human well-being. (See Box 5.1.) In each region, the dryland ecosystem shows higher infant mortality rates than the forest ecosystems, though the ratio of the two varies across regions. The ratio is highest in the former Soviet Union and lowest in Latin America. (See Table 5.5.)

The same comparisons can be performed for the MA subsystems. The results reveal even greater variation in patterns of human well-being than at the system level. As Figure 5.7 shows, infant mortality ranges from over 100 to under 10 in the MA subsystems, and some subsystems have a share of the world's population that is almost three times as large as their share of the world's GDP. At this level of resolution, the drylands do not emerge as clearly disadvantaged as they do in the system comparison. This is likely to be because the 56 MA subsystems are not distributed as evenly across geopolitical regions as the systems are, and therefore the strong effects of these geopolitical regions become more prominent in the subsystem analysis. For example, the MA subsystem with the highest infant mortality rate and the highest ratio of world population fraction to world GDP fraction is a forest subsystem: broadleaf, deciduous, open tree cover. Three quarters of this subsystem type are found within sub-Saharan Africa, where poverty rates in general are quite high, irrespective of ecosystem type.

It must be emphasized that none of these generalizations implies anything about causality. If infant mortality is high within a particular ecosystem, this does not mean that the ecosystem explains the high infant mortality. Rather, it indicates that the ecosystem is home to populations experiencing comparatively low levels of well-being and that these populations are therefore, other things being equal, potentially vulnerable to declines in ecosystem services.

Table 5.4. Human Well-being Indicators, by MA Subsystems. GDP and infant mortality rate estimates were not calculated for polar subsystems due to lack of appropriate resolution data.

MA System	Subsystem	Area (thousand sq. km.)	Population (million)	Population Density (people/ sq. km.)	GDP (bill. 2000 dollars)	GDP Per Capita (2000 dollars)	Infant Mortality Rate (deaths per thousand live births)
Coastal	coastal	6,020	1,022	170	9,148	8,956	41.5
Cultivated	agriculture/ two other land cover types	630	91	145	932	10,202	49.6
	agriculture/forest mosaic	4,459	294	66	3,017	10,272	48.2
	agriculture/other mosaic	5,922	508	86	3,001	5,903	62.2
	agriculture with forest	2,170	247	114	3,003	12,154	38.2
	agriculture with other vegetation	2,884	288	100	2,160	7,492	47.7
	cropland	8,270	3,013	243	8,924	4,433	55.3
	cropland/pasture	2,612	152	58	2,515	16,528	45.1
	forest with agriculture	1,601	97	61	1,969	20,235	15.9
	other vegetation with agriculture	6,659	405	61	2,299	5,677	65.2
	pasture	108	7.6	70	120	15,790	32.8
Dryland	dry subhumid	12,689	910	72	3,886	4,271	60.7
	semiarid	22,270	855	38	4,773	5,580	72.4
	arid	15,325	243	16	1,135	4,677	74.2
	hyper-arid	9,635	101	11	601	5,928	41.3
Forest	mosaic: tree cover/ other natural vegetation	2,409	53	22	217	4,137	59.3
	tree cover, broadleaved, deciduous, closed	6,526	348	53	3,312	9,503	58.9
	tree cover, broadleaved, deciduous, open	3,776	88	23	232	2,645	103.7
	tree cover, broadleaved, evergreen	12,210	266	22	1,436	5,394	60.3
	tree cover, burnt	298	0.3	1.0	2.4	8,238	15.9
	tree cover, mixed leaf type	3,182	51	16	953	18,843	12.4
	tree cover, needle-leaved, deciduous	3,804	3.5	0.9	28	8,127	27.2
	tree cover, needle-leaved, evergreen	9,032	370	41	5,184	14,013	28.5
	tree cover, regularly flooded, fresh	562	3.6	6.3	20	5,531	73.4
	tree cover, regularly flooded, saline (daily variation)	89	7.9	89	22	2,794	90.0

If the geopolitical regions are brought into the analysis explicitly, additional detail can be seen. Figure 5.8 (in Appendix A) shows that the differences across geopolitical regions are by and large more significant than the differences across ecosystem boundaries. Sub-Saharan Africa is less well off within each ecosystem than all other world regions, for example. There is also significant deviation from global averages within Asia: the Asian cultivated system contains 44% of the world's population, for example, but only 20% of the world's GDP.

By looking simultaneously at world geopolitical regions and the MA subsystems, the world is divided into 274 overlapping units. The basic patterns are seen in Figure 5.9 (in Appendix A). Well-being disparities across ecosystem types are lowest in the OECD countries and highest in Asia and sub-Saharan Africa. The very low variation within the OECD countries probably reflects the fact that high incomes and advanced infrastructures eliminate the kind of gross sensitivity to ecosystem effects that would influence infant mortality. This figure also shows the fundamentally

Table 5.4. *continued*

MA System	Subsystem	Area (thousand sq. km.)	Population (million)	Population Density (people/ sq. km.)	GDP (bill. 2000 dollars)	GDP Per Capita (2000 dollars)	Infant Mortality Rate (deaths per thousand live births)
Inland Water	50–100% wetland	2,157	13	6.3	454	33,623	6.1
	bog, fen, mire	2,305	7.4	3.2	98	13,243	25.1
	freshwater marsh, floodplain	4,606	403	87	1,123	2,789	68.5
	intermittent wetland/lake	4,946	128	26	558	4,355	68.2
	lake	9,538	522	55	5,390	10,329	45.5
	pan, brackish/saline wetland	619	11	18	59	5,279	59.8
	reservoir	649	37	57	278	7,504	47.1
	river	2,382	258	108	2,174	8,426	52.3
	swamp forest, flooded forest	1,902	21	11	82	3,993	76.7
Island	island state	3,949	375	95	1,787	4,767	36.0
	non-state island	3,125	230	74	5,242	22,794	11.8
Mountain	dry boreal/subalpine	997	10	10	46	4,436	85.2
	dry cool temperate montane	2,936	145	49	761	5,259	54.2
	dry subpolar/alpine	315	1.2	3.7	5.1	4,347	5.8
	dry subtropical hill	2,078	47	23	292	6,246	58.3
	dry tropical hill	467	16	33	56	3,595	73.8
	dry warm temperate lower montane	1,364	45	33	272	6,034	60.5
	humid temperate alpine/nival	1,605	4.4	2.8	33	7,381	51.1
	humid temperate hill and lower montane	3,888	373	96	3,101	8,321	42.8
	humid temperate lower/ mid-montane	1,591	91	57	1,232	13,512	29.8
	humid temperate upper montane and pan-mixed	5,143	54	10	357	6,659	39.1
	humid tropical alpine/nival	151	1.3	8.8	28	21,093	39.7
	humid tropical hill	942	57	60	198	3,487	55.4
	humid tropical lower montane	5,673	348	61	1,344	3,858	73.3
	humid tropical upper montane	212	16	74	99	6,253	35.1
	polar/nival	4,574	11	2.5	66	5,723	48.1
Polar	barrens and prostrate dwarf shrub tundra (includes rock/lichens and prostrate tundra)	2,628	0.5	0.2	8.5	18, 805	10.9
	forest tundra (includes low shrub tundra)	1,435	0.7	0.5	16	22,537	7.0
	graminoid, dwarf-shrub, and moss tundra	3,356	1.0	0.3	23	22,017	6.8
	ice	444	0.0	0.0	1.2		0.2
	lakes	595	4.0	6.7	48	12,019	10.7

different situation that sub-Saharan Africa is in with respect to well-being patterns. Almost all of its IMR values are higher than those found in any other region. Significant exceptions are the Asian dry boreal/subalpine subsystem and the Asian arid subsystem, where IMR values are much higher than the Asian average and well within the sub-Saharan African range.

5.3.4 Temporal Patterns

Most global socioeconomic indicators that are available in spatially disaggregated (sub-national) formats are not available in time series, and therefore it is difficult to say much about broad trends within the MA system boundaries. The exception to this generalization is

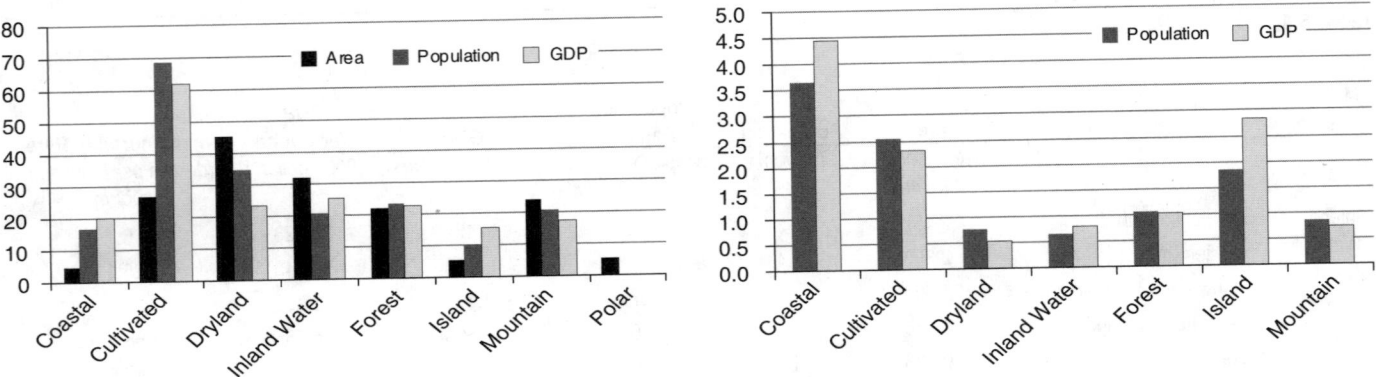

Figure 5.4. MA System Attributes as Percentage of World Total and Share of World Population and GDP as Ratio of Share of World Area (CIESIN 2004)

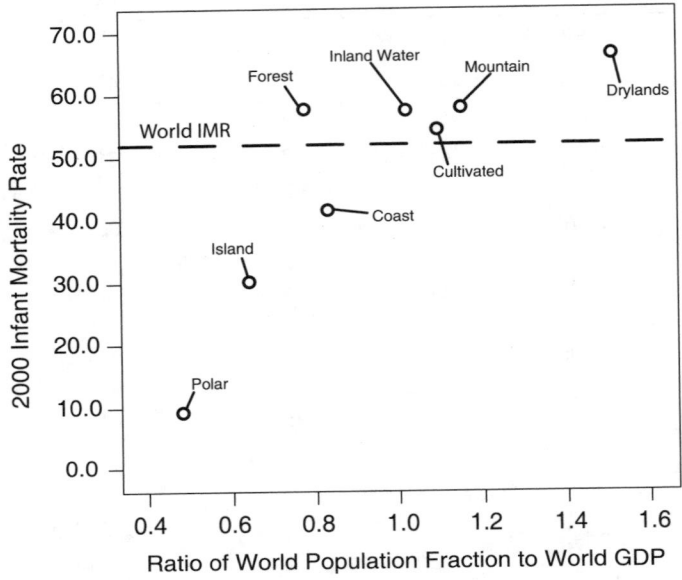

Figure 5.6. MA Systems and Relative Measures of Well-being

population. Because most countries carried out censuses both around 1990 and around 2000, and since these have been geo-referenced and integrated into a consistent grid, it is possible to estimate changes over the 1990–2000 period within the MA systems, as in Table 5.2.

Considering the global MA systems, the fastest growth rate during the 1990s occurred in the drylands, where population increased by 18.5%. The MA system with the greatest increase in total population is the cultivated system, where population increased by 506 million. If population growth is divided by land area, the highest value is observed in the coastal zone, where over the 1990s population grew by 23.3 people per square kilometer.

The fact that the highest growth rates are in an especially vulnerable ecosystem is significant. As the World Bank (2003) has pointed out, historically populations have migrated out of marginal lands to cities or to agriculturally productive regions; today the opportunities for such migration are limited due to poor economic growth in many African cities and much tighter immigration restrictions in wealthy countries.

Table 5.3 showed how the increased population in the drylands is distributed across world regions. More than half the people are located in Asia, but significant portions are also in Africa.

5.4 Sensitivity to Ecosystem Change

This section briefly reviews the state of knowledge about the degree to which the various dimensions of human well-being are sensitive to ecosystem change. In assessing the level of sensitivity, several issues arise that make an isolated understanding of the exact effect of a change in ecosystems on human well-being difficult. The affects of appropriating an ecosystem service today may have a different effect in the future on the appropriating social group. For example, the consumption of water today for irrigation will raise current crop productivity for the consuming population, but the impact on the future population is unknown, and so the net effect cannot be determined. Another difficulty in isolating the human well-being impact of ecosystem change is the distribution of ecosystem services. These may be derived from one geographical location but consumed in another one. Devising the net impact needs to take into account the costs and benefits to groups in all affected geographical locations.

5.4.1 Basic Material for a Good Life

The assessment of changes in the ecosystems on access to the basic needs for a good life has been one of the two dimensions of

BOX 5.1

Drylands and Human Well-being

Dryland ecosystems are characterized by extreme rainfall variability, recurrent but unpredictable droughts, high temperatures, low soil fertility, high salinity, grazing pressure, and fires. They reflect and absorb solar radiation, maintain balance in the functioning of the atmosphere, and sustain biomass and biodiversity. (This section is derived from World Bank 2003.)

Of the 500 million rural people who live on arid and dry semiarid land (see Table), most live in Asia and Africa, but there are also large pockets in Mexico and Northeastern Brazil. The low volume and extreme variability of precipitation limit the productive potential of this land for settled farming and nomadic pastoralism. Many ways of expanding agricultural production in the drylands—shifting cultivation from other areas, reducing fallow periods, switching farming practices, overgrazing pasture areas, cutting trees for fuelwood—result in greater environmental degradation.

Rural Population Living on Arid Lands (World Bank 2003)

Land Characteristics	Number of People
	(million)
Aridity only	350
Arid, slope	36
Arid, poor soil	107
Arid, slope, poor soil, forest	25
Total	518

The Southern Plains of North America, Africa's Sahel, and the inner Asian grasslands face similar climatic and soil characteristics but different political, financial, and institutional constraints. The result is differing patterns of resource management, with different impacts on human well-being.

The Southern Plains of North America

The European settlers in the Great Plains converted prime grazing land into intensive agricultural uses (monocropping, usually wheat). This pattern was badly suited to the lighter soils of the Southern Plains. Deep plowing dislodged soils, and monocropping mined soil nutrients. Large-scale farming in the 1920s pushed the expansion of wheat cultivation further onto native grasslands. By the next decade overgrazing, overplowing, and monocropping were exacerbated by the worst drought in U.S. history. An area of about 50 million hectares was affected each year in the Dust Bowl of the 1930s. (This section is based on Worster 1979.)

The response to the Dust Bowl included zoning laws for the most fragile areas, repurchases of submarginal private land, cash payments for leaving land fallow, and farm loans tied to approved land practices. In addition, there was planting of shelterbelts, adoption of soil and water conservation techniques such as the introduction of contour plowing, small dam and pond construction, mixed cropping, replanting of grasses, and state and federal protection of the remaining open grasslands.

Beginning in 1940, normal rainfall patterns resumed, while outmigration reduced the farm population and increased farm sizes (about 1 million people left the area between 1930 and 1970). But in the 1950s Dust Bowl II hit, followed in the 1970s by Dust Bowl III. Conservation practices had helped, but to achieve reliable production the United States needed to achieve a "climate-free" agriculture on the plains. In response, a striking feature has been the reliance on fossil fuel–intensive agricultural production with deep pumping of underground aquifers (up to 600 feet), and

heavy reliance on chemical fertilizers and mechanization. In the Texas High Plains Region, irrigated land area shrunk by 34% between 1974 and 1989 because the cost of overpumping the Ogallala aquifer exceeds the value of crops grown there. This vast aquifer is now being pumped faster than replenishment rates, with a net depletion rate of 3.62 million acre-feet (4.5 billion cubic meters) a year (Postel 1993).

The African Sahel

Throughout much of Africa the plowing and monocropping on fragile soils adopted in colonial times continued after independence (Swearingen and Bencherifa 1996). Inappropriate land use can rapidly lower soil quality, and intensive cultivation can deplete soil nutrients. Deforestation can cause erosion, washing away the layers of soil most suitable for farming. Two patterns are typical in Africa (note, however, that population growth in low density areas can be positive for resource management; see Tiffen et al. 1994):

- Growing populations convert high quality pastureland to grow cash crops. Herders lose the better grazing land, their security against drought. Migratory movements for herders are reduced, lower-quality land is more intensively grazed, and overgrazing leads to degradation.
- Poor subsistence farmers reduce fallow periods in order to grow more food to feed growing families. The reduction in fallow increases vulnerability to drought and without sufficient inputs, depletes soil nutrients. Degradation and soil erosion get worse.

In the Sahel, favorable rainfall from the 1950s to the mid-1960s attracted more people to the region. However, rainfall reverted to normal low levels after 1970, and by the mid-1970s many people and their livestock had died. The possibility that the Sahel could enter another period of favorable rainfall poses the risk of repeating the same tragedy as poor people are drawn back to the land. The Intergovernmental Panel on Climate Change (IPCC 2002) reports that Africa is highly vulnerable to climate change. Although the equatorial region and coastal areas are humid, the rest of the continent is dry subhumid to arid. Global warming scenarios suggest that soil moisture and runoff will be reduced in subhumid zones. Already, water storage has been reduced to critical levels in some lakes and major dams, with adverse repercussions for industrial activity and agricultural irrigation.

The poor quality of soils in the Sahel is another constraining environmental factor. Phosphorus deficiency, low organic content, and low water infiltration and retention capacity on much of African soil have been limiting factors in agriculture. Unlike climate variability, this problem can be addressed: soil quality can be augmented through careful management and soil nutrient supplementation. More difficult to address are the recurrent droughts.

The Asian Drylands

Population pressure on arable land in Asia is considerable and growing. Severe land degradation affects some 35% of productive land. The result has been to put more population pressure on the Inner Asian drylands. Most affected are Afghanistan, China, India, and Pakistan (FAO et al. 1994; ESCAP 1993), as well as Inner Asia's high steppe, the largest

(continues over)

BOX 5.1
continued

remaining pastureland in the world, which includes Mongolia, northwestern China, and parts of Siberia. Over thousands of years, these grasslands have been home to nomadic herders of horses, camels, goats, sheep, and cattle who practice elaborate systems of seasonal pasture rotation across wide stretches of land in response to climate fluctuations. Herd rotation has helped sustain the fertility and resilience of grassland ecosystems and improve the health of livestock (Ojima 2001).

Over the past decade, population pressures and competing uses on these fragile lands have made it hard to find the right balance between traditional land management and demand for higher agricultural productivity. Government policies that discouraged a nomadic lifestyle, herd movement, and temporary use of patchy grasses led to dependence on agricultural livelihoods and sedentary herds, which created greater pressure on local ecosystems and degraded fragile grasslands. The contrasting experiences of Mongolia and northwestern China illustrate some of the problems.

Mongolia has retained many traditional herding customs and customary tenure with land management as a commons. (This section is drawn from WRI 2000 and from Mearns 2001, 2002.) Herders rely on local breeds (which are stronger and more resilient) that graze year-round on native grasses. These customary practices were effectively supported by collective agriculture between the 1950s and 1980s. Policies allowed people and herds to move over large areas and provided the possibility of sustainable grasslands management under controlled-access conditions.

The economic transition since 1990 has not been conducive to sustainable management. Livestock mobility declined significantly. Many public enterprises closed. Having few alternatives, people turned to herding—often for the first time. The numbers of herders more than doubled from 400,000 in 1989 (17% of Mongolia's population) to 800,000 in the mid-1990s (representing 35%). Poverty also increased to 36% of the population by 1995 from a very low base in the 1980s. Herd size grew from the traditional 25 million head to about 30 million. Today, an estimated 10% of pastureland is believed to be degraded, causing noticeable increases in the frequency and intensity of dust storms.

This problem is considered manageable in Mongolia because population pressures are not very high. Rural population increased by about 50% from 1950 to 2000 (compared with a 700% increase in neighboring northwestern China).

As in Mongolia, the grasslands in China are state-owned. (This section is drawn from Mearns 2001, 2002.) But settled pastoralism and the conversion of grasslands to arable cultivation were more common in northwestern China than in Mongolia, beginning in the 1950s, when state-owned pastureland was allocated to "people's communes." The concentration of people in villages meant declining pasture rotation and expanding agriculture. Policies encouraged conversion of prime pasturelands into arable cropland, leading to salinization and wind erosion in some areas.

Common policies were applied to highly diverse circumstances, resulting in perverse outcomes and higher degradation in some places. Subsidies encouraged mixed farming systems, which put more pressure on fragile land than the traditional mobile pastoralism. Economic reforms in the early 1990s granted households nominal shares in the collective land pool. Shared areas were fenced off, making herd mobility more difficult. Subsidized inputs, income transfers, and deep pumping of underground aquifers encouraged a rapid increase in farming. From an estimated 3 million indigenous pastoralists in the 1950s in the Inner Mongolian part of northwestern China, farmers and livestock producers today number 20 million, and cattle doubled from 17 million head in 1957 to 32 million today.

China's western development plan shares two characteristics with the policies followed in the Southern Plains of the United States: intensification of agricultural production and creation of "climate-free" agriculture in the grasslands through irrigation from underground aquifers. The objective is to make the area a bread-and-meat basket to meet China's growing demands for protein-rich diets. But unlike the Southern Plains—where about 1 million farmers left between the 1930s and the 1970s, enabling reconsolidation of landholdings and conversion of vast grassland areas to protected areas—population pressures have continued to increase in China's grasslands. Poverty rates in these degraded and ecologically sensitive areas are well above the national average (25% in some provinces, compared with the national average of 6.3%). The frequency and intensity of dust storms are increasing. Estimates of areas degraded are 50–75%, compared with 10–15% in the grasslands of Mongolia.

Improving Prospects for Well-being in the Drylands

Agricultural research in China and India shows diminishing returns to investments in many high-potential areas, but investments in drylands can produce large returns in reducing poverty, even if yields are modest (Hazell 1998; Hazell and Fan 2000; Fan et al. 2000; Wood et al. 1999). Governments, researchers, and donor organizations are beginning to pay some attention to research and development on crop breeding varieties for people on marginal lands, but much more needs to be done by the public sector to replace antiquated crop varieties (UNEP 1992, 1997). In partnership with South African institutions, the International Maize and Wheat Improvement Center (a research center of the Consultative Group on International Agricultural Research) has developed two maize varieties for small farmers in South Africa's drought-prone, acidic, nutrient-depleted soils. Both varieties are drought-resistant, and one matures early, when farm food supplies are at their lowest. Trials from Ethiopia to South Africa have shown yields that are 34–50% higher than currently grown varieties (Ter-Minassian 1997; Rodden et al. 2002; Bardham and Mookherjee 2000). There are opportunities to achieve sustainable livelihoods in quite a few areas. But development decision-makers must recognize that the drylands are not homogeneous and cannot be made to function sustainably as non-drylands.

Some arid areas can take advantage of solar energy potential; others may have scenic value worthy of ecotourism development. The Mozambique Transfrontier Conservation Area Program and Burkina Faso's wildlife reserve development are two attempts in the direction of ecotourism that combine local and international cooperation. Research and innovations for appropriate service delivery—combined with policies that link human activities (farming, herding, and settlements) with natural processes (vegetation distribution, seasonal growing cycles, and watersheds)—can help sustain vulnerable dryland ecosystems while enhancing productivity to support growing populations.

Table 5.5. Ratio of Infant Mortality Rate in Drylands to Rate in Forests, by Region

Region	IMR in Drylands / IMR in Forests
	(ratio)
Asia	1.6
Former Soviet Union	2.6
Latin America	1.0
Northern Africa	1.2
OECD	1.4
Sub-Saharan Africa	1.1

human well-being that has been most thoroughly investigated, and there are numerous well-documented examples demonstrating that declining ecosystem services are capable of having serious negative consequences on incomes, food security, and water availability. Some studies have also suggested that the decline and in some cases even the collapses of several ancient civilizations—including the Mesopotamians, the ancient Greeks, the Mayans, the Maori, and the Rapanui of Easter Island—were associated with the overexploitation of biological resources (Deevey et al. 1979; Flenley and King 1984; van Andel et al. 1990; Ponting 1991; Flannery 1994; Diamond 1997; Redman 1999).

Economic theory shows clearly how continued improvements in income depend on growing levels of assets to be sustainable. If assets, or wealth, do not grow, incomes will eventually fall (Sachs et al. 2004). Some economists have attempted to quantify the natural resource components to assets, including such resource stocks as forests and minerals (Repetto et al. 1989; Vincent et al. 1997; Lange et al. 2003), and have sought to estimate the net social benefits of habitat conversion, taking into account both the narrow economic gains associated with conversion and the decline of broader social benefits. These studies support the generalization that private gains achieved by conversion are typically outweighed by the loss of public benefits, so that in overall soci-

etal terms, conversion of remaining intact habitat rarely makes net economic sense (Balmford 2002; Turner et al. 2003). For example, the early 1990s collapse of the Newfoundland cod fishery has cost tens of thousands of jobs, as well as at least $2 billion in income support and retraining (Beaudin 2001).

Recent reviews of such dynamics have concluded that many countries appear to be experiencing declines in net per capita assets when such resource components are taken into account (Dasgupta and Mäler 2004; Hamilton and Clements 1999). (See Box 5.2.) These findings suggest that some of the declines in well-being in sub-Saharan Africa may be in part attributable to declines in natural resource assets, and that other regions currently experiencing increases in well-being may be doing so at the expense of a declining resource base, which will create problems in the future.

5.4.2 Freedom and Choice

There are some direct connections between ecosystem services and the freedom and choice dimensions of well-being that are partly understood. The declining provision of fuelwood and drinking water as the result of deteriorating ecosystems, for example, has been shown to increase the amount of time needed to collect such basic necessities, which in turn reduces the amount of time available for education, employment, and care of family members. Such impacts are typically thought to be disproportionately experienced by women. However, the empirical foundation for this understanding is limited to a handful of isolated studies (e.g., Awumbila and Momsen 1995); little work based on comparative, cross-national data has been done.

The findings in the literature on common pool or common property resources is that when a resource is abundant relative to demand, there tend to be few rules about its use. If a resource is valuable and limited, however, then common property institutions develop (Ostrom 1990). Such institutions evolve according to the importance of the resource in question, the technology used to exploit it, and various social or political changes in that society. For many kinds of commons, such as forests, water, and grazing lands, resources may be controlled under government

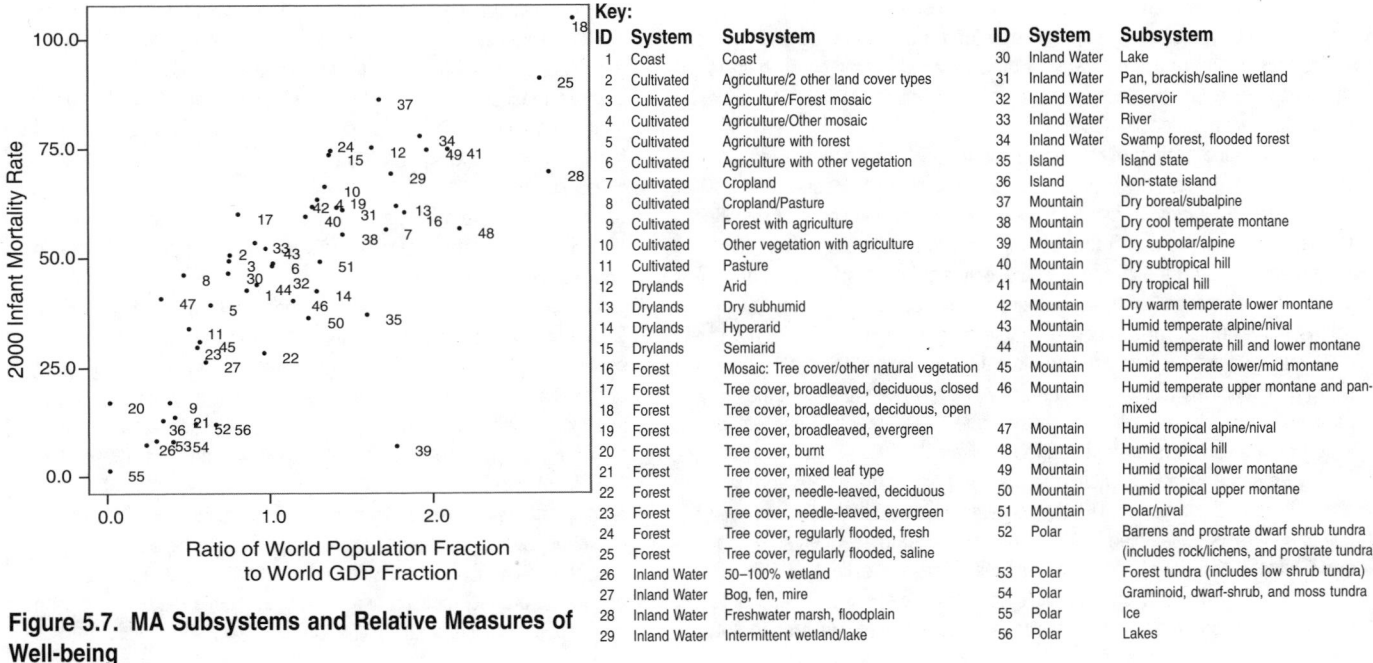

Figure 5.7. MA Subsystems and Relative Measures of Well-being

ID	System	Subsystem
1	Coast	Coast
2	Cultivated	Agriculture/2 other land cover types
3	Cultivated	Agriculture/Forest mosaic
4	Cultivated	Agriculture/Other mosaic
5	Cultivated	Agriculture with forest
6	Cultivated	Agriculture with other vegetation
7	Cultivated	Cropland
8	Cultivated	Cropland/Pasture
9	Cultivated	Forest with agriculture
10	Cultivated	Other vegetation with agriculture
11	Cultivated	Pasture
12	Drylands	Arid
13	Drylands	Dry subhumid
14	Drylands	Hyperarid
15	Drylands	Semiarid
16	Forest	Mosaic: Tree cover/other natural vegetation
17	Forest	Tree cover, broadleaved, deciduous, closed
18	Forest	Tree cover, broadleaved, deciduous, open
19	Forest	Tree cover, broadleaved, evergreen
20	Forest	Tree cover, burnt
21	Forest	Tree cover, mixed leaf type
22	Forest	Tree cover, needle-leaved, deciduous
23	Forest	Tree cover, needle-leaved, evergreen
24	Forest	Tree cover, regularly flooded, fresh
25	Forest	Tree cover, regularly flooded, saline
26	Inland Water	50–100% wetland
27	Inland Water	Bog, fen, mire
28	Inland Water	Freshwater marsh, floodplain
29	Inland Water	Intermittent wetland/lake

ID	System	Subsystem
30	Inland Water	Lake
31	Inland Water	Pan, brackish/saline wetland
32	Inland Water	Reservoir
33	Inland Water	River
34	Inland Water	Swamp forest, flooded forest
35	Island	Island state
36	Island	Non-state island
37	Mountain	Dry boreal/subalpine
38	Mountain	Dry cool temperate montane
39	Mountain	Dry subpolar/alpine
40	Mountain	Dry subtropical hill
41	Mountain	Dry tropical hill
42	Mountain	Dry warm temperate lower montane
43	Mountain	Humid temperate alpine/nival
44	Mountain	Humid temperate hill and lower montane
45	Mountain	Humid temperate lower/mid montane
46	Mountain	Humid temperate upper montane and pan-mixed
47	Mountain	Humid tropical alpine/nival
48	Mountain	Humid tropical hill
49	Mountain	Humid tropical lower montane
50	Mountain	Humid tropical upper montane
51	Mountain	Polar/nival
52	Polar	Barrens and prostrate dwarf shrub tundra (includes rock/lichens, and prostrate tundra)
53	Polar	Forest tundra (includes low shrub tundra)
54	Polar	Graminoid, dwarf-shrub, and moss tundra
55	Polar	Ice
56	Polar	Lakes

BOX 5.2

Economic Values Associated with Ecosystem Services

Many ecosystem services, such as the purification of water, regulation of floods, or provision of esthetic benefits, do not pass through markets. The benefits they provide to society, therefore, are largely unrecorded: only a portion of the total benefits provided by an ecosystem make their way into statistics, and many of these are misattributed (the water regulation benefits of wetlands, for example, do not appear as benefits of wetlands but as higher profits in water-using sectors). Moreover, for ecosystem services that do not pass through markets there is often insufficient incentive for individuals to invest in their maintenance (although in some cases common property management systems provide such incentives). Typically, even if individuals are aware of the services provided by an ecosystem, they are neither compensated for providing these services nor penalized for reducing them.

Nonmarketed ecosystem services do have economic value, and a number of techniques can be used to measure the economic benefits of the services and the costs of their degradation. (See Chapter 2.) A growing literature exists concerning measurements of the value of the nonmarketed ecosystem services, in order to provide better estimates of the total economic value of ecosystems (Pearce and Warford 1993). Some of the methods used to value nonmarketed services are still controversial, and in some cases estimates of the same services may differ by as much as several orders of magnitude. Moreover, even in the cases where the economic valuation methods are well established, the physical and ecological information necessary to value marginal changes in ecosystems may not be available. (For example, the calculation of the marginal cost of deforestation in terms of reduced water services requires accurate information on the relationship between forest cover loss and change in the timing and magnitude of river flows, and that biophysical information may not be available.)

Despite these challenges, even imperfect estimates of the economic costs and benefits of changes in ecosystem services can be useful for decision-making. Well-designed valuation studies tell us not only how much ecosystem services are worth, but inform resource management decisions with information about the economic benefits of alternative management options. Moreover, valuation studies help to identify who benefits from different ecosystem services and how various factors, including institutional ones, influence values. Information on beneficiaries and their willingness-to-pay under different institutional arrangements can help in the design of payment mechanisms or of mechanisms to turn economic values into actual financial flows.

The marketed portion of ecosystem services is often only a small portion of the total economic value of an ecosystem. In the case of forest ecosystems, for example, marketed services can include some provisioning services, such as timber production, and some cultural services, such as recreation. Even these services are only partially marketed: informal collections of fuelwood and non-timber products such as fruit and mushrooms, for example, are often not marketed (although local systems of rights or management may serve to maintain sustainable harvests). Regulating and supporting services have rarely been marketed, although some markets—such as for carbon sequestration—have recently been established to ameliorate the rate of change in climate services.

Figure A in Appendix A shows the results of one of the few comprehensive efforts to estimate the TEV of ecosystems on a significant scale, in this case forests in selected Mediterranean and nearby countries. Timber and fuelwood generally account for less than a third of estimated TEV, on average, and even this share is likely overestimated as it is easier to measure such provisioning services than other services. Recreation and

hunting benefits were imperfectly measured, but in European countries these benefits rival and sometimes exceed timber values. Watershed protection is an important benefit in Italy, Syria, Tunisia, Algeria, and Morocco and would likely have played an important role in several other countries as well had it been possible to better estimate its value. If forests were to be considered only for their timber, as is all too often the case, they would appear much less valuable than they actually are and would tend to be managed only for their extractive uses, such as timber harvest.

The estimated values of individual non-marketed ecosystem services are often substantial but are rarely included in resource management decisions, although incorporation of these values is now beginning to become somewhat more common. For example:

- *Recreational benefits of protected areas:* The annual recreational value of the coral reefs of each of six marine protected areas in the Hawaiian Islands in 2003 ranged from $300,000 to $35 million (van Beukering and Cesar 2004).
- *Water quality:* The net present value in 1998 of protecting water quality in the 360-kilometer Catawba River in the United States for a five-year period was estimated to be $346 million (Kramer and Eisen-Hecht 1999).
- *Nursery habitat for fisheries:* Deforestation of an 860-square-kilometer mangrove habitat in the state of Campeche, Mexico, at an average rate of 2 square kilometers a year in 1980–90 resulted in a loss in shrimp harvest each year of about 28.8 tons, amounting to a loss of approximately $279,000 in revenue (Barbier and Strand 1998). In contrast, a 1981 study of the marginal value of wetlands on the Gulf Coast of Florida for production of blue crab estimated the present value of only $7.40 per hectare (Lynne et al. 1981).
- *Water purification service of wetlands:* Approximately half of the total economic value of the Danube River floodplain in 1992—which included values associated with timber, cattle, fisheries, recreation, hunting, and the filtering of nutrients—could be accounted for in its role as a nutrient sink (Gren et al. 1995).
- *Native pollinators:* A study in Costa Rica found that forest-based pollinators increased coffee yields by 20% within 1 kilometer of forest (as well as increasing the quality of the coffee). During 2000–03, pollination services from two forest fragments (of 46 and 111 hectares) increased the income of a 1,100-hectare farm by $60,000 per year, a value commensurate with expected revenues from competing land uses (Ricketts et al. 2004).
- *Flood control:* Muthurajawela Marsh, a 3,100-hectare coastal peat bog in Sri Lanka, provides an estimated $5 million in annual benefits ($1,750 per hectare) through its role in local flood control (Emerton and Bos 2004).
- *Open space protection:* In an ex-urban area of Maryland in the United States, the marginal economic benefits per household of preserving neighboring open space range from $994 to $3,307 per acre of farmland preserved (Irwin 2002). And in Grand Rapids, Michigan, lots for residential homes that border forest preserves were found to sell at premiums of about $5,800–8,400 during the 1990s (19–35% of lot price) (Thorsnes 2002).
- *Biochemical resources:* Estimates of the economic value of species-rich ecosystems (such as neotropical forests) for bioprospecting for new pharmaceutical products range from $20 to $9,000 per hectare, depending on assumptions used in the economic models (Simpson et al. 1996; Rausser and Small 2000).

Relatively few studies have compared the TEV of ecosystems under alternate management regimes. The results of several that attempted to do so are shown in the Table. In each case where the TEV of sustainable management practices was compared to management regimes involving conversion of the ecosystem or unsustainable practices, the benefit of managing the ecosystem more sustainably exceeded that of the converted ecosystem, although the private benefits—that is, the actual monetary benefits captured from the services entering the market—would favor conversion or unsustainable management.

These studies are consistent with the understanding that market failures associated with ecosystem services lead to greater conversion of ecosystems than is economically justified. However, this finding would not hold at all locations. For example, the value of conversion of an ecosystem in areas of prime agricultural land or in urban regions often exceeds the total economic value of the intact ecosystem (although even in dense urban areas, the TEV of maintaining some "greenspace" can be greater than development of these sites). Similarly, in a study of the economic benefits of forest protection for maintaining water flows as "drought mitigation" for farmers in Eastern Indonesia, Pattanayak and Kramer (2001) found that where increased watershed protection mitigates droughts, the economic benefits can be sizable (as much as 10% of annual agricultural profits), but they found that forest cover did not necessarily increase baseflow for all households in a watershed or in all watersheds.

Comparisons of Economic Benefits of Retaining and Converting Ecosystems. (Values in dollars, rounded to two significant digits.) Each of the examples selected includes estimates of one or more regulating or cultural services in addition to provisioning services.

Ecosystem	Alternatives Compared	Services Included in TEV Calculations	Private Benefits	Total Economic Value	Source
Comparison of benefits of sustainably managed ecosystem to converted ecosystem					
Tropical forest, Cameroon	comparison of low-impact logging to small-scale farming or conversion to oil palm and rubber plantation	benefits from agricultural or plantation output, sedimentation control, flood prevention, carbon storage, and option, bequest and existence values; 10% discount rate over 32 years	Small-scale agriculture had greatest private benefits	across five study sites, average TEV of sustainable forestry was approximately $3,400 per hectare and that of small-scale farming $2,000 per hectare; across four of the sites, average TEV of conversion to oil palm plantation was $-1000 per hectare	Yaron 2001
Mangrove, Thailand	comparison of existing uses of mangrove system to conversion to shrimp farming	benefits from shrimp farming, timber, charcoal, NTFPs, offshore fisheries, and storm protection. 10% discount rate over 20 years	conversion to aquaculture had greatest private benefits	TEV value of intact mangroves was a minimum of $1,000 and possibly as high as $36,000 per hectare; TEV of shrimp farming was about $200 per hectare	Sathirathai and Barbier 2001
Wetland, Canada	comparison of intact wetlands to conversion to intensive farming	benefits of agriculture, hunting, angling, trapping; 4% discount rate over 50 years	conversion to agriculture had highest private benefits (in part due to substantial drainage subsidies)	TEV was highest for intact wetlands (average for three wetland types of approximately $5,800 per hectare) versus TEV of converted wetlands of $2,400 per hectare	Van Vuuren and Roy 1993
Tropical forest, Cambodia	comparison of traditional forest uses to benefits associated with commercial timber extraction	examined benefits associated with swidden agriculture and extraction of non-timber forest products (including fuelwood, rattan and bamboo, wildlife, malva nuts, and medicine) and ecological and environmental functions such as watershed, biodiversity, and carbon storage; 6% discount rate over 90 years	private benefits associated with unsustainable harvest practices exceeded private benefits of NTFP collection	total benefits were greatest for traditional uses, ranging from $1,300 to $4,500 per hectare (environmental services accounted for $590 per hectare while NTFPs provided between $700 and $3,900); private benefits for timber harvest ranged from $400 to $1,700 per hectare but after accounting for lost services the total benefits were $150 to $1,100 per hectare	Bann 1997
Comparison of benefits of establishing protected area to current use					
Coastal habitat, Jamaica	comparison of current management (includes destructive fishing, loss of mangroves, pollution) to establishment of Portland Bight Protected Area	benefits of fisheries, forestry, tourism, carbon sequestration, biodiversity, costal protection; 10% discount rate over 25 years	current overfishing has resulted in a decline of profits effectively to zero	total incremental benefits of establishment of protected area estimated to be $53 million ($28 per hectare) in the optimistic tourism scenario and $41 million ($22 per hectare) in the pessimistic tourism case; cost of protected area establishment and management would total $19 million ($10 per hectare) over the next 25 years, resulting in net benefits of $11 to $18 per hectare	Cesar et al. 2000
Marine protected areas, Hawaii	comparison of net benefits of protection of six existing MPAs with the costs associated with their protection	benefits associated with tourism, contribution to fisheries in adjacent areas, biodiversity, and amenity values; discount rate of 3%, period of 25 years		benefits for individual MPAs ranged from $15 million (Diamond Head) to $84 million (Hanauma), with management costs for these two MPAs of $1.1 million and $22 million respectively; the net benefit per hectare (benefits minus management costs) ranged from $144,000 (Diamond Head) to $17 million (Kahalu'u)	van Beukering and Cesar 2004

(continues over)

BOX 5.2
continued

The economic costs associated with damage to ecosystem services can be substantial:

- The early 1990s collapse of the Newfoundland cod fishery due to overfishing resulted in the loss of tens of thousands of jobs and has cost at least $2 billion (CAN$2.66 billion) in income support and retraining (Commission for Economic Cooperation 2001).
- In 1996, the external cost of U.K. agriculture associated with damage to water (pollution, eutrophication), air (emissions of greenhouse gases), soil (off-site erosion damage, carbon dioxide loss), and biodiversity was $2.6 billion (£1.566 billion at 1996 exchange rates)—9% of average yearly gross farm receipts for the 1990s (Pretty et al. 2000). Similarly, the cost of freshwater eutrophication in England and Wales was estimated to be $105–160 million per year in the 1990s, with an additional $77 million per year being spent to address those damages (Pretty et al. 2003).
- The largely deliberate burning in 1997 of approximately 50,000 square kilometers of Indonesian vegetation (about 60% of the total area burned from 1997 to 1998) affected around 70 million people (Schweithelm and Glover 1999). Some 12 million people required health care; overall economic costs—through lost timber and non-wood forest products, lost agriculture, reduced health, increased CO_2 emissions, lost industrial production, and lost tourism revenues—have been conservatively estimated at $4.5 billion (Ruitenbeek 1999; Schweithelm et al. 1999).
- The total damages for the Indian Ocean region over a 20-year time period (with a 10% discount rate) resulting from the long-term impacts of a massive 1998 coral bleaching episode are estimated to be between $608 million (if there is only a slight decrease in tourism-generated income and employment results) and $8 billion (if tourism income and employment and fish productivity drop significantly and reefs cease to function as a protective barrier) (Cesar and Chong 2004).
- The net annual loss of economic value associated with invasive species in the fynbos vegetation of the Cape Floral region of South Africa in 1997 was estimated to be $93.5 million (R455 million), equivalent to a reduction of the potential economic value without the invasive species of more than 40%. The invasive species have caused losses of biodiversity, water, soil, and scenic beauty, although they also provide some benefits, such as provision of firewood (Turpie and Heydenrych 2000).

Significant investments are often needed to restore or maintain nonmarketed ecosystem services. Examples include:

- In South Africa, invasive tree species threaten both native species and water flows by encroaching into natural habitats, with serious impacts for economic growth and human well-being. In response, the South African government established the Working for Water Programme. Between 1995 and 2001 the program invested $131 million (R1.59 billion at 2001 exchange rates) in clearing programs to control invasive species (van Wilgen 2004).

- The state of Louisiana in the United States has put in place a $14-billion wetland restoration plan to protect 10,000 square kilometers of marsh, swamp, and barrier islands in part to reduce storm surges generated by hurricanes (Bourne 2000).
- A plan to restore semi-natural water flows in the Everglades wetlands in the United States in part through the removal of 400 kilometers of dikes and levees is expected to cost $7.8 billion over 20 years (Kloor 2000).

In addition to efforts to measure the value of nonmarketed ecosystem services, recent years have also seen increasing efforts to devise mechanisms to bring these services into the market, thus improving incentives to conserve ecosystems (Pagiola et al. 2002). Examples include:

- *Markets for carbon sequestration:* Approximately 64 million tons of carbon dioxide equivalent were exchanged through projects from January to May 2004, nearly as much as during all of 2003 (78 million tons) (World Bank 2004). (See Figure B in Appendix A.) The value of carbon trades in 2003 was approximately $300 million. Some 25 percent of the trades (by volume of CO_2 equivalents) involve investment in ecosystem services (hydropower or biomass) (World Bank 2004). The World Bank has established a fund with a capital of $33.3 million (as of January 2005) to invest in afforestation and reforestation projects that sequester or conserve carbon in forests and agroecosystems while promoting biodiversity conservation and poverty alleviation. It is speculated that the value of the global carbon emissions trading markets could reach $44 billion in 2010 and involve trades totaling 4.5 billion tons of carbon dioxide or equivalent (http://www.pointcarbon.com/).
- *Markets for forest environmental services.* In 1997, Costa Rica established a nationwide system of payments for environmental services (Pago de Servicios Ambientales). Under this program, Costa Rica pays land users who conserve forests, thus helping to maintain environmental services such as downstream water flows, biodiversity conservation, carbon sequestration, and scenic beauty. Funds for the program come partly from earmarked taxes and partly from environmental service buyers, including the Global Environment Facility (biodiversity), Costa Rica's Office of Joint Implementation (carbon), and water users such as hydroelectric producers, municipal water utilities, and bottlers (watershed services). By 2001, over 280,000 hectares of forest had been incorporated into the program at a cost of about $30 million, with pending applications covering an additional 800,000 hectares. Typical payments have ranged from $35 to $45 a hectare per year for forest conservation (MA *Policy Responses,* Box 5.3; Pagiola 2002). Payments under Costa Rica's program do not reflect the values attached to the services provided so much as the costs associated with their provision. As a result, while this market mechanism provides for the cost-effective imposition of quantitative targets, it is not a typical market in the sense of private parties undertaking voluntary transactions in environmental services. Worldwide, the number of initiatives involving payments for ecosystem services is growing rapidly.

regulations, market mechanisms, or community-based institutions that develop among the users themselves. Even with these community institutions, in some cases ecosystem services have declined, although in other cases the development of effective partnerships or of participatory management institutions—with power-sharing between governments, local communities, or corporate resource owners—has reduced the further decline and even brought about improvement of ecosystem services.

In South Africa, for instance, data gathered over decades showed that invasive forestry trees not only threaten native species by encroaching into natural habitats, but also severely reduce stream flows, with serious impacts for economic growth and human well-being (van Wilgen et al. 1998). In response to this evidence, the South African government established its Working for Water Programme. With an annual budget now over $60 million, this simultaneously increases water availability, employs 20,000 skilled and semi-skilled workers, and addresses a major driver of biodiversity change (van Wilgen et al. 2002; Working for Water 2004). Similarly, in Madagascar continued upland deforestation by an estimated 50,000 slash-and-burn farmers has led to increased siltation and reduced water flows to 250,000 downstream rice farmers (Carret and Loyer 2003), and this played an important role in the Malagasy government's 2003 decision to triple the size of Madagascar's network of protected forests (J. Carret, personal communication).

5.4.3 Health

Connections between declining ecosystem services and human health are well documented and may be the best understood of the well-being impacts. As shown in Chapter 14, infectious disease impacts are rising as a consequence of land use change such as deforestation, dam construction, road building, agricultural conversion, and urbanization. Such effects can be observed in all regions of the world. The World Health Organization has attempted to quantify the environmental burden of disease through a modeling of the percentage of disability-adjusted life years lost as a consequence of such environmental drivers as unsafe water, air pollution, indoor smoke, and climate change. (See also Chapter 2.) Such analysis provides strong evidence that deteriorating water and air quality account for a large percentage of poor health outcomes in many locations, especially in developing countries (WHO 2002 Chapter 4).

The environmental impact on health, according to WHO analysis, is dependent on levels of poverty. Such impacts are highest in poor countries with high mortality rates, where unsafe water and indoor smoke from solid fuel use account for 9–10% of DALYs (WHO 2002:86).

5.4.4 Good Social Relations

There are clear examples of declining ecosystem services disrupting social relations. Indigenous societies whose cultural identities are tied closely to particular habitats or wildlife suffer if habitats are destroyed or wildlife populations decline. Such impacts have been observed in coastal fishing communities (see Chapter 19), in Arctic populations (see Chapter 25), in traditional forest societies, and among pastoral nomads (e.g., Mather 1999; Parkinson 1999). The 95% decline in the *Gyps* vulture populations of the Indian subcontinent since the mid-1990s (Cunningham et al. 2003) has led to divisions in the Parsi religion about how to dispose of their dead now that the traditional laying out of corpses in Towers of Silence is no longer practicable (Triveldi 2001).

Deterioration in ecosystems can also provide an opportunity for social relations when communities join together to form community-based institutions in response to degraded ecosystem services (Ostrom et al. 2002).

5.4.5 Security

The impact of declining ecosystem services on natural disasters is well understood and well documented, and this is further explored in Chapter 16. Although there has been much speculation about the relationship between ecosystem conditions and political violence (Myers 1994; Homer-Dixon 1999), scholars have been unable to demonstrate causal connections robustly (Gleditsch 1998).

5.5 Multiple Causal Mechanisms Link Ecosystem Change to Human Well-being

The causal connections between ecosystem change and changes in human well-being are not uniform. A variety of mechanisms link the two. This section delineates some distinct mechanisms and illustrates them with specific examples, in order to demonstrate the variety of pathways linking ecosystem change and human well-being. The presentation is not meant to imply that ecosystem changes are linked to human well-being in simple orderly mechanisms. Socioecological systems are dynamic and nonlinear, and their response to stress may not be predictable. They typically combine in complicated ways; yet drivers of behavior may be identified, and behavior may be bounded by political, social, technological, or economic limitations. Decision-making behavior about the use of different ecosystem services needs to consider trade-offs in human well-being.

The benefits and costs of declining ecosystem services are seldom evenly distributed. There are typically winners and losers, and this can lead to an improvement in well-being for one group at the expense of another. If mountain communities convert forests to agricultural lands, for example, that may reduce downstream ecosystem services to low-lying areas in the form of increased siltation and declining water quality. (See Chapter 24.) If wetlands are converted to human settlements in one area of a watershed, other communities in the watershed may experience diminished flood buffering capacities. (See Chapters 16 and 20.) When ecosystem change is linked to well-being change through this mechanism, some groups of people improve and other groups decline.

Another example is the growth of shrimp farming and the consequent damage of such aquaculture on mangroves. (See Chapter 19.) Stonich et al. (1997) have shown how, in Honduras, social conflict has increased between shrimp farm concession holders and those who are not concession holders but believe that shrimp farms are intruding on government-reserved natural resources.

The often-cited example of the Aral Sea provides a vivid illustration of how a government policy aimed at improving the national economy can generate large negative impacts on human well-being by focusing on the benefits of a single ecosystem service. Under Soviet rule, the small-scale independent irrigation systems were transformed into unsustainable large-scale collective irrigation systems for cotton production. With the Soviet Union's attention focused on cotton self-sufficiency, the long-term adverse effects of a rapid, large-scale expansion of inefficient irrigation systems, sole reliance on high-water demanding production systems, poor water distribution and drainage, and non-dose-related uses of fertilizers and pesticides around the Aral Sea were not considered high priority.

During the second half of the 1900s, these inefficient systems decreased water inflow into the sea to a mere trickle, shrunk the size of the sea by half, reduced the water level by 16 meters, and tripled its salinity (Micklin 1993). The government and the cotton producers were the initial winners. The Soviet Union met its need to be self-sufficient in cotton. However, there were many losers who suffered. Thirty-five million people have lost access to the lake for its water, fish, reed beds, and transport functions. The fishing industry around the Aral Sea has collapsed, with fishing ceasing in the 1980s.

More far-reaching environmental and ecological problems, such as dust storms, erosion, and poor water quality for drinking and other purposes have contributed to decreased human health in the area around the Aral Sea. Rates of anemia, tuberculosis, kidney and liver diseases, respiratory infections, allergies, and cancer have increased, and now far exceed the rest of the former Soviet Union and present-day Russia (Ataniyazova et al. 2001). High levels of reproductive pathologies (infertility, miscarriages, and complications during pregnancy and childbirth) have been observed in this region for more than 20 years. The rate of birth abnormalities, another serious consequence of pollution, is also increasing. One in every 20 babies is born with abnormalities, a figure approximately five times the rate in European countries (Ataniyazova et al. 2001; Ataniyazova 2003). (For more information about the Aral Sea and ecosystems, see Chapters 20 and 24.)

China has also witnessed the effects of ecosystem trade-offs, especially in connection with economic development and land use change in the north and west. Excessive collection of fuelwood, blind mining, harvesting of medicinal herbs, construction projects, rising populations, and farming have all contributed to China's desertification through deforestation, water mismanagement, and overgrazing. Excessive cutting of fuelwood has reduced the sand fixing vegetation in Qaidam Basin by one third. Illegal cultivation, harvesting, and exploitation have degraded 12 million hectares of rangeland, steppe, and grasslands from 1993 to 1996. The large-scale diversion of water from the upper reach of some rivers has caused the reduction and even the drying off of the water flow in lower reaches. Desertification has silted up rivers, raised riverbeds, and retreated lake surfaces (Asian Regional Network for Desertification Monitoring and Assessment, n.d.).

The spread of desertification in northern China is threatening villages and towns, thousands of kilometers of railway, highway, and irrigation canal, and a large number of reservoirs, dams, and hydraulic power stations due to sandstorms and dune movement. Since the early 1950s, there have been more than 70 heavy sandstorms and dust devils that have caused huge losses. In the sandstorm of May 1993 in northwest China, more than 12 million people were directly affected, 116 people were killed or missing, and more than 120,000 heads of livestock were lost. The direct economic loss was approximately $66 million (Asian Regional Network for Desertification Monitoring and Assessment, n.d.).

Finally, the Chesapeake Bay in the United States provides an example of major deterioration of an ecosystem through urbanization. Land use has changed considerably in the area. The population in the bay's watershed increased 34% between 1970 and 2000 (Chesapeake Bay Program 2003b), while the number of households grew by 67% (Chesapeake Bay Foundation Report, 2004). The low-density, single-use development pattern was accompanied by infrastructure such as roads, parking lots, and shopping malls, which had a negative impact on water quality. Agriculture also expanded, and nitrogen runoff from agricultural land has been the largest contributor to pollution to the bay (Horton 2003). Chemical runoff from urban centers has also gone up (Chesapeake Bay Program 2003a, 2003b). Increased levels of air

pollution have contributed to increased toxic pollution in the Chesapeake Bay.

Nitrogen pollution has depleted the oxygen levels of the waters (Cestti et al. 2003), which has resulted in increased dead zones, leading to a decline in the numbers of fish, crabs, and other aquatic life. The Chesapeake Bay has seen significant reductions in the water quality during the last decade. Declines in species like oysters and underwater grass have also accentuated the bay's capacity to naturally purify the water. Sustained high levels of pollution may result in groundwater contamination, which is the source of potable water for most people in the region.

The high pollution levels affect both income and health. The Chesapeake has been a major source of seafood, but declines in biodiversity in the bay have led to loss of income for the fishing communities who depended on it for their livelihoods (Horton 2003). The Water Quality Improvement Act requires all farms with sales of at least $2,500 to develop nutrient management plans (Joint Legislative Audit and Review Commission of the Virginia General Assembly 2003). Traditionally, farmers would rely on animal manure generated in poultry and dairy farms in the area. Since manure contains more phosphorus than nitrogen, the government advises farmers to add commercial nitrogen fertilizer. This has increased costs for farms. The animal farms, on the other hand, must now dispose of their manure as required by the Maryland Water Quality Improvement Act.

References

Asian Regional Network for Desertification Monitoring and Assessment, n.d.: Desertification Rehabilitation and Ecology Restoration in China Highlight. http://www.asia-tpn1.net/highlight.html.

Ataniyazova, O. A., 2003: Health and ecological consequences of the Aral Sea crisis. Paper presented at the session of Regional Cooperation in Shared Water Resources in Central Asia, the 3rd World Water Forum, Kyoto.

Ataniyazova, O.A., R.A. Baumann, A.K.D. Liem, U.A. Mukhopadhyay, E.F. Vogelaar, and E.R. Boersma, 2001: Perinatal exposure to environmental pollutants in the aral sea area, *Acta Paediatrica,* **90,** 801–8.

Awumbila, M., and J.H. Momsen, 1995: Gender and the Environment: Women's Time Use as a Measure of Environmental Change, *Global Environmental Change* **5(4),** 337–346.

Bann, C., 1997: An Economic Analysis of Tropical Forest Land Use Options, Ratanakiri Province, Cambodia, Economy and Environment Program for Southeast Asia, International Development Research Centre, Ottawa.

Barber, C.V., and J. Schweithelm, 2000: *Trial by Fire: Forest Fires and Forest Policy in Indonesia's Era of Crisis and Reform,* World Resources Institute, Washington, DC, 76 pp.

Barbier, E. B., and I. Strand, 1998: Valuing mangrove-fishery linkages: A case study of Campeche, Mexico, *Environmental and Resource Economics* **12,** 151–66.

Bardham, P.K., and D. Mookherjee, 2000: Capture and governance at local and national levels, *American Economic Review,* **90(2),** 135–39.

Baudrillard, J., 1998: *The Consumer Society: Myth and Structure,* Sage Publications, London, 208 pages.

Beaudin, M., 2001: Towards Greater Value: Enhancing Eastern Canada's Seafood Industry, Maritime Monograph Series, Canadian Institute for Research on Regional Development: Moncton, New Brunswick, 277 pages.

Balmford, A., et al., 2002: Economic Reasons for Conserving Wild Nature, *Science* **297,** 950–953.

Bourne, J., 2000: Louisiana's Vanishing Wetlands: Going, going. . . . *Science* **289,** 1860–1863.

Bruns, B., A. Mingat, and R. Rakotomalala, 2003: *Achieving Universal Primary Education by 2015: A Chance for Every Child,* World Bank, Washington, DC, 252 pages.

Carret, J., and D. Loyer, 2003: Madagascar Protected Area Network Sustainable Financing: Economic Analysis Perspective, Paper contributed to the World Park's Congress, Durban South Africa, September, Washington, DC: World Bank., 12 pages.

Cesar, H., M. Öhman, P. Espeut, and M. Honkanen, 2000: Economic Valuation of an Integrated Terrestrial and Marine Protected Area: Jamaica's Portland Bight. In: *Collected Essays on the Economics of Coral Reefs,* H. Cesar (ed.), CORDIO, Kalmar University, Kalmar, Sweden.

Cesar, H. and C.K. Chong, 2004: Economic valuation and socioeconomics of coral reefs: methodological issues and three case studies. In: *Economic Valuation and Policy Priorities for Sustainable Management of Coral Reefs,* Mahfuzuddin Ahmed, Chiew Kieok Chong, Herman Cesar (eds.), WorldFish Center, Malaysia, pp. 14–40.

Cestti, R., J. Srivastava, and S. Jung, 2003: *Agriculture Non-Point Source Pollution Control: Good Management Practices—The Chesapeake Bay Experience.* World Bank Working Paper 7, World Bank, Washington, DC, 54 pages.

Chesapeake Bay Foundation Report, 2004: Manure's impact on the rivers, streams and the Chesapeake Bay: Keeping manure out of the water.

Chesapeake Bay Program, 2003a: *Managing Storm Water on State, Federal and District-owned Lands and Facilities.* [online] Chesapeake Executive Council Directive No 01-1. Cited 22 October 2004. Available at http://www.chesapeakebay.net/pubs/stormwater_directive_120301.pdf.

Chesapeake Bay Program, 2003b: Urban Storm Water Runoff. [online] Chesapeake Bay Program, Maryland. Cited 22 October 2004. Available at http://www.chesapeakebay.net/stormwater.htm.

Cunningham, A.A., et al., 2003: Indian Vultures: Victims of an Infectious Disease Epidemic? *Animal Conservation* 6 189–197.

Commission for Economic Cooperation, 2001: *A North American Mosaic: State of the Environment Report,* Commission for Economic Cooperation, Canada.

Dasgupta, P. and I. Serageldin, 1999: *Social Capital: A Multifaceted Perspective.* World Bank, Washington DC, 424 pages.

Dasgupta, P., and K-G Mäler, 2000: Net national product, wealth, and social well-being. *Environment and Development Economics* 5, Parts 1&2:69–93, February & May 2000.

Deaton, A., 1997: *The Analysis of Household Surveys: A Microeconometric Approach to Development Policy.* Baltimore: Johns Hopkins University Press, 479 pages.

Deaton, A., 2003: How to monitor poverty for millennium development goals, *Journal of Human Development,* 4 (3), 353–378.

Deevey, E.S., D.S. Rice, M. Prudence, H.H. Vaughan, M. Breener, and M.S. Flannery, 1979: Mayan Urbanism—Impact on a Tropical Karst Environment. *Science* 206, 4416: 298–306.

Diamond, J. 1997: *Guns, Germs, and Steel: The Fates of Human Societies.* New York: Norton, 480 pages.

Emerton, L. and E. Bos, 2004: *Value: Counting Ecosystems as an Economic Part of Water Infrastructure,* World Conservation Union, Gland, Switzerland and Cambridge, UK. 88 pp.

ESCAP (Economic and Social Commission for Asia and the Pacific), 1993: *State of Urbanization in Asia and the Pacific.* United Nations, New York.

Fan, S., P. Hazzell, and S. Thorat, 2000: Government spending, agricultural growth and poverty in rural India, *American Journal of Agricultural Economics* 82(4):1038–51.

FAO (Food and Agriculture Organization of the United Nations), 2003: *The State of Food Insecurity in the World—2003.* FAO, Rome, 40pp.

FAO, UNEP, and UNDP (Food and Agriculture Organization, United Nations Environment Programme, and United Nations Development Programme), 1994. Land Degradation in South Asia: Its severity, causes and effects upon people. World Soil Resources Report, Rome.

Flannery, T., 1994: *The Future Eaters.* Reed Books: Australia.

Flenley, J.R. and S.M. King, 1984: Late Quarternary pollen records from Easter Island. *Nature* 307: 47–50.

Gleditsch, N. P., P. Wallensteen, M. Eriksson, M. Sollenberg, and H. Strand, 2002: Armed conflict 1946–2001: A new dataset, *Journal of Peace Research,* 39: 615–637.

Gleditsch, N. P. 1998: Armed conflict and the environment: A critique of the literature, *Journal of Peace Research* 35: 381–400.

Gren, I,K.-H. Groth and M. Sylven, 1995: Economic values of Danube floodplains, *Journal of Environmental Management* 45, 333–345.

Guha-Sapir, D. and R. Below, 2002: The quality and accuracy of disaster data: A comparative analyses of three global data sets, ProVention Consortium, *The Disaster Management Facility,* World Bank, Washington, DC.

Hamilton, K., and M. Clemens, 1999, Genuine saving in developing countries. *World Bank Economic Review,* 13:2, 333–56.

Hazell, P., 1998: *Why invest more in the sustainable development of less-favored lands?.* Report No. 20, International Food Policy Research Institute, Washington, DC Available at. http://www.ifpri.org/reports/0798RPT/0798RPTb.HTM

Hazell, P. and S. Fan, 2000: Should developing countries invest more in less-favored areas?: An empirical analysis of rural India. *Economic and Political Weekly* 35(17), pp. 1455–1464.

Horton, T., 2003: *Turning the tide: Saving the Chesapeake Bay.* Washington D.C.: Island Press, 386 pages.

Homer-Dixon, T., 1999: *Environment, Scarcity, and Conflict.* Princeton University Press, Princeton, 253 pages.

IPCC (Intergovernmental Panel on Climate Change), 2002: *Climate Change 2001: Impacts, Adaptation and Vulnerability.* McCarthy, J., Canziani, O., Leary, N., Dokken, D., White, K., eds. Cambridge: Cambridge University Press.

Irwin, E.G., 2002: The effects of open space on residential property values, *Land Economics* 78, 465–80.

Jaggers, K. and T. R. Gurr, 1995: Transitions to democracy: Tracking democracy's third wave with the polity III data, *Journal of Peace Research* 32 (November), 469–82.

Joint Legislative Audit and Review Commission of the Virginia General Assembly, 2003: *Implementation of the Chesapeake Bay Preservation Act.*

Kloor, K., 2000: Everglades restoration plan hits rough waters, *Science,* 288, 1166–1167.

Kramer, R.A., and J.I. Eisen-Hecht, 1999: *The Economic Value of Water Quality in the Catawba River Basin.* Nicolas School of the Environment, Duke University, USA.

Lange, G.M., R. Hassan, and K. Hamilton, 2003: *Environmental Accounting in Action: Case Studies from Southern Africa,* Cheltenham, UK: Edward Elgar Publishing Limited, 240 pages.

Lynne, G.D., P. Conroy, and F. J. Prochaska, 1981: Economic valuation of marsh areas for marine production processes, *Journal of Environmental Economics and Management* 8, 175–186.

Maddison A., 2001: *The World Economy: A Millennial Perspective.* OECD, Paris.

Mather, A., 1999: Society and the services of forests. In: *World Forests, Society and, Environment,* M. Palo and J. Uusivori (eds.), Kluwer Academic Press, pp. 86–89.

Mearns, R., 2001: *Contextual factors in the management of common grazing lands: Lessons from Mongolia and Northwestern China.* Proceedings of the XIX International Grassland Congress, São Pedro, Brazil, February 11–21, 2001. Piracicaba: Brazilian Society of Animal Husbandry.

Mearns, R., 2002: Taking stock: Policy, practice, and professionalism in rangeland development. Symposium on rangelands professionals and policy, February 2002, Kansas City, Missouri.

Micklin, P.P., 1993: The shrinking Aral Sea. *Geotimes,* 38(4), 14–18.

MA (Millennium Ecosystem Assessment), 2003: *Ecosystems and Human Well-Being: A Framework for Assessment.* Island Press, Washington, DC, 245 pp.

Millennium Project, 2004: Task Force 7, Achieving the Millennium Development Goals for water and sanitation: what will it take? Interim full report. Task Force on Water and Sanitation Millennium Project, February (available at http://www.unmillenniumproject.org/documents/tf7interim.pdf).

Modelski, G., and G. Perry III, 1991: Democratization in long perspective. *Technological Forecasting and Social Change,* 39(1–2), 22–34.

Modelski, G., and G. Perry, 2002: Democratization in long perspective. *Technological Forecasting and Social Change,* 69 (4), 359–376.

Myers, N., 1993: *Ultimate Security: The Environmental Basis of Political Stability.* Island Press, Washington, DC., 308 pages.

OECD (Organisation for Economic Co-operation and Development), 2001: *The Well-Being of Nations: The Role of Human and Social Capital.* OECD, Paris.

Ojima, D., 2001: Critical drivers of global environmental and land use changes in temperate Asia. Paper presented at the Open Symposium on Change and Sustainability of Pastoral Land Use Systems in Temperate and Central Asia. Ulaanbaatar, Mongolia.

Ostrom, E., 1990: *Governing the Commons: The Evolution of Institutions for Collective Action.* Cambridge University Press, Cambridge, 279 pp.

Ostrom, E., T. Dietz, N. Dolsak, P.C. Stern, S. Stonich, and E.U. Weber (eds.), 2002: *The Drama of the Commons.* National Academy Press, Washington, DC, 534 pp.

Pagiola, S., 2002: Paying for Water Services in Central America: Learning from Costa Rica. In: S. Pagiola, J. Bishop, and N. Landell-Mills, eds., *Selling Forest Environmental Services: Market-based Mechanisms for Conservation.* London: Earthscan.

Pagiola, S., N. Landell-Mills, and J. Bishop (eds.), 2002: *Selling Forest Environmental Services: Market-based Mechanisms for Conservation and Development.* London: Earthscan.

Parkinson, J., 1999: Indigenous people and forests. In: *World Forests, Society and Environment,* M. Palo and J. Uusivori (eds.), Kluwer Academic Press, 90–91.

Pearce, D.W., and J.W. Warford, 1993: *World Without End: Economics, Environment, and Sustainable Development,* Oxford: Oxford University Press.

Ponting, C., 1991: *A Green History of the World. The Environment and the Collapse of Great Civilizations.* Penguin, Harmondsworth.

Postel, S., 1993: Water and agriculture. In: *Water in Crisis, a Guide to the World's Fresh Water Resources,* P. Gleik (ed.), Oxford University Press, New York, 56–66.

Prescott-Allen, R., 2001: *The Wellbeing of Nations: A Country-by-Country Index of Quality of Life and Environment.* Island Press, Washington, 342 pp.

Pretty, J.N., C. Brett, D. Gee, R.E. Hine, C.F. Mason, et al., 2000: An assessment of the total external costs of UK agriculture, *Agricultural Systems,* **65,** 113–136.

Pretty, J.N., C.F. Mason, D.B. Nedwell, R.E. Hine, S. Leaf, and R. Dils, 2003: Environmental costs of freshwater eutrophication in England and Wales. *Environmental Science and Technology, 32,* 201–208.

Putnam, R., R. Leonardi, and R. Nanetti, 1993: *Making Democracy Work: Civic Traditions in Modern Italy.* Princeton University Press, Princeton.

Rausser, G.C., and A.A. Small, 2000: Valuing research leads: Bioprospecting and the conservation of genetic resources, *Journal of Political Economy,* **108,** 173–206.

Reddy, S.G., and T.W. Pogge, 2002: "How Not to Count the Poor," http://www.socialanalysis.org/, 73 pages.

Redman, C.L., 1999: *Human Impact on Ancient Environments,* Tucson: University of Arizona Press, 239 pages.

Repetto, R., W. McGrath, M. Wells, C. Beer, and F. Rossini, 1989: *Wasting Assets: Natural Resources and the National Income Accounts.* Washington, DC: World Resources Institute, 68 pages.

Ricketts, T.H., G.C. Daily, P.R. Ehrlich, and C.D. Michener, 2004: Economic value of tropical forest to coffee production. *Proceedings of the National Academy of Sciences,* **101(34),** 12579–12582.

Rodden, J., G.S. Eskeland, and J. Litvack, 2002: *Decentralization and the Challenge of Hard Budget Constraints.* MIT Press, Boston, 446 pages.

Ruitenbeek, J. 1999: Indonesia. In D. Glover & T. Jessup (eds.): *Indonesia's Fires and Haze. The Cost of Catastrophe.* Institute for Southeast Asian Studies, Singapore, 86–129.

Sachs, J.D., J.W. McArthur, G. Schmidt-Traub, M. Kruk, C. Bahadur, M. Faye, and G.McCord. 2004: "Ending Africa's Poverty Trap." Brookings Papers on Economic Activity no. 2: 117–216.

Schweithelm, J., and D. Glover, 1999: Causes and impacts of the fires. In D. Glover & T. Jessup (eds.): *Indonesia's Fires and Haze. The Cost of Catastrophe.* Institute for Southeast Asian Studies, Singapore, 1–13.

Schweithelm, J., T. Jessup, and D. Glover, 1999: Conclusions and policy recommendations. In D. Glover & T. Jessup (eds.): *Indonesia's Fires and Haze. The Cost of Catastrophe.* Institute for Southeast Asian Studies, Singapore, 130–143.

Sen, A., 1999: *Development as Freedom.* New York, Knopf.

Simpson, R.D., R.A. Sedjo, and J.W. Reid, 1996: Valuing biodiversity for use in pharmaceutical research, *Journal of Political Economy,* **104,** 163–185.

Stonich, S.C., J.R. Bort, and L.L. Ovares, 1997: Globalization of shrimp mariculture: The impact on social justices and environmental quality in Central America. *Society and Natural Resources* **10(2),** 161–179.

Swearingen, W. D., and A. Bencherifa (eds.), 1996: *North African Environment at Risk: State, Culture and Society in Arab North Africa.* Westview Press, Boulder, Colorado.

Ter-Minassian, T. (ed.), 1997: *Fiscal Federalism in Theory and Practice.* International Monetary Fund, Washington, DC.

Thorsnes, P., 2002: The value of a suburban forest preserve: Estimates from sales of vacant residential lots, *Land Economics* **78,** 426–441.

Tiffen, M., M. Mortimore, and F. Gichuki, 1994: *More People, Less Erosion: Environmental Recovery in Kenya.* John Wiley and Sons, Chichester.

Turner, R.K., J. Paavola, P. Cooper, S. Farber, V. Jessamy, and S. Georgiou, 2003: Valuing nature: lessons learned and future directions. *Ecological economics* **46:** 493–510.

Turpie, J., and B. Heydenrych, 2000: Economic consequences of alien infestation of the Cape Floral Kingdom's Fynbos vegetation. In: *The Economics of Biological Invasions,* C. Perrings, M. Williamson, and S. Dalmazzone (eds.), Edward Elgar, U.K., 152–182.

United Nations Millennium Project, 2004: A Global Plan to Achieve the Millennium Development Goals. www.unmillenniumproject.org/

UN (United Nations), 1953: *UN Statistics Yearbook,* United Nations, New York.

UNDP (United Nations Development Programme), 2003: *Human Development Report.* New York.

UNEP (United Nations Environment Programme), 1992: *World Atlas of Desertification.* Oxford University Press, New York.

UNEP, 1997: *World Atlas of Desertification.* Oxford University Press, New York.

van Andel, T.H., E. Zangger, and A. Demitrack, 1990: Land use and soil erosion in prehistoric and historic Greece; *Journal of Field Archaeology* **17:** 379–396.

van Beukering, P., and H. Cesar, 2004: *Economic Analysis of Marine Managed Areas in the Main Hawaiian Islands.* Cesar Environmental Economics Consulting, The Netherlands.

Van Wilgen, B.A., D.C. le Maitre, and R.M. Cowling, 1998: Ecosystem services, efficiency, sustainability and equity: South Africa's Working for Water Programme. *Trends in Ecology and Evolution* **13,9:** 378.

Van Wilgen, B.W., et al., 2002: Win-win-win: South Africa's Working for Water Programme. In S.M. Pierce, R.M. Cowling, T. Sandwith, and K. MacKinnon, eds., *Mainstreaming Biodiversity in Development: Case Studies from South Africa,* The World Bank, Washington, DC, 5–20.

Van Wilgen, B., 2004: The largest environmental programme in Africa continues to tackle alien invasives, *CSIR Technobrief* **11,** CSIR, South Africa (Online) Available at: www.csir.co.za/websource/ptl0002/pdf_files/technobrief/sep2004/workingforwater.pdf).

Vincent, J., et al., 1997: *Environment and Development in a Resource-Rich Economy: Malaysia under the New Economic Policy* (Cambridge, MA: Harvard Institute for International Development).

Wiebe, K. (ed.), 2003: *Land Quality, Agricultural Productivity, and Food Security.* Edward Elgar, Northampton, MA, 461 pages.

Wood, S., F. Nachtergaele, D. Nielsen, and A. Dai, 1999: Spatial aspects of the design and targeting of agricultural development strategies [online], International Food Policy Research Institute, Washington, DC. Cited October 22, 2004. Available at http://www.ifpri.org/divs/eptd/dp/papers/eptdp44.pdf.

Working for Water, 2004: Working for Water Annual Report 2003/4. Working for Water, Cape Town.

World Bank, 2003: *World Development Report 2003: Sustainable Development in a Dynamic World.* Oxford: Oxford University Press.

World Bank, 2004a: *World Development Indicators 2004.* World Bank, Washington, DC.

World Bank, 2004b: *State and Trends of the Carbon Market–2004.* World Bank, Washington, DC.

WHO (World Health Organization), 2002. *World Health Report.* Geneva.

WHO, 2004. *World Health Report.* Geneva.

Worster, D. 1979. *Dust Bowl: The Southern Plains in the 1930s.* Oxford University Press, New York.

WRI (World Resources Institute), 2000: *People and Ecosystems: The Fraying Web of Life.* Washington, DC: World Resources Institute, 389 pages.

Chapter 6

Vulnerable Peoples and Places

Coordinating Lead Authors: Roger E. Kasperson, Kirstin Dow

Lead Authors: Emma R.M. Archer, Daniel Cáceres, Thomas E. Downing, Tomas Elmqvist, Siri Eriksen, Carle Folke, Guoyi Han, Kavita Iyengar, Coleen Vogel, Kerrie Ann Wilson, Gina Ziervogel

Review Editors: Richard Norgaard, David Rapport

---

*This appears in Appendix A at the end of this volume.

Main Messages

Some of the people and places affected by changes in ecosystems and ecosystem services are highly vulnerable to the effects and are particularly likely to experience much of the damage to well-being and loss of life that such changes will entail. Indeed, many of these people and places are already under severe stress from environmental, health, and socioeconomic pressures, as well as new forces involved in globalization. Further threats arising from changes in ecosystems and ecosystem services will interact with these other on-going stresses to threaten the well-being of these groups while many others throughout the world benefit and prosper from human interactions with ecosystems.

The patterns and dynamics of vulnerability in coupled socioenvironmental systems are shaped by drivers operating at scales from the international to the local, all interacting with the specifics of places. The dominant drivers and patterns of vulnerability differ, depending on the threat or perturbation addressed, the scale of analysis selected, and not least the conceptual framework employed. While our existing knowledge of the sources and patterns of vulnerability is still incomplete, substantial progress is being made in this relatively new area of analysis, and vulnerability assessment is proving useful in addressing environmental management and sustainable development.

At a global level, various efforts over the past several decades have defined vulnerability indictors and indexes and have mapped relevant global patterns. Because they use different conceptual frameworks and consider vulnerability to different types of threats, these efforts largely identify different national-scale patterns of vulnerability. Examples in the chapter introduce major efforts to address vulnerability to environmental change broadly defined, as a dimension of environmental sustainability, in respect to climate change and natural hazards. Improvements in the state of knowledge and methodology development are needed generally to deepen understanding of these global patterns and their causes, although the topics of natural hazards, desertification, and food security have received more attention than others, due to the level of societal concern on these issues.

Trends in natural hazards reveal several patterns that are known with high confidence at the national level. The world is experiencing a worsening trend of human suffering and economic losses from natural disasters over the past several decades. In the last 40 years, the number of "great" disasters has increased by a factor of 4 while economic losses have increased by a factor of 10. The significance of these events to the social vulnerability of exposed human populations is of special concern. Even before the December 2004 tsunami, Asia was disproportionately affected, with more than 43% of all natural disasters and 70% of deaths occurring there over the last decade of the twentieth century. The greatest loss of life continues to be highly concentrated in developing countries as a group.

Desertification is another phenomenon that has received extensive attention. Vulnerability to desertification has multiple causes that are highly intermingled; like all vulnerability, it is the product of the interaction between environmental change and social and political systems. The driving forces of environmental change generally have a high patchiness, and effects vary widely with differences in social and geographic scales.

Food insecurity is a third primary area of concern in changes in ecosystem services. Multiple domains of vulnerability exist in food security regimes and livelihood systems. Production, economic exchanges, and nutrition are key elements, along with more-structural issues associated with the political economy. At this point in time, the more generalized, major contributions to

knowledge are emerging in the realms of better understanding of driving forces, interactions across biophysical scales and social levels, connections between ecosystems services and human well-being, and differential vulnerability at local levels. While many challenges remain in aggregating diverse case study findings, consistency is emerging around a number of themes:

- Socioeconomic and institutional differences are major factors shaping differential vulnerability. The linkages among environmental change, development, and livelihoods are receiving increasing attention in efforts to identify sources of resilience and increase adaptive capacity, but knowledge in this area is uneven in its coverage of environmental threats and perturbations as they act in relation to different ecosystems and livelihoods.

- Poverty and hazard vulnerability are often closely related, as the poor often lack assets and entitlements that allow them some buffer from environmental degradation and variability.

- The interactions of multiple forms of stress—economic, social, political, and physical—with environmental change can amplify and attenuate vulnerability abruptly or gradually, creating dynamic situations for assessment that have still to be fully captured in research methodologies. Major worldwide trends of population growth, urbanization, the spread of HIV/AIDS, economic development, and globalization are acting to shape patterns of vulnerability at national and local scales. The implications of these processes for climate change are still poorly understood.

The limitations of existing understanding point to the need for a variety of efforts to improve assessment and identify measures to reduce vulnerability. These include the need for a robust and consensual conceptual framework for vulnerability analysis, improved analysis of the human driving forces of vulnerability as well as stresses, clarification of the overlaps and interactions between poverty and vulnerability, the tracking of sequences of stresses and perturbations that produce cumulative vulnerability, the role of institutions in creating and mitigating vulnerability, the need to fill gaps in the knowledge base of global patterns of vulnerability, improved assessment methods and tools, and the need for interventions aimed at reducing vulnerability.

6.1 Introduction

The Third Assessment of the Intergovernmental Panel on Climate Change noted that over the past century average surface temperatures across the globe have increased by 0.6° Celsius and evidence is growing that human activities are responsible for most of this warming (IPCC 2001b). Human activities are also altering ecosystems and ecosystem services in myriad ways, as assessed in other chapters. While both positive and negative effects on human societies are involved, it is unrealistic to expect that they will balance out.

Many of the regions and peoples who will be affected are highly vulnerable and poorly equipped to cope with the major changes in ecosystems that may occur. Further, many people and places are already under severe stress arising from a panoply of environmental and socioeconomic forces, including those emanating from globalization processes. Involved are such diverse drivers of change as population growth, increasing concentrations of populations in megacities, poverty and poor nutrition, accumulating contamination of the atmosphere as well as of land and water, a growing dependence on distant global markets, growing gender and class inequalities, the ravages of wars, the AIDS epidemic, and politically corrupt governments. (See Chapter 3 for further discussion on drivers of change.) Environmental change will produce varied effects that will interact with these other stresses and multiple vulnerabilities, and they will take their toll particularly among the most exposed and poorest people of the world.

The most vulnerable human and ecological systems are not difficult to find. One third to one half of the world's population already lacks adequate clean water, and climate change—involving increased temperature and droughts in many areas—will add to the severity of these issues. As other chapters in this volume establish, environmental degradation affects all ecosystems and ecosystem services to varying degrees. Many developing countries (especially in Africa) are already suffering declines in agricultural production and food security, particularly among small farmers and isolated rural populations. Mountain locations are often fragile or marginal environments for human uses such as agriculture (Jodha 1997, 2002). Increased flooding from sea level rise threatens low-lying coastal areas in many parts of the globe, in both rich and poor countries, with a loss of life and infrastructure damages from more severe storms as well as a loss of wetlands and mangroves. (See Chapters 19 and 23.)

The poor, elderly, and sick in the burgeoning megacities of the world face increased risk of death and illness from growing contamination from toxic materials. Dense populations in developing countries face increased threats from riverine flooding and its associated impacts on nutrition and disease. These threats are only suggestive, of course, of the panoply of pressures that confront the most vulnerable regions of the world. It is the rates and patterns of environmental change and their interaction with place-specific vulnerabilities that are driving local realities in terms of the eventual severities of effects and the potential effectiveness of human coping mitigation and adaptation.

Research on global environmental change and on-going assessments in many locales throughout the world have greatly enriched our understanding of the structure and processes of the biosphere and human interactions with it. At the same time, our knowledge is growing of the effects that changes in ecosystems and ecosystem services have upon human communities. Nonetheless, the knowledge base concerning the vulnerabilities of coupled socioecological systems is uneven and not yet sufficient for systematic quantitative appraisal or validated models of cause-and-effect relationships of emerging vulnerability. Yet what we need to understand is apparent in the questions that researchers are addressing (Turner et al. 2003a): Who and what are vulnerable to the multiple environmental and human changes under way, and where? How are these changes and their consequences attenuated or amplified by interactions with different human and environmental conditions? What can be done to reduce vulnerability to change? How may more resilient and adaptive communities and societies be built?

In this chapter key definitions and concepts used in vulnerability analysis are first considered. Included in this is a clarification of what is meant by the terms "vulnerability" and "resilience." Several of the principal methods and tools used in identifying and assessing vulnerability to environmental change are then examined (but see also Chapter 2). Efforts to identify and map vulnerable places at the global scale are described, followed by three arenas—natural disasters, desertification, and food security—that have received substantial past analyses in vulnerability research and assessment. Several specific case studies that illustrate different key issues that pervade vulnerability assessments are presented and, finally, implications of our current knowledge for efforts to assess and reduce vulnerability and to build greater resilience in coupled socioecological systems are assessed.

6.2 Definitions and Conceptual Framework

The term vulnerability derives from the Latin root *vulnerare,* meaning to wound. Accordingly, vulnerability in simple terms means the capacity to be wounded (Kates 1985). Chambers (1989) elaborated this notion by describing vulnerability as "exposure to contingencies and stress, and the difficulty in coping with them." It is apparent from relating the notion of vulnerability to the broader framework of risk that three major dimensions are involved:

- exposure to stresses, perturbations, and shocks;
- the sensitivity of people, places, and ecosystems to stress or perturbation, including their capacity to anticipate and cope with the stress; and
- the resilience of exposed people, places, and ecosystems in terms of their capacity to absorb shocks and perturbations while maintaining function.

6.2.1 Conceptual Framework for Analyzing Vulnerability

A wide variety of conceptual frameworks have arisen to address the vulnerability of human and ecological systems to perturbations, shocks, and stresses. Here we draw on a recent effort of the Sustainability Science Program to frame vulnerability within the context of coupled socioecological systems (Turner et al. 2003a, 2003b). The framework seeks to capture as much as possible of the totality of the different elements that have been demonstrated in risk, hazards, and vulnerability studies and to frame them in regard to their complex linkages. (See Figure 6.1.)

The framework recognizes that the components and linkages in question vary by the scale of analysis undertaken and that the scale of the assessment may change the specific components but not the overall structure. It identifies two basic parts to the vulnerability problem and assessment: perturbation-stresses and the coupled socioecological system.

Perturbations and stresses can be both human and environmental and are affected by processes often operating at scales larger than the event in question (such as local drought). For example, globally induced climate change triggers increased variation in precipitation in a tropical forest frontier, while political strife elsewhere drives large numbers of immigrants to the frontier. The coupled socioecological system maintains some level of vulnerability to these perturbations and stresses, related to the manner in which they are experienced. This experience is registered first in terms of the nature of the exposure—its intensity, frequency, and duration, for instance—and involves measures that the human and environment subsystems may take to reduce the exposure. The coupled system experiences a degree of harm to exposure (risk and impacts), determined by its sensitivity. The linkage between exposure and impact is not necessarily direct, however, because the coupled system maintains coping mechanisms that permit immediate or near-term adjustments that reduce the harm experienced and, in some cases, changes the sensitivity of the system itself.

If perturbations and stresses persist over time, the types and quality of system resilience change. These changes are potentially irreversible, as the case of ozone depletion illustrates. Change may lead to adaptation (fundamental change) in the coupled system. The role of perception and the social and cultural evaluation of stresses and perturbations is important to both the recognition of stresses and the decisions regarding coping, adaptation, and adjustment. These decisions reflect local and regional differences in perceptions and evaluations. The social subsystem must be altered, or

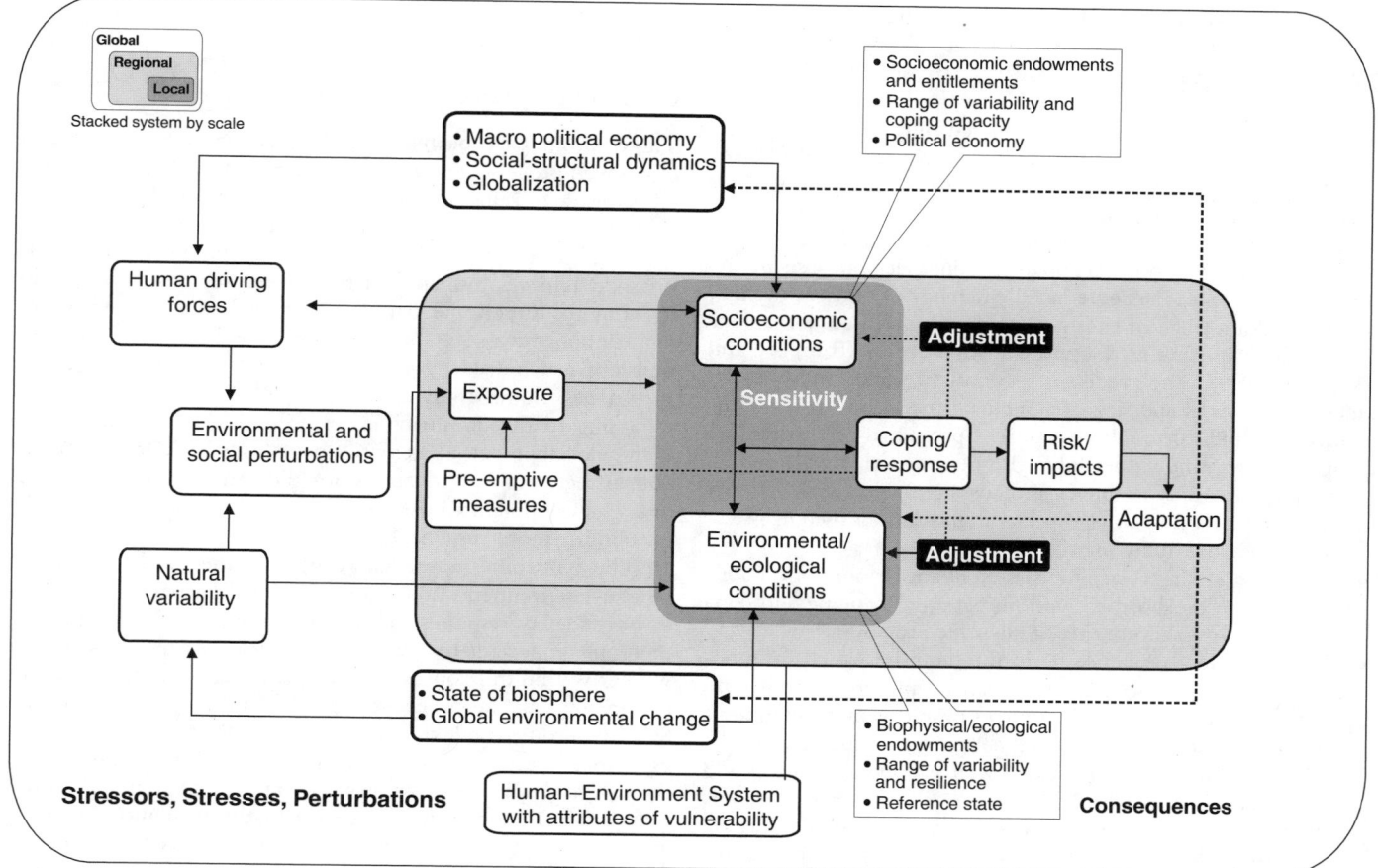

Figure 6.1. A Framework for Analyzing Vulnerability

it ceases to function (a place or region is abandoned, for example); the ecological subsystem changes in climate and vegetation. This process of more fundamental change, sometimes also referred to as "reorganization," may move the coupled socioecological system in a direction of greater sustainability, but perhaps at a cost to those depending on current patterns of ecosystem services. The MA *Policy Responses* volume addresses adjustments and adaptation in ecosystems and with respect to human well-being in greater detail. By definition, no part of a system in this vulnerability framework is unimportant.

6.2.2 The Concept of Resilience

The concept of resilience as applied to integrated socioecological systems may be defined as the amount of disturbance a system can absorb and still remain within the same state or domain of attraction, the degree to which the system is capable of self-organization (versus lack of organization or organization forced by external factors), and the degree to which the system can build and increase its capacity for learning and adaptation (Carpenter et al. 2001). Socioecological systems are complex adaptive systems that are constantly changing, and the resilience of such systems represents the capacity to absorb shocks while maintaining function (Holling 1995, 2001; Gunderson and Holling 2002; Berkes et al. 2002). When a human or ecological system loses resilience, it becomes vulnerable to change that previously could be absorbed (Kasperson and Kasperson 2001).

New insights have been gained during the last 10 years about the essential role of resilience for a prosperous development of

human society (Gunderson and Holling 2002). A growing number of case studies have revealed the tight connection between resilience, diversity, and the sustainability of socioecological systems (Berkes and Folke 1998; Adger et al. 2001).

Ecosystems with low resilience may still maintain function and generate resources and ecosystem services—that is, they may seem to be in good shape—but when subject to disturbances and stochastic events, they may reach a critical threshold and slide into a less desirable state. Such shifts may significantly constrain options for social and economic development, reduce options for livelihoods, and create environmental migrants as a consequence of the impact on ecosystem life-support.

In ecological systems, Lawton (2000) and Loreau et al. (2001) synthesized the evidence from many experiments and affirmed that the diversity of functionally different kinds of species affected the rates of stability and increased the reliability of ecosystem processes locally. Furthermore, a number of observations suggest that biodiversity at larger spatial scales, such as landscapes and regions, ensures that appropriate key species for ecosystem functioning are recruited to local systems after disturbance or when environmental conditions change (Peterson et al. 1998; Bengtsson et al. 2003). In this sense, biological diversity provides insurance, flexibility, and risk spreading across scales in the face of uncertainty and thereby contributes to ecosystem resilience (Folke et al. 1996). (See also Chapter 11.)

Ecological resilience typically depends on slowly changing variables such as land use, nutrient stocks, soil properties, and biomass of long-lived organisms (Gunderson and Pritchard 2002), which are in turn altered by human activities and socioeconomic driving forces (Lambin et al. 2001). The increase in social and economic vulnerability as a consequence of reduced resilience through land degradation and drought may cause losses of livelihood and trigger tension and conflict over critical resources such as fresh water or food (Homer-Dixon and Blitt 1998).

Increased vulnerability and fragility places a region on a trajectory of greater risk to the panoply of stresses and shocks that occur over time. Stressed ecosystems are often characterized by a "distress syndrome" that is indicated not only by reduced biodiversity and altered primary and secondary productivity but also by increased disease prevalence, reduced efficiency of nutrient cycling, increased dominance of exotic species, and increased dominance by smaller, shorter-lived opportunistic species (Rapport and Whitford 1999). The process is a cumulative one, in which sequences of shocks and stresses punctuate the trends, and the inability to replenish coping resources propels a region and its people to increasing vulnerability (Kasperson et al. 1995).

Key attributes of resilience in ecosystems, flexibility in economic systems, and adaptive capacity in institutions used in assessments include the following:

- Ecological resilience can be assessed by the amount of variability that can be absorbed without patterns changing and controls shifting to another set of keystone processes.
- Key sources of resilience lie in the requisite variety of functional groups; the accumulated financial, physical, human, and natural capital that provides sources for reorganization following disturbances; and the social networks and institutions that provide entitlements to assets as well as coping resources and social capital (Adger 2003).
- In an ecosystem, these key processes can be recognized as the processes that interact and are robust in an overlapping, redundant manner.
- When a system is disrupted, resilience is reestablished through regeneration and renewal that connect that system's present to its past.

Management can destroy or build resilience, depending on how the socioecological system organizes itself in response to management actions (Carpenter et al. 2001; Holling 2001; MA *Policy Responses*). There are many examples of management suppressing natural disturbance regimes or altering slowly changing ecological variables, leading to disastrous changes in soils, waters, landscape configurations, or biodiversity that did not appear until long after the ecosystems were first managed (Holling and Meffe 1996). Similarly, governance can disrupt social memory or remove mechanisms for creative, adaptive response by people in ways that lead to the breakdown of socioecological systems (McIntosh et al. 2000; Redman 1999). By contrast, management that builds resilience can sustain socioecological systems in the face of surprise, unpredictability, and complexity. Successful ecosystem management for human well-being requires monitoring and institutional and organizational capacity to respond to environmental feedback and surprises (Berkes and Folke 1998; Danter et al. 2000), a subject treated at the conclusion of this chapter.

6.3 Methods and Tools for Vulnerability Analysis

Many tools and methods exist for undertaking vulnerability analysis, as described in Chapter 2. This section describes several tools more specific to assessing vulnerability issues and outcomes. The vulnerability toolkit described here and in Chapter 2 is considerable, ranging from qualitative to quantitative methods, with various levels of integration among disciplines, and it is suitable for participation of stakeholders. Matching the types of analytical approaches in a toolkit to the characteristics of a specific assessment is a necessary step in scoping projects.

6.3.1 The Syndromes Approach

The syndromes approach aims to "assess and monitor a multitude of coupled processes taking place on different (spatial and temporal) scales with different specificities" (Petschel-Held 2002). The goal of the syndromes approach is to identify where intervention can help contribute to sustainable development pathways. In order to achieve this, similarities between regions are found by looking for functional patterns that are called "syndromes" (Schellnhuber et al. 1997). An assessment of these patterns of relationships is achieved by combining qualitative and quantitative approaches. Some 16 syndromes of global change are grouped according to the dominant logic: utilization of resources, economic development, and environmental sinks. The results enable critical regions to be identified for different syndromes, so that future development can set priorities for key areas necessary for establishing more-sustainable systems.

The syndromes approach recognizes the need to examine human-environment interactions, as global change is a function of how society responds to natural changes and vice versa. It is therefore important that the socioecological system is seen as a whole. Within this context, archetypal patterns are most relevant to representing the process of global change. For example, the Sahel syndrome (Lüdeke et al. 1999), characterizes a set of processes that result in the overuse of agriculturally marginal land. (Note that the names of syndromes represent an archetype rather than a specific location, event, or situation; for more detailed analysis of environmental change in the Sahel itself, see Chapter 22.)

The Sahel syndrome can be located in certain parts of the world and characterized by a number of factors. Its driving forces or core mechanisms include impoverishment, intensification of agriculture, and soil erosion, which in turn lead to productivity loss. Various factors might contribute to the disposition toward this syndrome, including socioeconomic dimensions, such as high dependence on fuelwood, and natural dimensions, such as aridity and poor soils. The core mechanisms can be quantitatively assessed to determine which areas of the world experience the syndrome most extensively and intensively. The syndromes approach is a transdisciplinary tool, drawing on both quantitative and qualitative assessments of dynamic patterns at a variety of scales, and by identifying patterns of unsustainable development, it can be used to target future development priorities aimed at enabling sustainable development.

6.3.2 Multiagent Modeling

Multiagent behavioral systems seek to model socioecological interactions as dynamic processes (Moss et al. 2001). Human actors are represented as software agents with rules for their own behavior, interactions with other social agents, and responses to the environment. Physical processes (such as soil erosion) and institutions or organizations (such as an environmental regulator) may also be represented as agents. A multiagent system could represent multiple scales of vulnerability and produce indicators of multiple dimensions of vulnerability for different populations.

Multiagent behavioral systems have an intuitive appeal in participatory integrated assessment. Stakeholders may identify with "their" agents and be able to validate a model in qualitative ways that is difficult to do for econometric or complex dynamic simulation models. However, such systems require significant computational resources (proportional to the number of agents), and a paucity of data for validation of individual behavior is a constraint.

6.3.3 Vulnerability and Risk Maps

The development of indicators and indices of vulnerability and the production of global maps are prominent vulnerability assessments techniques at the global level, although these approaches are still being developed to better capture the full concept of vulnerability. Global assessments using these techniques are described later in this chapter.

In order to bring conceptual understanding of vulnerability closer to their cartographic representations, vulnerability and risk mapping efforts are working to resolve several methodological challenges. Generally, risk maps are explicitly concerned with the human dimensions of vulnerability, such as the risks to human health and well-being associated with the impacts from natural hazards.

Given the common focus on human well-being at an aggregate level, vulnerability is quantified in terms of either single or multiple outcomes, such as water scarcity and hunger. Two exceptions are the hotspots of biodiversity (Myers et al. 2000) and the GLOBIO analysis (Nellemann et al. 2001), which are concerned with the vulnerability of biodiversity. For example, the hotspots of biodiversity identify areas featuring exceptional concentrations of endemic species and experiencing exceptional loss of habitat. The GLOBIO analysis relates infrastructure density and predicted expansion of infrastructure to human pressure on ecosystems in terms of the reduced abundance of wildlife. Limited progress, however, has been made as yet in integrating analyses of the vulnerability of human and ecological systems.

Many of the risk maps have been generated from remotely sensed data or information held in national data libraries. The maps are generally developed and displayed using a geographic information system. The analytical and display capabilities of GIS can draw attention to priority areas that require further analysis or urgent attention. Interactive risk mapping is presently in its infancy. The PreView project (UNEP-GRID 2003) is an interactive Internet map server presently under development that aims to illustrate the risk associated with natural disasters at the global level.

For the most part, risk maps have tended to be scale-specific snapshots at a particular time, rarely depicting cumulative and long-term risk. A challenge is linking global and local scales in order to relate indirect drivers (which operate at global, national, and other broad levels and which originate from societal, economic, demographic, technological, political, and cultural factors) to direct drivers (the physical expressions of indirect drivers that affect human and natural systems at regional or local scales). Temporally, risk maps generally depict short-term assessments of risk. The accuracy of these maps is rarely assessed, and risk maps are usually not validated empirically. Two exceptions are the fire maps and the maps of the risk of land cover change. The uncertainty that surrounds the input risk data needs to be explicit and should also be mapped.

A challenging problem for the effective mapping of risk is to move from solely identifying areas of stress or likely increased stress to mapping the resistance or sensitivity of the receptor system. This would highlight regions where the ability to resist is low or declining and the sensitivity of the receptor systems is high. The difficulty here lies in quantifying the ability to resist external pressures. Quantifying resistance, at least in ecological systems, is presently largely intractable as it requires information on the effects of different levels of severity of threats, which is

usually species-specific, as well as ways of integrating this information across assemblages of species or areas of interest.

A further challenge to risk mapping is the analysis of multiple and sequential stressors. Generally, single threats or stressors are analyzed and multiple stressors are rarely treated. The ProVention Consortium (2003) aims to assess risk, exposure, and vulnerability to multiple natural hazards. Possible limitations to undertaking a multiple hazard assessment of this kind include accounting for the different ways of measuring hazards (for example, in terms of frequency, intensity, duration, spatial extent), different currencies of measurement, varied data quality, and differences in uncertainty between varying hazard assessments.

Scale and how to represent significant variation within populations of regions are common challenges for global mapping exercises, with broad implications for vulnerability assessment (German Advisory Council on Global Change 1997). Political and social marginalization, gendered relationships, and physiological differences are commonly identified variables influencing vulnerability, but incorporating this conceptual understanding in global mapping remains a challenge. Global-scale maps may consider vulnerability of the total population, or they may consider the situation of specific groups believed to be particularly vulnerable. Because many indigenous peoples are less integrated into political and social support systems and rely more directly on ecosystem services, they are likely to be more sensitive to the consequences of environmental change and have less access to support from wider social levels.

Women and children are also often reported to be more vulnerable than men to environmental changes and hazards (Cutter 1995). Because the gendered division of labor within many societies places responsibility for routine care of the household with women, degradation of ecosystem services—such as water quality or quantity, fuelwood, agricultural or rangeland productivity—often results in increased labor demands on women. These increased demands on women's time to cope with loss of ecosystem services can affect the larger household by diverting time from food preparation, child care, and other beneficial activities. While women's contributions are critical to the resilience of households, women are sometimes the focus of vulnerability studies because during pregnancy or lactation their physiology is more sensitive and their ill health bears on the well-being of children in their care. Children and elderly people are also often identified as particularly vulnerable primarily because of their physiological status.

Measures of human well-being and their relationship to ecosystems services also often incorporate data on the sensitivity and resilience dimensions of vulnerability, expressed as assets, capabilities, or security. These measures are discussed in greater detail in Chapter 5.

6.4 Assessing Vulnerability

The causes and consequences of human-induced change in ecosystems and ecosystem services are not evenly distributed throughout the world but converge in certain regions and places. For some time, for example, Russian geographers prepared "red data maps" to show the locations of what they regarded as "critical environmental situations" (Mather and Sdasyuk 1991). The National Geographical Society (1989) created a map of "environmentally endangered areas" depicting areas of natural hazards, pollution sources, and other environmental stresses. Nonetheless, it is only in recent years that concerted efforts have been made to develop indices and generate maps that depict the global distribution of people and places highly vulnerable to environmental stresses.

As noted earlier, several challenges remain in developing indicators, indices, and maps that capture all the dimensions of vulnerability, but this section reviews major notable efforts that address vulnerability in the context of human security, as an aspect of environmental sustainability, and natural disasters and that point to environmental health issues addressed further in Chapter 14.

Although modest progress has occurred in identifying and mapping vulnerable places and peoples, the state of knowledge and methodology are still significantly limited. Few of the analyses presented here integrate ecological and human systems. They rarely treat multiple stresses, interacting events, or cumulative change. Indicators continue to be chosen without an adequate underlying conceptual framework and are typically not validated against empirical cases. For the most part, they are scale-specific and snapshots in time. Disaggregated data are lacking, and much remains to be done before a robust knowledge base at the global scale will exist.

In a demonstration project, the Global Environmental Change and Human Security Project of the International Human Dimensions Programme on Global Environmental Change (Lonergan 1998) mapped regions of ecological stress and human vulnerability, using an "index of vulnerability" composed from 12 indicators:

- food import dependency ratio,
- water scarcity,
- energy imports as percentage of consumption,
- access to safe water,
- expenditures on defense versus health and education,
- human freedoms,
- urban population growth,
- child mortality,
- maternal mortality,
- income per capita,
- degree of democratization, and
- fertility rates.

The criteria used in selecting indicators were that data were readily available, that the resulting "index" consisted of a small number of indictors, and that the indicators covered six major categories—ecological and resource indicators, economic indicators, health indicators, social and demographic indicators, political/social indicators, and food security indicators. Through cluster analysis, a vulnerability "index" was derived and then used to map estimated vulnerability patterns, such as one for Africa. (See Figure 6.2.)

The work of the Intergovernmental Panel on Climate Change (IPCC 2001a) has made clear that ongoing and future climate changes will alter nature's life-support systems for human societies in many parts of the globe. Significant threats to human populations, as well as some potential benefits, are involved. (See Box 6.1.) As the example on the Arctic region illustrates, changes that benefit some may harm others in the same area. (See also Chapter 25.)

But it is unrealistic to assume that positive and negative effects will balance out, particularly in certain regions and places. Many of the regions and human groups, the IPCC makes clear, will be highly vulnerable and poorly equipped to cope with the major changes in climate that may occur. Many people and places are already under severe stresses arising from other environmental degradation and human driving forces, including population growth, urbanization, poverty and poor nutrition, accumulating environmental contamination, growing class and gender inequalities, the ravages of war, AIDS/HIV, and politically corrupt governments. The IPCC points to the most vulnerable socioecological systems: one third to one half of the world's population lack adequate clean water; many developing countries are likely to suffer future declines in agricultural production and food security; sea level rise is likely to greatly affect low-lying coastal areas; small-island states face potential abandonment of island homes and relocation; and the poor and sick in growing megacities face increased risk for death and illness associated with severe heat and humidity.

In preparation for the World Summit on Sustainable Development in 2002, the Global Leaders for Tomorrow Environment Task Force (2002) of the World Economic Forum to created a global Environmental Sustainability Index. It has five major components developed from globally available national data, including one on reducing human vulnerability. (See Table 6.1.) While it would be desirable to display regional differences within countries, finer-scale information is not consistently available for many types of data.

Human vulnerability seeks to measure the interaction between humans and their environment, with a focus on how environmental change affects livelihoods. Two major issues are included in the vulnerability component (one of the five components in the overall index): basic human sustenance and environmental health. The index is based on five indicators: proportion undernourished in the total population, percentage of population with access to improved drinking water supply, child death rate from respiratory diseases, death rate from intestinal infectious diseases, and the under-five mortality rate. The standardized values for each indicator were calculated and converted to a standard percentile indicator for ease of interpretation. The indicators were unweighted. Country scores were then derived to demarcate global patterns, as shown in Table 6.2.

The United Nations Environment Programme (UNEP 2003) has also assessed definitions, concepts, and dimensions of vulnerability to environmental change in different areas of the world. In particular, it calls attention to the importance of environmental health in the vulnerability of different regions and places. It notes, for example, that every year thousands of people die from a range of disasters, but the fate of many of these people is never reported. The International Red Cross Federation (IFRC 2000) has shown that the death toll from infectious diseases (such as HIV/AIDS, malaria, respiratory diseases, and diarrhea) was 160 times the number of people killed in natural disasters in 1999. And this situation is becoming worse rapidly. It is estimated, for example, that over the next decade HIV/AIDS will kill more people in sub-Saharan Africa than died in all wars of the twentieth century.

The United Nations Development Programme, in *Reducing Disaster Risk: A Challenge for Development* (UNDP 2004), undertakes the formulation of a "disaster risk index," which it then uses to assess global patterns of natural disasters and their relationship to development. The Disaster Risk Index calculates the relative vulnerability of a country to a given hazard (such as earthquakes or floods) by dividing the number of people killed by the number of people exposed to the hazard. The analysts then compared the risk of the hazard (the number of people actually killed by the hazard in a country) with 26 indicators of vulnerability, selected through expert opinion. Analyzing a series of statistical analyses, a number of findings concerning the impact of development on disaster risk emerge:

- The growth of informal settlements and inner city slums has led to the growth of unstable living environments, often located in ravines, on steep slopes, along floodplains, or adjacent to noxious industrial and transport facilities.

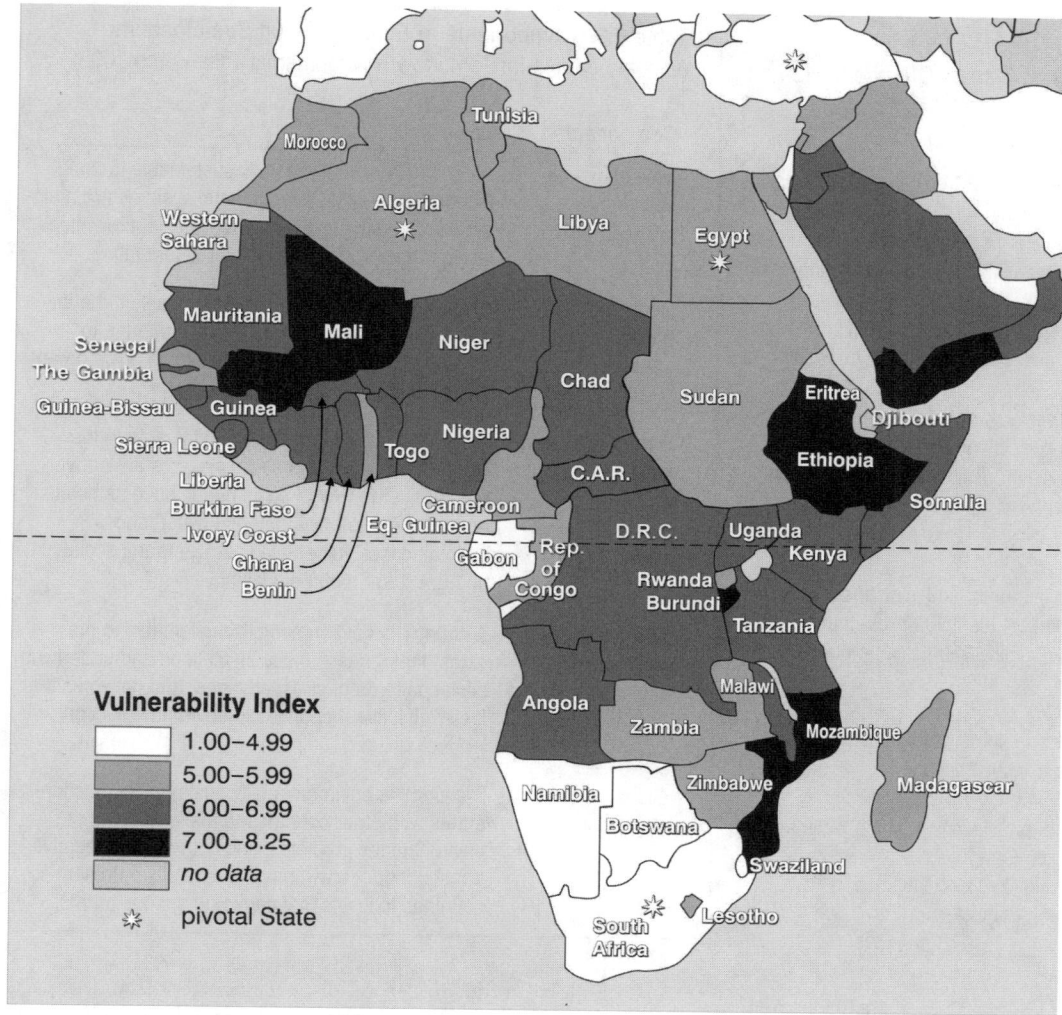

Figure 6.2. Vulnerability Index for African Countries (Lonergan 1998)

- Rural livelihoods are put at risk by the local impacts of global climate change or environmental degradation.
- Coping capacities for some people have been undermined by the need to compete in a globalizing economy, which presently rewards productive specialization and intensification over diversity and sustainability (UNDP 2004, p. 2).

6.5 Natural Hazards and Vulnerability

Natural hazards and disasters are products of both natural variability and human-environment interactions, and vulnerability to them has received substantial past attention. (See also Chapter 16.) The extremes of environmental variability are defined as hazards when they represent threats to people and what they value and defined as disasters when an event overwhelms local capacity to cope. Natural hazards offer a particularly dramatic view of the role of vulnerability in explaining patterns of losses among people and places. Indeed, research on this topic was the first realm to document the vast differences in the magnitude of losses among people and places experiencing the same types of events (White 1974). Since the 1970s, researchers have consistently reported greater loss of life among poorer populations and countries than in industrial countries, along with the inverse relationship for economic damage.

Natural hazards and disasters have always been a part of human history. Yet human relationships to hazards have evolved as the power of humans to shape natural landscapes and their biogeochemical processes has grown. Over the centuries, humans have changed from relatively powerless victims in the face of natural hazards and disasters to active participants shaping natural hazards and our vulnerability to them. Only recently has policy recognized that natural hazards are not "Acts of God" and begun to shift hazard management from a model of response and relief to an active engagement with mitigation, prevention, and integration of hazard management into development planning (ISDR 2002).

It is well established that the impacts of natural disasters continue to create uneven patterns of loss in populations around the world. The rising economic costs, the relative significance of those costs to the budgets of developing countries, the increasing numbers of people affected, and the decreasing loss of life demonstrate the dynamics of vulnerability across scales and experienced in local places.

6.5.1 Trends in Natural Hazards and Vulnerability

The best available data on a global scale (e.g., Swiss Re 2000; Munich Re 2003; CRED 2002) indicate that the world is witnessing a worsening trend of human suffering and economic loss

BOX 6.1
Threats and Potential Benefits of Climate Change to Human Societies (IPCC 2001a)

Threats

- Reduced potential crop yields in some tropical and sub-tropical regions and many mid-latitude regions
- Decreased water availability for populations in many water-scarce regions, particularly those with inadequate management systems
- An increase in the number of people exposed to vector-borne diseases (such as malaria) and waterborne diseases (such as cholera)
- Increases in the number of people dying from heat stress, particularly in large cities in developing countries
- A widespread increase in the risk of flooding for many human settlements throughout the world
- Severe threats to millions of people living on low-lying islands and atolls
- Threats to aboriginals living in Arctic and high mountains (for example, through the breakup of ice fields preventing people from reaching their traditional hunting and fishing grounds)

Potential Benefits

- Increased potential crop yields in some mid-latitude regions
- A potential increase in global timber supply from appropriately managed forests
- Increased water availability for populations in some water scarce regions (such as parts of South East Asia)
- Reduced winter mortality in mid- and high latitudes
- Improved marine transportation in the Arctic

Table 6.1. Components of Environmental Sustainability
(Global Leaders for Tomorrow Environmental Task Force 2002)

Component	Logic
Environmental systems	A country is environmentally sustainable to the extent that its vital environmental systems are maintained at healthy levels and to the extent to which levels are improving rather than deteriorating.
Reducing environmental stresses	A country is environmentally sustainable if the levels of anthropogenic stress are low enough to engender no demonstrable harm to its environmental systems.
Reducing human vulnerability	A country is environmentally sustainable to the extent that people and social systems are not vulnerable (in the way of basic needs such as health and nutrition) to environmental disturbances; becoming less vulnerable is a sign that a society is on a track to greater sustainability.
Social and institutional capacity	A country is environmentally sustainable to the extent that it has in place institutions and underlying social patterns of skills, attitudes, and networks that foster effective responses to environmental challenges.
Global stewardship	A country is environmentally sustainable if it cooperates with other countries to manage common environmental problems, and if it reduces negative transboundary environmental impacts on other countries to levels that cause no serious harm.

to natural disasters over recent decades.(Data available at the time this chapter was written do not include losses caused by the 2004 tsunami.) While the general trend is clear, the precise estimates vary somewhat, due to improvements in reporting over time, data gathering practices, and definitional differences across organizations. (See Chapter 16 for more detailed description of the limitations and variations among data sets.)

During the past four decades, the number of "great" catastrophes—when the ability of a region to help itself is distinctly overtaxed, making interregional or international assistance necessary—has increased about four times, while economic losses have increased over 10 times. (Munich Re 2000) (See Table 6.3.) This trend reflects the increasing economic costs of disasters, lives lost, and the unequal ability of nations to cope with the impacts. Natural disasters affected twice as many people in the 1990s as in the 1980s (CRED 2003). The annual average losses for all disasters over the 1990s were 62,000 deaths, 200 million affected, and $69 billion in economic losses (IFRC 2001). Although comprehensive global databases do not exist for smaller-scale natural hazard events, the significance of these more common events to the social vulnerability of exposed human populations is also a major concern (ISDR 2002; Wisner et al. 2004).

Throughout the twentieth century, three general observations can be drawn from global trends: the number of disasters has increased, economic losses from disasters have increased (primarily in industrial countries), and the ratio of loss of life to total population affected has decreased, although this decline has also been heavily concentrated in industrial societies. (See Figure 6.3 in Appendix A.)

The global trends in natural disaster occurrences and impacts suggest several important patterns of vulnerability among people and places at the same time that they mask considerable geographic variation. Asia is disproportionately affected, with more than 43% of all natural disasters in the last decade of the twentieth century. During the same period, Asia accounted for almost 70% of all lives lost due to natural hazards. In China alone, floods affected more than 100 million people on average each year (IFRC 2002).

Variation among types of natural hazards is also significant. Over the decade 1991–2000, the number of hydro-meteorological disasters doubled, accounting for approximately 70% of lives lost from natural disasters (IFRC 2001). Floods and windstorms were the most common disaster events globally, but not consistently the cause of greatest losses. Disasters causing the greatest number of deaths varied among regions, with floods causing the most deaths in the Americas and Africa, drought or famine the most in Asia, earthquakes the most in Europe, and avalanches or landslides narrowly exceeded windstorms or cyclones in Oceania. Chapter 16 provides a more comprehensive description of flood and fire hazards.

While the economic loss per event is much larger in industrial countries, the greatest losses still occur in developing nations in absolute numbers of lives as well as in relative impact as measured by percentage of GDP represented by disaster losses. (See Figure 6.4.)

Considering lack of resources and capacity to prevent or cope with the impacts, it is clear that the poor are the most vulnerable to natural disasters. Among the poorest countries, 24 of 49 face a

Table 6.2. Reducing Human Vulnerability: Country Scores (Global Leaders for Tomorrow Environmental Task Force 2002)

1.	Austria	85.1	49.	Colombia	71.7	97.	Zimbabwe	39.2
2.	Netherlands	85.1	50.	Trinidad and Tobago	71.4	98.	Namibia	38.5
3.	Sweden	85.0	51.	Jordan	70.9	99.	Gambia	37.3
4.	Canada	85.0	52.	Iran	70.7	100.	Laos	35.3
5.	Slovenia	85.0	53.	Kazakhstan	70.6	101.	Iraq	33.8
6.	Australia	84.9	54.	Tunisia	68.8	102.	Mongolia	32.8
7.	Finland	84.9	55.	Syria	68.1	103.	Myanmar (Burma)	32.6
8.	United Kingdom	84.8	56.	Mexico	67.2	104.	Ghana	32.3
9.	Norway	84.8	57.	Turkey	66.8	105.	Nepal	31.5
10.	Hungary	84.3	58.	Panama	66.2	106.	Bhutan	31.4
11.	Slovakia	84.3	59.	Brazil	66.0	107.	Senegal	30.6
12.	Switzerland	84.3	60.	Lithuania	64.8	108.	Sudan	29.5
13.	Ireland	83.9	61.	Algeria	64.2	109.	Gabon	25.6
14.	Iceland	83.6	62.	Bosnia and Herzegovina	63.7	110.	Congo	25.1
15.	Italy	82.7	63.	Romania	62.7	111.	Côte d'Ivoire	22.4
16.	New Zealand	82.2	64.	Libya	62.2	112.	Tajikistan	21.6
17.	France	82.2	65.	Egypt	62.1	113.	Benin	21.0
18.	Japan	82.1	66.	China	61.9	114.	Togo	18.3
19.	Denmark	82.0	67.	Jamaica	61.4	115.	Nigeria	18.2
20.	Greece	81.9	68.	Honduras	61.3	116.	Papua New Guinea	18.0
21.	South Korea	81.7	69.	Ecuador	61.2	117.	Uganda	15.4
22.	Uruguay	81.1	70.	Paraguay	60.7	118.	Cameroon	15.1
23.	Germany	80.9	71.	Morocco	60.4	119.	Burkina Faso	10.3
24.	Belgium	80.8	72.	Uzbekistan	60.3	120.	Kenya	10.2
25.	Spain	80.6	73.	Albania	59.8	121.	Tanzania	9.9
26.	Israel	80.4	74.	Thailand	58.9	122.	Mauritania	9.7
27.	United States	80.4	75.	North Korea	57.9	123.	Central African Rep.	9.4
28.	Chile	79.9	76.	Venezuela	57.8	124.	Mali	9.3
29.	Russia	79.7	77.	South Africa	57.7	125.	Cambodia	8.2
30.	Czech Republic	79.7	78.	Indonesia	57.5	126.	Guinea	8.1
31.	Belarus	79.3	79.	Philippines	56.4	127.	Madagascar	7.9
32.	Bulgaria	79.1	80.	Sri Lanka	56.3	128.	Haiti	7.9
33.	Costa Rica	79.1	81.	Kyrgyzstan	52.3	129.	Malawi	7.4
34.	Portugal	78.9	82.	Guatemala	52.3	130.	Zambia	6.9
35.	Poland	78.5	83.	Dominican Republic	51.5	131.	Burundi	6.4
36.	Moldova	77.3	84.	Peru	51.1	132.	Rwanda	6.1
37.	Croatia	76.6	85.	Botswana	51.0	133.	Mozambique	5.4
38.	Kuwait	76.5	86.	Armenia	51.0	134.	Niger	5.1
39.	Estonia	76.3	87.	Viet Nam	50.5	135.	Guinea-Bissau	5.1
40.	Saudi Arabia	76.2	88.	El Salvador	48.8	136.	Liberia	3.9
41.	Argentina	75.2	89.	Azerbaijan	47.6	137.	Chad	3.8
42.	United Arab Emirates	75.0	90.	Nicaragua	45.6	138.	Somalia	3.5
43.	Lebanon	74.8	91.	India	43.8	139.	Zaire	2.7
44.	Latvia	74.8	92.	Bolivia	43.5	140.	Ethiopia	2.4
45.	Macedonia	73.8	93.	Turkmenistan	42.0	141.	Sierra Leone	2.2
46.	Ukraine	73.6	94.	Pakistan	41.5	142.	Angola	1.9
47.	Malaysia	73.0	95.	Oman	41.0			
48.	Cuba	72.6	96.	Bangladesh	40.3			

high level of disaster risk; at least 6 countries have been affected by two to eight major disasters per year in the past 15 years, with long-term consequences for human development (UNEP 2002). Ninety percent of natural disaster–related loss of life occurs in the developing world. When countries are grouped according to the UNDP Human Development Index, socioeconomic differences are strongly reflected in disaster losses (IFRC 2001). For the 1990s, countries of low human development experienced about 20% of the hazard events and reported over 50% of the deaths and just 5% of economic losses. High human development countries accounted for over 50% of the total economic losses and less than 2% of the deaths.

In assessing the distribution of vulnerability, several limitations to existing research need to be considered. First, economic valua-tions do not reflect the difference in relative value of losses among wealthier and poorer populations or the reversibility of environ-mental damages incurred. Similarly, land degradation due to land-slides, flooding, or saline inundation from coastal events can diminish the natural capital resources of livelihoods, further com-pounding recovery challenges. The meaning of the economic value of these losses of ecosystem services is also difficult to capture and is seldom included in conventional economic assessments.

Second, because of the definitions of disaster used, local-scale disasters of significance to the affected community are often not reflected in these disaster statistics. If those losses were included, the figures on damages could easily be much higher.

Finally, there is the tendency to treat natural hazards in sepa-rate categories and to treat disasters as discrete, individual events.

Table 6.3. Great Natural Catastrophes and Economic Losses: Comparison of Decades, 1950–99 (Munich Re 2000)

Catastrophes and Losses	1950–59	1960–69	1970–79	1980–89	1990–99
Number	20	27	47	63	82
Economic losses (bill. 1998 dollars)	38.5	69.0	124.2	192.9	535.8
Insured losses (bill. 1998 dollars)	unknown	6.6	11.3	23.9	98.9

Note: Natural catastrophes are classified as "great" if the ability of the region to help itself is distinctly overtaxed, making interregional or international assistance necessary.

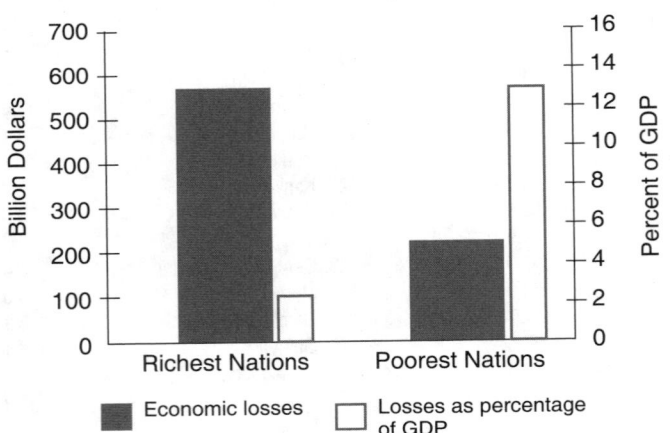

Figure 6.4. Disaster Losses, Total and as Share of GDP, in 10 Richest and Poorest Nations, 1985–99 (Abramovitz 2002)

This accounting practice limits insights into the consequences of threats from multiple hazards in one place and of sequences of disasters following upon one another. Over time, multiple and recurring hazards exacerbate vulnerability, and across scales, vulnerability is generally greater during the recovery period, when systems are already damaged. These patterns of differential impact affect efforts to cope with the impacts of environmental variability and degradation, as described earlier.

6.5.2 Explaining Vulnerability to Natural Hazards

Human-driven transformation of hydrological systems, population growth (especially in developing countries), and movements of people and capital into harm's way are major driving forces underlying the increasing numbers of disasters (Mitchell 2003). Conflict among people contributes further to vulnerability (Hewitt 1997). The causal reasons relate to basic characteristics of economy and political system but also to the perceptions, knowledge, and behavior of local managers and institutions (Hewitt 1997).

In some regions, significant environmental changes have resulted in the degradation of ecological services that mediated the effects of hydro-climatological events. Two common forms of ecological change—desertification and deforestation—can exacerbate the impacts of drought in some areas by reducing the moisture-holding capacity of the soil and contribute to increased flooding through reducing infiltration. (See Chapter 16.) In Honduras, de-

forestation contributed losses through increasing flooding as well as landslides following Hurricane Mitch in 1998. In other areas, efforts at river or flood control have reduced vulnerability to smaller hazard events, but increased losses when larger events overwhelmed dams, dykes, or levees and damaged the usually protected area.

The growth in numbers of people affected is a particularly important measure, as it provides an indication of the potential increase of exposure and sensitivity of people to environmental variability. The global annual average number of people affected has increased over the last decade, although the number of deaths due to disasters has declined. This shift highlights the potential for changes in pattern of vulnerability though adaptations. (See also MA *Policy Responses,* Chapter 11.) The greatest proportion of people affected resides in countries of medium human development, which include the large-population countries of Brazil, China, India, and Indonesia (IFRC 2001).

In addition to changing exposure, socioeconomic changes are shaping the overall patterns of vulnerability. First, while poverty is not synonymous with vulnerability, it is a strong indicator of sensitivity, indicating a lack of capability to reduce threats and recover from harm. The number of people living in poverty is increasing (UNDP 2002a). The greater number of people affected in medium human development countries may also reflect their experience with the additional challenges of transition, a situation somewhat akin to recovery, in which infrastructure and support systems, both physical and social, may be disrupted by the processes of change and be unable to contribute to reducing vulnerability.

Urbanization creates particular problems in disaster vulnerability. Due to the concentrations of people and complex infrastructure systems involved, the repercussions of an event in cities can spread quickly and widely, and the scale of resources needed for effective response is often challenging for national or international coordination. In many cases, these cities also draw in vast numbers of people seeking better lives, but they are often unable to keep up with the demand for planning, housing, infrastructure, and jobs. The informal housing that immigrants create is often located in marginal areas, such as hill slopes and floodplains, and accessible construction options cannot address the site limitations (Wisner et al. 2004). In 1950, just under 30% of the world's population (of 2.5 billion) lived in cities; by 2025 it is projected to be over 60% (of an estimated 8.3 billion) (UNDP 2002b). This rapid urbanization trend is particularly pronounced in countries with low per capita income. (See also Chapter 27.)

Globalization is contributing to natural hazard vulnerability as it is changing the sensitivity and coping options available (Adger and Brooks 1003; Pelling 2003). On an international scale, increasing connectedness is causing societies to become more dependent on services and infrastructure "lifelines." In such a connected world, the consequences of natural disaster reach far beyond the area physically damaged. It has been estimated that the possible extent of damage caused by a extreme natural catastrophe in one of the megacities or industrial centers of the world has already attained a level that could result in the collapse of the economic system of entire countries and may even be capable of affecting financial markets worldwide (Munich Re 2000, 2002). Globalization has also increased the risks faced by marginalized indigenous peoples; many of these are developmental effects that will become apparent over only the long term. Traditional coping mechanisms have come under severe pressure, and adaptation strategies, at one time effective, can no longer cope (Pelling 2003).

Data on global trends do not report on the social differentiation among victims, but case study evidence and other synthesis efforts indicate some social groups are continually disproportionately represented among those harmed the most (Wisner et al. 2004). These are often people who are marginalized within society, due to combinations of prejudice, lack of or ignored rights, and lack of access to social supports or personal resources or due to distance from concentrations of services and power. Indigenous peoples, such as the Inuit, Sami, and others from northern regions, represent the vulnerability of this type of situation well. (See Chapter 25 for further details). These circumstances often apply to poor people, women, children, elderly individuals, and ethnic minorities in affected areas. In addition, the elderly, children, women, and handicapped people are more likely to have physical limitations or special needs that reduce their ability to cope with disaster.

6.6 Desertification: Lessons for Vulnerability Assessment

Desertification—land degradation in drylands—has been a subject of interest for over 30 years, with numerous technical assessments and policy analyses, and it is a good example of changes in a coupled socioecological system that threaten livelihoods across large swaths of Earth. It is also a good example of understanding vulnerability. (See Downing and Lüdeke (2002) and Chapter 22 for more on drylands and desertification and a useful set of maps.)

Local to global studies of social vulnerability to desertification suggest at least three lessons for vulnerability from past experience:

- *Vulnerability is dynamic.* Desertification arises from the interactions of the environment and social, political, and economic systems—through the actions of stakeholders and the vulnerable themselves (Downing and Lüdeke 2002).
- *Vulnerability takes different forms at different scales.* Similar constellations of institutions have diverse effects at different social or geographic scales. The patchiness of driving forces, often represented in global scenarios, precludes developing a simple hierarchy from local vulnerability to global maps of desertification risks.
- *Vulnerability cannot be differentiated into different causes.* At the level of human livelihoods and systems, exposure to desertification is entangled with poverty, drought, water, food and other threats and stresses.

One example of the close coupling of social and environmental systems related to desertification is apparent in the syndromes approach developed by the Potsdam Institute for Climate Impact Research, which depicts the close linkages and components involved in the coupling. The basic idea behind syndromes is "not to describe Global Change by regions or sectors, but by archetypical, dynamic, co-evolutionary patterns of civilization–nature interactions" (Petschel-Held et al. 1999, p. 296). Syndromes are charted in dynamic process models that link state variables that change over time and between states. The scale is intermediate, reflecting processes that occur between household and national/macro scales. The typology of syndromes reflects expert opinion, modified over time based on modeling. Local case examples are used to generalize to mechanisms in the modeling and also to validate the syndrome results. Desertification is a case of several syndromes operating independently, reflecting the internal dynamics of places, resources, economies, and populations.

The syndrome approach illustrates how concepts of dynamic vulnerability might be implemented to understand multiple

stresses arising from the human use of ecosystem services. It takes the analysis one stage beyond purely biophysical explanations to examine linkages with human systems. The next steps might be integrated analysis at the level of different users of ecosystem services, and how they interact with each other in markets and in governance.

6.7 Food Insecurity

The arena of food security has been a third primary focus of vulnerability analysis. The severe famines in the 1980s in Africa saw the launch of dozens of famine early warning schemes. These implemented various designs, but all expanded beyond the simple monitoring of agricultural production. By the mid-1980s, Amartya Sen's entitlement theory (Sen 1981), which emphasized factors influencing the distribution of food as well as the absolute levels of available food, was widely circulated and implemented in food security monitoring. Attention to the socioeconomic failures that limit access to global food supplies became a substantial component of these efforts. More recently, more holistic approaches have sought to focus on livelihood security, to include food security, thereby widening the conceptual framing of vulnerability still further.

Much of the literature on food security focuses on human vulnerability; ecosystem services are limited to crop production, grazing for livestock, and to a lesser extent wild foods. While vulnerability assessment is maturing as an analytical tool, the need exists for assessments that are more dynamic and actor-oriented. An essential way forward in vulnerability analysis is to adopt a more precise terminology and nomenclature (see, e.g., the papers in Smith et al., 2003).

6.7.1 Methodology

Methodological lessons learned in vulnerability assessment over the past several decades reinforce the general messages of this chapter: food security is a relative measure that can be captured in various quantitative and semi-quantitative ways, but it is not an absolute condition that can be measured objectively. Food security is multidimensional and it integrates exposure to stresses beyond more narrow treatment of the production or availability of food. (See Chapters 8 and 18 for a further description of food provisioning services.) It is also clear that indicators of food security need to represent an explicit conceptual framework, such as that offered earlier in this chapter. The collation of indicators into profiles and aggregated indexes needs also to reflect the causal structure of food insecurity, going beyond the indiscriminate adding up of available indicators into a single index (see Downing et al. 2001).

A common feature of almost all food security (and livelihood) analyses is the recognition of multiple domains of vulnerability. Operational assessments commonly treat production, economic exchanges, and nutrition, while longer-term and more structural analyses include some measure of the political economy that underlies the more immediate dimensions of food security. Examples of operational assessments include India (MSSRF 2001) and Kenya (Haan et al. 2001). Figure 6.5 charts three domains of rural food insecurity for states in India.

A more heuristic illustration of the multiple dimensions of food security, related to climate change, is shown in Figure 6.6 (in Appendix A). The Figure is speculative, based on a subjective assessment of food security and climatic risks. Nevertheless, it clearly shows that global food production is of less concern than

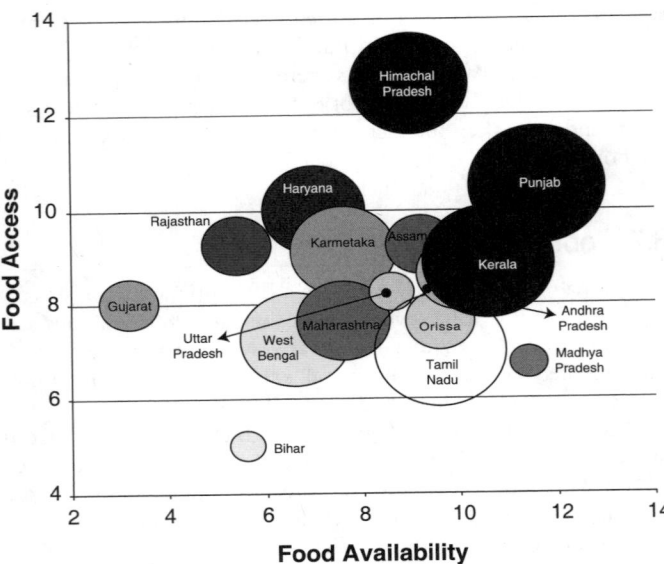

Figure 6.5. Food Insecurity Indicators of Rural India. Compiled at the state level, the MS Swaminathan Research Foundation aggregated food insecurity into three dimensions—food availability and production (x-axis), economic access (y-axis), and nutritional utilization (size of the circles, where larger is better-off). (MS Swaminathan Research Foundation 2001)

the impacts of droughts, which are already economically and socially significant for some livelihoods.

6.7.2 Wild Foods at the Local Scale

While a myriad of propositions regarding food security are possible, relating to different elements of causal structure, from the nature of the hungry themselves to the global political economy of food trade, here the case of wild foods and their role in food security is examined. (See also Chapters 5, 8, and 18 for food production and hunger issues.)

The most common approaches to food security are designed to balance consumption and production at the household level—including such indicators as expected yields of major foods (related to rainfall, soil quality, and pests, for instance), economic exchanges (such as terms of trade for agricultural sales or access to off-farm employment), hunting of wild foods, and some measures of entitlement through remittances from kin, official food relief, and relief work schemes. Set against the total of available food is the expected consumption, from meeting the FAO calorie standards to various levels of deprivation and starvation resulting in measurable effects on health. Aggregating to a regional or national level, such food balances guide policies for imports and exports, for targeted relief, and for declaration of a food crisis.

Notably absent from such food balances is the role of off-farm food collection—the gathering of wild foods either for consumption or sales. (See Chapter 8 for a more detailed description of the role of wild foods, including game, fish, and plants, in diets and for the underestimation in accounting in food balances.) In forest regions, these are called non-wood forest products and can be a major livelihood activity. Equally, few monitoring schemes include direct measures of ecosystem services such as charcoal sales, increased burdens of water shortages, or even effects of vegetation and land cover on livestock and pests. Nevertheless, for some marginal communities, such ecosystem services are essential and

particularly important for surviving food shortages (Ericksen 2003).

Investigations of two dryland sites in Kenya and Tanzania found that indigenous plants were an important source of raw material in the majority of coping mechanisms when alternative sources of food or income were required, such as when the harvest failed or sudden expenses had to be met. Such coping mechanisms included making use of trees for making and hanging beehives (flowering trees are also a source of nectar); of fuelwood for sale, burning bricks, or producing charcoal; of reeds, fibers, and wood for handicrafts such as mats or tools; and of fruit, vegetables, and tubers for food and sale. Indigenous plant-based coping mechanisms are particularly important for the most vulnerable, who have little access to formal employment or market opportunities, thus providing a crucial safety net in times of hardship. Wild fruits provide important nutrients to children during times when meals are reduced at home in many parts of Africa and South Asia (Brown et al. 1999), for example.

Such raw materials can often be acquired from communal land or from neighbors without cash transactions, and they are available at critical times of the year due to the climatic resilience of indigenous plants. In addition, the sale of livestock and poultry and engaging in casual labor, which are critical sources of cash during crises, often depend on ecosystem services, such as grazing land and fodder or forest products for fencing, construction, and other typical casual labor tasks. Table 6.4 shows the high percentage of households that depended on indigenous plant-based coping mechanisms in the Kenya and Tanzania site (Eriksen 2000), and Figure 6.7 illustrates the relative importance of indigenous foods. While the findings refer to a particular point in time (the 1996 drought), the widespread use of forest products as a source of food and income figures is consistent with findings from numerous other studies (Arnold 1995; Brown et al. 1999).

6.7.3 Global Influence on Local Food Balance

The literature on food security has a long tradition recognizing that local food balances are embedded in national economies and global flows of food trade and aid (for one representation, see Kates et al. 1988). A fictitious illustration captures the notion of global exposure:

During a drought, a farm household suffers a loss of yields in one of its fields of maize and beans. The field is primarily used for domestic consumption, cultivated by the women. Rainfall shortages are apparent with the delay in the onset of the rains—although the field is planted and later weeded by the women, the family does not apply expensive pesticides and fertilizers, expecting low returns during a poor season. Another field has a different problem. The head of the household acquired it as part of a community-based irrigation scheme

Table 6.4. Households That Depended on Indigenous Plant-based Coping Mechanisms in Kenya and Tanzania (Eriksen 2000)

Activities that Involve Use of Indigenous Plants	Share of Households, Kenya site	Share of Households, Tanzania site
	(percent)	
All use	94	94
Food use	69	54
Non-food use	40	42

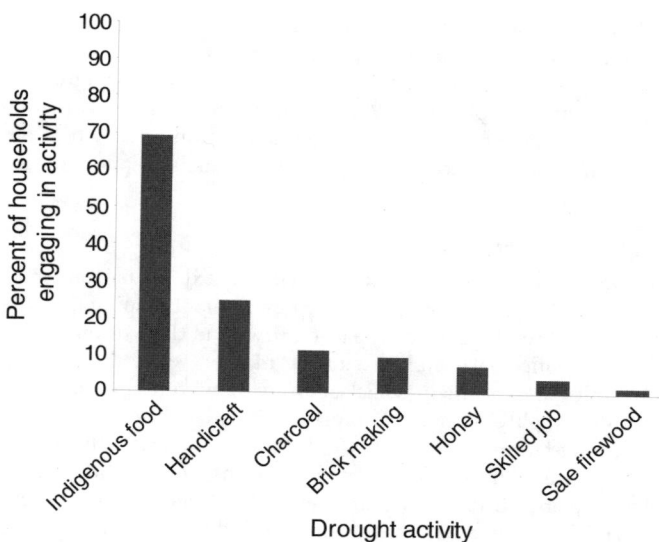

Figure 6.7. Use of Indigenous Plants in Mbitini, Kenya, by Activity during the 1996 Drought. "Skilled job" entailed tailoring, stone masonry/construction of houses, and woodcarving. Total number of households is 52. (Eriksen 2000)

that he joined a few years ago. He plants it this year with a cash crop of tomatoes and invests in fertilizer and pesticide. Halfway through the season, however, the drought restricts the availability of irrigation water. As a "junior" member of the scheme, his supply is reduced earlier than expected and his yields and quality are poor. When he tries to sell his crop to the local factory for processing into tomato juice, he discovers that there is a glut in the market due to a relaxation of import controls. Good conditions in a nearby country and export subsidies have produced a surplus, and the factory cannot afford to purchase local produce.

The fictitious example is not unrealistic—farmers have to contend with local conditions, with social, economic, and environmental relations in their community, and with the global and national food system. This global nature of vulnerability makes it impossible to clearly "bound" exposure, and it is often misleading to adopt a single spatial scale, as is often attempted in mapping vulnerability (as noted earlier regarding tools and methods).

6.7.4 Reporting Vulnerability

An essential way forward in food security analysis is to use at least a more precise terminology and nomenclature. A fairly simple scheme is proposed here, which makes clear four fundamental considerations that are not consistently reported: who is exposed, what the stresses are, what time frame is considered, and what consequences evaluated (see Downing et al. 2004). The notation below calls for reporting vulnerability (V) as specific to time frame (t); the sector, such as agriculture (s); the group, such as small-scale farmers, women farmers, or residents of peri-urban areas (g); and the consequences evaluated, such as food production, change in food purchasing power, nutritional levels or hunger (c).

$$^{t}V_{s,g}{}^{c}$$

(where s = sector, g = group, c = consequence, t = time frame and V = vulnerability)

For instance, an examination of climate change vulnerability in agriculture could offer greater utility to future comparisons and policy by specifying differences as follows:

- Climate change vulnerability (T = climate change, no other terms specified)
- Drought (T) vulnerability for food systems (s)
- Drought (T) vulnerability for smallholder (g) agriculturalists (s)
- Drought (T) vulnerability for smallholder (g) agriculturalists (s) at risk of starvation (c = health effects of reduced food consumption)

These four different statements about climate change vulnerability suggest the range of potential differences in assessment findings. The process of conducting a vulnerability assessment can be labeled VA. If the indicators are mapped, this is extended to a vulnerability assessment map, a VAM. A database of vulnerability indictors used in a VA (or VAM) can be labeled VI. Greater precision and analytical comparability could be gained by assigning a nomenclature to individual indicators (VI_x), such as:

t = time period (historical, present or specific projection)
g = group of people if specific to a vulnerable population
r = region (or geographic pixel)
★ = transformed indictors, as in standard scores

This basic set of relationships can be extended into a variety of assessment tools and facilitate comparison of case studies.

6.8 Exploring Vulnerability Concepts: Three Case Studies

The broad patterns of vulnerability apparent in the patterns and trends of natural disasters, the assessment of desertification, and the lessons from food security studies all demarcate important aspects of the sources and outcomes of stresses and perturbations on coupled socioecological systems. But it is well known that these interactions are highly place-specific. Thus it is useful to turn to particular cases to explore these issues in greater depth.

This section considers three specific examples. First, the situation of two types of resource-poor farmers in northeastern Argentina is examined, illustrating how vulnerability can take different forms with different types of farming systems. Second, we look at how shifting the scale of analysis or vulnerability and resilience yields quite different insights on the sources of vulnerability and the potential effectiveness of resilience-building strategies, using a case study from Southern Africa. Finally, efforts to reduce vulnerability and the challenges involved in assessing the benefits of different types of interventions are examined through a case study from the one of the poorest areas of India.

6.8.1 Resource-poor Farmers in Northeastern Argentina

The Misiones region, in a hilly area of northeastern Argentina, has a sub-tropical wet climate where about 60% of the original vegetation (sub-tropical forest) has now been replaced by agriculture, despite the fact that soils are fragile, ill-suited for continuous cropping, and subject to nutrient depletion and erosion (Rosenfeld 1998).

Subsistence farming is common in the region, and two major types of farmers can be distinguished. Both have a similar farm structure in terms of land, capital, and labor; both are very poor; and both types often cannot meet their basic needs (Cáceres 2003). But they have designed very different farming systems and developed contrasting strategies to interact with the wider context within which they operate. On the one hand, agroecological farmers have developed farming systems of very high diversity, use few external inputs, rely mostly on local markets, and are part of representative peasant farmer organizations. Tobacco growers, by contrast, manage less diverse agroecosystems, rely on external inputs provided by the tobacco industry, have a weak participation in local organizations, and are closely linked to external markets. (See Table 6.5.)

6.8.1.1 Agrobiodiversity

The number of domesticated animals and cultivated plants (agrobiodiversity) maintained by the two types of farmers is strikingly different. On average, agroecological farmers grow or raise three times as many species within a single farm as tobacco growers do. The total number of species in all surveyed farms is also very different: 97 species in the case of agroecological farmers and 41 species for tobacco growers. This indicates that agrecological farmers maintain a higher degree of heterogeneity among farms and a higher agrobiodiversity at the landscape-to-region level. Horticultural, aromatic, and medicinal crops and fruit trees are the most diverse categories both within and among farms.

Agrobiodiversity has a direct impact on food security (Altieri 1995). The more diverse farms are, the more likely they are to meet subsistence food needs. The opposite occurs in the case of farmers specialized in the production of commodities (such as tobacco), since most of the farm resources are allocated to a goal that does not strengthen local food security (Dewey 1979; Fleuret and Fleuret 1980). This situation is clearly observed in this case, where agroecological farmers grow more than three times as many species for food as tobacco growers do.

6.8.1.2 Technology

Agroecological farmers and tobacco growers also differ strongly in terms of farm technology. Although both draw on the same technological matrix (draft power and the use of fire to clear up land), the "final" technologies used in their farms are very different (Cáceres 2003). In order to produce their cash crop, tobacco growers rely on modern technology and a conventional approach to farming. This involves the use of high external input technology (chemical pesticides and fertilizers and high-yield seeds) and monocropping. Nearly all the inputs needed for tobacco production come from the market. Because tobacco growers have an extremely limited financial capacity, they rely on the credit provided by tobacco companies, which in turn buy the tobacco leaves from them, in a typical contract-farming relationship.

In contrast, the technology used by agroecological farmers rests mostly on the understanding and management of natural processes and cycles. Rather than relying on external inputs, they maximize the use of both local and agroecological knowledge and resources that are locally available. As a consequence of both their

traditions and the extension work of development agencies, the use of raised beds, composting practices, intercropping, biological pest control, and crop rotation is common among the agroecological farmers (Rosenfeld 1998). In order to gain access to this technology, these farmers do not need to develop a heavy reliance on the market, nor do they require the financial support of the agroindustry.

6.8.1.3 Scale Interactions

The socioeconomic and institutional context, in particular of markets and organization, is another key element shaping the vulnerability of rural societies. Tobacco growers in the Misiones have a less diversified relationship with the markets, since the tobacco companies are the main social actor with whom they interact. This is the highly asymmetrical relationship that typically develops in contract farming (Watts 1990). Tobacco growers are unable to make the most important farming decisions (such as how many tobacco plants to have, or which varieties), negotiate the quality and price of their tobacco with the agroindustry, or even decide which company to sell their product to.

The contacts of agroecological farmers with the agroindustry, on the other hand, are weak, and they are mostly linked to NGOs and governmental programs fostering rural sustainable development. Agroecological farmers have substantial control over the productive decisions concerning their farms and have developed a more diversified relationship with the market. They sell their production through different channels, of which the organic farmers' markets is the main one. In these markets, farmers and consumers meet once a week, when they set the price and other aspects of the commercial transactions. The wider range of products that agroecological farmers bring to the market also allows more spreading of commercial risks and thereby has a favorable impact on the stability of their cash flow.

The differences among these two types of farmers are even more noteworthy in terms of their participation in local organizations. Agroecological farmers not only relate with a higher number of organizations, they are also part of a larger number of grassroots representative organizations committed to peasant interests and civil rights. In contrast, tobacco growers are almost exclusively related to the Tobacco Growers' Association of Misiones, a highly bureaucratized organization that primarily represents tobacco company-interests (Schiavoni 2001). Yet the participation of tobacco growers in this organization is compulsory in order to be able to sell their tobacco to the agroindustry.

6.8.1.4 Synthesis: Differential Vulnerability

Agroecological farmers and tobacco growers share many key social and productive features. Both types of farmers and the envi-

Table 6.5. Differences between Agroecological Farmers and Tobacco Growers in Terms of Agrobiodiversity, Food Safety, Links with Markets, and Representative Organizations (P< 0.001, Mann-Witney U test for independent samples) (Cáceres 2003)

Variable	Agroecological Farmers			Tobacco Growers		
	Median	Minimum	Maximum	Median	Minimum	222
Total number of plant and animal species grown or raised on the farm	40	21	54	14	7	82
Number of species grown or raised for family consumption	28	18	42	10	4	11
Number of species sold in the market	5	3	10	2	1	1
Participation in organizations (number)	2	1	5	1	1	1

ronment in which they develop their farming strategies may be regarded as "vulnerable." However, as this case illustrates, factors shaping vulnerability can come together in a variety of ways that result in substantial variations in the magnitude and types of vulnerability, even among a group such as small-scale farmers, who are often assumed to be homogeneous.

Given these differences in vulnerability, the agroecological farmers appear less vulnerable overall than tobacco growers. Differences in agrobiodiversity, technology, and articulation to the wider context are the main factors underpinning this contrast. On the one hand, agroecological farmers appear to have developed more autonomous and resilient livelihood strategies. They manage more diverse and stable agroecosystems, produce more food, and show a stronger negotiating capacity within the political process. The strategy of tobacco growers, in contrast, depends far more on the agroindustry. They produce less food, have very limited negotiation power, and are more exposed to the control of tobacco companies and the fluctuations of tobacco prices and industry.

All this suggests that livelihood strategies used by different groups can dramatically increase or decrease their level of vulnerability. Since the articulation to the wider context is a key aspect in determining the vulnerability of poor farmers, the latter can change drastically due to external factors, no matter how "sensible" the within-farm decisions. This suggests that vulnerability involves the amplification and attenuation of a variety of conditions that depend on both internal and external circumstances, and that vulnerability changes over time with changing stresses or needs in households or with wider socioeconomic and political changes that increase or decrease access to various assets and opportunities.

6.8.2 Vulnerability and Resilience in Southern Africa: Perspectives from Three Spatial Scales

The southern African region is currently facing a suite of complex emergencies driven by a mix of factors, including HIV/AIDS, conflict, land tenure, governance, and lack of access to resources, coupled with climate risks—not least of which is the emergence of floods as a serious hazard (Mano et al. 2003; Vogel and Smith 2002; IPCC 2001a). Existing adaptive capacity is also, arguably, being increasingly eroded and undermined by such factors. The World Food Programme has recently estimated that around 14 million people in the region are in a heightened food insecurity situation (Morris 2002). Contributing factors emerging from this situation include, among others, low opening stocks of cereals from previous years, low grain reserves in some countries, low levels of preparedness for such food insecurity, and inappropriate and constraining policies that contributed to market failures (Mano et al. 2003).

This case examines the multiple roles of global environmental change as part of a complex suite of stressors (such as climate, governance, and health) and adaptation to such stressors in South Africa, using the 2002/03 famine situation in the Southern African Development Community as a backdrop. The theme of resilience and adaptation in the face of global change (Adger 2000) is analyzed at three spatial scales, moving from the regional (SADC) level to the district and community levels, focusing particularly on the role of information as a potential input into building sustainability. The greatest priority in such an investigation is less one of describing the problem than it is interactively crafting appropriate sustainable interventions. (No suitable "sustainable" interventions can be designed in isolation of the institutions and stakeholders involved.)

6.8.2.1 The SADC Region 2002/03 Season: Coping with Complex Environmental Stress

The contributions of various socioeconomic and political factors, often generated outside the region, have long been acknowledged to contribute to the complexities associated with climate stress and food insecurity facing Southern Africa (Benson and Clay 1998). Several of these myriad of factors usually become particularly important during a severe dry spell, flood, or other climate-driven event.

In response to the droughts of the 1970s, 1980s, and 1990s, international organizations, bilateral donors, African governments and NGOs established numerous early warning systems and enlarged institutional capacity to manage food security and risks (Moseley and Logan 2001). These entities have been actively undertaking efforts to reduce vulnerability to a number of risk factors in the region. A clear activity has been to examine current risks and threats primarily relating to drought-induced production deficits and to provide improved climate information to serve the agricultural sector (see, e.g., Archer 2003).

Another priority has been not only to increase the understanding of food provision and production but also to improve assessments of food procurement and access to food by households in the region (e.g., see Devereux 2000; Vogel and Smith 2002) and the factors (such as institutions, governance, and policy issues) that enhance or constrain access to food. The contributions of adverse synergies, including natural triggers (such as drought) and politics (such as civil stress) that have precipitated famines (Devereux 2000), have in some cases become more prevalent and endemic in sub-Saharan Africa.

A number of interesting adaptive measures have emerged from assessments undertaken of the 2002/03 famine in the region (see www.fews.net). Vulnerability assessments show, for example, that cereal production is sometimes not a key activity in procuring food in risk-prone households. Rather, it is food purchases and other inputs (remittances, gifts, and so on) that enable households to obtain food. Such insight on adaptation practices has only emerged from detailed food economy investigations. Such studies reveal and question the role of "food relief" as an intervention strategy in reducing the impacts of the crisis. Furthermore, the role of HIV/AIDS in aggravating the situation in several households is also emerging as a strong and negative factor (SADC FANR Vulnerability Assessment Committee 2003).

With the background of this regional scale, vulnerabilities to a similar suite of risks (including climate, management, and other factors) can be understood at the scale of South Africa and Limpopo Province. These case studies clearly show that, similar to the regional examples described earlier, a well-intentioned focus on early warning can do little to enhance resilience to risks if it is not coupled with a careful examination of the wider socioeconomic environment in which such activities operate (such as the policy environment, or institutional strengths and weaknesses), consistent with the northern Argentina case.

6.8.2.2 South Africa, 2002/03 Season—The National Scale

An unusually dry 2002/03 summer rainfall season caused widespread livestock mortality and water scarcity for growing crops in Limpopo, Mpumalanga, and North West Provinces in South Africa. In Limpopo, the provincial government requested 40 million rand in drought relief from the National Department of Agriculture, in addition to 6 million rand of provincial emergency funding that was made available (largely for subsidized fodder). Official estimates were that drought-related cattle mortalities exceeded 18,000.

A range of potentially valuable mechanisms to promote drought mitigation and risk reduction was, however, in place. Institutions and mechanisms included the Agricultural Risk Management Directorate, whose Early Warning Subdirectorate was substantively involved in improving awareness of early warning in the agricultural sector. The Early Warning Subdirectorate was established to improve forecast dissemination to smallholder farmers after forecasters and decision-makers realized that the information did not reach any further than provincial departments of agriculture (Archer and Easterling 2004). In addition, the National Agrometeorological Committee was established as a forum for reviewing updated seasonal outlook and provincial reports regularly throughout the season.

Essentially, the seasonal warning advisory was developed and disseminated at least to the provincial level in South Africa for the 2002/03 season. In spite of this, the adverse effects of climatic risk were substantial. Accepting that further investigation is required (and is planned), some preliminary observations on the 2002/03 season at the national scale in South Africa are possible.

As is well documented in a variety of case studies, forecasts, warnings, and information were in themselves insufficient to ensure action to improve resilience to environmental stress. In this case study, failures may have occurred in dissemination (for example, forecast information may not have been disseminated to extension officers or farmers). There may also have been failures in response capacity—even had farmers heard the seasonal warning, they may, for a variety of reasons, have been constrained in their ability to take anticipatory action (such as destocking). Last, there may have been weaknesses in institutional capability as well as weaknesses of "fit" and "interplay" between what institutions are providing and what is required (see, e.g., Folke et al. 1998; Berkes and Folke 1998; Orlove and Tosteson 1999; Raskin et al. 2002). Even with effective information dissemination, provincial, municipal, and local institutions may themselves be constrained in their ability to either recommend or support appropriate actions to improve resilience.

6.8.2.3 Vhembe District, Limpopo Province, 2002/03 Season

Results from research at the district and local level in Vhembe district of Limpopo Province show where gaps and weaknesses existed with regard to improved resilience to climatic risk in the 2002/03 season. It appears that this was the first season that the surveyed community (first surveyed in 2000/01) had exposure to seasonal forecast information. The Vhembe District Department of Agriculture and the District Department of Water Affairs and Forestry also received the forecast. Yet both at the community level and at the district institutional level, little response was apparent. Identifying the reasons for the lack of action is key to understanding the adverse drought effects at the national and provincial level described earlier.

First, it is clear that the forecast alone was insufficient, both for the needs of farmers and for district institutions. Both farmers and institutions explicitly asked for more guidance in terms of what actions might be appropriate in the light of the forecast or warning information. Farmers requested, for example, that when the seasonal forecast (or severe weather warning) was broadcast over the radio, the announcement needed to be coupled with an advisory. Such an advisory could include a wide range of general advice at various scales—at the district level, for instance, information on planting dates; at the farm level, very specific information on cultivars and planting. The District Department of Agriculture asked that the existing agricultural advisory be further developed and refined for local district conditions. The District

Department of Water Affairs and Forestry requested that the agricultural advisory be adapted for the water sector (and for other climate-sensitive sectors as well, such as health).

Second, farmers themselves may have been constrained in their ability to respond to information about climatic stress. The most commonly documented constraint on response capacity was resource limitation, including lack of access to credit, supplemental irrigation, land, and markets as well as lack of decision-making power (particularly in the case of women farmers) (Archer 2003). Further research in the area is seeking to understand the precise role of resource limitations and misdirected inputs (such as inappropriate irrigation infrastructure) in constraining both the ability to respond to forecasts and warnings and, more important, the ability to increase resilience and adaptive capacity.

There are also, however, encouraging signs in Vhembe district and at the national scale in South Africa of building adaptive capacity under conditions of climatic (and environmental) stress. Progress has been made in the dissemination of the forecast to district institutions and to the community level. And intermediary mechanisms described at the national scale (such as the programs under the Directorate of Agricultural Risk Management) show promise. There are signs that research on ways to improve adaptive capacity in South Africa is becoming increasingly well positioned to produce generalized recommendations that may inform policy.

6.8.2.4 Synthesis: Cross-scale Interactions and Multiple Stressors

The results from this case suggest that although gaps and weaknesses were evident in the ability of entities at different scales to decrease vulnerability to the emergence of multiple stressors, success stories were also apparent. In this example it is clear that the spatial scale is a valuable unit of analysis. The level of interplay, however, between scales of "intervention" is equally important (e.g., Orlove and Tosteson 1999).

This example illustrates the "misfit" between scales of research and intervention, between what is investigated and what is required. This example points to a greater understanding of these complex issues, particularly in a region undergoing complex shocks and stressors, and the deeper interrogation that is required of the range of institutional responses that may be needed to manage these systems effectively. The South African Weather Service, as the official national forecast producer, works with other forecast producers at the international and national levels to derive a multiple-sourced seasonal outlook, containing three-month rainfall and temperature forecasts. The forecast, looking specifically at the agricultural sector, is disseminated to the National Department of Agriculture and from there to provincial, district, ward extension, and finally farm level.

The process of sub-provincial dissemination of the forecast is still in progress. There are three areas of on-going activity to improve the system: the process of combining multiple source forecasts, the role of the National Disaster Management Centre in receiving forecasts and coordinating response in appropriate areas and sectors, and the sub-provincial receipt of, and response to, the forecasts.

At present, however, there remains a misfit between what is currently being provided by the forecast producers and the suggested requirements from the agricultural sector within the provincial levels. From the province down to ward level extension, suggested forecast information differs from the three-month temperature and rainfall forecasts provided from the national and international levels. Finer levels suggest information be provided on seasonal quality (such as information on intra-seasonal variability), advisories coupled to forecasts, retroactive forecast applications, and impact-specific interpretation of forecasts (Orlove and Tosteson 1999). To reiterate, the system is highly dynamic and should be seen as evolving. The key question remains how to best intervene to aid a system in building resilience to sustain socioecological systems under conditions of environmental stress and surprise.

6.8.3 The Benefits of Reducing Vulnerability in Bundelkhand, India

The Bundelkhand region in the central highlands of India consists of semiarid plateau land. Rising population, subsequent agricultural expansion, and increased demand for wood has led to rapid deforestation in the region, which together with poor land management practices and government-approved commercial logging has aggravated soil erosion and ecological degradation. Erratic rainfall coupled with soil erosion has further reduced soil productivity and contributed to crop failure, and the area is now highly degraded (EcoTech Services 1997). (This paper draws on Eco-Tech Services 1999; the study was carried out to support the Uttar Pradesh state government initiatives in the area, under a grant from the Government of the Netherlands.)

The region has some of the lowest levels of per capita income and human development in India. Illiteracy and infant mortality rates were high, and local inhabitants depended on rain-fed single-crop agriculture and small-scale livestock production. The forests that were the traditional source of livelihood have largely disappeared.

Lalitpur district lies at the heart of the Bundelkhand region. The main monsoon crops grown in the district are maize, gram, and groundnut, while the main winter crops are wheat, peas, and gram. Most people collect green fodder from their own land during kharif and feed harvest remains to the animals in rabi and summer. Harvest is sold as dry fodder. Most households use the same well through the year, and it takes approximately two hours per household to collect water each day. Nonavailability of potable water is a major problem across the district (EcoTech Services 1997).

6.8.3.1 Watershed Management

A technical plan for the Donda Nala watershed in Lalitpur district was drawn up, aimed at land treatment and drainage line treatment measures (EcoTech Services 1997). Land treatment measures sought to reduce the loss of topsoil and to augment rainwater retention and biomass production. Measures such as embankments, earthen gully (channel) plugs, and agroforestry were deemed applicable to cultivated land, while silvipasture was deemed applicable to uncultivated lands. Drainage treatments suggested by the plan included mechanical measures such as the construction of dams and surface water storage tanks. Long-term benefits envisioned from these measures were retention of topsoil and an increase in the moisture-retaining capacity of soil. The technical plan estimated that the high-grade lands in the watershed would show increased crop yields by about 50% in the first five years as a result of such improvements.

6.8.3.2 Quantifying Benefits

Benefits projected from the watershed management activities included increased productivity of land, improvement in the health of animals due to increased fodder availability, better access to drinking water, increased employment, lower rates of soil erosion, and stabilizing environmental degradation. For the economic analysis in the plan, the benefits were summarized as irrigation benefits, benefits from vegetative treatments, drinking water benefits, and employment benefits. (The assessment did not attempt to evaluate environmental and health benefits, which are more complex to quantify.)

Farmers realized benefits from cultivation in the form of increased profits. The incremental net profit was computed as the difference between current profits and potential future profits from cultivation. Assuming that prices would remain constant, profits in the future were estimated on the present value of future cultivation. It was estimated that the average annual incremental profit would be 3,910,700 rupees (or 1,450 rupees per acre) as a result of additional water on existing farmlands. It was estimated that there would be additional benefits due to cultivation on marginal lands due to a further 257 hectares coming under cultivation during monsoon and 90 hectares in winter. This value was estimated as 1,681,000 rupees.

Vegetative treatments led to increased biomass in the form of fodder, firewood, and timber. Locally accepted species were identified for long-term community-managed common land. The estimates from increased fodder availability were based on fodder collection amounts. The incremental production of dry fodder or crop residues was valued at the existing market rate and estimated at 777,800 rupees for the watershed as a whole. A detailed cost-benefit estimation of silvipastural treatments planned in the wastelands for a period of 30 years was also assessed to compute the net present value of the future stream of benefits. Some 420 hectares of land were to be covered under the afforestation plan.

The potential benefits from better access to drinking water were valued by using the opportunity cost of time saved in water collection for women. Three open wells were proposed in the villages of Agar, Dhurwara, and Ghisoli. These sought to enhance women's participation in the project and to benefit families who lacked easy access to drinking water. The new wells were typically located near a cluster so that these families would not have to go more than a quarter of a kilometer. The estimated cost of digging wells in the watershed was 304,065 rupees, and the total value of time savings was 45,090 rupees for the year. The value of this is projected to rise over time as daily wages increase.

Given the labor requirements for each type of project activity, the market and opportunity costs for labor were determined. The benefits were calculated from activity-specific labor components of the technical work plan. Total incremental benefits from employment were valued at the prevailing wage rate. The employment benefits disbursed in the first two years of project activities were estimated at 5,480,000 rupees.

The projected present value of the future stream of the total annual benefits from each of the estimated components provides the overall value for the stream of benefits accruing from the project. The average projected present value of benefits per hectare was 47,461 rupees as opposed to an average project activity cost of 7,500 rupees per hectare. (See Table 6.6.) Assuming a 30-year horizon, the present projected value of the estimated benefits were computed using a 12% discount rate. The net present value of total benefits worked out to be over 100 million rupees for the entire watershed.

6.8.3.3 Synthesis: Distributional Issues

Most of the village community of Lalitpur district consists of small farmers and landless people. While the benefits from additional employment and access to drinking water are projected to directly enhance their quality of life, benefits from irrigation and green fodder production (which are the major source of benefits) are

Table 6.6. Total Benefits for Donda Nala Watershed (EcoTech Services 1997)

Project Activity	Total Undiscounted Benefits	Total Discounted Benefits
	(Rs crores)	
Irrigation	16.5620	3.5799
Digging wells	0.1300	0.0281
Employment	0.5476	0.4132
Silvipasture	24.4177	6.0871
Forestry	5.5876	0.3949
Total benefits	**47.2449**	**10.5320**

likely to accrue to those with land or cattle. The benefits will reach poorer households only if the access to treated wastelands and to harvest can be assured.

6.9 Implications for Assessment and Policy

The discussion and cases in this chapter emphasize that the patterns and dynamics that shape the vulnerability of coupled socio-ecological systems are composed of a multitude of linkages and processes. As such, assessments of vulnerability need to be comprehensive, sensitive to driving forces at different scales, but also appreciative of the differences among places.

A number of observations relevant to attempts to assess and reduce vulnerability and to build resilience may be offered. First, conceptual frameworks of vulnerability have improved, representing human and biophysical vulnerabilities as a coupled socio-ecological system. However, the relationships across scales and the role of specific actors (as drivers of systems) are poorly represented in most frameworks, and the existing state of knowledge is still weak. Different components of the coupled socioecological system may have quite different vulnerabilities and may experience exposure to stresses and perturbations quite differently. Diverse impacts are likely as a result; broad frameworks should not be taken as reliable guides to local conditions. The term vulnerability is still used in disparate ways in many assessments; a clear nomenclature is required to make assessments more consistent and coherent.

Second, the driving conditions of vulnerability have been well characterized at least at a general level. Human alterations of ecosystems and ecosystem services shape both the threats to which people and places are exposed and their vulnerabilities to the threats. The same alterations of environment can have very different consequences, depending on the differential vulnerability of the receptor systems.

Third, poverty and hazard vulnerability are linked and often mutually reinforcing by creating circumstances in which the poor and those with limited assets have few options but to exploit environmental resources for survival. At the same time, poverty and vulnerability are overlapping but distinct conditions, and they require analysis to determine overlaps and interactions.

Fourth, vulnerability can also be increased by the interaction of stresses over time. In particular, sequences of stresses that erode coping capacity or lengthen recovery periods can have long-term impacts that still often are not adequately treated in many assessments. Capturing these dynamics of vulnerability in assessment is an ongoing challenge.

Fifth, socioeconomic and institutional differences are major contributors to patterns of differential vulnerability. The linkages among environmental change, development, and livelihood are attracting increasing attention as a nexus in building resilient communities and strengthening adaptive capacity, but existing knowledge is still uneven and not well developed.

Sixth, despite this general level of explanation, it is still difficult to document adequately the effects of different changes upon different human groups with precision. While environmental changes and natural disasters are affecting increasing numbers of people, the existing knowledge base of vulnerability and resilience is highly uneven, with much known about some situations and very little about others. Some of the most vulnerable peoples and places are those about which the least is known. New vulnerabilities may be realized in the future, as in the dramatic increase of flooding damages in Africa or the effects of HIV/AIDS as a compounding factor in livelihood security. Filling the major gaps is a high priority in improving current assessments.

Seventh, assessment methods are improving. Entering vulnerability assessments at different scales of analysis, and particularly the local scales of place-based assessments, holds potential for greater depth and understanding of the complexity and dynamics of changing vulnerability.

Finally, despite the limitations of theory, data, and methods, sufficient knowledge exists in most regions to apply vulnerability analysis to contemporary problems of ecosystem management and sustainable development in order to provide useful information to decision-makers and practitioners.

References

Abramovitz, J.N., 2001: Averting unnatural disasters. In: *State of the World 2001,* L.R. Brown et al., W.W. Norton & Company, New York, 123–142.

Adger, N., 2000: Social and ecological resilience: are they related? *Progress in Human Geography,* **24(3),** 347–364.

Adger, N., 2003: Social capital, collective action and adaptation to climate change. *Economic Geography* **79,** 387–404.

Adger, N. and N. Brooks, 2003: Does global environmental change cause vulnerability to disaster? In: *Natural Disasters and Development in a Globalizing World,* M. Pelling (ed.), Routledge, London, pp. 19–42.

Adger, W. Neil, M. Kelly, and N. H. Ninh, 2001: *Living with Environmental Change: Social Vulnerability, Adaptation, and Resilience in Vietnam.* Routledge, London, 336 pp.

Altieri, M. A. 1995: *Agroecology: The Science of Sustainable Agriculture.* Westview Press, Boulder, Colorado, 448 pp.

Archer, E.R.M., 2003: Identifying underserved end-user groups in the provision of climate information. *Bulletin of the American Meteorological Society* **84(11),** 1525–1532.

Archer, E.R.M. & W.E. Easterling, 2004: Constraints and Opportunities in the Application of Seasonal Climate Forecasts in a Resource-Limited Farming Community. (submitted to Natural Hazards)

Arnold, J.E.M., 1995: Socio-economic benefits and issues in non-wood forest products use. In: *Report of the International Expert Consultation on Non-Wood Forest Products, Non-Wood Forest Products Report 3,* FAO—Food and Agriculture Organization of the United Nations (ed.), Rome, pp. 89–123.

Bengtsson, J., P. Angelstam, T. Elmqvist , U. Emanuelsson, C. Folke, et al. 2003: Reserves, resilience and dynamic landscapes. *Ambio* 32(6), 389–396.

Benson, C. and E. Clay, 1998: The Impact of Drought on Sub-Saharan African Economies, *World Bank Technical Paper No. 401,* World Bank, Washington DC, USA.

Berkes, F. and C. Folke (eds.), 1998: *Linking Social and Ecological Systems. Management Practices and Social Mechanisms for Building Resilience.* Cambridge University Press, Cambridge, 476 pp.

Berkes, F., J. Colding, and C. Folke (eds.), 2002: *Navigating Social-Ecological Systems: Building Resilience for Complexity and Change.* Cambridge University Press, Cambridge, 416 pp.

Brown, K., L. Emerton, A. Maina, H. Mogaka, and E. Betser, 1999: *Enhancing the Role of Non-wood Tree Products in Livelihood Strategies of Smallholders in Semi-arid Kenya.* The Overseas Development Group/The School of Development Studies, University of East Anglia, Norwich.

Carpenter, S.R, B. Walker, J.M. Anderies, and N. Abel, 2001: From metaphor to measurement: Resilience of what to what? *Ecosystems* **4**, 765–781.

Cáceres, D., 2003: Los sistemas productivos de pequeños productores tabacaleros y orgánicos de la Provincia de Misiones. *Estudios Regionales,* **23**, 13–29.

Chambers, R., 1989: Editorial introduction: vulnerability, coping and policy. *IDS Bulletin* **2**(2), 1–7.

CRED (Centre for Research on the Epidemiology of Disasters), 2002: EM-DAT: The OFDA/CRED International Disaster Database—www.em-dat .net—Université Catholique de Louvain—Brussels – Belgium, Jan 2002–2003 [cited 02/02/2003]. Available from www.cred.be/emdat.

CRED (Centre for Research on the Epidemiology of Disasters), 2003: EM-DAT: The OFDA/CRED International Disaster Database—www.em-dat .net—Université Catholique de Louvain—Brussels – Belgium, Jan 2002–2003 [cited 02/02/2003]. Available from www.cred.be/emdat.

Cutter, S., 1995: The forgotten casualties: Women, children, and environmental change. *Global Environmental Change* **5**, 181–194.

Danter, K. J., D.L. Griest, G.W. Mullins, and E. Norland, 2000: Organizational change as a component of ecosystem management. *Society and Natural Resources* **13**(6), 537–547.

Devereux, S., 2000: *Famine in the Twentieth Century.,* IDS Working Paper, No.105. Institute of Development Studies, University of Sussex, UK.

Dewey, K. G., 1979: Agricultural development, diet and nutrition. *Ecology of Food and Nutrition* **8**(4), 265–273.

Downing, T.E., A. Patwardhan, E. Mukhala, L. Stephen, M. Winograd and G. Ziervogel, 2002: *Vulnerability assessment for climate adaptation.* Adaptation Planning Framework Technical Paper 3.

Downing, T.E., 2002: Linking sustainable livelihoods and global climate change in vulnerable food systems. *Die Erde* **133**, 363–378.

Downing, T.E. with R.E. Butterfield, S. Cohen, S. Huq, R. Moss, A. Rahman, Y. Sokona and L. Stephen, 2001: *Vulnerability Indices: Climate Change Impacts and Adaptation.* Policy Series 3. *UNEP,* Nairobi.

Downing, T.E. and M. Lüdeke, 2002: International desertification: Social geographies of vulnerability and adaptation. In: *Global Desertification: Do Humans Cause Deserts?* J.F. Reynolds and D.M. Stafford Smith (eds.), Dahlem University Press, Berlin, pp. 233–252.

Downing, T.E. and A. Patwardhan et al. 2004: Assessing Vulnerability for Climate Adaptation In *Adaptation Policy Frameworks for Climate Change,* B. Lim and E. Spangler-Sigfried (eds.), UNDP, Cambridge University Press, Cambridge, pp. 67–90.

EcoTech Services, 1997: *Technical Plan: Donda Nala Watershed, Jakhora Block, Lalitpur District,* Report of the Bundelkhand Integrated Watershed Management Project, Jhansi.

EcoTech Services, 1999: *Economic Analysis Study of the Donda Nala Watershed,* EcoTech Services India, New Delhi.

Eriksen, S., 2003: The role of indigenous plants in household adaptation to climate change: The Kenyan experience. In: *Climate change for Africa: Science, technology, policy and capacity building,* P.S. Low (ed.), Cambridge University Press, Cambridge, (in press).

Eriksen, S., 2000: *Responding to Global Change: Vulnerability and Management of Agro-ecosystems in Kenya and Tanzania.* Ph.D. Thesis, Climatic Research Unit, School of Environmental Sciences, University of East Anglia, Norwich, UK.

Fleuret, P. and A. Fleuret, 1980: Nutrition, consumption and agricultural change. *Human Organization* **39**(3), 250–260.

Folke, C., C.S. Holling, and C. Perrings, 1996: Biological diversity, ecosystems and the human scale. *Ecological Applications* **6**, 1018–1024.

Folke, C., L. Pritchard Jr., F. Berkes, J. Colding, and U. Svedin, 1998: *The Problem of Fit between Ecosystems and Institutions.* IHDP Working Paper No 2. IHDP, Bonn, Germany.

Franklin, S. and T.E. Downing, 2004: Vulnerability, Global Environmental Change and Food Systems. SEI Poverty and Vulnerability Programme/GEC-AFS Briefing Paper, Stockholm Environment Institute, Stockholm, Sweden, 4pp.

German Advisory Council on Global Change (WBGU), 1997: *World in Transition: The Research Challenge.* Springer Verlag, Berlin.

Global Leaders of Tomorrow Environment Task Force, 2002: *2002 Environmental Sustainability Index.* Yale Center for Environmental Law and Policy, New Haven, CT., 82 pp.

Gunderson, L.H. and Holling, C.S. (eds.), 2002: *Panarchy: Understanding Transformations in Systems of Humans and Nature.* Island Press, Washington, DC.

Gunderson, L.H. and L. Pritchard, (eds.), 2002: *Resilience and the Behavior of Large-Scale Ecosystems.* Island Press, Washington, DC, 240 pp.

Haan, N., G Farmer, and R Wheeler, 2001: *Chronic Vulnerability to Food Insecurity in Kenya: A WFP Pilot Study for Improving Vulnerability Analysis.* World Food Programme, Rome.

Hewitt, K., 1997: *Regions of risk: A geographical introduction to disasters.* Longman, Harlow.

Holling, C. S. and G.K. Meffe, 1996: Command and control and the pathology of natural resource management. *Conservation Biology* **10**: 328–337.

Holling, C.S., 1995: Engineering resilience vs. ecological resilience. In P. Schulze, ed., *Engineering within Ecological Constraints.* National Academy Press. Washington.

Holling, C.S., 2001: Understanding the complexity of economic, ecological, and social systems. *Ecosystems* **4**, 390–405.

Homer-Dixon, T. and J. Blitt, (eds.), 1998: *Ecoviolence: Links Among Environment, Population, and Security.* Rowman and Littlefield, Lanham, MD.

IFRC (International Federation of Red Cross and Red Crescent Societies), 2001: *World Disasters Report 2001: Focus on Recovery.* International Federation of Red Cross and Red Crescent Societies, Geneva.

IFRC, 2002: *World Disasters Report 2002: Focus on Reducing Risks.* International Federation of Red Cross and Red Crescent Societies, Geneva.

IPCC (Intergovernmental Panel on Climate Change), 2001a: *Climate Change 2001, Impacts, Adaptation, and Vulnerability,* Contribution of Working Group II to the Third Assessment Report of the Intergovernmental Panel on Climate Change, Cambridge University Press.

IPCC, 2001b: *Summary for Policy Makers: Working Group II: Climate Change 2001: Impacts and Adaptation.* Cambridge University Press, Cambridge.

ISDR (International Strategy for Disaster Reduction Secretariat), 2002: *Living with Risk: A Global Review of Disaster Reduction Initiatives.* ISDR, Geneva.

Jodha, N.S., 2002: *Life on the Edge: Sustaining Agriculture and Community Resources in Fragile Environments.* Oxford University Press, Oxford, 356 pp.

Jodha, N.S., 1997: Mountain Agriculture. In: B. Messerli and J.D. Ives, eds. *Mountains of the World: A Global Priority.* Parthenon, N.Y., pp. 313–336.

Kasperson, J.X. and R.E. Kasperson (eds.), 2001: *Global Environmental Risk.* Earthscan/ United Nations University Press, London, 574 pp.

Kasperson, J.X., R.E. Kasperson, and B.L. Turner, 1995: *Regions at Risk: Comparisons of Threatened Environments.* United Nations University Press, NY, 588 pp.

Kates, R. W., 1985: The interaction of climate and society. In: *Climate Impact Assessment SCOPE 27,* R. W. Kates, J. H. Ausubel and M. Berberian (eds.), Wiley, New York, pp. 3–36.

Kates, R. W., R. S. Chen, T. E. Downing, J. X. Kasperson, E. Messer, and S. R. Millman, 1988: *The Hunger Report: 1988.* World Hunger Center, Brown University, Providence, RI.

Lambin, E. and E. Bartholomé, 2003: Monitoring natural disasters and 'hot spots' of land cover change with SPOT VEGETATION data to assess regions at risks. [online] Cited May 2003. Available at http://www.geo.ucl.ac.be/ Disasters.htm.

Lambin, E.F, B.L. Turner II, H.J. Geist, S.B. Agbola, A. Angelsen, et al. 2001: Our emerging understanding of the causes of land-use and land-cover change. *Global Environmental Change* **11**, 261–269.

Lane, R., 1997: Oral histories and scientific knowledge in understanding environmental change: A case study in the Tumut Region, NSW. *Australian Geographical Studies* **35**(2), 195–205.

Lawton, J. H., 2000: Community ecology on a changing world. *Excellence in Ecology* Series no. 11. Ecology Institute, Oldendorf, Germany.

Lonergan, S., 1998: *The Role of Environmental Degradation in Population Displacement.* 2nd ed. Research Report 1. Victoria, B.C.: Global Environmental Change and Human Security Project, University of Victoria.

Loreau, M, S. Naeem, P. Inchausti, J. Bengtsson, J.P. Grime, et al. 2001: Biodiversity and ecosystem functioning: Current knowledge and future challenges. *Science* **294**, 804–808.

Lüdeke, M.K.B. and G. Petschel-Held, 1997: Syndromes of global change: An information structure for sustainable development. In: *SCOPE 58: Sustainability indicators – Report of the project on indicators for sustainable development,* B. Moldan and S. Billharz (eds.), Wiley & Sons, Chichester.

Lüdeke, M.K.B., O. Moldenhauer, and G. Petschel-Held, 1999: Rural poverty driven soil degradation under climate change: The sensitivity of the disposition toward the SAHEL Syndrome with respect to climate. *Env. Modeling Asses.* **4**, 315–326.

MSSRF (M S Swaminathan Research Foundation), 2001: *Food Insecurity Atlas of Rural India.* Chennai: MSSRF.

Mano, R., B. Isaacson, P. Dardel, 2003: *Identifying policy determinants of food security response and recovery in the SADC region: The case of the 2002 food emergency.* FANRPAN Policy Paper, FANRPAN.

Mather, J. and G. Sdasyuk, 1991: *Global Change: Geographic Approaches.* University of Arizona Press, Tucson.

McIntosh, R.J., J.A. Tainter and S.K. McIntosh, (eds.), 2000: *The Way the Wind Blows: Climate, History and Human Action.* Columbia University Press; New York, 448 pp.

Mitchell, J.K., 2003: European river floods in a changing world. *Risk Analysis* **23(3),** 567–574.

Morris, J.T., 2002: *Executive Summary of the First mission of the Special Envoy to Lesotho, Malawi, Mozambique, Swaziland, Zimbabwe and Zambia.* United Nations, Rome.

Moseley, W.G. and B.I. Logan, 2001: Conceptualising hunger dynamics: A critical examination of two famine early warning methodologies in Zimbabwe. *Applied Geography* **21,** 223–248.

Moss, S., C. Pahl-Wostl, and T.E. Downing, 2001: Agent based integrated assessment modelling: The example of climate change. *Integrated Assessment* **2(1),** 17–30.

Munich Re, 2000: *Topics 2000: Natural Catastrophes—the Current Position.* Munich, Germany. Available online at www.munichre.com.

Munich Re, 2003: *Topics: Annual Review of Natural Catastrophes 2002.* Munich, Germany. Available online at www.munichre.com.

Myers, N., R. Mittermeier, C. Mittermeier, G. da Fonseca, and J. Kent, 2000: Biodiversity hotspots for conservation priorities. *Nature* **403(24),** 853–858.

National Geographic Society, 1989: *Endangered Earth.* Washington DC, USA.

Nellemann, C., L. Kullerud, I. Vistnes, B. Forbes, T. Foresman, et al. 2001: *GLOBIO. Global Methodology for Mapping Human Impacts on the Biosphere.* UNEP/DEWA/TR.-1-3, UNEP-DEWA, Nairobi, Kenya.

Orlove, B. S., and J. L. Tosteson, 1999: The Application of Seasonal to Interannual Climate Forecasts Based on El Nino–Southern Oscillation (ENSO) Events: Australia, Brazil, Ethiopia, Peru, and Zimbabwe. *Working Paper WP99–3-Orlove.* Berkeley, CA: Institute of International Studies, Berkeley Workshop on Environmental Politics, http://repositories.cdlib.org/iis/bwep/WP99–3-Orlove.

Pelling, M., (ed.), 2003: *Natural Disasters and Development in a Globalizing World.* Routledge, London.

Peterson, G., C.R. Allen, and C.S. Holling, 1998: Ecological resilience, biodiversity, and scale. *Ecosystems* **1,** 6–18.

Petschel-Held, G., 2002: Requirements for socio-economic monitoring: The perspective of the millennium assessment and integrated assessments in general. Prepared for Social Monitoring: Meaning and methods for an Integrated Management in Biosphere Reserves. Biosphere Reserve Integrated Monitoring (BRIM) Series No.1.W. Lass and F. Reusswig, (eds.), 2002. Available at: www.pik-potsdam.de/~gerhard/homepage/pdfs/rome01.pdf

Petschel-Held, G., A. Block, M. Cassel-Gintz, J. Kropp, M.K.B. Lüdeke, O. Moldenhauer, F. Reusswig, and H. J. Schellnhuber, 1999: Syndromes of global change—A qualitative modeling approach to assist global environmental management. *Environmental Modeling and Assessment* **4,** 295–314.

Petschel-Held, G., and M.K.B. Lüdeke, 2001: Integration of case studies by means of qualitative differential equations. *Integrated Assessment* **2,** 123–138.

Planning Commission, Government of India, *Study on Bundelkhan.* Available at http://planningcommission.nic.in/.

ProVention Consortium, 2003: Disaster Risks Hotspots Program. [online]. Available at www.proventionconsortium.org.

Rapport, D.J. and W.G. Whitford, 1999: How Ecosystems Respond to Stress. *BioScience* **49 (3),** 193–203.

Raskin, P., T. Banuri, G. Gallopin, P. Gutman, A. Hammond, R.W. Kates, and R. Swart, 2002: *Great Transition: The Promise and Lure of Times Ahead.* SEI PoleStar Series Report no. 10. *Stockholm Environment Institute, Boston.*

Redman, C.L., 1999: *Human Impact on Ancient Environments.* University of Arizona Press, Tucson AZ, 288 pp.

Reynolds, F., and D.M. Stafford Smith, 2002: *Global Desertification: Do Humans Cause Deserts?* Dahlem University Press, Dahlem.

Robertson, M., P. Nichols, P. Horwitz, K. Bradby, and D. Macintosh, 2000: Environmental narratives and the need for multiple perspectives to restore degraded landscapes in Australia. *Ecosystem Health* **6(2),** 119–133.

Rosenfeld, A., 1998: *Evaluación de Sostenibilidad Agroecológica de Pequeños Productores (Misiones-Argentina).* Tesis de Maestría. Universidad Internacional de Andalucía, España.

SADC (Southern Africa Development Committee), FANR (Food, Agriculture and Natural Resources), and Vulnerability Assessment Committee, 2003: *Towards Identifying Impacts of HIV/AIDS on Food Insecurity in Southern Africa and Implications for Response: Findings from Malawi, Zambia and Zimbabwe.* Harare, Zimbabwe.

Schellnhuber, H.-J., A. Block, M. Cassel-Gintz, J. Kropp, G. Lammel, et al., 1997: Syndromes of global change. *GAIA,* **6(1),** 19–34.

Schiavoni, G., 2001: Organizaciones agrarias y constitución de categorías sociales. Plantadores y campesinos en el nordeste de Misiones (Argentina). *II Jornadas Interdisciplinarias de Estudios Agrarios y Agroindustriales.* Buenos Aires.

Sen, A., 1981: *Poverty and famines: An essay on entitlement and deprivation.* Clarendon Press, Oxford.

Smith, J., R.J.T. Klein, and S. Huq, 2003: *Climate Change, Adaptive Capacity and Development.* Imperial College Press, London.

Stephen, L. and T.E. Downing, 2001: Getting the scale right: A comparison of analytical methods for vulnerability assessment and household level targeting. *Disasters* **25(2),** 113–135.

Swiss Re, 2003: Natural catastrophes and man-made disasters in 2002. *Sigma* No. 2/2003. Available online at www.swissre.com.

Turner, B.L., II, R.E. Kasperson, P.A. Matson, J.J. McCarthy, R.W. Corell, et al., 2003a: A framework for vulnerability analysis in sustainability science. *Proceedings of the National Academy of Sciences* 100 (No. 14): 8074–8079.

Turner, B.L. II, P.A. Matson, J.J. McCarthy, R.W. Corell, L. Christiansen, et al. 2003b: Illustrating the coupled human-environment system: Three case studies. Proceedings of the National Academy of Sciences 100 (No. 14), 8080–8085.

UNDP (United Nations Development Programme), 2002a: *Human Development Report 2002.* Oxford University Press, New York and Oxford.

UNDP, 2002b: *The 2000 Revision and World Urbanization Prospects: The 2001 Revision.* Population Division of the Department of Economic and Social Affairs of the United Nations Secretariat, World Population Prospects 2001 [cited 18 December 2002]. Available from http://esa.un.org/unpp.

UNDP, 2004: *Reducing Disaster Risk: A Challenge for Development.* UNDP, New York, 146 pp.

UNEP (United Nations Environment Programme), 2002: *Global Environmental Outlook 3.* Earthscan, London.

UNEP, 2003: *Assessing Human Vulnerability to Environmental Changes: Concepts, Issues, Methods, and Case Studies.* UNEP, Nairobi, 57 pp.

UNEP-GRID, 2003: *Project of Risk Evaluation, Vulnerability, Information & Early Warning* (PreView). [online]. Available at http://www.grid.unep.ch/activities/earlywarning/preview/.

USDA (U.S. Department of Agriculture), 1998: Global desertification vulnerability. See: www.nrcs.usda.gov/technical/worldsoils/mapindex/desert.html

Vogel, C. and J. Smith, 2002: The politics of scarcity: Conceptualizing the current food security crisis in southern Africa. *South African Journal of Science* **98(7/8),** 315–317.

Watts, M., 1990: Peasants under contract: Agro-food complexes in the third world. In: *The food question. Profits versus people?* H. Bernstein, B. Crow, M. Mackintosh and C. Martin (eds.), Earthscan, London, pp. 149–162.

White, G. F. (ed.), 1974: *Natural Hazards: Local, National, Global.* Oxford University Press, London, 288 pp.

Wisner, B., P. Blaikie, T. Cannon, and I. Davis, 2004: *At Risk: Natural Hazards, People's Vulnerability, and Disasters.* Routledge, London. 464 pp.

Chapter 7

Fresh Water

Coordinating Lead Authors: Charles J. Vörösmarty, Christian Lévêque, Carmen Revenga
Lead Authors: Robert Bos, Chris Caudill, John Chilton, Ellen M. Douglas, Michel Meybeck, Daniel Prager
Contributing Authors: Patricia Balvanera, Sabrina Barker, Manuel Maas, Christer Nilsson, Taikan Oki, Cathy
    A. Reidy
Review Editors: Frank Rijsberman, Robert Costanza, Pedro Jacobi

---

★This appears in Appendix A at the end of this volume.

Main Messages

Global freshwater use is estimated to expand 10% from 2000 to 2010, down from a per decade rate of about 20% between 1960 and 2000. These rates reflect population growth, economic development, and changes in water use efficiency. Projections that this trend will continue have a high degree of certainty. Contemporary water withdrawal is approximately 3,600 cubic kilometers per year globally or 25% of the continental runoff to which the majority of the population has access during the year. If dedicated instream uses for navigation, waste processing, and habitat management are considered, humans then use and regulate over 40% of renewable accessible supplies. Regional variations from differential development pressures and efficiency changes during 1960–2000 produced increases in water use of 15–32% per decade.

Four out of every five people live downstream of, and are served by, renewable freshwater services, representing 75% of the total supply. Because the distribution of fresh water is uneven in space and time, more than 1 billion people live under hydrologic conditions that generate no appreciable supply of renewable fresh water. An additional 4 billion (65% of world population) is served by only 50% of total annual renewable runoff that is positioned in dry to only moderately wet conditions, with concomitant pressure on that resource base. Only about 15% live with relative water abundance.

Forest and mountain ecosystems serve as source areas for the largest amounts of renewable freshwater supply—57% and 28% of total runoff, respectively. These ecosystems each provide renewable water supplies to at least 4 billion people, or two thirds of the global population. Cultivated and urban ecosystems generate only 16% and 0.2%, respectively, of global runoff, but because of their close proximity to human settlements, they serve 4–5 billion people. Such proximity is also associated with nutrient and industrial water pollution.

From 5% to possibly 25% of global freshwater use exceeds long-term accessible supply. Overuse implies delivery of freshwater services through engineered water transfers or nonrenewable groundwater supplies that are currently being depleted. Much of this water is used for irrigation with irretrievable losses in water-scarce regions. All continents record overuse. In the relatively dry Middle East and North Africa, non-sustainable use is exacerbated, with current rates of freshwater use equivalent to 115% of total renewable runoff. In addition, possibly one third of all withdrawals come from nonrenewable sources, a condition driven mainly by irrigation demand. Crop production requires enormous quantities of fresh water; consequently, many countries that aim at self-sufficiency in food production have entrenched patterns of water scarcity. Alternatively, crops can be traded on global food markets, with some countries accruing substantial benefits from importing "virtual water" that would otherwise be required domestically to irrigate crops.

The water requirements of aquatic ecosystems in the context of expanding human freshwater use results in competition for the same resources. Changes in flow regime, transport of sediments and chemical pollutants, modification of habitat, and disruption of migration routes of aquatic biota are some of the key consequences of this competition. In many parts of the world, competition for fresh water has produced impacts that fully extend to the coastal zone, with effects including oxygen depletion, coastal erosion, and harmful algal blooms. Through consumptive use and interbasin transfers, several of the world's largest rivers (the Nile, the Yellow, and the Colorado in the United States) have been transformed into highly stabilized and in some cases seasonally nondischarging river channels.

The supply of fresh water continues to be reduced by severe pollution from anthropogenic sources in many parts of the world. Over the past half-century, there has been an accelerated release of artificial chemicals into the environment. Inorganic nitrogen pollution of inland waterways, for example, has increased substantially, with nitrogen loads transported by the global system of rivers rising more than twofold over the preindustrial state. Increases of more than tenfold are recorded across many industrialized regions of the world. Many anthropogenic chemicals are long-lived and transformed into by-products whose behaviors, synergies, and impacts are for the most part unknown as yet. As a consequence of pollution, the ability of ecosystems to provide clean and reliable sources of fresh water is impaired. Severe deterioration in the quality of fresh water is magnified in cultivated and urban systems (high use, high pollution sources) and dryland systems (high demand for flow regulation, absence of dilution potential).

The demand for reliable sources of fresh water and flood control has encouraged engineering practices that have compromised the sustainability of inland water systems and their provision of freshwater services. Prolific dam-building (45,000 large dams and possibly 800,000 smaller ones) has generated both positive and negative effects. Positive effects on human well-being have included flow stabilization for irrigation, flood control, drinking water, and hydroelectricity. Negative effects have included fragmentation and destruction of habitat, loss of species, health issues associated with stagnant water, and loss of sediments and nutrients destined to support coastal ecosystems and fisheries.

Water scarcity is a globally significant and accelerating condition for 1–2 billion people worldwide, leading to problems with food production, human health, and economic development. A high degree of uncertainty surrounds these estimates, and defining water scarcity merits substantial further analysis in order to support sound water policy formulation and management. Rates of increase in a key water scarcity measure—water use relative to accessible supply—from 1960 to present averaged nearly 20% per decade globally, with values of 15% to more than 30% per decade for individual continents. Inequalities in level of economic development, education, and governance result in differences in coping capacity for water scarcity.

The annual burden of disease from inadequate water, sanitation, and hygiene totals 1.7 million deaths and the loss of at least 50 million healthy life years. Some 1.1 billion people lack access to safe drinking water and 2.6 billion lack access to basic sanitation. Investments in drinking water supply and sanitation show a close correspondence with improvement in human health and economic productivity. Each person needs only 20 to 50 liters of water free of harmful contaminants each day for drinking and personal hygiene to survive, yet there remain substantial challenges to providing this basic service to large segments of the human population. Half of the urban population in Africa, Asia, and Latin America and the Caribbean suffers from one or more diseases associated with inadequate water and sanitation.

The state of freshwater resources is inadequately monitored, hindering the development of indicators needed by decision-makers to assess progress toward national and international development commitments. Substantial deterioration of hydrographic networks is occurring throughout the world, increasing the difficulty of making an accurate assessment of global freshwater resources. The same is true for groundwater monitoring, standard water quality monitoring, and freshwater biological indicators. New techniques make it possible to identify literally thousands of chemicals, including long-lived synthetic pharmaceuticals, in freshwater resources. But universal application of these techniques is lacking, and there are no systematic epidemiological studies to understand their impact on long-term human well-being.

Trade-offs in meeting the Millennium Development Goals and other international commitments are inevitable. It is *very certain* that the condition of inland waters and coastal ecosystems has been compromised by the conventional sectoral approach to water management, which, if continued, will jeopardize human well-being. In contrast, the implementation of the established ecosystem-based approaches adopted by the Convention on Biological Diversity, the Convention on Wetlands, the Food and Agriculture Organization, and others could substantially improve the future condition of water-provisioning services by balancing economic development, ecosystem conservation, and human well-being objectives.

7.1 Introduction to Fresh Water as a Provisioning Service

This chapter provides a picture of the recent history and contemporary state of global freshwater provisioning services. It documents a growing dependence of human populations on these services, which has resulted in a variety of activities aimed at stabilizing and delivering water supplies. So effective has been the ability of water management to influence the state of this resource, in terms of both its physical availability and chemical character, that anthropogenic signatures are now evident across the global water cycle. Much of this influence is negative due to overuse and poor management. The capacity of ecosystems to sustain freshwater provisioning services is thus strongly compromised throughout much of the world and may continue to remain so if historic patterns of managed use persist.

7.1.1 Fresh Water in the MA Context

Within the MA conceptual framework (see Chapter 1), water is treated as a service provided by ecosystems as well as a system (inland waters). Because the water cycle plays so many roles in the climate, chemistry, and biology of Earth, it is difficult to define it as a distinctly supporting, regulating, or provisioning service. Precipitation falling as rain or snow is the ultimate source of water supporting ecosystems. Ecosystems, in turn, control the character of renewable freshwater resources for human well-being by regulating how precipitation is partitioned into evaporative, recharge, and runoff processes. Together with energy and nutrients, water is arguably the centerpiece for the delivery of ecosystem services to humankind (Falkenmark and Folke 2003).

While recognizing the role of water in supporting and regulating services, the placement of this chapter among other provisioning services is done from a practical point of view, in part because water resources are the most tangible and well-documented aspect of this broader spectrum of freshwater services. This chapter assesses the condition and recent trends in global freshwater resources, examining the amount and condition of renewable and nonrenewable surface and groundwater supplies, changes in these supplies over time and into the near future, and the impacts on human well-being of changes in the service. Chapter 20 examines the role of inland water ecosystems that provide a multitude of services, including water, fish, habitat, cultural and aesthetic values, and flood prevention. Because fresh water is so essential to life on Earth, its assessment overlaps with services and ecosystem chapters across the MA.

Throughout this chapter reference is made to summary statistics on the fresh water associated with specific ecosystems. While ecosystems are strongly dependent on the water cycle for their very existence, at the same time these systems represent domains over which precipitation is processed and transferred back to the atmosphere as "green water" (through evapotranspiration drawn from soils and plant canopies in natural ecosystems and rain-fed agriculture). The remainder runs off as "blue water" which constitutes the renewable water supply that can pass to downstream users—both aquatic ecosystems and humans such as farmers who irrigate. These water flows can be tabulated across ecosystems to identify areas that are critical to human well-being as well as those that require particular attention in designing strategies for environmental protection. Box 7.1 defines key terms used in this analysis.

7.1.2 Setting the Stage

Prior to the twentieth century, global demand for fresh water was small compared with natural flows in the hydrologic cycle. With population growth, industrialization, and the expansion of irrigated agriculture, however, demand for all water-related goods and services has increased dramatically, putting the ecosystems that sustain this service, as well as the humans who depend on it, at risk. While demand increases, supplies of clean water are diminishing due to mounting pollution of inland waterways and aquifers. Increasing water use and depletion of fossil groundwater adds to the problem. These trends are leading to an escalating competition over water in both rural and urban areas. Particularly important will be the challenge of simultaneously meeting the food demands of a growing human population and expectations for an improved standard of living that require clean water to support domestic and industrial uses.

Meeting even the most basic of needs for safe drinking water and sanitation continues to be an international development priority. Some 1.1 billion people lack access to clean water supplies and more than 2.6 billion lack access to basic sanitation (WHO/UNICEF 2004). Reducing these numbers is a key development priority. By adopting the initial targets of the Millennium Development Goals, governments around the world have made a commitment to reduce by half the proportion of people lacking access to clean water supply and basic sanitation between 1990 and 2015.

The ministerial declaration from the 2nd World Water Forum in The Hague in 2000 captured the essence of the goals and challenges faced (see Box 7.2), including articulation of the importance of ecosystems in sustaining freshwater services. Water continues to rise in importance in major policy circles, with 2003 declared the International Year of Fresh Water, release of the first World Water Development Report (UN/WWAP 2003) by a collaboration of 24 U.N. agencies through the World Water Assessment Programme, and proclamation by the UN General Assembly of the International Decade of Action "Water for Life" in 2005–15.

Societies have benefited enormously through their use of fresh water. However, due to the central role of water in the Earth system, the effects of modern water use often reverberate throughout the water cycle. Key examples of human-induced changes include alteration of the natural flow regimes in rivers and waterways, fragmentation and loss of aquatic habitat, species extinction, water pollution, depletion of groundwater aquifers, and "dead zones" (aquatic systems deprived of oxygen) found in many inland and coastal waters. Thus, trade-offs have been made—both explicitly and inadvertently—between human and natural system requirements for freshwater services.

The challenge for the twenty-first century will be to manage fresh water to balance the needs of both people and ecosystems, so that ecosystems can continue to provide other services essential for human well-being. Human impacts on the capacity of ecosystems to continue delivering freshwater services are assessed in

BOX 7.1
Operational Definitions of Key Terms on Fresh Water

The global water cycle involves major transports that link Earth's atmosphere, land mass, and oceans, though the emphasis in this chapter is on the continental hydrologic cycle. The Figure here outlines the major fluxes of fresh water, which help to define the renewable supplies on which humans and ecosystems depend. The water cycle can be divided into a portion that is accessible to humans and that which is not. The portion of the global water cycle that is accessible to humans is shown in the diagram. The following nomenclature is used throughout this chapter.

Total Precipitation (P_t). This term is equivalent to the total sustainable water supply falling as rain and snow over the terrestrial portion of Earth. P_t represents the ultimate source of fresh water for recharge into soils, evaporation, and transpiration by plants in natural and cropped ecosystems, recharge into groundwaters, and, eventually, runoff and discharge through river corridors. For the purposes of this study, P_t represents climatic means, unless otherwise noted. P_t can be divided into precipitation that is accessible (P_a) or inaccessible (P_i) to humans on the land mass. Ocean precipitation is denoted as P_o.

Total Blue Water Flow (B_t). This term represents the global renewable water supply computed as surface and sub-surface runoff. "Total" here refers to "blue water" that is both accessible and inaccessible to humans. It is a subcomponent of P_t representing the net fresh water remaining after accounting for evapotranspiration (ET) losses to the atmosphere from the soils and vegetation of natural ecosystems and rain-fed agriculture, known as "green water" (G_t). Blue water represents the sustainable supply of fresh water that emanates from ecosystems and is then transferred through rivers, lakes, and other inland aquatic systems. These downstream ecosystems evaporate and consume water (C_{iws}) and reduce blue water flows. In basins occupied by humans, accessible blue water (B_a) is further reduced (B_a') through consumptive losses (C_a) from water resource management, such as irrigation.

Water Use (U_a). This represents water withdrawn or used by humans. U_a is derived from either accessible blue water flows (B_a) or nonrenewable sources, predominantly fossil groundwater mining, which constitutes a non-sustainable water use. Use is divided into domestic (D_a), industrial (I_a), and agricultural (A_a) applications, a part of which can be returned to inland water systems, though sometimes degraded in its quality in such return flows.

Water Consumption (C_a). The portion of water that is lost as net evapotranspiration after being withdrawn from an accessible supply source (U_a). Such losses are associated predominantly with irrigation, and emerge from both renewable and nonrenewable freshwater supplies. C_a is also referred to as irretrievable losses. While humans "consume" water directly for drinking, this is not termed water consumption but simply a component of domestic water use tabulated under U_a.

Non-sustainable Water Use (U_{an}). This is computed by comparing total water demand or withdrawals for human use (U_a) to the available renewable water supply (B_a). Where U_a exceeds B_a at the point of extraction, non-sustainable use is tabulated. For most parts of the planet, this will refer to the "mining" of groundwaters, especially in arid and semiarid areas, where recharge rates to the underground aquifer are limited. U_{an} can also embody the interbasin transport of fresh water from water rich to water poor areas.

Environmental Flows. These are the water requirements needed to sustain freshwater ecosystems.

Water Abundance and Scarcity. The conjunction of renewable freshwater supply, withdrawals, consumptive losses, and level of development can be used to define quantitative measures of water abundance or scarcity. The number of people supported on a unit of renewable freshwater flows (the "water crowding" index) will define thresholds of chronic water scarcity, as will use-to-supply ratios (U_a/B_t or U_a/B_a).

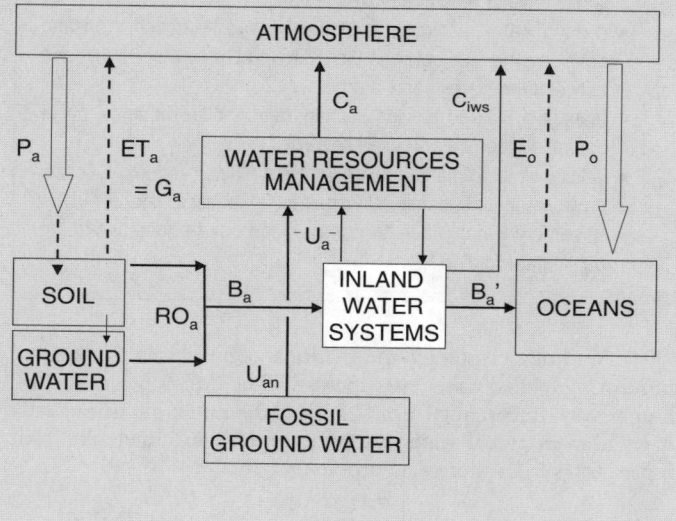

Chapter 20. Some options on balancing human and ecosystem water requirements are discussed in Chapter 7 of the MA *Policy Responses* volume.

Before describing the details of this chapter's assessment, a word is in order on the quality of information on which it is based. Monitoring the continental water cycle in a timely manner at the global scale using traditional discharge gauging stations—the mainstay of water resource assessment—continues to challenge the water sciences (IAHS 2001; NRC 1999; Kanciruk 1997). Data collection is now highly project–oriented, yielding often poorly integrated time series of short duration, restricted spatial coverage, and limited availability. In addition, there has been a legal assault on the open access to basic hydrometeorological data sets, aided in large measure by commercialization and fears surrounding piracy of intellectual property. Delays in data reduction and release (up to several years in some places) are also prevalent. Much information has yet to be digitized, and exists in difficult–to–use book and report formats.

Based on available global archives at the WMO Global Runoff Data Center, to which member states contribute voluntarily, there was arguably a better knowledge of the state of renewable surface water supplies in 1980 than today. Such statements apply to many parts of the world, including otherwise well monitored countries like the United States and Canada (IAHS 2001; Shiklomanov et al. 2002), though most marked declines are in the developing world. Our understanding of groundwater resources is even more limited, since well–log, groundwater discharge/ recharge, and aquifer property data for global applications are only beginning to be synthesized (Foster and Chilton 2003; UNESCO-IHP 2004). Information on water use and operation of infrastructure has never been assembled for global analysis (IAHS 2001; Vörösmarty and Sahagian 2000).

While remote sensing and models of the water cycle can be used to fill some data gaps, these approaches themselves produce a range of outputs arising from differences in their input data streams and detailed calculation procedures (e.g., Fekete et al.

BOX 7.2

Ministerial Declaration from the 2nd World Water Forum

The ongoing series of World Water Forums (Marrakech 1997, Hague 2000, Kyoto 2003, Mexico 2006), organized by the World Water Council and its partners, brings together a broad array of thousands of stakeholders to discuss strategies for sustainable development with respect to water. While there have been three such gatherings to date, outputs from the affiliated Ministerial Conference of the 2nd Forum are most relevant to the MA. This Ministerial Declaration captures the interconnections among ecosystem integrity, human actions affecting water supply, and human well-being. It is precisely these interactions that define the contemporary conditions and trends and that are suggestive of responses that foster water stewardship, sustainable water use, and progress toward development. These fundamental goals highlight the need for well-functioning ecosystems. They also reflect strongly the Millennium Development Goals:

- meeting basic human needs—that is, access to safe and sufficient water and sanitation, which are essential to health and human well-being;
- securing the food supply to enhance food security through a more efficient mobilization and use of water for food production;
- protecting ecosystems and ensuring their integrity through sustainable water resources management;
- sharing water resources to promote peaceful cooperation and develop synergies between the different uses of water within and between the states concerned;
- managing risks to provide security from floods, droughts, pollution, and other water-related hazards;
- valuing water to manage it in a way that reflects economic, social, environmental, and cultural values for all its uses; and
- governing water wisely to ensure good governance, including public participation.

2004). Without a sustained international commitment to baseline monitoring, global water assessments will be difficult to make and fraught with uncertainty. Box 7.3 gives the range of current estimates used in global water resource models, an uncertainty that in part arises from these data problems.

7.2 Distribution, Magnitude, and Trends in the Provision of Fresh Water

While it is true that there is an abundance of water across blue planet Earth, only a small portion of it exists as fresh water, and even a smaller fraction is accessible to humans. Nearly all water on Earth is contained in the oceans, leaving only 2.5% as fresh water. (See Table 7.1.) Of this small percentage, nearly three quarters is frozen, and most of the remainder is present as soil moisture or lies deep in the ground. The principal sources of fresh water that are available to society reside in lakes, rivers, wetlands, and shallow groundwater aquifers—all of which make up but a tiny fraction (tenths of 1%) of all water on Earth. This amount is regularly renewed by rainfall and snowfall and is therefore available on a sustainable basis.

Global averages fail to portray a complete picture of the world's water resource base, however. The basic climatology of the planet dictates that fresh water will be distributed unevenly around the globe, with abundant supplies across zones like the wet tropics and absolute water scarcity across the desert belts and in the rain shadow of mountains. For this assessment, both locally available runoff and water transported though river networks is considered (Vörösmarty et al. 2005). River corridor flows convey essential water resources to those living on the banks of large rivers, such as along the lower Nile. Figure 7.1 (in Appendix A) shows the broad range of sustainable water resources (blue water flows), which varies from essentially zero in many arid and semi-arid regions to hundreds and thousands of cubic kilometers per year as major river corridor flow. Such regional differences in the quantity of available fresh water establish the diverse patterns of water supply across the globe.

The supply of fresh water is conditioned by several additional factors, which amplify the patterns of abundance and scarcity. These factors include the distribution of humans relative to the supply of water (that is, access to water), patterns of demand, presence of water engineering to stabilize flows, seasonal and interannual climate variations, and water quality. The following sections assess the state of global freshwater supplies, demands (withdrawals or use), and water quality. The time domain covered here is the last several decades and into the near future of 2010–15.

7.2.1 Available Water Supplies for Humans

Estimates of global water supply are imprecise and complicated by several factors, including differences in data and methodologies used, loss of hydrographic monitoring capacity, alternative time frames considered, and distortions from land cover, climate, and hydraulic engineering that are increasingly a part of the water cycle. The renewable resource base expressed as long-term mean runoff has been estimated to fall between 33,500 and 47,000 cubic kilometers per year (Korzoun et al. 1978; L'vovich and White 1990; Gleick 1993; Shiklomanov and Rodda 2003; Fekete et al. 2002; Nijssen et al. 2001; Döll et al. 2002). Within-year variations also define the basic nature of water supply. At the continental scale, maximum-to-minimum runoff ratios vary between 2:1 and 10:1 (Shiklomanov and Rodda 2003), with individual rivers experiencing ratios far higher, such as in snowmelt-dominated basins or episodically flooded arid and semiarid river systems. These variations necessitate flow stabilization through hydraulic engineering for either protection (for example, from floods) or seasonal supply augmentation (for example, for dry-season agriculture or hydroelectricity).

Water supply can also be assessed from the standpoint of societal access to renewable runoff and river flow, from which humans can secure provisioning services. By one estimate (Postel et al. 1996), one third of global renewable water supply is accessible to humans, when taking into account both its physical proximity to population and its variation over time, such as when flood waves pass uncaptured on their way to the ocean. Such accessibility is considered as part of this assessment later in this chapter.

Groundwater plays an important role in water supply. It has been estimated that between 1.5 billion (UNEP 1996) and 3 billion people (UN/WWAP 2003) depend on groundwater supplies for drinking. It also serves as the source water for 40% of self-supplied industrial uses and 20% of irrigation (UN/WWAP 2003). For certain countries this dependency is even greater; for example, Saudi Arabia meets nearly 100% of its irrigation requirements through groundwater (Foster et al. 2000). Two important classes of groundwater can be identified. The first is renewable groundwater resources, closely linked to the cycling of fresh water, through which the ground is periodically replenished when sufficient precipitation is available to recharge soils or when floodplains become inundated. The second, fossil groundwater, is

BOX 7.3

Uncertainties in Estimates of Contemporary Freshwater Services, Use, and Scarcity

All entries are ranges in the units indicated and represent near-contemporary conditions.

Geographic Region	Renewable Water Supply[a]	Total Withdrawals	Mean Water Crowding	Mean Use-to-Supply (U_a/B_t) Ratio	Population with U_a/B_t Ratio Greater than 40%
	(cu. km. per year)		*(people/mill. m³/yr)*	*(percent)*	*(million)*
Asia	7,850–9,700	1,520–1,790	320–384	16–22	712–1,200
Former Soviet Union	3,900–5,900	270–380	48–74	6–8	56–110
Latin America	11,160–18,900	200–260	25–42	1–2	84–160
North Africa/Middle East	300–367	270–370	920–1,300	74–108	91–240
Sub-Saharan Africa	3,500–4,815	60–90	115–160	2–2	16–140
OECD	7,900–12,100	920–980	114–129	8–12	164–370
World Total	**38,600–42,600**	**3,420–3610**	**133–150**	**8–9**	**1,123–2,100**

[a] For the purpose of this intercomparison, supply is total supply (B_t). See also Box 7.1 and Table 7.2.

The ranges reported here are from three global-scale water resource models, two of which were used directly in the MA: University of New Hampshire (Vörösmarty et al. 1998a; Fekete et al. 2002; Federer et al. 2003) for the Condition and Trends Working Group assessment and Kassel University (Alcamo et al. 2003; Döll et al. 2003) used in the Scenarios Working Group. A third model from the University of Tokyo and Global Soil Wetness Project (Oki et al. 2001, 2003b; Dirmeyer et al. 2002) was also compared.

The global-scale correspondence for total supply, withdrawals, water crowding, and demand-to-supply ratio is high, but masks continental-scale differences. Such disparities can be large, as for water supply in Latin America, where large remote tropical river systems have proved difficult to monitor systematically. Substantial differences at the continental scale are noted for population living under severe water scarcity (use-to-supply >40%). The order-of-magnitude range apparent for sub-Saharan Africa can be linked in part to the distribution of sharp climatic gradients,that are difficult to analyze geographically. The result is also a function of the assumptions made regarding access to water. Because of such uncertainties, the current state-of-the-art in global models put 1–2 billion people at risk worldwide arising from high levels of water use. The MA models predict a much smaller range, from 2.0–2.1 billion.

Large uncertainties surround current estimates of water consumption by the largest user of water, agriculture. Recent estimates vary from 900 (Postel 1998) up to 2000 cubic kilometers per year (Shiklomanov and Rodda 2003). A value of 1200 cubic kilometers per year is reported in this assessment (Table 7.4).

Table 7.1. Major Storages Associated with the Contemporary Global Water System (Shiklomanov and Rodda 2003)

Type	Volume	Fraction of Total Volume	Fraction of Fresh Water
	(thous. cu. km.)	*(percent)*	*(percent)*
World ocean	1,338,000	96.5	–
Groundwaters	23,400	1.7	–
–Fresh	10,530	0.76	30.1
Soil moisture	16.5	0.001	0.05
Glaciers/permanent ice	24,100	1.74	68.7
Ice in permafrost	300	0.022	0.86
Lakes (fresh)	91	0.007	0.26
Wetlands	11.5	0.0008	0.03
Rivers	2.12	0.0002	0.006
Biological water	1.12	0.0001	0.003
Atmosphere	12.9	0.001	0.04
Total hydrosphere	1,386,000	100	–
Total fresh water	35,029	2.53	100

typically locked in deep aquifers that often have little if any long-term net recharge. Whenever this is extracted, it is functionally "mined," a particularly acute problem in arid regions, where replenishment times can be on the order of thousands of years (Margat 1990a, 1990b).

Establishing the contribution of groundwater to the global supply of freshwater inserts a substantial element of uncertainty into the overall assessment. Problems of poor data harmonization, incomplete and fragmentary inventories, and methodological difficulties are well documented (Revenga et al. 2000; UN/WWAP 2003; Morris et al. 2003). As a result, there is large uncertainty in estimates of fresh groundwater resources, ranging from 7 million to 23 million cubic kilometers (UN/WWAP 2003; Morris et al. 2003). While abundant, their use can be severely restricted by pollution (Foster and Chilton 2003) or by the cost of extracting water from aquifers, which rises progressively in the face of extraction rates exceeding recharge (Dennehy et al. 2002).

Another important water supply is represented by the widespread construction of artificial impoundments that stabilize river flow. Today, approximately 45,000 large dams (>15 meters high or between 5 and 15 meters high and a reservoir volume of more than 3 million cubic meters) (WCD 2000) and possibly 800,000 smaller dams (McCully 1996; Hoeg 2000) have been built for municipal, industrial, hydropower, agricultural, and recreational water supply and for flood control. Recent estimates place the volume of water trapped behind documented dams at 6,000–7,000 cubic kilometers (Shiklomanov and Rodda 2003; Avakyan and Iakovleva 1998; Vörösmarty et al. 2003). In drainage basins regulated by large reservoirs (>0.5 cubic kilometers) alone, one third of the mean annual flow of 20,000 cubic kilometers is stored (Vörösmarty et al. 2003). Assuming seasonal six-month low flows constitute roughly 40% of annual discharge (Shiklomanov and Rodda 2003), this impounded water represents a global potential to carry over an entire year's minimum flows.

Desalinization constitutes a renewable water supply using distillation and membrane techniques to withdraw salt from otherwise unusable water. While the technology continues to improve, desalinization remains the most costly means of supplying fresh water and is highly energy-intensive (Gleick 2000). Costs range between $1 and $4 per cubic meter, placing it well above the most expensive traditional sources (Gleick 2000). Despite this, in 2002 there were over 10,000 desalinization plants in 120 countries supplying more than 5 cubic kilometers per year, with a global market of $35 billion per year (UN/WWAP 2003). Collectively, these plants provide for much less than 1% of global freshwater use.

More than 70% of global installed desalinization capacity is in the oil-rich states of the Middle East and North Africa (UN/WWAP 2003). While its use may be difficult to justify for high-water-consumptive activities like irrigation, investments in desalinization technologies are likely to improve efficiency and bring down costs, creating a potentially important source at least for domestic drinking water (Gleick 2000), and the annual supply of desalinized water could double in 15 years (UN/WWAP 2003). The unresolved issue of adequately managing brine waste from the desalinization process to protect nearby coastal ecosystems requires special attention.

Finally, rainwater harvesting through traditional methods or modern technology is another way in which humans augment freshwater supply. Rainwater harvesting can directly increase the soil water content or be stored for later application as supplemental irrigation during dry periods. This is particularly important in places like India, which relies heavily on a short period of intense rainfall (WWC 2000). The groundwater authorities in India, for instance, have made it mandatory for multistoried buildings in New Delhi and several other states to have a rooftop rainwater harvesting system (Hindustan Times, Patna, September 2002). Rainwater harvesting can also be an appropriate technology for maintaining groundwater base flow and reducing flood peaks. (See MA *Policy Responses*, Chapter 7, for further discussion.)

7.2.1.1 Total Flows of Fresh Water

Ecosystems vary greatly in their exposure to precipitation and hence as source areas for renewable runoff that emerges as part of the hydrologic cycle. (See Table 7.2.) The proportional contribution of each ecosystem to global runoff is generally equivalent to the fraction of precipitation to which it is exposed. Forests therefore are associated with slightly more than half of global precipitation and yield about half of global runoff, while mountains represent one quarter of both global precipitation and runoff. Cultivated and island systems are the next most important source areas, each constituting about 15% of global runoff. All other systems contribute 10% or less. Paradoxically, dryland ecosystems, due to their large aerial extent, receive a nearly identical fraction of global precipitation as mountains do, yet because of substantial losses from the system due to evapotranspiration, they are a relatively minor contributor to global renewable water supply (<10%). Urban systems, because of their restricted extent (<<1% of land area), receive only 0.2% of global precipitation and provide the same very minor proportion of global runoff.

From a regional perspective, Latin America is most water-rich, with about one third of global runoff. Asia is next, with one quarter of global runoff, followed by OECD (20%), and sub-Saharan Africa and the former Soviet Union, each with 10%. The Middle East and North Africa is clearly driest and most water-limited, accounting for only 1% of global runoff.

7.2.1.2 Freshwater Flows Accessible to Humans

Ecosystems constitute the ultimate source areas for freshwater provisioning services. The accessibility of renewable water supply can be estimated through an index measuring the proportion of total annual renewable runoff generated locally that eventually flows through river corridors and encounters downstream human populations. The importance of upstream ecosystems as source areas for freshwater supply is demonstrated in Table 7.2. Cultivated, coastal, and urban systems, with sizable fractions of the global population, have from 90% to 100% of their renewable runoff accessible. Drylands also show high accessibility, likely reflecting the propensity of humans to settle near scarce freshwater resources. Mountains, forests, and inland waters each show 70–80% of total runoff as accessible to downstream populations. The exception is polar systems, which yield less than 20% of total runoff as accessible, reflecting their remote and generally uninhabited environment.

Populations served by accessible runoff emerging from individual ecosystems are typically in the billions. Cultivated systems, forests, inland waters, and mountains each serve at least 4 billion people. Four fifths of the world lives downstream of runoff from cultivated lands, followed by a nearly identical fraction downstream from forests. Inland waters and mountains provide water to two thirds of global population and drylands to one third. Remote islands and polar systems serve the fewest people. Runoff from urban systems, nearly all generated in close proximity to densely settled areas, serves nearly three quarters of the world's population.

The large fractions of total runoff expressed as accessible runoff indicate that, by and large, human society has positioned itself into areas with identifiable local sustainable water supplies or river corridor flows. A geographic distribution of human settlement thus is linked to the availability of fresh water (see also Meybeck et al. 2001). The global geography of accessible runoff, expressed in units of dependent population per unit of delivered flow, was shown in Figure 7.1. Mountains serve 3 times, forests 4 times, and inland waters 12 times as many people downstream through river corridors as they do through locally derived runoff. Urban areas nearly double the total service when tabulating downstream populations. Remaining ecosystems show more-limited importance in transferring precipitation as accessible runoff to downstream populations. For drylands, this is due to a lack of substantial quantities of runoff, while for coastal or island systems it is a consequence of short flow pathways to the ocean. Each of these systems still supplies 15–30% of global population with renewable and accessible runoff.

From a regional perspective, Latin America and Asia constitute the largest proportion (together nearly 60%) of global accessible runoff. And while the OECD, sub-Saharan Africa, and the former Soviet Union generate a large portion of the global runoff, substantial quantities are remote and inaccessible particularly in the former Soviet states (see also Postel et al. 1996). The Middle East and North Africa generates less than 1% of renewable accessible runoff.

Overall, the global fraction of total annual runoff that is accessible to humans is 75%, with slightly more than 80% of world

Table 7.2. Estimates of Renewable Water Supply, Access to Renewable Supplies, and Population Served by the Provision of Freshwater Services, Year 2000 Condition (computed based on methods in Vörösmarty et al. 2005; renewable water supply estimates from Fekete et al. 2002 from simulated water budgets using climatology data from 1950–96)

System[a] or Region	Area	Total Precipitation (P_t)	Total Renewable Water Supply, Blue Water Flows (B_t)	Renewable Water Supply, Blue Water Flows, Accessible to Humans[b] (B_a)	Population Served by Renewable Resource[c]
	(mill. sq. km.)		*thousand cubic kilometers per year*		(billion)
			[percent of global runoff]	*[percent of B_t]*	*[percent of world population]*
MA System					
Forests	41.6	49.7	22.4 [57]	16.0 [71]	4.62 [76]
Mountains	32.9	25.0	11.0 [28]	8.6 [78]	3.95 [65]
Drylands	61.6	24.7	3.2 [8]	2.8 [88]	1.90 [31]
Cultivated[d]	22.1	20.9	6.3 [16]	6.1 [97]	4.83 [80]
Islands	8.6	12.2	5.9 [15]	5.2 [87]	0.79 [13]
Coastal	7.4	8.4	3.3 [8]	3.0 [91]	1.53 [25]
Inland Water	9.7	8.5	3.8 [10]	2.7 [71]	3.98 [66]
Polar	9.3	3.6	1.8 [5]	0.3 [17]	0.01 [0.2]
Urban	0.3	0.22	0.062 [0.2]	0.062 [100]	4.30 [71]
Region					
Asia	20.9	21.6	9.8 [25]	9.3 [95]	2.56 [42]
Former Soviet Union	21.9	9.2	4.0 [10]	1.8 [45]	0.27 [4]
Latin America	20.7	30.6	13.2 [33]	8.7 [66]	0.43 [7]
North Africa/Middle East	11.8	1.8	0.25 [1]	0.24 [96]	0.22 [4]
Sub-Saharan Africa	24.3	19.9	4.4 [11]	4.1 [93]	0.57 [9]
OECD	33.8	22.4	8.1 [20]	5.6 [69]	0.87 [14]
World Total	**133**	**106**	**39.6 [100]**	**29.7 [75]**	**4.92 [81]**

[a] Note double-counting for ecosystems under the MA definitions.
[b] Potentially available supply without downstream loss.
[c] Population from Vörösmarty et al. 2000.
[d] For cultivated systems, estimates are based on cropland extent from Ramankutty and Foley 1999 within this MA reporting unit.

population (4.9 billion people) being served by these renewable and accessible water flows. However, while providing an estimate of long-term water supply, these figures overstate the effective availability of fresh water. Given that approximately 30% of annual runoff is uncaptured flood flow (Shiklomanov and Rodda 2003), the world's population has its access reduced from 75% to 53% of total runoff.

Globally, renewable freshwater services reflect the geographic distributions of both water supply and human populations. Four out of every five people live downstream of and are served by renewable freshwater services. (See Figure 7.2.) Thus, while the human population is generally well organized with respect to the availability of fresh water, 20% of humanity remains without any appreciable quantities of sustainable supply or must gain access to such resources through costly interbasin transfers from more water-rich areas. (See also Table 7.2.) These people are highly reliant on unsustainable water resources. For those with access to renewable supplies, a total of 65% of the world's population is served by the 50% of total annual renewable runoff that is positioned in dry to moderately wet conditions, with concomitant pressure on that resource base. Only 15% live with relative water abundance—that is, in conjunction with the remaining 50% of total runoff (represented by the high runoff-producing regions shown in the upper part of the curve in Figure 7.2). If uncaptured flood flow is incorporated into these calculations, for the 80% of

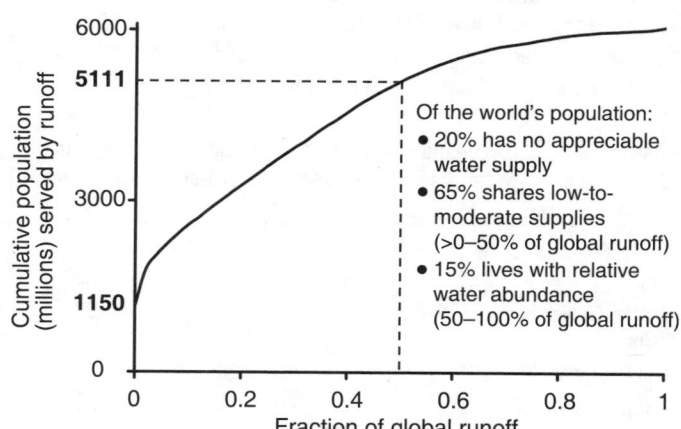

Figure 7.2. Cumulative Distribution of Population with Respect to Freshwater Services, 1995–2000. Fraction of runoff is ranked from low to high based on mean annual conditions. This distribution is also affected by seasonal variations in available runoff.

world population who reside in the lower half of the water availability spectrum in Figure 7.2 (65% plus 15% with no appreciable renewable freshwater flows), the effective supply is reduced from 50% to 35% of total runoff.

7.2.2 Water Use

Over the last few centuries, global water use has shown roughly an exponential growth and been linked closely to both population growth and economic development. There was a fifteenfold increase in global water withdrawals between 1800 and 1980 (L'vovich and White 1990), when population increased by a factor of four (Haub 1994). Since the 1900s, the overall increase has been sixfold (WMO 1997). Global consumptive water losses, primarily from evapotranspiration through irrigation, increased thirteenfold during this same period. A major, recent feature of human water use is the reduction in per capita use rates, dropping as of around 1980 from about 700 to 600 cubic meters per year, though the aggregate global withdrawal continues to increase (Gleick 1998; Shiklomanov and Rodda 2003).

While the general features of a historical rise in freshwater demands are clear, there are substantial uncertainties surrounding water use estimates, reflecting the current state of knowledge, assumptions (or lack thereof) on potential efficiency changes and reuse potential, number of years projected into the future, and interactions with market forces (Gleick 2000; Shiklomanov and Rodda 2003). The summary statistics from three global tabulations provided earlier, in Box 7.3, demonstrate the current degree of uncertainty.

Global water withdrawals today total about 3,600 cubic kilometers per year, with a wide range of use over individual continents. (See Table 7.3.) The largest user is Asia, accounting for nearly half of the world total, with OECD next, using about one third. The remaining continents each represent less than 10% of global use. Water use today is dominated by agricultural withdrawals (70% of all use), followed by industrial and then domestic applications. Withdrawals in agriculture are fundamentally defined by irrigation. In Asia, the Middle East and North Africa, and sub-Saharan Africa, agriculture accounts for 85–90% of all withdrawals. Driven by irrigation demand, overall withdrawals across MENA constitute 120% of renewable accessible supplies,

meaning that this region relies on nonrenewable supplies for food production. Agricultural water use in the former Soviet Union and the OECD is proportionally much lower, reflecting the water needs of other sectors in these industrial economies. In contrast, industrial water use is only 4% in sub-Saharan Africa, reflecting a low level of economic development.

Water lost from groundwater and surface water sources to the atmosphere through net evaporation (such as from irrigation, cooling towers, or reservoirs) is termed water consumption or irretrievable losses, which today represent a substantial fraction of water use. Contemporary irretrievable losses through irrigation, computed as the evapotranspiration component of agricultural withdrawals, are assessed here. (See Table 7.4.) Irretrievable losses from irrigation represent one third of all water use globally. The efficiency computed for irrigated agriculture (the ratio of water withdrawn to water consumed or lost through evapotranspiration on irrigated cropland) is on average 50% globally and varies from 25% (in Latin America) to 60% (in Asia). Additional losses from evaporation from reservoirs, irrigation ditches, and so on are difficult to estimate accurately but could total over 500 cubic kilometers per year (Postel 1998), thus indicating the conservative nature of the consumption estimates in Table 7.4. (See Box 7.3 earlier in this chapter for the range in current estimates of consumptive loss from irrigation.)

Non-sustainable water use could be a substantial component of total withdrawals. Earlier work based on documentary evidence showed approximately 200 cubic kilometers per year of global aquifer overdraft (Postel 1999; WWC 2000), though the estimate is regarded as highly uncertain (Foster 2000). This assessment of water supply and use (based on Vörösmarty et al. 2000, 2005; Fekete et al. 2002) using a geospatial framework (about 50-kilometer resolution) enables calculations to be made of the degree to which water withdrawal exceeds locally accessible supplies—in other words, non-sustainable water use (U_{an}). Worldwide, non-sustainable withdrawals can be computed using two endpoints: crop evaporative demands or water use statistics, which include both consumption and transport losses, some unknown fraction of which reenters the surface-groundwater system for potential reuse (Molden 2003). These endpoints give a calculated non-sustainable use of about 400–800 cubic kilometers per year. In terms of total freshwater withdrawals, 10–25% could represent nonrenewable use. When the earlier estimate of 200 cubic kilometers per year is also included, a large degree of uncertainty results, and from 5% to 25% of freshwater withdrawals could represent nonrenewable use.

Nevertheless, each of these estimates reflects a high dependence on existing water services, especially in areas where induced, chronic water stress necessitates costly water engineering remedies, groundwater depletion, or curtailment of water-using activities. Each continent shows a heavy reliance on such nonrenewable extraction, ranging up to one third of total use based on the high estimates. Asia and MENA show the greatest level of such dependence; OECD, the least. In MENA, 30% of all water use is from non-sustainable sources, and this use is equivalent to over one third of accessible renewable supplies.

Figure 7.3 (in Appendix A) shows the contemporary geography of such non-sustainable use and demonstrates the much larger impacts that arise at subcontinental scales. The summary in Table 7.4 may thus understate the true degree of this overconsumption locally. The spatial pattern of overuse is broadly consistent with previously reported regions of use exceeding supply, major water transfer schemes, or groundwater overdraft: Australia, western Asia, northern China, India, North Africa, Pakistan, Spain, Turkey, and the western United States (Muller 2000; Shah et al. 2000;

Table 7.3. Freshwater Services Tabulated as Withdrawals for Human Use over MA Regions and the World, 1995–2000 (WRI et al., 1998, updated using Shiklomanov and Rodda 2003, as in Vörösmarty et al. 2000; resampled to MA reporting units)

MA Geographic Region	Domestic Water Use D_a	Industrial Water Use I_a	Agricultural Water Use A_a	Total Use (Withdrawals) U_a
	(cu. km. per year)			
Asia	80	99	1,373	1,550
Former Soviet Union	34	115	188	337
Latin America	33	31	205	269
North Africa/ Middle East	22	15	247	284
Sub-Saharan Africa	10	4	83	97
OECD	149	489	384	1,020
Global Total	**328**	**753**	**2,480**	**3,560**

Table 7.4. Consumptive and Non-sustainable Freshwater Use over MA Regions and the World, 1995–2000. Renewable supplies calculated as for Table 7.2. Irrigated water consumption was computed over irrigation-equipped land (Döll and Siebert 2000) within the cropland domain depicted by Ramankutty and Foley (1999). Evapotranspiration losses from irrigated cropland (Vörösmarty et al. 1998; Federer et al. 2003) relative to available local runoff or, when available, river corridor flows determine non-sustainable use. See Figure 7.3 for geography of non-sustainable use.

Geographic Region	Consumptive Losses from Irrigated Agriculture (C_a)	Consumptive Losses from Irrigated Agriculture	Non-sustainable Water Use[a] (U_{an})	Non-sustainable Water Use	Non-sustainable Water Use as Share of Agricultural Water Use	Non-sustainable Water Use as Share of Total Water Use
	(cu. km./ year)	*(percent of agricultural use/total use)*	*(cu. km./year)*	*(percent of accessible renewable supplies)*	*(percent)*	*(percent)*
Asia	811	59 / 52	295–543	3–6	21–40	19–35
Former Soviet Union	78	41 / 23	20–58	1–3	11–31	6–17
Latin America	49	24 / 18	8–37	<0.1–0.4	4–18	3–14
North Africa/ Middle East	94	38 / 33	25–86	10–36	10–35	9–30
Sub-Saharan Africa	33	39 / 34	10–18	0.2–0.4	12–22	10–19
OECD	141	37 / 14	31–88	0.5–2	8–23	3–9
World Total	**1,210**	**49 / 34**	**391–830**	**1–3**	**16–33**	**11–23**

[a] Range represents crop demand alone (low estimate) versus reported withdrawals (high estimate, which includes delivery loss; Table 7.3). Recycling within river basins of irrigation withdrawals that are not consumed by crops reduces, to some unknown degree, the high estimate (see Molden 2003). Calculations assume a maximum 75-kilometer buffer around river corridors from which irrigation areas can secure fresh water.

Vörösmarty and Sahagian 2000; Dennehy et al.2002; EEA 2003; MDBC 2003; NLWRA 2004).

Non-sustainable use expressed as a proportion of irrigated agricultural withdrawals shows an even higher degree of dependency on nonrenewable supplies. Globally, about 15–35% of irrigation withdrawals are computed to be non–sustainable. Individual continental areas show percentages ranging from less than 10% to 40%, as in the case of Asia. Such high rates indicate an increasing degree of food insecurity. Given projections showing no major expansion in global cropland area (Bruinsma 2003), increasing pressure will be placed on irrigated cropland, which today provides nearly 40% of crop production (Shiklomanov and Rodda 2003; UN/WWAP 2003). By its very nature, this water use cannot persist indefinitely, and many regions of the world have well-documented cases of aquifer depletion and abandonment of irrigation, adding constraints to irrigated crop production arising from rising development costs, soil salinization, and competition for water required by sensitive ecosystems and commercial fisheries (Postel and Carpenter 1997; Postel 1998; Foster and Chilton 2003).

7.2.3 The Notion of Water Scarcity

The assessment thus far has shown a growing dependence of human society on accessible freshwater resources. To assess the state of these provisioning services more comprehensively, the supply of renewable water must be placed into the context of interactions with people and their use of water. A set of relative measures can be used in this regard.

One measure of dependence on fresh water is the population served per million cubic meters per year of accessible runoff (renewable supply). This is known as the "water crowding" index, with levels on the order of 600–1,000 people per million cubic meters per year (that is, 1,000–1,700 cubic meters per year supply per person) showing water stress, and above 1,000 people (that is, less than 1000 cubic meters per year per person) indicating extreme water scarcity (Falkenmark 1997). Another measure is the relative water use or water stress index (WMO 1997; UN/ WWAP 2003), expressed as the ratio of water withdrawals to supply. More sophisticated indicators are available that incorporate social and economic dimensions of water use (Raskin 1997; Sullivan et al. 2003), and these will be described in the section on water and human well-being. A major water scarcity indicator effort is under way through the World Water Assessment Programme (UN/WWAP 2003).

Worldwide, a substantial quantity of renewable freshwater supply—nearly 30,000 cubic kilometers per year—is accessible to humans. Thus contemporary use represents slightly more than 10% of annual supply. However, there is a substantial range in the share of accessible runoff used by humans across different continents as well as a rapidly changing picture over the last few decades. Time series of use indicate increasing pressures on the freshwater resource base.

Between 1960 and 2000, world water use doubled from about 1,800 to 3,600 cubic kilometers per year, a rate of about 17% per decade, with a slower (10%) increase projected to 2010. (See Table 7.5.) Individual continents show increases over the 1960–2000 timeframe from 15% up to 32% per decade. MENA has historically shown a great dependence on its freshwater supply, using well over half as early as 1960 and exceeding all renewable supplies shortly after 1980. Today its withdrawals represent 120% of accessible sustainable supply, and these are projected to rise to >130% by 2010. Asia, the former Soviet Union, and OECD countries show intermediate levels of use relative to supply over this period. In sub-Saharan Africa, substantial contributions of fresh water from river basins in the wet tropics coupled with rela-

Table 7.5. Indicators of Freshwater Provisioning Services and Their Historical and Projected Trends, 1960–2010. Water use, "water crowding" (population supplied per unit accessible renewable supply), and use relative to accessible supply, by region, are shown. These figures are based on mean annual conditions. The values for the relative use statistics shown rise when the sub-regional spatial and temporal distributions of renewable water supply and use are considered. (Population from Vörösmarty et al. 2000; demand estimates from WRI et al. 1998, updated using Shiklomanov and Rodda 2003, as in Vörösmarty et al. 2000; resampled to MA reporting units)

MA Geographic Region	Population	Water Use U_a	Water Crowding on Accessible Renewable Supply[a]	Use Relative to Accessible Renewable Supply[1] (U_a/B_a)
	(million)	*(km³/yr)*	*(people/mill. m³/yr)*	*(percent)*
Asia	1960: 1,490	1960: 860	1960: 161	1960: 9
	2000: 3,230	2000: 1,553	2000: 348	2000: 17
	2010: 3,630	2010: 1,717	2010: 391	2010: 19
Former Soviet Union	1960: 209	1960: 131	1960: 116	1960: 7
	2000: 288	2000: 337	2000: 160	2000: 19
	2010: 290	2010: 359	2010: 161	2010: 20
Latin America	1960: 215	1960: 100	1960: 25	1960: 1
	2000: 510	2000: 269	2000: 59	2000: 3
	2010: 584	2010: 312	2010: 67	2010: 4
North Africa/Middle East	1960: 135	1960: 154	1960: 561	1960: 63
	2000: 395	2000: 284	2000: 1,650	2000: 117
	2010: 486	2010: 323	2010: 2,020	2010: 133
Sub-Saharan Africa	1960: 225	1960: 27	1960: 55	1960: <1
	2000: 670	2000: 97	2000: 163	2000: 2
	2010: 871	2010: 117	2010: 213	2010: 3
OECD	1960: 735	1960: 552	1960: 131	1960: 10
	2000: 968	2000: 1,021	2000: 173	2000: 18
	2010: 994	2010: 1,107	2010: 178	2010: 20
World Total	**1960: 3,010**	**1960: 1,824**	**1960: 101**	**1960: 6**
	2000: 6,060	**2000: 3,561**	**2000: 204**	**2000: 12**
	2010: 6,860	**2010: 3,935**	**2010: 231**	**2010: 13**

[a] Renewable supply calculated as for Table 7.2, and refers to accessible blue water flows (B_a). Index uses full regional population.

tively poor water delivery infrastructure and restricted development mean that only 2% of renewable supply is tapped. In water-rich Latin America, relative use rates also remain low, at less than 5%.

The contemporary water crowding index is modest in almost all regions. Only MENA shows a value reflective of its well-known position as a highly water-scarce region. Over the last four decades there has been a sustained and substantial increase in the water crowding index with respect to accessible runoff, reflecting directly the impact of population growth. Worldwide, the number of people served per unit of supply has doubled during this period, at an average rate of 20% per decade. Several regions show even greater rates of increase—a tripling for MENA and sub-Saharan Africa and a more than doubling for Asia and Latin America. Globally, an additional 13% crowding in renewable supply is predicted between 2000 and 2010, with greatest regional increases expected in sub-Saharan Africa (30%) and MENA (20%). A slight slowing in rate of increase is noted globally, with near stability in the index for OECD and the former Soviet states.

Several cautionary notes are needed in interpreting these trends. The statistics are based on mean annual flows and access computed for 100% of individual continental and global populations. In the context of the 50% of continental runoff generated in dry to moderately wet climate zones (19,800 cubic kilometers per year) that serves the majority of global population, contemporary use represents nearly 20% of the mean annual supply. When seasonal variations in runoff are considered (reducing supplies to 13,900 cubic kilometers per year), withdrawals exceed 25% of the renewable resource. In addition, if dedicated instream uses of about 2,000 cubic kilometers per year for navigation, waste processing, and habitat management are considered (based on Postel et al. 1996), humans then use and regulate 40% or more of renewable accessible supplies.

Further, the crowding index does not take into account different countries' abilities to deal with water shortages. For example, high-income countries that are water-scarce may be able to cope to some degree with water shortages by investing in desalination or reclaimed wastewater. The study also discounts the use of fossil water sources because such use is unsustainable in the long term.

In addition, while the global numbers are well below the extreme scarcity threshold of 1,000 people per million cubic meters per year of renewable supply, they mask important local and regional differences and thus understate the true degree of stress (Vörösmarty et al. 2000, 2005). Prior assessments (Revenga et al. 2000) show that as of 1995 some 41% of the world's population, or 2.3 billion people, were living in river basins under water stress, with some 1.7 billion of these people residing in river basins under conditions of extreme water scarcity. From a river basin perspective, the Volta, Nile, Tigris and Euphrates, Narmada, and Colorado in the United States will show ongoing pressure through 2025 (Revenga et al. 2000). Another 29 basins will descend further into scarcity by 2025, including the Jubba, Godavari, Indus, Tapti, Syr Darya, Orange, Limpopo, Yellow, Seine, Balsas, and Rio Grande. Indicators based on mean annual conditions also mask important supply limits imposed by seasonal and inter-annual variability. For example, in India most of the annual water supply is generated as a result of the monsoons, which in many cases means both flooding downstream as well as seasonal drought.

Another measure of adequacy of the freshwater supply is the mean use-to-supply ratio. A set of thresholds for water stress was given by the United Nations in a recent global analysis that used this ratio based on mean annual conditions (WMO 1997): low (<10%), moderate (10–20%), medium/high (20–40%), and high (>40%). Using this classification and a grid-based approach necessary to capture the high degree of spatial heterogeneity (see Vörösmarty et al. 2000), the contemporary global-scale ratio is from low-to-moderate, as seen in Table 7.5, although entire continents are under a moderate (Asia, former Soviet Union, and OECD) to high (MENA) state of scarcity. This is in stark contrast to the situation in 1960, when uniformly low levels of scarcity were noted (with the exception of MENA). Globally, it has been shown that 2.5 billion people suffer from at least moderate levels of chronic water stress (Vörösmarty et al. 2000) and from 1–2 billion people suffer high levels of scarcity even when tabulations are made conservatively on total renewable supplies. Calculating the population at risk through a ratio based on accessible supplies would increase the overall exposure to stress.

Water scarcity as a globally significant problem is a relatively recent phenomenon, evolving only over the last four decades. Rates of increase in the relative use ratio from 1960 to the present averaged about 20% per decade globally, with values from 15% to more than 30% for individual regions. A slowing in the rate of increase in use is projected between 2000 and 2010, to 10% per decade globally. With anticipated population growth, economic development, and urbanization, a further increase in the relative use ratio for some continents is likely to remain high (MENA at 14% per decade, Latin America at 16%, and sub-Saharan Africa at 20%).

7.2.4 Environmental Flows for Ecosystems

In light of the expanding use of fresh water by humans and several indicators of growing water stress, an important issue emerges with respect to the sustainability of water provisioning services—that is, being able to continue providing water for human use while also meeting the water requirements of aquatic ecosystems so as to maintain their capacity to provide other services. "Environmental flows" refers to the water considered sufficient for protecting the structure and function of an ecosystem and its dependent species. These flow requirements are defined by both the long-term availability of water and its variability and are established through environmental, social, and economic assessment (King et al. 2000; IUCN 2003).

Determining how much water can be allocated to human uses or distorted through flow stabilization (such as dam construction) without loss of ecosystem integrity is central to an understanding of how freshwater ecosystems support human well-being through the range of provisioning, supporting, and regulating services. Assessment of water availability, water use, and water stress at the global scale has been the subject of on-going research. However, water requirements of aquatic ecosystems are only now being estimated globally and considered explicitly in these assessments (Smakhtin et al. 2003). Flow requirements can range globally from 20% up to 80% of mean annual flow, depending on the river type, its species composition, and the river health condition objectives sought (for instance, pristine, moderate modification from natural conditions, minimum flows), indicating the high degree of potential conflict with river regulation and human uses should the environment be preserved.

If human systems are viewed as being embedded within natural systems, human water use can expand to a "sustainability boundary" beyond which a substantial degradation of ecosystem services results (King et al. 2000; Postel and Richter 2003). Determining the location of the sustainability boundary is critical to successful management and rests on clearly defining what constitutes a degraded ecosystem. Environmental flows should consider both the quantity and timing of flow to maintain "naturally variable flow regimes" (Poff et al. 1997), whereby seasonal flow patterns are maintained with the aim of retaining the benefits provided by low and high flows. (See Figure 7.4.) Naturally low flows, for example, help exclude invasive species while high flows, especially floods, shape channels and allow the delivery of nutrients, sediments, seeds, and aquatic animals to seasonally inundated floodplains. High flows may also provide suitable migration and spawning cues for fish (Poff et al. 1997; Baron et al. 2002).

7.2.4.1 Global Trends in Water Diversion and Flow Distortion

While global trends in altered water regime are difficult to assemble with certainty due to incomplete information, they reflect an overall increase in regulation of the world's inland river systems (Revenga et al. 2000; Vörösmarty and Sahagian 2000). Tables 7.4 and 7.5 provided an indication of the scope of such changes. Water withdrawals show a doubling between 1960 and 2000, by which time irretrievable losses from irrigation alone totaled 34% of all global use.

One third of all rivers for which contemporary and predisturbed discharges could be compared in a compendium (Meybeck and Ragu 1997) showed substantial declines in discharges to the ocean. Long-term trend analysis (more than 25 years) of 145 major world rivers indicated more than one fifth with declines in discharge (Walling and Fang 2003). From 1960 to 2000 there was a near quadrupling of reservoir storage capacity and more than a doubling of installed hydroelectric capacity (Revenga et al. 2000). Worldwide, large artificial impoundments (storing each 0.5 cubic kilometers or more) now hold two to three months of runoff, capable of significant hydrograph distortion, with several major basins showing storage potentials of greater than a year's runoff (Vörösmarty et al. 2003). Much of this regulation occurred over the last 40 years.

Through consumptive use and interbasin transfers, several of the world's largest rivers (Nile, Yellow, Colorado) have been transformed into highly stabilized and in some cases seasonally nondischarging river channels (Meybeck and Ragu 1997; Kowalewski et al. 2000). In the case of the Yellow River, improved water management since 2000 has helped to restore flows (MWR 2004).

7.2.4.2 Recent History of Governance and Management for Environmental Flows

Over the last decade, policy solutions to developing environmental flows have taken several forms, depending on social and historical context, degree of scientific knowledge, water infrastructure,

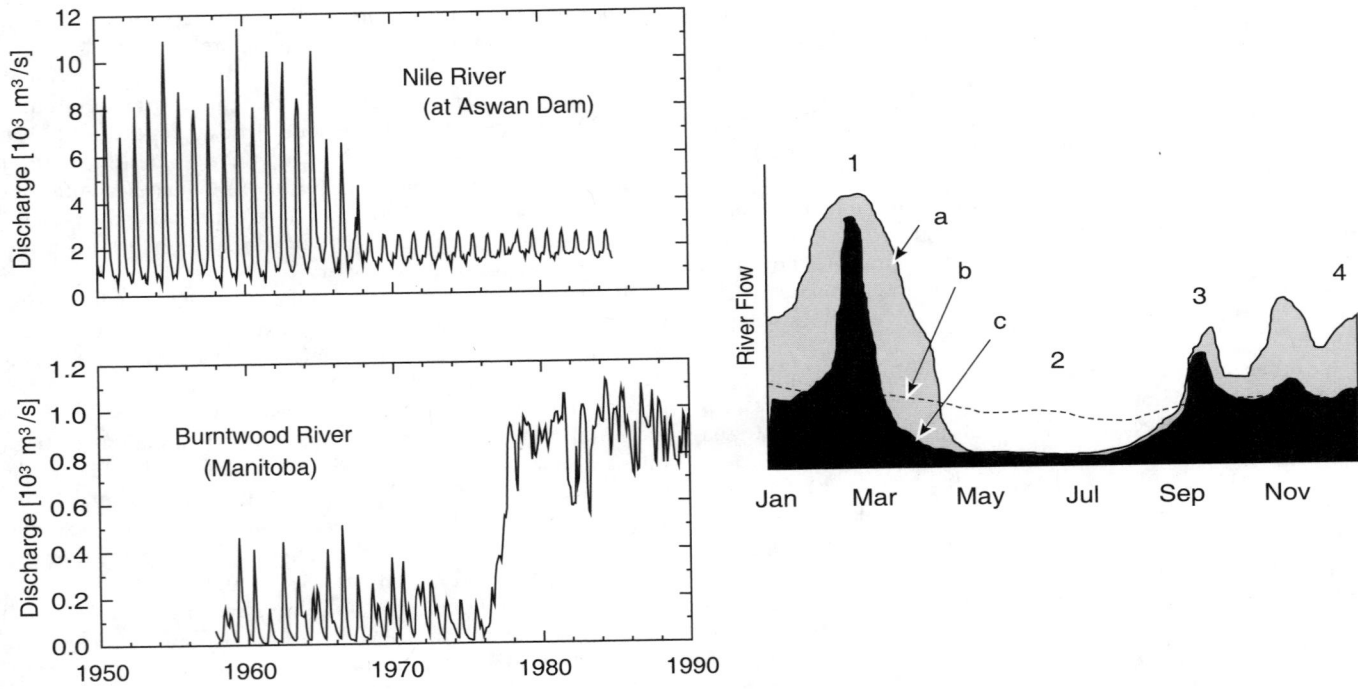

Figure 7.4. Managing for Environmental Flows: Contrasts among Natural, Reservoir-affected, and Reconstituted River Discharge Regimes. Observed alteration of natural flow regimes (left) arises from the provision of freshwater services, as through impoundment on the Nile River and interbasin transfer to optimize hydropower on the Burntwood River (Vörösmarty 2002). Environmental flow management attempts (right) to preserve key facets of the (a) natural flow regime in light of (b) typical 20th century flow distortion after damming. Condition (c) represents a partially "re-naturalized" flow regime, which retains important hydrologic characteristics: 1) peak wet season flood, 2) baseflow during the dry season, and 3) a "flushing" flow at the start of the wet season to cue life cycles, and 4) variable flows during the early wet season. Flow regime (b) shows many more negative effects than (c), even though both regulate similar volumes of water annually. (Right panel adapted from Tharme and King 1998)

and local ecosystem conditions. These approaches include managing the quantity and temporal pattern of water withdrawals or releases (Poff 2003; Postel and Richter 2003), developing water markets, and preemptively managing land use to protect watersheds.

Water allocation for environmental flows to sustain functioning freshwater ecosystems is practiced in parts of Australia, Europe, New Zealand, North America, and South Africa. However, there appears to be very little consideration of this matter anywhere in Asia, despite aggressive water extraction from many rivers during the dry season across the continent. But there is cause for cautious optimism. The calculation, adoption, and implementation of environmental flows are under consideration in other parts of the world. In addition, more than 2,000 river, lake, and floodplain restoration projects in at least 20 countries, particularly in Europe but also in Africa and Asia, are being carried out (DRRC 1998; UKRRC 2004; Richter et al. in prep.). Some key examples include the restoration of the Diawling delta in Mauritania (Hamerlynck and Duvail 2003), the Waza Logone floodplain in Cameroon (Loth 2004), the Danube and Rhine Rivers, and the South Florida Everglades—one of the largest ecosystem restoration projects ever attempted (Baron et al. 2002).

The shift toward management for natural flow regimes is also reflected by parallel shifts in public policy from laws favoring private interests and prior appropriations (as in much of the American West) to protecting water rights and environmental flows as part of the "public trust." In 1998, South Africa passed landmark legislation to aid decision-making on all or part of any significant

water resource (National Water Act 1998). One of the most progressive aspects of this act was establishment of a Reserve to support both essential human needs (water for drinking, food preparation, personal hygiene) and aquatic ecosystem integrity. Notably, this two-part Reserve—with human and environmental components—takes priority over other uses such as irrigation and industrial withdrawal. In Burkina Faso, a new water framework law (Loi d'Orientation sur L'eau), adopted in 2001, establishes the legal and institutional framework for promoting integrated basin management, equitable access, water for nature, and international cooperation. The legislation recognizes that "infrastructures which are built on a water course must maintain a minimal flow that guaranties aquatic life" (MEE 2001).

For many highly regulated river systems in North America (e.g., Colorado, Columbia, Missouri, Savannah), recent changes in dam operations and adaptive management plans are now fostering conditions that improve fish habitat, river-floodplain connectivity, and estuarine ecosystems, often at the cost of hydroelectric generation or navigability to barges (Postel and Richter 2003; Richter et al. in prep.). In addition, the decommissioning and removal of some dams has begun in the United States (Hart et al. 2002). In Australia, water allocation reforms have led to limits on future withdrawal (that is, a "water cap") in the Murray-Darling River basin, subsequent development of a water market where allocations are traded, and creation of incentives to increase water productivity and efficiency (Blackmore 1999; MDBC 2004). Similarly, water markets developed in Mexico, Chile, and some western states in the United States have been used to secure flows for ecosystems (Thobani 1997).

Watershed management strategies that integrate ecological principles have been used to prevent water supply crises from developing. An often-cited example is the New York City water supply management strategy, which includes protection of riparian habitat in the nearby source area of the Catskills Mountains, thus eliminating the need to construct a water filtration plant at an estimated cost of $6 billion. The ~400,000-hectare Pinelands National Reserve in nearby New Jersey is regulated under a Comprehensive Management Plan developed at the local, state, and federal level in 1978–79 (Good and Good 1984). The plan permits a wide spectrum of land use development categories, ranging from intensive development to full protection, and it successfully redirected human activities to areas deemed appropriate while protecting a large core area, which is ecologically sensitive, drought-prone, and nutrient-poor and which harbors a unique community of wildlife with a large number of endemic species (Walker and Solecki 1999; Bunnell et al. 2003). The benefits of maintaining high water quality are recognized outside the reserve through the delivery of relatively high-quality fresh water to an estimated 9 million people in New York City for less than if a water filtration plant were built. In addition, water discharged into Delaware Bay helps to support populations of anadromous fish and spawning horseshoe crabs, which in turn support large numbers of migrating shorebirds and local industries.

7.2.5 Water Quality

Summarizing patterns and trends in water quality, particularly at a global scale, encompasses an array of challenges that include basic definitional problems, a lack of worldwide monitoring capacity, and an inherent complexity in the chemistry of both natural and anthropogenic pollutants. From a management perspective, water quality is defined by its desired end use. Water for recreation, fishing, drinking, and habitat for aquatic organisms thus require higher levels of purity, whereas for hydropower, quality standards are much less important. For this reason, water quality takes on a broad definition as the "physical, chemical, and biological characteristics of water necessary to sustain desired water uses" (UN/ECE 1995).

Natural water chemistry is inherently highly variable over space and time (Meybeck and Helmer 1989; Meybeck 2003), and aquatic biota are adapted to this variability. With added pressure from human activities, the biogeophysical state of inland waters plus their variability is altered, often to the detriment of aquatic species (see Chapter 20), thereby compromising the sustainability of aquatic ecosystems. Many chemical, physical, biological, and societal factors affect water quality: organic loading (such as sewage); pathogens, including viruses in waste streams from humans and domesticated animals; agricultural runoff and human wastes laden with nutrients (such as nitrates and phosphates) that give rise to eutrophication and oxygen stress in waterways; salinization from irrigation and water diversions; heavy metals; oil pollution; literally thousands of synthetic and persistent engineered chemicals, such as plastics and pesticides, medical drug residues, and hormone mimetics and their by-products; radioactive pollution; and even thermal pollution from industrial cooling and reservoir operations.

Furthermore, despite important improvements in analytical methodologies (UN/ECE 1995; Meybeck 2002), the capacity to operationally monitor contemporary trends in water quality is even more limited than monitoring the physical quantity of water. In terms of the spatial coverage, frequency, and duration of monitoring, data currently available for global and regional-scale assessments are patchy at best, leading to oversimplified and sometimes misleading information. (See Table 7.6.)

Data abundance is generally associated with level of economic development: industrial countries show a higher level of data availability, while water quality in developing countries is less well monitored. Even when data from monitoring stations are available, they only provide a fragmented view of water quality issues for very local sections of rivers, necessitating potentially unreliable extrapolation to the rest of the basin (Meybeck 2002). For this reason, water quality assessments or trajectories are usually river- or station-specific. Even for the best-represented regions of the globe, a coherent time series of data is available for only the last 30 years or less, constraining the ability to clearly quantify trends in water quality.

Data comparability problems are yet another constraint on the utility of water quality data. Standardized protocols, in terms of sampling frequency, spatial distribution of sampling networks, and chemical analyses, are still not in place to ensure the production of comparable data sets collected in disparate parts of the world. The monitoring of groundwater supplies is even more problematic (Meybeck 2003; Foster and Chilton 2003); because ground-

Table 7.6. Data Assessment of Existing Monitoring Programs Worldwide. The entries relate to the quantity of available data, indicated by the number of + symbols. For the purposes of this assessment, data quantity is an aggregate measure of station network density, spatial coverage, frequency of data collection, and duration of monitoring programs. (Updated from Vörösmarty et al. 1997b)

Constituent	Industrial Countries	Rapidly Developing Countries	Other Developing Countries
Sediment			
Bedload	(+) 0	0	0
Total suspended (TSS)	+++	++	+
Carbon	+++	++	+
Dissolved Inorganic (DIC)	+++	++	+
Dissolved Organic (DOC)	++	+	0
Particulate Organic (POC)	+	0	0
Nitrogen			
Ammonium (NH_4)	+++	++	+
Nitrate (NO_3)	+++	++	+
Dissolved Organic (DON)	+	0	0
Particulate Organic (PON)	0	0	0
Phosphorus			
Phosphate (PO_4)	+++	++	+
Dissolved Organic (DOP)	0	0	0
Total (TP)	++	+	0
Metals			
Dissolved	++	+	0
Total	+	0	0
Particulate	+	0	0
Major dissolved constituents[a]	+++	++	+
Discharge	+++	++	+

[a] SO_4, Cl, Ca, Mg, K, Na, SiO_4, CO_3.

water is hidden from view, many pollution and contamination problems that affect supplies have been more difficult to detect and have only recently been discovered.

These many factors make it difficult to estimate the impact of changing water quality on global water supply. The following sections provide an overview assessment of trends in water quality that have bearing on the capacity of the contemporary water cycle to provide provisioning services for fresh water and on the sustainability of inland water systems. Other assessments specifically target water quality issues over selected regional-to-continental domains (e.g., AMAP 2002; Hamilton et al. 2004).

7.2.5.1 General Trends in Water Quality

The state of inland water quality illustrates the long-term and complex nature of human interactions with their environment. The earliest changes attributable to humans likely occurred in tandem with land use change in small to medium-sized catchments some 5,000 or 6,000 years ago in the Middle East and Asia, where water and sediment budgets were substantially altered (Wasson 1996; Vörösmarty et al. 1998b; Alverson et al. 2003; Meybeck et al. 2004). Water also has been considered since ancient times to be the preferred medium for cleaning, transporting, and disposing of wastes—establishing a tradition that today has substantially transformed the physical, biological, and chemical properties of global runoff.

A set of syndromes depicting riverine changes arising from anthropogenic pressures has been proposed (GACGC 2000; Meybeck 2003) through which society transforms inland fresh waters from a pristine state fully controlled by the natural Earth system to a modern condition in which humans provide many of the predominant controls. In most of the densely populated areas of the world, river engineering, waste production, and other human impacts have significantly changed the water and material transfers through river systems (Vörösmarty and Meybeck 1999, 2004) to the extent that this now likely exceeds the influence of natural drivers. This is true today in many parts of the Americas, Africa, Australasia, and Europe (Vörösmarty and Meybeck 1999, 2004).

The contrast between pristine and contemporary states can be dramatic and potentially global in scope. Changes to the global nitrogen cycle are emblematic of those in water quality more generally, through which high concentrations of people or major landscape disturbances (such as industrial agriculture) translate into a disruption of the basic character of natural water systems. In addition, modern changes often "reverberate" far downstream of the original point of origin. Compared with the preindustrial condition, loading of reactive nitrogen to the landmass has doubled from 111 million to 223 million tons per year (Green et al. 2004) or possibly 268 million tons (Galloway et al. 2004). (See also Chapter 12.) Model results show these accelerated loadings transformed into elevated freshwater transports through inland waterways to the coastal zone, doubling pre-disturbance rates from 21 million to 40 million tons per year (Green et al. 2004; Seitzinger et al. 2002). North America, continental Europe, and South, East and Southeast Asia show the greatest change. (See Figure 7.5 in Appendix A.)

Riverine transport of dissolved inorganic nitrogen (immediate precursors to nutrient pollution, algal blooms, and eutrophication) have increased substantially from about 2–3 million tons per year from the preindustrial level to 15 million tons today, with order-of-magnitude increases in drainage basins that are heavily populated or supporting extensive industrial agriculture. Rivers with high concentrations of inorganic nitrogen constitute a major global source for inorganic nitrogen, despite relatively modest contributions to aggregate water runoff. (See Figure 7.6.) While it is noteworthy that aquatic ecosystems "cleanse" on average 80% of their global incident nitrogen loading (Green et al. 2004; Howarth et al. 1996; Seitzinger et al. 2002; Galloway et al. 2004), the intrinsic self-purification capacity of aquatic ecosystems varies widely and is not unlimited (Alexander et al. 2000; Wollheim et al. 2001). As a result, sustained increases in loading from land-based activities are already reflected in the deterioration of water quality over much of the inhabited portions of the globe, they extend their impacts to major coastal receiving waters (e.g., Rabalais et al. 2002), and they are likely to continue well into the future (Seitzinger and Kroeze 1998).

While the stark contrast between pristine and contemporary states demonstrates the overall impact of anthropogenic influences on water quality, much of the contamination of fresh water has occurred over the last century. The main contamination problems 100 years ago were fecal and organic pollution from untreated human wastewater. Even though this type of pollution has decreased in the surface waters of many industrial countries over the last 20 years, it is still a problem in much of the developing world, especially in rapidly expanding cities (WMO 1997; UN/WWAP 2003). (See also Chapter 27.)

In developing countries, sewage treatment is still not commonplace, with 85–95% of sewage discharged directly into rivers, lakes, and coastal areas (UNFPA 2001; Bouwman et al. 2005), some of which are also used for water supply. Consequently, water-related diseases, such as cholera and amoebic dysentery, among others, claim millions of lives annually (WHO/UNICEF 2000). In Europe, organic pollution and contamination by toxic metals are probably now less than the levels observed between the 1950s and1980s, due to improved environmental regulation (Meybeck 2003). In the developing world, the riverine evolution is likely to be similar to that found in Europe, with a major lag corresponding to their different stages of industrialization, urbanization, and intensification of agriculture (Meybeck 2003).

New pollution problems from agricultural and industrial sources have emerged in industrial and developing countries and have become one of the biggest challenges facing water resources in many parts of the world (WMO 1997). In Western Europe and North America, on the one hand phosphorus contamination in waterways has been reduced considerably with the introduction

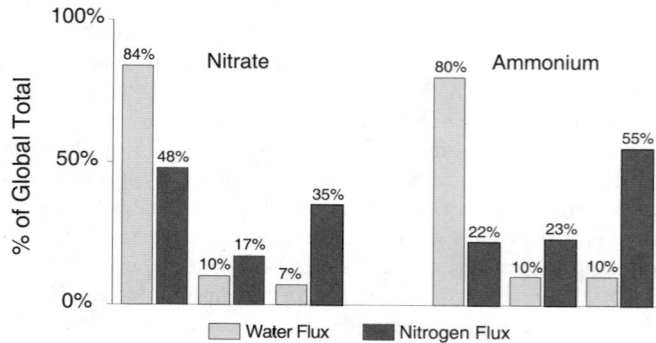

Figure 7.6. Global Summary of Inorganic Nitrogen Transport by Contemporary Rivers. Modern patterns of pollution from anthropogenic sources have created characteristically high-impact regions or "hotspots" that represent highly polluted river systems that today carry much greater quantities of nitrogen than their collective discharge would indicate. (Meybeck and Ragu 1997; Vörösmarty and Meybeck 2004)

of phosphate-free household detergents, investments in waste-water treatment plants, and to some degree modified agroecosystem management. On the other hand, residues of synthetic pharmaceuticals for humans and livestock are increasingly being discovered at low doses in rivers and lakes (Schiermeier 2003). There are indications that these residues can disturb the physiology of invertebrates, and it is still a matter of debate whether and, if so, to what degree these newly discovered pollutants may affect human physiology (Daughton and Ternes 1999; Jones et al. 2003).

Water contamination by pesticides has grown rapidly since the 1970s. In a medium-sized river basin like the Seine, over 100 different types of active molecules from pesticides can be found (Chevreuil et al 1998). Even if the use of xenobiotic substances is increasingly being regulated in Western Europe and North America, bans—when they exist—occur generally two to three decades after the first commercial use of the products. For example, DDT, atrazine (a common pesticide), and PCBs were in use for a long time before they were banned in parts of the industrial world. In general these bans take longer to implement in the developing world, so these products are still commercialized and used in some countries.

In the United States, PCB and DDT records in estuarine sedimentary archives peaked in the 1970s and are now markedly decreasing (Valette-Silver 1993). At the same time, persistent xenobiotics are widespread, with a recent study (Kolpin et al. 2002) finding traces of at least one drug, endocrine-disrupting compound, insecticide, or other synthetic chemical in 80% of samples from 139 streams in 30 states of the United States. The persistence of these products in continental aquatic systems can be high, and their degradation products can be more toxic than the parent molecules (Daughton and Ternes 1999). Because of the poor monitoring of the long-term effects of xenobiotics, the global and long-term implications of their use cannot be fully assessed.

7.2.5.2 Global Ranking of Water Quality Issues Based on Regional Assessment

A global water quality assessment, originally as part of the Dublin International Conference on Water and the Environment and in preparation for the Rio Summit (Meybeck et al. 1991) is summarized here. The original report determined a global ranking of key water quality issues based on U.N. Global Environmental Monitoring System data, the perceptions of local/regional scientists and managers, published reports and papers, and expert knowledge. Lakes, groundwater, and reservoir issues were considered, although as Siberia and northern Canada were not expressly covered in the 1991 report, these have been considered separately using the same approach (Meybeck 2003). Eleven variables were considered and ranked, the scoring of which ultimately reflects the aggregate impact of human pressures, natural rates of self-purification, and pollution control measures.

The results show that pathogens and organic matter pollution (from sewage outfalls, for example) are the two most pressing global issues (see Figure 7.7), reflecting the widespread lack of waste treatment. As water is often used and reused in a drainage basin context, a suite of attendant public health problems arises, thus directly affecting human well-being. At the other extreme, acidification is ranked #10 and fluoride pollution #11. The importance of the various issues varies between regions, however, and some of these globally low-ranked issues are particularly important in certain areas, such as acidification in Northern Europe, salinization in the Arabic peninsula, and fluoride in the Sahel and African Great Lakes (see maximum scores on Figure 7.7). Fluo-

ride and salinization issues are mostly due to natural conditions (rock types and climate), but mining-related salinization can also be found (for instance, in Western Europe). All other concerns directly arise through human influences. An annotated continental summary is given in Table 7.7.

Although these updated results correspond well to the state of water quality in the 1980–90s (Meybeck 2003), since the 1990s the situation in most developing countries and countries in transition is likely worse in terms of overall water quality. In Eastern Europe, Central and South populated Americas, China, India, and populated Africa, it is probably worse for metals, pathogens, acidification, and organic matter, while for the same issues Western Europe, Japan, Australia, New Zealand, and North America have shown slight improvements. Nitrate is still generally increasing everywhere, as it has since the 1950s. In the former Soviet Union there has been a slight improvement in water quality due to the economic decline and associated decrease in industrial activities. Eastern Europe has also seen some improvements, such as those in the Danube and the Elbe basins. A few rivers, such as the Rhine, have seen a stabilization of nitrate loads after 1995.

7.3 Drivers of Change in the Provision of Fresh Water

The drivers of change in the global water cycle and the system's capacity to generate freshwater provisioning services act on a variety of spatial and time scales. Throughout history, humans have pursued a very direct and growing role in shaping the character of inland water systems, often applied at local scales, but sometimes reflecting provincial or national policies on water. The collective significance of human influences on the hydrologic cycle may today be of global significance, but this has only recently begun to be articulated (Vörösmarty and Meybeck 2004).

Humans today control and use a significant proportion of the runoff—from 40% to 50% (Postel et al. 1996)—to which the vast majority has access. Given high numbers of people dependent on water provisioning services derived from ecosystems and the growing degree of water crowding, urbanization, and industrialization, the global water cycle is and will continue to be affected strongly by humans.

Water engineering to facilitate use by humans has fragmented aquatic habitats, interfered with migration patterns of economically important fisheries, polluted receiving waters, and compromised the capacity of inland water ecosystems to provide reliable, high-quality sources of water. Land cover changes have also altered the patterns of runoff and created sources of pollution, negatively affecting human health, aquatic ecosystems, and biodiversity. (See Chapter 20.) Due to a growing reliance on irrigated agriculture for domestic food production and international trade, freshwater services—in decline in many parts of the world through non-sustainable resource use practices—are directly linked to the global food security issue. (See Chapters 8 and 26.) Finally, natural climate variability and anticipated changes associated with greenhouse warming convey additional, major constraints on the provision of renewable freshwater services.

7.3.1 Population Growth and Development

Population growth is a major indirect driver of change in the provision of fresh water. Although freshwater supplies are renewed through a more or less stable global water cycle that produces precipitation in excess of evapotranspiration over the continents, the mean quantity of water supply available per capita is ever-decreasing due to population growth and expanding con-

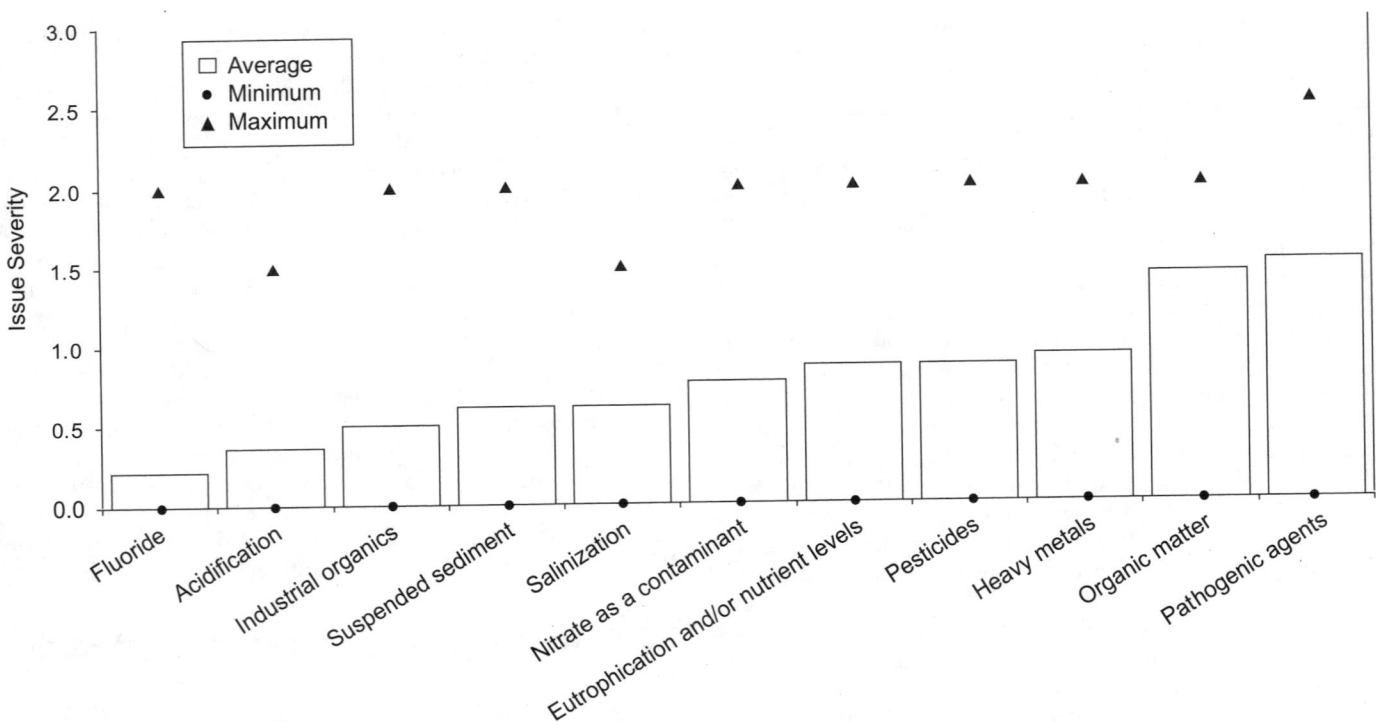

Figure 7.7. Ranking of Globally Significant Water Quality Issues Affecting the Provision of Freshwater Services for Water Resource End Uses. Averages show the general tendencies for specific pollutants, but a wide range is noted, with minima in all cases ranked zero and maxima often several times more severe than the mean condition. Although this ranking shows organic matter pollution and pathogens to be relatively more important at the global scale, information to quantify the degree to which water supplies are compromised by pollution is currently insufficient. Scores are as follows: 0: No problem or irrelevant; 1: Some pollution, water can be used if appropriate measures are taken; 2: Major pollution with impacts on human health and/or economic use, or aquatic biota; 3: Severe pollution—impacts are very high, losses involve human health and/or economy and/or biological integrity. (Based on expert opinion; Meybeck et al. 1991, updated by Meybeck 2003)

sumptive use (Shiklomanov and Rodda 2003). Human population doubled from 1960 until the present (Cohen 2003), and nearly 20 contemporary cities are home to 10 million people or more (Cohen 2003). Substantial flow stabilization and increased withdrawals have occurred across all regions, supporting an increase in the number of people sustained by the accessible, renewable water supply.

Continued growth in population will fuel increases in food production, which in the context of a stable cropland base (Bruinsma 2003) will require greater diversions of fresh water for irrigation or considerably more efficient use of water supplies. The same applies to industry and municipalities, amplifying current pressures on the global water supply. Economic development, technology, and lifestyle changes (such as increasing meat consumption) further define the functional availability of water in the context of declining per capita supplies. Over the twentieth century, water withdrawals increased by a factor greater than six—more than twice the rate of population growth (WMO 1997).

In addition to increased water demands, as mentioned in section 7.2.5, pollution from industry, urban centers, and agricultural runoff limits the amount of surface and groundwater available for domestic use and food production. Threats of water quality degradation are most severe in areas where water is scarce because the dilution effect is inversely related to the amount of water in circulation.

7.3.2 Managed Water Supplies

A broad array of water engineering schemes has enabled variability in the hydrologic cycle to be controlled and increasing amounts of water to be stored and withdrawn for human use. This technology refers to any sort of engineering used in the storage, management, and distribution of water, such as dams, canals, water transfers, irrigation ditches, levees, and so on. It also includes both traditional water harvesting techniques as well as modern production and treatment facilities like desalinization plants.

Global patterns of water management are not driven solely by investments in technology and large-scale engineering. Water is also managed through international trade, by way of the embodied or "virtual" water content of commodities exchanged. The agricultural sector, in particular, requires huge amounts of rainfall or irrigation water, much of which is lost to evapotranspiration, and in the case of irrigation there are also transit losses. Water input-to-crop output ratios, expressed on a weight-to-weight basis, vary from the hundreds to the thousands. Given enormous contrasts in local availability of fresh water, there is a potentially enormous comparative advantage in virtual water trade strategies that transport products from water-rich to water-poor areas.

This section first assesses the role of major engineering works in the provision of water and then considers the significance of virtual water trade of agricultural products in the global economy.

Table 7.7. Continental-scale Assessment of Major Water Quality Issues. The purpose of this table is to present a general overview. It does not capture fully large sub-regional differences that are known to occur. (Updated from Meybeck et al. 1991)

Continental Domain	Summary of Key Findings
Africa	Major sources of pollution in Africa, according to the 1992 assessment, are fecal contamination; toxic pollution downstream of major cities, industrial centers, and/or mining; and vector-borne diseases. The Nile Basin and Northern Africa show more contamination problems than other regions, but this also may be because of more information and monitoring stations in these regions, or more altered water flows that affect dilution potential in rivers.
Americas	In the United States and Canada, the major pollution problem is eutrophication from agricultural runoff and acidification from atmospheric deposition. Major problems also include persistent toxic water pollution from point and non-point sources. In South and Central America the major contaminant problems, except in the Amazon and Orinoco basins, where ecosystems are more intact and high flows foster dilution, are pathogens and organic matter, as well as industrial and mining discharges of heavy metals and pesticide and nutrient runoff.
Asia and the Pacific	Arid and semiarid regions tend to have different pollution problems than areas in the monsoon belt. In the Indian subcontinent the major problems are pathogens and contamination from organic matter. While these are prevalent in Southeast Asia as well, heavy metals, eutrophication, and sediment loads from deforestation are also critical in this sub-region. The Pacific Islands have higher levels of salinization than other regions in Asia, while still having problems with pathogens and organic matter, like much of the developing world. China has a combination of all pollution problems in its major watersheds. In the dry north, eutrophication, organic matter, and pathogens are major problems, while in the south in addition there is a large sedimentation problem. Finally, Japan, New Zealand, and Australia present similar pollution problems as other industrial nations, like the United States and Europe. Australia has particular problems with salinization due to agricultural practices, especially in the Murray-Darling Basin.
Europe	In the Nordic countries the major problem is acidification, while other contaminant levels are relatively low. In Western Europe eutrophication and nitrates pose the greatest challenge, while in Southern and Eastern Europe the major contaminants are organic matter and pathogens, nitrates, increasingly pesticides, and eutrophication.
Eastern Mediterranean and Middle East	Characterized by its arid climate, this area shows great demands and pressure on its scarce water resources. Industrial pollution and toxics are a problem in some locales, but overall salinization from over-abstraction is the key concern in this region.

7.3.2.1 The Role of Engineering on Water Supply

7.3.2.1.1 Dams and reservoirs

Humans have altered waterways around the world since historical times to harness more water for irrigation, industry, and domestic and recreational use. Dams have been a particularly significant driver of change, buffering against both spatial and temporal scarcity of water supplies and increasing the security of water and food supply over the past half-century. However, large engineering works that impound and divert fresh water have caused damage to key habitats and migratory routes of important commercial and subsistence fisheries (Revenga et al. 2000), as well as serious societal disruptions, including public health problems (as described later; see also Chapter 14) and forced displacements (WCD 2000).

Large dams are today the fundamental feature of water management across the globe (FC/GWSP 2004). Approximately 45,000 large dams (>15 meters in height) (WCD 2000) and possibly 800,000 smaller dams (McCully 1996; Hoeg 2000) are in place and an estimated $2 trillion has been invested in them over the last century. These facilities have served as important instruments for development, with 80% of the global expenditure of $32–46 billion per year focused on the developing world (WCD 2000).

Major stabilization of global river runoff from major engineering works expanded greatly between 1950 and 1990. (See Figure 7.8.) Currently the largest reservoirs—those with more than 0.5 cubic kilometers of storage capacity—intercept locally 40% of the water that flows off the continents and into oceans or inland seas (Vörösmarty et al. 2003). The volumetric storage behind all large dams represents from three to six times the standing stock of water held by natural river channels (Vörösmarty et al. 1997a, 2003; Shiklomanov and Rodda 2003). In addition, large reservoir construction has doubled or tripled the residence time of river water—that is, the average time that a drop of water takes to reach the sea, with the mouths of several large rivers showing delays on the order of many months to years (Vörösmarty et al. 1997a).

Such regulation has enormous impacts on the water cycle and hence aquatic habitats, suspended sediment, carbon fluxes, and waste processing (Dynesius and Nilsson 1994; Vörösmarty et al. 2003; Stallard 1998; Syvitski et al. 2005). Large dams, in particular, have been a controversial component of the freshwater debate. While contributing to economic development and food security, they also produce environmental, social, and human health impacts. A World Bank review (1996a) of the impacts and economic benefits of 50 large dams concluded that these projects showed proven economic and development benefits but had a mixed record in terms of their treatment of displaced people and environmental impacts. A further review by the World Commission on Dams on the performance of large dams showed considerable shortfalls in their technical, financial, and economic performance relative to proposed expectations, particularly irrigation dams, which often have not met physical targets, failed to recover cost, and have been less profitable than expected (WCD 2000).

In Pakistan, for example, the direct benefits from irrigation made possible by the Tarbela and Mangla dams are estimated at about $260 million annually, with the farmers who own irrigated land clearly benefiting from increased incomes (World Bank 1996a). However, the increased use of irrigation water has led to waterlogging and increased soil salinity in the Punjab area, with a

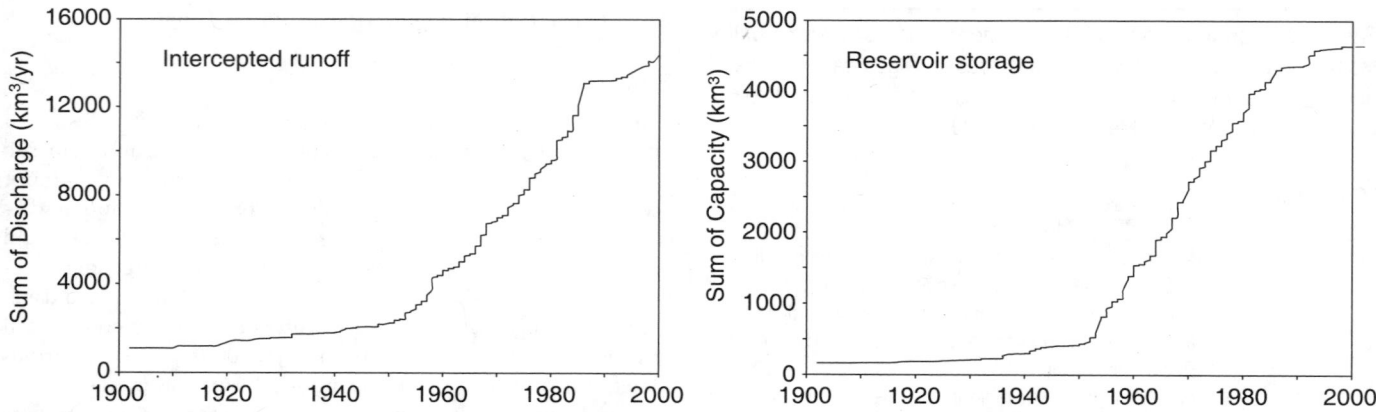

Figure 7.8. Time Series of Intercepted Continental Runoff and Large Reservoir Storage, 1900–2000. The series is taken from a subset of large reservoirs (>0.5 km³ maximum storage each), geographically referenced to global river networks and discharge. The years 1960–2000 have shown a rapid move toward flow stabilization, which has slowed recently in some parts of the world, due to the changing social, economic, and environmental concerns surrounding large hydraulic engineering works. (Vörösmarty and Sahagian 2000)

direct link to a decline in crop productivity, and an increase in malaria transmission (World Bank 1996b).

Hydroelectricity is another important benefit from dams. Total production of hydropower reached 2,740 terawatt-hours in 2001 or 19% of global electrical production, and many industrial (such as Norway and Iceland) and developing countries (Democratic Republic of Congo, Mozambique, Brazil, Honduras, Tajikistan, and Laos) rely on dams for more than 90% of their power production (UN/WWAP 2003). As with irrigation dams, in many circumstances the effectiveness of large dams for hydroelectricity generation has not been sufficient to meet the predicted benefits (WCD 2000), and they have caused loss of habitats and species as well as the displacement of millions of people (WCD 2000).

Flood control continues to be another major objective for building large dams. In Japan, for example, 50% of the population lives in flood-prone areas, and in the last 10 years floods have affected 80% of municipalities in the country. Japan is one of the top five dam-building countries in the world. Matsubara and Shimouke Dams on the Chikugo River in the Kyushu District in southern Japan, for instance, were built for flood control after a flood in 1953 inundated one fifth of the entire catchment, killing 147 people and destroying 74,000 households. These two dams successfully reduced peak flows in the river years later during a 1982 flood, saving lives and property (Green et al. 2000).

However, the effectiveness of large dams to replace the role of natural wetlands for flood mitigation is not well supported by scientific evidence. Wetlands and floodplains act as natural sponges; they expand by absorbing excess water in time of heavy rain and they contract as they release water slowly throughout the dry season to maintain streamflow. (See Chapter 20.) The large-scale conversion of floodplains and wetlands (some of it through dams) has resulted in declines in the natural mechanism for flood regulation. And while a handful of dams are being decommissioned in some countries (268 out of 80,000 in the United States, for example), an estimated 1,500 dams are under construction worldwide and many more are planned, particularly in the developing world (WWF and WRI 2004). River basins with the largest number of dams over 60 meters high planned or under construction include the Yangtze Basin in China with 46 large dams, the La Plata Basin in South America with 27, and the Tigris and Euphrates River Basin in the Middle East with 26 (WWF and WRI 2004).

The debate on cost, benefits, and performance of large dams continues, but given recent reviews (see WCD 2000), the traditional reliance on constructing such large operations for water supply is being called into question on environmental, political, and socioeconomic grounds (Gleick 1998; WCD 2000).

7.3.2.1.2 Interbasin transfers

Interbasin water transfers represent yet another form of securing water supplies that can greatly alleviate water scarcity. They include any canals, ditches, tunnels or pipelines that divert water from one river or groundwater system to another, typically from dammed reservoirs, and often represent massive engineering works involving both ground and surface waters. Changes to natural surface water hydrographs can be enormous and virtually instantaneous. The Great Man-Made River Project in Libya, for example, transports over 2 cubic kilometers of fossil groundwater a year through 3,500 kilometers of desert to huge coastal storage reservoirs that support 135,000 hectares of irrigable cropland, one third of the country's total (UN/WWAP 2003).

Two of the world's largest interbasin transfers are the 93% loss of flow (27 cubic kilometers per year) from the Eastmain River and a 97% gain of flow (53 cubic kilometers per year) in the La Grande River (Dynesius and Nilsson 1994), both in Canada. In total, the flow being diverted without return to its stream of origin in Canada alone totaled 140 cubic kilometers a year in the 1980s (Day and Quinn 1987), more than the mean annual discharge of the Nile River and twice the mean annual flow of Europe's Rhine River. The Farraka Barrage alone diverts over 9% of the Ganges River's historical mean annual flow and over 5% of the flow for the entire Ganges-Brahmaputra basin (Nilsson et al. 2005).

A gigantic diversion project is also under way in China, which proposes to move 40 cubic kilometers per year (MWR 2004) of water from southern China to the parched parts of northern China, thus connecting the Yangtze River with the Hai, Huai and Yellow Rivers. Three channels, two of which are over 1,000 kilometers long, will be needed for this transfer, which corresponds to 4% of the average flow of the Yangtze River (U.S. Embassy in China 2003). Developers plan to bring enough water to replenish groundwater aquifers in the north. This withdrawal from the Yangtze, even though it represents only a small fraction of the river's annual flow, will likely still have some effect on downstream ecosystems: sediment loads needed to maintain riparian and coastal wetlands will be reduced, and pollutants will be marginally less diluted, raising their concentration in the Yangtze River's lower reaches (U.S. Embassy in China 2003). In addition, as water flows north from one basin to another, the introduction of non-native species and the transfer of contaminants could affect native fauna in the receiving basins (Snaddon and Davies 1998; Snaddon et al. 1998; U.S. Embassy in China 2003).

Social effects of interbasin water transfers are complex. Populations in the recipient basin of water transfers gain water for irrigation, industry, and human consumption, all leading to indisputable economic and social benefits. However, those living in the basin of origin (and particularly those downstream of the diversion point) often lose precisely those same benefits (Boyer 2001), and many times they are displaced to other parts of the country, losing their homes and cultural heritage. While sometimes economic compensation is offered to people displaced by dams, the amounts usually do not cover the potential losses in terms of livelihoods, economic productivity, and cultural and historical heritage (WCD 2000).

Resettlement is an issue for water transfers as well as for dams, with many resettled communities suffering from a marginalized status, and cultural and economic conflicts with the population into which they are resettled. The central route of the Yangtze-to-Yellow water transfer in China, for example, will require the resettlement of 320,000 people, each of whom is supposed to receive the equivalent of $5,000 in compensation (U.S. Embassy in China 2003).

The trade-offs involved in interbasin transfer schemes include both direct societal costs and benefits, as well as those involving ecosystems services and biodiversity. Yet given increasing demands for water in the future, such transfers are likely to remain an important mechanism for alleviating regional water shortages (Nilsson et al. 2005).

7.3.2.2 Virtual Water in Trade

Virtual water, or VW, refers to the amount of fresh water used during the production process and thus "embodied" in a good or service (Allan 1993). While tabulations could be made for any product, VW has been explored mainly from the perspective of crop and livestock production and trade, given the predominance of agriculture in water use globally.

Operationally, VW in agriculture can be defined as the quantity of water used to support evapotranspiration in crops, which are then consumed domestically (as human food or animal feeds) or traded internationally. Additional water to process food products and to care for livestock can also be tabulated (Oki et al. 2003a), but VW estimates are fundamentally determined by irretrievable water losses through crops. There is a vast mismatch between the weight of agricultural commodities produced and the VW embodied in their production. For example, 1 kilogram of grain requires 1,000–2,000 kilograms (liters) of water, even under the most favorable of climatic conditions (Hoekstra and Hung 2002), producing 1 kilogram of cheese requires >5,000 kilograms of water, and 1 kilogram of beef requires an average of 16,000 kilograms of water (Hoekstra 2003).

Water has been transported in internationally traded products for hundreds of years, but the concept of trading VW has only recently begun to be considered as a mechanism to alleviate regional or global water security by exploiting the comparative advantage of water-rich or water-efficient countries (Allan 1996; Jaeger 2001). However, VW does not take into account the nature of food production systems and other factors, such as soil erosion, biodiversity impacts, or pollution. Moreover, for political and social reasons, countries may elect to be self-sufficient and independent in food production. For example, India, which is food self-sufficient in aggregate, serves as a net exporter of food and virtual water despite being water-stressed.

A substantial volume of VW trade in food commodities has nonetheless been taking place. (See Figure 7.9.) Worldwide, international VW trade in crops has been estimated at between 500 and 900 cubic kilometers per year, depending on tabulations made from the exporting or importing country perspective and the number of commodities considered (Oki et al. 2003a; Hoekstra and Hung 2002; Hoekstra 2003). (See Table 7.8.) An additional 130–150 cubic kilometers per year is traded in livestock and livestock products. For comparison, current rates of water consumption for irrigation total 1,200 cubic kilometers per year, and taking into account the use of precipitation in rain-fed agriculture as well, the total water use by crops has been estimated to range from 3,200 to 7,500 cubic kilometers per year, depending on whether allied agroecosystem evapotranspiration is included (Postel 1998; Rockström and Gordon 2001). The most important exporters of crop-related VW are the OECD and Latin America, though individual sub-regions, such as Western Europe, are net importers of VW. Asia (Central and South) is the largest importer of VW.

Of the top 10 virtual water exporters, 7 countries are in water-rich regions, while of the 10 largest importers of VW, 7 are highly water-short, indicating a general redistribution of VW from relatively wet to dry regions. However, the notable absence of clear-cut relationships linking the degree of domestic water scarcity to dependence on external VW supplies (Hoekstra and Hung 2002) suggests that an optimal redistribution of water through crop production and trade is yet to emerge. The consequences of food self-sufficiency thus entrench present-day patterns of water scarcity, as can decisions to pursue an aggressive export marketing strategy in the face of unsustainable water use. Future increases in water stress over the coming decades (Vörösmarty et al. 2000; Alcamo et al. 2000) and further integration of a global economy are likely to be powerful forces in adopting the notion of VW into food production and trade policies. An analysis of international trade in VW for Africa is provided in Box 7.4.

7.3.3 Land Use and Land Cover Change

Among the major processes influencing water quantity and quality at the river basin scale are changes in land use intensity and land cover change. (See Table 7.9.) Land use changes affect evapotranspiration, infiltration rates, and runoff quantity and timing. Particularly important for human well-being are contrasting reductions in the overall quantity of available runoff with some types of land cover change versus concentrated peaks of runoff associated with flooding under other land cover changes that can often be translocated far downstream through river networks (Douglas et al. 2005).

For example, expanding impervious areas due to urban expansion greatly increases the volume and rate of stormflow into receiving streams. Such changes also affect the water quality and biodiversity of freshwater ecosystems (Jones et al. 1997). Land use changes that compact soils and reduce infiltration are associated with deficiencies in groundwater recharge and dry period baseflow, the long-term global consequences of which are yet to be documented. Reduced infiltration can also lead to longer lifespans of pools with stagnant water, thus providing increased breeding opportunities for mosquitoes and other vectors of human disease.

The impact on local water budgets of changes from forest cover to pasture, agricultural, or urban land cover are well docu-

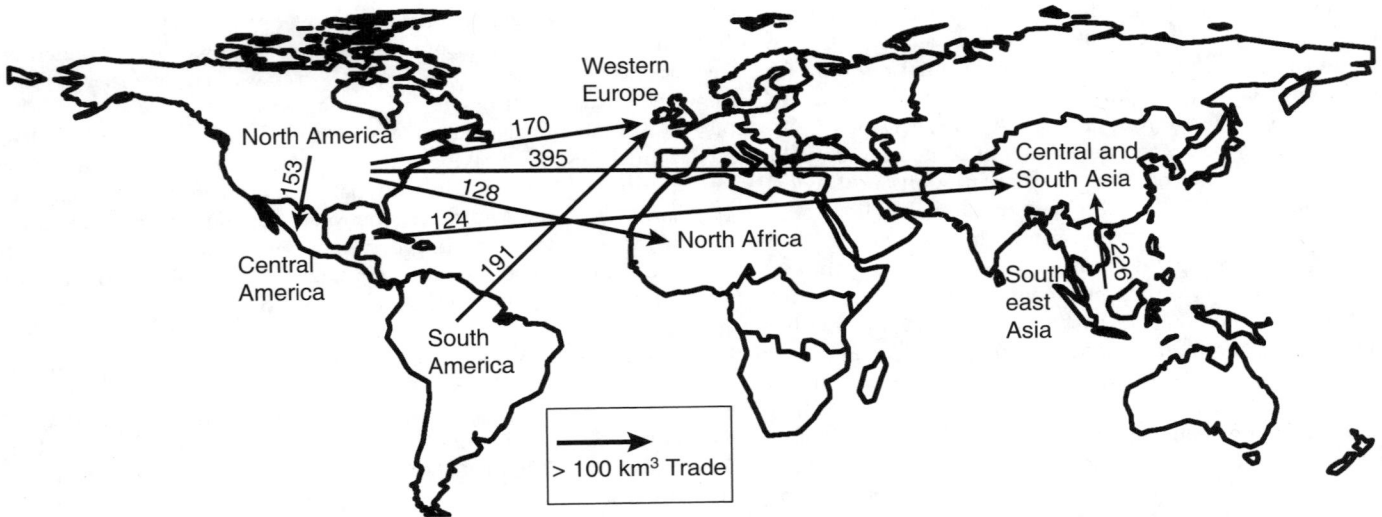

Figure 7.9. Net Inter-regional Trade in Major Crops Expressed as Embodied or "Virtual" Water Expended in Production of These Agricultural Commodities, 1995–99. The regions used differ from those used in the MA. Virtual water flows <100 km³ for the full period are not shown. Rain-fed and irrigated agriculture are considered, although estimates do not include transfer and drainage losses during irrigation. (Hoekstra and Hung 2003)

Table 7.8. Annual Transfer of Virtual and Real Water through Global Trade of Cereal and Meat Commodities, 2000. "Virtual" water in this table is estimated as the fresh water required by the importing country to produce the commodity, while "real" water is the fresh water expended by the exporter to produce the same commodity. Water equivalents are vastly greater than the actual weights traded, from 1000:1 to 3000:1 for cereals and >20,000:1 for beef. Through such trade there is a water-saving equivalent to approximately 20% of agricultural water withdrawals. (Oki et al. 2003a)

Commodity	Virtual Water Trade	Real Water Trade	Water "Saved"
	(cubic kilometers per year)		
Maize	130	50	80
Wheat	460	270	190
Rice	190	110	80
Barley	92	38	54
Cereal total	870	470	400
Beef	86	82	4
Pork	28	20	8
Chicken	37	25	12
Meat total	150	130	24

mented in the hydrological and ecological literature. While historically a large portion of the available information was generated for temperate and boreal areas of North America and Europe (Swank and Crossley 1988; Buttle et al. 2000), information is becoming available for selected sites in Amazonia, South Africa, and Australia, among others (Bruijnzeel 1990; Le Maitre et al. 1999). The global impact of 110,000 square kilometers per year net deforestation (FAO 1999) on runoff, however, has yet to be fully quantified.

Impacts of land use change patterns of weather and climate at different scales are only starting to be understood. (See Chapter 13.) Fragmenting a landscape alone can generate changes in local weather patterns (Avissar and Liu 1996; Pielke et al. 1997). At the continental level, land use changes can reduce recycling rates of water leading to reduced precipitation and distortions in the atmospheric circulation patterns that link otherwise widely separated regions of the globe (Chase et al. 1996; Costa and Foley 2000; Pitman and Zhao 2000). There has also been continental-to-global-scale acceleration in the loading of pollutants, including nutrients, onto the land mass associated with industrial agriculture, urbanization, and grazing. (See Chapters 12 and 15.) These inputs are translated into greatly elevated fluxes to and transport through inland water systems (Chapter 20), the effects of which pass in many cases fully to the coastal zone (Chapter 19).

Intensive agricultural and urbanized areas have expanded rapidly in the last 50 years. The current extent of cultivated systems provides an indication of the location of freshwater ecosystems that are likely to experience water quality degradation from pesticide and nutrient runoff as well as increased sediment loading (Revenga et al. 2000). (See Figure 7.10 in Appendix A.) Figure 7.11 (in Appendix A) shows, from a drainage basin perspective, the distribution and pattern of urban areas, as judged by satellite images of nighttime lights for 1994–95 (NOAA-NGDC 1998). Because more urbanized river basins tend to have greater impervious area as well as higher quantities of sewage and industrial pollution, this figure suggests the contemporary geography of pressures on freshwater systems arising from these classes of contaminants (Revenga et al. 2000).

The two Figures show contrasting patterns of modified land use across the world. Intensively cropped lands are concentrated in five areas: Europe, India, eastern China, Southeast Asia, and the midwestern United States, with smaller concentrations in Argentina, Australia, and Central America. Africa is striking for its lack of intensively cropped land, with the exception of small patches along the Nile, on the Mediterranean coast, and in South Africa. This reflects the minimal use of chemical inputs and the low level of agricultural productivity in most African countries. Figure 7.11 shows that highly urbanized watersheds are concentrated along the east coast of the United States, Western Europe, and Japan, with smaller concentrations in coastal China, India,

BOX 7.4

Virtual Water Content Associated with African Food Supply

The interplay between water availability and irrigation is critical in defining whether a country (or regions within a country) can be self-sufficient in food production and do so in a sustainable manner. This is especially true in Africa, where the climate and hydrology are highly unpredictable and as much as 40% of irrigation withdrawals in the driest regions are estimated to be non-sustainable (Vörösmarty et al. 2005). Africa also represents a flashpoint for future water scarcity and food security, with a large and rapidly growing population, enormous expanses of dry landscapes, extensive poverty, lack of investment in water infrastructure, and a lingering human health crisis.

Virtual water is the fresh water needed to produce crops embodying all evapotranspiration on rain-fed or irrigated cropland, plus any transit losses for irrigation (Raskin et al. 1995; Allan 1996). There are enormous throughputs of water within agroecosystems to satisfy the evaporative demands of crops, with ratios of >1,000-to-1 by weight for cereal products and >15,000-to-1 for beef (Hoekstra 2003). Thus, while food trade can be highly beneficial in simply economic terms, it could also help compensate for local water scarcity by exploiting the comparative advantage of water-rich countries to produce food (Allan 1996; Jaeger 2000).

The map of Africa (see Box 7.4 Figure A in Appendix A) shows the spatial distribution of annual virtual water production on rain-fed and irrigated croplands, computed from long-term average (1950–95) water balance terms. VW embodied in meat (beef, pork, and chicken) production was also estimated as the sum of VW in feed and fodder plus a portion of evapotranspiration that occurs over grazing lands, where it is assumed that 30% of net primary production and hence evapotranspiration could be used sustainably.

In Africa, much of the sustainable (rain-fed) agriculture occurs within the more humid regions of the continent, while most irrigated agriculture occurs in the semiarid and arid regions in northern and southern Africa and along the Sahel. At the continental scale, about 18% of total African VW is used for meat production, although this number is probably much higher because it is doubtful that all grazing land is used sustainably. Food imports (both crops and meat) represent over 20% of Africa's total VW consumption, illustrating a reliance on external sources for meeting the food needs of today's population. This reliance will likely continue to increase in the future, though some unknown fraction is intra-continental.

Globally, VW from crop production is computed to co-opt 14,600 cubic kilometers (20%) of the 66,400 cubic kilometers annual evapotranspiration. For Africa, crop production co-opts only 9% of annual evapotranspiration, a reflection of the fact that three quarters of Africa's cropland is located in arid and semiarid climates characterized by highly limited soil moisture stocks (Vörösmarty et al. 2005). The bar chart (see Box 7.4 Figure B in Appendix A) illustrates that while sub-Saharan Africa relies heavily on rain-fed agriculture (60–75% for South, East, West) and very little on irrigated agriculture (3–7%) for food production, North Africa has very little rain-fed crop production and obtains more than 60% of its within-region VW from irrigated agriculture. Much of this irrigation water is withdrawn from highly exploited river corridors, such as the Nile, as well as groundwater. To satisfy overall food demand, North Africa nearly doubles its available VW through food trade.

Table 7.9. Brief Overview of Hydrologic Consequences Associated with Major Classes of Land Cover and Use Change (Bosch and Hewlett 1982; Swank et al. 1988; Bruijnzeel 1990; Hornbeck and Smith 1997; Jipp et al. 1998; Swanson 1998; Bonnell 1999; Le Maitre et al. 1999; Buttle et al. 2000; Le Maitre et al. 2000; Zavaleta 2000; Zhang et al. 2001; Paul and Meyer 2001; Sun et al. 2001; Zoppou 2001; Tollan 2002)

Type of Land Use Change	Consequences on Freshwater Provisioning Service	Confidence Level
Natural forest to managed forest	slight decrease in available freshwater flow and a decrease in temporal reliability (lower long-term groundwater recharge)	likely in most temperate and warm humid climates, but highly dependent on dominant tree species
		adequate management practices may reduce impacts to a minimum
Forest to pasture/agriculture	strong increase in amount of superficial runoff with associated increase in sediment and nutrient flux	very likely at the global level; impact will depend on percentage of catchment area covered
	decrease in temporal reliability (floods, lower long-term groundwater recharge)	consequences are less severe if conversion is to pasture instead of agriculture
		most critical for areas with high precipitation during concentrated periods of time (e.g., monsoons)
Forest to urban	very strong increase in runoff with the associated increase in pollution loads	very likely at the global level with impact dependent on percent of catchment area converted
	strong decrease in temporal reliability (floods, lower long-term groundwater recharge)	stronger effects when lower part of catchment is transformed
		most critical for areas with recurrent strong precipitation events
Invasion by species with higher evapotranspiration rates	strong decrease in runoff	very likely, although highly dependent on the characteristics of dominant tree species
	strong decrease in temporal reliability (low long-term groundwater recharge)	scarcely documented except for South Africa, Australia, and the Colorado River in the United States

Central America, most of the United States, Western Europe, and the Persian Gulf (Revenga et al. 2000). While Figures 7.10 and 7.11 show the average composition of each large river basin in terms of intensively cultivated land or urban and industrial areas, they nonetheless hide within-basin differences that arise from highly localized patterns of crop production and urban point sources of pollution (Revenga et al. 2000).

The implications of these changes and the incomplete understanding of their consequences affect the manner in which humans interact with the water cycle. Integrated watershed management is the current paradigm for sustainable water use and conservation (Poff et al. 1997). It can yield important environmental and social benefits, as shown by a survey of 27 U.S. water suppliers that found the cost of water treatment in watersheds forested 60% or more was only half that of systems with 30% forest cover (Ernst 2004).

In practice, the integrated management approach is complex and difficult to implement because of limits to the understanding of interactions linking the physical and biotic processes that control water quantity and quality (Schulze 2004). Integrated management research typically has focused on local and short time scales and been limited to a very small portion of the world's watersheds. Most of the understanding of watershed dynamics and management principles comes from hydrological research on small watersheds and from studies at the local scale (Vörösmarty 2002). At present, the longest hydrological studies encompass only the last 20–40 years, but the recent application of GIS techniques facilitates reconstruction of past events to place the impact of contemporary land management into a longer-term perspective (Bhaduri et al. 2000).

One significant challenge to both scientific understanding and sound management is that multiple processes control water quantity, quality, and flow regime. The pattern and extent of cities, roads, agricultural land, and natural areas within a watershed influences infiltration properties, evapotranspiration rates, and runoff patterns, which in turn affect water quantity and quality. Additional challenges surround the fact that river basins extend across contrasting political, cultural, and economic domains (the Mekong River, for instance, flows through China, Laos, Thailand, Cambodia, and Viet Nam). Thus, there remains substantial uncertainty about the effects of management on different components of the hydrological cycle arising from the unique combinations of climatic, social, and ecological characteristics of the world's watersheds (Bruijnzeel 1990; Tollan 2002).

It is widely recognized that while much more information is needed to evaluate the impact of land use and cover change on freshwater provisioning services, integrated watershed management—despite its present degree of uncertainty—is both possible and would contribute significantly to improved management of water resources (Swanson 1998; Tollan 2002).

7.3.4 Climate Change and Variability

A major and natural characteristic of the land-based water cycle, and hence of water supply, is its variability over space and time. The large-scale patterns of atmospheric circulation dictate the world's climate zones and regional water availability. One particular concern arises from climate change, which in the past has shaped major shifts in the water cycle, such as changes in the Sahara from a much wetter region with abundant vegetation about 10,000 years ago to the desert of today (Sircoulon et al. 1999). A changing climate can modify all elements of the water cycle, including precipitation, evapotranspiration, soil moisture,

groundwater recharge, and runoff. It can also change both the timing and intensity of precipitation, snowmelt, and runoff.

Two issues are critical for water supply: changes in the average runoff supply and changes in the frequency and severity of extreme events, including both flooding and drought. Both of these changes have been difficult to articulate due to complexities in the processes at work as well as a non-uniform and, in many parts of the world, deteriorating monitoring network, as discussed in section 7.1.2.

Shiklomanov and Rodda (2003) present a study of continental-scale variations in water supply as represented in the observational record spanning 1921–88. They used data from a total of approximately 2,500 stations, maximizing, to the degree possible, length of record, suitably large river basins, and hydrographs reflecting near-natural conditions. The stations represented <10% of all available records and reflect great disparities in maximum length of record (from 5 to 178 years). (Statistically, the optimal record length for trend analysis is on the order of 30 years (Lanfear and Hirsh 1999; Shiklomanov et al. 2002), but detectability of a trend also depends on the relative lengths of the "base" (pre-change) and changed periods of record (Radziejewski and Kundzewicz 2004).)

Year-to-year variations over five continents were 10% or less (see Figure 7.12) but rose to as high as 35% when examining 27 climate-based subdivisions. Relatively dry periods occurred in the 1940s, 1960s, and late 1970s, with global runoff declining by up to 3,000 cubic kilometers a year. This is in contrast to relatively wet conditions in the 1920s, late-1940s to early 1950s, and mid-1970s. Though there are limitations to making such global statements, the overall conclusion with respect to renewable supplies of runoff is that despite some recent continental-scale trends (an increase in South America and decrease in Africa), there was no substantial global trend in renewable supplies of runoff over the 67 years tested. Labat et al. (2004) did, however, compute an increasing global trend in runoff. This was correlated to increasing global surface air temperature, amounting to 4% per degree Celsius over the last century, though with regional increases (Asia, North and South America) and decreases (Africa) or stability (Europe) over the last few decades.

Care must be exercised in interpreting such long-term trends, which are anticipated to be associated with climate change. Maps of trends presented by the IPCC (Houghton et al. 2001) show large-scale and spatially coherent increases as well as decreases in precipitation over multi-decadal periods that start in 1910, although these patterns shift depending on the time frame observed. A similar time dependency is evident in interpreting changes in rain-to-snow ratios across Canada, with a time frame of 1948–96 indicating completely opposite results than with a time series starting at 1960 and ending in 1990 (Mekis and Hogg 1999; Lammers et al. 2001).

The clearest signatures require long time periods and sufficient spatial integration units (that is, large drainage basins). Peterson et al. (2002), for example, found it impossible to detect a coherent trend in runoff without first aggregating the flow records from six large Eurasian rivers and over 65 years. Insofar as northern Eurasia is among the regions historically to show the clearest trends in climate warming and the general absence of other confounding effects such as land cover change and water engineering, these results point to the difficulty in assessing recent runoff trends.

Nonetheless, there is evidence that climate change may already be causing long-term shifts in seasonal weather patterns and the runoff production that defines renewable freshwater supply. Shifts toward less severe winters and earlier thaw periods in cold temperate climates that depend on snowfall and snowmelt result

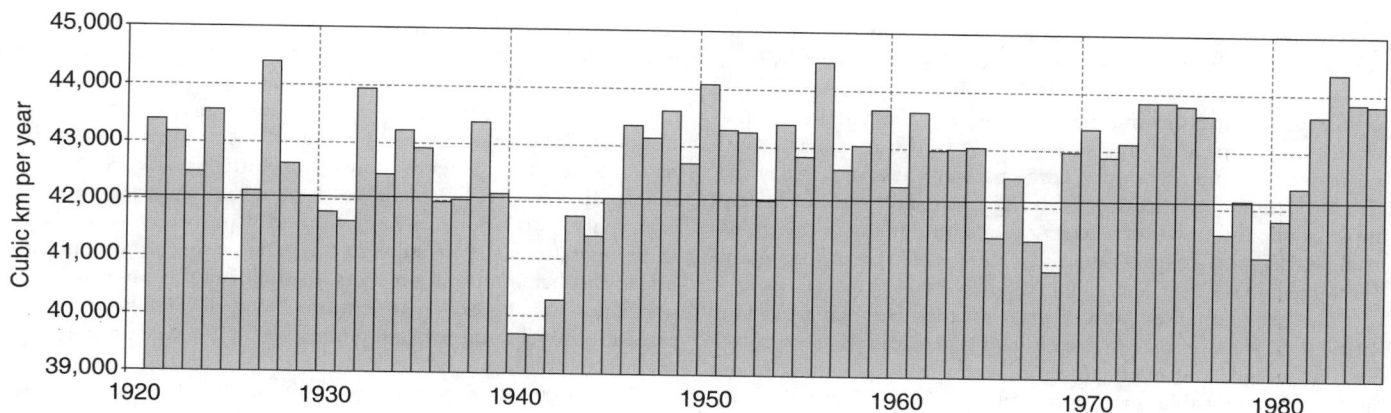

Figure 7.12. Time Series of Renewable Water Supply across the Global Landmass since 1920. The series is based on a subset of available discharge station records. (Shiklomanov and Rodda 2003)

in important changes in water availability (Dettinger and Cayan 1995; Hamlet and Lettenmaier 1999; WSAT 2000; Hodgkins et al. 2003). Multi-decade hydrological anomalies are apparent for Africa, with decreases on the order of 20% between 1951 and 1990 for both humid and arid zone basins that discharge into the Atlantic (Mahé 1993).

In the Sahel, persistent rainfall deficits could entrench desertification through a critical loss of water recycling between land and atmosphere, exacerbated by reduced soil infiltration when so-called hydrophobic soils are created in arid environments, and by soil compaction over poorly managed lands (Sircoulon et al. 1999). Such rainfall deficits also reduce replenishment of the groundwater resource, exacerbated by the decreased permeability of soils that favor storm runoff and flooding, even in the context of lower overall precipitation. In the transition zones between wet and dry regions across Africa, there is a highly uneven and erratic distribution of rainfall and river corridor flow (Vörösmarty et al. 2005). While this climate already produces chronic water stress, episodic droughts greatly increase the number of people at risk. Once each generation, the major sub-regions of the western Sahel, Horn of Africa, and SADCC region see a tripling in the number of people at risk from severe water stress (Vörösmarty et al. 2005).

An intensification of the water cycle, through more extreme precipitation in the United States (Karl et al. 1996; Karl and Knight 1998) and other parts of the world (Easterling et al. 2000; Houghton et al. 2001; Frich et al. 2002) has also been recorded. However, the effect of these increases on the rest of the hydrological cycle is only now being articulated.

In the United States, where sufficient records are available, Lins and Slack (1999) and Douglas et al. (2000) used stream gauging stations with 50 years of continuous records (from unregulated systems) to conclude that annual minimum and mean flows have increased. This was later confirmed by McCabe and Wollock (2002), who found statistically significant increases in annual moisture surplus (moisture that eventually becomes runoff) over the contiguous United States as a whole, but especially in the East. And while Yue et al. (2003) found similar increases in minimum and mean daily flows in northern Canada, they found the opposite to be true (significant decreases in minimum, mean, and maximum daily flows) in the southern part of the country.

Groisman et al. (2004) reported that warming in the northern half of the coterminous United States was related to a reduction in the extent of springtime snow cover and to the earlier onset of spring-like weather conditions and snow retreat. This has resulted in the increased frequency of cumulonimbus clouds and in a nationwide increase in very heavy precipitation. Warming in the southwest and northeast part of the country has led to greater summer dryness and increased fire danger. An interseasonal shift of precipitation from summer to fall in the Southeast was also noted.

The effect of increased precipitation extremes on floods is still debated (Douglas et al. 2000; Groisman et al. 2001; McCabe and Wolock 2002; Robson 2002; Milly et al. 2002) because flood response is influenced by many interacting factors, such as basin geology, terrain, and land cover as well as basin size and rainfall patterns. Also, the natural variability of flood flows can mask small changes in precipitation inputs.

Trends are also apparent in soil moisture distributed around the globe. Historical time series from more than 600 sites indicate a modest increase in growing period wetness for the majority of stations examined (Robock et al. 2000), contrary to the expectation (by general circulation models) of drier conditions in mid-continental areas due to climate change (e.g., Cubasch and Meehl 2000).

Taken together, these results indicate a high natural degree of variability and difficult-to-interpret shifts in runoff generation associated with historical climate change. The detection of such changes is complicated by the interactions among existing physical climate variations (that is, decadal and ENSO-type oscillations), land cover change, and water engineering, which for many parts of the world dominate the character of renewable water supplies.

7.3.5 Urbanization

During the twentieth century, the world's urban population increased almost fifteenfold, rising from less than 15% to close to half the total population (see Chapter 27), and by 2015 nearly 55% of the world will live in urban areas (UNPD 2003). In developing countries alone, the proportion of the population living in urban centers will rise from less than 20% in 1950 to 48% in 2015 (UNPD 1999, 2003). In fact, 60% of the fastest-growing cities with more than 750,000 people are located in the developing world, mostly in Asia (World Bank 2001). While 70% of the world's water use is for agriculture, the remaining withdrawals are for domestic household and other urban uses, including industry, and in many places these water resources are heavily polluted and limited by local shortages and distribution problems (UN-HABITAT 2003).

Urban residents bring with them a set of new challenges for water supply delivery, management, and waste treatment (WHO/UNICEF 2004; UN-HABITAT 2003). Because of the rapid rate of increase in cities around the world, water infrastructure is practically unable to keep apace, especially in the megacities with more than 10 million people. Large parts of these megacities lack the basic infrastructure for drinking water and sanitation, and most large cities in the developing world, and many in the industrial world, lack basic waste and storm water treatment plants (UN-HABITAT 2003).

The geographic location of many of these large and growing cities, such as close to coastal areas, and their rapid pace of growth has encouraged the overtapping of water resources that are not necessarily renewable, such as coastal aquifers. In Europe, for instance, nearly 60% of the cities with more than 100,000 people are located in areas where there is groundwater overabstraction (EEA 1995). Groundwater overexploitation is also evident in many Asian cities. Bangkok, Manila, Tianjin, Beijing, Chennai (formerly Madras), Shanghai, and Xian all have registered a decline in water table levels of 10–50 meters (Foster et al. 1998). These high levels of abstraction in many cases are accompanied by water quality degradation and land subsidence. For instance, the aquifer that supplies much of Mexico City had fallen by 10 meters as of 1992, with a consequent land subsidence of up to 9 meters (Foster et al. 1998).

Overabstraction is also an increasing problem with tourism-associated development, particularly in coastal areas. Groundwater overabstraction in such areas can reverse the natural flow of groundwater into the ocean, causing saltwater to intrude into inland aquifers. Because of the high marine salt content, even low concentrations of seawater in an aquifer are enough to make groundwater supplies unfit for human consumption (Scheidleder et al. 1999). Of 126 groundwater areas in Europe for which status was reported, 53 showed saltwater intrusion, mostly of aquifers used for public and industrial water supply (Scheidleder et al. 1999).

Unfortunately, the poor, mostly migrant workers from rural areas suffer most from reduced quality or quantity of water supply when they resettle to large cities. Poor residents of cities tend to concentrate in the outskirts, where safe drinking water and sanitation are less available, and they often depend on contaminated sources of water or intermediate water vendors who charge exorbitant prices.

In the context of these many problems, an emerging trend toward protecting water supplies for urban areas is noteworthy. A study of more than 100 of the world's largest cities, for example, found that more than 40% rely on runoff-producing areas that are fully or partially protected (Dudley and Stolton 2003). This reflects a growing recognition of the value of ecosystem services linked to sound watershed management approaches, as well as of the limits placed on urban water supply from polluted upstream source areas (UN-HABITAT 2003). The geography of downstream populations supported by upstream runoff-producing areas suggests the potential global importance of this management strategy. This is further demonstrated by Table 7.2, with data showing billions of people living downstream of particular MA ecosystems and their renewable freshwater flows.

7.3.6 Industrial Development

Industrial processes, which include withdrawals for manufacturing and thermoelectric cooling, today use about 20% of the total freshwater withdrawals, which has more than doubled between 1960 and 2000 (Shiklomanov and Rodda 2003). Even though this global use remains small in comparison to water used for agriculture, the current trend in shifting the manufacturing base from industrial to developing countries, due to globalization and international trade, is of concern for future water security. Much of the technology developed for industry is adapted to industrial nations, which are generally considerably more water-rich. When industrial plants are relocated to developing countries, many of which are water-poor or have limited water delivery services, these operations add pressure to the water resource base and increase conflict among water users. In addition, the environmental safeguards for effluent treatment are less well established or enforced in developing nations, adding to the scarcity problems by increasing pollution.

The most polluting industries, in terms of organic water pollutants, are those whose products are based on organic raw materials, such as food and beverage, paper and pulp, and textile plants (UN/WWAP 2003). Power station electric generation is the largest source of thermal water pollution. Estimates correlating water withdrawals for industrial use with population density by river basin show that many already water-stressed river basins are also centers for industrial production, such as in eastern China, India, and parts of Europe (UN/WWAP 2003).

Industrial emissions are released not only as thermal and chemical effluents into rivers and streams but also as gases and aerosols into the atmosphere. These can be transported for large distances and may end up deposited in other water bodies far from the emission source. Large areas of the continents show atmospheric deposition as the single most important source of nitrogen loading, with concomitant increases in pollutant transport through inland waterways (Green et al. 2004).

7.4 Consequences for Human Well-being of Changes in the Provision of Fresh Water

Water is essential for human well-being, but not all parts of the world receive the same amount or timing of available water supplies. Some areas contain abundant water throughout the year, others have seasonal floods and droughts, and still others have hardly any water at all. In river basins with high water demand relative to the available supply, water scarcity is a growing problem, as is water pollution. Water availability is already one of the major challenges facing human society, and the lack of water will be one of the key factors limiting development (WMO 1997; UN/WWAP 2003). The socioeconomic implications of delivering, using, managing, or buying water also have impacts on human well-being. This section begins with a brief overview of the benefits and required investments for water resource systems; examines the implications of freshwater scarcity, including treatment of pricing and equity issues; and concludes with descriptions of the consequences of too much water (flooding) and the connections between freshwater services and human health.

7.4.1 Freshwater Provision: Benefits and Investment Requirements

Over the long term, water use has generally increased geometrically, in line with population growth, increased food production, and economic development (L'vovich and White 1990), and during the last 40 years, there has been a doubling in the water used by society—from 1,800 to 3,600 cubic kilometers a year. In an aggregate sense, water is a required input generating value-added in all sectors of the economy, and trends in its use can be assessed from its ability to yield economic productivity. In the United States, for example, water productivity measured as GDP per cubic meter of freshwater use rose dramatically between 1960 and 2000 by about 25% per decade, to $18 per cubic meter (Postel and Vickers 2004), in response to shifts in regulation, technology, and restructuring of the economy.

Water provided for irrigation has a particularly important role, being responsible for 40% of global crop production (UN/WWAP 2003). And despite major challenges in conveying adequate drinking water and sanitation, more than 5 billion people are routinely provided with clean water and more than 3.5 billion have access to sanitation (WHO/UNICEF 2004). Further, with continued investments in water infrastructure, much of the world's population has benefited from allied improvements in public health, flood control, electrification, food security through irrigation, and associated economic development. From this standpoint, well-managed water resources have helped promote economic development, which is tied closely to improvements in many aspects of human well-being.

A good example is provided by a recent analysis (Hutton and Haller 2004) of the cost-effectiveness of different options to achieve MDG 7 (on access to safe water and basic sanitation). Of five scenarios tested, two considered Target 10—halving the proportion of people without sustainable access to safe water by 2015 and halving the proportion of people without sustainable access to improved sanitation. It was shown that for each dollar invested in both improved water supply and sanitation, a return of $3–34 can be expected. Among the health benefits of achieving the MDG drinking water target was a global reduction in diarrheal episodes of 10%. The economic benefits of simultaneously meeting the drinking water and sanitation MDG targets on households and the health sector amount to $84 billion per year, representing reduced health care costs, value of days gained from reduced illness, averted deaths, and time savings from proximity to drinking water and sanitation facilities for productive endeavor.

Because of the variability of the water cycle, economic benefits often accrue only after substantial investments in infrastructure and operations that stabilize and improve the reliability of water resources. Capital investments in water infrastructure totaled $400 billion in the United States over the last century (Rogers 1993). When the annual investment in water storage for irrigation globally during the 1990s of about $15 billion (WCD 2000) is tabulated, an important source of required capital can be seen, which can constitute a major fraction of agricultural investment for many developing countries (UN/WWAP 2003).

Worldwide, investments in dams have totaled $2 trillion (WCD 2000). World Bank lending for irrigation and drainage averaged about $1.5 billion per year from 1960 to 2000, although this continues to decrease from a peak of $2.5 billion in 1975 to its current rate of $500 million (Thompson 2001). Global costs for expanding irrigation facilities are estimated at $5 billion annually, but rehabilitation and modernization costs on existing irrigation works are estimated at an additional $10 billion or more per year (UN/WWAP 2003). Although projected funding for economic development and meeting the MDGs for the entire water sector is estimated to reach $111–180 billion a year, current investments in sanitation and water supply total from $10 billion to $30 billion annually (UN/WWAP 2003). Securing water resources is thus deeply embedded within development investment and planning but incompletely resolved. The private sector, with global revenues today standing at $300 billion annually (Gleick et al. 2002a), is a major player in providing potential investments, as described later.

7.4.2 Consequences of Water Scarcity

With population growth and the overexploitation and contamination of water resources, the gap between available water supply and water demand is increasing in many parts of the world. In areas where water supply is already limited, water scarcity is likely to be the most serious constraint on development, particularly in drought-prone areas. Earlier in this chapter we provided a quantification of water scarcity in physical terms. Here scarcity is mapped to issues relating directly to human well-being.

While decreased or variable water supply has sometimes presented itself as an opportunity to develop efficiency-enhancing responses (Wolff and Gleick 2002) and cooperation (Wolf et al. 1999; UN/WWAP 2003), more often it has spawned numerous development challenges, including increased levels of competition for water among people and between people and ecosystems; the use of non-sustainable supplies or development of costly alternatives; limits to economic growth, including curtailment of activities and required importation of food and other water-intensive commodities; pollution and public health problems; potential political and civil instability (Furlong and Gleditsch 2003; Miguel et al. 2004); and international disputes in transboundary river basins (Gleick 1998). These situations arise in part because society has typically managed ecosystems for one dominant service such as timber or hydropower without fully realizing the trade-offs being made in such management. This approach has led to the documented decline in freshwater ecosystem condition, with accompanying consequences for human well-being. The poor, whose livelihoods often depend most directly on ecosystem services, suffer most when ecosystems are degraded. (See Chapter 6.)

One of the problems thus far has been the difficulty of relating ecosystem condition to human well-being, particularly from the socioeconomic perspective. An emphasis on water supply, by developing more dams and reservoirs, coupled with weak enforcement of regulations, thus has limited the effectiveness of water resource management, particularly in developing regions of the world (Revenga and Cassar 2002).

As a consequence, policy-makers are now shifting from entirely supply-based solutions to demand management, highlighting the importance of using a combination of measures to ensure adequate supplies of water for different sectors, and slowly moving toward an integrated approach to water resources management (Schulze 2004) that is now linked directly to development initiatives (GWP 2000; UN-HABITAT 2003; Kakabadse-Navarro et al. 2004). Measures include improving water use efficiency, pricing policies, preservation of environmental flows, market incentives, privatization of water delivery, and public-private partnerships among others. (See MA *Policy Responses*, Chapter 7, for more on response measures in integrated water resource management.)

Human society has relied for decades on economic and social indicators for planning, but in virtually complete isolation of measures depicting the state and trends of ecosystem services. This section presents some of the latest findings relating water as an ecosystem service through social and economic indicators.

The need for integrated indicators or indices at the national or regional scale to help donors and decision-makers establish priorities in water resource management is widely acknowledged. Such metrics can also assist in monitoring progress toward sustainable development goals in a systematic manner. Many such tools have been proposed over the last several decades. For example, the Water Stress Index was developed in the 1970s to link population to water resources (Falkenmark 1997), and various other indices have been proposed, such as the Stockholm Environment Institute's Water Resources Vulnerability Index (Gleick et al. 2002b). New water scarcity indices capitalizing on geospatial data sets and high-resolution digital representations of river networks can define the climatic and hydrological sources of water-related stress (Vörösmarty et al. 2005).

One important indicator that combines physical, environmental, economic, and social information related to water availability and use is the Water Poverty Index. The WPI is similar to the Human Development Index but applicable to a more local scale, where the impacts of water scarcity are fundamentally expressed. It measures water stress at the household and community level and was designed to "aid national decision makers, at community and central government level, as well as donor agencies, to determine priority needs for intervention in the water sector" (Sullivan et al. 2003).

The WPI reflects an attempt to quantify inequities in water allocation and the inability of the poor to govern access to water. It has five key components, each based on a series of input variables that are weighed and aggregated into the overall index. When an element cannot be measured, proxy indicators are used in its place. The WPI relies in part on standardized data collected for other purposes, and thus can be used in comparative analysis of water stress across countries. For instance, to provide inputs on water management capacity the index uses Log GDP per capita, under-5 mortality rates, and a UNDP education index, all used previously in constructing the HDI. The five components of the WPI and some of its key variables are:

- *Water resources*: The physical characteristics of water availability and water quality. This component includes total water availability, its variability across time (seasonality), and its quality.
- *Access to water*: This includes not only the distance to water from dwellings but, more important, the time spent in collecting water, conflicts over water use, and access to sanitation.
- *Water use*: This represents withdrawals for domestic, agricultural, and industrial purposes. In many parts of the world, small-scale irrigation and livestock are key components of livelihood strategies and thus are tabulated as inputs.
- *Capacity to manage water*: This component is measured in terms of income, education level, membership in water users associations, and the burden of illness due to contaminated water.
- *Environmental integrity*: If the ecosystems that support water delivery are degraded, then provision of water per se plus the many services derived from freshwater systems will be jeopardized. This component evaluates the integrity of freshwater ecosystems based on the use of natural resources, crop losses reported in the previous five years, and household reports of land erosion. Overall, no variables of the actual condition of aquatic ecosystems are included, suggesting a component of the index that could benefit from revision.

The WPI has been tested internationally in 140 countries as well as at the local scale in South Africa, Tanzania, and Sri Lanka. Finland and Iceland were found to score the best, while Haiti and Ethiopia fared worst (Lawrence et al. 2003). The results of the local pilot analysis look promising, but the WPI would benefit from a better incorporation of ecosystem condition and capacity measures. Nevertheless, the WPI is a vehicle for understanding the complex relationship between water services and human well-being. Moreover, as the authors state, it constitutes "a systematic approach that is open and transparent to all" (Sullivan et al. 2003), allowing incremental improvements to the index to be made through community consensus.

There are also important gender-related issues associated with water poverty. Women and men usually have different roles in water and sanitation activities, and these differences are pronounced in rural areas across the developing world (Brismar 1997; UN/WWAP 2003). Women are most often the users, providers, and managers of water in rural households and the guardians of

household hygiene. In many parts of the world, women and girls can spend several hours a day carrying heavy water containers, suffering acute physical problems as a result (WEDO 2004). The inordinate burden of acquiring water also inhibits women's and girls' opportunities to secure an education and contribute to family income (WEDO 2004).

7.4.3 The Cost and Pricing of Water Delivery

Water users in most countries are generally charged but a small fraction of the actual cost of water abstraction, delivery, disposal, and treatment (Briscoe 1999; WHO/UNICEF 2000; Walker et al. 2000), and in some countries implicit and explicit water subsidies can reach up to 93% (Pagiola et al. 2002). Moreover, externalities associated with freshwater use, such as salinization of soils, degradation of ecosystems, and pollution of waterways, have been almost universally ignored, promoting current inefficiencies in use and threats to freshwater ecosystems. In general, those with access to abundant or underpriced water use it in a wasteful manner, while many, usually the poor, still lack sufficient access to water resources.

When water is in short supply or when it is polluted or unsafe to drink, the expense of delivering water services can rise dramatically or force curtailment in use. As scarcity increases, the cost of developing new freshwater resources also reflects the need to secure water from sources sometimes at great distances from the eventual user, often involving complex hydrological engineering (Hirji 1998; Rosegrant et al. 2002). Until recently there were few incentives in most countries to use water efficiently. However, increasing costs of water supply, dwindling supplies, and losses of aquatic habitat and biodiversity are increasingly providing incentives to value water as an economic good. In most countries, governments bear the burden of water delivery to users, but maintaining necessary infrastructure and expanding it to reach unserved users or improve the efficiency of water delivery is the exception rather than the norm (Pagiola et al. 2002). Inadequate funding results in a lack of new connections and unreliable service, with serious consequences for the poor, who usually incur higher costs when forced to obtain water from alternative sources (Pagiola et al. 2002).

Water can be priced in a number of different ways, and the past decade has shown the increasing application of several common methods, including flat fees, fixed fees plus volumetric charges, decreasing block rates, and increasing block rates. Some of these measures discourage waste, while others lead to overuse. (See MA *Policy Responses*, Chapter 7). This section surveys recent trends in the price and cost of water, reviews cost-recovery strategies, and assesses the impact on human well-being of privatization and public-private partnerships that deliver freshwater services.

7.4.3.1 The Price of Water and Recent Trends

There are enormous disparities in the price of fresh water supplied to end-users, reflecting a complex interplay among several factors, including proximity to natural sources of sufficient quantity and quality, level of economic development, investments—both public and private—in water infrastructure, and governance. A survey of urban households across the developing world showed water costs from both public and private sources varying by a factor of 10,000, from $0.00001 per liter (for piped supply in Calcutta) to as high as $0.1 per liter (through private water vendors) (UN-HABITAT 2003). Even municipal supplies can constitute a substantial fraction of monthly family expenditure—for example, up to 20% in informal settlements in Namibia (UN-HABITAT 2003). An analysis of urban areas in Asia showed that prices charged by informal water vendors are more than 100 times that from domestic connections (ADB 2001). In Benin, Burkina Faso, Kenya, Mauritania, and Uganda, household connection fees to piped water supplies exceeded per capita GDP by factors of up to 5:1, rendering these unaffordable (Collignon and Vezina 2000).

Cities also have seen a marked increase in the cost of financing new water supplies. In Amman, Jordan, during the 1980s groundwater sources were used to meet water needs at an incremental cost of $0.41 per cubic meter. As groundwater supplies declined, the city began to rely on surface water pumped from a site 40 kilometers away at an average incremental cost of $1.33 per cubic meter (Rosegrant et al. 2002). In another example, the real cost of water supply for irrigation in Pakistan more than doubled between 1980 and 1990 (Dinar and Subramanian 1997).

In Algeria, drought during 2000–02 forced cuts in the provision of water supply from municipal networks (access restricted to several hours every two to four days), despite large investments in water supply networks by the Algerian government since 1962 (UN-HABITAT 2003). The situation was further exacerbated by lack of maintenance of the network, with water losses through leaking pipes and underpricing of the resource use. Today, price increases and a major facilities upgrade are under way. Many African cities have exhausted and polluted local groundwater supplies, necessitating expensive transport of fresh water from distant suppliers (200 kilometers in the case of Dakar, Senegal) (UN-HABITAT 2003) or the need to invest in desalination, which is among the costliest methods of supplying fresh water (Gleick 2000; UN/WWAP 2003).

In addition to direct prices paid, additional costs are incurred by the poor provision of water services. Time spent in traveling to supplies, queuing, and transporting water can be a significant burden on household incomes for the poor. Compared with the late 1960s, households without piped water supply in Kenya, Uganda, and Tanzania today spend triple the time each day securing water, an average of over 90 minutes (UN-HABITAT 2003). Public taps are often in short supply, as in many Asian cities, where several hundred people are served by a single source (McIntosh and Yñinguez 1997). Further, the true costs associated with water delivery services are amplified by significant health burdens incurred when supplies are insufficient to meet basic needs. In the case of Lima, Peru, a major portion of household income (27%) is represented by medical costs and lost wages from water-related disease (Alcazar et al. 2000), while in Khulna, Bangladesh, an average of 10 labor days per month are lost due illness from poor water provision (Pryer 1993).

7.4.3.2 Cost Recovery

The fourth guiding principle of the Dublin Statement on Development Issues for the 21st Century (ICWE 1992) articulated that "water has an economic value" and "should be recognized as an economic good." At the same time, the statement argued that water should be available to all people at affordable prices. After much discussion and controversy, which continues to this day, the Ministerial Declaration from the 2nd World Water Forum (2002) established that "the economic value of water should be recognized and fully reflected in national policies and strategies by 2005" and that "mechanisms should be established by 2015 to facilitate the full cost pricing for water services, while the needs of the poor are guaranteed."

Supporters of full-cost water pricing argue that to improve efficiency, the set price of water needs to reflect the cost of supplying, distributing, and treating it. There is some evidence that this principle works. For instance, price increases for water in Bogor, Indonesia, reduced domestic consumption by 30% (Rosegrant et al. 1995). Proponents of full-cost water pricing also point out that most of the poor are not meeting their basic water needs today under current public management, usually because of lack of government capacity and resources. Consequently, poor communities are already paying higher prices through intermediate water vendors than if they were connected to a water delivery system.

But while pricing water to reflect its true cost is relatively simple in theory, the political and social obstacles are formidable. Opponents to the idea of full-cost water pricing claim that access to water is a fundamental human right. Water, like air, should therefore not be treated as an exchangeable, marketable commodity, because if market conditions rule, access to water becomes dependent on the ability to pay and not an inherent entitlement. In the eyes of many, establishing a price for water or privatizing its delivery puts many of the poorest, most marginalized people at risk of not getting enough water to meet basic needs.

The majority of OECD countries have adopted or are adopting, as an operating principle, the full-cost recovery concept in water management, although what should be covered under this "full cost" is still a matter of debate. Infrastructure costs, however, are not usually included (UN/WWAP 2003). As pricing was restructured and subsidies reduced during the 1990s and in the current decade in industrial countries to capture the full-cost recovery of water, the real price of water was increasing in 18 out of 19 countries surveyed (Australia being the only exception). In two thirds of OECD countries, over 90% of single-family homes are currently metered (OECD 1999).

The concept of full-cost water pricing in the developing world has been introduced with the support of local communities in situations where a more reliable service is assured. In Haiti, for example, shantytown residents with no connection to the water utility pay 10 times more for water from water vendors (water trucks) than those who are connected to the private water utility grid in nearby villages (Constance 1999). Residents connected to the grid have their water use metered and pay the corresponding fees.

7.4.3.3 Water Privatization

One of the most controversial trends in today's globalized economy is the increasing privatization of some water management and delivery services. In many countries, due to increasing costs of maintaining and expanding water networks and overstretched government budgets, private companies have been invited to take over some of the management and operations of public water systems. Private-sector investment in theory results in more financing for infrastructure as well as more-efficient operations and cost recovery, and the hope is that the public will benefit from a more stable and reliable water delivery system at a reasonable price.

Opponents to privatizing water services argue that putting private companies in charge of water will drive prices to the point that marginalized groups have no capacity to secure sufficient water even for their most basic of needs. In addition, because the profit motive fails to recognize environmental externalities, they argue that privatization will increase risk to the very ecosystems that help supply fresh water. The debate on public-private partnerships for water management was prominent on the agenda of the World Water Forum gatherings (in particular, at the 2nd Forum in The Hague and the 3rd Forum in Kyoto), as well as at the World Summit on Sustainable Development.

Despite trends toward privatization, at present over 80% of the world's investments in water, sanitation, and hydropower systems are by publicly owned bodies or international donors (Winpenny 2003). Therefore, the responsibility for providing water, over the short to medium term, will remain largely a public enterprise. Among industrial countries, there is much variation in the degree of privatization. In the United States in 2000, private companies provided only 15% of municipal water supply, although in the nineteenth century they provided nearly 95% (Gleick et al. 2002a). France, in contrast, shows more than half of all residents currently served by private companies (Gleick et al. 2002a).

In Latin America, Chile has been successful at delivering water through privatization, and nearly all houses in Santiago have access to clean water and sanitation. Despite exchange-rate fluctuations, a foreign company, Suez, has remained the water provider for Santiago and its region, investing over $1 billion in water infrastructure. Water in Chile has been priced at rates affordable by the middle classes, and stamps are given to poor people to guarantee near-universal access. Conversely, in Argentina, what looked like a positive trend did not withstand economic troubles. In 1993 privatization in Buenos Aires increased the share of residents served with water from 70% to 85%—an increase of 1.6 million people, with a concurrent drop in prices (Peet 2003). Exchange-rate fluctuations in many developing countries, such as the currency devaluation in Argentina in 2002, add challenges to successful implementation of privatization schemes. If the currency of a country devalues, the price paid for water will be worth much less, and the foreign firm could pull out of the market, leaving users without reliable service (Peet 2003).

South Africa is using a different pricing scheme to improve poor people's access to water and has made good progress in providing water to nearly two thirds of those who lacked access in 1994, when apartheid officially ended. Despite the relatively low cost of water, however, some rural residents opted to consume free—but contaminated—water from other sources. In February 2000, to improve public health for the poor, the government introduced a scheme to provide households with 6,000 free liters of water per month, enough to provide 25 liters per person per day, with charges for additional use.

Privatization can be executed in many ways, depending on the level of transfer from public to private hands. Full transfer of ownership and operations of water resource systems so far has been rare. The majority of cases embody the transfer of certain operational aspects, such as water delivery, but the ownership of the water resources usually remains with the state, thereby forming a public-private partnership.

These partnerships have been demonstrated over the last few years to capture the benefits of privatization without all of the risk (Blokland et al. 1999). They do not privatize all of the water assets, but they do give private actors control over some elements of the water rights, infrastructure, and distribution systems. Yet public entities typically maintain ownership over some or all of these systems. Public-private partnerships work best when strong regulatory controls exist. A typical arrangement in France, for example, delegates the operation, maintenance, and development of public potable water and sanitation to private companies, though public bodies retain ownership of the system (Barraque et al. 1994; Gleick et al. 2002c).

Experience has shown that a clear legal framework, where risks are decreased and the cost of capital decreases, would be necessary to enlist private-sector involvement (Winpenny 2003). More detailed analysis of privatization as a response option for the sustainable management of water resources and freshwater ecosystems is presented in Chapter 7 of the MA *Policy Responses* volume.

7.4.4 Consequences of Too Much Water: Floods

In addition to water scarcity, the accumulation of too much water in too little time in a specific area can be devastating to populations and national economies. (See Chapter 16.) According to the latest *World Disasters Report* (IFRC/RCS 2003), on average 140 million people are affected by floods each year, more than all other natural or technological disasters put together. Between 1990 and 1999, there were over 100,000 people reportedly killed by floods. The majority of these deaths were in Asia (56,000), followed by the Americas (35,000), Africa (9,000), and Europe (3,000) (IFRC/RCS 2000).

In addition to human lives, floods are a costly natural hazard in monetary terms, with more than $244 billion damage from 1990 to 1999, the most of any single class of natural hazard (IFRC/RCS 2000). Although this arises from potential changes in climate variability and extreme weather, humans also play an important role, settling and expanding into vulnerable areas (Kunkel et al. 1999; van der Wink et al. 1998).

While catastrophic flooding has negatively affected society for thousands of years, naturally occurring floods also provide benefits to humans through maintenance of ecosystem functioning such as sediment and nutrient inputs to renew soil fertility in floodplains, providing floodwaters to fish spawning and breeding sites and helping to define the dynamics to which coastal ecosystems are adapted. Although floods are primarily natural events, human activity influences their frequency and severity. By converting natural landscapes to urban centers, deforesting hillsides, and draining wetlands, humans reduce the capacity of ecosystems and soils to absorb excess water and to evaporate or transpire water back into the atmosphere, creating conditions that promote increased runoff and flooding.

There are, then, potentially costly consequences of upstream anthropogenic activities on hydrological function that place downstream populations at risk, sometimes affecting other nations, as in the case of more than 250 international river basins (Wolf et al. 1999). Douglas et al. (2005) reported on a simulation study suggesting that, in aggregate, a 32% conversion of forests to agriculture across the pan-tropics has led to a mean increase in annual basin yields of approximately 10%, with a concomitant rise in seasonal high flows. More than 800 million people live along floodplains in river basins containing some amount of tropical forest, and if the most threatened of the existing forests are converted to agriculture in the future, approximately 80 million of them could be at risk from the hydrologic impacts associated with these land conversions. Costa et al. (2003) present empirical evidence that large-scale savanna clearance in the Tocantins basin in Brazil (175,000 square kilometers) has been associated with increases of 24% in mean annual and 28% in wet season flows, independent of climate variations.

Nevertheless, there is some agreement that the most catastrophic floods in large basins result from storms so large and persistent that peak flows are unaffected by land cover (Calder 1999; Bruijnzeel 2004). Further, the proclivities of particular regions to landslides, soil erosion, and debris flows, as in the Himalayas, constitute the dominant source of risk (Gilmour et al. 1987; Hamilton 1987; Gardner 2002). Thus the costs and benefits of designing interventions to mitigate floods have their limits, and there may be little opportunity to escape potential vulnerabilities to flooding, given current patterns of human settlement in high-risk areas.

These findings should not suggest abandonment of good land stewardship, which yields fundamental benefits in sustaining ecosystem services. But they do argue for clearly identifying the source areas of hazard and designing response strategies to protect life and property. Even when specific and well-established policy goals for watershed protection are formulated, stakeholder interests and sustainable funding issues add to the challenge of designing effective upstream-downstream management strategies (Pagiola 2002).

Further information on the impact of natural hazards, including floods, on human well-being can be found in Chapter 16.

7.4.5 Consequences of Poor Water Quality on Human Health

Water is an essential resource for sustaining human health, and there is a basic per capita daily water requirement of 20 to 40 liters of water free from harmful contaminants and pathogens for the purposes of drinking and sanitation, which rises to 50 liters when bathing and kitchen needs are considered (Gleick 1996, 1998, 1999). Yet billions of people lack the services to meet this need, as documented earlier. Water-related diseases include four major classes: waterborne, water-washed, water-based, and water-related vector-borne infections (Bradley 1977). Threats to health also arise from chemical pollution.

7.4.5.1 Water-Related Diseases

As a whole, water-related diseases are a leading cause of morbidity and mortality in many parts of the developing world, with estimates ranging from 2 million to 12 million deaths per year (Gleick 2002), although monitoring and reporting remain poor in many countries. (See Table 7.10.) UN/WWAP (2003) reports 3.2 million deaths each year from water-related infectious disease, or about 6% of all deaths. The lack of access to safe water and to basic sanitary conditions also translates into the annual loss of 1.7 million lives and at least 50 million disability-adjusted life years. (The DALY is a summary measure of population health, calculated as the sum of years lost due to premature mortality and the healthy years lost due to disability for incident cases of the ill health condition. The DALY is not only an effectiveness indicator

in the economic evaluation of different intervention options but also a reflection of the impact of ill health on the income-generating capacity of the poor.)

The first three categories of water-related diseases are most clearly associated with lack of access to improved sources of drinking water, and in turn to ecosystem condition. Improved sanitation through the safe disposal of human waste is a major development objective that improves the health of those served directly by separating drinking water from wastewater. In developing countries, however, 90–95% of all sewage and 70% of industrial wastes are dumped untreated into surface waters (UNFPA 2001), placing both downstream populations as well as ecosystem functions at risk. (See Chapter 15.) The fourth category of water-related disease is associated with ecological conditions that favor disease vector breeding. These may be natural (such as those supporting malaria transmission by *Anopheles gambiae* mosquito across large parts of Africa south of the Sahara) or anthropogenic, through improperly planned irrigation systems, dams, and urban water systems. (See Chapter 14.)

Waterborne diseases are caused by consumption of water contaminated by human or animal waste and containing pathogenic parasites, bacteria, or viruses. They include the diverse group of diarrheal diseases as well as cholera, typhoid, and amoebic dysentery. These diseases occur where there is a lack of access to safe drinking water for basic hygiene, and most could be prevented by treating water before use. The World Health Organization estimates that there are 4 billion cases of diarrhea each year in addition to millions of other cases of illness associated with lack of access to safe water (WHO/UNICEF 2000). This translates into 1.7 million deaths per year, mostly among children under the age of five (WHO 2004). Morbidity and mortality from microbial contamination are orders of magnitude greater in developing countries than in the industrial world.

Water-washed diseases are caused by poor personal hygiene and skin or eye contact with contaminated water; their incidence is associated with the lack of access to basic sanitation and suffi-

---

Table 7.10. Selected Water-Related Diseases. Approximate yearly number of cases, mortality, and disability-adjusted life years. The DALY is a summary measure of population health, calculated as the sum of years lost due to premature mortality and the healthy years lost due to disability for incident cases of the ill-health condition. (WHO 2001, 2004)

Disease	Number of Cases	Disability-Adjusted Life Years (thousand DALYs)	Estimated Mortality (thousand)	Relationship to Freshwater Services
Diarrhea	4 billion	55,000[a]	1,700[a]	water contaminated by human feces
Malaria	300–500 million	46,500	1,300	transmitted by *Anopheles* mosquitoes
Schistosomiasis	200 million	1,700	15	transmitted by aquatic mollusks
Dengue and dengue hemorrhagic fever	50–100 million dengue; 500,000 DHF	616	19	transmitted by *Aedes* mosquitoes
Onchocerciasis (river blindness)	18 million	484	0	transmitted by black fly
Typhoid and paratyphoid fevers	17 million			contaminated water, food; flooding
Trachoma	150 million, 6 million blind	2,300	0	lack of basic hygiene
Cholera	140,000–184,000[b]		5–28[b]	water and food contaminated by human feces
Dracunculiasis (Guinea worm disease)	96,000			contaminated water

[a] Specifically attributable to unsafe water, sanitation, and hygiene from WHO (2002).

[b] The upper part of the range refers specifically to 2001 as reported in UN/WWAP 2003.

cient water for effective hygiene (Bradley 1977; Gleick 2002; Jensen et al. 2004). These include scabies, trachoma, and flea, lice, and tick-borne diseases. Trachoma alone is estimated to cause blindness in 6 million people (WHO/UNICEF 2000). In addition, the transmission of intestinal helminths (*Ascaris, Trichuris,* and hookworm) is linked to a lack of sanitation facilities and is estimated to account for a global annual loss of over 2 million DALYs.

Water-based diseases are those caused by aquatic organisms that spend part of their life cycle in the water and another part as parasites of animals. As parasites, they usually take the form of worms, using intermediate animal vectors such as snails to thrive, and then directly infecting humans either by boring through the skin or by being swallowed. They include Guinea worm infection, schistosomiasis (bilharzia), and a few other helminths (certain liver flukes of local importance in Southeast Asia, for instance, such as *Opisthorchis viverrini*) that infect humans through either direct contact with contaminated water or the consumption of uncooked aquatic organisms.

Although these diseases are not usually fatal, they prevent people from living normal lives and impair their ability to work. For instance, 200 million people worldwide are infected with schistosomiasis, of which 20 million suffer severe consequences, with an estimated global annual burden of 1.7 million DALYs (WHO 2004). The prevalence of water-based diseases often increases where dams are constructed, because stagnant water is the preferred habitat for aquatic snails, their most important intermediary hosts. For instance, the Akosombo Dam in Ghana, the Aswan High Dam on the Nile in Egypt, and the Diamma Dam at the mouth of the Senegal river have resulted in huge increases of local schistosomiasis prevalence. (See also Chapter 14.)

Water-related vector-borne diseases are caused by parasites that require a vector (such as insects) to develop and transmit the disease to humans. For example, *Anopheles* mosquitoes are the vectors for a protozoan parasite (*Plasmodium*) that causes malaria. These diseases are strongly ecosystem-linked, in contrast to the other three categories of water-related diseases, where water quality (and to some extent quantity) is the key determinant. Their distribution reflects the distribution of ecosystems suited to the propagation of the vectors.

Vector species, moreover, are highly diverse, so that detailed ecological requirements differ over wide ranges. Anopheline mosquitoes—vectors of malaria (1.3 million deaths a year and an annual burden of over 46 million DALYs), for example—breed in different types of freshwater ecosystems and brackish water coastal lagoons. *Aedes,* vectors of dengue and yellow fever, originally breeding in leaf axils of bromeliads, are cosmopolitan in human settlement areas, where they breed in small water pools. Urban filariasis vectors (*Culex* ssp.) breed in organically polluted water. And the blackfly vectors of onchocerciasis breed in oxygenated waters of rapids.

These vector-borne diseases are not typically associated with lack of access to safe drinking water but rather with water management practices in tropical and sub-tropical regions of the world. Several parasitic diseases endemic of tropical regions, such as Rift Valley Fever and Japanese encephalitis, spread easily with the presence of reservoirs, irrigation ditches and canals, and rice fields (WCD 2000). (See Chapter 14.) In all, more than 30 diseases have been linked to irrigation and paddy agriculture (WRI et al. 1998). Consequently, improved water management, drainage, and storage practices can help reduce the transmission risk, particularly in areas where anthropogenic conditions have led to the introduction of these diseases.

7.4.5.2 Chemical Pollution

Another set of diseases affecting industrial and developing nations alike arises in response to chemical pollution of water by heavy metals, toxic substances, and long-lived synthetic compounds. While evidence of the long-term impacts of chemical pollution can be detected even in the remote Arctic (AMAP 2002), the impacts on poor populations in developing countries are difficult to identify, given the lack of reliable and comprehensive records. However, exposure to chemical agents in water has been related to a range of chronic diseases, including cancer, lung damage, and birth defects. Many such diseases develop over several years, making the links between cause and effect difficult to establish. On a global scale, the burden of disease from chemical pollution is much lower than from microbial contamination and parasitic diseases, but in some highly polluted regions these risks can be substantial (WRI et al. 1998). Exposure to chemical pollutants can also compromise the immune system, rendering people more susceptible to microbial and viral infections. The cumulative and synergetic effects of long-term exposure to a variety of chemicals, especially at low concentrations, cannot be well quantified at present.

Naturally occurring inorganic pollutants constitute a class of chemical pollution with serious long-term health effects. Arsenic, which occurs naturally in some soils, for example, can become toxic when exposed to the atmosphere, as seen in areas with high water abstraction from underground aquifers (WRI et al. 1998). Arsenicosis is the result of arsenic poisoning from drinking arsenic-rich water over long periods of time and is a great concern in many countries, including Argentina, Bangladesh, China, India, Mexico, Thailand, and the United States (Bonvallot 2003). WHO estimated in 2001 that in Bangladesh alone, 35–77 million people—close to half the population—were exposed to drinking water from deep wells contaminated with high levels of arsenic (5–50 times the limit of 0.01 milligrams per liter recommended by WHO) (Bonvallot 2003). Arsenic is a carcinogen linked to skin, lung, and kidney cancer, although these diseases can go undetected for decades (WRI et al. 1998). In other parts of the world, high fluoride concentrations in drinking water have resulted in long-term effects that weaken the skeleton.

Chronic effects also arise from anthropogenic pollutants such as discharge from mining operations, pesticide runoff, and industrial sources. Long-term lead poisoning from old water pipes, for example, can cause significant neurological impairment (WRI et al. 1998). Mercury contamination can also originate from industrial discharge and runoff from mining activities, accumulating in animal tissue, particularly fish (WCD 2000).

Nutrient runoff is another concern from the standpoint of human health, especially in light of pandemic increase in loadings to inland water ecosystems, for example, of nitrogen (described earlier; see also Chapters 12 and 20). Although there is no global assessment of how many water bodies exceed the WHO guidelines on nitrate levels, most countries report that nitrates are one of the most common contaminants found in drinking water (WRI et al. 1998). Coastal and inland waters in regions with high levels of eutrophication have been observed to often propagate toxic algal blooms (toxic cyanobacteria) that can cause chronic disease. (See Chapter 19.) In China, for instance, the presence of cyanobacterial toxins in drinking water has been associated with elevated levels of liver cancer (WCD 2000). Excess nitrate in drinking water has also been linked to methaemoglobin anemia in infants, the "blue baby" syndrome (WRI et al. 1998).

Discharge from aquaculture facilities can also be loaded with pollutants, including high levels of nutrients from uneaten fish feed and fish waste, antibiotic drugs, and other chemicals, including disinfectants such as chlorine and formaline, antifoulants such as tributyltin, and inorganic fertilizers such as ammonium phosphate and urea (GESAMP 1997). These chemicals can significantly degrade the surrounding environment, particularly local waterways (GLFC 1999). The use of antibiotics and other synthetic drugs in aquaculture can also have serious health effects on people and ecosystems more broadly. The antibiotic chloramphenicol, for example, can cause human aplastic anaemia, a serious blood disorder that is usually fatal. While many countries have banned the use of chloramphenicol in food production, the level of enforcement varies considerably (GESAMP 1997; Health Canada 2004). A further risk from antibiotic use is the spread of antibiotic resistance in both human and fish pathogens. The U.S. Center for Disease Control and Prevention reported that certain antibiotic resistance genes in *Salmonella* might have emerged following antibiotic use in Asian aquaculture (Angulo 1999 as cited in Goldburg et al. 2001).

There is also evidence from studies on wildlife that humans may be at risk from persistent organic pollutants and residual material that has the ability to mimic or block the natural functioning of hormones, interfering with natural physiological processes, including normal sexual development (WRI et al. 1998). Certain chemicals such as PCBs, DDT, dioxins, and at least 80 pesticides are regarded as "endocrine disrupters," chemicals that may interfere with normal human physiology, undermining disease resistance and affecting reproductive health (WRI et al. 1998).

Finally, pharmaceutical products excreted by livestock or humans comprise a set of "emerging contaminants," whose impacts on human well-being, ecosystems, and species are not yet understood. These contaminants are hard to detect with current technologies, but their impact on wildlife are already observed in some parts of the world. In the United States, the first nationwide survey conducted in 1999 and 2000 found hormones in 37% of the streams surveyed and caffeine in more than half (Kolpin et al. 2002). Just recently, 42% of the sampled male bass in a relatively pristine stretch of the Potomac River in the United States were found to be producing eggs. The exact cause is still unknown, but it is hypothesized that it could be caused by chicken estrogen left over in poultry manure or perhaps human hormones discharged into the river with processed sewage.

7.4.5.3 Sanitation and Provision of Clean Water: Challenges for the Twenty-first Century

Providing "improved" clean water supply and sanitation to large parts of the human population remains a challenge (WHO/UNICEF 2004; United Nations Statistics Division 2004). (See Box 7.5 for definitions of improvement.) The most recently completed and comprehensive assessment of improved water and sanitation (WHO/UNICEF 2004) concluded that 1.1 billion people around the world still lack access to improved water supply and more than 2.6 billion lack access to improved sanitation, with strong geographic variations. (See Table 7.11 and Figures 7.13 and 7.14 in Appendix A.) Asia contains two thirds of all people who lack access to improved drinking water and three quarters of those who lack access to improved sanitation. Africa is next most prominent in terms of numbers still awaiting improvements in supply and sanitation. Other continents show much smaller numbers but may have relatively low rates of service, as in Oceania, with less than 50% served for both supply and sanitation.

There has been progressive improvement in the provision of sanitation since 1990 (see Table 7.12), recently prompted by the ambitious target for sanitation of the MDG environmental sus-

---

BOX 7.5
Defining Improved Water Supply and Sanitation

"Improved" water supply includes household connections, public standpipes, boreholes, protected dug wells, protected springs, and rainwater harvesting systems, but it does not include protected rivers or ponds, unprotected wells or springs, and unmonitored vendor-provided water (bottled water is not considered improved due to quantity limits arising from its high expense).

"Improved" sanitation technologies include connections to a public sewer, connections to a septic system, pour-flush latrines, simple pit latrines, and ventilated improved pit latrines. Excreta disposal systems are considered adequate if they are private or shared (but not public) and if they hygienically separate human excreta from human contact. "Not improved" sanitation systems are service or bucket latrines (where excreta are manually removed), public latrines, or open pit latrines.

---

tainability goal—namely, to halve by 2015 the proportion of people lacking such service in 1990. Worldwide, the goal was set to move coverage from 49% to 75%, and progress is nearly on track with the interim target for 2002 of 62% nearly attained. Of the nine regions analyzed, however, only four are on track or nearly, while five are behind schedule. The greatest challenge remains sub-Saharan Africa, which met only 4% of a targeted 17% improvement by 2002. Western Asia and Eurasia are less off their targets but have not moved forward. Overall, improvements in sanitation in rural areas have been significantly less than in urban areas, and there has even been a decline in the provision of sanitation in rural areas of Oceania and the former Soviet Union (WHO/UNICEF 2004).

The rapid and disorganized growth in cities and peri-urban areas in developing countries is likely to hinder progress toward improved water delivery and sanitation systems. In 2000 alone, 16 cities around the world became megacities, with more than 10 million inhabitants each, housing 4% of the world's population (United Nations 2002). Most of these megacities fall within regions already suffering from water stress (UN/WWAP 2003). In Africa, Asia, and Latin America, 25–50% of the population live in informal or illegal settlements around urban centers where no public services and no effective regulation of pollution and ecosystem degradation are available (UN-HABITAT 2003). Half of the urban population in Africa, Asia, and Latin America and the Caribbean suffers from one or more diseases associated with inadequate water and sanitation (UN/WWAP 2003).

Even if government or municipal authorities were inclined to expand water and sanitation services to informal urban settlements, the lack of formal land ownership, plot designation, and infrastructure make this very difficult and unlikely. In many countries, water and sanitation authorities are only allowed to provide services and connect households to the water grid if proof of land-ownership is provided (UN-HABITAT 2003). These problems are in addition to the basic inability of slum dwellers to pay for connection charges and monthly fees without subsidies. With urban populations expected to encompass 80% of the world's population by 2030 (UNPD 1999), the supply of water and sanitation to city dwellers is set to become one of the greatest challenges to development.

7.5 Trade-offs in the Contemporary Use of Freshwater Resources

This chapter has provided an assessment of the recent history and contemporary state of global freshwater provisioning services. It

Table 7.11. Access to Clean Water and Sanitation (WHO/UNICEF 2004)

Geographic Region[a]	Population Unserved by Improved Drinking Water Supply (million)	Unserved by Clean Drinking Water Supply (percent of region's population)	Population Unserved by Improved Sanitation (million)	Unserved by Improved Sanitation (percent of region's population)
Africa				
North	14	10	40	27
Sub-Saharan	288	42	438	64
Asia				
Western	22	12	39	21
South	237	16	933	63
Southeast	112	21	209	39
Eastern	302	22	756	55
Latin America and the Caribbean	59	11	134	25
Eurasia	20	7	48	17
Oceania	4	48	4	45
World Total	**1,060**	**17**	**2,600**	**42**

[a] According to WHO/UNICEF definition; does not correspond fully to MA reporting units.

Table 7.12. Regional Progress toward the MDG Sanitation Goal (WHO/UNICEF 2004)

Geographic Region[a]	Coverage in 1990	Coverage in 2002	Coverage Needed in 2002 to Remain on Track	Coverage Needed by 2015 to Achieve MDG Target
	(percent)			
Regions on track				
Eastern Asia	24	45	43	62
Southeast Asia	48	61	61	74
Regions nearly on track				
North Africa	65	73	74	82
Latin America and the Caribbean	69	75	77	84
Regions not on track				
South Asia	20	37	40	60
Sub-Saharan Africa	32	36	49	66
West Asia	79	79	84	90
Eurasia	84	83	88	92
Oceania	58	55	68	79
World Total	**49**	**58**	**62**	**75**

[a] According to WHO/UNICEF definition; does not correspond fully to MA reporting units.

has documented a growing dependence of human well-being on fresh water, which in turn has promoted a variety of engineering strategies aimed at delivering reliable freshwater supplies. So effective has been the ability of water management to influence the state of this resource that anthropogenic impacts are now evident across the global water cycle. Much of the human influence is negative due to overuse and poor management, which has resulted in human-induced water scarcity, widespread pollution, and habitat and biodiversity loss. The capacity of ecosystems to sustain freshwater provisioning services thus has been greatly compromised throughout much of the world and may continue to remain so if historic patterns of managed use persist.

Sector-specific decisions often drive the nature of human interactions with water, with often unintended or purposefully ignored externalities on ecosystems. There is no shortage of examples. Flow stabilization optimizing hydroelectricity can severely fragment and degrade aquatic habitats and lead to losses of economically important fisheries. Industrial development with poor effluent management can result in severe pollution, leading to the loss of aquatic ecosystem function and biodiversity. Connecting urban dwellers to water supply and sewerage systems without due attention to water treatment, as has been commonplace, results in the release of toxic compounds and waterborne diseases that affect downstream water users. In arid and semiarid regions, decisions to promote national food self-sufficiency can translate into great risk to downstream populations and costly infrastructure, as rivers that normally carry water and sediments nourishing coastal lands and floodplains are diverted onto croplands or stabilized behind dams.

Trade-offs are thus an unavoidable component of human-freshwater interactions. Trade-offs are also inevitable in meeting Millennium Development Goals and other international commitments. To demonstrate this, a heuristic analysis is presented here to explore how emphasis on a particular objective could influence the capacity to attain others. The analysis uses the contemporary setting as its starting point, which is then tracked with respect to the impact of five specific interventions. These correspond directly to major objectives embodied in the Kyoto Protocol (carbon mitigation), the MDGs (poverty alleviation, hunger reduction, improved water services), and the Conventions on Biological Diversity and Wetlands (pragmatic ecosystem maintenance applied to inland and coastal ecosystems).

A non-intervention case (current trends) is also considered, analyzing the implications of allowing contemporary trends to continue. A time frame of approximately 10–15 years is considered, allowing sufficient time for general patterns to emerge. This time frame also is associated with the first targets of the MDGs.

The interventions and their impacts are specifically viewed through the lens of freshwater services and ecosystem maintenance. Thus, for carbon mitigation the positive impacts of expanding hydropower to reduce carbon emissions are considered, together with the negative impacts of flow fragmentation that compromises the normal functions of inland freshwater and coastal ecosystems. To maximize relevancy to the international development agenda, the findings refer to poor countries alone. The interventions and key results are summarized in Table 7.13 and Figure 7.15. In each case the contemporary baseline is the starting point, given by the intermediate of three circles. Improvement is depicted by movement outward to the larger circle. Declining condition is represented by a move inward, and no appreciable change settles on the middle curve.

It is important to note that these experiments are not predictions but instead are thematic devices to demonstrate broad-scale effects that can be supported by findings in this chapter. Although the details could be argued legitimately one way or another, it is the basic character of the response that is sought. Furthermore, as will become apparent, it is the behavior of the full set of experiments rather than individual cases that becomes most instructive.

Current Trends in Figure 7.15 is the first case, representing no meaningful change in the pace at which human development is attained or interventions are made to reverse ongoing threats to ecosystem services. This scenario shows direct beneficiary effects on human well-being but also sustained and substantial declines in the condition of aquatic ecosystems. On the positive side, there is some alleviation of hunger through increased food production that relies on expanded irrigation and use of agrochemicals; continued improvement to health by way of drinking water and sanitation access; some progress toward reducing poverty; and an expansion of hydropower, which in some parts of the developing world (such as South America) is already an important source of energy, with some beneficiary effects on carbon mitigation.

At the same time, aquatic ecosystems and their biodiversity will be increasingly degraded in this scenario due to the combined forces of industrial, agricultural, and domestic sources of pollution, hydropower with associated flow fragmentation, and habitat destruction. Lack of environmental regulation and enforcement exacerbates the trend. Reduced and highly regulated water flows in rivers continue to decrease the transport of water and sediment to estuaries and coastal wetlands. Food provisioning services, in terms of natural inland and coastal fisheries, are in decline, and freshwater provisioning will continue to be placed in jeopardy by the dual threats of overuse and pollution.

Major supporting and regulating services also continue their decline due to loss of ecosystem function across both inland aquatic systems and their linked terrestrial ecosystems. Particularly relevant to fresh water are losses in flood control (from poor land management, erosion, loss of wetlands), in self-purification potential of waterways (from chronic and acute land-based sources of pollution), and in protection of human health (from inappropriate waste disposal). The links between ecosystem services and human well-being mean that these losses of natural services could ultimately compromise the attainment of important development goals.

While the value of controlling greenhouse gases or instituting the MDGs is almost universally accepted, results in Table 7.13 and Figure 7.15 suggest that pursuing each objective in isolation of other development goals or environmentally sound management principles will be counterproductive. Interventions in accordance with strategies being promoted through the Conventions on Biological Diversity and Wetlands, which stress protection and wise use of ecosystems and their services for sustainable development, yield several positive effects on human well-being. These improvements arise from a purposeful strategy of integrated environmental management, which links environmental stewardship directly to poverty alleviation, food security, and clean water targets (CBD 2004; Ramsar Convention 2004).

There is a growing recognition that maintaining biodiversity and ecosystem integrity will require compromise and trade-offs. A good example is the critical choice between providing water for crop production or for healthy rivers and wetlands. In areas where irrigation and storage reservoirs are upstream of sensitive ecosystems, both livelihoods and environmental integrity can be at stake. One possible strategy to accommodate potential losses in food production and income is by managing basin-wide improvements in water productivity for agriculture through new crop breeding, innovative technologies, and water reuse strategies (Molden 2003), all saving water and reducing the need for irrigation and flow stabilization.

While only qualitative in nature, these findings clearly demonstrate the consequences of optimizing one development goal or conservation objective over others. This assessment indicates that there would be substantial inconsistencies in the major development and sustainability strategies should they not become better integrated. The impacts of these conflicts on freshwater provisioning services and ecosystem functioning are likely to compromise the sought-after progress inherent in these same international commitments. The conjunction of several incongruous objectives will further exacerbate the deterioration of inland and coastal systems documented in Chapters 19 and 20.

It is very certain that the condition of inland waters and coastal ecosystems has been compromised by the conventional sectoral approach to water management, which, if continued, will constrain progress to enhance human well-being. In contrast, the ecosystem approach, as adopted by CBD, Ramsar, FAO, and others, shows promise for improving the future condition of water provisioning services, specifically by balancing the objectives of economic development, ecosystem needs, and human well-being.

Table 7.13. Major Objectives Optimized in Experiments to Discern the Compatibility of Development Goals and International Conventions. These objectives are considered in the context of freshwater provisioning services and protection of inland and coastal waters. General categories of responses are given, as depicted in Figure 7.13. Positive, intermediate, and negative effects are relative to contemporary condition. A time horizon of 10–15 years is considered.

Sectoral Intervention	Relevant International Commitment	Positive Effects	Intermediate or Small Effects	Negative Effects
Current trends (non-intervention)			some progress toward carbon mitigation, poverty reduction, hunger alleviation, and access to water services	persistent decline in health of inland and coastal ecosystems and their services (provisioning, regulating, supporting)
Carbon mitigation	Kyoto Protocol	reduced CO_2 emissions through increased reliance on hydropower assumed to override reservoir respiration and methane emission; progress on hunger reduction, water services, poverty reduction as under current trends	water storage for irrigation yields some reservoir fisheries for food; urban benefits of hydroelectricity; rural poverty alleviation effects small in relation to current trends	waterborne disease increases in tropical regions; dams fragment habitat and modify fluxes of constituents and water through inland waterways; loss of inland fisheries; erosion, nutrient imbalance in coastal systems due to upstream reservoir trapping
Hunger reduction	MDG 1, Target 2	major beneficial effects on nutrition	well-fed populations show increased health benefits and poverty reduction; consumptive losses from expanded irrigation mean less water for hydroelectricity; little effect on improved water/sanitation	expanded irrigation and impoundment storage means less available water for inland and coastal ecosystems
Improved water services (access to clean water and sanitation)	MDG 7, Target 10	improved health; increased productivity of labor reduces poverty	similar water quality as under current trends if waste treatment assumed (not the norm); no impact on carbon mitigation or hunger alleviation assumed	inland and coastal pollution from sewage, assuming no treatment
Poverty alleviation	MDG 1, Target 1	rising standards of living; increased availability of hydropower with benefits for carbon mitigation; increased food demands and availability	increased access to water services leads to improved health for those served; effect mitigated by increased pollution and water-related diseases for remaining poor	strong impacts on natural ecosystems from agricultural pollution; water diversions for crops and industrial production; river fragmentation from dams
Pragmatic ecosystem maintenance (inland and coastal wetlands)	Convention on Biological Diversity, Convention on Wetlands (Ramsar)	integrated management leads to protection of inland/coastal ecosystems with improved freshwater provision (quantity and quality)	land management improves carbon mitigation and crop productivity; food sources from aquatic systems; stable water supplies allow for some high-productivity irrigation and well-managed reservoirs (for C mitigation as well); improved water quality leads to better health; aggregate benefit from all factors reduces poverty	no single objective met fully; compromises among stakeholders inherent in such a multiobjective framework

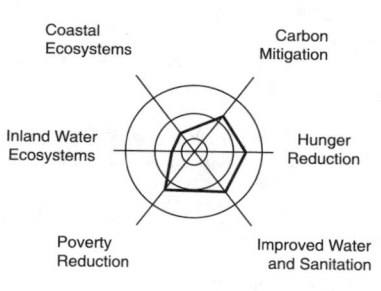
Current Trends

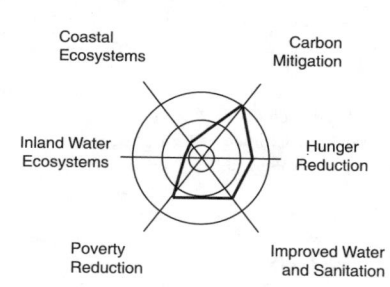
Kyoto Protocols: Carbon Mitigation

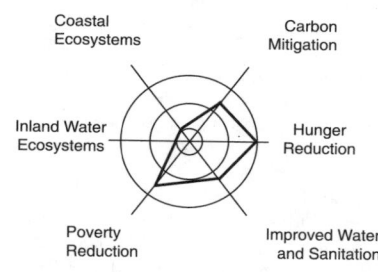
MDG: Hunger Reduction

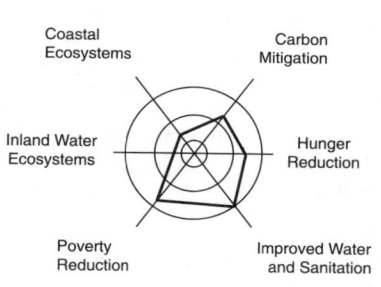
MDG: Improved Water Services

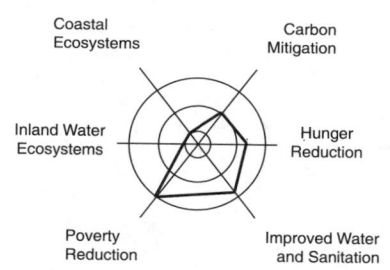
MDG: Poverty Alleviation

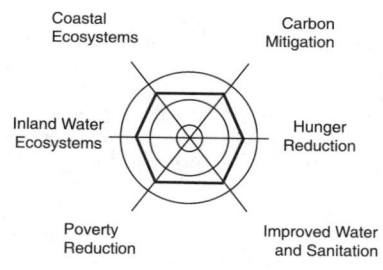
CBD, Ramsar: Pragmatic Ecosystem Maintenance

Figure 7.15. Trade-off Analysis, Depicting Major Interventions and Consequences on Condition of Ecosystems and Development Goals. Note that in the absence of integrated sustainable development and environmental protection plans, current trends and development-related interventions may compromise ecosystem functioning. Better balanced effects are noted by instituting strategies guiding the Convention on Biological Diversity and Convention on Wetlands (Ramsar). An approach balancing ecosystem protection and economic development could yield an aggregate net benefit to the entire suite of objectives. The contemporary starting point is the middle circle. Movement toward the outside circle indicates improvement while movement inward depicts negative trends. See text and Table 7.13 for further interpretation.

References

ADB (Asian Development Bank), 2001: *Water for All: The Water Policy of the Asian Development Bank,* Manila, Philippines.

Alcamo, J., T. Heinrichs, and T. Rösch, 2000: *World Water in 2025,* World Water Series Report #2, Center for Environmental Systems Research, University of Kassel, KasselGermany.

Alcamo, J., P. Döll, T. Henrichs, F. Kaspar, B. Lehner, et al., 2003: Global estimates of water withdrawals and availability under current and future "business-as-usual" conditions, *Hydrological Sciences Journal,* 48(3), pp. 339–48.

Alcazar, L., L.C. Xu, and A.M. Zuluaga, 2000: *Institutions, Politics, and Contracts: The Attempt to Privatize the Water and Sanitation Utility of Lima, Peru,* Policy research working paper, WPS 2478, World Bank, Washington, DC.

Allan, J.A., 1993: Fortunately there are substitutes for water: Otherwise our hydro-political futures would be impossible. In: *Priorities for Water Resources Allocation and Management,* ODA, London, UK, pp. 13–26.

Allan, J.A., 1996. The political economy of water: Reasons for optimism but long-term caution. In: *Water and Peace in the Middle East: Negotiating Resources in the Jordan Basin,* Tauris Academic Publications, London, UK, pp. 75–120.

Alexander, R.B., R.A. Smith, and G.E. Schwarz, 2000: Effect of stream channel size on the delivery of nitrogen to the Gulf of Mexico, *Nature,* **403,** pp. 758–61.

Alverson, K.D., R.S. Bradley, and T.F. Pedersen (eds.), 2003: *Paleoclimate: Global Change and the Future,* Springer, Berlin, Germany.

AMAP (Arctic Monitoring and Assessment Program), 2002: Arctic pollution 2002: Persistent organic pollutants, heavy metals, radioactivity, human health, changing pathways, AMAP, Oslo, Norway.

Angulo, F., 1999: Antibiotic use in aquaculture: Centers for Disease Control and Prevention memo to the record, 18 October 1999. Accessed 5 April 2001. Available at www.natlaquaculture.org/animal.htm.

Avakyan, A.B. and V.B. Iakovleva, 1998: Status of global reservoirs: The position in the late twentieth century, *Lakes and Reservoirs: Research and Management,* **3,** pp. 45–52.

Avissar, R. and Y. Liu, 1996: A three-dimensional numerical study of shallow convective clouds and precipitation induced by land-surface forcing, *Journal of Geophysical Research,* **101,** pp. 7499–18.

Baron, J.S., N.L. Poff, P.L. Angermeier, C.N. Dahm, P.H. Gleick, et al., 2002: Meeting ecological and societal needs for freshwater, *Ecological Applications,* **12(5),** pp. 1247–60.

Barraque, B., J.-M. Berland, and S. Cambon, 1994: *EUROWATER: Vertical report on France,* LATTS-ENPC, Paris, France.

Bhaduri, B., J. Harbor, B. Engel, and M. Grove, 2000: Assessing watershed-scale, long-term hydrological impacts of land-use change using a GIS-NPS model, *Environmental Management,* **26,** pp. 643–58.

Blackmore, D.J., 1999: The Murray-Darling Basin cap on diversions: Policy and practice for the new millennium, *National Water,* **(June),** pp. 1–12.

Blokland, M, O. Braadbaart, and K. Schwartz (eds), 1999: *Private Business, Public Owners: Government Shareholdings in Water Enterprises,* The Ministry of Housing, Spatial Planning, and the Environment, Nieuwegein, The Netherlands.

Bonvallot, V., 2003: L'arsenic quotidien, *Biofutur,* **232,** pp. 70–3.

Bosch, J.M. and J.D. Hewlett, 1982: A review of catchment experiments to determine the effect of vegetation changes on water yield and evapotranspiration, *Journal of Hydrology,* 55, pp. 3–23.

Bouwman, A.F., G. Van Drecht, J.M. Knoop, A.H.W. Beusen, and C.R. Meinardi, 2005: Exploring changes in river nitrogen export to the world's oceans, *Global Biogeochemical Cycles.* In press.

Boyer, D., 2001: Trade: The connection between environment and sustainable livelihoods, Working paper no. 5, Oxfam America, Boston, MA.

Bradley, D., 1977: Health aspects of water supplies in tropical countries. In: *Water, Wastes and Health in Hot Climates,* R. Feachem, M. McGarry, and D. Mara (eds.), John Wiley & Sons, London, UK, pp. 3–17.

Briscoe, J., 1999: The financing of hydropower, irrigation and water supply infrastructure in developing countries, *Water Resources Development,* **15,** pp. 459–91.

Brismar, A., 1997: *Freshwater and Gender: A Policy Assessment,* Stockholm Environment Institute, Stockholm, Sweden.

Bruijnzeel, L.A., 1990: *Hydrology of Moist Tropical Forests and Effects of Conversion: A State of Knowledge Review,* UNESCO International Hydrological Programme, International Institute for Aerospace Survey and Earth Sciences Programme on Geo-information for Environmentally Sound Management of Tropical Resources, and The International Association of Hydrological Sciences.

Bruijnzeel, L.A., 2004: Hydrological functions of tropical forests: Not seeing the soil for the trees? *Agriculture Ecosystems and Environment.* In press.

Bruinsma, J.E., 2003: *World Agriculture: Towards 2015/2030: An FAO Perspective,* Earthscan Publications Ltd, London, UK.

Bunnell, J.F., R.A. Zampella, R.G. Lathrop, and J.A. Bognar, 2003: Landscape changes in the Mullica river basin of the Pinelands National Reserve, New Jersey, USA, *Environmental Management,* **31(6),** pp. 696–708.

Buttle, J.M., I.F. Creed, and J.W. Pomeroy, 2000: Advances in Canadian forest hydrology, 1995–1998, *Hydrological Processes,* **14,** pp. 1551–78.

Calder, I.R., 1999: *The Blue Revolution: Land Use & Integrated Water Resources Management,* Earthscan, London, UK.

Calder, I., 2000: Land use impacts on water resources, FAO Land-Water Linkages in Rural Watersheds Electronic Workshop, Background paper no. 1, FAO, Rome, Italy.

Calder, I., 2004: Forests and water: Closing the gap between public and science perceptions, *Water Science and Technology,* 49(7), pp. 39–53.

Carpenter, S.R., 2003: *Regime Shifts in Lake Ecosystems: Pattern and Variation,* Excellence in Ecology 15, International Ecology Institute, Oldendorf/Luhe Germany.

CBD (Convention on Biological Diversity), 2004: *Convention on Biological Diversity Convention Text* [online]. Cited 1 November 2004. Available at www.biodiv.org/convention/articles.asp.

Chase, T.N., R.A. Pielke, T.G.F. Kittel, R. Nemani, and S.W. Running, 1996: Sensitivity of a general circulation model to global changes in leaf area index, *Journal of Geophysical Research,* **101(D3),** pp. 7393–408.

Chevreuil, M., D. Thevenot, P. Scribe, M. Blanchard, Y. Duclos et al., 1998: Micropolluants organiques: Une boîte de Pandore. In: *La Seine en Son Bassin,* M. Meybeck, G. de Marsily, and E. Fustec (eds.), Elsevier, Paris, France, pp. 439–81.

Cohen, M., 2002: Managing across boundaries: The case of the Colorado River Delta. In: *The World's Water: 2002–03,* P. Gleick et al. (eds.), Island Press, Washington, DC, pp. 133–47.

Collignon, B. and M. Vezina, 2000: *Independent Water and Sanitation Providers in African Cities,* UNDP–World Bank Water and Sanitation Program.

Constance, P., 1999: What price water? *IDB America* **26(7/8).** Available at http://www.iadb.org/idbamerica/Archive/stories/1999/eng/cone899.htm.

Costa, M.H. and J.A. Foley, 2000: Combined effects of deforestation and doubled atmospheric CO2 concentrations on the climate of Amazonia, *Journal of Climate,* **13,** pp. 35–58.

Costa, M.H., A. Botta, and J.A. Cardille, 2003: Effects of large-scale changes in land cover on the discharge of the Tocantins River, Southeastern Amazonia, *Journal of Hydrology,* **283,** pp. 206–17.

Daughton, C.G. and T.A. Ternes, 1999: Pharmaceuticals and personal care products in the environment: Agents of subtle change? *Environmental Health Perspectives,* **107** (Supplement 6), pp. 907–38.

Day, J.C. and F. Quinn, 1987: Dams and diversions: Learning from Canadian experience. In: *Proceedings of the Symposium on Interbasin Transfer of Water: Impacts and Research Needs for Canada,* National Hydrology Research Institute, Saskatoon, Canada, pp. 43–56.

Dennehy, K.F., Litke, D.W., and McMahon, P.B., 2002: The High Plains aquifer, USA: Groundwater development and sustainability. In: *Sustainable Groundwater Development,* K.M. Hiscock, M.O. Rivett, and R.M. Davison (eds.), Special Publications 193, Geological Society, London, pp. 99–119.

Dettinger, M.D. and D.R. Cayan, 1995: Large-scale atmospheric forcing of recent trends toward early snowmelt runoff in California, *Journal of Climate,* **8,** pp. 606–23.

Dinar, A. and A. Subramanian, 1997: *Water Pricing Experiences: An International Economy and Political Risks of Institutional Reform in the Water Sector,* World Bank working paper no. 386, Washington, DC.

Dirmeyer, P., X. Gao, and T. Oki, 2002: *GSWP-2: The Second Global Soil Wetness Project Science and Implementation Plan,* IGPO publication series no. 37, International GEWEX (Global Energy and Water Cycle Experiment) Project Office, 65 pp.

Döll, P. and S. Siebert, 2000: A digital global map of irrigated areas, *ICID Journal* **49(2),** pp. 55–66.

Döll, P., F. Kaspar, and B. Lehner, 2003: A global hydrological model for deriving water availability indicators: Model tuning and validation, *Journal of Hydrology,* **270,** pp. 105–34.

Douglas, E.M., R.M. Vogel, and C.N. Kroll, 2000: Trends in floods and low flows in the U.S.: Impact of spatial correlation, *Journal of Hydrology,* **240,** pp. 90–105.

Douglas, E.M., K. Sebastian, C.J. Vörösmarty, and S. Wood, 2005: The role of tropical forests in supporting biodiversity and hydrological integrity: A synoptic overview, World Bank working paper, Washington, DC.

Dudley, N. and S. Stolton, 2003: Running pure: The importance of forest protected areas to drinking water, World Bank/WWF, Gland, Switzerland/ Washington, DC.

Dynesius, M. and C. Nilsson, 1994: Fragmentation and flow regulation of river systems in the northern third of the world, *Science,* 266, pp. 753–62.

Easterling, D.R., J.L. Evans, P.Ya. Groisman, T.R. Karl, et al., 2000: Observed variability and trends in extreme climate events: A brief overview, *Bulletin of the American Meteorological Society,* **81(3),** pp. 417–25.

EEA (European Environment Agency), 1995: *Europe's Environment: The Dobris Assessment,* D. Stanners, and P. Bourdeau (eds.), EEA, Copenhagen, Denmark.

EEA, 2003: Europe's water: An indicator-based assessment, Topic report no. 1/2003, EEA, Copenhagen, Denmark, p. 93 and p. 97. Available at http://www.eea.eu.int.

Ernst, C., 2004: *Protecting the Source: Land Conservation and the Future of America's Drinking Water,* Trust for Public Land, Washington, DC.

Falkenmark, M., 1997: Society's interaction with the water cycle: A conceptual framework for a more holistic approach, *Hydrological Sciences Journal,* **42,** pp. 451–66.

Falkenmark, M. and C.Folke, 2003: Theme issue: Freshwater and welfare fragility: Syndromes, vulnerabilities and challenges, *Royal Society's Philosophical Transactions B Biology,* **358,** p. 1440.

FAO (Food and Agriculture Organization of the United Nations), 1999: State of the World's Forests, FAO, Rome, Italy.

FCGWSP (Framing Committee of the Global Water System Project), 2004: Humans transforming the global water system, *Eos: Transactions of the American Geophysical Union.* In press.

Federer, C.A., C.J. Vörösmarty, and B. Fekete, 2003: Sensitivity of annual evaporation to soil and root properties in two models of contrasting complexity, *Journal of Hydrometeorology,* **4,** pp. 1276–90.

Fekete, B.M., C.J. Vörösmarty, and W. Grabs, 2002: High resolution fields of global runoff combining observed river discharge and simulated water balances, *Global Biogeochemical Cycles,* **16(3),** Art. no. 1042.

Fekete, B.M., C.J. Vörösmarty, J. Roads, and C. Willmott, 2004: Uncertainties in precipitation and their impacts on runoff estimates, *Journal of Climate,* 17, pp. 294–304.

Foster, S.S. and P.J. Chilton, 2003: Groundwater: The processes and global significance of aquifer degradation, *Philosophical Transactions Of the Royal Society of London Series B Biological Sciences,* **358(1440),** pp. 1957–72.

Foster, S., A. Lawrence, and B. Morris, 1998: Groundwater in urban development: Assessing management needs and formulating policy strategies, World Bank technical paper no. 390, The World Bank, Washington, DC.

Foster, S., 2000: Groundwater at the World Water Forum, IAH (International Association of Hydrogeologists) News and Information. Cited 30 October 2004. Available at www.iah.org/articles/mar2000/art001.htm.

Foster, S., J. Chilton, M. Moench, F. Cardy, and M. Schiffler, 2000: Groundwater in rural development: Facing the challenges of supply and resource sustainability, World Bank technical paper no. 463, World Bank, Washington, DC.

Fraser, A., M. Meybeck, and E. Ongley, 1995: *Water Quality of World Rivers,* UNEP Environment Library report no. 14, UNEP, Nairobi, Kenya.

Frich, P., L.V. Alexander, P. Della-Marta, B. Gleason, M. Haylock, et al., 2002: Observed coherent changes in climatic extremes during the second half of the twentieth century, *Climate Research,* **19,** pp. 193–212.

Furlong, K. and N.P. Gleditsch, 2003: The boundary dataset, *Conflict Management and Peace Science,* **20,** pp. 93–117.

GACGC (German Advisory Council on Global Change), 2000: *World in Transition. Strategies for Managing Global Environmental Risks,* Annual Report 1998, Springer, Berlin, Germany.

Galloway, J.N., F.J. Dentener, D.G. Capone, E.W. Boyer, R.W. Howarth, et al., 2004: Global and regional nitrogen cycles: Past, present and future, *Biogeochemistry,* **70,** pp. 153–226.

Gardner, J.S., 2002: Natural hazards risk in the Kullu District, Himachal Pradesh, India, *Geographical Review,* **92,** pp. 282–306.

GESAMP (Group of Experts on the Scientific Aspects of Marine Environmental Protection), 1997: Towards safe and effective use of chemicals in coastal aquaculture, Reports and studies no. 65, GESAMP/ FAO, Rome, Italy, 40 p.

Gilmour, D.A., M. Bonell, and D.S. Cassells, 1987: The effects of forestation on soil hydraulic properties in the middle hills of Nepal: A preliminary assessment, *Mountain Research and Development,* **7,** pp. 239–49.

Gleick, P.H. (ed.), 1993: *Water in Crisis: A Guide to the World's Freshwater Resources,* Oxford University Press, London, UK.

Gleick, P.H., 1996: Basic water requirements for human activities: Meeting basic needs, *Water International,* 21, pp. 83–92.

Gleick, P.H., 1998: *The World's Water 1998–1999,* Island Press, Washington, DC.

Gleick, P.H., 1999: The human right to water, *Water Policy,* 1(5), pp. 487–503.

Gleick, P.H., 2000: *The World's Water 2000–2001,* Island Press, Washington, DC.

Gleick, P.H., 2002: *Dirty Water: Estimated Deaths from Water-Related Diseases 2000–2020,* Pacific Institute Research Report, Pacific Institute for Studies in Development, Environment, and Security, Oakland, CA.

Gleick, P.H., E.L. Chalecki, and A. Wong, 2002: Measuring water well-being: Water indicators and indices. In: *The World's Water: 2002–03,* P. Gleick et al. (eds.), Island Press, Washington, DC, pp. 87–112.

Gleick, P.H., G. Wolff, E.L. Chalecki, and R. Reyes, 2002a: The privatization of water and water systems. In: *The World's Water: 2002–03,* P. Gleick et al. (eds.), Island Press, Washington, DC, pp. 57–85.

Gleick, P. H., E.L. Chalecki, and A. Wong, 2002b. Measuring water well-being: Water indicators and indices. pp. 87–112. In: P. Gleick et al. (eds.) *The World's Water: 2002–03.* Washington, DC: Island Press.

Gleick, P.H., G. Wolff, E.L. Chalecki, R. Reyes, 2002c: *The New Economy of Water: The Risks and Benefits of Globalization and Privatization of Fresh Water,* Pacific Institute for Studies in Development, Environment, and Security, Oakland, CA.

GLFC (Great Lakes Fishery Commission), 1999: *Addressing Concerns for Water Quality Impacts from Large-Scale Great Lakes Aquaculture,* GLFC, Great Lakes Water Quality Board, International Joint Commission.

Goldburg, R.J., M.S. Elliott, and R.L. Naylor, 2001: *Marine Aquaculture in the United States: Environmental Impacts and Policy Options,* Pew Oceans Commission, Arlington, VA.

Good, R.E. and N.F. Good, 1984: The Pinelands National Reserve: An ecosystem approach to management, *BioScience,* **34(3),** pp. 169–73.

Green, C.H., D.J. Parker, and S.M Tunstall, 2000: *Assessment of Flood Control Options and Management,* WCD Thematic Review, Options Assessment: IV. Available at http://www.dams.org/docs/kbase/thematic/tr44main.pdf.

Green, P., C.J. Vörösmarty, M. Meybeck, J. Galloway, and B.J. Peterson, 2004: Pre-industrial and contemporary fluxes of nitrogen through rivers: A global assessment based on typology, *Biogeochemistry,* 68, pp. 71–105.

Groisman, P.Ya., R.W. Knight, and T.R. Karl, 2001: Heavy precipitation and high streamflow in the contiguous United States: Trends in the twentieth century, *Bulletin of the American Meteorological Society,* 82(2), pp. 219–46.

Groisman, P.Ya., R.W. Knight, T.R. Karl, D.R. Easterling, B. Sun, et al., 2004: Contemporary changes of the hydrological cycle over the contiguous United States: Trends derived from in situ observations, *Journal of Hydrometeorology,* 5, pp. 64–85.

GWP (Global Water Partnership), 2000: Integrated river basin management, Technical Advisory Committee background paper, Stockholm, Sweden.

Hamerlynck, O. and S. Duvail, 2003: *The Rehabilitation of the Delta of the Senegal River in Mauritania,* IUCN, Gland, Switzerland and Cambridge, UK, viii + 88 pp.

Hamilton, L.S., 1987: What are the impacts of deforestation in the Himalayas on the Ganges–Brahmaputra lowlands and delta? Relations between assumptions and facts, *Mountain Research and Development,* 7, pp. 256–63.

Hamilton, P.A., T.L. Miller, and D.N. Myers, 2004: Water quality in the nation's streams and aquifers: Overview of selected findings, 1991–2001, U.S. Geological circular 1265, USGS (US Geological Survey), Denver, CO.

Hamlet, A.F. and D.P. Lettenmaier, 1999:Effects of climate change on hydrology and water resources objectives in the Columbia River basin, *Journal of the American Water Resources Association,* 35, pp. 1597–624.

Hart, D.D., T.E. Johnson, K.L. Bushaw-Newton, R.J. Horwitz, A.T. Bednarek, et al., 2002: Dam removal: Challenges and opportunities for ecological research and river restoration, *BioScience,* **52(8),** pp. 669–81.

Haub, C., 1994: Population explosion. In: *The Encyclopedia of the Environment,* R. Eblen and W. Eblen (eds.), Hougton Mifflin, Boston, MA, p. 577.

Health Canada, 2004: Chloramphenicol in food products, Health Canada, Ottawa, Canada. Available at http://www.hc-sc.gc.ca/vetdrugs-medsvet/chloramphenicol_e.html#5.

Hirji, R., 1998: Inter-basin water transfers: Emerging trends, World Bank Report no. 22675 *Environment Matters,* **(Fall),** 4 pp.

Hoeg, K., 2000: Dams: Essential infrastructure for future water management, Paper presented at the Second World Water Forum for the International Commission on Large Dams, 17–22 March, The Hague, The Netherlands.

Hodgkins, G.A., R.W. Dudley, and T.G. Huntington, 2003: Changes in the timing of high river flows in New England over the 20th Century, *Journal of Hydrology,* **278,** pp. 244–52.

Hoekstra, A.Y. and P.Q. Hung., 2002: *Virtual Water Trade: A Quantification of Virtual Water Flows between Nations in Relation to International Crop Trade,* Value of Water Research report series no. 11, IHE (Institute for Water Education)-Delft, Delft, The Netherlands.

Hoekstra, A.Y., 2003: Virtual water: An introduction. In: *Virtual Water Trade: Proceedings of the International Expert Meeting on Virtual Water Trade,* A.Y. Hoekstra (ed.), Value of Water research report series no. 12, IHE (Institute for water Education)-Delft, Delft, The Netherlands. pp. 1–23.

Hornbeck, J.W., and R.B. Smith, 1997: A water resources decision model for forest managers, *Agricultural and Forest Meteorology,* **84,** pp. 83–8.

Houghton, J.T., Y. Ding, D.J. Griggs, M. Noguer, P.J. van der Linden, et al., (eds.), 2001: Climate change 2001: The scientific basis, Third Assessment Report, Intergovernmental Panel on Climate Change working group I, Cambridge University Press, Cambridge, UK.

Howarth, R.W., G. Billen, D. Swaney, A. Townsend, N. Jaworski, et al., 1996: Regional nitrogen budgets and riverine N and P fluxes for the drainages to the North Atlantic Ocean: Natural and human influences, *Biogeochem,* **35,** pp. 75–139.

Hutton, G. and L. Haller, 2004: *Evaluation of the Costs and Benefits of Water and Sanitation Improvements at the Global Level,* WHO/SDE/WSH/04.04, WHO, Geneva, Switzerland. Available at www.who.int/water_sanitation_health/wsh0404/en.

IAHS (International Association of Hydrologic Sciences), 2001: Global water data: A newly endangered species, *Eos: Transactions of the American Geophysical Union,* **82(5),** pp. 54, 56, 58.

ICWE (International Conference on Water and the Environment), 1992: *Development Issues for the 21st Century: The Dublin Statement and Report of the Conference,* 26–31 January, Dublin, World Meteorological Organization, Geneva, Switzerland.

IFRC/RCS (International Federation of the Red Cross and Red Crescent Societies), 2000: *World Disasters Report 2000: Focus on Public Health,* IFRC/RCS, Geneva, Switzerland.

IFRCRCS, 2003: *World Disasters Report 2003,* IFRC/RCS, Geneva, Switzerland.

IUCN (World Conservation Union), 2003: *Flow: The Essentials of Environmental Flows,* M. Dyson, G. Bergkamp, and J. Scanlon (eds.), Gland, Switzerland/Cambridge, UK.

Jaeger, C., 2001: The challenge of global water management. In: *Understanding the Earth System: Compartments, Processes, Interactions,* E. Ehlers and T. Krafft (eds.), Springer, Berlin, Germany, pp. 125–35.

Jensen, P.K., G. Jayasinghe, W. van der Hoek, S. Cairncross, and A. Dalsgaard, 2004: Is there an association between bacteriological drinking water quality and childhood diarrhoea in developing countries? *Tropical Medicine and International Health,* 9(11), pp. 1210–15.

Jones, K.B., K.H. Riitters, J.D. Wickham, R.D. Tankersley, Jr., R.V. O'Neill, et al., 1997: *An Ecological Assessment of the United States Mid-Atlantic Region: A Landscape Atlas,* US Environmental Protection Agency, Washington, DC.

Jones, O.A.H., N. Voulvoulis, and J.N. Lester, 2003: Potential impact of pharmaceuticals on environmental health, *Bulletin of the World Health Organization,* **81(10)** (October), Geneva.

Kakabadse-Navarro, Y., J. McNeely, and D. J. Melnick, 2004: Interim report of task force 6 on environmental sustainability, Millennium project, New York, NY. Available at www.unmillenniumproject.org.

Kanciruk, P., 1997: Pricing policy for federal research data, *Bulletin of the American Meteorological Society,* **78,** pp. 691–2.

Karl, T., R.W. Knight, D.R. Easterling, and R.G. Quayle, 1996: Indices of climate change for the United States, *Bulletin of the American Meteorological Society,* 77, pp. 279–92.

Karl, T.R. and R.W. Knight, 1998: Secular trends of precipitation amount, frequency, and intensity in the USA, *Bulletin of the American Meteorological Society,* **79(2),** pp. 231–41.

King, J.M, R.E. Tharme, and M.S. DeVilliers (eds.), 2000: *Environmental Flow assessments for Rivers: Manual for the Building Block Methodology,* Water Research Commission, Pretoria, South Africa.

Kolpin, D.W., Furlong, E.T., Meyer, M.T., Thurman, E.M., Zuagg, S.D. et al., 2002: Pharmaceuticals, hormones, and other organic wastewater contaminants in US streams, 1999–2000: A national reconnaissance, *Environmental Science and Technology,* **36,** pp. 1202–2111.

Korzoun, V.I., A.A. Sokolov, M.I. Budyko et al. (eds), 1978: *Atlas of World Water Balance and Water Resources of the Earth,* USSR Committee for the International Hydrological Decade, Leningrad, UNESCO, Paris, France.

Kowalewski, M., G.E. Avila Serrano, K.W. Flessa, and G.A. Goodfriend, 2000: Dead delta's former productivity: Two trillion shells at the mouth of the Colorado River, *Geology,* 28, pp. 1059–62.

Kunkel, K., K. Andsager, and D.R. Easterling, 1999: Trends in heavy precipitation events over the continental Unites States, *Journal of Climate,* **12,** pp. 2515–27.

Labat, D., Y. Goddéris, J.L. Probst, and J.L. Guyot, 2004: Evidence for global runoff increase related to climate warming, *Advances in Water Resources,* **27,** pp. 631–42.

Lammers, R.B., A.I. Shiklomanov, C.J. Vörösmarty, B.M. Fekete, and B.J. Peterson, 2001: Assessment of contemporary Arctic river runoff based on observational discharge records, *Journal of Geophysical Research – Atmospheres,* **106(D4),** pp. 3321–34.

Lawrence, P., J.R. Meigh, and C.A. Sullivan, 2003: *The Water Poverty Index: An International Comparison,* Keele Economic Research Papers 2003/18 and Centre for Ecology and Hydrology (CEH), Wallingford, UK.

Lanfear, K.J. and R.M. Hirsch, 1999: USGS study reveals a decline in long-record streamgauges, *Eos: Transactions of the American Geophysical Union,* **80,** pp. 605–7.

Le Maitre, D.C., D.F. Scott, and C. Colvin, 1999: A review of information on interactions between vegetation and groundwater, *Water SA,* **25,** pp. 137–52.

Le Maitre, D.C., D.B. Versfeld, and R.A. Chapman, 2000: The impact of invading alien plants on surface water resources in South Africa: A preliminary assessment, *Water SA,* **26,** pp. 397–408.

Lins, H.F. and J.R. Slack, 1999: Streamflow trends in the United States, *Geophysical Research Letters,* 26, pp. 227–30.

Loth, P. (ed.), 2004: *The Return of the Water: Restoring the Waza Logone Floodplain in Cameroon,* IUCN, Gland, Switzerland/ Cambridge, UK, xvi + 156 pp.

L'vovich, M.I. and G.F. White, 1990: Use and transformation of terrestrial water systems. In: *The Earth as Transformed by Human Action,* B.L. Turner, W.C. Clark, R.W. Kates, J.F. Richards, J.T. Mathews, et al. (eds.), Cambridge University Press, Cambridge, UK, pp. 235–52.

Mahé, G., 1993: *Les écoulements fluviaux sur la façade atlantique de l'Afrique: Etude des éléments du bilan hydrique et variations interannuelles: Analyse de situations hydrclimatiques moyennes et extremes,* Col. Etudes et Theses, ORSTOM (Office de la Recherche Scientifique et Technique de Outre-Mer/Institute of Research for Development and Cooperation), Paris, France.

Margat, J. 1990a: *Les Eaux Souterraines dans le Monde,* Orléans, Bureau de recherches géologiques et miniéres (BRGM), Département eau.

Margat, J. 1990b: Les Gisement d'Eau Souterraine, *La Recherche,* **221.**

McCabe, G. J. and D.M. Wolock, 2002: Trends and temperature sensitivity of moisture conditions in the conterminous United States, *Climate Research,* **20(1),** pp. 19–29.

McCully, P., 1996: *Silenced Rivers: The Ecology and Politics of Large Dams,* Zed Books, London, UK.

McIntosh, A.C. and C.E. Yñiguez, 1997: *Second Water Utilities Data Book,* Asian Development Bank, Manila, Philippines, 210 pp.

MDBC (Murray-Darling Basin Commission), 2003: Murray-Darling Basin Water Resources fact sheet, November, 7 pp. Available at www.mdbc.gov.au/publications/factsheets/water_resourcesver2.html.

MDBC, 2004: The cap. Available at www.mdbc.gov.au/naturalresources/the_cap/cap_brochure.pdf.

MEE (Ministry of the Environment and Water), 2001: Loi d'orientation relative à la gestion de l'eau : Loi n002–2001/AN du 8 février 2001 portant loi d'orientation relative à la gestion de l'eau, Ouagadougou, Burkina Faso.

Mekis, E. and W.D. Hogg, 1999: Rehabilitation and analysis of Canadian daily precipitation time series, *Atmosphere-Ocean,* **37,** pp. 53–85.

Meybeck, M. and R. Helmer, 1989: The quality of rivers: From pristine stage to global pollution, Palaeogeography, Palaeoclimatology, Palaeoecology *(Global and Planetary Change),* 1, pp. 283–309.

Meybeck, M. and A. Ragu, 1997: Presenting GEMS-GLORI, a compendium of world river discharges to the oceans, *International Association of Hydrological Science Publications,* **243,** pp. 3–14.

Meybeck, M., 2002: Riverine quality at the Anthropocene: propositions for global space and time analysis, illustrated by the Seine River, *Aquatic Sciences,* **64,** pp. 1–18.

Meybeck, M., R. Helmer, M. Dray, H. El Ghobary, A. Demayo, et al., 1991: Water quality: Progress in the implementation of the Mar del Plata Action Plan, UN Conference on Water, Dublin, WHO/UNEP.

Meybeck, M., P. Green, and C.J. Vörösmarty, 2001: A new typology for mountains and other relief classes: An application to global continental water resources and population distribution, *Mountain Research and Development,* **21,** pp. 34–45.

Meybeck, M., 2003: Global analysis of river systems: From Earth system controls to Anthropocene syndromes, *Philosophical Transactions of the Royal Society Of London Series B,* ?DOI 10.1098/rstb.2003, pp. 1379.

Meybeck, M., P. Green, and C.J. Vörösmarty, 2001: A new typology for mountains and other relief classes: An application to global continental water resources and population distribution, *Mountain Research and Development,* **21,** pp. 34–45.

Meybeck, M., C.J. Vörösmarty, R.E. Schulze, and A. Becker, 2004: Conclusions: Scaling relative responses of terrestrial aquatic systems to global changes. In *Vegetation, Water, Humans and the Climate,* P. Kabat, M. Claussen, P.A. Dirmeyer et al. (eds.), Springer, Berlin, Germany, pp. 455–64.

Miguel, E., S. Satyanath, and E. Sergenti, 2004: Economic shocks and civil conflict: An instrumental variables approach, *Journal of Political Economy,* **112(4),** pp. 725–53.

Milly, P.C.D., R.T. Wetherald, K.A. Dunne, and T.L. Delworth, 2002: Increasing risk of great floods in a changing climate, *Nature,* **415,** pp. 514–17.

Molden, D., 2003: Pathways to improving the productivity of water, In: *Issues of Water Management in Agriculture: Compilation of Essays,* Comprehensive assessment of water management in agriculture, Colombo, Sri Lanka, pp. 1–6.

Morris, B.L, A.R. Lawrence et al., 2003: *Groundwater and Its Susceptibility to Degradation: A Global Assessment of the Problem and Options for Management,* Early Warning and Assessment Reports Series, RS.03–3, UNEP, Nairobi, Kenya.

Muller, M., 2000: Inter-basin water sharing to achieve water security: A South African perspective, Presented in the World Water Forum in the Haag, 2000, Department of Water Affairs and Forestry, Pretoria, South Africa.

MWR (Ministry of Water Resources), 2004: Official Ministry of Water Resources of China. Cited 25 October 2004. Available at www.mwr.gov.cn/english1/20040823/39104.asp.

National Water Act (Act No. 36), 1998: South African National Water Act. Available at http://www-dwaf.pwv.gov.za/Documents/Legislature/nw_act/NWA.doc.

Nijssen, B., G.M. O'Donnell, D.P. Lettenmaier, D. Lohmann, and E.F. Wood, 2001: Predicting the discharge of global rivers, *Journal of Climate,* **14,** pp. 3307–23.

Nilsson, C., C. Reidy, M. Dynesius, and C. Revenga, 2005: *Fragmentation and Flow Regulation of the World's 292 Largest Rivers.* In review.

NLWRA (National Land and Water Resources Audit), 2004: Australian Natural Resources Atlas. Cited 30 October 2004. Available at audit.ea.gov.au/ANRA/water/water_frame.cfm?region_type = AUS&region_code = AUS&info = resourcesr.

NOAA-NGDC (National Oceanic and Atmospheric Administration–National Geophysical Data Center), 1998: Stable lights and eadiance calibrated lights of the world CD-ROM, View Nighttime Lights of the World data base, NOAA–NGDC, Boulder, CO. Available at http://julius.ngdc.noaa.gov:8080/production/html/BIOMASS/night.html.

NRC (National Research Council), 1999: *A Question of Balance: Private Rights and the Public Interest in Scientific and Technical Databases,* National Academy Press, Washington, DC.

OECD (Organisation for Economic Co-operation and Development), 1999: *The Price of Water: Trends in OECD Countries,* OECD, Paris, France.

Oki, T., Y. Agata, S. Kanae, T. Saruhashi, D. Yang et al., 2001: Global assessment of current water resources using total runoff integrating pathways, *Hydrological Sciences Journal,* 46, pp. 983–96.

Oki, T., M. Sato, A. Kawamura, M. Miyake, S. Kanae, et al., 2003a: Virtual water trade to Japan and in the world. In: *Virtual Water Trade: Proceedings of the International Expert Meeting on Virtual Water Trade,* AY. Hoekstra (ed.), Value of Water research report series no. 12, IHE (Institute for Water Education)-Delft, Delft, The Netherlands, pp. 221–33.

Oki, T., Y. Agata, S. Kanae, T. Saruhashi, and T. Musiake, 2003b: Global water resources assessment under climatic change in 2050 using TRIP, *IAHS Publication,* **280,** pp. 124–33.

Pagiola, S., 2002: Paying for water services in Central America: Learning from Costa Rica. In: *Selling Forest Environmental Services,* S. Pagiola, J. Bishop, and N. Landell-Mills (eds.), Earthscan, London, UK.

Pagiola, S., R. Martin-Hurtado, P. Shyamsundar, M. Mani, and P. Silva, 2002: *Generating Public Sector Resources to Finance Sustainable Development: Revenue and Incentive Effects,* World Bank technical paper no. 538, World Bank, Washington, DC.

Paul, M.J. and J.L. Meyer, 2001: Streams in the urban landscape, *Annual Review of Ecology and Systematics, 32,* pp. 333–65.

Peet, J., 2003: Priceless: A survey of water, *The Economist,* (July 17),.

Peterson, B.J., R.M. Holmes, J.W. McClelland, C.J. Vörösmarty, R.B. Lammers, et al., 2002: Increasing river discharge to the Arctic Ocean, *Science,* 298, pp. 2171–3.

Pielke, R.A., T.J. Lee, J.H. Copeland, J.L. Eastman, C.L. Ziegler, and C.A. Finley, 1997: Use of USGS-provided data to improve weather and climate simulations, *Ecological Applications,* 7, pp. 3–21.

Pitman, A. and M. Zhao, 2000: The relative impact of observed change in land cover and carbon dioxide as simulated by a climate model, *Geophysical Research Letters,* **27,** pp. 1267–70.

Poff, N.L., 2003: River flows and water wars: Emerging science for environmental decision making, *Frontiers in Ecology and the Environment,* **1(6),** pp. 298–306.

Poff, N.L., J.D. Allan, M.B. Bain, J.R. Karr, K.L. Prestegaard, et al., 1997: The natural flow regime: A paradigm for river conservation and restoration, *BioScience,* **47(11),** pp. 769–84.

Postel, S.L., G.C. Daily, and P.R. Ehrlich, 1996: Human appropriation of renewable fresh water, *Science,* **271,** pp. 785–88.

Postel, S.L. and S. Carpenter, 1997: Freshwater ecosystem services. In: *Nature's Services: Societal Dependence on Natural Ecosystems,* G.C. Daily (ed.), Island Press, Washington, DC, pp. 195–214.

Postel, S.L., 1998: Water for food production: Will there be enough in 2025? *BioScience,* **48,** pp. 629–35.

Postel, S.L., 1999: *Pillar of Sand: Can the Irrigation Miracle Last?* W.W. Norton, New York, NY, 313 pp.

Postel, S. and B. Richter, 2003: *Rivers for Life: Managing Water for People and Nature,* Island Press, Washington, DC.

Postel, S. and A. Vickers, 2004: Boosting water productivity. In: *State of the World 2004,* Worldwatch Institute, W.W. Norton, New York, NY.

Pryer, J., 1993: The impact of adult ill-health on household income and nutrition in Khulna, Bangladesh, *Environment and Urbanization,* **5,** pp. 35–49.

Rabalais, N.N., R.E. Turner, and D. Scavia, 2002: Beyond science into policy: Gulf of Mexico hypoxia and the Mississippi River, *BioScience,* **52,** pp. 129–42.

Radziejewski, M. and Z.W. Kundzewicz, 2004: Detectability of changes in hydrological records, *Hydrological Sciences Journal,* **49(1),** pp. 39–51.

Ramankutty, N. and J.A. Foley, 1999: Estimating historical changes in global land cover: Croplands from 1700 to 1992, *Global Biochemical Cycles,* **13(4),** pp. 997–1027.

Ramsar Convention on Wetlands, 2004: Convention text. Consulted 25 October 2004. Available at www.ramsar.org.

Raskin, P., E. Hansen, and R. Margolis, 1995: *Water and Sustainability: A Global Outlook,* Pole Star Series report no. 4, Stockholm Environmental Institute, Stockholm, Sweden.

Raskin, P. 1997. *Water Futures: Assessment of Long-range Patterns and Problems,* Background document to the Comprehensive Assessment of the Freshwater Resources of the World report, Stockholm Environmental Institute, Stockholm, Sweden.

Revenga, C., Brunner, J., Henniger, N., Kassem, K., and R. Payner, 2000: *Pilot Analysis of Global Ecosystems, Freshwater Systems,* World Resources Institute, Washington, DC.

Revenga, C. and A. Cassar, 2002: *Ecosystem management of water resources in Africa,* World Wide Fund for Nature (WWF) Global Network, World Wildlife Fund.

Richter, B.D., A.T. Warner, J. Meyer, and K. Lutz: A collaborative process for developing flow recommendations to restore river and estuary ecosystems, *River Research and Applications.* In prep.

Robock, A., K.Y. Vinnikov, G. Srinivasan, J.K. Entin, S.E. Hollinger et al., 2000: The global soil moisture data bank, *Bulletin of the American Meteorological Society,* **81(6),** pp. 1281–99.

Robson, A.J., 2002: Evidence for trends in UK flooding, *Philosophical Transactions of the Royal Society of London Series A- Mathematical, Physical, Engineering Sciences,* 1796, pp. 1327–43.

Rockström, J. and L. Gordon, 2001: Assessment of green water flows to sustain major biomes of the world: Implications for future ecohydrological landscape management, *Physical Chemistry of the Earth,* **26,** pp. 843–51.

Rogers, P., 1993: *America's Water: Federal Roles and Responsibilities,* Twentieth Century Fund, MIT Press, Cambridge, MA.

Rosegrant, M.W., R.G. Scheleyer, S.N. Yadav, and N. Satya, 1995: Water policy for efficient agricultural diversification: Market-based approaches, *Food Policy,* **20,** pp. 203–23.

Rosegrant, M.W., X. Cai, and S.A. Cline, 2002: *Worldwater and Food to 2025: Dealing with scarcity,* International Food Policy and Research Institute and International Water Management Institute, Washington, DC.

Scheidleder, J. Grath, G. Winkler, U. Stark, C. Koreimann, et al., 1999: *Groundwater Quality and Quantity in Europe,* S. Nixon (ed.), European Topic Centre on Inland Waters, European Environment Agency, Copenhagen, Denmark.

Schiermeier, Q., 2003: Studies assess risks of drugs in water cycle 2003, *Nature,* **424,** p. 5.

Seitzinger, S.P. and C. Kroeze, 1998: Global distribution of nitrous oxide production and N inputs in freshwater and coastal marine ecosystems, *Global Biogeochemical Cycles,* **12,** pp. 93–113.

Seitzinger, S.P., C. Kroeze, A.F. Bouwman, N. Caraco, F. Dentener, et al., 2002: Global patterns of dissolved inorganic and particulate nitrogen inputs in coastal ecosystems: Recent conditions and future projections, *Estuaries,* **25,** pp. 640–55.

Schulze, R.E., 2004: River basin responses to global change and anthropogenic impacts. In: *Vegetation, Water, Humans and the Climate,* P. Kabat, M. Claussen, P.A. Dirmeyer, et al. (eds), Springer, Heidelberg, Germany.

Shah, T., D. Molden, R. Sakthivadivel, and D. Seckler, 2000: *The Global Groundwater Situation: Overview of Opportunities and Challenges,* International Water Management Institute, Colombo, Sri Lanka.

Shiklomanov, A.I., R.B. Lammers, and C.J. Vorosmarty, 2002: Widespread decline in hydrological monitoring threatens Pan-Arctic research, *Eos: Transactions of the American Geophysical Union,* **83,** pp. 13, 16–17.

Shiklomanov, I.A. and J. Rodda, 2003: *World Water Resources at the Beginning of the 21st Century,* UNESCO, Paris, France.

Sircoulon, J., T. Lebel, and N.W. Arnell, 1999: Assessment of the impacts of climate variability and change on the hydrology of Africa. In: *Impacts of Climate Change and Climate Variability on Hydrological Regimes,* J.C. van Dam (ed.), UNESCO International Hydrology Series, Cambridge University Press, Cambridge, UK, pp. 67–84.

Smakhtin, V., C. Revenga, and P. Döll, 2004: Taking into account environmental water requirements in global-scale water resources assessments, Comprehensive assessment research report no. 2, International Water Management Institute, Colombo, Sri Lanka.

Snaddon, C.D. and B.R. Davies, 1998: A preliminary assessment of the effects of a small South African inter-basin water transfer on discharge and invertebrate community structure, *Regulated Rivers: Research and Management,* **14,** pp. 421–41.

Snaddon, C.D., M. Wishart, and B.R. Davies, 1998: Some implications of inter-basin water transfers for river functioning and water resources management in Southern Africa, *Aquatic Ecosystem Health and Management,* **1,** pp. 159–82.

Stallard, R.F., 1998: Terrestrial sedimentation and the carbon cycle: Coupling weathering and erosion to carbon burial, *Global Biogeochemical Cycles,* **12,** 231–57.

Sullivan, C.A., J.R. Meigh, A.M. Giacomello, T. Fediw, P. Lawrence, et al., 2003: The Water Poverty Index: Development and application at the community scale, *Natural Resources Forum,* **27,** pp. 189–99.

Sun, G., S.G. McNulty, J.P. Shepard, D.M. Amatya, H. Riekerk, et al., 2001: Effects of timber management on the hydrology of wetland forests in southern United States, *Forest Ecology and Management,* **143,** pp. 227–36.

Swank, W.T. and D.A.J. Crossley (eds.), 1988: *Forest Hydrology at Coweeta,* Springer Verlag, New York, NY.

Swank, W.T., L.W.J. Swift, and J.E. Douglass, 1988: Streamflow changes associated with forest cutting, species conversions, and natural disturbances. In: *Forest Hydrology and Ecology at Coweeta,* W.T. Swank and D.A.J. Crossley (eds.), Springer Verlag, New York, NY, pp. 297–312.

Swanson, R.H., 1998: Forest hydrology issues for the 21st century: A consultant's viewpoint, *Journal of the American Water Resources Association,* **34,** pp. 755–63.

Syvitski, J.P.M., C.J. Vörösmarty, A.J. Kettner, and P. Green, 2005: Impact of humans on the flux of terrestrial sediment to the global coastal ocean, *Science,* **308,** pp. 376–80.

Tharme, R.E, J.M. King, 1998: *Development of the Building Block Methodology for Instream Flow Assessments and Supporting Research on the Effects of Different Magnitude Flows on Riverine Ecosystems,* Water Research Commission in *Rivers for Life,* Postel and Richer, Cape Town, South Africa.

Thobani, M., 1997: Formal water markets: Why, when, and how to introduce tradable water rights, *The World Bank Research Observer,* **12,** pp. 161–79.

Thompson, R.L., 2001: The World Bank strategy to rural development with special reference to the role of irrigation and drainage, Keynote address on the occasion of the 52nd IEC meeting of the International Commission on Irrigation and Drainage, 16–21 September 2001, Seoul, South Korea.

Tollan, A., 2002: Land-use change and floods: What do we need most, research or management, *Water Science and Technology,* **45,** pp. 183–90.

UN (United Nations), 2002: World urbanization prospects: The 2001 revision, Department of Economic and Social Affairs, Population Division, UN, New York, NY. Available at http://www.unpopulation.org.

UNECE (United Nations Economic Commission for Europe), 1995: *State of the Art on Monitoring and Assessment: Rivers,* UNECE Task Force on Monitoring and Assessment, Draft Report V, RIZA, Lelystad, The Netherlands.

UNEP (United Nations Environment Programme), 1996: Groundwater: A threatened resource, UNEP Environment Library no. 15, UNEP, Nairobi, Kenya.

UNEP, 2002: *Global Environmental Outlook 3 (GEO-3),* Earthscan Publications, Ltd., London and Stirling, UK.

UNESCO-IHP (UN Educational, Scientific and Cultural Organization—International Hydrological Programme), 2004: The World-wide Hydrogeological Mapping and Assessment Programme (WHYMAP), A consortium mapping project of International Hydrological Program, International Geological Correlation Programme, International Association of Hydrogeologists, Commission for the Geological Map of the World, International Agricultural Exchange Association, Bundesanstalt fur Geowissenschaften und Rohstoffe—Germany). Available at www.bgr.de/b1hydro/index.html?/ b1hydro/fachbeitraege/a200401/e_whymap.htm.

UN-HABITAT, 2003: Water and sanitation in the world's cities: Local action for global goals, Earthscan, London, UK.

UNPD (UN Population Division), 1999: World population prospects: The 1998 revision, Vol. 1, UN, New York, NY.

UNPD, 2003: World population prospects: The 2002 revision, Highlights, ESA/P/WP.180, UN, New York, NY. Available at http://esa.un.org/unpp.

UN Statistics Division, 2004: *Millennium Indicators,* UN, New York, NY. Available at http://millenniumindicators.un.org/unsd/mi/mi_series_xrxx .asp?row_id = 669.

UNWWAP (United Nations World Water Assessment Program), 2003: Water for people: Water for life, UN World Water Development Report, UNESCO, Paris, France, 576 pp.

US Embassy in China, 2003: Update on China's south-north water transfer project, A June 2003 report, US Embassy, Beijing, China. Available at www .usembassy-china.org.cn/sandt/ptr/SNWT-East-Route-prt.htm.

van der Wink, G.R.M. Allen, J. Chapin, M. Crooks, W. Fraley, et al., 1998: Why the United States is becoming more vulnerable to natural disasters, *Eos: Transactions of the American Geophysical Union,* 79, pp. 533–7.

Valette-Silver, N.N.J., 1993: The use of sediment cores to reconstruct historical trends in contamination of estuarine and coastal sediments, *Estuaries,* **16,** pp. 577–88.

Vörösmarty, C.J.K. Sharma, B. Fekete, A.H. Copeland, J. Holden, et al., 1997a: The storage and aging of continental runoff in large reservoir systems of the world, *Ambio,* **26,** pp. 210–19.

Vörösmarty, C.J., R. Wasson, and J.E. Richey (eds.), 1997b: *Modeling the Transport and Transformation of Terrestrial Materials to Freshwater and Coastal Ecosystems: Workshop Report and Recommendations for IGBP Inter-Core Project Collaboration,* IGBP Report no. 39, International Geosphere-Biosphere Programme Secretariat, Stockholm, Sweden.

Vörösmarty, C.J., C.A. Federer, and A. Schloss, 1998a: Potential evaporation functions compared on U.S. watersheds: Implications for global-scale water balance and terrestrial ecosystem modeling, *Journal of Hydrology,* **207,** pp. 147–69.

Vörösmarty, C.J., C. Li, J. Sun, and Z. Dai, 1998b: Emerging impacts of anthropogenic change on global river systems: The Chinese example. In: *Asian Change in the Context of Global Change: Impacts of Natural and Anthropogenic Changes in Asia on Global Biogeochemical Cycles,* J. Galloway and J. Melillo (eds.), Cambridge University Press, Cambridge, UK, pp. 210–44.

Vörösmarty, C.J. and M.M. Meybeck, 1999: Riverine transport and its alteration by human activities, International Geosphere-Biosphere Programme *Global Change Newsletter,* **39,** pp. 24–9.

Vörösmarty, C.J., P. Green, J. Salisbury, and R. Lammers, 2000: Global water resources: Vulnerability from climate change and population growth, *Science,* **289,** pp. 284–8.

Vörösmarty, C.J. and D. Sahagian, 2000: Anthropogenic disturbance of the terrestrial water cycle, *BioScience,* **50,** pp. 753–65.

Vörösmarty, C.J., 2002: Global change, the water cycle, and our search for Mauna Loa, *Hydrological Processes,* **16,** pp. 1335–9.

Vörösmarty, C.J., M. Meybeck, B. Fekete, K. Sharma, P. Green, et al., 2003: Anthropogenic sediment retention: Major global-scale impact from the population of registered impoundments, *Global and Planetary Change,* **39,** pp. 169–90.

Vörösmarty, C.J. and M. Meybeck, 2004: Responses of continental aquatic systems at the global scale: New paradigms, new methods. In: *Vegetation, Water, Humans and the Climate,* P. Kabat, M. Claussen, P.A. Dirmeyer, et al. (eds.), Springer, Berlin, Germany, pp. 375–413.

Vörösmarty, C.J., E. Douglas, P. Green, and C. Revenga, 2005: Geospatial indicators of emerging water stress: An application to Africa, *Ambio,* **34(3),** pp. 230–6.

Walker, I., F. Ordonez, P. Serrano, and J. Halpern, 2000: *Pricing, Subsidies and the Poor: Demand for Improved Water Services in Central America,* Policy research working paper no. 2468, World Bank, Washington, DC.

Walker, R.T. and W.D. Solecki, 1999: Managing land use and land cover change: The New Jersey Pinelands biosphere reserve, *Annals of the Association of American Geographers,* 89(2), pp. 220–37.

Walling, D.E. and D. Fang, 2003: Recent trends in the suspended sediment loads of the world's rivers, *Global and Planetary Change,* **39,** pp. 111– 26.

Wasson, R.J., 1996: *Land Use and Climate Impacts on Fluvial Systems during the Period of Agriculture: Recommendations for a Research Project and Its Implementation,* PAGES workshop report, Series 96 – 2, International Geosphere-Biosphere Programme-PAGES (Past Global Changes), Bern, Switzerland, 51 pp.

WCD (World Commission on Dams), 2000: *Dams and Development: A New Framework for Decision-Making,* World Commission on Dams, Earthscan, London, UK.

WEDO (Women's Environment and Development Organization), 2004: Gender differences in water use and management. Available at www.wedo.org/ sus_dev/untapped3.htm.

WHO (World Health Organization), 2001: WHO, Geneva, Switzerland. Available at http://www.worldwaterday.org/2001/disease/index.html.

WHO, 2002: *World Health Report 2002: Reducing Risks, Promoting Healthy Life,* WHO, Geneva, Switzerland.

WHO, 2004: *World Health Report 2004: Changing History,* WHO, Geneva, Switzerland.

WHO/UNICEF (UN Children's Program), 2000: Global Water Supply and Sanitation Assessment 2000 Report, WHO, Geneva, Switzerland/ UNICEF, New York, NY. Available at http://www.who.int/docstore/water_sanitation_health/Globassessment/GlobalTOC.htm.

WHO/UNICEF, 2004: *Meeting the MDG Drinking Water and Sanitation Target: A Mid-term Assessment of Progress,* WHO, Geneva, Switzerland/ UNICEF, New York, NY, 33 pp.

Winpenny, J., 2003: Financing water for all: Report of the World Panel on Financing Water Infrastructure, World Water Council, Marseille, France, 54 pp.

WMO (World Meteorological Organization), 1997: *Comprehensive Assessment of the Freshwater Resources of the World,* UN, UNDP, UNEP, FAO, UNESCO, WMO, United Nations Industrial Development Organization, World Bank, SEI, WMO, Geneva, Switzerland.

Wolf, A., J. Natharius, J. Danielson, B. Ward, and J. Pender, 1999: International river basins of the world, *International Journal of Water Resources Development,* **15,** pp. 387–427.

Wolff, G. and P.H. Gleick, 2002: The soft path for water. In: *The World's Water 2002–2003: The Biennial Report on Freshwater Resources,* P. Gleick (ed.), Island Press, Washington, DC, pp. 1–32.

Wollheim, W.M., B.J. Peterson, L.A. Deegan, J.E. Hobbie, B. Hooker, et al., 2001: Influence of stream size on ammonium and suspended particulate nitrogen processing, *Limnology and Oceanography,* **46,** pp. 1–13.

World Bank, 1996a: *World Bank Lending for Large Dams: A Preliminary Review of Impacts,* Operations Evaluation Department, Precise number 125, 09/01/ 1996. Available at http://wbln0018.worldbank.org/oed/oeddoclib.nsf/0/ bb68e3aeed5d12a4852567f5005d 8d95?OpenDocument.

World Bank, 1996b: *Irrigation Investment in Pakistan,* Operations Evaluation Department, Precise Number 124, 09/01/1996. Available at http://wbln0018 .worldbank.org/oed/oeddoclib.nsf /View + to + Link + WebPages/601AF76 831E9A1C0852567F5005D8D81?OpenDocument.

World Bank, 2002. *Shrimp Farming and the Environment: Can Shrimp Farming Be Undertaken Sustainably?* Bangkok, Thailand: Network of Aquaculture Centres in Asia-Pacific. Online at: http://www.enaca.org/Shrimp/Publications.htm.

World Bank, 2001: World Development Indicators 2001, The World Bank, Washington, DC.

WRI (World Resources Institute)**, UNEP, UNDP, and World Bank,** 1998: *World Resources 1998–99: Environmental Change and Human Health,* Oxford University Press, New York, NY.

WSAT (Water Sector Assessment Team), 2000: *Water: The Potential Consequences of Climate Variability and Change for the Water Resources of the United States,* Lead author, P. Gleick, The Report of the Water Sector Assessment Team of the National Assessment of the Potential Consequences of Climate Variability and Change for the U.S. Global Change Research Program, USGS, Washington, DC.

WWC (World Water Council), 2000: *World Water Vision,* Earthscan, London, UK.

WWF and WRI (World Wide Fund for Nature and World Resources Institute), 2004: *Rivers at Risk: Dams and the Future of Freshwater Ecosystems,* WWF, Surrey, UK.

Yue, S, P. Pilon, and B. Phinney, 2003: Canadian streamflow trend detection: Impacts of serial and cross-correlation, *Hydrological Sciences Journal,* **48(1),** pp. 51–63.

Zhang, L., W.R. Dawes, and G.R. Walker, 2001: Reponse of mean annual evapotranspiration to vegetation changes at catchment scale, *Water Resources Research,* **37,** pp. 701–8.

Chapter 8

Food

Coordinating Lead Authors: Stanley Wood, Simeon Ehui

Lead Authors: Jacqueline Alder, Sam Benin, Kenneth G. Cassman, H. David Cooper, Timothy Johns, Joanne Gaskell, Richard Grainger, Sandra Kadungure, Joachim Otte, Agnes Rola, Reg Watson, Ulf Wijkstrom, C. Devendra

Contributing Authors: Nancy Kanbar, Zahia Khan, Will Masters, Sarah Porter, Stefania Vannuccini, Ulrike Wood-Sichra

Review Editors: Arsenio M. Balisacan, Peter Gardiner

----

*This appears in Appendix A at the end of this volume.

Main Messages

Despite the fact that food production per capita has been increasing globally, major distributional inequalities exist. Global food production has increased by 168% over the past 42 years. The production of cereals has increased by about 130%, but that is now growing more slowly. Nevertheless, an estimated 852 million people were undernourished in 2000–02, up 37 million from the period 1997–99. Of this total, nearly 96% live in developing countries. Sub-Saharan Africa, the region with the largest share of undernourished people, is also the region where per capita food production has lagged the most.

Rising incomes, urbanization, and shifting consumption patterns have increased per capita food consumption in most areas of the world. Food preferences, including those arising from cultural differences, are important drivers of food provision. As incomes have increased in regions such as East and Southeast Asia, so has demand for high-value products such as livestock and fish, but cereals are likely to remain the major single component of global diets and to occupy the predominant share of cultivated land.

A diverse diet, with sufficient protein, oils and fats, micronutrients, and other dietary factors is as important for well-being as access to and consumption of sufficient calories. Average daily energy (calorific) intake has declined recently in the poorest countries. Inadequate energy intake is exacerbated by the fact that poor people tend to have low-quality diets. The world's poorest rely on starchy staples for energy, which leads to significant protein, vitamin, and mineral deficiencies. Overconsumption is also a health problem. Nutritional status and children's growth rates improve with consumption of greater food diversity, particularly of fruits and vegetables.

A global epidemic of diet-related obesity and noncommunicable disease is emerging as increasingly urbanized people adopt diets that are higher in energy and lower in diversity in fruits and vegetables than traditional diets (known as the nutrition transition). Many countries now face the double burden of diet-related disease: the simultaneous challenges of significant incidence of acute, communicable diseases in undernourished populations and increasing incidence of chronic diseases associated with the overweight and obese.

An increasing number of people everywhere suffer from diseases caused by contaminated food. As the world eats more perishable foods such as meat, milk, fish, and eggs, the risk of food-borne illnesses is increasing. The relative health risks from food vary by climate, diet, income, and public infrastructure. Food of animal origin poses health risks particularly when it is improperly prepared or inadequately refrigerated. Microbial contamination is of special concern in developing countries. Non-microbial contaminants include metals and persistent organic pollutants. Other growing health concerns related to food production are diseases passed from animals to humans (zoonoses), toxin-containing animal wastes, and overuse of antibiotics in livestock production that may cause allergies or render human antibiotics less effective.

Local food production is critical to eliminating hunger and promoting rural development in areas where the poor do not have the capacity to purchase food from elsewhere. The number of food-insecure people is growing fastest in developing regions, where underdeveloped market infrastructures and limited access to resources prevent food needs from being satisfied by international trade alone. In these areas, local food production is critical to eliminating hunger and providing insurance against rising food prices. In addition, rural households gain income and employment from engaging in food provision enterprises. In sub-Saharan Africa, two thirds of the population relies on agriculture or agriculture-related activities for their livelihoods.

Maintaining a focus on raising the productivity of food production systems continues to be a priority for both global food security and environmental sustainability. While major cereal staples are likely to continue as the foundation of the human food supply, some doubts are being raised about our ability to reproduce past yield growth in the future—especially with regard to sustaining rates of yield growth in high-productivity systems that are already producing near the yield potential threshold, as well as in terms of the availability of land that is suitable for sustaining expanded food output needs.

Government policies are significant drivers of food production and consumption patterns, both locally and globally. Investments in rural roads, irrigation, credit systems, and agricultural research and extension serve to stimulate food production. Improved access to input and export markets boosts productivity. Opportunities to gain access to international markets are conditioned by international trade and food safety regulations and by a variety of tariff and non-tariff barriers. Selective production and export subsidies, including those embodied in the European Union's Common Agricultural Policy and the U.S. Farm Bill, stimulate overproduction of many food crops. This in turn translates into relatively cheap food exports that benefit international consumers at the expense of domestic taxpayers and has often undermined the ability of food producers in many poorer countries to enter international food markets.

The accelerating demand for livestock products is increasingly being met by intensive (industrial or so-called landless) production systems, especially for chicken and pigs, and especially in Asia. These systems have contributed to large increases in production: over the last decade, bovine and ovine meat production increased by about 40%, pig meat production rose by nearly 60%, and poultry meat production doubled. However, intensified livestock production poses serious waste problems and puts increased pressure on cultivated systems to provide feed inputs, with consequent increased demand for water and nitrogen fertilizer.

Per capita consumption of fish is increasing, but this growth is unsustainable with current practices. Total fish consumption has declined somewhat in industrial countries, while it has doubled in the developing world since 1973. Demand has increased without corresponding increases in supply productivity, leading to increases in the real prices of most fresh and frozen fish products at the global level. Pressure on marine ecosystems is increasing to the point where a number of targeted stocks in all oceans are near or exceeding their maximum sustainable levels of exploitation, and world fish catches have been declining since the late 1980s due to overexploitation. Inland water fisheries in the developing world are expanding slowly and will remain an important source of high-quality food for many of the world's poor, particularly in Africa and Asia; however, habitat modifications and water abstraction threaten the continued supply of freshwater fish. For the world as a whole, increases in the volume of fish consumed are made possible by aquaculture, which in 2002 is estimated to have contributed 27% of all fish harvested and 40% of the total amount of fish products consumed as food. Future growth of aquaculture will be constrained by development costs and by fishmeal and oil supplies, which are increasingly scarce.

Wild foods are locally important in many developing countries, often bridging the hunger gap created by stresses such as droughts and civil unrest. In addition to fish, wild plants and animals are important sources of nutrition in some diets, and some wild foods have significant economic value. In most cases, however, wild foods are excluded from economic analysis of natural resource systems as well as official statistics, so the full extent of their importance is improperly understood. In some cases, plants and animals are under pressure from unsustainable levels of harvesting, and there is a local need for conservation of wild food resources to satisfy the nutritional needs of those who do not have access to agricultural land or resources.

8.1 Introduction

The initial use and subsequent transformation of ecosystems for the purpose of meeting human food needs has been a vital, long-standing, and, for the most part, fruitful dimension of the human experience. The provision, preparation, and consumption of food are daily activities that for most societies represent an important part of their identity and culture. But while human ingenuity has transformed the specter of global famine into an unparalleled abundance of food, there are still too many people in the world for whom an adequate, safe, nutritious diet remains an illusion.

Before dealing squarely with the remaining inequities in food distribution and access, as well as the environmental damage often associated with the provision of food, the first and foremost fact is that our ability to provide sufficient food and to do so in increasingly cost-effective ways has been a major human and humanitarian achievement. It is all the more remarkable given that the past 50 years have seen the global population double, adding more mouths to be fed than existed on the planet in 1950. And according to most projections, it appears likely that growing food needs can be met in the foreseeable future, notwithstanding a growing list of technological, distributional, food safety, and health issues that require serious attention and action (Bruinsma 2003; Runge et al. 2003).

Figure 8.1 illustrates the trend in a number of key indicators of food provision. The most significant trend is the growth in food output from 1961 to 2003, increasing by over 160%, or 1.7% per year. As a consequence, average food production per capita also increased by around 25% during the period. Fueling this output growth in many parts of the world were long-term investments in the generation and distribution of new seeds and other farming technologies, and in infrastructure such as irrigation systems and rural roads. This allowed farm productivity to increase and marketing margins to decrease, reducing the price of many foods. Figure 8.1 shows that following significant spikes in the 1970s caused primarily by oil crises, there have been persistent and profound reductions in the price of food globally. It is *well established* that past increases in food production, at progressively

lower unit costs, have improved the health and well-being of billions of people, particularly the poorest, who spend the largest share of their incomes on food.

Despite rising food production and falling food prices, more than 850 million people still suffer today from chronic undernourishment, and the absolute number of hungry people is rising. In 1970 there were an estimated 959 million people suffering from hunger, or about one quarter of the world's population. By 1998 that number had been reduced to 815 million, but progress has been slow. And in sub-Saharan Africa, there are now many more hungry people than there were in 1970. There have also been recent declines in food security in South Asia and the transition economies. In 2000–02, the total number of 852 undernourished people globally was up 37 million from 1997–99. Of this total, 815 million people were in developing countries, up by around 38 million from the 777 million in 1997–99 (FAO 2001, 2004a). In industrial countries, approximately 1.6% of children under five are underweight (WHO 2004d).

This chapter provides insights into the structure and distribution of food provision, with particular emphasis on the relative contribution of various ecological systems. It examines trends in the core food sources (crops, livestock, and fisheries), some of the key linkages to ecosystems and ecosystem service provision, and the drivers of those trends. Examining the drivers of change is particularly important, since some are amenable to intervention so as to bring about improved outcomes, particularly with regard to greater provision of (or fewer trade-offs with) other ecosystem services. Finally, the chapter addresses linkages between human well-being and food access and use. The chapter does not dwell on the important issue of the specific ways in which food is cultivated or harvested, and how those ways affect ecosystem capacity and the provision of other services. These topics are the core focus of specific systems chapters, of which cultivated systems (Chapter 26), drylands (Chapter 22), inland waters (Chapter 20), coastal (Chapter 19), and marine (Chapter 18) systems are the ones most directly relevant. Key related service chapters are those on biodiversity (Chapter 4), fresh water (Chapter 7), and nutrient cycling (Chapter 12).

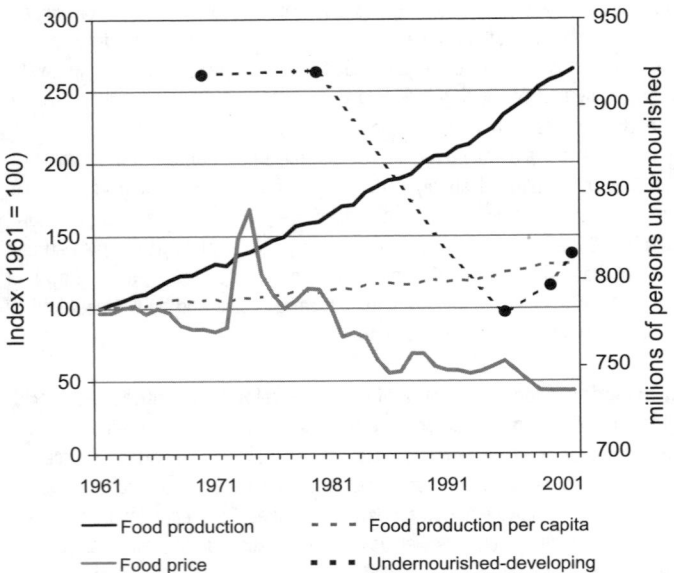

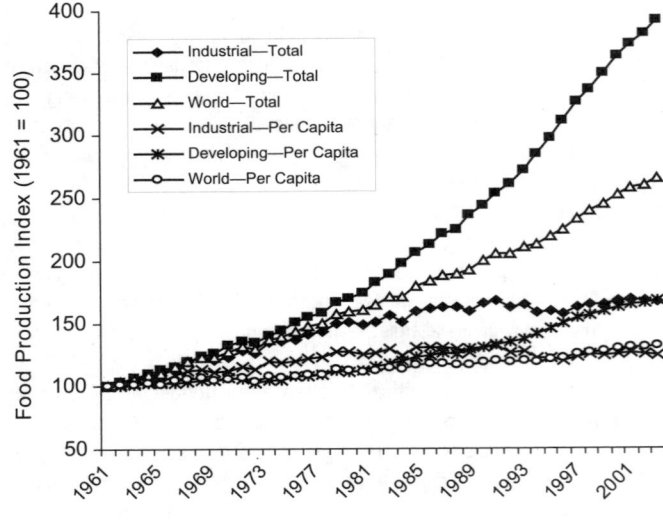

Figure 8.1. Trends in Key Indicators of Food Provision, 1961–2002 (FAOSTAT 2004; IMF various; FAO 2001, 2003, 2004a; World Bank 2004) Global Production, Prices, and Undernourishment, left. Food Supply in Industrial and Developing Countries, right.

8.2 Magnitude, Distribution, and Structure of Food Provision

This section has two main subsections. First, a contemporary perspective of the structure of food provision is presented, focusing both on major food groups and on a breakdown of food provision by system types. An assessment of the spatial distribution of global food production by value for crops, livestock, and fisheries is also presented. Second, a review in more depth is provided of the specific regional and food group trends for crops, livestock, and fisheries.

8.2.1 Structure and Distribution of Food Provision

The overall distribution of food production by MA system type and by major food group is presented in Table 8.1. Care must be taken in interpreting the Table for several reasons. First, the MA systems are neither mutually exclusive nor fully exhaustive of all terrestrial ecosystems. For example, a single cropland area might simultaneously be counted as belonging to several MA systems, since it is a *cultivated system* in a *dryland* area, situated in the *coastal* zone. Second, it is very difficult to obtain any reliable information on the quantity and value of wild sources of food (apart from commercial fisheries), even though they are extremely important in many parts of the world.

All crop production is considered to take place in cultivated systems. (See Chapter 26.) Dryland systems account for about 38% of total crop production, with forest and mountain ecosystems each accounting for about 25%, and coastal systems around 12%. Table 8.1 shows both annual and perennial crop production, irrigated and rain-fed production proportions, and an assessment of the food and feed utilization of crops. On average, about 53% of food crops find their way into food and 21% are used for feed.

The remaining 26% is categorized as used for seed, waste, or other industrial processing. Only a small share of perennial crop production is used for feed. However, a significant quantity of wild fisheries capture is used for feed—for aquaculture and, to a lesser extent, livestock. Aquaculture production is roughly split evenly between inland/fresh and coastal/brackish waters. Wild fish catches from freshwater systems are extremely difficult to estimate, as most go unreported. Some 63% of wild marine fish catches are from marine systems and 37% from coastal systems.

Figure 8.2 (in Appendix A) shows the spatial distribution of the total value of food production summarized in Table 8.1, indicating where the major calorie and protein sources of the world are concentrated. Figure 8.3 (in Appendix A) shows a detail for Asia, highlighting the importance of coastal zone systems in providing high values of both marine and terrestrial food sources. This dual pressure on coastal zones poses particular management challenges. (See Chapter 19.)

8.2.2 Distribution of and Trends in Domesticated and Wild Food Production

8.2.2.1 Domesticated Species

As domestication of plant and animal species favored for food production has evolved, the species base supporting food provision has been eroded. Of the estimated 10,000–15,000 edible plants known, only 7,000 have been used in agriculture and less than 2% are deemed to be economically important at a national level. Only 30 crops provide an estimated 90% of the world population's calorific requirements, with wheat, rice, and maize alone providing about half the calories consumed globally (Shand 1997; FAO 1998; FAOSTAT 2004).

There is a large potential for the improvement and greater use of neglected and underutilized species (FAO 1996; Naylor et al.

Table 8.1. The Global Structure of Food Provision by Food Category and MA Ecosystem (2000 production)

| Food/Feed Types | | Total Value | Share By Use | | Value by Selected MA System | | | | | | | |
			Food	Feed	Dryland	Forests	Cultivated Systems	Mountains	Polar	Inland Waters	Coastal	Marine
			(percent)		(billion 1989–91 dollars)							
Crops	Total	815	53.3	21.1	314	202	815	195			100	
	Irrigated	336			185	38	336	38				
	Rain-fed	479			129	165	479	157				
	Annual	663	49.3	23.0	254	164	663	151				
	Perennial	152	95.8	2.0	60	38	152	44				
Wild plants		n.a.										
Livestock	Total	576	83.0	15.7	294	98	242	150			35	
Wild meat		n.a.										
Fish	Total	158[a]					32		2	32	67	57
	Wild	93[a]	83.0	17.0					2	n.a.	34	57
	Aquaculture	65	100.0				32			32	33	n.a.
Aquatic plants	Wild	n.a.										
	Aquaculture	8					n.a.				8	
Total value of food production		**1,557**			**608**	**300**	**1,089**	**345**	**2**	**32**	**210**	**57**

Production values derived from 2000 production estimates weighted by 1989–91 global average international dollar prices for individual products in each food type group (FAOSTAT 2004; FAO Fishstat 2003; FAO 1997). The 1989–91 prices are the most recent set of complete and comparable prices covering all FAO crop and livestock products. Fisheries prices based on landed values by group of species. Production values by MA system and irrigated/rain-fed split derived by authors from GIS analysis of cropland, irrigated area, and pasture and livestock distribution. Non-food agricultural products were excluded from the analysis. Note that total value for each food group is not the sum of individual MA system values since MA systems overlap and not all MA systems are included in the table.

[a] Fisheries totals do not include wild inland water catches.
n.a. = data not available.

2004). In addition, along with traditional crop varieties, wild relatives of crop plants have been used to supply specific traits that have been introduced into crop plants using conventional breeding techniques, and, increasingly, using modern biotechnology (FAO 1998). There is also a large potential for the domestication and improvement of new crops, especially fruits, vegetables, and industrial (or cash) crops (Janick and Simon 1993), but the probability of developing new major staple crops is probably rather limited (Diamond 1999). With regard to livestock, of the estimated 15,000 species of mammals and birds, only some 30–40 (0.25%) have been used for food production, with fewer than 14 species accounting for 90% of global livestock production.

Since the origins of agriculture, farmers—and, more recently, professional plant and animal breeders—have developed a diverse range of varieties and breeds that contain a high level of genetic diversity within the major species used for food. For some crop species, there are thousands of distinct varieties (FAO 1998). Similarly, there are many breeds of livestock that originate from a single species. However, as larger and larger areas are planted with a smaller and smaller number of crop varieties, and as livestock systems are intensified, many of these varieties and breeds are at risk of being lost in production systems and increasingly are found only in ex situ collections. (See Chapter 26.) For example, FAO estimates that in Europe 50% of livestock breeds that existed 100 years ago have disappeared (Shand 1997).

Plant breeders have achieved yield increases through changing plant physiology and number of grains; increasing the oil, protein, and starch content of specific crops; shortening the maturity period for annual and perennial crops; and increasing drought resistance and nutrient use efficiency. Plant breeding per se has been complemented by deliberate programs of genetic enhancement or "base broadening" in order to incorporate genetic variation into plant breeders' stocks. Generally, there has been insufficient investment in such "pre-competitive" crop improvement activities (Simmonds 1993; FAO 1996; Cooper et al. 2000).

8.2.2.1.1 Crops

Over the 40 years from 1964 to 2004, the total output of crops expanded by some 144% globally, an average increase of just over 2% per year, always keeping ahead of global population growth rates. As shown in Table 8.2, output growth varied by region and over the period as a whole.

Despite a resurgence of crop output in the early to mid-1990s in response to both the decline in outputs from countries in tran-

sition and a surge in food prices, many middle-income and richer countries have seen a gradual slowing down in the growth of crop output in line with the deceleration of population growth and the attainment of generally satisfactory levels of food intake. Decelerating growth patterns in crop output have been most evident in industrial countries and in Asia more widely.

Output in the transition economies fell by about 30% between 1990 and 1995 from its fairly stable level in the mid to late 1980s. While output has since steadied around a lower level, a significant drop in average food energy intake and an increase in the incidence of malnutrition have been documented during the 1990s, as described elsewhere in this chapter.

In response to growing affluence and shifting dietary patterns that increased demand for both food and feed crops, growth of food output in Asia has been consistently high, at 3% a year or more since the early 1960s. The feed market is important not only for intensive livestock production, but increasingly for aquaculture, as seen in the rapid increase in soybean demand for carp cultivation in China.

While growth in overall crop output in sub-Saharan Africa has been relatively strong over the past two decades, beverage and fiber crops, predominantly for export, still represent a significant share of that production. Since food crop production has not grown as markedly, and population growth rates remain high, sub-Saharan Africa remains the only region in which per capita food production has not seen any sustained increase over the last three decades, and this has recently been in decline. In North Africa and the Middle East, growth in crop output has been both moderate and often erratic.

The past 40 years have also seen some considerable shifts in crop production, driven by changes in consumption. Figure 8.4 shows the trends in crop production by major crop group on a per capita basis. There have been four general trends exhibited by oilcrops; fruits and vegetables; cereals and sugar crops; and roots, tubers, and pulses.

Growth in output of oilcrops and vegetable oils between 1961 and 2001was consistently strong at just over 4% per year, largely propelled by a rapid growth in palm oil (8.2% per year), rapeseed oil (6.9% per year), and soybeans (4.1% per year). The principal commodities included in this category (and their global production quantities in million tons in 2001) include soybeans (177), oil palm (128), coconuts (52), groundnuts (36), and rapeseed (36). Cottonseed (37 million tons) is usually often part of this group, but it is excluded here as it is not considered a food product.

Food use of oil and vegetable oil crops, expressed in oil equivalent, grew from 6.3 kilograms per capita per year in 1964/66 to 11.4 kilograms in 1997/99. Demand has grown more in developing countries (5.0% per year) than in industrial ones (3.2%), stimulated by rising incomes and urbanization that have increased consumption of cooking oil, processed foods, and snacks. More than for any other crop (and excluding pastures), it is the global area expansion of oilcrops over the past 40 years that has driven cropland expansion. (See Box 8.1.)

Fruit and vegetable production grew in line with population during the 1960s and 1970s, when growing demand lead to increased per capita output. The principal commodities in this category, and their 2001 production in million tons, are tomatoes (106), watermelons (81), bananas (65), cabbages (61), grapes (61), oranges (60), apples (58), and dry onions (51). While per capita output growth was modest during the 1980s, it accelerated during the 1990s. Between 1961 and 2001, production of vegetables grew from 72 kilograms per capita on average per year to 126 kilograms, and that of fruits from 56 to 77 kilograms per year.

Table 8.2. Global and Regional Growth Rates in Crop Output (Bruinsma 2003)

Region	1969–99	1979–99	1989–99
	(percent per year)		
Sub-Saharan Africa	2.3	3.3	3.3
Near East/North Africa	2.9	2.9	2.6
Latin America and Caribbean	2.6	2.3	2.6
South Asia	2.8	3.0	2.4
East Asia	3.6	3.5	3.7
Developing countries	3.1	3.1	3.2
Industrial countries	1.4	1.1	1.6
Transition economies	−0.6	−1.6	3.7
World	**2.1**	**2.0**	**2.1**

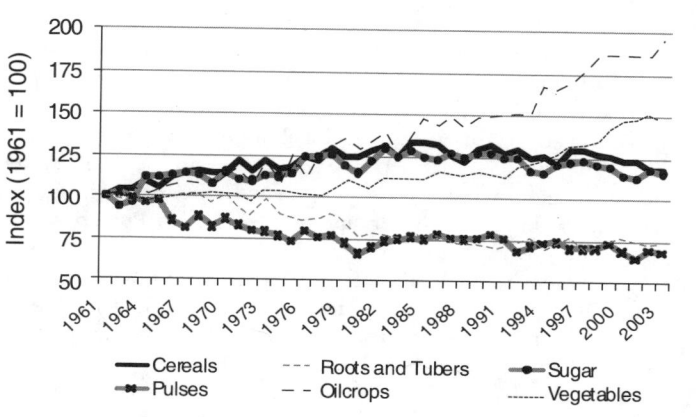
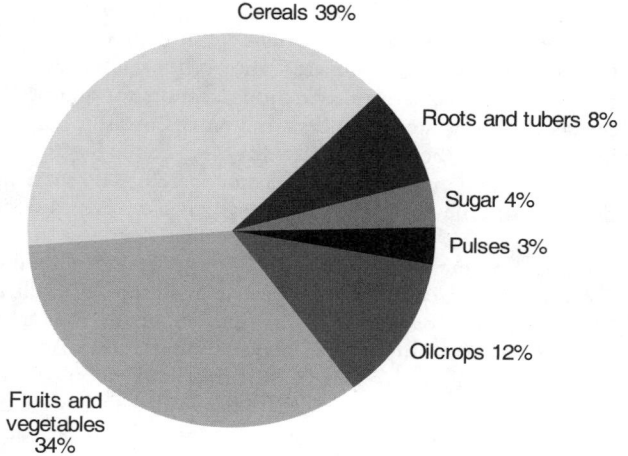

Figure 8.4. Aggregate Structure of Per Capita (1961–2003) and Total Food Crop Output by Group (2001–03 averages) (Calculated from FAOSTAT 2004)

BOX 8.1

Cropland Dynamics: The Case of Oil Crops (Bruinsma 2003: 101–03)

In 1961 the harvested areas of cereals and oilcrops stood at 648 million and 113 million hectares respectively. Over the past 40 years, output of oilcrops expanded dramatically, and by 2001 their harvested area stood at 233 million hectares compared to 674 million hectares for cereals. The harvested area of soybean alone expanded by some 50 million hectares. Most of the oilcrop expansion took place in land-abundant countries (Brazil, Argentina, Indonesia, Malaysia, the United States, and Canada).

The big four oil crops have been responsible for a good part of the expansion of cultivated land under all crops in developing countries and the world as a whole. In terms of harvested area, land devoted to the world's principal crops (cereals, roots and tubers, pulses, fibers, sugar crops, and oilcrops) expanded by 59 million hectares (or 6%) since the mid-1970s. (Increases in harvested area arise from a physical expansion of cultivated land, an expansion of land under multiple cropping (a hectare of arable land is counted as two if it is cropped twice in a year), or both. Therefore, the harvested area expansion under the different crops discussed here could overstate the extent to which physical area in cultivation has increased. This overstatement is likely to be more pronounced for cereals (where the arable area has probably declined even in developing countries) than for oilcrops, as the latter include also tree crops (oil and coconut palms and olive trees).)

A 105-million-hectare increase in harvested area in developing countries was accompanied by a 46-million-hectare decline in industrial countries and transition economies. The expansion of land under the big four oil crops was 63 million hectares—that is, they accounted for all the increase in world harvested area, and more than compensated for the drastic declines in the area under cereals in industrial countries and transition

economies. In these countries, the expansion of oilseed area (25 million hectares) substituted and compensated for part of the deep decline in the area sown to cereals. But in developing countries, it seems likely that it was predominantly new land that came under cultivation, as land under the other crops also increased.

These numbers illustrate the dramatic changes in cropping patterns that occurred, particularly in industrial countries, as a result of policies (such as EU support to oilseeds) and of changing demand patterns toward oils for food in developing countries and toward oilcakes/meals for livestock feeding everywhere. They also demonstrate that land expansion still can play an important role in the growth of crop production. The 200% increase in oilcrop output between 1974/76 and 1997/99 in developing countries was brought about by a 70% (50-million-hectare) expansion of land under these crops at the same time as land under other crops increased by an almost equal amount.

Particularly notable is the rapid expansion of the share of oil palm products (in terms of oil palm fruit) from Southeast Asia (from 40% of world production in 1974/76 to 79% in 1997/99) and the dramatically shrinking share from Africa (from 53 to 14%). Africa's share in terms of actual production of palm oil (9% of the world total, down from 37% in the mid-1970s) remained well below that of its share in oil palm fruit production. This denotes the failure to upgrade the processing industry, but also the potential offered by more-efficient processing technology to increase oil output from existing oil palm areas. The contrast of these production shares with the shares of land area under oil palm is even starker: Africa still accounts for 44% of the world total, three quarters of it in Nigeria.

Cereal and sugar crop production grew at an accelerated rate in the 1960s and 1970s, increasing their total per capita output by around 25% by 1980. The principal cereal crops, according to their 2001 production in million tons, are maize (615), paddy rice (598), wheat (591), barley (114), sorghum (60), millet (29), and oats (27).

Per capita cereal production peaked in the mid-1980s and has been in slow decline ever since. Sugar crop production followed broadly the same pattern as that for cereals. The per capita pro-

duction of roots, tubers, and pulses declined by around 25% between the early 1960s and the early 1980s, with pulses declining more rapidly at first. Since then production has roughly kept pace with population growth.

Overall, these trends suggest that higher-value cereals, fruits, and vegetables have tended to displace pulses and roots and tubers.

The cereal sector remains particularly important in several ways. Cereals provide almost half of the calories consumed directly by humans globally (48% in 2001) and will continue as the foundation of human food supply into the foreseeable future because of their high yields, nutrient density, and ease of cooking, transport, and storage compared with other staples such as root and starch crops. Cereal production accounts for almost 60% of the world's harvested crop area and an often disproportionately larger share of the usage of fertilizer, water, energy, and other agrochemical inputs. The cereal sector therefore is especially important from the perspective of ecosystem services and trade-offs between services both locally and globally. Chapter 7 in the MA *Scenarios* volume describes the technological and humanitarian successes of the cereal-based Green Revolution, as well as the subsequent and continuing controversy about the scale and longevity of its environmental and equity implications. At the heart of this debate lie many questions of trade-offs among ecosystem services and among elements of human well-being. One part of that debate has focused on the relative economic, social, and environmental costs of intensification versus expansion strategies for meeting global (cereal) food needs (Evenson and Gollin 2003; Conway 1997; Green et al. 2005) as well as on key assumptions regarding the scientific opportunities for improving future crop yield potential (Cassman 1999; Cassman et al. 2003).

Aggregate cereal consumption and production patterns are influenced by three major, codependent forces. The first force is a two-stage income effect in which cereal consumption increases in proportion with incomes as they grow from low levels, but a reversal in this behavior (technically, a reverse in the "income elasticity") is witnessed as incomes continue to rise and as basic energy and other dietary needs are met. At this stage most consumers tend to replace food staples like cereals with higher-value foods, such as animal protein and fruits and vegetables. The second force is urbanization, which often brings a shift in cereal preferences toward wheat and rice at the same time as an overall decline in the share of cereals in a more diverse diet. And the third force is the increasing role of coarse grains (maize, sorghum, millet) but also wheat and, to a lesser extent, rice as livestock feed. These forces, all at various stages of evolution in different parts of the world, have resulted in a net increase in per capita cereal consumption globally from 135 to 155 kilograms per year between 1961 and 2001, even though cereals now constitute a slightly lower proportion of total energy intake (down from 50% to 48%).

The trends are clearer if industrial- and developing-country groupings are distinguished. In industrial countries, per capita consumption of cereal as food fell from 148 to 130 kilograms per year (representing 38% and 31% respectively of dietary energy supply), while in developing countries per capita consumption increased from 129 to 162 kilograms per year (representing 59% and 53% respectively of DES). (See Box 8.2 for a description of trends in cereals for feed.)

Following a peak in food prices in 1996, there was strong growth in crop output in 1999 in both industrial and developing countries, but since then the general pattern of growth deceleration has resumed. In industrial countries, output actually declined in both 2001 and 2002. In the case of cereals, global output levels have stagnated since 1996, while grain stocks have been in decline. The area devoted to the major cereals has been decreasing at about 0.3% annually since the 1980s. These trends are likely to continue if real cereal prices continue to fall, causing farmers to switch to more profitable crops, such as vegetables and fruits. Loss of highly productive cereal-growing land is particularly acute in areas of rapid urban expansion, a common feature of development in many countries. Although there has been some cereal price recovery since 2001, prices still stand at some 30–40% lower than their peak in the mid-1990s (FAO 2004b).

Growth in the yield of the major cereals has been virtually constant for the past 35 years since the release of the first miracle varieties of wheat and rice and of the single-cross maize hybrids. And in many of the world's most important cereal production areas, there has even been a plateauing of yields in the past 15–20 years as average farm yields reached about 80–85% of the genetic yield potential (Cassman 1999). Such stagnation is evident in key rice-growing provinces in China, Java and other parts of Indonesia, Central Luzon in the Philippines, the Indian Punjab, Japan, and South Korea (Cassman et al. 2003), as well as for irrigated wheat in the Yaqui Valley of Mexico. However, yield growth rates will have to increase to meet future food demand unless more land area is devoted to cereal production. While in many low-productivity areas there is still considerable scope (and pressing need) for raising yields through the use of improved technologies and management practices, in high-productivity areas future yield growth will depend increasingly on raising genetic yield potential and more fine-tuned crop and soil management practices to allow consistent production near the yield potential ceiling.

Despite the potential contribution of genomics and molecular biology, as well as substantial research investments to improve photosynthesis during the 1970s and 1980s, there is as yet limited evidence that biotechnology approaches can help raise the yield potential ceiling. Indeed, there has been little progress toward increasing maximum net assimilation rates (photosynthesis minus respiration) in crop plants, and the determinants of yield potential are under complex genetic control that result in trade-offs between different options for increasing seed number, seed size, partitioning of dry matter among different organs, crop growth duration, and so forth (Denison 2003; Sinclair et al. 2004). Consideration of these issues has led to calls for caution in projecting forward past achievements in yield growth as a basis for assessing future food security, as well as for greater urgency in the key scientific challenges involved (Denison 2003; Cassman 2001).

8.2.2.1.2 Livestock

Livestock and livestock products are estimated to make up over half of the total value of agricultural gross output in industrial countries, and about a third of the total in developing countries, but this latter share is rising rapidly (Bruinsma 2003). The global importance of livestock and their products is increasing as consumer demand in developing countries expands with population growth, rising incomes, and urbanization. This rapid worldwide growth in demand for food of animal origin, with its accompanying effects on human health, livelihoods, and the environment, has been dubbed the "Livestock Revolution" (Delgado et al. 1999). Livestock production has important implications for ecosystems and ecosystem services, as it is the single largest user of land either directly through grazing or indirectly through consumption of fodder and feedgrains (Bruinsma 2003). Industrial livestock production, the most rapidly growing means of raising livestock, poses a range of pollution and human health problems. (See Chapter 26.) At the same time, livestock production can promote linkages between system components (land, crops, and water) and enables the diversification of production resources for poor farmers (Devendra 2000).

The overall annual growth rates for livestock product outputs are summarized by region and by time period in Table 8.3. The global growth rate is currently just over 2% per year and is declining over time, but this masks the true dynamics of the sector (and highlights the potential pitfalls of interpreting global-scale data), as there are large regional disparities. While growth rates in industrial countries, where people already enjoy adequate supplies of

BOX 8.2

The Growing Use of Crops as Feed (Delgado et al. 1999)

Crops are used both as feed inputs for intensive livestock systems and for direct or processed sources of food. Global use of cereals as feed increased at only 0.7% per year between 1982 and 1994 despite rapid increases in meat production. Growth rate in cereal use in industrial countries was negligible, while it increased by about 4% a year in the developing countries. Despite the higher growth rate, developing countries still use less than half as much cereal for feed as industrial countries do. During the early 1990s, concentrated cereal feed provided between 59% and 80% of the nutrition given to animals in the industrial world. By contrast, cereals accounted for only 45% of total concentrate feed in Southeast Asia, the developing region with the most intensive use of feed grains.

For the world as a whole, it is estimated that 660 million tons (in 1997) of mainly coarse grains, making up 35% of all cereal use, are fed to animals. Most of these are used in the United States and other industrial countries. Nevertheless, increasing amounts are being fed to intensive livestock in developing countries, as poultry and pig production increased. Over the last decade, the increase in cereal use for feed has been more gradual than expected, partly because of a reduction in intensive livestock production in the transition economies, partly because of high cereal prices in the EU, and partly because of increasing efficiency of feed conversion.

Poultry are very efficient feed converters, requiring only 2–2.5 kilograms of feed per kilogram of meat produced and even less per kilogram of eggs. Pigs require 2.5–4 kilograms of dry matter per kilogram of pig meat, while concentrate-fed ruminants require much more feed per kilogram of meat.

The use of cereals as feed has been fastest in Asia, where output growth has risen the most and land is scarce. In Other East Asia, Southeast Asia, and Sub-Saharan Africa, cereal use as feed grew faster than meat production, indicating that those regions are intensifying their use of feed per unit of meat output. Most of Asia, West Asia–North Africa, and Sub-Saharan Africa lack the capacity to produce substantial amounts of feed grain at competitive prices. The growing amounts of feed grains imported into these regions attest to this deficiency. Given that many developing countries cannot expand crop area, two possibilities remain: intensification of existing land resources and importation of feed. Because much of the gain from intensification will probably go toward meeting the increasing demand for food crops, substantially more feed grains will have to be imported by developing countries in the future.

Alternatives to crops in the way of feed include household waste products and crop residues. In developing countries, household food waste, such as tuber skins, stems, and leaf tops, has traditionally been an important feed source for backyard monogastric production in particular. But small-scale backyard operations are disappearing because of low returns to labor and increased competition from large-scale producers. Although each backyard operation is small, at the aggregate level such systems act as major transformers of waste into meat and milk. Because large operations are unlikely to find it cost-effective to collect small amounts of waste from many households, this source of animal feed may be underused in industrial systems.

Trends and Projections in the Use of Cereal as Feed. Figures are three-year moving averages centered on year shown. The 2020 projections are from the July 2002 version of the IMPACT model. (Delgado et al. 2003, calculated from data in FAOSTAT 2004)

Region	Total Cereal Use as Feed			
	1983	1993	1997	2020
	(million tons)			
China[a]	40–49	78–84	91–111	226
India	2	3	2	4
Other East Asia	3	7	8	12
Other South Asia	1	1	1	3
Southeast Asia	6	12	15	28
Latin America	40	55	58	101
Western Asia and North Africa	24	29	36	61
Sub-Saharan Africa	2	3	4	8
Developing world	128	194	235	444
Industrial world	465	442	425	511
World	**592**	**636**	**660**	**954**

[a] Ranges show high and low estimates based on data from various sources.

---

animal protein, have remained at just over 1% for the past 30 years, growth rates in developing countries as a whole have been high and generally accelerating. The trends in East Asia (and particularly China) are particularly strong, with livestock product growth rates of over 7% a year over the last 30 years, albeit from a low base. South Asia and the Middle East and North Africa have maintained long-term growth in livestock product output of over 3% per year.

As with crops, two regions have lagged behind in livestock production: the countries in transition and sub-Saharan Africa. The transition economies exhibit the same pattern as for crops—slow long-term shrinkage of output, followed by collapse in the early 1990s. Sub-Saharan Africa, faced with the world's highest stresses of poverty, hunger, and population growth (see Chapters 3, 6, and 7) and with continuing insecurity, particularly in pastoral areas within the subcontinent, has made slow progress; per capita livestock output has hardly increased at all in the past 30 years (Ehui et al. 2002).

With regard to the product structure of growth, Figure 8.5 presents the trends in growth of global output for each of the major livestock food product categories, expressed in per capita terms. Three broad groupings of trends are shown; for poultry meat; for pigmeat and eggs; and for bovine, mutton, and goat meat and milk. Poultry meat production has expanded almost ninefold, from some 2.9 to 11.2 kilograms per capita per year between 1961 and 2001. In developing countries, this entailed a production expansion from 1.0 to 7.7 kilograms per capita per year as population in those countries increased from 2.1 billion to 4.8 billion. In industrial countries, the equivalent figure was from 6.7 to 24 kilograms as population increased from 980 million to 1.3 billion. This quite remarkable growth in output has been achieved through rapid expansion of industrial ("landless") chicken rearing and processing facilities located in peri-urban areas throughout the world. (See Chapter 26.) These enterprises in turn depend on supplies of quality grain–based feedstuffs from national or international markets.

Table 8.3. Global and Regional Growth in Livestock Output
(Bruinsma 2003)

Region	1969–99	1979–99	1989–99
	(percent per year)		
Sub-Saharan Africa	2.4	2.0	2.1
Near East/North Africa	3.4	3.4	3.4
Latin America and Caribbean	3.1	3.0	3.7
South Asia	4.2	4.5	4.1
East Asia	7.2	8.0	8.2
Developing countries	4.6	5.0	5.5
Industrial countries	1.2	1.0	1.2
Transition economies	–0.1	–1.8	–5.7
World	**2.2**	**2.1**	**2.0**

While growth in the poultry meat sector has been relatively consistent since the early 1960s, the output of eggs and pork was slower both in its takeoff and in its subsequent growth, with higher and more sustained growth starting only in the early 1980s. Per capita production of both eggs and pork almost doubled between 1961 and 2001. Total production of eggs rose from 15.1 million to 57.0 million tons, and pork from 24.7 million to 91.3 million tons. In developing countries, annual per capita production of eggs and pork increased from 1.6 and 2.1 kilograms, respectively, in 1961 to 7.0 and 11.3 kilograms in 2001. In industrial countries, growth has been more modest, however, from 10.8 to 12.7 kilograms per capita in the case of eggs, and from 20.5 to 24.0 kilograms per capita in the case of pork during the same time period. Pig and poultry meat each now account for about a third of all meat produced worldwide, and more than one half of total pig production is in China.

Growth in milk (cattle and buffalo), beef, and mutton and goat meat production has, on the whole, kept pace with population growth rates, and average per capita global production has stayed relatively constant over the last 40 years. The 1961 global production of 344 million, 29 million, and 6 million tons of milk, beef, and mutton and goat meat, respectively, increased to 590

million, 59 million, and 11 million tons in 2001. Milk production has risen faster in developing than in industrial countries, from 32 to 50 kilograms per capita per year, but still lies far below the 264 kilograms per capita per year of industrial countries. Annual per capita production of beef increased in developing countries from 4.6 to 6.2 kilograms between 1961 and 2001, while in industrial countries, despite the large-scale switch to poultry meat, annual per capita beef production edged up from 19.6 kilograms in 1961 to 22.4 in 2001.

Looking back at the trends in the evolution of livestock systems, three points can be made: First, almost the entire expansion in output from poultry and pigs, globally, and from beef and milk cattle in industrial countries, has taken place in intensive, industrial production systems. Second, while providing food in relatively safe, reliable, and progressively cheaper ways, there have been many examples in both industrial and developing countries of a wide range of soil, water, and odor pollution problems, as well as potential large-scale health risks from the more intensive production of livestock. (See Chapter 26.) Third, the expansion of extensive beef production systems, primarily in South and Central America, has been associated with high rates of deforestation (Mahar and Schneider 1994; Kaimowitz 1996; Vosti et al. 2002).

Livestock productivity (output per head of livestock) continues to be higher in industrial than in developing countries, with the largest difference in the case of milk production, which is more than six times higher. In 2001, for example, milk yield was 3,075 and 480 kilograms per animal in industrial and developing countries, respectively. In general, sub-Saharan Africa and South Asia have the lowest output per animal compared with other parts of the world. In sub-Saharan Africa, milk production per animal has been declining since 1961, and in 2001, while production of beef per animal was about 65% of the world average, production of milk per animal was only 14% of the world average.

This low productivity level can be attributed to the types of production systems prevailing in sub-Saharan Africa. Generally, three phases of the income-herd relationship in smallholder producers can be distinguished, which coincide with the process of commercialization of the livestock sector: emergence, expansion, and contraction. Poor farmers raise few livestock, but as development begins to take place poor rural households are able to gradually expand their livestock holdings. The herd size gradually

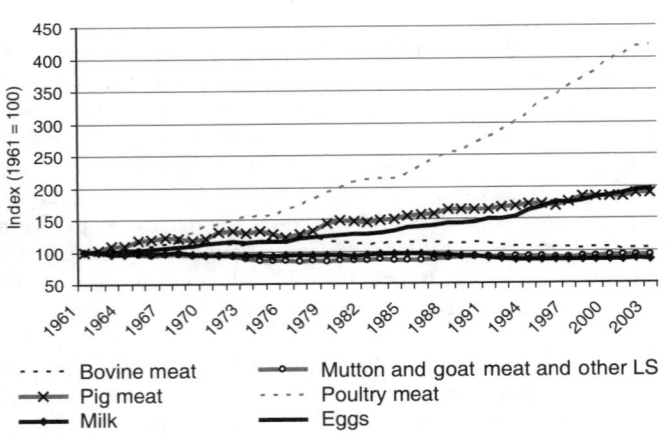

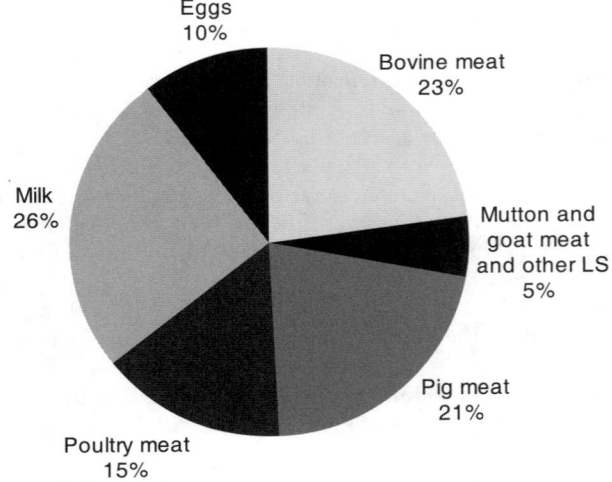

Figure 8.5. Aggregate Structure of Per Capita (1961–2003) and Total Livestock Output by Group (2001–03 averages) (Calculated from FAOSTAT 2004)

expands with further development, but there is a point of development in many rural economies after which most farmers choose to stop raising livestock. Beyond a certain income level, herd size for most households falls as productivity increases, and only a few specialized households evolve toward larger-scale commercial operations (McIntire et al. 1992).

To date, overall growth in livestock production has been sufficient to meet increases in demand without significant price increases, and relative to the long-term downward trend in prices for cereals, oils, and fats, the prices for livestock products have remained relatively stable. However, there are considerable differences between continents and countries in production and consumption, and international trade between surplus and deficit producers has increased. Developing countries, as a group, have become net importers of livestock products from industrial countries. Between 1990 and 2000, net imports of meat and milk to developing countries grew by more than 6% a year, while net imports of eggs declined by a little over 16%.

8.2.2.2 Wild Food Sources: Fisheries

Biodiversity provides a diverse range of edible plant and animal species that have been and continue to be used as wild sources of food, including plants (leafy vegetables, fruits, and nuts), fungi, bushmeat, insects and other arthropods, and fish (including mollusks and crustaceans as well as finfish) (Pimbert 1999; Koziell and Saunders 2001). Many types of wild food remain important for the poor and landless, especially during times of famine and insecurity or conflict, when normal food supply mechanisms are disrupted and local or displaced populations have limited access other forms of nutrition (Scoones et al. 1992). Even in normal times, these wild land-based foods are often important in complementing staple foods to provide a balanced diet, and plants growing as weeds may often be important in this respect (Johns and Staphit 2004; Cromwell et al. 2001; Satheesh 2000).

About 7,000 species of plants and several hundred species of animals have been used for human food at one time or another (FAO 1998; Pimbert 1999). Some indigenous and traditional communities use 200 or more species for food (Kuhnlein et al. 2001). The capacity of ecosystems to provide wild food sources is generally declining, as natural habitats worldwide are under increasing pressure and as wild plant and animal populations are exploited for food at unsustainable levels.

This section focuses on freshwater and marine fisheries, as they are globally significant sources of wild food, and it also covers aquaculture.

During the past century, the production and consumption of fish (including crustaceans and mollusks) has changed in important ways. Three trends are notable: average per capita consumption has increased steadily; the proportion of fish consumed at considerable distances from where it is harvested is growing; and an increasing number of fish stocks have been critically depleted by catch rates that exceed, often considerably, any commonly understood measure of maximum sustainable yield.

During the last four decades, the per capita consumption of fish as seafood increased from 9 to 16 kilograms per year. Table 8.4 shows fish production and utilization over the last half of the 1990s.

8.2.2.2.1 Trends in trade, commercialization, and intensification

Ninety percent of full-time fishers conduct low-intensive fishing (a few tons per fisher per year), often in species-rich tropical waters of developing countries. Their counterparts in industrial countries generally produce several times that quantity of fishing

output annually, but they are much fewer, probably numbering about 1 million in all (FAO 1999), and their numbers are declining. In industrial countries, fishing is seen as a relatively dangerous and uncomfortable way to earn an income, so as a result fishers from economies in transition or from developing countries are replacing local fishers in these nations.

Nearly 40% of global fish production is traded internationally (FAO 2002). Most of this trade flows from the developing world to industrial countries (Kent 1987; FAO 2002). Many developing countries are thus trading a valuable source of protein for an important source of income from foreign revenue, and fisheries exports are extremely valuable compared with other agricultural commodities. (See Figure 8.6.)

Although fish are consumed in virtually all societies, the levels of consumption differ markedly. Per capita consumption is generally higher in Oceania, Europe, and Asia than in the Americas and Africa. Small island countries have high rates of consumption; land-locked countries often low levels. Fish is eaten in almost all social strata, due to the large variety of fish species and products derived from them, ranging from the very exclusive and expensive and rare to the cheap and currently still plentiful.

8.2.2.2.2 Overfishing and sustainability

After 50 years of particularly rapid expansion and improving technological efficiency in fisheries, the global state of the resources is causing widespread concern. Between 1974 and 1999, the number of stocks that had been overexploited and were in need of urgent action for rebuilding increased steadily and by 1999 stood at 28% of the world's stocks for which information is available. While the percentage of overexploited stocks appears to have stabilized since the late 1980s, the latest information indicates that the number of fully exploited stocks has been increasing in recent years while the number of underexploited stocks has been decreasing steadily—from an estimated 40% in 1970 to 23% in 2004. The most recent information available from FAO suggests that just over half of the wild marine fish stocks for which information is available are fully or moderately exploited, and the remaining quarter is either overexploited or significantly depleted.

The Atlantic Ocean was the first area to be fully exploited and overfished, and fish stocks in the Pacific Ocean are almost all currently fully exploited. There still seems to be some minor potential for expansion of capture fisheries in the Indian Ocean and the Mediterranean Sea, although this may be due to environmental changes including eutrophication. Phytoplankton plumes near densely populated areas and riverine plumes have been associated with higher levels of fisheries productivity (Caddy 1993).

At the beginning of the twenty-first century, the biological capability of commercially exploited fish stocks was probably at a historical low. FAO has reported that about half of the wild marine fish stocks for which information is available are fully exploited and offer no scope for increased catches (FAO 2002). Of the rest, 25% are underexploited or moderately exploited and the remaining quarter are either overexploited or significantly depleted.

Although information on catches from inland fisheries is less reliable than for marine capture fisheries, it appears that freshwater fish stocks are recovering somewhat from depletion in the Northern Hemisphere, while the large freshwater lakes in Africa are fully exploited and in parts overexploited. Some fish species exhibit more dramatic threshold effects, appearing less able to recover than others.

Accentuating the ecological implications of the increase in capture fisheries production is an important trend in catch com-

Table 8.4. World Fishery Production and Utilization, 1996–2001

Production and Utilization	1996	1997	1998	1999	2000	2001[a]
			(million tons)			
Production						
Inland						
Capture	7.4	87.6	8.0	8.5	8.8	8.8
Aquaculture	15.9	17.5	18.5	20.2	21.4	22.4
Total inland	23.3	25.0	26.5	28.7	30.2	31.2
Marine						
Capture	86.0	86.4	79.2	84.7	86.0	82.5
Aquaculture	10.8	11.2	12.0	13.3	14.1	15.1
Total marine	96.9	97.5	91.3	98.0	100.2	97.6
Total capture	93.5	93.9	87.3	93.2	94.8	91.3
Total aquaculture	26.7	28.6	30.5	33.4	35.6	37.5
Total production	120.2	122.5	117.8	126.7	130.4	128.8
Utilization						
Human consumption	88.0	90.8	92.7	94.5	96.7	99.4
Non-food uses	32.2	31.7	25.1	32.2	33.7	29.4
			(kilograms)			
Per capita food fish supply	15.3	15.6	15.7	15.8	16.0	16.2

[a] Denotes projected data (Fisheries Centers, UBC).

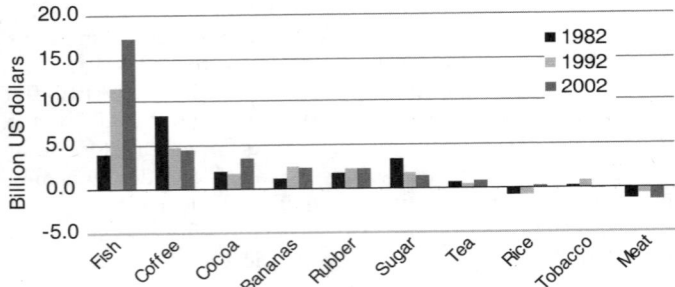

Figure 8.6. Developing-Country Net Exports of Fish and Selected Agricultural Commodities, 1982, 1992, and 2002 (FAO 2004)

---

position—over the past 30 years the average trophic level of fish landed from marine and freshwater ecosystems has declined. (See Box 8.3.) Trophic level decline is the progressive change in catch composition, in the case of marine systems, from a mixture of top predatory fish such as sharks and saithe, mid-trophic level fish such as cods and herrings, and a few lower trophic level animals such as shrimp to a catch of a few mid-trophic species such as whiting and haddock and many low-trophic species such as shrimp. This change is a result of three phenomena: the expansion of fisheries from benthic coastal production areas to the pelagic open ocean; the expansion of fisheries from the Northern Hemisphere (dominated by large shelves and bottom fish) to the Southern Hemisphere (dominated by upwelling systems and pelagic fish); and overfishing, possibly leading to a local replacement of depleted large predators by their smaller preys. This change in catch composition is sometimes called "fishing down marine food webs."

BOX 8.3
Trophic Level

One way to understand the structure of ecosystems is to arrange them according to who eats what along a food chain. (See Figure.) Each link along the chain is called a trophic level. Levels are numbered according to how far particular organisms are along the chain—from the primary producers at level 1 to the top predators at the highest level. Within marine systems, large predators such as sharks and saithe are at a high trophic level, cod and sardines are in the middle, and shrimp are at a low trophic level, with microscopic plants (mainly phytoplankton) at the bottom sustaining marine life (Pauly et al. 2003).

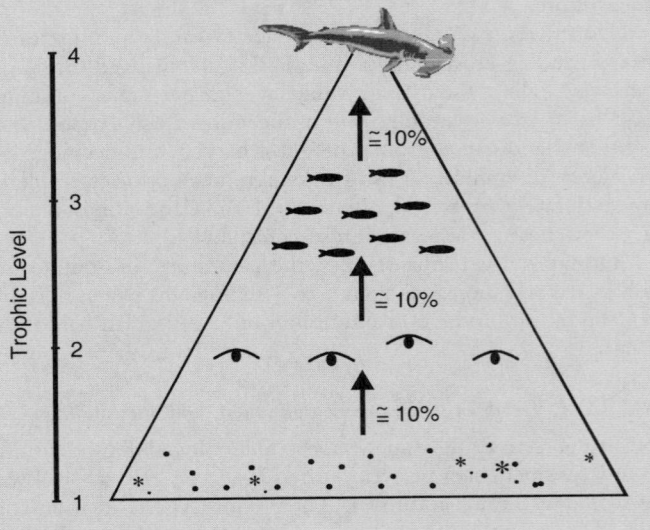

8.2.2.2.3 *Freshwater fisheries and food security*

Approximately 10% of wild harvested fish are caught from inland waters, likely a smaller proportion than in the early twentieth century. However, it is more difficult to measure freshwater fisheries catches than marine catches. They may be underreported by as much as a factor of two because informal fisheries activities, such as subsistence fisheries, are not accurately accounted for in national statistics (Coates 1995). Fish production from inland waters is almost entirely finfish, with negligible amounts of crustaceans or mollusks, except in localized areas. As shown in Figure 8.7, the mean tropic level of freshwater fisheries landings tends to be lower than that of marine catches.

The socioeconomic value of freshwater fish catches is especially high. Freshwater fish tend to be consumed in their entirety, with minimal wastage, providing key sources of protein for local communities. And in addition to their nutritional value, freshwater fisheries provide livelihoods for low-income and resource-poor groups. The high level of artisanal and informal activity, relying on labor-intensive catching methods, contributes to food security for vulnerable groups, including women and children.

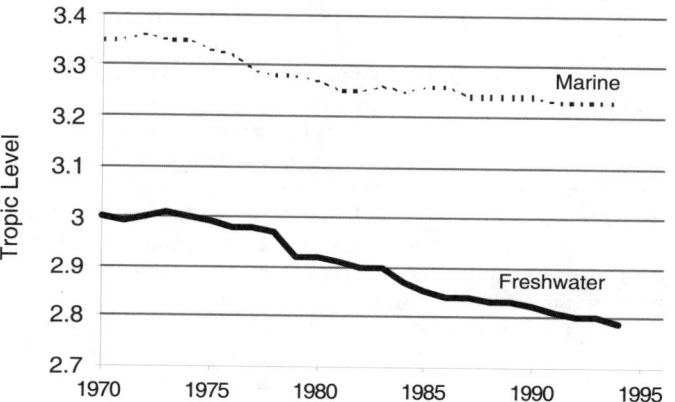

Figure 8.7. Decline in the Trophic Level of Fishery Catch, 1970–95 (Pauly et al. 1998)

8.2.2.2.4 *Aquaculture*

Although aquaculture is an ancient activity, it is only during the past 50 years that it has become a globally significant source of food. In 2002 it contributed approximately 27% of fish harvested and 40% (by weight) of all fish consumed as food. However, the variety of supply from aquaculture is well below that of capture fisheries: only five different Asian carp species account for about 35% of world aquaculture production, and inland waters currently provide about 60% of global aquaculture outputs.

The distinction between capture fisheries and aquaculture in fresh waters can be unclear. For example, extensive aquaculture in China includes catches from stocked rivers and lakes (which are substantial). While expanding aquaculture production can take the pressure off wild fisheries resources in some cases, in other cases the opposite is true (Naylor et al. 2000), as cultivation of carnivorous species can require large inputs of wild fish for feed. Overall, catches of wild fish for non-food uses are increasing faster than catches for food.

8.3 Food Provision and Biodiversity

This section reviews some of the key impacts of the provision of food on biodiversity. Since food provision involves the purposive management or exploitation of ecosystems to enhance food productivity, there are often trade-offs involved with other ecosystem services. In the past, when food production activities affected a smaller share of Earth's land and ocean bodies, and overall demand for ecosystem services supported by biodiversity was less than today, many of these trade-offs were not recognized or were not considered to be important. Now cultivated systems account for about 27% of the world's land surface and for a much higher share of habitable land (Wood et al. 2000).

The most direct impact of food provision on biodiversity has been through habitat conversion: around 43% of tropical and subtropical dry and monsoon forests and 45% of temperate broadleaf and mixed forests globally have been converted to croplands. Huge areas of the world are now planted to a small number of crop species or covered by modified pastures. In addition, rapid increases in coastal aquaculture have lead to the loss of mangrove ecosystems. Though future rates of conversion are expected to be much lower in absolute terms than historically, the major locations of agricultural expansion have frequently coincided with remnants of natural habitats with high biodiversity value (Myers et al. 2000). And the construction of roads and other infrastructure (such as irrigation canals), which are seen as key to promoting agricultural development and meeting the Millennium Development Goals, tends to dissect the landscape and to further limit the movement of wildlife and the dissemination of plant species.

Second, food provision affects wild biodiversity through its demand for inputs other than land, most notably water and nutrients, and through the pollution of ecosystems with pesticides and excess nutrients. Irrigated agriculture is a major user of fresh water (see Chapter 7), which, together with the direct loss of wetland habitats from conversion and the pollution of inland waters from excess nutrients, has a major negative impact on inland water biodiversity. (See Chapter 20.) As a consequence, wild fish populations in inland waters can be greatly reduced, often having the greatest negative impacts on the poor (Bene et al. 2003). Despite increases in water use efficiency, total water demand for agriculture is increasing and in many regions is projected to outstrip sustainable supplies over the coming decades. (See Chapter 7.)

Agriculture is the major consumer of reactive nitrogen, but only a fraction this is used in plant growth and retained in food products. The excess leads to biodiversity loss and reduced water quality in inland waters and coastal systems through eutrophication and to terrestrial plant diversity losses through aerial deposition. (See Chapters 12, 19, and 20.) Despite modest increases in nitrogen use efficiency, demand for fertilizer is projected to increase by 65% by 2050, leading to a doubling of current rates of N aerial deposition and N loading in waterways (Galloway et al. 2004).

Of the pesticides in widespread use, the most important effects on biodiversity are from persistent organic pollutants, since these have effects on large spatial and temporal scales. (See Chapter 25.) Many of the most persistent chemicals are being phased out through appropriate legislation and replaced by ones with fewer environmental impacts. However, the total use of pesticides is still increasing, and the poor regulatory environments in many countries mean that highly toxic chemicals continue to be used unsafely.

A third aspect of the impact of food provision on biodiversity concerns the effects within agricultural production systems and landscapes. Since agricultural landscapes (areas containing a significant share of cropland and pasture) now occupy 38% of Earth's land area, the maintenance of biodiversity within them is an important part of any overall strategy for biodiversity conservation. Even in relatively intensely farmed areas, cultivated crop production typically only covers a portion of the actual land areas, and much of the rest of the land can serve as habitat for wild species, if appropriately managed. However, in many agricultural landscapes wild biodiversity appears to be declining. For example, the pan-European bird index for farmland birds shows a declining trend since 1980 (see Chapter 26), in contrast to the situation for overall pan-European bird index.

One positive landscape-wide impact noted in sub-Saharan Africa, South Asia, and Southeast Asia is the trend of growing more trees in agricultural landscapes, for a wide variety of purposes. Trees stabilize and enhance soils, contribute in themselves to biodiversity, but also play host to a variety of birds and insects. Management practices can have major impacts on such biodiversity and the services that it provides for nutrient cycling, pest control, and pollination (Chapter 26), with positive spillovers for agricultural production.

The spread of invasive alien species is a fourth way that food provision affects biodiversity. While most of the world's major crops species are "alien" in the sense that their main production areas are outside their areas of origin (with notable exception of rice, the world's most important crop), none of the major crop plants are invasive. The greatest ecological risks probably arise from the spread of alien aquatic species. (See Chapter 20.) The introduction of the Nile perch in Lake Victoria, for example, led to the extinction of a large number of cichlid fish species.

Tilapia is the second most important fish species for aquaculture. Like carp, tilapia is vegetarian, and therefore tilapia-based aquaculture avoids many of the negative effects of carnivorous species. However, escapes into surrounding freshwater ecosystems may disrupt local species populations. Besides the direct use of alien species for food production, trade in food products is a major potential pathway for the introduction of pests and diseases, and most countries have quarantine systems to address this threat (FAO/NACA 2001).

Finally, when food provision is from wild sources, overexploitation and certain fishing practices can have major impacts on species composition. Overexploitation has been implicated as the leading threat to the world's marine fishes and has led to a decline in the average trophic level of catches, as described earlier. Overfishing affects not only the target species but also habitats, food webs, and non-target species. High-impact fishing (including bottom trawling, long-lining, gill netting, and dynamite fishing) causes damage to the biodiversity of sensitive habitats, such as cold-water reefs, tropical coral reefs, and seamounts, and to migratory seabirds (Pauly et al. 1998, 2003; Jackson et al. 2001). (See Chapter 18.)

Historically, many terrestrial species have become extinct due to hunting, and there are currently 250 mammal species, 262 bird species, and 79 amphibian species listed as threatened due to overexploitation for food (Baillie et al. 2004). In some groups of species and in some ecosystems, overexploitation is a particularly serious threat. In eastern and southeastern Asia, for example, almost all species of turtles and tortoises are in serious decline as a result of harvesting for human consumption and medicine, mainly in China (Baillie et al. 2004). In some cases overexploitation of plants, particularly medicinal plants, is also threatening many populations.

Food insecurity can have very severe consequences for local biodiversity. Famines, conflict, civil unrest, floods, and other natural disasters can decimate local food production and break food supply chains. In such cases, people are often forced to resort to exploitation of local wild plant and animal sources of food, often unsustainably.

8.4 Drivers of Change in Food Provision

The MA defines a driver as "any natural or human-induced factor that directly or indirectly causes a change in an ecosystem." (See Chapter 3.) In this section, that definition is limited to factors causing change in a specific ecosystem service: food provision.

Increased understanding of the drivers of change in food provision can generate insights into potential intervention opportunities for accelerating desired change and mitigating or adapting to less welcome trends. The discussion of drivers here is organized around two key dimensions. The first is the distinction recognized by the MA conceptual framework between indirect and direct drivers of ecosystem change. The second is the distinction between factors influencing food demand as opposed to those shaping food supply.

Assessing the impact of drivers for both demand and supply is particularly important in the case of food. The demand for food has long since outstripped the capacity of nature to provide it unaided, and for several millennia humans have transformed natural ecosystems for the singular purpose of obtaining more accessible, reliable, and productive sources of food to meet growing demands (Evans 1998; Smith 1995). The factors driving these changing demands must therefore be examined as a proper context for examining drivers of change in food provision.

Emerging patterns of food consumption provide early signals of the shifts in stresses on specific ecosystems in specific locations. In subsistence-oriented food production systems there is strong geographical coincidence of food consumption and ecosystem stress. In the increasingly globalized commodity trade and food industry sectors, the consumption-driven footprint of production on ecosystems might be several continents or oceans removed from the sites where consumption takes place.

Chapter 3 in this volume and Chapter 7 in the *Scenarios* volume contain information that is complementary to this section, particularly with regard to the treatment of indirect drivers such as technology, demographic trends, and economic growth. Chapter 26 in this volume also provides a brief summary with regard to the agricultural sector in exemplifying the important role of science and technology as a driver of change. That material is not repeated here, but appropriate cross-references are made.

Table 8.5 presents an assessment of the key indirect drivers of food provision, using separate grouping of drivers for food demand and supply. Table 8.6 presents the key direct (supply-side) drivers. For each driver the Tables provide a qualitative assessment of its rate of change and a judgment of its relevance in terms of influencing food provision. These variables are assessed both retrospectively over the past 50 years and for current and projected trends (up to 2015). Finally, to provide a slightly more nuanced perspective, these driver-specific assessments are provided for both industrial (In) and developing (Dg) regions in two adjacent rows. In the subsections that follow, every driver is not described in detail, but the key drivers where some relevant data exist are dealt with selectively. Trends in some important drivers are shown in Figure 8.8.

8.4.1 Indirect Drivers

8.4.1.1 Drivers of Food Demand

Eight factors are identified here that shape the demand for food. The first four of these (population growth, urbanization, economic growth, food prices) encompass the major demographic and economic trends that condition the demand for food and specific types of food in the aggregate. The remaining four factors (food marketing, food-related information, consumer attitudes to

Table 8.5. Indirect Drivers of Food Provision (compiled by authors from assessment of literature and evidence)

Drivers		Past 50 Years		Current Trends		Remarks/Examples
		Change	Relevance of Driver	Change	Relevance of Driver	
Demand factors						
Population growth and structure	In	+/++	med	–/+	low/med	Europe static/shrinking; North America still growing
	Dg	+++	high	+/+++	med/v. high	East Asia slow; SSA, WANA, SA highest growth rates
Urbanization	In	++	med	–/+	low	70–80% urbanized
	Dg	+++	med	++/+++	med/high	40% urbanized, 3%/yr growth, 80% of global urban total
Income growth	In	++	med/high	++	med/high	slow to medium long-term growth
	Dg	+/+++	high	–/+++	high	some negative, esp. SSA; strong growth: East Asia
Food prices	In	– –	med	–/o	low/med	well-integrated markets, productivity growth
	Dg	–	high	–/+	med/high	weaker markets, lower productivity growth
Food marketing: branding and advertising	In	++	med	+++	med	major diet changes are through switching brands/product
	Dg	+	low	+/++	med	less in poor rural areas, but increasing, e.g., radio, tv
Diet and health information	In	++	med	++/+++	med/high	increased information on the healthfulness or otherwise
	Dg	o/+	low	+/++	med	related to specific food types or food processing
Consumer concerns with production context	In	x	low	xx	low/med	concerns with environmental, food safety, child labor,
	Dg	o/x	low	o/x	low	equity, GMOs, animal welfare, etc. issues
Dietary (and lifestyle) preferences	In	o/x	low/med	o/x	low/med	largely consequence of marketing, diet, and health info
	Dg	x/xxx	med/high	xx	med/high	largely consequence of urbanization and income growth
Consumer demands for minimum produce grades, standards, labels	In	++	med/high	+++	high/v. high	most producers conform; contract farming on the rise
	Dg	o/+	low	o/+++	med/v. high	major challenge to poor smallholders
Supply factors						
Investments in infrastructure and institutions	In	++	med	+	med	industrial countries maintained investments in high stock
	Dg	–/+	high	–/+	very high	developing countries often underinvesting in low stock
Investments in science and technology	In	++	high	+/++	high	biotechnology: increasing, conventional: stable/decline
	Dg	o/+	high	–/+	very high	widening gap between industrial and developing R&D
Domestic price policies (e.g., producer subsidies, price controls)	In	++	med/high	+	med	powerful farm lobbies resist support reduction
	Dg	++	med/high	++	very high	policies often favor urban consumer
International trade regimes and regulations (e.g., WTO)	In	+	med	++	med	limited incentives for industrial-country concessions
	Dg	o/+	low/med	+/++	high	growing incentives developing countries to seek change
Regulatory environment for production practices	In	+	med	++	med/high	regulatory pressures on effluents, animal welfare, etc.
	Dg	o/+	low	+/++	med	less regulation/enforcement
Food industry integration and food retailing practices	In	+++	med	+/++	med	increased attention to on-farm standards and food safety
	Dg	o/+	med	+/++	high	increased incentives for smallholder collective marketing
Prices of produce and inputs	C/W	– –/+	high	o/++	med/high	prices increasing with scarcity of wild food sources
	In/Dg	–/–	high	–/–	high/v. high	real prices declining; raise productivity to compete
Access to information, technology and credit, markets	In	++	high	++	very high	growing ICT role; weather/price forecasts, credit
	Dg	o/+	high	+/++	very high	credit is a major constraint; ICT role growing fast
Level of market access/integration	In	++	high/v. high	+	high	more mature and integrated; lower transactions costs often
	Dg	–/+	high	–/++	very high	poor infrastructure, institutions; high costs
Insecurity and instability	In	o	very low	o	very low	not a significant issue; FSU a possible exception locally
	Dg	–/+	v. high)(loc)	–/++	v. high (loc)	critical loss of assets, resources

Key:

In – industrial-country grouping; Dg – developing-country grouping

Increases: + low; ++ medium; +++ high; decreases: – low, – – medium, – – – high; – –/+ indicates a range from – – to +

Change (no sign): x low, xx medium, xxx high, o no change.

C/W: cultivated/wild

ICT: information and communication technologies

Table 8.6. Direct Drivers of Food Provision (compiled by authors from assessment of literature and evidence)

Drivers		Past 50 Years		Current Trends		Remarks/Examples
		Change	Relevance of Driver	Change	Relevance of Driver	
Increasing climate variability and long-term climate change	In	+	low	+	low/med	current and projected changes often low/positive
	Dg	++	low/med	++	med/high	apparent increases in variability/extremes; high/neg
Area expansion of cropland, pasture, fishing grounds	In	+	low/med	–/+	low	little available unexploited area; some are in decline
	Dg	++	med/high	+/++	lmed/high	forest/habitat loss, urban growth loss
Intensification of production (e.g., seeds, irrigation, fertilizers)	In	+++	med/high	++	med	main source of growth in food output
	Dg	+/++	high	+/++	very high	mixed; sustainable increases critical for SSA
Degradation of underpinning resource stocks	In	++	med	o/+	med	overfishing of marine fisheries; agrobiodiversity loss
	Dg	– – –/+	very high	– – –/+	very high	major impact on soil degradation, wild food sources
Pest and disease incidence and adaptation	In	o/+	med/high	–/+	med	extensive (regulated) pesticide use; GM crops (US)
	Dg	o/++	high	o/++	high	greater pressures, less regulation; IPM increasing

Key:
In – industrial-country grouping; Dg – developing-country grouping.
Increases: + low, ++ medium, +++ high; decreases: – low, – – medium, – – – high; – –/+ indicates a range from – – to +
No change: o

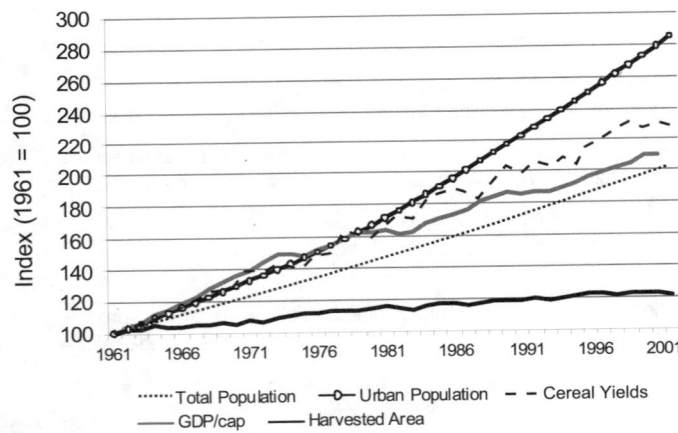

Figure 8.8. Trends in Selected Drivers of Food Provision Worldwide, 1961–2001 (FAOSTAT 2004; World Bank 2003)

production practices, and diet and lifestyle preferences) are often more subtle expressions of the interplay of the previous three drivers but are increasingly shaping the structure of food consumption. Finally, these trends are drawn together in the paradigm of the "diet transition" (Popkin 1993, 1998; WHO 2003b).

8.4.1.1.1 Population growth and age structure

Between 1961 and 2001 the major driver of growth in total food consumption was population growth. Global population doubled in this time period, from just over 3 billion to 6.1 billion, while the apparent consumption of calories per person increased on average by around 24% (FAOSTAT 2004). But from a global perspective the central role of population growth as a driver of food demand has started to decline, significantly in some regions of the world. Population growth rates, which peaked at 2.1% per year

in the late 1960s, had fallen to around 1.35% (or 78 million additional people) per year by the turn of the millennium (UNDP 2003). This still represents a daunting food security and humanitarian challenge, as approximately 90% of this increase is taking place in developing countries. Around half of the total population increase in developing countries will occur in sub-Saharan Africa and South Asia, where the incidence of hunger is already high and, according to estimates for 2000–02, increasing in absolute terms (FAOSTAT 2004; Bruinsma 2003; FAO 2004a).

By contrast, in Western Europe, transition countries, and East Asia, population growth is extremely low and in some cases negative. As a consequence of lower fertility rates and increased life expectancy, typical of richer countries, the average age of individuals in such countries is increasing. Conversely, countries with higher fertility rates, which are also often poorer, have younger age structures. (See Chapter 3.)

While population size and growth rates have direct consequences for food needs and the required resilience of food production systems, age structure has more subtle impacts. One such impact is that energy and diet diversity requirements are age- (and sex-) dependent and, for example, increase for mothers during pregnancy and lactation. Furthermore, there is evidence of differences in food consumption according to age in the United States (Blisard and Blaylock 1993) and Japan (Mori and Gorman 1999). For example, U.S. and Japanese studies found older people to be consuming more fruit, as well as eating more meals at home.

8.4.1.1.2 Urbanization

Urbanization has proceeded at such a pace that globally, urban dwellers will outnumber rural populations by around 2007. High-income countries currently have populations that are 70–80% urban, and the same pattern is being seen as development progresses elsewhere (such as in Latin America and the Caribbean). The proportion of those in developing countries who live in cities has doubled since 1960 from 22% to over 40%. This share is expected to grow to almost 60% by 2030 (UN 2004). Developing countries now account for around 80% of the world's urban population. In 2001, 13 of the world's 17 "megacities" were in developing countries, and by 2015 it is expected that figure will have risen to 17 (UN 2001). (See also Chapter 27.)

Urbanization affects many dimensions of food demand. First, food energy requirements of urban populations are generally less than those in rural areas because of more sedentary lifestyles (Clark et al. 1995; Delisle 1990). Urban consumers generally have higher incomes as well as access to a more diverse array of both domestic and imported foods. Urban lifestyles often mean less time at home, and more meals eaten away from home (Popkin 1993). As a consequence, urban consumers eat more processed and convenience foods. This raises issues of food cost, quality, and safety in terms of the use of appropriate inputs, especially safe water in food processing.

Empirical evidence shows that urban diets are more diversified and contain more micronutrients and animal proteins but with a considerably higher intake of refined carbohydrates as well as of saturated and total fats and lower intakes of fiber (Popkin 2000). Data for China, Indonesia, and Pakistan at two points in time show reduced consumption of cereals and roots and tubers and increased consumption of fruits and vegetables and meats among urban populations (Regmi and Dyck 2001). The greater diversity, including of fruits and vegetables, generally available to urban populations does not necessarily translate into increased diversity of consumption (Popkin et al. 2001; Johns and Shtapit 2004).

A widely observed trend in urban diets is the switch away from traditional staples (locally produced millet, sorghum, root crops, and plantains, for instance) and toward consumption of rice and wheat, even though cereal diet shares decline overall. Often the rice and wheat needs are met through imports. Rice is particularly attractive because its preparation is quick and simple relative to other cereals. Wheat gains popularity through increased consumption of bread, noodles, pasta, dumplings, and so on. Given the scale and speed of urbanization, particularly in developing countries, these dietary shifts amount to significant changes in the structure of food demand at the national and regional scale. While importing rice and wheat for urban markets requires foreign exchange and can undercut the market potential for domestic suppliers, it can increase the food security of urban populations (usually a politically important social group) by tapping into sources of food supply that are often more reliable and of higher quality than domestic sources. Because of domestic infrastructure constraints and related high transaction costs, it may also be more economical to import food even when local and foreign production costs are similar.

There are other impacts of urbanization apart from structural changes in food consumption. One is the loss of prime agricultural land as a consequence of urban expansion, often displacing food production into less productive land elsewhere. Another is the major changes in nutrient flows associated with the flow of food from rural to urban areas. Whereas organic matter residues were once recycled locally, this nutrient export from rural to urban areas can deplete soil nutrient content in the production areas and can concentrate nutrients in human wastes and other residues in and around cities. The latter incurs effluent treatment and disposal costs and often causes pollution of water courses or coastal waters. A good example is the depletion of soil fertility in the banana-growing regions of Uganda due to the high demand for cooking bananas in Kampala, which involves shipping complete stems for processing in the city. There are also significant other, often negative, effects associated with the continuous need for food transportation (such as the contribution of increased ex-

haust emissions to local particulate matter as well as to greenhouse gases).

8.4.1.1.3 *Income growth*

It is *well established* that income is the single most important factor determining the amount and quality of food consumption. The relative share of budget spent on food is significantly higher among the poor but decreases rapidly as incomes increase and basic nutritional needs are met (Engel's law) (Tomek and Robinson 1981). At higher levels of income, high-value, more nutritious, or more culturally prestigious foods, such as fresh seafood or imported specialty foods, replace less-valued food sources (as described earlier regarding the transition to high-value meat, fish, vegetables, and fruits in East Asia since the late 1980s). In particular, the extra demand for meat is driving the "Livestock Revolution."

The most widely used proxy of income derived from aggregate statistics is the national measure of GDP per capita. (See Chapter 2.) At a macroeconomic scale there is strong evidence of the association between national average energy supply (kilocalories per person per day) and national economic growth as measured by GDP. (See Figure 8.9.)

8.4.1.1.4 *Food advertising, information, consumer power, and changing food preferences*

The previous drivers have played and will continue to play key roles in shaping the overall quantity and structure of food consumption in generally predictable ways: more people, urban migration and urban lifestyles, higher incomes, and lower food prices. This second group of drivers acts to influence the food consumption decisions of individuals in more subtle ways. They are often factors that come increasingly into play as incomes rise and basic food needs are met, but far from exclusively so.

Consumers in industrial and developing countries, particularly in the urban environment, are exposed to advertising for food, and poor consumers are often the specific targets of food-related information and safety-net programs. Such advertising and information can directly alter food preferences. And as a variety of obstacles have slowly been removed, opportunities for providing consumers with information, particularly about food safety, nutri-

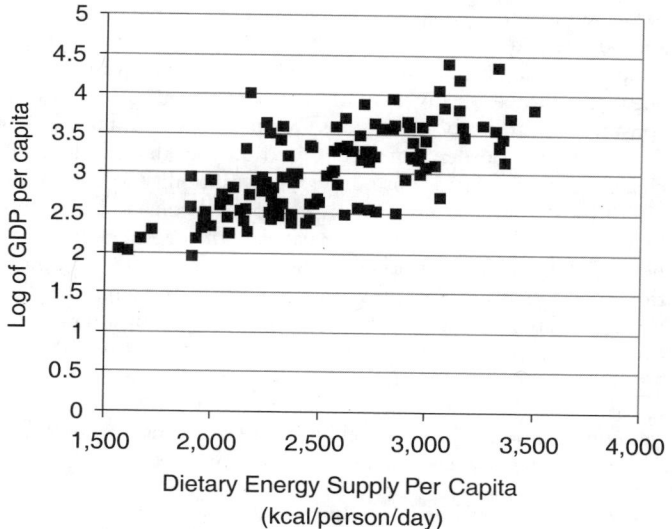

Figure 8.9. Association between National Average Dietary Energy Supply and GDP, Per Capita (Arcand 2001)

tion, and health have increased. At the same time, however, local and traditional knowledge that underpin traditional diets and cuisine is often in decline (Johns and Sthapit 2004).

Public information media—radio stations, newspapers, and television channels as well as public service access points such as clinics, schools, churches—are increasingly being used as cost-effective means of providing food and nutrition information to the public. Messages are varied, from direct commercial advertising to important public health information, such as the advantages of breast-feeding or the existence of contaminated food supplies locally. In richer countries, this phenomenon is most apparent with regard to the high levels of consumer interest and credence in information concerning health-related or weight-control-related attributes of food. While the latter are often passing fads, they do have significant impacts on food production. The current attention being given to low-carbohydrate diets for weight loss, for example, has had a measurable increase in demand for white meats, fish, and eggs and a measurable reduction in demand for wheat-flour based products and, in the United States, tomatoes (a large share of which are grown for tomato paste in pastas and pizzas).

In the transition to lifestyles more characteristic of industrial societies, retaining a strong traditional food system in which diet has recognized health, cultural, and ecological roles has allowed some countries to reduce the often concomitant increases in chronic noncommunicable diseases (Popkin et al. 2001). Asian and Mediterranean diets (Trichopolulou and Vaasiloppoulou 2000) provide the clearest examples. Traditional societies often see food, medicine, and health as interrelated. Food may have strong symbolic and religious value and is highly associated with cultural identity and social well-being (Etkin 1994; Johns 1990; Johns and Sthapit 2004).

Trade liberalization and the increased role of transnational food companies, urbanization, and migration, combined with the equalizing effects of rising incomes, has led to convergence between diets internationally. Yet cultural and religious factors appear to limit such convergence and to help retain dietary diversity (Bruinsma 2003).

8.4.1.1.5 The "diet transition"

Excessive consumption, particularly of some food types, has been associated with the growing phenomena of excessive body weight and obesity and with the associated health risks. Urbanization and socioeconomic changes have resulted in diets that are higher in energy and lower in diversity of fruits and vegetables than those consumed historically. As a consequence, many countries now face a "double burden" of diet-related disease: the simultaneous challenges of increased morbidity and susceptibility to communicable diseases among undernourished populations and increased incidence of chronic diseases associated with the overweight and obese (WHO 2003a; Ezzati et al. 2002). The pathway from traditional rural diets to those of increasingly urban and affluent societies and its attendant implications for nutrition and health has been dubbed the nutrition transition or the diet transition (Popkin 1993; Smil 2000; Receveur et al. 1997). (See Box 8.4.) The diet transition can be viewed as the integration of many of the individual consumption-related drivers just described, having both significant human and ecosystem health outcomes.

8.4.1.2 Drivers of Food Supply

8.4.1.2.1 Investments in agricultural research and development

It is *well established* that technological innovation is a major driver of increased agricultural productivity and in many cases is now the major source of increased productivity (Evenson et al. 1999; Acquaye et al. 2003; Roe and Gopinath 1998; Ruttan 2002). In turn, a major indirect driver of changes in food production systems has been the flow of innovations arising from investments in agricultural research and development. Worldwide, public investments in agricultural research nearly doubled in inflation-adjusted terms, from an estimated $11.8 billion (in 1993 dollars) in 1976 to nearly $21.7 billion in 1995. (See Table 8.7.) Developing countries account for just over half of the world's public agricultural research (Pardey and Beintema 2001). (See Table 8.8.)

Regional totals fail to reveal, however, that the public spending was concentrated in only a handful of countries. Just four countries—the United States, Japan, France, and Germany—accounted for two thirds of the $10.2 billion of public agricultural research done by rich countries in 1995. Similarly, China, India, and Brazil dominate the spending on agricultural research in developing countries. By the mid-1990s, about one third of the annual $33-billion investment in agricultural research worldwide was done by private firms, including those involved in providing farm inputs and processing farm products. The overwhelming majority ($10.8 billion, 94% of the global total) of this privately funded research was conducted in industrial countries (Pardey and Beintema 2001).

Chapter 7 of the *Scenarios* volume discusses the historical evolution of the impacts of investments in agricultural productivity. A meta-study of quantitative evidence on the economic payoffs from improved productivity attributable to agricultural research, which included 1,845 data points from evaluation studies published between 1950 and 1995, revealed that the mean average economic rate of return on investment was 30.4% per year, though the range was wide; there appeared not to have been, as was popularly believed, any observable decrease in the rate of return to research investment over time; nor was there any significant regional bias in payoffs—that is, the economic returns to research investments in sub-Saharan Africa were not statistically different from those in Asia (Alston et al. 2000). One persistent finding has been the importance in agricultural research of spillover of knowledge and technologies between different locations. Indeed, evaluation studies that have specifically taken account of knowledge and technology spillovers have shown that this has accounted for a large share, and in many cases more than half, of the overall economic benefits (Alston 2002). Nonetheless, local R&D is necessary to facilitate spillover, and the lower levels of local R&D in Africa as opposed to Asia help account for the former's lower level of productivity growth (Masters 2005).

8.4.1.2.2 Agricultural policy and trade: producer subsidies and import tariffs

One of the most important and controversial set of drivers conditioning food provision globally are agricultural production and trade policies, and especially the producer subsidy and tariff protection measures supported, in particular, by the European Union, the United States, and Japan. By subsidizing food production and exports, while keeping in place high import tariffs, particularly on semi-processed or processed foods, these OECD countries drive down food prices on the world market, undercutting the potential profitability of developing-country producers in their own markets while simultaneously limiting their export opportunities (Watkins and von Braun 2003).

In 2002, some $235 billion of the over $300 billion spent by OECD countries on their agricultural sectors (some six times the amount they allocate to overseas development aid) went to support agricultural producers. This support is paid for by higher

BOX 8.4

Diet and Nutrition Drivers: The "Diet Transition"

Popkin (1998) has described five broad nutrition patterns (see Figure), not restricted to particular periods of human history but presented as historical developments. "Earlier" patterns are not restricted to the periods in which they first arose, but continue to characterize certain geographic and socio-economic subpopulations.

Pattern 1: Collecting Food. This diet, which characterizes hunter-gatherer populations, is high in carbohydrates and fiber and low in fat, especially saturated fat. The proportion of polyunsaturated fat in meat from wild animals is significantly higher than in meat from modern domesticated animals. Activity patterns are very high and little obesity is found among hunter-gatherer societies.

Pattern 2: Famine. The diet becomes much less varied and subject to larger variations and periods of acute scarcity of food. During the later phases of this pattern, social stratification intensifies, and dietary variation increases according to gender and social status. The pattern of famine has varied over time and space. Some civilizations have been more successful than others in alleviating famine and chronic hunger. The types of physical activities changed, but there is little change in activity levels associated with this pattern.

Pattern 3: Receding Famine. The consumption of fruits, vegetables, and animal protein increases, and starchy staples become less important in the diet. Many earlier civilizations made great progress in reducing chronic hunger and famines, but only in the last third of the last millennium have these changes become widespread, leading to marked shifts in diet. However, famines continued well into the eighteenth century in portions of Europe and remain common in some regions of the world. Activity patterns start to shift and inactivity and leisure becomes a part of the lives of more people.

Pattern 4: Nutrition-related Noncommunicable Disease. A diet high in total fat, cholesterol, sugar, and other refined carbohydrates and low in polyunsaturated fatty acids and fiber, and often accompanied by an increasingly sedentary life, is characteristic of most high-income societies (and increasing portions of the population in low-income societies), resulting in increased prevalence of obesity and contributing to the degenerative diseases that characterize Omran's final epidemiologic stage.

Pattern 5: Behavioral Change. A new dietary pattern appears to be emerging as a result of changes in diet, evidently associated with the desire to prevent or delay degenerative diseases and prolong health. Whether these changes, instituted in some countries by consumers and in others also prodded by government policy, will constitute a large-scale transition in dietary structure and body composition remains to be seen. If such a new dietary pattern takes hold, it may be very important in increasing disability-free life expectancy.

The focus is increasingly on patterns 3 to 5, in particular on the rapid shift in much of the world's low- and moderate-income countries from pattern 3 to pattern 4, commonly termed the "diet transition" or "nutrition transition."

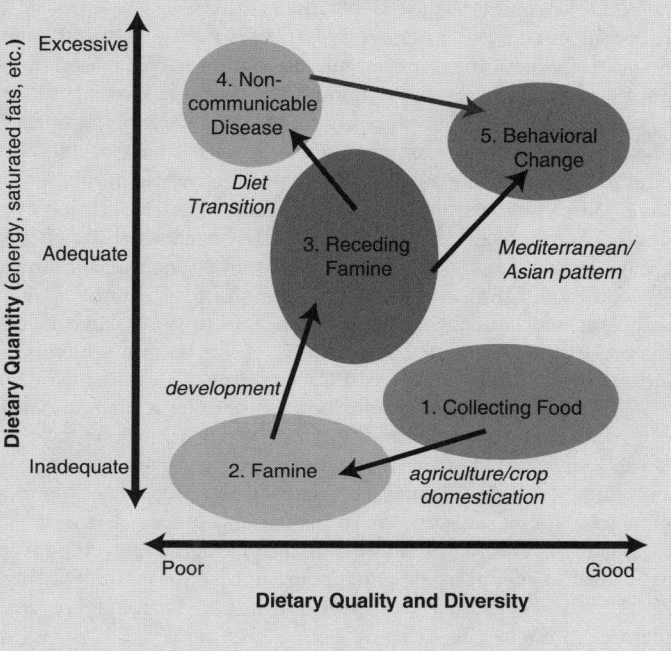

Table 8.7. Global Agricultural Research Spending, 1976–95.
Figures in parenthesis indicate number of countries. (Pardey and Beintema 2001)

Region	1976	1985	1995
	(thousand 1993 dollars)		
Developing countries (119)	4,738	7,676	11,469
Sub-Saharan Africa (44)	993	1,181	1,270
China	709	1,396	2,063
Asia and Pacific, excluding China (23)	1,321	2,453	4,619
Latin America and the Caribbean (35)	1,087	1,583	1,947
Middle East and North Africa (15)	582	981	1,521
Industrial countries (34)	7,099	8,748	10,215
Total (153)	**11,837**	**16,424**	**21,692**

Table 8.8. Public-Private Breakdown of Research Expenditures, Circa 1995 (Pardey and Beintema 2001)

Region	Public	Private	Total
	(thousand 1993 dollars)		
Developing countries	11,469	672	12,141
Industrial countries	10,215	10,829	21,044
Total	21,692	11,511	33,204

domestic food prices and by taxes ($100 billion in the EU, $44 billion in Japan, and $31 billion in the United States). And it represents around 31% of average farm income (18% in the United States and 36% in the EU). For individual commodities that OECD countries target for support (wheat, maize, cotton, dairy, beef, sugar, rice, and oilcrops), the levels of support can be much higher (OECD 2003a; Watkins and von Braun 2003). Government support in Japan has consistently represented around 85% of farmers' rice production revenues, while U.S. rice support has declined from around 50% in 2002 (OECD 2004). The economic losses to developing countries due to these policies has been estimated as some $24 billion a year in lost agricultural production and incomes of farm households and about $40 billion a year in lost access to markets in OECD countries (Diao et al. 2004).

Fisheries are another area of food production where subsidies have become controversial. In the early 1990s it was established that subsidies had probably contributed to an excessive buildup of fishing fleets worldwide during the preceding decade. (See Chapter 18.) Since then, it has proved difficult to limit the fishing of existing fleets, and fish stocks continue to be overexploited as illegal, unreported, and unregulated fishing spreads. Subsidies to the fishing industry are regulated through WTO agreements; however, the international community has agreed that existing WTO rules are not sufficient to "discipline" their use, and efforts are now being made to seek improved measures under the so-called Doha round of trade talks (Chang 2003).

From an ecosystem perspective, these market distortions have two major effects. In countries where subsidies are paid, food output increases to levels that would be uneconomic in the absence of subsidies, drawing proportionately more land, labor, and other resources into production and creating higher levels of agricultural pollution. While other factors such as increasing productivity have caused the net amount of agricultural land to grow more slowly or even to decline in some OECD countries, those effects would have been more significant in the absence of subsidies. Recently, the OECD reported that while nitrogen runoff, pesticide use, and agricultural greenhouse gas emissions have fallen since the mid-1980s in most OECD countries, they have increased in the United States (OECD 2003b).

The other major impact of producer subsidies is the reduction of production and income opportunities in other parts of the world where subsidies are not paid, although these impacts can be ambiguous. While it could be argued that reduced production incentives might conserve more habitat and reduce demands on local natural resources, it also can make poor people even poorer by limiting productivity enhancement incentives on existing lands, thus accelerating land degradation and further increasing pressure to convert more land.

International processes and agreements under the World Trade Organization have a major bearing on these drivers, particularly WTO's Agreement on Agriculture. This provides a framework for removing the "amber box" trade biases induced by producer subsidies and import tariffs, as well as for agreeing on a broader set of "green box" provisions dealing with support to producers for improving the environmental and landscape dimensions of farming through a range of "set-aside," "conservation," "countryside stewardship," or similar programs. The espoused intent of "green box" measures is to provide incentives to farmers to follow less polluting, more environmentally sensitive production practices. However, such provisions have become highly contentious as they are seen by many, including most developing countries, as a way of legitimizing existing rich-country producer support in another guise.

There are other ecosystem-relevant dimensions of the Agreement on Agriculture, including the notion of decoupling support to producers from the quantity of production so as to limit perverse incentives to overproduce food, use more agricultural land, use more potentially polluting agricultural inputs, and generate more wastes. The impacts of the WTO on agriculture are still emerging and developing, but its potential ramifications will likely grow, bringing with it a significant change in food production incentives globally. Through the WTO, for example, devel-oping countries are becoming increasingly effective at asserting demands for more liberal agricultural trade policies.

8.4.1.2.3 Food industry commercialization and integration

Just as agriculture has witnessed a gradual industrialization of the production process, so too have there been sweeping changes in food marketing, processing, and retailing practices. Even where industrialization of the production process has not taken place, the concentration and formalization of marketing is having significant repercussions on production decisions and on the need for improved smallholder collective action in order to stay engaged in food markets. Several forces at work are leading to the integration and formalization of food marketing and supply chains: growing economies of scope and scale in the transportation, processing, and retailing subsectors; falling food prices that have reduced profit margins and further encouraged consolidation in the post-harvest sector; a growing need to respond to consumer demands for specific type and quality of product that provides incentives to shorten the marketing "chain," such as vertical integration; a growing need for transparency and accountability in the certification of food sources, such as organic foods, and to satisfy appropriate (regulated or self–imposed) standards and food safety requirements; and the enormous expansion in the role of supermarkets, even in developing countries (Berdegue et al. 2003).

Thus an increasing share of food production is being contracted for before planting, with contracts that often involve producer obligations relating to minimum quantities, product quality, and delivery dates. Such stringent criteria are very difficult for smallholders to meet, and farmer associations and marketing groups are increasingly being formed to help respond to these needs. For example, in the United States there has been massive consolidation of farms due to the economic pressures to improve economies of scale for both production and marketing purposes (as well as, in recent times, to fully reap the benefit of production subsidies). In the 1920s and 1930s there were more than 6 million farms of around 40 hectares each. By the late 1990s, there were fewer than 2 million farms and they averaged 200 hectares each (Bread for the World 2003).

8.4.2 Direct Drivers

8.4.2.1 Climate Change and Climate Variability

Although there is a relatively rich literature on the potential impacts of long-term climate trends on future food production (see, for example, Rosenzweig and Parry 1996; Sombroek and Gommes 1996; Parry et al. 2004), evidence on the impacts of historical and recent climate change on food production is relatively sparse. Although climate is a major uncontrollable factor influencing food production (especially in areas of rain-fed agriculture), it is extremely difficult to reliably isolate the influence of climate from other factors such improved seeds, the use of irrigation, fertilizer, pesticides, crop and land management.

However, some data do exist. For example, based on over 80 years of crop yield and climate data in five central Corn Belt states in the United States, Thompson (1998) developed relationships describing the influence of monthly average temperature and total precipitation on corn yields. These suggest that 40 millimeters above normal precipitation in July would lead to a corn yield increase of 316 kilograms per hectare above the long-term average. It has also been suggested that U.S. corn and soybean yields could drop by as much as 17% for each degree that the growing season warms (Lobell and Asner 2003), although the level of certainty in such findings remains low (Gu 2003). Weather variability in China has been shown to have a measurable effect on year-to-year national grain output (Carter and Zhang 1998), and a study of the temperature and wine quality in the world's top 27 wine regions over the past 50 years reveals that rising temperatures have already affected vintage quality (Jones et al. 2004).

Global-scale cyclical weather patterns have also strongly influenced food production. This includes the impact of, in particular, the El Niño-Southern Oscillation and the North Atlantic Oscillation. Carlson et al. (1996) found that a negative Southern Oscillation Index (a measure of pressure difference in an ENSO event) can result in a corn yield that is 10% above trend line in U.S. Corn Belt states, and ENSO-based climate variability has significant impact on cereal production in Indonesia (Naylor et al. 2002), where year-to-year fluctuations in the August sea-surface temperature anomaly explain about half the interannual variance in paddy production during the main (wet) growing season. The North Atlantic Oscillation has also been shown to significantly correlate to vegetation productivity in northern Asia, with a surprisingly long lag time of one-and-a-half years (Wang and You 2004).

8.4.2.2 Area Expansion and Intensification

There have been two main direct drivers of growth in food production: the increase in the area extent of cultivation, grazing, or fishing and the intensity of production or exploitation within cultivated areas. Figure 8.8 showed increasing trends in both harvested area (area expansion) and cereal yields (a proxy of increased intensification). It is clear that, for crops, it is intensification rather than area expansion that has mainly driven increased food output. Over the past 40 years cropland area has expanded globally by some 15%—from 1.3 billion to 1.5 billion hectares (see Chapter 26), the area of pasture has grown some 11% from 3.14 billion to 3.48 billion hectares (FAOSTAT 2004), and practically all corners of the world's oceans are now accessible to the world's fishing fleet (given the capacity of modern fishing vessels to stay at sea for extended periods and the large amounts of catch in refrigerated holds).

While physical expansion in the area dedicated to food provision has been important in the past, rates of growth are now relatively low—and in some places in decline (for instance, in the European Union and Australia). This slowdown reflects both the slowing growth in global food demand and the more limited opportunities for area expansion. Just as with the large growth in crop yields per hectare of cropland, there has also been substantial increase in livestock production per animal; however, intensive livestock systems are dependent on inputs from a significant land area for feed production. (These trends are described in more detail in Chapter 26.)

Investments in agricultural research and the resulting flow of innovation have been key to the intensification process. Technical change and increased use of external inputs such as irrigation, fertilizer, and mechanical power contribute to changes in productivity—the formal means by which changes in intensification can be measured. Increased productivity can also come from the introduction of less capital-intensive food-feed systems, whereby both the main crops as well as the introduction of legumes can enhance the cropping system. The complementary advantages of both food and feed enable intensification of mixed crop-livestock systems and can increase total factor productivity (Devendra et al. 2001).

8.5 Food Provision and Human Well-being

The MA defines five dimensions of human well-being—basic material for a good life, security, health, good social relations, and freedom and choice (MA 2003)—into which the production and consumption of food maps in several ways. Food production, distribution, processing, and marketing provide employment and income to a large share of the world's population. About 2.6 billion people depend on agriculture for their livelihoods, either as actively engaged workers or as dependants (FAO 2004b). (The exact number of people dependent on food processing, distribution, and marketing for their livelihoods is not certain.) Food consumption contributes directly to health and is an important aspect of cultures and social relations, and indirectly supports improved security and freedom and choices.

There are several attributes of food that have a major bearing on its potential impact on human well-being—quantity and price, diversity and quality of its nutrient content, and safety. The actual impact of food depends on local food availability and the ability of consumers to gain access to and properly use it, as well as on individual food preferences. The dimensions of food availability, access, and utilization are integrated in the notion of food security, defined as "access by all people at all times to enough food for an active, healthy life" (Reutlinger and van Holst Pellekan 1986). For the poor, the price of food (as well as access to wild sources of food) is key to determining the value of incomes, and for many such people the relationship between wage rates and food prices is a critical determinant of human well-being. It is *well established* that a productive food and agriculture sector not only benefits individual farmers and food consumers but also provides a platform for economic growth (Mellor 1995; Hazell and Ramasamy 1991).

8.5.1 Health and Nutrition

The most direct and tangible benefit of food is its role in enabling individuals to pursue active, healthy, productive lives as a consequence of adequate nutrition. For these reasons, access to adequate, safe food has been recognized as a basic human right. The 1948 Universal Declaration of Human Rights proclaimed that "everyone has the right to a standard of living adequate for the health and well-being of himself and his family, including food." Nearly 20 years later, the International Covenant on Economic, Social and Cultural Rights (1966) developed these concepts more fully, stressing "the right of everyone to . . . adequate food" and specifying "the fundamental right of everyone to be free from hunger." These rights were specifically embodied in the 1989 International Convention on the Rights of the Child and have found further expression and practical interpretation through subsequent confirmation at the 1996 and 1999 World Food Summits (UN/SCN 2004).

Proper nutrition has many benefits for human, physical, and mental development. Better-fed children show improved educational performance, and better-fed adolescents and adults are able to lead more economically productive lives. Good nutrition also reduces neonatal and child mortality, helping to slow population growth by increasing birth intervals and reducing demand for large families. Well-nourished mothers are also more likely to survive childbirth themselves and to deliver healthier babies (ACC/SCN 2002).

Inadequate consumption of protein and energy as well as deficiencies in key micronutrients such as iodine, vitamin A, and iron are key factors in the morbidity and mortality of children and adults. An estimated 55% of the nearly 12 million deaths each year among children under five in the developing world are associated with malnutrition (UNICEF 1998). Malnourished children also have lifetime disabilities and weakened immune systems (UNICEF 1998). Moreover, malnutrition is associated with disease and poor health, which places a further burden on households as well as health care systems.

A commonly used measure of the impact of ill health on human well being is disability-adjusted life years. This is a measure of the burden of disease, reflecting the total amount of healthy life lost, to all causes, whether from premature mortality or from some degree of disability during a period of time. Table 8.9 summarizes the available evidence regionally and globally on the burden of nutrition-related health impacts measured in terms of DALYs.

While the human health consequences of undernourishment, since they arise largely out of conditions of poverty and marginalization, legitimately dominate global humanitarian concerns, health problems associated with overconsumption of (certain types of) food have grown alarmingly over the past 40 years. The health-related aspects of nutrition are examined in three steps. First, micronutrient deficiency is considered, since although this is most often associated with undernourishment, it can arise even in diets with adequate or excessive protein and energy content. Second, undernourishment-related aspects are looked at, largely acute and communicable diseases. And third, the evidence on overnourishment-related noncommunicable diseases is reviewed.

The constituents of good nutrition are fairly well understood, but dietary needs vary according to a range of context- and individual-specific characteristics, including climate, metabolism, and occupation. Guidelines indicate that a healthy adult diet includes a daily intake of 2,780 kilocalories for males and 2,235 kilocalories for females. Recommended daily allowances for protein are 37

grams for males and 29 grams for females (FAO/WHO/UNU 1985); however, these recommendations vary by country. Recommended daily intakes have also been established for a number of vitamins, minerals, and other micronutrients. For example, daily intakes have been established of 750 micrograms of vitamin A for both males and females, 130 micrograms of iodine for males, and 110 micrograms for females (FAO/WHO 2002). In addition, components of foods may have functions that include antibiotic, immuno-stimulation, nervous system action, anti-inflammatory, antioxidant, anti-glycemic, and hypolipidemic properties (Johns and Sthapit 2004).

Figure 8.10 presents a simplified schematic summary of the principal pathways of impact of food on human well-being, as mediated through nutrition. The following sections describe some of these pathways and, as far as available evidence permits, quantify the scale and intensity of specific nutrition, health, human development, and economic outcomes. But first the major trends in key measures of nutrition over the past 40 years are reviewed.

8.5.1.1 Diet Quantity: Dietary Energy Supply

The most commonly used proxy of food availability is the gross availability of food, as proxied by the average per capita availability of calories from food products, the dietary energy supply. Figure 8.11 shows levels and trends in DES over the period 1961 to 2001 for industrial and developing countries and for the world. Expansion of food production, coupled with declining real prices for food, increased DES by 24% globally over that period (an

Table 8.9. Regional and Global Burden of Nutrition-related Disease Risk Factors. The Table shows the estimated disease burden for each risk factor considered individually. These risks act in part through other risks and act jointly with other risks. Consequently, the burden due to groups of risk factors will usually be less than the sum of individual risks. The disability-adjusted life year is a measure of the burden of disease. It reflects the total amount of healthy life lost to all causes, whether from premature mortality or from some degree of disability, during a period of time. (Adapted from Ezzati et al. 2002; Ollila n.d.; and WHO 2002a)

Population/Risk Factor	Africa	Americas	East Mediterranean	Europe	Southeast Asia	Western Pacific	World
			(thousand)				
Total population	639,593	827,345	481,635	873,533	1,535,625	1,687,287	6,045,017
Childhood and maternal-undernutrition-related diseases							
			(DALYs as percent of regional and world population)				
Underweight	9.82	0.24	3.58	0.09	3.06	0.48	2.28
Iron deficiency	1.59	0.21	0.77	0.12	0.91	0.26	0.58
Vitamin A deficiency	2.57	0.04	0.61	0	0.42	0.03	0.44
Zinc deficiency	2.15	0.06	0.67	0.01	0.35	0.03	0.46
Malaria							0.74
HIV/AIDS							1.49
Respiratory infections							1.67
Iodine deficiency							0.04
Measles							0.4
Diarrhea							1.19
Other nutrition-related risks							
High blood pressure	0.69	0.78	1.02	2.22	0.98	0.83	1.06
High cholesterol	0.31	0.55	0.67	1.51	0.8	0.31	0.67
High BMI	0.23	0.89	0.6	1.35	0.27	0.35	0.55
Low fruit and vegetable intake	0.24	0.36	0.34	0.76	0.57	0.3	0.44
Diabetes							0.25

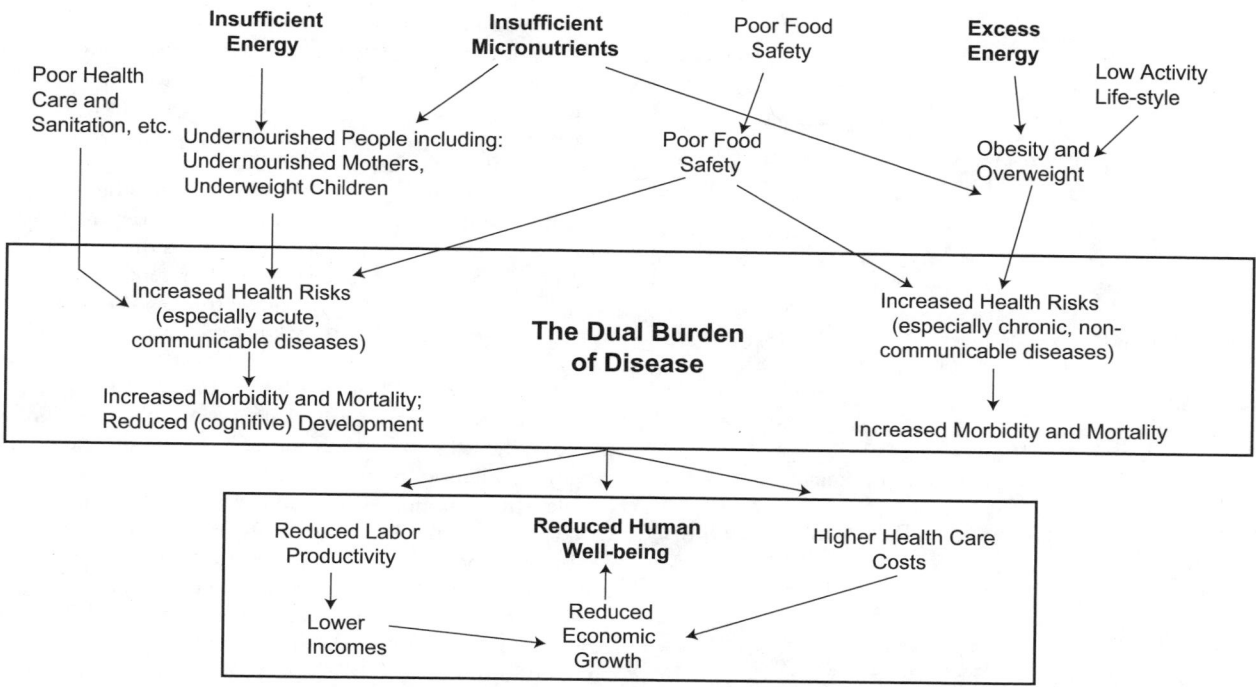

Figure 8.10. Key Linkages in the Nutrition, Health, and Economic Well-being Nexus

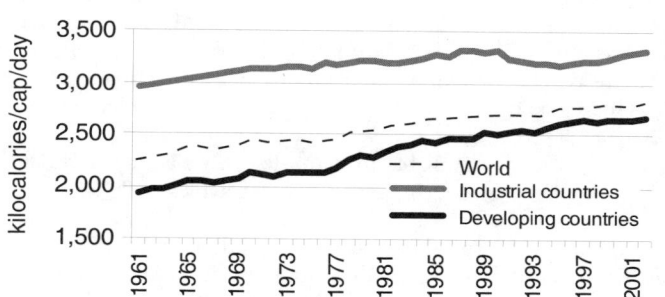

Figure 8.11. Dietary Energy Supply, 1961–2001 (FAOSTAT 2004)

11% and a 39% average increase, respectively, in industrial and developing countries). In 1961, with a world population of around 3 billion, the average DES was around 2,250 kilocalories per capita per day. By 2001, average DES had reached 2,800 kilocalories per capita despite a doubling of world population to 6.1 billion (FAOSTAT 2004).

But there have been, and remain, large geographic and socioeconomic differences in DES. Such disparities largely arise from differences in income, in the disposition of and access to food and to lands favorable for food production, and in dietary preferences and food utilization practices. Figure 8.12 depicts an obvious twin peak structure—one peak for developing countries and another for industrial ones. From the 1960s through to and especially during the 1980s, there was a progressive shrinking of the DES gap between these two groups (Wang and Taniguchi 2003). Although the average DES of developing countries increased from 1929 to 2,675 kilocalories per day over 40 years, it is still lower than the DES in industrial countries in 1961, some 2,947 kilocalories per day. And during those 40 years the figure in industrial countries increased to 3,285 kilocalories a day (FAOSTAT 2004).

The progress of individual countries varies more markedly. Over the same 40 years Indonesia's average DES grew by 68% from a very low 1,727 to 2,904 kilocalories per day, Uganda's

average figure barely increased from 2,318 to 2,398, while in the United States average DES grew by some 31%—from 2,883 to 3,766 kilocalories a day. With regard to overall shares of food consumption, industrial countries—with 24% of the world's population—consumed 29% of global calories, 34% of global protein, and 43% of global fat in 2001 (FAOSTAT 2004). (Trends in hunger are discussed in further detail later.)

Of particular concern in Figure 8.12, however, is that the trend of improvement was reversed during the 1990s, when a noticeable leftward shift (lower DES) became apparent in several parts of the distribution (Wang and Taniguchi 2003). The figure suggests that the nutritional status of at least some developing and middle-income countries worsened during the 1990s.

8.5.1.2 Dietary Quality and Diversity

There are many other dimensions of good nutrition besides dietary energy supply. In addition to quantity, key elements of good nutrition are diet quality and diversity. A healthy diet comprises both sufficient quantities and a proper mix of carbohydrates, proteins, fats, fibers, vitamins, and micronutrients, as well as components with health-mediating functional properties (Johns and Sthapit 2004). These can be derived from a diverse range of crop and livestock products as well as wild and cultivated fisheries products and other wild sources of food. (See Box 8.5.)

A handful of epidemiological studies from the United States and Europe, along with a few case studies from Africa and Asia, uphold the conventional wisdom concerning the benefits of a varied diet, particularly in fruits and vegetables. Nutritional status and child growth improve with consumption of greater food diversity, as do measures of functional properties of dietary components likely to play an important role (Johns and Sthapit 2004; Johns 2003).

Micronutrients are needed, as the term suggests, in only minuscule amounts; the consequences of their absence are severe, however, and contribute significantly to the burden of disease. Micronutrients enable the body to produce enzymes, hormones,

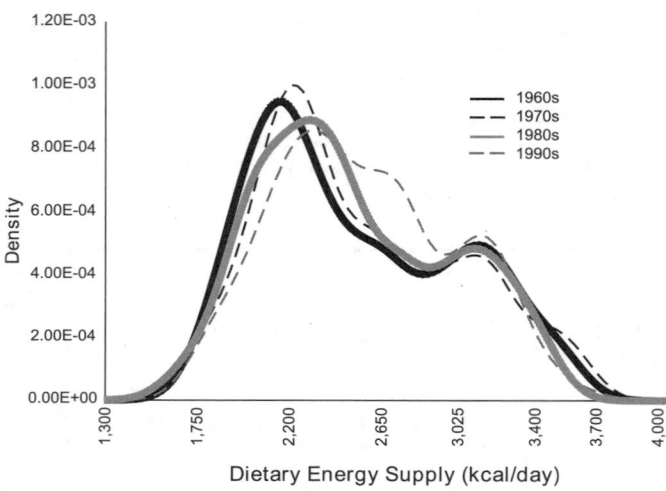

Figure 8.12. Distribution of Per Capita Dietary Energy Supply by Decade (Wang and Taniguchi 2003)

BOX 8.5

Biofortification (IFPRI/CIAT 2002)

It is now possible to breed plants for increased vitamin and mineral content, making "biofortified" crops a promising tool in the fight to end malnutrition and save lives. An estimated 3 billion people suffer the effects of micronutrient deficiencies because they lack money to buy enough meat, fish, fruits, lentils, and vegetables. Women and children in sub-Saharan Africa, South and Southeast Asia, and Latin America and the Caribbean are especially at risk of disease, premature death, and impaired cognitive abilities because of diets poor in crucial nutrients, particularly iron, vitamin A, iodine, and zinc.

We have the ability today to further improve and disseminate more widely iron-rich rice, quality protein maize, high-carotene sweet potato, and high-carotene cassava. The potential advantages of biofortification are that it does not require major changes in behavior by farmers or consumers, can directly address the physiological causes of micronutrient malnutrition, can readily be targeted to the poorest people, uses built-in delivery mechanisms, is scientifically feasible and cost-effective, and can complement other ongoing methods of dealing with micronutrient deficiencies.

The potential disadvantages are that biofortification may not benefit small-scale and poor farmers; it undermines dietary diversity and creates possible conflict with distinctive food cultures; nutrient traits (except (-carotene) are difficult to verify without advanced technology; public investment is needed where infrastructure, including formal seed systems and marketing , and economies are weak; evaluations will be time-consuming; and resistance to and ethical issues related to transgenic crops are found in some regions.

---

and other substances essential for proper growth and development. Different micronutrients interact: there is a correlation, for example, between iron deficiency and deficiency in other vitamins and minerals (WHO 2004b). In addition, the content of specific micronutrients in foods can affect the absorption of other minerals.

Iodine, vitamin A, iron, and zinc are the most important individual micronutrients in global public health terms; their lack represents a major threat to the health and development of populations the world over, particularly to preschool children and pregnant women in low-income countries. Vitamin A deficiency and iron deficiency (related to anemia) alone affect as many as 3.5 billion people (WHO 2004b). Micronutrients with high bioavailability that are provided from animal-source foods include minerals, calcium, iron, phosphorus, zinc, magnesium, and manganese, along with the vitamins thiamine B1, riboflavin B2, niacin, pyridoxine B6, and B12 (CAST 1999). Several other nutrients are known to be inadequate in developing countries, including B12, folate, and vitamin C. This section describes the four micronutrients most important for public health at present.

Vitamin A is derived from animal sources as retinol and from fruits and vegetables (such as dark leafy vegetables and yellow and orange non-citrus fruits) as carotene, which is converted into vitamin A in the body. Plant-derived vitamin A is more difficult to absorb than that from animals. Vitamin A deficiency significantly increases the risk of blindness and of severe illness and death from common childhood infections, particularly diarrheal diseases and measles. In communities where vitamin A deficiency exists, children are on average 50% more likely to suffer from acute measles.

Improvements in vitamin A status have been demonstrated to lead to a 23% reduction in mortality among children aged one to five (Rahmathullah 2003), so preventing between 1.3 million and 2.5 million deaths each year and saving hundreds of thousands of children from irreversible blindness. Vitamin A therapy is a standard treatment in children with measles infection in developing countries. Measles infection itself causes a transient immunosuppression. Measles as an acute catabolic disease is thought to use up body reserves of nutrients, making the child more vulnerable to disease in the first instance and unable to counter the effects of the disease. The risk of dying from diarrhea, malaria, and measles was increased by around 20–24% in Vitamin A–deficient children (Rice et al. 2003).

Iron is readily available in many foods, especially meat, fish, and poultry, as well as in some leafy vegetables such as spinach. WHO estimates that 4–5 billion people worldwide, many of them women of reproductive age and children under 12 (as many as half of all such women and children in developing countries), are affected by iron-deficiency-induced anemia. Iron deficiency is associated with malaria, intestinal parasitic infestations, and chronic infections. One strategy to mitigate iron deficiency has been to promote the use of home gardens with small animals. Other strategies include food fortification and the eradication of infections.

Iodine deficiency is the world's most prevalent—yet easily preventable—cause of brain damage. Iodine deficiency disorders jeopardize children's mental health. They affect over 740 million people, 13% of the world's population; 30% of the remainder are at risk. Chronic iodine deficiency causes goiter in adults and children. Serious iodine deficiency during pregnancy may result in stillbirths, abortions, and congenital abnormalities such as cretinism—a grave, irreversible form of mental retardation that has affected people living in iodine-deficient areas of Africa and Asia (Aquaron et al. 1993; Hsairi et al. 1994; Foo et al. 1994; Yusuf et al. 1994). Global rates of goiter, mental retardation, and cretinism are all falling, attributed in varying degrees to the increased use of iodized salt (WHO 2004b). Of far greater global and economic significance, however, is iodine deficiency disorder's less visible yet more pervasive level of mental impairment that lowers intellectual development.

The consequences of severe human zinc deficiency have been known since the 1960s, but only more recently have the effects of milder degrees of zinc deficiency, which are highly prevalent, been recognized. Trials have shown that zinc supplementation results in improved growth in children; lower rates of diarrhea, malaria, and pneumonia; and reduced child mortality. In total, about 800,000 child deaths per year and, through deaths and increased rates of infectious diseases in affected areas, some 1.9% of DALYs are attributed to zinc deficiency. The incidence of diarrhea is increased around 20% in zinc-deficient children and that of pneumonia around 10–40% (Black 2003).

8.5.1.3 Hunger

Undernutrition can be broadly categorized into protein energy undernutrition (the result of a diet lacking enough protein and calorie sources) and (specific) micronutrient deficiencies. PEM is probably the most important factor contributing to nutrition-related mortalities (Habicht 1992).

FAO estimates that 852 million people worldwide did not have enough food to meet their basic daily energy needs in 2000–02. This includes 9 million in industrial countries, 28 million in countries in transition, and 815 million in developing countries (FAO 2004a). Some 519 million hungry people live in Asia and the Pacific and 204 million in sub-Saharan Africa, around 60% and 24% respectively of the global total of undernourished people. Viewed as a share of regional population, this means that some 16% of Asians and 33% of sub-Saharan Africans are undernourished. The two most populous countries in the world—China and India—alone account for almost 43% of the global total of hunger, but the highest incidence rates, ranging from 40% to 55% of the population, are found in Eastern, Southern, and Central Africa (FAO 2004a).

The latest hunger estimates signal a disturbing reversal of trends reported since around 1970 of a gradual decline in both hunger incidence and the absolute number of hungry people. Between 1969–71 and 1995–97, the absolute number of hungry people in developing countries had fallen from around 959 million to 780 million people. During the period 1995–97 to 2000–02, however, while the proportion of undernourished in developing countries fell by 1%, the number of hungry increased by some 18 million people to a total of 815 million.

This trend reversal reported by FAO (2003) confirms the analysis of overall DES patterns globally undertaken by Wang and Taniguchi (2003). The regional trends in the number of undernourished in developing countries for the early to mid-1990s and from the mid-1990s to around 2000 are clearly shown in the upper panel of Figure 8.13. The progress in hunger reduction in the early 1990s occurred predominantly in China and India. But in the second half of the decade, progress in China slowed and in India reversed, while in the Near East and Central Africa the numbers of hungry increased throughout the 1990s. The large-scale humanitarian crisis existing in Central Africa has escalated unabated.

Chronic child hunger, in particular, is measured using anthropometric measures for height-for-age or stunting. Using cross-sectional data from 241 nationally representative surveys, de Onis et al. (2000) have shown that the prevalence of stunting in children has fallen in developing countries from 47% in 1980 to 33% in 2000 (by 40 million). Progress has not been uniform, however. Stunting has increased in East Africa but decreased in Southeast Asia, South Central Asia, and South America. Stunting has moderately improved in North Africa and the Caribbean. West and Central Africa show little progress. Despite the average decrease, child undernutrition remains a major public health problem.

Undernutrition has a huge global impact on morbidity and mortality due to infectious diseases. In spite of a general understanding that undernutrition increases susceptibility to infectious diseases, good estimates of etiological fractions for the influence

of hunger on infectious disease mortality predominantly exist for children under five, where being underweight confers about 50% of the mortality risk for the main infectious diseases in the developing world like diarrhea, malaria, pneumonia, and measles (Schelp 1998; Cebu Study Team 1992). Child growth (also in utero, resulting in low birth weight) has again a very strong effect on morbidity and mortality. Birth weight alone is the single most important predictor of mortality in early life.

Evidence suggests that PEM is linked with higher malaria morbidity/mortality (Caulfield et al. 2004). Both malaria and chronic hunger have effects on child growth. Undernutrition is highly prevalent in many areas in which morbidity and mortality from malaria is unacceptably high. The global burden of malaria is associated with various nutrient deficiencies as well as underweight status. Although the association is complex and requires additional research, improved nutritional status lessens the severity of malaria episodes and results in fewer deaths due to malaria. Deficiencies in vitamin A, zinc, iron, and folate as well as other micronutrients are responsible for a substantial proportion of malaria morbidity and mortality. It is recommended that nutrition programs should be integrated into existing malaria intervention programs.

Diarrheal diseases are an important cause of mortality and morbidity in children, leading to more malnutrition but also being a consequence of undernutrition, as susceptibility to diarrheal diseases is increased in malnourished children (Lanata and Black 2001). There is ample evidence that giving formula and the early introduction of solids increases susceptibility to diarrhea, which emphasizes the importance of access to clean water.

Disease affects a person's development from a very early age. Gastroenteritis, respiratory infections, and malaria are the most prevalent and serious conditions that can affect development in the first three years of life. It is estimated that children under the age of five in developing countries suffer from 3.5 episodes of diarrhea per year and between four and nine respiratory tract infections in their first two years of life (Mirza et al. 1997). Infections affect children's development by reducing their dietary intake, by causing a loss of nutrients, or by increasing nutrient demand as a result of fever.

Undernutrition also plays a significant role in morbidity among adults. The link between morbidity from chronic disease and mortality, on the one side, and a high body mass index, on the other side, has been recognized and analyzed in industrial countries primarily for the purpose of determining life insurance risk. These relationships have also been studied in developing countries. A study on Nigerian men and women has shown mortality rates among chronically energy-deficient people who are mildly, moderately, and severely underweight to be 40%, 140%, and 150% greater than rates among non-chronically energy-deficient people (ACC/SCN 2000).

There are also linkages between food provision and HIV/AIDS. Not only does good nutrition of afflicted individuals help maintain the quality of their life, but in a rural environment there are implications for the feasibility of continued engagement in food production activities. One challenge is to minimize demands on a household labor pool that is severely depleted by the incapacitation of the afflicted family member and the time devoted to care giving by other family members (Gillespie and Haddad 2002).

8.5.1.4 Obesity and Overweight

Obesity has become a global epidemic. At present over 1 billion adults are overweight, with at least 300 million considered clinically obese, up from 200 million in 1995 (WHO 2003a). Obesity

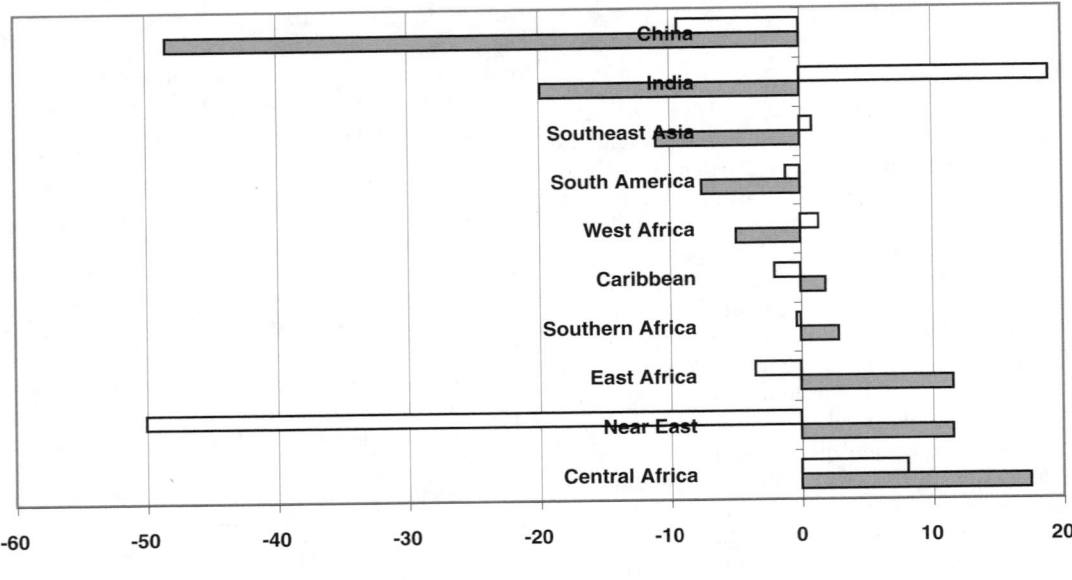

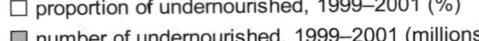

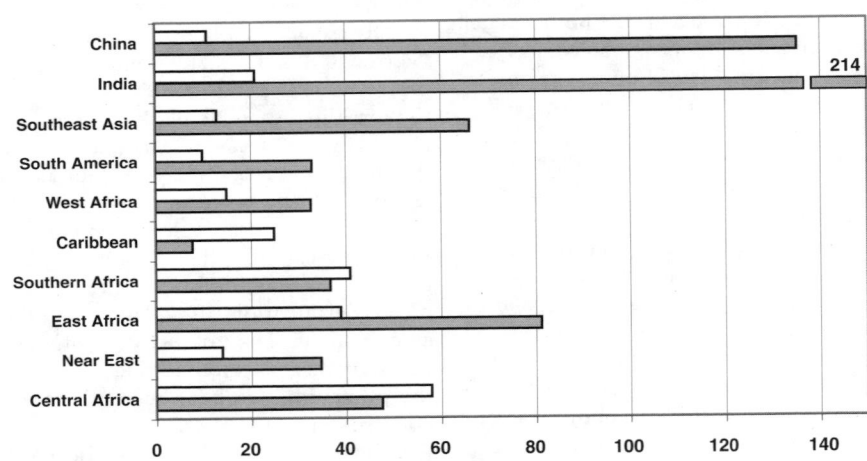

Figure 8.13. Status and Trends of Hunger, 1990–2001 (FAO 2003)

is now a major determinant of the global burden of chronic disease and disability. In many countries, significant incidence of both obesity and undernutrition may co-exist. (See Figure 8.14.) Obesity is a complex condition, with serious social and psychological dimensions, and can be found across almost all ages and socioeconomic groups (WHO 2003a). The prevalence of overweight and obesity is commonly assessed using the body mass index, which is defined as the weight in kilograms divided by the square of the height in meters (kilograms per square meter). A BMI of over 25 kilograms per square meter is defined as overweight, and a BMI of over 30 kilograms per square meter as obese (WHO 2003a).

Rising rates of obesity and overweight are due to both reduced physical activity and increased consumption of more energy-dense, nutrient-poor foods with high levels of sugar and saturated fats. Obesity rates often increase faster in developing countries

than in industrial ones (Chopra 2002; WHO 2003a). The underlying causes of these trends include urbanization, income growth, changing lifestyles, and globalization of and convergence of "western" diets, the "diet transition." This transition is generally associated with an epidemiological transition in which disease patterns shift over time so that infectious and parasitic diseases are gradually but not completely displaced, and noncommunicable diseases become the leading cause of death (Uusitalo et al. 2002). WHO reports that NCDs now account for 59% of the 57 million deaths annually and 46% of the global burden of disease. There is a *well-established* link between an unhealthy diet and several of the most important of these NCDs, including coronary heart disease, cerebrovascular disease, various cancers, diabetes mellitus, dental caries, and various bone and joint diseases (WHO 2003b). Low levels of physical activity exacerbate dietary causes of increased risk of NCDs (WHO 2003b).

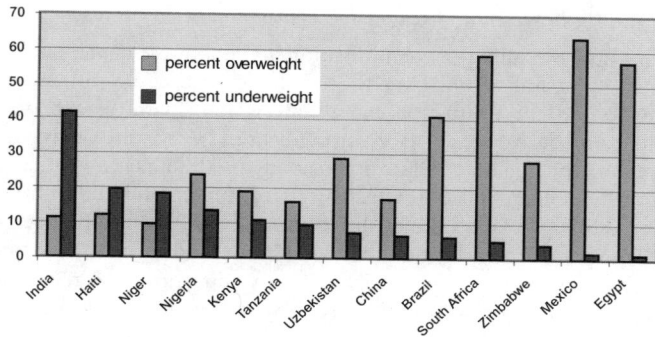

Figure 8.14. Double Burden of Undernutrition and Overnutrition among Women Aged 20–49 for Selected Developing Countries. A calculation of body mass index, BMI, can determine if a person is not eating enough, or is consuming too much. Many developing countries are facing both problems simultaneously. Note: data for each country are from the most recent year available. (Mendez et al. 2005)

Obesity rates have tripled or more since 1980 in some areas of North America, the United Kingdom, Eastern Europe, the Middle East, the Pacific Islands, Australasia, and China. Current obesity levels range from below 5% in China, Japan, and certain African nations to over 75% in urban Samoa. But even in relatively low prevalence countries like China, rates are almost 20% in some cities (WHO 2003a). A study by the Centers for Disease Control and Prevention in the United States found that almost 65% of Americans are overweight and around one quarter are obese (Flegal et al. 2002).

Child obesity is a growing concern and is already an epidemic in some areas. Based on an analysis of 160 nationally representative cross-sectional surveys from 94 countries, the global prevalence of overweight among pre-school children was assessed at 3.3%, representing, in developing countries, some 17.6 million children (de Onis et al. 2000). According to the U.S. Surgeon General, the number of overweight children in the United States has doubled and the number of overweight adolescents has tripled since 1980 (WHO 2003a).

Overweight and obesity lead to adverse metabolic effects on blood pressure, cholesterol, triglycerides, and insulin insensitivity (Rguibi and Belahsen 2004). Non-fatal health problems include respiratory chronic musculoskeletal problems, skin problems, and infertility. The more life-threatening problems are cardiovascular disease, conditions associated with insulin resistance such as type-2 diabetes, and certain types of cancers, especially the hormonally related and large-bowel cancers and gall bladder disease (WHO 2003b). Four out of the 10 leading global disease burden risk factors for NCDs identified by WHO (2002a) were diet-related: high blood pressure, high cholesterol, obesity, and insufficient consumption of fruits and vegetables.

An estimated 16.7 million deaths, or almost 30% of the global total, result from various forms of cardiovascular disease (WHO 2004c). Many of these are preventable by action on the major primary risk factors: unhealthy diet, physical inactivity, and smoking. Cancer accounts for about 7.1 million deaths annually (12.5% of the global total). Dietary factors account for about 30% of all cancers in western countries and approximately up to 40% in developing countries. Diet is second only to tobacco as a preventable cause (WHO 2004c). And up to 2.7 million lives could potentially be saved each year with sufficient global fruit and vegetable consumption (WHO 2004c). Fruits and vegetables as part

of the daily diet could help prevent major NCDs such as cardiovascular disease and certain cancers.

A number of concerns have been raised concerning the convergence of diets that accompanies globalization, in addition to the negative effects on health and nutrition. These are loss of cultural identity and increased resource use in food production and transportation (Bruinsma 2003). It is estimated, for example, that an average American diet requires about three times as much nitrogen fertilizer per capita as the average Mediterranean diet (Howarth et al. 2002).

The simplification of diets in terms of food sources as a consequence of the nutrition transition has also had implications on demand for certain types of food and hence on the structure of crop, livestock, and aquaculture production systems as well as the genetic diversity they embody (Johns 2003). Although an increasing number of processed foods are available, particularly in urban and higher-income markets, the species and intra-species diversity of food sources is narrowing (Johns 2003).

Conversely, agricultural intensification and the simplification of agricultural landscapes can limit the availability of and access to wild foods and to food plants growing as weeds that may be of nutritional importance, especially to landless poor people and to vulnerable groups within households (Scoones et al. 1992). Similarly, the decline of traditional fisheries (following commercial exploitation of coastal fisheries and damage to inland water ecosystems due to water extraction and diversion) can have severe negative nutritional consequences for poor artisanal fishers (DFID 2002).

8.5.2 Food Safety

Illness as a result of contaminated food is a widespread health problem and an important factor in reduced economic productivity. The global incidence of food-borne disease is difficult to estimate, but it has been reported that in 2000 alone 2.1 million people died from diarrheal diseases, and it is estimated that 70% of the 1.5 billion global episodes of diarrhea are due to biologically contaminated food (WHO 2002b). Additionally, diarrhea is a major contributor to malnutrition in infants and young children. In industrial countries, up to 30% of people reportedly suffer from food-borne diseases each year (WHO 2002b). In the United States, for example, around 76 million cases of food-borne diseases, resulting in 325,000 hospitalizations and 5,000 deaths, are estimated to occur each year (Mead et al. 1999). While the situation in developing countries is less documented, they bear the brunt of food safety problems due to the presence of a wide range of food-borne diseases, including those caused by parasites. For example, cholera is a major public health problem in developing countries (WHO 2002b).

Food contaminants can occur naturally or as a result of poor or inadequate production, storage, and handling. Hazardous agents include microbial pathogens, zoonotic disease agents, parasites, myco- and bacterial toxins, antibiotic drug residues, hormones, and pesticide residues. Genetically modified organisms and their potential to contain allergens or toxins have also begun to receive attention (Unnevehr 2003). Food safety risks vary with climate, diet, income level, and public infrastructure, and food-borne pathogens are more prevalent in developing countries, where food safety may be as important as food security for health and nutrition. Consumption of contaminated food may cause children to be stunted, underweight, and more susceptible to infectious diseases in childhood and later in life.

In addition to food safety risks associated with uncontrolled endemic diseases found in developing regions, there are also food safety risks that appear in highly developed production systems when animal concentrations are high, feeds contain contaminants, or meat and milk are improperly handled. There have been outbreaks of several zoonotic diseases (those naturally transmitted from animals to humans) recently in people, including avian influenza, severe acute respiratory syndrome, and Creutzfeldt-Jakob Disease. Of 1,709 human pathogens, 832 are zoonotic, and of 156 emerging diseases, 114 are zoonotic. Overall, zoonotic pathogens are more than three times more likely to be associated with emerging diseases than non-zoonotic pathogens. Examples include influenza, brucellosis, bovine spongiform encephalopathy, tuberculosis, and rabies (Kaufmann and Fitzhugh 2004).

Genetically modified foods are increasingly receiving attention in the food safety debate with regard to toxicity, allergenicity, stability of the modified genetic composition, nutritional effects associated with genetic modification, and unintended effects as a result of gene insertion (WHO 2005). Outcrossing (the movement of genes from GM plants into conventional crops or related species in the wild), as well as the mixing of crops derived from conventional seeds with those grown using GM crops, may have an indirect effect on food safety and food security if the gene products are toxic. The risk of seed mixing is real, as was shown when traces of a maize type that was only approved for feed use appeared in maize products for human consumption in the United States (WHO 2005).

GM foods currently available on the international market have passed risk assessments and are not likely to present risks for human health (WHO 2005). In the developing world, the approval and cultivation of GM crops is largely limited to soybean, maize, and cotton in Argentina, Brazil, China, India, Mexico, and South Africa (James 2004). Consumption of GM foods has not shown any adverse effects on human health in the general population in the countries where they have been approved. Nevertheless, food safety assessments are essential to GM approvals and thus need to be started early in the process of GM crop development (Unnevehr 2003).

Pesticide residues on food are also of growing concern. In the United States, for example, the Environmental Protection Agency has set maximum legal limits for pesticide residues on food commodities for sale domestically. In the most recent U.S. Food and Drug Administration studies, dietary levels of most pesticides were less than 1% of the acceptable daily intake established by the FAO and WHO (Bessin 2004). Consumption of pesticide residues in food remains a significant problem in most developing countries, however, especially for food that is produced and consumed locally (Dasgupta et al. 2002). Pesticides, especially organochlorides, are expected to increase in importance as a health concern, particularly in the context of multiple pesticide exposure. Although the long-term effects of pesticide exposure remain uncertain, evidence suggests that toxins may increase carcinogenic and neurotoxic health risks in susceptible sub-groups (Alavanja et al. 2004; Maroni and Fait 1993).

8.5.3 Household Economic Impacts

Consumption of food is also linked to cognitive development and labor productivity. Nutrition has a dynamic and synergistic relationship with economic growth through the channel of education, and the evidence shows that the causality works in both directions: better nutrition can lead to higher cognitive achievement and increased learning capacity, and thus to higher labor productivity as well as higher incomes. And higher levels of education lead to better nutrition.

8.5.3.1 Nutrition and Cognitive Development

This dual causality between nutrition and cognitive development is complex and varies over the life cycle of a family. In utero, infant, and child nutrition can all affect later cognitive achievement and learning capacity during school years, ultimately increasing the quality of education gained as a child, adolescent, and adult. Parental education affects in utero, infant, and child nutrition directly through the quality of care given (principally maternal) and indirectly through increased household income. Human capital development, primarily through education, has received merited attention as a key to economic development (Verner 2004), but early childhood nutrition has yet to obtain the required emphasis as a necessary facilitator of education and human capital development.

Despite the limited evidence demonstrating a causal link between poor nutrition and cognitive achievement, systematic evidence supports the argument that policy interventions in early childhood nutrition are crucially important for cognitive achievement, learning capacity, and, ultimately, household welfare. Specifically, available studies (Horton 1999) have shown that:

- PEM deficiency, as manifested in stunting, is linked to lower cognitive development and educational achievement;
- low birth weight is linked to cognitive deficiencies;
- iodine deficiency in pregnant mothers negatively affects the mental development of their children;
- iodine deficiency in children can cause delayed maturation and diminished intellectual performance; and
- iron deficiency can result in impaired concurrent and future learning capacity.

Children are most vulnerable to malnutrition in utero and before they reach three years of age, as growth rates are fastest and they are most dependent on others for care during this period. However, nutrition interventions, such as school feeding programs, among children of school age are also important for strengthening learning capacity.

Yet there remains significant uncertainty surrounding estimates of the monetary costs associated with the impact of hunger and malnutrition on school performance. Nevertheless, Behrman (2000) cites three studies suggesting that, by facilitating cognitive achievement, child nutrition and schooling can significantly increase wages. Micronutrients from animal-source foods are especially important in the health of women of reproductive age and in the cognitive development and school performance of children (Neumann and Harris 1999). Behrman (2000) concludes that while the link between health and educational attainment is not as robust as most studies suggest, and specific cost-benefit analysis is difficult to carry out, policies supporting nutrition make good sense, and the empirical basis for this is as sound as that of many other conventional assumptions in economics.

8.5.3.2 Nutrition and Labor Productivity

Much of the empirical work linking economic outcomes to nutrition to date has focused on agriculture, and it attempts to link farm output, profits, wages, or labor allocation choices to indicators of nutritional intake such as calories or to nutritional outputs such as weight-for-height, BMI, and height. Widely cited work by Strauss (1986) links the average calorie intake per adult in a household to the productivity of on-farm family labor in Sierra Leonean agriculture. For this sample, on average, a 50% increase in calories per consumer equivalent increased output by 16.5%, or 379 kilograms. For an increase of 50% in hours of family labor or in the area of cultivated land, this compares with an output response of 30% and 13%, respectively. Significantly, Strauss's findings show that the lower the calorie intake is, the more significant the output response is to increased calorie intake. For example, based on a daily intake of 1,500 calories per consumer equivalent, a mere 10% increase in calorie intake would increase output by nearly 5%.

Findings from Ethiopia, presented in Croppenstedt and Muller (2000), show that a 10% increase in weight-for-height and BMI would increase output and wages by about 23% and 27%, respectively. They also find that height, an indicator of a person's past nutritional experience, is a significant determinant of wages in Ethiopia, with a person who is 7.1 centimeters above the average height earning about 15% more wages. These findings have to be contrasted with the effect of other productivity-augmenting investments, such as education. Nutrition would appear to compare well with the 4% increase in cereal output attributed to an additional year of schooling in a rural Ethiopian household.

Poor nutritional status not only reduces a person's output, it may also prevent them from carrying out certain tasks. A study in Rwanda found that those who are poorly fed have to choose activities that are physically less demanding—and less well paid (Bhargava 1997). A low BMI and poor nutritional status may also limit productivity indirectly through absenteeism and reduced employment opportunities. Moreover, to carry out certain activities, undernourished people may have to put their muscle mass and heart rate under much greater strain than well-nourished people do, requiring more energy to produce the same output, which may lead to health problems in the long term.

There is also an increasing awareness of the role of micronutrients in people's nutritional status, as described earlier. For example, iron deficiency in adults negatively affects productivity as well as contributing to absenteeism. Basta et al. (1979) found that productivity among Indonesian rubber plantation workers with anemia was reduced by 20% compared with non-anemic workers. There is also some evidence that iodine deficiency during adulthood reduces productivity and work capacity (Hershman et al. 1986).

8.5.4 Macroeconomic Growth

There is clear evidence of an association between improvements in nutrition and improvements in macroeconomic growth, but the nature and strength of the association is sensitive to the context and time frame (Easterly 1999; Arcand 2001; Wang and Taniguchi 2003). Overall, inadequate nutrition is estimated to cause losses of between 0.23% and 4.7% per year in per capita GDP growth rates worldwide (Arcand 2001). Improved nutrition affects economic growth directly through its impact on labor productivity and indirectly through improvements in life expectancy. The reverse is also true, and analysis using 81 indicators found that a 1% increase in per capita GDP raises per capita calorie intake by approximately 540 kilocalories a day (Easterly 1999). Differences in economic growth explained 40% of cross-country differences in improved mortality rates over the last three decades (Pritchett and Summers 1996), while in a separate study covering some 65 countries between 1970 and 1995, half of the decline in child malnutrition from 1970 to 1995 could be attributed to income growth (Smith and Haddad 2000). Half of the economic growth that occurred in the United Kingdom and France in the eighteenth and nineteenth centuries has been attributed to improvements in nutrition and health (Fogel 1994).

More recent and comprehensive analysis of 114 countries over the period 1961–99 found that, on average, the GDP growth rate increased by 0.5% per capita for each 500 kilocalorie-per-day increase in dietary energy supply (Wang and Taniguchi 2003).

However, the results differed significantly between country groups and short-term and long-term perspectives. For some groups (East and Southeast Asia), GDP growth was up to four times higher, while in others (sub-Saharan Africa) GDP growth was absent or negative. Furthermore, short-term GDP impacts were also ambiguous (a mix of positive, neutral, and negative impacts). This is explained on the basis of a "nutrition trap" for some countries, particularly—and persistently—in sub-Saharan Africa. In such countries, significant increases in DES often translate into an expansion in short-term population growth rates that depress both per capita DES and GDP. To overcome this, nutrition and income need to attain levels at which further increases in DES do not promote population expansion, or DES increases need to be large and sustained (Wang and Taniguchi 2003).

8.5.5 Cultural Aspects

Food is not just an economic good and a requirement for good health, but also a centerpiece of culture. It is no coincidence that the word "company" stems from *com,* meaning together and *panis* meaning bread. Recalling the quasi-mystical status of the yam in much of African culture, Wole Soyinka, the Nigerian Nobel laureate for literature, speaking on the role of cultural leaders in changing attitudes and behaviors related to food and nutrition, said "Food is allied to culture in the most organic, interactive way, and one may be brought to the aid of, in enhancement of, or in celebration of the other" (IFPRI 2004).

Cultural aspects of food and acquired tastes (a type of "food bias") can be significant determinants of both physical and economic well-being. For example, in drought-prone southern Zambia, some drought-resistant crops, such as millet and sorghum, are not cultivated because southern Zambians have not acquired a taste for them despite the fact that these foods could provide food security during times of drought.

Food is so integral to the formation of cultural identities and global unity that 2004 was declared "The International Year of Rice" by FAO, to commemorate rice as a life force having "enormous impact on human nutrition and global food security" (FAO 2004). A grain of rice is compared with a grain of gold in Southeast Asia, and many Japanese perceive rice as the "heart" of their culture. Rice, along with many other foods, is used for consumption by people every day, as well as to celebrate religious and social holidays, festivals, and other special occasions.

Indeed, food rituals form the core of many religious rituals, and inadvertently affect markets and trade. Taboos associated with different types of food also set apart cultures, sects, and denominations. Not only do these taboos and rituals help promote social solidarity and maintain cultural boundaries, they also promote "food bias" and, thereby, contribute to the kinds of food trade a culture will initiate and promote. For example, religious beliefs associated with pigs in the Judaic and the Muslim traditions are not likely to initiate any pork trade in these regions. Similarly, Hinduism's taboo of beef, due to the sacred allegiance to the cow, can adversely affect the beef trade in certain regions of India. In North America, the turkey has become a cultural symbol associated with Thanksgiving, and more broadly is associated with Christmas, generating a large turkey trade during these seasons.

Recognizing that the enjoyment of wholesome food is important in the pursuit of happiness and, in part, as a backlash to the rapidly growing western/urban culture of fast food, a small but rapidly growing grassroots movement has been established that seeks to maintain the cultural significance of food and its consumption (Jones et al. 2003).

8.5.6 Distributional Dimensions

In low-income countries, most of which are characterized by a relatively large share of agriculture in the economy and a high proportion of rural population, food production has critical distributional impacts in the form, for example, of income or health inequalities. It has impacts on poverty alleviation, reduced inequalities in food consumption, improved nutrition and health, low commodity and food prices, and direct and indirect employment and income generation. The size and pattern of these impacts depend on a number of factors: the rate of growth of agricultural and food production across regions and types of crops, agriculture's share in the economy, the proportion of the population that is rural, asset (mainly land) distribution, rural infrastructure, well-functioning markets, availability of agricultural inputs and credit to the poorer farmers and in remote areas, and adequate research and extension (von Braun 2003).

Thus a whole set of factors combine to have a broad and equitable impact from increased food productivity. This explains the continued food distributional inequalities despite food production per capita increasing globally. Increased local food production remains critical to alleviating poverty and providing food security. In general, the experience of the last few decades shows that the higher the food (and agricultural) output (especially when it is due to higher labor and land productivity), the more equal the land distribution, the better the small farmer's access to inputs and markets, and the less suppression of agricultural prices, the greater is the positive impact on income and consumption distribution, poverty alleviation, and food security for the poor (von Braun 2003). Further, the more "distribution friendly" variables that are present, the stronger is their synergistic impact.

Livestock are also a significant source of income and consumption in low-income countries. Often they provide a supplementary source of income and income stability. However, in many cases livestock are a vital, even the sole, source of income for the poorest, the landless, pastoralists, sharecroppers, and widows. They are one of the few assets available to these groups. Livestock also allow the rural poor to exploit common property resources, such as open grazing areas, in order to earn incomes and reduce income variability, especially in semiarid areas.

References

ACC/SCN (Administrative Committee on Coordination/Standing Committee on Nutrition), 2002: *Nutrition: A Foundation for Development* (12 briefs). Geneva.

ACC/SCN, 2000: *Fourth report on the World Nutrition Situation: Nutrition Throughout the Life Cycle.* ACC/SCN in collaboration with the International Food Policy Research Institute, Geneva, 121 pp.

Acquaye, A.K.A., J.M. Alston, and P.G. Pardey, 2003: Post-war productivity patterns in U.S. agriculture: influences of aggregation procedures in a state-level analysis. *American Journal of Agricultural Economics,* **85(1),** 59–80.

Alavanja, M.C.R., J.A. Hoppin, and F. Kamel, 2004: Health effects of chronic pesticide exposure: Cancer and neurotoxicity. *Annual Review of Public Health,* **25,** 155–197.

Alston, J.M., 2002: Spillovers. *Australian Journal of Agricultural and Resource Economics,* 46(3), 315–346.

Alston, J.M., C. Chan Kang, M.C. Marra, P.G. Pardey, and T.J. Wyatt. 2000. *A Meta-Analysis of Rates of Return to Agricultural R&D: Ex Pede Herculem?* Research Report 113. International Food Policy Research Institute, Washington. 148 pp.

Aquaron, R., K. Zarrouck, M. Eljarari, R. Ababou, A. Talibia, and J.P. Ardissone, 1993: Endemic goiter in Morocco (Skoura-Toundoute areas in the high Atlas). *Journal of Endocrinological Investigation,* **16(1),** 9–14.

Arcand, J. 2001. *Undernourishment and Economic Growth: the Efficiency Cost of Hunger.* FAO Economic and Social Development Paper No. 147, FAO, Rome, 60 pp.

Baillie J.E.M., C. Hilton-Taylor, and S. Stuart (eds.), 2004: *2004 IUCN Red List of Threatened Species: A Global Species Assessment.* International Union for the Conservation of Nature, Gland, Switzerland, and Cambridge, UK, 217 pp.

Basta, S., S. Soekirman, D. Karyadi, and N.S. Scrimshaw, 1979: Iron deficiency anemia and the productivity of adult males in Indonesia. *American Journal of Clinical Nutrition,* **32,** 916–925.

Behrman, J., 2000. *Literature Review on Interactions between Health, Education and Nutrition and the Potential Benefits of Intervening Simultaneously in All Three.* International Food Policy Research Institute, Washington, DC, 33 pp.

Bene, C., A. Neiland, T. Jolley, S. Ovie, O. Sule, B. Ladu, K. Mindjimba, E. Belal, F. Tiotsop, and M. Baba, 2003: Inland fisheries, poverty, and rural livelihoods in the Lake Chad Basin. *Journal of Asian and African Studies,* **38(Part 1),** 17–51.

Berdegue, J.A., F. Balsevich, L. Flores, D. Mainville, and T. Reardon, 2003: *Food Safety and Food Security in Food Trade. Case Study: Supermarkets and Quality and Safety Standards for Produce in Latin America.* Focus 10. Brief 12 of 17. International Food Policy Research Institute, Washington, DC, 2 pp.

Bessin, R.T., 2004: Pesticide residues in foods: is food safety just a matter of organic vs. traditional farming? [online] Lexington: University of Kentucky Entomology. Cited February 18, 2005. Available at http://www.uky.edu/Agriculture/Entomology/entfacts/misc/ef009.htm.

Bhargava, A., 1997. Nutritional status and the allocation of time in Rwandese households. *Journal of Econometrics,* **77,** 277–295.

Black, R.E., 2003. Zinc deficiency, infectious disease and mortality in the developing world. *The Journal of Nutrition,* **133 (5),** 1485–1489.

Blisard, N. and J.R. Blaylock, 1993: *US Demand for Food: Household Expenditures, Demographics and Projections for 1990–2010,* Economic Research Service, United States Department of Agriculture, Technical Bulletin 1818, December 1993, Washington, DC.

Bread for the World Institute, 2003: *Agriculture in the Global Economy, Hunger 2003. 13th Annual Report on the State of the World's Hunger.* Washington DC, 157 pp.

Bruinsma, J. (ed.), 2003: *World Agriculture: Towards 2015/2030, An FAO Perspective.* Earthscan and FAO, London and Rome, 432 pp.

Caddy, J.F, 1993: Contrast between recent fishery trends and evidence for nutrient enrichment in two large marine ecosystems: the Mediterranean and the Black Seas. In: Kenneth Sherman, et al. (eds.), *Large Marine Ecosystems: Stress, Mitigation, and Sustainability,* American Association for the Advancement of Science, Washington, DC, pp.137–147.

Carlson, R.E., D.P. Todey, and S.E. Taylor, 1996: Midwestern corn yield and weather in relation to extremes of the Southern Oscillation. *Agronomy Journal/Production Agriculture,* **9,** 347–352.

Carter, C. and B. Zhang, 1998: The weather factor and variability in china's grain supply. *J Comparative Economic* **26,** 529.

Cassman, K.G., 2001. Crop science research to assure food security. In: J. Noesberger et al. (eds.), *Crop Science: Progress and Prospects.* CAB International, Wallingford, UK. pp. 33–51.

Cassman, K.G., 1999: Ecological intensification of cereal production systems: yield potential, soil quality, and precision agriculture. *Proceedings of the National Academy of Sciences,* **96,** 5952–5959.

Cassman, K.G., A. Dobermann, D.T. Walters, and H. Yang, 2003: Meeting cereal demand while protecting natural resources and improving environmental quality. *Annual Review of Environmental Resources,* **28,** 315–358.

CAST (Council for Agricultural Science and Technology), 1999: *Animal Agriculture and Global Food Supply.* Ames, Iowa, 92 pp.

Caulfield, L.E., S.A. Richard, and R.E. Black. *Undernutrition as an Underlying Cause of Malaria Morbidity and Mortality.* DCPP Working Paper No. 16, Center for Human Nutrition, Bloomberg School of Public Health, The Johns Hopkins University, Baltimore, MD, 38 pp.

Cebu Study Team, 1992: A child health production function estimated from longitudinal data. *Journal of Development Economics,* **38(2),** 323–351.

Chang, S.W., 2003: WTO Disciplines on fisheries subsidies: a historic step towards sustainability? *Journal of International Economic Law,* **6(4),** 879–921.

Chopra, M., 2002: Globalization and food: implications for the promotion of healthy diets. In: *Globalization, Diets and Non-communicable Diseases.* WHO, Geneva, 185 pp.

Clark, G., M. Huberman, and P. Lindert, 1995: A British food puzzle, 1770–1850, *Economic History Review,* **48 (2),** 215–37.

Coates, D., 1995: Inland Capture Fisheries and Enhancement: Status, Constraints and Prospects for Food Security. FISHAID Project, Port Moresby, Papua New Guinea. *Report presented at the International Conference on Sustainable Contribution of Fisheries to Food Security,* Kyoto, Japan, 4–9 December.

Conway, G., 1999: *The Doubly Green Revolution: Food for all in the 21st Century.* Cornell University Press, Ithaca, 335 pp.

Cooper, H.D., T. Hodgkin, and C. Spillane, 2000: *Broadening the Genetic Base of Crop Production.* Commonwealth Agricultural Bureaux International/ Food and Agriculture Organization/ International Plant Genetics Research Institute. Rome, 480 pp.

Cromwell, E., D. Cooper, and P. Mulvany, 2001: *Agricultural Biodiversity and Livelihoods: Issues and Entry Points for Development Agencies.* Overseas Development Institute, London, 52 pp.

Croppenstedt, A. and C. Muller, 2000: The impact of farmers' health and nutritional status on their productivity and efficiency: evidence from Ethiopia. *Economic Development and Cultural Change,* **48 (3),** 475–502.

Dasgupta S., C. Meisner, D. Wheeler, and Y. Jin, 2002: Agricultural trade, development and toxic risk. World Development, **30(8),** 1401–1412.

Delgado, C., M. Rosegrant, and N. Wada, 2003: Meating and milking global demand: stakes for small-scale farmers in developing countries. In: *The Livestock Revolution: A Pathway from Poverty? Record of a Conference Conducted by the ATSE Crawford Fund. Parliament House, Canberra, 13 August 2003.* A festschrift in honor of Derek E. Tribe.

Delgado C., M. Rosegrant, H. Steinfeld, S. Ehui, and C. Courbois, 1999: *Livestock to 2020: The Next Food Revolution. Food, Agriculture and the Environment* Discussion Paper 28. 2020 Vision. International Food Policy Research Institute. Washington, DC, 83 pp.

Delgado C., J. Hopkins, and V. Kelly, with P. Hazell, A. McKenna, P. Gruhn, B. Hoijati, J. Sil, and C. Courbois, 1998: *Agricultural Growth Linkages in sub-Saharan Africa.* International Food Policy Research Institute, Research Report No.107, Washington, DC, 154 pp.

Delisle, H., 1990: *Patterns of Urban Food Consumption in Developing Countries: Perspectives from the 1980s.* FAO, Rome, 92 pp.

Denison, R.F., 2003: Darwinian agriculture: When can humans find solutions beyond the reach of natural selection? *Quarterly Review of Biology* **78,** 145–168.

de Onis M., E.A. Frongillo, and M. Blossner, 2000: Is malnutrition declining? an analysis of changes in levels of child malnutrition since 1980. *Bulletin of the World Health Organization,* **78 (10),** 1222–1233. World Health Organization, Geneva.

Devendra, C., 2000: Animal production and rainfed agriculture in Asia: potential opportunities for productivity enhancement. *Outlook on Agriculture,* **29(3),** 161–75.

Devendra, C., C. Seville, and D. Pezo, 2001: Food-feed systems in Asia: review. *Asian-Australasian Journal of Animal Science,* **14(5),** 733–745.

DFID (Department for International Development, UK), 2002: Marine fisheries. Key sheets for sustainable livelihoods [online], cited February 28, 2005. Available at http://www.keysheets.org/green_8_marinefish.pdf.

Diamond, J., 1999: *Guns, Germs and Steel: The Fates of Human Societies.* W. W. Norton and Company, New York. 480 pp.

Diao, X., E. Diaz-Bonilla, S. Robinson, and D. Orden, 2005: *Tell Me Where It Hurts, and I'll Tell You Who to Call: Industrialized Countries' Agricultural Policies and Developing Countries,* MTID Discussion Paper No 84. International Food Policy Research Institute, Washington, DC.

Döll, P. and S. Siebert, 1999: *A Digital Map of Irrigated Areas.* Report No. A9901, Center for Environmental Systems Research, Germany, University of Kassel.

Easterly, W., 1999. Life during growth. *Journal of Economic Growth,* **4(3),** 239–276.

Ehui, S., S. Benin, T. Williams, and S. Meijer, 2002: *Food Security in sub-Saharan Africa to 2020.* Socio-economics and Policy Research Working Paper 49. International Livestock Research Institute (ILRI): Nairobi, Kenya, 60 pp.

Etkin, N.L. (ed.), 1994: *Eating on the Wild Side: The Pharmacologic, Ecologic, and Social Implications of Using Noncultigens.* University of Arizona Press, Tucson, AZ, 305 pp.

Evans, L.T., 1998: *Feeding the 10 Billion: Plants and Population Growth.* Cambridge University Press, Cambridge, UK, 247 pp.

Evenson, R.E. and D. Gollin, 2003: Assessing the impact of the green revolution. *Science,* **300,** 758–762.

Evenson, R.E., C. Pray, and M.W. Rosegrant, 1999: *Agricultural Research and Productivity Growth in India.* IFPRI Research Report No. 109. Washington, D.C., International Food Policy Research Institute, 103 pp.

Ezzati, M., A. Lopez, A. Rodgers, S. Vander Hoorn, C. Murray, and the Comparative Risk Assessment Collaborating Group, 2002: Selected major risk factors and global and regional burden of disease. *The Lancet,* 360 (November), 1347–1360.

FAO (Food and Agriculture Organization of the United Nations), 2004: International year of rice: rice is life. Online at: http://www.fao.org/rice2004/en/rice-us.htm.

FAO, 2004a: *The State of Food Insecurity in the World. Monitoring Progress towards the World Food Summit and Millennium Development Goals.* Rome.

FAO, 2004b: *The State of Food and Agriculture.* Food and Agriculture Organization of the United Nations, Rome, 208 pp.

FAO, 2003: *The State of Food Insecurity in the World. Monitoring Progress Towards the World Food Summit and Millennium Development Goals.* Rome. 37 pp.

FAO, 2002: *The State of World Fisheries and Aquaculture.* FAO, Rome, 150 pp.

FAO, 2001: *The State of Food Insecurity in the World. Monitoring Progress towards the World Food Summit and Millennium Development Goals.* Rome, 36 pp.

FAO, 1999: Fishery Information, Data and Statistics Unit Fisheries Circular No. 929, Revision 2. Number of fishers 1970–1997. Rome, 106pp.

FAO, 1998: *The State of the World's Plant Genetic Resources for Food and Agriculture.* Rome, 29 pp.

FAO, 1997: Computer Printout of FAOSTAT's International Commodity Prices 1989–91. Personal communication via Technical Advisory Committee, CGIAR. Rome.

FAO, 1996: *The Global Plan of Action for the Conservation and Sustainable Use of Plant Genetic Resources for Food and Agriculture.* Rome, 63 pp.

FAO FISHSTAT, 2003: Online at: http://www.fao.org/fi/statist/statist.asp.

FAO/ NACA (Network of Aquaculture Centers in Asia-Pacific), 2001: *Manual of Procedures for the Implementation of the Asia Regional Technical Guidelines on Health Management for the Responsible Movement of Live Aquatic Animals,* FAO Fisheries Technical paper 402/1. Rome. 106 pp.

FAOSTAT, 2004: Food and Agriculture Organization of the United Nations, Statistical Databases. Online at: http://faostat.fao.org

FAO/WHO (World Health Organization), 2002: *Human Vitamin and Mineral Requirements.* Report of a Joint Expert Consultation, Bangkok, Thailand. FAO, Rome, 303 pp.

FAO/WHO/UNU (United Nations University), 1985: *Energy and Protein Requirements.* Report of a Joint FAO/ WHO/ UNU Expert consultation. WHO Technical Report Series, **724,** 1–206.

Flegal, K.L., M.D. Carroll, C.L. Ogden, and C.L. Johnson, 2002: Prevalence and trends in obesity among US adults, 1990–2000. *Journal of the American Medical Association.* **288(14),** 1723–1727.

Fogel, R., 1994: Economic growth population theory and physiology: the bearing of long-term processes on the making of economic policy. *American Economic Review,* **84 (3),** 369–95.

Foley, J.A., M.H. Costa, C. Delire, N. Ramankutty, and P. Snyder, 2003: Green surprise? how terrestrial ecosystems could affect earth's climate. *Frontiers in Ecology and the Environment,* **1(1),** 38–44.

Foo, L.C., T. Zainab, G.R. Letchuman, M. Nafikudin, R. Arriman, P. Doraisingam, and A.K. Khalid, 1994: Endemic goiter in the Lemanak and Ai river villages of Sarawak. *Southeast Asian Journal of Tropical Medicine and. Public Health,* **25(3),** 575–8.

Galloway, J.N., F. Dentener, D. Capone, E. Boyer, R. Howarth, et al., 2004: Nitrogen cycles: past, present and future. *Biogeochemistry,* **70(2),** 153–226.

Gillespie, S. and L. Haddad, 2002: *Food Security as a Response to AIDS.* International Food Policy Research Institute, Annual Report Essay, *in Aids and Food Security,* IFPRI, Washington DC, pp. 10–16.

Green, R.E., S.J. Cornell, J.P.W. Scharlemann, and A. Balmford, 2005: Farming and the fate of wild nature. *Science* **307,** 550–555.

Gu, L., 2003. Comment on "climate and management contributions to recent trends in US agricultural yields." *Science* **300,** 1505b.

Habicht, J.P., 1992: Discussion: mortality, malnutrition and synergies: determinants and interventions. In: *Child health priorities for the 90's.* K. Hill for The Johns Hopkins University, Baltimore, MD, pp. 215–220.

Hazell, P.B.R. and C. Ramasamy (eds.), 1991: *The Green Revolution Reconsidered: The Impact of High-yielding Rice Varieties in South India.* Johns Hopkins University Press, Baltimore, MD, 286 pp.

Hershman, J.M., G.A. Melnick, and R. Kastner, 1986: Economic consequences of endemic goiter. In: *Towards the eradication of endemic goiter, cretinism and iodine deficiency.* J.T. Dunn, E.A. Pretell, C.H. Daza, and F.E. Viteri (eds.), Pan-American Health Organization, Scientific Publication No. 502, Washington, DC, pp. 96–106.

Horton, S., 1999: Opportunities for investment in nutrition in low-income Asia. *Asian Development Review,* **17 (1,2),** 246–273.

Howarth R.W., E. Boyer, W. Pabich, and J.N. Galloway, 2002: Nitrogen use in the United States from 1961–2000 and potential future trends. *Ambio* **31,** 88–96.

Hsairi, M., F. Ben Slama, C. Ben Rayana, R. Fakhfakh, B. Romdhane, et al., 2003: Prevalence of endemic goiter in the north western region of Tunisia. *Tunis Medicine,* **72(12),** 663–669.

IFPRI (International Food Policy Research Institute), 2004: *Ready for Action in Africa? 2020 Vision Initiative.* IFPRI Forum, Special Issue. May. Washington, DC, 8 pp.

IFPRI/CIAT (International Center for Tropical Agriculture), 2002: *Biofortification: Harnessing Agricultural Technology to Improve the Health of the Poor.* Future Harvest, Washington, DC, 4 pp.

Jackson, B.C.J., M.X. Kirby, W.H. Berger, K.A. Bjorndal, L.W. Botsford, et al., 2001: Historical overfishing and the recent collapse of coastal ecosystems. *Science* **293,** 692.

James, C., 2004: *Preview: Global Status of Commercialized Biotech/GM Crops: 2004.* ISAAA Briefs. No. 32. ISAAA, Ithaca, NY. 12 pp.

Janick, J. and J.E. Simon (eds.), 1993: *New Crops: Exploration, Research and Commercialization. American Society of Agronomy,* John Wiley & Sons, Inc., New York, 710 pp.

Johns, T., 2003. Plant biodiversity and malnutrition: simple solutions to complex problems. *African Journal of Food, Agriculture, Nutrition and Development.* **3**(1), 45–52.

Johns, T., 1990: *With Bitter Herbs They Shall Eat It: Chemical Ecology and The Origins of Human Diet and Medicine.* University of Arizona Press, Tucson, AZ, 356 pp.

Johns, T. and B.R. Sthapit, 2004: Biocultural diversity in the sustainability of developing country food systems. *Food and Nutrition Bulletin* **25,** 143–155.

Jones, G.V., M.A. White, and O.R. Cooper, 2004: Climate change and global wine quality. *Climatic Change* (in review).

Jones, P., P. Shears, D. Hillier, D. Comfort, and J. Lowel, 2003: Return to traditional values? a case study of slow food. British Food Journal, **105**(4–5), 297–304.

Kaimowitz, D., 1996: *Livestock and Deforestation in Central America in the 1980s and 1990s: A Policy Perspective.* Center for International Forestry Research CIFOR, Bogor, Indonesia, 95 pp.

Kaufmann, R. and H. Fitzhugh, 2004: Importance of livestock for the world's poor. In: *Perspectives in World Food and Agriculture.* C. Scanes and J. Miranowski (eds). Iowa State Press, 485 pp.

Kent, G., 1987: *Fish, Food and Hunger: The Potential of Fisheries for Alleviating Malnutrition.* Westview Press, London, 201 pp.

Koziell, I. and J. Saunders (eds.), 2001: *Living Off Biodiversity: Exploring Livelihoods and Biodiversity Issues in the Natural Resources Management,* International Institute for Environment and Development, London, 269 pp.

Kuhnlein, H.V., O. Receveur, and H.M. Chan, 2001, Traditional food systems research with Canadian indigenous peoples. *International Journal of Circumpolar Health* **60**(2), 112–112.

Lanata, C.F. and R.E. Black, 2001: Diarrheal diseases. In: *Nutrition and Health in Developing Countries.* R. Semba and M.W. Bloem (eds.) Humana Press Inc, Totowa, NJ, pp. 93–129.

Lobell, D. and G. Asner, 2003: Climate and management contributions to recent trends in US agricultural yields. *Science,* **299,** 1032.

Mahar, D. and R. Schneider, 1994. Incentives for tropical deforestation. some examples from Latin America. In: *The Causes of Tropical Deforestation.* K. Brown, D.W. Pearce, (eds.), UCL Press, London, pp.159–171.

Maroni, M. and A. Fait, 1993: Health—effects in man from long-term exposure to pesticides: a review of the 1975–1991 literature. *Toxicology,* **78,** 1–3.

Masters, W.A., 2005: Paying for prosperity: how and why to invest in agricultural R&D for development in Africa. *Journal of International Affairs* 58(2), 35–64.

McIntire, J., D. Bourzat, and P. Pingali, 1992: *Crop-livestock interactions in sub-Saharan Africa.* World Bank, Washington, DC, 246 pp.

Mead P.S., L. Slutsker, V. Dietz, L. F. McCaig, J.S. Bresee, et al., 1999: Food related illness and death in the United States. *Emerging Infectious Diseases,* **5**(5), 607–625.

Mellor, J.W. (ed.), 1995: *Agriculture on the Road to Industrialization.* Johns Hopkins University Press, Baltimore, MD, 358 pp.

Mendez, M.A., C.A. Monteiro, and B.M. Popkin, 2005: Overweight exceeds underweight among women in most developing countries. *American Journal of Clinical Nutrition,* **81,** 714 – 21.

Millennium Ecosystem Assessment, 2003: *Ecosystems and Human Well-being.* Island Press, Washington DC, 245 pp.

Mirza, M.N., L.E. Caulfield, R.E. Black, and W.M. Macharia, 1997: Risk factors for diarrheal duration. *American Journal of Epidemiology,* **146**(9), 776.

Mori, H. and W.D.Gorman, 1999: A cohort analysis of Japanese food consumption–old and new generations. *Senshu Economics Bulletin,* 34(2), 71–111.

Myers, N.A., R.A. Mittermeier, C.G. Mittermeier, G.A.B. da Fonseca, and J. Kent, 2000: Biodiversity hotspots for conservation priorities. *Nature,* 403, 853–858.

Naylor, R.L., W. Falcon, R. Goodman, M. Jahn, T. Sengooba, H. Tefera, and R. J. Nelson, 2004: Biotechnology in the developing world: a case for increased investments in orphan crops. *Food Policy 29,* 15–44.

Naylor, R., W. Falcon, N. Wada and D. Rochberg, 2002: Using El Nino-Southern Oscillation climate data to improve food policy planning in Indonesia. *Bulletin Indonesian Economic Studies,* **38,** 75.

Naylor, R., R. Goldburg, J. Primavera, N. Kautsky, M. Beveridge, et al., 2000: Effect of aquaculture on world fish supplies. *Nature,* 405, 1017–1024.

Neumann, C.G. and D.M. Harris, 1999: *Contribution of Animal Source Foods in Improving Diet Quality for Children in the Developing World.* [online] Commissioned paper for the World Bank, Washington, DC, cited 18 February 2005. Available at http://glcrsp.ucdavis.edu/projects/project_subpages/CNP_folder/CNPWB.pdf.

OECD (Organisation for Economic Co-operation and Development), 2004: Producer and Consumer Support Estimates OECD Database, 1986–2003. User Guide and Data Files. Online at: www.oecd.org/document/58/0,2340,en_2649_201185_32264698_1_1_1_1,00.html.

OECD, 2003a: *Agricultural Policies in OECD Countries: Monitoring and Evaluation.* Paris, 296 pp.

OECD, 2003b: *OECD Agricultural Outlook 2003/2008.* Paris, 212 pp.

Ollila, E., n.d.: Kansainvälinen terveyspolitiikka.[online] Cited February 28 2005. Available at http://www.valt.helsinki.fi/staff/jkoponen/b1104/eeva ollila220304.pdf.

Pardey, P.G. and N.M. Beintema, 2001: *Slow Magic: Agricultural R&D a Century after Mendel.* International Food Policy Research Institute Food Policy Report. Washington DC, 30 pp.

Parry, M.L., C. Rosenzweig, A. Iglesias, M. Lovermore, and G. Fischher, 2004: Effects of climate change on global food production under SRES emissions and socio-economic scenarios. *Global Environmental Change,* **14,** 53–67.

Pauly, D., J. Alder, E. Bennett, V. Christensen, P. Tyedmers, and R. Watson, 2003: The future for fisheries. *Science* **302,** 1359–1361.

Pauly, D, V. Christensen, J. Dalsgaard, R. Froese, and F. Torres, 1998: Fishing down marine food webs. *Science,* **279,** 860–863.

Pimbert, M., 1999: *Agricultural Biodiversity. In Cultivating Our Futures.* FAO, Rome.

Popkin, B., 2000: Urbanization and the Nutrition Transition. Policy Brief 7 of 10. Focus 3. In: *Achieving Urban Food and Nutrition Security in the Developing World.* J. Garret and M. Ruel (eds.), International Food Policy Research Institute, Washington, DC, 20 pp.

Popkin, B., 1998: The nutrition transition and its health implications in lower income countries. *Public Health Nutrition,* **1,** 5–21.

Popkin, B.M., 1993: Nutritional patterns and transitions. *Population and Development Review* **19**(1), 138–157.

Popkin B.M., S. Horton, and S. Kim, 2001: The nutrition transition and prevention of diet-related diseases in Asia and the Pacific. *Food Nutrition Bulletin,* **22,** 51–58.

Pritchett, L. and L.H. Summers, 1993: Wealthier is healthier. *Journal of Human Resources,* **31**(4), 842–868.

Rahmathullah, L., J.M. Tielsch, R.D. Thulasiraj, J. Katz, C. Coles, et al., 2003: Impact of supplementing newborn infants with vitamin A on early infant mortality: community based randomized trial in Southern India. *British Medical Journal,* **327**(7409), 254.

Ramankutty, N. and J. A. Foley, 1998: Characterizing patterns of global land use: an analysis of global croplands data. *Global Biogeochemical Cycles,* **12**(4), 667–685.

Receveur O., M. Boulay, and H.V. Kuhnlein, 1997: Decreasing traditional food use affects diet quality for adult Dene/Metis in 16 communities of the Canadian Northwest Territories. *Journal of Nutrition 12*(11), 2179–2186.

Regmi, A. and J. Dyck, 2001: Effects of urbanization on global food demand. In: *Changing Structure of Global Food Production and Trade,* A. Regmi (ed.), Economic Research Service, U.S. Department of Agriculture, Washington, DC, pp. 23–30.

Reutlinger, S. and J. van Holst Pellekan, 1986: *Poverty and Hunger: Issues and Options for Food Security in Developing Countries.* World Bank, Washington, DC, 82 pp.

Rguibi M. and R. Belahsen, 2004: Metabolic syndrome among Moroccan Sahraoui adult women. *American Journal of Human Biology,* **16,** 598–601.

Rice, A.L., K.P. West, and R.E. Black, 2003: Vitamin A deficiency. In: *Comparative Quantification of Health Risks: Global and Regional Burden of Disease Attributable to Selected Major Risk Factors.* M. Ezzati et al. (eds.), WHO, Geneva, 2248 pp.

Roe, T. and M. Gopinath, 1998: The ''miracle'' of U.S. agriculture. *Minnesota Agricultural Economist,* **69,** 1–4.

Rosenzweig, C. and M. Parry. 1996. Potential impact of climate change on world food supply. *Nature,* **367,** 133–138.

Runge, C.F., B. Senaur, P. Pardey, and M. W. Rosegrant, 2003: *Ending Hunger in Our Lifetime: Food Security and Globalization.* Johns Hopkins University Press, Baltimore, MD, 204 pp.

Ruttan, V., 2002: *Technology, Growth and Development: An Induced Innovation Perspective.* New York, Oxford University Press, 672 pp.

Satheesh, S.V., 2000: *Uncultivated Food and the Poor.* Proceedings of South Asian Workshop on Uncultivated Food and Plants. Policy Research for Development Alternative, Shaymoli, Dhaka, Bangladesh, pp.78–92.

Schelp, F.P., 1998: Nutrition and infection in tropical countries—implications for public health intervention: a personal perspective. *Nutrition,* **14(2),** 217–222.

Science Daily, 2003: Climate change in the vineyards: a taste of global warming [online]. Cited April 21, 2005. Available from Science Daily at http://www.sciencedaily.com/releases/2003/11/031104064327.htm.

Scoones, I., M. Melnyck, and J. Pretty (eds.), 1992: *The Hidden Harvest—A Literature Review and Annotated Bibliography.* International Institute for Environment and Development, London, 260 pp.

Shand, H., 1997: *Human Nature: Agricultural Biodiversity and Farm-based Food Security.* Rural Advancement Foundation International, Ottawa, 110 pp.

Simmonds, N.W., 1993: Introgression and incorporation. strategies for the use of crop genetic resources. *Biological Reviews,* **68,** 539–562.

Sinclair, T.R., L.C. Purcell, and C.H. Sneller, 2004: Crop transformation and the challenge to increase yield potential. *Trends in Plant Science,* **9,** 70–75.

Smil, V., 2000: *Feeding the World – A Challenge for the 21st Century.* The MIT Press, Cambridge, MA, 360 pp.

Smith, B.D., 1995 *The Emergence of Agriculture.* Scientific American Library, New York, 231 pp.

Smith, L. and L. Haddad, 2000: *Overcoming Child Malnutrition in Developing Countries: Past Achievements and Future Choices.* Food Agriculture and the Environment Discussion Paper 30. International Food Policy Research Institute, Washington, DC, 66 pp.

Sombroek, W.G. and R. Gommes, 1996: The climate change-agriculture conundrum. In *Global Climate Change and Agricultural Production.* F. Bazzaz and W. Sombroek, eds. West Sussex, Wiley, 345 pp.

Strauss, J., 1986: Does better nutrition raise farm productivity? *Journal of Political Economy,* 94(2), 297–320.

Thompson, L.M., 1998: Effects of changes in climate and weather variability on the yield of corn and soybean. *Agronomy Journal/Production Agriculture,* **1,** pp. 20–27.

Tomek, W.G. and K.L. Robinson, 1981: *Agricultural Product Prices.* Cornell University Press, Ithaca, NY, 367 pp.

Trichopoulou, A. and E. Vaasilopoulou, 2000: Mediterranean diet and longevity. *British Journal of Nutrition,* **84,** S209.

UN (United Nations), 2004: *World Urbanization Prospects: The 2003 Revision.* New York, 335 pp.

UN, 2001: *World Urbanization Prospects: The 2001 Revision.* New York, 190 pp.

UN, 1998: *World Urbanization Prospects: The 1998 Revision.* New York.

UNDP (United Nations Development Programme), 2003: *Human Development Report 2003.* Oxford University Press, New York, 384 pp.

UNICEF (United Nations Children's Fund), 1998: *State of the World's Children.* Oxford University Press, New York, 131 pp.

Unnevehr, L.J. (ed.), 2003: *Food Safety in Food Security and Food Trade.* 2020 Focus No. 10. International Food Policy Research Institute, Washington, DC, 38 pp.

UN/SCN (United Nations Standing Committee on Nutrition), 2004: *Nutrition for Improved Development Outcomes. 5th Report on The World Nutrition Situation.* WHO, Geneva, 143 pp.

Uusitalo, U., P. Pietinen, and P. Puska, 2002: Dietary transition in developing countries: challenges for chronic disease prevention. In: *Globalization, Diets and Non-communicable Diseases.* WHO, Geneva, pp.1–25.

Verner, D., 2004: *Education and Its Poverty-reducing Effects: The Case of Paraíba, Brazil,* Policy Research Working Paper 3321. World Bank, Washington DC.

Vosti, S., J. Witcover, and C. L. Carpentier, 2002: *Agricultural Intensification by Smallholders in the Western Brazilian Amazon.* International Food Policy Research Institute, Research Report 130, Washington, DC, 147 pp.

Wang, G. and L. You, 2004: Delayed impact of the North Atlantic Oscillation on biosphere productivity in Asia. *Geophysical Research Letters,* **31** (12), L12210, doi:10.1029/2004GL019766.

Wang, X. and K. Taniguchi, 2003: Does better nutrition enhance economic growth? the economic cost of hunger. In: *Nutrition Intake and Economic Growth. Studies on the Cost of Hunger.* K. Taniguchi and X. Wang (eds.), FAO, Rome, 126 pp.

Watkins, K. and J. von Braun, 2003: *Time to Stop Dumping On The World's Poor.* 2002–2003 International Food Policy Research Institute, Annual Report Essay. Washington, DC, 18 pp.

WHO (World Health Organization), 2005: *Modern Food Biotechnology, Human Health and Development: An Evidence-based Study.* Geneva. 84 pp.

WHO, 2004a: 20 questions on genetically modified foods [online]. WHO Geneva, cited February 28, 2005. Available at http://www.who.int/foodsafety/publications/biotech/20questions/en/.

WHO, 2004b: Micronutrient deficiencies [online]. WHO, Geneva, cited February 28, 2005. Available at http://www.who.int/nut/#mic.

WHO, 2004c: Chronic disease information sheets. Global strategy diet and physical activity [online]. WHO, Geneva, cited February 28, 2005. Available at http://www.who.int/dietphysicalactivity/publications/facts/en/.

WHO, 2004d: Global database on bodymass index [online]. WHO, Geneva, cited February 28, 2005. Available at http://www.who.int/nut/db_bmi.htm.

WHO, 2003a: Obesity and overweight. Global strategy on diet, physical activity and health [online] WHO, Geneva, cited February 28, 2005. Available at http://www.who.int/dietphysicalactivity/publications/facts/obesity/en/.

WHO, 2003b: *Diet, Nutrition and the Prevention of Chronic Diseases.* WHO Technical Report Series 916. Geneva, 160 pp.

WHO, 2002a: *Globalization, Diets and Non-communicable Diseases.* WHO, Geneva, 185 pp.

WHO, 2002b: Food safety and food borne illness [online] WHO, Geneva, cited February 28, 2005. Available at http://www.who.int/mediacentre/fact sheets/fs237/en/.

WHO, 2000: Collaborative study team on the role of breastfeeding on the prevention of infant mortality. Effect of breastfeeding on infant and child mortality due to infectious diseases in less developed countries: a pooled analysis. *The Lancet,* **355,** 451–455.

Wood, S., K. Sebastian and S.J. Scherr, 2000: *Pilot Analysis of Global Ecosystems: Agroecosystems.* A joint study by International Food Policy Research Institute and World Resources Institute. Washington, DC, 110 pp.

World Bank, 2004: *World Development Indicators* [CD-ROM] World Bank, Washington, DC.

World Bank, 2003: *World Development Indicators* [CD-ROM]. World Bank, Washington, DC.

Yusuf H.K., S. Quazi, M.N. Islam, T. Hoque, K.M. Rahman, et al., 1994: Current status of iodine-deficiency disorders in Bangladesh. *The Lancet,* **343(8909),** 1367–1368.

Chapter 9

Timber, Fuel, and Fiber

Coordinating Lead Author: R. Neil Sampson
Lead Authors: Nadia Bystriakova, Sandra Brown, Patrick Gonzalez, Lloyd C. Irland, Pekka Kauppi, Roger
   Sedjo, Ian D. Thompson
Contributing Authors: Charles V. Barber, Roland Offrell
Review Editors: Marian de los Angeles, Cherla Sastry

Main Messages

Global timber harvest has increased by 60% in the last four decades and will continue to grow in the near future, but at a slower rate. The growth rate of timber harvest has slowed in recent years and is likely to continue to grow more slowly in the foreseeable future. (In this chapter, "timber" is used to denote standing trees and their immediate products). The portion of harvested wood that was used for pulp increased threefold since 1961, reflecting a major shift in demand for timber products as pulpwood demand greatly outpaced the increased demand for sawn wood products.

Timber supplies for common industrial wood products appear to be ample for the near future, but there will be shortages of high-value species and premium quality woods due to past overharvesting. Timber production from both forest and agricultural plantations will increase over the near future, providing wood for pulp and common sawtimber. Premium woods from large and old trees of highly valuable species are scarce in most regions. They may be restored through protection and sustainable forest management in the longer term but will remain in short supply for the foreseeable future.

Plantations are providing an increasing proportion of timber products. In 2000, plantations were 5% of the global forest cover, but they provided some 35% of harvested roundwood, an amount anticipated to increase to 44% by 2020. The most rapid expansion will occur in the mid-latitudes, where yields are higher and production costs lower. Gains in production will also come from insect and disease-resistant trees, genetically improved trees with higher yields and improved fiber characteristics, improved planting techniques, and improved management. The net effect will increase the amount of products available to satisfy timber, fuel, and fiber needs, but with a reduced harvest from natural forests in most regions.

Major shifts in the location of timber production have resulted from a combination of globalization, economic stress, and changing national policies, and the future for sustainable timber supplies varies across areas and regions. In general, recent production shifts have been from north to south. In the northern boreal and temperate regions, forest-growing stocks have increased in the recent past, as overall annual growth has exceeded mortality and harvest. This trend will continue into the foreseeable future (*medium certainty*). In the low latitudes, deforestation and degradation continue to diminish natural forests. Contributing to this are examples of destructive, exploitive, and illegal logging practices. These combine with a complex of other drivers, including fuelwood extraction, agricultural expansion, development policies, and population pressures to contribute to forest degradation. The associated production losses are partially offset by plantation expansion, but premium species are seldom replaced, either in regrowing forests or in plantations.

International trade in forest products has increased at a rate much faster than the increase in production. The global value of timber harvested in 2000 was around $400 billion, and around one quarter of that entered into world trade, representing some 3% of total merchandise trade. In constant-dollar terms, global exports increased almost twenty-five-fold between 1961 and 2000. Five countries—the United States, Germany, Japan, the United Kingdom, and Italy—imported more than 50% of world imports in 2000, while Canada, the United States, Sweden, Finland, and Germany accounted for more than half of exports. During the past decade, China has increased its imports of logs and wood products by more than 50% and, if unabated, this rate of increase will put significant pressure on wood supplies in many regions, particularly Russia and Southeast Asia.

International moves toward sustainable forest management and forest certification have expanded rapidly in recent years. To date they have been used primarily in industrial countries, and only locally in developing countries, and they do not seem to be affecting timber production or trade significantly. They are, however, affecting forest management where certification is involved. Their broad future effectiveness remains uncertain, but they could become more important in some regions, such as Europe.

The global forestry sector annually provides subsistence and wage employment of 60 million work years, with 80% taking place in the developing world. While timber is processed mainly by industrial firms, the management of forests, including harvesting and transport, is dominated by individuals, families, and small companies. In the industrial world, the forestland is both privately (in Europe and the United States, for example) or publicly (as in Canada) owned and managed. In the developing world, most of the forest is a public resource. These ownership factors create very different and diverse opportunities for people to interact with and benefit from forest products and services. As production moves South in the near future, the trend indicates increasing forest employment opportunities in sub-tropical and tropical regions, with declining employment in temperate and boreal regions. Labor requirements per unit of output in all regions will continue to shrink due to technological change.

Illegal logging is a significant factor that skews timber markets and trade, particularly in tropical forest regions and countries with weak or transitional governments. It is estimated that up to 15% of global timber trade involves illegal activities, and the annual economic toll is around $10 billion. Addressing this problem will require a major effort by both governments and private industry.

On a global basis, the recorded value of fuelwood production per capita has fallen in recent years, and earlier concerns about a "wood energy crisis" have eased. Fuelwood is the primary source of energy for heating and cooking for some 2.6 billion people, and 55% of global wood consumption is for fuelwood. While populations in developing regions will continue to grow, estimates from FAO indicate that global consumption of fuelwood appears to have peaked in the 1990s and is now believed to be slowly declining. Accurate data on fuelwood production and consumption are difficult to collect, since much of it is produced and consumed by households in developing countries. It is also difficult to relate trends to ecosystem condition, since fuel is harvested from woody plants wherever they occur. Localized fuelwood shortages in Africa impose burdens on people who depend on it for home heating and cooking. The impact on people may be high prices in urban areas or lengthy and arduous travel to collect wood in rural areas.

The burning of wood fuel without appropriate smoke venting creates significant health hazards, and where wood shortages force poor families to shift to dung or agricultural residues for heating and cooking, the problems are exacerbated. An estimated 1.6 million deaths and 39 million disability-adjusted life years are attributed to indoor smoke pollution, with women and children most affected. Where adequate fuel is not available, the consumption of cooked food may decline, leading to adverse effects on nutrition and health. Human well-being can be enhanced in many developing regions through efforts to assure adequate and accessible fuel supplies, as well as to promote more-efficient stoves and more-effective smoke venting.

Wood and forest biomass, agricultural crops and residues, manure, municipal and industrial wastes, and many other non-fossil organic materials can produce renewable energy and fuel supplies through a variety of modern industrial processes. These renewable energy technologies are being rapidly developed throughout the world, but examples of full commercial exploitation are still fairly modest. Competition from low-cost and widely available fossil-based fuels currently limits the expansion of successful research and pilot projects into widespread commercial production. Biomass-based energy production, while likely to expand slowly in the near future, offers great promise for the mid- and long-term future, as the nonrenewable nature of fossil fuels, particularly natural gas and petroleum crude oil, begins to affect energy economics because of shortages and supply disruptions.

Global cotton production has doubled and silk production has tripled since 1961, accompanied by major regional shifts in the production areas. Production of other agricultural fibers such as wool, flax, hemp, jute, and sisal has declined. While cotton production has doubled, the land area on which it is harvested has stayed virtually the same. Major area expansion of cotton production in India and the United States has been offset by large declines in Pakistan and the former Soviet Union. These shifts have important impacts on limited water resources, since much of the cotton crop is irrigated, and on agricultural land use patterns, as cotton competes with food crops for arable land. There have also been significant impacts of the use of fertilizer and pesticides for the increased production of cotton. Silk production also experienced a major shift in production area, from Japan to China, due to lower labor costs.

There are still instances where species are threatened with extinction due to the trade in hides, fur, or wool, in spite of international efforts to halt poaching and trade. On the other hand, there are instances, such as for crocodiles, where international conservation efforts have restored species and established sustainable production of valuable commodities.

9.1 Introduction

This chapter on timber, fuel, and fiber covers a wide range of ecosystem services provided by an equally wide range of ecosystems. The major focus is on timber that is linked most closely with forest ecosystems and forest-related communities and industries, but it extends to fuel and fibers as well. Information about recent trends in supply and consumption of those products is presented and assessed in terms of the impact of that consumption on ecosystems and human well-being, and efforts are made to identify and assess the important drivers of change in each resource area. Trends in the condition of the ecosystems from which these services are derived are found in other chapters of this volume (particularly Chapters 21, 22, and 26).

Forests cover about 30% of the ice-free land area of Earth, but global forest cover has declined considerably through history (FAO 2001a; Williams 2003). (See also Chapter 21.) While there are copious statistics for the timber and wood products that enter into world trade, assessing the connection between production and forest ecosystems is complicated by the fact that the production statistics are reported by country rather than by biome or ecosystem.

The most complete and comparable statistics are maintained by the U.N. Food and Agriculture Organization, and its publications and databases are widely cited in this chapter. It should be noted that all statistics on wood production, whether for fuel or commercial purposes, have some limitations. Regional data are reported by FAO, but the regions are geographic or geopolitical, making their direct use in reflecting effects on forest ecosystems of limited value. For this assessment, FAO country data have been regrouped to approximate the dominant forest systems affected by the production trends reported. Table 9.1 lists the countries that were grouped into the seven major forested regions chosen for

Table 9.1. Regions Used in the Wood Products Analyses. The regions group countries with areas of closed forest by continent and climate type. (FAO 2003b)

Africa Tropical	Asia Temperate	Europe Temperate
Angola	Bhutan	Albania
Benin	China (PR)	Austria
Cameroon	Japan	Belarus
Central African Republic	Korea (DPR)	Belgium
Congo (Dem. Rep.)	Korea (Rep. of)	Bosnia and Herzegovina
Congo (Rep. of)	Mongolia	Bulgaria
Côte d'Ivoire	Nepal	Croatia
Equatorial Guinea	Russian Federation	Czech Republic
Ethiopia	Turkey	Denmark
Gabon		Estonia
Ghana	**Asia Tropical**	Finland
Guinea	Bangladesh	France
Guinea-Bissau	Brunei Darussalam	Germany
Liberia	Cambodia	Greece
Madagascar	India	Hungary
Mali	Indonesia	Ireland
Mozambique	Laos	Italy
Nigeria	Malaysia	Latvia
Senegal	Myanmar (Burma)	Liechtenstein
Sierra Leone	Papua New Guinea	Lithuania
Togo	Philippines	Luxembourg
Zambia	Sri Lanka	Macedonia (FYR)
	Thailand	Moldova
America Temperate	Viet Nam	Netherlands
Canada		Norway
United States		Poland
		Portugal
		Romania
America Tropical		Serbia and Montenegro
Belize		Slovakia
Bolivia		Slovenia
Brazil		Spain
Colombia		Sweden
Costa Rica		Switzerland
Dominican Republic		Ukraine
Ecuador		United Kingdom
El Salvador		
French Guiana		
Guatemala		**Southern Temperate**
Guyana		Argentina
Haiti		Australia
Honduras		Chile
Mexico		New Zealand
Nicaragua		South Africa
Panama		
Paraguay		
Peru		
Suriname		
Venezuela		

this assessment. Figure 9.1 (in Appendix A) indicates where these seven regions are.

In addition to timber, fuels derived from biomass are also considered in this chapter. More than 2 billion people worldwide rely on biomass for their main energy source (IEA 2003). Although wood and charcoal are the primary energy sources in many societies, particularly developing ones, the data for assessing the impacts of this use and consumption on ecosystem sustainability are largely unavailable. This is due to several factors, including the difficulty of monitoring the supply and consumption of wood fuels accurately, as much of it is done outside the market economy.

The last section of the chapter discusses some of the general trends in the production and consumption of important fibers derived from forest or agricultural crops, domestic animals, wildlife, and other sources. These products, each of which could be the subject of a major assessment, are treated sparingly due to limitations of space and time.

The ownership of forests varies on the basis of legal and cultural traditions and history. In some nations, central governments claim vast areas of forest but do little to enforce those claims. Poor administration and corruption allow illegal logging and other forms of resource extraction, including within protected forests. In other areas, highly developed property rights systems exist, and resource extraction is often tightly regulated as a result. More wood per hectare is generally harvested from private forests than public ones, because of more-intensive management and more-focused forest production objectives. While some private forests produce multiple benefits, many are managed mainly for timber. (See Table 9.2.)

For forests in protected areas, timber harvest may or may not be one of the management objectives. Even where forests are protected from harvesting in some protected areas, resource extraction (for commercial or local use) continues unabated owing largely to poor governance and corruption (WWF 2000; Khan et al. 1997; Smith et al. 2003; and see MA *Policy Responses*, Chapter 5).

In both temperate and tropical forests, many local communities have decision-making rights for the management of public forestlands. Such community forest management systems include joint forest management in India (TERI 2000) and nascent concession systems in Bolivia and Peru. In order to foster long-term maintenance of forest resources, communal forest management arrangements generally allocate all or part of the forest production

revenues to the local community. While the amount of forest products produced under such arrangements remains negligible as a fraction of total global output, community management can produce significant positive local economic impacts.

9.2 Timber and Related Products

While forests produce a wide range of services that are essential to human well-being, their major financial output consists of timber that can be used for a variety of manufacturing, building, fuel, and other materials. Timber is harvested from forest ecosystems (both intensively and extensively managed), forest and agricultural plantations, and trees outside forests. It should be noted that there is not always a clear distinction in the data and discussion of these issues between "planted forests" and "plantations." (See Chapter 21.)

9.2.1 Industrial Roundwood

According to FAO, industrial roundwood includes wood for the following commodities: sawlogs, veneer logs, pulpwood, other industrial roundwood (used for tanning, distillation, match blocks, poles, piling, posts, pitprops, and so on) and, in the case of trade, also chips and particles and wood residues. Although the title "industrial roundwood" suggests that this is wood harvested by industrial companies, the majority of the timber harvested in the world is still harvested by individuals, families, and small operations using a variety of traditional and modern methods. This use of the term "industrial" is better understood to mean wood that is produced for sale into commercial channels for processing or end use.

9.2.1.1 Production

Production of industrial roundwood increased between 1961 and 1980 in nearly all regions, but the increase, which was slowing gradually before 1980, slowed considerably between 1980 and 2000. (See Table 9.3.) Some of this decline in the rate of increase reflects the demise of the Soviet Union and the associated dramatic reduction in timber production. Russian and other former Soviet Union timber production, in excess of 300 million cubic meters a year in Soviet times, declined to one fourth of previous levels.

As the former centrally planned economy moved toward greater use of markets, the subsidies received by the transport sector were largely eliminated and with it, the subsidy that allowed

Table 9.2. Some Key Differences Affecting Forest Product Derivation from Public and Private Forests

Category	Public Forest	Private Forest
Management regime	extensive	intensive
Protected areas	common	uncommon
Land use	multiple use	dedicated use
Management costs (dollar per hectare)	low	high
Illegal logging	common	rare to nil
Planted forests	uncommon	common
Species planted	mostly endemic	mostly endemic
Plantation forests	uncommon	common
Species planted	mostly endemic	mostly exotic
Production per hectare	variable	generally high

Table 9.3. Industrial Roundwood Production, 1961–2000 (FAO 2003c)

Region	1961	1980	2000	1961–80	1980–2000
	(million cubic meters)			(percent change)	
Africa tropical	15	28	34	83	20
America temperate	335	478	625	43	31
America tropical	28	83	131	198	57
Asia temperate	282	336	239	19	−29
Asia tropical	34	101	86	199	−14
Europe temperate	286	352	377	23	7
Southern temperate	25	50	92	97	86
Other areas	13	19	25	50	36
Total	**1,018**	**1,446**	**1,610**	**42**	**11**

timber transport from the forests to distant processing facilities. In addition, the large declines experienced in Russian Federation GDP further depressed timber demand and production. A growing Russian economy together with new export markets, such as China, should revive some of the demand for Russian Federation wood products, and indeed timber production increased in the late 1990s. The transport subsidy is unlikely to return, however, and Russian timber production in the next decade will likely be substantially lower than in Soviet times.

9.2.1.1.1 *Illegal logging*

It should be noted that the FAO data (and Table 9.3) consist primarily of legally produced wood as reported to national governments, and so in regions where illegal logging is significant, the data may underreport the actual volume of timber harvested. It is difficult to estimate the magnitude of illegal removals, and quantitative estimates vary. But the amount of timber involved is both locally and globally significant. Contreras-Hermosilla (2002) estimates that as much as 15% of global timber trade involves illegalities and corrupt practices, amounting to annual losses in assets and revenues in excess of $10 billion. Curry et al. (2001) conclude that "illegal logging is rife in all the major tropical timber producing countries," citing studies showing that 70% of log production in Indonesia (50 million cubic meters per year) is derived from illegal sources; 80% of logging in the Brazilian Amazon during 1998 was illegal, and half of all timber in Cameroon is sourced through illegal logging (WRI 2000b).

Illegal logging takes many forms, including logging timber species protected by national or international law; logging in protected or prohibited areas; logging without authorization; extracting more timber than authorized; timber theft and smuggling; and fraudulent transfer pricing and other corrupt accounting practices. Its negative impacts include the impoverishment of forest landscapes, governments and local communities that are deprived of significant forest revenues, the strengthening of organized criminal enterprises, and inducements for the corruption of law enforcement and other officials (Contreras-Hermosilla 2002).

National and local governments, being unable to appreciably reduce illegal logging, often play down its extent. In Russia, for example, according to government figures illegal logging over the past decade has amounted to 1% of legally permitted logging. Independent NGO and scientific assessments, however, estimate illegal logging to constitute some 20% of the total, with illegal exports of particularly valuable species in border areas being double the legally permitted volumes (Sheingauz 2001). In other areas too, weak or poorly enforced legislation has been ineffective in protecting forests from illegal logging (for example, in the Congo basin) (Global Witness 2002).

9.2.1.1.2 *Forest plantations*

All the following statistics refer to forest plantations as opposed to planted forests, with the distinction being, for the latter, the application of planting as one of several normal silvicultural techniques used to reforest harvested areas. Plantations, in contrast, are intensively managed, regularly spaced stands, often monocultures, often of exotic species, and usually with the production of wood as the main product (FAO 2000c). Forest plantations covered 187 million hectares in 2000, with Asia accounting for 62% (FAO 2001a). This total represented a tenfold increase (from 17.9 million hectares) since 1980 (FAO 2001a). In 2000, the annual rate of new planting was estimated at 4.5 million hectares, of which 79% was in Asia, and much of this in China and India. (See Chapter 21 for a more complete description of trends in plantation forests.)

Yields from tropical plantations are high—often in the range of 10–30 cubic meters per hectare per year for *Eucalyptus* and *Pinus,* with some species on favorable soils reaching yields as high as 50–60 cubic meters per hectare. Given the amount of research that has gone into improving yield from planted stocks, these yields are likely to continue to increase. Because of high yields and increasing area, plantations provide a continuously increasing portion of the world's timber supply. According to FAO (2001a), plantations were only about 5% of global forest cover in 2000, yet provided some 35% of global roundwood, an amount anticipated to increase to 44% by 2020.

Plantation growth rates and wood quality are likely to improve with each successive rotation, as trees embodying new genetic improvements are increasingly being used to replace the old technology embodied in the prior harvest. Thus far the genetic improvements have come primarily through conventional breeding techniques. However, there are substantial efforts under way to apply modern biotechnology, including genetic engineering, to industrial trees and plantation forests (Sedjo 2004). The degree of acceptance of genetically engineered (transgenic) trees remains uncertain. But based on experience with genetically engineered crops, there is reason to believe that the adoption of transgenic technology for trees will be variable.

9.2.1.1.3 *Agricultural plantations*

Plantations of agricultural and industrial crops such as rubber, coconut, and oil palm are increasingly important sources of industrial wood for the forest industries of Asia. It is estimated that there are some 27.4 million hectares of these crops in the region. Changing technology and markets are resulting in the conversion of formerly wasted by-product logs to products such as plywood, particleboard, paper, and lumber for furniture and other uses (Durst et al. 2004).

Rubberwood (*Hevea brasiliensis*) is estimated to cover some 9 million hectares, primarily in Indonesia, Thailand, and Malaysia. Planted originally for the production of latex, the trees mature at about age 30 and are then replaced. Prior to the emergence of milling technology in the late 1970s, the wood was either used for energy or disposed of in open burning. Today, it is estimated that over 6.5 million cubic meters are processed into sawn lumber, largely for furniture (Durst et al. 2004). Rubberwood is also used for manufacturing a variety of wood panels. This production, which eases pressure on many formerly used native species, is currently the basis for exports of almost $1.5 billion annually from Malaysia and Thailand (Durst et al. 2004).

Coconut palm (*Cocos nucifera*) is another agricultural crop with emerging value for the forest products industry. Its primary use is for coconut oil, but after age 60, when yields begin to decline, the palms are generally felled and replaced. Removing the downed stems from the plantation is important for controlling disease, and selling them for lumber production increases plantation revenues.

It is estimated that, throughout Asia, palm wood production rates of around 5 million cubic meters per year will be available for at least 20–30 years. After that time, the existing plantations will be replaced by higher-yielding dwarf varieties that produce less valuable lumber, but which may be useful for products such as panels (Durst et al. 2004).

Oil palm (*Elaeis guineensis*) plantations cover an estimated 6.5 million hectares in Asia and are grown on a 25- to 30-year rotation (Durst et al. 2004). While this is a huge potential in terms of biomass, research and development are needed to overcome limitations in the use of this species for wood products. That research is under way, and if the experience in developing and using other formerly "unusable" species is any guide, increasing industrial use is likely in the future.

9.2.1.2 Consumption

Between 1961 and 1991, the market value of global wood consumption more than doubled in real terms, growing at 2.7% per year (FAO 1999). In terms of processed wood products, sawnwood increased by 20%, wood panels by 600%, and paper by 350% between 1961 and 1991. These increases were achieved, in large part, through improved production efficiencies, recovery of wood residues for use in wood panels, and paper recycling (FAO 1999).

An increasing supply of wood products has begun to flow into China as that economy continues to develop rapidly. China is now importing 107 million cubic meters of wood, which represents an increase of more than 50% during the period 1997–2003 (Sun et al. 2004). The demand for imported wood has developed in part as a result of policies limiting domestic harvest levels. Imports are made up of raw pulp and paper largely coming from Canada and Russia and sawn lumber and logs from Russia, Malaysia, and Indonesia (Sun et al. 2004). Imports from Russia have resulted in continued forest loss in there through rapid logging of mature forests. (See Chapter 21.) Demand from China is expected to continue to grow, resulting in increased social, environmental, and economic effects, particularly in Russia and Southeast Asia.

The substitution of a variety of other materials for wood has contributed to relatively slow growth in global timber consumption in recent years. Steel studs (with up to one-third recycled content) are replacing wood in some markets of industrial countries, and vinyl has largely replaced wood as a siding material in North America. Plastics have replaced wood in some specialty applications—for example, to a large extent in the European window frame market. The use of metal roofing is growing rapidly in many parts of the developing world and, along with other products, it has largely replaced wood as a roofing material.

As premium quality and specialty woods have become scarce due to high-grading and other forms of unsustainable harvesting, there has been considerable replacement of traditional wood products by new products. Solid hardwood furniture has been replaced in many situations by a thin veneer of hardwood glued to a manufactured panel. Solid boards for sheathing buildings have been replaced, first by plywood and then by composite oriented strandboard. Solid wood joists for flooring systems are being replaced by engineered I-joists fabricated from smaller pieces of wood from widely available second-growth trees that often do not produce the large and high-strength pieces needed for solid wood structural timbers. These I-joists offer a flatter floor, greater strength, and longer spans. These manufactured wood products are so far largely confined to markets in industrial countries.

In many parts of the world, newly engineered forms of wood flooring, often with sturdy artificial overlays, are gaining wide acceptance. Other new products include engineered panels for nonstructural uses such as furniture, flooring, and moldings. These are often constructed using sawdust, shavings, and other residuals from sawmills or plywood mills. Viet Nam has a plan to build as many as nine particleboard plants to support its growing furniture industry. Non-wood fibers are increasingly being tapped for these products; examples include panels from wheat straw in the North American wheat belt, and medium density fiberboard produced from oil palm stalks in Malaysia and Indonesia. Like

many new technologies, the agri-fiber panels have had difficulties with some early testing. However, the technology seems well adapted for fiber-short nations with growing economies, such as China.

In the paper industry, plastics and other materials have replaced some paper bags, packing papers, and paperboard. Increasing use of recycled fiber has caused roundwood use for paper in the United States to decline significantly since the mid-1990s. Markets for many paper products in North America are saturated, and low economic growth in areas like Europe and Japan, driven by sluggish economic growth and declining birth rates, contributes to the demand weakness.

9.2.1.3 Trade

While FAO (2003c) indicates that total roundwood harvest across the world has increased some 60% since 1961, the constant-dollar value of forest products exported has increased almost twenty-five-fold. (See Table 9.4.) Much of this is due to the increasing proportion of finished and semi-finished products in trade, as opposed to unprocessed logs or chips.

About one quarter of global timber production enters into international trade, according to FAO, and paper alone makes up about 2% of all global trade (IIED 1996). With a total value of around $100 billion in 1991 (FAO 1995), wood products constituted about 3% of total world trade. This trade is growing rapidly, particularly in industrial countries. By 1996–98, import value was estimated at almost $43 billion and export value was $135 billion (WRI 2000a). Five countries—the United States, Germany, Japan, the United Kingdom, and Italy—accounted for more than 50% of international imports in 2000, while Canada, the United States, Sweden, Finland, and Germany accounted for more than half of international exports (FAO 1999).

Much of the total global trade is within regions. Flows between the United States and Canada, between northern and central Europe, and within Southeast Asia dominate world trade (TMFWI 2001). Where these trade patterns have resulted in shifting timber production to more productive regions (as described later), total timber production has increased more rapidly than the area of forest harvested.

9.2.1.4 Employment

In recent decades there has been a decline in the labor inputs required by the forest products industry, where employment has experienced similar declines as many other extractive and manufacturing industries (U.S. Bureau of Labor Statistics 2004).

For example, from 1955 to the mid-1990s, labor productivity in logging in the United States increased an average of 1.45% annually (Perry 1999). Since output increased at about the same rate, however, employment levels remained relatively steady. Employment in forest product processing has declined in many countries in recent years due to increasing labor productivity. In the United States, for example, over the period 1997–2003 employment in the paper and paperboard production fell by one third while total production barely declined (U.S. Bureau of Labor Statistics 2004).

9.2.1.5 Future Availability

Estimates of future wood demand must be approached with some caution, as past assessments have regularly overestimated needs (Clawson 1979; Sedjo and Lyon 1990). More-recent estimates by FAO (1998b) suggest that the global wood demand by 2045/50 will be about 2 billion cubic meters a year and that supply should

Table 9.4. Forest Product Exports and Imports, 1961–2000 (Data from FAO 2003c)

Region	1961	1980	2000	1961–80	1980–2000
	(million 1996 dollars)			(percent change)	
Exports					
Africa tropical	0.2	1	2	590	50
America temperate	2.1	18	45	773	149
America tropical	0.1	1	4	1,214	239
Asia temperate	0.4	4	11	835	177
Asia tropical	0.2	4	10	2,145	113
Europe temperate	3.0	26	67	761	162
Southern temperate	0.1	1	5	2,039	278
Other areas	0.03	1	1	2,235	55
Total	6	57	145	839	156
Imports					
Africa tropical	–2	0.02	0.4	0.4	1,516
America temperate	258	1.6	8	30	426
America tropical	212	0.2	2	6	989
Asia temperate	0.4	13	33	3,094	152
Asia tropical	0.1	1	6	544	487
Europe temperate	3.7	33	68	776	108
Southern temperate	0.4	2	3	354	114
Other areas	0.4	4	7	1,069	63
Total	7	63	153		

be about the same, with Asia contributing about 700 million cubic meters and North America about 1 billion cubic meters.

Ultimately, the future availability of wood supplies is likely to be related more to growth and productivity of managed and planted forests than to the area of natural forests or gross forest stock. Production from natural forests will continue to be replaced by production from plantation forests, especially in tropical and sub-tropical forests (as described later). The same trend seems likely in many temperate forest areas. In the United States, for example, the harvest of softwood timber on non-plantation native forests is forecast to decline by 25% over the next 50 years, and standing stocks on non-plantation forests are expected to increase significantly (USDA Forest Service 2003).

Although the amount of forest cover is changing differently by region, other factors also contribute to timber supply and availability. (See Chapter 21.) The forest available for timber production has been declining in some parts of the world (in Europe, North America, Taiwan, and Japan, for instance) due to increasing emphasis on non-timber services such as nature conservation and wilderness. For example, some 21 million hectares of productive forestland have been set aside in conservation reserves (parks, wilderness, and so on) in the United States (USDA Forest Service 2001). This area has more than doubled since 1953, reflecting the increased emphasis on non-timber forest values in the latter half of the twentieth century (USDA Forest Service 2001).

In many areas of the world, suburban sprawl and the fragmentation of ownerships are having an effect on the availability of forests for harvesting, management, and recreational access. In addition, the resulting land use patterns create significant problems for land management and wildfire management in the rural-urban interface (Sampson and DeCoster 2000; Gordon et al. 2004). Large areas of land that meet the definition of forest for inventory purposes are not available for timber production due to the small

size of the ownerships or reduced access due to isolation from road networks by other ownerships (Tyrrell et al. 2004). Few national forest inventories effectively account for this factor.

In spite of increasing timber harvests, the reclassification of productive forest to non-timber use, and forest fragmentation, the forests available for timber production in Europe and North America have increased in terms of total standing timber stock. In Europe, it is estimated that annual growth in standing timber is around 700 million cubic meters while timber removals in 2000 were around 418 million cubic meters (FAO 2003b). In the United States, softwood sawtimber stocks in 1997 were 8% higher than in 1963, and hardwood sawtimber stocks had nearly doubled (up 95%) in the same period (USDA Forest Service 2001). With both regions tending to reduce harvesting intensity, particularly on publicly owned lands, the prospects for the near future are for continued growth in the standing stock of timber in both Europe and North America.

One potential factor that can deplete these timber stocks in the future is the threat of alien invasive species, which move more freely as a result of increased global trade. For example, recent infestations in central North America by several invasive insects have resulted in the mortality of large numbers of deciduous trees, especially maples and ashes. In the past, such invasives as Dutch elm disease (*Ophiostoma lumi* and *O. novo-ulmi*) have virtually eliminated individual tree species in some areas.

In addition to the volume of standing timber, the quality of the timber resource is also important. For example, North American eastern hardwood forests, while doubling in standing stock in recent decades, are often dominated by species such as red maple (*Acer rubrum*) that are not as valuable for timber as the original forests that were, in many areas, dominated by species such as oaks, hickories, and chestnuts. In many areas of the world, the premium quality trees and most valuable species have been unsustainably harvested, and supplies of those high-value woods are increasingly limited (FAO 2003b).

The uncertainties in this future prospect include political decisions affecting land use and access to the land for timber harvest, changing trade policies, and the uncertain effects of climate change on net forest growth rates. There are indications that the uncertainty of future access to forests for timber harvest is currently discouraging investment in the forest products industry in some regions.

9.2.2 Wood Pulp

Pulpwood is one component of industrial roundwood production, and the production of pulpwood from forest ecosystems is often tightly integrated with the production of other solid wood products. Pulpwood is derived from a variety of wood sources, ranging from the harvest of fast-growing young trees in plantations managed specifically for pulp production, to the small or lower-quality stems removed from managed forests to improve forest quality or health, to the shavings, trimmings, and other wood produced in the manufacture of sawn wood products. Pulpwood accounts for about a third of the roundwood harvested (including fuelwood) (IIED 1996).

In 1995, about 17% of the wood for paper came from primary forests (mostly boreal), 54% from regenerated forest, and 29% from plantations (IIED 1996). This is expected to change as increased output from plantation forests, many of which will mature in the next decade, reaches world markets.

The global production of wood pulp has almost tripled in the past 40 years. (See Table 9.5.) Regional experience is uneven, with the most rapid expansion occurring in the tropical regions, for a variety of reasons described later in the chapter. The rate of expansion has slowed significantly in recent years, and the current global market in pulp and paper products is marked by overcapacity in the industry and low prices. This has resulted in an unprecedented decline in U.S. pulp and paper production and capacity after 1999, for example (*Economist* 2000). Since late 2003, prices for pulp have rebounded, perhaps aided by high demands in the expanding Chinese economy. This current situation does not, however, change the outlook for further structural change and regional shifts in these industries, especially with increased production from plantations globally.

These factors suggest that the prospect for the near future is a continued leveling off of wood pulp production, with continued decline of major pulp mills in the mature production areas of Europe, Japan, and North America. Those impacts, if they occur, will not affect all regions similarly. The industry continues to shift both mill locations and wood supply contracts to regions of lower-cost production, such as South America and Asia. Such shifts are also tied to increased demand in Asia, especially China.

The rapid increase in wood pulp production over the past four decades was tied to several trends, such as increased population and literacy, leading to increased consumption of paper and paper products and increased use of packaging and packing materials as trade in manufactured consumer goods has grown. The slowdown of the rate of growth in wood pulp production, however, indicates a maturation of these markets and the impact of competing materials. Nevertheless, the International Institute for Environment and Development (1996) predicts continued increased consumption of paper globally, with most of the increase in Asia.

Even so, there appears to be little evidence to suggest that changes in forest ecosystem condition will materially affect the availability of wood pulp globally in the foreseeable future. In fact, the evidence suggests that the increased harvest of young plantations will continue to keep supplies ample and prices low. Since most of the end products of wood pulp (packaging, paper, and so on) are traded in market economies, it is unlikely that changes in the supply will affect consumer well-being, since these societies are generally less vulnerable to modest price changes, and world trade is active in most of these items.

The use of non-wood fiber in pulp supply is expected to remain low and concentrated in Asia. Non-wood materials made up 5.3% of global pulp in 1983 and 11.7% in 1994, and they are expected to reach 12–15% by 2010 (Pande 1998). Most of the non-wood fiber is used in "small-scale pulp mills" (less than 30 tons a day), many of which are currently being closed as a result of poor pollution controls and increasingly stringent standards (e.g., FAO 1998a). This industry has not been well supported by research and development of improved technology or pollution controls (Hunt 2001). Two scenarios are likely for the small-scale non-wood pulp sector. Either most small mills will eventually be replaced by larger regional mills that use wood fiber, or cost-effective pollution abatement will be developed, in which case the projected 10–15% of global production of paper from non-wood fibers may be realized.

9.2.3 Craft Wood

The use of wood as a basis for local crafts is common in forested and wooded regions. The products range across a wide variety—from musical instruments to special furniture, toys, and fixtures

Table 9.5. Wood Pulp Production, 1961–2000 (Data from FAO 2003c)

Region	1961	1980	2000	1961–80	1980–2000
		(million tons)		(percent change)	
Africa tropical	0	0.07	0.04		−41
America temperate	33	66	85	98	28
America tropical	0.5	4	8	673	128
Asia temperate	8	20	22	146	8
Asia tropical	0.05	1	7	1,292	804
Europe temperate	18	31	41	68	33
Southern temperate	1	4	9	253	117
Other areas	0.04	0.3	1	789	74
Total	**62**	**126**	**172**	**104**	**36**

and to artistic and decorative objects. The woods prized for many of these objects are often culled from commercial timber production lines.

While wood crafting is often a significant local economic or cultural activity, there are few data to address the importance of these wood products at the ecosystem, country, or regional level, and few reliable data at the global level to assess the condition and trends of the production of craft wood products. From the limited information available, Bali exports about $100 million of wood carvings a year, suggesting an important contribution to local economies. A study done by CIFOR (2002) indicated that the African craft wood industry is not sustainable, with a declining availability of preferred tree species in most African countries.

9.2.4 Carbon Sequestration in Wood Products

The international community has increased its focus on the issue of climate change due to the buildup of atmospheric carbon dioxide and other greenhouse gases. One topic receiving significant attention has been the ability of forests to sequester atmospheric CO_2 in the process of photosynthesis. (See Chapter 21.)

The harvesting of forest products from sustainably managed forests leads to medium-term carbon sequestration in the form of stable wood products that remain in use for decades or centuries (Skog and Nicholson 1998). Functions for quantifying the rate of retirement or turnover of a wide variety of forest products have been developed, allowing analysts to calculate the carbon fate of harvested wood products (Skog and Nicholson 1998).

At this time, there have been no carbon offset payments or credits recognized for wood product storage, and the national reporting guidelines under the United Nations Framework Convention on Climate Change have treated wood products as though the carbon contained was released in the same year that it was harvested (Houghton et al. 1997). This would be a reasonable assumption if the amount of wood products in use were stable, but it has been estimated that the total amount of wood stored in products was increasing globally at a rate equivalent to about 139 million tons of carbon per year in 1990 (Winjum et al. 1998). Continuing research and international negotiations are seeking practical ways in which this carbon sequestration can be appropriately documented and credited to countries under the UNFCCC and the Kyoto Protocol.

9.2.5 Drivers of Change in Production and Consumption of Timber Products

Increased production of wood products is associated with the growth in human population, improved literacy, increased industrial development, and the associated greater demand. Most of predicted future human population growth is expected to occur in developing countries and to place more demands on forest resources.

Nevertheless, as noted earlier, there are many factors affecting use of wood products that have resulted in reduced demand for some products, with changes in structural products and recycling; at the same time, other factors are leading to increased consumption, including increased literacy and the development of the Chinese economy. Still other factors are influencing how products are produced, where supplies come from, and which nations are importers and exporters of wood products. Aside from production and demand factors, climate change will have an impact on the forests of many nations in the future and will likely result in the increased use of plantations to sequester carbon to mitigate change (Gitay et al. 2001). In spite of difficulty in quantifying clear cause-and-effect relationships, some causal factors can

be identified and assessed, at least in a general manner. These are discussed briefly in this section. (See also MA *Scenarios,* Chapter 7.)

9.2.5.1 Globalization

The trend toward globalization of timber trade has significantly affected the forest products industries in recent decades. As markets become more globalized, companies tend to rely more on plantation forests than on natural ones, particularly in the southern temperate and Asian countries, where labor and materials tend to cost less. Technological advances have helped to enable long-distance operations. Information and communications technologies allow companies and traders to manage activities and processes across the world. Falling transportation costs, spurred in part by increased trade in manufactured commodities, help lower shipping costs for relatively low-value material, such as wood chips or other unprocessed fibers.

Market competition encourages firms to relocate production facilities or to buy production inputs in regions where lower labor costs, easier access to resources, higher timber yields, good governance, political stability, functioning logistics and service, availability of recycled fiber, or any combination of these factors exist that can bring costs down and increase profits. In 1999, for example, the production costs for bleached pulp in Chile was estimated at $330 per ton, compared with $420 per ton in northern industrial countries (*Financial Times* 1999).

During the 1970s and 1980s, world trade in wood chips emerged as a growing factor in the paper industry. At present a fleet of more than 100 specialized chip vessels operates under charter, hauling chips long distances. Increasingly, chip exports are being based on plantation forests in sub-tropical and tropical regions instead of primary forests or low-value secondary forests in temperate regions.

9.2.5.2 Supply Shifting

Shifting timber production is in part a function of globalization and the lower cost of fiber from plantations in regions of the Southern Hemisphere, but it is also a function of political decisions to protect forests from harvest (Berlik et al. 2002). When affluent regions use more wood products than they produce, the harvest is often shifted to regions that have less capability or desire to enforce oversight on such matters as environmental impact or worker safety and health. The result is greater negative impact than would arise if consumption were reduced, or production increased, in the affluent region (Berlik et al. 2002). Both depletion of supply and shifting technologies also result in changes among the species harvested.

High-quality veneer and sawn wood continue to be produced largely from primary or secondary natural-origin forests as opposed to plantations. A major exception is the management of plantations in northern and central Europe on long rotations for high quality sawlogs. Pulkki (1998) suggested that sustainable forest management techniques could ensure the long-term provision of valuable species from most tropical areas as well, but that is not a prospect for the near-term future. Teak offers a particularly good example of a high-quality hardwood species that can be grown to sawlog size in about 30–40 years in tropical areas. Little teak is now derived from natural origin forests (Pandey and Brown 2000), and the future of the species as a commercial product depends on plantations (Rasmussen et al. 2000).

Traditionally valued rare tropical hardwoods have seen shifting patterns of sourcing over the centuries as early supplies were depleted and replacement species emerged, often from other continents. Given the nature of the species, the trends in deforestation, and weak institutional capacity in producing areas, many valuable tropical species will decline markedly in availability in future decades (Adams and Castano 2002).

9.2.5.3 Increase in Plantations

The recent increase in plantations is forecast to continue into the foreseeable future, even if the rate of expansion proceeds more slowly than in the recent past. FAO estimates that the plantation share of roundwood production will grow from the current one third to almost half of total global production by 2040. Roundwood production from plantation forests is likely to provide 906 million cubic meters by 2045 compared with 331 million cubic meters in 1995 (FAO 2000c, 2001a).

There is considerable debate over the appropriate role of plantations in relation to sustainable forest management (UNFF 2003). Oliver (1999) estimates that plantations growing 12.5 cubic meters per hectare per year on 8% of the world's forested lands could satisfy current global wood consumption, easing pressure on natural forests. Others argue that the competition from plantations can render the long-term management of many forests uneconomical, thus increasing the incentives to exploit and abandon them or to sell them off for development or other land uses (TMFWI 2001). Investments in plantation forestry will only be made under good governance (Rice et al. 1997). Most plantation investment is in countries with stable political systems, high availability of land, and climate and soil conditions that promote high yields. Whether that concentration can continue for long without controversy over conflict with other land uses, traditional users, or environmental impact is highly uncertain (Durst and Brown 2002).

In the near future, the production from plantations will significantly increase due to the age structure of the current plantations. In 1995, it was estimated that some 55 million hectares of plantations were younger than 15, and some 22 million hectares were in the 0–5 year age class (Brown 2000). Many of those plantations will be coming into harvest age between 2005 and 2010, and to the extent that they reflect the enhanced yields being pursued through improved varieties, fertilization, and other management improvements, their impact on markets will be significant.

There is a large body of research available to enable plantation managers to improve yields from their lands by selecting specific stock or improved stock from various genetic improvements, depending on species. For example, Newton (2003) suggested that volumes in planted northern conifers can be improved by 7–26% at rotation age through genetic improvement, Matziris (2000) suggested an observed volume increase of 8–13% in Aleppo pine (*Pinus halepensis*) in Greece, and Li et al. (2000) estimated volume improvement in loblolly pine of 12–30% through tree improvement.

Planted forests supply less than 7% of wood used for fuels (FAO 2000c), but the opportunities to expand various forms of agroforestry in community settings, reducing travel times to obtain fuelwood and supplying livestock fodder, are immense and could have large impacts on the quality of life in many parts of the world (Sanchez 2002).

9.2.5.4 Mechanization and Utilization Technology

Mechanization of timber harvesting has reduced employment in the forestry sector, particularly in the tasks of felling trees and transporting logs. However, mechanization has been key to improving forest management and has led to reduced injury and mortality in the sector.

The output of forest products has risen faster than the production of industrial roundwood over the last 20 years. Conversion efficiency, recycling, and waste reduction are all contributing factors. In developing countries, however, efficiency is still low and improvements will increase fiber availability (Chen 2001). For example, while mills in the industrial world produce about 70% product from the wood received, mills in the developing world reach only 30% (WWF n.d.). Closing this "efficiency gap" through technology transfer and other means offers a significant opportunity for meeting future commodity demands without increasing pressure on forest ecosystems.

Engineered wood products are becoming increasingly common as a result of reductions in the availability of high-quality structural wood, competition from steel products, and cyclical wood prices. These products, derived as a result of new technologies, essentially turn low-quality wood and wood residues into products valuable for construction and furniture (Enters 2001). The use of engineered wood products in the North American market, for example, has grown at a rate of 20% per year since 1992, reaching more than 29 million cubic meters in 1997 and projected to rise to over 45 million cubic meters by 2005 (Taylor 2000). If the use of these technologies continues to spread, the pressure on some ecosystems and high-quality species will be eased.

9.2.5.5 Global Energy Sources and Costs

As nonrenewable fossil energy supplies decline and new sources of renewable energy are sought, the implications for the supply and consumption of forest products are significant. They are not, however, all pressing in one direction, so the potential effect is mixed and difficult to discern at this point.

To the extent that the cost of energy rises in response to fossil supply changes, forest product technologies that are energy-intensive will become less competitive. Whether this will affect some of the new engineered wood products is unclear. In addition, field production methods that rely on mechanization may suffer competitively against those that use more traditional means of growing, harvesting, and processing timber, and trade could be affected if transportation costs rise significantly.

On the other hand, increasing prices for fossil fuels are likely to encourage more-rapid development of biomass-based fuels, and their emergence could provide outlets for low-grade timber products that currently lack markets. On managed natural forests, in particular, the sale of low-grade timber could provide the financial support needed to invest in thinning, weeding, and other management practices that result in improved forest health and the production of higher-quality timber products. Such markets could also provide economic incentives for the establishment of agriculture or forest plantations devoted to woody crops grown specifically for energy production.

9.2.5.6 Sustainable Forest Management

Appropriate forest management can improve yields and permit long-term sustainable timber production from extensively managed forests. However, managing forests in a sustainable manner may create short-term reductions in timber harvest as managers seek to replace production goals with sustainability goals. (See Chapter 21.)

Although there are still significant differences in the national interpretations of sustainable forest management progress, FAO reports that in 2000 some 89% of forests in industrial countries were subject to a formal or informal management plan (FAO 2001a). Although reports from developing countries were less complete, about 6% of those forests were reported to be subject to a formal management plan covering at least five years. These efforts may be too new to assess the ultimate impact on timber harvest levels, and there is no international assessment of their effect at this time.

There is a clear efficiency gap in logging methods between forests in many tropical and sub-tropical areas and sustainably managed forests elsewhere. The result is considerable waste of wood, loss of forests, forest degradation, and soil damage in unsustainable operations. The Tropical Forest Foundation suggests that 50% less stem damage during operations would increase productivity on a given land base by 20%. The application of sustainable management practices in developing countries (that is, reduced impact logging) would go a long way toward improving long-term fiber yields and the effective and efficient use of extracted resources, while conserving forests as systems (FAO 2003b). Improved infrastructure, reduced corruption, and independent certification may all be required for improved forest management to become a reality in many developing nations.

9.2.5.7 Forest Certification

Forest certification has been developed as a means of helping demonstrate that specific forests are being managed in a sustainable manner under a defined set of standards as wood products are harvested (see MA *Policy Responses,* Chapter 8). It is described by proponents as a market-based instrument, with the implicit assumption that consumers will prefer certified forest products and be willing to pay a price premium sufficient to cover the cost of the certification process and the associated management techniques. Timber producers that have adopted certification have done so primarily to reduce pressure from environmental NGOs and to satisfy customers such as merchants and retailers who are at risk from NGO campaigning.

By 2000, several certification schemes had emerged, and some 80.7 million hectares, or about 2%, of the global forest area was certified (FAO 2001a). Virtually all of this land (92%) was in the industrial regions of North America and Europe. The movement has continued to expand rapidly, and by 2003 some 124.1 million hectares were reported certified on the Web sites of three large certification systems—the Programme for Endorsement of Forest Certification Schemes (formerly the Pan-European Forest Council), the Forest Stewardship Council, and the Sustainable Forestry Initiative—and the Canadian Standards Association added another 14.4 million hectares. By 2004, the estimate of certified forests in those four systems had grown to 164 million hectares (van Kooten et al. 2004). However, certification is still largely a phenomenon of industrial countries.

The implications of certification for the future production of timber remain uncertain. The costs of achieving certification are significant, and the extent to which those costs might create competitive imbalances is still uncertain. The hope that certified products would command a price premium that could cover the additional costs has not yet been realized, and it seems unlikely in the foreseeable future. It is possible that market pressures (particularly access to specialized markets in regions such as Europe) could create new and significant pressures for companies to undergo certification.

9.2.5.8 Deforestation

Deforestation and the loss of forests reduces the land's capacity to produce wood and other services. (See Chapter 21.) Loss of forests has a long history throughout Europe and western Asia, but during the last four decades the highest deforestation rates have been in tropical forests in Africa, Southeast Asia, and South America, where it is currently estimated that over 100,000 square kilometers per year are deforested (FAO 2001a).

Where deforestation continues in tropical forest regions, producers may be unable to maintain past production levels because forests are converted to other land uses or lands become too degraded to recover to forests. One example is in Africa, where producers in Ghana and Côte d'Ivoire are losing forest to large-scale conversion to coffee and cocoa plantations (TMFWI 2001). Certain forest types, particularly dry tropical forests, are especially subject to land use change.

9.2.5.9 Political and Economic Change

Political and economic changes occur regularly, and most create largely local or regional impacts on the production or consumption of timber products. Some, however, such as the collapse of the Soviet Union, the decision to eliminate timber harvest in major areas of China, or increased corruption (Smith et al. 2003) create impacts on the forest products industry around the globe. Given the increasing trend toward globalization and supply shifting described earlier, political decisions that once created mainly local or national impacts may now affect global supply and demand dynamics.

One dynamic that exists but is difficult to assess is the uncertainty associated with public policy. Investors who fear that public pressure will reduce access to timber supplies, or force new costs on production processes, may hesitate to make investments or may move their operations to locations perceived to be less risky. This raises the specter of extractive industries concentrating on the least restrictive areas, where public pressures for effective pollution controls or sustainable management are lower and the resulting environmental and social impacts are higher.

9.2.6 Environmental and Social Impacts of Timber Extraction

9.2.6.1 Environmental Impacts

Poorly planned or excessive timber harvesting can increase road access into remote forest areas, leading to a reduction in forest interior and increasing the "edge" effects associated with forest fragmentation (Kremsater and Bunnell 1999). This has resulted in wildlife population declines and reduced species richness (FWI/GFW 2002). Increased access to forests promotes illegal logging as well as poaching of wildlife resources and hunting of bushmeat. The trade in illicit bushmeat has become so widespread that in 2002 the signatories to the Convention of Biodiversity adopted a resolution and a program of action to deal with this issue at their Sixth Convention of the Parties, and the Convention on International Trade in Endangered Species of Wild Fauna and Flora has established a special working group to deal with this issue.

Environmental concerns over single-species plantation forests compared with managed natural forests include reduced biodiversity, degradation of soils, reduced water conservation, increased susceptibility to pest invasion, and impacts of genetic modifications on natural gene diversity (Nambiar and Brown 1997; Estades and Temple 1999; Lindenmayer and Franklin 2002; Thompson et al. 2002; Cossalter and Pye-Smith 2003; SCBD 2003). However, under many situations, when plantations are established on degraded lands and are appropriately managed they can increase local biodiversity through re-establishment of native species in the understory (Lugo et al. 1993; Allen et al. 1995).

9.2.6.2 Social Impacts

The importance of wood products to domestic industries, employment, building materials, and energy supplies can be estimated by several measures. Wood products facilities at all scales provide employment that can be important to local communities. Harvesting trees, hauling logs and products to market, and handling shipments at ports engage additional workers. In some instances, the cycle of work is dictated by weather, creating seasonal employment, for example in the Amazon region, where the flooding cycle halts logging and milling for months at a time.

FAO estimates that the global forestry sector provides subsistence and wage employment equivalent to 60 million work years, with 80% in developing countries (FAO 1999). Much of this involves people who work in an "informal" economy. One estimate is that for every job reported in the official surveys, there could be as many as two jobs that go unreported (TMFWI 2001). In addition, timber products and fuelwood are critical portions of many subsistence economies.

The number of people involved is significant, but the data for evaluation of trends and impacts on human well-being are scarce. For example, European enterprises with fewer than 20 employees are not included in formal employment surveys (EU 1997). Yet in the European Union it is estimated that over 90% of all firms have fewer than 20 employees (Hazley 2000). One database lists 7,000 Indian workers in furniture making, when it seems more likely that there could be several hundred thousand thus employed in this country of 1 billion people (TMFWI 2001). The FAO estimate of 60 million work years is likely to be an underestimate of the true figure.

The impacts of economic change and development strategies fall unevenly on different portions of society. Conversion of forests into timber production as a primary objective may reduce access and availability for non-timber resources and values, often at the expense of indigenous populations that were unable to profit from the increased industrial output. In many developing countries, forests are a primary source of energy, food, and medicine for some segments of society—often the poor. Forest sources of food are often most critical as buffer supplies to help subsistence societies through periods of crop shortages or seasonal famines, and medicines derived from forest plants are used by some three quarters of the world's people, with thousands of medicinal plants identified as important sources, particularly in tropical forest ecosystems. Where forests are converted to intensive timber production or plantations, the resulting loss of biological diversity is mainly of these non-timber species, with a resulting decline in well-being for the people dependent on them.

Public policy decisions that alter timber harvests also have important impacts on people. It is estimated that China's decision to restrict timber harvest due to concerns for the flooding caused by improper harvesting methods will reduce employment by up to 1 million jobs (*China Green Times* 2000). This is now having major impacts on regional wood markets as Chinese industry turns to international sources for logs and fiber.

Replacement of managed natural forests by plantations at large scales may affect local and indigenous communities, either through displacement or through the loss of the natural forest biodiversity that formerly provided sources of food, medicine, fuelwood, fodder, and small timber on which these communities depend for their livelihoods.

9.3 Non-wood Forest Products

A great number of non-wood forest products are of importance to people in virtually every forest ecosystem and elsewhere. These products contribute directly to the livelihoods of an estimated 400 million people worldwide and indirectly to those of more than a billion. They include foods, medicinal products, dyes, minerals, latex, and ornamentals among others. This section, however, considers only those that are fiber-based and serve as inputs to construction or craft purposes.

9.3.1 Bamboos

There are approximately 1,200 species of woody and herbaceous bamboos, the former being most important from the socioeconomic perspective (Grass Phylogeny Working Group 2001). Many woody bamboos grow quickly and are highly productive. For example, the shoots of *Bambusa tulda* elongate at an average rate of 70 centimeters per day (Dransfield and Widjaja 1995). Annual productivity values range between 10 and 20 tons per hectare per year, and bamboo stands may achieve a total standing biomass that is comparable to some tree crops (of the order of 20–150 tons per hectare) (Hunter and Junqi 2002). It is estimated that bamboo makes up about 20–25% of the terrestrial biomass in the tropics and sub-tropics (Bansal and Zoolagud 2002). A substantial amount of bamboo timber comes from plantations, although natural forests are also important.

Bamboos are multipurpose crops, with more than 1,500 documented uses. As a construction material, bamboo is widely used in all parts of the world where it grows, and because of its high strength-weight ratio, bamboo is the scaffolding material of choice across much of Asia. The tubular structure of the plant is optimally "engineered" for strength at minimum weight. In many places, its use is restricted almost exclusively for low-cost housing, usually built by the owners themselves. For this and other reasons, bamboo is often regarded as the "poor man's timber" and used as a temporary solution to be replaced as soon as improved economic conditions allow. However, architects' interest in working with bamboo has also led to this becoming a common building material for the wealthy. Modern manufacturing techniques allow the use of bamboo in timber-based industries to produce flooring, board products, laminates and furniture.

Despite its importance, very little is known about the worldwide distribution and resources of bamboo, especially in natural forests, although some preliminary regional assessments are available (Bystriakova et al. 2003, 2004). As a non-wood forest product, and one that is often harvested in non-forest settings, bamboo is not routinely included in forest inventories. According to FAO, statistical data on bamboo timber are only available for 1954 to 1971 (FAO 2001a). Today, very few countries monitor nonwood forest product supply and use at the national level, although program efforts in some countries (such as India) are beginning to occur.

Although reported figures on the area of bamboo forests are inconsistent, it is widely accepted that China is the richest country in the world in terms of bamboo resources. China's bamboo forests cover an estimated area of 44,000 to 70,000 square kilometers, mostly of *Phyllostachys* and *Dendrocalamus* species. Their standing biomass is estimated at more than 96 million tons (Feng 2001). Asia ranks first in bamboo production, and Latin America is second. It is estimated that bamboo in Latin America covers close to 110,000 square kilometers (Londoño 2001).

Worldwide, domestic trade and subsistence use of bamboo are estimated to be worth $8–14 billion per year. Global export of bamboo generates another $2.7 billion (INBAR 1999). Bamboo is increasingly being used as a substitute for wood in pulp and paper manufacturing, and currently India uses about 3 million tons of bamboo per year in pulp manufacture and China about 1 million tons, although China is set to increase the use of bamboo for paper to a target of 5 million tons per year.

In many countries in Asia, Africa, and Central and South America, bamboo products are used domestically and can be very significant in both household and local economies. In parts of Africa, for example, the majority of rural families depend entirely on raw bamboo for construction, household furniture, and fuel. Since the products are traded locally, statistics do not enter the national accounting systems. Thus the real value of bamboo products, as well as the impact of changing supplies on human well-being, is difficult to estimate.

9.3.2 Rattans

Rattan is a scaly, fruited climbing palm that needs tall trees for support. There are around 600 different species of rattan, belonging to 13 genera; the largest of which is *Calamus,* with some 370 species (Sunderland and Dransfield 2000). It is estimated that only 20% of the known rattan species are of any commercial value. The most important product of rattan palms is cane from the stem stripped of leaf sheaths. This stem is solid, strong, and uniform, yet highly flexible. The canes are used either in whole or round form, especially for furniture frames, or split, peeled, or cored for matting and basketry. Rattans require considerable treatment, including boiling and scraping to remove resins and dipping in insecticides and fungicides prior to drying. The range of indigenous uses of rattan canes is vast—from bridges to baskets, fish traps to furniture, crossbow strings to yam ties.

Rattans are almost exclusively harvested from the wild tropical forests of South and Southeast Asia, parts of the South Pacific (particularly Papua New Guinea), and West Africa. Much of the world's stock of rattan grows in over 5 million hectares of forest in Indonesia. Other Southeast Asian countries, such as the Philippines and Laos, have less rattan but have been relatively self-sufficient due to the appropriate size of their processing sector. No rattans grow naturally elsewhere, and even in these locations deforestation can lead to local extinction of rattans due to their dependence on mature forests.

In the last 20 years, the international trade in rattan has undergone rapid expansion. The trade is dominated by Southeast Asia, and by the late 1980s the combined annual value of exports of Indonesia, Philippines, Thailand, and Malaysia had risen to almost $400 million, with Indonesia accounting for 50% of this trade (Sunderland and Dransfield 2000). The net revenues from the sale of rattan goods by Taiwan and Hong Kong, where raw and partially finished products were processed, totaled around $200 million a year by the late 1980s (Sunderland and Dransfield 2000).

Worldwide, over 700 million people trade in or use rattan. Domestic trade and subsistence use of rattan are estimated to be worth $2.5 billion per year. Global exports of rattan generate another $4 billion (INBAR 1999).

9.3.3 Drivers of Change in Bamboo and Rattan Products

9.3.3.1 Increased Trade

Traditionally, bamboo was used domestically and supplies were extracted based on local requirements. Contemporary additional applications of bamboo have propelled it into new domestic and international markets, increasing profits and income for many participants in the sector. Bamboo generates substantial export income for several countries, such as China ($329 million in 1992) and the Philippines ($241 million in 1994) (INBAR 1999).

Indonesia has a clear advantage over other countries, with its overwhelming supply of wild and cultivated rattan (80% of the world's raw material), and rattan contributes about $300 million

to Indonesia's foreign exchange and is an important vehicle for rural development. It also raises the value of standing forests, as rattan is the most valuable of the non-wood forest products in the country, earning 90% of total export earnings from such products (INBAR 1999).

Much of Indonesian wild production was diverted from the international market to the domestic market as a ban on export of unprocessed rattan was phased in between 1979 and 1992. Other major rattan products manufacturers, such as the Philippines and China, are augmenting domestically produced supplies with imports from other rattan producers such as Myanmar, Papua New Guinea, and Viet Nam (often based on unsustainable harvesting), along with continued illicit supplies of Indonesian cane.

9.3.3.2 Depletion of Resource Base

Most bamboo-processing countries are facing a shortage of raw material. The causes of this range from overharvesting and conversion of bamboo forestland to settled agriculture or shifting cultivation. Restoring productive agricultural land to bamboo production is often difficult, as is seen in parts of Nepal, where farmers' concerns for food security are more pressing.

In Indonesia, Laos, and the Philippines, parts of the rattan resource base are becoming scarce. In Indonesia, large-diameter rattans are becoming scarce; in Laos, the rate of exploitation from accessible areas is unsustainably high; and the Philippines has recently become a rattan importer (INBAR 1999). In all cases, loss of forest cover is a main contributor to reduced supply of rattans. As a result of limited supplies of rattan in China and the Philippines, wood is now regularly used in place of large-diameter rattan as a main structural element in "rattan" furniture.

9.3.3.3 Biological Cycles

Many species of bamboo flower simultaneously at long intervals, then set seed and die. Where large areas of bamboo forest are involved, these area-wide disturbances are significant. India, for example, is currently experiencing a bamboo flowering that is expected to affect more than 10 million hectares and peak in 2007 (FAO 2004). The impacts are enormous. The amount of dead woody product vastly exceeds harvesting and storage capacity, leading to serious economic losses. Soils unprotected from erosion are damaged before vegetation is re-established, and local employment suffers. The huge influx of seed leads to a population explosion of rats that, when the seed supply is largely eaten, move out to compete with people for food in regional communities. The last flowering cycle in India (1911–12) led to serious famines, and research and policy measures are under way to mitigate the damage of the current flowering cycle (FAO 2004).

9.3.3.4 Plantations

While a few countries, such as China and India, have successfully promoted bamboo plantations, far more struggle with providing the needed technical and financial support. Potentially adverse effects of bamboo monoculture, such as possible depletion of soil, low biodiversity, and loss of genetic diversity, are largely uncertain.

Private-sector cultivation of rattan, from both large and small-scale plantations, has fallen below expectations and failed to respond to local raw material scarcities. The traditional rattan cultivation system in Kalimantan appears to be under threat, with reduced rattan garden establishment and some conversion of existing rattan gardens to other uses owing to low prices for the main cultivated species and new competing land use opportunities (Belcher 1999).

9.3.3.5 Technological Development and Substitution

Industrial use of bamboo has increased dramatically due to new developments in bamboo processing technology. Laminated bamboo board, bamboo mat plywood, bamboo particleboard, bamboo-fiber molds, floorings, and engineered timber (all called "composites") from bamboo fiber are currently available for building construction, architecture decorating, and other applications. In the pulp and paper industry, some bamboo species can be substituted for timber.

9.4 Fuel

In 2000, the world used approximately 1.8 billion cubic meters of fuelwood and charcoal (FAO 2003b). Energy use from fuelwood and charcoal accounts for 0.7–1.1 terawatts out of a global total energy use from all sources of 14.6 terawatts (Gonzalez 2001b). Although these statistics combine fuelwood and charcoal, this chapter discusses the two forms separately because of their different environmental and social impacts.

Although they account for less than 7% of world energy use, fuelwood and charcoal provide 40% of energy used in Africa and 10% of energy used in Latin America (WEC 2001), and 80% of the wood used in tropical regions goes to fuelwood and charcoal (Roda 2002). In Africa, 90% of wood use goes to fuelwood and charcoal, the highest of any region in the world (FAO 2003a). On the other hand, fuelwood and charcoal account for only 20% of wood use in temperate regions (Roda 2002).

The International Energy Agency projects that, by 2030, renewable energy sources will provide some 53% of residential energy consumption in developing countries as a whole, compared with 73% in 2000. In that projection, an estimated 2.6 billion people will continue to rely on traditional biomass for cooking and heating, and virtually all of that will be produced and consumed locally (IEA 2002a).

FAO (2001b) analyses since 1970 indicate that as certain regions in Asia and Latin America have industrialized—particularly China and Brazil—people have switched from fuelwood and charcoal to fossil fuels. Consequently, total global fuelwood use seems to have peaked somewhere around 2000, although total global charcoal use continues to rise (Arnold et al. 2003).

In recent projections of global energy use, the IEA indicates that with continuation of present government policies and no major technological breakthroughs, the use of combustible renewables and waste will grow by 1.3% a year, compared with an overall growth in energy use of 1.7% annually over the next decade (IEA 2002a). The projections reflect the conclusion that the growth of combustible renewables and waste will slow as people in developing countries (which presently use about 73% of world renewables) gain more disposable income and switch to using fossil fuels. A counter-trend may, however, result as the cost of fossil fuels rise and people are forced to use less convenient fuels. With the volatility of international oil prices, these trends will be highly irregular and difficult to predict.

9.4.1 Fuelwood

People harvest fuelwood by cutting or coppicing shrubs, by lopping branches off mature trees, or by felling whole trees. In many rural areas, local people prefer fuelwood from shrub species that will regenerate after coppicing (Gonzalez 2001a). Cooking and heating are the major end uses of fuelwood and charcoal. In some developing nations, wood and charcoal are important for commercial applications such as bakeries, street food, brick-making, smoking foods, and curing tobacco and tea, and fuelwood is an important source of income and employment in many rural areas.

Since developing societies tend to shift from wood fuels to other sources for home heating and cooking, the change in fuelwood production and consumption reflects both a change in economic condition and a change in ecosystem impact. Fuelwood is often produced and consumed largely outside the market system, in subsistence societies, and its value to human well-being is therefore not captured in unadjusted national economic statistics, such as GDP.

9.4.1.1 Production and Consumption

FAO provides the most consistently developed estimates over time for the use of fuelwood. (See Table 9.6.) It should be noted, however, that these data were developed from a fairly small sample, and in many countries the historical data were estimated from models based on population change and average consumption rates. These FAO country data have been grouped according to the forest regions established as being reflective of large forest biomes, but care must be taken in inferring that trends in use of fuelwood translate directly into impacts on forest ecosystems. While charcoal is largely produced from forests in the developing world, fuelwood is produced from woody plants wherever they are found on the landscape. Note, for example that the "other areas" category in Table 9.6, which reflects production outside the identified major forest areas, represents a significant amount of fuelwood production. To the extent that agroforestry grows in importance, more of these energy resources will be derived from agricultural and grazing systems.

The African tropical forest region experienced a near-doubling (91% increase) in rural population between 1960 and 2000, but recorded fuelwood value per capita declined, although the decline slowed only in the latter two decades. (See Table 9.7.) On the other hand, the Asian tropical region, which also experienced a doubling of rural populations (98% increase) in the period saw per capita fuelwood values decline at a far more rapid rate, particularly since 1980. This indicates that the dependence on wood as a rural energy source disproportionately declined in Asia relative to Africa (and other regions). This may indicate a shift away from wood fuel as Asian rural populations experienced more rapid development during the period. Nevertheless, FAO predicts an increased demand for wood in central and eastern Asia of about 25% by 2010 over the amount used in 1994 (RWEDP 1997).

Using FAO price data and adjusting for inflation to portray a constant-dollar value of fuelwood production, the average value of fuelwood produced per rural person has declined significantly (with the exception of the Southern temperate region) since 1980. While published prices may stem largely from urban economic transactions, rural populations were chosen for this comparison because of the importance of fuelwood to rural societies, particularly poor rural people.

FAO (2003a) projects that total fuelwood use in Africa will increase by 34% to 850 million cubic meters by 2020, but at a rate less than the rate of population growth. Local forest departments throughout Africa have continued to record locally severe problems of overharvesting to provide wood and charcoal for urban areas (Arnold et al. 2003; FAO 2003a). (See Box 9.1.) Declining local availability of fuelwood is also a problem in areas of India, Haiti, the Andes, and Central America, especially near large cities.

9.4.1.2 Future Availability

Although in the 1970s there was increasing concern that the exploding demand on fuelwood resources, driven by population increases,

Table 9.6. Fuelwood Production, 1961–2000 (Data from FAO 2003c)

Region	1961	1980	2000	1961–80	1980–2000
	(million cubic meters)			(percent change)	
Africa tropical	157	211	334	34	59
America temperate	48	93	75	92	–19
America tropical	154	192	247	24	29
Asia temperate	238	276	309	16	12
Asia tropical	498	521	513	5	–2
Europe temperate	100	61	57	–39	–7
Southern temperate	16	18	34	12	88
Other areas	113	160	220	42	38
Total	**1,325**	**1,532**	**1,791**	**16**	**17**

Table 9.7. Fuelwood Production Monetary Value Per Rural Person, 1961–2000 (Prices from FAO 2003c; CPI for real dollar adjustment from U. S. Department of Labor)

Region	1961	1980	2000	1961–80	1980–2000
	(2000 dollars per person)			(percent change)	
Africa tropical	17	101	60	478	–41
America temperate	48	797	43	1561	–95
America tropical	48	322	49	577	–85
Asia temperate	20	22	11	8	–50
Asia tropical	50	45	15	–9	–66
Europe temperate	27	29	12	7	–58
Southern temperate	67	48	102	–29	115
Other areas	9	39	21	321	–46
	286	**1,402**	**313**	**390**	**–78**

BOX 9.1
Fuelwood Supply Analysis in Southern Africa

A multiscale analysis of fuelwood availability in the Southern Africa region, done as part of the Southern Africa Millennium Assessment, demonstrates a method of identifying localized conditions that would otherwise be masked in a large-scale assessment (Scholes and Biggs 2004). The analysis utilizes a geographic model to compare the local biomass production rate to the local harvest rate. Where harvest exceeds production the stock will inevitably decline, and, despite some regrowth in the depleted area, the zone in which harvesting occurs expands until the effort required to transport the wood or charcoal exceeds its value.

At the scale of the entire Southern Africa region, much more wood is grown than is consumed as fuel (Scholes and Biggs 2004 Fig 7.1). Thus a regional analysis would lead to the conclusion that fuelwood supply is not a problem. A more fine-grained analysis, however, reveals several very clearly defined areas of local insufficiency that indicate unsustainable use:

- Western Kenya, southeast Uganda, Rwanda, and Burundi;
- Southern Malawi;
- the area around Harare in Zimbabwe and Ndola and Lusaka in Zambia;
- Lesotho; and
- locations in the former homelands in South Africa—in KwaZulu, Eastern Cape and Limpopo provinces, and around Gauteng.

SAfMA local studies confirm that fuelwood shortages are experienced at the last two locations, with the exception of the Gauteng spot. The generalized "rural Africa" model that predicts per capita woodfuel use clearly breaks down in this highly urbanized situation where electricity and coal are well established and relatively cheap energy sources. Conversely, SAfMA local studies in the Richtersveld and Gorongosa-Marromeu confirm that in areas indicated by the regional model to have a fuelwood sufficiency, this is indeed the case.

Checks with local experts and personal experience in the team confirmed that the fisrt three locations currently experience severe fuelwood deficiencies. Therefore, it seems that the regional-scale assessment correctly identified problem areas at a local scale. The authors attribute this to the fact that the underlying wood production models and fuel demand models are working at a resolution of 5 kilometers, slightly smaller than the typical radius of fuelwood depletion around population concentrations.

would have devastating effects on forest ecosystems, currently available evidence suggests that fuelwood demand has not become a major cause of deforestation (Arnold et al. 2003). There are local situations of concern, such as areas near settlements, where an income can be earned by cutting and carrying wood to buyers, but it appears that fuelwood supply is not important enough in most places, with the possible exception of some regions in Africa, to attract national policy intervention to reduce deforestation. The predicted supply for South and East Asia is well above the projected demand.

Recent estimates developed by FAO indicate that global consumption of fuelwood appears to have peaked in the late 1990s and is now believed to be slowly declining. Global consumption of charcoal appears to have doubled between 1975 and 2000, largely as a result of continuing population shifts toward urban areas (Girard 2002; Arnold et al. 2003). Since the production of charcoal entails a net loss of energy and is highly concentrated in forest areas of the developing world, increased charcoal consumption signals pressure on wood supplies.

9.4.1.3 Impacts on Human Well-being

Fuelwood is the main source of household energy for an estimated 2.6 billion people, and in urban areas of developing countries families may spend 20–30% of their income on wood and charcoal fuels (FAO 1999). In terms of impact on their well-being, the main problem for the ever-increasing number of urban dwellers may be price rather than availability of fuelwood.

In rural villages that rely on hand-gathered wood, local shortages may impose serious time constraints on women, whose task it usually is to collect fuelwood, as well as increased energy use and risk of injury associated with lengthy travel with heavy loads. While it is usually women who search for and carry wood on their heads and backs for rural use, many rural people load animals and carts to transport wood for sale in urban areas. Thus, urban demand translates into local economic opportunities. In areas where the demand exceeds the sustainable supply, this can result in serious impacts on local forests.

The lack of reliable and consistent data limits the ability to assess the impact of fuelwood trends on human well-being. People make fuel choices for a variety of reasons, including convenience, price, and reliability of supplies. For example, it has been found that price, availability, and ease of use are very important in affecting fuel choice among urban people, while the price of stoves and the level of pollution from the fuel did not seem to matter as much (Gupta and Kohlin 2003).

A 1996 survey of rural energy in six Indian states found that wood was becoming more scarce and difficult to obtain (ESMAP 2000). As a result, some poorer households were using less efficient fuels like straw and dung, while wealthier households were shifting up the "energy ladder" to purchase charcoal or fossil fuels. The most common response, however, was for households to increase their collection time to compensate for reduced availability and access (Arnold et al. 2003). This and other studies reinforce the conclusion that, in many rural areas, gathered supplies of fuelwood still constitute the main source of domestic energy, making these users more vulnerable to changes that affect their ability to get wood supplies. Reduced access may arise from resource shortages, from changes in land tenure (such as increased privatization), or increased distance to common property. In all circumstances, the result is a reduction in well-being for affected families.

FAO (2000a) estimates that about half the world's households cook daily with biomass fuels and that most of this cooking is done indoors with unvented stoves. Pollutants found in biomass smoke include suspended particulates, carbon monoxide, nitro-gen oxides, formaldehyde, and hundreds of other organic compounds such as polyaromatic hydrocarbons. In many parts of the world, for all or part of the year these pollutants are released from stoves in poorly ventilated kitchens and homes. Women, infants, and young children who spend more time in the home suffer the highest exposures.

Several studies have suggested that domestic smoke pollution is responsible for respiratory diseases, low birth weight, and eye problems. The evidence is overwhelming in the case chronic obstructive pulmonary disease in adults and acute respiratory infection in children. There is also evidence to suggest a relationship with perinatal conditions, blindness, tuberculosis, and lung disease. It has been suggested that domestic smoke pollution may also be related to asthma and cardiovascular disease (FAO 2000a), and in total an estimated 1.6 million deaths and 39 million disability-adjusted life years are attributed to indoor smoke, primarily in Africa, Southeast Asia, and the Western Pacific (WHO 2002).

There could be additional impacts in local areas experiencing a shortage of fuelwood. One is the increase in crop residues and dung that are used for cooking and heating in the absence of available wood. These fuels are less efficient and produce more smoke, and the burning of crop residues and dung reduces their availability for enhancing soil structure and fertility, leading to reduced food production. Where adequate fuels are not available, the consumption of cooked food may decline, leading to adverse effects on nutrition and health.

The conclusion to be drawn is that there are important human benefits to be gained from targeted efforts to improve fuelwood availability and accessibility in localities where it is now in short supply. The adoption of improved stoves, with higher efficiency and improved venting, would have important human well-being benefits across wide regions but has often proved more difficult than anticipated, for reasons of affordability and cultural acceptability.

9.4.2 Charcoal

Charcoal consists of the remnants of wood that has been subjected to partially anaerobic pyrolysis (decomposition under heat). Conversion of wood to charcoal creates a product with double the energy per unit mass that is less bulky and more convenient for transport, marketing, and sale than fuelwood. The major domestic end uses of charcoal are cooking and heating, often in the urban areas of developing countries where people are able to purchase, rather than gather, their home energy supplies. Charcoal is not as convenient as petroleum fuels, so as incomes rise people tend to shift from charcoal to coal, gas, or oil. Thus, charcoal consumption has tended to peak, then diminish, as development proceeds. Other large users of charcoal include light industrial users, such as blacksmiths and ceramic and brick makers, and Brazil alone produces approximately 6 million tons of charcoal annually for steel production (WEC 2001).

In Africa, there is a general trend to replace fuelwood with charcoal (Girard 2002). For example, in Bamako, Mali, the proportion of households that use charcoal has risen from nothing in 1975 to 50% in 1996, while the proportion using fuelwood has declined at the same rate (Girard 2002). This trend is expected to continue throughout Africa.

Converting wood to charcoal provides employment and has the advantage of using wood remnants and sawdust that are often otherwise wasted. Nevertheless, much of the charcoal is produced in low-efficiency "cottage industries," resulting in a net loss of energy of as much as two thirds of the energy contained in the original wood, although some of that lost energy is offset by the reduced energy required in transport to markets. Charcoal production, particularly if low-efficiency techniques are used, is a significant source of air pollutants and greenhouse gases.

Charcoal production is declining significantly in Europe and the southern temperate region. (See Table 9.8.) Where charcoal production is increasing to provide fuel for some urban areas, and where inefficient charcoal production methods are common, this may be a local concern for forest sustainability, although there is limited information available to assess these situations (Arnold et al. 2003). Certainly at the regional and global level, trends for charcoal production and use do not suggest broad threats to forest ecosystems.

9.4.3 Industrial Wood Residues

In many modern wood industries, residues that were formerly waste products now provide a portion of the electricity and heating needs of the mill or paper plant. For example, in the mid-1990s, it was estimated that the pulp and paper industry in the United States produced about 56% of its energy needs by burning the unused wood components removed in the pulping process (Klass 1998). In the United States, 98% of the bark, saw dust, and wood trimmings from sawmill operations, and the black liquor produced in the pulping process, are currently used as fuel or to produce other fiber products (Energy Information Administration 1994). Enters (2001) indicates that on average, only half the wood harvested in Asia is used and the rest is unused residue that goes to waste.

As industrial wood processing and paper-making residues have become increasingly used to generate energy, the main impact on local communities has been the associated reductions in air and water pollution. Historically, many mill communities tolerated smoke, chemical aerosols, and degraded stream reaches as a necessary part of maintaining the jobs and economic impact of the mill. Today, the communities that benefit from modern mills have fewer associated pollution burdens.

9.4.4 Biomass Energy

The world currently relies heavily on nonrenewable fossil energy sources such as coal, petroleum, and natural gas, and although long-term forecasts for declining supply of fossil fuels entail a high degree of uncertainty, Klass (1998) estimates that the gradual depletion of oil and natural gas reserves will become a major problem by 2050. As the availability of fossil fuels declines, the only renewable carbon resource large enough to substitute for or replace fossil resources for the production of fuels and electricity

is biomass. Policy implications include the opportunity to encourage more efficient and modern biomass systems through technological development and diffusion.

Industrial biomass includes energy systems generating electricity, heat, or liquid fuels from fuelwood, agricultural crops, or manure. In 2000, biomass other than fuelwood and charcoal may have provided 5% of global world energy (WEC 2001). Biogas produced from dung and other carbohydrate-based agriculture products like nonedible oil cake is another major source of energy in Asia, particularly in India and China (Deng 1995).

Current technologies for converting biomass into electricity and fuels include thermochemical and microbial processes such as combustion, gasification, liquefaction, and fermentation (Klass 2002). Biogas is most commonly produced using animal manure, mixed with water, stirred and warmed inside air-tight digesters that range in size from around 1 cubic meter for a small household unit to as large as 2,000 cubic meters for a commercial installation (Ramage and Scurlock 1996). The biogas can be burned directly for cooking and space heating or used as fuel in internal combustion engines to generate electricity.

Examples of thermochemical processes include wood-fueled power plants in which wood and woody wastes are combusted to produce steam that is passed through a turbine to produce electricity; the gasification of rice hulls by partial oxidation to yield fuel gas, which drives a gas turbine to generate electricity; and the refining of organic oils to produce diesel fuels. Another example is the alcoholic fermentation of corn to produce ethanol, which is then used in a variety of formulations in motor fuels (Klass 2002).

Soybeans and oil palms produce oil crops that can be processed directly into biodiesel. The combination of different biomass sources and conversion technologies can produce all the fuels and chemicals currently manufactured from fossil fuels. The major obstacle is the price competition from fossil fuels. While most analysts foresee the economic gap narrowing and reversing as fossil fuel prices rise in response to dwindling supplies, there are varying opinions as to when this may have a significant effect, with estimations up to the middle of the twenty-first century (Klass 2003), and some predicting that peak petroleum production may occur well within the first quarter (IEA 2002b).

The data on production of biomass-based liquid fuels and electricity are limited. One major source, the International Energy Agency (IEA 2003), pools estimates for all renewable sources, including energy generated from solar, wind, biomass, geothermal, hydropower and ocean resources, and biofuels and hydrogen derived from renewable resources.

Table 9.8. Charcoal Production 1961–2000 (Data from FAO 2003c)

Region	1961	1980	2000	1961–80	1980–2000
	(million cubic meters)			(percent change)	
Africa tropical	2.6	5.6	12.0	114	114
America temperate		0.5	0.9		70
America tropical	6.0	8.7	13.7	45	58
Asia temperate	0.3	0.3	0.5	8	43
Asia tropical	1.6	2.8	3.4	71	25
Europe temperate	0.3	0.4	0.2	24	−39
Southern temperate	0.6	0.6	0.3	4	−48
Other areas	2.9	4.7	8.3	60	76
Total	**14**	**24**	**39**	**64**	**67**

As noted earlier, fuels and electricity can be produced from almost any biomass resource, but commercial production has been limited. A few examples are power and steam production via the combustion of municipal solid wastes, of fuel gas recovered from landfills, and of biogas produced in wastewater treatment plants. Steam and hot water are produced in the gasification of wood and wood wastes to produce fuel gas for use in commercial buildings and the combustion of black liquor in the pulp and paper industry. Liquid fuels for internal combustion engines come from lipids and fuel oxygenates come from fermented grains. In the United States, production of fuel ethanol from corn has been commercialized, but it relies on federal subsidies and policies requiring the use of organic oxygenates in gasoline to reduce pollution in some areas of the country. A major research and development effort is in progress to displace corn with low-cost cellulosic feedstocks such as crop residues and non-merchantable wood produced through fuel reduction projects aimed at reducing the intensity and severity of forest fires (Sampson et al. 2001).

On a energy content basis, existing global standing biomass is estimated to be about 100 times the total annual consumption of coal, oil, and natural gas in the 1990s, and net annual production of biomass is 10 times annual energy consumption (Klass 1998). Incremental new biomass growth on carefully designed sustainable plantations that produce dedicated energy crops could eventually have large potential uses in meeting global energy demands.

There is considerable variation in the estimates of the biomass in agricultural wastes that might be available for energy production. For example, one study estimates that the potential amount of rice straw and husks available for energy might range from about 300 million to 1,900 million tons (Koopmans and Koppejan 1998). The range involves different assumptions about production as well as the extent to which available crop residues will be used for fuel, fodder, fertilizer, fiber, or feedstock. One of the issues in using crop residues for commercial energy production is that their use may depend on storage for prolonged periods after harvest. Also, the scale of biomass electric plants may exceed locally available feedstock supplies.

Taking into account the net primary productivity of the world's ecosystems and conventional energy technology, global biomass could provide energy at a theoretical rate of 9 terawatts (WEC 2001) to 26 terawatts (Holdren 1991), compared with the current rate of global energy use of 15 terawatts (Gonzalez 2001b). Some regional studies show significant supplies available. (See Table 9.9.) Realistic estimates of supply, however, need to be tempered by several factors:

- Much biomass, such as crop residues and logging wastes, are widely dispersed; making their collection for commercial use difficult and costly.
- The removal of organic material from producing crop and forestlands may compromise their ability to sustain productivity. Organic material returns are essential for maintaining soil quality, so only a portion of the waste biomass can be safely removed.
- The environmental impacts of increased biomass energy production, both positive and negative, need to be considered (Sampson et al. 1993).

An increasing role for biofuels in the world energy system would have significant local economic implications. Growing, harvesting, and transporting these fuels could provide new crop and employment opportunities for rural residents.

Because biofuels are produced from renewable sources, their use does not involve a net transfer of carbon dioxide into the atmosphere. As a result, where they replace fossil sources they can be counted as a positive benefit in attempts to address climate change.

To the extent that they become a significant force, the benefits accrue across all nations and societies. For example, the use of bagasse in Australia is estimated to reduce net emissions of CO_2 into the atmosphere by 226,000 tons per year (Ramage and Scurlock 1996).

The benefits of clean, renewable energy and fuels are evident, but the slow rate of their growth in relation to the growth in fossil fuel use reflects the difficult obstacles that biofuels face. An integrated, large-scale biomass energy industry has yet to emerge despite the major expenditures made to develop new technologies and scale them from research to production levels. In most of the industrial world, the lack of financing for first-time production facilities, the difficulty in assuring growers of adequate prices and producers of adequate supplies in the absence of market experience, and the lack of an energy infrastructure geared to dispersed, decentralized production facilities have all deterred industrial development of biomass fuel. The competition from fossil fuels has also contributed to the slow growth in biomass-based production of modern fuels, despite the steady advances from research and technological development.

9.4.5 Drivers of Change in the Use of Biomass Fuels

9.4.5.1 Fossil Fuel Availability

The near-term prospects for the future of biomass fuels remain one of slow growth. If fossil fuel prices continue to rise in the coming decades, the longer-term prospects (2030–50) for biofuels look very positive. The combination of drivers, including concern over global climate change, pollution, and fossil fuel depletion, appears poised to drive government policies and market forces toward an increased role for biomass-based modern energy sources.

9.4.5.2 Income and Development Levels

As incomes rise and development proceeds, people tend to shift from low-cost, heavy, or inefficient fuel sources to those that cost more but require less effort to obtain and use. If development efforts succeed in raising incomes and living standards, pressures on local ecosystems for fuelwood will diminish as people move up the energy ladder to other sources.

9.4.5.3 Technology Development and Transfer

Programs that successfully introduce more efficient cooking and heating stoves, modern renewable sources (such as solar, geothermal, and wind), or other energy innovations can reduce pressure on local sources of biomass fuels. The result can be improved human well-being by lowering the time and effort spent gathering fuel, lowering health impacts from smoke, and supporting improved diets.

Increased use of biomass for commercial energy production will require continued major investment in research, development, and technology transfer.

9.4.5.4 Resource Availability

Lack of accessible fuelwood supplies can be an important localized problem with serious impacts, particularly on rural or low-income people. This can be the result of an imbalance between population levels and local biomass production capability, as described in Box 9.1, in the absence of affordable or accessible energy options. Policy options may include efforts to increase local fuel production (through increased agroforestry), introduce technology innovation, or improve fuel transport.

Table 9.9. Consumption and Potential Supply of Biomass Fuels for 16 Asian Countries[a] (RWEDP 1997)

Consumption/Supply	1994			2010		
	Area (million hectares)	Mass (million tons)	Energy (petajoules)	Area (million hectares)	Mass (million tons)	Energy (petajoules)
Consumption						
Total woodfuels		646	9,688		812	12,173
Potential supply						
Sustainable woodfuel from forestland	416	670	10,047	370	629	9,440
Sustainable woodfuel from agricultural areas	877	601	9,021	971	692	10,381
Sustainable woodfuel from other wooded lands	93	54	810	81	47	708
Waste woodfuels from deforestation	(4)	606	9,083	(3)	438	6,566
Total potentially available woodfuels	1,382	1,931	28,962	1,420	1,806	27,095
50 percent of crop processing residues	877	219	3,458	971	322	5,105
Total potentially available biomass fuels		2,150	32,420		2,128	32,200

[a]Bangladesh, Bhutan, Cambodia, China, India, Indonesia, Lao PDR, Malaysia, Maldives, Myanmar, Nepal, Pakistan, Philippines, Sri Lanka, Thailand, and Viet Nam.

9.5 Fiber

9.5.1 Agricultural Plant Fibers

A wide variety of crops are grown for fiber production. Flax, hemp, and jute are generally produced from agricultural systems, while sisal is produced from the fiber contained in the leaves of the *Agave* cactus, which is widely cultivated in tropical and subtropical areas. (See Chapter 22.) Silk is a special case, produced by silkworms fed the leaves of the mulberry tree, grown in an orchard-like culture. The production of all the listed fibers except silk has declined in recent decades. (See Table 9.10.)

Competition from non-cellulosic fibers has increased significantly in recent years. (See Table 9.11.) According to the U.S. Department of Agriculture (whose data varies slightly from that of FAO in Table 9.10), total world fiber production has grown by 63% in the last two decades, while the proportion of natural (cellulosic) fibers has declined from almost two thirds to under one half (USDA-ERS 2003).

9.5.1.1 Cotton

Cotton is the single most important textile fiber in the world, accounting for over 40% of total world fiber production. It is an unusual crop, in that it is an oil crop grown for its fiber, which develops as elongated surface cells on the seedcoat. The cotton

seed itself constitutes about 65% of the harvested crop and contains about 17% oil and 24% protein (Gillham et al. 2003).

While some 80 countries around the world produce cotton, China, the United States, India, Pakistan, and the former Soviet Union have dominated global production since 1961, although their relative share of the global total has changed over time. (See Table 9.12.) Over 70% of the world's cotton is produced in the United States (above 30° north latitude), China, the former Soviet Union countries, and southern Europe (Gillham et al. 2003). The water and fertilizer requirements for high yields of cotton under intensive production are high, and it is this that leads to concentrated production in so few regions.

Cotton is produced on both irrigated and rain-fed cropland, and cotton demand has been the basis for major irrigation projects over the past century. In Uzbekistan, for example, major irrigation developments were constructed in the 1940s to convert the region into the primary cotton producer for the Soviet Union (Gillham et al. 2003). The resulting diversion of water, along with the intensive use of agrochemicals, resulted in disastrous environmental deterioration of the Aral Sea. (See Chapter 20.)

Global production of cotton has about doubled in the past 40 years, while the land harvested has stayed virtually the same (FAO 2003c). However, those global totals mask significant shifting of cotton growing. For example, a major area expansion in Pakistan has been offset by large declines in the rest of the world. FAO

Table 9.10. World Production of Selected Agricultural Fibers, 1961–2000 (FAO 2003c)

Item	1961	1980	2000	1961–80	1980–2000
	(thousand tons)			(percent change)	
Flax	697	620	522	−11	−16
Hemp	300	186	50	−38	−73
Jute and jute-like fibers	3,492	3,609	3,037	3	−16
Sisal	763	548	413	−28	−25
Silk, raw and waste production	33	69	107	111	56
Total	**5,284**	**5,032**	**4,129**	**−5**	**−18**

Table 9.11. World Textile Fiber Production, 1980–2000 (USDA-ERS 2003)

Item	1980 (thousand tons)	1980 (percent of year's total)	1990 (thousand tons)	1990 (percent of year's total)	2000 (thousand tons)	2000 (percent of year's total)
Rayon and acetate	3,243	10.6	2,758	7.0	2,216	4.4
Non-cellulosic fibers	10,479	34.2	14,899	37.7	26,137	52.4
Cotton	14,259	46.6	18,969	48.0	19,466	39.0
Wool (clean)	1,693	5.5	1,978	5.0	1,361	2.7
Silk	56	0.2	66	0.2	86	0.2
Flax	630	2.1	712	1.8	591	1.2
Hemp (soft)	258	0.8	165	0.4	57	0.1
Total Fibers	**30,618**	**100.0**	**39,548**	**100.0**	**49,914**	**100.0**

Table 9.12. Annual Production and Area Harvested of Cotton for Selected Countries and Rest of the World, 1961–2000 (FAO 2003c)

Country	1961	1980	2000	1961–80	1980–2000	1961–2000
Production		(thousand tons)			(percent change)	
China	800	2,707	4,417	238	63	452
India	884	1,292	1,641	46	27	86
Pakistan	324	714	1,825	120	155	463
United States	3,110	2,422	3,742	−22	55	20
Russia/former Sov. Un.	1,528	2,804	1,487	84	−47	−3
Other	2,815	3,966	5,505	41	39	96
Global Total	**9,461**	**13,905**	**18,618**	**47**	**34**	**97**
Area Harvested		(thousand hectares)			(percent change)	
China	3,868	4,915	4,041	27	−18	4
India	7,719	7,820	8,576	1	10	11
Pakistan	1,396	2,108	2,927	51	39	110
United States	6,327	5,348	5,285	−15	−1	−16
Russia/former Sov. Un.	2,335	3,147	2,545	35	−19	9
Other	10,216	10,981	8,482	7	−23	−17
Global Total	**31,861**	**34,319**	**31,856**	**8**	**−7**	**0**

data show some inconsistency in yields, with China's year 2000 yield of over 1 ton per hectare being significantly higher than the global average.

Although the rate of increase in cotton production has slowed since 1980, further growth in cotton production is set to continue through either additional planting or irrigation or through increased yields from improved varieties, management techniques, or pest protection. The reasons for declining production in some regions vary and include increased competition for available irrigation water, loss of productive soils to salinization, or declining markets and prices due to continued or increased competition from synthetic fibers.

In one major cotton-producing region—Uzbekistan—the area planted to cotton has declined steadily from a peak of 2.1 million hectares in 1987 to a reported 1.44 million hectares in 2000 (FAO 2003c). Since the demise of the Soviet Union, the ability of the region to trade cotton for food has diminished, and the need to become more self-sufficient in food is contributing to the decline in the area devoted to cotton production. In addition, the collapse of the economy contributed to a lack of fertilizer and other inputs due to the shortage of foreign exchange (Gillham et al. 2003).

Cotton plays a major role in the economies of many developing countries. In India, over 60 million people derive income from cotton and textiles. In Pakistan, textiles employ over one third of the industrial labor force, and in Uzbekistan 40% of the workforce relies on cotton (Gillham et al. 2003). In China, an estimated 50 million families grow cotton, illustrating that much of the world's cotton is produced by smallholders relying primarily on family labor. In these situations, the crop competes with food crops for available land, water, time, and energy, and strong markets or government policies that encourage expanded cotton production may create difficulty in meeting food production needs.

One of the major challenges in cotton production is the management of crop pests. The most widely known pest, the cotton bollworm (*Helicoverpa armigera*), causes millions of dollars worth of damage annually. One estimate suggests that India alone suffers over $300 million in annual damages from this pest (www.nri .org/work/bollworm.htm). This has led to major research efforts around the world to develop improved pest management techniques. One approach, genetically modified cotton, is being tested in many regions but has raised significant controversies.

Smallholders face significant competitive disadvantages in growing cotton, lacking the mechanical implements for timely operations, the inputs to raise yields or protect against pests, and the marketing ability to produce commercial amounts for sale. It has been estimated that it would take 75–150 smallholders, averaging a quarter to a half hectare each, to produce 100 bales of cotton—a common amount needed to attract a commercial contract (Gillham et al. 2003). Thus, many of the world's cotton producers need significant technical and marketing support to maintain cotton production as a viable agricultural option.

Significant changes in the supply of agricultural fibers can have an impact on those craftsworkers, artisans, and local producers who rely heavily on one or more of them for their livelihood. Slow gradual changes, which seem far more likely, will not be as disruptive, as they provide time for adaptation. Although changes in the production of cotton in any one region would not appear to have significant impact on the well-being of consumers, due to the extent to which the fiber is traded on world markets, such changes will affect local food supplies due to the competition of cotton production with food production.

While the controversy surrounding genetically modified organisms extends well beyond cotton, this crop is one where the issues are both current and particularly important. Media reports estimate that some 1.5 million hectare of GM cotton were planted in China in 2001 and some 100,000 hectares were grown in India in 2003 after the country approved testing in 2002 (Reuters 2002, 2004). To date, there are no official data on these crop varieties, but their use is growing rapidly, along with the associated controversies.

GM cotton is the result of genetic engineering that introduces genetic material from the Bt (*Bacillus thuringiensis*) organism into the cotton plant. This protein makes the crop more resistant to pests such as the cotton bollworm. Proponents of the technology point to evidence of increased yields and profits to growers, particularly small growers who lack the capital and equipment to control pests effectively. They also argue that the technology reduces the use of pesticides and lowers associated environmental impacts. Critics of the technology assert that early pest resistance is likely to vanish as pests evolve the capability to overcome the new defenses and that there are dangers of releasing genetic material into the environment that may not be subject to natural controls. They also express concerns that farmers may end up using more pesticide rather than less, as the need to control pests other than the target pests becomes more important.

These are issues of great importance to the future of many crops, and an adequate assessment of the technology and its implications is beyond the scope of this chapter. Such an assessment will be increasingly important as the world grapples with the implications of GMO crops, including cotton.

9.5.1.2 Silk

Silk has long been highly prized for the manufacture of fine cloth. It is produced primarily in Asia, where silkworm culture (sericulture) has been under way for centuries. Originally developed in China, silkworms and their host, mulberry trees, have been exported widely around the world. Although commercial sericul-

ture has been tested in many areas of Europe, northern Africa, and the Americas in the past, world production is now heavily centered in China, which accounted for 73% of reported world production in 2000 (FAO 2003c).

Silk production has tripled and the center of production has shifted from Japan to China over the last 40 years. In 1961, Japan produced 57% and China produced 20% of a total world supply of 32 million tons (FAO 2003c). By 2000, China was producing 73% and India was producing 14% of a world supply that had tripled to 110 million tons (FAO 2003c). China's silk production in 2000 (78 million tons) exceeded total world output in 1980 (68 million tons). The movement of silk production from Japan to China over the recent past appears to be linked primarily to lower labor costs in the very labor-intensive production process.

9.5.1.3 Flax, Hemp, Jute, and Sisal

FAO data contain statistics on several of the world's important fiber crops, including flax, hemp, jute and jute-like fibers, and sisal.

Flax is obtained from the stems of several varieties of *Linaceae usitatissimum*, an annual herb that has been cultivated since prehistoric times. The crop has been transported from its native Eurasia to all the temperate zones with cool, damp climates. It is also grown for oilseed production in many parts of the world and was the major source of cloth fiber (linen) until the growth of the cotton industry. Flax fiber cultivation on agricultural land involves dense plantings to prevent the annual plant from branching, then harvesting before maturity.

The total area devoted to flax production has declined from over 2 million hectares in 1961 to less than 450,000 hectares in 2000. The most significant decline was in the former Soviet Union, where the area devoted to flax went from over 1.6 million to about 200,000 hectares. During that same period, the most significant increase in production was reported from China, where production has grown fivefold to some 215,000 tons. In 2000, the three largest flax producers (China, France, and Russia) produced almost two thirds of total global output (FAO 2003c).

Hemp is the common name for *Cannabis sativa*, an annual herb that was native to Asia but is now widespread around the world due to its history of cultivation for bast fiber and drugs. The fiber, taken from the stem, was once widely used to produce various kinds of cordage, paper, cloth, oakum, and other products. Hemp production has declined dramatically since 1961, particularly in the former Soviet Union. In 2000, the two largest producers (China and North Korea) reported over half of total global production (FAO 2003c).

Jute is the common name for the tropical annuals of the genus *Corchorus*. Many species yield fiber, but the primary sources of commercial jute are two species (C. *capsularis* and C. *olitorius*) grown in the Ganges and Brahmaputra valleys of India. Jute is used primarily for coarse fabrics used in burlap, twine, and insulation. Total world production of these fibers declined between 1980 and 2000, particularly in China, Thailand, and Myanmar, and jute and jute-like fibers are now produced almost entirely in India and Bangladesh, where some 89% of total global production originated in 2000 (FAO 2003c).

Sisal is extracted from the leaves of the Agave cactus (*Agave sisalana* and *A. fourcroyides*), which is widely grown in dry tropical regions. The fibers are strong and used primarily for cordage, such as binding twine for hay bales. Over 70% of the sisal fiber production in 2000 was in Brazil and Mexico (FAO 2003c). The major decline between 1961 and 2000 was reported by Tanzania, where production fell from 200,000 to 20,000 tons in that period (FAO 2003c).

Fibers from *Musa* (banana and abaca), *Ciba, Patendra,* and *Bomba* species and coir from coconut palm are also used in many countries for local crafts, cloth, and other uses.

9.5.2 Wood Fibers

Fibers made of almost pure cellulose derived from wood pulp have been manufactured since the late 1800s. Rayon, the most widely known, was developed in France in the 1890s and was originally called "artificial silk" (Smith 2002). It has been commercially produced in the United States since 1910 (Fibersource 2004). In rayon production, purified cellulose is chemically converted into a soluble compound that is then passed through a "spinneret" to form soft filaments that are chemically treated or "regenerated" back into almost pure cellulose. The fibers are then used to produce cloth, cord, or other products. High-performance cords, such as those used in tires, were developed in the 1940s (Fibersource 2004).

At one time, rayon and cotton competed for similar end uses, but cotton's lower price gives it a competitive advantage. Rayon is moisture-absorbent, breathable, and easily dyed for use in clothing. It has moderate resistance to acids and alkalis and is generally not damaged by bleaches. As a cellulosic fiber, rayon will burn, but flame-retardant finishes can be applied. It is now manufactured primarily in Europe and Japan (Smith 2002), although production has declined almost 50% since 1980.

Lyocell is a more recently developed cellulosic fiber, which entered the consumer market in 1991 and was designated as a separate fiber group from rayon due to its unique properties and production processes. Lyocell is both biodegradable and recyclable, and virtually all of the chemicals used in production are reclaimed, making it a very environmentally friendly fiber (Smith 1999). Lyocell is stronger than cotton or linen both when dry and wet. These characteristics make it highly useful for a variety of clothing and similar uses. Since it is a manufactured fiber, the diameter and length of the fibers can be varied according to the desired end use, allowing the fiber to be substituted (or blended) for cotton- or silk-like appearances (Smith 1999). Industrial uses for lyocell include conveyor belts (due to its strength), cigarette filters, printers blankets, abrasive backings, carbon shields, specialty papers, and medical dressings (Smith 1999).

9.5.3 Animal Fibers

9.5.3.1 *Domestic Animals*

Animal skins and fibers such as wool and mohair are a staple in many societies' clothing and shelter. Most domesticated livestock provide multiple products such as milk, meat, and fiber. Ranching and herding occur largely in agricultural and dryland systems, and excessive grazing pressure is often cited as a driving force for degradation of those systems. As competition from synthetic fabrics has reduced the demand for wool in recent decades, wool production declined 16% between 1980 and 2000, after rising between 1960 and 1980. (See Table 9.13.) The number of live sheep declined 4.4% in that same period, but since the available FAO (2003c) data list all live animals together and do not differentiate those from which wool is harvested, the decline in wool animals is not clear.

The increase in hide production appears to reflect both increased population (associated with increased consumption of leather goods) and the growth of animal agriculture. FAO (2003c)

reports, for example, that the world population of live goats more than doubled between 1961 and 2000.

Skins and hides from domestic livestock are generally produced as a by-product of animals slaughtered for meat, so the trends in Table 9.13 are a reflection of the growth in animal agriculture, as well as increased demand. Wool production has stayed virtually unchanged over the last four decades, showing only a slight decline between 1961 and 2000. This appears to reflect the rough balance between increasing populations and reduced per capita wool usage as other fibers have replaced it in some markets.

9.5.3.2 *Wildlife*

Skins, furs, wools, and hairs from many species of wild mammals, reptiles, and even birds and fish are traded in the international market to make products ranging from clothing and accessories such as footwear, shawls, and wallets to ornaments and furnishings such as charms, rugs, and trophies. Consumers of these products range from local people in Southeast Asian communities using small pieces of tiger skin as magic amulets to ward off evil and illness, to the world's wealthy, wearing fashionable shahtoosh shawls made from the endangered Tibetan antelope.

The skins, hair, and furs from wild animals have been an important source of clothing and shelter for people throughout human history. In some cases today, this trade is putting further pressure on some of the world's most endangered species. For example, progress made over the years in stemming the demand for tiger bone medicine is being thwarted by what appears to be, in some countries, increased poaching of tigers for their skins. Despite their legal protection, the estimated illegal harvest of tens of thousands of Tibetan antelope annually for their wool has reduced populations to fewer than 75,000 animals, compared with an estimated 1 million at the beginning of the twentieth century. (See Box 9.2.)

For some species, the trade and use of skins and furs can be made sustainable. The revival of crocodilian populations in the wild is considered one of the great conservation success stories of the last quarter-century, demonstrating the effectiveness of the Convention on Trade of Endangered Species and sustainable use management programs. In 1969, all 23 species of crocodilians were threatened or had declining populations. Today, one third

---

BOX 9.2

Shahtoosh

The wool of the Tibetan Antelope (*Chiru Pantholops hodgsonii*), known as shahtoosh, is a valuable and widely traded commodity, despite the animal's protected status and a 23-year-old international trade ban. Unfortunately, the wool is not collected by combing or brushing the animal but by killing it, so that individual hairs can be plucked from the skin (www.traffic.org/25/wild4_3.htm).

Known as Chiru in its home range on the remote Qinghai-Tibetan Plateau of China, the Tibetan antelope lives at altitudes between 3,700 and 5,500 meters, with some animals venturing into the Ladakh region of India. More closely related to sheep and goats than to other antelope species, Tibetan antelope have developed a super-fine layer of hair to protect against the harsh plateau environment. IUCN classifies the Tibetan antelope as vulnerable to extinction.

Because items made from shahtoosh bring extraordinarily high prices, there is widespread poaching and smuggling of hides and wool. The harsh, remote region and the existence of well-armed and organized poaching gangs make law enforcement difficult and dangerous.

Table 9.13. World Production of Hides, Skins, and Greasy Wool, 1961–2000 (FAO 2003c)

Item	1961	1980	2000	1961–80	1980–2000
	(thousand tons)			(percent change)	
Cattle hides, fresh	4,070	5,655	7,389	39	31
Buffalo hides, fresh	322	488	811	52	66
Goat skins, fresh	261	390	840	49	115
Sheepskins, fresh	929	1,106	1,598	19	45
Total	**5,582**	**7,639**	**10,638**	**37**	**39**
Greasy wool	2,619	2,794	2,346		

of crocodilians can sustain a regulated commercial harvest and only four species are critically endangered.

The most likely cause of changes in wildlife-derived skins and fibers will be the ability of governments to control the poaching and trade in the skins and fibers from animals threatened with extinction. Where successful conservation efforts can result in a sustainably harvested supply, production will be maintained. Demand for particular animal products (and the resulting prices) may, in some instances, be driven more by fashion trends or cultural demands than by ecosystem conditions.

9.6 Sustainability of Timber, Fuel, and Fiber Services

While there are local and regional exceptions, the global production and consumption of most timber, fuel, and fiber goods over the last four decades has increased significantly, although the rate of increase has slowed during the past decade. In the process, the continents have become more interconnected through international trade, the value of which has grown much faster than global wood products output.

The impacts on forest ecosystems due to this increased production are difficult to generalize. In some cases, timber harvesting has directly contributed to degrading and deforesting forest ecosystems, most recently in tropical areas. This is particularly true where institutional controls are weak and where destructive and often illegal logging practices are common. In other situations, where modern forest technologies and effective governance occur, forest area and measures of condition are holding steady or improving in the face of increased production. While many negative impacts can be seen immediately, the full impacts on forests (either positive or negative) from particular harvesting practices may not be evident for many years.

For the near future, total global wood supplies are predicted to remain adequate for most market demands, if not in surplus. Increases in demand for forest products in the near future are likely to be met by increases in supply and are unlikely to create significant price increases that would create hardship on consumers. That does not apply to premium species or the high-quality woods that have been overharvested in the past. For the near future, those will be in short supply and will need to be replaced by other species or manufactured products or substitutes.

Fuelwood is a special case, largely because it is so important to the people who depend on it for heating and cooking. It is produced, harvested, and consumed locally in the regions where it is a critical factor in family well-being. Local assessments of fuelwood and its relationship to both ecosystems and communi-

ties are feasible and needed, as national or regional assessments of overall fuelwood adequacy mask critical community shortages.

In most parts of the world, changes in the production of the timber, fuel, and fiber in the near future (10–15 years) will be caused primarily by political, social, and economic forces rather than changes in the capacity of ecosystems to produce these services. Exceptions may be the capacity of local ecosystems to meet fuelwood demands in some rural subsistence economies, particularly in Africa, and the reduction in availability of some wildlife skins and fibers due to population declines. Where institutional capacity is weak, significant increases in industrial production of wood or fibers such as cotton can cause adverse impacts on local environments or disadvantage traditional users who relied on the ecosystem for food, shelter, medicines, or other nonindustrial products.

Due to increasing globalization of investments and trade, some of the more important policy impacts on the supply of forest products may come in the form of trade or transportation policies; subsidies or taxes for production, transportation, or manufacturing; economic development; or monetary policy. These policies are made outside the forestry and agricultural sectors, often for vastly different reasons of national interest.

The search for a sustainable future involves important challenges in the provision of timber, fuel, and fiber. Some of those challenges include:

- the skillful management of planted and natural forests to supply wood crops, employ local people, and support improvements in literacy, housing, nutrition, and health;

- the application of wise policy decisions that consider industrial production, environmental quality, and local communities and are supported by sufficient governance to achieve their objectives;

- the development and dispersal of science and technology to improve efficiencies in wood and fiber production and use, to protect important biodiversity, watershed, and social values, and to contribute to the alleviation of poverty; and

- reduced pressure on remaining natural forests to provide habitat for wild species and people whose future is threatened by the loss of those forest types.

These challenges include, but go beyond, science, management, technology, and laws. They are fundamental social issues that each society and nation will have to tackle.

In terms of the assessment made in this chapter, as noted earlier, global generalizations tend to mask local dynamics. Since policy is made locally, nationally, and regionally, the need to understand and assess local and regional conditions is a critical precursor to informed policy-making, and it is hoped that the global context illustrated here will contribute to those efforts.

References

Adams, M. and J. Castano, 2002: World timber supply and demand scenario, government interventions, issues and problems, *Proceedings of the International Conference on Timber Plantation Development,* Manila, 7–9 November, FAO, Rome, Italy.

Allen, R.B., K.H. Platt, and S.K. Wiser, 1995: Biodiversity in New Zealand plantations, *New Zealand Forestry,* **39(4),** pp. 26–9.

Arnold, M., G. Köhlin, R. Persson, and G. Shepherd, 2003: Fuelwood revisited: What has changed in the last decade? CIFOR occasional paper no. 39, CIFOR, Jakarta, Indonesia.

Bansal, A.K. and S.S. Zoolagud, 2002: Bamboo composites: Material of the future, *Journal of Bamboo and Rattan,* **1,** pp. 119–30.

Belcher, B., 1999: The bamboo and rattan sectors in Asia: An analysis of production-to consumption systems, INBAR working paper 22, International Network for Bamboo and Ratan, Beijing, China.

Berlik, M.M., D.B. Kittredge, and D.R. Foster, 2002: The illusion of preservation: A global environmental argument for the local production of natural resources, Harvard forest paper no. 26, Harvard University, Petersham, MA, 24 pp.

Brown, C., 2000: The global outlook for future wood supply from forest plantations, Working paper no. GFPOS/WP/03, FAO Forestry Policy and Planning Division, February, FAO, Rome, Italy.

Bystriakova, N., V. Kapos, C. Stapleton, and L. Lysenko, 2003: *Bamboo Biodiversity:* Information for planning, conservation and management in the Asia-Pacific region, United Nations Environmental Program, Nairobi, Kenya/ World Conservation Monitoring Center/INBAR, Beijing, China.

Bystriakova, N., V. Kapos, and L. Lysenko, 2004: *Bamboo* Biodiversity: Africa, Madagascar and the Americas, United Nations Environmental Program, Nairobi, Kenya/ World Conservation Monitoring Center, Cambridge, UK/ INBAR, Beijing, China.

Chen, X., 2001: Wood residues in China, Appendix 1. In: Trash or treasure? Logging and mill residues in Asia and the Pacific, T. Enters (ed.), RAP publication 2001/16, FAO, Bangkok, Thailand.

CIFOR (Center for International Forestry Research), 2002: Planning for woodcarving in the 21st century, InfoBrief April no. 1, CIFOR, Bogota, Indonesia.

Clawson, Marion, 1979: Forests in the long sweep of history, *Science,* **204(4398),** pp. 1168–74.

Contreras-Hermosilla, A., 2002: Law compliance in the forestry sector: an overview, WBI working paper, The World Bank, Washington, DC, 47 pp.

Cossalter, C. and C. Pye-Smith, 2003: *Fast-wood Forestry: Myths and Realities,* CIFOR, Jakarta, Indonesia.

Deng, K., 1995: Renewable energy benefits rural women in China, Chapter 7. In: *Energy as an Instrument of Socio-Economic Development,* J. Goldemberg and T. B. Johansson (eds.), United Nations Development Programme, New York, NY.

Dransfield, S. and E.A.Widjaja (eds), 1995: *Plant Resources of South-East Asia No 7: Bamboos,* Backhuys Publishers, Leiden, The Netherlands.

Durst, P.B., W. Killmann, and C. Brown, 2004: Asia's new woods, *Journal of Forestry,* **102(4),** pp. 46–53.

Durst, P.B. and C. Brown, 2002: Current trends and development of plantation forestry in Asia Pacific countries, *Proceedings of the International Conference on Timber Plantation Development,* Manila, 7–9 November, FAO, Rome, Italy.

Energy Information Administration, 1994: Manufacturing consumption of energy, US Department of Energy report DOE/EIA-0512(91), US Government Printing Office, Washington, DC.

Enters, T., 2001: Trash or treasure? Logging and mill residues in Asia and the Pacific, RAP publication 2001/16, FAO, Bangkok, Thailand.

ESMAP (Energy Sector Management Assistance Programme), 2000: Energy strategies for rural India: Evidence from six states, World Bank document, August, Joint United Nations Development Programme, New York, NY/ World Bank, Washington, DC.

Estades, C.F. and S.A. Temple, 1999: Deciduous-forest bird communities in a fragmented landscape dominated by exotic pine plantation, *Ecological Applications,* **9,** pp. 573–85.

EU (European Union), 1997: *Panorama of EU Industry 1997,* European Commission, Brussels, Belgium.

FAO (Food and Agriculture Organization of the United Nations), 1995: *State of the World's Forests,* FAO, Rome, Italy.

FAO, 1998a: Asia-Pacific forestry sector outlook study: Status, trends and prospects for non-wood and recycled wood in China, Working paper no. APF-SOS/WP/35, FAO, Rome, Italy.

FAO, 1998b: Global fibre supply model, FAO, Rome, Italy. Available at http://www.fao.org/DOCREP/006/X0105E/X0105E00.HTM.

FAO, 1999: *State of the World's Forests, 1999:* FAO, Rome, Italy.

FAO, 2000a: Wood energy, climate and health: International expert consultation, Field document no. 58, Regional Wood Energy Development Programme in Asia (GCP/RAS/154/NET), FAO, Bangkok, Thailand.

FAO, 2000b: Forest resources of Europe, CIS, North America, Australia, Japan and New Zealand, Geneva Timber and Forest study papers 17, United Nations, New York, NY/ Geneva, Switzerland.

FAO, 2000c: The global outlook for future wood supply from forest plantations, Working paper no. GFPOS/WP/03, FAO, Rome, Italy.

FAO, 2001a: Global Forest Resources Assessment 2000, FAO forestry paper 140, FAO, Rome, Italy. Available at www.fao.org.

FAO, 2001b: Past trends and future prospects for the utilization of wood for energy, Global Forest Products Outlook Study, FAO, Rome, Italy.

FAO, 2003a: African forests: A view to 2020, In: *Forestry Outlook Study for Africa,* FAO, Rome, Italy.

FAO, 2003b: *State of the World's Forests 2003,* FAO, Rome, Italy.

FAO, 2003c: FAOSTAT statistics database [online], FAO, Rome, Italy. Available at www.fao.org.

FAO, 2004: NWFP digest no. 4–04, FAO Forestry Department, FAO, Rome, Italy.

Feng, Lu, 2001: China's bamboo product trade: Performance and prospects, INBAR working paper no. 33, INBAR, Beijing, China.

Fibersource, 2004: Rayon fiber (viscose) fact sheet. Available at www.fibersource.com/f-tutor/rayon.htm.

FWI/GFW (Forest Watch Indonesia/Global Forest Watch), 2002: *The State of the Forest: Indonesia,* Bogor, Indonesia/ Washington, DC.

Gillham, F.E.M., T.M. Bell, T. Arin, G.A. Mathews, C. Le Rumeur, et al., 2003: Cotton production prospects for the next decade, World Bank technical paper no. 287. Available at http://www.icac.org.

Girard, P., 2002: Charcoal production and use in Africa: What future? *Unasylva,* **53 (211).**

Gitay, H., S. Brown, W. Easterling, B. Jallow, et al., 2001: Ecosystems and their goods and services. In: *Climate Change 2001: Impacts, Adaptation, and Vulnerability,* J. McCarthy, O. Canziani, N. Leary, D. Dokken, and K. White (eds.), Contribution of Working Group II to the third assessment report of the Intergovernmental Panel on Climate Change, Cambridge University Press, Cambridge, UK, pp. 235–342.

Global Witness, 2002: *Branching Out: Zimbabwe's Resource Colonialism in Democratic Republic of Congo,* Global Witness, London, UK.

Gonzalez, P., 2001a: Deserti?cation and a shift of forest species in the West African Sahel, *Climate Research,* **17,** pp. 217–28.

Gonzalez, P., 2001b: Energy use: Human. In: *Encyclopedia of Biodiversity,* S. Levin (ed.), Academic Press, San Diego, CA.

Gordon, J.C., R.N. Sampson, and J.K. Berry, 2004: The challenge of maintaining working forests at the WUI. In: *Forests at the Wildland-Urban Interface: Conservation and Management,* S.W. Vince, M.L. Duryea, E.A. Macie, and L.A. Hermansen (eds.), CRC Press, Boca Raton, FL, pp.15–23.

Grass Phylogeny Working Group, 2001: Phylogeny and subfamilial classification of the grasses (Poaceae), *Annals of the Missouri Botanical Garden,* **88,** pp. 373–457.

Gupta, G. and G. Kohlin, 2003: Preferences in urban domestic fuel demand: The case of Kolkata, India, Environmental Economics Unit, Department of Economics, Goteborg University, Goteborg, Sweden.

Hazley, C., 2000: *Forest-based and related industries of the European Union: Industrial districts, clusters and agglomerations,* ETLA (Elinkleinoelaman Tutkimuslaitos/Research Institute of the Finnish Economy), Helsinki, Finland.

Holdren, J.P., 1991: Population and the energy problem, *Population and Environment,* **12,** pp. 231–55.

Houghton, J.T., L.G., Meira Filho, B. Lim, K. Treaton, I. Mamaty, et al., (eds), 1997: Revised 1996 IPCC guidelines for national greenhouse gas inventories, Intergovernmental Panel on Climate Change. Available at http://www.ipcc-nggip.iges.or.jp/public/gl/invs1.htm.

Hunt, N., 2001: Review of PIRA non-wood fibre conference. Available at http://www.tappsa.co.za/html/review_of_pira.html.

Hunter, I.R. and W. Junqi, 2002: Bamboo biomass, INBAR working paper no. 36, INBAR, Beijing, China.

IEA (International Energy Agency), 2002a: *World Energy Outlook, 2002,* IEA, Paris, France.

IEA, 2002b: Renewables in global energy supply: An IEA fact sheet, International Energy Agency, November, IEA, Paris. France. Available at www.iea.org.

IEA, 2003: Renewables information: 2003, IEA, Paris, France. Available at www.iea.org.

IIED (International Institute for Environment and Development), 1996: *Towards a Sustainable Paper Path,* World Business Council for Sustainable Development, IIED, London, UK.

INBAR (International Network for Bamboo and Rattan), 1999: Socio-economic issues and constraints in the bamboo and rattan sectors: INBAR's assessment, INBAR working paper no. 23, Beijing, China.

Khan, M.L., S. Menon, and B.S. Bawa, 1997: The effectiveness of the protected area network on biodiversity conservation: A case study in Meghalaya state, *Biodiversity and Conservation,* **6,** pp. 853–68.

Klass, D.L., 1998: *Biomass for Renewable Energy, Fuels, and Chemicals,* Academic Press, San Diego, CA, 651 pp.

Klass, D.L., 2002: An introduction to biomass energy: A renewable resource. Available at http://www.bera1.org/about.html.

Klass, D.L., 2003: A critical assessment of renewable energy usage in the USA, *Energy Policy,* **31(2003),** pp. 353–67.

Koopmans, A. and J. Koppejan, 1998: Agricultural and forest residues: Generation, utilization and availability. In: *Proceedings of the Regional Expert Consultation on Modern Applications of Biomass Energy,* FAO Regional Wood Energy Development Programme in Asia, Report no. 36, FAO, Bangkok, Thailand.

Kremsater, L. and F.L. Bunnell, 1999: Edge effects: Theory, evidence and implications to management of western North American forests. In: *Forest Fragmentation: Wildlife and Management Implications,* J.A. Rochelle, L.A. Lehmann, and J. Wisniewski (eds.), Brill, Leiden, The Netherlands, pp. 117–53.

Li, B.L., S. McKeand, and R. Wier, 2000: Impact of forest genetics on sustainable forestry: Results from two cycles of loblolly pine breeding in the U.S., *Journal of Sustainable Forestry,* **10,** pp. 79–85.

Lindenmayer, D.B. and J.F. Franklin, 2002: *Conserving Forest Diversity: A Comprehensive Multi-scaled Approach,* Island Press, Washington, DC.

Londoño, X., 2001: Evaluation of bamboo resources in Latin America: A summary of the final report of the project 96–8300–01-4, INBAR, Beijing, China.

Lugo, A.E., J.A. Parrotta, and S. Brown, 1993: Loss in species caused by tropical deforestation and their recovery through management, *Ambio,* **22,** pp. 106–9.

Matziris, D.I., 2000: Genetic variation and realized genetic gain from Aleppo pine tree improvement, *Silvae Genetica,* **49,** pp. 5–10.

Nambiar, E.K.S. and A.G. Brown, (eds.), 1997: Management of soil, nutrients and water in tropical plantation forests, ACIAR monograph 43, Canberra, Australia.

Newton, P.F., 2003: Systematic review of yield responses of four North American conifers to forest tree practices, *Forest Ecology and Management,* **172,** pp. 29–51.

Oliver, C.D., 1999: The future of the forest management industry: Highly mechanized plantations and reserves or a knowledge-intensive integrated approach? *Forestry Chronicle,* **75,** pp. 229–45.

Pande, H., 1998: Non-wood fibre and global fibre supply, *Unasylva,* **49,**

Pandey, D. and C. Brown, 2000: Teak: A global overview, *Unasylva,* **51,**

Perry, I.W.H., 1999: Productivity trends in the natural resources ndustries: A cross cutting analysis. In: *Productivity in Natural Resource Industries,* R.D. Simpson, (ed.), Resources for the Future, Washington, DC.

Pulkki, R., 1998: Conventional versus environmentally sound harvesting: Impacts on non-coniferous tropical veneer log and sawlog supplies, *Unasylva,* **49.**

Ramage, J. and J. Scurlock, 1996, Biomass. In: *Renewable Energy-power for a Sustainable Future,* G. Boyle (ed.), Oxford University Press, Oxford, UK.

Rasmussen, J.N., A. Kaosa-ard, T.E. Boon, M.C. Diaw, K. Edwards et al., 2000: For whom and what? Principles, criteria and indicators for sustainable forest resources management in Thailand, DFLRI Report 6–2000, Danish Forest and Landscape Research Institute, Hoersholm, Denmark.

Reuters, 2002: China GMO cotton bad for environment: Greenpeace, Reuters News Service, 4 June.

Reuters, 2004: Monsanto-backed survey sees India GMO cotton bonus, Reuters News Service, 26 March.

Rice, R.E., R.E. Gullison, and J.W. Reid, 1997: Can sustainable management save tropical forests, *Scientific American,* **276,** pp. 44–9.

Roda, J.-M., 2002: Le point sur la place des bois tropicaux dans le monde, *Bois et Forêts des Tropiques,* **274,** pp. 79–80.

RWEDP (Regional Wood Energy Development Program in Asia), 1997: Regional study on wood energy today and tomorrow, Publication FD50, RWEDP, Bangkok, Thailand. Available at http://www.rwedp.org/public .html.

Sampson, R.N. and L.A. DeCoster, 2000: Forest fragmentation: Implications for sustainable private forests, *Journal of Forestry,* **98(3),** pp. 4–8.

Sampson, R.N., M.S. Smith, and S.B. Gann, 2001: Western forest health and biomass energy potential: A report to the Oregon Office of Energy, Oregon Office of Energy, Salem, Oregon, 53 pp. Available at http://www.energy .state.or.us/biomass/forest.htm.

Sampson, R. N., L.L. Wright, J.K. Winjum, J.D. Kinsman, J. Benneman, et al., 1993: Biomass management and energy, *Water, Air, and Soil Pollution,* **70(1–4),** pp. 139–59.

Sanchez, P.A., 2002: Soil fertility and hunger in Africa, *Science,* **295(5562),** pp. 2019–20.

SCBD (Secretariat to the Convention of Biological Diversity), 2003: Interlinkages between biodiversity and climate change, SCBD technical series no. 10, SCBD, Montreal, Canada.

Scholes, R.J. and R. Biggs (eds), 2004: *Ecosystem Services in Southern Africa: A Regional Assessment,* The Regional Scale Component of the Southern African Millennium Ecosystem Assessment (SAfMA), Council for Scientific and Industrial Research, Pretoria, South Africa. Available at www.millennium assessment.org.

Sedjo, R., 2004: Transgenic trees: Implications and outcomes of the Plant Protection Act, RFF discussion paper 04–10, Resources for the Future, Washington, DC.

Sedjo, R.A. and K.S. Lyon, 1990: The long-term adequacy of world timber supply, Resources for the Future, Washington, DC.

Sheingauz, A.S. (ed.), 2001: *Forest Complex of Khabarovsk Kray,* RIOTIP, Khabarovsk, Russia, 255 pp. [in Russian]

Skog, K. and G. Nicholson, 1998: Carbon cycling through wood products: The role of wood and paper products in carbon sequestration, *Forest Products Journal,* **48 (7/8),** pp. 75–83.

Smith, J., K. Obidzinski, Subarudi, and I. Suramenggala, 2003: Illegal logging, collusive corruption and fragmented governments in Kalimantan, Indonesia, *International Forestry Review,* **5(3),** pp. 293–302.

Smith, J.A., 2002: Rayon: The multi-faceted fiber, Ohio State University Extension fact sheet HYG-5538–02. Available at http://ohioline.osu.edu/hyg-fact/5000/5538.html.

Smith, J.A., 1999: Lyocell: One fiber, many faces, Ohio State University Extension fact sheet HYG-5572–99. Available at http://ohioline.osu.edu/hyg-fact/5000/5572.html.

Sun, X., E. Katsigiris, and A. White, 2004: Meeting China's demand for forest products: An overview of import trends, rorts of entry, and supplying countries, with emphasis on the Asia—Pacific Region, Forest Trends, Washington, DC/ Center for Chinese Agricultural Policy (CCAP), Beijing, China/ CIFOR, Bogota, Indonesia.

Sunderland, T.C.H., J. Dransfield, J, 2000: Species profiles rattans. In: Rattan: Current research issues and prospects for conservation and sustainable development, Expert Consultation on Rattan Development, Non-Wood Forest Products No. 14, FAO, Rome, Italy.

TERI (Tata Energy Research Institute), 2000: Case studies in joint forest management, TERI Press, New Delhi, India.

Taylor, R.E., 2000: Substitute products to lumber growing rapidly, Press release, Wood Markets Research Inc., Vancouver, BC, Canada.

Thompson, I., G. Patterson, S. Leiner, R. Nasi, C.N. de P. Pola, et al., 2002: Review of the status and trends of, and major threats to, forest biological diversity, Technical Series No. 7, Secretariat of the Convention on Biological Diversity, Montreal, Canada.

TMFWI (Tripartite Meeting on Forestry and Wood Industries), 2001: *Globalization and Sustainability: The Forest and Wood Industries on the Move,* Report for discussion at the tripartite meeting of the social and labour dimensions of the forestry and wood industries on the move, International Labour Organization, Geneva, Switzerland.

Tyrrell, M.L., M.H.P. Hall, and R.N. Sampson, 2004: *Dynamic Models of Land Use Change in the Northeastern USA,* GISF (Global Institute of Sustainable Forestry) research paper 003, Yale University School of Forestry & Environmental Studies/ State University of New York College of Environmental Science and Forestry (SUNY-ESF), 76 pp.

UNFF (United Nations Forum on Forests), 2003: The role of planted forests in sustainable forest management, Report of the expert group meeting, 25–27 March, New Zealand, UNFF, New York, NY. Available at http://www.maf-.govt.nz/mafnet/unff-planted-forestry-meeting/report-of-unff-meeting-nz .pdf.

US Bureau of Labor Statistics, 2004: Bureau of Labor Statistics data. Available at www.bls.gov.

USDA-ERS (US Department of Agriculture–Economic Research Service), 2003: *Cotton and Wool Situation Outlook Yearbook* (CWS 2003), USDA-ERS, Washington, DC.

USDA Forest Service, 2001: 1997 RPA final tables. Available at www.fs.fed .us.

USDA Forest Service, 2003: An analysis of the timber situation in the United States: 1952 to 2050, Gen. Tech. Rep. PNW-GTR-560, USDA Forest Service, Pacific Northwest Research Station, Portland, OR. Available at www .fs.fed.us/pnw/pubs/gtr560.

Van Kooten, G.G., H.W. Nelson, and I. Vertinsky, 2004: Certification of sustainable forest management practices, a global perspective on why countries certify, *Forest Policy and Economics.* In press.

WHO (World Health Organization), 2002: *The World Health Report 2002: Reducing the Risks, Promoting Healthy Life,* Geneva, Switzerland, 248 pp.

Williams, M., 2003: *Deforesting the Earth,* University of Chicago Press, Chicago, IL, 689 pp.

Winjum, J.K., S. Brown, and B. Schlamadinger, 1998: Forest harvests and wood products: Sources and sinks of atmospheric carbon dioxide, *Forest Science,* **44(2),** pp. 272–84.

World Energy Council (WEC), 2001: *Survey of Energy Resources,* 19th ed., WEC, London, UK.

WRI (World Resources Institute), 2000a: *A Guide to World Resources 2000–2001: People and Ecosystems: The Fraying Web of Life,* WRI, Washington, DC.

WRI, 2000b: An overview of logging in Cameroon, WRI, Washington, DC.

WWF (World Wide Fund for Nature/World Wildlife Fund), 2000: *Squandering Paradise? The Importance and Vulnerability of the World's Protected Areas,* N. Dudley (ed.), WWF, Gland, Switzerland.

WWF, n.d.: The forest industry in the 21st century. Available at http://www .wwf.de/imperia/md/content/pdf/waelder/nutzungneu/forest_industrie.pdf.

Chapter 10

New Products and Industries from Biodiversity

Coordinating Lead Author: Andrew J. Beattie
Lead Authors: Wilhelm Barthlott, Elaine Elisabetsky, Roberta Farrel, Chua Teck Kheng, Iain Prance
Contributing Authors: Joshua Rosenthal, David Simpson, Roger Leakey, Maureen Wolfson, Kerry ten Kate
Review Editor: Sarah Laird

---

*This appears in Appendix A at the end of this volume.

Main Messages

Bioprospecting is the exploration of biodiversity for new biological resources of social and economic value. It is carried out by a wide variety of industries that include pharmaceuticals, botanical medicines, crop protection, cosmetics, horticulture, agricultural seeds, environmental monitoring, manufacturing, and construction. There are between 5 million and 30 million species on Earth, each one containing many thousands of genes. However, fewer than 2 million species have been described, and knowledge of the global distribution of species is limited. History reveals that less than 1% of species have provided the basic resources for the development of all civilizations thus far, so it is reasonable to expect that the application of new technologies to the exploration of the currently unidentified and overwhelming majority of species will yield many more benefits for humanity.

Biodiversity is the fundamental resource for bioprospecting, but it is rarely possible to predict which genes, species, or ecosystems will become valuable for bioprospecting in the future. A wide variety of species—microbial, plant, and animal and their genes—have provided services, products, blueprints, or inspiration for products or the basis of industries. While species-rich environments such as tropical forests may be expected to supply many products in the long term, bioprospecting thus far has yielded valuable products from many diverse ecosystems, including temperate forests and grasslands, arid and semiarid lands, freshwater ecosystems, and montane and polar regions, as well as cold and warm oceans. In this context, the conservation of all biodiversity in all ecosystems would provide the most opportunities for bioprospecting in the future.

Well-regulated bioprospecting contributes to the joint goals of ecosystem conservation and social and economic development through partnerships and benefit-sharing. Bioprospecting can achieve multiple goals: generating revenues for protected areas, conservation projects, and local communities; building scientific and technological capacity to study and manage biodiversity; enhancing biodiversity science; raising awareness of the commercial and noncommercial importance of biodiversity; creating businesses dependent upon the sustainable management of resources; and, in rare instances, generating large profits for corporations and shareholders. These benefits may occur at local, regional, or national scales.

Market trends vary widely according to the industry and country involved, but many bioprospecting activities and revenues are expected to increase over the next decades. Several major new industries, such as bioremediation and biomimetics are well established and appear set to increase, while others have a less certain future. The current economic climate suggests that pharmaceutical bioprospecting is likely to increase, especially as new methods that use evolutionary and ecological knowledge enhance productivity.

Bioprospecting is one part of a package of economic activities that, when carefully implemented, use biodiversity in a way that contributes to the multiple objectives of the sustainable management of natural resources, poverty reduction, and economic development. Established biodiversity-based industries such as farming, forestry, grazing, and fisheries, along with local uses of biodiversity for foods, medicines, and fibers and for cultural activities and the development of new industries such as bioremediation, ecological restoration, and biomimetics, generate knowledge of and respect for the multiple benefits of biodiversity. While recent research clearly demonstrates the future resource potential of biodiversity, opportunities for bioprospecting industries in any given country will depend on many factors, ranging from the conservation status of its biodiversity to the trends in a variety of markets.

Global threats to biodiversity, and especially species losses, may affect the development of valuable new products for humanity, including medicines, industrial processes, and new crop varieties. The current global decline of biodiversity may affect bioprospecting in many ways. Serious undervaluation of such losses for bioprospecting result from a lack of recognition that a high proportion of commercially important species are either small or microscopic, and so losses go undetected. Other threats include loss of traditional knowledge, the impacts of some kinds of modern agricultural technologies, and depletion of natural resources.

Bioprospecting partnerships are increasingly supported by international and national laws and self-regulation measures, including codes of ethics, high-quality contracts, and transparent institutional policies that result in benefit-sharing. Recent international agreements include the 1992 Convention on Biological Diversity and the 2001 International Treaty on Plant Genetic Resources for Food and Agriculture. More than 100 countries have introduced or are developing laws and other policy measures that complement these international initiatives, regulating access to biological resources and benefit-sharing. Further, a range of documents developed by indigenous communities, researchers, professional associations, and bioprospecting companies has generated a significant shift in the ethical and legal framework within which bioprospecting operates. Nevertheless, serious issues remain, including achieving an appropriate balance between benefit-sharing and the creation of incentives for investment. These often-conflicting interests among potential partners may operate across local, national, and international scales.

10.1 Introduction

The number of species of use to humanity runs into many thousands, and those that form the basis of contemporary agriculture are well known—not least the major crops and domesticated animals that provide food (Baker 1978; Clutton-Brock 1999). Indigenous peoples use a very wide range of lesser-known species and often possess deep ecological knowledge that helps maintain the ecosystems in which they live (Myers 1983; Malaisse 1997).

In this context, it is widely assumed that the biological resources of the world have been thoroughly explored. Recent research shows that our knowledge of biodiversity is still very limited, however, and that the exploration of all types of organisms is likely to yield many more useful species for an unexpectedly wide variety of human needs and pursuits. To place this in perspective, flowering plants (angiosperms) have provided a wide variety of foods, drugs, cosmetics, fibers, and building materials. But it is now clear that this group of organisms, in its entirety, constitutes only a minor part of the total number of species on Earth and that vast resources remain in other species-rich groups such as the microbes and invertebrates (Wilson 1992; Torsvik et al. 2002; Crawford and Crawford 1998; Eisner 2003).

Some environments are also little explored. This is especially true of the oceans, where current exploration is revealing many new species every week and scientists expect to discover at least 2 million marine species over the next two or three decades. This may be an underestimate, however, as the number of species of marine nematode worms alone has been estimated at 1 million (Malakoff 2003). Even apparently well known groups such as the mammals and reptiles are revealing new species (Beattie and Ehrlich 2004), and recently a new family of frogs has been discovered in southern India—a major surprise, as amphibians have been studied intensively for decades (Hedges 2003).

The importance of the exploration of biodiversity for new products was recognized at the 1990 meeting of the International Society of Chemical Ecology in Goteborg, Sweden, in the Goteborg Resolution (Eisner and Meinwald 1990): "Natural products constitute a treasury of immense value to humankind. The current alarming rate of species extinction is rapidly depleting this treasury, with potentially disastrous consequences. The International Society of Chemical Ecology urges that conservation measures be mounted worldwide to stem the tide of species extinction, and that vastly increased biorational studies be undertaken aimed at discovering new chemicals of use to medicine, agriculture and industry. These exploratory efforts should be pursued by a partnership of developing and developed nations, in such fashion that the financial benefits flow in fair measure to all participants."

This chapter explores modern and emerging biodiversity-based products and industries and largely excludes traditional ones that have been developed throughout history. However, traditional uses of biodiversity are included when they have contributed to new ventures. The next section presents the multiple and disparate facets of bioprospecting across a wide range of industries. The third section discusses the variety of partnerships and benefit-sharing arrangements that have developed worldwide. The fourth section reviews the legal environment for bioprospecting, and the final section summarizes the major threats to the industry.

10.2 Overview of Industries Involved in Bioprospecting

Bioprospecting involves the use of a wide variety of species by a wide variety of industries (ten Kate and Laird 1999; Beattie and Ehrlich 2004). Some examples are provided in Table 10.1. The resource values of the species concerned to date have differed fundamentally in nature—in some cases it is the organism itself that provides the product, while in others the organism serves as a model or as inspiration for a copy, modified or otherwise. The examples given here are a small part of a much longer list and have been selected because they are either the subject of major ongoing investment or already a commercial reality.

Discovery is often achieved by considering where the desired product might have evolved naturally. Habitats or a group of organisms are then identified and explored. An example of the discovery and development of a product with self-cleaning properties is presented in Box 10.1.

Another example is the search for heat-tolerant industrial enzymes. As most enzymes are destroyed by heat, some industrial processes would be greatly enhanced if heat-tolerant enzymes were discovered. The question of where heat-tolerant enzymes would be expected to occur was pursued by exploring the microbial biodiversity of thermal springs. These habitats revealed microbes with heat-stable enzymes that are being applied to a variety of industrial processes, including paper and pulp manufacturing, biotechnology, commercial cleaning, and forensic science, with each generating important benefits or major revenues (e.g., see Moss et al. 2004). A possible new source of industrial enzymes of this type is the recently discovered bacterium *Pyrodictium,* which inhabits hydrothermal vents and can grow at temperatures between 85 and 121 degrees Celsius (Kashefi and Lovley 2003).

Other methods of drug discovery include combinatorial chemistry and rational drug design. While these have been developed independently of natural products, current thought is that natural products are likely to provide the best lead-molecules in the future (Chapman 2004; Ortholand and Ganesan 2004).

10.2.1 Pharmaceutical Bioprospecting

Interest in novel products from biodiversity has varied greatly in the last decade, with a general decline in pharmaceutical bioprospecting by major companies, although a resurgence is expected (Chapman 2004). Based on the knowledge that many important drugs, such as aspirin, were derived from natural products (Jack 1997)—that is, generated in the tissues of native species—the industry has at various times invested heavily in the exploration of species-rich communities such as rain forests and coral reefs in search of commercially profitable pharmaceuticals (Ismail et al. 1995; Bailey 2001).

Alarming levels of antibiotic resistance in many human pathogens is likely to provoke an increase in pharmaceutical bioprospecting, which remains a vital source of lead drug discovery (Wessjohann 2000; McGeer and Low 2003; Newman et al. 2003). Malaria, one of the world's most deadly diseases, has been treated historically with drugs derived from natural products—quinine, chloroquine, mefloquine, and doxycycline—and today the artemisinins derived from the Chinese herb Qinghao (*Artemesia annua*) are at the forefront of the battle against this parasite.

Some compounds from natural resources approved for marketing during the 1990s in the United States and various other countries are shown in Table 10.2. The probability that any single discovery actually reaches the marketplace remains low, however. For example, 75% of the drugs that entered phase 1 clinical trials in the Untied States in 1991 went on to phase 2, 36% entered phase 3, and only 23% received FDA approval. From another perspective, the probability of a drug being launched into the market was 5–10% during the pre-clinical research and development phase, 30% during phase 2A, 40% during phase 2B, 70% in phase 3, and 90% during the period of regulatory review (ten Kate and Laird 1999). This is because the conventional process of drug discovery has several distinct and increasingly expensive stages: acquisition of the natural material; extraction of the active compounds; primary screening against a range of human disease organisms; isolation and chemical characterization of the active compounds; secondary screening assaying the compounds in tissue cultures and experimental animals; structural chemistry and synthesis; pre-clinical development with a view to human trials; and clinical development, marketing, and distribution.

The magnitude of the resource was illustrated by Henkel et al. (1999), who provided a summary of the wide range of organisms from which drugs have been derived, including bacteria and fungi (both terrestrial and marine), plants, algae, and a variety of invertebrates, including worms, insects and mollusks. (See Figure 10.1 in Appendix A.)

Munro et al. (1999) demonstrated the importance of marine animals among diverse organisms screened for clinically significant cytotoxicity (such as is useful for anti-cancer drugs) and compared the relative importance of terrestrial versus marine organisms for this particular pharmaceutical activity. (See Figure 10.2.) They also showed the widespread distribution of this cytotoxicity among marine phyla, reminding us that many are relatively little known either to the general public or to the bulk of scientists. They include the Porifera (sponges), Bryozoa (sea mosses), Cnidaria (jellyfish), and Echinodermata (starfish and their relatives). (See Figure 10.3.)

Natural products are still important sources of novel compounds for pharmaceuticals. An average of 62% of new, small-molecule, nonsynthetic chemical entities developed for cancer research over the period 1982–2002 were derived from natural products. In antihypersensitive drug research, 65% of drugs currently synthesized can be traced to natural structures. This emphasizes the important role of many natural products as blueprints rather than the actual end points. Newman et al. (2003), who assembled these data, noted that they had not been able to identify a *de novo* combinatorial compound approved as a drug during

Table 10.1. Novel Products and Industries and the Organisms They Come From. The examples shown have either been the subject of major investment and research or have become commercial products. (Classification from Margulis and Schwartz 1998)

Category	Common Name	Phylum	Ecosystem of Origin
Products			
Antibiotics	ants, mollusks, plants, bacteria	Mandibulata, Mollusca, Anthophyta, Actinobacteria	terrestrial (e.g., temperate and tropical forests), marine
Antifreeze, cryoprotectants	fish, water bears	Craniata, Tardigrada	polar, marine, montane
Cold-active enzymes	fungi	Ascomycota	Antarctica
Self-cleaning surfaces/paints	various plants	Anthophyta	terrestrial (including wetlands)
Architectural design	termites	Mandibulata	mounds from tropical arid ecosystems
Fire detection devices	fire beetles	Mandibulata	temperate forest
Pest repellants	various insects	Mandibulata	terrestrial (including temperate forests and grasslands)
High-tensile fibers	spiders, moths	Chelicerata, Mandibulata	terrestrial (most ecosystems)
Surgical drugs	scorpions, wasps	Chelicerata, Mandibulata	terrestrial
Clinical drugs	leeches, fungi	Annelida, Basiodiomycota	terrestrial, aquatic
Fiber-optics	sponges	Porifera	marine
Industrial enzymes (textiles, pulp and paper)	primitive bacteria, fungi	Crenarchaeota, Ascomycetes, Basidiomycetes	terrestrial, aquatic, marine, extreme environments
Engineering materials, (ceramics, industrial crystals)	snails	Mollusca	marine
Model research organisms in science/medicine	slime moulds, round worms	Myxomycota, Nematoda	terrestrial, marine
Industrial adhesives	barnacles, velvet worms, gecko	Crustacea, Onychophora, Craniata	ocean, forest
Antifouling paints	sea moss, marine algae,	Bryozoa, Rhodophyta	marine coastal
Robotic and aeronautic design	fish, millipedes, bees, dragonflies, worms	Craniata, Mandibulata, Annelida	all ecosystems
Industrial pigments	single-cell algae	Dinomastigota Bacillariophyta Haptomonada	marine
Industries			
Nanotechnology	bacteria, viruses, algae	various	various (e.g., terrestrial, marine)
Biological mining	bacteria	various	terrestrial, aquatic
Biological control, crop protection (new developments)	many different groups	various: microbes, animal, plant	various
Biomonitoring (new developments)	many different groups	various: microbes, animal, plant	various (e.g., terrestrial, aquatic, marine)
Agriculture, horticulture (new developments)	mostly plants	Anthophyta	various (terrestrial)
Biomimetics	many different groups	various: plants, animals, microbes	various (e.g., terrestrial, marine)
Ecotourism	all groups	various	wide variety of tourism destinations
Bioremediation	mostly microbes	various (e.g., Proteobacteria)	various
Ecological restoration	mostly plants but invertebrates/ microbes being tested	various	various
Pharmaceuticals	many different groups	microbes, plants, animals	various
Botanical medicines	mostly higher plants	Anthophyta	various
Personal care/cosmetics	many different groups	various	various

this time frame, despite massive investment in this technique by pharmaceutical companies. (See Figures 10.4 and 10.5.) Some of the most striking examples of recent drug development based on natural products are the drugs that inhibit cell division. (See Table 10.3.)

The current assessment of bioprospecting by the large pharmaceutical companies is reflected in the focus of their research and development, where the major investment is in rational drug design and combinatorial chemistry (Olsen et al. 2002; Hijfte et al. 1999) rather than natural products. Such decisions have probably been based on three factors: recent advances in high through-put instrumentation, low "hit" rates from natural product exploration, and consequently the high risks of natural product investment. On the other hand, natural product bioprospecting is

BOX 10.1

The "Lotus Effect," an Example of Novel Products Commercialized through the Exploration of Biodiversity

Many important processes in nature occur at the interfaces between organisms and their environment. For example, the outermost barrier of plant leaves and shoots, the cuticle, can be regarded as an extracellular membrane deposited on the outer epidermal cell wall and is the necessary interface for plant-environment interactions. It is covered with epicuticular waxes that self-assemble into complex three-dimensional crystals. They are very important in repelling water, and this hydrophobicity occurs in extreme forms when the crystals generate the micro- and nano-roughness of about 0.2–5 μm. This leads to what is known as super-hydrophobicity, so that the leaf surface is never wetted. Water forms spherical droplets, due to surface tension, that rest on the outermost tips of the wax crystals.

After screening some 15,000 species by electron microscopy, it has been shown that micro- and nano-rough plant surfaces are self-cleaning. Dirt particles cannot adhere to the surface, and the contact area between them and the surface is extremely reduced, while at the same time the contact between the water droplets and the dirt particles is increased, resulting in greater adhesion to the water droplet. This super-hydrophobicity results in self-cleaning plant surfaces in the presence of rain, fog, or dew. The cleanliness originates from the combined effect of surface topography and hydrophobicity.

Research has shown it is possible to transfer this effect into biomimetic self-cleaning products, and in 1994 a patent process was initiated. In 1998 a European patent was granted, yielding the trademark Lotus-Effect®. Research and development involving 12 industrial companies led to more than 200 patents. In 1999 a facade paint named Lotusan® was successfully launched on the market, and there are now more than 350,000 buildings with self-cleaning coatings. The enormous range of industrial applications for these biomimetic surfaces comprise mainly external materials exposed to rain, such as the surfaces of buildings and vehicles. However, some special applications such as medical devices, pipelines, and textiles are being targeted. In the near future, architectural glass, awnings, and temporary spraycoats with Lotus-Effect® are expected on the market. Detailed information is available at www.lotus-effect.com.

the main activity of a variety of active small companies that sell their products to the larger ones that can afford the massive costs of drug development. Some contemporary researchers believe that natural product research is more likely to result in new lead discovery and that the great advantage of combinatorial chemistry is its capacity to take advantage of such leads. Chapman (2004) and Ortholand and Ganesan (2204) argue persuasively for this approach.

10.2.2 Ethnobotanical Bioprospecting

Historically, much corporate drug discovery has depended on indigenous knowledge delivered to modern science through ethnobotany. Over 50% of modern prescription medicines were originally discovered in plants, and plants continue to be the source of significant therapeutic compounds to this day (e.g., Pearce and Puroshothaman 1993; Cragg and Newman 2004). Many were developed because the plants were used in indigenous medicine, and some common drugs were first used only on a local scale. In Europe, for example, aspirin was first isolated from *Filipendula ulmaria* because it had long been used in folk medicine

to treat pain and fevers. When the Bayer company developed a synthetic derivative of salicylic acid called acetylsalicylic acid, they named it Aspirin—"a" for "acetyl" and "spirin" for *Spiraea*, the former Latin name for the genus. Another European folk cure that became a drug was derived from *Digitalis purpurea*, the leaves of which were first used to treat congestive heart failure. The active ingredients, digitoxin and diyoxin, remain an important treatment for heart ailments.

Farnsworth et al. (1985) showed that at least 89 plant-derived medicines used in the industrial world were originally discovered by studying indigenous medicine. Among the best known is quinine, used in South America to treat fever. This has been the single most effective cure for malaria. Quinine comes from the bark of trees of the genus *Cinchona* that grow in the Andean region. More recently, the drugs vincristine and vinblastine were discovered in the rosy periwinkle (*Catharanthus roseus*) from Madagascar. When the Eli Lilly company studied this plant, they found that the periwinkle had anti-cancer properties. Vincristine has given children with leukemia a likelihood of remission, and vinblastine has cured many people with Hodgkin's disease. Native American peoples used the mayapple (*Podophyllum peltatum*) to treat warts. Two important drugs have been derived from it: teniposide to treat bladder cancer and podophyllotoxin, from which a powerful anti-tumor agent has been synthesized. The biological origins of the top 150 prescription drugs in the United States are shown in Table 10.4.

Indigenous peoples generally have large pharmacopoeias, since plants are often the only source of medicine available to them. Ethnobotanical studies list a large number of plant species used medicinally (e.g., Cox and Balick 1994; Balick 1994; Peters et al. 1989; McCutcheon et al. 1992). The MA Sub-Global Mekong River Wetlands Assessment has identified 280 medically important plant species, 150 of which are in regular use. The ethnobotanical approach to drug discovery is more likely to succeed where people have lived in the same area over many generations and so have had more time to discover suitable medicines. Local medicines can be complex mixtures of chemicals, however, either from the whole plant or from several plant species, and their efficacy may be enhanced from interactions that take place in their preparation or consumption. Thus, when pharmacologists try to isolate individual chemicals from the plants they often do not achieve the same effect as the local preparation. This is one reason that many effective cures of indigenous peoples have not been developed by western medicine.

10.2.3 The Botanical Medicine Industry

Botanical medicines in commerce are generally whole plant materials as opposed to pharmaceuticals, which are often derived from specific biochemical compounds extracted from plants. Best-selling examples include ginkgo, St. John's wort, echinacea, garlic, ginseng, and various yeasts. (See Table 10.5.) The structure of this industry varies according to the particular medicines being produced, but typically there are several stages: collection from the wild or cultivation, followed by the purchase of materials by exporters, importers, wholesalers, brokers, or traders. Materials may then be tested for contamination, powdered, or extracted by processing companies or by manufacturers of the finished products. These may then be handled by specialized distributors before retailing to consumers.

Revenues from these products can be very large. For example, annual sales of medicinal ginkgo, garlic, evening primrose, and echinacea in Europe average $350 million (ten Kate and Laird 1999). The global sales of raw botanical materials by leading U.S.

Table 10.2. Some Compounds from Natural Sources Approved for Marketing in the 1990s in the United States and Elsewhere. These agents are either pure natural products, semi-synthetic modifications, or the pharmacophore is from a natural product. (From ten Kate and Laird 1999 with permission)

Generic	Brand name	Developer
In the United States and elsewhere		
Cladribine	Leustatin	Johnson & Johnson (Ortho Biotech)
Docetaxel	Taxotere	Rhône-Poulenc Rorer
Fludarabine	Fludara	Berlex
Idarubicin	Idamycin	Pharmacia & Upjohn
Irinotecan	Camptosar	Yakult Haisha
Paclitaxel	Taxol	Bristol-Myers Squibb
Pegaspargase	Oncospar	Rhône-Poulenc
Pentostatin	Nipent	Parke-Davis
Topotecan	Hycamtin	SmithKline Beecham
Vinorelbine	Navelbine	Lilly
Only outside the United States		
Bisantrene		Wyeth Ayerst
Cytarabine ocfosfate		Yamasa
Formestane		Ciba-Geigy
Interferon, gamma-la		Siu Valy
Miltefosine		Acta Medica
Porfimer sodium		Quadra Logic
Sorbuzoxane		Zeuyaku Kogyo
Zinostatin		Yamamouchi

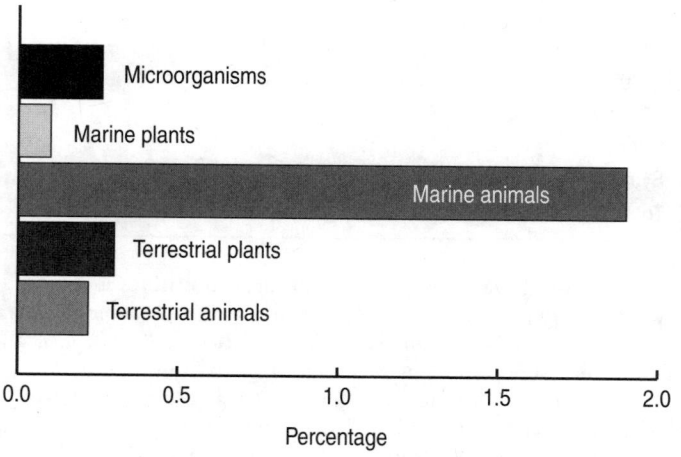

Figure 10.2. Distribution of Samples with Significant Cytotoxicity among Marine and Terrestrial Organisms (Munro et al. 1999)

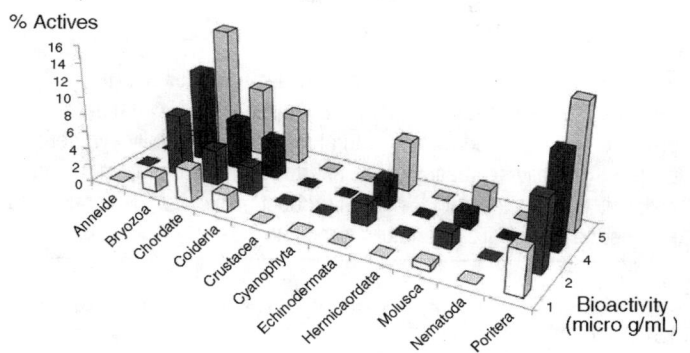

Figure 10.3. Distribution of Cytotoxicity among Marine Phyla (Munro et al. 1999)

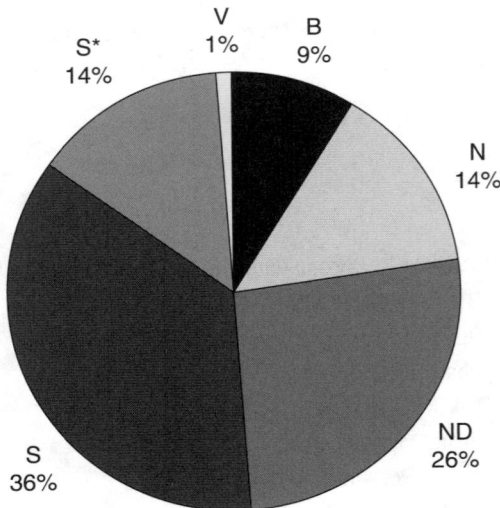

Figure 10.4. Sources of All New Chemical Entities, 1981–2002 (n = 1031). (Newman et al. 2003) B = biological, N = natural product, ND = derived from a natural product, S = totally synthetic drug, S* = totally synthetic but the pharmacophore was from a natural product, V = vaccine, NM = natural product mimic, i.e., designed from knowledge gained from a natural product.

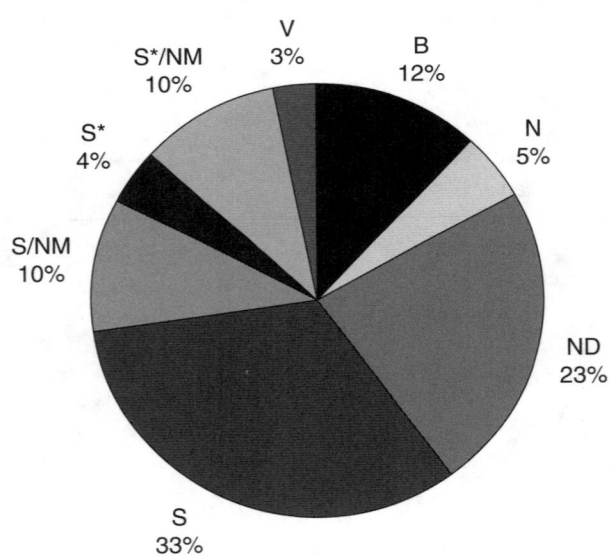

Figure 10.5. Sources of All Available Anti-cancer Drugs, 1940s–2002 (Newman et al. 2003) B = biological, N = natural product, ND = derived from a natural product, S = totally synthetic drug, S* = totally synthetic but the pharmacophore was from a natural product, V = vaccine, NM = natural product mimic, i.e., designed from knowledge gained from a natural product.

suppliers amount to approximately $1.4 billion (ten Kate and Laird 1999).

The "nutraceuticals" industry sells food ingredients or products believed to confer health or medical benefits. These include dietary supplements, individual nutrients, foods enhanced in various biotechnological ways, and fortified foods. Major products of this industry include dietary additives. Products include teas with added ginseng, probiotic yogurts, fruit juices fortified with calcium, and flour fortified with folic acid. Various companies in the

Table 10.3. Naturally Derived Microtubule Stabilizing Agents That Inhibit Cell Division and Are Useful against Cancer (Cragg and Newman 2003)

Name	Source	Status
Paclitaxel	*Taxus brevifolia* (plant) (made by semisynthesis)	clinical use
Docetaxel	semisynthesis from *Taxus* spp.	clinical use
Discodermolide	*Discodermia dissoluta* (marine) (made synthetically)	phase I in 2002
Eleutherobin	*Eleutherobia* sp (marine)	derivatives in preclinical development
Saracodictyin	*Sarcodictyon roseum* (marine)	derivatives in preclinical development
Epothilones	*Sorangium* sp. (terrestrial microbe)	naturally occurring compounds and derivatives in early clinical trials
Laulimalide	*Cacospongia mycofijiensis* (marine)	preclinical development
Dictyostatin-1	*Spongia* sp. and a deep-water Lithistid sponge (marine)	preclinical development
Jatrophane esters	*Euphorbia semiperfoliata* (plant)	preclinical development

Table 10.4. Biological Origins of Top 150 Prescription Drugs in the United States (World Resources Institute 2000, with permission, based on Grifo et al. 1997)

Origin	All Compounds	Natural Product	Semi-Synthetic	Synthetic	Share of Total
		(number)			(percent)
Animal	27	6	21	–	23
Plant	34	9	25	–	18
Fungus	17	4	13	–	11
Bacteria	6	5	1	–	4
Marine	2	2	0	–	1
Synthetic	64	–	–	64	43
Total	**150**	**26**	**60**	**64**	**100**

food industry have been interested in sugar substitutes such as the sweet-tasting proteins produced by plants such as *Dioscoreophyllum cumminisii*, *Thaumatococcus daniellii*, and *Richardella dulcifera*, all from West Africa, and *Capparis masaikai* from South China.

The nutraceuticals market for 1996 was estimated at $16.7 billion (ten Kate and Laird 1999), and interest is rapidly growing worldwide. Further information resources on botanical medicines include the herbage database at www.herbage.info.

10.2.4 The Personal Care and Cosmetics Industries

Personal care and cosmetics industries use wild harvested or cultivated products in a wide variety of products, including cosmetics, feminine hygiene, hair products, baby care, nail care, oral hygiene, deodorants, skin care, and fragrances. Different demo-

Table 10.5. Geographic Range of 1,464 Medicinal and Aromatic Plants in Trade in Germany, 1997 (ten Kate and Laird 1999, with permission)

Geographic Range	Medicinal and Aromatic Plant Species Found Only in the Region	Introduced Species	Total Medicinal and Aromatic Plant Species Found in the Region
Europe	16	71	605
Africa	63	16	343
Asia – temperate	248	13	849
Asia – tropical	90	10	318
Australasia	8	18	55
Pacific	1	1	13
North America	124	186	454
South America	106	25	207

graphics apply to their marketing; different products are directed to "prestige," "mass," and "alternative" markets.

The same agents and organizations as those that handle botanical medicines often handle the raw materials for these industries, which consist largely of dried plant products and oils from a wide variety of organisms. Many of the natural products of interest are derivatives of wild sources and include saponins, flavonoids, amino acids, anti-oxidants, and vitamins as well as various compounds from seaweeds, chitin from crustaceans, and fish oils. In some specialty markets, products are certified as being from organic or sustainable sources and others as "fair trade" products—that is, ethically or socially certified.

The value of this industry was estimated at $1.4 billion for the United States for 1996 and at $2.8 billion for 1997. Growth is also thought to be very rapid in markets as varied as Asia, Latin America, Europe, and Australia.

10.2.5 Biological Control and Crop Protection

Biological control is a more established biodiversity-based industry but it is currently expanding through new knowledge of biodiversity (Bellows 1999). Much biological control is for crop protection, using predators, parasites, or pathogens or their products to limit pests. (See Box 10.2.)

Biological control has also been used on many invasive species, however, including non-crop pests—chiefly animals and plants introduced outside their native environments that have successfully bred, often in vast numbers, uncontrolled by their natural enemies. (See Chapter 4 for a further discussion of invasive species.) In this context, biological control has been successfully implemented against rabbits in Australia, Europe, Argentina, and Chile; against cats in oceanic islands; and against various plant species that have escaped from gardens. The application of biological control as crop protection includes the control of soil pathogens; the protection of tree crops such as olives, citrus, coffee, bananas, and coffee; use in tree plantations, greenhouses, and grape vines; and the control of weeds in both terrestrial and aquatic environments as well as medical and veterinary pests.

Biocontrol is an industry that involves bioprospecting activity worldwide, particularly in developing alternatives to chemical pesticides that have severe environmental and occupational safety hazards. Bioprospecting for this industry requires study of the diversity both of the organism being controlled as well as the con-

trol agent(s). Biological control agents include plants, viruses, bacteria, fungi, insects, nematodes, and many other kinds of invertebrates and have been extremely successful in many parts of the world (Bellows and Fisher 1999). (See Box 10.3.)

Because nearly all these control agents are small organisms, however—many of them microscopic and with no appeal to the public—they have not been recognized as being in need of conservation action, even when threatened with extinction. In recent years, pests have been controlled by the simultaneous release of more than one control agent; for example, the so-called Bridal Creeper, a noxious vine, is controlled by a fungus, a beetle, and a leafhopper insect in Australia. None of the three control species were considered to have any economic importance until their utility in the biocontrol industry was demonstrated.

The biocontrol industry has had major problems when control agents introduced to areas where their own natural predators were absent have themselves become pests or plagues. In response, new and more rigorous screening protocols for new biological control agents have been developed so as to minimize the risks of such population explosions. These new developments have significantly increased the environmental safety and commercial viability of the biological control industry (Bellows 1999).

Box 10.3 illustrates the potential benefits of maintaining all species, most notably as a resource for the crop protection and biological control industries. In the specific case cited, sufficient weevil species were still extant in the host country, facilitating the discovery of the control agent. Such a situation has been repeated all over the world both for this and for the biological monitoring industry (Bellows and Fisher 1999).

The best data on the commercial value of biological control come from the crop protection markets, from which it is estimated that sales by the top 10 crop protection companies in 1997 totaled $25 billion. It is unclear what fraction of this total can be assigned to the use of species and species products, as they are used in many different ways, but there are also several billion more dollars spent on research and development. An alternative valuation method to crop protection sales figures has been suggested by Pimentel (1992), who examined the varied and multiple costs of pesticide use for food production in the United States. These costs included health bills of human applicators, veterinary costs, surface and groundwater contamination, pollinator losses, and administrative costs. The estimated total was $8 billion a year, much of which could be avoided by the replacement of chemical pesticides with biological control agents.

10.2.6 Biomimetics

Biomimetics is the generic name for a wide variety of biologically inspired technologies. The industries involved use the structures and materials of organisms as the models, blueprints, or inspiration for novel materials and manufactured products (Mann et al. 1989). Among the best-known examples are the shell and radula (teeth) of various mollusks that have informed the manufacture of high-tech ceramics and other materials, including car parts and industrial crystals. Another high-profile research program is the application of the properties of spider silk to the manufacture of novel high-tensile fibers (Beattie and Ehrlich 2004; Mann 2001; Craig 2003).

The most recent development is in the field of molecular biomimetics, which is providing much of the inspiration and design for nanotechnology (Sarikaya et al. 2003). The field has its own professional journal, *Biomimetics,* and the Darwinian theory of evolution by natural selection was in a sense reworded by Birchall (1989) for this context: "Biology does not waste energy manipu-

---

BOX 10.2

The Role of Biological Materials in Crop Protection (modified from ten Kate & Laird 1999 with permission)

Chemical Control

Natural products:
"Pure" natural products (isolated from nature and not changed chemically
Semisynthesized derivatives (modifications of compounds isolated from nature)

Synthetic compounds:
Synthetic analogue compounds built from templates originally discovered from natural products
"Pure" synthetic compounds (not based on a natural product lead)

Behavior-modifying chemicals:
Use of naturally derived and synthetic versions of signaling chemicals from living organisms, such as pheromones, to create insect traps and disrupt mating

Growth regulators:
Insect growth regulators: chemicals that interfere with the growth of pests
Plant growth regulators: chemicals such as gibberillic acid sprayed on to crops to increase the size and quality of fruit, to speed/delay ripening

Biological Control

Toxic protein-producing bacteria:
Over 30 recognized subspecies of the naturally occurring bacterium *Bacillus thuringiensis* (Bt) produce different insecticidal toxins

Baculoviruses:
Naturally occurring ("wild type") or modified viruses that, once ingested by insects, interfere with their metabolic processes and kill them

Fungal pesticides:
Fungal insecticides: three species of fungi are the main source of fungal insecticides—*Verticillium, Metarhizium* and *Beauvaria*
Fungal herbicides: fungi that affect the metabolic processes of weed species, killing them
Fungal fungicides and microsporidia

Bacterial pesticides:
Bacterial bactericides, bacterial fungicides, and bacterial insecticides

Natural predators, parasites, and parasitoids:
A wide variety of species, including many arthropods and nematodes, that seek out and kill pests

Crop Improvement

Genes for disease and pest resistance or herbicide tolerance:
Breeding resistance or tolerance into crop varieties, using either traditional methods or genetic engineering

---

BOX 10.3

Water Weeds and Weevils: The Importance of Individual Species to the Biological Control Industry

Salvinia is an aquatic fern from South America that has been accidentally carried beyond its natural area of distribution to localities where it has no natural enemies that regulate its growth. As a result, in many parts of the world it completely covers the surface of lakes, rivers, and canals with a deep layer of vegetation that prevents sunlight penetrating the water below. There are many damaging flow-on effects, such as the decline of light-dependent animals and plants, reduction in the oxygen content of the water, and overwhelming bacterial growth.

The Australian Commonwealth Scientific and Industrial Research Organisation searched for a biological control agent in its native environments in South America and a promising herbivorous weevil was discovered, cultured in large numbers and then released. Nothing happened. A second species of weevil that appeared to be very similar was tried, and this succeeded in reducing the weed with spectacular speed and success. (The photos in Appendix A show a lake before and several months after the introduction of the weevil.)

There were two interesting lessons from this experience. First, if you have seen one weevil you definitely have not seen them all! Second, you never know which species are going to be crucial. In this case, a tiny, little-known weevil from South America was worth millions of dollars by restoring water quality and enabling the return of fisheries and commercial waterway navigation throughout a large part of Australia. Biological control agents that fight crop pests are often equally obscure and yet have great value—not merely in terms of dollars but also in terms of human lives.

lating materials and structures that have no function and it eliminates those that do not function adequately and economically. The structures that we observe work and their form and microstructure has been developed and refined over millions of years . . . it is well, then, to look for fresh insights to biology at the wisdom encapsulated in the materials it uses."

Beattie and Ehrlich (2004) discuss a wide variety of examples. At present, the use of biomimetics is scattered throughout a variety of engineering, manufacturing, and construction industries, and it is difficult to identify its commercial worth or to predict its value in the future. Clearly, however, individual projects and products generate very large revenues.

10.2.7 Biomonitoring

Biological monitoring is an industry developing in response to the necessities of tracking down sources of pollution across large geographical areas. This would normally require vast resources in terms of conventional instrumentation, but the status of the environment can also be monitored by using organisms that routinely "sample" the environment, such as aquatic or marine filter-feeding animals. Provided that the species used are both widespread and common and that collection for lab testing does not compromise their populations, there is little need for instrumentation outside the analytical laboratory (Boyle 1987; Rosenberg and Resh 1993). Biomonitoring is also applied to the detection of pollutants in soils and may involve a range of selected test organisms, including bacteria, algae, earthworms, and nematodes. A selection of biomonitoring organisms is provided in Table 10.6.

10.2.8 Horticulture and Agricultural Seeds

The global horticulture industry is worth many billions of dollars, mainly based on cultivated plants. Although all horticultural spe-

Table 10.6. Types of Biomonitoring and the Organisms Used

Type	Common Name	Phylum
Freshwater	fish	Craniata
	insect larvae	Mandibulata (e.g., Ephemeroptera, Plecoptera, Trichoptera)
	mussels	Mollusca
	water fleas	Crustacea
	bristle worms	Annelida (Oligochaeta)
	mosses	Bryophyta
Soil	earthworms	Annelida (Oligochaeta)
	round worms	Nematoda
Marine	paddle worms	Annelida (Polychaeta)
	fish	Craniata
	sea squirts	Urochordata (Tunicata)
	bacteria	Proteobacteria
Air	bees, ants	Mandibulata (Hymenoptera)
	lichens	Mycophytophyta
Monitoring disturbance and rehabilitation	ants, butterflies, beetles, spiders	Mandibulata Chelicerata

cies are derived from wild species, current reliance on wild plant biodiversity is limited to a few areas, notably flowers harvested from native plants and genetic material taken from native plants to improve or establish new horticultural varieties. This is not to minimize the size or cost of the industrial effort developing new varieties, and the revenues they generate are large. For example, it is estimated that it can cost up to $5 million to develop a single new variety (ten Kate and Laird 1999).

The industry is made up of companies of all sizes—from large multinationals to small, local enterprises—and is backed up by specialized research organizations in many countries. These include universities and botanic gardens as well as commercial laboratories. The contemporary wild harvest may be small, as the commercial focus is generally on developing new varieties of familiar and popular plants, such as roses, hydrangeas, geraniums, and begonias. However, in some areas seed harvesting remains large-scale (and sometimes a problem, as described at the end of the chapter).

The development of new seed varieties for agriculture is a major use of plant biodiversity, some of it from wild, native plants, but much of it from the wealth of crop varieties that have been bred to adapt crops to a host of local conditions worldwide (Brush 2004). Contemporary technologies usually take advantage of very long development times. For example, the production of one wheat variety may involve thousands of plant breeding crosses and dozens of different individual lines, including wild ones, from many countries and over many centuries (see, e.g., Mujeeb-Kazi et al. 1996; Quick et al. 1996; Cassaday and Smale 2001). New varieties are developed through traditional plant breeding protocols, genetic engineering, or a combination of these two. The research budgets for agricultural biotechnology are estimated at $1.95 billion annually but only a small proportion of this involves the harvest of seeds from the wild.

Although most agricultural and horticultural varieties are derived from seeds collected from wild plants at some time in the past, the sources of almost all plant breeding materials today are seed banks maintained by major corporations, universities, botanic gardens, and regional, national, or international gene banks. There are many reasons for this, especially the time and expense required to take seeds from wild plants and breed them for compatibility with commercial varieties. There is serious concern about the growing genetic uniformity of crops, however, and about the availability of genes from wild ancestors and neglected varieties used to generate adaptive variety. With the burgeoning biotechnology industry developing ever more sophisticated genetically engineered crops, it is difficult to know what the demand will be for wild seeds in the future. Some of the key issues are summarized in Box 10.4.

10.2.9 Bioremediation

A more recent example of the potential for novel biodiversity-based industries is bioremediation. This industry is often associated with heavy industry and mining, especially in countries where the law requires restoration of abandoned industrial sites and mines (Crawford and Crawford 1998). Two common methods are applied by this industry: first, remediation by altering the site environment to allow resident, beneficial microorganisms to proliferate or, second, augmentation of the site by the addition of beneficial microbes. Both these methods explore microbial diversity not just for species tolerant to the pollutants concerned but those that metabolize them, either transforming them into less harmful derivatives or sequestering them from other species in the ecosystem. Success has often been hard to achieve, but in some

---

BOX 10.4

Some Reasons the Use of Wild Plant Resources May or May Not Increase (from ten Kate & Laird 1999 with permission)

Why demand for primitive germplasm may grow in the future:

- The need to improve resistance to disease requires access to more genes and the globalization of research and markets means that pests and disease are transferred faster, increasing demand for access.
- Breeders wish to broaden the genetic base of the material they use.
- The desire to move away from reliance on chemical pesticides to more biological approaches will require access to diverse genetic resources.
- Modern methods make it easier to use primitive materials.
- Public funds are drying up, so companies will need to access more diverse materials themselves.

Why demand for primitive germplasm may decrease in the future:

- It is becoming harder to gain access to materials from many countries.
- Fear of accidental infringement of patents and a reluctance to negotiate licenses and material transfer agreements will decrease the demand for access to cultivars.
- Modern methods mean there is more to be found in existing collections and less need to turn to primitive materials.
- Privatization and commercialization of research mean that the public institutions that were accessing materials are no longer doing so. It is less competitive for companies to work on unimproved materials.

cases highly toxic sites have been transformed into relatively benign areas suitable for other purposes (Flatman et al. 1994; Rittman and McCarty 2001). Microbial inoculation for the remediation of major oil spills is showing increasing promise (Mueller et al. 1992).

One example of fungal bioremediation illustrates that exploring biodiversity is key to the success of this technology and that the function of a microorganism in nature can be applied to a quite different twenty-first century commercial need. The major bioremediating fungi are white rot fungi, so-called because they degrade the dark-colored lignin in wood, leaving behind light-colored cellulose and giving WRF-decayed wood a bleached appearance. These fungi are common and diverse in forests worldwide and are a vast "bank" of chemical systems that have evolved to break down intractable materials such as wood. Without the chemical activity of WRF and other fungi known as brown rot fungi, the natural recycling of nutrients would be extremely slow.

The lignin-degrading systems of white rot fungi are now of major commercial importance because they also degrade many classes of pollutants such as PCBs, dibenzo dioxins, and dibenzo furans that have chemical properties in common with lignin (Joshi and Gold 1993). The industrial uses and potential of microbial biodiversity are further reviewed in Demain (2000), and the new technologies known as bioaugmentation, in which selected microorganisms are added to polluted substrates such as soils, are described fully by Mueller et al. (1992).

10.2.10 Ecological Restoration

Ecological restoration differs from bioremediation in that it attempts to recreate the ecosystem that once existed. In terrestrial situations, while some of the original structure is often achieved with rudimentary soils, some leaf litter, and elements of the vegetation, the restored species richness is generally reduced in comparison with the former ecosystem. Nevertheless, restoration is a much needed industry worldwide as a result of national and local government legislation requiring the repair of damaged ecosystems such as abandoned industrial and mining sites, eroded agricultural lands, and surface water degraded by a wide variety of human activities.

This demand has generated a new industry with active societies such as the Society for Ecological Restoration. The basic resource is biodiversity (Handel et al. 1994) and the species used are most often those from neighboring or comparable ecosystems that can be carefully harvested or grown offsite and moved to the restoration site (Harker et al. 2001; Whisenant 2001). Effective ecological restoration requires deep knowledge of species, their ecological functions, and their interactions with each other and the environment.

Agroforestry is discussed here not because it is used only for ecological restoration but because it has become such an important activity in this field. Simons and Leakey (2004) report as follows: "In recent years international aid to developing countries has developed a strong focus on poverty reduction. In parallel with this, the World Agroforestry Centre (formerly ICRAF) initiated its tree domestication program in the mid-1990s with a new focus on products with market potential from mainly indigenous species. With this came a shift from on-station formal tree improvement towards more active involvement of subsistence farmers in the selection of priority species for domestication and the implementation of the tree improvement process. In many regions in which ICRAF is active, farmers selected indigenous fruit trees as their top five priorities. Consequently, over the last decade a strategy for the domestication of indigenous trees producing high-value products of traditional and cultural significance has been developed."

This approach to improving the trees planted by farmers has many advantages (Leakey 1999):

- It has a clear poverty reduction focus, which has been endorsed by a review on behalf of the U.K. Department for International Development. The income derived from tree products is often of great importance to women and children.
- It has immediate impact by being implemented at the village level, thereby avoiding delays arising from constraints to the transfer of technology from the field station to the field that can be due to technical, financial, dissemination, and political difficulties.
- The approach being developed is focused on simple, low-cost, appropriate technology yielding rapid improvements in planting stock quality, based on selection and multiplication of superior trees that also produce fruits within a few years and at heights that are easily harvested.
- It builds on traditional and cultural uses of tree products of domestic and local commercial importance and meets local demand for traditional products.
- It promotes food and nutritional security in ways that local people understand, including promoting the immune system, which is especially important in populations suffering from AIDS.
- It can promote local-level processing and entrepreneurship, hence employment and off-farm economic development. These benefits can stimulate a self-help approach to development and empower poor people.
- It can be adapted to different labor demands, market opportunities, and land tenure systems and is appropriate to a wide range of environments.
- It builds on the rights conferred by indigenous knowledge and the use of indigenous species by the Convention on Biological Diversity and is a model for best practice.
- It builds on the commonly adopted farmer-to-farmer exchange of indigenous fruit tree germplasm as practiced in West and Central Africa—for example *Dacryodes edulis,* although native to southeast Nigeria and southwest Cameroon, is now found across much of the humid tropics of central Africa.
- It builds on the practice of subsistence farmers to plant, select, and improve indigenous fruits, such as marula (*Sclerocarya birrea*) in South Africa, where the yields of cultivated trees increased up to 12-fold and average fruit size is 29g, while trees in natural woodland are 21g (Shackleton et al. 2003).
- The domestication of new local cash crops provides the incentive for farmers to diversify their income and the sustainability of their farming systems.

Against these advantages there are possible disadvantages, such as reduced genetic diversity in wild populations as domesticated populations replace them. However, the implementation of some in situ or ex situ conservation of wild germplasm, together with the deliberate selection of relatively large numbers of unrelated cultivars, can minimize these risks. Indeed, the current situation, whereby each village develops its own set of cultivars, should maintain levels of regional intraspecific diversity.

To maximize the economic, social, and environmental benefits from domestication, it is crucial to develop post-harvest techniques for the extension of shelf life of the raw products, processing technologies to add value to them, and, of course, access to markets. Without this parallel preparation for increased commercialization, domestication will not provide all the benefits just described. The combination, however, has potential applications that extend beyond subsistence agriculture to agricultural diversification of farming systems. In tropical North Queensland, Australia, for example, this is linked, at least in part, to the development of an Australian "bush tucker" food industry supplying restaurants and supermarkets.

10.2.11 Ecotourism

Ecotourism is an industry in which tour operators prospect for localities rich in biodiversity or charismatic species. As returns on investment rely greatly on maintaining such attributes, it is in the interests of the industry to conserve many elements of biodiversity. Tourists are less likely, for example, to travel to rain forests or coral reefs that are degraded or to mountains and islands that have been deforested. In the majority of destinations, biodiversity is one of the main attractions for ecotourists. The Quebec Declaration on Ecotourism states that "ecotourism embraces the principles of sustainable tourism and contributes actively to the conservation of natural and cultural heritage and includes local and indigenous communities in its planning, development and operation, contributing to their well-being" (www.uneptie.org/pc/tourism/ecotourism/home.htm).

Ecotourism can generate large revenues, some of which goes to local communities, but it is often extremely difficult to make it sustainable, as the industry itself can also put major pressures on biodiversity: Local increases in the human population from both tourists and the employees that look after them, along with hotel, road, airport, and dock construction and off-road travel, can all contribute to local pollution, habitat fragmentation, lowering of water quality, and the influx of exotic plants and animals. In some cases, these activities lead to declines in the very animals the industry is built around (WRI 2000). Ecotourism often relies heavily on the presence of what are known as charismatic animals and plants, and it can be especially vulnerable to losses in biodiversity, which are occurring on a global scale and therefore threatening the sustainability of this industry. (See Chapter 17.)

10.2.12 Other Biodiversity-based Industries and Products

Other products and industries are also emerging, many of which depend on microbial diversity. These include a wide variety of microbially produced enzymes that contribute to treating industrial and agricultural wastes, driving diverse reactions in chemical engineering, processing wood and pulp, increasing the efficiency of textile manufacture, and industrial and domestic cleaning. Biological mining uses microbes that leach metals from low-grade ores and mine tailings. Biofuels, especially ethanol, are derived from a wide variety of plant species, and various microbial species generate biogas from landfills and waste dumps (ten Kate and Laird 1999; Beattie and Ehrlich 2004). Each of these applications is already a commercial reality, and almost any one of them may surpass the other biodiversity-based industries in commercial value in the future, assuming the development of appropriate markets.

10.3 Distribution and Value of the Resource

Biodiversity is global, and the long history of its use by humanity, together with the more recent history of bioprospecting, shows that important commercial species have been found in all parts of the world. Indeed, it appears impossible to predict in which ecosystems and therefore in which countries future products will be found.

At this point it is reasonable to ask if bioprospecting will be more profitable in species-rich areas of the world, particularly the sub-tropical and tropical forest areas and coral reefs. The evidence does not present a clear picture. Many biological resources have been derived from non-tropical areas, including some critical medicines such as aspirin and the drugs derived from the plant genera *Digitalis, Podophyllum,* and the Pacific Yew tree *Taxus brevifolia,* all of which are from temperate zones. Other products such as cryoprotectants and anti-freezes have come from cold-water fish and high-altitude arthropods. Materials for silk research and development, industrial adhesives, and mollusk-based ultra-structures have been derived from native species in a wide variety of non-tropical ecosystems, both terrestrial and marine. In addition, microbes for bioremediation and biological mining and species used in the biological control of agricultural pests and in biological monitoring have emerged from ecosystems at a wide variety of latitudes and altitudes.

With respect to ocean resources, a variety of drugs are derived from different non-tropical marine organisms such as tunicates, which have provided anti-tumor compounds currently undergoing clinical trials, and marine fungi that secrete powerful antibiotics (Rinehart 2000; Cueto et al. 2001). Tunicates and marine fungi inhabit the intertidal zones of oceans in many parts of the world, including temperate regions. Some industrial bioprospecting also takes place in extreme environments such as hot springs and the poles or at great oceanic depths, where the variety of species may be relatively low, but the species present are unique to those areas and harbor extremely valuable adaptations to high or low temperatures or to great pressures (see, e.g., Moss et al. 2003).

Much recent pharmaceutical bioprospecting has focused on species-rich ecosystems, especially tropical rain forests and coral reefs. Although there is a general trend of increasing species richness with lower latitudes and altitudes, these trends do not necessarily inform bioprospecting. Certainly, if the goal is to screen as many species as possible in the most cost-effective way, the use of species-rich ecosystems such as rain forests appears logical. In addition, evolutionary theory tells us that herbivory, especially by insects, is far more intense in the tropics than in the temperate zones. In this context, plant chemical defenses against herbivory are likely to be both stronger and more diverse in ecosystems such as tropical rain forests, and this may make some pharmaceutical bioprospecting more profitable in tropical than in temperate forests (Coley et al. 2003).

However, many areas of modern bioprospecting are even more target-orientated, asking, for example, where the desired product is most likely to have evolved. In this context, there is frequently no expectation that it has necessarily evolved in a species-rich ecosystem, but rather that it has evolved in response to a particular kind of natural selection. Thus cryoprotectants will have evolved in animals from extremely cold environments, and silk is a predatory device that has evolved in all kinds of terrestrial environments. It may be that certain kinds of biological resources emerge as being more frequent in species-rich ecosystems, but far more research is required to discover which ones. The current expectation is still that novel drugs are more likely to come from the tropics.

Various methods of valuing biodiversity were reviewed by Heywood et al. (1995), and further discussion of the valuation of ecosystem services can be found in Chapter 2. The wide variety of products, especially drugs that have been derived from ecosystems, may suggest that there are likely to be many more awaiting discovery, and therefore biodiversity is a vast source of future revenues. While the evidence presented in this chapter suggests that this is likely to be the case, more specific quantitative economic analysis may be required to further understand the likely returns on investment from bioprospecting. An appraisal of these is presented in Box 10.5.

The current reality is that there is no robust, reliable, and generally agreed way of assessing the commercial value of the novel biological resources of any given species, group of organisms (taxon), habitat, or ecosystem. The situation is exacerbated by the fact that we are still in the early stages of discovery, and many resources are assumed to be there but are as yet unknown. Species useful to society or industry can be obscure, microscopic, and from any habitat and the use may be derived from a gene, a product, a behavior, or a structure, thus values can be legitimately anticipated but their dimensions tend to emerge later.

There have been many attempts at establishing the commercial value of drugs derived from wild or cultivated species with a view to valuing the biodiversity from which they came. (See Table 10.7.) Farnsworth et al. (1985) estimated that 25% of prescriptions from community pharmacies in the United States during the period 1959–80 contained a compound derived from higher plants. The contribution of wild species has not diminished, as 57% all prescriptions in the United States for the period January–September 1993 contained an active compound derived from biodiversity (Grifo and Rosenthal 1997).

Even more recently, Laird and ten Kate (2002) reported on the findings by Newman and Laird (1999): "They found that natural products continue to be a major player in the sales of pharmaceutical agents: 10 of the 25 best-selling drugs in 1997, representing 42% of industry-wide sales, are either biological, natural products, or entities derived from natural products, with a total 1997 value of US$17.5 billion. The study also found that a significant portion—between 10% and 50%—of the ten top-selling drugs of each of the top 14 pharmaceutical companies are either natural products or entities derived from natural products."

Early estimates of the annual value of individual medicinal plant species ranged between $203 million and $600 million (Farnsworth and Soejarto 1985; Principe 1989). More recently, however, the drugs Taxol and Taxotere derived from a single species *T. baccata* yielded $2.3 billion in drug sales during 2000. While this suggests that individual species may be of very great value, it does not necessarily show that biodiversity conservation is a prerequisite for bioprospecting. For example, sales of the drug Navelbine derived from the rosy periwinkle (*C. reseus*) were worth $115.4 million in 2000, but this is a common tropical garden plant. Yet bioprospecting activities are valuable in several other ways, not least providing education and training, employment, and local and regional sources of revenues based on the harvesting, processing, manufacturing, distribution, and retailing of products.

Mendelsohn and Balick (1997) estimated the value of as-yet-undiscovered pharmaceuticals from plants in tropical forests at $109 billion. They noted that a severe constraint on this value was the high cost of finding the pharmaceuticals but that this cost could be reduced by ethnobotanical methods. Costs could be further reduced by ecologically driven discovery methods (Coley et al. 2003), especially if they were applied to a broader range of organisms including, for example, microbes and arthropods.

BOX 10.5

Some of the Principles and Problems of Valuing Biodiversity and Biological Resources

Several principles are important in considering the economic value of bioprospecting. First, economic values are determined on the margin. This means that values must be placed in the context of particular magnitudes of change. If the great majority of Earth's biodiversity were to be lost, the value of the lost opportunities for inventing and improving products would be astronomical. Less value would be foregone if fewer components of biodiversity were at risk.

Second, research and development is an inherently random process, and the outcomes are uncertain. The value to be assigned to a change in the biodiversity available for conducting research is related to the increase in the expectation of the outcome it affords.

Third, value is determined by scarcity. If there is a lot of something, a little more or less of it does not make much difference. Conversely, unique resources command large values because there are no substitutes for them.

These are illustrated by a thought experiment (modeled after Simpson et al. 1996). Suppose there are many species that might provide the source of a particular new product. Many analyses of the value of bioprospecting have focused on the expected reward to success: the probability of making a "hit" times the payoff from developing a successful product. However, the value of biodiversity on the margin—what we might label the value of the "marginal species"—is the incremental increase in the expected reward to success. It is the probability of making a "hit" times the payoff times the probability that none of the other species available for testing would have yielded the same success.

While commentators often emphasize the rewards accruing to success, other considerations may be more salient. As the number of species researched increases, the value of having more necessarily declines and, in the limit, vanishes. This can be explained as follows. If the probability that any one species chosen at random will yield a success is relatively high, it is unlikely that it will be necessary to test a large number of species in order to achieve a success. Conversely, if the probability of success in testing any one species is low, it is unlikely that two or more will prove redundant, but also unlikely that any will prove successful. Regardless of the likelihood of success in any given test, the value of the "marginal species" will be small when the number of species is large.

The same species may, of course, be tested for any of a number of different applications. Thus, in order to calculate the overall value of the "marginal species," one would have to sum the values in all potential applications, both current and anticipated. If there are relatively large numbers of species available for testing, comparably large numbers of potential applications would need to be identified for the value of the "marginal species" to be appreciable (although if new products complement one another, values may be greater; Craft and Simpson 2001). Moreover, not all species are equally attractive as potential research leads. Other things being equal, organisms that are "most different" from others will be more valuable. This is not because they are necessarily more likely to yield new products, but rather because they are more likely to yield new products in the event that other, more distantly related, organisms do not (Weitzman 1992).

Knowledge is also valuable. Researchers will test first those organisms most likely to yield a success and will be willing to pay more to do so (Rausser and Small 2000). The fact that some organisms are known to promise more leads means, necessarily, that others are considered less promising and less valuable. If promising prior information is available on the properties of species from better-known regions, the bioprospecting value assigned to the as-yet undescribed species of the world's remaining pristine ecosystems will be commensurately lower.

Table 10.7. Some Values for Plant-based Pharmaceuticals (Pearce and Moran 1994, adapted from Principe 1989)

Value	United States	OECD	World
	(billion 1990 dollars)		
Market value of trade in medicinal plants	5.7	17.2	24.2?
Market or fixed value of plant-based drugs on prescription	11.7 (1985) 15.5 (1990)	35.1 (1985)	49.8? (1985)
Market value of prescription and over-the-counter plant-based drugs	19.8 (1985)	59.4 (1985)	84.3 (1985)
Value of plant-based drugs based on avoided deaths:			
Anti-cancer only	120.0	360.0	
+ non-cancers	240.0 (1985)	720.0 (1985)	

Notes: Bracketed year indicates year estimate refers to.
Ratio of OECD to United States taken to be 3.
Value of statistical life taken to be $4 million in 1990 prices.
Lives saved taken to be 22,500–37,500 per annum in United States.
  Average is taken here (i.e., 30,000). Multiply OECD by 1.4 to get world estimates.

10.4 Recent Industry Trends

Most novel products are researched, developed, and produced in industrial countries, and there is a geographical mismatch between centers of biodiversity, which tend to be in the tropics, and centers of research and development, which are largely concentrated in the temperate zones (Barbier and Aylward 1996; Simpson and Sedjo 1996). With respect to pharmaceutical bioprospecting, while tropical/temperate partnerships have been formed and some developing countries are beginning to enter the industry independently, the prevailing situation is that the resources are currently considered most likely to be located in the tropical regions while the value creation in terms of development and manufacturing as well as consumption frequently takes place elsewhere.

The withdrawal of many of the largest pharmaceutical companies from bioprospecting during the last decade is based in part on the experience that large investments have yielded relatively few lead compounds for development. In recent years, several laboratories and some small companies, located in different parts of the world, have applied natural history knowledge and ecological and evolutionary criteria and theory to increase lead discovery.

For example, Coley et al. (2003) carried out pharmaceutical bioprospecting in the tropical forests of Panama. The theory of plant defense against attacks by herbivores predicts that older leaves and many other plant tissues are protected, at least in part, because cell walls are toughened by cellulose and lignin, neither of which is of medical importance. By contrast, young leaves must expand and so cannot be protected by such stiff, physical means but rather by repellent chemicals. The crucial inference is that young, expanding leaves will contain a greater variety of more active secondary metabolites than older leaves or other plant parts.

By focusing specifically on the collection of young leaves, the team has isolated a variety of novel molecules that they have tested for activity against three cancer cell lines, Chagas' disease, leishmaniasis, malaria, and HIV, and the research has identified some promising bioactive leads.

This approach, which exploits the vast databases of natural history together with ecological and evolutionary theory, has been given a variety of names, including ecologically driven drug discovery, the biorational approach, and hypothesis-driven drug discovery (Beattie and Ehrlich 2004; Coley et al. 2003). It is too early to assess the impact of these methods, except to say that such sampling of ecosystems for potential drugs is a major advance on the more traditional pharmaceutical protocols. This, in turn, is likely to increase the frequency of lead discovery and thus the value of the industry as well as its resource. In recent years, these methods of bioprospecting have been applied to many industries outside pharmaceuticals (see Table 10.8), including biological control, bioremediation, construction engineering, shipping, environmental monitoring, mining, industrial materials, manufacturing, and environmental restoration. These developments suggest a far greater role for bioprospecting in the future because they lead to more species being identified as useful for a much greater variety of human activities.

10.5 Benefit-sharing and Partnerships

Benefit sharing and the creation of partnerships within diverse bioprospecting industries can be both complex and time-consuming. Since many legal issues were largely clarified in the Convention on Biological Diversity, the protection of the rights of indigenous communities and source countries has often created tensions, with the investment sector concerned with altered levels of returns and profitability (Dalton 2004).

The chain of events leading to sales frequently involves multiple stages that include generating the appropriate knowledge, harvesting, processing, manufacturing, and distribution. Accordingly, the economics of each stage vary greatly, and assigning and protecting intellectual property is often an underlying factor. When agreements are reached, however, the types of benefits are varied and may include benefits to society such as increased food production, better health, and cleaner environments; benefits to the local suppliers such as employment, training, and capacity-building; and benefits to local, regional, national, or international corporations in the form of profits. Most current partnerships also emphasize the benefits of biodiversity conservation.

10.5.1 Examples of National and International Agreements and Partnerships on Ethnobotanical Bioprospecting

The CBD calls for fair and equitable sharing of benefits arising out of the utilization of genetic resources, including appropriate access to genetic resources. The application of the CBD has supported the intellectual rights of indigenous peoples. For example, scientists at Trivandrum Botanic Garden in India developed two medicines from plants used by tribal people, and royalties from the sale of these drugs now benefit the hill tribes that provided the original leads. A drug suitable for treating obesity is being developed by the CSIR in South Africa in association with a pharmaceutical company and the local San peoples. The intellectual property involved San knowledge of the plant, *Hoodia*, which when consumed in appropriate amounts retards hunger and hence helps through periods of drought. The development of the drug

Table 10.8. Status and Trends in Major Bioprospecting Industries

Industry	Current Involvement in Bioprospecting	Expected Trend in Bioprospecting	Social Benefits	Commerical Benefits	Biodiversity Resources
Pharmaceutical	tends to be cyclical	cyclical, posible increase	human health, employment	+++	P,A,M
Botanical medicines	high	increase	human health, employment	+++	mostly P
Cosmetics and natural personal care	high	increase	human health and well-being	+++	P,A,M
Bioremediation	variable	increase	environmental health	++	mostly M
Crop protection and biological control	high	increase	food supply, environmental health	+++	P,A.M
Biomimetics	variable	variable, possible increase	various (e.g., medicine)	++	P,A,M
Biomonitoring	variable	increase	environmental health	+	P,A,M
Horticulture and seed industry	low	steady	human well-being, food supply	+++	P
Ecological restoration	medium	increase	environmental health	++	P, A, M

Key: +++ = billion dollars   P = plants
     ++ = million dollars   A = animals
      + = profitable but amounts vary   M = microorganisms

and the fate of the indigenous intellectual property have been complex. (See Box 10.6.)

The World Health Organization estimates that some 3.5 billion people in the developing world depend mainly on plants for their primary health care. The development of botanical medicines for local peoples is therefore an important contemporary area of research. In Brazil the "Plants of the Northeast" program has stimulated a "Green Pharmacies" initiative in which local cures are tested for efficiency and toxicity and then manufactured by local people at affordable prices. Many countries, such as Thailand, India, Sri Lanka, Mexico, and China, have integrated traditional medicine into their national health care systems. Ethnobotanical bioprospecting has therefore contributed both to the enhancement of local medicine and to the search for modern drugs.

The economic importance of biodiversity was formally recognized in the CBD, which emphasized the conservation of biodiversity, sustainable use of its components, and fair and equitable sharing of the benefits arising out of the use of genetic resources. This has also been recognized by a wide variety of other agreements in many parts of the world. For example, the 1993 meeting of the Asian Coordinating Group for Chemistry expressed concern over the number of Asian plant samples being removed for study elsewhere. Some cases involved biopiracy, as the host country had not given permission for the plants to be exported. The issue of ownership was also debated at the Seventh Asian Symposium on Medicinal Plants, Spices and Other Natural Products in 1992 in Manila, and a code of ethics was published as the Manila Declaration, which said, in brief:

BOX 10.6

Pharmaceutical Bioprospecting and Commercialization: Two Case Histories

Hoodia gordonia. Scientists at South Africa's Council for Scientific and Industrial Research isolated the chemical entity extracted from *Hoodia gordonia* called P57 that suppresses appetite. This plant property has been used by the San people for generations, staving off hunger during prolonged hunting trips. P57 was patented in 1996. Phytopharm plc, a listed British company, was licensed in 1997 by the CSIR to undertake development and commercialization, but in August 1998 the company signed a licensing agreement with Pfizer Inc for this purpose. In mid-2003 Pfizer informed Phytopharm that it would be discontinuing the clinical development and returned the rights to Phytopharm, which is presently negotiating with another company to do the clinical development.

With international support, the South African San Council demanded recognition of their knowledge and a share of the benefits, and an agreement with the San was signed in March 2003. The CSIR will pay the San 8% of the milestone payments made by its licensee, Phytopharm, during clinical development over the next few years and will offer study scholarships to the San community. The San could earn 6% of all royalties if and when the drug is marketed, possibly in 2008, and $32,000 has already been paid. San milestone payments could reach $1.8 million, while the royalties could be $9.4 million annually during the years before the patent expires. South Africa, Namibia, Botswana, and Angola are all involved, so

income will go to the San Hoodia Benefit Trust that was established by the CSIR and the San.

Artemesia annua. This plant has been used by the Chinese to treat fevers for over 1,000 years. Extracts contain artemisinin that is effective against the malarial parasite, which is especially important where the parasite has evolved resistance to other drugs. It has several disadvantages, however, especially the difficulties of extraction and a short action time. In response, major programs have been established to generate superior derivatives, using the natural product as the blueprint.

One of the major players is the nonprofit Medicines for Malaria Venture established in 1999, which now funds and manages several projects, largely in partnership with the private sector, such as the Indian manufacturer Ranbaxy. This partnership has funded a research team at the University of Nebraska that has isolated a new class of synthetic endoperoxide antimalarials with superior properties to the original blueprint. In the future, other organizations, such as the National Institutes of Health in the United States, the Wellcome Trust in the United Kingdom, and the European & Developing Countries Clinical Trials Partnerships, aim to achieve far greater levels of participation by scientists and companies from developing countries.

- the biological resources of each region must be conserved,
- local scientists must be involved in research on local flora and fauna, and
- any commercial benefit arising from a regional resource must be shared equitably with the region.

This initiated a series of follow-up actions, including the 1994 meeting of ASOMPS in Malaysia, which produced the Melaka Accord that recognized the contribution of scientists from developing countries and sought legislation governing research into regional biological resources and sustainable development. The following year, Philippine Presidential Executive Order No. 247 was issued to regulate bioprospecting under two types of agreements: the Academic Research Agreement and the Commercial Research Agreement. And in 1996, Australia, Indonesia, Malaysia, the Philippines, and Thailand met in a UNESCO-funded a workshop in Kuala Lumpur that resulted in the Kuala Lumpur Guidelines on access to biological resources and sustainable development. This was followed by the Phuket Declaration issued by the International Conference on Biodiversity and Bioresources, which in line with the CBD stressed the protection of biodiversity, the sustainable utilization of biological resources, and the equitable sharing of commercial benefits. The 1998 meeting of ASOMPS in Hanoi urged the adoption of Philippine Executive Order No. 247 by other Southeast Asian nations. The state of Sarawak in Malaysia has passed The Sarawak Biodiversity Centre Ordinance to establish a Sarawak Biodiversity Centre and to regulate access to state biological resources.

While much pharmaceutical bioprospecting is still controlled by companies in industrial countries, there is a significant pharmaceutical industrial base emerging in developing ones as well. (See Table 10.9.) For example, Axxon Biopharm Inc. was established by a drug development program funded in part through grants from the International Cooperative Biodiversity Groups Program. The company seeks to commercialize nearly 100 leads in association with the Bioresources Development and Conservation Programme based in Nigeria. Flora Medicinal was established by Professor Jose Ribeiro da Silva in Rio de Janeiro and was recently acquired by Natura, a leading cosmetics and personal hygiene company.

Phyto Nova was established to research, develop, and market safe and affordable medicines for African wasting diseases, opportunistic infections, and other public health needs; to promote African traditional medicine, to scientifically validate natural products in order to ensure safety, efficacy and quality; and to ensure the sustainability of the supply of raw materials through conservation and local rural development. Centroflora, a Brazilian company, focuses on the production of organic, certified extracts from fruits and medicinal plants. The company, in association

with the Brazilian Institute of the Environment and Renewable Natural Resources, selects local communities to engage in the organic production of medicinal plants to ensure good management and quality control. Each of the companies just mentioned use ethnomedical knowledge as the basis of drug development. In line with traditional concepts, the line between food and medicine is indistinct, so that products known as phytonutrients, nutraceuticals, phytofoods, and phytocosmetics are also generated.

Another kind of partnership has been formed by the government of Sarawak and the U.S. company Medichem Research: Sarawak-Medichem Pharmaceuticals is a joint venture in which both parties share the risks and the rewards. The Instituto Nacional de Biodiversidad (InBio) in Costa Rica has developed a complex of partnerships with pharmaceutical companies from many parts of the world, including Merck & Co., Indena, Eli Lilly and Co., and Agrobiot S.A., and a wide variety of academic institutions and other organizations such as the Rockefeller Foundation and the MacArthur Foundation. All agreements include aspects of access, equity, technology transfer, and training of local scientists as well as the nondestructive use of biodiversity (Sittenfeld and Gamez 1993).

Furthermore, returns go to the Ministry of the Environment and Energy to help cover conservation costs, and 50% of royalties go to national conservation areas. Through 2000, InBio donated $400,000 to the Ministry for Environment and Energy for conservation, $790,000 to conservation areas, $713,000 to public universities, and $750,000 to its internal programs, notably those inventorying Costa Rican biodiversity (Laird and ten Kate 2002). Similarly, there have been major initiatives with respect to bioprospecting, benefit-sharing, and capacity building in Nigeria, Guinea, and Uganda (Carlson et al. 1997; Carlson et al. 2001).

The International Cooperative Biodiversity Groups explicitly use their drug discovery and bioinventory research process to generate enhanced research capacity, opportunities for sustainable economic activity, and incentives for conservation at each host-country site. The approach emphasizes equitable sharing of the benefits of both the research process and its discoveries. This experimental program is administered by the Fogarty International Center of the U.S. National Institutes of Health and supported by NIH, the National Science Foundation, and the U.S. Department of Agriculture.

In its first 10 years of ICBG (1993–2003), eight projects involved researchers from over 59 organizations in 12 countries on five continents. (See Table 10.10.) Investments by the U.S. Government agencies in these projects totaled approximately $29 million. Four major pharmaceutical companies, two agrochemical

Table 10.9. Examples of Pharmaceutical Developments in Biodiversity-rich Countries

Company	Country	Year	Number of Products	Number of Plant Species	Sales in 2002	Ethno-medical Leads	Benefit-sharing Policy
					(mill. dollars)		
Axxon Biopharm Inc.	United States/Nigeria	1999		10	not available	✓	✓
Centroflora	Brazil			21	21.5	✓	✓
Flora Medicinal	Brazil	1912	45	69		✓	
Phytonova Limited	South Africa	1999	4	3	0.2	✓	✓
Natura/Ekos	Brazil	2000	21	13	46.7	✓	✓
NuSkin	United States					✓	✓

Table 10.10. Projects of the International Cooperative Biodiversity Groups. Between 1993 and 2003, the ICBGs have comprised 17 projects working in 21 developing host countries, as well as the United States and the United Kingdom. In several cases partner institutions, particularly the pharmaceutical companies, have changed during the course of the project. As a result, those listed in any one group may include organizations that did not participate in the project at the same time. In addition to discovery of lead compounds for development of pharmaceuticals and agricultural agents, projects conduct research and training activities related to biological inventory, biodiversity conservation, benefit-sharing, and community development.

Years Active	Project Title, Country, Prospecting Focal Organisms	Principal Institutions Involved
1993–2008	Biodiversity Utilization in Madagascar and Suriname Suriname (1993–2003) and Madagascar (1998–2008) tropical plants (1993–2008), marine organisms (2003–08)	Virginia Polytechnic Institute and State University Missouri Botanical Garden; Conservation International; Madagascar National Centers for Pharmaceutical, Environmental and Oceanographic Research; Pharmaceutical Distribution Organization of Suriname (BGVS); Bristol-Meyers Squibb; Eisai Research Institute; Dow Agrosciences
1994–2000	Peruvian Medicinal Plant Sources of New Pharmaceuticals Peru tropical plants	Washington University (St. Louis) University of San Marcos (Peru); Peruvian Cayetano Heredia University; Searle-Monsanto Co.; Confederation of Amazonian Nationalities of Peru
1993–98	Chemical Prospecting in a Costa Rican Conservation Area Costa Rica arthropods	Cornell University Institute of Biodiversity (Costa Rica); University of Costa Rica; Bristol-Myers Squibb
1994–2003	Drug Development and Conservation of Biodiversity in West and Central Africa Nigeria and Cameroon tropical rainforest plants	Walter Reed Army Institute of Research Bioresources Development and Conservation Programme; Smithsonian Institution; University of Dschang (Cameroon); Pace University; University of Utah; International Center for Ethnomedical Drug Development (Nigeria)
1993–2003	Bioactive Agents from Dryland Biodiversity of Latin America Argentina, Chile, and Mexico arid lands plants, microorganisms	University of Arizona Argentine National Institute on Agricultural Technology; National University of Patagonia; Pontifical Catholic University of Chile; National Autonomous University of Mexico; Wyeth Pharmaceuticals; University of Illinois at Chicago; American Cyanamid Corp.
1998–2002	Drug Discovery and Biodiversity Among the Maya of Mexico Chiapas, Mexico temperate plants	University of Georgia College of the Southern Frontier; Molecular Nature Ltd. (UK)
1998–2008	Ecologically Guided Bioprospecting in Panama Panama rainforest plants (1998–2008), marine algae and invertebrates (2003–08)	Smithsonian Tropical Research Institute University of Panama; Oregon State University; National Secretariat for Science, Technology and Innovation (Panama); Gorgas Memorial Institute of Health Research; Monsanto; Novartis; Dow Agrosciences; Conservation International
1998–2008	Biodiversity of Viet Nam and Laos Viet Nam and Laos rainforest plants	University of Illinois at Chicago Purdue University; Research Institute for Medicinal Plants (Laos); Viet Nam National Institutes of Biotechnology, of Ecology and Biological Resources, and of Chemistry; Glaxo Smith-Kline; Bristol-Myers Squibb
2003–08	Conservation and Sustainable Use of Biodiversity of Papua New Guinea Papua New Guinea tropical plants, marine invertebrates	University of Utah Smithsonian Institution; University of Papua New Guinea; National Museum of Natural History (PNG); Wyeth Pharmaceuticals
2003–08	Building New Pharmaceutical Capabilities in Central Asia Uzbekistan, Kyrgyzstan temperate plants, terrestrial microbes	Rutgers University University of Illinois; Kyrgyz Agricultural Research Institute; Tashkent State Agrarian University; Diversa Corp.; Princeton University; Eisai Research Institute; Phytomedics, Inc.

Planning Grants 2003

Years Active	Project Title, Country, Prospecting Focal Organisms	Principal Institutions Involved
2003–05	Drug Development and Bio-cultural Diversity Conservation in the Pacific Islands Samoa, Tonga tropical plants, marine invertebrates, microorganisms	National Tropical Botanical Gardens, Hawaii AIDS Research Alliance; Samoan Ministry of Trade and Tourism; University of California; Anti-Cancer, Inc.; Diversa Corp.; Beth Israel Medical Center; Tongan Ministry of Agriculture and Forestry; Phenomenome Discoveries, Inc
2003–05	Potential Drugs from Poorly Understood Costa Rican Biota Costa Rica endophytic fungi, terrestrial microbes	Harvard Medical School National Institute of Biodiversity (Costa Rica)
2003–05	Drug Discovery and Biodiversity Conservation in Madagascar Madagascar plants, arthropods	State University of New York at Stony Brook University of Antananarivo, Madagascar; California Academy of Sciences; INDENA, Inc.; University of Eastern Piedmont, Italy
2003–05	New Drugs from Marine Natural Resources of Jamaican Reefs Jamaica coral reef organisms	University of Mississippi Discovery Bay Marine Laboratory; University of West Indies
2003–05	Studies of the Flora and Predator Bacteria of Jordan Jordan arid lands plants, bacteria	Research Triangle Institute Jordan University of Science and Technology; Virginia Polytechnic Institute and State University
2003–05	Biodiversity and Drug Discovery in the Philippines Philippines tropical plants, microorganisms, marine invertebrates	Michigan State University University of the Philippines
2003–05	Ecological Leads: Drugs from Reefs and Microbes in Fiji Fiji marine and freshwater organisms	Georgia Tech Research Corporation, School of Biology Scripps Institution of Oceanography; University of the South Pacific; South Pacific Applied Geoscience Commission; Bristol-Myers Squibb; Nereus Pharmaceuticals

companies, and two small biotech companies have at one time or another been affiliated with one or more projects. Private-sector support provided directly to the projects from partnering pharmaceutical companies, philanthropic foundations, and host-country governments totaled at least $2.5 million.

ICBG projects have variously focused on research and development with tropical, arid, or temperate plants, tropical arthropods, or endophytic and soil-associated microorganisms. Over 275,000 samples from more than 11,000 species of plants, 600 species of arthropods (mostly insects), and 500 species of microorganisms have been studied. Each group carries out assays on their own collections in multiple therapeutic areas, but almost all have an interest in cancer and malaria. Most groups also target a variety of infectious diseases, including parasitic and respiratory diseases that pose a high burden in the partner developing countries. Four have done work in agricultural areas, including veterinary medicines and insect, weed, nematode, and fungal pest control, predominantly through the industrial partners.

It is estimated that over 270 types of primary assays in 26 therapeutic and agricultural areas have been run over the 10-year life span of the project. Over 500 natural product compounds have been isolated that are active in one or more of the 22 therapeutic areas under study. Of these, approximately half are compounds new to science, but fewer than 50 have advanced to animal testing. Approximately 20 of these are currently considered active leads for drug development, although none has reached clinical trials to date.

Experiences from the first 10 years of the ICBG Program suggest that the industrial engines of innovation in drug discovery today are often small "biotech" companies rather than the pharmaceutical giants. This is especially true in natural products. Many large companies are no longer screening natural product samples because they take much longer to characterize and develop than synthetic molecules from their own libraries. The development of a natural product today usually involves a mix of large and small enterprises, none of which are well placed to undertake the entire task individually.

The pace of discovery of both taxonomically novel organisms and pharmacologically useful constituents has been shown to be higher today from marine and microbial sources than from the historically important plant kingdom, including tropical forests. And the low rate at which research on a newly collected organism leads to a commercial drug means that for the vast majority of projects the greatest benefits to development and conservation are likely to be gained from research, training, and technology transfer outcomes rather than from royalties on a marketed product.

10.5.2 Equity Considerations

Although bioprospecting research and development tends to be concentrated in industrial countries, the benefits to human well-being are often global. The principles for the treatment of intellectual property are well established (Rosenthal et al. 1999) and include protection of inventions using patents or other legal mechanisms; clear designation of the rights and responsibilities of all partners; sharing of benefits with the appropriate source-country parties; disclosure and consent of indigenous or other local stewards; information flow that balances proprietary, collaborative, and public needs; and respect for and compliance with relevant national and international laws, conventions, and other standards.

There is potential conflict between the routine scientific documentation of traditional medicines and the protection of indigenous intellectual property. For example, knowledge on the use of more than 1,100 medicinal plant species known to Malaysian

peoples, including the Iban, Bidayuh, Orang Ulu, Malay, Kadazan, and Orang Asli, is now in the public domain, so it is no longer possible to seek compensation for sharing knowledge. However, some organizations are considering whether indigenous knowledge in the public domain might be protected in some way—for example, through the deployment of indigenous knowledge databases or by citing local people as "discoverers" and co-owners of patents.

The CBD provides guidance on these issues. Article 8(j), for example, calls for a fair and equitable sharing of benefits with indigenous peoples when their ethnobotanical knowledge is used in drug research and development. Access to biological resources in some resource-rich countries is now regulated, including in the ASEAN countries of Malaysia, the Philippines, and Thailand and in the Andean Pact countries of Bolivia, Colombia, Ecuador, Peru, and Venezuela. Permit processing regulations include a formal application, contract negotiation, and publication of the contract. Negotiations such as these are not easy, and there has to be considerable good will on all sides for the views of the host countries and the research institutions and industrial organizations to be accommodated. Benefit-sharing agreements have also been created in industrial countries such as Australia, for example, between the Australian Institute of Marine Science and the State of Queensland (see www.aims.gov.au/pages/about/corporate/bsa -aims-qldgov.html).

At the global scale, the CBD provides guidelines with respect to:

- terms for prior informed consent and mutually agreed terms;
- the roles, responsibilities, and participation of stakeholders;
- relevant aspects relating to in situ and ex situ conservation and sustainable use;
- mechanisms for benefit-sharing, such as through technology transfer and joint research and development; and
- the means to ensure the respect, preservation, and maintenance of knowledge, innovations, and practices of indigenous and local communities embodying traditional lifestyles relevant to the conservation and sustainable use of biological diversity, taking into account work by the World Intellectual Property Organization.

An Ad Hoc Open-ended Working Group on Access and Benefit Sharing was established by the CBD Conference of the Parties in 2000. The mandate of the Working Group is to elaborate and negotiate an international regime on access and benefit sharing within the framework of the CBD.

10.6 The Legal Environment

Many significant changes in the legal and policy framework over the past decade have set the scene for better recognition of the rights of indigenous and local communities in transactions involving genetic resources and traditional knowledge. These changes include intergovernmental agreements, national measures, and the various codes, statements, and policies adopted by communities, researchers, and companies.

10.6.1 Intergovernmental Agreements

In recent years, states have agreed on a range of intergovernmental agreements that include provisions supporting the rights of sovereign nations to control access to their genetic resources and the rights of local and indigenous communities to control the use of their traditional knowledge systems and thus benefit from them. Some agreements, such as the CBD, the Convention to Combat Desertification, and the International Labour Organization's Convention No. 169 Concerning Indigenous Peoples (in 1989), are legally binding. Others, such as the 1994 United Nations Draft Declaration on the Rights of Indigenous Peoples, *Agenda 21* from the Earth Summit, and the Rio Declaration of 1992, are not legally binding but place a moral obligation on signature countries to conform with the provisions.

The CBD's voluntary Bonn Guidelines on Access to Genetic Resources and Benefit-sharing provide operational guidance for "users and providers" of genetic resources and information for governments that are drafting national laws as well as for governments, communities, companies, researchers, and other parties involved in such agreements. The scope of the guidelines includes "all genetic resources and associated traditional knowledge, innovations and practices covered by the CBD and benefits arising from the commercial and other utilization of such resources," with the exclusion of human genetic resources.

The guidelines describe steps in the access and benefit-sharing process, with sections on prior informed consent and mutually agreed terms as well as possible measures that countries and organizations should consider in response to their roles and responsibilities as providers and users of genetic resources and traditional knowledge. They outline recommendations for the participation of stakeholders and refer to incentive measures, accountability, national monitoring and reporting, verification, dispute settlement, and remedies. One appendix sets out suggested elements for material transfer agreements and another describes monetary and nonmonetary benefits that may be shared. The guidelines state that access and benefit-sharing systems should be based on an overall access and benefit-sharing strategy at the national or regional level. Given the complexity and uncertainty involved in access and benefit-sharing arrangements, such strategies can help communities and other groups to derive optimum benefits (ten Kate and Wells 2001).

Another recent development is the International Treaty on Plant Genetic Resources for Food and Agriculture, which has provisions on prior informed consent, benefit sharing, and farmers' rights. One important element of this treaty, which entered into force on 29 June 2004, is a multilateral system for access, for food and agriculture, to 35 crop genera and 29 forage species and associated benefit sharing. Its conditions for facilitated access to in situ plant genetic resources for food and agriculture allow for the protection of intellectual and other property rights. Benefits such as the exchange of information, access to and transfer of technology, and capacity building will be shared on a multilateral basis rather than with the specific provider of genetic resources.

Parties to this treaty agree that benefits should flow mainly to farmers involved in the conservation and sustainable use of plant genetic resources for food and agriculture, particularly in developing countries. The treaty encourages countries to take steps "to protect and promote Farmers' Rights," including protection of traditional knowledge and the right to participate in benefit sharing and in national decision-making. Communities may also benefit through involvement in conservation and sustainable use.

10.6.2 Intellectual Property Rights

At regional and national levels, there are various initiatives to apply and develop intellectual property law consistent with prior informed consent for access to genetic resources, prior approval for the use of traditional knowledge, and benefit sharing. Of interest in this area are the U.K. Commission on Intellectual Property Rights and Decision 486, "Common Intellectual Property Regime," of the Commission of the Andean Community, adopted in September 2000. The five Andean countries have at-

tempted to introduce provisions in harmony with both the World Trade Organization's Trade-Related Aspects of Intellectual Property Rights and the CBD. The decision provides that certain life forms shall not be considered inventions, that patent applications based on the region's genetic resources require a copy of an access contract, and that applications for a patent on an invention obtained or developed from traditional knowledge shall include a copy of a license from the community.

At the international level, there are discussions on the review and implementation of TRIPS (see, e.g., the Doha WTO Ministerial Declaration of November 20, 2001, paragraphs 17–19, and the TRIPS Council). The Intergovernmental Committee on Intellectual Property and Genetic Resources, Traditional Knowledge and Folklore of the World Intellectual Property Organization is considering intellectual property issues that arise in the context of access to genetic resources and benefit sharing, the protection of traditional knowledge, innovations and creativity, and the protection of expressions of folklore. For example, it is reviewing clauses related to IPRs in access and benefit-sharing agreements. WIPO is working on an electronic database of contract clauses and practices concerning access to genetic resources and benefit sharing. It is also considering elements of a sui generis system for the protection of traditional knowledge, and the Intergovernmental Committee has been considering ways to improve access to traditional knowledge for patent examiners so that patents are not improperly granted.

The African Model Law for the Protection of the Rights of Local Communities, Farmers and Breeders and for the Regulation of Access to Biological Resources aims to protect biodiversity and livelihood systems with a common tool (Ekpere 2001) and to guide African countries as they tailor national legislation and regional agreements dealing with the exchange of biodiversity knowledge, innovations, and practices.

A range of proposals has emerged concerning patents, from the meaning of "prior art," the scope of patents, and the test of "inventive step" to procedural requirements such as disclosure of country of origin and even proof of prior informed consent in patent applications. Indigenous groups have engaged with the patent system to challenge the granting of patents. For example, the Coordinating Body of Indigenous Organizations of the Amazon Basin, an umbrella organization that represents more than 400 indigenous groups in the region, joined with the U.S.-based Center for International Environmental Law to file a request before the U.S. Patent and Trademark Office asking it to re-examine a patent issued on a purported variety of *Banisteriopsis caapi,* or Ayahuasca—a plant that has a long traditional use in religious and healing ceremonies. The patent was annulled shortly thereafter but has subsequently been reinstated. Other forms of IPRs are also being investigated as a potential source of protection against expropriation of traditional knowledge. Geographical indications and trademarks have looked particularly promising (see Commission on Intellectual Property Rights 2002).

10.6.3 National Laws on Access to Genetic Resources and Traditional Knowledge

The CBD establishes the sovereign rights of states over their biodiversity but leaves parties a great deal of discretion on regulation and access. About 100 countries have introduced or are developing appropriate national legislation and other policy measures. The Philippines and Peru have also introduced legislation to regulate access to traditional knowledge, whether it is obtained in conjunction with genetic resources or not.

The CBD states that the right to determine access to genetic resources rests with government, but several national laws on this topic make such governmental consent contingent on prior informed consent and benefit-sharing agreements with the communities involved. The Philippines and the five countries of the Andean Community were in the vanguard of such legislation. The Philippines Executive Order 247 on Access to Genetic Resources requires the prior informed consent of indigenous communities for prospecting for biological and genetic resources within their ancestral lands and domains. And the Indigenous Peoples Rights Act of 1997 in the Philippines recognizes a wide range of rights held by the country's numerous indigenous groups, including land rights and a considerable measure of self-government within ancestral domains, including rights to "preserve and protect their culture, traditions and institutions."

The Andean Community's Decision 391 established a Common Regime on Access to Genetic Resources in Bolivia, Colombia, Ecuador, Peru, and Venezuela. It states that an applicant wishing access to genetic resources, their derivatives, or their "intangible component" (any knowledge, innovation, or individual or collective practice of actual or potential value associated with them) within the region must secure prior informed consent from, and share benefits with, the respective government, any supplier of an "intangible component," and, where appropriate, from the "owner, holder or administrator of the biological resource containing the genetic resource." To complement this, in 2002 Peru introduced a law protecting the collective knowledge of indigenous peoples related to biological resources.

The Indian Biological Diversity Act 2002 stipulates that no foreigner may obtain any biological resource occurring in India or knowledge associated thereto "for research or for commercial utilization or for bio-survey and bio-utilization" without prior approval of the National Biodiversity Authority, nor may foreigners apply for any intellectual property right for any invention based on a biological resource obtained from India without the Authority's approval. A National Biodiversity Fund will channel benefits received from foreign bioprospectors to "benefit-claimers," to conservation, and to development for the area from which the genetic resource or knowledge comes. Indian citizens and corporations must also give "prior intimation" to State Biodiversity Boards before obtaining any biological resource for commercial utilization or biosurvey, through which benefits will be shared at the state level. Local bodies are to constitute Biodiversity Management Committees to promote the conservation, sustainable use, and documentation of biodiversity within the area.

National legislation is also being drafted to cover issues of access and benefit sharing relating to the use of genetic resources that originate outside the country in question. The Norwegian government, for example, is proposing such legislation to cover the use in Norway of genetic material originating elsewhere.

10.6.4 Indigenous Peoples' Declarations, Codes, Research Agreements, and Policies

Complementing developments on national and international policy, a range of codes of ethics, research agreements, statements and declarations, and corporate and institutional policies have been developed by indigenous peoples, researchers, professional associations, and companies, marking a significant shift in the ethical context for bioprospecting partnerships. Although implementation often remains a challenge, these have helped to make equitable relationships between local communities or indigenous peoples and various outside groups more likely and have influenced the language incorporated into national and international law and contractual agreements.

Over the past 20 years, indigenous peoples organizations have issued a range of declarations and statements with clear demands in terms of bioprospecting. These demands include ownership and inalienable rights over their knowledge and resources; requirements for their prior informed consent; the right of veto over research and access to their land, knowledge, or resources; and benefit sharing. Such demands have led to calls for a moratorium on bioprospecting pending a legal framework for equitable partnerships.

Researchers have developed a number of codes of ethics and research guidelines through professional societies such as the International Society of Ethnobiology, the American Society of Pharmacognosy, and the Society for Economic Botany. These lay out general principles for research partnerships, obligations of the partners, and sometimes recommended guidelines for researcher behavior in the field. Various research organizations have developed institutional policies that establish general principles for their employees and associates.

An important example is the Principles on Access to Genetic Resources and Benefit-sharing for Participating Institutions, in which 28 botanic gardens and herbaria from 21 countries developed common standards on access to genetic resources and benefit sharing. The Limbe Botanic Garden in Cameroon and other institutions working with indigenous peoples and local communities have endorsed these principles, then developed in more detail their own policies to translate them into action. These policies address practical issues confronted by the institutions concerned, including their relationship with local communities (Laird and Mahop 2001; and see www.rbgkew.org/peopleplants/manual).

An interesting further example is the Micro-Organisms Sustainable Use and Access Regulation International Code of Conduct to facilitate access to microbial resources and to help partners with agreements when transferring them. Partners in the voluntary code include several countries and both nonprofit and commercial organizations.

A number of bioscience companies have also developed corporate policies setting out their approach to compliance with the CBD. These policies generally describe the scope of resources covered by the policy; the standard to which the company means to be held accountable (for example, absolute commitments or commitments to make reasonable or best efforts); how to obtain prior informed consent and ensure genetic resources and information are obtained legally; and commitments to obtain clear legal title to the materials and information acquired, to share benefits fairly and equitably, and to support conservation through environmentally sustainable sourcing. Some corporate policies describe the process followed to develop them and the indicators used to gauge success in their implementation (ten Kate and Laird 1999).

In the GlaxoSmithKline Policy Position on the CBD approved in February 2002, the company states that it is increasingly focused on drug discovery by screening synthetic chemical compounds, and thus has limited interest in collecting and screening natural material. Collecting programs have drawn to an end, and screening is no longer conducted in-house but by partners in countries such as Brazil and Singapore. However, the policy supports the principles enshrined in the CBD when conducting relevant activities. The document does not address prior informed consent from local communities per se, but it states that the company has always undertaken only to work with organizations and suppliers with the expertise and legal authority to collect samples and to ensure that governments in developing countries are informed of and consent to the nature and extent of any collecting program.

10.7 Threats to and Impacts of Bioprospecting

A number of threats to biodiversity were discussed in Chapter 4, and each of these is also a threat to the sustainability of bioprospecting. The loss of biodiversity directly removes the resource base for bioprospecting, and declines in abundance of elements of biodiversity can reduce the ability and increase the costs of sampling. In addition to these main threats, the loss of traditional knowledge and modern agricultural practices have also contributed to declines in the potential for bioprospecting industries.

Bioprospecting itself also has had impacts on biodiversity, and many legal agreements now specify the need for sustainability with respect to issues such as harvesting from the wild. Sometimes, however, these issues are less relevant because the species of interest for bioprospecting are removed from the wild in such small numbers. For example, an individual termite under investigation for pharmaceutical analysis most likely involves a sample of a few hundred individuals from a single colony containing millions of individuals. Similarly, a bacterium taken from a gram of soil is cultured in the laboratory. At the other end of the spectrum, however, large quantities of species or products such as bark may be required for some pharmaceutical research and development, and special conservation measures may be required.

10.7.1 Biodiversity Loss

The current and future ability of countries, regions, and localities to generate novel products and industries is likely to be threatened by the loss of the basic resource, biodiversity, at all levels: genes, populations, species, and ecosystems. There is abundant evidence that such losses are widespread (see Chapter 4 and Balmford et al. 2003), and there is little sign that the losses are slowing, except in circumstances specifically aimed at biodiversity protection, such as the establishment of effective protected areas. (See Chapter 5 of MA *Policy Responses* volume.)

It is ironic that the recent explosion of new techniques in the biological, chemical, and physical sciences that has generated a vastly improved capacity to understand and use biodiversity has been accompanied by a global decline in this very resource. The loss of biodiversity may not only lead to a loss of commercial opportunity but may also compromise ecosystem function (see Chapter 11; Loreau et al. 2002; Coleman and Hendrix 2000). While there is much debate over exactly how many species are becoming extinct each year, it is abundantly clear that a very high proportion of species are losing their constituent populations at an alarming rate (Hughes et al. 1997; Ehrlich and Daily 1993).

In some forested regions there is a direct conflict of interest between logging on the one hand and human health and bioprospecting on the other. In Eastern Amazonia, for example, where native plants provide most of the medicines used locally, the removal of trees that supply medicinal leaves, fruits, bark, or oils has critically diminished the supply of medicines required by both the rural and the urban poor (Shanley and Luz 2003). Short-term, low-value commodities gained by logging may be matched by the sustainable use of non-timber forest products (Emery and McLain 2001) and, in rare instances, superseded by the high-value products that could be gained by bioprospecting.

For example, the pharmaceutically important tree species *T. brevifolia* was considered worthless to the timber companies logging the forests where it grew, but its pharmaceutical value has been far greater than that of the timber species around it. Another

pharmaceutically important plant species, *Calophyllum lanigerum*, was first collected from the forests of Sarawak, but when teams returned to the original collection area for more specimens they found it had been logged and the remnant populations showed less activity (Laird and ten Kate 2002).

While global threats to biodiversity may one day affect bioprospecting, not least for pharmaceuticals (Cragg and Newman 1999; Grifo and Rosenthal 1997), there are few documented cases in which bioprospecting has been compromised by the loss of a natural community or an individual species. Given the many examples in this chapter, however, the indiscriminate loss of species or of the communities where they reside is likely to be a major threat to bioprospecting, even when their values are currently unknown or even suspected.

Many species vital for crop protection and hence large commercial revenues, for example, have been discovered in the habitat of the pest species only after intensive and prolonged research. The weevils responsible for the pest control in Australian lakes described earlier, for instance, were virtually unknown until they were needed. The same story applies to hundreds more species used to protect crops worth billions of dollars (Bellows and Fisher 1999; ten Kate and Laird 1999). Thus while the potential threat to bioprospecting through the loss of biodiversity appears very large, the actual consequences of such losses to the industry at present are very small.

10.7.2 Loss of Traditional Knowledge

Losses of traditional knowledge of biological resources in recent centuries has been well documented (see Chapter 17), and it is very likely that much local knowledge of medicines has been lost to humanity in general and to pharmaceutical prospecting in particular (Laird 2002). The current situation has been reviewed by Maffi (2001), and a growing literature on the issue (e.g., Mathooko 2001 and other publications from the International Society of Ethnobiology) documents global losses in traditional knowledge of biological sources worldwide, especially as older generations are unable, for various reasons, to pass on their wisdom to the next generations.

10.7.3 Modern Agricultural Methods

The losses of crop genetic diversity due to modern agricultural methods have been well documented (WCMC 1992; Groombridge and Jenkins 2002). In China, for example, only 10% of the 10,000 wheat varieties present in 1949 were available in the 1970s, while in Mexico only 20% of maize varieties planted in the 1930s remain and in the United States only 15–20% of apple, cabbage, maize, pea, and tomato varieties grown in the nineteenth century are available today.

The environmental effects of genetically modified crops remain unclear. But modern agricultural methods more broadly, including the removal of native vegetation, creation of larger fields, and increased use of irrigation, have resulted in biodiversity declines and losses on a large scale. In areas maintained for agricultural production, therefore, profitable bioprospecting is less likely. This may apply even to soil microorganisms, which are more specialized and less diverse in agricultural systems, although these have been shown to be restored through various agroforestry practices (Leakey 1999). Despite the lower levels of biodiversity in agricultural systems, particularly those managed under modern agricultural methods, it may be that fragments of original ecosystems in the midst of broadscale agriculture may harbor crop relatives or genetic systems of commercial value because of their adaptations to the regions or systems of interest.

10.7.4 Overharvesting

Overharvesting is a serious issue in some regions, especially when it involves plants for botanical medicines and pharmaceuticals. The situation is frequently exacerbated by poverty, especially when harvesting wild plants is both the sole source of income and the sole source of medicine for local use (Edwards 2004). Recent studies show both the dramatic effects of overharvesting, such as the declining recruitment of Brazil Nut seedlings (Peres et al. 2003), and the more subtle effects, such as the progressive shortening of available cane length in rattan populations (Siebert 2004). To avoid this kind of situation, companies such as Shaman Pharmaceuticals have instituted sustainable harvesting protocols for wild plants, such as *Croton lechleri,* a drug development candidate (King et al. 1997).

Some marine species have also been overharvested for natural products research (Farrier and Tucker 2004; Benkendorff 2002). In particular, cone shells of the molluscan family Conidae are prized for their highly variable toxins (conotoxins) for application to many areas of medicine, including pain control, cancer treatment, and microsurgery. Widespread harvesting of these animals for medical research, in addition to their collection for the sale of their shells to tourists, has led to the threatened extinction of many species throughout the tropics. Chivian et al. (2003) sum up the situation as follows: "With up to 50,000 toxins, cone shells may contain the largest and most clinically important pharmacopoeia of any genus in nature. To lose them would be a self-destructive act of unparalleled folly."

Food or medicinal species may be overharvested when used for export or for consumption in large urban areas (Cunningham 1993; Ferreira 1995; Malaisse 1997; MacKinnon 1998). When resources have already been degraded or reduced, as in forest systems by activities such as logging, the effects of overharvesting can be faster and more significant to local markets. While forest degradation tends to reduce the availability of medicinal resources, a few useful species thrive in the secondary growth that follows timber extraction (Shanley and Luz 2003). In South Africa, Namibia, and Botswana, wild devil's claw (*Harpagophytum procumbens*) is widely collected for its analgesic and anti-inflammatory properties, and trade data show that about 700 metric tons are collected each year. While this is actively managed in Botswana and elsewhere, and it is a protected plant across its range, the vast distances involved make enforcing regulations against overharvesting for this species extremely difficult.

The impacts of overharvesting have been recognized in many areas, and measures have been introduced to reduce levels of wild harvest. For example, Mayapple *Podophyllum* are now cultivated commercially in the United States. Another strategy, exemplified by the search of more sustainable sources of Taxol, has been to harvest from different species in different parts of the world following sustainability agreements, thereby reducing the pressure on individual populations of species.

References

Bailey, F., 2001: *Bioprospecting: Discoveries Changing the Future.* The Parliament of the Commonwealth of Australia, Government Publishing, Canberra.
Baker, H., 1978: *Plants and Civilization.* Wadsworth, California.
Balick, M.J. 1994: Ethnobotany, drug development and biodiversity conservation—exploring the linkages. In: *Ethnobotany and the Search for New Drugs,* Chadwick, D.J. and Marsh, J. (eds.). Ciba Foundation Symposium 185, John Wiley and Sons, New York.
Balmford, A., R.E. Green, and M. Jenkins, 2003: Measuring the changing state of nature. *Trends in Ecology and Evolution,* 18: 326–330.
Barbier, E.B. and B.A. Aylward, 1996: Capturing the Pharmaceutical Value of Biodiversity in a Developing Country. *Environmental and Resource Economics,* 8:157–181.
Beattie, A.J. and P.R. Ehrlich, 2004: *Wild Solutions: How Biodiversity is Money in the Bank.* Second Edition. Yale University Press, New Haven.
Bellows, T.S. 1999. Whither Hence, Prometheus? The Future of Biological Control. In: *Handbook of Biological Control* (Bellows and Fisher, eds.), Academic Press, New York.
Bellows, T.S. and T.W. Fisher, 1999: *Handbook of Biological Control.* Academic Press, New York.
Benkendorff, K., 2002: Potential Conservation Benefits and problems Associated with Bioprospecting in the Marine Environment. In: *A Zoological Revolution: Using Native Fauna to Assist in Its Own Survival,* Lunney, D. and C. Dickman, (eds.). Royal Zoological Society of New South Wales, Mosman, Australia. pp 90–100.
Birchall, J.D., 1989: The importance of the study of minerals to materials technology. In: *Biomineralization: Chemical and Biochemical Perspectives,* S. Mann, J. Webb, and R.J.P. Williams (eds.), Verlagsgesellschaft, Hamburg, pp. 491–509.
Boyle, T.P., 1987: *New Approaches to Monitoring Aquatic Systems.* American Society for Testing Materials Publication, Philadelphia, USA.
Brush, S.B., 2004: *Farmer's Bounty: Locating Crop Diversity in the Contemporary World.* Yale University Press, New Haven.
Carlson, T., M. Iwu, S.R. King, C. Obialor, and A. Ozioko, 1997: Medicinal Plan Research in Nigeria: An Approach for Compliance with the Convention of Biological Diversity. *Diversity,* 13: 29–33.
Carlson, T., B. Foula, J. Chnnock, S.R. King, G. Abdourahmaue, et al. 2001: Case Study on Medicinal Plant research in Guinea: Prior Informed Consent, Focused Benefit Sharing and Compliance with the Convention on Biological Diversity. *Economic Botany* 55: 478–491.
Cassaday, K. and Smale, M. 2001: Benefits from giving and receiving genetic resources. *Plant Genetic Resources Newsletter* 127: 1–10.
Chapman, T. 2004: The Leading Edge. *Nature* 430:109–115.
Chivian, E., C.M. Roberts, and A.S. Bernstein, 2003: The threat to cone snails. *Science,* 302:391.
Clutton-Brock, J., 1999: *A Natural History of Domesticated Animals.* (2nd Ed.) Cambridge University Press, Cambridge **Coley,** P.D., Heller, M.V., Aizprua, R., Arauz, B., Flores, N., et al. 2003: Using ecological criteria to design plant collection strategies for drug discovery. *Frontiers in Ecology and Environment,* 1: 421–428.
Coleman, D.C. and P.F. Hendrix, 2000: *Invertebrates as Webmasters in Ecosystems.* CABI Publishing, New York.
Cox, P.A. and Balick, M.J. 1994: The ethnobotanical approach to drug discovery. *Scientific American,* June 1994, p 2–7.
Craft, A.B. and R.D. Simpson, 2001: The social value of biodivertsity in new pharmaceutical product research. *Environmental and Resource Economics,* 18: 1–17.
Cragg, G.M. and D.J. Newman, 1999: Discovery and development of antineoplastic agents from natural sources. *Cancer Invest.,* 17: 153–163.
Cragg, G.M. and D.J. Newman. 2004. A tale of two tumor agents: Topoisomerase 1 and Tubulin. The Wall and Wani contribution to cancer chemotherapy. *Journal of Natural Products.* 67: 232–244.
Craig, C.L., 2003: *Spiderwebs and Silk.* Oxford University Press, Oxford.
Crawford, R.L. and D.L. Crawford, 1998: *Bioremediation: Principles and Applications.* Cambridge University Press, Cambridge.
Cueto, M., P.R. Jensen, C. Kauffman, W. Fenical, E. Lobkovsky, and J. Clardy, 2001: Pestalone, a New Antibiotic Produced by a Marine Fungus in Response to Bacterial Challenge. *Journal of Natural Products,* 64: 1444–1446.
Cunningham, A.B. 1993: Guidelines for equitable partnerships in new natural products development: Recommendations for a code of practice. In: *Ethics, Ethnobiological Research and Biodiversity: Research Report,* WWF International, Gland, Switzerland.
Dalton, R. 2004: Bioprospects less than golden. *Nature* 429: 598–600.
Demain, A.L., 2000: Small bugs, big business: The economic power of the microbe. *Biotechnology Advances,* 18: 499–514.
Edwards, R. 2004: No remedy in sight for herbal ransack. *New Scientist.* January, p. 10.
Ehrlich, P.R. and G.C. Daily, 1993: Population extinction and saving biodiversity. *Ambio,* 22: 64–68.
Eisner, T., 2003: *For the Love of Insects.* Belknap Harvard, Cambridge Massachusetts.
Eisner, T. and J. Meinwald, 1990: The Goteborg Resolution. *Chemoecology,* 1:38.
Ekpere, J.A. 2001: *The African Model Law: The Protection of the Rights of Local Communities, Farmers and Breeders, and for the Regulation of Access to Biological Resources.* An Explanatory Booklet. OAU, the Gaia Foyndation, Institute of Sustainable Development, Ethiopia.

Emery, M.R. and R.J. McLain, 2001: *Non-Timber Forest Products.* Food Products Press, Haworth Press, New York.

Farnsworth, N.R. and D.D. Soejarto, 1985: Potential consequences of plant extinction in the United States on the current and future availability of prescription drugs. *Economic Botany,* 39:231–240.

Farnsworth, N.R., O. Akerele, A.S. Bingel, D.D. Soejarto, and Z. Guo, 1985: *Medicinal Plants in Therapy.* World Health Organization **63**:965–981.

Farrier, D. and L. Tucker, in press: Access to Marine Bioresources: Hitching the Conservation Cart to the Bioprospecting Horse. *The International Journal of Marine and Coastal Law.*

Ferreira, A., 1995: Saving the Mopane Worm. *Food Insects Newsletter,* **8**:8.

Flatman, P.E., D.E. Jerger, and J.E. Exner, 1994: *Remediation Field Experience.* CRC Press, London.

Grifo, F. and J. Rosenthal (eds.), 1997: *Biodiversity and Human Health.* Island Press, Washington D.C. pp. 379.

Grifo, F., D. Newman, A.S. Fairfield, B. Bhattacharya, and J.T. Grupenhoff, 1996: The Origins of Prescription Drugs, in: Grifo, F. and Rosenthal, J. (eds.) *Biodiversity and Human Health.* Island Press, Washington DC.

Groombridge, B. and M.D. Jenkins, 2002: *World Atlas of Biodiversity UNEP,* University of California Press, Berkeley.

Handel, S.N., G.R. Robinson, and A.J. Beattie, 1994: Biodiversity resources for restoration ecology. *Restoration Ecology,* **2**: 230–241.

Harker, D., G. Libby, K. Harker, S. Evans, and M. Evans, 2001: *Landscape Restoration Handbook.* Lewis Publishers, Boca Raton.

Hedges, S.B., 2003: The coelacanth of frogs. *Nature,* **425**:669–670.

Henkel, T., R.M. Brunne, H. Muller, and F. Reichel, 1999: Statistical Investigation into the Structural Complementarity of Natural Products and Synthetic Compounds. *Angew. Chem. Int. Ed.* **38**: 643–647.

Heywood, V.H., 1995: *Global Biodiversity Assessment.* UNEP and Cambridge University Press, Cambridge.

Hijfte, L.V., G. Marciniak, and N. Froloff, 1999: Combinatorial Chemistry, Automation and Molecular Diversity: New Trends in the Pharmaceutical Industry. *Journal of Chromatography,* **725**: 3–15.

Hughes, J.B., G.C. Daily, and P.R. Ehrlich, 1997: Population diversity: its extent and extinction. *Science,* **278**: 689–692.

Ismail, G., M.Mohamed, and L. Bin Din, 1995: *Chemical Prospecting in the Malaysian Forest,* Pelanduk Publications, Malaysia.

Jack, D.B. 1997: One Hundred Years of Aspirin. *The Lancet,* **350**: 437–445.

Joshi, D. and M.H. Gold, 1993: Degradation of 2,4,5-trichlorophenol by the lignin-degrading basidiomycete *Phanerochaete chrysosporium. Applied Environmental Microbiology,* 59:1779–1985.

Kashefi, K. and D.R. Lovley, 2003: Extending the upper temperature limit for life. *Science,* 301:934.

King, S.R., E. Meza, F. Ayala, L.E. Forero, M. Pena, V. Zak, and H. Bastien. 1997: *Croton lechleri* and sustainable harvest and management of plants in pharmaceuticals, phytomedicines and cosmetics industries. In: *International Symposium on Herbal Medicine,* D.S. Wozniak, S. Yuen, M. Garrett, and T.M. Shuman, (eds.). International Institute for Human Resources Development, San Diego State University, San Diego, USA.

Laird, S.A., 1993: Contracts for biodiversity prospecting. In: *Biodiversity Prospecting,* W.V. Reid et al. (eds.), WRI, Washington DC pp. 99–130.

Laird, S.A., 2002: *Biodiversity and Traditional Knowledge.* Earthscan, London.

Laird, S.A. and K. ten Kate, 2002: Linking Biodiversity Prospecting and Forest Conservation. In: *Selling Forest Environmental Services,* S. Pagiola, J. Bishop, and N. Landell-Mills (eds.). Earthscan, London. pp. 151–172.

Laird, S.A. and Mahop, T. 2001: The Limbe Botanic and Zoological Gardens Policy on Access to Genetic Resources and Benefit-Sharing. Limbe, SW Province, Cameroon.

Leakey, R.R.B. 1999. Agroforestry for biodiversity in farming systems. In: *Biodiversity in Agroecosystems,* W.W. Collins and C.O. Qualset (eds.) pp. 127–145, CRC Press, New York.

Leakey, R.R.B., 2001a: Win: Win land use strategies for Africa. 1. Building on experience with agroforests in Asia and Latin America. *International Forestry Review.* 3: 1–10.

Leakey, R.R.B., 2001b: Win: Win land use strategies for Africa. 2. Capturing economic and environmental benefits with multi-strata agroforests. *International Forestry Review.* 3:11–18.

Loreau, M, S. Naeem, and P. Inchausti, 2002: *Biodiversity and Ecosystem Functioning.* Oxford University Press, Oxford.

MacKinnon, K. 1998. Sustainable use as a conservation tool in the forests of SE Asia. In: *Conservation of Biological Resources,* E.J. Milner–Gulland and R. Mace (eds.) Blackwell Science, Oxford. pp. 174–192.

Maffi, L. (ed.), 2001: *On Biocultural Diversity Linking Language, Knowledge, and the Environment.* Smithsonian Institution Press, Washington.

Malaisse, F., 1997: *Se Nourir en Foret Claire Africaine.* Les Presses Agronomique de Gembloux, Gembloux, Belgium.

Malakoff, D., 2003: Scientists counting on census to reveal marine biodiversity. *Science,* 302:773.

Mann, S., J. Webb, and R.J.P. Williams, 1989: *Biomineralization: Chemical and Biochemical Perspectives.* VCH Verlagsgesselschaft, Weinheim.

Mann, S. 2001. *Biomineralization.* Oxford University Press, Oxford.

Mathooko, J. M., 2001: *The Status and Future of African Traditional Ecological Knowledge in the Sustainability of Aquatic Resources.* IUCN.

McCutcheon, A.R., S.M. Ellis, R.E.W. Hancock, and G.H.N. Towers, 1992: Antibiotic screening of medicinal plants of the British Columbian native peoples. *Journal of Ethnopharmacology* 37: 213–223.

McGeer, A. and D.E. Low, 2003: Is resistance futile? *Nature Medicine,* **9**: 390–392.

Mendelsohn, R. and M.J. Balick, 1997: Valuing undiscovered pharmaceuticals in tropical forests. *Economic Botany* 51: 328.

Moss, D., S.A. Harbison, and D.J. Saul. 2003. An easily automated, closed tube forensic DNA extraction procedure using a thermostable proteinase. *International Journal of Legal Medicine.* 117:340–349.

Mueller, J.G., S.M. Resnick, M.E. Shelton, and P.H. Pritchard, 1992: Effect of inoculation on the biodegradation of weathered Pridhoe Bay crude oil. *Journal of Industrial Microbiology,* 10: 95–102.

Mujeeb-Kazi, A., R.L. Villareal, L.A. Gilchrist, and S. Rajaram, 1996: Registration of five wheat germplasm lines resistant to *Helminthosproium* leaf blight. *Crop Science* 36: 216.

Munro, M.H.G., J.W. Blunt, E.J. Dumdei, S.J.H. Hickford, R.E. Lill, S. Li, C.N. Battershill, and A.R. Duckworth, 1999: The discovery and development of marine compounds with pharmaceutical potential. *Journal of Biotechnology,* 70: 15–25.

Myers, N., 1983: *A Wealth of Wild Species: Storehouse for Human Welfare.* Westview Press, Boulder, Colorado.

Newman, D.J. and S.A. Laird, 1999: The influence of natural products on 1997 pharmaceutical sales figures. In: *The Commercial Uses of Biodiversity,* K. ten Kate, and S.A. Laird (eds). Earthscan, London.

Newman, D.J., G. M. Cragg and K.M. Snader. 2003. Natural products as sources of new drugs over the period 1981–2002. *Journal of Natural Products.* 66:1022–1037.

Olsen, N., T. Swanson, and R. Luxmoore, 2002: *Biodiversity and the Pharmaceutical Industry,* UNEP.

Ortholand, J.Y. and A. Ganesan, 2004: Natural Products and Combinatorial Chemistry: Back to the Future. *Current Opinion in Chemical Biology* 8:271–280.

Pearce, D. and S. Puroshothaman, 1993: *Protecting Biological Diversity: The Economic Value of Pharmaceutical Plants.* The Centre for Social and Economic Research in the Global Environment, Discussion Paper 92–97, University College, London.

Pearce, D. and D. Moran, 1995: *The Economic Value of Biodiversity.* Earthscan, London.

Peres, C.A., and 16 authors. 2003. Demographic threats to the sustainability of Brazil Nut exploitation. *Science.* 302:2112–2114.

Peters, C.P., A.H. Gentry, and R.O. Mendelsohn, 1989: Valuation of an Amazonian rainforest. *Nature* 339:655–656.

Pimentel, D., 1992: Environmental and economic costs of pesticide use. *Bioscience,* 42:750–760.

Principe, P. P., 1989: The economic significance of plants and their constituents as drugs. *Economic and Medical Plant Research,* 3:1–17.

Quick, J.S., K.K. Nkongolo, F.B. Peairs and J.B. Rudolph, 1996: Registration of Russian wheat aphid-resistant wheat germplasm CORWA1. *Crop Science* 36:217.

Rinehart, K.L., 2000: *Antitumour Compounds from Tunicates.* John Wiley & Sons, New York.

Rosenberg, D.M. and V.H. Resh, 1993: *Freshwater Biomonitoring and Benthic Macro-Invertebrates.* Chapman and Hall, London.

Rosenthal, J.P., D. Beck, A. Bhat, J. Biswas, L. Brady, K. Bridboard, S. Collins, G. Cragg, J. Edwards, A. Fairfield, M. Goyylieb, L.A. Gschwind, Y. Hallock, R. Hawks, R. Hegyeli, G. Johnson, G.T. Keusch, E.E. Lyons, R. Miller, J. Rodman, J. Roskoski, and D. Siegel-Causey, 1999: Combining high risk science with ambitious social and economic goals. *Pharmaceutical Biology* 37, Supplement, pp 6–21.

Rittman, B.E. and P.L. McCarty, 2001: *Environmental Biotechnology: Principles and Applications.* McGraw Hill, Boston.

Sarikaya, M., C. Tamerler, A. Jen, K. Schulten, and F. Baneyx, 2003: Molecular biomimetics: nanotechnology through biology. *Nature Materials,* **2**:577–585.

Shackleton, C.M., J. Botha, and P.L. Emanuel, 2003: Productivity and abundance of *Sclerocarya birrea* subsp. *caffra* in and around rural settlements and protected areas of the Bushbuckridge lowveld, South Africa. Forests, *Trees and Livelihoods* **13**: 217–232.

Shanley, P. and L. Luz, 2003: The impacts of forest degradation on medicinal plant use and implications for health care in Eastern Amazonia. *Bioscience,* **53**: 573–590.

Siebert, S.F. 2004. Demographic effects of collecting Rattan cane and their implications for sustainable harvesting. *Conservation Biology.* **18**:424–431.

Simons, A.J. and R.R.B. Leakey, 2004: Tree domestication in tropical agroforestry. *Agroforestry Systems* **61**: 167–181.

Simpson, R.D. and R.A. Sedjo, 1996: *Investments in Biodiversity Prospecting and Incentives for Conservation.* Resources for the Future, Discussion Paper 96–14, Washington DC.

Simpson, R.D., R.A. Sedjo, and J.W. Reid, 1996: Valuing biodiversity for use in pharmaceutical research. *Journal of Political Economy,* **104**: 163–185

Sittenfeld, A. and R. Gamez, 1993: *Biodiversity prospecting,* by INBio. In: *Biodiversity Prospecting,* W.V. Reid et al.(eds.), WRI, Washington DC, pp69–97.

ten Kate, K. and S.A. Laird. 1999. *The Commercial Use of Biodiversity: Access to Genetic Resources and Benefit-Sharing.* Royal Botanic Gardens, Kew and European Communities, Earthscan Publications Ltd, London.

ten Kate, K. and A. Wells, 2001: *Preparing a National Strategy on Access to Genetic Resources and Benefit-Sharing. A Pilot Study.* Royal Botanic Gardens, Kew and UN Development Programme/UNEP Biodiversity Planning Support Programme.

Torsvik, V., L. Ovreas, and T.F. Thingstad, 2002: Prokaryotic diversity—magnitude, dynamics and controlling factors. *Science,* **296**: 1064–1066.

Weitzman, M.L., 1992: On diversity. *Quarterly Journal of Economics,* **107**: 363–406.

Wessjohann, L.A., 2000: Synthesis of natural-product-based compound libraries. *Current Opinion in Chemical Biology,* **4**:303–309.

Whisenant, S.G., 2001: *Repairing Damaged Wildlands.* Cambridge University Press, Cambridge.

Wilson, E.O. 1992: *The Diversity of Life.* Belknap Press of Harvard University Press, Cambridge, Massachusetts.

WCMC (World Conservation Monitoring Centre). 1992. *Global Biodiversity: Status of the Earth's Living Resources.* Chapman and Hall, London

WRI (World Resources Institute), 1994: *World Resources: A Guide to the Global Environment.* Oxford University Press, Oxford.

WRI in collaboration with UNEP (United Nations Environment Programme), UNDP (United Nations Development Programme), and World Bank. 2000: *World Resources 2000–2001: People and Ecosystems: The Fraying Web of Life.* Washington DC: WRI.

Chapter 11
Biodiversity Regulation of Ecosystem Services

Coordinating Lead Authors: Sandra Díaz, David Tilman, Joseph Fargione
Lead Authors: F. Stuart Chapin III, Rodolfo Dirzo, Thomas Kitzberger, Barbara Gemmill, Martin Zobel, Montserrat Vilà, Charles Mitchell, Andrew Wilby, Gretchen C. Daily, Mauro Galetti, William F. Laurance, Jules Pretty, Rosamond Naylor, Alison Power, Drew Harvell
Contributing Authors: Simon Potts, Claire Kremen, Terry Griswold, Connal Eardley
Review Editors: Gerardo Ceballos, Sandra Lavorel, Gordon Orians, Stephen Pacala, Jatna Supriatna

Main Messages

Biodiversity, including the number, abundance, and composition of genotypes, populations, species, functional types, communities, and landscape units, strongly influences the provision of ecosystem services and therefore human well-being (*high certainty*). Processes frequently affected by changes in biodiversity include pollination, seed dispersal, climate regulation, carbon sequestration, agricultural pest and disease control, and human health regulation. Also, by affecting ecosystem processes such as primary production, nutrient and water cycling, and soil formation and retention, biodiversity indirectly supports the production of food, fiber, potable water, shelter, and medicines.

Species composition is often more important than the number of species in affecting ecosystem processes (*high certainty*). Thus, conserving or restoring the composition of communities, rather than simply maximizing species numbers, is critical to maintaining ecosystem services. Changes in species composition can occur directly by species introductions or removals, or indirectly by altered resource supply due to abiotic drivers (such as climate) or human drivers (such as irrigation, eutrophication, or pesticides).

Although a reduction in the number of species may initially have small effects, even minor losses may reduce the capacity of ecosystems for adjustment to changing environments (*medium certainty*). Therefore, a large number of resident species, including those that are rare, may act as "insurance" that buffers ecosystem processes in the face of changes in the physical and biological environment (such as changes in precipitation, temperature, or pathogens).

Productivity, nutrient retention, and resistance to invasions and diseases tend to increase with increasing species number in experimental ecosystems that have been reduced to a small number of species (10 or fewer). This is known with *high certainty* for experimental herbaceous ecosystems and with *low certainty* for natural ecosystems, especially those dominated by long-lived species. In natural ecosystems these direct effects of biodiversity loss may often be masked by other environmental changes that are caused by the factors that resulted in the loss of biodiversity (such as eutrophication or climate change). Nevertheless, human activities that cause severe reductions in species number can directly impair these ecosystem services.

Preserving interactions among species is critical for maintaining long-term production of food and fiber on land and in the sea (*high certainty*). The production of food and fiber depends on the ability of the organisms involved to successfully complete their life cycles. For most plant species, this requires interactions with pollinators, seed disseminators, herbivores, or symbionts. Therefore, land use practices that disrupt these interactions will have a negative impact on these ecosystem services.

Intended or accidental changes in the composition of ecological communities can lead to disproportionately large, irreversible, and often negative alterations of ecosystem processes, causing large monetary and cultural losses (*high certainty*). In addition to direct interactions, the maintenance of ecosystem processes depends on indirect interactions, whose disruption can lead to unexpected consequences. These consequences can occur very quickly; for example, in a wide range of terrestrial, marine, and freshwater ecosystems, the introduction of exotic species by humans has altered local community interactions. Alternatively, these consequences may be manifest only after a long time. For example, the intraspecific genetic diversity of certain plant species decreases when the populations of their animal pollinators or dispersers are reduced.

Invasion by exotic species, facilitated by global trade, is a major threat to the biotic integrity of communities and the functioning of ecosystems. Empirical evidence suggests that areas of high species richness (such as hot spots) are more susceptible to invasion than species-poor areas. On the other hand, within a given habitat the preservation of its natural species pool appears to decrease its susceptibility to invasions. On the basis of our present theoretical knowledge, however, we still cannot predict with accuracy whether a certain organism will become a serious invader in a given ecosystem.

The extinction of local populations, or their reduction to the point that they become functionally extinct, can have dramatic consequences in terms of regulating and supporting ecosystem services. Local extinctions have received little attention compared with global extinctions, despite the fact that the former may have more dramatic ecosystem consequences than the latter. Before becoming extinct, species become rare and their ranges contract. Therefore their influence on ecosystem processes decreases, even if local populations persist for a long time, well before the species becomes globally extinct. We do not have sufficient knowledge to predict all the consequences of these local extinctions. However, because they tend to be biased toward particular organisms that depend on prevailing land uses and types, rather than occurring at random, we can anticipate some of the most obvious impacts.

The properties of species are more important than species number in influencing climate regulation (*medium certainty*). Climate regulation is influenced by species properties via effects on sequestration of carbon, fire regime, and water and energy exchange. The traits of dominant plant species, such as size and leaf area, and the spatial arrangement of landscape units are particularly important in climate regulation. The functional characteristics of dominant species are thus a key element determining the success of mitigation practices such as afforestation, reforestation, slowed-down deforestation, and biofuel plantations.

The diversity of landscape units also influences ecosystem services (*high certainty*). The spatial arrangement of habitat loss, in addition to its amount, determines the effects of habitat loss on ecosystem services. This is because the effects of habitat loss on remaining habitat fragments are greater on the edges than in their cores. Thus, fragmentation of habitat has disproportionately large effects on ecosystem services. These effects are best documented in the case of carbon sequestration and pollination in the tropics.

Maintenance of genetic and species diversity and of spatial heterogeneity in low-input agricultural systems reduces the risk of crop failure in a variable environment and reduces the potential impacts of pests and pathogens (*high to medium certainty*). Agroforestry systems, crop rotations, intercropping, and conservation tillage provide opportunities to protect crops and animals from pests and diseases while maintaining yields without heavy investment in artificial chemicals.

Global change drivers that affect biodiversity indirectly also affect biodiversity-dependent ecosystem processes and services. Among these global change drivers, a major threat to biodiversity-dependent human well-being is large-scale land use change, especially the intensification and extensification associated with large-scale industrial agriculture (*high certainty*). This threat is most obvious for those human groups that are already vulnerable because their livelihoods rely strongly on the use of natural and seminatural ecosystems. These include subsistence farmers, the rural poor, and traditional societies.

A considerable amount of new research is needed to understand the role of different components of biodiversity in the provision of ecosystem

services. Although the available evidence clearly points to the key importance of the maintenance of the genetic, species, and landscape diversity of ecosystems in order to preserve the ecosystem services they provide, important knowledge gaps remain to be filled. These are particularly obvious in the case of high-diversity ecosystems, ecosystems dominated by long-lived plants, and trophic levels other than plants.

11.1 Introduction

Biodiversity refers to the number, abundance, and composition of the genotypes, populations, species, functional types, communities, and landscape units in a given system. Biodiversity is both a response variable that is affected by changes in climate, resource availability, and disturbance (see Chapter 4) and a factor with the potential to influence the rate, magnitude, and direction of ecosystem processes. This chapter focuses on this second aspect—the effects of biodiversity on ecosystem processes and the ecosystem services that humans obtain from them.

Ecosystem services are broadly defined as the benefits provided by ecosystems to humans; they contribute to making human life both possible and worth living (Daily 1997; MA 2003). Biodiversity affects numerous ecosystem services, both indirectly and directly. Some ecosystem processes confer direct benefits on humanity, but many of them confer benefits primarily via indirect interactions.

This chapter focuses on regulating and supporting ecosystem services (see Chapter 1) that result from interactions between two or more species or genotypes. The regulating ecosystem services addressed in this chapter include pollination, seed dispersal, climate regulation, carbon sequestration, and pest and disease control. (See Figure 11.1.) Biodiversity also provides supporting ecosystem services, which are necessary for the production of all other—more direct—ecosystem services. For example, by influencing primary production and nutrient and water cycling, biodiversity indirectly supports the production of food, fiber, and shelter. The enormous value of biodiversity per se and its importance in the provision of cultural ecosystem services are described in detail in Chapters 4, 10, and 17. Here the focus is on how biodiversity affects the quantity and temporal stability of the supply of those services.

Consideration of all components of biodiversity—genotypes, species, functional traits and types, communities, and landscape units—is essential in order to understand its role in ecosystem processes and thus in the provision of ecosystem services. Although traditionally the focus has been mainly on species number, there is now broad consensus that functional diversity—the value, range, and relative abundance of organismal traits present in a community—is the most important component of biodiversity influencing ecosystem functioning (Díaz and Cabido 2001; Loreau et al. 2001; Hooper et al. 2005). Recent scientific literature on the functional role of biodiversity has generated conflicting results that are sometimes difficult to interpret. However, some basic points of agreement have emerged that are relevant to land use and conservation policies.

Most of the current evidence and theory described early in this chapter deal with direct interactions among terrestrial plants. Although a growing number of studies incorporate other ecosystem processes, most of what we know about biodiversity effects on ecosystem functioning refers specifically to the production of plant biomass (the tissues formed using the solar energy captured by photosynthetic plants). However, there is growing empirical evidence suggesting that the influence of interactions between

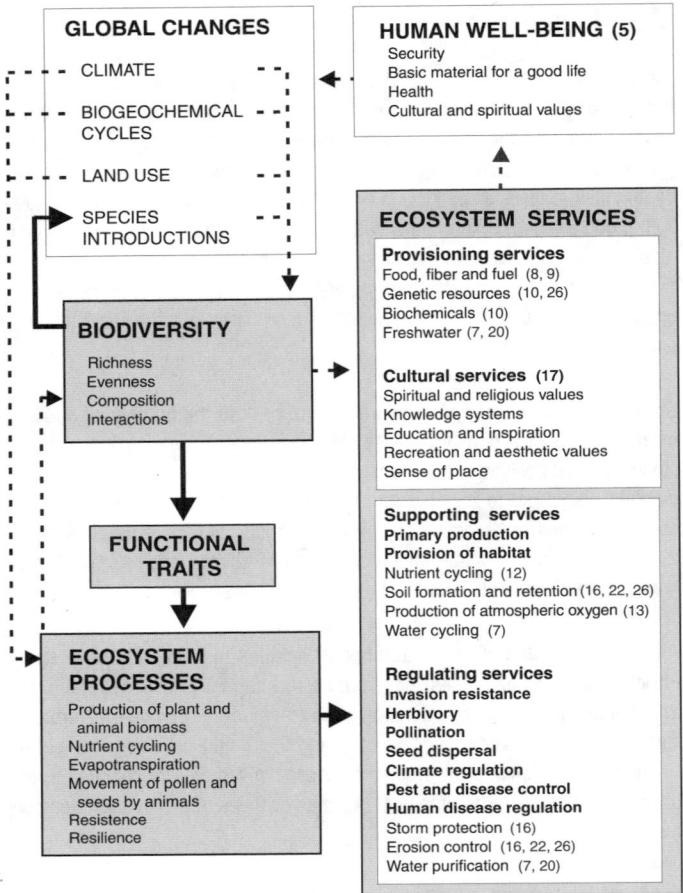

Figure 11.1. Biodiversity as Response Variable Affected by Global Change Drivers and as Factor Modifying Ecosystem Processes and Services and Human Well-being (modified from Chapin et al. 2000). Solid arrows indicate the links that are the focus of this chapter. Regulating services are the benefits obtained from the regulation of ecosystem processes. Supporting services are those that are necessary for the production of all other services. Ecosystem services in bold are developed in detail in this chapter. Other services are not addressed here because the role of biodiversity in regulating them in ecosystem processes is minor or uncertain, or because they are developed in detail elsewhere in this volume (relevant chapter number is indicated in parentheses).

plants and microorganisms and between plants and animals as well as the influence of indirect interactions on ecosystem properties are both important and widespread. Because these intertrophic and indirect interactions have received much less attention in the literature, they are emphasized in this chapter. The effects of terrestrial biodiversity on ecosystem services is discussed first, followed by a similar discussion of the effects of marine biodiversity on ecosystem processes provided by oceans and coastal areas, a topic whose importance has been recognized only recently.

Two major aspects of ecosystem functioning form the focus of this analysis: resource dynamics at given point in time (which includes processes such as primary and secondary production, nutrient cycling, and water dynamics) and long-term stability of such processes in the face of environmental variability or directional change.

11.2 Terrestrial Biodiversity Effects on Supporting Services

Region-to-region differences in ecosystem processes are driven mostly by climate, resource availability, and disturbance—not by differences in species richness. In most ecosystems, changes in the number of species are the consequence of changes in major abiotic and disturbance factors, so that the ecosystem effect of species richness (number of species) per se is expected to be both comparatively small and very difficult to isolate. For example, variation in primary productivity depends strongly on temperature and precipitation at the global scale and on soil resources and disturbance regime at the region-to-landscape-to-local scales. Factors that increase productivity, such as nutrient addition, often lead to lower species richness because more productive species outcompete less productive ones. In nature, therefore, high species diversity and high productivity are often not positively correlated (Grime 1979, 2001; Gough et al. 1994; Waide et al. 1999).

When elaborating management recommendations, it is extremely important to bear these considerations in mind and to interpret the conclusions of experiments within the right context. Here again, if taken uncritically, results from synthetic assemblages may lead to misleading recommendations to land managers (see also Fridley 2001; Schmid 2002; Hodgson et al. 2005).

Similarly, artificial increases of species richness in naturally species-poor areas (such as moorlands, boreal forests, or desert shrublands) may not result in any substantial "improvement" in ecosystem services. In natural ecosystems, low species richness does not necessarily imply impaired ecosystem properties and services. In most synthetic-community experiments, the assemblage of experimental communities occurs by random draws of species from a species pool decided by the experimenter, and the abundances of different species are artificially even, at least at the initial stages. This strongly differs from the process of community assembly that occurs in nature, in which the local assemblage is the product of a "filtering" process exerted by the environment on the regional species pool (Zobel 1997; Díaz et al. 1998).

Extinction in natural ecosystems similarly tends to be biased to certain organisms (Vitousek et al. 1997; Grime 2002; see also Chapter 4) rather than being a random process. In most natural assemblages, any unused resources are likely to be quickly used by members of the species pool, even in communities with low species richness (Zobel 1997; Hodgson et al. 1998). Some low-species-richness combinations in synthetic communities, on the other hand, are artificially maintained and could not persist in nature (Hodgson et al. 1998; Lepš 2005). This does not necessarily invalidate the results of experiments based on artificial assemblages; however, it is crucial to understand that low species richness in these experiments and in real ecosystems stem from different causes. Low diversity in nature tends to occur at either very high or very low productivity and to result from different processes in each situation. Most often, it results from strong abiotic constraints at low productivity and high biotic constraints at high productivity (see, e.g., Huston 1999; Pärtel et al. 2000).

Nevertheless, biodiversity can directly affect supporting services such as primary productivity, soil formation, and nutrient cycling, which in turn influence provisioning services such as genetic resources and production of food, timber, fuel, and fiber. The mechanisms by which that happens, and the empirical evidence accumulated to date are discussed in the remainder of this section.

In general, the most important component of plant biodiversity influencing ecosystem services is functional composition. And, other things being equal, a greater number of resident species should result in greater production, higher nutrient retention, and enhanced resistance to invasion, at least in experimental studies using a low number of resident species. There is also some indication that a high number of species within each functional type (a group of organisms that responds to the environment or affects ecosystem processes in a similar way) should lead to more stability in the face of perturbations than a low number of species, although direct experimental evidence of this is limited.

Productivity of agroecosystems may not increase if individual agricultural fields are planted with higher species richness, because under intensive modern agriculture a single strain of a single crop species is likely to give the highest yield or the highest profit in a given field (Swift and Anderson 1993). The situation may be different, however, in agroecosystems managed using approaches that incorporate biodiversity (such as the "agricultural diversification" or "integrated pest management" paradigms described later), where species composition and resource levels are under much looser control by the land managers, and thus recruitment from the natural species pool plays a more relevant role (Swift and Anderson 1993; Fridley 2001).

In high-biodiversity agriculture, a larger number of species may provide "ecological insurance" against crop failures, which is especially important to poor farmers disconnected from market insurance systems. (See Chapters 6 and 26.) Changes in species numbers may have only subtle short-term effects in some ecosystems but may directly influence their capacity for long-term adjustment in the face of a changing environment.

Major changes in species composition due to direct introduction or removal of species, or caused indirectly by changing relative abundances via altered resource supply (such as irrigation or eutrophication), can shift the functional trait composition of ecosystems and therefore deeply modify their derived services. Therefore, the preservation of the integrity, in terms of size and composition, of the regional species pool is a key factor in maintaining the rate, magnitude, and long-term persistence of those ecosystem processes that support ecosystem services. The regional species pool is defined as the set of species occurring in a certain region that is capable of coexisting in the target community.

11.2.1 Ecosystem Resource Dynamics, with Emphasis on Primary Production

The relationship between plant species richness and ecosystem production (including both total biomass achieved and the rate at which that biomass is achieved—that is, productivity) and the efficiency of resource use is probably the single most tested—and most debated—aspect of the relationship between biodiversity and ecosystem functioning. One important source of controversy among authors stems from the different results that emerge from studies at different spatial scales, where the controlling processes may differ (Fridley 2001; Loreau et al. 2001; Hooper et al. 2005). Specifically, some studies have focused on the relationship between species richness within a single habitat, whereas others have compared patterns among different habitats. For example, when ecosystems developed in different habitats are compared, soil fertility is a strong determinant of primary production and plant species diversity.

Two synthesis articles published in the last few years provide a good overview of the state of the art in this topic (Loreau et al. 2001; Hooper et al. 2005). Some of the main empirical findings and underlying theoretical issues related to the role of biodiversity in regulation ecosystem resource dynamics in general, and to primary production in particular, are summarized here.

Experimental manipulation of species richness in greenhouse (Naeem et al. 1995; Symstad et al. 1998) and large-scale field experiments (Tilman et al. 1996, 1997a, 2001; Hector et al. 1999, 2001) has shown a positive relationship between plant species richness and primary production, especially at low number of species (see Schläpfer and Schmid 1999; Schwartz et al. 2000; Hector 2002; Schmid et al. 2002; Tilman et al. 2002b). (See Table 11.1.)

In some experimental studies, total plant biomass has experimentally been shown to be greater, on average, and levels of soil nitrate—the limiting resource—lower (less leaching) at higher levels of plant species richness (Tilman et al. 1996, 1997a). Tilman et al. (2001) found that both species richness and functional type composition were significant controllers of productivity, and that no low-richness plot was as productive as many higher-richness combinations of species were. However, other experimental studies have found that ecosystem processes are more strongly linked to plant species and functional type composition than to species richness (Hooper and Vitousek 1997; Tilman et al. 1997a; Wardle

et al. 1997a, 1999; Crawley et al. 1999; Lavorel et al. 1999; Kenkel et al. 2000; Paine 2002; Hooper and Dukes 2004). The influence of plant species richness or composition on soil processes such as decomposition and microbial activity is less well understood (Wardle et al. 2004).

For forest ecosystems, most data come from observational field surveys that compare natural, managed, or old forest plantations with different tree species richness. Single-species tree stands and adjacent two-species stands have been the most studied (reviewed by Cannell et al. 1992; Kelty et al. 1992). In general, two-species forests are more productive than stands dominated by a single species, but they are not necessarily more productive than the best monoculture.

Inhibitory and enhancing effects are also common. In the United Kingdom, for example, studies of two-species combinations of a pool of four species have shown that mixtures that con-

Table 11.1. Main Components of Biodiversity Involved in Supporting and Regulating Ecosystem Services Addressed in This Chapter. Bullets indicate importance and/or degree or certainty (•••> ••> •). The mechanisms and shape of the relation between the provision of ecosystem services and diversity remain highly speculative in many cases. In the cases of most saturating curves, the level at which diversity effects saturate for different ecosystem services is poorly known. Biodiversity also contributes to provisioning and cultural ecosystem services in important ways.

Ecosystem Services	Main Components of Biodiversity Involved	Mechanisms That Produce the Effect	How the Provisioning of Service Scales to Diversity
Supporting services			
Amount of primary production	••• functional composition of plant assemblage	faster-growing, bigger, more efficient, more locally adapted plants will produce more biomass	complex relationship; processes depend on identity of dominant species, not species richness
		in low-diversity systems, coexisting plants with very different (complementary) resource use strategies will take up more resources	saturating curve
	•• species richness of plant assemblage	a larger species pool is more likely to contain groups of complementary species and individual species that are highly productive, both of which should lead to higher productivity of the community	saturating curve
Stability of primary production	••• genetic diversity	large genetic variability within a crop species buffers production against losses due to diseases and environmental change	saturating curve
	••• species richness	polycultures (more than one species cultivated together) maintain production over a broader range of conditions	saturating curve
	••• functional composition of plant assemblage	life history, resource use strategy, and regeneration strategy of dominant plants determine resistance and resilience of ecosystem functioning against perturbations	complex relationship; stability depends on identity of dominant species, not species richness saturating curve; subordinate species can totally or partially compensate for functions of dominants
Provision of habitat	••• habitat diversity, including spatial distribution, size and shape of landscape units	connectivity, landscape heterogeneity, and large landscape units are necessary for migrating species and species that need large foraging areas	complex relationship, likely to be different for different kinds of organisms
	••• functional composition of vegetation	some vertebrates need a complex vegetation structure for breeding and roosting	complex relationship; stability depends on identity of dominant species, not species richness
	•• species richness	the more species at each trophic level, the more species herbivores, predators, and/or pathogens are provided a resource base	saturating curve
Regulating services			
Invasion resistance	••• species composition	some key native species are very competitive or can act as biological controls to the establishment and naturalization of aliens	complex relationship; processes depend on identity of dominant species, not species richness
	••• arrangement of landscape units	landscape corridors (e.g., roads, rivers, extensive crops) can facilitate the spread of aliens	complex relationship; size and nature of suitable corridors likely to be different for different organisms
	•• species richness and diversity	all else being equal, species-rich communities are more likely to contain highly competitive species and fewer vacant niches, and therefore to be more resistant to invasions	decreasing curve, often exponential decay to zero in experimental studies

Table 11.1. *continued*

Ecosystem Services	Main Components of Biodiversity Involved	Mechanisms That Produce the Effect	How the Provisioning of Service Scales to Diversity
Pollination	••• functional composition of pollinator assemblage	loss of specialized pollinators leads to a reduction of number and quality of fruits produced and plant genetic impoverishment	complex relationship; processes depend on identity of dominant species, not species richness
	•• species richness of pollinator assemblage	lower pollinator species richness leads to a reduction of number and quality of fruits produced and plant genetic impoverishment	linear relationship for co-evolved pollination systems; saturating curve or linear relationship for generalist pollination systems
	•• arrangement and size of landscape units	large landscape units and/or connectivity among them maintain plant genetic pool and number and quality of fruits	saturating curve
Climate regulation	••• arrangement and size of landscape units	size and spatial arrangement of landscape units over large areas influence local-to-regional climate, by lateral movement of air masses of different temperature and moisture	threshold for effect is patch size (landscape diversity) of about 10 km diameter, depending on wind speed and topography
	•• functional composition of vegetation	height, structural diversity, architecture, and leaf seasonal patterns modify albedo, heat absorption, and mechanical turbulence, thus changing local atmospheric temperature and air circulation patterns	linear relationship between albedo and heating; albedo depends on structural diversity and on the plant functional types that dominate the canopy
Carbon sequestration	••• arrangement and size of landscape units	carbon loss is higher at forest edges; as forest fragments decline in size, a larger proportion of the total landscape is losing carbon	nonlinear relationship; as patches get larger, changes in carbon sequestration should saturate (the edges become a smaller proportion of total area); conversely, as patches get smaller, carbon loss increases exponentially with degree of fragmentation
	•• functional composition of vegetation	fast-growing, fast-decomposing, short-leaved, small-sized plants retain less carbon in their biomass than slow-growing, slow-decomposing, long-leaved, large-statured plants	saturating relationship with plant size; linear relationship with surface area of landscape units; note that the diversity has to do with the column to the left; in some cases the shape of this relationship is not related to diversity
	• species richness of vegetation	high species richness can slow down the spread of pests and pathogens, which are important agents of disturbance and carbon loss from ecosystems	saturating curve
Pest and disease control in agricultural systems	••• genetic diversity of crops	reduces density of hosts for specialist pests, and thus their ability to spread	saturating curve, but substantial effects are achieved with only a few species
	•• high richness of crop, weed, and invertebrate species	similar to genetic diversity, but also increases habitat for natural enemies of pest species	saturating curve in general, but some weed or invertebrate species may lead to a complex relationship
	•• spatial distribution of landscape units	natural vegetation patches intermingled with crops are the habitat of many natural enemies against insect pests	saturating curve as the size and number of natural vegetation patches increase; saturation point likely to be different for different groups of natural enemies

tain the pine *Pinus sylvestris* are always more productive than any monospecific stand due to the nursing effect of pines on other species with no detriment to itself, whereas mixtures of Norway spruce (*Picea abies*) and alder (*Alnus glutinosa*) have lower productivity than monospecific stands of either species (Brown 1992). These results suggest that, for temperate forests, species identity and combination might be more important than tree species richness per se. Forest productivity seems to be improved in two-species stands when there is complementarity in resource use (for instance, early and late successional species, shade-tolerant and shade-intolerant species, or different duration of the growing season).

Few studies have compared primary production of forests over a wide range of species richness. The Forest Inventory and Analysis database in the United States shows a positive correlation between tree species richness and stand productivity (Caspersen and Pacala 2001). However, the lack of environmental description hinders the interpretation of this association, especially because for two-species mixtures it is known that whether mixtures are more productive than pure stands depends on site conditions. In the western Mediterranean basin, productivity has also been compared across a range of forests with different tree species richness (Vilà et al. 2003). Here, monospecific pine forests have lower wood production than mixed (two- to five-species) forests. However, the species-rich forests are associated with humid climates, certain bedrock types, and early successional stages, which may be the cause of higher productivity.

The production of leaf litter may be greater in forests with two or more tree species than in monospecific forests, but whether there was a positive effect beyond two-species mixtures depended on the species and functional identity of the dominant tree species (Vilà et al. 2004). Similarly, the effect of tree species interactions on decomposition, a key process in nutrient cycling, seems to be species- and mixture-specific (Fyles and Fyles 1993), depending on the leaf litter quality of the trees and the associated microbial detritivore community (Blair et al. 1990; Wardle et al. 1997). Species-specific effects do not necessarily contradict the possibility of diversity effects. However, the lack of long-term monitoring with a reasonable number of species limits the ability to assess the relative importance of composition versus number of tree species. Therefore, general conclusions about the causal links between species richness and ecosystem processes in forests cannot yet be made.

On the other hand, an increasing number of reports indicate that the functional components of biodiversity (value, range, and relative abundance of plant traits) play an important role in ecosystem resource dynamics. Significant associations of ecosystem processes with plant functional composition and richness have been found more consistently than associations with species richness (Díaz and Cabido 2001). Considerable evidence, both from experiments and from nonmanipulative field studies of plant communities, shows that not all species are equally important to ecosystem functioning. Some are particularly crucial due to their traits or relative abundance.

In particular, the relative distribution of plant biomass among species is highly inequitable in most communities, with a minority of species (dominants) contributing most of the total biomass. The traits of the dominant plant species are usually the key drivers of an ecosystem's processing of matter and energy (see Hobbie 1992; Aerts 1995; Chapin et al. 1996; Aerts and Chapin 2000; Lavorel and Garnier 2002 for reviews). Therefore, at any given time the relative roles of species and functional type richness in ecosystem functioning tend to be small compared with the effect of the most dominant species. In these situations, the loss or introduction of dominant plant species may lead to much more important shifts in ecosystem functioning than those of other plant species, irrespective of changes in species richness (Lepš et al. 1982; Hooper and Vitousek 1997; McGrady-Steed et al. 1997; Wardle et al. 1997a; Mikola 1998; Symstad et al. 1998; Grime et al. 2000).

Even in situations where the average effect of the loss of randomly chosen plant species is a decrease in ecosystem productivity and nutrient use, large variations in ecosystem processes exist depending on which species or functional types are lost. The literature on invasive species provides dramatic examples of major ecosystem changes brought about by very small changes in species richness, usually the addition of a single species. (See Table 11.2.)

Biodiversity can influence ecosystem processes via at least two qualitatively different but not mutually exclusive mechanisms. One is the "niche complementarity effect" or "niche differentiation effect." Because the range of functional types is likely correlated with species number, species-rich communities may achieve more efficient resource use in a spatially or temporally variable environment than in species-poor communities (Tilman 1999; Loreau 2000). Complementary interactions, which are caused by differences among species in their resource and environmental needs, allow combinations of species to obtain more resources and produce more biomass than could any single species. Typical examples of resource-use complementarity are plant species with shallow and deep roots, warm-season and cool-season grasses, and diurnal and nocturnal pollinators or predators. In species-poor situations, increasing species richness would add novel traits, which will allow a more complete use of available resources. Positive interactions between species, such as facilitation and mutualism (increased availability of nitrogen to grasses as a consequence of the presence of nitrogen-fixing legumes, for example), may also enhance biomass production.

The second mechanism that can explain the positive effects of species diversity on ecosystem processes is the "sampling effect" (Aarssen 1997; Huston 1997; Tilman et al. 1997b), also called the "selection probability effect" (Loreau 1998): the greater the number of species initially present in an ecosystem, the higher the probability of including a species that performs particularly well under these conditions. Because any given species has a greater chance of being present at higher species richness, communities with higher number of species would be more likely to contain "better-performing" species (bigger, faster-growing, more tolerant to the prevailing conditions, more likely to have facilitative effects on other species, and so on) and thus to function "better" than species-poor communities. The sampling effect emphasizes the effects of a single dominant species and its greater chance of being present ("sampled") in communities with more species.

The niche complementarity effect and the sampling effect are not mutually exclusive, and their relative importance varies among ecosystems, depending on the environmental conditions. For example, the niche complementarity effect should be most relevant in areas of high spatial heterogeneity of environmental conditions and resource availability, whereas the sampling effect should be most relevant in small habitat patches, in early successional communities, and in areas with high resource availability (Fridley 2001). Differences among species are central to both mechanisms (Díaz and Cabido 2001). This is because the traits of the dominant plants have a strong influence on local ecosystem functioning (sampling effect) and because the greater the differences among coexisting species in terms of traits, the more likely they are to be complementary (rather than overlapping) in their resource use (niche complementarity effect).

Most biodiversity studies have focused on plant biomass, in part because of its importance in the production of food and fiber. However, much less is known about biodiversity effects on other important ecosystem processes, such as nutrient cycling, secondary production, or water dynamics. Another difficulty in generalizing from past biodiversity studies is that most empirical findings and theoretical developments are derived from a focus on herbaceous plant communities, where results are expressed rapidly. Further progress in the understanding of the role of biodiversity on ecosystem processes and services will depend on widening the scope of investigation toward other ecosystem processes, vegetation types, and trophic levels.

11.2.2 Ecosystem Stability, with Emphasis on Primary Production

For continued delivery of ecosystem services, both rate and magnitude of ecosystem processes and their stability over long periods of time, especially in the face of environmental variability, matter. Stability of an ecosystem is defined as its capacity to persist in the same state. Ecosystem stability is often divided into two components: resistance and resilience. Resistance is the capacity of a system to remain in the same state in the face of perturbation. Resilience is the rate at which a system returns to its former state after being displaced from it by a perturbation (Lepš et al. 1982). Temporal variability in community composition, including that associated with the invasion by non-native species, is an inverse measure of resistance.

Ecological theory predicts a positive relationship between species richness and the stability of ecosystems. Species-rich communities should have greater interspecific variation in responses to perturbation or environmental variation, and therefore variation in ecosystem services should be less than in species-poor communities (Tilman 1996; Doak et al. 1998; Yachi and Loreau 1999; Lehman and Tilman 2000). In addition, when species compete, the number of feedback loops in a competitive community in-

Table 11.2. Ecological Surprises Caused by Complex Interactions. Voluntary or involuntary introductions of species often trigger unexpected alterations in the normal provision of ecosystem services by terrestrial, freshwater, and marine ecosystems. Thus the introductions or deletions can have consequences opposite the intended management goals and can affect ecosystem services negatively. In all cases, the community and ecosystem alterations have been the consequence of indirect interactions among three or more species.

Study Case	Nature of the Interaction Involved	Ecosystem-service Consequences	Source
Introductions			
Top predators			
Introduction of brown trout (*Salmo trutta*) in New Zealand for angling	trophic cascade, predator increases primary producers by decreasing herbivores	negative — increased eutrophication	Flecker and Townsend 1994
Introduction of bass (*Cichla ocellaris*) in Gatun Lake, Panama	trophic cascade, top predator decreases control by predators of mosquito larvae	negative — decreased control of malaria vector	Zaret and Paine 1973
Introduction of pine marten (*Martes martes*) in the Balearic Islands, Spain	predator of frugivorous lizards (main seed dispersers)	negative — decreased diversity of frugivorous lizards due to extinction of native lizards on some islands; changes in dominant shrub (*Cneorum tricoccon*) distribution because marten replaced the frugivorous-dispersing role	Riera et al. 2002
Introduction of Artic fox (*Alopex lagopus*) in the Aleutian archipelago	predator of seabirds that transport large quantities of nutrient-rich guano from productive ocean waters to land	negative — reduced transport of nutrient from ocean to land; reduced soil fertility, nutrient status of plants, primary productivity and induced compositional shifts from productive grass-sedge to less productive shrub-forb communities	Croll et al. 2005
Intraguild predators			
Potential egg parasitoid (*Anastatus kashmirensis*) to control gypsy moth (*Lymantria dispar*)	hyperparasitism (parasitoids that may use parasitoids as hosts)	negative — disruption of biological control of pests; introduced parasitoid poses risk of hyperparasitism to other pest-regulating native parasitoids	Weseloh et al. 1979; see other examples in Rosenheim et al. 1995
Gambusia and *Lepomis* fish in rice fields to combat mosquitoes	intraguild predator (adult fish feed on juveniles as well as on mosquito larvae)	opposed to goal — decreased control of disease vector (mosquito)	Blaustein 1992
Intraguild preys			
Opposum shrimp (*Mysis relicta*) in Canadian lakes to increase fish production	intraguild prey depletes shared zooplankton preys	opposed to goal — decreased salmonid fish production	Lasenby et al. 1986
Apparent competitors			
Rats (*Rattus* spp) and cats (*Felis catus*) in Steward Island, New Zealand	rats induce high cat densities and increase predation on endangered flightless parrot (*Strigops habroptilus*)	negative — reduced diversity	Karl and Best 1982 see Müller and Bordeur (2002) for more examples
Herbivores			
Zebra mussel (*Dreissena polymorpha*) in Great Lakes, United States	zebra mussel reduces phytoplankton and outcompetes native bivalves	negative — reduced diversity positive — increased water quality	Benson and Boydstun 1995 Lodge 2001
Mutualists			
Myna bird (*Acridotheres tristis*) for worm pest control in Hawaiian sugarcane plantations	myna engages in the dispersal of the exotic woody weed *Lantana camara*	negative — increased invasion by *Lantana* produced impenetrable thorny thickets, reduced agricultural crops and pasture carrying capacity, and sometimes increased fire risk; displaces habitat of native birds	Pimentel et al. 2000
Ecosystem engineers			
Earthworm (*Pontoscolex corethrurus*) in Amazonian tropical forests converted to pasture	dramatically reduces soil macroporosity and gas exchange capacity	negative — reduced soil macrofaunal diversity and increased soil methane emissions	Chauvel et al. 1999
C4 perennial grasses *Schizachyrium condesatum, Melinis minutiflora* in Hawaii for pasture improvement	increases fuel loads, fuel distribution, and flammability	negative — increased fire frequency affecting fire-sensitive plants; reduced plant diversity; positive feedback for further invasion of flammable exotic species on burned areas	D'Antonio and Vitousek 1992
N-fixing firetree (*Myrica faya*) in Hawaii	increases soil N levels in newly formed N-poor volcanic soils	negative — increased fertility, increased invasion by other exotics, reduced regeneration of native *Metrosideros* tree, alteration of successional patterns	Vitousek et al. 1987

(continues over)

creases with species richness. Community biomass is stabilized because a decline in abundance of one species allows its competitors to increase, partially compensating for the initial decrease. In total, theoretical analyses suggest that increased species richness should slightly destabilize the production by individual species but more greatly stabilize production by the entire community (May 1973; Doak et al. 1998; Lehman and Tilman 2000).

Experimental manipulations provide weak evidence to support these theoretical predictions. In well-controlled laboratory experiments, species-rich communities were more resistant to perturbation (Naeem and Li 1997; McGrady-Steed et al. 1997). Also, African comparative field data suggested that greater species richness led to greater ecosystem stability (McNaughton 1993). Year-to-year variation in total community biomass (an inverse

Table 11.2. *continued*

Study Case	Nature of the Interaction Involved	Ecosystem-service Consequences	Source
Deletions/harvesting			
Top predators			
Sea otter (*Enhydra lutris*) harvesting near extinction in southern California	cascading effects produce reductions of kelp forests and the kelp-dependent community	negative — loss of biodiversity of kelp habitat users	Dayton et al. 1998
Pollution-induced reductions in predators of nematodes in forest soils	heavy metal bioaccumulation produces reductions in nematophagous predators and increases herbivorous nematodes	negative — disruption of forest soil food webs and increases in belowground herbivory; decrease in forest productivity	Parmelee 1995
Intraguild predators			
Declining populations of coyote (*Canis latrans*) in southern California	releases in raccoons (*Procyon lotor*) and feral house cats	negative — threat to native bird populations	Crooks and Soulé 1999
Overharvesting of seals and sea lions in Alaska	diet shifts of killer whales increased predation on sea otters	negative — conflict with other restoration programs; failure of reintroduction of sea otters to restore kelp forest ecosystems	Estes et al. 1998
Keystone predators			
Harvesting of triggerfish (*Balistaphus*) in Kenyan coral reefs	triggerfish declines release sea urchins, which outcompete herbivorous fish	negative — increased bioerosion of coral substrates; reduced calcium carbonate deposition	McClannahan and Shafir 1990
Herbivores			
Voluntary removal of sheep and cattle in Santa Cruz Is., United States, for restoration	release of the exotic plant component from top-down control	opposite to goal — explosive increases in exotic herbs and forbs and little recovery of native plant species	Zavaleta et al. 2000
Overfishing in the Caribbean, reducing herbivorous and predatory fish and reducing fish biomass	lack of fish grazers allowed macroalgae to outcompete coral following disturbances	negative — coral cover was reduced from 52% to 3%, and macroalgae increased from 4% to 92%	Hughes 1994
Ecosystem engineers			
Voluntary removal of exotic tamarisk (*Tamariscus* sp.) for restoration of riparian habitats in Mediterranean deserts	long-established tamarisk has replaced riparian vegetation and serves as habitat to endangered birds	opposite to goal — reduction in biodiversity; structural changes in riparian habitats	Zavaleta et al. 2000

measure of resistance) in a Minnesota grassland in the United States was greater in plots with lower species richness (Tilman 1996). However, these grassland plots differed in species richness mainly because of different rates of nitrogen addition, not because of direct experimental control of species richness (e.g., Givnish 1994; Huston 1997); therefore, additional field experiments are required to confirm these findings.

The evidence for a positive effect of biodiversity on stability is stronger in the case of resistance and weaker in the case of resilience (Schmid et al. 2002). Both components of ecosystem stability are strongly influenced by key traits of the dominant species, which explains why the effect of species life history on the stability characteristics of an ecosystem usually outweighs the effects of species richness (Lepš et al. 1982; Sankaran and McNaughton 1999; Osbornová et al. 1990; Grime et al. 2000).

In addition, there can be trade-offs between the traits that favor resistance and those that favor resilience (Lepš et al. 1982; McGillivray et al. 1995). For example, the dominance of short-lived, fast-growing, nutrient-demanding plants, with high output of persistent seeds, leads to high resilience and low resistance. These systems, such as annual grasslands, change very easily in the face of a perturbation but return to their initial condition relatively quickly. On the other hand, communities dominated by long-lived, slow-growing, stress-tolerant plants that allocate much energy to storage and defense tend to be more resistant and less resilient. These systems, such as mature forests in relatively dry climates, are resistant to environmental perturbations, but when

they are finally displaced away from their initial condition, they recover very slowly. Management alternatives that simultaneously try to maximize both resistance and resilience are therefore not likely to succeed.

Although the resistance/resilience characteristics of an ecosystem can be explained to a large extent by the functional traits of the most abundant species, less abundant species also contribute to the long-term preservation of ecosystem functioning. For example, subordinate and rare plants, despite their often negligible role in resource dynamics, can be crucially important in maintaining species richness of higher trophic levels (species further up the food chain) (Lepš et al. 1998; Lepš 2005). Subordinate and rare species can increase in abundance under changing environmental conditions, providing a source of colonizers or acting as positive or negative "filters" to the establishment of other species (Grime 1998; Fukami and Morin 2003; Magurran and Henderson 2003). The key role of some less abundant species, often mediated by complex and indirect interactions, is addressed in more detail later in the chapter.

The presence of multiple species, abundant or rare, within each functional type increases functional redundancy and may have important implications for ecosystem stability (Walker 1995; Grime 1998; Walker et al. 1999; Hooper et al. 2005). Functional redundancy occurs when several species in a community carry out the same process, such as nitrogen fixation. It is important because the larger the number of functionally similar species in a community, the greater the probability that at least some species will survive changes in the environment and maintain the functional properties of the ecosystem (Walker 1992; Chapin et al. 1996; Naeem and Li 1997). If there is no functional redundancy (that is, species richness is low in any given functional type), the loss of a single species could result in the elimination of an entire functional type (for instance, all the nitrogen-fixers, all the woody deciduous species, all the scavengers, or all the nocturnal pollinators), which would have a larger impact on ecosystem functioning than randomly deleting the same number of species from a variety of functional types.

Direct empirical support for this idea is still scarce, but species assigned to the same functional type have been reported to differ in their tolerances to frost (Gurvich et al. 2002), warming (Chapin et al. 1996), drought (Buckland et al. 1997), disturbance (Cowling et al. 1994; Walker et al. 1999), and changes in soil and atmosphere composition (Dormann and Woodin 2002). This suggests that the effect of species loss should depend on the number and composition of the species remaining, with the largest changes occurring when the last member of a functional type is lost. Thus the effect of species loss on stability cannot always be easily predicted (Díaz et al. 2003).

11.3 Terrestrial Biodiversity Effects on Regulating Services

11.3.1 Invasion Resistance

Invasions of species beyond their native range constitute a global driver of change of major concern for the conservation of natural and managed areas. Invasive species threaten biodiversity (Wilcove et al. 1998), change ecosystem functioning (Levine et al. 2003), and have economic costs (OTA 1993; Pimentel et al. 2000). For example, the economic costs of invasive exotic (alien) species in the United States are estimated in the tens of billions of dollars, the majority of which is due to crop losses and the application of herbicides and pesticides to reduce exotic weeds and pests. In addition, millions of dollars are spent annually in the United States to control numerous invasive species, including purple loosestrife (*Lythrum salicaria*, $45 million), Australian Melaleuca tree (*Melaleuca quenquenervia*, $3–6 million), feral pigs (*Sus scrofa*, $500,000), brown tree snake (*Boiga irregularis*, $4.6 million), fire ant (*Solenopsis invicta*, $200 million), gypsy moth (*Lymantria dispar*, $11 million), Dutch elm disease (*Ophiostoma ulmi*, $100 million), and aquatic weeds (several species, $100 million) (Pimentel et al. 2000).

Invasive species can have important negative impacts on ecosystem services and human well-being (OTA 1993; Pimentel et al. 2000): weeds and pests reduce agricultural yields; invasive eels reduce freshwater fisheries; invasive termites damage homes and other infrastructure; aquatic weeds clog waterways used for transportation and recreation; invasive mussels clog water pipes, threatening the flow of water used in such tasks as cooling power plants; invasive grasses increase fire frequency and intensity, threatening homes and other infrastructure. Conversion of native communities to invasive-dominated communities also has aesthetic and cultural impacts.

Trends in species introductions (Levine and D'Antonio 2003; Padilla and Williams 2004; Ruiz et al. 2000; Ribera Siguan 2003) and modeling predictions (Sala et al. 2000) strongly suggest that biological invasions will continue to increase in number and impact. In addition, human impacts on environmental characteristics required by native species (via eutrophication, pollution, nonsustainable harvesting, and so on) suggest that biotic resistance to

invasions may decrease and that the number of communities dominated by invasive species will increase.

Invasibility—the overall susceptibility to invasion—depends on a region's climate and environmental properties and on the interaction between the invader and the recipient community (Lonsdale 1999; Hooper et al. 2005). The presence and abundance of invaders in an ecosystem are functions of both invasibility of the system and of the supply of invading species or propagules. Here we focus on the resistance to invasions that may be afforded by species already present in a community—biotic resistance (Elton 1958). Biotic resistance is defined as the ability of resident species to inhibit the establishment, growth, survival, and reproduction of invasive species. Biotic resistance may vary from habitat to habitat and over time due to changes in the identity, composition, and diversity of the species in the community.

In general, the available evidence and theoretical predictions suggest that higher species richness and functional type richness can increase the resistance of a community against invasion by exotic species. In addition, some individual species may be particularly important in conferring invasion resistance to a community. Therefore, all else being equal, maintaining native species assemblages should diminish the ability of exotic species to become invasive, and it is most likely that the loss of biodiversity from a particular habitat will decrease the invasion resistance of this habitat.

The location on the landscape where exotic species are most likely to invade can also be predicted. Numerous studies have found a positive correlation between native and exotic species richness across habitats (Rejmánek 1996; Levine et al. 2002; Stadler et al. 2000; Lonsdale 1999; Stohlgren 2003 and references therein), where high native species richness is not the cause of high richness of exotic species. Rather, these studies suggest that the factors that promote the richness and coexistence of native species, such as benign climate, intermediate levels of disturbance, and habitat heterogeneity, also promote the richness and coexistence of exotic species (Levine and D'Antonio 1999; Byers and Noonburg 2003). These results have major conservation implications, because they suggest that hot spots for diversity are particularly at risk of invasion by introduced species, and that the loss of native species (from communities of low or high native species richness) is expected to increase invasibility.

A number of mutually compatible mechanisms have been proposed to explain the effect of biodiversity on invasion resistance (Mack et al. 2000). For all hypothesized mechanisms, it is the traits of the resident species, not merely the species richness, that determine the invasibility of a system (Foster et al. 2002; Prieur-Richard et al. 2002; Dunstan and Johnson 2004).

One proposed mechanism for high species richness inhibiting invasibility is the "niche hypothesis," which suggests that communities that are relatively impoverished in numbers of native species cannot provide biological resistance to exotic species because there are unused resources in the system (sometimes referred to as a vacant niche). Diverse communities will resist invaders because they reduce resource availability and increase competition. Consistent with the niche hypothesis, the loss of biodiversity has been shown to reduce invasion resistance in experiments in which biodiversity and community composition have been manipulated while holding the habitat conditions constant (e.g. Stachowicz et al. 1999; Dukes 2002; Naeem et al. 2000; Hector et al. 2001; Kennedy et al. 2002; Fargione et al. 2003; van Ruijven et al. 2003).

Several types of biodiversity loss decrease invasion resistance, including losses of species richness, of functional richness, and of particular species. Loss of biodiversity may reduce competition and provide increased space and resources for invading species. For example, reduced species richness in a grassland experiment led to increases in resource availability (both light and soil nitrogen) and caused higher levels of invasion (Knops et al. 1999). Invasion resistance in diverse stands has been associated with the closeness of neighbors (Kennedy et al. 2002) and with increasing temporal stability (reducing fluctuations of open space, for example) (Stachowicz et al. 1999). At this local scale, species invasion seems limited not only by species richness, but also the richness of functional types (grasses, herbs, and shrubs) (Symstad 2000). Overall, these studies suggest invading species can be most successful when they make use of resources that are incompletely used by the resident community (for example, brown trout) (Fargione et al. 2003).

The loss of biodiversity is most likely to result in unused resources in habitats that already have low functional redundancy, and for communities, such as oceanic or habitat islands, in which functional redundancy is also limited at the regional species pool level (that is, few species can disperse there naturally) (Rejmánek 1996). Thus, certain communities are susceptible to invasion because of a lack of competition from endemic species occupying one or more niches—a lack of biotic resistance. For example, the higher success of invasion by vertebrates in oceanic islands compared with corresponding continental areas is partially explained by the lack of native vertebrates that could act as predators or competitors (Brown 1989). Some stressful environments may have low species richness and in some cases low functional redundancy, but invasion is constrained by environmental conditions.

Natural enemies (pathogens, parasites, and herbivores) are important agents of biotic resistance to invasion. Invaders benefit from escaping their specialized natural enemies left behind in their region of origin, but they may be inhibited by the accumulation of natural enemies in the invaded range (Maron and Vilà 2001). Naturalized plant species that have accumulated more pathogen species native to their new habitat are less frequently listed as noxious weeds, whereas naturalized plant species that escape a greater proportion of their native pathogen species are more frequently listed as this, implying that associations with these pathogens help keep them from becoming pests in their native range (Mitchell and Power 2003).

It has also been shown that invasive animal species have fewer parasites in invaded than in native ranges (Torchin et al. 2003). Many invasive animal species have become pests only after losing their native parasites, which suggests a possible role for parasite species richness in controlling invasive species.

Similarly, invasive species may be successful because they accumulate fewer root pathogens than rare species (Klironomos 2002). In the Netherlands, weeds invading across an experimental gradient of plant species richness were found to be significantly reduced by the presence of a plant species, *Leucanthemum vulgare,* that acted as a host to parasitic nematodes, which then acted to control invading weeds (van Ruijven et al. 2003). Generalist native herbivores can also reduce the growth, seed set, and survival of introduced plants, but the evidence that they hinder the spread of invasive exotic plants is scarce (Maron and Vilà 2001). The natural enemies hypothesis is an integral part of the conceptual basis for biological control, in that specialized enemies are identified and introduced to control pests. Numerous examples of successful biological control demonstrate the importance of natural enemies in controlling invasive species (Hajek 2004; see also Chapter 10).

There is a general consensus that invasions flourish in areas disturbed by human activities (Hobbs and Huenneke 1992). Disturbances can be defined as events that create available space for the germination of propagules, increase the availability of resources, and reduce competition with colonizing species, such as changes in land use resulting in soil erosion or changes in water courses. Temporary increases in the availability of resources can reduce competition and increase the establishment and expansion of plant populations (Davis et al. 2000; Davis and Pelsor 2001). The ability of a biotic community to resist invaders may thus depend on its susceptibility to disturbances that create resource pulses.

Disturbance-induced invasions are more common when the disturbance in question does not have a long evolutionary history in an area. For example, livestock tends to favor invasion by exotic plants in areas where large herbivores have only recently been introduced (Milchunas et al. 1988; Díaz et al. 1999). There is some theoretical and empirical evidence suggesting that increased species and functional type richness can increase invasion resistance by decreasing both average resource availability and resource fluctuations (Prieur-Richard and Lavorel 2000). In addition, disturbances may interact with each other, with the highest rates of invasion occurring after multiple disturbances (such as biomass removal, fire, or soil disturbance) (Petryna et al. 2002).

Impacts of invasive species include altering the local environment in directions that are more favorable for them but less favorable to native species. Specifically, invading species may alter geomorphic processes (soil erosion rates, for instance, or sediment accretion), biogeochemical cycling, hydrological cycles, or fire or light regimes (Macdonald et al. 1996; Levine et al. 2003). For example, invading trees in the fynbos of upland South Africa reduce stream flow from mountain catchment areas, altering the hydrological regime of the whole area. In the fynbos biome, there are over 1500 threatened plant species and over 50% are threatened by the spread of introduced trees and shrubs, which prevent germination and growth of native species (Le Maitre 1996). Similarly, in Great Britain (Usher 1987), in the mixed oak (*Quercus petraea*) and holly (*Ilex aquifolium*) woodlands, the introduced species *Rhododendron ponticum* is thought to inhibit woodland regeneration both by casting a dense shade and by forming an impenetrable leaf litter layer on the ground.

Many invasive species also enhance the frequency and intensity of fires, to which many native species are not adapted. For example, numerous invasive grasses produce a great deal of flammable standing dead material and many resprout quickly after fires, giving them a competitive advantage over native species (D'Antonio and Vitousek 1992). However, some invasive species may have positive effects on native species. For example, some native species may benefit from preying upon invasive species, such as the endangered Hawaiian hawk (*Buteo solitarius*), which benefits from preying on the now-established invasive rat (*Rattus norvegicus*) (Klavitter et al. 2003).

Another hypothesis is that invasive species may exhibit positive feedbacks on subsequent invaders, either through mutualistic interactions or by modifying ecosystem properties (Simberloff and Von Holle 1999; D'Antonio and Vitousek 1992). For example, although it is thought that native species benefit more from the presence of mycorrhizal fungi than exotic plants do and that plant invaders will often be non-mycorrhizal (Klironomos 2002; Bever 2003), the intentional introduction of "improved" ectomycorrhizal fungi to increase crop or forest plantation production has altered the invasibility of many systems. These fungi may form mutualisms with invasive species and replace indigenous flora and fungi (Richardson et al. 2000). Another example of an established invader promoting subsequent invasions via mutualistic interactions is introduced honeybees that provide reliable pollination to invading plants.

Although we have focused on the inhibitory effects of native species on invaders, it is also possible that some native species may benefit invaders. For example, generalist herbivores disperse the seeds of the exotic plants they consume over long distances, having more of a facilitating than an inhibiting effect on exotic plant invasion (Maron and Vilà 2001), especially in regions with a short evolutionary history of grazing by ungulates.

11.3.2 Direct and Indirect Interactions between Species

Many ecosystem processes and the services they provide depend on obligate or facultative interactions among species. Direct interactions between plants and fungi, plants and animals, and indirect interactions involving more than two species are essential for ecosystem processes such as transfer of pollen and many seeds, transfer of plant biomass production to decomposers or herbivores, construction of habitat complexity, or the spread or suppression of plant, animal and human pathogens. Because of this, interactions between different trophic levels are among the most important processes by which biodiversity regulates the provision of ecosystem services, as illustrated in Figure 11.1 (see also Chapin et al. 2000a). Although experimental evidence is growing (e.g. van der Putten et al. 2001; Haddad et al. 2001), most of the examples come from the dramatic community and ecosystem effects of the introduction or removal of only one or a small number of species. There is clearly still insufficient information to determine whether there are general principles that describe how biotic linkages between different trophic levels and indirect interactions affect various ecosystem processes. Nevertheless, the available studies suggest that the integrity of these interactions is important for maintaining ecosystem processes and that threats to them via habitat destruction and fragmentation (see Box 11.1) are likely to result in losses of ecosystem service.

11.3.2.1 Interactions between Plants and Symbiotic Microorganisms

The interactions between plants and symbiotic microorganisms, such as mycorrhizal fungi, endophytic fungi, and nitrogen-fixing microorganisms, can greatly influence ecosystem processes and have considerable impacts in the provision of ecosystem services by natural and agricultural ecosystems. These interactions are complex and can tip the balance between different plant-community members, with various consequences for the provision of plant-related ecosystem services.

The effects of mycorrhizal fungi on plant communities are both profound and widespread. Arbuscular mycorrhizal fungi form symbiotic relationships with approximately 80% of the land plants on Earth (Smith and Read 1997), in which the mycorrhizal fungus receives benefits from the plant in the form of carbon and provides various benefits to the plant, such as phosphorus absorption (Jakobsen et al. 2002) and resistance to pathogens (Klironomos 2000).

The abundance, species composition, and richness of AMF communities influence the productivity, composition, and species richness of plant communities. This is because AMF have different effects on different plant species, ranging from mutualism to parasitism (Sanders 1993; van der Heijden et al. 1998a; Moora et al. 2004; Rillig 2004), and therefore benefit some species more than others (Grime et al. 1987; Gange et al. 1993; Hartnett et al. 1993; Moora and Zobel 1996; Wilson et al. 2001; van der Heijden et al. 2003). It is likely that AMF enhances plant species di-

versity when they favor less abundant species, but decreases in plant diversity are likely when AMF favor dominant plant species (Urcelay and Díaz 2003).

The presence and species composition of the AMF community can even alter the relationship between plant species richness and productivity. In the absence of AMF, the relationship between plant species richness and productivity is positive and linear, whereas in the presence of AMF, the relationship is positive but asymptotic (Klironomos et al. 2000). The effects of different AMF species can also differ considerably.

Increasing AMF species richness can result in more-efficient exploitation of soil phosphorus and an increase in the size of the plant nutrient pool. Van der Heijden et al. (1998b) found that increased AMF species richness led to a significant increase in the amount of soil phosphorus captured by the plant community.

Much less is known about the effects of the richness and composition of ectomycorrhizal fungi communities on ecosystem processes (Dahlberg 2001). EMF are common in nutrient-limited forest ecosystems and can play a critical role in tree nutrition and carbon balance, supplying soil resources to their plant host in exchange for sugars (Smith and Read 1997). The effects of EMF on plants appear to be species-specific, such that the loss of EMF species richness could, in theory, reduce plant species richness and productivity (e.g., Timonen et al. 1997; Baxter and Dighton 2001). However, no relationship has been found between ecosystem productivity and EMF species richness (Gehring et al. 1998), although more research is clearly needed before general conclusions can be drawn.

Based on the limited available evidence, it is likely that other fungal groups also play important functional roles. For example, systemic fungal endophytes (fungi that live inside aboveground plant tissues and receive nutrition and protection from the host) change the performance, herbivore resistance, biomass allocation, and final biomass of plant individuals and may thus also have a considerable effect on competitive interactions among plants (Clay and Holah 1999; Matthews and Clay 2001; Pan and Clay 2002). The presence of fungal endophytes may also inhibit the activity of other microbial organisms like AMF (Chu-Chou et al. 1992; Guo et al. 1992) or soil invertebrates (Bernard et al. 1997), with possible indirect effects on plant community diversity or productivity. Toxic alkaloids in the leaf litter of endophyte-infected plants could inhibit decomposition, slowing rates of nutrient cycling (Bush et al. 1997). Pathogenic fungi in the root zone may influence plant distribution and competition by favoring certain species (de Rooij-van der Goes et al. 1998; Packer and Clay 2000). There are currently no results from biodiversity-ecosystem functioning experiments explicitly considering endophytes and soil fungal pathogens, but the complex relationships described above suggest that fungal diversity may play an important role in the structure and functioning of ecosystems.

Ecosystem productivity and carbon accumulation may be enhanced by nitrogen-fixing microorganisms. These include both nitrogen-fixing bacteria in symbiotic relationships with plants (especially, but not exclusively, legumes), and free-living microorganisms. As in the case of AMF, not only the presence but also the identity of the symbiotic nitrogen-fixing bacteria is important, since different genotypes may have different effects on host plant species (Thrall et al. 2000).

The input of nitrogen to soils from nitrogen-fixing plants is crucial in the productivity and successional dynamics of many natural ecosystems and can have important positive and negative impacts on ecosystem services (Walker and Vitousek 1991; Doyle 1994; Fridley 2001). Some of the positive effects of biodiversity on plant biomass production have been attributed, at least in part,

BOX 11.1

Impacts of Habitat Fragmentation on the Links between Biodiversity and Ecosystem Processes

Throughout the world, habitat fragmentation is one of the most critical threats to biodiversity and ecosystem services, such as pollination, seed dispersal, herbivory, and carbon sequestration (Dirzo 2001a; Laurance et al. 2002; Dirzo and Raven 2003). In the tropics, for example, millions of hectares of forest are destroyed each year (Whitmore 1997; Achard et al. 2002), typically leaving small islands of forest surrounded by a sea of pastures, crops, and scrubby regrowth. In other areas, such as the eastern United States and much of Europe, forests have been fragmented for centuries. Hence, the fragmented landscape is rapidly becoming one of the most ubiquitous features of our planet.

Habitat fragments are ecologically different from intact habitat, and they are often biologically depauperate. This occurs for several reasons. First, habitat destruction is often nonrandom. Humans tend to clear areas overlaying productive, well-drained soils and to avoid areas with steep or strongly dissected topography. Consequently, habitat remnants are often confined to areas with poor soils, rugged topography, and low species richness. Second, because they are limited in area, habitat fragments contain only a fraction of the habitat diversity found in a particular area (Wilcox 1980). Third, small fragments usually have higher local extinction rates than large fragments because they contain smaller, more vulnerable populations (MacArthur and Wilson 1967; Stratford and Stouffer 1999). Fourth, habitat fragments are influenced by edge effects, which are ecological changes associated with the artificial, abrupt margins of habitat fragments (Janzen 1986; Laurance et al. 2002; Hobbs and Yates 2003). Edge effects can be remarkably varied, altering physical gradients, species distributions, and many ecological and ecosystem processes. (See Figure.) Fifth, fragments that are isolated tend to support fewer species than do those that are near other habitat areas (MacArthur and Wilson 1967; Lomolino 1984).

By altering species richness, relative abundance, and composition, habitat fragmentation also indirectly affects many ecosystem processes. Smaller fragments often become hyper-disturbed, leading to progressive changes in floristic composition (Laurance 1997; Hobbs and Yates 2003).

Distances to Which Various Edge Effects Penetrate the Interiors of Fragmented Rainforests in Central Amazonia. Some edge effects, such as microclimatic alterations like higher vapor pressure deficits and lower soil-moisture content, penetrate only limited distances (<50 m) into forest fragments. Other edge effects, however, such as elevated wind disturbance, can penetrate hundreds of meters into fragments. As a result, even large fragments can be substantially altered by wind disturbance.

to the presence of nitrogen-fixing legumes in the assemblages (e.g. Hector et al. 1999). Nitrogen-fixing symbiotic relationship is at the very basis of intercropping (Vandermeer 1989) and renders high economic profit in many semi-natural pasture and agroforesty ecosystems. However, the invasion by trees with nitrogen-fixing symbiotic microorganisms has had dramatic consequences in some naturally nutrient-poor ecosystems (such as the firetree, *Myrica faya,* in Hawaii, where a nitrogen-fixing tree was previously absent).

11.3.2.2 Interactions between Plants and Animals

The provision of ecosystem services by plants and animals is inextricably linked. This is because animals interact with plants directly by eating them or by moving their pollen and seeds across the landscape. This has short-term consequences on ecosystem processes and also long-term evolutionary consequences (such as evolution of plant chemical defenses and floral and fruit structures). Major ecosystem services are supported by the direct interactions between plants and animals, such as herbivory, pollination, and seed dispersal. Animals and plants can also influence each other indirectly by changing each other's habitat and resource availability (such as the provision of nesting sites by plants to animals or the increased availability of soil nutrients to plants due to physical disturbance caused by animals).

11.3.2.2.1 Herbivory

Herbivory—the consumption of plant tissues or fluids by animals—is ubiquitous in ecosystems and often has a dramatic impact on ecosystem processes. This section describes the ecosystem effects of the interactions between wild herbivores and plants. The ecosystem effects of domestic herbivores are addressed in Chapters 8 and 22.

The consequences of herbivory for ecosystem services and human well-being go far beyond its widely recognized role in terms of impact on the production of plant biomass (such as food, wood, and fiber). This is because herbivores consume an important portion of the world's primary production, and in many cases they stimulate plant biomass production and nutrient cycling and favor stability (by decreasing the amount of standing dead biomass, for instance, and thus the probability of high-temperature fires). Herbivory has also played a key role in the development of plant functional biodiversity over evolutionary time (Dirzo 2001b).

Herbivory tends to be an antagonistic interaction in which plant performance (yield, reproduction, and survival) is often negatively affected. However, we now know that the impacts of herbivory move along a gradient from negative, neutral (compensation), and even positive (overcompensation) effects on plants (Strauss and Agrawal 1999). For example, mammals may speed

New trees regenerating near forest edges tend to be disturbance-loving pioneer and secondary species rather than old-growth, forest-interior species (Viana et al. 1997; Laurance et al. 1998b). Large canopy and emergent trees, which contain a high proportion of forest biomass, are particularly vulnerable to fragmentation (Laurance et al. 2000). As the biomass from the dead trees decomposes, it is converted into greenhouse gases such as carbon dioxide and methane. This loss of living biomass is not offset by increased numbers of lianas and small successional trees in fragments (Laurance et al. 2001), which have lower wood densities and therefore store less carbon than the old-growth species they replace (Nascimento and Laurance 2004). In fragmented forests worldwide, millions of tons of atmospheric carbon emissions may be released each year by this process. Edge-related losses of biomass increase sharply once fragments fall below 100–400 hectares in area, depending on fragment shape (Laurance et al. 1998b).

In addition to reduced carbon storage, the rate of carbon cycling is also altered in fragmented habitats. In undisturbed forests, carbon can be stored for very long periods in large trees, some of which can live for more than 1,000 years (Chambers et al. 1998). In forest fragments, however, the residence times for carbon will decrease as smaller, short-lived plants replace large old-growth trees (Nascimento and Laurance 2004). The dynamics of this cycle can have major effects on carbon storage in vegetation and soils and on the rate of input of organic material into tropical rivers and streams (Wissmar et al. 1981).

There is limited understanding of the ecosystem consequences of the effects of fragmentation on complex interspecific interactions. Many species can be negatively affected by secondary or "ripple effects" in fragmented habitats (Terborgh et al. 1997). For example, plants that rely on specialized pollinators can experience reduced fecundity in fragments if their key pollinators disappear (Aizen and Feinsinger 1994), although exotic pollinator species can sometimes compensate for the loss of native pollinators (Hobbs and Yates 2003). Moreover, the loss of seed dispersers has dramatically affected the life cycle of plants worldwide (Chapman and Chapman 1996; Terborgh and Wright 1994; Wright and Duber 2001), and scientists have shown that in areas affected by fragmentation or with heavy poaching, the number of seeds dispersed decreases (e.g., Wright and Duber 2001; Wright et al. 2000). Reduced dispersal may in turn decrease the genetic diversity of plant populations, since seeds are one of the main vectors of gene flow between populations (Pacheco and Simonetti 2000). Small, fragmented plant populations may show increased inbreeding, reduced genetic fitness, and increased susceptibility to environmental stress (Heschel and Paige 1995).

In addition, the rapid loss of large predators (wolves, bears, and tigers, for example) in many fragmented landscapes can lead to a phenomenon known as mesopredator release (Soulé et al. 1988), in which medium-sized omnivores (coyotes, raccoons, coatis, and opossums, for instance) that were formerly controlled by the large predators undergo population explosions. These omnivores may then decimate vulnerable species, such as nesting birds (Crooks and Soulé 1999) and large-seeded trees (Asquith et al. 1997).

In summary, from the point of view of biodiversity-mediated ecosystem services, habitat fragments are not simply reduced versions of nonfragmented habitats. Rather, they are often fundamentally altered in terms of their species composition and ecosystem functioning. By reducing biodiversity, habitat fragmentation affects a number of regulating processes, such as herbivory, pollination, seed dispersal, and carbon storage. When fragmented, forests may have a diminished capacity to provide natural products such as certain fruits, fibers, game, and pharmaceuticals; they may experience drastically altered fire regimes that can affect local communities, livestock, and croplands; and they may have a reduced capacity for capturing and storing atmospheric carbon in its living vegetation.

up biomass production by plants by removing dead parts or by eliminating apical dominance leading to the proliferation of secondary branches. This gradient of herbivore impacts and plant responses, together with the lack of information for most ecosystems, makes it difficult to understand the role of herbivores in ecosystem processes or the effect of their biodiversity on that role (e.g., van der Putten et al. 2001; Wardle et al. 2003). However, as this section illustrates, although negative impacts on plants are common, ecosystem services and even biodiversity maintenance depend on this biotic interaction.

Approximately 50% of the total species richness is accounted for by phytophagous insects and their food plants (Strong et al. 1984; Heywood and Watson 1995). Most of this biodiversity is concentrated in the tropics (Dirzo and Raven 2003), where herbivory rates, largely by insects, are also higher (Coley and Barone 1996).

The loss of herbivores can affect species throughout the community in ways that can be difficult to predict. In both tropical and nontropical ecosystems, herbivory by mammalian vertebrates can be high, and it appears that, regardless of species richness, removal of large herbivores can have profound effects on ecosystem diversity and functioning, including terrestrial, marine, and freshwater ecosystems (Pimm 1980). In the tropics, for instance, loss of mammalian herbivores resulting from hunting and habitat deterioration may reduce herbivory and seed dispersal, resulting in patches with high density and low species richness of seedlings. The high density and low species richness of these patches increases the abundance of insect herbivores and their parasites (Chapman and Chapman 1996; Dirzo 2001b; Wright 2003). The consequences of species loss depend on both the magnitude and type of animals that are removed and also on the potential for the remaining animals to ecologically compensate in the absence of those lost. In most systems, current knowledge is insufficient to predict the effect of herbivore loss reliably.

In grasslands and rangelands where native or domestic ungulates have been present over evolutionary time, these herbivores assist with nutrient cycling and buffer against disturbances. Herbivores open up the vegetation by eating and trampling it. Also, their feces and urine decompose faster than plant litter. As a consequence, nutrients, especially nitrogen, are recycled faster. In areas where ungulates have been present over evolutionary time, the loss or voluntary suppression of grazing leads to considerable accumulation of standing and dead biomass. In some areas, this increases fire frequency and intensity, with negative consequences for plant and soil communities (Collins et al. 1998; Perevolovsky and Seligman 1998). Biomass accumulation has also been reported to favor rodent outbreaks, because the tall, dense canopy provides a refuge from predators (Noy-Meir 1988).

In addition, herbivores can change the characteristics of their host plants over ecological and evolutionary time. Herbivory has selected for adaptive responses by plants, including physical and chemical traits, such as the omnipresent plant secondary metabolites (tannins, alkaloids, cardiac glycosides, non-protein amino acids, and so on). These traits have in turn selected for adaptive responses by animals, including detoxification mechanisms. Such adaptive and counter-adaptive responses lead to coevolutionary changes that, beyond their academic importance, have important practical ramifications in terms of, for example, biological control and pharmacology (see Chapter 10).

The role of herbivores in supporting ecosystem services related to the maintenance of genetic resources and food production has been underestimated. The high economic losses of crops caused by insect pests would suggest, at first glance, that herbivory reduces ecosystem services, particularly if considering only the economic value. This interaction can have direct ecological and economic benefits, however, when herbivores operate as effective control agents of potential weeds (reviewed later in this section).

In addition, the impacts of herbivores on wild plants have led to the evolution of defensive mechanisms, particularly secondary plant metabolites, which are of great actual or potential importance for humans. For instance, about 25% of the currently prescribed drugs have their origin in defensive plant secondary compounds (Dirzo and Raven 2003), which in turn are believed to have arisen as a result of the interactions between plants and herbivores over evolutionary time.

The potential benefit of many other metabolites, still poorly investigated, is considerable. For example, the metabolite dihydromethyldihydroxypyrrolydine (DMDP) is produced in the foliage of the tropical liana *Omphalea diandra*. This and other related species in the genus are strongly protected against phytophagous insects, except the highly specialized caterpillars of the moth *Urania fulgens,* which, in turn, sequester the metabolite in their bodies. Remarkably, this metabolite plays some role in blocking the activity of HIV, has negative effects on bruchid beetles that attack stored grains in the tropics, and has shown some activity against cancer and diabetes (Dirzo and Smith 1995).

11.3.2.2.2 Pollination

Pollination, the transfer of pollen between flowers, without which many plants cannot achieve sexual reproduction, is an interaction between animals and plants that is essential for the provision of plant-derived ecosystem services. Worldwide, there is increasing realization of the extent to which both wild plant communities and agricultural systems depend on pollination services (Buchmann and Nabhan 1996; Allen-Wardell et al. 1998). (See Box 11.2.)

Because many fruits and vegetables require pollinators, pollination services are critical to the production of a considerable portion of the vitamins and minerals in the human diet. When agroecosystems are managed in a way that reduces a diverse assemblage of native pollinators, crops are at risk of suffering yield losses (Kremen et al. 2002).

Estimates of the annual monetary value of pollination vary widely, from $120 billion per year for all pollination ecosystem services (Costanza et al. 1997), to $200 billion per year for the role of pollination in global agriculture alone (Richards 1993). The range of these numbers reflects the lack of common methods for valuing the services provided by nature in general (see Chapter 2) and pollinators in particular. Recent research in coffee ecosystems in Costa Rica (Ricketts et al. 2004) however have shown that for stingless bee pollinators, which nest only in the forest, the

services provided by adjacent forest patches contribute to 20% greater coffee yields within one kilometer of the forest, and 7% overall to the income of the coffee farms.

Existing evidence indicates that species richness and composition of pollinators are linked with plant reproduction and establishment and thus with all the supporting, regulating, and provisioning services that stem from terrestrial vegetation. The direct impact of losing effective pollinators is primarily on plant reproductive success and fruit production. Most pollination systems are "somewhat generalized" (Waser et al. 1996), in that most flowers attract and can be pollinated by a range of pollinators that often vary under different climatic conditions. Therefore flowers usually will continue to be visited even if the most effective pollinators have been eliminated. Because some pollinators are much more effective than others, however, less pollen may be deposited, or it may be deposited at the wrong place on the plant, or the visits may occur at times when the flower is less receptive to receiving pollen. Rarely will plants completely fail to produce seed when their most effective pollinator is removed; they are more likely to produce less seeds or fruit of reduced viability or quantity.

Previously, low fruit production in plants was widely attributed to nutrient limitation, but increasingly studies have pointed to pollen limitation as a cause of fruiting failure (Burd 1994; Johnson and Bond 1994). The contribution of pollination to crop yields is beginning to garner attention on the scientific agenda and to be considered an essential agricultural input for optimal production. Pollination is now increasingly recognized as a key component of biodiversity and sustainable livelihoods, and an International Pollinators Initiative has been formed to address pollinator conservation.

Adequate richness and density of pollinators also influence plant genetic diversity and thus indirectly affect supporting ecosystem services related to it. Threats to pollination services may lead to genetic impoverishment of species. Pollination is the means by which genes are exchanged in a population. Where the number of individuals of a given species is low (as a result of habitat fragmentation, for instance, or selective harvesting) pollinators may carry fewer pollen grains to each flower visited (Kearns et al. 1998; Kunin 1992). In self-compatible species (where individuals may fertilize themselves), this "pollination deficit" leads directly to increased inbreeding, reduced genetic fitness, and increased susceptibility to environmental stress (Heschel and Paige 1995), as seen often in small fragmented plant populations.

Although most pollination systems tend to be generalized, the greatest risks of reproductive failure or genetic impoverishment occur in highly specialized pollination systems, where the suite of effective pollinators is the smallest. Specialized pollination systems occur most commonly in desert ecosystems (Ollerton and Kamner 2002; Waser et al. 1996). The greatest richness of bee species, for example, occurs in arid and semiarid environments such as Israel and the American Southwest (O'Toole 1993). Closely related *Acacia* tree species in Tanzania drylands flower at different times of the day, thereby reducing the opportunities for sharing pollinators (Stone et al. 1998).

Nevertheless, at a local level, whether in moist tropical systems or desert systems, there is a strong linkage between effective pollination systems and biodiversity. Larger individual fruit of more uniform shape and better seed production generally correlate with a greater number of visits from pollinators (Alderz 1966). Since changing weather conditions may favor some pollinators over others, having the largest suite of potential pollinators is the best

BOX 11.2
Global Status of Pollinators

Approximately 80% of Angiosperms, including many important agricultural species, are pollinated by animals (the rest are wind- or water-pollinated or are self-compatible). Worldwide, the number of flower-visiting species is estimated to be about 300,000 (Nabhan and Buchmann 1997). Bees (Hymenoptera: Apidae) account for 25,000–30,000 species (O'Toole and Raw 1991) and together with flies, butterflies and moths, wasps, beetles, and some other insect orders encompass the majority of pollinating species (Buchmann and Nabhan 1996). Vertebrate pollinators include bats, non-flying mammals (monkeys, rodents, lemurs, and so on), and birds.

The challenges of identifying declines in pollinators are considerable given the rarity of many species, the lack of baseline data, and high spatial and temporal variation in pollinator populations (Williams et al. 2001). Evidence is generally either direct, from isolated case studies showing declines of specific taxa in a particular place or time, or indirect, from studies of pollinator abundance across gradients of human disturbance. If, as seems to be the case, pollinator populations are reduced in areas with human disturbance, and the area affected by that disturbance is increasing, we can expect pollinator populations to decline over time.

Direct evidence of pollinator declines has been reported in at least one region or country on every continent except Antarctica, which has no pollinators. However, no consistent assessment is available at the continental level, though efforts are currently under way on at least two continents.

Marked declines of bumblebees (*Bombus* spp.) have been reported for the United Kingdom (Williams 1986), Belgium (Rasmont 1988), and eastern Germany (Peters 1972) and for native solitary bee species in Germany (Westrich 1989) and in the United Kingdom (Falk 1991). Changes have been attributed to habitat loss resulting from agricultural intensification. Day (1991) compiled information on the status of bees from several national *Red Data Lists,* identifying more than 400 listed species from north of the Alps but virtually none from the Mediterranean. Although several case studies from Poland, Lithuania, Turkey, Russia, and Ukraine are available, data are insufficient to draw conclusions about general trends in these countries (Banaszak 1995). Similarly, a widespread pollinator decline may be occurring in North America (Buchmann and Nabham 1996; Allen-Wardell et al. 1998), but conclusive data are not yet available (see special section in *Conservation Ecology* 5:1 (2001)).

Honeybee (*Apis mellifera*) colonies, both managed and wild, have undergone marked declines in both the United States and some European countries. The number of managed honeybee colonies in the United States has dropped from 5.9 million in the 1940s to 1.9 million in 1996 (Ingram et al. 1996; USDA National Agricultural Statistics Service 1997), and most feral colonies have also been lost (Kearns et al. 1998). In the European Union, honeybee colonies are reported to have declined by 16% between 1985 and 1991, with losses expected to increase (Williams et al. 1991). A major cause of honeybee declines is parasitic mites (*Varroa jacobsoni* and *Acarapsis woodi*). The range expansion of Africanized honeybees in the United States is also predicted to decrease managed honeybee colonies, largely because beekeepers fear liability lawsuits (Allen-Wardell et al. 1998).

The related Himalayan cliff bee (*Apis laboriosa*) has declined significantly. In a regional study, all but one censused cliff showed declines in number of colonies or total loss across a 15-year period (Ahmad et al. 2003). Bee population characteristics may show changes before population declines can be detected. For example, the most abundant orchid bee in lowland forest in Panama, *Euglossa imperialis,* frequently has high levels of sterile males resulting in low effective population sizes (Zayed et al. 2003). Recent research points to reduced genetic diversity in specialist bees compared with generalists (Packer et al. 2005).

Butterfly (Lepidoptera) populations have decreased in Europe, based on local and national studies in the United Kingdom, the Netherlands, and Germany. Comparison with historical records (1970–82) showed that half of British resident butterflies have disappeared from over 20% of their range and that a quarter have declined by more than 50% (Asher et al. 2001). Swaay and Warren (1999) report in the *Red Data Book of European Butterflies* that many European butterflies are under serious threat because of changing land use and agricultural intensification.

Mammalian and bird pollinators also show strong declines. Nabhan (1996) notes that 45 species of bats, 36 species on non-flying mammals, 26 species of hummingbirds, 7 species of sunbirds, and 70 species of passerine birds are of global conservation concern. The black and white ruffed lemur of Madagascar, an important pollinator of the island's celebrated Traveler's Palm, is highly threatened (Buchmann and Nabhan 1996). Lower visitation rates by bats and reduced fruit set occurred on a dry forest tree, *Ceiba grandiflora,* in disturbed habitats (Quesada et al. 2003).

Pollinator biodiversity is sensitive to a number of factors, many of them related to land use. Given that these drivers are widespread and often increasing, the indirect evidence indicates that declines in pollinators may also be increasing. In order to persist in agroecosystems, pollinators need local floral diversity and nesting sites. Large monocultures fail to provide these. For example, cultivated orchards surrounded by other orchards have significantly fewer bees than orchards surrounded by uncultivated land (Scott-Dupree and Winston 1987). On melon farms in the western United States, wild bee communities become less diverse and abundant as the proportion of natural habitat surrounding farms declines (Kremen et al. 2004). The most important species for crop pollination became locally extinct throughout large parts of the landscape. In addition, all species declined along this gradient, so more resistant species could not compensate for the loss of more sensitive species (Kremen 2004).

The implications for pollinator services are evident: only farms near natural habitats sustained communities of pollinators sufficiently large to provide needed levels of pollination (Kremen et al. 2002). Distance from natural habitat affected pollinator communities and services in a similar way on coffee farms in Costa Rica (Ricketts et al. 2004; Ricketts 2004). The sizes, shapes, and interdigitation patterns of natural habitat in an agricultural landscape may profoundly affect the persistence of pollinators.

Globally important threats to plant-pollinator systems, while based on land use practices, are driven by a number of forces of varying scales and points of origin. These include forces driving agricultural intensification and consequent habitat loss and fragmentation of wild ecosystems, climate change, use of environmental chemicals, diseases and parasites of pollinator populations, changing fire regimes, introduction of alien plants, and competition with introduced pollinators. Each of these forces may introduce what appear to be only marginal impacts, but effects can cascade through the ecosystem in ways that may have serious repercussions for pollinator populations.

For example, the introduction of domesticated livestock to grassland ecosystems may depress pollinators if the livestock pressure exceeds the levels of grazing to which the resident pollinator populations are adapted. Intensively managed livestock tend to trample pathways and water edges that otherwise serve as nesting sites and water access points for wasp and bee pollinators (Gess and Gess 1993). Changes in the herbaceous layer of vegetation, due to grazing and the introduction of tall, fire-tolerant grasses, may lead to hotter fires, which destroy the dead wood that several groups of bees use as nesting sites (Vinson et al. 1993).

insurance policy for reproductive success and consistent gene flow between plant individuals (Kremen et al. 2002).

Poor reproduction observed in several rare plants has been linked to the loss of specialized pollinators. Examples are populations of members of the Scrophularaceae plant family in South Africa (Steiner 1993) and bird-pollinated vines in Hawaii (Lord 1991). The high degree of mutualism seen in some pollination interactions is illustrated by plants such as figs, yuccas, and food plants that are both pollinated by and serve as brood sites for the larval stage of many lepidopteran pollinators. Highly specialized relationships occur between fig tree species (considered keystone species for the maintenance of several vertebrate populations in the forest) (Terborgh 1986) and their pollinators, fig wasps, making them particularly dependent on the pollinators (Wiebes 1979). Some geographical regions of the world may have a higher occurrence of specialized pollination systems than others. South Africa, for example, has hundreds of plant species that rely on long-tongued flies for pollination. Many of these plant species rely on a single long-tongued fly species (Johnson 2004).

Key pollinators for one plant species may also provide pollination services to other plants at other times of the year. For example, Sampson (1952) noted that grazing livestock may destroy or alter riparian vegetation that serves as a key resource to pollinators at certain times of the year, thus reducing the ability of those pollinators to carry out pollination services not only on the riparian vegetation but on other plants flowering at different seasons. There is a concern that pollinator declines could, through such interconnectedness, ultimately affect multiple trophic levels (Allen-Wardell et al. 1998), yet understanding of these complex and diffuse relationships is still very incomplete. A growing body of research, however, is investigating the interactions among members of "pollination webs," similar to the complex interactions that define food webs (Memmott 1999).

Human well-being and plant reproductive success are bound together by the need for a large and diverse suite of pollinators to assure continued and reliable delivery of effective pollination services. Pollination services generally cannot be reduced to a focus on a single "service provider." The world's agricultural community is presently largely relying on the domesticated honeybee, *Apis mellifera,* to provide a complex and variable service, and that specific provider is faced with a number of disease and parasite challenges. A matrix of healthy natural ecosystems, interspersed and adjacent to human settlements and agricultural fields, can provide significant insurance that pollination services remain intact.

11.3.2.2.3 Seed dispersal

The movement of seeds away from the parent plant is an essential process in plant population and community dynamics. This is achieved in various ways, including wind, water, or explosion of fruit capsules. Most plants, however, including those directly used and managed by humans, depend on seed dispersal by animals. The seeds of a large proportion of woody plants are dispersed by animals (about 80–95% in the tropics and about 30–60% in temperate forests) (Jordano 1992). Many herbaceous plants also rely on animals for their seed dispersal, but the literature on these links and on their ecosystem-service importance is much sparser than that for woody species.

Seeds can be dispersed by animals that eat the fruit and discard the seeds (frugivores) or by seed eaters. In the latter case, most seeds do not survive consumption, but the survival of a small proportion of them is enough to ensure the perpetuation of plant populations. Fruit-eating animals include insects and vertebrates, ranging from ants to elephants, although in tropical forests a variety of frugivorous birds and mammals are the main vertebrate dispersal agents (e.g., Leighton and Leighton 1984). Species that are important for forest regeneration include those of birds, bats, monkeys (Julliot 1996), opossums (Medellin 1994), fish (Goulding 1980), and ants (Horvitz and Beattie 1980; van der Pijl 1982). Flying seed dispersers (bats and birds) are the main vectors that promote forest regeneration in human-disturbed forests by carrying seeds from adjacent habitats to disturbed areas (Gorchov et al. 1993; Silva et al. 2002).

The removal of a frugivore species may have severe effects on several plant species. Most seed dispersal systems can be characterized as generalized (many animals disperse several species of fruits) (Jordano 1987). However, even in generalized seed dispersal systems each animal species deposits seeds in a distinct pattern that affects plant distribution (Jordano and Schupp 2000). One single species of animal may operate as the disperser of several plant species. For instance, agoutis (medium-size rodents; *Dasyprocta* spp.), are the main seed dispersal agent of several large-seeded plants in tropical ecosystems and thus influence the floristic diversity of the understory (Asquith et al. 1999).

In a similar manner to pollination, reduced dispersal also may decrease the genetic diversity of plant populations, since seeds are one of the main vectors of gene flow between populations (Pacheco and Simonetti 2000). The reduction of frugivore populations may have disproportionately large effects. For example, when an animal population is reduced, its resource use shifts to the most preferred items, such that the least preferred resources are used little if at all. Not eating the fruits may have negative impacts on the populations of plants with animal-dispersed fruits. Thus reductions of animal populations (rather than extinction) may be sufficient to dramatically change the ecosystem services provided by frugivores (Redford and Feinsinger 2001).

The value of seed dispersal is hard to estimate, but many tree crops of high economic importance depend on the seed dispersal services of animals. Conversely, the persistence of large enough populations of wild vertebrates strongly depends on the availability of fruits of such crops. Several trees whose crops have an important role in local and export economies depend on seed dispersal by wild vertebrates. Examples include the Brazil nut (*Bertholletia excelsa*), which represents a multimillion-dollar business, and the açai palm (*Euterpe oleracea*) (Baider 2000). Also, several cosmetics are based on nuts or seeds from tropical forests.

Several tree species, such as figs and palms, are also some of the most important keystone species in the tropics (Terborgh 1986; Galetti and Aleixo 1998), because they serve as food sources during periods of fruit scarcity. Monkeys, tapirs, peccaries, and several bird species rely on keystone fruit species in Neotropical forests, and empirical evidence suggests that the structure of vertebrate communities could collapse if these keystone plant species are removed from the forest (Terborgh 1986). The overharvesting of Brazil nuts, açai palm, and Araucaria pine seeds (*Araucaria angustifolia*) in many areas—including inside protected areas—is threatening not only the plant populations but also the animals that depend on their seeds, such as peccaries, toucans, and other large-bodied frugivores (Galetti and Aleixo 1998; Solórzono-Filho 2001; Baider 2000; Moegenburg 2002).

11.3.2.3 Predation and Food Web Interactions

Indirect interactions among species are widespread in nature and refer to the effects of one species on a second species mediated by a third species. For example, a predator may increase abundances of some plant species by reducing the abundance of herbivores. It is difficult to predict the effects of changes to these interactions because the indirect links are often poorly understood. Even if such interactions are known to exist, their strength, and hence their effects, typically vary with environmental conditions (Berlow et al. 1999). However, if these interactions are disrupted, disproportionately large, and often unexpected, alterations in ecosystem properties and services may occur.

Because indirect interactions are often not immediately obvious, and because the loss or addition of organisms with certain traits can trigger positive feedback (self-accelerating) processes in ecosystems, the introductions or removal of species can cause "ecological surprises." Human alterations of the species composition of natural ecosystems can be unintended—such as mortality due to pollution, accidental species introductions, and extinctions caused by habitat losses—or deliberate, as when actions are taken in pursuit of some management goals—such as sustained exploitation, increased production, improved provision of ecosystem services, conservation, restoration, or increased attraction of tourists. Both types of interventions can disrupt ecosystem functioning and alter the provision of ecosystem services. Although some accidental changes have improved the provision of some services, highly undesirable effects are by far more common, or at least more commonly reported. These often involve very important monetary, environmental, and cultural costs.

Not all ecosystems are equally likely to yield unexpected or unwanted results. Rich, complex food webs have higher functional redundancy and more indirect interactions, many of which are weak. A system with many weak interactions may be more resistant to environmental change or loss of individual species. In highly interconnected food webs, however, the effect of changes in richness of one functional group are less predictable and may affect the abundance of other species or the richness of other functional groups through a complex set of direct and indirect interactions (e.g., Buckland and Grime 2000). Trophically simple systems such as temperate freshwater communities have responded particularly strongly to the deletion of high trophic level species. For example, high-latitude lakes have simple food structures and low functional redundancy and therefore are highly vulnerable to food-chain disruption (Schindler 1990). Strong interactors can also cause destabilization (Luckinbill 1979). Modeling approaches also suggest that increasing diversity can increase food-web stability under the condition that most of the interactions within the food web are weak (McCann and Hastings 1997). Stability increases with dietary breadth and number of alternative prey (Fagan 1997; Morin 1999).

The traits of introduced or removed species strongly affect whether these changes will result in unexpected or unwanted disruptions. The structure and organization of communities are often dependent on a few interactions, changes in which have disproportionately large effects relative to their abundance. Therefore, preserving functional diversity and interactions may be more important than maintaining species richness per se. Introduction or removal of species for which there are few functional analogues are likely to produce the strongest effects. Specifically, introduction of species with traits not found in species already present can produce large-scale alterations of ecosystem processes and structure (such as the introduction of exotic N-fixing trees and C_4 grasses in Hawaii).

The removal of a top predator often induces increases in herbivores and thus reductions in plants, altering community structure and ecosystem properties. Predators, by preferentially eating a competitively dominant prey, may facilitate increases in abundance of other species (Paine 1969). Removal of such keystone predators can greatly reduce prey diversity, because the dominant competitor may seriously reduce populations of species. Such effects of predator removals have been extensively documented in numerous terrestrial and aquatic ecosystems (Pace et al. 1999). Ecosystems in which such effects, known as "trophic cascades," are particularly likely include physically homogeneous habitats with few consumers, food-limited predators and herbivores, systems in which predators strongly suppress herbivores, and nutrient-enriched systems. For example, in whole-lake experiments, nutrient enrichment strongly promotes trophic cascading (Carpenter et al. 1995). Trophic cascades may be smaller when mid-level omnivorous consumers compensate for the activities of the suppressed herbivores. For example, the exclusion of fish in Venezuelan rivers did not produce cascading effects, as these fish eat both insects and algae (Flecker 1996).

As mentioned earlier, interactions among species can vary spatially and temporally and may range from strongly positive to strongly negative. For example, in dry woodlands, shrubs may facilitate the establishment of tree seedlings during dry years, but they also provide habitat for beetles (*Tenebrionidae*) that eat the seedlings (Kitzberger et al. 2000). Therefore, the effects of shrubs on the trees vary in space and time. In wetter years, direct facilitation becomes less important, and net effects may become dominated by indirect negative effects caused by the herbivores.

11.3.2.4 Ecosystem Engineers

Ecosystem engineers are organisms that directly or indirectly modulate the availability of resources other than themselves to other species by causing physical changes in the biotic or abiotic materials of the environment. They may dramatically modify the composition and functioning of an ecological community (Jones et al. 1994). On land, woody plants dominate the physical structure of the habitat. Deforestation causes massive changes in habitat structure and leads to loss of species at several trophic levels (Holling 1992). Some species modify an ecosystem's disturbance regimes because they have traits that affect probabilities of disturbance. For instance, highly flammable grasses induce high fire frequency, which in turn alters community composition and ecosystem functioning (D'Antonio and Vitousek 1992; Levine et al. 2003). In hurricane-prone tropical forests, deeper-rooted trees may be less likely to fall during high winds, thereby altering understory communities (Lawton and Jones 1995). Examples of animal ecosystem engineers are numerous. Beavers damming streams, termites building mounds, and elephants killing trees are all examples of animals that modify the structure of their habitat. These modifications can strongly affect the hydrology, productivity, and the provisioning of ecosystem products (such as fish in beaver ponds).

11.3.3 Biodiversity Effects on Climate Regulation

Certain components of biodiversity, such the characteristics of the dominant species and the distribution of landscape units, influence the capacity of terrestrial ecosystems to sequester carbon and regulate climate at the local, regional, and global scales. (See also Chapter 13.) Indirect feedback to global climate may accrue because plants sequester carbon in biomass (decreasing carbon release to the atmosphere). Climate may also be altered by plants through changes in albedo, evapotranspiration, temperature, and fire regime. Changes in land use, over large land surface areas, will change how biodiversity affects climate. Equally important are the functional traits of dominant plant species and the spatial arrangement of landscape units. Thus biodiversity needs to be explicitly considered in climate change mitigation practices such as afforestation, reforestation, slowed-down deforestation, and bio-fuel plantations.

11.3.3.1 Biophysical Feedbacks

The functional traits and structural complexity of plant canopies influence water and energy exchange through their effects on albedo (Chapter 13). Albedo is the proportion of incoming radiation that is reflected by the land surface back to space. Complex canopies trap more reflected radiation, thereby reducing albedo. In dense vegetation, albedo is determined by the properties of the dominant plant functional types, with albedo decreasing from grasses to deciduous shrubs and trees to conifers (Chapin et al. 2002). In open-canopied ecosystems, which account for 70% of the ice-free terrestrial surface (Graetz 1991), all individuals contribute to albedo, and more biologically diverse—and hence more structurally complex—communities have lower albedo (Thompson et al. 2004). For example, the increase in shrub density in Arctic tundra in response to regional warming (Sturm et al. 2001) has reduced regional albedo and increased regional heating (Chapin et al. 2000b). (See Chapter 25.)

Greater structural diversity of the canopy increases the efficiency of water and energy exchange, which influences water use efficiency of vegetation and runoff to streams. Complex canopies generate mechanical turbulence that mixes within-canopy air with the bulk atmosphere and therefore increases the efficiency with which water, heat, and CO_2 are exchanged between the ecosystem and the atmosphere. (See Chapter 13.) Mechanical turbulence depends on the structural diversity of the vegetation—the number, size, and arrangement of roughness elements such as trees or shrubs. Changes in structural diversity are particularly important when they add individuals that are taller or more wind-resistant than the surrounding vegetation. Even a low density of trees (less than 100 trees per hectare, for example) in a savanna or woodland substantially increases turbulent exchange with the atmosphere (Thompson et al. 2004).

The functional composition of vegetation (for instance, the structural complexity, phenology, or height) influences not only the total quantity of energy absorbed and exchanged with the atmosphere but also the partitioning of this energy flux among three pathways: latent heat flux (evapotranspiration) as a result of evaporation of water at the surface and its condensation in the atmosphere, sensible heat flux (heat associated with a temperature increase of the air), and ground heat flux (the heat conducted into the ground) (Oke 1987).

Forests transmit a larger proportion of their energy to the atmosphere as latent heat (evapotranspiration) than grasslands do because of their deeper roots and greater leaf area (Chapin et al. 2002). They therefore have a net moistening effect on the atmosphere (Shukla et al. 1990), which becomes a moisture source for downwind ecosystems. In the Amazon, for example, 60% of precipitation comes from water transpired by upwind ecosystems. Species with traits that enhance stand-level evapotranspiration, such as high stomatal conductance, therefore enhance the regional precipitation derived from a given moisture source. Since water is the resource that most strongly limits global plant production (Chapin et al. 2002; Gower 2002), these properties also contribute substantially to global productivity. In the boreal forest, post-fire deciduous stands have higher albedo and stomatal conductance than pre-fire conifer stands (Baldocchi et al. 2000) and therefore have a net cooling effect on climate. Because increasing temperatures will increase fire frequency, this may act as a one of the few potential negative feedbacks to high-latitude warming (Chapin et al. 2000c).

Large-scale changes in landscape patterns have effects on regional climate. The diversity of patches on a landscape exerts an additional impact on biophysical coupling between land and atmosphere and therefore on local-to-regional climate. Large patches (more than 10 kilometers in diameter) that have lower albedo and higher surface temperature than neighboring patches create cells of rising warm air above the patch (convection); this air is replaced by cooler moister air that flows laterally from adjacent patches (advection). Climate models suggest that these landscape effects substantially modify local-to-regional climate. In Western Australia, the replacement of native heath vegetation by wheatlands increased regional albedo. As a result, air tended to rise over the dark heathland, drawing moist air from the wheatlands to the heathlands. The net effect was a 10% increase in precipitation over heathlands and a 30% decrease in precipitation over croplands (Chambers 1998). Most vegetation changes generate a climate that favors the new vegetation, making it difficult to return the vegetation to its original state.

11.3.3.2 Carbon Sequestration

Biodiversity affects carbon sequestration primarily through its effects on species traits, particularly traits related to growth (which governs carbon inputs) and woodiness, a key determinant of carbon turnover rate within the plant. As described earlier, species diversity can enhance productivity through temporal and spatial niche diversification and through increasing the probability of including productive species in the community. Species differences in productivity result from a wide range of plant traits, including growth rate, allocation patterns, phenology, nutrient use efficiency, resource requirements, traits that influence access to resource pools (such as root depth or symbioses with mycorrhizae or N-fixing microorganisms), and traits that influence conditions that limit growth (such as temperature or moisture) (Lambers and Poorter 1992). Woodiness is particularly important in enhancing carbon sequestration because woody plants tend to contain more carbon, live longer, and decompose more slowly than smaller herbaceous plants.

Plant species also strongly influence carbon loss via decomposition and their effects on disturbance. Decomposition is influenced by traits linked to leaf litter quality (carbon quality and nutrient concentrations, for example), effects on soil environment (temperature, moisture, oxygen, and so on), carbon exudation rate from roots, and interactions with other species (Eviner and Chapin 2004). For example, wood decomposes more slowly than herbaceous material, and slow-growing plants characteristic of low-resource environments produce leaves that decompose more slowly than those of more rapidly growing plants (Cornelissen 1996; Pérez-Harguindeguy et al. 2000), enhancing carbon sequestration.

In general, the suite of traits that promotes rapid growth and high productivity also leads to rapid decomposition. Thus there is a tradeoff among traits that promote short-term carbon accumulation versus long-term carbon storage. Plant traits also influence the probability of disturbances such as fire, wind-throw, and human harvest, which temporarily change forests from accumulating carbon to releasing it (Valentini et al. 2000; Schulze et al. 2000). In addition to the effects of plant traits on carbon gain and loss from ecosystems, other forms of diversity can be important by influencing the spread of pests and pathogens, which are important agents of disturbance and carbon loss from ecosystems (see next section).

Landscape diversity and spatial pattern also influence carbon loss from ecosystems. In particular, the edges of forest fragments are often places of high plant mortality because the radically altered environment at forest edges kills trees via wind throw and desiccation (e.g., Hobbs 1993 for Australia; Chen et al. 1992 for western North America; Laurance et al. 1998a for Amazonia). Elevated tree mortality leads to a decline of living biomass near forest edges (Laurance et al. 1997) and an increase in decomposition (Laurance et al. 2000). The net effect is a decline in carbon storage at the edges of forest fragments. As forest fragments decline in size, a larger proportion of the total landscape loses carbon. Another potential effect of habitat fragmentation is the alteration of natural fire regimes, either by reducing the frequency and extent of fires (for example, when fires are suppressed in the surrounding matrix) (Baker 1994) or by increased burning in ecosystems that are highly vulnerable to fire (as in tropical rainforests) (Gascon et al. 2000; Cochrane and Laurance 2002; see also Chapter 16).

11.3.4 Pest and Disease Control in Agricultural Systems

Both the diversity of natural enemies and the landscape diversity may influence pest and disease control in agricultural systems. Yields of desired products from agroecosystems may be reduced by attacks of herbivores above and below ground, fungal and microbial pathogens, and competition with weeds. Modern agriculture has focused on reducing biodiversity in order to generate monocultures of the most profitable species or genetic variety. Landscape diversity (such as the intermixing of crop and noncrop patches) and crop rotation can also reduce the need to breed for new pest and disease resistance and to discover new pesticides.

However, biodiversity may enhance pest resistance in agricultural systems through both ecological and evolutionary processes. Because of the high population densities and short life cycles of many weeds and pests, resistance to synthetic biocides typically evolves rapidly, necessitating continuing costly investments to develop and employ new synthetic biocides. Most improvements in crop resistance to herbivores, pathogens, and weeds are transitory. Use of biodiversity can reduce the frequency with which biocides need to be applied and, hence, the selective pressure and rate at which resistance evolves (Palumbi 2001).

Biodiversity-based techniques that reduce or eliminate the need for biocides can be based on the species richness of crop plants or natural enemies (pathogens or parasites). Techniques that use crop plant biodiversity to reduce or eliminate application of biocides include intercropping of genetically different strains of a single crop species, intercropping of crop plants of different species, and crop rotation. Techniques that encourage populations of predators, parasites, and pathogens of the species that attack crop plants include no-till or low-till soil management and planting of other plant species that either repel crop predators or attract them away from the crop.

11.3.4.1 Techniques Based on Crop Plant Biodiversity

The productivity of agricultural systems with high crop genetic diversity or species richness tends to be more stable over time than that of low-diversity systems, in part due to improved pest and disease control (Power and Flecker 1996; Power 1999) (see also the earlier section on ecosystem stability). Traditional agricultural systems often include substantial planned genetic and species diversity (Pretty 2002). In contrast, the low diversity of most commercial monoculture systems often results in large crop losses from a pest complex that is less diverse but more abundant than that in more diverse systems. Indeed, a low-diversity global strategy of food production could potentially be destabilized by pests and disease (Tilman et al. 2002a).

A large proportion of global food production is accounted for by just three crops: wheat, rice and maize. The relative scarcity of outbreaks of diseases on these three crops is a testament to the success of plant breeding, cultivation practices, and the use of agrochemicals. Because of the rapid evolution of biocide-resistant organisms, however, these successes may not be sustainable in the long term. For example, within about one or two decades of the introduction of each of seven major herbicides, herbicide-resistant weeds were observed. Insects also frequently evolve resistance to insecticides within a decade. Resistant strains of bacterial pathogens appear within one to three years of the release of many antibiotics for livestock (Palumbi 2001).

By the beginning of the twenty-first century, some 2,645 cases of resistance of species to biocides had been recorded in insects and spiders, involving more than 310 pesticide compounds and 540 different insect species (www.cips.msu.edu/resistance/; www.cips.msu.edu/resistance/rmdb/background.html). During the 1990s, there was a 38% increase in compounds to which one or more arthropod species were resistant, and a 7% increase in arthropod species that are resistant to one or more pesticides.

Increased genetic diversity of crops nearly always decreases pathogen-related yield losses. Recently rice blast, a major and costly fungal pathogen of rice, was controlled in a large region of China by planting alternating rows of two rice varieties (Zhu et al. 2000). This tactic increased profitability and reduced the use of a potent pesticide. The use of mixtures of different crop varieties has been shown to effectively retard the spread or evolution of fungal pathogens of grains (Ngugi et al. 2001; Mundt 2002). There is some evidence that these approaches may also be useful for the control of plant viruses (Power 1991; Matson et al. 1997; Hariri et al. 2001).

High crop species richness enhances the ecosystem services derived from agriculture and often improves the stability of production over time by reducing the incidence of herbivores, pathogens, and weeds. In monoculture plantations of rubber trees, sugarcane, or cacao, the larger and more isolated the plantation, the greater the impact of herbivores on the plants of the agroecosystem (Harper 1977; Strong 1974). Plantations of cacao and rubber tend to have considerably lower levels of herbivory when adjacent to natural, diverse forests. In a review of reported tests of herbivore density in polycultures compared with monocultures, 52% of 287 herbivore species occurred at lower densities in polycultures compared with only 15% that occurred at higher density (Andow 1991). Sometimes even growing a mixture of two crops is enough for broad pest control; for example, in the Philippines, intercropping maize and peanuts helps to control the maize stemborer (Conway 1997). Numerous studies indicate that increasing crop species richness commonly decreases the severity of weed infestations (Liebman and Staver 2001). This is because greater crop species richness often increases the overall usage of available resources by the crops, leaving fewer leftover resources on which weeds can subsist.

Plant species richness also tends to suppress the spread of viral infection in crops: 89% of plant viruses with a known transmission mechanism are transmitted by plant-feeding insects (Brunt et al. 1996). Greater plant species richness reduces the abundance of their insect vectors, and so the majority of viruses that are transmitted by insects tend to be found at lower densities in polycultures than monocultures (Power and Flecker 1996). The richness of crop species in an agroecosystem has a much less predictable effect on the prevalence of microbial pathogens that do not rely on insect vectors, such as most fungi (Matson et al. 1997).

Fungal diseases are usually but not always less severe in polycultures than monocultures (Boudreau and Mundt 1997). Variations occur because the effects of intercropping on disease dynamics depend on a variety of factors, including microclimate effects and the spatial scale of pathogen dispersal (Boudreau and Mundt 1994; Boudreau and Mundt 1997). Crop diversification can alter microclimate in ways that either encourage or inhibit pathogen growth, depending on the characteristics of the pathogen, plants, and local environment (Boudreau and Mundt 1997). Long-distance aerial dispersal is an important survival strategy for fungal and fungus-like pathogens that cause crop diseases, such as rusts (*Uredinales*), powdery mildews (*Erysiphales*), and downy mildews (members of the protist family *Peronosporaceae*) (Brown and Hovmøller 2002). Therefore, deployment of increased crop species richness at larger spatial scales may be necessary to reduce their spread. This idea is supported by studies of increased crop genetic diversity (Zhu et al. 2000; Wolfe 2000), but untested with species diversity.

11.3.4.2 Techniques Based on the Biodiversity of Natural Enemies of Crop Predators, Parasites, and Pathogens

The species richness of natural enemies of pests increases with that of crops (Andow 1991). Compared with monocultures, species-rich agroecosystems are likely to have higher predation and parasitism rates and higher ratios of natural enemies of herbivores, all of which may contribute to lower pest densities. The spraying of biocides is much more likely to wipe out the organisms that control the pests than the pests themselves or to so reduce their predator populations that the resurgence of pests can cause considerable damage before control is reestablished (Naylor and Ehrlich 1997). Traditional subsistence systems that rely on diverse agroecosystems, such as the Javanese home garden or the milpa farming system in Mexico, typically support natural enemies of pests, such as spiders, ants, and assassin bugs (see the Javanese rice paddy case study in Chapter 26). However, the positive impacts of increased species richness on natural pest control are not universal (Altieri and Schmidt 1986).

Natural pest control services are likely to be detrimentally affected by loss of species richness (Schläpfer et al. 1999). However, in only a few cases has the role of natural enemy species richness in controlling pests been tested explicitly. Species richness of parasitoids increases parasitism rates in the armyworm caterpillars in some but not all locations in the United States (Menalled et al. 1999). Perhaps the most comprehensive understanding of the importance of predator species richness comes from spiders. There are indications of complementarity of function among spider species—that is, they catch prey using different methods, occupy different microhabitats, or are active at different times or seasons. Because of this, increasing spider species richness leads to higher and less variable predation rates and increased food web stability (Marc and Canard 1997; Riechert et al. 1999; Sunderland and Samu 2000; and see section 11.2.1 for general discussion of diversity and functional complementarity).

Recent theoretical evidence suggests that the species richness of predators and parasites of herbivorous insects may be important for the control of some types of insect pests, whereas composition, the presence of a particular predator or parasite species, may be more important than species richness for other types of pest (Wilby and Thomas 2002b), though this is yet to be rigorously tested in the field. Understanding whether and when natural enemy species richness will increase pest control is an important goal of contemporary agroecological science.

Mixtures of two or more plant species have also been developed to manipulate the density of pests and their natural enemies. For example, two kinds of plants are sometimes cultivated together with maize to control stem-borers: a plant that repels the insects and another that attracts them. This strategy has also been shown to be helpful in suppressing the parasitic weed *Striga* (Khan et al. 2000). Natural plant compounds that have been used in traditional farming systems can be useful in controlling pests in many agricultural settings. Examples include the neem tree (*Azadirachta indica*)—a natural insecticide source that has been used against rice pests in India for decades—and a variety of other plants such as the custard apple (*Annona* sp.), turmeric (*Curcuma domestica*), Simson weed (*Datura stranonium*), and chili peppers (*Capsicum frutescens*) (Pretty 1995).

11.3.4.3 Integrated Pest Management and Low-till Cultivation Systems

Integrated pest management, an approach that combines traditional agricultural systems with modern techniques, includes the promotion of natural pest controls through enhanced biodiversity of crops and natural enemies of crop pests, parasites, and pathogens (as just described), the development of host-plant resistance, and the use of pesticides when absolutely necessary. IPM can be highly successful in mitigating pest pressure in regions where farmer training programs and information services are adequate (Conway 1997; Naylor and Ehrlich 1997). However, despite cases of notable success with IPM, such as the control of the brown planthopper in Indonesian rice systems (Kenmore 1984), relatively few crops are managed widely with IPM techniques on a global scale. Because of favorable pricing policies for pesticides in many locations and the knowledge-intensive nature of IPM, this practice has yet to significantly reduce the amount of pesticides applied in agriculture worldwide.

A central component of IPM is a low-till cultivation system, which maintains a permanent or semi-permanent organic cover on the soil, consisting of either a growing crop or dead organic matter in the form of a mulch or green manure. Low-till cultivation provides habitat for natural enemies to control insect pests and increases local genetic, species, and landscape diversity, as well as enhancing soil stability, organic matter content, and carbon sequestration (Sánchez 1994; Swift 1999; Pretty and Ball 2001; Lal 2004). However, this practice often relies on the heavy use of herbicides to control weeds that might otherwise be controlled by tillage and thus can have strong negative impacts on plant biodiversity (Pretty 2002). No-till with no or minimum use of herbicides is also a viable option, at least for small farms (Petersen et al. 2000; Ekboir 2002).

11.3.4.4 Summary on Biodiversity and Natural Pest Control

To summarize, the maintenance of natural pest control services is strongly dependent on biodiversity. This service benefits food security, rural household incomes, and national incomes of many countries. (See also Boxes 11.3 and 11.4.) In many cases, perhaps the majority of them, increased crop genetic diversity and species richness at different trophic levels lead to more efficient natural control of pests and diseases in agricultural systems. However, further research is required to elucidate the ecological mechanisms of pest and disease control in order to understand both the successes and failures of reduced-input agricultural systems.

Nonetheless, the available evidence suggests that conserving the genetic diversity of crops and crop relatives at a global scale and deploying that diversity locally will protect and enhance natu-

---

BOX 11.3

Biodiversity and the Multifunctionality of Agricultural Systems

Modern agricultural methods brought spectacular increases in productivity (Conway 1997; Pretty 2002; Tilman et al. 2002a). Large-scale agriculture, however, brings simplification and a loss of biological diversity and thus reduces the potential of agriculture to provide ecosystem services other than food production. Worldwide, a third of the 6,500 breeds of domesticated mammals and birds are under immediate threat of extinction owing to their very small population size. Over the past century, it is believed that 5,000 animal breeds have already been lost. The situation is most serious in the already industrialized farming systems, with half of breeds at risk in Europe and a third at risk in North America. Asia, Africa, and Latin America have approximately 20% of their breeds at risk (FAO/UNEP 2000; Blench 2001). There is strong evidence that more genetic diversity keeps options open for both breeders and farmers in the face of a changing environment.

Unlike many other economic sectors, agriculture is inherently multifunctional. It jointly produces much more than just food, fiber, or oil, having a profound impact on many elements of local, national, and global economies and ecosystems (FAO 1999; see also Chapters 10 and 17). These impacts can be negative or positive. For example, an agricultural system that depletes organic matter or erodes soil while producing food imposes costs that others must bear; but one that sequesters carbon in soils and keeps both planned and unplanned species richness high enhances eco-

system services other than food and fiber production. For centuries, traditional agricultural systems have contributed to ecosystem services such as regulation of water supply, soil fertility, and plant and animal pathogens and pests; storm protection and flood control; and carbon sequestration. In contrast, industrialized agriculture has become progressively more expensive in terms of energy (inorganic fertilizers, pumped irrigation, and mechanical power) (Pretty 1995), human and environmental health (toxic contamination, soil erosion and salinization, eutrophication of land and water) (Conway and Pretty 1991; Pretty 1995; Altieri 1995; EEA 1998), and social impact (rural uprooting, poverty, and economic inequity) (Pretty 2002; see also Chapter 6).

Sustainable agricultural systems that substitute goods and services derived from nature for externally derived fertilizers, pesticides, and fossil fuels enhance the provision of ecosystem services and human well-being in several ways. First, they increase the energy-efficiency of food production (Pretty 1995; Pretty and Ball 2001) (see also Chapters 8 and 26), thus decreasing the externals costs to society as a whole. Second, they enhance the provision of human health (see also Chapter 14). Third, by protecting genetic, species, and landscape diversity, they enhance the provision of biodiversity-linked regulating and supporting ecosystem services derived from it.

---

ral pest control services that provide economic and food production benefits. Moreover, high-biodiversity agriculture has cultural and esthetic value and can reduce many of the externalized costs of irrigation, fertilizer, pesticide, and herbicide inputs associated with monoculture agriculture (Pretty et al. 2000 2001).

11.3.5 Biodiversity Effects on Human Disease Regulation

Human health, particularly risk of exposure to many infectious diseases, may depend on the maintenance of biodiversity in natural ecosystems. (See Chapter 14.) Over 60% of human pathogens are naturally transmitted from animals to humans (Taylor et al. 2001). Many of these are transmitted by arthropod vectors from wildlife species, creating the potential for ecological processes to affect human disease risk. A greater richness of wildlife species might be expected to sustain a greater number of pathogen species that can infect humans. However, evidence is accumulating that greater wildlife species richness may decrease the spread of wildlife pathogens to humans. The effect of biodiversity on disease risk is also expected to depend on the details of interactions between the wildlife host and arthropod vector species. Unfortunately, such data are lacking for most such diseases.

Spread of one disease for which there is data, Lyme disease, seems to be decreased by the maintenance of the biotic integrity of natural ecosystems. Lyme disease is the most common vector-transmitted disease of humans in North America, and thousands of cases occur annually in Europe and Asia as well (Ostfeld and Keesing 2000a). Where it has been studied in eastern North America, the ticks that transmit the disease primarily acquire the pathogen from the white-footed mouse, *Peromyscus leucopus* (Barbour and Fish 1993). Therefore, ecological processes that reduce the number of ticks feeding on mice have the potential to reduce disease transmission to humans.

A greater number of small mammal species could reduce the number of ticks feeding on mice either by reducing mouse abundance through competition or by attracting ticks that would oth-

erwise have fed on mice. Modeling analyses of data collected in southeastern New York State suggests that the current level of mammal biodiversity decreases disease risk to humans by up to 50% relative to realistic scenarios of decreased biodiversity (Schmidt and Ostfeld 2001; LoGiudice et al. 2003; Ostfeld and LoGiudice 2003). In a complementary analysis of large-scale geographic gradients in mammal biodiversity, states in the eastern U.S. inhabited by more species of small mammals reported fewer cases of Lyme disease per capita (Ostfeld and Keesing 2000a).

In another survey, Lyme disease risk was over four times greater in forest fragments less than 2 hectares in area than in larger fragments that typically harbor a greater number of mammal species (Allan et al. 2003). In these latter two studies, Lyme disease risk also appeared to be a function of other variables correlated with mammal species richness, such as climate, geographic location, and the presence and abundance of specific mammal species. Together, these results strongly suggest that current biodiversity of small mammals supports public health by reducing peoples' risk of contracting Lyme disease, but that this ecosystem service is being eroded by habitat fragmentation.

Risks of other infectious diseases might also depend on biodiversity, although data to fully understand such links are sparse and inconsistent. Lyme disease is epidemiologically representative of emerging diseases in general. Vector-transmitted diseases are over twice as likely as other diseases to be emerging diseases, and 75% of emerging human diseases are naturally transmitted from animals to humans (Taylor et al. 2001). Therefore, biodiversity might be important in controlling many emergent diseases. Whether the same biological processes that appear to control Lyme disease risk also control risk of other vector-borne pathogens remains largely untested, however. Whether biodiversity can also decrease the risk of wildlife pathogens that do not require arthropod vectors for transmission to humans is even less well understood. Thus, the available data indicate that human health is supported as an ecosystem service by biodiversity in some cases, but the generality of this service is poorly known. (See also Chapter 14.)

BOX 11.4

Putting a Monetary Value on High-biodiversity Agricultural Landscapes

How much are traditional high-diversity agricultural landscapes worth? It is relatively easy to assess the negative costs of unsustainable agriculture in terms of abatement and treatment costs following pollution, increased sediment deposition into dams, the socioeconomic costs of rural uprooting, and so on. It is much more difficult to calculate the value of both the positive direct contributions of agricultural systems containing highly planned and unplanned biodiversity and the indirect effects on supporting and regulating ecosystem services. Environmental economists have developed methods for assessing people's stated preferences for environmental goods through hypothetical markets (see Chapter 2), which permits an assessment of their willingness to pay for nature's goods and services or to accept compensation for losses (Stewart et al. 1997; Hanley et al. 1998, Brouwer et al. 1999).

A variety of these assessment methods suggests that traditional agricultural landscapes are highly valued. Although it is impossible to precisely quantify this, several proxies can be used, including how much governments are willing to pay farmers to produce certain habitats or landscapes, how often the public visits the countryside, and how much they spend when they get there (Willis et al. 1993; Foster et al. 1997; Stewart et al. 1997; Hanley et al. 1998).

U.K. government programs have attempted to preserve and restore some of the habitat and other positive countryside attributes that were lost during intensification. The annual per household benefit of these areas, using a variety of valuation methods (including contingent valuation, choice experiments, and contingent ranking), varies from £2–30 to £380. If we take the range of annual benefits per household to be £10–30 and assume that this is representative of the average households' preferences for all landscapes produced by agriculture, then this suggests national benefits of the order of £200–600 million per year. Expressed on a per hectare basis, annual benefits are £20–60 per hectare of arable and pasture land.

Another study compared paired organic and nonorganic farms, and concluded that organic agriculture produces £75–125 per hectare of positive externalities each year, with particular benefits for soil health and wildlife (Cobb et al. 1998).

Another proxy measure of how much we value landscapes can be made based on actual visits made to the countryside. Each year in the United Kingdom, day and overnight visitors make some 433 million visit-days to the countryside and another 118 million to the seaside (Pretty et al. 2003). The average spent per day or night varies from nearly £17 for U.K. day visitors to £58 for overseas overnight visitors. This indicates that the 551 million visit-days to the countryside and seaside result in expenditure of £14 billion per year. This is three and a half times greater than the annual public subsidy of farming. While none of these estimates are definitive, in total they clearly indicate that the landscape is highly valued by society.

Should farmers receive public support for the ecosystem services they produce in addition to food? Should those that pollute or otherwise decrease the provision of ecosystem services to the public have to pay for restoring them? The external costs and benefits of agriculture raise important policy questions for both industrial and developing countries. Three categories of policy instruments are available: advisory and institutional measures, regulatory and legal measures, and economic instruments. In practice, effective pollution control and supply of desired public goods requires a mix of all three approaches, together with integration across sectors. Regulatory and legal measures can be used to internalize external costs: those who decrease the ecosystem-service potential of the environment below a set standard are subject to penalties. Economic instruments can also be used to make sure that those who damage the environment bear the costs of the damage and also as a reward for good behavior. A variety of economic instruments are available for achieving internalization, including environmental taxes and charges, tradable permits, and targeted use of public subsidies and incentives.

11.4 Biodiversity Effects on the Provision of Marine Ecosystem Services

The ocean covers approximately 70% of Earth's surface area and contains nearly 99% of its habitable volume, so ecosystem services disrupted in the ocean will have large global consequences. The services provided by ocean ecosystems include global materials cycling, transformation and detoxification of pollutants and wastes, support of coastal recreation and tourism, and support of world fisheries and aquatic ecosystems. (See Chapter 18.) All these services are affected by the diversity of life in the ocean, although quantification of many of the links between biodiversity and marine ecosystem services has only occurred recently (Peterson and Lubchenco 1997). Marine biodiversity provides many of the same types of services as those of terrestrial biodiversity just described, with the exceptions of pollination and seed dispersal.

11.4.1 Invasion Resistance

In several marine ecosystems, decreases in the richness of native taxa were correlated with increased survival and percentage cover of invading species. This suggests that, as in terrestrial plant ecosystems, invasion resistance is enhanced by the integrity of the native species pool. For example, diverse systems use resources such as available space more completely. In experimentally assembled benthic (sea floor) communities, decreasing the richness of native taxa was correlated with increased survival and percent cover of invading species. Open space was the limiting resource for invaders, and a higher species richness buffered communities against invasion through increasing temporal stability (such as reducing fluctuations of open space) (Stachowicz et al. 1999). High biodiversity is also expected to contribute to community resilience by creating insurance through functional redundancy (Stachowicz et al. 2002). Although there are few studies of the effects of biodiversity in marine ecosystems, the available evidence suggests that marine systems may possess similar mechanisms of invasion resistance as found in terrestrial systems.

11.4.2 Direct and Indirect Interactions between Marine Species

11.4.2.1 Interactions between Plants and Symbiotic Microorganisms

Coral reefs and the ecosystem services they provide are seriously threatened by a hierarchy of anthropogenic threats. (See Chapter 19.) As one of the most species-rich communities on Earth, coral reefs are responsible for maintaining a vast storehouse of genetic and biological diversity. Substantial ecosystem services are provided by coral reefs, such as habitat construction, nurseries and spawning grounds for fish, nutrient cycling and carbon and nitrogen fixing in nutrient-poor environments, wave buffering, sediment stabilization, and tourism. Reef-related fisheries constitute approximately 9–12% of the world's fisheries. Coral reefs support the pelagic food web by exporting nutritional material such as mucous, wax esters, and dissolved organic matter. The total economic value of reefs and associated services is estimated as $503 million in Australia and as $900 million in the Caribbean (Moberg and Folke 1999).

Corals require a symbiosis with zooxanthellan algae, which provide carbon, and ecosystem services can be maintained only if the interaction between corals and their obligate symbiotic algae is preserved. The interaction with zooxanthellae is strain-specific and changes with temperature and biogeographic region, light environment, and depth (Baker et al. 2004). High temperatures, such as experienced globally as a result of the 1998 El Niño events, disrupt the symbiosis, make corals less resilient to other stresses, and can lead to massive coral mortality (Hughes et al. 2003). Thus there is a direct causal link between climate warming and disruption of a critical biological interaction that can trigger collapse of an entire reef system, with consequent loss of ecosystem services that are provided. (See Chapter 19.)

11.4.2.2 Ecosystem Engineers and Herbivory

Macroalgae and corals modify wave action regimes and allow sediment stabilization, greatly affecting intertidal community diversity (Lawton and Jones 1995). Corals are threatened by a variety of human impacts, and many kelp macroalgae communities are threatened by overgrazing. The effects of overgrazing may be reversible. For example, recovery of sea otter (*Enhydra lutris*) populations after decades of overhunting on the coast of western North America has promoted the reestablishment of structure-forming kelp forests and its associated community as a result of the reestablishment of the predation of herbivorous sea urchins by otters (Dayton et al. 1998; Springer et al. 2003).

11.4.2.3 Predators and Food Webs

Overfishing reduces the capacity of the marine system to continue to provide ecosystem services by impoverishing and threatening marine biodiversity, particularly top predators (Myers and Worm 2003). The loss of a top predator is likely to have effects on their prey and other species throughout the food web. Removal of fish with key characteristics from the ecosystem may result in loss of resilience and a change in the ecosystem from one equilibrium state to another (e.g., Sutherland 1974; Hughes 1994). For example, recent declines in great whales, a preferred food of killer whales, caused the killer whales to shift to sea otters. Rapid decimation of otters by killer whales took predation pressure off a keystone herbivore, urchins, which then overgrazed kelp beds and transformed them into crustose algal–dominated communities called "urchin barrens" (Springer et al. 2003).

As in terrestrial and aquatic communities, there are many examples of how biodiversity, particularly the loss of populations of individual species, influences ecosystem processes and the provisioning of ecological services. In the rocky intertidal zone, for example, most primary productivity is contributed by a few strong interacting species (Paine 2002). A loss of biodiversity that includes those species has a large effect on primary productivity. Similarly, Duarte (2000) found a strong link between ecosystem functioning and biodiversity in seagrass beds worldwide, with the caveat that ecosystem processes depend on particular members of a community rather than on species numbers.

Some species may have a disproportionately large effect relative to their abundance (Power et al. 1996). For example, the main predators of large commercial fish species are not larger fish, but rather small jellyfish that feed on fish larvae (Purcell 1989). In addition, species loss in species-rich communities is more likely to be compensated for by increases of functionally similar species, as described early in the chapter in the section on ecosystem stability.

Many species interactions vary spatially and temporally from strongly positive to strongly negative. For example, predatory whelks (*Nucella emarginata* and *N. canaliculata*) in intertidal communities consume mussels (*Mytilus trossulus*), but also influence them indirectly through their effects on barnacles (*Balanus glandula*), habitat facilitators of mussels. These spatially and temporally fluctuating interactions have important consequences on community structure and ecosystem organization (Berlow 1999).

11.4.3 Biodiversity Effects on Climate Regulation

The major importance of marine biodiversity in climate regulation appears to be via its effect on biogeochemical cycling and carbon sequestration. The ocean, through its sheer volume and links to the terrestrial biosphere, plays a huge role in cycling of almost every material involved in biotic processes. (See Chapter 12.) Of these, the anthropogenic effects on carbon and nitrogen cycling are especially prominent.

Biodiversity influences the effectiveness of the biological pump that moves carbon from the surface ocean and sequesters it in deep waters and sediments (Berner et al.1983). Some of the carbon that is absorbed by marine photosynthesis and transferred through food webs to grazers sinks to the deep ocean as fecal pellets and dead cells. The efficiency of this trophic transfer and therefore the extent of carbon sequestration is sensitive to the species richness and composition of the plankton community (Ducklow et al. 2001). Some phytoplankton in the southern ocean, for example, are more palatable than others, so an increase in their abundance increases grazing, the formation of fecal pellets, and the export of carbon to depth. (See Chapter 25.)

The biodiversity of marine sediments can play a key role in ecosystem processes. Sedimentary habitats cover most of the ocean bottom and therefore constitute the largest single ecosystem on Earth in terms of spatial coverage. Although only a small fraction of benthic organisms that reside in and on sediments have been described and few estimates of total species numbers and biogeographic pattern have been attempted, there is sufficient information on a few species to suggest that sedimentary organisms have a significant impact on major ecological processes (Snelgrove et al. 1997). Benthic organisms contribute to the regulation of carbon, nitrogen, and sulfur cycling, to water column processes, to pollutant distribution and fate, to secondary production and transport, and to the stability of sediments. Linkages between groups of organisms and the level of functional redundancy of marine sediment biodiversity is poorly known, and there are very few empirical studies (e.g., Bellwood et al. 2004).

11.4.4 Biodiversity Effects on Pollution and Human Disease Regulation

The marine microbial community provides critical detoxification services—filtering water, reducing effects of eutrophication, and degrading toxic hydrocarbons. Very little is known about how many species are necessary to provide detoxification services, but these services may critically depend on one or a few species. For example, American oysters in Chesapeake Bay on the U.S. East Coast were once abundant but have sharply declined, and with them their filtering ecosystem services (Lenihan and Peterson 1996). Reintroduction of large populations of filtering oysters may significantly improve water clarity in the bay (Jackson et al.

2001). The process of degrading toxic hydrocarbons, such as those in an oil spill, into carbon and water requires oxygen. Nutrient pollution can generate oxygen deprivation and thereby significantly reduce the ability of marine microbes to detoxify hydrocarbons (Peterson and Lubchenco 1997).

11.5 Biodiversity, Ecosystem Services, and Human Well-being: Challenges and Opportunities

The message emerging from the evidence assessed in this chapter is clear: the loss of biodiversity can reduce the provision of ecosystem services essential for human well-being. Knowledge of the links between biodiversity and ecosystem processes is still incomplete, but existing evidence suggests that a precautionary approach may be prudent and that research should be targeted to assist with the development of appropriate management interventions.

The biggest challenges are posed by the limited understanding of the ways in which biodiversity regulates ecosystem functioning at local and regional scales and the intrinsic difficulty of predicting unexpected, accelerated, and some times irreversible changes triggered by alterations of local and regional biodiversity by human intervention. Global extinctions are serious and irreversible, but alteration of the functional composition of local communities, the extinction of local populations, or their reduction to levels that do not allow them to play strong ecosystem roles (functional extinctions) are of major concern.

The vast majority of supporting and regulating ecosystem services provided by biodiversity are delivered at the local to regional scale. Often, when the functioning of a local ecosystem has been pushed beyond a certain limit by direct or indirect biodiversity alterations, the ecosystem service losses may persist for a very long time. In this sense, modern industrial agricultural practices based on the reduction of local biodiversity to one or a very small group of desired species is a major threat to the maintenance of supporting and regulating ecosystem services.

The evidence presented in this assessment suggests than in many instances biodiversity conservation is an economically sound way of improving human well-being. Conserving and managing biodiversity sustainably can maintain a number of ecosystem services whose importance is only now starting to be recognized without necessarily compromising the delivery of economic products from ecosystems.

The idea that there is an unavoidable trade-off between biodiversity and the economic output of ecosystems ignores the external costs of intensive ecosystem exploitation. When these considerable costs are taken into account, including those of lost supporting and regulating services—in the case of agricultural intensification, for instance, external costs are related to pollution and related health hazards, erosion, and carbon emissions resulting from the burning of fossil fuel by machinery and the production of pesticides—the net benefits of intensive exploitation are substantially reduced.

Thus by minimizing external costs and maximizing nonprovisioning ecosystem services, management practices that incorporate biodiversity may represent a cost-effective option. This is particularly important for the less-favored sectors of society, such as local indigenous communities and subsistence farmers, who normally bear the largest burden of those external costs. The recognition of both the external costs and the value of supporting and regulating ecosystem services can provide a solid basis for developing appropriate schemes of biodiversity management.

References

Aarssen, L.W., 1997: High productivity in grassland ecosystems: effected by species diversity or productive species? *Oikos,* **80,** 183–184.

Aerts, R. and F.S. Chapin, 2000: The mineral nutrition of wild plants revisited: A re-evaluation of processes and patterns. *Advances in Ecological Research,* **30,** 1–67.

Aerts, R., 1995: The advantages of being evergreen. *Trends in Ecology and Evolution,* **10,** 402–407.

Ahmad, F., S.R. Joshi, and M.B. Gurung, 2003: The Himalayan cliff bee *Apis laboriosa* and the honey hunters of Kaski. Indigenous honeybees of the Himalayas 1; 52 pp. International Centre for Integrated Mountain Development, Kathmandu, Nepal.

Alderz, W.C., 1966: Honeybee visit numbers and watermelon pollination. *Journal of Economic Entomology,* **59,** 28–30.

Allan, B.F., F. Keesing, and R.S. Ostfeld, 2003: Effect of forest fragmentation on Lyme disease risk. *Conservation Biology,* **17,** 267–272.

Allen-Wardell, G., Bernhardt, T., Bitner, R., Burquez, A., Cane J., et al. 1998: The potential consequences of pollinator declines on the conservation of biodiversity and stability of crop yields. *Conservation Biolog,y* **12,** 8–17.

Altieri, M.A. and L.L. Schmidt, 1986: Cover crops affect insect and spider populations in apple orchards. *California Agriculture,* **40,** 15–17.

Altieri, M.A., 1995: *Agroecology—The Science of Sustainable Agriculture.* Westview Press, London, 448 pp.

Andow, D.A., 1991: Vegetational diversity and arthropod population response. *Annual Review of Entomology,* **36,** 561–586.

Asher, J., M. Warren, R. Fox, P. Harding, G. Jeffcoate, and S. Jeffcoate, 2001: The Millennium Atlas of Butterflies in Britain and Ireland. OUP.

Asquith, N.M., J. Terborgh, A.E. Arnold, and C.M. Riveros, 1999: The fruits the agouti ate: Hymenaea courbaril seed fate when its disperser is absent. *Journal of Tropical Ecology,* **15,** 229–235.

Asquith, N.M., S.J. Wright, and M.J. Clauss, 1997: Does mammal community composition control recruitment in neotropical forests? Evidence from Panama. *Ecology,* **78,** 941–946.

Baider, C., 2000: Demografia e ecologia de dispersão de frutos de Bertholletia excelsa Humb. & Bonpl. (Lecythidaceae) em castanhas silvestres a Amazônia Oriental. Instituto de Biociências. São Paulo, USP: 231.

Baker, A.C., C.J. Starger, T.R. McClanahan, and P.W. Glynn, 2004: Corals' adaptive response to climate change. *Nature,* **430,** 741–741.

Baker, W.L., 1994: Restoration of landscape structure altered by fire suppression. *Conservation Biology,* **8,** 763–769.

Baldocchi, D., F.M. Kelliher, T.A. Black, and P.G. Jarvis, 2000: Climate and vegetation controls on boreal zone energy exchange. *Global Change Biology,* **6** (Suppl. 1), 69–83.

Barbour, A.G. and D. Fish, 1993: The Biological and Social Phenomenon of Lyme Disease. *Science,* **260,** 1610–1614.

Baxter, J.W. and J. Dighton, 2001: Ectomycorrhizal diversity alters growth and nutrient acquisition of grey birch (*Betula populifolia*) seedlings in host-symbiont culture conditions. *New Phytologist,* **152,** 139–149.

Bellwood, D.R., T.P. Hughes, C. Folke, and M. Nystrom, 2004: Confronting the coral reef crisis. *Nature,* **429,** 828.

Benson, A.J. and C.P. Boydstun, 1995: Invasion of the zebra mussel in the United States. In: *Our living resources: a report to the nation on the distribution, abundance and health of U.S. plants, animals, and ecosytems,* E.T. LaRoe, et al. (eds.), U.S. Department of Interior, National Biological Service, Washington D.C., pp. 445–446.

Berlow, E.L., S.A. Navarette, C.J. Briggs, M.E. Power, and B.A. Menge, 1999: Quantifying variation in the strengths of species interactions. *Ecology,* **80,** 2206–2224.

Berlow, E.L., 1999: Strong effects of weak interactions in ecological communities. *Nature,* **398,** 330–334.

Bernard, E.C., K.D. Gwinn, C.D. Pless, and C.D. Williver, 1997: Soil invertebrate species diversity and abundance in endophyte-infected tall fescue pastures. In: *Neotyphodium/Grass Interactions,* C.W. Bacon and N.S. Hill (eds.), Plenum Press, New York, pp. 125–135.

Berner, R., A.C. Lasaga, and R.M. Garrels, 1983: The carbonate-silicate geochemical cycle and its effect on atmospheric carbon dioxide over the past 100 million years. *American Journal of Science,* **283,** 641–683.

Bever, J.D., 2003: Soil community feedbacks and the coexistence of competitors: conceptual frameworks and empirical tests. *New Phytologist,* **157,** 465–473.

Blair, J.M., R.W. Parmelee, and M.H. Beare, 1990: Decay rates, nitrogen fluxes, and decomposer communities of single- and mixed species foliar litter. *Ecology,* **71,** 317–321.

Blench, R., 2001: Why conserve livestock biodiversity? In: *Living Off Biodiversity,* I. Koziell and J. Saunders (eds.), IIED, London.

Boudreau, M.A. and C.C. Mundt, 1994: Mechanisms of alteration in bean rust development due to intercropping, in computer-simulated epidemics. *Ecological Applications,* **4,** 729–740.

Boudreau, M.A. and C.C. Mundt, 1997: Ecological approaches to disease control. In: *Environmentally Safe Approaches to Crop Disease Control,* N.A. Rechcigl and J.E. Rechcigl (eds.), CRC Press, Boca Raton, pp. 33–62.

Brown, A.H.F., 1992: Functioning of mixed-species stands at Gisburn, N.W. England. In: *The Ecology of Mixed-Species Stands of Trees,* M.G.R. Cannell, D.C. Malcolm and P.A. Robertson (eds.), Blackwell Scientific Publications, London, pp. 125–150.

Brown, J.H., 1989: Patterns, modes and extents of invasions of invasions by vertebrates. In: *Biological Invasions: A Global Perspective,* J.A. Drake, H. Mooney, F. di Castri, R. Graves, F. Kruger, M. Rejmanek and M. Wiliamson (eds.), John Wiley & Sons, pp. 85–109.

Brown, J.K.M. and M.S. Hovmøller, 2002: Epidemiology—Aerial dispersal of pathogens on the global and continental scales and its impact on plant disease. *Science,* **297,** 537–541.

Brunt, A.A., K. Crabtree, M.J. Dallwitz, A.J. Gibbs, L. Watson, and E.J. Zurcher (eds.), 1996 onwards: *Plant Viruses Online: Descriptions and Lists from the VIDE Database.* Available at http://image.fs.uidaho.edu/vide.

Buchmann, S.L. and G.P. Nabhan, 1996: The forgotten pollinators. Island Press, Washington, D.C., 312 pp.

Buckland, S. and J. Grime, 2000: The effects of trophic structure and soil fertility on the assembly of plant communities: a microcosm experiment. *Oikos,* **91,** 336–352.

Buckland, S.M., J.P. Grime, J.G. Hodgson, and K. Thompson, 1997: A comparison of plant responses to the extreme drought of 1995 in northern England. *Journal of Ecology,* **85,** 875–882.

Bush, L.P., H.H. Wilkinson, and C.L. Schardl, 1997: Bio-protective alkaloids of grass-fungal endophyte symbioses. *Plant Physiology,* **114,** 1–7.

Byers, J.E. and E.G. Noonburg, 2003: Scale dependent effects of biotic resistance to biological invasion. *Ecology,* **84,** 1428–1433.

Cannell, M.G.R., D.C. Malcolm, and P.A. Robertson, 1992: *The Ecology of Mixed Species Stands of Trees.* Blackwell Scientific Publications, Oxford, 312 pp.

Carpenter S.R., D.L. Christensen, J.J. Cole, K.L. Cottingham, X. He, J.R. Hodgson, J.F. Kitchell, S.E. Knight, M.L. Pace, D.M. Post, D.E. Schindler, and N. Voichick, 1995: Biological control of eutrophication in lakes. *Environmental Science and Technology,* **29,** 784–786.

Caspersen, J.P. and S.W. Pacala, 2001: Successional diversity and forest ecosystem function. *Ecological Research,* **16,** 895–904.

Chambers, S., 1998: *Short- and Long-Term Effects of Clearing Native Vegetation for Agricultural Purposes.* PhD., Flinders University of South Australia.

Chapin, F.S., III, A.D. McGuire, J. Randerson, R. Pielke, Sr., D. Baldocchi, S.E. Hobbie, N. Roulet, W. Eugster, E. Kasischke, E.B. Rastetter, S.A. Zimov, and S.W. Running, 2000c: Arctic and boreal ecosystems of western North America as components of the climate system. *Global Change Biology,* **6,**1–13.

Chapin, F.S., III, H.L. Reynolds, C.M. D'Antonio, and V. M. Eckhart, 1996: The functional role of species in terrestrial ecosystems. In: *Global Change and Terrestrial Ecosystems,* B. Walker and W. Steffen (eds.), Cambridge University Press, Cambridge, pp.403–428.

Chapin, F.S., III, P.A. Matson, and H.A. Mooney, 2002: *Principles of Terrestrial Ecosystem Ecology.* Springer-Verlag, New York, 472 pp.

Chapin, F.S.III, E.S. Zavaleta, V.T. Eviner, R.L. Naylor, P.M. Vitousek, H.L. Reynolds, D.U. Hooper, S. Lavorel, O.E. Sala, S.E. Hobbie, M.C. Mack, and S. Díaz, 2000a: Consequences of changing biodiversity. *Nature,* **405,** 234–242.

Chapin, F.S.III, W. Eugster, J.P. McFadden, A.H. Lynch, and D.A. Walker, 2000b: Summer differences among arctic ecosystems in regional climate forcing. *Journal of Climate,* **13,** 2002–2010.

Chapman, C.A. and L.J. Chapman, 1996: Frugivory and the fate of dispersed and non-dispersed seeds of six African tree species. *Journal of Tropical Ecology,* **12,** 491–504.

Chen, J., J.F. Franklin, and T.A. Spies, 1992: Vegetation responses to edge environments in old-growth Douglas-fir forests. *Ecological Applications,* **2,** 387–396.

Chu-Chou, M., B. Guo, Z.-Q. An, J.W. Hendrix, R.S. Ferriss, M.R. Siegel, C.T. Dougherty, and P.B. Burrus, 1992: Suppression of mycorrhizal fungi in fescue by the *Acremonium coenophialum* endophyte. *Soil Biology and Biochemistry,* **24,** 633–637.

Clay, K. and J. Holah, 1999: Fungal endophyte symbiosis and plant diversity in successional fields. *Science,* **285,** 1742–1744.

Cobb, D., R. Feber, A. Hopkins, and L. Stockdale, 1998: *Organic Farming Study.* Global Environmental Change Programme Briefing 17, University of Sussex, Falmer.

Cochrane, M.A. and W.F. Laurance, 2002: Fire as a large-scale edge effect in Amazonian forests. *Journal of Tropical Ecology,* **18,** 311–325.

Coley, P.D. and J.A. Barone, 1996: Herbivory and plant defenses in tropical forests. *Annual Review of Ecology and Systematics,* **27,** 305–335.

Collins, S.L., A.K. Knapp, J.M. Briggs, J.M. Blair, and E.M. Steinauer, 1998: Modulation of diversity by grazing and mowing in native tallgrass prairie. *Science,* **280,** 745–747.

Conway, G.R. and J.N. Pretty, 1991: *Unwelcome harvest. Agriculture and pollution.* Earthscan Publications Ltd, London 645 pp.

Conway, G.R., 1997: *The Doubly Green Revolution.* Penguin, London, 352 pp.

Cornelissen, J.H.C., 1996: An experimental comparison of leaf decomposition rates in a wide range of temperate plant species and types. *Journal of Ecology,* **84,** 573–582.

Costanza, R., R. d'Arge, R. de Groot, S. Farber, M. Grasso, B. Hannon, K. Limburg, S. Naeem, R.V. O'Neill, J. Paruello, R.G. Raskin, P. Sutton, and M. van den Belt, 1997: The value of the world's ecosystem services and natural capital. *Nature,* **387,** 253–260.

Cowling, R.M., P.J. Mustart, H. Laurie, and M.B. Richards,1994: Species diversity; functional diversity and functional redundancy in fynbos communities. *South African Journal of Science,* **90(6),** 333–337.

Crawley, M.J., S.L. Brown, M.S. Heard, and G.R. Edwards, 1999: Invasion-resistance in experimental grassland communities: species richness or species identity? *Ecology Letters,* **2,**140–148.

Crooks, K.R. and M.E. Soulé, 1999: Mesopredator release and avifaunal extinctions in a fragmented system. *Nature,* **400,** 563–566.

D'Antonio, C.M. and P.M. Vitousek, 1992: Biological invasions by exotic grasses, the grass/fire cycle, and global change. *Annual Review of Ecology and Systematics,* **23,** 63–88.

Dahlberg, A., 2001: Community ecology of ectomycorrhizal fungi: an advancing interdisciplinary field. *New Phytologist,* **150,** 555–562.

Daily, G.C., 1997: *Nature's Services: Societal Dependence on Natural Ecosystems.* Island Press, Washington D.C., 392 pp.

Davis, M.A. and M. Pelsor, 2001: Experimental support for a resource-based mechanistic model of invasibility. *Ecology Letters,* **4,** 421–428.

Davis, M.A., J.P. Grime, and K. Thompson, 2000: Fluctuating resources in plant communities: A general theory of invasibility. *Journal of Ecology,* **88,** 528–534.

Day, M.C., 1991: Towards the conservation of Aculeate Hymenoptera in Europe. Convention on the Conservation of European Wildlife and Natural Habitats. Council of Europe Press, Strasbourg, *Nature and Environment Series* **51.**

Dayton, P.K., M.J. Tegner, P.B. Edwards, and K.L. Riser, 1998: Sliding baselines, ghosts, and reduced expectations in kelp forest communities. *Ecological Applications,* **8,** 309–322.

de Rooij-van der Goes, P.C.E.M., B.A.M. Peters, and W.H. van der Putten, 1998: Vertical migration of nematodes and soil-borne fungi to developing roots of *Ammophila arenaria* (L.) link after sand accretion. *Applied Soil Ecology,* **10,** 1–10.

Díaz, S. and M. Cabido, 2001: Vive la différence: plant functional diversity matters to ecosystem functioning (review article). *Trends in Ecology and Evolution,* **16,** 646–655.

Díaz, S., M. Cabido and F. Casanoves, 1998: Plant functional traits and environmental filters at a regional scale. *Journal of Vegetation Science,* **9,** 113–122.

Díaz, S., M. Cabido and F. Casanoves, 1999: Functional implications of trait-environment linkages in plant communities. In: *Ecological Assembly Rules: Perspectives, Advances, Retreats,* E. Weiher and P.A. Keddy (eds.), Cambridge University Press, Cambridge, pp. 338–362.

Díaz, S., A.J. Symstad, F.S. Chapin III, D.A. Wardle, and L.F. Huenneke, 2003: Functional diversity revealed by removal experiments. *Trends in Ecology and Evolution,* **18,** 140–146.

Dirzo, R., 2001a: Tropical forests. In: *Global Biodiversity in a Changing Environment.* Chapin, F.S.,O.E. Sala, and E. Huber-Sannwald, (eds.), Springer, New York. pp. 251–276.

Dirzo, R., 2001b: Plant-mammal interactions: lessons for our understanding of nature, and implications for biodiversiy conservation. In: *Ecology: Achievement and Challenge,* M.C. Press, N.J. Huntly, and S. Levin, S. (eds.), Blackwell, London. pp. 319–335.

Dirzo, R. and P.H. Raven, 2003: Global state of biodiversity and loss. *Annual Review of the Environment and Resources,* **28,** 137–167.

Dirzo, R. and N.G. Smith, 1995: Potential human benefits of the Urania/Omphalea coevolution. Box 6.2–1: In: *Biodiversity and Ecosystem Functioning:*

Ecosystem Analyses. Section 6.2 of Global Biodiversity Assesment. UNEP. Cambridge University Press.

Doak D.F., D. Bigger, E.K. Harding, M.A. Marvier, R.E. O'Malley, and D. Thomson, 1998: The statistical inevitability of stability-diversity relationships in community ecology. *The American Naturalist,* **151,** 264–276.

Dodd, M.E., J. Silvertown, K. McConway, J. Potts, and M. Crawley, 1994: Stability in the plant communities of the Park Grass Experiment: the relationships between species richness, soil pH and biomass variability. *Philosophical Transactions of the Royal Society of London B,* **346,** 185–193.

Dormann, C.F and S.J. Woodin, 2002: Climate change in the Arctic: Using plant functional types in a meta-analysis of field experiments. *Functional Ecology* 16 :4–17.

Doyle, J., 1994: Phylogeny of the Legume family—An approach to understanding the origins of nodulation. *Annual Review of Ecology and Systematics,* **25,** 325–349.

Duarte, C.M., 2000: Marine biodiversity and ecosystem services: an elusive link. *Journal of experimental marine Biology and Ecology,* **250,** 117–131.

Ducklow, H.W., D.K. Steinberg, and K.O. Buesseler, 2001: Upper ocean carbon export and the biological pump. *Oceanography,* **14,** 50–58.

Dukes, J. K., 2002: Species composition and diversity affect grassland susceptibility and response to invasion. *Ecological Applications,* **12,** 602–617.

Dunstan, P.K. and C. Johnson, 2004: Invasion rates increase with species richness in a marine epibenthic community by two mechanisms. *Oecologia,* **138,** 285–292.

Ekboir, J. (ed.), 2002: *CIMMYT 2000–2001 World Wheat Overview and Outlook: Developing No-Till Packages for Small-Scale Farmers.* Mexico, DF: CIMMYT.

Elton, C.S., 1958: *The Ecology of invasions by animals and plants.* Chapman & Hall, London, 181 pp.

Estes, J.A., M.T. Tinker, T.M. Williams, and D.F. Doaks, 1998: Killer whale predation on sea otters linking oceanic and nearshore ecosystems. *Science,* **282,** 473–476.

Eviner, V.T. and F.S. Chapin, 2003: Functional matrix: A conceptual framework for predicting multiple plant effects on ecosystem processes. *Annual Review of Ecology Evolution and Systematics.* **34:** 455–485.

Fagan, W.F., 1997: Omnivory as a stabilizing feature of natural communities. *The American Naturalist,* **150,** 554–567.

Falk, S., 1991: *A review of scarce and threatened bees, wasp and ants of Great Britain. Research and Survey in Nature Conservation no. 35.* Nature Conservancy Council, UK.

Fargione, J., C. Brown, and D. Tilman, 2003: Community assembly and invasion: An experimental test of neutral versus niche processes. *Proceedings of the National Academy of Sciences of the United States of America,* **100,** 8916–8920.

Flecker, A.S. and C.R. Townsend, 1994: Community wide consequences of trout introduction in New Zealand streams. *Ecological Applications,* **4,** 798–807.

Flecker, A.S., 1996: Ecosystem engineering by a dominant detritivore in a diverse tropical stream. *Ecology,* **77,** 1845–1854.

Foster, B.L., V.H. Smith, T.L. Dickson, and T. Hildebrand, 2002: Invasibility and compositional stability in a grassland community: relationships to diversity and extrinsic factors. *Oikos* 99, 300–307.

Foster, V., I.J. Bateman, and D. Harley, 1997: Real and hypothetical willingness to pay for environmental preservation: a non-experimental comparison. *Journal of Agricultural Economics,* **48(1),** 123–138.

Fridley, J.D., 2001: The influence of species diversity on ecosystem productivity: how, where, and why? *Oikos* **99,** 514–526.

Fukami, T. and P. Morin, 2003: Productivity-biodiversity relationships depend on the history of community assembly. *Nature,* **424,** 423–426.

Fyles, J.W. and I.H. Fyles, 1993: Interaction of Douglas-fir with red alder and salal foliage litter during decomposition. *Canadian Journal of Forest Research,* **23,** 358–361.

Galetti, M. and A. Aleixo, 1998: Effects of palm heart harvesting on avian frugivores in the Atlantic rain forest of Brazil. *Journal of Applied Ecology,* **35,** 286–293.

Gange, A.C., V.K. Brown, and G.S. Sinclair, 1993: Vesicular- arbuscular mycorrhizal fungi- a determinant of plant community structure in early succession. *Functional Ecology,* **7,** 616–622.

Gascon, C., G.B. Williamson, and G.A.B. Fonseca, 2000: Receding forest edges and vanishing reserves. *Science,* **288,** 1356–1358.

Gehring, C.A., T.C. Theimer, T.G. Whitham, and P. Keim, 1998: Ectomycorrhizal fungal community structure of pinyon pines growing in two environmental extremes. *Ecology,* **79,** 1562–1572.

Givnish, T.J., 1994: Does diversity beget stability? *Nature,* **371,** 113–114.

Gorchov, D.L., F. Cornejo, C. Ascorra, and M. Jaramillo, 1993: The role of seed dispersal in the natural regeneration of rain forest after strip-cutting in the Peruvian Amazon. In: *Frugivory and Seed Dispersal: Ecological and Evolutionary Apects,* T.H. Fleming and A. Estrada (eds.), Kluwer Academic Publisher, Dordrec.

Gough, L., J.B. Grace, and K.L. Taylor, 1994: The relationship between species richness and community biomass—the importance of environmental variables. *Oikos,* **70,** 271–279.

Goulding, M., 1980: *The Fishes and the Forest.* University of California Press, Berkley.

Gower, S.T., 2002: Produtivity of terrestrial ecosystems. In: *Encyclopedia of Global Change,* H.A. Mooney and J. Canadell, (eds.), Blackwell Scientific, Oxford, pp. 516–521.

Graetz, R.D., 1991: The nature and significance of the feedback of change in terrestrial vegetation on global atmospheric and climatic change. *Climatic Change,* **18,** 147–173.

Grime J.P., V.K. Brown, K. Thompson, G J. Masters, S.H. Hillier, I.P. Clarke, A.P. Askew, D. Corker, and J.P. Kielty, 2000: The response of two contrasting limestone grasslands to simulated climate change. *Science,* **289,** 762–765.

Grime, J.P. 2001: *Plant strategies, vegetation processes, and ecosystem properties,* John Wiley & sons, Chichester, UK. 417 pp.

Grime, J.P., 1979: *Plant Strategies and Vegetation Processes.* J. Wiley, Chichester.

Grime, J.P., 1998: Benefits of plant diversity to ecosystems: immediate, filter and founder effects. *Journal of Ecology,* **86,** 902–910.

Grime, J.P., 2002: Declining plant diversity: Empty niches or functional shifts? *Journal of Vegetation Science,* 13, 457–460.

Grime, J.P., J.M.L. Mackey, S.H. Hillier, and D.J. Read, 1987: Floristic diversity in a model system using experimental microcosms. *Nature,* **328,** 420–422.

Guo, B.Z., J.W. Hendrix, Z.-Q. An, and R.S. Ferriss, 1992: Role of Acrimonium endophyte of fescue on inhibition of colonization and reproduction of mycorrhizal fungi. *Mycologia,* **84,** 882–885.

Gurvich, D.E., S. Díaz, V. Falczuk, M. Pérez- Harguindeguy, M. Cabido, and P.C. Thorpe, 2002: Foliar resistance to simulated extreme climatic events in contrasting plant functional and chorological types. *Global Change Biology,* **8,** 1139–1145.

Haddad, N.M; D. Tilman, J. Haarstad, M. Ritchie, and J.M. Knops, 2001: Contrasting effects of plant richness and composition on insect communities: A field experiment. *American Naturalist,* **158,** 17–35.

Hajek, A.E., 2004: Natural enemies : an introduction to biological control. Ann E. Hajek Publisher, Cambridge, UK ; New York : Cambridge University Press, 378 pp.

Hanley, N., D. MacMillan, R.E. Wright, C. Bullock, I. Simpson, D. Parrison, and R. Crabtree, 1998: Contingent valuation versus choice experiments: estimating the benefits of environmentally sensitive areas in Scotland. *Journal of Agricultural Economics,* **49,** 1–15.

Hariri, D., M. Fouchard, and H. Prud'homme, 2001: Incidence of Soil-borne wheat mosaic virus in mixtures of susceptible and resistant wheat cultivars. *European Journal of Plant Pathology,* **107,** 625–631.

Harper, J.L., 1977: *Population Biology of Plants.* Academic Press, London.

Hartnett, D.C., B.A.D. Hetrick, G.W.T. Wilson, and D.J. Gibson, 1993: Mycorrhizal influence on intra- and interspecific neighbour interactions among co-occurring prairie grasses. *Journal of Ecology,* **81,** 787–795.

Hector, A., 2002: Biodiversity and the functioning of grassland ecosystems: multi-site comparisons. In: *The Functional Consequences of Biodiversity,* A.P. Kinzig, S.W. Pacala, and D. Tilman, (eds.), Princeton University Press, Princeton, pp. 71–95.

Hector, A., B. Schmid, C. Beierkuhnlein, M.C. Caldeira, M. Diemer, et al., 1999: Plant diversity and productivity experiments in European grasslands. *Science,* **286,** 1123–1127.

Hector, A., K. Dobson, A. Minns, E. Bazeley-White, and J.H. Lawton, 2001: Community diversity and invasion resistance: an experimental test in a grassland ecosystem and a review of comparable studies. *Ecological Research,* **16,** 819–831.

Heschel, M.S. and K.N. Paige, 1995: Inbreeding depression, environmental stress, and population size variation in scarlet gilia (*Ipomopsis aggregata*). *Conservation Biology,* **9,** 126–33.

Heywood, V.H. and R. Watson, eds. 1995. Global biodiversity assessment. Cambridge University Press, Cambridge, UK.

Hobbie, S.E., 1992: Effects of plant species on nutrient cycling. *Trends in Ecology and Evolution,* **7,** 336–339.

Hobbs, R. J. and C. J. Yates. 2003. Impacts of ecosystem fragmentation on plant populations: generalizing the idiosyncratic. *Australian Journal of Botany,* **51,** 471–488.

Hobbs, R.J. and L.F. Huenneke, 1992: Disturbance, diversity and invasion: implications for conservation. *Conservation Biology,* **6,** 324–337.

Hobbs, R.J., 1993: Effects of landscape fragmentation on ecosystem processes in the Western Australian wheatbelt. *Biological Conservation,* **64,** 193–201.

Hodgson, J.G, K. Thompson, P.J. Wilson, and A. Bogaard, 1998: Does biodiversity determine ecosystem function? The Ecotron experiment reconsidered. *Functional Ecology,* **12,** 843–848.

Hodgson, J.G., G. Montserrat-Marti, J. Tallowin, K.Thompson, S. Díaz, et al., 2005: How much will it cost to save grassland diversity? *Biological Conservation,* **122,** 263–273.

Holling, C.S., 1992: Cross-scale morphology, geometry, and dynamics of ecosystems. *Ecological Monographs,* **62,** 447–502.

Hooper, D.U., 1998: Effects of plant composition and diversity on nutrient cycling. *Ecological Monographs,* **68,** 121–149.

Hooper, D.U. and J.S. Dukes, 2004: Overyielding among plant functional groups in a long-term experiment. *Ecology Letters,* **7,** 95–105.

Hooper, D.U. and P.M. Vitousek, 1997: The effects of plant composition and diversity on ecosystem processes. *Science,* **277,** 1302–1305.

Hooper, D.U., F.S. Chapin III, J.J. Ewel, A. Hector, P. Inchausti, et al., 2005. Effects of biodiversity on ecosystem functioning: a consensus of current knowledge and needs for future research. *Ecological Monographs,* **75,** 3–35.

Horvitz, C.C. and A.J. Beattie, 1980: Ant dispersal of Calathea (Marantaceae) seeds by carnivorous ponerines (Formicaridae) in a tropical rain forest. *American Journal of Botany,* **67,** 321–326.

Hughes, T.P. 1994. Catastrophes, phase shifts, and large-scale degradation of a Caribbean coral reef. *Science,* **265,** 1547–1551.

Hughes, T.P., A.H. Baird, D.R. Bellwood, M. Card, S.R. Connolly, et al., 2003: Climate change, human impacts, and the resilience of coral reefs. *Science,* **301,** 929–933.

Huston, M.A., 1997: Hidden treatments in ecological experiments: re-evaluating the ecosystem function of biodiversity. *Oecologia,* **110,** 449–460.

Huston, M.A., 1999: Local processes and regional patterns: appropriate scales for understanding variation in the diversity of plants and animals. *Oikos,* **86,** 393–401.

Ingram M., G.C. Nabhan, and S.L. Buchmann, 1996: Impending pollination crisis threatens biodiversity and agriculture. *Tropinet* **7:** 1.

Jackson, J.B.C., M.X. Kirby, W:H: Berger, K.A. Bjorndal, L.W. Botsford, B.J. Bourque, R.H. Bradbury, R. Cooke, J. Erlandson, J.A. Estes, et al., 2001: Historical overfishing and the collapse of coastal ecosystems. *Science,* **293,** 629–638.

Jakobsen, I., S.E. Smith, and F.A. Smith, 2002: Function and diversity of arbuscular mycorrhizae in carbon and mineral nutrition. In: *Mycorrhizal Ecology,* M.G.A. Van der Heijden, and I.R. Sanders (eds.), Springer, Berlin, pp. 75–92.

Janzen, D.H., 1986: The eternal external threat. In: *Conservation Biology: The Science of Scarcity and Diversity,* M.E. Soulé (ed.) Sinauer, Sunderland, Massachusetts, pp. 286–303.

Johnson, S.D. 2004: An overview of plant-pollinator relationships in southern Africa. *International Journal of Tropical Insect Science,* **24,** 45–54.

Johnson, S.D. and W.J. Bond, 1994: Evidence for widespread pollen limitation of fruiting success in Cape wildflowers. *Oecologia,* **109,** 530–534.

Jones, C.G., J.H. Lawton, and M. Shachak, 1994: Organisms as ecosystem engineers. *Oikos,* **69,** 373–386. Organisms as ecosystem engineers. *Oikos,* **69,** 373–386.

Jordano, P., 1987: Patterns of mutualistic interactions in pollination and seed dispersal: connectance, dependence asymmetries, and coevolution. *The American Naturalist,* **129,** 657–677.

Jordano, P., 1992: Fruits and frugivory, In: *Seeds: The Ecology of Regeneration in Plant Communities,* M. Fenner (ed.), Commonwealth Agricultural Bureau International, Wallingford,

Jordano, P. and E.W. Schupp, 2000: Seed disperser effectiveness: The quantity component and patterns of seed rain for Prunus mahaleb. *Ecological Monographs,* **70,** 591–615.

Julliot, C., 1996: Seed dispersal by red howling monkeys (*Alouatta seniculus*) in the tropical rain forest of French Guiana. *International Journal of Primatology,* **17,** 239–258.

Karl, B.J. and H.A. Best, 1982: Feral cats on Stewart Island: their food and their effects on kakapo. *New Zealand Journal of Zoology,* **9,** 287–294.

Kearns, C.A., D.W. Inouye, and N.M. Waser, 1998: Endangered mutualisms: the conservation of plant-pollinator interactions. *Annual Review of Ecology and Systematics,* **29,** 83–112.

Kelty, M.J., B.C. Larson, and C.D. Oliver (eds.), 1992: *The Ecology and Sylviculture of Mixed-Species Forests.* Kluwer Academic Publishers, Dordrecht.

Kenkel, N.C., D.A. Peltzer, D. Baluta, and D. Pirie, 2000: Increasing plant diversity does not influence productivity: empirical evidence and potential mechanisms. *Community Ecology,* **1,** 165–170.

Kenmore, P.E, F.O. Carino, C.A. Perez, V.A. Dyck, and A.P. Gutierrez,1984: Population regulation of the brown planthopper within rice fields in the Philippines. *Journal of Plant Protection in the Tropics,* **1,** 19–37.

Kennedy, T.A., S. Naeem, K.M. Howe, J.M.H. Knops, D. Tilman, and P.B. Reich, 2002: Biodiversity as a barrier to ecological invasion. *Nature,* **417,** 636–638.

Khan, Z.R., J.A. Pickett, J. van den Berg, and C.M. Woodcock, 2000: Exploiting chemical ecology and species diversity: stem borer and *Striga* control for maize in Africa. *Pest Management Science,* **56,** 1–6.

Kitzberger, T., D.F. Steinaker, and T.T. Veblen, 2000: Effects of climatic variability on facilitation of tree establishment in northern Patagonia. *Ecology,* **81,** 1914–1924.

Klavitter J.L, J.M Marzluff, and M.S Vekasy, 2003: Abundance and demography of the Hawaiian hawk: Is delisting warranted? *Journal of Wildlife Management* **67,**165–176.

Klironomos, J.N, J. McCune, M. Hart, and J. Neville, 2000: The influence of arbuscular mycorrhizae on the relationship between plant diversity and productivity. *Ecology Letters,* **3,** 137–141.

Klironomos, J.N., 2000: Host-specificity and functional diversity among arbuscular mycorrhizal fungi. In: *Microbial Biosystems: New Frontiers,* C.R. Bell, M. Brylinsky, and P. Johnson-Green (eds.), Atlantic Canada Society for Microbial Ecology, Halifax, pp. 845–851.

Klironomos, J.N., 2002: Feedback with soil biota contributes to plant rarity and invasiveness in communities. *Nature,* **417,** 67–70.

Knops, J.M.H., D. Tilman, N.M. Haddad, S. Naeem, C.E. Mitchell, J. Haarstad, M.E. Ritchie, K.M. Howe, P.B. Reich, E. Siemann, and J. Groth, 1999: Effects of plant species richness on invasions dynamics, disease outbreaks, insect abundances, and diversity. *Ecology Letters,* **2,** 286–293.

Kremen, C. 2004. Pollination services and community composition: does it depend on diversity, abundance, biomass or species traits? In: Solitary Bees: *Conservation, Rearing and Management for Pollination.* Freitas, B.M. and Pereira, J.O.P. (eds), Beberibe, Ceara, Brazil. Pp. 115–123

Kremen, C., N. M. Williams, R. L. Bugg, J. P. Fay, and R. W. Thorp, 2004: The area requirements of an ecosystem service: crop pollination by native bee communities in California. *Ecology Letters,* **7:**1109–1119.

Kremen, C., N.M. Williams, and R.W. Thorp, 2002: Crop pollination from native bees at risk from agricultural intensification. *Proceedings of the National Academy of Science,* **99,** 16812–16816.

Kunin, W.E., 1992: Density and reproductive success in wild populations of Diplotaxis erucoides (*Brassicaceae*). *Oecologia,* **91,** 129–133.

Lal, R., 2004: Soil carbon sequestration impacts on global climate change and food security. *Science* **304:**1623–1627.

Lambers, H. and H. Poorter, 1992: Inherent variation in growth rate between higher plants: A search for physiological causes and ecological consequences. *Advances in Ecological Research,* **23,** 187–261.

Laurance, W.F., D. Perez-Salicrup, P. Delamonica, P.M. Fearnside, S. D'Angelo, A. Jerozolinski, L. Pohl, and T.E. Lovejoy, 2001: Rain forest fragmentation and the structure of Amazonian liana communities. *Ecology* **82:**105–116.

Laurance, W.F., L.V. Ferreira, J.M. Rankin-de Merona, S.G. Laurance, R. Hutchings, and T.E. Lovejoy, 1998b: Effects of forest fragmentation on recruitment patterns in Amazonian tree communities. *Conservation Biology,* **12:** 460–464.

Laurance, W.F., T.E. Lovejoy, H.L. Vasconcelos, E.M. Bruna, R.K. Didham, P.C. Stouffer, C. Gascon, R.O. Bierregaard, S.G. Laurance, and E. Sampiao, 2002: Ecosystem decay of Amazonian forest fragments: a 22-year investigation. *Conservation Biology* **16:**605–618.

Laurance, W.F., 1997: Hyper-disturbed parks: edge effects and the ecology of isolated rainforest reserves in tropical Australia. In: *Tropical Forest Remnants: Ecology, Management, and Conservation of Fragmented Communities,* W.F. Laurance and R.O. Bierregaard (eds.), University of Chicago Press, Chicago, pp. 71–83.

Laurance, W.F., H.L. Vasconcelos, and T.E. Lovejoy, 2000: Forest loss and fragmentation in the Amazon: implications for wildlife conservation. *Oryx,* **34,** 39–45.

Laurance, W.F., L.V. Ferreira, J.M. Rankin-de Merona, and S.G. Laurance, 1998a: Rain forest fragmentation and the dynamics of Amazonian tree communities. *Ecology,* **79,** 2032–2040.

Laurance, W.F., S.G. Laurance, L.V. Ferreira, J. Rankin-de Merona, C. Gascon, and T.E. Lovejoy, 1997: Biomass collapse in Amazonian forest fragments. *Science,* **278,** 1117–1118.

Lavorel, S. and E. Garnier, 2002: Predicting changes in community composition and ecosystem functioning from plant traits: Revisiting the holy grail. *Functional Ecology,* **16,** 545–556.

Lavorel, S., C. Roschette, and J.-D. Lebreton, 1999: Functional groups for response to disturbance in Mediterranean old fields. *Oikos,* **84,** 480–498.

Lawton, J.H. and C.G. Jones, 1995: Linking species and ecosystems: organisms as ecosystem engineers. In: *Linking Species and Ecosystems,* C.G. Jones and J.H. Lawton (eds.), Chapman and Hall, New York, pp. 141–150.

Le Maitre, D.C., B.V. Wilgen, R. Hapman, and D. McKelly, 1996: Invasive plants and water resources in the Western cape province, South Africa: modelling the consequences of a lack of management. *Journal of Applied Ecology,* **33,** 161–172.

Lehman, C.L. and D. Tilman, 2000: Biodiversity, stability, and productivity in competitive communities. *The American Naturalist,* **156,** 534–552.

Leighton, M. and D.R. Leighton, 1984: Vertebrate responses to fruiting seasonality within a bornean rainforest. In: *Tropical Rainforests: Ecology and Management,* S.L. Sutton, T.C. Whitmore, and A.C. Chadwick (eds.), Blackwell Science Publishing, Oxford.

Lepš, J., 2005: Diversity and ecosystem function. In: *Vegetation Ecology,* E. van der Maarel (ed.), Blackwell, Oxford, pp. 199–237.

Lepš, J., J. Osbornová and M. Rejmánek, 1982: Community stability, complexity and species life-history strategies. *Vegetatio,* **50,** 53–63.

Lepš, J., K. Spitzer and J. Jaro, 1998: Food plants, species composition and variability of the moth community in undisturbed forest. *Oikos,* **81,** 538–548.

Levine, J.M. and C.M. D'Antonio, 2003: Forecasting biological invasions with increasing international trade. *Conservation Biology,* **17,** 322–326.

Levine, J.M., M. Vilà, C.M. D'Antonio, J.S. Dukes, K. Grigulis, and S. Lavorel, 2003: Mechanisms underlying the impact of exotic plant invasions. *Philosophical Transactions of the Royal Society of London,* **270,** 775–781.

Levine, J.M. and C.M. D'Antonio, 1999: Elton revisited: a review of evidence linking diversity and invasibility. *Oikos,* **87,** 15–26.

Levine, J.M., T. Kennedy, and S. Naeem, 2002: Neighborhood scale effect of species diversity on biological invasions and their relationship to community patterns. In: *Biodiversity and Ecosystem Functioning,* M. Loreau, S. Naeem and P. Inchausti (eds.), Oxford University Press, Oxford, pp. 114–124.

Liebman, M. and C.P. Staver, 2001: Crop diversification for weed management. In: *Ecological Management of Agricultural Weeds,* M. Liebman, C.L. Mohler and C.P. Staver (eds.), Cambridge University Press, Cambridge, UK, pp. 322–374.

Lodge, D., 2001: Lakes. In: *Global Biodiversity in a Changing Environment,* F.S. Chapin, O.E. Sala and E. Huber-Sannwald (eds.), Springer Verlag, New York, pp. 277–314.

LoGiudice, K., R.S. Ostfeld, K.A. Schmidt, and F. Keesing, 2003: The ecology of infectious disease: effects of host diversity and community composition on Lyme disease risk. *Proceedings of the National Academy of Sciences, USA,* **100,** 567–571.

Lonsdale, W.M., 1999: Global patterns of plant invasions and the concept of invasibility. *Ecology,* **80,** 1522–1536.

Lord, J.M., 1991: Pollination and seed dispersal in Freycinetia baueriana, a dioecious liane that has lost its bat pollinator. *New Zealand Journal of Botany,* **29,** 83–86.

Loreau, M., 2000: Biodiversity and ecosystem functioning: recent theoretical advances. *Oikos,* **91,** 3–17.

Loreau, M.,1998: Separating sampling and other effects in biodiversity experiments. *Oikos,* **82,** 600–602.

Loreau, M., S. Naeem, P. Inchausti, J. Bengtsson, J.P. Grime, et al., 2001: Ecology—Biodiversity and ecosystem functioning: Current knowledge and future challenges. *Science,* **294,** 804–808.

Luckinbill, L.S., 1979: Regulation, stability, and diversity in a model experimental microcosm. *Ecology,* **60,** 1098–1102.

MA (Millennium Ecosystem Assessment), 2003: Ecosystem and Human Well-Being. Island Press, Washington, Covelo, London.

Macdonald, I.A.W., L.L. Loope, M.B. Usher, and O. Hamann, 1996: Wildlife conservation and the invasion of nature reserves by introduced species: a global perspective. In: *Biological Invasions. A Global Perspective,* J.A. Drake, H.A. Mooney, F. Di Castri, R.H. Groves, F.J. Kruger, M. Rejmanek, and M. Williamson (eds.), J. Wiley, Chichester, pp. 215–255.

Mack, R.N., D. Simberloff, W.M. Lonsdale, H. Evans, M. Clout, and F.A. Bazzaz, 2000: Biotic invasions: causes, epidemology, global consequences, and control. *Ecological Applications,* **10,** 689–710.

Magurran, A. and P. Henderson, 2003: Explaining the excess of rare species in natural species abundance distributions. *Nature,* **422,** 714–716.

Marc, P. and A. Canard, 1997: Maintaining spider biodiversity in agroecosystems as a tool in pest control. *Agriculture Ecosystems and Environment,* **62,** 229–235.

Maron, J.L. and M. Vilà, 2001: When do herbivores affect plant invasions? Evidence for the natural enemies and biotic resistance hypotheses. *Oikos,* **95,** 361–373.

Matson, P.A., W.J. Parton, A.G. Power, and M.J. Swift, 1997: Agricultural intensification and ecosystem properties. *Science,* **227,** 504–509.

Matthews, J.W. and K. Clay, 2001: Influence of fungal endophyte infection on plant-soil feedback and community interactions. *Ecology,* **82,** 500–509.

May, R.M., 1973: *Stability and Complexity in Model Ecosystems.* Princeton University Press, Princeton, 235 pp.

McCann, K. and A.P. Hastings, 1997: Re-evaluating the omnivory-stability relationship in food webs. *Proceedings of the Royal Society of London B, Biological Sciences,* **264,** 1249–1254.

McGillivray, C.W., J.P. Grime, S.R. Band, R.E, Booth, B. Campbell, G.A.F. Hendry, S.H. Hillier, J.G. Hodgson, R. Hunt, A. Jalili, et al., 1995: Testing predictions of the resistance and resilience of vegetation subjected to extreme events. *Functional Ecology,* **9,** 640–649.

McGrady-Steed, J., P.M. Harris, and P.J. Morin, 1997: Biodiversity regulates ecosystem predictability. *Nature,* **390,** 162–165.

McNaughton, S.J., 1993: Biodiversity and function of grazing ecosystems. In: *Biodiversity and Ecosystem Function,* E.-D. Schulze and H.A. Mooney (eds.), Springer-Verlag, Berlin, Germany, pp. 361–383.

Memmott, J. 1999: The structure of a plant-pollinator food web. *Ecology Letters,* **2,** 276–280.

Menalled, F.D., P.C. Marino, S.H. Gage, and D.A. Landis, 1999: Does agricultural landscape structure affect parasitism and parasitoid diversity? *Ecological Applications,* **9,** 634–641.

Mikola, J., 1998: Effects of microbivore species composition and basal resource enrichment on trophic-level biomasses in an experimental microbial-based soil food web. *Oecologia,* **117,** 396–403.

Milchunas, D.G, O.E. Sala, and W.K. Lauenroth, 1988: A generalized model of the effects of grazing by large herbivores on grassland community structure. *The American Naturalist* **132**:87–106.

Mitchell, C.E. and A.G. Power, 2003: Release of invasive plants from fungal and viral pathogens. *Nature,* **421,** 625–627.

Moberg, F and C. Folke. 1999: Goods and services associated with coral reef ecosystems. *Ecological Economics,* **29,** 215–233.

Moegenburg, S. M., 2002: Harvest and management of forest fruits by humans: implications for fruit-frugivore interactions. In: *Seed dispersal and frugivory: ecology, evolution and conservation.* D. Levey, W.R. Silva and M. Galetti (eds). Oxon, CABI Publishing. 479–494.

Moora, M. and M. Zobel, 1996: Effect of arbuscular mycorrhiza on inter- and intraspecific competition of two grassland species. *Oecologia,* **108,** 79–84.

Moora, M., M. Öpik, R. Sen, and M. Zobel, 2004: Rare vs. common *Pulsatilla* spp. seedling performance in soils from contrasting native habitats. *Functional Ecology,* **18,** 554–562.

Morin, P., 1999: Productivity, intraguild predation, and population dynamics in experimental food webs. *Ecology,* **80,** 752–760.

Muller, C.B. and J. Bordeur, 2002: Intraguild predation in biological control and conservation biology. *Biological Control,* **25,** 216–223.

Mundt, C.C., 2002: Use of multiline cultivars and cultivar mixtures for disease management. *Annual Review of Phytopathology,* **40,** 381–410.

Myers, R.A. and B. Worm, 2003: Rapid worldwide depletion of predatory fish communities. *Nature,* **423,** 280–283.

Nabhan, G.C., 1996: *Global list of threatened vertebrate wildlife species serving as pollinators for crops and wild plants. Forgotten Pollinators Campaign.* Arizona-Sonora Desert Museum, Arizona.

Nabhan, G.P. and S.L. Buchmann, 1997: Services provided by pollinators. In: *Nature's Services* (ed. Daily G.), pp. 133–150. Island Press, Washington D.C.

Naeem S., J.M.H. Knops, D. Tilman, K.M. Howe, T. Kennedy, and S. Gale, 2000: Plant diversity increases resistance to invasion in the absence of covarying extrinsic factors. *Oikos,* **91,** 97–108.

Naeem, S. and S. Li., 1997: Biodiversity enhances ecosystem reliability. *Nature,* **390,** 507–509.

Naeem, S., L.J. Thompson, S.P. Lawler, J.H. Lawton, and R.M. Woodfin, 1995: Empirical evidence that declining species diversity may alter the performance of terrestrial ecosystems. *Philosophical Transactions of the Royal Society of London B,* **347,** 249–262.

Naylor, R.L. and P.R. Ehrlich, 1997: Natural pest control services and agriculture. In: *Nature's Services: Societal Dependence on Natural Ecosystems,* G.C. Daily (ed.), Island Press, Washington D.C., pp. 151–174.

Ngugi, H.K., S.B. King, J. Holt, and A.M. Julian, 2001: Simultaneous temporal progress of sorghum anthracnose and leaf blight in crop mixtures with disparate patterns. *Phytopathology,* **91,** 720–729.

Noy-Meir, I., 1988: Dominant grasses replaced by ruderal forbs in a vole year in undergrazed mediterranean grasslands in Israel. *Journal of Biogeography,* **15,** 579–587.

O'Toole, C., 1993: Diversity of native bees and agroecosystems. In: *Hymenoptera and Biodiversity,* J. LaSalle and I.D. Gauld (eds.), CAB Int., Wallingford, UK, pp. 169–196.

Oke, T.R. 1987: *Boundary Layer Climates.* Methuen, London, UK.

Ollerton, J. and L.L. Kamner, 2002: Latitudinal trends in plant-pollinator interactions: are tropical plants more specialised? *Oikos,* **98,** 340–350.

Osbornová, J., M. Kovárová, J. Lepš, and K. Prach, (eds.), 1990: *Succession in Abandoned Fields. Studies in Central Bohemia, Czechoslovakia.* Geobotany 15, Kluwer, Dordrecht.

Ostfeld, R.S. and F. Keesing, 2000: Biodiversity and disease risk: the case of Lyme disease. *Conservation Biology,* **14,** 722–728.

Ostfeld, R.S. and K. LoGiudice, 2003: Community disassembly, biodiversity loss, and the erosion of an ecosystem service. *Ecology,* **84,** 1421–1427.

OTA (Office of Technology Assessment), 1993: *Harmful Non-Indigenous Species in the United States.* Publication no. OTA-F-565, US Government Printing Office, Washington, DC.

O'Toole, C. and A. Raw, 1991: Bees of the World. – Sterling Publishing Co, New York, USA.

Pace, M.L., J.J. Cole, S.R. Carpenter, and J.F. Kitchell, 1999: Trophic cascades revealed in diverse ecosystems. *Trends in Ecology and Evolution,* **14,** 483–488.

Pacheco, L.F. and J.A. Simonetti, 2000: Genetic structure of a mimosoid tree deprived of its seed disperser, the Spider Monkey. *Conservation Biology,* **14,** 1766–1775.

Packer, L., A. Zayed, J.C. Grixti, L. Ruz, R.E. Owen, F. Vivallo, and H. Toro, 2005: Conservation genetics of potentially endangered mutualisms: Reduced levels of genetic variation in specialist versus generalist bees. *Conservation Biology* **19,** 195–202.

Packer, A. and K. Clay, 2000: Soil pathogens and spatial patterns of seedling mortality in a temperate tree. *Nature,* **404,** 278–281.

Padilla, D.K. and S.L. Williams. 2004: Beyond ballast water: aquarium and ornamental trades as sources of invasive species in aquatic ecosystems. *Frontiers in Ecology and the Environment,* **2,** 131–138.

Paine, R.T. 2002: Trophic Control of Production in a Rocky Intertidal Community. *Science,* **296,** 736–739.

Paine, R.T., 1969: A note on trophic complexity and community stability. *The American Naturalist,* **103,** 91–93.

Palumbi, S.R. 2001: Evolution—Humans as the world's greatest evolutionary force. *Science,* **293,** 1786–1790.

Pan, J.J. and K. Clay, 2002: Infection by the systemic fungus Epichloe glyceriae and clonal growth of its host grass Glyceria striata. *Oikos,* **98,** 37–46.

Pärtel, M., M. Zobel, J. Liira, and K. Zobel, 2000: Species richness limitations in productive and oligotrophic plant communities. *Oikos,* **90,** 191–193.

Pérez-Harguindeguy, N., S. Díaz, J.H.C. Cornelissen, F. Vendramini, M. Cabido, and A. Castellanos, 2000: Chemistry and toughness predict leaf litter decomposition rates over a wide spectrum of functional types and taxa in central Argentina. *Plant and Soil,* **218,** 21–30.

Peters, G.,1972: Ursachen für den Rückgang der seltenen heimishchen Hummelarten (Hym., Bombus et Psithyrus). *Entomologische Berichte* **1972:** 85–90.

Petersen, P, J.M. Tardin, and F. Marochi, 2000: Participatory development of non-tillage systems without herbicides for family farming: the experience of the center-south region of Paraná. *Environ. Dev. and Sust.,* **1,** 235–252.

Peterson, C.H. and J. Lubchenco, 1997: Marine Ecosystem Services. In: *Nature's Services: Societal Dependence on Natural Ecosystems,* G. Daily (ed.), Island Press, Washington, DC, Pp. 177–195.

Petryna, L., M. Moora, C. Nunez, J.J. Cantero, and M. Zobel, 2002: Are the invaders disturbance-limited? Management for conservation of mountain grasslands in Central Argentina. *Applied Vegetation Science,* **5,** 195–202.

Pimentel, D., L. Lach, R. Zuniga, and D. Morison, 2000: Environmental and economic costs of nonindigenous species in the United States. *BioScience,* **50,** 53–65.

Pimm, S.L., 1980: Food web design and the effects of species deletions. *Oikos,* **35,** 139–149.

Power, A. and A.S. Flecker, 1996: The role of biodiversity in tropical managed ecosystems. In: *Biodiversity and Ecosystem Processes in Tropical Forests,* G.H. Orians, R. Dirzo and J.H. Cushman (eds.), Springer-Verlag, New York, pp. 173–194.

Power, A., 1999: Linking ecological sustainability and world food needs. *Environment, Development and Sustainability,* **1,** 185–196.

Power, A.G., 1991: Virus spread and vector dynamics in genetically diverse plant populations. *Ecology,* **72,** 232–241.

Power, M.E., D. Tilman, J.A. Estes, B.A. Menge, W.J. Bond, L.S. Mills, G. Daily, J C. Castilla, J. Lubchenco, and R.T. Paine, 1996: Challenges in the quest for keystones. *BioScience,* **46,** 609–620.

Pretty J., C. Brett, D. Gee, R. Hine , C. F. Mason, J.I.L. Morison, H. Raven, M. Rayment, and G. van der Bijl, 2000: An assessment of the total external costs of UK agriculture. *Agricultural Systems,* **65,** 113–136.

Pretty J., C. Brett, D. Gee, R.E. Hine, C.F. Mason, J.I.L. Morison, M. Rayment, G. van der Bijl, and T. Dobbs, 2001: Policy challenges and priorities for internalising the externalities of agriculture. *Journal of Environmental Planning and Management,* **44,** 263–283

Pretty, J.N. and A. Ball, 2001: *Agricultural Influences on Emissions and Sequestration of Carbon and Emerging Trading Options.* CES Occasional Paper 2001–03, University of Essex, Colchester.

Pretty, J.N., 1995: *Regenerating Agriculture.* Earthscan, London, and National Academy Press, Washington, DC, 320 pp.

Pretty, J.N., 2002: *Agri-Culture: Reconnecting People, Land and Nature.* Earthscan, London, 261 pp.

Pretty, J.N., C.F. Mason, D.B. Nedwell, and R.E. Hine, 2003: Environmental costs of freshwater eutrophication in England and Wales. *Environmental Science and Technology* **37,** 201–208

Prieur-Richard, A. H., S. Lavorel, A. Dos Santos, and K. Grigulis. 2002. Mechanisms of resistance of Mediterranean annual communities to invasion by *Conyza bonariensis:* Effects of native functional composition. *Oikos,* **99,** 338–346.

Prieur-Richard, A.-H. and S. Lavorel, 2000: Invasions: the perspective of diverse plant communities. *Austral Ecology,* **25,** 1–7.

Purcell, J.E., 1989: Predation of fish larvae and eggs by the hydromedusa Aequorea Victoria at a herring spawning ground in British Columbia. *Canadian Journal of Fisheries and Aquatic Sciences,* **46,** 1415–1427.

Quesada, M., K. E. Stoner, V. Rosas-Gerrero, C. Palacios-Guevara, and J. A. Lobo, 2003: Effects of habitat disruption on the activity of nectarivorous bats (Chiroptera: Phyllostomidae) in a dry tropical forest: implications for the reproductive success of the neotropical tree *Ceiba grandiflora. Oecologia* **135:** 400–406.

Redford, K.H. and P. Feinsinger, 2001: The half-empty forest: sustainable use and the ecology of interactions. In: *Conservation of Exploited Species,* J.D. Reynolds, G.M. Mace, K.H. Redford, and J.G. Robinson (eds.), Cambridge University Press, Cambridge, pp. 370–400.

Rejmánek, M., 1996: Species richness and resistance to invasions. In: *Biodiversity and Ecosystem Processes in Tropical Forests.* Ecological Studies 122, G.H. Orians, R. Dirzo and J.H. Cushman (eds.), Springer, Berlin, pp. 153–172.

Ribera Siguan, M.A., 2003: Pathways of biological invasions of marine plants. In: *Invasive Species: Vectors and Management Strategies,* G.M. Ruiz and J.T. Carlton (eds.), Island Press, Washington, DC.

Richards, K.W., 1993: Non-Apis bees as crop pollinators. *Rev. Suisse Zool.,* **100,** 807–822.

Richardson, D.M, N. Allsopp, C.M. D'Antonio, S.J. Milton, and M. Rejmánek, 2000: Plant invasions-the role of mutualisms. *Biological Review,* **75,** 65–93.

Ricketts, T. H., 2004: Tropical forest fragments enhance pollinator activity in nearby coffee crops. *Conservation Biology* **18:**1–10.

Ricketts, T.H., G.C. Daily, P.R. Ehrlich, and C.D. Michener, 2004: Economic value of tropical forest production to coffee production. *Proceedings of the National Academy of Sciences USA,* 2004, **101,** 12579–12582.

Riechert, S.E., L. Provencher, and K. Lawrence, 1999: The potential of spiders to exhibit stable equilibrium point control of prey: Tests of two criteria. *Ecological Applications,* **9,** 365–377.

Rillig, M., 2004: Arbuscular mycorrhizae and terrestrial ecosystem processes. *Ecology Letters,* **7,** 740–754.

Ruiz, G.M., P.W. Fofonoff, J.T. Carlton, M.J. Wonham, and A.H. Hines, 2000: Invasion of coastal marine communities in North America: Apparent patterns, processes, and biases. *Annual Review of Ecology and Systematics,* **31,** 481–531.

Sala, O.E., F.S. Chapin, J.J. Armesto, E. Berlow, J. Bloomfield, et al., 2000: Global Biodiversity Scenarios for the Year 2100. *Science,* **287,** 1770–1774.

Sampson, A.W., 1952: *Range Management, Principles and Practices.* Wiley, New York.

Sanchez, P., 1994: Tropical soil fertility research: towards the second paradigm. *World Congress of Soil Science,* 15, Transactions, Mexico, Vol. 1, pp. 65–88.

Sanders, I.R., 1993: Temporal infectivity and specificity of vesicular-arbuscular mycorrhizas in co-existing grassland species. *Oecologia,* **93,** 349–355.

Sankaran, M. and S. McNaughton, 1999: Determinants of biodiversity regulate compositional stability of communities. *Nature*, **401**, 691–693.

Schindler, D.W., 1990: Experimental perturbations of whole lakes as tests of hypotheses concerning ecosystem structure and functioning. *Oikos*, **57**, 25–41.

Schindler, D.W., S.R. Carpenter, J.J. Cole, J.F. Kitchell, and M.L. Pace, 1997: Influence of food web structure on carbon exchange between lakes and the atmosphere. *Science*, **277**, 248–251.

Schläpfer, F. and B. Schmid, 1999: Ecosystem effects of biodiversity: A classification of hypotheses and exploration of empirical results. *Ecological Applications*, **9**, 893–912.

Schläpfer, F., B. Schmid, and I. Seidl, 1999: Expert estimates about effects of biodiversity on ecosystem processes and services. *Oikos*, **84**, 346–352.

Schmid, B., 2002: The species richness-productivity controversy. *Trends in Ecology and Evolution*, **17**: 113–114.

Schmid, B., J. Roshi, and F. Schlapfer, 2002: Empirical evidence for biodiversity-ecosystem functioning relationships. In: *The Functional Consequences of Biodiversity*, A.P. Kinzig, S.W. Pacala and D. Tilman (eds.), Princeton University Press, Princeton, pp. 120–245.

Schmidt, K.A. and R.S. Ostfeld, 2001: Biodiversity and the dilution effect in disease ecology. *Ecology*, **82**, 609–619.

Schulze, E.-D., C. Wirth, and M. Heimann, 2000: Climate change: Managing forests after Kyoto. *Science*, **289**, 2058–2059.

Schwartz, M.W., C.A. Brigham, J.D. Hoeksema, K.G . Lyons, M.H. Mills, and P.J. van Mantgem, P.J., 2000: Linking biodiversity to ecosystem function: implications for conservation ecology. *Oecologia*, **122**, 297–305.

Scott-Dupree, C.D. and M.L. Winston, 1987: Wild bee pollinator diversity and abundance in orchard and uncultivated habitats in the Okanagan Valley, British Columbia. *Canadian Entomologist*, **119**, 735–745.

Shukla, J., C. Nobre, and P. Sellers, 1990: Amazon deforestation and climate change. *Science*, **247**, 1322–1325.

Silva, W.R., P. Marco Jr., E. Hasui, and V.S.M. Gomes, 2002: Patterns of fruit-frugivore interactions in two Atlantic forest bird communities of south-eastern Brazil: implications for conservation, In: *Seed Dispersal and Frugivory: Ecology, Evolution and Conservation*, D.J. Levey, W.R. Silva and M. Galetti (eds.), CABI Publishing, Oxon, pp. 423–436.

Simberloff, D. and B. Von Holle, 1999: Positive interactions of nonindigenous species: invasional meltdown? *Biological Invasions*, **1**, 21–32.

Smith, S.E. and D.J. Read, 1997: *Mycorrhizal Symbiosis*. Academic Press, London.

Snelgrove P, T.H. Blackburn, P.A. Hutchings, D.M. Alongi, J.F. Grassle, H. Hummel, G. King, I. Koike, P.J.D. Lambshead, N.B. Ramsing, and V. Solis-Weiss, 1997. *AMBIO*, **26**, 578–583.

Solórzano-Filho, J.A., 2001: Demografia, fenologia e ecologia da dispersão de sementes de Araucaria angustifolia (Bert.) Kuntze (Araucariaceae) numa população relictual em Campos de Jordão, SP. São Paulo, Universidade de São Paulo: 155.

Springer, A.M., J.A. Estes, G.B. van Vliet, T.M. Williams, D.F. Doak, E.M. Danner, K.A. Forney, and B. Pfister, 2003: Sequential megafaunal collapse in the North Pacific Ocean: An ongoing legacy of industrial whaling? *Proceedings of the National Academy of Sciences of the United States of America*, **100**, 12223–12228.

Stachowicz, J.J., H. Fried, R.B. Whitlatch, and R.W. Osman, 2002: Biodiversity, invasion resistance and marine ecosystem function: reconciling pattern and process. *Ecology*, **83**, 2575–2590

Stachowicz, J.J., R.B. Whitlatch, and R.W. Osman, 1999: Species diversity and invasion resistance in a marine ecosystem. *Science*, **286**, 1577–1579.

Stadler, J., A. Trefflich, S. Klotz, and R. Brandt, 2000: Exotic plant species invade diversity hot spots: the alien flora of northwestern Kenya. *Ecography*, **23**, 169–176.

Steiner, K.E., 1993: Has Ixianthes (Scrophulariaceae) lost its special bee? *Plant Systematics and Evolution*, **185**, 7–16.

Stewart, L., N. Hanley, and I. Simpson, 1997: *Economic valuation of the agri-environment schemes in the UK*. Report to HM Treasury and the Ministry of Agriculture, Fisheries and Food. Environmental Economics Group, University of Stirling, Stirling.

Stohlgren, T.J., 2003: The rich get richer: patterns of plant invasion in the United States. *Frontiers in Ecology and the Environment*, **1**, 11–14.

Stone, G.N., P.G. Willmer, and J.A. Rowe, 1998: Partitioning of pollinators during flowering in an African Acacia community. *Ecology*, **79**, 2808–2827.

Stratford, J.A. and P.C. Stouffer, 1999: Local extinctions of terrestrial insectivorous birds in a fragmented landscape near Manaus, Brazil. *Conservation Biology*, **13**, 1416–1423.

Strauss, S.Y. and A.A. Agrawal, 1999. The ecology and evolution of plant tolerance to herbivory. *Trends in Ecology and Evolution*, **14**, 179–185.

Strong, D.R., 1974: Rapid asymptotic species accumulation in phytophagous insect communities: the pests of cacao. *Science*, **185**, 1064–1066.

Strong, D.R., J.H. Lawton, and T.R.E. Southwood, 1984: *Insects on Plants: Community Patterns and Mechanisms*. Blackwell Scientific, Oxford.

Sturm, M. C. Racine, and K. Tape, 2001: Climate change—Increasing shrub abundance in the Arctic. *Nature*, **411**, 546–547.

Sunderland, K. and F. Samu, 2000: Effects of agricultural diversification on the abundance, distribution, and pest control potential of spiders: a review. *Entomologia Experimentalis Et Applicata*, **95**, 1–13.

Sutherland, J., 1974: Multiple stable points in natural communities. *American Naturalist*, **108**, 859–873.

Swift, M., 1999: Integrating soils, systems, and society. *Nature and Resources*, **35**, 12–20.

Swift, M.J. and J.M. Anderson, 1993: Biodiversity and ecosystem function in agricultural systems. In: *Biodiversity and Ecosystem Function*, E.D. Schulze and H.A. Mooney (eds.), Springer Verlag, Berlin, Germany, pp. 15–41.

Symstad, A.J., 2000: A test of the effect of functional group richness and composition on grassland invasibility. *Ecology*, **81**, 99–109.

Symstad, A.J., D. Tilman, J. Wilson, and J. Knops, 1998: Species loss and ecosystem functioning: effects of species identity and community composition. *Oikos*, **81**, 389–397.

Taylor, L.H., S.M. Latham, and M.E.J. Woolhouse, 2001: Risk factors for human disease emergence. *Philosophical Transactions of the Royal Society of London B*, **356**, 983–989.

Terborgh, J. and S.J. Wright, 1994: Effects of mammalian herbivores on plant recruitment in two neotropical forests. *Ecology*, **75**, 1829–1833.

Terborgh, J., 1986: Community aspects of frugivory in tropical forests. In: *Frugivores and Seed Dispersal*, A. Estrada and T.H. Fleming (eds.), Dr. W. Junk Publishers, Dordrecht, The Netherlands, pp. 371–384. The American Naturalist 132:87–106

Terborgh, J., L. Lopez, J. Tello, D. Yu, and A.R. Bruni, 1997: Transitory states inrelaxing ecosystems of tropical land-bridge islands. In: *Tropical Forest Remnants: Ecology, Management, and Conservation of Fragmented Communities*, W.F. Laurance and R.O. Bierregaard, Jr. (eds.), University of ChicagoPress, Chicago, Illinois, pp. 256–274.

Thompson, C., J. Beringer, F.S. Chapin III, and A.D. McGuire, 2004: Structural complexity and land-surface energy exchange along a vegetation gradient from arctic tundra to boreal forest. *Journal of Vegetation Science*, **15**, 397–406.

Thrall, P.H., J.J. Burdon, and M.J. Woods, 2000: Variation in the effectiveness of symbiotic associations between native rhizobia and temperate Australian legumes: interactions within and between genera. *Journal of Applied Ecology*, **37**, 52–65.

Thrush, S.F. and P.K. Dayton, 2002: Disturbance to marine benthic habitsts by trawling and dredging: implications for marine biodiversity. *Annual Review of Ecology and Systematics*, **33**, 449–47.

Tilman, D., 1996: Biodiversity: Population versus ecosystem stability. *Ecology*, **77**, 350–363.

Tilman, D.,1999: The ecological consequences of changes in biodiversity: A search for general principles. *Ecology*, **80**, 1455–1474.

Tilman, D., K. Cassman, P. Matson, R. Naylor, and S. Polasky, 2002a: Agricultural sustainability and intensive production practices. *Science*, **418**, 671–677.

Tilman, D., J. Knops, D. Wedin, and P. Reich, 2002b: Experimental and observational studies of diversity, productivity, and stability. In: *The Functional Consequences of Biodiversity*, Kinzig, A.P. S.W. Pacala, and D. Tilman (eds.), Princeton University Press, Princeton, pp. 42–70.

Tilman, D., J. Knops, D. Wedin, P. Reich, M. Ritchie, and E. Siemann, 1997a: The influence of functional diversity and composition on ecosystem processes. *Science*, **277**, 1300–1302.

Tilman, D., C.L. Lehman, and K.T. Thomson, 1997b: Plant diversity and ecosystem productivity: theoretical considerations. *Proceedings of National Academy of Sciences USA*, **94**, 1857–1861.

Tilman, D., P.B. Reich, J. Knops, D. Wedin, T. Mielce, and C. Lehman, 2001: Diversity and productivity in a long-term grassland experiment. *Science*, **294**, 843–845.

Tilman D, D. Wedin, and J. Knops, 1996: Productivity and sustainability influenced by biodiversity in grassland ecosystems. *Nature*, **379**, 718–720.

Timonen, S., H. Tammi, and R. Sen, 1997: Outcome of interactions between genets of two Suillus spp. and different Pinus sylvestris L. genotype combinations: identity and distribution of ectomycorrhizas and effects on early seedling growth in N-limited nursery soil. *New Phytologist*, **137**, 691–702.

Torchin, M.E., K.D. Lafferty, A.P. Dobson, V.J. McKenzie, and A.M. Kuris, 2003: Introduced species and their missing parasites. *Nature,* **42,** 8–630.

Urcelay, C. and S. Díaz, 2003: The mycorrhizal dependence of subordinates determines the effect of arbuscular mycorrhizal fungi on plant diversity. *Ecology Letters,* **6,** 388–391.

USDA National Agricultural Statistics Service, 1997: 1996 Honey production report. U. S. Department of Agriculture, Washington D.C.

Usher, M.B., 1987: Invasibility and wildlife conservation: invasive species on nature reserves. *Philosophical Transactions of the Royal Society of London B,* **314,** 695–710.

Valentini, R.G., G. Matteucci, A.J. Dolman, E.D. Schulze, C. Rebmann, et al., 2000: Respiration as the main determinant of carbon balance in European forests. *Nature,* **404,** 861–864.

van der Heijden, M.G.A., A. Wiemken, and I.R. Sanders, 2003: Different arbuscular mycorrhizal fungi alter coexistence and resource distribution between co-occurring plant. *New Phytologist,* **157,** 569–578.

van der Heijden, M.G.A., J.N. Klironomos, M. Ursic, P. Moutoglis, R. Streitwolf-Engel, T. Boller, A. Wiemken, and I.R. Sanders, 1998b: Mycorrhizal fungal diversity determines plant biodiversity, ecosystem variability and productivity. *Nature, 396,* 69–72.

van der Heijden, M.G.A., T. Boller, A. Wiemken, and I.R. Sanders, 1998a: Different arbuscular mycorrhizal fungal species are potential determinants of plant community structure. *Ecology,* **79,** 2082–2091.

van der Pijl, L. 1982: *Principles of Dispersal in Higher Plants.* 3rd Edition. Springer-Verlag, New York.

van der Putten, W.H. and B.A.M. Peters, 1997: How soil-borne pathogens may affect plant competition. *Ecology,* **78:**1785–1795

van der Putten, W.H., L.E.M. Vet, J.A. Harvey, and F.L. Wackers, 2001: Linking above- and belowground multitrophic interactions of plants, herbivores, pathogens, and their antagonists. *Trends in Ecology and Evolution,* **16,** 547–554.

van Ruijven, J., G.B. De Deyn, and F. Berendse, 2003: Diversity reduces invasibility in experimental plant communities: the role of plant species. *Ecology Letters,* **6,** 910–918.

Vandermeer, J., 1989: *Ecology of Intercropping.* Cambridge University Press, Cambridge.

Viana, V.M., A.A. Tabanez, and J. Batista, 1997: Dynamics and restoration of forest fragments in the Brazilian Atlantic moist forest. In: *Tropical Forest Remnants: Ecology, Management, and Conservation of Fragmented Communities,* W.F. Laurance, and R.O. Bierregaard (eds.), University of Chicago Press, Chicago, pp. 351–365.

Vilà, M, J. Vayreda, C. Gracia, and J.J. Ibáñez, 2003: Does tree diversity increase production in pine forests? *Oecologia,* **135,** 299–303.

Vilà, M., J. Vayreda, C. Gracia, and J. Ibáñez. 2004. Biodiversity correlates with regional patterns of forest litter pools. *Oecologia,* **139,** 641–646.

Vinson, S.B., G.W. Frankie, and J. Barthell, 1993: Threats to the diversity of solitary bees in a neotropical dry forest in Central America. In: *Hymenoptera and Biodiversity,* J. LaSalle and I.D. Gauld (ed.), CAB Int., Wallingford, UK, pp. 53–82.

Vitousek, P., H. Mooney, J. Lubchenco, and J. Melillo, 1997: Human domination of Earth's ecosystems. *Science,* **277,** 494–499

Vitousek, P., L. Walker, L. Whiteaker, D. Mueller-Dombois, and P. Matson, 1987: Biological invasion by *Myrica faya* alters ecosystem development in Hawaii. *Science,* **238,** 802–804

Waide, R.B., M.R. Willig, C.F. Steiner, G. Mittelbach, L. Gough, S.I. Dodson, G.P. Juday, and R. Parmenter, 1999: The relationship between productivity and species richness. *Annual Review of Ecology and Systematics,* **30,** 257–300.

Walker, B., 1995: Conserving biological diversity through ecosystem resilience. *Conservation Biology,* **9,** 747–752.

Walker, B.H., 1992: Biodiversity and ecological redundancy. *Conservation Biology,* **6,** 18–23.

Walker, B.H., A. Kinzig, and J. Langridge, 1999: Plant attribute diversity, resilience and ecosystem function: the nature and significance of dominant and minor species. *Ecosystems,* **2,** 95–113.

Walker, L. and Vitousek P. M., 1991: an invader alters germination and growth of a native dominant tree in Hawaii. *Ecology,* **72,** 1449–1455.

Wardle, D., K. Bonner, and K. Nicholson, 1997: Biodiversity and plant litter: Experimental evidence which does not support the view that enhanced species richness improves ecosystem function. *Oikos,* **79,** 247–258.

Wardle, D., K. Bonner, G. Barker, G. Yeates, K. Nicholson, R. Bardgett, R. Watson, and A. Ghani, 1999: Plant removals in perennial grassland: Vegetation dynamics, decomposers, soil biodiversity, and ecosystem properties. *Ecological Monographs,* **69,** 535–568.

Wardle, D.A., G.W. Yeates, W. Williamson, and K.I. Bonner, 2003: The response of a three trophic level soil food web to the identity and diversity of plant species and functional groups. *Oikos,* **102,** 45–56.

Wardle, D.A., R.D. Bardgett, J.N. Klironomos, H. Setälä W.H. van der Putten, and D.H. Wall, 2004: Ecological linkages between aboveground and belowground biota. *Science,* **304,** 1629–1633.

Waser, N.M., L. Chittka, M. Price, N.M. Williams, and J. Ollerton, 1996: Generalization in pollinations systems and why it matters. *Ecology,* **77,** 1043–1060.

Whitmore, T.C., 1997: Tropical forest disturbance, disappearance, and species loss. In: *Tropical Forest Remnants: Ecology, Management, and Conservation of Fragmented Communities,* W.F. Laurance and R.O. Bierregaard, Jr. (eds.), University of Chicago Press, Chicago, Illinois, pp. 3–12.

Wiebes, J.T., 1979: Co-evolution of figs and their insect pollinators. *Annual Review of Ecology and Systematics,* **10,** 1–12.

Wilby, A. and M.B. Thomas, 2002b: Natural enemy diversity and natural pest control: patterns of pest emergence with agricultural intensification. *Ecology Letters,* **5,** 353–360.

Wilcove, D.S., D. Rothstein, J. Dubow, A. Phillips, and E. Losos, 1998: Quantifying threats to imperiled species in the United States. *BioScience,* **48,** 607–615.

Wilcox, B.A., 1980: Insular ecology and conservation. In: *Conservation Biology: An Evolutionary-Ecological Perspective,* M.E. Soulé and B.A. Wilcox (eds.), Sinauer Associates, Sunderland, Massachusetts, pp. 95–117.

Williams, I.H., S.A. Corbet, and J.L. Osborne, 1991: Beekeeping, wild bees and pollination in the European Community. *Bee World,* **72,** 170–180.

Williams, N.M., R.L Minckley, and F.A. Silveira, 2001: Variation in native bee faunas and its implications for detecting community change. *Conservation Ecology,* **5,** 57–89.

Williams, P.H., 1986: *Bumble bees and their decline in Britain.* Central Association of Beekeepers, UK.

Willis, K., G. Garrod, and C. Saunders, 1993: *Valuation of the South Downs and Somerset Levels Environmentally Sensitive Areas.* Centre for Rural Economy, University of Newcastle upon Tyne.

Wilson, G.W.T., D.C. Hartnett, M.D. Smith, and K. Kobbeman, 2001: Effects of mycorrhizae on growth and demography of tallgrass prairie forbs. *American Journal of Botany,* **88,** 1452–1457.

Wolfe, M.O., 2000: Crop strength through diversity. *Nature, 406,* 681–682.

Wright, S.J. and H.C. Duber, 2001: Poachers and forest fragmentation alter seed dispersal, seed survival, and seedling recruitment in the palm Attalea butyraceae, with implications for tropical tree diversity. *Biotropica, 33,* 583–595.

Wright, S.J., 2003: The myriad consequences of hunting for vertebrates and plants in tropical forests. *Perspectives in Plant Ecology, Evolution and Systematics,* **6,** 73–86.

Wright, S.J., H. Zeballos, I. Domínguez, M.M. Gallardo, M.C. Moreno, and R. Ibáñez, 2000: Poachers alter mammal abundance, seed dispersal, and seed predation in a Neotropical forest. *Conservation Biology,* **14,** 227–239.

Yachi, S. and M. Loreau, 1999: Biodiversity and ecosystem functioning in a fluctuating environment: the insurance hypothesis. *Proceedings of the National Academy of Science USA,* **96,** 1463.

Zhu, Y., H.R. Chen, J.H. Fan, Y.Y. Wang, Y.Li, J.B. Chen, J.X. Fan, S.S. Yang, L.P. Hu, et al., 2000: Genetic diversity and disease control in rice. Nature, **406,** 718–722.

Zobel, M., 1997: The relative role of species pools in determining plant species richness: an alternative explanation of species coexistence? *Trends in Ecology and Evolution,* **12,** 266–269.

Chapter 12

Nutrient Cycling

Coordinating Lead Authors: Patrick Lavelle, Richard Dugdale, Robert Scholes
Lead Authors: Asmeret Asefaw Berhe, Edward Carpenter, Lou Codispoti, Anne-Marie Izac, Jacques Lemoalle,
    Flavio Luizao, Mary Scholes, Paul Tréguer, Bess Ward
Review Editors: Jorge Etchevers, Holm Tiessen

Main Messages

An adequate and balanced supply of elements necessary for life, provided through the ecological processes of nutrient cycling, underpins all other ecosystem services. The cycles of several key elements—phosphorus, nitrogen, sulfur, carbon, and possibly iron and silicon—have been substantially altered by human activities over the past two centuries, with important positive and negative consequences for a range of other ecosystem services and for human well-being.

In preindustrial times, the annual flux of nitrogen from the atmosphere to the land and aquatic ecosystems was 90–130 teragrams (million tons) per year. This was more or less balanced by a reverse "denitrification" flux. **Production and use of synthetic nitrogen fertilizer, expanded planting of nitrogen-fixing crops, and the deposition of nitrogen-containing air pollutants have together created an additional flux of about 200 teragrams a year, only part of which is denitrified.** The resultant N accumulation on land and in waters has permitted a large increase in food production, but at the cost of increased emissions of greenhouse gases and a frequent deterioration in freshwater and coastal ecosystem services, including water quality, fisheries, and amenity value.

Phosphorus is also accumulating in ecosystems at a rate of 10.5–15.5 teragrams per year, which compares with the preindustrial rate of 1–6 teragrams of phosphorus a year, mainly as a result of the use of mined P in agriculture. Most of this accumulation is occurring in soils, which may then be eroded into freshwater systems, causing deterioration of ecosystem services. This tendency is likely to spread and worsen over the next decades, since large amounts of P have accumulated on land and their transport to water systems is slow and difficult to prevent.

Sulfur emissions have been progressively reduced in Europe and North America but not yet in the emerging industrial areas of the world: China, India, South Africa, and the southern parts of South America. A global assessment of acid deposition threats suggests that tropical ecosystems are at high risk. Human-induced alteration of the iron and silicon cycles is less well understood, but it is believed, with *medium certainty*, to be a significant factor in altering the productivity of the ocean. This may be a significant benefit to the service of carbon sequestration.

Human actions, many associated with agriculture, have increased the "leakiness" of ecosystems with respect to nutrients. Tillage often damages soil structure, and pesticides may decrease useful nontarget organisms, increasing nutrient leaching. Simplification of the landscape and destruction of riparian forests, wetlands, and estuaries allow unbuffered flows of nutrients between terrestrial and aquatic ecosystems. Specific forms of biodiversity are critical to the performance of the buffering mechanisms that ensure the efficient use and cycling of nutrients in ecosystems.

In contrast to the issues associated with nutrient oversupply, there remain large parts of Earth, notably in Africa and Latin America, where harvesting without nutrient replacement has led to a depletion of soil fertility, with serious consequences for human nutrition and the environment.

12.1 Introduction

Nutrients comprise the 22 or so chemical elements known to be essential for the growth of living organisms. (See Table 12.1.) The list varies somewhat because some elements are only needed in very specific groups of organisms or specific circumstances.

This chapter deals mainly with nitrogen, sulfur, phosphorus, and carbon—all elements needed in relatively large quantities (the so-called macronutrients), and with cycles that have been substantially altered by human activities. Emerging issues related to iron and silicon will also be addressed.

Broadly speaking, nutrients can occur in gaseous form (such as N_2, CO_2), mineral form (such as apatite, the main P-containing mineral), inorganic ionic form (NH_4^+, NO_3^-, SO_4^{2-}, H_2PO4^-), and organic form (bound into various C-based compounds in living or dead organisms or their products). Nutrients are mostly taken up by plants in the ionic form and by animals in organic forms through consumption of living or dead tissues; microorganisms in general may use nutrients in any mineral or organic form, with sometimes high degrees of specialization at the guild or species level. The interconversion between forms is mediated by the ecosystem.

Nutrient cycling describes the movement within and between the various biotic or abiotic entities in which nutrients occur in the global environment. These elements can be extracted from their mineral or atmospheric sources or recycled from their organic forms by converting them to the ionic form, enabling uptake to occur and ultimately returning them to the atmosphere or soil. Nutrient cycling is enabled by a great diversity of organisms and leads to creation of a number of physical structures and mechanisms that regulate the fluxes of nutrients among compartments. These structures and processes act as buffers to limit losses and transfers to other ecosystems, as described later. Nutrients are distributed among a large number of living or dead compartments, and their relative abundance among these compartments is typical of certain ecosystems. For example, this is the case in terrestrial ecosystems, where nutrients may be greatly concentrated in living biomass (such as tropical rainforests) or in humus and soil organic matter (such as tundra ecosystems) (Lavelle and Spain 2001).

Fertility is the potential of the soil, sediment, or water system to supply nutrient elements in the quantity, form, and proportion required to support optimum plant growth (implicitly, in the context of ecosystem services, for human benefit). The largest flux of nutrients is their release from organic materials, as a result of decomposition by microbial communities. This flux may not be measurable, as part of it may be immediately reincorporated in microbial biomass. Microbial activity depends primarily on the availability of a food source and on regional and local climatic, edaphic, or hydrological factors. Locally, biological parameters such as the chemical composition of the organic material (which depends in turn on the plant community that produced it) and the soil invertebrates present act as proximate determinants.

The maintenance of fertility is a supporting service for the production of food, timber, fiber, and fuel. It is also necessary for ecological processes such as succession and for the persistence and stability of ecosystems. In systems intensively managed by humans, such as cultivated systems, the inherent fertility of ecosystems is supplemented through fertilization and management practices, such as the use of N-fixing plants, the acceleration of microbial processes by tillage, the addition of suitable organic inputs to the soil, and, in many parts of the world, biomass burning.

Fertilization is the input of nutrients to a system by humans, deliberately or as an unintended consequence of other activities. It includes supplements of N, P, S, potassium, calcium, magnesium, and micronutrients for agriculture; atmospheric N and S deposition; and the effects of elevated CO_2 in the atmosphere. Iron fertilization in the ocean has been performed experimentally (Coale et al. 1996; Moore et al. 2001).

It is possible to have an excessive supply of nutrients. In aquatic systems, this condition is known as eutrophication. Eutrophication, usually resulting from leaching of nutrients from soils managed for agriculture, is a form of nonvoluntary fertilization of

Table 12.1. Major Elements Needed for Plant Growth and Their Concentrations in Plants, the Upper Meter of Soil, and Ocean Water
(Fortescue 1980; Bohn et al. 1979)

Element	Content in Elemental Form (µg/g)			Major Forms	Biological Function/*Source*
	Plant	**Soil**	**Ocean**		
Macronutrients (>0.1 % of dry mass)					
C	454,000	20,000	28	CO_2	in organic molecules; photosynthesis/*atm.CO_2; OM*
O	410,000	490,000	857,000	O_2	in organic molecules; cellular respiration/*water; OM*
H	55,000	650	108,000	H_2O	in organic molecules/*water; OM*
N	30,000	1,000	0.5	NO_3^- or NH_4^+	in proteins, nucleic acids, and chlorophyll/*biol. fix of N_2; OM mineralization; atm. deposition*
K	14,000	10,000	380	K^+	principal positive ion inside cells; control of stomatal aperture; enzyme activity/*OM mineralization; weathering*
P	2,300	800	0.07	$H_2PO_4^-$ or HPO_4^{2-}	in nucleic acids, phospholipids, and electron carriers in chloroplasts and mitochondria/*OM mineralization; weathering*
Ca	18,000	10,000	400	Ca^{2+}	in adhesive compounds in cell walls; control of membrane permeability; enzyme activation/*weathering; OM mineralization*
Mg	3,200	6,000	1,350	Mg^{2+}	component of chlorophyll; enzyme activation; ribosome stability/*weathering; OM mineralization*
S	3,400	500	885	SO_4^{2-}	component of proteins and many coenzymes/*OM mineralization; atm. deposition*
Cl	2,000	100	19,000	Cl^-	in photosynthesis/*OM mineralization; atm. deposition*
Micronutrients (<0.2 % of dry mass)					
Fe	140	40,000	0.01	Fe^{2+} or Fe^{3+}	needed for synthesis of chlorophyll; component of many electron carriers
Mn	630	800	0.002	Mn^{2+}	in photosynthesis; enzyme activation
Mo	0.05	3	3	MoO_4^{--}	in nitrogen metabolism, required for nitrogen fixation
Cu	14	20	0.003	Cu^{2+}	enzyme activation; component of electron carriers in chloroplasts
Zn	100	50	0.01	Zn^{2+}	enzyme activation; protein synthesis; hormone synthesis
Bo	50	10	4.6	$H_2Bo_3^-$	involved in sugar transport
Ni				Ni^{2+}	nitrogen metabolism cofactor
Si	1,000	330,000	3	$Si(OH)_4$	support tissues
Co	0.5	8	0.0003	Co^{2+}	required by N-fixing plants
Na	1,200	7,000	10,500	Na^+	beneficial to higher plants
Se	0.05	0.01		$H_2SeO_3^-$	beneficial to higher plants
I	0.005	5		I^-	beneficial to higher plants

Note: *OM* = Dead organic matter

inland and coastal waters. Increase in nutrients in sewage effluent is another form of cultural eutrophication related to increases in human population.

Fertility can also be decreased by human activities, through soil erosion, nutrient mining (harvesting of nutrients at a rate in excess of their rate of replenishment), alteration of the soil biota or structure, salinization through poorly managed irrigation, acidification through inappropriate fertilization, deposition of acidifying pollutants, or excessive use of N-fixing crops.

Life on Earth is closely regulated by the efficient cycling and availability of nutrients. Human manipulation of this service has greatly affected all ecosystems. Climate regulation is affected by decomposition and nutrient cycling at regional or continental scales through the release of greenhouse gases and carbon sequestration in ecosystems. When nutrient cycling is impaired, the aesthetic and recreational value of freshwater and marine ecosystems may decrease significantly.

Nutrient cycling requires a large number of different organisms from diverse functional groups. It is a prime example of "functional biodiversity" in action. Conversely, dysfunctions in nutrient cycling, leading, for example, to eutrophication, have severe negative effects on biodiversity.

Nutrient cycling occurs everywhere, being a necessary part of the function of all ecosystems, but at widely varying rates. The specific ecosystem service of fertility is unevenly distributed around the land and oceans of the world. The most fertile soils—for instance, the deep, dark, loamy soils of the Eurasian steppes, the North American prairies and the South American pampas—have mostly been converted from grasslands to intensive agroecosystems over the past two centuries. Agriculture is currently expanding into areas with soils of inherently lower fertility, such as the old, red soils of tropical Africa, South America, and Southeast Asia (Wood et al. 2000).

The world's oceans contain vast reservoirs of nutrients. However, these are found primarily at depths below about 200 meters, where there is insufficient light for (net) photosynthesis (Dugdale 1976). As a consequence, high levels of marine primary production require the uplifting of nutrients from deeper water into the euphotic zone (Summerhayes et al. 1995). Along the west coast of the continents, equator-ward trade winds drive surface water away from the coast to be replaced by water rich in nutrients from the sub-euphotic zone. These coastal upwelling systems constitute only about 1% of the ocean surface but contribute about 50% of the world's fisheries (Ryther 1969), due to not only high new production rates but also to a short food chain in which much of the phytoplankton production is eaten directly by fish. The central oceans and seas, especially in tropical regions, are generally low in nutrients and productivity.

Human activities have resulted in large-scale changes in nutrient cycles over the last two centuries, which have occurred at an accelerated rate since about 1950. Specifically, shifts in land use patterns, increased fertilization associated with high-yield crops, and lateral transfer of nutrients across ecosystem boundaries have dramatically changed the rate, pathways, and efficiency of nutrient cycling. Traditional small-scale, low-input cultivation practices generally lead to nutrient depletion when fallow periods are shortened. On the other hand, the sustainability of soil fertility under large-scale, high-input intensive agriculture is still in question, given that this form of agriculture is only a few decades old.

The increase in demand for food, fuel, and fiber during the last 50 years has led to supplementation of the natural nutrient supply in agroecosystems by artificial sources, and the global annual nitrogen and phosphorus input to ecosystems has more than doubled in this period (Vitousek et al. 1997; Smil 2000; Falkowski et al. 2000).

All over the world, the complex regulation mechanisms in natural systems and biological controls operated by plants and invertebrates across many scales have been severely impaired, wherever food production has increased through the input of additional nutrients and use of tillage. Large-scale additions of nutrients in agroecosystems can no longer be retained and recycled locally. Undesired transfers from terrestrial to aquatic ecosystems have become a serious and growing problem worldwide (Howarth et al. 2000). At the same time, deposition from the atmosphere of nutrients, originating from industry, agriculture, biomass fires, and wind erosion, is spreading unprecedented quantities of N, P, and possibly Fe and Si to downwind ecosystems over large regions (Brasseur et al. 2003).

In contrast, severe nutrient depletion is observed in soils in some regions, notably in sub-Saharan Africa (Sanchez 2002). In these areas, particularly where the inherent soil fertility is low for geological and ecological reasons and fertilizer inputs are limited by economic constraints, the nutrient stock in croplands is decreasing. (See Tables 12.2 and 12.3.) This is having a serious impact on human food security in the region.

Some of the important questions addressed in this chapter include: What will be the consequences for ecosystems and human well-being of the expected 30% increase in the human contribution to fixed N and other nutrients over the next 30 years? (See also Chapter 9 of the *Policy Responses* volume). Is the increasing frequency and extent of eutrophication observed in fresh and marine waters likely to be stopped or reversed? How will these changes in the natural nutrient cycling system affect other ecosystem services?

12.2 Important Issues Common to All Nutrient Cycles

12.2.1 Ecosystem Stoichiometry: The Balance between Nutrients

Living organisms tend to contain a relatively constant proportion of elements, especially carbon, nitrogen, and phosphorus. In

Table 12.2. Total Nutrient Balance in Africa (Henao 2002)

Country	Year (Average)			
	1981–85	1986–90	1991–95	1996–99
	(NPK–kg/ha)			
Algeria	9.2	2.9	34.7	45.3
Angola	−33.9	−27.9	−36.5	−50.1
Benin	−55.7	−63.9	−57.9	−53.3
Botswana	−6.8	−15.8	−11.8	1.9
Burkina Faso	−46.6	−51.1	−56.2	−55.6
Burundi	−65.4	−62.7	−93.0	−87.4
Cameroon	−34.5	−44.5	−51.5	−54.2
Cape Verde	−91.5	−52.8	−42.0	−41.7
Central Africa	−54.3	−64.5	−62.3	−66.9
Chad	−56.7	−59.0	−58.8	−62.6
Comoros	−92.6	−87.5	−86.7	−89.3
Congo DR	−61.3	−64.4	−63.8	−63.2
Congo PR	−78.0	−80.5	−85.6	−76.5
Djibuti	−104.6	−79.1	−99.5	−101.8
Egypt	19.0	7.9	−49.7	−54.0
Equatorial Guinea	−63.9	−65.6	−68.8	−61.4
Eritrea			−48.1	−51.3
Ethiopia	−74.4	−69.8	−62.7	−63.6
Gabon	−47.0	−54.7	−64.4	−67.9
Kenya	−72.7	−72.6	−74.3	−60.5
Lesotho	−31.5	−41.4	−25.1	−52.1
Libya	71.6	69.2	101.8	29.6
Madagascar	−65.4	−72.2	−73.2	−75.1
Malawi	−71.5	−64.6	−62.3	−83.8
Morocco	−36.9	−63.6	-49.5	−59.4
Mozambique	−44.5	−43.1	−43.3	−61.6
Namibia	−60.5	−69.8	−47.4	−52.0
Reunion	208.1	199.6	149.1	23.9
Rwanda	−151.5	−136.3	−128.4	−123.8
Seychelles	−98.7	−107.0	−68.1	−55.4
Somalia	−78.7	−87.9	−61.9	−58.8
South Africa	27.0	−12.0	3.1	−18.6
Sudan	−61.5	−57.7	−59.2	−58.3
Swaziland	−78.4	−85.0	−89.2	−86.1
Tanzania	−64.5	−55.6	−63.4	−66.2
Tunisia	−13.6	−6.7	−23.2	−28.0
Uganda	−70.8	−74.5	−76.9	−74.9
Zambia	−16.0	−41.2	−21.7	−47.3
Zimbabwe	−35.6	−51.4	−27.1	−45.8

Table 12.3. Total Nutrient Balance in Latin America and in Central America and the Caribbean (Henao 2002)

Country	Year (Average)			
	1981–85	1986–90	1991–95	1996–99
	(NPK–kg/ha)			
Argentina	−109.1	−108.8	−105.4	−98.9
Belize	−189.6	−106.3	−125.5	−143.7
Bolivia	−97.4	−105.1	−132.7	−142.9
Brazil	−67.7	−72.3	−79.7	−79.5
Chile	−54.7	−21.1	24.5	101.7
Colombia	−87.7	−55.3	−68.3	−66.0
Costa Rica	−50.4	−22.7	−18.8	63.2
Dominican Rep	−133.6	−85.8	−83.6	−70.0
Ecuador	−68.5	−76.4	−85.4	−63.1
El Salvador	−80.5	−63.9	−83.5	−78.6
French Guiana	109.6	−24.8	−86.6	−69.4
Guatemala	−91.7	−77.8	−88.5	−96.1
Guyana	−150.0	−108.4	−137.9	−132.0
Honduras	−133.7	−132.1	−136.8	−72.9
Jamaica	−120.2	−76.5	−91.2	−90.7
Mexico	−33.2	−27.2	−47.1	−47.4
Nicaragua	−105.5	−76.8	−93.9	−92.8
Panama	−118.6	−74.1	−89.1	−67.5
Paraguay	−88.7	−98.9	−116.2	−117.1
Peru	−97.3	−59.2	−80.2	−63.8
Suriname	−97.2	−121.7	−151.9	−83.5
Trinidad & Tobago	−110.9	−163.0	−131.8	−98.5
Uruguay	−35.9	−33.9	−35.8	−2.6
Venezuela	12.1	113.3	6.3	−29.2

freshwater microalgae, the elemental contents expressed as molar stoichiometric ratios are C (125), N (19), and P (1) (Reynolds 1990), the same order of magnitude as the Redfield ratio (106:16:1) proposed for suspended particulate matter in oceans (Redfield 1958). The average proportions for land plants are about 200:13:1 (although these can be much wider for woody plants) because the need for structural tissue drives up the C content. Since the proportions of various functional groups of plants are also relatively fixed, ecosystems too have broad proportionality in their nutrient content. This has major implications: for instance, it means that the cycles are inextricably linked through shared organic molecules produced by living organisms. A perturbation to one cycle is a perturbation to all.

In response to an unbalanced composition of nutrients in their food, organisms tend to either eliminate excess nutrients or develop strategies to better capture nutrients that limit their growth. For example, in large water bodies, the efficiency of remineralization is higher for P than for N. This causes a shift in the N/P ratio of the available nutrients. This shift may be corrected by natural nutrient inputs, or further modified by inputs from human activities in the watershed. The result may be twofold: first, a new phytoplankton community may develop, with species more aligned to the new N/P nutrient ratio and concentration, and, second, the C/N/P composition of the phytoplankton may change, affecting its nutritional value for the primary consumers and energy transfer efficiency in the food chain.

Up to the 1980s, N was considered the nutrient that limits primary production of the world oceans. The "North Atlantic paradigm"—the view that oceanic primary production is under control of dissolved inorganic N—was established on the basis of studies mostly conducted in the North Atlantic Ocean, where most of the marine stations were situated. However, this view of the world oceans' biogeochemistry has proved to be incorrect. Multidisciplinary studies in a wider range of locations and situations have revealed the important role of other nutrients (Dugdale et al. 1995). (See Boxes 12.1 and 12.2.) For instance, silicon is essential for diatoms, the workhorses of the marine ecosystem, and may be quickly depleted as an excess of N and P becomes available. Phosphorus is required for all phytoplankton, including for N fixation (described later), and so does not have a fixed stoichiometric ratio. Interaction of the biota with the great nutrient cycles in the oceans makes it necessary to consider each situation with regard to potential nutrient limitation.

12.2.2 The Role of Organic Matter

Organic matter (that is, C-based compounds of biological origin) plays several pivotal roles in determining nutrient availability to plants. The largest labile stock of nutrients in soils, sediments, and waters is typically contained in organic compounds. Since uptake by plants is almost exclusively in the inorganic form, the biologically mediated process of organic matter decomposition is crucial to nutrient availability (Parton et al. 1988).

Soil and sediment organic matter is a heterogeneous mix of particles of partly decayed plant and animal tissues; the living biomass of microbes (principally fungi and bacteria); amorphous, decomposition-resistant high molecular weight C polymers known as humic substances; and a small quantity of simpler, less recalcitrant organic molecules, such as carbohydrates, amino acids, and lipids.

BOX 12.1

Iron as an Essential Fertilizer in Oceans

A number of trace metals are required for plant growth. Most appear to be present in sufficient quantities. However, iron is a special case for ocean organisms as it is required in many enzyme systems, including those in photosynthesis and in the processing of inorganic and gaseous nitrogen. A shortage of Fe has been shown to influence the uptake kinetics of $Si(OH)_4$—reducing, for example, the maximal uptake rate (Leynaert et al. 2001). Open ocean Fe enrichment experiments found Fe to affect the initial slope of the light versus carbon uptake curve in primary production (Barber et al. 1996).

The distribution of Fe in the oceans is poorly known due in part to the difficulty of obtaining uncontaminated samples, as the techniques for obtaining clean samples were developed only recently. However, the vertical profiles of Fe often follow the same pattern as NO_3 (Martin 1992). Since the NO_3 profiles are the result of biological production and regeneration, it appears that biological processes to a large extent control the distribution of Fe also. Input of Fe from the atmosphere is an important source of Fe, and some regions of the open ocean may experience enhanced productivity from the occurrence of remote dust events.

Silicon: A Key Element Linking Land to Ocean Ecosystems

Silicon, like iron, has its major sources on land but is a critical element in marine ecosystems. Dust atmospheric deposits allow transport from land to oceans and other continents, with significant effects on transfers of carbon to the deep ocean and neoformation of clay minerals that regulate nutrient cycling in terrestrial ecosystems.

Sustained C storage in the ocean requires a mechanism known as the "biological pump"—the export of particulate organic C to the deep ocean (Buesseler 1998). $Si(OH)_4$ is required for building the frustule of diatoms, the most productive and fastest growing of the phytoplankton (Smetacek 2000). Diatoms contribute as much as 75% of the annual primary production in coastal upwellings and Antarctic waters and about 40 % of the global marine annual primary production (Nelson et al. 1995) Where Si is in short supply, the plankton community is dominated by small-bodied algae, whose primary production simply cycles within the surface waters. To sink into the deep ocean waters, and thus be sequestered for useful periods of time, C must be in larger, heavier bodies, such as diatoms (Dugdale et al. 1995). Large areas on the fringe of the Southern Ocean are apparently Si-limited at certain times of the year. If we are to better understand and model the marine cycle of C, Si has to be taken into consideration by marine biogeochemists (Tréguer and Pondaven 2000; Ridgwell et al. 2002).

During the last decade, the Si budget of the world ocean and of key marine regions has been revised (Tréguer et al. 1995; Nelson et al. 1995; Tréguer and Pondaven 2000; De Master 2002). Our present best estimate for the annual production of the biogenic Si deposited in diatom frustules is 240 (40 Tmol per year for the world ocean (Nelson et al. 1995). The estimates of the net export of biogenic Si range between 120 and 129 teramoles (10^{12} moles) per year (Tréguer et al. 1995; Ridgwell et al. 2002).

In Amazonia, recycling of Si by plants allows neosynthesis of soil clay minerals that accumulate at the soil surface forming thick microporous horizons of microcrystalline kaolinite where aluminum oxide horizon should have normally formed (Lucas et al. 1993). A similar mechanism has been observed in Mexican paramo with a perennial grass cover (Dubroeucq et al. 2002). In that case, allophane is formed instead of imogolite, an aluminium-rich colloid that is toxic at high concentrations. Soil fauna then redistribute these minerals in the upper part of the soil profile.

Organic matter provides energy for all microbial and faunal activities and thus allows them to build the microaggregate structures that control soil hydraulic properties and serve to further conserve organic matter. For example, deposition of straw on the soil surface of degraded soils of the Sahelian region attracts termites that feed on this resource and significantly improve water infiltration and storage in the galleries and porous constructions (Mando et al. 1997). The amorphous polymers act as reserves of nutrients, which are sequestered in their chemical structures for periods of centuries to millennia. Since its surface bears a significant electrical charge, organic matter (along with clay) is the main location on which the cationic and anionic forms of plant nutrients are retained prior to uptake, without being leached out of the soil (Jenkinson and Rayner 1977; Schlesinger 1997; Lavelle and Spain 2001).

Agricultural practices and the policies that accompany their evolution may have a significant impact on the role and dynamics of organic matter in the provision of soil ecosystem services. For example, the prohibition against burning sugarcane crop residues in some tropical countries will probably have an impact on the storage of C in soils. Pasture rotation practiced in some areas aims to reconstitute organic stocks by the annual crops by inserting perennial grass lays that will increase organic matter stocks (Franzluebbers et al. 2000).

12.2.3 Nutrient Retention in Ecosystems: Buffers and Safety Nets at All Scales

In natural ecosystems, regulation of nutrient cycling operates at different scales of time and space, allowing the flow of nutrients released by microbial activities to adjust to plant demand, thus limiting losses to other parts of the ecosystems or to different ecosystems. In natural ecosystems, this "synchrony" between release of nutrients and their use by microorganisms and plants is determined by complex interactions among physical, chemical, and biological processes. It is rarely achieved to a comparable degree in agroecosystems, which as a result lose nutrients to aquatic ecosystems or to the atmosphere (Cadisch and Giller 1997).

Regulation of nutrient fluxes occurs in biological structures ("self-organizing systems," according to Perry 1995), which can be observed at seven different scales, ranging from microbial aggregates (Scales 1 and 2), through ecosystems (Scales 3 and 4) to landscapes (Scales 5 and 6) and the entire biosphere (Scale 7) (Lavelle and Spain 2001; Lavelle et al. 2004), as follows.

Scale 1, microbial communities in microbial aggregates and biofilms—At this scale, highly diverse microbial populations may control some transformations, whereas others that rely on a low number of species are vulnerable. For example, this is the case for microorganisms that control nitrification (the transformation of ammonia into nitrate, a form usable by plants), a critical step in N cycling. The functionality of these communities is rarely impaired by ecosystem degradations in spite of decreases in microbial diversity reported in some cases. The most sensitive function may be nitrification, since it is operated by a relatively small, diverse group of microbes. The risk to this function seems to be very limited, although it is speculated that threshold effects might be observed in some conditions.

Scale 2, microbial loops involving microbes and their micro-predators in soil and sediment aggregates, leaf packs, and freshwater systems or water columns—These are rarely severely impaired, although changes in soil nematode communities, and in microbial abundance and composition, have been observed in response to aggressive land use practices such as agricultural intensification (Bongers 1990; Yeates 1994). When larger-scale regulating systems are impaired, the diversity and abundance of the micro-predator food web (protozoa, nematodes, and acari in soils; specific invertebrate microplancton in fresh water and seas) may affect parameters of nutrient cycling, such as C and N mineralization (DeRuiter et al. 1993), through changes in their regulation of microbial activities. An index of soil "maturity" has been proposed to evaluate the functionality of this compartment (Bongers 1990). Clear changes in its communities have been observed, although effects on nutrient cycling have been reported only in laboratory experiments.

Scale 3, physical structures of animals and roots that have an impact on sediments and soils by their bioturbation effects—Mutually beneficial interactions with microflora and indirect effects of organisms that create the soil architecture by producing solid bricks of associated soil particles (called aggregates) and pores of different sizes and shapes (such as galleries, burrows, and interaggregate voids) allow regulations of nutrient cycling at different nested scales of time and space (Lavelle and Spain 2001).

For example, digestion of soil organic matter in earthworm guts is largely performed by the microflora that was ingested with the soil and further stimulated during the gut transit. In the case of the tropical geophagous earthworm *Millsonia anomala,* 90% of the energy thus assimilated by the worm is further spent on mechanical activities—the transit of huge amounts of soil (up to 20 times the weight of the worm daily; 1,000 milligrams of dry soil per hectare per year for an earthworm community). Soil that has transited through the earthworm guts is deposited in the soil and at the surface in the form of globular casts that are the stable aggregates that constitute most of the soil in the upper horizon. These aggregates are microsites where C-sequestration and nutrient conservation are active. Porosity created by these biological activities participates in water infiltration and storage and in soil aeration (Lavelle and Spain 2001).

Diversity of plants and soil invertebrates may influence nutrient cycling and prevent nutrient losses at the ecosystem scale, although the generality of this process has not yet been demonstrated nor have underlying mechanisms been clearly identified (Chauvel et al. 1999; Tilman and Downing 1994). The functionality of the nutrient buffer provided by biodiversity can be evaluated through indicators of soil and sediment quality based on invertebrate communities (Velasquez 2004; Ruiz 2004). Direct and indirect effects of soil bioturbators on nutrient cycling are reasonably well understood, and their use as indicators of soil quality is now well established.

Scale 4, ecosystems as mosaics of physical domains of different species of plants and animals—Excess nutrients released in a patch may diffuse away and be absorbed in adjacent nutrient-depleted patches. This feature is deliberately manipulated in most agroforestry and agricultural systems with plants that are associated with high nutrient needs (such as an annual crop of cereals) planted in patches or bands to capture excess nutrients (such as legume shrubs). Secondary successions in forests may also rely on different timing and rates of accumulation of nutrients in adjacent patches (Bernier and Ponge 1994). Research has proved the value of having a mosaic of patches at different stages of the succession (that is, a combination of patches with young, mature, and senescent groups of trees), although models suggesting optimal combinations of these patches do not exist.

Scale 5, landscapes or seascapes—At this scale, nutrients released in excess from nutrient-rich ecosystem patches may be absorbed by adjacent nutrient-poor patches. For instance, riparian forests may absorb nutrients leaking from crop areas and thus prevent them from entering a river system. On the River Seine in France, 25–55% of the N coming from below the root zone of agricultural land or from aquifers is retained or eliminated by riparian forest or wetlands before reaching surface waters (Billen and Garnier 1999). More generally, it is estimated that wetlands on average intercept 80% of N flowing from terrestrial systems (although figures vary due to temperature and size of the area; see Chapter 7). Again, there is ample empirical evidence of the efficiency of these buffers, although predictive models do not exist (Krug 1993; Haycock et al. 1993).

Scale 6, river basins, oceanic biogeochemical provinces, terrestrial biomes—At this scale, climatic factors (such as ambient temperature or the level of water in terrestrial systems) regulate the overall rates of biological activities and chemical transformations. Organic matter produced in warm surface layers of aquatic bodies descends to cold, deep strata, where consumers are not active due the lack of light. Specific mechanisms of thermal convection and deep ocean currents driven by temperature-, salinity-, and density-induced gradients move nutrient-rich deep ocean water to the surface, where potential consumers are active.

Scale 7, the biosphere—Global atmospheric and oceanic circulation created by the redistribution of energy at the Earth's surface determines distribution of nutrients through various mechanisms, including deposition of atmospheric dust and nutrient conveyer belts associated with major currents in the oceans. The variation in nutrient concentrations in different basins of the world's oceans has its origin in the circulation of the water between the ocean basins, driven by changes in heat input at the ocean surface and changes in salinity from variations in river inputs to the different basins. The macro-circulation of the oceans has been described as a "conveyer belt" (Broecker and Peng 1982), wherein deep water formed in the north Atlantic sinks and makes its way through the Antarctic and then north into the Pacific and Indian Oceans. As it flows, it receives organic particles formed by primary production and fecal pellets formed by grazing by zooplankton that are traveling in the opposite direction. This counter-current exchange mechanism results in increasing nutrient concentrations in the deep water as it travels westward from the Atlantic Ocean.

The scale structure is hierarchical, with each scale comprising the sum of the elements present in smaller scales. Effects at small scales are accumulative, and their effects are additive at larger scales. Impairment of services at one scale is likely to be caused, at least in part, by changes at a smaller scale, and it is at these smaller scales that interventions are most effectively targeted in order to tackle the causes of large-scale trends. Interventions at larger scales generally are only able to mitigate the effects of change in the short term.

The existence of different levels of buffers at nested scales also has the potential to reduce the vulnerability of nutrient cycling services, since different options exist to support the services. Damage occurring at lower scales, such as accidental loss of soil fauna that aerate soil (earthworms, for instance), may not have immediate or lasting effects, since soil structures created by these invertebrates may last long enough to enable populations to be restored if the disturbance is not too severe.

However, sometimes disturbances have multiple effects that accumulate over time and space, causing the system to collapse. Highly intensive land management for crop production, for example, can have detrimental effects at Scales 2–5, reducing soil fauna and root diversity (Scales 2 and 3) through use of pesticides and tillage (Lavelle and Spain 2001), homogenizing the ecosystem at plot scale (Scale 4) through use of monocultures, and simplifying the landscape composition through the use of fewer crop varieties (Scale 5). Precise data on the degree and duration of impacts that will lead to irreversible effects, such as soil and sediment erosion or replacement of native by invasive species of invertebrates, and other severe consequences for biodiversity and the vulnerability of the system are currently lacking. At present, only indicators of soil quality allow us to compare different sites and states and to evaluate trends.

12.2.4 Soil Erosion

Soil erosion is one of the key processes underlying land degradation and desertification. Erosion affects nutrient cycling and reduces the fertility of the soil through a reduction in the pool of available nutrients. Soil erosion results in drastic modifications to the structure as well as the biological and chemical properties of the soil matrix. The resulting dust and sediments have off-site impacts that may be as large, or larger, than the loss of production sustained on the eroded site. Persistently high rates of erosion affect more than 1.1 billion hectares of land worldwide (Berc et al. 2003; Jacinthe and Lal 2001), redistributing 75 billion tons of soil (Pimente et al. 1995) with 1.5 to 5.0% C-content (Lal 2001).

When nutrient-rich topsoil is transported from one place to another, nutrients are redistributed over the landscape. In the process, some is lost to riverine systems and eventually to the ocean, with major off-site economic and human well-being consequences, such as silting of reservoirs and eutrophication of lakes. On the other hand, dust deposits transported in the high atmosphere from Africa to North America are likely to stimulate the formation of new clay minerals in highly weathered soils and contribute (in other regions) to fertilization of nutrient-deficient marine areas. (See Figure 12.1.)

Erosion reduces the potential to sequester atmospheric CO_2 in soils by reducing the primary productivity. Exposure of subsurface layers to near-surface environments results in acidification of carbon-containing layers in most soils (Lal 2003). Soil erosion is the main way in which stable, mineral-associated soil organic C is translocated in large quantities into aquatic systems (Starr et al. 2000). The organic matter can end up in anoxic environments, where it is better protected than on land, and the accompanying nutrient fluxes may increase aquatic primary productivity. Thus not all C in eroded soil is returned to the atmosphere immediately, but the net effect is likely to be carbon emissions.

12.2.5 Input and Output Processes

Ecosystem nutrient balance is the net result of inputs minus outputs. Negative and positive balances are ultimately unsustainable. The magnitude and duration of nutrient imbalance that can be tolerated is determined by an ecosystem's buffering capacity. This is roughly indexed to the size of the nutrient stocks, divided by the normal net flux, which gives the turnover time.

Input of nutrients to ecosystems occurs through five processes. First, weathering from geological sources generally produces relatively small quantities of nutrients over long periods of time but is nevertheless an important input mechanism that sustains the levels of P, potassium, iron, aluminum, sodium, and silicon in natural ecosystems. The nature and composition of bedrock, in interaction with the climate, largely determines the flux. In all cases, this flux decreases with the age of the weathering surface. The rate at which anions are liberated through weathering determines the long-term capacity of ecosystems to absorb acid deposition, an ecosystem service of major importance in regions downwind of anthropogenic sources of NO_x and SO_2.

Second, atmospheric input of nutrients can occur through wet or dry deposition of elements previously released to the atmosphere by fires (biomass or combustion of fossil fuels), intensive farming practices (such as pig farms or cattle feedlots), and wind erosion. Atmospheric inputs have been substantially increased by human activities.

Third, biological processes include the fixation of atmospheric C (CO_2) through photosynthesis, and atmospheric N (N_2) through biological N fixation.

Fourth, nutrients can be released from the biomass of mobile organisms that enter an ecosystem and suffer mortality. This also occurs through the lateral transfer of nutrients, primarily in water flows. The burgeoning human trade in agricultural and forest products is now a significant pathway of nutrient transfer glob-

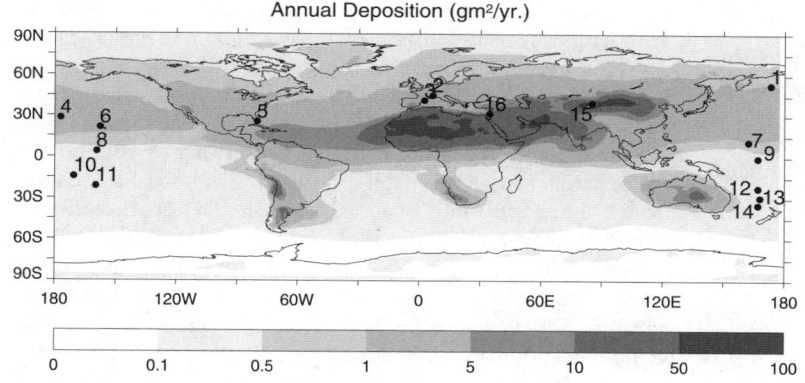

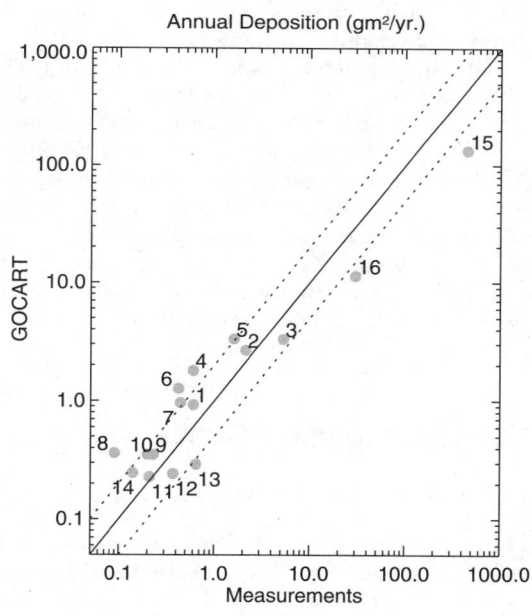

Figure 12.1. Global Dust Deposition. Estimates of global dust emissions can be derived from dust cycle models combine with remote sensing. The output of one such model, GOCART, is shown here, along with validation data for 16 locations. The estimates from a variety of models range from about 1,000 TG/yr (Tegen and Miller 1998) to 3,000 Tg/yr (Mahowald et al. 1999), for particles with radii of less than 10μm. The often-cited wider range of global dust emissions of 60–3,000 Tg/yr (Duce 1995) is partly based on such model estimates, and partly on extrapolations from very limited regional measurements. A central value of 2,000 Mt/yr global dust deposition appears reasonable. It excludes the small-scale transport of larger soil particles, which do not remain airborne for long. To obtain estimates of nutrient deposition from these, the dust deposition can be multiplied with the crustal abundance of the elements: O : 46.6%; Si : 27.7%; Al : 8.1%; Fe : 5.0%; Ca 3.6%; Na : 2.8%; K : 2.6%; and Mg 2.1%. These averages neglect regional variability, which is not well determined for dust aerosol. For the nutrients deposited with the dust, the bioavailability of the substances in the dust must also be considered, which in the case of iron is estimated to range between 1 and 10% (e.g., Fung et al. 2000). Thus the upper limit of biologically active Fe transport from land to the oceans as dust is 2,000 x 0.05 x 0.1 = 10 TgFe/yr, and the lower limit is probably around 0.5 TgFe/yr. (Tg = 10^{12}g)

ally—some of it from the nutrient-depleted developing world to the nutrient-saturated industrial world and, on a smaller scale, between rural and urban ecosystems.

Fifth, direct anthropogenic inputs occur through fertilization practices used in intensive agriculture and through the release of human sewage and livestock wastes.

The output of nutrients from ecosystems also involves five processes. Soil erosion is the main mechanism whereby nutrients are transported in large quantities from terrestrial to aquatic ecosystems. Although a natural process of soil rejuvenation, erosion is typically accelerated to many times above the long-term "natural" rate in systems where cultivation, overgrazing, and vegetation clearance are practiced. The essential nutrients that are most affected by erosion are C, P, K, and N.

Second, leaching is the vertical flow of water in the soil profile that transports significant amounts of nutrients in solution from the soil system into groundwater and thence laterally to rivers, lakes, and oceans. Leaching losses of nutrients are highest in cultivated or disturbed systems.

Third, gaseous emissions of CO_2, CH_4, and CO (among other gases) to the atmosphere result from the decomposition of organic matter, including digestion by animals, and the vastly accelerated decomposition that occurs in fires. Processes related to the conversion between inorganic forms of N lead to emissions of N_2, N_2O, NO, and NH_3. Phosphorus has no significant gaseous forms in most ecosystems. Anthropogenic activities, such as ploughing, fertilization, fossil fuel burning, flooding, drainage, deforestation, and changes to fire regimes, have altered the amounts and proportions of emissions of nutrients to the atmosphere. This is the ultimate underlying cause of contemporary climate change and air quality deterioration.

The fourth output source is the emigration of fauna or the harvest of crop, forest, fish, or livestock. As noted earlier, export from one ecosystem generally means import to another.

Fifth, the effective permanent removal of nutrients from the biosphere only occurs at a slow rate and through a small number of processes. For instance, for the atmospheric concentration of CO_2 to stabilize at levels that will not cause dangerous climate changes, anthropogenic carbon emissions must drop, within the next few centuries, to a level determined by the long-term sequestration sinks to a few teragrams (million tons) of carbon per year (Prentice et al. 2001).

12.3 Global Nutrient Cycles

12.3.1 The Global Nitrogen Cycle

There have been extremely significant changes in the global nitrogen cycle in the last two centuries, and N inputs to the global cycle have approximately doubled in this time. (See Figures 12.2 here and 12.3 in Appendix A.) Although most aspects of this balance have been assessed with a high level of accuracy (fertilizer inputs, atmospheric deposition, inland N fixation), a large uncertainty remains regarding the extent of N fixation in oceans, as explained later. Three processes are primarily responsible for the increased flows of N in the global cycle.

First, industrial-era combustion, especially of fossil fuels, has increased the emission of reactive N gases (NO_y) to the atmosphere, where they participate in the production of tropospheric ozone (the main harmful component of air pollution) before depositing, as a gas, as nitric acid dissolved in precipitation, or as dry aerosols on land or sea. Because of the reactivity of the gases, the impact is restricted to a region of up to about 1,000 kilometers

downwind of the source. Worldwide deposition of N as wet or dry deposits, in the form of NO_3^- or NH_4^+ ions, is especially concentrated in regions with farm cattle production. Highest deposition rates reach 50 kilograms per hectare per year, with local maximums of up to 100 kilograms in Europe, North America, China, and India, as well as Southwest America, Colombia, and a few regions in Africa (see Chapter 9). In other places, the recent rise in deposition (compared with the beginning of the twentieth century) has remained limited.

Initially, N deposition stimulates net primary productivity, since N is the most widely limiting nutrient in terrestrial ecosystems (largely because it is so readily lost by gaseous emission or leaching). Once the capacity of the recipient ecosystem for N inputs is reached (a point known as N saturation, indicated by a sudden increase in nitrates in water draining from the system), the excess is leached into adjacent rivers, lakes, and coastal zones, causing eutrophication. This can lead to biodiversity loss in both terrestrial and aquatic ecosystems, and, in severe cases, net primary production may decline (Schulze et al. 1989). However, given the moderate rise in N use predicted, the minor contribution to plant growth that Nadelhoffer et al. (1999) have calculated should remain limited. According to Frink et al. (2001), this should pose little hazard to biodiversity.

Second, the invention of the Haber-Bosch process for converting atmospheric N_2 to ammonia laid the foundation for the exponential growth in N fertilizer use in the second half of the twentieth century. It enabled the high-yielding crops of the Green Revolution, which brought about a large increase in the production of relatively cheap food and improved the well-being of millions of people. However, less than half of the applied N fertilizer finds its way into the crop plant. The remainder leaches into water bodies or returns to the atmosphere, most benignly as N_2, but some as the powerful and long-lived greenhouse gas N_2O, which is also involved in stratospheric ozone depletion. The atmospheric concentration of N_2O has been rising by roughly 0.8 parts per trillion per year (0.25%) during the industrial era, largely through this mechanism. In 1998, it averaged 314 ppt, up from a preindustrial level of 270 ppt (Prather et al. 2001). (See also *Policy Responses* volume, Chapter 9.) The present trend indicates a plateau may have been reached, although some models predict future increases. Whichever model is correct, no decrease is expected, since human population seems to have already exceeded the maximum number that can be supported without chemical fertilizers.

Third, the natural process of biological N fixation has been harnessed for agricultural purposes. Worldwide plantings of N-fixing crops, such as soybeans, now capture about 40 teragrams of nitrogen a year, an ecosystem service worth several billion dollars annually in avoided fertilizer costs and contributing substantially to human nutrition. The negative consequences are ultimately similar to those resulting from industrial N fixation: increased emissions of N_2O and leaching of N from the land into water bodies once organic N has been mineralized. In addition, severe acidification of soils following repeated harvests of N-fixing crops can occur, unless balancing quantities of cations are added (Pate 1968). Increase in N fixation is still considered important for sustained agricultural production, and is set to continue.

In marine ecosystems, estimates of N fixation by organisms vary by more than tenfold, ranging from less than 30 to more than 300 teragrams per year (Vitousek et al. 1997). There is speculation, supported by some evidence, that biological N fixation through the cyanobacterium *Trichodesmium* has increased during the modern era as a result of increased iron fertilization by windborne dust (Moore et al. 2001).

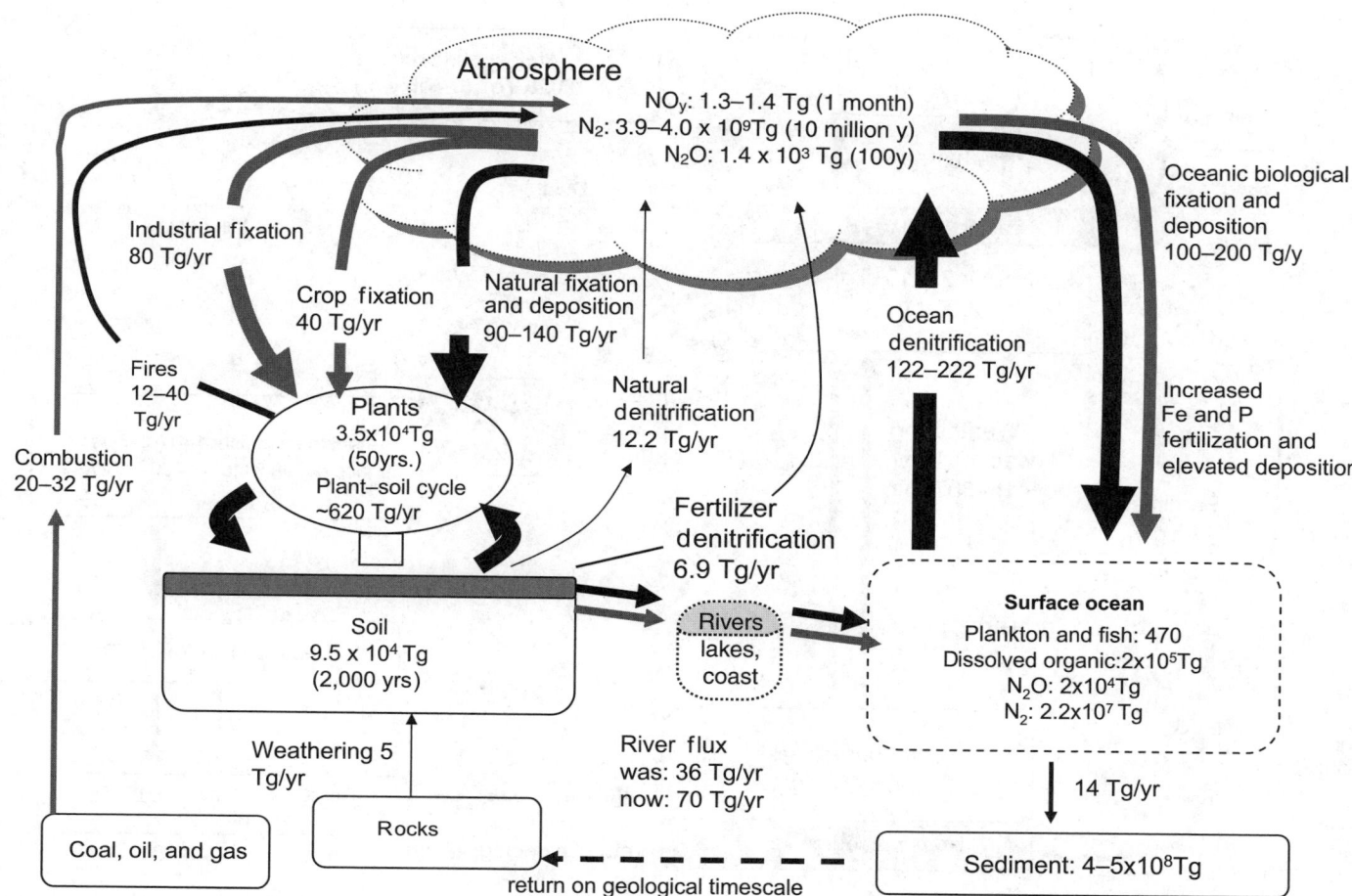

Figure 12.2. Key Pools, Fluxes, and Turnover Times in the Global Nitrogen Cycle as Modified by Human Activity. The turnover times are represented in parentheses. The size of the pools (i.e., the stocks in a particular form) are all expressed as TgN, which equals 10^{12}g of nitrogen in elemental form, or a million tons. The fluxes or flows between pools are in TgN/yr. Pools where the flows in and out are small relative to the size of the stock have a slow turnover time—in other words, they change slowly, but are also slow to mend once altered. The solid arrows represent the basic "natural" (i.e., preindustrial) cycle. The grey arrows and boxes represent the new or enhanced flows and accumulations caused by human activity. The width of the arrows is approximately proportional to the flux rates. It is clear that the addition of a major new flux from atmosphere to land, by way of industrial and crop nitrogen fixation, has created an imbalance leading to increased flow to the ocean, in the process contributing to eutrophication of rivers and lakes. Some of the nitrogen oversupply leads to increased emissions of N_2O and NO_y, which are increasing in the atmosphere, contributing to global warming, tropospheric pollution, and stratospheric ozone depletion. (Reeburgh 1997; Prather et al. 2001; Brasseur et al. 2003)

Part of the accumulated N is eliminated in gaseous forms through the denitrification process. Although it resembles a natural mitigation of eutrophication, nitrous oxides produced from fertilizer denitrification account for 6.9 teragrams of N per year, representing a flux of 56% increase in total denitrification from terrestrial ecosystems. Furthermore, 20–32 teragrams in gaseous forms are released to the air each year through combustion, creating atmospheric pollution. (See Chapter 13.)

The overall N budget on Earth has thus been significantly modified. In preindustrial times, the annual flux of nitrogen from the atmosphere to the land and aquatic ecosystems was estimated at 90–140 teragrams of N per year. This was more-or-less balanced by a reverse "denitrification" flux. Production and use of synthetic nitrogen fertilizer, expanded planting of nitrogen-fixing crops, and deposition of nitrogen-containing air pollutants together create an additional flux of about 210 teragrams a year, only part of which is denitrified (Vitousek et al. 1997). This 210-teragram increase can be attributed to chemical fertilizers (80), biomass burning (40), N-fixation in legume crops (40), fossil fuel combustion (20), land clearing (20), and wetland drainage (10).

12.3.2 The Global Phosphorus Cycle

The lithosphere is the ultimate source of all phosphorus in the biosphere. (See Figure 12.4.) Paradoxically, while apatite (the naturally occurring phosphate rock) is one of the most easily weathered primary minerals, P is amongst the least biologically available major nutrients. This is because the forms of phosphorus in the biosphere are poorly soluble, immobile, or otherwise inaccessible.

As a result, P occurs in sufficient supply in young, arid, and neutral soils, although with some exceptions, depending on the nature of the parent material. On the other hand, P often co-limits (with N) plant and animal production on old, highly weathered soils, such as those that dominate tropical Africa, South America, and Australia. Since NH_4^+ and NO_3^- are both more readily leached out of soils than phosphate, freshwater and some

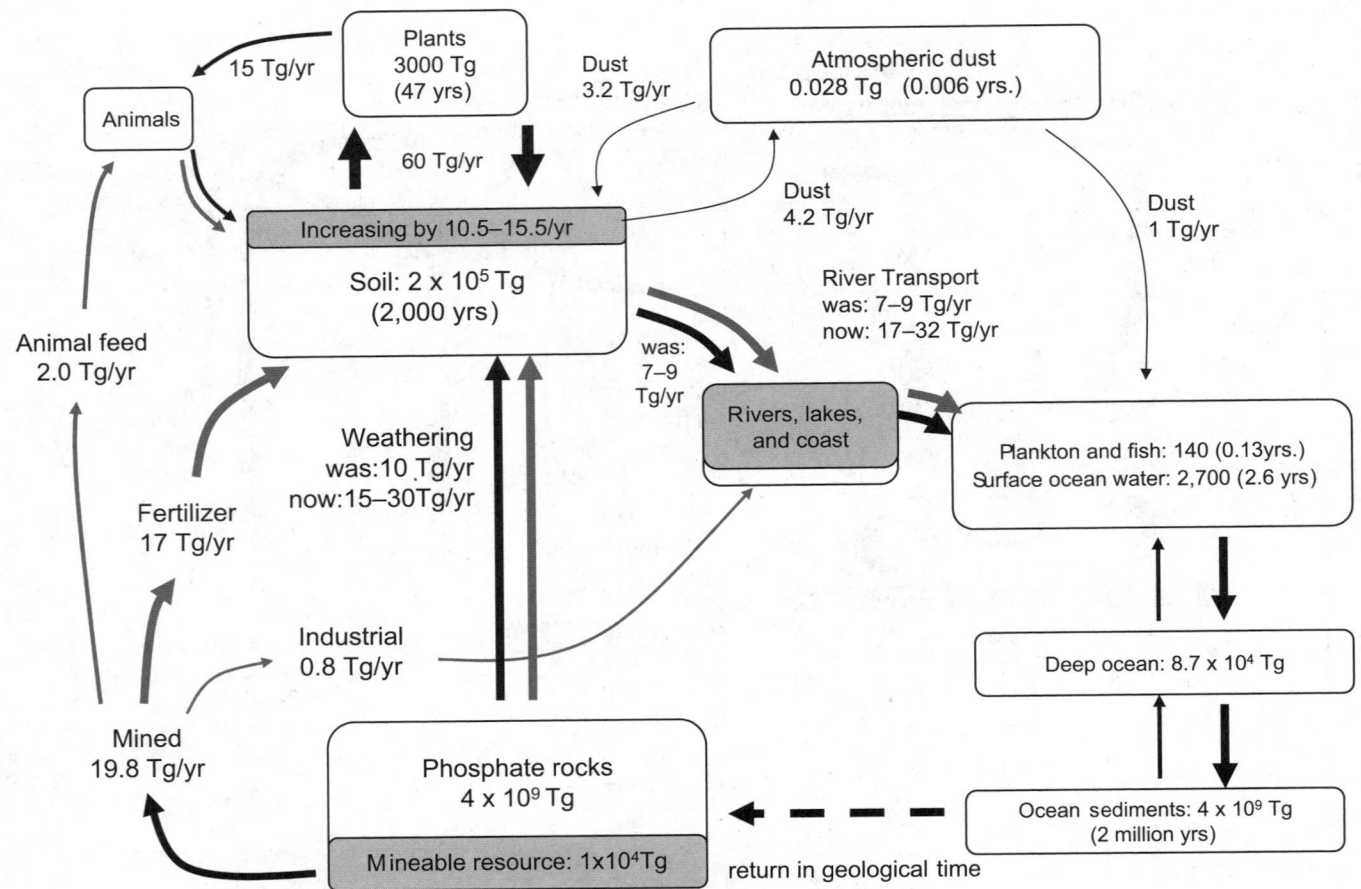

Figure 12.4. Schematic Diagram of Main Parts of Global Phosphorus Cycle. Pools are in TgP, fluxes in TgP/yr, and turnover times are in years. The grey arrows are the fluxes that are entirely or partly anthropogenic. P is building up in terrestrial soils as a result of P fertilizers, and is leaking into rivers, lakes, and coastal waters, where it is the main driver of eutrophication. (Reeburgh 1997; Carpenter et al. 1999)

coastal ecosystems are typically more responsive to increases of P than of N, making P the principal driver of eutrophication in lakes and estuaries. Phosphorus is transported principally adsorbed onto or absorbed into soil particles rather than in solution. In addition, where intensive animal husbandry is practiced, P can be lost in large quantities in surface runoff. As such, phosphorus in fact does not strictly cycle (other than in very long geological time frames) but follows a one-way path from terrestrial to aquatic systems. Return from marine systems to land in the form of bird guano, although sometimes locally important, is a very minor flux in total.

The availability of P in landscapes where it is scarce is greatly enhanced by biological processes. Specialized symbiotic fungi, known as mycorrhizae, transfer P from inaccessible forms to the plant and help to reduce leakage of P from the system. There is widespread empirical evidence that low P availability constrains biological N fixation (Smith 1992) contributing to the co-limitation just mentioned. However, the mechanism resulting in this constraint remains poorly understood (Vitousek et al. 2002).

Figure 12.4 shows the contemporary and preindustrial global P cycles. The contemporary cycle is not in balance. As a result of the large P inputs from the lithosphere, mainly through phosphate mining, and accelerated weathering as deep soil is exposed as a result of surface erosion, P is accumulating in terrestrial ecosystems in both the industrial and developing world (with some conspicuous exceptions, such as most of sub-Saharan Africa). The main mechanism by which the P leaves the land and enters fresh-

water ecosystems is soil erosion. Agricultural P is the principal driver of eutrophication. P concentrated in sewage effluents and animal and industrial wastes, including P-containing detergents, makes a relatively small global contribution (Bennett et al. 2001), although it may be important locally. For example in the United States 36% of sewage sludge is applied to the land, with the rest going to landfills, incineration, or other "surface disposal" methods.

Because the amount of P accumulated on land is large, and the processes of release are relatively slow but hard to prevent, this problem is highly likely to grow substantially in the coming decades. However, the rapid increase in no-till agricultural systems in several parts of the world and an increasing trend to incorporate buffer strips or hedgerows in agricultural landscapes may help to mitigate the problem significantly in some areas. For example, in the United States in 1989, 3% of cropland was no-till; in 1998 it was 16.3% (20 million hectares). In Brazil, 3% of cropland was no-till in 1990 and in 1998 it was 25% (10 million hectares). In Argentina, the figure was 2% in 1990 and 28% (6 million hectares) in 1998. And in Australia, the no-till area jumped from 0.1% in 1990 to 50% (10 million hectares) in 1998. (Eutrophication is dealt with extensively later in this chapter.)

12.3.3 The Global Sulfur Cycle

In many respects the sulfur cycle parallels the nitrogen cycle, except for a significant input from the lithosphere via volcanic activity and the absence of a biological process of S fixation from the atmosphere to the land or water. (See Figure 12.5.) The main human perturbation to the global S cycle is the release of SO_x (SO_2 plus a small amount of SO_3) to the atmosphere as a result of burning S-containing coal and oil and the smelting of sulfite ores.

SO_x gas impairs respiration in humans at high concentrations and is moderately toxic to plants. Other S gases, such as H_2S and mercaptans (sulfur-containing organic chemical substances), are not very toxic but are highly offensive to human olfaction even at low concentrations. Consequently, S-containing gases are usually vented from tall smokestacks in order to be widely diluted at ground level. Coupled with the simultaneous removal of ash particles from the smoke, this contributed greatly to the emergence of the "acid rain" deposition problem in the twentieth century (Smil 1997). Sulfuric acid is one of the major components of acid deposition, along with nitric acid, carbonic acid, and various organic acids. In preindustrial times it was largely neutralized by the simultaneous deposition of alkaline ash. High increase in S deposition (22–47 teragrams per year) has led to a situation where this compensation is no longer operating.

In the atmosphere SO_x forms SO_4^{2-}, a crystalline aerosol that acts as a powerful nucleus for cloud condensation and helps to retard climate change. The sulfate dissolves in rainwater, forming dilute sulfuric acid, and is deposited on Earth's surface in wet or dry form, the proportions of which depend on the prevailing climate.

Damage to ecosystems results not so much from the direct effects of acid, SO_x, or SO_4^{2-} on plants (S, in small doses, is a fertilizer), but from the direct effect of leaching of SO_4^{2-} from soil in drainage. To maintain electrical neutrality of drainage, cations, principally Ca^{2+}, are lost; the resulting acidification brings Al^{3+} and H^+ into solution (Galloway 2003). The Al^{3+} ion impairs nutrient absorption, especially phosphorus uptake, by the roots of all but a very specialized group of plants. It is also highly detrimental to aquatic organisms and ecosystems. Soil, river, and lake acidification is extremely difficult and expensive to remedy. The buffering capacity of the ecosystem, which is related to soil depth, soil chemistry, and weathering rate, provides an ecosystem service worth billions of dollars, both in avoided damage and mitigation actions. The capacity of this service is, however, finite and easily exceeded.

As a result of severe human health problems associated with SO_2 in urban smog and concerns regarding ecosystem health in areas exposed to high loadings of acid deposition, sulfur has been progressively reduced or eliminated from industrial, domestic, and transport sector emissions in Europe and North America. This has been achieved by switching from high-S coal and oil to lower-S fuels and by installing flue-gas desulfurization equipment. This has been so successful that S deposition has declined to such an extent that sulfur is now becoming a limiting nutrient in many parts of Europe. The result is that N deposition, resulting from vehicular, industrial, and agricultural emissions, has now become the major

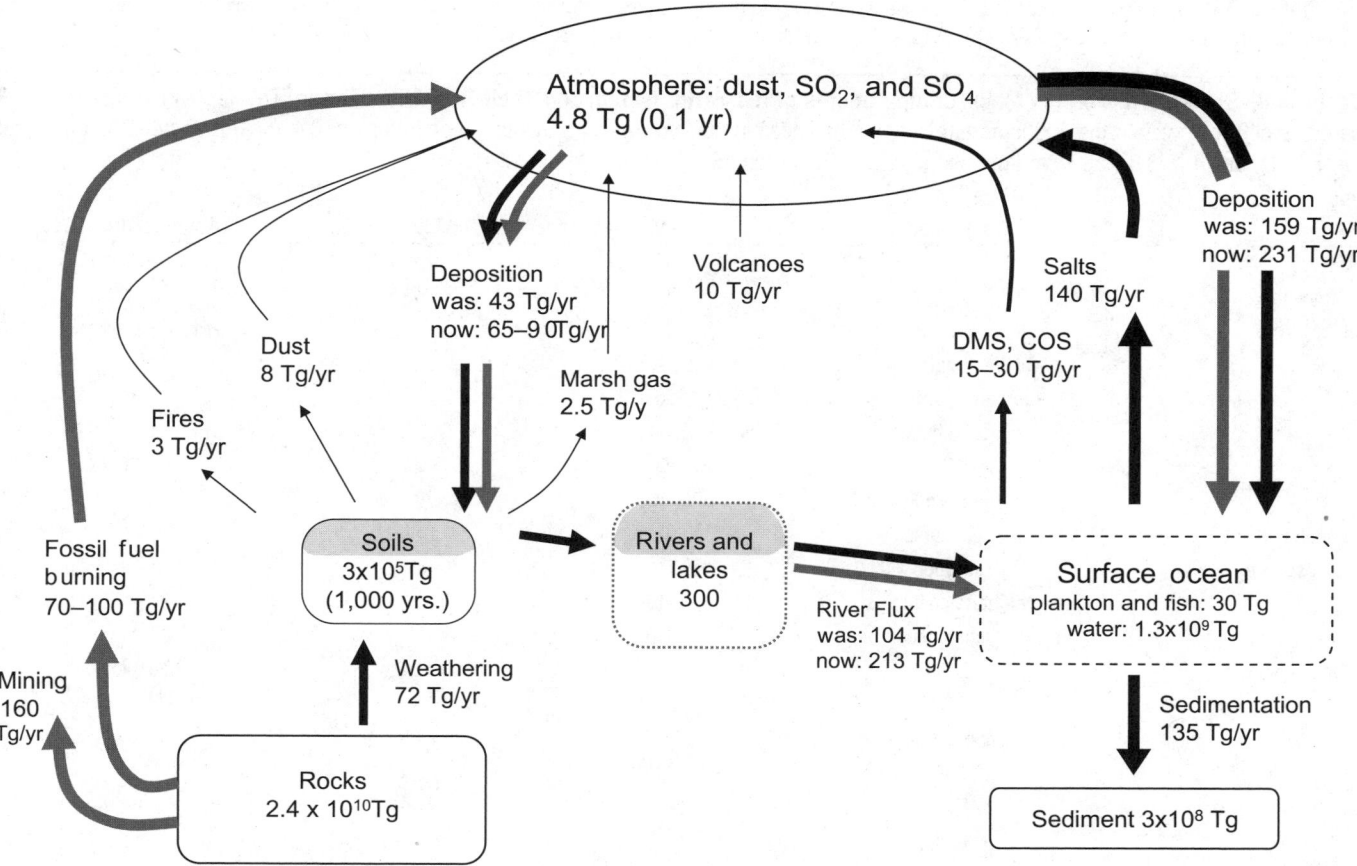

Figure 12.5. Main Pools and Fluxes in Global Sulfur Cycle. Pools are in TgS, fluxes in TgS/yr, and turnover times in years. The solid arrows represent the preindustrial cycle, while the grey arrows and boxes show the human additions, circa 2000. Emissions of S from burning fossil fuels and smelting ores are the main causes of increased S deposition, which accumulates in soils, leading to acidification of the soils and particularly of poorly buffered freshwater bodies draining from them. (Reeburgh 1997; Brasseur et al. 2003)

component of acid deposition in these regions. A perverse consequence of this is that global warming will increase by about 0.3 watts per square meter (about 10%) as the anti-greenhouse effects of sulfate aerosols diminish (Ramaswamy et al. 2001).

However, sulfur emission reduction is not widely practiced in the emerging industrial areas of the world: China, India, South Africa, and the southern parts of South America. A global assessment of acid deposition threats (Kuylenstierna et al. 2001), based on a combination of emission locations, wind transport patterns, and the buffering capacity of soils in the receiving regions, suggests that tropical ecosystems in the developing world are at high risk of acidification. (See Figure 12.6 in Appendix A.)

12.3.4 The Global Carbon Cycle

The global carbon cycle has been assessed recently and comprehensively by the Intergovernmental Panel on Climate Change (Prentice et al. 2001) because of its centrality to the issue of global climate change. This chapter will not repeat that work nor the discussion of C-climate interactions covered in Chapter 13 on air quality and climate. Suffice to say here that the global C cycle is currently out of balance (see Table 12.4), principally as a result of the burning of fossil fuels, but also due to the conversion of high C-density natural ecosystems, such as forests and grasslands, to lower C-density agroecosystems. It should be noted that the C cycle has been perturbed by about 13% relative to its preindustrial state, compared with figures of 100% or more for the N, P, and S cycles (Falkowski et al. 2000).

There is an important interconnectivity between the cycles of N, P, S, Fe, and Si and the C cycle (Mackenzie et al. 2002).

Human actions have significantly perturbed those cycles, and this has knock-on effects on the C cycle due to considerations of stoichiometry and the co-limitation or regulation of key processes. Mechanisms such as N fertilization and the sequestration of C in deep oceans through the "biological pump" contribute an uncertain but large proportion of the ~ 4 billion tons per year carbon storage service that the biosphere currently provides (Prentice et al. 2001). Some key processes, such as N and CO_2 fertilization of terrestrial ecosystems, are highly likely to reach saturation, possibly during this century (Scholes et al. 2000).

The global fixation of carbon through photosynthesis has been suggested as a general index of the health of both terrestrial and aquatic ecosystems, all other conditions of temperature, moisture, and nutrient supply being equal (Schlesinger 1997). Global indices of this type (such as the Normalized Differential Vegetation Index) are now available and show an increase in biomass in the high northern hemisphere latitudes, consistent with climate change, and no overall trend (but high interannual variability) in the sub-tropical deserts. They also reveal large areas of algal blooms in coastal areas, especially those that are semi-enclosed, such as the China Sea, indicating eutrophication.

12.4 Consequences of Changes to Nutrient Cycles

Alterations of nutrient cycling include situations of nutrient excess, leading to eutrophication of soils and water bodies, and nutrient deficiency linked to soil exhaustion and some specific natural situations in oceans.

Table 12.4. Stocks of Carbon in Major Compartments of the Earth System and Their Residence Times. The Table indicates change in carbon stocks per year in the 1990s as a result of imbalances in the cycle caused by human emissions of 6.4 ± .6 PgC from the burning of fossil fuels, and 1.4–3.0 PgC from land use change. (Reeburgh 1997; Prentice et al. 2001)

	Components	Stock	Residence Time	Accumulation Rate
		(PgC)	(years)	(Pg/year)
Atmosphere	CO_2	750	3–5	3.2 ± 0.2
Land biota	plants	550–680	50	land net uptake
Soil	peat	360	>10^5	1.4 ± 0.7
	inorganic carbonates	1220		
	microbial biomass	15–30	<10	
	POC	250–500	<10^2	
	amorphous polymers	600–800	10^2–10^5	
Lakes and rivers	sediments	150	10^{-1}–10^3	?
Lithosphere	kerogen	15 x 10^6	>>10^6	
	methane clathrates	11 x 10^3	–	
	limestone	60 x 10^6	–	
Ocean: surface	DOC	40	–	oceanic net
	POC	5	–	uptake 1.7 ± 0.5
	living biomass	2	10^{-1}–10^1	
Ocean: deep	DIC	38,000	~2 x 10^3	
	DOC	700	5 x 10^3	
	POC	20–30	10^1–10^2	
	sediments	150		

Notes: 1 Pg = 10^{15}g = 1 billion tons
Kerogen consists of coal, oil, gas, and other lower-grade fossil carbons such as lignite and oil shales.
Clathrates are a solid form of methane hydrates found at depth in ocean sediments.
DIC: dissolved inorganic C; DOC: dissolved organic C; POC: particulate organic C.

12.4.1 Nutrient Excess in Fresh and Marine Waters

As noted earlier, a major consequence of fertilizer inputs and atmospheric deposits, and of impairment of buffers and regulatory mechanisms at all scales, is eutrophication of aquatic systems—both fresh and saline waters. (See Chapters 19 and 20 for more on coastal and inland water systems.)

12.4.1.1 Eutrophication of Aquatic Ecosystems

Eutrophication is the fertilization of surface waters by nutrients that were previously scarce (Carpenter et al. 1999). In the 1960s it became obvious that a change was occurring in many lakes and reservoirs, especially in industrial countries, resulting from an increase in their nutrient load. This was a consequence of human activity, with increased inputs of urban and industrial wastewater and agricultural runoff containing mainly C, N, and P. The problem is now apparent in many coastal areas as well. Eutrophication is regarded as the most widespread water quality problem in many countries (OECD 1982; NRC 1992; Nixon 1995; Carpenter et al. 1998; Howarth et al. 2000). (See also Box 12.3.)

Eutrophication leads to many changes in the structure and function of aquatic ecosystems and thus the services they provide. The symptoms of eutrophication are well known: increase in phytoplankton, benthic, and epiphytic algae and bacterial biomass; shifts in composition to bloom-forming algae, which may be toxic or inedible; development of rooted macrophytes and macroalgae along the shores; anoxia (oxygen depletion) in deep waters; increased incidence of fish and shellfish mortality; decreases in water transparency; taste, odor, and water treatment problems; and in coastal areas, coral mortality. Such characteristics are detrimental to many water uses, including for drinking, fisheries, and recreation. Although possible, remediation measures are costly, and mostly consist of reducing the inputs of nutrients to tolerable levels.

Most attention has focused on P inputs rather that on N or C, because P is often the element limiting the growth of aquatic biota in temperate freshwater environments. Phosphorus is less abundant than N in fresh waters relative to plant needs, and its concentrations are reduced to very low levels by uptake during the growing season. It is therefore P that often regulates the extent of algal and other plant development in the aquatic environment.

BOX 12.3

Case Study of Eutrophication of Inland Waters: Lake Victoria, East Africa

Observations in 1990–91 in the center of Lake Victoria have been compared with data from 1960–61 (Hecky et al. 1994). The results indicated a stronger stratification, with less oxygen in a greater part of the deep zone of the lake. This eutrophication process results mainly from an increase of human activities in the watershed, with increased inputs of nutrients linked with urbanization, deforestation, and cultivation. Some shift in the phytoplankton population has also occurred. The increase in nutrients first led to an increase in diatoms. The resulting depletion in dissolved silicon promoted a shift in the diatom species and finally the establishment of blue-green alga (Cyanophycea) (Verschuren et al. 1998). Introduction of alien fish species such as *Lates niloticus* (Nile perch) and four species of *Oreochromis* (tilapia) also contributed to a strong modification of the trophic food web, with probable feedback on the phytoplankton community, as well as the fish community, with a strong decrease in the number of the endemic cichlids species (Lévêque and Paugy 1999).

A debate took place during the 1970s on the possible reduction or ban of P in detergents as a means of reducing the eutrophication of lakes. The pro-P argument relied on the small fraction of P load resulting from the detergents as compared with other urban or agricultural sources (Lee and Jones 1986). While phosphorus use has been totally banned in some countries (in Switzerland and some states of the United States), and partial restrictions occur in others, the use of high-P detergents is unrestricted in a number of other countries.

Agricultural production in some countries is not keeping pace with the increase in food demand. (See Chapter 8 for more on food production). Although in sub-Saharan Africa and Latin America a significant part of the required increase in cereal production could result from an increase in the cultivated area, the increase would transform land that is currently supplying other ecosystem services. Increase in food production is also possible from higher yields in existing agricultural areas (Pinstrup-Anderson 1999). This will require the development of more-sustainable practices making a better use of natural biological processes (Swift and Woomer 1994) and an increased use of fertilizers, which could be mitigated by an improvement in the efficiency of their use.

In Western Europe or North America, the quantity of fertilizer applied today is the same as it was in 1970, whereas the average yield of wheat, for example, has more than doubled. Assuming a slowdown in the growth of world population and crop production and an improvement in fertilizer use efficiency, it is forecast that total fertilizer use will have to increase from the present level of 140 million tons $N + P_2O_5 + K_2O$ to 167–199 million tons per year by 2030, meaning annual growth rates of 0.7–1.3%. This compares with an annual rate of increase over the past 30 years of 2.4% (FAO 2000). A significant part of the increase in P fertilizers may enter the aquatic reservoir if conservation practices and erosion control are not implemented on a very large scale.

12.4.1.2 Eutrophication and Carbon Sequestration

Elevated net primary production associated with eutrophication would seem to favor carbon sequestration. However, when recycling of C through decomposer chains prevails on sedimentation in rivers and lakes, these systems behave as sources of CO_2 since their CO_2 partial pressure (pCO_2) is higher than that of the atmosphere. In a survey of pCO_2 in the surface water of 1,835 lakes around the world, 87 % of the samples were supersaturated (mean pCO_2 : 1,034 (atm) (Cole et al. 1994). In highly polluted aquatic systems, the pCO_2 may reach higher values, such as 5,700 (atm in the upper estuary of the Scheldt River, which receives industrial and urban wastes from France, Belgium, and the Netherlands (Frankignoulle et al. 1996).

The net global C budget of inland waters has been approached from a different point of view. From their measurements on lakes, Cole et al. (1994) have estimated the potential release of CO_2 from lakes at about 0.14 billion tons of carbon per year, which is about half as large as riverine transport of organic and inorganic C to the ocean. This indicates that terrestrial systems, among which forests are usually considered as C sinks, export some organic matter that later contributes to CO_2 emission. From considerations of primary production, Dean and Gorham (1998) estimated that lakes are currently accumulating organic C at an annual rate of about 42 teragrams a year. Most of the C in all but the most oligotrophic of these lakes is primary production in the lakes themselves. The sediments of reservoirs accumulate an additional 160 teragrams annually, and peatlands contribute 96 teragrams annually. These three C pools collectively cover less than 2% of Earth's surface and constitute a C sink of about 300 teragrams a year.

12.4.1.3 Marine Dead Zones

Low oxygen conditions in coastal marine waters are primarily the result of enrichment in nitrogen with consequent enhanced growth of phytoplankton. These nitrogen-fed phytoplankton sink to the sea floor when they die, and the organic matter is regenerated by bacterial activity, consuming oxygen. At low oxygen concentrations, most marine life is unable to survive, leading to the designation of "marine dead zones." The size of these reaches up to 70,000 square kilometers (Brian et al. 2004), and they have been reported off South America, Japan, China, Australia, New Zealand, and the west coast of North America. The number of such areas has doubled every decade as more and more artificially produced nitrogen fertilizers are used in agriculture and as human population increases result in increased nitrogen containing sewage effluent.

The hypoxic conditions are seasonal in some regions, such as the Gulf of Mexico (fed by nutrients from the Mississippi River) and off the coast of Oregon, where low oxygen conditions have appeared in recent years. The Oregon events are related to upwelling conditions when north winds displace surface waters and deep waters rise as replacements. These deep waters have high nutrients, but in recent years have also had low oxygen content. Benthic populations, such as crabs, are quickly affected by such conditions and can provide useful indicators for early warning of low oxygen.

The occurrence of dead zones and their size appears to be a function of nitrogen inputs, which continue to grow in most parts of the world. Denitrification (which returns nitrate to the atmosphere as N_2 and N oxides) takes effect only when oxygen concentrations are already low and so becomes effective only after the dead zone phenomenon has occurred. In the future, this phenomenon may be as important as overfishing in the decline of fisheries. (See Chapter 18 for more on marine fisheries.)

12.4.2 Nutrient Deficiencies

A significant proportion of agricultural soils, mainly located in developing countries, are suffering nutrient deficiencies. Similar situations may occur in coastal and marine systems as a result of shortage of water flow from terrestrial systems, or naturally in the high nutrient, low chlorophyll zones, where deficiencies in Si and Fe limit primary production.

12.4.2.1 Agricultural Soils

The fertility of any soil will decline if the nutrient content of the harvest removed from the system (as grain, timber, livestock, and so on) exceeds the nutrient input from natural and anthropogenic sources. In general, the nutrient balances in the industrial world are positive, especially for N, as crops use less than half of the applied fertilizer, leading to the eutrophication problem just described. In large areas of South America (Wood et al. 2000) and Africa (Smaling et al. 1997; Sanchez 2002), on the other hand, the nutrient balance is negative, leading to declining soil fertility. In the case of South America, the magnitude of the imbalance appears to be decreasing as incomes rise and farmers can afford more fertilizer. In Africa, the cost of fertilizer to low-income farmers is usually prohibitive.

The situation is exacerbated by two factors: ecological features and farmers' perception of risk in many of the poorest developing regions. First, much of the agricultural population in nutrient-deficient areas lives on soils derived from basement rocks, on very old, stable land surfaces. The P and base cation content of these soils is inherently low; as a result, natural biological N fixation is also low. Because the soils are sandy and low in organic C, they lose N through leaching. Their low N status leads them to be burned frequently (since the grass that grows on them is too low in N for cattle to digest in the wintertime), causing further N loss. Nitrogen is the key component of protein, and it is precisely these areas that show a steady decline in per capita protein consumption, to levels well below the recommended daily intake. The same areas also show high levels of stunted growth associated with malnutrition (Scholes and Biggs, in press).

Where farmers perceive that there is a risk of not achieving an acceptable level of yield at harvest (perhaps because of drought) to cover their input costs, they are often unwilling to invest in fertilizers to replace nutrients like P and K removed in the harvested produce.

12.4.2.2 High Nutrient, Low Chlorophyll Regions of the Ocean

Large parts of the ocean are characterized by the presence of adequate N and P in the euphotic zone but low phytoplankton biomass and low primary and new production (Minas et al. 1986). The best known of these regions are between the coast of Ecuador out to the Galapagos Islands, the equatorial Pacific out to the dateline, the northeast Pacific, and portions of the Southern Ocean.

Open ocean fertilization experiments have shown the first of these to be limited by Fe (Coale et al. 1996). Responses to Fe additions have been observed in the Southern Ocean as well, but with limited effects on phytoplankton productivity and growth rates. The equatorial Pacific is chronically low in $Si(OH)_4$ with some secondary effects due to the relatively low Fe concentrations (Ku et al. 1995). The Southern Ocean north of the polar front has been identified as a low $Si(OH)_4$ region (Dugdale et al. 1995). The origin of this condition is now understood as the result of a seasonal drawdown of $Si(OH)_4$ by diatoms, which proceeds southward over the course of the austral summer. Drawdown of NO_3 is small compared to the uptake of $Si(OH)_4$, a condition that may be related to low Fe (Takeda 1998). However, unusually high $Si(OH)_4$ uptake by diatoms can result from other processes that slow their growth (Claquin et al. 2002).

The interest in high nutrient, low chlorophyll regions has been sparked by the possibility of increasing their productivity through fertilization, with relatively small quantities of Fe and/or Si. This has been suggested as an option for slowing the increase in atmospheric CO_2. However, the scientific understanding of the consequences of full-scale implementation of such an action remains insufficient for adequate assessment.

12.4.3 Threats to the Global Marine Nutrient Cycling System

Global marine nutrient cycling is beginning to be understood in spite of the complex interactions among physical, chemical, and biological processes that occur across the moving boundaries of water masses. Several processes require examination, as each may be a significant source of increased or decreased services in the future.

First, N fixation seems to be enhanced when Fe and P are added (Beherenfeld and Kolber 1999; Bidigare and Ondrusek 1996; Sanudo Wilhelmy et al. 2001; Karl et al. 1997, 2002). Since the expansion of the Sahara in the early 1970s, the dust load has increased nearly fourfold (Prospero and Nees 1986). Furthermore, Tegen and Fung (1995) have shown that about half the dust reaching the Equatorial Atlantic is due to disturbed soil conditions, and this dust contains more Fe than undisturbed desert dust. Increased N availability is likely to increase C fixation and its further sequestration in sediments and would therefore constitute a positive contribution to mitigation of climate change.

Second, denitrification is considered a useful process in the elimination of part of the N burden in estuaries and coastal areas, whereas it may limit production in N-limited areas, as N fixation is severely limited by Fe and Si availability (Naqvi et al. 2000; Codispoti and Christensen 1985). This is thought to be the current state of the world's oceans. If so, this would imply that oceanic biological processes (as distinguished from the physical dissolution of CO_2 into water due to increased partial pressures of CO_2 in the atmosphere) are actually adding C to the atmosphere instead of removing it (e.g., Falkowski 1997).

Third, increased NH_4^+ from terrestrial effluents inhibits the ability of diatoms to use NO_3^-. In San Francisco Bay, a long-term decline in productivity has been ascribed to increased NH_4 concentrations of anthropogenic origin (Karl et al. 2001). However, the Anamnox process (oxidization of NH_4 into free N by reaction with nitrite) is likely to mitigate the problem in anoxic areas (Devol 2003).

Fourth, decreased productivity is linked to reduced inputs of Si and Fe from terrestrial sources. Changes in the input of $Si(OH)_4$ from rivers as a result of damming and changes in farm practices may also have an impact on the productivity of marine diatoms (Leynaert et al. 2001). When $Si(OH)_4$ concentrations fall below about 2 μM, other algal groups are able to outcompete diatoms, potentially leading to the dominance of less desirable or even toxic bloom species.

12.5 Monitoring and Assessment

Given the large diversity of mechanisms, pools, and fluxes involved in nutrient cycling, decision-makers face a large number of different, although closely related issues. (See Table 12.5.) For example, eutrophication of fresh water, which is mainly a consequence of excessive nutrient inputs and mismanagement of wastes at the landscape scale, is aggravated by atmospheric N deposits and by the reduction of plant cover and biodiversity at the plot and landscape scales. Once the pools, fluxes, or mechanisms relevant to the particular issue have been identified, adequate indicators or descriptors of their sizes or intensity can be identified and monitored. (See Table 12.6.)

Although alterations in nutrient cycling are generally observed at scales from plot to landscape and region, the mechanisms involved are generally operating at much smaller scales. Efforts to improve the general quality of nutrient cycling could improve from fine-tuning the inputs to the needs of cultivated plants as much as possible, in order to limit the risk of leakage from terrestrial ecosystems to groundwater and to freshwater and marine systems, and through greater attention to the state of the various buffering systems described earlier.

In practice, the best indicators of nutrient cycling efficiency will be measured at the plot level (diversity and cover of plants and indicators of quality based on soil fauna communities) and at the landscape level (density and distribution of buffering zones such as riparian forests, diversity of land use types in mosaics). Links between diversity at these two scales and adequate nutrient cycling have been observed in a number of studies (Wardle et al. 1999; Niklaus et al. 2001; Reich et al. 2001; Tilman et al. 2001). Cover crops, diversification of land use types in time (rotations) and space, no-tillage practices at the plot scale, and the use of hedgerows, riparian forests, and wetlands have all been shown to improve the tight cycling of nutrients (Nair 1993; Palm 1995; Entry and Emmingham 1996; Inamdar et al. 1999; Fassbender et al. 2000).

However, with a few exceptions, the mechanisms that link measurements of landscape features and plant and invertebrate diversities and so on to nutrient cycles have not been established satisfactorily, nor have thresholds for degradation of their diversity been established. This is in part due to a lack of extensive datasets that would allow statistical relationships to be established. However, the apparently large time lag between the loss of biological diversity and the impairment of soil functions might obscure significant statistical relationships (Lavelle et al. 2004).

12.6 Implications of Altered Nutrient Cycles for Human Well-being

Nutrient cycling and soil or water fertility play key roles in terrestrial and aquatic systems, and these roles benefit many segments of society. Substantial human benefits are derived directly and indirectly from nutrient cycling and fertility services.

Principally, millions of people earn their living from the production of commodities, such as food and other products, yielded by terrestrial and aquatic systems. The livelihoods of many rural households in developing countries in particular are highly dependent on direct harvesting of wild foods and non-food products. (See Chapters 18 and 26 for more on livelihoods based on food production.) More generally, benefits are derived directly from the use of noncommercialized ecosystem products generated by the fertility service of terrestrial and aquatic systems, such as biodiversity and its services. Thus nutrient cycling and fertility are essential for supporting the supply of farmed and wild products and the benefits people derive from their consumption and use.

Various approaches and methods have been developed to value nutrient cycling and fertility services. These methods can be grouped into two broad categories: benefit-based and cost-based approaches. (See Chapter 2.) The most commonly used benefit-based method is the production function approach that measures the marginal contribution of fertility to the total value of generated products after accounting for the marginal contribution of all other inputs and factors used in the production process. Examples of studies that have used such an approach to quantify relationships include yield reduction (productivity loss) versus land degradation (Aune and Lal 1995; Lal 1995). The most researched land degradation factors are soil erosion and nutrient mining in the United States (Hertzler et al. 1985; Burt 1981; Pierce et al. 1984), Mali (Bishop and Allen 1989), Zimbabwe (Grohs 1994), and Ethiopia (Sutcliffe 1993).

In principle, this method is applicable to both commercialized and noncommercial products of terrestrial and aquatic systems. However, it is relatively easier to apply to situations where products are commercially exploited and marketed and where the value of products directly harvested from the wild for noncommercial use is usually calculated using prices and values of similar products exchanged in the market. But the accuracy and realism of figures derived from the production function approach is questionable. A fundamental assumption of this method is that the market prices of agricultural outputs reflect their marginal costs of production; in most economies, market distortions (such as U.S. and European farming subsidies) ensure that market prices are not equal to marginal costs.

Among the most commonly used cost-based approaches is the replacement cost method. This uses the value of commercial fer-

Table 12.5. Problems in Ecosystem Functioning Linked to Nutrient Cycling Dysfunctions and Indicators to Assist Diagnostic.
Numbers of scales in the first column relate to classification of scales made in section 12.2.3.

Scales	Dysfunction Observed	Possible Diagnostic	Indicators
Terrestrial			
1–2: Micro	plant growth limited by unexpected nutrient deficiencies	microbial diversity modified	diversity in microfoodwebs communities (micro- and meso-fauna)
		beneficial or key microorganisms (N-fixers; mycorrhizae; white rot fungi) absent	nodulation in legumes
	fungal or bacterial diseases	impairment of soil structure preventing optimal function of foodweb controls	Maturity Index (Bongers 1990)
3–4 Ecosystem	soil compaction and erosion	abundance and functional diversity of plants and soil natural tillers (earthworms, termites, ants) insufficient to allow adequate regulation of microbial communities and soil structure	synthetic indices of soil quality (Vélasquez 2004; Ruiz-Camacho 2004)
	deficient water infiltration		
	C losses		
5: Landscape	groundwater and adjacent freshwater systems loaded with nitrates and phosphates	leakage due to dysfunction of regulations at scales 1—4 not mitigated by the presence of suitable buffer zones for nutrient absorption and refugia for useful soil fauna	landscape metrics given by analysis of composition and structure of landscape
	erosion	insufficient C supply for faunal activities	
6: Global	soil nutrient depletion at large scales	unbalanced nutrient budgets for terrestrial biomes inadequate management options	erosion, indices of soil quality crop production
		lack of formation and financial support	indicators of human development
Fresh Water			
1–2: Micro	algal blooms	eutrophication (nutrient excess loading)	OECD trophic levels (based on total P, chlorophyll, transparency in lake water)
3–4: Ecosystem	changes in fish communities	aquatic habitat destruction; increased input of suspended solids from land erosion	in rivers, changes in water regime; natural habitat diversity; invertebrates and fish communities structure
	fish kills	eutrophication; excess nutrient and organic matter load, pollution by pesticides or other humanmade substances	low oxygen in rivers and anoxia in deep layers of lakes
			in lakes, changes in hydrological seasonality
5: Landscape	decrease in inland fisheries	decrease in water quality and in self-epuration capacity; increased control on river flows (dam construction, canals, regulated river, etc.)	decrease in aquatic biodiversity and in habitat diversity changes in riparian vegetation
		pollution (nutrients and chemicals)	total N and P in water
6: Global	decrease in availability of fresh water (drinking water)	alteration of the self-epuration capacity	total N and P in water
Marine and Coastal			
1–2: Micro	algal blooms fish kills	changes in N-cycling microbial communities due to sewage effluents (nutrients and heavy metals)	changes in microbial communities
3–4: Ecosystem	algal blooms fish kills	organic matter loading producing hypoxia	low oxygen concentration
	decreases in productivity; decreased fish catches	reduced water and nutrient flows from terrestrial ecosystems (damming, irrigation, industrial use of water)	low chlorophyll concentration
5: Landscape	eutrophication (algal blooms fish kills)	habitat disturbance (landfill, mangrove/seagrasses clearing/disruption of estuaries)	habitat structure in coastal environments
		impaired regulation of populations of primary producers by fishes due to overfishing	decreasing trend in fish catches
6: Global	eutrophication in non-directly affected areas	nutrient overloading of major current systems	nutrient load in water

Table 12.6. Biophysical Structures That Allow Regulation of Nutrient Cycling at Different Scales, Processes Involved, and Indicators of Their Functionality

Scale	Biophysical Structures (self-organizing systems)	Process	Indicators of Service	Human Alterations	Indicator of Alteration
Micro	soil + sediments: microaggregates waters: water column	micro-foodweb effects	soils: maturity index (Bongers 1990) all: diversity in microfood-webs communities (micro- and meso fauna)	organic and chemical contamination	fertilizer and pesticide application
Ecosystem	biogenic structures created by roots and bioturbators	habitat creation modified access to resources regulation of hydraulic properties and organic matter dynamics	physical structure of soils and sediments bioindicators of soil quality (Ruiz-Camacho 2004, Vélasquez 2004) bioindicators of fresh water quality (AFNOR 1992)	organic and chemical contamination physical disruption (tillage)	fertilizer and pesticide application description of tillage and other practices
	herbivory, foodweb effects	acceleration of nutrient release from living plant biomass CH$_4$ production in cattle guts	nutrient balances at plot level monitoring at plot level	conversion of natural and crop systems to pastures	grazing pressure
	saprophagy, comminution incorporation in stable aggregates	changes in decomposition rates organic matter sequestered	monitoring decomposition rates invertebrate communities (bioindicators of soil quality) distribution of organic matter in particle size fractions	organic and chemical contamination	land use practices
	quality of organic matter produced by primary producers	influence rates and pathways of nutrient cycling and release	chemical analysis of plant material: C: nutrient ratios, phenolic compounds	change in plant cover	N—fixing primary producers; ligneous vs. herbaceous
	texture and structure of soils and sediments; variations in ecosystem and landscape	regulation of nutrient storage and release	texture, nature of clay minerals, effective calcium carbonate porosity, aggregation	fertilization sedimentation and dust deposits physical disruption, compaction of soils and sediments	external inputs (fertilizers, deposits, importation) texture, formation of dunes (desert ecosystems) tillage, grazing pressure

(continues over)

tilizers needed to maintain stable levels of productivity (or natural fertility) as a proxy for the value of this ecosystem service. However, the replacement cost method generally underestimates the marginal contribution of natural fertility to production, as it assumes perfect substitutability between natural and manufactured fertility. It is known that this is not the case, although economic theory struggles to suggest the replacement value of natural processes for which no human-made process is a full substitute (see Freeman 1993 for a further discussion of this issue).

As an illustrative approximation, the nitrogen made available through cycling and nonagricultural biological nitrogen fixation on land totals around 5,000 teragrams a year. Hence, at a global average price of nitrogen fertilizer ($1,100 per ton of nitrogen), its annual replacement value is $5.5 trillion. Even if only the cy-cling on agricultural fields is considered, the total annual replacement value is around $1.1 trillion. Similarly, the estimate for phosphorus is 60 teragrams a year on land, with a annual replacement value of $240 billion, and of $48 billion for agricultural lands only. As marine systems are not deliberately fertilized, such an approach does not apply, but nutrient cycling is essential in support of coastal and ocean fisheries.

More-recent approaches to the valuation of the nutrient cycling service, such as the Habitat Equivalent Analysis, although they have not yet been applied in a comprehensive manner, would certainly provide much lower figures (Huguenin et al. in press). Other measures of the value of soil nutrients stocks include user cost or resource rent estimated as the dynamic price or production value lost to future generations from a unit of the nutri-

Table 12.6. *continued*

Scale	Biophysical Structures (self-organizing systems)	Process	Indicators of Service	Human Alterations	Indicator of Alteration
Landscape	vertical distribution in soils and water columns	transfers and losses	monitoring concentrations along soil profiles and water columns and sediments	intensification of agriculture and forestry	changes in landscape composition and structure (remote sensing)
				urbanization	
	mosaic of ecosystems (patches, ecotones, and buffering zones)	buffering zones regulating transfer of nutrient between domains (terrestrial to aquatic)	composition of landscape/seascape as assessed by remote sensing techniques (fragmentation, homogeneity, shape etc. indices)	pollution	fertilizer application
					detergent use
					N—fixation in agriculture and forestry
		transfer from continent to rivers and oceans	eutrophication		population density
			erosion rates, stream discharge		socioeconomic indicators (education, wealth)
Global	homeostatic interactions among atmosphere, lithosphere, and biosphere on Earth	transport across oceans (conveyer belts) and continents (dust depositions)	remote sensing measurement of dust clouds	climate change	global climatic parameters
			ocean stream monitoring		

ents' stock depleted today (Hertzler et al. 1985; Brekke et al. 1999; Nakhumwa 2004).

Nutrient cycling and fertility also generate noncommercial but nevertheless important benefits for various groups in society. They increase the risk-buffering capacity of agroecosystems—that is, the capacity of these systems to respond to environmental, climatic, and economic risks by adapting to these stresses without decreasing their productive capacities. The specific ecological mechanisms through which this occurs have not been fully documented, but it appears that belowground biodiversity plays an important role in this regard (Lavelle et al. 2004). Small-scale farmers in developing countries are particularly exposed and vulnerable to these forms of risk (Izac 2003). It is particularly difficult, both conceptually and empirically, to calculate the economic value of this enhanced capacity to manage risks. This enhanced capacity can also be thought of as a contribution to the long-term maintenance of the productive capacity of a system or to its sustainable use, so future generations will also benefit from this.

There are other issues related to the negative effects of service impairment that occur outside the immediate area where fertility is declining. For instance, soil erosion is particularly important and creates on-site and off-site (externality) effects, and economic evaluation of soil erosion is therefore not complete without the inclusion of costs and benefits for both on- and off-site effects. These may include the cost of dredging silted dams and of restoring impaired inland and coastal aquatic systems, in addition to the direct cost of fertilizer applications. Generally only anecdotal evidence is available concerning the costs of impaired aquatic systems, although estimates of the benefits and costs of nutrient loading for specific fisheries have been provided (UNEP 2002)

The historical changes in nutrient cycling and fertility analyzed in this chapter affect human well-being in two different ways. The first is directly, through changes in the perceived benefits or values of these ecosystem services from an anthropogenic perspective. Some of these changes in well-being are relatively easy to quantify in monetary terms; others are much more difficult to assess in this way, but nevertheless affect extremely important dimensions of human livelihoods. The second way in which human well-being is affected by these changes is through specific benefits that are currently not explicitly valued by society but that may have significant future ecological and welfare implications, such as the capacity of organisms involved in the marine nitrogen cycle to respond to global environmental changes.

In the case of terrestrial systems, negative trends in fertility and nutrient cycling have historically been remedied by greater and greater applications of fertilizers (inorganic and organic). However, such remedies can mask the effects of these trends only to a degree and can also result in increased costs of production. World fertilizer consumption increased rapidly in the 1970s and 1980s but slowed down in the 1990s, partly due to environmental legislation in industrial countries. (See Chapter 26.)

In the case of coastal and marine systems, high anthropogenic nutrient loads have resulted in anoxia and loss of fisheries, changes in the composition of inorganic nitrogen, declining productivity in some estuaries, and changes in species composition of the phytoplankton. (See Chapter 19.) These undesirable effects are directly related to agricultural fertilization and animal husbandry practices and to human population increases. Although many industrial countries are able to mitigate some of these effects through sewage treatment practices and in some cases control of agricultural practices, such mitigation is not universal.

Many countries have coped with the decreasing profitability of agriculture, caused in part by the necessity to supplement the natural nutrient cycle with fertilizers, by providing direct public subsidies to primary producers. Although low crop productivity across sub-Saharan Africa cannot be attributed entirely to increased costs of production, it is clear that deterioration in ecosystem services related to fertility has had a direct impact on productivity and subsequent rural livelihoods. Claims that fertility depletion is the single most important cause of poverty in sub-

Saharan Africa have not been supported by empirical evidence and failed to consider the myriad of other factors that also contribute to the highest rural poverty in the world. It is, however, undeniable that a net loss in nutrient cycling services that are essential for primary production has a negative effect on the financial situation of the rural poor.

One consequence of the changes documented in this chapter is the loss of sustainability, or resilience, in production systems. This is mainly, but not uniquely, through a loss of nutrient stocks, sequestered carbon, and biodiversity. This consequence is particularly important for future generations of farmers and particularly so in cases where threshold levels of irreversibility seem to have been reached. In these cases, this generation's running down of natural capital is either increasing costs of production for future generations of farmers or, in some cases, is closing off future production options altogether.

References

AFNOR (Association Française de Normalisation), 1992: Détermination de l'indice biologique global normalisé (IBGN). NF T 90–350, Paris.

Aune, J.B. and Lal, R., 1995: The Tropical Soil Productivity Calculator. A model for assessing effects of soil management on productivity', in R. Lal and B.A. Stewart, eds., *Soil Management: Experimental Basis for Sustainability and Environmental Quality.* Advances in Soil Science, Boca Raton: CRC Press, Lewis Publishers, pp. 499–520.

Barber, R.T., M.P. Sanderson, S.T. Lindley, F. Chai, J. Newton, et al., 1996: "Primary productivity and its regulation in the equatorial Pacific during and following the 1991–1992 El Nino." *Deep Sea Res II* **43**: 933–969.

Behrenfeld, M.J., and Z.S. Kobler, 1999: Widespread iron limitation of phytoplankton in the South Pacific Ocean. *Science,* **283**:840–843.

Bennet, E.M., S.R. Carpenter and N.F. Caracao, 2001: Human impact on erodable phosphorus and eutrophication: a global perspective. *Bioscience* 51 227–234.

Berc, J., R. Lawford, J. Bruce, L. Mearns, D. Easterling, et al., 2003: *Conservation Implications of Climate Change: Soil Erosion And Runoff From Cropland: A Report from the Soil and Water Conservation Society.* Soil and Water Conservation Society, Ankeny, Iowa.

Bernier, N., and J.F. Ponge, 1994: Humus form dynamics during the sylvogenetic cycle in a mountain spruce forest. *Soil Biol. Biochem.* **26**:183–220.

Bidigare, R.R. and M.E. Ondrusek, 1996: "Spatial and temporal variability of phytoplankton pigment distributions in the central equatorial Pacific Ocean." *Deep Sea Res Part 2* **43(4–6)**: 809–834.

Billen, G. and J. Garnier, 1999: Nitrogen transfers through the Seine drainage network: a budget based on the application of the 'Riverstrahler' model. *Hydrobiologia* **410**: 139–150.

Bishop, J. and J. Allen, 1989: *The On-Site Costs of Soil Erosion in Mali,* Working Paper No.21. The World Bank, Environmental Department, Washington, DC.

Bohn, H.L., B.L. McNeal, and G.A. O'Connor, 1979: *Soil Chemistry.* John Wiley & Sons.

Bongers, T., 1990: The maturity index: an ecological measure of environmental disturbance based on nematode species compostion. *Oecologia* **83**:14–19.

Brasseur, G.P., P. Artaxo, L.A. Barrie, R.J. Delmas, I. Galbally, et al., 2003: An integrated view of the causes and impacts of atmospheric changes. In: Brasseur, G.P., R.G. Prinn, A.A.P. Pszenny (eds.) *Atmospheric Chemistry in a changing world.* IGBP Series, Springer: Berlin pp. 207–230.

Brekke, K.R., Iversen, V., and Aune, J.B., 1999: Tanzania's Soil Wealth. *Environment and Development Economics* **4(1999),** pp 333–356.

Broecker, W.S. and T.-H. Peng, 1982: *Tracers in the Sea.* Eldigeo Press, Palisades, New York.

Buesseler, K.O., 1998: The decoupling of production and particulate export in the surface ocean *Global Biogeochem. Cycles.* **12**: 297–310.

Burt, O.R., 1981: 'Farm Level Economics of Soil Conservation in the Palouse Area of the Northwest.' *American Journal of Agricultural Economics* **63(1):** 83–92.

Casciotti, K.L., D.M. Sigman, and B.B. Ward, 2003: Linking diversity and stable isotope fractionation in ammonia-oxidizing bacteria. *Geomicrobiology Journal* **20**:4.

Cadisch, G., and K. E. Giller, eds. 1997: *Driven by Nature.* CAB International, Wallingford, U.K.

Carpenter, E.J., J. Montoya, J. Burns, M. Mulholland, A. Subramanian, and D.G. Capone, 1999: Extensive bloom of N_2 fixing symbiotic association in the tropical Atlantic Ocean. *Marine Ecology Progress Series.* **185**:273–283.

Chauvel, A., M. Grimaldi, E. Barros, E. Blanchart, T. Desjardins, M. Sarrazin, and P. Lavelle 1999: Pasture degradation by an Amazonian earthworm. *Nature* **389**:32–33.

Cinderby, S. H.M. Cambridge, R. Herrera, W.K. Hicks, J.C.I. Kuylenstierna, F. Murrary, and K. Olbrich, 1998: *Global Assessment of Ecosystem Sensitivity to Acidic Deposition.* Stockholm Environment Institute, Sweden.

Claquin, P., V. Martin-Jezequel, J.C. Kromkamp, M.J. Veldhuis, and G.W. Kraay, 2002: "Uncoupling of silicon compared with carbon and nitrogen metabolisms and the role of the cell cycle in continuous cultures of Thalasiosira pseudonana (Bacillariophyceae) under light, nitrogen and phosphorus control." *J.Phycol.* **38**: 922–930.

Coale, K., K.H. Coale, K.S. Johnson, S.E. Fitzwater, R.M. Gordon, et al., 1996: "The IronEx-II experiment produces massive phytoplankton blooms in the Equatorial Pacific." *Nature* **383**: 495–501.

Codispoti, L. A., and J. P. Christensen, 1985: Nitrification, denitrification and nitrous oxide cycling in the eastern tropical South Pacific Ocean. *Mar. Chem.* **16**: 277–300.

Cole, J.J., N.F. Caraco, G.W. Kling, and T.K. Kratz, 1994: C dioxide supersaturation in the surface waters of lakes. *Science* **265**: 1568–1570.

Dean, W.E. and Gorham, E., 1998: Magnitude and significance of carbon burial in lakes, reservoirs, and peatlands. *Geology,* Vol. **26**, no. **6**, pp. 535–538.

De Master D.J., 2002: The accumulation and cycling of biogenic silicon in the Southern Ocean: revisiting the marine silicon budget. *Deep-Sea Research Part II,* **49**: 3155–3168.

DeRuiter, P. C., J. A. Vanveen, J. C. Moore, L. Brussaard, and H. W. Hunt. 1993: Calculation of nitrogen mineralization in soil food webs. *Plant and Soil* **157**:263–273.

Devol, A. H. 2003: Solution to a marine mystery. *Nature* **422**: 575–576.

Dubroeucq, D., D. Geissert, I. Barois, and M-P. Ledru, 2002: Biological and mineral features of Andisols in the Mexican volcanic highlands. *Catena* **49**, 183–202.

Duce, R.A., 1995: Sources, distributions, and fluxes of mineral aerosols and their relationship to climate. In: Charlson, R., Heintzenberg, J. (Eds.), *Aerosol Forcing of Climate.* Wiley, New York, pp. 43–72.

Dugdale, R.C., 1976: Nutrient Cycles, p. 467, *In* J. J. Walsh, ed. *The Ecology of the Seas.* W.B.Saunders, Philadelphia.

Dugdale, R.C., F.P. Wilkerson, and H.J. Minas, 1995: The role of a siliconte pump in driving new production. *Deep-Sea Res. (I Oceanogr. Res. Pap.)* **42**:697–719.

Entry, J. A., and W. H. Emmingham, 1996: Nutrient content and extractability in riparian soils supporting forests and grasslands. *Applied Soil Ecology* **4**:119–124.

Falkowski, P.G., 1997: Evolution of the nitrogen cycle and its influence on the biological sequestration of CO_2 in the ocean. *Nature* **387**: 272–275.

Falkowski, P., R.J. Scholes, E. Boyle, J. Canadell, D. Canfield, et al., 2000: The global C cycle: A test of our knowledge of Earth as a system. *Science* **290**, 291–296.

FAO (Food and Agriculture Organization), 2000: *Fertilizer Requirements in 2015 and 2030.* Rome.

Fassbender H.W., L. Alpizar, J. Heuveldop, H. Fölster, and G. Enrique, in press: Modeling agroforestry systems of cacao (*Theobroma cacao*) with laurel (*Cordia alliodora*) and poro (*Erythrina poeppigiana*) in Costa Rica. III. Cycles of organic matter and nutrients. *Agroforestry Systems:* 49–62.

Fortescue, J.A.C., 1980: *Environmental Geochemistry: a Holistic Approach.* Springer-Verlag,, New York, NY.

Frankignoulle, M., I. Bourge, and R. Wollast, 1996: Atmospheric CO_2 fluxes in a highly polluted estuary (the Scheldt). *Limnol. Oceanogr.* **41**: 365–369.

Franzluebbers, A.J., J.A. Stuedemann, H.H. Schomberg, and S.R. Wilkinson, 2000: Soil organic C and N pools under long-term pasture management in the Southern Piedmont USA. *Soil Biology and Biochemistry* **32**:469–478.

Frink, C.R., P.E. Waggoner, and J.H. Ausubel, 2001: Nitrogen on the Land: Overcoming the Worries. Lifting fertilizer efficiency and preserving land for non farming. *Pollution Prevention review,* **11(3):** 77–82

Freeman, A.M. III, 1993: *The Measurement of Environmental and Resource Values* (1st and 2nd eds.). Resources for the Future.

Fung, I.Y., S. Meyn, and I. Tegen, 2000: Iron supply and demand in the upper ocean. *Global Biogeochemical Cycles* **14**, 281–295.

Galloway, J.N. 2003: Acid deposition: S and N cascades and elemental interactions. In *Interactions in the major Biogeochemical Cycles.* (eds.) Melillo, J.M., Field, C.B. and Moldan, B. **SCOPE 61,** Island Press, Washington. pp 259–272.

Garnier, J., J. Némery, G. Billen, and S. Théry, 2004: Nutrient dynamics and control of eutrophication in the Marne River system: modelling the role of exchangeable phosphorus. *Journal of Hydrology.*

Grantham, B.A. F. Chan, K.J. Nielsen, D.S. Fox, J.A. Barth, et al., 2004: Upwelling-driven nearshore hypoxia signals ecosystem and oceanographic changes in the northeast Pacific. *Nature* 429, 749–754

Green, P., C.J. Vörösmarty, M. Meybeck, J. Galloway, and B.J. Peterson, 2004: Pre-industrial and contemporary fluxes of nitrogen through rivers: A global assessment based on typology. *Biogeochemistry,* **68(1),** 71–105.

Grohs, F., 1994: Economics of Soil Degradation, Erosion and Conservation: A Case Study of Zimbabwe. Wissenschaftsverlag Vauk Kiel KG, Kiel, Germany.

Haycock N.E., G. Pinay, C. Walker, 1993: Nitrogen retention in river corridors: European perspective. *Ambio* 22:340–346.

Hecky, R.E., F.W.B. Bugenyi, P. Ochumba, J.F. Talling, R. Mugidde, et al., 1994: Deoxygenation of the deep water of Lake Victoria, East Africa. *Limnol. Oceanogr.* **39**: 1476–1481.

Hertzler G., C.A. Ibanez-Meier, and R.W. Jolly, 1985: User Costs of Soil Erosion and their Effect on Land Agricultural Prices. Costate Variables and Capitalised Hamiltonian. *American Journal of Agricultural Economics,* **67(5)** December: 948–1009.

Howarth, R.W., D. Anderson, J. Cloern, C. Efring, C. Hopkinson, et al., 2000 Nutrient pollution of coastal rivers, bays and seas. *Issues in Ecology* **7**: 15.

Huguenin, M.T., C.G. Legett, and R.W. Paterson, in press: Economic valuation of soil fauna. *European Journal of Soil Biology.*

Hutchinson, G.E., 1957: A treatise on limnology. 1. *Geography, Physics and Chemistry.* John Wiley and Sons, New York, 1015 p.

Inamdar, S.P., R.R. Lowrance, L.S. Altier, R.G. Williams, and R.K. Hubbard, 1999: Riparian Ecosystem Management Model (REMM): II. Testing of the water quality and nutrient cycling component for a Coastal Plain riparian system. *Transactions of the Asae* 42:1691–1707.

Izac, A-M.N., 2003: Economic aspects of soil fertility management and agroforestry practices. In, Trees, Crops and Soil Fertility. *Concepts and Research Methods,* eds. G. Schroth and F. L. Sinclair, pp. 13–38. CABI, Wallingford, UK.

Jacinthe, P.A., and R. Lal, 2001: A Mass Balance Approach to Asssess CO_2 Evolution During Erosional Events. *Land Degradation & Development* 12:329–339.

Jenkinson D S., and Rayner J. H., 1977: The turnover of soil organic matter in some of the Rothamsted classical experiments. *Soil Science* **123**:298–305.

Karl, D.M., R. Letelier, L. Tupas, J. Dore, J. Christian, and D. Hebel, 1997: The role of nitrogen fixation in biogeochemical cycling in the subtropical North Pacific Ocean. *Nature* **388**: 533–538

Karl. D.M., R R. Bidigare, and R.M. Letelier, 2001: Long-term changes in plankton community structure and productivity in the North Pacific Subtropical Gyre: The domain shift hypothesis. *Deep-Sea Research II* 48: 1449–1470.

Karl, D., A. Michaels, B. Bergman, D. Capone, E. Carpenter, et al., 2002: Dinitrogen fixation in the world's oceans. *Biogeochemistry* 57/58: 47–98.

Krug, A., 1993: Drainage history and land use patterns of a Swedish river system. Their importance for understanding N and P load. *Hydrobiologia* **251**:285–296

Ku, T.L., S. Luo, M. Kusakabe, and J.K.B. Bishop, 1995: 228Ra-derived nutrient budgets in the upper equatorial Pacific and the role of "new" silicate in limiting productivity. *Deep Sea Res II* 42: 479–497.

Kuylenstierna, J.C.I., H. Rodhe, S. Cinderby, and K. Hicks, 2001: Acidification in developing countries: ecosystem sensitivity and the critical load approach on a global scale. *Ambio* 30, 20–28.

Lal, R., 1995: Erosion-Crop Productivity Relationship for Soils of Africa. *Soil Sci. Soc. Am. J.* 59: 661–667.

Lal, R., 2001: Soil Conservation for C-sequestration, *In* D. E. Stott, et al., eds. Sustaining the Global Farm. Selected Papers from the 10th International Soil Conservation Organization Meeting held May 24–29, 1999 at Purdue University and the USDA-ARS National Soil Erosion Research Laboratory.

Lal, R., 2003: Soil Erosion and the global C budget. *Environment International* 29:437–450.

Lavelle, P. and A.V. Spain, 2001: *Soil Ecology.* Kluwer Scientific Publications, Amsterdam, 691p.

Lavelle, P., D. Bignell, M. Austen, P. Giller, G. Brown, et al., 2004: Vulnerability of ecosystem services at different scales: role of biodiversity and implications for management. In D.H. Wall (ed.), *Sustaining Biodiversity and Functioning in Soils and Sediments.* Island Press, Washington, DC.

Lee, G.F. and R.A. Jones, 1986: Detergent bans and eutrophication. *Environmental Science and Technology,* **20,**330–331.

Leynaert, A., P. Treguer, C. Lancelot, and M. Rodier, 2001: Silicon limitation of biogenic silica production in the Equatorial Pacific. *Deep Sea Res I* 48: 639–660.

Lévêque, C. and Paugy, D., 1999: Les poissons des eaux continentales africaines. Diversité, écologie, utilisation par l'homme. Paris, IRD Editions, 521 p.

Lucas, Y., F.J. Luizâo, A. Chauvel, J. Rouiller, and D. Nahon, 1993: The relation between biological activity of the rain forest and mineral composition of the soils. *Science* **260,** 521–523.

Mackenzie, F.T., L.M. Ver, and A. Lerman, 2002: Century-scale nitrogen and phosphorus controls of the C cycle. *Chemical Geology,* **190**: 13–32.

Mahowald, N., K.E. Kohfeld, M. Hansson, Y. Balkanski, S. Harrison, et al., 1999. Dust sources and deposition during the Last Glacial Maximum and current climate: a comparison of model results with paleodata from ice cores and marine sediments. *Journal of Geophysical Research* **104**, 15895–16436.

Mando, A., and R. Miedema, 1997: Termite-induced change in soil structure after mulching degraded (crusted) soil in the Sahel. *Appl Soil Ecol* 6:241–249.

Martin, J., 1992: Iron as a limiting factor in oceanic productivity, *in* Falkowski and Woodhead (eds), *Primary Productivity and Biogeochemical Cycles in the Sea.* Plenum Press, New York.

Minas, H.J., M. Minas, and T.T. Packard, 1986: Productivity in upwelling areas deduced from hydrographic and chemical fields. *Limnology and Oceanography* **31**:1182–1206.

Moore, J.K., S.C. Downey, D.M. Glover, and I.Y. Fung, 2001: Iron cycling, and nutrient-limitation patterns in surface waters of the World Ocean. *Deep-Sea Res. II* 49:463–507.

Nadelhoffer, K.J., et al., 1999: Nitrogen Deposition Makes a Minor Contribution to C Sequestration in Temperate Forests. *Nature* 398:145–148.

Nair, P. K. R. 1993: *An Introduction to Agroforestry.* Kluwer Academic Publishers/ICRAF, Dordrecht.

Nakhumwa, T. 2004: Dynamic costs of soil degradation and determinants of adoption of soil conservation technologies by smallholder farmers in Malawi. Unpublished PhD Thesis, University of Pretoria, Pretoria.

Naqvi, S.W.A., D.A. Jayakumar, P.V. Narvekar, H. Nalk, V.S. Sarma, et al., 2000: Increased marine production of N_2O due to intensifying anoxia on the Indian continental shelf. *Nature* 408: 346–349.

Nelson D.M., P. Treguer, M.A. Brzezinski, A. Leynaert, and B. Queguiner, 1995: Production and dissolution of biogenic silicon in the ocean: revised global estimates,comparison with regional data and relationship with biogenic sedimentation. *Global Biogeochemical Cycles,* **9**: 359–372.

Niklaus, P.A., P.W. Leadley, B. Schmid, and C. Korner, 2001: A long-term field study on biodiversity x elevated CO_2 interactions in grassland. *Ecological Monographs* 71:341–356.

Nixon, S.W. 1995: Coastal marine eutrophication: A definition, causes and future concerns. *Ophelia* **41,** 199–219.

NRC (National Research Council), 1992: *Soil and water quality: An agenda for agriculture.* National Academic Press, Washington DC.

OECD (Organisation for Economic Co-operation and Development), 1982: *Eutrophisation des eaux. Méthodes de surveillance, d'évaluation et de lutte.* Paris, 164 pp.

Palm, C.A., 1995:Contribution of agroforestry trees of nutrient requirements of intercropped plants. *Agroforestry systems* 30:105–124.

Parton W. J., J.W.B. Stewart, and C.V. Cole, 1988: Dynamics of C, N, P and S in grasslands soils: a model. *Biogeochemistry* 5:109–131.

Pate, J.S., 1968: Physiological Aspects of Inorganic and Intermediate Nitrogen Metabolism (with special reference to the legume, *Pisum arvense* L.). In *Recent Aspects of Nitrogen Metabolism in Plants* E.J. Hewitt and C.V. Cutting, (eds.) Academic Press, London, pp 219–240.

Perry, D. A., 1995: Self-organizing systems across scales. *TREE* 10:241–245.

Pierce F.J., W.E. Larson, R.H. Dowdy, and W.A.P. Graham, 1984: Soil Productivity in the Corn Belt: An assessment of erosion's long-term effects. *Journal of Soil and Water Conservation* 39:131–136.

Pimente, D., C. Harvey, P. Resosudarmo, K. Sinclair, D. Kurz, et al., 1995: Environmental and Economic Costs of Soil Erosion and Conservation Benefits. *Science* **267**:1117–1123.

Pinstrup-Anderson, P., R. Panya-Lorch, and M.W. Rosegrant, 1999: *World food prospects: critical issues for the early twenty-first century.* International policy Research Institute. Washington DC.

Prather, M., D. Ehhalt, F. Dentener, R. Derwent, E. Dlugokencky, et al., 2001: Atmospheric chemistry and greenhouse gases. In: *Climate change 2001: The scientific basis. Volume 1 of the Third Assessment Report of the Intergovernmental Panel on Climate Change.* Cambridge University Press. Pp 239–288.

Prentice, I.C., G.D. Farquhar, M.J.R. Fasham, G.L. Goulden, M. Heimann, et al., 2001: The C cycle and atmospheric CO_2. In: *Climate change 2001: The scientific basis. Volume 1 of the Third Assessment Report of the Intergovernmental*

Panel on Climate Change. Intergovernmental Panel on Climate Change Cambridge University Press. Pp 183–237 .

Prospero, J.M. and R.T. Nees, 1986: Impact of the North African Drought and El Nino on mineral dust in the Barbados trade winds. *Nature* 320:735–738.

Ramaswamy, V., O. Boucher, J. Haigh, D. Hauglustaine, J. Haywood, et al., 2001: Radiative forcing of climate change. In: *Climate change 2001: The scientific basis. Volume 1 of the Third Assessment Report of the Intergovernmental Panel on Climate Change.* Cambridge University Press. Pp 349–416.

Redfield A.C., 1958: The biological control of chemical factors in the environment. *American Scientist, 46*: 205–221.

Reeburgh, W.S., 1997: Figures Summarizing the Global cycles of biogeochemically important elements. *Bull. Ecol. Society of America* 78(4):260–267.

Reich, P.B., J. Knops, D. Tilman, J. Craine, D. Ellsworth, et al., 2001: Plant diversity enhances ecosystem responses to elevated CO2 and nitrogen deposition (vol 410, pg 809, 2001). *Nature* 411:824–826.

Reynolds, C.S., 1990: *The Ecology of Freshwater Phytoplankton.* Cambridge University Press, Cambridge, UK, 384 p.

Ridgwell A., A. Watson, and D. Archer, 2002: Modeling the oceanic Si inventory to perturbation, and consequences for atmospheric CO_2. *Global Biogeochemical Cycles.* **16**:4.

Ruiz-Camacho, N., 2004: Mise au point d'un système de bioindication de la qualité du sol base sur l'étude des peuplements de macro-invertébrés. 270p. Université Paris VI. Doctorate. Université Paris VI, Paris.

Ryther, J.H., 1969: Photosynthesis and fish production in the sea. *Science* **166**:72.

Sanchez, P., 2002: Soil fertility and hunger in Africa. *Science* **295(5562):** 2019–2020.

Sanudo-Wilhelmy, S.A., A. Kustka, D.G. Capone, D. Hutchins, C. Gobler, et al., 2001: Phosphorus limitation of N_2 fixation in the central Atlantic Ocean. *Nature* 411:66–69.

Scholes, R.J., E.D. Schulze, L.F. Pitelka, and D.O. Hall, 2000: Biogeochemistry of terrestrial ecosystems. In: *The Terrestrial Biosphere and Global Change: Implications for Natural and Managed Ecosystems.* B.H. Walker, W. Steffen, J. Canadell, and J. Ingram (eds). IGBP Book Series 4, Cambridge University Press. Pp 271–303.

Scholes, R.J. and R. Biggs (in press): Report of the southern African regional scale Millennium Ecosystem Assessment. CSIR, Pretoria

Schlesinger, W.H., 1997: *Biogeochemistry: Analysis of Global Change.* Second ed. Academic Press.

Schulze E,D., 1989: Air pollution and forest decline in a spruce (*Picea abies*) forest. *Science* 244:776–783

Smaling E.M.A., M.S. Nandwa, and B.H. Janssen, 1997: Soil Fertility in Africa Is at Stake. In: Replenishing Soil Fertility in Africa. *Am. Soc. of Agr. and Soil Sci. Soc. of Am. Special Publ* 51: 48–61.

Smetacek, V., 2000: The giant diatom dump. *Nature* 406:574–575

Smil, V., 1997: *Cycles of Life: Civilisation and the Biosphere.* Scientific American Library, New York. 221 pp.

Smil, V., 2000: Phosphorous in the Environment: Natural Flows and Human Interferences. *Annual Reviews of Energy and Environment* 25:53–88.

Smith, V.H., 1992: Effects of nitrogen:phosphorus supply ratios on nitrogen fixation in agricultural and pastoral systems *Biogeochemistry* 18: 19–35.

Starr, G.C., R. Lal, R. Malone, D. Hothem, L. Owens, and J. Kimble, 2000: Modeling soil C transported by water erosion processes. *Land Degradation & Development* **11**:83–91.

Summerhayes, C.P., K.-C. Emeis, M.V. Angel, R.L. Smith, and B. Zeitzschel, 1995: Upwelling in the Ocean: Modern Processes and Ancient Records, p. 422, *In* B. Zeitzschel, ed. Upwelling in the Ocean: Modern Processes and Ancient Records. John Wiley and Sons, Chichester.

Sutcliffe, J.P. 1993: Economic Assessment of Land Degradation in Ethiopian Highlands: A case study. Addis Ababa, Ethiopia: National Conservation Strategy Secretariat, Ministry of Planning and Economic Development.

Swift, M.J. and P. Woomer, 1994: *The Biological Management of Tropical Soil Fertility.* Wiley-Sayce, New York.

Takeda, S. 1998: Influence of iron availability on nutrient consumption ratio of diatoms in oceanic waters. *Nature* 393: 774–777.

Tegen, I. and I.Y. Fung, 1995: Contribution to the mineral aerosol load from land surface modification. *J. Geophys. Res.,* **100**: 18,707–18,726.

Tegen, I. and R. Miller, 1998: A GCM study on the interannual variability of soil dust aerosol. *Journal of Geophysical Research* 103, 25975–25995

Tilman, D. and J. A. Downing. 1994: Biodiversity and stability in grasslands. *Nature* 367:363–365.

Tilman, D., P.B. Reich, J. Knops, D. Wedin, T. Mielke, and C. Lehman, 2001: Diversity and productivity in a long-term grassland experiment. *Science* **294**:843–845.

Tréguer P., D.M. Nelson, A. J. Van Bennekom, D.J. DeMaster, A. Leynaert, and B. Queguiner, 1995: The balance of silicon in the world ocean: a re-estimate. *Science, 268*: 375–379.

Tréguer, P. and P. Pondaven, 2000: Global Change: silica control on carbon dioxide. *Nature, 406*: 358–359.

UNEP (United Nations Environment Programme), 2002: *Global Environment Outlook 3.* Earthscan.

Velasquez, E. 2004: Bioindicadores de calidad del suelo basados en comunidades de macrofauna y su relacion con variables fisicas e quimicas del suelo. 250p. Universidad Nacional de Colombia, Cali.

Verschuren, D., D.N. Edgington, H.J. Kling, and T.C. Johnson, 1998: Silicon depletion in Lake Victoria: Sedimentary signals at offshore stations. *J.-Great-Lakes-Res.* 24: 118–130.

Vitousek, P.M., J.D. Aber, R.W. Howarth, G.E. Likens, P.A. Matson, et al., 1997: Human alteration of the global nitrogen cycle—sources and consequences. *Ecological Applications* 7:737–750.

Vitousek, P.M., K. Cassman, C. Cleveland, T. Crews, C.B. Field, et al., 2002: Towards an ecological understanding of biological nitrogen fixation. *Biogeochemistry* **57/58**:1–45.

Wardle, D.A., K.I. Bonner, G.M. Barker, G.W. Yeates, K.S. Nicholson, et al., 1999: Plant removals in perennial grassland: vegetation dynamics, decomposers, soil biodiversity, and ecosystem properties. *Ecological Monographs* 69:535–568.

Wood, S., K. Sebastian, and S.J. Scherr, 2000: *Agroecosystems.* Pilot Analysis of Global Ecosystems. IFPRI and WRI, Washington DC.

Yeates, G.W. and A.F. Bird, 1994: Some observations on the influence of agricultural practices on the nematode faunae of some south Australian soils. *Fundamental and Applied Nematology* 17:133–145.

Chapter 13
Climate and Air Quality

Coordinating Lead Authors: Jo House, Victor Brovkin

Lead Authors: Richard Betts, Bob Constanza, Maria Assunçao Silva Dias, Beth Holland, Corinne Le Quéré, Nophea Kim Phat, Ulf Riebesell, Mary Scholes

Contributing Authors: Almut Arneth, Damian Barratt, Ken Cassman, Torben Christensen, Sarah Cornell, Jon Foley, Laurens Ganzeveld, Thomas Hickler, Sander Houweling, Marko Scholze, Fortunat Joos, Karen Kohfeld, Manfredi Manizza, Denis Ojima, I. Colin Prentice, Crystal Schaaf, Ben Smith, Ina Tegen, Kirsten Thonicke, Nicola Warwick

Review Editors: Pavel Kabat, Shuzo Nishioka

---

*This appears in Appendix A at the end of this volume.

Main Messages

Ecosystems, both natural and managed, exert a strong influence on climate and air quality. Ecosystems are both sources and sinks of greenhouse gases, aerosol precursors, and pollutants. Their physical properties affect heat and water fluxes, influencing temperature and precipitation—altering, for example, the reflection of solar radiation (albedo) and the flow of water through plants to the atmosphere (evapotranspiration), where it becomes available for rainfall. Thus ecosystems provide the following atmospheric "services":

- warming (for example, sources of greenhouse gases and reduction of albedo by boreal forests compared with bare soil and snow);

- cooling (for example, sinks of greenhouse gases, sources of aerosols that reflect solar radiation, and evapotranspiration);

- water recycling and regional rainfall patterns (for example, evapotranspiration and sources of cloud condensation nuclei);

- atmospheric cleansing (for example, sinks for pollutants such as tropospheric ozone, ammonia, NO_x, sulfur dioxide, and methane);

- pollution sources (for example, particulates from biomass burning, NO_x, carbon monoxide, and precursors of tropospheric ozone); and

- nutrient redistribution (for example, source of nitrogen deposited elsewhere and reduction of erosion and nutrient-rich airborne dust compared with bare soil).

Changes in ecosystems have made a large contribution to the changes in radiative forcing (the cause of global warming) between 1750 and the present. The main drivers are deforestation, fertilizer use, and agricultural practices. Ecosystem changes account for about 10–30% of the radiative forcing of carbon dioxide from 1750 to present and a large proportion of the radiative forcing due to methane and nitrous oxide. Ecosystems are currently a net sink for carbon dioxide and tropospheric ozone, while they remain a net source of methane and nitrous oxide. Management of ecosystems has the potential to significantly modify concentrations of a number of greenhouse gases, although this potential is small in comparison to IPCC scenarios of fossil fuel emissions over the next century (*high certainty*). Ecosystems influence the main anthropogenic greenhouse gases in several ways:

- **Carbon dioxide**—Preindustrial concentration, 280 parts per million; concentration in 2000, 370 ppm. About 40% of the emissions over the last two centuries and about 20% of the CO_2 emissions during the 1990s originated from changes in land use and land management, primarily deforestation. Terrestrial ecosystems have been a sink for about a third of cumulative historical emissions and a third of the 1990s total (energy plus land use) emissions. The sink may be explained partially by afforestation/reforestation/forest management in North America, Europe, China, and other regions, and partially by the fertilizing effects of nitrogen deposition and increasing atmospheric CO_2. Ecosystems were on average a net source of CO_2 during the nineteenth and early twentieth century and became a net sink sometime around the middle of the last century (*high certainty*).

- **Methane**—Preindustrial concentration, 700 parts per billion; concentration in late 1990s, 1750 ppb. Natural processes in wetland ecosystems account for 25–30% of current CH_4 emissions, and about 30% of emissions are due to agriculture (ruminant animals and rice paddies).

- **Nitrous oxide**—Preindustrial concentration, 270 ppb; concentration in late 1990s, 314 ppb. Ecosystem sources account for about 90% of current N_2O emissions, with 35% of emissions from agricultural systems, primarily driven by fertilizer use.

- **Tropospheric ozone**—Preindustrial, 25 Dobson Units; late 1990s, 34 DU. Several gases emitted by ecosystems, primarily due to biomass burning, act as precursors for tropospheric ozone. Dry deposition in ecosystems accounts for about half the tropospheric ozone sink. The net global effect of ecosystems is a sink for tropospheric ozone.

Land cover changes between 1750 and the present have increased the reflectivity to solar radiation (albedo) of the land surface (*medium certainty*), partially offsetting the warming effect of associated CO_2 emissions. Deforestation and desertification in the tropics and sub-tropics leads to a reduction in regional rainfall (*high certainty*). The biophysical effects of ecosystem changes on climate depend on geographical location and season. With *high certainty*:

- Deforestation in seasonally snow-covered regions leads to regional cooling during the snow season due to an increase in surface albedo and leads to warming during summer due to reduction in evapotranspiration.

- Large-scale tropical deforestation (hundreds of kilometers) reduces regional rainfall, primarily due to decreased evapotranspiration.

- Desertification in tropical and sub-tropical drylands leads to decrease in regional rainfall due to reduced evapotranspiration and increased surface albedo.

Biophysical effects such as this need to be accounted for in the assessment of options for mitigating climate change. For example, the warming effect of reforestation in seasonally snow-covered regions due to albedo decrease is likely to exceed the cooling effect of additional carbon storage in biomass.

Ecosystems are currently a net sink for several atmospheric pollutants, including tropospheric ozone (which causes respiratory problems and plant damage), CO_2 (which leads to ocean acidification, with negative effects on calcifying organisms such as corals and coccoliths), and ammonia (which contributes to health problems, eutrophication of lakes, and acidification of N-saturated ecosystems). Fertilizer use has led to increased ecosystem emissions of N gases (which contribute to acid rain and eutrophication of lakes). The net effect of ecosystems on acid rain and stratospheric ozone depletion are small compared with industrial emissions.

Vegetation burning, both natural and human-induced, is a major cause of air pollution. Particulates, tropospheric ozone, and carbon monoxide are toxic to humans at levels reached as a result of biomass burning. In the 1990s, biomass burning was responsible for about a quarter of global carbon monoxide emissions, just under half of particulate aerosol emissions, and a large but poorly quantified fraction of tropospheric ozone precursor emissions. Smoke plumes cause changes in plant productivity (generally decreases), changes in rainfall (generally decreases), and economic losses due reduced visibility (affecting transport, for example).

The self-cleansing ability of the atmosphere is fundamental to the removal of many pollutants and is affected by ecosystem sources and sinks of various gases. Removal of pollutants involves chemical reactions with the hydroxyl radical. OH concentration and hence atmospheric cleansing capacity has declined since preindustrial times but likely not by more than 10%. The net contribution of ecosystem changes to this decline is currently unknown. The reactions are complex, but generally emissions of NO_x and hydrocarbons from biomass burning increase tropospheric ozone and OH concentrations, and emissions of CH_4 and carbon monoxide from wetlands, agricultural practices, and biomass burning decrease OH concentration.

The most important ecosystem drivers of change in climate and air quality in the past two centuries have been deforestation (net CO_2 emissions, net increase in surface albedo, and rainfall reduction), agricultural practices (increasing emissions of CH_4, N_2O, and other N gases), and biomass burning (emissions of toxic pollutants). Wetland draining has decreased CH_4 emissions but increased emissions of CO_2 and N_2O. The net short-term (20–100 year) effect on radiative forcing is cooling (*medium certainty*), while the long-term effect is probably warming (*low certainty*). Management of drylands to increase vegetation cover reduces soil carbon loss, reduces dust emissions, and increases water recycling. Loss of species richness has probably not had significant impacts on climate and air quality in the recent past, but shifts in functional types—such as trees versus grasses, deciduous versus evergreen trees, or calcifying versus non-calcifying plankton—could alter the biological storage of carbon and trace gas emissions in the future.

Ecosystem interactions with the atmosphere are highly nonlinear, with many feedbacks and thresholds that, if passed, may lead to abrupt changes in climate and land cover. Human-induced land cover changes may become irreversible due to ecosystem-climate feedbacks. For example, in the Sahara-Sahel region, two alternative land cover types are theoretically sustainable: savanna and desert. If a threshold in loss of vegetation cover contributing to rainfall reduction is crossed, the desert state becomes self-sustaining (*low certainty*). The complexity and incomplete understanding of the feedbacks make it hard to predict thresholds and their future changes.

13.1 Introduction

Living matter builds bodies of organisms out of atmospheric gases such as oxygen, carbon dioxide and water, together with compounds of nitrogen and sulphur, converting these gases into liquid and solid combustibles that collect the cosmic energy of the sun. After death, it restores these elements to the atmosphere by means of life's processes . . . Such a close correspondence between terrestrial gases and life strongly suggest that the breathing of organisms has primary importance in the gaseous system of the biosphere; in other words, it must be a planetary phenomenon.

(Vernadsky 1926)

The composition of the atmosphere and the climate we experience are products of the co-evolution of the biosphere, atmosphere, and geosphere over billions of years (Vernadsky 1926; Zavarzin 2001). The climate and the concentration of various gases in the atmosphere are determined by the flow of energy (radiation, heat) and materials (such as water, carbon, nitrogen, trace gases, aerosols) between the atmosphere, ocean, soils, and vegetation. These interlinking components are referred to collectively as the Earth System to stress their inter-dependence. (See Box 13.1.) Lovelock and Margulis (1974) proposed the Gaia Hypothesis: that biospheric feedbacks regulate the climate within a range suitable for life. Although not universally accepted, Gaia remains an inspirational idea in Earth System science (Lenton 1998; Watson 1999; Kirchner 2003).

Ecosystems alter atmospheric chemistry, providing both sources and sinks for many atmospheric constituents that affect air quality or that affect climate by changing radiative forcing. In this chapter we refer to these as "biogeochemical effects." Ecosystems further influence climate through the effects of their physical properties on water fluxes (such as rainfall) and energy balance

(such as temperature). (See Figure 13.1.) We refer to these as "biophysical effects."

The ability of ecosystems to modify climate and air quality and thereby provide a service to humans occurs both through natural processes and as a result of ecosystem management. For example, the conversion of carbon dioxide and water to oxygen by ecosystems billions of years ago could be considered the fundamental ecosystem service, enabling evolution and maintenance of a breathable atmosphere. Not all impacts of ecosystems on the atmosphere and climate are beneficial to human well-being, and the effects often depend on the location and magnitude of the impact. A change in magnitude can change the sign—for example, a small temperature increase may help some people at some locations by, say, extending the crop-growing season or potential area, but a large temperature increase is detrimental to the majority of people in the majority of locations through, for instance, damage to crops and human health (IPCC 2001d).

This chapter assesses all major effects of ecosystems on climate and air quality, be it "good" or "bad" for human well-being. The impacts of climate and air quality on ecosystems and human well-being, in contrast, are dealt with in detail by other assessments (e.g., IPCC 2001b, 2001d; WHO 2002; WMO 2003; Brasseur et al. 2003a; Emberson et al. 2003) and are not the focus of the MA other than as drivers of ecosystem change (which are summarized in Chapter 3 and in MA *Scenarios,* Chapter 7, in several sections later in this chapter, and in relevant sections of other chapters). For more detailed reviews of the science behind global climate change see IPCC (2001a); see also Kabat et al. (2004) and Kedziora and Olejnik (2002) on biophysical mechanisms and impacts and Brasseur et al. (2003a) on atmospheric chemistry.

During the Quaternary period (approximately the past 2.5 million years), the Earth System has shown a persistent pattern of glacial-interglacial cycles during which the climate and atmospheric composition varied between fairly consistent bounds, as shown by ice core measurements (Petit et al. 1999, EPICA community members 2004). These quasi-periodic cycles are triggered primarily by variations in Earth's orbit. The associated changes in climate and in carbon dioxide, methane, and other atmospheric constituents are controlled by mechanisms involving both terrestrial and ocean ecosystems (IPCC 2001a; Prentice and Raynaud 2001; Steffen et al. 2004; Joos and Prentice 2004). However, the balance of these mechanisms is not well understood, and this implies uncertainties in predicting future changes, especially on time scales of centuries or longer.

Burning fossil fuels, changes in land cover, increasing fertilizer use, and industrial emissions over the past two centuries have propelled the Earth System outside the boundaries of the natural system dynamics of the Quaternary period (*high certainty*). The current concentration of carbon dioxide and methane are unprecedented in the last 420,000 years and possibly in the last 20 million years, and the rate of increase is unprecedented in at least the last 20,000 years (IPCC 2001a, 2001d; see also MA *Scenarios,* Chapter 7). The increase in temperature in the twentieth century was the largest of any 100 years in the last 1,000 years (IPCC 2001a, 2001d). The Intergovernmental Panel on Climate Change concluded that "there is newer and stronger evidence that most of the warming observed over the last 50 years is attributable to human activities" (IPCC 2001a, 2001d).

Human intervention in global biogeochemical cycles has triggered a chain of biogeochemical and biophysical mechanisms that will continue to affect both atmospheric chemistry and climate on time scales from years to millennia (*high certainty*). Even if emissions ceased today, past emissions would continue to have an impact in the future related to the lifetime of the emitted gas in the

---

BOX 13.1

The Earth System, Thresholds, and Feedbacks

The "Earth System" has several interacting components: the atmosphere, ocean, terrestrial and marine biosphere, cryosphere (ice, including permafrost), the pedosphere (soils), and humans. These components are tightly linked with each other. Ecosystems are an integral part of the Earth System; they provide different services to the climate system via numerous physical and chemical mechanisms that control fluxes of energy (radiation, heat), water, and atmospheric constituents.

The Earth System is highly nonlinear: climate, air quality, and ecosystem distribution across the planet may change quite abruptly in response to smooth changes in external forcing (as occurred, for example, during the last deglaciation about 15,000 years ago). Current theories support the possibility of multiple stable states (regimes of a particular balance of components that are resistant to change) and abrupt transitions between these different states or regimes (as with desert and vegetation in the Sahara/Sahel region). These transitions are reinforced by positive (amplifying) feedback loops between components of the Earth System, whereby a small change in one component can cause changes in other components that continue to push the system away from its previous state and toward a new one. Conversely, negative (stabilizing) feedbacks can maintain stable states by preventing the system moving beyond certain thresholds.

Numerous examples of feedbacks between ecosystems, climate, and atmospheric constituents are mentioned throughout this chapter.

- **Climate–greenhouse gases, positive feedback.** Warming enhances emissions of CH_4, N_2O, and tropospheric ozone precursors (NO_x and VOCs) (*very certain*). Warming reduces inorganic ocean uptake of CO_2, increases soil emissions of CO_2, and has been predicted to reduce carbon storage in terrestrial and ocean ecosystems; the net result is an increase in atmospheric CO_2 (*high certainty*). Increasing concentration of greenhouse gases causes further warming.

- **CO_2 fertilization–CO_2 uptake, negative feedback.** Increased atmospheric concentration of CO_2 has a fertilizing effect on plants, increasing uptake of CO_2 and reducing atmospheric concentrations (*high certainty*).

- **Taiga-tundra albedo–temperature, positive feedback.** Afforestation of snow-covered regions due to northward movement of forest boundaries in a warmer world, or due to tree planting, reduces albedo, leading to further warming, less snow, and further reduced albedo (*high certainty*).

- **Tropical rainforest–precipitation, positive feedback.** Large-scale reduction in tropical rainforest cover reduces regional precipitation, potentially causing further forest loss and precipitation reduction (*high certainty*).

- **Sahara-Sahel vegetation–precipitation, positive feedback.** Decreased vegetation cover increases albedo, reduces soil-atmosphere water recycling, and reduces monsoon circulation, decreasing precipitation, all of which further suppress vegetation cover (*high certainty*).

- **DMS-cloudiness, negative feedback.** Emissions of DMS by ocean ecosystems (and of VOCs by terrestrial ecosystems) increase cloud condensation nuclei, cooling Earth, reducing photosynthesis and emissions of DMS (and VOCs), and increasing thermal stability, reducing cloud formation (*medium certainty*).

- **Tropical forest–tropospheric ozone, positive feedback.** High levels of tropospheric ozone have a deleterious effect on vegetation, compromising further uptake of tropospheric ozone (*medium certainty*).

- **Pollution–reduction in cleansing capacity, positive feedback.** For example, CH_4 in the atmosphere reduces OH concentration and atmospheric cleansing capacity, increasing lifetime and atmospheric concentrations of CH_4 (*high certainty*).

---

atmosphere, atmospheric chemistry, and inertia in different parts of the Earth System (such as the uptake and mixing of carbon dioxide in the ocean and the response of sea level rise to temperature and ice melting) (IPCC 2001d Figure 5.2).

A summary of the main ecosystem effects on climate and air quality, the drivers, and the impacts on human well-being that are discussed in this chapter is presented in Figure 13.2. Changes in climate or air quality are often simultaneously affected by several atmospheric constituents. Likewise, a particular atmospheric constituent can affect both climate and air quality. Furthermore, ecosystem drivers (such as deforestation, biomass burning, and agricultural practices) often simultaneously affect biogeochemical and biophysical properties, and their effects can work in the same or opposite directions. Thus it is often not possible to quantify cause and effect. Each atmospheric constituent and vegetation property considered in this chapter is summarized in Table 13.1, along with its magnitude and distribution, main drivers, and impacts.

13.2 Biogeochemical Effects of Ecosystems on Climate: Greenhouse Gases and Aerosols

Many atmospheric constituents determine the radiative forcing of Earth's climate (IPCC 2001a, 2001d). (Radiative forcing is the change in net vertical irradiance (radiation or energy) of the tropopause (upper troposphere), with an increase in radiative forcing implying an increase in global temperature. Global warming potential is an index, relative to CO_2, describing the radiative properties of greenhouse gases based on their effectiveness at absorbing long-wave radiation, and the time they remain in the atmosphere. For more detailed explanations of these terms, see IPCC (2001a, 2001d).)

Key atmospheric compounds that have an ecosystem source or sink include:

- greenhouse gases that absorb long-wave radiation from Earth's surface, leading to warming—carbon dioxide, methane, tropospheric ozone (formed from precursors NO_x, methane, and volatile organic compounds), and nitrous oxide; and

- aerosols, of which some types (such as sulfate aerosols) reflect solar radiation leading to cooling, while others (such as black carbon) trap radiation leading to warming.

Ecosystems have played a significant role in past radiative forcing (see Figure 13.3A) and in current sources and sinks of greenhouse gases and aerosols (see Figure 13.3B). The net biochemical contribution of ecosystems to historical radiative forcing has been to increase global warming, accounting for about 10–30% of the radiative forcing of CO_2 from 1750 to the present (Brovkin et al. 2004) and a large proportion of the warming due to CH_4 and N_2O, while reducing tropospheric ozone forcing. Ecosystems are currently a net sink for CO_2 and tropospheric ozone, while they remain a net source of CH_4 and N_2O.

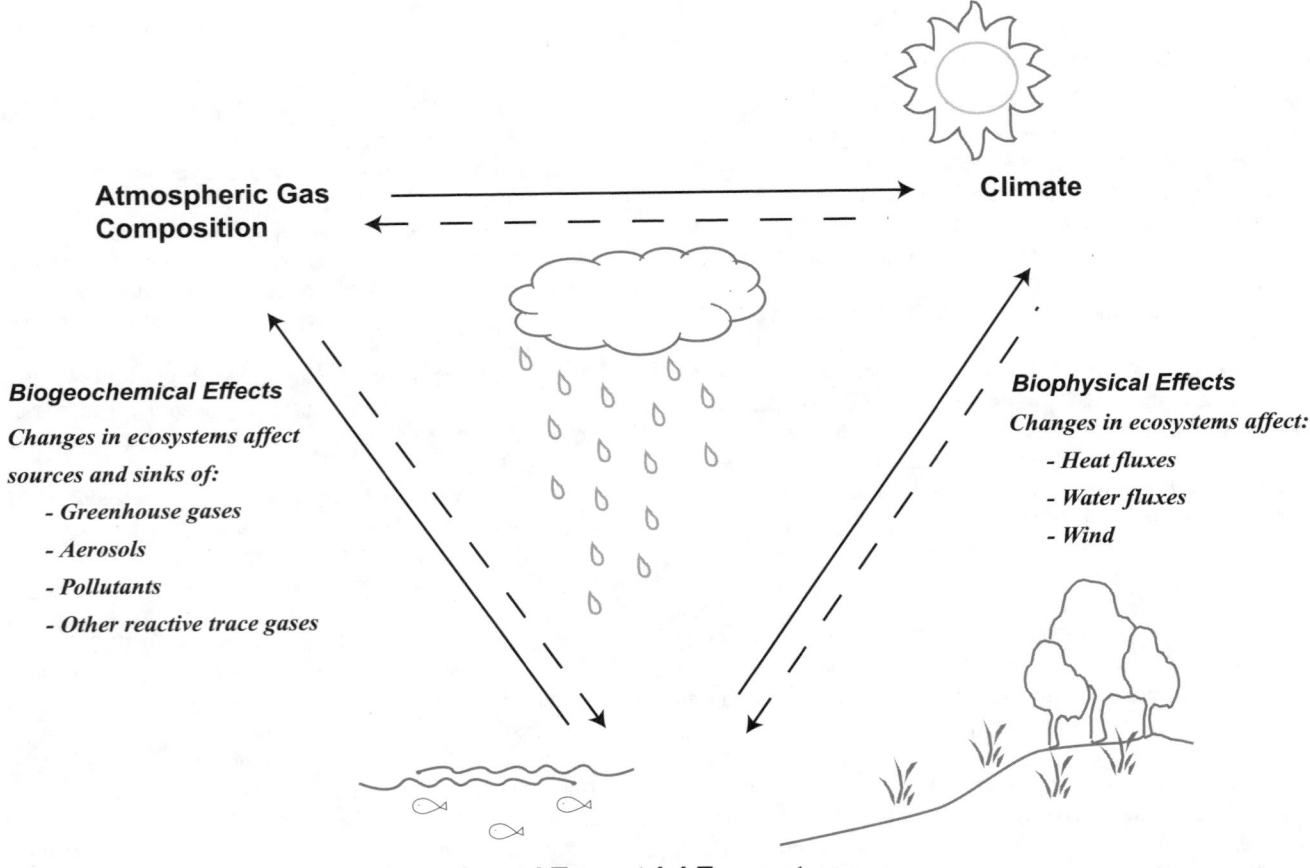

Atmospheric Gas Composition

Climate

Biogeochemical Effects

Changes in ecosystems affect sources and sinks of:
- *Greenhouse gases*
- *Aerosols*
- *Pollutants*
- *Other reactive trace gases*

Biophysical Effects

Changes in ecosystems affect:
- *Heat fluxes*
- *Water fluxes*
- *Wind*

Aquatic and Terrestrial Ecosystems

Figure 13.1. Ecosystem Effects on the Atmosphere and Climate. Ecosystems, the concentration of many atmospheric gases/species, and the climate all strongly interact. However, it is the effect of ecosystems on atmospheric air quality (as sources and sinks of pollutants) and on the climate (both directly due to biophysical properties of vegetation, and indirectly as a source and sink of greenhouse gases and aerosols) that is the focus of this chapter, as indicated by the solid arrows.

13.2.1 Carbon Dioxide

Increasing carbon dioxide concentration has had more impact on historical radiative forcing than any other greenhouse gas. In addition, CO_2 has a fertilizing effect on most land plants, while rapid injection of CO_2 into the atmosphere causes acidification of the global ocean, with negative implications for calcifying organisms. Ecosystems are both a source and sink for CO_2. Management of ecosystems for carbon storage is currently regarded as an important ecosystem service by policy-makers, therefore more information is included for this gas than for others. A summary of pertinent details of the carbon cycle is presented in this section; for more details, see Prentice et al. (2001), Kondratyev et al. (2003), and Field and Raupach (2004).

Carbon, the basic building block of all animal and plant cells, is converted to carbohydrates by the process of plant photosynthesis. Terrestrial plants capture CO_2 from the atmosphere; marine plants (phytoplankton) take up carbon from seawater, which exchanges CO_2 with the atmosphere. Plant, soil, and animal respiration returns carbon to the atmosphere, as does burning biomass. Burning fossilized biomass (fossil fuels) also returns carbon, captured by plants in Earth's geological history, to the atmosphere.

CO_2 is continuously exchanged between the atmosphere and the ocean; it dissolves in surface waters and is then transported into the deep ocean (the "solubility pump"). It takes roughly one year for CO_2 concentration in surface waters to equilibrate with the atmosphere, but subsequent mixing between surface waters and deep waters, which drives the ongoing uptake of increased atmospheric CO_2, takes decades to centuries. Some of the dissolved carbon that is taken up by marine plants sinks in the form of dead organisms, particles, and dissolved organic carbon (the "biological pump"). A small amount remains in ocean sediments; the rest is respired at depth and eventually recirculated to the surface. The biological pump acts as a net sink for CO_2 by increasing the concentration of CO_2 at depth, where it is isolated from the atmosphere for decades to centuries, causing the concentration of CO_2 in the atmosphere to be about 200 parts per million lower than it would be in the absence of marine life (Sarmiento and Toggweiler 1984; Maier-Reimer et al 1996).

It has been widely assumed that ocean ecosystems are at steady state at present, although there is now much evidence of large-scale trends and variations (Beaugrand et al. 2002; Chavez et al. 2003; Richardson and Schoeman 2004). Changes in marine ecosystems, such as increased phytoplankton growth rate due to the fertilizing effect of iron in dust (see section 13.4.4.3) and shifts in species composition due to ocean acidification (see section 13.4.2.1) or for other reasons, have the potential to alter the oceanic carbon sink. The net impact of changes in ocean biology on global CO_2 fluxes is unknown.

CO_2 concentration varied within consistent bounds of 180 to 300 ppm during glacial-interglacial cycles. Prior to the industrial

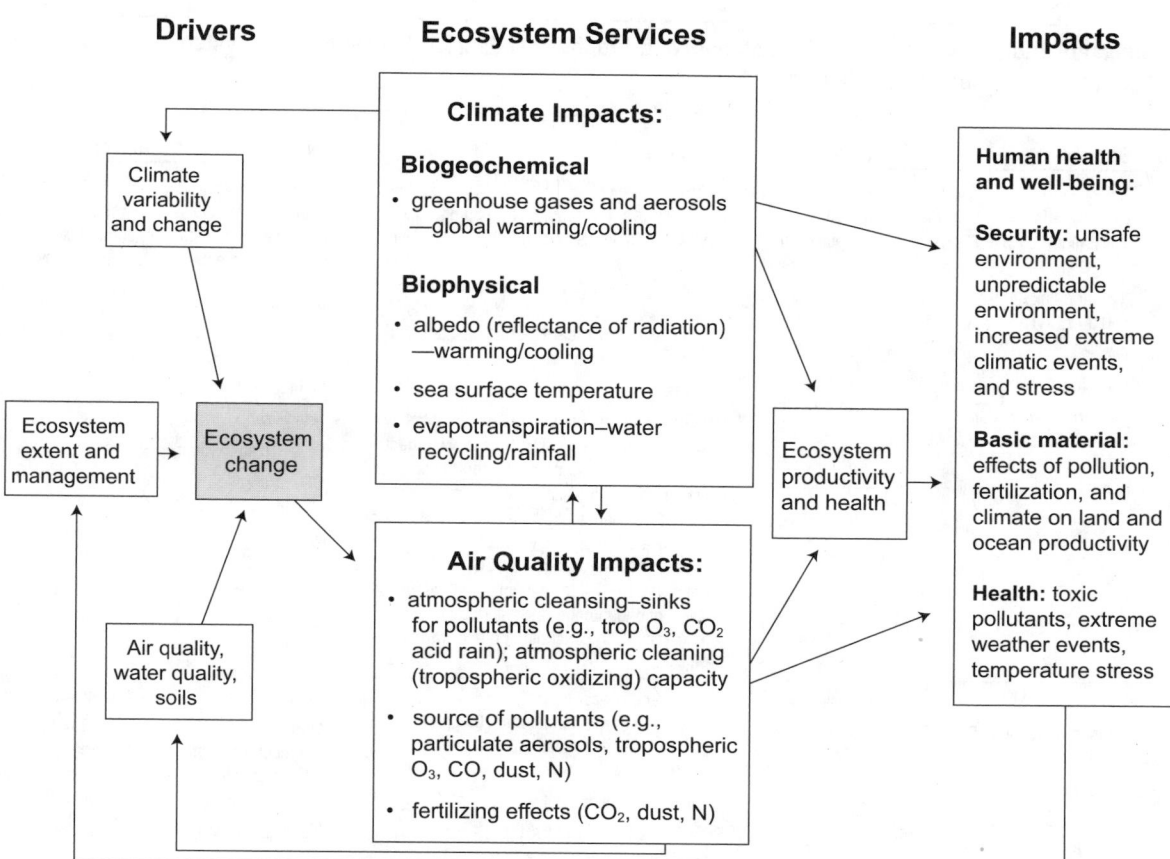

Figure 13.2. Ecosystem Effects on Climate and Air Quality: Services, Drivers, and Impacts on Human Well-being. This figure summarizes the services, drivers, and impacts discussed in this chapter. Arrows represent a direct impact on human health and well-being of the two ecosystems services of this chapter (climate regulation and regulation of atmospheric composition). Note: not all arrows are shown, for simplicity (e.g., direct impacts of climate change on atmospheric composition through changes in atmospheric chemistry).

revolution (that is, before 1750), CO_2 concentration was about 280 ppm, and since then it has risen rapidly, reaching 370 ppm in 2000 (MA *Scenarios,* Chapter 7). It has been estimated that about 40% of CO_2 emissions over the last two centuries came from land use change (primarily deforestation), while 60% came from fossil fuel burning (DeFries et al. 1999). (See Figure 13.4.) About 40% of total CO_2 emissions have remained in the atmosphere.

Oceans are estimated to have taken up approximately a quarter, an amount that can be fully accounted for by the solubility pump. This means that terrestrial ecosystems took up about a third of all emissions (Prentice et al. 2001, House et al. 2002) through a combination of ecosystem processes whose relative importance is still not firmly established but that probably include growth of replacement vegetation on cleared land (e.g., Dixon et al. 1994; Houghton et al. 1998; McGuire et al. 2001; Goodale et al. 2002); agricultural and forest management (e.g., Spiecker et al. 1996; Houghton et al. 1999); other land management practices, such as fire suppression leading to woody encroachment (e.g., Houghton et al. 1999); and fertilizing effects of elevated CO_2 and nitrogen deposition (e.g., Lloyd 1999; Holland et al. 1997).

Analyses of historical atmospheric CO_2 concentrations preserved in ice cores and more recent atmospheric measurements suggest that the land was a net source of CO_2 during the nineteenth and early twentieth centuries (that is, emissions exceeded uptake), and that land changed to a net sink around the 1940s (Bruno and Joos 1997; Joos et al. 1999). Model analyses with reconstructed land use and environmental data indicate a later change from source to sink (1960s to 1970s), due to decreasing deforestation in the tropics, forest regrowth in North America and Asia (Houghton 2003; McGuire et al. 2001; Ramankutty and Foley 1999; Brovkin et al. 2004), and increased uptake of CO_2 by extant ecosystems (McGuire et al. 2001). However, there are uncertainties in modeling the magnitude of changes in the terrestrial carbon budget resulting from several sources: differences in land cover data sets; the lack of systematic global inventory data for vegetation and soil carbon density; and poor quantification of N and CO_2 fertilization effects and climate impacts on ecosystems (Prentice et al. 2001; House et al. 2003).

Measured fluxes of CO_2 during the 1980s and 1990s are shown in Table 13.2. During this period, ecosystems were a net CO_2 sink. Model results indicate that, during the 1990s, terrestrial ecosystems accounted for about 20% of the total emissions (land plus fossil fuels) but were a sink for about a third of the total emissions. Figure 13.5 (in Appendix A) shows a reconstruction of the spatial distribution of ocean and terrestrial net fluxes in the latter half of the 1990s, based on atmospheric measurements (Rödenbeck et al. 2003). These net fluxes are not broken down into source/sink terms or their underlying drivers. Information on regional fluxes due to different drivers assessed by different methods is reviewed in House et al. (2003). Generally, areas of deforestation and forest degradation in the tropics are losing carbon, while areas of afforestation and forest growth in North America and Europe are gaining carbon (House et al. 2003).

Table 13.1. Summary Table of Atmospheric Constituents and Biophysical Factors Affected by Ecosystems: Trends, Drivers, and Impacts on Ecosystem and Human Well-being[a]

Atmospheric Constituents	Sources	Sinks	Trends	Ecosystem Drivers	Impacts
CO_2	*Ecosystem:* land use change (mainly deforestation) ≈ 1.6 (0.5 to 3.0) PgC/yr (net land flux uptake of 1.2 ± 0.9 PgC/yr) *Other:* fossil fuel 6.3 ± 0.4 PgC/yr [note: numbers updated since IPCC 2001a, b; see Table 13.2]	*Ecosystem:* terrestrial uptake (photosynthesis) ≈ 2.8 (0.9 to 5.0) PgC/yr *Other:* ocean uptake (dissolution and mixing) 1.9 ± 0.7 PgC/yr	*Atmospheric concentration:* Increased from preindustrial 280 ppm to 370 ppm (2000). Increased by 3.2 ± 0.1 PgC/yr during 1990s. Average annual rate of increase rising. Projected to continue rapid increase due to fossil fuel burning and long atmospheric lifetime: ≈ 250 years, but a small amount persists for much longer. *Ecosystem:* Terrestrial source until around middle of last century, then increasing sink. Sink likely to decline due to limited management opportunities, saturation of CO_2 fertilization effect, and climate impacts. Ocean ecosystems show evidence of large-scale trends and variations, but the net impact of these changes on CO_2 fluxes is unknown. Non-biological uptake by the ocean will continue, but the rate will decline with increasing CO_2 concentration and warmer climate.	– climate change – land use and land management: deforestation, afforestation, reforestation, forest management, agricultural management – biomass burning – N fertilization – Fe fertilization (dust) – CO_2 fertilization effects	– climate: positive radiative forcing (heating) – ocean acidification: reduced growth of oceanic calcifying organisms including corals, potential negative impacts on fish production – "fertilizing" effect on plants
CH_4	*Ecosystem:* peatlands/wetlands 92–237 TgC/yr ruminants 80–115 TgC/yr rice 25–100 TgC/yr termites 20 TgC/yr oceans 10–15 TgC/yr biomass burning 23–55 TgC/yr *Other:* energy 75–110 TgC/yr, landfills 35–73 TgC/yr waste treatment 14–25 TgC/yr methane hydrates 5–10 TgC/yr	*Ecosystem:* soil uptake 30 TgC/yr *Other:* tropospheric OH reactions 506 TgC/yr stratospheric loss 40 TgC/yr	*Atmospheric concentration:* Increased from preindustrial 700 ppb to 1,745 ppb in 1998. Increased 7.0 ppb/yr during 1990s. Atmospheric lifetime: 8.4 years. Growth rate peaked in 1981 at 17 ppb/yr but is highly variable from year to year. *Ecosystem:* Increasing terrestrial source; sink relatively small contribution to the overall trend; growth rate slowed 1990–96 partly due to decreased northern wetland emissions rates from anomalously low surface temperatures and reduction in OH from strat. O_3 depletion; removal rates increased 1990–2000 by +0.5%/yr.	– climate change – land use and land management: agricultural practices, wetland draining – biomass burning – flooding	– climate: positive radiative forcing (greenhouse gas, heating) – tropospheric ozone formation – stratospheric ozone formation – tropospheric oxidizing capacity reduction
CO	*Ecosystem:* mostly tropical sources vegetation 150 TgC/yr oceans 50 TgC/yr biomass burning 700 TgC/yr *Other:* oxidation of: – CH_4 800 TgC/yr – VOCs 430 TgC/yr fossil/domestic fuel 650 TgC/yr	*Ecosystem:* dry (surface) deposition 190 TgC/yr (Hauglustaine et al. 1998) *Other:* OH reaction 1920 TgC/yr (Hauglustaine et al. 1998)	*Atmospheric concentration:* Preindustrial concentration unknown, 1998 concentration 80 ppb. Increasing 6 ppb/yr during 1990s. Atmospheric Lifetime: 0.08–0.25 years. Slowly increasing till late 1980s, then decreased possibly due to catalytic converters decreasing automobile emissions, increased again late 1990s. Increase may be mostly in Northern Hemisphere (Haan et al. 1996), which already contains twice as much CO as Southern Hemisphere.	– biomass burning – ecosystem uptake – land use and land management: vegetation cover	– human health: hypoxia, neurological problems, cardiovascular disease – tropospheric ozone precursor – tropospheric oxidizing capacity reduction, removal of OH – indirect climate impacts: reacts with other greenhouse gases
N_2O	*Ecosystem:* tropical soils 4.0 TgN/yr temperate soils 2.0 TgN/yr agricultural soils 4.2 TgN/yr ocean 3 TgN/yr cattle/feedlots 2.1 TgN/yr biomass burning 0.5 TgN/yr *Other:* industrial 1.3 TgN/yr atmosphere (NH_3 oxid.) 0.6 TgN/yr	*Ecosystem:* N_2O uptake by soils and conversion to N_2 (relatively small) *Other:* stratospheric reactions that deplete ozone 12.3 TgN/yr	*Atmospheric concentration:* Increased from preindustrial 270 ppb to 314 ppb in 1998. Increased by 0.8ppb/yr during 1990s. Increase rate slower in 1990s than 1980s. Atmospheric lifetime: 120 years. *Ecosystem:* Terrestrial and oceanic sources, exponential rise in concentration since preindustrial. Agricultural emissions increased fourfold from 1900 to 1994 (Kroeze et al. 1999).	– climate change: emission higher in wetter soils – land use and management: acceleration of the global N cycle due to fertilizer use and agricultural N fixation, animal production – biomass burning – N deposition – atmospheric NO_x pollution	– climate: positive radiative forcing (greenhouse gas, heating) – stratospheric ozone depletion

Table 13.1. *continued*

Atmospheric Constituents	Sources	Sinks	Trends	Ecosystem Drivers	Impacts
NO_x (NO and NO_2) (precursors of nitrate)	*Ecosystem:* soils (mostly tropical) 13–21 TgN/yr biomass burning 7.1 TgN/yr *Other:* fossil fuel 33.0 TgN/yr aircraft 0.7 TgN/yr lightning 5.0 TgN/yr stratosphere <0.5 TgN/yr	*Ecosystem:* canopy uptake of soil emissions 4.7–8 TgN/yr and of wet and dry deposition *Other:* reaction with OH to form nitric acid (HNO_3), which collects on aerosols (dry deposition) or dissolves in precipitation (wet deposition)	*Atmospheric Concentration:* Difficult to quantify because of the tremendous spatial and vertical variability. Atmospheric Lifetime: <0.01–0.03 years. Nitrate concentrations declined recently due to emission controls. *Ecosystem:* Difficult to quantify trend, largely stable.	– climate change: warming increases emissions – land use and management: tropical deforestation reduces soil emissions, but reduces canopy uptake more so net emissions increase. Acceleration of global N cycle due to fertilizer use, etc. – biomass burning	– human health: direct respiratory effects and respiratory effects of aerosols – tropospheric ozone precursor – tropospheric oxidizing capacity increase – acid rain formation – fertilization of plants (deposition) – eutrophication of lakes (deposition and nitrate leaching)
NH_3	*Ecosystem:* domestic animals 22 TgN/yr fertilizer use 9 TgN/yr crops (+ decomposition) 4 TgN/yr natural soils 2 TgN/yr oceans 8 TgN/yr biomass burning 6 TgN/yr	*Ecosystem:* direct soil and plant uptake wet and dry deposition (affected by vegetation cover) *Other:* reaction with OH (very small percentage)	*Atmospheric Concentration:* Documentation of trends is challenging because of the relatively short atmospheric lifetime. Atmospheric lifetime: 1 day–1 week. *Ecosystem:* Rise in agricultural sources exponential, much of the growth occurring since 1950. Main source areas Europe and North America (fertilizer use) and India (cattle).	– land use and management: acceleration of global N cycle due to fertilizer use, agricultural intensification/management; land cover change – biomass burning	– human health: hypoxia, pfisteria, respiratory effects – eutrophication of lakes (deposition and nitrate leaching) – acid neutralization and production – aerosol/particulate formation
$SO_2/SO_4/H_2S$ DMS (dimethylsulfide) (precursors of sulfate aerosols)	*Ecosystem:* biomass burning [SO_2] 2.2 Tg/yr land biota [H_2S] 1.0 Tg S/yr marine plankton [DMS] 24 Tg S/yr *Other:* fossil fuel emissions [SO_2] 76 Tg S/yr volcanoes [SO_2] 9.3 Tg S/yr	*Ecosystem:* direct soil and plant uptake wet and dry deposition (affected by vegetation cover) ecosystems are a sink for about 30% of SO_2 emissions and sulphate aerosols *Other:* reaction with OH	*Atmospheric concentration:* Sulfate concentration in 1960s four times that of preindustrial, but declined recently due to stringent emissions regulations. Patchy distribution around source areas — polluted regions North America, Europe, and China. SO_2 emissions declining in North America and Europe, rising in South and East Asia. Emissions are projected to decrease substantially over the next century. *Ecosystem:* Mostly stable.	– climate change – land use and management; land cover change – biomass burning	– human health: respiratory effects of aerosols – climate: negative radiative forcing (cooling) – indirect climate impacts: cloud condensation nuclei – acid rain – reduced NPP though reduced solar radiation
VOCs (volatile organic compounds)	*Ecosystem:* vegetation (mostly tropical): isoprene 220 TgC/yr terpene 127 TgC/yr acetone 50 TgC /yr methanol 70–350 Tg/yr biomass burning 33 TgC/yr *Other:* fossil fuel 161 TgC/yr	*Ecosystem:* direct soil and plant uptake wet and dry deposition (affected by vegetation cover) *Other:* reaction with OH	*Atmospheric concentration:* Difficult to quantify because of the tremendous spatial and vertical variability. Atmospheric lifetime: < 1 day to >1 week. Emissions probably increased due to increasing use of gasoline and other hydrocarbon products. *Ecosystem:* Deforestation has probably decreased natural emissions.	– climate change (warming increases emissions) – land use and management: forest cover change, agricultural management, use of fertilizers – biomass burning – N deposition	– human health: aerosol precursor (terpene), respiratory effects – indirect climate impacts: cloud condensation nuclei – tropospheric ozone formation (isoprene) – tropospheric oxidizing capacity (increase) – organic acid formation – acid rain
Aerosols: organic matter	*Ecosystem:* biomass burning 54 Tg/yr biogenic (VOC oxidation, plant debris, humic matter and microbial particles) 56 Tg/yr *Other:* fossil fuel 28 Tg/yr	wet and dry deposition (affected by vegetation cover)	*Ecosystem:* Biogenic aerosols increasing due to increase in oxidizing agents, e.g., NO_3 and O_3, possible three- to fourfold increase since preindustrial times (Kanakidou et al. 2000). Not much is known about emissions from plants, microbes, and humic matter, but likely to be strongly affected by land use change.	– land use and management – biomass burning	– human health: respiratory effects – climate impacts: negative radiative forcing (cooling) – indirect climate impacts: cloud condensation nuclei – reduced NPP though reduced solar radiation

(continues over)

Table 13.1. *continued*

Atmospheric Constituents	Sources	Sinks	Trends	Ecosystem Drivers	Impacts
Aerosols: black carbon (0–2 μm)	*Ecosystem:* biomass burning 5.7 Tg/yr *Other:* fossil fuel 6.6 Tg/yr	wet and dry deposition (affected by vegetation cover)	*Atmospheric concentration:* Concentration and trend uncertain.	– land use and management: vegetation cover – biomass burning	– climate: positive radiative forcing (heating) – human health: respiratory effects – reduced NPP though reduced solar radiation
Aerosols: dust	*Ecosystem:* mineral (soil) dust 2000 Tg/yr (1–2 μm 300 Tg/yr); mainly from desert/dryland areas, <10 % from disturbed soil surfaces (Tegen et al. 2004) *Other:* industrial dust (>1 μm) 100 Tg/yr	wet and dry deposition (affected by vegetation cover)	*Atmospheric concentration:* Concentration and trend uncertain. *Ecosystem:* Decrease in North American emissions since dust bowl years due to changes in management. Emissions from Sahara/Sahel increased significantly since 1960s, possibly due to changing wind patterns and desertification. Chinese desert areas and loess plateau variable, trend and causes not clear. Climate change impacts–model results inconclusive.	– climate variability – climate change – land use and management: reduction in land cover, agricultural management – desertification	– human health: respiratory effects, irritation – climate impacts: radiative forcing net sign and magnitude unclear as reflects incoming radiation and traps outgoing radiation – fertilizing effects of iron in ocean and phosphate on land in some regions, increasing productivity, indirect climate effect as CO_2 sink – reduced NPP though reduced solar radiation – visibility reduction
Tropospheric ozone	*Ecosystem:* ecosystem precursors VOCs (isoprene), NO_x, CH_4, CO primarily from biomass burning in the tropics *Other:* transport from stratosphere 475 Tg O_3/yr precursors in urban pollution	*Ecosystem:* dry deposition: 620–1178 Tg O_3/yr *Other:* stratosphere/troposphere exchange: 400–1440 Tg O_3/yr	*Atmospheric Concentration:* Increased from preindustrial conc. 25 DU (Dobson units) to 34 DU (370 Tg O_3) in late 1990s. Increased from 1970 to 1980 but no clear trend from 1980 to 1996. Difficult to quantify due to the high reactivity and spatial and temporal variability of sources, but satellite measurements may improve quantification. Models predict increasing tropospheric O_3 driven regionally by increasing emissions of pollutants. Atmospheric lifetime: 0.01–0.05 years. Concentrated over areas of urban pollution and biomass burning.	– emissions of key precursor trace species including VOCs, CH_4, NO_x, CO – climate change: warming increases concentration – biomass burning	– human health: UV exposure – climate: positive radiative forcing (heating) – troposphere oxidizing capacity – stratospheric ozone production
OH radical	reactions between tropospheric ozone, non-methane hydrocarbons, and NO_x in UV light	reactions with many reduced compounds, especially CO and CH_4	Probably declining since preindustrial but not by more than 10%.	– emissions of key precursors and sinks	– reduced OH leads to reduced tropospheric oxidizing capacity

Biophysical Surface Properties	Non-Forests	Forests	Trends	Ecosystem Drivers	Impacts
Surface albedo	snow-free: 0.16 (tall grasslands) to 0.6 (sand desert) snow-covered: 0.5 to 0.8	snow-free: 0.11 (tropical evergreen) to 0.2 (deciduous) snow-covered: 0.2 to 0.25	Increase in mid-latitudes due to deforestation until middle of twentieth century, now decrease due to regrowth in some mid-latitude areas. Increase in tropics.	– land use and management: primarily forest cover	– climate: radiative forcing (higher albedo = more reflection = cooling)
Water fluxes (evapotranspiration)	up to 5 mm/day	up to 10 mm/day	Evapotranspiration decrease, especially in tropics, due to deforestation.	– land use and management: primarily forest cover	– climate: direct impacts on radiative forcing and indirect impacts via clouds – hydrological cycle
Surface roughness	up to 0.1m	1.0–2.5m	Decrease, especially in tropics, due to deforestation.	– land use and management: primarily forest cover	– climate: atmospheric circulation (wind)

[a] All numbers relate to the 1990s and are from IPPC 2001b unless otherwise stated.

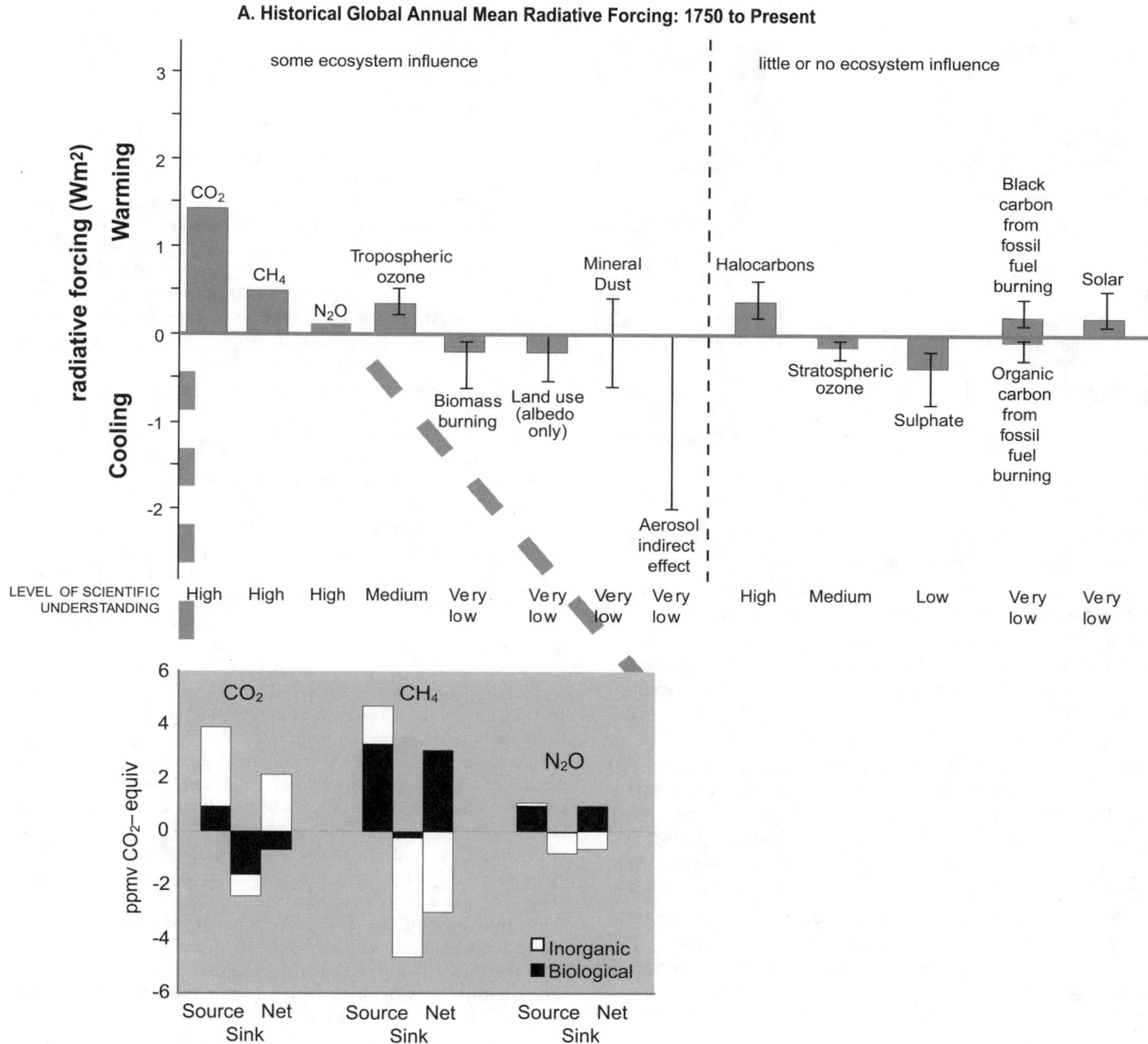

Figure 13.3. Contribution of Ecosystems to Historical Radiative Forcing and Current Greenhouse Gas Emissions (Adapted from IPCC 2001a, 2001b). Figure A is the radiative forcing caused by changes in atmospheric composition, alteration in land surface reflectance (albedo), and variation in the output of the sun for the year 2000 relative to the conditions in 1750. The height of the bar represents a best estimate, and the accompanying vertical line a likely range of values. We have separated factors with a significant ecosystem influence from those without. The indirect effect of aerosols shown is their effect on cloud droplet size and number, not cloud lifetime. Some of the radiative components are well mixed over the globe, such as CO_2, thereby perturbing the global heat balance. Others represent perturbations with stronger regional signatures because of their spatial distribution, such as aerosols. Radiative forcing continues to be a useful tool to estimate to a first order, the relative climate impacts such as the relative global mean surface temperature response due to radiatively induced perturbations, but these global mean forcing estimates are not necessarily indicators of the detailed aspects of the potential climate responses (e.g., regional climate change). Figure B is the relative contribution of ecosystems to sources, sinks, and net change of three of the main greenhouse gases. These can be compared by conversion into CO_2–equivalent values, based on the global warming potential (radiative impact times atmospheric lifetime) of the different gases. For CH_4 and N_2O, a 100-year time scale was assumed; a shorter time scale would increase the relative value compared with CO_2 and a longer time scale would reduce it. Ecosystems are also a net sink for tropospheric ozone, but it is difficult to calculate emissions in CO_2–equivalent values.

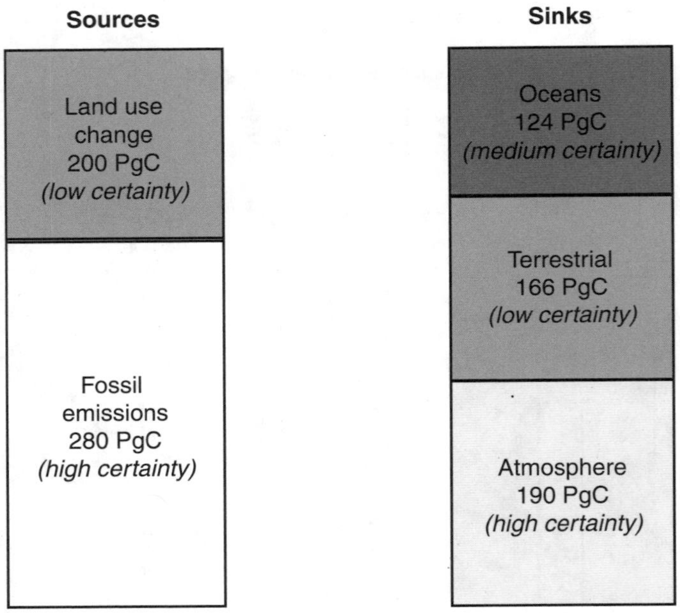

Sources

Sinks

Land use change
200 PgC
(low certainty)

Fossil emissions
280 PgC
(high certainty)

Oceans
124 PgC
(medium certainty)

Terrestrial
166 PgC
(low certainty)

Atmosphere
190 PgC
(high certainty)

Figure 13.4. Carbon Sources and Sinks over the Last Two Centuries. Total carbon losses to the atmosphere due to historical land use change have been estimated at around 200 PgC (DeFries et al. 1999): two thirds to three quarters of this loss was due to conversion of forestland to cropland or other land uses; other carbon losses included degradation of grasslands and shrublands and the conversion of non-forestland to cropland. Fossil fuel emissions from pre-industrial times to 2000 are estimated as 280 PgC (Marland et al. 2000, update in Prentice et al. 2001), but the atmospheric increase during the same period was only 190 PgC. About 124 PgC, or $\approx$26% of the total emissions, were taken up by the oceans primarily as a result of chemical and physical processes (dissolution and mixing) (House et al. 2002 based on Gruber 1998; Sabine et al. 1999, 2002; Prentice et al. 2001; Langenfelds et al. 1999; Manning 2001). The remaining 166 PgC or ~34% was absorbed by the land biosphere. In this analysis, historical land use change is responsible for about 40% of the observed growth in atmospheric CO_2; Brovkin et al. (2004) estimate a range of 25–49%, with the lower end of this range being more likely.

Concern about global warming and the implementation of the Kyoto Protocol has led to carbon uptake for climate regulation being considered as an important ecosystem service (MA *Policy Responses,* Chapter 13). Forests can be managed as a sink (including preventing deforestation and promoting afforestation, reforestation, and improved forest management). This approach implies that once the forests stop growing, they must be protected to avoid loss of most of the carbon store (and to encourage long-term storage in soil carbon pools). Alternatively, forest biomass can be used to produce long-lived products that store carbon (such as furniture), or as a substitute for materials that are energy-intensive to produce (such as aluminum and plastics), or as biomass fuels that are used instead of fossil fuels.

In these ways, an area of forest can continue to offset CO_2 emissions indefinitely and may provide other services at the same time. The potential trade-offs with other environmental and socioeconomic values are also relevant—for example, biodiversity maintenance. (See Chapter 9 and MA *Policy Responses,* Chapter 13). The prospect of carbon trading under the Clean Develop-

ment Mechanism, along with public and industry awareness of climate change issues, is already promoting small-scale forest activities with a view to carbon sequestration and the production of modern biomass fuels.

Forest degradation resulting from overexploitation can result in substantial carbon losses. For example, about 0.5 megagrams of carbon per hectare per year is being lost in Southeast Asia (Kim Phat et al. 2004) by this mechanism. Forest fragmentation leads to increased rates of big tree mortality, decomposition, and fire, causing carbon losses greater than deforestation in some areas (Nascimento and Laurance 2004). In central Amazonia, model estimates for the first half of this century suggest annual fragmentation losses of 4–5 megagrams of carbon per hectare per year. Timber removal from extant Amazonian forests is likely to account for a loss equal to about 10–20 megagrams of carbon per hectare per year, and fires for double that amount (Nepstad et al. 1999). Agroforestry systems have an annual sequestration capacity of 0.2–0.3 megagrams of carbon per hectare per year, and about 400 million hectares of degraded land are potentially suitable for agroforestry systems globally (IPCC 2000). One North Indian agroforestry system sequestered up to 19.6 megagrams of carbon per hectare per year (Singh et al. 2000).

Agricultural management alters the amount of carbon contained in soils and also affects other greenhouse gases such as nitrous oxide and methane (e.g., Rosenberg et al. 1998; Paustian et al. 2000; West and Post 2002; Witt et al. 2000). Soil carbon is lost when land is cleared and tilled, but it can be regained through low-tillage agriculture and other methods (Lal et al. 2004). It remains uncertain how much carbon contained in soil eroded from agricultural landscapes is delivered to the atmosphere in policy-relevant timescales (Renwick et al. 2004). Actions taken to promote carbon sequestration in soils would reduce dust emissions, which have a fertilizing effect on the ocean, thus reducing the uptake of CO_2 by the ocean and diminishing the effectiveness of this response strategy to an unknown degree (Ridgwell et al. 2002). Complex interactions of this type highlight the need to consider all implications of management options (see MA *Policy Responses,* Chapter 13). In the lowland tropics and sub-tropical areas where rice is now grown continuously throughout the year (continuous flooding), soil carbon has been seen to increase (Bronson et al. 1997; Cheng 1984). With intensive management typical of these high-yield continuous rice systems, net carbon sequestration rates of 0.7–1.0 megagrams of carbon per hectare per year have been measured (Witt et al. 2000). (See Chapter 26.)

Management of marine ecosystems to increase oceanic carbon sequestration ("ocean engineering," for instance via iron fertilization) has some theoretical potential, but very little is known about potential ecological and geochemical risks of such an endeavor. Due to its comparatively low estimated cost (probably in the range of a few dollars per ton of CO_2), the approach is financially attractive but the capacity is too small to slow down anthropogenic CO_2 increase significantly. Early upper limit calculations show that the potential for reducing atmospheric CO_2 is limited to a few tens of parts per million (Peng and Broecker 1991; Joos et al. 1991), in agreement with the most recent high-resolution ice core data of dust deposition and atmospheric CO_2 (Röthlisberger et al. 2004). More recent calculations put the maximum potential at around 1.0 petagrams (1.0 billion tons) of carbon per year for a maximum of 100–150 years, although it is likely to be much smaller, say less than 0.2 petagrams of carbon per year (Matear and Elliott 2004; Caldeira et al. 2004).

Future trends in atmospheric CO_2 are likely to depend more on fossil fuel emissions than on ecosystem change. Although land use management can have a significant impact on CO_2 concentra-

Table 13.2. Annual Fluxes of Carbon Dioxide over the Last Two Decades. Positive values represent atmospheric increase (or ocean/land sources); negative numbers represent atmospheric decrease (ocean/land sinks). Land and ocean uptake of CO_2 can be separated using atmospheric measurements of oxygen (O_2) in addition to CO_2 because biological processes on land involve simultaneous exchange of O_2 with CO_2, while ocean uptake does not (ocean ecosystems are assumed to be in equilibrium for the purposes of this calculation). While this technique allows quantification of the net contribution of terrestrial ecosystems, breaking this down into sources and sinks requires modeling. The CO_2 released due to land use change during the 1980s has been estimated as the range across different modeling approaches. The difference between the modeled land use change emissions and the net land-atmosphere flux can then be interpreted as uptake by terrestrial ecosystems (the "residual terrestrial sink"). (IPCC data from Prentice et al. 2001)

Source of Flux	IPCC		Update[a]	
	1980s	1990s	1980s	1990s
	(gigatons of carbon equivalent per year)			
Atmospheric increase	+3.3 ± 0.1	+3.2 ± 0.1		
Anthropogenic emissions (fossil fuel, cement)	+5.4 ± 0.3	+6.3 ± 0.4		
Ocean-atmosphere flux	−1.9 ± 0.6	−1.7 ± 0.5	−1.8 ± 0.8	−1.9 ± 0.7
Land–atmosphere flux:	−0.2 ± 0.7	−1.4 ± 0.7	−0.3 ± 0.9	−1.2 ± 0.9
—Land use change[b]	+1.7 (+0.6 to +2.5)	incomplete	+1.3 (+0.3 to 2.8)	+1.6 (+0.5 to +3.0)
—Residual terrestrial sink	−1.9 (−3.8 to +0.3)	incomplete	−1.6 (−4.0 to −0.0)	−2.8 (−5.0 to −0.9)

[a] Same data as Prentice et al. 2001 but including a correction for the air-sea flux of oxygen caused by changes in ocean circulation (Le Quéré et al. 2003). Other estimates of the air-sea oxygen correction have given slightly different results for the 1990s, mostly due to the fact that direct observations of heat change in the ocean have not yet been compiled for after 1998 (Keeling and Garcia 2002; Plattner et al. 2002; Bopp et al. 2002).

[b] The IPCC estimated range for the land use change flux is based on the full range of Houghton's bookkeeping model approach (Houghton 1999) and the CCMLP ecosystem model intercomparison (McGuire et al. 2001). The update is based on the full range of Houghton (2003) and DeFries et al. (2002); the CCMCP analysis only extended to 1995.

tions in the short term (Prentice et al. 2001), the maximum feasible reforestation and afforestation activities over the next 50 years would result in a reduction in CO_2 concentration of only about 15–30 ppm by the end of the century (IPCC 2000). Even if all the carbon released so far by anthropogenic land use changes throughout history could be restored to the terrestrial biosphere, atmospheric CO_2 concentration at the end of the century would be about 40–70 ppm less than it would be if no such intervention had occurred (Prentice et al. 2001, House et al. 2002). Conversely, complete global deforestation over the same time frame would increase atmospheric concentrations by about 130–290 ppm (House et al. 2002).

This compares with the projected range of CO_2 concentrations in 2100, under emissions scenarios developed for the IPCC, of 170–600 ppm above 2000 levels, mostly due to fossil fuel emissions (Prentice et al. 2001). The ability of the land and ocean to take up additional increments of carbon decreases as the CO_2 concentration rises, primarily due to the finite buffering capacity and rate at which ocean water can take up CO_2, as well as the saturation of the CO_2 fertilization response of plant growth (Cox et al. 2000; Prentice et al. 2001). Global warming is predicted to have a strong positive feedback on the carbon cycle due, for example, to increases in soil organic matter decomposition and a reduction in inorganic ocean uptake due to reduced CO_2 solubility and ocean stratification at higher temperatures, as described later.

13.2.2 Methane

Methane is a greenhouse gas and an energy source (natural gas). It is involved in many atmospheric chemistry reactions, including formation of tropospheric and stratospheric ozone and the reduction of atmospheric cleansing capacity (see section 13.4.1). The major source of methane is microorganisms living in a variety of anaerobic environments such as flooded wetlands and rice paddies, the guts of termites and ruminant animals, the ocean, landfill sites, and waste treatment plants. Other soil bacteria re-oxidize CH_4, preventing much of the CH_4 produced in anaerobic (wet) soils from reaching the atmosphere and accounting for a small but significant sink of atmospheric CH_4 in soils remote from methane sources.

The current atmospheric concentration of CH_4 is more than twice that of preindustrial times, as Table 13.1 indicated (see also MA *Scenarios,* Chapter 7). The growth rate peaked in 1981 and has declined since (Prather et al. 2001), with no increase in the concentration between 1999 and 2002 (Dlugokencky et al. 2003). The observed values are subject to high interannual variability due to changes in sources, sinks, atmospheric transport, atmospheric chemistry, and climate variability. Emissions from ecosystems account for about 70% of total emissions (Prather et al. 2001), with about 30% from wetlands and 30% from agriculture. The spatial distribution of emissions from natural wetlands and agriculture (based on modeling) is shown in Figure 13.6. While northern wetlands are rather well studied with respect to magnitude and drivers of CH_4 emissions, the lack of data from tropical wetlands is a major knowledge gap.

Agriculture (ruminant animals and rice paddies) is the most important anthropogenic driver of CH_4 emissions. In pastoral countries such as Bolivia, Uruguay, and New Zealand, CH_4 emissions by ruminant livestock are responsible for over 40% of all greenhouse gas emissions (when expressed as CO_2 equivalents)

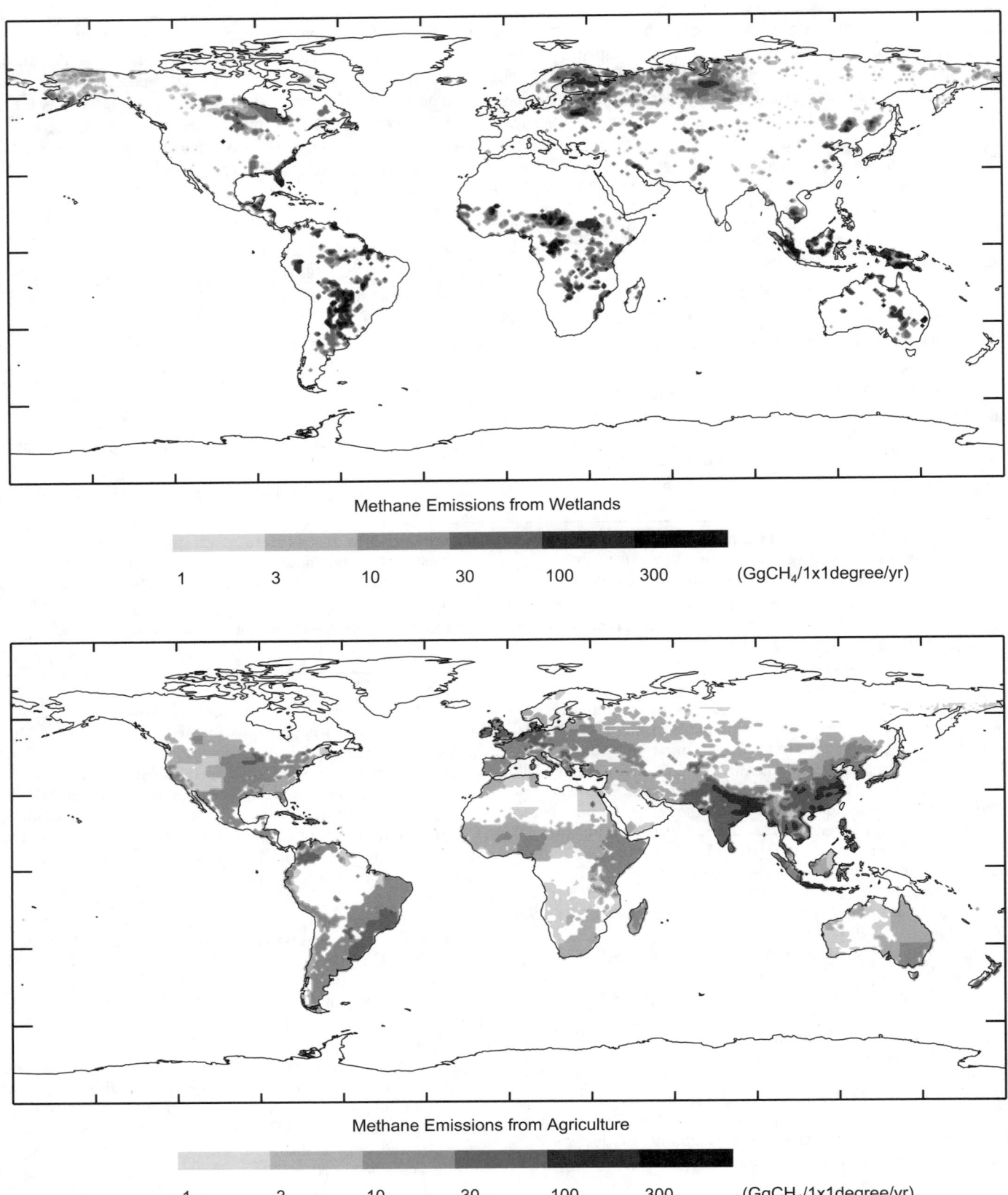

Methane Emissions from Wetlands

1   3   10   30   100   300   (GgCH$_4$/1x1degree/yr)

Methane Emissions from Agriculture

1   3   10   30   100   300   (GgCH$_4$/1x1degree/yr)

Figure 13.6. Map of Methane Fluxes from Wetlands and Agriculture Models. The top figure shows the spatial location of methane fluxes from natural wetlands according to Walter and Heimann (2000). The total emissions are probably too high; a maximum of about 160–170 GgCH$_4$/yr may be more realistic (Houweling et al. 2000). The bottom figure shows emissions from agriculture sources (rice paddies, animals and animal wastes) according to the model EDGAR (v3.2) (Olivier et al. 1994).

(UNFCCC at ghg.unfccc.int). Control of this methane source through changing diet is being examined as a potential control on global warming. Methane emissions from lowland, irrigated rice systems are affected by various management practices. For example, the use of green manures can substantially increase CH_4 emissions compared with the use of inorganic N fertilizers (Wassmann et al. 2000a, 2000b). (More detailed discussion of rice paddies can be found in Chapter 26.)

Wetland and peatland soils under waterlogged or seasonally frozen conditions tend to store carbon fixed by plants during photosynthesis. This is because in waterlogged soils, decomposition of plant material is slower than in aerated soils and is accompanied by a relatively slow release of CH_4 compared to the faster CO_2 efflux of aerobic soils. Wetland draining for agriculture, for forestation, or due to water extraction leads to a decrease in CH_4 production but a rapid increase in CO_2 release from soils that are often very rich in organic matter. The current consensus is that, while some areas are carbon sinks and some are carbon sources, most wetlands—in particular, the northern peat-forming ones—have a carbon balance close to zero (Callaghan et al. 2004).

CH_4 has a higher global warming potential than CO_2 (23:1 on a 100-year time horizon), although CO_2 is much longer-lived in the atmosphere. In the short term (20–100 years), CH_4 emissions have a more powerful effect on the climate per unit mass than CO_2. Thus as most wetlands are currently sources of CH_4, they are also net sources of radiative forcing (*medium certainty*) (Friborg et al. 2003). Draining of wetlands leads to a decrease in radiative forcing (cooling) in the short term, but in the long term the opposite may be true (*low certainty*) (IPCC 2001a; Christensen and Keller 2003).

The climate change impacts on peatlands have different feedbacks depending on the location and extent of global warming. Some regions, such as northern Alaska, are predicted to experience—and are indeed already experiencing—soil drying, leading to net losses of carbon as CO_2 and decreases in CH_4 emissions (Chapter 25). In the subarctic, however, where recent decadal warming has led to the loss of permafrost with thermokarst erosion and a wetting of the soils as a consequence, a significant increase in CH_4 emissions has been seen (Christensen et al. 2004), with the overall effect of an increase in radiative forcing. In tropical regions, the key issue is what changes will occur to seasonal flooding—draining will lower the impact on radiative forcing as the CH_4 emissions are very high, yet the carbon store in peat is insufficient for emissions of CO_2 to overwhelm this in the long term.

Preliminary model results of changes in wetland areas under future climate (e.g., Cox et al. 2000) suggest that the northern wetlands will increase their carbon storage and subsequent CH_4 emissions while tropical ones will lose significant amounts of carbon and decrease CH_4 emissions. Furthermore, rising soil temperature and enhanced microbial rates could increase methane emissions, amplifying global warming by 3.5–5% by the end of this century (Gedney and Cox 2003; Gedney et al. 2004).

13.2.3 Nitrous Oxide

Ecosystems are a source of nitrogen in various gaseous forms, each with different effects on climate and air quality. They are also a sink for atmospheric nitrogen, taking up N_2 directly from the atmosphere (nitrogen fixation) or reactive nitrogen after deposition (wet deposition–rained out, or dry deposition). For a detailed explanation of the nitrogen cycle and the role of ecosystems, see Chapter 12. The nitrogen cycle has been profoundly altered by use of synthetic fertilizers. (See Chapter 26.)

Nitrous oxide is a powerful, long-lived greenhouse gas (GWP 296:1 on a 100-year time horizon), which (unlike other N oxides) is unreactive in the troposphere. Atmospheric concentrations of N_2O have increased since preindustrial times from 270 ppb to 314 ppb (MA *Scenarios*, Chapter 7). Ecosystem sources—primarily soil microorganisms in an array of environments—account for about 90% of N_2O emissions and a small fraction of N_2O uptake (Prather et al. 2001). Enhanced ecosystem N_2O emissions are mainly driven by increased fertilizer use, agricultural nitrogen fixation, and atmospheric nitrogen deposition (Nevison and Holland 1997; Prather et al. 2001). Wetland draining also increases N_2O emissions. Fertilizer use and nitrogen deposition are projected to increase substantially in the tropics (Matson et al. 1999; Prather et al. 2001).

13.2.4 Tropospheric Ozone

Besides being a greenhouse gas, tropospheric ozone is a toxic pollutant. It is highly reactive in the atmosphere and also helps maintain the atmospheric cleansing capacity. It is formed in the atmosphere in the presence of light from precursors: mainly volatile organic compounds (the most important VOC being isoprene), nitrogen oxides (NO and NO_2, collectively denoted as NO_x), CH_4, and carbon monoxide. Biomass burning is an ecosystem source of precursors, but urban pollution sources dominate, with very high concentrations of ozone mostly appearing downwind of urban areas. In addition to being a source of tropospheric ozone precursors, ecosystems account for about half the total sink for tropospheric ozone, through dry deposition (Prather et al. 2001). Ecosystems are thus currently a net sink for tropospheric ozone. Deforestation reduces this sink (see Figure 13.7): NO_x soil emissions decline, but canopy uptake declines more. Where forests are replaced with agriculture, NO_x emissions increase further.

The concentration, sources, and sinks of tropospheric ozone are difficult to quantify due to its high reactivity and the spatial and temporal variability of sources and sinks. Most surface measuring stations show an increase from 1970 to 1980, but no clear trend from 1980 to 1996. Models predict increasing tropospheric ozone in the future, driven regionally by increasing emissions of its precursors (Prather et al. 2001).

13.2.5 Aerosols

Ecosystems are sources and sinks for a variety of aerosols (or aerosol precursors) that directly affect radiative forcing, causing warming or cooling depending on their properties (such as reflectivity) and location (such as height or the underlying surface) (Penner et al. 2001). Many aerosols affect cloud formation, which in turn affects radiative forcing in complex ways. The net effect of clouds on radiative forcing remains uncertain, but increasing the aerosol load probably, on average, causes cooling. The net effect of aerosols on climate is matter of intensive investigation, and while the field is progressing rapidly, more research is needed before firm conclusions can be drawn.

Sulfur compounds (sulfur dioxide, hydrogen sulfide, and dimethyl sulphide) contribute to the formation of sulfate aerosols with a negative radiative forcing (climate cooling) (MA *Scenarios*, Chapter 7). Industrial sources dominate, despite declines in some regions due to controls and legislation (Rodhe 1999; Penner et al. 2001). Ecosystems are a sink for about 30% of SO_2 emissions and for sulfate aerosols. Dimethyl sulfide, emitted by marine phytoplankton when they die or are eaten, contributes to cloud formation. DMS emissions are quite variable in relation to phytoplankton species and mode of release. Global mean DMS concentration in surface waters is fairly well known, but regional emissions are

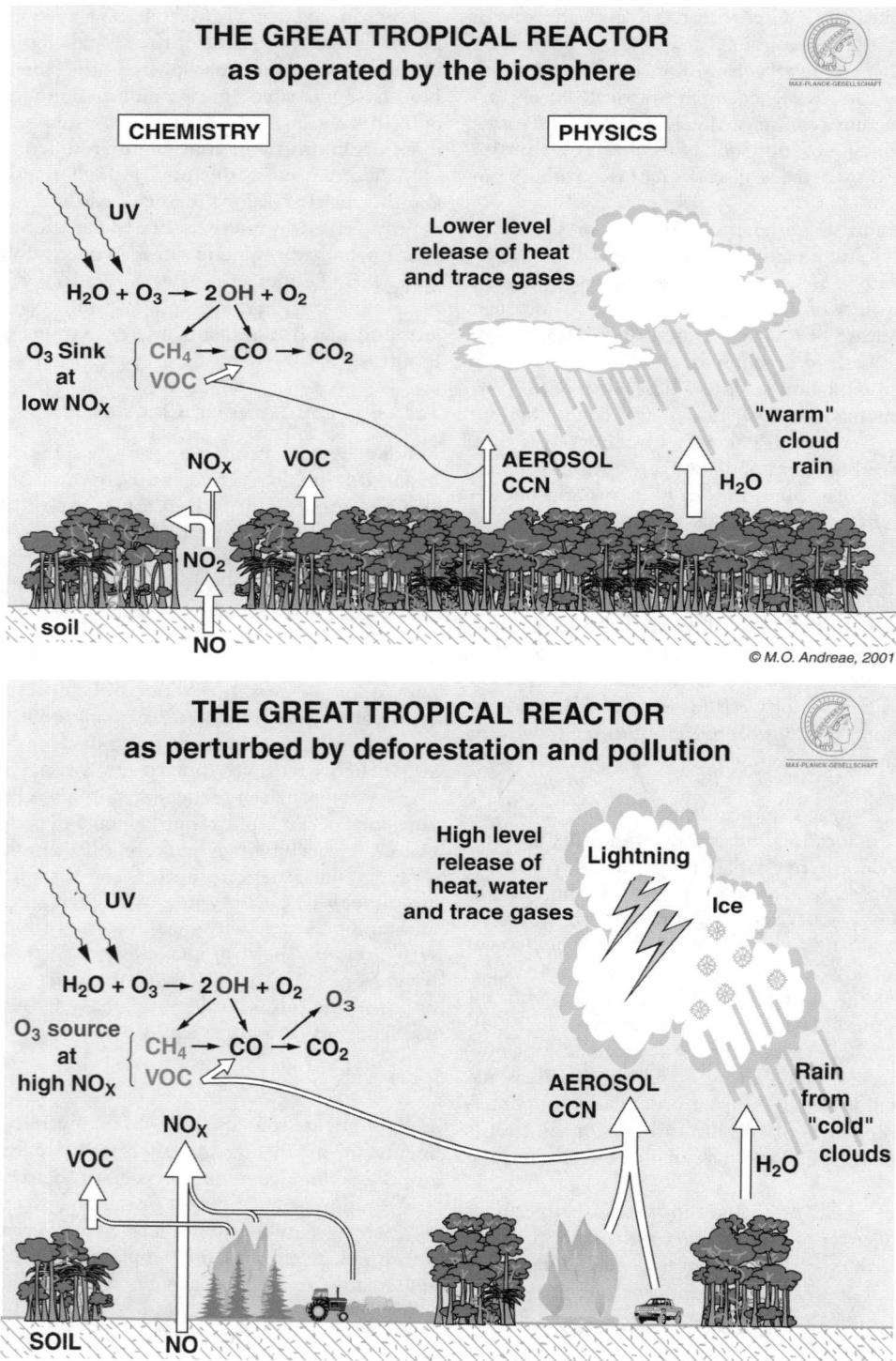

Figure 13.7. The Tropical Reactor—Biochemical and Biophysical Interactions in the Tropics (Andreae 2001). The top figure shows the natural biochemical and biophysical fluxes and interactions over intact forest. The bottom figure shows the fluxes when the natural forest area is subject to deforestation and pollution. The land changes from a net sink of tropospheric ozone to a net source when the canopy does not trap NO_x emissions from soil, and emissions of NO_x from agriculture increase. Aerosols from vegetation fires and pollutions sources (e.g., cars) act as cloud condensation nuclei leading to higher storm clouds. Overall rainfall is reduced due to a reduction or evapotranspiration (water recycling through vegetation).

more uncertain. Climate-related change in surface ocean stratification is bound to affect phytoplankton species distribution and succession, with likely consequences for DMS production and potentially also for cloud formation (Kiene et al. 1996).

Carbonaceous aerosols of many forms are emitted by ecosystems and can cause warming or cooling depending on their composition, size, shape, and location. IPCC (2001a) concluded that biomass-burning aerosols have a net cooling effect (indoor biomass fuel burning was underrepresented in this estimate). But these aerosols can also reduce cloudiness, which is thought to enhance climate warming (Penner et al. 2003). Black carbon (soot from incomplete combustion of biomass and fossil fuels) has been suggested to have a large warming effect on the climate, due in part to various feedbacks (Hansen and Sato 2001; Jacobson 2002), although the magnitude of this effect is disputed (Penner et al. 2003; Roberts and Jones 2004). Air quality impacts of biomass-burning aerosols are dealt with later in this chapter.

There are huge uncertainties regarding emissions and trends of biogenic aerosols (VOC oxidation products, plant debris, humic matter, and microbial particles), thus IPCC (2001a) did not estimate their net radiative warming impacts. Their contribution could be significant in densely vegetated regions of the tropics, and it is under threat as a result of land use change.

Mineral dust is entrained into the atmosphere from sparsely vegetated soils. Dust scatters solar radiation and absorbs terrestrial radiation. Its net impact on radiative forcing is uncertain, but updates since IPCC (2001a) infer that the net effect is cooling (Kaufman et al. 2001). Its trends and drivers are described later in this chapter.

13.3 Biophysical Effects of Ecosystems on Climate

13.3.1 Surface Properties and Climate Processes Affected by Ecosystems

Ecosystems affect climate through the alteration of energy and water fluxes in the lowest atmosphere, or planetary boundary layer (about 1–2.5 kilometers above Earth's surface) (Pitman et al. 2004). Within this layer, vertical profiles of temperature and humidity depend strongly on the partitioning of energy between sensible heat and latent heat. Over land, this partitioning is largely controlled by ecosystems. Over bare, dry land, the energy is transported via sensible heat, resulting in relatively high surface air temperatures. Vegetation canopies transpire water extracted from the root zone, increase the upward latent heat flux, and cool the surface air (Avissar et al. 2004). Modification of the fluxes of water and energy by ecosystems has significant regional effects on precipitation, temperature, and wind. Globally averaged impacts are small, complex, and hard to detect against the background of natural climate variability and anthropogenic climate change. Key physical properties and processes affected by ecosystems are summarized here:

- **Surface albedo** is the fraction of solar radiation reflected back into the atmosphere from Earth's surface. Higher albedo means that more energy leaves the planetary boundary layer (net cooling of the atmosphere). Vegetation traps radiation, generally reducing albedo compared with, for instance, snow cover or bare ground in dry lands. In agricultural regions, tillage usually decreases albedo since bare soil in moist climates is generally darker (less reflective) than plant canopies. Forests are very effective at trapping radiation by multiple reflection within the canopy; this effect is particularly strong in snow-covered regions where trees extend above the snow, while short vegetation such as crops and pastures are covered by snow. (See Figure 13.8 in Appendix A.) Phytoplankton modify ocean surface albedo, with different types either reducing (Frouin and Lacobellis 2002) or increasing it (Brown and Yoder 1994; Balch et al. 1991).

- **Transpiration** is the flux of water from the ground to the atmosphere through plants, controlled by the opening and closing of tiny pores in the leaf's surface called stomata. The volume of water transpired is determined by vegetation rooting depth, leaf area, soil moisture, temperature, wind, and stomatal conductance (which is biologically regulated). Transpiration drives the hydrological cycle—recycling rain water back to the atmosphere to be rained out elsewhere. Thus terrestrial ecosystems mediate the service of water recycling. Through transpiration and precipitation, water evaporated over the ocean is transported into the interior of continents. A part of the rainfall escapes immediate recycling and forms river runoff; thus the presence of vegetation reduces the fraction of rainfall going into runoff. (See Figure 13.9.) As runoff is part of the freshwater flux into the surface ocean, changes in terrestrial ecosystems can in principle affect ocean dynamics. Transpiration cools the surface during the daytime and increases air humidity in the near-surface atmospheric layer. Increased concentration of water vapor (a greenhouse gas) leads to reduced fluctuations in the diurnal temperature cycle by increasing the night temperatures. Photosynthesis is tightly coupled to transpiration, but while increased atmospheric CO_2 concentration in the future is likely to enhance photosynthesis, it may tend to reduce transpiration due to reduced stomatal conductance (*medium certainty*).

- **Cloud formation** has strong but complex effects on global and regional climate (Stocker et al. 2001). Evapotranspiration determines the availability of water vapor for the formation of clouds. Clouds alter the radiation balance (low clouds are cooling while high cirrus clouds are warming), air circulation, and precipitation. Vegetation also affects cloud formation via changes in surface albedo and roughness. Some of the atmospheric constituents with ecosystem sources act as cloud condensation nuclei: in particular DMS emitted by marine plankton, VOCs emitted by some types of vegetation, and some aerosols emitted during biomass burning. Increased concentrations of CCNs produce more and smaller cloud droplets, making clouds more reflective and persistent; this has a cooling effect on Earth. In addition, such clouds tend to rise in the atmosphere, delaying the onset of rain; increasing ice formation, rainfall intensity, and lightning; creating more violent convective storms; and altering energy balances and air circulation (Andreae et al. 2004). The net effect on the total rainfall within a given area is unknown.

Both marine and terrestrial biota naturally regulate CCN concentrations to remain at fairly low levels (Charlson et al. 1997; Williams et al. 2002). Increased DMS and VOC emissions increase CCNs, which reduces radiation and cools the planet; this in turn reduces photosynthesis and emissions of DMS and VOCs and increases thermal stability, thus reducing the probability of cloud formation in a negative feedback loop. The natural regulation mechanism is becoming overwhelmed by anthropogenic emissions of aerosols and deforestation. Some aerosols, such as soot particles, absorb sunlight, which cools the surface and heats the atmosphere, reducing cloud formation.

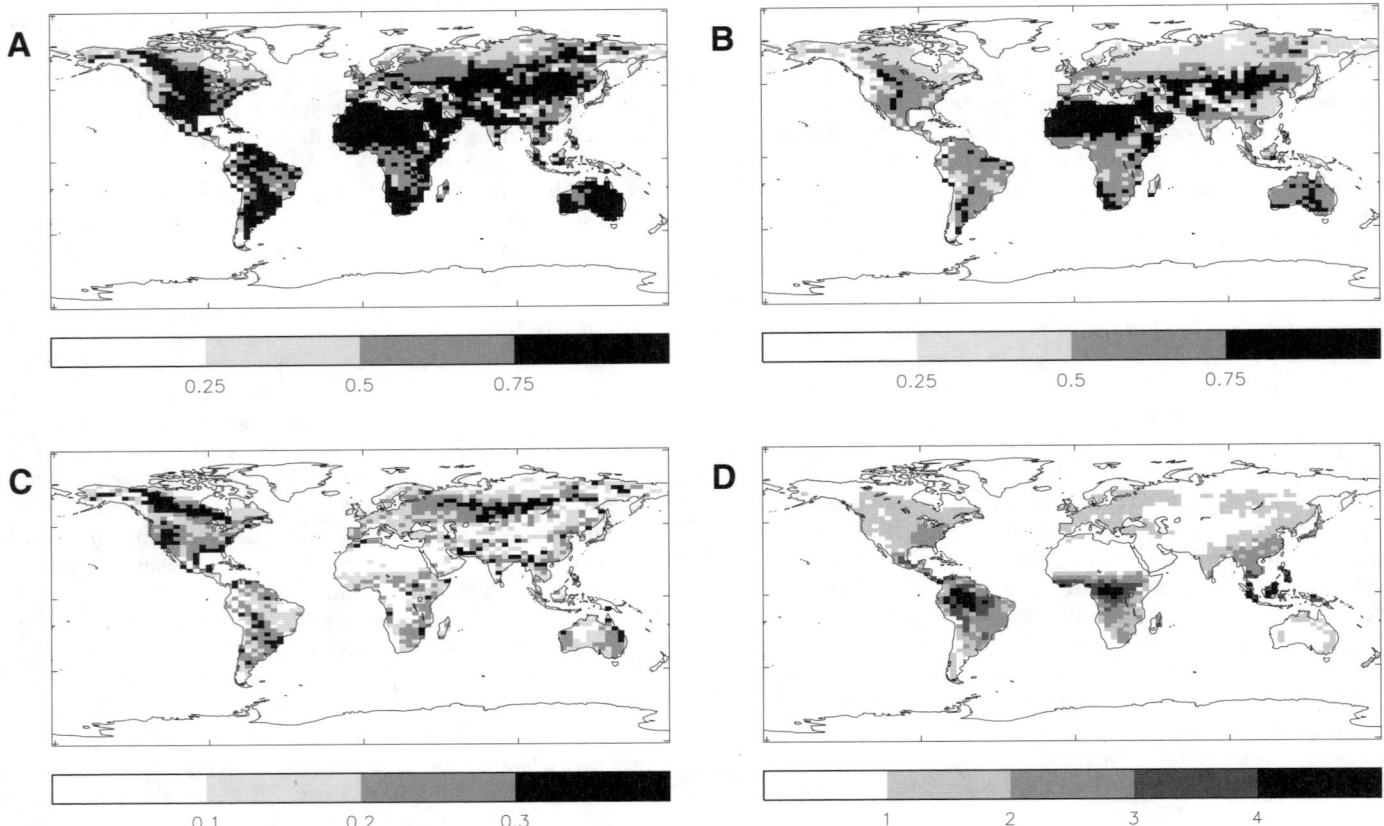

Figure 13.9. The Influence of Terrestrial Vegetation on Water Recycling (in mm/day). A general circulation model of climate has been used to simulate the ratio of land evapotranspiration to precipitation (Betts 1999) with (A) present-day vegetation and (B) all vegetation removed leaving bare soil. The difference between the two (C) illustrates the general increase in recycling of water back into the atmosphere via evapotranspiration when vegetation in present. For reference, (D) shows the absolute rate of evapotranspiration simulated with present-day vegetation.

- **The aerodynamic properties of the surface** (roughness length) modify the strength and direction of the surface wind. On land, the height and cover of surface vegetation are the main determinants of roughness length.
- **Sea surface temperature** is warmed by phytoplankton that trap radiation within the surface layer. A warmer surface reduces vertical mixing (ocean stratification) and ice cover, with potential feedbacks on regional circulation (Sathyendranath et al. 1991; Miller et al. 2003). Stratification reduces the flow of nutrients, feeding back on phytoplankton growth and composition, CO_2 uptake, and climate.

13.3.2 Biophysical Hotspots

The biophysical impacts of land use on climate are region- and season-specific and are often confounded by other climate drivers (for example, greenhouse gases and aerosols). On the global scale, biophysical effects of historical land cover changes are limited. Most model results indicate a slight biophysical cooling, partly offsetting the biogeochemical warming due to CO_2 emissions from land cover change (Betts 1999; Matthews et al. 2004; Brovkin et al. 2004). Regionally, the biophysical climate effects of ecosystem change can be substantial.

13.3.2.1 Tropical Forests

Tropical forests in South America, Africa, and Southeast Asia are being cleared to make land available for agriculture. (See Chapters 21 and 26.) The change in vegetation from forest to pasture or crops increases surface albedo (leading to cooling of the atmosphere), decreases roughness, and reduces evapotranspiration (reducing rainfall and leading to local warming) (Henderson-Sellers et al. 1993). Measurable impacts can be expected when the area of deforestation is on a scale of a few hundred kilometers. Trees can get access to deep soil water and have been observed to maintain evapotranspiration through the dry season at levels equal to the wet season. Thus extensive deforestation generally leads to decreased regional rainfall (see Box 13.2), although under some conditions rainfall can be higher over areas with partial deforestation compared with areas with no deforestation (Durieux et al. 2003; Avissar et al. 2002). Since forest existence crucially depends on rainfall, the relationship between tropical forests and precipitation forms a positive feedback, which, under certain conditions, theoretically leads to the existence of two steady states: rainforest and savanna (Sternberg 2001; Oyama and Nobre 2003), although some models suggest only one stable climate-vegetation state in the Amazon (Claussen 1998).

Biophysical impacts of tropical deforestation are different in the wet and dry seasons. In the dry season, non-forest areas become hot and dry during the daytime, and air flows from forests to non-forest areas, enhancing thermal turbulence. These conditions favor the formation of shallow rain clouds over non-forest areas. During the wet season, evaporation in forest and pasture are about the same but the forest reflects less radiation due to lower albedo, leading to warming of the planetary boundary layer.

Case Study: Deforestation and Rainfall Impacts in the Amazon (Maria Silva Dias, personal communication, 2002)

The state of Rondônia in the southwest Amazon was opened for colonization in the early 1970s by developers, following government incentives. Afonso Andrade was one of the first farmers, clearing an area of about 2,000 hectares for farming cattle and crops like rice, beans, and corn for local consumption. During the first five years, when Andrade's property was mostly surrounded by forest, he would plant a brown bean crop at the end of the wet season. Even if rain was scarce, the seeds would germinate and the crops would grow because dew was very abundant during the nighttime, and "the soil would be wet early in the morning." The flow of moisture from the neighboring forest during daytime would sustain the dew formation during nighttime. As deforestation proceeded, however, the forest was further away in successive years, the atmosphere and the soil became drier during the dry season, and the forest moisture was diluted into a larger area. Now it is very hard to get a crop going during the dry season without irrigation.

Warming generates more thermal turbulence, which favors the formation of clouds and rainfall over forest areas (Nobre et al. 2004). Cloud formation is further affected by aerosols from biomass burning and other sources (Williams et al. 2002).

DeFries et al. (2002) found that projected future land cover change, which is likely to occur mostly in the sub-tropics and tropics, will have a warming effect on climate, driven mostly by decrease in evapotranspiration. Increasing atmospheric CO_2 decreases stomatal conductance in many species. If this effect occurs on a large scale, it will reduce latent heat flux and therefore increase land surface temperature (Sellers 1996). In some cases, agricultural leaf area index (the area of green leaf per unit area of ground) is higher than forest leaf area index, which will reduce the effect of decreased stomatal conductance on large-scale transpiration (Betts et al. 1997). Decreased canopy conductance may contribute to a decrease in precipitation in regions where water recycling by vegetation is important—for example, Amazonia (Betts et al. 2004).

13.3.2.2 Boreal Forests

The presence of forests in boreal regions reduces the albedo of the land surface compared with short tundra vegetation. Solar radiation is trapped within the forest canopy, causing warming (Betts and Ball 1997). This effect is particularly accentuated during the snowy season, when short vegetation becomes fully covered with snow, which strongly reflects solar radiation back to the atmosphere (Harding and Pomeroy 1996; Hall et al. 2004). Increased air temperature leads to earlier snow melt. The treeline boundary is limited by temperature, so the relationship between forest and air temperature forms a positive taiga-tundra feedback. This biophysical mechanism plays a substantial role in Earth System dynamics; for example, a reduction of forest cover may have helped trigger the onset of the last glaciation (Gallimore and Kutzbach 1996; de Noblet et al. 1996), while enhanced forest cover has contributed to the regional warming during the mid-Holocene (Foley et al. 1994).

Boreal deforestation leads to spring cooling and extension of the snow season due to albedo changes. During the growing season, trees have a denser, more productive canopy than herbaceous plants, and therefore they transpire more water, cooling surface air (Pielke et al. 1998). This hydrological effect is of primary importance during the summer, so deforestation leads to summer warming. Model results suggest that for deforestation in most boreal forest areas, the cooling effect of albedo changes dominates over the hydrological warming effect on annual average surface temperature (Chalita and Le Treut 1994; Betts 1999; Brovkin et al. 1999). A sea ice–albedo feedback enhances the cooling effect of boreal deforestation (Bonan et al. 1992; Brovkin et al. 2003).

In the recent past, deforestation in the temperate and boreal regions has likely led to a biophysical cooling that partially offset the warming effect of associated CO_2 emissions. In the future, reforestation of regions permanently covered by snow in winter is likely to lead to an increase in global temperature, as the biophysical warming due to albedo changes outweighs the biogeochemical cooling due to uptake of CO_2 in forest stands (Betts 2000; Claussen et al. 2001). This would be counter to the aims of carbon sequestration schemes in these regions. It is also expected that warming at the high northern latitudes will be substantially amplified through the taiga-tundra feedback (Brovkin et al. 2003), although changes in permafrost, forest fire frequency, and outbreaks of pests complicate projections for vegetation cover dynamics in the boreal and polar regions. (See also Chapters 21 and 25.)

13.3.2.3 Sahel/Sahara

In the Sahel region of North Africa, vegetation cover is almost completely controlled by rainfall. The rainy season lasts between two and four months, during which rainfall is highly unpredictable: rainy days can be followed by weeks with no rainfall. When vegetation is present, it quickly recycles this water, as Figure 13.9 illustrated, generally increasing regional precipitation and, in turn, leading to a denser vegetation canopy (Dickinson 1992; Xue and Shukla 1993). This positive feedback between vegetation cover and precipitation amplifies rainfall variability in the Sahel (Zeng et al. 1999). Model results suggest that land degradation leads to a substantial reduction in water recycling and may have contributed to the observed trend in rainfall reduction in the region over the last 30 years (Xue et al. 2004). Combating degradation should maintain or restore the water recycling service, increasing precipitation and contributing to human well-being in the region. (See Chapter 22.)

The Sahara Desert, to the north of the Sahel, is another important example of ecosystem-climate interactions. While rainfall there is too low to support much vegetation at present, it was not so in the past. During the mid-Holocene, about 9,000–6,000 years ago, Sahelian vegetation was greatly extended to the north (Prentice and Jolly 2000). Changes in Earth's orbit increased summer insolation, enhancing monsoon circulation and increasing moisture inflow into the region, and this effect was greatly enhanced by the reduced albedo of the vegetation itself (Kutzbach et al. 1996; Braconnot et al. 1999; Claussen et al. 1999; Joussaume et al. 1999). A rather abrupt collapse in west Saharan rainfall and vegetation cover occurred about 5,500 years ago (deMenocal et al. 2000). This abrupt change is consistent with the existence of alternative stable states in the climate-vegetation system (Claussen 1998; Brovkin et al. 1998). (See Figure 13.10.)

The Sahara desert today differs from other sub-tropical deserts in its exceptionally high albedo. Net cooling of the atmosphere leads to a horizontal temperature gradient and induces a sinking motion of dry air, suppressing rainfall over the region (Charney 1975). Low precipitation reduces the vegetation cover, increasing bare ground with high albedo. This positive feedback maintains desert conditions. On the other hand, if precipitation increases there is more vegetation, the albedo is lower, surface temperature

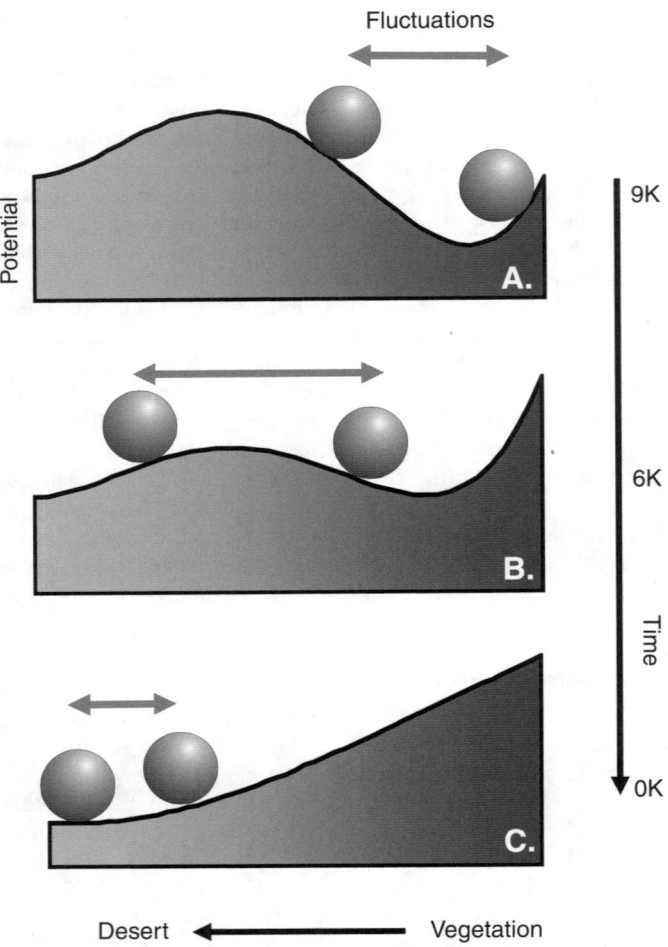

Figure 13.10. Changes in Stability of the Climate-Vegetation System in the Sahara/Sahel Region According to Climate Model Simulations of the Last 9,000 Years (Renssen et al. 2003). Hypothetically, strong positive feedbacks between precipitation and vegetation in the Sahara/Sahel region can lead to the existence of two steady states, desert and "green" (savanna-like). Balls and arrows in the figure indicate the maximum range of the fluctuations induced by large-scale atmospheric and oceanic variability. A) About 9,000 years ago, the system fluctuated in the vicinity of the green state, while the desert state with a shallow potential minimum was also stable. B) About 6,000 years ago, the potential became equal for both states and the system fluctuated between desert and green states. C) Later, the green state lost stability. Desert is the only steady state at present, and precipitation variability is reduced in comparison with the two-state system (A and B).

is higher, and the gradient in temperature between the land and ocean increases—amplifying monsoon circulation and upward air motion over the desert, resulting in increased summer rainfall.

A shift from "green" to "desert" state and vice versa is strongly influenced by externally induced fluctuations in rainfall (Wang and Eltahir 2000; Renssen et al. 2003). In the future, global warming may increase moisture in the Sahara/Sahel region (Brovkin et al. 1998), but it is unlikely that Sahara greening will reach mid-Holocene levels (Claussen et al. 2003).

13.3.2.4 Wetlands

Evapotranspiration of water from wetlands during the day leaves the air above the surface heavy with water vapor, preventing a loss of energy. In sub-tropical regions this is often sufficient to hold night temperatures above freezing point. When wetlands are drained, this ecosystem service is lost. Marshall et al. (2003) showed that the likelihood of agriculturally damaging freezes in southern Florida has increased as a result of the conversion of natural wetlands to agriculture. In January 1997, a rare freeze in southern Florida caused losses to vegetable and sugarcane crops that exceeded $300 million.

13.3.2.5 Cultivated Systems

Intensified agriculture generally increases leaf area index because crops are bred for maximum ability to intercept light. Therefore the effects of past extension of agriculture may, to a certain extent, be reversed by intensification, with increased leaf area during the summer season being a move back toward denser vegetation with characteristics closer to a natural forested state (Gregory et al. 2002). Where seasonal crops replace evergreen vegetation, this will not be the case.

13.4 Effects of Ecosystem on Air Quality

Ecosystems effect the concentrations of many atmospheric compounds that have a direct deleterious effect (for example, pollution) or a beneficial effect (for example, fertilization) on human well-being. Ecosystems are often both sources and sinks for various trace gases that undergo complex atmospheric reactions, simultaneously affecting several aspects of air quality in different ways. It is therefore often hard to quantify the current net effect of ecosystems or of ecosystem change on a particular aspect of air quality. This section concentrates on main net effects, complementing the summary information for each atmospheric constituent that was presented in Table 13.1.

13.4.1 OH and Atmospheric Cleansing Capacity

Some reactive gases released or absorbed by ecosystems are involved in maintaining the ability of the atmosphere to cleanse itself of pollutants through oxidation reactions that involve the hydroxyl radical. (See Box 13.3.) The reactions are complex but, generally, emissions of NO_x and hydrocarbons from biomass burning increase tropospheric ozone and OH concentrations. CH_4 and CO are removed by OH, so emissions of these gases from wetlands, agriculture, and biomass burning decrease OH concentration. Net effects of deforestation are uncertain.

A proxy for estimating atmospheric cleansing capacity (or tropospheric oxidizing capacity) is the concentration of OH, but this is hard to measure directly because of its short lifetime (on the order of seconds). Model estimates, based on measurements of compounds destroyed by OH, indicate a decline in OH concentration since preindustrial times, but the change has probably been less than 10% (Prinn et al. 2001; Jöckel et al. 2003). There is some concern that the pursuit of a hydrogen energy economy could further reduce oxidizing capacity (Schulze et al. 2003; Warwick et al. 2004). A threshold-dependent collapse in atmospheric cleansing capacity would have major implications for air quality (Brasseur et al. 1999). The fundamental importance of tropospheric oxidizing capacity to air quality and atmospheric chemistry means that improving the understanding of OH in the atmosphere, and of the role of ecosystems in regulating OH, is a focus for intensive scientific research.

13.4.2 Pollution Sinks

13.4.2.1 CO₂ and Ocean Acidification

Increased atmospheric CO_2 concentrations are having adverse effects on certain ocean ecosystems, and since terrestrial ecosystems are currently a net sink for CO_2 (as described earlier), this could be considered an ecosystem service of benefit to human well-being. The component of anthropogenic CO_2 from fossil fuels and land use change that has dissolved in the ocean has increased the acidity of the ocean to a degree that is unprecedented in recent geological history (Caldeira and Wickett 2003). The global mean surface ocean pH has decreased by 0.12 units since preindustrial times. A further decrease of 0.25 units will occur during this century if atmospheric CO_2 concentration rises to 750 ppmv (Wolf-Gladrow et al. 1999).

Increased acidity changes carbonate chemistry in the surface ocean, with negative impacts on ocean organisms such as corals, coccolithophores, and foraminifera that make their bodies from calcium carbonate. The rate of calcification (growth) of the organisms will decrease (Riebesell et al. 2001), with possible consequences for ecosystem services such as tourism and fish production (Guinotte et al. 2003). A rise in the atmospheric CO_2 concentration to double the preindustrial level may reduce global calcification by as much as 30% (Kleypas et al. 1999). The negative feedback of declining growth of calcifying organisms on the atmospheric CO_2 increase is likely very small (Heinze 2004).

13.4.2.2 Tropospheric Ozone

Sources, sinks, and trends in tropospheric ozone concentration were described earlier in this chapter. Tropospheric ozone can have adverse effects even at relatively low concentrations. Harmful concentrations occur in urban areas but also in the region of vegetation fire events and of high NO_x emissions from fertilizer use, particularly when atmospheric conditions trap air, such as in valleys or temperature inversions. While ecosystems in some regions are a source of ozone precursors, globally they are a net sink. Ozone is destroyed by reacting with plant tissues.

Adverse human health effects of tropospheric ozone include impacts on pulmonary and respiratory function and the aggrava-

tion of pre-existing respiratory diseases such as asthma, resulting in increased excess mortality (Thurston and Ito 1999; WHO 2000). Current tropospheric ozone concentrations in Europe and North America cause visible leaf injury and reduced yield of some crops and trees (Mclaughlin and Percy 1999; Braun et al. 1999; Mclaughlin and Downing 1995; Ollinger et al. 1997). It is estimated that for wheat there is a 30% yield reduction for a seasonal seven-hour ozone daily mean of 80 nanomoles per mole, a concentration level that has been found in parts of the United States and Europe (Fuehrer 1996). Economic losses in the Untied States alone may amount to several billion dollars annually. Fowler et al. (1999) estimate that the proportion of global forests exposed to potentially damaging ozone concentrations will increase from about 25% in 1990 to about 50% in 2100.

13.4.2.3 Acid Rain and Acid Regulation

Acid deposition of SO_2, sulfate, NO_x, and nitrate has increased mainly due to industrial emissions (Satake et al. 2000; Rodhe et al. 1995). Biomass burning emissions also affect the chemical composition of rainfall over large areas, while soil emissions of NO_x are as important as industrial emissions in tropical areas, with large increases driven by increasing fertilizer use. Ecosystems are also a sink for NO_x and sulfur compounds. The net effect of ecosystems is probably a sink for compounds that contribute to acid rain. Industrial sources of acid deposition have caused damage to sensitive ecosystems, especially in northern Europe and parts of northeastern North America (Emberson et al. 2003; see also MA *Scenarios,* Chapter 7). Industrial emissions are now declining in these areas due to pollution control, but they continue to rise in other areas, such as Southeast Asia.

Ammonia is the only important gaseous alkaline component in the atmosphere. NH_3 neutralizes many of the acid compounds emitted, forming sulfate and nitrate aerosols. NH_3 in high concentrations is harmful to human health, causes eutrophication of lakes, and can also contribute to acidification in N-saturated ecosystems. Although the quantity of NH_3 volatilized from fertilized fields and animal feedlots can be extremely large (Bouwman et al. 1997), ecosystems as a whole are a net sink for NH_3 (Dentener and Crutzen 1994; Holland et al. 1997).

13.4.2.4 Stratospheric Ozone

Decreases in stratospheric ozone and subsequent increases in ultraviolet (UV-B) radiation have adverse effects on human and animal health, plant growth and mortality, and marine organisms. The net effect of ecosystems on stratospheric ozone is very small compared with industrial emissions, but the effect is probably to reduce the quantity of stratospheric ozone-destroying compounds. As with acid rain, compounds that destroy stratospheric ozone have greatly increased as a result of industrial activities. There are natural ecosystem sources of methyl halides (marine ecosystems and biomass burning), and ecosystems are the primary source of N_2O, which also destroys stratospheric ozone when oxidized in the atmosphere. However, ecosystems are also sinks for halocarbons. As industrial emissions have been controlled by the Montreal Protocol and fallen dramatically, concentrations of most ozone-depleting gases in the atmosphere are now at or near a peak (WMO 2003), while concentrations of N_2O continue to rise.

13.4.3 Pollution Sources

13.4.3.1 Biomass Burning—Vegetation Fires

Vegetation fires are a common natural phenomenon in many regions and vegetation types. Fires maintain vegetation diversity, productivity, and nutrient cycling (although too-frequent fire can lead to impoverishment of nutrients). Fire is also a common land management tool, particularly in the tropics, where it is used to clear land (shifting agriculture, for example, or disposing of residues) (see Box 13.4), to maintain grasslands for cattle grazing, to prevent encroachment of weeds, and to prevent destructive canopy fires and catastrophic wildfires. Prevention of uncontrolled wildfire is an integral component of land use policies: balancing the benefits of controlled use of fire with minimizing the many adverse effects of uncontrolled fires. Several studies have suggested increased fire risk during the twenty-first century (e.g., Flannigan and van Wagner 1991; Price and Rind 1994; Stocks et al. 1998; Mouillot et al. 2002).

Combustion of plant biomass in vegetation fires (or as an energy source, see Chapter 9) produces a mixture of compounds, including greenhouse gases, toxic pollutants, and reactive gases. Toxic pollutants are mostly the result of incomplete combustion; biomass burning is often inefficient, varying with biomass type and load, fire intensity, and weather conditions. Air pollution from biomass burning is associated with a broad spectrum of acute and chronic health effects (Schwela et al. 1999; WHO 2002; Brasseur et al. 2003b). Emissions associated with burning biomass for energy (mostly fuelwood and dung) have been linked to high levels of indoor pollution and major health effects (see Table 13.3); we do not consider this an ecosystem source, however, but rather one arising from use of ecosystem resources. Estimates of the emissions of biomass combustion products in vegetation fires are subject to large uncertainties because of the difficulties inherent in estimating the amount of biomass burned (Andreae and Merlet 2001).

13.4.3.1.1 Particulates

Particles small enough to be inhaled into the lungs, typically less than 10 micrometers (PM10), are associated with the most serious effects on humans, including respiratory disease, bronchitis, reduced lung function, lung cancer, and other cardiopulmonary sources of mortality and morbidity. Studies do not show threshold concentrations below which effects are not observed (WHO 2000, 2002). Particulate levels in plumes associated with large-scale tropical fires can be 2 to 15 times those observed in urban areas (Brauer and Hisham-Hashim 1998), although exposure levels are generally less than those from indoor air pollution.

During the peak of the burning season, the number of particles in the air is an order of magnitude higher than during the rest

---

BOX 13.4

Case Study: Biomass Burning in and around the Amazon Basin during the Dry Season

Farmers prepare their crop and pastureland by burning to clear weeds; developers burn the forest to open new farmland. The alternative to burning crop and pasture land is to use herbicides or machinery to cut and mulch the weeds. Herbicides are expensive for the local farmer. The machinery is expensive to buy and operate (but could be bought by cooperatives). Burning is cheap and effective in the short term (two to three years), although it becomes costly to the crop yield in the medium term (three to five years). In areas with poor soils, open areas have been abandoned and a secondary forest is growing back. In areas with less poor soils, fertilizers have been introduced to compensate for the lack of nutrients. In a few places the culture of not burning is beginning to be adopted by more educated farmers.

---

of the year. Solar radiation reaching the surface is reduced by about 10–30%, lowering both the surface temperature and the light available for plant growth. In China, the effect of atmospheric aerosols and regional haze from all pollution sources is reducing wheat yields by 5–30% (Chameides et al. 1999). In some circumstances, for example in cloudless conditions in areas with high solar radiation (Cohan et al. 2002), plant production could be increased by haze cover. Particulates tend to reduce rainfall but increase the likelihood of intense storms (Andreae et al. 2004). Rainfall reduction has a positive feedback effect, making further fires more likely in, for example, the Amazon (Koren and Kaufman, 2004), but intense rains storms can contribute to putting out fires.

13.4.3.1.2 Tropospheric ozone

Biomass burning emits tropospheric ozone precursors (VOCs, NO_x, CH_4, and CO). The impacts of tropospheric ozone were described earlier. High levels of tropospheric ozone can accumulate at a regional scale during the biomass-burning season (e.g., Swap et al. 2003). The interannual trends and seasonal cycle of tropospheric ozone concentrations correspond to the seasonal cycle and extent of biomass burning in tropical Africa, Latin America, and Asia (Thompson and Hudson 1999). Ozone concentrations can reach values that are larger than those observed in the most polluted urban areas of the world. Values of 100–120 ppb have been measured during the dry season in the southwest Amazon Basin; background values during the rainy season are around 10–15 ppb (Kirchhoff et al. 1996; Cordova et al. 2004).

13.4.3.1.3 Carbon monoxide

Potential health effects of CO include hypoxia, neurological deficits, neurobehavioral changes, and increases in daily mortality and cardiovascular diseases. CO toxicity is mostly associated with indoor pollution from biomass fuel burning, but some studies show effects of CO even at very low concentrations (WHO 2000; Schwela et al. 1999). Dangerous levels are only occasionally observed during vegetation fires, although fatalities caused by excessive carbon monoxide concentrations alone or in combination with other pollutants have been reported, as in China in 1987 (Schwela et al. 1999).

13.4.3.1.4 Other compounds

Volatile organic compounds, including benzene, toluene and xylene, and polynuclear aromatic hydrocarbons, have been identified in fire smoke plumes (Muraleedharan et al. 2000; Radojevic 2003) and are known or suspected carcinogens, mutagens, and teratogens with the potential to cause serious long-term effects. Volatized heavy metals can also pollute the environment. In 1992, severe wildfires spread into the 30-kilometer exclusion zone around the Chernobyl Power Plant in Belarus, burning radioactively contaminated vegetation and increasing the level of radioactive cesium in aerosols 10 times (Dusha-Gudym 1996).

13.4.3.2 Nitrogen Pollution

Elevated deposition of nitrogen compounds is driven by emissions of various N gases. Ecosystem sources include increased fertilizer use, rice paddies, ruminant animals, and biomass burning. Ecosystem N deposition adds to the burden of nitrate leaching to groundwater due to fertilizer use in agricultural ecosystems, causing changes in the functioning and stability of many sensitive ecosystems (for example, heathlands and bogs), particularly in the most populated parts of the world. In heavily affected areas (the Netherlands, for instance), the drinking water standard for nitrate has been exceeded, and lakes and coastal waters suffer from eutrophication. Episodes of high wet deposition of nitrogen are suspected of occasionally affecting even remote marine ecosystems. Changes in the nitrogen cycle and its impacts are dealt with in detail in Chapter 12.

13.4.3.3 Dust Pollution

Agricultural intensification coupled with increasing population density and climate variability in many areas of Sahelian Africa have led to soil degradation and greater soil exposure to wind erosion, increasing the sources of atmospheric dust in recent years (N'Tchayi 1994; Nicholson 1998). An increase in local dust storms is widely considered to be related to ill health (fever, coughing, sore eyes) and has been implicated in meningococcal meningitis epidemics in the region (Molesworth 2002). Dust emanating from this region and the Sahara has been implicated in respiratory problems as far away as the United States (Prospero 2001). Dust storms cause a strong reduction in visibility, resulting in serious disruptions in ground and air traffic. These conditions not only occur in the dust source regions themselves, they can also be problematic downwind. For example, visibility in Beijing is often adversely affected by dust storms originating in the Gobi Desert in springtime (Sugimoto et al. 2003; Shimuzu et al. 2004).

The major dust source regions are deserts, but semiarid regions where vegetation is sparse and soil surfaces disturbed by human activities also contribute to the atmospheric burden of dust. Soil disturbance has been estimated to account for up to 10% of total dust emissions (Tegen et al. 2004). Long-term increases in dust over the Atlantic are possibly associated with desertification of northern Africa (Prospero and Lamb 2003). There are many land use and climate drivers and feedbacks that are likely to affect dust production in the future, with climate drivers dominating. While both the magnitude and the direction of change in dust are uncertain, some models suggest that dust production could decline in a warming world due to increased vegetation in arid and semiarid regions (Mahowald and Luo 2003; Tegen et al. 2004).

13.4.4 Fertilizing Effects

Carbon dioxide, nitrogen gases (NO_x, NH_3), and nutrients in dust particles can all have fertilizing effects on terrestrial plants, potentially increasing production of services such as food and timber. Fertilization from these sources is one possible contributory mechanism for the increasing terrestrial CO_2 sink in recent years. Estimates of the magnitude of this effect in the past and future are limited by incomplete understanding of soil carbon dynamics, plant nutrient relationships, and plant physiology (Oren 2001; Finzi et al. 2002; Hungate et al. 2003; Zak et al. 2003; Norby et al. 2004; Nowak et al. 2004). Nitrogen and dust deposition on the ocean also have the potential to increase phytoplankton production, and the supply of iron in dust is thought to be a major control on the strength of biological carbon uptake in the ocean (Martin et al. 1991). Ecosystems are currently a net sink for CO_2 and a net source of NO_x, while vegetation cover reduces dust emissions (as described earlier in this chapter).

13.4.4.1 Carbon Dioxide

Carbon dioxide in the atmosphere and ocean is necessary to support plant photosynthesis. Most marine plants are not limited by CO_2 but by nutrients and light. On the other hand, most terrestrial plants are limited to some extent by CO_2 supply, and thus rising atmospheric concentrations can have a fertilizing effect, enhancing productivity both directly (Farquhar et al. 1980) and indirectly though stomatal closure and improvements in water use

efficiency (Drake et al. 1997; Farquhar 1997; Körner 2000). The strength of the response in terrestrial plants depends on the photosynthetic pathway. Theoretically, those with a C_3 pathway (trees, cold climate plants, most nontropical grasses, and most agricultural crops, including wheat and rice) respond more strongly than those with a C_4 pathway (most tropical grasses, some desert shrubs, and some crops, including maize and sugarcane), although field experiments suggest a more complex picture (Owensby et al. 1993; Polley et al. 1996; Porter and Navas 2003; Nowak et al. 2004).

Experimental doubling of CO_2 in Free-Air CO_2 Enrichment systems produces an average aboveground biomass increase of 17% for C_3 and C_4 agricultural crops and a 20% increase in agricultural yield under conditions of ample N and water, but there is a wide range of responses among individual studies (Kimball et al. 2002). Increases are generally greater in conditions of low water availability. Trees in open top chambers have shown an enhancement of the annual wood mass increment of about 27% (Norby et al. 1999). This strong response of trees to elevated CO_2 has been confirmed in Free-Air CO_2 Enrichment experiments in young forest plantations (DeLucia et al. 1999; Hamilton et al. 2002; Nowak et al. 2004). However, the response of mature forests may be different from that of young forests for various reasons (Norby et al. 1999; Curtis and Wang 1998).

13.4.4.2 Nitrogen Deposition

Nitrogen limitation to plant production is widespread. (See Chapter 12.) There has been a rapid increase in reactive N deposition over the past 50 years (Vitousek et al. 1997; Holland et al. 1999). There is much field evidence that N deposition increases NPP (e.g., Chapin 1980; Vitousek and Howarth 1991; Bergh et al. 1999; Spieker et al. 1996) and soil carbon storage (Fog 1988; Bryant et al. 1998). When the nitrogen saturation limit is reached, as is thought to have happened in highly polluted areas of Europe, plants can no longer process the additional nitrogen and may suffer from deleterious effects of associated pollution (Shulze et al. 1989; Aber et al. 1998; see also Chapter 7). N addition leads to changes in plant species composition and an overall reduction in diversity. In general, the impacts are most pronounced in nutrient-poor systems, where N deposition enhances growth of the most responsive species, which often outcompete and eliminate rare species that occupy N-deficient habitats (Mooney et al. 1999).

13.4.4.3 Dust as a Fertilizer

Nutrients in dust particles (especially phosphate and iron) can act as fertilizers when deposited on oceans and land (Piketh et al. 2000). Fertilization of the ocean from expanded desert sources, and the resulting increase in ocean ecosystem CO_2 uptake, is thought to be one of the drivers of change in glacial-interglacial atmospheric CO_2 concentrations (Watson et al. 2000; Ridgwell et al. 2002; Bopp et al. 2003). Ice core records indicate decreased dust input in the ocean may account for up to about a quarter of the 80-ppm atmospheric CO_2 increase at the last glacial-interglacial transition (Rothlisberger et al. 2004). A recent synthesis (Piketh et al. 2000) suggests that aerosols derived from the southern African continent are increasing carbon uptake downwind in the Indian Ocean. Changes in dust sources, transport, and deliberate iron fertilization of the ocean could affect marine productivity, future CO_2 uptake (Mahowald and Luo 2003; Tegen et al. 2004), and marine N fixation and N_2O release (Denman et al. 1996), but the magnitude and associated impacts are uncertain. (Sources and drivers of dust are described earlier; see also Chapter 22.)

13.5 Climate Variability and Climate Feedbacks

13.5.1 Interannual Variability

Biological activities are dependent on climatic conditions, and therefore biogenic emissions of gases often show a seasonal cycle and interannual climate variability linked to natural climate variations. In some cases the interannual variability is large—for example, when biological processes are limited by water availability. This variability in turn affects biological sources and sinks of atmospheric compounds and biophysical properties of vegetation. For instance, during recent El Niño events atmospheric CO_2 increase has doubled or tripled compared with other times, partly due to reductions in land uptake caused by the effects of high temperatures, drought, and fire on terrestrial ecosystems in the tropics (Tian et al. 1998; Clark et al. 2004).

Warming enhances emissions of VOCs (which increase by 20–30% per degree Celsius) (Guenther et al. 2000), CH_4, and N gases, which generally tend to increase tropospheric ozone and OH (although these reactions also depend on water and light availability). Changes in OH concentration further affect the variability in concentration of some atmospheric compounds such as CH_4. Effects of interannual variability on biogenic sources and sinks are one of a number of processes that affect CH_4 and tropospheric ozone (Dlugokencky et al. 1998; Warwick et al. 2002; Sudo and Takahashi 2001).

Enhanced biomass associated with the La Niña phase of the El Niño-Southern Oscillation, followed by droughts in the El Niño phase, produce above average biomass burning emissions from savannas in southern Africa (Swap et al. 2003). In 1997–98, fires associated with an exceptional drought caused by ENSO devastated large areas of tropical rain forests worldwide (Siegert et al. 2001). Emissions of associated gases such as CO_2, CO, CH_4, and other trace gases have been correlated with large biomass burning events in tropical and boreal regions (Langenfelds et al. 2002).

13.5.2 Climate Feedbacks on Ecosystems, Climate, and Air Quality

Changes in global and regional climate, partially bought about by ecosystem change, can in turn lead to further changes in ecosystem sources and sinks of gases and biophysical properties. Temperature and moisture changes will cause a variety of changes in sources, sinks, and chemical reactions in the atmosphere, the net effect of which is uncertain and may differ from place to place. The predominant climate feedbacks operate though changes in the carbon cycle and CO_2 emissions, with a strong positive feedback predicted under future climate change. Methane emissions from wetlands are expected to increase under some conditions (such as permafrost melting) and to decrease in others (such as the drying of northern and tropical soils), as described earlier. Nitrous oxide emissions are generally higher in wetter soils. Increased emissions of tropospheric ozone precursors, NO_x and VOCs, occur under warmer conditions.

Where climate change causes shifts in vegetation there will be regional biophysical effects—for example, due to the northward shift of boreal forests and enhanced vegetation cover in the Sahara (Brovkin et al. 2003; Claussen et al. 2003). Vegetation loss could lead to positive climate feedbacks though biophysical effects—for instance, feedback on local drying from Amazon dieback (Betts et al. 2004)—and could lead to increased dust emissions. Changes in

water availability affect transpiration, with drought reducing water recycling and rainfall.

On the land, warming increases the rate of decomposition of soil organic matter, thereby reducing carbon storage in soil. Although soil warming experiments have shown an increased rate of decomposition for the first one to three years only (Jarvis and Linder 2000; Oechel et al. 2000; Luo et al. 2001; Rustad et al. 2001), this represents the burning-off of the labile (easily decomposed) component only. The larger pool of more chemically stable soil organic matter is potentially vulnerable to warming over longer time scales (Cramer et al. 2001; Joos et al. 2001; Knorr et al. 2005). The effect of global warming on vegetation cover is highly uncertain but likely also to affect atmospheric CO_2. One coupled climate-carbon cycle model has predicted a dieback of tropical forests in South America, which, along with increased soil organic matter decomposition and subsequent carbon loss, would lead to an additional 200 ppm increase in atmospheric CO_2 (Cox et al. 2000). Another model predicted a smaller feedback (Friedlingstein et al. 2001).

In the oceans, sea surface temperature increase and changes in the global water cycle tend to increase vertical stratification (layering) and to slow down global ocean circulation. Warming reduces the solubility of CO_2 in the ocean. Stratification slows the mixing into deep layers of excess carbon in the surface water. Stratification further reduces nutrient input into the surface zone and leads to a prolonged residence time of phytoplankton at the surface, near light. Models indicate the net effect is reduced phytoplankton productivity (Bopp et al. 2001; Joos et al 1999). Models estimate that the combined effect of warming and circulation changes on ocean physics and biology will reduce the oceanic CO_2 uptake by 6–25% in 1990–2050, thus providing a positive climate feedback (Maier-Reimer et al. 1996; Sarmiento et al. 1998; Matear and Hirst 1999; Joos et al. 1999, Bopp et al. 2001; Plattner et al. 2001).

Changes in ocean circulation, pH, and temperature are also likely to have additional effects on ocean biology that have not been quantified in these models and that may induce further CO_2 feedbacks. These include changes in the community structure, net production, and bio-calcification. The effect of bio-calcification is estimated to increase the ocean carbon sink by less than 2.5% (Riebesell et al. 2001). The quality and magnitude of biological changes will vary over space and time and is highly uncertain. While the combined inorganic and biological changes tend to reduce global uptake of anthropogenic carbon, the global net effect on carbon uptake of the ocean biological changes alone is unknown. Altered size and timing of phytoplankton blooms can also potentially reduce fish production (Chavez et al. 2003; Beaugrand et al. 2002; Platt et al. 2003).

13.6 Impacts of Changes in Climate and Air Quality on Human Well-being

13.6.1 Impacts of Changes in Climate on Human Well-being

According to the IPCC (2001d), "The earth's climate system has demonstrably changed on both global and regional scales since the pre-industrial era, with some of these changes attributable to human activities. . . . Projected climate change will have beneficial and adverse effects on both environmental and socioeconomic systems, but the larger the change and rate of change in climate, the more the adverse effects dominate." Changes in climate are linked to all aspects of human well-being as defined by the Millennium Ecosystem Assessment (MA 2003). This section provides a summary of the detailed results presented by IPCC (2001a, 2001b, 2001c, 2001d) unless otherwise stated.

13.6.1.1 Security

An increase in frequency and severity of floods and droughts has been noted in some areas. A fourfold increases in economic losses for catastrophic weather events from the 1980s to the 1990s (average annual global loss $40 billion in the 1990s) has been partly linked to regional climatic factors and partly to socioeconomic factors. IPCC projections include increasing ecological shocks and stress as well as vulnerability to them, alongside a reduction in the ability to predict and plan for the weather.

13.6.1.2 Health

Many vector-, food-, and water-borne infectious diseases are known to be sensitive to changes in climatic conditions, as are production of spores and pollens and the climatically related production of photochemical air pollutants. Floods increase risk of drowning, diarrhea, respiratory diseases, water-contamination diseases, hunger, and malnutrition. Heat waves in Europe and America have been associated with a significant increase in urban mortality. For example, during the European heat wave of 2003, almost 15,000 additional deaths were estimated to have occurred in France, mostly in elderly people (WHO 2004). Warmer wintertime temperatures can also result in reduced wintertime mortality in cold climates.

Indirect climate effects on human health include changes in water quality, air quality, food availability and quality, population displacement, and economic disruption. Poor understanding of the role of socioeconomic and technological factors in shaping and mitigating health impacts, and the difficulty in separating climate variability impacts from climate change impacts, means that current estimates of the potential health impacts of global warming are based on models with *medium* to *low certainty.*

The World Health Organization (WHO 2002) has estimated that global warming was responsible in 2000 for approximately 2.4% of worldwide diarrhea, 6% of malaria in some middle-income countries, and 7% of dengue fever in some industrial countries. (See Table 13.3.) These factors contribute to the estimated mortality of 154,000 deaths and 5.5 million disability-adjusted life years, mostly in Southeast Asia and Africa. (Such estimates are of high uncertainty, however, due to the difficulties in establishing direct causality.)

Overall, global warming is projected to increase threats to human health, particularly in lower-income populations predominantly within tropical and sub-tropical countries: thermal stress effects amplified with higher projected temperature increases; expansion of areas of potential transmission of malaria and dengue; greater increases in deaths, injuries, and infections from floods and storms; and water quality degraded by higher temperatures and salinization, with changes modified by changes in water flow volume.

13.6.1.3 Basic Material for a Good Life

The impacts of global warming include changes in species distributions, population sizes, the timing of reproduction or life-cycle events, and the frequency of pest and disease outbreaks (IPCC 2001b, 2001d). Growing season has lengthened by one to four days in the Northern Hemisphere during the last 40 years, with earlier onset of life-cycle events (such as flowering, migration, and breeding). Coral reef bleaching has increased in frequency. (See Chapter 19.)

The productivity of ecological systems is highly sensitive to climate change, and projections of change in productivity range from increases to decreases. Models of cereal crops indicate that in some temperate areas, potential yields increase for small increases in temperature but decrease with larger temperature changes. In most tropical and sub-tropical regions, potential yields are projected to decrease for most projected increases in temperature. An increase in frequency of disturbance by fire and insect pests is projected.

Stratification of the ocean at warmer temperatures may reduce phytoplankton productivity and thus fish production (Platt et al. 2003). A further increase in frequency and extent of coral reef bleaching is projected, along with loss of coastal wetlands and erosion of shorelines. Diversity in ecological systems is expected to be affected by climate change and sea level rise, with an increased risk of extinction of some vulnerable species. Projected climate change would exacerbate water shortages and water-quality problems in many water-scarce areas of the world but would alleviate it in others. Some systems—including coral reefs, glaciers, mangroves, boreal and tropical forests, polar and alpine systems, prairie wetlands, and temperate native grasslands—are particularly vulnerable to climate change because of limited adaptive capacity and may undergo significant and irreversible damage (IPCC 2001b, 2001d).

13.6.1.4 Good Social Relations

The impacts just described may compound the risk of conflict over natural resources.

Tropical and dryland regions are likely to incur more detrimental impacts than temperate and cold regions (IPCC 2001b, 2001d). People in poor countries are most vulnerable due to lower adaptive capacity. Climate change is expected to have negative impacts on development, sustainability, and equity (IPCC 2001b, 2001d; Toth 1999). The aggregated market sector effects are estimated to be negative for many developing countries for all magnitudes of global mean temperature increase studied and are estimated to be mixed for industrial countries for up to a few degrees Celsius warming and negative for warming beyond that point.

The global value of the climate regulation services of ecosystems was estimated by Costanza et al. (1997) to be $2 trillion per year, of which $800 billion was attributed to the biological role of ecosystems, principally carbon storage in forests and changes in greenhouse gas emissions and albedo from converting grasslands to agriculture. The remainder was due to nonbiological oceanic uptake of CO_2. This global value is a synthesis of published estimates of ecosystem service values for several different biomes, using a range of valuation techniques. Extrapolating from the biome values to a global aggregate is likely to underestimate the true total value because these are partial valuations in several ways. First of all, not all biomes were represented in the available literature. Second, not all processes (biochemical and biophysical) and feedbacks that generate ecosystem climate services were considered. For example, increasing loss of forests might alter other ecosystems so dramatically as to change their function in the carbon cycle, such as altering temperature in the oceans and net ocean uptake of CO_2.

Damages from reductions in carbon sequestration capacity may be nonlinear, with damages increasing more than proportionally to forest loss. The unit demand for an ecosystem service is likely to increase rapidly as its supply diminishes; in other words, there is reason to expect that the marginal value of forests for climate control may increase with forest loss. In this case, ag-

Table 13.3. Attributable Mortality and Disability-Adjusted Life Years from Environmental Risk Factors, 2000.[a] The risk factors and measured adverse outcomes of exposure are as follows: unsafe water, sanitation, and hygiene–diarrhea; urban air pollution–cardiovascular mortality, respiratory mortality, lung cancer, mortality from acute respiratory infections in children; indoor smoke from solid fuels–acute respiratory infections in children, chronic obstructive pulmonary disease, lung cancer; climate change–diarrhea, flood injury, malaria, malnutrition, dengue fever, cardiovascular mortality, population movement. (WHO 2002)

	World	Africa	North America[b]	South and Central America	Eastern Mediterranean	Europe	Southeast Asia	Western Pacific
				(thousand)				
Mortality								
Unsafe water, sanitation, and hygiene	1,730	608	1	54	270	18	699	77
Urban air pollution	799	32	28	35	59	107	164	373
Indoor smoke from solid fuels	1,619	392	0	26	118	21	559	503
Climate change[c]	154	54	0	0	21	0	74	3
DALYs								
Unsafe water, sanitation, and hygiene	54,158	18,636	61	2,045	8,932	736	19,727	4,018
Urban air pollution	7,865	485	200	360	727	859	1,852	3,386
Indoor smoke from solid fuels	38,539	12318	6	773	3,572	544	15,227	6,097
Climate change[c]	5,517	1,893	3	94	768	17	2,572	170

[a] Uncertainty ranges (range of coefficient of variation): water and indoor air pollution 0 to 4.9; urban air pollution 10 to 14.9; climate change >15.

[b] North America: United States, Canada, and Cuba.

[c] Climate change impacts are modeled effects on disease, flood risk, and food production for modeled climate in year 2000 compared with mean climate in 1961–90.

gregating the marginal valuation methods used may underestimate the economic value of total forest climate control services. While the direct use of the Costanza et al. (1997) service values is problematic in many policy spheres, which need the marginal values, the review by Balmford et al. (2002) of the relative values of intact and human-modified ecosystems suggests that in general terms, the loss of nonmarketed services associated with ecosystem loss or conversion frequently exceeds the (marketed) benefits.

13.6.2 Impacts of Changes in Air Quality on Human Well-being

Impacts of air pollution on human health can be dramatic, as exemplified by the "Asian/atmospheric brown cloud" and the smoke haze generated by 1997–98 fires in Indonesia. Health effects can also be more subtle and are increasingly widespread. Industrial pollution is not the concern of this chapter, but some of its effects were included in Table 13.3 for illustration. Ecosystem emissions, particularly those resulting from biomass burning, can add to the burden of industrial pollution and affect human well-being in nonurban areas, while ecosystem sinks can reduce the negative impacts of industrial air pollution. Some ecosystem air quality effects described earlier in this chapter are summarized below according to the Millennium Ecosystem Assessment (MA 2003) definition of well-being:

13.6.2.1 Security

Vegetation fires can cause damage to property and life, with effects of smoke on transport and effects of toxic pollutants on health. The health implications of changes in clean air are outlined in a following section.

13.6.2.2 Access to Resources

Some pollutants with ecosystem sources and sinks are deleterious to ecosystem health, affecting production of resources such as food and timber. For example, ecosystems are currently a net sink for tropospheric ozone and compounds that contribute to acid rain as well as for CO_2 (ocean acidification and impacts on marine organisms). Agricultural ecosystems are a net source of nitrogen compounds that contribute to acid rain and eutrophication of lakes, decreasing agricultural and fish production. On the other hand, some of these N compounds have a fertilizing effect, increasing plant productivity up to the point where the ecosystems become saturated with that nutrient.

13.6.2.3 Health: Clean Air

Biomass burning is a source of particulates, tropospheric, ozone and CO, all of which have harmful respiratory effects. Smoke pollution generated by vegetation fires occasionally reaches levels with major public health and economic impacts—usually when wildfires or land management fires get out of hand under extreme weather conditions. Vegetation fires particularly enhance the risk of acute respiratory infections in childhood, a major killer of young children in developing countries, and affect the health of women already exposed to high levels of indoor air pollution (Schwela et al. 1999).

Few epidemiological studies investigate short-term and long-term implications of vegetation fires for human health. The health impacts of burning biomass (mainly fuelwood or dung) as an energy source in indoor cookstoves has been studied in more detail. (See also Chapter 9.) Exposures are far more concentrated and chronic than for vegetation fires, but since many of the compounds emitted are the same, it is useful to note the impacts for comparison. WHO has estimated that 1.6 million deaths and 39 million DALYs worldwide were attributable to indoor biomass burning, with women and children particularly at risk (WHO 2002). Ecosystems are a net sink for tropospheric ozone, reducing the impacts of urban air pollution. Dust adversely affects respiration and is reduced by vegetation cover.

13.6.2.4 Good Social Relations

Ecosystem reduction of air pollution, such as acid rain and ocean acidification, can limit damage to ecosystems valuable for aesthetic, cultural, religious, recreational, or educational purposes. On the other hand, the detrimental impacts of some wildfires on economies, human health, and safety have consequences comparable in severity to other major natural hazards and could lead to transboundary conflicts. In addition to the air quality impacts of fire mentioned already, wildfire can lead to the destruction of ecologically or economically important resources (such as timber and biodiversity), adding to rapid environmental changes and degradation. Smoke plumes can cause visibility problems, resulting in accidents and economic loss including closure of airports and marine traffic.

Fires can be catastrophic in areas that have been long protected from them, allowing a buildup of fuel, and where human settlement has extended into forest areas. In the 1980s and 1990s, the most serious pollution problems were noted in the Amazon Basin and in Southeast Asia. Land use fires and uncontrolled wildfires in Indonesia and neighboring countries in 1991, 1994, and 1997 created regional smog layers that lasted for several weeks. Box 13.5 provides examples of impacts and losses of particular fire events. Advances in satellite data and atmospheric transport models are expected to improve monitoring, evaluation, and early warning systems to prevent fires or manage impacts. (See Chapter 16 for more information on fire impacts other than air quality.)

13.7 Synthesis: Effects of Ecosystem Change and Management on Climate and Air Quality Services

Table 13.4 provides a synthesis of the different biochemical and biophysical effects of each MA ecosystem type on climate and air quality. This section addresses the most pertinent types of ecosystem change and management due to the scale of their impacts or their relevance to the MA.

13.7.1 Changes in Ecosystem Cover and Management

The largest effects of ecosystems on air quality and climate due to human-induced changes in land cover and management have been associated with deforestation, agricultural management (fertilizer use, cattle, and irrigation), and biomass burning. Deforestation and agricultural practices are mainly driven by population growth, urbanization, and economic development and are modified by policies and subsidies. Chapter 7 in the *Scenarios* volume (and Chapter 3 here, more briefly) describes many of these direct and indirect driver and linkages. While industrial countries have been responsible for most of the industrial impacts on climate and air quality in the past, management of tropical ecosystems in particular has played a role and will likely continue to have a significant impact in the future.

13.7.1.1 Forest Cover and Management

Change in forest cover has had a larger impact on global and regional climate than any other ecosystem driver. Deforestation

---

BOX 13.5

Recent Major Fire Episodes and Losses

- Regional haze episodes caused by forest fires occurred in SE Asia on several occasions during the 1990s (Radojevic 2003). Measurements in Brunei in 1998 during a particularly severe haze episode caused mainly by local fires recorded many compounds, including VOCs (such as benzene and toluene), aldehydes, cresol, phenol, acetic acid, polynuclear aromatic hydrocarbons, heavy metals, and levels of particulates exceeding air quality guidelines (Muraleedharan et al. 2000). In 1994, the fires burning in Indonesia caused the visibility to drop to as low as 500 meters in Singapore. During the 1997 South East Asian smog episode, when particle levels in some areas were up to 15 times higher than normal, the Malaysian government was close to evacuating the 300,000 inhabitants of the city of Kuching (Brauer and Hisham-Hashim 1998), and the loss of an aircraft and 234 human lives in Sumatra was partially attributed to air traffic control problems caused by the smog.

- On the Indonesian islands of Kalimantan and Sumatra during 1997–98, an estimated 9 million hectares of vegetation burned. Some 20 million people in Indonesia alone suffered from respiratory problems, mainly asthma, upper respiratory tract illness, and skin and eye irritation during the episode, with nearly four times as many acute respiratory illnesses as normal reported in South Sumatra (Heil and Goldammer 2001). A first assessment of costs of damages caused by the fire episode on 4–5 million hectares was $4.5 billion (short-term health damages; loss of industrial production, tourism, air, ground and maritime transportation; fishing decline; cloud seeding and fire-fighting costs; losses of agricultural products, and timber; and direct and indirect forest benefits) (EEPSEA 1998).

- The fires burning in Mexico during the 1998 episode forced the local government to shut down industrial production in order to decrease additional industrial pollution during the fire-generated smog. Daily production losses were about $8 million (Schwela et al. 1999).

- In 2002, forest and peat fires in the Moscow region resulted in the worst haze seen in Moscow in 30 years. This has caused severe cardiovascular and respiratory problems among the population of Moscow, especially among children (GFMC 2003).

---

has been a major source of CO_2, only partially offset by reforestation, afforestation, and forest management activities and by the fertilizing effects of N and CO_2. Immediately after deforestation, tropical soils are a source of N_2O, although emissions decline to original level or below after 15–20 years. Tropical forests are also an important source of VOCs; therefore deforestation reduces VOC emissions, although this will depend on the emission rate of the replacement vegetation. Deforestation reduces the sink for tropospheric ozone and N gases (Ganzeveld and Lelieveld 2004).

Forest cover affects albedo, particularly in boreal snow-covered regions. Deforestation increases albedo (cooling). Model results suggest that historical deforestation has led to a cooling of the land surface (Betts 2001; Govindasamy et al. 2001) comparable to the warming caused by CO_2 emissions resulting from the same deforestation (Brovkin et al. 2004), and that this biophysical effect of historical deforestation is necessary to explain the observed climate during the second half of the nineteenth century (Crowley 2000; Bauer et al. 2003). Loss of forest cover profoundly affects the water cycle, reducing water recycling and local rainfall, but the net hydrological effect of deforestation is less certain, especially on a global scale (Rind 1996).

Table 13.4. Summary of Important Ecosystem Fluxes and Biophysical Properties, by Ecosystem Type

Biome	Major Biochemical Impacts	Major Biophysical Impacts

Cultivated systems	CO_2 source: conversion to cropland, management
	   sink: management (e.g., low tillage)
	CH_4 source: rice paddies, ruminant animals, termites
	   sink: upland soils
	N_2O source: soils, cattle/feedlots, fertilizer use
	NO_x source: soils
	NH_3 source: cattle, feedlots, fertilizer, plants, soils
	VOCs source: oxygenated VOCs (e.g., methanol, ethanol, acetone)
	dust source: disturbed soil surfaces and reduced vegetation cover
		albedo: increase when forest conversion to cropland, decrease in case of irrigation, decrease where leaf area index higher than natural vegetation
		transpiration: decrease in case of forest conversion to cropland, increase for irrigated systems

Dryland systems (including savannas and grasslands)	CO_2 source: biomass burning, devegetation,
	   sink: woody encroachment
	CH_4 source: biomass burning, ruminants, termites
	   sink: upland soils
	CO source: biomass burning
	N_2O source: soils
	NO_x source: soils
	NH_3 source: plants, animal waste, soils
	VOCs source: plants, biomass burning
	S source: biomass burning
	particulates source: biomass burning
	tropospheric O_3 source: biomass burning
	CO source: biomass burning
	dust source: devegetation, degradation, and erosion
		albedo: increase in case of desertification
		surface runoff: increase in case of desertification

13.7.1.2 Agriculture

Agriculture is a significant source of greenhouse gases, accounting for about 5% of total CO_2 emissions (Rosenberg et al. 1998), about a quarter of CH_4 emissions (rice paddies and ruminant animals) (Praether et al. 2001), and about a third of N_2O emissions (agricultural soils and cattle/feedlots) (Praether et al. 2001). Agricultural management can reduce carbon loss or promote storage to some extent (Lal et al. 2004; Renwick et al. 2004). The use of nitrogen fertilizers profoundly alters the nitrogen cycle, leading to increased emissions of N gases that, in addition to contributing to global warming, contribute to acid rain and eutrophication of lakes, increase the atmospheric cleansing capacity, destroy stratospheric ozone, and may cause respiratory and other health problems. Dust is lost from cultivated and denuded soil surfaces. Agricultural crops often have a higher leaf area index than natural vegetation, reducing albedo. Irrigation increases water recycling, raising latent heat flux and cooling the surface.

13.7.1.3 Wetlands

Wetland draining for agriculture, forestry, or water extraction leads to a decrease in CH_4 production and an increase in CO_2 and N_2O emissions, probably with a net decrease in radiative forcing on short time-frames (20–100 years), but in the longer term the opposite may be true (*low certainty*) (IPCC 2001a; Christensen and Keller 2003). The same will be true for wetlands experiencing drying due to global warming, such as Northern Alaska (now and in the future), and for tropical seasonally flooded areas (if they dry in the future) (*low certainty*). Where climate change has led to loss of permafrost, the net effect is increased CH_4 emissions that will likely continue in the future.

13.7.1.4 Dryland Management and Degradation

Management of drylands to increase vegetation cover and reduce soil erosion increases carbon storage, reduces dust sources, and increases rainfall recycling. Potential impacts are significant given the very large areas involved. Drylands store more carbon in soils than in biomass and are thus more vulnerable to carbon loss through soil erosion, but with good potential for increasing belowground carbon storage (IPCC 2000). (See Chapter 22.)

13.7.1.5 Biomass Burning

Biomass burning is a major source of toxic pollutants, greenhouse gases, and reactive gases—causing major health and visibility problems and contributing to global warming. Greenhouse gases emitted during fires are CO_2, CH_4 (5–10% of all sources), N_2O, and tropospheric ozone precursors: NO_x (just over 10% of all sources), CO (a quarter of all sources), and VOCs. Aerosols from biomass burning have a net cooling effect. Fire suppression reduces emissions from burning and encourages woody plant biomass to increase and act as a carbon sink. (In the United States, for example, this may have amounted to a sink of 0.2 petagrams of carbon per year during the 1980s (Houghton et al. 1999).) However, fire suppression can also increase the risk of future, catastrophic wildfires (Schwela et al. 1999). Pollutants include particulates, precursors of tropospheric ozone, and CO, plus a number of trace gases and compounds (such as polynuclear aromatic com-

Table 13.4. *continued*

Biome	Major Biochemical Impacts	Major Biophysical Impacts

Forest and woodland systems	CO_2 source: deforestation
	  sink: afforestation, reforestation, forest management
	CH_4 source: biomass burning, termites
	  sink: upland soils
	CO source: biomass burning, decomposition
	N_2O source: soils
	NO_x source: soils
	  sink: canopy
	NH_3 source: plants, animal waste, soils
	VOCs source: plants, biomass burning
	S source: biomass burning
	particulates source: biomass burning
	tropospheric O_3 source: biomass burning
	CO source: biomass burning
		albedo: increase in case of deforestation, decrease due to afforestation
		
		transpiration: decrease in case of deforestation, increase for afforestation

Urban systems	CO_2 source: biomass (fuel) burning? CH_4 source: landfill, biogas N_2O source: landfills VOCs source: landfills S source: landfills tropospheric O_3 source: indoor biomass fuel burning CO source: indoor biomass fuel burning particulates: indoor biomass fuel burning	"heat island" effect albedo: increase with expansion and vegetation replacement transpiration: decreases with expansion and vegetation replacement
Inland water systems	CH_4 source: intermittent flooding of vegetation (remineralization)	freshwater incursions to ocean and effects on ocean circulation

Coastal systems	CO_2 sink: biological pump
	  source: upwelling
	  net balance unknown
	CH_4 source: remineralization
	N_2O source: denitrification	

Marine systems	CO_2 sink: biological and solubility pumps N_2O source: remineralization CH_4 source: remineralization DMS source: plankton	phytoplankton blooms—reduced albedo (warming sea surface temperatures)
Polar systems	CO_2 source: permafrost melting CH_4 source: permafrost melting	reduced ice cover due to warmer surface and longer growing season—decreased albedo and further warming
Mountain systems	CO_2, CH_4 and N_2O production under snowpack can constitute a significant proportion of the annual trace gas budget	reduced ice cover due to warmer surface and longer growing season—decreased albedo and further warming, shift in treeline
Island systems	Depends on land cover as above, no specific impacts	deforestation changes in wind patterns alters ocean upwelling, warming sea surface temperatures
Wetlands	CH_4 source: anaerobic respiration—decreased by draining CO_2 source: peatland burning, aerobic respiration after draining CO source: peatland burning N_2O source: soils	reduced evapotranspiration in case of draining

pounds, aldehydes, organic acids, sulfur dioxide, and methyl halides (stratospheric ozone depletion)).

13.7.2 Changes in Biodiversity, Invasive Species, and Disease

Change in species diversity in the strict sense is thought not to have a large influence on climate and air quality, although climate, climate change, and air quality conditions have a large influence on biodiversity. (See Chapters 4 and 11.) Changes in the relative abundance of different functional types (such as needle-leaved versus deciduous trees, shrubs versus grasses, and diatoms versus cocolithophorids), however, may have substantial impacts on sources and sinks of gases and on other ecosystem properties (e.g., Riebesell et al. 2001; Scherer-Lorenzen et al. 2005). Loss of biodiversity could further affect the adaptability and resilience of ecosystems and their ability to migrate with changing climate (Schulze and Mooney 1993; Tilman et al. 1997; Nepstad et al. 1999; Loreau et al. 2001; Kim Phat et al. 2004).

Furthermore, the loss of particular species could have a substantial impact on ecosystem functioning. Such "keystone species" or "ecosystem engineers" (Jones et al. 1994) may not necessarily be identified in advance, which makes preventive mitigation policy difficult. For a review of climate-biodiversity interactions, see Gitay et al. (2002) as well as Chapters 4 and 11. Some examples of drivers of change in functional type abundance are provided here, including climate change, species invasions, and disease.

Encroachment of invasive woody species is generally an additional sink of CO_2, changes biophysical properties (increases LAI, increases transpiration, and reduces albedo), and may reduce biodiversity—for example, Mesquite (*Prosopis* sp.) invasions in Texas in the United States (Archer et al. 2001; Dugas et al. 1996; Gibbens 1996).

The "fertilizing" effect of increased CO_2 levels benefit some species (most trees) more than others (such as grasses) (Nowak et al. 2004), giving them a competitive edge. For example, Smith et al. (2000) showed that elevated CO_2 increased the success of an invasive C_4 grass species in the Mojave Desert, potentially reducing biodiversity and altering ecosystem function. As with invasive tree species, this functional shift from grass to trees will also affect biophysical properties, biodiversity, and sources/sinks of various trace gases. Trees are not always the winners; for example, lianas respond more strongly than trees to the fertilizing effects of increased atmospheric CO_2 concentration (Condon et al. 1992; Granados and Koner 2002; Phillips et al. 2002.). Lianas enhance tree mortality and suppress tree growth, which could ultimately reduce carbon storage in forests.

The chestnut blight in the United States around the 1900s caused a switch from chestnut trees, which do not emit isoprene (a VOC involved in tropospheric ozone formation), to oaks, which do, approximately doubling the biomass of isoprene-emitting species (Lerdau and Keller 1997).

Regime shifts of marine pelagic ecosystems, which have occurred in Arctic waters since the mid-1980s, have caused major breakdown in fishery production. Diatom-dominated phytoplankton communities were replaced by extensive coccolithophorid blooms in the Barents Sea and the eastern Bering Sea (Smyth et al. 2004), causing massive changes in ecosystem structure. With coccolithophores being predominant producers of DMS, the observed regime shifts are likely to have altered the sulfur cycle and cloud formation in these areas, affecting air quality and water recycling.

Certain types of marine organisms (calcifiers—coccolithophorids, foaminifera, and corals) form shells of calcium carbonate ($CaCO_3$). This process releases CO_2 to the ambient seawater, countering part of the photosynthetic CO_2 drawdown (that is, the drawdown of CO_2 by coccolithophorids is much smaller than that of non-calcifying phytoplankton). A shift in species composition away from coccolithophorids, for example due to changes in ocean acidity, would increase the ocean's CO_2 storage capacity. On the other hand, biogenic particles containing $CaCO_3$ and SiO_2 sink faster than other particles, which implies that the plankton (coccolithophorids and diatoms) producing these two minerals should increase the drawdown of carbon from surface to depth in the ocean. Thus shifts in composition of the marine ecosystems have the potential to influence the oceanic carbon sink (Francois et al. 2002; Klaas and Archer 2002), but at present we cannot quantify the probability, extent, or direction of the likely future changes or their consequences for climate.

References

Aber, J., W. McDowell, K. Nadelhoffer, A. Magill, G. Berntson, et al., 1998: Nitrogen saturation in temperate forest ecosystems—Hypotheses revisited. *Bioscience,* **48(11),** 921–934.

Andreae, M.O. and P. Merlet, 2001: Emission of trace gases and aerosols from biomass burning. *Global Biogeochemical Cycles,* **15(4),** 955–966.

Andreae, M.O., D. Rosenfeld, P. Artaxo, A.A. Costa, G.P. Frank, K.M. Longo, and M.A.F. Silva-Dias, 2004: Smoking rain clouds over the Amazon. *Science,* **303(5662),** 1337–1342.

Andreae, M. O., 2001: The Biosphere: Pilot or Passenger on Spaceship Earth?, In: *Contributions to Global Change Research,* Heinen, D., S. Hoch, T. Krafft, C. Moss, P. Scheidt, and A. Welschhoff (eds.), German National Committee on Global Change Research, Bonn, pp. 59–66.

Archer, S., T.W. Boutton, and K.A. Hibbard, 2001: Trees in grasslands: biogeochemical consequences of woody plant expansion. In: *Global Biogeochemical Cycles and their Interrelationship with Climate,* E.D. Schulze, S.P. Harrison, M. Heimann, E.A. Holland, J. Lloyd, C. Prentice, and D.S. Schimel (eds.), Academic Press, San Diego, pp. 115–137.

Avissar, R., P.L.S. Dias, M. Dias, and C. Nobre, 2002: The large-scale biosphere-atmosphere experiment in Amazonia (LBA): Insights and future research needs. *Journal of Geophysical Research-Atmospheres,* **107(D20),** 8086, doi:10.1029/2002JD002704.

Avissar, R., C.P. Weaver, D. Werth, R.A. Pielke Sr., R. Rabin, A.J. Pitman, and M.A. Silva Dias, 2004: Regional climate. In: *Vegetation, Water, Humans, and the Climate: A New Perspective on an Interactive System,* P. Kabat, M. Claussen, P.A. Dirmeyer, J.H.C. Gash, L. Bravo de Guenni, M. Meybeck, R.A. Pielke, C.J. Vörösmarty, R.W.A. Hutjes, and S. Lütkemeier (eds.), Springer, New York, USA, pp. 21–32.

Balch, W.M., P.M. Holligan, S.G. Ackleson, and K.J. Voss, 1991: Biological and optical properties of mesoscale coccolithophore blooms in the Gulf of Maine. *Limnology and Oceanography,* **36(4),** 629–643.

Balmford, A., A. Bruner, P. Cooper, R. Costanza, S. Farber, et al., 2002: Economic reasons for conserving wild nature. *Science,* **297(5583),** 950–953.

Bauer, E., M. Claussen, V. Brovkin, and A. Huenerbein, 2003: Assessing climate forcings of the Earth system for the past millennium. *Geophysical Research Letters,* **30(6),** 1276, doi:10.1029/2002GL016639.

Beaugrand, G., P.C. Reid, F. Ibanez, J.A. Lindley, and M. Edwards, 2002: Reorganization of North Atlantic marine copepod biodiversity and climate. *Science,* **296(5573),** 1692–1694.

Bergh, J., S. Linder, T. Lundmark, and B. Elfving, 1999: The effect of water and nutrient availability on the productivity of Norway spruce in northern and southern Sweden. *Forest Ecology and Management,* **119(1–3),** 51–62.

Betts, A.K. and J.H. Ball, 1997: Albedo over the boreal forest. *Journal of Geophysical Research-Atmospheres,* **102(D24),** 28901–28909.

Betts, R.A., 1999: Self-beneficial effects of vegetation on climate in an Ocean–Atmosphere General Circulation Model. *Geophysical Research Letters,* **26(10),** 1457–1460.

Betts, R.A., 2000: Offset of the potential carbon sink from boreal forestation by decreases in surface albedo. *Nature,* **408,** 187–190.

Betts, R.A., 2001: Biogeophysical impacts of land use on present-day climate: near-surface temperature and radiative forcing. *Atmospheric Science Letters,* **2(1–4),** 39–51 (doi:10.1006/asle.2001.0023).

Betts, R.A., P.M. Cox, M. Collins, P.P. Harris, C. Huntingford, and C.D. Jones, 2004: The role of ecosystem-atmosphere interactions in simulated Amazonian precipitation decrease and forest dieback under global climate warming. *Theoretical and Applied Climatology* [online] http://www.springeronline.com doi:10.1007/s00704-004-0050-y.

Betts, R.A., P.M. Cox, S.E. Lee, and F.I. Woodward, 1997: Contrasting physiological and structural vegetation feedbacks in climate change simulations. *Nature,* **387(6635),** 796–799.

Bonan, G.B., D. Pollard, and S.L. Thompson, 1992: Effects of boreal forest vegetation on global climate. *Nature, 359(6397),* 716–718.

Bopp, L., K.E. Kohfeld, C. Le Quéré, and O. Aumont, 2003: Dust impact on marine biota and atmospheric CO_2 during glacial periods. *Paleoceanography,* **18,** 10.1029/2002PA000810.

Bopp, L., C. Le Quéré, M. Heimann, A.C. Manning, and P. Monfray, 2002: Climate-induced oceanic oxygen fluxes: Implications for the contemporary carbon budget. *Global Biogeochemical Cycles,* **16(2),** doi:10.1029/2001GB 001445.

Bopp, L., P. Monfray, O. Aumont, J.L. Dufresne, H. Le Treut, et al., 2001: Potential impact of climate change on marine export production. *Global Biogeochemical Cycles,* **15(1),** 81–99.

Bouwman, A.F., D.S. Lee, W.A.H. Asman, F.J. Dentener, K.W. VanderHoek, and J.G.J. Olivier, 1997: A global high-resolution emission inventory for ammonia. *Global Biogeochemical Cycles,* **11(4),** 561–587.

Braconnot, P., S. Joussaume, O. Marti, and N. de Noblet, 1999: Synergistic feedbacks from ocean and vegetation on the African monsoon response to mid-Holocene insolation. *Geophysical Research Letters,* **26(16),** 2481–2484.

Brasseur, G., R.G. Prinn, and A.A.P. Pszenny (eds.), 2003a: *Atmospheric Chemistry in a Changing World: An Integration and Synthesis of a Decade of Tropospheric Chemistry Research.* Global Change: The IGBP Series, Springer, Berlin, Germany, 300 pp.

Brasseur, G.P., P. Artaxo, L.A. Barrie, R.J. Delmas, I. Galbally, et al., 2003b: An integrated view of the causes and impacts of atmospheric changes. In: *Atmospheric Chemistry in a Changing World,* G.P. Brasseur, R.G. Prinn, and A.A.P. Pszenny (eds.), Springer Verlag, Heidelberg, Germany, 207–230.

Brasseur, G.P., J.J. Orlando, and G.S. Tyndall, 1999: *Atmospheric Chemistry and Global Change.* Oxford University Press, New York, USA, 688 pp.

Brauer, M. and J. Hisham-Hashim, 1998: Indonesian fires: Crisis and reaction. *Environmental Science and Technology, 32,* 404A-407A.

Braun, S., B. Rihm, C. Schindler, and W. Flückiger, 1999: Growth of mature beech in relation to ozone and nitrogen deposition: An epidemiological approach. *Water Air and Soil Pollution,* **116(1–2),** 357–364.

Bronson, K.F., K.G. Cassman, R. Wassmann, D.C. Olk, M. van Noordwijk, and D.P. Garrity, 1997: Soil carbon dynamics in different cropping systems in principal ecoregions of Asia. In: *Management of Carbon Sequestration in Soil,* R. Lal, J.M. Kimble, R.F. Follett, and B.A. Stewart (eds.). 1, CRC Press, Boca Raton, FL, USA, pp. 35–57.

Brovkin, V., M. Claussen, V. Petoukhov, and A. Ganopolski, 1998: On the stability of the atmosphere-vegetation system in the Sahara/Sahel region. *Journal of Geophysical Research-Atmospheres, 103(D24),* 31613–31624.

Brovkin, V., A. Ganopolski, M. Claussen, C. Kubatzki, and V. Petoukhov, 1999: Modeling climate response to historical land cover change. *Global Ecology and Biogeography,* **8(6),** 509–517.

Brovkin, V., S. Levis, M.F. Loutre, M. Crucifix, M. Claussen, A. Ganopolski, C. Kubatzki, and V. Petoukhov, 2003: Stability analysis of the climate-vegetation system in the northern high latitudes. *Climatic Change,* **57(1–2),** 119–138.

Brovkin, V., S. Sitch, W. von Bloh, M. Claussen, E. Bauer, and W. Cramer, 2004: Role of land cover changes for atmospheric CO2 increase and climate change during the last 150 years. *Global Change Biology,* **10,** 1253–1266.

Brown, C.W. and J.A. Yoder, 1994: Coccolithophorid blooms in the global ocean. *Journal of Geophysical Research-Oceans, 99(C4),* 7467–7482.

Bruno, M. and F. Joos, 1997: Terrestrial carbon storage during the past 200 years: A Monte Carlo analysis of CO_2 data from ice core and atmospheric measurements. *Global Biogeochemical Cycles,* **11(1),** 111–124.

Bryant, D.M., E.A. Holland, T.R. Seastedt, and M.D. Walker, 1998: Analysis of litter decomposition in an alpine tundra. *Canadian Journal of Botany-Revue Canadienne De Botanique,* **76(7),** 1295–1304.

Caldeira, K., M.G. Morgana, D. Baldocchi, P.G. Brewer, C.-T.A. Chen, G.-J. Nabuurs, N. Nakicenovic, and G.P. Robertson, 2004: A portfolio of carbon management options. In: *Toward CO_2 Stabilization: Issues, Strategies, and Consequences,* C.B. Field and M.R. Raupach (eds.), Island Press, Washington, DC, UK, p. 527.

Caldeira, K. and M.E. Wickett, 2003: Oceanography: Anthropogenic carbon and ocean pH. *Nature, 425(6956),* p. 365.

Callaghan, T.V., L.O. Björn, Y. Chernov, F.S. Chapin III, T.R. Christensen, et al. 2004: Arctic tundra and polar desert ecosystems. In: *Impacts of a Warming Climate—Arctic Climate Impact Assessment,* ACIA, (ed.), Cambridge University Press, Cambridge, UK, 144v pp. http://www.acia.uaf.edu

Chalita, S. and H. Le Treut, 1994: The albedo of temperate and boreal forest and the Northern Hemisphere climate: a sensitivity experiment using the LMD GCM. *Climate Dynamics,* **10(4–5),** 231–240.

Chameides, W.L., H. Yu, S.C. Liu, M. Bergin, X. Zhou, et al., 1999: Case study of the effects of atmospheric aerosols and regional haze on agriculture: An opportunity to enhance crop yields in China through emission controls? *Proceedings of the National Academy of Sciences of the United States of America,* **96(24),** 13626–13633.

Chapin, F.S., 1980: The mineral-nutrition of wild plants. *Annual Review of Ecology and Systematics,* **11,** 233–260.

Charlson R.J., J.E. Lovelock, M.O. Andreae, and S.G. Warren, 1987: Oceanic phytoplankton, atmospheric sulphur, cloud albedo and climate, *Nature,* **326,** 655–661, 1987

Charney, J.G., 1975: Dynamics of deserts and drought in the Sahel. *Quarterly Journal of the Royal Meteorological Society,* **101(428),** 193–202.

Chavez, F.P., J. Ryan, S.E. Lluch-Cota, and M. Niquen, 2003: From anchovies to sardines and back: Multidecadal change in the Pacific Ocean. *Science,* **299(5604),** 217–221.

Cheng, Y.-S., 1984: Effects of drainage on the characteristics of paddy soils in China. In: *Organic Matter and Rice,* International Rice Research Institute, Los Banos, Philippines, pp. 417–430.

Christensen, T.R., T.R. Johansson, H.J. Akerman, M. Mastepanov, N. Malmer, T. Friborg, et al., 2004: Thawing sub-arctic permafrost: Effects on vegetation and methane emissions. *Geophysical Research Letters,* **31(4),** L04501, doi:10. 1029/2003GL018680.

Christensen, T.R. and M. Keller, 2003: Element interactions and trace gas exchange. In: *Interactions of the Major Biogeochemical Cycles: Global Change and Human Impacts,* J. Melillo, C.B. Field, and B. Moldan (eds.). SCOPE 61, Island Press, Washington, DC, USA.

Clark, D.A., 2004: Tropical forests and global warming: slowing it down or speeding it up? *Frontiers in Ecology and the Environment,* **2(2),** 73–80.

Claussen, M., 1998: On multiple solutions of the atmosphere-vegetation system in present-day climate. *Global Change Biology,* **4(5),** 549–559.

Claussen, M., V. Brovkin, and A. Ganopolski, 2001: Biogeophysical versus biogeochemical feedbacks of large-scale land cover change. *Geophysical Research Letters,* **28(6),** 1011–1014.

Claussen, M., V. Brovkin, A. Ganopolski, C. Kubatzki, and V. Petoukhov, 2003: Climate change in northern Africa: The past is not the future. *Climatic Change,* **57(1–2),** 99–118.

Claussen, M., C. Kubatzki, V. Brovkin, A. Ganopolski, P. Hoelzmann, and H.J. Pachur, 1999: Simulation of an abrupt change in Saharan vegetation in the mid-Holocene. *Geophysical Research Letters,* **26(14),** 2037–2040.

Cohan, D.S., J. Xu, R. Greenwald, M.H. Bergin, and W.L. Chameides, 2002: Impact of atmospheric aerosol light scattering and absorption on terrestrial net primary productivity. *Global Biogeochemical Cycles,* **16(4),** 1090, doi:10.1029/ 2001GB001441.

Condon, M.A., T.W. Sasek, and B.R. Strain, 1992: Allocation patterns in two tropical vines in response to increased atmospheric CO_2. *Functional Ecology,* **6(6),** 680–685.

Cordova, A.M., K. Longo, S. Freitas, L.V. Gatti, P. Artaxo, A. Procópio, M.A.F. Silva Dias, and E.D. Freitas, 2004: Nitrogen oxides measurements in an Amazon site and enhancements associated with a cold front. *Atmospheric Chemistry and Physics Discussions,* **4,** 2301–2331.

Costanza, R., R. d'Arge, R. de Groot, S. Farber, M. Grasso, B. Hannon, K. Limburg, S. Naeem, R.V. O'Neill, J. Paruelo, R.G. Raskin, P. Sutton, and M. van den Belt, 1997: The value of the world's ecosystem services and natural capital. *Nature,* **387(6630),** 253–260.

Cox, P.M., R.A. Betts, C.D. Jones, S.A. Spall, and I.J. Totterdell, 2000: Acceleration of global warming due to carbon-cycle feedbacks in a coupled climate model. *Nature,* **408(6813),** 184–187.

Cramer, W., A. Bondeau, F.I. Woodward, I.C. Prentice, R.A. Betts, et al., 2001: Global response of terrestrial ecosystem structure and function to CO2 and climate change: results from six dynamic global vegetation models. *Global Change Biology,* **7(4),** 357–373.

Crowley, T.J., 2000: Causes of climate change over the past 1000 years. *Science,* **289(5477),** 270–277.

Curtis, P.S. and X.Z. Wang, 1998: A meta-analysis of elevated CO_2 effects on woody plant mass, form, and physiology. *Oecologia,* **113(3),** 299–313.

de Noblet, N.I., I.C. Prentice, S. Joussaume, D. Texier, A. Botta, and A. Haxeltine, 1996: Possible role of atmosphere-biosphere interactions in triggering the last glaciation. *Geophysical Research Letters,* **23(22),** 3191–3194.

DeFries, R.S., L. Bounoua, and G.J. Collatz, 2002: Human modification of the landscape and surface climate in the next fifty years. *Global Change Biology,* **8(5),** 438–458.

DeFries, R.S., C.B. Field, I. Fung, G.J. Collatz, and L. Bounoua, 1999: Combining satellite data and biogeochemical models to estimate global effects of human-induced land cover change on carbon emissions and primary productivity. *Global Biogeochemical Cycles,* **13(3),** 803–815.

DeLucia, E.H., J.G. Hamilton, S.L. Naidu, R.B. Thomas, J.A. Andrews, et al., 1999: Net primary production of a forest ecosystem with experimental CO_2 enrichment. *Science,* **284,** 1177–1179.

deMenocal, P., J. Ortiz, T. Guilderson, J. Adkins, M. Sarnthein, L. Baker, and M. Yarusinsky, 2000: Abrupt onset and termination of the African Humid Period: rapid climate responses to gradual insolation forcing. *Quaternary Science Reviews,* **19(1–5),** 347–361.

Denman K,, E Hofmann, and H. Marchant, 1996: Marine biotic responses to environmental change and feedbacks to climate. In *Climate change 1995: Contribution of Working Group I to the Second Assessment of the Intergovernmental Panel on Climate Change,* J.T. Houghton, L.G. Meira Filho, B.A. Callender, N. Harris, A. Kattenberg, and K. Maskell (eds.), Cambridge University Press, Cambridge, 485–516.

Dentener, F. and P.J. Crutzen, 1994: A global 3D model of the ammonia cycle. *Journal of Atmospheric Chemistry,* **19,** 573–602.

Dickinson, R.E., 1992: Changes in land use. In: *Climate System Modeling,* K.E. Trenberth (ed.), Cambridge University Press, Cambridge, UK, 698–700.

Dixon, R.K., S. Brown, R.A. Houghton, A.M. Solomon, M.C. Trexler, and J. Wisniewski, 1994: Carbon pools and flux of global forest ecosystems. *Science,* **263,** 185–190.

Dlugokencky, E.J., S. Houweling, L. Bruhwiler, K.A. Masarie, P.M. Lang, J.B. Miller, and P.P. Tans, 2003: Atmospheric methane levels off: Temporary pause or a new steady-state? *Geophysical Research Letters,* **30(19),** 1992, doi:10.1029/2003GL018126.

Drake, B.G., M.A. Gonzales-Meler, and S.P. Long, 1997: More efficient plants: a consequence of rising atmospheric CO_2? *Annual Reviews of Plant Physiology and Plant Molecular Biology,* **48,** 609–639.

Dugas, W.A., R.A. Hicks, and R.P. Gibbens, 1996: Structure and function of C_3 and C_4 Chihuahuan Desert plant communities. Energy balance components. *Journal of Arid Environments,* **34(1),** 63–79.

Durieux, L., LA.T.Machado, and H. Laurent, 2003: The impact of deforestation on cloud cover over the Amazon arc of deforestation, *Remote Sensing of Environment,* **86,** 132–140

Dusha-Gudym, S.I. 1996: The effects of forest fires on the concentration and transport of radionuclides. In: *Fire in ecosystems of boreal Eurasia,* J.G. Goldammer and V.V. Furyaev (eds.),. Kluwer Academic Publishers, Dordrecht, 476–480.

EEPSEA (Economy and Environment Program for Southeast Asia), 1998: *The Indonesian Fires and Haze of 1997: The Economic Toll. Economy and Environment Program for SE Asia.* World Wide Fund for Nature (WWF), Singapore, http://www.idrc.org.sg/eepsea/specialrept/specreptIndofire.htm.

Emberson, L., M. Ashmore, and F. Murray (eds.), 2003: *Air Pollution Impacts on Crops and Forests: A Global Assessment. Air Pollution Reviews, Vol. 4,* Imperial College Press, London, UK, 388 pp.

EPICA community members, 2004: Eight glacial cycles from an Antarctic ice core. *Nature,* **429(6992),** 623–628.

Farquhar, G.D., 1997: Carbon dioxide and vegetation. *Science,* **278,** 1411.

Farquhar, G.D., S. von Caemmerer, and J.A. Berry, 1980: A biochemical model of photosynthetic CO_2 assimilation in leaves of C_3 plants. *Planta,* **149,** 78–90.

Field, C.B. and M.R. Raupach, 2004: *The Global Carbon Cycle: Integrating Humans, Climate, And The Natural World.* Vol. SCOPE Series No. 62, Island Press, Washington, DC, 526 pp.

Finzi, A.C., E.H. DeLucia, J.G. Hamilton, D.D. Richter, and W.H. Schlesinger, 2002: The nitrogen budget of a pine forest under free air CO_2 enrichment. *Oecologia,* **132,** 567–578.

Flannigan, M.D. and C.E. van Wagner, 1991: Climate change and wildfire in Canada. *Canadian Journal of Forest Research-Revue Canadienne De Recherche Forestiere,* **21(1),** 66–72.

Fog, K., 1988: The effect of added nitrogen on the rate of decomposition of organic-matter. *Biological Reviews of the Cambridge Philosophical Society,* **63(3),** 433–462.

Foley, J.A., J.E. Kutzbach, M.T. Coe, and S. Levis, 1994: Feedbacks between climate and boreal forests during the Holocene epoch. *Nature,* **371(6492),** 52–54.

Fowler, D., J.N. Cape, M. Coyle, C. Flechard, J. Kuylenstierna, K. Hicks, D. Derwent, C. Johnson, and D. Stevenson, 1999: The global exposure of forests to air pollutants. *Water Air and Soil Pollution,* **116(1–2),** 5–32.

Francois, R., S. Honjo, R. Krishfield, and S. Manganini, 2002: Factors controlling the flux of organic carbon to the bathypelagic zone of the ocean. *Global Change Biology,* **16,** doi:10.1029/2001GB001722.

Friborg, T., H. Soegaard, T.R. Christensen, C.R. Lloyd, and N.S. Panikov, 2003: Siberian wetlands: Where a sink is a source. *Geophysical Research Letters,* **30(21),** 2129, doi:10.1029/2003GL017797.

Friedlingstein, P., L. Bopp, P. Ciais, J.L. Dufresne, L. Fairhead, H. LeTreut, P. Monfray, and J. Orr, 2001: Positive feedback between future climate change and the carbon cycle. *Geophysical Research Letters,* **28(8),** 1543–1546.

Frouin, R. and S.F. Lacobellis, 2002: Influence of phytoplankton on the global radiation budget. *Journal of Geographic Research,* **107(D19),** 5-1-5-10.

Fuehrer J., 1996: The critical level for effects of ozone on crops and the transfer to mapping. In: *Critical levels for ozone in Europe.* L. Karanlampi and L. Skarby (eds), UN-ECE workshop report. Department of Ecology and Environmental Science University of Koupio Finnland.

Gallimore, R.G. and J.E. Kutzbach, 1996: Role of orbitally induced changes in tundra area in the onset of glaciation. *Nature,* **381,** 503–505.

Ganzeveld, L., and J. Lelieveld, 2004: Impact of Amazonian deforestation on atmospheric chemsitry, *Geophysical Research Letters,* **31,** L06105, doi:10.1029/

Gedney, N. and P.M. Cox, 2003: The sensitivity of global climate model simulations to the representation of soil moisture heterogeneity. *Journal of Hydrometeorology,* **4,** 1265–1275.

Gedney, N., P.M. Cox, and C. Huntingford, 2004: Climate feedback from wetland methane emissions. *Geophysical Research Letters.*

GFMC (Global Fire Monitoring Centre), 2003: The wildland fire season 2002 in the Russian Federation, *International Forest Fire News,* **28** (January–June).

Gibbens, R.P., R.A. Hicks, and W.A. Dugas, 1996: Structure and function of C_3 and C_4 Chihuahuan Desert plant communities. Standing crop and leaf area index. *Journal of Arid Environments,* **34(1),** 47–62.

Goodale, C.L., M.J. Apps, R.A. Birdsey, C.B. Field, L.S. Heath, et al., 2002: Forest carbon sinks in the northern hemisphere. *Ecological Applications,* **12(3),** 891–899.

Govindasamy, B., P.B. Duffy, and K. Caldeira, 2001: Land use changes and northern hemisphere cooling. *Geophysical Research Letters,* **28(2),** 291–294.

Granados, J. and C. Körner, 2002: In deep shade, elevated CO_2 increases the vigor of tropical climbing plants. *Global Change Biology,* **8(11),** 1109–1117.

Gregory, P.J., J.S.I. Ingram, R. Andersson, R.A. Betts, V. Brovkin, et al., 2002: Environmental consequences of alternative practices for intensifying crop production. *Agriculture Ecosystems & Environment,* **88(3),** 279–290.

Gruber, N., 1998: Anthropogenic CO_2 in the Atlantic Ocean. *Global Biogeochemical Cycles,* **12(1),** 165–191.

Guenther, A. C. Geron, T. Pierce, B. Lamb, P. Harley and R. Fall, 2000: Natural emissions of non-methane volatile organic compounds, carbon monoxide, and oxides of nitrogen from North America, *Atmospheric Environment,* **34,** 2205–2230.

Guinotte, J.M., R.W. Buddemeier, and J.A. Kleypas, 2003: Future coral reef habitat marginality: temporal and spatial effects of climate change in the Pacific basin. *Coral Reefs,* **22(4),** 551–558.

Hall, F.G., A.K. Betts, S. Frolking, et al., 2004: The Boreal climate. In: *Vegetation, Water, Humans, and the Climate: A New Perspective on an Interactive System,* P. Kabat, M. Claussen, P.A. Dirmeyer, J.H.C. Gash, L. Bravo de Guenni, M. Meybeck, R.A. Pielke, C.J. Vörösmarty, R.W.A. Hutjes, and S. Lütkemeier (eds.), Springer, New York, USA, pp. 93–114.

Hamilton, J.G., E.D. DeLucia, K. George, S.L. Naidu, A.C. Finzi, and W.H. Schlesinger, 2002: Forest carbon balance under elevated CO_2. *Oecologia,* **131,** 250–260.

Hansen, J.E. and M. Sato, 2001: Trends of measured climate forcing agents. *Proceedings of the National Academy of Sciences of the United States of America,* **98(26),** 14778–14783.

Harding, R.J. and J.W. Pomeroy, 1996: Energy balance of the winter boreal landscape. *Journal of Climate,* **9(11),** 2778–2787.

Heil, A. and J.G. Goldammer, 2001: Smoke-haze pollution: A review of the 1997 episode in Southeast Asia. *Regional Environmental Change,* **2(1),** 24–37.

Heinze, C., 2004: Simulating oceanic $CaCO_3$ export production in the greenhouse. *Geophysical Research Letters,* **31(16),** L16308, doi:10.1029/2004GL 020613.

Henderson-Sellers, A., R.E. Dickinson, T.B. Durbidge, P.J. Kennedy, K. McGuffie, and A.J. Pitman, 1993: Tropical deforestation: modelling local- to regional-scale climate. *Journal of Geophysical Research-Atmospheres*, **98(D4)**, 7289–7315.

Holland, E.A., B.H. Braswell, J.F. Lamarque, A. Townsend, J. Sulzman, et al., 1997: Variations in the predicted spatial distribution of atmospheric nitrogen deposition and their impact on carbon uptake by terrestrial ecosystems. *Journal of Geophysical Research-Atmospheres*, **102(D13)**, 15849–15866.

Holland, E.A., F.J. Dentener, B.H. Braswell, and J.M. Sulzman, 1999: Contemporary and pre-industrial global reactive nitrogen budgets. *Biogeochemistry*, **46(1)**, 7–43.

Houghton, R.A., 1999: The annual net flux of carbon to the atmosphere from changes in land use 1850–1990. *Tellus Series B-Chemical and Physical Meteorology*, **51(2)**, 298–313.

Houghton, R.A., 2003: Revised estimates of the annual net flux of carbon to the atmosphere from changes in land use and land management 1850–2000. *Tellus Series B-Chemical and Physical Meteorology*, **55(2)**, 378–390.

Houghton, R.A., 2003: Why are estimates of the terrestrial carbon balance so different? *Global Change Biology*, **9(4)**, 500–509.

Houghton, R.A., E.A. Davidson, and G.M. Woodwell, 1998: Missing sinks, feedbacks, and understanding the role of terrestrial ecosystems in the global carbon balance. *Global Biogeochemical Cycles*, **12(1)**, 25–34.

Houghton, R.A. and J.L. Hackler, 1999: Emissions of carbon from forestry and land-use change in tropical Asia. *Global Change Biology*, **5(4)**, 481–492.

Houghton, R.A., J.L. Hackler, and K.T. Lawrence, 1999: The US carbon budget: Contributions from land-use change. *Science*, **285(5427)**, 574–578.

House, J.I., I.C. Prentice, and C. Le Quere, 2002: Maximum impacts of future reforestation or deforestation on atmospheric CO_2. *Global Change Biology*, **8(11)**, 1047–1052.

House, J.I., I.C. Prentice, N. Ramankutty, R.A. Houghton, and M. Heimann, 2003: Reconciling apparent inconsistencies in estimates of terrestrial CO_2 sources and sinks. *Tellus Series B-Chemical And Physical Meteorology*, **55(2)**, 345–363.

Hungate, B.A., J.S. Dukes, M.R. Shaw, Y. Luo, and C.B. Field, 2003: Nitrogen and climate change. *Science*, **302(5650)**, 1512–1513.

IPCC (Intergovernmental Panel on Climate Change), 2000: *Land Use, Land-Use Change and Forestry*, Watson, R.T., I.R. Noble, B. Bolin, N.H. Ravindranath, D.J. Verado, and D.J. Dokken (eds.), Cambridge University Press, Cambridge, UK, 377 pp.

IPCC, 2001a: *Climate Change 2001: The Scientific Basis. Contribution of Working Group I to the Thirsd Assessment Report of the Intercovernmental Panel on Climate Change*, Houghton, J. T., Y. Ding, D. J. Griggs, M. Noguer, P. J. van der Linden, X. Dai, K. Maskell, and C. Johnson (eds.), Cambridge University Press, Cambridge, UK, 881 pp.

IPCC, 2001b: *Climate Change 2001: Impacts, Adaptation, and Vulnerability. Contribution of Working Group II to the Thirsd Assessment Report of the Intercovernmental Panel on Climate Change*, McCarthy, J. J., O.F. Canziani, N.A. Leary, D.J. Dokken, and K.S. White (eds.), Cambridge University Press, Cambridge, UK, 1032 pp.

IPCC, 2001c: *Climate Change 2001: Mitigation. Contribution of Working Group III to the Third Assessment Report of the Intergovernmental Panel on Climate Change*, Metz, B., O Davidson, R. Swart, and J. Pan (eds.), Cambridge University Press, Cambridge, UK, 752 pp.

IPCC, 2001d: *Climate Change 2001: Synthesis Report*, (Watson, R. T. and Core Writing Team (eds.), Cambridge University Press, Cambridge, UK, 398 pp.

Jacobson, M.Z., 2002: Control of fossil-fuel particulate black carbon and organic matter, possibly the most effective method of slowing global warming. *Journal of Geophysical Research-Atmospheres*, **107(D19)**, 4410, doi:10.1029/2001JD001376.

Jarvis, P. and S. Linder, 2000: Botany—Constraints to growth of boreal forests. *Nature*, **405(6789)**, 904–905.

Jöckel, P., C.A.M. Brenninkmeijer, and P.J. Crutzen, 2003: A discussion on the determination of atmospheric OH and its trends. *Atmospheric Chemistry and Physics Discussions*, **2(4)**, 1261–1286.

Jones, C.G., J.H. Lawton, and M. Shachak, 1994: Organisms as ecosystem engineers. *Oikos*, **69(3)**, 373–386.

Joos, F., R. Meyer, M. Bruno, and M. Leuenberger, 1999: The variability in the carbon sinks as reconstructed for the last 1000 years. *Geophysical Research Letters*, **26(10)**, 1437–1440.

Joos, F. and I.C. Prentice, 2004: A paleo perspective on the future of atmospheric CO2 and climate. In: *The Global Carbon Cycle*, SCOPE Series No. 62, Island Press, Washington, DC, USA, pp. 165–186.

Joos, F., I.C. Prentice, S. Sitch, R. Meyer, G. Hooss, G.-K. Plattner, S. Gerber, and K. Hasselmann, 2001: Global warming feedbacks on terrestrial carbon uptake under the Intergovernmental Panel on Climate Change (IPCC) emission scenarios. *Global Biogeochemical Cycles*, **15(4)**, 891–907.

Joos, F., J.L. Sarmiento, and U. Siegenthaler, 1991: Estimates of the effect of Southern Ocean iron fertilization on atmospheric CO_2 concentrations. *Nature*, **349(6312)**, 772–775.

Joussaume, S., K.E. Taylor, P. Braconnot, J.F.B. Mitchell, J.E. Kutzbach, et al., 1999: Monsoon changes for 6000 years ago: Results of 18 simulations from the Paleoclimate Modeling Intercomparison Project (PMIP). *Geophysical Research Letters*, **26(7)**, 859–862.

Kabat, P., M. Claussen, P.A. Dirmeyer, J.H.C. Gash, L.B. de Guenni, M. Meybeck, C.J. Vörösmarty, R.W.A. Hutjes, and S. Lütkemeier (eds.), 2004: *Vegetation, Water, Humans and the Climate: A New Perspective On An Interactive System. Global Change: The IGBP Series*, Springer, Berlin, Germany, 566 pp.

Kaufman, Y.J., D. Tanre, O. Dubovik, A. Karnieli, and L.A. Remer, 2001: Absorption of sunlight by dust as inferred from satellite and ground-based remote sensing. *Geophysical Research Letters*, **28**, 1479–1482.

Kedziora A., J. Olejnik, 2002: Water balance in Agricultural Landscape and options for its management by change of plant cover structure of landscape. In: *Landscape ecology in agroecosystems management*. L. Ryszkowski (ed.). CRC Press, Boca Raton, London, New York, Washington D.C., 57–110.

Keeling, R.F. and H.E. Garcia, 2002: The change in oceanic O_2 inventory associated with recent global warming. *National Academy of Science USA*, **99**, 723–727.

Kiene, R.P., P.T. Visscher, M.D. Keller, and G.O. Kirst (eds), 1996: *Biological and Environmental Chemsitry of DMSP and Related Sulfonium Compounds*. Plenum Press, New York.

Kim Phat, N., W. Knorr, and S. Kim, 2004: Appropriate measures for conservation of terrestrial carbon stocks—analysis of trends of forest management in Southeast Asia. *Journal of Forest Ecology and Management*, **191**, 283–299.

Kimball, B.A., K. Kobayashi, and M. Bindi, 2002: Responses of agricultural crops to free-air CO_2 enrichment. *Advances in Agronomy*, **77**, 293–368.

Kirchhoff, V.W.J.H., J.R. Alves, F.R. daSilva, and J. Fishman, 1996: Observations of ozone concentrations in the Brazilian cerrado during the TRACE A field expedition. *Journal of Geophysical Research-Atmospheres*, **101(D19)**, 24029–24042.

Kirchner, J.W., 2003: The Gaia hypothesis: conjectures and refutations. *Climatic Change*, **58(1–2)**, 21–45.

Klaas, C. and D.E. Archer, 2002: Association of sinking organic matter with various types of mineral ballast in the deep sea: Implications for the rain ratio. *Global Biogeochemical Cycles*, **16(4)**, 1116, doi:10.1029/2001GB001765.

Kleypas, J.A., R.W. Buddemeier, D. Archer, J.P. Gattuso, C. Langdon, and B.N. Opdyke, 1999: Geochemical consequences of increased atmospheric carbon dioxide on coral reefs. *Science*, **284(5411)**, 118–120.

Knorr, W., I.C. Prentice, J.I. House, and E.A. Holland, 2005: Long-term sensitivity of soil carbon turnover to warming. *Nature*, **433**, 298–301.

Kondratyev, K.Y., V.F. Krapivin, and C.A. Varotsos, 2003: *Global Carbon Cycle and Climate Change*. Springer—Praxis Publishing, Berlin, Germany, 368 pp.

Koren, I., Y.J. Kaufman, L.A. Remer, and J.V. Martins, 2004: Measurement of the effect of Amazon smoke on inhibition of cloud formation. *Science*, **303(5662)**, 1342–1345.

Körner, C., 2000: Biosphere responses to CO_2 enrichment. *Ecological Applications*, **10(6)**, 1590–1619.

Kroeze, C., A. Mosier, and L. Bouwman, 1999: Closing the global N_2O budget: A retrospective analysis 1500–1994. *Global Biogeochemical Cycles*, **13(1)**, 1–8.

Kutzbach, J., G. Bonan, J. Foley, and S.P. Harrison, 1996: Vegetation and soil feedbacks on the response of the African monsoon to orbital forcing in the early to middle Holocene. *Nature*, **384(6610)**, 623–626.

Lal, R., M. Griffin, J. Apt, L. Lave, and M.G. Morgan, 2004: Ecology: Managing soil carbon. *Science*, **304(5669)**, 393–393.

Langenfelds, R.L., R.J. Francey, B.C. Pak, L.P. Steele, J. Lloyd, C.M. Trudinger, and C.E. Allison, 2002: Interannual growth rate variations of atmospheric CO_2 and its delta[13]C, H_2, CH_4, and CO between 1992 and 1999 linked to biomass burning. *Global Biogeochemical Cycles*, **16(3)**, 1048, doi:10.1029/2001GB001466.

Langenfelds, R.L., R.J. Francey, L.P. Steele, and M. Battle, 1999: Partitioning of the global fossil CO_2 sink using a 19-year trend in atmospheric O_2. *Geophysical Research Letters*, **26(1897)**, 1900.

Lenton, T.M., 1998: Gaia and natural selection. *Nature*, **394(6692)**, 439–447.

Lerdau, M. and M. Keller, 1997: Controls on isoprene emission from trees in a subtropical dry forest. *Plant Cell and Environment*, **20(5)**, 569–578.

Lloyd, J., 1999: The CO_2 dependence of photosynthesis, plant growth responses to elevated CO_2 concentrations and their interaction with soil nutrient status, II. Temperate and boreal forest productivity and the combined effects of increasing CO_2 concentrations and increased nitrogen deposition at a global scale. *Functional Ecology*, **13(4)**, 439–459.

Loreau, M., S. Naeem, P. Inchausti, J. Bengtsson, J.P. Grime, et al. 2001: Biodiversity and ecosystem functioning: Current knowledge and future challenges. *Science,* **294,** 804–808.

Lovelock, J.E. and L. Margulis, 1974: Atmospheric homoestasis by and for the biosphere: the Gaia hypothesis. *Tellus,* **26,** 2–10.

Luo, Y.Q., S.Q. Wan, D.F. Hui, and L.L. Wallace, 2001: Acclimatization of soil respiration to warming in a tall grass prairie. *Nature,* **413(6856),** 622–625.

Mahowald, N.M., and C. Luo, 2003: A less dusty future?, *Geophysical Research Letters,* **30(17),** 1903, doi:10.1029/2003GL017880

Maier-Reimer, E., U. Mikolajeewicz, and A. Winguth, 1996: Future ocean uptake of CO_2—Interaction between ocean circulation and biology. *Climate Dynamics,* **2,** 63–90.

Manning, A.C., 2001: *Temporal variability of atmospheric oxygen from both continuous measurements and a flask sampling network: Tools for studying the global carbon cycle.* University of California, San Diego, La Jolla, California, USA, 190 pp.

Marshall, C.H., R.A. Pielke, and L.T. Steyaert, 2003: Crop freezes and land-use change in Florida. *Nature,* **426(6962),** 29–30.

Martin, J.H., R.M. Gordon, and S.E. Fitzwater, 1991: The case for iron. *Limnology and Oceanography,* **36,** 1793–1902.

Matear, R.J. and B. Elliott, 2004: Enhancement of oceanic uptake of anthropogenic CO_2 by macronutrient fertilization. *Journal of Geophysical Research-Oceans,* **109(C4),** C04001, doi:10.1029/2000JC000321.

Matear, R.J., and A.C. Hirst, 1999: Climate change feedback on the future oceanic CO2 uptake. *Tellus Series B,* **51,** 722–733.

Matson, P.A., W.H. McDowell, A.R. Townsend, and P.M. Vitousek, 1999: The globalization of N deposition: ecosystem consequences in tropical environments. *Biogeochemistry,* **46(1),** 67–83.

Matthews, H.D., A.J. Weaver, K.J. Meissner, N.P. Gillett, and M. Eby, 2004: Natural and anthropogenic climate change: Incorporating historical land cover change, vegetation dynamics and the global carbon cycle. *Climate Dynamics,* **22(5),** 461–479.

McGuire, A.D., S. Sitch, J.S. Clein, R. Dargaville, G. Esser, et al., 2001: Carbon balance of the terrestrial biosphere in the twentieth century: Analyses of CO_2, climate and land use effects with four process-based ecosystem models. *Global Biogeochemical Cycles,* **15(1),** 183–206.

McLaughlin, S. and K. Percy, 1999: Forest health in North America: Some perspectives on actual and potential roles of climate and air pollution. *Water Air and Soil Pollution,* **116(1–2),** 151–197.

McLaughlin, S.B. and D.J. Downing, 1995: Interactive effects of ambient ozone and climate measured on growth of mature forest trees. *Nature,* **374(6519),** 252–254.

MA (Millennium Ecosystem Assessment), 2003: *Ecosystems and Human Well-being: A Framework for Assessment.* Island Press, Washington, DC, USA, 245 pp.

Miller, A. J., M.A. Alexander, G.J. Boer, F. Chai, K. Denman, et al., 2003: Potential feedbacks between Pacific Ocean ecosystems and interdecadal climate variations. Bulletin of the American Meteorological Society, 84, 617–633.

Molesworth, A.M., L.E. Cuevas, A.P. Morse, J.R. Herman, and M.C. Thomson, 2002: Dust clouds and spread of infection. *Lancet,* **359,** 81–82.

Mooney, H.A., J. Canadell, F.S. Chapin III, J.R. Ehleringer, C. Körner, R.E. McMurtrie, W.J. Parton, L.F. Pitelka, and E.D. Schulze, 1999: Ecosystem physiology responses to global change. In: *Implications of Global Change for Natural and Managed Ecosystems. A Synthesis of GCTE and Related Research.,* B.H. Walker, W.L. Steffen, J. Canadell, and J.S.I. Ingram (eds.). IGBP Book Series No. 4, Cambridge University Press, Cambridge, UK, pp. 141–189.

Mouillot, F., S. Rambal, and R. Joffre, 2002: Simulating climate change impacts on fire frequency and vegetation dynamics in a Mediterranean-type ecosystem. *Global Change Biology,* **8(5),** 423–437.

Muraleedharan, T.R., M. Radojevic, A. Waugh, and A. Caruana, 2000: Chemical characterization of the haze in Brunei Darussalam during the 1998 episode. *Atmospheric Environment,* **34(17),** 2725–2731.

N'Tchayi, G.M., J. Bertrand, M. Legrand, and J. Baudet, 1994: Temporal and spatial variations of the atmospheric dust loading throughout West Africa over the last 30 years. *Annales Geophysicae-Atmospheres Hydrospheres and Space Sciences,* **12(2–3),** 265–273.

Nascimento, H.E.M. and W.F. Laurance, 2004: Biomass dynamics in Amazonian forest fragments. *Ecological Applications,* **14(4),** S127–S138

Nepstad, D.C., A. Veríssimo, A. Alencar, C. Nobre, E. Lima, P. Lefebvre, P. Schlesinger, C. Potter, P. Moutinho, E. Mendoza, M. Cochrane, and V. Brooks, 1999: Large-scale impoverishment of Amazonian forests by logging and fire. *Nature,* **398(6727),** 505–508.

Nevison, C. and E. Holland, 1997: A reexamination of the impact of anthropogenically fixed nitrogen on atmospheric N_2O and the stratospheric O_3 layer. *Journal of Geophysical Research-Atmospheres,* **102(D21),** 25519–25536.

Nicholson, S.E., C.J. Tucker, and M.B. Ba, 1998: Desertification, drought, and surface vegetation: An example from the West African Sahel. *Bulletin of the American Meteorological Society,* **79(5),** 815–829.

Nobre, C.A., M.A. Silva Dias, A.D. Culf, J. Polcher, J.H.C. Gash, J.A. Marengo, and R. Avissar, 2004: The Amazonian climate. In: *Vegetation, Water, Humans, and the Climate: A New Perspective on an Interactive System,* P. Kabat, M. Claussen, P.A. Dirmeyer, J.H.C. Gash, L. Bravo de Guenni, M. Meybeck, R.A. Pielke, C.J. Vörösmarty, R.W.A. Hutjes, and S. Lütkemeier (eds.), Springer, New York, USA, pp. 79–92.

Norby, R.J., K. Kobayashi, and B.K. Kimball, 2001: Rising CO_2: future ecosystems—Commentary. *New Phytologist,* **150(2),** 215–221.

Norby, R.J. and Y. Luo, 2004: Evaluating ecosystem responses to rising atmospheric CO2 and global warming in a multi-factor world. *New Phytologist,* **162(2),** 281–293

Norby, R.J., S.D. Wullschleger, C.A. Gunderson, D.W. Johnson, and R. Ceulemans, 1999: Tree responses to rising CO_2 in field experiments: implications for the future forest. *Plant Cell and Environment,* **22(6),** 683–714.

Nowak, R.S., D.S. Ellsworth, and S.D. Smith, 2004: Functional responses of plants to elevated atmospheric CO_2—do photosynthetic and productivity data from FACE experiments support early predictions? *New Phytologist,* **162(2),** 253–280.

Oechel, W.C., G.L. Vourlitis, S.J. Hastings, R.C. Zulueta, L. Hinzman, and D. Kane, 2000: Acclimation of ecosystem CO_2 exchange in the Alaskan Arctic in response to decadal climate warming. *Nature,* **406(6799),** 978–981.

Olivier, J.G.J., A.F. Bouwman, C.W.M. Vandermaas, and J.J.M. Berdowski, 1994: Emission database for Global Atmospheric Research (EDGAR). *Environmental Monitoring and Assessment,* **31(1–2),** 93–106.

Ollinger, S.V., J.D. Aber, and P.B. Reich, 1997: Simulating ozone effects on forest productivity: Interactions among leaf-, canopy-, and stand-level processes. *Ecological Applications,* **7(4),** 1237–1251.

Oren, R., D.S. Ellsworth, K.H. Johnsen, N. Phillips, B.E. Ewers, C. Maier, K.V.R. Schäfer, H. McCarthy, G. Hendrey, S.G. McNulty, and G.G. Katul, 2001: Soil fertility limits carbon sequestration by forest ecosystems in a CO_2-enriched atmosphere. *Nature,* **411(6836),** 469–472.

Owensby, C.E., P.I. Coyne, J.M. Ham, L.M. Auen, and A. Knapp, 1993: Biomass production in a tallgrass prairie ecosystem exposed to ambient and elevated levels of CO_2. *Ecological Applications,* **3(4),** 644–653.

Oyama, M.D. and C.A. Nobre, 2003: A new climate-vegetation equilibrium state for Tropical South America. *Geophysical Research Letters,* **30(23),** 2199, doi:10.1029/2003GL018600.

Paustian, K., J. Six, E.T. Elliott, and H.W. Hunt, 2000: Management options for reducing CO_2 emissions from agricultural soils. *Biogeochemistry,* **48(1),** 147–163.

Peng, T.-H. and W.S. Broecker, 1991: Dynamic limitations on the Antarctic iron fertilization strategy. *Nature,* **349(6306),** 227–229.

Penner, J.E., M. Andreae, H. Annegarn, L. Barrie, J. Feichter, D. Hegg, A. Jayaraman, R. Leaitch, D. Murphy, J. Nganga, and G. Pitari, 2001: Aerosols, their direct and indirect effects. In: *Climate Change 2001: The Scientific Basis. Contribution of Working Group I to the Third Assessment Report of the Intergovernmental Panel on Climate Change, IPCC, Geneva, Switzerland,* J.T. Houghton (ed.), Cambridge University Press, Cambridge, UK, pp. 289–348.

Penner, J.E., S.Y. Zhang, and C.C. Chuang, 2003: Soot and smoke aerosol may not warm climate. *Journal of Geophysical Research-Atmospheres,* **108(D21),** 4657, doi:10.1029/2003JD003409.

Petit, J.R., J. Jouzel, D. Raynaud, N.I. Barkov, J.M. Barnola, et al., 1999: Climate and atmospheric history of the past 420,000 years from the Vostok ice core, Antarctica. *Nature,* **399(6735),** 429–436.

Phillips, O.L., R.V. Martínez, L. Arroyo, T.R. Baker, T. Killeen, et al., 2002: Increasing dominance of large lianas in Amazonian forests. *Nature,* **418(6899),** 770–774.

Pielke, R.A., R. Avissar, M. Raupach, A.J. Dolman, X.B. Zeng, and A.S. Denning, 1998: Interactions between the atmosphere and terrestrial ecosystems: influence on weather and climate. *Global Change Biology,* **4(5),** 461–475.

Piketh, S.J., P.D. Tyson, and W. Steffen, 2000: Aeolian transport from southern Africa and iron fertilization of marine biota in the South Indian Ocean. *South African Journal of Science,* **96(5),** 244–246.

Pitman, A.J., H. Dolman, B. Kruijt, R. Valentini, and D. Baldocchi, 2004: The climate near ground. In: *Vegetation, Water, Humans, and the Climate: A New Perspective on an Interactive System,* P. Kabat, M. Claussen, P.A. Dirmeyer, J.H.C. Gash, L. Bravo de Guenni, M. Meybeck, R.A. Pielke, C.J. Vörösmarty, R.W.A. Hutjes, and S. Lütkemeier (eds.), Springer, New York, USA, pp. 9–20.

Platt, T., C. Fuentes-Yaco, and K.T. Frank, 2003: Spring algal bloom and larval fish survival. *Nature, 423(6938),* 398–399.

Plattner, G.K., F. Joos, T. F. Stocker, and O. Marchal, 2001: Feedback mechanisms and sensitivities of ocean carbon uptake under global warming. *Tellus,* Ser. B., **53,** 564–592.

Plattner, G.K., F. Joos, and T.F. Stocker, 2002: Revision of the global carbon budget due to changing air-sea oxygen fluxes. *Global Biogeochemical Cycles,* **16(4),** 1096, doi:10.1029/2001GB001746.

Polley, W.H., H.B. Johnson, H.S. Mayeux, and C.R. Tischier, 1996: Are some of the recent changes in grassland communities a response to rising CO_2 concentrations? In: *Carbon Dioxide, Populations and Communities,* C. Körner and F.A. Bazzaz (eds.), Academic Press, San Diego, USA, pp. 177–195.

Porter, H. and M.-L. Navas, 2003: Plant growth and competition at elevated CO_2: on winners, losers and functional groups. *New Phytologist,* **157(2),** 175–198.

Prather, M., D. Ehhalt, F. Dentener, R. Derwent, E. Dlugokencky, E. Holland, I. Isaksen, J. Katima, V. Kirchhoff, P. Matson, P. Midgley, and M. Wang, 2001: Atmospheric chemistry and greenhouse gases. In: *Climate Change 2001: The Scientific Basis. Contribution of Working Group I to the Third Assessment Report of the Intergovernmental Panel on Climate Change, IPCC, Geneva, Switzerland,* J.T. Houghton (ed.), Cambridge University Press, Cambridge, UK, pp. 239–287.

Prentice, I.C., G.D. Farquhar, M.J.R. Fasham, M.L. Goulden, M. Heimann, et al. 2001: The carbon cycle and atmospheric carbon dioxide. In: *Climate Change 2001: The Scientific Basis. Contribution of Working Group I to the Third Assessment Report of the Intergovernmental Panel on Climate Change,* 1st ed. J.T. Houghton, Y. Ding, D.J. Griggs, M. Noguer, P.J. van der Linden, X. Dai, K. Maskell, and C.A. Johnson (eds.), Cambridge University Press, Cambridge, UK, 185–225.

Prentice, I.C., D. Jolly, and BIOME 6000 Participants, 2000: Mid-Holocene and glacial-maximum vegetation geography of the northern continents and Africa. *Journal of Biogeography,* **27(3),** 507–519.

Prentice, I.C. and D. Raynaud, 2001: Paleobiogeochemistry. In: *Global Biogeochemical Cycles in the Climate System,* E.-D. Schulze, M. Heimann, S.P. Harrison, E. Holland, J. Lloyd, I.C. Prentice, and D.S. Schimel (eds.), Academic Press, San Diego, USA, pp. 87–93.

Price, C. and D. Rind, 1994: The impact of a 2xCO$_2$ climate on lightning-caused fires. *Journal of Climate,* **7(10),** 1484–1494.

Prinn, R.G., J. Huang, R.F. Weiss, D.M. Cunnold, P.J. Fraser, et al., 2001: Evidence for substantial variations of atmospheric hydroxyl radicals in the past two decades. *Science,* **292(5523),** 1882–1888.

Prospero, J.M., 2001: African dust in America. *Geotimes,* **46(11),** 24–27.

Prospero, J.M. and P.J. Lamb, 2003: African droughts and dust transport to the Caribbean: Climate change implications, *Science,* **302(5647),** 1024–1027.

Radojevic, M., 2003: Chemistry of forest fires and regional haze with emphasis on Southeast Asia. *Pure and Applied Geophysics,* **160(1–2),** 157–187.

Ramankutty, N. and J.A. Foley, 1999: Estimating historical changes in global land cover: Croplands from 1700 to 1992. *Global Biogeochemical Cycles,* **13(4),** 997–1027.

Renssen, H., V. Brovkin, T. Fichefet, and H. Goosse, 2003: Holocene climate instability during the termination of the African Humid Period. *Geophysical Research Letters,* **30(4),** 1184, doi:10.1029/2002GL016636.

Renwick, W.H., S.V. Smith, R.O. Sleezer, and R.W. Buddemeier, 2004: Comment on "Managing Soil Carbon" (II). *Science,* **305(5690),** 1567c–1567c.

Richardson, A.J. and D.S. Schoeman, 2004: Climate impact on plankton ecosystems in the Northeast Atlantic. *Science,* **305(5690),** 1609–1612.

Ridgwell, A.J., M.A. Maslin, and A.J. Watson, 2002: Reduced effectiveness of terrestrial carbon sequestration due to an antagonistic response of ocean productivity. *Geophysical Research Letters,* **29(6),** 1095, doi:10.1029/2001GL 014304.

Riebesell, U., I. Zondervan, B. Rost, and R.E. Zeebe, 2001: Effects of increasing atmospheric CO_2 on phytoplankton communities and the biological pump. *Global Change Newsletter,* **47,** 12–15.

Rind, D., 1996: The potential for modeling the effects of different forcing factors on climate during the past 2000 years. In: *Climate Variations and Forcing Mechanisms of the Last 2000 Years,* P.D. Jones, R.S. Bradley, and J. Jouzel (eds.), Springer, Berlin, 563–581.

Roberts, D.L. and A. Jones, 2004: Climate sensitivity to black carbon aerosol from fossil fuel combustion. *Journal of Geophysical Research-Atmospheres,* **109(D16),** D16202, doi:10.1029/2004JD004676.

Rödenbeck, C., S. Houweling, M. Gloor, and M. Heimann, 2003: CO_2 flux history 1982–2001 inferred from atmospheric data using a global inversion of atmospheric transport. *Atmospheric Chemistry and Physics,* **3,** 1919–1964.

Rodhe, H., 1999: Human impact on the atmospheric sulfur balance. *Tellus Series A-Dynamic Meteorology and Oceanography,* **51(1),** 110–122.

Rodhe, H., P. Grennfelt, J. Wisniewski, C. Ågren, G. Bengtsson, K. Johansson, P. Kauppi, V. Kucera, L. Rasmussen, B. Rosseland, L. Schotte, and G. Selldén, 1995: Acid reign '95?—Conference Summary Statement. *Water Air And Soil Pollution,* **85,** 1–14.

Rosenberg, N.J., C.V. Cole, and K. Paustian, 1998: Mitigation of greenhouse gas emissions by the agriculture sector—New technologies, policies and measures offer potential to mitigate emissions while improving productivity and ecosystem health: An introductory editorial. *Climatic Change,* **40(1),** 1–5.

Röthlisberger, R., M. Bigler, E.W. Wolff, F. Joos, E. Monnin, and M.A. Hutterli, 2004: Ice core evidence for the extent of past atmospheric CO_2 change due to iron fertilisation. *Geophysical Research Letters,* **31(16),** L16207, doi:10.1029/2004GL020338.

Rustad, L.E., J.L. Campbell, G.M. Marion, R.J. Norby, M.J. Mitchell, A.E. Hartley, J.H.C. Cornelissen, and J. Gurevitch, 2001: A meta-analysis of the response of soil respiration, net nitrogen mineralization, and above ground plant growth to experimental ecosystem warming. *Oecologia,* **126(4),** 543–562.

Sabine, C.L., R.M. Key, K.M. Johnson, F.J. Millero, A. Poisson, J.L. Sarmiento, D.W.R. Wallace, and C.D. Winn, 1999: Anthropogenic CO_2 inventory of the Indian Ocean. *Global Biogeochemical Cycles,* **13(1),** 179–198.

Sarmiento, J.L. and J.R. Toggweiller, 1984: A new model for the role of the ooceans in determining atmospheric pCO_2. *Nature,* **308,** 621–624

Sarmiento, J.L., T.M.C. Hughes, R.J. Stouffer, and S. Manabe, 1998: Simulated response of the ocean carbon cycle to anthropogenic climate warming. *Nature,* **393,** 245–249.

Satake, K., S. Kojima, T. Takamatsu, J. Shindo, T. Nakano, et al. 2001: Acid rain 2000—Conference Statement; looking back to the past and thinking of the future. *Water Air and Soil Pollution,* **130,** 1–16.

Sathyendranath, S., A.D. Gouveia, S.R. Shetye, P. Ravindran, and T. Platt, 1991: Biological control of surface temperature in the Arabian Sea. *Nature,* **349(6304),** 54–56.

Scherer-Lorenzen, M., C. Körner, E.D. Schulze (eds.), 2005: *Forest Diversity and Function: Temperate and Boreal Systems.* Springer, Berlin. pp.400

Schulze, E.D., W. Devries, M. Hauhs, K. Rosen, L. Rasmussen, C.O. Tamm, and J. Nilsson, 1989: Critical loads for nitrogen deposition on forest ecosystems. *Water Air and Soil Pollution,* **48(3–4),** 451–456.

Schultz, M.G., T. Diehl, G.P. Brasseur, and W. Zittel, 2003: Air pollution and climate-forcing impacts of a global hydrogen economy. *Science,* **302(5645),** 624–627.

Schulze, E.D. and H.A. Money (eds.), 1993: *Biodiversity and Ecosystem Function.* Springer-Verlag, Berlin, Germany.

Schwela, D., J.G. Goldammer, L.H. Morawska, and O. Simpson, 1999: *Health Guidelines for Vegetation Fire Events—Guideline Document.* Published on behalf of United Nations Environment Programme, Nairobi. World Health Organisation, Geneva, Switzerland, Singapore,197 pp.

Sellers, P.J., L. Bounoua, G.J. Collatz, D.A. Randall, D.A. Dazlich, et al. 1996: Comparison of radiative and physiological effects of doubled atmospheric CO2 on climate. *Science,* **271(5254),** 1402–1406.

Shimizu, A., N. Sugimoto, I. Matsui, K. Arao, I. Uno, T. Murayama, N. Kagawa, K. Aoki, A. Uchiyama, and A. Yamazaki, 2004: Continuous observations of Asian dust and other aerosols by polarization lidars in China and Japan during ACE-Asia. *Journal of Geophysical Research-Atmospheres,* **109(D19),** D19S17, doi:10.1029/2002JD003253.

Siegert, F., G. Ruecker, A. Hinrichs, and A.A. Hoffmann, 2001: Increased damage from fires in logged forests during droughts caused by El Niño. *Nature,* **414(6862),** 437–440.

Singh, T.P., V. Varalakshmi, and S.K. Ahluwalia, 2000: Carbon sequestration through farm forestry: case from India. *Indian Forester,* **126(12),** 1257–1264.

Smith, S.D., T.E. Huxman, S.F. Zitzer, T.N. Charlet, D.C. Housman, J.S. Coleman, L.K. Fenstermaker, J.R. Seemann, and R.S. Nowak, 2000: Elevated CO_2 increases productivity and invasive species success in an arid ecosystem. *Nature,* **408(6808),** 79–82.

Smyth, T.J., T. Tyrrell, and B.J. Tarrant, in press: Time-series of coccolithophore activity in the Barents Sea from twenty years of satellite imagery. *Geophysical Research Letters.*

Spiecker, H., K. Mielikäinen, M. Köhl, and J.P. Skovsgaard, 1996: *Growth Trends in European Forests.* European Forest Institute Research Report No. 5, Springer Verlag, Berlin-Heidelberg, Germany.

Steffen, W., A. Sanderson, J. Jäger, P.D. Tyson, B. Moore III, P.A. Matson, K. Richardson, F. Oldfield, H.-J. Schellnhuber, B.L. Turner II, and R.J. Wasson (eds.), 2004: *Global Change And The Earth System: A Planet Under Pressure.* Global Change: The IGBP Series, Springer, Berlin, Germany, 336 pp.

Sternberg, L.D.L., 2001: Savanna-forest hysteresis in the tropics. *Global Ecology and Biogeography,* **10(4),** 369–378.

Stocker, T.F., G.K.C. Clarke, H. Le Treut, et al. 2001: Physical Climate Processes and Feedbacks. In: *Climate Change 2001: The Scientific Basis. Contribution of Working Group I to the Third Assessment Report of the Intergovernmental Panel on Climate Change,* 1st ed. J.T. Houghton, Y. Ding, D.J. Griggs, M. Noguer, P.J. van der Linden, X. Dai, K. Maskell, and C.A. Johnson (eds.), Cambridge University Press, Cambridge, UK, pp. 417–470.

Stocks, B.J., M.A. Fosberg, T.J. Lynham, L. Mearns, B.M. Wotton, Q. Yang, J.Z. Jin, K. Lawrence, G.R. Hartley, J.A. Mason, and D.W. McKenney, 1998: Climate change and forest fire potential in Russian and Canadian boreal forests. *Climatic Change,* **38(1),** 1–13.

Sudo, K. and M. Takahashi, 2001: Simulation of tropospheric ozone changes during 1997–1998 El Niño: Meteorological impact on tropospheric photochemistry. *Geophysical Research Letters,* **28(21),** 4091–4094.

Sugimoto, N., I. Uno, M. Nishikawa, A. Shimizu, I. Matsui, X.H. Dong, Y. Chen, and H. Quan, 2003: Correction to "Record heavy Asian dust in Beijing in 2002: Observations and model analysis of recent events" (vol 30, art no 1640, 2003). *Geophysical Research Letters,* **30(16),** 1834, doi:10.1029/2003GL018215.

Swap, R.J., H.J. Annegarn, J.T. Suttles, M.D. King, S. Platnick, J.L. Privette, and R.J. Scholes, 2003: Africa burning: A thematic analysis of the Southern African Regional Science Initiative (SAFARI 2000). *Journal of Geophysical Research-Atmospheres,* **108(D13),** 8465, doi:10.1029/2003JD003747.

Tegen, I., M. Werner, S.P. Harrison, and K.E. Kohfeld, 2004: Relative importance of climate and land use in determining present and future global soil dust emission. *Geophysical Research Letters,* **31(5),** L05105, doi:10.1029/2003GL019216.

Thompson, A. M., and R. D. Hudson, 1999: Tropical tropospheric ozone (TTO) maps from Nimbus 7 and Earth Probe TOMS by the modified-residual method: Evaluation with sondes, ENSO signals, and trends from Atlantic regional time series, *Journal of Geophysical Research,* **104,** 26,961–26,975.

Thurston, G.D. and K. Ito, 1999: Epidemiological studies of ozone exposure effects. In: *Air pollution and health,* H. Koran, and S. Holgate (eds.) Academic Press.

Tian H.Q., J.M. Melillo, D.W. Kicklighter, 1998: Effect of interannual climate variability on carbon storage in Amazonian ecosystems, *Nature,* **396(6712),** 664–667

Tilman, D., J. Knops, D. Wedin, P. Reich, M. Ritchie, and E. Siemann, 1997: The influence of functional diversity and composition on ecosystem processes. *Science,* **277,** 1300–1302.

Toth, F.L., 1999: *Fair weather? Equity Concerns in Climate Change.* EarthScan, London, UK, 212 pp.

Vernadsky, V.I., 1926: *Biosfera.* Nauka, Leningrad, Russia,146 pp.

Vernadsky, V.I., 1998: *The Biosphere.* Copernicus Books, New York, USA, 192 pp.

Vitousek, P.M., J.D. Aber, R.W. Howarth, G.E. Likens, P.A. Matson, D.W. Schindler, W.H. Schlesinger, and D.G. Tilman, 1997: Human alteration of the global nitrogen cycle: Sources and consequences. *Ecological Applications,* **7(3),** 737–750.

Vitousek, P.M. and R.W. Howarth, 1991: Nitrogen limitation on land and in the sea: how can it occur ? *Biogeochemistry,* **13(2),** 87–115.

Walter, B.P. and M. Heimann, 2000: A process-based, climate-sensitive model to derive methane emissions from natural wetlands: Application to five wetland sites, sensitivity to model parameters, and climate. *Global Biogeochemical Cycles,* **14(3),** 745–765.

Wang, G.L. and E.A.B. Eltahir, 2000: Biosphere-atmosphere interactions over West Africa. II: Multiple climate equilibria. *Quarterly Journal of the Royal Meteorological Society,* **126(565),** 1261–1280.

Warwick, N.J., S. Bekki, K.S. Law, E.G. Nisbet, and J.A. Pyle, 2002: The impact of meteorology on the interannual growth rate of atmospheric methane. *Geophysical Research Letters,* **29(20),** 1947, doi:10.1029/2002GL015282.

Warwick, N.J., S. Bekki, E.G. Nisbet, and J.A. Pyle, 2004: Impact of a hydrogen economy on the stratosphere and troposphere studied in a 2-D model. *Geophysical Research Letters,* **31(5),** L05107, doi:10.1029/2003GL019224.

Wassmann, R., H.U. Neue, R.S. Lantin, L.V. Buendia, and H. Rennenberg, 2000a: Characterization of methane emissions from rice fields in Asia. I. Comparison among field sites in five countries. *Nutrient Cycling in Agroecosystems,* **58(1–3),** 1–12.

Wassmann, R., H.U. Neue, R.S. Lantin, K. Makarim, N. Chareonsilp, L.V. Buendia, and H. Rennenberg, 2000b: Characterization of methane emissions from rice fields in Asia. II. Differences among irrigated, rainfed, and deepwater rice. *Nutrient Cycling in Agroecosystems,* **58(1–3),** 13–22.

Watson, A.J., 1999: Coevolution of the Earth's environment and life: Goldilocks, gaia and the anthropic principle. In: *James Hutton: Present and Future,* G.Y. Craig and J.H. Hull (eds.). Geological Society Special Publication No. 150, Geological Society, London, UK, pp. 75–88.

Watson, A.J., D.C.E. Bakker, A.J. Ridgwell, P.W. Boyd, and C.S. Law, 2000: Effect of iron supply on Southern Ocean CO_2 uptake and implications for glacial atmospheric CO_2. *Nature,* **407,** 730–733.

Watson, A.J. and J.E. Lovelock, 1983: Biological homeostasis of the global environment—the parable of Daisyworld. *Tellus Series B-Chemical and Physical Meteorology,* **35(4),** 284–289.

West, T.O. and W.M. Post, 2002: Soil organic carbon sequestration rates by tillage and crop rotation: A global data analysis. *Soil Science Society of America Journal,* **66(6),** 1930–1946.

WHO (World Health Organization), 2000: *Air Quality Guidelines for Europe.* WHO Regional Publications, European Series, No. 91, Geneva, Switzerland, 273 pp.

WHO, 2002: *Tha World Health Report 2002: Reducing the Risks, Promoting Healthy Life.* World Health Organisation, Geneva, Switzerland, 248 pp.

WHO, 2004. *Heat-waves: Risks and Responses.* Health and Global Environmental Change Series, No. 2. WHO Regional Office for Europe, Copenhagen, Denmark, 124pp.

Williams, E., D. Rosenfeld, N. Madden, J. Gerlach, N. Gears, et al., 2002: Contrasting convective regimes over the Amazon: Implications for cloud electrification. *Journal of Geophysical Research-Atmospheres,* **107(D20),** 8082, doi:10.1029/2001JD000380.

Witt, C., K.G. Cassman, D.C. Olk, U. Biker, S.P. Liboon, M.I. Samson, and J.C.G. Ottow, 2000: Crop rotation and residue management effects on carbon sequestration, nitrogen cycling and productivity of irrigated rice systems. *Plant and Soil,* **225(1–2),** 263–278.

WMO (World Meteorological Organization), 2003: *Scientific Assessment of Ozone Depletion: 2002.* Global Ozone Research and Monitoring Project Report No. 47 [online], Geneva, 498 pp., Available at http://www.unep.org/ozone and http://www.unep.ch/ozone.

Wolf-Gladrow, D.A., U. Riebesell, S. Burkhardt, and J. Bijma, 1999: Direct effects of CO_2 concentration on growth and isotopic composition of marine plankton. *Tellus Series B-Chemical and Physical Meteorology,* **51(2),** 461–476.

Xue, Y. and P.A. Dirmeyer, 2004: The Sahelian climate. In: *Vegetation, Water, Humans, and the Climate: A New Perspective on an Interactive System,* P. Kabat, M. Claussen, P.A. Dirmeyer, J.H.C. Gash, L. Bravo de Guenni, M. Meybeck, R.A. Pielke, C.J. Vörösmarty, R.W.A. Hutjes, and S. Lütkemeier (eds.), Springer, New York, USA, pp. 59–78.

Xue, Y. and J. Shukla, 1993: The influence of land surface properties on Sahel climate. Part 1: desertification. *Journal of Climate,* **6,** 2232–2245.

Zak, D.R., W.E. Holmes, A.C. Finzi, R.J. Norby, and W.H. Schlesinger, 2003: Soil nitrogen cycling under elevated CO_2: A synthesis of forest FACE experiments. *Ecological Applications,* **13(6),** 1508–1514.

Zavarzin, G.A., 2001: Stanovlenie Biosfery. *Vestnik Rossijskoi Akademii Nauk,* **71(11),** 988–1001, (in Russian).

Zeng, N., J.D. Neelin, K.-M. Lau, and C.J. Tucker, 1999: Enhancement of interdecadal climate variability in the Sahel by vegetation interaction. *Science,* **286(5444),** 1537–1540.

Chapter 14

Human Health: Ecosystem Regulation of Infectious Diseases

Coordinating Lead Authors: Jonathan A. Patz, Ulisses E.C. Confalonieri
Lead Authors: Felix P. Amerasinghe, Kaw Bing Chua, Peter Daszak, Alex D. Hyatt, David Molyneux, Madeleine Thomson, Dr. Laurent Yameogo, Mwelecele-Malecela-Lazaro, Pedro Vasconcelos, Yasmin Rubio-Palis
Contributing Authors: Diarmid Campbell-Lendrum, Thomas Jaenisch, Hassane Mahamat, Clifford Mutero, David Waltner-Toews, Christina Whiteman
Review Editors: Paul Epstein, Andrew Githeko, Jorge Rabinovich, Philip Weinstein

---

*This appears in Appendix A at the end of this volume.

Main Messages

According to the World Health Organization, infectious diseases still account for close to one quarter of the global burden of disease. Major tropical diseases, particularly malaria, meningitis, leishmaniasis, dengue, Japanese encephalitis, African trypanosomiasis, Chagas disease, schistosomiasis, filariasis, and diarrheal diseases still infect millions of people throughout the world (*very certain*). .

The magnitude and direction of altered disease incidence due to ecosystem changes depend on the particular ecosystems, type of land use change, disease-specific transmission dynamics, and the susceptibility of human populations. Anthropogenic drivers that especially affect infectious disease risk include destruction or encroachment into wildlife habitat, particularly through logging and road building; changes in the distribution and availability of surface waters, such as through dam construction, irrigation, or stream diversion; agricultural land use changes, including proliferation of both livestock and crops; deposition of chemical pollutants, including nutrients, fertilizers, and pesticides; uncontrolled urbanization or urban sprawl; climate variability and change; migration and international travel and trade; and either accidental or intentional human introduction of pathogens (*medium certainty*).

There are inherent trade-offs in many types of ecosystem changes associated with economic development, where the costs of disease emergence or resurgence must be weighed against a project's benefits to health and well-being. Such trade-offs particularly exist between infectious disease risk and development projects geared to food production, electrical power, and economic gain. To the extent that many of the risk mechanisms are understood, disease prevention or risk reduction can be achieved though strategic environmental management or measures of individual and group protection (*high certainty*).

Intact ecosystems play an important role in regulating the transmission of many infectious diseases. The reasons for the emergence or reemergence of some diseases are unknown, but the main biological mechanisms that have altered the incidence of many infectious diseases include altered habitat, leading to changes in the number of vector breeding sites or reservoir host distribution; niche invasions or interspecies host transfers; changes in biodiversity (including loss of predator species and changes in host population density); human-induced genetic changes of disease vectors or pathogens (such as mosquito resistance to pesticides or the emergence of antibiotic-resistant bacteria); and environmental contamination of infectious disease agents (*high certainty*).

Disease/ecosystem relationships that best illustrate these biological mechanisms include the following examples with *high certainty* (unless stated otherwise):

- Dams and irrigation canals provide ideal habitat for snails that serve as the intermediate reservoir host species for schistosomiasis; irrigated rice fields increase the extent of mosquito breeding areas, leading to greater transmission of mosquito-borne malaria, lymphatic filariasis, Japanese encephalitis, and Rift Valley fever.

- Deforestation alters malaria risk, depending on the region of the world. Deforestation has increased the risk of malaria in Africa and South America (*medium certainty*).

- Natural systems with intact structure and characteristics generally resist the introduction of invasive human and animal pathogens brought by human migration and settlement. This seems to be the case for cholera,

kala-azar, and schistosomiasis, which have not become established in the Amazonian forest ecosystem (*medium certainty*).

- Uncontrolled urbanization of forest areas has been associated with mosquito-borne viruses (arboviruses) in the Amazon, and lymphatic filariasis in Africa. Tropical urban areas with poor water supply systems and lack of shelter promote transmission of dengue fever.

- There is evidence that habitat fragmentation, with subsequent biodiversity loss, increases the prevalence of the bacteria that causes Lyme disease in North America in ticks (*medium certainty*).

- Zoonotic pathogens (complete natural life cycle in animals) are a significant cause of both historical diseases (including HIV and tuberculosis) and newly emerging infectious diseases affecting humans (such as SARS, West Nile virus, and Hendra virus).

- Intensive livestock agriculture that uses subtherapeutic doses of antibiotics has led to the emergence of antibiotic strains of *Salmonella, Campylobacter,* and *Escherichia coli* bacteria. Overcrowded and mixed livestock practices, as well as trade in bushmeat, can facilitate interspecies host transfer of disease agents, leading to dangerous novel pathogens, such as SARS and new strains of influenza.

Human contact with natural ecosystems containing foci of infections increases the risk of human infections. Contact zones between systems are frequently sites for the transfer of pathogens and vectors (whenever indirect transmission occurs) to susceptible human populations such as urban-forest borders (malaria and yellow fever) and agricultural-forest boundaries (hemorrhagic fevers, such as hantavirus) (*high certainty*). The different types and subtypes of systems (natural, cultivated, and urban) may contain a unique set of infectious diseases (such as kala-azar or plague in drylands, dengue fever in urban systems, and cutaneous leishmaniasis in forest systems), but some major diseases are ubiquitous, occurring across many ecosystems (such as malaria and yellow fever) (*very certain*).

Tropical developing countries are more likely to be affected than richer nations in the future due to their greater exposure to the vectors of infectious disease transmission and environments where they occur. Such populations have a scarcity of resources to respond to and plan environmental modifications associated with economic activities (*high certainty*). However, international trade and transport leave no country entirely unaffected.

The following diseases (*high certainty*) are ranked as high priority for their large global burden of disease and their high sensitivity to ecological change:

- malaria across most ecological systems;

- schistosomiasis, lymphatic filariasis, and Japanese encephalitis in cultivated and inland water systems in the tropics;

- dengue fever in tropical urban centers;

- leishmaniasis and Chagas disease in forest and dryland systems;

- meningitis in the Sahel;

- cholera in coastal, freshwater, and urban systems; and

- West Nile virus and Lyme disease in urban and suburban systems of Europe and North America.

14.1 Introduction

This chapter focuses on infectious diseases whose incidence has been shown or is suspected to be related to anthropogenic ecological change. Mechanisms of change occur through a variety of ways, including altered habitats or breeding sites for disease vectors or reservoirs, niche invasions, loss of predator species, biodiversity change, host transfer, and changes in (intermediate) host population density.

Infectious diseases stemming from health infrastructural deficiencies, such as poor sanitation and lack of adequate vaccine coverage, as well as those linked to specific sociocultural factors, such as airborne and sexually transmitted diseases, are not covered in this chapter, even though these lead to a large global burden of disease. Readers should refer to Chapter 5 of this volume and Chapter 12 of *Policy Responses* for an assessment of noninfectious disease and related health topics.

Ecosystems affect human health in many ways, either directly or indirectly. Many major pharmaceuticals, including aspirin, digitalis, quinine, and tamoxifin, originated from plants. (See Chapter 10.) Intact ecosystems protect against mortality and injuries from floods and mudslides. (See Chapter 16.) Human health depends on access to food, clean water, clean air, and sanitation. (See also Chapters 8, 7, 13, and 27.) Watershed conditions influence water quality and wetlands help remove toxins from water—increased soil runoff following deforestation, for example, has led to mercury contamination of Amazonian fish. (See Chapter 20.) Toxic algal blooms threaten food safety. (See Chapter 19). Watershed protection has been used to offset the cost of drinking water treatment facilities. (See MA, *Policy Responses,* Chapter 7.) Finally, a broad range of noninfectious disease health risks and prevention strategies are detailed in Chapter 16 of *Policy Responses.*

Infectious diseases account for 29 of the 96 major causes of human morbidity and mortality listed by the World Health Organization, representing 24% of the global burden of disease (WHO 2004). As numerous reports address the health effects of poor sanitation and drinking water treatment, this chapter focuses more specifically on diseases with known links to anthropogenic ecological change. Table 14.1 and Figure 14.1 (in Appendix A) show the current extent and distribution of infectious and parasitic diseases around the globe.

The incidence of many of these diseases is not declining. According to WHO (WHO 2002), African trypanosomiasis, dengue, and leishmaniasis are emerging and expanding and do not yet have a standardized control program in place. In addition, malaria, schistosomiasis, and tuberculosis persist even though active control programs have been established. Comparison of disease-burden figures published in WHO's latest *World Health Report* (2004) with the same statistics from the previous report (WHO 2002) shows that malaria, meningitis, leishmaniasis, dengue, and Japanese encephalitis are increasing. Tropical diseases with essentially no change include diarrheal diseases, trypanosomiasis, Chagas disease, schistosomiasis, and filariasis. However, onchocerciasis (river blindness) shows a declining trend.

While this chapter summarizes known links between ecological degradation and altered infectious disease transmission or emergence, natural systems can also be a source of pathogens, and destruction of an ecosystem may, in some cases, reduce the prevalence of disease in an area. Destroyed ecosystems have led to the disappearance of foci of disease, but this has resulted more from economic development rather than from any planned disease control. Yet environmental modification has been, for millennia, a key means for controlling disease vectors—from the drainage of swamps in Rome to reduce mosquitoes to deforesta-

tion in Zimbabwe to protect cattle from trypanosomiasis. At this point in history, however, the scale of ecological change may be leading to disease emergence or reemergence, and this is the issue to which the assessment in this chapter is directed.

14.1.1 Historical Perspective on Infectious Diseases and Development

Over the millennia, people have used and changed the habitable environment. Ten thousand years ago, agriculture and large settlements developed. Several of today's most pervasive diseases originally stemmed from domestication of livestock. Tuberculosis, measles, and smallpox, for example, emerged following the domestication of wild cattle. Infectious agents or pathogens of vertebrate mammals that infect humans as incidental hosts are called zoonotic, and the resultant diseases are zoonoses. Many pathogens that are currently passed from person to person (anthroponotic), including some influenza viruses and HIV, were formerly zoonotic but have diverged genetically from their ancestors that occured in animal hosts. Many diseases thought to be caused by noninfectious agents, including genetically based and chronic diseases, are now known to be influenced or directly caused by infectious agents (UNEP in press).

In the last two centuries, the spread of industrial and post-industrial change, rapid population growth, and population movements have quickened the pace and extensiveness of ecological change. New diseases have emerged even as some pathogens that have been around for a long time are eradicated or rendered insignificant, such as smallpox. Environmental and ecological change, pollutants, the widespread loss of top predators, persistent economic and social crises, and international travel that drives a great movement of potential hosts have progressively altered disease ecology, affecting pathogens across a wide taxonomic range of animals and plants (Epstein 1995).

14.1.2 Ecology of Infectious Diseases

Intact ecosystems maintain a diversity of species in equilibrium and can often provide a disease-regulating effect if any of these species are either directly or indirectly involved in the life cycle of an infectious disease and occupy an ecological niche that prevents the invasion of a species involved in infectious disease transmission or maintenance. Disease agents with much of their life cycle occurring external to the human host, such as water- and vector-borne diseases, are subjected to environmental conditions, and it is these diseases for which most linkages to ecosystem conditions have been found (Patz et al. 2000).

Infectious diseases are a product of the pathogen, vector, host, and environment. Thus, understanding the nature of epidemic and endemic diseases and emerging pathogens is essentially a study of the population biology of these three types of organisms, as well as of environmental factors. In addition to ecologically mediated influences on disease, changes in the level of infectious diseases can themselves disrupt ecosystems (such as bird populations or predator-prey relationships altered by West Nile virus) (Daszak et al. 2000; Epstein et al. 2003).

Recent interest in infectious disease threats to public health has focused on emerging and reemerging pathogens. From a scientific perspective, looking at emerging infectious diseases is useful, as they display different adaptive mechanisms of evolution that have been "successful" in leading to the survival or even increased spread of a microorganism. In a narrow sense, the study of the ecology of emerging infectious diseases tries to understand (and possibly also predict) the mechanisms that lead to the ability

Table 14.1. Burden of Infectious and Parasitic Disease in 2003, by WHO Region and Mortality Stratum. Mortality stratum is a way of dividing up the WHO regions, which are based on geography, into units which are more similar in terms of health performance (i.e., separating Australia, Japan, and New Zealand out from China, the Philippines, and others in the Western Pacific Region, and Canada, the United States, and Cuba from the rest of the Americas, where health status is poorer). They are based on WHO estimates of adult and child mortality, with some arbitrary threshold to group them into different classes (see second column in the table). The data are based on nationally reported health statistics, although there is sometimes some estimation by WHO if national statistics are poor or non-existent. (WHO 2004)

Region	Mortality Stratum	Total DALYs[a]	DALYs from Infectious and Parasitic Diseases	Infectious and Parasitic Diseases as Share of Total
		(thousand)	*(thousand)*	*(percent)*
Africa	high child, high adult	160,415	75,966	47.4
	high child, very high adult	200,961	111,483	
Americas	very low child, very low adult	46,868	1,228	2.6
	low child, low adult	81,589	6,719	8.2
	high child, high adult	17,130	3,944	23.0
Southeast Asia	low child, very low adult	62,463	10,598	17.0
	high child, high adult	364,110	78,355	21.5
Europe	very low child, very low adult	51,725	891	1.7
	low child, low adult	37,697	2,040	5.4
	low child, high adult	60,900	2,734	4.5
Eastern Mediterranean	low child, low adult	24,074	1,529	6.4
	high child, high adult	115,005	30,881	26.9
Western Pacific	very low child, very low adult	16,384	322	2.0
	low child, low adult	248,495	23,349	9.4
Total		**1,487,816**	**350,039**	**23.5**

[a] Disability-adjusted life year: years of healthy life lost, a measure of disease burden for the gap between actual health of a population compared with an ideal situation where everyone lives in full health into old age. (WHO World Health Report 2004)

to switch hosts and establish in a new host—from the perspective of a given pathogen (as described later in this chapter).

Definitions of the term "emerging" are given early in the literature (e.g., Krause 1981; Lederberg et al. 1992). Emerging diseases are those that have recently increased in incidence, impact, or geographic or host range (Lyme disease, tuberculosis, West Nile virus, and Nipah virus, respectively); that are caused by pathogens that have recently evolved (such as new strains of influenza virus, SARS, or drug-resistant strains of malaria); that are newly discovered (Hendra virus or Ebola virus); or that have recently changed their clinical presentation (hantavirus pulmonary syndrome, for instance). Many authors vary in their definitions of "recent," but most agree that emerging infectious diseases are those that have developed within the last 20–30 years (Lederberg et al. 1992). "Reemerging" diseases are a subclass of emerging diseases that historically occurred at significant levels but that became less significant due to control efforts and only recently increased in incidence again, such as dengue fever and cholera.

14.1.3 Trade-offs

While preservation of natural ecosystems can prevent disease emergence or spread, there are recognized trade-offs between ecological preservation and human disease. Malaria control efforts, for example, which relied heavily on the insecticide DDT, caused enormous damage to wetland systems and beyond. (See MA *Policy Responses,* Chapter 12.)

Probably the best documented examples of trade-offs involving ecosystem change, development, and disease are associated with water supply projects for agriculture or electrical power. (See Box 14.1.) Dams and irrigation systems were one of the most visible symbols of water resources development and management in the twentieth century. Irrigation systems are estimated to consume 70–80% of the world's surface freshwater resources and produce roughly 40% of its food crops. (See Chapter 7.) The pace of irrigation development has increased rapidly over the past half-century, in order to meet the increasing food requirements of human populations. But irrigation and dam construction can also increase transmission of diseases such as schistosomiasis, Japanese encephalitis, and malaria.

Such trade-offs also have an important temporal aspect. For example, draining wetlands can reduce mosquito breeding sites for immediate benefit, but the wetland services of filtering, detoxifying, or providing species habitat will be lost.

14.1.4 Ecosystem Services Relevant to Human Health

The relationships between ecological systems, their services, human society, and infectious diseases are complex. (See Figure 14.2.) The primary drivers of ecosystem changes are linked to population growth and economic development. These changes trigger several ecological mechanisms that can often increase the risk of infectious disease transmission or can change conditions of vulnerability, such as malnutrition, stress and trauma (in floods and storms, for example), immunosuppression, and respiratory ailments associated with poor air quality.

BOX 14.1

Trade-offs: Dams—Food and Power versus Disease

From 1930 to 1970, dams were synonymous with economic development, consuming an estimated $3 trillion in global investments but providing food security, power, local employment, and the expansion of physical and social infrastructure such as roads and schools (WCD 2000). Apart from their direct benefits, dams also are recognized to mediate indirect benefits, both economic and social, that are often ignored in quantifications of economic benefits focused on crop production. These include the use of water for horticulture, livestock farming, fisheries, and domestic purposes.

Case studies done for the World Commission on Dams show services and benefits ranging from irrigation and electricity generation for domestic and industrial purposes to flood protection, tourism, fisheries, local employment, and water supply. At the household level it is accepted that irrigation projects, and implicitly large irrigation dams, have contributed to greater food security and improved nutrition. The magnitude of the impact at the national level is less clear. In India (one of the largest builders of irrigation dams), for instance, estimates of total food increase attributable to new land brought under irrigation range from 10% to 30% and nutrition levels have improved by 14% over the past 25 years. Over the past 50 years, the country has achieved a marginal per capita increase in food availability and a decrease in the proportion of rural population below the poverty line (people without the capacity to purchase their own food requirements). However, the absolute number of people below the poverty line has increased by approximately 120 million people (WCD 2000).

The financial and economic profitability of large dam projects also presents a mixed picture. The WCD's evaluation of 14 large dam projects showed a shortfall of roughly 5% in the average economic internal rate of return, between the appraisal estimate and the evaluation estimate, with 4 projects falling below the 10% rate of return that is deemed acceptable in a developing-country economic context. The WCD concluded that irrigation dam projects have "all too often" failed to deliver on the economic profitability promised, even when defined narrowly in terms of direct project costs and benefits. However, added to the direct costs are additional costs in terms of adverse economic, social, environmental, and health impacts, such as the loss of thousands of hectares of tropical forests and their associated flora and fauna, the suffering of physically and livelihood-displaced communities, the loss of downstream fisheries and agricultural productivity, and impaired health due to water-related diseases.

Diseases such as malaria, schistosomiasis, onchocerciasis, lymphatic filariasis, and Japanese encephalitis have at various times scourged humankind in different parts of the world, and together with diarrheal and intestinal diseases caused by microbial agents and helminths they continue to be a serious threat to human health. There is an extensive literature on disease outbreaks or increased endemicity occurring in the aftermath of large-scale water resources development over the past 50–75 years (for example, Surtees 1975; Service 1984, 1989; Gratz 1987; Hunter et al. 1993; Jobin 1999). However, there is a serious lack of comparative burden of disease estimations and economic cost estimations relating to these disease outbreaks or increased endemicity.

There is a wide spectrum of human disturbances to ecosystems and their services that may change disease risk via biological mechanisms described in the next section. Of course, human activities not associated with environmental modifications may also have a role in the production of infectious diseases, both in their emergence or resurgence. This is the case of the infectious processes associated with human behavior, such as those transmitted by direct contact (such as AIDS or skin infections), airborne infections, and some food-borne infections.

Ecosystem changes can mediate the influence of anthropogenic activities in changing the epidemiological patterns of human infectious diseases by reducing or increasing disease incidence. (See Box 14.2.) Most ecological systems have a unique set of infectious diseases; however, some diseases, such as malaria, are more ubiquitous and can be found across ecological systems such as drylands, forests, and wetlands, although with somewhat different dynamics.

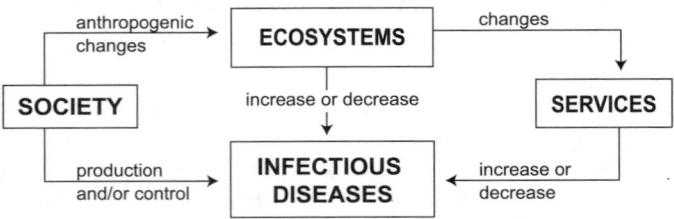

Figure 14.2. Relationships between Society, Ecosystem Services, and Human Infectious Diseases

14.2 Trends and Drivers of Changes in Disease Risk

Most emerging diseases are driven by human activities that modify the environment or otherwise spread pathogens into new ecological niches (Taylor et al. 2001). Examples of direct anthropogenic drivers that affect disease risk include wildlife habitat destruction, conversion, or encroachment, particularly through deforestation and reforestation; changes in the distribution and availability of surface waters, such as through dam construction, irrigation, and stream diversion; agricultural land use changes, including proliferation of both livestock and crops; deposition of chemical pollutants, including nutrients, fertilizers, and pesticides; uncontrolled urbanization; urban sprawl; climate variability and change; migration and international travel and trade; and either accidental or intentional human introduction of pathogens. (See also Chapter 3.)

These anthropogenic drivers of ecosystem disturbance can lead to specific changes in ecosystems that may or may not lead to disease emergence via mechanisms that are more directly relevant to life cycles or transmission of infectious diseases. There is concern that the extent of ecosystem changes in recent decades and the multiple ways in which habitats and biodiversity are being altered are increasing the odds that infectious diseases will be affected at some level. The specific biological mechanisms altering disease incidence, emergence, or reemergence are described here and, by way of illustration, disease case studies in this chapter are organized according to these biological mechanisms.

The relationship between infectious diseases and ecological changes is shown in Figure 14.3.

Disturbance or degradation of ecosystems can have biological effects that are highly relevant to infectious disease transmission. The reasons for the emergence or reemergence of some diseases

BOX 14.2

The Resilience of the Amazon Forest in Preventing the Establishment of Infectious Diseases

The Amazon forest system in Brazil has been the subject of successive cycles of occupation and development since the last quarter of the nineteenth century, starting with the rubber boom up to the 1980s, when road building and the expansion of cattle ranching also became important drivers. These developments attracted human migrations from other parts of the country, either for temporary work or permanent settlement (Confalonieri 2001). Most of the migrants came from northeastern Brazil, which is endemic for diseases like kala-azar, schistosomiasis, and Chagas disease.

As for the introduction of schistosomiasis, both infected humans and snail intermediate hosts have been found in the region (Sioli 1953). However, the foci of the disease were established only in the periphery of a few major cities, in snail breeding sites created by humans, such as pools, ponds, and channels. No foci of transmission of the disease were created among the riverine populations, some including infected human migrants. The main reason was the absence of the snail species *Biomphalaria spp*, which were not able to develop probably due to the characteristics of the fresh water of the natural systems, which do not have the appropriate mineral salts necessary for the formation of the shell of the snails (Sioli 1953).

A similar situation has been observed with kala-azar, which is endemic in rural areas of the Brazilian northeast, involving humans, sand flies,

dogs, and wild canids as reservoir animals. So far the disease has become established only in two geographically restricted areas of the Brazilian Amazon: in a savanna area in the northern part and in a periurban setting in the central part (Confalonieri 2000). In both situations it seems that the dogs are the only reservoir species involved in the transmission cycle, and the pathogen did not pass to wild populations of vertebrates (Guerra 2004; Silveira et al. 1997).

In the early 1990s, cholera entered Brazil through the Peruvian border and moved down the Amazon River and its tributaries, where a few small outbreaks occurred. When the "cholera wave" reached the major cities in the Amazon region, hundreds or even thousands of people were affected. In the "rural"/riverine areas of western Brazilian Amazon, the disease affected people only for a few months and vanished without any significant control measures being implemented. The suspected major reasons for this, in addition to the low human population density in the area, was that the left margin tributaries of the Amazon River were unsuitable for the survival of *Vibrio cholerae,* due especially to their low pH.

These case studies are good example of the "nonreceptiveness" of natural systems—and the people living on them—to the introduction of alien pathogens or parasites and their invertebrate carriers, which did not evolve in these environments due to a natural resilience. This has important implications both for public health and for conservation.

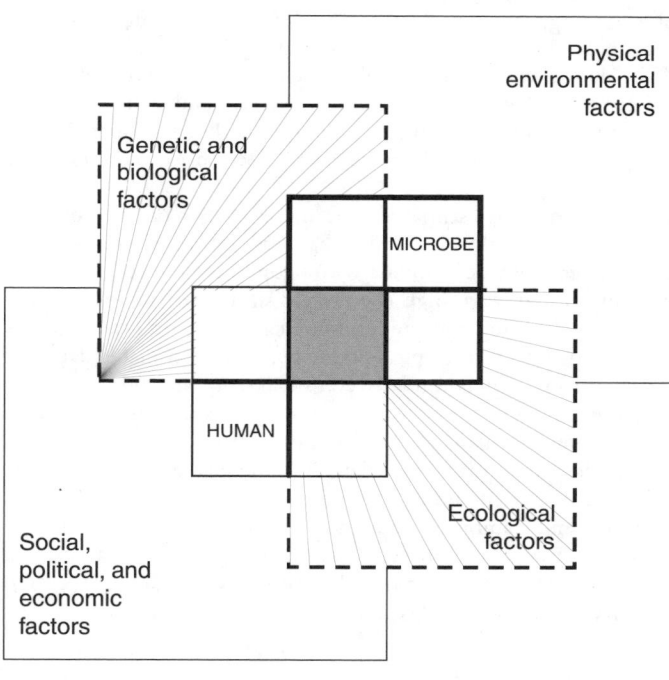

Figure 14.3. Convergence Model for the Emergence of Infectious Diseases, Combining Multiple Causes of Disease Emergence or Reemergence (Institute of Medicine 2003)

are unknown, but the following mechanisms have been proposed and have altered the incidence of many diseases (Molyneux 1997; Daszak et al. 2000; Patz et al. 2004):

- altered habitat leading to changes in the number of vector breeding sites or reservoir host distribution;
- niche invasions or transfer of interspecies hosts;

- biodiversity change (including loss of predator species and changes in host population density);
- human-induced genetic changes in disease vectors or pathogens (such as mosquito resistance to pesticides or the emergence of antibiotic-resistant bacteria); and
- environmental contamination by infectious disease agents (such as fecal contamination of source waters).

While the rest of this chapter is organized according to these mechanisms (see Table 14.2), it is important to recognize that in many instances these often overlap or act in combination, sometimes resulting in non-linear or synergistic effects on disease transmission. It should be noted that emerging or resurging diseases may occur across many ecosystems even though they have been grouped here according to the major ecosystem in which they are prevalent.

14.2.1 Altered Habitat/Breeding Sites: Effect on Infectious Disease Transmission

Disturbance of habitats due to alterations in land cover or climatic change is considered to be the largest factor altering the risk of infectious diseases—for example, by affecting breeding sites of disease vectors or the biodiversity of vectors or reservoir hosts. Examples of diseases emerging or resurging due to habitat change that occur across many ecosystems are described here for cultivated, drylands, forest, urban, and coastal systems.

14.2.1.1 Irrigation and Water Development in Cultivated Systems

According to the FAOSTAT database (apps.fao.org/default.jsp), the global extent of irrigated agricultural land increased from 138 million hectares in 1961 to 271 million hectares in 2000. In 1950, there were an estimated 5,000 large irrigation and multipurpose dams in the world, which has now increased to more than 45,000. These provide water for 30–40% of irrigated agricultural land and

Table 14.2. Mechanisms of Disease Emergence and Examples of Diseases across Ecosystems

Mechanisms	Ecosystems				
	Cultivated Systems	Dryland Systems	Forest Systems	Urban Systems	Coastal Systems
Habitat alteration	schistomiasis Japanese encephalitis malaria	hantavirus Rift Valley fever meningitis	malaria arboviruses (e.g., yellow fever) onchocerciasis	lymphatic filariasis Dengue fever malaria	cholera
Niche invasion or host transfer	Nipah virus BSE (mad cow) SARS influenza		HIV (initially)	leishmaniasis	
Biodiversity change	leishmaniasis	onchocerciasis	rabies onchocerciasis	lyme disease	
Human-driven genetic changes	antibiotic-resistant bacteria		chagas disease	chagas disease	
Environmental contamination of infections agents	cryptosporidiosis leptospriosis			leptospirosis	diarrheal diseases

generate 19% of global electricity supplies. (See Chapter 8 for more on water impoundment and offtake for irrigation.)

Inevitably, this growing trend in water resources development has resulted in qualitative and quantitative changes in natural biodiversity and in changed levels of interaction between humans, vectors, and disease agents. Large-scale human resettlement, which usually occurs in conjunction with irrigation development, has resulted in exposure to disease in nonimmune populations. Furthermore, multiple cropping has changed movement patterns of temporary agricultural labor, increasing risks of disease transmission and dissemination, while overcrowding and low nutritional status has increased susceptibility to infection. Water resources development has often been accompanied by invasions of new disease-carrying vectors or population changes in existing vectors, as well as similar changes in disease agents, which have increased the risks of disease (Bradley and Narayan 1988).

Global statistics on major microbial and parasitic diseases associated with water resource development show differential degrees of mortality and morbidity due to different diseases. Comparisons in terms of disability-adjusted life years (DALYs) show that infectious and parasitic diseases contribute 23.5% (340 million DALYs) of the total global burden of disease.

Unfortunately, DALY estimates only exist for global and regional scales at present, and it is not possible to use these to make direct spatial or temporal comparisons of disease burdens associated with large-scale water resources developments. Such information would be needed to evaluate properly the impact of water resource developments on human health. A similar caveat applies to economic indicators that point to significant impacts. For instance, malaria alone is estimated to reduce the economic growth rate in seriously affected countries by 1.3% per year and to cost these countries billions of dollars a year (Malaney et al. 2004). Once again, however, such figures cannot be disaggregated to determine the economic costs associated with water resources development in particular.

Small dams, below 15 meters in height, are far more numerous than large dams and are closely linked to agriculture. They have been advocated by the international financing community as manageable and practical solutions to land and water conservation. They serve more purposes than large dams: for example, a small multipurpose project may provide water for domestic supplies, fishing, cattle, and irrigation while providing flood control. Although there is no accurate estimate of the number of small dams in the world, it is likely that their collective volume is greater than that of large dams. In Nigeria and Zimbabwe, for instance, the shore length of small dams has been estimated to be 8–10 times that of the large reservoirs (Jewsbury and Imevbore 1988).

Similarly, it is estimated that small dams have an equal or greater impact on human health than large dams. There is usually a high degree of water contact with people and animals, so disease transmission rates are high (Hunter et al. 1993). However, there have been relatively few epidemiological studies on disease trends around small dams in tropical countries, despite available data that show a strong association between small dams and substantial increase in disease. For instance, intense transmission of diseases such as schistosomiasis, onchocerciasis, malaria, lymphatic filariasis, and dracunculosis are associated with small dams in many African countries, including Cameroon, Kenya, Ghana, Mali, Rwanda, and Zambia (Hunter et al. 1993). As in the case of large water resource developments, comparative health statistics in terms of DALYs or other useful parameters such as years of life lost or years lived with disability are yet to be determined.

14.2.1.1.1 Schistosomiasis

Irrigation canals are known to provide ideal habitat for the snails that serve as an intermediate reservoir host for schistosomiasis. For example, reduced salinity and increased alkalinity of water associated with irrigation development along the Senegal River have been shown to increase fecundity and growth of freshwater snails (Southgate 1997).

A literature review of the association between schistosomiasis and the development of irrigation projects along the Tana River in Kenya has been provided by Mutero (2002). Two species of the genus *Schistosoma* occur in Africa, namely *S. haematobium* and *S. mansoni*. Clinical signs for the two infections are blood in the urine and blood in the stools, respectively. *S. haematobium* has the highest prevalence along the lower Tana, where the now largely abandoned Hola irrigation scheme is situated. In 1956, when the scheme began, there were no snail vectors of *S. haematobium* due to a lack of suitable habitat because of the scheme's elevation above the river. A decade later, there was a 70% prevalence of urinary schistosomiasis among local children, which rose to 90% by 1982 due to poorly maintained irrigation channels.

14.2.1.1.2 Mosquito-borne diseases and tropical rice irrigation

There are many examples worldwide of vector-borne disease problems linked to water resources development (see reviews of Bradley 1977; Mather and That 1984; Service 1984). (See Box 14.3.) Tropical rice irrigation systems, in particular, have been linked to vector-borne diseases such as malaria and Japanese encephalitis (reviews by Lacey and Lacey 1990; Amerasinghe 2003). Major ecological impacts of irrigated rice include an enormous increase in the extent of mosquito breeding surface and an increase in the availability of habitat where multiple cropping occurs. These factors may selectively favor some species and displace or change the relative dominance of certain species or genotypes.

With malaria, in particular, the result can be marked changes in disease equilibrium, which may increase or decrease depending on the transmission capability of a particular mosquito species (Akogbeto 2000). The varied epidemiology of malaria in different cultivated systems in Africa is aptly reviewed by Ijumba and Lindsay (2001), who coined the phrase "paddies paradox" to describe situations where irrigation increases vector populations but may or may not increase malaria.

This anomaly has been largely attributed to differences in socioeconomic and ecological environments inside and outside irrigation schemes. For instance, villages surrounded by irrigated rice fields in Kenya showed a 30- to 300-fold increase in the number of the local malaria vector, *Anopheles arabiensis,* compared with those without rice irrigation, yet malaria prevalence was significantly lower in these villages (0–9% versus 17–54%) (Mutero et al. 2003). The most plausible explanation for this appeared to be the tendency of *An. arabiensis* to feed more on cattle than people in irrigated villages.

BOX 14.3

Irrigation: Infectious Disease Case Studies from Sri Lanka and India

Two of the most recently documented examples of ecosystem disturbance and mosquito-borne disease are from South Asia, and they provide contrasting examples of the aggravation of health problems resulting from irrigation development in a tropical environment (Sri Lanka) and a more northern desert region (India).

In Sri Lanka, the Accelerated Mahaweli Development Project developed 165,000 hectares of land (much of it forested land previously unoccupied by humans) for irrigated rice cultivation, resulting in the settlement of 1 million persons into the malaria-endemic lowland dry zone of the country between 1980 and 1990. Varied impacts of mosquito-borne diseases were observed in the irrigated rice systems. Malaria increased two- to fivefold in all systems within the first two to three years of settlement (Samarasinghe 1986). *Plasmodium falciparum* infections increased from the normal 5% of infections to 24% in some regions. Some systems have remained highly malarious 10–15 years later, but the disease burden has decreased in other areas. Upstream impacts also were recorded, with outbreaks of malaria in villages along the banks of the Mahaweli River in normally non-malarious hill country areas—a consequence of decreased water flow, pooling, and vector breeding as a result of water impoundment at upstream dams (Wijesundera 1988).

The major vector of malaria in Sri Lanka is *Anopheles culicifacies,* but in areas of the Mahaweli Project it was observed that additional species such as *An. annularis* and *An. subpictus* also played a significant role in transmission, as their populations increased due to the availability of suitable breeding habitats (Amerasinghe et al. 1992; Ramasamy et al. 1992).

On top of the malaria burden came Sri Lanka's first major epidemic of Japanese encephalitis in System-H of the Mahaweli in 1985–86 (more than 400 cases and 76 deaths), followed by a second epidemic in 1987–88 (more than 760 cases and 138 deaths). The catalyst appears to have been the promotion of smallholder pig husbandry in a misguided attempt to generate supplementary income among farmers. In a rice irrigation system where *Culex tritaeniorhynchus* and other *Culex* vectors of Japanese encephalitis were breeding prolifically, the outcome was catastrophic.

The Mahaweli represents a complex of gross physical ecosystem disturbance in terms of forest clearing, dam, reservoir and canal construction, and the maintenance of standing or flowing water virtually throughout the year, which erased the normal trend of wet and dry periods. Added to this was biological disturbance in terms of the replacement of a diverse natural forest flora and fauna by the introduction of a virtual crop monoculture (rice), and a dominant large mammal population (humans, often from areas nonendemic to diseases such as malaria and Japanese encephalitis), together with fellow-traveler species (garden plants, vegetables, fruit trees, livestock, domestic pets, poultry, rodents, and so on). Whether overall plant and animal biodiversity was diminished or not is debatable, but it is clear that the natural biodiversity was replaced by a crop-related one that afforded opportunities for disease causing–organisms and their vectors to have an impact.

The development of irrigated agriculture in the Thar Desert, Rajasthan, in northwestern India provides another telling example of ecosystem disturbance exacerbating disease burden. Here, the major change was the provision of surface water to a desert area.

The Thar Desert was traditionally only mildly prone to malaria, but in the last six decades it has undergone drastic change in physiography and microclimate concomitant with irrigation development. Contrasting trends in the balance between *An. culicifacies* (an efficient vector) and *An. stephensi* (a poor vector) have occurred in the Thar Desert, with *An. stephensi* constituting 94% of the two species in desert areas, but the opposite situation (overwhelmingly more *An. culicifacies*) holding in irrigated areas. As a result, the prevalence of malaria in the irrigated areas has increased almost fourfold between the 1960s and today, with several epidemics in the past 15 years. As in Sri Lanka, this has been accompanied by a high incidence of *P. falciparum* infections in the irrigated areas, rising from 12% in 1986 to 63% by 1994 (Tyagi 2002).

Although excessive rainfall triggered by the El Niño Southern/Oscillation has probably contributed to malaria epidemics in the Thar Desert, Tyagi (2002) relates most of the recent epidemics to the phenomenon of "inundative vectorism"—the sudden ushering of one or more vector species in prodigiously high densities in virgin areas such as a recently irrigated desert). Malaria in the Thar Desert is now effectively transmitted in three ways: in the irrigated area it is transmitted in tandem by the native *An. stephensi* and invader *An. culicifacies*; in the dryland areas, by *An. stephensi*; and in the non-command flood-prone southern areas, mainly by *An. culicifacies*.

In rural India during the 1990s, "irrigation malaria" was responsible for endemic transmission in a population of about 200 million people (according to Sharma 1996). This has been attributed to poorly maintained irrigation systems, illegal irrigation, water seepages, poor drainage, and a rise in water tables associated with irrigation that created conditions suitable for the breeding of the major vector *An. culicifacies* and slow running streams that favor another vector, *An. fluviatilis*.

Japanese encephalitis is confined to Asia and is almost always associated with rice ecosystems. Region-wide, with an estimated 50,000 cases per year and 20% fatality and disability rates, the disease takes a considerable social and economic toll (Hoke and Gingrich 1994). The primary vector, *Culex tritaeniorhynchus,* occurs throughout Asia and breeds abundantly in flooded rice fields, as does another important vector, *C. vishnui* (India, Thailand, Taiwan). Other vectors, such as *C. gelidus* (in Indonesia, Sri Lanka, Thailand, and Viet Nam), *C. fuscocephala* (in Malaysia, Thailand, Taiwan, and Sri Lanka), and *C. annulus* (in Taiwan), breed in a variety of habitats, some of them associated with irrigated rice.

The transmission cycle of Japanese encephalitis involves an amplifying host, which is usually the domestic pig (in some instances, birds of the heron family (Ardeidae) are also involved) (Hoke and Gingrich 1994). Thus, irrigated riceland communities in which pig husbandry is traditionally carried out are likely sites for the disease, as both vector and amplifying host are brought together. The disease is endemic in irrigated ricelands in Thailand and China. Explosive outbreaks of Japanese encephalitis in new irrigation systems have been reported from the Terai region of Nepal and in Sri Lanka (Joshi 1986; Peiris et al. 1992). Extensive use of synthetic nitrogenous fertilizers in Indian rice fields has been blamed for significant increases in the populations of this disease's vectors; elevated nitrogen in the rice field water increases the density of mosquito larvae (Victor and Reuben 2000; Sunish and Reuben 2001).

Another disease linked to irrigation developments is lymphatic filariasis. Commonly called "elephantiasis," filariasis is caused by a mosquito-borne helminth. Outbreaks often occur from rising water tables following water project developments. In Ghana, for instance, rates of infection, worm load, annual bites per person, and annual transmission potential have been found to be higher in irrigated areas than in communities without irrigation (Appawu et al. 2001). This was also confirmed by another observation, where opening irrigation channels during the dry season resulted in a significant increase of filariasis vectors (Dzodzomenyo et al. 1999). On the other hand, conversion of swamps into rice fields on Java Island, Indonesia, resulted in a decrease of breeding sites for vectors and therefore a decrease in disease transmission (Oemijati et al. 1978).

14.2.1.1.3 African trypanosomiasis (African sleeping sickness)

Tsetse flies are widely distributed in West and Central Africa and parts of East Africa; they are the vectors of animal and human trypanosomes that cause African sleeping sickness. They are a highly adaptable group of generalized vectors feeding on available hosts throughout their distribution and are associated with the ability to adapt rapidly to changing habitats and vegetation. For instance, in Kenya in 1964 an outbreak of sleeping sickness was associated with the spread of flies from the natural shoreline habitats of Lake Victoria to vegetation within settlements characterized by thickets of *Lantana camara*; flies were feeding on humans and cattle, with cattle acting as the reservoir host of the parasite. More recently, during the 1980s an epidemic in Busoga, Uganda, was the result of civil unrest and abandonment of traditional ag-

ricultural practices and crops (coffee and cotton), followed by the spread of *L. camara* along village edges (UNEP in press).

In West Africa the behavior of tsetse flies in Southeast Nigeria and Côte d'Ivoire peridomestic populations has been closely associated with villages with a high population of domestic pigs and, again, *Lantana* as well as other vegetation (coconuts, yams, and bananas). In West Africa, *Glossina palpalis* and *G. tachinoides* appear to feed preferentially on pigs, where they act as a "dilution host"—reducing the risk of sleeping sickness to humans, as do cattle in Uganda (UNEP in press).

14.2.1.1.4 Rodent-borne hemorrhagic viruses

These infections are caused by different species of arenaviruses, with wild rodents of the genera *Calomys, Sigmodon, Akodon,* and *Zygodontomys* acting as their natural hosts and reservoirs. They have been especially recorded in Argentina (Junin virus), Bolivia (Machupo virus), and Venezuela (Guanarito virus) (Simpson 1978; Salas et al. 1991; Maiztegui 1975; de Manzione et al. 1998).

These infections often occur in outbreaks involving a few dozen to thousands of cases, mainly in rural populations, and humans become infected through contact with the urine and feces of infected rodents. In the specific case of Junin virus infections there is an additional occupational component: agricultural workers risk excess exposure during the harvesting of corn.

Human infection of the three diseases occurs in both villages and the wider countryside, primarily due to the contact between susceptible human hosts and the naturally infected rodent species in agro-ecosystems. In short, these viral infections are linked to the expansion of agriculture into natural systems in South America.

14.2.1.2 Habitat Change and Disease in Drylands and Grasslands

As with other systems, drylands have specific components and services that are relevant to human health issues. (See Box 14.4.) These can be grouped in two general categories: those that are part of natural systems (either biological or nonbiological) that can pose risks to human health and those associated with social interventions to promote human livelihoods in dry environments.

Water availability is the major limiting factor in drylands—not only for the survival of wild species and for agricultural and livestock production systems (see Chapter 22), but also for human health. Water scarcity and the poor quality of available water can increase the risk of transmission of pathogens associated with poor hygiene practices, leading to food-borne and water-borne diseases, which are major health problems in impoverished communities living in these areas. The greatest impacts are most often experienced by the more vulnerable social segments, such as high morbidity and mortality rates in children due to diarrheas. These infections are aggravated by the already existing chronic physical health problems common in impoverished communities, such as malnutrition, as well as a lack of adequate medical care in what are often marginalized communities.

Climate extremes such as droughts can have severe impacts on human health via different pathways, both direct and indirect. Direct effects are basically associated with the exacerbation of water scarcity as well as with food deprivation, resulting in famines. Droughts also have indirect effects that are mediated by social and demographic mechanisms such as individual and population stress and migrations. Movements of rural communities deprived of water and food in extreme situations can become an important determinant of the spatial redistribution of endemic infectious diseases. This is the case of kala-azar in northeastern

---

BOX 14.4

Meningitis in West Africa and Its Connection to Ecology, Overgrazing, and Dust Clouds

No animal reservoirs of infections are involved in the transmission of meningococcal meningitis in Africa. This is an airborne disease, transmitted from person to person through aerosols that are inhaled. The disease occurs in high endemic levels in Africa only in the so called meningitis belt, which is a large dry area in the Sahel. The specific environmental mechanisms involved in determining the biological vulnerability of the human population to this infection is not well known, but it is generally assumed that the low relative humidity is an important factor in decreasing the resistance of the upper respiratory tract. The role of dust storms is also being investigated (Molesworth et al. 2002).

The countries of Sahelian Africa sandwiched between the hot dry Saharan Desert and the moist humid forests of the Guinea Coast are among the poorest in the world. The region includes significant portions of Senegal, Mauritania, Mali, Burkina Faso, Niger, Chad, Sudan, and Eritrea, which roughly correspond to the zone receiving between 200 and 600 millimeters of annual rainfall. In this region, approximately 90% of the population depends on subsistence agriculture. As a result of rapid demographic change, climate variability, and economic drivers, farmers have expanded their cropping areas to increase production.

Agricultural intensification (often resulting in overgrazing) coupled with higher population densities in many areas has led to fallow periods that are insufficient to recuperate the soil. This has resulted in the loss of topsoil due to wind erosion, thus increasing the sources of atmospheric dust. Wind erosion may cause three types of agricultural damage: sedimentation at undesired places, crop damage, and soil degradation (Stark 2003).

Wind erosion has resulted in an increase in local dust storms, widely considered to be related to ill health during the dry season. For instance, in a detailed study of farmers' perceptions on the causes and consequences of wind erosion in Niger, its impact on health (fever, coughing, and sore eyes) was of greater concern than its contribution to crop damage or loss of topsoil (Bielders 2001). Dramatic increases in atmospheric dust in Sahelian West Africa have been noted in recent years (Ben Mohamed 1986; N'tchayi 1994; Nicholson 1998); atmospheric dust emanating from this region and the Sahara has been implicated in respiratory problems many thousands of miles away (Prospero 2001).

Dust storms have also been implicated in changes in the spatial and temporal dynamics of meningococcal meningitis epidemics in the region (Molesworth 2002). Key factors that have been identified as determinants of areas at risk of epidemic meningitis are land cover and absolute humidity (Molesworth 2003). The identification of these determinants has significant implications for directing essential monitoring and intervention activities and health policy. It also provides a basis for monitoring the impact of climate variability and environmental change on epidemic occurrence in Africa.

Brazil. In the major droughts of the early 1980s and 1990s, massive intraregional migrations of people from endemic rural areas to cities in search of subsistence and governmental assistance created new foci of the disease. This has resulted in outbreaks of infectious diseases at the periphery of major cities (Confalonieri 2001).

Besides natural biotic and abiotic factors, anthropogenic environmental modifications in dryland areas can also create conditions for the emergence of health problems related to development projects in these areas. This is the case of the efforts to increase food production in drylands by designing schemes for the provision of water, such as irrigation, and the building of dams has resulted in an increase in the incidence of tropical diseases, as mentioned earlier.

Wild fauna and insect ecology is also important for some infectious diseases located in dryland systems, including plague in the scrublands of southwestern United States, northeastern Brazil, and Peru and in regions of India and southern Africa. Another focal infection is kala-azar (*Leishmania donovani*), which is present mostly in arid areas such as Sudan (Thomson et al. 1999), Mediterranean countries, and South America.

14.2.1.2.1 Hantavirus

Rodent-borne hantavirus occurs both in arid grasslands (for example, in North America), as well as agricultural systems (particularly in South America and Asia). One of the best-known outbreaks occurred in the spring and summer of 1993, when acute respiratory distress with a high fatality rate was diagnosed among previously healthy individuals in the Four Corners region of the southwestern United States (Engelthaler et al. 1999). The disease, hantavirus pulmonary syndrome (HPS), was traced to infection by a previously unrecognized hantavirus. The virus (Sin Nombre virus) was found to be maintained and transmitted primarily within populations of a common native field rodent, the deer mouse Peromyscus spp. Transmission to humans is thought to occur through contact with virus in secretions and excretions of infected mice.

Recent studies have now shown that the El Niño effects during 1991–92 helped boost the reservoir populations of rodents in the region (Engelthaler et al. 1999; Glass et al. 2000). Unseasonal rains during the usually dry summer months in 1992 produced favorable environmental conditions in the spring and summer of 1992 that led to the outbreak of HPS. Parmenter and colleagues reported that populations of deer mice at the Long-Term Ecological Research station approximately 90 kilometers south of Albuquerque, New Mexico, were ten- to fifteenfold higher during the HPS outbreak period than the previous 20–year average (Parmenter et al. 1993). Glass and colleagues (2000) further showed the potential of using remotely sensed data to monitor conditions and identify high-risk areas up to a year in advance of anticipated disease outbreaks.

14.2.1.2.2 Rift Valley Fever

Extensive Rift Valley Fever outbreaks were not reported until 1951, when an estimated 20,000 people were infected during an epidemic among cattle and sheep in South Africa. Outbreaks were reported exclusively from sub-Saharan Africa until 1977–78, when 18,000 people were infected and 598 deaths were reported in Egypt (CDC 2002).

All known Rift Valley Fever virus outbreaks in East Africa from 1950 to May 1998, and probably earlier, followed periods of abnormally high rainfall (Woods et al. 2002). Analysis of these records and of Pacific and Indian Ocean sea surface temperature anomalies, coupled with vegetation data from satellites, showed that accurate predictions of Rift Valley Fever outbreaks in East Africa could be made up to five months in advance. Concurrent near-real-time monitoring with such data may identify actual affected areas (Linthicum et al. 1999).

Dams and irrigation can increase the breeding sites of the Rift Valley Fever vector, exacerbating the effect of extreme rainfall. (See Figure 14.4.) Several ecological changes have been reported in the epidemic region in Mauritania following dam construction, irrigation, and heavy rainfall (Jouan et al. 1990). Environmental factors, such as hydroelectric projects, and low-grade transmission among domestic animals could have enhanced the disease's survival and subsequent outbreaks (Lefevre 1997).

14.2.1.3 Habitat Change and Disease in Forest Systems

The major vector-borne diseases are focused in the tropics. There is a significant overlap between the distribution of the majority of important vectors of human and animal diseases and the biological richness of tropical rain forest ecosystems, woodland savannas, and boundaries of these ecosystems. (See Box 14.5.) It is the degradation of these ecosystems, the behavior and ecology of the vectors at the forest edge, the impact of deforestation on the interactions between humans with vectors, and reservoir hosts at the interface that determine the epidemiology of human infective agents. Additional factors are the behavior and degrees of immunity of local or migrant populations, their interaction with and the behavior of reservoir hosts, and the availability and effectiveness of surveillance systems and quality of local health care (UNEP in press).

14.2.1.3.1 Malaria

Deforestation, with subsequent changes in land use and human settlement patterns, has coincided with an upsurge of malaria or its vectors in Africa (Coluzzi et al. 1979, 1984, 1994), Asia (Bunnag et al. 1979), and Latin America (Vittor et al. in press; Tadei et al. 1998). (See Box 14.6.)

The capacity of different *Anopheles* mosquitoes to transmit malaria varies between species. Anopheline species themselves also occupy a variety of ecological niches. *An. darlingi* in South America, *An. gambiae* in Africa, and *An. dirus* in Southeast Asia are the predominant and highly effective vectors in their respective regions.

When tropical forests are cleared for human activities, they are typically converted into agricultural or grazing lands. This process is usually exacerbated by construction of roads, causing erosion and allowing previously inaccessible areas to become colonized (Kalliola and Paitan 1998) and anopheline mosquitoes to

invade. Cleared lands and culverts that collect rainwater are far more suitable breeding sites for malaria-transmitting anopheline mosquitoes than forest (Tyssul Jones 1951; Marques 1987; Charlwood and Alecrim 1989). Forest-dwelling *Anopheles* species either adapt to newly changed environmental conditions or disappear from the area, which offers other anophelines a new ecological niche (Povoa et al. 2001).

14.2.1.3.2 Forest Arboviruses in the Amazon

A wide variety of arboviruses occurs in the Amazon forest, a consequence of the extreme diversity of both arthropod vectors and wild vertebrates.

Thirty-two arbovirus types have been associated with human disease in the Brazilian Amazon region. Of these, four are important in public health because of their link to epidemics: the Oropouche virus (family *Bunyaviridae*), dengue and yellow fever viruses (*Flaviviridae*), and Mayaro virus (*Togaviridae*). It is noteworthy that Oropouche and dengue viruses are associated with human epidemics in urban areas while Mayaro and yellow fever occur in rural areas. All arboviruses (except dengue) that have been isolated in the Brazilian Amazon are maintained within complex cycles in the forest, where many species of blood-sucking arthropods act as vectors and several wild vertebrates act as reservoir hosts.

There have been historical changes in the Amazonian environment due to natural cyclical processes such as climate variability and as the result of human economic and geopolitical activities. The latter, which includes deforestation, construction of dams and highways, and mining, can disrupt to a greater extent the fragile equilibrium of the forest ecosystem, with impacts in the dynamics of virus transmission (Dégallier et al. 1989; Shope 1997; Vasconcelos et al. 2001).

Dam construction has been associated with the emergence of several different arboviruses, some of them responsible for human disease while others were previously unknown (Vasconcelos et al. 2001b).

Comparative studies carried in the 1970s and 1980s in Altamira and Tucurui municipalities prior, during, and after the construction of the Tucurui Dam in the State of Pará, Brazil, showed that inadequate management of the environment can cause an increase in the occurrence of a known virus or the appearance of a new one (Pinheiro et al. 1977; Dégallier et al. 1989). Examples

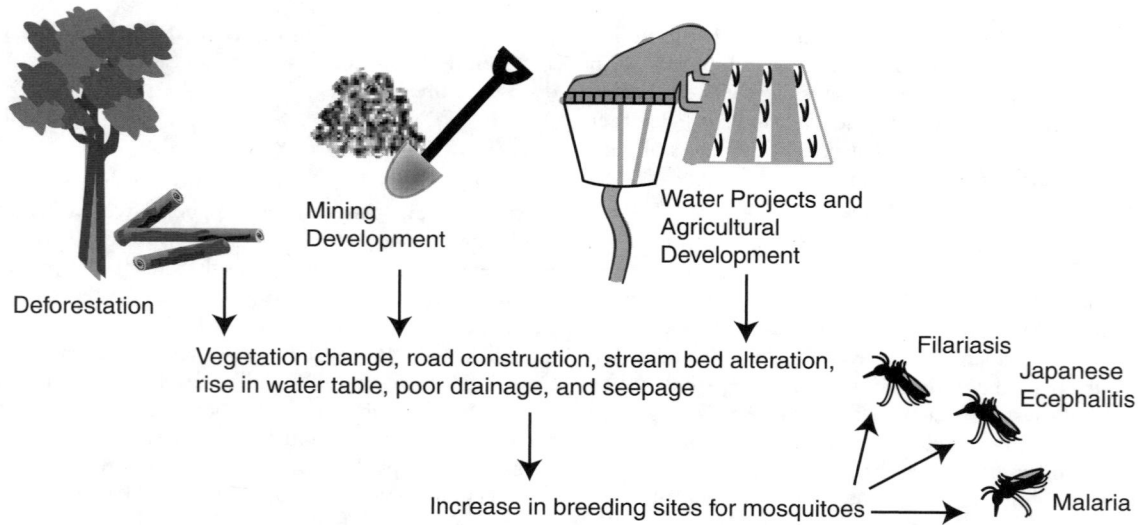

Figure 14.4. Habitat Change and Vector-borne Diseases

BOX 14.5

Reemergence of Onchocerciasis Related to Deforestation in Africa

Prior to the 1970s, onchocerciasis (also known as river blindness) was a neglected disease. Its devastating effects were largely borne by rural populations of West Africa living near the fast-flowing rivers of the Sahel. When the Onchocerciasis Control Programme was started in 1974, some of West Africa's richest riparian lands were uninhabited. In villages sited in river valleys near to the white water rapids that form the major breeding sites of the blackfly vector, it was not unusual to find 60% of adults afflicted with the disease and 3–5% blind. As a direct consequence of the disease, communities were forced to abandon their villages en masse. Today, 30 years and $600 million after the program was first launched, the disease has been controlled through one of the most successful public health campaigns in history (Benton 2002).

The filarial worm that causes the disease (*Onchocerca volvulus*) is transmitted in West Africa solely by blackflies, which are members of the *Simulium damnsum* species complex. Understanding the spatial and temporal distribution of the vectors of onchocerciasis has been key to their successful control (Boakye et al. 1998), given that the *S. damnosum* species complex comprises many distinct sibling species with varying capacities to transmit *O. volvulus*. The savanna species *S. sirbanum* and *S. damnosum* s.str. were identified early on in the program as the species associated with the blinding form of the disease. The distribution of the different members of the *S. damnosum* species complex is generally related to vegetation zones, forest, and savanna, but seasonal changes in their distribution occurs on an annual cycle as the monsoon winds and their accompanying rainfall aid dispersal and result in enhanced river flow and the creation of breeding sites. According to Baker (Baker et al. 1990), members of the *S. damnosum* complex move average distances of 15–20 kilometers daily and may migrate over a total distance of 400–500 kilometers.

From the beginning of the control program, vector control has been dogged by "reinvasion"—the long-range northwards movement of female blackflies with the potential to reestablish breeding grounds in savanna areas during the wet season and to carry *O. volvulus* parasite back into vector-cleared areas. Flies are known to be carried from more southerly permanent breeding sites on a generally northeasterly tack across West Africa by the associated winds. In the late 1980s, a reverse form of migration was also noted in which savanna flies, both *S. sirbarnum* and *S. damnosum* s.str., appeared to have extended their range into areas pre-

viously only inhabited by forest species (Thomson et al. 1996). The migrations of savanna species of *S. damnosum* s.l. into the forest zones was considered a serious threat to the health of populations living in the forested area, as it suggested that the blinding, savanna form of the disease may spread to the forested areas.

In 1988 a Task Force of the Onchocerciasis Control Programme in Sierra Leone determined that breeding sites there were the source of the annual invasion of savanna species of *S. damnosum* that were actively recolonizing the controlled breeding sites of Mali each rainy season. The Task Force also found that the savanna fly *S. sirbarnum* was widely distributed throughout the country, including many forested areas (Baker 1989). This has led to the speculation on the possible role of deforestation and rainfall decline on the distribution of different species of *S. damnosum* s.l. (Walsh et al. 1993), a role that was later confirmed by a detailed cytotaxonomic study of *S. damosum* larvae found breeding in a deforested area in Ghana (Wilson et al. 2002).

Deforestation in West Africa has been implicated in the southward movement of the savanna species of *S. damnosum* s.l. in the region—the most significant vectors of the blinding form of onchocerciasis (Wilson 2002; Thomson and Connor 1996). This has important implications for the newly developed African Programme for Onchocerciasis Control; unlike its predecessor, the Onchocerciasis Control Programme that successfully controlled savannah species of *S. damnosum* s.l. using insecticides, the new program is heavily dependant on the widespread distribution of the micro-filaricidal drug ivermectin. Should current control measures fail, then emergence of the blinding disease in deforested areas can be expected. Curiously, forest cover has further implications to the success of new program, as the recent distribution of ivermectin in forested areas has resulted in a number of deaths—thought to be related to the presence of yet another filarial worm, *Loa loa*. This parasite, which has emerged from obscurity, is now considered a major impediment to the success of the control program, and mapping its potential distribution has become a priority activity and may also be relevant to the control of other filarial worms in the region, such as lymphatic filariasis (Thomson and Connor 2000).

The exact effect of land cover changes, especially deforestation, on the composition of the vector population differs from place to place. However, as a net effect an increase of the more virulent vector and therefore an increase of morbidity from onchocerciasis can be noted.

---

of arboviruses that emerged or reemerged in the Brazilian Amazon region and the factors responsible are shown in Table 14.3.

Uncontrolled urbanization or colonization near forest areas has been typically associated with the emergence of Oropouche fever and Mayaro fever viruses (Vasconcelos et al. 2001a). The Oropouche virus has been responsible for at least 500,000 infections in the last 40 years in the Amazon, and it is spreading.

Current modifications to the forest ecosystem are likely to result in the spread of infections if vectors and reservoirs find better ecological conditions. This mechanism may explain the large epidemics of Oropouche fever virus in the Brazilian Amazon from 1960 to 1990, as well as in other Latin American countries, especially Peru and Panama (Pinheiro et al. 1998; Watts et al. 1998; Saeed et al. 2000). Alternatively, if the ecological changes are detrimental to the nonhuman carriers of the virus, they will probably disappear due to the absence of the basic elements necessary for their survival. This could explain the absence of many previously identified virus species that are no longer found despite continued surveillance (Vasconcelos et al. 2001a).

Deforestation for agricultural expansion has been the most important factor associated with the spread of yellow fever in Africa and its reemergence in Brazil (in the state of Goiás), although climatic extremes are also important, for instance Brazil in 2000 (Vasconcelos et al. 2001b). Yellow fever is maintained by a sylvatic cycle between primates and mosquitoes in the forest. Evidence from epidemics in Côte d'Ivoire (1982), Burkina Faso (1983), Nigeria (1986 and 1987), and Mali (1987) and other, long-term, studies have established a clear link between deforestation and yellow fever's increasing area of endemicity (Cordellier 1991). However, it should be noted that yellow fever epidemics occur in urban and dry savanna areas as well and can be transmitted by different mosquito species more adapted to these environments.

14.2.1.4 Habitat Change and Infectious Diseases in Urban Systems

Uncontrolled urbanization has many adverse human health consequences, primarily due to health infrastructure problems and

---

BOX 14.6

Gold Mining and Malaria in Venezuela

Bolívar state in southern Venezuela near the border with Brazil and Guyana covers an area of 24 million hectares (the size of the United Kingdom), 70% of which is forested. It is presently affected by deforestation mainly associated with logging, agriculture, dam construction, and gold mining. Before 1980, malaria was occasionally reported from this state among the indigenous population. Malaria in this area was classified by Gabaldon (1983) as difficult to control since the vector, *An. darlingi,* was found to bite mostly outside houses (thus harder to spray with insecticides). Also, the human populations were mainly Amerindians with seminomadic habits in remote areas (Gabaldon 1983).

During the late 1950s, construction was started on a road to connect Ciudad Bolívar, capital of Bolivar state, to Boa Vista, capital of Roraima state in Brazil. The road construction brought workers from all over the country and opened new opportunities for people seeking land for agriculture, gold and diamond mining, and forest exploitation. In the 1980s, the boom of gold and diamonds attracted a wave of migrants from different parts of the country as well as illegal migrants from Brazil, Guyana, Colombia, and Dominican Republic, among others. Malaria started to increase steadily, reaching the highest peak of over 30,000 cases in 1988, with over 60% cases due to *P. falciparum* (MSDS 2000). Malaria has since become endemic-epidemic in Bolívar state.

Studies carried out in several villages spanning 1999–2000 showed that only three species of anophelines were caught on human landing catches: *An. darlingi, An. marajoara,* and *An. neomaculipalpus* (Moreno et al. 2002). In general, biting densities were low (fewer that two bites per person per hour). Nevertheless, there was a strong correlation between *An. darlingi* density and malaria incidence (P<0.001). In this area, studies on breeding places showed that up to 13 species of anophelines were present, with *An. triannulatus* being the most abundant. The most productive breeding sites in terms of anophelines species diversity and density were the abandoned mine dug outs, which vary in size from a few meters in diameter to several kilometers, followed by lagoons created by flooding of streams on artificial and natural depressions of terrain (Moreno et al. 2000).

Table 14.3. Probable Factors in Emergence of Arbovirus in Brazilian Amazon Region and Association with Human Disease (Vasconcelos et al. 2001)

Virus	Probable Factors for Emergence	Disease in Humans
Dengue	poor mosquito control; increased urbanization in tropics	yes, epidemic
Guaroa	flooding of dam[a]	yes, sporadic cases
Gamboa	flooding of dam[a]; migrating birds	not yet
Mayaro	deforestation	yes, limited outbreak
Oropouche	deforestation; increase of colonization and urbanization in Amazon	yes, epidemic
Triniti	flooding of dam[a]	not yet
Yellow fever	urbanization in tropics; deforestation; lack of widespread immunization	yes, epidemic
Anopheles A viruses[b]	flooding of dam[a]	not yet
Changuinola viruses[c]	flooding of dam[a]; deforestation; use of subsoil	not yet

[a] During construction of dam in Tucurui, Pará State, millions of hematophagous insects were obtained in a few days, from which several strains of previously known and new arboviruses were obtained

[b] In this serogroup of family *Bunyaviridae*, the new viruses Arumateua, Caraipé, and Tucurui were isolated, and there was also an increase in circulation of Lukuni and Trombetas viruses.

[c] In this serogroup of family *Reoviridae*, 27 new arboviruses were isolated in Tucurui, 4 in Carajas (use of subsoil), and 8 in Altamira (deforestation for several purposes) from phlebotominae sandflies.

---

overcrowded, unsanitary conditions. This section focuses on infectious diseases linked to urbanization.

14.2.1.4.1 Lymphatic filariasis

Lymphatic filariasis is one of the most prevalent tropical diseases, with some 120 million people infected, primarily in India and Africa, where there has been no decline in the incidence of the disease for the past decade. The disease is reported to be responsible for 5 million DALYs lost annually, ranking third after malaria and tuberculosis (WHO/TDR 2002).

Distribution and transmission of the lymphatic filariasis are closely associated with socioeconomic and behavioral factors in endemic populations. In Southeast Asia, urban Bancroftian filariasis *Wuchereria bancrofti* infection, a filarial nematode, is related to poor urban sanitation, which leads to intense breeding of *C. quiquefasciatus,* its principal mosquito vector (Mak 1987). In Sri Lanka, *C. quinquefasciatus* was not originally present in a forested environment, but rapidly invaded as soon as forest clearing began and settlement expanded (Asmerasinghe in press). In urban East Africa, filariasis is also transmitted by *C. quinquefasciatus,* but in rural areas across tropical Africa it is spread by the same anopheline mosquito species that transmit malaria (Manga 2002).

14.2.1.4.2 Dengue fever

In terms of morbidity and mortality, dengue fever—caused by a virus that has four serotypes or genetic variants—is the most important human viral disease carried by mosquitoes. It is caused by a flavivirus and is endemic in about 100 countries and found in all continents except Europe (WHO/TDR 2004). Around 80 million cases are reported every year, of which 550,000 people need hospital treatment and about 20,000 die. Although dengue is primarily a tropical disease, it has become a great concern in countries with temperate climates because of an increased number of imported cases, resulting from increased air travel and the introduction of *Aedes albopictus,* an exotic vector adapted to a cold climate (Kuno 1994).

However, the primary vector of large dengue epidemics is *Ae. aegypti,* a day-biting mosquito that prefers to feed on humans and that breeds at sites typically found in the urban environment: discarded tires, cans, and other trash that accumulates rainwater; water storage devices in houses; flower pots and even plants that collect water. This species has made extraordinary evolutionary adjustments to coexist with human beings since its origins in Africa as a forest species feeding principally on wild animals (rodents and so on) and laying eggs in tree holes containing rainwater.

One subspecies—*Ae. aegypti aegypti*—evolved to become highly adapted to indoor or peridomestic environs, breeding in artificial containers, and followed humankind on its journeys and migrations throughout the globe (Monath 1994).

The spread of the disease is associated with the geographical expansion of the vector species, favored by current housing and water supply conditions as well as garbage collection practices in developing countries. Therefore, uncontrolled urbanization, which is frequently associated with population growth and poverty (resulting in substandard housing and inadequate water and waste management systems), plays a major role in creating the conditions for dengue epidemics. Analysis of associations between social and economic variables, such as income and education level, in residential urban areas and the incidence of dengue infection has shown that low-income neighborhoods have the highest levels of infection (Costa and Natal 1998). Moreover, communities with high infestations of *Ae. aegypti*, especially if they are near forests that maintain yellow fever virus, face a significant risk of yellow fever epidemics.

14.2.1.4.3 Other diseases linked to urbanization

Soil disturbance from building construction in arid environments can encourage coccidioidomycosis, a fungal pneumonia. Increased aridity and eventual desertification due to increasing global temperatures may increase the potential for infection. Coccidioidomycosis is spread by dust, and disease outbreaks are often preceded by increased rain, followed by dry periods, and especially in the wake of soil disturbances. A well-documented outbreak followed the 1994 Northridge, California, earthquake, when 317 cases resulted (Schneider et al. 1997). In a study in Greece, factors thought to contribute to the extraordinary increases in cases were a drought lasting five to six years, abundant rain in 1991 and 1992, construction of new buildings, and arrival of new susceptible residents to the endemic areas (Pappagianis 1994)

Leishmaniasis has been associated with urban settlements in forested regions and is widespread in 22 countries in the New World and in 66 in the Old World (WHO, TDR 2004). There are two major types of leishmaniasis: cutaneous and visceral (kala-azar). Cutaneous leishmaniasis is originally a forest disease but may adapt to urban settings with some vegetation and cycling in dogs.

Uncontrolled urbanization or colonization near forest areas has been typically associated with the emergence of Oropouche fever and Mayaro fever viruses (Vasconcelos et al. 2001a). The relationship between malaria and urbanization in two cities is presented in Box 14.7.

14.2.1.5 Ecological Change and Infectious Disease in Coastal and Freshwater Systems

Cholera and severe forms of gastroenteritis are caused by *Vibrio cholerae* and *V. parahaemolyticus*. In tropical areas, cases are reported year-round. In temperate areas, cases are mainly reported in the warmest season. The seventh cholera pandemic is currently spreading across Asia, Africa, and South America. In 1992, a new strain or serogroup (*V. cholerae O139*) appeared and has been responsible for epidemics in Asia. During the 1997/98 El Niño, excessive flooding caused cholera epidemics in Djibouti, Somalia, Kenya, Tanzania, and Mozambique. Warming of the African Great Lakes due to climate change may create conditions that increase the risk of cholera transmission in the surrounding countries.

Pathogens are often found in coastal waters, and transmission occurs through consumption of shellfish or brackish water or

---

BOX 14.7
Malaria and Urbanization in Brazil

The transmission of malaria in urban areas in the Americas occurs when urbanization invades the habitat of vectors. These phenomena have been observed in cities of Brazil such as Belém (Pará state) located near the mouth of the Amazon River and in Manaus (Amazonas state).

In a recent study comparing malaria transmission and epidemiology over 60 years in Belém, Póvoa et al. (2002) showed that the incriminated vectors in the 1940s, *An. aquasalis* and *An. darlingi*, are still currently important vectors. The anopheline species diversity has increased from 2 in the 1930s to 6 in the 1940s to 10 in the 1990s. *An. darlingi* was eliminated from Belém in the 1960s and was absent for approximately 20 years, probably due to the destruction of breeding sites in forested areas as urbanization increased. During this period the reported malaria cases were attributed to human immigration from rural areas into Belém (Marques 1986) or people from Belém traveling to endemic areas during the holidays and returning with malaria parasites in their blood (Souza 1995). Nevertheless, *An. aquasalis* was present in coastal areas influenced by tides.

The population of Belém increased from 206,331 in 1942–43 to 934,322 in 1980–89 and then to 1,367,677 in 1996. The number of malaria cases went from 363 in 1942–43 to 1,197 in 1980–89 and 2,716 in 1996. Approximately 80% of the original forest has been destroyed (Povoa et al. 2002), but the expansion of the city (particularly the District of Daent) toward the forested protected area and low coastal swampy areas provided new larval habitats, resulting in a higher mosquito diversity and closer proximity between human dwellings and mosquito habitat, which increases the transmission of malaria.

A similar situation has occurred in Manaus, where the process of urbanization eliminated *An. darlingi* from the city by 1976 (Tadei et al. 1998). Nevertheless, the accelerated expansion of the suburbs into the surrounding jungle reestablished the contact between the human population and the principal vector, *An. darlingi*, which by 1988 colonized the city again and triggered malaria epidemics. Vectors are also adapting to new circumstances; for example, mosquitoes have varied their feeding time in areas where humans have intervened (Tadei et al. 1998). At present, malaria cases continue to be reported from the periurban areas of Manaus.

---

through bathing. Coastal waters in both industrial and developing countries are frequently contaminated with untreated sewage (see Chapter 19), and warmth encourages microorganism proliferation. The presence and increased transmission of *Vibrio spp.* (some of which are pathogens that cause diarrhea), such as *V. vulnificus*, a naturally occurring estuarine bacterium, has been associated with higher sea surface temperatures. Phytoplankton organisms respond rapidly to changes in environmental conditions and are therefore sensitive biological indicators of the combined influences of climate change and soil and water pollution. Algal blooms are associated with several environmental factors, including sunlight, pH, ocean currents, winds, sea surface temperatures, and runoff (which affects nutrient levels), as described in Chapters 12 and 19.

V. cholerae and other gram-negative bacteria can be harbored in many forms of algae or phytoplankton. *V. cholerae* can assume a noncultivable but viable state, returning to a cultivable, infectious state with the same conditions that promote algal blooms (Colwell 1996). Some species of copepod zooplankton apparently provide an additional marine reservoir for *V. cholerae*, facilitating its long-term persistence in certain regions, such as in the estuaries of the Ganges and Brahmaputra in India. According to this theory, the seasonality of cholera epidemics may be linked to the seasonality of plankton (algal blooms) and the marine food chain. Studies using remote sensing data of chlorophyll-containing phytoplankton have shown a correlation between cholera cases and sea surface temperatures in the Bay of Bengal. Interannual variability in cholera incidence in Bangladesh is also linked to El Niño/Southern Oscillation and regional temperature anomalies (Lobitz et al. 2000), and cholera prevalence has been associated with progressively stronger El Niño events spanning a 70-year period (Rodo et al. 2002).

14.2.2 Niche Invasion or Interspecies Host Transfer Effects on Infectious Disease Transmission

The emergence of many diseases has been linked to the interface between tropical forest communities, with their high levels of biodiversity, and agricultural communities, with their relatively homogenous genetic makeup but high population densities of humans, domestic animals, and crops. For instance, expanding ecotourism and forest encroachment have increased opportunities for interactions between wild nonhuman primates and humans in tropical forest habitats, leading to pathogen exchange through various routes of transmission (Wolfe et al. 2000). (See Box 14.8.)

14.2.2.1 Cultivated Systems and Niche Invasion

Intensive farming practices have seen the emergence of several devastating herd and flock diseases, including Nipah virus, bovine spongiform encephalopathy, foot and mouth disease, severe acute respiratory syndrome, and avian influenza.

14.2.2.1.1 Nipah virus in Malaysia

The emergence of many diseases can be viewed as a pathogen invading a new or recently vacated niche. For example, Nipah virus emerged in Malaysia in 1999, causing over 100 human deaths (Chua et al. 2000). This highly pathogenic virus (with a case fatality rate greater than 40 %) was previously unknown as a human pathogen, and emerged from its natural reservoir hosts (fruit bats) via domestic animal (pig) amplifier hosts.

The ecological changes that caused this virus to invade a new niche appear to be a complex series of anthropogenic alterations to fruit bat habitat and agriculture set in a background of increasing ENSO-caused drought (Daszak et al. 2001; Field et al. 2001; Chua et al. 2002). First, the virus appears not to be able to move directly from bats to humans, so the development of a pig industry in Malaysia has been a crucial driver of emergence. Second, fruit bat habitat has been largely replaced in peninsular Malaysia by crop plants such as oil palm. Third, deforestation in Sumatra coupled with increasing amplitude, intensity, and duration of ENSO-driven drought has led to repeated, significant seasonal haze events that cover Malaysia. These events reduce the flowering and fruiting of forest trees that are the natural food of fruit bats and may have changed the fruit bats' migration patterns and ability to find food.

These interrelated events are theorized to have coincided just prior to the Nipah virus outbreaks, when the large mid-1990s

---

BOX 14.8

The Importance of Ecological Change and Zoonotic Diseases

Zoonotic pathogens are the most significant cause of emerging infectious diseases affecting humans in terms of both their proportion and their impact. Some 1,415 species of infectious organisms are known to be pathogenic to humans; 61% of these are zoonotic and 75% of those considered as emerging pathogens are zoonotic (Taylor et al. 2001). More important, zoonotic pathogens cause a series of EIDs with high case fatality rates and no reliable cure, vaccine, or therapy (such as Ebola virus disease, Nipah virus disease, and hantavirus pulmonary syndrome). They also cause diseases that have the highest incidence rates globally (such as AIDS). AIDS is a special case, because it is caused by a pathogen that jumped host from nonhuman primates and then evolved into a new virus. Thus it is essentially a zoonosis (Hahn et al. 2000) and is thought to have infected the highest number of human individuals of any disease in history.

Viruses such as Junin, Machupo, and Guanarito hemorrhagic fever agents in Argentina, Bolivia, and Venezuela, transmitted to humans through the urine of wild rodents, came through the expansion of agricultural practices to new areas; hantaviruses, initially recognized in Korea, came through the urine of infected rodents and were then identified in Asia and Europe. The Junin virus, which causes a hemorrhagic fever, also emerged when the production of vegetables that serve as food source for rodents increased in Argentina in 1957.

Because of the key role of zoonoses in current public health threats, wildlife and domestic animals play a key role in the process by providing a "zoonotic pool" from which previously unknown pathogens may emerge (Daszak et al. 2001). The influenza virus is an example, which causes pandemics in humans after periodic exchange of genes between the viruses of wild and domestic birds, pigs, and humans. Fruit bats are involved in a high-profile group of EIDs that include rabies and other

lyssaviruses, Hendra virus, Menangle virus (Australia), and Nipah virus (Malaysia and Singapore). This has implications for further zoonotic disease emergence. A number of species are endemic to remote oceanic islands, and these may harbor enzootic and potentially zoonotic pathogens (Daszak et al. 2000).

The current major infectious disease threats to human health are therefore emerging and reemerging diseases, with a particular emphasis on zoonotic pathogens jumping host from wildlife and domestic animals. A common, defining theme for all EIDs (of humans, wildlife, domestic animals, and plants) is that they are driven to emerge by anthropogenic changes to the environment. Because threats to wildlife habitat are so extensive and pervading, it follows that many of the currently important human EIDs (such as AIDS and Nipah virus disease) are driven by anthropogenic changes to wildlife habitat such as encroachment, deforestation, and others. This is essentially a process of natural selection in which anthropogenic environmental changes perturb the host-parasite dynamic equilibrium, driving the expansion of those strains suited to the new environmental conditions and driving expansion of others into new host species (Cunningham et al. in press). The selection process acts on the immense pool of varied pathogen strains circulating within the population (c.f. the "zoonotic pool," Morse 1993).

Thus, very few EIDs are caused by newly evolved pathogens, although notable exceptions include drug-resistant pathogens, newly re-assorted influenza strains, and pathogens with point mutations that increase their virulence (such as canine parvovirus, Parrish et al. 1985). Even in these examples, it is possible that the new strains were already present in the pathogen population. For example, recent work shows that drug-resistant strains of some common microbes circulate within rodent populations in areas outside normal contact with antibiotics (Gilliver et al. 1999).

ENSO event was associated with a drop in fruit production and the appearance of *Pteropus vampyrus* (the key Nipah virus reservoir) for the first time at the index farms, where Nipah virus was initially identified (Chua et al. 2002).

14.2.2.1.2 Bovine spongiform encephalopathy ("mad cow" disease)

The background of the BSE epidemic is well known. In order to improve the protein content in the diet of cattle, ground-up sheep and cattle remains—including of the brain and spinal cord—were fed to cattle. This practice was claimed as economically rational because it turned a waste product into a valuable food. But from an ecological perspective it was anything but rational; cattle are normally vegetarian, and are certainly not cannibals (Prusiner 1997).

While it was originally argued that this practice is harmless, a similar disease in sheep (scrapie) was known for centuries, and one in humans (kuru), transmitted through the ritual cannibalism of human brains, was already known in New Guinea. In time, the causal agent of BSE, an unusual protein called a prion, was transmitted to humans, causing a devastating, rapidly progressive, still untreatable brain disease, called new variant Creutzfeldt-Jakob disease. So far, the size of this human epidemic has been modest, but transmission is probably still occurring through blood transfusions (Llewelyn et al. 2004) and surgical instruments that cannot be sterilized. As well as these human health effects, the BSE epidemic had an immense economic and psychological cost to farmers, as millions of cattle were slaughtered prematurely.

14.2.2.1.3 Severe acute respiratory syndrome

SARS gained international attention during an outbreak that began sometime during November 2001 in China. Wet markets, a known source of influenza viruses since the 1970s, were found to be the source of the bulk of the infections (Webster 2004). (Live-animal markets, termed "wet markets," are common in most Asian societies and specialize in many varieties of live small mammals, poultry, fish, and reptiles (Brieman et al. 2003).) The majority of the earliest reported cases of SARS were of people who worked with the sale and handling of wild animals. The species at the center of the SARS epidemic are palm civet cats (*Paguna larvata*), raccoon dogs (*Nyctereutes procuyoinboides*), and Chinese ferret badgers (*Melogale moschata*) (Bell at al. 2004). As of July 2003, there had been 8,096 cases and 774 deaths reported worldwide (WHO 2004).

14.2.2.1.4 Avian influenza

Avian influenza virus has caused fatalities in humans, highlighting the potential risk that this type of infection poses to public health (Capua and Alexander 2004). Genetic reassortment within a person coinfected with human and avian strains of influenza virus could potentially link the high transmissibility associated with human-adapted viruses with the high rates of mortality observed in the avian cases, thus triggering a potentially devastating pandemic (Ferguson et al. 2004). The "Spanish flu" pandemic of 1918–19 was the largest infectious disease event in recorded history, killing over 20 million people. Recently a single gene coding for the viral haemagglutinin protein from the 1918 pandemic influenza A strain was identified and produced extraordinary virulence in a mouse model (Kobasa et al. 2004). Such highly virulent recombinant viruses will continue to pose a threat through agricultural practices.

14.2.2.2 Forest Systems and Interspecies Host Transfer

14.2.2.2.1 Bushmeat hunting and disease emergence

The global trade in bushmeat is quite extensive. In Central Africa, 1–3.4 million tons of bushmeat are harvested annually (Fa and Peres 2001). Also, in West Africa, a large share of protein in the diet comes from bushmeat; in Côte d'Ivoire, for example, 83,000 tons are eaten each year (Feer 1993) and in Liberia, 75% of meat comes from wildlife (105,000 tons consumed a year) (Fa and Peres 2001). The bushmeat harvest in West Africa includes significant numbers of primates, so the opportunity for interspecies disease transfer between humans and non-human primates is not a trivial risk.

Contact between humans and other animals can provide the opportunity for cross-species transmission and the emergence of novel microbes into the human population. Road building is linked to the expansion of bushmeat consumption that may have played a key role in the early emergence of human immunodeficiency virus types 1 and 2 (Wolfe et al. 2000). Simian foamy virus has been found in bushmeat hunters (Wolfe et al. 2004), and workers collecting and preparing chimpanzee meat have become infected with Ebola (WHO 1996). The initiation of a local epidemic of monkeypox (an orthopoxvirus similar to smallpox), which continued for four generations of human-to-human contact, has been attributed to the hunting of a red colobus monkey (Jezek et al. 1986). Also, there are other specific human activities that pose risks similar to those of wildlife hunting and butchering. For example, in the Tai forest in Côte d'Ivoire, a researcher performing a necropsy on a chimpanzee contracted Ebola (Le Guenno et al. 1995).

The Taxonomic Transmission Rule states that the probability of successful cross-species infection increases the closer hosts are genetically related (chimpanzees are closer genetically to humans, for example, than birds or fish are), since related hosts are more likely to share susceptibility to the same range of potential pathogens (Wolfe et al. 2000). Surveillance of nonhuman primates—reservoirs or sources for microbial emergence—can flag emerging pathogens (Wolfe et al. 1998).

14.2.2.2.2 Diseases transferred from human hosts to wildlife

Cross-species transmission also increases the probability that endangered nonhuman primate species and other wildlife will come into contact with human pathogens. The parasitic disease Giardia was introduced to the Ugandan mountain gorilla, *Gorilla gorilla beringei,* by humans through ecotourism and conservation activities (Nizeyi et al. 1999). Gorillas in Uganda also have been found with human strains of *Cryptosporidium* parasites, presumably from ecotourists (Graczyk at al 2001). The invasion of human tuberculosis into the banded mongoose (Alexander et al. 2001) represents another case of a human pathogen invading a free-ranging wildlife species.

These host transfer and emergence events not only affect ecosystem function, they could possibly result in a more virulent form of a human pathogen circling back into the human population from a wildlife host and may also allow the development of other transmission cycles to develop outside human-to-human contact.

14.2.3 Biodiversity Change Effects of Infectious Disease Transmission

Biodiversity change includes issues of species replacement, loss of key predator species, and variation in species population density. (See Chapter 4.) The equilibrium among predators and prey, hosts, vectors, and parasites is an essential service provided by ecosystems. (See Chapter 11.) It plays the role of controlling the emergence and spread of infectious diseases, although it is only recently that this protective function of biodiversity has been acknowledged (Chivian 2001).

14.2.3.1 *Biodiversity Change at the Forest/Urban System Interface*

Lyme disease can be used as a model system to illustrate some effects of biodiversity change on infectious disease transmission. In eastern U.S. oak forests, studies on the interactions between acorns, white-footed mice (*Peromyscus leucopus*), moths, deer, and ticks have linked defoliation by gypsy moths with the risk of Lyme disease (Jones et al. 1998). Most vectors feed on a variety of host species that differ dramatically in their function as a reservoir—that is, their probability of transmitting the infection from host to vector. Increasing species richness has been found to reduce disease risk (Schmidt and Ostfeld 2001), and the involvement of a diverse collection of vertebrates in this cases may dilute the impact of the main reservoir, the white-footed mouse (Ostfeld and Keesing 2000a).

Habitat fragmentation also plays a part in disease emergence; mouse populations that are isolated in fragments seem to fluctuate, unregulated by biotic interactions. Moreover, predators tend to be absent in small woodlots, and probable competitors occur at lower densities in these areas than in more continuous habitat. Therefore, habitat fragmentation causes a reduction in biodiversity within the host communities, increasing disease risk though the increase in both the absolute and relative density of the primary reservoir.

The same conclusions may apply to a number of other diseases, including cutaneous leishmaniasis, Chagas disease, human granulocytic ehrlichiosis, babesiosis, plague, louping ill, tularemia, relapsing fever, Crimean Congo hemorrhagic fever, and LaCrosse virus (Ostfeld and Keesing 2000b).

14.2.3.2 *Biodiversity Change in Forest Systems: Variation in Population Density*

Most of the human cases of rabies acquired from bat bites in Brazil result from isolated attacks, usually by sick bats that drop from their resting places or fly into houses and are unwisely grasped by a person. But over the last few decades several reports have been published of outbreaks of vampire bat (*Desmodus rotundus*) attacks on humans in Latin America, occasionally transmitting rabies (Thomas and Haran 1981; Schneider and Burgoa 1995).

Recently compiled data show that vampire bats were the second most frequent species transmitting rabies to humans in Brazil (25% of all cases) in 1993, second only to dogs (Schneider and Burgoa 1995); this was also true previously in Mexico (Anonymous 1991). In the majority of vampire bat attacks, the underlying motive for the attacks seemed to be the same: bat populations deprived of their abundant and readily obtained primary food sources (animals) sought alternative hosts (humans) to feed on. In rural areas, it has happened when animals such as cattle or pigs were rapidly eliminated from an area (MacCarthy 1989; Schneider 1991; Lopez 1992; Costa et al. 1993).

Similar attacks can occur when wild vertebrates are reduced following establishment of human settlements in remote areas (Almansa and Garcia 1980; Schneider 1991). For example, massive attacks by bats have occurred in the gold mining camps in the Amazon, where wild animal species that served as food sources for the bats were depleted due to overhunting or were chased away by the noise produced by water pumps and airplanes (Uieda et al. 1992; Coelho 1995; Schneider et al. 1996; Confalonieri 2001).

In this context, bat attacks on humans are a direct consequence of human-induced environmental modifications caused by the elimination of native species of animals in natural landscapes, with a consequent host shift and an increase in the likelihood of rabies transmission.

14.2.4 Human-induced Genetic Changes of Disease Vectors or Pathogens

One of the key properties of microbes that have successfully managed a "host transfer," migrating into a new ecological niche, is the potential for mutability. This mutagenic potential of the microbe is exploited once selection pressure (through ecological change) is exerted on the microbe. For example, Brault et al. (2004) have associated genetic shifts in the Venezuelan encephalitis virus with epidemics in animal hosts in southern Mexico due to extensive deforestation and habitat changes, where the subsequent replacement of mosquito vector species is exerting new evolutionary pressures on the virus.

14.2.4.1 *Resistant Bacteria from the Use of Antibiotics in Animal Feed*

Intensive animal production (both agriculture and aquaculture) has many impacts on ecosystems and human well-being. (See Chapters 19 and 26.) Antibiotics are routinely used for prophylaxis and growth promotion in high-production livestock agriculture rather than being used sparingly for medical purposes. Such subtherapeutic levels exert selective pressure on the emergence of resistant bacteria. *Campylobacter* bacteria sampled from pigs in South Australia, for example, show widespread resistance (60–100%) to antibiotics such as erythromycin, ampicillin, and tetracycline, and *E. coli* strains showed widespread resistance to multiple antibiotics (Hart et al. 2004). Livestock have also been shown to be reservoirs of drug-resistant *Salmonella* bacteria (Busani et al. 2004) and other *E. coli* that are resistant even to newer-generation antibiotics, like cephalosporins (Shiraki et al. 2004).

Salmonella enteritidis likely stemmed from antibiotic prophylaxis in the poultry industry that led to removal of *S. gallinarum* and *S. pullorum* during the 1960s and the creation of a vacant niche in the gut, which was subsequently filled by *S. enteritidis*. These three pathogens share a common surface antigen; thus, flock immunity prior to removal of the former two species would have prevented the latter from becoming established.

14.2.4.2 *Genetic Changes in Disease Vectors at Forest/Urban Interface*

Chagas disease is a deadly disease transmitted by triatomine beetles in South and Central America. The niche and trophic relationships of these insects have direct epidemiological importance, and pesticide resistance is changing the ecology and transmission of this disease.

Only a few cases of typical insecticide resistance have been reported (Busvine 1970; Cockburn 1972), but important niche adaptations to houses and domestic areas seem to have taken place, particularly for the *Triatoma infestans* beetle (Dujardin et al. 1997a, 1997b; Gorla et al. 1997; Noireau et al. 1997; Panzera et al. 1997; Schofield et al. 1997). Domiciliary adaptation generally involves genetic simplification (Schofield 1994; Dujardin et al. 1998; Rabinovich et al. 2001), including the loss of genetic material that may make triatomines highly susceptible to chemical control (Borges et al. 1999) and may cause simplification of some specific characteristics, such as a reduction in body size and sexual dimorphism (Steindel 1999). Specific markers can differentiate populations that have survived chemical control from invasive populations (Costa 1999).

Triatomine "ecological successions" (replacement of some species by other species) follows control programs and environmental changes such as deforestation (Costa 1999). For instance, *T. sordida* has progressively replaced *T. infestans* because of elimination of the latter from houses. And *T. pseudomaculata*, which is predominantly peridomiciliary (around dwellings), has replaced *T. brasiliensis* in some areas treated with insecticides (Diotaiuti et al. 1998). The Brazilian Amazon, where Chagas disease was always considered endemic in wild animals, presently has at least 18 triatomine species reported, 10 of which are infected by *Trypanosoma cruzi,* the deadly human form of the disease (Coura et al. 2002; Teixeira et al. 2000).

14.2.5 Environmental Contamination of Infectious Agents of Diseases

14.2.5.1 Cryptosporidiosis

Cryptosporidium, a protozoan that completes its life cycle within the intestine of mammals, sheds high numbers of infectious oocysts that are dispersed in feces. One hundred and fifty-two species of mammals have been reported to be infected with *C. parvum,* which causes diarrhea and gastrointestinal illness. *C. parvum* infections of humans were first reported in 1976 and have since been reported from 90 countries. It is highly prevalent in ruminants and readily transmitted to humans. *Cryptosporidium* oocysts are very small (~3 microns) and are difficult to remove from water; another study found 13% of treated water in the United States still contained *Cryptosporidium* oocysts, indicating some passage of microorganisms from source to treated drinking water (LeChevallier and Norton 1995).

In a survey of farms in the state of Pennsylvania, 64% returned at least one bovine stool sample that was positive for *C. parvum,* and all samples were positive at 44% of farms. All cattle had full access to water courses that could be contaminated with these parasites (Graczyk et al. 2000). Environmental factors such as land use, climate extremes, and inadequate water treatment are now recognized as contributing factors in the spread of cryptosporidiosis (Rose et al. 2002).

The extent of the problem of *C. parvum* in tropical environments where water supplies are less well controlled is unknown. However, the impact of changing climate patterns and increased frequency of severe weather events on watershed and water storage facilities is likely to increase the level of oocyst contamination (Graczyk et al. 2000).

14.2.5.2 Leptospirosis

Leptospirosis is a bacterial waterborne disease disseminated by mammals, which shed the pathogen in their urine. Leptospirosis occurs across the globe. In tropical areas, the disease usually occurs as outbreaks during the rainy season, due to the flooding of densely populated low-lying areas infested with domestic rats, which breed prolifically because of poor garbage collection practices. In Rio de Janeiro, for example, epidemics have always been associated with seasonal increases in rainfall (Roberts et al. 2001). Outbreaks in rural areas are not as common as in urban areas.

14.2.6 Synergies between Malnutrition and Infectious Diseases

Malnutrition, as a consequence of environmental degradation, has a huge global impact on morbidity and mortality due to infectious diseases. For children under five years of age in the developing world, being underweight equates to about half the mortality risk of the main infectious diseases such as diarrhea, malaria, pneumonia, and measles. (See Chapter 8 for more on the health impacts of malnutrition). (See also Box 14.9.)

BOX 14.9

Nomadic Lifestyle, Land Cover Change, and Infectious Diseases

Nomads often face a different spectrum of health problems than nonnomadic populations. Their adaptation to a specific environment—often one that supports human settlement only on a transient basis—makes them sensitive indicators of land cover changes. Important inequalities between nomadic and nonnomadic populations exist not only with regard to the disease spectrum, but even more important with regard to access to health care (Sheik-Mohamed and Velena 1999). Nomadic lifestyle interferes with issues of compliance and adherence, but health care as an instrument of control used by settled populations is also an issue. Primary health care for nomads would have to be adapted to their life-style and needs. Continuous drought, for example, may force nomads into living conditions that threaten their health as well as their whole lifestyle (Loutan and Lamotte 1984). Becoming sedentary carries a major risk of mortality and infectious disease for nomadic populations in the African rain forest.

Nomadic populations live in the semiarid regions of the African continent or in the rain forest regions of Africa, South America, and Asia. Two big groups that can be distinguished are pastoralists (in the semiarid regions) and hunters/collectors (in rain forests). Kalahari Bushmen or Australian Aborigines would also be part of the hunters/collectors group.

For nomadic populations in the semiarid regions of Africa, water supply and management is a crucial issue. Water is also an essential part in many land cover changes and is of course essential to health. Child mortality in the Sahel region of Africa was found to be higher, and general access to health care was found to be limited. Infectious disease prevalence in pastoralists differs from settled populations with a profile that is directed toward diseases that have a reservoir in cattle, such as tuberculosis or brucellosis, and those that require long-term treatment, such as some sexually transmitted diseases and again tuberculosis (Niamir-Fuller and Turner 1999).

Pygmies in the Central African rain forest have been protected against some infectious diseases (including HIV1) due to their partial isolation. However, they are fully susceptible when coming into contact with carriers. Major land cover changes like deforestation in the Central African basin increase contact between different populations. Infections that require long-term treatment, like leprosy or tuberculosis, are difficult to target in nomadic populations in the rain forest.

14.3 New Tools and Methods for Assessment of Ecosystem Change and Human Disease

Table 14.4 summarizes the published literature on the link between ecological change and human disease. New tools applied across these ecologically linked diseases will improve understanding of these linkages and of the emergence of infectious diseases. Time-series analysis, geographic information systems, and spatial analysis (which encompass a range of technologies and approaches, including digital mapping, analysis of remotely sensed imagery, spatial statistics, ecological niche modeling, and the use of global positioning systems), for example, have proved useful in studying diseases emerging from land use change. (See also Chapter 3.) There have also been a number of significant reviews of tools that have been applied specifically to public health (Beck et al. 2000; Thomson and Connor 2000; Hay 2000).

Table 14.4. Infectious Diseases and Mechanisms of Potential Changing Incidence as Related to Ecosystem Changes

Disease	Cases per Year	DALYs[a] (thousand)	Emergence Mechanism	Anthropogenic Drivers	Geographical Distribution	Expected Variation from Ecological Change	Confidence Level
Malaria	350 million	46,486	niche invasion; vector expansion	deforestation; water projects	tropical (America, Asia, and Africa)	+ + + +	+ + +
Lymphatic filariasis	120 million	5,777	habitat alteration	water projects; urbanization	tropical America and Africa	+	+ + +
Schistosomiasis	120 million	1,702	intermediate host expansion	dam building; irrigation	America; Africa; Asia	+ + + +	+ + + +
Dengue fever	80 million	616	vector expansion	urbanization; poor housing conditions	tropical	+ + +	+ +
HIV	42 million	84,458	host transfer	forest encroachment; bushmeat hunting; human behavior	global	+	+ +
Onchocerciasis	18 million	484	habitat alteration	spillways of dams	Africa; tropical America	+ +	+ + +
Chagas disease	16–18 million	667	habitat alteration	deforestation; urban sprawl and encroachment	Americas	+ +	+ + +
Leishmaniasis	12 million	2,090	host transfer; habitat alteration	deforestation; agricultural development	tropical Americas; Europe and Middle East	+ + + +	+ + +
Meningitis	223,000 (in 2002)	6,192	habitat alteration; dust storms	desertification	Saharan Africa	+ +	+ +
Hantavirus	200,000	–	variations in population density of natural food sources	climate variability		+ +	+ +
Rabies	35,000 deaths	1,160	biodiversity loss, altered host selection	deforestation and mining	tropical	+ +	+ +
Trypanosomiasis	30,000–500,000	1,525	habitat alteration	deforestation	Africa	+ + +	+ +
Japanese encephalitis	30,000–50,000	709	vector expansion	irrigated rice fields	Southeast Asia	+ + +	+ + +
Rift Valley fever	27,500 (Kenya 1998)		heavy rains	climate variability and change; dam building	Africa; Middle East	+ + +	+ +
Lyme disease	23,763 (U.S. 2002)		depletion of predators; biodiversity loss; reservoir expansion	habitat fragmentation	North America and Europe	+ +	+ +
SARS	8,098[e]		host transfer	intensive livestock operations mixing wild and domestic animals	global	+	+

Ecological niche modeling is an approach used in biogeography to predict the distributional range of species from existing occurrence data (Anderson et al. 2003). Using genetic algorithms as a decision tool in a GIS containing layers of environmental information (such as topography, climate, and vegetation), epidemiological and spatial risk stratification can be achieved from data on the location of vectors or pathogens. This approach has been successfully used in the case of Chagas disease and for vectors of leishmaniasis and filovirus infections (Peterson et al. 2002, 2004a, 2004b).

Increasingly in recent years, meteorological satellite data have been used to help model the spatial and seasonal dynamics of disease transmission and develop early warning systems (Connor et al. 1998). These relatively low cost and easy-to-use data sources have become familiar to public health services in Africa. For example, environmental data indicating areas at risk of malaria epidemics are beginning to be routinely incorporated into the WHO/UNICEF-supported disease surveillance software Health-Mapper that is widely used by ministries of health (WHO 2001).

Table 14.4. *continued*

Disease	Cases per Year (thousand)	DALYs[a]	Emergence Mechanism	Anthropogenic Drivers	Geographical Distribution	Expected Variation from Ecological Change	Confidence Level
West Nile virus and other encephalitides	5,483 (U.S. average 2002–04)	–	niche invasion	international travel; climate variability	Americas; Eurasia	+ +	+
BSE	133[d]		host transfer	intensive livestock farming	Europe	+	+
Cholera	[b]	[c]	sea surface temperature rising	climate variability and change	global (tropical)	+ + +	+ +
Cryptosporidiosis	[b]	[c]	contamination by oocystes	poor watershed management where livestock exist	global	+ + +	+ + + +
Coccidioidomycosis	–	–	disturbing soils	climate variability	global	+ +	+ + +
Ebola	–		forest encroachment; bushmeat hunting	forest encroachment	Africa	+	+
Guanarito; Junin; Machupo	–	–	biodiversity loss; reservoir expansion	monoculture in agriculture after deforestation	South America	+ +	+ + +
Oropouche/ Mayaro virus in Brazil	–	–	vector expansion	forest encroachment; urbanization	South America	+ + +	+ + +
Leptospirosis	–	–	habitat alteration	agricultural development; urban sprawl	global (tropical)	+ +	+ + +
Nipah/Hendra viruses	–		niche invasion	industrial food production; deforestation; climate abnormalities	Australia; Southeast Asia	+ + +	+
Salmonellosis	–		niche invasions	antibiotic resistance from using antibiotics in animal feed	global	+	+

Key:

+	low
+ +	moderate
+ + +	high
+ + + +	very high

[a] Disability-adjusted life year: years of healthy life lost, a measure of disease burden for the gap between actual health of a population compared with an ideal situation where everyone lives in full health into old age. (WHO World Health Report 2004)

[b] and [c] Diarheal disease (aggregated) deaths and DALYs respectively: 1,798 ˘ 1,000 cases and 61,966 ˘ 1,000 DALYs.

[d] Human cases from 1995 to 2002.

[e] From November 2002 to July 2003, probable cases of SARS reported to the World Health Organization.

References

Akogbeto, M., 2000: Lagoonal and coastal malaria at Cotonou: entomological findings. *Sante,* Jul-Aug., **10(4)**, 267–275.

Alexander, K.A., E. Pleydell, M. Williams, J.F.C. Nyange, A.L. Michel, 2001: The zoonotic importance of mycobacterium tuberculosis: a potential threat to free-ranging wildlife populations? *Journal of Emerging Diseases,* **8**, 598.

Almansa, J.C. and R.C. Garcia, 1980: Incidencia del murciélago hematófago *Desmodus rotundus* sobre los indígenas Yanomami de Venezuela, Donana, *Acta Vediebrata,* **7**, 1113–1117.

Amerasinghe, F.P., 2003: Irrigation and mosquito-borne diseases. *Journal of Parasitology,* **89** (Suppl.), S14–S22.

Amerasinghe, P.H., F.P. Amerasinghe, R.A. Wirtz, N.G. Indrajith, W. Soma-pala, L.R. Pereira, and A.M.S. Rathnayake, 1992: Malaria transmission by indoor resting *Anopheles subpictus* Grassi in a new irrigation project in Sri Lanka. *Journal of Medical Entomology,* **29**, 577–581.

Anderson, R.P., D. Lew, and A.T. Peterson, 2003: Evaluating predictive models of species' distributions: criteria for selecting optimal models. *Ecological Modelling* **162**: 211–232.

Appawu, M.A., S.K. Dadzie, A. Baffoe-Wilmot, and M.D. Wilson, 2001: Lymphatic filariasis in Ghana: entomological investigation of transmission dynamics and intensity in communities served by irrigation systems in the Upper East Region of Ghana. *Tropical Medicine & International Health* **6**: 511–516.

Baker, R. 1989: Unpublished report to OCP, Entomological Investigations in Sierra Leone, April–August 1988.

Baker, R., P. Guillet, A. Seketeli, P. Poudiougo, D. Boakye, M.D. Wilson, and Y. Bissan, 1990: Progress in controlling the reinvasion of windborne vectors into the western area of the onchocerciasis control program in West-Africa.

Philosophical Transactions of the Royal Society of London Series B-Biological Sciences, **328,** 731–750.

Beck, L.R., B.M. Lobitz, and B.L. Wood, 2000: Remote sensing and human health: new sensors and new opportunities. *Emerging Infectious Diseases, Calif State Univ, Monterey Bay, CA USA NASA, Ames Res Ctr, Moffett Field, CA 94035 USA,* **6,** 217–227.

Bell, D., S. Robertson, P.R. Hunter, 2004: Animal origins of SARS corona virus: possible links with international trade in small carnivores. *Philosophical Transactions of the Royal Society of London, Biological Sciences.* July, **359** (1447), 1107–1114.

Brieman, R.F., M.R. Evans, W. Preiser, J. Maguire, A. Schnur, A. Li, H. Bekedam, J.S. MacKenzie, 2003: Role of china in the quest to define and control acute respiratory syndrome. *Emerging Infectious Diseases;* **9 (9),** p 1037.

Ben Mohamed, A. and J.P. Frangi, 1986: Results from ground-based monitoring of spectral aerosol optical-thickness and horizontal extinction—some specific characteristics of dusty Sahelian atmospheres. *Journal of Climate and Applied Meteorology,* **25,** 1807–1815.

Benton, B., J. Bump, A. Seketeli, and B. Liese, 2002: Partnership and promise: evolution of the African river-blindness campaigns. *Annals of Tropical Medicine and Parasitology,* **96,** 5–14.

Bielders, C.L., S. Alvey, and N. Cronyn, 2001: Wind erosion: The perspective of grass-roots communities in the Sahel. *Land Degradation & Development,* **12,** 57–70.

Boakye, D.A., C. Back, G.K. Fiasorgbor, A.P.P. Sib, and Y. Coulibaly, 1998: Sibling species distributions of the *Simulium damnosum* complex in the West African Onchocerciasis control Programme area during the decade 1984–93, following intensive larviciding since 1974. *Medical and Veterinary Entomology,* **12,** 345–358.

Borges, E. C., H. H. R. Pires, S. E. Barbosa, C. M. S. Nunes, M. H. Pereira, A. J. Romanha and L. Diotaiuti, 1999: Genetic variability in brazilian triatomines and the risk of domiciliation. *Memórias do Instituto Oswaldo Cruz,* **94** (Suppl. 1), 371–373.

Bradley, D., 1977: The health implications of irrigation schemes and man-made lakes in tropical environments. *Water, Wastes and Health in Hot Climates.* R. Feachem, M. Mcgarry and D. Mara (eds.), John Wiley, London, pp. 18–29.

Bradley, D. and R. Narayan, 1988: Epidemiological patterns associated with agricultural activities in the tropics with special reference to vector-borne diseases. *Effects of Agricultural Development on Vector-Borne Diseases.* FAO Publication, No. AGL/MISC/12/87, pp.35–43.

Brault, A.C., A.M. Powers, D. Ortiz, J.G. Estrada-Franco, R. Navarro-Lopez, S.C. Weaver, 2004: Venezuelan equine encephalitis emergence: enhanced vector infection from a single amino acid substitution in the envelope glycoprotein. *PNAS, USA,* **101(31),** 11344–11349.

Bunnag, T., S. Sornmani, S. Phinichpongse, C. Harinasuta, 1979: Surveillance of water-borne parasitic infections and studies on the impact of ecological changes on vector mosquitoes of malaria after dam construction. *SEAMEO-TROPMED Seminar: environmental impact on human health in Southeast and East Asia, 21st.* Tokyo: Tsukuba.

Busvine, J. R., 1970: Chagas disease control and the possibilities of resistance in triatomids. *Observations on a Visit to South America.* A World Health Organization Report.

Busani L., C. Graziani, A. Battisti, A. Franco, A. Ricci, et al., 2004: Antibiotic resistance in Salmonella enterica serotypes Typhimurium, Enteritidis and Infantis from human infections, foodstuffs and farm animals in Italy. *Epidemiol Infect,* **132,** 245–51.

Capua, I., and D.J. Alexander, 2004: Human health implications of avian influenza viruses and paramyxoviruses. *Eur J Clin Microbiol Infect Dis,* **23,** 1–6.

Charlwood, J.D. and W.A. Alecrim, 1989: Capture-recapture studies with the South American malaria vector *Anopheles darlingi,* Root. *Annals of Tropical Medicine and Parasitology,* 83(6), 569–576.

Chivian, E., 2001: Species loss and ecosystem disruption: the implications for human health. *Canadian Medical Association Journal,* Jan, **164(1),** 66–69.

Chua, K.B., W.J. Bellini, P.A. Rota, B.H. Harcourt, A. Tamin, et al. 2000: Nipah virus: a recently emergent deadly paramyxovirus. *Science,* **288,** 1432–1435.

Chua, K.B., B.H. Chua, and C.W. Wang, 2002: Anthropogenic deforestation, El Nino and the emergence of Nipah virus in Malaysia. *Malaysian J. Pathol.* **24,** 15–21.

Cockburn, J. M., 1972: *Laboratory investigations bearing on possible insecticide resistance in traitomid bugs.* World Health Organization, WHO/VBC/72. 359 pp: 1–9.

Coelho, G.E., 1995: *Aspectos relacionados às agressões humanas da raiva, no Estado de Roraima, Brasil.* Monografia, Univ. Brasília, Faculdade de Ciências da Saúde, Brasília, DF, 29 pp.

Coluzzi, M., 1984: Heterogeneities of the malaria vectorial system in tropical Africa and their significance in malaria epidemiology and control, *Bull World Health Organ.,* **62**(Suppl), 107–13.

Coluzzi, M., 1994: Malaria and the Afrotropical ecosystems: impact of man-made environmental changes, *Parassitologia,* **36(1–2),** 223–7.

Coluzzi, M, A. Sabatini, V. Petrarca, M.A. Di Deco, 1979: Chromosomal differentiation and adaptation to human environments in the *Anopheles gambiae* complex, *Trans R Soc Trop Med Hyg,* **73(5),** 483–97.

Colwell RR, 1996. Global climate and infectious disease: the cholera paradigm. *Science* 274: 2025–31.

Confalonieri, U.E.C., 2000: Environmental Change and Human Health in the Brazilian Amazon. *Global Change and Human Health,* **1(2),** 174–83.

Confalonieri, U.E.C., 2001: Global environmental change and health in brazil: review the present situation and proposal for indicators for monitoring these effects. In: *Human Dimensions of Global Environmental Change,* D.J. Hogan and M.T. Tomasquin (eds.). Brazilian Perspectives, Acad. Brasil. Cien., Rio de Janeiro, pp. 43–78.

Connor, S.J., M.C. Thomson, S.P. Flasse, and A.H. Perryman, 1998: Environmental information systems in malaria risk mapping and epidemic forecasting. *Disasters,* **22,** 39–56.

Cordellier, R, 1991: The epidemiology of yellow fever in Western Africa. *Bull World Health Organ,* **69,** 73–84.

Costa, J., 1999: The synanthropic process of chagas disease vectors in Brazil, with special attention to triatoma brasiliensis Neiva, 1911 (Hemiptera, Reduviidae, Triatominae) Population, Genetical, Ecological, and Epidemiological Aspects. *Memórias do Instituto Oswaldo Cruz,* **94** (Suppl. 1), 239–241.

Costa, A.I. and D. Natal, 1998: Geographical distribution of dengue and socio-economic factors in an urban locality in southeastern Brazil. *Rev Saude Publica* **32:** 232–6.

Costa, J.M., R.F. Bonito, and A.S. Nishioka, 1993: An outbreak of vampire bat bite in a Brazilian Village. *Trop. Med. Parasitol,* **44,** 1219–1220.

Coura, J.R., A.C. Junqueira, O. Fernandes et al., 2002: Emerging Chagas Disease in Amazonian Brazil. *Trends in Parasitology,* **18(4):**171–176.

Craig, M.H., R.W. Snow, and D. le Sueur, 1999: A climate-based distribution model of malaria transmission in sub-Saharan Africa. *Parasitology Today,* **15,** 105–111.

Cunningham, A.A, P. Daszak, and J.P. Rodríguez. In press. Pathogen pollution: defining a parasitological threat to biodiversity conservation. *Journal of Parasitology.*

Daszak, P. and A.A. Cunningham, 1999: Anthropogenic change, biodiversity loss and a new agenda for emerging diseases. *Journal of Parasitology,* Aug., **85(4),** 742–746.

Daszak, P, A. Cunningham, and AD. Hyatt, 2000: Emerging infectious diseases of wildlife-threats to biodiversity and human health. *Science,* **287**(5452).

Daszak, P., A.A. Cunningham, and A.D. Hyatt, 2001: Anthropogenic environmental change and the emergence of infectious diseases in wildlife. *Acta Tropica,* **78(2),** 103–116.

Dégallier, N., A.P.A. Travassos da Rosa, J-P. Hervé, P.F.C. Vasconcelos, J.F.S. Travassos da Rosa, G.C. Sá Filho, F.P. Pinheiro, 1989: Modifications of arbovirus eco-epidemiolgy in Tucurui, Pará, Brazilian Amazonia, related to the construction of a hydroelectric dam. *Arbovirus Research in Australia, Proceedings 5th Symposium,* Queensland Institute of Medical Research, Brisbane, pp.124–135.

de Manzione, N., R.A. Salas, H. Paredes, O. Godoy, L. Rojas, et al., 1998: Venezuelan hemorrhagic fever: clinical and epidemiological studies of 165 cases. *Clin. Infect. Dis,* **26,** 308.

Diotaiuti, L., E.C. Borges, E.S. Lorosa, R.E. Andrade, F.F.C. Carneiro, O.F. Faria Filho and C.J. Schofield, 1998: Current Transmission of Chagas Disease in The State of Ceará, Brazil. *Memórias do Instituto Oswaldo Cruz,* **93** (Suppl. II), 66.

Dujardin, J. P. , H. Bermúdez and C. J. Schofield, 1997a: The use of morphometrics in entomological surveillance of sylvatic foci of *Triatoma infestans* in Bolivia. *Acta Tropica,* **66,** 145–153.

Dujardin, J. P., H. Bermúdez, C. Casini, C. J. Schofield and M. Tibayrenc. 1997b: Metric differences between sylvatic and domestic *Triatoma infestans* (Hemiptera: Reduviidae) in Bolivia. *Journal of Medical Entomology,* **34(5),** 544–551.

Dujardin, J., M. Muñoz. T. Chávez, C. Ponce, J. Moreno and C. J. Schofield, 1998: The Origin of *Rhodnius prolixus* in Central America. *Medical and Veterinary Entomology,* **12,** 113–115.

Dzodzomenyo, M., S.K. Dunyo, C.K. Ahorlu, W.Z. Coker, M.A. Appawu, E.M. Pederson, and P.E. Simonsen, 1999: Banocroftian filariasis in an irrigation project community in southern Ghana. *Trop. Med. Int. Health,* Jan., **4(1),** 13–18.

Engelthaler, D.M., D.G. Mosley, J.E. Cheek, C.E. Levy, K.K. Komatsu, et al., 1999: Climatic and environmental patterns associated with hantavirus pulmonary syndrome, Four Corners region, United States. *Emerg Infect Dis* **5**: 87–94.

Epstein, P.R., E. Chivian, and K. Frith, 2003: Emerging diseases threaten conservation. *Environmental Health Perspectives,* Volume **111**, Number 10, August 2003: A506–A507.

Epstein, P.R. 1995: Emerging diseases and ecosystem instability: new threats to public health. *American Journal of Public Health,* **85(2).**

Fa, J.E. and C.A. Peres, 2001: Game vertebrate extraction in African and Neotropical forests: an intercontinental comparison, *in* Reynolds, J.D., Mace, G.M., Redford, K.H., and Robinson, J.G., *Conservation of exploited species:* Cambridge, UK, Cambridge University Press, p. 203–241.

Feer, F., 1993: The potential for sustainable hunting and rearing of game in tropical forests, in Hladik, C.M., Hladik, A., Linares, O.F., Pagezy, H., Semple, A., and Hadley, M. (eds.), *Tropical Forests: People and Food: biocultural interactions and applications to development*: Pearl River, N.Y., Parthenon Publishing Group, p. 691–708.

Ferguson, N.M., C. Fraser, C.A. Donnelly, A.C. Ghani, and R.M. Anderson, 2004: Public health risk from the avian H5N1 influenza epidemic. *Science,* **304,** 968–9.

Field, H., P. Young, J.M. Yob, J. Mills, L. Hall, and J. Mackenzie, 2001: The natural history of Hendra & Nipah viruses. *Microbes and Infection,* **3,** 307–314.

Gabaldon, A., 1983: Malaria eradication in Venezuela: doctrine, practice, and achievements after twenty years. *Am. J. Trop. Med. Hyg.,* **32,** 203–211.

Gilliver, M.A., M. Bennett, M. Begon, S.M. Hazel, and A. Hart, 1999: Antibiotic resistance found in wild rodents. *Nature,* **401,** 233–234.

Glass G., J. Cheek, J.A. Patz, T.M. Shields, T.J. Doyle, et al., 2000: Predicting high risk areas for Hantavirus Pulmonary Syndrome with remotely sensed data: the Four Corners outbreak, 1993. *J Emerg Infect Dis* 2000;**6**: 239–246.

Gorla, D. E., J.P. Dujardin, and C.J. Schofield, 1997: Biosystematics of old world triatominae. *Acta Tropica,* **63,** 127–140.

Graczyk, T.K., B.M. Evans, C.J. Shiff, H.J. Karreman, and J.A. Patz, 2000: Environmental and geographical factors contributing to watershed contamination with cryptosporidium parvum oocysts. *Environ. Res.,* Mar., **82(3),** 263–271.

Graczyk, T.K., A.J. DaSilva, M.R. Cranfield, J.B. Nizeyi, G.R. Kalema, N.J. Pieniazek. 2001: Cryptosporidium parvum genotype 2 infections in freeranging mountain gorillas (Gorilla gorilla beringei) of the Bwindi Impenetrable National Park, *Uganda. Parasitol Res,* **87,** 368–70.

Gratz, N.G., 1987: The impact of rice production on vector-borne disease problems in developing countries. In: *Vector-Borne Disease Control through Rice Agroecosystem Management.* International Rice Research Institute, Los Banos, Philippines, pp. 7–12.

Guerra, J.A., M.L. Barros, N.F. Fe, M.V. Guerra, E. Castellon, M.G. Paes, and I.A. Sherlock, 2004: Visceral leishmaniasis among Indians of the State of Roraima, Brazil: clinical and epidemiological aspects of the cases observed from 1989 to 1993. *Rev Soc Bras Med Trop* 37: 305–11.

Hahn, B.H., G.M. Shaw, K.M. de Cock, and P.M. Sharp, 2000: Aids as a zoonosis: Scientific and public health implications. *Science,* **287,** 607–614.

Hart, W.S., M.W. Heuzenroeder, and M.D. Barton, 2004: Antimicrobial resistance in Campylobacter spp., Escherichia coli and enterococci associated with pigs in Australia. *J Vet Med B Infect Dis Vet Public Health* **51,** 216–21.

Hay, S. I., 2000: An overview of remote sensing and geodesy for epidemiology and public health application. *Advances in Parasitology,* 47, 1–35.

Hoke, C.H. and J.B. Gingrich, 1994: Japanese Encephalitis. In: *Handbook of Zoonoses. 2nd Edition. Section B: Viral.* G.W. Beran (ed.). CRC Press, Boca Raton, pp. 59–70.

Hunter, J.M., L. Rey, K.Y. Chu, E.O. Adekolu-John, and K.E. Mott, 1993: *Parasitic diseases in water resources development: the need for intersectoral negotiation.* World Health Organization, Geneva, 152 pp.

Ijumba, J.N. and S. Lindsay, 2001: Impact of irrigation on malaria in Africa: paddies paradox. *Medical and Veterinary Entomology,* **15,** 1–11.

Institute of Medicine, 2003: *Microbial Threats to Health: Emergence, Detection, and Response.* National Academy Press, Washington.

Jewsbury, J.M. and A.M.A. Imevbore, 1988: Small dam health statistics. *Parasitology Today,* **4,** 57–58.

Jezek Z., I. Arita, M. Mutombo, C. Dunn, J.H. Nakano, and M. Szczeniowski, 1986: Four generations of probable person-to-person transmission of human monkeypox. *Am J Epidemiol* 123: 1004–12.

Jobin, W., 1999: *Dams and Disease: Ecological design and health impacts of large dams, canals and irrigation systems.* E. & Fn. Spon. (Taylor & Francis Group), London & New York, 580 pp.

Jones, C.G., R.S. Ostfeld, M.P. Richard, E.M. Schauber, and J.O. Wolff, 1998: Chain reactions linking acorns to gypsy moth outbreaks and Lyme disease risk. *Science,* **279,** 1023–1026.

Joshi, D.D., 1986: Japanese encephalitis in Nepal. *JE HFRS Bulletin,* **3**:15.

Jouan, A., Adam F, Coulibaly I, Riou O, Philippe B, et al., 1990: Epidemic of Rift Valley fever in the Islamic republic of Mauritania. Geographic and ecological data. *Bull. Soc. Pathol. Exot.,* **83(5),** 611–620.

Kalliola, R, and S. Flores Paitan, eds., 1998: *Geoecologia y Desarollo Amazonico: Estudio integrado en la zona de Iquitos, Peru.* Sulkava: Finnrklama Oy.

Kobasa, D., A. Takada, K. Shinya, M. Hatta, P. Halfmann, et al., 2004: Enhanced virulence of influenza A viruses with the haemagglutinin of the 1918 pandemic virus. *Nature,* 431, 703–7.

Krause, R.M., 1981: *The Restless Tide: The Persistent Challenge of the Microbial World.* National Foundation for Infectious Diseases. Washington, DC.

Kuno G., and R.E. Bailey 1994: Cytokine responses to dengue infection among Puerto Rican patients. *Mem Inst Oswaldo Cruz* **89**: 179–82.

Lacey, L.A. and C.M. Lacey, 1990: The medical importance of riceland mosquitoes and their control using alternatives to chemical insecticides. *Journal of the American Mosquito Control Association,* **6**(Suppl):1–93.

LeChevallier, M.S., and W.D. Norton, 1995: Giardia and Cryptosporidium in raw and finished water. *J. Am. Water Works Assoc.,* **87,** 54–68.

Lederberg, J., R.E. Shope, and S.C. Oakes, 1992: *Emerging Infections: Microbial Threats to Health in the United States.* Institute of Medicine, National Academy Press. Washington, DC.

Lefevre, P.C., 1997: Current status of rift valley fever. What lessons to deduce from the epidemics of 1977 and 1987?. *Med. Trop.,* **57**(Suppl. 3), 61–64.

Le Guenno B., P. Formentry, M. Wyers, P. Gounon, F. Walker, and C. Boesch, 1995: Isolation and partial characterisation of a new strain of Ebola virus. *Lancet* 345: 1271–4.

Linthicum, K.J., A. Anyamba, C.J. Tucker, P.W. Kelley, M.F. Myers, and C.J. Peters, 1999: Climate and satellite indicators to forecast Rift Valley fever epidemics in Kenya. *Science,* **285,** 397–400.

Llewelyn, C. A., Hewitt, P. E., Knight, R. S. G., Amar, K., Cousens, S., Mackenzie, J., Will, R. G., 2004: Possible transmission of variant Creutzfeldt-Jakob disease by blood transfusion. *The Lancet,* **363**(9407), 417–21.

Lobitz, B., L. Beck, A. Huq, B. Wood, G. Fuchs, A.S. Faruque, and R. Colwell, 2000: Climate and infectious disease: use of remote sensing for detection of Vibrio cholerae by indirect measurement. *Proc Natl Acad Sci USA,* **97,** 1438–4.

Lopez, A. et al., 1992: Outbreak of human rabies in the Peruvian jungle. *Lancet,* **339,** 1408–1411.

Loutan L. and J.M. Lamotte, 1984: Seasonal variation in nutrition among a group of nomadic pastoralists in Niger. *The Lancet* **8383** (1), 845 – 947

Mac Carthy, T., 1989: Human depredation by vampire bats (Desmodus rotundus) following a hog cholera campaign. *Am. J. Trop. Med. Hyg.,* **40,** 1320–1322.

Maiztegui, J.I., 1975: Clinical and epidemiologic patterns of Argentine haemorrhagic fever. *Bull. WHO,* **50**:567–575.

Mak, J.W., 1987: Epidemiology of lymphatic filariasis. *Ciba Found Symp,* 127, 5–14.

Marques, A.C., 1986: Migrations and the dissemination of malaria in Brazil. *Mem. Inst. Oswaldo Cruz,* **81**(Suppl II), 17–30.

Marques, A.C., 1987: Human migration and the spread of malaria in Brazil. *Parasitol Today,* **3,** 166–70.

Mather, T.H. and T.T. That, 1984: *Environmental management for vector control in rice fields.* FAO Irrigation and Drainage Paper No. 41, 152 pp.

Molesworth, A.M., L.E. Cuevas, A.P. Morse, J.R. Herman, and M.C. Thomson, 2002: Dust clouds and spread of infection. *Lancet,* **359,** 81–82.

Molesworth, A.M., L.E. Cuevas, S.J. Connor, A.P. Morse, and M.C. Thomson, 2003: Environmental risk and meningitis epidemics in Africa. *Emerging Infectious Diseases,* **9**(10), 1287–1293.

Molyneux, D.H., 1997: Patterns of change in vector-borne diseases. *Ann Trop Med Parasitol* 91:827–39.

Monath, T.P., 1994: Dengue: the risk to developed and developing countries. *Proc Natl Acad Sci* USA **91**: 2395–400.

Moreno, J., Y. Rubio-Palis, and P. Acevedo, 2000: Identificación de criaderos de Anopheles en un área endémica del estado Bolívar. *Bol. Dir. Malariol. San. Amb.,* **40,**122–129.

Moreno, J and Y. Rubio-Palis, E. Pérez, V. Sánchez, and E. Páez, 2002: Evaluación de tres métodos de captura de anofelinos en un área endémica del estado Bolívar, Venezuela. *Entomotropica,* **17(2),** 157–165.

Morse, S.S. 1993: Examining the origins of emerging viruses. In: *Emerging Viruses.* S. S. Morse (ed.) Oxford University Press, New York, pp. 10–28.

Mutero, C.M., 2002: Health impact assessment of increased irrigation in the Tana River Basin, Kenya. In: *The Changing Face of Irrigation in Kenya. Opportu-*

nities for Anticipating Change in Eastern and Southern Africa. International Water Management Institute (IWMI), Colombo.

Mutero, C.M., C. Kabutha, V. Kimani, L. Kabuage, G. Gitau, et al. 2003: A transdisciplinary perspective on the links between malaria and agroecosystems in Kenya, *Acta Tropica* **89**: 171–186

Niamir-Fuller, M. and M.D. Turner, 1999: A review of recent literature on pastoalism and transhumance in Africa. In *'Managing Mobility in Africa Rangelands—The legitimization of transhumance'* (Niamir-Fuller M., Ed.), pp. 18–46, Intermediate Technology Publications Ltd, London UK.

Nicholson, S.E., C.J. Tucker, and M.B. Ba, 1998: Desertification, drought, and surface vegetation: An example from the West African Sahel. *Bulletin of the American Meteorological Society,* **79**, 815–829.

Nizeyi, J.B., R. Mwebe, A. Nanteza, M.R. Cranfield, G.R. Kalema, and T.K. Graczyk, 1999: Cryptosporidium sp. and Giardia sp. infections in mountain gorillas (Gorilla gorilla beringei) of the Bwindi Impenetrable National Park, Uganda. *J Parasitol* **85**: 1084–8.

Ostfeld, S.R. and F. Keesing, 2000a: Biodiversity and disease risk: the case of Lyme disease. *Conservation Biology,* **14(3)**, 722–728.

Ostfeld, S.R. and F. Keesing, 2000b: The function of biodiversity in the ecology of vector-borne zoonotic diseases. *Can. J. Zool.,* **78**, 2061–2078.

Panzera, F., S. Hornos, J. Pereira, R. Cestau, D. Canale, L. Diotaiuti, J. P. Dujardin and R. Pérez, 1997: Genetic variability and geographic differentiation among three species of triatomine bugs (Hemiptera-Reduviidae). *American Journal of Tropical Medicine and Hygiene,* **57(6)**, 732–739.

Pappagianis D., 1994: Marked increase in cases of coccidioidomycosis in California: 1991, 1992, and 1993. *Clin Infect Dis* **19** Suppl 1:S14-S18.

Parrish, C.R., P.H. O'Connell, J.F. Evermann, and L.E. Carmichael, 1985: Natural variation in canine parvovirus. *Science,* **230**,1046–1048.

Parmenter, C.A., T.L. Yates, R.R. Parmenter, and J.L. Dunnum, 1999: Statistical sensitivity for detection of spatial and temporal patterns in rodent population densities. *Emerg Infect Dis* **5**: 118–25.

Patz, J.A., T.K. Graczyk, N. Geller, and A.Y. Vittor. 2000: Effects of environmental change on emerging parasitic diseases. *Int J Parasitol,* **30**, 1395–1405.

Patz, J.A., P. Daszak, G.M. Tabor, A.A. Aguirre, M. Pearl, et al., 2004: Unhealthy Landscapes: Policy Recommendations on Land Use Change and Infectious Disease Emergence. *Environ Health Perspect,* **101**:1092–98.

Peiris, J.S.M., F.P. Amerasinghe, P.H. Amerasinghe, C.B. Ratnayake, S.H.P.P. Karunaratne, and T.F. Tsai, 1992: Japanese encephalitis in Sri Lanka—the study of an epidemic: vector incrimination, porcine infection and human disease. *Transactions of the Royal Society of Tropical Medicine and Hygiene,* **86**, 307–313.

Peterson, A.T., V. Sanchez-Cordero, C.B. Beard, and J.M. Ramsey, 2002: Ecologic niche modeling and potential reservoirs for Chagas disease, Mexico. *Emerg Infect Dis* **8**: 662–7.

Peterson, A.T., R.S. Pereira, and V.F. Neves, 2004a: Using epidemiological survey data to infer geographic distributions of leishmaniasis vector species. *Rev Soc Bras Med Trop* **37**: 10–4.

Peterson, A.T., J.T. Bauer, and J.N. Mills, 2004b: Ecologic and geographic distribution of filovirus disease. *Emerg Infect Dis* **10**: 40–7.

Pinheiro, F.P., G. Bensabath, A.P.A. Travassos da Rosa, R. Lainson, J.J. Shaw, et al. 1977: Public health hazards among works along the Trans-Amazon Highway. *J. Occupat. Med.,* **7**, 490–497.

Pinheiro, F.P., A.P.A. Travassos da Rosa & P.F.C. Vasconcelos, 1998: An overview of Oropouche fever epidemics in Brazil and neighbour countries. In: Travassos da Rosa, A.P.A., P.F.C. Vasconcelos & J.F.S. Travassos da Rosa (ed.). *An overview of arbovirology in Brazil and neighbouring countries.* Instituto Evandro Chagas, Belém, pp.186–192.

Povoa, M.M., J.E. Conn, C.D. Schlichting, J.C.Amaral, M.N. Segura, et al., 2003: Malaria vectors and the re-emergence of Anopheles darlingi in Belém, Pará, Brazil. *J. Med. Enotmol.* Jul; **40** (4): 379–86.

Povoa, M., R. Wirtz, R. Lacerda, M. Miles, D. Warhurst, 2001: Malaria vectors in the municipality of Serra do Navio, State of Amapa, Amazon Region, Brazil. *Mem Inst Oswaldo Cruz,* Feb;**96(2)**, 179–84.

Prospero, J.M., 2001: African dust in America. *Geotimes,* **46**, 24–27.

Prusiner, S.B., 1997: Prion diseases and the BSE crisis. *Science,* **278**, 245–51.

Rabinovich J., N. Schweigmann, V. Yohai, and C. Wisnivesky-Colli, 2001: Probability of Trypanosoma cruzi transmission by Triatoma infestans (Hemiptera: Reduviidae) to the opossum Didelphis albiventris (Marsupialia: Didelphidae). *Am J Trop Med Hyg* **65**: 125–30.

Ramasamy, R., R. de Alwis, A. Wijesundera, and M.S. Ramasamy, 1992: Malaria transmission at a new irrigation project in Sri Lanka: the emergence of Anopheles annularis as a major vector. *American Journal of Tropical Medicine and Hygiene,* **47**, 547–553.

Roberts, L., U.E.C. Confalonieri, and J.L. Aaron, 2001: Too Little, Too Much: How the Quantity of Water Affects Human Health. Chapter in, Aaron J.L. and Patz J.A. (eds), *Ecosystem Change and Public Health: A Global Perspective.* Johns Hopkins University Press, Baltimore.

Rodo, X., M. Pascual, G. Fuchs, A.S.G. Faruque, 2002: ENSO and cholera: a nonstationary link related to climate change? *Proc. Natl. Acad Sci. USA.* Oct, 99(20), 12901–6.

Rose, J.B., D.E. Huffman, and A. Gennaccaro, 2002: Risk and control of waterborne cryptosporidiosis. *FEMS Microbiol Rev* 26: 113–23.

Saeed, M.F., H. Wang, M. Nunes, P.F.C. Vasconcelos, S.C. Weaver, R.E. Shope, D.M. Watts, R.B. Tesh, A.D.T. Barrett, 2000: Nucleotide sequences and phylogeny of the nucleocapsid gene of Oropouche virus. *J. Gen. Virol.,* 81, 743–748.

Salas, R., N. de Manzione, R.B. Tesh, R. Rico-Hesse, R.E. Shope, A. Betancourt, O. Godoy, R. Bruzual, M.E. Pacheco, B. Ramos, et al., 1991: Venezuelan haemorrhagic fever. *Lancet,* **338** (8774) 1033–6.

Schmidt, K.A. and R.S. Ostfeld, 2001: Biodiversity and the dilution effect in disease ecology. *Ecology,* **82(3)**, 609–619.

Schneider, M.C., 1991: *Situación Epidemiológica de la Rabia Humana Transmitida por Murciélagos en el Brasil.* In: Reunión de Consulta sobre la Atención a Personas Expuestas a la Rabia Transmitida por Vampiros, Washington, DC, Abril 1–5, 1991. Organización Panamericana De La Salud, pp. 63–82.

Schneider, M.C. and C.S. Burgoa, 1995: Algunas consideraciones sobre la rabia humana transmitida por murciélago. México, *Rev. Salud. Publ.,* **37**, 354–362.

Schneider, M.C., C. Santos-Burgoa, J. Aron, B. Munoz, S. Ruiz-Velazco, and W. Uieda, 1996: Potential force of infection of human rabies transmitted by vampire bats in the Amazon region of Brazil. *Am. J. Trop. Med. Hyg.,* **55(6)**, 680–684.

Schneider E., R.A. Hajjeh, R.A. Spiegel, R.W. Jibson, E.L. Harp, et al. 1997: A coccidioidomycosis outbreak following the Northridge, Calif, earthquake. *Jama-Journal of the American Medical Association* 277: 904–908.

Schofield, C. J., 1994: *Triatominae, Biology and Control.* Eurocommunica Publications, UK.

Schofield, C. J. and J. P. Dujardin. 1997: Chagas disease vector control in Central America. *Parasitology Today* **13(4)**, 141–144.

Service, M.W., 1984: Problems of vector-borne diseases and irrigation projects. *Insect Science and its Application,* **5**, 227–231.

Service, M.W., 1989: Rice, a challenge to health. *Parasitology Today,* **5**, 162–165.

Sharma, V.P., 1996: Re-emergence of malaria in India. *Indian Journal of Medical Research,* **103**, 26–45.

Sheik-Mohamed, A. and J.P. Velema, 1999: Where health care has no access: the nomadic populations of sub-Saharan Africa. *Tropical Medicine and International Health,* **4(10)**, 695–707.

Shiraki, Y., N. Shibata, Y. Doi, and Y. Arakawa, 2004: Escherichia coli producing CTX-M-2 beta-lactamase in cattle, Japan. *Emerg Infect Dis* **10**, 69–75.

Shope, R.E., 1997: Emergence of arbovirus diseases following ecological modifications: epidemiological consequences. In: *Factors in the emergence of diseases,* Saluzzo JF, B. Dodet (ed.). Elsevier, Paris, pp.19–22.

Silveira, F.T, J.J. Shaw, C.N. Bichara, et al., 1997: Leishamniose Tegumentar Americana, in R.N.Q. Leão (org.), Doenças Infecciosas e Parasitárias. Enfoque Amazônico. CEJUP, Belém, Pará, pp. 631–44.

Simpson, D.I.H., 1978: Viral haemorrhagic fevers of man. *Bull. WHO,* **56(6)**, 819–832.

Sioli H., 1953: Schistosomiasis and limnology in the Amazon region. *Am J Trop Med Hyg* 2: 700–7.

Southgate, V.R., 1997: Shistosomiasis in the Senegal River Basin: before and after the construction of dams at Diama, Senegal and Manantali, Mali and future prospects. *J Helminthol,* **71(2)**, 125–32.

Souza, J.M., 1995: Malaria autóctone na grande Belém: problema dos poderes constituídos (Estado e Município da Grande Belém), ao lado das empresas e associações comunitárias das áreas atingidas. *O Plasmodio,* 2:1–2.

Stark, G., 2003: "Causes and consequences and control of wind erosion in Sahelian Africa: a review." *Land Degradation & Development* **14**: 95–108.

Steindel, M., 1999: Trypanosoma cruzi interaction with its vectors and vertebrate hosts. *Memórias do Instituto Oswaldo Cruz,* **94** (Suppl. 1), 243–245.

Sunish I.P. and R. Reuben, 2001: Factors influencing the abundance of Japanese encephalitis vectors in ricefields in India—I. Abiotic. *Medical and Veterinary Entomology* **15**: 381–392.

Surtees, G., 1975: Mosquitoes, arboviruses and vertebrates. In: *Man-made lakes and human health,* N.F. Stanley and M.P. Alpers (eds.). London Academic Press, pp. 21–34.

Tadei, W.P., B.T. Thatcher, J.M.M. Santos, V.M. Scarpassa, I.B. Rodríguez, and M.S. Rafael, 1998: Ecologic observations on anopheline vectors of malaria in the Brazilian Amazon. *Am. J. Trop. Med. Hyg.*, **59**, 325–335.

Taylor, L.H., S.M. Latham, and M.E.J. Woolhouse, 2001: Risk factors for human disease emergence. *Phil. Trans. Soc. Lond. B.*, **356**, 983–989.

Teixeira, A.R.L., P.S. Monteiro, J.M. Rebelo et al., 2000: Emerging Chagas Disease: trophic network and cycle of transmission of Trypanosoma cruzi from palm trees in the Amazon. *Emerging Infectious Diseases* **7(1)**:100–112.

Thomas, J.G. and H.J. Haran, 1981: Vampire bat bites seen in humans in Panamá: their characterization, recognition and management. *Milit. Med.*, **146**, 410–412.

Thomson, M.C. and S.J. Connor, 2000: Environmental information systems for the control of arthropod vectors of disease. *Medical and Veterinary Entomology*, **14**, 227–244.

Thomson, M.C. J. B. Davies, R. J. Post, M. J. Bockarie, P. A. BeechGarwood, and J. Kandeh, 1996: The unusual occurrence of savanna members of the *Simulium damnosum* species complex *(Diptera: Simuliidae)* in southern Sierra Leone in 1988. *Bulletin of Entomological Research*, **86**,(3) 271–280.

Thomson, M.C., D.A. Elnaiem, R.W. Ashford, and S.J. Connor, 1999: Towards a kala azar risk map for Sudan: mapping the potential distribution of *Phlebotomus orientalis* using digital data of environmental variables. *Trop Med Int Health*, **4**, 105–13.

Tyagi, B.K., 2002: *Malaria in the Thar Desert: facts, figures and future.* Agrobis, Índia, 165 pp.

Tyssul Jones, T.W., 1951: Deforestation and epidemic malaria in the wet and intermediate zones of Ceylon. *Indian J Malariology*, 5(1), 135–61.

Uieda, W., G.E. Coelho, and I. Menegola, 1992: Morcegos Hematófagos IV. Agressões humanas no Estado de Roraima e considerações sobre sua incidência na Amazônia. *Resumo*, Seminário Nacional de Raiva, S. Paulo, 7 – 11 Dez, 1992, Inst. Biológico, S. Paulo (mimeo).

UNEP (United Nations Environment Programme) (in press). Sustaining Life: How Human Health Depends on Biodiversity. Oxford University Press – Interim executive summary available at: www.med.harvard.edu/chge/bio .html.

Vasconcelos, P.F.C., A.P.A. Travassos da Rosa, S.G. Rodrigues, E.S. Travassos da Rosa, N. Dégallier, J.F.S. Travassos da Rosa, 2001: Inadequate management of natural ecosystem in the Brazilian Amazon region results in the emergence and reemergence of arboviruses. *Cadernos de Saúde Pública*, **17**(Supl.), 155–164.

Victor, T.J. and R. Reuben, 2000: Effects of organic and inorganic and inorganic fertilizers on mosquito populations in rice fields of southern India. *Med Vet Entomol.*, **14**, 361–368.

Vittor A.Y., R. Gilman, J. Tielsch, G.E. Glass, T.M. Shields, W.S. Lozano, V.P. Cancino, and J.A. Patz, (in press): The impact of deforestation on malaria risk: association between human-biting rate of the major South American malaria vector, Anopheles darlingi, and the extent of deforestation in the Peruvian Amazon. *J Am Trop Med Hyg*.

Walsh, J.F., D.H. Molyneux, and M.H. Birley, 1993: Deforestation: effects on vector-borne disease. *Parasitology* **106** Suppl: S55–75.

Watts, D.M., G. Ramirez, C. Cabezas, M.T. Wooster, C. Carrillo, M. Chuy, E.J. Gentrau, C.G. Hays, 1998: Arthropod-borne viral diseases in Peru. In: Travassos da Rosa, A.P.A., P.F.C. Vasconcelos, J.F.S. Travassos da Rosa (ed.). *An overview of arbovirology in Brazil and neighbouring countries*. Instituto Evandro Chagas, Belém, pp. 193–218.

WCD (World Commission on Dams), 2000: *Dams and Development: A New Framework for Decision-Making.* Earthscan, UK, 404 pp.

Webster, R.G., 2004: Wet Markets- a continuing source of severe acute repiratory syndrome and influenza. *Lancet*, 363 (9404); 234.

WHO (World Health Organization), 1996: Ebola haemorrhagic fever in Gabon—Update 3 from 26 April 1996. [online] World Health Organization, Geneva. Available at http://www.who.int/csr.don/1996_04_26b/en/.

WHO, 2001: *Malaria Early Warning Systems, Concepts, Indicators and Partners, "A framework for field research in Africa".* WHO, Geneva.

WHO, 2002: *World Health Report, 2002: Reducing Risks, Promoting Healthy Life.* WHO, Geneva.

WHO, 2003: Summary of probable SARS cases with onset of illness from 1 November 2002 to 31 July 2003. [online] World Health Organization, Geneva. Cited September 26, 2003. Available at http://www.who.int/csr/sars/country/table2003_09_23/en/.

WHO, 2004: *World Health Report, 2004: Changing History.* WHO, Geneva.

WHO/TDR, 2002: *Disease burden and epidemiological trends.* [online] World Health Organization. Cited February 2002. Available at http://www.who .int/tdr/diseases/lymphfil/direction.htm#burden.

Wijesundera, M. de S., 1988: Malaria outbreaks in new foci in Sri Lanka. *Parasitology Today*, **4**, 147–150.

Wilson, M.D., et al., 2002: Deforestation and the spatio-temporal distribution of savannah and forest members of the *Simulium damnosum* complex in southern Ghana and south-western Togo. *Transaction of The Royal Society of Tropical Medicine and Hygiene*, **96**, 632–639.

Wolfe, N.D., A.A. Escalante, W.B. Karesh, A. Kilbourn, A. Spielman, A.A. Lal, 1998: Wild primate populations in emerging infectious disease research: The missing link? *Emerging Infectious Diseases*, Apr-Jun, **4(2)** 149–158.

Wolfe, N.D., M.N. Eitel, J. Gockowski, P.K. Muchaal, C. Nolte, A.T. Prosser, J.N. Torimiro, S.F. Weise, D.S. Burke, 2000: Deforestation, hunting and the ecology of microbial emergence. *Global Change and Human Health*, 1(1), 10–25.

Wolfe N.D., W.M. Switzer, J.K. Carr, V.B. Bhullar, V. Shanmugam, et al., 2004: Naturally acquired simian retrovirus infections in central African hunters. *Lancet* 363:932–7(2004).

Woods, C.W., A.M. Karpati, T. Grein, N. McCarthy, P. Gaturuku, et al., 2002: An Outbreak of Rift Valley Fever in Northeastern Kenya, 1997–98. *Emerging Infectious Diseases*, Feb:**8** (2).

Chapter 15

Waste Processing and Detoxification

Coordinating Lead Authors: Kenneth R. Hinga, Allan Batchelor
Lead Authors: Mohamed Tawfic Ahmed, Oladele Osibanjo
Contributing Authors: Noëlle Lewis, Michael Pilson
Review Editors: Naser Faruqui, Angela Wagener

Main Messages

Ecosystem processes may act to reduce concentrations of substances that are directly or indirectly harmful to humans. This capacity is finite, has been exceeded locally in many places, and is exceeded across some whole regions. The cases of stratospheric ozone depletion by chlorofluorocarbons and climate change due to greenhouse gas buildup demonstrate that the capacity to alter the environment in harmful ways through by-products of human activities has now reached global proportions.

Humankind produces a large variety of wastes that are introduced into the environment either by accident or by design. Wastes are by-products of human activity and include human excrement wastes, agricultural wastes, energy and manufacturing wastes, industrial and consumer chemical products, medical and veterinary products, and transportation emissions.

All cases of wastewater-borne disease, human health impairment due to contaminants, environmental degradation due to waste discharges, and contaminant-caused impacts on biota are failures in waste management. The management of wastes is an important function of human societies and essential to the promotion of human well-being. The mismanagement, or neglect of management, for any of the many waste types leads to impairment of human health, to economic losses, to aesthetic value loss, and to damages to ecosystems, biodiversity, and ecosystem function.

Deferring waste management actions until the problems become large is not an effective management approach. The costs of trying to reverse damages to waste-degraded ecosystems or remove toxins from the environment, if possible at all, can be extremely large and burdensome on society.

The capacity of an environment to adsorb wastes without damage to human well-being or to ecosystems depends on the ecosystems' ability to detoxify, process, or sequester (that is, isolate from the biosphere) waste contaminants. The no-damage limit or capacity for assimilation is highly variable. This variability is a result of the properties of different contaminants and the differing characteristics of specific ecosystems. Loading limits also depend on human judgments as to what is an acceptable level of human health risk or alteration of an ecosystem.

The levels of a waste chemical that are harmful to the ecosystem may be either much higher or much lower than for human health. For example, by the time human health effects are noticed, there may already have been substantial changes in the ecosystem and vice versa. Further, the sensitivity of different organisms within ecosystems may be very different for a particular contaminant. Different waste contaminants may have local, regional, or global impacts.

Some waste contaminants (such as metals and salts) cannot be converted to harmless materials and will remain in the environment permanently. With continued introduction into an environment, the concentrations of such contaminants will continue to increase. Many other contaminants (such as organic chemicals and pathogens) can be degraded to harmless components. And depending on the properties of the contaminant and its locations in the environment, degradation can occur at relatively fast or extremely slow rates. The more slowly a contaminant is detoxified (that is, the more persistent it is), the greater the possibility that harmful concentrations of the contaminant will be reached in the environment either locally or globally.

Some wastes, such as nutrients and organic matter, are normal components of natural ecosystem processes but can reach harmful levels due to human activities. Inputs of these materials may sufficiently exceed natural rates so that ecosystem functions are modified or impaired. Nutrients and organic matter, when applied at appropriate rates and locations, are an important resource for the improvement of agricultural soils.

The fate and effects of chemicals introduced into the environment can be predicted (in most cases with useful accuracy), so as to allow informed waste management decisions to be made. Due to the great number of possible contaminants, as well as the continuous generation of new compounds, and to the complex interactions within ecosystems, our understanding of the consequences of some contaminants is incomplete.

The problems associated with wastes and contaminants are in general growing worldwide. Some wastes are produced in nearly direct proportion to population size, such as sewage wastes. Other wastes and contaminants reflect the affluence of society. An affluent society uses and generates a larger volume of waste-producing materials such as domestic trash and home-use chemicals. Where there is significant economic development, loadings of certain wastes are expected to increase faster than population growth. The generation of some wastes, such as industrial ones, does not necessarily increase with population or development state. These wastes may often be reduced through regulation aimed at encouraging producers to clean discharges or to seek alternate manufacturing processes.

It is not possible at this stage to state whether the intrinsic waste detoxification capability of Earth will increase or decrease with a changing environment. The detoxification capabilities of individual locations change with changing conditions. However, it is certain that at high waste-loading rates, the intrinsic capability of ecosystems can be overwhelmed such that wastes build up in the environment to the detriment of human well-being and the loss of ecosystem biodiversity and functions.

15.1 Introduction

This chapter addresses two major topics: the general characteristics and patterns of waste production and the capacity of ecosystems to detoxify or otherwise adsorb human-produced wastes. The first is addressed by showing some trends in major categories of wastes, attempting to indicate in particular the relationships between population and level of development and waste production. The second topic is complex due to the wide variety of types of wastes produced by human activities and to the complex set of ecosystem processes that determine the fate and effects of wastes in various ecosystems. The chapter shows how ecosystem processing of wastes and waste loadings interact to determine the damages of particular wastes.

The term "wastes" in this chapter refers to materials for which there is no immediate use and that may be discharged into the environment. It includes materials that might otherwise be useful if they were not in the environment, such as oil after an oil spill or pesticides once they are no longer at their site of application. The term "pollutant" is not generally used here. A waste in the environment is a contaminant but not necessarily a pollutant in the sense of loss of use of ecosystem services.

Although there is no attempt here to systematically assemble and estimate the damage to humans and ecosystems done by past waste releases, a few examples of the types, issues involved, and magnitude of damages of wastes that exceeded ecosystems' capacity are provided in order to illustrate the importance of managing wastes.

The intent of this chapter is to lay down a foundation for decisions regarding wastes and to establish the importance of waste management for human well-being at both the local and the global scale.

15.1.1 Humanity's Many Types of Wastes

Human activities discharge many types of materials into the environment:

- *Industrial by-products resulting from the production of durable goods, pharmaceuticals, and other manufactured goods* used by society: Some of these materials are novel to natural systems (xenobiotic), which means they did not exist on Earth before being manufactured by humans. Ecosystems may be very ineffective in detoxifying novel chemicals, and so they may be particularly persistent and thus accumulate in the environment.
- *Nondegradable wastes, which cannot be broken down to harmless materials*: These can only be diluted. This category includes metals and salt wastes. Typically, salts are not a problem once they reach the ocean, but metals cause problems in any ecosystem.
- *Pesticides*: These may indiscriminately kill even beneficial insects, and persistent pesticides may accumulate in organisms and have harmful effects on other organisms in the food web.
- *Fertilizers,* the most important of which, quantitatively, are nitrogen and phosphorus compounds.
- *Excrement, or sewage wastes,* rich in organic matter and in nitrogen and phosphorus (plant fertilizers or nutrients) and carrying pathogens: The organic matter in sewage can remove oxygen from aquatic systems, and the nutrients may stimulate plant growth and alter ecosystem structure and function.
- *Natural materials,* which are often released at rates that greatly increase environmental concentrations, thereby often harming ecosystems and human health: Included here are toxic metals, salt wastes, acid wastes, reactive nitrogen, carcinogenic polycyclic aromatic hydrocarbons (found in smoke and exhaust), and petroleum products.
- *The by-products of day-to-day human activities*: This category includes materials made from paper, plastics, glass, metals, and products such as household chemicals, and pharmaceuticals, which become wastes after they are used and, in one form or another, end up in the environment.

Table 15.1 provides a more detailed listing of the major categories of wastes introduced into the environment. The individual materials within a category may have different behaviors in the environment and represent very different types and levels of risks. It should be recognized that some of the categories in the table include a great number of different types of materials. For example, the Registry of Toxic Effects of Chemical Substances (MDL 2003) lists over 150,000 chemicals. Another compendium, *The Merck Index* (O'Neill et al. 2003), lists over 10,000 individual biologically active pharmaceuticals, chemicals, and biological compounds. Further, thousands of new compounds are being synthesized and manufactured each year.

Wastes may be classified as either "point source" or "nonpoint source." Point source wastes are those discharged from a specific facility or at a specific location. Typical of these facilities are industrial operations or sewage processing facilities where the discharge location is readily identifiable, often a single pipe or smokestack. (It should be noted that a sewage treatment plant is not the actual source, but a collecting point for wastes.) The other major category, non-point source wastes, includes urban runoff, acid rain, and agricultural runoff. The distinction between point and non-point sources is somewhat arbitrary, however, as the place in time and space at which a large number of small point sources become non-point source is subjective.

Different types of wastes are also often mixed together, making management of wastes more difficult. For example, the mixing of industrial effluents with domestic sewage impairs possible beneficial uses of waste waters. Domestic sewage, properly treated to kill pathogens, may make suitable fertilizer and soil enhancers for agricultural purposes. However, metals or chemicals often contaminate centrally collected sewage, precluding the use of the sewage for such beneficial purposes. Many household chemicals such as pesticides, pharmaceuticals, and cleaners may also render domestic sewage unsuitable. For example, irrigation with industrial waste waters has been associated with enlarged livers, cancers, and malformation rates in areas in China (Yaun 1993) and with cadmium poisoning in Japan (WHO 1992).

Reuse of domestic waste waters is easier when grey waters (those, for example, from washing) are not mixed with black waters (those containing excrement). The grey waters can be used especially for small-scale irrigation with low risk of disease transmission (Faruqui et al. 2004). If mixed wastes (such as municipal wastes) are incinerated, special technology is required to prevent potentially harmful materials (such as noncombustible metals) from entering the incinerator's air emissions and causing air pollution. The ash of incinerated mixed waste may also have high levels of contaminants, requiring careful disposal.

15.1.2 Types of Damage Caused by Wastes

Wastes can cause harm in many different ways. It is convenient to consider three general different types of harm:

- direct impairment of human health;
- damage to ecosystems or organisms that creates economic losses; and
- damage to organisms in an ecosystem, with loss of biodiversity.

15.1.2.1 Impact of Wastes on Human Health

There are many examples of human health problems associated with wastes:

- Pathogens in sewage wastes transmit diseases. Such pathogens include cholera, typhoid, shigella, and viruses, causing diseases such as diarrhea, polio, meningitis, and hepatitis. It is estimated that 1.8 million children in developing countries (excluding China) died from diarrheal disease in 1998, caused by microorganisms, mostly originating from contaminated food and water (WHO 2003) (some reported food cases may not be from wastes but from direct contamination by other humans and poor hygiene in food preparation). Worldwide, a lack of suitable sanitary waste treatment is estimated to cause 12 million deaths per year (Davidson et al. 1992).
- For metals, the severe health effects of mercury being discharged into the environment were learned in the painful lesson of Minimata Bay Disease (actually not an infectious disease, but mercury poisoning) first identified in the late 1950s, where nearly 3,000 people were stricken with disease or died after mercury was dumped into the environment. Lead is another metal of high concern. Lead has entered the environment as an additive in gasoline and paints. For example, lead exposure in Mexico has resulted in 40–88% of the children in various communities having blood levels of lead higher than exposure guidelines (Romieu et al. 1994, 1995). Lead exposure can reduce growth and cause learning disabilities and neurological problems.
- A number of persistent organic pollutants have become of enough concern to stimulate the generation of international conventions to stop their use (such as the Stockholm Conven-

Table 15.1. Major Categories of Wastes and Contaminants Listed by Source

Category	Types of Wastes	Character of Source	Extent of Impact
Industrial sources			
Energy producers Coal, oil, and gas, production of coking coal Nuclear plants	metals, PAH, fixed nitrogen, waste heat, fly ash, spent fuel, CO_2	point source	local to regional to global
Manufacturing and chemical wastes	wide variety of types; often synthetic chemicals, solvents, and/or metals	point source	local to regional
Mining	metal-contaminated water and soils, acidified water	point source	local to regional
Transportation accidents	oil spills and chemical spills	point source	local to regional
Waste incineration	particulates, PAH, dioxins, fixed nitrogen, phthalates	point source	local to regional
Agricultural sources			
Livestock production systems	pathogens, including species-jumping bacteria/viruses, organics, nutrients, salts; pharmaceuticals, including antibiotics	non-point source	local to regional
Cropping systems	herbicides, fungicides, and insecticides; nonusable plant materials, nitrogen, phosphorus	point and non-point sources	local
Land preparation and rangeland management	PAH, particulates (from set fires)	point and non-point sources	local to regional
Human habitation sources			
Sewage	pathogens, fertilizers, organic matter, residual pharmaceuticals	point and/or non-point sources	local to regional
Heating source emissions	PAH, particulates	non-point source	local to regional
Consumer hazardous materials	cleaners, paints, automotive fluids, pesticides, fertilizers, batteries, cells	point and/or non-point sources	local
Trash	organics, leachates containing nutrients, salts, metals, plastics, glass	point and/or non-point sources	local
Transportation, including shipping, aviation, and automotive sources	PAH, reactive nitrogen, lubricating oils, coolants, lead	non-point source	regional

tion on Persistent Organic Pollutants and the Aarhus Protocol on Persistent Organic Pollutants). For example, the concentrations of polychlorinated biphenyls in fish from certain waters are great enough to warrant official advisories against eating the fish. PCBs have been found in human tissues and human milk throughout the world (Jensen and Slorach 1991).

- It is estimated that air pollution in China causes more than 50,000 premature deaths and 400,000 new cases of chronic bronchitis a year (Harrison and Pearce 2000; UNEP 2000). Air pollution may aggravate asthma, and in the United States approximately 600 children die annually from asthma and 150,000 are hospitalized (CDC 1995). (See also Chapters 13 and 27 for a further assessment of air pollution.)

Humans are exposed to wastes by drinking water containing hazardous substances from waste residues, by ingesting foods with waste residues, by inhaling airborne wastes, or by through-the-skin exposure after physical contact with waste residues. Drinking contaminated water is a common route of exposure for many wastes that are leached into surface or groundwaters. Examples of food exposure routes include agricultural pesticide residues on foods, contamination during food preparation, and bacteria accumulated by edible clams that filter and concentrate particles from water.

Inhalation is a route of exposure for airborne wastes, such as the polycyclic aromatic hydrocarbons in smoke and automotive exhausts, or for pesticides carried downwind of the site of application. Through-the-skin (dermal) exposure is most often associated with worker exposure to materials they are handling, but it is also a route for wastes, such as for residential use pesticides, and for parasites. The dermal and inhalation exposure routes highlight significant overlaps between worker safety issues and waste issues.

15.1.2.2 Economic Losses

Detriment to ecosystems or specific organisms by the introduction of harmful wastes can lead to significant economic losses:

- When a body of water becomes anoxic from eutrophication through excess nutrient inputs, it can no longer support commercial, subsistence, or sport fisheries. Worldwide, there are now at least 146 areas in the coastal environment where low oxygen concentrations occur either chronically, seasonally, or episodically (UNEP 2004).
- A major economic loss has resulted from the use of the antifoulant agent tributyltin used in paints for ship and boat hulls.

Pleasure boats are often kept in high-density moorings and docks in close proximity to where oysters are grown. The species and race used for the oyster aquaculture industry in Europe is about 100 times more sensitive to TBT than are typical aquatic test organisms. As the pleasure boating community adopted TBT-based paints as the hull paints of choice in the 1970s, oyster farming operations throughout Europe experienced huge decreases in yield. For example, in Arcachon Bay, France, annual oyster production dropped from around 13,000 tons to about 3,000 tons between 1977 and 1983. Further, surviving oysters grew deformed, rendering them unsuitable for market. Economic losses for Arcachon Bay over this period were estimated at 800 million francs (Alzieu 1990). Banning the use of TBT paints on small vessels alleviated the immediate threat to the European oyster industry.

- Discharges of wastes into the environment can also cause economic losses even where there may be no clear detriment to the local organisms. For example, in U.S. coastal waters about 15% of potential commercially harvestable shellfish beds are closed due to water quality problems, primarily from sewage discharges (NOAA/ORCM 2003).
- When human health is affected by wastes such that a person is unable to work, there are direct economic consequences.

15.1.2.3 Damage to Ecosystems and Loss of Biodiversity

Wastes present in the environment can harm organisms. Terrestrial organisms may be exposed to wastes through the food chain, through inhalation, or through direct contact. Aquatic organisms may also take up wastes directly from the water.

- Where sufficiently high levels of contamination occur, organisms may receive acutely toxic exposures and die in mass. For example, fish kills have resulted from industrial discharges or runoff of agricultural pesticides (Heileman and Siung-chang 1990). Deaths of aquatic and bird wildlife are often the immediate and visible result of large oil spills, although there may also be long-term residual toxicity resulting from large oil spills (Peterson et al. 2003).
- More subtle and less evident are wastes that change the behavior or the biology of organisms but are not in themselves lethal. A well-known example was the widespread use of DDT and its effects on high-trophic-level birds, which did not die but could not successfully reproduce. Currently, there is concern that other chemicals, such as hormone disrupters, may affect organisms at concentration levels below those causing effects evident in standard toxicity tests. Some of these may disrupt the hormone systems of some organisms, affecting their behavior or reproductive physiology and reducing their capability to survive or reproduce (NTP 2001; deFur et al. 1999).
- Ecosystems can be damaged by changes in their chemical composition. For example, acid rain—a result of sulfur and fixed nitrogen emissions from power plants, motor vehicles, and agriculture—can alter the chemistry of soil. This stresses vegetation and alters the species composition of lakes and streams, sometimes to the point of making the lake unable to support fish life (Driscoll et al. 2001; Galloway 2001).

15.1.2.4 Different Thresholds for Human Health, the Ecosystem, and Economic Loss

The thresholds for effects in the three general types of damage—human health, economic losses, and damage to ecosystems—may be quite different. Acceptable waste concentration limits to protect human health may overwhelm some ecosystems and result in economic losses.

- The effective pesticide DDT has a low acute (short-term) toxicity to humans. From a human perspective, its use was relatively safe and is still used inside households for malaria control, and although there could be significant sublethal human health effects as well, it was primarily DDT's effects on the reproductive capabilities of birds that made its use unacceptable.
- The antifoulant paint ingredient TBT is apparently harmless to mammals (including humans) at environmental concentrations that are lethal to oysters. Although the use of TBT caused economic losses in Europe and harmed some other mollusk populations as well, ingestion of TBT-contaminated organisms was not considered or regulated to address a human health problem.
- The effects of low levels of many carcinogenic chemicals (such as PAH) or ionizing radiation are of great concern for human health where each individual is valued, yet are usually not risks to the health of wildlife populations.

Given the different sensitivity of different species, including humans, to chemical residues, a chemical exposure standard set to reduce human health impacts may not be sufficient to protect wildlife, or vice versa. As such, individual evaluations are required to fully understand the consequences of varying waste concentrations on human health and ecosystems.

15.2 Trends in Waste Production

The amounts of wastes released into the environment in many ways depend on choices made by governments, organizations, and individuals. Trends for some categories of wastes by themselves are not necessarily predictive of future trends, as waste-generating behaviors may change. And the future magnitude of waste problems depends on how wastes are managed in relation to the capacity of ecosystems to detoxify wastes.

In order to understand the actions required for adequate waste management, recognizing the role of ecosystems in waste detoxification, it is useful to examine relationships between waste production and different segments of the human population. The production of some wastes is closely related to population size, while some production is more related to human practices in agriculture and manufacturing, which often correlate with a nation's state of development.

15.2.1 Population-proportional Wastes

The amounts of human excrement wastes (feces and urine) produced are essentially proportional to human population size. Although there may be some differences in the composition and per capita production of excrement depending upon the nutrition status of the population (for example, the nitrogen content of sewage depends on the amount of protein in the diet), the overall variability in composition is relatively small. The total amount of human excrement wastes will increase in proportion to population growth.

The urbanization of populations (see Chapter 27) and the general increase in coastal populations (see Chapter 19) will tend to concentrate the excrement production into relatively limited areas. If damage to ecosystems from the nutrients and organic matter in human sewage is to be prevented, and if sewage-carried pathogens are to be reduced or eliminated, efforts to manage human excrement wastes must increase in proportion to the population size. Where the oxygen demand (mostly due to the decay of carbon compounds in the sewage) and nutrient fertilizers in excrement wastes are not removed from sewage, especially in waste-receiving waters near large urban areas, the loadings to the environment can easily exceed the capacity of ecosystems to absorb them without causing harm to the environment. Pathogen destruction and the removal of oxygen demand can be managed very effectively by modern sewage treatment. With additional treatments (and often significant expense), the amount of nitrogen and phosphorus discharges may be significantly reduced. (See later description in this chapter, Chapter 12, and MA *Policy Responses*, Chapter 10.)

15.2.2 Development-related Wastes

The amount of consumer or municipal waste produced on a per capita basis has a relationship with the development status of countries. (See Figure 15.1.) In general, poverty reduction would be expected to increase the production of consumer waste even without increasing populations. However, policies and practices may also significantly affect the amount and types of municipal wastes produced as societies develop.

The density and character of solid municipal wastes also differs at different states of development. (See Figure 15.2.) In the United States and Europe, for example, solid waste has a large fraction of light materials, such as carton boxes, paper bags, and plastic bags, whereas in many developing countries there is a larger fraction of higher density solid waste, such as gravels, glass, food wastes, and unusable metals.

There are three different types of agricultural wastes: fertilizers, pesticides, and organic wastes (such as manure). The use patterns of these are different and typically depend on a country's state of development. In industrial nations, fertilizers were applied in increasing amounts per unit area until the late 1980s, when application declined. (See Figure 15.3.) Recognizing the contribution of agricultural runoff to degraded water quality and eutrophication and the expense of unnecessary overfertilization, improved farming practices were introduced in many countries,

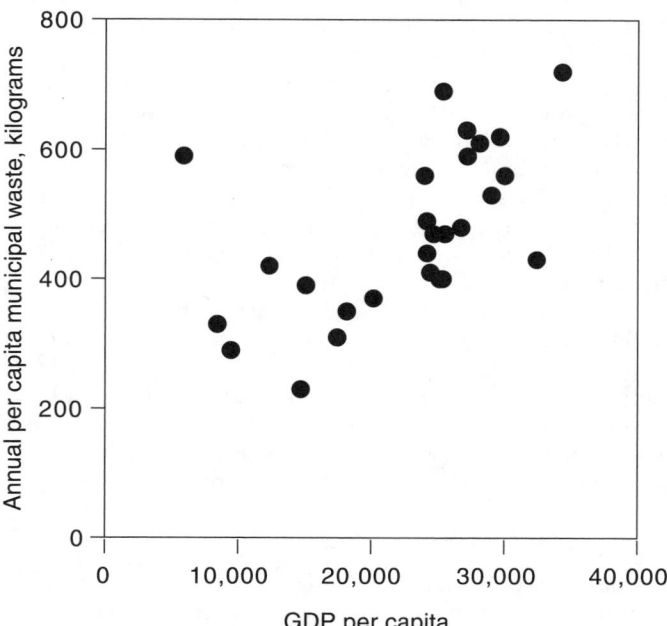

Figure 15.1. Municipal Waste Production in OECD Countries by per Capita GDP (Harrison and Pearce 2001; UNDP 2003)

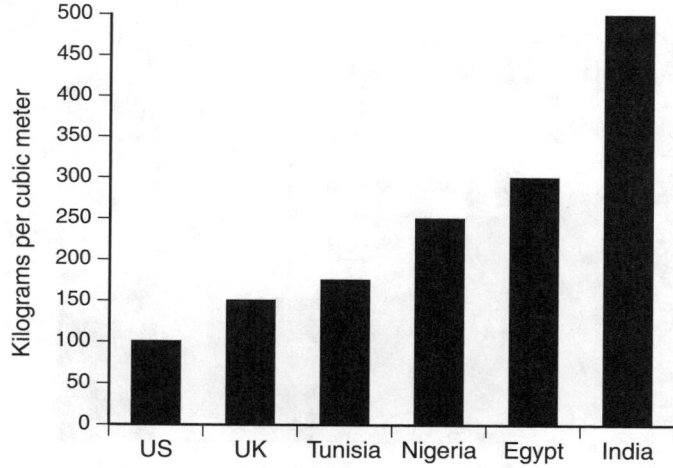

Figure 15.2. Density of Municipal Solid Waste in Selected Countries (UNEP 1996)

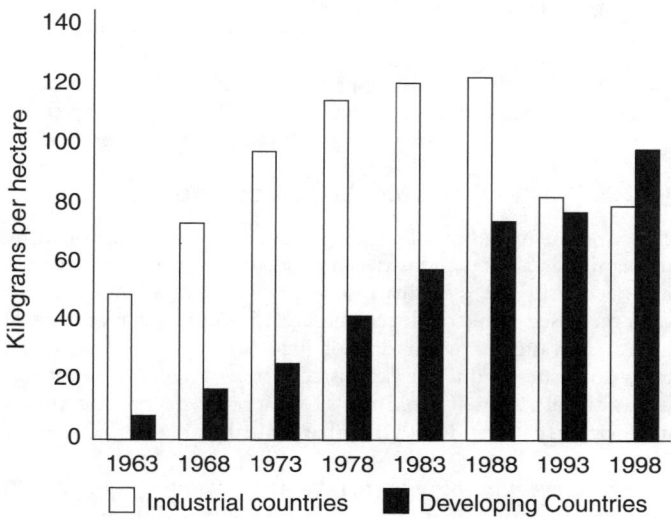

Figure 15.3. Fertilizer Use Trends in Industrial and Developing Countries, 1963–98 (Harrison and Pearce 2001)

which reduced the need for more fertilization yet maintained production. In contrast, in many developing countries the application of fertilizer on an area basis is still increasing and has surpassed the rate in industrial countries. There is, however, considerable variation between industrial and developing countries: industrial ones use more pesticides overall. (See Figure 15.4). Hidden within this overall trend are some additional important trends:

- Industrial countries are quicker to move to newer, less toxic (to humans), more environmentally suitable systems and products.

- Developing countries are often still using older, less expensive, more toxic pesticides at significant risk to farm workers, their families, and others in the vicinity of farms (Goldman and Tran 2002), with poverty exasperating such problems.

- Many developing countries have stockpiled wastes that are likely to eventually breach containment and be a source of potential harm if not destroyed or disposed of properly (Goldman and Tran 2002), again a problem exasperated by poverty, which reduces the capacity to deal with such problems.

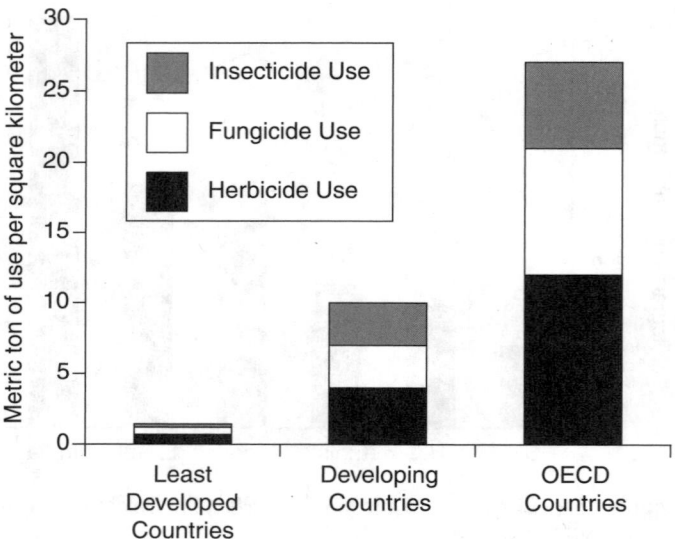

Figure 15.4. Pesticide Use in Countries at Different Levels of Development (Goldman and Tran 2002)

- Many individuals applying pesticides, particularly in developing countries, lack appropriate training. This leads to pesticide poisonings, unnecessary use, and misapplication of pesticides.

15.2.3 Wastes Controlled by Regulations

The amount of industrial wastes produced can be controlled by regulation and good industrial manufacturing practices, and it is not necessarily the case that discharge of industrial wastes grows with increased industrial levels. Figure 15.5 shows that releases of industrial wastes to water do not increase in proportion to the gross domestic product of various countries. The trend is not significantly different if looking only at the waste produced as a function of the fraction of GDP that is attributable to the industry sector.

The prevention of industrial discharges can be required by a strong regulatory framework and can be achieved through treatment of waste streams (or pollution control) and through pollution prevention (often called "green chemistry") through modification or selection of manufacturing processes to reduce or eliminate the production of wastes in the manufacturing process while providing the same products (Greer 2000).

15.2.4 Wastes Controlled by a Combination of Factors

Some wastes do not fit into a single one of the categories just described. For example, emissions from cars and trucks include carbon monoxide, reactive forms of nitrogen, soot, PAH, and carbon dioxide. The number of vehicles in any country depends on population size, the wealth of the population, and the policies in place. The emissions from an individual vehicle depend on regulation and enforcement of emission standards and on the kinds of vehicles (that is, the size and engine capacity and type) in use.

15.3 The Necessity of Waste Management

The production of wastes is a normal function of all living organisms, and individuals, groups of organisms, and societies depend on the capacity of ecosystems to detoxify such wastes. Without

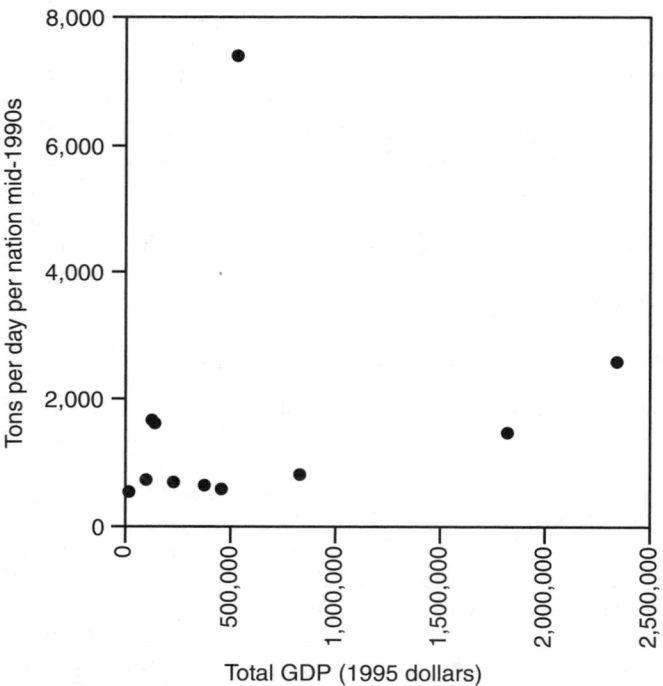

Figure 15.5. National Releases of Organic Water Pollutants in Selected Countries. The nations included (in order of lowest to highest GDP) are Ukraine, Indonesia, Russia, India, Brazil, China, United Kingdom, France, Germany, Japan, and United States. (Harrison and Pearce 2000; WRI 2003)

an ability to manage wastes, organisms cannot live indefinitely. A lesson often taught in introductory biology classes provides a useful illustration. Bacteria inoculated into a rich culture medium will at first grow and multiply. But eventually, as waste products from bacterial metabolism build up in the medium, the bacteria can no longer multiply and may die. The death happens even though there may still be plenty of food for the bacteria. They are poisoned by their wastes.

Higher organisms require systems to process wastes and expend significant energy to deal with wastes. For example, a human body has kidneys, liver, large intestine, and bladder that are primarily committed to waste processing. Further, the circulatory system is just as important for transporting wastes to be processed or eliminated as it is for carrying oxygen and nutrients to cells. Waste management is an important and necessary function of higher organisms.

Similarly, a portion of society's energy must be committed to the important but unglamorous task of waste processing in order to promote health and well-being. In a human-dominated Earth, the practice of placing wastes out of reach is no longer a long-term solution. The dumping of wastes, such as at sea, or the transport of wastes from one country to another does not necessarily prevent the wastes from causing human detriment.

15.3.1 Local Scale

On the scale of a large city, sewage treatment and solid waste management are particularly important functions. The level of commitment necessary to accomplish these tasks may be illustrated by examination of the expenditures in cities where sewage and solid waste services are considered adequate. In an examination of budgets for some major U.S. cities, for example, the combined expenditures of solid waste disposal and sewage treatment range from about half the costs of police and fire protection (New York and San Diego) to equal to those expenditures (Detroit and Houston). Establishing effective waste management functions is an extremely important development goal, as the poor are particularly susceptible to detrimental exposure to wastes (Goldman and Tran 2002; see also description later in this chapter).

15.3.2 National Scale

Regulatory management of wastes is generally conducted through national governments and through state or provincial authorities. Insufficient regulatory management often leads to human health impairment, economic loss, or ecosystem degradation. The costs of failure to prevent waste problems can be very high, especially those for cleaning up contaminated sites. For example, the U.S. Environmental Protection Agency estimated (EPA 1998) that it would require $32.9 billion in public funds to remediate the 5,664 listed contaminated sites in the United States at that time, which is in addition to the considerable private funds spent on these projects. Another example is the costs of remediation in U.S. coastal waters, where it is estimated that upgrading sewage treatment plants to mitigate nitrogen-caused low oxygen problems in estuaries will cost up to $20 billion for the Chesapeake Bay and about $10 billion for Long Island Sound (Boesch 1996). In addition to such remediation costs, the full cost of failure in waste management includes broader human, ecosystem, and economic costs.

15.3.3 Global Scale

Some contaminants are of global significance, particularly persistent chemicals subject to long-range transport. Accordingly, a number of international conventions and protocols have come into force or been signed to manage such chemicals. Notable among these are:

- the 1979 Geneva Convention on Long-Range Transboundary Air Pollution and its eight protocols;
- the 1985 Vienna Convention for the Protection of the Ozone Layer and its Montreal Protocol on Substances that Deplete the Ozone Layer (plus amendments);
- the 1989 Basel Convention on the Control of Transboundary Movements of Hazardous Wastes and Their Disposal;
- the 1992 United Nations Framework Convention on Climate Change and its Kyoto Protocol;
- the 1998 Rotterdam Convention on the Prior Informed Consent Procedure for Certain Hazardous Chemicals and Pesticides in International Trade; and
- the 2001 Stockholm Convention on Persistent Organic Pollutants.

In addition, there are further efforts to manage chemicals on a global basis, such as the U.N.'s Strategic Approach to International Chemicals Management initiative.

15.4 The Capacity of Ecosystems to Detoxify or Use Wastes

The ability of an ecosystem to reduce waste concentrations depends on both the properties of the waste and the properties of the ecosystem. In order to understand the potential harm of a waste or to determine how much capacity an ecosystem has to assimilate a particular waste, it is necessary to examine the ecosystem processes that are responsible for the fates of wastes. Although there are thousands of types of materials in wastes, ecosystem processes are relatively few in number.

Table 15.2 lists the ecosystem processes that may act to reduce the concentrations or impacts of wastes in an environment over time. There are two fundamentally different types of processes:

- *Processes that act to change wastes into less toxic forms ("detoxification")*: Some types of wastes may be completely destroyed by processes in the environment or used in the environment in a way that renders them harmless. It should be recognized that while processes that alter waste materials are often on the path to detoxification, the initial alteration of a waste may not reduce the potential of the waste to do harm.
- *Processes that move and transport wastes*: These reduce concentrations of waste by diluting them into larger areas or larger volumes of water. However, some waste transport processes may also concentrate wastes into "hot spots" of relatively high waste concentrations.

Different types of waste have different properties and interact with or are subject to different environmental processes. Table 15.3 lists the general detoxification characteristics for different types of waste materials.

15.4.1 Detoxification Processes

15.4.1.1 Microbial Degradation

Organic wastes can usually be broken down or even consumed as food by some organisms. When degraded, the waste may be made less toxic and its harmful effects reduced. Remineralization is the complete breakdown of an organic chemical (for example, into its basic components such as carbon dioxide, nitrogen, phosphorus, and water), such as may occur by microbial digestion. Remineralization completely destroys toxicity inherent in such waste. Most remineralization is conducted by microbes (primarily bacteria and fungi).

The ability of microbes to metabolize and remineralize different wastes is highly variable. Remineralization depends on the type and number of microbes in the total community that are capable of degrading a particular waste. The number of waste-degrading microbes in turn depends on prior exposure to the waste (or similar types of waste) through chronic or single-event exposures. Degradation of the wastes may increase after exposure as a result of an increase in number of waste-degrading microbes or the induction of appropriate enzyme systems.

The rates at which microbes can degrade wastes are influenced by temperature and thus may change seasonally. The presence of oxygen often exerts a strong influence on degradation rates, and wastes may persist for very long periods of time in sediments that have little or no oxygen, even though microbes in oxygen-rich environments may readily degrade the same waste. Marine and aquatic sediments and water-saturated soils in wetlands are often low in oxygen. Hence, the water availability in a specific location, acting through changes in the amount of saturated sediments and soils or even through changes in soil moisture, may have significant effects on contaminant degradation rates.

Some wastes break down in anaerobic conditions. Specifically, the removal of chlorine atoms from PCB molecules, a necessary first step in their remineralization, seems to occur (though slowly) in anaerobic conditions (e.g., Mondello et al. 1997).

High concentrations of waste contaminants may also be toxic to the bacteria and fungi that could otherwise degrade the contaminant were it present in lower concentrations. This restricts the breakdown of a contaminant to the less contaminated areas of a site, slowing the overall degradation.

Table 15.2. Processes Involved in Waste Processing and Detoxification

Processes	Factors Affecting Processes
Contaminant movements in soils	
Sorption to sediments (may reduce bioavailability)	chemical nature of soils; especially organic content and cation exchange capacity, basicity of compound
Leaching by precipitation water flow (transport from and through surface sediments)	porosity of soils, rate and pattern of rainfall, vegetation cover, slope of soil surface, polarity of compound
Groundwater transport	clay content, cation exchange capacity, porosity, hydraulic gradient, polarity of compound
Volatilization and dust transport	vapor pressure of chemical, vegetation cover, moisture content of soils
Soil erosion	vegetation cover, soil moisture content, soil cohesion
Biological transports	bioturbation, bioconcentration, and animal movements and migration
Contaminant movements in water bodies	
Mixing or dilution	volume of receiving waters, stratification
Advection and dispersion	velocity of water, turbulence regime
Water residence time	water input rate, tidal height, salinity distribution
Particle and sediment interactions and scavenging to sediments	solubility of contaminant (Kow), particle deposition rate
Sediment-water exchange	diffusion rate, bioturbation rate, water level fluctuations
Sequestering of chemicals (acid volatile sulfides in sediments)	oxidation state of sediments, availability of sulfides
Scavenging to (accumulation at) sea–air interface	solubility of contaminant
Volatilization	vapor pressure of contaminant, wind speed and water surface roughness
Aerosol formation	breaking waves
Precipitation from solution	solubility limits
Biological transports	uptake by organisms, settling of organic materials, food chain transfers
Contaminant movements in air	
Wind transport	wind speed, particle size and density
Wet deposition	precipitation patterns and solubility
Dry deposition and adsorption	surface area, air turbulence, "strictness" of contaminant
Processes responsible for alteration and destruction of contaminants	
Direct photolysis	strength and wavelength of light
Oxidation/reduction	properties of chemicals, oxidation state of media
Acid, base, and neutral hydrolysis	properties of chemicals, moisture availability
Microbial transformations	type of chemical bonds (e.g., chlorinated/holgenated vs. natural compounds); native community of microbial degraders; microbial degradation rates may depend upon waste concentration, temperature, the presence or absence of oxygen, pH, prior exposure, availability of co-metabolites, and influence of synergism
Radioactive decay	rate of decay intrinsic to radionuclide
Die-off (of pathogens and indicator organisms)	temperature and other conditions in environment, presence of organisms to ingest pathogens, presence of vector organisms

Chemicals that can be used as food by microbes tend to break down more quickly than poor food-quality chemicals. Some organic chemicals are especially slow to break down in the environment. For many of these, their resistance to microbial degradation results from the presence of many attached chlorine or bromine atoms. While there are many naturally occurring organic compounds incorporating chlorine or bromine—which degrade, albeit slowly, naturally—many of those synthesized by the chemical industry are quite resistant to microbial degradation. The chlorine-carbon bonds in these synthetic chemicals are not naturally abundant, and organisms do not have enzymes effective at degrading these chemicals. One group of such resistant chemicals with known toxic effects are the persistent organic pollutants, some of which have been banned from further production by international treaties (as described earlier in this chapter).

The time required for the concentration of POPs in a contaminated area to decrease measurably is typically measured in decades (Wania and Mackay 1995). For example, results from the NOAA Mussel Watch Program show that PCBs (one group of POPs) in Delaware Bay are decreasing at a rate of only about 5% per year (a half-life of about 13 years). This decrease is partly due to slow degradation of the PCBs but also to their dilution and spreading. On a national scale, the average PCB concentration in mussels in the United States is only decreasing at a rate of 2% per year (a half-life of 34 years). Another example is dioxin contamination in Viet Nam, which was a result of defoliant use between 1962 and 1970. Concentrations of dioxin in human milk in Viet Nam have been decreasing with a half-life of about 4 years (calculated from data in Schecter et al. 1995). This decrease is a result of both degradation and slow dispersion of the dioxin. The slow rate of loss, coupled with the high initial exposures, results in dioxin concentrations in human milk some 30 years after the contamination ceased that are still 10 times higher than in nonexposed areas (Schecter et al. 1995).

Where large amounts of waste are present, such as in an oil spill, the degradation rates of the wastes may be limited by factors needed for microbial growth such as nitrogen and phosphorus. If bacteria cannot increase in number, the degradation rates of compounds will remain slow. The physical nature of spills may also inhibit microbial breakdown. If the spilled waste remains in large clumps, or patches, the microbes cannot reach or degrade the interior of the clump, and only the surface of the clump is subject to microbial decay, greatly slowing the digestion of the waste.

For many organic wastes, and particularly large (high molecular weight) compounds, it is important to distinguish between an initial alteration of the chemicals in the waste (a relatively small change in the form of the waste) and complete remineralization. A small alteration of a parent compound may result in persistent and/or toxic daughter products that may be of as much concern as the parent waste chemical.

With such a complex array of factors that can influence biodegradation, it is not usually a simple matter to predict the persistence of a degradable waste in an environment. However, experience with wastes has been gained and models have been developed that allow such predictions to be made with good confidence. In addition, understanding of the process affecting waste degradation has allowed the development of bioremediation techniques that create the right conditions to accelerate biodegradation in some contaminated areas (Alexander 1994) (see MA *Policy Responses*, Chapter 10).

15.4.1.2 Pathogen Die-off

Some human pathogens are rendered harmless in the environment. Many bacteria and some viruses that only grow in condi-

Table 15.3. Detoxification Characteristics for Different Types of Waste Materials. For the types of waste that cannot be destroyed by natural environmental processes, as opposed to dilution or sequestration, no time frame for detoxification is given. Half-life is the time it takes for the concentration of a particular material to reduce by half. The use of half-life is a measure that implies first-order kinetics. It should be recognized that detoxification is usually a biological process that will have much more complex kinetics than first-order. Nevertheless, half-life is a useful simplifying approach.

Type of Waste	Characteristics of Waste and Major Processes for Detoxification	Time Frames for Detoxification, Where Applicable
Airborne sulfur dioxide and sulfates	Component of acid rain. Can be buffered by some soils until buffering capacity of soils is exceeded.	
Airborne oxides of nitrogen	Component of acid rain. Contributor to eutrophication. (see box on eutrophication.)	
Polycyclic aromatic hydrocarbons (PAH)	Many PAH are carcinogenic. PAH may be photodegraded and remineralized by microbes. Many PAH have a high affinity to adsorb onto particles and soils.	Rate of photodegredation of PAH depends upon specific configuration of individual PAH. In air or sunlit waters, half-lives of PAH may range from hours to days. Lower molecular weight PAH can be readily degraded in aerobic soils and sediments, with typical half-lives in days to weeks. Higher molecular weight PAH degrade much more slowly, with half-lives in weeks to months. In anaerobic conditions, PAH may persist indefinitely.
Toxic metals, especially mercury, lead, and cadmium	Cannot be degraded. May be bound to other material (i.e., sulfides) in certain conditions so that toxicity of metals is not exhibited if there is sufficient binding material. Other binding sites include organics, and they may form precipitates such as hydroxides and oxy-hydroxides that may render them less bio-available and in the process reduce their toxicity.	Rate of dilution depends upon environmental setting, while precipitation processes are pH/redox regulated. Many metal-contaminated sites may remain contaminated for decades to centuries.
Halogenated organic compounds, especially DDT and its metabolites, PCB, PCT, dieldrin, and short-chained halogenated aliphatic compounds	Some can be photodegraded. May be degraded by microbes. Many have high affinity to adsorb onto particles and soils. One route of degradation is the volatilization of chemicals from sediments or soils to the atmosphere, where it is degraded by strong UV light.	Due to having halogen-carbon bonds that only occur in trace amounts in nature, microbes do not generally have an ability to degrade these compounds readily. However, general detoxification enzymes within microbes can degrade some of these chemicals, albeit slowly. For example, half-lives of PCBs in contaminated sediments are typically decades.
Petroleum hydrocarbons	Petroleum hydrocarbons are a mixture of primarily alkane compounds and PAH. The alkanes are readily degraded by microbes in the right conditions. PAH are described above. Where petroleum hydrocarbons are introduced into the environment in large quantity, such as an oil spill, the oil may remain in large drops so that bacteria cannot get to most of the oil. In large spills, there may not be sufficient nitrogen and phosphorus in the local environment to permit hydrocarbon-degrading microbes to multiply.	Lighter hydrocarbons can degrade with half-lives of hours to days when found in low concentrations and in aerobic conditions. Heavier components of oil—i.e., that left over after the light hydrocarbons are degraded or volatilized—can persist indefinitely (such as tarballs). Light or heavy hydrocarbons in anaerobic conditions will degrade very slowly if at all. In addition, high concentrations of hydrocarbons may initially exhibit localized toxic effects, causing a delay before degradation proceeds.
Toxins of biological origin from algae, fungi, and bacteria	Presumably these compounds are subject to microbial and/or photodegradation.	Rates of degradation probably rapid, with half-lives in hours or days. The toxicity of a water body disappears soon after the organisms producing the toxins are no longer present.

(continues over)

tions found in the human body lose viability in the relatively harsh conditions of the environment. Both viruses and bacteria may be ingested and utilized for food (therefore remineralized) by other organisms in the environment without detrimental effects. However, other pathogens, such as vector-borne diseases, complete part of their life cycles outside of the human body and are maintained by their in hosts in the environment.

15.4.1.3 Photochemical Degradation

In the atmosphere, and to a lesser extent in surface waters, some compounds are altered by interaction with ultraviolet light or by chemically reactive compounds produced by ultraviolet light, such as ozone or hydroxyl radicals. Some limited groups of organic chemicals have structures (chromophores) that adsorb light energy and are activated into a particularly reactive state, altering their structure. This structural change may not reduce the toxicity and may even create more toxic daughter products, but it may be a first step toward remineralization, which can be followed by microbial degradation. The rates of alteration by photochemical degradation may vary widely between chemicals that have only slightly different structures. Reactive compounds in the atmosphere, especially ozone, may also act to alter the structure of airborne wastes.

15.4.1.4 Sequestration of Wastes

Some waste materials may be sequestered in the environment in such a way that they are not biologically available and do not exhibit toxicity. The sequestration of certain metals in marine and

Table 15.3. *continued*

Type of Waste	Characteristics of Waste and Major Processes for Detoxification	Time Frames for Detoxification, Where Applicable
Nitrogen compounds	Nitrogen exists in different forms and is an essential plant nutrient. The presence of elevated nitrogen concentrations in water can lead to proliferation of algae and other plant species. This can reduce the value of the resource and can affect human health. For example, nitrate levels over 20mg/l are considered the safe upper limit in potable water.	Rates of transformation are dependent on the form. Dentrification can be rapid (hours) in the presence of bio-available carbon and acceptable pH and temperature conditions. Assimilation into plant biomass is light-, temperature-, and plant species-dependent and may be hours to days.
Ammonia toxicity	Ammonia is oxidized to nitrate in aerobic environments. Local pH and salinity conditions can influence the level of toxicity by influencing the ionized and un-ionized fractions.	Rate of oxidation is temperature and oxygen-dependent, with typical half-lives of hours to days.
Phosphorus	An important plant nutrient that is associated with fertilizers and detergents. Phosphates bind with some metals, for example iron and aluminum salts, and can from insoluble salts of calcium and magnesium under high pH conditions in oxidizing conditions. Reducing conditions can result in the release of preciously bound phosphorus. Phosphorus may be removed by plant uptake.	Minutes to hours if chemical precipitation/release processes are involved; days to months in the case of biological uptake being species-dependent.
Organic loading	Most organic matter can be remineralized to carbon dioxide by microbes. Some organic matter is resistant and will degrade much more slowly.	Rate of oxidation is temperature and oxygen-dependent, with typical half lives of hours to days for readily degradable organics and months to years for resistant materials.
Acid wastes	High acid wastes (i.e., low pH wastes) can result in the mobilization of metals that may increase toxicity. Acidity can be buffered in both soils and water until the buffering capacity is exceeded. The presence of sulfur can influence acidity, for example the oxidation of sulfur compounds can lead to a reduction in pH while sulfate reduction through the generation of alkalinity can reduce the acidity, which can be reflected in an increase in the pH.	Hours to weeks if conditions are favorable. Chemical processes are generally more rapid than biologically mediated processes.
Pathogens	Some pathogens may lose viability in environment or may be utilized as food by other organisms. Some pathogens are endemic in the environment and do not lose viability.	Survival half-lives of non-endemic pathogens is typically hours to days.
Radionuclides	Natural radioactive decay to stable isotopes (see metals for associated processes).	Half-lives of most medical-use radionuclides range from hours to weeks. Half-lives for the bulk of radioactive wastes from power plants range from decades to millennia.
Solid salts of alkali metals and alkaline earth metals	Cannot be destroyed.	
Waste heat	Heat loss occurs through dilution and radiation of heat energy to the atmosphere.	

aquatic sediments by acid volatile sulfides is an example. These metals are bound into a mineral form that is not biologically available, and as long as there are sufficient sulfides to bind all the metals, no toxicity is exhibited (Ditoro et al. 1992). Where the concentrations of metals exceed that of sulfides, the sediments exhibit toxicity. Sequestration may be reversible. If conditions are altered, the sequestration may break down and the wastes returned to toxic forms. In the case of acid volatile sulfides, the metal sequestration takes place in anoxic sediments, and if the sediments become oxidized through disturbance or ecosystem change, the metals may be released from their bound state and become available and exhibit toxicity. Some chemicals may also be sequestered within soil or sediment organic matter.

Humans often manage wastes by sequestration or immobilization, such as in managed landfills or high-level radioactive waste storage sites. As with natural sequestration, sufficient time may bring about a change in conditions and release wastes back to the environment.

15.4.1.5 Incineration

Organic wastes are often remineralized by incineration. Efficient incineration can completely destroy the toxicity of an organic or pathogenic waste. Inefficient incineration, however, results in the production of carcinogenic polycyclic aromatic hydrocarbons, dioxins, and furans. If a metal-containing waste is burned, the metals may be emitted to the atmosphere, where they can enter food chains and cause environmental damage and detriment to human health. Incineration in large amounts, especially of sulfur-containing materials (such as some coals) can add acidity to the atmosphere, resulting in acid rain and ecosystem damage. A variety of engineering techniques are available to capture the contaminants and reduce the emissions.

15.4.2 Waste Transport Processes

As noted earlier, there are a number of processes that tend to disperse materials in the environment—for example, pesticide leaching into groundwater, runoff with rainwater to water courses, and evaporation or volatilization into the atmosphere, where it may be carried by winds. Soil and wind erosion may also carry pesticides sorbed to soil particles into water bodies and to downwind areas.

Reduction of waste concentrations in a single body of water is best understood as a result of two processes: dispersion (dilution by mixing into larger volumes of water) and advection (water moving downstream). Both these processes reduce the concentration of the waste at its point of entry in the ecosystem. These also apply to contaminants in air.

Dispersion and advection have been counted on for millennia to manage human wastes. If the wastes can be detoxified by the environment, and the loading rates do not exceed the capacity of the environment to process them without undesirable change to the ecosystem, such an approach can be effective. However, with increasing human population densities and waste loads, coupled with many types of wastes that do not degrade rapidly, such a simple approach is rarely adequate. Even wastes that are diluted into the entire atmosphere of the planet have reached concentrations above acceptable levels. Examples are the upper-atmosphere ozone-depleting chemicals chlorofluorocarbons and methyl bromide, which are now the subject of coordinated efforts to reduce their emissions through the Vienna Convention and the Montreal Protocol.

While there is a natural tendency for all things to disperse and move from areas of high concentration to low concentration, there are also processes that effectively act to concentrate some wastes and create relative hot spots. The next two sections examine these processes in more detail.

15.4.2.1 Partitioning and Scavenging

One particularly important process that influences the fate of chemicals, especially in aquatic systems, is that of chemical partitioning. Many waste chemicals have low solubility in water. Even when present in water at concentrations below their absolute solubility, low-solubility chemicals may be strongly attracted to particles. If a solid surface is available, such as small particles in the water, the chemical will tend to adsorb or attach to the particle. The strength of this affinity is referred to as a partition coefficient (which for organic chemicals correlates with the octanol/water partition coefficient, or Kow). Some authors express this general concentration mechanism as "solvent switching" (MacDonald et al. 2002). The particle affinity of metals depends on other factors, especially the oxidation state of the metal.

Particles in the water column tend to settle to the bottom. This may occur in standing water or in areas of low flow, such as upstream of a dam in a river. The adsorption of chemicals to particles and the subsequent settling of particles acts to transport (or scavenge) chemicals from the water column to sediments. In this fashion, relatively high concentrations of chemicals can build up in sediments even in areas that originally had dilute concentrations of waste in the water. The sediments act as a reservoir for wastes, often to the detriment of benthic organisms. And depending on the persistence of the waste, the sediments may remain contaminated even if the inputs are stopped. If the sediments are disturbed by high water flows such as floods or major storms or by mechanical means such as dredging or construction projects, the sediments may become "new" sources for wastes.

15.4.2.2 Biological Uptake and Trophic-level Concentration

Many waste chemicals may have a strong affinity for biological tissue, particularly for the lipid (fatty) tissues in organisms. An aquatic organism may move large amounts of water across its metabolic surfaces (gills or equivalent) in order to obtain oxygen. This provides an opportunity for chemicals (especially organic compounds such as the POPs) to be taken up into the organism, concentrating the chemical by another type of solvent switching. As with sediment partitioning, the concentrations in the organism may become much higher than in the water. The magnitude of this effect is usually described as the "bioconcentration factor."

The concentrations of wastes can increase in organisms further up the food chain. This process is often called "bioamplification" and may be viewed as a "solvent depletion process" (MacDonald et al. 2002). (Biomagnification is another term often used in the same sense as bioamplification, although biomagnification is also sometimes used in the same sense as bioconcentration. The use of these terms is not standard.) When a higher trophic level organism ingests a food that has been contaminated (by bioaccumulation), it may digest most of the food but none of the waste. The waste then becomes much more concentrated in the gut, exposing the higher-trophic-level organism to high concentrations of waste. Where upper-trophic-level organisms cannot metabolize or excrete the waste, the waste increases in concentration in bodies of organisms in successively higher trophic levels, and thus bioamplifies. In this fashion, persistent chemicals that were originally present at low, nonharmful levels in the environment build up to harmful levels in the tissues of higher-trophic-level organisms, including humans (MacDonald et al. 2002).

15.5 Determining the Capacity of an Ecosystem to Assimilate Wastes

The assimilative capacity of an ecosystem to adsorb waste may be defined as the amount and rate of a given waste that can be added to an ecosystem before some specified level of detrimental effect is reached. Deciding on a safe or acceptable level is not usually a simple matter, as is clear from the complex set of processes and the wide array of possible waste types just described. Further, the "acceptable level" incorporates human value judgments, which may be different for different people and may vary over time.

The human value judgments include such considerations as:

- the level of risk that is acceptable;
- whether environmental standards are based upon some absolute level or whether risks are balanced against benefits;
- the costs of mitigating the effects of wastes;
- the manner in which noneconomic properties of an ecosystem are valued; and
- the allocation of benefits, or risks, to different sectors of Earth's population, and between the present and the future.

Hence the capacity of an ecosystem to assimilate wastes is not usually determined by purely scientific or objective study. Science may be able to clearly describe the consequences of any particular waste loading, but it is the application of human values to the consequences that are responsible for setting a "safe" or acceptable limit.

15.5.1 Safe Levels of Exposure

The concept of "safe" is reasonably clear for pathogen wastes. If the disease organism is present in water or food, it is not safe to drink the water or eat the food. In practice, however, it is often not easy or routine to directly detect the presence of disease organisms in the environment. A separate test or procedure might be necessary for each species of pathogen, and the culturing of pathogens for test purposes is often difficult and may be dangerous. The presence of nonharmful organisms that are found in human wastes and that are relatively easy to measure are usually used to indicate the possible presence of disease organisms. This approach assumes that pathogenic organisms do not survive longer in the environment than do the indicator organisms, which may not always be the case. Still, the use of indicator organisms has been effective (but not perfect) in protecting human health for many decades. The common use of fecal bacteria for indicators is not useful in predicting the presence of parasitic worms.

Chemical wastes are often divided into two categories: those having threshold effects and those having no threshold effects. For threshold chemicals, there are assumed to be no detrimental effects below a certain level of human exposure, although above that threshold, detrimental effects may be found. For no-threshold chemicals, primarily carcinogens, it is usually assumed that there is detriment at all levels of exposure, no matter how small. For these, the probability of damage is greater with greater levels of exposure.

Establishing a safe level for a contaminant that has a distinct threshold effect is conceptually straightforward, but there may be considerable technical complexities in arriving at a standard, such as how to apply results of laboratory animal studies to humans. In practice, maximum exposure limits are usually set well below the threshold level for the observed effect in order to account for uncertainties in the measurement of the threshold, individual sensitivity, and the possibility of greater than estimated exposures to a waste for some persons. There is always some concern that a chemical may have subtle detrimental effects at low levels of exposure that are hard to identify as being caused by the chemical.

For non-threshold carcinogenic chemicals and for ionizing radiation, the risk increases with exposure. The effects are an increase in probability that the exposed person could get a cancer during his or her lifetime. As cancers may arise from different causes, the cause of a particular case of cancer may not be unequivocally attributable to a particular exposure, and the effects are observed in population statistics of how many persons contract cancer of a particular type. The setting of a "safe" level for a carcinogen requires a value judgment of the level of risk that persons are willing to accept, plus the knowledge of the probability of a cancer developing in an individual from a given dose (that is, the dose-response relationship). An example might be the standard for acceptable concentration of a chemical in drinking water. This standard may be based on a value of an individual's lifetime risk being no more than one chance in a million (usually used by U.S. EPA) or one in 100,000 (usually used by WHO) of contracting cancer from that chemical if they have a lifetime exposure to the chemical at the level of the standard.

The level of exposure that is deemed acceptable varies considerably from chemical to chemical depending on its toxicity. For example, the FAO/WHO *CODEX Alimentarius* standards for maximum residue levels range from 0.05 to 25 milligrams per kilogram for different pesticides on apples.

Where contaminants occur together and have similar mechanisms of detrimental effects, these effects may be additive, acting together so the threshold for effect may be reached at levels that would have no effect for individual chemicals. There is also concern that the effects of some contaminants may be synergistic in that exposure to a combination of two or more contaminants will lead to a much greater effect than would be expected from the simple addition of the effects of the individual contaminants. Finally, the effects of some contaminants may be antagonistic in that together there is less effect than would be expected from the addition of multiple contaminants (Yang 1994).

15.5.2 Predicting and Managing the Risks of Wastes

Determination of human health impairment from exposure to wastes can be very complex. For each waste or each component of a mixed waste stream, an evaluation of the impairment depends on a knowledge of the fate of the wastes in the environment, how much exposure humans will get from the waste, and how much detriment will occur from a given level of exposure.

In evaluating the detrimental effects of a waste or actions taken to reduce exposure and detriment from wastes, it is usually desirable to evaluate the total number of exposed persons in the population—the cumulative human-health impairment. For example, two differing waste management options might protect all individuals from exposures that exceeded standards for individual protection, but the options may differ in the total dose to the population.

Toxicology and risk assessment are developed areas of science, and various national and international bodies have used appropriate techniques to determine health risks and set appropriate standards and guidelines. Professional organizations working on these issues include the Environmental Mutagen Society, Genetic Toxicology Association, Society of Environmental Contamination and Toxicology, Society for Occupational and Environmental Health, and many others (see, for example, www.health.gov/environment/ehpcsites.htm). Although the scientific community has the ability in most cases to predict (with useful accuracy) the fate and effects of chemicals, due to the great number of possible contaminants (as well as the continuous generation of new compounds) and the fact that the specific conditions within a specific ecosystems may not be known, our understanding of the consequences of some contaminants is incomplete.

15.6 Drivers of Change in Waste Processing and Detoxification

At this stage it is not possible to state, on a global scale, whether the capacity of ecosystems will increase or decrease in response to climate change. The change in average local or global temperatures of a few degrees is not thought likely to have much effect on the distribution of waste materials in the environment (MacDonald et al. 2002) or on a temperature-dependent microbial degradation rate.

As described earlier, the capacity for an environment to assimilate wastes is highly dependent upon local conditions. The bacteria and other decomposing organisms that detoxify susceptible chemicals or reuse nutrient wastes are highly dependent upon local conditions such as oxygen availability, moisture, and temperature. Hence, changes in local climate may have significant effects of waste assimilation capacity of different ecosystems. The conditions at some locations may allow the microbial community to be better able to process certain types of wastes. Other locations may suffer a reduced inherent ability to detoxify.

Changing climatic conditions may also have an effect on the susceptibility of organisms to wastes. Organisms living near their physiological temperature limits may be particularly sensitive to stress from waste contaminants. In such cases, small temperature changes may cause significant differences in the effects of contaminants.

A safe generalization for virtually all types of wastes is that the ability of environments to detoxify them can be overwhelmed at high waste loading rates. If waste production increases with growing populations and improved development of nations proceeds faster than waste management efforts, then it seems likely that there will be an increasing number and size of locations on Earth where the detoxification capabilities of the ecosystem will be overwhelmed and waste concentrations will build up to the detriment of human well-being to damage ecosystems with a loss of biodiversity.

It is hard to predict with a high degree of confidence which environments will be most subject to impacts by different types of wastes. Nevertheless, Table 15.4 is an assessment of the likelihood of ecosystems receiving different types of wastes and contaminants. Table 15.5 lists a number of the driving forces acting at the local level and the effects those drivers will have on waste processing and detoxification.

15.7 Selected Waste Issues

This section describes some key waste issues. A great number of waste types and issues could be cited as examples, and it is not possible for this chapter to attempt to detail the entire scope of waste types and issues in similar detail. The selection here is merely representative of this range, and should not be interpreted as suggesting that these wastes are the most important to consider for future waste management actions.

15.7.1 Consumer Household Wastes and Hazardous Materials

15.7.1.1 Household Trash, a Mixed-waste Stream

A multitude of products are used by individual consumer households or in nonindustrial commercial settings. Many of these products generate trash and require disposal. These include food wastes, paper products, plastics, and metals and may also contain harmful chemicals. Depending on local practice, wastes may be combined or attempts may be made to keep the different materials separate. Food wastes, when kept separate, may be used as compost and for soil enrichment. Some paper products, plastics, glass, and metals can be recycled. What is not separated and reused must be disposed of. Once different types of materials are mixed (usually by the individuals or households generating the trash), it is much more difficult to separate them for reuse. Polyethylene wrappers, bags, or sheets are a major problem in Africa as they litter urban areas, are not biodegradable, and do not burn readily.

Common practices are to either landfill (dump) or incinerate the wastes. Both options have drawbacks. Landfilling permanently takes up space, often in short supply in areas of high population density. Improperly designed and managed landfills may attract vermin, may contaminate groundwater, become visual blights, and emit objectionable odors. Although portions of the organic materials in landfills may decay and generate the green-house gas methane (which could be collected and used as a resource), much of the disposed materials does not degrade and will effectively persist permanently.

Incineration generates air pollution, with the amount of pollution dependent upon the investment in technologies to reduce or capture emitted metals and other hazardous materials before they leave the smokestack (NRC 1999). Where open burning is used, the simplest form of incineration, there is no opportunity for recapture of harmful materials in the smoke. Clearly, keeping the different types of wastes separate and reusing those materials is beneficial relative to either landfilling or incineration. However, most community recycling programs reduce the total volume of the domestic waste stream by only 30–50%.

The amount of household refuse is likely to increase with increasing affluence of persons, as indicated earlier in Figure 15.1.

There can also be differences in societal practices that would deviate from this trend. Another example of development-dependent use of materials is provided by the per capita use of paper in different countries. (See Figure 15.6).

In parts of the developing world, municipal waste collection is often very ineffective, and much of the wastes are dealt with by a network of urban wastepickers (Furedy 1990, 1994). In most cases, wastepicking is driven by poverty. Picking through wastes of more-affluent people provides access to resources of clothing, fuel, housing, and even jobs for the poorest. The waste stream can be mined for the raw materials to support small-scale industries, but the conditions for the waste pickers are usually very unhealthy (Yhdego 1991). There have been some efforts to organize wastepickers and improve waste collection efficiency, recycling, and conditions for the workers (Furedy 1992; LIFE 1995; Poerbo 1991).

15.7.1.2 Consumer-use Hazardous Materials

Consumers use many products that contain hazardous materials. A partial list of such products sorted into general categories includes the following:

- *Household cleaners*: bleach, ammonia, disinfectants, drain opener, furniture polish and wax, oven cleaner, spot remover
- *Laundry products*: laundry detergent, fabric softener, bleach, perchloroethylene
- *Lawn and garden products*: fertilizer, pesticides, herbicides, gasoline, oil
- *Home maintenance products*: paint, paint thinner, stains, varnish, adhesives, caulk
- *Pesticides*: insecticide, mothballs, pet spray and dip, rat and mouse poison, weed killer, disinfectant, flea collars, insect repellant
- *Health and beauty products*: hairspray, hair remover, fingernail polish, fingernail polish remover, hair coloring products, cosmetics, medications
- *Automotive products*: antifreeze, brake fluid, car wax and cleaners, gasoline, oil filters, transmission fluid, windshield washer fluid, lead-acid batteries, tires
- *Other*: charcoal briquettes, lighter fluid aerosol cans, art and craft materials, lighter fluid, pool chemicals, shoe polish, batteries, electronic components, light bulbs

Many of these products represent potential waste streams and potential impacts on the environment. In addition, some of them may have detrimental effects on sewage treatment systems, especially individual household septic systems. Consumer use may be a significant fraction of chemical use not usually associated with individual consumers. For example, nonagricultural uses of pesticides in the United States represents about 30% of total pesticide sales (Donaldson et al. 2002). At least one pesticide is used in 77% of U.S. households, with most households using multiple types of pesticides (Donaldson et al. 2002). Pesticides are used in homes for nuisance insect control, protection of structures, and (in some countries) for control of insect-borne diseases. The risks of pesticide use in the home are relatively large. Between 1981 an 1990, on average 20,000 pesticide exposures a year were reported in emergency rooms throughout the United States, with 82% of those reportedly due to exposure in the home (Blondell 1990).

Further, many of these products may end up in domestic wastes, as indicated. Where toxic materials are placed in landfills, they represent an additional hazard as they may leach into groundwater and render it unsafe for consumption.

15.7.2 Persistent Organic Pollutants

POPs are a category of waste of special concern because of their longevity and biological effects. One definition of POPs is pro-

Table 15.4. Assessment of the Likelihood of Ecosystems Receiving Different Types of Wastes and Contaminants

Contaminant	Dryland		Inland Waters		Coastal Waters		Marine	Mountains	Polar	Forest	Urban
	Industrial Country	Developing Country	Industrial Country	Developing Country	Industrial Country	Developing Country					
Airborne sulfur dioxide and sulfates	XXX	XXX	XXX	XXX	–	–	–	XXX	XXX	XXX	XX
Airborne oxides of nitrogen	XXX	XXX	XXX	XXX	XX	XX	X	X	–	X	XX
Polycyclic aromatic hydrocarbons	XX	X	XXX	XX	XXX	XXX	X	–	–	–	XX
Toxic metals, especially mercury, lead, and cadmium	XXX	XX	XXX	XXX	XXX	XXX	X	X	X	X	XX
Halogenated organic compounds, especially DDT and its metabolites, PCB, PCT, dieldrin, and short-chained halogenated aliphatic compounds	XXX	XXX	XXX	XXX	XX	XX	X	X	X	X	X
Petroleum hydrocarbons	XXX	XX	X	X	XXX	XXX	X	–	XX	X	X
Toxins of biological origin from algae, fungi, and bacteria	–	–	XXX	XXX	XX	XX	X	–	–	–	X
Eutrophication	–	–	XXX	XXX	XXX	XXX	X	–	–	–	X
Ammonia toxicity	–	XX	XXX	XXX	XX	XX	–	–	–	X	X
Organic loading	–	–	XXX	XXX	XXX	XXX	–	–	–	·	
Acid wastes	XX	XX	XX	XX	–	–	X	X	X		
Pathogens	X	X	XXX	XXX	X	XX	–	–	–	–	X
Selected indicators of water quality: biological/chemical oxygen demand, dissolved oxygen, pH, coliform bacteria	X	XXX	XX	XXX	XXX	XXX	X	–	–	–	–
Selected radionucleides	X	X	X	X	X	X	X	–	–	–	–
Solid salts of alkali metals and alkaline earth metals	XXX	XXX	XXX	XXX	–	–	–	–	–	–	–
Other substances that have caused significant local environmental problems in the past, such as arsenic, boron, elemental phosphorus, selenium, and fluoride	XXX	XX		XXX	XXX	XXX	–	–	–	–	–
Waste heat			XXX	XXX	XXX	XXX	–	–	–	–	–
Domestic refuse (mixed wastes)	XXX	XX	X	XX	X	X	X	–	–	–	XXX

Key:

xxx = highly probable

xx = moderately probable

x = somewhat probable

– = not likely or relevant

vided in the 1998 Aarhus Protocol on Persistent Organic Pollutants of the 1979 Geneva Convention on Long-Range Transboundary Air Pollution: "Persistent organic pollutants (POPs) are organic substances that: (i) possess toxic characteristics; (ii) are persistent; (iii) bioaccumulate; (iv) are prone to long-range transboundary atmospheric transport and deposition; and (v) are likely to cause significant adverse human health or environmental effects near to and distant from their sources."

Further production of these chemicals will be prohibited under international treaty. The Aarhus Protocol and the 2001 Stockholm Convention on Persistent Organic Pollutants ban the production of an initial list of POPs (12 under Stockholm and 16 under Aarhus), including pesticides, industrial chemicals, and industrial by-products, and include provisions for adding other chemicals to the restricted lists. The Aarhus Protocol has 36 signatories and went into force in October 2003. The Stockholm Convention has 151 signatories and entered into force in May 2004.

A common feature of POPs is that they are all heavily chlorinated compounds. Although there are numerous naturally produced chlorine and bromine compounds (Gribble 2003), they occur at low concentrations in ecosystems, and there are few naturally occurring enzymes and metabolic pathways efficient at breaking down these compounds. POPs have a low solubility in water and a high affinity

Table 15.5. Principal Drivers of Change Summarized from Individual Chapters, with a Provisional Assessment of the Implications for Detoxification

System	Driver of Change	Some Considerations for Detoxification
Cultivated systems	change in vegetation (seasonal contribution to inputs)	increase in soil loss with negative consequences for waste/contaminant attenuation due to loss of part of the resource
	change in exposure (erosion) temperature interruption in nutrient flow	the soil through erosion becomes a waste/contaminant in its own right causing accretion in and loss of wetlands; a change in the nutrient status of receiving water bodies due to P bound to soil and an increase in turbidity can reduce photosynthetic capacity of water, affecting nitrogen/phosphorus cycling
	changes in moisture regime (irrigation) removal of biomass	possible greater ranges in temperature that affect moisture content (increase in evaporation losses) and rates of particular transformation processes by influencing metabolic rates
	introduction of inorganic nutrients	cropping reduces return nutrient flows due to active removal of vegetation
	introduction of herbicides, pesticides	change in nature of nutrients from organically derived to inorganic due to fertilization
	salinization	direct addition of herbicides/pesticides to crops and transfer of these to other ecosystems either by aerosol or elution
	aquaculture in coastal areas floodplains (accreting systems) cultivation of floodplains	introduction of irrigation water can result in salinization of soil, if irrigation scheduling and drainage are not managed, etc.; irrigation return flows can result in an increase in salinity and nutrient flows into receiving waters
	change in soil structure	introduction of ponds and fish farming can result in increase in waste loads in affected waters; the ponds themselves may assist in internal transformation processes afforded by the extended hydraulic retention periods
		regulated flows due to damming can result in dessication and loss of floodplain functionality, with sediments trapped in dams (rather than on floodplains); should the trapped sediments be subjected to anoxic/anaerobic conditions, previously bound phosphorus, for example, could be released and contribute to eutrophication
		cultivation of floodplains can result in loss of changes in carbon supply and have implications for nutrient cycling and retention
		change in soil structure and properties, i.e., loss of carbon, changes in carbon exchange capacity, pH, etc. can have implications for contaminant removal
Dryland systems	changes in vegetation cover	soil erosion, loss of capacity to buffer due to there simply being less soil
	increased nutrient load (livestock) use of wetlands (water)	draining of wetlands for cultivation reduces opportunities for sediment trapping and, depending on site-specific circumstances, may actually contribute to sediment loss
	water retention (dams) change in exposure (erosion)	reduction in P removal, related to sediment loss and dentrification due to loss of anaerobic environments associated with wetland loss
	temperature interruption in nutrient flow	the storage of water in impoundments increases opportunity for contaminant degradation/immobilization due to increased opportunities for sedimentation and extended detention periods
	changes in moisture regime (irrigation) removal of biomass	the loss of floodplains to agriculture as a result of regular flows will reduce their value in terms of sediment accreting and nutrient trapping/release systems
	introduction of inorganic nutrients introduction of herbicides, pesticides	water retention–appearance of systems with symptoms of eutrophication is likely to increase as a consequence of nutrient retention, cycling, and internal generation (i.e., C fixation as algae)
	salinization	capacity of drylands to successfully reduce pesticides/herbicides is likely to be compromised by a reduction in area due to conversion to cropping and higher applied loads
	aquaculture in coastal areas floodplains (accreting systems)	the risks of salinization both as a result of changes in land use as well as increasing manufacturing and support activities will increase with a further deterioration of dryland systems
	cultivation of floodplains fire	the increase in aquaculture ponds in drylands in coastal areas will improve the capacity of the system to trap wastes/contaminants, but on the downside this could be to the detriment of people because of bioconcentration/transfer from sediment to harvestable biota
Inland waters	loss of wetlands (cropping)	see above for drylands
	loss of wetlands (irreparable alteration, i.e. reclamation for structures)	reduction in capacity to detoxify due to direct loss of habitat, increasing loads, and changes in nature of contaminants
	loss of wetlands water abstraction loss of water changes in land use (e.g., mining)	damming increases capacity to transform wastes, including nutrients, with consequences on water quality; detention times, days to weeks, can lead to eutrophication
	loss of wetlands through modification of hydrology (e.g., high peaks from urban areas, constrictions due to culverts, etc.)	
	abstraction for human use/urban and agriculture storage	
	deterioration due to waste loads	
	deterioration due to increased introduction of waterborne sanitation	
	increase in baseflows	
	salinization	

(continues over)

Table 15.5. *continued*

System	Driver of Change	Some Considerations for Detoxification
Coastal	interruption of coastal processes by stabilizing rivers direct loss of habitat (urban/harbor) change in structure (e.g., prawn ponds) deterioration due to waste loads deterioration due to increased introduction of waterborne sanitation increase in baseflows stabilization of flows transportation	see urban increasing nutrient loads, both from local as well as regional origin, may lead to impairment of water quality as the capacity of the system to transform the wasted could be compromised; this is already apparent in the presence of anoxic zones, for example, in the Gulf of Mexico
Marine	reduction in biomass transportation (increase in globalization and trade) dumping accidental spills	increased risk of point source waste load due to increase in global shipping trade increase in nutrient load overriding dilution effect
Forest	changes in land use changes in species composition changes in runoff characteristics changes in water quality (sediment) forest loss acidification fire	see cultivation, inland waters
Mountains	land use changes, deforestation, cropping systems industrial use deforestation, overgrazing, and inappropriate cropping practices lead to irreversible losses of soil and ecosystem function	as above
Polar	land use changes atmospheric composition changes infrastructure development, mining ecotourism	increase in waste loads with temperature being limiting factor limited capacity for microbial degradation due to low temperature
Urban	population expansion increase in impervious surfaces simplification of environments hydrololgy (rapid runoff, contained) nutrient load contaminant types loss of habitat due to direct transformation increase in demand for externally sourced resources, e.g., power, water, food, gas, petroleum waste sludge management	increase in nutrient and other wastes reduced "natural" capacity but potential to replace with engineered systems, constrained by finances and political will, different for industrial and developing countries value systems, immediate survival as opposed to high levels, e.g., aesthetics contaminant types and loads reflect socioeconomic status increased conversion of all other ecosystems, e.g., feedlots, energy provision, landfills, industrial development, to support change in lifestyles

for tissue, especially the fats in tissue, so they tend to accumulate in organisms through bioaccumulation from water or food. As these compounds are not metabolized and are only slowly excreted by mammals and birds, the concentrations of POPs tend to bioamplify, so that organisms at high trophic levels, including humans, are particularly susceptible to detrimental effects.

POP-based pesticides were widely used and are still applied. For example, about 2.6 million tons of DDT were used from 1950 to 1993, while the figure for toxaphene during the same period was 1.33 million tons (Voldner and Li 1995). The remarkable efficacy of this class of insecticides, especially in controlling insect-borne diseases such as malaria, coupled with the intensification of agricultural systems led to their production and wide use. However, use of most of these compounds was curtailed in

most countries in the 1970s after their toxicity was demonstrated. While primarily active on nerve conduction chemistry, these substances are acknowledged carcinogens and suspected teratogens, immunotoxins, and hormone disrupters (Guillette et al. 1994; Zahm and Ward 1998; Holladay and Smialowicz 2000; Solomon and Schettler 2000). In many parts of the world, DDT is still used for malaria control and chlordane is still the chemical of choice for termite control. A serious problem facing many developing and transition countries is the issue of stocks and reservoirs of obsolete, discarded, and banned POP pesticides and PCBs. There is an estimated 120,000 tons of obsolete stock of pesticides in Africa (FAO 2001; Tanabe 1988; Goldman and Tran 2002).

Polychlorinated biphenyls have been used mainly in electrical transformers, capacitors, hydraulic fluids, adhesives, plasticizers,

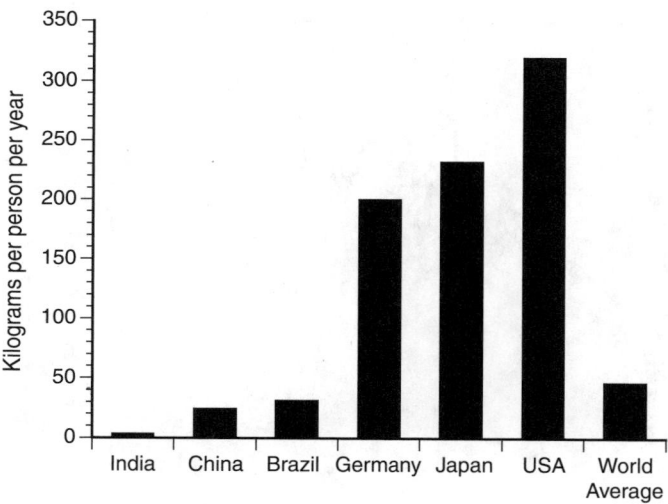

Figure 15.6. Paper Production per Capita in Selected Nations
(Brown et al. 1998)

heat transfer fluids, lubricants, cutting oils, and fire-retardants. Global production of PCBs to date has been estimated to be 1.68 million tons (AMAP 2000; UNEP/GEF 2003). PCBs enter the aquatic environment from industrial effluent or urban waste discharges (GESAMP 2001). Dioxins are unintentional products from combustion sources and manufacturing processes such as municipal solid waste incineration, energy production, motor vehicles, smelting, bleaching of paper pulp, and the manufacture of some pesticides and herbicides. Dioxins and the chlorinated furans are also produced naturally during volcano eruption and forest fires (Gullett and Touati 2003).

Volatilization is the principal means of wide dissipation of POPs in the environment. All POPs have a certain vapor pressure or a tendency for the chemical to become a gas and enter the atmosphere. Further, some POPs are initially emitted directly into the atmosphere. POPs may be dispersed over great distances in the air and can now be found anywhere on the globe. They have even reached the polar regions in large quantities through a process called global distillation, where these substances preferentially settle out in the colder air of the polar regions (Wania 2003 and see Chapter 25). POPs also have a strong affinity to associate with particulate matter. In aquatic systems, POPs partition or bind to particles and they sink to bottom sediments, which are their main environmental repository. These legacy POPs can seep back into the water and atmosphere many years or decades after emissions themselves have ceased.

15.7.3 Hydrocarbons

Petroleum hydrocarbons remain the dominant energy source in industrial and most developing countries (UNEP 1994), and yet they are associated with a range of environmental issues. The devastating effects on marine organisms of large oil spills from shipping, oil exploration, and production, for example, are well known (NRC 2003). Management of spilled oil in waters has largely been by attempts to contain and recover oil through use of booms and mechanical skimming and reprocessing of the oil recovered. Unfortunately, in large oil spills most is not recoverable. Unrecovered oil eventually decays in the environment due to microbial degradation, but different fractions of petroleum hydrocarbons are degraded at different rates. The lighter fractions can be degraded relatively rapidly, typically within days to weeks

in temperate conditions, except where they are mixed into sediments or soils where there is no oxygen (as described earlier). The heavier fractions are very persistent, requiring weeks to months to be degraded, even under conditions with sufficient oxygen. Hydrocarbons that get mixed into soils and sediments where there is little oxygen persist for years to decades.

One sub-group of hydrocarbons that can cause significant problems are the polycyclic aromatic hydrocarbons. The primary environmental source of these is a residue of combustion processes, though they are also found to some extent in petroleum oils. PAHs—both the parent compounds and alkylated homologues—are carcinogens (and mutagens and teratogens), and though subject to microbial degradation, they are the most environmentally persistent of the hydrocarbons. PAHs emitted into the atmosphere will deposit on land in dry and wet deposition and may accumulate in aquatic sediments through scavenging (GESAMP 2001).

Recent studies have established that land-based sources are the major input of hydrocarbons into the marine environment (GESAMP 2001). The major land-based sources identified are urban runoff, refinery effluents, municipal wastes, and used lubricating oils. Used lubricating oils are often contaminated with metals and high concentrations of PAHs, which makes them particularly hazardous. Pathways through which used oil get into surface water sources include oil dumped down drains that discharges into surface waters and oil poured on the ground and washed into groundwater or surface water.

15.7.4 Salinization

Salinization is a process resulting in an increasing concentration of salts in soils, water, or both. A number of activities can be linked to increases in the salt concentration of surface waters and soils. These include soil disturbance by activities that expose the soil and subsoil to weathering processes and subsequent leaching of salts the nature of which are determined by the composition of the parent material; the use of water for flushing human and industrial wastes; water recycling; water losses due to evaporation or evapotranspiration, such as in irrigation systems; discharge of saline groundwater; and changes in land cover, such as the replacement of trees with grasses, permitting already saline groundwaters to approach the surface due to a reduction in evapotranspiration by deep-rooted trees.

An example of the effects of changes in land use is shown in Figure 15.7, which illustrates increasing conductivity and sulfate concentrations (measures of salinity) in response to open cast (strip) mining in one of the principal coal mining areas in South Africa. The strip mining exposes the soil and subsoil profiles to weathering processes, resulting in the dissolution of minerals, in particular pyrite.

Salts do not degrade, and their concentrations can only be influenced by dilution through the use of sufficient water to move them to where they have less influence—in the ocean, for example. When water with even a low salt content is added continuously to an ecosystem, and when the water is allowed to evaporate or be lost through plant evapotranspiration, the salt remains and accumulates. For example, irrigation water with a salt content of 0.3 grams per liter applied at a rate of 10,000 cubic meters per hectare per year transfers 3,000 kilograms of salt per hectare per year into the soil (Oosterbaan 2003; see also Chapter 22).

Salinization of agricultural soils affects plant growth by restricting the uptake of water by the roots through their high osmotic pressures and by interfering with a balanced absorption of

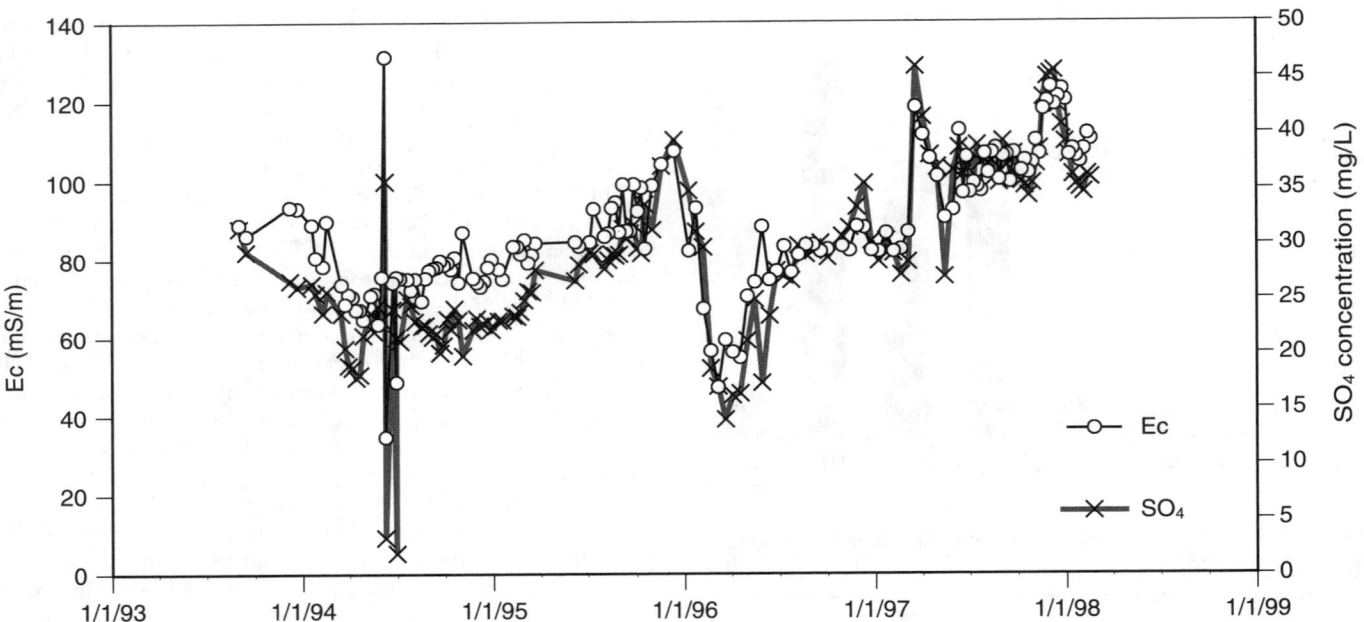

Figure 15.7. Salinization in the Olifant River, South Africa, 1993–98. Changes in electrical conductivity and sulphates measured in the Olifant River showing decreasing quality, attributed largely to coal mining activities. The Ec is a measure of the salt in the water. (S. African Dept. of Water Affairs and Forestry 1999 gauging station B3H017Q01)

essential nutritional ions by the plants. Different plants have different levels of sensitivity, so salinization can cause a shift in local plant communities.

These processes can have local economic consequences. A study on the effects of salinity changes in an irrigation area associated with the Vaal River in South Africa suggested that income of a specific group of farmers could be reduced by up to 84% as a result of crop changes and reductions in yields (Du Preez et al 2000; Viljoen and Armour 2002).

Estimates of the extent of salinization in the world run in the hundreds of millions of hectares, primarily in the more arid regions (Brinckman 1983; see also Chapter 22). The costs of salinization are high, with an estimate provided by the Australian Museum of AUS$500 million a year in Victoria alone (AM 2003).

The economic impact of salinization of surface waters runs beyond agriculture alone; Table 15.6 provides an example of the costs of increasing salinity on various sectors of society in a South African case study (Urban Econ 2000). The costs and factors considered for the identified sectors include:

- *in households*: maintenance costs (water heaters and plumbing), accelerated replacement costs, costs of soaps and detergents as well as the current anticipated salinity level;
- *in agriculture*: cost of water supply, maintenance costs, plant, chemicals;
- *in mining*: evaporative cooling circuits, service water circuits, metallurgical plant circuits, flotation circuits, electrolytic processes, irrigation water circuits, probabilistic distribution of management behavior, cost of water supply, cost of chemicals, cost of water treatment, cost of discharge, costs of scaling, salinity ratio;
- *in industry*: cost of water supply, maintenance costs, cost of chemicals, cost of plant replacement, current anticipated salinity, salinity threshold where expenditure patterns change; and
- *in water services*: cost of water supply, maintenance costs, chemical costs, plant replacement costs, current or anticipated salinity, salinity threshold where expenditure patterns change.

The greatest direct cost implications occur at the household sector, attributed to the fact that this constitutes the largest group of treated water users in the economy, compared with sectors that use predominantly untreated water.

With increasing pressure on surface waters, particularly in arid areas, there is likely to be an increase in the demand for water recycling and reuse (IWA 2004). This will undoubtedly lead to an increase in salinization of water, with associated consequences for downstream users.

15.7.5 Wetlands: Natural Defense Mechanism for Pollution Abatement

Depending on their type, wetlands can improve water quality, provide flood control, provide habitat for young of commercially valuable fish, provide habitat for many types of wildlife, help prevent erosion, and help reduce waterborne disease (see Chapters 7, 14, 19, and 20). Wetlands represent one of the major mechanisms to treat and detoxify a variety of waste products, and there have been many efforts to construct artificial wetlands to obtain these wetland functions.

15.7.5.1 Wetland Processes

One of the key functions that wetlands perform is to reduce concentrations of nitrogen in water. The close proximity of aerobic and anaerobic conditions often found in wetlands create a suitable environment for denitrification to take place. Denitrification converts nitrate, readily used by plants, to nitrogen gas, generally unavailable to plants, thereby reducing eutrophication (as described in the next section). Some wetlands have been found to reduce the concentration of nitrate by 90%, and artificially constructed wetlands have been developed specifically to treat nitrogen-rich sewage effluents.

Wetlands also act as a filter or trap for many waterborne wastes, including metals, organic chemicals, and pathogens. Metals and many organic compounds are adsorbed to the sediments

Table 15.6. Summarized Direct Costs of Salinity Changes on Selected Economic Sectors in South Africa, 1995. Negative values indicate no additional costs due to salinization. The percentages represent the increase in total operating costs due to salinization. For example, the net costs of mining were estimated to increase by 3.17% at 600 milligrams per liter. (Urban Econ 2000)

Sector	Salinity (mg/l TDS)						Contribution at 600 mg/l
	200	400	600	800	1,000	1,200	
	(milligrams per liter)						*(percent)*
Mining	−7.309	−2.212	0.844	4.863	10.209	17.816	3.17
Business and services	−1.843	0.487	1.211	1.707	2.209	2.697	4.55
Manufacturing 1	−0.145	0.028	0.086	0.123	0.160	0.198	0.32
Manufacturing 2	−2.825	0.294	1.351	1.993	2.635	3.278	5.07
Agriculture	0.000	0.000	0.439	0.439	0.427	0.503	1.65
Households (suburban)	−35.12	−11.71	11.70	35.12	58.53	81.95	43.94
Households (townships)	−27.93	−9.309	9.309	27.93	46.54	65.16	34.94
Households (informal)	−5.081	−1.694	1.694	5.081	8.469	11.85	6.36
Total	**−80.25**	**−24.11**	**26.64**	**77.25**	**129.22**	**183.46**	**100.00**

in the wetlands, and the relatively slow passage of water through many wetlands provides time for pathogens to lose their viability or be consumed by other organisms in the ecosystem. For easily degraded chemicals, their adsorption to sediments and slow passage through wetlands provides time for their degradation. However, for metals and persistent organic chemicals, wetlands become permanent traps, as these wastes either do not degrade or degrade very slowly. Many metals are held in sediments by precipitation with sulfides or as surface oxides (Barnes et al 1991).

Although metals and persistent organic chemicals can build up to high enough concentrations to have detrimental effects on the wetland functions (for example, the impairment of denitrification by metals (Slater and Capone 1984)), moderate waste loadings can generally be tolerated by wetlands without loss of services. Moderate loadings of plant nutrients lead to enrichment (analogous to fertilization of crops or lawns), while severe loadings will lead to a major loss in wetland productivity, structure, and function through eutrophication. Unfortunately, the threshold between where loadings are tolerated and where they will do damage to wetlands is not easily determined and depends on the specific conditions in each wetland.

15.7.5.2 Wetlands and Human Activities

The impact of human activities on wetlands has been drastic, and it is speculated that some 50% of world wetlands have been lost (see Chapter 20), with the greatest changes occurring in industrial countries in the first half of the twentieth century (BEST 2001). Wetlands have been drained for agricultural purposes or filled to create lands suitable for construction. Quite often wetlands have also been used as the end point of wastewater effluent, receiving large volumes of industrial, municipal, and agricultural wastewater. The impact of these effluents on the quality, density, and structure of living communities is substantial, as indicated by diversity indices (Patric 1976; Stevenson 1984). Wastewater effluent can significantly affect phytoplankton species diversity and community structure (Sullivan 1984; Gab Allah 2001), thereby reducing the wetland's capacity to detoxify wastes and to re-oxygenate water. The impairment of wetlands' ability to detoxify of wastes is affecting the quality of groundwater particularly, for example, in northeastern Egypt (Gab Allah 2001), where elevated levels of salinity and high concentrations of heavy metals in groundwater are reported.

15.7.6 Eutrophication

The addition of nitrogen or phosphorus (both essential, and often limiting, in plant growth) can have very undesirable effects on freshwater and marine systems (see Chapters 19 and 20). In freshwater systems, phosphorus is usually in shortest supply. Additions of phosphorus can stimulate large blooms of algal types usually not found in abundance. In some cases, dense filamentous algal mats form, altering the environment to the exclusion of other species and reducing biodiversity. The increased levels of algae sink to the bottom and are broken down by bacteria and other organisms. This decay of the plant material takes up oxygen from the water, and with the decay of enough plant material, the bottom water can become anoxic. The link between phosphorus additions and the undesirable effects on lakes and rivers has been clearly established through both direct experimentation with whole lakes (Schindler et al. 1971; Schindler 1973) and through cross-lake comparisons (Vollenweider 1976).

In coastal and marine systems, nitrogen is usually the limiting nutrient. Additions of any of the various forms of reactive nitrogen usually stimulate plant growth. Reactive nitrogen is nitrogen in the forms of nitrate, nitrite, ammonia, or organic nitrogen that is readily biologically available. Nitrogen gas (N_2 or di-nitrogen), which constitutes 79% of the atmosphere, cannot be used by most plant species. Figure 15.8 shows how the rate of plant production in several coastal marine ecosystems is very strongly influenced by the input of reactive nitrogen nutrients.

As with lakes, the increase in plant material, when settled to the bottom of a bay or estuary, can lead to loss of oxygen and the exclusion of all higher organisms, including the fish, clams, crabs, shrimp, and other valuable harvestable seafood. Eutrophication of coastal waters often stimulates microscopic forms of algae (phytoplankton), which in turn causes a decrease in light penetration to areas that would normally support seagrass beds, an environment that provides a valuable nursery for many desirable species. There is no simple relationship between N loading rates and effects, because the effects depend on the depth, circulation, temperatures, and other characteristics of each system.

Eutrophication in coastal waters has also been linked to the increased prevalence of large blooms of toxic phytoplankton, or red tides (see Chapter 19). The toxins in red tide species may be accumulated in marine organisms and cause significant toxicity to

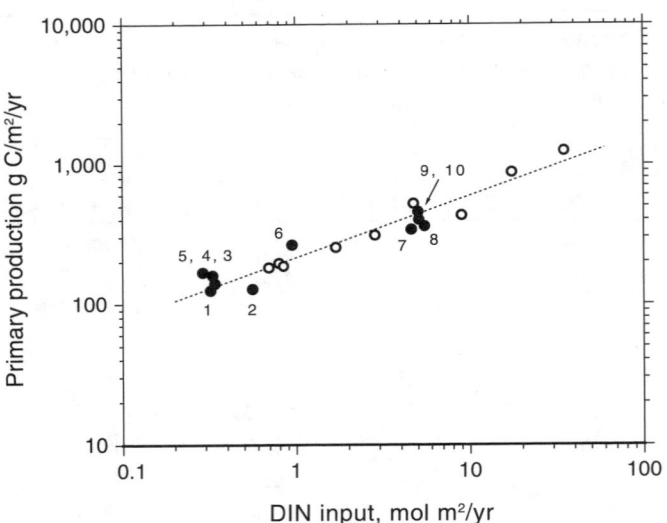

Figure 15.8. Marine Plant Production and Nitrogen Loadings.
The increase in phytoplankton production by increased dissolved inorganic nitrogen inputs. Open circles are data from the MEL Marine enclosures. Numbered systems are: 1. Scotian Shelf; 2. Sargasso Sea; 3. North Sea; 4. Baltic Sea; 5. N. Central Pacific; 6. Tomales Bay; 7. Continental Shelf, New York; 8. Outer-Southeast U.S. Continental Shelf; 9. Peru Upwelling; 10. Georges Bank. (Nixon and Pilson 1983)

humans. It should also be noted that some nuisance algal species are not correlated with eutrophication (for examples, see Keller and Rice 1989; Gobler and Sanudo-Wilhelmy 2001; Gobler et al. 2002).

The amount of fixed nitrogen and phosphorus cycling through ecosystems has been greatly increased by human activities (see Chapter 12). Phosphorus is usually mined from mineral deposits. A major source of nitrogen is an industrial process (the Haber process) that converts nitrogen gas from the atmosphere into ammonia. The vast majority of nitrogen fertilizers in use today are from this process. Reactive nitrogen is also generated during combustion processes in industrial, heating, and automotive combustion. Combustion sources produce enough reactive nitrogen in areas of high activity, such as Western Europe and the eastern United States, to elevate atmospheric concentrations of reactive nitrogen to up to 10 times natural concentrations.

The combination of synthetic nitrogen in agricultural fertilizers, nitrogen-rich human sewage inputs, and deposition of atmospheric reactive nitrogen to watersheds has raised the global average river concentrations of nitrogen to five times preindustrial levels, and rivers in heavily populated areas and industrialized areas have nitrogen concentrations some 25 times higher than natural levels (Meybeck 1982). The delivery of this nitrogen to coastal waters has, on a global average, doubled nitrogen concentrations in coastal waters, certainly one of the largest chemical alterations of Earth's ecosystems. The relative importance of the agricultural, combustion, and sewage inputs of nitrogen varies between different watersheds (Hinga et al. 1991), with all three sources being important in some systems. Management of nitrogen eutrophication may require addressing all three sources.

Low levels of eutrophication probably do not adversely affect aquatic ecosystems or impair human well-being. Indeed, as in agricultural systems, fertilization of coastal systems may lead to greater harvests of seafood (Nixon 2003). However, the very large

loadings of fertilizers that occur with intense agricultural and industrial activity and dense human population (especially in the coastal zone) have pushed many aquatic systems into conditions that exceed the capability of the system to adsorb the nutrients without detrimental effects. These systems are marked with anoxic conditions and the loss of biodiversity (including coral reefs) and harvestable species. (See Chapters 19 and 20.)

References

Alexander, M. 1994: *Biodegradation and Bioremediation.* Academic Press, San Diego.

Alzieu, C., 1990: Environmental problems caused by TBT in France: Assessment, Regulations. Paper presented at the *3rd International Organotin Symposium,* Monaco, 17–20 April, 1990.

AM (Australian Museum), 2003: Salinisation—One of our Biggest Environmental Problems. Available at http://www.amonline.net.au/factsheets/salinisation.htm.

AMAP (Arctic Monitoring and Assessment Programme), 2000: *PCB in the Russian Federation: Inventory and Proposals for Priority Remedial Actions.* 2003–12–03, Arctic Monitoring and Assessment Council, Center for International Projects, Moscow

Barnes, L.J., J. Janssen, J.H. Sherren, R.O. Versteegh, and O. Scheere, 1991: A new process for the removal of sulphate and heavy metals from contaminated water extracted by a geohydrological control system. Paper presented at the *Chem. E. Research Event. Research and Technology Symposium and Recruitment Fair.* Rugby Inst. Chem. Eng., Rugby, U.K. 33–36 pp.

BEST (Board on Environmental Studies and Toxicology), 2001: *Compensating for Wetland Loss under the Clean Water Act.* National Academy Press, Washington, D.C.

Boesch, D.F., 1996: Science and management in four U.S. coastal ecosystems dominated by land-ocean interactions. *Journal of Coastal Conservation,* **2,** 103–114.

Brinckman, R., 1983: *Land Reclamation and Water Management.* ILRI publication 27, International Institute for Land Reclamation and Improvement, Wageningen.

CDC (Centers for Disease Control and Prevention), 1995: Asthma—United States, 1982–1992. *Morbidity and Mortality Weekly Report,* **43,** 952–955.

Davidson, J., D. Meyers, and M. Chakraborty, 1992: *No Time to Waste—Poverty and the Global Environment.* Oxfam, Oxford, 217 pp.

deFur, P., M. Crane, C.G. Ingersoll, and L.J. Tattersfield (eds.), 1999: *Endocrine Disruption in Invertebrates: Endocronology, Testing, and Assessment.* Society of Environmental Toxicology and Chemistry, Pensacola.

Ditoro, D.M., J.D. Mahony, D.J. Hansen, K.J. Scott, A.R. Carlson, and G.T. Ankley, 1992: Acid volatile sulfide predicts the acute toxicity of cadmium and nickel in sediments. *Environ. Sci. Technol.,* **26,** 96–101.

Driscoll, C.T., G.B. Lawrence, A.J. Bulger, T.J. Butler, C.S. Cronan, et al. 2001: *Acid Rain Revisited: advances in understanding since the passage of the 1970 and 1990 Clean Air Act Amendments.* Science Links Publication Vol 1, no 1, Hubbard Brooks Research Foundation. Hanover, New Hampshire, USA.

Du Preez, C., M. Strydom, P. le Roux, J. Pretorius, L. van Rensburg, and A. Bennie, 2000: *Effect of water quality on irrigation farming along the lower Vaal River: The influence on soils and crops.* WRC Report No 740/1/00, Water Research Commission, Pretoria South Africa.

FAO (Food and Agriculture Organization), 2001: *Inventory of Obsolete, Unwanted and/or Banned Pesticide Stocks in Africa and the Near East.* Rome.

Faruqui, N.I., C.A. Scott, and L. Raschid-Sally, 2004: Confronting the realities of wastewater use in irrigated agriculture: Lessons learned and recommendations. In: *Wastewater Use in Irrigated Agriculture,* C.A. Scott, N.I. Faruqui, and L. Raschid-Sally (eds.), CABI Publishing, Wallingford, 208.

Gab Allah, M., 2001: Environmental impact of wastewater effluent on the phytoplankton communities in great Bitter Lake, Suez Canal, Egypt. *Az. J. Microbiology,* **53**(199–214).

Galloway, J.N., 2001: Acidification of the World: Natural and Anthropogenic. *Water, Air, & Soil Pollution,* **130**(1–4), 17–24.

GESAMP (Joint Group of Experts on the Scientific Aspects of Marine Environmental Protection), 2001: *Protecting Oceans from Land-Based Activities; Land-Based Sources and Activities Affecting the Quality and Uses of the Marine, Coastal and Associated Freshwater Environment.* Rep. Stud. GESAMP 71, 162 pp.

Gobler, C. and S. Sanudo-Wilhelmy, 2001: Effects of organic carbon, organic nitrogen, inorganic nutrients, and iron additions on the growth of phytoplankton and bacteria during a brown tide bloom. *Marine Ecology Progress Series,* **209,** 19–34.

Gobler, C.J., M.J. Renaghan, and N.J. Buck, 2002: Impacts of nutrients and grazing mortality on the abundance of *Aureococcus anophagefferens* during a New York brown tide bloom. *Limnology and Oceanography,* **47(1),** 129–141.

Goldman, L. and N. Tran, 2002: *Toxics and Poverty: The Impact of Toxic Substances on the Poor in Developing Countries.* The World Bank, Washington, DC.

Greer, L.E., 2000: Anatomy of a Successful Pollution Reduction Project. *Environmental Science and Technology,* **34(1),** 254A–261A.

Gribble, G.W., 2003: Diversity of naturally produced organohalogens. *Chemosphere,* **52,** 289–297.

Guillette, L.G., Jr., T.S. Gross, G.R. Masson, H.F. Matter, H.F. Percival, and A.R. Woodward, 1994: Developmental Abnormalities of the Gonad and Abnormal Sex Hormone Concentrations in Juvenile Alligators from Contaminated and Control Lakes in Florida. *Envrionmental Health Perspectives,* **102,** 680–688.

Gullett, B.K. and A. Touati, 2003: PCDD/F emissions from forest fire simulations. *Atmospheric Environment,* **37,** 803–813.

Harrison, P. and F. Pearce, 2001: *AAAS Atlas of Population and Environment.* American Association for the Advancement of Science, Washington, DC.

Heileman, L.I. and A. Siung-Chang, 1990: An analysis of fish kills in coastal and inland waters of Trinidad and Tobago, West Indies, 1976–1990. *Caribbean Marine Studies,* **1(2),** 126–136.

Hinga, K.R., A.A. Keller, and C.A. Oviatt, 1991: Atmospheric deposition and nitrogen inputs to coastal waters. *Ambio,* **20,** 256–260.

Holladay, S.D. and R.J. Smialowicz, 2000: Development of the Murine and Human Immune System: Differential Effects of Immunotoxicants Depend on Time of Exposure. *Environmental Health Perspectives,* **108 (Suppl. 3),** 463–473.

IWA (International Water Association), 2004: First International Conference on On-Site Waste Water Treatment and Recycling, February 12–14, Perth, Australia.

Jensen, A.A. and S.A. Slorach, 1991: *Chemical Contaminants in Human Milk.* CRC Press, Inc., Boca Raton.

Keller, A.A. and R.L. Rice, 1989: Effects of nutrient enrichment on natural populations of the brown tide phytopolankton *Aureococcus anophagefferens* (Chrysophyceae). *Journal of Phycology,* **25,** 636–646.

MacDonald, R., D. Mackay, and B. Hickie, 2002: Contaminant amplification in the environment. *Environmental Science and Technology,* **36,** 456a–462a.

MDL (MDL Information, Inc.), 2003: The Registry of Toxic Effects of Chemical Substances (RTECS). Available at http://www.cdc.gov/niosh/rtecs.html

Meybeck, M., 1982: Carbon, nitrogen and phosphorus transport by world rivers. *American Journal of Science,* **282,** 401–450.

Mondello, F.J., M.P. Turcich, J.H. Lobos, and B.D. Erickson, 1997: Identification and modification of biphenyl dioxygenase sequences that determine the specificity of polychlorinated biphenyl degradation. *Appl. Environ. Microbiol.,* **63,** 3096–3101.

Nixon, S.W., 2003: Replacing the Nile: Are anthropogenic nutrients providing the fertility once brought to the Mediterranean by a great river? *Ambio,* **32(30–39).**

Nixon, S.W. and M.E.Q. Pilson, 1983: Nitrogen in estuarine and coastal marine ecosystems. In: *Nitrogen in the Marine Environment,* E.J. Carpenter and D.G. Capone (eds.), Academic Press, New York.

NOAA/OCRM (National Oceanographic and Atmospheric Administration, Office of Ocean and Coastal Resource Management), 2003: Polluted Runoff. 2003. Available at http://www.ocrm.nos.noaa.gov/pcd/6217.html.

NRC (National Research Council), 2003: Oil in the Sea III, Inputs, Fates, and Effects, Washington, D.C.

NTP (National Toxicology Program), 2001: *National Toxicology Program's Report of the Endocrine Disruptors Low-Dose Peer Review.* U.S. Department of Health and Human Services, National Institute of Environmental Health Services, National Institutes of Health, Research Triangle Park, August, 487 pp.

O'Neil, M.J., A. Smith, P.E. Heckleman, J.R. Obenchain, J.A.R. Gallipeau, and M.A. D'Arecca (eds.), 2003: *Merck Index: An Encyclopedia of Chemicals, Drugs, & Biologicals, 13th Edition.* Merck Publishing Group, Whitehouse Station, New Jersey.

Oosterbaan, R.J., 2003: Soil Salinity. Available at http://www.waterlog.info/salinity.pdf.

Patric, 1976: The effect of invasion rate, species pool and size of area on the structure of the diatom community. *Proceedings of the National Academy of Sciences Philadelphia,* **58(1335–1342).**

Peterson, C.H., S.D. Rice, J.W. Short, D. Esler, J.L. Bodkin, B.E. Ballachey, and D.B. Irons, 2003: Long-term ecosystem response to the Exxon Valdez oil spill. *Science,* **302(5653),** 2082–2086.

Romieu, I., T. Carreon, L. Lopez, C. Rios, and M. Hernandez-Avila, 1995: Environmental urban lead exposure and blood levels in children of Mexico City. *Environmental Health Perspectives,* **103,** 1036–1040.

Romieu, I., E. Palazuelos, M. Hernandez-Avila, C. Rios, I. Munoz, and C. Jimenez, 1994: Sorces of lead exposure in Mexico City. *Environmental Health Perspectives,* **102,** 384–389.

Schecter, A., L.C. Dai, L.T. Thuy, H.T. Quynh, D.Q. Minh, H.D. Cau, P.H. Phiet, N.T. Nguyen, J.D. Constable, and R. Baughman, 1995: Agent orange and the Vietnamese: The persistence of elevated dioxin levels in human tissues. *American Journal of Public Health,* **85,** 516–522.

Slater, J. and D.G. Capone, 1984: Effects of metals on nitrogen fixation and denitrification in slurries of anoxic saltmarsh sediment. *Marine Ecology Progress Series,* **18,** 89–95.

Solomon, G.M. and T. Schettler, 2000: Endocrine disruption and potential human health implications. *Canadian Medical Association Journal,* **163,** 1471–1476.

Stevenson, R.J., 1984: Epilithic and epipelic diatoms in the Sanduskie River, with emphasis on species diversity and water pollution. *Hydrobiologia,* **114,** 161–176.

Sullivan, R.J., 1984: Mathematical expression of diatom results: Are these pollution indicies valad and useful? In: *8th International Diatom Symposium,* M. Ricard (ed.), Koeltz Scientific Books, Koenigstein, 775–779.

Tanabe, S., 1988: PCB problems in the future: foresight from current knowledge. *Environmental Pollution,* **50,** 5–28.

UNEP (United Nations Environment Programme), 1994: *Environmental Data Report 1993–1994,* Oxford, UK.

UNEP, 2000: *Global Environment Outlook.* Nairobi, Kenya.

UNEP, 2004: *Geo Yearbook 2003.* Nairobi, Kenya.

UNEP/GEF (Global Environment Facility), 2003: *Regionally Based Assessment of Persistent Toxic Substances Global Report.* Nairobi, Kenya.

Urban Econ, 2000: The economic cost effects of salinity: integrated report. WRC Report No: TT 123/00, Water Research Commission, Pretoria, South Africa.

U.S. EPA (Environmental Protection Agency), 1998: *Progress Toward Implementing Superfund, fiscal Year 1998, Report to Congress.* EPA/540/R/02/001, Washington, DC.

Viljoen, M. and R. Armour, 2002: *The economic impact of changing water quality on irrigated agriculture in the lower Vaal and Riet Rivers.* WRC Report No 947/1/02, Water Research Commission, Pretoria South Africa.

Voldner, E.C. and Y.F. Li, 1995: Global usage of selected persistent organochlorines. *Science of the total Environment,* **160/161,** 201–210.

Vollenweider, R.A., 1976: Advances in defining critical loading levels for phosphorus in lake eutrophication. *Memorie dell'Instituto di Idrobiologia,* **33,** 53–83.

Wania, F., 2003: Assessing the potential of persistent organic chemicals for long-range transport and accumulation in polar regions. *Environmental Science and Technology,* **37,** 1344–1351.

Wania, F., and D. Mackay 1995: A global distribution model for persistent organic chemicals. Science of the Total Environment 160/161: 211–232.

WHO (World Health Organization), 1992: *Environmental Health Criteria 135: Cadmium-Environmental Aspects.* Geneva, Switzerland. 156 pp.

WHO, 2003: General information related to microbiological risks in food. [Online] 2003. Available at http://www.who.int/foodsafety/micro/general/en.

WRI (World Resources Institute), 2003: EarthTrends: the Environmental Information Portal. Cited 12/09/2003. Available at http://earthtrends.wri.org/text/index.htm.

Yang, R.S. 1994: *Toxicology of Chemical Mixtures.* Academic press, San Diego.

Yuan, Y., 1993: Etiological study of high stomach cancer incidence among residents in wastewater irrigation areas. *Environmental Protection Science,* **19(10),** 70–73.

Zahm, S.H. and M.H. Ward, 1998: Pesticides and Childhood Cancer. *Environmental Health Perspectives,* **106 (Suppl. 3),** 893–908.

Chapter 16

Regulation of Natural Hazards: Floods and Fires

Coordinating Lead Author: Lelys Bravo de Guenni
Lead Authors: Manoel Cardoso, Johan Goldammer, George Hurtt, Luis Jose Mata
Contributing Authors: Kristie Ebi, Jo House, Juan Valdes
Review Editor: Richard Norgaard

---

*This appears in Appendix A at the end of this volume.

Main Messages

Floods and fires are natural processes occurring in the biosphere that have affected the planet for millennia. As a result of human development and the growth of public and private wealth, these events are having increasing impacts in term of costs and benefits for ecological and human systems.

It is clear that extreme events result in differentiated impacts across ecological and human systems, including impacts on human well-being (such as economic condition, health, happiness, and so on). Different individuals and groups experience different costs and benefits at various spatial and temporal scales.

Under certain circumstances, ecosystem conditions may serve to alleviate the impacts of an extreme event on human systems. For example, the economic impacts of a particular flood event on a community depend largely on the amount of unoccupied floodplains available for flood waters.

Ecosystem conditions may increase or reduce both costs and benefits at various temporal and spatial scales. Examples of these services include sediment and nutrient deposition in floodplains and deltas and natural fire regimes that help to sustain and rejuvenate ecosystems through periodic burning.

Ecosystem conditions that serve to modulate the impacts of extreme events on human well-being have experienced various changes as a result of a variety of drivers that affect ecosystems. The relationships between drivers and their impacts on ecosystem services and extreme events on human well-being are varied and highly complex and are not well understood.

Our knowledge of how ecosystems ameliorate or accentuate the impacts of extreme events on human well-being is limited for a variety of reasons. First, in many parts of the world understanding of the frequency and magnitude of extreme events is not well developed. Second, the collection of trend data on the impacts of extreme events on human well-being has been uneven and the data are often of poor quality. Third, in those cases where good trend data exist they are typically focused on first-order economic impacts and ignore other measures related to human well-being. Fourth, few rigorous quantitative studies have been performed that properly analyze the various factors that explain temporal and spatial patterns of extreme events or trends in impacts. Fifth, very few studies have sought to identify and quantify the services provided by ecosystems in the context of regulation of extreme events such as floods and fires.

Available studies on extreme events, their impacts on human well-being, and the roles of ecosystem services do, however, allow several qualitative assertions to be made:

- Humans are increasingly occupying regions and localities that are exposed to extreme events, establishing settlements, for example, on coasts and floodplains, close to fuelwood plantations, and so on. These actions are exacerbating human vulnerability to extreme events. Many measures of human vulnerability show a general increase, due to growing poverty, mainly in developing countries.

- Impacts of natural hazards are increasing in many regions around the world. Annual economic losses from extreme events increased tenfold from the 1950s to 1990s. From 1992 to 2001, floods were the most frequent natural disaster (43% of the 2,257 disasters), and floods killed 96,507 people and affected more than 1.2 billion people over the decade. A large number of damaging floods occurred in Europe in the last decade.

Material flood damage recorded in Europe in 2002 was higher than in any previous year.

- Human vulnerability is usually the primary factor that explains trends in impacts, and, in general, it overwhelms any positive effect on human well-being.

- Interactions of modern human activities with ecosystems have contributed to increasing human vulnerability and to the impact of extreme events on human well-being.

- Appropriate management of ecosystems can be an important tool to reduce vulnerability and can contribute to the reduction of negative impacts of extreme events on human well-being.

16.1 Introduction

The impacts of natural disasters have changed dramatically over the past 30 years. While the number of people reported affected by natural disasters increased from just over 700 million in the 1970s to nearly 2 billion in the 1990s, deaths from natural disasters fell from nearly 2 million in the 1970s to just under 800,000 in the 1990s. The reasons behind these statistics are complex and need further analysis. However, the drop in fatalities can be attributed in part to better disaster preparedness, although most deaths occur in developing countries, where disaster preparedness measures are less well developed.

This chapter examines the role of ecosystems in the context of the impacts of floods and fires on human systems. Floods and fires are considered natural hazards—that is, natural processes or phenomena occurring in the biosphere that may become damaging for human as well as for natural systems. The outcome of a natural hazard becomes a natural disaster as the result of the interaction of human or ecosystem vulnerability and the extent and severity of the damage to the human group or ecosystem receiving it. Many other natural hazards exist that could have been included in the assessment, such as droughts, tropical cyclones, volcanic eruptions, and earthquakes, for example. Only floods and fires are considered here, however, because they are most strongly subject to feedback processes and most directly influenced by human activities such as urbanization and environmental degradation. Deforestation, for example, has a direct effect on the incidence and magnitude of flood events.

This chapter focuses on the roles that ecosystems play in modulating the effects of extreme events, and in particular in protecting human well-being from the impacts of floods and fire. Protection can be defined as contributing to the prevention of harm as well as to the receipt of benefits. It is well understood that extreme events result in differentiated impacts to human well-being, with resulting costs and benefits to people at different spatial and temporal scales (Pielke Jr. 2000a, 2000b). For example, the tropical cyclone that devastates a neighborhood is a disaster to those who lose their homes but also an opportunity for aquifer recharge, especially after a long drought period. In the context of flooding, the impacts of a particular event may be reduced by the presence of unoccupied floodplains that allow flood waters to pass through unpopulated areas. Benefits from flooding may occur through the transport of sediments and nutrients to the coastal zone, although the consequences of this are often negative.

It is of course important to recognize that natural events do not always bring benefits to human well-being; some may result in harms or costs. For example, flood conditions may foster the spread of disease or disease vectors harmful to humans, such as

West Nile virus or hantavirus. Thus, understanding valuation of ecosystem services in the context of extreme events is complicated due to the interacting benefits and costs. Logically, if ecosystems play an intervening role between extreme events and human well-being, then ecosystem services must also provide differentiated impacts on humans.

Understanding the role of ecosystem services in protecting human well-being is further complicated by the diverse array of factors that contribute to protection of and harm to human well-being in the context of extreme events. Such impacts result from the effects of the extreme event itself as well as the vulnerability of human systems. Many of these effects can be considered independent of the intervening role of ecosystems, such as when a powerful tropical cyclone strikes a highly urbanized area. In this case, the economic losses are unavoidable, although human losses can be prevented or minimized.

Ecosystem conditions and their services can play a role in modulating both the event and the human systems that create conditions of vulnerability. This is also true for natural systems. As an example, a perennial grass from a Mediterranean ecosystem might become more vulnerable to extreme wildfires and rainfall events under severe erosion and land degradation conditions (De Luis et al. 2004). In the case of flooding, local or regional ecosystem conditions, such as increased deforestation, may contribute to the magnitude or scope of particular flooding events, setting the stage for increased vulnerability. Human vulnerability is conditioned by the characteristics of local ecosystems, social systems, and human modifications to them.

Figure 16.1 depicts in a highly simplified manner the various inter-relationships of extreme events, human and ecological systems, and the resulting effects (both benefits and costs) to human well-being. Highlighted in red is the role of ecosystem services in contributing to benefits and costs related to human well-being in the context of extreme events. The arrows represent the differentiated impacts of benefits and costs for both ecosystems and human systems. These impacts are modulated by the ecosystem regulating services.

16.1.1 Ecosystem Services in the Context of Regulation of Extreme Events

This section describes the mechanisms by which ecosystems regulate floods and fires, producing benefits to both ecosystems and humans. Table 16.1 describes the mechanisms by which main ecosystems regulate natural hazards.

These mechanisms the role of natural hazards in preserving biodiversity patterns and key biophysical processes in the bio-

sphere (Wohlgemuth et al. 2002). These natural phenomena play an important role in the natural cycle of matter and energy. Fires, for example, are part of the natural behavior of the biosphere, and floods are an efficient mechanism for natural transport of dissolved or suspended material (Pielke Jr. 2000b).

In the case of humans, ecosystem regulation can provide protection from the adverse consequences of natural hazards to human well-being.

16.1.1.1 The Role of Wetlands, Floodplains, and Coastal Ecosystems in the Regulation of Floods

The preservation of natural areas is important for flood attenuation. For example, some natural soils (not affected by human activities) have a large capacity to store water, facilitate transfer of groundwater, and prevent or reduce flooding. The capacity to hold water is dependent on soil texture (size of soil particles and spaces between them) and soil structure (nature and origin of ag-

Table 16.1. Key Role of Ecosystems in Regulating Extreme Events

Ecosystem	Role in Flood Regulation	Role in Fire Regulation
Cultivated	crop cover provides flood protection, conditioned on good management	part of the management of some cropping systems, e.g., sugar cane, timber, etc.
Dryland	protection through vegetation cover; recharge of aquifers	biodiversity issues: adaptation mechanisms to fire
Forest	protection from floods providing flood attenuation and soil loss prevention	part of the natural system; reducing wood fuel accumulation; biodiversity issues
Urban	move people away from flood-prone areas, conditioned on good urban planning	move people away from natural fire-prone areas; scale benefits from more effective fire prevention and control
Inland Waters	provide mechanisms for flood attenuation potential (wetlands, lakes, etc.)	wildfires control, e.g., pit fires control by wetlands
Coastal	benefits from sediment transport to the coastal zone; flood protection provided by coastal ecosystems (barrier beaches, mangroves, etc.)	not applicable
Marine	benefits from nutrient transport to the oceans	not applicable
Polar	discharge regulation to oceans in the Arctic system (freshwater provision to Arctic oceans)	not applicable
Mountains	regulating flood-related events (slope stability)	main source of wood fuel
Islands	benefits from sediment transport to oceans through floods from the mainland; aquifer recharge as main source of fresh water	not applicable

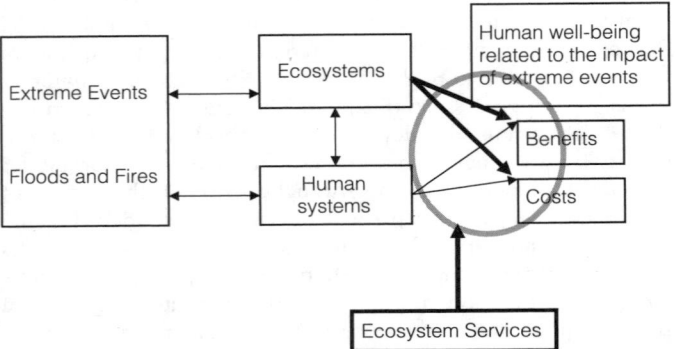

Figure 16.1. Inter-relationships between Extreme Events, Ecosystems, and Social Systems

gregates and pores). For instance, clay soils have a larger capacity to hold water than sandy soils due to pore size.

Other intact areas, such as barrier beaches, offer natural flood protection. Important inland water components, such as wetlands and lakes, are key agents of flood attenuation through energy dissipation of runoff peaks. Gosselink et al. (1981) determined that the forested riparian wetlands adjacent to the Mississippi in the United States during pre-settlement times had the capacity to store about 60 days of river discharge. With the subsequent removal of wetlands through canalization, leveeing, and drainage, the remaining wetlands have a current storage capacity of less than 12 days discharge—an 80% reduction of flood storage capacity. The extensive loss of these wetlands was an important factor contributing to the severity and damage of the 1993 flood in the Mississippi Basin, the most severe flooding in recent U.S. history (Daily et al. 1997). Similarly, the floodplain of the Bassee River in France performs a natural service by providing an overflow area when the Seine River floods upstream of Paris. A valuation analysis that highlights the economic need to conserve this natural environment is presented by Laurans (2001).

Wetlands, however, should not be viewed as single units; when examined collectively on a catchment scale, they are all interlinked. Although an isolated wetland may perform a significant flood control function, effective control is more often the result of the combined effect of a series of wetlands within a particular catchment area (Verry and Boelter 1978). Thus, when considered singly, upstream wetlands may appear to have an insignificant effect on flood attenuation. Although upstream wetlands are often numerous, and their cumulative effect may be considerable, most flood control benefits of wetlands are derived from floodplain wetlands (Bullock and Acreman 2003).

Coastal areas, including coastal barrier islands, coastal wetlands, coastal rivers floodplains, and coastal vegetation, all play an important role in reducing the impacts of floodwaters produced by coastal storm events and have been characterized by Leatherman et al. (1995) as a "movable boundary" that continuously responds to fluvial and deltaic fluxes.

Storm regulation in the coastal zone can be achieved by preserving natural buffers such as coral reefs, mangrove forests, sandbars, and so on. When such buffers have been destroyed, management agencies will sometimes try to restore damaged habitats or construct artificial wetlands.

16.1.1.2 *The Role of Ecosystems in Regulating Fires*

Fire regulation is defined here as the capacity of ecosystems to maintain natural fire frequency and intensity. Fires occur due to a combination of fuel, flammability, ignition, and spread conditions, which are generally related to climate and land cover. For example, precipitation is closely related to fuel load and flammability. Persistent high precipitation allows for increases in fuel by favoring vegetation growth, but leads to conditions that may be too wet to allow fires. Persistent low precipitation, on the other hand, limits plant growth and fuel supply. High temperature and wind speed generally enhance flammability and fire spread. High flammability combined with high fuel load increase the likelihood of intense fires. Temperate regions suffer from catastrophic fires after periods of fuel buildup, whereas tropical forests can store large amounts of fuel, but humid conditions naturally prevent flammability.

In addition to climate and land cover, factors related to economy, education, and technology are also important for fire regulation. For example, economic and educational conditions that determine the use of agricultural techniques other than fires may

reduce accidental fires (Nepstad et al. 1999a). Fire regulation can also benefit from investments in systems for monitoring fire risk and activity (Prins et al. 1998) and from disaster-relief mechanisms in cases of catastrophic fires (FEMA 2004). In these cases, one of the most important tools for regulating fires is the use of satellite data.

Remote sensing data can provide fire occurrence information, help improve preparedness, and aid decision-making on fire regulation (Dwyer et al. 2000; Justice et al. 2003; Grégoire et al. 2003). These data are usually reported as "fire pixels" on satellite maps where active burning is highlighted. Several satellite-based systems are in use to monitor fire activity (e.g., Setzer and Malingreau 1996; Prins et al. 1998; Giglio et al. 2000; Justice et al. 2002). For example, the Moderate Resolution Imaging Spectroradiometer (MODIS) Fire Products reports fire activity globally at spatial resolutions as small as 1 kilometer square and temporal resolutions of up to four times per day. (See Figure 16.2 in Appendix A.)

Satellites can also provide information on recent patterns of the area affected by fires, based on comparisons between repeated images. For example, GBA2000 (Grégoire et al. 2003) and GLOBSCAR (Global Burn Scars) (Kempeneers et al. 2002) are inventories of the area burned worldwide in 2000.

When analyzing fire patterns from remote sensing, it is important to consider that some factors can interfere with fire detection. For example, fires occurring at different times to the detection times (Ichoku et al. 2003) or under clouds or forest canopies may not be detected and hence under-recorded (Setzer and Malingreau 1996); exposed soils and solar reflections may be misinterpreted as fires (Giglio et al. 1999); and low spatial resolution can impair the identification of small burned areas (Simon 2002).

Thus to properly interpret satellite fire patterns and improve detection algorithms, comparisons between satellite- and field-based fire data (ground-truthing) are essential. While more studies are needed, there are indications that satellites tend to report fewer active fires than actually occur on the ground and that frequent (sub-daily) temporal sampling is as important as high spatial resolution.

16.1.2 Hazard Regulation and Biodiversity

It is increasingly accepted that a large pool of species is required to sustain the structure and functioning of ecosystems, especially in landscapes with intensive land use. In this context, a group of related hypotheses (Loreau et al. 2001) proposes that biodiversity may act as an insurance or buffer against changing environmental conditions. However, the suggestion that biodiversity controls ecosystem processes, including stability, needs to be analyzed in a much wider array of conditions, since not all ecosystems respond equally to environmental disturbance and since some perturbations may be better buffered by biodiversity than others (Roy 2001).

In the face of major natural disturbances such as fire and flood and an increased probability of their occurrence due to global change, the buffering capacity attributed to biodiversity may play a key factor in ecosystem recovery. The existence of a minimum species richness could help ensure the presence of species or functional groups with dominant roles in biomass production and regulation of nutrient fluxes. (See also Chapters 11 and 12.) A number of observations suggest that after a disturbance or when environmental conditions change, biodiversity richness at larger spatial scales, such as landscapes and regions, can ensure that appropriate key species for ecosystem functioning are recruited to local systems. (See also Chapter 4.)

The importance of natural disturbance processes in the maintenance of landscape diversity has been documented for many protected areas of the world. An example of this is the effect of fires in Yellowstone National Park. In 1988, fires burned a total of 45% (400,000 ha) of the park (Christensen et al. 1989). Reconstructions of the history of fires in Yellowstone suggest that the last time a fire of this magnitude occurred was in the early 1600s. Although plant species richness was still increasing at the scale of sampling plots (less than 10 square meters) five years after the 1988 fires, overall species richness at the landscape level had been unaffected by fires. However, relative abundance varied among species (Turner et al. 1997).

Large-scale disturbance are very important in structuring ecosystems. Changes in this disturbance regime—for example, as a result of climate change—could have substantial implications from the ecological point of view (Turner et al. 1997).

The role of biological diversity in providing insurance, flexibility, and risk distribution across various scales within the context of ecosystem resilience and under conditions of uncertainty is discussed in Chapter 4.

16.2 Magnitude, Distribution, and Changes in the Regulation of Natural Hazards

This section presents the ecosystem components and mechanisms that provide regulating services for floods and fires and their magnitude, distribution, and variability. First we describe key ecosystem components providing regulation of floods and fires at a global scale. Then we present evidence of the historical changes that have occurred in the capacity of ecosystems to provide flood and fire regulation. Changes in the occurrence and magnitude of these natural hazards and their impacts on human well-being provide an indirect measure of the ability of ecosystem conditions to provide regulation services. Unfortunately, data on the direct relationship between the magnitude (frequency and intensity) of natural hazards and changes in the ecosystem conditions as a direct consequence of anthropogenic actions are practically non-existent. Therefore, incidence and trends in natural hazard impacts on ecological and social systems are used as proxy measures of the regulation capacity of ecosystems.

16.2.1 Ecosystem Conditions That Provide Regulatory Services

16.2.1.1 Flood Regulation

Wetlands, floodplains, lakes, and reservoirs are the main providers of flood attenuation potential in the inland water system. Geographic distribution of wetlands is described in different databases (such as Matthews from GISS, University of Kassel, and Ramsar). The University of Kassel in Germany has developed a map of the global distribution of wetlands, lakes, and reservoirs. (See Figure 20.1 in Chapter 20.) Wetlands cover an estimated 6.6% of the global land area (excluding Antarctica and Greenland), and lakes and reservoirs alone cover 2.1% (Lehner and Döll 2004).

Flood attenuation potential can be estimated by the "residence time" of rivers, reservoirs, and soils. (See Figure 16.3 in Appendix A.) Residence time is defined as the time taken for water falling as precipitation to pass through a system. The longer the residence time, the larger the buffering capacity to attenuate peak flood events. Larger rivers, such as the Congo and the Amazon, have a greater attenuation capacity than smaller rivers. Data for populations living in different zones of water residence time are presented in Figure 16.4. Nearly 2 billion people live in areas

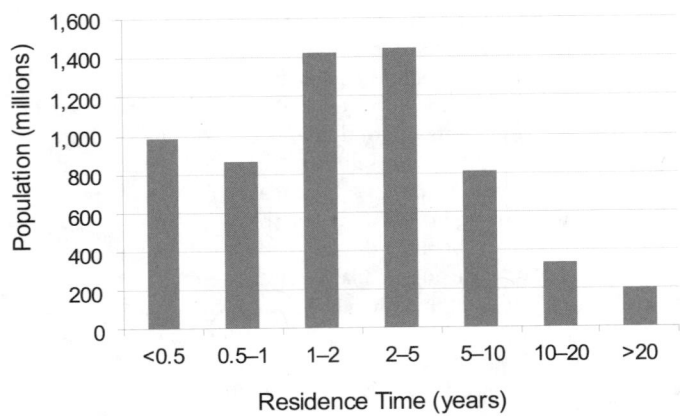

Figure 16.4. Population by Basin for Different Residence Time in Lakes, Reservoirs, Rivers, and Soils (WSAG, University of New Hampshire 2004)

with a residence time of one year or less and are thus located in areas of high flood risk with low attenuation potential. Most of these people live in northern South America, highly populated regions of northern India and South East Asia, Central Europe, and the Southwest coast of Africa.

16.2.1.2 Fire Regulation

Several ecosystem conditions are related to fire regulation—principally the amount of vegetation, and therefore fuel, in the system. Climate, land cover, and land use are important factors linked to these conditions. Climate variability is a dominant factor affecting large wildfires in the western United States, although current fire management practices focus on reducing fuel availability (McKenzie et al. 2004). Regions with climates that present distinct dry and wet seasons have potentially more fires. Under these climates, vegetation growth during the wet season can increase fuel load, and flammable conditions during the dry season can lead to more frequent and intense fires. In addition to seasonality in precipitation, temperature and wind speed are important. Hot conditions and intense winds may promote high fire frequency and intensity by favoring fuel dryness and fire spread.

Land cover and land use are important because they can affect fuel load, flammability, number of ignition events, and spread conditions. For example, forests with deep root systems may take longer to become flammable during dry periods than vegetation with shallow roots. Soils that have low water-holding capacity may lead to flammable conditions after short dry periods. Fire suppression reduces sources of ignition but may favor fuel buildup and the likelihood of more intense fires when they do occur. Land use practices such as pasture maintenance can lead to higher number of trigger events, but the use of firebreaks can help to lower the spread of fire.

16.2.2 Impacts of Ecosystem Changes on Underlying Capacity to Provide a Regulating Service

Direct quantification of the underlying capacity of ecosystems to regulate floods and fires is difficult, and few studies have sought to identify and quantify changes in these regulating services. This section looks at the direct consequences of a reduced capacity to regulate floods and fires, therefore, and presents evidence for changes in flood incidence and flood-related damages as well as changes in fire activity.

16.2.2.1 Evidence for Changes in Flood Incidence and Flood-related Damages

A damaging flood is defined as a flood in which individuals or societies suffer losses related to the event. In almost all cases a damaging flood results from a combination of physical and societal processes (Pielke Jr. 2000b). All types of floods, such as riverine or coastal floods, sudden snow melt floods, or floods after heavy, intense rainfall, have become more destructive in recent years (IFRC 2001). Moreover, projections show that this tendency is likely to become even more pronounced in the future. From 1950 to 1990, annual economic losses from extremes events increased steadily (Swiss Re 2003), largely due to several indirect socioeconomic drivers, including population increase and accumulation of wealth in vulnerable areas, and to some extent direct drivers, such as climate and climate change.

Hydrological variables (precipitation and stream flow, for example) show strong spatial and temporal variability. Occasionally, they take on extremely high values (heavy precipitation occurs, perhaps), with substantial impacts on ecosystems and human society. Thus floods have been a major issue since the beginning of civilization (floods of the Tigris and Euphrates were documented in ancient Babylonian and Sumerian texts, for instance) and continue to be so.

According to data compiled by the International Red Cross (IFRC 2001), floods account for over two thirds of the average of 211 million people a year affected by natural disasters. Every year extreme weather and climate events cause significant morbidity and mortality worldwide. From 1992 to 2001, floods were the most frequent natural disaster (43% of the 2,257 recorded disasters), killing 96,507 people and affecting more than 1.2 billion people (OFDA/CRED 2002). In the Americas, floods accounted for 45% of all deaths from disasters (IFRC 2003). Table 16.2 presents the total number of deaths due to floods and wild fires, by continent, during 1990–99. Asia is the most affected continent in terms of human-related flood losses.

In 2003, flood events did not occur to the same extent as they had in (northern) summer 2002; rather, heat waves dominated, especially in Europe. However, a Swiss Reinsurance Company study found that economic losses from "catastrophes" in 2003 were on the order of $70 billion (Swiss Re 2003). Natural disasters, including floods, accounted for $58 billion. Swiss Re also estimated that "natural and human disasters" claimed 60,000 human lives in 2003. Another study put economic losses at $60 billion and deaths due to extreme events at 75,000 (Munich Re 2003). Significant floods did occur in Nepal, France, Pakistan, and China during the (northern) summer of 2003 (Munich Re 2003).

The majority of large floods have occurred in Asia during the last few decades, but few countries have been free of damaging floods. In many countries at least one destructive flood (including storm surges) has occurred since 1990 (Kundzewicz and Schellnhuber 2004). Pielke and Downton (2000) found an annual increase of 2.93% in the total flood damage in the United States over the period 1932–97. This increase has been attributed to both climate factors, such as increasing precipitation, and socioeconomic factors, such as increasing population and wealth.

Many damaging floods have also occurred in Europe in the last decade. Material damage in 2002 was higher than in any previous year. For instance, the floods in Central Europe in August 2002 caused damage totaling nearly 15 billion euros (Kundzewicz and Schellnhuber 2004). Several destructive floods also occurred in other parts of the world in 2002, including China, Russia, and Venezuela. An evaluation of the damages, by affected area, is presented in Table 16.3.

It is also important to emphasize that damaging floods have occurred in arid and semiarid regions, such as summer floods in Arizona (Hirschboeck 1987).

The number of flood events for each continent and decade since the beginning of last century is shown in Figure 16.5. The data source for this figure is the EM-DAT global disaster database from OFDA/CRED International Disaster Database at the University of Louvein in Belgium (OFDA/CRED 2002). Only events that are classified as disasters are reported in this database. (An event is declared as a disaster if it meets at least one of the following criteria: 10 or more people reported killed; 100 or more people reported affected; international assistance was called; or a state of emergency was declared (OFDA/CRED 2002).) Figure 16.5 shows a clear increase in the number of floods since the 1940s for every continent and a roughly constant rate of increase for each decade. However, it should be noted that although the number has been increasing, the actual reporting and recording of floods have also increased since 1940, due to the improvements in telecommunications and improved coverage of global information.

Regional changes in the timing of floods have been observed in many areas. The maximum daily flow of the River Elbe in Germany (3,000 cubic meters per sec) has been exceeded eight times in winter (most recently in 1940) and only three times in summer (last in 2002) (Mudelsee et al. 2003). Thus, severe winter

Table 16.3. Examples of Floods during the Summer of 2002 (Dartmouth Flood Observatory)

Location and Continent	Duration	Affected Region	Damage
	(days)	(sq. km.)	(dollars per sq. km.)
C. Europe, Europe	18 (August)	252,300	79,270
S. Russia, Asia	12 (June)	224,600	1,945
W. Venezuela, South America	11 (July)	224,900	13
NW China, Asia	10 (June)	252,000	1,587
NW China, Asia	8 (July)	127,600	287

Table 16.2. Deaths Due to Floods and Wild Fires, by Continent, 1990–99 (IRFC World Disaster Report 2001)

Phenomenon	Oceania	United States and Canada	Rest of Americas	Europe	Africa	Asia	Total
Floods	30	363	35,235	2,839	9,487	55,916	**103,870**
Wild Fires	8	41	60	127	79	260	**575**

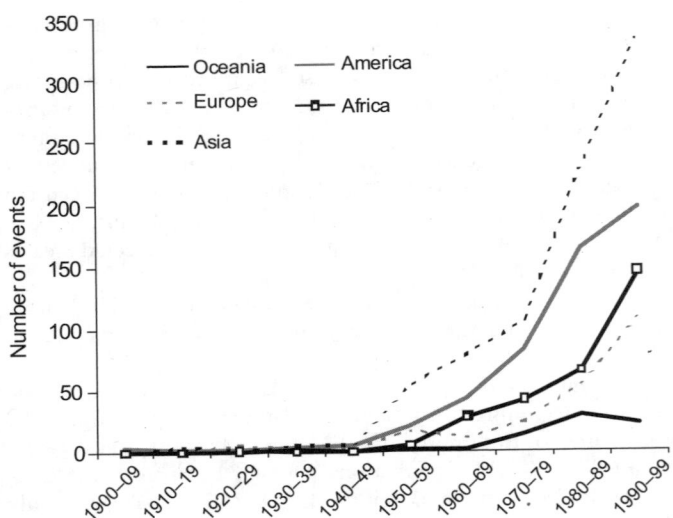

Figure 16.5. Number of Recorded Flood Events by Continent and Decade in Twentieth Century (OFDA/CRED 2002)

floods of the Elbe have not occurred in the last 64 years. Also, as indicated by Kundzewicz et al. (2004), intense and long duration summer precipitation occurred in Central Europe in 1997 and 2002 leading to very damaging floods.

Flood processes are controlled by many factors, climate being one of them. Other nonclimatic factors include changes in terrestrial systems (that is, hydrological and ecological systems) and socioeconomic systems. In Germany, for instance, flood hazards have increased (Van der Plog et al. 2002) partly as a result of changes in engineering practices, agricultural intensification, and urbanization (direct and indirect drivers).

During the 1960s and 1970s, more than 90% of natural disasters in the United States were the result of climate extremes (Changnon and Easterling 2000), and it has been estimated that about 17% of all the urban land in the United States is located within the 100-year (the return period) flood zone (Pielke and Downton 2000). Likewise, in Japan, about 50% of the population and 70% of infrastructure are located on floodplains, which account for only 10% of the land area. On the other hand, in Bangladesh the percentage of flood-prone areas is much higher, and inundation of more than half of the country is not uncommon. For instance, about two thirds of the country was inundated in the 1998 flood (Mirza 2003).

16.2.2.2 *Evidence for Changes in Fire Activity*

It is difficult to make a complete assessment of the changes in global fire activity because of the diverse methodologies employed and the limited availability of data. At large scales, recent fire activity patterns (during the last 10 years) based on remote sensing data from the satellites are relatively well known (Dwyer et al. 2000; Justice et al. 2002). Data at the local scale are less common and generally only available for specific regions where fire occurrence has been documented from ground-based observations (Schelhaas and Schuck 2002; NIFC 2004). In some cases, these records cover many years (10 to 100) and extend our knowledge on fire patterns further into the past. Tree rings have also been used in the analysis of forest fires for the last few centuries. For longer time scales, fire patterns have been determined from analyses of plant materials in sediments (Clark 1997), which can provide evidence of fires that occurred hundreds to millions of years ago.

Active fire detections from satellite data made from April 1992 to March 1993 showed that most fire activity occurs from July to August and from November to January (Dwyer et al. 2000). Seventy percent of the detected fires were located in the tropics, with 50% of all fires in Africa. Globally, fires affected 19% of savanna areas, 20% of broadleaf forests, 20% of crops and pasture areas, 4% of shrublands, and 4% of grassland areas (Dwyer et al. 2000).

Information on the global area burned in 2000 is available from two satellite-based inventories: the GBA2000 (Grégoire et al. 2003) and the GLOBSCAR (Kempeneers et al. 2002). Based on GBA2000, fires affected ~350 × 10⁶ ha in 2000; 81% of this area occurred in woodlands and scrublands, 16.5% in grasslands and croplands, 1.5% in coniferous and mixed forests, and 1.2% in broadleaf forests (Bartholomé et al. 2003). Most of the fire activity occurred from July to September and from December to January. Figure 16.6 (in Appendix A) illustrates the global pattern of burned areas based on GBA2000.

According to GLOBSCAR, the global burned area in 2000 was ≈200 × 10⁶ ha (Kempeneers et al. 2002). Of this total, 59% was located in Africa, 11% in Asia, 9% in Australia, 8% in Europe, 7% in South America, and 6% in North America. Most of the fires occurred during June-August and November-January.

Despite the potential to provide large-scale information, satellite-based global fire data sets covering long periods of time are rare (Lepers 2003). However, for tropical regions data on major fires from January 1997 to December 2000 are available from the ATSR World Fire Atlas Project. Based on these data, Lepers (2003) determined that 36% of exceptional or frequent fire events occurred in South America, followed by Africa (30%) and Asia (20%), mostly during El Niño conditions. In Asia and in Central and South America, these fires appear to be associated with areas experiencing deforestation and forest degradation (Lepers 2003). (See Figure 16.7 in Appendix A.)

Independent studies have also shown that high fire activity in tropical regions can be driven by changes in land use. For example, higher fire activity was related to conversion of peat swamp forests in Indonesia (Page et al. 2002). In Amazonia, large-scale fire patterns mirror large-scale patterns of deforestation (Skole and Tucker 1993), road construction (Prins et al. 1998; Laurence et al. 2001; Cardoso et al. 2003), and logging activities (Cochrane et al. 1999a; Nepstad et al. 1999). Consequently, increases in fire activity are likely to occur in response to future development programs in these regions, if the current relationships between fire and land use continue to hold (Cardoso et al. 2003).

At the local scale, important data sets on fire patterns include inventories of areas burned. These data are commonly reported as aggregated state/country statistics. For example, fires in the United States burned an average of 4.1 million hectares a year from 1960 to 2002 (NIFC 2004). In Europe, from 1961 to 1999 forest fires and other woodland burned an average of nearly 450,000 hectares annually (Schelhaas and Schuck 2002). Forest fires in Australia affected on average 360,000 hectares a year from 1956 to 1971 and 480,000 hectares from 1983 to 1996 (Gill and Moore 2002). These are annual averages, but there is significant variability between years.

Differences in methodologies and data availability make it difficult to compare figures from different regions and time periods. Yet the trends in these data sets are informative. For example, records available for extended periods show a general long-term reduction in the area burned. In the United States, the area burned has declined by more than 90% since 1930 (Hurtt et al. 2002) and in Sweden, the area burned fell from ~12,000 to ~400 hectares a year between 1876 and 1989 (Pyne and Goldammer 1997). In both countries, fires were reduced due to changes in land use. In Europe as a whole, however, the total extent and the interannual variability of the area of burnt forest are higher for the period 1975–2000 than for the 1960s, due, it is presumed, to changes in land use and climate (Schelhaas and Schuck 2002).

Data on major fire events from OFDA/CRED indicate a global increase in the number of major fire events after 1960. (See Figure 16.8.) According to CRED, changes were greatest in North America, where the number of major fires increased from ~10 during the 1980s to ~45 during the 1990s. CRED data sets only include data on major disasters (as defined by OFDA/CRED 2002) and are compiled from several sources, including U.N. agencies, nongovernmental organizations, insurance companies, research institutes, and press agencies.

Evidence of past fire activity is provided by sediment analyses. Indicators include materials directly affected by or the products of fires, such as charcoal (Sanford et al. 1985), or materials such as pollen that can be used to determine the presence of specific plant species associated with fire regimes (Camill et al. 2003). Data from sediment records are especially important in providing long-term trends in fire activity at large scales. For example, data from sea sediments show that sub-Saharan Africa had low fire activity until about 400,000 years ago and indicate that humans had a significant influence on the occurrence of fire in the Holocene (Bird and Cali 1998). Analyzes from lake sediments show that natural fires have influenced forests in Amazonia for the last 7,000 years, including impacts on current patterns of forest structure (Turcq et al 1998). Data from soils also show that fires have disturbed lowland forests in the Amazonian region for the past 6,000 years (Sanford et al. 1985). In Minnesota, fires have been shown to result from shifts in vegetation caused by climate changes during the Holocene (Camill et al. 2003). An important common result from these studies is the link between high fire activity, dry climates, and fuel accumulation.

16.3 Causes of Changes in the Regulation of Floods and Fires

The complex relationships between direct and indirect drivers are the main causes of change in the regulation of all natural hazards,

including floods and fires. Understanding the patterns of distribution of such drivers may be partially achieved by describing which ecosystems are undergoing the most rapid and largest transformations. Such areas are likely to have the most reduced capacity to regulate natural hazards.

Increased attention has been paid to extreme weather and climate events over the past few years due to increasing losses associated with them. For a more accurate understanding of weather impacts, however, it is necessary to acknowledge the actual and potential impacts of changes in both climate and society (Pielke Jr. et al. 2003). For instance, there are examples where economic impacts of hurricanes have increased dramatically during prolonged periods of rather benign hurricane activity (Pielke and Landsea 1998), and it is thought that a major factor conditioning hurricane losses in the United States is the significant increase in wealth and population in the coastal areas (Landsea et al. 1999).

Human occupancy of the floodplains and the presence of floodwaters produce losses to individuals and society. Different pressures have caused increases in population density in flood-prone areas, especially poverty, which has been responsible for the growth of informal settlements in susceptible areas around megacities in many developing countries.

An immediate question that emerges from the increases in flood damage is the extent to which a rise in flood hazard and vulnerability can be linked to climate variability and change. This has been treated extensively in the Third Assessment Report of the Intergovernmental Panel on Climate Change and recently reviewed by Kundzewicz and Schellnhuber (2004). However, climate change is just one of the many drivers affecting the regulation of floods. A more detailed discussion about drivers affecting flood regulation is given in the next section.

Changes in fire regulation can affect many other ecosystem services—supporting, provisioning, regulating, and cultural services. Supporting services affected by changes in the fire regulation capacity of ecosystems include nutrient cycling (for example, through fluxes and changing stocks of carbon and nitrogen) and primary production (a decrease due to vegetation removal, for instance, or increase by soil fertilization). Provisioning services affected include food production (increased by short-term fertilization, decreased through long-term nutrient losses) and availability of genetic resources (decreased by removal of flora and fauna). Regulating services affected include climate (changes in surface albedo and greenhouse gas emissions) and disease regulation (increase in the likelihood of respiratory diseases due to reduction of air quality). Finally, cultural services affected include recreation opportunities and aesthetic experiences (short-term effects on the landscape in protected areas, for example, or airport closings due to reduced visibility caused by smoke).

16.3.1 Drivers Affecting the Regulation of Floods

In the MA framework a driver is defined as any natural or human-induced factor that directly or indirectly causes a change in an ecosystem, affecting its capacity to provide a service. Evidently, climate variability, climate change, and natural or anthropogenic land cover changes constitute physical and biological factors that directly affect the regulation of floods in terms of both processes and magnitude through changes in extreme rainfall events and peak runoff magnitude. If extreme events become more frequent and intense, the capacity of the system to provide the regulation could be affected. However, the distribution of the impacts of extreme events is not uniform across the world; their impact is greater in poorer and more vulnerable regions.

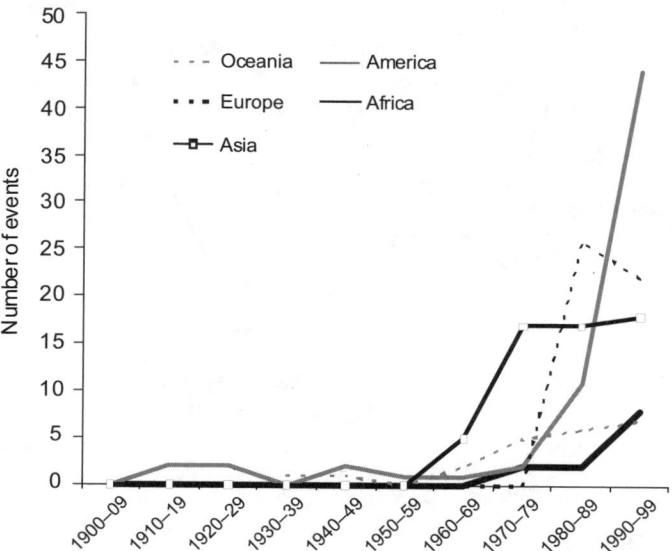

Figure 16.8. Number of Recorded Wild Fires by Continent and Decade in Twentieth Century (OFDA/CRED)

Precipitation is a critical factor in causing floods, and its characteristics—such as intensity, location, and frequency—appear to be changing. During the twentieth century precipitation increased by 0.5–1.0% per decade in many areas in the mid- and high latitudes of the Northern Hemisphere (IPCC 2001). Moreover, increases in intense precipitation have been reported even in regions where total precipitation has decreased (Kabat et al. 2002). It is difficult to generalize, however, as some regions have shown decreases in both total and intensity of precipitation. Although intense rainfall is a sufficient condition to increase flood hazard, there are a number of other nonclimatic factors that exacerbate flood hazard (Kundzewicz and Schellnhuber 2004).

A conceptual framework for understanding the human- and human-influenced processes contributing to damaging floods is presented in Figure 16.9. The broad integrative framework presented by Pielke and Downton (2000) may be used to understand the role of the drivers (direct or indirect) and policies in determining actual and potential flood outcomes.

Lepers (2003) presented the most rapidly changing areas in terms of deforestation and forest degradation, which are predominately in the tropics. Nearly half of the affected forests are in South America, with 26% in Asia and 12% in Africa. Concurrently, the tropical regions of these three continents have had higher flood incidence and higher relative vulnerability (people killed per million exposed per year) (UNDP 2004).

Figure 16.10 (in Appendix A) shows the number of floods between 1980 and 2000 (OFDA/CRED data set), together with the area of deforestation and forest degradation during the same period. Since there were multiple data sets used to describe deforested areas, the map indicates the number of data sets where deforestation was effectively identified (see Lepers 2003 for more details). The number of flood events is reported as total number of floods by country during the period 1980–2000. Tropical regions, mainly in Asia and South America, have been severely affected by floods and suffered an important loss in forest area and increasing forest degradation, as shown in Figure 16.10.

There are also physical, biological, and anthropogenic factors that influence the regulation of floods, such as land use change and human encroachment of natural areas, such as wetlands, forest, and vegetation in coastal zones. These perturbations of natural areas with the subsequent reduction of ecosystem functions affect flood attenuation potential and soil water storage capacity. It is clear that land conversion, deforestation, and loss of ecosystems such as wetlands have increased in the last decades (see Chapters 20, 21, and 28), but the impact of these drivers is unevenly distributed across different regions. Water extraction and diversion also constitute a direct anthropogenic driver that affects regulation of floods.

The indirect drivers that strongly influence the regulation of floods are population growth, the level of economic, scientific, and technological development, a lack of governance, and population settlement preferences. Impacts vary between countries but are manifest in changes in exposure of human populations to extreme events, in the increase or decrease of GDP (depending on the region), in the capacity of governments to respond to flood-related disasters, and in the commercial and recreational activities developed along major river floodplains. The impacts are more intensely felt in densely populated areas in poorer countries, where farming activities, property, and commercial activities are most at risk.

16.3.2 Drivers Affecting the Regulation of Fires

Direct and indirect drivers can change the capacity of ecosystems to provide fire regulation. For example, climate change is a natu-

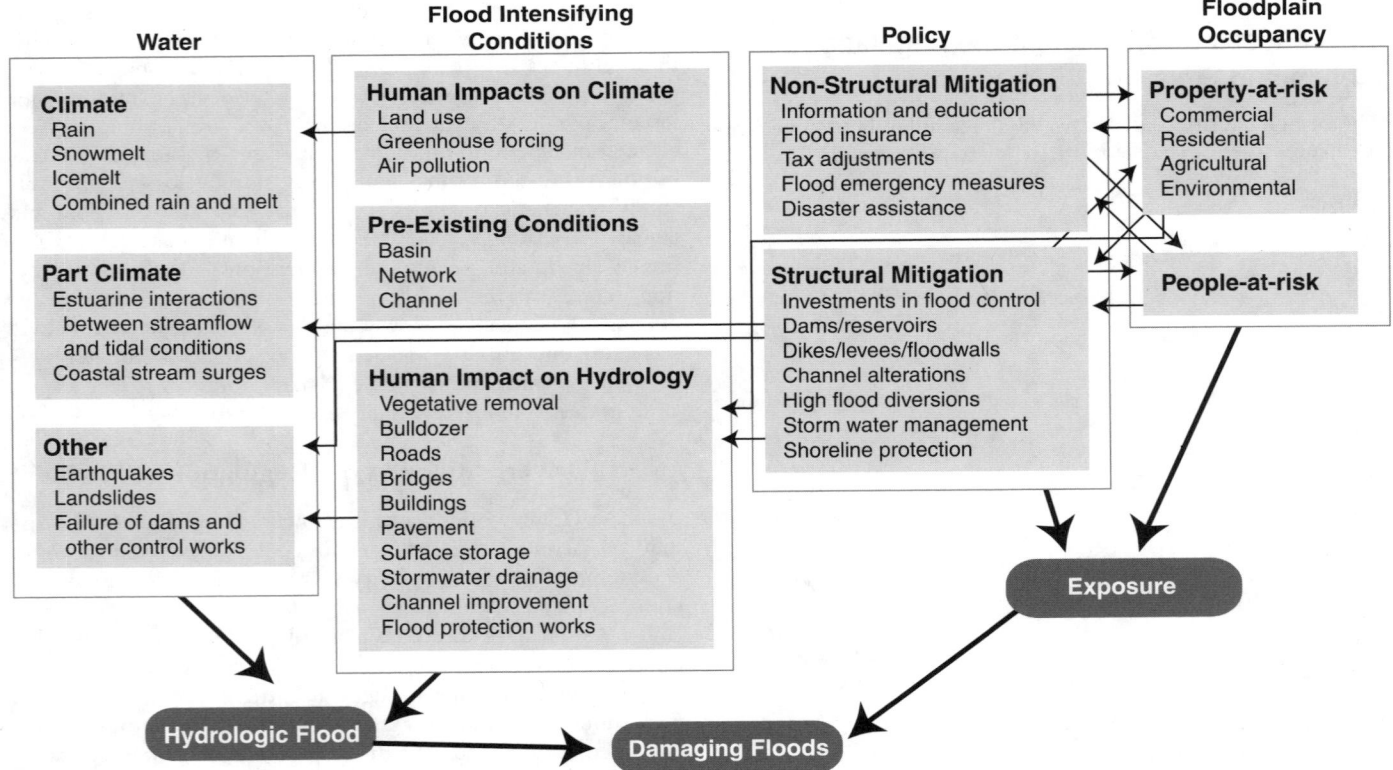

Figure 16.9. Factors Contributing to Damaging Floods (Pielke and Downton 2000)

ral and direct driver that can significantly affect fire regulation. Changes in precipitation variability may affect fire frequency and intensity. Low precipitation is associated with low fire frequency and intensity due to low biomass and fuel conditions. Similarly, high precipitation is associated with low fire frequency and intensity due to extremely wet conditions (Cardoso and Hurtt 2000).

Regions with climates that present distinct dry and wet seasons have potentially more fires. Under these climates, fuel load can increase through vegetation growth during the wet season and the high flammability during the dry season, leading to more frequent and intense fires. In addition to varying precipitation patterns, hotter conditions and more intense winds may promote a higher frequency and intensity of fires by favoring fuel dryness and fire spread. Climate effects may also depend on the state of the ecosystem. Forest trees with deep root systems may resist reduced precipitation longer before becoming flammable compared with trees with shallower roots, such as savannas species (Nepstad et al. 1994). And ecosystems with soils that have low water-holding capacity may become flammable after even short dry periods.

Unfortunately, it is not easy to predict future climate conditions. Although there is evidence that the global average temperature is increasing (IPCC 2001), other climatic data, particularly at a regional, sub-continental scale, are less clear.

Land use and land cover are important direct anthropogenic drivers of fire regulation. Included here are land management, land clearance and agriculture, housing development, logging, harvesting and reforestation, and fire suppression schemes. These can change fuel load, flammability, number of ignition events, and spread conditions. For example, fire suppression seeks to reduce the sources of ignition and short-term fire activity. However, long-term fire suppression may lead to high fuel loads and increase the likelihood of catastrophic fires, as demonstrated in the United States. Fire suppression also leads to lower trigger events, but intentional and accidental fires related to land management lead to a higher number of trigger events. Firebreaks lower spread, but actions that provide fire corridors (such as reforestation) may increase fire spread. Logging and harvesting may reduce fuel loads if biomass is completely removed or may increase fuel loads when biomass debris is left. Changes in land use and land cover have increased in developing regions in the tropics, leading to more fires (Page et al. 2002). In other regions, such as North America, however, land management has reduced fire frequency but increased fuel loads and the likelihood of more intense fires (Hurtt et al. 2002).

Indirect drivers of fire regulation are mainly linked to human activities. These drivers include demographic, economic, sociopolitical, and scientific and technological drivers. (See Chapter 3.) Demographic factors such as changes in population distribution may lead to different land use and land cover patterns, which are direct drivers of fire regulation. Changes in population may also alter property exposure to fire. Increase of population in fire-prone areas can potentially lead to higher exposure to fires, to increased risk of accidental fires, and to economic and human losses that are generally related to the level of preparedness for extreme events in the affected region.

The degree of economic development is also important. Higher levels of economic development generally improve fire regulation. For example, access to machines and technology, such as tractors, offers an alternative to the use of fires as a land clearance tool for land management. Educational programs on fire risks and effects and to alternatives to the use of fire as a land use management tool also tend to improve fire regulation. In addition, higher levels of economic development allow for the use of

fire risk and activity monitoring systems and for the development of programs for disaster relief in case of extreme events.

Sociopolitical drivers, such as environmental policy, aid in fire regulation through many processes, including development of legislation and mechanisms of law enforcement and fire research, which leads to better knowledge of fires and how to fight them.

Scientific and technological drivers, such as the monitoring of fire activity, also enhance fire regulation by providing tools that help forecast and track fire activity. Satellite-based data collection systems, for example, provide additional information on fire frequency and behavior, as well as fire times and position. These can help support decisions on fire regulation resources, can improve law enforcement, and can contribute to fire research.

16.4 Consequences for Human Well-being of Changes in the Regulation of Natural Hazards

In this section the consequences and impacts of extreme events on several aspects of human well-being are presented. These impacts result from the reduced capacity of ecosystems to regulate the magnitude and intensity of these events and from the ability of ecosystems and human populations to cope with the consequences of such events—their exposure, sensitivity, and resilience. The latter are all components of "vulnerability," which is a complex and multidimensional concept, intrinsic to ecological and social systems. The degree of vulnerability of an ecosystem or human group will define the impacts of extreme events on society and ecosystems. Vulnerability is discussed extensively in Chapter 6. In this section the vulnerability concept is treated as a tool to understand the differentiated impacts of extreme events on different societal groups.

16.4.1 Security and the Threat of Floods and Fires

16.4.1.1 *Floods*

Many factors are likely to contribute to the increase of those affected by disasters. One of these is the vulnerability profile of the population. As more people move into urban areas and slum settlements they are increasingly living in susceptible (disaster-prone) regions. Migrants to big cities are normally at greatest risk, since they occupy the most hazard-prone locations on unstable slopes and flood-prone areas. However, urbanization is not necessarily a factor increasing disaster risk if it is managed in an appropriate and adequate way. Local authorities and their interactions with public, private, and civil society may play an important role in urban risk reduction to bridge the gap between national and international risk management players and local communities (UNDP 2004). But this implies a high level of municipal governance.

Environmental degradation increases the negative effects of extreme events on society. While disaster preparedness measures do help save lives, the failure to reduce risks more broadly may be contributing to the higher numbers of disaster-affected people. Unfortunately, data are not uniformly collected, but more information would likely show an increase in the number of affected people, although the definition of "affected" varies (IFRC 2003).

16.4.1.2 *Fires*

The detrimental impacts and consequences of some wildfires on economies, human health, and safety are comparable in severity to other major natural hazards. Unlike the majority of the geological and hydro-meteorological hazards, however, vegetation fires represent a natural and human-caused hazard that can be predicted, controlled, and, in many cases, prevented through the application of appropriate policies. Most countries already have laws, regulations, and action plans for management of forest fires and response procedures in place. These are often insufficient, however, or inadequately implemented.

Loss estimates have been used to quantify the impacts of fires on society. Globally, such losses are difficult to quantify, but data do exist from areas where fire occurrence is well documented. For example, U.S. federal agencies spent an average of $768 million a year in fire suppression between 1994 and 2002 (NIFC 2004). From 1995 to 1999, the annual costs of prescribed fires increased from $20 million to $99 million (NIFC 2004). In 2003, fires in California burned over 750,000 hectares, caused 24 fatalities, and destroyed over 3,000 residences (FEMA 2004).

An analysis of national fire policies concluded that mitigation policies were generally weak, rarely based on reliable data of forest fire extent, causes, or risks, and did not involve landowners most likely to be affected (ECE/FAO 1998). Furthermore, they were sometimes the result of ill-conceived forest management policies, particularly policies aimed at total fire exclusion that led to fuel accumulation and catastrophic fire outbreaks. Future fire policies will need to find a balance between the various ecological, agricultural and energy benefits of biomass burning and environment and health problems.

16.4.2 The Effects of Natural Hazards on Economies, Poverty, and Equity

Economic indicators such as GDP are highly affected by natural disasters, especially in the most vulnerable areas. Thus poorer nations experience lower economic losses but a relatively higher drop in GDP following a natural disaster, while the opposite trend is observed in wealthier nations (MunichRe 2001).

In a recent UNDP report on natural disasters, mortality from floods was inversely correlated with GDP per capita. There was also a negative correlation between deaths from flooding and local density of population (UNDP 2004). This fact, although contradictory, might be explained because of the higher expected mortality in rural and remote areas with limited health care and low disaster preparedness. The most important factors contributing to high risk from floods were low GDP per capita, low density of population, and a high number of exposed people.

In the same study, 147 countries with populations exposed to floods were identified. India, Bangladesh, Pakistan, and China were at the top of the list of countries with high absolute and relative populations exposed to floods. This is a consequence of the large populations living along floodplains and low-lying coasts in this part of the world. Other countries, such as Bhutan and Nepal and the Central American and Andean states, also have large absolute and relative populations exposed to floods because of their mountainous topography and important population centers located on river floodplains (UNDP 2004).

Rural areas with low population density and poor health coverage that are prone to flooding are also more vulnerable to flood-related diseases due to their lower flood evacuation abilities. Disaster reduction measures for these areas can also offer opportunities for these communities, such as involving women in maintaining local social networks focused on risk reduction. In Cox's Bazar (Bangladesh), women's involvement in disaster preparedness activities organized through local networks—including education, reproductive health, and micro-enterprise groups—has resulted in a significant reduction in the number of women killed following tropical cyclones (UNDP 2004).

16.4.3 Impacts of Flood and Fires on Human Health

Natural hazards adversely affect human health both directly and indirectly. The direct physical effects that occur during or after flooding include mortality (mostly from flash floods), injuries (such as sprains or strains, lacerations, contusions), infectious diseases (respiratory illnesses, for example), poisoning (from carbon monoxide, say), diseases related to the physical and emotional stress caused by the flood, and such other effects as hypothermia from loss of shelter. The number of deaths associated with flooding is closely related to the local characteristics of floods and to the behavior of victims (Malilay 1997).

Indirect effects of floods can also cause human injury and disease, such as waterborne infections and vector-borne diseases, acute or chronic effects of exposure due to chemical pollutants released into floodwaters, and food shortages. Studies have also observed increased rates of the most common mental disorders, such as anxiety and depression, following floods (Hajat et al. 2003). These and other psychological effects may continue for months or even years. The physical and health impacts of floods in the United Kingdom have been the subject of longitudinal studies (Tapsell et al. 2003). Flooding events caused adverse physical effects in about two thirds of vulnerable people and adverse mental and physiological effects in more than three quarters. The physical effects lasted about 12 months on average, while the psychological impacts lasted at least twice as long.

Impacts can also be categorized according to when they occurred relative to the event (during the impact phase, immediate post-impact phase, or during the recovery phase). For example, injuries are likely to occur in the aftermath of a flood disaster, as residents return to dwellings to clean up damage and debris.

Although natural hazards cause a significant number of injuries and diseases, the available morbidity data are limited, which restricts our understanding of both the impacts and the possible effective response options.

Campbell-Lendrum et al. (2003) used comparative risk assessment methods to estimate the current and future global mortality burden from coastal flooding due to sea level rise and from inland flooding and mass movement caused by an increase in the frequency of extreme precipitation. As quantitative estimates were not available, longer-term health impacts due to population displacement, economic damage to public health infrastructures, increased risk of infectious disease epidemics, and mental illness were excluded from their analyses, although these impacts are likely to be greater than the acute impacts. Despite this, the global model developed has been shown to be relatively accurate when tested against more detailed assessments at a national level.

The model has other limitations, however. In the absence of detailed data on the relationships between intensity of precipitation, the likelihood of a flood or mudslide disaster, and the magnitude of health impact variables and their effects, it was assumed that flood frequency was proportional to the frequency with which monthly rainfall exceeded the 1-in-10-year limit (that is, upper 99.2% confidence interval) of the baseline climate. It was also assumed that determinants of vulnerability were distributed evenly throughout the population of a region, so that the change in relative risk of health impacts was proportional to per capita change in risk of experiencing such an extreme event. Changes in the frequency of coastal floods were defined using published models that estimate change in sea level for various climate scenarios. These changes were applied to topographical and population distribution maps to estimate the regional change in incidence of exposure to flooding. The regional changes were summed to estimate potential worldwide impacts. The model did not account for changes in the frequency of storm surges.

From the model, it was estimated that the impact of climate change on flooding in 2000 amounted to192,000 disability-adjusted life years, with the Eastern Mediterranean, Latin American and Caribbean, and the Western Pacific regions suffering the largest burdens of flood-related disease (Campbell-Lendrum et al. 2003). Potentially large changes in flood-related mortality were estimated under various climate change scenarios. Subgroups vulnerable to adverse health effects of floods include the elderly, those with prior health problems, the poor, and those with dependents (especially children) (Hajat et al. 2003).

Thus, floods and other extreme weather events should be considered multiple stressors that include the event itself, the disruptions and problems of the recovery period, and the worries or anxieties about the risk of recurrence of the event (Penning-Rowsell and Tapsell 2004). The perceived risk of recurrence can include a perceived failure on the part of relevant authorities to alleviate risk or provide adequate warnings. These sources of stress and anxiety, along with pre-existing health conditions, can have significant impacts on the overall health and well-being of flood victims. Medical authorities, social services departments, insurance companies, and other organizations need to provide better post-event social care for people affected by extreme weather events.

The population at risk, policy-makers, and emergency workers may undertake activities to reduce health risks before, during, and after a flood event. Traditionally, the fields of engineering and urban planning aim to reduce the harmful effects of flooding by limiting the impact of a flood on human health and economic infrastructure. Mitigation measures may reduce, but not eliminate, major damage. Early warning of flood risk and appropriate citizen response has been shown to be effective in reducing disaster-related deaths (Noji 2000). From a public health point of view, planning for floods during the inter-flood phase aims at enabling communities to respond effectively to the health consequences of floods and allows the local and central authorities to organize and effectively coordinate relief activities, including making the best use of local resources and properly managing national and international relief assistance. In addition, medium to long-term interventions may be needed to support populations affected by flooding.

References

Bartholomé, E., A.S. Belward, F. Achard, S. Bartalev et al., 2003: *Use of Data from the VEGETATION Instrument for Global Environmental Monitoring: Some Lessons from the GLC 2000 and the GBA 2000 Projects.* Institute for Environment and Sustainability, Joint Research Centre, European Commission.

Bird, M.I. and J.A. Cali, 1998: A million-year record of fire in sub-Saharan Africa. *Nature*, 394, pp. 767–79.

Bullock, A. and M. Acreman, 2003: The role of wetlands in the hydrological cycle. *Hydrology and Earth System Science*, 7(3), pp. 358–89.

Campbell-Lendrum, D.H., C.F. Corvalan, and A. Pruss-Ustun, 2003: How much disease could climate change cause? In: *Climate Change and Human Health: Risks and Responses*, A.J. McMichael, D. Campbell-Lendrum, C.F. Corvalan, K.L. Ebi, A. Githeko, J.D. Scheraga, and A. Woodward (eds.), WHO/WMO/UNEP.

Christensen, N.L., J.K. Agee, P.F. Brussard, J. Hughes, D.K. Knight et al., 1989: Interpreting the Yellowstone fires of 1988. *Bioscience*, 39, pp. 678–685.

Cochrane, M.A. and M.D. Schulze, 1999: Fire as a recurrent event in tropical forests of the eastern Amazon: Effects on forest structure, biomass, and species composition. *Biotropica*, 31(1), pp. 1–16.

Cochrane, M.A., A. Alencar, M.D. Schulze, C.M. Souza, D.C. Nepstad, P. Lefebvre, and E.A. Davidson, 1999a: Positive Feedbacks in the Fire Dynamic of Closed Canopy Tropical Forests. *Science*, 11, pp. 1832–35.

Clark, J.S., 1997: An introduction to sediment records of biomass burning. In: *Sediment Records of Biomass Burning and Global Change*, J.S. Clark, H. Cachier,

J.G. Goldammer, and B. Stocks (eds.), NATO ASI Series, Series I, *Global Environmental Change*, 51, pp. 1–5.

Daily, G.C., S. Alexander, P.R. Ehrlich, L. Goulder, J. Lubchenco et al., 1997: *Ecosystem Services: Benefits Supplied to Human Societies by Natural Ecosystems.* Island Press, Washington, DC.

Daly, C., D. Bachelet, J.M. Lenihan, R.P. Neilson, W. Parton, and D. Ojima, 2000: Dynamic simulation of tree-grass interactions for global change studies. *Ecological Applications*, 10, pp. 449–69.

De Luis, M., J. Raventos, J. Cortina, J.C. Gonzalez-Hidalgo, and J.R. Sanchez, 2004: Fire and torrential rainfall: Effects on the perennial grass *Brachypodium retusum. Plant Ecology*, 173(2), pp. 225–32.

Dwyer, E., S. Pinnock, J.M. Gregoire, and J.M.C. Pereira, 2000: Global spatial and temporal distribution of vegetation fire as determined from satellite observations. *Int. J. Remote Sensing*, 21(6–7), pp. 1289–302.

ECE/FAO (Economic Commission for Europe/Food and Agriculture Organization of the United Nations), 1998: *Forest fire statistics 1994–1996,* United Nations, Geneva, 19 pp.

FEMA (Federal Emergency Management Agency), 2004: The California Fires Coordination Group—A Report to the Secretary of Homeland Security.

Giglio, L., J.D. Kendall, and C.O. Justice, 1999: Evaluation of global fire detection algorithms using simulated AVHRR infrared data. *Int. J.Remote Sensing*, 20(10), pp. 1947–85.

Giglio, L., J.D. Kendall, and C.J. Tucker, 2000: Remote sensing of fires with the TRMM VIRS. *Int. J. Remote Sensing*, 21(1), pp. 203–7.

Gill, A.M. and P.H.R. Moore, 2002: Fire situation in Australia. *Int. Forest Fires News*, 26, pp. 2–8.

Gosselink, J.G., W.H. Conner, J.W. Day, and R.E. Turner, 1981: Classification of wetland resources: land timber, and ecology. In: *Timber Harvesting in Wetlands*, B.D. Jackson and J.L. Chambers (eds.), Division of Continuing Education, Louisiana State University, Baton Rouge.

Green, P.A., C.J. Vörösmarty, M. Meybeck, J.N. Galloway, B.J. Peterson, and E.W. Boyer, 2004: Pre-industrial and contemporary fluxes of nitrogen through rivers: A global assessment based on typology. *Biogeochemistry*, 68(1), pp. 71–105.

Grégoire, J.M., K. Tansey, and J.M.N. Silva, 2003: The GBA2000 initiative: Developing a global burnt area database from SPOT-VEGETATION imagery. *Int. J. Remote Sensing*, 24(6), pp. 1369–76.

Hajat, S., K.L. Ebi, S. Kovats, B. Menne, S. Edwards, and A. Haines, 2003: The human health consequences of flooding in Europe and the implications for public health: A review of the evidence. *Applied Environmental Science and Public Health*, 1, pp. 13–21.

Hirschboeck, K., 1987: Hydroclimatically-defined mixed distribution in partial duration flood series. In: *Hydrologic Frequency Modelling*, V.P. Singh (ed.), D. Reidel, Norwell, MA.

Hurt, G.C., S. W. Pacala, P.R. Moorcroft, J. Caspersen, E. Shevliakova, R. Houghton, and B. Moore, 2002: Projecting the future of the US carbon sink. *Proceedings of the National Academy of Sciences of the United States (PNAS)*, 99(3), pp. 1389–94.

IFRC (International Federation of Red Cross and Red Crescent Societies), 2001: World Disasters Report 2001. Available at http://www.ifrc.org/publicat/wdr2001.

IFRC, 2003: World Disasters Report 2001. Available at http://www.ifrc.org/publicat/wdr2003.

IPCC (Intergovernmental Panel on Climate Change), 2001: Climate Change 2001: The Scientific Basis. Contribution of Working Group I to The Third Assessment Report of the Intergovernmental Panel on Climate Change. J.T. Houghton, Y. Ding, D.J. Griggs, M. Noguer, P.J. van der Linden, X. Dai, K. Maskell, and C.A. Johnson (eds.), Cambridge University Press, Cambridge.

Justice, C.O., L. Giglio, S. Korontzi, J. Owens, J.T. Morisette, D. Roy, J. Descloitres, S. Alleaume, F. Petitcolin, and Y. Kaufman, 2002: The MODIS fire products. *Remote Sensing of Environment*, 83, pp. 1–2, 244–62.

Kabat P., R.E. Schulze, M.E. Hellmuth, and J.A. Veraart (eds.), 2002: *Coping with impacts of climate variability and climate change in water management: a scoping paper.* DWC-Report No. DWCSSO-01(2002), International Secretariat of the Dialogue on Water and Climate, Wageningen.

Landsea, C.W., R.A. Pielke Jr., A.M. Mestas-Nuñez, and J.A. Knaff, 1999: Atlantic basin hurricanes: Indices of climatic changes. *Climate Change*, 42, pp. 89–129.

Laurans, Y., 2001: Economic valuation of the environment in the context of justification conflicts: development of concepts and methods through examples of water management in France. *International Journal of Environment and Policy*, 15(1), pp. 94–115.

Leatherman, S.P., R. Chalfont, E.C. Pendelton, T.L. McCnadless, and S. Funderburk, 1995: *Vanishing lands: Sea level, society and the Chesapeake Bay.* University of Maryland, Annapolis, 47 pp.

Lehner, B. and P. Döll, 2004: Development and validation of a global database of lakes, reservoirs and wetlands. *Journal of Hydrology,* **296,** pp. 1–4, 1–22.

Lepers, E., 2003: Synthesis of the Main Areas of Land-cover and Land-use change. Report for the Millennium Ecosystem Assessment.

Loreau, M., S. Naeem, P. Inchausti, J. Bengtsson, J.P. Grime, A. Hector, D.U. Hooper, M.A. Huston, D. Rafaelli, B. Schmid, D. Tilman, and D.A. Wardle, 2001: Biodiversity and ecosystem functioning: current knowledge and future challenges. *Science,* **294,** p. 804–8.

Malilay, J., 1997. Floods. In: *The Public Health Consequences of Disasters,* E.Noji (ed.), Oxford University Press, New York.

McKenzie, D., Z. Gedalof, D.L. Peterson, and P. Mote, 2004: Climate change, wildfire and conservation. *Conservation Biology,* **18(4),** pp. 890–902.

Mirza, M.M.Q., 2003: The three recent extreme floods in bangladesh: A hydro-meteorological analysis. In: *Floods Problem and Management in South Asia,* M. Mirza, A. Dixit, and A: Nishat (eds.), Kluwer Academic Publisher, Netherlands.

MunichRe, 2003: Annual Review: Natural Catastrophes. Available at http://www.munichre.com/publications/302–03202_en.pdf?rdm = 54724.

MunichRe, 2001: *Topics 2000: Natural Catastrophes: The Current Position.* Special Millennium Issue, Munich.

Nepstad, D.C., C.R. de Carvalho, and S. Vieira, 1994: The role of deep roots in the hydrological and carbon cycles of Amazonian forests and pastures. *Nature,* **372(6507),** pp. 666.

Nepstad, D.C., A. Verissimo, A. Alencar, C. Nobre, E. Lima, P. Lefebvre, P. Schlesinger, C. Potter, P. Moutinho, E. Mendoza, M. Cochrane, and V. Brooks, 1999: Large-scale impoverishment of Amazonian forests by logging and fire. *Nature,* **98,** pp. 505–8.

Nepstad, D.C., A.G. Moreira, and A.A. Alencar, 1999a: *Flames in the Rain Forest: Origins, Impacts and Alternatives to Amazonian Fires.* The Pilot Program to Conserve the Brazilian Rain Forest, Brasilia, Brazil.

NIFC (National Interagency Fire Center), 2004: Wildland Fire Statistics, Boise, ID.

Noji, E.K., 2000: The public health consequences of disasters. *Prehospital Disaster Med,* 15, pp. 147–57.

OFDA/CRED (Office of U.S. Foreign Disaster Assistance/Centre for Research on the Epidemiology of Disasters), 2002: EM-DAT: The OFDA/CRED International Disaster Database. Available at http://www.cred.be/emdat.

Page, S.E., F. Siegert, J.O. Rieley, H.D. Boehm, A. Jaya, and S. Limin, 2002: The amount of carbon released from peat and forest fires in Indonesia during 1997. *Nature,* 420(6911), pp. 61–5.

Penning-Rowsell, E. and S.M. Tapsell, 2004: Extreme weather and climate events and public health responses. WHO/EEA Workshop, Bratislava, Slovakia, 9–10 February.

Pielke Jr., R.A. and M. Downton, 2000: Precipitation and damaging floods: Trends in the United States, 1932–97. *J. of Climate,* **13,** pp. 3625–37.

Pielke Jr., R.A., 2000a: Risk and vulnerability assessment of coastal hazards. In: *The Hidden Costs of Coastal Hazards: Implications for Risk Assessments and Mitigation.* The H. John Heinz III Center for Science, Economics and the Environment, Island Press, Washington DC.

Pielke Jr., R.A., 2000b: Floods impacts on society. In: *Floods, Vol.I.* D.J. Paker (ed.), Routledge, London.

Pielke Jr., R.A., J. Rubera, C. Landsea, M.L. Fernández, and R. Klein, 2003: Hurricane vulnerability in Latin America and The Caribbean: Normalized damage and loss potentials. *Natural Hazards Review,* **4(3),** pp. 101–14.

Pielke Jr., R.A. and C.W. Landsea, 1998: Normalized U.S. hurricane damage, 1925–95. *Wea. Forecasting,* **13,** pp. 621–31.

Prins, E.M., W.P. Menzel, and J.M. Feltz, 1998: Characterizing spatial and temporal distributions of biomass burning using multi-spectral geostationary satellite data. Ninth Conference on Satellite Meteorology and Oceanography, 25–29 May, Paris, France.

Pyne, S.J. and J.G. Goldammer, 1997: The culture of fire: an introduction to anthropogenic fire history. In: *Sediment Records of Biomass Burning and Global Change,* J.S Clark, H. Cachier, J.G. Goldammer, and B. Stocks (eds.), NATO ASI Series, Series I, *Global Environmental Change,* **51,** pp. 71–114.

Roy, J. 2001: How does biodiversity control primary productivity? In: *Global terrestrial productivity,* J. Roy, B. Saugier, and H.A. Mooney (eds.), Academic Press, San Diego.

Sanford, R.L., J. Saldarriaga, K.E. Clark, C. Uhl, and R. Herrera, 1985: Amazon rain forest fires. *Science,* **227,** pp. 53–5.

Schelhaas, M.J. and A. Schuck, 2002: Forest fires in Europe, 1961–98. *Int. Forest Fire News,* **27,** pp. 76–80.

Setzer, A.W. and J.P. Malingreau, 1996: AVHRR monitoring of vegetation fires in the tropics: Towards a global product. In: *Biomass Burning and Global Change,* J.S. Levine (ed.), MIT Press, Cambridge.

Simon, M., 2002: GLOBSCAR Products Qualification Report. Technical Note GLBS/ESA/QR, European Space Agency.

Skole, D. and C. Tucker, 1993: Tropical deforestation and habitat fragmentation in the Amazon: Satellite data from 1978 to 1988. *Science,* **206,** pp. 1905–9.

Swiss Re, 2003: Natural catastrophes and man–made disasters in 2002. Sigma No. 2/2003. Available on line at www.swissre.com.

Tapsell, S.M., S.M. Tunstall, and T.Wilson, 2003: *Banbury and Kidlington Four Years after the Flood: An Examination of the Long-term Health Effects of Flooding.* Enfield, Flood Hazard Research Centre, Middlesex University, UK.

Turcq, B., A. Sifedine, L. Martin, M.L. Absy, F. Soubles, K. Suguio, and C. Volkmer-Ribeiro, 1998: Amazonia rainforest fires: Alacustrine record of 7000 years. *Ambio,* 27(2), pp. 139–42.

Turner, M.G., V.H. Dale, and E.H. Everham III, 1997: Fires, hurricanes, and volcanoes: Comparing large disturbances. *Bioscience,* **47(11),** pp. 758–68.

UNDP (United Nations Development Programme), 2004: Reducing disaster risk: A challenge for development. Bureau for Crisis Prevention and Recovery. Available at http://www.undp.org/bcpr.

van der Plog, R.R., G. Machulla, D. Hermsmeyer, J. Ilsemann, M. Gieska, and J. Bachmann, 2002: Changes in land use and the growing number of flash floods in Germany. In: *Agricultural Effects on Ground and Surface Waters: Research at the Edge of Science and Society,* J. Steenvorden, F. Claessen, and J. Willems (eds.), IAHS Publ. No. 273, pp. 317–22.

Verry, E. S. and D.H. Boelter, 1978: Peatland hydrology. In: *Wetlands Functions and Values: The State of Our Understanding,* P. Gresson, J.R. Clark, and J.E. Clark (eds.), American Water Resources Association, Minneapolis, MN.

Chapter 17

Cultural and Amenity Services

Coordinating Lead Authors: Rudolf de Groot, P.S. Ramakrishnan
Lead Authors: Agnes van de Berg, Thaya Kulenthran, Scott Muller, David Pitt, Dirk Wascher, Gamini
Wijesuriya
Contributing Authors: Bas Amelung, Nesa Eliezer, Aspara Ram Gopal, Mechtild Rössler
Review Editors: Xu Jianchu, Hebe Vessuri

Main Messages

Human culture is strongly influenced by ecosystems, and ecosystem change can have a significant impact on cultural identity and social stability. Human cultures, knowledge systems, religions, heritage values, social interactions, and the linked amenity services (such as aesthetic enjoyment, recreation, artistic and spiritual fulfillment, and intellectual development) have always been influenced and shaped by the nature of the ecosystem and ecosystem conditions in which culture is based. At the same time, humankind has always influenced and shaped its environment. Rapid loss of culturally valued ecosystems and landscapes lead to social disruptions and societal marginalization, now occurring in many parts of the world.

To achieve conservation and sustainable use of ecosystems, "traditional" and "formal" knowledge systems need to be linked. There is an emerging need and opportunity for building bridges between these two systems to improve the quality of human life. The complex relationships that exist between ecological systems and cultural systems can be understood only by linking our formal knowledge system, based on a hypothetical-deductive approach and inductive reasoning to understand ecosystems, with the traditional knowledge system, derived from societal experiences and perceptions. Our understanding of the tangible benefits derived from traditional ecological knowledge, such as medicinal plants and local species of food, is relatively well developed. However, our knowledge of the linkages between ecological processes and social processes, and their tangible and intangible benefits (such as spiritual and religious values), and of the influence on sustainable natural resource management at the landscape level needs to be strengthened.

Loss of traditional knowledge systems has many direct and indirect effects on ecosystems and human welfare. The loss of traditional knowledge has a direct effect on the depletion of fauna and flora and the degradation of the habitats and ecosystems generally. Traditional is knowledge is largely oral, and there is significant loss every time an old person dies without leaving a record of what they know. Equally significant is the loss of languages—the vehicles by which cultures are communicated and reproduced. It is estimated that more than 5,000 linguistic groups contain the traditional knowledge of humankind, many of which may disappear by 2020. TK is a key element of sustainable development, particularly in relation to plant medicine and agriculture, which may offer solutions and cures to pandemics such as AIDS and cancer as well as to many other health problems that are emerging with globalization.

The importance of cultural services and values is not currently recognized in landscape planning and management. These fields could benefit from a better understanding of the way in which societies manipulate ecosystems and then relate that to cultural, spiritual, and religious belief systems. This realization is reflected in the emphasis placed by many international organizations, such as UNEP, UNESCO, FAO, IUCN, and WWF, in recognizing "cultural landscapes," "cultural agro-ecosystems," World Heritage Sites, and Biosphere Reserves. The so-called ecosystem approach implicitly recognizes the importance of a socioecological system approach, and policy formulations should empower local people to participate in managing natural resources as part of a cultural landscape, integrating local knowledge and institutions.

In planning and managing ecosystems, a balance must be found between cultural and amenity services. Due to changing cultural values and perceptions, there is an increasing tendency to create landscapes with high amenity values (for aesthetic and recreational use, for example) at the expense of traditional landscapes with high cultural and spiritual values. The remaining traditional landscapes require urgent protection in order to create diversified landscape systems that contribute to strengthening buffering mechanisms and

that reduce the vulnerability of ecosystems and human society to environmental change.

Better information is needed on the economic importance of cultural and amenity services. Many cultural and amenity services are not only of direct and indirect importance to human well-being (in terms of improved physical and mental health and well-being), they also represent a considerable economic resource; for example, tourism generates approximately 11% of global GDP and employs over 200 million people. Approximately 30% of these revenues are related to cultural and nature-based tourism. In planning ecosystem use or conversion, these values have not been fully taken into account in the analysis of trade-offs. The costs of the loss of ecosystem services and the benefits of their continued availability should be shared more equitably among all stakeholders.

17.1 Introduction

17.1.1 Nature of the Service

Human cultures have always been influenced and shaped by the nature of the ecosystem (e.g., Ramakrishnan 1998). At the same time, humankind has always influenced and shaped its environment to enhance the availability of certain valued services. While there are specific cultural "services" that ecosystems provide (such as aesthetic enjoyment, recreation, spiritual fulfillment, and intellectual development), it is quite artificial to separate these services or their combined influence on human well-being. For example, a jogger in Central Park in New York City obtains a recreational benefit from that ecosystem through aesthetic enjoyment and physical exercise while simultaneously perhaps gaining spiritual benefits from watching a swan land in the lake. Similarly, a farmer in India may have a strong spiritual and religious connection to the local ecosystem and actively protect sanctuaries of forests. As a result, sophisticated health care systems associated with traditional knowledge of herbs often maintained in these forests may develop, and the cultural identity of the local society is maintained through close association with that local ecosystem.

Recognizing that different types of spiritual, intellectual, and physical links between human cultures and ecosystems are inseparable, this chapter seeks to explore the dimensions of the human-ecosystem relationship for the main types of cultural and amenity services provided by ecosystems and landscapes. Based on various literature sources (e.g., De Groot 1992; De Groot et al. 2002; Ramakrishnan et al. 2002; Van Droste et al. 1999) the following six categories have been distinguished:

- cultural identity (that is, the current cultural linkage between humans and their environment;
- heritage values ("memories" in the landscape from past cultural ties);
- spiritual services (sacred, religious, or other forms of spiritual inspiration derived from ecosystems);
- inspiration (the use of natural motives or artifacts in arts, folklore, and so on);
- aesthetic appreciation of natural and cultivated landscapes; and
- recreation and tourism.

Although cultural services are one of the four main service categories identified by the Millennium Ecosystem Assessment, they cannot be treated independently: cultural and amenity services depend especially on supporting and regulating services; at the same time, the expression of cultural services influences the way ecosystems are viewed in terms of their other services (for instance, fish have a food value but may also have a spiritual value, and fishing may be a traditional way of life).

Throughout this chapter, care has been taken to give a balanced representation of the main "worldviews" regarding human-nature relationships, ranging from those of the more traditional and indigenous societies to those of highly industrialized ones. There are striking differences in the way cultural and amenity services are perceived, experienced, and valued by different cultures, which can often be related to differences in the ecosystem conditions in which they originated and the way societies have changed ecosystem conditions and evolved with their environment. The dynamic nature of human-environment interactions leads to continuous changes in the perception and appreciation of cultural and amenity services and greatly contributes to cultural diversification.

17.1.2 Key Questions and Cross-cutting Issues

This chapter addresses how ecosystem changes affect cultural and amenity services and thereby human well-being. For each cultural service considered, three main issues are addressed: current status and dependence on ecosystem condition; observed changes in the availability of ecosystem services, causes for change, and future trends; and the effects on human well-being of changes in the availability of ecosystem services.

Thus for each service, a brief overview is given of its nature, its magnitude and distribution, and its dependence on ecosystem condition, illustrated by means of quantitative data where available on the ecosystem properties providing the service (such as landscape and biodiversity features) and with reference to the systems chapters in this volume (Chapters 18–27). It should be noted that the availability of cultural and amenity services is partly determined by the physical and biotic environment (such as the presence of landscape features with scenic, inspirational, or sacred values), and partly by culture. Thus similar environmental features (species, forests, soil, waterfalls, and so on) will be valued differently by different societies, depending on the cultural background and the way societies have shaped their environment during the course of their development.

In addition, changes in ecosystems in the recent past (since about 1960) and how these have influenced the capacity to provide cultural and amenity services (either positively or negatively) are described, along with predicted trends for the next 10 years. The direct causes for these changes will be briefly described, with reference to the proximate drivers or indirect causes described in Chapter 3.

The importance of a service to human well-being can be described by many different indicators (improved physical and psychological health, for instance, or income). (See Chapter 5.) Where available, examples are given of the economic importance of cultural and amenity services, including monetary data (with reference to Chapter 2, regarding methods and tools for economic valuation of ecosystem services). The consequences of changes in cultural and amenity services for human welfare are discussed near the end of the chapter.

17.1.3 Knowledge Systems

Cultural and amenity services are entirely determined by human perceptions of their environment. Human perceptions, in turn, are the product of the knowledge system of which the individual or community is a part. All knowledge systems, whether "traditional" or "formal" (or however labeled), reflect the history of ideas as much as some objective body of "facts." (The neutral term traditional is used here; other equivalent terms are local or indigenous, which tend to be much more location-specific. In contrast, "formal" knowledge is often referred to as "scientific."

One challenge is to validate the former and integrate it into the latter, to the extent possible.) Fundamental is the social context in which the traditional knowledge system of thousands of cultures has evolved. (See Box 17.1.) Important in this social construction is the idea of key paradigms (or mythologies), which even if not scientifically tested in the sense of being based on experiment and verification, are logical and provide insight in understanding how systems, including ecosystems, function (Berger and Luckmann 1966).

While formal knowledge in ecology has largely been a prerogative of natural scientists, analyzing natural phenomena through hypothetico-deductive methods and inductive reasoning, traditional knowledge evolves locally in different communities through an experiential approach, with differences in the way each creates knowledge. Except for some instances involving direct economic values, such as non-timber forest products that may have food, fiber, or medicinal value, the origin and meaning of this knowledge has not been properly documented (Berkes 1999), and there is significant loss every time an old (knowledgeable) person dies without leaving a record of knowledge and experience.

The loss of traditional knowledge has a direct effect on the depletion of fauna and flora and the degradation of the habitats and ecosystems generally. For example, in the transmigration program in Indonesia the traditional knowledge of the transmigrant is of no value under the changed ecological situation, leading to adoption of wrong technologies and ending up in land degradation (Whitten et al. 1987).

Equally significant is the loss of languages, which are the main vehicles by which cultures are communicated and reproduced (in addition to the reflection of human-nature relationships in dance, other art forms, rituals, and architecture, such as in Stonehenge and the Pyramids). It is estimated that there are more than 5,000 indigenous linguistic groups, representing over 350 million people, which contain most of humankind's traditional knowledge. Many of these linguistic groups may disappear by 2020 (United Nations 2004), which is an important obstacle to finding pathways for more sustainable ecosystem management (Berkes et al. 2000). It is also true that much of the traditional knowledge that existed in Europe (such as knowledge on medicinal plants) has gradually eroded due to rapid industrialization during the past century (Hughes 1998).

17.2 Distribution, Magnitude, and Trends in Cultural and Amenity Services

17.2.1 Cultural Identity

Throughout human evolution, human societies have developed in close interaction with the natural environment, which has shaped their cultural identity, value systems (Balee 1989), and economic well-being. However, since the human-nature relationship is influenced by factors such as ownership, ethics, religion, and so on (Hanna and Jentoft 1996), it varies widely across cultures, evolving in both space and time. For instance, for many traditional forest dwellers in the tropics, shifting agriculture is a way of life; for those living in the savanna grasslands of tropical Africa, nomadic pastoralism is a major activity (with limited shifting agriculture), while others living under more extreme climatic conditions, such as the peoples of the Tibetan and central Asian highlands, tend to be nomadic pastoralists and those living in coastal areas and the Arctic regions tend to be depend on fishing. This variety of lifestyles and livelihoods, which are "dictated" by

---

BOX 17.1

Traditional Knowledge Systems

Many traditional societies (including indigenous and tribal) with extended association with nature and natural resources have accumulated empirical knowledge about the natural resources around them, especially food and medicines (National Academy of Sciences 1975; Berlin 1992; Hladik et al. 1993). Many such societies also have accumulated traditional wisdom based on the intrinsic realization that humans and nature form part of an indivisible whole and therefore should live in partnership with each other. This ecocentric view is widely reflected in their reverential attitudes toward plants, animals, rivers, and Earth, often concretized in iconography and imagery of the sculptural forms, a way of transmitting the timeless truths of human-nature ethics (Vatsayan 1993).

Traditional ecological knowledge, although it may have a strong element of the "formal," stands apart in that it is largely derived through societal experiences and perceptions accumulated through a process of trial and error during interactions with nature and natural resources. This implies that while "formal" emphasizes universality of the knowledge created by the given methodology, TEK has a certain degree of location-specificity, but with a strong human element that emphasizes social emancipation (Elzinga 1996). Traditional knowledge enables society to relate to

a value system that they understand and appreciate and therefore participate in the process of the quality of life they cherish.

The dichotomy between the universality of formal knowledge and the location-specific nature of TEK hides two distinct elements: the difference between scientific knowledge and common sense (which concerns all societies) and the difference between cultural patterns of thought embedded in the formal knowledge and non-western approaches of the natural and social world. It should be added, however, that below the considerable location-specific diversity, TEK often has undeniable universal characteristics.

In any case, we need to move beyond this perceptional divergence and arrive at generalizations across locations, after validation from an eco-science perspective where required, in order to integrate the two knowledge systems and use them for ecosystem management. For example, traditional systems of medicine such as Ayurveda, which is well developed throughout India, are now getting linked with cultural tourism in this part of the world, which is tending to be of global value. This is in addition to hundreds of ethnic medical practices spread across the world. Similarly, a whole variety of lesser-known plants of food value have not been integrated into our food production systems (National Academy of Sciences 1975).

---

different ecosystem conditions, led to different knowledge systems and to cultural diversification.

17.2.1.1 Current Status and Dependence on Ecosystem Condition

Language, knowledge, and the environment have been intimately related throughout human history. Local and indigenous languages are the repositories of traditional knowledge about the environment and its systems, its management, and its conservation, which in the contemporary context needs analysis and validation. (See Figure 17.1.) (Ramakrishnan 2001; Ramakrishnan et al. 2004).

Approximately two thirds of the world's languages are linked to forest-dwellers; indeed, almost 50% of all languages are spoken in tropical/sub-tropical moist broad-leaved forest biomes (see www.terralingua.org). Furthermore, nearly 24% of all languages are spoken in tropical and sub-tropical grassland, savanna, and shrubland biomes. But just as with species, the world is now undergoing a massive extinction crisis of languages and cultures. At present, the greatest losses are occurring in high-risk situations, such as where languages are not officially recognized and people are marginalized by rapid industrialization, globalization, depopulation, poor health, low literacy, or considerable ecosystem degradation. Especially threatened are the languages of indigenous peoples, who number 350 million, representing over 5,000 linguistic groups in 70 countries, according to a special UNESCO meeting in New York in May 2004 (see www.unesco.org/culture/indigenous).

External forces, especially national and international development policies, are dispossessing traditional peoples of their land, resources, and lifestyles, forcing them to subsist in highly degraded environments. People who lose their linguistic and cultural identity may lose an essential element in a social process that commonly teaches respect for nature and understanding of the natural environment (Ramakrishnan et al. 1998). Many traditional societies view culture and environment as complementary, and efforts aimed at maintaining cultural identity also often promote environmental conservation (Stevens 1997). The concept of "cultural

landscapes" (described in the following section) is an example of traditional societies co-evolving with their environment. (See Box 17.2.)

17.2.1.2 Observed Change, Causes of Change, and Future Trends

Human societies are not immune to changes in their environment. The continuing overconsumption of natural resources is resulting in erosion of time-tested and value-based institutions in many societies. Among the most powerful forces that influence both local cultures and ecosystems are various government policies and the expansion of national, regional, and international markets that stimulate privatization of land and aim to "fix" populations in a particular space, leading to a loss of traditional lifestyles (as with pastoralists and nomadic peoples).

For example, central government policies in Somalia in the 1970s and 1980s sought to "settle" semi-nomadic groups so they could be better "controlled" and provide taxes to government. Another example is government policies that are driven by international market forces determining coffee prices, which in the Western Ghat region in southern India resulted in the extension of coffee plantations into dried zones that are ecologically unsuitable for production, leading eventually to abandonment of the plantations and forest degradation (Ramakrishnan et al. 2002).

The rapid decline in traditional value systems and changing values among the younger generation are linked phenomena that are widespread. Human societies, traditional or otherwise, always tend to perceive the landscape around them as a carved-out cultural landscape. Indeed, now there is a renewed interest even about urban landscapes that could be made self-sustaining to the extent possible through urban agriculture (sometimes referred to as "urbaculture"), a variety of city-based gardens, "bioshelters," green corridors or greenways, and so on (Burel and Baudry 2003).

In the mountain regions of both the developing and the industrial world, there is an increasing realization that the lost cultural landscape should be conserved where they exist or redeveloped where they are already lost (Ramakrishnan et al. 2003; Maurer and Holl 2003). Particularly in the developing-

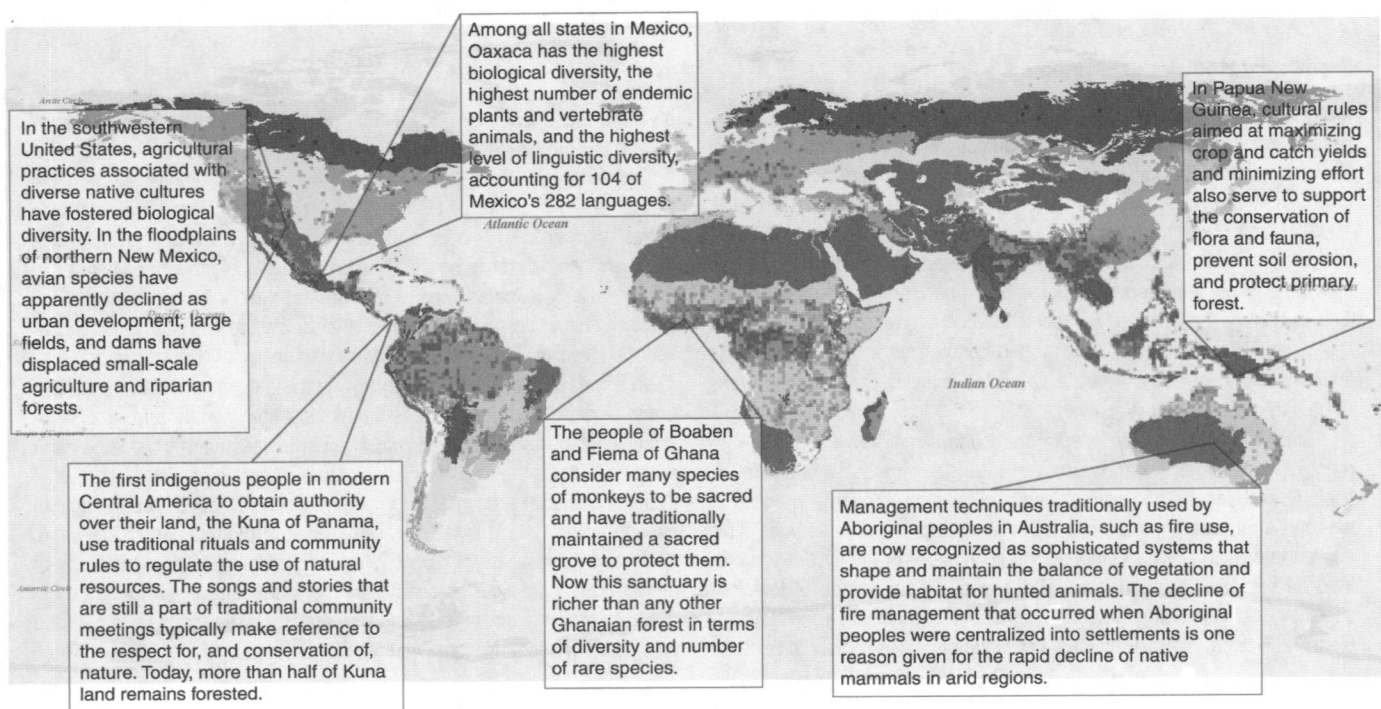

In the southwestern United States, agricultural practices associated with diverse native cultures have fostered biological diversity. In the floodplains of northern New Mexico, avian species have apparently declined as urban development, large fields, and dams have displaced small-scale agriculture and riparian forests.

Among all states in Mexico, Oaxaca has the highest biological diversity, the highest number of endemic plants and vertebrate animals, and the highest level of linguistic diversity, accounting for 104 of Mexico's 282 languages.

In Papua New Guinea, cultural rules aimed at maximizing crop and catch yields and minimizing effort also serve to support the conservation of flora and fauna, prevent soil erosion, and protect primary forest.

The first indigenous people in modern Central America to obtain authority over their land, the Kuna of Panama, use traditional rituals and community rules to regulate the use of natural resources. The songs and stories that are still a part of traditional community meetings typically make reference to the respect for, and conservation of, nature. Today, more than half of Kuna land remains forested.

The people of Boaben and Fiema of Ghana consider many species of monkeys to be sacred and have traditionally maintained a sacred grove to protect them. Now this sanctuary is richer than any other Ghanaian forest in terms of diversity and number of rare species.

Management techniques traditionally used by Aboriginal peoples in Australia, such as fire use, are now recognized as sophisticated systems that shape and maintain the balance of vegetation and provide habitat for hunted animals. The decline of fire management that occurred when Aboriginal peoples were centralized into settlements is one reason given for the rapid decline of native mammals in arid regions.

Figure 17.1. Links between Language, Culture, and the Natural Environment: Some Examples (Map produced by Terralingua in partnership with the Conservation Biology Institute; data on the world's languages made available by SIL International; www.terralingua.org)

BOX 17.2

Some Examples of Evolving Human–Nature Relationships

- To Naskapi Indians of Labrador, ownership means shared identity (Henriksen 1986). With deep respect for the harsh environment in which they live, the dependence on nature and natural resources is reflected in the ethnobiological knowledge they possess.
- For others, like the Bushman of Australia, this linkage is reflected in the ritual acts used to kill animals (Campbell 1996).
- For the Lake Racken fishing community concerned with crayfish management, the way in which the formal knowledge system is contextualized with traditional knowledge represents a recent adaptation to combat acidification problems in the lake (Olsson and Folke 2001).
- Combining traditional knowledge with the formal in a complementary fashion, the livelihood needs of the Inuit and Cree communities in the Hudson Bay area of Canada were harmonized in the context of the impact of hydroelectric dams, for effective co-management of natural resources in the area (Fenge 1997).
- In the Great Fish River Valley in South Africa, local Xhosa people place great cultural and utilitarian value on key resource patches such as mountains, forests in various stages of succession, and a variety of grazing lands. In many cases the diversity of resource patches is the consequence of people interacting with the land, where, through a variety of induced disturbances, these resource patches are created. The different types of resource patches provide different kinds of resources, thus satisfying villagers' basic needs. These include both practical, physical needs as well as cultural and spiritual needs (see MA *Multiscale Assessments,* South Africa).

country context, where rural poor abound, the developmental paradigm based on high-energy input monoculture of crops is increasingly debated (Ramakrishnan 2001). Thus, for instance, are we satisfied with having patches of protected biodiversity in the form of nature reserves, placed as islands in a vast ocean of monocultures, or are we looking for more heterogeneity in our landscapes, so that biodiversity is not merely restricted to nature reserves? The latter approach will provide greater resilience to the biosphere by strengthening the internal buffering mechanisms against uncertainties in the environment (see, e.g., Holling 1995).

17.2.1.3 Consequences of Change for Human Well-being

The observed estrangement of people from their land and traditional way of life leads to overexploitation and degradation of ecosystems, which in turn leads to poverty and loss of cultural identity. (For a more in-depth discussion, see Rutten 1992.) Unless ecosystem management is firmly rooted in the local cultural ethos, it can affect the livelihood concerns of large numbers of people, particularly marginalized societies in the developing world, causing social disruptions and ecological degradation. There is an increasing danger of culture-specific land use systems being gradually wiped out, without any viable alternatives in place. If this trend continues, apart from ecological catastrophes, large-scale social disruptions could occur, as is already evident among many traditional societies (United Nations 2004).

For a new perspective to emerge, and to ensure that human well-being and cultural identity remain linked to ecosystem services, there needs to be a reconciliation between ecology, economics, and ethics. The challenge, therefore, lies in learning lessons from the past and in developing an adaptive management strategy that is economically sound and specific to the socioecological system in question.

17.2.2 Cultural Heritage

A large part of our cultural heritage is associated with ecosystems and landscapes with special features that remind us of our historic roots, both collectively and individually (such as special, usually old trees, the remains of traditional cultivation systems, or historic artifacts). These ecosystems and landscape elements give us a sense of continuity and understanding of our place in our natural and cultural environment and are increasingly valued as expressed by the designation of cultural landscapes and sites with special historic interest.

17.2.2.1 Current Status and Dependence on Ecosystem Condition

Cultural landscapes are complex socioeconomic expressions of (mainly) terrestrial ecosystems that have co-evolved under the influence of biophysical factors (such as climate, relief, soil type, water availability, and so on) as well as of human societies at different levels of their cultural, social, and technological development. In many places in the world, long-standing traditions in agri-, silvi-, viti-, and aqua-cultural ecosystem management have contributed to the development of a wide range of productive and characteristic landscapes on cultivated systems. (See also Chapter 26.)

Often this ecosystem management is based on traditional ecological knowledge, sociocultural practices, or religious beliefs, and human perception therefore has a strong influence on defining landscapes. This is echoed by Ellis et al. (2000), whose hierarchical landscape classification system builds upon ecotopes that are defined as "the smallest homogeneous ecosystem units within landscapes." Thus, both natural and cultural features are taken into account when proposing the following definition: "Cultural landscapes are spatially defined units whose character and functions are defined by the complex and region-specific interaction of natural processes with human activities that are driven by economic, social and environmental forces and values" (Wascher 2004)

Hence, sustainable cultural landscapes should offer both high heritage values and (relatively) stable ecosystem functions. Ideally, these objectives should be reached on the basis of efficient resource management (wise use), seeking synergy between ecosystem processes and cultural interferences (the latter including economic interests). Table 17.1 illustrates the linkages between cultural landscapes and associated ecosystem functions.

Table 17.1 and several examples illustrate the large variety among cultural landscapes and heritage services in terms of scale and character. In the Netherlands, the historic *slagen* (long stretched land parcels) landscape *Krimpenerwaard* is a specific type of *polder* landscape situated in the "Green Heart" of the country. The Green Heart–*polder* is located between Amsterdam, Rotterdam, and the Hague and is a land reclamation system based on a systematic drainage process that determines the characteristic structural and functional landscape patterns of the area. Its characteristic features include long and narrow access roads; straight, parallel drainage ditches in regular sequences linking up with naturally meandering water courses in right-angle patterns; land segregations; blind alleys; and numerous parallel ditches.

In Portugal and Spain, *montado* and *dehesa* landscapes consist of open evergreen forests of cork and holm oaks *Quercus* spp., or open oak savanna, with tree densities ranging from 20 to 60 trees per hectare in an irregular pattern, with relatively open understory or partially closed by shrub encroachment. Despite its use for cork production and multi-functionality with regard to other agricultural management regimes (such as grazing and small-scale crop-

land), the *montado* and *dehesa* landscapes are also valued for their biological diversity, heterogeneity, and cultural interest due to their strong identity and recreation potential (Ferreira et al. 2003).

Many cultural landscapes, such as the River Ganges and parts of the Himalayas, are defined by their religious significance and are of great importance to a large portion of the world's population, as described later in this chapter.

Thus it is clear that maintenance of cultural heritage is an important service of especially semi-natural and cultivated ecosystems and landscapes. Many European countries have therefore developed specific policies and legislation for the conservation of cultural landscapes, and many private organizations are engaged in their care. In the United Kingdom, for instance, the National Trust owns or manages 200 historic houses, 230 gardens, and 25 industrial monuments plus 240,000 hectares of beautiful countryside and 550 miles of coast. At the global level, initiatives have also emerged to conserve landscapes directly—through, for example, the World Heritage Convention (UNESCO 1972; Rössler 2000). (See Box 17.3.)

Within the European Union, national agricultural legislation typically set objectives for the protection and restoration of landscapes and to provide public access to these landscapes. In addition to regulations and voluntary agreements, many OECD countries adopt economic incentives for agricultural landscape conservation and restoration (see Table 17.2), such as through area payments and management agreements, which can be interpreted as a rough approximation of the "willingness to pay" for the maintenance of cultural and heritage values.

Other initiatives target field-based collaborative management at the local and regional levels, including transboundary regions. For instance, the Collaborative Management Working Group within IUCN's Commission on Environmental, Economic and Social Policy promotes and supports field-based co-management initiatives, draws lessons and methods from experience, and supports the development of participatory mechanisms for the management of natural resources through local capacity building (knowledge, skills, attitudes, and institutions) and the elaboration of national, regional, and global policies. Projects address a number of topical areas such as the co-management of protected areas and agricultural landscapes and the involvement of local communities in ecosystem conservation, with an emphasis on poor communities in particularly harsh and fragile ecosystems, such as arid lands, mountains, and coastal areas. (See also MA *Policy Responses,* Chapter 14.)

17.2.2.2 Observed Changes, Causes of Change, and Future Trends

In cultural landscapes, ecosystem processes are mainly driven by human land use changes. Because these have taken place over the entire history of human civilization, it is difficult to introduce objective, widely accepted points of reference. Compared with early cultivation history, however, modern forms of land management and reclamation appear to have more erosive effects on the character and processes of traditional cultural landscapes. Dominant trends include decreasing landscape diversity, altered hydrological systems (drainage and irrigation), intensification of land use, and landscape fragmentation, all of which have affected human social structures, ecosystem functions, and heritage values. Even protected sites, including many of those designated under the World Heritage Convention, are at risk of losing their status due to various internal and external pressures. African, Arab, and Asian UNESCO sites appear to be at higher risk than those in Europe or in North and Latin America. (See Figure 17.2.)

Table 17.1. Examples of Cultural Landscapes, by Biome, with Selected Ecosystem Functions

Biome	Cultural Landscapes	Some Examples of Ecosystem Functions	Ecosystem State and Characteristics
Humid tropical	Salina landscape (Densu Delta, Ghana)	habitat for thousands of wetland species 20 communities with fishing being their primary activity million-dollar salt industry	Ramsar wetland: 6,700 hectares tidal influences extend upstream for some 10km heavily populated with urban estate development
Semiarid tropical	Arnhem land/ dreamland (Australia)	revitalization of native flora and fauna through patch fire management preventing disastrous wild fires tourism main income due to attractivity of Kakadu and Litchfield National Parks	eucalypt grassy woodlands and open tropical savannas pastoral or Aboriginal land management major threats are changes in the fire regime, feral carnivores, cattle grazing, and mining
Humid temperate	hedgerow landscapes (e.g., France, United Kingdom, Germany)	protection against soil erosion wood production grassland farming habitat/corridors for native species acting as natural pest control recreation	regionally distinctive types, regarding patterns, plant compostions, materials, and management threats: agricultural intensification and abandonment
Warm Mediterranean	Dehesa (Spain) and Montada (Portugal)	cork is key export business openland pig farming and transhumance (local products) high biodiversity hunting grounds micro-climate	characteristic pattern of evergreen forest in variable densities of native cork oaks threats: extensification and abandonment, fires, irrigation projects, tree diseases
Semiarid boreal	prairie pothole land- scape (Canada)	farmland hunting ("duck factory") biodiversity	mosaic of 4 million small wetlands; 51 percent of all North American breeding ducks threats: agricultural activities (pesticides, nutrients)
Warm desert	farm-based wildlife landscapes (Namibia)	wildlife-based rural development biodiversity (including elephant and endangered black rhinoceros) tourism	75 percent of wildlife is found in these landscapes threats: hunting
Cold desert	Ladakh landscape	unique architecture makes use of local materials such as mud, stone, and wood and of indige- nous construction techniques (Gupta 2000)	cold high-altitude desert rainshadow region, cut off Himalaya monsoon clouds chemical reactions in rocks carved fantastic ("lunar") landscapes

Four basic driving forces are considered to affect cultural land- scapes: polarization of land use (intensification, extension, aban- donment of land, and simplification of land use, which in turn is driven by national and international policies that stimulate mono- cultures and cash crops); policy responses (site protection, agri- environmental measures, planning schemes, and so on); infra- structure, urbanization, tourism, resource extraction, and energy facilities; and climate change and its effects on ecological, land use, and demographic systems.

During recent years there has been increasing public demand for cultural landscape and associated amenity goods and services linked to rising disposable incomes, more leisure time, and other factors. Public and policy-driven shifts toward greater land use diversification, small-scale developments, and more environmen- tally friendly land management have also occurred. Increasing awareness of these issues, especially in Europe and Japan, favors multifunctional landscapes that provide humans with food and raw materials, drinking water, space for recreation, a sense of identity, and heritage values (Wascher 2000).

17.2.2.3 Consequences of Change for Human Well-being

Cultural landscapes include living societies as an integral part of their landscape units. From a socioecological viewpoint, these in-

terconnections are significant for ensuring a sustainable livelihood for traditional societies, such as the shifting agricultural societies in the tropics (Ramakrishnan 2001) and in many Central and East European countries, and loss of these cultural landscapes can have many social and economic consequences. (See Box 17.4.)

A review of the past 30 years of implementation of the World Heritage Convention reveals a broad interpretation of the heri- tage concept. The inclusion of cultural landscapes, and in particu- lar those associated with natural elements rather than material cultural evidence (which may be insignificant or even absent), has changed the perception and the practice of the convention. This evolution in the interpretation of the World Heritage Convention represents a growing recognition of the wealth and complexity of numerous values (including intangible ones) associated with protected areas, and in particular with sites of outstanding ecolog- ical or cultural value. Experience has shown that an inclusive ap- proach is crucial for the designation and management of World Heritage sites, for the benefit of the people living in and around them, of the conservation community, and of humanity as a whole (Rössler 2000).

17.2.3 Spiritual Services

Most people feel the need to understand their place in the uni- verse, and they search for spiritual connections to their environ-

World Heritage Cultural Landscapes

The Convention Concerning the Protection of the World Cultural and Natural Heritage (known as the World Heritage Convention), adopted by the General Conference of UNESCO in 1972, established a unique international instrument that recognizes and protects both the cultural and natural heritage of outstanding universal value (Rössler 2000). The World Heritage Convention's definition of heritage provided an innovative and powerful opportunity for the protection of cultural landscapes as "works of man or the combined works of nature and man."

Although there is still debate about the criteria for selecting World Heritage Sites and the type of management imposed on them, the impact of the inclusion of cultural landscapes in the implementation of the World Heritage Convention was considerable in many ways, such as for the recognition of intangible values and of the heritage of local communities and indigenous people; for the importance of protecting biological diversity by maintaining cultural diversity within cultural landscapes; for the management and traditional protection ensuring the conservation of the nominated cultural properties or cultural landscapes; and for the interpretation, presentation, and management of the properties.

Many cultural landscapes have been nominated and inscribed on the World Heritage List since the 1992 landmark decision to include them in the list. (See Figure.) Often they are associative cultural landscapes, which may be physical entities or mental images embedded in a people's spirituality, cultural tradition, and practice.

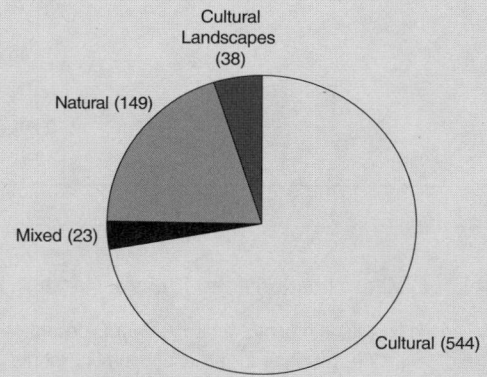

Distribution of 754 World Heritage properties located in 129 State Parties

Cultural Landscapes (38)
Natural (149)
Mixed (23)
Cultural (544)

World Heritage sites generally are cornerstones in national and international conservation strategies. This far-reaching concept faces new challenges in the future, including:

- creating new institutional networks between international instruments, but also protected area agencies, to fully explore the links between the different categories and protection systems—such a complementary relationship might be formalized through a close link between the World Heritage Convention and other international agreements such as the European Landscape Convention;
- enhancing new partnerships, as recommended by the Venice celebration on 30 years of the World Heritage Convention; and
- enlarging the circle in sharing information about protected area systems and cultural landscapes, in particular on achievements, success stories, and model cases.

One topic to be explored is how World Heritage sites can serve as cornerstones for sustainable local and regional development.

ment both through personal reflection and more organized experiences (as part of religious rules, rituals, and traditional taboos, for example). Ecosystems provide an important measure for this orientation in time and space, which is reflected by spiritual values placed on certain ecosystems (such as "holy" forests), species (sacred plants and animals, for instance), and landscape features (such as mountains and waterfalls). (See Box 17.5.)

17.2.3.1 Current Status and Dependence on Ecosystem Condition

The initial impetus among early civilizations and contemporary traditional societies (those living close to nature and natural resources) for biodiversity conservation seems to have arisen out of religious belief systems. The most common element of all religions throughout history has been the inspiration they have drawn from nature (*physis*), leading to a belief in non-physical (usually supernatural) beings (Frazier 1922). The idea of "unity" between humans and nature is present in all major religions and influences the management of ecosystems and our attitude toward species. The concept of *Sarvabhutadaya* in Buddhism implies that humans are an integral part of the ecosystem, with a sense of compassion and fellowship—that we give back what we have taken from the biosphere. In the Bible and the Koran, reference is made to the importance of nature as a source of life for humans and their fellow-creatures.

Thus belief systems are a fundamental aspect of people's culture that strongly influences their use of natural resources. The concept of the "scared grove" (ecosystem) that traditionally served as an area for religious rituals to appease nature-linked deities (the Wind, Water, Fire, Sun, and so on) as well as a site of worship for ancestral spirits could be viewed as symbolic of the spiritual services derived from nature. Traditional societies all over the world have institutionalized sacred landscapes and ecosystems in a variety of ways, large and small, as part of their belief systems. (See, for example, *Places of Peace and Power* at www.sacredsites.com and *The Sacred Mountains Foundation* at www.sacredmountains.com.) Sacred groves, once strictly protected for cultural and religious reasons, now often remain as islands of biodiversity in an otherwise degraded landscape and are widespread across the globe. (See Box 17.6.)

Perhaps because of their awe-inspiring landscape characteristics, mountains, for instance, have been linked to all major religions in all continents and are sacred to nearly 1 billion people (Wijesuriya 2001; Berbaum 1997). Examples include Mount Kaila (Himalayas), Adams Peak (Sri Lanka), and the Sierra Nevada de Santa Marta (Colombia). There are also sacred or culturally valued species that stand out as a class apart. Sometimes these have restrictions on their usage (see Box 17.7), but in any case such species have implications for management of natural ecosystems with community participation, as described at the end of this chapter.

In addition to the more formalized spiritual ties between humans and nature, there are many other examples of the spiritual importance of ecosystems and species, such as the classic work by Aldo Leopold (1949) on land ethics and the feeling of spiritual enlightenment that many people experience when viewing wildlife (whales, for instance) or "inspiring" landscapes.

17.2.3.2 Observed Changes, Causes of Change, and Future Trends

Changes in geographic religious spheres of interest (such as the advent of Christianity in Europe), industrialization and urbanization, and many other social, political, and institutional changes

Table 17.2. Landscape Conservation Schemes and Funding for Selected Countries, 1998. The share of total expenditure on biodiversity, habitats, and landscape as a percentage of the total producer support estimate for 1998 was as follows: Canada: < 1%; Norway: 20%; Poland: < 1%; Switzerland: 4%; and EU: < 1% (the percentage for EU is higher than this, however, as only 9 member states are included in this calculation, while the PSE covers 15 member countries). (OECD 2001)

Scheme	Objective	Area	Share of agricultural area	Funding
		(thousand hectares)	*(percent)*	*(thousand 1998 dollars)*
Austria				
Mountains and less favored areas	landscape	1,214	35	238,301
Finland				
Supplementary protection	landscape	173	6	37,594
Greece				
Maintainance of landscape elements	landscape	..	..	5,594
Japan				
Yusuhara village	landscape	..	..	31/hectare
Netherlands				
Landscape conservation subsidy	landscape	..	..	623
Landscape and farmyard planting	landscape	0.15	< 1	1,246
Landscape elements (province)	landscape	..	..	2,928
Norway				
Area and cultural landscape	landscape	1,050		524,165
Preservation of buildings	architecture	370		
Local management of areas	landscape	50	15	1,590
Portugal				
Maintaining traditional farming	landscape	439	11	46
Sweden				
Conserving biodiversity and cultural heritage	nature and culture	1,583	51	140,242

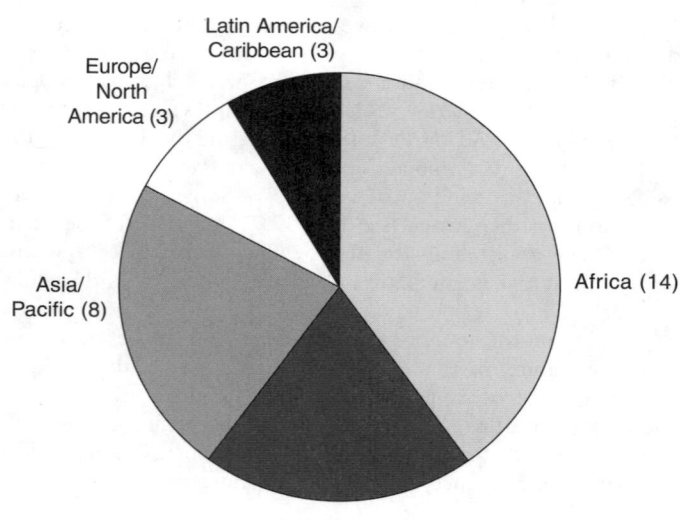

Figure 17.2. Regional Distribution of World Heritage Sites in Danger, 2004

Latin America/Caribbean (3)
Europe/North America (3)
Asia/Pacific (8)
Arab States (7)
Africa (14)

BOX 17.4

Loss of Ecosystem Functions and Cultural Heritage Values

- Farming in the limestone hills of Southwest Cyprus became economically less rewarding, resulting in the abandonment (shrub growth) and destruction of traditional landscape elements (Dower 2000).
- In *dehesa* landscapes (Spain), the planting of conifers (*Pinus pinaster*) and exotic broad-leaved trees (*Eucalyptus ssp*) brought about the most radical change, entirely replacing major parts of *dehesa* landscapes with large single-species plantations.
- Over the last 30 years, Cinque Terre (Liguria, Italy) is dramatically losing its traditional landscape character: approximately 85% of the terraces built and maintained over 1,000 years have fallen into disrepair and been abandoned (Stovel 2002).

over time (including the education system), spurred by economic development, led to the decline of many traditional belief systems in many parts of the world. This had a large impact on the exploitation of natural resources and the way ecosystems have been managed. The impact of culture-linked change in natural ecosystems is expressed through the rapid changes seen in the perception of societies toward culturally valued ecosystems and landscapes, notably "sacred groves." Destruction of these sacred ecosystems

BOX 17.5

Spiritual Traditions Linked to Nature and Natural Resources

- Pre-Columbian societies in the Americas held the widespread view that Earth and all her creatures are sacred and that therefore permission had to be sought before the resources could be used, or else the spirits of those resources would seek revenge (Hughes 1998).
- For the enlightened sages of the eastern tradition, the forest is a world of wisdom, peace, and spirituality. The term "Aaranya," in the Sanskrit language of antiquity, comes from Aa for "no" and Ranya for "war," meaning a place of nonviolence (Saraswati 1998).
- A strong feeling of human participation in the universal order pervades the Vedas, the ancient scriptures of the Hindu religion, which is an oral tradition of wisdom, at least 5,000 years old (Vannucci 1993).
- The concept of the Cosmic Tree (the Tree of Life) represents the center of the Universe in the eastern culture and is part of many traditional belief systems.
- The cosmologies of American Indians, Australian aboriginals, New Zealand Maori, and many others are intimately connected with the land (Carmichael 1994) and extend to cover all elements of nature such as mountains, rivers, plants, animal, fish, and even human beings (Matunga 1994; Wijesuriya 2001).

BOX 17.6

Sacred Landscapes and Groves around the World (Hughes and Chandran 1988)

- In Africa, possibly the original home of humankind, sacred groves still exist in the sub-Saharan region. For the Kikyus of East Africa, cutting trees, breaking branches, gathering firewood, burning grass, and hunting animals are prohibited from groves that have the sacred Mugumu tree. These are still common in Ghana, Nigeria, Zimbabwe, and South Africa, often under the control of the local tribal leader. In Egypt, it was an ancient practice to have a sacred grove along with a sacred lake. Egyptians conserved many sacred species such as Palm and Persea (*Mimusops laurifolius, M. shcimperi*, called ished in Egyptian).
- Siberians used the groves for the rites of Shamanism. The nomadic Ostyaks and Voguls of the Ob river basin protected them very strictly, considering even eagles alighting in a grove as sacred.
- Chinese, Japanese, and Koreans have many groves linked with Buddhist temples. Shifting agriculture-based hill people of the Yunnan province in China have designated sacred woodlands. Balinese in Indonesia have "monkey forests," which are fragments of the ancient rain forest dedicated to the Hindu monkey God, Hanuman.
- Australian aborigines have groves dedicated to ancestral spirits of the ancient "Dreamtime," when the landscape was shaped. Maoris of New Zealand call the sacred sites Waahi Tapu, which include trees and forests, among many other natural features.
- Europe had thousands of sacred groves in ancient times, such as Mt. Atlas in Greece, with its sacred forests, and the Celts, Slavs, and Germans all worshipped in groves and regarded the Oak as the most divine tree.
- The Maya people cultivated certain trees like Cacao (*Theobroma cacao*) for a valuable drink for Mayan priests and royalty, and its seeds were widely used as currency in Mesoamerica. Tribes such as the Ojibwas and Utes reserved certain sections of the forest where hunting was prohibited, except when in great need.

(for timber, for instance, or through warfare) started in the fifth century BC during the Persian invasion in Greece. And with the advent of Christianity, most of the sacred groves and sacred sites in Mediterranean Europe were eliminated, being considered "pagan" (Hughes and Chandran 1988). Similarly, in the northeastern hill area of India, only a few scattered sacred groves now remain where formerly each Khasi village had its own (Ramakrishnan 1992).

In more recent times, there has been a growing interest in protecting the value systems of indigenous communities through initiatives such as natural heritage and cultural heritage conservation, human rights, and so on—as in *Akwé: Kon Voluntary Guidelines for the Conduct of Cultural, Environmental and Social Impact Assessments Regarding Developments Proposed to Take Place on, or which are Likely to Impact on, Sacred Sites and on Lands and Waters Traditionally Occupied or Used by Indigenous and Local Communities* from the Secretariat of the Convention on Biological Diversity and in the IUCN working group on Cultural Values of Protected Areas. (See Box 17.8.)

17.2.3.3 Consequences of Change for Human Well-being

The world is passing through an "emerging systems" view of life, mind, and consciousness and human evolution, which could have profound consequences for our social and political structures (Capra 1982). On a spiritual dimension, slow gradual changes in value systems and cultural values have already started happening. The traditional wisdom, embedded in the concept of sacred species, ecosystems, and landscapes and its revival in the contemporary context of biodiversity conservation (such as World Heritage Sites) is worth noting. Rather than taking a merely mechanistic view of Earth processes, where humans are continually struggling for unlimited material progress through economic growth mediated by technological innovations, a greater appreciation of interconnections between ecological and social systems is emerging.

17.2.4 Inspirational Services

Natural and cultivated systems inspire an almost unlimited array of cultural and artistic expressions, including books, magazines, film, photography, paintings, sculptures, folklore, music and dance, national symbols, fashion, and even architecture and advertisement. Consciously or subconsciously, representations of natural (and cultivated) ecosystems in art, writings, and so on remind us of our ties with nature (and our cultural heritage) and shape our views and appreciation of the represented ecosystems and species.

17.2.4.1 Current Status and Dependence on Ecosystem Condition

Five main types of inspirational services are distinguished and briefly described here: verbal art and writings inspired by nature, the performing arts, fine arts, design and fashion, and the media in general.

Many literary and oratory works use nature as a source of inspiration. Poet-naturalist Henry David Thoreau spent a year living in a simple cabin at Walden Pond in Concord, Massachusetts, in 1845, which resulted in *Walden,* his eulogy on nature and its spiritual dimension—long considered a classic of the genre. Naturalist John Muir believed that "wilderness mirrors divinity, nourishes humanity, and vivifies the spirit," while Ralph Waldo Emerson, in his first essay "Nature," published in 1836, claimed that spirit is present behind and throughout nature (Enger and Smith 1995). Since then, many writers have had a strong inspirational impact, such as Aldo Leopold's *A Sand County Almanac*

BOX 17.7

Sacred Species (Ramakrishnan et al. 1998)

- The Bodhi (Pipal tree; *Bot. Ficus religiosa*) is sacred to Buddhists. The tree that provided shelter for the Buddha to attain enlightenment is in Bodhgaya in India (recently declared as a World Heritage Site). Its sapling was sent to Sri Lanka in the third century BC and is still surviving, thus qualifying as the oldest recorded tree. It is one of the most sacred places of the Buddhists in Sri Lanka, and the Na tree (National Tree) is sacred as it is extensively used for temple building and supports associated bird diversity.
- The Sacred Lotus (*Nelumbo nucifera Gaertn*), an icon of Buddhism (associated with Buddhist heaven) and Hinduism (also associated with the energy centre of the human body) and the national symbol of India, is revered for its sanctity, for its multipurpose medicinal properties, and for numerous uses of the whole plant, all over Asia.
- Prevalent in the Mediterranean region, the sacred value is attached to species like oak, olive, apple, and may even extend right up to the Central Himalayan region, where oaks (*Quercus spp*) are culturally valued keystone species in an ecological sense, acting as a trigger for ecosystem/landscape rehabilitation.
- *Ocimum sanctum* (locally known as Tulsi) is an important multipurpose medicinal plant, which is not only worshipped as a Goddess incarnate but also put on an elevated platform in the entrance to Hindu homes.

BOX 17.8

Global Concern for Protecting Biodiversity-linked Spiritual Values

The ILO Convention on Indigenous and Tribal Peoples, 1989, though signed by only 14 state parties, suggests the need to uphold indigenous and tribal peoples' right to recognition and retention of customary law and practices, with special reference to control over land and resources, with many more new initiatives.

The World Heritage Convention in 1972 recognized that culture and nature are complementary and started listing both natural and cultural products of "outstanding universal values" and developed the concept of the cultural landscape, thus recognizing the spiritual links maintained with nature by different cultures. Other conventions, declarations, and initiatives in this direction are the UNESCO Man and the Biosphere program, the International Decade of the World's Indigenous People, The World Conference on Science for the Twenty-First Century, the UNESCO Recommendation on the Safeguarding of Traditional Culture and Folklore, and Agenda 21. The most comprehensive document on this aspect is the United Nations Draft Declaration on the Rights of Indigenous Peoples, which was drafted for approval in 2004.

This trend of changing attitudes toward recognizing the culture-nature link is complemented in the area of cultural heritage conservation as well. Over the last three decades, increasing interest in indigenous cultures brought major changes in recognizing intangible values. The concept of cultural landscape now encompasses all items, both natural and human-created artifacts, such as historical and religious monuments, as items of intangible value (Wijesuriya 2001), with many national governments adopting legislation to protect interests of traditional societies—such as the Native American Graves Protection and Repatriation Act in the United States in 1990, the Archaeological Resources Protection Act in the United States in 1979, the Historical Place Trust Act of New Zealand, and the Burra Charter of Australia.

(1949) on land ethics, *Silent Spring* by Rachel Carson (1962), and the poem by Samuel Taylor Coleridge entitled "To Nature" (Farrel 1992).

The performing arts—dance, song, drama, theatre, and so on—have entertained and delighted people for thousands of years. For example, Indian classical art forms seek to uplift the human spirit to a higher level of awareness, an awareness that is both inward as well as outward. This is expressed by a verse from the Sanskrit work *Abhinaya Dharpana* that signals a student's initiation into the world's oldest existing dance forms, *Bharathanatyam,* a classical dance style predominant in South India. About 66% of the 500 hand gestures in Bharathanatyam relate to ecosystems. Wetlands and water have also inspired music, such as "Swan Lake" from Tchaikovsky and the "Water Music Suite" from Handel. (See also Figure 17.3.)

Dance can be a powerful medium to address environmental and development issues. For instance, dance was one of the prime movers that instilled nationalism among the masses during the freedom movement in India in the 1930s and 1940s. Dance and song through the media of film, photography, and records or CDs can be used to inspire the needed intergenerational movement for conservation of ecosystems. Examples include the "Dance for the Earth and its People" promoted by the IUCN/WCPA Task Force on Cultural and Spiritual Values of Protected Areas.

The fine arts, expressed through crafts, painting, and sculpture, have always made extensive use of ecosystems as a source of inspiration. For instance, Vietnamese stone-crafted turtles and lotus incense holders, block prints narrating the lotus plant's life-history, bamboo grove candle holders, and woven scenes of rice fields on fabric are inspired by the prevailing rice fields, the ponds and lakes, and the bamboo groves and forests of Viet Nam today. And the motifs of baby carrier baskets of a Borneo tribe include tigers, dragons, and human faces that serve to protect the baby and nourish his or her soul to attain the proper social and spiritual level (Heidi Munan of Borneo, personal communication). Exam-

ples from the industrial world include the work of the famous French Impressionist painters Claude Monet and Camille Corot in the 1800s, who used landscapes as their source of inspiration (for example, Monet's *Water Lilies* and Corot's *Souvenir de Morte-fontaine*).

Designs and fashion have for generations captured the beauty of the natural world and reproduced them onto items of utilitarian use—from crockery to home furnishings and clothing, such as the china of Royal Doulton and Noritake, the daily-worn *molas* of the indigenous Kuna women of Panama, the fabrics of Laura Ashley, and the Kanchivaram saris of India. In the latter case, the artist who sees nature, the weaver who interprets it, and the woman who wears a sari all become one in their wonder of and homage to the beauty of nature. In the industrial world, many industrial and architectural designs and many national symbols—the bald eagle in the United States, for instance—also use nature as an example and source of inspiration.

Radio, films, videos, television, the Internet, photography, and advertising all use nature as a source of inspiration to make programs and sell products. The National Geographic, Discovery, and Animal Planet Channels on television in the United States are examples of this, as is the ARKIVE initiative in the United Kingdom, which attempts to maintain photographs, videos, and sound recordings of species so that they may remain available even if these species become extinct (see www.arkive.org). Over the past 50 years our emotional and economic dependence on this service has grown constantly and we are now "consuming"

Figure 17.3. Bavarian State Ballet Performance in Wetland

this inspirational service of nature through media, often without being aware of it.

17.2.4.2 Observed Changes, Causes of Change, and Future Trends

Urbanization and the increasing influence of the global market economy have strongly influenced the inspirational ties between humans and nature. The continued degradation of cultural landscapes and pristine ecosystems have led to changing perceptions regarding what is considered valuable in terms of providing inspiration to culture and art. Thus, even degraded ecosystems inspire the creation of songs, drama, dance, films, and photography, although they are not only used to show the beauty of, for example, eroded sand dunes but are often used as examples to warn of the dangers of the changes in our environment. The numbers of products of inspirational services depicting ecosystem degradation are potential indicators of the effect changes in these ecosystem services has on human welfare.

On the other hand, positive trends can be observed. For example, since about 2001, eco-textiles of banana and pineapple linen have started to appear in Southeast Asia (at the World Eco-Fiber and Textile Forum 2001 in Kuching, Malaysia, for instance), along with craft products such as handbags, rugs, and cushions made of jute, *mengkuang,* and *pandan* (traditional Malaysian and Southeast Asian natural fibers). And in Panama, there is a growing interest in the *molas* (stitched textile designs produced by the Kuna people).

Consumer and purchasing choices will change through the changed values placed on the various inspirational services, and it is expected that the early years of this century will see a marked increase in the use of natural dyes and cultivated fibers for indige-

nous crafts and functional items. In many parts of the world, women will play a vital role in the choice and purchase of consumer products, since they are the primary managers of their homes and the primary purchasers of a family's needs.

17.2.4.3 Consequences of Change for Human Well-being

The ability to experience and express inspiration from natural, semi-natural, and cultivated ecosystems is important for the well-being of many, if not all, people. As one writer once put it "without nature, life would be very dull indeed" (van Dieren and Wagenaar Hummelink 1979). Determining the consequences of the loss of inspirational services caused by a loss in quality and quantity of valued ecosystems is difficult, however. The gradual change from direct and participative experience of nature (through all senses) to its virtual representation through the media and the impact of this change on human well-being is hard to describe, let alone quantify.

Various measures of the dependence of human society on inspirational services have been suggested. These include the number of people engaged in various art activities, the number of people growing and harvesting the raw material used to create fashion and art, the quality and variety of natural resources used for art activities, the variety and numbers of art pieces created, and the price people are prepared to pay for products based on these services. In principle, these indicators could be used to measure the effect of changes in inspirational services on human health (physical and emotionally) and income caused by ecosystem change.

17.2.5 Aesthetic Services

Natural environments are an important source of aesthetic pleasure for people all over the world. The high aesthetic value of nature is reflected in many areas of human behavior, such as the use of plants and flowers as decorative elements in interiors, the use of computer screensavers depicting natural environments, and the demarcation of "scenic routes."

To most people, the fact that nature is beautiful is so obvious and self-evident that they rarely take time to think about it. Likewise, scientists have for a long time neglected this topic because there was no need to prove that nature is beautiful or to explain this phenomenon. Scientific interest in this topic was raised only when it became clear that aesthetic values of nature were being threatened by the ongoing human demand for expansion and that these deserved protection in their own right. In the United States, for example, the National Environmental Policy Act of 1969, which required federal agencies to take into consideration the impacts of large-scale interventions on the natural environment, constituted an important impetus for systematic scientific inquiry into the aesthetic quality of nature.

17.2.5.1 Current Status and Dependence on Ecosystem Condition

Three general findings about aesthetic services are worth noting: people's preference for natural over built environments, people's preference for park-like settings, and the existence of individual differences in preferences for wild versus cultivated landscapes. With a few exceptions (e.g., Chokor and Mene 1992; Yu 1995), nearly all studies have focused on industrial countries, which are the focus therefore of this section. However, as will be noted, one of the most remarkable findings of environmental perception research is the overwhelming similarity in aesthetic preferences between people from different subgroups and with different backgrounds (Kaplan and Kaplan 1989). Thus there is no indication that the assessment presented here would be highly different for developing countries.

A great number of studies in environmental aesthetics have shown that people display, in general, a strong preference for natural over built environments (see reviews by Ulrich 1983; Kaplan and Kaplan 1989; Hartig and Evans 1993). In samples of European and North American adults, for example, photographs of natural scenes consistently receive higher ratings for scenic beauty than photographs of urban scenes do (e.g., Stamps 1996). (See Figure 17.4.) In fact, this preference is so strong that even plain grassland is generally considered equally or more beautiful than any built environment, including pretty townscapes such as the monumental buildings along the river Seine in Paris (Ulrich 1983).

People's preference for natural over built environments can also be inferred from behavioral indicators, such as the higher prices paid for real estate surrounded by trees or adjacent to parks (e.g., Luttik 2000) and the higher number of recreational stays in natural areas. The latter observation is substantiated by the finding that aesthetic pleasure has consistently been found to be one of the most important motivations for outdoor recreation. (See the section on recreation and ecotourism.)

The preference for natural over built environments has been observed across all times and cultures. Even very early urban people apparently took aesthetic pleasure in nature, as is indicated by the gardens of the ancient Egyptian nobility, the walled gardens of Persian settlements in Mesopotamia, and the gardens of merchants in medieval Chinese cities (Ulrich 1993). Consequently, several researchers have proposed that people's preference for nature may be the result of an ancient evolutionary history (Ulrich 1983; Kaplan 1987). In particular, they have suggested that modern humans prefer nature because evolution has made contact with natural environments an innate source of restoration and well-being. The promise of restoration stimulates people to seek out contact with non-threatening natural environments that contain resources and opportunities that are necessary for survival.

In corroboration with this assumption, numerous studies have demonstrated that contact with nature may enhance restoration from stress and increase health and well-being (e.g., Hartig et al. 2003; Ulrich 1983; Ulrich et al. 1991; Van den Berg et al. 2003). For example, Ulrich (1984) has shown that patients who were recovering from gall bladder surgery had shorter postoperative hospital stays and required fewer injections of painkillers when they were given a room with a natural view than when they were

in one looking out at a brick wall. Likewise, Hartig and colleagues (2003) have shown that fatigued individuals who walked through natural environments showed more positive changes in mood state, ability to concentrate, and physiological stress levels than fatigued individuals who walked through built environments.

Aesthetic preference for different types of natural environments is strongly dependent on the environment's ecological condition. In general, people prefer natural settings that are healthy, lush, and green. Verdant vegetation is preferred over arid landscapes (Abello and Bernaldez 1986), and forests with sick trees receive much lower preference ratings than healthy forests (Ulrich 1986). These findings are often interpreted as evidence that aesthetic quality is identical to ecological quality. However, it is necessary to distinguish aesthetic values and preferences associated with traditional knowledge systems from those from formal knowledge systems. Although there are some areas in which aesthetic quality and ecological quality may overlap, these two values may diverge strongly in other areas, and aesthetic (traditional knowledge) values need to be considered in their own right and must not be confounded with ecological (formal knowledge) values.

Although people prefer nearly all natural environments to urban environments, this does not mean that they find all natural environments equally beautiful. Certain natural environments are consistently judged as more beautiful than others. Kellert's (1993) review of the environmental perception literature states that European, North American, and Asian populations consistently prefer park-like settings. Most of these studies used rankings of photos or slides. Among the characteristics of park-like settings that people prefer are depth, (half-)openness, uniform grassy coverings, presence of water, absence of threat, and scattering of trees.

Like the general preference for natural over built environments, the preference for park-like natural landscapes has also been explained as a genetic disposition that impels modern humans to seek out the natural settings that, for early humans, were most likely to offer primary necessities of food, water, security, and exploration (Heerwagen and Orians 1993;). Thus it appears that our aesthetic judgments of natural settings are still to a large extent based on implicit assessments of their survival value, even though most of us are no longer directly dependent on nature for our primary supplies.

In addition to the general preference tendencies just described, there are important individual differences in aesthetic preferences for natural landscapes across different times and cul-

Figure 17.4. Preference for Natural over Built Environments. Numerous studies in environmental esthetics have shown that natural environments are generally considered more beautiful than urban environments. This "love for nature," or biophilia, has been explained as an adaptive genetic mechanism that stimulates people to seek out environments that are beneficial for their health. In line with this assumption, experimental studies have demonstrated that contact with natural environments is associated with greater health benefits than contact with urban environments, especially greater and more complete recovery from stress. (Photos from Van den Berg et al. 2003)

tures. For instance, historical analyses have revealed that the appreciation of wilderness in the western world has changed dramatically over the centuries. Until late in the seventeenth century, wild, uncultivated land was generally regarded with indifference and hostility (Nash 1973). But the Romantic Era artists and intellectualists of the eighteenth century began to describe wild places in terms of divinely endowed beauty and order (Thacker 1983), and public perceptions began to change. Since then, more and more people have adopted a positive attitude toward wilderness.

Negative perceptions of wilderness continue to exist in certain groups and cultures, however, even in modern times. Indeed, empirical investigations of modern people's landscape preferences indicate that differences between groups and cultures can nearly always be interpreted in terms of differences in the preferred degree of "wildness" in natural landscapes (Kaplan and Kaplan 1989; Van den Berg 1999). In particular, farmers and low-income groups have been found to prefer managed natural landscapes with a high degree of human influence, while urbanites and high-income groups have been found to prefer wild natural landscapes with a low degree of human influence.

17.2.5.2 Observed Changes, Causes of Change, and Future Trends

The general preference for natural over built environments appears to be relatively stable across different times and cultures. Yet there are reasons to believe that the strength of this preference may vary depending on the degree of stress and mental overload. In particular, Staats et al. (2003) have found that the preference for nature over the city was twice as strong in individuals who were asked to imagine that they suffered from stress and attentional fatigue. These findings suggest that nature becomes more important to people as their levels of stress and mental exhaustion increase.

Urbanization, industrialization, and globalization mean that life is becoming more stressful for people all over the world. Particularly in developing countries, rapid and uncontrolled urban expanse may lead to increased levels of stress and stress-related diseases. These higher levels of stress may result from environmental factors, such as noise and air pollution, but also from social factors, such as unemployment and poverty (World Resources Institute 1996). Thus, it can be expected that people's preference for natural over built environments will become stronger with increasing urbanization. Paradoxically, while the appreciation of nature can be expected to increase with increasing urbanization, the supply of nature and access to natural settings tend to decrease with urban expansion, thereby underlining the importance of green spaces in and near cities.

While the effects of urbanization on the appreciation of nature may apply to all types of nature, regardless of its aesthetic or ecological value, it can also be expected that urbanization will specifically affect the popularity of wilderness settings. As pointed out earlier, preference for wilderness tends to be higher among urban residents. These findings suggest that the popularity of wilderness environments may increase as more and more people start to live in urban areas. At the same time, a lack of recognition of the aesthetic value of wilderness can lead to less value being attributed to wilderness areas in parts of the world where people still live in or near the wilderness. Taken together, these developments may eventually lead to a situation in which the majority of the world population longs for a wilderness that no longer exists.

17.2.5.3 Consequences of Change for Human Well-being

Contact with nature has been related to a large number of health and economic benefits, including decreased levels of stress, mental fatigue, and aggression (restorative effects) (e.g., Hartig et al. 2003); decreased need for health care services and decreased levels of aggression and criminality due to restorative effects of contact with nature (Kuo and Sullivan 2001; Ulrich 1984); increased health due to increased levels of activity stimulated by the presence of attractive nature in the nearby work and living environment (Taylor et al. 1998); increased social integration due to the function of urban natural settings as social meeting places (Kweon et al. 1998); improved motoric development in children who regularly engage in outdoor activities (Fjortoft 1997); increased worker productivity and creativity in offices with plants or views of nature (Lohr et al. 1996); economic benefits for society due to enhanced employability, reduced criminal behavior, and lower substance abuse by disadvantaged youth who participate in wilderness programs (Russel et al. 1998); and increased value of real estate property in natural surroundings (Anderson and Cordell 1988; Luttik 2000).

Most of these benefits apply to all types of nature, including plants, green spaces, and agricultural areas, and are not necessarily dependent on the ecological value of an area. Contact with ecologically valuable nature, such as wilderness areas, may provide the individual with additional benefits, such as increased self-confidence and personal growth, which may be of crucial importance to certain groups, such as youth-at-risk (teenagers from disrupted families, for instance) (Fredrickson and Anderson 1999). However, contact with wilderness may also evoke fears and increase the risk of hazards and diseases (such as Lyme disease or accidents), in particular for people who are unfamiliar with wilderness environments and their potential threats and dangers (Bixler and Floyd 1997).

Based on the benefits just described, it can be expected that a decline in aesthetic services due to a reduction in the availability of and access to natural areas for urban residents may have important detrimental effects on public health, societal processes, and economics.

17.2.6 Recreation and Tourism

Many ecosystems have important value as a place where people can come for rest, relaxation, refreshment, and recreation. Through the aesthetic qualities and almost limitless variety of landscapes, natural and cultural environments provide many opportunities for nature-based recreational activities, such as walking, bird-watching, camping, fishing, swimming, and nature study. With increasing numbers of people, affluence, and leisure time, the demand for recreation in natural areas and cultivated landscapes will most likely continue to increase in the future.

17.2.6.1 Current Status and Dependence on Ecosystem Condition

Travel and tourism have been interrelated throughout human history via ancient roots related to play, ritual, and pilgrimages. Tourism has been referred to as both "a sacred journey and a profane vision quest" (Graburn 1976). Some anthropologists have even suggested that tourism is preeminently a "secular ritual," and that in many contemporary societies it fulfills some of the functions once met by sacred rituals (Graburn 1983). The driving agents of this host-visitor interaction can be recreation and enjoyment, the search for knowledge, religious pilgrimages, and so on. The World Tourism Organization, the most comprehensive collector of data on tourism, distinguishes several types of tourism, including cultural tourism, rural tourism (agri-tourism), and nature tourism (including ecotourism and adventure tourism) and, secondarily, "sun-and-beach tourism" and "fitness, wellness and health tourism."

Cultural tourism is a form of experiential tourism based on the search for and participation in new and deep cultural experiences of an aesthetic, intellectual, emotional, or psychological nature (Reisinger 1994). Cultural landscapes and heritage services are important attractions for people wanting to experience other cultures and religions. The Ganges River–based cultural and sacred landscape system in India, for example, is visited every year by millions of people, being sacred for close to a billion people of the Indian subcontinent. Similarly, the Demajong landscape of the Tibetan Buddhists in the Eastern Himalayan State of Sikkim, India, and the Koyasan landscape in Japan are equally important for Buddhists living in that part of the world. More than 1 million people visit Koyasan annually.

Rural tourism can be interpreted in a number of ways. Over the last decade, the concept has come to encompass more and more activities. For instance, Bramwell and Lane (1994) included activities and interests in farms, adventure, sport, health, education, arts and heritage, and even natural sites. Pedford (1996) added aspects of living history such as rural customs and folklore, local and family traditions, values, beliefs, and common heritage. And Turnock (1999) further broadened the view of rural tourism to embrace all aspects of leisure appropriate in the countryside (cultural landscapes). The growing overlap of cultural tourism and rural tourism led MacDonald and Jolliffe (2003) to integrate the two concepts into "cultural rural tourism."

The World Conservation Union (IUCN) defined ecotourism in 1996 as tourism that "is environmentally responsible travel and visitation to relatively undisturbed natural areas, in order to enjoy and appreciate nature (and any accompanying cultural features—both past and present) that promotes conservation, has low negative visitor impact and provides for beneficially active socio-economic involvement of local populations." It is estimated that in 1997 nature tourism, including ecotourism, accounted for approximately 20% of total international travel (WTO 1998) and that nature travel is increasing between 10% and 30% a year (WRI 1990).

17.2.6.2 Observed Changes, Causes of Change, and Future Trends

There is evidence of rapid growth of nature- or ecotourism (Skayannis 1999), demonstrated in the surging growth of international arrivals to the countries with high biodiversity. (See Table 17.3.) Travel and tourism was one of the few industries identified in *Agenda 21* as having the potential to make a positive contribution to healthier national economies as well as a healthier planet. Tourism is now the primary economic development strategy for many developing nations, as demonstrated in 1996 when all the presidents of Central America at a summit in Nicaragua declared their intentions to make tourism the primary revenue source for the region (UN 51/197 1996). Similar sentiments have been expressed throughout the world.

Research indicates that nature tourism has experienced a surge in demand that has far exceeded supply (Diamantis 1998). Tourism is a well-recognized agent of change, and the rapid expansion of recreation and tourism planning in recent years has led to the need for managing its impacts. Yet the cultural phenomenon of societies protecting special areas for visitors has been common for centuries. Indeed, in many cases it was the increasing arrivals of travelers to special sites that were the impetus for site designation and protection (Eagles et al. 2001).

Well-planned and well-managed tourism has proved to be one of the most effective tools for long-term conservation of biodiversity when the right conditions, such as market feasibility, social and physical carrying capacity, management capacity at local level, and clear and monitored links between tourism development and conservation, are present. For example, a study of nature-based tourism in southern Africa in 2000 estimated the aggregate value to be $3.6 billion per year, which represented approximately half the total income from foreign travel in the region (the other half was contributed mostly by business travel and visits to family and friends) (MA *Multiscale Assessments,* South African Assessment). (See also Box 17.9.)

Sustainable tourism, in the context of development, has been defined as "all forms of tourism development, management and activity, which maintain the environmental, social and economic integrity and well being of natural, built and cultural resources in perpetuity" (FNNPE 1993). In the years since the concept of sustainable tourism was first defined, a consensus has formed on the basic objectives and targets. Sustainable tourism should contribute to the conservation of biodiversity and cultural diversity; should contribute to the well-being of local communities, enhancing social equity and respect for the rights and sovereignty of local communities and indigenous people; should include an interpretation/learning experience; should involve responsible action on the part of tourists and the tourism industry; should be appropriate in scale; should require the lowest possible consumption of nonrenewable resources; should respect physical and social carrying capacities; should involve minimal repatriation of earned revenue; and should be locally owned and operated (through local participation, ownership, and business opportunities, particularly for rural people).

Now more than ever, the protection of natural and cultural areas is intimately connected to the tourism industry. High growth and demand have greatly influenced the management trends of protected areas, with the interaction between humans and the environment as one of the main factors. These effects in the protected-area tourism management industry include linking sustainable use and conservation, increasing travel to protected areas, moving toward self-regulation in the tourism industry, acknowledging the important financial aspects of tourism to protected areas, and acknowledging the importance of the sociocultural aspects of sustainable tourism (Eagles et al. 2001).

17.2.6.3 Consequences of Change for Human Well-being

It is important to note that in countries without large mineral resources, tourism is often the major source of foreign income (WTTC 1999). (See also Box 17.10.) It is useful to compare income from nature-based tourism to that generated from the other main sectors based on ecosystem services: agriculture, forestry, fisheries, and the provision of water. Assuming that nature-based tourism is half of all tourism, and excluding the manufacturing sector knock-on effects of agriculture, forestry, and fisheries, the contribution by nature-based tourism is nearly equal to the other natural resource sectors combined (WTO 1998; WTTC 1999). It is important to note that these other sectors are growing slowly (1–3% a year) while tourism is growing rapidly (5–15% a year).

Thus, the balance of policy drivers in relation to natural resources is likely to shift over the next few decades, from being strongly influenced by the needs of agriculture, forestry, and fishing to being more influenced by considerations of conservation and aesthetics. The dominance of industries based on nonrenewable resources, such as mining and oil extraction, must in the long term decline, but it is likely to remain high over the next quarter-

Table 17.3. Examples of Hotspots of Countries with High Biodiversity and Tourism Growth of More than 200 Percent (Conservation International 2003, based on data from WTO)

Hotspot/Country	International Arrivals			Growth 1990–2000	
	1990	1995	2000	Number	Increase
	(thousand people)			*(thousand people)*	*(percent)*
Indo-Burma					
Laos	14	60	300	286	2,043
Myanmar	21	117	208	187	890
Viet Nam	250	1,351	2,140	1,890	756
Succulent Karoo/Cape Floristic Region					
South Africa	1,029	4,684	6,001	4,972	483
Caribbean					
Cuba	327	742	1,700	1,373	420
Brazilian Cerrado/Atlantic Forest					
Brazil	1,091	1,991	5,313	4,222	387
Mesoamerica					
Nicaragua	106	281	486	380	358
El Salvador	194	235	795	601	310
Guinean Forests					
Nigeria	190	656	813	623	328
Tropical Andes					
Peru	317	541	1,027	710	224
Madagascar and Indian Ocean Islands					
Madagascar	53	75	160	107	202
Eastern African Mountains and Coastal Forests					
Tanzania	153	285	459	306	200

BOX 17.9

Inter-American Development Bank Lessons on Tourism Development with Conservation (Conservation International 2003)

The Brazilian state of Bahia harbors one of the most threatened conservation hotspots, the Atlantic rain forest. The $400-million PRODE-TUR I project, funded by the IDB from 1994 to 2001, improved and expanded eight international airports, built and improved over 800 kilometers of highways and access roads, provided water and sewage infrastructure, and attracted more than $4 billion in private tourism investment. Its negative impacts on the environment, though, became clear to IDB officers: uncontrolled settlement of people looking for jobs, private building in environmentally sensitive areas, encroachment on rain forests and mangroves, and impacts on coastal reefs and other coastal ecosystems.

Intense pressure from local and international NGOs and community groups, supported by bank officials, ultimately overcame the initial resistance from investor groups and development-oriented government officers to allocate funds for conservation. The result was the conservation of 22 historical heritage sites and the beginning of efforts to conserve over 70,000 hectares of coastal ecosystems and protected areas, including the creation of the new Serra do Conduru State Park. These lessons are being applied to new IDB projects in the region.

BOX 17.10

Economic Importance of Cultural and Nature-based Tourism

The economic importance of global travel and tourism is indicated by a few figures on the sector as a source of jobs and national income; about 30% of these revenues are related to cultural and ecotourism. Global travel and tourism:

- generates 11% of global GDP (WTTC), growing at 7.5% per year (Carsten Loose, personal communication);
- employs 200 million people or 7.6 % of total employment for the world (WTTC);
- transports nearly 700 million international travelers per year—a figure that is expected to double by 2020 (WTTC);
- accounts for 36% of trade in commercial services in industrial economies and 66% in developing economies (WTO);
- accounts for 36% of trade in commercial services in industrial economies and 66% in developing economies (WTO);
- constitutes 3–10% of GDP in advanced economies and up to 40% in developing economies (WTO);
- generated $464 billion in tourism receipts in 2001 (WTO); and
- is one of the top five exports for 83% of countries and the main source of foreign currency for at least 38% of countries (WTO).

century. A key trade-off is between the social benefits that such sectors offer now and the long-term benefits that may be afforded by nature-based tourism.

Management is frequently the weak link in the connection between tourism and the environment (Valentine 1992). Tourism provides both benefits and hazards, and the monitoring and controlling of impacts is necessary in order to mitigate the negative impacts from uncontrolled visitation, both ecologically and socioculturally (such as prostitution and the spread of diseases); to prepare for the expected rapid increase in visitor arrivals as well as rapid increase in the value of pristine lands; to move beyond past relationship failures between host ecosystems, visitors, local cultures, foreign developers, governments, indigenous groups, and scientists; and to allow crucial economic and natural science contributions to community and indigenous self-determination and resource conservation within rapidly changing environments.

Responsiveness to the relationships between cultures, biodiversity, and tourism is important to the objectives of the Convention on Biological Diversity—that is, the conservation of biological and cultural diversity and the sustainable use of the components of biodiversity—while intimately linked to issues of equity as well. The CBD Guidelines (Decision VII/14) on Biodiversity and Tourism Development are the most recent, comprehensive, and multilateral effort toward more sustainable tourism development. Its coordinating framework represents one of the best opportunities to improve global human well-being by strengthening protected area management systems (public, private, or indigenous); by increasing the value of sound ecosystems through generating income, jobs, and business opportunities in tourism and related business networks; by sharing information, capacity building, and public notification; and by allowing people to internalize the benefits of the biodiversity that has been a part of their historical, natural, and cultural heritage.

17.3 Drivers of Change in Cultural and Amenity Services

Changes in ecosystem characteristics are determined by direct and indirect drivers (see also Chapter 3), which in turn can affect sociocultural, spiritual, and recreational activities. The consensus now seems to be that complex interactions between the indirect and direct drivers—including market forces (both national and international), taxes and subsidies, consumption patterns, population migration and resettlement, land ownership, autonomic cultural rights, participation in decision-making, poverty, and the problem of invasive species (to mention but a few)—lead to land degradation and loss of ecosystem services (Lambin et al. 2001).

Issues such as population and poverty, which are often assumed to be ultimate drivers of ecosystem transformation, are now recognized as much more complex, even under diverse socioecological-economic-political situations as found in India, China, and the United States (Indian National Science Academy et al. 2001). For example, conversion of Mediterranean mixed cultivation systems, such as traditional olive cultivation combined with livestock grazing, into intensive cropping systems is the consequence of agricultural policies and subsidies, which in turn lead to increased mobility that causes, for example, landscape fragmentation. These changes may lead to both negative and positive effects in terms of the real or perceived availability and value of cultural and amenity services. In many parts of the world, for instance, so-called cultural landscapes are highly valued for aesthetic or historic reasons but from an ecological point of view are highly degraded (for instance, the heath landscapes in the Netherlands, a succession

stage that is artificially maintained by preventing natural forest regrowth).

The problem of invasive species, an important global change phenomenon, is becoming a major issue in the maintenance of cultural and amenity services in different parts of the world. A global synthesis and a recent international initiative on invasive species (Drake et al. 1989), suggest that invasions by exotic species have a strong impact on land transformations and land degradation and affect traditional livelihoods. For the poorer sections of rural society, particularly in the developing tropics, the adverse impact of invasions can be critical because these communities depend on natural ecosystems for socioeconomic as well as cultural and spiritual well-being (Ramakrishnan 1991).

In spite of the disruptions caused to ecosystem characteristics, humans have both learned to appreciate changes in ecosystems (such as conversion of natural ecosystems into landscapes that, over time, have developed cultural-historic values) and intentionally transformed natural systems into landscapes with special cultural, spiritual, or amenity values (such as urban parks, sacred landscapes, and recreational sites). However, changes in value systems (a loss of religious beliefs, say, or cultural identity) have also led to the loss of previously valued sacred or historic landscapes.

17.4 Consequences for Human Well-being of Changes in Cultural and Amenity Services

The importance of a service to human well-being can be described by many different indicators, including environmental safety (low risk of natural disasters, provision of clean water, and so on), economic security (employment and income), health (physical and psychological), and social aspects (cultural identity, traditional knowledge, social networks, and so on). (See Chapter 5 for further details.)

As described in previous sections, natural and cultivated systems provide many cultural and amenity services that contribute significantly to the general well-being of humans. Inspirational, aesthetic, and recreational services of ecosystems are important not only for their therapeutic value (physically and mentally) and other human well-being aspects but also for their considerable economic value. Changes in ecosystem conditions will always change the availability of ecosystem services and hence affect human well-being (either positively or negatively). Part of these changes in well-being can be measured by economic valuation methods, including monetary data. (See Chapter 2 for methods and tools for economic valuation of ecosystem services.) It is beyond the scope of this chapter to substantially expand on this, but three types of impacts of changes in ecosystem-based cultural services on human well-being are briefly discussed.

17.4.1 Cultural Identity and Social Values

As described earlier, population growth and economic development in many parts of the world have led to changes in traditional land use, cultural values, and spiritual ties between human society and their surrounding ecosystems. In most cases, this has meant that economic gains, including increased use of amenity services (such as tourism) has led to the loss of cultural identity and heritage values. Recently, a reverse in the trend has become noticeable, where cultural identity and heritage values are being rediscovered and restored while simultaneously bringing economic benefits to the region. A good example of this is the long-standing and evolving interest of UNESCO, as part of its World Heritage Centre on culturally valued natural landscape systems (Rossler 2000; UNESCO 2003). Also, the emerging interest of

FAO on the Globally Important Ingenious Agricultural Heritage Systems is indicative of this growing interest in conserving and sustainably developing cultural landscapes with economic benefits to local communities and society at large (Ramakrishnan 2003).

17.4.2 Human Health

The loss of cultural ties between people and ecosystems often leads to a loss of cultural identity, causing increased social disruption and stress that in turn causes a whole array of mental and physical health effects. Similarly, a loss of opportunities to enjoy the inspirational, aesthetic, and recreational benefits of natural and cultural landscapes has negative mental and physical health effects. And the loss of traditional knowledge systems can have negative health effects, notably through plant medicine that could help humankind deal with pandemics like AIDS, cancer, and other health problems in a globalizing world.

17.4.3 Material Well-being

Many of the changes described in this chapter have considerable economic and financial consequences. On the one hand, more modern and large-scale land use systems, increased tourism, and so on bring higher financial revenues. On the other hand, social disruption and negative health effects lead to higher costs (to prevent and combat crime, diseases, environmental problems, and so on). The problem is that the higher revenues usually accrue to a small number of specific stakeholders (landowners, for instance, and tourist companies) while the costs in terms of loss of cultural identity and reduced health and income are felt by society as a whole (and usually the more vulnerable people), including future generations. The challenge is to find a balance between the maintenance of cultural and amenity services and values and the (sustainable) development of their full economic potential. "Diversity in use" seems to be the key here: scientific evidence is mounting that if all services and associated values are properly taken into account, multifunctional use of ecosystems is not only environmentally and socioculturally more sustainable but also economically more beneficial than single-function use (e.g., Balmford et al. 2002).

17.5 Lessons Learned

17.5.1 Landscape Management and Sustainability Issues: The Ecosystem Approach

Current international conservation initiatives are increasingly based on the "ecosystem approach" (see CBD Decisions V/6 and VII/11) and the "eco-region" approach of the World Wide Fund for Nature, although there have been both older and more recent attempts to take a more integrative socioecological system approach to managing natural resources. The cultural and amenity values of landscape are one important dimension in this integrated method.

The concept of Biosphere Reserves (in which humans are viewed as an integral part of the component ecosystems), the concept of UNESCO's "cultural" World Heritage Sites, and the recently initiated Globally Important Ingenious Agricultural Heritage Systems of FAO are indicative of the importance attached to the cultural and spiritual dimensions of the issues. Emphasis is also being directed toward conservation linked with sustainable use of these systems, viewing them not merely as ecosystems in a biophysical sense but more appropriately as constantly evolving "socioecological systems."

17.5.2 Cultural Basis for Landscape Management

Landscape planning and management needs to be based on a better understanding of the way in which societies manipulate ecosystems and to consider cultural, spiritual, and religious belief systems. Human societies understand and interact with landscapes through a cultural lens, and traditional knowledge has played an important role in mediating a sustainable relationship between biophysical and human systems. (See Box 17.11.) This is an area of evolving interest, with possible linkages between traditional and formal knowledge systems to create landscape management institutions and practices, though this still remains somewhat problematic.

Traditional beliefs, practices, and knowledge are often embedded in shared territory, common property rights, and lifestyles. In November 2002, the Convention on Wetlands (Ramsar 1971) adopted Resolution VIII.19 on 'Taking into account cultural values in the management of sites' and is now working in various parts of the world for its implementation. The purpose of this resolution is twofold: to reconnect people with nature, by strengthening traditional cultural links, and to promote an integrated perception of the natural and cultural heritage of wetland sites, which can attract visitors and provide benefits to local communities. As the examples in Box 17.12 illustrate, the motivations for conservation range from spiritual to utilitarian, and in many situations they could potentially play a significant role in fostering sustainability.

17.5.3 Traditional Technologies

The term "technology" is taken here to represent the composite of all protocols, processes, practices, and institutions that are applicable to the management of natural resources, documented or transmitted through oral tradition. As with TEK, there are many examples of such technologies underpinning the development of complex societies that have long-lasting relationships with landscapes.

For example, the development of plant cultivars as part of landscape organization by traditional societies in South America dates back to at least 10,000 BP (Pearsall 1992). Pre-Hispanic cul-

---

> **BOX 17.11**
>
> ### Some Examples of Landscape Management and Traditional Knowledge
>
> Throughout Africa, natural resource management practices are traditionally linked to religious sanctions. The rules and regulations are implemented through living authorities, often with pragmatic objectives that are relevant to conservation issues too. In the Miombo woodlands in Southern Africa, for instance, it is prohibited to cut fruit trees or trees growing around "sacred" water springs. Sacred groves, often occurring on hills or in river valleys, are protected for ceremonial reasons, as an abode for departed souls, as a source for natural water springs, or as a source for medicinal plants and other non-timber forest products (Clarke et al. 1996), but they also perform critical ecological functions.
>
> Buddhist monks are prohibited from doing any harm to trees and animals by the code of conduct known as *Vinaya* (the discipline). Locations related to Buddha, the place of birth, enlightenment, and death, are recognized and protected as places of worship. Sri Pada (or Adams Peak) in Sri Lanka, a landscape of rich diversity, is considered by the Buddhists, Christians, Hindus, and Muslims as a place of worship and is protected (Wijesuriya 2001).

tures managed complex ecosystems and conserved biodiversity, which through an extended historical process of cultural adaptation reached a surprising degree of stability. These included the grazing systems of native *Camelidae* in the Punas, complex lacustrine agricultural systems of the Mexican Chinapas, Zenu hydraulic society in the Caribbean lowlands of Colombia, and the shifting agricultural systems that permitted maintenance of diversity in the Amazonian and Mayan forests (Monasterio 1994). Similarly, *Miombo* savanna landscape management practices are typical of what is found in many parts of tropical Africa (Campbell 1996), with a long history of connectivity between people and the ecosystem, where the traditional bush-fallow rotational *Chitemane* system of agriculture is linked with livestock husbandry.

17.5.4 Adaptive Management Strategies

Adaptive management—an interactive process of "learning by doing"—is founded on the premise that natural systems are dynamic and complex and that information on which to base decision-making is inevitably incomplete. Specific management strategies and actions are therefore approached as experiments that can be reviewed and adapted based on the information gained from monitoring systems on the strengths and weaknesses of these strategies (Holling 1978; Lee 1999; Borrini-Feyerabend et al. 2001). This has been promoted as the approach of choice by a number of international bodies (IUCN 1999). Tools that enable local perspectives and voices to be articulated in planning processes can help bridge this gap, notably methods such as Participatory Rural Appraisal, Participatory Learning and Action, and Participatory Assessment, Monitoring and Evaluation (Zanetell and Knuth 2002). An important proviso is that the application of such methods should not be mechanical or ultimately substitute for the development of collaborative relationships and ongoing communication that underpin real knowledge sharing between stakeholders (Poffenberger 2000).

The integration of social and cultural dimensions of resource management within an adaptive management framework requires an integrative approach by practitioners at the level of knowledge, worldview, and practice. The integration of traditional and formal knowledge systems through "knowledge partnerships" involves a creative blending of technical and local perspectives to achieve a balanced approach to managing landscapes. (Jiggins and Roling 2002; Zanetell and Knuth 2002). An example of successful co-management is the collaboration between the Inuvialuit and the government in the Northwest Territories of Canada (Harriet Kuhnlein, 2003, personal communication).

Combining the reductionistic, formal perspective of knowledge with a more "traditional" and more holistic perspective toward natural resource management is likely to yield better results, although the proportionality of these two elements will differ depending on the socioecological systems being dealt with (Ramakrishnan 2001). Cases where success has been realized by combining the two knowledge systems can act as "field laboratories" for scientific research and as reference points for monitoring environmental change brought about through appropriately designed technologies derived from an integration of formal and traditional knowledge systems.

References

Abello, R.P. and F.G. Bernaldez, 1986: Landscape preference and personality, *Landscape and Urban Planning,* **13,** pp. 19–28.

Balee, W., 1989: The culture of Amazonian forests, *Advances in Economic Botany,* 7, pp. 1–21.

Balmford, A., A. Bruner, P. Cooper, R. Costanza, S. Farber, et al., 2002: Economic reasons for conserving wild nature, *Science,* **297,** pp. 950–3.

Berbaum, E., 1997: The spiritual and cultural significance of mountains. In: *Mountains of the World,* B. Messerli and J.D. Ives (eds.), Parthenon Publications, Carnforth, Lancashire, UK, pp. 39–60.

Berger, P. and T. Luckmann, 1966: *The Social Construction of Reality,* Doubleday, New York, NY.

Berkes, F., 1999: *Sacred_Ecology: Traditional Ecological Knowledge and Resource Management,* Taylor & Francis, London, UK.

Berkes, F., J. Colding, and C. Folke, 2000: Rediscovery of traditional ecological knowledge as adaptive management, *Ecological Applications,* **10(5),** pp. 1251–62.

Berlin, B., 1992: *Ethnobiological Classification: Principles of Categorization of Plants and Animals in Traditional Societies,* Princeton University Press, Princeton, NJ.

Bixler, R.D. and M.F. Floyd, 1997: Nature is scary, disgusting and uncomfortable, *Environment and Behavior,* **29,** pp. 443–67.

Borrini-Feyerabend, G., M. Taghi Farvar, M.T. Nguinguiri, and V.A. Ndangang, 2001: *Comanagement of Natural Resources: Organising, Negotiating and Learning-by-doing,* The World Conservation Union (IUCN), Gland, Switzerland/ GTZ (Gesellschaft für Technische Zusammenarbeit), Eschborn, Germany.

Bramwell, B. and B. Lane, 1994: *Rural Tourism and Sustainable Rural Development,* Channel View, London, UK.

Burel, F. and J. Baudry, 2003: *Landscape Ecology: Concepts, Methods and Applications,* Science Publications, Enfield, NH, 362 pp.

Campbell, B. (ed.), 1996: *The Miombo in Transition: Woodlands and Welfare in Africa,* Centre for International Forestry Research, Bogor, Indonesia, 266 pp.

Capra, F., 1982: *The Turning Point,* Flamingo, London, UK.

Carmichael, D.L., J. Hubert, B. Reeves, and A. Schanche (eds), 1994: *Sacred Sites, Sacred Places,* Routledge, London, UK.

Carson, R., 1962: *Silent Spring,* Houghton Mifflin, Boston, MA, 304 pp.

Chokor, B.A. and S.A. Mene, 1992: An assessment of preference for landscapes in the developing world: Case study of Warri, Nigeria, and environs, *Journal of Environmental Management,* **34,** pp. 237–56.

Clarke, J., W. Cavendish, and C. Coote, 1996: Rural households and Miombo woodlands: Use, value and management. In: B. Campbell (ed.), *The Miombo in Transition: Woodlands and Welfare in Africa,* Centre for International Forestry Research, Bogor Indonesia, pp. 101–35.

Conservation International, 2003: *Tourism and Biodiversity: Mapping Tourism's Global Footprint,* Conservation International, Washington, DC.

De Groot, R.S., 1992: Functions of nature: Evaluation of nature in environmental planning, management and decision-making, *Wolters Noordhoff BV,* Groningen, Netherlands, 345 pp.

De Groot, R.S., M. Wilson, and R. Boumans, 2002: A typology for the description, classification and valuation of ecosystem functions, foods and services. In: The dynamics and value of ecosystem services: Integrating economic and ecological perspectives, *Ecological Economics,* **41(3)** (Special issue).

Diamantis, D., 1998: Consumer behavior and ecotourism products, *Annals of Tourism Research,* **25(2),** pp. 515–28.

Dower, M., 2000: In: *The Face of Europe: Policy Perspectives for European Landscapes,* D.M. Wascher (ed.), Report on the implementation of the PEBLDS (Pan-European Biological and Landscape Diversity Strategy) Action Theme 4 on European Landscapes, Published under the auspice of the Council of Europe, European Center for Nature Conservation, Tilburg, The Netherlands, 60 pp.

Drake, J.A., H.A. Mooney, F. di Castri, R.H. Groves, F.J. Kruger, et al., (eds.), 1989: *Biological Invasion: A Global Perspective,* SCOPE 37, John Wiley, New York, NY, 525 pp., pp. 281–300.

Eagles, P., M. Bowman, and T. Chang-Hung, 2001: *Guidelines for Tourism in Parks and Protected Areas of East Asia,* IUCN, Gland, Switzerland, 99 pp.

Ellis, E.C., R.G. Li, L.Z. Yang, and X. Cheng, 2000: Long-term change in village-scale ecosystems in China using landscape and statistical methods, *Ecological Applications,* **10,** pp. 1057–73.

Elzinga, A., 1996: Some reflections on post-normal science. In: *Culture, Perceptions and Environmental Problems: Interscientific Communication on Environmental Issues,* M. Rolen (ed.), Swedish Council for Planning and Coordination of Research, Stockholm, Sweden, pp. 32–46.

Enger, E.D. and B.F. Smith, 1995: Environmental science: A study of interrelationships, 5th ed., William C. Brown Publishers, pp. 21–22.

Farrel, K., 1992: *Art and Nature: An Illustrated Anthology of Nature Poetry,* The Metropolitan Museum of Art, New York, NY/Bulfinch Press, Medford, OR/Little Brown and Company, Boston, MA/Toronto, ON, Canada/London, UK, pp. 11, 50, 132.

Fenge, T., 1997: Ecological change in the Hudson Bay bioregion: A traditional ecological knowledge perspective, *Northern Perspectives,* **25,** pp. 2–3.

Ferreira, A.P., T. Pinto Correi, and R. Mata Olmo, 2003: Montado/Dehesas: Case Study. In: *Learning from Transfrontier Landscapes,* D. Wascher et al., Landscape Europe, Wageningen, The Netherlands. In press.

Fjortoft, I., 1997: The natural environment as a playground for children: The effects of outdoor activities on motor fitness of pre-school children, Paper presented at conference Urban Childhood Conference, Trondheim, Norway, 9–12 June 1997.

FNNPE (Federation of Nature and Natural Parks of Europe), 1993: *Loving Them to Death? Sustainable Tourism in Europe's Nature and Natural Parks,* Eupen, Belgium.

Fredrickson, L.M. and D.H. Anderson, 1999: A qualitative exploration of the wilderness experience as a source of spiritual inspiration, *Journal of Environmental Psychology,* **19,** pp. 21–39.

Graburn, N.H.H (ed.), 1976: *Ethnic and Tourist Arts,* University of California Press, Berkeley, CA.

Graburn, N.H.H., 1983: Tourism and prostitution: Articles review, *Annals of Tourism Research,* **19,** pp. 437–56.

Gupta, D., 2000: *Strategies for Protection and Management of Cultural Landscapes in India,* Iron Bridge Institute, University of Birmingham, Shropshire/Birmingham, UK.

Hanna, S. and S. Jentoft, 1996: Human use of the natural environment: An overview of social and economic dimensions. In: *Rights to Nature: Ecological, Economic, Cultural and Political Principles of Institutions for the Environment,* S. Hanna, C. Folke, and K-G. Maeler (eds.), Island Press, Washington, DC, pp. 35–55.

Hartig, T. and G.W. Evans, 1993: Psychological foundations of nature experience. In: *Behavior and Environment: Psychological and Geographical Approaches,* T. Gärling and R.G. Golledge (eds.), Elsevier Science Publishers, Amsterdam, The Netherlands, pp. 427–57.

Hartig, T., G.W. Evans, J.D. Jamner, D.S. Davis, and T. Gärling, 2003: Tracking restoration in natural and urban field settings, *Journal of Environmental Psychology,* **23,** pp. 109–23.

Heerwagen, J.H. and G.H. Orians, 1993: Humans, habitats and aesthetics. In: *The Biophilia Hypothesis,* S.R. Kellert and E.O. Wilson (eds.), Island Press, Washington, DC.

Henriksen, G., 1986: Rettigheter, oppgaver og ressurser: Tre forutsetninger for likeverd I etnisk sammensatte statssamfunn. In: *Identitet og Livsutfoldelse,* R. Erke and A. Hogmo (eds.), Universitets-forlaget, Oslo, Norway, pp. 116–28.

Hladik, C.M., A. Hladik, O.F. Linares, H. Pagezy, A. Semple, and M. Hadley, 1993: *Tropical Forests, People and Food: Biocultural Interactions and Applications to Development,* UNESCO-MAB (Man and Biosphere) book series 13, UNESCO, Paris, France/ Parthenon Publishing, Carnforth, Lancashire, UK, 852 pp.

Holling, C.S., (d.), 1978: *Adaptive Environmental Assessment and Management,* Wiley for United Nations Environment Programme, Chichester, UK.

Holling, C.S., 1995: Sustainability: The cross-scale dimension. In: *Defining and Measuring Sustainability: The Biophysical Foundations,* M. Munasinghe and W. Shearer (eds.), UN University, Tokyo, Japan/ The World Bank, Washington, DC, pp. 65–75.

Hughes, J.D., 1998: Sacred groves of the ancient Mediterranean area: Early conservation of biological diversity. In: *Conserving the Sacred: For Biodiversity Management,* P.S. Ramakrishnan, K.G. Saxena, and U.M. Chandrashekara, (eds.), UNESCO, Paris, France/Oxford & IBH Publishing, New Delhi, India, pp. 101–21.

Hughes, J.D. and M.D.S. Chandran, 1998: Sacred groves around the earth: An overview. In: *Conserving the Sacred: For Biodiversity Management,* P.S. Ramakrishnan, K.G. Saxena, and U.M. Chandrashekara, (eds.), UNESCO, Paris, France/ Oxford & IBH Publishing, New Delhi, India, pp. 69–86.

INSA/ CAS/ USNAS (Indian National Science Academy/ Chinese Academy of Sciences/ and US National Academy of Sciences), 2001: *Growing Populations, Changing Landscapes: Studies From India, China And The United States,* M.G. Wolman, P.S. Ramakrishnan, P.S. George, S. Kulkarni, P.S. Vashishtha, et al., (eds.), INSA/ CAS/ USNAS, National Academy Press, Washington, DC, 299 pp.

IUCN (World Conservation Union), 1999: Sustainable use within an ecosystem approach, IUCN submission to the 5th Meeting of the Subsidiary Body for Scientific, Technical and Technological Advice to the Convention on Biological Diversity, Geneva, Switzerland.

Jiggins, J. and N. Roling, 2002: Adaptive management: Potential and limitations for ecological governance of forests in a context of normative pluriformity. In: *Adaptive Management: from Theory to Practice,* J.A.E. Oglethorpe (ed.), IUCN, Gland, Switzerland, pp. 93–104.

Kaplan, S., 1987: Aesthetics, affect and cognition: Environmental preference from an evolutionary perspective, *Environment and Behavior,* **19,** pp. 3–32.

Kaplan, S. and R. Kaplan, 1989: *The Experience of Nature: A Psychological Perspective,* Cambridge University Press, New York, NY.

Kellert, S.R., 1993: The biological basis for human values of nature. In: *The Biophilia Hypothesis,* S.R. Kellert and E.O. Wilson (eds.), Island Press, Washington, DC.

Kuo, F.E. and W.C. Sullivan, 2001: Aggression and violence in the inner city: Impacts of environment via mental fatigue, *Environment & Behavior,* **33(4),** Special Issue on Restorative Environments, pp. 543–71.

Kweon, B.C., W.C. Sullivan, A.R. Wiley, 1998: Green common spaces and the social integration of inner city older adults, *Environment and Behavior,* **30(6),** pp. 832–58.

Lambin, E.F., B.L. Turner II, H.J. Geist, S. Agbola, A. Angelsen, et al., 2001: The causes of land-use and land-cover change: Moving beyond the myths, *Global Environmental Change: Human and Policy Dimensions,* **11,** pp. 261–9.

Lee, K.N., 1999: Appraising adaptive management, *Conservation Ecology,* **3,** pp. 3. Available at http://www.consecol.org/vol3/iss2/art3.

Leopold, A., 1949: *A Sand County Almanac,* Oxford University Press, New York, NY 226 pp.

Lohr, VI, C.H. Pearson-Mims, and G.K. Goodwin, 1996: Interior plants may improve worker productivity and reduce stress in a windowless environment, *Journal of Environmental Horticulture,* **14(2),** pp. 97–100.

Luttik, J., 2000: The value of trees, water and open space as reflected by house prices in the Netherlands, *Landscape and Urban Planning,* **48,** pp. 161–7.

MacDonald, R. and L. Jolliffe, 2003: Cultural rural tourism: Evidence from Canada, *Annals of Tourism Research,* **30(2),** pp. 307–22.

Matunga, H., 1994: Waahi tapu: Maori sacred sites. In: *Sacred Sites, Sacred Places,* D.L. Carmichael, J. Hubert, B. Reeves, and A. Schanche (eds.), Routledge, London, UK, pp. 217–26.

Maurer, M. and O. Holl, (eds.), 2003: *Natur als Politikum,* RLI-Verlag, Vienna, Austria, 633 pp.

Monasterio, M., 1994: Traditional prehistoric ecotechnologies for the management of biodiversity in Latin America, *Biology International Special Issue,* **32,** pp. 12–22.

Nash, R., 1973: *Wilderness and the American Mind,* Revised ed., Yale University Press, New Haven, CT.

OECD (Organisation for Economic Co-operation and Development), 2001: Environmental Indicators for Agriculture, Volume 3: Methods and Results, OECD, Paris, France.

Olsson, P. and C. Folke, 2001: Local ecological knowledge and institutional dynamics for ecosystem management: A study of crayfish management in the Lake Racken watershed, Sweden, *Ecosystems,* **4,** pp. 85–104.

Pearsall, D.M., 1992: The origin of plant cultivation in South America. In: *The Origins of Agriculture: An International Perspective,* C.W. Cowan and P.J. Watson (eds.), Smithsonian Institution Press, Washington, DC, pp. 173–205.

Pedford, J., 1996: Seeing is believing: The role of living history in marketing local heritage. In: *The Marketing of Tradition,* T. Brewer (ed.), Hisarlink Press, Enfield Lock: pp. 13–20.

Poffenberger, M., 2000: Local knowledge in conservation. In: *Beyond Fences: Seeking Social Sustainability in Conservation,* G. Borrini-Feyerabend and D. Buchan(eds.), IUCN, Gland, Switzerland/ Cambridge, UK, pp. 41–3.

Ramakrishnan, P.S., 1991: *Ecology of Biological Invasion in the Tropics,* National Institute of Ecology and International Science Publications, New Delhi, India, 195 pp.

Ramakrishnan, P.S., 1992: *Shifting Agriculture and Sustainable Development: An Interdisciplinary Study from North-Eastern India,* UNESCO-MAB (Man and Biosphere) Series, Paris, France/ Parthenon Publishing, Carnforth, Lancashire, UK, 424 pp. (Republished by Oxford University Press, New DelhiIndia, 1993).

Ramakrishnan, P.S., 1998: Ecology, economics and ethics: Some key issues relevant to natural resource management in developing countries, *International Journal of Social Economics,* 25, PP. 207–25.

Ramakrishnan, P.S., 2001: *Ecology and Sustainable Development,* National Book Trust, New Delhi, India, 198 pp.

Ramakrishnan, P.S., 2003: Globally important ingenious agricultural heritage systems (GIAHS): Global change and mountain biosphere management. In: *Global Change Research in Mountain Biosphere Reserves,* C. Lee and T. Schaaf (eds.), Proceedings of the International Workshop, Entlebuch Biosphere Reserve, 10–13 November, pp. 132–46.

Ramakrishnan, P.S., K.G. Saxena, and U.M. Chandrashekara (eds.), 1998: *Conserving the Sacred for Biodiversity Management,* UNESCO, Paris, France/ Oxford & IBH Publishing, New Delhi, India, pp. 101–21.

Ramakrishnan, P.S., R.K. Rai, R.P.S. Katwal, and M. Mehndiratta, (eds.), 2002: *Traditional Ecological Knowledge for Managing Biosphere Reserves in South* *and Central Asia,* UNESCO, Paris, France/ Oxford & IBH Publishing, New Delhi, India, 536 pp.

Ramakrishnan, P.S., K.G. Saxena, S. Patnaik, and S. Singh, 2003: *Methodological Issues in Mountain Research: Socio-Ecological System Approach,* UNESCO, Paris, France/ Oxford & IBH Publishing, New Delhi, India, 283 pp.

Ramakrishnan, P.S., R. Boojh, K.G. Saxena, U.M. Chandrashekara, D. Depommier, et al., 2004: *One Sun, Two Worlds: An Ecological Journey,* UNESCO, Paris, France/ Oxford & IBH Publishing, New Delhi, India. In press.

Ramsar, 1971: *Convention on Wetlands of International Importance Especially as Waterfowl Habitat,* RAMSAR U.N.T.S. no. 14583, vol. 996 (1976), Ramsar, p. 243.

Reisinger, Y., 1994: Tourist–host contact as a part of cultural tourism, *World Leisure and Recreation,* **36(2).**

Rössler, M., 2000: World heritage cultural landscapes. In: The George Wright Forum, *The Journal of the George Wright Society,* Special issue: Landscape stewardship: New directions in conservation of nature and culture, **17(1),** pp. 27–34.

Russell, K.C., J.C. Hendee, and S. Cooke, 1998: Social and economic benefits of a U.S. Wilderness Experience Program for Youth-at-Risk in the Federal Job Corps, *International Journal of Wilderness,* **4(3),** pp. 32–8.

Rutten, M., 1992: *Selling Wealth to Buy Poverty: The Process of the Individualization of Land Ownership among the Maasai Pastoralists of Kajiado District, Kenya,* Verlag Breitenbach Publishers, Germany.

Skayannis, P., 1999: *Planning Tourism, Development and the Environmental Protection in the Coastal Area of Magnesia: Local Interests and Expectations,* Association of European Schools of Planning annual conference, 7–10 July, Bergen, Norway, AESOP, University of Groningen, Groningen, The Netherlands.

Staats, H., A. Kieviet, and T. Hartig, 2003: Where to recover from attentional fatigue: An expectancy-value analysis of environmental preference, *Journal of Environmental Psychology,* **23,** pp. 147–57.

Stamps, A.E., 1996: People and places: Variance components of environmental preferences, *Perceptual and Motor Skills,* **82,** pp. 323–34.

Stevens, S., 1997: *Conservation Through Cultural_Survival: Indigenous Peoples and Protected Areas,* Island Press, Washington, DC.

Stovel, H., 2002: Cultural landscapes: A new approach for heritage conservation, Paper written in conclusion of the ITUC02 Course of ICCROM (International Organization for Conservation of Cultural Heritage), Rome, Italy.

Taylor, A.F., A. Wiley, F.E. Kuo, W.C. Sullivan, 1998: Growing up in the inner city: Green spaces as places to go, *Environment and Behavior,* **30(1),** pp. 3–27.

Thacker, C., 1983: *The Wildness Pleases: The Origins of Romanticism,* Croom Helm, London, UK.

Turnock, D., 1999: Sustainable rural tourism in the Romanian Carpathians, *Geographical Journal,* **165,** pp. 192–99.

Ulrich, R.S., 1983: Aesthetic and affective response to natural environment. In: *Human Behavior and Environment: Advances in Theory and Research,* I. Altman and J.F. Wohlwill (eds.), Volume 6, Plenum Press, New York, NY, pp. 85–125.

Ulrich, S.R., 1986: Human responses to vegetation and landscapes, *Landscape and Urban Planning,* **13,** pp. 29–44.

Ulrich, S.R., 1993: Biophilia, biophobia and natural landscapes. In: *The Biophilia Hypothesis,* S.R. Kellert and E.O. Wilson, (eds.), Island Press, Washington, DC.

Ulrich, R.S., R.F. Simons, B.D. Losito, E. Fiorito, M.A. Miles, et al., 1991: Stress recovery during exposure to natural and urban environments, *Journal of Environmental Psychology,* **11,** pp. 201–30.

UN (United Nations), 1996: Decision 51/197, A/RES/51/197 87th plenary meeting of the General Assembly, 17 December 1996, UN, New York, NY.

UN, 2004: *Permanent Forum on Indigenous Issues: Report of the 3rd Session,* ECOSOC Report E/2004/43, UN, New York, NY.

UNESCO (United Nations Educational, Scientific and Cultural Organization), 1972: *Convention Concerning the Protection of the World Cultural and Natural Heritage,* Adopted by the General Conference at its seventeenth session, Paris, 16 November 1972, UNESCO, Paris, France.

UNESCO, 2003: The sacred in an interconnected world, *Museum International,* 218 (September) (Special Issue), Paris, France.

US NAS (National Academy of Sciences), 1975: *Underexploited Tropical Plants with Promising Economic Value,* National Academy of Sciences, Washington, DC, 189 pp.

Valentine, P.S., 1992: Review: Nature-based tourism. In: *Special Interest Tourism,* B. Wetler and C.M. Hall (eds.), Belhaven, London, UK.

Van den Berg, A.E., 1999: *Individual Differences in the Aesthetic Evaluation of Natural Landscapes,* Dissertatiereeks Kurt Lewin Instituut 1999–4, Rijksuniversiteit Groningen, Groningen, The Netherlands.

Van den Berg, A.E., S.L. Koole, and N.Y. Van der Wulp, 2003: Environmental preference and restoration: (How) are they related? *Journal of Environmental Psychology,* **23,** pp. 135–46.

Van Dieren, W. and M.G. Wagenaar Hummelink, 1979: *Nature's Price: the Economics of Mother Earth,* Marion Boyars Ltd., London, UK/ Boston, MA, 193 pp.

Vannucci, M., 1993: *Ecological Readings in the Veda,* D.K. Print World, New Delhi, India, 216 pp.

Vatsayan, K., 1993: *Prakriti,* Indira Gandhi National Centre for the Arts, New Delhi, India, 64 pp.

von Droste, B.M.R. and S. Titchen (eds.), 1999: *Linking Nature and Culture,* Report of the Global Strategy Natural and Cultural Heritage Expert Meeting, 25–29 March 1998, Amsterdam, The Netherlands, UNESCO/ Ministry for Foreign Affairs/ Ministry for Education, Science, and Culture, The Hague, The Netherlands.

Wascher, D.M. (ed.), 2000: *The Face of Europe: Policy Perspectives for European Landscapes,* Report on the implementation of the PEBLDS (**Pan-European Biological and Landscape Diversity Strategy**) Action Theme 4 on European Landscapes, published under the auspice of the Council of Europe, ECNC European Center for Nature Conservation, Tilburg, The Netherlands, 60 pp.

Wascher, D.M., 2004: Landscape indicator development: Steps towards a European approach. In: *The New Dimensions of the European Landscape,* R. Jongman (ed.), Proceedings of the Frontis workshop on the future of the European cultural landscape Wageningen, The Netherlands 9–12 June 2002, Wageningen UR Frontis Series Nr. 4, Kluwer Academic Publishers, Dordrecht, The Netherlands, pp. 237–52.

Whitten, A.J., H. Haernman, H.S. Alikodra, and M. Thohari, 1987: *Transmigration and the Environment in Indonesia,* IUCN, Gland, Switzerland.

Wijesuriya, G., 2001: Protection of sacred mountains: Towards a new paradigm. In: *Proceedings of the UNESCO Thematic Expert Meeting on Asia-Pacific Sacred Mountains,* World Heritage Centre UNESCO, Agency for Cultural Affairs of Japan and Wakayama Prefectural Government, Japan, pp. 129–46.

WRI (World Resources Institute), 1990: *World Resources Report,* WRI, New York, NY/ Basic Books, Inc., New York, NY.

WRI, 1996: *World Resources, 1996–97: The Urban Environment,* WRI, Washington, DC, 400p.

WTO (World Tourism Organization), 1998: *Tourism Market Trend: Americas 1988–1997,* WTO, Madrid, Spain.

WTTC (World Travel and Tourism Council), 1999: Agenda 21 for the travel and tourism industry: Towards environmentally sustainable development progress, Report no. 2, Summary 4 pp. Available at http://www.wttc.org.

Yu, K., 1995: Cultural variations in landscape preference: Comparisons among Chinese sub-groups and Western design experts, *Landscape and Urban Planning,* **32,** pp. 107–26.

Zanetell, B.A. and B.A. Knuth, 2002: Knowledge partnerships: Rapid rural appraisal's role in catalyzing community-based management in Venezuela, *Society and Natural Resources,* **15,** pp. 805–25.

Chapter 18

Marine Fisheries Systems

Coordinating Lead Authors: Daniel Pauly, Jacqueline Alder
Lead Authors: Andy Bakun, Sherry Heileman, Karl-Hermann Kock, Pamela Mace, William Perrin, Kostas Stergiou, Ussif Rashid Sumaila, Marjo Vierros, Katia Freire, Yvonne Sadovy
Contributing Authors: Villy Christensen, Kristin Kaschner, Maria-Lourdes Palomares, Peter Tyedmers, Colette Wabnitz, Reg Watson, Boris Worm
Review Editors: Joseph Baker, Patricia Moreno Casasola, Ariel Lugo, Avelino Suárez Rodríguez, Lingzis Dan Ling Tang

---

★This appears in Appendix A at the end of this volume.

Main Messages

All oceans are affected by humans to various degrees, with overfishing having the most widespread and the dominant direct impact on food provisioning services, which will affect future generations. Areas beyond the 50 meters depth are mainly affected directly by fishing and indirectly by pollution. Fish are also directly affected by coastal pollution and degradation when their life cycle takes them into coastal habitats. Recent studies have demonstrated that global fisheries landings peaked in the late 1980s and are now declining despite increasing fishing effort, with little evidence that this trend is reversing under current practices. Fishing pressure is so strong in some marine systems that the biomass of some targeted species, especially larger high-value fish and those caught incidentally (the "bycatch"), has been reduced to one tenth or less of the level that existed prior to the onset of industrial fishing. In addition, the average trophic level of global landings is declining, which implies that we are increasingly relying on fish that originate from the lower part of marine food webs.

Industrial fleets are fishing with greater efficiency, further offshore, and in deeper waters to meet the global demand for fish. Until a few decades ago, depth and distance from coasts protected much of the deep ocean fauna from the effect of fishing. However, recent large investments in fishing capacity and navigation aids have led to fleets that now cover the world's ocean, including polar and deep, low-productivity areas, where catches are affecting easily depleted populations of long-lived species. The biomass of large pelagic fish in these areas taken by longlines, purse seines, and drift nets has also plummeted. Studies on available data have shown that deep-sea fisheries that collapsed in the 1970s have not recovered.

Overfishing has negative impacts on marine biodiversity. The lowered biomasses and fragmented habitats resulting from the impacts of fishing have led to local extinctions, especially among large, long-lived, slow-growing species with narrow geographical ranges. In addition, the ability of the component ecosystems and their embedded species to withstand stresses resulting from climate change and other human impacts will be reduced, though direct demonstration of this effect may not be evident in many systems for some decades.

Destructive fishing practices have long-term impacts on marine habitats. Destructive practices such as trawling, dynamiting, and dredging change the structure of marine ecosystems, with consequential changes in their capacity to provide services, such as food provisioning and income generation. Long-term losses in species and habitats through destructive fishing ultimately reduce the biodiversity of these affected systems, resulting in a further loss of services such as coastal protection. Some systems may recover and improve the availability of some services and products fairly quickly; other more vulnerable systems, such as cold-water corals and seamounts, may take hundreds of years to recover.

The implementation of no-take marine reserves combined with other interventions, such as controls on fishing capacity, would be a more proactive response to fisheries management than current reactive approaches. Marine reserves can contribute to better fisheries management—helping to rebuild stocks through enhanced recruitment and spill-over effects, maintaining biodiversity, buffering marine systems from human disturbances, and maintaining the ecosystems that fisheries rely on.

Aquaculture is not a solution to the problem of declining wild-capture fisheries. Good governance and effective management of wild-capture fishing are likely to be more successful approaches. Farmed species such as salmon and tuna, which use fishmeal, may in fact contribute to the problem since much of the fishmeal and oil currently used in the aquaculture industry is derived from wild-caught small pelagic fish. In some countries, such as Chile, small pelagic fish that were once a source of cheap protein for people are now largely diverted for fishmeal.

The supply of wild marine fish as a cheap source of protein for many countries is declining. Per capita fish consumption in developing countries (excluding China) has declined from 9.4 kilograms per person in 1985 to 9.2 kilograms in 1997. In some areas, fish prices for consumers have increased faster than the cost of living. Fish products are heavily traded, and approximately 50% of fish exports are from developing countries. Exports from developing countries and the Southern Hemisphere presently offset much of the demand shortfall in European, North American, and East Asian markets.

The proposed future uses of marine systems pose significant policy challenges. Ocean ranching of marine organisms, bioprospecting, seabed mining, and carbon sequestration in deep ocean waters are foreseeable uses of marine systems. However, the potential impacts of these activities are not well known. In some cases no or only limited field studies have been conducted to test the theoretical basis for the activity. Policies will need to deal with the uncertainty of potential impacts and the limited understanding of marine biodiversity. National and regional ocean policies that incorporate zoning for various uses within an integrated ecosystem-based management framework are likely to be needed. Such policies might include marine protected areas that can contribute to the restoration of species and habitats and thus form part of a precautionary strategy for guarding against management errors.

18.1 Introduction

Most of Earth—70.8%, or 362 million square kilometers—is covered by oceans and major seas. Marine systems are highly dynamic and tightly connected through a network of surface and deep-water currents. The properties of the water generate different density layers, thermoclines, and gradients of light penetration in marine systems, which result in productivity varying vertically. Tides, currents, and upwellings break this stratification and, by forcing the mixing of water layers, enhance primary production.

One widely accepted classification divides marine systems into four biomes (Longhurst et al. 1995; Longhurst 1998): the coastal boundary zone, trade-winds, westerlies, and polar. (See Figure 18.1.) These biomes are subdivided into a total of 57 biogeochemical provinces with distinct seasonal patterns of surface nutrient enrichment, which determine primary production levels and, ultimately, fisheries yield. The provinces of the coastal boundary zone biome largely overlap with the large marine ecosystems of K. Sherman and collaborators (see Watson et al. 2003), and hence those are implicitly included here. For practical reasons, we also refer to the U.N. Food and Agriculture Organization's classification that has been used to report on global fisheries statistics since 1950 and that divides the world's oceans up into 18 FAO statistical areas (FAO 1981).

The coastal boundary zone that surrounds the continents is the most productive part of the world ocean, yielding about 90% of marine fisheries catches. Overall, coastal and marine fisheries landings averaged 82.4 million tons per year during 1991–2000, with a declining trend now largely attributed to overfishing. The other three biomes are less productive, and their deep waters are exploited mainly for their large pelagic fish. The four biomes are described in detail in the next section.

In this assessment, the marine system is defined as the marine waters from the low-water mark to the high seas that support marine capture fisheries and deepwater (>50 meters) habitats.

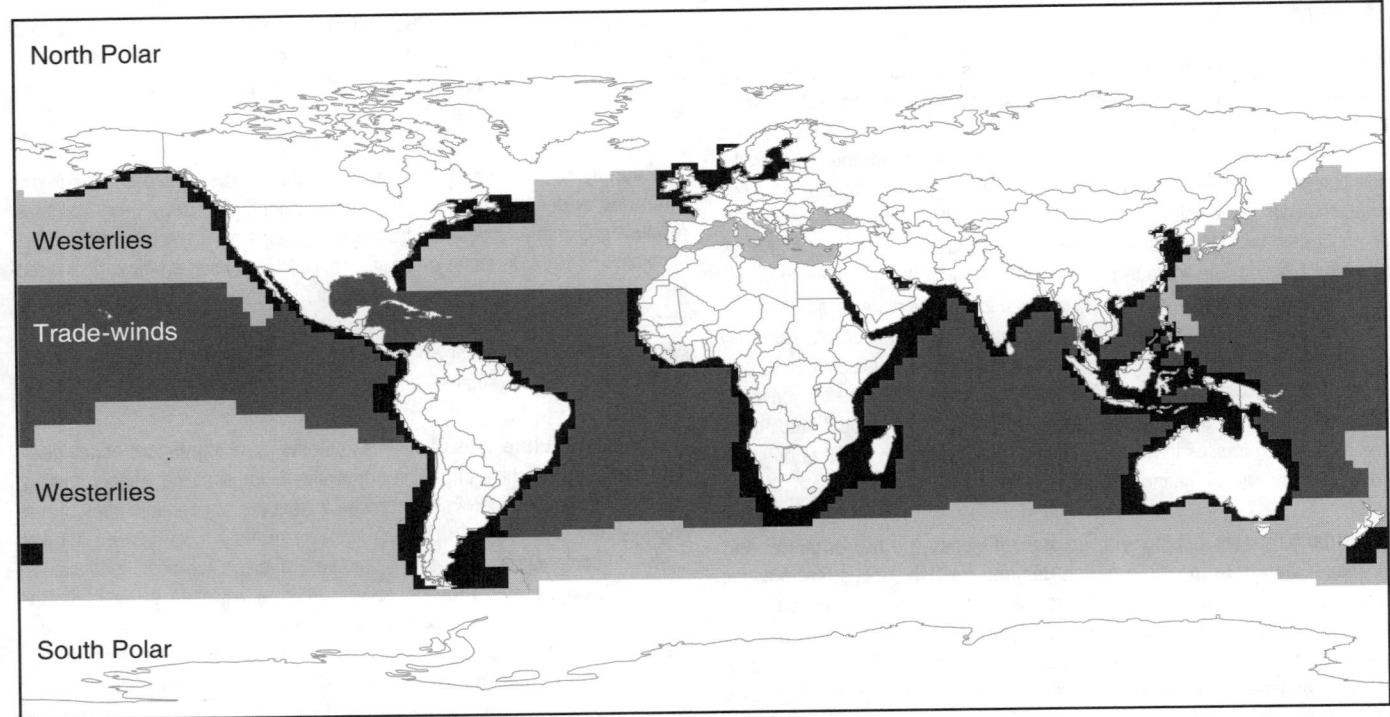

Figure 18.1. Classification of World's Oceans. Four "Biomes" were identified: Polar, Westerlies, Trade-winds, and Coastal Boundary (Longhurst et al. 1995; Longhurst 1998). The Coastal Boundary is indicated by a black border around each continent. Each of these Biomes is subdivided into Biogeochemical Provinces. The BGP of the Coastal Boundary Biome largely overlaps with LMEs identified by K. Sherman and coworkers (see Watson et al. 2003).

This definition spatially overlaps with coastal systems, which are bounded inland by land-based influences within 100 kilometers or 100-meters elevation (whichever is closer to the sea) and seaward by the 50-meter depth contour. Chapter 19 focuses on coastal habitats and coastal communities, however. It does not overlap conceptually with this chapter, which focuses on the condition and trends of fisheries resources in marine ecosystems for the following reasons:

- Living marine resources and their associated ecosystems outside of coastal areas (as defined by the MA), which maintain the food provisioning services of marine systems, have been affected over the last 50 years mostly by fishing.

- Our level of understanding of fisheries and the availability of information needed to assess the impact of fisheries are much better than for other human activities in marine systems. However, studies on biodiversity changes in marine systems are lagging behind our understanding of fisheries systems or terrestrial biodiversity changes. And overall, our understanding of long-term impacts and their interactions with other activities (current and future) is very limited.

- Chapter 19 describes the condition and trends of marine habitats and significant marine animals from the high-water mark to the 50-meter bathymetric line. Thus it discusses in detail the condition and trends of shallow inshore coastal habitats such as coral reefs, mangroves, and seagrasses, as well as important fauna such as seabirds, turtles, and marine mammals. Since most human uses of marine systems (tourism, gas and oil extractions, and so on) occur in the coast, they are discussed in detail there. On the other hand, the impact of human use, especially fishing, on deeper-water systems such as shelves, slopes, seamounts, and so on are discussed in this chapter.

- Chapter 4, on biodiversity, includes many non-fisheries aspects of marine biodiversity not covered here. Chapter 12, on

nutrient cycling, discusses the cycling of carbon, nitrogen, and phosphorus and the changes in these cycles in marine systems. And Chapter 13, on air quality and climate, highlights possible changes, including acidification, carbon sequestration, and fluxes in marine systems, over the short term.

Nevertheless, this chapter touches on various aspects of marine ecosystems such as marine biodiversity as they relate to fisheries and deepwater habitats, and some activities such as tourism and transportation are also mentioned. But there is currently insufficient information available to assess which activities have relatively more impact than others in marine systems.

Marine ecosystems are diverse—some are highly productive, and all are important ecologically at the global scale and highly valuable to humankind. The major ecosystem services (as described in Chapter 1) derived from marine ecosystems are summarized in Table 18.1.

Marine systems play significant roles in climate regulation, the freshwater cycle, food provisioning, biodiversity maintenance, energy, and cultural services, including recreation and tourism. They are also an important source of economic benefits, with capture fisheries alone worth approximately $81 billion in 2000 (FAO 2002); aquaculture worth $57 billion in 2000 (FAO 2002); offshore gas and oil, $132 billion in 1995; marine tourism, much of it in the coast, $161 billion in 1995; and trade and shipping, $155 billion in 1995 (McGinn 1999). There are approximately 15 million fishers employed aboard decked and undecked fishing vessels in the marine capture fisheries sector. About 90% of these fishers work on vessels less than 24 meters in length (FAO n.d.).

More than a billion people rely on fish as their main or sole source of animal protein, especially in developing countries. (See Table 18.2.) Demand for food fish and various other products from the sea is driven by population growth, human migration

Table 18.1. Percentage of Animal Protein from Fish Products, 2000 (FAO 2003)

Region	Share of Animal Protein from Fish Products
	(percent)
Asia (excluding Middle East)	27.7
Oceania	24.2
Sub-Saharan Africa	23.3
Central America and Caribbean	14.4
North America	11.5
South America	10.9
Europe	10.6
Middle East and North Africa	9.0

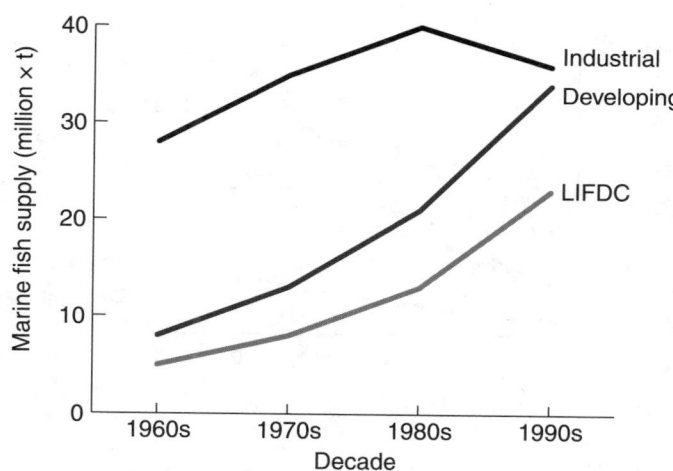

Figure 18.2. Average Domestic Marine Fish Supply, Lesser-Income Food-Deficit Countries, 1961–99 (FAO 2002)

Table 18.2. Summary of Ecosystem Services Provided by Different Marine System Subtypes

Direct and Indirect Services	Inner Shelf	Outer Shelves, Edges, Slopes	Seamounts and Mid-ocean Ridges	Deep Sea and Central Gyres
Food — human	**	**	**	**
Food — animal	**	**		
Fiber, timber, fuel	*	*		*
Medicines, other services	*			
Biodiversity	**	**	*	*
Biological regulation				
Nutrient cycling and fertility	*	*	*	*
Atmospheric and climate regulation	*	*		*
Human disease control				
Waste processing				
Flood/storm protection				
Employment	**	**	*	*
Cultural and amenity	**			
Key: * some importance ** very important				

toward coastal areas, and rising incomes that increase demand for luxury seafood.

Detailed data on fisheries catches—that is, food provisioning, the major ecosystem service considered here—are available since 1950 for (groups of) species for all FAO areas and maritime countries of the world. (See www.fao.org for tabular statistics and www.seaaroundus.org for spatially disaggregated statistics.) These show that catches increased more rapidly than the human population through the 1950s and 1960s, leading to an increase in available seafood. (See Figure 18.2.) This period also saw the depletion of many local stocks, but this was masked by the global increase of landings. The first fisheries collapse with global impact on prices of fishmeal and its substitutes was the Peruvian anchoveta, in 1971/72, which fell from an official catch of 12 million tons annually in the 1972–73 season (in reality, probably 16 million tons annually; see Castillo and Mendo 1987) to 2 million tons in 1973 (Tsukayama and Palomares 1987), ushering in two decades of slow growth and then stagnation in global fish catches. (See Figure 18.3.)

18.2 Condition and Trends of Marine Fisheries Systems

18.2.1 Global Trends

The mid-twentieth century saw the rapid expansion of fishing fleets throughout the world and an increase in the volume of fish landed. These trends continued until the 1980s, when global marine landings reached slightly over 80 million tons per year; then they either stagnated (China included; FAO 2002) or began to slowly decline (Watson and Pauly 2001). However, regional landings peaked at different times throughout the world, which in part masked the decline of many fisheries.

Indeed, the world's demand for food and animal feed over the last 50 years has resulted in such strong fishing pressure that the biomass of some targeted species, such as the larger, higher-valued species and those caught incidentally (the "bycatch"), has been reduced over much of the world by a factor of 10 relative to levels prior to the onset of industrial fishing (Christensen et al. 2003;

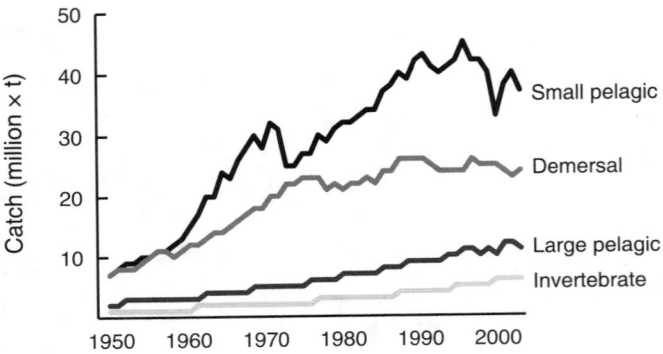

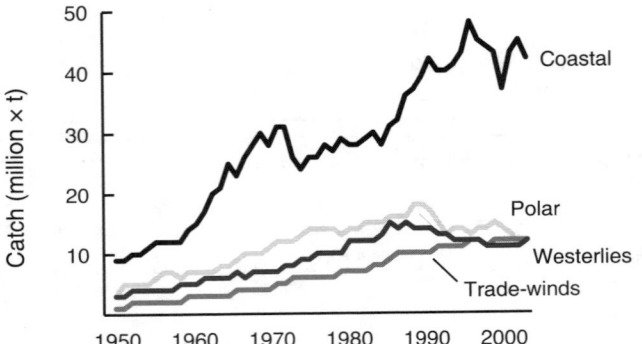

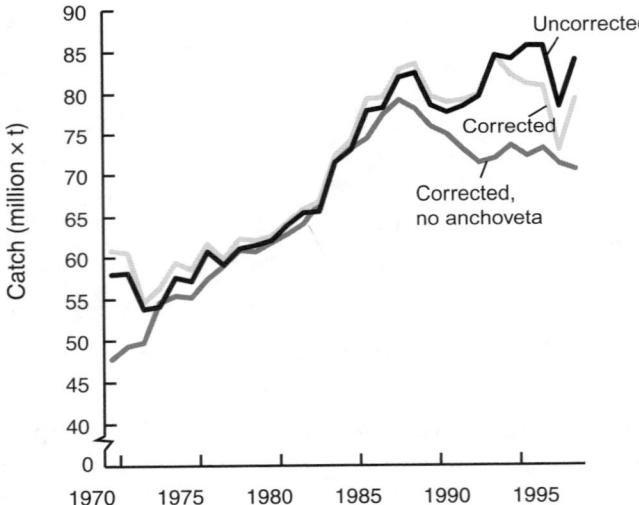

Figure 18.3. Estimated Global Fish Catches, 1950–2001, by Target Group and Biome, with Adjustment for Over-reporting from China. Note: bottom graph is the total landings, adjusted and not adjusted for China. (Watson and Pauly 2001)

fully exploited, and moderately exploited (FAO 2002). The FAO analysis lists the status of commercially important stocks where there is sufficient information. While the information presented is simple and many people use it to reflect the state of fisheries globally, they have the potential to provide an overoptimistic estimate of the state of fisheries. First, the figures presented only consider stocks currently exploited and exclude those that were fished either to extinction or abandoned over the last 50–100 years. Second, the reporting is based on over 1,500 stocks, with an assessed ''stock'' actually representing species distributed over large areas—that is, the aggregate of many stocks that are at varying states of exploitation. Moreover, the ''stocks'' presented do not represent the thousands of stocks that are fished by small-scale fishers that are not assessed or included in official statistics. For example, there are thousands of coral reef fish stocks that are fished by small-scale fishers in areas such as Indonesia and the Philippines, which are severely overfished but not a part of the FAO global analysis.

Until a few decades ago, depth and distance from coasts protected much of the deep-ocean fauna from the effect of fishing (Figure 18.5). However, fleets now fish further offshore and in deeper water with greater precision and efficiency, compromising areas that acted as refuges for the spawning of many species of commercial interest to both industrial and artisanal fleets (Kulka et al. 1995; Pauly et al. 2003). (See Figure 18.6.) Investments in the development of fishing capacity have led to fleets that cover the entire world's oceans, including polar and deep-sea areas and the low-productivity central gyres of the oceans. Trawl catches particularly target easily depleted accumulations of long-lived species, and the biomass of large pelagic fish has also plummeted (Worm and Myers 2003).

Not only are once inaccessible areas of the ocean increasingly being fished, they are also increasingly exploited for other ecosystem services. The marine realm is seen by many as the next frontier for economic development, especially for gas and oil and other energy sources (wind, gas hydrates, and currents), seabed mining (such as polymetallic nodules), bioprospecting, ocean dumping, aquaculture, and carbon sequestration. Worm et al. (2003) have identified pelagic ''hotspots'' of biodiversity (see Figure 18.7 in Appendix A), while Bryant et al. (1998) identified key coral reef areas. These and other hotspots, which may play a key role in supporting ecosystem services such as biodiversity, will be negatively affected by these developments unless they are appropriately managed.

The gas and oil industry is worth more than $132 billion annually, and the potential for further development is considered high (McGinn 1999). Current levels of development in the deeper ocean environments are low, but future rises in the price of carbon-based fuels could make the extraction of crude oil and gas further offshore financially feasible.

Mining in shallow offshore coastal areas for gold, diamonds, and tin is already under way and there is little doubt that the technology can be developed for deeper mining for a range of minerals, including manganese nodules, cobalt, and polymetallic sulfides, given appropriate economic incentives (Wiltshire 2001). It is assumed that institutional constraints in the future may ensure that activities in these deep-sea environments are conducted in a sustainable way and with minimal impact. This raises a number of questions regarding the nature and scale of the impacts on the little-known deep-ocean habitats. Mineral extraction may include ship- or platform-based processing of the extracted product, which has the potential to pollute the adjacent ecosystems. While bioprospecting does not have the magnitude of physical impacts that oil and mineral extraction does, there is nevertheless the po-

Myers and Worm 2003). In addition, with fleets now targeting the more abundant fish at lower trophic levels (see Figure 18.4), it would be expected that global catches should be increasing rather than stagnating or decreasing, as is actually occurring. Indeed, this by itself indicates the extent that fishing has affected marine ecosystems.

Changes in trophic levels of global and regional catches are considered a better reflection of trends in fisheries than the proportion of fish stocks that are reported as depleted, overexploited,

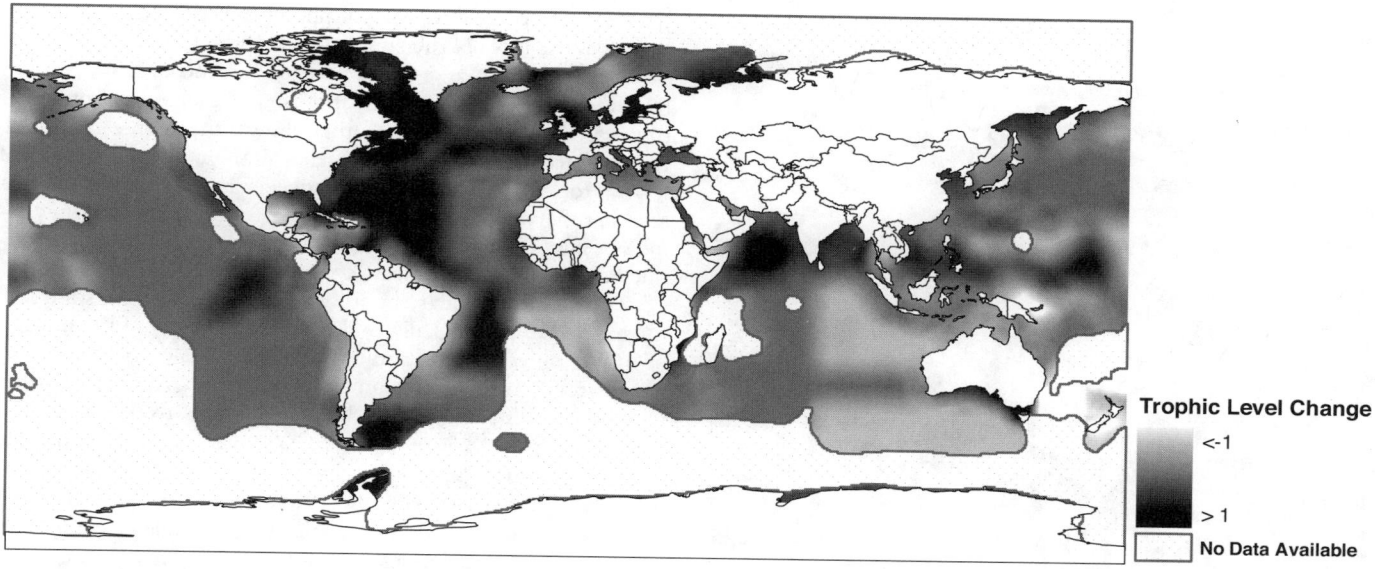

Figure 18.4. Trophic Level Change, 1950–2000

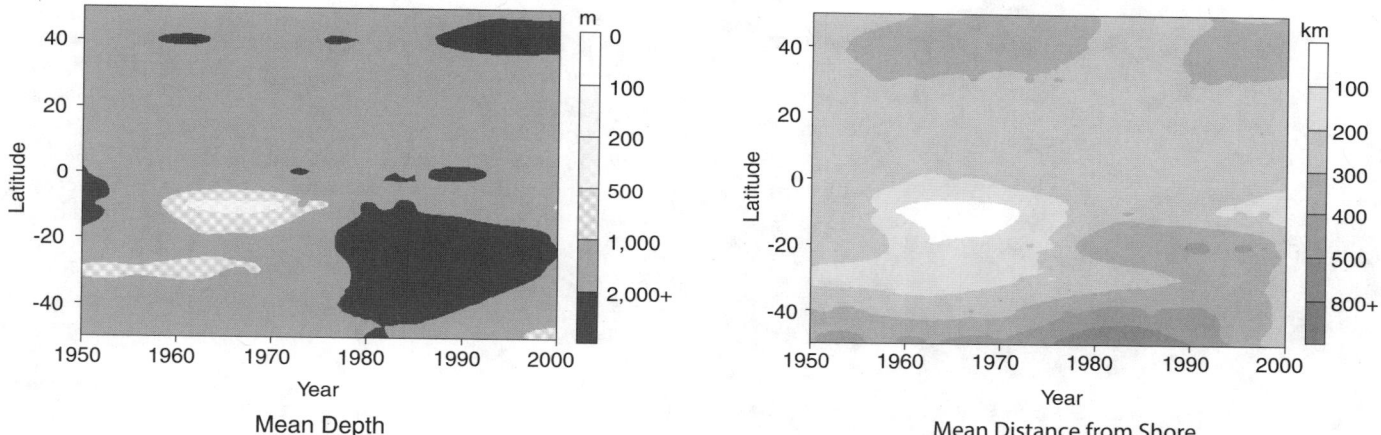

Figure 18.5. Mean Depth of Catch and Mean Distance of Catch from Shore since 1950. Both panels show that fisheries catches increasingly originate from deep, offshore areas, especially in the Southern Hemisphere.

tential for overexploitation (see Chapter 10), resulting in impacts similar to those described for overfishing.

Ocean biomes are also strongly affected by humans. While the coastal biome is heavily affected by coastal development and land-based pollution sources (see Chapter 19), the three other biomes have been affected by a variety of actions such as oil spills, over-hunting of marine mammals, seabird mortalities, and ocean dumping of waste. (See Tables 18.3 and 18.4.) For instance, the estimated 313,000 containers of low-intermediate emission radio-active waste dumped in the Atlantic and Pacific Oceans since the 1970s pose a significant threat to deep-sea ecosystems should the containers leak (Glover and Smith 2003), which seems likely over the long term. Other examples include seabird populations that have been seriously affected by fishing and oil pollution, such as the estimated 14,000 seabirds killed each year by the Alaskan longline groundfish fishery between 1993 and 1997 (Stehn et al. 2001) and the chronic pollution along the coast of Chubut (Argentina) that has significantly increased Magellanic penguin (*Spheniscus megellanicus*) mortality (Gandini et al. 1994).

Knowledge of the effects of persistent organic and inorganic pollutants on marine fauna, including reproductive effects, is lim-

ited, and we know even less about how these pollutants interact with fisheries impacts. Similarly, non-fishery factors and their impacts on habitats, primary productivity, and other ecosystem features are often described in the literature. However, their joint and cumulative effect on ecosystems is usually not assessed, limiting comparative analyses. Other human activities, as well as climate (due to natural variation and anthropogenic sources), influence marine systems, but their effects cannot usually be clearly separated from the impact of fishing. However, this should not detract from the urgent need to implement sustainable fisheries practices.

18.2.2 Coastal Boundary Zone Biome

The coastal boundary zone biome (10.5% of the world ocean) consists of the continental shelves (0–200 meters) and the adjacent slopes—this is, from the coastlines to the oceanographic front usually found along the shelf edges (Longhurst 1998). The 64 large marine ecosystems listed in Sherman and Duda (1999), which serves as a conceptual framework for an increasing number of multisectoral projects, largely match the biogeochemical prov-

1965

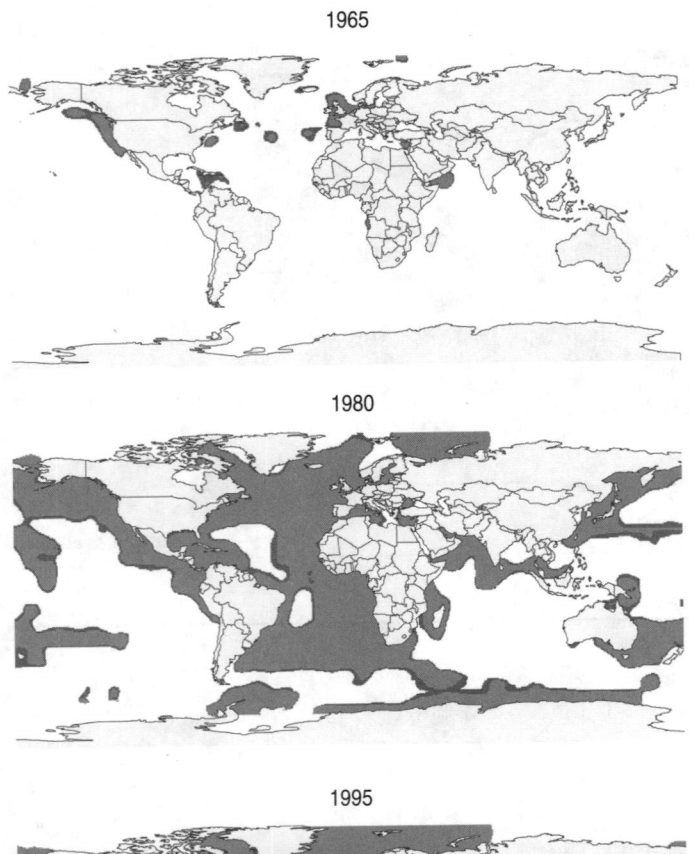

1980

1995

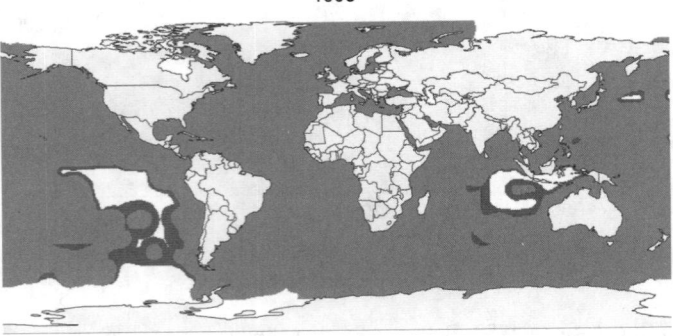

Figure 18.6. Year of Maximum Catch, 1965, 1980, and 1995. Solid lines indicate the current year and shading indicates that the maximum catch has already been reached. (SAUP 2005)

inces of Longhurst's coastal boundary biome and hence are implicitly considered here.

This biome fully includes the coastal systems (0–50 meters) covered in Chapter 19, the outer shelves (50–200 meters), and most of the continental slopes (200–1,000 meters). The coastal component of this biome is the first to have been accessed by fisheries, and it provides the bulk (53% in 2001; R. Watson, Sea Around Us Project, unpublished data) of the world's marine fisheries catches. The major processes that lead to ecosystem services such as food provisioning and biodiversity from this biome are described here.

Marine food webs are based largely on primary production by microscopic algae, the phytoplankton. This production occurs in the lighted, upper layers of the ocean, especially in the coastal zone, and is intensified by processes that lift nutrient-laden water from deeper layers. Most of this production is then either grazed by herbivorous zooplankton (mainly copepods) or falls to the sea bottom in form of detritus aggregates known as marine snow,

which is formed of decomposed phytoplankton and zooplankton as well as the feces of the zooplankton attacked by bacteria while on the way down and consumed by benthic organisms upon reaching the sea floor. Little marine snow reaches the bottom of tropical seas due to, among other things, the higher metabolic rates of bacteria in warm waters. Hence, there is less benthos and fewer ground fish to catch in the deeper reaches of tropical seas than in otherwise comparable temperate or polar seas and upwelling systems. This creates a limit to the expansion of deep-sea benthic fisheries in tropical areas (Longhurst and Pauly 1987).

The coastal boundary biome is the most significant source of marine fish landed globally, and it also bears many of the impacts of fishing on ecosystems and of other human activities. Most depleted stocks are in this biome. In the North Atlantic, this has resulted in substantial marine biomass declines over the last 100 years (see Figure 18.8 in Appendix A) as well as the mean trophic level of the catch (see Figure 18.9) over the last 50 years. The majority of bottom-trawling fleets operate in this zone, affecting large areas of the seabed on a continual basis while catching both target and nontarget species. Holmes (1997) suggested that trawling destroyed seabed habitat and this contributed to fish declines in heavily trawled areas. A century of trawling in the North Sea has reshaped part of the seabed and changed the structure of the ecosystem (Malakoff 1998). Areas of the greatest decline in landings are in the coastal boundary biome.

Fishing pressure in this biome is not just attributed to excessively large industrial fleets but also to small-scale or recreational fishers, whose landings have a minor impact when viewed individually but who collectively can significantly deplete local resources, as described later. Coastal habitats such as coral reefs and similar biogenic bottom structures (for example, soft corals and sponge beds) are degraded where destructive fishing methods such as explosives, poisoning, and trawling are used by small-scale fishers (Cesar et al. 2003). Such fishing practices have particular impacts on coral reef habitats and the ability of damaged reefs to recover.

Coral reef fisheries are overexploited in many reef systems around the world (Christensen and Pauly 2001; Jameson et al. 1998). Although many fisheries have become unsustainable due to the scale of high-technology improvements to boats and fishing gear, even small changes of technology can shift the balance toward unsustainability. In Pacific Islands, for instance, spearfishers' catch of large humphead wrasse (*Cheilinus undulates*) used to be limited due to their reliance on snorkeling gear. However, scuba diving equipment has recently given fishers access to wider areas of reef and let them use other methods such as cyanide and dynamite both day and night, decimating humphead wrasse as well as populations of other large fish that are sold to the live fish market (Birkeland and Friedlander 2002).

The trading of fish sourced in the coastal boundary biome has undermined food security in coastal communities of the developing world. The demand for fish in local, regional, or international markets can, though increased prices, promote overfishing when demand from a luxury market largely exceeds the supply and fisheries management is ineffective. Eight of the top 40 food-deficient countries are also major fish producers and exporters (Kurien 1998). Much of the fish from the coastal boundary biome is exported to industrial countries, often to service the national debt of developing nations. The export of captured marine fish from developing countries has removed a cheap source of protein from their people in some cases. Senegal, for instance, which is a significant exporter of marine products, also has a protein deficit among its rural population because the growth of export-oriented

Table 18.3. Summary of Human Disturbances at the Deep-sea Floor, in Pelagic Waters, and on Continental Slopes (Deep-sea floor from Glover and Smith 2003)

Human Use	Temporal Scale	Knowledge of Impacts/Severity/Spatial Scale	Estimated Importance in 2025
Past impacts			
Deep-sea floor			
Dumping of oil/gas structures	isolated incidents (now banned)	good/low/regional	low
Radioactive waste disposal	1950s–90s	good/low/local	low
Lost nuclear reactors	1960s onwards	good/low/local	low
Dumping of munitions	1945–76 (now banned)	poor/low/local	low
Pelagic waters and continental slopes			
Dumping of wastes	until 1980s (now regulated)	low/low/low	moderate
Present impacts			
Deep-sea floor			
Deep-sea fisheries	1950s onward	good/high/regional	high (overfished)
Collateral damage by trawling	1950s onward	good/high/regional	high
Deep-sea oil and gas drilling	1990s onward	poor/moderate/basin	moderate
Dumping of bycatch causing food falls	1990s onward	poor/moderate/basin	moderate
Research and bioprospecting at vents	1960s onward	good/low/local	very low
Underwater noise	1960s onward	poor/low?/local	probably low for benthos
Pelagic waters and continental slopes			
Fishing	until 1950s very limited; steadily increasing, especially since 1960s	good/good/global	high (some unsustainable or highly variable)
Transportation — oil spills	increasing accidental oil spills until 1990s; decreasing	good/low/isolated	moderate
Transportation — other pollution (oil from bilges, litter, ballast)	despite regulations, still occurring	low/low/basin	moderate
Sewage discharge	1990s	good/good/local	high
Mining — minerals, gas and oil (slopes only)	1980s onward	good/good/local	high
Future impacts			
Deep-sea floor			
Polymetallic nodule mining	10–20 year time scale	poor/very high/regional-basin	high
CO$_2$ sequestration	10–30 year time scale	poor/very high/local-regional	high
Dumping of sewage sludge	5–10 year time scale	good/moderate/local-regional	moderate
Dumping of dredge spoil	5–10 year time scale	poor/low/local	moderate
Climate change	50–100 year time scale	poor/very high/basin-global	low
Manganese crust mining	unknown	poor/high/local	low
Polymetallic sulphide mining	unknown	poor/high/local	low
Methane hydrate extraction	unknown	poor/moderate/regional	low
Pelagic waters and continental slopes			
Dumping of dredge spoils	< 5 years	good/poor/local	high
Aquaculture	5–10 years	low/low/local	moderate-high
Tourism — mega cruise liners or offshore structures	unknown	low/low/low	low

fisheries disrupted domestic supplies of cheap, small pelagic fish (UNEP 2002a).

In contemporary literature on fisheries economics, it is accepted that once fisheries cease to be open access (by instituting some form of property rights), they can be managed sustainably to ensure the holder of quotas the maximum discounted eco-nomic rent, which is the highest present value of the sum of all future flows of resource rent from a given fishery (Hannesson 2000; Arnason and Gissurarson 1999). However, many authors have challenged this view because whether the stock is managed sustainably will also depend on a number of other factors, including the price-cost ratio of landing a unit weight of fish and the

Table 18.4. Summary of Specific Deep-sea Habitats (Based on Baker et al. 2001)

Deep-sea Habitat	Current Condition/Threats	Potential Threats
Hydrothermal vents	limited disturbances – currently due to limited research undertaken on vents; low number of species but high endemism and high abundance	high potential for biotechnology, mining, energy, and tourism
Seamounts	few of the more than 10,000 seamounts have been studied; high endemism on studied seamounts; some seamounts are heavily exploited for fisheries resources, trawling damages benthic habitats	mining of ferromanganese oxide and polymetallic sulfides
Deep-sea trenches	highly unique "hadal" fauna, much of it associated with soft sediments and holothurians; high endemism; supports diverse and abundant bacterial community; no known disturbances	research, biotechnology, and waste disposal
Deep-water corals	limited knowledge, speculation that they are more widespread than currently known; high diversity except for fish and mollusks compared with tropical reefs; colonies growing on gas and oil platforms; damaged by trawling but spatial extent unknown; gas and oil platforms can damage corals	biotechnology
Polymetallic nodules	primarily inhabited by hard substrate epifauna with foraminifera dominating by abundance and coverage; diverse fauna; limited disturbance resulting from current research and feasibility studies	high for mineral exploration including spoils disposal, especially if prices increase; mining also has potential to have an impact on pelagic communities
Cold seep and pockmarks	limited knowledge; high endemism; limited disturbances except for Gulf of Mexico (trawling and oil exploitation) or areas of research	biotechnology and mineral exploitation
Gas hydrates	limited knowledge and disturbance but scientific studies emerging in Gulf of Mexico, offshore India, and Japan; important bacterial diversity	source of methane gas for energy, but potential to exacerbate climate change
Submarine canyons	high diverse flora and fauna including commercially important species such as lobsters; important nursery areas for marine species; some areas affected by fishing and oil exploitation	gas and oil developments

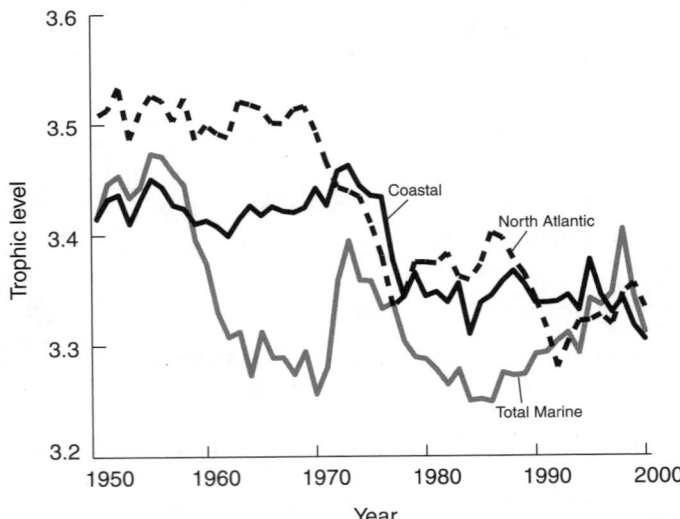

Figure 18.9. Changes in Trophic Level in North Atlantic and Coastal Areas at Less Than 200 Meters Depth, and Total Marine Landings, 1950–2000 (SAUP 2005)

discount rate applied to calculate the discounted rent. If both this ratio and the discount rate are very high, it may well be economically rational for the quota holder to deplete the fish stocks. In fact, it will be optimal in a strictly economic sense, as doing so may provide the maximum economic rent from a resource (Clark 1973; Heal 1998; Sumaila 2001; Sumaila and Walters 2004). Current economic models for fisheries, including those based on property rights, need to better accommodate underlying biological constraints.

18.2.3 Trade-winds Biome

The trade-winds biome (covering 38.5% of the world ocean) lies between the northern and southern sub-tropical convergences, where a strong water density gradient hinders nutrient recycling between deep layers and upper surface layers. The resulting low levels of new primary production make these zones the marine equivalent of deserts. Therefore, fisheries in this biome rely mainly on large pelagic fish, especially tunas, capable of migrating over the long distances that separate isolated food patches. In the eastern tropical Pacific, a major portion of the tuna purse-seine catch results from exploitation of a close association with pelagic dolphins, which suffered severe depletion in the 1970s due to incidental kills in the tuna purse seines (Gerrodette 2002). Between 1990 and 2000, 1.5 to 3.5 million Northeastern spotted dolphins (*Stenella attenuata*) were incidentally captured annually in tuna seine nets (Archer et al. 2002).

One exception to the general low productivity of the trade-winds biome is around islands and seamounts, where physical processes such as localized upwelling allow for localized enrichment of the surface layer. Above seamounts, these processes also lead to the retention of local production and the trapping of advected plankton, thus turning seamounts into oases characterized by endemism and, when pristine, high fish biomass.

Exploitation of the demersal resources of seamounts usually occurs in the form of intense trawling pulses, mainly by distant water fleets, which reduce biomass to extremely low levels, reduce diversity in the associated pelagic systems, and destroy biogenic bottom structures and their associated benthic diversity. Similar exploitation occurs along ocean ridges, such as in the North Atlantic and the Central Indian Ocean, where poorly documented bottom trawl fisheries developed in the 1990s outside of any regulatory regimes.

Overall, the trade-winds biome contributed 15% of the world's marine fisheries catch in 2001. Of this, 34% consisted of large pelagic fish and the rest were largely deep demersal species.

18.2.4 Westerlies Biome

In the westerlies biome (35.7% of the world's oceans), seasonal differences in the depth of the mixed layer result from seasonality in surface irradiation and wind stress, inducing strong seasonality of biological processes, including a spring bloom of phytoplankton. The fisheries of this biome, mainly targeting tuna and other large pelagic species, are similar to those of the trade-winds biome.

The westerlies and trade-winds biomes are also inhabited by an enormous number of small mesopelagic fish that aggregate during the day at depths of 500–1,000 meters, forming a dense layer of fish and invertebrates, especially squid, and that migrate upward every night to feed on zooplankton at the surface layer (vertical migration). Their aggregate biomass, almost 1 billion tons (Gjøsaeter and Kawaguchi 1984), has often been described as a potential resource enabling further fisheries development. However, mesopelagic fish rarely occur in fishable concentrations, and their bodies tend to contain large amounts of wax esthers, which render their flesh unpalatable to humans.

Overall, the westerlies biome contributed 15% of the world's marine fisheries catch in 2001. Of these, 9% were large pelagic fish with the rest consisting of small pelagic fish (40%), demersal fish (23%), and squid (11%). The marine environment in this biome is relatively unaffected by human use other than fishing. However, the large pelagic fish, such as tuna and shark, are strongly exploited.

18.2.5 Polar Biome

The polar biome covers only 15% of the world ocean and accounts for 15% of global marine fish landings. Its vertical density structure is determined by low-salinity waters from spring melting of ice. The bulk of annual primary production occurs in ice-free waters during a short intense summer burst. However, primary production under lighted ice occurs over longer periods, especially in Antarctica.

The Arctic fisheries along the north coast of Siberia, Alaska, and Canada (FAO Area 18) are poorly documented, and the few thousand tons of landings reported for this area by FAO are likely to be underestimates. The Arctic marine system is important for the well-being of indigenous people living in the area. For instance, marine mammals, such as whales and seals, are an important source of food and are of significant cultural value. However, high levels of persistent organic pollutants in their blubber pose a health concern. (See also Chapter 25.) Climate change has the potential to have a significant impact on the people of this area,

since the ice forms a fundamental part of subsistence, shelter, travel, safety, and culture in the region. Oil and gas exploitation pose another set of issues for inhabitants of the Arctic (through social changes, for instance) and the ecosystem (through impacts on marine mammals, habitat damage or changes, oils spills and contamination).

The Antarctic krill, *Euphausia superba,* consumes the primary production from both open waters and under the ice and then serves as a food source for a vast number of predators, notably finfish, birds (including penguins), and marine mammals. As in the Arctic, the marine mammal populations of Antarctica were largely decimated before the middle of the twentieth Century. There is also a relatively small direct fishery for krill (about 150,000 tons per year) in Antarctica, which may expand if krill proves a suitable feed for salmon or other forms of farming.

The development of fisheries in the southern polar biome demonstrates the fragility of fish stocks, marine mammals, and seabirds in terms of the impacts of exploitation by humans in just over 30 years. The distant water fleet of the former Soviet Union began exploiting this biome in the mid-1960s when ships began to deplete stocks of Marbled notothenia (*Notothenia rossii*), Mackerel icefish (*Champsocephalus gunnari*), and Gray notothenia (*Lepidonotothen squamifrons*) in different areas of the South Indian, South Atlantic, and South Pacific Oceans. (See Figure 18.10.) In all these areas, the same catch trends emerged: within a few years of opening fisheries, catches would peak and then rapidly decline to a small fraction of their original biomass. This operating mode of distant water fleets, including those of Russia, Chile, Argentina, France, and the United Kingdom, continued until the beginning of the 1990s.

The formation of the Commission for the Conservation of Antarctic Marine Living Resources in 1982 brought the first conservation measures for stocks of Marbled notothenia (in 1985). Other stocks remained unmanaged until the 1990s, when dra-

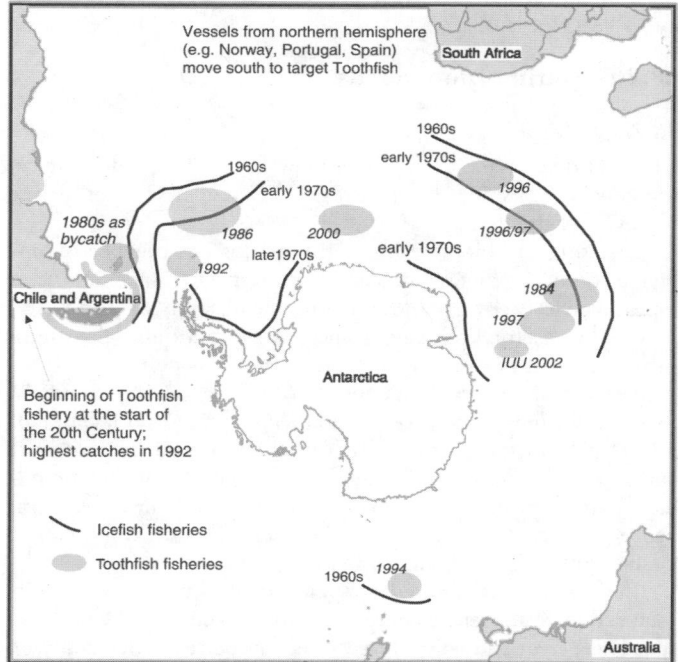

Figure 18.10. Growth of Fishing for Icefish and Toothfish in the Southern Polar Biome, 1900 to Present (Sabourenkov and Miller 2004; Kock 1991, 2001)

matic declines in stocks of Mackerel icefish at South Georgia and Kerguelen were recognized. Assessment, management, and control of the fisheries became much more stringent (Constable et al. 2000), and Mackerel icefish around South Georgia recovered sufficiently to allow limited commercial catches from the mid-1990s onward (CCAMLR 2002). However, Kerguelen stocks have remained at a very low level (Duhamel and Claudet 2002).

In the second half of the 1980s, the Soviet Union developed a longline fishery on Patagonian toothfish *Dissostichus eleginoides* (Kock 1991, 1992). The same declining catch trend for these long-lived, slow-growing species emerged for the stock around the Prince Edward Islands, which was reduced to very low levels within a few seasons. As a side effect, a large numbers of seabirds (such as albatrosses) became hooked on lines during the process of setting and hauling (Kock 2001). The situation became more aggravated when longline fishing was extended to virtually all grounds in the northern part of the southern polar biome from 1996/97 onward, and concern over the sustainability of stocks grew. Nevertheless, illegal, unregulated, and unreported fishing on the highly prized Patagonian toothfish increased dramatically from 1997 onward, and it is estimated that 80–90% of the current catch is taken illegally (CAMLR 2002).

While it was possible to reduce IUU fishing around South Georgia to low levels, fishing pressure by IUU vessels remained high on other fishing grounds, notably in the Southern ocean, in FAO Area 58, despite considerable efforts by France and Australia to improve surveillance around the territories under their control. Commission members are working in closer cooperation with countries to assist with the apprehension of IUU vessels (as, for example, in the *Viarsa* incident; see www.intrafish.com).

New fisheries are still being developed in the southern polar biome despite the lessons learned about the vulnerability of the local fish (including the Patagonian toothfish) to high levels of exploitation. Thus, New Zealand started an exploratory fishery on Antarctic toothfish (*Dissostichus mawsoni*) in the Ross Sea in the 1998/99 season (CCAMLR 2002) and has increased catches every year since.

18.2.6 Marine Biodiversity

18.2.6.1 Global Trends

This section provides a brief overview of marine biodiversity in the context of fishing. Chapter 4 gives a broader view of marine biodiversity trends.

Assessing the biodiversity of the oceans has not been completed, both in general terms and with respect to specific system types, such as rocky grounds on continental slopes (Carlton et al. 1999). The factors influencing species distributions and patterns of species richness are only just emerging for widespread habitats such as soft sediments (MacPherson 2002; Gray 2002). Similarly, methods for measuring diversity and its patterns are evolving rapidly, so that in the future, if such methods are put into practice and applied in research and monitoring activities, our understanding of the condition and trends of marine biodiversity will improve significantly (Price 2002; MacPherson 2002; Warwick and Clarke 2001; Warwick and Turk 2002).

Information on commercially important or threatened species required for management purposes is quite limited. New non-fish species are frequently discovered and described in association with fisheries surveys or, more recently, environmental impact assessments. However, recent initiatives, such as the Census of Marine Life, are increasing the rate at which new knowledge on marine life is becoming available, although understanding of most taxa other than fish is very limited and reflects a failure to seek any

systematic understanding of fisheries systems. For some groups important in fisheries catches, notably finfish and cephalopods, online databases that provide information on all species described so far do exist (see www.fishbase.org and www.cephbase.org). Also, a fair understanding of the factors influencing the distribution (depth, temperature, and so on) and population dynamics of major commercial species is available.

It is widely assumed that marine fish and invertebrates are somewhat less susceptible to extinction than most other marine as well as terrestrial and freshwater organisms. However, recent advances in methodology allowed studies that have questioned this assumption (Dulvey et al. 2003; Hutchings 2000). Although few marine species are known to have become globally extinct in the last century, there are numerous instances of extirpations of marine fish species—for example, the European sturgeon (*Acipenser sturio*) in the North Sea and the Green wrasse (*Anampses viridis*) in Mauritius.

Recent analyses suggest that marine extinctions may have been underestimated because of low detection abilities and a generally poor understanding of the conservation status of species that live in the marine realm (Dulvey et al. 2003; Carlton et al. 1999). Moreover, given that a major cause of declines or local extinctions in marine populations is overexploitation and that exploitation is rapidly increasing in scope and volume, there is a real likelihood that extirpations will increase (Dulvey et al. 2003). Other factors such as environmental degradation and climate change, alone or in combination with exploitation, can also play a role in local extinctions. As the first step toward global extinction, extirpations (local extinctions) cannot be dismissed as unimportant or irrelevant to a species' status and have significant impacts on the provision of ecosystem services. (See Chapters 4 and 11.)

The now rapidly fading notion that marine fish and invertebrates are inherently more resilient to impacts on their populations than other wildlife was based on a number of unfounded assumptions about their biology, particularly in the case of species that release large numbers of eggs into the open water. The key assumption was that high fecundity combined with a seemingly high dispersal capacity of the eggs or larvae, high recruitment variability, and wide-ranging distributions minimizes the risk of extirpations or extinctions even under heavy fishery exploitation. However, scientific support for this assumption is lacking or poor. Indeed, there is now an emerging consensus that marine fish are no more resilient to extirpations or extinctions than any other wildlife species of similar size (Roberts and Hawkins 1999; Hutchings 2000; Sadovy 2001).

18.2.6.2 Ecosystem and Habitat Diversity

The number of species present and their relative abundance is an important aspect of biodiversity and is threatened in marine systems. Overfishing and destructive fishing methods have an impact on marine ecosystems by changing community structure and altering trophic and other interactions between ecosystem components and by directly modifying habitats, notably when trawlers erode biogenic bottom structures (Pandolfi et al. 2003). By removing important components of the ecosystem, such as algal feeding fish in coral reef systems, overfishing results in altered ecological states that may be impossible to restore to former conditions. (See Chapter 19.)

A number of generalities can be drawn from the literature on biodiversity. One is that biological production declines with increasing "trophic level" (the number of feeding levels that organisms are removed from phytoplankton and other primary producers; see Chapter 8 for an explanation of the trophic level concept). In fisheries, most catches occur at around trophic level 3, which consists of small fish (such as sardines and herrings) feeding on herbivorous zooplankton (zooplanktivorous fish), and around trophic level 4, which consists of fish that prey on the zooplanktivorous fish (such as cods and tunas). Many fish, however, have intermediate trophic levels, as they tend to feed on a wide range of food items, often feeding on zooplankton as juveniles and other fish when adults. (See www.fishbase.org for diet composition data and trophic level estimates on thousands of fish species and the corresponding references.)

Biomass energy is transferred up the food web, with transfer efficiencies between trophic levels ranging, in marine ecosystems, from about 5% to 20%, with 10% a widely accepted mean. Thus the productivity of the large, high trophic–level fish that are traditionally targeted will always be lower than that of lower trophic–level fish. This has led to suggestions that fisheries yields should be increased by deliberately "fishing down the food chain" (Sprague and Arnold 1972)—that is, by exploiting species located at lower trophic levels more intensively. But this is already occurring throughout the world's oceans as a result of the decline in catches of the large, slow-growing high trophic–level fish, which are gradually being replaced, in global landings, by smaller, shorter-lived fish, at lower trophic levels (Pauly et al. 1998). Unfortunately, fishing down marine food webs does not necessarily lead to increased catches (see earlier Figures 18.3 and 18.9). Indeed, globally both the landings and their mean trophic levels are currently falling under the pressure of fisheries (Pauly et al. 1998; Watson and Pauly 2001); what seems to be increasing worldwide is the abundance of jellyfish, which are increasingly exploited throughout the world and exported to East Asia.

The deep ocean bottom contains some of the least explored areas of the world, with only 0.0001% of the deep seabed subject to biological investigations thus far (WWF/IUCN, 2001; Gray 2002). Nevertheless, studies have revealed a wealth of diverse habitats in the deep sea, which include seamounts, cold-water coral reefs, hydrothermal vents, deep-sea trenches, submarine canyons, cold seeps and pockmarks, and gas hydrates and polymetallic nodules. Of those, seamount ecosystems and cold-water coral reef communities are particularly threatened by high-impact fishing methods, such as bottom trawling (Thiel and Koslow 2001; Freiwald et al. 2004).

Scientific exploration of seamounts is minimal, with only approximately 300 of them sampled biologically, out of what is believed to be tens of thousands worldwide (ICES, 2003; see also seamounts.sdsc.edu). As mentioned previously, seamounts increase the biological productivity of waters surrounding them. The tops and upper flanks of seamounts also tend to be biological hotspots, with potentially high species diversity and endemism. Marine mammals, sharks, tuna, and cephalopods all congregate over seamounts to feed, and even seabirds have been shown to be more abundant. Suspension feeders, such as corals, dominate seamount benthic fauna. Seamounts may also act as "stepping stones" for transoceanic species dispersal (WWF/IUCN, 2001).

Our knowledge of cold-water coral diversity is also limited, and new reefs are still being discovered. For example, the largest known cold-water reef—35 kilometers long and 3 kilometers wide—was discovered off the Norwegian coast in June 2002 (Freiwald et al. 2004). There are few quantitative studies of fauna associated with cold-water corals, but it is known that they provide habitat for high diversity of associated species. More than 800 species have been recorded in the *Lophelia pertusa* reefs in the northeast Atlantic, and 3,000 species of fish and mollusks have been identified on deepwater reefs in the Indo-West Pacific region (WWF/IUCN, 2001).

The biggest threat to deep-sea coral reefs comes from trawling activities. WWF (2002) suggest that 30–50% of the deep-water corals along the Norwegian coast have already been lost due to bottom trawling, marine pollution, and oil and gas exploration.

Inconsistent and opportunistic sampling in deep and isolated areas, where cold waters and deep-sea corals are located, hampers efforts to study these habitats, and it is likely that global assessments will underestimate the biodiversity of these areas. More is known about local habitats and local extinctions in warm waters, such as the loss of the sawfish (*Pristis pectinata*) in Mauritania (UNEP 2002b) and the Chinese bahaba (*Bahaba taipingensis*) in Hong Kong (Sadovy and Cheung 2003).

18.2.6.3 Species Diversity

The lowered biomass and fragmented habitats resulting from overexploitation of marine resources is likely to lead to numerous extinctions, especially among large, long-lived, late-maturing species (Sadovy and Cheung 2003; Sadovy et al. 2003a; Denney et al. 2002).

Fishing is thus one of the major direct anthropogenic forces that has an impact on the structure, function, and biodiversity of the oceans today. Climate change will also have impacts on biodiversity through changes in marine species distributions and abundances. In the coastal biome, other factors, including water quality, pollution, river and estuarine inputs, have large impacts on coastal and marine systems. (See Chapter 19.) Historical overfishing and other disturbances have caused dramatic decreases in the abundance of large predatory species, resulting in structural and functional changes in coastal and marine ecosystems and the collapse of many marine ecosystems (Jackson et al. 2001).

One well-documented example is that of the historic fishing grounds ranging from New England to Newfoundland and Labrador, which once supported immense cod fisheries but which have now been almost completely replaced by fisheries targeting invertebrates, the former prey of these fish (providing a classic example of fishing down marine food webs). The system that once supported cod has almost completely disappeared, fueling fears that this species will not rebuild its local populations, even though fishing pressure has been much reduced (Hutchings and Ferguson 2000; Hutchings 2004; Lilly et al. 2000). However, some collapsed stocks have been able to recover once fishing pressure is removed: the North Sea herring fishery collapsed due to overharvest in the late 1970s but recovered after a four-year closure (Bjørndal 1988). On a much smaller scale, but nevertheless widespread throughout the tropics, coral reef areas have been degraded by a combination of overfishing, pollution, and climate variability. (See Chapter 19.)

18.2.6.4 Genetic Diversity

An important component of biodiversity is genetic diversity (FSBI 2004). Even for those marine groups that are taxonomically well documented, relatively little is known about the subdivision of species into populations with distinct genetic (and sometimes morphological) features, which are of evolutionary importance and of potential human use. Lack of knowledge about appropriate conservation units can lead to inadvertent overexploitation of distinct populations and to their extirpations (Taylor 1997; Taylor and Dizon 1999); where recovery is possible, it may take decades or centuries, as in the case of some populations of large species of whale (Clapham et al. 1999). In some cases, genetic diversity may be irretrievably lost due to a "bottleneck" effect caused by overexploitation, as with the northern elephant seal (*Mirounga angustirostris*) population, which was nearly exterminated by early commercial sealing (Bonnell and Selander 1974; Stewart et al. 1994).

18.3 Drivers of Change in Marine Fisheries Systems

There are two direct drivers and several indirect drivers of changes in marine ecosystems. The climate, due to its natural variability and increasingly because of greenhouse gas emissions, drives a number of changes affecting marine ecosystems, while government policy primarily drives change through the effect on investment in fisheries, with direct drivers such as overfishing resulting from government subsidies. Economic factors, including an increase in demand reflected in an increase in price and food preferences, also affect fisheries, with population growth exacerbating most of these.

18.3.1 Climate Change

Climate change is a direct driver in marine systems (McLean et al. 2001) and its potential impacts are described later, in the section on choices and trade-offs. Changing wind patterns and sea temperatures have an impact on various oceanographic processes, including upwellings (for example, Benguela) and surface currents (as in the Gulf Stream) (McLean and Tsyban 2001). (See also Chapters 12 and 13.) These currents may slow down, shift spatially or disappear altogether, resulting in changes in population abundance and distribution for many marine species.

There may be local extirpations, but global extinctions in the oceans are unlikely to result from climate change alone. Recent results from monitoring sea temperatures in the North Atlantic suggest that the Gulf Stream may be slowing down and affecting abundance and seasonality of plankton that are food for larval fish (Richardson and Schoeman 2004). Declining larval fish populations and ultimately lower adult stocks of fish will affect the ability of overexploited stocks to recover (Beaugrand et al. 2003).

Climate-induced changes in their physical characteristics (such as currents and circulation patterns) and their chemical characteristics (such as nutrient availability) will affect marine ecosystems directly. These impacts include sea surface temperature–induced shifts in the spatial distribution of some species and compositional changes in biodiversity, particularly at high latitudes. A poleward shift of marine production due mainly to a longer growing season at high latitudes is anticipated. While a complete shutdown of the North Atlantic circulation is unlikely, it cannot be ruled out, even in the foreseeable future (IPCC 2003).

A poleward shift of marine production due mainly to a longer growing season at high latitudes is anticipated (IPCC 2003). Recent findings show that warming in the Northern Hemisphere will cause a northern shift of distribution limits for various species through improved growth and fecundity in the north and lower growth or even extinctions in the south of this range. Such shifts may seriously affect fishing activities in the North Sea (Portner et al. 2001) and other productive areas of the world's oceans. However, current knowledge of the impacts of climate change in marine ecosystems is still poor, and literature on the subject is scarce.

Current scenarios of global climate change include projections of increased upwelling and consequent cooling in temperate and sub-tropical upwelling zones. Such cooling could disrupt trophic relationships and favor less complex community structures in these areas (Aronson and Blake 2001; Barret 2003). Marine export production may be reduced (estimated at − 6%), although regional changes may be either negative or positive (from − 15%

zonal average in the tropics to + 10% in the southern polar biome) (Bopp et al. 2001).

18.3.2 Subsidies

There are two forms of subsidies: direct financial support in industrial countries (price supports, for instance) and indirect support, notably in the form of open access policies that allow resource rents to be spent on excess capacity. The latter happens when the surplus of funds from a fishery after all costs have been paid are spent on purchasing additional capacity.

Financial subsidies are considered to be one of the most significant drivers of overfishing and thus indirect drivers of change in marine ecosystems. In most cases, government subsidies have resulted in an initial increase of overall effort (number of fishers and size of fleet), which translates into increased fishing pressure and overexploitation of a number of species. While it appears that the number of fishing vessels (see Figure 18.11) and fishers stabilized in the late 1990s, other subsidies, e.g. cheap fuel subsidies, can keep fleets operating even when fish are scarce. Without such subsidies, many of these fisheries would cease to be economically viable (Munro and Sumaila 2002).

Subsidies also play a role in fisheries expansion. Globally, the provision of subsidies to the fisheries industry has been variously quantified at $20 billion to over $50 billion annually, the latter roughly equivalent to the landed value of the catch (Christy 1997). More conservative estimates are provided by Milazzo (1998) and by an OECD (2000) study, recently reanalyzed and scaled to the North Atlantic by Munro and Sumaila (2002). The latter suggested an annual subsidy of $2.5 billion for a part of the world ocean that contributes about one sixth of the world catch.

The subsidies given to fisheries vary between countries and range from unemployment benefits in Canada to tax exemption in the United States and payment of fees to gain access to foreign fishing grounds by the European Union (Kaczynski and Fluharty 2002). For instance, in 1997 Canada provided over $198 million in unemployment benefits to its fishing sector; the United States gave $66 million in tax exemptions, and the European Union provided subsidies of $155 million to obtain access to other countries fishing grounds. Each of these have the effect of either reducing the cost of fishing or increasing the net revenues fishers obtain, and hence they lead to more fishing than would have been the case without the subsidies.

Over half the subsidies in the North Atlantic have negative effects on fleet development (Munro and Sumaila 2002). This, perhaps surprisingly, includes decommissioning subsidies, which

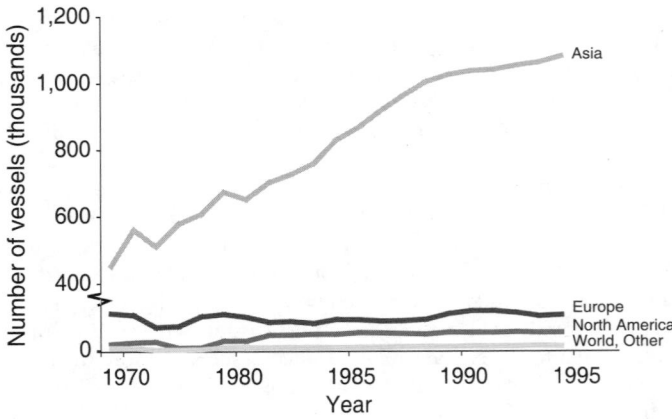

Figure 18.11. Trend in Fishing Vessels, 1976–2000 (FAO 2003)

have been shown under most circumstances to have the effect of helping to modernize fleets, thereby bringing about an increase in their catching powers.

18.3.3 Demand and Fish Prices

Overfishing drives ecosystem change, including changes in biodiversity, as described earlier. The growing demand and correspondingly increase in prices has contributed to overfishing. Marine products are in demand as a luxury food as well as for subsistence in many coastal communities, and as feed for aquaculture and livestock. It is the relatively high prices for these products combined with subsidies (plus the use of coastal systems as disposal sites for their waste products) that makes aquaculture a feasible industry.

It has been reported that bluefin tuna (*Thunnus thynnus*) have sold on the Tokyo market for as much as 20 million yen for a single fish (Japan Times 2001). Other fishery products such as eel larvae (*Anguilla* spp) and large prawns are also extremely high priced commodities. Such very high prices generate extreme pressures for overexploitation that are sometimes nearly impossible to counter through local management measures. As such items become increasingly scarce, they increasingly assume the status of luxury foods. The result is that increasing scarcity, rather than causing a relaxation of pressure on the remaining remnants of the resource populations, may act to increase the incentives to harvest the remaining individuals. For example, the Chinese bahaba (*Bahaba taipingensis*) is highly sought after for its swimbladders used in traditional medicine. Consequently, this fish—which fetches $20,000–64,000 per kilogram (see Figure 18.12)—has been exploited to critically low levels (Sadovy Cheung 2003).

18.3.4 Shifting Food Preferences and Consumption

It could be argued that human population growth and the resulting need for inexpensive food have been driving fisheries expansion. However, human population growth did not drive excess fleet capacity from Northern Europe and Japan into the southern oceans. Human population growth did not stimulate people in countries not accustomed to eating fish to shift toward a heavy consumption of seafood, as seen in China where income growth and urbanization has fueled fish consumption (Delgado et al. 2003). In industrial countries, such as the United States, fish is no longer a cheap source of protein compared with other sources. The price of fish has increased in real terms while the price of red meat has lost half its value over the last 20 years (Delgado et al. 1999).

Factors driving overfishing other than human population growth are also at work. One of these is increase in incomes and

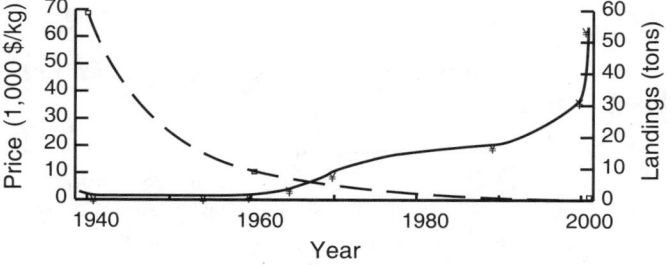

Figure 18.12. Estimated Annual Landings of *Bahaba Taipingensis* (Giant Yellow Croaker) and Swimbladder Market Price in Hong Kong, 1940–2000 (in constant 2000 US Dollars) (Sadovy and Cheung 2003)

therefore fish consumption in various countries that previously did not appear in international markets, such as China (Ahmed et al. 1999). Another factor is the consumption of fish promoted as part of a "healthy" diet and changing food preferences in many industrial countries.

18.3.5 Technological Change

Historically, the global expansion of fisheries has been driven by successive waves of technological innovation, much of it developed for naval warfare following the Industrial Revolution, two World Wars, and the cold war. These innovations included the invention of steam and diesel engines, the onboard manufacturing of ice, and blast freezing, all of which expanded the range of industrial fishing vessels. This expansion was followed by the incorporation of an enormous array of electronic devices facilitating fish detection, including radar and acoustic fish finders on fishing vessels, culminating in the introduction of GPS technology and detailed seabed mapping at the end of the cold war. These technologies, while improving the safety of people working at sea, also allowed fishers to aim for specific places with high fish abundances, places that once were protected by the depths and vastness of the oceans.

18.3.6 Illegal Fishing

The profits of fisheries that choose to operate outside of national and international laws and conventions can be very high. In some areas there is a lack of surveillance, enforcement, and monitoring due to high operational costs. In other areas corruption and cheating are tolerated due to the economic conditions or social obligations within a country. Managers recognize illegal, unreported, and unregulated fishing as a global problem, and recent initiatives, such as FAO's International Plan of Action, will help formulate strategies to deal with the problem. Regional fishery management organizations, such as the International Commission for the Conservation of Atlantic Tunas, are dealing with nonmembers and members who do not comply with management measures through the use of economic sanctions. "Name and shame" strategies are being used by NGOs to force companies and governments to comply with international management measures (for example, the Coalition of Legal Toothfish Operators, see www.colto.org/vessels.htm).

Large-scale cheating in some fisheries that were supposedly regulated internationally has led to extreme depletion of some living marine resources. An especially egregious example is the misreporting of Soviet whale catches to the International Whaling Commission (Brownell and Yablokov 2002). Between 1947 and 1972, some 90,000 more whales were taken in the Southern Hemisphere than was reported, including more than 3,000 endangered southern right whales that were supposedly fully protected at the time. Similar violations occurred in the North Pacific and Indian Oceans, and it was only with the introduction of international on-board observers that these practices ceased. It is now widely agreed that independent surveillance is an essential part of any fishery management and enforcement plan.

18.3.7 Globalization

Fish represent the fastest-growing food commodity entering international trade (Preston 1997). Accordingly, fish and fish products are an extremely valuable source of foreign exchange to many countries, in some cases providing as much as half of their total available foreign exchange income. For example, in Guinea-Bissau fishing agreements with the EU finance more than 45% of the government's annual fisheries operations budget, though this country only receives a very small fraction of the value (<10%) of the fish taken by European fleets (Kaczynski and Fluharty 2002).

Stocks of bluefin and other large tuna species around the world are being strained by fishing pressure driven by the extremely high prices such fish fetch in the Japanese luxury fish markets. Traditional local fish foods are, in many cases, no longer available to local consumers due to their inability to match the prices available by shipping the products elsewhere. An example is Senegal, where exports have disrupted local supplies of fish (UNEP 2002a). Consequently, highly nutritious fish foods produced in poorer regions of the world are increasingly being eaten by more economically advantaged populations in distant areas of North America, Europe, and East Asia. Of particular concern is the East and Southeast Asian market for shark fins that is threatening many shark species around the world, which are already under pressure from being a significant part of the bycatch of many pelagic fisheries.

One benefit of globalization is the improved quality of fish, because most importing countries demand that exporting facilities meet Hazard Analysis and Critical Control Point standards, which require exporting countries to follow safe food processing and handling standards. The associated benefits have been mainly to industrial countries, however. In developing countries, benefits have been limited to companies that can afford the required investment (Atta-Mills et al. 2004) or to the few local fishers able to participate in "boutique" fisheries for live fish, seahorses, and aquarium fish, which are low volume but a high-price export product (Erdmann and Pet-Soede 1996; Tomey 1997; Sadovy and Vincent 2002; Alder and Watson in press).

Export fisheries have also influenced the aquaculture industry, especially for salmon and shrimp, which are bred to meet the demand from industrial countries for luxury high-value seafood. For example, salmon (much of it farmed) was the leading fish export commodity of the EU in 1998 (Smith and Taal 2001). Countries such as Thailand that are the leading producers of shrimp (much of it from aquaculture) are often the leading exporters.

Increasing exports have contributed to the expansion of fishing fleets (facilitated by subsidies) leading to overcapacity and overexploitation as seen in the development of the pollock industry in the 1980s in Alaska (St Clair 1997). Depending on the fishery, this can lead to habitat destruction through trawling and biodiversity loss through, for example, turtles caught in shrimp trawls, albatross and sharks caught by longlines, and other bycatch in various fisheries (Hall et al. 2000).

Globalization clearly has the benefit of supplying foreign exchange to developing countries and potentially decreasing national debt. But this benefit has been at the cost of domestic supplies of fish resources, resulting in increasing domestic prices; in India, for example, the cost of fish has increased faster than the cost-of-living index and other meats (Kurien 1998) and has decreased food security.

18.3.8 Other Drivers

Habitat changes in coastal systems are a major driver of fisheries declines. (See Chapter 19.) Other factors of lesser apparent importance are invasive species, pollution, and disease. Human impacts, especially exploitation, are increasing. Moreover, persistent and widespread misconceptions about the ability of marine fish populations to withstand and recover from fishing continue to undermine initiatives to address the root causes of these problems (Roberts and Hawkins 1999; Hutchings 2004). Habitat loss or damage is caused by a range of fishing practices (from bottom

trawling to the use of dynamite), by pollution, or possibly by global warming, as in the case of extensive bleaching of coral reefs. Even well-intended attempts to remediate declines in fisheries through stocking can be problematic as hatchery operations have an impact on the genetic structure of wild stocks.

Two additional processes have effects similar to subsidies. One is the rapid increase in the demand for fish, reflected in increased prices of fish products, which in the last 50 years have increased three to four times faster than the consumer price index (Delagado and Courbois 1999). The other is the low price of fuel, which keeps numerous, otherwise bankrupt fisheries afloat in many countries. Moreover, due to the decline in stock abundance, the catch and edible protein per amount of fuel burned has decreased over time (Tydemers 2004). Indeed, fisheries are probably the only sector of the economy that has decreasing fuel efficiency (compared with, say, trucking, aviation, or manufacturing). Obviously, this growing dependence of the fishing industry on cheap fuel makes it highly vulnerable to fuel price increases, as well as to implementation of the Kyoto protocol or similar agreements that would tax industries for their energy intensity (Pauly et al. 2003).

18.4 Choices, Trade-offs, and Synergies within the System

Marine systems are still considered a new frontier for development by some people (McNutt 2002), and therefore a number of choices and trade-offs over fisheries will need to be made in the future. History has shown that once humans exhaust resources on land they look to the sea for alternatives. In repeating history, coastal environments are becoming degraded (for loss of coral reefs, see Chapter 19) and biodiversity is declining, beginning with the loss of large predators at high trophic levels (Pauly et al. 1998; Myers and Worm 2003). Now areas deeper and further offshore are increasingly exploited for fisheries and other resources such as oil and gas.

Marine fish resources often have value and benefits beyond that of food security. Some species are of considerable cultural importance (salmon are an important part of aboriginal culture in the Northeast Pacific, for instance), while others generate substantial income from tourism (especially dive tourism) and recreation (Rudd and Tupper 2002). Yet others may be important keystone species within their community, with a loss even at local levels cascading throughout the ecosystem. (See Chapter 11.) These trade-offs need to be considered when allocating resource access. Nevertheless, there are some uses of marine systems that have minimal impacts and that can be developed in tandem with other uses such as tourism, well-managed recreational fisheries, and bioprospecting. (See Table 18.5.)

18.4.1 Environmental Impacts of Capture Fishing versus Other Uses

Contrary to the coastal systems, where many uses are mutually incompatible, few other economic activities in the marine realm directly preclude fishing. In fact, the major problem for fishers is other fishers. Thus, for example, by modifying habitats, trawlers affect the yield of other fishers who do not use such destructive gear.

Three different classes of multiple uses and synergies can be identified:

- relationships between fisheries and other sectors, such as aquaculture and coastal development;

Table 18.5. Trade-offs and Synergies in Marine Ecosystems

		Extraction		Conservation		Aquaculture		Other
	Fishing	Bioprospecting	Mining, Gas and Oil	Tourism	Biodiversity	Growout	Farm	Shipping
Fishing		minor trade-offs if the levels of bioprospecting not excessive and fishing sustainable—however, if aquaculture has genetic impact then the story changes; the need for biologically active products may force managers to improve the management of exploited fisheries	few trade-offs as seen except in the immediate vicinity; some gas and oil facilities have provided refuges for fish stocks and therefore a hedge against overexploitation	major trade-offs since people enjoy seeing wildlife, especially diving, and lobby for their protection (e.g., seahorses), but can have social consequences if not managed properly; may protect some species if they are valued by the tourism industry	major and varied trade-offs, if destructive fishing such as trawling takes place, then food provisioning is traded off against biodiversity, forgoing biodiversity over a range of ecosystems whereas longlining forgoes seabird biodiversity; few synergies but could provide a niche for new species that are bioactive or species to move into the niche	minor trade-offs since argument is that fish would have been caught anyway and therefore just ensures the economic value is realized, but this could affect fisheries since it takes away the potential for wild-capture fisheries; possibility to improve coastal communities economically since they do not have to spend so much time fishing	major trade-offs especially if genetic dilution takes place or diseases introduced; while it may provide more high-quality fish it does not necessarily provide same total fish tonnage; can reduce the price of wild capture fish, making fishing financially difficult; in developing countries, often export-oriented and therefore risks the food security of the country; possibility of maintaining some species that are at risk of overexploitation	few trade-offs or synergies
Bioprospecting			few trade-offs except in the immediate vicinity unless areas of mineral, gas, and oil exploitation also contain organisms of high bioactivity, then trade-offs are needed; may provide refuges, as noted above	few trade-offs if the bioprospecting is done with minimal impact or small footprint; strong synergism in that the bioprospecting could form the basis of ecotourism	few trade-offs (see tourism); strong synergism since maintaining biodiversity will maintain bioactivity	minor trade-offs unless aquaculture introduces diseases that threaten the populations of bioactive species; farms could be used to grow out biologically active species	major trade-offs if genetic dilution occurs as well as the introduction of diseases; if produced on large scale, could threaten the livelihood of small-scale collectors; provides the facilities for mass production	few trade-offs or synergies

(continues over)

• relationships between fisheries and top predators or charismatic fauna (marine mammals, seabirds, turtles); and

• competition within the fisheries sector.

Generally, fisheries do not appear to be affected to a large extent by other extractive activities, such as oil or seabed mining, at least relative to the wide impact of the fisheries themselves.

The issue of competition with humans does not arise with marine turtles, which along with marine mammals and seabirds are key indicator species for problems and changes in the marine environment. (See Chapter 19 for more on marine wildlife.)

It has been proposed that marine mammals directly or indirectly compete with fisheries (commercial and artisanal) for resources targeted by fisheries. This perceived competition has been used to justify annual sustainable harvests of marine mammals during the last decade (Lavigne 2002) and also to justify the resumption of whaling in many international fora (Holt 2004). Though competition may occur at small local scales, this issue warrants much further investigation. A recent analysis of global trophic overlap between marine mammals and fisheries indicated that there is limited competition in the Northern Hemisphere on a large scale, while competition between the two is low in most other areas of the world (Kaschner 2004; Kaschner and Pauly 2004). Moreover, the analysis suggested that, overall, fisheries are more likely to adversely affect marine mammal species, particularly those with restricted ranges, than vice versa (Kaschner 2004).

Examples of marine mammals adversely affecting humans do exist. However, such impacts are far less severe and mostly fisheries-related, such as when killer whales take fish from the catch of longline fisheries in Alaska. Reducing the competition between higher vertebrates and fisheries is likely to involve both technological changes in the way fishing gears are deployed and the creation of suitably large marine reserves (as described later).

The third group of interactions occurs between fishers and fleets. Essentially, these interactions are shaped by the fact that each fisher looks for exclusive access to the resource. In fact, the technological improvements that characterize modern fisheries, and that enable access to resources deeper and further offshore, are a response to competition between fishers. This competition, which drives the technological development of fisheries, has over time eliminated the refuges, such as depth and distance offshore, that naturally protected fisheries resources (Pauly et al. 2002).

The case study in Box 18.1 documents an example of how European Union subsidies for technological development and fleet improvements gave Mediterranean trawlers access to fish populations in previously out-of-reach areas of the deep sea.

A significant proportion of world fish stocks and catches is overexploited or depleted (Watson and Pauly 2001; FAO 2002) and the marine habitats that many of the world's fish stocks rely on at some stage of their life cycle are being degraded. (See Chapter 19.) The combination of overfishing and degradation or conversion of habitats, which contribute to the loss of biodiversity and food provisioning, occurs almost everywhere. In developing countries this is aggravated by export-driven fisheries that overexploit their resource base and that divert food away from the do-

Table 18.5. *continued*

	Extraction		Conservation		Aquaculture		Other
	Bioprospecting	Mining, Gas and Oil	Tourism	Biodiversity	Growout	Farm	Shipping
Mining, Gas and Oil			major trade-offs since most tourists seeking natural experience, not high infrastructure and possible pollution; onshore infrastructure could facilitate offshore tourism (e.g., NW of W. Australia)	major local impacts if spill takes place, minor if footprint small and pollution contained; platform provides a niche for new species or species to move into the area	major trade-offs— while the risk of spills on the farms is low, if it does happen, financial and ecological impacts extremely high, therefore generally they do not coexist; possibly onshore infrastructure synergies	see growout	few trade-offs or synergies
Tourism				minor trade-offs, tourism can have an impact on biodiversity at the local scale from overfishing, collecting, etc. (e.g., Red Sea overfishing to meet restaurant trade); tourism provides an incentive to maintain biodiversity since that is one of the attractions to the area	the offshore infrastructure and the concept of penned fish may not appeal to many tourists; the associated pollution with aquaculture facilities as well as with tourism facilities (e.g., human diseases in shellfish); limited	see growout	
Biodiversity						trade-off in terms of the introduction of diseases into wild populations; alterations to population structure; localized habitat changes; declining food supply for other species that consume small pelagics/krill; possibly maintenance of species at risk	genetic dilution; same as growout

mestic market. (See Figure 18.13.) As a result, the fishing sector has declined as a source of employment in many industrial countries.

18.4.1.1 Food and Protein

Overfishing affects human well-being through declining food availability in the long term, since fewer fish are available for consumption and the price of fish increases (Alder and Sumaila 2004). Due to declines in coastal habitats, fishers are forced to go further offshore and for longer periods of time, resulting in reduced food security (Alder and Christanty 1998). In Canada, the collapse of the cod fishery resulted in severe unemployment (see Figure 18.14), compounded by restrictions on subsistence fishing (Neis et al. 2000).

While fish is a healthy, luxury food in high demand by the industrial world, it is still a significant and cheap source of protein for many countries in the developing world. However, per capita consumption of fish in the latter is much lower than in the industrial world. (See Figure 18.15.) Therefore declines in the availability of cheap fish protein either through overfishing, habitat changes, or shifting trade practices contribute to reduced food security in countries such as Ghana, Senegal, and Chile (Atta-Mills et al. 2004; Alder and Sumaila 2004).

The developing world produces just over 50% of the value of fish that is traded globally, and much of the fish caught in the

developing world is exported to industrial countries (FAO 2002). Fishing to meet export demands should, theoretically, provide funds to allow the import of cheaper fish and protein products, reduce the government debt, and supply cheap protein to the local population. However, the benefits to developing countries from trade in fish will not be realized if the funds generated are not reinvested in the economy. That this is not always the case may contribute to the fact that some of the major fish-exporting countries are also the least developed. (See Table 18.6.)

The industrial world, in particular the United States, the EU, and Japan, have been able to buffer against declines in fish availability and increases in prices because they have been able to purchase or otherwise get access to high-quality fish. Indeed, the per capita consumption of fish by industrial countries was 21.7 kilograms in 1997, compared with 9.2 kilograms in the developing world (excluding China, although if China is included, the per capita consumption for the developing world rises to 14 kilograms due to China's massive consumption of locally produced farmed freshwater fish) (Delgado et al 2003).

18.4.1.2 Livelihoods

Fisheries and fish products provide direct employment to nearly 27 million people (FAO 2002). Globally, the bulk of people employed in fisheries are poor, and many are without alternative

BOX 18.1

Subsidy-driven Removal of a Natural Reserve

The Mediterranean Sea has a long history of fisheries exploitation with a variety of gears. Yet the landings of many species, including highly valued demersals, have increased in the last decades (Anonymous 2002; Fiorentini et al. 1997). A large part of the landings consist of the juvenile of demersal species (Stergiou et al. 1997; Lleonart 1999; Lloret et al. 2001), such as hake, *Merluccius merluccius*—one of the six most important species, in terms of both landings and landed value, in the Mediterranean Sea (Fiorentini et al. 1997). Hake landings rose from about 5,000 tons in the late 1940s in the eastern and western Mediterranean to about 15,000 and 40,000 tons in the mid-1990s in two areas of the Mediterranean (Fiorentini et al. 1997).

The increase in landings is not only the result of increasing fishing mortality (that is, increasing effort and technological developments), but also due to recent development of new fisheries and expansion of fishing grounds (mainly by fishing in deep waters). Increasing landings are also related to the increase in phytoplankton production due to eutrophication of Mediterranean waters (e.g., Caddy et al. 1995; Fiorentini et al. 1997; Tserpes and Peristeraki 2002). For some demersal species, increasing landings might have been mitigated by the fact that a large part of the spawning population was not available to fishing gears because of:

- life history—with the onset of maturation, many demersal fish species (e.g., *M. merluccius* (Abello et al. 1997), *Pagrus pagrus* (Labropoulou et al. 1999), and *Pagellus erythrinus* (Somarakis and Machias 2002)) migrate to deep bottom areas that were not accessible to trawlers—and
- the geomorphology of the Mediterranean Sea deep areas provides natural refuges (low-take zones), allowing mature individuals to contribute to recruitment and stock size maintenance (Caddy 1990, 1993, 1998; Abella et al. 1997).

In the last several years, however, the modernization of trawlers and small-scale vessels, driven mainly by EU subsidies, allowed fishing to expand in deeper waters and to operate during harsh weather conditions. As a result, deepwater fisheries, practiced for instance with bottom or vertical longlines, were developed, heavily exploiting the large mature hake as well as of other species (e.g., *Pagellus bogaraveo*). Given the absence of appropriate management procedures (Lleonert 1999; Stergiou et al. 1997), this will lead to recruitment overfishing and the fisheries will eventually collapse, adding another data point to the records on the negative effect of subsidies.

sources of work and sustenance. In addition, fish and fishing are enormously important to the cultural life of many coastal communities, and often define the "quality of life" of people with a cultural tradition of harvesting the sea (Johannes 1981).

The global consequences of exploiting marine resources are numerous and significant. Overcapacity in the global fleet implies that both labor and physical capital are wasted (Mace 1997) and could be used more beneficially in other sectors of the economy, where they are most needed. The huge deficits detract from human well-being in sectors such as education and health care. The Common Fisheries Policy of the European Community, for example, allows for vessels to be decommissioned to reduce effort in some countries while simultaneously subsidizing others to increase their fishing capacity (Alder and Lugten 2003).

18.4.1.3 Habitats

Some coastal habitats have been converted to other uses, such as mangroves for coastal aquaculture ponds or cage culture of high-valued species such as penaeid shrimp, salmon, or tuna. This conversion can affect wild-capture fisheries, which use these coastal habitats for part of their life cycle. It also has sometimes caused displacement of fishers, loss of revenues, and social unrest. Coastal residents often no longer have access to cheap protein or sources of income. (See Figure 18.16) In addition growing juvenile fish taken from wild populations, in the case of tuna, in pens or cages can also be a significant source of pollution for the area in which they are located.

Area closures and the halt of destructive fishing have improved fisheries, especially in coral reefs (Roberts et al. 2002). Overall, however, overfishing and habitat destruction continue throughout the world (Jackson et al. 2001; Jennings et al. 2001).

Ultimately, overfishing has a significant impact on most marine systems. Habitat degradation due to pollution, infrastructure development, and so on contributes to the further degradation of the ecosystem and declining human well-being (mercury and PCBs in Baltic Sea fish, for example) or impedes recovery of the marine or coastal system (as when the conversion of mangroves affects important nursery habitat for some species of fish). Considering the serious nature of many non-fisheries impacts, attention should be given to addressing these impacts and within the context of integrated coastal management. (See Chapter 15 of the MA *Policy Responses* volume).

Offshore areas, especially the deep sea, vents, and seamounts, are at risk of being exploited beyond recovery. It is difficult to say how much direct impact the crossing of such thresholds will have on human well-being. Nevertheless, it will ultimately affect ecosystem services. The lack of good information on these systems, past experiences regarding the impact of fishing, and lack of good international management frameworks add to the uncertainty of the impacts on these systems.

Climate change, El Niño/Southern Oscillation, and hydrological conditions will have a significant impact on some marine habitats, especially coastal areas (Chapter 19) and in polar marine and coastal systems (Chapter 25) (Chavez et al. 1999). Climate change will affect marine systems but may not have as severe an impact on open-ocean and deep-water systems (McLean et al. 2001).

18.4.1.4 National Economies and Foreign Exchange

Many areas where overfishing is a concern are also low-income, food-deficit countries. For example, the exclusive economic zones of Mauritania, Senegal, Gambia, Guinea Bissau, and Sierra Leone in West Africa all accommodate large distant water fleets, which catch significant quantities of fish. (See Figure 18.17 in Appendix A.) Much of it is exported or shipped directly to Europe, while compensation for access is often low compared with the value of the product landed. These countries do not necessarily benefit through increased fish supplies or increased government revenue when foreign distant water fleets access their waters. In some countries, such as Côte d'Ivoire, the landings of distant water fleets can lower the price of fish, which affects local small-scale fishers.

Although Ecuador, China, India, Indonesia, and the Philippines, for example, do not provide access to large distant water fleets, these low-income, food-deficit countries are major exporters of high-value fish products such as shrimp and demersal fish. As shown in the West African example, several countries in the region export high-value fish, which should provide a significant

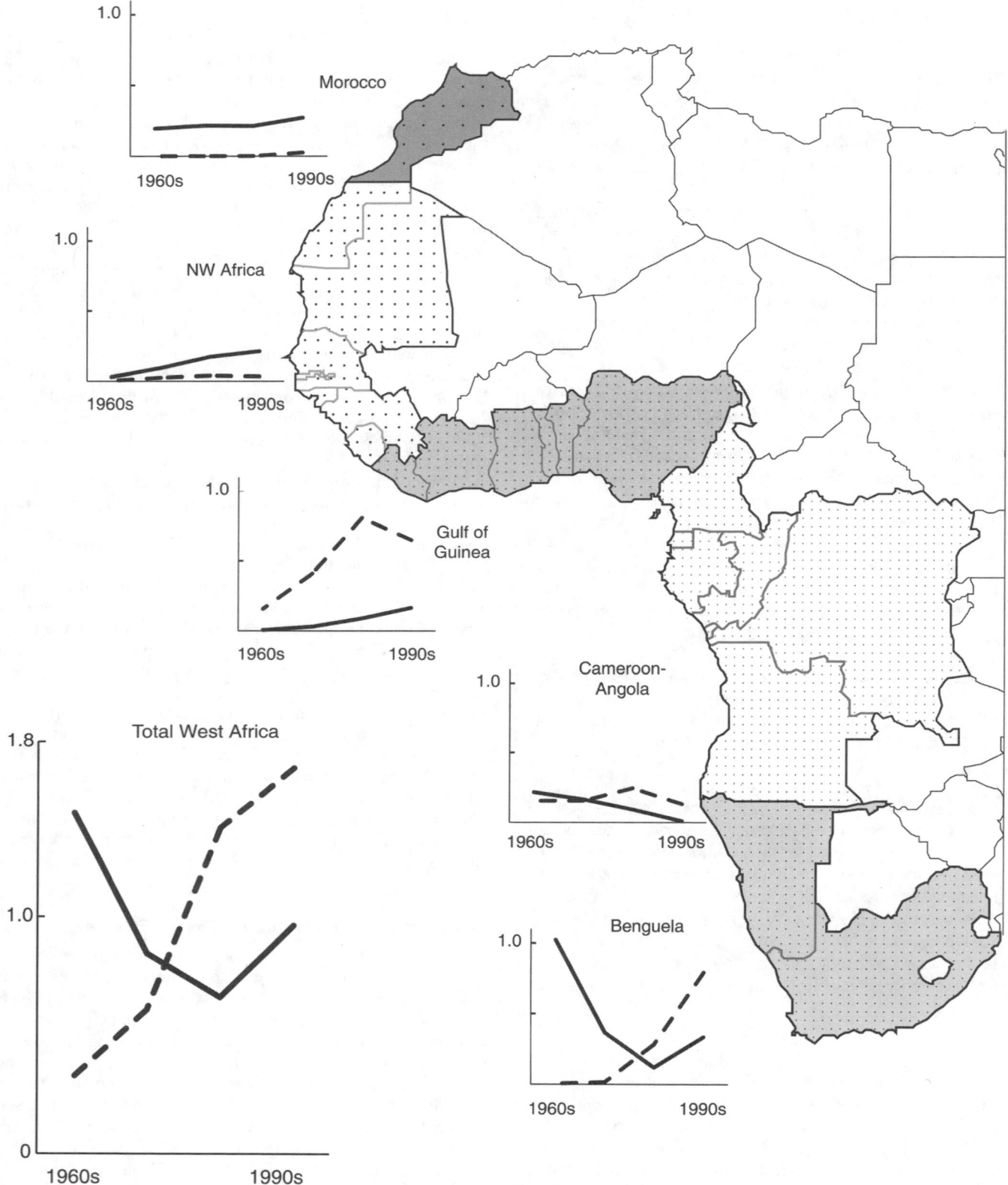

Figure 18.13. Trend in Imports and Exports from Fisheries in Western Africa, 1960s to 1990s. All graphs are in million tons. Dashed lines are imports and solid lines are exports. (Alder and Sumaila 2004)

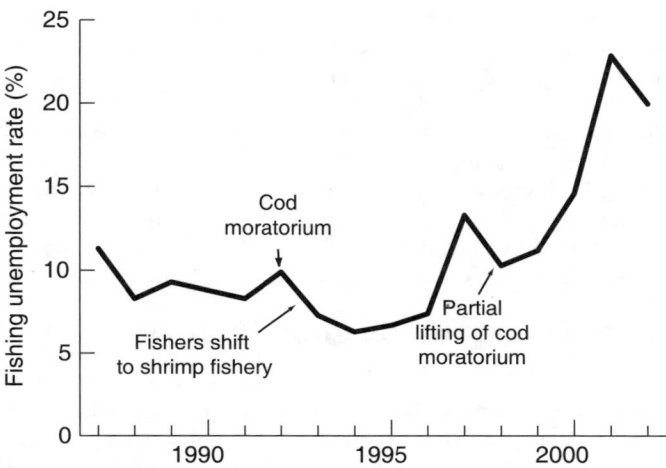

Figure 18.14. Unemployment in the Newfoundland Fishing Sector, 1987–2002 (Stats Canada 2003)

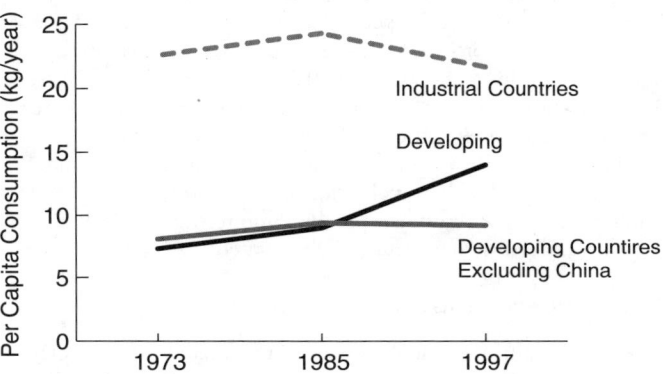

Figure 18.15. Trend in Per Capita Fish Consumption, 1973–97 (Delgado et al. 2003)

Table 18.6. Major Exporters of Marine Products, 2000. Those shown in bold are lower-income food-deficit countries. (FAO 2003)

Country/Area	Export Value (billion dollars)	Country/Area	Export Value (billion dollars)
Thailand	**4.4**	Australia	1.0
China	**3.7**	**Morocco**	**1.0**
Norway	3.6	Japan	0.8
United States	3.1	Argentina	0.7
Canada	2.8	Mexico	0.7
Denmark	2.8	New Zealand	0.7
Chile	1.8	**Ecuador**	**0.6**
Taiwan	1.8	Sweden	0.5
Spain	1.6	Belgium	0.5
Indonesia	**1.6**	Singapore	0.5
South Korea	1.5	Faeroe Islands	0.4
Viet Nam	1.5	**Philippines**	**0.4**
India	**1.4**	Italy	0.4
Russian Federation	1.4	Ireland	0.3
Netherlands	1.4	**Bangladesh**	**0.3**
United Kingdom	1.3	Portugal	0.3
Iceland	1.2	South Africa	0.3
Peru	1.1	Panama	0.3
Germany	1.1	Greenland	0.3
France	1.1	Senegal	0.3

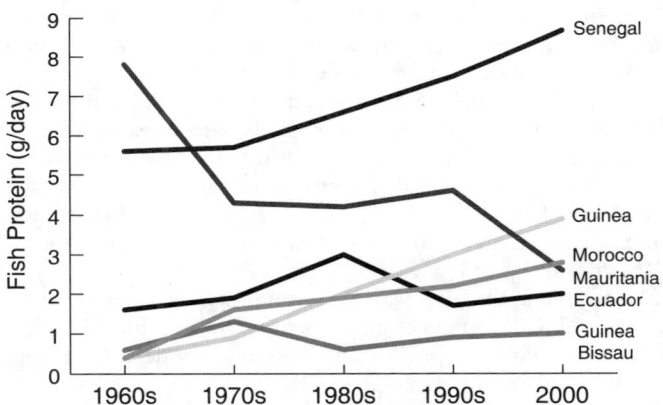

Figure 18.16. Available Fish-Based Protein of Selected Lesser-Income Food-Deficit Countries, by Decade since 1960 (FAO 2003)

national economic gain so that cheaper forms of protein can be imported. In countries such as Ghana, however, the value of exports is often less than the value of imported fish, and the volume of imported fish does not meet the domestic demand for fish (Atta-Mills et al 2004).

Fish products are heavily traded, and exports from developing countries and the Southern Hemisphere presently offset much of the demand shortfall in European, North American, and Northeast Asian markets (Ahmed et al. 1999). Given the global extent of overfishing, however, it is likely that the global decline in marine fisheries landings, which already affects the poorer consumers in developing countries, will also catch up with consumers in industrial countries (Garcia and Newton 1997).

18.4.2 Unintentional Trade-offs

There are many unintended consequences of policies. However, it is worth noting that the economics of natural resources, including fisheries, usually only consider market forces and not the underlying biological constraints or environmental costs. For example, building an access road to a fishing village with the intention of connecting it to various cultural, social, and economic amenities (markets) has been shown in many cases to lead to massive increase in fishing effort and a collapse of local resources, the exploitation of which was previously regulated by the absorptive capacity of the local market.

Fish aggregating devices, which are large floating structures, are constructed and use materials so that fish, especially pelagic fish, are attracted to them. They were seen as a cost-effective means to increase fish catches without affecting other aspects of the marine ecosystem or juvenile tuna stocks. However, monitoring of catches by these devices has shown that juvenile tunas are also attracted, which may, by increasing the effort on tuna juveniles and other "by-catch" species, have an impact on tuna stocks. Indeed, the mean size of caught fish is diminishing in both yellowfin tuna (*T. albacares*) and bigeye tuna (*T. obesus*) (Bromhead et al. 2003).

Perverse incentives in fisheries are well studied by economists. They include size limits for landed fish, which encourage undersized bycatch to be discarded at sea, and decommissioning schemes that lead to fleet modernization and "technology stuffing." Alder and Lugten (2002) discuss this in the context of the EU Common Fisheries Policy.

18.5 Choices, Trade-offs, and Synergies with Other Systems

The ocean is the ultimate receptacle for discharges from the land and coast. While land use has significant and frequently obvious impacts on terrestrial and coastal systems, impacts on deeper marine systems are not as evident. However, indirect impacts such as air pollution deposition and water pollution are being detected in marine systems (such as the POPs found in high Arctic marine mammals). Marine systems play an important role in climate regulation, the water cycle, and coastal processes, especially in the recycling of nutrients. El Niño events highlight the widespread geographic and ecosystem impact of perturbations of marine systems on pelagic fish stocks, coastal areas, and land systems.

18.5.1 Climate Change

Climate change is affecting the global distribution of heating and cooling of the ocean surface, which is a major factor in determining the large-scale patterns of ocean current flow and will have an impact on the marine ecosystem, including fisheries, in a number of ways (Mclean and Tsyban 2001).

Most marine fish and many other marine organisms have complex life cycles, featuring at least one planktonic early life stage. Consequently, reproductive habits and behaviors may be specifically adapted to current conditions, which will transport passively drifting larvae to appropriate nursery grounds. Bakun (1996) has shown that for the Brazilian sardine (*Sardinella brasiliensis*), enrichment of nearshore waters, the presence of a retention mechanism, and concentration of small fish in nearshore waters must all occur at the right place and time for successful recruitment. Climate change, acting through changes in sea temperature and especially wind patterns, will disturb and displace fisheries. Disruptions in current flow patterns in marine and estuarine systems, including changes to freshwater inputs as predicted under climate change, may cause great variations in reproductive success (Welch et al. 1998; Francis and Mantua 2003).

Research in the Northeast Pacific indicates that alternating climate "regimes" can affect marine ecosystem structure. Haigh et al. (2001) coupled a planktonic ecosystem model to a general ocean circulation model for the Pacific Ocean (18° S to the Bering Strait) to study the 1976 "regime shift." The results showed clear geographical patterns in primary production and significant changes in the levels of primary production and standing stocks of phytoplankton and zooplankton before and after 1976. This implies that fish and other marine species will establish new geographic distributions as they seek new areas with the optimal temperature and food supplies. For example, it can be expected that North Pacific cold temperate species will move northward and will be replaced, along the coast of British Columbia, by warm temperate species (Welch et al. 1998). A similar set of replacements has been occurring in the last two decades in Europe, notably herring and pilchards in the Bay of Biscay, the English Channel, and the North Sea.

The capacity of fish species to adapt rapidly in reaction to such environmental changes is not clear at the present time. There are some hypothetical mechanisms available that might conceivably facilitate quite rapid adaptive adjustments (e.g., Bakun 2001). However, these adjustments imply high populations (biomass) ca-

pable of producing numerous and varied offspring, some of which will be adapted to the newly opened niches. Also, oceanographic conditions may be changing too rapidly. Given the low biomasses for most exploited stocks, the prospect of rapid environmental change is doubly problematic for marine fisheries and marine biodiversity preservation.

Finally, it should be noted that the changes in physical features predicted by various global and regional simulation models under likely CO_2 emission scenarios fall well outside the parameter range so far observed under natural marine regimes (IPCC 2003).

18.5.2 Interactions with Coastal Systems

One of the major interactions between marine fisheries and coastal systems is through the life history of the large number of marine species that use coastal areas, especially estuaries, mangroves, and seagrasses, as nurseries. (See Chapter 19.) Thus, coastal habitat modifications and coastal pollution, as well as inshore fishing, can adversely affect offshore fisheries by reducing the supply of recruits to the offshore adult stocks. The scale of this problem varies between continents, however. For example, a large fraction of the fish species occurring along the continental shelf of eastern North America produce juveniles that are dependent on estuarine habitats, while fish along the coasts of western Australia, northwestern Africa, or southwestern South America appear to depend less on coastal systems.

The presence of juveniles in coastal systems and of adults further offshore means that small-scale inshore and industrial offshore fisheries effectively compete for the same resources, even though they are apparently geographically separated.

18.5.3 Interactions with Island Systems

Small islands throughout the Pacific, Indian, and Atlantic Oceans have coastal and shelf areas whose physical and biological features differ from those typical of the biogeochemical province in which the islands are located. In many cases, these islands, whose nearshore fisheries have often reduced the limited coastal resources, are attempting to get access to pelagic offshore resources (see Chapter 23) (Petersen 2003). This has led to a new set of conflicts between island states and industrial countries with distant water fleets.

From the perspective of island states, the oceanic resource—even in its present depressed state—far exceeds the coastal fish population they initially relied on. The United Nations Convention on the Law of the Sea and various other treaties discussed later in this chapter have begun to provide mechanisms to address some of these issues of perception and equity.

18.6 User Rights and Protection Status of Marine Ecosystems

Marine systems are often described as "commons" (for everyone's use) and their overexploitation as a tragedy (Hardin 1976). While this may hold true for the open ocean, complex property rights exist in many coastal areas. The property rights in question can be traditional (aboriginal), historical or local, and commercial (that is, the government sells the right to gain access to resources). The boundaries between these rights are frequently unclear and in some cases generate conflicts.

Some of the major groups of fishers involved in exploiting marine resources are described briefly in this section, along with major issues, outlooks, and prospects.

The introduction of UNCLOS and other jurisdictional restrictions has had minimal impact on countries or companies that can afford to "buy" access to fishing grounds. Other countries find themselves having to fish further offshore or purchase cheap, low-quality fish. This situation is clearly demonstrated in West Africa, where Europe has extended its fishing grounds from Northwest Africa to the whole of the West coast. Countries such as Nigeria and Ghana have seen a reduction in their traditional fishing grounds, intensification of small-scale fishing in coastal waters since they can not afford to gain access to foreign EEZs, and increases in imports of cheap, low-quality fish (Atta-Mills et al. 2004).

The term "aboriginal" refers to the descendents of the original inhabitants of countries whose population now largely consists of recent immigrants (such as in the Americas, Australia, and New Zealand). Aboriginal peoples are often marginalized, especially with regard to access to natural resources. In some areas, their existence is not even legally recognized. Other areas recognize aboriginal rights based on the original occupancy of the territory. Such rights are now recognized in Canada's Constitution and Supreme Court decisions, for instance (Haggan and Brown 2002). Similar rights were granted to New Zealand Maoris, who now own one third of the country's fishing quotas. Examples of aboriginal groups with strong fishing traditions are the First Nations along the Pacific coast of Canada and the United States, the Maoris of New Zealand, and the Torres Strait Islanders of Northern Australia. In some situations, their rights can be a source of conflict with non-aboriginal fishers, whose behaviors and activities are governed by additional or different laws and regulations (Tomlinson 2003).

Aboriginal fishing methods tend to be used only in the coastal zone, but their impact can extend much farther, as the target species often have broad offshore distributions. Aboriginal fisheries are often subsistence fisheries, though not necessarily so. Indeed, the term "subsistence fisheries" is frequently a misnomer, since much of what is caught by subsistence fisheries is actually sold. True subsistence fisheries without subsequent sale of harvested products are now rare (an example is shellfish gathering in Kwazulu Natal, South Africa; see Branch et al. 2002).

Small-scale or artisanal fishers exist both in industrial and developing countries, and consequently their definition varies. However, competition between small-scale and large-scale fisheries is a common issue in all countries. Competition can be either direct (such as fixed gears versus trawlers targeting shrimps in tropical waters, as described in the next section) or indirect (such as coastal traps versus trawlers targeting different age classes of the same cod stock off Newfoundland). There is also competition for political and financial support as well as markets. The resulting conflicts, often couched as "equity" issues and aggravated by large-scale fleets, have tended to paralyze regulatory agencies in both industrial and developing countries. Community-based property rights may be one way of dealing with this issue, as seen in the Philippines, where local communities have jurisdiction over inshore waters within 15 kilometers of the coast (Rolden and Sievert 1993). However, synergies can also emerge with new markets, infrastructure improvements (processing and transport), and technology transfers occurring. Figure 18.18 illustrates the respective role and impacts of small-scale versus large-scale fisheries in one case study.

In most of the industrial world, such as the European Union and United States, the large-scale sector receives the bulk of government subsidies. Consequently, the economic efficiency of these fisheries is questionable. Another characteristic of large-scale fleets is that once local waters are depleted, they lead the expansion further abroad, often requesting subsidies to gain access to the resources of other countries, as with EU fleets off West Africa (Kaczynski and Fluharty 2002). However, the industrial fishing sector is declining as a source of employment in the industrial world. EU member countries and Japan, for example, are finding it difficult to recruit young workers into the industry and are turning to developing countries in Africa and Asia to crew their vessels (FAO 2002; Morales 1993).

Recreational fishing was considered relatively benign until recently, mainly because information about its impact has been limited. FAO's first estimate of global recreational catches was put at only 0.5 million tons (Coates 1995), but recent estimates of over 1 million tons are probably more accurate (Coleman et al. 2004). Indeed, recent studies in British Columbia have shown that for inshore fisheries, the catch from the recreational sector can exceed the commercial sector (Forrest 2002). Recreational fishing is an important economic activity in some countries; in the United States it is worth approximately $21 billion a year; in Canada, $5.2 billion a year (Department of Fisheries and Oceans 2003); and in Australia, $1.3 billion a year (Henry and Lyle 2003). Although recreational fisheries often promote catch and release practices, the mortality associated with these practices is not well known and may be high (Wilde and Pope 2003).

There are major conflicts between commercial fishers, especially small-scale or artisanal ones, and recreational rod-and-line and underwater spear fishers in many countries (in Portugal, for instance). Large numbers of so-called recreational fishers in fact fish for a living and are not licensed or regulated in any way. In southern Europe, there is evidence that spear fishing has had a significant impact on inshore fish communities (to depths of 30–40 meters), with a decrease in abundance and mean size of many species, such as groupers and large sea breams. In addition, pressure on fish populations may have contributed to the disappearance of some small-scale fishing gear (such as small-hook longlines) due to the decline in catch rates and profitability (K. Erzini, Universidade do Algarve, personal communication 2004).

Marine tourism is a growing industry, principally in the marine wildlife tours sector. Similarly, coral reef tourism has increased in visitation levels and value, with a current net present value estimated at $9 billion (Cesar et al. 2004). The Great Barrier Reef attracts 1.6 million visitors each year and generates over $1 billion annually in direct revenue (Harriott 2002).

18.6.1 Competition between User Groups—Equity and Access Rights

Given the involvement of a number of user groups, management of fisheries and marine natural resource allocation requires the consideration of rights and equity among stakeholders. There is a strong tendency among fisheries economists to view individual transferable quotas and another form of "rights-based fishing" as the solution to mismanagement of fisheries (Hannesson 2000; Arnason and Gissurarson 1999). Others argue that the citizens of each country in question are the implicit owners of coastal and shelf fishery resources and that exploiting these resources can only be granted as a privilege for which payment is due, for example, through annual auctions (Macinko and Bromley 2002). The solution to these access and equity issues lies within the sociopolitical situation of each country rather than within the realm of science.

Shallow waters of tropical continental-shelf ecosystems are characterized by relatively high fish densities and the presence of high-value species, such as penaeid shrimps (Longhurst and Pauly 1987). The incursion of trawlers, which target shrimps for export, into shallow shelf ecosystems (10–100 meters) where artisanal

FISHERY / BENEFITS	LARGE-SCALE	SMALL-SCALE
Number of fishers employed	about 2 million	over 12 million
Annual catch of marine fish for human consumption	about 29 million tons	about 24 million tons
Capital cost of each job on fishing vessels	$30,000 – $300,000	$25 – $2,500
Annual catch of marine fish for industrial reduction to meal and oil, etc.	about 22 million tons	Almost none
Annual fuel oil consumption	14–19 million tons	1–3 million tons
Fish caught per ton of fuel consumed	2–5 tons	10–20 tons
Fishers employed for each $1 million invested in fishing vessels	5–30	500–4,000
Fish and invertebrates discarded at sea	16–40 million tons	None

Figure 18.18. Comparison of Small-scale and Large-scale Sub-sectors in Norwegian Fisheries, 1998. Schematic illustration of the duality of fisheries prevailing in most countries of the world. This largely reflects the misplaced priorities of fisheries "development" but also offers opportunity for reducing fishing mortality on depleted resources while maintaining social benefits. The solution here is to reduce mainly the large-scale fisheries. (David Thompson in Alverson et al. 1994)

fisheries operate results in open competition between the two fisheries for the same resource (Pauly 1997).

Catching and discarding undersized fish is another source of conflict between the two fisheries. Shrimp trawl fisheries, particularly for tropical species, have been found to generate more discards than any other fishery type while accounting for small fraction of global marine fish landings (Kelleher 2004). Since many discarded species are of commercial value to small-scale fisheries, bitter conflicts between the two fisheries often ensue, leading to conflicts between fishers using different gear and political infighting over resource allocation and bycatch removal quotas (Alverson et al. 1994; Kelleher 2004; Pauly 1997).

Artisanal and industrial fisheries also experience conflicts in markets where they use the same inputs such as fuel and gear or they catch the same or similar species of fish (Panayotou 1982). Industrial fisheries may bid up the prices of fishing inputs or their massive landings may depress fish prices, making small-scale fishers increasingly uncompetitive. Industrial fisheries have access to low-interest institutional credit and loans and government subsidies, whereas small-scale fishers usually only have access to informal credit at interest rates many times higher than the institutional rates (Panayotou 1982).

Institutional support is often skewed in favor of the large-scale fishers because of their apparently higher efficiency and greater contribution to economic growth; their use of fewer landing areas, which allows economies of scale in the provision of infrastructure and the delivery of assistance programs; their political and economic power; and a general bias in favor of large-scale fisheries under an open-access regime. In contrast, small-scale fishers are geographically dispersed and lack political and economic power and hence do not generally benefit from institutional support (Bennett et al. 2001).

Several management interventions and conflict mitigation processes can minimize conflicts. These range from outright banning of certain gears—as occurred in western Indonesia in 1980, for example, when trawling was banned (Sardjono 1980)—to clarifying property rights and gear and area restrictions. In some cases, the role and recognition of traditional and local ecological knowledge are increasing as management responsibilities are transferred to local communities and stakeholders. For example, traditional and local knowledge is used to manage a bait fishery in the Solomon Islands (Johannes et al. 2000). Protection of the rights of small-scale fishers is also recognized explicitly in the Convention on Biological Diversity decision VII/5 and the FAO Code of Conduct for Responsible Fisheries.

18.6.2 Effectiveness of International Instruments in Managing Shared Stocks

While there are more than 100 fisheries access agreements (multilateral and bilateral) currently used to manage access to marine resources, few are monitored or evaluated for their effectiveness, equitable access, and sharing of economic benefits. The European Union has initiated a monitoring program for the EU's Common Fisheries Policy, and other regional fisheries bodies are considering monitoring programs, but none have been developed to date (FAO 2001a).

A recent study of international fisheries instruments for the North Atlantic suggested that global and regional treaties covering the area are of limited effectiveness; the treaties themselves are not found to be weak, but many governments appear unwilling to act on commitments made under them (Alder and Lugten 2002). The study showed that the 14 relevant instruments have a strong North-South gradient of decreasing implementation, as measured by adherence to reporting requirements, issuance of reports, and other formal requirements. On the other hand, there was no such gradient in the state of the stocks, and it could therefore be argued that these instruments are of limited effectiveness even when completely implemented. Consequently, reinforcing international agreements and ensuring they are implemented is likely to contribute significantly to solutions to the present problem of overfishing. This will require that these bodies free themselves from the influence (dominant participation) of the very industry they are intended to regulate, as documented in a detailed study of the U.S. Fisheries Management Councils (Okey 2003).

Various legal instruments, such as the Marine Mammal Protection Act in the United States, explicitly require protection of the food and habitat of charismatic marine fauna. International instruments such as the International Whaling Commission and the Convention on International Trade in Endangered Species of Wild Fauna and Flora restrict harvesting of species or limit bycatch.

A recent joint study between the CBD and the UNCLOS found that whereas the provisions of these two conventions are complementary and mutually supportive regarding the conservation and sustainable use of marine and coastal biodiversity, an important legal gap exists with respect to commercially oriented activities relating to marine genetic resources in areas beyond national jurisdiction. (Each country's national legislation could address the issues within its EEZ.) The gap is yet to be addressed by the international community, and it would seem particularly important given the increasing importance of genetic resources in these areas and the threat posed to them by various activities, such as bio-prospecting by multinational pharmaceutical companies, that may be carried out without due regard to conservation and equity principles.

The implementation of North Atlantic agreements is among the best funded and supported administratively, and yet fisheries in the North Atlantic continue to decline (Alder and Lugten 2002). The refusal of EU-member governments to set sustainable harvest levels or to take a precautionary approach to the Common Fisheries Policy is a clear example of a well-funded management system supported by sound science with overfished stocks. Agreements in the South Atlantic, and for the most part in the rest of the world, are largely similar in nature to North Atlantic legal instruments. However, there are substantially fewer agreements elsewhere, and the level of support for them is much less, although some of the instruments of the southern polar biome are exceptions in terms of support. Globally, there is potential for instruments to assist in finding solutions to fisheries sustainability problems, but the lack of political commitment to their use to effect change prevents significant progress toward such solutions.

18.6.3 Effectiveness of Marine Protected Areas

Marine protected areas with no-take reserves at their core can reestablish the natural structures that have enabled earlier fisheries to maintain themselves. (See also Chapter 4.) MPAs are not a recent concept. Historically, many fisheries were sustained because a portion of the target population was not accessible. Most targeted fisheries were offshore or in areas adjacent to lands with low human populations and therefore subject to relatively low threat. However, modern fishing technology for mapping the seabed and for finding and preserving fish (artificial ice and blast freezing) expanded the reach of fishing fleets.

A number of recent studies have demonstrated that MPAs can help in managing fisheries (Roberts et al. 2002). Most of these studies have covered spatially small areas and primarily in tropical shelf systems, although emerging studies from temperate areas, such as New Zealand and Chile, have also demonstrated MPA effectiveness. However, other studies have found that MPAs have not delivered the expected benefits of protecting species and their habitats (Hilborn et al. 2004; Edgar and Barrett 1999; Willis et al. 2003). In many cases failure was due to either not including MPAs as part of a broader coastal management system or a lack of management effectiveness, funding, or enforcement. In the Gulf of Mexico, for example, the establishment of MPAs merely shifted fishing effort to other areas and increased the vulnerability of other stocks and endangered species (Coleman et al. 2004). Knowledge on the size and location of MPAs that can act as effective buffers against the impacts of fishing requires further research.

It has been widely and repeatedly demonstrated that marine protected areas, particularly no-take marine reserves, are essential to maintain and restore biodiversity in coastal and marine areas (COMPASS and NCEAS 2001). Their wide-scale adoption is inhibited by the perception that biodiversity is unimportant relative to fishers' access to exploitable resources. Therefore, the proponents of marine reserves have been saddled with the additional task of demonstrating that setting up no-take reserves will increase fisheries yields in the surrounding areas, as well as determining the appropriate size and siting of marine reserves that are needed to at least sufficiently offset the loss of fishing grounds. This requirement, combined with initiatives by recreational fishers asserting rights to fish, has effectively blocked the creation of marine reserves in many parts of the world.

Thus while the cumulative area of marine protected areas is now about 1% of the world's oceans, only about one tenth of that—0.1% of the world's oceans—is effectively a no-take area. This gives an air of unreality to suggestions that 20% and an optimum of 30–50% of the world's ocean should be protected from fishing to prevent the loss of some species now threatened with extinction and to maintain and rebuild some currently depleted commercial stocks (National Research Council 2001; Roberts et al. 2002; Airame et al. 2003; Agardy et al. 2003). Even the more modest CBD target of 10% MPA coverage by 2012 will be hard to reach.

One approach to resolving this dilemma is to take an adaptive management approach so that the use of MPAs within a suite of fisheries management options can be assessed and modified as new information emerges and lessons learned are shared (Hilborn et al. 2004). This avoids unrealistic expectations on the improved performance of MPAs. Any approach to the use of MPAs in managing marine ecosystems would also benefit enormously from including performance monitoring and enforcement programs to address some of the management problems that have traditionally hindered effectiveness (Coleman et al. 2004).

If properly located and within a context of controlled fishing capacity, no-take marine reserves enhance conventional fisheries management outcomes. They may, in some cases, reduce catches in the short term, but they should contribute significantly to improving fishers' livelihoods as well as biodiversity over the mid to long term. Marine reserves generally perform this way in inshore shelf systems (such as reefs); many case studies, as shown in Saba Marine Park (Netherlands Antilles), Leigh Marine Reserve (New Zealand), and Sumilon Island Reserve (Philippines), are described in detail in Roberts and Hawkins (2000) to support this. However, understanding of the effectiveness of marine reserves in managing fisheries in deeper oceanic areas is more limited. Further, the protection and monitoring of these deep-sea areas and other undamaged areas may, in line with the precautionary principle, avoid the need for mitigation or restoration of the systems later, when costs are likely to be higher (and in some cases restoration may not be viable).

Already, the demand for fish resources has pushed fishing fleets into international waters, and as other resources become scarcer in national waters (such as gas, oil, minerals, and carbon sinks), conflicts over the best use of these common resources and spaces will increase. Hence the growing call for ocean zoning, including the creation of no-take zones that would reestablish the reserves that were once in place due to vessels lacking the technology to gain access to deeper, offshore areas, which in the past has protected exploited species.

18.6.4 Effectiveness of Fisher Mobilization

Recently, small-scale and artisanal fishers, with the assistance of organizations such as the International Collective in Support of Fishworkers, have established local groups to lobby government and industry. The mobilization of these groups is taking place globally in industrial and developing countries alike. In India (Mishra 1997), Chile (Phyne and Mansilla 2003), the Philippines, and Japan (Kurien 2004), for instance, fisher unions and collectives have used their numbers to lobby government and industry for improved resources access, better working conditions, and social benefits. In addition, these groups have provided mechanisms to enable women, who play a number of roles within the fishing sector (including sometimes also fish capture), to participate more in fisheries management. However, considerable resources and the legal basis to demand change are required, and in some countries mobilization of fishers is discouraged or hindered by governments or cultural norms (Alauddin and Hamid 1999).

18.7 Sustainability and Vulnerability of Marine Fisheries

Fishing has grown to become a much larger business than it was in the 1950s. Modern, highly capitalized fleets now range the oceans of the world, in some cases competing for a limited common resource with small-scale fishers and local communities. Overcapitalization of the global fishing industry is, in fact, an enormous problem. FAO estimated that on a global basis world fisheries operated at an annual deficit of up to $20 billion in the 1990s (Milazzo 1998). Thus, present-day commercial fisheries actually represent a large net burden on other economic sectors. This unfortunate situation is at least to some degree due to the lack of a sound scientific basis to correctly gauge the productivity of the resources and to effectively manage impacts in the face of large-amplitude variability in both physical and biological aspects of ocean ecosystems.

In ecological terms, many marine fish appear to be vulnerable to anything above low levels of fishing pressure because of their biological characteristics, including their late age of sexual maturation, association with specialized or limited habitats that might also be vulnerable or widely damaged by fishing activities (such as trawling), and vulnerable life-history strategies such as aggregation spawning (Ehrhardt and Deleveaux 2001). Distinct genetic units may be susceptible to irrecoverable declines if they are self-recruiting or exhibit Allee-type effects (effects of birth rate declines at low population densities), whereby they cannot recover below some lower threshold of population numbers or density (Stephens and Sutherland 1999).

18.7.1 Natural Population Variability and Sustainability

Analysis of multidecadal time series fishery data has shown that many important fish populations are highly variable (e.g., Csirke and Sharp 1984; Schwartzlose et al. 1999). This is often in response to changes in ocean climate and ecosystem structure. This variability greatly compounds the difficulties in properly managing fisheries exploitation (Bakun and Broad 2002). For example, a level of fishing pressure calculated to be sustainable may cause unexpected collapse of a population whose productivity has decreased naturally. Moreover, even when evidence of declining numbers is available, the fishing industry may cite natural variability to argue against the results and question the need for action and may continue with unsustainable levels of exploitation.

Sustainability in fisheries, as evidenced by long-term data series showing no downward trend in catches, is generally a matter of fishers lacking either the technical means to catch more fish or the outlets (the large markets) to sell more than a small fraction of the fished population (Pauly et al. 2002). Increasing demand relative to the productivity of local resources and the technology to gain access to the entire distributional range of a species will tend to deplete the resource.

In a number of case studies, responses to fisheries management problems have mitigated or reversed the impact of fisheries. For instance, the introduction of community-based management of reef areas in the Philippines has resulted in increased fish landings that ultimately improved the well-being of those communities (Russ and Alcala 1994). Also, more effective enforcement measures for Namibian fisheries and the nationalization of the fishery sector contributed to better socioeconomic conditions for many coastal communities (Erastus 2002). In general, relatively small and often single-species fisheries can be restored, as has occurred in the Peruvian hake (*Merluccius gayi peruanus*) fishery (Instituto de Mar del Peru 2004).

However, there are also a number of spectacular failures—cod (*Gadus morhua*) in Newfoundland and orange roughy (*Hoplostethus atlanticus*) in New Zealand are two often-cited examples. A combination of increasing fishing efficiency through technology, expansion by Canadian fishers into fishing grounds previously used by foreign fleets, and an apparent but misleading increase in stock density as the overall biomass collapsed in the 1980s partly account for the repeated underestimation of the problem facing cod (Walters and Maguire 1996). Similarly, early in the development of orange roughy fisheries, there were management failures as the stocks were harvested at much higher rates than considered sustainable (Francis et al. 1995 in Clark 2001). Also, responses or interventions that mitigate or reverse negative effects of the large-scale, multispecies fisheries, such as the demersal fisheries within EU waters, managed by the EU Common Fishery Policy have been implemented since the early 1970s. But these were ineffective due to lack of political will to decrease effort, poor data quality for specific fisheries and countries, and limited enforcement (Alder and Lugten 2002).

Some fisheries have been certified as sustainable by the Marine Stewardship Council and the Marine Aquarium Council. Only seven fisheries, including for Thames (U.K.) herring (*Clupea harengus*) and Western Australian rock lobster (*Panulirus cygnus*), have been MSC-certified since the program began in 1997, however, and most are small-scale with no impact at regional or global scales. Nevertheless, such schemes are important in communicating to consumers the need to manage fisheries sustainably and in educating them about the source and harvest techniques of the seafood products they purchase. Similarly, the Marine Aquarium Council's certification scheme has had a positive impact on the aquarium trade (GEF 2004).

Conventions that have been effective in sustainably managing fisheries, such as the Pacific Salmon Treaty and the North Pacific Anadromous Fish Convention, have had small-scale impacts since they are focused on a particular species or set of species. Broader bodies, such as the International Commission for the Conservation of Atlantic Tunas, face greater challenges in managing the fisheries in their charge sustainably. Some fisheries, such as those under the EU's Common Fisheries Policy and those in Canada and the United States (cod and Atlantic salmon (*Salmo salar*), for instance), have not been managed sustainably despite sound scientific evidence being available suggesting that reduced fishing effort would significantly assist the situation.

18.7.2 Thresholds

It has not so far been possible to predict the critical thresholds beyond which a fish stock will collapse, and the major stock collapses of the last few decades have been a surprise, even to those involved in monitoring and managing these stocks. One well-known example is Newfoundland's northern cod (*G. morhua*). Almost the same scenario was reenacted 10 years later, in 2001, in Iceland, which very nearly lost its cod stock (Marine Research Institute Reykjavik 2002), in spite of the Icelandic government's commitment to sound fisheries management. Because of the unpredictability of these thresholds, precautionary approaches such as those involving marine protected areas and reductions in fishing effort (and in fishing mortality) are likely to safeguard against such thresholds being reached.

18.7.3 Areas of Rapid Change

Within 10–15 years of their arrival at a new fishing ground, new industrial fisheries usually reduce the biomass of the resources they exploit by an order of magnitude, (Myers and Worm 2003). This is well illustrated by the Gulf of Thailand demersal fisheries (Eiamsa-Ard and Amornchairojkul 1997), by orange roughy fisheries around various seamounts around New Zealand (Koslow et al. 2000), by the Antarctic fisheries discussed in Chapter 25, by the live reef fish trade (Sadovy et al. 2003b), and by a multitude of others. This process is often accelerated by encouragement from governments to "diversify" fisheries, often resulting in fleet overcapacity and a drive to exploit new or "unconventional" species. This level of biomass reduction renders the species in question extremely vulnerable to subsequent exploitation and other perturbations, notably those likely to result from climate change.

Many areas of the coastal zone have undergone rapid changes directly due to coastal use and indirectly through upstream changes and land use. The consequences of this have been significant habitat loss, declining coastal environments, and reduced fish landings, as discussed in Chapter 19.

18.8 Management Interventions in Marine Systems

Any management initiative aiming to reduce the impacts of marine resource use or to strengthen sustainable use and conservation of biodiversity need to be addressed at different levels and by various means, and to involve local communities, including indigenous peoples. In the case of industrial fisheries, industry needs to be included as well. There are numerous examples from around the world of local involvement resulting in recovery of ecosystems and social benefits—the Philippines (Alcala and Russ 1994), Chile (Castilla 2000), and Brazil (Ferreira and Maida 2001), for example. There are also regional initiatives, as shown by the Regional Fishery Management Councils in the United States, that develop management plans to rebuild stocks. All sectors of society can be a part of the solution—governments by enforcing mandates and ensuring compliance with appropriate environmental codes, industry by operating responsibly, NGOs by providing capacity and training where needed, and consumers by demanding goods and services that are provided at minimal impact on marine ecosystems.

A number of international instruments can also be used to manage fishing and its impacts on the marine environment, domestically and internationally. Various fishing instruments have been discussed throughout this chapter. FAO's Code of Conduct for Responsible Fisheries includes approaches for avoiding or mitigating the impacts of fishing on other components of ecosystems and is one of the few fisheries-specific instruments that includes fishing impacts on other species.

Other instruments, such as the Convention on International Trade in Endangered Species of Wild Fauna and Flora, can be used indirectly to manage the impact of fishing on threatened species through the development of species management strategies. CITES is an international agreement among 167 countries to cooperatively manage international trade in species of conservation concern, to ensure that trade does not threaten their survival in the wild.

The recent listing of seahorses and two other species of shark on CITES Appendix II now requires countries to determine that proposed exports will not be detrimental to the survival of the species in the wild (known as a nondetriment finding) as a precondition to permitting export. From a practical perspective, this means that countries that are parties to CITES need to implement management strategies to ensure that the export of seahorses and listed shark species (and therefore catch levels and methods) does not threaten the sustainability of their fisheries and wild populations. The Convention on the Conservation of Migratory Species of Wild Animals (known as the Bonn Convention) and its Agreement on the Convention of Albatrosses and Petrels, as well as FAO's International Plan of Action for Seabirds, are other examples of international management interventions in marine systems.

The continuation of present fisheries trends, including the buildup of fishing capacities, suggests a serious risk of losing more fisheries. Interventions such as the halting of destructive fishing practices by developing alternative technologies or financial incentives, reducing fishing effort, and establishing MPAs are needed to reverse the current trends. There is likely to be no single most suitable intervention; rather, a mix of interventions is likely to be the most effective approach. The composition of that mix will require an adaptive approach to management of marine ecosystems.

18.8.1 Integrating Management of Sectors in Marine Areas

No global or regional framework for integrated management of the oceans exists. Internationally, some activities on or in the high seas are managed through a range of organizations. For example, the International Seabed Authority manages seabed mining through UNCLOS, and the International Maritime Organization and the conventions it administers (such as the International Convention for the Prevention of Pollution from Ships, known as MARPOL) manages marine transportation and pollution, including dumping of waste at sea. These organizations have worked with industry on measures to reduce the impact of hazardous and damaging activities. Such measures have included, for example, the introduction of double-hulled oil tankers to lower the risk of oil spills. However, there is no integrated approach to managing ocean use, which has resulted in concern over some issues, such as bioprospecting, high-seas MPAs, and the management of marine biodiversity. Not all issues can be addressed within UNCLOS and changes to the convention or the introduction of a new legislative framework (including zoning plans) may be needed to overcome current impediments to making and managing the needed trade-offs for equitable and sustainable use of ocean space and deep-sea resources.

National ocean policies based on sustainable development principles have been successful frameworks for integrating the management of the various marine sectors. Despite countries declaring their exclusive economic zones over the last 25 years, few countries have actually formulated or implemented comprehensive ocean policies. It is only in the last five years that we have

seen the introduction of such policies in Australia, Canada, the United States, and the Netherlands (Alder and Ward 2001), along with, most recently, the Pacific Islands Regional Ocean Policy (South Pacific Regional Environment Program 2003).

UNEP has a comprehensive Regional Seas Programme that includes 13 regions and 140 countries with a focus on tackling the sources of degradation of marine and coastal systems. The program has a broad mandate and can integrate various sectors, including transportation and oil development, and initiatives at the regional scale to address pollution problems. It can also play a key role in establishing MPAs crossing multiple borders, as seen in the successful Mediterranean network of specially protected areas that conserve critical areas through MPAs, reserves, and refuges.

The European Community's marine strategy is an example of the challenges in managing marine ecosystems on a regional basis (EC 2002). Other regional programs, such as OSPAR for the North-East Atlantic and HELCOM for the Baltic, also integrate various marine sectors to address a range of marine issues.

18.8.2 Integrating Coastal Management and Ocean Policy

Oceanic and coastal ecosystems are tightly linked. While integrated coastal management is now well entrenched in many countries (see Chapter 19), the development of ocean policies lags behind. Where ocean policies are in place, there is recognition of the need to take an ecosystem-based approach and to ensure that coastal management plans and ocean policies are harmonized. Depending on the legislative basis and jurisdictional issues, coastal management may be embedded within ocean policy, as demonstrated in Canada's Oceans Act.

18.8.3 Marine Protected Areas

Attempting to maintain fisheries for depleted or collapsed fish populations (through subsidies, for example) is economically and ecologically damaging. As provided for in the 1982 Law of the Sea Convention, the 1995 UN Fish Stock Agreement, and the 1995 FAO Code of Conduct, declining populations must be rebuilt and marine ecosystem productivity must be restored as far as possible. No-take marine reserves can make an important contribution when they are part of an overall policy to maintain fisheries. No-take marine reserves are important for providing places where critical life stages, such as adult spawners and juveniles, can find refuge and for providing additional insurance to more conventional fisheries management while safeguarding against local extinctions.

Extinction is a gradual process, but many species of commercial fish species already have severely depleted populations. Avoiding the loss of threatened species must involve not only allowing their biomass to rebuild, but also rebuilding the ecosystems in which they live. MPAs may also contribute to reducing the impacts of global climate change by increasing biomass and widening age structure so that the population of fish stands a greater chance of withstanding wider fluctuations in the environment.

Moreover, the implementation of no-take marine reserves combined with other interventions, such as controls on fishing capacity, would be a more proactive response to fisheries management than current reactive approaches. Small reserves may be effective in protecting sedentary organisms, since they do not move or only move small distances. But marine reserves intended for the conservation or sustainable use of fish, marine mammals, seabirds, and large species need to be larger and particularly appropriately located to take into account the life characteristics of such species. Some species can spend various parts of their life cycle in different habitats (larval stages in estuaries, juvenile stages in coastal seagrass meadows, and adult stages in the open ocean, for instance), and marine reserves need to be strategically located to account for these differences. In areas left open for fishing, however, explicit consideration needs to be given to the food required to maintain recovered populations, and the scientific tools (such as ecosystem models) exist to perform the required accounting for biomass. Populations of small fish and invertebrates presently not exploited should not be viewed as a latent resource that should be developed. They are the remaining food basis for marine mammals, seabirds, and large fish that need to be sustained and rebuilt (as described in the next section).

The immediate and urgent need to manage risks to the marine biodiversity of seamounts and cold-water coral reefs through measures such as the elimination of destructive fishing practices, such as bottom trawling, has been highlighted in a number of recent international fora. These include the fourth meeting of the United Nations Open-ended Informal Consultative Process on Oceans and the Law of the Sea (2003), the 2003 World Parks Congress (recommendation 5.2.3 and the Congress document on emerging issues), the Eighth Meeting of the Convention on Biological Diversity Subsidiary Body on Scientific, Technical and Technological Advice (SBSTTA recommendation VII/5) (2003), the 2003 Defying Ocean's End conference, the 10th Deep-Sea Biology Symposium (2003), and the 2nd International Symposium on Deep-Sea Corals (2003). These meetings have resulted in initiatives to protect cold-water reefs, as through a marine protected area established by Norway to conserve the Tisler Reef along the Norwegian-Swedish border.

18.8.4 Iron Enrichment of Ocean Waters for Carbon Sequestration and Increased Fish Yields

There is growing interest in experiments wherein large areas of low productivity oceanic waters are fertilized using micronutrients, principally iron (Boyd et al. 2000). One major reason for these experiments is to investigate the potential for sequestration of atmospheric carbon, whereby carbon is taken up by the cells of primary producers (the growth of marine phytoplankton populations is often limited by the availability of iron in the water), which would then sink to the bottom as marine snow (Cole 2001). The net effect of this process on carbon sequestration is not clear, however, because localized algal outbursts can also lead to anoxia and the production of methane, a powerful greenhouse gas.

It has been suggested that the enhanced primary production resulting from fertilization could also lead to increased fish yields and could even help alleviate dietary iron deficiencies in some parts of the world (Jones 2001). However, the success of such schemes is extremely doubtful, given the wide range of evolved adaptations that are required for planktivorous species such as anchovies to thrive in a highly productive variable environment, such as the Peruvian upwelling system (Bakun 1996).

18.8.5 Selected Examples of Human Responses to the Sustainability Challenge

While the problem of overfishing has been globally identified at high policy level as a key fisheries issue since the mid-1940s, the broad-scale recognition that it is also an important component of the global environment issue only emerged in the early 1990s in the wake of the U.N. Conference on Environment and Development (Garcia 1992). Even though the word "overfishing" was in common usage long before then (applying to "isolated" instances), there was still a general belief that the oceans had a an enormous capacity to provide fish and invertebrates for direct human consumption and for reduction products such as fishmeal and fish oils used in intensive food production systems (see, e.g., Idyll 1978; Pike and Spilhaus 1962).

In theory, the 1982 United Nations Convention on the Law of the Sea should have been an international instrument for wise use of the oceans: it espouses the right and need for coastal nations to monitor and manage their fish stocks. In retrospect, however, UNCLOS exacerbated overfishing problems in at least two important ways. First, as it gave coastal nations the ability to declare a 200-mile EEZ, many national governments saw this as an opportunity to greatly augment their fishing industries as sources of more secure employment and export earnings. While a few industrial countries managed to achieve some of the expected benefits by testing and adopting new management measures (such as limited entry and fishing rights), most others simply failed to realize them because of uncontrolled, anarchic, and subsidized development of chronic overcapacity (FAO 1993).

The subsidies injected into the industry in the 1980s and 1990s resulted in immense global overcapacity of fishing fleets, perhaps the most important problem—and apparently the hardest one to resolve—faced today in marine resource management (Mace 1997). For example, the EU's subsidies for the construction of fishing vessels in the late 1980s and early 1990s resulted in overcapacity, and until recently attempts to reduce this capacity were not very effective (Alder and Lugten 2002). Subsidies in the form of employment insurance to Newfoundland fishers were one of the factors that contributed to overcapacity in the industry and the ultimate collapse of the cod fishery (Brubaker 2000).

Second, the UNCLOS requirement that coastal nations without sufficient fishing capacity to exploit the resources within their respective EEZs should make these resources available to other nations has ultimately proved detrimental to many developing nations. Although these countries do receive some form of reimbursement from other nations acquiring access, this has frequently resulted in payments that are significantly less than the value of the resource (particularly the long-term value; see Kaczynski and Fluharty 2002), in underreporting of foreign catches, in depletion of developing nations' deep-sea resources, and in depletion of the coastal resources that supported local fishing communities.

The change in the perception that there was considerable potential to increase the exploitation levels of marine resources, including fisheries, which began in the early 1990s, was influenced by several factors: the globalization of information (making it more rapidly and more widely available), the awareness-raising work of the World Commission on Environment and Development and UNCED, and the active mobilization of the media by the NGO environmental movement. Fishing capacity was widely perceived to be getting out of hand. Opportunities for further expansion of fisheries were diminishing as an increasing portion of the globe was being surveyed and assessed, albeit yielding few new areas for fishing or new fisheries. Deepwater species were mostly found to have very low productivity. Many environmental organizations such as the World Wide Fund for Nature that had previously focused mainly activities on terrestrial systems and marine mammals began to expand into fisheries. Legislation was tightened in many nations, and work began on several binding and nonbinding international instruments, some related to ecosystems in general and others specifically on marine systems and fisheries.

Subsequently, there has been a marked increase in national and international instruments that are gradually changing people's perceptions of the sustainability of current practices. For marine systems, the most important such instruments range from the 1992 Rio Declarations formulated at UNCED to several FAO International Plans of Action, including the International Plan of Action for the Management of Fishing Capacity (FAO 2001b) and the International Plan of Action to Deter, Prevent and Eliminate Illegal, Unreported and Unregulated Fishing (FAO 2001b). (See also MA, *Policy Responses, Chapter 6.*) All of these embody some facet of the "precautionary approach," which has been instrumental in shifting attention to the benefits of conservative harvest strategies rather than risk-prone management (Mace 2001).

Concerted efforts to implement these agreements have been launched in many regions, with results ranging from unanticipated levels of success, to moderate success, and to failure. In terms of successes, the most notable accomplishments appear to have been those where national legislations have been modified to better accomplish the spirit and intent of the international instruments.

For example, the most recent amendment to the Magnuson-Stevens Fisheries Conservation and Management Act in the United States (popularly known as the Sustainable Fisheries Act) embodied the precautionary approach (without mentioning it by name) as exemplified in Annex 2 on the implementation of UNCLOS (UN 1995), in which the fishing mortality associated with maximum sustainable yield is suggested as a limit reference point to be avoided rather than a target that is routinely exceeded. This has resulted in considerable reductions in fishing mortality for some previously overfished U.S. fisheries, and several have rebuilt to levels not recorded in three or more decades.

One of the most dramatic examples is the density of Georges Bank scallops, which increased eighteenfold in the seven years following exceptional recruitment and implementation of management measures for bottom fishing that resulted in substantial reductions in fishing mortality during the fishing closure. Other U.S. examples include mid-Atlantic scallops (*Haliotis* spp), Georges Bank haddock (*Melanogrammus aeglefinus*), Georges Bank yellowtail flounder (*Limanda ferruginea*), Atlantic stripped bass (*Morone saxatilis*), Atlantic Acadian redfish (*Sebastes fasciatus*), Pacific chub mackerel (*Scomber japonicus*), and Pacific sardine (*Sardinops sagax*) (Murawski et al. 2000; NEFSC 2002). Yet while these and many other stocks in the United States and elsewhere are slowly but steadily rebuilding (NMFS 2004), the biomass of most of them is generally still well below historic levels. Closures of seamount fisheries on the Chatam Rise in New Zealand have also resulted in the slow rebuilding of orange roughy biomass (Clark 2001). Globally, however, these success stories represent a small proportion of the many overfished stocks.

Other examples of successful national initiatives include Australia's implementation of the Environment Protection and Biodiversity Act in 1999, which requires several fisheries to be assessed for their ecological sustainability. Regionally, the ICCAT has been able to work toward compliance of management measures for nonmember countries through tuna import and export restrictions by member countries.

Notable failures in fisheries management are exemplified by attempts to reduce harvest overcapacity while maintaining that exploitation of natural resources is a human right. This "birthright" is hard to dispute: all humans ought to have the "right" to a healthy life with adequate nutrition. And for some people, fish are one of the few available sources of protein. Solutions to this dilemma lie in assistance that improves the health, education, and non-fisheries-related employment opportunities of coastal communities as well as empowering coastal communities to manage resources sustainably.

Even relatively rich industrial nations that routinely pay their farmers not to produce crops seem unable to resolve the fish harvesting overcapacity problem, however. The European Union, for example, appears to have been largely unsuccessful with the capacity reduction plan in its Common Fisheries Policy, which has been in place for more than 15 years. With a few exceptions, substantial overcapacity reductions have been recorded in cases where individual transferable quotas, which allocated a quota of the catch to license holders as well as allowing those individuals to sell part or all of their quota, have been implemented (for instance, ITQs in New Zealand, Iceland, and some U.S. fisheries, such as Atlantic surf clams (*Spisula solidissima*) and ocean quahogs (*Arctica islandica*), South Atlantic wreckfish (*Polyprion americanus*), and Pacific halibut (*Hippoglossus stenolepis*) and sablefish (*Anoplopoma fimbria*)).

Although the emphasis in recent years has been on unsustainable fishing practices, fisheries represent only one of many human influences on marine ecosystems. In coastal marine systems in particular, coastal development—with concomitant problems of local pollution and habitat destruction—is very important. (See Chapter 19.) Non-fisheries human influences such as marine debris and oil slicks are also important on the high seas. As a result, as described earlier, several nations are attempting to develop legislation and policies to facilitate integrated management of marine systems—that is, coordinated management of all alternative uses of the ocean. Such uses include harvesting marine species for food and other purposes, aquaculture, research, oil and gas exploration, ocean mining, dredging, ocean dumping, energy generation, ecotourism, marine transportation, and defense. To date, it has proved difficult to integrate the management of all these activities because the authorities regulating these activities are usually independent of one another (Sissenwine and Mace 2003).

In Australia, the Environment Protection and Biodiversity Act requires management agencies to demonstrate that the fishery or fisheries are ecologically sustainable, with defined benchmarks to quantitatively assess sustainability. Since the introduction of this Act in 1999, more than 35 fisheries have been assessed (www.ea.gov.au/coasts/fisheries/assessment/index.html). The private sector has responded similarly to the introduction of the Marine Stewardship Council's accreditation scheme for fisheries, which is supported by major fish buyers, such as Unilever, and conservation organizations, such as WWF.

With increased public awareness of the limits of marine systems, binding and nonbinding international instruments in place, tightened national legislations, and a few success stories that emphasize the positive benefits of conservation, the future of the oceans may appear brighter than it did a decade ago. However, this is not the time for complacency, particularly considering the likely increase in pressure on natural resources that will result from the world's growing human population and rising incomes.

References

Abella, A.J., J.F. Caddy, and F. Serena, 1997: Do natural mortality and availability decline with age? An alternative yield paradigm for juveniles fisheries, illustrated by the hake *Merluccius merluccius* fishery in the Mediterranean. *Aquatic Living Resources,* **10,** 257–269.

Agardy, T., P. Bridgewater, M.P. Crosby, J. Day, P.K. Dayton, R. et al., 2003: Dangerous targets? Unresolved issues and ideological clashes around marine protected areas. *Aquatic Conservation-Marine and Freshwater Ecosystems,* **13(4),** 353–367.

Ahmed, M., C. Delgado, and S. Sverdrup-Jensen, 1999: The growing need for fisheries policy research in developing countries. In: *Fisheries Policy research in developing countries: issues, priorities and needs,* M. Ahmed, C. Delgado, S. Sverdrup-Jensen, and R.A.V. Santos (eds.). ICLARM Conference Proceedings 60, 1–4.

Airame, S., J.E. Dugan, K.D. Lafferty, H. Leslie, D.A. McArdle, and R.R. Warner, 2003: Applying ecological criteria to marine reserve design: A case study from the California Channel Islands. *Ecological Applications,* **13(1),** S170–S184.

Alauddin, M. and M.A. Hamid, 1999: Shrimp culture in Bangladesh with emphasis on social and economic aspects. In: *Towards Sustainable Shrimp Culture in Thailand and the Region,* P.T. Smith (ed.), Australian Centre for International Agricultural Research (ACIAR), Canberra (Australia), 53–62.

Alcala, A.C. and G.R. Russ, 1994: Sumilon Island Reserve: 20 years of hopes and frustrations. *NAGA, The ICLARM Quarterly,* **17(3),** 8–12.

Alder, J. and L. Christanty, 1998: Taka Bonerate: Developing a strategy for community-based management of marine resources. In: *Living through Histories. Culture, History and Social Life in South Sulawesi,* K. Robinson and M. Paeni (eds.), Department of Anthropology, Research School of Pacific and Asian Studies, the Asutralian National University, Canberra (Australia). Published in association with The National Archives of Indonesia (Arsip Nasional Republik Indonesia), 229–248.

Alder, J. and T. Ward., 2001: Australia's ocean policy—sink or swim? *Journal of Environment and Development,* **10,** 266–289.

Alder, J. and G. Lugten, 2002: Frozen fish block: how committed are North Atlantic States to accountability, conservation and management of fisheries? *Marine Policy,* **26(5),** 345–357.

Alder, J. and R.U. Sumaila, 2004: Western Africa: a fish basket of Europe past and present. *Journal of Environment and Development,* **13,** 156–178.

Alder, J. and R. Watson, in press: Globalization and its effects on fisheries. In: *Proceedings of the Workshop Globalization: Effects on fisheries, 12–14 August 2004, Quebec (Canada),* W. Taylor, M. Schechter, and L. Wolfson (eds.), Cambridge University Press, NY (USA).

Alverson, D.L., M.H. Freeberg, J.G. Pope, and S.A. Murawski, 1994: *A Global Assessment of Fisheries Bycatch and Discards.* FAO Fisheries Technical Paper 339, Food and Agriculture Organization of the United Nations (FAO), Rome (Italy), 233 pp.

Anonymous, 2002: *Patterns and propensities in Greek fishing effort and catches.* Report to the EU (DGXIV) Project 00/018, European Commission, Brussels (Belgium).

Archer, F., T. Gerrodette, and A. Jackson, 2002: *Preliminary estimates of the annual number of sets, number of dolphins chased, and number of dolphins captured by stock in the tuna purse-seine fishery in the eastern tropical Pacific, 1971—2000.* Southwest Fisheries Science Center (SWFSC), National Marine Fisheries Service, NOAA, La Jolla, CA (USA), 26 pp.

Arnason, R. and H. Gissurarson (eds.), 1999: *Individual Transferable Quotas in Theory and Practice: Papers Exploring and Assessing the Radical Reorganization of Ocean Fisheries in the Final Decades of the 20th Century.* The University of Iceland Press, Reykjavik (Iceland).

Aronson, R.B. and D.B. Blake, 2001: Global climate change and the origin of modern benthic communities in Antarctica. *American Zoologist,* **41,** 27–39.

Atta-Mills, J., U.R. Sumaila, and J. Alder, 2004: The decline of a regional fishing nation: The case of Ghana and West Africa. *Natural Resources Forum,* **28(1),** 13–21.

Baker, C.M., B.J. Bett, D.S.M. Billett, and A.D. Rogers, 2001: An environmental perspective. In: *The status of natural resources on the high seas,* WWF/IUCN (ed.), WWF/IUCN, Gland (Switzerland), 1–67.

Bakun, A., 1996: *Patterns in the Ocean: Ocean Processes and Marine Population Dynamics.* California Sea Grant College System, University of California, La Jolla, CA (USA), 323 pp.

Bakun, A., 2001: 'School-mix feedback': a different way to think about low frequency variability in large mobile fish populations. *Progress in Oceanography,* **49(1–4),** 485–511.

Bakun, A. and K. Broad, 2003: Environmental 'loopholes' and fish population dynamics: comparative pattern recognition with focus on El Niño effects in the Pacific. *Fisheries Oceanography,* 12:458–473.

Barrett, P., 2003: Palaeoclimatology—Cooling a continent. *Nature,* **421(6920),** 221–223.

Beaugrand, G. and P.C. Reid, 2003: Long-term changes in phytoplankton, zooplankton and salmon related to climate. *Global Change Biology,* **9,** 801–817.

Beaugrand, G., K.M. Brander, J.A. Lindley, S. Souissi, and P.C. Reid, 2003: Plankton effect on cod recruitment in the North Sea. *Nature,* **426(6967),** 661–664.

Bennett, E., A. Neiland, E. Anang, P. Bannerman, A.A. Rahman, et al. 2001: *Towards a better understanding of conflict management in tropical fisheries: evidence from Ghana, Bangladesh and the Caribbean.* CEMARE Research Paper 159, Centre for the Economics and Management of Aquatic Resources (CEM-

ARE), Department of Economics, University of Portsmouth, Portsmouth (UK), 22 pp.

Birkeland, C. and A. Friedlander, 2002: *The importance of refuges for reef fish replenishment in Hawai'i.* Hawai'i Audubon Society, Honolulu, HI (USA), 19 pp.

Bjørndal, T. 1988: The Optimal Management of North Sea Herring, *Journal of Environmental Economics and Management,* 15, 9–29.

Bonnell, M.L. and R.K. Selander, 1974: Elephant Seals—Genetic-Variation and near Extinction. *Science,* **184(4139),** 908–909.

Bopp, L., P. Monfray, O. Aumont, J.L. Dufresne, H. Le Treut, G. Madec, L. Terray, and J.C. Orr, 2001: Potential impact of climate change on marine export production. *Global Biogeochemical Cycles,* **15(1),** 81–99.

Boyd, P.W., A.J. Watson, C.S. Law, E.R. Abraham, T. Trull, et al. 2000: A mesoscale phytoplankton bloom in the polar Southern Ocean stimulated by iron fertilization. *Nature,* **407(6805),** 695–702.

Branch, G.M., J. May, B. Roberts, E. Russell, and B.M. Clark, 2002: Case studies on the socio-economic characteristics and lifestyles of subsistence and informal fishers in South Africa. *South African Journal of Marine Science-Suid-Afrikaanse Tydskrif Vir Seewetenskap,* **24,** 439–462.

Bromhead, D., J. Foster, R. Attard, J. Findlay, and J. Kalish, 2003: *A review of the impact of fish aggregating devices (FADs) on tuna fisheries: final report to Fisheries Resources Research Fund.* Bureau of Rural Sciences (BRS), Canberra (Australia), 122 pp.

Brownell, R.L.J. and A.V. Yablokov, 2002: Illegal and pirate whaling. In: *Encyclopedia of Marine Mammals,* W.F. Perrin, B. Würsig, and J.G.M. Thewissen (eds.), Academic Press, San Diego, CA (USA), 615–617.

Brubaker, E., 2000: Unnatural Disaster: How Politics Destroyed Canada's Atlantic Groundfisheries. In Political Environmentalism, Anderson, T. (ed.), Hoover Institution Press, Stanford (USA). Cited December 2004, Available at http://www.environmentprobe.org/enviroprobe/pubs/ev551.htm

Bryant, D., L. Burke, J.W. McManus, and M. Spalding, 1998: *Reefs at Risk: A map based indicator of threats to the world's coral reefs.* World Resources Institute (WRI), Washington, D.C. (USA), 56 pp.

Burke, L., Y. Kura, K. Kassem, C. Ravenga, M. Spalding, and D. McAllister, 2001: *Pilot Assessment of Global Ecosystems: Coastal Ecosystems.* World Resources Institute (WRI), Washington, D.C. (USA), 94 pp.

Caddy, J.F., 1990: Options for the regulation of Mediterranean Demersal fisheries. *Natural Resource Modelling,* **4(4),** 427–475.

Caddy, J.F., 1993: Towards a comparative evaluation of human impacts on fishery ecosystem of enclosed and semi-enclosed seas. *Review of Fisheries Science,* **1(1),** 57–95.

Caddy, J.F., 1998: *Issues in Mediterranean Fisheries Management: Geographical Units and Effort Control.* Studies and Reviews No. 70. General Fisheries Council for the Mediterranean, Food and Agriculture Organization of the United Nations (FAO), Rome (Italy), 63 pp.

Caddy, J.F., R. Refk, and T. Dochi, 1995: Productivity Estimates for the Mediterranean—Evidence of Accelerating Ecological Change. *Ocean & Coastal Management,* **26(1),** 1–18.

Carlton, J.T., J.B. Geller, M.L. Reaka-Kudla, and E.A. Norse, 1999: Historical extinctions in the sea. *Annual Review of Ecology and Systematics,* **30,** 515–538.

Castilla, J.C., 2000: Roles of experimental marine ecology in coastal management and conservation. *Journal of Experimental Marine Biology and Ecology,* **250(1–2),** 3–21.

Castillo, S. and J. Mendo, 1987: Estimation of Unregistered Peruvian Anchoveta (*Engraulis ringens*) in Official Catch Statistics, 1951–1982. In: *The Peruvian Anchoeta and Its Upwelling Ecosystem: Three Decades of Change,* D. Pauly and I. Tsukayama (eds.), ICLARM, Manilla (Philippines), 109–116.

CCAMLR, 2002: *Statistical Bulletin.* Commission for the Conservation of Antarctic Marine Living Resources (CCAMLR), Hobart (Australia), 155 pp.

Cesar, H., L. Burke, and L. Pet-Soede, 2003: *The Economics of Worldwide Coral Reef Degradation.* ICRAN, Cambridge (UK) and WWF Netherlands, Zeist (Netherlands).

Chavez, F.P., P.G. Strutton, C.E. Friederich, R.A. Feely, G.C. Feldman, et al. 1999: Biological and chemical response of the equatorial Pacific Ocean to the 1997–98 El Nino. *Science,* **286(5447),** 2126–2131.

Christensen, V. and D. Pauly, 2001: Coral reef and other tropical fisheries. In: *Encyclopedia of Ocean Sciences,* J.H. Steele, S.A. Thorpe, and K.K. Turekian (eds.). 1, Academic Press, San Diego, CA (USA), 534–538.

Christensen, V., S. Guénette, J.J. Heymans, C.J. Walters, R. Watson, D. Zeller, and D. Pauly, 2003: Hundred-year decline of North Atlantic predatory fishes. *Fish and Fisheries,* **4(1),** 1–24.

Christy, F.T., 1997: Economic Waste in Fisheries: Impediments to Change and Conditions for Improvement. In: *Proceedings from the 20th American Fisheries Society Symposium: Global Trends—Fisheries Management, 14–16 June 1994,*

Seattle, WA (USA), E.K. Pikitch, D.D. Huppert, and M.P. Sissenwine (eds.), American Fisheries Society, Bethesda, MD (USA), 28–39.

Clapham, P.J., S.B. Young, and R.L. Brownell, 1999: Baleen whales: conservation issues and the status of the most endangered populations. *Mammal Review,* **29(1),** 35–60.

Clark, C.W., 1973: Economics of Overexploitation. *Science,* **181(4100),** 630–634.

Clark, M., 2001: Are deepwater fisheries sustainable?—the example of orange roughy (*Hoplostethus atlanticus*) in New Zealand. *Fisheries Research,* **51(2–3),** 123–135.

Coates, D., 1995: Inland capture fisheries and enhancement: status, constraints and prospects for food security. Paper presented at the *International Conference on sustainable Contribution of Fisheries to Food Security.* Organized by the Government of Japan in collaboration with the Food and Agriculture Organization of the United Nations (FAO), Kyoto (Japan), 82 pp.

Cole, K., 2001: Open Ocean iron Fertilization for Scientific Study and Carbon Sequestration. Paper presented at the *First National Conference on Carbon Sequestration,* 14–17 May. NETL Publications, Washington, D.C. (USA).

Coleman, F.C., P.B. Baker, and C.C. Koenig, 2004: A review of Gulf of Mexico marine protected areas: Successes, failures, and lessons learned. *Fisheries,* **29(2),** 10–21.

COMPASS and NCEAS (National Center for Ecological Analysis and Synthesis), 2001: Scientific Consensus Statement on Marine Reserves and Marine Protected Areas. Cited December 2003. Available at http://www.nceas.ucsb.edu/Consensus

Constable, A.J., W.K. de la Mare, D.J. Agnew, I. Everson, and D. Miller, 2000: Managing fisheries to conserve the Antarctic marine ecosystem: practical implementation of the Convention on the Conservation of Antarctic Marine Living Resources (CCAMLR). *Ices Journal of Marine Science,* **57(3),** 778–791.

Csirke, J. and G.D. Sharp, (eds), 1984: *Reports of the Expert Consultation to Examine the Changes in Abundance and Species Composition of Neritic Fish Resources,* San Jose, Costa Rica, 18–29 April 1983. FAO Fish. Rep. Ser. No.291 (1). FAO, Rome (Italy), 102 pp.

Delgado, C.L., M. Rosegrant, H. Steinfeld, S. Ehui, and C. Courbois, 1999: *Livestock to 2020: The Next Food Revolution.* Food, Agriculture, and the Environment Discussion Paper 28, International Food Policy Research Institute, Washington, D.C. (USA); Fisheries and Agriculture Organization of the UNited Nations (FAO), Rome (Italy); International Livestock Research Institute, Nairobi (Kenya), 83 pp.

Delgado, C.L., N. Wada, M.W. Rosegrant, S. Meijer, and M. Ahmed, 2003: *Fish to 2020: Supply and Demand in Changing Global Markets.* International Food Policy Research Institute (IFPRI), Washington, D.C. (USA) and WorldFish Center, Penang (Malaysia), 226 pp.

Denney, N.H., S. Jennings, and J.D. Reynolds, 2002: Life-history correlates of maximum population growth rates in marine fishes. *Proceedings of the Royal Society of London Series B-Biological Sciences,* **269(1506),** 2229–2237.

DFO (Fisheries and Oceans Canada), 2004: 2000 Survey of Recreational Fishing in Canada. [online] Cited November 2004. Available at http://www.dfo-mpo.gc.ca/communic/statistics/recreational/canada/2000/index_e.htm.

Duhamel, G. and J. Claudet, 2002: *Preliminary analysis on the Kerguelen shelf icefish* Champsocephalus gunnari *stock from 1996/97 to 2000/01: no evidence in the recovery.* WG-FSA-02/65, Commission for the Conservation of Antarctic Marine Living Resources (CCAMLR), Hobart (Australia).

Dulvey, N.K., Y. Sadovy, and J.D. Reynolds, 2003: Extinction vulnerability in marine populations. *Fish and Fisheries,* **4(1),** 25–64.

EC (European Commission), 2002: *Towards a strategy to protect and conserve the marine environment.* Communication from the Commission to the Council and the European Parliament, Brussels (Belgium), 35 pp.

Edgar, G.J. and N.S. Barrett, 1999: Effects of the declaration of marine reserves on Tasmanian reef fishes, invertebrates and plants. *Journal of Experimental Marine Biology and Ecology,* **242(1),** 107–144.

Ehrhardt, N.M. and V. Deleveaux 2001: Report on the 199–2001 Nassau grouper stock assessments in the Bahamas. Government of the Bahamas, Nassau.

Eiamsa-Ard, M. and S. Amornchairojkul, 1997: The marine fisheries of Thailand, with emphasis on the Gulf of Thailand trawl fishery. In: *Status and Management of tropical coastal fisheries in Asia,* G. Silvestre and D. Pauly (eds.), ICLARM Conference Proceedings 53, 85–95.

Erastus, A.N., 2002: *The Development of the Namibianisation Policy in the Hake Subsector, 1994–1999.* NEPRU Working Paper No. 82, The Namibian Economic Policy Research Unit (NEPRU), Windhoek, Namibia.

Erdmann, M.V. and L. Pet-Soede, 1996: How fresh is too fresh? The live reef food fish trade in Eastern Indonesia. *NAGA, The ICLARM Quarterly,* **19(1),** 4–8.

FAO (Food and Agriculture Organization of the United Nations), 1981: *Atlas of the living resources of the seas.* Fisheries and Agriculture Organization of the United Nations, Rome (Italy), 23 pp and 73 maps.

FAO, 1993: *Marine Fisheries and the Law of the Sea: a decade of change. A special chapter (revised) of the State of Food and Agriculture.* Fisheries Circular No. 853, Food and Agriculture Organization of the United Nations, Rome (Italy), 66 pp.

FAO, 2001a: *Report of the Second Meeting of FAO and Non-FAO Regional Fishery Bodies or Arrangements, 20–21 February.* FAO Fisheries Report No. 645, Food and Agriculture Organization of the United Nations, Rome (Italy).

FAO, 2001b: *International Plan of Action to Prevent, Deter and Eliminate Illegal, Unreported and Unregulated Fishing.* Food and Agriculture Organization of the United Nations, Rome (Italy).

FAO, 2002: *The State of World Fisheries and Aquaculture 2002.* Food and Agriculture Organization of the United Nations, Rome (Italy).

FAO, 2003: FAOSTAT2 Statistics Database. [online] Food and Agriculture Organization of the United Nations. Cited November 2004.

FAO, n.d.: Number of fishers doubled since 1970. [online] Food and Agricultural Organization of the United Nations. Cited 30 November 2003. Available at http://www.fao.org/fi/highligh/fisher/c929.asp.

Ferreira, B.P. and M. Maida, 2001: Fishing and the Future of Brazil's Northeastern Reefs. *InterCoast,* **38,** 22–23.

Fiorentini, L., J.F. Caddy, and J.I. de Leiva, 1997: *Long and short-term trends of Mediterranean fishery resources.* Studies and Reviews. General Fisheries Council for the Mediterranean no. 69, Food and Agriculture Organization of the United Nations (FAO), Rome (Italy), 72 pp.

Forrest, R., 2002: Estimating the sports catch in northern BC. In: *Restoring the past to salvage the future: report on a community participation workshop in Prince Rupert, BC,* T.J. Pitcher, M.D. Power, and L. Wood (eds.), Fisheries Centre, University of British Columbia (UBC), Vancouver (Canada), Fisheries Centre Research Reports 10(7)38–40.

Francis, R.I.C.C., M.R. Clark, R.P. Coburn, K.D. Field, and P.J. Grimes, 1995: *Assessment of the ORH 3B orange roughy fishery for the 1994± 1995 fishing year.* NZ Fish. Assess. Res. Doc. 95/4, National Institute of Water & Atmospheric Research (NIWA), Wellington (New Zealand), 43 pp.

Francis, R.C. and N.J. Mantua, 2003: Climatic influences on salmon populations in the NE Pacific. In Assessing Extinction Risk for West Coast Salmon. NMFS-NWFSC-56. Seattle, USA, 37–76pp.

Freiwald, A., J.H. Fosså, A. Grehan, T. Koslow, and J.M. Roberts, 2004: *Cold-water coral reefs. Out of sight—no longer out of mind.* UNEP-WCMC, Cambridge (UK), 86 pp.

FSBI (Fisheries Society of the British Isles), 2004: *Effects of fishing on biodiversity in the North Sea.* Briefing Paper 3, Cambridge (UK), 13 pp.

Gandini, P., P.D. Boersma, E. Frere, M. Gandini, T. Holik, and V. Lichtschein, 1994: Magellanic Penguins (*Spheniscus-Magellanicus*) Affected by Chronic Petroleum Pollution Along Coast of Chubut, Argentina. *Auk,* **111(1),** 20–27.

Garcia, S.M., 1992: Ocean fisheries management. The FAO programme. In: *Ocean management in global change,* P. Fabbri (ed.), Elsevier Applied Science, 381–418.

Garcia, S.M. and C. Newton, 1997: Current Situation, trends and prospects in world capture fisheries. In: *Proceedings of the 20th American Fisheries Society Symposium: Global Trends -Fisheries Management, 14–16 June 1994, Seattle, WA (USA),* E.K. Pikitich, D.D. Huppert, and M. Sissenwine (eds.), American Fisheries Society, Bethesda, MD (USA), 3–27.

GEF (Global Environment Facility), 2004: *Philippines and Indonesia: Marine Aquarium Market Transformation Initiative (MAMTI).* The World Bank, Jakarta, Indonesia, 172 pp.

Gerrodette, T., 2002: Tuna-Dolphin issue. In: *Encyclopedia of marine mammals,* W.F. Perrin, B. Würsig, and J.G.M. Thewissen (eds.), Academic Press, San Diego, CA (USA), 1269–1273.

Gjøsaeter, J. and K. Kawaguchi, 1980: *A review of the world resources of mesopelagic fish.* FAO Fisheries Technical Paper 193, Fisheries and Agriculture Organization of the United Nations, Rome (Italy), 151 pp.

Glover, A.G. and C.R. Smith, 2003: The deep-sea floor ecosystem: current status and prospects of anthropogenic change by the year 2025. *Environmental Conservation,* **30(3),** 219–241.

Gray, J.S., 2002: Species richness of marine soft sediments. *Marine Ecology Progress Series,* **244,** 285–297.

Haggan, N. and P. Brown, 2002: Aboriginal fisheries issues: the west coast of Canada as a case study. In: *Production systems in fishery management,* D. Pauly and M.L.D. Palomares (eds.), Fisheries Centre, University of British Columbia (UBC), Vancouver (Canada), Fisheries Centre Research Reports 10(8) 17–21.

Haigh, S.P., K.L. Denman, and W.W. Hsieh, 2001: Simulation of the planktonic ecosystem response to pre- and post-1976 forcing in an isopycnic model of the North Pacific. *Canadian Journal of Fisheries and Aquatic Sciences,* **58(4),** 703–722.

Hall, M.A., D.L. Alverson, and K.I. Metuzals, 2000: Bt-Catch: problems and solutions. *Marine Pollution Bulletin,* 41:204–219.

Hannesson, R., 2000: A note on ITQs and optimal investment. *Journal of Environmental Economics and Management,* **40(2),** 181–188.

Hardin, G., 1976: Naturalist at Large—Fishing Commons. *Natural History,* **85(7),** 9-&.

Harriott, V.J., 2002: *Marine tourism impacts and their management on the Great Barrier Reef.* CRC Reef Research Centre Technical Report No. 46, CRC Reef Research Centre, Townsville (Australia), 41 pp.

Heal, G.M., 1998: *Valuing the Future: Economic Theory and Sustainability.* Columbia University Press, New York, NY (USA), 224 pp.

Henry, G.W. and M. Lyle, 2003: *The national recreational and indigenous fishing survey.* FRDC Project No. 99/158, Australian Government Department of Agriculture, Fisheries and Forestry, Canberra (Australia), 190 pp.

Hilborn, R., K. Stokes, J.J. Maguire, T. Smith, L.W. Botsford, et al., 2004: When can marine reserves improve fisheries management? *Ocean & Coastal Management,* **47(3–4),** 197–205.

Holmes, B., 1997: Destruction follows in trawlers' wake. *New Scientist,* **14 June.**

Holt, S., 2004: Sharing our seas with whales and dolphins. *FINS- Newsletter of ACCOBAMS,* **1(1),** 2–4.

Hutchings, J.A., 2000: Collapse and recovery of marine fishes. *Nature,* **406(6798),** 882–885.

Hutchings, J.A., 2004: Evolutionary biology—The cod that got away. *Nature,* **428(6986),** 899–900.

Hutchings, J.A. and M. Ferguson, 2000: Temporal changes in harvesting dynamics of Canadian inshore fisheries for northern Atlantic cod, *Gadus morhua. Canadian Journal of Fisheries and Aquatic Sciences,* **57(4),** 805–814.

ICES (International Council for the Exploration of the Sea), 2003: Seamounts—Hotspots of Marine Life. [online] Cited November 2004. Available at http://www.ices.dk/marineworld/seamounts.asp.

Idyll, C.P., 1978: *The Sea Against Hunger—Harvesting the Oceans to Feed a Hungry World.* Thomas Y. Crowell Company, New York, NY (USA), 222 pp.

Instituto de Mar del Peru, 2004: II Panel Internacional de expertos para la evaluacion de la merluz peruana [online] Cited November 2004. Available at http://www.imarpe.gob.pe/imarpe/panel_2_merlu_03marzo_2004.php

IPCC (Intergovernmental Panel on Climate Change), 2003: *Climate Change 2001: The Scientific Basis. Contribution of Working Group I to the Third Assessment Report.* J.T. Houghton, Y. Ding, D.J. Griggs, M. Noguer, P.J. van der Linden, X. Dai, K. Maskell, C.A. Johnson (eds.). Cambridge University Press, Cambridge (UK), 892 pp.

Jackson, J.B.C., M.X. Kirby, W.H. Berger, K.A. Bjorndal, L.W. Botsford, et al., 2001: Historical overfishing and the recent collapse of coastal ecosystems. *Science,* **293(5530),** 629–638.

Jameson, S.C., M.V. Erdmann, G.R. Gibson Jr., and K.W. Potts, 1998: *Development of Biological Criteria for Coral Reef Ecosystem Assessment.* Atoll Research Bulletin No.450, United States Environmental Protection Agency, Office of Science and Technology, Health and Ecological Criteria Division; Smithsonian Institution, Washington, D.C. (USA), 102 pp.

Japan Times, 2001: One tuna fetches 20 million yen at Tsukiji. January 6.

Jennings, S., J.K. Pinnegar, N.V.C. Polunin, and K.J. Warr, 2001: Impacts of trawling disturbance on the trophic structure of benthic invertebrate communities. *Marine Ecology-Progress Series,* **213,** 127–142.

Johannes, R.E., 1981: Words of the Lagoon: Fishing and Marine Lore in the Palau District of Micronesia. University of California Press, Berkley (USA), 245 pp.

Johannes, R.E., M.M.R. Freeman, and R.J. Hamilton, 2000: Ignore Fishers' Knowledge and Miss the Boat. *Fish and Fisheries,* **1,** 257–271.

Jones, I.S.F., 2001: The global impact of ocean nourishment. Paper presented at the *First National Conference on Carbon Sequestration.* NETL Publications, Washington, D.C. (USA).

Kaczynski, V.M. and D.L. Fluharty, 2002: European policies in West Africa: who benefits from fisheries agreements? *Marine Policy,* **26(2),** 75–93.

Kaschner, K., 2004: *Modelling and mapping of resource overlap between marine mammals and fisheries on a global scale.* Zoology, Ph.D., University of British Columbia (UBC), Vancouver (Canada).

Kaschner, K. and D. Pauly, 2004: *Competition between marine mammals and fisheries: food for thought.* Report for the Humane Society by the Fisheries Centre, UBC, Vancouver (Canada), 28 pp.

Kelleher, K., 2004: Discards on the world's marine fisheries: An update. FAO Fisheries Technical Paper No. 470, FAO, Rome (Italy), 131 pp.

Kock, K.H., 1991: The state of exploited fish stocks in the Southern Ocean—a review. *Archiv für Fischereiwissenschaft,* **41(1),** 1–66.

Kock, K.H., 1992: *Antarctic Fish and Fisheries.* Cambridge University Press, Cambridge (UK), 359 pp.

Kock, K.H., 2001: The direct influence of fishing and fishery-related activities on non-target species in the Southern Ocean with particular emphasis on longline fishing and its impact on albatrosses and petrels—a review. *Reviews in Fish Biology and Fisheries,* **11(1),** 31–56.

Koslow, J.A., G.W. Boehlert, J.D.M. Gordon, R.L. Haedrich, P. Lorance, and N. Parin, 2000: Continental slope and deep-sea fisheries: implications for a fragile ecosystem. *Ices Journal of Marine Science,* **57(3),** 548–557.

Kulka, D.W., J.S. Wroblewski, and S. Narayanan, 1995: Recent Changes in the Winter Distribution and Movements of Northern Atlantic Cod (*Gadus-Morhua Linnaeus,* 1758) on the Newfoundland-Labrador Shelf. *Ices Journal of Marine Science,* **52(6),** 889–902.

Kurien, J., 1998: Does international trade in fishery products contribute to food security? [online] Proceedings of the FAO E-mail conference on Fish Trade and Food Security 19 October—12 December 1998. Food and Agriculture Organization of the United Nations, Fish Utilization and Marketing Service. Available at http://www.tradefoodfish.org/articles.php?pageid=art&article=article01.

Kurien, J., 2004: *Responsible Fish Trade and Food Security: Toward understanding the relationship between international fish trade and food security.* Conducted jointly by the Food and Agriculture Organisation of the United Nations (FAO) and the Royal Norwegian Ministry of Foreign Affairs, Rome (Italy), 121 pp.

Labropoulou, M., A. Machias, and N. Tsimenides, 1999: Habitat selection and diet of juvenile red porgy, *Pagrus pagrus* (Linnaeus, 1758). *Fishery Bulletin,* **97(3),** 495–507.

Lavigne, D.M., 2002: Harp seal—*Pagophilus groenlandicus.* In: *Encyclopedia of Marine Mammals,* W.F. Perrin, B. Würsig, and J.G.M. Thewissen (eds.), Academic Press, San Diego, CA (USA), 560–562.

Lilly, G.R., D.G. Parsons, and D.W. Kulka, 2000: Was the increase in shrimp biomass on the northeast Newfoundland shelf a consequence of a release in predation pressure from cod? *Journal of Northwest Atlantic Fisheries Science,* **27,** 45–61.

Lleonart, J., 1999: Precautionary approach and Mediterranean fisheries. Paper presented at the *Precautionary approaches to local fisheries and species introductions in the Mediterranean. CIESM Workshop Series 7,* 23–26 september, Kerkennah (Tunisia), 15–24 pp.

Lloret, J., J. Lleonart, I. Sole, and J.M. Fromentin, 2001: Fluctuations of landings and environmental conditions in the north-western Mediterranean Sea. *Fisheries Oceanography,* **10(1),** 33–50.

Longhurst, A. and D. Pauly, 1987: *Ecology of Tropical Oceans.* Academic Press, San Diego, CA (USA), 407 pp.

Longhurst, A., S. Sathyendranath, T. Platt, and C. Caverhill, 1995: An Estimate of Global Primary Production in the Ocean from Satellite Radiometer Data. *Journal of Plankton Research,* **17(6),** 1245–1271.

Longhurst, A.R., 1998: *Ecological Geography of the Sea.* Academic Press, San Diego, CA (USA), 398 pp.

Mace, P.M., 1997: Developing and Sustaining World Fisheries Resources: The State of the Science and Management. In: *Proceedings of the 2nd World Fisheries Congress: Developing and Sustaining World Fisheries Resources—The State of Science and Management:,* D.A. Hancock, D.C. Smith, A. Grant, and J.P. Beumer (eds.), CSIRO Publishing, Collingwood (Australia), 1–20.

Mace, P.M., 2001: A new role for MSY in single-species and ecosystem approaches to fisheries stock assessment and management. *Fish and Fisheries,* **2(1),** 2–32.

Macinko, S. and D.W. Bromley, 2002: *Who Owns America's Fisheries.* Island Press Publication Services, Washington, D.C. (USA), 48 pp.

MacPherson, E., 2002: Large-scale species-richness gradients in the Atlantic Ocean. *Proceedings of the Royal Society of London Series B-Biological Sciences,* **269(1501),** 1715–1720.

Malakoff, D., 1998: Fisheries science—Papers posit grave impact of trawling. *Science,* **282(5397),** 2168–2169.

McGinn, A.P., 1999: *Worldwatch Paper No. 145: Safeguarding the Health of Oceans.* Worldwatch Institute, Washington, D.C. (USA), 87 pp.

McLean, R.F. and A. Tsyban, 2001: Coastal Zones and Marine Ecosystems. In: *Climate Change 2001: Working group II: Impacts, Adaptation and Vulnerability,* J.J. McCarthy, O.F. Canziani, N.A. Leary, D.J. Dokken, and K.S. White (eds.), Published for the Intergovernmental Panel on Climate Change, Cambridge University Press, Cambridge (UK), 345–382.

McNutt, M.K., 2002: Developing the Ocean Opportunities and Responsibilities: A new era of ocean exploration will feature intelligent vehicles that track fish populations and "drifters" that constantly monitor ocean temperature and current. *The Futurist,* Jan–Feb, 38–43.

Milazzo, M.J., 1998: *Subsidies in World Fisheries. A Reexamination.* WorldBank Technical Paper No. 406, The WorldBank, Washington, D.C. (USA), 86 pp.

Mishra, B.K., 1997: Fisheries Co-operatives in India. *Co-op Dialogue,* **7(2),** 28–31.

Morales, H.L., 1993: Deep Sea, Long Hours: The Condition of Fishworkers on Distant Water Vessels. *Samudra,* **7,** 11.

Munro, G.R. and U.R. Sumaila, 2002: The impact of subsidies upon fisheries management and sustainability: the case of the North Atlantic. *Fish and Fisheries,* **3(4),** 233–250.

Myers, R.A. and B. Worm, 2003: Rapid worldwide depletion of predatory fish communities. *Nature,* **423(6937),** 280–283.

NEFSC (Northeast Fisheries Science Center), 2002: *Assessment of 20 Northeast Groundfish Stocks through 2001. A Report of the Groundfish Assessment Review Meeting (GARM), Northeast Fisheries Science Center, Woods Hole, Massachusetts, October 8–11, 2002.* Northeast Fisheries Science Center Reference Document 2–16, Woods Hole, MA (USA), 544 pp.

Neis, B., R. Jones, and R. Ommer, 2000: Food Security, Food Self-Sufficiency, and Ethical Fisheries Management. In: *Just Fish: Ethics and Canadian Marine Fisheries,* H. Coward, R. Ommer, and T. Pitcher (eds.). Social and Economic Papers 23, Institute of Social and Economic Research, Memorial University of Newfoundland, St John's (Canada), 154–173.

NMFS (National Marine Fisheries Service), 2004: *NMFS Strategic Plan for Fisheries Research.* Technical Memorandum NMFS F/SPO-61, U.S. Department of Commerce, National Oceanic and Atmospheric Administration, Silver Spring, MD (USA), 148 pp.

NRC (National Research Council), 2001: *Marine Protected Areas: Tools for Sustaining Ocean Ecosystem.* National Academy Press, Washington, D.C. (USA), 288 pp.

OECD (Organisation for Economic Co-operation and Development), 2000: *Transition to Responsible Fisheries Economic and Policy Implications.* Paris (France), 268 pp.

Okey, T.A., 2003: Membership of the eight Regional Fishery Management Councils in the United States: are special interests over-represented? *Marine Policy,* **27(3),** 193–206.

Panayotou, T., 1982: *Management concepts for small-scale fisheries: Economic and social aspects.* FAO Fisheries Technical Paper 228, Fisheries and Agriculture Organization of the United Nations (FAO), Rome (Italy), 53 pp.

Pandolfi, J.M., R.H. Bradbury, E. Sala, T.P. Hughes, K.A. Bjorndal, et al. 2003: Global trajectories of the long-term decline of coral reef ecosystems. *Science,* **301(5635),** 955–958.

PISCO (Partnership for Interdisciplinary Studies of Coastal Oceans), 2002: The Science of Marine Reserves. Cited December 2004. Available at http://www.piscoweb.org

Pauly, D., 1997: Small-scale fisheries in the tropics: marginality, marginalization, and some implications for fisheries management. In: *Proceedings of the 20th American Fisheries Society Symposium: Global Trends -Fisheries Management, 14–16 June 1994, Seattle, WA (USA),* E.K. Pikitich, D.D. Huppert, and M. Sissenwine (eds.), American Fisheries Society, Bethesda, MD (USA), 40–49.

Pauly, D., V. Christensen, J. Dalsgaard, R. Froese, and F. Torres Jr., 1998: Fishing down marine food webs. *Science,* **279,** 860–863.

Pauly, D., J. Alder, E. Bennett, V. Christensen, P. Tyedmers, and R. Watson, 2003: The future for fisheries. *Science,* **302(5649),** 1359–1361.

Pauly, D., V. Christensen, S. Guénette, T.J. Pitcher, U.R. Sumaila, C.J. Walters, R. Watson, and D. Zeller, 2002: Towards sustainability in world fisheries. *Nature,* **418(6898),** 689–695.

Petersen, E., 2003: The catch in trading fishing access for foreign aid. *Marine Policy,* **27(3),** 219–228.

Phyne, J. and J. Mansilla, 2003: Forging linkages in the commodity chain: The case of the Chilean salmon farming industry, 1987–2001. *Sociologia Ruralis,* **43(2),** 108–+.

Pike, S.T. and A.F. Spilhaus, 1962: *Marine Resources.* National Academy of Sciences, National Research Council, Washington, D.C.

Portner, H.O., B. Berdal, R. Blust, O. Brix, A. Colosimo, et al., 2001: Climate induced temperature effects on growth performance, fecundity and recruitment in marine fish: developing a hypothesis for cause and effect relationships in Atlantic cod (*Gadus morhua*) and common eelpout (*Zoarces viviparus*). *Continental Shelf Research,* **21(18–19),** 1975–1997.

Preston, G.L., 1997: *Review of Fishery Management Issues and Regimes in the Pacific Island Region.* South Pacific Regional Environment Programme, South Pacific Commission, South Pacific Forum Agency, Noumea (New Caledonia), 72 pp.

Price, A.R.G., 2002: Simultaneous 'hotspots' and 'coldspots' of marine biodiversity and implications for global conservation. *Marine Ecology Progress Series,* **241,** 23–27.

Richardson, A. J. and D. S. Schoeman, 2004: Climate impact on plankton ecosystems in the Northeast Atlantic. Science 305:1609–1612.

Roberts, C.M. and J.P. Hawkins, 1999: Extinction risk in the sea. *Trends in Ecology & Evolution,* **14(6),** 241–246.

Roberts, C. and J. P. Hawkins. 2000: Fully-protected marine reserves: a guide. WWD Endangered Seas Campaign, WWF, Washington D. C. (USA) and University of York, York (UK), 131 pp.

Roberts, C.M., C.J. McClean, J.E.N. Veron, J.P. Hawkins, G.R. Allen, et al., 2002: Marine biodiversity hotspots and conservation priorities for tropical reefs. *Science,* **295(5558),** 1280–1284.

Rolden, R. and R. Sievert, 1993: Coasta Resources Management: A Maunual for Government Officials and Community Organizers, Department of Agriculture, Manilla (Philippines)

Rudd, M.A. and M.H. Tupper, 2002: The impact of Nassau grouper size and abundance on scuba diver site selection and MPA economics. *Coastal Management,* **30(2),** 133–151.

Sadovy, Y., 2001: The threat of fishing to highly fecund fishes. *Journal of Fish Biology,* **59,** 90–108.

Sadovy, Y. and W.L. Cheung, 2003: Near extinction of a highly fecund fish: The one that nearly got away. *Fish and Fisheries,* **4(1),** 86–99.

Sadovy, Y.J. and A.C.J. Vincent, 2002: Ecological Issues and the Trades in Live Reef Fishes. In: *Coral reef fishes: Dynamics and diversity in a complex ecosystem,* P.F. Sale (ed.), Academic Press, San Diego, CA (USA), 391–420.

Sadovy, Y., M. Kulbicki, P. Labrosse, Y. Letourneur, P. Lokani, and T.J. Donaldson, 2003a: The humphead wrasse, *Cheilinus undulatus:* synopsis of a threatened and poorly known giant coral reef. *Reviews in Fish Biology and Fisheries,* **13(3),** 327–364.

Sadovy, Y.J., T.J. Donaldson, T.R. Graham, F. McGilvray, G.J. Muldoon, M.J. Phillips, M.A. Rimmer, A. Smith, and B. Yeeting, 2003b: *While Stocks Last: The Live Reef Food Fish Trade.* Asian Development Bank (ADB), Manila (Philippines), 169 pp.

Sardjono, I., 1980: Trawlers banned in Indonesia. *ICLARM Newsletter,* **3(4),** 3.

SAUP (Sea Around Us Project), 2005: Sea Around Us database. [online] Cited 5 February. Available at http://www.searoundus.org.

Schwartzlose, R.A., J. Alheit, A. Bakun, T.R. Baumgartner, R. Cloete, et al., 1999: Worldwide large-scale fluctuations of sardine and anchovy populations. *South African Journal of Marine Science,* **21,** 289–347.

Sherman, K. and A.M. Duda, 1999: Large marine ecosystems: An emerging paradigm for fishery sustainability. *Fisheries,* **24(12),** 15–26.

Sissenwine, M.P. and P.M. Mace, 2003: Governance for responsible fisheries: an ecosystem approach. In: *Responsible fisheries in the marine ecosystem,* M. Sinclair and G. Valdimarsson (eds.), Fisheries and Agriculture Organization of the United nations (FAO), Rome (Italy) and CAB International, Wallingford (UK), 363–380.

Smith, M. and C. Taal, 2001: *EU Market Survey 2001: Fishery Products.* Centre for the Promotion of Imports from Developing Countries (CPI), Rotterdam (Netherlands), 108 pp.

South Pacific Regional Environment Program, 2003: Pacific Islands Regional Ocean Policy. [online] Cited November 2004. Available at http://www.spc.int/piocean/policy/oceanpolicy.htm.

Sprague, L.M. and J.H. Arnold, 1972: Trends in the use and prospects for the future harvest of world fisheries resources. *J. Amer. Oil. Chemist. Soc.,* 49: 345–350

St Clair, J., 1997: Fishy Business. *In These Times,* May 26, 14–16.

Stehn, R.A., K.S. Rivera, S. Fitzgerald, and K.D. Whol, 2001: Incidental catch of Seabirds by Longline Fisheries in Alaska. In: *Seabird Bycatch: Trends, Roadblocks and Solutions,* E.F. Melvin and J.K. Parrish (eds.), Annual Meeting of the Pacific Seabird Group, February 26–27, 1999, Blaine Washington. University of Alaska Sea Grant, AK-SG-01–01, Fairbanks, AK (USA), 204.

Stephens, P. A., and W. J. Sutherland, 1999: Consequences of the Allee effect for behaviour, ecology and conservation. Trends in Ecology and Evolution 14: 401–405.

Stergiou, K.I., E.D. Christou, D. Georgopoulos, A. Zenetos, and C. Souvermezoglou, 1997: The Hellenic Seas: physics, chemistry, biology and fisheries. *Oceanography Marine Biology: an Annual Review,* **35(415–538).**

Stewart, B.S., P.K. Yochem, H.R. Huber, R.L. DeLong, R.J. Jameson, et al., 1994: History and present status of the northern elephant seal population. In: *Elephant seals: Population ecology, behavior and physiology,* B.J. LeBoeuf and R.M. Laws (eds.), University of California Press, Berkeley, CA (USA), 29–48.

Sumaila, R.U., 2001: Generational Cost Benefit Analysis for Evaluating Marine Ecosystem Restoration. In: *Fisheries Impacts on North Atlantic Ecosystems: Evaluations and Policy Exploration,* T.J. Pitcher, U.R. Sumaila, and D. Pauly (eds.), Fisheries Centre Research Reports 9(5) 3–9.

Sumaila, U.R. and C. Walters, 2005: Intergenerational discounting: a new intuitive approach. *Ecological Economics,* **52,** 135–142.

Taylor, B.L., 1997: Defining "population" to meet management objectives for marine mammals. In: *Molecular Genetics of Marine Mammals.,* A.E. Dizon and a.W.F.P. S. J. Chivers (eds.). Society for Marine Mammalogy Special Publication 3, The Society for Marine Mammalogy, Lawrence, KA (USA), 49–65.

Taylor, B.L. and A.E. Dizon, 1999: First policy then science: why a management unit based solely on genetic criteria cannot work. *Molecular Ecology,* **8(12),** S11–S16.

Thiel, H. and J.A. Koslow (eds.), 2001: *Managing Risks to Biodiversity and the Environment on the High Sea, Including Tools Such as Marine Protected Areas—Scientific Requirements and Legal Aspects.* Proceedings of the Expert Workshop held at the International Academy for Nature Conservation Isle of Vilm (Germany), 27 February—4 March. German Federal Agency for Nature Conservation, Bonn (Germany), 214 pp.

Tomey, W.A., 1997: Review of Developments in the World Ornamental Fish Trade: Update, Trends and Future Prospects, in K.P.P. Namibar and T. Singh (eds.), *Sustainable Aquaculture: Proceedings of the INFOFISH-AQUATECH '96 International Conference on Aquaculture,* Kuala Lumpur, Malaysia.

Tomlinson, Z., 2003: Abrogation or regulation? How Anderson v. Evans discards the Makah's treaty whaling right in the name of conservation necessity. *Washington Law Review* 78 (4): 1101–1129

Tserpes G. and P. Peristeraki, 2002: Trends in the abundance of demersal species in the southern Aegean Sea. *Scientia Marina,* **66 (Suppl 2),** 243–252.

Tsukayama, I. and M.L.D. Palomares, 1987: Monthly catch and catch composition of Peruvian anchoveta (*Engraulis ringens*) (Northern-Central stock, 4–14° S), 1953 to 1982. In: *The Peruvian anchoveta and its upwelling ecosystem: three decades of change,* D. Pauly and I. Tsukayama (eds.), ICLARM, Manila (Philippines), ICLARM Stud. Rev. 15. 89–108.

UN (United Nations), 1995: Agreement for the Implementation of the Provisions of the United Nations Convention on the Law of the Sea of 10 December 1982 Relating to the Conservation and Management of Straddling Fish Stocks and Highly Migratory Fish Stocks. United Nations Conference on Straddling Fish Stocks and Highly Migratory Fish Stocks. [online] Cited November 2004. Available at http://www.un.org/Depts/los/fish_stocks_conference/fish_stocks_conference.htm.

UNEP (United Nations Environment Programme), 2002b: *Environmental impact of trade liberalization and trade-linked measures in the fisheries sector.* National Oceanographic and Fisheries Research Centre, Nouadhibou (Mauritania).

UNEP, 2002a: *Integrated Assessment of Trade Liberalization and Trade-Related Policies: A Country Study on the Fisheries Sector in Senegal.* United Nations Environment Programme, 81 pp.

Walters, C. and J.J. Maguire, 1996: Lessons for stock assessment from the northern cod collapse. *Reviews in Fish Biology and Fisheries,* **6(2),** 125–137.

Warwick, R.M. and K.R. Clarke, 2001: Practical measures of marine biodiversity based on relatedness of species. In: *Oceanography and Marine Biology, Vol 39.* 39, 207–231.

Warwick, R.M. and S.M. Turk, 2002: Predicting climate change, effects on marine biodiversity: comparison of recent and fossil molluscan death assemblages. *Journal of the Marine Biological Association of the United Kingdom,* **82(5),** 847–850.

Watson, R. and D. Pauly, 2001: Systematic distortions in world fisheries catch trends. *Nature,* **414(6863),** 534–536.

Watson, R., D. Pauly, V. Christensen, R. Froese, A. Longhurst, et al., 2003: Mapping natural ocean regions and LMEs. In: *Large Marine Ecosystems of the World: Trends in Exploitation, Protection, and Research,* G. Hempel and K. Sherman (eds.), Elsevier, New York, NY (USA), 375–397.

Welch, D.W., Y. Ishida, and K. Nagasawa, 1998: Thermal limits and ocean migrations of sockeye salmon (*Oncorhynchus nerka*): long-term consequences of global warming. *Canadian Journal of Fisheries and Aquatic Sciences,* **55(937–948).**

Wilde, G.R., K.L. Pope, and R.E. Strauss, 2003: Estimation of fishing tournament mortality and its sampling variance. *North American Journal of Fisheries Management,* **23(3),** 779–786.

Willis, T.J., R.B. Millar, R.C. Babcock, and N. Tolimieri, 2003: Burdens of evidence and the benefits of marine reserves: putting Descartes before des horse? *Environmental Conservation,* **30(2),** 97–103.

Wiltshire, J., 2001: Future Prospects for the Marine Minerals Industry. *Under-Water Magazine,* May/June.

Worm, B., H.K. Lotze, and R.A. Myers, 2003: Predator diversity hotspots in the blue ocean. *Proceedings of the National Academy of Sciences of the United States of America,* **100(17),** 9884–9888.

WWF (World Wide Fund for Nature), 2002: Koster/Yttre Hvaler—A Potential MPA. [online] WWF Germany. Cited November 2004. Available at http://www.ngo.grida.no/wwfneap/Publication/briefings/Hvaler.pdf.

WWF/IUCN (World Conservation Union), 2001: *The status of natural resources on the high-seas.* Gland (Switzerland).

Chapter 19

Coastal Systems

Coordinating Lead Authors: Tundi Agardy, Jacqueline Alder
Lead Authors: Paul Dayton, Sara Curran, Adrian Kitchingman, Matthew Wilson, Alessandro Catenazzi, Juan Restrepo, Charles Birkeland, Steven Blaber, Syed Saifullah, George Branch, Dee Boersma, Scott Nixon, Patrick Dugan, Nicolas Davidson, Charles Vörösmarty
Review Editors: Joseph Baker, Patricia Moreno Casasola, Ariel Lugo, Avelino Suárez Rodríguez, Lingzis Dan Ling Tang

★This appears in Appendix A at the end of this volume.

Main Messages

Coastal ecosystems—coastal lands, areas where fresh water and salt water mix, and nearshore marine areas—are among the most productive yet highly threatened systems in the world. These ecosystems produce disproportionately more services relating to human well-being than most other systems, even those covering larger total areas. At the same time, these ecosystems are experiencing some of the most rapid environmental change: approximately 35% of mangrove area has been lost or converted (in those countries for which sufficient data exist, which encompass about half of the area of mangroves) and approximately 20% of coral reefs have been destroyed globally in the last few decades, with more than a further 20% being degraded. Coastal wetland loss in some places has reached 20% annually (*high certainty*).

Coastal systems are experiencing growing population and exploitation pressures; nearly 40% of the people in the world live within 100 kilometers of the coast. Demographic trends suggest coastal populations are increasing rapidly, mostly through migration, increased fertility, and tourist visitation to these areas (*high certainty*). Population densities on the coasts are nearly three times that of inland areas. Communities and industries increasingly exploit fisheries, timber, fuelwood, construction materials, oil, natural gas, sand and strategic minerals, and genetic resources. In addition, demand on coastal areas for shipping, waste disposal, military and security uses, recreation, aquaculture, and even habitation are increasing.

Coastal communities aggregate near the types of coastal systems that provide the most ecosystem services; these coastal subtypes are also the most vulnerable. Within the coastal population, 71% live within 50 kilometers of estuaries; in tropical regions, settlements are concentrated near mangroves and coral reefs. These habitats provide protein to a large proportion of the human coastal populations in some countries; coastal capture fisheries yields are estimated to be worth a minimum of $34 billion annually. However, many of these habitats are unprotected or marginally protected; as a result, ecosystems services in many areas are at risk (*medium certainty*).

Human pressures on coastal resources are compromising many of the ecosystem services crucial to the well-being of coastal economies and peoples. Coastal fisheries have depleted stocks of finfish, crustaceans, and mollusks in all regions (*high certainty*). Illegal and destructive fisheries often cause habitat damage as well as overexploitation. Large-scale coastal fisheries deprive coastal communities of subsistence and are causing increasing conflicts, especially in Asia and Africa. Demands for coastal aquaculture have been on the rise, partly in response to declining capture fisheries, but the doubling of aquaculture production in the last 10 years has also driven habitat loss, overexploitation of fisheries for fishmeal and fish oil, and pollution. Overexploitation of other resources such as mangroves for fuelwood, sand for construction material, seaweeds for consumption, and so on also often undermine the ecological functioning of these systems.

The greatest threat to coastal systems is development-related loss of habitats and services. Many areas of the coast are degraded or altered, such that humans are facing increasing coastal erosion and flooding, declining water quality, and increasing health risks. Port development, urbanization, resort development, aquaculture, and industrialization often involve destruction of coastal forests, wetlands, coral reefs, and other habitats. Historic settlement patterns have resulted in centers of urbanization near ecologically important coastal habitats: 58% of the world's major reefs occur within 50 kilometers of major urban centers of 100,000 people or more, while 64% of all mangrove forests and 62% of all major estuaries occur near such centers. Dredging, reclamation, engineering works (beach armoring, causeways, brid-

ges, and so on) and some fishing practices also account for widespread, usually irreversible, destruction of coastal habitats (*medium certainty*).

Degradation is also a severe problem, because pressures within coastal zones are growing and because such zones are the downstream recipients of negative impacts of land use. Worldwide, human activities have increased sediment flows in rivers by about 20%, but reservoirs and water diversions prevent about 30% of sediments from reaching the oceans, resulting in a net reduction of sediment delivery to coasts of roughly 10% (*high certainty*). The global average for nitrogen loading has doubled within the last century, making coastal areas the most highly chemically altered ecosystems in the world, with resulting eutrophication that drives coral reef regime shifts and other irreversible changes to coastal ecosystems. Nearly half the people living along coasts have no access to sanitation and thus face decreasing ecosystem services and increasing risks of disease. Mining and other industries cause heavy metal and other toxic pollution. Harmful algal blooms and other pathogens, which affect the health of both humans and marine organisms, are on the rise, in part because of decreased water quality. Invasions of alien species have already altered marine and coastal ecosystems, threatening ecosystem services.

The health of coastal systems and their ability to provide highly valued services is intimately linked to that of adjacent marine, freshwater, and terrestrial systems, and vice versa. Land-based sources of pollutants are delivered by rivers, from runoff, and through atmospheric deposition, and these indirect sources account for the large majority (77%) of pollutants (*high certainty*). In some areas, especially drylands, pollution in coastal zones contaminates groundwater. Another linkage occurs between expanding desertification and pollution of coral reef ecosystems caused by airborne dust. Destruction of coastal wetlands has similarly been implicated in crop failures due to decreased coastal buffering leading to freezing in inland areas (*medium certainty*).

Sub-national sociological data suggest that people living in coastal areas experience higher well-being than those living in inland areas, but the acute vulnerability of coastal ecosystems to degradation puts coastal inhabitants at greater relative risk. The world's wealthiest populations occur primarily in coastal areas (per capita income being four times higher in coastal areas than inland), and life expectancy is thought to be higher in coastal regions, while infant mortality is thought to be lower (*medium certainty*). However, many coastal communities are politically and economically marginalized and do not derive the economic benefits from coastal areas. Wealth disparity has denied many coastal communities access to resources. Access issues have in turn led to increased conflict, such as between small-scale artisanal fishers and large-scale commercial fishing enterprises. Regime shifts and habitat loss have led to irreversible changes in many coastal ecosystems and losses in some ecosystem services. Finally, given the fact that many degraded coastal systems are near thresholds for healthy functioning (*medium certainty*), and that coastal systems are simultaneously vulnerable to major impacts from sea level rise, erosion, and storm events, coastal populations are at risk of having their relatively high levels of human well-being severely compromised.

Trade-offs occur not only within coastal ecosystems, but also between the different uses of coastal systems and inland areas. In general, the choice to exploit coastal resources results in a reduction of other services; in some cases, overexploitation leads to loss of most other services (*medium certainty*). Within the coastal system, choices that result in irreversible changes, such as conversion of coastal habitat for industrial use, urbanization, or other coastal development, often bring short-term economic benefits but exact longer-term costs, as regulating and provisioning services are permanently lost. Choices made outside coastal areas, such as the decision to divert water for agriculture and thus reduce the flow of fresh water to estuaries, are cause for particular concern because virtually none of the benefits accrue to the coastal sector. Estuaries and coral reefs are the most threatened of all coastal ecosystems, precisely because impacts are both direct (originating from activity within the ecosystem), and indirect (originating in watersheds and inland areas).

Management of coastal systems to maximize the supply of services has been inadequate, but some negative trends are slowing and degradation can be halted with policy reform and by scaling up small successes to broader-scale initiatives. Effective coastal area management requires the integration of management across many sectors that have traditionally been separated. Because coastal systems are strongly affected by activities both in and outside of coastal regions, watershed management is a necessary element of effective coastal management. Integrated coastal management, marine protected area networks that effectively protect the most ecologically important habitats, and comprehensive ocean zoning all hold great promise. Restoration of some coastal habitats such as marshlands and mangrove is being undertaken. Other success stories do exist, but such successes have generally been small-scale, and scaling up has proved difficult. Business as usual will not avert continued degradation, associated loss of services, and declining human well-being in certain portions of society, such as coastal communities in developing countries and much of the low- to middle-income populace of industrial countries (*high certainty*).

19.1 Introduction

Coastal and marine ecosystems are among the most productive, yet threatened, ecosystems in the world; included in this category are terrestrial ecosystems, areas where fresh water and salt water mix, and nearshore coastal areas and open ocean marine areas. For the purpose of this assessment, the ocean and coastal realm has been divided into two major sets of systems: "coastal systems" inshore and "marine fisheries systems."

Coastal systems are places where people live and where a spate of human activity affects the delivery of ecosystem services derived from marine habitats; marine fisheries systems are places that humans relate to and affect mainly through fisheries extraction. Continental shelf areas or large marine ecosystems span both coastal and marine systems and provide many key ecosystem services: shelves account for at least 25% of global primary productivity, 90–95% of the world's marine fish catch, 80% of global carbonate production, 50% of global denitrification, and 90% of global sedimentary mineralization (UNEP 1992).

These shelf areas contain many different types of coastal systems, including freshwater and brackish water wetlands, mangrove forests, estuaries, marshes, lagoons and salt ponds, rocky or muddy intertidal areas, beaches and dunes, coral reef systems, seagrass meadows, kelp forests, nearshore islands, semi-enclosed seas, and nearshore coastal waters of the continental shelves. Many of these coastal systems are highly productive; Table 19.1 illustrates the relative productivity of some of these coastal ecosystems compared with selected terrestrial ecosystems.

In this assessment, the inland extent of coastal ecosystems is defined as the line where land-based influences dominate up to a maximum of 100 kilometers from the coastline or 50-meter elevation (whichever is closer to the sea, as per Small and Nicholls 2003) and with the outward extent as the 50-meter depth contour. Marine ecosystems begin at the low water mark and encompass the high seas and deepwater habitats. (See Figure 19.1.)

The resulting definition of coastal systems is geographically constrained and departs from many earlier assessments. The nar-

Table 19.1. Relative Productivity Estimates for Select Coastal and Terrestrial Ecosystems (based on Odum and Barrett in press)

Ecosystem Type	Mean Net Primary Productivity (kilograms per sq. meter per year)	Mean Biomass per Unit Area (kilograms per sq. meter)
Swamp and marsh	2.0	15
Continental shelf	0.36	0.01
Coral reefs and kelp	2.5	2
Estuaries	1.5	1
Tropical rain forest	2.2	45

rower band of coastal zone is a terrestrial area dominated by ocean influences of tides and marine aerosols, and a marine area where light penetrates throughout. This narrow definition was chosen for two reasons, relating to inshore and offshore boundaries: first, it focuses on areas that truly rely on and affect coastal ecosystems and it omits areas that may be near the coast but have little connection to those ecosystems (such as areas in valleys behind coastal mountain ranges); second, the "watery" portion of the coastal zone to 50 meters depth captures shallow water ecosystem like coral reefs but avoids deeper portions of the continental shelves in which fisheries impacts are paramount above all others (which are treated extensively in Chapter 18).

The heterogeneous ecosystems embodied in these coastal systems are dynamic, and in many cases are now undergoing more rapid change than at any time in their history, despite the fact that nearshore marine areas have been transformed throughout the last few centuries (Vitousek et al. 1997). These transformations have been physical, as in the dredging of waterways, infilling of wetlands, and construction of ports, resorts, and housing developments, and they have been biological, as has occurred with declines in abundances of marine organisms such as sea turtles, marine mammals, seabirds, fish, and marine invertebrates (Jackson et al. 2001; Myers and Worm 2003). The dynamics of sediment transport and erosion deposition have been altered by land and freshwater use in watersheds; the resulting changes in hydrology have greatly altered coastal dynamics. These impacts, together with chronic degradation resulting from land-based and marine pollution, have caused significant ecological changes and an overall decline in many ecosystem services. (Known rates of change and degradation in coastal subtypes are described later in this chapter.)

Dependence on coastal zones is increasing around the world, even as costs of rehabilitation and restoration of degraded coastal ecosystems is on the rise. In part, this is because population growth overall is coupled with increased degradation of terrestrial areas (fallow agricultural lands, reduced availability of fresh water, desertification, and armed conflict all contributing to decreased suitability of inland areas for human use). Resident populations of humans in coastal areas are rising, but so are immigrant and tourist populations (Burke et al. 2001). At the same time, wealth inequities that result in part from the tourism industry decrease access to coastal regions and resources for a growing number of people (Creel 2003). Nonetheless, local communities and industries continue to exploit coastal resources of all kinds, including fisheries resources; timber, fuelwood, and construction materials; oil, natural gas, strategic minerals, sand, and other nonliving natural resources; and genetic resources. In addition, people increasingly

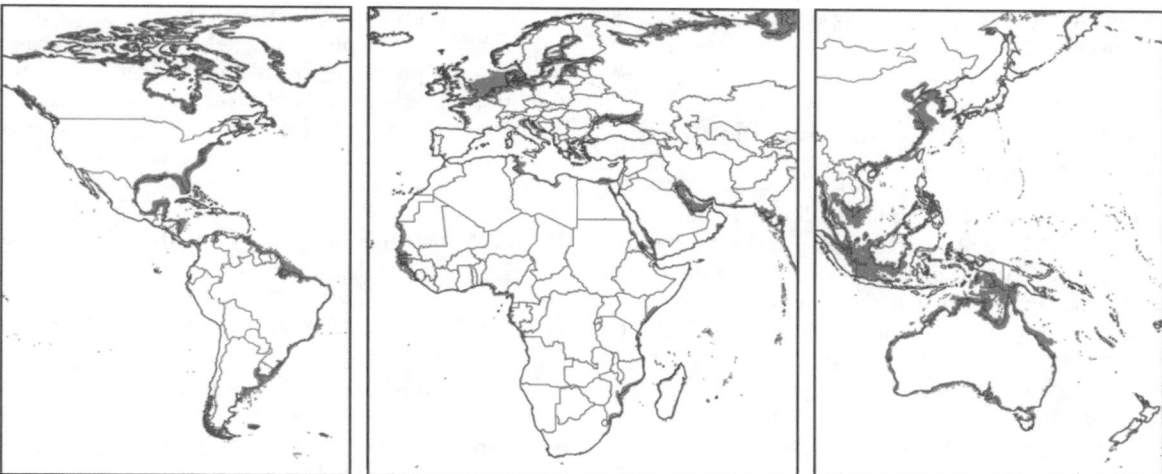

Figure 19.1. Coastal and Marine Systems Delimitation

use ocean areas for shipping, security zones, recreation, aquaculture, and even habitation. Coastal zones provide far-reaching and diverse job opportunities, and income generation and human well-being are currently higher on the coasts than inland.

Despite their value to humans, coastal systems and the services they provide are becoming increasingly vulnerable (*high certainty*). Coastal systems are experiencing growing population and exploitation pressures in most parts of the world. Though the thin strip of coastal land at the continental margins and within islands accounts for less than 5% of Earth's land area, 17% of the global population lives within the coastal systems as defined in this chapter, and 39% of global population lives within the full land area that is within 100 kilometers of a coast (CIESIN 2000). Population density in coastal areas is close to 100 people per square kilometer compared with inland densities of 38 people per square kilometer in 2000. Though many earlier estimates of coastal populations have presented higher figures (in some cases, near 70% of the world population was cited as living within the coastal zone), previous estimates used much more generous geographic definitions of the coastal area and may be misleading (Cohen 1995; Tibbetts 2002). That we have used a narrower definition and refined the coastal population numbers downwards in no way implies that coastal systems have lesser importance to humans—on the contrary, this assessment underlines the central extent to which human well-being is linked to the health and productivity of coastal systems.

Human pressures on coastal resources compromise the delivery of many ecosystem services crucial to the well-being of coastal peoples and national economies. Coastal fisheries, like many more offshore fisheries, have severely depleted stocks. (See Chapter 18.) These depletions not only cause scarcity in resource availability, they also change the viability of coastal and marine food webs, affecting the delivery of other services such as coastal protection (Dayton et al. 1995, 2002).

Biological transformations are also coupled to physical transformations of the coastal zone. Habitat alteration is pervasive in the coastal zone, and degradation of habitats both inside and outside these systems contributes to impaired functioning. Similarly, human activities far inland, such as agriculture and forestry, affect coastal ecosystems when fresh water is diverted from estuaries or when land-based pollutants enter coastal waters (nearly 80% of the pollutant load reaching the oceans comes from terrestrial sources).

These chemical transformations affect the functioning of coastal systems and their ability to deliver services. Thus, changes to ecosystems and services occur as a function of land use, freshwater use, and activities at sea, even though these land-freshwater-marine linkages are often overlooked.

Larger forces are also at play. Coastal areas are physically vulnerable: many areas are now experiencing increasing flooding, accelerated erosion, and seawater intrusion into fresh water; these changes are expected to be exacerbated by climate change in the future (IPCC 2003). Such vulnerabilities are currently acute in low-lying mid-latitude areas, but both low-latitude areas and polar coastlines are increasingly vulnerable to climate change impacts. Coral reefs and atolls, salt marshes, mangrove forests, and seagrasses will continue to be affected by future sea level rise, warming oceans, and changes in storm frequency and intensity (*high certainty*) (IPCC 2003). The ecosystems at greatest risk also support large numbers of people; thus human well-being is at risk from degradation of coastal systems.

In general, management of coastal resources and human impacts on these areas is insufficient or ineffective, leading to conflict, decreases in services, and decreased resilience of natural systems to changing environmental conditions. Inadequate fisheries management persists, often because decision-makers are unaware of marine resource management being ineffective, while coastal zone management rarely addresses problems of land-based sources of pollution and degradation (Agardy 1999; Kay and Alder in press). Funds are rarely available to support management interventions over the long term.

At the same time, the incidence of disease and emergence of new pathogens is on the rise, and in many cases coastal degradation has human health consequences as well (NRC 2000; Rose et al. 2001). Episodes of harmful algal blooms are increasing in frequency and intensity, affecting both the resource base and people living in coastal areas more directly (Burke et al. 2001; Epstein and Jenkinson 1993).

Effective measures to address declines in the condition of coastal systems remain few and far between and are often too little, too late. Restoration of coastal habitats, although practiced, is generally so expensive that it remains a possibility only on the small scale or in the most industrialized countries. Education about these issues is lacking. The assessment in this chapter aims to contribute to a better understanding of the condition of coastal ecosystems and the consequence of changes in them, and thereby to help decision-makers develop more appropriate responses for the coastal environment.

19.2 Coastal Systems and Subtypes, Marine Wildlife, and Interlinkages

Total global coastlines exceed 1.6 million kilometers and coastal ecosystems occur in 123 countries around the world (Burke et al. 2001). The MA coastal system includes almost 5% of the terrestrial surface area of Earth. Coastal systems are a complex patchwork of habitats—aquatic and terrestrial. Figure 19.2 illustrates the heterogeneity of the habitats, human communities, and interconnected systems commonly referred to as the coastal zone. The diversity of habitat types and biological communities is significant, and the linkages between habitats are extremely strong (IOC 1993).

Scaling is a very important consideration in deciding how to treat the varied set of habitats in coastal systems, since investigations at fine scales will not reveal the global situation, and investigations at coarse scales will inevitably exclude important detail (O'Neill 1988; Woodmansee 1988). Thus, for the purposes of this discussion, the coastal system is divided into eight subtypes, relying in part on former classification systems (e.g., Allee et al. 2000)

and in part on the model set forth in other chapters of the MA. Each subtype is described separately, including discussions of the services each provides, and is then assessed in terms of current condition and trends in the short-term future. In subsequent sections in which we discuss drivers of change, trade-offs, management interventions, and implications for human well-being, the coastal system is treated as a single unit.

19.2.1 Coastal Subtypes: Condition and Trends, Services and Value

19.2.1.1 Estuaries, Marshes, Salt Ponds, and Lagoons

Estuaries—areas where the fresh water of rivers meets the salt water of oceans—are highly productive, dynamic, ecologically critical to other marine systems, and valuable to people. Worldwide, some 1,200 major estuaries have been identified and mapped, yielding a total digitized area of approximately 500,000 square kilometers. (See Figure 19.3.)

There are various definitions of an estuary. One commonly accepted one is "a partially enclosed coastal body of water which is either permanently or periodically open to the sea and within which there is a measurable variation of salinity due to the mixture of sea water with freshwater derived from land drainage" (Hobbie 2000). Other definitions accommodate the fact that the

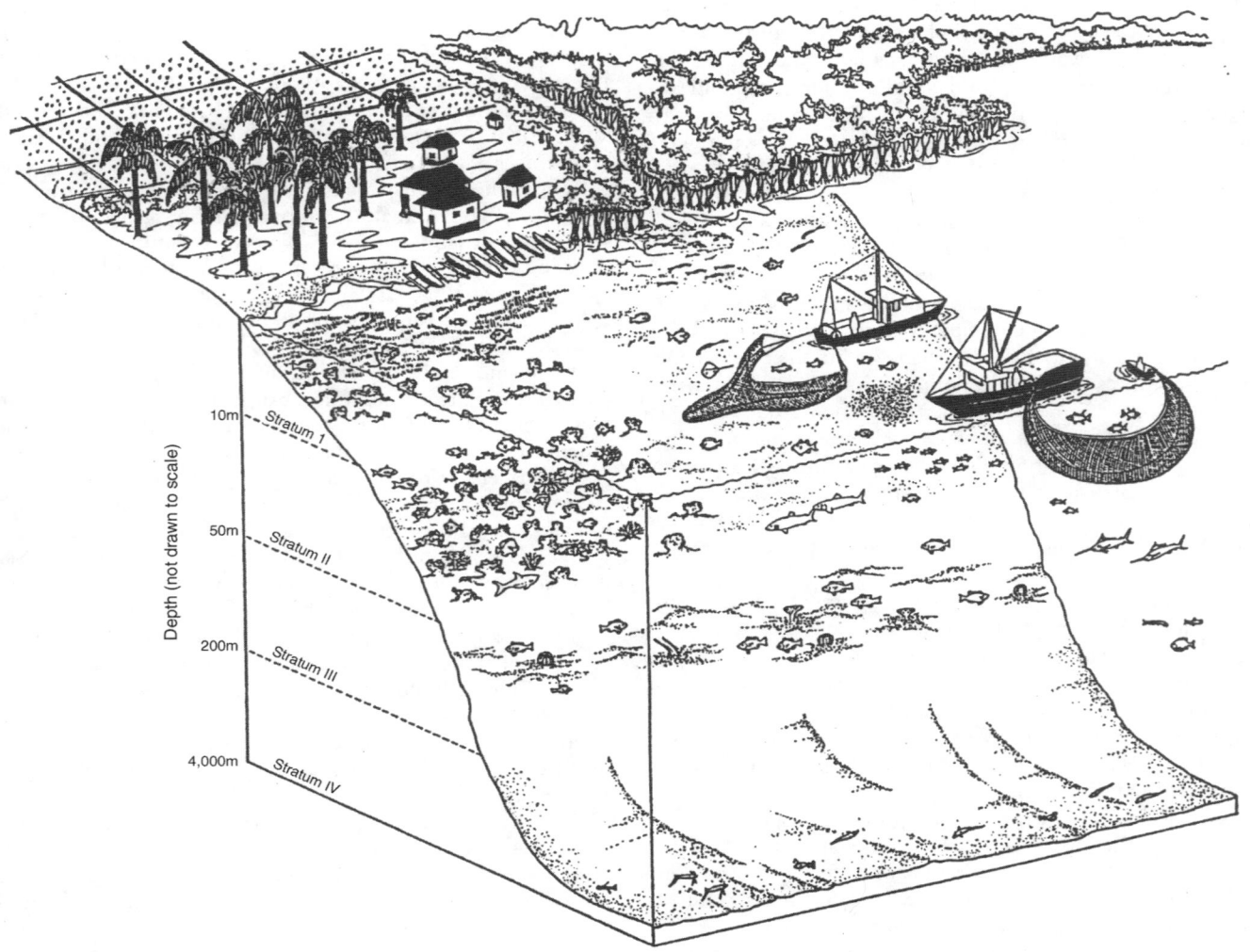

Figure 19.2. Schematic of Coastal System (Pauly et al. 1998)

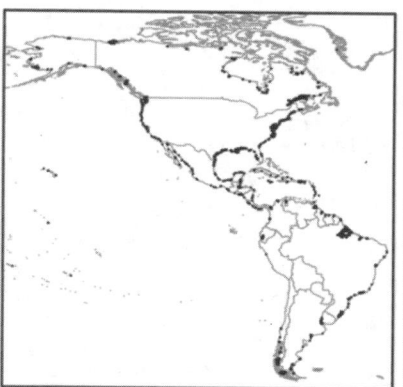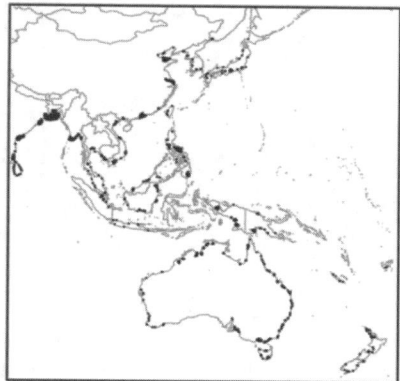

Figure 19.3. Distribution of World's Major Estuaries (UNEP-WCMC 2003b)

range of estuarine organisms is often larger than suggested by a "biophysical" definition. Coastal marshes and lagoons are essentially extensions of true estuaries and are included in estuarine analysis and assessment. Mangroves are also often found in estuaries, but their importance to coastal communities warrants a separate detailed discussion, which is given in the next section.

Regardless of location or latitude, estuaries, marshes, and lagoons play a key role in maintaining hydrological balance, filtering water of pollutants, and providing habitat for birds, fish, mollusks, crustaceans, and other kinds of ecologically and commercially important organisms (*high certainty*) (Beck et al. 2001; Levin et al. 2001). The 1,200 largest estuaries, including lagoons and fiords, account for approximately 80% of the world's freshwater discharge (Alder 2003).

Of all coastal subtypes, estuaries and marshes support the widest range of services and may be the most important areas for ecosystems services. One of the most important processes is the mixing of nutrients from upstream as well as from tidal sources, making estuaries one of the most fertile coastal environments (Simenstad et al. 2000). There are many more estuarine-dependent species than estuarine-resident species, and estuaries provide a range of habitats to sustain diverse flora and fauna (Dayton 2003). Estuaries are particularly important as nursery areas for fisheries and other species, and form one of the strongest linkages between coastal, marine, and freshwater systems and the ecosystem services they provide (Beck et al. 2001).

Freshwater wetlands close to the coast form a salinity gradient and play a key role in maintaining freshwater flows. These areas are also under pressure for conversion to other uses, as well as for fish production. Many of these freshwater wetlands have been lost, and those that remain are under threat from coastal development, with pollution exacerbating threats. The European Union Habitats Directive has declared the conservation of coastal freshwater wetlands a priority (Ledoux et al. 2003).

An array of anthropogenic impacts has degraded, altered, or eliminated these ecosystems in many areas. The main threats include the loss or destruction of large areas of an estuary's watershed; eutrophication; effects of non-nutrient pollutants such as pesticides, herbicides, and bacteria; overfishing; invasions of exotic species; and, most important, habitat conversion within estuaries themselves. There has been a substantial loss of estuaries and associated wetlands globally (Levin et al. 2001). In California, for example, less than 10% of natural coastal wetlands remain, while in the United States more generally, over half of original estuarine and wetland areas have been substantially altered (Dayton 2003). In Australia, 50% of estuaries remain undamaged, although these

are away from current population centers (Dayton 2003). Of the world's major estuaries, 62% occur within 25 kilometers of urban centers having 100,000 or more people.

Estuaries, especially those in proximity to urban centers, are often subjected directly and indirectly to trade-offs between development and conservation. Alterations such as infilling, dredging, channeling, installation of harbor works including seawalls and groins affect estuaries directly. Altering soft bottom habitat to hard bottom in the process often affects estuaries indirectly by creating conditions for new assemblages of species, and facilitating range expansions of invasive species (Ruiz and Crooks 2001). The resulting ecosystems may have losses in some ecosystem services and biodiversity. In New Zealand, invasive species have displaced commercially important mussel beds, causing significant economic losses for many mussel farmers (NOAA News Online 2003).

Figure 19.4 shows the interplay among urbanization, port development, and estuary loss worldwide. (See also Box 19.1.) Changes to freshwater flows through river impoundment and diversion are indirect trade-offs—worldwide, human activities have increased sediment flows in rivers by about 20% but reservoirs and water diversions prevent about 30% of sediments from reaching the oceans, resulting in a net reduction of sediment delivery to coasts of roughly 10% (*high certainty*) (Syvitski et al. 2005; Vörösmarty et al. 2003). Delivery of ecologically important nutrients is also impeded by freshwater diversion in watersheds, affecting not only coastal ecology but also marine fisheries yields. In the Nile Delta region of the Mediterranean, fish yields dropped significantly following the construction of the Aswan Dam (Nixon 2003). Although biomass levels rebounded from increasing nutrient input through human sewage, species composition was altered, and fish caught from the polluted waters of the Nile estuary continue to have human health impacts.

Poor management of watersheds often leads to degradation of estuaries. Agricultural and grazing practices that destroy natural riparian habitats have resulted in floods and burial of the natural estuarine habitats under silt and enriched sediment (Teal and Teal 1969). Urbanization of watersheds interrupts natural flows of both fresh water and nutrients, and it increases pollution. Agricultural inputs often result in excessive nutrient loading, which in turn causes large coastal areas to become eutrophied, hypoxic, or even anoxic (Boesch et al. 2001; D'Avanzo et al. 1996). An extreme example is the massive dead zone (up to 15,000 square kilometers) in the Gulf of Mexico (Turner and Rabalais 1994). Eutrophication is pervasive close to most of the world's large estuaries and all centers of human population, and the resulting ecosystem

Figure 19.4. Distribution of Major Ports and Estuaries (UNEP-WCMC 2003b; GDAIS 2004)

BOX 19.1
Case Study of the Paracas National Reserve

The Paracas National Reserve (335,000 hectares) is located along the Peruvian Pacific Coast, 250 kilometers south of the capital city, Lima. The reserve represents the best example of Pacific sub-tropical coastal desert on the South American continent (Rodriguez 1996). It includes relics of the coastal desert plant communities (*lomas*). Paracas is one of the most biologically productive marine areas in the world, serving as a home for nearly 300 fish species, over 200 migratory bird species (60 of which migrate between Peru and the United States), and marine mammals and reptiles. The reserve also provides food for human populations in local communities and numerous coastal cities, providing about 60% of the seafood consumed by the people of Lima, which is home to 8 million people.

Historically, the arid coasts near Paracas gave rise to numerous pre-Colombian cultures, including the Paracas culture, and their villages built up "a life of unexpected richness in the arid dunes" (Stone-Miller 1995). The allochthonous subsidies from the sea may explain the apparent contrast between the aridity of the habitat (Paracas is a Quechuan word meaning "sand falling like rain") and the richness of the Paracas culture. Today the industrial effluents from fish meal and fish oil factories reaching the Paracas Bay cause massive deaths of fish and marine invertebrates. Overfishing and overcollecting of invertebrates has reduced the food source of numerous seabirds and marine mammals, whose populations have been declining continuously since the middle of the last century. Currently, a fractionation plant to process natural gas is being built in the buffer zone of the protected area within the Paracas Bay, adding another source of environmental risk to an already vulnerable and degraded marine ecosystem.

The Paracas National Reserve is an important source of income for local fishers. Overfishing and overcollecting might have serious social and economic consequences in the town bordering the reserve, where most of the economy is centered on sea products. An economic valuation of Independencia Bay (Cuadros Dulanto 2001), a 25-by-9 kilometer bay in the southern part of the reserve, calculated its direct use value as $17.42 million. Fish and seafood accounted for 98% of this, whereas guano accounted for 1.4% and algae 0.4%. The value of indirect use, calculated through a model accounting for carbon sequestration by phytoplankton, was of $181,124 per year. Potential, existence, and biodiversity values were estimated to be $9.5 million, $2.7 million, and $29.8 million, respectively.

changes are difficult (though perhaps not impossible) to reverse once algae take over benthic habitats or cause shifts in trophic structure.

Estuarine systems are among the most invaded ecosystems in the world, with exotic introduced species causing major ecological changes (Carlton 1989 and 1996). Often introduced organisms change the structure of coastal habitat by physically displacing native vegetation (Grosholz 2002; Harris and Tyrrell 2001; Murray et al. 2004). For example, San Francisco Bay in California has over 210 invasive species, with one new species established every 14 weeks between 1961 and 1995 (Cohen and Carlton 1995, 1998). Most of these bioinvaders were bought in by ballast water of large ships or occur as a result of fishing activities (Carlton 2001). The ecological consequences of the invasions include habitat loss and alteration, altered water flow and food webs, the creation of novel and unnatural habitats subsequently colonized by other exotic species, abnormally effective filtration of the water column, hybridization with native species, highly destructive predators, and introductions of pathogens and disease (Bax et al. 2003; Ruiz et al. 1997).

Salt ponds and salinas are formed when evaporation causes constrained marine waters to become hypersaline. Some are naturally formed and others are artificial, such as salt pans and shrimp ponds. In effect, these subtypes are the biophysical opposites of estuaries, yet these coastal features provide key feeding areas for coastal birds and have their own unique biological communities. In the Red Sea region, these salt flats contribute nitrogen to adjacent mangroves (Potts 1980; Saifullah 1997b). Many of these features are seasonal or ephemeral and provide certain services only during certain times of year. Salt ponds and salt flats are often converted for other uses.

Salt marshes and coastal peat swamps (see Chapter 20) have also undergone massive change and destruction, whether they are within estuarine systems or along the coast. Salt marsh subsidence has occurred in part due to restricted sediment delivery from watersheds. Peat swamps in Southeast Asia have declined from 46–100% in countries monitoring changes (MacKinnon 1997). Coastal birds using estuaries and salt marshes both are indicators of ecosystem condition and provide many of the aesthetic ecological services of coastal systems (Benoit and Askins 2002); shorebird diversity and abundance has declined dramatically in the last few decades (International Wader Study Group 2003). Changes in relative sea level have affected and continue to affect salt marsh productivity and functioning, especially the ability of marshes to accumulate and retain sediments (Adam 2002). Relative sea level is a function of absolute sea level, changes in land level due to plate tectonics, and sediment delivery levels. Since sea level is rising due to climate change and land subsidence, and since freshwater diversion impedes delivery of sediments to estuarine systems (Vörösmarty and Meyback 1999), salt marshes will continue to be degraded and lost (Cahoon et al. 1999). The greatest threat may be to salt marshes in the tropics, which are relatively poorly studied (Adam 2002).

In many parts of the world, freshwater wetlands occur inland along the gradient of coastal ecosystems that begins offshore and moves inland through estuaries and salt marshes. Such coastal freshwater wetlands include herbaceous wetlands (marshes) and arboreal wetlands (swamps). Freshwater wetlands are discussed in detail elsewhere in this volume (Chapters 7 and 20), but it should be noted that the provision of ecosystem services by coastal systems can be highly dependent on the condition of these freshwater wetlands, and many have been and continue to be degraded by coastal development, changes to hydrology, and pollution.

19.2.1.2 Mangroves

Mangroves are trees and shrubs found in intertidal zones and estuarine margins that have adapted to living in saline water, either continually or during high tides (Duke 1992). Mangrove forests are found in both tropical and sub-tropical areas (see Figure 19.5 in Appendix A), and global mangrove forest cover currently is estimated as between 16 million and 18 million hectares (Valiela et al. 2001; Spalding et al. 1997). The majority of mangroves are found in Asia.

Mangroves grow under a wide amplitude of salinities, from almost fresh water to 2.5 times seawater strength; they may be classified into three major zones (Ewel et al. 1998) based on dominant physical processes and geomorphological characters: tide-dominated fringing mangroves, river-dominated riverine mangroves, and interior basin mangroves. The importance and quality of the various goods and services provided by mangroves varies among these zones (Ewel et al. 1998). Fringe forests provide protection from typhoons, flooding, and soil erosion; organic matter export; animal habitat; and a nursery function. Riverine mangroves also provide protection from flooding and erosion, as well as sediment trapping, a nursery function, animal habitat, and the harvest of plant products (due to highest productivity). Basin forests provide a nutrient sink, improve water quality, and allow the harvest of plant products (due to accessibility).

These forests thus provide many ecosystem services, playing a key role in stabilizing land in the face of changing sea level by trapping sediments, cycling nutrients, processing pollutants, supporting nursery habitats for marine organisms, and providing fuelwood, timber, fisheries resources. They also buffer land from storms and provide safe havens for humans in the 118 coastal countries in which they occur (Spalding et al. 1997). Mangroves have a great capacity to absorb and adsorb heavy metals and other toxic substances in effluents (Lacerda and Abrao 1984). They can also exhibit high species diversity. Those in Southeast Asia, South Asia, and Africa are particularly species-rich, and those in association with coral reefs provide food and temporary living space to a large number of reef species. In some places mangroves provide not only nursery areas for reef organisms but also a necessary nursery ground linking seagrass beds with associated coral reefs (Mumby et al. 2004). Removal of mangrove can thus interrupt these linkages and cause biodiversity loss and lower productivity in reef and seagrass biomes.

Mangroves are highly valued by coastal communities, which use them for shelter, securing food and fuelwood, and even as sites for agricultural production, especially rice production. Due to their function as nurseries for many species, fisheries in waters adjacent to mangroves tend to have high yields; annual net values of $600 per hectare per year for this fishery benefit have been suggested (Giesen et al. 1991). In addition, an annual net benefit of $15 per hectare was calculated for medicinal plants coming from mangrove forests, and up to $61 per hectare for medicinal values (Bann 1997). Similarly large economic benefits are calculated for shoreline stabilization and erosion control functions of mangroves (Ruitenbeek 1992).

Many mangrove areas have become degraded worldwide, and habitat conversion of mangrove is widespread (Farnsworth and Ellison 1997). Much of the coastal population of the tropics and sub-tropics resides near mangroves; 64% of all the world's mangroves are currently within 25 kilometers of major urban centers having 100,000 people or more. Mangroves have been converted to allow for aquaculture and for agriculture, including grazing and stall feeding of cattle and camels (which in Pakistan, for instance, is the second most serious threat to mangrove ecosystems (Saifullah 1997a)). Mangrove forests are also affected by removal of trees for fuelwood and construction material, removal of invertebrates for use as bait, changes to hydrology in both catchment basins or nearshore coastal areas, excessive pollution, and rising relative sea levels (Semesi 1992, 1998).

Along with conversion to agriculture, salt pans, and urban and industrial development, an important cause of loss is the aquaculture industry, typically through conversion of mangrove wetlands to shrimp or prawn farms. This destruction is particularly wasteful and costly in the long term, since shrimp ponds created out of mangrove forest lose their productivity over time and tend to become fallow in 2–10 years (Stevenson 1997). Historically, abandoned shrimp ponds are rarely restored, but new policy directives and a shift in the aquaculture industry is helping to make aquaculture less destructive and more prone to supporting restoration or regrowth in some parts of the world.

Estimates of the loss of mangroves from countries with available multiyear data (representing 54% of total mangrove area at present) show that 35% of mangrove forests have disappeared in the last two decades—at the rate of 2.1%, or 2,834 square kilometers, per year (Valiela et al. 2001). In some countries, more than 80% of original mangrove cover has been lost due to deforestation (Spalding et al. 1997). In summary, the current extent of mangroves has been dramatically reduced from the original extent in nearly every country in which data on mangrove distribution have been compiled (Burke et al. 2001). The leading human activities that contribute to mangrove loss are 52% aquaculture (38% shrimp plus 14% fish), 26% forest use, and 11% freshwater diversion (Valiela et al. 2001). Restoration has been successfully attempted in some places, but this has not kept pace with wholesale destruction in most areas.

19.2.1.3 Intertidal Habitats, Deltas, Beaches, and Dunes

Rocky intertidal, nearshore mudflats, deltas, beaches, and dunes also provide ecosystem services such as food, shoreline stabilization, maintenance of biodiversity (especially for migratory birds), and recreation.

Rocky intertidal habitats display interesting patterns of biological regulation and have been the location of much of the research that provided the foundation for our knowledge of predator-prey interactions, keystone species, and other biological regulation (Foster et al. 1988; Paine 2002; Sebens 1986). (See Chapter 11 for more on biological regulation in coastal systems.) The rocky intertidal habitats of temperate areas are highly productive and, in some cases, an important source of food for humans (Murray et al. 1999b). Food and bait collection (including mollusks and sea-weeds) and human trampling have substantially depleted many of the organisms in these habitats. In the United States, the rocky intertidal zone has undergone major transformation in the last few decades: the California mussel *Mytelus californianus* has become very rare, the seastar *Pisaster* sp. is now almost never seen, and the once abundant black abalone (*Haliotis cracherodii*) can no longer be found in southern California (Dayton 2003). In addition, dozens of formally abundant nudibranch species are now rare (Tegner and Dayton 2000). Similar trends have been observed elsewhere in the world (Dayton 2003). Along the Yellow Sea coast, China has lost around 37% of habitat in intertidal areas since 1950, and South Korea has lost ~43% since 1918 (Birdlife International 2004a).

Intertidal mudflats and other soft-bottom coastal habitats play pivotal roles in ocean ecology, even though research and public interest have not historically focused on these habitats. Soft-bottom coastal habitats are highly productive and can be extraordinarily diverse (Levin et al. 2001), with a species diversity that may rival that of tropical forests (Gray 1997). Mudflats are critical habitat for migrating shorebirds and many marine organisms, including commercially important species like the horseshoe crab (*Limulus polyphemus)* and a variety of clam species. Unfortunately, mudflats are commonly destroyed during port development or maintenance dredging (Rogers et al. 1998), and coastal muds in many areas are highly contaminated by heavy metals, PCBs, and other persistent organic pollutants, leading to mortality and morbidity in marine species and to human health impacts.

Coastal deltas are extremely important microcosms where many dynamic processes and human activity converge. The IPCC has identified "deltas, estuaries, and small islands" as the coastal systems most vulnerable to climate change and sea level rise (IPCC 2003). Deltas are high population and human land use areas and are dynamic and highly vulnerable. They are also experiencing significant global changes as a class in themselves, aside from their overlap with the categories of mangrove, marshes, and wetlands (discussions of which do not capture all the dynamic influences in deltas).

Beaches and sandy shores also provide ecological services and are being altered worldwide. Sandy shores have undergone massive alteration due to coastal development, pollution, erosion, storms, alteration to freshwater hydrology, sand mining, groundwater use, and harvesting of organisms (Brown and McLachlan 2002). Disruptions to the sand balance in many locations is causing the total disappearance of beaches and with it the loss of ecological services, such as the provision of food to migratory birds, provision of nesting habitat, delivery of land-based nutrients to the nearshore coastal system, and provision of both food and recreational space to humans. Removal of beach wrack (seaweeds cast up on beaches) near urban centers and tourism resorts also alters habitat and services.

Dune systems occur inland of the intertidal zone but are commonly found in conjunction with beaches and sandy shores. These habitats are often highly dynamic and mobile, changing their form in both the short and long term. Although dune systems are not as productive exporters of nutrients as many other coastal systems, they act as sediment reserves, stabilize coastlines, provide areas for recreation, and provide breeding and feeding sites for seabirds and other coastal species. Dunes support high species diversity in certain taxonomic groups, including endangered bird, plant, and invertebrate species. Encroachment in dune areas often results in shoreline destabilization, resulting in expensive and ongoing public works projects such as the building of breakwaters or seawalls and sand renourishment. In the United

States alone, coastal erosion of dunes and beaches costs $500 million in property losses annually (The Heinz Center 2000). Not only are such projects costly, they also have cascading impacts throughout the coast and nearshore areas.

19.2.1.4 Coral Reefs and Atolls

Coral reefs exhibit high species diversity and endemism and are valued for their provisioning, regulating, and cultural services (McKinney 1998). Reef-building corals occur in tropical coastal areas with suitable light conditions and high salinity and are particularly abundant where sediment loading and freshwater input is minimal. The distribution of the world's major coral reef ecosystems is shown in Figure 19.6 (in Appendix A). Reef formations occur as barrier reefs, atolls, fringing reefs, or patch reefs, and many islands in the Pacific Ocean, Indian Ocean, and Caribbean Sea have extensive reef systems occurring in a combination of these types. Coral reefs occur mainly in relatively nutrient-poor waters of the tropics, yet because nutrient cycling is very efficient on reefs and complex predator-prey interactions maintain diversity, productivity is high. However, with a high number of trophic levels the amount of primary productivity converted to higher levels is relatively low, and reef organisms are prone to overexploitation.

Reefs provide many of the services that other coastal ecosystems do, as well as additional services: they are a major source of fisheries products for coastal residents, tourists, and export markets; they support high diversity that in turn supports a thriving and valuable dive tourism industry; they contribute to the formation of beaches; they buffer land from waves and storms and prevent beach erosion; they provide pharmaceutical compounds and opportunities for bioprospecting; they provide curios and ornamentals for the aquarium trade; and they provide coastal communities with materials for construction and so on (Ahmed et al. 2004).

The fine-tuned, complex nature of reefs makes them highly vulnerable to negative impacts from overuse and habitat degradation—when particular elements of this interconnected ecosystem are removed, negative feedbacks and cascading effects occur (Nystrom et al. 2000). Birkeland (2004) describes ecological ratcheting effects through which coral reefs are transformed from productive, diverse biological communities into depauperate ones, along with similar cascading effects caused by technological, economic, and cultural phenomena. Coral reefs are one of the few marine environments displaying disturbance-induced phase shifts: a phenomenon in which diverse reef ecosystems dominated by stony corals dramatically turn into biologically impoverished wastelands overgrown with algae (Bellwood et al. 2004).

Most tropical reefs occur in developing countries, and this is where the most intensive degradation is occurring (Burke et al. 2002). Of all the world's known tropical reef systems, 58% occur within 25 kilometers of major urban centers having populations of 100,000 or more. Coral reefs are at high risk from many kinds of human activity, including coastal construction that causes loss of habitat as well as changes in coastal processes that maintain reef life; coastal constructions that change physical processes; destructive fishing and collecting for the marine ornamental trade; overfishing for both local consumption and export (Chapter 18); inadequate sanitation and poor control of run-off leading to eutrophication; dumping of debris and toxic waste; land use practices leading to siltation; oil spills; and degradation of linked habitats such as seagrass, mangrove, and other coastal ecosystems (Wilkinson 2000, 2002). In 1999, it was estimated that approximately 27% of the world's known reefs had been badly degraded or destroyed in the last few decades (Wilkinson 2000), although the latest estimates are of 20% of reefs destroyed (Wilkinson 2004) and more than a further 20% badly degraded or under imminent risk of collapse.

Of all the world's ecosystems, coral reefs may be the most vulnerable to the effects of climate change (Hughes et al. 2003). Although the mechanisms are not clear, warming seawater triggers coral bleaching, which sometimes causes coral mortality. Corals bleach when the symbiotic zooxanthellae that live in the tissue of the coral polyps and catalyze the reactions that lead to calcium carbonate deposition are changed or expelled. Bleaching does not automatically kill corals, but successive bleaching events in close proximity, or prolonged bleaching events, often do lead to mass mortality (Pandolfi et al. 2003). However, it has been estimated that approximately 40% of the reefs that were seriously damaged in the 1998 coral bleaching events are either recovering well or have fully recovered (Wilkinson 2004).

Climate change also has other detrimental impacts on coral. For example, rising carbon dioxide levels change the pH of water, reducing calcium carbonate deposition (reef-building) by corals. Climate change also facilitates the spread of pathogens leading to the spread of coral diseases. It has been suggested that climate change will reduce the world's major coral reefs in exceedingly short time frames—one estimate suggests that all current coral reefs will disappear by 2040 due to warming sea temperatures (Hughes et al. 2003), and it is not known whether the reefs that take their place will be able to provide the same level of services to humans and the biosphere.

Coral reefs are highly degraded throughout the world, and there are likely to be no pristine reefs remaining (Hughes et al. 2003; Pandolfi et al. 2003; Gardner et al. 2003). Historical analysis of conditions suggests that reef degradation, involving the decline of large animals, then smaller animals and reef-building species, precedes the emergence of bleaching and disease (Pandolfi et al. 2003). This suggests that overfishing, combined with pollution from land-based sources, predisposes reefs to be less resilient to disease and the effects of climate change. Such pollution includes increases in turbidity resulting from sediments washing into nearshore waters or from release during dredging, which results in significantly lower light levels reach corals, disrupting photosynthesis in algal symbionts and reducing calcification rates (Yentsch 2002). The coral reefs of the Caribbean Sea and portions of Southeast Asia have suffered the greatest rates of degradation and are expected to continue to be the most threatened (Gardner et al. 2003).

19.2.1.5 Seagrass Beds or Meadows

Seagrass is a generic term for the flowering plants that usually colonize soft-bottom areas of the oceans from the tropics to the temperate zones (some seagrass can be found on hard-bottom areas but the ones occupied are usually small). In estuarine and other nearshore areas of the higher latitudes, eelgrass (*Zostera* spp.) forms dense meadows (Deegan et al. 2001). Further toward the tropics, manatee and turtle grass (*Thalassia testudinum* and *Syringodium filiforme*) cover wide areas. Along with mangroves, seagrass is thought to be a particularly important in providing nursery areas in the tropics, where it provides crucial habitat for coral reef fishes and invertebrates (Gray et al. 1996; Heck et al. 1997). Seagrass is highly productive and an important source of food for many species of coastal and marine organisms in both tropical and temperate regions (Gray et al. 1996). It also plays a notable role in trapping sediments and stabilizing shorelines.

Seagrass continues to play an important ecological role even once the blades of grass are cut and carried by the water column. Drift beds, composed of mats of seagrass floating at or near the surface, provide important food and shelter for young fishes (Kulczycki et al. 1981), and the deposit of seagrass castings and macroalgae remnants on beaches is thought to be a key pathway for nutrient provisioning to many coastal invertebrates, shorebirds, and other organisms. For instance, nearly 20% of the annual production of nearby seagrass (over 6 million kilograms dry weight of beach cast) is deposited each year on the 9.5-kilometer beach of Mombasa Marine Park in Kenya, supporting a wide variety of infauna and shorebirds (Ochieng and Erftemeijer 2003).

Tropical seagrass beds or meadows occur both in association with coral reefs and removed from them, particularly in shallow, protected coastal areas such as Florida Bay in the United States, Shark Bay and the Gulf of Carpentaria in Australia, and other geomorphologically similar locations. Seagrass is also pervasive (and ecologically important) in temperate coastal areas such as the Baltic Seas (Fonseca et al. 1992; Green and Short 2003; Isakkson et al. 1994). The distribution of these major seagrass beds is shown in Figure 19.7 (in Appendix A).

Human impacts, including dredging and anchoring in seagrass meadows, coastal development, eutrophication, hypersalinization resulting from changes to inflows, siltation, habitat conversion for the purposes of algae farming, and climate change, are all causing widespread damage to seagrasses globally (Duarte 2002). Increased nutrient inflows into shallow water coastal areas with limited flushing (prime areas for seagrass growth) can cause algal and epifaunal encrustation of seagrass blades (Duarte 1995), limiting their ability to photosynthesize and in extreme cases smothering the meadows altogether (Deegan et al. 2001; Short and Wyllie-Echeverria 1996). Major losses of seagrass habitat have been reported from the Mediterranean, Florida Bay, and Australia (Duarte 2002). Present losses are expected to accelerate, especially in Southeast Asia and the Caribbean (Burke et al. 2001; Duarte 2002), as eutrophication increases, algal grazers are overfished, and coastal development increases.

19.2.1.6 Kelp Forests

The productivity of kelp forests rivals that of the most productive land systems (Dayton 2003). These temperate ecosystems have a complex biological structure organized around large brown algae, supporting a high diversity of species and species interactions. Kelp support fisheries of a variety of invertebrate and finfish, and the kelp itself is harvested for food and additives. Kelp forests are remarkably resilient to natural disturbances such as wave impacts, storm surges, and other extreme oceanographic events (Dayton 2003).

Kelp forests and other macroalgae provide specialized nursery habitats for some species. For instance, the upper layers of kelp provide nursery habitat for young rockfish and other organisms. Kelp communities consist of several distinct canopy types supporting many herbivores. Most important among these are sea urchins, which are capable of destroying nearly all fleshy algae in most kelp systems, and the spines of the red sea urchin (*Strongylocentrotus franciscanus*) provide crucial nursery habitat for other sea urchin species (Tegner and Dayton 1977). Factors affecting the abundance of sea urchins are thus important to the integrity of kelp ecosystems (Dayton 2003).

Unfortunately, the biological communities of many kelp forests have been so destabilized by fishing that they retain only a fraction of their former diversity (Tegner and Dayton 2000). It is likely that no kelp systems exist in their natural condition (Dayton

2003), and there have been enormous system responses to human impact. Fishing impacts (see Chapter 18) can cause cascading effects, reducing diverse kelp forests to much simplified sea urchin–dominated barren grounds. Such "urchin barrens" are exactly as the name implies: devoid of many normal forms of life and dominated by urchins. Urchin barrens are or were prevalent in the northwest Atlantic (Labrador to Massachusetts), the Aleutian Islands, southern California, the Chilean coast, Japan, New Zealand, and Australia.

Removal of predators plays a key role in these regime shifts, some of which regularly oscillate between states, while others remain in the barren state for long periods of time. For example, in the Atlantic Ocean large fish such as halibut (*Hippoglossus hippoglossus*), wolfish (*Anarichus latifrons*), and cod (*Gadus spp.*), which are the key predators of sea urchins, have been largely removed from the system, causing sea urchin populations to explode (Tegner and Dayton 1977; Dayton et al. 1998). Following this, directed exploitation and disease led to a collapse of the urchin populations, but kelp forests have not fully recovered and continue to be vulnerable to waves of exotic species (Dayton 2003).

In other places, kelp communities are tied to sea otter populations. When sea otters were decimated in the Aleutian Islands through hunting, kelp forests were destroyed by booming populations of sea urchins. Following protection of sea otters, the kelp forests temporarily recovered, but the barrens returned in the 1990s when the otters began declining again (Estes et al. 1998). The health of kelp forests is thus strongly related to the health of the predator populations.

19.2.1.7 Other Benthic Communities: Rock and Shell Reefs, Mud Flats, Coastal Seamounts, and Rises

Although public interest in coastal biodiversity has tended to focus on coral reefs, many other coastal systems harbor vast amounts of species (Gray 1997; Gray et al. 1997). Within estuaries, for instance, oyster reefs are considered important nursery areas, not just for oysters but also for a wide range of fish species, other mollusks, crabs, and other fauna. Rock reefs, for example, provide rich nursery habitat for fisheries, such as those that occur in the extensive banks inshore from the upwelling areas of the northern Gulf of Guinea in West Africa (Binet and Marchal 1993), as well as in temperate areas such as in the Mediterranean Sea. Mud flats in the intertidal area and on banks are also productive habitats that exhibit surprising species diversity.

Hard-bottom habitats below the photic zone tend to be dominated by sponges, corals, bryozoans, and compound ascidians. Most of these temperate, non-reef-building corals are found in deeper waters beyond the coastal limit, although their ecosystem dynamics and the threats facing them are similar to many coastal systems. Human-induced disturbances can cause major ecological damage and compromise biodiversity, regardless of whether these communities occur more inshore or offshore. Bottom trawling and other fishing methods that rake the benthos have destroyed many of these communities already (Dayton 2003; Jennings and Kaiser 1998). These impacts on biodiversity sometimes result in permanent losses when endemic or restricted species are wiped out. (See the section on biodiversity in Chapter 18.)

About 70% of Earth's seafloor, including that located within the MA coastal system, is composed of soft sediment (Dayton 2003). Although soft-sediment habitats do not always appear as highly structured as some terrestrial or marine reef habitats, they are characterized by extremely high species diversity. There is now strong evidence of fishing effects on seafloor communities that have important ramifications for ecosystem function and re-

silience (Dayton 2003; Rogers et al. 1998). Given the magnitude of disturbance by trawling and dredging and the extension of fishing effort into more vulnerable benthic communities (Chapter 18), this type of human disturbance is one of the most significant threats to marine biodiversity (Dayton 2003). Sponge gardens in soft substrates face particular threat from bottom trawling, since the soft substrate is easily raked by heavy trawling gear.

In places, the ocean floor's soft sediment is interrupted by highly structured seamounts with highly diverse communities of organisms (Dayton 1994). These underwater mountains or volcanoes are usually found far offshore and are thought to be crucial for many pelagic fish species, not just as sites for breeding and spawning, but also as safe havens for juvenile fishes seeking refuge from open ocean predators (Johannes et al. 1999). Since the vast majority of large seamounts occur in deeper marine waters, they are discussed in detail in Chapter 18. However, smaller seamounts occur in conjunction with coral reefs and elsewhere in the coastal zone, and they contribute significantly to coastal fisheries production and biodiversity maintenance. Because their high species diversity is concentrated into a relatively small, localized area, and because of their occasionally high endemism, seamounts are extremely vulnerable to fishing impacts. (See Chapter 18.)

Other benthic habitats that might be expected to fall into this subtype are not discussed in this assessment, such as the fjords of Norway and non-kelp-dominated rocky slopes and banks. Cold water corals of the temperate deeper waters are discussed in Chapter 18. Some of these habitats provide ecosystem services important to humankind, and some are also being degraded, but these habitats are either so specialized as to make generalizations impossible, or assessment information is lacking at the global scale.

19.2.1.8 Semi-enclosed Seas

A semi-enclosed sea is legally defined as "a gulf, basin or sea surrounded by two or more States and connected to another sea or the ocean by a narrow outlet or consisting entirely or primarily of the territorial seas and exclusive economic zones of two or more coastal States" (Convention on Law of the Sea, Article 122). Although this is a geopolitical, not an ecological, definition, and despite the fact that large portions of semi-enclosed seas thus defined fall outside the MA category of "coastal," these areas are described here as another coastal subtype. (Chapter 18 mentions these systems in regard to fisheries as well.)

Notable examples of semi-enclosed seas include the Mediterranean, Black, Baltic, and Red Seas and the Gulf of Aden. Semi-enclosed seas can be intercontinental (such as the Mediterranean Sea), intracontinental (such as the Black and Baltic Seas), or marginal (such as the North and Bering Seas). Gulfs with restricted openings such as the Gulf of California in Mexico and the Gulf of Thailand could also be considered "semi-enclosed." These systems all share similar attributes: they tend to be highly productive (primarily due to exogenous inputs from lands nearby), often have high species diversity and endemism, are heavily used by the countries and communities that border them, and are often at high risk from pollution.

Perhaps more than open ocean systems, semi-enclosed seas are directly linked to human well-being. Many of the world's great civilizations sprung up along the shorelines of semi-enclosed seas, which have historically provided food, trade routes, and waste processing services to burgeoning human populations. Today most semi-enclosed seas of the world are highly valued as tourism and recreational venues, adding to their value in continuing to provide food and other services (Sheppard 2000). Yet they are becoming highly degraded due to demands placed on them and their physical configuration.

Freshwater inflows to semi-enclosed seas have been severely curtailed in most areas, robbing them of recharging waters and nutrients. A particularly acute case of this degradation has occurred in the Gulf of California, which now receives only a trickle of water through the now dry, but once very fertile, delta of the Colorado River (GIWA 2003). At the same time, water reaching these basins is often of poor water quality due to land-based sources of pollution such as agricultural and industrial waste (GESAMP 2001). Such degradation is highly prevalent in semi-enclosed seas with major river drainages, such as the Black Sea (Bakan and Büyükgüngör 2000), Baltic Sea (Falandysz et al. 2000; Kautsky and Kautsky 2000), and even large parts of the Mediterranean Sea (Cognetti et al. 2000). The limited flushing and long recharge times in semi-enclosed seas means that pollutants are not as quickly diluted as in the open sea, and eutrophication and toxics loading are often the result.

Virtually all semi-enclosed seas have undergone dramatic transformation as the consequence of coastal development, ever-increasing fishing pressures, declines in freshwater input, and pollutant loading. The pollution that enters semi-enclosed seas from drainage basins is a significant source of degradation in these physically constrained coastal areas, especially in regions with major river basins and high rainfalls (for instance, see Cognetti et al. 2000 on the Adriatic Sea and Bakan and Büyükgüngör 2000 on the Black Sea). In the Bosporus region of Turkey, sewage pollution has been implicated in the decline of many fish species. However, land-based sources of pollution can also be a problem in arid and semiarid regions, as evidenced by the extensive local degradation of coral reefs in the Red Sea caused by seepage and runoff of untreated sewage into nearshore waters (Sheppard 2000).

Negative synergies often act together to bring about cataclysmic change in ecosystem condition in relatively short amounts of time. The Black Sea, which once supplied much of Europe with fisheries products, has undergone a slow but chronic environmental degradation in the last century as industrial pollution from major rivers, including the Danube, Dniester, and Dnieper, as well as more coastally based pollution, contaminated the waters. Overfishing and wetlands destruction occurred during roughly the same period, but intensified even as the health of the sea began to falter. When an Atlantic ctenophore, *Mnemiopsis leidyi*, was introduced through ship ballast water sometime in the 1980s, the voracious predator eagerly preyed on the struggling biota, causing the loss of over two dozen major fisheries (Zaitsev and Mamaev 1997). In recent years, the anoxic layer of this basin has expanded and moved upwards, making restoration of the sea to its once-vibrant state difficult.

19.2.2 Marine Wildlife

The world's oceans and coasts are home to many hundreds of species of marine mammals, turtles, crocodiles, and seabirds—some common, others rare; some with global distributions, others with narrow coastal distributions. Those with wide-ranging distributions demonstrate the connectivity of ecosystems and the need for holistic approaches to management of coastal and marine systems. Several species are threatened, either because they have not recovered from earlier exploitation (such as the Northern right whale, *Eubalaena glacialis*) or because they continue to suffer excessive mortality, mainly through incidental catches or as bycatch of fishing (such as the vaquita, *Phocoena sinus,* a dolphin

endemic to the northern Gulf of California (D'Agrosa et al. 2000) and albatrosses (Stehn et al. 2001)).

Other human activities also threaten marine wildlife. Recent studies have found strong correlations between mass strandings of some marine mammals, such as beaked whales (family Ziphiidae), and military low frequency sonar exercises (Piantadosi and Thalmann 2004). More widespread is the threat of incidental catch in fisheries. Bycatch is currently recognized as a significant threat to conservation of small cetaceans (Dawson et al. 1998) and seabirds (Tasker et al. 2000).

19.2.2.1 Turtles and Crocodiles

None of the 23 known crocodile species have gone extinct despite local extirpations and multiple threats to their habitats as well as interactions with humans (Webb 1999). Although some species of crocodile are still threatened with extinction, others have increased in number and through appropriate management plans are being harvested sustainably (Ross 1998).

Marine turtles, along with marine mammals and seabirds, are key indicator species for problems and changes in the marine environment. The overall situation of the seven marine turtle species found worldwide is no better than that of most marine mammals. Human-related impacts—particularly habitat destruction, direct harvest of adults and eggs, international trade, bycatch, and pollution—are seriously threatening the survival of marine turtles. All seven species of turtles are listed under the Convention on International Trade in Endangered Species of Wild Fauna and Flora Appendix I, thereby restricting international trade in turtles or turtle-derived products between parties to the convention. According to the *IUCN Red List,* three of the seven species are critically endangered with extinction, three are endangered, and the status of the Australian flatback (*Chelonia depressa*) remains unknown due to insufficient information.

Although survival of marine turtles is threatened on a global scale, at the regional scale different turtle subpopulations show different growth trajectories. However, this may be a reflection of data availability. For example, information about turtle populations in Africa has been lacking until recently (Fretey 2001) and is still largely incomplete.

Green turtle (*C. mydas*) populations are particularly at risk in the Indo-Pacific, primarily due to high levels of directed take of adults, juveniles, and eggs. Leatherback turtle populations (*Dermochelys coriacea*) are especially at risk in the Eastern Pacific. It has been estimated that the number of leatherback turtles in that region has decreased from just under 100,000 adult females in 1980 to fewer than 3,000 adult females in 2000 (Spotila et al. 2000). Conservative estimates are that longline and gill-net fisheries were responsible for the mortality of at least 1,500 female leatherbacks per year in the Pacific during the 1990s (Spotila et al. 1996).

Similarly, leatherbacks and loggerhead turtles (*Caretta caretta*) at sea suffer from high rates of mortality due to unsustainable levels of bycatch in various fisheries (notably longline fisheries). Should these levels be sustained, Eastern Pacific leatherback turtles are anticipated to become extinct in the next few decades (Crowder 2000). In many parts of the world, however, direct harvest (as occurs for the hawksbill, *Eretmochelys imbricata*) and incidental capture of marine turtles in inshore fisheries represent a greater source of mortality than bycatch in longline fisheries (Seminoff 2002; Kaplan 2001).

In addition to mortalities experienced at sea, habitat loss and destruction of nesting beaches and important foraging grounds have contributed to marine turtle population declines (WWF 2003). Turtle products, such as jewelry made from hawksbill shells, also threaten marine turtles. Thousands of turtles die from eating or becoming entangled in nondegradable debris each year. Trash, particularly plastic bags, causes mortality for species like the leatherback, which cannot distinguish between floating bags and jellyfish prey. Pollution has also been linked to increased incidence of fibropapilloma disease, which kills hundreds of turtles annually (Herbst et al. 2004). However, the greatest recent historical losses in turtle populations occurred as a result of early European colonization of the Americas, when trade in turtle products helped finance further exploration and settlement, as occurred in the Caribbean (Carr 1979; Jackson et al. 2001).

19.2.2.2 Marine Mammals

Marine mammals are affected and frequently threatened by fisheries and other human activities (Northridge 2002). In the past, the main threats were large-scale whaling and sealing operations focused initially on the waters of northern Europe and Asia. Operations soon extended to Antarctica and reduced populations to small fractions of their former abundances (Perry et al. 1999) or extirpated them completely, as with the now extinct Atlantic grey whale (Mitchell and Mead 1977) or the Caribbean monk seal (Kenyon 1977; Gilmartin and Forcada 2002). While many of the pinniped (seals, sea lions, and walrus) species appear to have recovered quite successfully from former exploitation levels, recovery of some of the heavily depleted whale species has been slow, making them more susceptible to other emerging threats, such as bycatch in commercial fisheries or climate change (Clapham et al. 1999).

In recent decades, incidental entanglement in fishing gear, chemical and acoustical pollution, habitat degradation, climate change, and ship strikes are regarded as the most serious human-related threats for marine mammals, although impacts of these are highly variable for different species.

Small cetaceans such as dolphins are probably most threatened by bycatch (Northridge 2002; Kaschner 2003)—in some cases, to the verge of extinction, such as the vaquita (D'Agrosa et al. 2000). And worldwide estimated mortalities across all species add up to several hundred thousands every year (Read et al. 2003). Although entanglement in fishing gear is generally not fatal for the larger baleen whales, it may seriously affect the ability of an animal to feed and may potentially result in starvation (Clapham et al. 1999).

Increasing levels of chemical pollution and marine debris in the marine environment are likely having impacts on most marine mammal species through ingestion of pollution and floating plastic debris or entanglement (Merrick et al. 1987). Various health problems in marine mammals have been associated with high levels of accumulated pollutants that have been found in many species of predatory marine mammals (Aguilar and Borrell 1994).

Pinniped species combined represent the most abundant group of marine mammals in terms of population size. However, a high proportion of pinniped species are restricted to polar waters, and this group is most likely to be negatively affected by climate change (Harwood 2001). Currently, almost a quarter of all pinniped species are listed as endangered or vulnerable in the *IUCN Red List*.

19.2.2.3 Waterbirds

Many waterbirds are dependent on coastal systems (see Chapter 20 for a more detailed assessment of waterbird status and trends), and waterbirds themselves are important in the delivery of a number of coastal ecosystem services, including nutrient cycling, recreation, food provisioning, and cultural values. Coastal systems

are vital for both shorebirds and seabirds, which use coastal areas for breeding, foraging and resting. There are 336 species of seabirds (Schreiber and Burger 2002). Some species, notably gulls, have increased because of widespread discarding of bycatch. Others have strongly declined in recent decades, both due to the reduction of their food base by fisheries and because they are caught as bycatch of pelagic fisheries.

Shorebirds are declining worldwide: of populations with a known trend, 48% are declining in contrast to just 16% increasing (International Wader Study Group 2003). For shorebirds in Africa and Western Eurasia, three times as many populations are decreasing as are increasing, although the trend status of the majority of populations seems not to have changed significantly over the last 10–20 years. Overall, 45 (34%) of African-Eurasian migratory shorebird populations are regarded as of conservation concern due to their decreasing or small populations (Stroud et al. 2004). Similarly, 54% of shorebird populations occurring in North America are in a significant or persistent decline, with only 3% increasing significantly and as many as 80% of populations in this region showing evidence of declines (Morrison et al. 2001). However, shorebird trend status in other regions is poorly known and has not been reassessed since the 1980s.

Information on trends in shorebirds and seabirds is highly variable geographically. For shorebird (wader) flyways in Africa-Eurasia, trend information is available for 93% of populations using the coastal East Atlantic flyway and 76% using the Black Sea/Mediterranean flyway. Only 35% of populations on the West Asia/East Africa flyway have good trend information, and the status of resident African populations is particularly poorly known (only 30%) (Stroud et al. 2004). While fewer seabirds than inland waters species have become extinct, a much larger proportion (41.8%) of extant seabirds are globally threatened. (See Chapter 20.) The decline in seabirds is occurring in all parts of the world and across major habitat types. The most threatened families are albatrosses (90.5% of species globally threatened), penguins (58.8%), petrels and shearwaters (42.9%), and frigate birds (40%).

Land use change and habitat loss and degradation seem to continue to be drivers of shorebird declines. For example, the decline of certain long-distance East Atlantic flyway populations (while other populations on the same flyway are stable or increasing) has been attributed to their high dependency on deteriorating critically important spring staging areas, notably the international Wadden Sea, that are being affected by commercial shellfisheries. Similar situations are reported from other flyways and key spring staging areas such as Delaware Bay in the United States and the Yellow Sea coast. Maintaining the ecological character of such staging areas is increasingly recognized as vital for the survival of Arctic-breeding species, yet many remain under threat (Baker et al. 2004; Davidson 2003).

For seabirds, direct drivers of declines are likely to be different from those of coastal and freshwater waterbirds. For example, for albatrosses—the seabirds showing the most dramatic current population declines—it is highly certain that the main driver is adult mortality caused by pelagic (longline) fisheries in southern oceans (BirdLife International 2004b).

For sea- and shorebirds, climate change is considered to be additional to the drivers of land use change and habitat loss and degradation. For example, changes in the non-breeding distribution of coastal wintering shorebirds in western Europe have been attributed to rising mid-winter temperatures (Austin et al. 2000), and seabird breeding failures in the North Sea in 2004 have been linked to a northward shift in plankton distribution driven by rising sea temperatures (Edwards and Richardson 2004).

Any effects of climate change on waterbirds are generally considered to be additive to the impacts of direct drivers such as habitat degradation. However, it is predicted that reduction in areas of Arctic tundra breeding habitat will contribute to population declines in high-Arctic breeding species (*medium certainty*). Similar shifts in distribution in several other parts of the world are well known and occur in relation to El Niño events (*medium certainty*).

19.2.3 Summary and Linkages with Other Systems

Coastal ecosystems are diverse, highly productive, ecologically important on the global scale, and highly valuable for the services they supply. (See Table 19.2.) Dividing the coastal system into separate subtypes and discussing each one independently obscures the fact that these habitats and the ecological processes within them are highly linked, with water mediating many of these linkages. While it is true that all habitats are ultimately connected in the marine environment, some habitats are more intimately connected than others.

Coral reefs provide a good example of this interconnectedness (Hatcher et al. 1989). The internal interconnectedness of coral reefs has historically been emphasized, giving the impression of self-contained entities: very productive ecosystems with nutrients essentially locked up in the complex biological community of the reef itself. Many of the most ecologically crucial habitats for reef organisms are actually not on the coral reef itself, however, but rather in seagrass beds, mangrove forests, and seamounts sometimes far from the reef (Birkeland and Frielander 2002; Mumby et al. 2004). Thus the coral reef ecosystem depends on these essential linked habitats as well. Currents and the mobile organisms themselves provide the linkages among the reefs, nursery habitats, and places where organisms move to feed or breed.

One of the strongest links between coastal subsystems is that between areas that act as nursery grounds for fish species. The majority of the world's marine fishery species are caught or reared in continental shelf waters, and many of these species spend at least some part of their life histories in the nearshore coastal habitats (Sherman 1993, cited in Burke et al. 2001). When nursery areas are lost due to habitat conversion, freshwater diversion from estuaries, or degradation, fisheries even outside the nursery area can be significantly affected (Deegan and Buchsbaum 2001; Lenanton and Potter 1987). Loss of nursery areas has been implicated in the collapse of some fisheries in North America, North Africa, and elsewhere (Chambers 1992; Deegan 1993).

Nursery areas and other habitats crucial for fisheries production can also be ecologically "lost" when degraded by seemingly natural (or, in any case, biotic) events. Harmful algal blooms, for instance, can be devastating to eggs and larvae of fish and can thus cause loss of nursery services. Often the population growth of such harmful algae is spurred by eutrophic conditions—the result of agricultural, sewage, aquacultural, or fish processing wastes overcoming the assimilative capacity of the coastal environment.

The ocean and coastal habitats are not only connected to each other, they are also inextricably linked to land. (See Table 19.3.) Fresh water is one specific mediator here: rivers and streams bring nutrients as well as pollutants to the ocean, groundwater flows to coastal systems, and the ocean gives some of these materials back to land via the atmosphere, tides and seiches, and other pathways, such as the deposition of anadromous fish (salmon carcasses, for instance) after spawning. The salinization of aquifers from marine intrusion, usually due to excessive freshwater extraction) is another factor. Seawater to freshwater linkages also occur; in experimental settings, polluted coastal water has been shown to contaminate freshwater aquifers (Jones 2003). But the atmosphere also provides a linkage, and land-sea-air interactions sometimes create complex feedback mechanisms between impacts on one habitat type and consequent impacts on another. For example, in

---

Table 19.2. Summary of Ecosystem Services and Their Relative Magnitude Provided by Different Coastal System Subtypes. The larger circles represent higher relative magnitude.

Direct and Indirect Services	Estuaries and Marshes	Mangroves	Lagoons and Salt Ponds	Intertidal	Kelp	Rock and Shell Reefs	Seagrass	Coral Reefs
Food	●	◉	•	●	•	●	•	•
Fiber, timber, fuel	●	●	•					
Medicines, other	•	•	•		●			•
Biodiversity	●	●	◉	⬤	◉	●	◉	●
Biological regulation	●	●	◉	•		•		◉
Freshwater storage and retention	•		•					
Biochemical	•	•		•				•
Nutrient cycling and fertility	●	●	◉	•		•		●
Hydrological	•							
Atmospheric and climate regulation	◉	◉	◉	•		•		◉
Human disease control	◉	◉	◉	•		•		•
Waste processing	●	●	◉			•	•	•
Flood/storm protection	◉	◉		•	•	◉		●
Erosion control	•	●				•	•	◉
Cultural and amenity	●	•	◉	●	•	◉	◉	●
Recreational	●	•	•	●	•			●
Aesthetics	◉	•	◉	●				●

Table 19.3. Fluxes from Land to Sea and from Sea to Land, Differentiating between Natural and Anthropogenic Factors (Modified from Kjerfve et al. 2002)

Factor	Land to Sea	Sea to Land
Natural	river discharge	energy and debris from hurricanes
	groundwater	
	sediment	cold water and nutrients from upwelling
	nutrients and minerals	
	humics and organics	wave action
	storm debris	salt and salt aerosols
	earthquake debris	sand
	volcanic debris	nutrients through carcasses, guano
Anthropogenic	sediment (increase from land use and decrease from dams)	oil and chemical spills
		chronic input of oil and chemicals
	nutrients and organic matter from agriculture and sewage	sewage from ships
		ballast water with exotic organisms
	coliform bacteria	debris from ships
	herbicides and pesticides	brackish infiltrations of groundwater reservoirs by water extraction
	heavy metals	
	oil and chemicals	pharmaceuticals

Florida in the United States, the loss of coastal wetlands and their buffering capability may have caused severe freezes affecting inland agricultural lands in recent winters, costing millions of dollars in failed crops (Marshall et al. 2003).

Coastal systems serve as a major sink for sediments and are major sites of nutrient-sediment biogeochemical processes. Water quality in river systems plays a crucial role in the sustainability of coastal aquatic habitats, food webs, and commercial fisheries that serve as a major protein source for humans (Burke et al. 2001). The transport of sediment and biotically active materials (nutrients and toxic substances) to the coastal zone through long-distance river transport ultimately links the continental landmass to the oceans (Vörösmarty and Meybeck 1999). (See Box 19.2.) Thus coastal issues need to be addressed from a system perspective involving the whole catchment scale and the coupling of human and natural systems.

The cross-habitat movement of nutrients, detritus, prey, and consumers exerts major effects on populations and food webs in practically all habitats and can sustain communities of abundant consumers even in places with little or no primary productivity (Polis et al. 1997). This relationship is particularly strong in the coastal system, especially where highly productive oceanic waters meet relatively unproductive, dryland habitats (Polis and Hurd 1996).

The Pacific coast of Peru is one of the best examples of this, where high- and low-productivity systems are juxtaposed: highly productive marine waters associated with upwelling of the Humboldt current are next to one of the world's most arid areas, the Atacama desert. The system of the Humboldt current has a primary productivity rate that makes it one the world's richest marine areas (Arntz and Fahrbach 1996), whereas the desert it faces receives less than 5 millimeters of rainfall a year. Other examples are the Namib Desert facing the Benguela upwelling system, the

Banc d'Arguin region of Mauritania, and the coasts and islands of the Gulf of California in Mexico.

The movement of nutrients from the ocean to land can occur in two different pathways. The first is the guano pathway, which includes the accumulation of seabird excrement. This pathway is likely to be significant only for islands and rocky shores where sea birds congregate in large numbers. The second is the detritus/scavenger pathway, with a significant amount of biomass entering the terrestrial system through algal or seagrass mats and through animal carcasses washing ashore. Fish or mammals may also become vectors of marine-derived energy and nutrients by migrating over large distances. River otters and sea lions have been shown to enrich terrestrial vegetation with marine-derived nitrogen in coastal environments.

Perhaps the best-known example is anadromous Pacific salmon (*Oncorhynchus kisutch, O tshawytscha, O. nerka*), the carcasses of which fertilize forests (Helfield and Naiman 2003) and provide a valuable source of nutrients for scavengers in the sites where they congregate to spawn (Ben-David et al. 1998b). In regions these salmon carcasses seem to be a keystone nutrient resource for scavengers, populations of such scavengers are greatly affected by reductions in anadromous fish stocks (Willson and Halupka 1995).

The idea that marine resources are also a key resource to human populations is verified archeologically. Moseley (1975) proposed a "marine hypothesis" to demonstrate that the paradigm of agricultural economy as being the foundation of civilizations does not hold for ancient populations in coastal Peru. He proposed instead that the enormous productivity of the upwelling system caused the rise of Andean civilizations. Numerous archeologists have challenged this hypothesis, noting that other sources of food had to be available for populations exposed to high variability in marine productivity. However, there is no doubt that marine productivity accounted for a large part of the diet in several major coastal civilizations.

In the Atlantic, for example, cod was said to fuel the immigration and growth of New England and Canadian maritime population centers, and in Europe herring is thought to have underpinned the mercantile expansion. The declining availability of marine resources has affected large portions of these populations even today. More recently, it has been surmised that declining availability of coastal and freshwater fish for subsistence fishers in West Africa has driven the increase in the illegal bush meat trade. This trade, in turn, has imperiled many endangered species in the region and is thought to contribute to outbreaks of primate-borne and other viruses in human populations (Brashares et al. 2004).

Ocean climate in one region may affect land and coastal systems in another, and in complex ways. For instance, it is now surmised that the warming of the Indian Ocean has caused the recent droughts of the Sahel, directly affecting millions of people through increased crop failure and decreased health (Giannini 2003), while the increased desertification of the Sahel region may have caused mortality of corals half a world away through the transport and subsequent deposition of Saharan dust. (See MA Caribbean Sub-global Assessment.)

Thus, negative impacts on coastal ecosystems, whether on land, in areas of fresh or brackish water, or in the sea itself, have enormous ramifications for the health and productivity of other terrestrial and marine systems, in addition to affecting coastal systems and their provisioning of ecosystem services. As human population pressures continue to grow, these declines in coastal ecosystem services will increase the strain on coastal communities and have negative impacts on human well-being in coastal systems.

BOX 19.2
Trends in Sediment Loads into Coastal Zones

Fluvial systems evolve along with the landscape, and the sediment load observed today is influenced by the geologic history of these paleo-systems. Therefore it remains difficult to determine the sediment flux of unaffected rivers, given the natural variability within fluvial systems. While there is no accepted estimation for the paleo-flux of sediment to the coastal zones (Syvitski 2001, 2003), Milliman and Syvitski (1992) argued that the modern 20 gigatons a year global flux value may have been 50% smaller about 2,000 years ago, before human impact was great.

A recent study of the annual sediment load records for the world's rivers shows many examples of nonstationary behavior (Walling and Fang 2003). Simple trend analysis of this database indicates that about 50% of the sediment load records showed evidence of statistically significant upward or downward trends, with the majority evidencing declining loads. In about 50% of rivers, the sediment load records showed no evidence of significant trends. In some rivers, loads are declining as a result of dam construction and the implementation of soil and water conservation and sediment control programs. In other systems, loads are increasing due to land clearance and land use change and intensification, along with other forms of catchment disturbance and increased runoff as a result of increased precipitation and runoff. The results suggest that the dominant trends in sediment flux to the global coastal zone are either stability or a decrease. This analysis has not included rivers from other areas of the world, such as Africa, Southeast Asia, and South America.

Under this picture of the world's decreasing sediment load, less river sediment discharge alters the sedimentation-erosion equilibrium within the coastal zone. Coarse-grained bed load is normally taken to represent 10% or less of the total sediment discharge delivered to the coast. Hence, it has been assumed that a decrease of approximately 5% of the total sediment flux represents the critical threshold, beyond which the coastal system is likely to show evidence of significant deterioration (such as coastal erosion). This level of change results in mangrove siltation and severe erosion of coastal ecosystems and beaches (Lacerda et al. 2002). Thus river sediment flux plays an important role in the sediment budget of the coast.

Dramatic and virtually instantaneous changes are recorded in water fluxes measured at river discharge monitoring stations before and after impoundment. The Nile River has experienced, as many river systems worldwide, reduced flows and distortion of runoff due to water use for irrigation (Nixon 2003). (See Figure.)

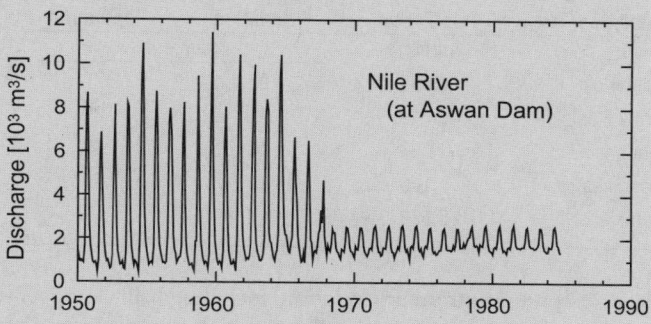

Compared with the past five decades, both river discharge and sediment load will probably decrease for some large fluvial systems 30–40% in the next 50 years (Vörösmarty et al. 1997; Vörösmarty and Meybeck 1999) and decrease to 50% in the next 100 years as a result of human activities and dam construction (Yang et al. 1998). Thus general erosion in the coastal zone, including estuaries, deltas, and associated beach systems, seems to be inevitable.

The future discharge of sediment to the coastal oceans will continue to be controlled by humans and climate change. Determining the balance between increasing sediment loads (from land use, engineering, climate change, and climate variability) and decreasing sediment loads from reservoirs, engineering, climate change, and climate variability is of utmost importance for sound coastal zone and resource management (Syvitski 2003).

19.3 Coastal Systems and Human Communities

19.3.1 Humans in the Coastal System: Demographics and Use of Services

Humans are a natural element within coastal systems and have been so for thousands of years. However, the balance of nature in these systems has become altered. While human dependence on coastal systems has greatly increased in the last centuries, the impacts on the ecology of these habitats have become so severe that their productivity and functioning have been altered, mostly in the last few decades. It is increasingly difficult for coastal systems to accommodate the increased collective demands of growing populations and markets.

Coastal populations are not spread evenly throughout the coastal zone. Using night light analysis, Small and Nicholls (2003) graphically demonstrated the concentration of habitation on the world's coasts. Quantitative analysis of newer population data has shown that there has been a decrease in the rate at which interior populations are increasing relative to coastal populations. Coastal population densities are nearly three times that of inland areas: in 2000, population density in coastal areas was 99.6 people per square kilometer, while in inland areas density was 37.9 people

per square kilometer (Kay and Alder in press). At the turn of the millennium, half of the world's major cities (those with more than 500,000 people) were found within 50 kilometers of a coast. Growth in these cities since 1960 was significantly higher than in inland cities of the same size (Kjerfve et al. 2002).

Not only are population pressures high relative to those in many other ecosystems worldwide, but the bulk of those pressures stress many of the most ecologically important and valuable ecosystems within coastal zones. Some 71% of the world's coastal people live within 50 kilometers of an estuary, 31% live within 50 kilometers of a coral reef system, 45% live within 50 kilometers of mangrove wetlands, and 49% live within 50 kilometers of seagrass ecosystems (See Table 19.4.) This is not accidental, of course—these habitats and the ecosystem services they provide present many of the "pull" factors that resulted in initial settlement along a coast as well as subsequent migration to it. Historically, settlements first inhabited the sheltered areas near estuarine bays (many with associated mangrove and seagrass) and reef-protected coasts and only later expanded to other coastal areas.

Conversely, 58% of the world's major coral reef systems occur within 25 kilometers of urban centers with more than 100,000 people; 62% of major estuaries occur near such urban centers, and 64% of major mangrove forests are found near major centers. This

Table 19.4. Share of World and Coastal Populations Living within 50 Kilometers of Estuaries, Coral Reefs, Mangroves, and Seagrass. Based on spatially referenced population data; due to overlap of some habitat types, figures do not add up to 100 percent. (CIESN 1995)

Subtypes	Population	Share of World Population	Share of Coastal Population
	(million)	*(percent)*	
Estuaries	1,599	27	71
Coral Reefs	711	12	31
Mangroves	1,030	18	45
Seagrass	1,146	19	49
Total	**5,596**		

means that pressures from urbanization, including habitat conversion as cities and their areas of influence grow, are affecting the majority of these key coastal habitats. In fact, analysis of areas of recent rapid land cover change shows that all but three of the world's cities showing greatest rates of change and highest population occur in coastal areas, in both the tropics as well as higher latitudes. (See Chapter 28 for more on this work.)

By all commonly used measures, the human well-being of coastal inhabitants is on average much higher than that of people in inland communities. Of the world's total GNP of approximately $44 trillion, 61% comes from coastal areas within 100 kilometers of a coastline. Whereas per capita GNP in 1999 averaged only $4,018 across all inland areas, in the 100-kilometer coastal area it was nearly four times as much—at $16,035. Figure 19.8 shows that the concentration of global wealth as measured by GNP occurs primarily in coastal regions, although concentrations of wealth also occur in some inland areas (especially in the United States and Europe). Infant mortality and life expectancy indices are also thought to be relatively better in coastal areas. This situation partly explains why rates of population increase are highest in coastal areas.

Nonetheless, many coastal communities are at risk. There are considerable physical risks associated with living in some coastal areas; low-lying atolls, for example, are at risk of catastrophic events such as hurricanes, cyclones, tsunamis, and storm surge

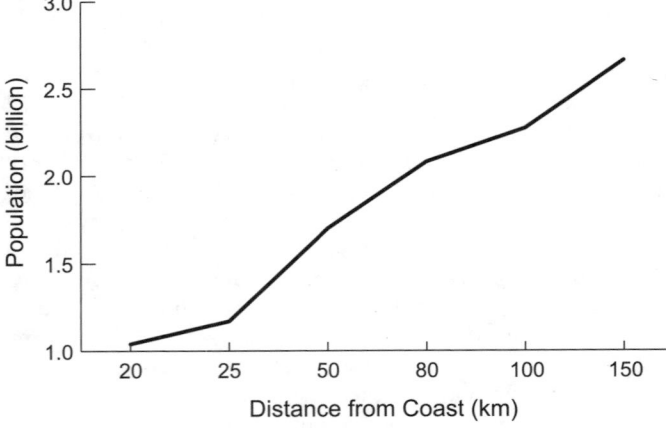

Figure 19.8. Population Density by Distance from Coast (CIESIN 2003)

flooding, as well as losses incurred from both sudden and chronic shoreline erosion. Figure 19.9 illustrates potential global vulnerability to erosion by highlighting areas where soil types and slope make shorelines prone to erosion and inundation from storm events. Many of these risks are increasing with climate change–driven changes to meteorology. And some countries, such as the United Kingdom, are developing contingency plans to cope with such changes (see www.foresight.gov.uk/fed.html). Some areas are prone to flooding because of relative changes in sea level—the average global sea level rise is projected at 1–2 millimeters per year over the next century (Church et al. 2001). This is an especially acute problem in small island nations, atoll communities, and low-lying flood-prone areas like much of Bangladesh.

Coastal communities are also at risk because the coastal ecosystems they exploit and rely on are stressed—and many are nearing ecological breaking point or thresholds (Birkeland 2004; Dayton 2003). Technological advances that allow greater access to resources, including boat design, navigation, fishing gear, and oil exploration methods and equipment, have pushed the use of many coastal resources beyond sustainable limits. Such advances have also increased the conflicts between large-scale industries and small-scale local users, such as subsistence fishers (Curran and Agardy 2002). Poorly planned or executed development has already compromised the ability of many coastal ecosystems to provide regulating services such as maintenance of hydrological balance, nutrient fluxes, and shoreline stabilization (Kay and Alder in press). Thus the relatively high levels of human well-being experienced by many coastal communities are at risk of declining as ecosystems continue to be degraded, lost, or rendered unproductive.

Human communities are also at risk from the health implications of these degraded ecosystems. Cholera and other waterborne diseases are on the rise in coastal countries (Anderson et al. 2001) and may be related to eutrophication-driven algal blooms (Colwell and Spira 1992; Islam et al. 1990). Cholera affects human well-being directly by increasing human morbidity and mortality rates, but it also has severe economic impacts in coastal countries (Rose et al. 2001). For instance, tuna coming from countries having incidences of cholera must be quarantined; this restriction affects many of the major tuna-producing and -exporting countries.

Algal blooms (including red tides) have caused neurological damage and death in humans through consumption of affected seafood (Rose et al. 2001). There are significant health impacts from swimming and bathing in water contaminated by fecal coliform and other pathogens; approximately half the people living in coastal areas have no access to sanitation, and even where sewage treatment exists it is often inadequate, with the result that coastal areas become polluted (UNEP 2002). In a particularly severe outbreak in Italy in 1989, harmful algal blooms cost the coastal aquaculture industry $10 million and the Italian tourism industry $11.4 million (UNEP 1992). Ciguatera, a tropical fish disease causing severe illness and sometimes mortality in humans who consume affected fish, is on the rise, both in terms of the number of cases and number of affected areas.

Human health effects are also caused by pollution of nearshore waters, whereby humans eat fish or other marine products that contain heavy metals, PCBs, and other toxins that have bioaccumulated in the food chain (Verlaan 1997). UNEP and the Water Supply and Sanitation Council estimate the global economic costs related to pollution of coastal waters is $16 billion annually (www.wsscc.org), much of which is due to human health impacts.

Changes in coastal systems also affect the well-being of those living there and elsewhere in more subtle ways. The destruction

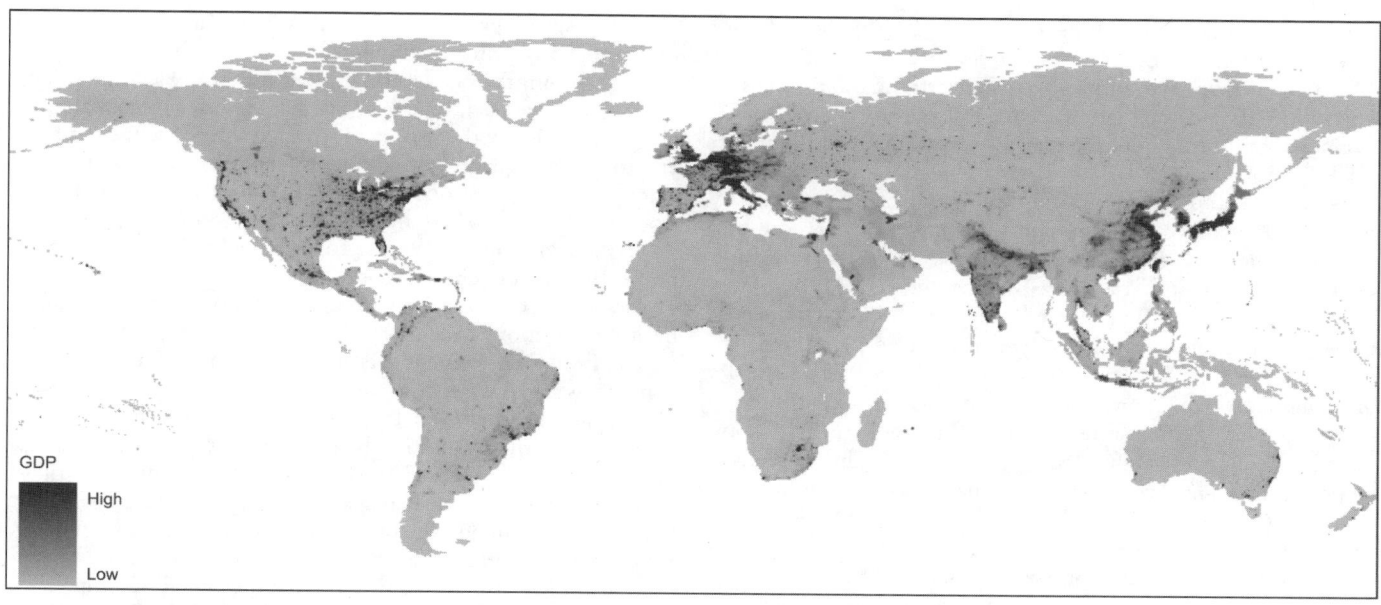

Figure 19.9. Relative Levels of GDP (CIESIN 2003; World Bank 2004)

of places that create opportunities for recreation, that are spiritually or culturally important, or that could potentially increase our knowledge and respect for the natural world entail costs that are more difficult to quantify. Surveys everywhere show that humans maintain strong spiritual connections to the sea and care about its condition, even if they live far inland with no direct reliance on coastal areas for obtaining food or employment, for example.

19.3.2 The Value of Coastal System Services

Coastal ecosystems provide a wide range of services to human beings (Wilson et al. 2004). These include regulation and supporting services such as shoreline stabilization, nutrient regulation, carbon sequestration, detoxification of polluted waters, and waste disposal; provisioning services such as supply of food, fuelwood, energy resources, and natural products; and amenity services such as tourism and recreation. These services are of high value not only to local communities living in a coastal zone (especially in developing countries), but also to national economies and global trade (Peterson and Lubchenco 1997).

In addition to the production of marketable goods and services, such as commercial fisheries and tourism, coastal systems provide services such as nutrient recycling, support for terrestrial and estuarine ecosystems, habitat for plant and animal life, and the satisfaction people derive from simply knowing that a beach or coral reef exists (Wilson et al. 2004). While estimating exchange-based values of marketed services in this case is relatively straightforward, as observable trades exist from which to measure value (Freeman 1993), estimating the economic value of coastal services not traded in the marketplace is more difficult (Freeman 1993; Bingham et al. 1995). However, such analysis often reveals social costs or benefits associated with coastal ecosystem services that otherwise would remain hidden or unappreciated. Market values and nonmarket values are discussed separately in this section.

Studies of specific regions and biomes give us some idea of the enormous economic value of coastal habitats (Balmford et al. 2002). The Wadden Sea in northern Europe, for instance, has provided up to one quarter of the North Sea catch of plaice, sole, shrimp, dab, and herring (De Groot 1992). Coral reef–based fisheries are also valuable: those in Southeast Asia generate $2.4

billion per year (Burke et al. 2001). Although it is widely cited that coral reefs contribute about one quarter of the annual total fish catch in developing countries, providing food to about 1 billion people in Asia alone, the empirical evidence to support such statements is not strong. However, the value of reef fisheries in this region is undeniably significant: Cesar et al. (2003) estimated net benefit streams of reef-dependent fisheries in Asia at over $2 billion.

In principle, a global picture of the potential economic value associated with the coastal zone can be built up via the aggregation of a number of existing valuation studies. For example, in a preliminary estimate of the total economic value of ecosystem services provided by global systems, Costanza et al. (1997) showed that while the coastal zone covers only 8% of the world's surface, the goods and services provided by it are responsible for approximately 43% of the estimated total value of global ecosystem services: $12.6 trillion (in 1997 dollars). While controversial (Pimm 1997; Pearce 1998), this preliminary study made it abundantly clear that coastal ecosystem services do make significant contributions to human well-being at a global scale. Furthermore, it demonstrated the need for additional research and indicated that coastal areas are among the ecosystems most in need of additional study (Costanza 2000).

19.3.2.1 Market Coastal Values

Coastal ecosystems are among the most productive in the world today, rivaling even tropical rainforests in terms of their overall productivity of raw materials and goods used by humans (Primavera 1991; Spurgeon 1992; Barbier 1993). As the following examples show, many coastal regions are valued through market activities that directly support humans—such as fishing, hunting, fuelwood and woodchip extraction, harvesting ornamental materials, and the extraction of medical resources.

Coastal systems generate a variety of seafood products such as fish, mussels, crustaceans, sea cucumbers, and seaweeds (Moberg and Folke 1999; Ronnback 1999). Many commercially important marine species, like salmon, shad, grouper, snapper, bluefish, striped bass, and invertebrates (such as shrimp, lobster, crabs, oysters, clams, mussels), use coastal nursery habitats. Capture fisheries in coastal waters alone account for $34 billion in yields annually. (See Chapter 18.) Given this level of economic productivity, it is perhaps not surprising that overfishing and intensive aquaculture have caused serious ecological and social problems in coastal regions throughout the world (Primavera 1991; Primavera 1997; Jackson et al. 2001).

Valuation studies of food directly or indirectly supplied by coastal systems have predominantly focused on the economic value of fishery products (Batie and Wilson 1978; Lynne et al. 1981; Farber and Costanza 1987; Buerger and Kahn 1989; Rivas and Cendrero 1991; Bennett and Reynolds 1993; Ruitenbeek 1994; Kaoru et al. 1995; Deb 1998; Gilbert and Janssen 1998; Ronnback 1999; Barbier 2000; Sathirathai and Barbier 2001). Most often, the market price of seafood products is used as a proxy when calculating the value of ecosystem goods provided by coastal systems. For example, the annual market value of seafood supported by mangroves has been calculated to range from $750 to $16,750 (in 1999 dollars) per hectare (Ronnback 1999). High-value species are harvested from coral reefs to meet live fish demand in restaurants, mainly in Asia. (See Chapter 18.)

Coastal areas also provide the foundation for the mariculture (marine aquaculture) industry, which uses coastal space or relies on wild stock to produce valuable fisheries products, from tiger prawns to bluefin tuna. Human reliance on farmed fish and shellfish is significant and growing. Global annual per capita consumption of seafood averages 16 kilograms, and one third of that supply currently comes from aquaculture (Lubchenco 2003). Globally, aquaculture is the fastest-growing food-producing sector, with production rates doubling in weight and value from 1989 to 1998 (Goldburg et al. 2001). Much of that growth has occurred in the shrimp and salmon farming industries.

Besides food and raw materials, at least three other types of marketable goods are provided by coastal systems: genetic, medical, and ornamental resources. For example, coral reefs have been shown to be an exceptional reservoir of natural bioactive products, many of which exhibit structural features not found in terrestrial natural products (Carte 1996). The pharmaceutical industry has discovered several potentially useful substances among the seaweeds, sponges, mollusks, corals, sea cucumbers, and sea anemones of reefs (Carte 1996; Moberg and Folke 1999). (See Chapter 10 for more on bioprospecting in coastal systems.) Furthermore, many coastal products are collected not only as food but also to sell as jewelry and souvenirs. Mother-of-pearl shells, giant clams, and red coral are collected and distributed as part of a worldwide curio trade (Craik et al. 1990). The marine aquarium market is now a multimillion-dollar industry trading in live reef-dwelling fishes that are collected and shipped live from coral reef communities (Moberg and Folke 1999; Wabnitz et al. 2003).

19.3.2.2 Nonmarket Coastal Values

In addition to marketable goods and products, landscape features and ecological processes within the coastal zone also provide critical natural services that contribute to human well-being and have significant economic value (Farber and Costanza 1987). As the data just cited suggest, much of what people value in the coastal zone—natural amenities (open spaces, attractive views), good beaches for recreation, high levels of water quality, protection from storm surges, and waste assimilation/nutrient cycling—is provided by key habitats within coastal systems. In Thailand, the conversion of mangroves to shrimp aquaculture ponds reduced the total economic value of the intact mangroves by 70% in less than a decade (Balmford et al. 2002).

Open space, proximity to clean water, and scenic vistas are often cited as a primary attractor of residents who own property and live within the coastal fringe (Beach 2002). Hedonic pricing techniques have been used to show that the price of coastal housing units varies with respect to characteristics such as ambient environmental quality (proximity to shoreline, for example, or water quality) (Johnston et al. 2001). For example, Leggett and Bockstael (2000) use hedonic techniques to show that water quality has a significant effect on property values along the Chesapeake Bay in the United States. They use a measure of water quality—fecal coliform bacteria counts—that has serious human health implications and for which detailed, spatially explicit information from monitoring is available. The data used in this analysis consist of sales of waterfront property on the western shore of the Chesapeake Bay between 1993 and 1997 (Leggett and Bockstael 2000). The authors consider the effect of a hypothetical localized improvement in observed fecal coliform counts on a set of 41 properties. The projected increase in property values due to the hypothetical reduction in coliform bacteria totaled approximately $230,000. Extending the analysis to calculate an upper limit benefit for 494 properties, it is estimated that the benefits of improving water quality at all sites would be around $12.145 million (Leggett and Bockstael 2000).

Stretches of beach, rocky cliffs, estuarine and coastal marine waterways, and coral reefs provide numerous recreational and scenic opportunities. Boating, fishing, swimming, walking, beachcombing, scuba diving, and sunbathing are among the leisure activities that people enjoy worldwide and thus represent significant economic value (Farber 1988; King 1995; Kawabe and Oka 1996; Ofiara and Brown 1999; Morgan and Owens 2001). Both travel cost and contingent valuation methods are commonly used to estimate this value. (See Chapter 2 for more on these valuation techniques.) For example, the Chesapeake Bay estuary has also been the focus of considerable research on nonmarket recreational values associated with coastal systems. When attempting to estimate the monetary worth of water quality improvements in Chesapeake Bay, Bockstael et al. (1989) focused on recreational benefits because it was assumed that most of the increase in well-being associated with such improvements would accrue to recreational users. The authors estimated the average increases in economic value for beach use, boating, swimming, and fishing with a 20% reduction in total nitrogen and phosphorus being introduced into the estuary. Using a combination of the two valuation methods, the annual aggregate willingness to pay for a moderate improvement in the Chesapeake Bay's water quality was estimated to be in the range of $10–100 million (in 1984 dollars) (Bockstael et al. 1989).

Global tourism has been deemed the world's most profitable industry, and coastal tourism is one of its fastest-growing sectors. Much of this tourism centers on aesthetically pleasing landscapes and seascapes, intact healthy coastal ecosystems with good air and water quality, opportunities to see diverse wildlife, and so on. For instance, much of the economic values of coral reefs—with net benefits estimated at nearly $30 billion each year—is generated from nature-based and dive tourism (Cesar et al. 2003). The demand for biologically rich sites to visit increases the value of intrinsically linked habitats such as mangroves and seagrass beds. Temperate bays and estuaries can similarly generate tourism revenues of similar orders of magnitude.

The link between tourist visits and the revenues from and condition of the coastal system has not been analyzed at the global level, but local case studies point to a strong correlation between value and condition. In the United States alone, reef ecosystems with their nursery habitats support millions of jobs and billions of dollars in tourism each year. For example, reef-based tourism generated over $1.2 billion in the Florida Keys alone, while in Hawaii, reefs generate some $360 million per year, with annual gross revenues generated from just one half-square-mile coral reef reserve exceeding $8.6 million (Birkeland 2004).

As these reefs decline in biodiversity and ecosystem health, these nature-based tourism industries stand at risk (Cesar and Chong 2004). In Jamaica and Barbados, for instance, destruction of coral reefs resulted in dramatic declines in visitation; loss of revenue streams subsequently led to social unrest and even further tourism declines (MA Sub-global Assessment on Caribbean Sea). Similarly, "willingness to pay" studies in the Indian Ocean suggest that the health of coral reefs is an important factor for tourists: they were willing to pay, on average, $59–98 extra per holiday to experience high-quality reefs (Linden et al. 2002). And in Florida, reef degradation is rapidly changing the structure of the tourism market, from high-value, low-volume tourism toward larger numbers of budget travelers (Agardy 2004).

Recreational fishing is also a major industry in many parts of the world, and it primarily targets marine or anadromous fishes in coastal ecosystems. The estimated revenue generated by coral reef–based recreational fisheries reaches several hundred million dollars annually (Cesar et al. 2003). The coastal zone also supplies nonmarket values associated with both recreational and commercial fisheries by providing some of the most productive habitat refugia in the world (Gosselink et al. 1974; Turner et al. 1996). Eelgrass, salt marsh, and intertidal mud flats all provide a variety of services associated with their nursery functions (Gosselink et al. 1974; Turner et al. 1996).

As already noted, improvements in the condition of these habitats may ultimately lead to measurable increases in the production of market goods such as fish, birds, and wood products. In other cases, however, ecological productivity itself can represent a unique class of values not captured by traditional market-based valuation methods. (See Box 19.3.) Instead, these values represent an increase in the production of higher trophic levels brought about by the increased availability of habitat, though analysis must be careful not to risk double counting some aspects of value or measuring the same benefits in different ways.

The seas and coasts are also of great spiritual importance to many people around the world, and such values are difficult to quantify. While the depth and breadth of these values are as diverse as the cultures that are found worldwide, there is the common theme of a cultural or spiritual connection. For example, the Baju peoples of Indonesia (Sather 1997) and the aboriginal people of the Torres Strait in Australia have a culture intimately connected to oceans, while many of the native peoples of North America have similar strong ties to coastal systems. Even systems on which we place low economic value today may be of importance and value tomorrow because they support species that may turn out to have pharmaceutical value or because they support species or habitat types that may become rare and endangered in the future. This gives them high option value associated with an individual's willingness to pay to safeguard the option to use a natural resource in the future, when such use is not currently planned. Non-use values are representative of the value that humans bestow upon an environmental resource, despite the fact they may never use or even see it.

In summary, ecosystem services are critical to the functioning of coastal systems and also contribute significantly to human well-being, representing a significant portion of the total economic value of the coastal environment. The best available market and nonmarket data suggest that substantial positive economic values

---

BOX 19.3
Examples of Productivity Analyses

In an example of coastal wetland productivity analysis, Johnston et al. (2002) used a simulation model based on biological functions that contribute to the overall productivity of the food web in the Peconic Estuary System in Suffolk County, New York, in the United States. Based on habitat values for fin and shellfish, birds, and waterfowl, an average annual abundance per unit area of wetland habitat in the estuary system was estimated by summing all relevant food web values and habitat values for a year (Johnston et al. 2002). The value of fish and shellfish was based on commercial harvest values. The marginal value of bird species usage of the habitat was based on the benefits human receive from viewing or hunting waterfowl. Using these values as input data, the simulation model resulted in annual marginal asset values for three wetland types: eelgrass ($1,065 per acre per year); salt marsh ($338 per acre per year); and intertidal mud flat ($67 per acre per year).

Farber and Costanza (1987) estimated the marginal productivity of a coastal system in Terrebonne Parrish, Louisiana, in the United States by attributing commercial values for several species to the net biomass, habitat, and waste treatment of the wetland ecosystem. Arguing that the annual harvest from an ecosystem is a function of the level of environmental quality, the authors chose to focus on the commercial harvest data for five different native species—shrimp, blue crab, oyster, menhaden, and muskrat—to estimate the marginal productivity of wetlands. The annual economic value (marginal product) of each species was estimated (in 1983 dollars) as shrimp, $10.86 per acre; blue crab, $0.67 per acre; oyster, $8.04 per acre; menhaden, $5.80 per acre; and muskrat pelts, $12.09 per acre. Taken together, the total value of marginal productivity of wetlands in Terrebonne Parrish was estimated at $37.46 per acre.

---

can be attached to many of the marketed and nonmarketed services provided by coastal systems.

19.4 Projections of Trends, Areas of Rapid Change, and Drivers of Change

19.4.1 Projections of Trends and Areas of Rapid Change

Coastal habitat loss is likely to continue and possibly accelerate as increasing and sometimes conflicting demands for coastal space and resources rise (*high certainty*). Coastal systems and the habitats within them are rapidly becoming degraded around the globe; many have been lost altogether. Sometimes the changes are natural (such as hurricanes and naturally occurring climate variation), but more often than not the impacts are human-induced. These anthropogenic impacts are direct, such as the filling in of wetlands, or indirect, such as the diversion of fresh water from estuaries or land-based sources of pollution. Habitat is lost, usually permanently, when coastal development and marine resource use is destructive or unsustainable.

The greatest factor leading to loss of coastal habitats is conversion of wetlands, including marshes, seagrass beds, mangrove forests, beaches, and even mudflats to make way for coastal development. In the Philippines, for instance, 210,500 hectares of mangrove—40% of the country's total mangrove cover—were lost to aquaculture from 1918 to 1988 (UNESCO 1993). By

1993, only 123,000 hectares of mangroves were left—equivalent to a loss of 70% in roughly 70 years (Nickerson 1999; Primavera 2000). Transportation infrastructure claims much coastal land and will continue to do so as roads are widened, ports and airports are expanded, and so on. Climate change–induced sea level rise will likely exacerbate rates of habitat loss due to development, especially in vulnerable areas such as atolls, deltas, and floodplains (Nicholls 2004). Habitat conversion and loss is thus expected to continue, at least until all available natural habitat is used up or until policy reform stems the tide of habitat loss.

Exploitation beyond sustainable levels is likely to continue and even increase in rate for many resources (*high certainty*). Coastal ecosystems will likely continue to be used for both commercial and artisanal fisheries, and if current trends continue many of these stocks will be depleted to commercial and ecological extinction. The drivers behind coastal resource overexploitation may be direct, such as consumption, or they may be indirect, such as marginalization, perverse subsidization, political corruption, and socioeconomic condition (Myers and Kent 2001). (See Chapter 18.)

Some members of the biological community in coastal habitats have special roles to play in maintaining ecological interactions; the removal of keystone species, for example, can cause large-scale ecological havoc (Kaufman and Dayton 1997). The removal of fish and invertebrates that graze algae living on seagrasses can destroy seagrass beds when heavy algal mats subsume the seagrass meadows. Human activities also affect coastal ecology indirectly by causing the alteration and degradation of distant habitat and by causing mortality of species within the habitat (Keough and Quinn 1998). This threat is often unseen, noticed only once the cumulative effects of degradation has altered or destroyed these ecosystems.

One of the most severe anthropogenic impacts on coastal areas in the near future will likely be through continued interference with hydrology and water flows to the coast (Pringle 2000) (*medium certainty*). Diversion of fresh water from estuaries and riparian-zone conversion of land for agriculture, human use, and hydro-electric generation causes the hypersalinization of estuarine areas and renders them unable to fulfill these important ecological functions and services (Diop et al. 1985; Weinstein and Kreeger 2000). Reduced water delivery to coasts also lowers sediment delivery and greatly accelerates rates of deltaic loss and coastal erosion. For instance, the damming of the Nile caused severe erosion and exacted high costs due to the need for shoreline protection, as well as loss of fertility of agricultural lands in the floodplain. Fisheries in the Nile Delta region of the Mediterranean have also been altered and yields decreased, at least in part due to silicate depletion and changes in phytoplankton communities away from diatoms.

Although there are many specific, often quantified benefits derived from the use and diversion of water in river basins, such hydrological changes are expected to cause rapid change to many estuaries, deltas, and semi-enclosed seas worldwide in coming years, with largely unknown consequences. (See Box 19.4.)

The next few decades will see large increases in rates of eutrophication and prevalence of hypoxic or dead zones as levels of nutrient inputs and wastes rise and as ocean waters warm (*high certainty*). Some 77% of the pollutant load reaching the coastal ecosystems currently originates on land, and 44% of this comes from improperly treated wastes and runoff (Cicin-Sain et al. 2002). These figures are expected to rise if population growth continues to outpace proper sanitation and if agricultural and other runoff remains unregulated. The result will be increased rates of eutrophication through the addition of large quantities of fertilizers, sewage, and other non-natural nutrients, which will change the processes occurring in these ecosystems (NRC 2000). Eutrophied conditions are evident in virtually all coastal waters near areas of human habitation, being especially acute in areas where coastal wetlands and their filtering function have been destroyed. High nutrient concentrations are expected to have particularly large impacts on the ecology of semi-enclosed and other seas in arid areas (Beman et al 2005).

Since nutrient production through agricultural waste and human sewage are expected to increase in the future, and since wetland loss will likely occur at current or higher rates, eutrophication will undoubtedly increase worldwide (*medium certainty*). Numerous river basin and coastal zone studies (in the Baltic region, for instance, the Mississippi River and Gulf of Mexico, the North Sea, the Northern Adriatic, and the Black Sea) have shown that elevated levels of nutrients, coastal eutrophication, toxic phytoplankton blooms, and bottom-water hypoxia are a consequence of human settlement and industrialization. It has been estimated that fluvial fluxes of inorganic N and P to the world oceans have increased severalfold over the last 150–200 years. In certain regions, such as in Western Europe, N and P levels are ten- to twentyfold over pre-industrial levels (Meybeck and Ragu 1997; Vörösmarty and Meybeck 1999).

With *high certainty,* pollutant levels are expected to increase in the near future, despite effective controls on some substances in some areas. River loadings of biotically active elements, metals, hormones, antibiotics, and pesticides are known to have increased severalfold since the beginning of the industrial era, and levels of these toxins are expected to continue to increase. Pollutants not only affect water quality, and with it many provisioning and amenity services, but are also implicated in large-scale failures of fish farming operations. These failures are extremely costly (white spot syndrome in shrimp cost India $200 million over three years, and it nearly caused the collapse of the shrimp farming industry in Ecuador in 1999), and they can affect both ecosystem health at the farming site and human health where the product is consumed. Human health effects from all forms of pollutants have not been comprehensively quantified, but coastal pollutant-related human mortality and morbidity are undoubtedly on the rise (Verlaan 1997).

The geographically largest impacts to coastal systems will be caused by global climate change, and since rates of warming are generally expected to increase in the near future, projected climate change–related impacts are also expected to rise (IPCC 2003). Warming of the world's seas degrades coastal ecosystems and affects species in many ways: by changing relative sea level faster than most biomes can adapt; by stressing temperature-sensitive organisms such as corals and causing their death or morbidity (in corals, this is most often evidenced by coral bleaching); by changing current patterns and thus interfering with important physiobiotic processes; and by causing increased incidence of pathogen transmission. Coral reefs may be the most vulnerable, having already evidenced rapid change, and some projections predict the loss of all reef ecosystems during this century (Hughes et al. 2003). Climate change also alters the temperature and salinity of estuary and nearshore habitats, making them inhospitable to species with narrow temperature tolerances. Warming can also exacerbate the problem of eutrophication, leading to algal overgrowth, fish kills, and dead zones (WRI 2000). (See Figure 19.10 for the location of major hypoxic areas in coastal systems.) Finally, warming is expected to further increase the transmission rates of pathogens and hasten the spread of many forms of human and nonhuman disease.

BOX 19.4

Water Diversion in Watersheds versus Water and Sediment Delivery to Coasts

The degree to which river water and sediment reach the coastal zone depends on other human activities, such as the construction of structures for water diversion, flood control, power generation, and recreation. Reservoirs and irrigation channels can retain a large proportion of the fluvial sediment discharge (Farnsworth and Milliman 2003). According to Vörösmarty et al. (1997, 2003), the 663 dams with large reservoirs (greater than 0.5 cubic kilometers maximum storage capacity) globally store about 5,000 cubic kilometers of water (approximately 15% of the global river water discharge). Also, global large reservoirs intercept more than 40% of global water discharge, and approximately 70% of this discharge maintains a sediment trapping efficiency of more than 50%.

Further analysis of the recent history of anthropogenic sediment retention by large dams (Vörösmarty et al. 2003) indicates that between 1950 and 1968, there was an increase from 5% to 15% in global sediment trapping, another increasing trend to 30% by 1985, and stabilization thereafter. As much as 25% of the current sediment load from the land to the coastal zones is trapped behind reservoirs. The trapping effect of fresh water discharge and suspended sediment by 45,000 dams analyzed in this study has dramatic impacts on water and sediment destined for the global coastal zone and inland seas. (See Figure.) Assuming that the global natural sediment discharge is between 18 and 20 gigatons per year, then the combined impact of all large dams will be on the order of 4–5 gigatons per year. Therefore, modern dam construction reduces the global sediment flux to the world's coastal zones by 25–30%.

According to Syvitski (2003), by decreasing sediment loads to the river through damming, coastal erosion is increased, and coastal marine ecosystems frequently deteriorate. Many dramatic examples of river control and utilization and their impacts on coastal systems have been recognized. After the Aswan Dam was completed in 1964, for example, the productive fishery collapsed and was reduced by 95%, and the delta subsided rapidly. The fishery remained unproductive for 15 years. It began a dramatic recovery during the 1980s, coincident with increasing fertilizer use and thus a flux of nutrients, expanded agricultural drainage, and increasing human population and sewage collection systems (see Nixon 2003). Similarly, after the Colorado River in the United States was dammed, sediment and nutrient discharge decreased dramatically and the shrimp catch in Baja California collapsed. After completion of the Kotri Barrage on the Indus River in Pakistan in 1956, fish catch decreased by a factor of three. And in China's Sea of Bohai, when the sediment discharge of the Yellow River was reduced the shrimp fishery decreased by 85% and the percentage of high-quality fish dropped by an order of magnitude.

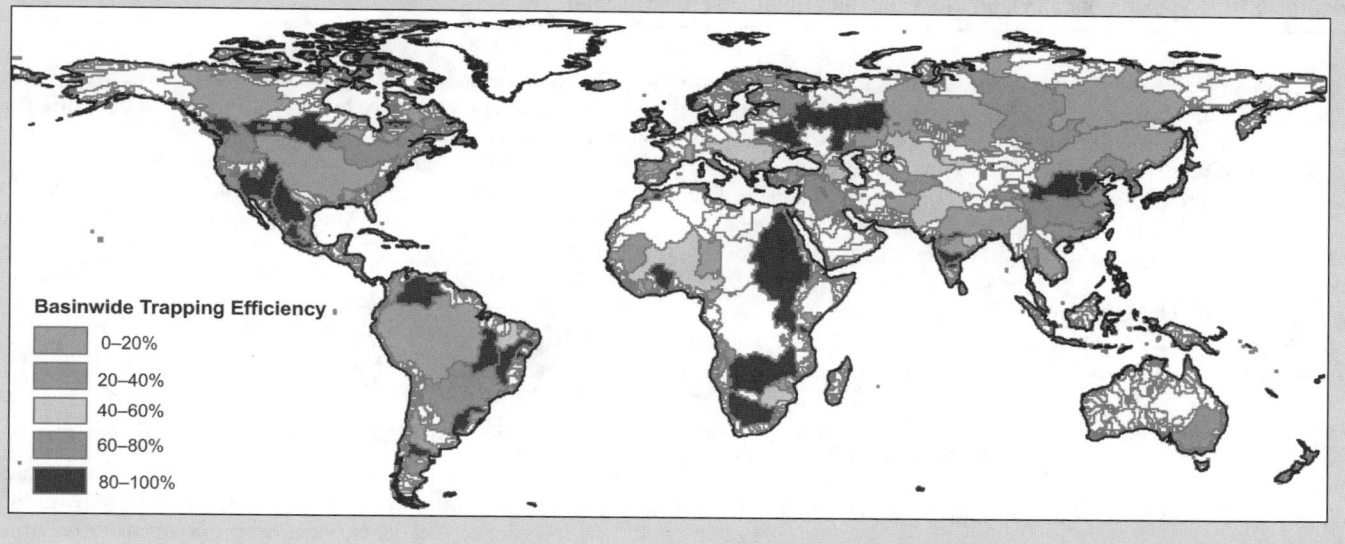

Basinwide Trapping Efficiency
- 0–20%
- 20–40%
- 40–60%
- 60–80%
- 80–100%

Climate change–related sea level rise will cause continued inundation of low-lying areas, especially where natural buffers have been removed (Church et al. 2001). Sea level rise is due to thermal expansion of ocean waters and melting of land based–ice, and both expansion and ice melts are expected to increase (IPCC 2003). In most if not all cases, global climate change impacts act in negative synergy with other threats to marine organisms and can be the factor sending ecosystems over the threshold levels of stability and productivity. In limited cases, new habitats may be created. Changes in weather patterns modeled in some extreme scenarios of climate change—including increased precipitation in some areas, abrupt warming at the poles, and increased frequency and intensity of storm events—would affect oceanic circulation (perhaps even leading to the collapse of thermohaline circulation) and currents as well as the ability of organisms to live or reproduce.

Different coastal subtypes, habitats, and even taxonomic groups will be affected by these direct and indirect impacts to greater or lesser degrees. Coral reefs may be the most vulnerable of all coastal subtypes (*medium certainty*), since multiple threats affect systems and since tolerances for corals and related reef species are generally of a narrow range. Estuaries are also vulnerable because these systems are directly subject to impacts from land (Gosselink et al. 1974; Turner et al. 1996) and water. Semi-enclosed seas are more vulnerable to degradation than open ocean basins—and because more isolated coastal waters have higher endemism, biodiversity is at greater risk in these areas.

Looking ahead 10–50 years suggests that some geographic areas of the world are expected to show particularly high rates of change and loss of certain ecosystem services. Southeast Asia, with its burgeoning population growth, limited land area, and largely ineffective controls on fisheries, pollution, and coastal development, is expected to continue to be an area of extremely rapid coastal change with losses in food provisioning, biodiversity, nutrient cycling, and storm protection services (*high certainty*). Small islands will continue to suffer dramatic alterations to their coastal

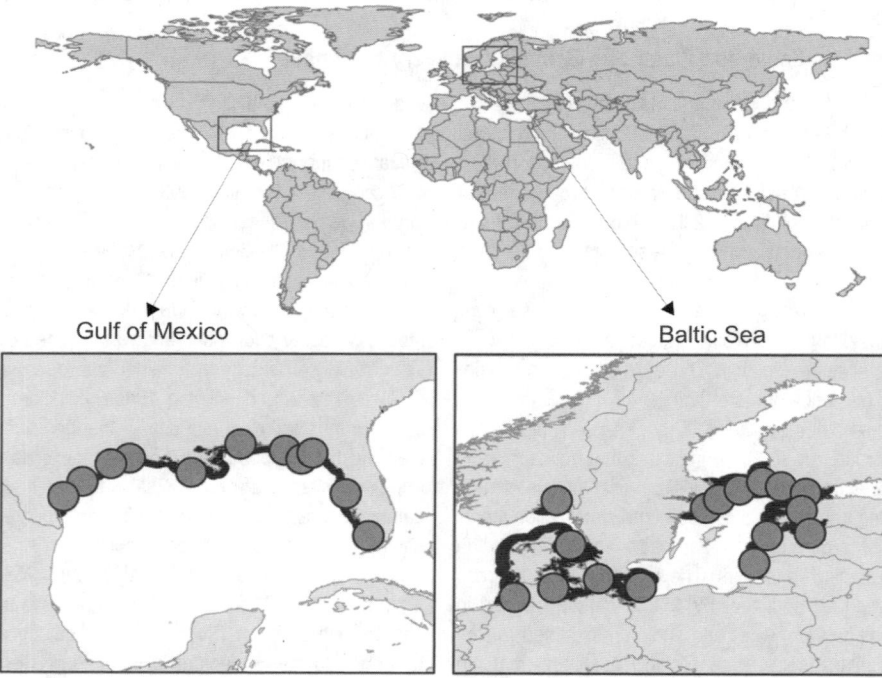

Figure 19.10. Hypoxic Zones in Gulf of Mexico and Baltic Sea (UNEP 2004)

environments, especially in the Pacific Ocean, Indian Ocean, and Caribbean Sea, where archipelagos of small islands support large numbers of residents and tourists but where monitoring and enforcement of regulations is difficult due to the distances between islands and limited resources. The areas of greatest change in land use that are situated in the coastal zone, such as those in the Middle East region, will also suffer rapid coastal change in the coming years.

The continued degradation of coastal ecosystems is paradoxical. Despite the value of coastal areas in supporting the tourism industry, for instance, coastal tourism development often uses habitats such as estuaries, mangroves, marshes, and atoll lagoons for waste disposal, degrading these areas and reducing their capacity to provide ecosystem services such as waste processing and coastal protection. Tourism development also results in conversion of habitat to accommodate infrastructure, resulting in loss of dune systems, wetlands, and even coral reefs. Damming damages estuaries and reduces fisheries yields, even if there are benefits of freshwater diversion for increasing food supply in terrestrial systems.

The costs of such trade-offs are significant, especially since the economic value of coastal developments that are put at risk by loss of protective and regulating services are high. A relatively new and rapidly growing form of coastal development that severely affects coastal ecosystems is uncontrolled building of shrimp ponds and other aquaculture sites (Lubchenco 2003). Dredging of waterways, as well as sand and coral mining, also cause habitat loss. Urbanization has enormous impact on the coasts, both in developing countries where displaced landless people often take up residence in urban shanties and in industrial countries where urban and suburban sprawl threaten natural habitats and ecosystem services. Finally, humans increasingly cause the loss of coastal habitats through destructive fishing practices such as blast fishing (the use of underwater explosives) and trawling (dragging of weighted nets along the sea floor).

For some degraded coastal habitats, such as mangroves, marshes, and areas of seagrass, it may be possible to regain ecosystem services through restoration, but the prohibitively high costs prevent restoration being an effective policy for other habitat types. Some ecosystems under the right conditions may recover or regenerate without intervention, but in most ecosystems active and expensive restoration may be necessary. Toxin loadings, pathogens, and alien species invasions will further stress coastal ecosystems and may impede natural recovery and managed restoration; human well-being will suffer as a consequence unless significant improvements to coastal management are systematically made across wide regions of the globe (*high certainty*).

19.4.2 Drivers of Change in the Coastal System

As noted previously, population growth is highest in coastal countries, and population densities within the coastal system are high. Urban areas are often concentrated on the coast: half of all major cities (with populations above 500,000 inhabitants) are located in coastal systems. Population doubling rates are highest in coastal areas.

However, the link between sheer population number and environmental quality is not clear-cut. Some authors argue that a direct link exists between the number of people and the quality of the environment or loss of diversity, regardless of consumption patterns (McKee et al. 2004). Others argue that the number of households is better correlated to the environmental impact or ecological footprint left by humans (Liu et al. 2003). In the coastal zone, however, neither population numbers nor household numbers tell the full story. Patterns of consumption and other human behaviors greatly influence the ecological footprint left by communities, and migration and its effects often spell the difference between sustainable and unsustainable use (Creel 2003; Curran and Agardy 2002). Local resource use and migration patterns are also affected by local and international markets.

In many industrial countries, urban sprawl is a major driver behind coastal ecosystem impacts and habitat loss. In the United

States, for example, it is the pattern of growth, which includes runaway land consumption, dysfunctional suburban development patterns, and exponential growth in automobile use, rather than population growth itself in the coastal zone that has affected ecosystems and their services (Beach 2002).

National and local economies influence the ability of countries to manage resource use and lessen impacts on ecosystem services. Industrial countries with strong economies have the ability to put resource management programs in place, undertake pollution mitigation and ecological restoration, and support surveillance and enforcement. However, wealthy countries also tend to be proportionately greater consumers, and their large-scale industries often threaten the environment (Creel 2003). Agribusiness and other large-scale industries often have a disproportionately large voice in democratic governments, since they can underwrite extensive lobbying on their behalf (Speth 2004), and subsidization can also steer such industries away from sustainability (Myers and Kent 2001).

Even individual wealth can have a negative impact on the environment. Expensive chemicals are generally available only to industry or the wealthy (such as tributyltin, used to prevent fouling of ship hulls, which has harmed marine species and caused changes in sex in exposed organisms), while in the industrial world, improved access to drugs threatens coastal systems, since antibiotics and hormones (especially ethinyloestradiol, a synthetic estrogen used in birth control pills) find their way into streams and rivers and eventually into coastal systems (Colburn et al. 1996). Since the magnitude of the impact of these chemicals on coastal ecology and on human health is not fully understood, there has been little impetus to implement mitigation measures to prevent pollutants from entering streams, rivers, sewers, and estuaries.

Foreign markets and globalization have been major drivers behind degradation of coastal ecosystems and diminishing services. Globalization causes greater mobilization of fishers and other users, greater flow of information and access to resources, increased fishery or other trade-related pollution and habitat loss, and loss of rights and representation of local peoples, leading to marginalization (Alder and Watson 2004). Access to markets and growing consumer demand (for both legal and illegal goods) increase pressures on resources and can lead to overexploitation and habitat loss.

For instance, conversion of habitat for aquaculture drives much of the loss of habitat and services in coastal South America and Southeast Asia. Although in Latin America, habitat conversion is undertaken primarily by large international corporations, in Thailand and Viet Nam there is a more balanced mixture of small- and large-scale farms. Production is geared completely toward export markets. The growth in this industry has little or nothing to do with population growth or local demands for sources of food. In Ecuador, more than 50,000 hectares of mangrove forest has been cleared to make shrimp ponds since 1969, representing a 27% decline in mangrove cover. During the same period, shrimp ponds have gone from zero to over 175,000 hectares. While there has been some recent reforestation in Ecuador (representing approximately 1% increase in a four-year period), this may be more to do with increasing market competition with Southeast Asian producers.

In Thailand, both primary conversion of mangroves and wetlands and secondary conversion of rice, rubber, and other agricultural crops to shrimp farms has occurred. Ten years of observations of shrimp farm production in Thailand (Lebel et al. 2002) suggests that once shrimp farms are established, the resulting sedimentation, salinization, and changed tidal influences may seriously im-

pede natural or planned regeneration of coastal forests or tidal basin species and may alter animal communities in waterways and wetlands. An analysis of shrimp farm production also demonstrates the multitude of linkages via the vital flow of water between human-based, land-based, coast-based, and marine-based systems. (See Box 19.5.)

The aquaculture-driven conversion of coastal habitat in Asia presents lessons about understanding drivers of ecosystem change in all coastal habitats. While it is necessary to separate threats to ecosystems in order to assess their impact, it is equally important to note that most coastal areas are facing multiple threats simultaneously, and many have experienced chronic impacts over long periods of time. Table 19.5 presents a typology of drivers of change in coastal systems and ecosystem services.

In a set of systems as complex and diverse as coastal systems, however, it is more germane to discuss drivers behind certain classes of impacts separately, rather than speaking of coastal ecosystem degradation more generally. Arguably, the greatest impacts on coastal systems worldwide are caused by the conversion of habitat for the purposes of coastal development (wetlands infilling, dredging of bays and harbors for port development, and so on) and through certain kinds of resource use (mangrove harvest, destructive fishing, and the like). These changes cause major if not total losses in ecosystem services and are largely irreversible.

For this reason, much attention has been paid to population growth in the coastal zone and the ways in which population drives habitat loss. Certainly this is true in poorly developed areas, where mangrove remains an important source of fuelwood and competition for increasingly scarce fisheries forces fishers to use unsustainable techniques. However, population is not the only driver behind habitat loss, and a confluence of chronic negative impacts may eventually lead to as debilitating a loss of ecosystem services as the more visible loss of habitat caused by growing populations.

19.5 Trade-offs, Synergies, and Management Interventions

19.5.1 Trade-offs, Choices, and Synergies

A central concern in coastal management is one of making trade-offs between ecosystem services in allocating increasingly scarce resources among society's members. Decision-makers face questions such as, Should this shoreline be cleared and stabilized to provide new land for development, or should it be maintained in its current state to serve as wildlife habitat? Should that wetland be drained and converted to agriculture, or should more wetland area be created to provide nutrient filtration services? Should this coral reef be mined for building materials and the production of lime, mortar, and cement, or should it be sustained to provide renewable seafood products and recreational opportunities?

To choose from among competing options, it is often necessary to compare the value that various groups in society receive from any improvement in a given coastal ecosystem with the value these groups give up to degrade the same system. Given this, a key question comes down to, What gets counted and how? Unfortunately, there are usually very few (if any) studies that can provide decision-makers with information on the full range of values provided by coastal ecosystem services, which is needed to evaluate specific trade-offs.

The wide variety of habitats, resources, and ecosystem services provided by coastal systems, and the strong interlinkages between these various components and processes suggest that complicated choices and difficult trade-offs exist whenever any form of coastal

BOX 19.5

Four Pathways to Coastal Ecosystem Degradation and Poverty through Shrimp Production in Thailand (Lebel et al. 2002 and references therein)

Shrimp production and sales are one of the fastest-growing food commodity markets in the world. Farmed shrimp production in the world market went from 815,250 tons in 1999 to 1.6 million tons in 2002. The industry has been growing at the rate of 10–20% a year over the last five years. The increase in production results from the spread of shrimp farm production along coastal ecosystems and tidal plains around the world. The area farmed for shrimp is approximately 1.2 million hectares. This number does not include the estimated 250,000 hectares of abandoned shrimp farms around the globe.

In 2000, the vast majority of shrimp farms were located in Pacific rim countries, along the coasts of South, East, and Southeast Asia. Approximately 89% of shrimp farmland is located in Asia. Thailand is the leading exporter of shrimp, with 25% of the world market, and the growth in Thai shrimp farms has been dramatic. In 1995, there were 19,700 farms covering about 65,000 hectares; by 2003 there were 35,000 farms encompassing 80,000 hectares.

Shrimp farm production has garnered attention from environmentalists because the building of shrimp farms is often linked to mangrove forest clearing. But extensive work by Lebel and colleagues in Thailand over the last decade suggests that the relationship between shrimp farm production and coastal ecosystem degradation is much more complex than popular science would predict. Lebel describes four pathways through which shrimp farm production there degrades coastal ecosystems and affects coastal communities' livelihoods. The complex interaction between shrimp farm production for global markets, coastal ecosystems, farmer/fisher livelihoods, and unsustainable and short-lived capacity of individual shrimp ponds spurred both a dramatic deterioration of coastal ecosystems over the last decade and recent, aggressive exploration of inland freshwater systems as a substitute.

The first pathway is sedimentation. Artificial shrimp ponds must be dredged and cleaned regularly. Typically, farmers empty the sludge into nearby coastal creeks or river basins. The resulting sedimentation has several effects. The filling of creeks and river tidal basins deters small fisher navigation through the creeks and tidal basins, disrupts nesting and breeding grounds of coastal dependent marine and shore species, and diminishes coastal fisheries. The buildup of sediment also diminishes the flushing and nutrient cycling role of tidal surges, further depleting the quality of coastal creeks and tidal basins. Sedimentation results from the building of fish ponds, which are usually bulldozed, the regular dredging of the ponds, and the more frequent pumping of pond water into coastal creeks. Although regulations stipulate that all larger farms must operate post-production water treatment ponds, many do not.

Salinization, the second pathway, is the result of three processes. First, the standing water in ponds and the resulting evaporation yields a buildup of salt in the pond water. Second, because tidal surges are minimized, the salt buildup is not naturally flushed out to sea. Third, in some locations saline water is pumped or trucked inland because local water

sources are too fresh. The impacts appear to be most serious in areas without immediate access to coastal inlets or the ocean. Here, wastewater effluent from ponds is dumped into canals and waterways previously for irrigation of rice, orchards, and rice-sugar palms systems. Productivity declines and some tree species die, making it hard for non-shrimp farming land uses to persist in an area. These "off-site" or landscape effects have often been a source of sharp conflict between shrimp farmers and other farmers. Over the past couple of years, techniques for rearing shrimp under freshwater conditions have greatly improved and spread.

Although the building of shrimp farms typically takes place on private land, the residual creeks, wetlands, and shoreline are frequently understood to be public lands, so public accessibility can be the third pathway of degradation. Because of the value of shrimp, however, shrimp farm producers limit access to once-public natural resources that are near ponds. Guards actively dissuade local fishers, farmers, and hunters from using these public goods. Those who need access to these public resources are those in most need of the livelihood income generated from gathering freshwater clams, plants and greens, and small fish.

Impoverization is the fourth pathway identified by Lebel et al. The establishment of shrimp farms improves wage-based employment opportunities for many local residents, and the presence of shrimp farms in an area provides opportunities for short-term work in sorting, pond cleaning, equipment maintenance, and reselling of inputs. Where factories are present, there can also be substantial wage-earning opportunities, primarily for young women. As a result, the poorest members of coastal villages, who used to define their livelihoods as coastal fishers, now work as low-wage earners in shrimp farm production. In addition, during the early to middle years of the growth in shrimp production, many small fishers and farmers established their own farms, despite the high cost of entry. Many of these smaller producers quickly lost their ponds, however, as the economies of scale, the high incidence of disease outbreaks, and the fluctuation in prices precluded success for those with few assets in reserve. Currently many owners of coastal land rent their property to large corporate shrimp farm producers. Rental contracts typically include clauses that abrogate a renter's responsibility for payment if the shrimp harvest fails.

In both cases, coastal residents—whether low-wage workers on shrimp farms or owners of land rented to shrimp farm producers—are increasingly vulnerable to global market price fluctuations that affect the profitability of local shrimp ponds, as well as the rapidly deteriorating quality of the coastal ecosystem. The latter not only increases the likelihood of disease outbreaks and shrimp pond abandonment, it also precludes possible alternative coastal-based resource livelihoods once the shrimp farm economy collapses or moves elsewhere. Recently, as a result of the deterioration of coastal sites and limited new sites for expansion along the coast, corporate research and development has focused on shifting shrimp production into freshwater systems.

development or protection takes place. For example, the choice to cut down mangrove forest to build a seaside resort will not only involve opportunity costs in reducing mangrove availability to local people, it will also have an impact on other uses of the coastal zone, such as fishing, and will dramatically reduce ecological services such as storm buffering, maintaining water and sediment balances, water purification, nutrient delivery, biodiversity maintenance, and provisioning of nursery areas for coastal fishery species. Similarly, the decision to protect a key habitat via establishment of a marine protected area will mean that access to resources will be restricted, and it may incur additional costs such as overexploitation of resources outside the protected area, as well as the costs of protected area management. Thus decisions over the management of coastal systems need to consider the various trade-offs inherent in alternative management practices (Brown et al. 2001).

Often the trade-offs are related to who has access to resources or who benefits from coastal development (Creel 2003). For ex-

Table 19.5. Drivers of Change in Coastal Ecosystems

Direct Drivers	Indirect Drivers
Habitat Loss or Conversion	
Coastal development (ports, urbanization, tourism-related development, industrial sites)	population growth, poor siting due to undervaluation, poorly developed industrial policy, tourism demand, environmental refugees and internal migration
Destructive fisheries (dynamite, cyanide, bottom trawling)	shift to market economies, demand for aquaria fish and live food fish, increasing competition in light of diminishing resources
Coastal deforestation (especially mangrove deforestation)	lack of alternative materials, increased competition, poor national policies
Mining (coral, sand, minerals, dredging)	lack of alternative materials, global commons perceptions
Civil engineering works	transport and energy demands, poor public policy, lack of knowledge about impacts and their costs
Environmental change brought about by war and conflict	increased competition for scarce resources, political instability, inequality in wealth distribution
Aquaculture-related habitat conversion	international demand for luxury items (including new markets), regional demand for food, demand for fishmeal in aquaculture and agriculture, decline in wild stocks or decreased access to fisheries (or inability to compete with larger-scale fisheries)
Habitat Degradation	
Eutrophication from land-based sources (agricultural waste, sewage, fertilizers)	urbanization, lack of sewage treatment or use of combined storm and sewer systems, unregulated agricultural development, loss of wetlands and other natural controls
Pollution: toxics and pathogens from land-based sources	lack of awareness, increasing pesticide and fertilizer use (especially as soil quality diminishes), unregulated industry
Pollution: dumping and dredge spoils	lack of alternative disposal methods, increased enforcement and stiffer penalties for land disposal, belief in unlimited assimilative capacities, waste as a commodity
Pollution: shipping-related	substandard shipping regulations, no investment in safety, policies promoting flags of convenience, increases in ship-based trade
Salinization of estuaries due to decreased freshwater inflow	demand for electricity and water, territorial disputes
Alien species invasions	lack of regulations on ballast discharge, increased aquaculture-related escapes, lack of international agreements on deliberate introductions
Climate change and sea level rise	insufficient controls on emissions, poorly planned development (vulnerable development), stressed ecosystems less able to cope
Overexploitation	
Directed take of low-value species at high volumes exceeding sustainable levels	population growth, demand for subsistence and market goods (food and medicinal), industrialization of fisheries, improved fish-finding technology, poor regional agreements, lack of enforcement, breakdown of traditional regulation systems, subsidies
Directed take for luxury markets (high value, low volume) exceeding sustainable levels	demand for specialty foods and medicines, aquarium fish, and curios; lack of awareness or concern about impacts; technological advances; commodification
Incidental take or bycatch	subsidies, bycatch has no cost
Directed take at commercial scales decreasing availability of resources for subsistence and artisanal use	marginalization of local peoples, breakdown of traditional social institutions

ample, conflicts frequently occur between large-scale commercial fisheries and small-scale (local) artisanal or subsistence fishing (see Chapter 18), or between tourism resort development and local communities who frequently receive little if any of the derived profits (nor even national economies in some instances). Zoning areas can reduce trade-offs and allow a suite of benefits to be derived from the same ecosystem, whether this occurs through smaller-scale marine protected areas (Brown et al. 2001; Villa et al. 2001) or through other coastal planning efforts such as those being installed throughout the Great Barrier Reef in Australia (Day 2002). Ocean zoning is also slowly becoming accepted as a

problem-solving tool in much the same way that land use zoning evolved slowly (and simultaneously) in many regions of the world decades ago. Zoning plans and permitting procedures for development that is potentially environmentally harmful are most effective when taking into account the costs of losing the ecosystem processes and services that these areas provide (U.S. Oceans Commission 2004).

Environmental impact studies that take into account the full value of the most important coastal areas where ecological processes are concentrated help decision-makers understand and quantify the trade-offs to be made when coastal development, environmental degradation through waste discharge, or exploitation of coastal areas occurs (Bocksteal et al. 1989; Brown et al. 2001). However, such studies require the kind of detailed assessment information that is lacking in many coastal areas and countries.

Some choices, when made in concert with others, will have an exponentially larger impact on ecosystem services than merely the additive effect of individual choices (a synergetic effect). For instance, if a management authority authorizes the development of coastal hotels that do not have sewage treatment facilities and at the same time authorizes fisheries on reefs nearby, the combined effect of increased nutrient pollution and decreased abundances of grazing fishes leads to algal overgrowth of the reefs and, in extreme cases, a regime shift from coral reefs to algal reefs (Birkeland 2004; McManus et al. 2000). Recovery from such alternate states is very difficult to achieve—and since the alternate state (algal reef) may not be as attractive to tourists, the resort business may well falter (Moberg and Ronnback 2003). Thus decision-makers who weigh not only the immediate costs and benefits from development but also the longer-term ones make better and often economically more viable choices.

Long time frames are extremely important to keep in mind. Many of the impacts humans have on coastal systems are small-scale, but when these become chronic, the cumulative impact may be quite large. In coastal systems that are downstream of recipients of terrestrial environmental degradation and sites of more immediate and direct degradation, threats to ecosystem health are multiple and especially cumulative. In these cases, decisions about resource and space utilization that are viewed holistically, with the long term in mind, are likely to have better outcomes for society.

19.5.2 Management Interventions

The story of human impacts on coastal ecosystems is a complex one involving not only a large number of diverse drivers acting simultaneously but also cumulative effects over time. Unfortunately, effective responses to such impacts on natural systems have typically only emerged after changes have taken effect, and management of coastal areas remains largely reactive.

Complex problems require comprehensive solutions. Integrated management of watersheds, land use planning, and impact assessment are key to protecting coastal ecosystems (Sorenson 1997). For this reason, tackling the issues of loss and degradation of coastal areas by addressing single threats to these environments has not proved effective in the past. The holistic approach—looking at how human activities affect coastal ecosystems, identification of key threats, and implementation of management that is integrated across all sectors—is a relatively new focus and is likely to produce much more effective decision-making. Effective management of these crucial areas means coordinated pollution controls, development restrictions, fisheries management, and scientific research.

Resource use that is managed in a way that considers the impacts that resource removal has on all linked ecosystems and human well-being has proved to be more effective than sectoral or single-species management (Kay and Alder 2004). Fisheries management agencies and conservationists are promoting ecosystem-based fisheries management—management that looks at multispecies interactions and the entire chain of habitats these linked organisms need in order to survive and reproduce (Agardy 2002). Due to the linkages between marine fisheries production and coastal ecosystem condition, the protection of coastal habitats figures very prominently in ecosystem-based fisheries management (Pauly et

al. 2002). However, truly holistic integrated management of coastal areas also requires complementary watershed management and land use planning to ensure that negative impacts do not reach coastal areas from outside the coastal realm.

Significant strides have been made in coastal management in the last few decades, in both the industrial and the developing world. Many of the world's 123 coastal countries have coastal management plans and legislation, and new governance arrangements and regulations are being developed every year (Burke et al. 2001). In 1993, it was estimated that there were 142 coastal management initiatives outside the United States and 20 international initiatives (Sorensen 1993). By 2000, there were a total of 447 initiatives globally, including 41 at the global level (Hildebrand and Sorensen 2001). This dramatic increase in activity was attributed both to initiatives that had started since 1993 and to the improved ability to find information on coastal management initiatives though the use of the Internet (Kay and Alder in press). The latest survey estimates that there are 698 coastal management initiatives operating in 145 nations or semi-sovereign states, including 76 at the international level (Sorensen 2002).

Yet even countries with well-developed coastal zone plans that have been in place for decades struggle with overexploitation of resources, user conflicts, habitat loss, and indirect degradation of ecosystems from activities occurring sometimes hundreds of kilometers away from the coastal zone itself. Management has not kept pace with degradation, as the number of management interventions worldwide has only increased two- or threefold over the last decade, while degradation of many habitats like coral reefs and mangroves has increased significantly more in the same time (Kay and Alder in press).

Some key coastal habitats such as mangrove forests, marshes, and seagrass meadows can be, and are being, restored once degraded. The science of mangrove restoration is relatively advanced, especially in the new world where natural species diversity is low and where replanting a few species can restore the ecosystems and most services quickly (Kaly and Jones 1998). Marshlands are also easily restored, as long as major alterations to hydrology have not taken place. Such restoration initiatives are risky, however, since it has yet to be shown that the full range of ecosystem services can be supported by artificially reconstructed wetlands (Moberg and Ronnback 2003; NRC 1992). Coral reef transplantation, though technologically possible, can only be practiced at a small scale and has had limited success (Moberg and Ronnback 2003). Furthermore, the costs of such restoration can be enormous, as the $7.8-billion price tag for the restoration of the Everglades cord grass system in Florida in the United States attests. In fact, most full-scale restoration (habitat reconstruction) is practiced in highly industrialized countries that are able to finance the high costs over the long time frames needed.

Management interventions to deal with pollution in coastal areas have largely failed. One method of mitigation is to conserve, reconstruct, or construct new wetlands that act as filters of these pollutants before the compounds enter the coastal environment. Another is to encourage land use practices such as buffer strips in agriculture and forestry to prevent the runoff of fertilizers, sediments, and so on. Municipal waste and storm runoff is sometimes controlled to limit hydrocarbons and other toxic inputs, and regulations regarding dredging operations help control the release of pollutants deposited into coastal sediments. However, no country has succeeded in comprehensively limiting pollution of the nearshore environment, despite the large number of initiatives and regulations in place.

One reason for these failures is that neither the status of coastal habitats nor the full values of coastal systems are known in many parts of the world. Effective management of coastal systems and the evaluation of trade-offs and choices requires both information and awareness. Education plays a key role in supplying both, and although education about ecology has generally improved in recent decades, education on marine systems is underfunded and underdeveloped (Kay and Alder in press). Further applied multidisciplinary research on ecosystem function, sustainable yields, and economic valuation of coastal ecosystems is also needed (Lubchenco 1998). Research focused on fundamental questions about ecosystem function, impacts, and efficacy of management measures will aid decision-makers in mitigating loss and degradation of these habitats. Fully protected areas help in this regard because they provide crucial control sites to test management interventions and allow for baseline monitoring. Better economic valuations—particularly quantitative estimates of marginal benefits—are also required to understand fully the importance of coastal systems.

Individual sites are sometimes recognized for their valuable services, and management interventions are put in place to conserve these habitats and the species within them through marine protected areas (NRC 2001). These may be small fisheries reserves in which resource extraction is prohibited, or they may occur in the context of larger multiple-use areas. Increasingly, marine protected areas are being established in networks in order to safeguard key areas of the coastal and marine environment over a geographically large area (Agardy 1999; Murray et al. 1999a; Pauly et al. 2002). A prime example of this is the network of reserves encompassed by the newly rezoned Great Barrier Reef Marine Park in Australia (Day 2002).

In order for marine protected areas to succeed in meeting the objectives of conserving habitats and protecting fisheries and biodiversity, their management seeks to address all the direct threats to marine and coastal areas. In most habitats, these threats are multiple and cumulative over time. Thus protected areas that address only one of these threats will usually fail to conserve the ecosystem or habitats and the services they provide (Agardy 1997).

Marine and coastal protected areas already dot coasts around the world, and the number of protected areas continues to increase. The last official count of coastal and marine protected areas, in 2003, yielded 4,116 (Spalding et al. 2003), a marked increase over the 1,308 listed in 1995 (Kelleher et al. 1995), though this is a significant underestimate because unconventional protected areas that do not fit the IUCN categories for protected areas are typically not counted.

By far the bulk of these protected areas occur in coastal zones, and many include both terrestrial and aquatic components. However, even with the large number of individual sites, coverage accounts for less than 1% of the world's oceans. Many marine protected areas occur in relatively close proximity to human settlements—in fact, nearly 10% of the world lives within 50 kilometers of a marine protected area, and over 25% of the worldwide coastal population lives within 50 kilometers of one. (See Table 19.6.)

Management effectiveness of most marine protected areas remains questionable, and many of these areas have no operational management or enforced legislation at all. It is *well established* that marine protected area tools are not being used to their fullest potential anywhere in the world (Agardy et al. 2003). Nonetheless, there are good examples of effective marine management, such as the Great Barrier Reef Marine Park. And examples such as this highlight how even a protected area that begins with relatively modest protection measures can be strengthened over time (Lawrence et al. 2002).

Table 19.6. Share of World and Coastal Populations Living Close to a Coastal Marine Protected Area

Category	Within 50 Kilometers of MPA	Within 100 Kilometers of MPA	Within 150 Kilometers of MPA
	(percent)		
World population	9	19	26
Coastal population	25	51	70

Tenure of marine areas and some forms of traditional use can also be effective coastal conservation interventions, even when these patterns of sustainable use of marine and coastal resources occur outside of conventional protected areas (Curran and Agardy 2002; Young 2004). Common property and common property management regimes have evolved in many coastal communities and have in some cases been shown to be much more effective than conventional, top-down methods in keeping resource use to sustainable limits (Agardy 1997; Curran and Agardy 2002). Legitimizing such traditional uses remains an issue in many coastal countries, and recently nongovernmental organizations have begun to liaise with governments to help codify use rights for local communities.

An analysis of the efficacy of coastal and marine protected areas, sustainable traditional use regimes, and common property management regimes highlights the fact that all such local action must be supplemented by effective management at much larger scales (Agardy 1999). Indeed, the interlinkages between terrestrial environments, fresh water, coastal systems, and the marine realm prevent local interventions from succeeding unless the larger context is addressed. Coastal zone management at the provincial, state, or national level can help scale up management efforts, as can zoning initiatives (Norse in press). Coastal management is a particularly important facet of national policy-making, as most coastal zones exist wholly within the exclusive economic zones of individual nations (Sorenson 1997).

A relatively recent movement in this direction is the coupling of coastal zone management with catchment basin or watershed management, as has occurred under the European Water Framework Directive and projects undertaken under the LOICZ (Land–Sea Interactions in the Coastal Zone) initiative. Such freshwater–marine system coupling has resulted in lower pollutant loads and improved conditions in estuaries. However, due to the fluid nature of the marine system and the large-scale interconnectivities, even larger-scale integrated management initiatives are required for effective management of coastal and marine systems over the long term.

Several international instruments provide a framework for such larger-scale regional cooperation, including the United Nations Convention on the Law of the Sea (UNCLOS 1982), U.N. Regional Seas Conventions and Action Plans, the Global Programme of Action for the Protection of the Marine Environment from Land-based Activities (GPA 1995), the Jakarta Mandate on the Conservation and Sustainable Use of Marine and Coastal Biological Diversity (CBD 1995), the RAMSAR Convention, Chapter 17 of *Agenda 21* (UNCED 1992), and Paragraph 29 of the Implementation Plan of the World Summit on Sustainable Development (WSSD 2002). While some of these international agreements pertain more directly to marine systems (as discussed in Chapter 18), all carry obligations or give guidance to parties on management of coastal areas. Yet while many international agreements and policies promote the idea of ecosystem-based management, the practical application of the concept is still being developed.

Global treaties and multilateral agreements can bridge some of the gaps that occur between small-scale interventions on the ground and large-scale coastal problems, but most of these international instruments have not been effective in reversing environmental degradation (Speth 2004). For shared coastal and marine resources, it may well be that regional agreements will prove more effective, especially when such agreements capitalize on better understandings of costs and benefits accruing from shared responsibilities in conserving the marine environment.

Large marine ecosystems have been put forward as a logical way to frame such agreements (Duda and Sherman 2002; Kimball 2001). Each of the world's 64 LMEs averages 200,000 square kilometers and is characterized by distinct bathymetry, hydrology, productivity, and trophically dependent populations (Sherman 1993). The LME concept was originally applied in the fisheries context under CCAMLR to take into account predator/prey relationships and environmental factors affecting target stocks; thus Antarctica was the site of the first truly ecosystem-based approach to fisheries management (Griffis and Kimball 1996). Several recent international instruments refer to LMEs, and the geographic units serve as the basis for some global assessments, such as GIWA (UNEP's Global International Waters Assessment). In many parts of the world, however, the political constituency for nations to cooperate to conserve large-scale ecosystems is lacking, though this situation may well be improving (Wang 2004).

Coastal ecosystems are crucial elements of the global environment, supporting not only marine food webs but also providing key services for humankind. To stave off the dramatic losses in coastal habitats that are now occurring worldwide, valuing these habitats and communicating their value to the public is crucial. And because in many parts of the world migration dramatically undermines regulation of coastal resource use, migration patterns and the drivers behind them merit investigation to provide the foundation for migration policies. Coastal systems are so complex, and the impacts humans have on them so varied, that coastal ecosystem services will only be successfully protected when the entire spectrum of threats and integrated responses to them are addressed. As human dependence on coastal services grows, management will continue to be challenged to manage the coastal environment more effectively.

19.6 Coastal Systems and Human Well-being

The coastal systems of the world are crucially important to humankind and are under ever-increasing threat from activities within and outside the coastal zone. Provisioning, regulating, supporting and cultural services have all been affected by human use and indirect impacts on coastal habitats, and some habitat types are close to being degraded to the point that important services will be lost altogether. Diminishing services caused by poor choices threaten the well-being of not only coastal communities, but coastal nations and the global community as well.

Many of these impacts affect rural communities in developing nations, especially where livelihoods are closely tied to availability of coastal resources. However, coastal degradation affects people in industrial countries as well, and has an impact on suburban and urban human well-being. For instance, according to a new study by the European Commission, a fifth of the coastline of the newly enlarged European Union is eroding away from human-induced causes, in a few cases as much as 15 meters (49 feet) shoreline

erosion inland a year (European Commission 2004). Such erosion threatens homes, roads, and urban infrastructure and the safety of individuals, as well as affecting biodiversity.

Resource overexploitation and coastal degradation undermine subsistence use of coastal ecosystems. Small rural populations are not the only ones to suffer from overexploitation and mismanagement, however—national economies are affected as well. For instance, potential net benefit streams from coral reefs include fisheries, coastal protection, tourism, and biodiversity values are estimated to total $29.8 billion annually (Cesar et al. 2003). Much of these revenues are at risk from ever-accelerating rates of coastal degradation. When the negative impacts from overfishing are coupled with inadequate environmental management that allows increases in pollutant levels and stresses coral reef health, the consequences can be a full-fledged ecosystem collapse or regime shifts to alternate (and less desirable) states (Birkeland 1997).

Many coastal communities, especially in poorer developing countries, are trapped in what has been called "a vicious cycle of poverty, resource depletion and further impoverishment" (Cesar et al. 2003). As in many other coastal and marine ecosystems, marginalization of fishers is largely responsible for "Malthusian" or exponentially increasing rates of overfishing (Pauly 1997). This phenomenon is not unique to coral reefs, of course, but once coral reefs are destroyed, restoration is extremely difficult, and the costs brought about by loss of services such as coastal protection continue to be incurred for long periods thereafter (Moberg and Ronnback 2003).

Pollution puts coastal inhabitants at great risk—directly, by affecting human health, and indirectly, by degrading the resource base on which many of them depend. Poor sanitation affects not only slum dwellers. For instance, South Asian waters are highly polluted throughout the region, partly as a result of 825 million people who live without basic sanitation services (UN System-Wide Earthwatch, cited in Creel 2003). Pathogens are spread more quickly and reach greater numbers of people in coastal ecosystems that have become degraded. Chronic exposure to heavy metals and other bioaccumulating pollutants may not cause death in large numbers of people, but their cumulative effect can lead to reproductive failure and significantly decreased well-being. Food security is also greatly compromised in degraded coastal ecosystems.

Yet even when people are made aware of the importance of coastal ecosystems, they still may not be able to stop the kinds of activities that destroy or degrade these areas unless alternative resources or livelihoods are made available to them. For instance, boat-builders of the coastal and island communities of East Africa have little choice but to harvest mangrove for boat construction from key nursery habitats, which support the very fisheries on which their boat-building industry is based (Agardy 1997). Few alternative materials for boat building exist, except when conservation projects have expressly built in alternatives and training on how to use them. In areas in which resource extraction is moving beyond ecologically sustainable limits or the removal of the resource causes major physical changes to the habitat, the search for alternatives is particularly crucial.

A "business as usual" approach is projected to lead to continued loss of habitats and species, with attendant changes to ecosystem services and negative impacts on many coastal-dependent industries and coastal communities. Degradation will result in future choices of either accepting loss of ecosystem services or investing in costly restoration programs that are not guaranteed to reinstate the full range of services. Connectivity of systems and the large spatial scale of impacts will mean that local-scale or site-specific conservation and management investments will be increasingly at risk as overall coastal and marine conditions deteriorate. Changes in species distribution and abundance in response to climate change, resource use, and pollution may render many protected areas ineffective.

Yet enough is known to change the current approach and begin to systematically develop strategic plans for more effective protection and more sustainable use of coastal ecosystems (Kay and Alder, in press). Coastal areas could be zoned to allow appropriate uses in various areas, reduce user conflicts, and limit the impacts of detrimental trade-offs. Marine protected areas could well serve as starting points for such zoning measures, as well as acting as small-scale models for integrating coastal and marine management across all sectors (Agardy 2002).

In all parts of the world, it will be crucial to find ways to involve local communities in planning management interventions and zoning schemes in order to better safeguard resources, coastal areas, and human well-being. At the same time, ecological linkages between systems must be maintained in order to continue the delivery of services. Effective management for sustainable use of coastal systems will best be achieved by applying an ecosystem-based, whole-catchment approach that addresses land use upstream and the use of marine resources far out to sea. Multilateral, regional initiatives and agreements could help foster an integrated and comprehensive approach and may well lessen the costs of management through economies of scale. Regional cooperation schemes would facilitate a scaling up of management interventions that have to date been on too small a scale, and thereby help abate declines in coastal services and related human well-being.

References

Adam, P., 2002: Saltmarshes in a time of change. *Environmental Conservation,* **29(1),** 39–61.

Agardy, T., 1997: *Marine Protected Areas and Ocean Conservation.* RG Landes Company and Academic Press, Austin, TX (USA), 244 pp.

Agardy, T., 1999: Creating havens for marine life. *Issues in Science and Technology,* **16(1),** 37–44.

Agardy, T., 2002: An environmentalist's perspective on responsible fisheries: The need for holistic approaches. In: *Responsible Fisheries in the Marine Ecosystem,* M. Sinclair and G. Valdimarson (eds.), Food and Agriculture Organization of the United Nations (FAO) and CAB International, Rome (Italy) and Wallingford (UK), 65–85.

Agardy, T., 2004: America's coral reefs: Awash with problems. *Issues in Science and Technology,* **20(2),** 35–42.

Agardy, T., P. Bridgewater, M.P. Crosby, J. Day, P.K. Dayton, et al. 2003: Dangerous targets? Unresolved issues and ideological clashes around marine protected areas. *Aquatic Conservation-Marine and Freshwater Ecosystems,* **13(4),** 353–367.

Aguilar, A. and A. Borrell. 1994. Assessment of organochlorine pollutants in cetaceans by means of skin and hypodermic biopsies. In: *Nondestructive Biomarkers in Vertebrates* (M.C. Fossi and C. Leonzio eds.). Lewis Publishers Inc., p. 245–267.

Ahmed, M., C.K. Chong, and H. Cesar, 2004: *Economic Valuation and Policy Priorities for Sustainable Management of Coral Reefs. International Consultative Workshop,* Worldfish Center, Penang (Malaysia), 222 pp.

Alder, J., 2003: *Distribution of estuaries worldwide.* Sea Around Us Project, UBC, Vancouver, B.C. (Canada).

Alder, J. and R. Watson, 2004: Globalization and its effects on fisheries. In: *Proceedings of the Workshop Globalization: Effects on fisheries, 12–14 August 2004, Quebec (Canada),* W. Taylor, M. Schechter, and L. Wolfson (eds.), Cambridge University Press, NY (USA).

Allee, R., M. Dethier, B. Brown, L. Deegan, R.G. Ford, et al. 2000: *Marine and Estuarine Ecosystem and Habitat Classification.* NOAA Technical Memorandum NMFS-F/SPO-43, Silver Spring, MD (USA).

Anderson, D.M., P. Andersen, V.M. Bricelj, J.J. Cullen, and J.E. Rensel, 2001: *Monitoring and Management Strategies for Harmful Algal Blooms in Coastal Waters.* APEC #201-MR-01.1, Asia Pacific Economic Programme, Singapore, and Intergovernmental Oceanographic Commission Technical Series No. **59,** Paris (France).

Anonymous, 2004: The Global Maritime Boundaries Database. [CD-ROM] General Dynamics Advanced Information Systems. Cited November 2004. Available

Arntz, W.E. and E. Fahrbach, 1996: *El Niño: experimento climático de la naturaleza.* Fondo de Cultura Económica, Ciudad de México (México).

Austin, G.E., I. Peachel, and M.M. Rehfisch, 2000: Regional trends in coastal wintering waders in Britain. *Bird Study,* **47,** 352–371.

Bakan, G. and Büyükgüngör, 2000: The Black Sea. In: *Seas at the Millenium: An Environmental Evaluation,* C. Sheppard (ed.). Volume 1. Regional Seas: Europe, The Americas and West Africa, Elsevier Science Ltd., Oxford (UK) and Pergamon Press, Amsterdam (Netherlands), 285–305.

Baker, A.J., P.M. González, T. Piersma, L.J. Niles, I. de Lima Serrano do Nascimento, et al. 2004: Rapid decline in red knots: fitness consequences of decreaed refueling rates and late arrival in Delaware Bay. *Proceedings of the Royal Society of London B,* 271, 875–882.

Balmford, A., A. Bruner, P. Cooper, R. Costanza, S. Farber, et al. 2002: Ecology—Economic reasons for conserving wild nature. *Science,* 297(5583), 950–953.

Bann, C., 1997: The economic valuation of mangroves: An manual for researchers. [online] Cited November 2004. Available at http://web.idrc.ca/uploads/user-S/10305674900acf30c.html.

Barbier, E.B., 1993: Sustainable Use of Wetlands—Valuing Tropical Wetland Benefits—Economic Methodologies and Applications. *Geographical Journal,* 159, 22–32.

Barbier, E.B., 2000: Valuing the environment as input: review of applications to mangrove-fishery linkages. *Ecological Economics,* 35(1), 47–61.

Batie, S.S. and J.R. Wilson, 1978: Economic Values Attributable to Virginia's Coastal Wetlands as Inputs in Oyster Production. *Southern Journal of Agricultural Economics,* 111–118.

Beach, D., 2002: *Coastal Sprawl: The Effects of Urban Design on Aquatic Ecosystems in the United States.* Prepared for the Pew Oceans Commission, Arlington, VA (USA).

Beck, M.W., K.L. Heck, K.W. Able, D.L. Childers, D.B. Eggleston, et al. 2001: The identification, conservation, and management of estuarine and marine nurseries for fish and invertebrates. *Bioscience,* 51(8), 633–641.

Bellwood, D.R., T.P. Hughes, C. Folke, and M. Nystrom, 2004: Confronting the coral reef crisis. *Nature,* 429(6994), 827–833.

Beman J.M., K.R. Arrigo, and P.A. Matson, 2005: Agricultural runoff fuels large phytoplankton blooms in vulnerable areas of the ocean. *Nature,* 434:211–214.

Ben-David, M., R.T. Bowyer, L.K. Duffy, D.D. Roby, and D.M. Schell, 1998b: Social behavior and ecosystem processes: River otter latrines and nutrient dynamics of terrestrial vegetation. *Ecology,* 79(7), 2567–2571.

Bennett, E.L. and C.J. Reynolds, 1993: The Value of a Mangrove Area in Sarawak. *Biodiversity and Conservation,* 2(4), 359–375.

Benoit, L.K. and R.A. Askins, 2002: Relationship between habitat area and the distribution of tidal marsh birds. *Wilson Bulletin,* 114(3), 314–323.

Binet, D. and E. Marchal, 1993: The Large Marine Ecosystem of Shelf Areas in the Gulf of Guinea: Long-Term Variability Induced by Climatic Changes. In: *Large Marine Ecosystems: Stress, Mitigation, and Sustainability,* K. Sherman, L. Alexander, and B. Gold (eds.), American Association for the Advancement of Science, Washington, D.C. (USA), 104–118.

Bingham, G., R. Bishop, M. Brody, D. Bromley, E. Clark, et al. 1995: Issues in Ecosystem Valuation—Improving Information for Decision-Making. *Ecological Economics,* 14(2), 73–90.

BirdLife International, 2004a: Threatened Birds of the World 2004. [CD-ROM]. Available at http://www.birdlife.net/datazone/search/species_search .html.

BirdLife International, 2004b: State of the World's Birds 2004: indicators for our changing world. BirdLife International, Cambridge (UK), 73 pp.

Birkeland, C., 1997: Implications for resource management. In: *Life and Death of Coral Reefs,* C. Birkeland (ed.), Chapman and Hall, New York, NY (USA), 411–435.

Birkeland, C., in press: Ratcheting down the coral reefs. *BioScience.*

Birkeland, C. and A. Friedlander, 2002: *The importance of refuges for reef fish replenishment in Hawai'i.* Hawai'i Audubon Society, Honolulu, HI (USA), 19 pp.

Bocksteal, N.E., K.E. McConnell, and I.E. Strand, 1989: Measuring the benefits of improvements in water quality: the Chesapeake Bay. *Marine Resource Economics,* 6, 1–18.

Boesch, D.F., R.H. Burroughs, J.E. Baker, R.P. Mason, C.L. Rowe, and R.L. Siefert, 2001: *Marine Pollution in the United States: Significant Accomplishments, Future Challenges.* Prepared for the Pew Oceans Commission, Arlington, VA (USA).

Brashares, J.S., P. Arcese, M.K. Sam, P.B. Coppolillo, A.R.E. Sinclair, and A. Balmford, 2004: Bushmeat Hunting, Wildlife Declines, and Fish Supply in West Africa. *Science,* **306,** 1180–1183.

Brown, A.C. and A. McLachlan, 2002: Sandy shore ecosystems and the threats facing them: some predictions for the year 2025. *Environmental Conservation,* **29(1),** 62–77.

Brown, K., W.N. Adger, E. Tompkins, P. Bacon, D. Shim, and K. Young, 2001: Trade-off analysis for marine protected area management. *Ecological Economics,* **37(3),** 417–434.

Buerger, R. and J.R. Kahn, 1989: New York value of Chesapeake striped bass. *Marine Resource Economics,* **6(1),** 19–25.

Burke, L., L. Selig, and M. Spalding, 2002: *Reefs at Risk in Southeast Asia.* World Resources Institute (WRI), Washington, D.C. (USA), 72 pp.

Burke, L., Y. Kura, K. Kassem, C. Ravenga, M. Spalding, and D. McAllister, 2001: *Pilot Assessment of Global Ecosystems: Coastal Ecosystems.* World Resources Institute (WRI), Washington, D.C. (USA), 94 pp.

Cahoon, D.R., J.W. Day Jr., and D.J. Reed, 1999: The influence of surface and shallow subsurface soil processes on wetland elevation: a synthesis. *Current Topics in Wetland Biogeochemistry,* **3,** 72–88.

Carlton, J.T., 1989: Man's role in changing the face of the oceans: biological invasions and implications for conservation of near-shore marine environments. *Conservation Biology,* **3,** 265–273.

Carlton, J.T., 1996: Marine Bioinvasions: The alteration of marine ecosystems by nonindigenous species. *Oceanography,* **9(1),** 36–43.

Carlton, J.T., 2001: *Introduced Species in U.S. Coastal Waters: Environmental impacts and Management Priorities.* Prepared for the Pew Oceans Commissions, Arlington, VA (USA), 36 pp.

Carr, A.F., 1979: *The Windward Road: Adventures of a Naturalist on Remote Caribbean Shores.* University Press of Florida, Gainesville, FL (USA), 258 pp.

Carte, B.K., 1996: Biomedical potential of marine natural products. *Bioscience,* 46(4), 271–286.

CBD (Convention on Biological Diversity), 1995: Jakarta Mandate on Marine and Coastal Biological Diversity. [online] Cited November 2004. Available at http://www.biodiv.org/programmes/areas/marine/.

Cesar, H. and C.K. Chong, 2004: Economic Valuation and Socioeconomics of Coral Reefs: Methodological Issues and Three Case Studies. In: *Economic Valuation and Policy Priorities for Sustainable Management of Coral Reefs,* M. Ahmed, C.K. Chong, and H. Cesar (eds.), WorldFish Center, Penang (Malaysia).

Cesar, H., L. Burke, and L. Pet-Soede, 2003: *The Economics of Worldwide Coral Reef Degradation.* ICRAN, Cambridge (UK) and WWF Netherlands, Zeist (Netherlands).

Chambers, J.R., 1992: Coastal degradation and fish population losses. In: *Stemming the Tide of Coastal Fish Habitat Loss. Marine Recreational Fisheries Symposium, 7–9 March 1991, Baltimore, MD (USA),* R.H. Stroud (ed.), National Coalition for Marine Conservation, Savannah, GA (USA), 45–51.

Church, J.A., J.M. Gregory, P. Huybrechts, M. Kuhn, K. Lambeck, et al. 2001: Changes in Sea Level. In: *Climate Change (2001). The Scientific Basis. Contribution of Working Group 1 to the Third Assessment Report of the Intergovernmental Panel on Climate Change.,* J.T. Houghton, Y. Ding, D.J. Griggs, et al. (eds.), Cambridge University Press, Cambridge (UK), 639–694.

Cicin-Sain, B., P. Bernal, V. Vandeweerd, S. Belfiore, and K. Goldstein, 2002: *Oceans, Coasts and Islands at the World Summit on Sustainable Development and Beyond. Integrated Management from Hilltops to Oceans.,* Center for the Study of Marine Policy, Newark, DE (USA).

CIESIN (Center for International Earth Science Information Network), 2003: Gridded population of the world (GPW) Version 3 beta. [online]. Available at http://quin.unep-wcmc.org/MA/index.cfm (with username and password).

CIESIN and World Bank, 2004: Unpublished Data. Described in "Natural Disaster Hotspots: A Global Risk Analysis" by Maxx Dilley, Robert S. Chen, Uwe Deichmann, Arthur L. Lerner-Lam, Margaret Arnold. 2005. World Bank, Washington, DC.

Clapham, P.J., S.B. Young, and R.L. Brownell, 1999: Baleen whales: conservation issues and the status of the most endangered populations. *Mammal Review,* **29(1),** 35–60.

Cognetti, G., C. Lardicci, M. Abbiati, and A. Castelli, 2000: The Adriatic Sea and the Tyrrhenian Sea. In: *Seas at the Millenium: An Environmental Evaluation,* C.R.C. Sheppard (ed.). Volume 1. Regional Seas: Europe, The Americas and West Africa, Elsevier Science Ltd., Oxford (UK) and Pergamon Press, Amsterdam (Netherlands), 267–284.

Cohen, A.N. and J.T. Carlton 1995: *Nonindigenous Aquatic Species in a United States Estuary: A Case Study of the Biological Invasions of the San Francisco Bay and Delta.* A report for the United States Fish and Wildlife Service, Washington D.C. (USA).

Cohen, A.N. and J.T. Carlton, 1998: Accelerating invasion rate in a highly invaded estuary. *Science,* **279(5350),** 555–558.

Cohen, J.E., 1995: *How many people can the Earth support?* W. W. Norton & Company, New York (USA) and London (USA), 532 pp.

Colburn, T., D. Dumanoski, and J.P. Myers, 1996: *Our Stolen Future: Are We Threatening Our Fertility, Intelligence and Survival. A Scientific Detective Story.* Dutton Press, New York (USA), 306 pp.

Colwell, R.R. and W.M. Spira, 1992: The ecology of *Vibrio cholerae.* In: *Cholera: Current Topics in Infectious Disease,* D. Barua and W.B.I. Greenough (eds.), Plenum Medical Book Company, New York, NY (USA), 107–127.

Costanza, R., 2000: The Ecological, Economic and Social Importance of the Oceans. In: *Seas at the Millennium: An Environmental Evaluation,* C.R.C. Sheppard (ed.). Volume 3—Global Issues and Processes, Elsevier Science Ltd., Oxford (UK) and Pergamon Press, Amsterdam (Netherlands), 393–403.

Costanza, R., R. dArge, R. deGroot, S. Farber, M. Grasso, et al. 1997: The value of the world's ecosystem services and natural capital. *Nature,* **387(6630),** 253–260.

Craik, W., R. Kenchington, and G. Kelleher, 1990: Coral reef management. In: *Ecosystems of the World: Coral Reefs,* Z. Dubinsky (ed.). 25, Elsevier, New York, NY (USA), 453–467.

Creel, L., 2003: *Ripple Effects: Population and Coastal Regions.* Making the Link: Population Reference Bureau, 8 pp.

Crowder, L., 2000: Leatherback's survival will depend on an international effort. *Nature,* **405(6789),** 881–881.

Cuadros Dulanto, M.H., 2001: Valoración económica total de la biodiversidad en Bahía Independencia, Reserva Nacional de Paracas. In: *Valoración económica de la diversidad biológica y servicios ambientales en el Perú,* M. Glave and R. Pizarro (eds.), Irg/Biofor, Lima (Peru).

Curran, S.R. and T. Agardy, 2002: Common property systems, migration and coastal ecosystems. *Ambio,* **31(4),** 303–305.

D' Avanzo, C., J.N. Kremer, and S.C. Wainright, 1996: Ecosystem production and respiration in response to eutrophication in shallow temperate estuaries. *Marine Ecology-Progress Series,* **141(1–3),** 263–274.

D'Agrosa, C., C.E. Lennert-Cody, and O. Vidal, 2000: Vaquita bycatch in Mexico's artisanal gillnet fisheries: Driving a small population to extinction. *Conservation Biology,* **14(4),** 1110–1119.

Davidson, N.C., 2003: Declines in East Atlantic wader populations: Is the Wadden Sea the problem? *Wader Study Group Bulletin,* 101/102, 9–10.

Dawson, S.M., A. Read, and E. Slooten, 1998: Pingers, porpoises and power: Uncertainties with using pingers to reduce by catch of small cetaceans. *Biological Conservation,* **84(2),** 141–146.

Day, J.C., 2002: Zoning—lessons from the Great Barrier Reef Marine Park. *Ocean & Coastal Management,* **45(2–3),** 139–156.

Dayton, P.K., 1994: Community landscape: Scale and stability in hard bottom marine communities. In: *Aquatic Ecology: Scales, Patterns and Processes,* P.S. Giller, A.G. Hildrew, and D.G. Raffaelli (eds.), Blackwell Press, Oxford (UK), 289–332.

Dayton, P.K., 2003: The importance of the natural sciences to conservation. *American Naturalist,* **162(1),** 1–13.

Dayton, P.K., S. Thrush, and F. Coleman, 2002: *Ecological Effects of Fishing in Marine Ecosystems of the United States.* Prepared for the Pew Oceans Commission, Arlington, VA (USA).

Dayton, P.K., S.F. Thrush, M.T. Agardy, and R.J. Hofman, 1995: Environmental-Effects of Marine Fishing. *Aquatic Conservation-Marine and Freshwater Ecosystems,* **5(3),** 205–232.

Dayton, P.K., M.J. Tegner, P.B. Edwards, and K.L. Riser, 1998: Sliding baselines, ghosts, and reduced expectations in kelp forest communities. *Ecological Applications,* **8(2),** 309–322.

De Groot, R.S., 1992: *Functions of Nature: evaluation of nature in environmental planning, management and decision-making.* Wolters Noordhoff BV, Groningen (Netherlands), 345 pp.

Deb, A.K., 1998: Fake blue revolution: environmental and socio-economic impacts of shrimp culture in the coastal areas of Bangladesh. *Ocean & Coastal Management,* **41(1),** 63–88.

Deegan, L.A., 1993: Nutrient and Energy-Transport between Estuaries and Coastal Marine Ecosystems by Fish Migration. *Canadian Journal of Fisheries and Aquatic Sciences,* **50(1),** 74–79.

Deegan, L.A. and R.N. Buchsbaum, 2001: The Effect of Habitat Loss and Degradation on Fisheries. In: *The decline of fisheries resources in New England: Evaluating the impact of overfishing, contamination, and habitat degradation,* R.N. Buchsbaum, W.E. Robinson, and J. Pederson (eds.), University of Massachusetts Press, Amherst (Netherlands).

Deegan, L.A., A. Wright, S.G. Ayvazian, J.T. Finn, H. Golden, R.R. Merson, and J. Harrison, 2002a: Nitrogen loading alters seagrass ecosystem structure and support of higher trophic levels. *Aquatic Conservation-Marine and Freshwater Ecosystems,* **12(2),** 193–212.

Deegan, L.A., A. Wright, S.G. Ayvazian, J.T. Finn, H. Golden, R.R. Merson, and J. Harrison, 2002b: Nitrogen loading alters seagrass ecosystem structure and support of higher trophic levels. *Aquatic Conservation: Marine and Freshwater Ecosystems,* **12,** 193–212.

Diop, E.S., J.P. Barusseau, and J.L. Saos, 1985: Mise en évidence du fonctionnement inverse de certains estuaires tropicaux. Conséquences géomorphologiques et sédimentologiques (Saloum et Casamance, Sénégal). *Revue Americaine de Sedimentologie,* **32:** 543–552.

Duarte, C.M., 1995: Submerged aquatic vegetation in relation to different nutrient regimes. *Ophelia,* **41,** 87–112.

Duarte, C.M., 2002: The future of seagrass meadows. *Environmental Conservation,* **29(2),** 192–206.

Duda, A.M. and K. Sherman, 2002: A new imperative for improving management of large marine ecosystems. *Ocean & Coastal Management,* **45(11–12),** 797–833.

Duke, N.C., 1992: Mangrove Floristics and Biogeography,. In: *Tropical Mangrove Ecosystems,* A.I. Robertson and D.M. Alongi (eds.), American Geophysical Union, Washington, D.C. (USA), 63–100.

EC (European Commission), 2004: *Living with coastal erosion in Europe: Sediment and Space for Sustainability. Guidelines for implementing local information systems dedicated to coastal erosion management. Information system functionalities.* Service contract B4–3301/2001/329175/MAR/B3. Brussels (Belgium).

Edwards, M. and A.J. Richardson, 2004: Impact of climate change on marine pelagic phenology and trophic mismatch. *Nature,* **430,** 881–884.

Epstein, P.R. and J.R. Jenkinson, 1993: Harmful algal blooms. *Lancet,* **342,** 1108.

Estes, J.A., M.T. Tinker, T.M. Williams, and D.F. Doak, 1998: Killer whale predation on sea otters linking oceanic and nearshore ecosystems. *Science,* **282(5388),** 473–476.

Ewel, K.C., R.R. Twilley, and J.E. Ong, 1998: Different kinds of mangrove forests provide different goods and services. *Global Ecology and Biogeography,* **7(1),** 83–94.

Falandysz, J., A. Trzosińska, P. Szefer, J. Warzocha, and B. Draganik, 2000: The Baltic Sea, especially southern and eastern regions. In: *Seas at the Millennium: An Environmental Evaluation,* C.R.C. Sheppard (ed.) Volume 1, Regional Seas: Europe, The Americas and West Africa, Elsevier Science Ltd., Oxford (UK) and Pergamon Press, Amsterdam (Netherlands), 99–120.

Farber, S., 1988: The Value of Coastal Wetlands for Recreation—an Application of Travel Cost and Contingent Valuation Methodologies. *Journal of Environmental Management,* **26(4),** 299–312.

Farber, S. and R. Costanza, 1987: The Economic Value of Wetlands Systems. *Journal of Environmental Management,* **24(1),** 41–51.

Farnsworth, E.J. and A.M. Ellison, 1997: The global conservation status of mangroves. *Ambio,* **26(6),** 328–334.

Farnsworth, K.L. and J.D. Milliman, 2003: Effects of climatic and anthropogenic change on small mountainous rivers: the Salinas River example. *Global and Planetary Change,* **39(1–2),** 53–64.

Fonseca, M.S., W.J. Kenworthy, and G.W. Thayer, 1992: Seagrass beds: nursery for coastal species. In: *Stemming the Tide of Coastal Fish Habitat Loss. Marine Recreational Fisheries Symposium, 7–9 March 1991, Baltimore, MD (USA),* R.H. Stroud (ed.), National Coalition for Marine Conservation, Savannah, GA (USA), 141–147.

Foster, M.S., A.P. De Vogelaere, C. Harrold, J.S. Pearse, and A.B. Thum, 1988: *Causes of spatial and temporal patterns in rocky intertidal communities of central and northern California.* Memoirs of the California Academy of Sciences 9. California Academy of Sciences, San Francisco, CA (USA), 45 pp.

Freeman III, A.M., 1993: *The Measurement of Environmental and Resource Values: Theory and methods.* Resources for the Future, Washington, D.C. (USA).

Gardner, T.A., I.M. Cote, J.A. Gill, A. Grant, and A.R. Watkinson, 2003: Long-term region-wide declines in Caribbean corals. *Science,* 301(5635), 958–960.

GDAIS (General Dynamics Advanced Information Systems), 2004: The Global Maritime Boundaries Database. Herndon, USA. [Distributed as CD-ROM]

GESAMP (Group of Experts on the Scientific Aspects of Marine Environmental Protection), 2001: Protecting the oceans from land-based activities—Land-based sources and activities affecting the quality and uses of the marine, coastal and associated freshwater environment. IMO/FAO/UNESCO-IOC/WMO/WHO/IAEA/UN/UNEP Joint Group of Experts on the Scientific Aspects of Marine Environmental Protection, GESAMP No. **71,** 162 pp.

Giannini, A., R. Saravanan, and P. Chang, 2003: Oceanic Forcing of Sahel Rainfall on Interannual to Interdecadal Time Scales. Science, **302,** 1027–1030.

Giesen, W., M. Baltzer, and R. Baruadi, 1991: *Integrating Conservation with Land-Use Development in Wetlands of South Sulawesi.* Asian Wetland Bureau, Bogor (Indonesia).

Gilbert, A.J. and R. Janssen, 1998: Use of environmental functions to communicate the values of a mangrove ecosystem under different management regimes. *Ecological Economics,* **25(3),** 323–346.

Gilmartin, W.G. and J. Focada, 2002: Monk Seals—*Monachus monachus, M. tropicalis* and *M. schauinslandi.* In: *Encyclopedia of Marine Mammals,* W.F. Perrin, B. Würsig, and J.G.M. Thewissen (eds.), Academic Press, San Diego, CA (USA), 756–759.

GIWA (Global International Waters Assessment), 2003: [online] Cited November 2004. Available at www.giwa.net.

Goldburg, R.J., M.S. Elliot, and R.L. Naylor, 2001: *Marine Aquaculture in the United States: Environmental Impacts and Policy Options.* Prepared for the Pew Oceans Commission, Arlington, VA (USA).

Gosselink, J.G., E.P. Odum, and R.M. Pope, 1974: *The Value of the Tidal Marsh.* Publication LSC1-SG-74–03, Louisiana State University Centre for Wetland Resources, Baton Rouge, LA (USA).

GPA (Global Programme of Action), 1995: Washington Declaration on the Protection of the Marine Environment from Land-Based Activities. [online] Cited November 2004.

Gray, C.A., D.J. McElligott, and R.C. Chick, 1996: Intra- and inter-estuary differences in assemblages of fishes associated with shallow seagrass and bare sand. *Marine and Freshwater Research,* **47(5),** 723–735.

Gray, J.S., 1997: Marine biodiversity: Patterns, threats and conservation needs. *Biodiversity and Conservation,* **6(1),** 153–175.

Gray, J.S., G.C.B. Poore, K.I. Ugland, R.S. Wilson, F. Olsgard, and O. Johannessen, 1997: Coastal and deep-sea benthic diversities compared. *Marine Ecology-Progress Series,* **159,** 97–103.

Green, E.P. and F.T. Short, 2003: *World Atlas of Seagrasses.* University of California Press, Berkeley, CA (USA), 304 pp.

Griffis, R.B. and K.W. Kimball, 1996: Ecosystem approaches to coastal and ocean stewardship. *Ecological Applications,* **6(3),** 708–712.

Grosholz, E., 2002: Ecological and evolutionary consequences of coastal invasions. *Trends in Ecology & Evolution,* **17(1),** 22–27.

Harris, L.G. and M.C. Tyrrell, 2001: Changing community states in the Gulf of Maine: synergism between invaders, overfishing and climate change. *Biological Invasions,* **3,** 9–21.

Harwood, J., 2001: Marine mammals and their environment in the twenty-first century. *Journal of Mammalogy,* **82(3),** 630–640.

Hatcher, B., R. Johannes, and A. Robinson, 1989: Review of the research relevant to the conservation of shallow tropical marine ecosystems. *Oceanography and Marine Biology,* **27,** 337–414.

Heck, K.L.J., D.A. Nadeau, R. Thomas, and 50–54, 1997: The nursery role of seagrass beds. *Gulf of Mexico Science,* **15(1),** 50–54.

Helfield, J.M. and R.J. Naiman, 2003: Effects of salmon-derived nitrogen on riparian forest growth and implications for stream productivity: Reply. *Ecology,* **84(12),** 3399–3401.

Herbst, L., A. Ene, M. Su, R. Desalle, and J. Lenz, 2004: Tumor outbreaks in marine turtles are not due to recent herpesvirus mutations. Current Biology, **14(17),** R697-R699.

Hildebrand, L. and J. Sorensen, 2001: Draining the Swamp and beating away the alligators: Baseline 2000. *Intercoast Network,* 20–21.

Hobbie, J.E., 2000: *Estuarine Science: A Synthetic Approach to Research and Practice.* Island Press, Washington, D.C. (USA), 540 pp.

Hughes, T.P., A.H. Baird, D.R. Bellwood, M. Card, S.R. Connolly, et al. 2003: Climate change, human impacts, and the resilience of coral reefs. *Science,* **301(5635),** 929–933.

International Wader Study Group, 2003: Waders are declining worldwide. Paper presented at the *2003 International Wader Study Group,* 26–28 September, Cádiz (Spain).

IOC (Intergovernmental Oceanographic Commission),1993: *Assessment and monitoring of large marine ecosystems.* UNESCO, Paris (France).

IPCC (Intergovernmental Panel on Climate Change), 2003: *Climate Change 2001: The Scientific Basis. Contribution of Working Group I to the Third Assessment Report.* J.T. Houghton, Y. Ding, D.J. Griggs, M. Noguer, P.J. van der Linden, X. Dai, K. Maskell, C.A. Johnson (eds.). Cambridge University Press, Cambridge (UK), 892 pp.

Islam, M.S., B.S. Drasar, and D.J. Bradley, 1990: Long-term persistence of toxigenic *Vibrio cholerae* 01 in the mucilaginous sheath of a blue-green alga, *Anabaena variabilis. The Journal of tropical medicine and hygiene,* **93(2),** 133–139.

Jackson, J.B.C., M.X. Kirby, W.H. Berger, K.A. Bjorndal, L.W. Botsford, et al. 2001: Historical overfishing and the recent collapse of coastal ecosystems. *Science,* **293(5530),** 629–638.

Jennings, S. and M.J. Kaiser, 1998: The effects of fishing on marine ecosystems. *Advances in Marine Biology,* **34,** 201–314.

Johannes, R.E., L. Squire, T. Graham, Y. Sadovy, and H. Renguul, 1999: *Spawning aggregations of Groupers (Serranidae) in Palau.* Marine Conservation Research Series Publication No **1,** The Nature Conservancy., 144 pp.

Johnston, R.J., J.J. Opaluch, T.A. Grigalunas, and M.J. Mazzotta, 2001: Estimating amenity benefits of coastal farmland. *Growth and Change,* **32(3),** 305–325.

Johnston, R.J., T.A. Grigalunas, J.J. Opaluch, M. Mazzotta, and J. Diamantedes, 2002: Valuing estuarine resource services using economic and ecological models: the Peconic Estuary System study. *Coastal Management,* **30(1),** 47–65.

Jones, N., 2003: Sea water 'pumps' pollutants into coastal aquifers. *New Scientist,* 17 May, 21.

Kaly, U.L. and G.P. Jones, 1998: Mangrove restoration: A potential tool for coastal management in tropical developing countries. *Ambio,* **27(8),** 656–661.

Kaoru, Y., V.K. Smith, and J.L. Liu, 1995: Using Random Utility-Models to Estimate the Recreational Value of Estuarine Resources. *American Journal of Agricultural Economics,* **77(1),** 141–151.

Kaplan, I.C., 2001: Pacific Leatherback and Loggerhead Turtle Populations: Estimating the Relative Importance of Longline Effects vs. Other Anthropogenic Mortality. Paper presented at the *International Tuna Conference,* 24–27 May, Lake Arrowhead, CA (USA).

Kaschner, K., 2003: *Review of small cetacean bycatch in the ASCOBANS area and adjacent waters—current status and suggested future actions.* on the Conservation of Small Cetaceans of the Baltic and North Seas (ASCOBANS-UN), Bonn (Germany).

Kaufman, L.S. and P.J. Dayton, 1997: Impacts of marine resource extraction on ecosystem services and sustainability. In: *Nature's Services: Societal Dependence on Natural Ecosystems,* G. Daily (ed.), Island Press, Washington, D.C. (USA), 275–293.

Kautsky, L. and N. Kautsky, 2000: The Baltic Sea, including Bothnian Sea and Bothnian Bay. In: *Seas at the Millenium: An Environmental Evaluation,* C.R.C. Sheppard (ed.). Volume 1, Regional Seas: Europe, The Americas and West Africa, Elsevier Science Ltd., Oxford (UK) and Pergamon Press, Amsterdam (Netherlands), 121–133.

Kawabe, M. and T. Oka, 1996: Benefit from improvement of organic contamination of Tokyo Bay. *Marine Pollution Bulletin,* **32(11),** 788–793.

Kay, R. and J. Alder, in press: *Coastal Planning and Management.* 2nd edition ed. EF&N Spoon, London (UK).

Kelleher, G., C. Bleakley, and S. Wells, 1995: *A global representative system of marine protected areas.* Vol. 1, Great Barrier Reef Marine Park Authority, the World Bank, the World Conservation Union (IUCN), World Bank, Washington, D.C. (USA).

Kenyon, K.W., 1977: Caribbean monk seal extinct. *Journal of Mammalogy,* **58,** 97–98.

Keough, M.J. and G.P. Quinn, 1998: Effects of periodic disturbances from trampling on rocky intertidal algal beds. *Ecological Applications,* **8(1),** 141–161.

Kimball, L.A., 2001: *International Ocean Governance. Using International Law and Organizations to Manage Resources Sustainably.* IUCN, Gland (Switzerland) and Cambridge (UK), 124 pp.

King, O.H., 1995: Estimating the value of marine resources: A marine recreation case. *Ocean & Coastal Management,* **27(1–2),** 129–141.

Kjerfve, B., W.J. Wiebe, H.H. Kremer, W. Salomons, J.I.C. Marshall Crossland, N. Morcom, N. Harvey, and J.I.O. Marshall Crossland, 2002: *Caribbean Basins: LOICZ Global Change Assessment and Synthesis of River Catchment/ Island-Coastal Sea Interactions and Human Dimensions; with a desktop study of Oceania Basins.* LOICZ-IPO, Texel (Netherlands), 174 pp.

Kulczycki, G.R., R.W. Virnstein, and W.G. Nelson, 1981: The Relationship between Fish Abundance and Algal Biomass in a Seagrass—Drift Algae Community. *Estuarine Coastal and Shelf Science,* **12(3),** 341–347.

Lacerda, L.D. and J.J. Abrao, 1984: Heavy metal accumulation by mangrove and saltmarsh intertidal sediments. *Revista Brasileira de Botanica,* **7,** 49–52.

Lacerda, L.D., H.H. Kremer, B. Kjerfve, W. Salomons, J.I. Marshall, and C.J. Crossland, 2002: *South American Basins: LOICZ Global Change Assessment and Synthesis of River Catchment—Coastal Sea Interaction and Human Dimensions.* LOICZ-IPO, Texel (Netherlands), 212 pp.

Lawrence, D., R. Kenchington, and S. Woodley, 2002: *The Great barrier Reef. Finding the Right Balance.* Melbourne University Press, Melbourne (Australia), 296 pp.

Lebel, L., N.H. Tri, A. Saengnoree, S. Pasong, U. Buatama, and L.K. Thoa, 2002: Industrial transformation and shrimp aquaculture in Thailand and Vietnam: Pathways to ecological, social, and economic sustainability? *Ambio,* **31(4),** 311–323.

Ledoux, L., J.E. Vermaatb, L. Bouwerb, W. Salomonsb, and R.K. Turnera, 2003: ELOISE research and implementation of the EU policy in the coastal zone. [online] Cited November 2004. Available at http://130.37.129.100/ english/o_o/instituten/IVM/research/eloise/pdf/btbch1.pdf.

Leggett, C.G. and N.E. Bockstael, 2000: Evidence of the effects of water quality on residential land prices. *Journal of Environmental Economics and Management,* **39(2),** 121–144.

Lenanton, R.C.J. and I.C. Potter, 1987: Contribution of Estuaries to Commercial Fisheries in Temperate Western Australia and the Concept of Estuarine Dependence. *Estuaries,* **10(1),** 28–35.

Levin, L.A., D.F. Boesch, A. Covich, C. Dahm, C. Erseus, et al. 2001: The function of marine critical transition zones and the importance of sediment biodiversity. *Ecosystems,* **4(5),** 430–451.

Liu, J.G., G.C. Daily, P.R. Ehrlich, and G.W. Luck, 2003: Effects of household dynamics on resource consumption and biodiversity. *Nature,* **421(6922),** 530–533.

Lubchenco, J., 1998: Entering the century of the environment: A new social contract for science. *Science,* **279(5350),** 491–497.

Lubchenco, J., 2003: The Blue Revolution: A Global Ecological Perspective. *World Aquaculture Magazine,* **34(4),** Guest Editorial.

Lynne, G.D., P. Conroy, and F.J. Prochaska, 1981: Economic Valuation of Marsh Areas for Marine Production Processes. *Journal of Environmental Economics and Management,* **8(2),** 175–186.

MacKinnon, J., 1997: *Protected Area Systems Review of the Indo-Malayan Realm.* Asian Bureau for Conservation, UNEP-World Conservation Monitoring Centre, Cambridge (UK).

Marshall, C.H., R.A. Pielke, and L.T. Steyaert, 2003: Wetlands: Crop freezes and land-use change in Florida. *Nature,* **426(6962),** 29–30.

McKee, J.K., P.W. Sciulli, C.D. Fooce, and T.A. Waite, 2004: Forecasting global biodiversity threats associated with human population growth. *Biological Conservation,* **115(1),** 161–164.

McKinney, M.L., 1998: Is marine biodiversity at less risk? Evidence and implications. *Diversity and Distributions,* **4(1),** 3–8.

McManus, J.W., L.A.B. Menez, K.N. Kesner-Reyes, S.G. Vergara, and M.C. Ablan, 2000: Coral reef fishing and coral-algal phase shifts: implications for global reef status. *Ices Journal of Marine Science,* **57(3),** 572–578.

Merrick, R.L., T.R. Loughlin, and D.G. Calkins, 1987: Decline in Abundance of the Northern Sea Lion, *Eumetopias jubatus,* in Alaska, 1956–86. *Fishery Bulletin,* **85(2),** 351–365.

Meybeck, M., 1976: Total mineral dissolved transport by world's major rivers. *Hydrological Science Bulletin,* **21,** 265–284.

Meybeck, M. and R.G.D.G.P. (19791, 215–246., 1979: Concentration des eaux fluviales en elements majeurs et apports en solution aux oceans. *Revue de Geologie Dynamique et de Geographie Physique,* **21(3),** 215–246.

Meybeck, M. and A. Ragu, 1997: Presenting Gems Glori, a compendium of world river discharge to the oceans. In: *Scientific Assembly of the International Association of Hydrological Sciences (IAHS),* 23 April–3 May 1997, Rabat (Morocco), B. Webb (ed.). IAHS Publication **243,** 3–14.

Milliman, J.D. and J.P.M. Syvitski, 1992: Geomorphic Tectonic Control of Sediment Discharge to the Ocean—the Importance of Small Mountainous Rivers. *Journal of Geology,* **100(5),** 525–544.

Mitchell, E. and J.G. Mead, 1977: The history of the gray whale in the Atlantic ocean. Paper presented at the *2nd Conference on the Biology of Marine Mammals,* 12–15 December. Society of Marine Mammalogy, San Diego, CA (USA), 12 pp.

Moberg, F. and C. Folke, 1999: Ecological goods and services of coral reef ecosystems. *Ecological Economics,* **29(2),** 215–233.

Moberg, F. and P. Ronnback, 2003: Ecosystem services of the tropical seascape: interactions, substitutions and restoration. *Ocean & Coastal Management,* **46(1–2),** 27–46.

Morgan, C. and N. Owens, 2001: Benefits of water quality policies: the Chesapeake Bay. *Ecological Economics,* **39(2),** 271–284.

Morrison, R.I.G., Y. Aubrey, R.W. Butler, G.W. Beyersbergen, G.M. Donaldson, et al. 2001: Declines in North American shorebird populations. *Wader Study Group Bulletin,* **94,** 34–38.

Moseley, M.E., 1975: *The Maritime Foundations of Andean Civilization.* Cummings Publications, Menlo Park, CA (USA), 131 pp.

Mumby, P.J., A.J. Edwards, J.E. Arias-Gonzalez, K.C. Lindeman, P.G. Blackwell, et al. 2004: Mangroves enhance the biomass of coral reef fish communities in the Caribbean. *Nature,* **427(6974),** 533–536.

Murray, S.N., J.A. Zertuche-Gonzalez, and L. Fernandez, in review: *Invasive seaweeds: Status of knowledge and economic policy considerations for the Pacific Coast of North America.* Center for Environmental Cooperation, Montreal (Canada).

Murray, S.N., T.G. Denis, J.S. Kido, J.R. Smith, and 40:100–106., 1999a: Human visitation and the frequency and potential effects of collecting on rocky intertidal populations in southern California marine reserves. *California Oceanic Cooperative Fisheries Investigations (CalCOFI) Reports*, **40**, 100–106.

Murray, S.N., R.F. Ambrose, J.A. Bohnsack, L.W. Botsford, M.H. Carr, G.E. et al. 1999b: No-take reserve networks: Sustaining fishery populations and marine ecosystems. *Fisheries*, **24(11)**, 11–25.

Myers, N. and J. Kent, 2001: *Perverse Subsidies: How Misused Tax Dollars Harm the Environment and the Economy*. Island Press, Washington, D.C. (USA), 277 pp.

Myers, R.A. and B. Worm, 2003: Rapid worldwide depletion of predatory fish communities. *Nature*, **423(6937)**, 280–283.

Nicholls, R.J., 2004: Coastal flooding and wetland loss in the 21st century: changes under the SRES climate and socio-economic scenarios. *Global Environmental Change-Human and Policy Dimensions*, **14(1)**, 69–86.

Nickerson, D.J., 1999: Trade-offs of mangrove area development in the Philippines. *Ecological Economics*, **28(2)**, 279–298.

Nixon, S.W., 2003: Replacing the Nile: Are anthropogenic nutrients providing the fertility once brought to the Mediterranean by a great river? *Ambio*, **32(1)**, 30–39.

NOAA (National Oceanic and Atmospheric Adminstration), 2003: Invasive Marine Species found on Georges Bank. [online] Cited November 2004. Available at http://www.noaanews.noaa.gov/stories2003/s2125.htm.

Norse, E.A., 2005: Ending the Range Wars on the Last Frontier: Zoning the Sea. In: *Marine Conservation Biology*, E.A. Norse and L.B. Crowder (eds.), Island Press, Washington, D.C. (USA).

Northridge, S.P., 2002: Fishing Industry, Effects of. In: *Encyclopedia of marine mammals*, W.F. Perrin, B. Würsig, and J.G.M. Thewissen (eds.), Academic Press, San Diego, CA (USA), 442–447.

NRC (National Research Council), 1992: *Restoration of Aquatic Systems: Science, Technology, and Public Policy*. National Academy Press, Washington, D.C. (USA), 576 pp.

NRC, 2000: *Clean Coastal Waters)*. National Academy Press, Washington, D.C. (USA).

NRC, 2001: *Marine Protected Areas: Tools for Sustaining Ocean Ecosystem*. National Academy Press, Washington, D.C. (USA), 288 pp.

Nystrom, M., C. Folke, and F. Moberg, 2000: Coral reef disturbance and resilience in a human-dominated environment. *Trends in Ecology & Evolution*, **15(10)**, 413–417.

Ochieng, C.A. and P.L.A. Erftemeijer, 2003: The seagrasses of Kenya and Tanzania. In: *World Atlas of Seagrasses*, E.P. Green and F.T. Short (eds.), University of California Press, Berkeley, CA.

Ofiara, D.D. and B. Brown, 1999: Assessment of economic losses to recreational activities from 1988 marine pollution events and assessment of economic losses from long-term contamination of fish within the New York Bight to New Jersey. *Marine Pollution Bulletin*, **38(11)**, 990–1004.

O'Neill, R.V., 1988: Hierarchy theory and global change. In: *Scales and Global Change*, T. Rosswal, R.G. Woodmansee, and P.G. Risser (eds.), John Wiley & Sons, New York, NY (USA), 29–44.

Paine, R.T., 2002: Trophic control of production in a rocky intertidal community. *Science*, **296(5568)**, 736–739.

Pandolfi, J.M., R.H. Bradbury, E. Sala, T.P. Hughes, K.A. Bjorndal, et al. 2003: Global trajectories of the long-term decline of coral reef ecosystems. *Science*, **301(5635)**, 955–958.

Pauly, D., 1997: Small-scale fisheries in the tropics: marginality, marginalization, and some implications for fisheries management. In: *Proceedings of the 20th American Fisheries Society Symposium: Global Trends-Fisheries Management, 14–16 June 1994, Seattle, WA (USA)*, E.K. Pikitich, D.D. Huppert, and M. Sissenwine (eds.), American Fisheries Society, Bethesda, MD (USA), 40–49.

Pauly, D., V. Christensen, S. Guenette, T.J. Pitcher, U.R. Sumaila, C.J. Walters, R. Watson, and D. Zeller, 2002: Towards sustainability in world fisheries. *Nature*, **418(6898)**, 689–695.

Pearce, D., 1998: Auditing the Earth: The Value of the World's Ecosystem Services and Natural Capital. *Environment*, **40(2)**, 23–28.

Perry, S.L., D.P. DeMaster, and G.K. Silber, 1999: The status of endangered whales: An overview. *Marine Fisheries Review (Special issue)*, **61(1)**, 1–6.

Peterson, C.H. and J. Lubchenco, 1997: On the value of marine ecosystem services to society. In: *Nature's Services: Societal Dependence on Natural Ecosystems*, G. Daily (ed.), Island Press, Washington, D.C. (USA), 177–194.

Pimm, S.L., 1997: The value of everything. *Nature*, **387(6630)**, 231–232.

Polis, G.A. and S.D. Hurd, 1995: Extraordinarily high spider densities on islands: flow of energy from the marine to terrestrial food webs and the absence of predation. *Proceedings of the National Academy of Sciences USA*, **92**, 4382–4386.

Polis, G.A. and S.D. Hurd, 1996: Allochthonous input across habitats, subsidized consumers, and apparent trophic cascades: Examples from the ocean-land interface. In: *Food Webs: Integration of Patterns and Dynamics*, G.A. Polis and K.O. Winemiller (eds.), Chapman & Hall, New York, NY (USA), 275–285.

Potts, M., 1980: Blue-Green-Algae (Cyanophyta) in Marine Coastal Environments of the Sinai Peninsula—Distribution, Zonation, Stratification and Taxonomic Diversity. *Phycologia*, **19(1)**, 60–73.

Primavera, J.H., 1991: Intensive Prawn Farming in the Philippines—Ecological, Social, and Economic-Implications. *Ambio*, **20(1)**, 28–33.

Primavera, J.H., 1997: Socio-economic impacts of shrimp culture. *Aquaculture Research*, **28(10)**, 815–827.

Primavera, J.H., 2000: Development and conservation of Philippine mangroves: institutional issues. *Ecological Economics*, **35(1)**, 91–106.

Pringle, C.M., 2000: Threats to US public lands from cumulative hydrologic alterations outside of their boundaries. *Ecological Applications*, **10(4)**, 971–989.

Read, A.J., P. Drinker, and S.P. Northridge, 2003: *By-Catches Of Marine Mammals In U.S. Fisheries and a First Attempt to Estimate the Magnitude of Global Marine Mammal By-Catch*. International Whaling Commission (IWC)—Scientific Committee Meeting, 16–19 June 2003, Berlin (Germany), 12 pp.

Rivas, V. and A. Cendrero, 1991: Use of Natural and Artificial Accretion on the North Coast of Spain—Historical Trends and Assessment of Some Environmental and Economic Consequences. *Journal of Coastal Research*, **7(2)**, 491–507.

Rogers, S.I., M.J. Kaiser, and S. Jennings, 1998: Ecosystem effects of demersal fishing: A European perspective. In: *Effects of fishing gear on the sea floor of New England.*, E.M. Dorsey and J. Pederson (eds.), Conservation Law Foundation, Boston, MA (USA), 68–78.

Ronnback, P., 1999: The ecological basis for economic value of seafood production supported by mangrove ecosystems. *Ecological Economics*, **29(2)**, 235–252.

Rose, J.B., P.R. Epstein, E.K. Lipp, B.H. Sherman, S.M. Bernard, and J.A. Patz, 2001: Climate variability and change in the United States: Potential impacts on water- and foodborne diseases caused by microbiologic agents. *Environmental Health Perspectives*, **109**, 211–221.

Rose, M.D. and G.A. Polis, 1998: The distribution and abundance of coyotes: The effects of allochthonous food subsidies from the sea. *Ecology*, **79(3)**, 998–1007.

Ross, J.P. (ed).1998: Crocodiles: Status Survey and Conservation Action Plan, 2nd edition, IUCN Gland Switerland

Ruitenbeek, H.J., 1992: *Mangrove management: an economic analysis of management options with a focus on Bintuni Bay, Irian Jaya, Indonesia*. 90, Environmental Reports No. 8, Environmental Management Development in Indonesia Project (EMDI), Jakarta (Indonesia) and Halifax (Canada).

Ruitenbeek, H.J., 1994: Modeling Economy Ecology Linkages in Mangroves—Economic Evidence for Promoting Conservation in Bintuni Bay, Indonesia. *Ecological Economics*, **10(3)**, 233–247.

Ruiz, G.M. and J.A. Crooks, 2001: Biological invasions of marine ecosystems: patterns,effects, and management. In: *Waters in Peril*, L. Bendell-Yound and P. Gallagher (eds.), Kluwer Academic Publications, Dordrecht (Netherlands), 1–17.

Ruiz, G.M., J.T. Carlton, E.D. Grosholz, and A.H. Hines, 1997: Global invasions of marine and estuarine habitats by non-indigenous species: Mechanisms, extent, and consequences. *American Zoologist*, **37(6)**, 621–632.

Saifullah, S.M., 1997a: Management of the Indus Delta Mangroves. In: *Coastal Zone Management Imperative for Maritime Developing Nations*, B.U. Haq, S.M. Haq, G. Kullenberg, and J.H. Stel (eds.), Kluwer Academic Publishing, Amsterdam (Netherlands), 333–347.

Saifullah, S.M., 1997b: Mangrove ecosystem of Red Sea coast (Saudi Arabia). *Pakistan Journal of Marine Sciences*, **6**, 115–124.

Sather, C., 1997: *The Bajau Laut: Adaptation, History, and Fate in a Maritime Fishing Society of South-Eastern Sabah*. Oxford university Press, Oxford (UK), 359 pp.

Sathirathai, S. and E.B. Barbier, 2001: Valuing mangrove conservation in southern Thailand. *Contemporary Economic Policy*, **19(2)**, 109–122.

Schreiber, E.A. and J. Burger, 2002: *Biology of Marine Birds*. CRC Press, Florida, FL (USA).

Sebens, K.P., 1986: Spatial Relationships among Encrusting Marine Organisms in the New-England Subtidal Zone. *Ecological Monographs*, **56(1)**, 73–96.

Semesi, A.K., 1992: Developing Management Plans for the Mangrove Forest Reserves of Mainland Tanzania. *Hydrobiologia*, **247(1–3)**, 1–10.

Semesi, A.K., 1998: Mangrove management and utilization in Eastern Africa. *Ambio,* **27(8),** 620–626.

Seminoff, J., 2002: *2002 IUCN Red List Global Status Assessment. Green turtle (Chelonia mydas).* Marine Turtle Specialist group (MTSG), IUCN/SSC Red List Programme, 87 pp.

Sheppard, C.R.C., 2000: The Red Sea. In: *Seas at the Millennium: An Environmental Evaluation,* C.R.C. Sheppard (ed.). Volume 2. Regional Chapters: The Indian Ocean to the Pacific, Elsevier Science Ltd., Oxford (UK) and Pergamon Press, Amsterdam (The Netherlands), 35–45.

Sherman, K., 1993: Large Marine Ecosystems as Global Units for Marine Resources Management: An Ecological Perspective. In: *Large Marine Ecosystems: Stress, Mitigation, and Sustainability,* K. Sherman, L.M. Alexander, and B.D. Gold (eds.), American Association for the Advancement of Science Press, Washington, D.C. (USA), 3–14.

Short, F.T. and S. WyllieEcheverria, 1996: Natural and human-induced disturbance of seagrasses. *Environmental Conservation,* **23(1),** 17–27.

Simenstad, C.A., S.B. Brandt, A. Chalmers, R. Dame, L.A. Deegan, R. Hodson, and E.D. Houde, 2000: Habitat-Biotic Interactions. In: *Estuarine Science: A Synthetic Approach to Research and Practice,* J.E. Hobbie (ed.), Island Press, Washington, D.C. (USA), 427–455.

Small, C. and R.J. Nicholls, 2003: A global analysis of human settlement in coastal zones. *Journal of Coastal Research,* **19(3),** 584–599.

Sorensen, J., 1993: The International Proliferation of Integrated Coastal Zone Management Efforts. *Ocean & Coastal Management,* **21(1–3),** 45–80.

Sorensen, J., 1997: National and international efforts at integrated coastal management: Definitions, achievements, and lessons. *Coastal Management,* **25(1),** 3–41.

Sorensen, J., 2002: Baseline 2000 Background Report: The Status of Integrated Coastal Management as an International Practice (Second Iteration). [online] Cited November 2004. Available at http://www.uhi.umb.edu/b2k/baseline 2000.pdf.

Spalding, M., S. Chape, and M. Jenkins, 2003: State of the World's Protected Areas. [online] Cited November 2004. Available at http://valhalla.unep-wcmc.org/wdbpa/sowpr/Introduction.pdf.

Spalding, M.D., F. Blasco, and C.D. Field, 1997: *World Mangrove Atlas.* The International Society for Mangrove Ecosystems, Okinawa (Japan), 178 pp.

Speth, J.G., 2004: *Red Sky at Morning: America and the Crisis of the Global Environment.* Yale University Press, New Haven, CT (USA), 304 pp.

Spotila, J.R., R.D. Reina, A.C. Steyermark, P.T. Plotkin, and F.V. Paladino, 2000: Pacific leatherback turtles face extinction. *Nature,* 405(6786), 529–530.

Spotila, J.R., A.E. Dunham, A.J. Leslie, A.C. Steyermark, P.T. Plotkin, and F.V. Paladino, 1996: Worldwide population decline of *Dermochelys coriacea:* Are leatherback turtles going extinct? *Chelonian Conservation Biology,* **2,** 209–222.

Spurgeon, J.P.G., 1992: The Economic Valuation of Coral Reefs. *Marine Pollution Bulletin,* **24(11),** 529–536.

Stehn, R.A., K.S. Rivera, S. Fitzgerald, and K.D. Whol, 2001: Incidental catch of Seabirds by Longline Fisheries in Alaska. In: *Seabird Bycatch: Trends, Roadblocks and Solutions,* E.F. Melvin and J.K. Parrish (eds.), Annual Meeting of the Pacific Seabird Group, February 26–27, 1999, Blaine Washington. University of Alaska Sea Grant, AK-SG-01–01, Fairbanks, AK (USA), 204.

Stevenson, N.J., 1997: Disused shrimp ponds: Options for redevelopment of mangroves. *Coastal Management,* **25(4),** 425–435.

Stone-Miller, R., 1995: *Art of the Andes: From Chavin to Inca.* Thames & Hudson, London (UK), 224 pp.

Stroud, D.A., N.C. Davidson, R. West, D.A. Scott, L. Haanstra, O. Thorup, B. Ganter, and S. Delany, (compilers on behalf of the International Wader Study Group) 2004: *Status of migratory wader populations in Africa and Western Eurasia in the 1990s.* Vol. **15,** *International Wader Studies,* 1–259 pp.

Syvitski, J., 2001: Supply and flux of sediment along hydrological pathways: Anthropogenic influences at the global scale. *LOICZ Newsletter,* **20,** 4–7.

Syvitski, J.P.M., 2003: Supply and flux of sediment along hydrological pathways: research for the 21st century. *Global and Planetary Change,* **39(1–2),** 1–11.

Syvitski, J.P.M., C.J. Vörösmarty, A.J. Kettner, and P. Green. 2005. Impact of humans on the flux of terrestrial sediment to the global coastal ocean. *Science* **308:** 376–380.

Tasker, M.L., C.J. Camphuysen, J. Cooper, S. Garthe, W.A. Montevecchi, and S.J.M. Blaber, 2000: The impacts of fishing on marine birds. *Ices·Journal of Marine Science,* **57(3),** 531–547.

Teal, J. and M. Teal. 1969. *Life and Death of a Saltmarsh.* Audubon/Ballantine Books, NY. 274 pp.

Tegner, M.J. and P.K. Dayton, 1977: Sea-Urchin Recruitment Patterns and Implications of Commercial Fishing. *Science,* 196(4287), 324–326.

Tegner, M.J. and P.K. Dayton, 2000: Ecosystem effects of fishing in kelp forest communities. *Ices Journal of Marine Science,* **57(3),** 579–589.

The Heinz Center, 2000: *Evaluation of Erosion Hazards.* The John Heinz III Center for Science, Economics and the Environment, Washington, D.C. (USA), 205 pp.

Tibbetts, J., 2002: Coastal cities—Living on the edge. *Environmental Health Perspectives,* **110(11),** A674-A681.

Turner, R.E. and N.N. Rabalais, 1994: Coastal Eutrophication near the Mississippi River Delta. *Nature,* **368(6472),** 619–621.

Turner, R.K., S. Subak, and W.N. Adger, 1996: Pressures, trends, and impacts in coastal zones: Interactions between socioeconomic and natural systems. *Environmental Management,* **20(2),** 159–173.

U.S. Commission on Ocean Policy, 2004: *An Ocean Blueprint for the 21st Century.,* The U.S. Commission on Ocean Policy, Washington, D.C. (USA).

UNCED (United Nations Conference on Environment and Development), 1992: Agenda 21, Chapter 17. Protection of the Oceans, all kinds of Seas, including enclosed and semi-enclosed seas, and coastal areas and the Protection, Rational Use and Development of their Living Resources. [online] Cited November 2004. Available at http://www.oceanlaw.net/texts/agenda21 .htm.

UNCLOS (United Nations Convention on the Law of the Sea), 1982: [online] Cited November 2004. Available at http://www.un.org/Depts/los/convention_agreements/texts/unclos/closindx.htm.

UNEP (United Nations Environment Programme), 1992: *The world environment 1972–1992: Two decades of challenge.* Chapman & Hall, New York, NY (USA), 884 pp.

UNEP, 2002: Oceans and Coastal Areas. Coastal Threats. [online] Cited November 2004. Available at http://earthwatch.unep.net/oceans/coastalthreats .php.

UNEP, 2004: *Geo Yearbook 2003.* United Nations Environment Programme, Nairobi (Kenya), 76 pp.

UNEP-WCMC (World Conservation Monitoring Centre), 2003a: Mangrove database. [CD-ROM]. Available at Cambridge (UK).

UNEP-WCMC, 2003b: Estuaries database. [CD-ROM]. Available at Cambridge (UK).

UNEP-WCMC, 2003c: Seagrass database. [CD-ROM]. Available at Cambridge (UK).

UNEP-WCMC, 2003d: Coral Reef database. [CD-ROM]. Available at Cambridge (UK).

UNESCO, 1993: Coasts—Managing Complex Systems. 2004. Available at http://www.unesco.org/csi/intro/coastse.pdf.

Valiela, I., J.L. Bowen, and J.K. York, 2001a: Mangrove forests: One of the world's threatened major tropical environments. *BioScience,* **51(10),** 807–815.

Valiela, I., J.L. Bowen, and J.K. York, 2001b: Mangrove Forests: One of the World' Threatened Major Tropical Environments. *BioScience,* **51(10),** 807–815.

Verlaan, P.A., 1997: The Importance of Coastal Management to Human Health: Toward a Sustainable World. In: *International Perspectives on Environment, Development and Health,* G.S. Shahi, B.S. Levy, A. Binger, T. Kjellstrom, and R.S. Lawrence (eds.), Springer Publishing Company, Inc., New York, NY (USA).

Villa, F., L. Tunesi, and T. Agardy, 2001: Zoning marine protected areas through spatial multiple-criteria analysis: the case of the Asinara Island National Marine Reserve of Italy. *Conservation Biology,* **16(2),** 515–526.

Vitousek, P.M., H.A. Mooney, J. Lubchenco, and J.M. Melillo, 1997: Human domination of Earth's ecosystems. *Science,* **277(5325),** 494–499.

Vörösmarty, C.J. and M.M. Meybeck, 1999: Riverine transport and its alteration by human activities. *IGBP Global Change Newsletter,* **39,** 24–29.

Vörösmarty, C.J., M. Meybeck, B. Fekete, and K. Sharma, 1997: The potential impact of neo-Castorization on sediment transport by the global network of rivers. In: *Human Impact on Erosion and Sedimentation,* D. Walling and J.-L. Probst (eds.), IAHS Press, Wallingford (UK), 261–272.

Vörösmarty, C.J., M. Meybeck, B. Fekete, K. Sharma, P. Green, and J.P.M. Syvitski, 2003: Anthropogenic sediment retention: major global impact from registered river impoundments. *Global and Planetary Change,* **39(1–2),** 169–190.

Wabnitz, C., M. Taylor, E. Green and T. Razak, 2003: From Ocean to Aquarium; The global trade in marine ornamental species. UNEP-WCMC, Cambridge, UK.

Walling, D.E. and D. Fang, 2003: Recent trends in the suspended sediment loads of the world's rivers. *Global and Planetary Change,* **39(1–2),** 111–126.

Wang, H.L., 2004: Ecosystem management and its application to large marine ecosystems: Science, law, and politics. *Ocean Development and International Law,* 35(1), 41–74.

Webb, G., 1999: Sustainable use of marine crocodiles and turtles. international Union for the Conservation of Nature (IUCN). 2004. Available at http://www.iucn.org/themes/ssc/susg/docs/newsletter/october.pdf.

Weinstein, M.P. and D.A. Kreeger, 2000: *Concepts and controversies in tidal marsh ecology.* Kluwer Academic Publishers, Dortrecht (Netherlands).

Wilkinson, C., 2000: Executive Summary. In: *Status of Coral Reefs of the World: 2000,* C. Wilkinson (ed.), Australian Institute of Marine Science (AIMS), Townsville (Australia), 7–21.

Wilkinson, C., 2002: *Status of Coral Reefs of the World: 2002.* Australian Institute of Marine Science (AIMS), Townsville (Australia), 378 pp.

Wilkinson, C., 2004: *Status of of Coral Reefs of the World: 2004.* Australian Institute of Marine Science (AIMS), Townsville (Australia).

Willson, M.F. and K.C. Halupka, 1995: Anadromous Fish as Keystone Species in Vertebrate Communities. *Conservation Biology,* **9(3),** 489–497.

Wilson, M.A., R. Costanza, R. Boumans, and S. Liu, in press: Integrated Assessment and Valuation of Ecosystem Goods and Services provided by Coastal Systems. *Biology and the Environment: Proceedings of the Royal Irish Academy.*

Woodmansee, R.G., 1988: Ecosystem processes and global change. In: *Scales and Global Change: Spatial and Temporal Variability in Biospheric and Geospheric Processes,* P.G. Risser, R.G. Woodmansee, and T. Rosswall (eds.). SCOPE Report 35, John Wiley & Sons, London (UK) and New York (USA), 11–27.

World Bank, 2004: World Development Indicators 2004. Washington, DC.

WRI (World Resources Institute), 2000: *People and ecosystems: The fraying web of life.* Washington, DC.

WSSD (World Summit on Sustainable Development), 2002: Plan of Implementation: Paragraph 29. 2004.

WWF (World Wide Fund for Nature), 2003: Marine turtles: Global voyagers threatened with extinction. WWF International. Available at http://www.panda.org/downloads/species/finalmarineturtlebrochurepdf.pdf.

Yang, S.L., Q.Y. Zhao, and I.M. Belkin, 2002: Temporal variation in the sediment load of the Yangtze river and the influences of human activities. *Journal of Hydrology,* **263(1–4),** 56–71.

Yang, Z.S., J.D. Milliman, J. Galler, J.P. Liu, and X.G. Sun, 1998: Yellow River's Water and Sediment Discharge Decreasing Steadily. *EOS,* **79(48),** 589–592.

Yentsch, C.S., C.M. Yentsch, J.J. Cullen, B. Lapointe, D.A. Phinney, and S.W. Yentsch, 2002: Sunlight and water transparency: cornerstones in coral research. *Journal of Experimental Marine Biology and Ecology,* **268(2),** 171–183.

Young, E., 2004: Taboos could save the seas. *New Scientist,* 17 April, 9.

Zaitsev, Y. and V. Mamaev, 1997: *Biological Diversity in the Black Sea. A Study of Change and Decline.* United Nations Publications, New York, NY (USA), 208 pp.

Chapter 20

Inland Water Systems

Coordinating Lead Authors: C. Max Finlayson, Rebecca D'Cruz

Lead Authors: Nickolay Aladin, David Read Barker, Gordana Beltram, Joost Brouwer, Nicholas Davidson, Laurie Duker, Wolfgang Junk, Michael D. Kaplowitz, Henk Ketelaars, Elena Kreuzberg-Mukhina, Guadalupe de la Lanza Espino, Christian Lévêque, Alvin Lopez, Randy G. Milton, Parastu Mirabzadeh, Dave Pritchard, Carmen Revenga, Maria Rivera, Abid Shah Hussainy, Marcel Silvius, Melanie Steinkamp

Contributing Authors: Vyascheslav Aparin, Elena Bykova, Jose Luis García Calderón, Nikolay Gorelkin, Ward Hagemeijer, Alex Kreuzberg, Eduardo Mestre Rodríguez, Iskander Mirabdullaev, Musonda Mumba, Igor Plotnikov, Vladislav Talskykh, Raisa Toryannikova

Review Editors: Robert Costanza, Pedro Jacobi, Frank Rijsberman

---

*This appears in Appendix A at the end of this volume.

Main Messages

Inland water habitats and species are in worse condition than those of forest, grassland, or coastal systems (*medium certainty*). It is *speculated* that 50% of inland water habitats were lost during the twentieth century. It is *well established* that for many ecosystem services, the capacity of inland water systems to produce these services is in decline and is as bad as or worse than that of other systems. More than 50% of inland waters (excluding lakes and rivers) have been lost in parts of North America, Europe, and Australia, but on a global scale there is insufficient information on the extent of specific inland water habitats, especially those of a seasonal or intermittent nature, to substantiate the extent of habitat loss.

In addition to the loss of inland water systems, degradation is widespread. As with habitat loss, it has not, on the whole, been possible to quantify this with great confidence at a continental scale, although many site-specific instances have been well documented. The species biodiversity of inland waters is among the most threatened of all ecosystems, and in many parts of the world it is in continuing and accelerating decline. Global climate change is expected to exacerbate the loss and degradation of many inland water systems and the loss or decline of their species; however, projections about the extent of such loss and degradation or decline are not yet well established.

The loss and degradation of inland water systems have been driven directly by many pressures, acting individually and synergistically or cumulatively. The direct drivers of loss and degradation of inland waters are well known and documented and include changes in land use or cover due to vegetation clearance, drainage, and infilling, especially connected to expansion of agriculture; the spread of infrastructure, whether for urban, tourism and recreation, aquaculture, agriculture, or industrial purposes; the introduction and spread of invasive species; hydrologic modification; overharvesting, particularly through fishing and hunting; pollution, salinization, and eutrophication; and global climate change, which is expected (*high certainty*) to lead to even further degradation and to exacerbate existing pressures. While it is known that cumulative and synergistic effects between multiple pressures occur, there is insufficient quantitative analysis to readily tease out the relative individual and combined effects and their importance.

Agricultural development has historically been the principal cause of the loss of inland water systems worldwide (*high certainty*). It is estimated that by 1985, 56–65% of suitable inland water systems had been drained for intensive agriculture in Europe and North America, 27% in Asia, and 6% in South America. The construction of dams and other structures along rivers has resulted in fragmentation and flow regulation of almost 60% of the large river systems in the world. In many countries, the construction of large dams is still a controversial issue. Water pollution and eutrophication are widespread and in many countries have led to the degradation of many inland water systems. In addition to direct adverse effects on biodiversity, pollution has reduced the capacity of inland waters to filter and assimilate waste. Threats of water quality degradation are most severe in areas where water is scarce (dryland systems). Toxic substances and artificial chemicals are increasingly being released into waterways, with uncertainty about their long-term effects on ecosystems and humans. In recent years the devastation caused by invasive species has been increasingly recognized worldwide.

The decline of inland water systems has placed the ecosystem services derived from these systems and human well-being at increasing risk. Provisioning services from inland waters, such as fish, are essential for human well-being, with estimates of more than 50 million people involved directly in inland fisheries. At present almost 50% of the world depends on rice as a staple food item; this is expected to increase, and by 2020 some 4 billion people will depend on rice. Supporting and regulating services are critical to sustaining vital ecosystem functions.

Flow regulation within and between inland waters and links between surface and groundwater are critical ecosystem services that have been degraded on a global scale. The disruption of natural flooding regimes has devastated many riverine habitats and led to decreased sediment transport and a loss of flood buffering and nutrient retention. Flooding can cause severe hardship to humans, with the 1998 floods in China causing an estimated $20 billion worth of damage, but it is also essential for maintaining sediment-based fertility of floodplains and supporting fish stocks in large rivers.

In addition, inland waters have significant aesthetic, artistic, educational, cultural, and spiritual values, and they provide invaluable opportunities for recreation by many communities and, increasingly, for tourism. The economic value of these services is known for many local habitats but not necessarily well quantified economically nor recognized by policy-makers and given priority within resource development and conservation agencies in most countries.

Trade-offs between services from inland waters have been considerable, yet poorly considered. Alteration of rivers through infrastructure has improved transportation, provided flood control and hydropower, and boosted agricultural output by making more land and irrigation water available. At the same time, rivers have been disconnected from their floodplains and other inland water habitats, water velocity in riverine systems has decreased, in some places rivers have been converted to a chain of connected reservoirs, and groundwater recharge has been reduced. In other places, infrastructure has increased the likelihood of flooding by diverting water and increasing flows. These changes have, in turn, affected the migratory patterns of fish species and the composition of riparian habitat, opened up paths for exotic species, changed coastal ecosystems, and contributed to an overall loss of freshwater biodiversity and inland fishery resources. Irrigation has led to increased food production in drylands but in many cases is unsustainable without extensive public capital investment as waterlogging and pollution (especially eutrophication and salinization) degrade the system and other services and encourage the introduction or spread of human disease vectors.

The assessment of the extent and change of inland water systems at a continental level is compromised by the inconsistency and unreliability of data (*high certainty*). Estimates of the extent of inland water systems vary from 530 million to 1,280 million hectares. The extent and distribution of inland waters is unevenly or even poorly known at the global and regional scales, partly due to confusion over definitions as well as difficulties in delineating and mapping habitats with variable boundaries as a result of fluctuations in water levels. Larger wetlands, lakes, and inland seas have been mapped along with major rivers; there are some 10,000 lakes that are over 1 square kilometer, and peatlands are estimated to cover more than 400 million hectares. Smaller habitats that are critical for many communities are not well mapped or delineated.

On the whole, available information focuses on the broader regional or global scales. This introduces uncertainty into many assessments and necessitates caution when attempting to make comparisons between data sets, especially when collected at different spatial scales. Innovative tools for effective assessment of the status and trends of inland water systems and their species, especially in those parts of the world where data are lacking, inadequate, or in need of updating, are required.

20.1 Introduction

Inland water systems encompass habitats such as lakes and rivers, marshes, swamps and floodplains, small streams, ponds, and cave waters. These have a variety of biological, physical, and chemical characteristics. As coastal wetlands (such as estuaries, mangroves, mudflats, and reefs) are considered in Chapter 19, the broad definition of wetland adopted by the Convention on Wetlands in 1971, which includes inland, coastal, and marine habitats, is not used in this chapter. All inland aquatic habitats, however—whether fresh, brackish, or saline—as well as inland seas are considered.

As there is no clear boundary between inland and coastal ecosystems, this delineation is indicative only and is not strictly applied where there are strong interactions between the biodiversity, services, and pressures that affect inter-connected habitats. Rice fields, aquaculture ponds, and reservoirs are included in this chapter's analysis. The supply of fresh water and its regulation, both in terms of water quality and flow, are considered in Chapters 7, 15, and 16. Groundwater as a system is addressed here, recognizing that important links occur with many surface-water habitats.

This chapter provides a brief description of the services provided by inland waters, together with the condition and trends of their habitats and species. More detailed information on the specific services derived from inland waters (such as water supply and waste processing) is found in other Chapters. Where information is available, the drivers of change in the condition of these habitats and their species are related to the condition of the services and any subsequent effects on human well-being. Trade-offs and responses to changes in the habitats and species are also presented (with further information being provided in the MA *Policy Responses* volume).

Inland water systems have a temporal dimension—varying from perennial to ephemeral—and a dynamic dimension, including flowing systems (rivers), standing waters (lakes and ponds), and systems with at times large seasonal fluctuations in water depth—with some being waterlogged and others flooded permanently, seasonally, intermittently, or even episodically. The term wetland is often used to define all inland aquatic systems, such as lakes, rivers, or lagoons. At other times it is used to describe a narrower group of habitats that represent a variety of shallow, vegetated systems, such as bogs, marshes, swamps, and floodplains. Extensive information on wetland definition and delineation is available (e.g. Finlayson and van der Valk 1995; Mitsch et al. 1994), but the failure to consider fully the different dimensions and definitions that have been used around the world has resulted in confusion and inaccurate analyses on the extent and condition of these systems (Finlayson and Spiers 1999). In this chapter, the terms "inland water systems" or "inland waters" are used wherever possible unless specific habitat types are unambiguously referred to in the source material.

The extent and distribution of inland waters is poorly and unevenly known at the global and regional scales due to differences in definitions as well as difficulties in delineating and mapping habitats with variable boundaries due to fluctuations in water levels (Finlayson et al. 1999). In many cases, comprehensive documentation at the regional or national levels also does not exist. The larger habitats, such as lakes and inland seas, have been mapped along with the major rivers, but for many parts of the world smaller wetlands are not well mapped or delineated. As a consequence, assessment of the extent of and change in inland water habitats at the continental level is compromised by the inconsistency and unreliability of the data. The most recent attempt to ascertain the extent and distribution of inland water systems (Lehner and Döll 2004) is shown in Figure 20.1 (in Appendix A). As with previous estimates, these data contain many inaccuracies and gaps. For example, intermittently inundated habitats are not

included, and there are many inaccuracies because of problems of scale and resolution.

20.2 Services Derived from Inland Water Systems

With the exception of the provision of fresh water, comprehensive global analyses of services provided by inland waters have not been undertaken, nor has the link between the condition and trend of the biodiversity, including habitats, and the provision of ecosystem services been strongly made at this scale. As such, the knowledge base of the true value of inland water systems is poorly known (see, e.g., Finlayson et al. 1999; Tockner and Stanford 2002; Malmqvist and Rundle 2002). Assessments of inland waters have not always considered inland saline waters, which are particularly widespread and important in many arid regions of the world.

A generalized list of services provided by or derived from inland waters has been compiled from a number of sources. (See Table 20.1.) The value of these services has been estimated at

Table 20.1. Ecosystem Services Provided by or Derived from Inland Water Systems

Services	Comments and Examples
Provisioning	
Food	production of fish, wild game, fruits, grains, etc.
Freshwater[a]	storage and retention of water for domestic, industrial, and agricultural use
Fiber and fuel	production of logs, fuelwood, peat, fodder
Biochemical	extraction of materials from biota
Genetic materials	medicine, genes for resistance to plant pathogens, ornamental species, etc.
Biodiversity	species and gene pool
Regulating	
Climate regulation	greenhouse gases, temperature, precipitation, and other climatic processes; chemical composition of the atmosphere
Hydrological flows	groundwater recharge and discharge; storage of water for agriculture or industry
Pollution control and detoxification	retention, recovery, and removal of excess nutrients and pollutants
Erosion	retention of soils
Natural hazards	flood control, storm protection
Cultural	
Spiritual and inspirational	personal feelings and well-being
Recreational	opportunities for recreational activities
Aesthetic	appreciation of natural features
Educational	opportunities for formal and informal education and training
Supporting	
Soil formation	sediment retention and accumulation of organic matter
Nutrient cycling	storage, recycling, processing, and acquisition of nutrients
Pollination	support for pollinators

[a] See also Chapter 7 for commentary on how this is variously considered a provisioning or regulating service.

$2–5 trillion annually (Costanza et al. 1997; Postel and Carpenter 1997). This wide range represents major differences in methods, reliability, and accuracy of the economic data and differences in the definition and area of habitats being assessed. Despite ongoing discussion about methods and data quality, it is *well established* that these systems are highly valued and extremely important for people in many parts of the world. It is *speculated,* but not well documented globally, that the loss and degradation of inland water systems has resulted in an immense loss of services.

As Chapter 7 deals solely with the critical service of the provision of fresh water, this service is not considered further here. Other chapters that contain information on services provided by or derived from inland water systems include those on food (Chapter 8), nutrient cycling (Chapter 12), waste processing and detoxification (Chapter 15), regulation of natural hazards (Chapter 16), cultural and amenity services (Chapter 17), and cultivated systems (Chapter 26).

Table 20.2 contains a summary of estimates for the global average value of services derived from or provided by inland and coastal water systems (referred to generically as wetlands in the source documents). The figures presented are average global values based on sustainable use levels and taken from two synthesis studies—Schuyt and Brander (2004), calibrated for the year 2000, and Costanza et

al. (1997), calibrated for 1994—that together cover more than 200 case studies. Most of the data are derived from Schuyt and Brander (2004), except for the aesthetic information service and climate regulation. The total economic value of 63 million hectares of wetland around the world would, according to this data, amount to about $200 billion a year (which is a conservative estimate, since for many services no economic data were available). Costanza et al. (1997) arrived at a figure of $940 billion, mainly due to higher estimates for several services (flood control at $4,539 per hectare per year, for example, water treatment at $4,177 per hectare, and water supply at $3,800 per hectare).

Despite such figures becoming available, the importance of services derived from inland waters (such as fresh water, fish, and groundwater recharge) is often taken for granted or treated as a common good, with the real value only being recognized after the services have been degraded or lost. This is demonstrated particularly well in semiarid and arid regions, with Lake Chad in western Africa being very illustrative, as it has a multiplicity of valuable services (see Table 20.3), which are mostly in decline. This situation is common globally, particularly where population pressures are high and services have been overexploited or inland water systems have been inappropriately managed.

In addition, while the value of particular services may be low in terms of global economic analyses, it can be extremely high locally. This is evident in the preliminary analyses undertaken in 10 inland water systems in the Zambezi in southern Africa, where multiple services—including subsistence agricultural crops, fish production, natural products, and livestock grazing—were estimated to be worth $123 million per year (Seyam et al. 2001). Such analyses (other examples in Lupi et al. 2002; Emerton et al. 1998) are fraught with assumptions, but they do illustrate the relative worth of the main services to local populations.

An analysis of global ecosystems has established with some confidence that for a standardized set of services, the capacity of inland water habitats (referred to as freshwater habitats in the source material) to produce these services is in decline and is as bad as or worse than the other systems considered (Revenga et al. 2000; WRI et al. 2000). (See Figure 20.2 in Appendix A.) This conclusion has been supported by global analyses of the condition of inland water systems, such as those undertaken for large lakes and inland seas (Beeton 2002), flowing waters (Malmqvist and Rundle 2002), floodplains (Tockner and Stanford 2002), temperate freshwater wetlands (Brinson and Ines Malvarez 2002), tropical wetlands (Junk 2002), and salt lakes (Williams 2002).

20.2.1 Hydrological Regulation

It is well recognized that some inland waters serve as important storage sites, accumulating water during wet periods and providing a reserve of water during dry periods by maintaining base flow in adjacent rivers (e.g., Revenga et al. 2000; Malmqvist and Rundle 2002). Similarly, it is increasingly known that some inland waters, such as lakes and marshes, attenuate floods by retaining water or storing it in the soil and therefore reducing the need for engineered flood control infrastructure (Abramovitz 1996). While it has been known for many years that aquatic vegetation attenuates surface flows, the considerable value of this service is not often widely and accurately assessed in economic terms. (See Chapter 16.) In contrast, figures on the cost of flood damage are readily available after this function has been lost or seriously eroded by unsustainable development; for example, the 1998 flash floods in China caused an estimated economic loss of $20 billion (Qu 1999).

While the damage caused by floods is often discussed, it must also be recognized that natural floods provide an essential service

Table 20.2. Total Economic Value of Ecosystem Services Provided by Wetlands (Costanza et al. 1997; Schuyt and Brander 2004)

	Average Value (dollars per hectare per year)
Provisioning services (products obtained from wetlands)	**601**
Fishing	374
Hunting	123
Water supply	45
Raw materials (thatch, timber, fodder, fertilizer, etc.)	45
Fuelwood	14
Other (genetic, medicinal, and ornamental resources)	?
Cultural services (nonmaterial benefits obtained from wetlands)	**1,373**
Aesthetic information	881
Recreation and tourism	492
Other (e.g., artistic, spiritual, historic, or scientific information)	?
Regulating services (benefits obtained from ecosystem processes)	**1,086**
Flood control/water regulation	464
Water treatment	288
Nursery function	201
Climate regulation	133
Other (e.g., sediment control, biological control)	?
Supporting services (ecosystem functions necessary to maintain all other services)	**214**
Habitat/refugia for biodiversity	214
Other (e.g., primary products, soil formation, nutrient/biogeochemical cycling)	?
Total value	**3,274**

Table 20.3. Change in Ecosystem Services Derived from Lake Chad (White et al. 2004)

Ecosystem Service	Services in Lake Chad	Change in Services	Trend
Provisioning services			
Food – plant crops	rice, maize, cowpeas, wheat, cotton, millet, groundnuts, cassava	increase in production of food crops	up
Food – aquatic plants	spirulina for commercial production	loss of commercial plant production	down
Food – fish	harvesting for local diet and for trade	less fish for food and for trade	down
Food – milk	milk from livestock	decrease in milk available from Kuri population	down
Food – meat	meat from cattle	decrease in meat available from Kuri population	down
Fuel – wood	timber and fuelwood from floodplain forests	recession of floodplain and drying out of habitat for floodplain forests	down
Genetic resources	laboratory for genetic studies on endemic livestock breeds	decrease in availability of genetic material	down
Biochemicals	mineral resources used as salt and in preparation of soap and medicines	less deposition of mineral-contributing production of trade goods	down
Fresh water – surface water	water for domestic and agricultural use	decrease in volume and surface area	down
Fresh water – groundwater	groundwater recharge provides water supply	decrease in groundwater recharge	down
Regulating services			
Climate regulation	precipitation and temperature control	decrease in precipitation	down
Water regulation	seasonal fluctuations replenish farmland and feeding areas for fish	less replenishing of farmland for crops and fish feeding areas	down
Erosion control	aquatic vegetation holds sediment	less erosion control	down
Water purification	suspended solids reduced as water spreads across floodplain	increase in deposition of suspended solids	down
Storm protection	holds storm water, provides flood control	less flood protection	down
Cultural services			
Cultural diversity	permanent residents and seasonal herders	loss of cultural diversity	down
Spiritual/religious values	locally grown spirulina used as a treatment to ward off sorcerers	less growth of algae used in traditional treatment	down
Knowledge systems	traditional village-based systems of fisheries management	breakdown of village systems without fish resource base	down
Cultural heritage values	historical cultural landmarks of ancient Sao people	possibly no change	stable
Recreation	habitat for game species, local hunting reserve	fewer game species	down
Supporting services			
Soil formation	retains sediment, adding to islands, banks, polders	less soil formation and island/bank-building	down
Nutrient formation	fertile soil from alluvial deposits	less nutrient cycling with seasonal water fluctuations	down
Primary production	abundant aquatic vegetation for wildlife	less aquatic vegetation; more grazing habitat for domestic livestock	mixed
Habitat	habitat for endemic cattle, rich avifauna, endemic fish, diverse mammals	less habitat for domestic and wild aquatic species	down

to millions of people. For example, the livelihoods of many people depend on floods to replenish the soil and nutrients of the floodplains used in flood-recession agriculture and for grazing and to clean and renew streams and sandbanks to permit fish passage for migration and the enhancement of fish production, as on the Pongolo floodplain in southern Africa (Heeg and Breen 1982). Floods also replenish sediment in coastal areas. Despite a lack of reliable quantitative evaluations, the importance of hydrological regulation by inland water systems is widely recognized around the world (Mitsch and Gosselink 2000).

20.2.2 Sediment Retention and Water Purification

Over the past few decades the valuable role that plants and substrates play in many inland waters by trapping sediments, nutri-

ents, and pollutants has been *well established* (see Chapter 15) and is illustrated in many analyses (see reviews of the condition of inland water systems cited earlier). Wide-scale vegetation clearing has caused erosion to increase, filling many shallow water bodies with sediment and disrupting the transport of sediment to coastal areas. Excessive amounts of sedimentation due to land disturbance are a global problem and have severely degraded many coastal-marine habitats, especially coral reefs close to shore. (See Chapter 19.) It is speculated that soil retention in inland waters would have ameliorated the impacts of excess sedimentation on coastal systems.

In addition to retaining sediments, the vegetation in some inland water systems, such as lakes and swamps, can remove high levels of nutrients, especially phosphorus and nitrogen, commonly associated with agricultural runoff, which could otherwise result in eutrophication of receiving ground, surface, and coastal waters. (See Chapter 12.) For example, cypress swamps in Florida in the United States can remove 98% of the nitrogen and 97% of the phosphorus that would otherwise have entered the groundwater (Brown 1981), and vegetation along the edge of Lake Victoria, East Africa, was found to have a phosphorus retention of 60–92% (Arcadis Euroconsult 2001). Inland water systems can also export nutrients, and although the general conditions under which these systems retain or export nutrients are known (e.g., Richardson and Vepraskas 2001; Mitsch and Gosselink 2000), they are often not investigated sufficiently on a site-specific basis.

The capacity of many wetland plants to remove pollutants derived from chemical or industrial discharges and mining activities is *well established* and increasingly used as a passive treatment process. (See Chapters 10 and 15.) The floating water hyacinth (*Eichhornia crassipes*) and large emergent species (such as some *Typha* and *Phragmites* species), for example, have been used to treat effluents from mining areas that contain high concentrations of heavy metals such as cadmium, zinc, mercury, nickel, copper, and vanadium. In West Bengal, India, water hyacinth is used to remove heavy metals, while other aquatic plants remove grease and oil, enabling members of a fishing cooperative to harvest one ton of fish a day from ponds that receive 23 million liters of polluted water daily from both industrial and domestic sources (Pye-Smith 1995). In another example, the Nakivubo papyrus swamp in Uganda receives semi-treated effluent from the Kampala sewage works and highly polluted storm water from the city and its suburbs. It has been established that during the passage of the effluent through the swamp, sewage is absorbed and the concentrations of pollutants are considerably reduced, at an estimated value of $2,220–3,800 per hectare per year (Emerton et al. 1998). These examples are indicative of the extremely important role that these habitats can play in removing pollutants from wastewater effluents.

It is also *well established* that not all inland water systems can assimilate all types and amounts of waste. Excessive loads of domestic sewage or industrial effluent can degrade inland water systems, with a consequent loss of biota and services. (See Chapter 15.) The environmental problems associated with waste from mining operations are a good example of the limited waste-processing capacity of these ecosystems. Recent examples of this problem include the failures of engineered waste containment structures, as occurred in 1999 in southern Spain, where more than 5 million cubic meters of heavy metal-laden sludge flowed into the Guadiamar River and part of the Coto Doñana wetlands (Bartolome and Vega 2002), and the discharge in 2000 of 100,000 cubic meters of cyanide and heavy metal-contaminated wastewater from the Baia Mare mine in Romania, which affected the Tisza, Szamos, and Danube Rivers (WWF 2002). In both these cases, species and ecosystem services from wetlands were severely affected by the excessive and toxic waste loads (Bartolome and Vega 2002; WWF 2002).

20.2.3 Recharge/Discharge of Groundwater

The issues of groundwater supply, use, and quality have received far less attention around the world than surface waters, even in industrial countries. Our understanding of groundwater resources is more limited as sufficient data, such as covering groundwater discharge/recharge and aquifer properties, for global applications are only beginning to be synthesized (Foster and Chilton 2003; UNESCO-IHP 2004). While many wetlands exist because they overlie impermeable soils or rocks and there is, therefore, little or no interaction with groundwater, numerous wetlands are fed largely by groundwater, and recharge of the aquifer occurs during flooding periods. It is well known though that many groundwater resources are vulnerable to a variety of threats, including overuse and contamination. (See Chapter 7.)

The importance of groundwater for human well-being is *well established*; between 1.5 billion (UNEP 1996) and 3 billion people (UN/WWAP 2003) depend on groundwater supplies for drinking. It also serves as the source water for 40% of industrial use and 20% of irrigation (UN/WWAP 2003). Many people in rural areas depend entirely on groundwater.

Because of unsustainable withdrawals, parts of India, China, West Asia, the former Soviet Union, the western United States, and the Arabian Peninsula, among other regions, are experiencing declining water tables, limiting the amount of water that can be used and raising the costs of getting access to it (Postel 1997; UNEP 1999). Overpumping of groundwater can lead to land subsidence, as has been recorded in megacities such as Mexico City, Manila, Bangkok, and Beijing (Foster et al. 1998). In coastal areas, lowering water tables can cause the underground intrusion of saline water, rendering these freshwater sources unusable for human consumption. In 9 of 11 European countries, for example, especially along the Mediterranean coast, where groundwater over-exploitation is reported, saltwater intrusion has become a serious problem. The main cause is groundwater overabstraction for public water supply (EEA 2003).

A common outcome of groundwater overabstraction is dryland salinization, which renders the soil unusable for cropping (see Chapter 22), often exacerbated by irrigation practices, such as those that have affected about 40% of the dryland area in West Asia (Harahsheh and Tateishi 2000). Salinity and waterlogging have affected 8.5 million hectares or 64% of the total arable land in Iraq, while 20–30% of irrigated land has been abandoned due to salinization (Abul-Gasim and Babiker 1998). In Azerbaijan, some 1.2 million hectares (about a third of total irrigated area) has been affected by salinization, and much of it has been abandoned (State Committee on Ecology and Control of Natural Resources Utilization 1998).

The ability of inland waters to recharge groundwater has been *well established*. For example, in monetary terms a 223,000-hectare swamp in Florida has been valued at $25 million per year for its role in storing water and recharging the underlying aquifer (Reuman and Chiras 2003). And the Hadejia-Nguru wetlands in northern Nigeria, in addition to supporting fishing, agriculture, and forestry, play a major role in recharging aquifers that are used by local people for domestic water supplies—a service estimated as being worth $4.8 million per year (Hollis et al. 1993).

20.2.4 Climate Change Mitigation

Inland water systems play two critical but contrasting roles in mitigating the effects of climate change: the regulation of greenhouse gases (especially carbon dioxide) and the physical buffering of climate change impacts. Inland water systems have been identified as significant storehouses (sinks) of carbon as well as sources of carbon dioxide (such as boreal peatlands), as net sequesters of organic carbon in sediments, and as transporters of carbon to the sea. Although covering an estimated 3–4% of the world's land area, peatlands are estimated to hold 540 gigatons of carbon (Immirizy and Maltby 1992), representing about 1.5% of the total estimated global carbon storage and about 25–30% of that contained in terrestrial vegetation and soils (Joosten and Clarke 2002; Lévêque 2003). Many wetlands also sequester carbon from the atmosphere through photosynthesis and act as traps for carbon-rich sediments from watershed sources. It is likely that one of the most important roles of wetlands may be in the regulation of global climate change through sequestering and releasing a major proportion of fixed carbon in the biosphere (Mitsch and Wu 1995).

Inland waters also contribute to the regulation of local climates. Possibly the most widely publicized example is that of the Aral Sea, where a combination of desiccation and pollution have altered the local climate, with dire effects on human health (as described later in the chapter). Similarly, the burning and degradation of peatland in Southeast Asia have degraded the atmosphere and affected the health of a large but possibly indeterminate number of people if the long-term effects on livelihoods as a consequence of the land degradation are considered. Getting accurate measurements of such effects and the number of people actually affected by changes in local climates is likely to prove difficult in some instances due to an absence of data and the dispersed nature of some effects or the population affected.

20.2.5 Products from Inland Water Systems

Inland water systems are a major source of products that can be exploited for human use, including fruit, fish, shellfish, deer, crocodile and other meats, resins, timber for building, fuelwood, peat, reeds for thatching and weaving, and fodder for animals. Many of these products are exploited at subsistence, cottage industry, or the larger commercial scale in most parts of the world.

Arguably the most important product derived from inland waters in terms of human well-being on a global scale is fish and fishery products. An estimated 2 million tons of fish and other aquatic animals are consumed annually in the lower Mekong Basin alone, with 1.5 million tons originating from natural wetlands and 240,000 tons from reservoirs. The total value of the catch is about $1.2 billion (Sverdrup-Jensen 2002). In recent years, the production of fish from inland waters has become dominated by aquaculture operations, mainly carp in China for domestic consumption and salmon, tilapia, and perch mainly for export to other countries (Kura et al. 2004). (See Chapters 8, 19, and 26.) In fact, since 1970 aquaculture has become the fastest-growing food production sector in the world, increasing at an average rate of 9.2% per year—an outstanding rate compared to the 2.8% rate for land-based farmed meat products (FAO 2004). In parallel, consumption of freshwater fish has shown the greatest increase over recent years, especially in China, where per capita consumption of fish increased nearly tenfold between 1981 and 1997 (Delgado et al. 2003; Kura et al. 2004).

Inland fisheries are of particular importance in developing countries, as fish is often the only source of animal protein to which rural communities have access (Kura et al. 2004). A large proportion of the recorded inland fisheries catch comes from developing countries, and the actual catch is thought to be several times the official 2001 figure of 8.7 million tons, as much of the inland catch is underreported (FAO 1999; Kura et al. 2004). Indeed, FAO considers its data on freshwater harvests so uncertain that it declined to give a comprehensive analysis of inland trends in its latest report on the state of world fisheries and aquaculture (FAO 2004).

Most of the above-mentioned increase in freshwater fish consumption has occurred in Asia, Africa, and more moderately in South America. In 1999, China accounted for 25% of the annual catch, India 11%, and Bangladesh 8% (Fishstat 2003). In North America, Europe, and the former Soviet Union, landings of fish have declined, whereas in Oceania they have remained stable (FAO 1999). Despite this increase in landings, maintained in many regions by fishery enhancements, such as stocking and fish introductions, the greatest overall threat for the long-term sustainability of inland fishery resources is the loss of fishery habitat and the degradation of the terrestrial and aquatic environments (FAO 1999). Historical trends in commercial fisheries data for well-studied rivers show dramatic declines over the twentieth century, mainly from habitat degradation, invasive species, and overharvesting (Revenga et al. 2000).

The Great Lakes of North America, shared by the United States and Canada, illustrate the value of inland fishery in these countries. These lakes have supported one of the world's largest freshwater fisheries for more than 100 years, with the commercial and sport fishery now collectively valued at more than $4 billion annually (Great Lakes Information Network 2004). The fishery consists of a mix of native and introduced species, some of which are regularly restocked. The fishery declined due to the combined effects of overfishing, pollution, and the introduction of invasive species. Recent years have seen a major resurgence in Lake Erie's fish production as some populations have recovered, and in Lake Ontario as a new fishery has been developed. However, this has only occurred in response to considerable expenditure in support of fish stocking and administration, including facilitating cooperation between governmental agencies; the real overall cost of recovering the fishery may not be accurately known (Dochoda 1988).

Another critical product derived from wetlands is rice—the staple food for nearly 50% of the world's peoples, mainly in Asia (FAO 2003). The world per capita rice consumption in 1990 was 58 kilograms per year of milled rice. This represents 23% of the average world per capita caloric intake and 16% of the protein intake (International Rice Research Institute 1995). In Asia alone, more than 2 billion people obtain 60–70% of their calories from rice and its derived products (FAO 2003). Rice is also the most rapidly growing source of food in Africa and is of significant importance to food security in an increasing number of low-income food-deficit countries. It is estimated that by 2020, 4 billion people—more than half the world's population—will depend on rice as a staple of their diet (International Rice Research Institute 1999).

Peatlands in particular, as a diverse group of habitats, provide many useful products. Peat soil has been mined extensively for domestic and industrial fuel, particularly in Western Europe but also in South America, and peat mining for use in the horticulture industry is a multimillion-dollar industry in Europe (Finlayson and Moser 1991; Maltby et al. 1996; Joosten and Clarke 2002). Whereas peat mining can be destructive in terms of the biodiversity values of the affected areas, there is an increased emphasis on sustainable practices through improved planning, water regulation, and post-mining restoration (Joosten and Clarke 2002). Peatlands also provide foods in the form of berries and mushrooms, and sometimes timber, all of which can be locally important. The tropical peat swamp forests of Southeast Asia, for example, have been an important source of tropical hardwood and are also a source of products that contribute significantly to the economy of local communities, including fish, fruit, latex, and tannins (Rieley et al. 1996). In all regions of the world there are indigenous people whose livelihoods and cultures are sustained by peatlands.

20.2.6 Recreation and Tourism

It is extremely apparent that the aesthetics as well as the diversity of the animal and plant life of many inland water systems has attracted tourism. Many inland water sites are protected as National Parks, World Heritage Sites, or wetlands of international importance (that is, Ramsar sites) and are able to generate considerable income from tourist and recreational uses. In some locations tourism plays a major part in supporting rural economies, although there are often great disparities between access to and involvement in such activities. The negative effects of recreation and tourism are particularly noticeable when they introduce inequities and do not support and develop local economies, especially where the resources that support the recreation and tourism, such as wetlands, are degraded.

The income generated by recreation and tourism can be a significant component of local and national economies. Recreational fishing can generate considerable income: some 35–45 million people take part in recreational (both inland and marine) fishing in the United States, spending $24–37 billion each year on their hobby (Thomsen 1999; Ducks Unlimited 2002). In 2001, freshwater fishing (including the Great Lakes) alone generated more than $29 billion from retail sales and more than $82.1 billion in total economic output (including contributions to household income and taxation revenues) (American Sportfishing Association 2001). The total economic value of such activities extends far beyond the direct expenditure and includes, for example, contributions to local property markets and taxation revenues.

The value of recreation and tourism from inland water systems is widely recognized in many other parts of the world, but not necessarily as well quantified (Finlayson and Moser 1991). There are many inland waters with significant recreational value for which a monetary value cannot easily be given because visitors use the area without direct payment. Employing economic valuation techniques, such as willingness to pay and other methods (see Chapter 2), to investigate the value that users ascribe to a wetland is becoming the topic of increased research and documentation. The recreational value of the Norfolk Broads wetlands in the United Kingdom, for instance, was estimated at $57.3 million per year for people living relatively close to the Broads and $12.9 million per year for those living further away (Barbier et al. 1997).

In considering such analyses, the cost of repairing any degradation or providing facilities for visitors must also be taken into account. It is well known, for example, that overuse of popular fishing or camping spots around lakes or along rivers can lead to severe degradation and result in the demise of such activities and the loss of income-generation opportunities.

Although not strictly speaking a "recreational" function, the educational value of wetlands is closely related: there are many wetland education centers and programs around the world that involve the general public and schoolchildren in practical activities in their local wetland environments; these activities span the border between education and recreation. Approximately 160,000 people a year visit a 40-hectare wetland complex in the heart of London; created from a series of reservoirs, it offers 30 lakes and marshes, boardwalks, hides, and pathways as well as an exhibition center that educates visitors on the functions and values of inland water ecosystems, biodiversity issues, and other environmental matters in an essentially recreational setting (Peberdy 1999).

20.2.7 Cultural Value

Inland waters are closely associated with the development of human culture—notably, for example, in the Indus, Nile, and Tigris-Euphrates valleys (Finlayson and Moser 1991)—and many major cities are built near rivers. In some cultures inland waters may have deep religious significance for local people. In Tibet, for example, pre-Buddhist belief identified various lakes as sacred,

making them objects of worship as well as ensuring their protection from pollution and other harm. As Buddhism took over, these beliefs remained, albeit in a modified form, and certain lakes in Tibet are still sacred to the people, with strict regulations on their exploitation (Dowman 1997).

Cranes have a near-sacred place in the earliest legends of the world and have featured prominently in art and folklore for millennia. For example, the Brolga (*Grus rubicunda*) figures prominently in some indigenous Australian folklore and culture, and in Northeast Asia cranes are revered as symbols of longevity and peace (Wetlands International 1999). At anything but a local scale, however, cultural values are a relatively poorly documented service of inland waters, despite the many instances where wetlands have significant religious, historical, archaeological, or other cultural values for local communities. (See Chapter 17.)

20.3 Condition of Inland Water Systems

The information base for assessing the condition of inland water systems globally is widely documented and summarized in many reports (e.g., Finlayson et al. 1992; Finlayson and Moser 1991; Moser et al. 1993, 1996; Whigham et al. 1993; Mitsch 1994; McComb and Davis 1998). But it is as widely documented that at a global or continental scale there are large gaps in information (Finlayson and Spiers 1999; Darras et al. 1999; Revenga et al. 2000; Brinson and Ines Malvare 2002; Junk 2002; Malmqvist and Rundle 2002; Williams 2002). The information in this section on the biodiversity of inland water systems is largely based on analyses undertaken by Revenga and Kura (2003), which made use of global and regional-scale datasets while identifying the inadequacy of many information sources and the difficulties of gaining access to others.

20.3.1 Extent and Change of Inland Water Systems

Estimates of the extent of wetlands at a global level vary from 530 million to 1,280 million hectares, but it is *well established* that this is a clear underestimate (Spiers 1999; Finlayson et al. 1999). Estimates of the global extent of wetlands are highly dependent on the definitions for wetlands used in each inventory, the type of source material available, the methodology used, and the objectives of the investigation. The 1999 *Global Review of Wetland Resources and Priorities for Wetland Inventory* estimated wetlands extent from national inventories as approximately 1,280 million hectares (Finlayson et al. 1999), which is considerably higher than previous estimates derived from remotely sensed information.

Nevertheless, the GRoWI figure is considered an underestimate, especially for the Neotropics. For example, Ellison (2004) contends that in central America the savannas should be classed as seasonal wetlands rather than grasslands; it is not known if this is the case in other savanna landscapes. Analyses of wetland inventory in Mexico (CNA-INUBAN 1999) and Brazil (Maltchik 2003) similarly illustrated the poor state of knowledge covering wetland classification and inventory. Another limitation of the data used in GRoWI is that for certain wetland types (such as intermittently flooded inland wetlands, peatlands, artificial wetlands, seagrasses, and coastal flats), data were incomplete or not readily accessible (Finlayson et al. 1999).

Even so, the data collated by Finlayson et al. (1999) suggest that the largest area of wetlands is in the Neotropics (32%), with large areas also in Europe and North America. But note that figures provided by Lehner and Döll (2004) (see Table 20.4) suggest that Asia may contain a greater and Europe a lesser area of wetlands. Table 20.4 presents the two best available estimates from

Table 20.4. Estimates of Inland (Non-marine) Wetland Area
(Finlayson et al. 1999; Lehner and Döll 2004)

Region	1999 Global Review of Wetland Resources	2004 Global Lakes and Wetlands Database
	(million hectares)	
Africa	121–125	131
Asia	204	286
Europe	258	26
Neotropics	415	159
North America	242	287
Oceania	36	28
Total area	**~ 1,280**	**~ 917**

Note: Not all wetland types are equally represented in the underlying national inventory data. Some countries lack information on some types of wetlands.

wetlands extent: the GRoWI assessment (Finlayson et al. 1999) and the WWF/Kassel University Global Lakes and Wetlands Database (GLWD) (Lehner and Döll 2004).

Mapping exercises have been undertaken for inland waters, but the level of detail varies from region to region. The most recent global map, with a 1-minute resolution, was produced by combining various digital maps and data sources (Lehner and Döll 2004), but it still suffers from the problems of definition and scale outlined by Finlayson et al. (1999). Problems with the scale and resolution of data sources for inventory have been shown for northern Australia, where estimates of the area of inland water systems from 10 data sources varied from 0 to 98,700 square kilometers (Lowry and Finlayson 2004).

Inventories of major river systems, including data on drainage area, length, and flow volume are available, but there is considerable variability between estimates, based on the method and definitions used. Information on river flow volume and discharge, for example, varies considerably depending on the water balance model applied and the different time periods or locations for the measurement of discharge (see Revenga and Kura 2003).

Information on the estimated 5–15 million lakes across the globe is also highly variable and dispersed (WWDR 2003). Large lakes have been mapped reasonably well, but issues of scale also occur with smaller lakes being more difficult to map. Nevertheless, there is no single repository of comprehensive lake information, which makes assessment of these water bodies difficult and time-consuming. A high proportion of large lakes—those with a surface area over 500 square kilometers—are found in Russia and in North America, especially Canada, where glacial scouring created many depressions in which lakes have formed. Tectonic belts, such as the Rift Valley in East Africa and the Lake Baikal region in Siberia, are the sites of some of the largest and most "ancient" lakes, all of which have highly diverse species assemblages. Some of the largest lakes are saline, with the largest by far being the Caspian Sea (422,000 square kilometers). There are many saline lakes occurring on all continents and many islands; given the impermanence of many, it is difficult to derive accurate values for their number worldwide.

Reservoirs are also widespread; the number of dams in the world has increased from 5,000 in 1950 to more than 45,000 at present (WCD 2000). These reservoirs provide water for 30–40%

of irrigated agriculture land and generate 19% of global electricity supplies (WCD 2000).

Peatlands are known to occur in at least 173 countries throughout most parts of the world, from Arctic systems through temperate to tropical regions (Joosten 1992). Their total area has been estimated as approximately 400 million hectare, of which the vast majority are in Canada (37%) and Russia (30%), which together with the United States account for over 80% of the global peatland resource. The largest area of tropical peatland is in Indonesia (6–7% of the global area). Peatlands are estimated to store 30% of Earth's surface soil carbon (Joosten and Clarke 2002). The global area of paddies has been estimated as 1.3 million square kilometers (130 million hectares) (Aselmann and Crutzen 1989), of which almost 90% is in Asia, but it is likely that this figure is now out of date. Information on other human-made wetlands is variable and even lacking for some countries.

Groundwater systems have received slightly increased attention in recent years. These systems vary in size, from the small-scale alluvial sediment along rivers to extensive aquifers such as the 1.2 million square kilometers of the Guarani aquifer located across parts of Argentina, Brazil, Paraguay, and Uruguay (Daniel-opol et al. 2003). Groundwater systems have many connections and interactions with surface waters, although many of these are not well understood. Some aquifers are better known for their biodiversity values, such as the karst systems of Slovenia that cover some 8,800 square kilometers and are increasingly known for their high species biodiversity (see Box 20.1), while others are not known at all.

The loss and degradation of inland waters have been reported in many parts of the world (Finlayson et al. 1992; Mitsch 1998; Moser et al. 1996), but there are few reliable estimates of the actual extent of this loss. Dugan (1993) speculated that on a global scale the loss of wetlands was about 50%, but he did not provide supporting evidence, and as reliable estimates of the extent of wetlands (and particularly of intermittently inundated wetlands in semiarid lands) are lacking, it is not possible to ascertain the extent of wetland loss reliably.

The information available on the distribution of inland waters is on the whole better for North America than for many other parts of the world. The overall area of wetlands in North America includes 2.72 million hectares in Mexico (CNA-IBUNAM 1999), 127–168 million hectares in Canada (Wiken et al. 1996; Moore and Wiken 1998; National Wetlands Working Group 1988; Warner and Rubec 1997), and 43 million hectares in the conterminous United States (Dahl 2000).

As with the information on the distribution of wetlands, data on their conditions and trends are on the whole better for the United States than that for many other parts of the world. The United States is one of the few countries that systematically monitors change in wetlands extent. From the mid-1970s to mid-1980s, wetland losses (excluding lakes and rivers) in that country amounted to about 116,000 hectares per year (Dahl and Johnson 1991). This rate of loss decreased by 80% to a loss of approximately 23,700 hectares a year from 1986 to 1997, with 98% of these losses being from forested and freshwater wetlands, mostly from conversion or drainage for urban development and agricultural purposes (Dahl 2000). As of 1997, an estimated 42.7 million hectares remains out of the 89 million hectares of wetlands present in the United States at the time of European colonization (Dahl 2000).

The overall decline in the rate of loss observed in the United States is attributed primarily to wetland policies and programs that promote restoration, creation, and enhancement of wetlands, as well as incentives that deter the draining of wetlands. Between

BOX 20.1
Biodiversity of Karsts in Slovenia (Information supplied by G. Beltram from multiple sources)

Approximately 8,800 square kilometers or 44% of the surface area of Slovenia is carbonate bed-rock, known as karst. It is very permeable and supports many caves and fissures. Over many centuries the karst areas have been greatly modified by humans, with eventual replacement of the deciduous forests by dry, rocky pastures and meadows, small arable fields, dry stone walls, and karst pools. Logging, grazing, forest fires, and strong winds have further degraded the karsts through soil erosion and exposed a stark and bare-stone landscape. In the last 50 years or so, further change has occurred as local people abandoned the agricultural land and as shrubs and trees invaded meadows and arable fields—changing the vegetation structure and cultural significance of these areas (Beltram and Skoberne 1998). Pollution and habitat destruction are also problematic.

The subterranean caves and fissures within the karst are important for biodiversity as well as human use. The subterranean fauna, particularly aquatic stygobiontic species, is very rich and includes about 800 endemic fauna taxa. Many species also have extremely small distributions. The fauna in the caves is varied, with many species not found elsewhere—for example, the cave vertebrate (*Proteus anguinus*), tubeworm (*Marifugia cavatica*), mollusk (*Kerkia kusceri*), cnidarian (*Velkovrhia enigmatica*), and water fleas (*Alona sketi* and *A. stochi*), as well as a number of stygobiontic snails (*Gastropoda*) and epizoic turbellearians (*Temnocephalida*). Additionally, the crustacean fauna, including amphipods, copepods, and isopods, is extremely rich.

The karst landscape has a strong cultural heritage dating back centuries. Significant settlements were constructed near natural springs, and natural and human-made pools were used for watering domestic animals. With the demise of agricultural activities, some caves have become popular tourist destinations. Groundwater from the karst is a very important source of domestic water supply for almost half of Slovenia's inhabitants, making recharge of this source a very important function of the landscape.

are poorly known or assessed in most parts of the world, with very fragmentary information available; and that every group of organisms considered, including aquatic plants, invertebrate, and vertebrate animal species, contained examples of extinct, critically endangered, endangered, and vulnerable taxa.

Although small in global area compared with marine and terrestrial ecosystems, inland water systems are relatively species-rich (McAllister et al. 1997). (See Table 20.5.) Marine systems contain over six times as many known species as inland waters, but cover over two thirds of the globe, compared with inland water systems, which occupy less than 1% as much area. Over three quarters of known species are terrestrial, but these systems have similar relative species richness to inland water systems. Inland water systems also support a disproportionately large number of species of some taxonomic groups. For instance, some 40% of known species of fish inhabit inland waters (more than 10,000 species out of 25,000 species globally), and about 25–30% of all vertebrate species diversity is concentrated close to or in inland waters (Lévêque et al. in press). There are about 100,000 described species of freshwater fauna worldwide (Lévêque et al. in press). Half of these are insects (see Table 20.6), about 12,000 are crustaceans, 5,000 are mollusks, and some 20,000 are vertebrate species. It is anticipated that the number of aquatic animals will be far higher than current estimates as more species from inland waters are described every year—about 200 new fish species are described each year (Lundberg et al. 2000).

Because many inland wetlands are geographically isolated, levels of endemism of freshwater species are particularly high, especially in ancient lakes, such as the Great East African Lakes (Tanganyika, Malawi, and Victoria), Lake Baikal, Lake Biwa, and Lake Ohrid, which have been isolated for millions of years (Lévêque et al. in press).

The *IUCN Red List,* a widely used indicator for assessing the conservation status of plants and animals, does not comprehensively assess inland water species. For instance, only a very small proportion of the species in most freshwater taxa, such as aquatic plants, mollusks, crustaceans, and insects, have been assessed. However, among the taxa that have been comprehensively assessed, such as amphibians and birds, a high number of species are threatened with extinction (IUCN 2003), as described further later in this section.

Another global measure of the status of animal species is the Living Planet Index developed by WWF and UNEP-WCMC (Loh and Wackernagel 2004). The LPI provides a measure of the trends in more than 3,000 populations of 1,145 vertebrate species around the world. It is an aggregate of three separate indices of

1986 and 1997, the country had a net gain of about 72,870 hectares of upland wetlands, mostly due to Federal protection and restoration programs and an increase in the area of lakes and reservoirs by 47,000 hectares due to creation of new impoundments and artificial lakes (Dahl 2000).

20.3.2 Status of Inland Water Species

Data on the condition and trends of freshwater species are for the most part poor at the global level, although some countries have reasonable inventories and indicators of change of inland water species (such as Australia, Canada, New Zealand, South Africa, and the United States) (Revenga and Kura 2003). This does not mean that data are not available; there are considerable data on freshwater species and populations, but they are not easily accessible. For example, there are many extensive records in museums and universities around the world, but these are often not centrally located or electronically archived.

Revenga and Kura (2003) assessed the level of knowledge of the distribution and condition of inland water biodiversity at the global level. Key conclusions from this assessment indicated that fish and waterbirds are by far the best studied groups of inland water species, although with considerable regional differences; that aquatic plants, insects, freshwater mollusks, and crustaceans

Table 20.5. Relative Species Richness of Different Ecosystems (McAllister et al. 1997)

Ecosystems	Habitat Extent	Species Diversity	Relative Species Richness [b]
	(percent of world)	(percent of known species)[a]	
Freshwater	0.8	2.4	3.0
Marine	70.8	14.7	0.2
Terrestrial	28.4	77.5	2.7

[a] Does not add up to 100 because 5.3% of known symbiotic species are excluded.

[b] Calculated as the ratio between species diversity and habitat extent

Table 20.6. Current State of Knowledge of Global Species Richness of Inland Water Animal Groups (Lévêque et al. in press; Revenga and Kura 2003; IUCN et al. 2004)

Phylum	Described Species
	(number)
Porifera (sponges)	197
Cnidaria (hydra, freshwater jelly fish)	30
Nemertea (ribbon worms)	12
Plathelminthes (flatworms)	c. 500
Gastrotrichia	c. 250
Rotifers	1,817
Nematods (microscopic worms)	3,000
Annelids (segmented worms)	c. 1,000
Bryozoa (moss animals)	70—75
Mollusks (mussels, snails, slugs, etc.)	c. 6,000
Crustaceans (crabs, crayfish, etc.)	c. 12,000
Arachnids (spiders, etc.)	5,000
Insects	> 50,000
Vertebrates	
Fish	13,400
Amphibians	3,533
Reptiles	c. 250
Birds	c. 1,800
Mammals	c. 122

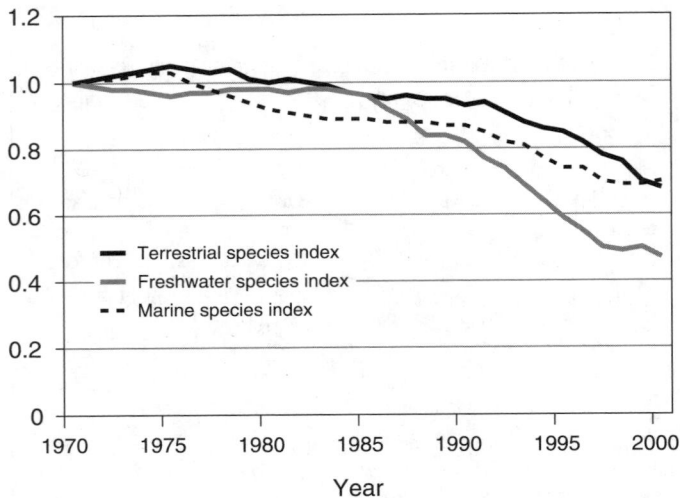

Figure 20.3. Living Planet Index for Terrestrial, Inland Water, and Marine Systems (Loh and Wacknernagel 2004)

change in freshwater, marine, and terrestrial species. The 2004 freshwater species population index, which took into account trend data for 269 temperate and 54 tropical freshwater species populations (93 of which were fish, 67 amphibians, 16 reptiles, 136 birds, and 11 mammals), shows that freshwater populations have declined consistently and at a faster rate than other species groups assessed, with an average decline of 50% between 1970 and 2000. Over the same period, both terrestrial and marine fauna decreased by 30%. (See Figure 20.3.) While the index has a bias in the available data toward North America and Europe, and particularly toward bird populations, data collection and collations have been undertaken each year to extend the veracity of this index, with initial indications of a continuing decline remaining constant (Loh and Wackernagel 2004).

20.3.2.1 Aquatic Plants and Fungi

The definition of aquatic plants has been and is still debated, but in general plants that tolerate or require flooding for a minimum duration of time are considered wetland plants. Aquatic macrophytes include angiosperms (flowering plants), macroalgae, pterophytes (pteridophytes, ferns), and bryophytes (mosses, hornworts, and liverworts). Gymnosperms (conifers, cycads, and their allies) do not have strictly aquatic representatives, but include a number of tree species that tolerate waterlogged soil, such as the bald cypress.

It is estimated that up to 2% (250 species) of pterophytes and 1% (2,500 species) of angiosperms are aquatic, but their geographic distribution, diversity patterns, or conservation status have not been summarized globally, although information exists for specific regions. Some families of aquatic angiosperms show highest diversity at tropical latitudes, while others show higher diversity in temperate regions, making it evident that the typical latitudinal gradient found in terrestrial species diversity does not apply. Locally, many macrophyte species may be threatened or lost, especially in lakes undergoing eutrophication; as an example, many isoetid species are widespread but increasingly threatened throughout their range (e.g. Nichols and Lathrop 1994). The maximum diversity of bryophytes is found in highly oceanic regions, where cool or temperate and consistently moist climate conditions persisted over geological time (Groombridge 1992). Bryophytes found in lowland aquatic environments, including pools and reservoirs, tend to have very restricted distribution and therefore are rare and threatened (Revenga and Kura 2003).

A relatively small fraction (over 600 species) of fungi are considered freshwater species, but recent studies indicate that there are many more freshwater fungi to be discovered in temperate and tropical regions; the total number estimated at between 1,000 and 10,000 (Palmer et al. 1997). Information on the conservation status of aquatic fungi is very limited.

20.3.2.2 Invertebrates

Information on aquatic invertebrate species diversity is fragmentary, with a few descriptive global overviews of particular taxa and some more detailed regional inventories. The conservation status of aquatic invertebrates has not been comprehensively assessed, except for regional assessments of certain taxonomic groups such as Odonata (dragonflies and damselflies; see Box 20.2) (Clausnitzer and Jodicke 2004), freshwater mollusks (mostly mussels), and freshwater crustaceans (Master et al. 1998; IUCN 2003). Assessments of the status of known mollusk species have been conducted for a limited number of taxa and regions, including the Mekong, which has a very diverse freshwater mollusk fauna (Dudgeon 2002a, 2002b, 2002c); Lake Biwa, Japan, which has 73% of the freshwater mussel species described in the country, of which 43% are endemic; and Lakes Baikal, Tanganyika, and Titicaca. (See Box 20.3.)

IUCN (2003) reports 130 freshwater species of aquatic insects, 275 species of freshwater crustacean, and 420 freshwater mollusks as globally threatened, although no comprehensive global assessment has been made of all the species in these groups. For the

---

BOX 20.2

Status of Odonata (Dragonflies and Damselflies)
(Clausnitzer and Jodicke 2004)

A recent global review of the threat status of dragonflies and damsel-flies in 22 regions covering most of the world (except for parts of Asia) found that there are many more dragonfly species now regarded as threatened than are listed in the *IUCN Red List*, which currently lists 130 species (*medium certainty*). It is important to note, however, that the criteria and categories used to assess conservation status are not harmonized across regions.

In Australia, for example, there are 4 species currently on the *Red List* as globally threatened but another 25 are considered to be in critical condition and an additional 30% of species are data-deficient in regional assessments. In North America, 25 species (6%) are of conservation concern; in the Neotropics, 25 species are considered globally threatened and a further 45 species are considered of high conservation priority, with many others being data-deficient. In eastern Africa, 90 species are considered appropriate for globally threatened status. And in southern Africa, although 2 species are currently recognized as globally threatened, a further 11 are now considered threatened in regional assessments. In Madagascar, 2 species are currently recognized as globally threatened, but because of high diversity and endemism, a large number—111 species, 64% of the fauna—are of conservation concern, although all species are data-deficient. In the Western Indian Ocean Islands, 3 species are recognized as globally threatened and 33 are now regarded as critical; in Sri Lanka, no species are currently on the *Red List*, although 47 species (all endemic) are regarded as threatened with extinction. Finally, in Europe 6 species are *Red-listed,* although two of these are now considered out of danger, and a further 22 species are of concern as their populations are declining. And in Turkey, Iran, and the Caucasus, there are 5 species on the *Red List* and 27 regarded as critical.

In most areas assessed, habitat loss and degradation of wetlands (and forests) were considered the major drivers of declines in Odonate species, often associated with overabstraction and pollution of water as well as the impacts of alien invasive species.

BOX 20.3

Endemism of Mollusks in Inland Waters

Twenty-seven areas of special importance for freshwater mollusk endemism worldwide have been identified, in three types of wetlands:

- ancient lakes: Baikal, Biwa, and Tanganyika, where 70%, 52%, and 64% of molluscan species are endemic respectively;
- lower river basins: 93% of the total freshwater mollusk species found in the Mobile Bay region of the Alabama-Tombigbee River basin in the United States are endemic; another notable center of endemism is found in the Lower 500 kilometers of the Mekong River basin, where 92% of molluscan species are endemic; and
- springs and underground aquifers (Australia, New Caledonia, the Balkans, western United States, Florida, and the Cuatro Cienegas basin in Mexico).

Inland Waters	Gastropods	Bivalves	Total
	%	%	%
Ancient lakes			
Baikal	78	52	73
Biwa	50	56	52
Sulawesi	c. 80	25	c. 76
Tanganyika	66	53	64
Malawi	57	11	46
Victoria	46	50	48
Ohrid	76		
Titicaca	63		
Major river basins			
Mobile Bay Basin	93	54	78
Lower Uruguay River and Rio de la Plata	48	21	37
Mekong River (lower 500 km)	92	13	73
Lower Congo basin	25	n/a	
Lower Zaire basin	25	n/a	

United States, one of the few countries to assess freshwater mollusks and crustaceans comprehensively, 50% of known crayfish species and two thirds of freshwater mollusks are at risk of extinction, and at least one in 10 freshwater mollusks are likely to have already gone extinct (Master et al. 1998).

20.3.2.3 Freshwater Fish

Most global and regional overviews of freshwater biodiversity include more information on fish than any other taxa (Cushing et al. 1995; Gopal et al. 2000; Groombridge and Jenkins 1998; Maitland and Crivelli 1996). A number of regional overviews of the status of freshwater fish are available, yet many of the existing overviews underestimate the number of species, as there are still many species to be described and assessed. There is, therefore, a high level of uncertainty about the status of fish in many inland waters. Estimates of the number of freshwater fish in Latin America vary from 5,000 to 8,000; in tropical Asia and Africa, there are an estimated 3,000 species on each continent (Revenga and Kura 2003), although these figures are almost certainly under-estimates. The Mekong River alone is considered to have 1,200–1,700 species (WRI et al. 2003). (See Box 20.4.) North America is estimated to have more than 1,000 species, and Europe and

Australia have several hundred species each (Revenga and Kura 2003).

With respect to their conservation status, estimates are that in the last few decades more than 20% of the world's 10,000 described freshwater fish species have become threatened or endangered or are listed as extinct (Moyle and Leidy 1992). The *IUCN Red List* (2004) classifies 648 freshwater ray-finned fish species as globally threatened. However, the coverage for freshwater fish is highly biased to particular regions for which more data are available, such as North and Central America or the African Rift Valley Lakes. For example, of the ray-finned fishes listed as threatened in the *IUCN Red List,* 122 are found in the United States and 85 in Mexico, partially reflecting the high level of knowledge in these two countries (IUCN 2003).

In the 20 countries for which assessments are most complete, an average of 17% of freshwater fish species are globally threatened. In addition, there are a few well-documented cases that show clearly this level of threat. The most widely known is the apparent disappearance of between 41 and 123 haplochromine cichlids in Lake Victoria (Harrison and Stiassny 1999), although taxonomic questions remain a problem in accurately assessing this group of fish. In Europe (including the former Soviet Union),

---

The vertebrate fauna of the Mekong River basin is difficult to quantify due to the incomplete state of the inventory and taxonomic effort. Many published figures are considered to be underestimates. The Mekong River Commission (1997) estimated that in the Laotian, Vietnamese, Cambodian, and Thai part of the basin there were some 830 mammal species, 2,800 bird species, 1,500 fish species, 250 amphibians, and 650 reptiles.

Many of these species are threatened. For example, among the mammals this includes the fishing cat (*Prionailurus viverrinus*), the hairy-nosed otter (*Lutra sumatrana*), the smooth-coated otter (*Lutrogale perspicillata*), and the Oriental small-clawed otter (*Aonyx cinerea*). A high proportion of bird species are in decline (Dudgeon 2002), particularly those that rely on sandbars and large river stretches for breeding or feeding. These include the Plain Martin (*Riparia paludicola*) and the now extinct White-eyed River Martin (*Pseudochelidon sirintarae*). Two crocodile species occur, although the population of the estuarine crocodile (*Crocodylus porosus*) is likely very low. A number of aquatic and semi-aquatic turtles, snakes, and lizards occur, many of which are hunted for subsistence or sold for food and medicine in local markets. A substantial illegal market also exists for many wildlife products.

The fish fauna is considered to be diverse, although this has not been well documented. There are an estimated 700 freshwater fish species in Cambodia. The diversity at a family level seems to be high, with some 65 families in the Cambodian and 50 in the Laotian parts of the basin. Fish introductions have occurred with the Nile tilapia (*Oreochromis niloticus*) and mosquito fish (*Gambusia affinis*), now considered as pests. A number of large native species have declined—the giant catfish (*Pangasianodon gigas*), river catfish (*Pangasius sanitwongsei*), thicklip barb (*Probarbus* labeamajor), and the giant barb (*Catlocarpio siamensis*) are now rare (Mattson et al. 2003).

---

there are 67 threatened species of freshwater fish, including sturgeons, barbs, and other cyprinids (IUCN 2003).

20.3.2.4 Amphibians

Amphibians are found in many types of freshwater habitats—from ponds, streams, and wetlands to leaf litter, trees, underground, and vernal (temporary) pools. Although some amphibians thrive in cold or dry conditions, the group reaches its highest diversity and abundance in warm, humid climates.

The recent Global Amphibian Assessment (IUCN et al. 2004) lists 5,743 known species of frogs, toads, salamanders, and caecilians, of which 3,908 species depend on fresh water during some stage of their life cycle, while the rest do not require fresh water to breed or develop. The study also shows nearly one third (1,856 species) of the world's amphibian species are threatened with extinction, a large portion of which (964 species) are freshwater—a far greater level of threat than for birds (12% of all species) and mammals (23% of all species). In addition, at least 43% of all species are declining in population, indicating that the number of threatened species can be expected to rise in the future. In contrast, less than 1% of species show population increases.

The rate of decline in the conservation status of freshwater amphibians is far worse than that of terrestrial species. As amphibians are excellent indicators of the quality of the overall environment, this underpins the notion of the current declining condition of freshwater habitats around the world.

Species associated with flowing water were found to have a higher risk of extinction than those associated with still water. For species of known status (that is, excluding those that are data-deficient), as many as 42% are globally threatened and as many as 168 amphibian species may already be extinct—at least 34 amphibian species are known to be extinct, while another 134 species have not been found in recent years and are possibly extinct. Salamanders and newts have an even high level of threat (46% globally threatened or extinct) than frogs and toads (33%) and Caecilians (2%, although knowledge of these is poor, with only one third assessed).

The largest numbers of threatened species occur in Latin American countries such as Colombia (208 species), Mexico (191 species), and Ecuador (163 species). However, the highest levels of threat are in the Caribbean, where more than 80% of amphibians are threatened in the Dominican Republic, Cuba, and Jamaica, and 92% in Haiti (IUCN et al. 2004). The major threat to amphibians is habitat loss, but a newly recognized fungal disease is seriously affecting an increasing number of species. Those species dependent on flowing water (usually streams) have a much higher likelihood of being threatened than those that use still water (often temporary rain-fed pools or other small freshwater pools). (See Figure 20.4 in Appendix A.) Basins with the highest number of threatened freshwater amphibians include the Amazon, Yangtze, Niger, Parana, Mekong, Red and Pearl in China, Krishna in India, and Balsas and Usumacinta in Central America. All these basins have between 13 and 98 threatened freshwater species.

20.3.2.5 Reptiles

There are around 200 species of freshwater turtles throughout the warm temperate and tropical regions of the world; information on the distribution of these species is available through the World Turtle Database (emys.geo.orst.edu), which contains maps of all the known localities of every freshwater (and terrestrial) turtle species. Of the 200 species of freshwater turtles, 51% of the species of known status have been assessed as globally threatened, and the number of critically endangered freshwater turtles more than doubled in the four years preceding 2000 (van Dijk et al. 2000). Of 90 species of Asian freshwater turtles and tortoises, 74% are considered globally threatened, including 18 species that are critically endangered and one, the Yunnan box turtle, which is already extinct (van Dijk et al. 2000).

Crocodiles, alligators, caimans, and gharials are widespread throughout tropical and sub-tropical aquatic habitats. Of the 23 species of crocodilians, which inhabit a range of wetlands including marshes, swamps, rivers, lagoons, and estuaries, 4 are critically endangered, 3 are endangered, and 3 are vulnerable (IUCN 2003). The other species are at lower risk of extinction, but depleted or extirpated locally in some areas (Revenga and Kura 2003). The most critically endangered crocodilian is the Chinese alligator, which is restricted to the lower reaches of the Yangtze River; it is estimated that only 150 individuals remain in the wild (IUCN/SSC Crocodile Specialist Group 2002). The major threats to crocodilians are habitat degradation and overexploitation (Revenga and Kura 2003).

There are several species of freshwater snakes in the world. The wart or file snakes (*Acrochordidae*) are adapted to aquatic life, with two species occurring in freshwater habitats (Uetz and Etzold 1996); there is little information on their conservation status. In addition, there are many semi-aquatic snakes, with some being considered vulnerable (IUCN 2003).

20.3.2.6 Birds

Waterbirds (bird species that are ecologically dependent on wetlands), particularly migratory waterbirds, are relatively well studied, with time series data available for some populations in North America and Southern and Northwest Europe for up to 40 years. Global information on waterbird population status and trends is compiled and regularly updated (Wetlands International 2002).

Detailed information and review of status for waterbird species has been compiled in North America (Morrison et al. 2001; Brown et al. 2001; U.S. Fish and Wildlife Service 2004) and for the Western Palaearctic and Southwest Asia (e.g., Delany et al. 1999). For African-Eurasian waterbird populations, comprehensive analyses have been compiled for Anatidae (ducks, geese, and swans) (e.g., Scott and Rose 1996) and waders (Charadrii) (Stroud et al. 2004). In East Asia, Bamford et al. (in press) have collated and reviewed the current status and trends of waders, while information for Gruidae (cranes) and Anatidae is available from Miyabayashi and Mundkur (1999) and the Asia-Pacific Migratory Waterbird Conservation Committee (2001). Networks of both large and small wetlands along migratory flyways are of key importance as resting and feeding sites. In semiarid landscapes, many waterbirds migrate in response to periodic and regionalized flooding that produces a temporally dispersed array of habitats (Roshier et al. 2001). The wetlands in the Sahel region of Africa provide a good example. (See Box 20.5.)

In all regions, population sizes of waterbirds are better known than population trends. Trends have been estimated for half of all waterbird populations and almost three quarters of European populations, but for only one third of populations in the Neotropics, and many trends are yet to be statistically quantified. The status of sedentary populations is much less well known that that of migratory ones.

Many waterbird species are globally threatened (Davidson and Stroud 2004), and the status of both inland and marine/coastal birds is deteriorating faster than those in other habitats (*high certainty*). Of the 35 bird families whose species are wholly or predominantly coastal/marine or inland wetland–dependent, 20% of the 1,058 species for which assessment data exist are currently globally threatened or extinct. Of these, 42 species—half of which

are island-endemic rails—are extinct and 41 species (4%) are critically endangered. There are globally threatened species in 60% of these families. The percentage of globally threatened waterbirds (including seabirds) is shown in Figure 20.5.

The status of birds continues to deteriorate in all parts of the world and across all major habitat types (*high certainty*). Waterbirds dependent on freshwater ecosystems, especially those using marine and coastal ecosystems, have deteriorated in status faster than the average for all threatened species (see Figure 20.6), but similarly to other migratory bird species.

Shorebirds are declining worldwide: of populations with a known trend, 48% are declining (Stroud et al. 2004). Other waterbirds have as bad or worse global status as shorebirds, including

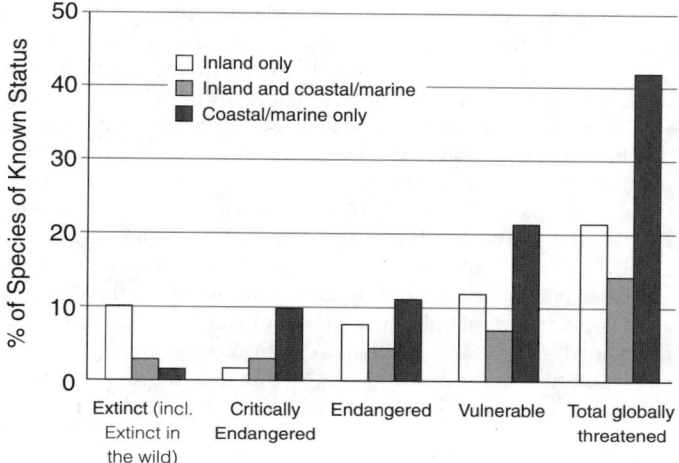

Figure 20.5. Percentage of Globally Threatened Waterbirds, Including Seabirds, in Different Threat Categories. Each waterbird family is allocated as either depending on only inland wetlands, depending on only coastal/marine systems, or depending on both inland and coastal/marine systems. (BirdLife International 2004)

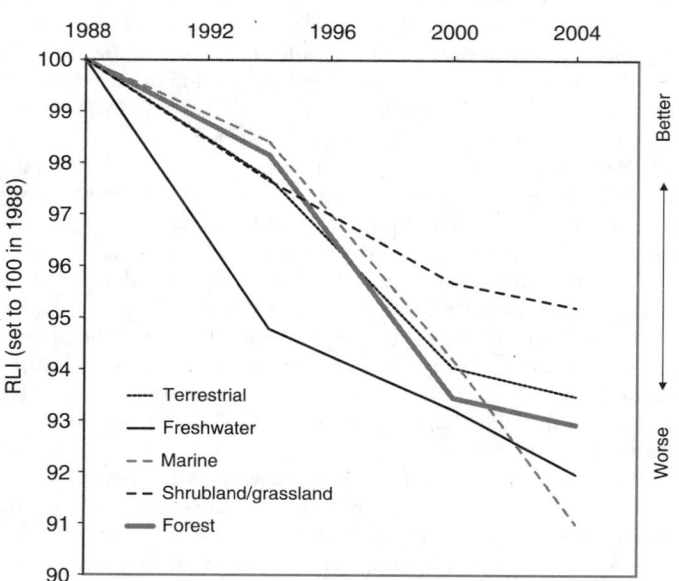

Figure 20.6. Red List Indices for Birds in Marine, Freshwater, and Terrestrial Ecosystems and for Birds in Forest and Shrubland/Grassland Habitats (Butchart et al. in press)

BOX 20.5

Sahel Wetlands (Information supplied by J. Brouwer: www.iucn.org/themes/cem)

The Sahel area of Africa is an important area for migratory birds, situated between the Sahara desert to the north and the more humid savanna and forests to the south. It is semiarid, with 200–600 millimeters of rainfall per year, and comprises the wetlands of the Senegal River, the Inner Delta of the Niger in Mali, the Hadejia-Nguru wetlands in northern Nigeria, Waza-Logone in northern Cameroon, and Lake Chad, as well as thousands of smaller, isolated wetlands.

In Niger, an estimated 1.1 million waterbirds are present during January-February, with 750,000 on the smaller, isolated wetlands. As in other semiarid regions, the waterbirds depend on a network of wetlands that are variously wet and dry, spatially and temporally. The water chemistry and vegetation composition also varies between wetlands, providing a diversity of habitats that is essential for the many birds that migrate through the region.

The wetlands are also used extensively by local people, as they are highly productive and important for grazing, fishing, and market gardening. As the human population has increased, so has pressure on these wetlands—land use is becoming more intensified, and many wetlands are threatened with change, which in turn is likely to adversely affect the waterbird populations that depend on these.

divers (67% of populations of known trend decreasing), cranes (47%), rails (50%), skimmers (60%), darters (71%), ibis and spoonbills (48%), storks (59%), and jacanas (50%). Only gulls (18%), flamingos (18%), and cormorants (20%) appear to have a relatively healthy status. A similar picture emerges for at least one region, Africa-Eurasia, where the status of some waterbird families is even worse than their global status. In this region, only grebes and gulls (9% decreasing) appear to have a relatively healthy status.

20.3.2.7 Mammals

Although most mammals depend on fresh water for their survival, and many feed in rivers and lakes or live in close proximity to freshwater ecosystems, as exemplified by many large mammals in Africa, only a few are considered aquatic or semi-aquatic mammals. Revenga and Kura (2003) provide an analysis of the status of aquatic and semi-aquatic mammals, including water otters, shrews, demans, tenrecs, marsh and swamp rabbits, aquatic rodents such as beavers, muskrats, nutria, and the capybara. Otters, seals (such as the Lake Baikal seal (*Phoca sibrica*)), manatees, river dolphins, and freshwater porpoises are among the most threatened mammals in the world. For example, of the five species of Asian freshwater cetaceans, four are threatened with extinction and one species, the Irrawaddy River dolphin, is data-deficient (IUCN 2003). Some 37% of inland water-dependent mammals are globally threatened, compared with 23% of all mammals (Revenga et al. in press). This includes otters (50% of species of known status threatened), seals (67% threatened), manatees (100% threatened), river dolphins and porpoises (100% threatened), and wetland-dependent antelopes (29% threatened) (Revenga et al. in press).

20.4 Drivers of Change in Inland Water Systems

Analyses over the past two decades have identified a suite of drivers of change in inland water systems (e.g. Ellison 2004; Revenga and Kura 2003; Beeton 2002; Brinson and Ines Malvarez 2002; Junk 2002; Malmqvist and Runndle 2002; Tockner and Stanford 2002; Finlayson et al. 1992; Finlayson and Moser 1991; Moser et al. 1993, 1996; Whigham et al. 1993; Mitsch 1994; McComb and Davis 1998; Williams 2002). These reviews have focused mainly on biophysical pressures that are currently directly affecting adversely, or are likely to in the future, the ecological condition of inland water systems. The direct drivers of change in inland water systems are presented diagrammatically in Figure 20.7 (in Appendix A) (Ratner et al. 2004). The importance of also addressing the indirect drivers of wetland change has been increasingly recognized—for example, in Australia (Finlayson and Rea 1999) and most emphatically in the Mediterranean (Hollis 1992).

The direct drivers of wetland and riverine loss and degradation include changes in land use or cover due to vegetation clearance, drainage, and infilling; the spread of infrastructure whether for urban, tourism and recreation, aquaculture, agriculture, industrial, or even military purposes; the introduction and spread of invasive species; hydrologic modification to inland waters; overharvesting, particularly through fishing and hunting; pollution, salinization, and eutrophication; and, more recently, global climate change. These issues have been explored in many site-based analyses and comprehensive databases, and inventories for some exist or are being developed, while others are only now being assessed in a systematic manner (Kira 1997; Finlayson et al. 1999; Jorgensen et al. 2001; Revenga and Kura 2003).

In some cases many drivers operate together. For example, Finlayson et al. (1993) provide an analysis of the effects of multiple drivers on wetland habitats in the lower Volga, Russia, and further

information on the multiple drivers and changes in the Caspian Sea is given in Box 20.6. Too often, though, these pressures are addressed in isolation and without an adequate information base; climate change is expected to exacerbate the problems. While the effect of such drivers on inland water systems is known with *medium*-to-*high certainty,* management responses are often undermined by an absence of sufficient information. The same drivers that affect surface waters, especially those associated with agricultural, urban, and industrial development, have also contributed to the degradation of groundwater systems (Danielopol et al. 2003).

20.4.1 Physical Change, Including Drainage, Clearing, and Infilling

Outside Western Europe and North America (including Mexico), there is very little systematic information available on the extent of loss of inland waters. The loss of wetlands worldwide has been speculated at 50% of those that existed in 1900 (Dugan 1993)—a figure that includes inland wetlands and possibly mangroves, but not large estuaries and marine wetlands. Although the accuracy of this figure has not been established due to an absence of reliable data (Finlayson et al. 1999), it is *well established* that much of the loss of wetlands has occurred in the northern temperate zone during the first half of the twentieth century.

BOX 20.6

Caspian Sea (Adapted from many sources, including www.grida.no/soe.cfin?country = caspian_*sea* and www.caspiamenvironment.org)

The Caspian Sea is the largest inland water body and is surrounded by Azerbaijan, Iran, Kazakhstan, Russia, and Turkmenistan. It is a major economic asset to the region, being rich in hydrocarbon deposits and many species of fish, crustaceans, and shrimp. Some waterbirds and the Caspian seal are also commercially hunted. The Volga River in the northwest provides about 80% of the annual 300 cubic kilometers of freshwater inflow. Evaporation is more than one meter per year, while salinity ranges from fresh to highly saline in the Kara Bogaz Gol, a small basin along the Turkmen coastline. The sea supports a diverse range of habitats and species, including many endemic aquatic taxa.

The sea is under great pressure from desertification and deforestation, river regulation, urbanization and industrial development, agricultural and aquacultural development, and pollution. The water is polluted and eutrophic and has been invaded by many non-native species. There are growing fears of further contamination from oil and gas developments. The comb jelly fish (*Mnemiopsos leidyi*) has invaded and spread throughout the Caspian more rapidly than it did in the Black Sea. (See Chapter 19.)

The value of the caviar industry has been greatly affected by an 80% decrease in sturgeon landings between 1985 and 1995. This has been caused by reduced access to breeding grounds due to the construction of large dams along the inflowing rivers, pollution, overfishing, and conversion of surrounding habitat to rice cultivation. Fluctuating sea levels over many decades have also resulted in major changes to the aquatic flora and vegetation of the Volga delta and riparian forests of the Samur delta. Mass mortalities of Caspian seals, one of only two freshwater species, have been reported, likely as a consequence of pollution by heavy metals and persistent organic pollutants.

The health and lifestyle of many people in the region have been adversely affected by changes to the sea and the surrounding landscape. This has included health effects of pollution as well as reduced access to resources and basic food commodities.

Since the 1950s, many tropical and sub-tropical wetlands, particularly swamp forests, have increasingly been lost or degraded. (See Box 20.7.) In South America, peatlands linked with the Andean paramos ecosystems, also called the high mountain water towers, are increasingly targeted for agriculture, including the practices of drainage and burning (Hofstede et al. 2003). A recent inventory of Patagonian peatlands (Blanco and Balze 2004) identified agriculture and forestry as the main causes of peatland disturbance, with peat mining (mainly for use in agriculture and horticulture) as a third, but increasing, threat.

It is highly certain that clearing or drainage for agricultural development is the principal cause for wetland loss worldwide. By 1985 it was estimated that 56–65% of available wetland had been drained for intensive agriculture in Europe and North America, 27% in Asia, 6% in South America and 2% in Africa—a total of 26% loss to agriculture worldwide (OECD 1996). In China, some of the most extensive peatland areas (> 5,000 square kilometers) occur at 3,500-meters elevation on the Tibetan Plateau, the source of the Yellow and Yangtze Rivers. Large networks of drainage canals were constructed there in the 1960s and

1970s to increase the area for livestock grazing, leading to a dramatic drop in peatland area and a subsequent degradation and loss of the peat, desertification, and loss of water retention capacity (UNDP/GEF/GOC 2003).

Conversion of peatlands for intensive agriculture has been a common feature in most parts of the world for many centuries, particularly in Europe, but also more recently in the highlands of the Andes, China, and parts of Africa. The most dramatic loss of peatlands to agriculture has been in some of the countries with a rich peatland heritage, such as Finland, the Netherlands, Estonia, Denmark, and the United Kingdom. The Netherlands (once one third peatland) lost virtually all (>99%) of its natural peatlands over the last two centuries (Brag et al. 2003; Joosten 1994).

Irrigated agriculture is the leading driver in water withdrawals worldwide, resulting in large changes in river flows (Revenga et al. 2000; see also Chapter 7)—flows that are essential in sustaining ecosystem services and species. The global extent of irrigated agricultural land has increased from 138 million hectares in 1961 to 271 million hectares in 2000, and it currently accounts for an estimated 40% of total food production even though it represents only 17% of global cropland area (Wiseman et al. 2003). (See Chapter 26.) Its negative impacts on inland waters tend to be disproportionate to the irrigated land area involved.

Well-documented examples include the biodiversity losses and human health impacts seen in the Aral Sea in Central Asia and the impacts of water diversions on the wetlands in the Murray-Darling Basin of Australia (Kingsford and Johnson 1998; Lemly et al. 2000). The important Turkish wetland bird–breeding site, Kus Cenneti, is currently being adversely affected by low flows due to diversions during the bird breeding season, which is also the main irrigation period (De Voogt et al. 2000). Numerous detrimental changes in the ecological condition of the Hadejia-Nguru wetland complex in Nigeria have also been associated with the Kano River Irrigation Project (Lemly et al. 2000).

The Aral Sea in Central Asia represents one of the most extreme cases in which water diversion for irrigated agriculture has caused severe and irreversible environmental degradation of an inland water system. (See Box 20.8 here and Figure 20.8 in Appendix A.) The volume of water in the Aral basin has been reduced by 75% since 1960, due mainly to large-scale upstream diversions of the Amu Darya and Syr Darya flow for irrigation of close to 7 million hectares of land (UNESCO 2000; Postel 1999). This loss of water, together with excessive chemical inputs from agricultural runoff, has caused a collapse in the fishing industry, a loss of species diversity and wildlife habitat, and an increase in human pulmonary and other diseases in the area resulting from the high toxicity of the salt concentrations in the exposed seabed (Postel 1999; WMO 1997).

There are many other well-documented examples where diversion of water for agriculture has caused a decline in the extent and degradation of inland water systems and their species richness. In the majority of the cases, the most affected people are the poor, who depend on freshwater resources (whether from wetlands, rivers, and lakes) not only for drinking water but as a source of food supply, especially animal protein, and of income from fisheries, reed harvesting, and so on. Lake Chad provides an example where major ecosystem change has occurred (see Figure 20.9 in Appendix A) as a consequence of both human-induced and natural changes, with subsequent loss of many species and ecosystem services as the lake shrank from about 25,000 square kilometers in surface area to one twentieth its size over 35 years at the end of the twentieth century. A drier climate and high agricultural demands for water in more recent years are the primary reasons for Lake Chad's degradation (Coe and Foley 2001).

BOX 20.7
Southeast Asian Peatlands

In Southeast Asia, most of the once-extensive tropical peat swamp forests have been heavily degraded, and large extents have been lost over the last four decades. The main cause of this has been logging for timber and pulp. This started with selective logging of forests, but it has increasingly been replaced by clear-felling. Over the last two decades, this has been exacerbated by the conversion of peat swamp forests to agriculture, particularly oil palm plantations. The peatlands of Malaysia and Indonesia are especially threatened by persistent changes. Drainage and forest clearing threatens their stability and makes them susceptible to fire. Attempts have been made to harness the deeper peat soils, often with a high rate of failure and resulting in one of the environmental disasters of the last century, with millions of hectares of peatlands burned and emitting large amounts of CO_2 into the atmosphere.

In 1997, during a drought linked with the El Niño-Southern Oscillation, land clearing and subsequent uncontrolled fires severely burned about 5 million hectares of forest and agricultural land on the Indonesian island of Borneo (Glover and Jessup 1999; Wooster and Strub 2002). The amount of carbon released into the atmosphere from these fires reached an estimated 0.8–2.6 billion tons (Page et al. 1997). BAPPENAS-ADB (1999) reported an estimated 156.3 million tons from the 1997/98 Indonesia peat fires, based on an estimate of only 750,000 hectares being burned. Revised estimates by Tacconi (2003) of the actual area burned brings the total to 442 million tones or 27% of global emissions from land use change in 1989–95. In economic terms (using $7 per ton), this would amount to over $3 billion. A noxious, yellow haze covered the region for several months, which had a serious economic and health impact—some 200,000 people were hospitalized with respiratory, heart, and eye and nose irritations. There are ongoing concerns for the health of the 70 million people in six countries affected by the haze.

Early economic assessments place the damage to timber, agriculture, and other benefits derived from the forests at $4.5 billion in addition to the actual cost of fighting the fires (Glover and Jessup 1999). The fires compound the loss of peatlands through clearing and failed attempts to cultivate large areas for rice, such as has occurred in large areas in Kalimantan (Rieley and Page 1997).

BOX 20.8

Aral Sea (Information supplied by Elena Kreuzberg-Mukhina, Nikolay Gorelkin, Alex Kreuzberg, Vladislav Talskykh, Elena Bykova, and Vyacheslav Aparin, and taken from Micklin 1993; Beeton 2002; UNEP 2002)

The degradation of the Aral Sea as a consequence of the expansion of the area under cotton and abstraction of water for large-scale irrigation is well known. The hydrological change has included the construction of at least nine water reservoirs and 24,000 kilometers of channels, with 40% of the annual water inflow of 80–100 cubic kilometers withdrawn for irrigation. The consequences for the Aral Sea have been enormous, and although estimates of the extent of change vary, the sea is now only about 20% of its former volume. The surface area has been reduced by a half or two thirds, the water level is some 16–22 meters lower, and the salinity has increased somewhere between three and twelve times. The shoreline has retreated 100–150 kilometers and exposed something like 45,000 square kilometers of former seabed, creating a salty desert and more than 100 million tons of salty dust.

The sea now has three separate entities: the Small Sea, with an area of 3,000 square kilometers, a volume of 20 cubic kilometers, and a salinity of 18–20 grams per liter; the eastern part of the Large Sea, with an area of 9,150 square kilometers, a volume of 29.5 cubic kilometers, and a salinity of 120 grams per liter; and the western part of the Large Sea, with an area of 4,950 square kilometers, a volume of 79.6 cubic kilometers, and a salinity of 80 grams per liter.

These changes have caused a collapse of the fishing industry, and many plant and animal species have been lost. Only a few of the former 34 fish species survive, with some becoming extinct, such as the Aral sturgeon, Aral trout, Chu sharpray, Tukestan dace, and Kessler's loach.

Waterbirds have similarly been drastically affected, with a loss of breeding and stopover habitats for migratory species, such as those in the deltas of the Amu Darya and Syr Darya. This has seen a decline in habitat for breeding mute swan, Dalmatian and Great White pelican, and Pygmy cormorant, among others. New habitats have been created through the construction of irrigation areas, but these do not compensate adequately for the losses.

The local climate has been dramatically affected. For example, the average humidity has decreased from around 40% to 30%, leading to increased desertification, with subsequent loss of pasture productivity and impacts on human well-being. The latter is also associated with the pollution of the water and increased occurrence of dust storms. The consequences of the management decisions for the Sea have been drastic, but some at least were foreseen, and deliberate trade-offs were made in favor of economic outcomes. In 1995, the cost of making a net water saving of 12 cubic kilometers was estimated as $16 billion, but the prospects for funding were limited, and so it is likely that current conditions will prevail, with the continuing demise of the aquatic ecosystem and human well-being.

However, with the collapse of the agricultural industry in the region in recent years, the demand for water has decreased to some extent, and partly alleviated the situation. (For further information on the human well-being consequences of changes to the Aral Sea, see Chapter 5.)

20.4.2 Modification of Water Regimes

Water regimes of inland waters have been modified by humans for centuries, with the last 50 years in particular witnessing large-scale changes in many parts of the world, often associated with drainage and infilling activities as described earlier (Brinson and Ines Malvarez 2002; Junk 2002; Malmqvist and Rundle 2002; Tockner and Stanford 2002; see also reviews cited earlier). Modifications include construction of river embankments to improve navigation, drainage of wetlands for agriculture, construction of dams and irrigation channels, and the establishment of inter-basin connections and water transfers. (See Table 20.7 and Boxes 20.9 and 20.10; see also Chapter 7.) These changes have improved transportation, provided local flood control and hydropower, boosted fisheries, and increased agricultural output by making more land and irrigation water available. At the same time, physical changes in the hydrological cycle have resulted in the disconnection of rivers from their floodplains and wetlands, caused seasonal changes in water flows, increased the likelihood and severity of flooding (see Chapter 16), disrupted links with groundwater systems, and enabled saline water to intrude on freshwater systems in many coastal regions.

Further, these changes have also altered the flow velocity in rivers—transforming some to large lakes, such as the Kariba lake

Table 20.7. Alteration of Freshwater Systems Worldwide (Revenga and Kura 2003)

Alteration	Pre–1990	1900	1950–60	1985	1996–98
Waterways altered for navigation (km)	3,125	8,750	–	> 500,000	
Canals (km)	8,750	21,250	–	63,125	–
Large reservoirs[a]					
Number	41	581	1,105	2,768	2,836
Volume (sq. km.)	14	533	1,686	5,879	6,385
Large dams (>15 m high)	–	–	5,749	–	41,413
Installed hydro capacity (megawatts)	–	–	< 290,000	542,000	−660,000
Hydro capacity under construction (megawatts)	–	–	–	–	−126,000
Water withdrawals (cu. km. per year)	–	578	1,984	−3,200	−3,800
Wetlands drainage[b] (cu. km.)	–	–	–	160,000	–

– ` Data not available.

[a] Large reservoirs are those with a total volume of 0.1 cubic kilometers or more. This is only a subset of the world's reservoirs.

[b] Includes available information for drainage of natural bogs and low-lying grasslands as well as disposal of excess water from irrigated fields.

Danube River

Engineering structures have inexorably altered the Danube River, one of the major rivers of central and eastern Europe. Since 1950, hundreds of artificial lakes have been constructed along the Danube and its tributaries to provide storage and release of water for flood control, hydropower, navigation, irrigation, and domestic and industrial water supply. The construction of dikes and reservoirs has led to the loss of floodplain zones, with important loss of habitats and modification of the Danube's hydrological and sediment regimes.

Structures built along the first 1,000 kilometers of the river have formed an almost uninterrupted artificial waterway through a chain of 59 hydropower dams. The delta has also been changed with the construction of polders, canals, dikes, and fish farms, which along with eutrophication have led to major ecological changes in the river (IUCN 1992). These changes have altered the nature of the river and the delta, negatively affecting both services and the biodiversity, such as the extent of fisheries and a reduction or even loss in some places of the filtering capacity provided by reed beds and riparian vegetation. Further adverse change is expected with construction of the Bystroye navigation canal through the delta.

South American Wetlands and Rivers

The construction of hydroelectricity schemes poses a major threat to wetlands in South America. In Brazil, rapidly rising energy demands have stimulated ambitious plans to build dams on nearly all major rivers, except the main stream of the Amazon (Junk and Nunes de Mello 1987; World Energy Council 2003). However, many rivers have low gradients, and in such cases dams inundate large areas and provide little energy; for example, the Balbina reservoir on the Uatuma River in the Brazilian Amazon covers 2,300 square kilometers and produces <10 megawatts per square kilometer. There are likely to be significant socioeconomic trade-offs from the construction of these reservoirs to produce hydroelectricity (Fearnside 1989).

In Brazil, rapidly expanding agriculture, mainly for soybean production, has increased demand for inexpensive transport along the rivers. Waterways (*hidrovias*) have been constructed or are under construction (Brito 2001), which involves straightening sinuous stretches of the river channels, dragging, removing obstacles such as logs and rocky outcrops, and placing signals for ship traffic. Environmental impact analyses are lacking in most cases. In 2000, the Brazilian government stopped plans to construct a *hidrovia* through the Pantanal of Mato Grosso. This project would have threatened one of the largest wetlands in the world (Ponce 1995; Hamilton 1999). Plans have not yet been abandoned by private enterprises, however, and infrastructure construction is proceeding.

ian zones, opened up access to exotic species, and contributed to an overall loss of freshwater biodiversity and inland fishery resources (Revenga et al. 2000), as well as led to alterations to upstream and downstream habitats. Dams also affect the magnitude and timing of water flow and sediment transport of rivers, often for long distances downstream. The Aswan High Dam in Egypt, for example, has led to reduced sediment transport for more than 1,000 kilometers downstream (McAllister et al. 1997). A further example of the downstream effects of dams is illustrated in the Indus delta, where rapidly accelerating mangrove loss as a result of reduced freshwater flows has seriously jeopardized the livelihoods of 135,000 people who rely on mangrove products to a total economic value of $1.8 million a year for fuelwood and fodder, as well as a coastal and marine fisheries sector that generates domestic and export earnings of almost $125 million annually (Iftikhar 2002).

Other examples of large-scale drivers of change in inland water systems are those affecting the Dead Sea (ILEC and UNEP 2003) and the Mesopotamian marshlands in Iraq (Partow 2001; UNEP 2002). Lying in the heart of the Syrian-African rift valley at the southern outlet of the Jordan River, the Dead Sea—417 meters below sea level—is the world's saltiest large water body. It is severely threatened by excessive water withdrawals in the north and dams and industrial development in the south as a result of ever-increasing industry, agriculture, and tourism. The annual flow of the Jordan River was approximately 1,370 million cubic meters in the 1950s, while today the total river discharge to the Dead Sea is about 300 million cubic metes a year. As a result, the level of the lake is dropping by about one meter each year (ILEC and UNEP 2003).

The Mesopotamian marshlands have also been severely affected in recent decades. These covered an area of 15,000–20,000 square kilometers before being reduced by drainage and dam construction along the Tigris and Euphrates Rivers (Partow 2001). Now they cover less than 400 square kilometers. The capacity of dams along these rivers currently exceeds the annual discharge of both rivers, drastically reducing the supply of downstream floodwaters that were so important in delivering sediments and nutrients to the marshland. Further, in the early 1990s drainage schemes were used to divert large amounts of water from the marshlands—an event that was made easier by the upstream damming. (See Figure 20.10 in Appendix A.)

There are now more than 45,000 large dams (more than 15 meters high) (WCD 2000), 21,600 of which are in China. This represents a 700% increase in the water stored in river systems compared with natural river channels since 1950 (Vörösmarty et al. 1997). Water storage and sediment retention from dams have had enormous impacts on suspended sediment and carbon fluxes, as well as on the waste processing capacity of aquatic habitats (Vörösmarty et al. 1997). The construction of large dams has doubled or tripled the residence time of river water (Revenga et al. 2000), with enormous impacts on suspended sediment and carbon fluxes, waste processing, and aquatic habitat, and has resulted in fragmentation of the river channels. Revenga et al. (2000) found 37% of 227 river basins around the world were strongly affected by fragmentation and altered flows, 23% moderately affected, and 40% unaffected. (See Figure 20.11 in Appendix A.) Strongly or moderately fragmented systems are widely distributed globally. Small dams can also have major affects on the ecological condition of inland water systems (Oritz Rendan 2001), and many inland surface and groundwater systems have also been affected by modifications at smaller scales.

The extent of recent change is illustrated by figures collated for Asia and South America. In Asia, 78% of the total reservoir

in southern Africa; creating a chain of connected deep reservoirs, such as those along the Volga River, Russia; leading to channelization, such as that along the Mississippi and Missouri Rivers in the United States; or significantly reducing flows to floodplains and downstream habitats, including deltas such as the Indus in Pakistan. Similarly, converting wetlands for agricultural purposes without completely destroying them, as with much of sub-Saharan agriculture or paddy (rice), still results in hydrological change.

Modifications to water regimes have drastically affected the migration patterns of birds and fish and the composition of ripar-

volume has been constructed in the last decade, and in South America almost 60% of all reservoirs have been built since the 1980s (Avakyan and Iakovleva 1998). The debate about the construction of dams is ongoing (WCD 2000)—weighing up, for example, the benefits against the potential adverse consequences of constructing further dams in the upper Mekong in China (Dudgeon 2003).

The effects of modification of flow regimes on fish migrations have been reviewed by Revenga and Kura (2003). The direct impacts of dams on diadromous fish species such as salmon are now *well established*. Indirect impacts of flow alteration, such as the reduction of floods and loss of lateral connections on floodplains, are also important. In many instances the construction of reservoirs has resulted in the disappearance of fish species adapted to river systems and the proliferation of species adapted to lakes, many of which were non-native. Examples include the decline of the sturgeon and the caviar industry in rivers such as the Volga in Russia (Finlayson et al. 1993). In West Africa, a sharp decline of *Mormyridae* (an elephant-nosed fish family of Osteoglossiformes) was observed in Lakes Kainji and Volta after the inundation of their preferred habitats as a result of dams (Lévêque 1997).

Cases of adverse impact on the structure of riparian vegetation and morphology from dams, embankments, and canals are also widely reported (Nilsson and Berggren 2000). In tropical Asia, change in flooding patterns due to river modification has affected riverine and wetland-dependent mammal populations, such as marshland deer and the Asian rhino in Thailand, India, and China, and diadromous fish stocks, such as sturgeons in China (Dudgeon 2000c). Similar cases have been reported by Pringle et al. (2000) for North and South America.

20.4.3 Invasive Species

The introduction of some non-native (alien) invasive species has contributed to species extinction in some freshwater systems (see Malmqvist and Rundle 2002; Tockner and Stanford 2002; see also reviews cited earlier). The problems caused by invasive species are very much a global concern (Mooney and Hobbs 2000). The spread of exotic species in inland waters is increasing with the spread of aquaculture, shipping, and global commerce. (See Boxes 20.11 and 20.12.) Examples include the pan-tropical weeds salvinia (*Salvinia molesta*) and water hyacinth (*Eichhornia crassipes*) that originated in South America but which are now widely distributed across the tropics. The cane toad (*Bufo marinus*), bullfrog (*Rana catesbeiana*), European domestic pig (*Sus scrofa*), carp (*Cyprinus carpio*), and zebra mussel (*Dreissena polymorpha*) are examples of animals that have become established outside of their native range and disrupted the inland water systems that they have invaded.

Species such as the water hyacinth, Canadian pondweed (*Elodea canadensis*), and mimosa (*Mimosa pigra*) have spread around the globe (Sculthorpe 1967; Gopal 1987; Walden et al. 2004) and remain largely unaffected by extensive and expensive control or management programs. Canadian pondweed illustrates the dilemma caused by invasive species. It is the first documented example of the explosive growth of an aquatic weed that originated in North America and in the late nineteenth century invaded the waterways of Europe. In becoming established, it grew rapidly, reproducing vegetatively, and reached maximum population densities within a period of a few months to four years. These population densities were maintained for up to five years and then declined to levels that were not considered a nuisance. The reasons for the rapid increase and subsequent decline were not deter-

---

BOX 20.11

Invasive Species and European Rivers (Information supplied by H. Ketelaars)

For many centuries canals have been constructed between rivers and other water bodies in Europe, through which species actively migrated or were aided by shipping traffic, either in ballast water or outside on the hull of ships. The Volga-Baltic Waterway, reconstructed in 1964, connecting the Caspian basin with the Baltic region, is one example that has enabled translocation of many aquatic species, such as copepods, rotifers, the onychopod *Bythotrephes longimanus,* and several fish species to the Volga basin. The Main-Danube Canal, officially opened in 1992, is another that has allowed many Ponto-Caspian invertebrate species to reach the Rhine basin and from there to disperse to other basins, mainly in ballast water.

Intentional introductions of aquatic species have occurred mainly in the past two centuries. The North American amphipod *Gammarus tigrinus* was deliberately introduced in 1957 in the German rivers Werra and Weser because the local gammarid fauna had disappeared due to excessive chloride pollution. The mysid *Mysis relicta* has been introduced in many Scandinavian lakes to stimulate fish production. Three North American introduced crayfish species have established themselves in many European waters and introduced "crayfish plague" (*Aphanomyces astaci*), which has almost eliminated the native crayfish (*Astacus astacus*).

At least 76 non-European freshwater fish have been introduced into European fresh waters, with approximately 50 establishing self-sustained populations. When introductions between areas within Europe are also considered, the number of introduced species is more than 100. The numerically most important families are cyprinids and salmonids, of which grass carp (*Ctenopharyngodon idella*), silver carp (*Hypophthalmichthys molitrix*), rainbow trout (*Oncorhynchus mykiss*), and brook char (*Salvelinus fontinalis*) are now widely distributed in Europe. Only a few introductions have resulted in the spread of economically important species.

---

mined, and this paradox still affects efforts to manage invasive species in inland waters.

Efforts to determine which of the many species that are introduced into new environments have not been fully successful have done little other than illustrate that only a small proportion of introduced species are likely to flourish and become serious problems (Williamson 1996; Manchester and Bullock 2000). While many species have features that enable them to take advantage of changed ecological conditions, there are likely as many factors that would limit establishment and growth.

Water hyacinth is an example of a widespread alien species that has caused considerable economic and ecological damage in inland water systems around the world (Gopal 1987). It is believed to be indigenous to the upper reaches of the Amazon basin, was spread in the mid-nineteenth century throughout much of the world for ornamental purposes, and now has a pan-tropical distribution. The plant spreads quickly to new rivers and lakes, clogging waterways and infrastructure, reducing light and oxygen in freshwater systems, and causing changes in water chemistry and species assemblages that affect fisheries. Water hyacinth control and eradication has become one of the top priorities for many environmental government agencies, with biological control being increasingly successful. (See Chapter 10.)

Many fish species have been spread beyond their native ranges, often as an important component of aquaculture (FAO

BOX 20.12

North America's Great Lakes and Invasive Species

Alien invasive species have threatened the Great Lakes ever since Europeans settled in the region. And since the 1800s, more than 140 exotic aquatic organisms of all types—including plants, fish, algae, and mollusks—have become established in the Great Lakes. The rate of introduction of exotic species has increased with human activity in the watershed. More than one third of the organisms have been introduced in the past 30 years, a surge coinciding with the opening of the St. Lawrence Seaway (Great Lakes Information Network 2004).

Approximately 10% of the Great Lakes' non-indigenous species have had significant impacts, both economic and ecological. For example, the sea lamprey has cost millions of dollars in losses to recreational and commercial fisheries and millions of dollars in control programs. Alewife fish littered beaches each spring and altered food webs, thereby increasing water turbidity before salmonids such as chinook salmon (themselves exotic) were stocked as predators and the foundation of a new recreational fishery.

Since 1991, the Great Lakes Panel on Aquatic Nuisance Species has worked to prevent and control the occurrence of aquatic alien invasive species in the Great Lakes, although efforts have not been totally successful.

1999). Introductions are usually done to enhance food production and recreational fisheries or to control pests such as mosquitoes and aquatic weeds. Introduced fish, for example, account for 96% of fish production in South America and 85% in Oceania (Garibaldi and Bartley 1998). The introduction of non-native fish, however, has had severe ecological costs. A survey of 31 studies of fish introductions in Europe, North America, Australia, and New Zealand found that in 77% of the cases, native fish populations were reduced or eliminated following the introduction of non-native fish. In 69% of cases, the decline followed the introduction of a single fish species, with salmonids responsible for the decline of native species in half of these (Ross 1991). The introduction of salmonids is attracting increased attention, as they have reduced the genetic diversity of wild stocks. In Canada and the United States, 68% of the recorded extinctions of 27 species and 13 sub-species of fish were due in part to the introduction of alien species (Miller et al. 1989). Similarly, in Mexico and Colombia, introduced fish pose a major risk to native fish and fisheries stock (Contreras-Balderas 2003; Alavrado and Gutiérrez 2002).

Fish introductions to tropical Asia and Latin America over the last 150 years have occurred mainly either to enhance food production (carps and tilapias, for instance) or for recreational purposes (piscivorous fish such as trout and bass for sport fishing) (Revenga and Kura 2003). The impacts of these species on the native fish fauna and ecosystems have not been well documented, although Fernando (1991) reports that introduced fish were not found to have caused severe damage to indigenous species except for some incidents in Latin America where piscivores were introduced.

In recent decades, tilapia species have been established and become a substantial contributor to inland fisheries in Mexico, the Dominican Republic, northeast Brazil, and Cuba (where as much as 90% of the fishery is tilapia species) (Fernando 1991). Although this has not resulted in the collapse of native fish stocks in most cases, it does indicate a significant shift in the composition and structure of biological communities in those systems.

In tropical Asia, herbivores and omnivores, such as Indian, Chinese, and common carps, account for the majority of introductions (Revenga and Kura 2003). Except in China, these temperate species of carps have not contributed much to fishery yield in the tropics. In comparison, tilapias have had a similar effect here as in Central America, boosting capture fishery in Sri Lanka and Thailand and aquaculture in Philippines, Taiwan, and Indonesia (Fernando 1991). In China, the world's largest producer of inland fisheries, carp contributes significantly to fisheries production. Although research on the impact of introduced species on the native aquatic ecosystems of China is limited, a few well-documented cases exist, such as Dianchi Lake in Yunnan Province, and Donghu Lake in Hunan Province, where it has been shown that indigenous and endemic fish species assemblages have significantly changed and many of their populations have declined (Xie et al. 2001).

Further information on invasive aquatic species can be found on the Global Invasive Species Program web site. Analyses of the economic costs of non-native invasive species are becoming more common, as shown by the fishery examples just mentioned, as are risk assessments of important species (e.g. Finlayson et al. 2000; van Dam et al. 2000, 2002a). The importance of alien invasive species in inland waters is likely to increase in response to global change (van Dam et al. 2002b).

20.4.4 Fisheries and Other Harvesting

Inland fisheries are a major source of protein for a large part of the world's population. People in Cambodia, for example, obtain roughly 60–80% of their total animal protein from the fishery resources of the Tonle Sap alone (MRC 1997). In some landlocked countries, this percentage is even higher; for example, in Malawi about 70–75% of the total animal protein for both urban and rural low-income families comes from inland fisheries (FAO 1996).

Global production of fish and fishery products from inland waters in 2002 amounted to 32.6 million tons (FAO 2004)—8.7 million tons from wild capture fisheries and the rest (23.9 million tons) from inland aquaculture. There is little dispute that major increases in the harvest of freshwater fish have occurred over the last two decades, mostly in the developing world, but as these statistics show, much of this increase is the product of aquaculture operations and enhancement efforts such as fish stocking and the introduction of non-native fish species in lakes and rivers.

Increased freshwater harvests, however, do not indicate healthy freshwater fish stocks or healthy aquatic ecosystems. In fact, FAO's last major assessment of inland fisheries (1999) reported that most inland capture fisheries that rely on natural reproduction of the stocks are overfished or are being fished at their biological limit and that the principal factors threatening inland capture fisheries are fish habitat loss and environmental degradation. In addition, one of the limitations in monitoring the state and condition of inland fish stocks is that the catch from inland fisheries is believed to be underreported by a factor of two or three, due to the large volume of harvest that is consumed locally, and remains unrecorded (FAO 1999). Asia and Africa are the two leading regions in inland capture fisheries, accounting for 90% of the catch in 2002 (FAO 2004). China alone accounts for at least one quarter of the inland catch, followed by India (9% of the catch), Bangladesh (8%), and Cambodia (4%) (FAO 2004).

Aquaculture continues to grow more rapidly than any other animal food-producing sector, at an average rate of 8.9% per year since 1970—a much higher rate than that for capture fisheries (1.2%) or terrestrial farmed meat products (2.8%) (FAO 2004). Most aquaculture production (58%) comes from the freshwater environment, the main producer by far being China. Between

1970 and 2000, inland water aquaculture production in China increased at an average annual rate of 11%, compared with 7% for the rest of the world (FAO 2004). However, many aquaculture operations, depending on their design and management, can and have contributed to habitat degradation, pollution, introduction of exotic species, and the spread of diseases through the introduction of pathogens (Naylor et al. 2000; see also Chapter 26).

Many other species of vertebrates are also harvested from inland waters, some in large numbers—such as turtles, waterbirds, crocodiles, and frogs. Overharvesting, whether for food, medicinal purposes, or recreation, has become a problem in many countries, and many species are locally or regionally threatened. For example, the increase in the harvesting and trade of freshwater turtles in South and Southeast Asia is causing severe declines in species populations, putting some of these species at risk of extinction (van Dijk et al. 2000). Because of the increase in trade, 11 proposals to list turtle species under Appendix II of CITES were accepted by consensus at the CITES Conference of the Parties in November 2002 (CITES 2002).

20.4.5 Water Pollution and Eutrophication

It is *well established* that nutrient concentrations have increased substantially in rivers throughout the world (Heathwaite et al. 1996; Revenga et al. 1998), resulting in eutrophication, harmful algal blooms, and high levels of nitrate in drinking water (Malmqvist and Rundle 2002). (See Chapters 7, 12, 15, and 19.) Many specific examples are available for inland water systems (e.g. Malmqvist and Rundle 2002; Tockner and Stanford 2002). For instance, the agricultural sector contributes an average of 50% of the total load of nitrogen and phosphorus to the Danube River in Europe, domestic sources contribute about 25%, and industry or atmospheric deposition 25%. Hazardous substances of particular concern are pesticides, ammonia, PCBs, polyaromatic hydrocarbons, and metals (IUCN 1992). (See also Chapter 15.) Industry and mining are responsible for most of the direct and indirect discharges of hazardous substances into the Danube and Volga Rivers in Europe, while transport is an important source of oil pollution (IUCN 1992; Popov 1992). Microbiological contamination by pathogenic bacteria, viruses, and protozoa is an important water quality problem in many regions of the world (see Chapters 7 and 15), and diffuse discharges from agriculture are important sources of micro-pollutants for both surface and groundwaters.

Jorgensen et al. (2001) notes that eutrophication is the most widespread problem in lakes and reservoirs and also one of the most difficult to abate. Cyanobacteria blooms have increased and are a major problem in inland and coastal waters worldwide. The problem of increased eutrophication from land-based activities is well shown for the Mississippi River in the United States, with problems along the length of the river and in the coastal zone—the so-called dead zone in the Gulf of Mexico.

It is *well established* that pollution from point sources such as mining has had devastating impacts on the biota of inland waters in many parts of the world. For example, the release of stored tailings (mine wastes) from the Aznalcollar mine some 50 kilometers from the Doñana National Park in Spain illustrated the problems that can occur to both ecological and socioeconomic activity of the areas downstream (Bartolome and Vega 2002). Following a spillage in 1998, an estimated 5.5 million cubic meters of acidic, metal-enriched water and 1.3–1.9 million tons of toxic tailings were spread over 4,600 hectares of downstream habitats—with fatal consequences for much of the biota in the affected area and a consequent disruption to the tourism industry. The cost of re-

moving the tailings and contaminated soil reached about 3.8 billion euros.

In developing countries, an estimated 90% of wastewater is discharged directly to rivers and streams without any waste processing treatment, and in some locations both surface and groundwater have been so polluted that they are unfit even for industrial uses (WMO 1997). Threats of water quality degradation are usually most severe in areas where water is scarce due to the reduced capacity for waste dilution. These threats are exacerbated by industrial and agricultural practices that channel waste products into inland waters, including caves and other underground water.

Meybeck (2003) provides an overview of water pollution problems for inland waters. (See Table 20.8.) In industrial countries, fecal contamination has been largely eliminated, while new problems, particularly from agriculture run-off, are increasing everywhere. In other countries this is not the case, and fecal contamination is a major problem. In developing countries, urban and industrial pollution sources are increasing faster than related wastewater treatment.

Contamination by pesticides has increased rapidly since the 1970s, with many different substances being involved. In the Seine basin, in France, for example, more than 100 different active molecules are known to occur (Chevreuil et al 1998). The use of persistent chemicals is now increasingly regulated in Western Europe and North America. Records of PCBs and DDT in sedimentary archives peaked in the 1970s and are now markedly decreasing (Valette-Silver 1993). The persistence of these products can be high, however, and their degradation products can be more toxic than the parent molecule. (See Chapter 15.) Additionally, it is extremely difficult to assess and address the effects of multiple chemicals together in inland waters, both in the short and the long term.

Toxic substances are known to be a serious and increasing threat in developing countries as land use in watersheds changes. Chemical pollution from urban domestic and industrial sources and from pesticides is increasing in many key lake watersheds such as Lake Baikal and the African Great Lakes (Ntakimazi 1992;

Table 20.8. Major Water Quality Issues in Inland Water Systems at the Global Scale (Maybeck 2003)

Issue	Rivers	Lakes	Reservoirs	Groundwaters
Pathogens	•••	•	•	••
Suspended solids	••	na	•	na
Decomposable organic matter	•••	•	••	•
Eutrophication	•	••	•••	na
Nitrate	•	0	0	•••
Salinization	•	0	•	•••
Trace metallic elements	••	••	••	••
Organic micropollutants	•••	••	••	•••
Acidification	•	•	••	0

Key: ••• severe or global deterioration observed
     •• important deterioration
     • occasional or regional deterioration
     0 rare deterioration
     na not applicable

Hecky and Bugenyi 1992). Lake Baikal water, fish, and seals all contain measurable levels of organochlorine compounds (Kucklick et al. 1994). Concentrations of chlorinated organic compounds have declined in some North American Great Lakes fish species but remain high for all fish species in Lakes Michigan and Ontario (Rowan and Rasmussen 1992). There have also been recent discoveries of endocrine-disrupting toxics in pulp wastewater that have caused abnormal male sexual organs to develop in alligators, feminization of male fish and turtles, and masculinization of female fish (Mathiessen and Sumpter 1998; Mathiessen 2000). Further information on human health and toxic substances can be found in Chapter 14.

20.4.6 Climate Change

It is arguable whether or not climate change has already affected inland waters and their species, but it is anticipated (*medium certainty*) that it will directly or indirectly affect the biota and services provided by inland waters (van Dam et al. 2002b; Gitay et al. 2002). As climate change will increase the pressure on habitats that are already under severe pressure from other drivers just described and will interact in a synergistic manner, it is considered briefly here.

The certainty with which with which we can attribute cause and effect of climate change is undermined by the extent of our data and existing knowledge; in all but a few cases the data are inadequate. We are, however, highly confident that many inland waters are vulnerable to climate change. Particularly vulnerable are those at high latitudes and altitudes, such as Arctic and sub-Arctic bog communities, or alpine streams and lakes (Gitay et al. 2002; IDEAM 2002), as well as those that are isolated (Pittock et al. 2001) or are low-lying and adjacent to coastal wetlands (Bayliss et al. 1997). Groundwater systems will also suffer as climate change affects recharge of aquifers (Danielopol et al. 2003).

The major expected impacts to inland waters include warming of rivers, which in turn can affect chemical and biological processes, reduce the amount of ice cover, reduce the amount of dissolved oxygen in deep waters, alter the mixing regimes, and affect the growth rates, reproduction, and distribution of organisms and species (Gitay et al. 2002). It is *very certain* that sea level rise will affect a range of freshwater systems in low-lying coastal regions. For example, low-lying floodplains and associated swamps in tropical regions could be displaced by salt-water habitats due to the combined actions of sea level rise and larger tidal or storm surges (Bayliss et al. 1997; Eliot et al. 1999). Plant species not tolerant to increased salinity or inundation could be eliminated, while salt-tolerant mangrove species could expand from nearby coastal habitats. Changes in the vegetation will affect both resident and migratory animals, especially if these result in a major change in the availability of staging, feeding, or breeding grounds for particular species (Boyd and Madsen 1997; Zockler and Lysenko 2000).

The most apparent faunal changes will probably occur with migratory and nomadic bird species that use a network of wetland habitats across or within continents. The cross-continental migration of many birds is at risk of being disrupted due to changes in habitats (see references in Walther et al. 2002). Reduced rainfall and flooding across large areas of arid land will affect bird species that rely on a network of habitats that are alternately or even episodically wet and fresh or drier and saline (Roshier et al. 2001). Responses to these climate-induced changes will be affected by fragmentation of habitats or disruption or loss of migration corridors or even by changes to other biota, such as increased exposure to predators by wading birds (Butler and Vennesland 2000), as a consequence of adaptation to and mitigation of climate change (Gitay et al. 2002).

It is anticipated with *medium certainty* that fish species distribution will move toward the poles, with cold-water fish being further restricted in their range, and cool and warm-water fish expanding in range. Aquatic insects, on the other hand, will be less likely to be restricted, given that they have an aerial life stage. Less mobile aquatic species, such as some fish and mollusks, will be more at risk because is it thought that they will be unable to keep up with the rate of change in freshwater habitats (Gitay et al. 2002). Climate change may also affect the wetland carbon sink, although the direction of the effect is uncertain due to the number of climate-related contributing factors and the range of possible responses (Gitay et al. 2002). Any major change to the hydrology and vegetative community of a wetland will have the potential to affect the carbon sink. Vegetation changes associated with the water drawdown in northern latitudes, for example, result in increased primary production, biomass, and slower decomposition of litter, causing the net carbon accumulation rate to remain unchanged or even increase. Other aspects of climate change, such as longer and more frequent droughts and the thawing of permafrost, will have negative effects on the carbon balance in peatland.

The extent of change in inland waters as a consequence of climate change should not be addressed in isolation of other drivers of change, as many of the adverse effects of the above-mentioned drivers of change will be exacerbated by climate change (Gitay et al. 2002; van Dam et al. 2002). Further, the affects of climate change will be felt across many of the services delivered by inland waters; as an example, a sensitivity projection for Canada's river regions in response to climate change indicates that there will be an increase in flood and river erosion that will affect the use and value of rivers for recreation, conservation, fisheries, water supply, and transportation (Ashmore and Church 2001).

20.5 Trade-offs, Synergies, and Management Interventions for Inland Water Systems

Management of inland waters worldwide has been regularly based on decision-making mechanisms that have not included sufficient consideration of the wider implications or outcomes of specific actions or responses (see Finlayson et al. 1992; Finlayson and Moser 1991; Whigham et al. 1993; Mitsch 1994; Jaensch 1996; McComb and Davis 1998; Ali et al. 2002). The assessment and case studies provided in this chapter illustrate the outcomes of management decisions that have not considered the trade-offs between services provided by inland waters. These decisions have often resulted in the degradation of inland waters, and the loss or decline in the multiple services they provide, in favor of a smaller number of services, such as the supply of fresh water for drinking or irrigation or the supply of hydroelectricity or transport routes. The case studies cited earlier of the Aral and Caspian Seas illustrate the adverse effects that such sectorally based decisions can have. More multisectorally based responses and decisions are required if we are to reverse the loss and degradation of inland waters and the decline in the services that they deliver. Further information on management responses is provided in the MA *Policy Responses* volume in chapters on biodiversity (Chapter 5), nutrient management (Chapter 9) and freshwater (Chapter 7).

The past loss and degradation of inland waters and their services is increasingly being recognized through international conventions and treaties as having exceeded the value gained through such actions. In response, the Ramsar Wetlands Convention has provided leadership and worked collaboratively with other organizations, both informally and through formal agreements, such as the joint work plans agreed with the Convention on Biological Diversity, to develop more multisectoral approaches to stop and reverse the loss and degradation of wetlands. This includes working collaboratively to address the Millennium Development Goals (see MA *Policy Responses,* Chapter 19) and to reduce the rate of loss of biodiversity by 2010 (see Chapter 4 in this volume and MA *Policy Responses,* Chapter 5). Many other international collaborative efforts and initiatives are underway, some linked with and many others independent of the Ramsar Convention.

The Mediterranean wetland program (MedWet) is one collaborative, multisectoral initiative that is formally linked with the Ramsar Convention and has resulted in strident calls and actions to not only halt the loss and degradation of wetlands but to reverse their consequences. The program has evolved considerably since the initial declaration of intent was made in Grado, Italy, in February 1991 (Anon 1992). The declaration contained a recommendation that all supranational organizations, Mediterranean governments, NGOs, and concerned individuals adopt the following goal: to stop and reverse the loss and degradation of Mediterranean wetlands.

It further recommended a number of actions that should be included in a strategy to support this goal:

- identification of priority sites for wetland restoration and rehabilitation and the development and testing of techniques for their complete rehabilitation;
- evaluation of existing and proposed policies to determine how they affect wetlands;
- increased institutional capacity to conserve and effectively manage wetlands through vigorous education and training programs;
- integrated management of all activities concerning wetlands, their support systems, and the wider area surrounding them carried out by properly funded and well-staffed multidisciplinary bodies with active participation of representatives of government, local inhabitants, and the scientific and nongovernmental community;
- open consultation and free flow of information when managing wetlands; and
- adoption and enforcement of national and international legislation for better management.

The declaration was not received with enthusiasm by some key sectors; however, the individual recommendations have since been repeated or extended in many fora and with widespread acceptance, the most recent being in the Chilika Statement agreed at the Asian Wetland Symposium 2005, Bhubaneswar, India, in February 2005 (www.wetlands.org/news&/docs/AWS_Declaration .pdf).

In the early 1990s, the concept of replacing lost wetlands received increasing support (e.g., see Finlayson and Larsson 1991; Finlayson et al. 1992; Hollis et al. 1992), and more attention is now directed toward wetland restoration worldwide (see Eiseltova 1994; Eiseltova and Biggs 1995; Zalidis et al. 2002).

However, current rates of restoration are inadequate to offset the continued rate of wetland loss in many regions. Given this situation, the Ramsar Convention on Wetlands has proposed a series of guidelines to assist in reversing the loss of wetlands. These cover the current thinking and agreement on priority topics for management of inland waters, but due to political considerations many are not as prescriptive as requested by some parties, especially when dealing with indirect drivers of change, such as trade and population growth. The current guidance covers these topics:

- wise use of wetlands;
- national wetland policies;
- laws and institutions;
- river basin management;
- participatory management;
- wetland communication, education, and public awareness;
- designation of Ramsar sites;
- management of wetlands;
- international cooperation;
- wetland inventory;
- impact assessment;
- water allocation and management;
- coastal management; and
- peatlands.

One of the key barriers in developing management responses to prevent further loss and degradation of wetlands is the unwillingness to undertake effective actions. Sufficient knowledge is generally now available to know what actions are required to stop further loss and degradation and when these are most likely to be effective. (The general reviews cited at the start of section 20.3 provide guidance to a wealth of useful information.) There is also inadequate adoption and understanding of "ecosystem approaches" for managing inland waters, including the precautionary principle, as espoused by the Convention on Biological Diversity and the Convention on Wetlands (Ramsar Convention Secretariat 2004). The World Commission on Dams (WCD 2000) illustrated some of the contradictory issues faced in managing inland waters. Ongoing debate about the allocation of water for environmental outcomes in rivers and associated wetlands illustrates the trade-offs that have long been inherent features in water management, especially at a river basin scale. Further dialogue is required to ensure the delivery of water allocations from dams to support a wider range of services than has generally been the case.

The extent of loss and degradation of wetlands, and trade-offs in services have resulted in an increasing number of large and small restoration projects, driven by legislation and public attitude, particularly in North America and Europe, and increasingly in Australia. The cost and complexity of large-scale restoration are shown by the plan for the restoration of the Everglades, USA (CERP 1999). A comprehensive plan containing more than 60 components has been prepared to restore, protect and preserve the water resources of central and southern Florida, including the Everglades wetlands. The plan has important environmental and economic benefits and is anticipated to cost US$7.8 billion over 30 years. The responses to the accidental release of tailings (mine wastes) from the Anzacollar mine site upstream of the Donana wetlands in Spain in 1998 also illustrate the complexity of large-scale restoration programmes involving both environmental and economic issues (Gallego Fernandez and Garcia Nove 2002; G. Schmidt personal communication). The removal of the waste, treatment of contaminated water, acquisition of contaminated land and rehabilitation cost the regional and central governments and the European Union some E208 million; the mining company spent a further E79 million and suffered an operational loss of E17 million; with another E81 million from the European Union being allocated for inter-related rehabilitation, including re-establishing some of the separately altered hydrological features of this important wetland.

In response to the complex nature of many management issues for inland waters, a good deal of effort has been invested in developing collaborative and integrated management structures that address common interests and differences between agencies or states over the services provided by shared inland water systems. (See Iza 2004 for an analysis of international agreements for the conservation of freshwater ecosystems.) In some instances, integrated and comprehensive strategies and action plans have been developed in support of active interventions, regionally and locally.

The Mediterranean wetland initiative, for example, is a successful mechanism for the conservation and wise use of wetlands throughout the Mediterranean region through local and regional actions and international cooperation (Papayannis 2002). More specific thematic initiatives or action plans cover the management of invasive species (e.g., Wittenberg and Cock 2001; McNeeley et al. 2001), the reintroduction and maintenance of biodiversity (e.g. Bibby et al. 1992), or the integration of development with conservation (e.g., Davies and Claridge 1993).

In some cases, specific technical methods suitable for local application have been developed (see Zalidis et al. 2002 for information on wetland restoration in the Mediterranean), with increasing recognition that integrated collation, collection, and use of data and information are essential aspects of an effective management mechanism, whether they are focused on local or regional issues. This recognition has resulted in the development of models that provide a basis for standardized inventory, risk assessment and evaluation, and monitoring, such as that proposed for an Asian wetland inventory, and integrated analyses incorporating community consultation and communication (Finlayson et al. 2002; Finlayson 2003; Ramsar Convention Secretariat 2004).

There has in recent years been increased interest in the development of mechanisms to encourage and support the capacity of local communities to contribute to the management of inland waters, particularly where local knowledge and experience can be constructively used (Ramsar Convention Secretariat 2004). Recognition of the beneficial outcomes that can occur when local people are involved in the management of inland waters and their services now underpins efforts by the Ramsar Convention to encourage best management practices. This concept is implicit in the guidance provided by the Convention covering policy and legal instruments, economic and social interactions, and technical tools (Ramsar Convention Secretariat 2004).

The challenge for the Convention and others is to ensure that such instruments and tools are used effectively and, as required, improved. This can be done within an adaptive management regime, noting that this necessitates active learning mechanisms, the involvement of key stakeholders, and the balancing of vested interests. All too often, however, the involvement of local communities has not occurred or has not been effective at resolving conflict between users and resource managers (Carbonell et al. 2001), or indeed, between competing users of sites listed as Wetlands of International Importance.

The Ramsar concepts of wise use and ecological character can be used to guide management interventions for wetlands (Ramsar Convention Secretariat 2004). Wise use of wetlands has been defined by the Convention as "their sustainable utilisation for the benefit of humankind in a way compatible with the maintenance of the natural properties of the ecosystem." "Sustainable utilisation" is in turn defined as "human use of a wetland so that it may yield the greatest continuous benefit to present generations while maintaining its potential to meet the needs and aspirations of future generations." "Ecological character" is defined under the Convention as "the sum of the biological, physical, and chemical components of the wetland ecosystem, and their interactions, which maintain the wetland and its products, functions, and attributes."

A suggested redefinition of this is under discussion, which would ensure that ecosystem services (referred to as products, functions, and attributes, in the definition above) are considered

as central components of ecological character and not just dependent on the ecological components and processes. Such a definition could read "the combination of the ecosystem components, processes, and services that characterize the wetland." Wise use could similarly be redefined to reflect the emphasis on ecosystem services and human well-being. These redefinitions further emphasize the close match between the Ramsar concepts and the conceptual framework of the Millennium Ecosystem Assessment, with the latter being more explicit about the emphasis on ecosystem services and human well-being.

20.6 Inland Water Systems and Human Well-being

It is *well established* that the services provided by inland waters are vital for human well-being and poverty alleviation (Dugan 1990; Revenga and Kura 2003; Finlayson et al. 1992; Finlayson and Moser 1991; Whigham et al. 1993; Mitsch 1994; McComb and Davis 1998; Lundqvist and Gleick 2000). A list of the services provided by inland waters was provided at the start of this chapter in Table 20.1. The benefits of these services to human well-being, and hence the consequences of reduced availability and supply for human well-being, are discussed in more detail in individual chapters that cover services derived from inland water systems—in particular, fresh water (Chapter 7), food (Chapter 8), nutrient cycling (Chapter 12), waste processing and detoxification (Chapter 15), regulation of natural hazards (Chapter 16), and various cultural and amenity services (Chapter 17). An analysis of human well-being and its relationship to ecosystem services is provided in Chapter 5. This section provides a brief assessment of specific examples of the relationship between the degradation of inland waters and human well-being. (See Table 20.9.)

The ecosystem services of inland water systems provide a basis for human well-being for people who live in close proximity to the system as well as those who live much further away. As human well-being is strongly affected by the extent to which people are able to meet their most basic needs (water, food, shelter, and health) in a secure manner, the sustainable use of inland waters for ensuring human well-being is vital. (See Box 20.13, as well as Box 20.8 earlier in the chapter.) This is well illustrated by the infrastructure and trade networks that have been developed to supply, for example, drinking water, food, and energy from lakes and reservoirs that can be located far from densely populated urban areas. It is also *well established* that in both rural and urban areas the poor are likely to suffer most when the availability and quality of water and food is reduced, whether due to failures in the infrastructure and trade networks or the demise of the systems themselves. (See Chapters 6, 7, and 8.) The impacts on human well-being of degraded supporting and regulating services from inland waters is often not recognized as readily, but it can be as significant as changes to provisioning services—for example, a reduction in the capacity of a wetland to filter water or to detoxify wastes can have significant consequences for human health, even if food provision remains adequate.

It is also known with *high certainty* that maintenance of an adequate flow of good-quality water is needed to maintain the health of inland water systems as well as estuaries and deltas. The reverse is also true: healthy inland water systems generate and maintain adequate flows of good-quality water. As the supporting services of inland waters are the result of interactions among the ecological components within the system and those in the catchment, human well-being is inexorably linked to the maintenance of the ecological character of inland water systems. Because of

Table 20.9. Summary of Critical Changes in Inland Water Systems and Services and Their Impacts on Human Well-being (WWDR 2003)

Major Drivers of Change in Inland Waters	Major Impacts on Services Derived from Inland Waters	Function(s) at Risk	Major Impacts on Human Well-being	Vulnerable People or Places
Population and consumption growth	increases water abstraction and acquisition of cultivated land through inland water drainage; increases requirements for all other activities, with consequent risks	virtually all ecosystem functions, including habitat, production, and regulation functions	increased health risks reduced quality and quantity of water	urban communities
Infrastructure development (dams, dikes, levees, diversions, interbasin transfers, etc.)	loss of integrity alters timing and quantity of river flows, water temperature, nutrient and sediment transport, and thus delta replenishment; blocks fish migrations; increases mosquito breeding	water quantity and quality, habitats, floodplain fertility, fisheries, delta economies	increased agricultural productivity reduced food security reduced economic opportunities increased health risks	downstream communities
Land conversion	eliminates key components of aquatic environment; loss of functions, integrity, habitat, and biodiversity; alters runoff patterns; inhibits natural recharge; fills water bodies with silt	natural flood control, habitats for fisheries and waterfowl, recreation, water supply, water quantity and quality	reduced household security loss of productive land increased release of carbon dioxide into the atmosphere reduced recreational, cultural, historical, or religious values	
Overharvesting and exploitation	depletes living resources, ecosystem functions (leading to fire and drought), and biodiversity (groundwater depletion, fisheries collapse)	food production, water supply, water quality, and water quantity	reduced food security reduced economic opportunities (e.g., tourism) increased risk of natural disasters	communities living adjacent to and dependent on inland water resources
Introduction of exotic species	outcompetes native species, alters production and nutrient cycling, loss of biodiversity	food production, wildlife habitat, recreation	reduced food security (e.g., reduced genetic variety and resilience)	
Release of pollutants to land, air, or water	pollution of water bodies alters chemistry and ecology of rivers, lakes, and wetlands	water supply, habitat, water quality, food production	reduced quality of water reduced food security reduced household security	
Climate change	greenhouse gas emissions produce dramatic changes in runoff and rainfall patterns, loss of coastal areas to sea level rise, increased erosion of shorelines, degradation of water quality by rising temperatures, changes in water flow volume, increased salt-water intrusion, increased water demand for irrigation, increased flood damage, increased drought frequency	shoreline protection, water quality, dilution capacity, transport, flood control	reduced household security reduced quality and quantity of water reduced productive land	

the complexities of these interactions, management of supporting services is likely to be best served by a holistic river basin approach, within which the resource base is assessed and managed in an integrated manner (Hollis 1998; Ramsar Convention Secretariat 2004). Implementation of a river basin or ecosystem approach implies stakeholders' acceptance that there may need to be trade-offs between them for access to the services provided by the river and its associated habitats.

It is widely accepted that the loss and degradation of inland waters has reduced their natural ability to buffer or ameliorate the impacts of floods (see Chapter 16) and hence threaten the security

of individuals and entire communities. For example, in Southern Africa in 1999 and 2000, devastating floods affected more than 150,000 families (Mpofu 2000); degradation of wetlands such as the Kafue in Zambia, damming of rivers, deforestation, and overgrazing led to a reduced absorption of excess water and magnified the impact of the floods (Chenje 2000; UNDHA 1994). The same applies to floodplains where increasing human habitation, drainage of wetlands, and river canalization have severely restricted the capacity to buffer floods in many places and increased people's vulnerability to flooding. (See Chapters 7 and 16.) In Central Europe in 2002, extreme flooding as a consequence of unusually high rainfall was exacerbated by physical alterations along the rivers and changes in the water retention capacity of the riparian zone and upper catchment. Floods and droughts also typically affect the poorest people most severely, as they often live in vulnerable areas and have few financial resources for avoidance, mitigation, or adaptation. (See Chapters 6 and 16.) Few countries have been free of damaging floods during the last few decades (Kundzewicz and Schellnhuber 2004).

Although largely eliminated in wealthier nations, water-related diseases are among the most common causes of illness and mortality affecting the poor in developing countries. The extent of water pollution and its link with human health in many countries is well known. The World Health Organization has estimated that there are 4 billion cases of diarrhea each year in addition to millions of other cases of illness associated with a lack of access to clean water. (See Chapter 7.) Water-borne diseases that result in gastrointestinal illness (including diarrhea) are caused by consuming contaminated water. (See Chapter 14.)

Perhaps less recognized as a major influence on human well-being, but as potentially debilitating to people, are actions that degrade inland water systems and result in a reduction in water supply or encourage the spread and abundance of disease vectors. (See Chapters 5 and 14.) Schistosomiasis, for example, has been spread by the construction of dams and large lakes in many countries, and interference with the hydrology of wetlands has exacerbated the incidence of mosquito-borne diseases. In 2000, the estimated mortality due to water sanitation hygiene–associated diarrheas and some other water sanitation–associated diseases

(schistosomiasis, trachoma, intestinal helminth infections) was 2,213,000. There were an estimated 1 million deaths due to malaria, and more than 2 billion people were infected with schistosomes and soil-transmitted helminths, of whom 300 million suffered serious illness. The majority of those affected by water-related mortality and morbidity are children under five (WWDR 2003). Since many illnesses are undiagnosed and unreported, the true extent of these diseases is unknown (Gleick 2002).

Water-related diseases that are exacerbated by the degradation of inland waters (see Chapter 14) include those caused by the ingestion of water contaminated by human or animal feces or urine containing pathogenic bacteria or viruses, such as cholera, typhoid, amoebic and bacillary dysentery, and other diarrheal diseases; diseases passed on by intermediate hosts such as aquatic snails or insects that breed in aquatic ecosystems, such as dracunculiasis, schistosomiasis, and other helminths as well as dengue, filariasis, malaria, onchocerciasis, trypanosomiasis, and yellow fever; and diseases that occur when there is insufficient clean water for washing and basic hygiene or when there is contact with contaminated water, such as scabies, trachoma, typhus, and flea-, lice-, and tick-borne diseases.

In addition to disease from inland waters, water-borne pollutants have a major effect on human health, often through their accumulation in the food chain. Many countries now experience problems with elevated levels of nitrates in groundwater from the large-scale use of organic and inorganic fertilizers. Excess nitrate in drinking water has been linked to methemoglobin anemia in infants, the so-called blue baby syndrome. Arsenicosis, the effect of arsenic poisoning when drinking arsenic-rich water over a long period, is also known and is a particularly severe problem in Bangladesh and Western Bengal, where some 35–77 million inhabitants are exposed to excessively high levels of arsenic in water drawn from wells (Bonvallot 2003). On the whole, though, it is still extremely difficult to quantify the cumulative effects of long-term exposure to a variety of chemicals at what seem like low concentrations. (See Chapter 15.)

There is increasing evidence from wildlife studies that humans are at risk from a number of chemicals that mimic or block the natural functioning of hormones, interfering with natural body processes, including normal sexual development. (See Chapter 15.) Chemicals such as PCBs, DDT, dioxins, and those from at least 80 pesticides are regarded as "endocrine disrupters," which may interfere with human hormone functions, undermining disease resistance and reproductive health. Pharmaceuticals in the environment represent an emerging environmental issue, with many being only partially removed by conventional wastewater treatment and therefore being deposited into a variety of receiving waters. The presence of these compounds in inland waters is considered harmful for humans even though the extent of harm remains uncertain. It is certain, though, that the degradation of inland water systems reduces the potential of these systems to mitigate the effects of pollutants through detoxification and waste processing and results in an overall reduction in human well-being.

It is expected that continued degradation of inland water systems will result in further reduction in human health, especially for vulnerable people in developing countries where technological fixes and alternatives are not as readily available. The evidence that the degradation of inland waters results in a loss of services and reduction in human health is incontrovertible, and yet degradation continues at a global scale. Conserving and using sustainably the services derived from inland waters is an ongoing challenge for society, as is reducing the negative downstream consequences of inland waters degradation. Failure to reduce and reverse the loss and degradation of inland water systems will further undermine human well-being. The problem of continued loss and degradation is both environmental and social—it is *well established* that the loss and degradation of inland waters has and continues to reduce the ecosystem services available for people.

References

Abramovitz, J. N. 1996: Imperiled Waters, Impoverished Future: The Decline of Freshwater Ecosystems. Worldwatch Paper 128. Worldwatch Institute, Washington, DC, U.S.A.

Abul-Gasim, S. and M. Babiker, 1998: Iraq's Food security: the sand dunes fixation project. *Desertification Control Bulletin, 33,* 2–10

Ali, A., Md Rawi, C.S., Mansor, M., Ramakrishna, S. & T. Mundkur, (eds) 2002: The Asian wetlands: bringing partnerships into good wetland practices, Proceedings of the Asian Wetland Symposium 2001 "bringing partnerships into good wetland practices', Penang, Malaysia, August 2001, Penebit Universiti Sains Malaysia, Pulau Pinang, Malaysia

Alvarado, F. H and B.F. Gutiérrez, 2002: *Especies hidrobiológicas continentales introducidas y transplantadas y su distribución en Colombia..* Ministerio del Ambiente de Colombia, Convención Ramsar, Corporación Autónoma Regional del Valle del Cauca. Bogotá. 176 p.

American Sportfishing Association. 2001: Economic impact of sportfishing in the US by type of fishing [online] Available at: www.asafishing.org/asa/statistics/economic_impact/economic_impact_table. html

Anon 1992: *A Strategy to Stop and Reverse Wetland Loss and Degradation in the Mediterranean Basin.* IWRB and Regione Friuli-Venezia Giulia, Trieste, Italy. 40p.

Arcadis Euroconsult, 2001: Buffering Capacity of Wetlands Study (BCWS): Final Report Vol. 1 Main Report. Lake Victoria Environmental Management Project (LVEMP), United Republic of Tanzania/World Bank.

Aselmann, I. and Crutzen, P.J. 1989: Global distribution of natural freshwater wetlands and rice paddies, and their Net Primary Productivity, seasonality and possible methane emissions. *Journal of Atmospheric Chemistry* **8,** 307–358

Ashmore, P. and Church, M. 2001: *The Impact of Climate Change on Rivers and River Processes in Canada.* Geological Survey of Canada Bulletin 555. Ottawa, Canada.

Asia-Pacific Migratory Waterbird Conservation Committee, 2001: Asia-Pacific Migratory Waterbird Strategy: 2001 – 2005. Wetlands International – Asia Pacific, Kuala Lumpur, Malaysia, 67 pp.

Avakyan, A.B. and V.B. Iakovleva, 1998: "Status of Global Reservoirs: The Positioning the Late Twentieth Century. *Lakes & Reservoirs: Research and Management* **3**: 45–52.

BAPPENAS-ADB (National Development Planning Agency and Asian Development Bank), 1999: *Causes, Extent, Impact and Costs of 1997/1998 Fires and Drought.* Final report, Annex 1 and 2, Planning for Fire Prevention and Drought Management Project. Jakarta, Indonesia.

Barbier, E. B., M.C. Acreman, and D. Knowler, 1997: *Economic valuation of wetlands: A guide for policy makers and planners.* Ramsar Convention Bureau, Gland, Switzerland, 127 pp.

Bartolome, J. and I. Vega, 2002: *Mining in Donana: learned lessons.* WWF Spain, Madrid. 25 pp.

Bayliss, B., K. Brenman, I. Elliot, M. Finlayson, R. Hall, T. House, B. Pidgeon, D. Walden and P. Waterman,1997: *Vulnerability Assessment of Predicted Climate Change and Sea Level Rise in the Alligator Rivers Region, Northern Territory Australia.* Supervising Scientist Report 123, Supervising Scientist, Canberra, Australia 134 pp.

Beeton, A.M., 2002: Large freshwater lakes: present state, trends, and future. *Environmental Conservation* **29 (1)**: 21–38.

Beltram, G. and P. Skoberne, 1998: *Agricultural and Forestry Practices for Sustainable Use of Biological and Landscape Diversity in Slovenia.* In: M. Küttel and ‘G. Thélin (eds.). Nature for East and Best, Congress, Basel 22–26 1997, Proceedings. SAEFL, Bern, Switzerland. pp. 91–102.

Bibby, C.J., N.J. Collar, M.J. Crosby, M.F. Heath, C. Imboden, T.H. Johnson, A.J. Long, 1992: *Environmental Conservation* 30(2): 104–130

Blanco, D.E., de la Balze, V.M. (eds) 2004: *Los Turbales de la Patagonia: Bases para su inventario y la conservación de su biodiversidad.* Publicación No. 19. Wetlands International. Buenos Aires, Argentina.

Bonvallot, V. 2003: . L'arsenic quotidien. Biofutur, **232,** 70–73.

Boyd, H. and J. Madsen, 1997: Impacts of global change on artic-breeding bird populations and migration. Pp. 201–217. In: Global change andArtic Terrestrial Ecosystems, W.C. Oechel, T. Callaghan, T. Gilmanov, J.I. Holten, B. Maxwell, U. Molau and B. Sveinbjoernsson (eds), Global Change and

Arctic Terrestrial Ecosystems. *Ecological Studies* 124. Springer, New York. p. 201–217.

Brag, O., Lindsay, R., Risager, M., Silvius, M. and Zingstra, H. 2003: *Strategy and Action Plan for Mire and Peatland Conservation in Central Europe.* Wetlands International, Wageningen, The Netherlands.

Brinson, M.M. and A. Ines Malvarez, 2002: Temperate freshwater wetlands: types, status, and threats. *Environmental Conservation* 29 (2): 115–133.

Brito, M., 2001: Eixos amazônicos de integração e desenvolimento—obras e empreendimentos. In *Biodiversidade na Amazônia Brasileira,* ed. A. Verríssimo, A. Moreira, D. Sawyer, I. dos Santos, L.P. Pinto, pp. 321–326. Brasília, Brasil: Estação Liberdade & Instituto Socioambiental.

Brown, S., 1981: A comparison of the structure, primary productivity, and transpiration of cypress ecosystems in Florida. *Ecol. Monogr.* **51,** 403–27.

Brown, S., C. Hickey, B. Harrington and R. Gill, R. (eds), 2001: *United States Shorebird Conservation Plan,* 2nd Edition. Manomet Center for Conservation Sciences, Manomet, USA. [online] Available at www.manomet.org/USSCP/files.htm

Butchart, S.H.M., A.J. Stattersfield, L.A. Bennun, S.M. Shutes, H.R. Akça-kaya, J.E.M. Baillie, S.N., Stuart, C. Hilton-Taylor and G.M. Mace, 2004. Measuring global trends in the status of biodiversity: Red List Indices for birds. Philosophical Transactions, LoS Biology, 2, 2294–2304.

Butler, R.W. and R.G. Vennesland, 2000: "Integrating climate change and predation risk with wading bird conservation research in North America." *Waterbirds* 23(3), 535–540.

Carbonell, M., N. Nathai-Gyan, and C.M. Finlayson, 2001: *Science and local communities: strengthening partnerships for effective wetland management.* Ducks Unlimited Inc, Memphis, USA. 93 pp.

CERP, 1999. Comprehensive Everglades Restoration Plan 1999. Available online at: http://www.evergladesplan.org/about/rest_plan.cfm

Chenje, M. (ed.), 2000: *State of the Environment Zambezi Basin 2000: Summary.* Maseru, Lusaka and Harare, SADC, IUCN, ZRA and SARDC.

Chevreuil, M., D. Thevenot, P. Scribe, M.Blanchard, Y. Duclos, M. Garmouna, et al. 1998: Micropolluants organiques: une boîte de Pandore. In *La Seine en son bassin,* M. Meybeck, G. de Marsily and E. Fustec (eds), Elsevier, Paris, France. p. 439–481.

Clausnitzer, V. and R. Jodicke, 2004: Guardians of the watershed: Global status of dragonflies: critical species, threat and conservation. *International Journal of Ordonatology* Vol **7 (2).** ISSN 1388–7890.

CNA-IBUNAM (Comisión Nacional del Agua-Instituto de Biología de la UNAM), 1999: Inventario de Cuerpos de Agua Continentales y Costeros, Vegetación. Comisión Nacional del Agua. Gerencia de Saneamiento y Calidad del Agua.

Coe, M.T. and Foley, J.A. 2001. Human and natural impacts on the water resources of the Lake Chad Basin. *Journal of Geophysical Research (Atmospheres)* **106,** 3349–3356.

Contreras-Balderas, S., 2003: *Las Presas del Noreste.* In: De la lanza-Espino G. y García-Calderón J.L. (Eds.) Lagos y Presas de México AGT Editor, Mexico. p. 484–499.

Costanza, R., R. d'Arge, R. de Groot, S. Farber, M. Grasso, B. Hannon, K. Limburg, S. Naeem, R.V. O'Neill, J. Paruelo, R.G. Raskin, P. Sutton, P. & M. van den Belt, 1997: The value of the world's ecosystem services and natural capital. *Nature* 387: 253–260.

Cushing, C.E., K.W. Cummins, and G.W. Minshall, 1995: *Ecosystems of the World 22: Rivers and Stream Ecosystems.* Elsevier Science Publishers, Amsterdam, The Netherlands.

Dahl, T. E. and C. E. Johnson, 1991: *Status and Trends of Wetlands in the Conterminous United States, Mid-1970's to Mid-1980's.* Washington, DC: U.S. Department of the Interior, Fish and Wildlife Service.

Dahl, T.E., 2000: *Stats and trends of wetlands in the conterminous United States 1986–1997.* U.S. Department of the Interior, Fish and Wildlife Service, Washington, D.C. 82 pp.

Danielopol, D.L., Griebler, C., Gunatilaka, A. and Notenboom, J. 2003: Present state and future prospects for groundwater ecosystems. *Environmental Conservation* 30, 104–130.

Darras, S., M. Michou and C. Sarrat, 1999: *A first step towards identifying a global delineation of wetlands.* IGBP-DIS Working Paper # 19, Toulouse, France. 64 pp.

Davidson, N.C. and D.A. Stroud, D.A. 2004: African-Eurasian Flyways: current knowledge, population status, and future challenges. Proceedings of the Waterbirds around the World conference, Edinburgh, April 2004. Her Majesty's Stationary Office, London

Davies, J. and G.F. Claridge (eds), 1993: *Wetland benefits: the potential for wetlands to support and maintain development.* AWB Publication No 87, Kuala Lumpur,

Malaysia, IWRB Special Publication No 21, Slimbridge, UK, and Wetlands of the Americas Publication No 11, Ottawa, Canada. 45 pp.

De Voogt, K., G. Kite, P. Droogers, and H. Murray-Rust. 2000. Modelling water allocation between a wetland and irrigated agriculture in the Gediz Basin, Turkey. *Water Resources Development,* **16,** 639–650.

Delany, S., C. Reyes, E. Hubert, S. Pihl, E. Rees, L. Haanstra and A. van Strien, 1999: Results from the International Waterbird Census in the Western Palearctic and Southwest Asia, 1995 and 1996.. Wetlands International Publication No. 54. Wageningen, The Netherlands. p. 1–178

Delgado, C.L., N. Wada, M.W. Rosegrant, S. Meijer and M. Ahmed, 2003: *Outlook for Fish to 2020: Meeting Global Demand.* International Food Policy Research Institute & WorldFish Centre, 28 pp.

Diaz-Carreras, Juan H., (n.d.): Comprehensive Everglades Restoration Plan [online] Available at: http://www.dep.state.fl.us/lands/conference/mateials/Diaz%20-%20Fixing%20Everg lades.pdf

Dochoda, M., 1988: Great Lakes Fishery Commission. [online] Available at: http://www.glfc.org/fishmgmt/fsecon.htm

Dowman, K., 1997: *The Sacred Life of Tibet.* Thorson (Harper Collins), London, 324 pp.

Ducks Unlimited, 2002: Recreation [online] Available at: www.ducks.org/conservation/improve_recreation.asp.

Dudgeon, D., 2000: "Large-scale hydrological changes in tropical Asia: prospects for riverine biodiversity." *BioScience* 50(9): 793–806.

Dudgeon, D., 2000a: Large-scale hydrological alterations in tropical Asia: prospects for riverine biodiversity. *BioScience* **50:** 793–806.

Dudgeon, D., 2000b: The ecology of tropical Asian rivers and streams in relation to biodiversity conservation. *Annual Review of Ecology & Systematics* **31:** 239–263.

Dudgeon, D. 2000c: Riverine wetlands and biodiversity conservation in tropical Asia. In: *Biodiversity in Wetlands: Assessment, Function and Conservation* B. Gopal, W.J. Junk & J.A. Davis (eds). Backhuys Publishers, The Hague: 35–60.

Dudgeon, D., 2003: The contribution of scientific information to the conservation and management of freshwater biodiversity in tropical Asia. *Hydrobiologia* **500** (Special Issue): 295–314.

Dugan, P. (ed.), 1993: *Wetlands in Danger.* World Conservation Atlas. (Oxford University Press: New York.)

Dugan, P.J. (ed.), 1990: *Wetland Conservation: A Review of Current Issues and Required Action.* IUCN, Gland, Switzerland, 94 pp.

Eiseltova, M., 1994: *Restoration of lake ecosystems.* IWRB Publication 32, Slimbridge, UK. 182 pp.

Eiseltova, M. and J. Biggs (eds), 1995: *Restoration of stream ecosystems—an integrated catchment approach.* IWRB Publication 37, Slimbridge, UK. 170 pp.

Eliot I, C.M. Finlayson and P. Waterman, 1999: Predicted climate change, sea level rise and wetland management in the Australian wet-dry tropics. *Wetlands Ecology and Management* **7,** 63–81.

Ellison, A.M., 2004: Wetlands of Central America. *Wetlands Ecology and Management* **12,** 3–55.

Emerton, L., P. Iyango, Luwum and A. Malinga, 1998: *The Present Economic Value of Nakivubo Urban Wetland, Uganda.* IUCN—The World Conservation Union, Eastern Africa Regional Office, Nairobi and National Wetlands Programme, Wetlands Inspectorate Division, Ministry of Water, Land and Environment, Kampala, 30 pp.

European Environment Agency 2003: Europe's water: An indicator-based assessment. Topic report No 1/2003, p. 93 and p. 97

FAO (Food and Agriculture Organization of the United Nations), 1996: Malawi Fishery Country Profile. FAO Fisheries Department [online] Available at: http://www.fao.org/fi/fcp/malawie.htm.

FAO, 1999: Inland Water Resources and Aquaculture Service, Fishery Resources Division, *Review of the State of World Fishery Resources: Inland Fisheries, 1999.* FAO Fisheries Circular, No. 942.

FAO, 2003: *Rice and Us.* [online] International Year of Rice 2004 Fact Sheet. Available at http://www.fao.org/rice2004/en/rice-us.htm

FAO, 2004: The State of the World Fisheries and Aquaculture 2004. FAO Fisheries Department, Rome, Italy.

Fernando, C.H., 1991: Impacts of fish introductions in tropical Asia and America. *Canadian Journal of Fisheries and Aquatic Sciences* 48 (Suppl. 1): 24–32.

Finlayson C.M. & M. Moser (eds), 1991: *Wetlands: A global perspective.* Facts on File, Oxford.

Finlayson C.M., T. Hollis and T. Davis (eds), 1992: *Managing Mediterranean wetlands and their birds.* Proceedings of an IWRB International Symposium, Grado, Italy, February 1991. IWRB Special Publication No 20, Slimbridge, UK. 285 pp.

Finlayson C.M. & A.G. van der Valk, 1995: *Classification and inventory of the world's wetlands:* Advances in Vegetation Science 16, Kluwer Academic Press, Dordrecht, The Netherlands.

Finlayson C.M. & N. Rea, 1999: Reasons for the loss and degradation of Australian wetlands. *Wetlands Ecology & Management* **7,** 1–11.

Finlayson C.M. & A.G. Spiers (eds), 1999: *Global Review of Wetland Resources and Priorities for Wetland Inventory.* Supervising Scientist Report 144 & Wetlands International Publication 53, Supervising Scientist, Canberra, 520 pp.

Finlayson C.M., N.C. Davidson, A.G. Spiers and N.J. Stevenson, 1999: Global wetland inventory—Status and priorities. *Marine and Freshwater Research* 50, 717–727.

Finlayson, M., van Dam, R., Walden, D. and Storrs, M. 2000: Risk Assessment for Managing the Tropical Weed, *Mimosa pigra.* In *Assessment and management of alien species that threaten ecosystems, habitats and species.* Abstracts of keynote addresses and posters presented at the sixth meeting of the Subsidiary Body on Scientific, Technical and Technological Advice, held in Montreal, Canada, from 12 to 16 March 2001. Montreal ,CBD Technical Paper no. 1. p. 17–20.

Finlayson C.M., G.W. Begg, J. Howes, J. Davies, K. Tagi and J. Lowry, 2002: *A manual for an inventory of Asian wetlands (version 1.0).* Wetlands International Global Series 10, Wetlands International, Kuala Lumpur, Malaysia. 72 pp.

Finlayson C.M., 2003: *Integrated inventory, assessment and monitoring of tropical wetlands. Environmental Monitoring of Tropical and Subtropical Wetlands,* Proceedings of a Conference, Maun, Botswana, 4–8 December 2002. Okavango Report Series No 1, HOORC, Maun, Botswana. Pp 16–44.

Finlayson, C.M., Y.S. Chuikow, R.C. Prentice and W. Fischer, 1993: *Biogeography of the lower Volga, Russia: an overview.* IWRB Special Publication No. 28, Slimbride, UK. 16 pp.

Finlayson, C.M. and T. Larsson (eds), 1991: *Wetland Management and Restoration,* Proceedings of a workshop, Sweden, 12–15 September 1990, Swedish Environmental Protection Agency Report 3992, pp. 17–23.

Foster, S., A. Lawrence, and B. Morris. 1998. Groundwater in Urban Development: Assessing Management Needs and Formulating Policy Strategies. World Bank Technical Paper No. 390. The World Bank, Washington, DC, U.S.A

Gallego Fernandez, J.B. and F. Garcia Nove, 2002: Restoration of tidal marshes in Guadalquivir river estuary, south-west Spain. In: *Restoration of Mediterranean Wetlands,* G.C. Zalidis, T.L. Crissman and P.A. Gerakis (eds), Hellenic Ministry of the Environment, Physical Planning and Public Works, Athens, and Greek Biotope/Wetland Centre, Thermi, Greece. pp. 213–228.

Garibaldi, L. and D. M. Bartley, 1998: The database on introductions of aquatic species (DIAS). Food and Agriculture Organization of the United Nations (FAO) Aquaculture Newsletter (FAN), no. 20. Available on the Web at http://www.fao.org/fi/statist/fisoft/dias/index.htm.

Gitay, H., Suarez, A., Watson, R., Ansimov, O., Chapin, F.S., Victor Cruz, R., Finlayson, M. et al 2002. Climate change and biodiversity. IPCC Technical Paper V, Intergovernmental Panel on Climate Change, Geneva, Switzerland. 77 pp.

Gleick, 2002: *Dirty Water: Estimated Deaths from Water-Related Diseases 2000–2020.* Pacific Institute for Studies in Development, Environment, and Security [online] Available at www.pacinst.org, page 12/12

Glover, D. and T. Jessup (eds), 1999: *Indonesia's fires and haze: the cost of catastrophe.* Institute of South East Asian Studies, Singapore and IDRC, Ottawa, Canada. 160 pp.

Gopal, B., 1987: *Water Hyacinth.* Aquatic Plant Studies 1, Amsterdam and New York: Elsevier Science.

Gopal, B., W.J. Junk, and J.A. Davis (eds), 2000: *Biodiversity in wetlands: assessment, function, and conservation,* volume 1. Leiden, The Netherlands: Backhuys Publishers.

Great Lakes Information Network, 2004: Fish and fisheries of the Great Lakes region—overview. [online] Available at http://www.great-lakes.net/envt/flora-fauna/wildlife/fish.html1#over

Groombridge, B., 1992: Global Biodiversity: Status of the Earth's Living Resources. Chapman and Hall, London, U.K.

Groombridge, B. and M. Jenkins, 1998: *Freshwater Biodiversity: a preliminary global assessment.* World Conservation Monitoring Centre—World Conservation Press, Cambridge, UK. vii + 104 pp + 14 Maps.

Hamilton, S.K., 1999: Potential effects of a major navigation project (Paraguay-Paraná Hidrovia) on inundation in the Pantanal floodplains. *Regulated Rivers: Research and Management,* **15,** 289–299.

Harahsheh, H. and R. Tateishi, 2000: *Environmental GIS Database and Desertification Mapping of West Asia.* Paper presented at the Workshop of the Asian Region Thematic Programme Network on Desertification Monitoring and Assessment, Tokyo, 28–30 June 2000.

Harrison, I. J. and M. J. Stiassny, 1999: The Quiet Crisis: A Preliminary Listing of the Freshwater Fishes of the World that Are Extinct or 'Missing in Action'. In *Extinctions in Near Time* R.D.E. MacPhee (ed.), Kluwer/Plenum Publishers, New York, U.S.A. 271–331.

Heathwaite, A.L., Johnes, P.J. and Peters, N. 1996: Trends in nutrients. *Hydrological Processes* 10, 263–293.

Hecky, R. and Bugenyi, F. 1992: Hydrology and chemistry of the African Great Lakes and water quality issues: problems and solutions. *Mitteilungen Internationale Vereinigung für Limnologie* 23, 45–54.

Heeg, J. and C.M. Breen (eds) 1982: *Man and the Pongolo floodplain.* Pretoria, South Africa: South African National Scientific Programs Report 56, CSIR, Pretoria, South Africa.

Hofstede, R. Segarra, P. and Mena V. 2003: *Los Páramos del Mundo.* Proyecto Atlas Mundial de los Páramos. Global Peatland Initiative/NC-IUCN/EcoCiencia. Quito, Peru.

Hollis G.E., J. Patterson, T. Papayannis and C.M. Finlayson, 1992: Sustaining wetlands: Policies, programs and partnerships. In: *Managing Mediterranean wetlands and their birds,* C.M. Finlayson, G.E. Hollis and Davis, T.J. (eds), IWRB Special Publication No. 20, Slimbridge, UK, pp. 281–285.

Hollis, G.E. 1992: The causes of wetland loss and degradation in the Mediterranean. In: *Managing Mediterranean wetlands and their birds,* C.M. Finlayson, G.E. Hollis and Davis, T.J. (eds), IWRB Special Publication No. 20, Slimbridge, UK, pp. 83–90.

Hollis, G.E., W.M. Adams and M. Aminu-Kano (eds), 1993: *The Hadejia-Nguru Wetlands—Environment, Economy and Sustainable Development of a Sahelian Floodplain Wetland.* IUCN, Gland, Switzerland, 244 pp.

Hollis, G.E. 1998: Future wetlands in a world short of water. In: *Wetlands for the future,* AJ McComb & JA Davis (eds), Gleneagles Publishing, Adelaide, 5–18.

IDEAM (Instituto de Hidrología, Metereología y Estudioa Ambientales), 2002: Páramos y ecosistemas alto andinos de Colombia en condición hotspot and global climatic tensor. Ministerio del Media Ambiente, Programa de las Naciones Unidas para el Desarrollo, Bogota, Colombia. 387 pp.

Iftikhar, U., 2002: Valuing the economic costs of environmental degradation due to sea intrusion in the Indus Delta', in IUCN, Sea Intrusion in the Coastal and Riverine Tracts of the Indus Delta—A Case Study. IUCN—The World Conservation Union Pakistan Country Office, Karachi.

ILEC and UNEP (International Lake Environment Committee Foundation and United Nations Environment Programme), 2003: *World Lake Vision: A Call to Action.* 39 pp. ISBN 4–9901546–0–6.

Immirzi, C.P. and E. Maltby, 1992: The global status of peatlands and their role in carbon cycling. In: *Wetlands and climate change. Feasibility investigation of giving credit for conserving wetlands as carbon sinks.* Patterson, J. 1999 (ed), Wetlands International Special Publication 1, p 35.

International Rice Research Institute, 1995: *World Rice Statistics 1993–94,* International Rice Research Institute, Manila, Philippines, p. xv.

International Rice Research Institute, 1999: *The Facts of Rice: Population, Poverty, and Food Security in Asia.* International Rice Research Institute Annual Report 1998–1999.

IUCN (World Conservation Union), 1992: *Environmental Status Reports: Volume 4—Conservation Status of the Danube Delta.* Gland, Switzerland. 107 pp.

IUCN, 2003: *2003 IUCN Red List of Threatened Species.* [online] Available at: http://www.redlist.org

IUCN, Conservation International, and NatureServe 2004. Global Amphibian Assessment. www.globalamphibians.org

Iza, A., 2004. International Water Governance: Conservation of Freshwater Ecosystems. Volume 1 International Agreements—Compilation and Analysis. IUCN, Gland Switzerland and Cambridge, U.K. 324 pp.

Jaensch, R. (ed), 1996: Wetland Conservation in the Pacific Islands Region, Proceedings of the Regional Workshop on Wetland Protection and Sustainable Use in Oceania, June 1994, Port Moresby. Wetlands International-Asia Pacific, Canberra.

Joosten, H. (compiler) 1992. *Estimated peatland per country.* Global Peatland Initiative, Wageningen, The Netherlands.

Joosten, H. 1994: Turning the tides: experiences and perspectives of mire conservation in the Netherlands. In: *Mires and Man: Mire Conservation in a Densely Populated Country- The Swiss Experience,* A. Grünig (ed.), 300–310, Swiss Federal Institute for Forest, Snow and Landscape Research, Birmensdorf, Switzerland.

Joosten, H. and D. Clarke. 2002: *Wise Use of Mires and Peatlands—Background and Principles Including a Framework for Decision-making.* International Mire Conservation Group and International Peat Society. NHBS, Totnes, Devon.

Jorgensen, S.E., R. de Bernardi, T. Ballatore, and V. Muhandiki, 2001: *Lake Watch 25: changes in the world's lakes.* Draft report. International Lake Environment Committee, Kusatsu, Japan.

Junk, W.J. & J.A.S. Nunes de Mello, 1987: Impactos ecológicos das represas hidreétricas na bacia amazônica brasileira. *Tübinger Geographische Studiern,* **95,** 367–385.

Junk, W., 2002: Long-term environmental trends and the future of tropical wetlands. *Environmental Conservation* 29 (4): 414–435

Kaufman, L., 1992: "Catastrophic Change in Species-Rich Freshwater Ecosystems: The Lessons from Lake Victoria." *Bioscience* 42 (11): 846–858.

Kingsford, R.T., and W. Johnson. 1998. Impact of water diversions on colonially-nesting waterbirds in the Macquarie Marshes of arid Australia. *Colonial Waterbirds* **21,**159–170.

Kira, T., 1997: Survey of the state of world lakes. In: *Guidelines for lake management, volume 8: the world's lakes in crisis,* Chapter 7, Jørgensen, S.E. and S. Matsui (eds), Kusatsu, Japan: International Lake Environment Committee and United Nations Environment Programme.

Kucklick, J., Bidleman, T., McConnell, L., Walla, M. & Ivanov, G. 1994: Organochlorines in water and biota of Lake Baikal, Siberia. *Environmental Science & Technology* 28: 31–7.

Kundzewicz, Z. W, and H.J. Schellnhuber, 2004: Floods in the IPCC perspective, *Natural Hazards* **31**: 111–128.

Kura, Y., C. Revenga, E. Hoshino and G. Mock 2004. Fishing for Answers. World Resources Institute, Washington D.C., USA. 138 p.

Lehner, B. and P. Döll, 2004: Development and validation of a global database of lakes, reservoirs and wetlands. *Journal of Hydrology* **296/1–4**:1–22.

Lemly, A.D., R.T. Kingsford, J.R. Thompson. 2000. Irrigated agriculture and wildlife conservation: conflict on a global scale. *Environmental Management.* **25,** 485–512.

Lévêque, C. 1997: *Biodiversity dynamics and conservation: the freshwater fish of tropical Africa.* Cambridge, UK: Cambridge University Press.

Lévêque, C. 2003. Ecology: From Ecosystem to Biosphere. Science Publishers Inc. Plymouth, U.K. 472 p.

Lévêque, C., Balian,E.V., Martens, K (in press): An assessment of animal species diversity in continental water systems. *Hydrobiologia.*

Loh, J, and M. Wackernagel (eds), 2004: *The Living Planet Report 2004.* World Wide Fund for Nature. Gland, Switzerland.

Lowry, J. and C.M. Finlayson 2004. A review of spatial data sets for wetland inventory in northern Australia. Supervising Scientist Report 178, Supervising Scientist, Darwin, Australia. 18 p.

Lundberg, J.G., M. Kottelat, G.R. Smith, M.L.J. Stiassny, and A. C. Gill, 2000: So many fishes, so little time: an overview of recent ichthyological discovery in continental waters. *Annals of the Missouri Botanical Garden* **87**: 26–62.

Lundqvist, J. and P. Gleick, 2000: Comprehensive Assessment of the Freshwater Resources of the World: Sustaining our Waters into the 21st Century, Stockholm Environment Institute.

Lupi, F., M.D. Kaplowitz and J.P. Hoehn, 2002: The economic equivalency of drained and restored wetlands in Michigan. *American Journal of Agricultural Economics* **84,** 2355–1361.

Maitland, P.S. and A.J. Crivelli, 1996: *Conservation of Freshwater Fish,* MedWet–Tour du Valat, Arles (France), 94 pp.

Malmqvist, B. and S. Rundle, 2002: Threats to the running water ecosystems of the world. *Environmental Conservation* 29 (2): 134–153.

Maltby, E., C.P. Immirzi and R.J. Safford (eds) 1996: Tropical Lowland Peatlands of Southeast Asia. IUCN, Gland, Switzerland.

Maltchik, L., 2003: Three new wetlands inventories in Brazil. In: *Interciancia* Vol: **28 (7):** 421–423.

Manchester, S.J. & J.M. Bullock, 2000: The impacts of non-native species on UK biodiveristy and the effectiveness of control. *Journal of Applied Ecology* 37: 845–864.

Master, L. L., S. R. Flack, and B. A. Stein (eds), 1998: Rivers of Life: Critical Watersheds for Protecting Freshwater Biodiversity. The Nature Conservancy, Arlington, Virginia, U.S.A.

Mathiessen, P. 2000: Is endocrine disruption an significant ecological issue? *Ecotoxicology* 9, 21–24.

Mathiessen, P.and J.P. Sumpter, 1998: Effects of estrogenic substances in the aquatic environment. In *Fish Ecotoxicology,* T. Braunbeck, D. Hinton and B. Streit (eds), Basel, Switzerland: Birkhauser Verlag. p. 319–335.

Mattson, N.S., K. Buakhamvongsa, N. Sukamasavin, T. Nguyen and V. Ouk. 2002: Cambodia Mekong giant fish species: on their management and biology. MRC Technical Paper No.3. Mekong River Commission, Phnom Penh. 29 p.

McAllister, D.E., A.L. Hamilton and B. Harvey, 1997: Global freshwater biodiversity:striving for the integrity of freshwater ecosystems. *Sea Wind— Bulletin of Ocean Voice International* **11(3)**: 1–140.

McComb, A. J., and J.A. Davis (eds), 1998: *Wetlands for the Future.* (Gleneagles Publishing, Adelaide.)

McNeely, J.A., H.A. Mooney, L.E. Neville, P.J. Schei, and J.K. Waage (eds.), 2001: *Global strategy on invasive alien species.* IUCN, Cambridge, U.K., in collaboration with the Global Invasive Species Programme. 48 pp.

Meybeck, M. 2003: Global analysis of river systems: from Earth system controls to Anthropocene syndromes. Philosophical Transactions of the Royal Society London, B. **358**, 1935–1955.

MRC (Mekong River Commission), 1997: *Greater Mekong Sub-Region State of the Environment Report.* Bangkok, Thailand: Mekong River Commission and United Nations Environment Programme.

Miller, R. R., J. D. Williams and J. E. Williams, 1989: Extinctions of North American Fishes During the Past Century. *Fisheries* 14 (6): 22–38.

Mitsch, W. J., R.H. Mitsch, and R.E. Turner (eds), 1994: Wetlands of the Old and New Worlds: ecology and management. In: *Global Wetlands: Old World and New World,* W. J. Mitsch (ed.) pp.3–56, Elsevier, Amsterdam.

Mitsch, J. and X. Wu, 1995: Wetlands and global change. In: *Advances in Soil Science, Soil Management and Greenhouse Effect ,* Lal, R., J. Kimble, E. Levine, B.A. Stewart (eds.), CRC Lewis Publishers, Boca Raton, FL, pp. 205–230.

Mitsch, W.J., 1998: Protecting the world's wetlands: threats and opportunities in the 21st century. In: *Wetlands for the Future,* A.J. McComb and J.A. Davis (ed.), Contributions from INTECOL's V International Wetlands Conference, Perth, Australia, Gleneagles Press, Adelaide, SA, Australia, pp. 19–31.

Mitsch, W.J. (ed.) 1994: Global Wetlands: Old World and New. Elsevier, Amsterdam, The Netherlands. 967 pp.

Mitsch, W. J. and J.G. Gosselink, 2000: Wetlands. 3rd edn. John Wiley, New York, U.S.A. 920 pp.

Miyabayashi, Y. and T. Mundkur, 1999: *Atlas of key sites for Anatidae in the East Asian Flyway.* Wetlands International—Japan, Tokyo and Wetlands International—Asia Pacific, Kuala Lumpur, Malaysia. 148 pp.

Mooney, H.A. and R.J. Hobbs (eds), 2000: *Invasive species in a changing world.* Island Press, Washington D.C. 457 pp.

Moore, H. and E. Wiken, 1998: *Wetland Conservation Analysis.* Canadian Council on Ecological Areas (CCEA) Technical Report. Ottawa, Ontario.16 pgs.

Morrison, R.I.A.G., Y. Aubry, R. Butler, G.W. Beyersbergen, G.M. Donaldson, C.L. Grotto-Trevor, P.W. Hicklin, V.H. Johnson and R.K. Ross, 2001: Declines in North America shorebird populations. *Wader Study Group Bulletin* **94**: 34–38.

Moser, M., R.C. Prentice and J. van Vessem (eds), 1993: *Waterfowl and wetland conservation in the 1990s: A global perspective.* Proceedings of an IWRB Symposium, St Petersburg Beach, Florida, USA, 12–19 November 1992. IWRB Special Publication No 26, Slimbridge, UK.

Moser, M., C. Prentice and S. Frazier, 1996: A global overview of wetland loss and degradation. In: *Papers, Technical Session B, Vol. 10/12B, Proceedings of the 6th Meeting of the Conference of Contracting Parties, Brisbane, Australia, 19.27 March 1996.* pp. 21–31, Ramsar Convention Bureau: Gland, Switzerland.

Moyle, P.B. and R.A. Leidy, 1992: Loss of Biodiversity in Aquatic Ecosystems: Evidence from Fish Faunas. In: *Conservation Biology: The Theory and Practice of Nature Conservation, Preservation and Management,* P.L. Fiedler and S.K. Jain (eds.), New York. New York: Chapman and Hall. pp. 127–169

Mpofu, B., 2000: *Assessment of Seed Requirements in Southern African Countries Ravaged by Floods and Drought 1999/2000.* SADC Food Security Programme, Food, Agriculture and Natural Resources. [online] Available at http://www.sadc-fanr.org.zw/sssd/mozcalrep.htm

National Wetlands Working Group (NWWG), 1988: *Wetlands of Canada. Ecological Land Classification Series, No.24.* Canada Committee on Ecological Land Classification; Sustainable Development Branch, Environment Canada. ISBN No. 0–921317–13–1. Polyscience Publications. Montreal, Quebec 452 pgs.

Naylor, R.L., R.J. Goldburg, J.H. Primavera, N.Kautsky, M.C.M. Beveridge, J. Clay, C. Folke, J. Lubchenco, H. Mooney and M. Troell., 2000: "Effect of Aquaculture on World World Fish Supplies." *Nature* 405: 1017–1024.

Nichols, S. and R.C. Lathrop, 1994: Cultural impacts on macrophytes in the Yahara lakes since the late 1800s. *Aquatic Botany* 47: 225–247.

Nilsson, C. and K. Berggren, 2000: Alterations of riparian ecosystems caused by river regulation. *BioScience* **50(9)**: 783–792.

Ntakimazi, G. 1992: Conservation of the resources of the African Great Lakes: why? An overview. *Mitteilungen Internationale Vereinigung für Limnologie* 23, 5–9.

OECD (Organisation for Economic Co-operation and Development), 1996: *Guidelines for aid agencies for improved conservation and sustainable use of tropical and subtropical wetlands.* Paris, France.

Ogutu-Ohwayo, R. 1999: The Nile perch in Lake Victoria: the balance between benefits and negative impacts of aliens. In *Invasive species and biodiversity management,* O.T. Sandlund, P.J. Schei and A. Viken (eds.). Kluwer Academic Publishers, Dordrecht, The Netherlands. P. 43–67.

Ortiz G., 2001: *Administración del Agua. Aplicación de Instrumentos de Política Hidráulica*: En Escenarios Alternativos. Instituto Mexicano de Tecnología del Agua, México; 227 p.

Page, S.E., F. Siegert, J. O'Reiley, H-D von Boehem, A. Jaya and S. Limin, 1997: The amount of carbon released from peat and forest fires in Indonesia during 1997. *Nature* **420**, 61–65.

Palmer, M. A., A.P. Covich, B.J. Finlay, J. Gilbert, K.D. Hyde, R.K. Johnson, T. Kairesalo, S. Lake, C.R. Lovell, R.J. Naiman, C. Ricci, F. Sabater, and D. Strayer, 1997: Biodiversity and ecosystem processes in freshwater sediments. *Ambio* **26(8)**: 571–577.

Papayannis, T., 2002: *Regional action for wetlands: the Mediterranean experience,* MedWet/Tour du Valat, Arles (France), 96 p.

Partow, H., 2001: *The Mesopotamian Marshlands: demise of an ecosystem.* UNEP, Nairobi, Kenya. 46 pp.

Peberdy, Kevin, 1999: *The Wetland Centre—A Showcase.* Proceedings of the *"WATERFOWL" INFORMATION NETWORK* International Conference, The Royal Veterinary College, London, 16 & 17 September 1999.

Pittock, B., D. Wratt *et al.,* 2001: Chapter 12. Australia and New Zealand. In: *Climate Change 2001: Impacts, Adaptations, and Vulnerability.* Contribution of Working Group II to the Thirds Assessment Report of the International Panel on Climate Change. McCarthy, J.J., Canziani, O.F., Leary, N.A., Dokken, D.J., White, K.S. (eds). pp. 591–640. Cambridge University Publication Press.

Ponce, V.M., 1995: *Impacto Hidrológico e Ambiental da Hidrovia Paraná-Paraguai no Pantanal Matogrossense—Um Estudo de Referência.* San Diego, California, USA: San Diego State University.

Popov, V.1992: Agricultural landse and water pollution in the lower Volga. In Wetland Conservation and Management in the Lower Volga, Russia, Finlayson, C.M. (ed.), 18–22, IWRB Special Publication No 18, Slimbridge, UK.

Postel, S. and S. Carpenter, 1997: Freshwater Ecosystem Services In: *Nature's Services: Societal Dependence on Natural Ecosystems.* Island Press, Washington DC. pp. 195–214.

Postel S., 1997: *Last Oasis: Facing Water Scarcity,* 1997. W.W. Norton & Company, New York.

Postel, S., 1999: *Pillar of Sand: Can the Irrigation Miracle Last?* New York, W.W. Norton & Company, 313 pp.

Pringle, C.M., M.C. Freeman and B. J. Freeman, 2000: Regional effects of hydrologic alterations on riverine macrobiota in the New World: tropical-temperate comparisons. *BioScience* 50(9): 807–823.

Pye-Smith, C., 1995: Salvation from Sewage in Calcutta Marshes. *People & the Planet,* **4(1),** 20–22.

Qu, J., 1999: Environmental Protection Knowledge (China's Red Flag Publishing House, Beijing), p. 354 (in Chinese). In: *China's Forest Policy for the 21st Century*: P. Zhang, G. Shao, G. Zhao, D.C. Le Master, G.R. Parker, J.B. Dunning, Jr., and Q. Li, *Science* Vol. **288**, 2135–2136.

Rabuor, C.O. and J.J. Polovina, 1995: An analysis of the multigear, multispecies fishey in the Kenyan waters of Lake Victoria. *NAGE, the ICLARM Quarterly. April, 1995*: 34–37.

Ramsar Convention Secretariat, 2004: Ramsar Handbooks for the Wise Use of Wetlands. 2nd Edition. Ramsar Convention Secretariat, Gland, Switzerland.

Ratner, B.D., Dong Thanh Ha, Mam Kosal, Ayut Nissapa, Somphanh Chanphengxay 2004. Undervalued and Overlooked: Sustaining Rural Livelihoods through better Governance of Wetlands. Studies and Review Series, World Fish Centre, Penang, Malaysia.

Reuman, S.C. & D.D. Chiras, 2003: *The Population Explosion in The 2003 World Book Encyclopedia.* Published by World Book Inc. 22 volume set with 14,000 pages.

Revenga, C. and Y. Kura, 2003: *Status and Trends of Biodiversity of Inland Water Ecosystems.* Secretariat of the Convention on Biological Diversity, Montreal, Technical Series no. 11. 120 pp.

Revenga, C., J. Brunner, N. Henninger, K. Kassem and R. Payne, 2000: *Pilot Analysis of Global Ecosystems: Freshwater Systems,* World Resources Institute, Washington D.C. 83 pp

Revenga, C., S. Murray, J. Abramovitz, and A. Hammond, 1998: *Watersheds of the World: Ecological Value and Vulnerability.* World Resources Institute, Washington, DC, 200 pp.

Revenga, C., I. Campbell, R. Abell, P. de Villiers and M. Bryer, in press: Prospects for Monitoring Freshwater Ecosystems towards the 2010 targets. In *Beyond extinction rates: monitoring wild nature for the 2010 target,* Philosophical Transactions, Biological Sciences, the Royal Society, London, U.K.

Richardson, J. L. and M. J. Vepraskas (eds), 2001: Wetland Soils: Genesis, Hydrology, Landscapes, and Classification. Lewis Publishers, Boca Raton, Florida, U.S.A.

Rieley, J.O., A.A. Ahmad-Shah and M.A. Brady. 1996: The nature and extent of tropical peat swamps. In: *Tropical Lowland Peatlands of Southeast Asia.* Maltby, E., Immirzi, C.P. & Safford, R.J. (eds) IUCN, Gland, Switzerland.

Rieley, J.O. and S.E. Page (eds) 1997. Biodiversity and Sustainability of Tropical Peatlands. Samara Publishing Ltd, Cardigan, U.K.

Roshier, D.A., P.H. Whetton, R.J. Allan, and A.I. Robertson, 2001: "Distribution and persistence of temporary wetland habitats in arid Australia in relation to climate". *Austral Ecology, 26,* 371–384.

Ross, S. T., 1991: Mechanisms Structuring Stream Fish Assemblages: Are There Lessons From Introduced Species? *Environmental Biology of Fishes* 30: 359–368.

Rowan, D. & Rasmussen, J. 1992: Why don't Great Lakes fish reflect environmental concentrations of organic contaminants?—an analysis of between lake variability in the ecological partitioning of PCBs and DDT. *Journal of Great Lakes Research* 18, 724–41.

Schuyt, K. and L. Brander 2004: The economic value of the World's wetlands. WWF Living Waters: Conserving the Source of Life. Gland, Switzerland. 31 pp.

Scott, D.A. and P.M. Rose, 1996: Atlas of Anatidae Populations in Africa and Western Eurasia. Wetlands International Publication No. 41. Wetlands International, Wageningen, The Netherlands.

Sculthorpe, C.D., 1967: *The biology of aquatic vascular plants.* Edward Arnold, London.

Seyam, I.M., A.Y. Hoekstra, G.S. Ngabirano and H.H.G. Savenije, 2001: The value of freshwater wetlands in the Zambezi basin. Value of Water Research Report Series No. 7 IHE Delft, The Netherlands.

Spiers, A.G., 1999: Review of international/continental wetland resources. In: *Global Review of Wetland Resources and Priorities for Inventory.* C. M. Finlayson and A. G. Spiers (eds) Supervising Scientist Report No. 144, 63.104. Canberra, Australia.

State Committee on Ecology and Control of Natural Resources Utilization 1998: National Environmental Action Plan, Azerbaijan Republic. Baku, Azerbaijan.

Stroud, D.A., N.C. Davidson, R. West, D.A. Scott, L. Hanstra, O. Thorup, B. Ganter and S. Delany (compilers) 2004: Status of migratory wader populations in Africa and Western Eurasia in the 1990s. *International Wader Studies* **15**: 1–259

Sverdrup-Jensen, S., 2002: *Fisheries in the Lower Mekong Basin: Status and perspectives.* MRC Technical Paper No. 6, Mekong River Commission, Phnom Penh. 103 pp. ISSN: 1683–148.

Thomsen, J.B., 1999: Looking for hotspots. World Conservation, 2, 6–7. [online] Available at: www.iucn.org/bookstore/bulletin/1999/wc2/content/hotspots.pdf.

Tockner, K. and J.A. Stanford, 2002: Riverine flood plains: present state and future trends. *Environmental Conservation* 29 (3): 308–330.

Uetz, P. and T. Etzold, 1996: The EMBL/EBI Reptile Database, *Herpetological Review* **27 (4)**

UNDHA (United Nations Department of Humanitarian Affairs), 1994: *First African Sub-Regional Workshop on Natural Disaster Reduction,* Gaborone, 28 November to 2 December 1994. Gaborone.

UNDP/GEF/GOC (United Nations Development Programme, Global Environment Facility, and Government of China), 2003: Management Plan for Ruoergai National Nature Reserve. UNDP/GEF/GOC Wetland Biodiversity Conservation and Sustainable Use in China, January 2003.

UNEP (United Nations Environment Programme), 1996: *Groundwater: A Threatened Resource.* UNEP Environment Library No. 15, Nairobi, Kenya.

UNEP, 1999: *GEO-2000.* United Nations Environment Programme. London and New York, Earthscan.

UNEP, 2002: *Vital Water Graphics—An Overview of the State of the World's Fresh and Marine Waters.* UNEP, Nairobi, Kenya. ISBN: 92–807–2236–0.

UNEP, 2003: *Post-Conflict Environmental Assessment: Afghanistan.* Report by the United Nations Environmental Programme. ISBN 92–1–158617–8

UNESCO (United Nations Educational, Scientific, and Cultural Organization), 2000: *Water Related Vision for the Aral Sea Basin for the Year 2025.* Paris, France: UNESCO.

U.S. Fish and Wildlife Service, 2004: *Waterfowl population status 2004.* Department of the Interior, Washington, D.C. 54 pp.

Valette-Silver, N. N. J. 1993: The use of sediment cores to reconstruct historical trends in contamination of estuarine and coastal sediments. Estuaries **16,** 577–588.

van Dam, R.A., Walden, D.J. and Begg, G.W. 2002a: A preliminary risk assessment of cane toads in Kakadu National Park. Supervising Scientist Report 164, Supervising Scientist, Darwin, Australia.

van Dam, R., Walden, D., Begg, G. and Finlayson., M 2000: Ecological Risk Assessment of the Cane Toad, Bufo marinus, in Karadu National Park, Australia. In *Assessment and management of alien species that threaten ecosystems, habitats and species.* Abstracts of keynote addresses and posters presented at the sixth meeting of the Subsidiary Body on Scientific, Technical and Technological Advice, held in Montreal, Canada, from 12 to 16 March 2001. Montreal ,CBD Technical Paper no. 1. p. 21–24.

van Dam, R., H. Gitay, M. Finlayson, N.J. Davidson and B. Orlando, 2002b: *Climate Change and Wetlands: Impacts, Adaptation and Mitigation.* Background document (DOC.SC26/COP8–4) prepared by for the 26th Meeting of the Standing Committee of the Convention on Wetlands (Ramsar Convention) held in Gland, Switzerland, 3–7 December 2002.

van Dijk, P.P., B. L. Stuart, and A. G.J. Rhodin, 2000: Asian Turtle Trade: Proceedings of a Workshop on Conservation and Trade of Freshwater Turtles and Tortoises in Asia Chelonian Research Monographs, No. 2, Chelonian Research Foundation in association with WCS, TRAFFIC, WWF, Kadoorie Farm and Botanic Gardens US Fish and Wildlife Service. Chelonian Research Foundation Lunenburg, Massachusetts, U.S.A. 164pp.

Vörösmarty, C.J., K.P. Sharma, B.M. Fekete, A.H. Copeland, J. Holden, J. Marble, and J.A. Lough, 1997: The Storage and Aging of Continental Runoff in Large Reservoir Systems of the World. *Ambio* 26(4): 210–219.

Walden, D., R. van Dam, M. Finlayson, M. Storrs, J. Lowry and D. Kriticos 2004: *A risk assessment of the tropical wetland weed* Mimosa pigra *in northern Australia.* Supervising Scientist Report 177, Supervising Scientist, Darwin NT.

Walther, G.R., E. Post, P. Convey, A. Menzel, C. Parmesan, T.J.C. Beebeef, J.M. Fromentine, O. Hoegh-Guldberg and F. Bairlein, 2002: Ecological responses to recent climate change. *Nature* **416,** 388–395.

Warner, B. G. and Rubec, C.D.A. (eds) 1997: *The Canadian Wetland Classification System,* Second Edition. National Wetlands Working Group. Wetlands Research Centre, University of Waterloo. Waterloo, Ontario. 68 pp.

Wetlands International, 1999: *Atlas of Key Sites for Cranes in the North east Asian Flyway.* Simba Chan (ed).

Wetlands International, 2002: *Waterbird Population Estimates. 3rd Edition.* Consultation Draft available on-line at: http://www.wetlands.agro.nl.

Whigham, D.F., RE Goode and J. Kvet (eds) 1993: *Wetland Ecology and Management—Case Studies,* Kluwer Academic Publishers, Dordrecht, The Netherlands.

White, R.P., C. Revenga, J. Nackoney and E. Hoshino 2004: *Wetlands , drylands, and ecosystem services.* Report Prepared by World Resources Institute for the Millennium Ecosystem Assessment, 43 p.

Wiken, Ed B., D. Gauthier, I. Marshall, K. Lawton, and H. Hirvonen, 1996: *A perspective on Canada's ecosystems. An overview of terrestrial and marine ecozones.* Canadian Council on Ecological Areas. Occasional paper 14.

Williams, W.D., 2002: Environmental threats to salt lakes and the likely status of inland saline Ecosystems in 2025. *Environmental Conservation* 29 (2): 154–167

Williamson, M., 1996: *Biological invasions.* London: Champan & Hall.

Wiseman, R., D. Taylor and H. Zingstra (eds) 2003: Proceedings of the workshop on agriculture, wetlands and water resources: 17th Global Biodiversity Forum, Valencia, Spain. November 2002. National Institute of Ecology and International Scientific Publications, New Delhi, India. 122 p.

Witte, F., T. Goldschmidt, J. Wanink, M. van Oijen, K. Goudswaard, E. Witte-Mass and N. Bouton, 1992: The Destruction of an Endemic Species Flock: Quantitative Data on the Decline of the Haplochromine Cichlids of Lake Victoria. *Environmental Biology of Fishes* 34:1–28.

Wittenberg, R. and M.J.W. Cock (eds), 2001: *Invasive alien species: a toolkit of best prevention and management practices.* CAB International, Wallingford, UK. 228 pp.

Wooster, M.J. and N. Strub, 2002: Study of the 1997 Borneo fires: quantitative analysis using global area coverage (GAC) satellite data. *Global Biogeochemical Cycles* 16.

WCD (World Commission on Dams), 2000: *Dams and Development: A new Framework for Decision-making.* The Report of the World Commission on Dams. London, U.K.: Earthscan Publications Ltd.

World Energy Council, 2003: *The Hydroelectric Power Option in Brazil Environmental, Technological and Economic Aspects.* London, U.K. URL: www.world-energy.org

WMO (World Meteorological Organization), 1997: *Comprehensive Assessment of the Freshwater Resources of the World.* Stockholm, Sweden: WMO and Stockholm Environment Institute.

WRI (World Resources Institute), United Nations Development Programme, United Nations Environment Programme, and World Bank, 2000. *World Resources 2000–2001: People and Ecosystems: The Fraying Web of Life.* World Resources Institute, Washington DC.

WWDR (World Water Development Report) 2003: *Water for People, Water for Life.* United Nations World Water Assessment Programme. UNESCO/Berhahn Boks, Paris, France, 575 pp. Available at www.unesco.org/water/wwap/wwdr/index.shtmlcheck publisher – web

WWF (World Wide Fund for Nature), 2002: *The Ecological Effects of Mining Spills in the Tisza River System in 2000.* WWF International Danube-Carpathian Programme, Vienna, Austria.

Xie, Y., Z. Li, W.P. Gregg and D. Li, 2001: Invasive species in China—an overview. *Biodiversity and Conservation* 10: 1317–1341.

Zalidis, G.C., T.L. Crissman and P.A. Gerakis (eds), 2002: *Restoration of Mediterranean Wetlands,* Hellenic Ministry of the Environment, Physical Planning and Public Works, Athens, and Greek Biotope/Wetland Centre, Thermi, Greece. Pp. 213–228.

Zockler, C. and I. Lysenko, 2000. *Water Birds on the Edge. First circumpolar assessment of climate change impact on Arctic breeding water birds.* WCMC Biodiversity Series No. 11, World Conservation Monitoring Centre—World Conservation Press, Cambridge, U.K.

Chapter 21

Forest and Woodland Systems

Coordinating Lead Authors: Anatoly Shvidenko, Charles Victor Barber, Reidar Persson
Lead Authors: Patrick Gonzalez, Rashid Hassan, Petro Lakyda, Ian McCallum, Sten Nilsson, Juan Pulhin, Bernardt van Rosenburg, Bob Scholes
Review Editors: Marian de los Angeles, Cherla Sastry

BOXES

FIGURES

TABLES

---

*This appears in Appendix A at the end of this volume.

Main Messages

Forest ecosystems are extremely important refuges for terrestrial biodiversity, a central component of Earth's biogeochemical systems, and a source of ecosystem services essential for human well-being. The area and condition of the world's forests has, however, declined throughout recent human history. In the last three centuries, global forest area has been reduced by approximately 40%, with three quarters of this loss occurring during the last two centuries. Forests have completely disappeared in 25 countries, and another 29 countries have lost more than 90% of their forest cover. Although forest cover and biomass in Europe and North America are currently increasing following radical declines in the past, deforestation of natural forests in the tropics continues at an annual rate of over 10 million hectares per year—an area larger than Greece, Nicaragua, or Nepal and more than four times the size of Belgium. Moreover, degradation and fragmentation of many remaining forests are further impairing ecosystem functioning.

Information about the world's forest is limited and unevenly distributed. The *Global Forest Resources Assessment 2000* (FRA-2000) done by the Food and Agriculture Organization of the United Nations reports that only 22 out of 137 developing countries possess a series of time-series inventories, 28 countries have no inventory, and 33 have only partial inventories. Further, 34 countries only have national forest inventories from before 1990, while only 43 have inventories completed after 1990. More than half the inventories used to compile FRA-2000 were either more than 10 years old or incomplete. Forest information is also inadequate—and sometimes statistically unreliable—for many industrial countries. Despite the proliferation of new remote sensing technologies, the reliability of remote sensing products remains uncertain.

Forests, particularly those in the tropics, provide habitat for half or more of the world's known terrestrial plant and animal species. This biodiversity is essential for the continued health and functioning of forest ecosystems, and it underlies the many ecosystem services that forests provide.

Forests and woodlands play a significant role in the global carbon cycle and, consequently, in accelerating or decelerating global climate change. Forests contain about 50% of the world's terrestrial organic carbon stocks, and forest biomass constitutes about 80% of terrestrial biomass. Forests contribute over two thirds of global terrestrial net primary production. Slowing forest loss and restoring forest cover in deforested areas could thus help mitigate climate change.

More than three quarters of the world's accessible freshwater comes from forested catchments. Water quality declines with decreases in forest condition and cover, and natural hazards such as floods, landslides, and soil erosion have larger impacts.

The provisioning services obtained from forests have substantial economic value. Forests annually provide over 3.3 billion cubic meters of wood (including 1.8 billion cubic meters of fuelwood and charcoal), as well as numerous non-wood forest products that play a significant role in the economic life of hundreds of millions of people. The combined economic value of "non-market" (social and ecological) forest services may exceed the recorded market value of timber, but these values are rarely taken into account in forest management decisions.

The rural poor are particularly dependent on forest resources. As many as 300 million people, most of them very poor, depend substantially on forest ecosystems for their subsistence and survival. The 60 million indigenous people who live in forest areas are especially dependent on forest resources and the health of forest ecosystems. Although use of forest resources on its own is often insufficient to promote poverty alleviation, forest loss and degradation has significant negative consequences on human well-being.

Forests play important cultural, spiritual, and recreational roles in many societies. For many indigenous and otherwise traditional societies, forests play an important role in cultural and spiritual traditions and, in some cases, are integral to the very definition and survival of distinct cultures and peoples. Forests also continue to play an important role in providing recreation and spiritual solace in more modernized, secular societies, and forests and trees are symbolically and spiritually important in most of the world's major religious traditions.

Forest loss and degradation are driven by a combination of economic, political, and institutional factors. The main direct drivers of tropical deforestation are agricultural expansion, high levels of wood extraction, and the extension of roads and other infrastructure into forested areas. Indirect drivers include increasing economic activity and associated market failures, a wide range of policy and institutional weaknesses and failures, the impacts of technological change, low public awareness of forest values, and human demographic factors such as population growth, density, and migration. While temperate and boreal forest cover has stabilized and even increased, the quality of these forests is still threatened by air pollution, fire, pest and disease outbreaks, continued fragmentation, and inadequate management. Climate change threatens forests in all biomes.

Many forests are used almost to their full potential to provide fiber and fuel. By 2020, demand for wood and woodfuel is expected to grow considerably. This growth in demand is likely to stimulate the establishment of more industrial plantations, more-careful management of natural forests, and technological improvements in the efficiency of wood use. However, the establishment of plantations often results in trade-offs with services other than fiber production and with biodiversity.

Many developing countries have not effectively used forest resources in support of development efforts. Widespread corruption in the forestry sector has resulted in valuable forest resources frequently being seized and controlled by political and economic elites. The poor have often seen access to forest resources diminish and have not widely shared in the benefits of forest resource exploitation.

The paradigm of sustainable forest management has been widely embraced at national and international policy levels, but it has not yet been implemented to the point where it is appreciably mitigating the negative trends affecting the world's forests. SFM provides an increasingly sophisticated set of policies and tools for setting forest management on a more sustainable trajectory. Implementing SFM, however, requires overcoming many of the same economic, political, and institutional hurdles that drive deforestation and forest degradation. In addition, forest management would benefit from anticipating and incorporating resilience to the present and likely future impacts of climate change on forest ecosystems. Past policies aspired to control change in forest ecosystems assumed to be stable. The new imperative is to develop policies to manage the capacity of forest to cope with, adapt to, and shape changes. Responding to this imperative requires new information and new knowledge, including advanced science and technology, more effective national and global systems for forest inventory and monitoring, involvement of people in decision-making about forest management and use, and strengthened dialogue and cooperation with decision-makers in other sectors.

21.1 Introduction

Forest ecosystems, which for the purposes of this chapter include woodlands with an interrupted tree canopy, serve important ecological functions and provide wood and numerous other products that contribute significantly to human well-being at local, national, and global levels. The diverse ecosystem services provided by forests include the conservation of soil and water resources, positive influences on local climate, the mitigation of global climate change, the conservation of biological diversity, improvement of urban and peri-urban living conditions, the protection of natural and cultural heritage, subsistence resources for many rural and indigenous communities, the generation of employment, and recreational opportunities. Research indicates that forests supply about 5,000 different commercial products (Chiras 1998), and the forestry sector contributes about 2% of global GDP (FAO 1997). The centrality of forests for humanity has been acknowledged internationally in recent environmental agreements and processes including the United Nations Framework Convention on Climate Change, the Convention on Biological Diversity, the United Nations Convention to Combat Desertification, and the United Nations Forum on Forests.

While it is clear that the value of forest ecosystem services is very high, there are many gaps in scientific understanding and few practical solutions to reconciling the conflicts that arise from the competing values that different user groups ascribe to different forest services. Interests of landowners, local communities, governments, and the private sector vary and frequently conflict in both spatial and temporal terms. The time horizon for using individual forest services is substantially different, for example, for forest-dependent indigenous communities and large logging companies.

About 8,000 years ago, forest covered an estimated 6.2 billion hectares of the planet—about 47% of Earth's land surface (Billington et al. 1996). Peoples of the preagricultural era likely had significant impacts on these forest ecosystems. Some aboriginal tribes are thought to have caused numerous extinctions (such as of North American and Australian megafauna) (Flannery 1994, 2001), and forest has been actively cleared and manipulated in composition through fire and other means for thousands of years (Williams 2003).

From today's perspective, however, preagricultural impacts on overall forest cover appear to have been slight. Since that time, the planet has lost about 40% of its original forest (*high certainty*), and the remaining forests have suffered varying degrees of fragmentation and degradation (Bryant et al. 1997; Matthews et al. 2000; Ball 2001; Wade et al. 2003). Most of this loss has occurred during the industrial age, particularly during the last two centuries, and in some cases much more recently. Some analyses have yielded substantially smaller estimates. Richards (1990), for example, estimates global loss of forests to have been only about 20%.

Much of the progress of human civilization has been made possible by the conversion of some forest areas to other uses, particularly for agricultural expansion. However, this process has resulted in many trade-offs with forest ecosystem services, many of which have not been recognized.

Extensive biodiversity loss—including losses of genetic, species, and habitat diversity—has been one result of the shrinking of the world's forests (Myers 1996; McNeely et al. 1995; Reid and Miller 1989). There is also evidence that forest genetic resources as a whole have declined in quality, especially in areas where high-quality timber has been selectively extracted (Rodgers 1997; Kemp and Palmberg-Lerche 1994). Forest loss has also had a negative impact on the provision of ecosystem services, such as regulation of hydrological cycles. The negative impacts of deforestation appear most directly at the local level, where communities lose access to timber, fuelwood, and bushmeat or suffer increased flooding and landslides. Impacts have also been felt on a

much larger scale as well. The widespread salinization of land and rivers in Australia, for example, is the result of extensive woodland clearing and the subsequent introduction of European agriculture (McFarlane et al. 1992; MDBC 1999).

Public awareness of the importance of forests and public concern over forest loss has grown substantially over the past several decades. Numerous international bodies such as the World Commission on Environment and Development (WCED 1987) and the World Commission on Forests and Sustainable Development (WCFSD 1999) have voiced concern about the deepening forest crisis, and the theory and practice of making the transition to "sustainable forest management" is a topic of intensive national and international debate. (See Box 21.1.)

The forest issue was a contentious topic at the 1992 U.N. Conference on Environment and Development, which sought—and ultimately failed—to reach agreement on an international convention on forests. Instead, UNCED adopted a nonbinding set of "Forest Principles," which gave life to a series of U.N. forest initiatives: the Intergovernmental Panel on Forests (1995–97), the Intergovernmental Forum on Forests (1997–2000), the United Nations Forum on Forests (2000–05), and the Collaborative Partnership on Forests (2000). These bodies have provided an important international "soft law" forum to debate global forest policy and have catalyzed a considerable amount of technical work on forest management and policy. It is nevertheless unclear to what extent this international forest policy architecture influences the government and private-sector decisions that actually affect forests on the ground (Chaytor 2001; Bass 2003). Current international and national processes addressing forest management are discussed in detail in MA *Policy Responses,* Chapter 8.

Reliable and comprehensive data and information are essential for determining forest conditions and trends and for development of national and international forest policies. As a whole, information on the world's forests has improved over the past few decades. This is partly a result of the emergence of new technologies such as remote sensing, but it is also due to improving data collection in some countries and to the efforts of scientific researchers and international institutions.

FAO holds the mandate within the U.N. system to compile, analyze, and supply global forest information, and the organization has steadily improved its capacities in this regard, resulting most recently in the *Global Forest Resources Assessment 2000,* discussed at length in this chapter. The UNEP World Conservation Monitoring Centre is another institution that has developed extensive information sources on the conservation aspects of forests. National reporting requirements under environmental conventions such as UNFCCC, CBD, and UNCCD are also a relatively new source of forest information. Many important contributions have been made by the scientific community as well (see, e.g., Wade et al. 2003). Finally, a number of international nongovernmental organizations such as IUCN–World Conservation Union, the World Wide Fund for Nature, and the World Resources Institute have developed their capacities to compile, analyze, and disseminate high-quality forest information.

Despite this progress, available information on the world's forests still contains many gaps and shortcomings. National data for some countries are not reliable, and the overall state of knowledge about the condition and trends of forests in many regions is incomplete. In addition, improvements in forest information have not been accompanied by effective sustainable forest management. Instead, increasing information capacities have provided an ever-more detailed picture of forest decline and its impacts on human well-being.

BOX 21.1
Defining and Measuring "Sustainable Forest Management"

It is well established that "sustainability" means satisfying present needs without compromising future options, but it is not obvious what this means in practical terms for forest management. The concept of "sustained yield" forest management for timber—based on the concept of equilibrium between growth and timber harvest that can be sustained in perpetuity (Thang 2003)—has evolved in line with the broader view of sustainable development articulated by the World Commission on Environment and Development (WCED 1987) and endorsed by the 1992 UNCED Forest Principles, subsequent international processes, as well as many national forest policies. The MA thus differentiates "sustained yield management"—the management and yield of an individual resource or ecosystem service—and "sustainable management," which refers to the goal of "ensuring that a wide range of services from a particular ecosystem is sustained" (MA 2003).

Although sustainable forest management is now widely accepted as the overriding objective for forest policy and practice, it is not easy to define. The problem is that "what is defined as sustainable forestry will vary greatly over space and time as society's needs and perceptions evolve." The Center for International Forestry Research therefore adopted a broad definition in which sustainable forest management means "maintaining or enhancing the contribution of forests to human well-being, both of present and future generations, without compromising their ecosystem integrity, i.e., their resilience, function and biological diversity" (Sayer et al. 1997). Further specification can only be accomplished through the elaboration of SFM criteria and both quantitative and qualitative indicators to measure progress in meeting those criteria. Accordingly, the Intergovernmental Panel/Forum on Forests process identified the development of SFM criteria and indicators as a high priority for international and national action, and nine regional processes involving 149 countries have been launched since 1992 to develop and implement SFM C&I (ECOSOC 2004).

Each of these processes is developing its own distinctive set of C&I to measure progress toward SFM in particular regions and forest biomes (Anonymous 1994, 1995; ITTO 1998; CCFM 2003). CIFOR, meanwhile, has developed a "C&I Toolbox" for the forest management unit level, which includes a generic C&I template (CIFOR C&I Team 1999). The template elaborates C&I within the framework of six SFM objectives:

- policy, planning, and institutional framework are conductive to sustainable forest management;
- ecosystem integrity is maintained;
- forest management maintains or enhances fair intergenerational access to resources and economic benefits;
- concerned stakeholders have acknowledged rights and means to manage forests cooperatively and equitably;
- the health of the forest actors, cultures, and the forest is acceptable to all stakeholders; and
- yield and quality of forest goods and services are sustainable.

The experience of CIFOR and the many regional processes attempting to develop and implement operational SFM C&I make two things clear. First, there is no one, neat definition of SFM that can be applied everywhere, although there are a number of core common elements. Second, "SFM is to a great extent a social issue. . . . In other words, forest policy must be part of comprehensive economic policy as expressed through agricultural policy, land use and population policy, tax codes, forest and recycling policy and other approaches to managing demand and supply" (Funston 1995; see also Folke et al. 2002). Kaimowitz (2003) notes that past efforts to promote sustainable forest management did not focus enough on macroeconomic, agricultural, infrastructure, finance, and energy policies that slowed progress in implementation of the SFM paradigm. In this respect, Canadian initiatives on SFM (such as partnership in forests, models forests, and so on) are a promising tool for implementing sustainable forest management (e.g., Collate 2003; Weaver 2003).

Despite negative trends in some regions, the world's forests still demonstrate considerable vitality and resilience and retain the potential to meet growing human needs—if, that is, they are managed more sustainably. The recent history of boreal forests, for example, has demonstrated their strong regeneration capacity despite high levels of natural and anthropogenic disturbance. The total area of closed forests in Russia has registered a net increase of about 80 million hectares over the past 40 years, even though about 55 million hectares were clear-cut during this period (Shvidenko and Nilsson 2002). Studies also show that tropical forests can regenerate when agriculture in an area is abandoned (such as around the ancient city of Angkor in Cambodia, on Mexico's Yucatan peninsula, and in old sugarcane fields in Venezuela) (Richards 1996; Hamilton 1976).

A number of countries and regions have undergone periods of extensive forest loss but then developed solid legislative, economic, and social backgrounds for the transition to sustainable forest management. For example, Europe lost 50–70% of its original forest cover, mostly during the early Middle Ages, and North America lost about 30%, mostly in the nineteenth century (WRI et al. 1996; Chiraz 1998; UNECE 1996). Forest policies and economic development in the twentieth century in these regions, however, have encouraged forest restoration and plantation development, restoring a significant part of the forest cover in both Europe and North America. Yet many forests in these regions

continue to decline in quality, are becoming increasingly fragmented, and suffer the impacts of industrial pollution.

In many parts of the developing world, deforestation continues to accelerate in tandem with poverty and high levels of population growth. For these regions, the transition to sustainable forest management is a much greater challenge. And the stakes for the global community are much higher: if tropical developing countries must wait until they reach the levels of economic development—and deforestation—of Europe before making this transition, a large percentage of known terrestrial species may become extinct in the meantime due to the disproportionate number of species found within their forests (Rodrigues et al. 2003).

This chapter assesses the condition of the world's forest ecosystems and trends in the services they provide for human well-being. It begins with a presentation of some key definitions and a brief discussion of some of the data and methodological issues the authors confronted in compiling and assessing the information presented. The chapter then reviews forest and woodland extent, condition, and changes. Subsequent sections assess the services provided by forest ecosystems and the direct and indirect drivers of changes in forest and woodland cover and condition. Finally, the chapter reviews the implications of these changes for human well-being. This chapter should be read in conjunction with Chapter 9 of this volume, on timber, fuel, and fiber, and Chapter

8 of the *Policy Responses* volume, on wood, fuelwood, and non-wood forest products.

21.2 Definitions, Methods, and Data Sources

The choices of definitions, methods, and data sources made for this chapter have a profound influence on the presentation of statistics and findings on global forest conditions and trends. This section therefore discusses the choices made, the rationale behind them, and the strengths and limitations of the definitions, methods, and sources used. These should be borne in mind when reviewing the data presented in subsequent sections.

21.2.1 Definitions of Forest and Woodland

There is no single, agreed definition of "forest," due to varying climatic, social, economic, and historical conditions. The situation is complicated by the fact that for many governments, "forest" denotes a legal classification of areas that may or may not actually have tree cover.

A variety of definitions of forest are in use. For example, the *Global Biodiversity Outlook* (Secretariat of the Convention on Biological Diversity 2001) defines forests as "ecosystems in which trees are the predominant life forms" and notes that a more precise definition than this remains surprisingly elusive because trees occur in many different ecosystems, at different densities, and in different forms. Most definitions refer to canopy or crown cover, which is essentially the percentage of ground area shaded by the crowns of the trees when they are in full leaf. The U.N. Framework Convention on Climate Change process has adopted a nationally defined threshold of between 10% and 40% canopy closure. A number of remote sensing products of the last decade (MODIS, GLC-2000) have introduced other approaches (see edcddac.usgs.gov/glcc/glcc.html, glcf.umiacs.umd.edu/data/lanfcover/data.shtml, www.gvm.sai.jrc.it/glc2000/defaultGLC2000.htm, duckwater.bu.edu/lc/dataset). Estimates of forest or woodland area thus vary widely depending on the definitions used. The precise definitions employed should therefore be borne in mind when comparing forest cover data provided by different institutions.

This chapter mainly follows the definition of forest used by FAO's *Global Forest Resources Assessment 2000* (FAO 2000, 2001c). The FAO definition covers ecosystems that are dominated by trees (defined as perennial woody plants taller than 5 meters at maturity), where the tree crown cover (or equivalent stocking level) exceeds 10% and the area is larger than 0.5 hectares (FAO 2000, 2001b, 2001c). The term includes forests used for production, protection, multiple use, or conservation, as well as forest stands on agricultural land (such as windbreaks and shelterbelts of trees with a width of more than 20 meters) and plantations of different types. It also includes both naturally regenerating and planted forests. The term excludes stands of trees established primarily for agricultural production, such as fruit tree plantations, and trees planted in agroforestry systems (but rubber and cork oak stands are included). Billions of trees outside the forest in cities, along roads and rivers, on farms, and so on are not included in the two categories just described.

The threshold of 10% is crucial in this definition. In many countries, "forest" is typically defined as areas with substantially higher levels of canopy closure, for example 30–40%, depending on age, in Russia (FFSR 1995) and 60% in South Africa (Scholes 2004). In the classification of forests introduced by UNEP-WCMC, all forest classes have a minimum threshold of 30% except for the class including sparse trees and woodlands, for which canopy closure is from 10% to 30% (UNEP-WCMC 2004). Another controversial feature of the FAO definition is its inclusion of "temporarily unstocked areas" (clear-cuts, burnt areas, and so on) as forest. This means that a country may have logged or burned most of its forest, but—unless it converts the area to another officially noted productive land use—it will appear to have retained the same forest area as before (WRM 2002; Wunder 2003). These definitional issues generate some problems with analysis of FRA-2000 data and the conclusions that flow from that analysis.

Nonetheless, the FAO definition has been adopted because it is the first consistent definition of forests to be applied globally. A global assessment such as this one obviously requires a consistent global definition of "forest" and a global dataset that adheres to that definition. The strengths and limitations of FRA-2000 are summarized in Box 21.2.

FRA-2000 defines "closed forests" as those with a canopy cover of more than 40% (and it is this class of forest that is incorporated into the system maps and analysis throughout this volume). "Open forests" have a canopy cover of between 10% and 40%. "Fragmented forests" (which are not quantitatively defined by FRA-2000) refer to mosaics of forest patches and non-forestland. Closed forests, open forests, and fragmented formerly closed forests, as a rule, are ecologically substantially different from one another.

In this chapter, "woodland" refers to the type of land cover characterized by trees and shrubs: "other wooded land." Other wooded land, or OWL, is defined by FRA-2000 as land with a tree crown cover (or equivalent stocking level) of 5–10% of trees able to reach a height of 5 meters at maturity, a crown cover of more than 10% of trees not able to reach a height of 5 meters at maturity (such as dwarf or stunted trees), or shrub and bush cover of more than 10%. OWL excludes areas with the tree, shrub, or bush cover just specified but of less than 0.5 hectares and width of 20 meters, as well as land predominantly used for agricultural practices (FAO 2000, 2001c). Trees growing in areas that do not meet the forest and OWL definitions are excluded (FAO 2001c). Such trees are included in assessments of "trees outside forests."

Plantations are defined by FRA-2000 as "forests established by planting or/and seeding in the process of afforestation or reforestation, and consisting of introduced species or, in some cases, indigenous species." There is a substantial difference between plantations in the tropics and those in temperate and boreal countries. Broadly, there are two different types of plantations: short-rotation, fast-growing species plantations (such as *Eucalyptus* and *Pinus*) and plantations of long-rotation, slow-growing species of valuable hardwoods. In the tropics, important hardwood plantation species include teak (*Tectona grandis*) and rosewood (*Dalbergia spp.*). Common hardwood plantation species in the temperate zone include oak (*Quercus spp.*), ash (*Fraxinus spp.*), poplar (*Populus spp.*), and walnut (*Juglans spp.*). This is not, however, a hard-and-fast distinction, since some medium-rotation softwood saw-log plantations (in South Africa, for instance, and New Zealand) also produce valuable timber.

21.2.2 Variations in National-level Forest Information

Thirty years ago, it was noted that "more is known about the surface of the moon than about how much of the world's surface is covered by forests and woodlands" (Persson 1974). Since then, the quantity and quality of available information has improved in some countries but has declined in others and overall remains inadequate. Information about the status of forest inventories in

---

BOX 21.2

Strengths and Limitations of the *Global Forest Resources Assessment 2000* (FAO 2001c; Matthews 2001; R. Persson, personal communication, 2004)

FRA-2000 is the most comprehensive, globally consistent assessment of global resources available and is the basis for the assessment presented in this chapter. The definitional and methodological choices made by FRA-2000, however, substantially affect the conclusions of the assessment and are therefore important to understand and take into account.

FRA-2000 presents new estimates of global forest cover in both 2000 and 1990. The *1990 FAO Global Forest Resources Assessment* (FRA-1990) used different crown-cover thresholds for industrial (20%) and developing (10%) countries that hindered consistent global analyses and comparisons. FRA-2000 uses a consistent threshold of 10% for all countries and has adjusted the FRA-1990 estimate of forest cover using the 10% global threshold as well. As a result, the FRA-2000 estimate of 1990 forest cover—the baseline from which changes in forest cover are calculated—has been revised upward, to 3.95 billion hectares from 3.44 billion hectares, a 15% increase over the original estimate made in 1990, with the biggest revisions occurring in industrial countries.

In many regions of the world, the use of 10% crown cover as a minimum threshold conflicts with scientific definitions of "forest" as a vegetation type as well as with traditional use and understanding of the term. While the need for a consistent global definition of "forest" is clear, the rationale for setting the threshold at such a low percentage is contested by many, and a number of other FRA methodological decisions remain questionable. The definition of plantations as "forest," for example, affects estimates of net forest loss in the tropics and obscures the actual rates of natural forest loss.

While most industrial countries have relatively good forest cover data, there are serious problems with the way forest cover data are reported in Canada and Russia. Because these two countries account for more than 65% of all forests in industrial countries, these national methodological inconsistencies skew results for the entire temperate and boreal region.

FRA-1990 used mathematical models to compensate for poor data availability in developing countries. The FRA-2000 analysis, however, is based on national forest inventory data supplemented by remote sensing information and expert opinion. While many national data used by FRA-2000 were obsolete or incomplete, the remote sensing survey used to supplement national data relied on images covering only 10% of total tropical forest area, focusing on the same randomly selected 117 sites surveyed in 1990. Deforestation, however, is not randomly distributed—it is highly concentrated along roads and rivers (Stokestad 2001)—and it is therefore arguable that a 10% sampling rate is insufficient to identify accurately how much forest survives intact and how much is being lost.

---

different countries can be found in the forest resources assessment publications of FAO (FAO 1982, 1993, 1995b, 1999a, 2000, 2001a, 2001b, 2001c). (The text of this section is largely based on Janz and Persson 2002.)

Most industrial countries have some kind of forest inventory. In 1990, 18 of 34 of these countries (containing 76% of the forest area in industrial countries) derived their forest area information from sampling-based national forest inventories, some of which were quite old. In the other 16 countries, information had been compiled by aggregating local inventories, which were usually carried out for forest management purposes (FAO 1995a). Such information contains unknown errors and is usually biased, as the aggregation method is known to produce significant underestimates of volumes and increment. A good example can be taken from Germany, where at the end of the 1980s a sampling-based national forest inventory was introduced that reported a stock-per-hectare increase from 155 cubic meters per hectare in 1985 to 298 cubic meters in 1990 (ECE/FAO 1985, 1992). Overall, it has been noted that the situation in several industrial countries (particularly in the former Soviet Union) is less than satisfactory for national and international forest policy development and implementation (FAO 2001d).

For developing countries, the quality of forest resources information also varies greatly. FRA-1990 reported that all but seven developing countries had at least one estimate of forest cover dating from between 1970 and 1990, usually based on remote sensing. Only 25 out of 143 countries had made more than one assessment. On average, the figures supplied to FAO were about 10 years old (FAO 1995a). FRA-2000 (FAO 2001d) reports that only 22 countries (of 137) have repeated inventories, 28 countries have no inventory, and 33 have a partial forest inventory; 34 have an inventory from before 1990 and 43 have one from after 1990. More than half the inventories used by FAO were either more than 10 years old or incomplete. Very few developing countries have up-to-date information on forest resources, and fewer have a national capacity for generating such information.

Knowledge of area and condition of forests in many countries does not seem to be improving, and in many cases it is actually declining. Additional issues of concern relate to the quality rather than the quantity of forestry-related information:

- There is often a strong interest in new technologies and the production of "showpiece" maps. Modern forest inventories are sometimes only forest mapping exercises, which do not contain all the information needed for sustainable forest management.

- There is a great deal of reliance on remote sensing technologies. However, remote sensing cannot provide information in many areas for which there is a need for better information, such as forest ownership and tenure, protection status, purpose and success rate of plantations, biodiversity, and production and consumption of forest-derived services.

- Few forest inventories are undertaken as part of regular monitoring schemes—most are one-time undertakings. As a result, comparable inventory information for different time periods is frequently unavailable.

The demand for forest information is increasing (for example, in response to the Intergovernmental Panel/Forum on Forests Proposals for Action and to report to international conventions), but there is no corresponding allocation of resources. The funds available for forest inventories are actually decreasing in many countries due to budget cuts and structural adjustment policies. For example, 20–30 years ago forest inventories in many countries were supported financially by international agencies, including FAO. Today hardly a single inventory of this type is being supported financially.

The trend in reliability of national data on forests for countries of Africa, Asia, and Latin America and the Caribbean can be estimated by classification of countries as having low (L), medium (M), or high (H) quality data (Persson 1974; FAO 1993; FAO 2001c). National inventory methods have been used as the main criterion for data quality classes. The ratio L:M:H (in percent to total amount of countries included in the survey) was 27:63:10 in 1970, 23:56:21 in 1990, and 11:25:64 in 2000. While there is an improving trend, progress has been slow, despite the increasing capabilities and utilization of remote sensing and other modern techniques. And some countries, such as Gabon and Côte d'Ivoire, have been covered by extensive inventories in the past but are now placed in the "low" category. Indeed, across Africa the number of countries in the "low" category increased between 1970 and 1990. The trend is more positive in Asia, but some of this improvement may due to the use of rapid assessment remote-sensing inventories.

21.2.3 Data Collection Methodology for FRA-2000

The FAO FRA process aims to collect statistical information on forests directly from countries. Information on temperate and boreal zones in industrial countries is collected through questionnaires. The national figures are then adjusted to fit FRA forest definitions. The data in FRA for developing countries, however, result from a dialogue between FRA and the countries, which included a number of steps: countries are requested to supply information; independent information (such as remote sensing) is used to corroborate the information received; estimates and outputs (partly by countries) are produced; and validation by and dialogues with countries are held. For FRA-2000, over 1,500 national and international reports were analyzed, and the information obtained was reclassified to fit FRA definitions in consultation with the providing countries.

It is evident, however, that official national statistics have many shortcomings following from weaknesses in national inventory methods and the varying political, administrative, and economic conditions of individual countries. In its analysis of the reliability of FRA-2000 data, FAO has pointed out that global results cannot be more accurate than national data and that all gaps and uncertainties of countries' statistics inevitably affect the FRA-2000 conclusions (FAO 2001c:350–51). Improvement in the information provided by international assessments such as FRA requires improvement in the information supplied at the national level. By using national statistics, however, FRA-2000 allows for ongoing improvement in the assessment process, whereby new information can be incorporated as it becomes available.

21.2.4 Global Forest Mapping Methodologies

In an effort to provide spatial definition of forests, the MA used two global maps produced by FRA-2000: the FRA-2000 global forest cover map (see Figure 21.1 in Appendix A) and the FRA-2000 global ecological zone map (see Figure 21.2 in Appendix A). In this assessment, these two datasets have been combined with a global continent map (ESRI 1998), in order to demonstrate forest class area geographically by continent and ecoregion. For the global summary statistics of the MA, the forest system was calculated from >40% forest cover classes of the Global Land Cover 2000 dataset.

The GLC2000 land cover database has been chosen as a core dataset for the MA—in particular, as a main input dataset to define the boundaries between systems such as forest, grassland, and cultivated systems. GLC2000 used the VEGA2000 dataset, providing a daily global image from the vegetation sensor onboard the SPOT4 satellite from November 1999 to December 2000. The GLC2000 dataset identified globally a total of 10 tree cover types: broadleaf evergreen; broadleaf deciduous (closed); broadleaf de-

ciduous (open); needle-leaved evergreen; needle-leaved deciduous; mixed leaf type; regularly flooded fresh; regularly flooded saline; mosaic: tree cover/other natural vegetation; and burnt. Forests were identified as having a minimum of 15% tree cover and 3 meters height. Closed forests were defined as having more than 40% tree cover. When aggregated, forest areas in the GLC2000 compare spatially well with the forest areas defined in the FRA2000. (For further details of GLC2000, see Chapter 2.)

The forest map in Figure 21.1, developed using coarse resolution satellite imagery, relied mainly on the Global Land Cover Characteristics Database. Source data for the forest map were drawn from the 1995–96 dataset and consisted of five calibrated advanced very high resolution radiometer bands and a normalized difference vegetation index (FAO 2001c). Results of an accuracy assessment showed that overall map accuracy is approximately 80%. Closed forests are more accurately mapped than the average accuracy, with open and fragmented forests less accurately mapped and other wooded lands least accurately mapped (FAO 2001c).

The ecological zone map in Figure 21.2 was developed using national and regional maps of potential vegetation and climate data. A globally consistent classification was adopted, based on the Köppen-Trewartha climate system in combination with natural vegetation characteristics. A total of 19 global ecological zones have been defined and mapped, ranging from the evergreen tropical rain forest zone to the boreal tundra woodland zone (FAO 2001c).

Although this chapter uses some estimates of forest area derived from these two maps, it should be noted that FRA-2000 only uses these maps to indicate the spatial distribution of forests and does not use data derived from the maps in its statistics on forest extent and cover.

Remote sensing methods are becoming an important tool for improving data and knowledge on the world's forests in the future. New and planned satellite sensors appear to be very promising in this regard, and several global initiatives (GTOS, GOFC-GOLD) are focusing on their further development. Experiences over the last decade, however, illustrate a number of problems with the estimation of forest cover and extent from space. (See Box 21.3.)

21.3 Condition and Trends in Forest and Woodland Systems

21.3.1 Forest Area

FRA-2000 estimates the total area of global forests at 3,869 million hectares (0.6 hectares per capita), or about 30% of the world's land area (see Table 21.1), with closed forests accounting for 3,335 million hectares. (Table 21.1, unlike FRA-2000, divides Russian forests into their European and Asian parts based on national statistics in order to give a more accurate assessment of the relative areas of forest on the European and Asian continents.) This can only be taken as an approximate estimate, however, due to the methodological problems that the FAO faced with respect to data weaknesses and inconsistencies among countries, as described in the preceding section. As noted there, national data for many developing countries are scarce and unreliable, and inconsistencies also exist for some industrial countries. Data for Canada, for example, are aggregated from provincial sources and report only "productive forestland," while "unproductive forests" are reported by FRA-2000 as other wooded land, even though many of them meet the FAO definition of forest. This anomaly resulted

BOX 21.3
Forest in Recent Global Land Cover Assessments Using Remote Sensing Methodologies

During the past decade, a number of attempts have been made to estimate forest area in the framework of global land cover assessments using various remote sensing methodologies. The major features of four satellite-based 1-kilometer land cover products in wide use by the international scientific community (McCallum et al. 2004) are compared and analyzed here.

The International Geosphere-Biosphere Program product (version 2.0) is derived from advanced very high resolution radiometer data from April 1992 to March 1993. This methodology employed a multi-temporal unsupervised classification of a normalized difference vegetation index with post-classification refinement using multi-source data. In total, 17 land cover classes were considered (USGS 2003).

The University of Maryland product used the IGBP AVHRR dataset, utilizing all five AVHRR channels as well as the NDVI, to derive 41 multi-temporal metrics from monthly composites to represent the phenology of global vegetation. UMD used a supervised classification tree method, resulting in a total of 14 land classes (Hansen and Reed 2000).

Global Land Cover 2000 is based on daily mosaics of four spectral channels and NDVI from VEGETATION-SPOT 4 imagery acquired from November 1, 1999, to December 31, 2000; data from other sensors have been used to solve specific problems. A total of 22 classes were produced (JRC 2003).

The MODIS-Terra product utilized monthly composites of eight input parameters from October 2000 to October 2001. The classification, which resulted in 20 land cover classes corresponding to IGBP classes, was produced using a supervised approach with a decision tree algorithm (MODIS 2002).

Land classifications differed between some of these remote sensing methodologies. McCallum et al. (2004) carried out a comparison of these four products, applying physiognomic aggregation of different land classes (cf. Hansen and Reed 2000) using the IGBP classification (with 17 classes) as a base. Differences in estimated areas of the same classes are significant. (See Figure.)

All these remote sensing methodologies contain forest classes. Comparisons of the satellite-derived data and FRA-2000 are presented in the

accompanying Table. Area estimates by all four remote sensing methodologies are less than those by FRA-2000, averaging −26.4% and varying from −13.5% (GLC 2000) to −43.6% (UMD, although in this case the large difference is additionally affected by incompatible classifications). The four remote sensing estimates, when compared for each aggregated IGBP class, vary from 12% to 74%. The reasons for the significant underestimates of forest area by these remote sensing methodologies compared with FRA-2000 are not completely clear. The remote sensing underestimates can probably be explained by the coarse resolution of the remote sensing technology used, fragmentation of forests in many regions, lack of satisfactory ground truth data for proper validation and verification of the remote sensing data, and the use of different classifications. By contrast, in this chapter's attempt to use the FAO global forest map the total area of the world's forests was estimated to be 4,356 million hectares—that is, about 12% more than the total provided by the FRA-2000 data.

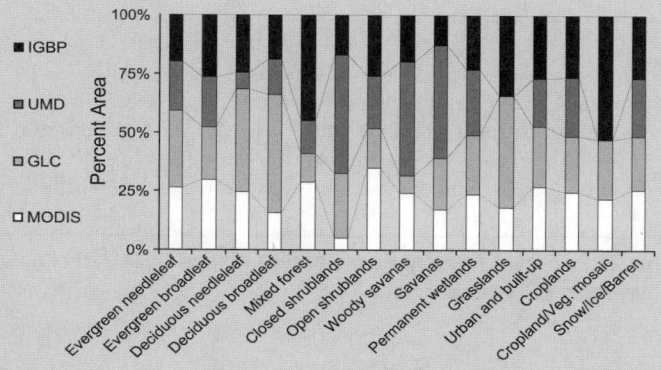

Comparison of Forest Classes Derived from Four Global Land Cover Projects Using the IGBP Classification Terminology. The areas are presented as the percentage of an individual class to the total area of all forest classes inside of each product. (McCallum et al. 2004)

Forest Area: Comparison of Four Global Remote Sensing Land Cover Products and FRA2000 Using IGBP Classes 1–5 (McCallum et al. 2004)

IGBP Forest Classes	GLC-2000	IGBP	MODIS	UMD	Average	Range of Variation from Average
			(million hectares)			*(percent)*
1 – Evergreen needleleaf	943	480	598	521	636	33.2
2 – Evergreen broadleaf	1,281	1,342	1,502	1,108	1,308	12.4
3 – Deciduous needleleaf	377	193	201	56	207	63.6
4 – Deciduous broadleaf	627	221	172	174	298	73.9
5 – Mixed forests	320	981	696	323	580	55.2
Total	3,548	3,217	3,168	2,182	3,029	19.5
			(percent)			
Difference with FRA2000	−13.5	−21.6	−23.7	−46.8	−26.4	

in underreporting of more than 170 million hectares (40%) of Canadian forestland. With this and similar adjustments, the global forest cover corresponding to the FAO definition would probably increase by about 5%.

Forests are not distributed evenly across the globe, as Figures 21.1 and 21.2 indicate. Although average forest cover on all con-

tinents except Antarctica exceeds 20%, vast territories are either completely bereft of forests or have negligible forest cover. FRA-2000 estimated that 56 countries have an average of only 3.9% forest cover. On the other hand, the six biggest forest countries—Russia, Brazil, Canada, the United States, China, and Australia—contain about 56% of the world's forests.

Table 21.1. Forest Area by Region in 2000 (FAO 2001c, modified for Europe and Asia; see explanation in text)

Region	Land Area	Forest Area — Natural Forests	Forest Area — Plantation	Forest Area — Total	Forest Coverage
		(million hectares)			(percent)
Africa	2,978	642	8	650	22
Asia	4,362	1,105	120	1,225	28
Europe	983	334	28	362	37
North and Central America	2,137	532	18	549	26
Oceania	849	194	3	198	23
South America	1,755	875	10	886	51
World total	**13,064**	**3,682**	**187**	**3,869**	**30**

Note: In this chapter, Russian forests are divided into European and Asian parts based on national statistics.

The area of the world's forests estimated using satellite-based methods is 4,356 million hectares (see Table 21.2), which is close to the recent estimate made by the UNEP World Conservation Monitoring Centre—4,540 million hectares (UNEP-WCMC 2002). The total area of closed forests is estimated at 2,860 million hectares (about two thirds of the total), with major areas in Asia, North America, and South America. The different values derived from FRA2000 and satellite data are mostly due to the different national thresholds of closed forests (in Russia, for instance, forests are classified as closed if canopy closure is more than 25%).

The threshold used for national inventories substantially changes estimates of forest area, as is clearly illustrated in Figure 21.3. Nonetheless, estimates of the extent of the world's forests are of the same order of magnitude. The satellite-based total area of global forests is only 11% more than FRA-2000 data in Table 21.2. The total area of global forests plus OWL is estimated to be 5,576 million hectares (about 42% of Earth's land area), which is very close to the corresponding FRA-2000 estimate of 5,532 million hectares.

The estimate in Table 21.2 for the area of tropical closed forests (1,229 million hectares) is in the range of previous estimates, such as those of IUCN at 1,140 million hectares (Collins et al. 1991; Sayer et al. 1992; Harcourt and Sayer 1996), project TREES at 1,165 million hectares (Mayaux et al. 1998), and Achard et al. (2002) at 1,116 million hectares for 1997, without Central America and Oceania. The FRA-1990 estimate was 1,298 million hectares (FAO 1996).

Of the total area of 1,494 million hectares of open and fragmented forests in Table 21.2, more than half (53%) is situated in tropical ecoregions and about 22% is in the boreal zone. A significant part of these forests in the tropics consist of sparse forests in dryland areas and degraded forests. By contrast, the majority of

Table 21.2. Forest Area by Biome and Continent (FRA 2000 Forest Cover Map; FRA 2000 Global Ecological Zone Map; global continents derived from ESRI world map)

Biome	Africa	Asia	Europe	North and Central America	Oceania	South America	Total
				(million hectares)			
Closed forests							
Polar	0	2	1	3	0	0	6
Boreal	0	495	156	295	0	0	945
Temperate	0	97	114	237	13	7	471
Sub-tropical	2	91	11	85	19	6	212
Tropical	274	222	0	77	46	609	1,229
Subtotal	*277*	*908*	*282*	*696*	*77*	*622*	*2,862*
Open and fragmented forests							
Polar	0	5	3	6	0	0	15
Boreal	0	158	46	109	0	0	313
Temperate	0	71	101	47	4	5	226
Sub-tropical	6	49	18	36	14	18	141
Tropical	344	133	0	26	30	264	798
Subtotal	*350*	*415*	*168*	*225*	*48*	*287*	*1,494*
Total forests							
Polar	0	7	4	9	0	0	21
Boreal	0	653	202	404	0	0	1,258
Temperate	0	168	215	284	17	12	697
Sub-tropical	8	140	29	121	33	24	353
Tropical	618	355	0	103	76	873	2,027
Total	**627**	**1,323**	**450**	**921**	**125**	**909**	**4,356**

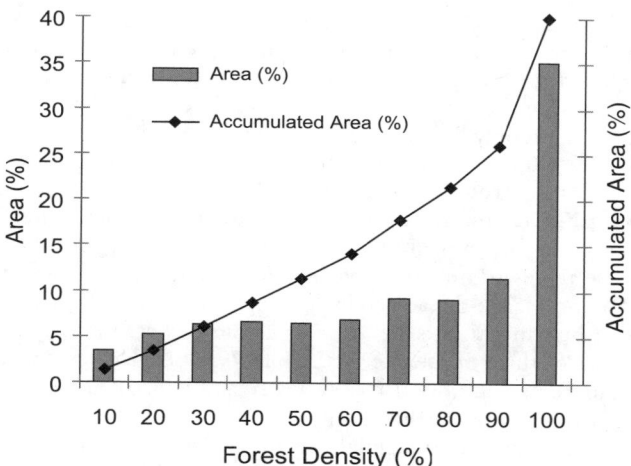

Figure 21.3. Distribution of World's Forests by Canopy Closure

boreal open forests are naturally sparse woodlands in the transition belt of the taiga-tundra ecotone.

The estimate derived from satellite data of the global area of other wooded land is 1,220 million hectares (see Table 21.3), which is 15% lower than the FRA-2000 statistical estimate. OWL plays a significant environmental and protective role in many regions, particularly in the arid tropics and sub-tropics, where it is also known as open savanna woodland. In the boreal ecoregion, areas of OWL are relatively small (about 11% of the total) and often represented by shrubs such as dwarf pine (*P. pumila*) and dwarf birches in Northern Eurasia. Although OWL has a low commercial value, these woodlands have a large and mostly unrecorded value to local people, provide soil and water protection services, and harbor biodiversity across vast landscapes.

Data derived from the FRA-2000 statistical tables and the FRA-2000 global maps are not completely consistent, and making specific area comparisons between these two sources is rather difficult. This is largely due to the different compilation methods used. In particular, differences between the two methods for North America and Oceania are noticeable. In the case of North America, the reported forest cover for Canada and the United States both appear to refer to productive or commercial forest cover only, as mentioned earlier. Therefore, the FRA-2000 forest cover map identifies a greater forest cover than is reported in the country statistics. In the case of Oceania, the forest map underesti-

mates the amount of forest cover compared with the statistics. One likely reason for this is the lack of good satellite imagery (due to persistent cloud cover) preventing the mapping of several Pacific islands (FAO 2001c). In addition, differences in classification between open and fragmented forest and other wooded land may be a factor.

21.3.2 Distribution by Aggregated Forest Types (Ecological Zones)

Different global classifications of the world's forests by forest type are generally largely incompatible. The classification by UNEP-WCMC includes 26 aggregated forest types—15 in tropical forests and 11 in non-tropical biomes (UNEP-WCMC 2004). Based on criteria equivalent to Köppen-Trewartha climatic groups, FRA-2000 considered five domains (biomes)—tropical, subtropical, temperate, boreal, and polar, which are divided in 20 global ecological zones (FAO 2001c, Table 47–2). The latter classification is used in this section, and distribution estimates are based on the remote sensing data sources rather than the FRA-2000 statistical datasets.

Three quarters of the world forests is located in two biomes—tropical (46%) and boreal (29%). Tropical rain forest is the most extensive forest type in the world, constituting 26% of global forest area and nearly 60% of tropical forest area. Most rain forests are in South America (582 million hectares), Africa (270 million hectares), and Asia (197 million hectares). Tropical rain forests are closed-canopy evergreen broadleaf forests that generally require continual temperatures of at least 25 Celsius and annual rainfall of at least 1,500 millimeters (Richards 1996). Tree diversity in tropical rain forests is very high, with often more than 100 tree species per hectare.

Tropical moist deciduous forests cover some 510 million hectares. They develop in areas with a dry season of three to five months, and they vary from closed forests to open savanna forests, depending on dry-season length, human pressures, and fire regimes. Only about one third of these forests are closed primary forest areas; the rest are open and fragmented forests, including significant areas of secondary forest created by disturbances such as agricultural clearing and fire. The soils are in general better than in rain forests areas, and human population pressure is therefore higher. In Asia, these forests contain commercially important species like teak (*Tectona grandis*) and sal (*Shorea robusta*). In tropical dry forests the dry season is longer than in the moist deciduous (open tropical) forests. Remaining areas of tropical dry forests are relatively small, consisting mostly of open forest.

Table 21.3. Area of Other Wooded Land by Biome and Continent (FRA 2000 Forest Cover Map; FRA 2000 Global Ecological Zone Map; global contents derived from ESRI world map)

Biome	Africa	Asia	Europe	North and Central America	Oceania	South America	Total
				(million hectares)			
Polar	0	5	3	6	0	0	15
Boreal	0	45	9	36	0	0	90
Temperate	0	34	22	3	3	4	67
Sub-tropical	22	78	6	7	25	27	164
Tropical	492	38	0	5	46	289	871
Total other wooded land	*514*	*207*	*39*	*67*	*74*	*319*	*1,220*
Total forest	*627*	*1,323*	*450*	*921*	*125*	*909*	*4,356*
Total	**1,141**	**1,530**	**489**	**988**	**199**	**1,228**	**5,576**

The temperate and boreal forests occur from the sub-tropics to the arid steppes and sub-Arctic, with the northernmost growing at 72°30′ in central Siberia (Abaimov et al. 1997) at an annual average temperature of −15° to −17° Celsius. They are mostly distributed in 55 industrial countries (in Europe, the former Soviet Union, North America, Australia, Japan, and New Zealand). Detailed and reasonably reliable information concerning these forests is available (FAO 2000). The total area of forest in these countries was estimated to be 1,914 million hectares, supplemented by an additional 795 million hectares of other wooded lands. Thus the total area of forest and other wooded land is estimated to be 2,478 million hectares, which accounts for 47% of global tree cover. More that one third (38%) of the total in these zones is located in the former Soviet Union, 29% in North America, 9% in Europe, and 25% in Australia, Japan, and New Zealand. On average, these countries have 1.3 hectares of forest per capita—about double the global average, although there is great variation between countries (from nearly none in Malta and Azerbaijan to 6 hectares per capita in Russia and 31 hectares per capita in Australia). These statistics do not include China, which has significant areas of temperate and boreal forests—30% and 8% respectively of the country's total forest area of 163.5 million hectares.

Countries of these biomes accounted for in the *Temperate and Boreal Forest Resources Assessment* contain 47% of predominantly coniferous forest (mostly genera *Pinus, Picea, Larix, Abies*), 26% of predominantly broadleaf forest (many genera, including *Populus, Betula, Quercus, Fraxinus, Tilia*), and 27% of mixed coniferous and broadleaf forests. Other forests types (bamboos, palms, and so on) cover small areas in Japan. Coniferous forests serve as a major source of global industrial wood, and the broadleaf forests include a number of high-value commercial species. (See Chapter 9.)

21.3.3 Wood Volume and Biomass

Wood volume, woody biomass, and total live biomass are important indicators of the potential of forests to provide various products and services, including carbon sequestration. Based on available information from 166 countries (about 99% of the world's forest area), FRA-2000 estimated the total global standing volume (aboveground volume of all standing trees, living or dead, with diameter at breast height over 10 centimeters) to be 386 billion cubic meters and the global aboveground woody biomass to be 432 billion tons (dry matter), which gives average values of 100 cubic meters and 109 tons per hectare, respectively. IPCC (2000) estimated the total carbon stock of vegetation in forest to be 359 billion tons of carbon. These data vary greatly over continents and countries. Average standing volume, for example, varies from about 60 cubic meters per hectare in Oceania and Asia to 125 cubic meters in North and South America, while the ratio of aboveground biomass (tons) to standing volume (cubic meters) varies from about 0.5 in Europe to 1.6 in South America (FAO 2001c).

21.3.4 Extent of Natural Forests

There are numerous ways of characterizing the degree of "naturalness" of forests—old growth, ancient, intact, frontier, natural, secondary, modified, and so on—and although there are no consistent, agreed definitions and information with which to classify forests in this manner is poor, FAO has defined natural forests as forests composed of indigenous trees regenerated naturally (FAO 2000c, 2002b).

Although FRA-2000 considered all forests except plantations to be "natural" (FAO 2001c), FRA-2005, which is currently under preparation, considers four classes of decreasing "naturalness"—primary forest, modified natural forest, semi-natural forest, and forest plantations (FAO 2004). Forest inventories as a rule do not characterize forests by their degree of naturalness, however, and so only limited assessments are available.

One attempt to inventory the extent of natural forests by the World Resources Institute identified the global extent of "frontier forests"—remaining large, intact natural forest ecosystems big enough to maintain all of their biodiversity (Bryant et al. 1997). These represent only 40% of the planet's remaining forests, and 39% of these are threatened by logging, agricultural clearing, and other human activities. Seventy-six countries were found to have lost all of their frontier forests, while 70% of what remains lies within three countries (Brazil, Canada, and Russia), and only 3% lies within the temperate zone.

This chapter uses a simplified three-class approach to forest naturalness, limiting the classification to "natural" (self-regenerating, generally multi-species, mixed age stands of native species, with a natural disturbance regime); "semi-natural" (some degree of human intervention in regeneration, species selection, and disturbance); and "anthropogenic" (established or significantly transformed by humans). Using regional expertise and some published sources (e.g., Vorob'ev et al. 1984; Bryant et al. 1997; Atlas 2002) it can be tentatively estimated that about 70% of the world's forests can be considered to be natural, 20% semi-natural, and 10% anthropogenic (half of which are plantations).

21.3.5 Trees Outside of Forests

Trees outside of forests, or TOF, occur in many formations, such as shelterbelts, shade and other elements of agroforestry, roadside plantings, village and urban plantings, orchards, and individual trees on farms and other private land. Although there are no consistent global data on the coverage or extent of TOF, FRA-2000 provides a global review of this, acknowledging the diversity of the multiple functions and benefits (FAO 2001c). For example, about 70% of the land area of Java (Indonesia) has trees but only 23% of this is classified as forest (Persson 2003).

TOF provide important services, including contributing to food security, particularly for rural populations (Auclair et al. 2000; Glen 2000; Klein 2000). In many Asian countries, particularly those with low forest cover, TOF supply the majority of fuelwood (Arnold et al. 2003). For example, more than 75% of fuel production comes from non-forestland (mostly from TOF) in Bangladesh, India, the Philippines, and Thailand, although with significant variation among countries (Bhattarai 2001). Shelterbelts are an important component of agroforestry landscapes in many countries of the northern hemisphere (see, e.g., Yukhnovsky 2003). Quantitative data on TOF are scarce and not comparable, however, since they are mostly limited to regional and national case studies (FAO 2003b), although some counties (such as France, the United States, India, and Bangladesh) have initiated efforts to gather national-scale quantitative information on TOF (FAO 2001c).

21.3.6 Distribution of People in Forest Areas

The current distribution of people living in and adjacent to forest and woodland areas is the result of a long historical process of social and economic development. Significant factors influencing population distribution include topography, degree and direction of landscape transformation, and forest types. Currently, about three quarters of humanity lives in three ecological zones classified as aggregated forest ecoregions (needle-leaved evergreen, closed broadleaf deciduous, and broadleaf evergreen), although a far smaller number actually live in or adjacent to forested areas (CIESIN 2000).

As a rule, more-densely populated regions have less natural forest and more plantations than less populated regions (Persson ·2003). Typical examples are China and India, with a combined population of about 2.3 billion and forest area of just 228 million hectares (FAO 2001c). Based on U.N. population statistics, UNEP-WCMC has derived detailed information on the ratio of forest area to people at both the national level and for 12 large regions in 1996 (UNEP-WCMC 2004). The overall global number was 0.7 hectares per person, with a large variation between regions—from 0.07 for Middle East to 5.6 for Russia and 6.5 for Australasia—and by ecological zones. In the tropics, the highest ratio (1.85 hectares per person) was in rain forest areas and the lowest (0.24 hectares per person) in dry deciduous forests (FAO 1993).

Tropical rain forests typically have low human population densities. This is largely because rain forest soils are frequently low in nutrients and therefore unsuitable for continuous agriculture. Although many rain forest areas can support traditional forms of extensive rotational ("shifting") cultivation, and have done so for millennia, this form of agriculture is unable to support high human population densities. In areas with good soils (such as volcanic or sedimentary soils), rain forests have long since been converted to agricultural landscapes.

Forests are a significant source of employment. Global recorded forest-based employment is about 47 million full-time equivalents, 17 million of whom are in the formal sector (ILO 2001; Blombaeck and Poschen 2003). Labor force trends and dynamics vary among countries and regions, but in general forestry sector employment is decreasing. (See Chapter 9.) The forestry sector labor force in Europe and the former Soviet Union, for example, is expected to decrease by 7% during the coming decade (ECE 2003).

21.4 Changes in Global Forest Area and Condition

21.4.1 Parameters of Change

Four basic change processes determine trends in global forest area and are defined for this chapter as follows:

* *Deforestation* is the conversion of forests to another land use or the long-term reduction of the tree canopy cover below 10%.
* *Afforestation* is the establishment of forest plantations on land that, until then, was not classified as forest. It implies transformation from non-forest to forest.
* *Reforestation* is the establishment of forests plantations on temporarily unstocked lands that were considered as forest in the recent past.
* *Natural expansion of forests* means the expansion of forests through natural succession on land that, until then, was under another land use (such as forest succession on land previously used for agriculture). It implies a transformation from non-forest to forest.

Net changes in forested area are a superimposition of these four major processes. While net changes are important to monitor, it is also important to disaggregate exactly what is being lost and what is being gained. A focus on net changes—for example, plantation establishment offsetting natural forest loss, and gains in forest cover in industrial countries offsetting forest losses in tropical developing countries—may obscure the severity of natural forest losses in tropical regions.

Forest degradation and *forest improvement* describe changes in forest condition, but not changes in an area's land use or land cover status. FRA-2000 defined these as changes within the forest, which negatively (forest degradation) or positively (forest improvement) affect the structure or function of the stand or site and thereby lower (degrade) or increase (improve) the capacity to supply ecosystem services (FAO 2001c). As previously noted, though, there is little consensus among definitions of forest degradation and deforestation. Some logged areas, for example, are severely degraded to the point of being virtually devoid of trees and previous ecological characteristics and functions, and many would argue that such areas should be counted as effectively deforested, irrespective of their formal legal or management status.

21.4.2 Changes in Global Forest Cover

Clearing of forests for other land uses, particularly agriculture, has accompanied human development for the whole of documented human history. Historically, deforestation has been much more intensive in temperate regions than in the tropics, and Europe is the continent with the least amount of original forests remaining. As a whole, clearance prior to the industrial era was a slow and steady process over a long period of time. In the more recent past, many countries and regions experienced much higher rates of forest conversion, and many currently industrialized countries experienced deforestation rates in the nineteenth century very similar to those now occurring in many tropical developing countries.

The relationship between agricultural expansion and forest decline has been analyzed, and the following preliminary conclusions emerge: agricultural land is expanding in about 70% of countries, declining in 25%, and is static in 5%; forest area is decreasing in two thirds of countries where agricultural land is expanding, but expanding in the other one third of those countries; and forests are expanding in 60% of countries where agricultural land is decreasing and are declining in 36% of this group of countries (FAO 2003b). A complicated combination of economic and social development factors, levels of agricultural productivity and urbanization, climatic and geographical peculiarities, and countries' previous histories determine rates of deforestation in particular places.

Significant deforestation in tropical forests has been documented for 1990–2000. The total loss of natural tropical forests is estimated for this period at 15.2 million hectares per year (FAO 2001c). Taking into account relatively small natural expansion of tropical forests (+1.0 million hectares a year) and plantations that have been developed at +1.9 million hectares annually, the net change in tropical forest area was estimated by FRA-2000 to be −12.3 million hectares. In contrast, during this period a net increase of forest area was observed in temperate and boreal zones (+2.9 million hectares a year, of which +1.2 million hectares were forest plantations and +1.7 million were due to the change in area of natural forests). In total, then, the net change in global forest area is estimated at −9.4 million hectares per year. (See Table 21.4 and Figure 21.4.)

The net annual change in forest area for 1980–90 was estimated to be −13 million hectares (FAO 1995b) (including losses of 6.1 million hectares per year in tropical moist forests and 3.8 million hectares per year in tropical dry forests), and −11.3 million hectares for 1990–95 (FAO 1997). This would indicate that net global forest loss has slowed down since the 1980s (FAO 2001c). However, much of this is due to increases in plantation forestry, and although the global net change in forest area was lower in the 1990s than in the 1980s, the rate of loss of natural forests remained at approximately the same level.

Table 21.4. Forest Area Changes, 1990–2000, in Tropical and Non-tropical Areas (FAO 2001c)

| Domain | Natural Forest | | | | | Forest Plantations | | | Total Forest |
| | Losses | | | Gains | | Gains | | | |
	Deforestation	Conversion to Forest Plantation	Total Loss	Natural Expansion	Net Change	Reforestation	Afforestation	Net Change	Net Change
	(million hectares per year)								
Tropical	−14.2	−1.0	−15.2	+1.0	−14.2	+1.0	+0.9	+1.9	−12.3
Non-tropical	−0.4	−0.5	−0.9	+2.6	+1.7	+0.5	+0.7	+1.2	+2.9
Global	−14.6	−1.5	−16.1	+3.6	−12.5	+1.5	+1.6	+3.1	−9.4

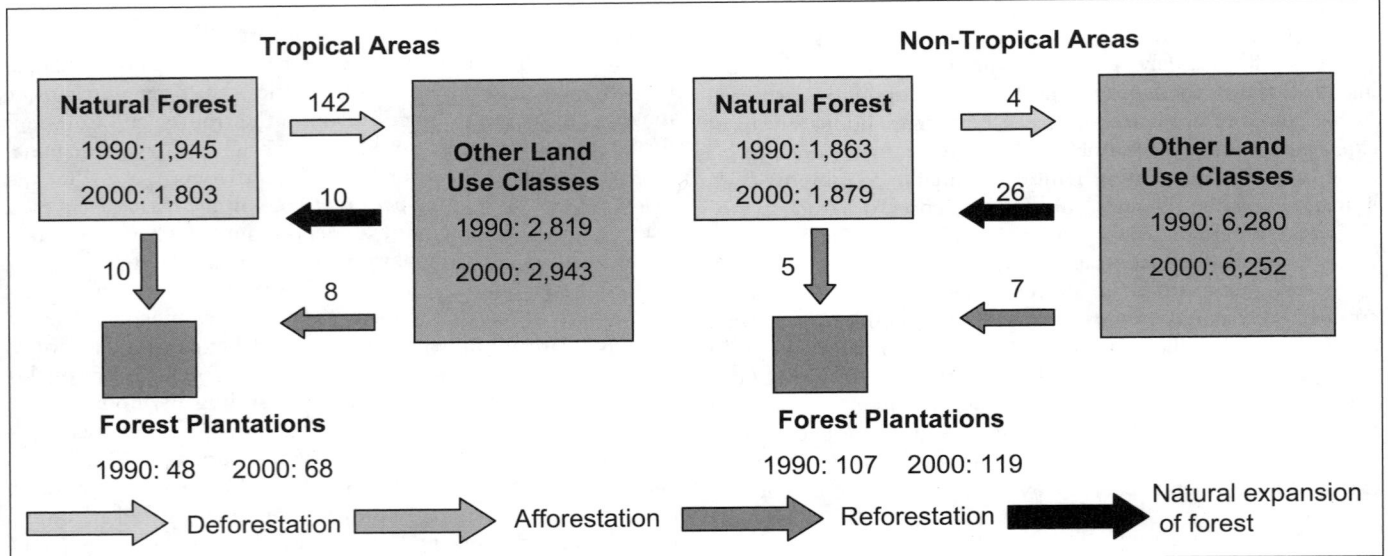

Figure 21.4. Major Change Processes in World's Forest Area, 1990–2000 (in million hectares) (FRA 2000; FAO 2001c)

Matthews (2001), however, reached a different conclusion, finding that in absolute terms, more tropical forest was lost in the 1990s than in the 1980s. According to this estimate, net deforestation rates have increased in tropical Africa, remained constant in Central America, and declined only slightly in tropical Asia and South America. The certainty of this estimate is unknown.

It is likely that deforestation in developing countries has continued since 2000 at practically the same rate as during the 1990s, about 16 million hectares per year, corresponding to 0.84% for the 1990s and 0.80% since 2000. The difference in these estimates is definitely within the uncertainty limits of the techniques used. However, both national inventories and remote sensing data often do not adequately record the regrowth of secondary forests in many areas. If better data on this were available, they would likely reduce the net area change in forest cover for many regions (see, e.g., Faminov 1997).

Recent remote sensing surveys of individual biomes and forest types have reported different, often lower, rates of deforestation than those reported in FRA-2000. The research program TREES (Tropical Ecosystem Environment observation by Satellite) estimated annual losses of humid tropical forests on three major continents between 1990 and 1997 at 5.8 ± 1.4 million hectares with a further 2.3 ± 0.7 million hectares of visibly degraded forests (Achard et al. 2002). This is about one fifth less than the estimates provided from the sources just discussed. However, estimated un-

certainties of forest cover were substantial ($1,150 \pm 54$ million hectares and $1,116 \pm 53$ million hectares for 1990 and 1997 respectively).

On the other hand, a study by DeFries et al. (2002) found that the rate of tropical deforestation actually increased by about 10% from the 1980s to the 1990s, in contrast to the 11% reduction reported by FRA-2000, and supporting Matthews (2001). This is not surprising, since methods vary among different surveys. It must therefore be realized that coarse-resolution remote sensing data still cannot provide detailed reliable information about changes in forest area. Rather, existing published figures provide only estimates of the order of magnitude of forest cover change (DeFries et al., 1995, 2000; Holmgren and Turesson 1998; McCallum et al. 2004).

Trends in deforestation and net changes in forest area vary across regions, although there are many commonalities within major biomes across regions. The major areas with rapid deforestation are currently in the tropics. Africa accounts for over 50% of net recent global deforestation, although the continent hosts only 17% of the world's forests. Ten tropical countries (six of them in Africa) had net annual change in forest areas of more than 3%, and four countries (three in Africa plus Nicaragua) had change rates of 2.5–3.0% between 1990 and 2000 (FAO 2001c). Net change of forest area by continent is presented in Figure 21.5.

Quantitative data on the dynamics of other wooded land are weak. Many national sources reported substantial transformation

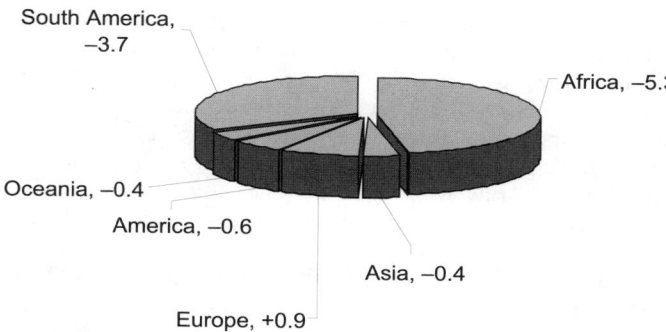

South America, −3.7

Africa, −5.3

Oceania, −0.4

America, −0.6

Asia, −0.4

Europe, +0.9

Figure 21.5. Net Change in Forest Area by Continent (in million hectares per year) (FRA 2000; FAO 2001c)

and decline of woodlands, in particular in dry regions, due to extensive conversion for agriculture and excessive harvesting by rural communities (Hassan 2002). Open savanna woodlands in South Africa, for example, have lost about half of their original extent of approximately 42 million hectares (Low and Rebello 1996).

21.4.3 Forest Plantations

Development of forest plantations can have significant impacts on the dynamics of forest areas in some regions. Forest plantations covered 187 million hectares in 2000, with significant regional variation—62% are in Asia and only 17 % in Europe. Ten countries account for 79% of global forest plantation area, and six of these account for 70%. Globally, broadleaf-species account for 40% of forest plantation area, with *Eucalyptus* the principal genus (10% of the global total); coniferous species account for 31%, with *Pinus* constituting 20% of the global total. Genus is not specified in FAO statistics for the remaining 29% (FAO 2001c).

According to national data supplied to FRA-2000, the tropical forest plantation estate has increased from 17.8 million hectares in 1980 to 43.6 million hectares in 1990 and about 70 million hectares in 2000 (Brown 2000; FAO 2001c). However, these and other data on the growth and extent of plantations are not fully comparable due to different definitions, incomplete statistics from different countries, and different approaches to estimation of coverage. For example, Europe and North America are not included in the FAO figures for 1980 and 1990 (FAO 1995a). According to official national data, the annual increase in plantations is 4.5 million hectares globally, of which 3 million hectares are estimated to be successful (FAO 2001c). About 90% of new plantations are in Asia and South America. Although plantations constitute only 5% of global forest cover, they were estimated to supply about 35% of global roundwood in 2000, and it is expected that this figure will increase to 44% by 2020 (ABARE-Jaakko Pöyry 1999; see also Chapter 9).

Other estimates of the rate of increase of plantations are also available. Pandey (FAO 1995a), for example, claimed that the total area of plantations for 1990 in tropics should be reduced by one third (although even this analysis is likely to have overestimated the extent of plantations in some countries such as India) due to peculiarities of the system of accounting for and estimation of plantations. Persson (1995) estimated the planted area in 1990 for all countries in the range of 148–173 million hectares, pointing out that plantation data for China are uncertain, and he estimated annual forest plantation increase in the tropics to have been about 0.5–1 million hectares for the 1970s, 1–1.5 million hectares for the 1980s, and about 2 million hectares for 1990–95. According to Persson (1995), plantations covered nearly 100 million hectares in 1970 and 120 million (100–30 million) hectares in 1980.

Based on socioeconomic analysis, Trexler and Haugen (1995) estimate that the total area of plantation in the tropics is likely to grow by 66.8 million hectares from 1995–2045, including 37.8 million hectares in Asia, 24.5 million hectares in Latin America, and 4.6 million hectares in Africa. Nilsson and Schopfhauser (1995) estimated the global availability of lands suitable for plantations and agroforestry at 345 million hectares. The reliability of these estimates is difficult to assess, however, because they do not consider specifics of many local socioeconomic and social processes, including the potential for expanded plantation establishment to cause social conflicts.

It can be seen from this short review that data on plantations are very uncertain and often contradictory. There are a number of reasons for this. The number of countries assessed for FRA-1980, FRA-1990, and FRA-2000 plantation estimates varied considerably, ranging from only 76 tropical developing countries in 1980 to all 213 countries in 2000. Plantation area is often overestimated if it is calculated from the number of plants produced or planted rather than from actually reforested or afforested areas. The area actually planted is often less than the planned area of plantation, which is often the reported area. Loss of plantations is often not included in national reporting, while the officially planted area is added each year. And finally, there is an inherent bias to exaggerate the success of plantation establishment.

Globally, 48% of forest plantation trees are destined for industrial end-use, 26% for nonindustrial uses, and 26% for unspecified uses (FAO 2001c). Industrial plantations provide raw material for commercial wood and paper products and can generate significant local employment opportunities. (See Chapter 9.) For example, some 1.5 million hectares of plantations in South Africa provide 1.63% of the global supply of pulp, 0.76% of paper, and 0.3% of sawn timber (Bethlehem and Dlomo 2003). Nonindustrial plantations are established to provide soil and water conservation, combat desertification, maintain biodiversity, absorb carbon, supply fuelwood, and rehabilitate fragile and degraded lands. During the last two decades, the major trend has been an increase in plantations established for industrial purposes, which have increased by about 25% since 1980 (FAO 2001c).

Forest plantations have potentially high productivity. On average, mean annual increments of *Eucalyptus* and *Pinus* are in the range of 10–20 cubic meters per hectare per year, but some species (e.g., *E. grandis, E. saligna, P. caribea*) can reach an MAI of up to 50–60 cubic meters, while *Araucaria* and *Acacia* can attain an MAI of up to 20–25 cubic meters per hectare per year. MAI for *Pinus, Picea*, and *Larix* plantations on the best sites in temperate and southern boreal zones can reach 12–15 cubic meters per hectare per year (Webb et al. 1984; Wadsworth 1997; Sagreev et al. 1992). The length of the rotation period for plantations varies from 5–10 to 30 years for major tropical species to 100–200 or more years for major boreal species. Along with the high MAI, the rotation period substantially affects the capacity of plantations to provide carbon sequestration services.

Many plantations do not in practice achieve these high potential growth rates. A number of studies (e.g., Nilsson 1996; McKenzie 1995; White 2003) have concluded that it is seldom possible to achieve high productivity in large-scale plantations, that insufficient forest management results in low survival rates and poor plantation condition, that monocultural plantations increases risks of pest and disease outbreaks, that production costs are often substantially underestimated, that knowledge about plantation growth and yield is poor, and that reliable and operative monitoring systems on plantation condition and dynamics only exist in a few countries. Many of these risks can be overcome where good management practices have been applied.

Plantations have been criticized for their environmental and social impacts—particularly in the tropics, where plantations have replaced natural forests, degrading water and soil resources, and resulting in negative impacts on local and indigenous communities who lose access to lands that formerly supplied them with subsistence resources and livelihoods (e.g., Carrere and Lohmann 1996; Carrere 1999; White 2003). Kanowski (2003) notes that in addition to the suboptimal performance of some plantations, many plantations have been established without appropriate consideration or recognition of trade-offs that were made with other forest services and with the rights and interests of various stakeholders. In Indonesia, for example, the timber plantation program has been a significant driver of natural forest loss, and the establishment of plantations (both for timber and oil palm) was a significant driving force behind the forest and land fires that beset Indonesia during 1997 and 1998 (Barber and Schweithelm 2000). A number of studies have also highlighted the risk of invasive alien species that can escape from plantations (e.g., Richardson 1998; Allen et al. 1997; De Wit et al. 2001).

Development of forest plantations can generate significant social conflicts. For example, in dryland areas plantation species may use more water than the natural vegetation, resulting in less recharge of groundwater and a reduction in streamflow available for other uses (Carrere and Lohmann 1996). Plantations can have social impacts because they employ fewer people than would find jobs on the agricultural land that they may replace and they can increase the price of farmland. They may also influence the viability of agro-enterprises if too many people in an industry sell their land for plantations. Cossalter and Pye-Smith (2003) evaluated such concerns for "fast wood" (fast-growing, short rotation species grown for charcoal, pulp, and wood-fiber panel products) plantations, which make up a relatively small but rapidly growing segment of global plantations. They concluded that the impacts of fast-wood plantations depend largely on their management. When poorly planned and executed, fast-wood plantations can cause significant social and environmental problems, but when well planned and executed, they can deliver not just large quantities of wood but a range of other environmental and social benefits.

Similar issues are raised, if not so acutely, by longer-rotation softwood and hardwood plantations. The long-established teak plantations on Java, for example, have been a perennial source of social conflict between local communities and the state forestry corporation that manages them (Peluso 1992). Although fast-wood plantations in the tropics appear to be the type most often responsible for negative environmental and social impacts (Cossalter and Pye-Smith 2003), they are nevertheless also expected to increase the fastest relative to other types of plantations. This is because increasing globalization of the markets for pulp and fiber exerts strong pressure in favor of the lowest-cost producers, based on the interaction of land, labor, and capital costs, combined with productivity. The trend is therefore toward short crop rotations in locations that can provide the highest productivity and the lowest costs (Kanowski 1997).

21.5 Services Provided by Forests and Woodlands

The 1992 U.N. Forest Principles identified the multifunctional and multiservice purpose of the world's forests: "Forest resources and forest lands shall be managed and used sustainably to fulfill social, economic, ecological, cultural and spiritual needs of present and future generations" (Forest Principles 1992). The services provided by forests and woodlands are numerous and diverse on all spatial and temporal levels, and include provisioning, regulating, cultural, and supporting services. Some national classifications account for as many as 100 different kinds of forest services, such as delivery of industrial and fuelwood, water protection and regulation, ecotourism, and spiritual and historical values (e.g., Sheingauz and Sapozhnikov 1988; Mather 1999). (See Figure 21.6.) These various forest services relate to each other in many different ways, ranging from synergistic to tolerant, conflicting, and mutually exclusive. The multiservice paradigm of forest management is therefore quite clear in theory but is often very difficult to implement, as it frequently requires difficult choices and trade-offs.

Market approaches can only be used to estimate the value of a few forest services, mostly the ones related to provisioning and that enter formal markets, although markets are also developing for carbon and biodiversity (Scherr et al. 2004). There is no consistent methodology, and usually insufficient information, to estimate credible values for many other forest services. (See Chapter 2.) One recent (and controversial) estimate of the annual value of forest ecosystem services totaled $4.7 trillion, roughly 15% of the global GNP (Costanza et al. 1997). An estimate for the value of Mexico's forests is some $4 billion a year (Abdger et al. 1995). The annual total annual loss to Indian society as a result of forest degradation is estimated at about $12 billion (Joshi and Singh 2003). Ricketts et al. (2004) showed that during 2000–03, pollination services from two forests with a total area of about 150 hectares translated into $60,000 a year for a Costa Rican coffee firm due to increased coffee yield (by 20%) and quality.

Approaches such as these do provide at least an order-of-magnitude insight into the importance of forests for people (Agarwal 1992). Many researchers successfully apply monetary methods to "nonmarket" and often "nontraditional" services. The concept of total economic value (Pearce 1990; see also Chapter 2) has become one of most widely used frameworks for identifying and categorizing forest benefits (Emerton 2003). TEV aims to account comprehensively for all forest services, estimating direct values (such as timber, fuelwood, NWFPs, grazing and fodder, and recreation), indirect values (including watershed protection, erosion control, macro-climate regulation, and carbon sequestration), option values (considering future economic options in all affected sectors, such as industrial, agricultural, pharmaceutical, and recreational) and existence values (landscape, aesthetic, heritage, cultural, religious, ritual, and so on). In spite of substantial progress in the theory, conceptual basis, and methodology of TEV during the last two decades (Bishop 1999; Lette and de Boo 2002), forest valuation studies often remain a purely academic exercise and rarely have an impact on practical planning and management (Emerton 2003).

The loss and degradation of natural forest as described in the preceding section has been accompanied by a decline in supply of many forest services. These impacts are felt most acutely by rural communities living in or near forests, who suffer a decline in livelihood resources and well-being (Byron and Arnold 1999), although urban dwellers are also affected. For example, based on the comparison of satellite images of 448 U.S. urban areas, over the last 10 years American cities have lost 21% of their forested areas, the damage of which has been estimated to be over $200 billion, although no estimates of the benefits provided by the land use change were calculated (ENN–Reuters 18 September 2003 on American Forests).

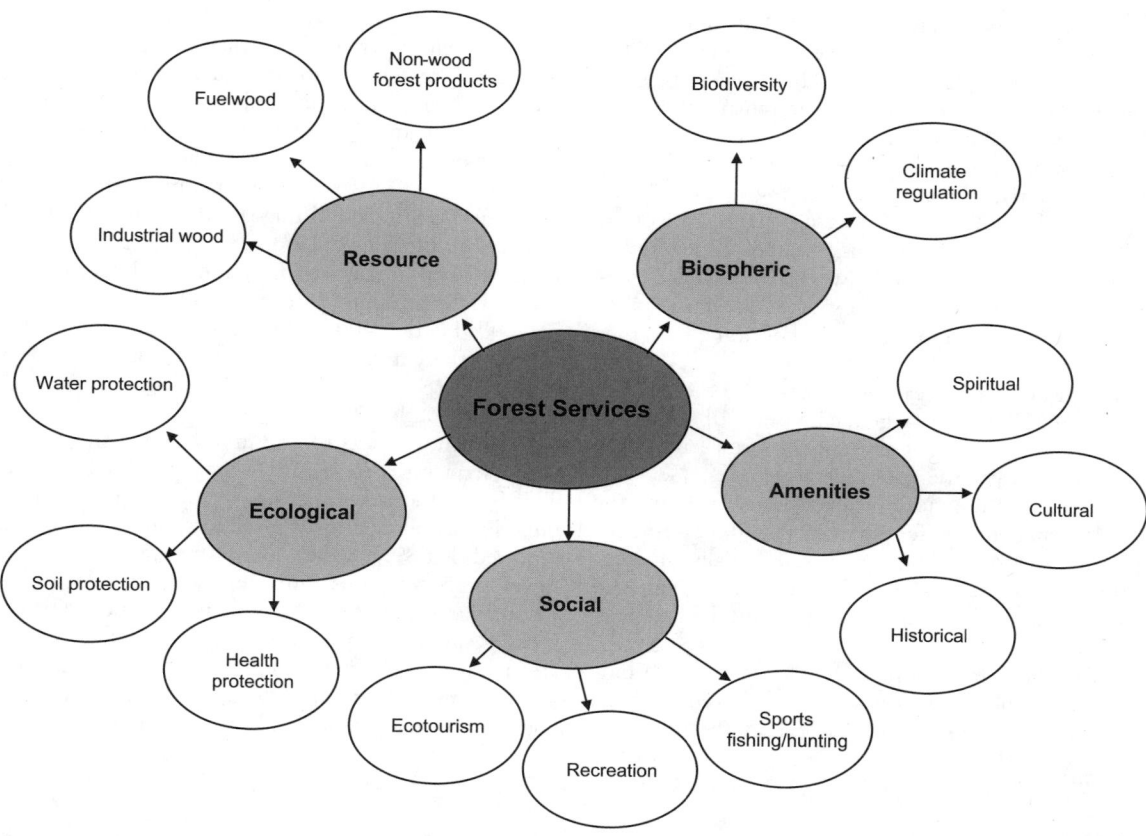

Figure 21.6. Major Classes of Forest Services

21.5.1 Biodiversity

Forests are an important repository of terrestrial biodiversity across three important dimensions: structural diversity (such as areas of forests, natural and protected forests, species mixture, and age structure), compositional diversity (numbers of total floral/faunal species, for example, and endangered species), and functional diversity (the impact of major processes and natural and human-induced disturbances) (Noss 1990; Paumalainen 2001).

Tropical forests cover less than 10% of Earth's land area but harbor between 50% and 90% of Earth's terrestrial species (WRI et al. 1992). The ancient tropical forests of Malaysia, for example, are home to 2,650 tree species, 700 species of birds, 350 species of reptiles, 165 species of amphibians, 300 species of freshwater fish, and millions of invertebrate species (Isik et al. 1997). Other types of forests are not as species-rich as tropical ones but are relatively species-rich ecosystems within their own contexts. Even boreal forests, which harbor only a small number of indigenous tree species (fewer than 100 in Northern Eurasia, for instance) (Atrokhin et al. 1982), have a high diversity at the ecosystem level, and some of their major tree species exhibit high adaptability to extreme climatic conditions. Larch forests, for example, grow at annual average temperature from $+8°$ to $-17°$ Celsius (Sherbakov 1975).

The importance of forest biodiversity for both its existence value as a major component of global biodiversity and its utilitarian value as the source of innumerable biological resources used by people has been recognized by the Convention on Biological Diversity and numerous other bodies and studies (e.g., Heywood et al. 1995; WRI et al. 1992). More recently, studies have shown that biodiversity is also an essential factor in sustaining ecosystem

functioning and hence the ecosystem services that forests provide (Naeem et al. 1999). Biodiversity thus provides the underpinning for many of the other forest services discussed in this section. It can also be viewed as a vast storehouse of information from which future services can be derived.

Considerable information on forest-related biodiversity has become available over the past decade (e.g., Heywood 1995; Secretariat of the Convention on Biological Diversity 2001; Groombridge and Jenkins 2002), but consistent global assessments and monitoring are still difficult due to data insufficiency and incompatibility, different standards and definitions, and geographical and thematic gaps in available assessments. Efforts to assess the nature and distribution of biodiversity rely on the selection of particular subsets of species, species assemblages, or environmental features that can be used as surrogates to measure biodiversity as a whole (Margules et al. 2002). A recent global analysis of gaps in protection of biodiversity within the global network of protected areas, for example, used recently completed surveys on the global spatial distribution of over 11,000 species of mammals, amphibians, and threatened bird species as surrogates (Rodrigues et al. 2003).

Forest decline threatens biodiversity at all levels. IUCN estimates that 12.5% of the world's species of plants, 44% of birds, 57% of amphibians, 87% of reptiles, and 75% of mammals are threatened by forest decline (IUCN 1996, 1997). *The World List of Threatened Trees* (Oldfield et al. 1998) indicates that more than 8,000 tree species (9% of the total) are currently threatened with extinction.

It is difficult to say with precision the extent to which forest habitat loss results in population or species extinctions, because our knowledge of forest biodiversity is so incomplete. Nonetheless, it is clear that deforestation, particularly in the tropics, is having extremely negative impacts on biodiversity. Fifteen of the 25 biodiversity "hotspots" originally identified by Myers (1997)—areas with high levels of plant endemism and high levels of habitat loss and threat that between them contain the remaining habitat of 44% of all plant species and 35% of all vertebrate species worldwide—contain tropical forests. These areas once covered nearly 12% of Earth's land surface, but their remaining natural habitat has been reduced to only 1.4% of that surface—that is, 88% of the hotspots' original natural habitat has disappeared. Brooks et al. (2002) concluded that habitat loss in the world's biodiversity hotspots has left extremely large numbers of species threatened, with a high probability of extinction in the absence of immediate conservation action.

Development of protected area systems has been the primary strategy for conserving biodiversity generally (see MA *Policy Responses,* Chapter 5), and significant amounts of forest have come under protected status over the past several decades. (See Box 21.4.) Given the multiple functions of forests, however, and the impracticality of placing enough forests in protected areas to conserve the full range of forest biodiversity substantially, maintenance of the diversity of forest-dependent species in managed forests (such as logging concessions) is also an important strategy (Sayer et al. 1995).

Modification of forest management practices to include biodiversity conservation objectives may not generally require large additional investments, at least in tropical forests (Johns 1997).

Some of the simple but important measures that can be taken to this end include retention of small refuge areas and the maintenance of riparian buffer strips at the level of the management unit, and distributing logged and unlogged areas in an appropriate way across the broader landscape. There is also a growing awareness among tropical ecologists that secondary forests recovering from alternative land uses may play an important role in conserving biodiversity (Brown and Lugo 1990; Dunn 2004). It is generally accepted that forest plantations, particularly even-aged and single-species plantations, are less favorable as habitat for a wide range of taxa in different regions of the world (Allen et al. 1997; Davis et al. 2000; Hartley 2002; Humphrey et al. 2002; IUFRO 2003), although there may be some exceptions, such as in degraded landscapes or areas with low original forest cover (Brockie 1992; Kwok and Corlett 2000).

Managing forests to conserve biodiversity (and other non-wood services) requires that a management regime be in place. This is not the case for many countries, particularly developing ones. Eighty-three non-tropical (including all industrial) countries reported that 89% of their forests are managed, although data for developing countries indicate that only 123 million hectares (about 3% of forest area) are managed under formal long-term plans. Regional variation is very high: about 1% of the total forest area in Africa, about 25% in Asia, 85% in Oceania, 55% in North and Central America, and 3% in South America are managed according to such plans. While areas in the tropics seem slightly

---

BOX 21.4

Forest Protected Areas

A proportion of forests in most countries have protected status. Recent statistics from the UNEP World Conservation Monitoring Centre (UNEP-WCMC 2002) reveal that around 10.4%—470 million hectares—of the world's 4,540 million hectares of forests are under various forms of protection. Countries use a variety of systems for classifying their protected areas. Some have very detailed classifications (up to 20–25 forest protected area categories), although many use some variant of the simpler international IUCN classification system. A CD-ROM published by UNEP-WCMC and the Center for International Forestry Research (Iremonger et al. 1997) contains a detailed analysis of forest protected areas by ecological zone, country, and region. About 7.8% of the world's forests are included in areas that are protected to the level of IUCN categories I-VI.

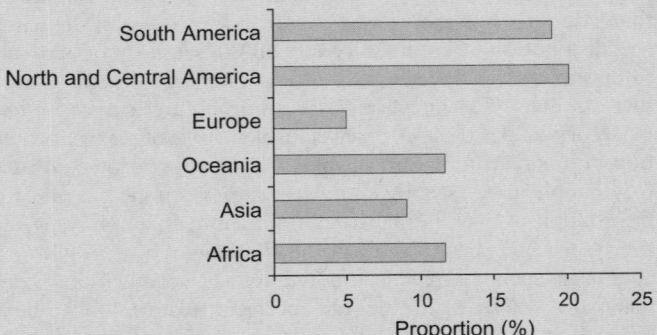

Percentage of Forest Area in Protected Status by Continent (FRA 2000)

According to the global map of protected forest areas produced by FAO and UNEP-WCMC as a part of FRA-2000, protected areas cover 479 million hectares—12.4% of the world's forests. Some 15.2% of tropi-

cal forests are protected, 11.3% of sub-tropical forests, 16.3% of temperate forests, and 5% of boreal forests. The highest percentage of protected forests is found on the American continent (about 20%) and the lowest in Europe (5%). (See Figure.) These data are not, however, completely consistent (cf., e.g., FFSR 1999; WDPA 2003).

Biodiversity conservation is now widely recognized as the most important objective of protected areas (although many current protected areas were not designed with this in mind), and this objective generates specific size and distribution requirements. Many studies have shown, for example, that conservation of a great deal of biodiversity is not viable in small fragmented areas. Experience in some countries, however, yields examples where even small protected areas can meet some biodiversity conservation objectives (Isaev 1991). This is particularly the case where small protected areas are surrounded by sustainably managed forests. Some countries have established specific categories of protected areas for particular purposes, such as water or soil conservation. TBFRA-2000, for example, reported such designations for soil protected in most industrial countries such as Russia (90.8 million hectares), Kazakhstan (9.3 million hectares) and Greece (2.5 million hectares).

Formal designation of protected areas, however, does not guarantee their effective management. In many developing countries, protected areas often exist on paper but lack active management and are in fact subject to illegal logging and wildlife poaching, agricultural encroachment, and settlement. In some cases they may be completely or partially treeless. As a result, experts and policy-makers frequently debate how best to divide limited financial and technical resources between designating new protected areas, strengthening management of existing protected areas, and establishing novel forms of community-based conservation—working with local and indigenous communities outside of formal state-run protected areas (Barber et al. 2004).

underestimated, the data for some large non-tropical forest countries should be used with caution due to different national definitions of "managed forests" (FAO 2001c; Shvidenko 2003).

21.5.2 Soil and Water Protection

The global condition of the world's soils and hydrological systems is not well known, but it is considered to be far from satisfactory. (See Chapters 7, 20, and 22.) From 1945 to 1990, a vast area (1.2 billion hectares) of land is estimated to have suffered moderate to extreme soil degradation, and degraded areas accounted for 17% of Earth's vegetated lands (Oldeman et al. 1990; WRI 1992). The major causes for this extensive degradation are a number of current practices in agriculture, forest management, and grazing.

In many regions, forest is a major stabilizing component of natural landscapes, providing protection of soil and water, households, and fields and reducing or preventing floods and landslides. In the Ukraine, for example, soils on 11% of the territory are in good condition, 18% satisfactory, 22% in conflict, 25% pre-critical, and 24% in critical ecological condition. The relative conditions in different areas are strongly correlated with the extent of forest cover, which varies from 26% to 3% across the country (Yukhnovsky 2003). Levels of soil erosion in the tropics may be 10–20 times higher on areas cleared of forests, due to construction of roads, skidder tracks, and log landings during mechanical logging, than in undisturbed natural forests, and this is particularly the case in mountainous and other areas characterized by fragile soils (Wiersum 1984; Dickinson 1990; Baharuddin and Rahim 1994; Douglas 1996; Chomitz and Kumari 1998).

Regulation of hydrological cycles and processes is one of the important services provided by forests at large scales. Globally, forests' hydrological functions have been claimed to include increasing precipitation and decreasing evaporation; regulating the total and redistribution of surface and belowground runoff; smoothing out the seasonal course of river discharges; increasing total annual river runoff; protecting landscapes against soil erosion and landslides, in particular in mountains; preventing and mitigating the consequences of floods; maintaining water quality; protecting river banks against destruction (abrasion); and preventing siltation of reservoirs (e.g., Protopopov 1975, 1979; Rakhmanov 1984; Rubtsov 1990; Pielke et al. 1998; Bruijnzel 2004).

While forests play an undeniably important hydrological role across the globe, the specifics of this role vary substantially among biomes, landscapes, and forest types. Differences between tropical, temperate, and boreal zones, for example, can be great (e.g., Hamilton and King 1983; Bruijnzeel 1989; Versfeld et al. 1994; Sandström 1995). Research in tropical forest areas indicates that the roles of forests in watershed hydrology may have been overestimated in some cases: in arid areas, for example, trees evaporate more water than other vegetation types, and there is little evidence that forests attract precipitation (with the exception of cloud forests); forests reduce runoff but are not always effective at flood prevention, since tropical forest soils become rapidly saturated in tropical rainstorms. Some studies do conclude, however, that forests promote an even seasonal and annual flow, particularly in the dry season (Hamilton and King 1983; Dhawan 1993; Cossalter and Pye 2003; Kaimowitz 2004).

Because of the important role that forests play in protecting watersheds, many countries grant protected status to forests that serve this purpose. Such forests often include protective belts along rivers, lakes, artificial water reservoirs, and other bodies of water, and forests on steep slopes (Dudley and Stolton 2003; De-Philip 2003).

Many of the world's major rivers begin in mountain highlands, and more than half the planet relies on fresh water flowing from these areas. (See Chapter 24.) One third of the world's 105 largest cities obtain their water supply from forested watersheds. However, 42% of the world's main river basins have lost more than 75% of their original forest cover, and there is a clear relationship between population density and forest loss in the river basin (Revenga et al. 1998). The Yangtze watershed in China (home to some 400 million people) has lost 85% of its original forest cover, and in 1998 severe flooding along the Yangtze killed more than 3,800 people and caused over $20 billion in damage (Eckholm 1998). Many countries in the arid zone face an acute water deficit. In South Africa, for example, 54% of available runoff is currently used, and the level of use is expected to increase by 75% during the next 25 years. Twenty-eight countries experienced shortages of fresh water in 1998, and this number is expected to increase to 56 countries by 2026 (Versfeld et al. 1994; Sandström 1995).

Water and soil protection services of forests depend critically on the area and spatial distribution of forests over landscapes. It has been suggested that in the temperate zone, the minimal forest coverage that provides significant protection of landscapes over large territories varies from 7% to 30%, depending on region, climate, vegetation type, specific landscape features, and other factors (Protopopov 1975; Shvidenko at al. 1987). In semiarid and arid conditions, under a system of protective tree shelterbelts and trees outside of forest over the area to be protected, significant levels of protection for water and soil services can be attained with forest cover of 3.5% to 5–6%, as in the steppe zone of Ukraine and Russia, for example (Pilipenko et al. 1998).

21.5.3 Protection of Fragile Ecosystems: Forests in Mountains, Drylands, and Small Islands

Forests play a specific and very important environmental and social role in fragile ecosystems and landscapes, such as mountains, drylands, and small island ecosystems, particularly at the local level. People in these areas often have a high dependence on forest services. (See Chapters 22, 23, and 24.)

Forests in mountains have local and regional value as regulators of water supplies, centers of biological diversity, providers of forest products, and stabilizers of land against erosion. Due to the generally greater precipitation in mountains and the high ability of montane (in particular cloud) forests to capture atmospheric water, mountains play an extremely important role in maintenance of hydrological cycles affecting large territories. The alpine catchment of the Rhine River, for example, occupies only 11% of the river basin but supplies 31% of the annual flow and more than 50% of the summer flow (Price 1998). In semiarid and arid areas, over 90% of river flow comes from the mountains.

Trees and forests in dryland areas provide fuelwood, small roundwood (poles for building houses and fences), non-wood forest products (foods, medicinal products, bushmeat, fodder, and so on), and diverse regulating and cultural services. Their most critical functions in many dryland areas are soil conservation, shade, and shelter against wind.

The forest cover of 52 small island states and territories is insignificant in global terms—only 0.2% of global forest area in 1995 (FAO 2003b). But forests and trees on these islands are extremely important for the well-being of the local populations, the conservation of biological diversity, and the maintenance of environmental conditions both on land and in surrounding marine ecosystems. (See Chapter 23.) They play an important role in protecting watersheds, maintaining water supply, and protecting the marine environment.

Mangroves and other tidal forests are highly productive ecosystems that are important feeding, breeding, and nursery backgrounds for numerous commercial fish and shellfish, including most commercial tropical shrimp. (See Chapter 19.) FAO (2003b) has reported decreases in the extent of mangrove forest from 1980 to 2000 at an annual rate exceeding 1%.

Forests on small islands are extremely important for coastal protection against the strong winds, high rainfall, and storm surges of cyclones and hurricanes, and they serve as sediment traps for upland runoff sediments (Roennbaek 1999). Biodiversity conservation is another crucial service provided by forests on small islands. High endemism is an intrinsic feature of small island biodiversity: about 30% of higher plant species, 20–25% of birds, and 25–50% of mammals are island endemics (WRI et al. 1996).

21.5.4 Fiber, Fuel, and Non-wood Forest Products

Wood is currently the most economically important forest product. During 1996–2000, about 3.3 billion cubic meters of wood were harvested annually from the world's forests, and roundwood production has steadily increased by approximately 0.8% per year (FAO 2002a). By region, the largest annual increases during this time were observed in Europe (4.8%) and Oceania (3.5%). Only Asia experienced a substantial decrease in harvested wood (–1.2% per year), which is explained by dramatic decreases in three countries: Malaysia (–33%), Indonesia (–16%, due to economic and political disturbances during this period), and China (–8%, due to the drastic national measures taken in 1998 to restrict harvesting). A significant proportion of harvested wood and wood products is traded internationally. For a detailed assessment of wood production, see Chapter 9.

In the temperate and boreal zone, 63% of forests are classified as available for wood supply. The average growing stock of forest available for wood supply is between 105 and 145 cubic meters per hectare, with considerable variation among countries (from less than 50 to over 250 cubic meters for some European countries with strong silvicultural traditions). On average, the growing stock of forest available for wood supply increased by about 640 million cubic meters a year during the last decade, mostly in Europe and North America, due to forest management and global change (in particular, a longer growing period in the boreal zone, increased temperature and precipitation, elevation of the atmospheric concentration of CO_2, and increasing nutrient deposition) (e.g., Myneni et al. 2001; Ciais et al. 2004).

Removal as a percentage of mean annual increment is an indicator of sustainability of wood supply. For the temperate and boreal region as a whole, this figure is estimated to be 52.6%, with strong regional variations between North America (78.6%), Europe (59%), and the former Soviet Union (16.8%). Conifers are used more intensively (62.5%) than broadleaf forests (42.2%). Annual felling in the temperate and boreal domain (1,632 million cubic meters, of which over half is in the United States and Canada and 28% in Europe) is, however, substantially higher than the level of removals (1,260 million cubic meters)—implying a high level of harvest loss (FAO 2002a).

The total area under timber harvesting schemes in 43 selected countries accounting for approximately 90% of the world's tropical forests is estimated to be about 50 million hectares (55% in tropical forests of Asia and Oceania, 33% in Latin America, and 12% in Africa). Annually harvested area is estimated to be about 11 million hectares (29% in Africa, 54% in Asia and Oceania, and 17% in Latin America), although harvesting intensity is highly variable by country, ranging from 1 to 34 cubic meters per hectare (FAO 2001c).

Accessibility of forests is also an important factor for assessing the sustainability of wood supply. Approximately 51% of the world's forests are within 10 kilometers of major transportation infrastructure, including big rivers (from 38% in South America to 65% in Africa), and 78% are within 50 kilometers. Boreal and tropical forests are more remote than others. Some 14% of the world's forests were considered unavailable for wood supply, as they are located either in protected areas (12.4%) or in inaccessible mountain areas (FAO 2001c).

The importance of plantations as a source of timber is likely to continue to increase. For example, it is expected that forest plantations in China (currently about 47 million hectares) will provide up to 150 million cubic meters of wood annually (Jiang and Zhang 2003).

Fuelwood meets about 7% of energy demand worldwide, including about 15% in developing countries and 2% in industrial countries (WEC 1999). Globally, about 1.8 billion cubic meters of wood is used annually for fuel (including charcoal production). However, there is a large amount of variation in these figures, and more than 70% of energy needs in 34 developing countries and more than 90% in 13 countries (of which 11 are in Africa) are met through fuelwood. Woodfuel constitutes about 80% of total wood use in developing countries, where about one third of the total forest plantations were established primarily for that purpose. More than 60% of these plantations are in Asia and 25% are in Latin America. Plantations currently supply 5% of woodfuel, although it is estimated that woodfuel supply from plantations will grow 3.5-fold by 2020 (FAO 2001c).

Estimates of the potential of the world's forests to meet most of the world's demand for fibers and fuel in the future vary considerably and are significantly affected by economic accessibility and protection status of forests. Hagler (1995) estimated that only 2.1 billion hectares of forest are usable for fiber and fuel and that this forest area can sustain a long-term harvest of 3.7 billion cubic meters of wood per year. This study did not, however, consider current and potential wood supply from trees outside forests. Nilsson (1996) estimated that by 2020 the world demand for industrial roundwood and fuelwood (including charcoal) will be 2.4 billion and 4.25 billion cubic meters, respectively. However, a forecast by Broadhead et al. (2001) for fuelwood is only about 1.9 billion cubic meters for the same year (including charcoal).

According to these analyses, the world's forests are very close to exhausting their fiber and fuel potentials, and intensive measures will be needed to satisfy the deficit projected for 2020. Nilsson (1996) argues that even if the high forecasts are accurate, the deficit will not occur in reality due to market mechanisms, which are likely to achieve equilibrium between supply and demand. Broadhead et al. (2001) argue that fuelwood demand has in fact already peaked. More detailed information on this issue may be found in Chapter 9, where considerable evidence is presented in support of the conclusion that a global shortage of wood, per se, is not likely to occur in the near future, although there are likely to be significant regional disparities, such as unsatisfied market demand for large-dimension timber of high quality. Overall, land use changes and policy decisions will likely have a greater impact on forest ecosystems than timber harvest.

No doubt, wood has a great future. As construction material (25% of the annual global wood harvest), wood outperforms steel and concrete on an environmental basis (CWC 1999). Wood is a renewable resource and can also be recycled or reused. Development of new wood-processing technologies and products, environmental scrutiny, new applications, and new markets are some of trends that are expected to influence wood supply and demand over the next few decades (Roche et al. 2003; Pisarenko and Strakhov 2004), and rapid urban growth in developing countries has substantially increased the demand for industrial wood and fuel (Scherr et al. 2002).

Non-wood forest products (defined as goods of biological origin other than wood, derived from forests, other wooded land, and trees outside the forests) (FAO 1999b) include a tremendous diversity of items—some of which enter formal markets, but many that do not (FAO 2001b, 2001c, 2001f; UNECE 1998). They can be classified in a number of broad categories according to their end use: edible products; fodder for domestic animals; medicines; perfumes and cosmetics; colorants; ornamentals; utensils, handicrafts, and construction materials; and exudates like gums, resins, and latex. Overall, they play an important role in the daily life and well-being of hundreds of millions of people worldwide as well as in the national economies of many countries.

At least 150 NWFPs are of major significance in international trade, and the annual export value of these products was estimated at $11 billion in 1994. China is the leading exporter of NWFPs, followed by India, Indonesia, Viet Nam, Malaysia, the Philippines, and Thailand (Iqbal 1995). NWFPs provide subsistence, employment, and income, particularly for the rural poor, and support small, household-based enterprises, especially in developing countries (e.g., Arnold 1998; Ciesla 1998). The most reliable estimates indicate that from 200 million to 300 million people earn much of their subsistence income from nonindustrial forest products (Byron 1997). From 150 million to 200 million people belonging to indigenous groups in over 70 countries, mostly in tropics, depend on NWFPs to sustain their way of life, including their culture and religious traditions (CIDA 1998).

Edible NWFPs—vegetables, fruits, nuts, seeds, roots, mushrooms, spices, bushmeat, bee products, insects, eggs, nests, and so on—are particularly important in tropical and sub-tropical regions. For example, bushmeat and fish provide more than 20% of the protein in 62 least developed countries (Bennet and Robinson 2000). And in rural areas of many countries, a significant relationship exists between food security and the degree of contribution of NWFPs to households (Odebode 2003). From 8% to 46% of indigenous tree species serve as a source for food and fodder in the Pacific region (Siwatibau 2003).

Many edible products are increasingly exported, including honey and beeswax from Africa, bamboo from China (1.6 million tons of fresh shoots exported in 1999), and wild edible mushrooms, mostly morel mushrooms (a trade with a total annual value of $50–60 million). In the mid-1990s, the cost of importing of edible NWFPs to three main markets—Europe, the United States, and Japan—was estimated at about $2.5 billion (Iqbal 1995).

Fodder is of great importance in many regions, particularly in the arid and semiarid zones and in animal-based production systems. In many developing countries, 30–40% of domestic animals depend on forests for grazing and fodder (FAO 2001c).

About three quarters of the people in developing countries use traditional medicines, and the ratio of traditional healers to western-trained doctors reaches 150:1 in some African countries (FAO 2001c). Medicinal plant species (mostly from the forest) used by local populations and as trade products number in the thousands, and some 4,000 commercially important medicinal plant species are used in Southeast Asia alone. The value of the world trade in medicinal plants in 1992 was on the order of $171 million (Iqbal 1995). Medicinal plant exports are economically important for some countries, such as Morocco and Egypt, which

export from 7,000 to 15,000 tons of medicinal plants annually (Lange and Mladenova 1997).

Forest plants are also widely used in the development of modern medicines for heart disease, cancers, leukemia, and HIV/AIDS. According to one survey, 90% of the most-prescribed pharmaceuticals in the United States contain compounds of forest origin (Lyke 1995). This is particularly remarkable in light of the fact that only 5–15% of higher plant species have been investigated for the presence of bioactive compounds (ten Kate and Laird 1999).

Rattan is the most important internationally traded NWFP. There are more than 600 species of rattan, some 10% of which are commercially used. Bamboo (more than 500 species) is the most commonly used NWFP in Asia, where about 20 million tons are produced annually. (See Chapter 9.) The average annual value of the world trade in bamboo ware is on the order of $36 million (FAO 2001c).

Global estimates of the total monetary value of NWFPs are very approximate and express an order of magnitude rather than documented market prices, particularly for subsistence uses. A number of studies (Myers 1997; UN-CSD/IPF-CSD 1996; Michie et al. 1999) have attempted to estimate the value of the subsistence use of NWFPs, arriving at figures ranging from $90 billion to $120–150 billion. This aggregate figure includes valuation of fodder and grazing ($40–50 billion); edible products ($20–25 billion); traditional medicines derived from plants, insects, and animals ($35–40 billion); and non-wood construction materials, such as thatch grass and bamboo, and other similar items ($25–35 billion).

21.5.5 Carbon Sequestration

Forests play an important role in the global carbon cycle and consequently in regulating the global climate system. Two main features of forests define this role. First, the world's forests accumulate a major part of the planet's terrestrial ecosystem carbon. Second, forests and wetlands are the two major land cover classes that are able to provide long-term sequestration of carbon. Accumulation of carbon in wood and soils results in a more significant share of total net primary productivity being stored in the long term than in other land cover classes and can represent as much as 10–15% of NPP (Field and Raupach 2004; Shvidenko and Nilsson 2003).

Estimates for the carbon stock in the world's forest ecosystems vary in the range of 352–536 billion tons of carbon (Dixon et al. 1994; Houghton 1996; Brown 1998; Saugier et al. 2001). The IPCC estimate of carbon content for three major forest biomes (covering 4.17 billion hectares) is 337 billion tons in vegetation and 787 billion tons in the top 1 meter layer of soils (IPCC 2001a). FRA-2000 estimated the aboveground tree biomass at 422 billion tons of dry matter (or 5.45 kilograms of carbon per square meter). The estimate by Kauppi (2003), based on FRA-2000, is 300 billion tons of carbon for tree biomass of forest ecosystems. Previously reported estimates—8.6 kilograms carbon per square meter by Dixon et al. (1994), 10.6 kilograms by Houghton (1999), and 6.6 kilograms by Kauppi (2003)—significantly overestimated densities. Based on analysis of all available sources and taking into account the above analysis of global forest area, it is estimated here that the total biomass of forest ecosystems is likely to include 335–365 billion tons of carbon (a priori confidence interval 0.9).

Forest carbon stocks and fluxes, and the major drivers of their dynamics, have been quantified for certain globally important forest areas (Zhang and Justice 2001; Houghton et al. 2001b; Fung et al. 2001; DeFries et al. 2002; Dong et al. 2003; Birdsey and Lewis 2003; Baker et al. 2004). Four major processes define whether forests serve as a net carbon sink or source: net primary productivity, decomposition (heterotrophic respiration), natural and human-induced disturbances (including harvest and consumption of forest products), and transport of carbon to the lithosphere and hydrosphere. The rate of accumulation of carbon over a whole ecosystem and over a whole season (or other period of time) is known as the net ecosystem productivity. In a given ecosystem, NEP is positive in most years and carbon accumulates, even if only slowly. However, major disturbances such as fires or extreme events that cause the death of many components of the biota release greater-than-usual amounts of carbon. The average accumulation of carbon over large areas or long time periods is called net biome productivity.

Productivity of forests varies significantly by continents, ecological zones, and countries, and no consistent global inventory of forest net primary productivity exists. Current estimates are based on potential (but not actual) forest cover and do not adequately take disturbances into account. This results in overestimation of biomass by 40–50% and overestimates of NPP by 25–35% for some large regions of the planet (Haberl 1997; Shvidenko et al. 2001).

Based on current understanding of the terrestrial vegetation global carbon cycle, NPP is estimated at 60 billion tons of carbon per year (e.g., Melillo et al. 1993; Goldweijk et al. 1994; Alexandrov et al. 1999), decomposition at 50 teragrams of carbon per year, net ecosystem productivity at 10 billion tons of carbon per year, and net biome productivity at 1 billion tons of carbon per year (0.7 ± 1.0 billion tons during 1988–99). The proportion of global NPP provided by forests is different in different climate zones and remains rather uncertain. Factors that influence the net uptake of carbon by forests include the direct effects of land use and land cover change (such as deforestation and regrowth), harvest and forest management, and the response of forest ecosystems to CO_2 fertilization, nutrient deposition, climatic variation, and disturbances.

Deforestation in the tropics has the greatest impact on the carbon cycle of any land use and land cover change. It is reported that land use change (mostly deforestation) is the source of 1.6 ± 0.8 billion tons of carbon per year (Houghton et al. 1999, 2001), although other estimates of net mean annual carbon fluxes from tropical deforestation and regrowth were 0.6 (0.3–0.8) and 0.9 (0.5–1.4) billion tons for the 1980s and 1990s (DeFries et al. 2002). Dixon et al. (1994) estimated that global forests were a net source of 0.9 ± 0.4 billion tons of carbon in the 1990s, including large sources in the low-latitude forests (1.6 ± 0.4 billion tons a year) and net sinks in mid-latitude (0.26 ± 0.09 billion tons a year) and high-latitude (0.48 ± 0.1 billion tons a year) forests.

Inversion studies using atmospheric-transport models indicate that land in the temperate and boreal latitudes of the Northern Hemisphere was a sink for 0.6–2.7 billion tons of carbon a year during the mid-1980s to mid-1990s, although patterns of spatial distribution of this sink are rather contradictive (Fan et al. 1998; Bousquet et al. 2000; Rayner et al. 1999; Battle et al. 2000; Prentice et al. 2001), and there is substantial interannual variation of forest NBP, which can reach three- to fivefold for large regions. Goodale et al. (2002) estimated that northern forests and woodlands provided a total sink for 0.6–0.7 billion tons of carbon a year during the early 1990s, consisting of 0.21 tons in living biomass, 0.08 tons in forest product, 0.15 tons in dead wood, and 0.13 tons in soil organic matter.

Russian forests, which account for about two thirds of total boreal forests, experienced severe disturbances during this period,

which resulted in an estimated annual carbon sink for 1988–92 of 0.11 billion tons of carbon (Goodale et al. 2002). Later it has been shown that the forest sink in Russia during this period was minimal over the last four decades: the annual average NBP of Russian forests has been estimated at 0.43 billion tons of carbon per year from 1961 to 1998 (Shvidenko and Nilsson 2003). Canadian forests served as a net carbon sink before the 1980s but became a carbon source as the result of increased disturbances and changes in the age class distribution (Kurz and Apps 1999).

Recently disturbed and regenerated forests usually lose carbon from both soil and remnant vegetation, whereas mature undisturbed forests maintain an overall neutral carbon balance (Apps et al. 2000). The rate of carbon sequestration depends upon age, site quality, species composition, and the style of forest management. Mature and over-mature boreal forests in many cases actually serve as a net carbon sink (Schulze et al. 1999), which probably relates to accumulation of carbon in forest soils and uneven-aged forest structure.

The post-Kyoto international negotiation process envisages an important role for forests in current and future efforts to mitigate climate change. Forest management operations that simultaneously improve the condition and productivity of forests and stabilize natural landscapes are able to increase the carbon stock of forest ecosystems and ensure its persistence. These activities include afforestation and reforestation, thinning, improving forest protection, increasing efficiency of wood processing, and use of wood for bioenergy. Numerous studies show significant potential of the world's forests in this respect. Implementation of special carbon management programs in Russia, for example, allows for sequestration of 200–600 teragrams of carbon annually during the next 100 years in a globally competitive carbon market (Shvidenko et al. 2003). The ability of forests to sequester carbon effectively takes on special significance since the Kyoto Protocol entered into force in 2005. Implementation of successful carbon management will require improvements in national forest policies, legal instruments, monitoring and reporting in many countries, and general progress in the transition of world forestry to sustainable forest management.

Plantations are also increasingly established as a response to climate change. A number of countries already have programs to establish forest plantations for carbon sequestration. In Costa Rica, for example, reforested conservation areas are credited with income for the carbon sink and watershed protection services they provide (Chichilnisky and Heal 1998). By 2000, about 4 million hectares of plantations worldwide were established with funding for carbon sequestration. However, despite much progress in the post-Kyoto Protocol international negotiation process, some important political and economic questions concerning the use of forestry and land use change for mitigating climate change remain to be resolved. The protocol allows carbon sequestered by afforestation or reforestation after 1990 to be counted as an offset for emissions under certain circumstances. Some observers (e.g., Schulze et al. 2002) fear that this might offer incentives to fell older, natural woodland (for which no offsets are available) and replace them with plantations. However, the accounting and verification procedures, such as those agreed in the Marrakesh Accords to the protocol, are designed to eliminate such perverse incentives.

21.5.6 Sociocultural Values and Services

Forests are highly valued for a host of social, cultural, and spiritual reasons. Forests and people have co-developed, with people shaping the physical nature of most forests (including those we today consider "natural") and the forest, in turn, exerting a powerful influence over human cultures and spiritual beliefs (Laird 1999; Posey 1993; UNESCO 1996). For many indigenous and traditional societies, forests are sacred and sometimes supernatural places, linked to both religious beliefs and the very identity of some communities and peoples (Parkinson 1999). The widespread existence of "sacred groves" in many societies is a physical manifestation of this spiritual role and has contributed to forest conservation. (See Chapter 17.)

Forests provide spiritual and recreational services to millions of people through forest-related tourism. Nature-based tourism has increased more rapidly than the general tourism market, evolving from a niche market to a mainstream element of global tourism, with annual growth rates estimated to be in the range of 10–30%. (See Chapter 17.) Although it is difficult to estimate with any precision what proportion of regular tourism has been redefined as "nature-based" or how many "nature-based tourists" are drawn to destinations because they are forested, it is nevertheless evident that forests, woodlands, and the species they support are a significant element of many ecotourism destinations—from the national parks of North America to the megafauna–rich savannas of Africa.

21.5.7 Services Provided by Agroforestry Systems

Although forests and woodlands can be a substantial component of agroforestry systems, trees outside forests are also a crucial component of these systems. Services provided by agroforestry systems vary between different climate regions and include woody and non-woody forest products for commercial and subsistence use; maintenance of soil fertility via organic matter input to the soil, nitrogen fixation, and nutrient recycling (Szott and Palm 1996; Buresh and Tian 1998); reduction of water and wind erosion (Beer et al. 1998; Yukhnovsky 2003); conservation of water via greater infiltration (Bharati et al. 2002); enhanced carbon capture (Lopez et al. 1999); and maintenance and management of biodiversity in agricultural landscapes (Beer et al. 2003).

21.5.8 Discussion

While it is clear that the value of forest services is very high, there are many gaps in scientific understanding and few practical solutions to reconciling the conflicts that arise from the competing values that different user groups ascribe to different forest services. Interests of landowners, local communities, governments, and the private sector vary and frequently conflict in both spatial and temporal terms. The time horizon for using individual forest services is substantially different, for example, for forest-dependent indigenous communities and large logging companies.

There are many similarities in the importance and use of forest services in industrial and developing countries, as well as clear geographical, national, and user group differences. For example, the relative importance of wood production has been ranked as "high" and "medium" by 78% and 89% of respondents in United States and France, respectively, but estimates for grazing were 33% and 4%, and for nature protection 50% and 100% (Agarwal 1992).

Expert estimates presented in Tables 21.5 and 21.6 indicate, to some extent, current understanding of the relative importance of different forest services for tropical and non-tropical forests. Although it is not easy to predict future trajectories of changes for these estimates, demands on forests as sources of both fiber and other services will undoubtedly grow significantly. Two central factors of global change, however, will likely be determinative: the extent to which development challenges are met and poverty is reduced in many parts of the world (IIASA and FAO 2002) and

the extent to which the direct and indirect impacts of climate change on the capacity of forests to provide services might exceed the resilience of forest ecosystems in many regions.

21.6 Drivers of Change in Forest Ecosystems

Understanding the drivers of change in forest condition at different spatial and temporal scales is a complicated task. As a rule, such changes are the result of interactions among many factors—social, ecological, economic, climatic, and biophysical. (See Chapter 3.) Rapid population growth, political instability, market forces, institutional strengths or weaknesses, natural and human-induced disturbances, and many other factors may be important. Biophysical factors, such as a region's history of landscape transformation (Mertens and Lambin 2000), the high sensitivity of forest soils to machinery used for logging (Protopopov 1979), or the high flammability of boreal forests (Kasischke and Stocks 2000) can also play a significant role (McConnell 2004).

21.6.1 Tropical Forest Ecosystems

Forest degradation and conversion to other land uses are the two main processes of change occurring in natural tropical forest ecosystems. Numerous studies have attempted to ascertain the direct and indirect drivers of tropical deforestation and the relationships among them, and broadly conclude that in many situations it is impossible to isolate a single cause due to the complex socioeconomic processes involved, and the diverse circumstances in which it occurs, which often obscures underlying patterns (Walker 1987; Roper 1996). Despite this complexity, it is clear that tropical deforestation is caused by a combination of direct and indirect drivers, that these drivers interact with each other, often synergistically, and that the specific combinations of drivers vary between regions of the globe, countries, and even between localities within countries.

The assessment of tropical deforestation provided by Geist and Lambin (2001, 2002) and further elaborated in Lambin et al. (2003) is presented here. It provides a comprehensive review and synthesis of recent literature and draws on analysis of 152 subnational case studies.

21.6.1.1 Direct Drivers

Direct drivers of tropical deforestation are human activities or immediate action at the local level, such as agricultural expansion, that originate from intended land use and directly affect forest cover (Geist and Lambin 2002). These direct drivers can be broadly categorized into those related to agricultural expansion, wood extraction, and infrastructure extension.

Agricultural expansion includes shifting cultivation (both traditional swidden agriculture and the more destructive "slash-and-burn" cultivation); permanent agriculture, which may be at large or small scales and, in the latter case, for either commercial or subsistence purposes; pasture creation for cattle ranching; and sponsored resettlement programs with the objective of converting forest to agriculture, estate crops, or timber plantations.

Wood extraction includes commercial wood extraction (state-managed or private logging concessions), fuelwood extraction and charcoal production for both domestic and industrial uses, and polewood extraction for both domestic and urban uses. Most timber extraction in tropical regions is done without effective management, and logging often inflicts a great deal of damage on the remaining forest stand (Verissimo et al. 2002; Schneider et al. 2002), although technologies of reduced impact logging have been successful on an experimental scale (Sist et al. 1998; Ceder-

Table 21.5. Major Services Provided by Tropical Forests and Woodlands to Various User Groups (Based on regional expert estimates)

User Group	Freshwater Yield	Fuel	Timber and Pulp	NWFP	Biodiversity	Amenities	Carbon Storage
Local communities	5	5	3	4	2	4	2
Loggers	2	4	5	2	1	2	2
Downstream users							
Cities	4	3	4	3	2	4	2
Agriculture	5	4	3	4	3	3	1
Industry	3	2	5	1	0	1	1
Timber traders	1	3	5	3	0	0	1
National	5	4	4	3	4	4	3
Global	3	4	3	4	5	3	3

Key:
5 – crucial   2 – moderately important
4 – very important   1 – sporadic use
3 – important   0 – not used

Table 21.6. Major Services Provided by Temperate Forests and Woodlands to Various User Groups (Based on regional expert estimates)

User Group	Freshwater Yield	Fuel	Timber and Pulp	NWFP	Biodiversity	Amenities	Carbon Storage
Local communities	5	5	3	4	2	4	2
Loggers	2	4	5	2	1	2	2
Downstream users							
Cities	4	3	4	3	2	4	2
Agriculture	5	4	3	4	3	3	1
Industry	3	2	5	1	0	1	1
Timber traders	1	3	5	3	0	0	1
National	5	4	4	3	4	4	3
Global	3	4	3	4	5	3	3

Key:
5 – crucial   2 – moderately important
4 – very important   1 – sporadic use
3 – important   0 – not used

gen 1996; Mårn and Jonkers 1981; Applegate et al. 2004) (See also MA *Policy Responses,* Chapter 8.) Illegal logging is also a major concern in many tropical countries. (See Chapter 9.) Illegal logging drives harvesting above planned legal limits, thereby impairing efforts at sustainable forest management, and is a powerful element of organized crime (e.g., Curry et al. 2001; Tacconi et al. 2003). According to assessments by international institutions such as the World Bank and WWF, about 70 countries have substantial problems with illegal logging, leading to annual losses of government income exceeding $5 billion and total economic losses of about $10 billion (Pisarenko and Strakhov 2004).

Infrastructure extension includes transport infrastructure (roads, railroads, and rivers); market infrastructure (such as sawmills and food markets); settlement expansion; and a variety of resource extraction, energy, and industrial infrastructure (such as hydropower, oil exploration, mining, and electrical grids).

Agricultural expansion is by far the most important direct driver of deforestation (in as much as 96% of cases studied) (Geist and Lambin 2002), and higher prices for agricultural products are a key indirect driver (Angelsen and Kaimowitz 1999). There is considerable regional variation in the kinds of agricultural expansion affecting tropical forests. Slash-and-burn clearing in Asia, for example, is more prevalent in uplands and foothills, whereas in

Latin America, it is mainly limited to lowland areas. Pasture creation for cattle ranching is a major direct driver of forest loss in mainland South America, but much less so in other regions.

Similar regional variation exists for commercial wood extraction, which was a factor in 67% of cases studied, but varied from being a direct driver of deforestation in 78% of Asian cases to 40% of Latin American cases and 26% of African cases. This is not surprising, since significant industrial logging for the international tropical timber trade now occurs only in seven Asian countries (Indonesia, Malaysia, Myanmar, Cambodia, Laos, Papua New Guinea, and the Solomon Islands), although many other countries have commercial logging operations for domestic and international markets (FAO 2002a). In some cases, large timber corporations have taken advantage of weaker or more corrupt governments (Forests Monitor 2001), which have ceded large tracts of forests to logging firms—for instance, 75% of Cameroon's forest area (WRI 2000b) and 50% of the forest area of Gabon (WRI 2000a).

By contrast to the relative importance of commercial logging in Asia, fuelwood gathering for domestic use was found to be a direct driver in 53% of African cases but only 33% of Asian and 18% of Latin American cases.

Infrastructure expansion was found to be a direct driver in 72% of cases overall, varying from 47% in Africa to 66% in Asia and 83% in Latin America. In particular, road extension was found to be one of the main specific direct drivers of tropical deforestation, especially in Latin America. The extension of roads, rail, and water transport now leaves 65% of forests in Africa 10 kilometers or less from a transportation line (FAO 2001c). By contrast, the development of private enterprise infrastructure (dams, mines, oil exploration) appears to be a minor direct driver of tropical deforestation globally, although it is important in some regions (such as hydropower development in Southeast Asia and oil development in the Peruvian, Ecuadorian, and Colombian Amazonian lowlands).

Tropical deforestation can rarely be explained by a single direct driver. In the Geist and Lambin assessment, single direct drivers only explained 6% of the cases. In particular, agricultural expansion in tandem with infrastructure development and/or logging are the most frequent combinations of direct drivers ("tandems") causing deforestation. The infrastructure-agriculture tandem explained more than one third of the cases and was relatively evenly distributed across regions. In 90% of these cases, the extension of road networks caused extension of permanently cropped land and cattle pasture, thereby resulting in deforestation. The logging-agriculture tandem explained only 10% of all cases in the study but was an important direct driver of deforestation in Southeast Asia and parts of China: the leading specific driver in most Asian cases is commercial, chiefly state-run logging activities, leading to the expansion of cropped land.

21.6.1.2 Indirect Drivers

Indirect drivers of deforestation are fundamental social processes, such as human population dynamics or agricultural policies, that underpin the direct drivers and either operate at the local level or have an indirect impact from the national or global level (Geist and Lambin 2002). These indirect drivers fall into five broad categories: economic, policy and institutional, technological, cultural/sociopolitical, and demographic. Each of these is complex even at the level of a general typology. (See Table 21.7.) They are of course even more complex in particular countries and contexts, and, like direct drivers, indirect drivers rarely function alone.

Economic factors, particularly those related to economic development through a growing cash economy, are highly important across many regions. Many cases are characterized by the marginalization of farmers who have lost their resource entitlements, combined with development brought about through public or private investments (Geist and Lambin 2002).

Institutional factors are also frequently important and are closely tied to economic drivers. These may involve formal pro-deforestation policies and subsidies (for colonization, agricultural expansion, or logging, for instance) as well as "policy failures" such as corruption or forestry sector mismanagement. Property rights issues, although much discussed in the deforestation literature, were only a major indirect driver in the cases Geist and Lambin analyzed for Asia and tended to have an ambiguous effect on forest cover: both tenurial insecurity (such as open access conditions and denial of indigenous land rights) and the legalization of land titles (enhanced tenurial security) were reported to influence deforestation in a similar manner. While property rights issues may not be the most dominant factors driving deforestation, it is widely recognized that clear property rights are a fundamental basis for instituting sustainable forest management. (See Box 21.5.)

Among demographic factors, only in-migration of colonists to sparsely populated forest areas appeared to be significant; population increase due to high fertility rates has not been a primary driver of deforestation at a local scale or over a few decades. Population increases are always combined with other factors (Geist and Lambin 2002).

21.6.1.3 Summary of Drivers

In summary, while it is possible to identify with some certainty the factors underlying tropical deforestation in a general sense, it is very difficult to pinpoint a uniform set of drivers and their relative contributions that can be said to apply generally at a global or even regional level. In a separate review of 140 models analyzing the causes of tropical deforestation, Angelsen and Kaimowitz (1999) raised significant doubts about many conventional hypotheses in the debate about deforestation; they found that more roads, higher agricultural prices, lower wages, and a shortage of off-farm employment generally lead to more deforestation, although how technical change, agricultural input prices, household income levels, and tenure security affect deforestation remains unknown. The role of macroeconomic factors such as population growth, poverty reduction, national income, economic growth and foreign debt was also found to be ambiguous. Moreover, the study found that the "win-win" hypothesis that economic growth and removal of market distortions will benefit both people and forests is not well supported by the available evidence. Rather, economic liberalization and currency devaluations tend to yield higher agricultural and timber prices that, in general, will promote deforestation (Angelsen and Kaimowitz 1999).

21.6.2 Temperate and Boreal Forest Ecosystems

Contrary to the situation in tropical forests, an important feature of forest dynamics in temperate and boreal zones is natural reforestation and expansion of forests. This process has been typical for the entire boreal zone during the last 40 years, and in Northern Eurasia this was due largely to the great restoration potential of boreal forests and the suppression of fire from the 1960s to the mid-1990s (Shvidenko and Nilsson 2002). Data for North America are less available, but fragmented satellite observations suggest that reforestation and forest expansion has been common for the entire circumpolar zone. Indeed, many temperate counties have initiated programs of reforestation and improvement of existing forests (UNECE/FAO 2003), resulting in increased net forest cover in temperate and boreal forest ecosystems.

Drivers of increasing forest cover in temperate industrial countries include the intensification of agriculture and agricultural overproduction, resulting in set-aside policies; loss of soil fertility; the increasing value of forests' amenity services; climate protection and watershed protection uses; and growing public understanding of the environmental values of forests.

In Europe, many forests were cleared centuries ago to allow agricultural expansion. Some of that agricultural land has become uneconomic to farm (see Chapter 26); meanwhile, the values of other forest services (amenity, conservation and protection, timber) have increased. Thus, the economically optimal land use has changed over the last century and trees have been either replanted or allowed to regenerate naturally. A number of countries in Europe have developed national polices aimed at conversion of some agricultural and marginal land into forest. And in Russia, the economic situation and social changes during the past decade have led to abandonment of over 30 million hectares of arable land, which is regenerating naturally into forest, trees, and bushes (Kljuev 2001).

Forest quality, however, has not necessarily improved across the temperate and boreal zones. Indeed, forests in Europe showed

Table 21.7. Generalized Typology of the Indirect Drivers of Tropical Deforestation (Adapted from Geist and Lambin 2001)

Economic change (economic growth, development, commercialization)	market growth and commercialization	rapid market growth (especially exports), rise of cash economy, increasing commercialization, incorporation into global economy
		increased market accessibility (especially of semi-urban and urban markets)
		lucrative foreign exchange earnings
		growth of demand for forest-related consumer goods due to rise in well-being
	specific economic structures	large individual (mostly) speculative gains
		poverty and related factors
		economic downturn or crisis
		indebtedness, heavy foreign debt
	urbanization and industrialization	urbanization; growth of urban markets
		industrialization: rapid expansion of new basic, heavy, and forest-based or forest-related industries
	special economic parameters	comparative advantage due to cheap, abundant production factors in resource extraction and use
		artificially low-cost production conditions (e.g., through subsidies)
		price increases or decreases for cash crops, fuel, land
Policy and institutional factors	formal policies	taxation, charges, tariffs, prices
		credits, subsidies, licenses, concessions, logging bans
		economic development (e.g., agriculture, land use policy, infrastructure)
		finance, investment, trade
		population (including migration and resettlement)
		other forestry sector policies
	informal policies (policy climate)	corruption and lawlessness
		growth or development coalitions
		bureaucratic mismanagement and poor performance
		clientelism, vested (private) interests
		role of civil society (e.g., NGOs)
	property rights regimes	insecure tenure and resulting open access in forest areas
		privatization of public lands
		state assertion of control over private, communal, or customary lands
		inequality in land access, ownership, and control
Technological change	agro-technological change	land use intensification
		land use extensification
		other changes (landholding, production orientation, etc.)
	technological applications in the wood sector	damage and waste due to poor logging performance
		waste in wood processing, poor industry performance
		lack of cheap technological alternatives to fuelwood; poor industrial and domestic furnace performance
	other production factors in agriculture	low level of technological inputs
		land-related factors (landlessness, land scarcity)
		labor-related factors (limited availability)
		capital-related factors (no credit, limited irrigation)
Cultural/socio-political factors	public attitudes, values, and beliefs	public unconcern or lack of (public and political) support for forest protection and sustainable use; low educational levels; frontier mentality; dominance of other public values (e.g., modernization, development)
		unconcern about the welfare of others and future generations; low perception of public citizenship and responsibilities
		beliefs about how environmental change affects other things that individuals value
	individual and household behavior	unconcern by individuals about the environment as reflected in increasing levels of demands, aspirations, and consumption, commonly associated with commercialization and increased income
		situation-specific behavior of actors: rent-seeking, nonprofit orientation, extent of adherence to traditional resource use modes
Human population dynamics	population growth, density, spatial distribution, and life cycle features (e.g., age, gender structure)	

BOX 21.5

Influence of Property Rights on Forest Cover Change and Forest Management

From the colonial period until recently, governments have legally owned most forests. The tradition of government ownership originated in medieval Europe and was transported to most colonies and adopted by imperial states in the sixteenth and seventeenth centuries (White and Martin 2002). Except for in the United States, Mexico, China, and Papua New Guinea, government ownership of forests spread throughout Africa, the Americas, and South and East Asia as new governments took rights from native peoples and centralized the control and management of forest resources in public forest agencies. Currently, about half (51%) of forests and other wooded lands are in public ownership in Europe (without Russia) and the rest is privately owned. National variation of ownership in temperate and boreal countries is significant: in a number of countries (such as Canada, Russia, Ukraine, and Bulgaria), forests and OWL are almost exclusively owned publicly; in others, forests are owned privately (for example, 92% in Portugal, 82% in Austria, and 80% in Sweden) (UNECE/FAO 2003).

By 1982, over 80% of the closed forests in developing countries were public land (FAO 1982). A 2002 study (White and Martin 2002) estimated that about 77% of the world's forest are owned and administered by governments based on national laws, at least 4% are reserved for communities, at least 7% are owned by local communities, and approximately 12% are owned by individuals. (See Table.)

In general, governments in countries with large amounts of forest have traditionally opted to transfer access rights and management authority to large-scale private industry through logging concessions. Gillis (1992) estimated that in 1980 about 90% of all industrial roundwood was derived from logging concessions. Data from 16 countries in Africa, the Americas, and Southeast Asia for which concession information is available reveal that 396 million hectares (44.2% of the total forest area) are under concessions. In some of these areas, particularly in Southeast Asia, the access and use rights granted to forest concessions have contributed to the massive exploitation of forest resources and the marginalization of forest-dependent communities (Broad 1995; Kummer 1992).

In the last decade or so, some governments have introduced reforms in forest ownership policies in favor of community access and ownership. These reforms were propelled by at least three factors: government recognition of the claims of indigenous and other local communities; growing evidence of the capacity of local communities to carry out sustainable forest management, due to their traditional management practices and their direct stake in forest sustainability; and the increasing realization that governments and public forest agencies have often not been good mangers of public forests (White and Martin 2002). Currently, small farms, communities, and indigenous people own or have usufruct rights over one fifth of forests in developing countries. India's Joint Forest Management Programme can be cited as one generally positive example of implementation: over 35,000 village organizations now participate in the program, covering 18% of all state forests where 147 million people live in and around forests (Forest Trends 2002), although the process is not simple and the results have not all been positive (Arnold 2001).

Social conflicts often accompany the process of change and redistribution of property rights, in particular use of lands of indigenous communities for industrial forestry and agricultural purposes, including forcible or illegal seizure of land. Reservation of indigenous territories is considered an important tool for conserving natural forests in many countries, particularly in the tropics. Recognized indigenous territories constitute 20% of the Brazilian Amazon, for example. Conflicts, however, between indigenous peoples in these territories and newcomers—such as illegal farmers associated with the Landless Rural Workers Movement—are quite common.

There is no single, "correct" forest property rights regime for all cases. Each country must find its own balance among public, private, and community rights. Whatever particular balance a country strikes, however, forest property rights need to be clear and enforceable. Formal legal establishment of property rights does not guarantee their effective implementation or enforcement. In many developing countries (and some countries in transition), forest property rights are legally mandated but are not implemented due to weak enforcement capacity or corruption.

Estimated Distribution of Forest Ownership for Selected Categories (White and Martin 2002)

	Public		Private	
	Administered by Governments	Reserved for Community and Indigenous Groups	Community/ Indigenous	Individual/Firm
Categories		*(percent of total)*		
Global forest estate	77	4	7	12
Developing countries	71	8	14	7
Developed countries	81	1	2	16
Tropical countries	71	6	13	10
Top 17 megadiverse countries	65	6	12	17
Top five roundwood products	80	7	6	7

a continuous deterioration from 1986 to 1995 due to air pollution, with the proportion of healthy trees falling from 69% in 1988 to 39% in 1995. Results for 1995–2001 show stabilization at a high level of damage, with almost a quarter of the sample trees rated as damaged due to air pollution (EC-UN/ECE 2002). For example, sulfur from the world's biggest source of sulfur emissions, Norilsk in northern Siberia (about 2 million tons of sulfur dioxide per year), caused tree mortality and degradation of more than 2 million hectares of surrounding forest tundra landscapes during the last four decades (Nilsson et al. 1998; Bruce et al. 2004).

Air pollution induces changes in tree physiology, phenology, and biochemical cycling. Among air pollutants affecting forest health, sulfur, nitrogen, heavy metals, and ozone are the most pervasive, although the complexity of forest decline in relation to air pollution suggests that decline in condition has been due to the combined impacts of eutrophication, acidification, and climate change (Nelleman and Thomsen 2001; see also Chapter 25).

The impacts of pollution on forests are not confined to industrial countries. Although anthropogenic emissions of sulfur dioxide have recently declined in most industrial countries in Europe and North America, emissions have increased in a number of countries of Asia, Africa, and Central and South America. Emissions of nitrogen oxides due to human activities remain constant or have increased over vast regions. (See Chapter 13.)

Pest outbreaks also seriously affect the quality of temperate and boreal forest ecosystems. Between 2000 and 2003, harmful forest insect outbreaks in Canada and Siberia affected more than 20 million hectares of boreal forests. The area affected by bark beetles in British Columbia increased during 2002–03, doubling to 4.2 million hectares (Berg and Henry 2003), from which the expected loss of timber is estimated to be CAN\$20 billion, in addition to the increased risk of catastrophic fires. In northern Siberia, more than 10 million hectares of larch forests were defoliated by Siberian silkworm in 2001 and 2002 (MNR 2003). The main underlying cause of these increases in natural disturbances in the boreal zone was the extremely hot and dry summers and mild winters that occurred between 1998 and 2003 (e.g., Ivanov 2003).

21.6.3 Fires in Forest Ecosystems

Fire is a crucial disturbance factor affecting tropical, temperate, and boreal forests. In many regions (the boreal zone, for instance, and savannas), fire is an essential and ecologically important process that organizes structure and functioning of forest ecosystems and substantially affects flows of energy and matter. For many other forest ecosystems, however, fire is a negative factor that severely damages forests and can lead to long-term degradation (FAO 2001e; WGWF 2003).

The incidence and severity of forest fires appears to have accelerated over the past few decades (Kasischke and Stocks 2000). (See also Chapter 16.) Until recently, for example, fire in tropical evergreen forests had a negligible distribution and impact. However, tropical rain forest conversion to rangeland and agricultural systems, slash-and-burn practices, and landscape fragmentation, exacerbated by the El-Niño Southern Oscillation, have resulted in the dramatic increase of wildfires in tropical rain forests during the last two decades (Muller-Dombois and Goldammer 1990; WGWF 2003; Mutch 2003).

The El Niño–driven fires of 1997–98 burned more than 20 million hectares in Latin America and Southeast Asia. The burnt area in Kalimantan (Indonesian Borneo) alone was about 6.5 million hectares, of which 3.2 million hectares was forest or forest areas that had recently been severely degraded or converted to plantations and other agricultural uses (Tacconi 2003). The complete economic, social, and ecological consequences of these fires have not been quantified, although some studies have yielded at least partial estimates of lost wood and impacts on wildlife and human health (e.g., Barber and Schweithelm 2000; WWF-Indonesia and EEPSEA 1998). The cost of carbon loss from the forests due to the 1998 fires in Latin America is roughly estimated at \$10–15 billion, and severe respiratory health problems together with widespread transport disruption were estimated to cost \$9.3 billion (WGWF 2003).

Increased fire activity has also been observed in other forest biomes. During the last two decades, forest fires in boreal North America (Canada and Alaska) have burned an average of 3 million hectares annually (national statistics available from the Global Fire Monitoring Center, www.fire.uni-freiburg.de). Apart from the influence of weather, shortcomings in forest fire management contributed to this increase. Human activities since 1900 have altered forest structure and fuel loadings to such an extent that they have eliminated the natural fire regime on over half the land area (260 million hectares) of the conterminous United States (Schmidt et al. 2002). In 2002, about 3 million hectares of U.S. forests burned, causing the deaths of 21 firefighters. In Russia, about 15 million hectares of forest burned in 2003. In that same year, forest fires destroyed 5% (386,000 hectares) of Portugal's forest and killed 20 people (the average annual burned area during the previous decade was about 50,000 hectares (Baptista and Carvalho 2002)), and the official estimate of economic damage of fire in 2002 was about \$1 billion.

Although an inventory of the global fire situation was prepared as part of FRA-2000 (FAO 2001d), available national information is incomplete, and the certainty of data for many regions in unknown. The satellite-based Global Burned Area Product for the year 2000 reported the global burned area of terrestrial vegetation to be 351 million hectares (JRC 2000). The reliability of this estimate is not known, however, due to the coarse resolution of the remote sensing data used and the absence of ground-truthing for many large regions. Nevertheless, the main conclusion is evident: forest fires have become a global factor negatively affecting the condition and functioning of terrestrial biota, and experiences over the past decade show that the risk and threats of forest fires are widespread across the globe.

21.6.4 Climate Change and Forests

During the last 30 years the world has experienced significant temperature increases, particularly in northern high latitudes (IPCC 2001a). (See Chapter 25.) The climatic scenarios considered by the Third Assessment Report of the IPCC projects the increase in global annual average surface temperature by the year 2100 to be 1.4–5.8° Celsius higher than the mean over the period 1990 to 2001. In some regions, this projected warming will generate a climate not experienced in recent evolutionary history. Western North America, for example, could be 2–5° Celsius above the range of temperatures that have occurred over the past 1,000 years, and vast regions in Siberia could be warmer by 6–10° Celsius. Moreover, temperatures are projected to continue to increase beyond 2100 even if atmospheric concentrations of greenhouse gases were to be stabilized by that time (Houghton et al. 2001).

As a whole, precipitation patterns are also predicted to increase, although this is mostly expected in winter precipitation, and many regions will face either a very small change or a decrease in summer precipitation. In particular, the latter is expected in regions of dry forests and woodlands. Finally, climate variability, such as the frequency of extreme events and occurrence of dry and hot periods, are expected to increase substantially (IPCC 2001a, 2001b).

These dramatic changes will be accompanied by the "fertilization effect" of increasing CO_2 concentration and nutrient deposition, which may substantially affect the state, functioning, and dynamics of the world's forests (Chapin et al. 2004). Although there is a lack of knowledge on the adaptive capacity of tree species, it is likely that an increase of temperature of a few degrees may accelerate productivity of forests, but any further increase will affect forest ecosystems in a clearly negative way (Walker et al. 1999). In spite of the fact that many experiments with leaves, shoots, and tree seedlings indicate a significant increase of productivity due to CO_2 fertilization, these effects on forests will be saturated in a short time (Scholes et al. 1999). There are also experimental data that do not support CO_2 fertilization models (Pacala 2004).

In many regions, adaptation of some forests, such as those on peat and wetlands covered by trees and shrubs, may be practically impossible. Melting of permafrost at high latitudes will cause dramatic changes in hydrological regimes of huge areas (Chapin et al. 2004). Satellite-based measures of the greenness of the boreal forest zone indicate a lengthening of the growing season over the past two decades (Nemani et al. 2003). In dry forests, net decreases in available soil moisture will decrease forest productivity. Many of these regions are also affected by El Niño/La Niña and other climatic extremes, and significant increases in land degradation and impoverishment of forests are likely (IPCC 2001a, 2001b).

Tropical montane cloud forests are especially vulnerable to climate change (Markham 1998). Various lines of evidence show that these have already been affected by climate change (Bubb et al. 2004), either through declines in the species they support (Pounds et al. 1999) or through rising cloudbanks (Still et al. 1999), which are a consequence of both climate change and regional land use change (e.g., see Lawton et al. 2001).

However, the degree to which changes in climate have already affected (e.g., see Walther et al. 2002) and continue to affect (Aber et al. 2001) productivity indicators of forests and their ability to supply services varies across space and time. This is because of the varying life cycles of forests, where climate changes within the former life cycle have a more immediate effect on regeneration success following disturbances (Price et al. 1999a, 1999b); differing values placed on forests by society (Spittlehouse 1997); disagreement on whether impacts of climate change are positive or negative (Körner and Arnone III 1992); and the varying priorities of governments for addressing other impacts (Spittlehouse 1997).

The impacts of changing climate also vary among different measures of ecosystem productivity. For example, because short growing seasons and the sum of active temperatures are the main factors limiting growth in boreal and alpine forests (Stewart et al. 1998), a projected increase in temperature may lead to higher net primary productivity values in most of these forest stands (Bugmann 1997, but see possible limitations in Barber et al. 2000), while net ecosystem productivity values will show decreases due to increased decomposition (Valentini et al. 2000, but see Giardina and Ryan 2000). On the other hand, should higher temperatures together with lower summer precipitation values occur in these forest stands, then harsher summer drought conditions may decrease NPP values as a result of lowered photosynthetic rates associated with reduced stomatal conductance (Sellers et al. 1997). Such a scenario will lead to a further decrease in NEP values due to decomposition.

The possible negative climatic effects caused by drought could be partly or fully mitigated by elevated CO_2 levels. High CO_2 levels have been found to be associated with increased photosynthetic rates and increased water use efficiency of various forest species; this could potentially lead to increased forest productivity. Evidence for this, however, is still inconclusive (Kirschbaum and Fischlin 1996). In humid evergreen tropical forest in Costa Rica, annual growth in the period 1984–2000 varied inversely with the annual means of daily minimum temperature because of increased respiration at night (Clark et al. 2003). On the other hand, a network of Amazon forest inventory plots shows a carbon accumulation rate of 1 ton of carbon per hectare per year since 1979 (Baker et al. 2004). Tree recruitment and mortality have increased significantly in the Amazon in the past two decades, with recruitment consistently exceeding mortality (Phillips et al. 2004).

Finally, warmer and drier conditions will result in increased forest and woodland fires (Laurance and Williamson 2001; Kasischke and Stocks 2000), leading to reduced transpiration and increased carbon emissions and thus creating a positive feedback whereby more-frequent and severe fires result in complete deforestation (Cochrane et al. 1999). It has been estimated that in 1997–98 net forest fire emissions (from biomass and soil losses) may have released carbon that was equivalent to 41% of worldwide fossil fuel use (Houghton et al. 2001). Therefore drier conditions will clearly add pressure to both ecosystem services (such as negative net biome productivity) (Apps et al. 2000) and the economic potential of these natural resources (see Dixon et al. 1994 for the economic importance of forests) and will also affect human health due to smoke-related impacts as a result of forest fires (Cochrane 2003).

Because climate change alters the spatial and temporal patterns of temperature and precipitation (the two most fundamental factors determining the distribution and productivity of vegetation), climate change will cause geographical shifts in the ranges of individual species and vegetation zones. In West Africa, lower rainfall and higher temperature due to climate change and desertification have shifted the Sahel, Sudan, and Guinean vegetation zones 25–30 kilometers southwest toward areas of higher rainfall and lower temperature in the period of about 1945–93 (Gonzalez 2001). In New Mexico in the United States, a 1954–58 drought caused a permanent 2-kilometer shift of xeric piñon-juniper woodland into mesic ponderosa pine forest (Allen and Breshears 1998), and some climate modeling shows extensive latitudinal and altitudinal shifts of vegetation zones across North America and Siberia (Iverson and Prasad 1998; Pan et al. 1998; Bachelet et al. 2001).

The dynamic nature of the environment within which sustainable forest management must take place means that simple representation of the more tangible forest elements, such as productivity indicators, in static areas (such as protected areas) is unlikely to be sufficient for long-term protection and hence sustainability (Stewart et al. 1998; Rodrigues et al. 2000; Hannah et al. 2002). Consequently, flexible forest and woodland management will be needed to adapt to some of the effects of future climate change, which will certainly have widespread affects on forest and woodland systems (Dixon et al. 1994; Cohen and Miller 2001; Spittlehouse and Stewart 2003).

21.7 Human Well-being and Forests and Woodland Systems

Forests and woodlands supply essential services to human well-being across the world, and human-forest interactions manifest themselves in many direct and indirect ways, each depending variously on the amount of forest, its condition, and its distribution over the landscape.

More than 1.7 billion people live in the 40 nations with critically low levels of forest cover, in many cases hindering prospects for sustainable development. The number of people living in low-forest-cover nations will probably triple by 2025, reaching 4.6 billion, and 13 additional countries will experience forest resources scarcity (Gardner-Outlaw and Engelman 1999). Human population growth has drastically shrunk the forest-to-people ratio from 1.2 hectares per capita in 1960 to 0.6 hectares per capita at present. By 2025, the ratio is predicted to decline further, to 0.4 hectares per capita (Gardner-Outlaw and Engelman 1999).

The expected decline in the per capita availability of forests in developing countries generates additional problems for sustainable development. In many parts of the developing world, direct harvesting of forest products by rural families contributes to more than 50% of total consumption and other household needs (Cavendish 2000; Hassan 2002; Hassan et al. 2002; Godoy and Bawa 1993; Kusters and Belcher 2004; Peters et al. 1989; Sheil and Wunder 2002; Sunderland and Ndoye 2004). This large group of people is particularly vulnerable to the negative impacts of declining forest cover.

Diminishing access to forest products significantly affects human well-being in developing countries. Inadequate supplies of paper could emerge as a significant impediment to development during this century, and 80% of the world's population has yet to achieve the level of paper use deemed necessary to meet basic needs for literacy and a minimal level of education and communication (Gardner-Outlaw and Engelman 1999). About 2.4 billion people use energy derived from biomass, mostly from forests and woodlands (Arnold et al. 2003; see also Chapter 9), and most of the 240 million poor people in forested regions in developing countries depend heavily on forests and trees for their livelihoods (World Bank 2003).

Development of modern forest industries can generate local employment and thereby improve the standard of living of forest communities. Still, a significant proportion of wood harvested in tropical countries is exported as unprocessed logs. For instance, 38–48% of unprocessed roundwood was imported from net export African member countries of the International Tropical Timber Organization during 1995–2000 (Buttoud et al. 2002), and domestic demand for wood products in these countries remains very low, at around 0.1 cubic meters per person a year for timber.

Population-related pressure on forests is greatest in countries where per capita forest cover is low, and although forests are often protected or planted as population pressures increase, this is usually only in high-productivity zones (such as Bangladesh, Java, parts of Kenya, and India) but rarely in areas of low productivity (Persson 2003). Many countries in the developing world are facing local woodfuel and NWFP scarcity, and the situation is expected to become more acute (FAO 2003a). Because it is often women and children who search for fuel and edible forest products and so on, such shortages have particularly negative impacts on these sectors of the population.

Of course, not all deforestation is necessarily undesirable, and many areas of forest have been lost after the negative consequences of such loss have been carefully considered and weighed against the benefits. For many countries, past and present, converting some forestlands to agricultural, infrastructure, industrial, and urban uses has been a necessary and accepted mechanism for economic development and progress. Unfortunately, deforestation, particularly in the tropics, has often resulted in conversion to unsustainable land uses and has not delivered the anticipated benefits to economic development.

It is projected that tropical deforestation will likely continue unabated through 2020 and that demand for fuelwood will continue to rise in Africa and some other regions of the developing world, watershed protection will continue to deteriorate, and countries will not likely improve efforts to implement sustainable forest management (FAO 2003a; Kaimowitz 2003). National forest services generally remain underequipped to counter these trends. A survey of government expenditures in 24 African countries in 1999 showed that forest expenditures averaged 82¢ per hectare, of which international financing accounted for 37¢ (FAO 2003a). Most national funding goes to staff salaries, while the international component generally goes to investments in material and information systems.

The condition of forests in individual countries and the well-being of forest-dependent peoples are closely tied to economic development levels and trends. Russia is an interesting and informative example, as the severe economic situation of the last 15 years has led to large-scale decline of the forest sector. The production of major forest products decreased between 1988 and 1998 by three- to fivefold (Bourdin et al. 2000), dramatically affecting the well-being of about 3 million people in regions where forests are a major source of employment and subsistence. Many hundreds of forest settlements now suffer from unemployment and a lack of basic living conditions; subsistence farming, gathering mushrooms and wild berries and fruits, fishing, and poaching have become major sources for subsistence in many forest regions. This situation is heavily influenced by an inadequate forest policy, although in recent years there has been a slow but evident restoration of the Russian forest sector, driven largely by market mechanisms (Shvidenko 2003).

The extent and distribution of forests are important at all spatial levels, from the local to the continental. Even if a country as a whole has a sufficient amount of forested area in the aggregate, forest cover in particular regions or landscapes may still be insufficient to meet the demand for services. Redistribution of forest cover over a landscape is difficult, however, and requires long-term, consistent policies at the national level. Improving the condition of forests and their contribution to human well-being is an important and urgent task, both nationally and internationally. Recent history, such as international efforts working with the Tropical Forestry Action Plan (FAO 1985; Winterbottom 1990), clearly shows both how difficult it is to achieve sustainable forest management in the contemporary world and that many problems remain to be solved in order to realize the potential benefits that forests and woodlands have to offer.

References

Abdger, W.N., K. Brown, R. Cervigni, and D. Moran, 1995: Total economic value of forests in Mexico. *Ambio* 24(5), 286–296.

Aber, J, R.P. Neilson, S. McNulty, J.M. Lenihan, D. Bachelet and R.J. Drapek, 2001: Forest processes and global environmental change: predicting the effects of individual and multiple stressors. *BioScience,* **51,** 735–751.

Achard, F., H.D. Eva, H.-J. Stibig, P. Mayaux, J. Gallego, T. Richards, and J.-P. Malingreau. 2002. Determination of deforestation rates of the world's humid tropical forests. *Science,* **297,** 999–1002.

Agarwal, A. (ed), 1992: *The Price of Forests.* Proceedings of a seminar on the Economics of the Sustainable Use of Forest Resources. Center for Science and Environment, New Delhi, India, 364 pp.

Alexandrov, G.A., T. Oikawa, and G.Esser, 1999: Estimating terrestrial NPP: what the data say and how they may be interpreted? *Ecological Modeling,* **117,** 361–369.

Allen, C.D. and D.D. Breshears, 1998: Drought-induced shift of a forest–woodland ecotone: Rapid landscape response to climate variation. *Proceedings of the National Academy of Sciences of the USA,* **95,** 14 839–14 842.

Allen, D.G, J.A. Harrison, R. Navarro, B.W. van Wilgen and M.W. Thompson, 1997. The impact of commercial afforestation on bird populations in Mpumalanga province, South Africa—Insights from bird atlas data. Biological Conservation 79: 173–185.

Angelsen, A. and D. Kaimowitz, 1999: Rethinking the Causes of Deforestation: Lessons from Economic Models. *The World Bank Research Observer,* Vol. 14 No. 1. February, pp. 73–98.

Anonymous, 1994: *European Criteria and Most Suitable Quantitative Indicators for Sustainable Forest Management.* Adopted by the first expert level follow-up meeting of the Helsinki Conference. Geneva, 19 pp.

Anonymous, 1995: *Criteria and Indicators for Conservation and Sustainable Management of temperate and Boreal Forests.* The Montreal Process, Canadian Forest Service, Hull, Quebec, 25 pp.

Applegate, G., F.E. Putz, and L.K Snook, 2004: *Who Pays and Who Benefits from Improved Tinber Harvesting Practices in the Tropics.* CIFOR, Bogor, 26 pp.

Apps, M.J., J.S. Bhatti, D. Halliwell, H. Jiang, and C. Peng, 2000: Simulated carbon dynamics in the boreal forest of central Canada under uniform and random disturbance regimes. In (Eds.) R. Lal, J.M. Kimble, B.A. Stewart. In: *Global Climate Change and Cold Regions Ecosystems,* R. Lal, J.M. Kimble, and B.A. Stewart (eds.). Advances in Soil Science, Lewis Publishers, Boca Raton, pp. 107–121.

Arnold, M. 2001: Forestry, Poverty and Aid. CIFOR Occasional Paper No. 33. CIFOR, Bogor, 17 pp.

Arnold, J.E.M., 1998: Socio-economic benefits and issue in non-wood forest product use. Non-Wood Forest Products, Food and Agricultural Organization of the United Nations, Rome, pp. 89–145.

Arnold, M., G. Kohlin, R. Persson, and G. Shepherd, 2003: *Fuelwood Revisited: What has Changed in the Last Decade?* CIFOR Occasional Paper No 39, Bogor, 35 pp.

Atlas, 2002: Atlas of Russia's Intact Forest Landscapes. Biodiversity Conservation Center, Greenpeace Russia, International Socio-Ecological Union, World Resource Institute, Moscow, 186 pp.

Atrokhin, V.G., K.K.Kalutskiy and F.T. Tjurikov, 1982: *Tree Species of the World. Volume 3. Tree Species of the USSR,* Forest Industry, Moscow, 264 pp. [in Russian].

Auclair, D., R. Prinsley and S. Davis, 2000: *Trees on Farms in Industrialized Countries: Silvicultural, environmental and Economics Issues.* Kuala Lumpur, Malaysia, IUFRO.

Bachelet, D., R.P. Neilson, J.M. Lenihan and R.J. Drapek. 2001: Climate change effects on vegetation distribution and carbon budget in the United States. *Ecosystems* **4,** 164–185.

Baharuddin, K., and N. Abdul Rahim, 1994: Suspended sediment yield resulting from selective logging practices in a small watershed in Peninsular Malaysia. *J. Tropical Forest Science* **7,** 286–295.

Baker, T.R., O.L. Phillips, Y. Malhi, S. Almeida, L. Arroyo, et al. 2004. Increasing biomass in Amazonian forest plots. *Philosophical Transactions of the Royal Society of London B,* pp. 353–365.

Ball, J.B., 2001: Global Forest Resources: History and Dynamics. The Forest Handbook. Oxford, Blackwell Science, 3–22 pp.

Baptista, M. and J. Carvalho. 2002: Fire situation in Portugal. *International Forest Fire News,* No 27, pp. 65–67.

Barber, C.V., K.R. Miller and M. Boness (Eds.), 2004: *Securing Protected Areas in the Face of Global Change. Issues and Strategies.* Gland, Switzerland and Cambridge, UK, IUCN, 234 pp.

Barber, C.V. and J. Schweithelm, 2000: *Trial by Fire. Forest Fires and Forestry Policy in Indonesia's Era of Crisis and Reform.* World Resources Institute, Washington DC, 84 pp.

Barber, V. A., G. P. Juday, and B. P. Finney, 2000: Reduced growth of Alaskan white spruce in the twentieth century from temperature-induced drought stress. *Nature,* **405,** 668–673.

Bass, S.M.J., 2003. International Commitments, Implementation and Cooperation. In *Congress Proceedings,* Volume C. Proceedings of the 12th World Forestry Congress, September 21 to 28, 2003 Quebec, Canada, pp. 257–264.

Battle, M., M.L. Bender, P.P. Tans, J.W.C. White, J.T. Ellis, T.Convey and R.J. Francey, 2000: Global carbon sinks and their variability inferred from atmospheric O2 and δ^{13}C. *Science,* **287,** 2467–2470.

Beer, J., R.Muschler, D. Kass, and E. Somariba, 1998: Shade management in coffee and cacao plantations. *Agroforestry Systems* **38,** 139–164.

Beer, J., C.A.Harvey, M.Ibrahim, J.M. Harmand, E. Somariba and F. Jimenz, 2003: Service functions of agroforestry systems. In *Congress Proceedings,* Volume B. Proceedings of the 12th World Forestry Congress, September 21 to 28, 2003, Quebec, Canada, pp. 417–424.

Bennet, E.L., and J.G. Robinson, 2000: *Hunting of Wildlife in Tropical Forests, Implications for Biodiversity and Forest Peoples.* Environment Department Papers, No 76, Washington DC, World Bank, 56 pp.

Berg, E.E., and J.D. Henry, 2003: The history of spruce bark beetle outbreaks in the Kluane region as determined from the dendrochronology of selected forest stands Online at: http://alaska.fws.gov/nwrO/kenai/biology/index .html.

Bethlehem, L., and M. Dlomo, 2003: Forests, economics, and the development agenda. Proceedings of the 12th World Forestry Congress, Volume C, Quebec, Canada, pp. 3–9.

Bharati, L., K.H. Lee, T.M. Isenhart, and R.C. Schulz. 2002. Soil-water infiltration under crops, pasture, and established riparian buffer in Midwestern USA. *Agroforestry Systems.* 56(3): 249–257.

Bhattarai, T., 2001: Trees Outside Forests: The Woodfuel Production Context. In *Fuelwood. Crisis or Balance,* G. Köhlin (ed.). Proceedings of the Workshop at School of Economics and Commercial Law, 6–8 June, Marstrand, Göteborg University, Göteborg, pp 280–285.

Billington, C., V. Kapos, M.S.Edwards, S. Blyth, and S. Iremonger, 1996: *Estimated Original Forest Cover Map—a First Attempt.* WCMC, Cambridge, UK.

Birdsey, R.A. and G. M. Lewis. 2003. *Carbon in U.S. Forests and Wood Products, 1987–1997: State-by-State Estimates.* U.S. Department of Agriculture, Forest Service, Newtown Square, PA, 42 pp.

Bishop, E.B., 1999: *Valuing Forests: A Review of Methods and Applications in Developing Countries.* International Institute for Environment and Development, Environmental Economics Programme, London.

Blombaeck, P., and P. Poschen, 2003: Decent work in forestry? Enhancing forestry work and forest based livelihoods. In *Congress Proceedings,* Volume A. Proceeding of the 12th World Forestry Congress, September 20 to 28, 2003, Quebec, Canada, pp. 231–240.

Bourdin, N.A., V.M. Shlikov, V.A. Egornov and V.V.Sakhanov, 2000: *Forest Complex (State, Problems, Perspectives).* Moscow, 473 pp. [in Russian].

Bousquet, P., P. Peylin, P. Ciais, C. Le Quere, P.Friedlingstein, and P. Tans, 2000: Regional changes in carbon dioxide fluxes of land and oceans since 1980. *Science,* **290,** 1342–1346.

Broad, R., 1995: The political economy of natural resources: case studies of the Indonesian and Philippine forest sectors. *Journal of Developing Areas,* **29(3),** 317–339.

Broadhead, J., J. Bahdon and A.Whiteman, 2001: Woodfuel consumption modeling and results. Annex 2 in ''Past trends and future prospects for the utilization of wood for energy'', Working Paper No: GFPOS/WP/05, Global Forest Products Outlook Study, FAO, Rome, 111 pp.

Brockie, R., 1992: *A Living New Zealand Forest.* David Bateman, Auckland, 172 pp.

Brooks, T.M., R.A. Mittermeier, G.A.B. da Fonesca, A.B. Rylands, W.R. Constant, et al. 2002: Habitat loss and extinction in the hotspots of biodiversity. Conservation Biology 16 (4), 909–923.

Brown, C., 2000: The Global Outlook for Present and Future Wood Supply from Forest Plantation. Working Paper No GFPOS/WP/03, FAO, 145 pp.

Brown, S., 1998: Present ad future role of forests in global climate change. In: *Ecology Today: An Anthology of Contemporary Ecological Research,* B. Goalp, P.S. Pathak and K.G. Saxena (eds), International Scientific Publications, New Delhi, pp. 59–74.

Brown, S. and A.E. Lugo, 1990: Tropical secondary forests. Journal of Tropical Ecology 6:1–32.

Bruce, F., N. Fresco, A.Shvidenko, K. Danell, and F. Stuart Chapin, III, 2004: Geographic variations in anthropogenic drivers that influence the vulnerability and resilience of social-ecological systems. *Ambio,* 33(6), 376–381.

Bruijnzeel, L.A., 1989: Hydrology of moist tropical forests and effects of conversion: A state of knowledge review, UNESCO-IHP, Humid tropics programme, Paris.

Bruijnzeel, L.A., 2004: Hydrological Functions of tropical forests: not seeing the soil for the trees? *Agriculture, Ecosystems and Environment* 104, 185–228.

Bryant, D., D. Nielson, and L. Tangley, 1997: *The Last Frontier Forests.* World Resources Institute, Washington DC, 43 pp.

Bubb, P, I. May, L. Miles and J. Sayer. 2004. *Cloud Forest Agenda.* UNEP-WCMC, Cambridge, UK.

Bugmann, H. 1997. Sensitivity of forests in the European Alps to future climatic change. *Climate Research,* 35–44.

Buresh, R.J. and G. Tian, 1998: Soil improvement by trees in sub-Saharan Africa. *Agroforestry Systems* **38,** 51–76.

Buttoud, G., P. Lefakis and J. Bakouma, 2002: Processing in Africa. *ITTO Tropical Forest Update,* 12, No 2, 15–18.

Byron, N., 1997: Challenges and opportunities policy options for the forestry sector in the Asia-Pacific region. In: FAO Asia-Pacific Forestry Sector Outlook Study Working Papers, Bangkok, 64 pp.

Byron, R.N. and J.E.M. Arnold, (1999): What futures for the people of the tropical forests? *World Development,* **27 (5),** 789–805.

Byron, N., 2003: Reinvigorating world forestry. In: *Congress Proceedings,* Volume C. Proceedings of the 12th World Forestry Congress, September 20 to 28, 2003, Quebec, Canada, pp. 10–17.

Carrere, R, 1999: *Ten Replies to Ten Lies.* World Rainforest Movement, Montevideo, Uruguay. Plantations Campaign Briefing Paper, 11 pp.

Carrere, R. and L. Lohmann, 1996: *Pulping the South: Industrial tree plantations and the world paper economy.* Zed Books, London, UK, 280 pp.

Cavendish, W., 2000: Empirical regularities in the poverty-environment relationship of rural households: evidence from Zimbabwe. *World Development* 28(11): 1979–2003.

CCFM (Canadian Council of Forest Ministers), 2003: Defining Sustainable Forest Management in Canada. Criteria and Indicators 2003. Ottawa, 20 pp.

Chapin, F.S. III, G. Peterson, F. Berkes, T.V. Callaghan, P. Angelstam, et al. 2004: Resilience and vulnerability of northern regions to social and environmental change. *Ambio,* **33(6),** 344–349.

Chaytor, B., 2001. *The Development of Global Forest Policy: Overview of Legal and Institutional Frameworks.* IIED, London, UK.

Chichilnisky, G. and G.M. Heal, 1998: Economic returns from the biosphere. Nature 391 (February), 629–630.

Chiras, D., 1998: *Environmental Science: A Systems Approach to Sustainable Development.* Waldsworth Publishing Company, Belmont, California, USA.

Chomitz, K. and K. Kumari, 1998: The domestic benefits of tropical forests: A critical review. *The World Bank Research Observer,* **13 (1),** 13–35.

Ciais, P., I. Janssens, A. Shvidenko, C. Wirth, Y. Malhi, et al. 2004: The potential for rising CO2 to account for the observed uptake of carbon by tropical, temperate and boreal forests. In: H. Griffiths and P.J. Jarvis (eds.) The Carbon Balance of Forest Biomes, Garland Science/BIOS Scientific Publishers, pp. 109–149.

CIDA (Canadian International Development Agency), 1998: *Deforestation: Tropical Forests in Decline.* Quebec, Canada, 50 pp.

CIESIN (Center for International Earth Science Information Network), 2000: Gridded Population of the World (GPW), Version 2, Columbia University, Palisades, NY: CIESIN. Dataset available at http://sedac.ciesin.columbia .edu/plue/gpw.

Ciesla, W.M., 1998: *Non-wood Forest Products from Conifers.* NWFP 12, Food and Agriculture Organization of the United Nations, Rome, 124 pp.

CIFOR C&I Team (Center for International Forestry Research), 1999. *The CIFOR Criteria and Indicators Generic Template.* Bogor, Indonesia.

Clark, D.A., S.C. Piper, C.D. Keeling, and D. B. Clark, 2003: Tropical rain forest tree growth and atmospheric carbon dynamics linked to interannual temperature variation during 1984–2000. *Proceedings of the National Academy of Sciences of the USA,* 100, 5852–5857.

Cochrane, M.A., 2003: Fire science for rainforests. *Nature,* **421,** 913–919.

Cochrane, M.A., A. Alencar, M.D. Schulze, C.M. Souza Jr., D.C. Nepstad, et al. 1999: Positive feedbacks in the fire dynamic of closed canopy tropical forests. *Science,* **284,** 1832–1835.

Cohen, S. and Miller, K., 2001: Climate Change 2001: *Impacts, Adaptations and Vulnerability.* In: J.J. McCarthy, O.F. Canziani, N.A. Leary, D.J. Dokken and K.S. White (eds.), IPCC. Cambridge, Cambridge University Press, pp. 735–800.

Collins, N.M., J.A. Sayer, and T.C.Whitemore, 1991: *The Conservation Atlas of Tropical Forests: Asia and the Pacific.* Macmillan, London, UK, 256 pp.

Cossalter, C. and C. Pye-Smith, 2003: *Fast-Wood Forestry. Myths and Realities.* Center for International Forestry Research, Bogor, Indonesia, 50 pp. London, UK and Telapak Indonesia, Bogor, Indonesia, 36 pp.

CWC (Canadian Wood Council), 1999: Life cycle analysis for residential buildings. Environmental Bulletin, 27 pp.

Davis, A.J., H. Huijbregts, and J. Krikken, 2000: The role of local and regional processing in shaping dung beetle communities in tropical forest plantations in Borneo. *Global Ecology and Biogeography Letters,* **9,** 281–292.

De Wit, M., D. Crookes and B.W. van Wilgen. 2001. Conflicts of interest in environmental management: Estimating the costs and benefits of a tree invasion. Biological Invasions 3: 167–178.

DeFries, R., M. Hansen, and J. Townshend, 1995: Global discrimination of land cover types from metrics derived from AVHRR pathfinder data. *Remote Sensing of Environment,* **54(3),** 209–222.

DeFries, R., M.C. Hansen, J.R.G. Townshend, A.C. Janetos, and T.V. Loveland, 2000: A new global 1-km data set of percentage tree cover derived from remote sensing. *Global Change Biology,* **6,** 247–254.

DeFries, R,S., R. A. Houghton, M.C. Hansen, C.B. Field, D.Skole, and J. Townshead, 2002: Carbon emissions from tropical deforestation and regrowth based on satellite observations for the 1980s and 1990s. *Proceedings of the National Academy of Sciences USA,* **99(22),** pp.14256–14261.

DePhilip, M., 2003. "The Giving Rivers: Protected areas and environmental flows in North America." Contribution by Michele DePhilip of the Nature Conservancy to the Workshop on "Building Broader Support for Protected Areas", 5th World Congress on Protected Areas, Durban, South Africa. September.

Dhawan, B.D., 1993: Coping with floods in Himalayan rivers. Economic and Political weekly, May 1, 1993, pp. 849–853.

Dickinson, A., M.B. Amphlett, ans P. Bolton, 1990: Sediment discharge measurements Magat catchment. Summary Report 1986–1988. Report No OD 1222. Hydraulics Research, Wallingford, UK, 97 pp.

Dixon, R.K., S. Brown, R.A.Houghton, A.M. Solomon, M.C. Trexler, and J.Wisniewski, 1994: Carbon pools and flux of global forest ecosystems. *Science,* **263,** 185–190.

Dong, J., R.K. Kaufmann, R.B. Myneni, C.J. Tucker, P.E. Kauppi, J. Liski, W. Buermann, V. Alexeyev, M.K. Hughes, 2003. Remote sensing estimates of boreal and temperate forest woody biomass: carbon pools, sources, and sinks. *Remote Sensing of Environment,* **84,** 393–410.

Douglas, I., 1996: The impact of land-use changes, especially logging, shifting cultivation, mining and urbanization on sediment yields in humid tropical southeast Asia: a review with special reference to Borneo. *Int. Assoc. Hydrol. Sci. Publ.* **236,** pp. 463–471.

Dudley, N. and S. Stolton. 2003. *Running Pure. The importance of forest protected areas to drinking water.* Washington DC: World Bank/WWF Alliance for Forest Conservation and Sustainable Use.

Dunn, R.R., 2004: Recovery of Faunal Communities During Tropical Forest Regeneration. *Conservation Biology* **18(2):** 302–309.

ECE (Economic Commission for Europe) 2003: *Employment Trends and Prospectus in the European Forest Sector.* European Forestry Outlook Studies Report, Geneva, 45 pp.

ECE/FAO (Food and Agriculture Organization), 1985: *Forest Resources of the ECE Region.* ECE/TIM/27. Geneva, Switzerland, 222 pp.

ECE/FAO, 1992: *1990 Forest Resources Assessment.* Vol. 1. ECE/TIM/62. New York, 348 pp.

Eckholm, E., 1998: Stunned by floods, China hastens logging curbs, *The New York Times,* 27 September 1998.

ECOSOC (Economic and Social Council), 2004. Criteria and indicators of sustainable forest management. Report of the Secretary-General. UN Forum on Forests, Fourth Session, 3–14 May, 2004, Geneva. E/CN.18/2004/11, 21 pp.

ESRI (Environmental Systems Research Institute), 1998: *World Countries, 1998.* Redlands, California, USA. Available at http://www.esri.com.

Emerton, L., 2003: Tropical forest valuation: Has it all been a futile exercise? In Congress Proceedings, Volume A. Proceedings of the 12th World Forestry Congress, Quebec, Canada, 21 to 28 September 2003, pp. 103–110.

Faminov, M., 1997: The disappearing Amazon rainforest problem. International Association of Agricultural Economists, Canadian Newsletter, http:// www.oac.uoguelph.ca/www/Agec/IAAE/Art_Faminov01.htm

Fan, S., M. Gloor, J. Mahlman, S. Pascala, J. Sarmiento, T. Takahashi and P. Tans, 1998: A large terrestrial carbon sink in North America implied by atmospheric and oceanic carbon dioxide data and models. *Science,* **282.** 442–446.

FAO (Food and Agriculture Organization of the United Nations), 1982: *Tropical Forest Resources.* FAO Forestry Paper 30, Food and Agriculture Organization of the United Nations, Rome, 106 pp.

FAO, 1993: *Forest Resources Assessment 1990—Tropical Countries.* FAO Forestry Paper 112, Food and Agriculture Organization of the United Nations, Rome, 88 pp.

FAO, 1985. *Tropical Forestry Action Plan.* Food and Agriculture Organization of the United Nations, Rome, 159 pp.

FAO, 1995a. *Forest Resources Assessment 1990. Global Synthesis.* FAO Forestry Paper 124, Food and Agriculture Organization of the United Nations, Rome, 102 pp.

FAO, 1995b. *Forest Resources Assessment 1990—Tropical Forest Plantation Resources,* by D. Pandey. FAO Forestry Paper 128, Food and Agriculture Organization of the United Nations, Rome, Rome, 90 pp.

FAO, 1996: *Forest Resources Assessment 1990. Survey of Tropical Forest Cover and Study of Change Processes Based on Multi-Data High Resolution Satellite Data.* FAO Forestry Paper 130, Food and Agriculture Organization of the United Nations, Rome, 170 pp.

FAO, 1997: *State of the World's Forests 1997.* Food and Agricultural Organization of the United Nations, Rome, 200 pp.

FAO, 1999a: *State of the World's Forests 1999.* Food and Agriculture Organization of the United Nations, Rome, 154 pp.

FAO, 1999b: Towards a harmonized definition of non-wood forest products. *Unasylva,* **198,** pp. 63–64.

FAO, 2000: *Forest Resources of Europe, CIS, North America, Australia, Japan and New Zealand.* Main Report. ECE/TIM/SP/17, UN, NEW York and Geneva, 457 pp.

FAO, 2001a: Forestry country profiles. Available on the Internet www.fao.or g/forestry/fo/country/nav_world.jsp.

FAO, 2001b. Forest Resources Assessment homepage. Available on the Internet www.fao.org/forestry/fo/fra/index/jsp.

FAO, 2001c: *Global Forest Resources Assessment 2000—Main Report.* FAO Forestry Paper 140, Food and Agriculture Organization of the United Nations, Rome, 482 pp.

FAO, 2001d: *The Global Forest Resources Assessment 2000. Summary Report.* Information note. COFO2001/INF.5, 21 pp.

FAO, 2001e: *FRA Global Forest Fire Assessment 1990–2000.* FAO Working Paper 55, Food and Agriculture Organization of the United Nations, Rome, 495 pp.

FAO, 2001f: *Non-wood Forest Products of Africa—a Regional and National Overview.* Working Paper FOPW/01/1, Food and Agriculture Organization of the United Nations, Rome.

FAO, 2002a: *Forest Products.* FAO Yearbook. FAO Forestry Series, No 35, Food and Agriculture Organization of the United Nations, Rome, 243 pp + sup l.

FAO, 2002b: *Expert Meeting on Harmonizing Forest-related Definitions for Use by Various Stakeholders* Proceedings of a Workshop, 23–25 January 2002, Rome, Food and Agriculture Organization of the United Nations, Rome, 193 pp.

FAO, 2002c: *Second Expert Meeting on Harmonizing Forest-related Definitions for Use by Various Stakeholders,* 11–13 September 2002, Rome, Food and Agriculture Organization of the United Nations, Rome, 394 pp.

FAO, 2003a: *Forestry Outlook Study for Africa.* Food and Agriculture Organization of the United Nations, Rome, 66 pp.

FAO, 2003b: *State of the World's Forests 2003.* Food and Agriculture Organization of the United Nations, Rome, 151 pp.

FAO, 2004. *Specification of National Reporting Tables for FRA 2005.* FAO Forest Resources Assessment Programme. Working Paper 81, Food and Agriculture Organization of the United Nations, Rome, 37 pp.

FFSR, 1995: *Manual for Forest Inventory in Russian Forest Fund.* Part 1. Organization of Forest Inventory. Field Works. Federal Forest Service of Russia, Moscow, 174 pp. [in Russian].

FFSR (Federal Forest Service of Russia), 1999: *Forest Fund of Russia (State by January 1 1998).* Moscow, 649 pp. [in Russian].

Field, C.B. and M.R. Raupach, 2004: The Global Carbon Cycle. Integrating Humans, Climate and the Natural World. SCOPE 62, Island Press, 526 pp.

Flannery, T., 1994. *The Future Eaters.* Reed Books, Chatswood, N.S.W., Australia, 423 pp.

Flannery, T., 2001. *The Eternal Frontier: An Ecological History of North America and Its Peoples.* Atlantic Monthly Press, Boston, Massachusetts, 432 pp.

Folke, C., S. Carpenter, T. Elmqvist, L. Gunderson, C.S. Holling, et al. 2002: *Resilience and Sustainable Development: Building Adaptive Capacity in a World of Transformation.* The Environmental Advisory Council to the Swedish Government, Stockholm, 52 pp.

Forest Principles, 1992: *The UN Conference on Environment and Development, Rio-de-Janeiro* [Non-legally Binding Authoritative Statement on Principles for a Global Consensus on the Management, Conservation and Sustainable Development of All Types of Forests].

Forest Trends, 2002: *Strategies for Strengthening Community Property Rights over Forests: Lessons and Opportunities for Practitioners.* Washington, DC, 42 pp.

Fung, J., A. Chen, C. Peng, S. Zhao and L. Ci, 2001: Changes in forest biomass carbon storage in China between 1949 and 1998. *Science,* 292: 2320–2322.

Funston, M., 1995. *Sustainable Forest Management.* USDA Forest Service, Sustainable Development Interdeputy Area Team Discussion Paper No. 2. May. Available at http://www/fs/fed/us./land/sustain_dev/susdev2.html.

Gardner-Outlaw T. and R. Engelman, 1999: *Forest Futures.* Population Action International, 67 pp.

Geist, H.J. and E.F. Lambin, 2001. *What Drives Tropical Deforestation? A meta-analysis of proximate and underlying causes of deforestation based on sub-national case study evidence.* LUCC International Project Office, Louvain-la-Neuve, Belgium. LUCC Report Series No. 4, 116 pp.

Geist, H.J. and Lambin, E.F. 2002. Proximate causes and underlying driving forces of tropical deforestation. *BioScience,* **52,** 143–149.

Giardina, C. P. and M. G. Ryan, 2000: Evidence that decomposition rates of organic carbon in mineral soil do not vary with temperature. *Nature,* **404,** 858–861.

Gillis, M., 1992: Forest concession management and revenue policies. Chapter 7 in N. P. Sharma (ed.) *Managing the world's Forests: Looking for Balance Between Conservation and Development.* Inowa, Kendall/Hunt, pp. 139–175.

Glen, W.M., 2000: Trees outside the forest assessment in the Sudan: a contribution to the Forest Resource Assessment 2000 report. FAO, Rome (draft paper).

Godoy, R., and K. S. Bawa, 1993: The economic value and sustainable harvest of plants and animals from the tropical forests: assumptions, hypotheses and methods. *Economic Botany* 47:215–219.

Goldweijk, K.K., J.G. van Minnen, G.J.J Kreileman, .M. Bloedbeld, and R. Leemans, 1994: Simulating the carbon flux between the terrestrial environment and the Atmosphere. *Water, Air and Soil Pollution,* 76, 99–230.

Goodale, C.L., M.J. Apps, R.A. Birdsey, C.B. Field, L.S. Heath, et al. 2002: Forest carbon sink in the Northern Hemisphere. *Ecological Applications,* 12(3), 891–899.

Gonzalez, P., 2001: Desertification and a Shift of Forest Species in the West African Sahel. Climate Research, 17, 217–228.

Groombridge, B. and M.D. Jenkins, 2002. *World Atlas of Biodiversity. Earth's Living Resources in the 21st Century.* University of California Press, Berkeley. Published in association with the UNEP World Conservation Monitoring Centre, 340 pp.

Haberl, H., 1997: Human appropriation of Net Primary Productivity as an environmental indicator: Implications for sustainable development. *Ambio* **26 (3),** 143–146.

Hagler, R.W., 1995: The Global Wood Fiber Balance: What It Is; What It Means. Proceedings TAPPI Global Fiber Supply Symposium, Chicago, Oct. 5–6, TAPPI Press, Atlanta, Georgia, USA.

Hamilton, L.S., 1976: Tropical rainforest use and preservation: A study of problems and practices in Venezuela. International Series No. 4, Sierra Club, San Francisco, California, 72 pp.

Hamilton, L.S., and P.N. King, 1983: *Tropical Forested Watersheds. Hydrologic and Soils Response to Major Uses or Conversion.* Westview Press, Boulder, Colorado, USA, 168 pp.

Hannah, L., G. Midgley, and D. Millar, 2002: Climate change-integrated conservation strategies. *Global Ecol. Biogeog.* 11, 485–495.

Hansen, M.C., and B. Reed, 2000: A comparison of the JGBP DISCover and University of Maryland 1km global land cover products. *International Journal of Remote Sensing,* 21, 1365–1373.

Harcourt, C.S., and J.A. Sayer, 1996: *The Conservation Atlas of Tropical Forests: the Americas.* Macmillan, London, UK, 335 pp.

Hartley, M.J., 2002: Rationale and methods for conserving biodiversity in plantation forests. *Forest Ecology and Management,* 155, 81–95.

Hassan, R.M. (ed), 2002: *Accounting for Stock and Flow Values of Woody Land Resources,* Centre for Environment Economics and Policy in Africa, University of Pretoria, 144 pp.

Hassan, R., P. Mbuli and C. Dlamini, 2002: *Natural Resource Accounts for the State and Economic Contribution of Forests and Woodland Resources in Swaziland.* CEEPA Discussion Paper, No. 2, Centre for Environmental Economics and Policy in Africa, University of Pretoria, 66 pp.

Heywood, V. (ed.), 1995: *Global Biodiversity Assessment.* United Nations Environment Programme, Nairobi, Kenya, 1140 pp.

Holmgren, P. and T.Turesson, 1998: Satellite Remote Sensing for Forestry Planning—A Review. Scandinavian Journal of Forest Research 13, pp. 90–110.

Houghton, R.A., 1996: Land-use change and terrestrial carbon: the temporal record. In: *Forest Ecosystems, Forest Management and the Global Change,* M.J. Apps and D.T. Price (eds.), NATO ASI Series, Vol. 1 40, Springer Verlag, Berlin, Heidelberg, pp. 117–134.

Houghton, R.A., 1999: The annual net flux of carbon to the atmosphere from changes in land use 1850–1990. *Tellus,* **51B,** 298–313.

Houghton, R.A., K.T. Lawrence, J.L. Hackler, and S. Brown. 2001. The spatial distribution of forest biomass in the Brazilian Amazon: a comparison of estimates. *Global Change Biology,* 7, 731–746.

Humphrey, J.W., R. Ferris, M.R. Jukes, and A.J. Peace, 2002: The potential contribution of conifers plantations to the UK Biodiversity Action Plan. *Botanical Journal of Scotland,* 54, 49–62.

IIASA and FAO (International Institute for Applied Systems Analysis and Food and Agriculture Organization), 2002: Global agro-ecological assessment for agriculture in the 21th century, by G. Fisher, M. Shah, H. van Velthuizen and F.O. Nachtergaele. Laxenburg, Austria and Rome, 155 pp.

ILO (International Labour Organization), 2001: Social and Labour Dimensions of the Forestry and Wood Industries on the Move. Report for the ILO Forestry and Wood Industries Meeting, September 2001.

IPCC (Intergovernmental Panel on Climate Change), 2001a: *Intergovernmental Panel on Climate Change. Climate Change 2001: The Scientific Basis.* Cambridge University Press, Cambridge, UK, 881 pp.

IPCC, 2001b: *Intergovernmental Panel on Climate Change. Climate Change 2001: Impacts, Adaptation, and Vulnerability.* Cambridge University Press, Cambridge, UK, 1032 pp.

IPCC, 2000: *Land Use, Land-Use Change, and Forestry.* Special Report of the Intergovernmental Panel on Climate Change.Cambridge University Press, Cambridge, UK, 377 pp.

Iremonger, S., C. Ravilious, and T. Qiunton, 1997: A statistical analysis of global forest conservation. In: *A Global Overview of Forest Conservtion.* CD-ROM, CIFOR and WCMC, Cambridge.

Iqbal, M., 1995: *Trade Restrictions Affecting International Trade in Non-wood Forest Products.* NWFP 8, FAO, Rome.

Isaev, A.S. (ed.), Forestry at the Beginning of 21th Century. Ecology, Moscow, 332 pp. [in Russian].

Isik K., F. Yaltirik, and A. Akesen A., 1997: Forests, biological diversity and the maintenance of the natural heritage. In Proceedingas of the XI World Forestry Congress, Volume 2, Antalya, Turkey, pp 3–19.

IUCN (World Conservation Union–IUCN), 1996: *1996 IUCN Red List of Threatened Animals.* Gland, Switzerland.

IUCN, 1997: *1996 IUCN Red List of Threatened Plants.* Gland, Switzerland.

IUFRO (International Union of Forest Research Organizations), 2003: Science and Technology—Building the Future of the World's Forests. Planted Forests and Biodiversity. IUFRO Occasional Paper 15, Vienna, Austria, 50 pp.

Ivanov, B.I. (ed), 2003: The Review of Conditions and Tendencies in Climate Change in Yakutia, Russian Academy of Sciences, Yakutsk, Russia, pp. 56. [in Russian].

Iverson, L.R. and A.M. Prasad. 1998. Predicting abundance of 80 tree species following climate change in the eastern United States. *Ecological Monographs* **68,** 465–485.

Janz, K and R. Persson, 2002: How to Know More About Forests? Supply and Use of Information for Forest Policy. CIFOR Occasional Paper, No.36. CIFOR, Bogor, 28 pp.

Jiang, Z., and S.Y. Zhang, 2003: China's plantation forests for sustainable wood supply and development. Proceedings of the 12th World Forestry Congress, Quebec, Canada, Volume B, 327–331.

Johns, A.G., 1997: *Timber Production and Biodiversity Conservation in Tropical Rain Forests.* Cambridge University Press, Cambridge, 256 pp.

Joshi, M. and P.P. Singh, 2003: Applying natural resource economics to forest degradation: Lessons from India. Proceedings of the 12th World Forestry Congress, Quebec, Canada, Volume A, p.127.

JRC (Joint Research Centre), 2000: Global Burnt Area 2000 (GBA2000) dataset. Available at http://www.gvm.jrc.it/fire/gba2000/index.htm.

JRC, 2003: Global Land Cover 2000 database. Database available at http://www.gvm.jrc.it/glc2000.

Kaimowitz, D., 2003: From Rio to Johannesburg and beyond: Forest conservation and rural livelihoods in the global South. In *Congress Proceedings,* Volume A. Proceedings of the 12th World Forestry Congress, Quebec, Canada, Septembn 21 to 28, 2003 Quebec, Canada, pp. 10–15.

Kanowski, P., 1997. *Afforestation and Plantation Forestry.* Resource Management in Asia-Pacific Project, Australian National University, Canberra, Australia. Working Paper No. 6.

Kanowski, P., 2003. Challenges to Enhancing the Contributions of Planted Forests to Sustainable Forest Management. Paper presented at UNFF Intersessional Experts Meeting on the Role of Planted Forests in Sustainable Forest Management, 24–28 March 2003, Wellington, NZ.

Kasischke, E.S. and B.J.Stock (eds), 2000: *Fire, Climate Change and Carbon Cycling in the Boreal Forest.* Ecological Studies 138, Springer Verlag, Berlin-Heidelberg New York, 461 pp.

Kauppi, R.E., 2003: New, low estimate for carbon stock in global forest vegetation based on inventory data. *Silva Fennica,* **37(4),** 451–457.

Kemp, R.H. and C. Palmberg-Lerche. 1994. "Conservation of genetic resources in forest ecosystems." In: Readings in sustainable forest management. Rome: FAO Forestry Paper. pp. 101–117.

Kirschbaum, M.U.F. and A. Fischlin, 1996: Climate change impacts on forests. In: *Climate Change 1995—Impacts, Adaptations and Mitigation of Climate Change: Scientific-Technical Analysis,* R.T. Watson, M.C. Zinyowera, R.H. Moss, and D.J. Dokken (eds.). Contribution of Working Group II to the Second Assessment Report of the IPCC. Cambridge, Cambridge University Press, pp. 94–126.

Klein, C., 2000: On large-area inventory and assessment of trees outside the forest. *Unasylva,* 200(51), 3–10.

Kljuev, N.N., 2001: Russia and its regions. External and internal threats. Nauka, Moscow, 214 pp. [in Russian].

Körner, C. and J.A. Arnone III, 1992: Responses to elevated carbon dioxide in artificial tropical ecosystems. *Science,* **257,** 1672–1675.

Kummer, D., 1992: *Deforestation in the Post War Philippines.* Quezon City: Ateneo de Manila University Press.

Kurz, W.A. and M.J. Apps, 1999: A 70-year retrospective analysis of carbon fluxes in the Canadian forest sector. *Ecological Applications,* **9,** 526–547.

Kusters, K., and B. Belcher (editors), 2004: *Forest products, livelihoods and conservation: case studies of non-timber forest product systems. Volume 1: Asia.* CIFOR, Bogor, Indonesia.

Kwok, H.K. and R.T. Corlett, 2000: The bird communities of a natural secondary forests and a *Lophostemon confertus* plantation in Hiog Kong. *Forest Ecology and Management,* **13,** 227–234.

Laird, S.A., 1999. Forests, Culture and Conservation. In Posey, D.A. (Ed.), 1999. *Cultural and Spiritual Values of Biodiversity.* United Nations Environment Programme, Nairobi, Kenya.

Lambin, E.F., J.H. Geist, and E. Lepers, 2003: Dynamics of land-use and land cover change in tropical regions. *Annual Reviews of Environment Resources,* **28,** 205–241.

Lange, D., and M. Mladenova, 1997: Bulgarian model for regulating the trade in plant material. In: *Medicinal Plants for Forest Conservation and Health Care,* FAO Non-Wood Forest Products Series II, Rome, pp. 149–156.

Laurance, W.F. and B.Williamson, 2001: Positive feedbacks among forest fragmentation, drought, and climate change in the Amazon. *Conservation Biology,* **15,** 1529–1535.

Lawton, R.O., U.S. Nair, R.A. Pielke Sr, and R.M. Welch, 2001: Climatic impact of lowland deforestation on nearby montane cloud forests. *Science* 294:584–587.

Lette, H., and H. de Boo, 2002: Economic Valuation of Forests and Nature: A Support Tool for Effective Decision-making. International Agricultural Centre (IAC) Wageningen and Natural Reference Centre for Agriculture, Nature Management and Fisheries (EC-LNV), Ede.

Lopez, M., A.Schloenvoigt, M. Ibrahim, C. Klein, and M. Kanninen, 1999: Cuantificacion del carbono almacenado en el suelo de un sistema silvopastorali en la zone Atlantica de Costa Rica. *Agroforesteria en las Americas* **6(23),** 51–53

Low, A.B. and A.G. Rebelo, 1996: *Vegetation of South Africa, Lesotho and Swaziland.* Department of Environment and Tourism, Pretoria.

Lyke, J., 1995: *Forests and Human Health.* International Forestry Issue Brief 13, US Forest Service, Washington DC.

Margules, C., R.L. Pressey & P.H. Williams. 2002. "Representing biodiversity: data and procedures for identifying priority areas for conservation." *J. Biosci.* (*Suppl. 2*) 27, 309–326. Indian Academy of Sciences.

Mårn, H.M., and W. Jonkers, 1981: Logging damage in tropical high forest.—In tropical forests. Source of energy Through Optimization and diversification. Universati Pertanian Malaysia, pp. 27–38.

Mather, A., 1999: Society and the services of forests. Chapter 8 in: *World Forests, Society and Environment,* M.Palo and J. Uusivori (eds.), Kluwer Academic Press, 86–89.

Matthews, E., 2001. *Understanding the FRA 2000.* Washington DC: World Resources Institute. Forest Briefing No. 1, 11 pp.

Matthews, E., R. Payne, M. Rohweder, and S. Murray, 2000: *Pilot Analysis of Global Ecosystems (PAGE): Forest Ecosystems.* World Resources Institute, Washington, DC, 100 pp.

Mayaux, P., F. Achard, and J.-P. Malingreau, 1998: Global tropical forest area measurements derived from coarse resolution satellite imagery: a comparison with other approaches. *Environmental Conservation* **25,** 37–52.

McCallum, I., S. Nilsson, M. Obersteiner, A. Shvidenko, 2004: A decade of Earth observation using global land cover datasets. International Journal of Earth Observations and Geoinformatics (submitted).

McConnell, W.J., 2004: Forest cover change—tales of the unexpected, *Global Change NewsLetter,* No 57, March, pp. 8–11.

McFarlane, D.J., George, R.J., and P. Farrington, 1992. Changes in the hydrological cycle. In Hobbs, R.J. and Saunders, D.A. (eds.) *Reintegrating Fragmented Landscapes,* Springer-Verlag, New York, pp. 146–186.

McKenzie, C.R., 1995: *Global Supply Outlook—South Asia.* Proceedings TAPPI Global Fiber Supply Symposium, Chicago, Oct. 5–6, TAPPI Press, Atlanta, Georgia, USA.

McNeely, J.A., Gadgil, C. Leveque, C. Padoch and K. Redford, 1995: "Human Influences on Biodiversity." In Heywood, V. (Ed.) 1995. Global Biodiversity Assessment. Cambridge, UK: Cambridge University Press.

MDBC (Murray-Darling Basin Ministerial Council), 1999: *The Salinity Audit of the Murray-Darling Basin. A 100-year Perspective.* Canberra.

Melillo, J.M., A.D. McGuire, D.W. Kicklighter, et al., 1993: Global climate change and terrestrial net primary production. *Nature,* **363,** 234–240.

Mertens, B., and E.F. Lambin, 2000: Land-cover-change trajectories in South Cameroon. Annals of the Association of American Geographers, **90(3),** 467–494.

Michie, B., C. Chandrasekharan, and P.Wardle, 1999: Production and trade in forest goods. Chapter 7 in: *World Forests, Society and Environment,* M.Palo and J. Uusivori (eds.), Kluwer Academic Press, 78–85.

MA (Millennium Ecosystem Assessment), 2003: *Ecosystems and Human Well-being. A Framework for Assessment.* Island Press, Washington DC. 245 pp.

MNR (Ministry of Natural Resources), 2003: Results of activities of MNR of the Russian Federation in 2002 and tasks of development of natute's protection and nature's resource complex in 2003. Moscow, 60pp. + Suppl. [in Russian].

MODIS (Moderate Resolution Imaging Spectroradiometer), 2002: Terra Land Cover Type 96 Day L3 Global 1 km ISIN Grid V003. Database available at : http://duckwater.bu.edu/lc/mod12ql.html.

Muller-Dombois, and J.G. Goldammer, 1990: Fire in tropical ecosystem and global environmental change. In: Fire in the Tropical Biota Ecosystems. Processes and Global Challenges, J.G.Goldammer (ed.). Eological Studies 84, Springer Verlag, pp. 1–10.

Mutch, R.W., 2003. Fire situation in Brazil. *International Forest Fire News,* No 28, pp. 45–50.

Myers, N., 1996: The world's forests: Problems and potentials. *Environmental Conservation,* 23(2), 156–168.

Myers, N., 1997: The world forests and their ecosystem services. In: *Nature's Services: Societal Dependence on Natural Ecosystems,* G. Daily (ed.), Island Press, Washington, DC, pp 215–235.

Myneni, R.B., J. Dong, C.J. Tucker, R.K. Kaufman, P.E. Kauppi, et al. 2001: A large carbon sink in the woody biomass of northern forests. *Proceedings of the National Academy of Sciences of the USA (PNAS),* **98(26),** pp. 14784–14789.

Naeem, S., F.S. Chapin III, R. Costanza, P. Ehrlich, F. Golley, et al. 1999. Biodiversity and Ecosystem Functioning: Maintaining Natural Life Support Processes. *Issues in Ecology* No. 4, Fall 1999. Ecological Society of America.

Nelleman, C. and M.G. Thomsen, 2001: Long-term changes in forest growth: Potential effects of nitrogen deposition and acidification. Water, Air, and Soil Pollution **128,** 197–205.

Nemani R.R., C.D. Killing, H. Hashimoto, W.M. Jolly, S.C. Piper, C.J. Tucker, R.B. Myneni, and S.W. Running, 2003: Climate-driven increases in global net primary production from 1982 to 1999. Science 300, 1560–1563.

Nilsson, S., 1996: *Do We Have Enough Forests?* IUFRO Occasional Paper, No 5. IUFRO Secretariat, Vienna, 71 pp.

Nilsson, S., 2000: Challenges for the boreal forest zone and IBFRA. In: Disturbance in Boreal Forest Ecosystems: Human Impacts and Natural Processes, S.G. Conard (ed.). Proceedings of the IBFRA 1997 Annual Meeting, General Technical Report NC-209, USDA, Forest Service, North Central Research Station, pp. 1–16.

Nilsson, S. and W. Schopfhauser, 1995: The carbon-sequestration potential of a global afforestation program. *Climatic Change,* 30, 267–293.

Nilsson, S.,Blauberg, K., Samarskaja, E., and V. Kharuk, 1998: Pollution stress of Siberian forests. In R. Linkov and R. Wilson, Air Pollution in the Ural Mountains, Kluwer Academic Publishers, pp. 31–54.

Noss, R.F., 1990: Indicators for monitoring biodiversity: A hierarchical approach. Essay, *Conservation Biology* 4: 355–364.

Odebode, S.O., 2003: Contribution of selected non-timber forest products to household food security in Osun State, Nigeria. Proceedings of 12th World Forestry Congress, Quebec, Canada, September 21 to 28, Volume A, Forest for People, p. 55.

Oldeman, L.R., R.T.A. Hakkeling, W.G. Soembroek, 1990: *World Map of the Status of Human-*Induced Soil Degradation: An Explanatory Note. International Soil Reference and Information Centre, Wageningen, The Netherlands.

Oldfield, S.F., Lusty, C., Mackinven, A., The World List of Threatened Trees. World Conservation Press, Cambridge, UK.

Pacala, S., 2004: Forest inventory data falsify ecosystem models of CO_2 fertilization. Available at http://www.eeb.princetin.edu/FACULTY/Pacala/Pacala.html

Palo, M. and J. Uusivuori (eds.), 1999: *World Forests, Society and Environment.* Kluver Academic Publishers, Dordrecht/London/Boston, 404 pp.

Pan, Y.D., J.M. Melillo, A.D. McGuire, D.W. Kicklighter, L.F. Pitelka, et al. 1998: Modeled responses of terrestrial ecosystems to elevated atmospheric CO2: A comparison of simulations by the biogeochemistry models of the Vegetation/Ecosystem Modeling and Analysis Project (VEMAP). *Oecologia,* **114,** 389–404.

Parkinson, J., 1999: Indigenous people and forests. In: *World Forests, Society and Environment,* M.Palo and J. Uusivori (eds.), Kluwer Academic Press, pp. 90–91.

Paumalainen, J., 2001: *Structural, Compositional and Functional Aspects of Forest Biodiversity in Europe.* ECE/TIM/DP/22, New York and Geneva, 88 pp.

Pearce, D.W., 1990: An Economic Approach to Saving the Tropical Forests. Discussion Paper 90–06, London Environmental Economic Centre, London.

Peluso, N. 1992. *Rich Forests, Poor People: Resource Control and Resistance in Java.* University of California Press, Berkeley, 321 pp.

Persson, R. 1974. *World Forest Resources. Review of the World's Forest Resources in the Early 1970:s.* Royal College of Forestry, Department of Forest Survey, Research Notes Nr 17. Stockholm, 274 pp.

Persson, R., 2003: Assistance to forestry. Experiences and potential for improvement. CIFOR, Bogor, 120 pp.

Persson, R., 1995: Den globala skogssituationen 1990 (The global forestry situation 1990). Report Nr 2. Swedish University of Agricultural Sciences, Umeå, Sweden, 338 pp.

Peters, C. M., A. H. Gentry, and R. O. Mendelsohn, 1989: Valuation of an Amazonian rainforest. *Nature* 339:655–656.

Phillips, O.L., T.R. Baker, L. Arroyo, N. Higuchi, T. Killeen, et al. 2004: Pattern and process in Amazon tree turnover, 1976–2001. *Philosophical Transactions of the Royal Society of London B* 359: 381–407.

Pielke, R.A., R. Avissar, M. Raupach, A.J. Dolman, X. Zeng, and S. Denning, 1998: Interactions between the atmosphere and terrestrial ecosystems: influence on weather and climate. *Global Change Biology* **4,** 461–475.

Pilipenko, A.I., and Y.J.Yukhnovskt, 1998: Backgrounds of optimal parameters of shelterbelts. Science Herald of the Ukrainian National Agrarian University, Vol 10, pp.236–248, Kiev, Ukraine [in Ukrainian]

Pimm, S.L., G.J. Russell, J.L. Gittleman and T.M. Brooks, 1995. The future of biodiversity. *Science* 269: 347–350.

Pisarenko, A., and V. Strakhov, 2004: Forest Management in Russia. From Utilization to Management, Jurisprudence, Moscow 551 pp. [in Russian].

Posey, D., 1993: Indigenous knowledge in the conservation and use of world forests. In: *World Forests for the Future. Their Use and Conservation,* K. Ramakrishna and G.W. Woodwell (eds). Yale University Press, New Haven and London, pp. 59–77.

Pounds, J.A., M.P.L Fogden and J.H. Campbell. 1999. Biological responses to climate change on a tropical mountain. *Nature* 398:611–615.

Prentice, I.C., G.D. Farquhar, M.J.R. Fasham, M.L. Goulden, M. Heimann, et al. 2001. In: T.J. Houghton and D. Yihui (eds) *Climate Change 2001: the Scientific Basis. The Intergovernmental Panel on Climate Change (IPCC),* Third Assessment Report. Cambridge University Press, Cambridge, UK, pp.183–237.

Price, M.F., 1998. Mountains: globally important ecosystems. *Unasylva,* 49(195): 3–12.

Price, D.T., D.H. Halliwell, M.J. Apps M.J. and C.H. Peng, 1999a: Adapting a patch model to simulate the sensitivity of Central-Canadian boreal ecosystems to climate variability. *Journal of Biogeography,* **26,** 1101–1113.

Price, D.T., C. Peng C., M.J. Apps and D.H. Halliwell, 1999b: Simulating effects of climate change on boreal ecosystem carbon pools in central Canada. *Journal of Biogeography,* 1237–1248.

Protopopov, V.V., 1975: *Environmental Role of Dark-Coniferous Forests.* Nauka, Novosibirsk, 327 pp. [in Russian].

Protopopov, V.V., 1979: *Environmental Role of Forests of the Basin of Lake Baikal,* Nauka, Novosibirsk, 256 pp. [in Russian].

Paumalainen, J., 2001: *Structural, Compositional and Functional Aspects of Forest Biodiversity in Europe.* ECE/TIM/DP/22, New York and Geneva, 88 pp.

Rakhmanov, V.V., 1984: *Hydro-Climatic Role of Forests.* Forest Industry, Moscow, 241 pp [in Russian].

Ramakrishna, K. and G. Woodwell, 1993: *World Forests for the Future: Their Use and Conservation.* Yale University Press, New Haven and London, 156 pp.

Rayner, P.J., I.G. Enting, R.J. Francey, and R. Lanenfelds, 1999: Reconstructing the recent carbon cycle from atmospheric CO_2, $\delta^{13}C$, and O_2/N_2 observations. *Tellus,* **51B,** 213–232.

Reid, W.V., 1992: How many species will there be? In Whitmore, T.C., and J.A. Sayer, Tropical Deforestation and Species Distinction, Chapman and Hall ,London, pp. 55–74.

Reid, W.V. and K.R. Miller. 1989. Keeping Options Alive: The Scientific Basis for Conserving Biodiversity. Washington DC: World Resources Institute, 105 pp.

Revenga, C., S. Murray, J. Abramovitz and A. Hammond, 1998: *Watersheds of the World: Ecological Value and Vulnerability.* World Resources Institute and Worldwatch Institute, 1998, 200 pp.

Repetto, R., and M. Gillis (eds.), 1988: Public Policies and the Misuse of Forest Resources. Cambridge University Press, Cambridge, UK, 432 pp.

Rhoades, C.C., G.E.Eckert, and D.C. Coleman, 1998: Effect of pasture trees on soil nitrogen and organic matter: implications for tropical montane forest restoration. *Restoration Ecology* **6(3),** 262–270.

Richards, J.F., 1990: Land transformation. In Turner, B.L, W.C. Clark, R.W. Kates, J.F. Richards, J.T. Mathews and W.B. Meyer (Eds): *The Earth as Transformed by Human Action,* Cambridge University Press, UK, pp. 163–180.

Richards, P.W. 1996. *The Tropical Rain Forest.* Second edition. Cambridge University Press, Cambridge, UK, 576 pp.

Ricketts, T.H., G.C.Daily, P.R. Ehrlich, and C.D.Michener, 2004: Economic value of tropical forest to coffee production. *PNAS* 101:34, 12579–12582. August 24.

Richardson, D.M. 1998. Forestry trees as invasive aliens. *Conservation Biology* **12,** 18–26.

Roche de la, I. A., O'Connor, and P. Tetu. 2003: Wood products and sustainable construction. Proceedings of 12th World Forestry Congress, Quebec, Canada, September 21 to 28 2003, Volume A Forests for People, 135–143.

Rodgers, W.A. 1997. "Patterns of loss of forest biodiversity—A global perspective". Proceedings of the 11th World Forestry Congress, Antalya Turkey, 13–22 October 1997.

Rodrigues, A.S.L., R.D. Gregory and K.J. Gaston K.J., 2000: Robustness of reserve selection procedures under temporal species turnover. *P. Roy. Soc. Lond. B Bio.*, **267**, 49–55.

Rodrigues, A.S.L., S.J. Andelman, M.I. Bakarr, L. Boitani, T.M. Brooks, et al. 2003. *Global Gap Analysis: Towards a Representative Network of Protected Areas.* Washington DC: Center for Applied Biodiversity Science, Conservation International, Washington, DC.

Rubtsov, V.M., A. A. Derjugin, Y.N. Salmina, and V.I.Gurzev, 1990: Water-regulating Role of Taiga Forests. Agropromizdat, Moscow, Russia, 224 pp. [in Russian]

Sagreev, V.V., V.I. Sukhikh, A.Z. Shvidenko, N.N. Gusev N.N., and A.G. Moshkalev, 1992: *All-Union Reference Data for Forest Inventory.* Kolos, Moscow, 495 pp [in Russian].

Saugier, B., J. Roy and H. Mooney, 2001: Estimations of global terrestrial productivity: converging toward a single number? In: *Terrestrial Global Productivity,* J. Roy, B. Saugier, and H. Mooney (eds.). Physiological Ecology, Academic Press, pp. 543–557.

Sayer, J.A., C.S. Harcourt, and N.M. Collins, 1992: *The Conservation Atlas of Tropical Forests:* Africa. Macmillan, London, UK, 228 pp.

Sayer, J.A., and S. Iremonger, 1999: The state of the world's forest biodiversity. Chapter 12 in: *World Forests, Society and Environment,* M.Palo and J. Uusivori (eds.), Kluwer Academic Press, pp. 129–134.

Sayer, J.A.,P.A. Zuidema, and M. Rijks, 1995: Managing for biodiversity in humid tropical forests. *Commonwealth Forestry Review,* **74,** pp. 282–287.

Sayer, J.A., J.K. Vanclay, and N. Byron, 1997. *Technologies for Sustainable Forest Management: Challenges for the 21st Century.* Occasional Paper No 12, CIFOR, Bogor, Indonesia, 12 pp.

Schmidt, K.M., J.P. Menakis, C.C. Hardy, W.J. Hann, and D.L. Bunnell, 2002: *Development of Coarse-scale Spatial Data for Wildland Fire and Fuel Management.* U.S. Department of Agriculture, Forest Service, Fort Collins, Colorado, USA.

Schneider, R., E. Arima, A. Verissimo, P. Barreto, and C. Souza, Jr., 2002: *Sustainable Amazon: Limitation and Opportunities for Rural Development,* Word Bank Technical Paper No 515, Environment Series, World Bank, Washington DC.

Scherr, S., A. White, and D. Kaimowitz, 2002: *Making Markets Work for Forest Communities.* Washington, DC, Forest Trends and CIFOR.

Scherr, S., A. White, and A. Khare, 2004. For Services Rendered; The current status and future potential of markets for the ecosystem services provided by tropical forests. ITTO Technical Series No. 21, Yokohama, Japan. 72 pp.

Scholes, R.J., E.-D. Schulze, L.F. Pitelka and D.O. Hall, 1999: Digeochemistry of terrestrial ecosystems. In: The Terrestrial Biosphere and Global Change: Implications for natural and managed ecosystems. B.H. Walker, W.L. Steffen, J. Canadell, and J.S.I. Ingram (eds) Cambridge University Press, pp 88–105.

Scholes, R.J., and R.Biggs, 2004: *Ecosystem Services in Southern Africa: A Regional Assessment,* Council for Scientific and Industrial Research, Pretoria, South Africa.

Schulze, E.-D., J. Lloyd, F.M. Kelliher, C. Wirth, C. Rebmann, et al. 1999: Productivity of forests in the Eurosiberian boreal region and their potential to act as a carbon sink—a synthesis. *Global Change Biology,* 5, 703–722.

Schulze, E.-D., R. Valentini, and M.-J. Sanz. 2002. The long way from Kyoto to Marrakesh: Implications of the Kyoto Protocol negotiations for global ecology. *Global Change Biology,* **8,** 505–518.

Secretariat of the Convention on Biological Diversity, 2001. *Global Biodiversity Outlook.* Montreal, Canada, 182 pp.

Sheingauz, A.S., and A.P. Sapozhnikov, 1988: Classification of functions of forests. Forest Science, 3, pp. 7–16 [in Russian].

Sheil, D., and S. Wunder, 2002: The value of tropical forest to local communities: complications, caveats and cautions. *Conservation Ecology* 6(2):9. [online] URL: http://www.consecol.org/vol6/iss2/art9.

Sherbakov, I.P., 1975: Forest Cover of North East of the USSR. Nauka, Novosibirsk, Russia, 344 pp. [in Russian].

Shvidenko, A., 2003: Russian forests at the beginning of the third millennium: Status and trends. In: *Congress Proceedings,* Volume B. Proceedings of the 12th World Forestry Congress, 21 to 28 September, 2003, Quebec, Canada, pp. 27–34.

Shvidenko, A.Z., A.A. Strochinsky, Yu.N. Savich, and S.N. Kashpor, 1987: *Standard and Reference Data for Inventory of Forests of Ukraine and Moldova.* Ukrainian Ministry of Forest, Kiev, 559 pp. [in Russian].

Shvidenko, A.Z., S. Nilsson, V.S. Stolbovoi, V.A. Rozhkov and M. Gluck, 2001: Aggregated estimation of basic parameters of biological production and the carbon budget of Russian terrestrial ecosystems: 2. Net primary production, Russian Journal of Ecology, 32, No 2, 83–90.

Shvidenko, A. and S. Nilsson, 2002: Dynamics of Russian forests and the carbon budget in 1961–1998: An assessment based on long-term forest inventory data. *Climatic Change* 55, 5–37.

Shvidenko, A. and S. Nilsson, 2003: A synthesis of the impact of Russian Forests on the global carbon budget for 1961–1998, *Tellus,* **55B,** 391–415.

Shvidenko, A., E. Vaganov and S. Nilsson, 2003: Biospheric role of Russian forests at the beginning of the third millennium: carbon budget and Kyoto Protocol. *Siberian Ecological Journal* **6,** 549–658 [in Russian].

Sist, P., D. Dykstra and R. Fimbel, 1998: Reduced-Impact Logging Guidelines for Lowland and Hill Dipterocarp Forests In Indonesia. CIFOR Occasional Paper No. 15. CIFOR, Bogor.

Siwatibau, S., 2003: Forests, trees and human needs in Pacific communities. Proceedings of the 12th World Forestry Congress, Volume A, Quebec, Canada, pp.29–36.

Spittlehouse, D.L., 1997: Forest management and climate change. In: *Responding to Global Climate Change in British Columbia and the Yukon,* E. Taylor and B. Taylor (eds.), Environment Canada, Vancouver, BC, pp. 24–1–24–8.

Spittlehouse, D.L. and Stewart, R.B. 2003. Adaptation to climate change in forest management. *Journal of Ecosystems and Management* (in review).

Stewart, R. B., Wheaton, E. and Spittlehouse, D., 1998: Climate change: implications for the boreal forests. In: *Emerging Air Issues for the 21st Century: the Need for Multidisciplinary Management,* Proceedings of a Speciality Conference, Sep. 22–24, 1997, Calgary, AB, Legge, A.H. and L.L. Jones (eds.), Air and Waste Management Assoc., Pittsburg, PA, pp. 86–101.

Still, C., P.N. Foster and S.N Schneider. 1999. Simulating the effects of climate change on tropical montain cloud forests. *Nature* 398: 608–610.

Sun, X., E. Katsigirls, and A. White, 2004: Meeting China's Demand for Forest Products: An Overview of Import Trends, Ports of Entry, and Supplying Countries, With Emphasis on the Asia-Pacific Region, Washington, DC, Forest Trends, Chinese Center for Agricultural Policy and Center for International Policy Research.

Sunderland, T. C. H., and O. Ndoye (editors), 2004: *Forest products, livelihoods and conservation: case studies of non-timber forest product systems. Volume 2: Africa.* CIFOR, Bogor, Indonesia.

Szott, L.T., and C.A. Palm, 1996: Nutrient stocks in managed and natural humid tropical follows. **186,** 293–309.

Tacconi, L., 2003: Fires in Indonesia: Causes, Costs and Policy Implications. CIFOR Occasional Paper No. 38, Bogor, 24 pp.

Tacconi, L., M. Boscolo, and D. Brack, 2003: *National and International Policies to Control Illegal Forest Activties.* Center for International Forestry Research, Bogor, Indonesia. Report prepared for the Ministry of Foreign Affairs, Government of Japan, 27 pp.

Thang, H.C., 2003: Current perspectives on sustainable forest management and timber certification. In: Congress Proceedings, Volume A—Forests for People. Proceedings of the 12th World Forestry Congress, 21 to 28 September 2003, Quebec, Canada, pp. 307–314.

Ten Kate, K. and S.A. Laird, 1999. *The Commercial Use of Biodiversity. Access to Genetic Resources and Benefit Sharing.* Earthscan, London, UK, 416 pp.

TNC, 2003: The Third National Communication of Russia to the UN FCCC. Moscow, 54 pp [in Russian].

Trexler, M.C. and C. Haugen, 1995: *Keeping it Green: Tropical Forestry Opportunities for Mitigating Climate Change.* World Resource Institute and EPA, Washington, DC, 52 pp.

UN-CSD/IPF, 1996: *Report of the Secretary General to the Session of IPF in Geneva,* March 1996.

UN/ECE (United Nations Economic Commission for Europe), 1996: Long-Term Historical Changes in the Forest Resource. Geneva Timber and Forestry Paper No 10, ECE/TIM/SP/10. United Nations Economic Condition for Europe and FAO, Timber Section, Geneva, Switzerland.

UN/ECE, 1998: *Non-Wood Goods and Services of the Forests.* Geneva Timber and Forest Study report No15, Geneva.

UN/ECE, 2002: *Forest Conditions in Europe. 2001 Executive Report.* Geneva and Brussels, 29 pp.

UN/ECE/FAO (Food and Agriculture Organization), 2003: *State of Europe's Forests 2003.* The MCPFE Report on Sustainable Forest Management in Europe, Vienna, Austria, 126 pp.

UNEP-WCMC (United Nations Environment Programme World Conservation Monitoring Centre), 2004: Forest Information Center. Available online at www.unep-wcmc.org/forest/data.

UNEP-WCMC, 2002: [online] World Conservation Monitoring Centre. Available online at www.unep-wcmc.org/forest/world.htm (Revision date: 12 December 2002).

UNESCO, 1996. *Sacred Sites—Cultural Integrity, Biological Diversity.* Programme proposal, November 1996, Paris.

USGS (United States Geological Survey), 2003: Global Land Cover Characteristics (GLCC) Data Base, Version 2.0. Database available at http://edcdaac.usgs.gov/glcc/glcc.html.

Valentini, R., G. Matteucchi, H. Dolman, E-D. Schulze, C. Rebmann, et al. 2000: Respiration as the main determinant of carbon balance in European forests. *Nature,* 404, 861–864.

Verissimo, A., Cochrane, M., Souza Jr., C., and R. Salomao. 2002. Priority areas for establishing national forests in the Brazilian Amazon. Conservation Ecology 6:4, URL http://www.consecol.org/vol6/iss1/art4.

Versfeld, D.B., B.W. van Wilgen, J.M. Bosch and F.J. Kruger, 1994: The impact of forestry on water resources in South Africa. Report to the Ministry of Water affairs and forestry, Division of Forest science and Technology. CSIR/WNNR, Pretoria.

Vorob'ev, G.I., K.D. Mukhamedshin, and L.M. Devjatkin, 1984: *World Forestry and Forest Management,* Forest Industry, Moscow, 351 pp. [in Russian].

WDPA (World Commission on Protected Areas), 2003: World Database on Protected Areas. CDRom.

Walker, R.T., 1987. Land Use Transition and Deforestation in Developing Countries. *Geographical Analysis.* Vol. 19 (1), pp. 18–30.

Walker, B.H., W.L. Steffen, J. Canadell, and J.S.I. Ingram (eds), 1999: The terrestrial biosphere and global change: implication for natural and managed ecosystems. Synthesis Volume. IGBP Book Series 4, Cambridge University Press, 450 pp.

Walther, G-R , E. Post, P. Convey, A. Menzel, C. Parmesan, et al. 2002: Ecological responses to recent climate change. *Nature,* 416, 389–395.

Wadsworth, F.H., 1997: *Forest Production for Tropical America.* USDA Forest Service Agr. Handbook 710, USDA Forest Service, Washington, DC.

WBGU (German Advisory Council on Global Change), 1998: Die Anrechnung biologischer Quellen und Senken im Kyoto-Protokoll: Fortschritt oder Ruckschlag fur den globalen Umweltschutz. Sonergutachten 1998, Bremerhaven, Germany, 76 pp.

WCED (World Commission on Environment and Development), 1987: *Our Common Future.* Oxford, UK: Oxford University Press. 393 pp.

WCFSD (World Commission on Forests and Sustainable Development), 1999: *Our Forests . . . Our Future.* Summary Report, 40 pp. Available at the Internet http://:www.whrc.org.

Webb, D.B., P.J. Wood, J.P. Smith, and G.S. Henman, 1984: *A Guide to Species Selection for Tropical and Subtropical Plantations.* Tropical Forestry Paper No 15, Commonwealth Forestry Institute, Oxford, UK.

Weaver, J.W., 2003: Partnership in our forests. The Canadian approach to sustainability. In: Congress Proceedings, Volume C. Proceedings of the 12th World Forestry Congress, September 21 to 28, Quebec, Canada, pp. 34–40.

WEC (World Energy Council), 1999: *The Challenge of Rural Energy Poverty in Developing Countries.* FAO/WEC, London.

WGWF (Working Group on Wildland Fire), 2003: An overview of vegetation fires globally. A background paper presented to the International Wildland Fire Summit, Sydney, Australia, 8 October 2003. Prepared by the Working Group on Wildland Fire, UN International Strategy for Disaster Reduction and the Global Fire Monitoring Center. *International Forest Fire News,* No 29, pp. 40–55 [available at http://www.fire.uni-freiburg.de/iffn/iffn_29/IWFS-6-Background.pdf].

White, A., 2003. Forest Plantations: Good for What and for Whom? *Asian Timber.* May/June.

White, A. and A. Martin, 2002: *Who Owns the World's Forests? Forest Tenure and Public Forests in Transition.* Forest Trends and Center for International Environmental Law, Washington, DC, 27 pp.

Wiersum, K.F., 1984: Surface erosion under various tropical agroforestry systems. In: O'Loughlin, Cl.L., Pearce, A.J. (Eds.) Effects of Forest Land Use on Erosion and Slope Stability. IUFRO, Vienna, pp. 231–230.

Williams, M., 2003. *Deforesting the Earth: From Prehistory to Global Crisis.* University of Chicago Press, Chicago, USA, 689 pp.

Winterbottom, R., 1990. Taking Stock: The Tropical Forestry Action Plan After Five Years. World Resources Institute, Washington DC.

World Bank, 2003: Sustainable Development in a Dynamic World. World Development Report 2003, Washington, DC.

WRI (World Resources Institute), 1992: *World Resources 1992–1993. A Guide to the Global Environment.* Oxford University Press, New York and Oxford, 382 pp.

WRI, 2000a: *A First Look At Logging In Gabon.* WRI, Washington, DC, 72 pp.

WRI, 2000b: *An Overview of Logging in Cameroon.* WRI, Washington, DC, 56 pp.

WRI, 2001: *People and Ecosystems: The Frying Web of Life.* World Resources Institute, Washington, DC.

WRI, IUCN and UNEP, 1992. *Global Biodiversity Strategy.* Washington, DC, 260 pp.

WRI/UNEP/World Bank, 1996. *World Resources 1996–1997.* New York, Oxford University Press, 400 pp.

WRM (World Rainforest Movement), 2002. The FAO Forest Assessment: Concealing the Truth. Bulletin No. 61, August 2002.

Wunder, S. 2003: Forests without trees? A note on problematic forest definitions and change assessment. Proceedings of 12th World Forestry Congress, Quebec City, Canada, September 21 to 28, 2003, Volume B-Forests for the Planet, Quebec, Canada, p. 68.

WWF-Indonesia and EEPSEA (World Wide Fund for Nature-Indonesia and Economy and Environment Program for Southeast Asia). 1998. The Indonesian Fires and Haze of 1997: The Economic Toll. Singapore and Jakarta.

Yukhnovsky, V.Yu., 2003: *Agro-Forest Landscapes of Plain Ukraine: Optimization, Standards, Ecological Aspects.* Kiev, 272 pp [in Ukrainian].

Zhang, Q. and C.O. Justice, 2001: Carbon emissions and sequestration potential of Central African ecosystems. *Ambio,* **30,** 351–355.

Chapter 22

Dryland Systems

Coordinating Lead Authors: Uriel Safriel, Zafar Adeel

Lead Authors: David Niemeijer, Juan Puigdefabregas, Robin White, Rattan Lal, Mark Winslow, Juliane Ziedler, Stephen Prince, Emma Archer, Caroline King

Contributing Authors: Barry Shapiro, Konrad Wessels, Thomas Nielsen, Boris Portnov, Inbal Reshef, Jillian Thonell, Esther Lachman, Douglas McNab

Review Editors: Mohammed El-Kassas, Exequiel Ezcurra

Main Messages

Drylands cover about 41% of Earth's land surface and are inhabited by more than 2 billion people (about one third of world population). Drylands are limited by soil moisture, the result of low rainfall and high evaporation, and show a gradient of increasing primary productivity, ranging from hyper-arid, arid, and semiarid to dry subhumid areas. Deserts, grasslands, and woodlands are the natural expression of this gradient.

Dryland populations on average lag far behind the rest of the world on human well-being and development indicators (*high certainty*). The current socioeconomic condition of dryland peoples, about 90% of whom are in developing countries, lags significantly behind that of people in other areas.

Existing water shortages in drylands are projected to increase over time due to population increase, land cover change, and global climate change. From 1960 to 2000, global use of fresh water (drylands included) expanded at a mean rate of 25% per decade. The availability in drylands is projected to decline further from the current average of 1,300 cubic meters per person per year (in 2000), which is already below the threshold of 2,000 cubic meters required for minimum human well-being and sustainable development. This increased water stress will lead to reduced productivity of croplands and availability of fresh water, resulting in further adverse impacts on human well-being in drylands. There is a high degree of certainty that global climate change, land use developments, and land cover changes will lead to an accelerated decline in water availability and biological production in drylands.

Transformation of rangelands and other silvipastoral systems to cultivated croplands is leading to significant, persistent decrease in overall dryland plant productivity. Extreme reduction of rangeland vegetation cover through grazing of forage and collection of fuelwood exposes the soil to erosion. Transformation of rangelands to cultivated systems (approximately 15% of dryland grasslands, the most valuable dryland range, were converted between 1950 and 2000), in combination with inappropriate dryland irrigation and cultivation practices has led to soil salinization and erosion. These processes reduce the provision of water-related services, which affects the provision of many other significant dryland services and goods, culminating in persistent reduction of primary production.

Among dryland subtypes, ecosystems and populations of semiarid areas are the most vulnerable to loss of ecosystem services (*medium certainty*). Population density within drylands decreases with increasing aridity from 10 persons per square kilometer in the hyper-arid drylands to 71 persons in dry subhumid drylands. Conversely, the sensitivity of dryland ecosystems to human impacts that contribute to land degradation increases with increasing aridity. Therefore, the risk of land degradation is greatest in the median section of the aridity gradient (mostly the semiarid drylands), where both sensitivity to degradation and population pressure (expressed by population density) are of intermediate values.

It is thought that some 10–20% of the world's drylands suffer from one or more forms of land degradation (*medium certainty*). Despite the global concern aroused by desertification, the available data on the extent of land degradation in drylands (also called desertification) are extremely limited. In the early 1990s, the Global Assessment of Soil Degradation, based on expert opinion, estimated that 20% of drylands (excluding hyper-arid areas) was affected by soil degradation. A recent MA commissioned desk study (Lepers 2003) based on regional data sets (including hyper-arid drylands) derived from literature reviews, erosion models, field assessments and remote sensing found much lower levels of land degradation in drylands. Coverage was not complete, but the main areas of degradation were estimated to cover 10% of

global drylands. Most likely the true level of degradation lies somewhere between the 10% and 20% figures. To identify precisely where the problems occur and the true extent of degradation will require a more in-depth follow-up to these exploratory studies.

Desertification, which by definition occurs only in drylands, causes adverse impacts on non-dryland ecosystems (*high certainty*). Desertification has both direct and indirect impacts on non-dryland ecosystems and peoples. For example, dust storms resulting from wind soil erosion, driven by degradation of the dryland vegetation cover, may affect people and ecosystems elsewhere. Similarly, transport of sediments, pesticides, and nutrients from dryland agricultural activities affects coastal ecosystems. Droughts and loss of land productivity are considered predominant factors in the migration of people from drylands to other areas (*medium certainty*).

Traditional and other current management practices contribute to the sustainable use of ecosystem services. Many existing practices help prevent desertification. These include enhanced and traditional water harvesting techniques, water storage and conservation measures, reuse of safe and treated wastewater for irrigation, afforestation for arresting soil erosion and improving ground water recharge, conservation of agrobiodiversity through diversification of crop patterns, and intensification of agriculture using technologies that do not increase pressure on dryland services. Policies that involve local participation and community institutions, improve access to transport and market infrastructures, and enable land users to innovate are essential to the success of these practices.

Alternative livelihoods have a lower impact on dryland ecosystem services. These livelihoods still depend on the condition of drylands services but rely less on vulnerable services and make use of the competitive advantages drylands can offer over other systems. They can include dryland aquaculture for production of high-value food and industrial compounds, controlled-environment agriculture (such as greenhouses) that requires relatively little land, and tourism-related activities.

Depending on the level of aridity, dryland biodiversity is relatively rich, still relatively secure, and is critical for the provision of dryland services. Of 25 global "biodiversity hotspots" identified by Conservation International, 8 are in drylands. The proportion of drylands designated as protected areas is close to the global average, but the proportion of dryland threatened species is lower than average. At least 30% of the world's cultivated plants originated in drylands and have progenitors and relatives in these areas. A high species diversity of large mammals in semiarid drylands supports cultural services (mainly tourism); a high functional diversity of invertebrate decomposers in arid drylands supports nutrient cycling by processing most arid primary production; a high structural diversity of plant cover (including microphyte diversity of soil crusts in arid and semiarid areas) contributes to rainfall water regulation and soil conservation, hence to primary production and its generated diversity of the dryland wild and cultivated plants.

22.1 Introduction

This chapter describes the current condition of dryland systems with respect to the services they provide and the drivers that determine trends in their provision. Within the context of the mounting global concern caused by land degradation in drylands (defined as desertification in the text of the United Nations Convention to Combat Desertification), the chapter assesses desertification as a persistent reduction in the services provided by dryland ecosystems, leading to unsustainable use of the drylands and their impaired development. The chapter also explores options for the sustainable use of drylands and points to human and societal responses that have succeeded or failed.

"Desertification" means land degradation in arid, semiarid, and dry subhumid areas resulting from various factors, including climatic variations and human activities. Land degradation means reduction of or loss in the biological or economic productivity and complexity of rain-fed cropland, irrigated cropland, range, pasture, forest, or woodlands resulting from land uses or from processes arising from human activities and habitation patterns (UNCCD 1992). Though this definition excludes the hyper-arid drylands, this chapter explores land degradation in all global drylands, including the hyper-arid areas.

22.1.1 Definition and Subtypes of Dryland Systems

Drylands are characterized by scarcity of water, which constrains their two major interlinked services—primary production and nutrient cycling. Over the long term, natural moisture inputs (that is, precipitation) are counterbalanced by moisture losses through evaporation from surfaces and transpiration by plants (evapotranspiration). This potential water deficit affects both natural and managed ecosystems, which constrains the production of crops, forage, and other plants and has great impacts on livestock and humans.

Drylands are not uniform, however. They differ in the degree of water limitation they experience. Following the UNEP terminology, four dryland subtypes are recognized in this assessment—dry subhumid, semiarid, arid, and hyper-arid—based on an increasing level of aridity or moisture deficit. The level of aridity typical for each of these subtypes is given by the ratio of its mean annual precipitation to its mean annual evaporative demand, expressed as potential evapotranspiration. The long-term mean of this ratio is termed the aridity index.

This chapter follows the *World Atlas of Desertification* (Middleton and Thomas 1997) and defines drylands as areas with an aridity index value of less than 0.65. The UNCCD, although excluding the hyper-arid dryland from its consideration, adopted the classification presented in the *World Atlas,* which is based on a global coverage of mean annual precipitation and temperature data collected between 1951 and 1980. The temperature data, together with the average number of daylight hours by month, were used to obtain a global coverage of corrected Thornthwaite's potential evapotranspiration values (Middleton and Thomas 1997). Aridity index values lower than 1 indicate an annual moisture deficit, and the *World Atlas* drylands are defined as areas with AI ≤ 0.65—that is, areas in which annual mean potential evapotranspiration is at least ~1.5 greater than annual mean precipitation.

Using index values, the four dryland subtypes can be positioned along a gradient of moisture deficit. Together, these cover more than 6 billion hectares, or 41.3% of Earth's land surface. (See Table 22.1.) Though the classification of an area as a dryland subtype is determined by its aridity index, which relates to the mean values of precipitation, it is important to remember that these areas do experience large between-year variability in precipitation.

Dryland subtypes can also be described in terms of their land uses: rangelands, croplands, and urban areas. (See Table 22.2.) Rangelands and croplands jointly account for 90% of dryland areas and are often interwoven, supporting an integrated agropastoral livelihood.

Drylands occur on all continents (between 63° N and 55° S; see Figure 22.1) and collectively comprise nearly half of the global landmass. The rest of the land area is primarily taken up by polar

and by forest and woodland systems (the latter overlapping with the dryland system; see Box 22.1).

Drylands are not spread equally between poor and rich countries: 72% of the global dryland area occurs within developing countries and only 28% within industrial ones. Furthermore, the proportion of drylands occupied by developing countries increases with aridity, reaching almost 100% for the hyper-arid areas. (See Figure 22.2.) Consequently, the majority of dryland peoples live in developing countries (that is, from 87% to 93%, depending on how the former Soviet Union countries are categorized), and only 7–15% reside in industrial countries. (See Figure 22.3.)

22.1.2 Ecosystems in Drylands

Although there are only four dryland subtypes, there are a greater number of dryland ecosystems within the subtypes. These are aggregated into large, higher-order units known as biomes, which are characterized by distinctive life forms and principal plant species (such as tundra, rainforest, grassland, or desert biomes). Whereas the MA dryland subtype boundaries are determined by two climatic factors (precipitation and evaporation), many environmental factors are used to delineate the boundaries of the different biomes. Many different systems of biome classification are presently used. Five well-recognized classification systems of terrestrial biomes identify 12–17 biomes within drylands, depending on the scheme adopted. (See Chapter 4.)

This chapter uses the classification of the World Wide Fund for Nature that designates terrestrial biomes as "terrestrial ecoregions." Each ecoregion delineates large land units containing a distinct assemblage of ecosystems, with boundaries approximating the extent of natural ecosystems prior to major land use change (Olson et al. 2001). These are further aggregated into four "broad" dryland biomes—desert, grassland, Mediterranean (mainly scrubland), and forest (mainly woodland)—that successively replace each other along the aridity gradient (see Figure 22.4), with decreased aridity leading to an increase in plant cover, stature, and architectural complexity. However, there is no exact match between the four dryland subtypes and the four broad dryland biomes, such that forest and grassland, for example, may occur at different areas of the same dryland subtype. The number of broad biomes that may occur within a dryland subtype increases with reduced aridity, and the diversity of biomes peaks in the semiarid subtype, which also covers the largest area of the various subtypes.

The presence of different biomes within each dryland subtype demonstrates that biological species respond not only to overall moisture deficit but also to other environmental variables, such as soils and geomorphological and landscape features. Furthermore, a greater degree of species richness and diversity of ecosystem services is observed as aridity declines. Although dryland services are provided by the biomes' ecosystems, the MA opted to report on ecosystems, or simply "systems," unified primarily by their range of aridity. This approach is justified for two reasons. First, it bypasses the many inherent differences in biome classification systems. Second, it better reflects current trends, as many dryland ecosystems have been and continue to be transformed into more simplified, cultivated ecosystems whose functioning is overwhelmingly dominated by the moisture deficit.

22.2 Ecosystem Services

The MA categorized ecosystem services into supporting, provisioning, regulating, and cultural services. (See Chapter 1.) The

Table 22.1. Statistical Profile of the Dryland System (Area from Deichmann and Eklundh 1991; global area based on Digital Chart of the World data (147,573,196.6 sq. km; year 2000 population from CIESIN 2004)

Subtypes	Aridity Index	Current Area Size (mill. sq. km.)	Current Area Share of Global (percent)	Dominant Broad Biome	Current Population Total (thousand)	Current Population Share of Global (percent)
Hyper-arid	<0.05	9.8	6.6	desert	101,336	1.7
Arid	0.05–0.20	15.7	10.6	desert	242,780	4.1
Semiarid	0.20–0.50	22.6	15.2	grassland	855.333	14.4
Dry subhumid	0.50–0.65	12.8	8.7	forest	909,972	15.3
Total		**60.9**	**41.3**		**2,109,421**	**35.5**

Table 22.2. Land Uses in Drylands (MA core data)

	Rangelands[a] Area (sq. km)	Rangelands[a] Share of Dryland Subtype (percent)	Cultivated Area (sq. km)	Cultivated Share of Dryland Subtype (percent)	Urban Area (sq. km)	Urban Share of Dryland Subtype (percent)	Others[b] Area (sq. km)	Others[b] Share of Dryland Subtype (percent)
Dry subhumid	4,344,897	34	6,096,558	47	457,851	4	1,971,907	16
Semiarid	12,170,274	54	7,992,020	35	556,515	2	1,871,146	8
Arid	13,629,625	87	1,059,648	7	152,447	1	822,075	5
Hyper-arid	9,497,407	97	55,592	0.6	74,050	1	149,026	2
Total	**39,642,202**	**65**	**15,203,818**	**25**	**1,240,863**	**2**	**4,814,155**	**8**

[a] Rangeland figures are based on available data on rangelands in drylands of developing countries (Reid et al. 2004; Thornton et al. 2002) and estimates for rangeland areas in the remaining drylands based on the assumption of uniformity in the rangeland's share of each dryland subtype.

[b] Inland water systems in drylands (3%) and other areas unaccounted for by the assessed land uses (5%).

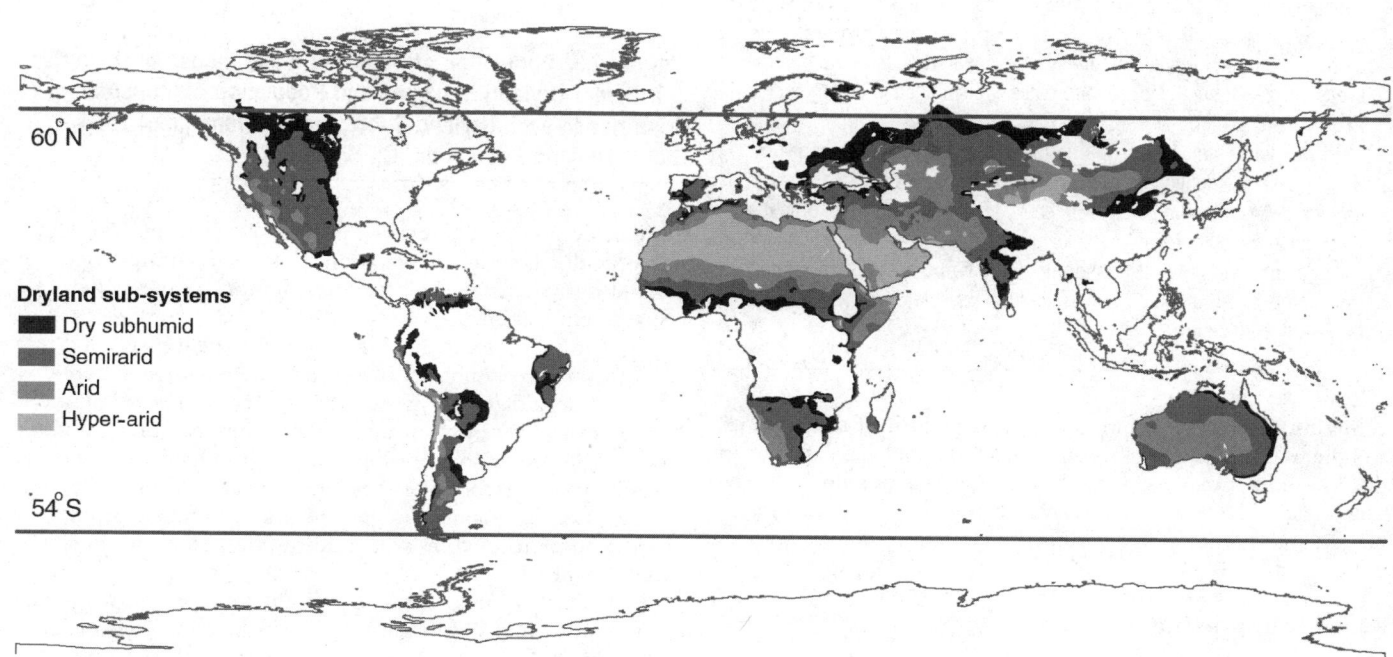

Figure 22.1. Dryland Systems and Subtypes

Forests in Drylands

This is not an oxymoron—forests do occur in drylands. Eighteen percent of the area of the dryland system is occupied by the forest and woodland system, though the probability of encountering forests in drylands decreases with their aridity. Australia is a good example, as seen in the Figure (which is a magnification of a section of Figure 20.1).

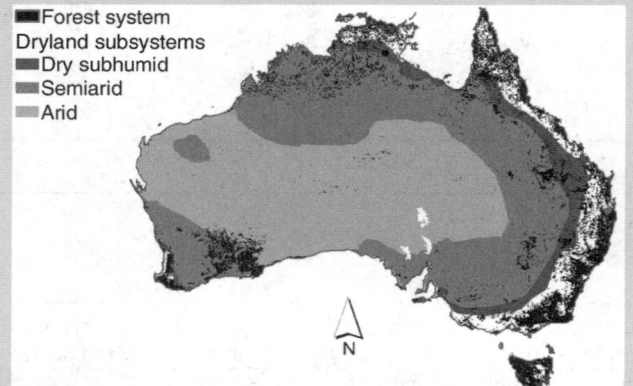

In general, aridity increases inland, and the forest and woodland system prevails along the coasts. The dry subhumid dryland subtype that is adjacent to the forest and woodland system has the greatest amount of overlap between the two systems, as compared with other drylands subtypes. But the distribution of forest in dry subhumid areas is patchier than it is within the forest and woodland system outside the drylands. Forests also occur in the much wider semiarid zone of Australia but are there mostly confined to the less dry seaward direction.

Forests occur in the dry subhumid subtype in Africa but are very scattered and rare in the semiarid zone. In China and India, with dry subhumid areas wider than in Australia, forests penetrate deep into dry subhumid areas. In Europe, where many dry subhumid areas are surrounded by non-drylands, forests are scattered all over the dryland areas. In the Americas, forests are patchily distributed in dry subhumid and semiarid regions. If forests do occur in the relatively humid range of the drylands and seem well adapted to these dryland conditions, why is their distribution patchy and not contiguous? Do the dryland forest patches occur in patches of locally less arid conditions, or is the patchiness a result of human exploitation? Answers to these questions may be critical for evaluating the use of the carbon sequestration service of these two dryland subtypes, which constitute nearly a quarter of Earth's surface area.

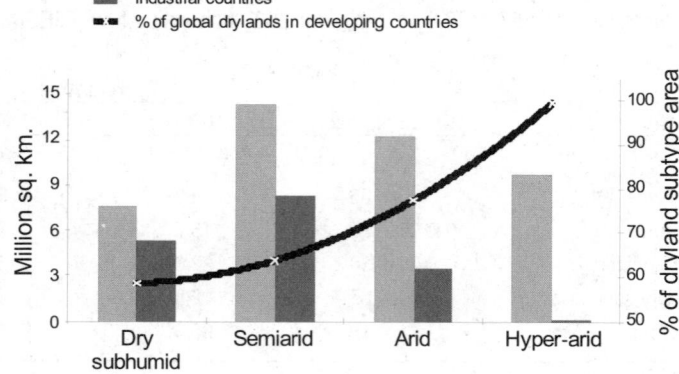

Figure 22.2. Dryland Subtypes and Socioeconomic–Political Status. The relative share of developing and industrial countries in the global drylands, by area and percentage taken up by developing countries. (MA core data)

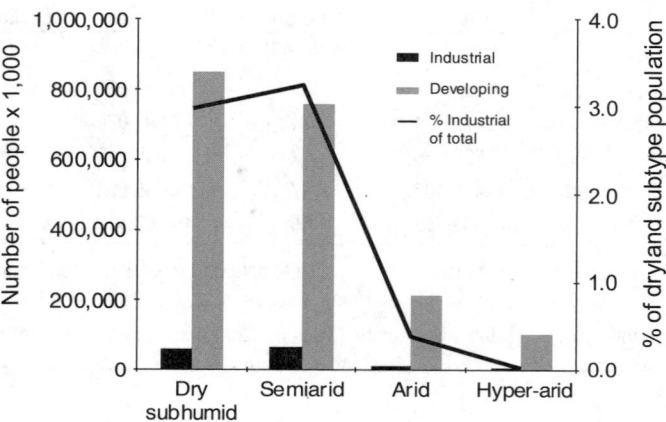

Figure 22.3. Population of Developing and Industrial Countries in Different Dryland Subtypes and Population of Industrial Countries in each Dryland Subtype as Percentage of Total Global Dryland Population (CIESIN 2004)

condition of dryland-significant services in each of these groups and the trends in their provision to dryland peoples are assessed in this section. For many services, however, data about global condition and trends are not readily available, and only generic information about processes governing the condition of these services is provided.

22.2.1 Supporting Services

22.2.1.1 Soil Development: Formation and Conservation

Though primary production in drylands is constrained by water, it is soil properties that determine how much of the rainfall will be stored and subsequently become available during dry periods. The availability of moisture in soil is also an important factor in nutrient cycling, a requisite for primary production. Therefore, soil formation and soil conservation are key supporting services of dryland ecosystems, the failure of which is one of the major drivers of desertification.

The slow process of soil formation, in which plants and microorganisms are intimately involved, is frequently countered by faster soil degradation expressed through erosion or salinization. Hence the services of soil formation and conservation jointly determine the rate of soil development and its quality. The rate of soil formation (hundreds to thousands of years) (Rust 1983) and its degree of development (depth of soil, infiltration depth, and organic content) decline with aridity (Nettleton and Peterson 1983; Sombroek 1990).

In hyper-arid areas, surfaces are often capped with mineral crusts that reduce infiltration and help generate soil-eroding flashfloods. In many arid drylands, dispersed plant clumps are often embedded in a matrix of apparently bare soil covered by a thin crust of photosynthetic cyanobacteria, with mosses and lichens added in semiarid drylands (Büdel 2001). The crusts reduce water penetration and thus channel runoff, sediments, nutrients, and seeds to the plant clumps, which then become active sites of

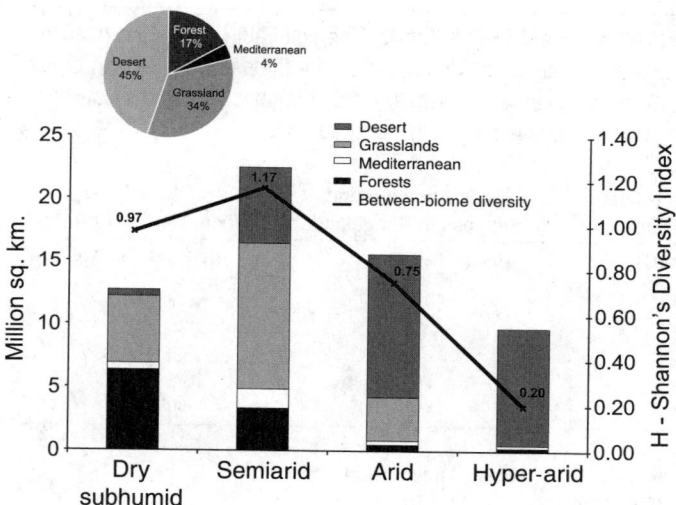

Figure 22.4. Global Area Covered by Dryland Subtypes and Their Broad Biomes. The line, points, and figures stand for the broad biome diversity index (Shannon's H); the pie chart shows the aggregated percentages for broad biomes by dryland subtypes.

soil formation and organic matter decomposition (Puigdefabregas et al. 1999). These crusts are therefore instrumental in soil development in and around the clumps and in soil conservation in the surrounding matrix (Aguiar and Sala 1999). However, they develop slowly and are sensitive to trampling or air pollution. Dry subhumid soils, on the other hand, are protected from erosion by multilayered, structurally complex vegetation (Poesen et al. 2003) that permits high water infiltration and storage, as well as water extraction by the same vegetation (Puigdefabregas and Mendizabal 1998).

22.2.1.2 Nutrient Cycling

This service supports the services of soil development and primary production through the breakdown of dead plant parts (thus enriching the soil with organic matter) and the regeneration of mineral plant nutrients. Unlike non-drylands, where soil microorganisms are major players in nutrient cycling, invertebrate macro-decomposers are the most important in drylands, and their significance increases with aridity. The significance of microorganisms, such as microbes and fungi, declines with aridity due to their strict moisture dependence. In addition, the role of large herbivores in nutrient cycling in arid and hyper-arid areas is limited due to lack of drinking water sources. Therefore macro-decomposers such as termites, darkling beetles (*Tenebrionidae*), and other invertebrates (many of which are soil dwellers) that are less water-sensitive become important for nutrient cycling in drylands. These organisms "prepare" the litter for microbial activity and increase the soil infiltration capacity.

When arid drylands are used as rangelands, however, most of the primary production takes place through the livestock rather than the macro-decomposers. Due to the high metabolic needs of mammalian herbivores (compared with cold-blooded macro-decomposers), much of the organic carbon consumed by the livestock does not return to enrich the soil but is respired or extracted from the ecosystem as meat, hair, milk, and other animal products. In addition, while most of the excreta of macro-decomposers is deposited within the soil, much of the livestock waste on the surface is volatilized and removed from the soil nutrient pool. Thus, livestock have the potential to gradually deplete the range-

land nitrogen reserve and further exacerbate the nutrient limitations for primary production in arid and hyper-arid rangelands (Ayal et al. 2005). However, this depletion may be partially mitigated by biological nitrogen fixation and by dust deposition (Shachak and Lovett 1998).

When drylands are used for crops, tillage and excessive use of pesticides can reduce the role of soil-dwelling macro-decomposers. This, together with low root biomass of annual crops, can impair nutrient cycling and reduce soil organic carbon and its associated nutrients. Whereas wastes from subsistence cropping systems in drylands are locally recycled and only small proportions of crop products are exported, cropping systems that export products lose nutrients, hence their fertilizer inputs are high.

22.2.1.3 Primary Production

The net primary production for global drylands, based on satellite observations at an 8-kilometer and 10-day resolution from 1981 to 2000, was 703 ± 44 grams per square meter (Cao et al. 2004), significantly lower than the values for the MA's cultivated system ($1,098 \pm 48$ grams) and the forest and woodland system (869 ± 34 grams). But averaging over all the dryland subtypes masks the effect of the aridity gradient.

The NPP of rangelands is mostly generated by the natural dryland plant community, in contrast to cultivated drylands, where the NPP is generated by agricultural crops and is often elevated due to two imported inputs—irrigation water and fertilizers. Whereas the NPP of monitored rangeland sites was 40–90% higher in non-dryland than in dryland countries, in the same years the yield of wheat in the croplands was more than three times greater in the non-dryland countries. (See Table 22.3.) Further analysis that includes wheat yield data for more countries and more years suggests a relative advantage of cultivation in non-drylands, but at the same time highlights the significance of socioeconomic conditions, which apparently determine the amount of resources that can be mobilized for promoting the service of primary production. On average, industrial dryland countries produced wheat yields nearly as high as those produced by non-dryland developing countries. The yield of developing dryland countries was low compared with that of non-dryland countries; even among industrial countries, non-dryland ones did much better than dryland ones.

In order to distinguish between the relatively low NPP of drylands that is due to their inherent moisture deficit and the additional decline in primary production due to land degradation, rain use efficiency—the ratio of NPP to rainfall—can serve as a measure of primary production service condition (Le Houerou 1984; Le Houerou et al. 1988; Pickup 1996): it separates out reduced NPP due to reduced rainfall from declines in NPP driven by land degradation (Prince et al. 1998) and also separates increased NPP due to increased rainfall from the effects of added irrigation and fertilizer use.

Several global MA systems, drylands included, show a trend of NPP increase for the period 1981 to 2000 (Cao et al. 2004). The slope of the linear regression for drylands (regression coefficient = 5.2) does not significantly differ from those of cultivated and forest and woodland systems. However, high seasonal and interannual variations associated with climate variability occur within this trend on the global scale. In the drylands, this variation in NPP was negatively correlated with temperature and positively with precipitation (Cao et al. 2004)—two drivers that are expected to further affect dryland primary production through global anthropogenic climate change.

Table 22.3. Primary Production Expressed in NPP of Rangelands and Wheat Yield in Croplands. The first Table shows the relations between aboveground biomass in monitored rangelands[a] and total wheat yield in each country's croplands[b] for the same year,[c] with dryland countries compared with non-dryland ones. The second Table is a comparison of mean annual wheat yields for selected dryland countries (in which most of the area is categorized as dryland) and temperate non-dryland countries, industrial and developing.[b]

Area	Country (year NPP measured)	Mean Annual Rainfall in Rangeland	Mean Aboveground Biomass in Rangeland	Wheat Yield of Country
		(millimeters)	*(grams per sq. meter)*	*(tons per hectare)*
Dryland countries[d]	Mongolia (1990)	280	100	1.3
	Kazakhstan (1978, 1992)	351	83	1.3
Non-dryland countries[e]	Sweden (1968)	537	141	4.3
	United Kingdom (1972)	858	188	4.2

Area	Country	Mean of Yields 1994–2003	Mean Annual Yield for Country Categories
		(tons per hectare)	
Dryland, developing	Kazakhstan	0.9	1.3
	Morocco	1.2	
	Iran	1.8	
Dryland, industrial	Australia	1.8	2.0
	Israel	1.8	
	Spain	2.5	
Non-dryland, developing	Uruguay	2.2	2.2
	Belarus	2.3	
	Bangladesh	2.1	
Non-dryland, industrial	Japan	3.6	5.7
	Sweden	5.9	
	United Kingdom	7.7	

[a] Data from NPP in grasslands database of Oak Ridge National Laboratory: http://daac.ornl.gov/NPP/html_docs/npp_site.html.

[b] Data from FAOSTAT: http://apps.fao.org/faostat.

[c] Except for Kazakhstan, where latest NPP are from 1978 and first-wheat yield data are from 1992.

[d] NPP measured in cold temperate steppes of both countries (modified Bailey ecoregion classification).

[e] NPP measured in rangelands within humid temperate forests (modified Bailey ecoregion classification).

22.2.2 Regulating Services

22.2.2.1 Water Regulation

Water is the limiting resource for dryland biological productivity, and thus water regulation is of major significance. This regulation determines the allocation of rainfall for primary production (enrichment of soil moisture); for irrigation, livestock watering, and domestic uses (storage in groundwater and surface reservoirs); and for the occurrence of flashfloods and their associated damages (soil erosion, reduced groundwater recharge, excessive clay and silt loads in downstream water bodies). Vegetation cover modulates the water regulation service, and its efficiency in intercepting rainfall determines the fraction available for human use. In rangelands, vegetation removal and livestock trampling can increase soil water erosion through disintegration of the biological soil crust. Similarly, in croplands tillage increases the risk of sealing and crusting (Hoogmoed 1999). Water regulation may be augmented by landscape management (terraces, small dams, and so on), which slows down surface runoff, thereby promoting water infiltration and flood avoidance.

22.2.2.2 Climate Regulation

Dryland ecosystems regulate their own local climate to some extent as their vegetation cover determines the surface reflectance of solar radiation as well as water evaporation rates. Drylands are also involved in regulation of the global climate, through local carbon sequestration by their vegetation. Both these phenomena are described here in some detail.

22.2.2.2.1 Regulation of local climate through surface reflectance and evaporation

The vegetative cover of drylands depends on inputs of incident solar radiation and rainfall. Conversely, the outputs from drylands, the fraction of the incident radiation reflected by the surface (the albedo), and the fraction of soil water transpired and evaporated (evapotranspiration) drive atmospheric energy- and water-balance processes. The provision of this service becomes apparent when it has either been degraded, as in the Sahel drought (Xue and Dirmeyer 2004), or enhanced, as in the rainfall patterns in Israel (Steinberger and Gazit-Yaari 1996) and the U.S. Great Plains (Barnston and Schickedanz 1984). Vegetation cover in drylands can either reduce albedo, resulting in increased surface and near-surface temperatures, or shade the surface leading to low surface temperatures. Both contrasting effects may lead, through different paths, to an identical effect on rainfall. (See Figure 22.5.)

The overexploitation of vegetation (Xue and Dirmeyer 2004) and the removal of the crust by trampling in arid and semiarid drylands (Warren and Eldridge 2001) lead to increased albedo, lower surface temperatures, lower convective activity, and reduced rainfall (Charney et al. 1975). Albedo may also increase due to surface dust cover, a result of dust storms promoted by greater surface exposure after vegetation removal (Williams and Balling 1995). Thus, the conservation of vegetation cover promotes the service of local climate regulation directly through its effect on albedo and indirectly through arresting dust generation.

A degraded vegetation cover also leads to reduced shade, increased surface and near-surface temperatures, and a rapid decrease in soil moisture, which leads to reduced evaporation. This reduction in overall evaporation links to reduced rainfall generation—low evaporation, reduced water flux into the atmosphere, a decrease in the amount of energy used to evaporate or transpire

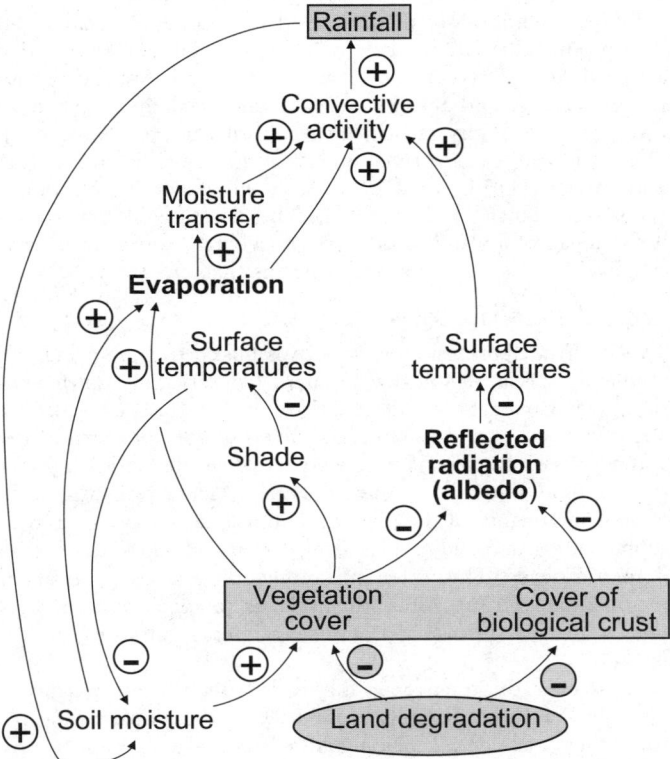

Figure 22.5. The Service of Climate Regulation in Drylands. The central grey box—the components of biodiversity involved in service provision—maintenance of soil moisture (bottom left) and modulation of rainfall (top). In bold—the major alternative/complementary function involved in the effect of live vegetation cover on rainfall; successive multiplication of signs along each trajectory generates an increase in rainfall (+) when service is ameliorated and a decrease in rainfall (–) when land is degraded. Land degradation (grey circle) degrades the service through affecting surface temperature; when surface temperatures increase along the albedo trajectory, it decreases along the evaporation trajectory; this trend is reversed when land is not degraded.

water, and reduced convective heating all combine to produce less rainfall (Williams and Balling 1995).

In climate modulated both by albedo and by evaporation, lower rainfall further reduces soil moisture and vegetation cover and induces further degradation in service provision. The prevalence of the climate regulation service is demonstrated by a few amelioration cases, such as the 10–25% increase in rainfall in the northern Negev of Israel attributed to reduced albedo resulting from controlled grazing, afforestation, and irrigated agriculture in a semiarid region (Otterman et al. 1990).

22.2.2.2.2 *Regulation of global climate through carbon sequestration*

Carbon sequestration (the uptake of atmospheric CO_2 by ecosystems and transformation into plant biomass) controls atmospheric CO_2 concentrations, which regulate the global climate through the "greenhouse effect." Part of the sequestered carbon is emitted back to the atmosphere through the respiration of plants and decomposers, but what is left—the live and the dead above- and belowground plant parts—constitutes the addition to the organic carbon reservoir. Some of the plant litter converts into recalcitrant humus, thereby enhancing the soil organic carbon pool and the formation of secondary carbonates through precipitation.

Plant biomass per unit area of drylands is low (about 6 kilograms per square meter) compared with many terrestrial ecosystems (about 10–18 kilograms). But the large surface area of drylands gives dryland carbon sequestration a global significance. Whereas organic carbon (in aboveground vegetation and soil) declines with aridity, inorganic soil carbon increases as aridity increases. (See Table 22.4.) Altogether, total dryland soil organic and inorganic carbon reserves comprise, respectively, 27% and 97% of the global soil organic and soil inorganic global carbon reserves.

22.2.2.3 *Pollination and Seed Dispersal*

Cases of tight associations between dryland plants and pollinators are known (such as Agava and its pollinator) (Arizaga et al. 2000), but the extent to which changes in land use in the drylands affect the pollination service and the dependence of dryland plant species on pollination has not been fully explored. Seeds of many dryland plant species are dispersed by fruit-eating birds, often prior to or after their cross-desert seasonal migration (for instance, in the Mediterranean basin) (Izhaki et al. 1991). Domestic and wild mammalian herbivores disperse seeds attached to their fur or through consuming them and then defecating, which promotes dispersal and enhances the chance of germination (as in African acacia trees) (Ward 2003). Livestock and other animals may also transfer seeds from improved pasture lands to neighboring non-managed rangelands (CGIAR 1997). Thus the services of pollination and seed dispersal are of significance, but assessment of their condition, importance, and trends requires more attention.

22.2.3 Provisioning Services

22.2.3.1 *Provisions Derived from Biological Production*

22.2.3.1.1 *Food and fiber*

The major dryland cereals and legumes (together with vegetable and fruits) constitute the main crops and basic food for 800 million farmers in drylands (CGIAR 1997). A large part of the dryland population depends on crop and livestock production as a livelihood and contributes significantly to the gross domestic product and trade. Livestock are raised mostly in rangelands or in agropastoral systems, and they constitute a major source of protein and income. Wool is provided by livestock and wild mammals

Table 22.4. Estimates of Dryland Carbon Reserves

| | Biotic[a] | Soil | | Totals | Share of Global |
		Organic[b]	Inorganic[c]		
		(gigatons of carbon)			*(percent)*
Hyper-arid and arid	17	113	732	862	28
Semiarid and dry subhumid	66	318	184	568	18
Total in drylands	**83**	**431**	**916**	**1,430**	**46**
Global totals[d]	**576**	**1,583**	**946**	**3,104**	
		(percent)			
Share of global	14	27	97		

[a] Adapted from IPCC 2001.

[b] Means of data of Eswaran et al. 2000 and Allen-Diaz et al. 1996 adapted to the dryland subtype classification by J. Puigdefabregas.

[c] Adapted from Eswaran et al. 2000.

[d] Means of values assembled from various sources by Jonathan Adams, Oak Ridge National Laboratory, http://www.esd.ornl.gov/projects/qen/carbon2.html.

such as guanacos (*Lama guanicoe*) and vicunas (*Vicugna vicugna*) in South America (Fernandez and Busso 1997).

Fiber is produced by both croplands and rangelands. For instance, cotton (*Gossypium* spp.) and sisal (*Agave sisalana*) are widely cultivated, while timber and silk are produced on a smaller scale. Fiber, vegetable oil, vegetables, fruits, and nuts provisioned by dryland ecosystems are also exported to non-dryland countries. The food provision service of drylands may be impaired by soil erosion (in rain-fed croplands, a long dry season with no plant cover challenges the soil conservation service), salinization (in irrigated croplands with poor drainage), and nutrient depletion (the removal of commodity crops challenges the nutrient cycling service). (For more on the condition and trends of food provision, see Chapter 8.)

22.2.3.1.2 Woodfuel

Most woodfuel (the collective term for fuelwood, charcoal, and other wood-derived fuels) is provided by trees or bushes inhabiting natural dryland ecosystems that are also often used as range. Hence the exploitation of this service is often a trade-off with the provision of forage. Overexploitation for woodfuel harvesting impairs the soil conservation service, and it leads to soil erosion and hindered vegetation regeneration. This downward spiral of service degradation encourages reforestation and afforestation for woodfuel provisioning, using drought and salinity-tolerant tree species and strains (Sauerhaft et al. 1998). Fuelwood is used predominantly at the household level, for cooking and heating (Amous 1997), and may constitute a sizable proportion of the energy consumed in many dryland countries—for example, 57% in Senegal in 1999 (IEA 2001).

22.2.3.1.3 Biochemicals

Many species of dryland plants are used by dryland peoples for medicinal and cosmetic purposes and as spices, which highlights the significance of dryland plant biodiversity. However, excessive exploitation puts many of these species at risk of extinction and contributes to soil loss and consequent erosion. Attempts to cultivate such species in order to reduce the pressure on natural ecosystems often fail, because the production of the active compounds by these plants is rather low under stress-reduced cultivation regimes. The adaptations of dryland plants to varying and

extreme conditions are often derived from unique biochemicals they produce that are the key to environmental tolerance or that act to deter herbivores and parasites. Further investigation into the generation and activity of these chemicals helps promote drought- and salinity resistance in cultivated crops (Wang et al. 2002) and can lead to development of novel medicines, such as anti-cancer (Haridas et al. 2001) and anti-malarial compounds (Golan-Goldhirsh et al. 2000). Biochemicals are also manufactured as part of dryland aquaculture, providing a source of alternative livelihoods, as described later in this chapter.

22.2.3.2 Freshwater Provisioning

The freshwater provisioning service is linked to supporting and regulating services—soil development (conservation and formation), water regulation, and, to a lesser extent, climate regulation. Vegetation cover and its structural diversity control much of the water provisioning service. This vegetation depends on water provisioning, but it is also instrumental in generating and maintaining the quality of the service. The resultant water is used to support rangeland and cropland vegetation and also livestock and domestic needs. The water provision service is also critical for maintaining wetlands within the drylands, to enable these ecosystems to provide a package of services of great significance in drylands.

However, the total renewable water supply from drylands is estimated to constitute only around 8% of the global renewable water supply (about 3.2 trillion cubic meters per year) (Vörösmarty et al. 2005), and only about 88% of this is accessible for human use. Thus, almost one third of the people in the world depend on only 8% of the global renewable water resources, which makes per capita availability in drylands just 1,300 cubic meters per year. It is substantially less than the average global availability and even lower than the 2,000 cubic meters regarded as a minimum by FAO (FAO 1993).

To mitigate this shortage, exacerbated by the large within- and between-years variability in rainfall, a variety of practices have been developed. From the least to the most technology-laden ones, these are:

- watershed management, including conservation and rehabilitation of degraded vegetation cover for generating and capturing surface runoff for deep storage in the soil (protecting it from evaporation) (Oweis 2000);

- floodwater recharge and construction of dams and weirs for minimizing impact of floods and water loss;
- irrigation, to circumvent the temporal variability in provision (often based on extraction from aquifers, with frequent over-pumping leading to salinization)—however, the transportation of water from other ecosystems that may be severely affected and the salinization of the irrigated drylands often make this option unsustainable;
- mining of nonrenewable fossil aquifers (which are quite common in drylands), for cultivation that is otherwise impossible;
- treatment of wastewater, mainly from urban sources, and reusing it for irrigation—a promising practice provided that concerns about adverse impacts on human health, crops, soils, and groundwater can be overcome (Karajeh et al. 2000); and
- desalination of brackish water and seawater for all uses (which is safe and uses renewable sources but has a high energy demand and is relatively costly, and the accumulated brine often poses a salinization risk).

These interventions are critical for relieving pressure on the water systems of drylands.

22.2.4 Cultural Services

22.2.4.1 Cultural Identity and Diversity

Dryland peoples identify themselves with the use of their surrounding ecosystem and create their own unique ecosystem-inspired culture. (See Chapter 17.) Drylands have high cultural diversity, in keeping with the ecosystem diversity along the aridity gradient. One expression of this is that 24% of global languages are associated with the drylands' grassland, savanna, and shrubland biomes. Typical to drylands are the diverse nomadic cultures that have historically played a key role in development of dryland farming systems (Hillel 1991). Ecosystem functions and diversity generate cultural identity and diversity that in turn conserve ecosystem integrity and diversity. A negative feedback loop is therefore expected between land degradation and cultural degradation in drylands.

22.2.4.2 Cultural Landscapes and Heritage Values

The term "cultural landscape" is a socioeconomic expression of the biophysical features of ecosystems that mutually contribute to the development of a characteristic landscape, and it signifies a heritage value. (See Chapter 17.) In drylands, the heritage value can be nurtured either by landscapes that reflect the human striving for "conquering the desert" or by ones reflecting aspirations to "live with the desert." Transformation of rural to urban ecosystems is an expression of changed livelihoods that modify the landscape and its cultural values and often degrade cultural heritages. Actions to conserve outstanding Cultural Heritage Sites that are cultural landscapes are under way (UNESCO 2004), and 21 such sites have been identified, of which 8 are in drylands.

22.2.4.3 Servicing Knowledge Systems

Dryland ecosystems also contribute to human culture through both formal ("scientific") and traditional knowledge systems. (See Chapter 17.) Drylands have generated significant contributions to global environmental sciences. Arid Cultural Heritage Sites (such as Lake Turkana National Park and Ngorongo Conservation Area) have generated knowledge of paleo-environments and of human evolution (UNESCO 2004); studies of desert organisms have revealed adaptations to extreme environmental stresses (e.g., Schmidt-Nielsen 1980); and studies of desert ecology have in-spired modern community and ecosystem ecology (e.g. Noy-Meir 1973, 1974; Rosenzweig 1995).

Dryland traditional knowledge has co-evolved with the cultural identity of dryland peoples and their environment and its natural resources and has generated many unique systems of water harvesting, cultivation practices, climate forecasting, and the use of dryland medicinal plants. The degradation of this knowledge in many cases has often led to adoption of unsustainable technologies. The exploration, conservation, and integration of dryland traditional knowledge with adapted technologies have been identified as priority actions by the Committee of Science and Technology of the UNCCD (ICCD 2000).

22.2.4.4 Spiritual Services

Many groves, tree species, and individual trees have spiritual significance to dryland peoples due to their relative rarity, high visibility in the landscape, and ability to provide shade. In ancient times in the Middle East and North Africa, spiritually significant social and religious activities took place under tree canopies. The sites of individual trees have been used for anointing rulers, hosting legal hearings, burial of community and religious dignitaries, and religious rituals, and individual trees themselves have become sacred and named after deities. For instance, the Hebrew names of *Quercus* and *Pistacia* (the dominant species of the eastern Mediterranean shrubland and woodland biomes)—*Alon* and *Ela*—derive from the words for God and Goddess respectively. Protected from grazing and cutting, these sacred trees have reached dimensions far larger than they ever attain in their natural climax community (Zohary 1973: 505–07). These sacred groves often conserve islands of indigenous ecosystems in a transformed landscape and contribute to a unique cultural landscape. Similar services are also provided by other drylands, such as the religious, ceremonial, and historical sites of Native Americans (Williams and Diebel 1996) and aboriginal Australians. (See Chapter 17.) In hyper-arid drylands, trees are far rarer, and indigenous nomadic people do not generally identify individual trees as sacred, although they can have spiritual values.

22.2.4.5 Aesthetic and Inspirational Services

There are outstanding literary and historical examples for inspiration generated by dryland landscapes (such as the Old and New Testaments). The stark contrast between inland wetlands and surrounding dryland areas, linked with the significance of water bodies to the well-being of dryland people, could have generated the association of the Mesopotamian marshlands with the Garden of Eden (Hamblin 1987). Dryland ecosystems are also a source of inspiration for non-dryland people. The 1950s Walt Disney film "The Living Desert" brought desert ecosystems and biodiversity to the attention of millions prior to the television era and was declared "culturally significant" in the year 2000 by the U.S. Library of Congress.

The popular conception of dryland peoples among non-dryland groups is one of human struggle against harsh natural conditions producing rich cultures nurtured by strong moral values, as well as naive romantic notions of life in the desert (Fernandez and Busso 1997). However, while the media has largely promoted the conservation of desert heritage in recent years, others have responded by trying to "green" desert areas or make them "bloom," which has often resulted in an aesthetically appealing landscapes of oasis-like patches of agricultural land set in sharply contrasting surrounding desert (Safriel 1992)—but with a loss of dryland biodiversity.

22.2.4.6 Recreation and Tourism

Large, sparsely populated, low-pollution arid and hyper-arid areas provide attractive holiday destinations for many. There are significant constraints to dryland tourism, however, including the general remoteness and isolation, which increases the cost of travel; lack of recreation amenities and security; the harsh climate, which means residential facilities have high energy demands; the high water demand of tourists, which places already scarce water under extra pressure; and often a direct competition with other livelihoods over the use of natural resources and energy. These issues are being addressed through various approaches, including treatment and reuse of local wastewater (Oron 1996), construction and architectural solutions for passive cooling and heating (Etzion et al. 1999), and the use of the dryland-abundant solar energy as a power source (Faiman 1998).

Drylands are also attractive for cultural tourism associated with historical and religious sites, for coastal tourism (such as Mediterranean beaches), and for health-related tourism (such as the Dead Sea). Dryland biodiversity is also a major draw for ecotourism. Paradoxically, this is because most drylands are devoid of woodlands and dense high vegetation and hence free of obstructions to view wildlife. For instance, African savanna safaris are generally designed around a few "charismatic" large mammal species and mass seasonal migrations of large herbivores, and many tourists pack resorts along the route of the spectacular seasonal trans-Saharan bird migration. The significance of the dryland cultural service to tourism is demonstrated by Kenya, where 90% of tourists visit a game park (White et al. 2000). Finally, although ecotourism generates income for dryland peoples, it often causes habitat degradation, as described later (White et al. 2000).

22.2.5 Biodiversity and the Provision of Dryland Services

22.2.5.1 Dimensions, Structure, and Composition of Dryland Biodiversity

Species richness declines with decreasing primary productivity (Rosenzweig and Abramsky 1993) and vice versa (Tilman et al. 2001); hence dryland species richness should decrease with aridity. Indeed, the low number of flowering plant species in the hyper-arid subtype rapidly increases with reduced aridity (in agreement with an increase of between-"broad biome" diversity) and peaks in the dry subhumid subtype. However, contrary to expectation, species diversity declines in non-dryland temperate humid areas. But as might be expected, it is nearly half that of tropical areas. (See Figure 22.6.) The significance of biodiversity for each of the major dryland biomes is discussed in this section. (See also Chapter 4.)

22.2.5.1.1 Deserts

Some 7,000 terrestrial amphibian, reptile, bird, and mammal species live in the desert biome. This covers 25% of global terrestrial fauna of these groups—22% of which also live in other biomes and 3% are found exclusively in deserts. For comparison, the richest terrestrial biome—tropical and sub-tropical moist broadleaf forests—supports around 70% of global terrestrial fauna, 28% of which are species endemic to that biome. Thus species richness of desert vertebrates is as much as a third of the most vertebrate-rich biome on Earth, signifying that desert biodiversity may be quite high in spite of harsh conditions. Because functional groups in deserts may each have only a few species, however, the redundancy in service provision is low, and human pressure may reduce it further (Huenneke 2001). Indeed, in spite of the remoteness

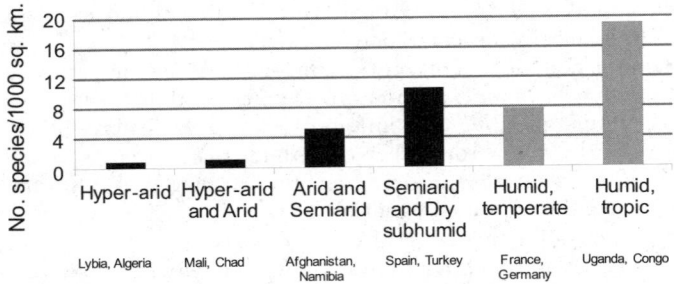

Figure 22.6. Number of Species of Flowering Plants in Selected Countries across the Aridity Gradient (per 1,000 sq. km). Each column represents a mean of two countries. Selected dryland countries are at least 95% dryland, either of one subtype, or two of roughly similar dimensions. Grey columns indicate non-dryland countries. (WRI 2004; CIA 2004)

and isolation, human impact on deserts in the form of settlements and infrastructure is mounting, and there has been a 6% loss of habitat between 1950 and 1990.

22.2.5.1.2 Grasslands

Grasslands (the temperate grasslands, savannas, and shrubland biome and the tropical and sub-tropical grasslands, savannas, and shrubland biome) occur in the semiarid and the dry subhumid dryland subtypes, and their biodiversity is richer than that of deserts (12% and 28% respectively of the global terrestrial vertebrate fauna are found in these two biomes). Much is known of the functioning of natural grasslands, many of which function as rangelands: plant diversity increases productivity (Tilman et al. 2001), and communities with many functional groups generate higher production than those with fewer groups. Yet within a grassland community a few abundant species account for a large fraction of grassland ecosystem functions, whereas the many rare species account for a small fraction of its functions (Sala et al. 1996; Solbrig et al. 1996).

It is important to note that the relationships between diversity and function are not linear, and a threshold in species richness has been identified below which ecosystem function declines and above which it does not change (Vitousek and Hooper 1993). Unfortunately, many natural grasslands have been transformed to croplands and most dryland cultivated lands are in these biomes. This transformation continues, and some 15% and 14% of the natural habitats in the semiarid and dry subhumid subtypes were transformed between 1950 and 1990.

22.2.5.1.3 Mediterranean forests, woodlands, and shrublands biome

The Mediterranean biome, comprising xeric woodlands and shrublands, occurs within semiarid and dry subhumid areas with a Mediterranean climate and is subjected to intensive human impact, especially in the Mediterranean basin, resulting in plant adaptations to clearing, grazing, fires, and drought (Davis et al. 1996). Species richness is high (Mooney et al. 2001), with the Mediterranean basin supporting 25,000 vascular plants (10% of global species), of which 60% are endemic; 10% of the global vertebrates species inhabit the Mediterranean biome.

The biome's biodiversity is threatened by its small geographic coverage, fragmentation, high human population density, abandonment of traditional practices, tourism, continued habitat conversion (2.5% of Mediterranean habitat was lost between 1950 and 1990), and invasive alien species (Mooney et al. 2001). Agriculture, grazing, and frequent fires have decreased dryland forests, converting many to grasslands (Solbrig et al. 1996). At the same time, the abandonment of rangelands in the Mediterranean basin has influenced secondary succession, which has eliminated open-habitat species and reduced diversity. Consequently, many endemic and rare species are currently restricted to protected areas surrounded by degraded or altered landscapes that act as a barrier to migration in response to environmental change.

22.2.5.2 The Role of Biodiversity in the Provision of Dryland Ecosystem Services

22.2.5.2.1 Involvement of biodiversity in packages of services

There are many dryland species that are directly involved in the provision of a range of ecosystem services. One such example is African acacia (Ashkenazi 1995), which provides for soil development and conservation (roots, canopy, and litter), forage (leaves and pods eaten by livestock), fuelwood (dead twigs), and food (edible gums). It is also involved in nutrient cycling (symbiosis with nitrogen-fixing bacteria) and generates cultural services (as described earlier). And it supports other biodiversity: a large number of animal species depend on it for shelter, shade, nest sites, and food. (Often this is of mutual benefit: wild and domestic mammals disperse the seeds, thus determining the spatial distribution of the species.)

The numerous dryland plant species of different growth forms jointly provide a package of services through their ground cover and structure, which provide the drylands' most important services of water regulation and soil conservation as well as forage and fuelwood provision and climate regulation. In arid and semiarid areas, clumps of bushes and annuals embedded in the matrix of a biological soil crust—which consists of an assemblage of several species of cyanobacteria (that provide the added benefit of nitrogen fixation), microalgae, lichens, and mosses—jointly generate soil conservation and water regulation (Shachak and Pickett 1997). In many arid and semiarid areas, this biodiversity of "vegetation cover" and biological soil crusts is linked to a diversity of arthropod species that process most of the living plant biomass, constituting the first link of nutrient cycling.

22.2.5.2.2 Involvement of dryland biodiversity in a single service

Individual species can also be important providers of a single service, such as individual dryland plant species serving as a "biogenetic resource" for cross-breeding and improvement of domesticated species to which they are genetically related. These species are either the progenitors of currently cultivated species that were domesticated millennia ago (Higgs and Jarman 1972; Harlan 1977) or they are relatives of those progenitors. It is estimated that 29–45% of the world's currently cultivated plants originated from drylands (FAO 1998). The progenitors and wild relatives of these originally dryland-cultivated plants (such as wheat, barley, rye, millet, cabbage, sorghum, olive, and cotton) are an important component of dryland biodiversity. However, only a few of them inhabit dryland protected areas and enjoy active in situ conservation, and much of their potentially useful genetic diversity has not yet been fully screened and may be under threat due to habitat loss (Volis et al. 2004).

Assemblages of dryland species can also jointly generate a single service, such as populations of large mammalian herbivores—from antelopes to elephants—providing for the cultural service of ecotourism, especially in the eastern and southern grasslands and savannas of Africa. These now occur mainly in protected areas and in ranches, and their management for the sustainable provi-

sion of this service is a scientific, legal, and sociopolitical challenge.

22.2.5.2.3 Trends in the involvement of dryland biodiversity in service provision

Evidence for human impacts on specific dryland biodiversity components that affect service provision is associated with livestock grazing. Livestock often impair the service of forage provision when prime forage species are replaced by non-palatable, often invasive species, leading to replacement of the grassland vegetation by encroaching bush or the reduction of the litter-decomposing termite populations, which impairs nutrient cycling, primary production, and carbon sequestration (Zeidler et al. 2002; Whitford and Parker 1989). Human-induced climate change may also alter the primary production and other dryland services, since plants of the three photosynthetic pathways (C3, CAM, or C4) co-occur in drylands and are expected to respond differently to climate change and to elevated atmospheric CO_2 (Huenneke and Noble 1996).

The loss of biodiversity from drylands is not likely to affect all services uniformly. Rather, primary production and the provisioning services derived from it, as well as water provision, will be more resilient than recreation and ecotourism. (See MA *Scenarios*, Chapter 10.) This is based on the observation that ecosystem services performed by top predators will be lost before those performed by decomposers. The service of supporting biodiversity (by generating and maintaining habitats of required value and ample size) is expected to be degraded faster than the service of provisioning biological products.

The direct threats to the service of supporting biodiversity include not only land degradation but also habitat loss and fragmentation, competition from invasive alien species, poaching, and the illegal trade in biodiversity products. Indirect threats include the losses of the drylands-specific "keystone" species (Paine 1966) and "ecosystem engineer" species (which modify the dryland environment for the benefit of other species) (Jones et al. 1994). Finally, not only losses but also addition of species may impair service provision. For example, Eucalyptus tree species introduced to southern Africa have invaded entire catchments of natural vegetation, causing large-scale changes in water balance and depriving water from lower catchments (Van Wilgen et al. 1998).

22.2.6 Integration: Services, Biodiversity, Livelihoods, and Aridity

Figure 22.7 highlights the interrelationships between major ecosystem services, between services and biodiversity, and between services and the livelihoods they support across the aridity gradient. Water regulation is the overarching dryland service, and its effect cascades through the interrelated supporting services of soil conservation and nutrient cycling to primary production and water provision. Whereas the service of water provision is the most significant one supporting the farming livelihoods prevailing in the dry subhumid and the semiarid subtypes, the primary production-dependent service of forage provision is the most significant service for pastoralists. Other primary production-dependent services are the provision of biochemicals and fuelwood, which serve both farmers and pastoralists but also generate independent, alternative livelihoods based on medicinal plants and biomass-generated energy. Forage, fuelwood, and biochemicals are goods produced by a diversity of plant species, which are both a product and a generator of primary production.

The structural diversity of the vegetation cover is the most significant dryland biodiversity component, since it is instrumen-

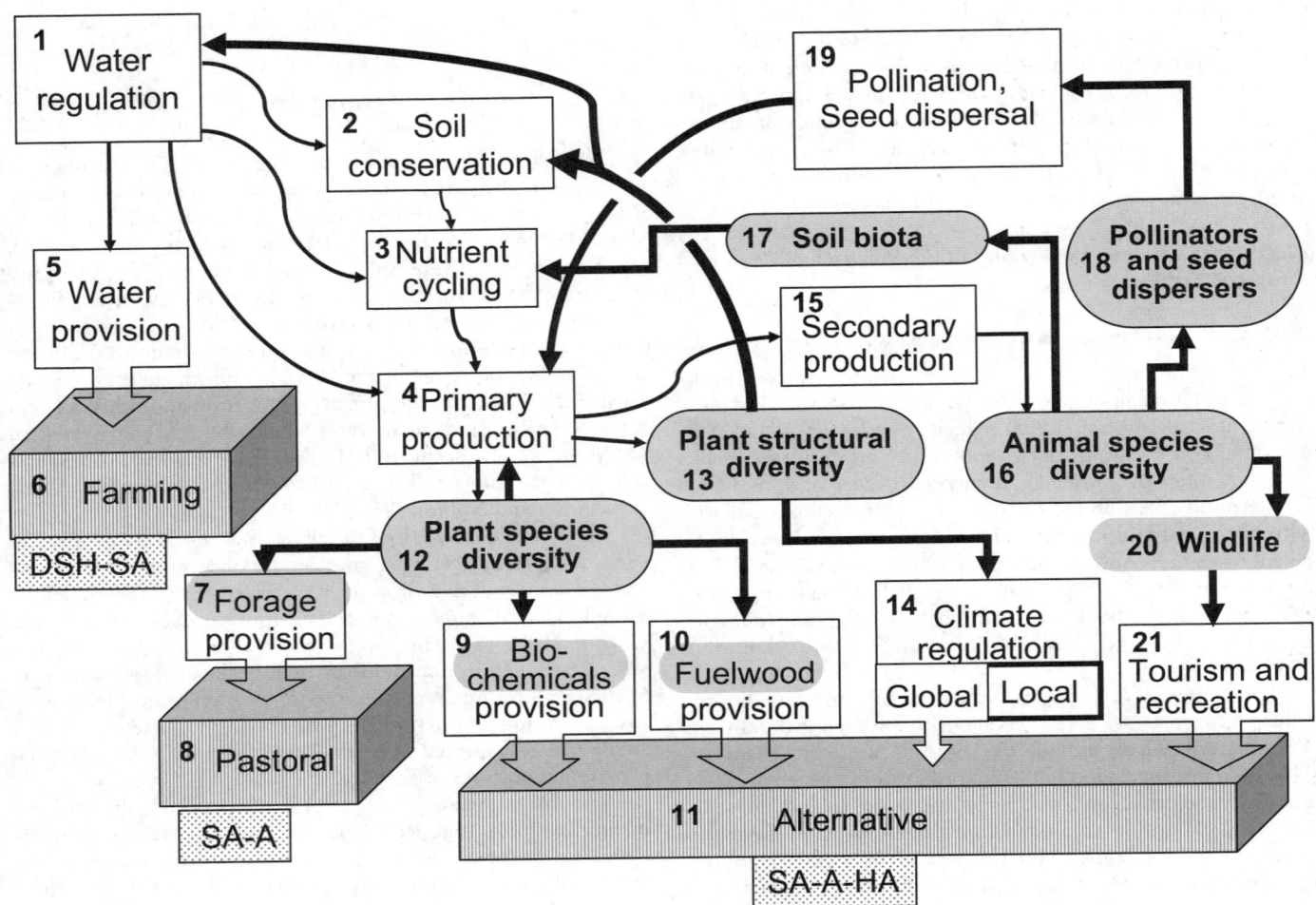

Figure 22.7. Linkages between Services, Biodiversity, Livelihoods, and Dryland Subtypes. Rectangular boxes—services; rounded boxes—components of biodiversity; three-dimensional boxes—livelihoods; dotted rectangles—dryland subtypes: DSH, SA, A, and HA—dry subhumid, semiarid, arid, and hyper-arid, respectively; thin arrows—direct effects of services; thick arrows—involvement of biodiversity. Follow the boxes in the order of their numbers, which streamlines with the text in section 22.2.6.

tal in the water regulation service. Plant structural diversity is also involved in soil conservation and water regulation, as well as in climate regulation. Plant biodiversity supports animal biodiversity since secondary production directly depends on primary production, and the diversity of animals is regulated by plant structural diversity through provisions of a diversity of habitats and shelters for animals. Critical components of dryland animal biodiversity are the diversity of surface and soil decomposers (supporting nutrient cycling) and larger wildlife (supporting the cultural service of tourism and recreation).

While pastoralists control the exploitation of rangeland services through managing stocking rates and livestock species composition, cropland cultivation constitutes more-intensive management (tillage, irrigation, and so on) that often taxes biodiversity and does not always generate the expected increased provision of services. While cultivation intends to increase service provision, it sometimes impairs the provision of all three supporting services and affects the water regulation service. One way out may then be turning to alternative livelihoods, as described later. Whereas the prevalence of traditional livelihoods decreases with aridity, existing and emerging alternative livelihoods, each supported by a different service, are expected to augment or even replace traditional livelihoods and hence may increase with aridity.

22.3 Condition and Trends in Dryland Systems

This section assesses the issue of greatest concern—land degradation in the drylands. The condition of service provision to each of the three major dryland livelihoods—pastoral, farming, and "alternative"—is also considered. There is an overarching trend toward further water scarcity, exacerbated by aridity, with reductions of 11%, 13%, 15%, and 18% in the annual per capita supplies of surface runoff augmented by flows from outside each subtype expected for the period 2000–10 in the dry subhumid, semiarid, arid, and hyper-arid subtypes respectively (Vörösmarty et al. 2005).

22.3.1 Land Degradation

22.3.1.1 How Much of the Drylands Is Degraded?

The critical drivers of change in drylands are those leading to a degraded condition of primary production. The process by which this service is degraded, compared with its provision prior to human pressures, is called "land degradation." Land degradation in drylands is termed "desertification" and can be viewed as an expression of a persistent decline in the ability of a dryland ecosystem to provide goods and services associated with primary productivity. Thus, an indicator of the condition of drylands is the degree of "land degradation" or desertification. (See Box 22.2.)

Despite the global importance of desertification, the available data on the extent of land degradation in drylands are limited. To date, there are only two studies with global coverage and both have considerable weakness. But in the absence of anything better they have been widely used as a basis for national, regional, and global environmental assessments.

The best known study is the Global Assessment of Soil Degradation (Oldeman et al. 1991). Intended as an exploratory study, it did not include any remote sensing or field measurements and was based on expert opinion only. A more detailed assessment—Soil Degradation in South and Southeast Asia—also relied heavily on expert opinion (Middleton and Thomas 1997). A more thorough, measurement-based global follow-up has not been conducted. Additionally, these studies only considered soil degradation and placed a strong emphasis on erosion, which is extremely hard to measure. These studies also formed the basis of the data and maps presented in the *World Atlas of Desertification* (Middleton and Thomas 1997). A reported 20% of the world's drylands (excluding hyper-arid areas) suffer from soil degradation, mainly caused by water and wind erosion, which is presented as the prime cause for 87% of the degraded land (Middleton and Thomas 1997; Oldeman and Van Lynden 1997; Lal 2001a).

The second study with global coverage is that of Dregne and Chou (1992), which covers both soil and vegetation degradation. It was based on secondary sources, which they qualified as follows: "The information base upon which the estimates in this report were made is poor. Anecdotal accounts, research reports, travelers' descriptions, personal opinions, and local experience provided most of the evidence for the various estimates." This study reported that some 70% of the world's drylands (excluding hyper-arid areas) were suffering from desertification (soil plus vegetation degradation).

Recognizing the lack of adequate data on land degradation, the MA commissioned a desk study (Lepers 2003; Lepers et al. 2005) that compiled more-detailed (and sometimes overlapping) regional data sets derived from literature review, erosion models, field assessments, and remote sensing. This study found less alarming levels of land degradation (soil plus vegetation) in the drylands (including hyper-arid regions). Achieving only partial coverage, and in some areas relying on a single data set, it estimated that

only 10% of global drylands were degraded. This includes 17% of drylands in Asia degraded, but in the Sahel region in Africa—an area reported as highly degraded by the Global Assessment (Oldeman et al. 1991) and by Dregne and Chou (1992)—few localities with degradation were found. The global number of people who live on lands determined by Lepers (2003) as degraded is about 20 million, much lower than the 117.5 million people living on lands defined as degraded by GLASOD.

All these assessments have their weaknesses. Due to the poor quality of the information sources, Dregne and Chou's numbers are most likely an overestimation. For example, they report figures as high as 80–90% for both rangeland and cropland degradation in the drylands of individual countries (such as Kenya and Algeria). Such high levels are hard to reconcile with data from the FAOSTAT database, which show that over the last 40 years the average per hectare yields of major cereals cultivated in Kenya and Algeria have increased 400–600 kilograms. Most likely the true level of degradation lies somewhere between the figures reported by GLASOD and those of Lepers (2003). This implies that there is *medium certainty* that some 10–20% of the drylands are suffering from one or more forms of land degradation. And the livelihoods of millions of people, whether they actually reside in the degraded areas or just depend on them, are affected by this degradation, including a large portion of the poor in drylands.

Even if the most conservative estimate of 10% is used, however, a total land area of over 6 million square kilometers is affected by desertification, an area roughly twice the size of India, the seventh largest country in the world. It should be borne in mind, however, that to determine the true extent of degradation and identify precisely where the problems occur will require a more in-depth follow-up to the three exploratory studies discussed here, combining analysis of satellite data with extensive ground-truthing. (See Box 22.3.) The ongoing Land Degradation Assessment in Drylands project, an international U.N. initiative to assess the status of land degradation in the drylands, is an appropriate response to this daring challenge.

22.3.1.2 Land Use, Land Degradation, and the Aridity Gradient

Human populations and land uses change across the aridity gradient. The best way to express the effect of the gradient on dryland peoples and their land use is through estimating the amount of water per capita available in each dryland subtype. As expected,

---

BOX 22.2
Desertification as Land Degradation

The UNCCD defines desertification as land degradation in the drylands (" 'Desertification' means land degradation in arid, semi-arid and dry sub-humid areas"), yet the two terms are often used as if they are distinct (for example, "Land degradation and desertification in desert margins," in Reich et al. 2000). The UNCCD also defines "land" by its primary productivity service (" 'land' means the terrestrial bio-productive system") and "land degradation" as an implicit loss of provision of this service (" 'land degradation' means reduction or loss . . . of the biological or economic productivity").

The definition of biological productivity and economic benefit depends on users' priorities. Transforming woodland to cropland may decrease biological productivity and degrade the economic benefit of firewood production, for example, but increase the economic benefit of food production. With respect to the mechanisms of land degradation, changes in the properties of the land (soil, water, vegetation) do not correspond linearly to changes in productivity. Loss of productivity can also be attributed to

non-human-induced factors such as rainfall variability and human factors such as low labor input. Thus a range of interacting variables that affect productivity should be addressed in order to assess land degradation objectively and unambiguously.

Commonly considered degradation processes are vegetation degradation, water and wind erosion, salinization, soil compaction and crusting, and soil nutrient depletion. Pollution, acidification, alkalization, and waterlogging are often important locally (Oldeman 1994; Lal 2001a; Dregne 2002). Field experiments, field measurements, field observations, remote sensing, and computer modeling are carried out to study these processes. The higher the aggregation level in each of these study approaches, the more problematic each of the methods becomes, either because of upscaling issues or because of questionable extrapolations and generalizations (Stocking 1987; Scoones and Toulmin 1998; Matthews 2000; Mazzucato and Niemeijer 2000b; Lal 2001a; Warren et al. 2001; Dregne 2002).

BOX 22.3

Satellite Remote Sensing and Desertification

Attempts to map desertification are often unsatisfactory. A key requirement for mapping desertification is that the term is defined in a way that leads to objective and practical measurement criteria. Earth-observing instruments carried on satellites (Prince 1999) routinely map land surface variables that respond to desertification, such as albedo, surface temperature, and vegetation cover—all with appropriate spatial resolution and regular global coverage. Unfortunately, factors that are not related to desertification also affect these properties; for example, AVHRR data have been used to monitor interannual changes of vegetation cover in dry regions (Tucker et al. 1991; Nicholson et al. 1998), but these are frequently caused by rainfall fluctuations, not desertification (Prince 2002).

Persistent reduction in productivity is an expression of desertification (Prince 2002), and it is routinely measured using satellite derived vegetation indices (such as the normalized difference vegetation index). NDVI measures the amount of solar radiation absorbed by the vegetation, from which simple methods can be used to estimate net primary production (Prince 1991). Prince (2002) has suggested that a persistent reduction of NPP below its potential, a reduction that does not disappear during wetter periods, could identify areas that may be experiencing desertification, a measure that is both practical and based on the underlying mechanism.

In the absence of human impacts the NPP is set by climate, soils, vegetation productive capacity, and growing season weather conditions.

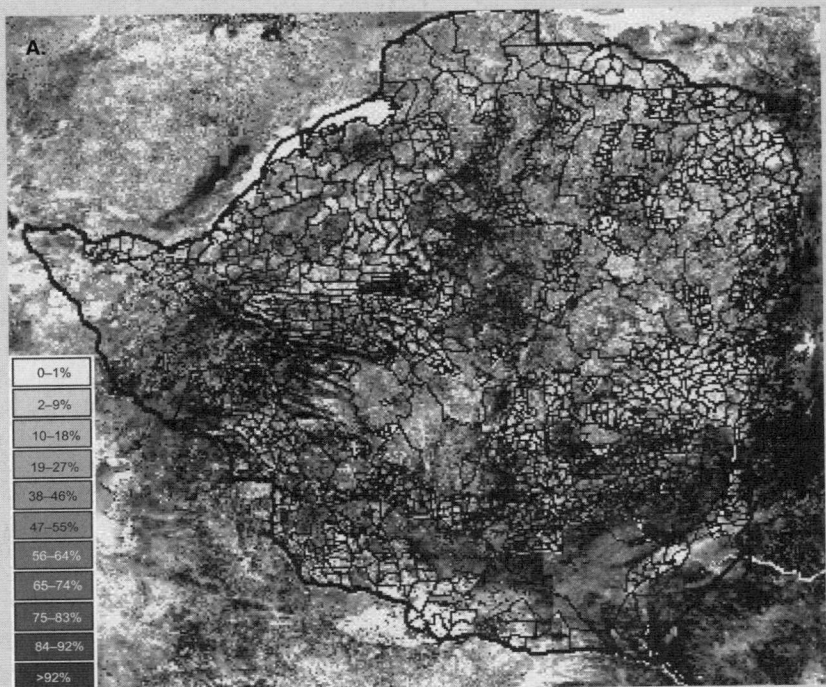

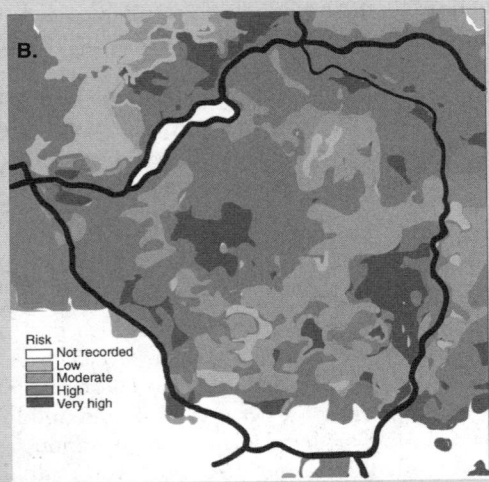

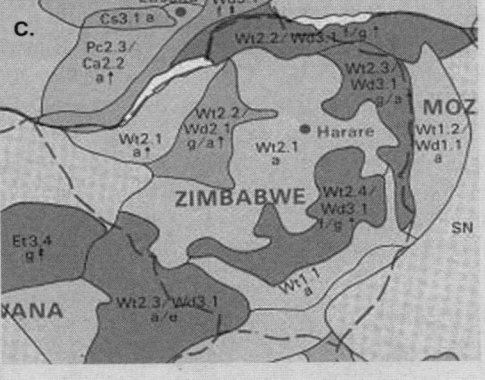

Figure A. Maps of Desertification in Zimbabwe. Map A. Local NPP Scaling (LNS) map. Gray scale indicates the LNS value, from white (LNS = 0–1%, i.e., desertified) to dark gray (LNS>92%, near potential productivity). Note correspondence with fourth level administrative boundaries (See Map D) and the correlation of low LNS with the overcrowded communal lands that are known to be seriously degraded. LNS was calculated using NPP estimated from SPOT VEGETATION 1 sq.km. satellite NDVI data, 1998–2002. Map B. Desertification Risk Map (Eswaran and Reich 2003). This map is a combination of the NRCS global desertification map and a global population density map. Note that the spatial resolution is much less than in Map A and the values of "risk" unfortunately cannot be validated. Map C. Zimbabwe, excerpted from the Global Assessment of Soil Degradation (GLASOD) map (Oldeman et al. 1990). Letter symbols: Wt-loss of topsoil by water erosion; Wd-terrain deformation/mass movement by water erosion; Pc-physical deterioration by compaction/crusting; f-deforestation and removal of the natural vegetation; g-overgrazing; a-agricultural activities; e-overexploitation of vegetation for domestic use. Key to numeric symbols: x.y, where x indicates degree of degradation (1-light; 2-moderate; 3-strong; 4-extreme) and y indicates percentage of mapping unit affected (1-less than 5%; 2–5 to 10%; 3–10 to 25%; 4-more than 50%); ↑-medium rate. As in Map B, the resolution is low and hard to validate on the ground. Map D. Zimbabwe land tenure classes. Note the correspondence of communal lands (white) and commercial farms (dark) to low and high LNS in Map A, respectively.

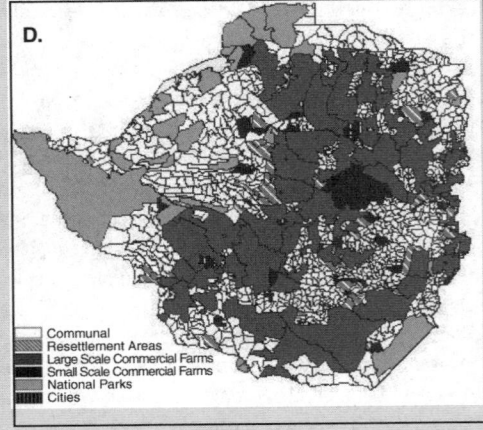

continued

Unfortunately these cannot be measured at an adequate resolution in most desertification-prone regions. An alternative, however, is to use the NPP maps themselves and to employ a statistical method to estimate the NPP of non- or less-degraded areas (Prince 2004). A large region can be classified into homogeneous areas that consist of land having the same climate, soils, and vegetation structure, in which only the human impacts vary. The NPP of the grid cells (pixels) measured by the satellite instrument that fall in each area can be normalized, and the highest NPP values can be used to estimate the potential NPP. All other pixels in that area can then be represented as a percentage of the potential NPP for the same area.

In the example (Figure A), Zimbabwe was stratified into regions in which the principal natural characteristics and existing vegetation types are uniform, based on maps of land cover and rainfall. The regions varied in area but were typically 1,000–10,000 square kilometers. Following the earlier outlined procedure, a percentage reduction from potential NPP was calculated for each pixel. This method is known as the Local NPP scaling method. Other methods are possible, based on closer study of desertification and NPP; under development is Local NPP Ranking, which depends on the recent observation (Wessels et al. 2004) that the NPP of some long overused land in South Africa differs from non-degraded land by a constant proportion. (See Figure B.)

Satellite data are able to detect changes in the productivity of the vegetation; hence, future synoptic primary production mapping is likely to identify regions with persistent reductions in NPP, indicating changes that need closer investigation and ground verification. The definition of desertification used here is land that has suffered a shift to a reduced NPP, even when rainfall is not limiting or is equally limiting on desertified and non-desertified land. This newly emerging procedure detects shifts to reduced NPP relative to the potential NPP. Its validity, utility, and practicality, however, have yet to be demonstrated for global-scale desertification monitoring.

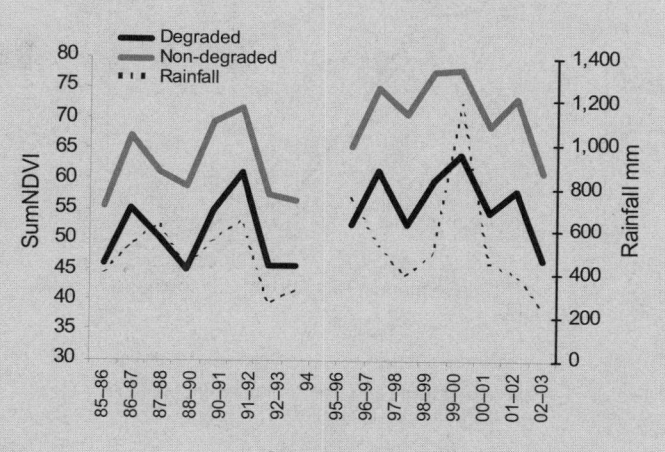

Figure B. Interannual Variation in Net Primary Production of Neighboring Non-degraded and Degraded Areas within the Same Land Potential Class in South Africa. Primary production was estimated using satellite measurements of the Normalized Difference Vegetation Index summed over the growing season. SNDVI is generally positively related to average rainfall. The SNDVI in the degraded areas relative to the non-degraded was reduced by a near constant proportion. Also, in wetter years, the SNDVI in degraded sites exceeded that of non-degraded sites in dry years (Wessels et al. 2004)

human population decreases with aridity. (See Figure 22.8.) It is also expected that annual runoff (as well as surface flows from non-drylands into drylands) will decrease with aridity. However, water supply per person also decreases with aridity (Vörösmarty et al. 2005 Figure 20.8b). Namely, the rate of decrease in water supply with aridity is greater than the rate of decline of the human population with aridity. This suggests that as aridity increases, the ability of the ecosystem to provide the water required by the local population decreases. Thus not only does water supply become low with increased aridity, but there is a mismatch between the supply and the number of people, and the increased aridity-linked gap between supply and demand creates water scarcity. This provides added insight into the decline of human population, cropping, and urbanization with aridity.

However, the most pronounced decrease of population and cropland occur from the semiarid to arid drylands. This may be partly due to the larger spatial extent of the global semiarid dryland compared with other subtypes. On the other hand, use of rangeland peaks in the arid subtype; drier areas are not attractive to livestock, and in areas with higher productivity pastoralism gives way to farming. Thus other than urban areas (2%), most of the world's dryland area is divided by land cover and land use between rangelands (65%) and croplands (25%), although some of these are actually interwoven rangeland and croplands, supporting a mixed, integrated agropastoral livelihood. Both pastoralism and farming and their combination are often implicated as drivers of degradation.

It is likely that the sensitivity of dryland ecosystems to human impact increases with aridity—a little human pressure may not destabilize a dry subhumid ecosystem, but it will degrade the productivity of a semiarid one. And even less pressure will destabilize an arid dryland, in which the capacity of the inherent resilience mechanisms is lower than in less arid drylands. On the other hand, human population pressures and the associated pressure of livestock decrease with aridity, as Figure 22.8 indicates.

Much of the drylands have traditionally been used as rangelands, but with the increase in dryland human populations, a gradual transformation of rangelands to croplands has occurred. Although a large proportion of the hyper-arid dryland is used as range, the value of this range is low and so is livestock density. In the dry-subhumid, the extent of rangeland is low since much of it has been converted to cropland, but grazing pressure is reduced, possibly due to the higher potential profitability of croplands. (Note that human density is greater than livestock density only in this subtype).

Unlike sensitivity to human pressure, which increases with aridity, and human pressure, which decreases with aridity (from 70.7 persons per square kilometer in the dry subhumid to 10.4 persons in the hyper-arid drylands), land degradation—at least as presented by GLASOD—follows a hump-shaped curve, with a maximum in the arid and semiarid drylands. (See Figure 22.9.)

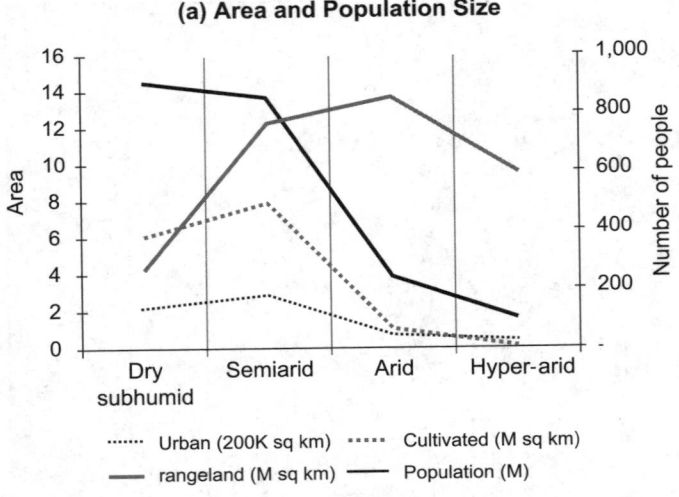

(a) Area and Population Size

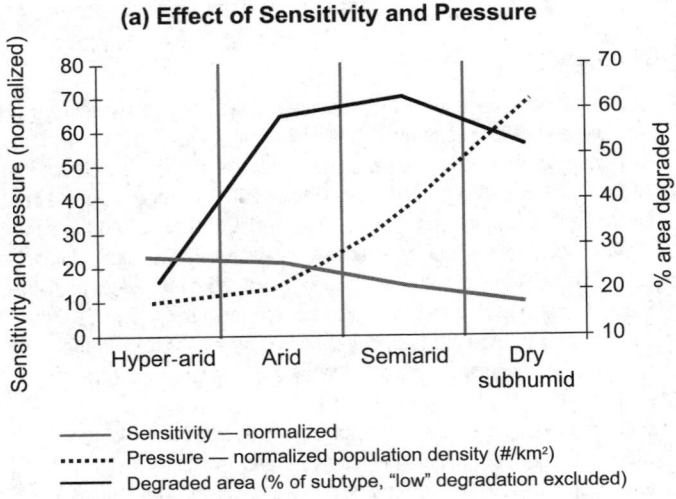

(a) Effect of Sensitivity and Pressure

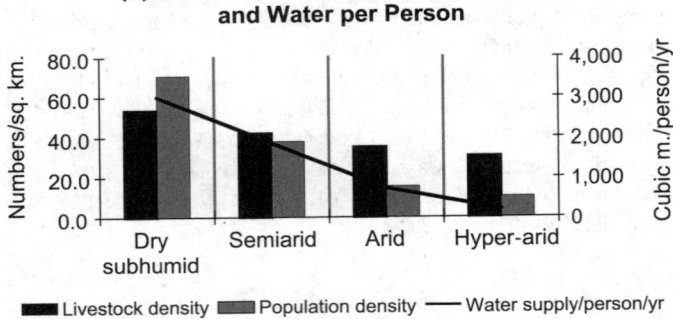

(b) Densities of Livestock and People, and Water per Person

Figure 22.8. Land Use, Human and Livestock Populations, and Water Availability across the Aridity Gradient. Land use and human population size (a) note the different scale for size of urban area). Rangeland figures are based on available data on rangelands in drylands of developing countries (Reid et al. 2004; Thornton et al. 2002) and estimates of rangeland areas in the remaining drylands based on the assumption of uniformity in the rangeland's share of each dryland subtype. Human population densities (b) (CIESIN 2004) and livestock averages of mean densities for developing countries only (Thornton et al. 2002). Line-water supply per person: total runoff generated by a dryland subtype and augmented by inflows (e.g., rivers) from other subtypes or other MA systems, divided by the number of people living in the subtype but taking into account the position of humans along river corridors, in areas of higher (or lower) runoff, etc. Thus, the points represent population-weighted means in terms of the flows per person based on the populations served. (Fekete et al. 2002; Vörösmarty et al. 2005)

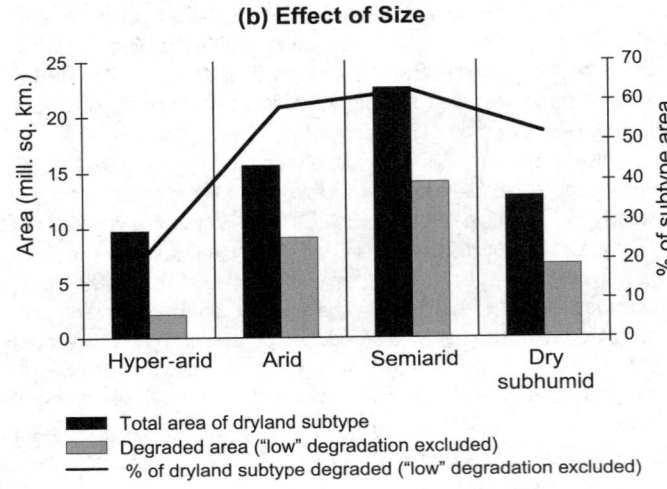

(b) Effect of Size

Figure 22.9. Dryland Degradation across the Aridity Gradient. Effect of aridity (a) and effect of subtype global size (b). Sensitivity to human pressure is 1-median of Aridity Index; sensitivity and pressure are normalized, lowest values set to 10; land degradation is from GLASOD (1990), excluding "low" degradation category, which may be hard to distinguish from "no degradation." (Population density data from CIESIN 2004)

This distribution of land degradation fits a model in which degradation is a function of the product of sensitivity and pressure: when sensitivity is linear but pressure increases exponentially with aridity, the degradation curve is biased to the lower aridity section. The peak is closer to the semiarid than to the arid section of the gradient, and the value for the dry-subhumid subtype is higher than that for the hyper-arid subtype.

These relationships between degradation, sensitivity, and pressure emerge when sensitivity is expressed as an inverse function of aridity and when pressure is a function of population density. The peak in percentage degradation coincides with the peak in

global dryland size, however—the semiarid dryland has the largest global extent and the highest degradation percentage. Thus the most extensive degradation occurs in the central section of Figure 22.9, which also happens to be the most extensive global dryland subtype—the semiarid drylands. This subtype and the arid ecosystems subtype—and especially the transition between the two—have medium sensitivity and are driven by a medium anthropogenic pressure, a combination that generates the highest vulnerability and may result in desertification.

22.3.2 Condition and Trends of Rangelands

Dryland rangelands support approximately 50% of the world's livestock and also provide forage for wildlife (Allen-Diaz et al. 1996). Global data on the extent of rangelands within drylands are available only for developing countries. Based on Reid et al. (2004), rangelands occupy 69% of the drylands of the developing world. Their greatest extent is in the semiarid subtype (14 million square kilometers) and the proportion of dryland they occupy increases with aridity—from 34% in the dry-subhumid subtype to 54% in the semiarid, 87% in the arid, and up to 97% in the hyper-arid subtype. Data on livestock numbers in drylands are available only for industrial countries, where densities are calculated according to land units, mostly administrative units (Thornton et al. 2002).

Combining data from Reid et al. (2004) and Thornton et al. (2002), it is possible to estimate livestock densities per dryland sub-type as well as for just the rangelands within those dryland subtypes. Livestock densities for rangelands within the subhumid, semiarid, and arid subtypes are relatively uniform (32–35 animals per square kilometer of rangeland) but drop to a low 15 animals in rangelands within the hyper-arid subtype. The combined densities of sheep, goats, and cattle per dryland subtype area unit steadily decline with aridity (53 animals per square kilometer in dry subhumid to 31 animals in the hyper-arid subtype). The number of animals per unit area of a dryland subtype is greater than those per rangeland unit area, especially in the hyper-arid and the dry subhumid subtypes. This suggests that many animals (especially cattle) do not range freely in the dry subhumid areas, probably due to competition with cultivation, and are kept in fertile areas within the low-productivity hyper-arid rangelands, such as desert oases and along desert rivers—areas not classified as pastures.

22.3.2.1 Semiarid Rangelands

Most of the arid drylands are used as rangeland, but during the second half of the nineteenth century large-scale commercial stockbreeding spread over the semiarid drylands of North and South America, South Africa, and Australia. Both the type of herbivore and the grazing management applied (including fire prevention) were new to these semiarid ecosystems. The resulting disturbance created a "transition trigger" that, combined with drought events, led to a progressive dominance of shrubs over grass (Scholes and Hall 1996), a process that may have been facilitated by increasing atmospheric CO_2 levels that differentially favored C3 shrubs over C4 grasses (Archer et al. 1995; Biggs et al. 2002). The transition of grasslands to shrublands (see Figure 22.10) is widely reported around the world. This transition generated a mosaic of plant clumps within a "matrix" devoid of much vegetation, which encourages surface runoff, topsoil erosion, and exposure of rocky surfaces (Abrahams et al. 1995; Safriel 1999).

Eventually, the degradation due to overstocking and range mismanagement led to a decline in livestock numbers after peaking at the beginning of the twentieth century—40% loss in New Mexico (Fredrickson et al. 1998), 45% loss in western New South Wales (Mitchell 1991) and 60% loss in Prince Albert District *karoo* (Milton and Dean 1996)—and to substitution of sheep for cows, as they are better adapted to graze on tussock grasses.

22.3.2.2 Dry Subhumid Rangelands

Two opposing trends in the global area of temperate grasslands are evident: expansion at the expense of woodland and contraction due to the encroachment of cultivation. Examples of the first process are the expansion at the expense of forests in North America following European colonization (Walter 1968) and, in the Caucasus, a 27% increase in the last century at the expense of oak forest (Krenke et al. 1991). Grasslands' natural primary production generates high biomass of many species of herbaceous forage plants, and the second transformation involved redirecting the grasslands' primary production service for maximizing the production of seeds of a few species of domesticated grasses. The

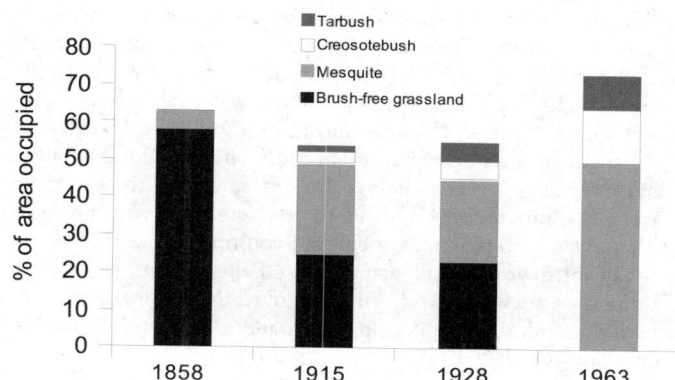

Figure 22.10. Transition of Grassland to Shrubland due to Stock-breeding in Semiarid Rangelands. The figure shows the percent of area occupied by dense (≧55% of perennial plant composition) brush cover of the major shrubs at various dates on the Jornada Plain of the Jornada Experimental Range in New Mexico. (USDA-ARS 2003)

conversion of grasslands to grain crops occurred some millennia ago in the loess region of north-central China, in the last three centuries in the Russian Federation (162 million hectares), and during the last century in the United States (121 million hectares) (Ramankutty and Foley 1999). However, though temperate grasslands are fairly resilient to grazing, fire, and tillage that maintains the topsoil (Lavrenko and Karamysheva 1992), more-intensive cultivation dramatically reduces the provision of supporting services.

Population increase is the main driver of the tropical savanna-cropland transition, and the dry subhumid African savannas north of the equator represent barely 14% of the original cover and are retreating at an estimated rate of 0.15% (or 340 square kilometers) per year (WRI 2004). Where human population in tropical savanna is relatively low but soil fertility high, fire is excluded, and grazing is heavy, however, range degradation occurs due to topsoil erosion driven by bush encroachment (Scholes and Hall 1996).

Rangelands of the Mediterranean xerophytic shrubland and woodland are relatively resilient to human impact. The Mediterranean landscape is of a fine-grained spatial heterogeneity, with mosaics of cultivated fields and stands of different vegetation succession stages that result from the combined effect of fire, grazing, and cropping. This dynamic mosaic of land uses maintains the soil conservation service, though frequent and extensive alterations may reduce it (Naveh 1991).

22.3.3 Condition and Trends of Cultivated Drylands

22.3.3.1 Oases in Hyper-arid and Arid Drylands

In the hyper-arid and arid drylands (the desert biome), most cultivation is either in oases or in croplands, where crops are irrigated by fluvial, ground, or local water sources or by a combination of these. Traditionally, oases were carefully managed by combining crops and water regulations. Since the middle of the twentieth century, oases have borne increasing demographic and investment pressures resulting in larger water abstraction, which has led to soil salinization and huge loss of surrounding vegetation through overgrazing and fuelwood exploitation followed by soil erosion.

Two examples are indicative of these trends. Recent indirect drivers of change in the Maghrebian oasis (northern border of the

Sahara) have been population increase, policies to settle nomadic populations, investments generated by migrants working abroad, and transformation of self-sufficiency to open-market economies in countries of the Maghreb. The expansion of cultivated areas irrigated by water from deep aquifers and a lack of drainage (de Haas et al. 2001) led to spreading shallow water tables and subsequent soil salinization (Mtimet and Hachicha 1995). In the Nile delta, traditional water management of the summer flood kept the water table at 5–7 meters below the surface during the low flow season. But nineteenth- and twentieth-century water management policies and practices, including construction of the Aswan dam, enabled year-round irrigation and the growth of out-of-season crops, such as cotton. These reduced the water table depth, however, resulting in soil salinization and a dramatic decline in crop yield (Ruf 1995).

22.3.3.2 Semiarid and Dry Subhumid Agriculture

Where massive agricultural encroachment into temperate grasslands has occurred, and where drylands are exposed to freezing and thawing and suffer a relatively high recurrence of drought and strong winds, topsoil is at risk of being lost and with it all the provisioning services. This happened in the Chinese loess plateau south of the Yellow River bend millennia ago and in North American central plains in the first quarter of the twentieth century. Once the topsoil is lost to wind, the underlying layers—rich in calcium carbonate and poor in organic matter—develop crusts that impede infiltration and foster further soil loss through gully erosion (Mainguet 1996). In the United States, temperate grassland soils can lose up to 50% of their original carbon within the first 50 years of cultivation (Conner et al. 2002). In parallel to carbon loss, conversion of grasslands to croplands can result in loss of fertility, increased soil erosion, and decreased water quality through larger sedimentation and non-point chemical pollution by salts, nutrients, and pesticides.

When agricultural encroachment of tropical savannas occurs, there is a risk of soil impoverishment through erosion and nutrient depletion (Stoorvogel and Smaling 1990; Drechsel et al. 2001) unless nutrients and water are carefully managed. Subsequent erosion may leave the subsurface soil horizon with its limited water infiltration capacity exposed, resulting in further soil loss by gully erosion. However, if land husbandry practices are properly adjusted, the increased crop production required to meet a growing population does not necessarily have to lead to nutrient depletion (Scoones 2001; Niemeijer and Mazzucato 2002; Mortimore and Harris 2004).

The long, traditional agricultural use of Mediterranean shrublands and woodlands in the Mediterranean basin has prevented soil loss through sophisticated terrace systems used across hill slopes. But the pulse of rural population increase that started at the end of the nineteenth century occurred too fast to allow widespread land conditioning, and it triggered extensive soil losses. The outcome was large upstream areas of exposed rock (Roquero 1990) and accretion pulses in river deltas (Hoffmann 1988).

22.3.4 Condition and Trends of Alternative Livelihoods

Rather than grappling with drylands' low biological productivity, dryland peoples have explored ways to exploit other ecosystem functions that can serve them better, even better than if they exploited the same functions in non-drylands. Such emerging "alternative livelihoods" do not depend on traditional land uses and are generally undemanding on land and natural resource use. Alternative livelihoods have minimal dependence on land primary productivity for producing subsistence products; do not impair the provision of other services of dryland ecosystems; generate more income per investment from local dryland resources, compared with the traditional, land biological production-dependent livelihoods; and provide people with a competitive edge over others who follow the same practices outside drylands. Five major alternative livelihoods are discussed, in order of decreasing dependence on land primary productivity and hence decreasing desertification risk and increasing likelihood of sustainability.

22.3.4.1 Dryland Afforestation

Silviculture and rain-fed horticulture are not very common in drylands and depend on labor-intensive construction and maintenance of runoff-harvesting structures (Evenari et al. 1982; Droppelmann et al. 2000). However, dryland silviculture and horticulture provide better soil protection than agriculture because tree canopies are denser and tree root system deeper and more extensive than those of agricultural crops and because trees continue to provide soil conservation after harvest, unlike agricultural crops, where very little is left to cover the soil after harvest.

Dryland afforestation for firewood production is a livelihood that depends on the biological production service. Unlike grazing or cropping, afforestation provides a superb soil conservation service for an area greater than that occupied by the forest itself since flashfloods are not generated by dryland forests. Dryland afforestation will qualify as an alternative livelihood if it generates more income than a traditional dryland land use and, even better, if it can generate more income than non-dryland silviculture.

Carbon sequestration by forests and their contribution to above- and belowground carbon reserves and the recent emergence of "carbon trading" under the Clean Development Mechanism may make the required difference. This is because most non-dryland ecosystems with good provision of the biological productivity service are already either cultivated or afforested. On the other hand, though the global drylands are less efficient than non-drylands in carbon sequestration, their potential for further carbon sequestration is high and has not yet been developed, while non-dryland capacity is already close to the maximum. Experiences in Israel have shown that a mean annual addition to the carbon reserve of drylands of 150 grams of carbon per square meter per year is possible, generated mainly during winter (rather than during summer in non-drylands), thus storing twice as much carbon as the adjacent nonforested rangeland (Gruenzweig et al. 2003).

Dryland afforestation only counters desertification in those cases where it is used as an alternative for unsustainable cropping and grazing practices. It is not a suitable alternative for rangelands, agroforestry systems, and areas under natural plant cover, as these offer equal or better protection against desertification. Also, in the case of degraded croplands, the benefits are quickly lost if trees are watered through irrigation (with its potential for salinization and water resource depletion) rather than through runoff harvesting.

22.3.4.2 Controlled Environment's Cash Crop Agriculture

There are dryland agriculture practices that can qualify as alternative livelihoods by employing plastic covers in agrotechnology. The plastic cover allows nearly full light penetration and at the same time offers options for locally manipulating many other crop-relevant environmental factors (Arbel et al. 1990). This practice ranges from covering individual rows of low-stature crops, with no additional intervention and for only a part of the growing season (mainly in the least dry drylands), to covering plots within "growth houses" or "greenhouses," within which several of the internal environmental conditions are artificially manipulated to the extent that the crop is virtually separated from the outer dryland environment (mainly in the driest drylands). The plastic enclosure reduces evapotranspiration and thus maximizes water use efficiency (Pohoryles 2000); it reduces insecticide use and makes CO_2 fertilization feasible.

Often the crops are grown on artificial substrates, with nutrients supplied and water provided by irrigation that incorporates fertilizer application ("fertigation"), and pollination may be provided by commercially raised and marketed pollinators (BioBee 2000). Together these enable intensification of biological productivity while economizing on land resources. This approach uses the dryland's abundant incident light and winter warmth, but it is otherwise nearly independent of local ecosystem services. Provided that it does not deplete local water resources and that salinity is controlled and not allowed to leach into groundwater, this practice does not generate desertification.

However, this livelihood requires investment in infrastructure, such as energy for ventilation and cooling, making it more suitable for industrial countries. The intensification of crop growth generates more yield per unit of investment, but the crops need to have high market value—namely, cash crops. The production of cash crops in drylands may be more profitable than in the non-drylands, on account of two physical/climatic features of most (but not all) drylands: high irradiation due to relatively infrequent cloud cover and higher ambient winter temperatures relative to those in the nearest non-dryland areas. Indeed, the gross value added and the cash generated per unit area from that part of the hyper-arid dryland of Israel in which intensive greenhouse agriculture is widely practiced is higher than those of all other types of Israeli agriculture, including those of the least dry areas of the country (Portnov and Safriel 2004).

22.3.4.3 Aquaculture in Drylands

Dryland aquaculture is inherently advantageous to dryland agriculture because although aquatic organisms live in water they do not transpire it, so water losses from aquaculture are predominately from evaporation rather than raised evapotranspiration. Furthermore, many more aquatic species than terrestrial crop species are tolerant of salinity and even thrive in it. Thus, dryland aquaculture can prosper on fossil aquifers (quite common in drylands) whose high salinity greatly curtails their use by dryland agriculture.

When dryland aquaculture borrows the technology of dryland greenhouses (as just described), water conservation is even greater than it is in agricultural greenhouses due to zero transpiration of aquatic organisms. At the same time, dryland aquaculture does not compete for water with dryland agriculture due to the divergent salinity tolerances of terrestrial plants and aquatic organisms (Kolkovsky et al. 2003). Since dryland aquaculture is always more economic on land than dryland agriculture, land use as well as water use efficiencies are high. Thus dryland aquaculture, like dryland controlled-environment cash crop agriculture, does not depend on local ecosystem services and need not cause desertification.

Dryland aquaculture is based on aquatic animals and plants (mostly micro-algae) or some combination of these. The productivity of aquatic animals is not light- and CO_2-dependent, hence the costs of feeding the animals are greater than those of fertilizing the plants. There is, however, an added cost of water filtration due to the enrichment of the water by the surplus organic load of animal feed and animal excretions. This cost can be reduced by integrating animal and plant aquaculture, in which algae thrive on the animal waste-enriched water, or the enriched water can be used for irrigation of crops. Plant aquaculture is advantageous on animal aquaculture in that feeding is not required and organic load is not a problem. Also, given that most aquatic plants are either very small or unicellular, their growth is much faster than that of terrestrial plant crops, and the ratio of harvested to nonharvested biomass of the crop is much higher than that of terrestrial plants.

Dryland aquaculture of both plants and animals is more advantageous than aquaculture elsewhere due to the abundance of light for aquatic plants (Richmond 1986) and of winter warmth for both plants and animals (Kolkovsky et al. 2003). An added benefit is the higher availability and hence the lower price of land in drylands than in non-drylands and the reduced competition with agriculture on land in the drylands. Most of the products of dryland aquaculture are cash crops, such as ornamental fish, high-quality edible fish and crustaceans, and industrially valuable biochemicals produced by micro-algae, such as pigments, food additives, health food supplements, and pharmaceutical products.

22.3.4.4 Urban Livelihoods

Though "dryland development" and "rural development" are often used synonymously, dryland cities as an alternative to dryland villages may be a sustainable option for settling more people in drylands because the cities consume, and hence affect, fewer land resources than dryland farming and pastoral livelihoods do. However, this depends on the potential of dryland cities to provide livelihoods as well as living conditions that are advantageous compared with those provided by other cities.

A combination of appropriate building materials, architectural design (Etzion et al. 1999), and urban planning (Pearlmutter and Berliner 1999) can provide living conditions in drylands that are as comfortable as and much cheaper than those provided by non-dryland cities. This is because drylands are endowed with two climatic features that are highly conducive to "passive" (energy-saving) climate control. The very low air humidity in the driest drylands makes summer evaporative cooling very efficient and cost-effective, and the low dryland cloud overcast means that solar radiation (aided by appropriate positioning, dimensions, and technological design of glass windows) provides efficient and cost-effective winter warming (Etzion and Erell 2000).

Thus the use of fossil fuels for cooling, and of fossil or biomass fuels for warming, can be much lower in driest dryland cities than elsewhere. Furthermore, fossil fuels can be nearly completely replaced by solar energy-generated power (Faiman 1998) due to the high year-round intense solar radiation coupled with the low overcast of many drylands. Given the potential (though rarely realized) advantages of living in dryland cities and their relatively low impact on dryland services, a policy of encouraging urban livelihoods in appropriately designed and functioning dryland cities could significantly contribute to sustainable dryland management. Dryland tourism may be one such livelihood.

22.3.4.5 Dryland Tourism

Dryland tourism is driven by the increasing affluence, free time, and mobility of a relatively large segment of the global population coupled with the growing craving for uncongested, unpolluted, pastoral, pristine landscapes. Drylands offer many unique scenic, wildlife, biodiversity, historical, cultural, and spiritual services. Hence employment in the tourist industry may become an increasingly important alternative dryland livelihood for both rural and urban dryland people.

Though urban and tourism-related dryland livelihoods are economic on land use, their impact on dryland water resources requires attention. Given the large water demand of dryland agriculture, the per capita water demand of a dryland city is likely to be lower than that of a rural dryland village. However, the tourist industry is a significant consumer of water. Irrespective of dryland urban versus dryland rural development, the growing demand versus the diminishing supply of renewable water in drylands has catalyzed the improvement of technologies for recycling and reuse of wastewater and for water desalination in drylands (NRC 1999). These are also helping to address the water demand incurred by the dryland tourist industry.

22.3.5 Condition and Trends of Dryland Biodiversity

22.3.5.1 Species Endangerment and Extinction in the Drylands

Most available information on species threat status and extinctions is listed by country, and no assessment has been made on overall species status for drylands, let alone for specific dryland subtypes. However, to gain some insight on the situation we selected three relatively large countries (each more than half a million square kilometers in size) that are virtually 100% drylands and are geographically isolated from each other as to minimize species identity: Kazakhstan in Central Asia (semiarid and arid), Mali in north equatorial Africa (arid and hyper-arid), and Botswana in south of equatorial Africa (semiarid and dry subhumid). The combined number of terrestrial threatened and non-threatened vertebrate species for these countries totals 1,593 species (IUCN 2004), occurring over an area of more than 4.5 million square kilometers. Only one species is known to be globally extinct (a mammal in Kazakhstan), and 69 species (4.3%) are threatened (see Table 22.5), which appears to be much lower than global species endangerment (12–53% for different groups; see Chapter 4).

No correlation exists between the species richness of each group and its proportion of threatened species, and it is likely that threat status is more related to body size. The dryland birds and mammals of these countries are more prone to extinction than amphibians and reptiles are, and though there are many more bird than mammal species, the highest proportion of threatened species is among mammals. It is likely that this is because, on average, mammals are larger than other vertebrates, have larger home ranges, and hence are more affected by habitat loss or are subject to greater hunting pressure or persecution by humans.

Analyzing a single country—Israel—by dryland subtype reveals 3% of the desert biome vertebrate and plant species (hyper-arid and arid subtypes combined) are threatened, compared with 7.7% of the Mediterranean biome species (semiarid and dry subhumid subtypes), which suggests that, at least in Israel, threats decline with aridity. These figures also include locally and globally extinct species: five freshwater fish and one amphibian species lost in Israel are globally extinct. Furthermore, 57% of the breeding birds associated with wetlands and inland waters in Israel have become locally extinct or are threatened (compared with only 27% of non-wetland birds) (Nathan et al. 1996). This is in agreement with the global situation, whereby species of freshwater habitats are at greatest risk of extinction (see Chapter 4), and it demonstrates the intense pressures on wetlands in drylands.

22.3.5.2 Conservation of Biodiversity in Drylands

Protected areas occupy 8% of global drylands, which is close to the global average for all systems (10.6%). The fraction of each subtype area within protected areas declines with aridity: 9% of dry subhumid drylands, 8% of the semiarid drylands, and 7% of arid drylands, although the figure for the hyper-arid subtype is the highest at 11%, just above the global average. The high degree of protection provided in the dry subhumid subtype has two explanations. First, greater political attention is focused on the conservation of less-arid areas and, second, due to the high population pressure in dry subhumid areas, there is a greater awareness of conservation needs there. At the same time, the negligible population pressure and low competition for dryland services make hyper-arid drylands ideal for designation of protected areas, which explains the high degree of protection there. Protected areas in drylands are managed for three different reasons: to support biodiversity, to promote the provision of their cultural services, and to promote biodiversity's role in provisioning all other services.

Of the 25 global "biodiversity hotspots"—terrestrial areas where at least 0.5% or 1,500 of the world's 300,000 plant species are endemics and with habitat loss expressed in the decimation of 70% or more of their primary vegetation (Myers et al. 2000)—8 are in drylands. These are the Succulent Karoo and Cape Floristic Province, both in southwestern Africa; the Brazilian Cerrado; Central Chile; the California Floristic Province; the Mediterranean Basin; the Caucasus; and parts of southwest Australia. (Note, however, that most of these hotspots represent the Mediterranean type biome only.) Of the 134 terrestrial "ecoregions" (200 global ecoregions defined by Olson and Dinerstein 1998) identified as priority conservation targets, 24 are within drylands. Almost 30% of the global Centers of Plant Diversity (WWF/IUCN 1994) contain, at least partially, drylands. Using the IUCN Protected Area classification, dryland protected areas that are managed to support biodiversity, and with the highest degree of protection and least access to people, occur mainly in the semiarid drylands and occupy 11.4% of all dryland protected areas.

Table 22.5. Endangerment of Vertebrate Species in Three Dryland Countries. Data for all IUCN categories of endangerment combined. (Data from IUCN 2004)

	All Species		Amphibians	Reptiles	Birds	Mammals
	Total	Threatened	Threatened within Each Group			
	(number)	*(number and percent)*	*(percent)*			
Mali	340	20 (6)	0	1	3	9
Kazakhstan	462	35 (8)	7	4	4	10
Botswana	349	14 (4)	0	0	4	4
Total	**1,593**	**69 (4)**	**2**	**1**	**3**	**8**

In protected areas maintained primarily for the provision of cultural services, human exploitation and occupation are restricted, but visitors are welcome and tourism is encouraged and hence they offer potential economic benefits to local people. In total, these account for 44.7% of all dryland protected areas combined and are most common in the arid subtype. Many of these are national parks or (officially non-protected) private game farms, which provide many of the same cultural services. In many of the national and private parks, especially in Africa, management is geared to maximizing the number of large game species for visitors, which may lead to trampling and overgrazing due to overpopulation of game.

It is increasingly difficult to set aside formally protected areas that largely exclude human populations, especially in dryland countries with high poverty levels. In some regions, historically protected areas have been associated with oppressive political regimes and the exclusion of local inhabitants, where wildlife conservation goals have been put ahead of human needs. This image of protected area use has been challenged over the past decade, especially through increased efforts to promote community-based natural resource management. Protected areas managed for their provisioning services, which allow for controlled exploitation of ecosystem goods, account for 43.9% of dryland protected areas and are most common in the semiarid drylands.

22.4 Drivers of Change

22.4.1 Conceptual Framework of Dryland Drivers: The Desertification Paradigm and Its Counterpart

The overarching change in drylands is land degradation defined as a persistent decrease in provisioning of ecosystem services—also frequently termed desertification, as described earlier. A number of phenomena are tied in through various links, interactions, and feedbacks, which jointly make up the "desertification paradigm." This section assesses the validity of the desertification paradigm by presenting the relevant direct and indirect drivers, including both biophysical and socioeconomic drivers. More-recent research has led to the development of an alternative "counter-paradigm" concerning drylands processes; this approach identifies interventions that prevent the occurrence of desertification. Between the two paradigms there is general agreement on the role of the direct biophysical drivers discussed in the next section. Where the common "desertification paradigm" and the emerging "counter-paradigm" differ is in the proposed role of indirect drivers. This is the subject of the subsequent two sections.

22.4.1.1 Direct Bioclimatic Drivers

The relatively low productivity and low soil moisture content of drylands are often exacerbated by the uni-modal pattern of rainfall, resulting in a long period during which soil moisture falls. Low and infrequent precipitation patterns and radiation-induced evaporation jointly and directly drive a linear sequence of biophysical processes. In this cycle, low soil moisture leads to low plant productivity, poor soil development, and high runoff, resulting in a high susceptibility of drylands to soil erosion.

The same two bioclimatic factors that are drivers of vulnerability to erosion are also drivers of vulnerability to salinization. Low rainfall does not leach the surface salts into deeper soil layers, down below the root zone of plants, hence the dryland topsoil is vulnerable to salinization. These natural conditions make dryland ecosystems vulnerable to land degradation, while both drought and human activities can further exacerbate existing vulnerabilities.

22.4.1.2 The Desertification Paradigm

The desertification paradigm holds that bioclimatic drivers and anthropogenic drivers that traditionally maintain dryland ecosystems in a stable state become drivers of change, pushing the transition from sustainable exploitation of ecosystem goods and services to a new ecosystem state of much lower level of service provision. Extremes of direct bioclimatic drivers, rainfall fluctuations leading to droughts, and extensive, intensive, and frequent fires, when coupled with indirect anthropogenic drivers, jointly become drivers of change that through an intricate chain of processes lead to a downward spiral of productivity ending in irreversible land degradation—that is, desertification.

Human population growth in drylands, which increased 18.5% between 1990 and 2000, has been the highest of any MA system (CIESIN 2004). Increased aspirations for raised standards of living are driving an increase in the exploitation of ecosystem services, often accompanied by an increased use of labor and new technologies. This is expressed as a proliferation of livestock and expansion of agriculture and through the adoption of intensive farming practices. The adverse impacts are further amplified when intensification of human activities coincides with droughts, which temporarily but drastically reduce soil and plant productivity.

Cropland in drylands has lower productivity than "wetter" croplands. Farmers attempt to address this through supplements such as additional water, fertilizers, and pesticides. If they cannot afford these, rain-fed dryland croplands are left fallow, but land shortages reduce fallow length, often leaving the land insufficient time to recover. Irrigation increases productivity, maintains vegetation cover, and helps protect soil from erosion. However, dryland irrigation accelerates soil salinization due to the often high salinity of available irrigation water and high evaporation. Salinization in croplands directly affects plant growth, dramatically impairing the provision of the soil's productivity service. Over time, soil erosion and salinization reduce productivity to the point where cropland has to be abandoned. In a similar way, rangelands may be degraded due to overgrazing.

The indirect anthropogenic drivers of change in drylands are diverse and act on several scales. They include demographic drivers, such as local population growth or immigration resulting from regional population growth; economic drivers, such as local and global market trends; and sociopolitical drivers, such as local and regional land tenure policies as well as scientific and technological innovations and transfer. According to the desertification paradigm, these combined drivers intensify pressure on drylands (in areas already in use and "virgin" territory) in anticipation of increased provision of ecosystem services. However, this pressure for increased productivity frequently fails and, worse, can lead to decreased productivity.

The impact of reduced land productivity is manifest through reduced income, malnutrition, and poor health, culminating in famine and increased mortality rates. People frequently abandon degraded land in order to avoid this impact and either intensify the use of other intact but lower-quality land or transform more rangeland to cropland, practices that may delay but not avoid further land degradation. Since alternative livelihood opportunities are few, migration from rural to urban areas and transfrontier migration often follows. These migrations often create environmental refugees, a situation that exacerbates poverty and urban sprawl and can bring about internal and transboundary social, ethnic, and political strife. These may encourage foreign intervention, which has the potential for destabilizing local, regional, and even global political and economic systems. This chain of processes driven by anthropogenic drivers of change leads to a downward spiral of productivity loss and increasing poverty.

Furthermore, the paradigm implies that since soil degradation and vegetation degradation are linked to increased aridity as part of a negative feedback loop, desertification is practically irreversible, and its inevitability increases with aridity (Cleaver and Schreiber 1994). And since desertification takes place mainly where agriculture is the major source of local livelihoods, agricultural practices are often blamed for desertification and the associated decline in the provision of ecosystem services and rise of poverty. Finally, the paradigm claims that investments are usually required to make dryland agriculture sustainable, but these are generally in short supply due to poverty. Thus poverty is not only a result of desertification but a cause of it.

22.4.1.3 The Counter-paradigm

According to the counter-paradigm, the drivers, processes, and events described in the desertification paradigm do exist, but the chain of events that leads to desertification and the chain-reaction cycle of reduced ecosystem productivity and poverty are far from inevitable. This section identifies the conceptual weaknesses of the desertification paradigm and presents the counter-paradigm approach.

The "desertification narrative" dates back to the 1920s and 1930s, when concerns about a presumed extension of the Sahara began to be raised and when claims of harmful African farming practices coincided with reports about the American "Dust Bowl" experience (Anderson 1984; Swift 1996). So even before the term "desertification" was coined, the narrative or paradigm already existed (Swift 1996). Although large-scale drought-related famines in drylands are not a new phenomenon (Nicholson 1979), however, there has always been a recovery after each drought rather than the irreversible collapse implied by the desertification paradigm. This suggests that the desertification paradigm does not fully describe what is happening on the ground. In particular, as described in this section, there are problems with the knowledge and understanding of dryland systems that form the basis for the theoretical foundation of the paradigm; the evidence for degradation on which the paradigm builds; and its assumptions about human response to changes in the natural and socioeconomic environment.

22.4.1.3.1 Inherent instability of drylands

The desertification paradigm assumes that natural dryland ecosystems are in a stable state that can be disrupted by population growth induced by overcultivation and overgrazing leading to a degraded condition of the service of primary production. Based on this assumption, measured deviations from a theoretical natural stable state are considered degradation, so human and livestock populations above calculated carrying capacities indicate degradation (Leach et al. 1999).

The notion of this "balance of nature" has been increasingly challenged by twentieth century ecologists, however, and an increasing number of studies in the last two decades have shown that dryland systems exhibit large variability in space and time and that many of them are far better described in terms of non-equilibrium systems (Ellis and Swift 1988; Behnke et al. 1993; Leach et al. 1999). A lot of this work has focused on African savanna systems, but the principles are likely to apply to other dryland regions. The basic premise is that irregular droughts prevent the establishment of a stable equilibrium between plants and livestock (Ellis and Swift 1988). Droughts, disease, and social upheaval have had similar disruptive impacts on many dryland farming systems, introducing elements of non-equilibrium systems there as well. Misinterpreting these systems as in equilibrium leads

to an overestimation of land degradation because each (temporary) shift in the balance of vegetation versus human and livestock populations is interpreted as a sign of degradation.

22.4.1.3.2 Problems defining and detecting degradation of an unstable system

Recent remote sensing studies (see Box 22.4), using multiyear analyses of rain-use efficiencies and vegetation indices, have revealed that the widely claimed land degradation in the Sahel may have been a temporary phenomenon caused by the droughts of the 1980s (Nicholson et al. 1998; Prince et al. 1998; Eklundh and Olsson 2003). In contrast, traditional desertification assessments have tended to be simple snapshots in time that have extrapolated observations and measured rates of change linearly into the future. To some extent that may be appropriate for equilibrium systems undergoing a transition, but such linear extrapolations are not justifiable for non-equilibrium systems that undergo continuous change (Niemeijer 1996).

While the understanding of the basic principles of the individual biophysical processes underlying the desertification paradigm is essentially correct, the methods of assessing and quantifying the processes have been problematic. Evidence for degradation has been based on the assessment of vulnerabilities using national, regional, and continental soil surveys and models of carrying capacity, as well as experimental plot studies, expert opinion, and nutrient balance models. While each method is sound in its own right, findings cannot simply be extrapolated in time and space to map out an essentially dynamic and spatially heterogeneous phenomenon such as desertification (Stocking 1987; Mazzucato and Niemeijer 2000b; Scoones 2001). To take just one example, erosion, which is rightfully a major concern in the desertification paradigm, is extremely difficult to measure accurately, let alone to extrapolate to large areas given the high spatial variability of rainfall, geomorphology, and soils in most dryland environments. Due to difficulties in measuring and extrapolating wind and water erosion accurately, landscape-level erosion figures based on plot measurements can lead to overestimation of the actual erosion by a factor of 10 to 100 times (Stocking 1987, 1996; Warren et al. 2001).

22.4.1.3.3 Responses to keep ahead of degradation

The desertification paradigm is grounded in simplistic, mechanistic thinking about human responses to the dryland environment and the processes of desertification. This section presents the crux of the counter-paradigm. An understanding of the dynamism of human responses helps explain why degradation estimates based on carrying capacity concepts of the desertification paradigm can be somewhat misleading.

It is important to note that the paradigm has its root in the environmental sciences (soil science, agronomy, and to a lesser degree forestry) and is strongly influenced by Malthusian thinking on the population-environment nexus. The sciences that deal with human behavior have been largely ignored; as Swift (1999) noted: "Soil science has been brilliantly informed by reductionist physics and chemistry, poorly informed by biology, ecology and geography and largely uninformed by the social sciences." The premise of the desertification paradigm is that land users, in response to their needs, increase pressure on the land in unsustainable ways, leading to decreasing productivity and a downward spiral of poverty and further degradation.

However, there is increasing evidence that these negative feedback loops need not occur. Rather, dryland populations, building on long-term experience with their dynamic environments as well as active innovation, can stay ahead of degradation

BOX 22.4

How Much of the Dryland Is Degraded?

The top Figure shows the severity of soil degradation as it was reported in the late 1980s by experts for the Global Assessment of Soil Degradation (Oldeman et al. 1991). Darker tones indicated more severe land degradation. A band of high to very high soil degradation severity can be seen across the Sahel, as well as similar degradation severity in northern and southern Africa.

The bottom Figure shows a map of the trend in primary production between 1982 and 2000. It was based on temporal analysis of satellite imagery (as described below). Dark tones indicate an increase in vegetation productivity, medium tones a no-change situation, light tones a decrease in productivity over time, and white areas are non-drylands and hyper-arid areas left out of the analysis. This Figure suggests a vegetation recovery since the Sahelian droughts of the 1970s and 1980s. Such a recovery would not be expected in areas of severe soil degradation, because soil degradation reduces the capacity of the soil to absorb and store rainwater, reduces soil nutrients for plant growth, and creates less suitable conditions for seed germination.

It is difficult to reconcile the "greening" of Africa (bottom Figure) in often the very same parts of dryland Africa reported as severely degraded in the 1980s (top Figure). What is more, areas of high degradation in the top Figure do not necessarily coincide with similarly located and shaped areas of decreasing vegetation productivity in the bottom Figure. While these Africa-wide results are preliminary, they are corroborated by more detailed studies for the Sahel region (Nicholson et al. 1998; Prince et al. 1998; Eklundh and Olsson 2003). These results suggest that desertification may be much less pronounced than the GLASOD map suggests, but more detailed analysis of the remote sensing material will be needed, especially in relation to the correlation between rainfall dynamics and vegetation cover response.

Methods: For the bottom Figure, 18 years of continental normalized difference vegetation index data (1982–2000, 1994 excluded) from the NOAA Pathfinder program were analyzed and compiled into a single image. Yearly vegetation productivity (calculated as the sum of NDVI values across the year: iNDVI) was calculated, resulting in a temporal linear regression of productivity over time for each pixel. The result was subsequently smoothed to improve legibility. Hence the image illustrates the general trend in vegetation productivity. Assuming that many areas in the drylands of Africa in the analyzed period experienced an increase in rainfall, some of the slightly darker tones may indicate a no-change situation, and medium-toned areas may indicate a decrease in vegetative response to rainfall. (Nielsen and Adriansen 2005; Nielsen in prep)

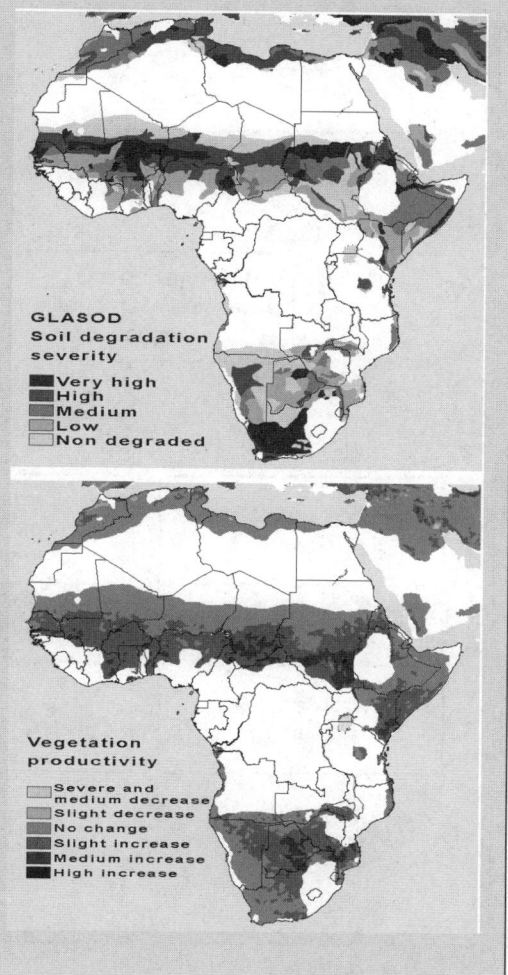

GLASOD
Soil degradation severity

■ Very high
■ High
■ Medium
■ Low
□ Non degraded

Vegetation productivity

□ Severe and medium decrease
□ Slight decrease
■ No change
■ Slight increase
■ Medium increase
■ High increase

---

by intensifying their agricultural practices and enhancing pastoral mobility in a sustainable way (Prain et al. 1999; Niemeijer and Mazzucato 2002; Mortimore and Harris 2004). In these scenarios, population growth does not lead to degradation and poverty but to a Boserupian-style intensification and improved environmental management. (Esther Boserup (1965) analyzed different trends of technological development of countries and continents over centuries and concluded that population growth provides the impetus for technological change. She found that the increased need for food and land scarcity caused by population growth was commonly countered by an intensified use of technologies in which more labor was used in conjunction with land improvement technologies.)

There is, for example, a mounting body of evidence that in the African Sahel region, once considered the centerpiece of the desertification paradigm, land users are achieving higher productivity by both intensifying and improving their land management practices—capitalizing on improved organization of labor, more extensive soil and water conservation, increased use of mineral fertilizer and manure, and new market opportunities (Scoones 2001; Niemeijer and Mazzucato 2002; Tiffen and Mortimore 2002; Mortimore and Harris 2004).

These studies have shown that yields per hectare and food output per capita and livestock sales are largely determined by policies and market opportunities within the constraints posed by the natural environment. It is also argued that population growth

is not the overriding driver of either desertification or sustainable land management, but that the impact of population growth is largely determined by the rate of change and the way in which people adjust to their increasing numbers, mediating the effect on the environment and their own well-being through adaptations of local informal institutions, technological innovations, income diversification, and livelihood options and strategies (Mazzucato and Niemeijer 2002; Mortimore and Harris 2004).

The message of the counter-paradigm is that the interacting direct and indirect drivers combined with the local situation can create a range of different outcomes and that raising a general alarm based on questionable scientific evidence in the end is much less effective than identifying individual problem areas where large influxes of refugees or other complicating factors have led to an unsustainable local response. It is also crucial to distinguish between problems originating from the natural harsh and unpredictable conditions of dryland ecosystems and problems caused by unsustainable management of the environment, since the remedies will often be different.

Figure 22.11 shows the interrelationship between the two paradigms. The desertification paradigm focuses only on the negative interactions (left side of figure), whereas the counter-paradigm allows for both negative (left side of figure) and positive interactions (right side), depending on how humans respond to the direct and indirect biophysical and anthropogenic drivers. The counter-paradigm offers a much more flexible approach in

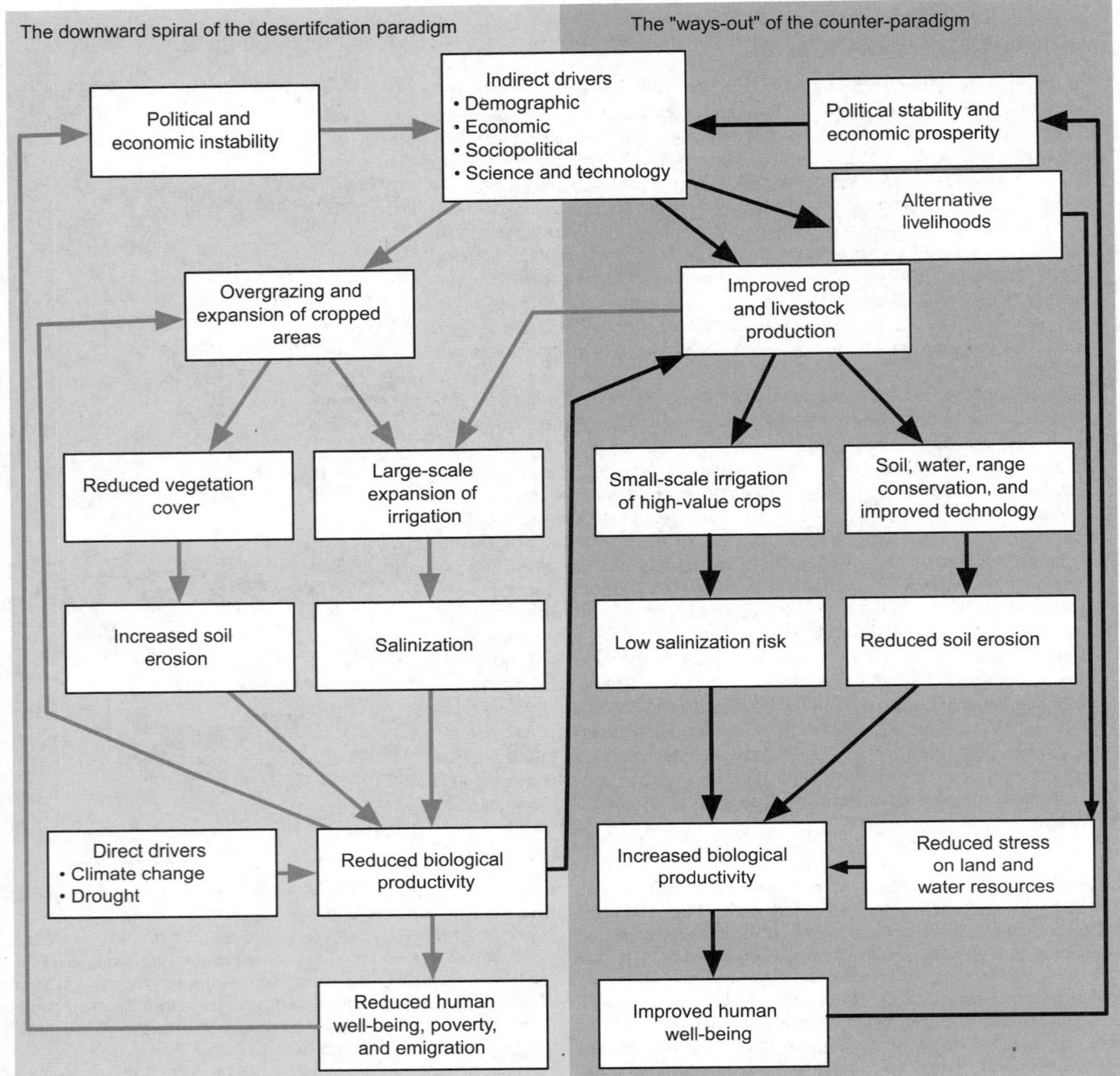

Figure 22.11. Desertification Paradigm and Counter-paradigm. The schematic shows the inter-relationships between the two paradigms: The desertification paradigm focuses only on the negative interactions (left side of figure), leading to a downward spiral of desertification. The right side shows the counter-paradigm, which entails developments that can help avoid or reduce desertification. In the counter-paradigm, land users respond to stresses by improving their agricultural practices on currently used land. Both development pathways occur today in various dryland areas.

that it allows for multiple sustainable development pathways and does not impose a single, intervention-based development model as the only way out.

22.4.2 Socioeconomic and Policy Drivers

22.4.2.1 Policy Drivers—Successes and Failures

Top-down policies with minimal participation of local communities often lead to land degradation, whereas policies encouraging

participation and local institutions can induce a sustainable intensification of primary production. For example, a study of eight countries in West Asia and North Africa (Hazell et al. 2002) revealed that past agricultural policies favoring only rich farmers promoted agricultural growth that led to environmental degradation. On the other hand, policies that emphasize risk-reducing strategies, that secure property rights, and that take into account both technical and socioeconomic constraints do ensure adequate incentives for participation in resource management and can thus avoid degradation (Sanders et al. 1996; Pender et al. 2001; Hazell et al. 2002). Community-based land use decision-making and social networks also contribute to the success of non-degrading agriculture in drylands (Mazzucato et al. 2001).

There are also examples for the Boserupian "induced innovation" triggered by increasing needs and sustained by the community's investments in agricultural infrastructure, markets, and conservation practices; in the Machakos (dry subhumid Kenya), demand from the nearby growing city of Nairobi since the 1930s has motivated farmers to restore degraded lands by terracing, penning, and feeding livestock; manuring croplands; planting high-value trees; adopting labor-saving plows and new maize varieties; and creating small-scale water harvesting and irrigation structures. In Gombe (semiarid to dry subhumid north-central Nigeria), farmers repeatedly changed their livelihoods base from livestock to sorghum/groundnut, to cotton, and to maize as market access and prices shifted and as fertilizer and new cultivation technologies became available through government programs since the 1940s (Tiffen and Mortimore 2002). In these two cases, market demand increased returns on agricultural production, which stimulated development toward a more labor-intensive system in which labor-intensive soil and water conservation practices paid off through higher yields and better prices at the market.

However, there are other situations where induced innovation has not occurred and higher population has increased pressure on the land, which has not been compensated for by more investment in soil and water conservation measures (Lopez 1998; Kates and Haarmann 1992). A comparison of these successes and failures suggests that critical drivers for successful indigenous dryland rehabilitation include access to technologies that can increase labor and land productivity faster than population growth and access to markets (Pender and Kerr 1998; Pender et al. 2001).

Next to intensification, diversification away from a small list of regular crops or mono-cropping systems provides economic sustainability as well as maintenance of service provision. Trees interspersed with crops provide year-round ground cover and livestock feed, protecting soils from the degradation spiral. A greater diversity of crops also provides a more consistent stream of employment and income to the farm year-round, reducing the incentive for out-migration and hedging against short-term drought risk. Higher-value products provide farmers with the income needed to reinvest in soil nutrient replacement, in turn building soil organic matter. In some areas, such practices have traditionally been an integral part of the system; in other areas, they are currently being stimulated through development interventions.

22.4.2.2 Demographic Drivers

Population is an important driver; however, population growth projections for drylands can be confounded by the impact of socioeconomic and health problems, like HIV/AIDS. For example, in Botswana, which is mostly semiarid and where one in three adults is reported to be HIV-positive, a 20% decline in population is predicted between 2000 and 2050 (UNFPA 2003). This situation could also apply to other dryland countries in Africa. Rapid demographic changes—increases or decreases—make planning resource management more problematic.

Migrating populations can be a source of additional pressure on dryland environments and on resource management when livestock temporarily concentrate at key resources such as water points. Under these circumstances, conflicts over water often arise between nomads and farmers (as in the dry subhumid part of Tanzania) (Mbonile 2005). A transition between migration as a temporary livelihood strategy to permanent migration creates additional pressure on drylands, as described later.

Nomadism can be described as a rangeland management practice that over the centuries has proved to be sustainable and within the carrying capacity of drylands. Sedentarisation of nomads and other policies and infrastructure that promote farming in rangelands at the lower limit of cropping viability can act as drivers of land degradation. The concentration of human and livestock populations in particular areas can reduce the ability of nomads to adjust their socioeconomic activities in the face of stresses such as droughts, and a Convention on Nomadic Pastoralism to protect pastoralists' rights and empower them economically, socially, and politically has been suggested (CENESTA 2002). Sedentarization under the Tribal Grazing Land Policy in Botswana has not yet caused large-scale environmental degradation, but it has reduced the resource base and options for both environmental and societal resilience to natural environmental variability (Thomas et al. 2000). Similarly, in Kenya's semiarid Laikipia District, sedentarization of the previously nomadic population in a dryland wetland placed the people in an escalating human-wildlife conflict (Thenya 2001).

22.4.2.3 Land Tenure Policies

Land tenure practices and policies in drylands can also act as indirect drivers of land degradation. When farmers and herders lose control or long-term security over the land they use, the incentives for maintaining environmentally sustainable practices are lost. Problems of water scarcity, groundwater depletion, sedimentation, and salinity can all be symptoms of deeper policy and institutional failures, including a lack of well-defined, secure, tradable property rights (Ahmad 2000). According to this argument, it is essential that people perceive that they have secure ownership over local natural resources for management to be effective. However, security of tenure need not imply systems of private property rights. For instance, long-established collective and community-based management of village tank systems has been more effective than the current proliferation of privately owned boreholes (Gunnell and Krishnamurthy 2003).

22.4.2.4 Water Policies

Water policies are relevant to many provisioning and supporting services in dryland areas. These policies include allocation systems, pricing, government investments in water resource development, and priorities in conservation measures. Water allocation for irrigation has caused degradation in some dryland areas where flows in semiarid rivers used for irrigation, such as the River Ord in Western Australia, are highly variable and unpredictable. Therefore the proportionate water release strategies have been found to be unsuitable and to cause detrimental effects to the riverine ecosystem (Dupe and Pettit 2002). Irrigation policy decisions also depend on other factors, such as water availability and pricing and anticipated crop prices (Norwood and Dumler 2002).

Increasing water scarcity and degradation of quality are also linked to water sharing between upstream-downstream riparian users (Lundqvist 1999). A frequent policy focus on the aggregate availability of water—more specifically, the ratio between the number of people in a country or region and the amount of water that is naturally available—hides how much of that water can and is withdrawn and used by different people (Lundqvist 1999). Therefore, a shift from the mindset of resource development to one of resource management and conflict resolution is more useful (Shah et al. 2001).

Institutional reforms such as pricing of water have been slow to materialize due in part to strong political interest groups resisting policy changes in the water sector (Ahmad 2000). The National Water Act of South Africa is an example of legislative innovation attempting to address these issues (Kamara and Sally 2003).

22.4.2.5 Governance Approaches

Central or local government investments in infrastructure and accessibility to credit can also influence sustainability or vulnerability of dryland livelihoods as well as determine human well-being in these areas. Large-scale government-driven projects can facilitate the sustainable development of drylands, as seen in developed dryland areas in Israel, California, and Australia. But if inappropriately designed, implemented or managed, they can lead to desertification, as in the Aral Sea region. Around the Mongolian-Chinese borderland, increasing cooperation with the Chinese government through development projects has led to both economic benefits for the Mongol population and to changes and homogenization of land forms, increasing sand dune encroachment and vulnerability (Jiang 2004; Brogaard and Zhao 2002).

The failure of African governments to devolve power to affected people and to link environmental degradation to economic policy has been seen by some as a significant drawback in combating desertification and drought (Darkoh 1998). As a result of these failings, many programs lack local support or are undermined by conflicting trade and agricultural policies pursued by governments.

22.4.2.6 Economic Drivers: Local Markets and Globalization

International, national, and local market dynamics and private and public-sector financial flows are treated as indirect drivers of pastoral and agricultural practices of dryland people, driving either sustainable use of dryland natural resources or desertification. Regarding globalization, the increasing focus on raising production for exports in Ghana (mostly semiarid) and Mexico (more than half of the country arid to dry subhumid), for instance, has led to increasing degradation (Barbier 2000). The negative impacts from increased access to markets challenge the conclusion of Zaal and Oostendorp (2002) that much of the explanation for the successful intensification in the Kenyan Machakos may be attributed to access to enlarged markets.

As far as local markets are concerned, these drive livestock management decisions and determine the effects that land degradation and droughts have on human well-being (Turner and Williams 2002). Local markets for off-farm labor also influence farm-level resources and resource management decisions, particularly regarding the use of fertilizers and land improvements (Lamb 2003; Pender and Kerr 1998).

22.4.3 Biophysical Indirect and Direct Drivers

22.4.3.1 Water Use

Freshwater resources like lakes, rivers, and aquifers are essential to the transition of rangelands to croplands by providing fresh water for irrigation. The intensity of this driver decreases with distance of the dryland from the source. The proximity of fresh water generates interventions in water transportation infrastructure, which accelerate the use of the provisioning services. These interventions can cause a degradation of several dryland services.

22.4.3.2 Global Climate Change

Anthropogenically induced global warming has been detected in the last 50 years (IPCC 2001). Dryland-specific and comprehen-sive information and predictions for the dryland system are not readily available, but it can be inferred that global warming has driven climate changes that have also affected many drylands. These may include the 0.3% rainfall decrease per decade during the twentieth century between 10° N to 30° N; the 2–4% increase in the frequency of heavy precipitation events in mid-latitudes of the Northern Hemisphere over the latter half of the last century; the more frequent, persistent, and intense warm episodes of the El Niño-Southern Oscillation phenomenon since the mid-1970s; and the increased frequency and intensity of droughts in parts of Asia and Africa in recent decades. These trends are expected to continue, whereas precipitation will either decrease or increase in different regions (IPCC 2001).

The combination of global climate change induced by anthropogenic emissions of greenhouse gases and the fact that carbon dioxide is both the most significant greenhouse gas and an important ingredient for primary production constitutes a potential driver of dryland services. The service of biological productivity, on which so many dryland peoples directly depend, is most sensitive to this combined driver.

The water deficit by which dryland primary production is constrained is caused by low precipitation but also high evaporative demand of the dryland atmosphere, which makes plants lose water each time they open their stomata to let in carbon dioxide, a raw material for primary production. With increased CO_2 concentration in the air, plants shorten the time of stomatal opening, thus reducing water losses, or they maintain transpiration rate but increase overall production, made possible by the increased CO_2 concentration. Furthermore, rainfall may locally increase due to climate change, which too may promote primary production. However, increased temperatures may be above the optimum for dryland plants and may also increase evaporation from soil surfaces, hence reducing soil moisture and even negating possible increases in rainfall. Should plant cover decline, the service of water regulation and hence also primary production will be disrupted. Modeling projects decreases in grain and forage quality in the drylands (IPCC 2001).

Climate change is also likely to drive changes in the water provision service through reduction of water quality and due to increased solubility of minerals with the temperature increase. Since global climate change is expected to increase the intensity of rainstorms, this together with the reduced plant cover will increase the incidence of flashfloods. These increased freshwater flows (Mirza et al. 1998) may offset the water quality degradation but increase soil erosion. Also, the projected higher frequency of dry spells might encourage dryland farmers to increase water withdrawals for irrigation. Since sea level rise induced by global warming will affect coastal drylands through salt-water intrusion into coastal groundwater, the reduced water quality in already overpumped aquifers will further impair primary production of irrigated croplands.

Rangelands will be affected too, by projected changes in grassland/shrubland boundaries due to climate change driving changes in plant community composition (Sala et al. 2000). On the other hand, dryland scrublands and woodlands, used mostly for livestock grazing, will be affected by a greater frequency and extent of fire (Howden et al. 1999). Climate change will also increase habitat fragmentation and thus detrimentally affect dryland biodiversity (Neilson et al. 1998).

Overall, climate change is expected to exacerbate desertification (Schlesinger et al. 1990). Furthermore, it is conceivable that it might amplify the potential negative effects of an existing management regime on the services of interest, increase the risks of land degradation, and raise the cost of intervention and reversal

(Fernandez et al. 2002). However, climate change is expected to have different effects on the various dryland subtypes; Canziani et al. (1998) suggested that since plants and animals of fluctuating environments are better adapted to environmental change, the adaptability of biodiversity to climate change will increase with aridity, since the drier dryland subtypes are also environmentally less stable than the less dry subtypes.

22.4.3.3 Floods

Drylands are characterized by low, unpredictable, and erratic precipitation. The expected annual rainfall typically occurs in a limited number of intensive, highly erosive storms. This produces overland flows that usually develop into violent floods. These floods can be a major driver of soil erosion and soil loss, and the dry spells between storms increase the risk of crop failure. However, these floodwaters can also replenish freshwater resources, deposit fertile minerals and organic debris, and recharge groundwater or the soil profile.

The prevalence of flash floods in drylands typically leads to a number of responses from farmers directed at storage of runoff and flood water, mainly for increasing crops and forage. These include using catchments of up to several hectares (Pacey and Cullis 1986) with or without mechanical or chemical treatment to reduce infiltrability (UNEP 1983); creating micro-catchments of several square meters around a single bush or tree; cultivating wadis that are naturally flooded following rainstorms; spreading the water over extensive tracts to reduce the kinetic energy and enhance infiltration; constructing diversion channels, stone or earth bunds, and even wood bunds to irrigate farmlands of hundreds or even a few thousands hectares; and combinations of several of these techniques (Reij et al. 1988; van Dijk 1995; Niemeijer 1999). Runoff farming is suitable in arid and semiarid areas where direct rainfall is too low for cropping, but in dry subhumid areas it would lead to extensive periods of waterlogging, causing yield reduction. In many of the drier areas floodwaters are also used to recharge wells and fill basins used for drinking water for livestock and humans or for some dry season gardening.

22.4.3.4 Fires

Natural and induced fires are drivers of land cover, soil condition, and biodiversity, especially in the dry subhumid and semiarid dryland subtype. With respect to soil condition, nitrogen and organic carbon are largely lost to the atmosphere or converted to inert forms (charcoal) by fire, while soil erodibility increases for a period after the event. Thus highly recurrent fires can lead to soil degradation.

Historically, land use and management changes have modified the temporal and spatial patterns of fire occurrence and intensity, with strong consequences on soil fertility and the composition of the vegetation it supports. In general, traditional land users maintained a fine-grained spatial pattern of small fires with low on-site recurrence, such as the aboriginal fire management in Australia (Griffin and Friedel, 1984a, 1984b, 1985), which could be extrapolated to other semiarid and dry subhumid dryland.

Dryland fires are often controlled by grazing and browsing of either wild herbivores or livestock. The twentieth-century commercial agricultural and stock breeding systems as well as wildlife management regulations led to widespread fire prevention together with overstocking of rangelands. The outcome has been a new pattern of larger patches of higher-intensity fires, which is claimed to be one of the triggers for grass-shrub transition (Scholes and Hall 1996). World carbon emissions from savanna burning are estimated at 0.87 billion tons of carbon per year

(Scholes and Hall 1996). In the northern Mediterranean basin, the burned area has been increasing at an annual rate of 4.7% since 1960 due to vegetation regrowth after agricultural abandonment (Le Houerou 1992).

22.5 Trade-offs, Synergies, and Interventions

This section compares and contrasts the major options available for drylands management. Each category of options is assessed in terms of trade-offs, as far as gains and losses of services with regard to their impact on human well-being; synergies, where one type of management option leads to multiple benefits; and vulnerability (losses greater than gains and no synergies) and sustainability (losses are equal to or smaller than gains).

22.5.1 Traditional Dryland Livelihoods

22.5.1.1 Woodland–Rangeland–Cropland Trade-offs and Synergies

Historically, dryland livelihoods have been based on a flexible combination of hunting, gathering, cropping, animal husbandry, and fishing. Archeological records and anthropological studies have revealed shifts in livelihood strategies over time in the same location and often involving the same cultures. As a consequence, land use changes both in time and space as an adaptation to new economic possibilities, in response to environmental or climatic changes, or as a result of war or drought-induced migration (Robbins 1984; Berry 1993; Niemeijer 1996). Land use changes are thus both responses to changes in the provision of ecosystem goods and services and drivers of changes in this provision.

Population increase drives a growing tension between pastoral rangeland and cultivated land use. This can lead to intercultural conflicts and service degradation as herders and farmers claim access to and use of the same land (van Driel 2001). Depending on annual rainfall, supplemental or full irrigation may be introduced for conversion to cultivated systems, often requiring capital investment by governments or farmers. In the long run, trade-offs between the two land uses can also lead to a tighter cultural and economic integration, with herders cultivating more land, farmers holding more livestock, and an increased exchange of services (Breusers et al. 1998; Mazzucato and Niemeijer 2003).

Woodlands are a source of wild fruits, edible plants, and wildlife that can be important sources of food and off-farm income—vital during years of drought. Woodlands also often have cultural and religious significance for the local population, which protects them to some degree against overexploitation. When woodlands are transformed to croplands, the tree volume decreases. In many traditional systems, however, certain trees and shrubs are not removed because their fruits, leaves, or other products are used for consumption or medication or are traded, as described earlier. The species composition changes and the soil cover decreases following clearance, especially during the dry season, when no crops are grown. The transformation negatively affects regulation and cultural services and reduces biodiversity, but at the same time it increases food production and creates multiple livelihood opportunities.

Woodlands that are increasingly used by livestock (and often managed with fire) may develop into rangelands with a reduced tree cover and increased grass or shrub cover. Over time, the species composition changes as a result of grazing, browsing, and fire (more fire-resistant species become dominant, for instance). For herders, economic productivity increases, but for hunters and gatherers the changes in species composition and the reduced habitat for wildlife negatively affects their livelihood. On a larger scale, the disappearance or transformation of woodlands reduces the service of supporting biodiversity by eliminating corridors for migration and refuge from predation and disturbances.

In semiarid and dry-subhumid areas, different land uses meet and there is the greatest potential for both trade-offs and synergies. Afforestation, silvipastoral, and agropastoral systems develop in response to population growth, environmental changes, and economic and political developments. Natural biodiversity is replaced by agrobiodiversity, where different species and landraces of livestock and crops are introduced. Synergies are found in mixed farming practices, where a single farm household combines livestock rearing and cropping (Slingerland 2000).

Synergies are also found where different households or communities engage in either livestock herding or farming and trade food and services. Such interactions can decrease livestock pressure on rangelands through fodder cultivation and provision of stubble to supplement livestock feed during forage scarcity. At the same time, cultivated systems benefit from manure provided by livestock. Many West African farming systems are based on this kind of integration of pastures and farmland (Prudencio 1993; Steenhuijsen 1995; Mazzucato and Niemeijer 2000a; van Driel 2001). Growing crops in the most fertile areas and grazing livestock on the less fertile land can optimally exploit spatial and temporal variability in service provision.

When used infrequently and controlled, fire plays an important role in the management of most pastoral and cropping systems. Pastoralists use controlled fire to get fresh shoots that are more digestible for their livestock. Farmers use fire to clear new land or old fallows for cultivation and in some cases also to remove the remaining crop residue at the onset of the wet season. In both cases, the use of fire promotes nutrient cycling essential for maintaining the productivity of rangeland and cultivated land. Although fire provides a temporary boost to the provisioning services, carbon and nitrogen are lost to the atmosphere, and the excessive and improper use of fire can also lead to land degradation.

22.5.1.2 Use of Water Resource for Cultivated Drylands

The relative scarcity of dryland water dictates trade-offs in land use and often creates competition and conflicts between different riparian users, as well as upstream-downstream conflicts. For example, dryland-crossing rivers provide drinking water, irrigation, and navigational uses for multiple countries. Water abstracted for irrigation often conflicts with the downstream needs of wetland areas of coastal or inland deltas. On a more local scale, farmers and pastoralists may compete for use of water between irrigation and livestock use. These competitive uses of water resources pose political, economic, and ecological conflicts and trade-offs that sometimes go back to ancient time.

Effective water harvesting and water conservation practices reduce runoff and erosion and increase crop performance, but they also reduce downstream water supply. This reduction in runoff and erosion generally reduces flooding and siltation of fields and water ways. More significant trade-offs need to be considered when runoff farming and water harvesting include intentional clearing of the catchment areas, by removing vegetation and sometimes using artificial coatings to seal the soil. This helps increase the amount of runoff that can be harvested from the catchment, thereby increasing water availability in the run-on area for crop production or livestock watering. However it leads to a loss of ecosystem services for the catchment area itself. Such catchment clearing is labor-intensive and not common in traditional

runoff farming and soil and water conservation practices, except in the arid and hyper-arid regions (Pacey and Cullis 1986; Reij et al. 1996; Bruins et al. 1987).

22.5.2 Drylands and Other Systems

Six MA-defined systems overlap with the dryland system, some of which overlap with each other. Cultivated systems (44% of dryland area), inland water systems (rivers and wetlands, 28%), coastal systems (9%), mountain systems (32%), forests (12%), and urban systems (2%) are all embedded in drylands. These systems generate services for the drylands in which they are embedded, and often feedback loops and even synergies develop between the drylands and the systems embedded in them. The trade-offs and synergies between cultivated systems in drylands and other land uses are discussed in previous sections. The linkages of dryland systems with coastal and mountain systems are important and are discussed in the related chapters. Inland waters and urban systems are assessed in this section.

22.5.2.1 Inland Waters and Wetlands

Inland water systems—rivers, canals, lakes, and wetlands—are an integral component of the dryland system and relate to provision of many services, including freshwater provision and supporting biodiversity. Rivers in drylands often feed dryland freshwater lakes (such as the Aral Sea and Lake Chad), landlocked salty lakes (the Dead Sea and Salt Lake in Utah), or end up in dryland deltaic (such as the Mesopotamian marshlands) or landlocked marshes (the Okavango, for example). There is a major trade-off and potential for conflict here: increasing abstraction of water from rivers is essential for agricultural production but it reduces the quantity and quality of water reaching lakes and marshes, leading to reduction in surface area and increase in water salinity.

For example, the Aral Sea lost about 60% of its 68,000 square kilometers between 1960 and 1998, a figure expected to rise to about 70% by 2010, and its salinity increased from 10 to 45 grams per liter (DFD 1996). This resulted in the total collapse of the 44,000 annual tonnage of the lake-based fishing industry and in wind transport of the salt and pesticide-laden soil particles to other parts of the Aral Sea basin, with severe health effects on the population (DFD 1996).

The large-scale abstraction of water for irrigation is directly expressed in shrinkage of wetlands; for example, the Mesopotamian marshlands lost 89% of 20,000 square kilometers between 1970 and 2000, which also affected fisheries of the Persian Gulf (UNEP 2001). Similarly, Lake Chad lost 95% of its 25,000-square-kilometer area from 1963 to 1997 due to irrigation appropriation and climate (USAID/FEWS 1997; ITAP 2003). Such changes could have a significant impact on the 180 fish species in Lake Chad, which are the second most important source of household income in the Nigerian drylands (Sarch and Birkett 2000).

Water abstraction also affects the provision of other services of dryland wetlands, including nutrient cycling, primary production, soil formation away from wetlands, provision of food (both animal and wild food plants) (Brouwer and Mullié 1994), provision of fuelwood and biochemicals, climate regulation through evaporative cooling, and removal of several pollutants from water. Many dryland wetlands are critical for survival of cross-desert migratory birds. Wetlands also provide cultural services (spiritual services and tourism, for instance, as described earlier).

22.5.2.2 Urban Systems

The proportion of the global population living in urban areas is expected to increase following a historical trend, with the urban fraction increasing to around 52% by 2010 (see Chapter 27) and to 60% by 2030 (UN 2002). This projection implies that nearly all the population growth over the next three decades will occur in urban areas. Such growth has consequences for drylands, depending on whether it is undertaken with proper planning and provision of services, infrastructure, and facilities. In hyper-arid drylands, a much larger fraction of the population is urbanized. (See Figure 22.12.) This may be a result of concentration of livelihood opportunities and better living conditions in otherwise harsh settings (CIESIN 2004). Overall urban population density increases with decreasing aridity, accompanied by a decreasing per capita income (expressed by GNP, calculated as dollars per capita). Such correlations with aridity can be linked to the reduction of provision of services with increasing aridity. On the other hand, the "ecological footprint" or impacts of dryland urban centers on adjoining rangelands and cultivated systems cannot yet be assessed.

22.5.2.3 Systems Away from Drylands—Dryland Dust

Drylands also affect non-dryland areas, indirectly at the sociopolitical level (through environmental refugees, for example, or immigration), and in various direct ways, such as the dust particles carried by winds from the drylands. Dust from the Gobi desert is carried to the Pacific coasts of America, and Saharan dust is carried to the Caribbean islands (Prospero and Nees 1986) and the Amazon basin (Swap et al. 1992). Chemical contaminants and bacterial and fungal spores adhere to the surface of these dust particles, which can be hazardous to people and are suspected to have already affected organisms of the Caribbean coral reefs (Smith et al. 1996). It is hypothesized that this and other recently emerging coral reef degradation episodes are coincidental with the desertification-driven increased frequencies of Saharan dust storms (Shinn et al. 2000).

22.5.3 Climate Change and Carbon Sequestration

During the last century, global drylands have experienced anthropogenically induced climate changes that are predicted to continue and even to accelerate during the present century, as noted earlier. Drylands ecosystems contribute carbon emissions to the atmosphere (0.23–0.29 billion tons of carbon a year) as a result of desertification and related vegetation destruction, through increased soil erosion and a reduced carbon sink (Lal 2001b). This latter effect is expected to intensify with climate change, but if

they are properly managed, dryland systems have the potential to function as a carbon sink.

Lal (2001b) estimated the potential of dryland ecosystems to sequester up to 0.4–0.6 billion tons of carbon a year if eroded and degraded dryland soils were restored and their further degradation were arrested. Furthermore, Lal also pointed out that through active ecosystem management, such as reclamation of saline soils and formation of secondary carbonates, carbon sequestration can be further enhanced. This will add sequestration of 0.5–1.3 billion tons of carbon a year; similar magnitudes of potential carbon sink capacity of dryland ecosystems have been estimated by Squires et al. (1995) on a global scale. This restoration and enhancement of dryland condition, if undertaken at a global scale, could have a major impact on the global climate change patterns.

A significant change in the direction of national and local policies would be needed to implement such restoration and enhancement in the carbon sequestration service. Knowledge gaps also need to be filled by collecting information on credible rates of the extent and severity of soil degradation at different spatial scales; biotic and soil carbon pools and fluxes; the impact of land use changes and desertification on the carbon sequestration dynamics; and the cost-benefit ratio of soil improvement and carbon sequestration practices for small landholders and subsistence farmers in dryland ecosystems.

However, there are also numerous hidden costs of enhanced soil carbon sequestration that must be considered (Schlesinger 1999). Such enhancements require the addition of mineral or organic fertilizer (especially nitrogen and phosphorus) and water, which would need significant capital investment. An incentive to further enhance the natural condition of the service even with the costs involved and at the expense of other services is that drylands can be instrumental in counter balancing the increased anthropogenic emissions of carbon dioxide to the atmosphere. In the context of the Clean Development Mechanism proposed by the Kyoto Protocol, developing countries can attract investments from industrial countries and multinational industries. Such projects are expected to generate income through emerging international carbon trading (World Bank 2003a, 2003b), which may offset the expenses. The long-term impact of such projects in developing countries depends on the future of the Kyoto Protocol and the Clean Development Mechanism.

22.6 Human Well-being in Dryland Systems

22.6.1 Indicators of Human Well-being in Drylands

In general, the human well-being of dryland peoples is lower than that of people in other MA systems (see Chapter 5). Dryland peoples have the highest infant mortality rates, and their economic condition (as expressed by the GNP per capita) is the lowest. (See Figure 22.13.) Though the two factors may be linked, the question remains as to what drives the relative and absolute low human well-being in drylands.

The MA defines human well-being as a composite of the basic materials for a good life, freedom and choice, health, good social relations, and security (MA 2003) and implies that these are directly or indirectly linked to the availability of ecosystem services. One hypothesis is therefore that HWB in drylands is low because the natural rate of provision of ecosystem services is inherently low. Hence, the relatively low rate of water provision not only reduces biological productivity on which most dryland peoples depend, it also restricts people's access to clean drinking water and adequate sanitation, thus worsening their health.

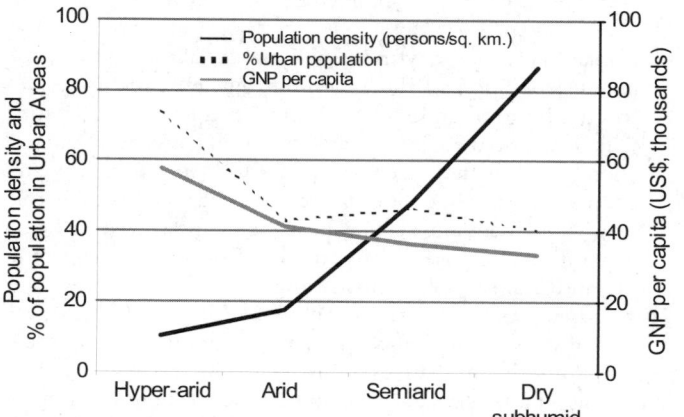

Figure 22.12. Impacts of Urbanization and Population Density on Income Levels in Drylands (CIESIN 2004)

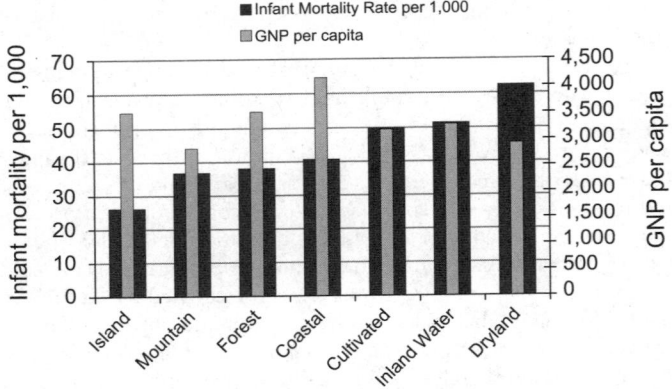

Figure 22.13. Comparison of Infant Mortality Rates and GNP per Capita across the MA Systems in Asia (CIESIN 2004)

Another hypothesis is that land degradation in drylands further reduces the natural rate of service provision, which either drives or exacerbates HWB in the drylands. The degree to which these dryland communities are reliant on dryland ecosystems to sustain them (Webb and Harinarayan 1999; Nyariki et al. 2002) and the extent to which their relatively low HWB is linked to services' failure can be assessed by extending the analysis to subtypes of drylands, with the underlying assumption that the level of aridity is associated with the quality of ecosystem services as well as to the level of stress or degradations observed in drylands. (See Figure 22.14.)

This correlation manifests itself when observable well-being indicators are measured. The hunger rate in children under the age of five and infant mortality rates in drylands demonstrate a clear linkage to the level of aridity. It can be argued that semiarid areas are worse off in terms of human well-being as a result of a high degree of sensitivity and high degree of pressure, which also generate the highest degree of land degradation. The region-to-region variability is also significant, with drylands in sub-Saharan Africa and Asia lagging well behind drylands in the rest of the world, whereas the GNP per capita in OECD dryland countries exceeds that of dryland countries in other regions almost by an order of magnitude. This is not surprising, considering that economic performance relates to many other governance and macro- and microeconomic factors. Thus the economic status of dryland societies is not entirely linked to the low availability of basic ecosystem services such as water and biological productivity. This section assesses the contribution of ecosystem services versus socioeconomic factor to components of HWB in the drylands.

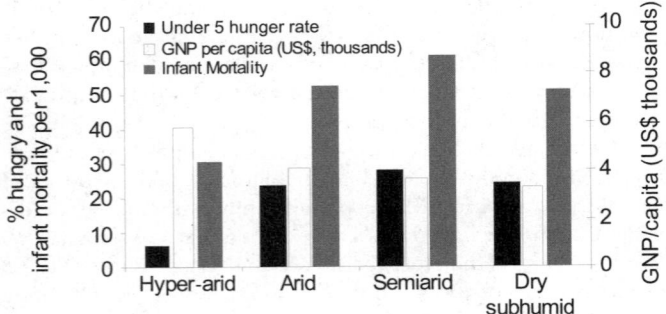

Figure 22.14. Human Well-being Statistics by Dryland Subtypes (OECD countries excluded)

22.6.2 Human Well-being Components in Drylands

The MA conceptual framework (MA 2003) broadly identifies five key components of human well-being.

- *Basic materials for a good life.* The low biological production of drylands, constrained by water, is the ecosystem condition that limits the provision of basic materials for a good standard of living. This also limits the livelihood opportunities in drylands and often leads to practices, such as intensified cultivation, that cannot be serviced due to low and further impaired nutrient cycling and water regulation and provision, requiring adjustments in management practices or the import of nutrients and water provided by services of other ecosystems.

- *Health.* The two key factors contributing to poor health in drylands are malnutrition and limited access to clean drinking water, again reflecting their low biological production and water provision. In Asia, for example, the fraction of children under the age of five facing hunger is 36% in drylands compared with 15% in the forest and woodland system (averaged for the different subtypes of each of these systems) (WHO 2004). However, poor health is also exacerbated by poor health-related infrastructures.

- *Good social relations.* The quality of social relations can be gauged in terms of social strife (wars and political upheavals) and refugees. Environmental refugees leave their homes due to environmental degradation and lack of viable livelihoods. And the sale of stock, wage labor, borrowing of cash for food, and the sale of valuables all precede their migration (Black 2001). Other categories include people displaced for political reasons that may affect the availability of services in drylands to which they have been relocated. Thus demography and sociopolitical drivers, more than the direct condition of ecosystem services, contribute to the quality of social relations.

- *Security.* Food security is an essential element of human well-being in drylands and is related to socioeconomic marginalization, lack of proper infrastructure and social amenities, and often the lack of societal resilience. Climatic events like prolonged droughts and excessive floods also drive insecurity in drylands. But sociopolitical drivers like land tenure practices that relate to sharing and conservation of natural resources or that generate land cover change that may limit traditional pastoral livelihood opportunities can greatly affect food security. Snel and Bot (2002) proposed a list of socioeconomic indicators of land degradation, including indicators to reflect insecurity caused by land degradation. Other security attributes, however, can be only indirectly related to the condition of the ecosystem.

- *Freedom and choice.* Dryland people are not affected just by the unique condition of the ecosystem but are also further restricted by local and global political and economic factors, which can exert major limitations on their freedom and life choices. With the exception of OECD countries, dryland peoples are mostly politically marginalized; that is, their role in political decision-making processes is perceived as being insignificant. Consequently, market factors that determine dryland farmers' decision-making and the effects on their well-being are often critical (Turner and Williams 2002).

This discussion demonstrates that provisioning of ecosystem services is not the only driver of low HWB in drylands. The lack of a strong HWB-service correlation can be attributed to a number of exogenous factors, including political marginality, slow growth of health and education infrastructure, facilities and services, and so on. What needs to be assessed, though, is whether or not there are thresholds in the provision of services that either independently or together with the exogenous factors cause human well-being to drop to a level that crosses the poverty threshold.

22.6.3 The Relative Dependence of Human Well-being on Ecosystems and Socioeconomic Drivers

Natural climatic fluctuations can move the level of service provision to the point of crossing a threshold from a relatively high, stable rate of provision to a lower yet relatively stable state of provision. This can be a temporary transition, but a prolonged drought or overexploitation can exert a persistent impact, resulting in a permanently reduced rate of service provision that does not regain the previous level even when the impact is removed (e.g., Puigdefabregas 1998).

The grassland/scrubland transition is an example of a dryland ecosystem that moves from one relatively stable state with a high provision of forage to a different relatively stable state with a lower provision of forage due to human impact. Even when the impact is removed, the lower-service state persists, as described earlier. Whether such a threshold is associated with a threshold in socioeconomic and political drivers and a resulting significant change in human well-being is not certain. The Sahel drought of the 1980s, for example, produced a devastating reduction in HWB, yet the region reverted to its former state once the prolonged drought terminated. Thus it is likely that the issue of concern is not that of thresholds but of vulnerability to a persistent ecosystem change, coupled with concomitant, resultant, or driving socioeconomic and political changes.

The relative contribution of the condition of the dryland ecosystem and of socioeconomic drivers to poverty attributes—such as per capita rural cereal production, the time spent by household members collecting water and fuelwood, the quantity of annual household consumption that is derived from common lands, the percent of children under five who are underweight, wasted, or stunted, and the percent of the rural population below the poverty line (Shyamsundar 2002)—has not yet been adequately assessed.

22.6.4 Responses to Improve Human Well-being in Drylands

The section explores the response interventions that can mitigate or reverse the effect of degraded ecosystem and service condition on the well-being of dryland people. Experience has shown that locally appropriate interventions can introduce dynamics of sustainability and can improve human well-being by interrupting the vicious circle of poverty leading to overexploitation of services and to environmental degradation and desertification.

Traditionally, dryland societies have been able to cope with their harsh environment through livelihood adaptations over extended periods. A number of commentators have explored the "failure" of dryland farmers in sub-Saharan Africa and elsewhere to develop as quickly as farmers in non-dryland regions during the Green Revolution (Singh 2004). Analyses conducted at the smallholder level have concluded that while overall progress has been compromised by widespread insecurity, significant achievements have been made by dryland farmers due to social resilience, the evolution of knowledge over time, and successful farmer adaptation (Bird and Shepherd 2003; Mortimore 2003). These achievements are contrasted with the failures of non-dryland farmers at comparable poverty levels (Mehta and Shah 2003). This section assesses the relative role of traditional practices and modern technologies in building capacities and public participation for improving HWB in the drylands.

22.6.4.1 Traditional Knowledge

Traditional response options combine tested approaches for resource management based on insights into the local natural and socioeconomic environment with continuous experimentation to deal with changes in that environment (Prain et al. 1999). Such approaches have enabled communities to live in the harsh dryland environments for millennia. In oases, for example, traditional management approaches are based on the appropriate usage of the physical and geomorphological factors. These include production and distribution systems for water management, architectures that regulate micro-climate, cultivation of salinity-tolerant fruit species such as the date palm, waste recycling systems, and sand dune stabilization techniques.

Traditional methods of water harvesting allow the replenishment of the resource and its long-term availability. They make effective use of local topographic and soil characteristics. In some cases, horizontal underground tunnels drain water from the surface of the groundwater table. Vertical ventilation shafts augment supplies by capturing night-time humidity. Locally adapted architectural innovations are also used to facilitate water conservation by condensing atmospheric water, including stone heaps, dry walls, little cavities, and depressions in the soil, thus allowing the plants to overcome periods of high drought.

Contrary to the literal implication of the term, however, "traditional" knowledge is not static but evolves over time, incorporating elements of local experimentation and integrating new ideas and technologies brought from outside or observed during seasonal or temporary migrations (Mazzucato and Niemeijer 2000b). Local knowledge can significantly contribute to human well-being because it has the benefit of integrating the multiple constraints posed by the natural and social environment in ways that are often lacking with introduced technologies. Where local approaches are failing, it is important to distinguish cases where the technologies themselves are fine but need a more enabling political or socioeconomic environment from those cases where a technological solution is needed in the form of improved local technologies or the introduction of new technologies.

22.6.4.2 Adaptation of New Technologies

The gradual introduction, trial, and development of new technologies have allowed considerable progress to be made in some dryland farming communities (Chapman et al. 1996). The adaptation of new techniques is entirely dependent upon the skill and environmental awareness of dryland farmers (Twomlow et al. 1999). Integration of new technologies with tested management approaches is a measure that can improve human well-being at various levels. Three such examples are:

- Integrated water resource management approaches include consideration of the full extent of water resources available within a catchment. Even in hyper-arid areas, opportunities can often be identified to use additional water sources, such as rainwater, seasonal floodwaters, and wastewaters, to supplement over-reliance on groundwater or variable rainfalls. Cumulative uses of water in a catchment are also considered in order to prevent overuse of water by multiple users over time and the resulting depletion of aquifers.
- Integrated water and farming approaches highlight the importance of considering water management in relation to other factors in the dryland production system. For example, the potential to economically irrigate a crop of turnips to provide

an additional source of nutritional value for lactating dairy cows is considerably affected by sowing dates, soil type, and insect damage (Jacobs et al. 2004).

- Farming systems approaches (e.g., Singh 1998) look into the whole production and farming system for synergies among its components, such as arable cropping, livestock management, alternative land use systems, and management of village commons or degraded lands.

22.6.4.3 Capacity Building and Public Participation

Dynamically evolving local and "traditional" practices into which adapted new technologies are integrated have the potential to build capacity for improving HWB in drylands. However, this potential is not always realized. For example, the Indian National Demonstration Programme, which was coupled to training and visiting system (beginning in the 1960s), was unsuccessful in the drylands of India (Singh 1998). This was in contrast to the Green Revolution that occurred elsewhere, with the instant adoption of the technology package over large areas in high-capacity irrigated regions. Singh (1998) concludes that this was because the dryland farmers were less willing to take risks and invest in the new technologies since they were poorer to start with and suffered from water-related constraints and uncertainties in the production process and because the demonstration approach could not convince them that the technologies were appropriate to their needs. Following such experiences, more-recent approaches to capacity development in drylands have adopted a more participatory approach to capacity development, working with farmers as they develop their knowledge, according to their needs and perceptions, rather than demonstrating to them what they could or should do.

Similarly, participatory learning approaches have been used to facilitate the transfer of technologies. In Australia, for example, increasing use is being made of on-farm experiments for capacity development (e.g., Foale et al. 2004; Lawrence et al. 2000). Public participation in resource-use decision-making in drylands is increasingly seen as a key to demand-management for scarce water resources (Kulkarni et al. 2004) and biodiversity conservation (Solh et al. 2003). Participation in dryland decision-making is generally structured according to the prevailing system of land and water rights. For instance, "community-based" projects in Sudano-Sahelian West Africa have often attempted to improve resource management by spatially delimiting appropriate land uses, strengthening the community's exclusionary powers, and clarifying specific claims to village resources. However, such moves can also increase social conflicts (Turner 1999). Thus, further experimentation and experience in merging "traditional" with advanced knowledge as a tool for capacity building for increasing HWB in the drylands may still be required.

22.6.5 Services, Degradation, and Human Well-being

This chapter is an attempt to present "drylands" and "dryland peoples" not as homogenous entities but as a continuum of ecosystems and their human inhabitants arranged along a global aridity gradient in which life and livelihoods are constrained by water. The magnitude of this constraint determines the makeup of the suite of services provided by the ecosystems and, accordingly, the land uses by people and their respective livelihoods. The chapter also examines the mutual interactions between ecosystems and people across this gradient and explores the degree to which the services provided by the ecosystems are used sustainably or are overexploited.

The findings are that both sustainable use and overexploitation occur, but they depend more on socioeconomic drivers than on the degree of water constraint and the resulting dimensions and qualities of the provided ecosystem services. Thus the greatest pressure on ecosystem services takes place at intermediate aridity and not, as might be expected, in the least arid drylands where population density is highest, or in the most arid areas, where population is lowest. The high overexploitation of services is inferred by physical, biological, and social phenomena—soil erosion and salinization, reduced biodiversity and biological productivity, and reduced income expressed by reduced human well-being—and it is reflected by the highest rate of infant mortality and hunger among children. It can therefore be suggested that where aridity is intermediate, on average, there is a mismatch between the rate of service provision and the intensity of exploitation. But there are deviations from this average, expressed in sustainable use of the ecosystems.

Two interlinked drivers are involved in generating this sustainability. The first one is a selective use of services depending on their divergent condition in different areas. For example, in areas where the quality and provision of cultural services are high, local people choose livelihoods served by cultural services (such as ecotourism in the African savanna). And in areas where the quality of provisioning services is high, people choose livelihoods served by these services (such as food crops in natural or managed desert oases). The second factor that promotes sustainability is the adaptation of sociocultural institutions and practices to the prevailing natural condition of the services and the implementation of policies that recognize the natural constraints and create economic instruments that provide for sustainability of the use of services, combined with the promotion of good human well-being.

This assessment also highlights the significance of dryland biodiversity as a whole rather than just individual, selected species in the provision of every single dryland service. It is therefore implied that livelihoods and human well-being depend on biodiversity just as they depend on services. Thus in addition to protected areas, most dryland management, land uses, and livelihoods that maintain biodiversity in drylands will contribute significantly to the well-being of dryland peoples. Also, the chapter suggests that some natural attributes of the drylands have the potential, already realized in some places, to provide dryland peoples with a competitive edge economically (through "alternative livelihoods," for example). Together with policies based on sociocultural and socioeconomic considerations, all dryland livelihoods—pastoral, farming, and "alternative"—can contribute to alleviation of the current high relative poverty and low human-well being of dryland peoples.

References

Abrahams, A., A. J. Parsons, and J. Wainwright, 1995: Effects of vegetation change on inter-rill runoff and erosion, Walnut Gulch, southern Arizona. *Geomorphology,* **13,** 37–48.

Ahmad, M., 2000: Water pricing and markets in the Near East: policy issues and options. *Water policy,* **2,** 229–242.

Aguiar, M. R. and O. E. Sala, 1999: Patch structure, dynamics and implications for the functioning of arid ecosystems. *Trends in Ecology and Systematics,* **14(7),** 273–277.

Allen-Diaz, B., F. S. Chapin, S. Diaz, M. Howden, J. Puigdefabregas, and M. Stafford Smith, 1996: Rangelands in a changing climate: Impacts, adaptations and mitigation. In: *Climate Change 1995—Impacts, Adaptation and Mitigation,* W. T. Watson, M. C. Zinyowera, R. H. Moss, and D. J. Dokken, (eds.), pp. 131–158.

Amous, S. 1997: *The Role of Wood Energy in Africa.* [online] Forestry Department, Food and Agriculture Organization of the United Nations, Rome,

Italy. Cited 29 October 2004. Available at http://www.fao.org/docrep/x2740e/x2740e00.htm.

Anderson, D., 1984: Dispersion, dust bowl, demography, and drought: The colonial state and soil conservation in East Africa during the 1930s. *Journal of the Royal African Society,* **83 (332),** 321–343.

Arbel, A., I. Segal, O. Yekutieli, and N. Zamir. 1990: Natural ventilation of greenhouses in desert climate, *Acta Horticulturae,* **281,** 167–174.

Archer, S., D. S. Schimel, and E. A. Holland, 1995: Mechanisms of shrubland expansion: land use, climate or CO_2? *Climatic Change* **29,** 91–99.

Arizaga, S., E. Ezcurra, E. Peters, F. Ramírez de Arellano, and E. Vega, 2000: Pollination ecology of *Agave macroacantha* (Agavaceae) in a Mexican tropical desert. II. The role of pollinators, *American Journal of Botany,* **87**(7), 1011–1017.

Ashkenazi, S., 1995: *Acacia trees in the Negev and the Arava, Israel.* Hakeren Hakayemet LeIsrael, Jerusalem.

Ayal, Y., G.A. Polis, Y. Lubin, and D.E. Goldberg, 2005: How can high animal biodiversity be supported in low productivity deserts? The role of macrodetrivory and physiognomy. In: *Biodiversity in Drylands,* M. Shachak and R. Wade, (eds.), Cambridge University Press, Cambridge: pp. 15–29.

Barbier, E. R., 2000: Links between economic liberalization and rural resource degradation in the developing regions, *Agricultural economics,* **23**(3), 299–310.

Barnston, A.G. and P.T. Schickedanz, 1984: The effect of irrigation on warm season precipitation in the southern Great Plains. *Journal of Climate and Applied Meteorology,* **23,** 865–888.

Behnke, R.H., I. Scoones, and C. Kerven, (eds.), 1993: *Range ecology at disequilibrium: New models of natural variability and pastoral adaptation in African savannas.* ODI, London.

Berry, S., 1993: *No Condition is Permanent: The Social Dynamics of Agrarian Change in Sub-Saharan Africa,* The University of Wisconsin Press, Madison, Wisconsin.

Biggs, T. H., J. Quade, and R. H. Webb, 2002: $\delta^{13}C$ values of soil organic matter in semiarid grassland with mesquite (*Prosopis*) encroachment in southeastern Arizona. *Geoderma* **110,** 109–130.

BioBee, 2000: BioBee Biological Systems, Kibbutz Sde Eliyahu. [online] Cited 29 October 2004. Available at http://www.seliyahu.org.il/eBees.htm.

Bird, K. and A. Shepherd, 2003: Livelihoods and Chronic Poverty in Semi-Arid Zimbabwe. *World Development,* **31**(3), 591–610.

Black, R., 2001: Environmental refugees: myth or reality? New Issues in Refugee Research Working Paper No. 34 Journal of Humanitarian Assistance [online] Cited 27 May 2004. Available at: http://www.jha.ac/articles/u034.pdf.

Boserup, E., 1965: *The conditions of agricultural growth: The economics of agrarian change under population pressure,* Allen and Unwin, London.

Breusers, M., S. Nederlof, and T. van Rheenen, 1998: Conflict or symbiosis? Disentangling farmer-herdsman relations: the Mossi and Fulbe of Central Plateau, Burkina Faso. *Journal of Modern African Studies,* **36,** 357–380.

Brogaard, S. and X.Y. Zhao, 2002: Rural reforms and changes in land management and attitudes: A case study from Inner Mongolia, China. *Ambio,* **31 (3),** 219–225

Brouwer, J. and W.C. Mullié, 1994: Potentialités pour l'agriculture, l'élevage, la pêche, la collecte des produits naturels et la chasse dans les zones humides du Niger. In: *Atelier sur les zones humides du Niger.* Proceedings of a workshop, 2–5 November 1994, La Tapoa/Parc du W, Niger. P.Kristensen (ed.). IUCN-Niger, pp. 27–51. [English version: The potential of wetlands in Niger for agriculture, livestock, fisheries, natural products and hunting]

Bruins, J.H., M. Evenari, and A. Rogel, 1987: Run-off farming management and climate. In: Progress in desert research, L. Berkofsky, and M.G. Wurtele, (eds.), Rowman & Littlefield, Totowa, New Jersey, pp. 3–14.

Büdel, B., 2001: Synopsis: Comparative Biogeography and Ecology of Soil-Crust Biota, In: *Biological Soil Crusts: Structure, Function and Management,* J. Belnap and O. L. Lange (eds.), Springer-Verlag, Berlin, pp. 141–154.

Cao, M., S. D. Prince, J. Small and S. J. Goetz, 2004: Remotely Sensed Interannual Variations and Trends in Terrestrial Net Primary Productivity 1981–2000. *Ecosystems,* **7,** 232–242.

Canziani, O.F., S. Díaz, E. Calvo, M. Campos, R. Carcavallo, et al. 1998: Latin America. In: *The Regional Impacts of Climate Change: An Assessment of Vulnerability. Special Report of IPCC Working Group II* [Watson, R.T., M.C. Zinyowera, and R.H. Moss (eds.)]. Intergovernmental Panel on Climate Change, Cambridge University Press, Cambridge, United Kingdom and New York, NY, USA, pp. 187–230.

CENESTA (Centre for Sustainable Development & Environment), 2002: Workshop on rangeland management and pastoralism in arid lands in Iran. [online] Cited 12 June 2002. Available at: http://www.cenesta.org/projects/FoodSovereignty/PastoralismWorkshopReport.pdf

CGIAR (Consultative Group on International Agricultural Research), 1997: Consultative Group on International Agricultural Research. [Online] *World's Dryland Farmers Need New Agricultural Technology—"Green Revolution" Never Reached Them.* CGIAR Newsletter 4 (4), September 1997, Cited 15 May 2004. Available at: http://www.worldbank.org/html/cgiar/press/dryland.html.

Chapman A.L., J.D. Sturtz, A.L. Cogle, W.S. Mollar, and R.J. Bateman, 1996: Farming systems in the Australian semi-arid tropics—A recent history. *Australian Journal of Experimental Agriculture,* **36 (8),** 915–928.

Charney, J.G., P.H. Stone, and W.J. Quirk, 1975: Drought in the Sahara: A biogeophysical feedback mechanism. *Science* **187,** 434–435.

CIA (Central Intelligence Agency), 2004: The World Fact Book. [online] Cited 29 October 2004. Available at http://www.cia.gov/cia/publications/factbook/geos/cg.html

CIESIN, 2004: [Online]Alpha version 3 of Gridded Popualtion of the World. Available at http://beat.sedac.ciesin.Columbia.edu/gpw

Cleaver, K.M. and G.A. Schreiber, 1994: *Reversing the spiral: the population, agriculture, and environment nexus in sub-Saharan Africa. Directions in development.* The World Bank, Washington, D.C.

Conner, R., A. Seidl, L. Van Tassell, and N. Wilkins, 2002: *United States Grasslands and Related Resources: An Economic and Biological Trends Assessment.* The National Cattlemen's Beef Association, The Nature Conservancy and Ducks Unlimited, 152 pp.

Darkoh, M.B.K., 1998: The nature, causes and consequences of desertification in the drylands of Africa. *Land Degradation and Development,* **9 (1),** 1–20.

Davis, G.W., D.M. Richardson, J.E. Keeley, and R.J. Hobbs, 1996: Mediterranean-type ecosystems: the influence of biodiversity on their functioning. In: *Functional roles of biodiversity: a global perspective,* H.A. Mooney, J.H. Cushman, E. Medina, O.E. Sala, and E.D. Schulze, (eds.), SCOPE/UNEP, John Wiley and Sons, Chichester, pp. 151–183.

de Haas, H., A. Bencherifa, L. de Haan, H. El Ghanjou, A. El Harradji, et al. 2001: *Migration, agricultural transformations and natural resource exploitation in the oases of Morocco and Tunisia.* Final Scientific Report, IMARROM project, IC18-CT97–0134 (EC, DGXII, INCO-DC), 1–297, Amsterdam, University Amsterdam.

DFD (German Remote Sensing Data Center), 1996: The Aral Sea Homepage, [online], Available at http://www.dfd.dlr.de/app/land/aralsee/

Drechsel, P., L. Gyele, D. Kunze, and O. Cofie, 2001: Population density, soil nutrient depletion, and economic growth in sub-Saharan Africa. *Ecological Economics,* **38,** 251–258.

Dregne, H.E., 2002: Land degradation in the drylands. *Arid Land Research and Management,* **16**(2), 99–132.

Dregne, H.E. and N.Chou, 1992: Global desertification and costs. In: *Degradation and restoration of arid lands,* H.E. Dregne (ed.), Texas Tech University, Lubbock, pp. 249–282.

Droppelmann, K. J., J.E. Lehmann, J. Ephrath, and P. R. Berliner, 2000: Water use efficiency and uptake patterns in a runoff agroforestry system in an arid environment. *Agroforestry Systems,* **49,** 223–243.

Dupe, R. G., and N.E. Pettit, 2002: Ecological perspectives on regulation and water allocation for the Ord River, Western Australia. *River Research and Applications,* **18 (3),** 307–320.

Eklundh, L., and L. Olsson, 2003: Vegetation index trends for the African Sahel 1982–1999. *Geophysical Research Letters,* **30**(8), 1430.

Ellis, J.E. and D.M. Swift, 1988: Stability of African Pastoral Ecosystems: Alternate Paradigms and Implications for Development. *Journal of Range Management* **41 (6),** 450–459.

Eswaran, H., P.F. Reich, J.M. Kimble, F.H. Beinroth, E. Padamnabhan, and P. Moncharoen, 2000: Global carbon stocks. In: *Global Climate Change and Pedogenic Carbonates,* R. Lal, J.M. Kimble, H. Eswaran and B.A. Stewart (eds.), CRC/Lewis Publishers, Boca Raton, FL: 15–25.

Eswaran, H. and P. Reich, 2003: *World Soil Resources Map Index.* United States Department of Agriculture, Natural Resources Conservation Service. http://www.nrcs.usda.gov/technical/worldsoils/mapindx/#regional, 2003. Accessed 3 June 2003.

Etzion, Y. and E. Erell., 2000: Controlling the transmission of radiant energy through windows: A novel ventilated reversible glazing system. *Building and Environment,* **35,** 433–444.

Etzion, Y., D., Pearlmutter, E. Erell, and I. Meir, 1999: Adaptive architecture: Low-energy technologies for climate control in the desert. In: *Desert Regions,* B. A. Portnov and Paul A. Hare, (eds.), Springer, Berlin, pp. 291–305.

Evenari, M., L. Shanan, and N. Tadmor, 1982: *The Negev. The Challenge of a Desert.* Harvard University Press, Cambridge.

Faiman, D., 1998: Solar energy in arid frontiers: designing a photovoltaic power plant for Kibbutz Samar, Israel, In: *The Arid Frontier,* H.J. Bruins and H. Lithwick, (eds.), Kluwer, Dordrecht, pp. 321–336.

FAO (Food and Agriculture Organization of the United Nations), 1993: *The State of Food and Agriculture 1993.* Food and Agriculture Organization of the United Nations, Rome.

FAO, 1998: *The State of the World's plant genetic resources for food and agriculture.* Food and Agriculture Organization of the United Nations, Rome.

Fekete, B.M., C.J. Vörösmarty, and W. Grabs, 2002: High resolution fields of global runoff combining observed river discharge and simulated water balances. *Global Biogeochemical Cycles,* **16(3),** art. no. 1042.

Fernandez, R.J., E.R.M. Archer, A.J. Ash, H. Dowlatabadi,, P.H.Y. Hiernaux, et al. 2002: *Degradation and Recovery in Socio-ecological Systems: a view from the household/farm level.* Chapter 17 in Reynolds, J.F., and D.M. Stafford Smith, eds. Report of the 88th Dahlem Workshop on *An Integrated Assessment of the Dimensions of Global Desertification.* Berlin: Dahlem University Press

Fernandez, O.A. and C.A. Busso, 1997: Arid and Semi-arid Rangelands: Two Thirds of Argentina, Rangeland Desertification Report No. 200: 41–60. [online] Cited 29 October 2004. Available at: http://www.rala.is/rade/rala report/Fernandez.pdf

Foale, M.A., M.E. Probert, P.S. Carberry, D. Lack, S. Yeates, et al. 2004: Participatory research in dryland cropping systems—monitoring and simulation of soil water and nitrogen in farmers' paddocks in Central Queensland. *Australian Journal of Experimental Agriculture,* **44 (3),** 321–331.

Fredrickson, E., K. M. Havstad, R. Estell, and P. Hyder, 1998: Perspectives on desertification: southwestern United States. *Journal of Arid Environments,* **39(2),** 191–208.

GLASOD (Global Assessment of Soil Degradation), 1990: International Soil Reference and Information Centre, Wageningen, Netherlands, and United Nations Environment Programme, Nairobi, Kenya. [online] Cited 29 October 2004. Available at http://lime.isric.nl/index.cfm?contentid=158

Golan-Goldhirsh, A., P. Sathiyamoorthy, H. Lugasi-Evgi, Y. Pollack, and J. Gopas, 2000: Biotechnological potential of Israeli desert plants of the Negev. *Acta Horticulturae,* **523,** 29–37.

Griffin, G. F. and M. H. Friedel, 1984a: Effects of fire on Central Australian rangelands, I. Fire and fuel characteristics and changes in herbage and nutrients. *Australian Journal of Ecology,* **9,** 381–393.

Griffin, G. F. and M. H. Friedel, 1984b: Effects of fire on central Australian rangelands. II. Response of tree and shrub populations, *Australian Journal of Ecology,* **9,** 395–403.

Griffin, G. F. and M. H. Friedel, 1985: Discontinuous change in central Australia: some implications of major ecological events for land management. *Journal of Arid Environments,* **9**: 63–80.

Gruenzweig, J. M., T. Lin, E. Rotenberg, A. Schwartz, and D. Yakir, 2003: Carbon sequestration in arid-land forest. *Global Change Biology,* **9,**791–799.

Gunnell Y. and A. Krishnamurthy, 2003: Past and present status of runoff harvesting systems in dryland peninsular India: A critical review. *Ambio,* **32 (4),** 320–324.

Hamblin, D.J., 1987; Has the garden of eden been located at last? *Smithsonian Magazine,* 18. No. (2) [online] Cited 29 October 2004. Available http://www.ldolphin.org/eden/

Haridas, V., M. Higuchi, G. S. Jayatilake, D. B. K. Mujoo, M. E. Blake, et al. 2001: Avicins: Triterpenoid saponins from *Acacia victoriae* (Bentham) induce apoptosis by mitochondrial perturbation. *Proceedings of the National Academy of Sciences,* **98,** 5821–5826.

Harlan, J.R., 1977: Plant and Animal Distribution in Relation to Domestication. In: *The Early History of Agriculture,* J. Hutchinson, J.G.G. Clark, E.M. Jope, and R. Riley, (Eds.), Oxford University Press, Oxford, pp. 13–25.

Hazell, P., T. Ngaido, and N. Chaherli, 2002: *Policy and institutional options for agricultural growth, poverty alleviation, and environmental sustainability in the dry areas of West Asia and North Africa.* Presentation at a Workshop on Agriculture, Environment and Human Welfare in West Asia and North Africa, ICARDA, Aleppo, Syria, May, 2002. IFPRI, Washington, D.C. Mimeo.

Higgs, E.S. and M.R. Jarman, 1972: The Origins of Animal and Plant Husbandry. In: *Papers in Economic Prehistory,* (Ed.), E.S. Higgs, Cambridge University Press, Cambridge, pp. 3–13.

Hillel, D.J., 1991: *Out of the earth: civilization and the life of the soil.* The Free Press, New York.

Hoffmann, G., 1988: *Holozänstratigraphie und Küstenlinienverlagerung an der Andalusischen Mittelmeerküste.* 2. Berichte aus dem Fachbereich Gewissenschaften der Universitas Bremen.

Hoogmoed, W., 1999: Tillage for soil and water conservation in the semi-arid tropics, *Tropical Resource Management Papers,* **24.** Wageningen University and Research Center, Wageningen.

Howden, S.M., P.J. Reyenga, H. Meinke, and G.M. McKeon, 1999: Integrated Global Change Impact Assessment on Australian Terrestrial Ecosystems: Overview Report. Working Paper Series 99/14, CSIRO Wildlife and Ecology, Canberra, Australia, 51 pp.

Huenneke, L.F., 2001: Deserts. In: F.S. Chapin III, O.E. Sala, and E. Huber-Sannwald, (eds.), *Global biodiversity in a changing environment—scenarios for the 21st century,* Springer Verlag, New York, pp. 201–222.

Huenneke, L.F. and I. Noble, 1996: Ecosystem function of biodiversity in arid ecosystems. In *Functional roles of biodiversity: a global perspective* H.A., Mooney, J.H. Cushman, E. Medina, O.E. Sala, and E.D. Schulze, (eds.), SCOPE/UNEP. John Wiley and Sons, Chichester. pp. 99–128.

ICCD (Convention to Combat Desertification), 2000: Traditional Knowledge: Report of the ad hoc Panel, ICCD/COP (4)/CST/2 [On line] Cited 28 October 2004. Available at http://www.unccd.int/cop/officialdocs/cop4/pdf/cst2eng.pdf.

IEA (International Energy Agency) 2001: *Energy Balances of Non-OECD Countries (2001 Edition),* Organisation for Economic Co-operation and Development, Paris.

IPCC (Intergovernmental Panel on Climate Change), 2001: *Climate Change 2001: Working Group I: The Scientific Basis, Summary for Policymakers,* Cambridge University Press, Cambridge.

ITAP (International Technical Advisory Panel), 2003: Restoration Planning Workshop, *Building a Scientific Basis for the Restoration of the Mesopotamian Marshlands.* Convened by Eden Again Project and The Iraq Foundation, 68 pp.

IUCN (World Conservation Union), 2004: 2004 IUCN Red List of Threatened Species[Online]Cited 24 November 2004. Available at www.redlist.org

Izhaki, I, P.B. Walton and U.N. Safriel, 1991: Seed shadows generated by frugivorous birds in an eastern Mediterranean scrub. *Journal of Ecology,* **79,** 575–590.

Jacobs, J.L., G.N. Ward, and G. Kearney, 2004: Effects of irrigation strategies and nitrogen fertiliser on turnip dry matter yield, water use efficiency, nutritive characteristics and mineral content in western Victoria. *Australian Journal of Experimental Agriculture,* **44 (1),** 13–26 2004.

Jiang, H., 2004: Cooperation, land use, and the environment in Uxin Ju: The changing landscape of a Mongolian-Chinese borderland in China. *Annals of the Association of American Gegoraphers,* **94** (1), 117–139.

Jones, C. G., J. H. Lawton, and M. Shachak, 1994: Organisms as Ecosystem engineers, *Oikos,* **69,** 373–386.

Kamara, A. and H. Sally, 2003: Water for food, livelihoods and nature: simulations for policy dialogue in South Africa. *Physics and chemistry of the Earth,* **28(20–27),** 1085–1094.

Karajeh, F., A. Saporov, V. Petrunin and T. Nugaeva, 2000: Use of treated wastewater from Almaty for fee-crop irrigation. In: *New Approaches to Water Management in Central Asia,* Adeel, Z. (Ed.), UNU Desertification Series No. 3, United Nations University, Tokyo.

Kates, R. and V. Haarmann, 1992: Where the poor live: Are the assumptions correct? *Environment,* **34,** 4–28.

Kolkovsky, S., G. Hulata, Y. Simon, R. Segev and A. Koren, 2003: *Integration of Agri-Aquaculture Systems—The Israeli Experience,* In; Integrated Agri-Aquaculture Systems, A Resource Handbook for Australian Industry Development, G.J. Gooley and F.M. Gavine (eds.), Rural Industries Research and Development Corporation, RIRDC Publication, Kingston, ACT, Australia, pp. 14–23.

Krenke, A. N., G. M. Nikolaeva, and A.B. Shmarin, 1991: The effects of natural and anthropogenic changes on heat and water budgets in the central Caucasus, USSR. *Mountain Research and Development,* **11,** 173–182.

Kulkarni H., P.S.V. Shankar, S.B. Deolankar and M. Shah, 2004: Groundwater demand management at local scale in rural areas of India: a strategy to ensure water well sustainability based on aquifer diffusivity and community participation. *Hydrogeology Journal,* **12 (2),** 184–196 .

Lal, R., 2001a: Soil degradation by erosion. *Land Degradation & Development,* **12,** 519–539.

Lal, R. 2001b: Potential of desertification control to sequester carbon and mitigate the greenhouse effect. *Climatic Change,* **51,** 35–72.

Lamb, R.L., 2003: Fertilizer use, risk, and off-farm labor markets in the semi-arid tropics of India. *American Journal of Agronomic Economy,* **85 (2),** 359–371.

Lavrenko, E. M. and Z.V. Karamysheva, 1992: Developing answers and learning in extension for dryland nitrogen management. Steppes of the former Soviet Union and Mongolia, In: *Natural Grassland,* R.T. Coupland (ed.), Elsevier, Amsterdam, pp. 3–59.

Lawrence D.N., S.T. Cawley and P.T. Hayman, 2000: Developing answers and learning in extension for dryland nitrogen management. *Australian Journal of Exmperimantal Agriculture,* **40 (4),** 527–539.

Leach, M., R. Mearns, and I. Scoones, 1999: Environmental entitlements: dynamics and institutions in community-based natural resource management. *World Development,* **27 (2),** 225–247.

Le Houerou, H. N. 1992: Vegetation and Land Use in the Mediterranean Basin by the Year 2050: A Prospective Study. In: *Climatic Change and the Mediterranean* J. D. M. L. Jeftic, and G. Sestini (eds.), London: Edward Arnold (Hodder & Stoughton), pp. 175–231.

Le Houerou, H.N., 1984: Rain use efficiency: a unifying concept in arid-land ecology. *Journal of Arid Environments,* **7,** 213–247.

Le Houerou, H.N., R.L. Bingham, and W. Skerbek, 1988: Relationship between the variability of primary production and variability of annual precipitation in world arid lands. *Journal of Arid Environments,* **15,** 1–18.

Lepers, E., 2003: Synthesis of the Main Areas of Land-cover and Land-use Change. Millennium Ecosystem Assessment, Final Report. Available at www.geo.ucl.ac.be/LUCC/lucc.html.

Lepers, E., E.F. Lambin, A.C. Janetos, R. DeFries, F. Achard, et al. 2005: A synthesis of rapid land-cover change information for the 1981–2000 period. *BioScience,* **55 (2),** 19–26.

Lopez, R. E. 1998: *Where development can or cannot go: the role of poverty-environment linkages.* 1997 Annual World Bank Conference on Development Economics. The World Bank, Washington, D.C.

Lundqvist, J., 1999: *Rules and roles in water policy and management—classification of rights and obligations.* In: Proceedings of the SIWI/IWRA Seminar "Towards Upstream/Downstream Hydrosolidarity", Stockholm. pp. 61–67.

MA (Millennium Ecosystem Assessment), 2003: *Ecosystems and Human Well-being; A Framework for Assessment.* Island Press, Washington DC.

Mainguet, M., 1996: Aridite, secheresse et degradation dans les aires sechs de Chine. *Secheresse,* **1(7),** 41–50.

Matthews, E., 2000: *Understanding the FRA 2000.* Forest Briefing, 1, World Resources Institute, Washington, D.C.

Mazzucato, V. and D. Niemeijer, 2000a: The Cultural Economy of Soil and Water Conservation: Market Principles and Social Networks in Eastern Burkina Faso. *Development and Change,* **31(4),** 831–855.

Mazzucato, V. and D. Niemeijer, 2000b: *Rethinking soil and water conservation in a changing society: A case study in eastern Burkina Faso,* Tropical Resource Management Papers, 32. Wageningen University, Wageningen.

Mazzucato, V., D. Niemeijer, L. Stroosnijder, and R. Röling, 2001: *Social networks and the dynamics of soil and water conservation in the Sahel.* SA Gatekeeper Series 101, International Institute for Environment and Development, London.

Mazzucato, V., and D. Niemeijer, 2002: Population growth and the environment in Africa: Local informal institutions, the missing link. *Economic Geography,* **78(2),** 171–193.

Mazzucato, V., and D. Niemeijer, 2003: Why do Savings Institutions Differ within the Same Region? The Role of Environment and Social Capital in the Creation of Savings Arrangements in Eastern Burkina Faso. *Oxford Development Studies,* **31** (4), 519–529.

Mbonile, M. J., 2005: Migration and intensification of water conflicts in the Pangani Basin, Tanzania. *Habitat International,* **29,** 41–67.

Mehta, A. K,. and Shah, A., 2003: Chronic Poverty in India: Incidence, Causes and Policies. *World Development,* **31,** 491–511

Middleton, N. and D. Thomas, 1997: *World Atlas of Desertification,* Arnold, London.

Milton, S. J. and R.J. Dean, 1996: *Karoo Veld. Ecology and management.* ARC Range and Forage Institute, Lynn East, South Africa.

Mirza, M.Q., R.A. Warrick, N.J. Ericksen, and G.J. Kenny, 1998: Trends and persistence in precipitation in the Ganges, Brahmaputra and Meghna Basins in South Asia. *Hydrological Sciences Journal,* **43,** 845–858.

Mitchell, P. B., 1991: Historical perspectives on some vegetation and soil changes in semi-arid New South Wales. *Vegetatio* **91,** 169–182.

Mooney, H.A., K. M.T. Arroyo, W.J. Bond, J. Canadell, R.J. Hobbs, S. Lavorel, and R.P. Neilson, 2001: Mediterranean-climate ecosystems. In: *Global biodiversity in a changing environment—scenarios for the 21st century,* F.S. Chapin III, O.E. Sala, and E.Huber-Sannwald, (eds.), Springer Verlag, New York, pp. 157–199.

Mortimore, M., 2003: Long-term Change in African Drylands: Can Recent History Point Towards Development Pathways? *Oxford Development Studies,* **31(4),** 503–518

Mortimore, M., and F. Harris, 2004: Do small farmers' achievements contradict the nutrient depletion scenarios for Africa? *Land Use Policy,* In press.

Mtimet A., and M., Hachicha, 1995: Salinisation et hydromorphie dans les oasis tunisiennes—*Sécheresse* 6, n° 4.

Myers N., R.A. Mittermeier, C.G. Mittermeier, G.A.B da Fonseca, and J. Kent, 2000: Biodiversity hotspots for conservation priorities. *Nature,* **403,**853–845.

Nathan, R., U.N. Safriel, and H. Shirihai, 1996: Extinction and vulnerability to extinction at distribution peripheries: An analysis of the Israeli breeding avifauna. *Israel Journal of Zoology,* **42,** 361–383.

Naveh, Z., 1991: The role of fire in Mediterranean vegetation. *Botanika Chronika,* **10,** 385–405.

Neilson, R.P., I.C. Prentice, B. Smith, T. Kittel, and D. Viner, 1998: Simulated changes in vegetation distribution under global warming. In: *The Regional Impacts of Climate Change: An Assessment of Vulnerability. Special Report of IPCC Working Group II,* Watson, R.T., M.C. Zinyowera, and R.H. Moss (eds.), Cambridge University Press, Cambridge, United Kingdom and New York, NY, USA, pp. 441–446.

Nettleton, W. D. and Peterson, F. F, 1983: Aridisols. In: *Pedogenesis and Soil Taxonomy,* L. P. Wilding, N. E Smeck, G. F. Hall (eds.), Elsevier Publishers B.V, Amsterdam, pp. 165–215.

Nicholson, S.E., 1979: The methodology of historical climate reconstruction and its application to Africa. *Journal of African History,* **20 (1),** 31–49.

Nicholson, S.E., C.J. Tucker, and M.B. Ba, 1998: Desertification, drought, and surface vegetation: An example from the West African Sahel. *Bulletin of the American Meteorological Society,* **79(4),** 1–15.

Nielsen, T.T., in prep: Long term trends in African vegetation productivity. *Global Change.*

Nielsen, T.T. and H.K. Adriansen, 2005: Government policies and land degradation in the Middle East. *Land Degradation and Development,* **16,** 151–161.

Niemeijer, D., 1996: The Dynamics of African Agricultural History: Is it Time for a New Development Paradigm? *Development and Change,* **27 (1),** 87–110.

Niemeijer, D., 1999: Environmental dynamics, adaptation, and experimentation in indigenous Sudanese water harvesting. In: *Biological and cultural diversity: The role of indigenous agricultural experimentation in development,* G. Prain, S. Fujisaka, D.M. Warren, (eds.), IT Studies in Indigenous Knowledge and Development, Intermediate Technology Publications, London, pp 64–79.

Niemeijer, D. and V. Mazzucato, 2002: Soil degradation in the West African Sahel: How serious is it? *Environment,* **44(2),** 20–31.

Norwood, C. A., and T.J. Dumler, 2002: Transition to dryland agriculture: Limited irrigated vs. dryland corn. *Agronomy Journal,* **94** (2), 310–320.

Noy-Meir, I., 1973: Desert ecosystems: Environment and producers. *Annual Review of Ecology and Systematics,* **4,** 25–51.

Noy-Meir, I., 1974: Desert ecosystems: Higher trophic levels. *Annual Review of Ecology and Systematics,* **5,** 195–214.

NRC (National Research Council), 1999: *Water for the Future: The West Bank and Gaza Strip, Israel and Jordan.* National Academy Press, Washington D.C.

Nyariki, D.M., S.L. Wiggins, and J.K. Imungi, 2002: Levels and causes of household food and nutrition insecurity in drylnd Kenya. *Ecology of Food and Nutrition,* **41** (2), 155–176.

Oldeman, L.R., 1994: The Global Extent of Soil Degradation. In *Soil Resilience and Sustainable Land Use,* D.J. Greenland and I. Szabolcs (eds.), CAB International, Wallingford, pp 99–118.

Oldeman, L.R., R.T.A. Hakkeling, and W.G. Sombroek, 1990: *World map on status of human-induced soil degradation (GLASOD).* UNEP/ISRIC, Nairobi, Kenya.

Oldeman, L.R., R.T.A Hakkeling, and W.G. Sombroek, 1991: *World map of the status of human-induced soil degradation: an explanatory note,* Second revised edition, International Soil Reference and Information Centre/United Nations Environment Programme, Wageningen/Nairobi.

Oldeman, L.R., and G.W.J. van Lynden, 1997: Revisiting the GLASOD methodology. In *Methods for Assessment of Soil Degradation,* (ed.) R. Lal, W.H. Blum, C. Valentine, and B.A. Steward, New York, CRC Press, pp. 423–439.

Olson, D. M., and E. Dinerstein, 1998: The Global 200: a representation approach to conserving the Earth's most biologically valuable ecoregions. *Conservation Biology,* **12,** 502–515.

Olson, D.M., E. Dinerstein, E.D. Wikramanayake, N.D. Burgess, G.V.N. Powell, et al., 2001: Terrestrial ecoregions of the world: a new map of life on earth. *BioScience,* **51,** 933–938.

Oron, G., 1996: Management modeling of integrative wastewater treatment and reuse systems. *Water Science and Technology,* 33 **(10–11),** 95–105.

Otterman, J., A. S. Manes, P. Rubin, P. Alpert, and D. O. C. Starr, 1990: An increase of early rains in southern Israel following land-use change? *Boundary-Layer Meteorology,* **53,** 333–351.

Oweis, T.Y., 2000: Coping with increased water scarcity in dry areas: Increased water productivity. In: *New Approaches to Water Management in Central Asia,* Z. Adeel, (ed.), UNU Desertification Series No. 3, United Nations University, Tokyo.

Pacey, A., and A. Cullis, 1986: *Rainwater harvesting. The collection of rainfall and runoff in rural areas.* Intermediate Technology Publications, London.

Paine, R.T., 1966: Food web complexity and species diversity, *American Naturalist,* **100,** 65–75.

Pearlmutter, D. and P. Berliner, 1999: Urban microclimate in the desert: planning for outdoor comfort under arid conditions, In: *Desert Regions,* B.A. Portnov and A. Paul Hare, (eds.), Springer, Berlin, pp. 279–290.

Pender J.L. and J.M. Kerr, 1998: Determinants of farmers' indigenous soil and water conservation investments in semi-arid India. *Agricultural Economy,* **19** (1–2), 113–125.

Pender, J., S. J. Scherr, and G. Durón, 2001: Pathways of development in the hillsides of Honduras: causes and implications for agricultural production, poverty, and sustainable resource use. In: *Tradeoffs or Synergies? Agricultural Intensification, Economic Development and the Environment,* D.R., Lee, and C. B Barrett, (eds.), Wallingford, UK: CAB International.

Pickup, G., 1996: Estimating the effects of land degradation and rainfall variation on productivity in rangelands: an approach using remote sensing and models of grazing and herbage dynamics. *Journal of Applied Ecology,* **33,** 819–832.

Poesen, J., J. Nachtergaele, G. Verstraeten, and C.Valentin, 2003: Gully erosion and environmental change: importance and research needs. *Catena* **50(2),** 91–133.

Pohoryles, S., 2000: Program for efficient water use in Middle East agriculture, In: *Water for Peace in the Middle East and Southern Africa,* Green Cross International, The Hague, pp. 18–38.

Portnov, B.A. and U.N. Safriel, 2004: Combating desertification in the Negev: dryland agriculture *vs.* dryland urbanization. *Journal of Arid Environment,* **56,** 659–680.

Prain, G., S. Fujisaka, and M.D. Warren, (eds.), 1999: *Biological and agriculutral diversity: The role of indigenous agricultural experimentation in development.* Intermediate Technology Publications, London, 218 pp.

Prince, S.D., 1991: A model of regional primary production for use with coarse-resolution satellite data. *International Journal of Remote Sensing,* **12,** 1313–1330.

Prince, S.D., 1999: What practical information about land-surface function can be determined by remote sensing? Where do we stand? In: *Integrating hydrology, ecosystem dynamics, and biogeochemistry in complex landscapes,* J. D. Tenhunen and P. Kabat (eds), Dahlem Workshop Reports, John Wiley & Sons Ltd., Chichester, pp 39–60.

Prince, S.D., 2002: Spatial and temporal scales of measurement of desertification. In: *Global desertification: do humans create deserts?* M. Stafford-Smith and J. F. Reynolds (eds.), Dahlem University Press, Berlin, pp. 23–40.

Prince, S.D., 2004: Mapping desertification in Southern Africa In: *Land Change Science: Observing, Monitoring, and Understanding Trajectories of Change on the Earth's Surface.* G. Gutman, A. Janetos, C. O. Justice, et al., (eds.), Kluwer, Dordrecht, pp. 163–184.

Prince, S.D., E. Brown de Colstoun, and L.L. Kravitz, 1998: Evidence from rain-use efficiencies does not indicate extensive Sahelian desertification. *Global Change Biology,* **4(4),** 359–374.

Prospero, J.M. and R.T. Nees, 1986: Impact of the North African drought and El Niño on mineral dust in the Barabados Trade Winds. *Nature,* **320,** 735–738.

Prudencio, C.Y., 1993: Ring Management of Soils and Crops in the West African Semi-Arid Tropics: The Case of the Mossi Farming System in Burkina Faso, *Agriculture, Ecosystems and Environment,* **47 (3),** 237–264.

Puigdefabregas, J., 1998: Ecological impacts of global change on drylands and their implications for desertification. *Land Degradation & Development,* **9,** 393–406.

Puigdefabregas, J., A., Sole, L. Gutierrez, G. del Barrio and M. Boer, 1999: Scales and processes of water redistribution in drylands: results from the Rambla Honda field site in southeast Spain. *Earth Science Reviews,* **48,** 39–70.

Puigdefabregas, J. and T. Mendizabal, 1998: Perspectives on desertification: western Mediterranean. *Journal of Arid Environments,* **39,** 209–224.

Ramankutty, N. and J.A. Foley, 1999: Estimating historical changes in global land cover: Croplands from 1700 to 1992. *Global Biogechemical Cycles,* **13[4],** 997–1027.

Reich, P., H. Eswaran, S. Kapur, S. and E. Akca, 2000: Land Degradation and Desertification in Desert Margins, [online] International Symposium On Desertification/2000-Konya, Cited 29 October 2004. Available at http://www.toprak.org.tr/isd/isd

Reid, R.S., P.K. Thornton, G.J. McCRabb, R.L. Kruska, F. Atieno, and P.G. Jones, 2004: Is it possible to mitigate greenhouse gas emissions in pastoral ecosystems of the tropics? *Development and Sustainability,* **6,** 91–109.

Reij, C., P. Mulder, and L. Begemann, 1988: *Water harvesting for Plant Production.* World Bank Technical Paper, 91, World Bank, Washington, D.C.

Reij, C., I., Scoones, and C. Toulmin, 1996: *Sustaining the Soil: Indigenous Soil and Water Conservation in Africa.* Earthscan, London.

Richmond, A., 1986: Halotolerant microloage: A future crop for arid lands. in: *Progress in Desert Research,* L. Berkofsky and M.G. Wurtele (eds.), Rowman & Littlefield, Totowa, New Jersey, pp. 67–86.

Robbins, L.H., 1984: Late Prehistoric Aquatic and Pastoral Adaptations West of Lake Turkana, Kenya. In: *From Hunters to Farmers: The Causes and Consequences of Food Production in Africa,.* J.D. Clark, and S.A. Brandt, (eds.), University of California Press, Berkeley, London, pp. 206–211.

Roquero, C., 1990: Mediterranean soils behavior in relation to soil erosion. In: *Strategies to combat desertification in Mediterranean Europe,* J.L.Rubio and R.J. Rickson (eds.), Commission of the European Communities, Luxembourg, pp. 40–76.

Rosenzweig, M. L., 1995: Species Diversity in Space and Time. Cambridge University Press, Cambridge.

Rosenzweig, M.L. and Z.Abramsky, 1993: How are diversity and productivity related? In: *Species diversity in ecological communities: historical and geographical perspectives,* R.E. Ricklefs and D. Schluter, (eds), University of Chicago Press, Chicago, pp 52–65.

Ruf, T, 1995: Histoire hydraulique et agricole et lutte contre la salinisation dans le delta du Nil. *Secheresse,* **4(6),** 307–318.

Rust, R. H., 1983: *Alfisols,* In: *Pedogenesis and Soil Taxonomy,* L. P. Wilding, N. E. Smeck, and G. F. Hall, (eds.), Elsevier Publishers D.V., Amsterdam, pp. 253–281.

Safriel, U.N, 1992: The regional and global significance of environmental protection, nature conservation and ecological research in Israel. In *Judaism and Ecology:* Aubrey Rose (ed.) Cassell, London, .pp. 91–99.

Safriel, U.N., 1999: The concept of sustainability in dryland ecosystems. in: T. W. Hoekstra and M. Shachak (eds.), *Arid Lands Management—Toward Ecological Sustainability,* Urbana: University of Illinois Press, pp. 117–140.

Sala, O.E., W.K. Laurenroth, S.J. McNaughton, G. Rusch, and X. Zang, 1996: Biodiversity and ecosystem function in grasslands. In: *Functional roles of biodiversity: a global perspective,* H.A. Mooney, J.H. Cushman, E. Medina, O.E. Sala and E.D. Schulze (eds.), SCOPE/UNEP, John Wiley and Sons, Chichester, pp. 129–149.

Sala, O.E., F.S. Chapin III, J.J. Armesto, E. Berlow, J. Bloomfield, et al. 2000: Global Biodiversity Scenarios for the Year 2100, *Science,* **287,** 1770–1774.

Sanders, J. H., B. I. Shapiro and S. Ramaswamy, 1996: *The Economics of Agricultural Technology Development in Sub-Saharan Africa.* John Hopkins University Press, USA.

Sauerhaft, B., P.R. Berliner, and T.L. Thurow, 1998: The fuelwood crisis in arid zones: runoff agriculture for renewable energy production. In: *The Arid Frontier,* H.J. Bruins, and H. Lithwick, (eds.), Kluwer, Dordrecht, pp. 351–364.

Sarch, M-T, and C. Birkett, 2000: Fishing and Farming at Lake Chad: Responses to Lake-level Fluctuations. *Geographical Journal,* **166,** 156–172.

Schlesinger, W.H., 1999: Carbon sequestration in soils. *Science,* 284: 2095.

Schlesinger, W.H., J.F. Reynolds, G.L. Cunningham, L.F. Huenneke, W.M. Jarrell, R.A. Virginia, and W.G. Whitford, 1990: Biological feedbacks in Global Desertification. *Science,* **247,** 1043–1048.

Schmidt-Nielsen, K., 1980: *Desert Animals: Physiological Problems of Heat and Water.* Dover Publications, New York.

Scholes, R. J. and D. O. Hall, 1996: The carbon budget of tropical savannas, woodlands and grasslands. In: *Global Change: Effects on Coniferous Forests and Grasslands,* A. I. Breymeyer, D. O. Hall, J. M. Melillo, and G. I. Agren, (Eds.), John Wiley & Sons Ltd., pp. 71–100.

Scoones, I. (ed.), 2001: *Dynamics and diversity: Soil fertility and farming livelihoods in Africa.* Earthscan, London, 244 pp.

Scoones, I. and C. Toulmin, 1998: Soil nutrient budgets and balances: what use for policy? *Agriculture, Ecosystems & Environment,* **71,** 255–267.

Shachak, M., and G.M. Lovett, 1998: Atmospheric deposition to a desert ecosystem and its implication for management. *Ecological Applications,* **8,** 455–463.

Shachak, M. and S.T.A. Picket, 1997: Linking ecological understanding and application: patchiness in a dryland system. In: *The Ecological Basis of Conservation,* S.T.A. Pickett, R.S. Ostfeld, M. Shachak and G.E. Likens (eds.), Chapman & Hall, New York, pp. 108–119.

Shah, T., A. D. Roy, A. S. Qureshi, and J. X. Wang, 2001: Sustaining Asia's groundwater boom: An overview of issues and evidence. [Online] Cited 29 October 2004. Available at: http://www.water-2001.de/supporting/Asia_Groundwater_Boom.pdf.

Shinn, E.A., G.W. Smith, J.M. Prospero, P. Betzer, M.L. Hayes, V. Garrison, and R.T. Barber, 2000: African Dust and the Demise of Caribbean Coral Reefs. *Geophysical Research Letters.* **27(19)**, 3029–3032.

Shyamsundar, P., 2002: *Poverty—Environmental Indicators.* World Bank Environmental Economics Series Paper No. 84.

Singh, H.P., 1998: Sustainable development of the Indian desert: The relevance of the farming systems approach. *Journal of Arid Environment,* **39 (2):** 279–284.

Singh, R., 2004: Simulations on direct and cyclic use of saline waters for sustaining cotton-wheat in a semi-arid area of north-west India. *Agricultural Water Management,* **66,**153–162.

Slingerland, M., 2000: *Mixed farming: Scope and constraints in West African Savanna,* Tropical Resource Management Papers, 34, Wageningen University, Wageningen.

Smith, G.T., L.D. Ives, I.A. Nagelkerken, and K.B. Ritchie, 1996: Caribbean sea fan mortalities. *Nature,* **383,** 487.

Snel, M. and A. Bot, 2002: Draft paper: Some suggested indicators for Land Degradation Assessment of Drylands, LADA e-mail conference 9th of October—4th of November 2002 [online] Cited 20 May 2004. Available at: http://www.fao.org/ag/agl/agll/lada/emailconf.stm.

Solbrig, O.T., E. Medina, and J.F. Silva, 1996: Biodiversity and tropical savanna properties: a global view. In: *Functional roles of biodiversity: a global perspective*: H.A., Mooney, J.H.Cushman, E. Medina, O.E. Sala, and E.D. Schulze, (eds.), SCOPE/UNEP, John Wiley and Sons, Chichester, pp. 186–211.

Solh M, A. Amri , T. Ngaido and J. Valkoun, 2003: Policy and education reform needs for conservation of dryland biodiversity. *Journal of Arid Environment,* **54 (1),** 5–13.

Sombroek, W. G., 1990: Aridisols of the World, occurrence and potential. In: *Characterization, Classification and Utilization of Aridisols,* In: J. M. Kimble, and W. D. Nettleton (eds.), Proceedings of the Fourth International Soil Correlation Meeting (ISCOM IV), Lincoln, NE, USDA, Soil Conservation Service. Part A: Papers, pp. 121–128.

Squires, V., E.P. Glenn and A.T. Ayub (eds.) 1995: Combating Global Climate Change by Combating Land Degradation, Proceedings of a Workshop held in Nairobi, Kenya, 4–8 September 1995, UNEP, Nairobi, Kenya.

Steenhuijsen Piters de, B., 1995: *Diversity of fields and farmers, Explaining yield variations in northern Cameroon.* Dissertation, Landbouwuniversiteit Wageningen, Wageningen.

Steinberger, E. H. and N. Gazit-Yaari, 1996: Recent changes in the spatial distribution of annual precipitation in Israel. *Journal of Climate,* **9,** 3328–3336.

Stoorvogel J.J., and E.M.A. Smaling, 1990: *Assessment of soil nutrient depletion in Sub-Saharan Africa: 1983–2000.* Report 28, The Winand Staring Centre for Integrated Land, Soil and Water Research, Wageningen, The Netherlands.

Stocking, M., 1987: Measuring land degradation. In: *Land degradation and society,* P. Blaikie and H. Brookfield (eds.), Methuen & Co. Ltd, London, pp. 49–63.

Stocking, M., 1996: Soil erosion: breaking new ground. In: *Challenging Received Wisdom in African Environmental Change,* M. Leach, R. Mearns, (Eds.), The Lie of the Land: James Currey/ International African Institute, London, pp. 140–154.

Swap, R., M. Garstang, S. Greco, R. Talbot, and P. Kallberg, 1992: Saharan Dust in the Amazon Basin. *Tellus,* **44B,** 133–149.

Swift, J., 1996: Desertification: narratives, winners and losers. In: *The Life of the Land: Challenging Received Wisdom in African Environmental Change,* M. Leach and R. Mearns (eds.), James Currey/ International African Institute, London, pp. 73–90.

Swift, M.J., 1999: Integrating soils, systems and society. *Nature & Resources.* **35 (4),** 12–20.

Thenya, T., 2001: Challenges of conservation of dryland shallow waters, Ewaso Narok swamp, Laikipia District, Kenya. *Hydrobiologia,* **458,** 107–119.

Thomas, D.S.G., D. Sporton, and J. Perkins, 2000: The environmental impact of livestock ranches in the Kalahari, Botswana: Natural resource use, ecological change and human response in a dynamic dryland system. *Land Degradatoin and Development,* **11 (4):** 327–341.

Thornton, P.K., R.L. Kruska, N. Henninger, P.M. Krisjanson, R.S. Reid, et al., 2002: *Mapping Poverty and Livestock in the Developing World.* ILRI (International Livestock Research Institute), Nairobi, Kenya, 124 pp. [online] Cited 29 October 2004. Available at (http://www.ilri.cgiar.org/InfoServ/Webpub/fulldocs/mappingPLDW/media/5.htm)

Tiffen, M. and M. Mortimore, 2002: Questioning desertification in dryland sub-Saharan Africa. *Natural Resources Forum,* **26(3),** 218–233.

Tilman, D., P.B. Reich, J. Knops, D. Wedin, T. Mielke, and C. Lehman, 2001: Diversity and Productivity in a Long-Term Grassland Experiment. *Science,* **294,** 843–845.

Tucker, C.J., H.E. Dregne, and W.W. Newcomb, 1991: Expansion and contraction of the Sahara desert from 1980 to 1990. *Science,* **253:** 299–301.

Turner M.D., 1999: Conflict, environmental change, and social institutions in dryland Africa: Limitations of the community resource management approach. *Society and Natural Resources,* **12 (7),** 643–657.

Turner, M.D. and T.O. Williams, 2002: Livestock market dynamics and local vulnerabilities in the sahel. *World Development,* **30 (4),** 683–705.

Twomlow S, C. Riches, D. O'Neill, P. Brookes and J. Ellis-Jones, 1999: Sustainable dryland smallholder farming in sub-Saharan Africa, *Annals of Arid Zone,* **38 (2),** 93–135

UNEP (United Nations Environment Programme), 1983: *Rain and stormwater harvesting in rural areas.* Water Resources Series, 5, Tycooly International Publishing Limited, Dublin.

UNEP, 2001: *The Mesopotamian Marshlands: Demise of an Ecosystem.* Early Warning and Assessment Technical Report TR.01–3, Prepared by H. Partow, UNEP, Nairobi, Kenya. 46pp.

UNESCO, 2004: World Heritage List. [Online] Cited 29 October 2004. Available at http://whc.unesco.org/pg.cfm?cid = 31.

UN (United Nations), 2002: *World Urbanization Prospects: The 2001 Revision.* United Nations, New York.

UNFPA (United Nations Population Fund) 2003: [online] "Saving Women's Lives", Smith College Lecture by Thoraya Ahmed Obaid, Executive Director, UNFPA, 26 March 2003, Cited 29 October 2004. Available at http://www.unfpa.org/news/news.cfm?ID = 271&Language = 1,.

USAID/FEWS (U.S. Agency for International Development/Famine Early Warning System), 1997: [online] *Lake Chad—Untapped Potential.* FEWS Special Report 97–4, Available at http://www.fews.org/fb970527/fb97sr4.html.

USDA-ARS (U.S. Department of Agriculture-Agricultural Research Service), 2003: Jornada Experimental Range, Las Cruces (NM),[online] Cited 29 October 2004. Available at http://usda-ars.nmsu.edu/JER/brush-invasion.html.

Van Dijk, J.A., 1995: Taking the Waters: Soil and Water Conservation among Settling Beja Nomads in Eastern Sudan. *African Studies Centre Research Series,* **4.** Avebury, Aldershot.

Van Driel, A., 2001: *Sharing a valley: The changing relations between agriculturalists and pastoralists in the Niger Valley of Benin,* Research Report, 64, African Studies Centre, Leiden.

Van Wilgen, R.M., B.W, Cowling, and D.C. Le Maitre, 1998: Ecosystem services, efficiency, sustainability and equity: South Africa's Working for Water Programme, *Trends in Ecology & Evolution,* **13,** 378.

Vitousek, P.M. and D.U. Hooper, 1993: Biological diversity and terrestrial ecosystem biogeochemistry. In: *Ecosystem Function of Biodiversity,* E. D. Schulze, and H.A. Mooney, (eds.), Springer, Heidleberg, pp. 3–14.

Volis, S., Y. Anikster, L. Olsvig-Whittaker and S. Mendlinger, 2004: The influence of space in genetic-environmental relationships when environmental heterogeneity and seed dispersal occur at similar scale. *American Naturalist,* **163 (2),** 312–327.

Vörösmarty, C.J., E.M. Douglas, P.A. Green, and C. Revenga, 2005: Geospatial indicators of emerging water stress: An application to Africa. *Ambio,* **34,** 230–236.

Walter, H. 1968: *Die Vegetation der Erde in öko-physiologischer Betrachtung.* Gustav Fischer Verlag, Stuttgart.

Wang, S., W. Zheng, J. Ren, and C. Zhang, 2002: Selectivity of various types of salt-resistant plants for K^+ over Na^+. Journal of Arid Environments, 52, 457–472.

Ward, K.D., 2003: Three-way interactions between Acacia, large mammalian herbivores and bruchid beetles—a review. *African Journal of Ecology,* **41**(3), 257–265.

Warren, A., S. Batterbury, and H. Osbahr, 2001: Soil erosion in the West African Sahel: a review and an application of a "local political ecology" approach in South West Niger. *Global Environmental Change,* **11,** 79–95.

Warren, S.D. and D.J. Eldridge, 2001: Biological soil crusts and livestock in arid ecosystems: are they compatible? In: *Biological Soil Crusts: Structure, Function, and Management,* J. Belnap and O.L. Lange (eds.), Springer, Berlin, pp. 401–415.

Webb P, and A. Harinarayan, 1999: A measure of uncertainty: the nature of vulnerability and its relationship to malnutrition. *Disasters,* **23 (4),** S. 292–305.

Wessels, K.J., S.D. Prince, P.E. Frost, and D. van Zyl, 2004: Assessing the effects of human-induced land degradation in the former homelands of northern South Africa with a 1 km AVHRR NDVI time-series. *Remote Sensing of Environment,* **91,** 47–67.

White, R.P, S. Murray, and M. Rohweder, 2000: *Pilot Analysis of Global Ecosystems (PAGE): Grassland Ecosystems.* World Resources Institute, Washington, DC, 69 pp.

Whitford, W.G. and L.W. Parker, 1989: Contributions of soil fauna to decomposition and mineralization processes in semiarid and arid ecosystems. *Arid Soil Research and Rehabilitation,* **3,** 199–215.

WHO (World Health Organization), 2004: Global Database on Child Growth and Malnutrition. [Online] Cited 22 July 2004. Available at: http://www.who.int/nutgrowthdb/

Williams, M.A.J. and R.J. Balling, 1995: *Interactions of Desertification and Climate.* Edward Arnold Press, London.

Williams, J.R. and P.I. Diebel, 1996: The economic value of the prairie. In: *Prairie Conservation: Preserving North America's Most Endangered Ecosystem,* F.B. Samson and F.I. Knopf, (ed.), Island Press, Washington, DC, pp 19–35.

World Bank, 2003a: *BioCarbon Funds,* World Bank, Washington, DC.

World Bank, 2003b: *ProtoType Carbon Funds,* World Bank, Washington, DC.

WRI (World Resources Institute), 2004: EarthTrends, the Environmental Information Portal [Online], Drylands, People, and Ecosystem Goods and Services: A Web-based Geospatial Analysis (pdf version), by R.P. White and J. Nackoney. Cited 29 October.

WWF/IUCN (World Wide Fund for Nature/World Conservation Union), 1994: *Centres of Plant Diversity: a guide and strategy for their conservation.* IUCN Publications Unit, 3 volumes, Cambridge, UK.

Xue, Y. and P.A. Dirmeyer, 2004: *The Sahelian climate.* In: *Vegetation, Water, Humans, and the Climate: A New Perspective on an Interactive System,* P. Kabat, M. Claussen, P.A. Dirmeyer, J.H.C. Gash, L. Bravo de Guenni, et al. (eds.), Springer, New York, pp. 59–78.

Zaal, F., and R.H. Oostendorp, 2002: Explaining a miracle: Intensification and the transition towards sustainable small-scale agriculture in dryland Machakos and Kitui Districts, Kenya *World Development,* **30 (7),** 1271–1287.

Zeidler, J., S. Hanrahan, and M. Scholes, 2002: Landuse intensity affects range condition in arid to semi-arid Namibia. *Journal of Arid Environments,* **52 (3),** 389–403.

Zohary, M., 1973: *Geobotanical Foundations of the Middle East.* Gustav Fischer Verlag, Sttutgart.

Chapter 23

Island Systems

Coordinating Lead Authors: Poh Poh Wong, Eduardo Marone, Paulo Lana, Miguel Fortes
Lead Authors: Dorian Moro, John Agard, Luís Vicente
Contributing Authors: Jillian Thönell, Paola Deda, Kalemani Jo Mulongoy
Review Editors: Joseph Baker, Ariel Lugo, Avelino Suárez Rodríguez

---

*This appears in Appendix A at the end of this volume.

Main Messages

The coastal, marine, and inland ecosystems of islands provide valuable regulating, provisioning, and cultural services to more than 500 million people. Many small islands have a strong traditional dependence on marine and coastal biodiversity for their food, tools, industry, medicine, transport, and waste disposal. With increasing human population pressures through high migration and reproductive rates, island systems face several serious issues both in the immediate and the near future.

Islands systems, in spite of size, category, climate, and social conditions, share a commonality, identified here as the "isola effect." This represents the physical seclusion of islands as isolated pieces of land exposed to different kinds of marine and climatic disturbances and with a more limited access to space, products, and services when compared with most continental landmasses. In addition, subjective issues such as the perceptions and attitudes of islanders themselves on their conditions and their future on the island are incorporated into the "isola effect."

Coastal fisheries, a particularly important and traditional source of food, protein, and employment on many islands, are seriously depleted. Overfishing has already deprived island communities of subsistence fishing and caused conflicts in many tropical islands across Asia. Island states and their exclusive economic zones comprise 40% of the world's oceans and earn significant foreign exchange from the sale of offshore fishery licenses, but this situation cannot last forever.

Watershed modification on islands has had a negative impact on water resources in terms of water quality and quantity as well as flow regime. Despite limited coverage on some islands, forested watersheds are critical regulators of island hydrology. Without adequate freshwater resources, small islands depend on desalinated or imported water. Island water supply is often threatened by pollution, particularly from poorly treated sewage.

The natural land cover of island systems has changed drastically under the pressure of growing human populations and consequent exploitation of the landmass. On some islands, the impact has exceeded critical thresholds, particularly along the coastal fringe. Anthropogenic changes range from deforestation for cropland to urbanization and the abandonment of degraded land. All these have immediate repercussions on habitat destruction and loss of biodiversity. One conspicuous effect of natural and anthropogenic actions in the coastal zone threatening islands systems is the erosion of soft coastlines (sandy and muddy beaches).

Island systems are highly dependent on outside sources for food, fuel, and even employment, which together increase the economic fragility of many islands. At the same time, island resources are increasingly affected by globalization and trade liberalization. It is questionable whether regional or international groupings of islands, such as the small island developing states, can respond adequately to such pressures.

Energy constraints are particularly critical in island systems. The usually limited size of islands, their constrained capacity to provide ecosystem services (in spite of type or size), and often their distance from large-scale energy supply systems are key factors to explain why energy issues are an important factor in island systems. However, oceans—through currents, tides, waves, and thermal and salinity gradients—offer a source of new renewable forms of energy that remain underexplored.

Low-lying island systems are under threat from climate change and predicted sea level rise. These in turn are expected to have serious conse- quences on flooding, coastal erosion, water supply, food production, health, tourism, and habitat depletion. The sea level rise would be severe or devastating to millions of people living on low-lying islands and atolls. The projected changes in temperature and rainfall could disrupt terrestrial and marine ecosystems on most islands, especially small ones. Increased flooding and coastal erosion will have serious consequences for the tourism industry. The incidence of dengue fever has been correlated to the Southern Oscillation Index, and extremes in rainfall are likely to exacerbate diarrheal illnesses. Islands need to develop appropriate coastal assessments and management so as to adapt to these changes in a sustainable manner.

The coastal systems of islands, such as coastal forests, dunes, mangroves, coral reefs, and seagrass meadows, are being altered through agriculture, aquaculture, coastal urban sprawl, industrialization, and resort development. In addition, these changes produce further stresses on the island systems, such as the production of sewage, solid waste, and water pollution. These alterations exacerbate the fragility of island systems.

23.1 Introduction

This chapter presents an overview of the status of and trends in island ecosystem services. Its aim is to recognize the services that islands, as a composite of ecosystems, provide to human well-being, to discuss the drivers of change on island systems, and to offer a critical assessment of these drivers in light of the trade-offs between change and ecosystem services.

23.1.1 Overview

Islands differ in their geological and geomorphologic settings and geography, in their physical, biological, climatic, social, political, cultural, and ethnic characteristics, and in their stage of economic development. Yet they share several characteristics that not only unify them as a distinct category but underscore their overall vulnerability in the context of sustainable development (Maul 1993; Leatherman 1997).

As islands and island chains are associated with specific geophysical settings, they are strongly influenced by the surrounding ocean and atmosphere. Their geophysical conditions may result in economic strength for some, but they can pose hazards that threaten the economic viability of others. Some islands are densely populated, and islanders rely on the sea for sustenance and economic viability, but many islands do not have the material or human resources to address the issues that are central to human well-being.

Islands generally have a distinct character or uniqueness, though these component characteristics are difficult to define (Granger 1993). One prevalent idea is that many islands can be defined by their small size and isolation in relation to the mainland, thus justifying a special and differential treatment to small island countries. On the other hand, Srinivasan (1986) has presented an alternate view that "many of the alleged problems of small economies are either not peculiar to small economies or can be addressed through suitable policy measures" and that "causes of economic and social stagnation in some of these economies cannot be attributed to their smallness."

This chapter addresses the environmental issues relevant to the ecosystem services that fit the technical and legal descriptions pertaining to islands and that are unique to these island systems. This uniqueness is here called the "isola effect," which takes into account the particular physical seclusion of islands as isolated pieces of land exposed to different kinds of marine and climatic disturbances and with a more limited access to space, products, and services when compared with most continental landmasses. In addition, subjective issues such as the perceptions and attitudes of islanders themselves on their conditions and on their future on the island are incorporated into the "isola effect."

Some emphasis is placed on small island developing states, as they are most at risk from projected global changes. Like other small islands, SIDS share limited physical size; generally limited natural resources; high susceptibility to climatic changes and natural hazards such as tropical cyclones (hurricanes) and associated storm surges, droughts, tsunamis, and volcanic eruptions; and relatively thin fresh water supplies that are highly sensitive to sea level changes. The vulnerability of islands or island economies can be attributed not only to smallness itself but also to the disproportionate impacts of natural disasters. Smaller island systems tend to have limited fertile soils and an unreliable water supply, and natural hazards can seriously affect their economic base. Ecological constraints such as sea level rise, salinization of coastal aquifers, and changes in rainfall distribution are exacerbated in such systems and are expected to increase with climate change and its anticipated oceanic impacts.

Small island developing states, in particular, experience even more specific challenges and vulnerabilities arising from the interplay of socioeconomic and environmental factors, such as small populations and economies, weak public- and private-sector institutional capacities, remoteness from and dependence on international markets, high cost of transportation, limited diversification in production and exports, export concentration, and income volatility and vulnerability due to exogenous economic shocks, leading to greater volatility than in other countries (Nurse et al. 2001; CBD 2004).

23.1.2 Definitions and Categorization

Islands can be defined and categorized in a number of ways, each useful for some purposes, but no single definition or categorization fits all needs. Most available definitions on islands tend to incorporate the size factor (Granger 1993). Additional threshold criteria include remoteness and morphology, population size, and gross domestic product, but their validity remains questionable.

Islands are usually defined as pieces of land surrounded by water, formally smaller than Greenland, which has 2.2 million square kilometers (Gorman 1979). As such, they can include independent island states, archipelagic states, and islands associated with large countries.

Combinations of area and population have been proposed to define islands, such as 13,000–20,000 square kilometers with fewer than 1.0–1.2 million people. The UNESCO Man and the Biosphere Programme consider "small islands" to be 10,000 square kilometers or less in surface area and to have 500,000 or fewer residents (Hess 1990). Additional criteria, including surface area, GNP, and population size, were used to define Pacific island systems as "small," "very small," and "micro" (Lillis 1993). The Commonwealth Secretariat uses a threshold of 1.5 million people to define "smallness," which is accepted by UNESCO. Of the developing countries and territories with populations below 1 million, 60 are islands and only 16 are not (UN 2002).

From a biological perspective, many islands are small biotopes, with terrestrial (nonmigratory) fauna that have been sufficiently isolated from continents that there have been little if any movements or genetic exchanges with continental populations, leading to local adaptation and endemism (Rosenzweig 1995; Vicente 1999). For migratory species, however, island landmasses are not difficult to reach, and island ecosystems can provide critical habitat for species that are not genetically unique to that island (such as

breeding beaches for marine turtles and stopover sites for migrant birds).

Islands can be categorized by physical aspects such as latitude (tropical, temperate, or Arctic), underlying geology or island structure (continental islands and oceanic islands, with the latter subdivided into volcanic islands and carbonate islands), hydrology (a runoff basin area), altitude (high versus low islands) and habitat (suitable habitats for an organism that are surrounded by unsuitable areas, such as mountaintops, lakes, caves, or host plants), land area, or human population or by some political (such as former colonial affiliation) or economic index (GDP). Islands can also be grouped by sociocultural categories—either at the centre or at the periphery of a culture, an economy, or some other national or regional designation. Human colonization patterns provide another distinction. In the Caribbean, for instance, few of the original peoples remain, whereas the peoples of Pacific have been there for at least 2,000 years.

According to Article 121 of Part VII of the International Convention on the Law of the Sea, an island is a naturally formed piece of land surrounded by water on all sides, emerging above the surface of the sea at the highest tide, capable of sustaining human habitation or economic life on its own, and with dimensions that are smaller than that of a continent.

Conceptually, the MA defines islands as lands isolated by surrounding water and with a high proportion of coast to hinterland. The degree of isolation of an island from the mainland in terms of natural and social aspects leads to the "isola effect."

For mapping and statistical purposes, the MA uses the ESRI ArcWorld Country Boundary dataset, which contains nearly 12,000 islands, including islands belonging to the Association of Small Island States and in the Small Island Developing States Network. In this chapter, populated islands with more than 0.15 square kilometers of surface area, up to the size of Greenland, are considered islands. In addition, islands had to be separated from the mainland by at least a distance of 2 kilometers, and only when the isolation or the perceptions of the islanders could be verified.

23.1.3 Insularity

Insularity is a distinguishing feature of islands and is influenced by their size to some extent. For example, islands cannot materially modify their macro-climate because of their size, with the exception of the largest, such as Greenland, New Guinea, Borneo, Sumatra, Hispaniola, Madagascar, and Sri Lanka (Granger 1993). Island systems have highly coupled terrestrial and marine ecosystems due to their large ratios of coastline lengths to land area. In such contexts, the impacts of natural or anthropogenic changes are much more immediately visible than for larger continental systems (Brookfield 1990). Moreover, islands with limited areas have limited capacity to buffer or trade off natural hazards or anthropogenic disturbance.

Although insularity is clearly increased by geographic, socioeconomic, and political isolation (Granger 1993), sociocultural factors are probably more important in defining the insular characteristics of islands. The more powerful the links with the outside world, the less pronounced will be insularity, no matter the size of the island.

The perception that islanders have of themselves can be explored further to refine the uniqueness and peculiarities of islands. Human sciences consider "islands" as places where the inhabitants see themselves as islanders. The German Brockhaus Encyclopedia's definition of islands includes not only the conventional idea of a piece of land surrounded by water on all sides, but also the idea that water, and especially the sea, permeates the whole of the island—physically and culturally—and that the island is submitted to some kind of insular marine condition leading to the "isola effect."

23.2 Condition and Trends in Island Ecosystem Services

Island systems provide important ecosystem services, such as biodiversity, fisheries, energy, fresh water, vegetation cover, traditional ecological knowledge, and tourism. Insularity leads to an obvious strengthening of the linkages between island ecosystem services and people. Over time, these linkages have been further affected by human pressure, which has contributed to an increase in the vulnerability of island ecosystems and to a reduction of species diversity (Baldacchino 2004).

23.2.1 Island Biodiversity

23.2.1.1 Isolation

Surrounded by water, which functions as a barrier to terrestrial animal and plant dispersion, islands provide a clear example of ecological isolation where biodiversity issues assume critical importance through endemism. (See also Chapter 4.) The size, distance, and period of isolation from large landmasses often culminate in high levels of adaptive specialization and thus high levels of endemism. Isolation, as a by-product of biogeographic insulation, is a key factor of evolutionary change, for it allows the genetic reservoir of a population to become distinct from that of other populations. Island isolation has usually led to a high level of plant and animal specializations associated with high endemism, and this is especially true for small isolated oceanic islands (Whittaker 1998; Dullo et al. 2002).

The very nature of isolation is, however, important to humans and biodiversity, for it is their isolation that has often excluded threatening processes from causing the extinction of many species. For example, the red fox in continental Australia has had a devastating impact on native mammals, yet its absence from many of Australia's offshore islands allows mammals that are endangered or extinct on the adjacent mainland to persist (Algar et al. 2002).

23.2.1.2 Dispersal, Speciation, and Extinction in Islands

The isolation of oceanic islands lends itself to another important phenomenon for nonmigratory species: low or no dispersal. Island species, particularly those on small islands, evolved in competition with a relatively low number of other species under the influence of natural selective forces peculiar to insular conditions (CBD 2004). Therefore, the flora and fauna have reduced competitive ability, small populations, and narrow distributional ranges compared with continental areas (Dullo et al. 2002). This is not necessarily true for continental islands (that is, within the influence of continents or large landmasses), which have a more complete fauna that fills the available habitats and closely resembles that of the adjacent mainland portion of the continent (Gibbons 1990). This is the case, for instance, for recently formed forests in Puerto Rico, which are composed of both native and alien species (Lugo and Helmer 2004).

The difficulties in dispersal and the isolation of populations in small isolated islands is due to the barrier that the sea presents to the dispersal of terrestrial species. The number of species on an island is therefore a consequence of area, the distance from continental landmasses, impoverishment, and in some cases competition with species that have become established earlier. The surface of a given region allows us to predict, with a high level of certainty, the number of present species based on the species–area relationship of MacArthur and Wilson (1967). Nevertheless, other authors, having analyzed the influence of constraints other than area, have found alternative variables that help to explain regional predictions of the number or richness of species for particular taxonomic groups, such as the number of vascular plants, vegetation height, or number of soil types (Case 1975; Dueser and Brown 1980).

Rapid speciation is frequent, and morphological and physiological change and adaptation are inevitable in many insular taxa. Adaptive radiation will proceed to the extent that new niches become available, particularly following disturbance events, such as land clearing. General mechanisms that promote population "smallness" (*sensu* Caughley 1994) include local catastrophes such as fire, anthropogenic change, direct or indirect killing, introduced predators and competitors, and introduced diseases. Each of these parameters will affect the biota that has become specialized to an insular ecosystem, increasing the probabilities of extinction. Island ecosystems are especially sensitive to disturbances and vulnerable to extinction, which can occur at rates that often exceed those of continental systems. As such, islands have been sites of concentrated extinction: of the 724 known animal extinctions in the last 400 years, about half were of island species, and at least 90% of bird species that became extinct in that period were island dwellers (CBD 2004).

23.2.1.3 Islands as Biodiversity Hotspots

Since island species tend to be concentrated in small areas, the contribution of islands to biodiversity is out of proportion to their land area, and many of them are considered biodiversity "hot spots" in global terms (Mittermeier et al. 1998). Although islands constitute less than 7% of the land surface of the world, one in six of Earth's known plant species occur on oceanic islands (Fisher 2004). In addition, endemism is typically high on islands. For instance, more than 80% of vascular plants of Saint Helena and the Hawaiian Islands are endemic (Rosabal 2004). High altitudinal ranges coupled with aridity (among Mediterranean islands) or tropical climes (among equatorial islands) further encourage endemism.

23.2.1.4 Island Biodiversity and Human Well-being

The health and wealth of island ecosystems and the conservation of biodiversity have important implications for the ecological, social, and economic well-being of island populations. Island systems provide habitats for plant, animal, and microbial species inhabiting both marine and terrestrial environments. Together with geological features, these habitats have particular value due to their high endemism or their absence on nearby mainland areas, making islands important refuges for many species.

Marine and coastal biodiversity still remain essential for many islanders, particularly those living in traditional societies, to meet their daily needs for food, tools, industry, medicine, transport, and waste disposal, in spite of new technologies and lifestyles. This is the case for many of the Pacific islands, including the Marshall Islands, Kiribati, and Tuvalu, which together contain some of the highest coastal biodiversity in the world (UNEP 2004c). Biodiversity is a particularly essential component of food security in small, isolated islands.

23.2.2 Fisheries

For many islands, and especially small oceanic islands and island states, fish provide an almost indispensable source of animal protein. In the Philippines, some 1,500 coastal communities (70% of the population) account for 40–60% of the national fish capture (www.oneocean.org). Traditional methods do not generally deplete the fish stocks. However, modern fishing methods, pollution, and the impacts of natural hazards have meant that the limits of sustainable fishing have been reached on many small islands or areas of larger islands. (See also Chapter 18.)

The small islands of the Pacific, Caribbean, and Indian oceans have narrow coastal shelves surrounded by deep waters. A simple fishing pressure index based on estimates of the number of people actively fishing (according to FAO) per kilometer of coastline suggests that fishing pressure is greatest in the China-Philippines area. (See Figure 23.1 in Appendix A.) Overfishing in the near shore of these islands has led artisanal fishers to venture further offshore for access to pelagic resources such as the large tunas. This has led to encounters and conflict with the already well established industrial factory ships of more industrialized countries or other island states fishing in these waters using longlines or purse seines. These conflicts over marine resources are increasingly being arbitrated through the provisions of the United Nations Convention on the Law of the Sea.

The importance of fisheries can be illustrated by reference to the Caribbean, though there are similarities with other regions. The FAO database shows that in both Central America and the Caribbean, about 500,000 people were actively fishing in the 1990s (or less than 0.1% of a total the population of 145 million), with fish protein contributing about 7% of total protein consumption. The export value of fish and fisheries products increased from $400.6 million in 1976 to $1.6 billion in 2000. Per capita annual consumption of fish in the Caribbean is approximately 15 kilograms, which is approximately three times as much as in the United States.

Consumption in several SIDS in the Caribbean is higher than local production and has to be satisfied by imports. These are very high in the insular states and account for the majority of the fish supplied for human consumption—such as in Haiti (70%), Jamaica (78%), and Martinique (80%). The composition of imports in the small island states is dominated by dried, salted, and smoked fish but fresh, chilled, and frozen products are also imported, mainly by countries with a tourism industry. The FAO database shows that exports of fish from the Caribbean (mostly for the U.S. market) have also been growing steadily and in 2000 were valued at approximately $1.2 billion. Export products are dominated by high-value commodities such as shrimp *Penaeus* sp., spiny lobster (*Panulirus* sp.), tunas (*Thunnus* sp.), snappers (Lutjanidae), and queen conch (*Strombus gigas*), which command premium prices on the international market.

Perhaps one of the most important roles of fisheries is the employment opportunities they offer for thousands of people in a region where the high levels of unemployment continue to be a major concern. The fisheries sector provides stable full-time and part-time direct employment for more than 200,000 people and indirect employment for another approximately 100,000 in the secondary sector (processing and marketing), boat building, net making, and other support industries. In addition, it is estimated that each person in the fisheries industry has five dependents, making the total number of people who depend on fisheries for their livelihood approximately 1.5 million. Those engaged in fishing often have low levels of formal education, limited access to capital, and limited occupational and geographical mobility.

Further information on fisheries can be found in Chapter 18.

23.2.3 Fresh Water

The issue of freshwater resources on islands involves many of the same problems facing developing countries in general, including inadequate human and financial resources. However, islands also have unique physical, demographic, and economic features, including relatively limited surface areas and natural resource bases (arable land, fresh water, mineral resources, conventional energy sources), greater sensitivity to natural disasters (typhoons, hurricanes, cyclones, earthquakes, volcanoes), and an isolation from mainland systems—all of which contribute to the vulnerability of their water resources. These all lead to the impact of the surrounding sea being more pronounced for small islands than for large islands and mainland areas.

23.2.3.1 Physical Conditions

Fresh water is scarce in many small islands, which mainly rely on rainfall harvesting, surface reservoirs and flows, or groundwater lens floating on top of the salt water for the majority of their resources. Severe water shortages are often experienced on atolls and raised limestone islands where there are no rivers. The amount of fresh water available on islands is dependant on rainfall, and this varies according to the geographic location of the island and its climate conditions. Natural events, such as El Niño, can result in a shift of expected rainfall patterns so that islands that normally have abundant rain, such as some of the central Pacific islands, may also experience periods of drought.

Vanuatu, for instance, experienced major droughts in 1978 and 1983; Samoa had the same experience in 1971 and 1989; and Fiji in 1987, 1992, and 1997. The 1987 Fiji drought was one of the worst in a century, beginning in the 1986 dry season and extending through the 1986/87 wet season.

The El Niño event in 1997/98 brought some of the worst droughts on record to the Northern Mariana Islands, Guam, the Marshall Islands, Nauru, Papua New Guinea, Fiji, Tonga, Samoa, and American Samoa. The Marshall Islands received slightly over two inches of rain from January to March 1998, just 8% of the norm. After more than four months of the El Niño–caused drought, the Marshall Islands government declared the country a disaster area. Desalination plants were sent to Majuro and Ebeye, the two main urban centers, while smaller water makers were installed on ships to provide fresh water to the outer islands. From August 1997 to March 1998, the highlands of PNG experienced one of the worst droughts on record, creating a national crisis and the need for an airlift of emergency food and water supplies (Lean 2004).

One of the main natural sources of fresh water on islands and in coastal areas is groundwater reservoirs. Water balance is not easy to determine, and average groundwater recharge normally requires in the region of 20–25% of rainfall, which is not easily retained on islands. Although there are several technologies available, islands cannot expect to develop their groundwater resources easily. Overpumping of groundwater through bore holes can lead to salination problems, which can have serious consequences for food production and human well-being. Also, salination may be enhanced by natural hazards, such as sea level rise, that cause higher penetration of sea water into the freshwater aquifers.

In several SIDS, freshwater shortage is amplified by the lack of effective water delivery systems and waste treatment, coupled with increasing human populations and expanding tourism, both of which may result in the overabstraction of water, contamination through poor sanitation and leaching from solid waste, and the use of pesticides and fertilizers (Bridgewater 2004).

23.2.3.2 Desalination

Desalination of the surrounding seawater to provide a source of fresh water is an option that has been explored by a number of islands, but expensive existing technologies mean this it is still a very costly way of supplementing the freshwater supply. Technology to implement reverse osmosis that leads to seawater desalination is proving a useful alternative to improving freshwater supply (Veza 2001). Ocean thermal energy conversion plants are also being proposed for island states not only to generate energy from thermal gradients but to help with the desalination process. (See Chapter 7 of MA *Policy Responses* for more on desalination).

23.2.3.3 Water Quality

Lack of safe drinking water and sanitation is one of the major causes of disease and death worldwide (WHO 2001). On islands, particularly small islands with rugged interiors (such as islands off the east coast of Peninsular Malaysia), people tend to be concentrated on the more gently sloping lands along a coastline. The resulting high population densities can cause problems for the safety of water supplies, which can easily become polluted by poor sanitation facilities or by facilities that are sited too close to the source. Also, the increased use of pesticides and fertilizers and leaching from solid waste disposal sites pose additional pollution hazards to ground and surface water on many islands. (See also Chapter 15.)

23.2.4 Forestry and Vegetation Cover

The extent of forest cover varies greatly among islands (Dulloo et al. 2002; CBD 2004). The forest cover of SIDS represents less than 1% of the forest area of the world. Insular Africa has 0.006%, insular Asia 4%, insular Caribbean 0.15%, and insular Oceania 0.9% of the total forest surface of the world (FAO 2001). In spite of the relatively small area of forest cover in global terms, some tropical and sub-tropical islands have significant forest cover and are characterized by comparatively short distances between upland and coastal forest areas.

Forest is estimated to cover a total of 75 million hectares, or about 63% of the combined land area of 41 SIDS, compared with the world average of 29.6% (CBD 2004). Under such conditions, island forests are critical regulators of freshwater supply for consumption, irrigation, and industrial uses. Forests also contribute directly to food security through the provision of food and animal products. Also on many tropical islands, mangroves are an important source of fuelwood and household products, provide a nursery for many marine fish and invertebrate species, and protect the coast from erosion.

For many of the larger islands, such as Borneo, forests also contribute significantly to the national economy and to the international trade in wood and non-timber forest products (Wilkie et al. 2002). In addition, forest cover buffers against natural hazards and anthropogenic disturbance. The prevention of erosion by forest cover has a direct impact on the health of coastal and marine systems by reducing the sediment load. Forests also play a buffering role, particularly in small tropical islands, against the impacts of tropical storms, hurricanes, and cyclones combined with high rainfall levels (Wilkie et al. 2002). On Hainan Island, where rubber plantations have replaced the local forests, it has been found that the plantations can still have an important hydrological ecoservice function (Jiang and Wang 2003).

Although the overall rate of island deforestation appears to have slowed down in the last decade, annual deforestation on islands is almost three times the world average rate (0.8% compared with 0.3%) (FAO 1999). The main causes of deforestation include conversion for agricultural use and for infrastructure development such as roads, ports, housing, and tourism development (CBD 2004).

Regarding global biodiversity, loss of forests in island systems often has more serious impacts than forest loss in continents due to intensified interactions of various activities within a limited geographic space and to the loss of endemic species and rare ecosystems (FAO 1999).

23.2.5 Cultural Services

23.2.5.1 Traditional Ecological Knowledge

The term traditional ecological knowledge commonly refers to the knowledge that indigenous peoples have about their environment, which is used to sustain themselves and to maintain their cultural identity. TEK covers a wide range of subjects, from agriculture, fishing, plants, and forests to general aspects of culture. (See Box 23.1 and Chapter 17.) Local cultivators, fishers, and other resource users often have a profound knowledge of the highly varied environments that could be better tapped in assessing the potential use of locally available resources and the sustainable development opportunities of their environments.

TEK is an integral part of the dynamics of island systems and the islanders who live there. For example, Indonesia has strong traditional medicine and many varieties are practiced, the oldest being the *Jamu* system of herbal medicine (Erdelen et al. 1999). Some 10% of Indonesia's total flora is estimated to have a medical value, and some 40 million Indonesians depend directly on biodiversity (Erdelen et al. 1999).

In addition, many stories and beliefs of islanders show the role of traditional villages and communities in improving the marine environment. For example, the Balinese believe in the harmony between God, communities, and nature for coastal management, and the traditional village has many roles, including protecting the coastal region from destruction by outsiders, promoting availability of knowledge to communities, assessing problems caused by populations, and maintaining healthy natural resources for the next generation (sustainable development) (Sudji 2003).

The greatest use of TEK on islands relates to sustainable use and management within customary inshore fishing grounds, for example in Fiji (Veitayaki 2004), in the customary prohibition on the use of resources (*ra'ui*) in Rarotonga in the Cook Islands, and in the village reserves in Samoa in the Pacific (MacKay 2001). Traditional ecological knowledge and customary sea tenure are also integrated into the conservation management of bumphead parrotfish (*Bolbometopon muricatum*) in Roviana Lagoon in the Solomon Islands (Aswani and Hamilton 2004).

TEK has also been of direct benefit in the protection of reefs from adverse impacts from commercial and recreational fisheries, scuba diving, snorkeling, aquarium fish collection, and onshore development (Calamia 1996). It has helped ensure sustainable development of the intertidal zone, with a focus on shellfish gathering and marine tenure in the atoll communities of western Kiribati, Micronesia, which are under pressure from population growth, urbanization, extractive technologies, and expanding market opportunities (Thomas 2001). While it is possible to integrate local and scientific knowledge in fisheries (Mackinson 2001), policy-makers and managers find TEK generally unsystematic and its generally unstructured nature makes its difficult to use in regional and national decision-making.

Traditional ecological knowledge has also served as a foundation for the conservation of trees outside forests in small island states of the Pacific Ocean (Thaman 2002). In addition, the knowledge of traditional agroforestry systems and associated traditional knowledge can serve as basis for addressing deforestation, forest degradation, agro-deforestation, and loss of diversity. In Pagbilao, Philippines, ecological knowledge has been shown to

BOX 23.1

Traditional Knowledge That Is Important to Environmental Management (Unit B5 1998)

Agriculture
- The many different varieties of crop plants and their utilization
- The best places, conditions, and times for planting, caring for, and harvesting crops
- Food storage techniques
- Control of crop sicknesses, insects, and other pests
- Management of agricultural land, both seasonally and from year to year; planting sequences or rotations; periods of fallow to allow the land to recover; techniques for soil improvement
- Control of erosion and wind damage
- Identification or classification of soils
- Water management and irrigation, including complex systems of aqueducts and irrigated terraces
- Controls on land use and access to land

Fishing
- Fishing methods and materials
- Knowledge of fish species and their behavior, migration, and reproduction
- Best fishing locations, times, and techniques for each species
- Controls on fishing: limited access to fishing areas, taboo areas or seasons, catch restrictions
- Changes in fishing resources, effects of overfishing, "how things used to be"

Animals and Hunting
- Behavior of species and hunting or trapping methods

- Controls or limitations on hunting: taboo areas, special times for hunting and restrictions to special occasions or special ranks

Plants and the Forest
- Useful trees and the qualities and uses of their woods
- Techniques for cutting and hauling trees from the forest
- Edible plants and plant parts (nuts, leaves, bark, roots, and so on)
- Medicinal plants and their uses
- Genetic resources, varieties, or special features of plants; loss of varieties
- Changes in the forest, loss of forest cover (where the forest used to be)

General
- Traditional names for and classifications of species and communities
- Calendars related to the weather, to celestial bodies (solar and lunar cycles, appearance or movement of stars), or to association with natural events such as the flowering or fruiting of trees or the migration of birds
- Weather patterns and prediction, cycles of rain and drought, changes in climate
- Natural catastrophes, cyclones, tsunamis, floods; signs and warnings; effects and areas affected
- Changes in the environment, past extent of the forest and agricultural areas, former locations and populations of villages
- Environmental knowledge: who possessed it, how it was used and transmitted

improve economic assessment of the mangroves (Ronnback and Primavera 2000), and in the Marshall Islands and atolls, traditional knowledge of medicinal plants serves as an inexpensive way to maintain human health (Nandwani and Dasilva 2003).

TEK has provided new biological and ecological insights, is useful in resource management and environmental assessment, has been used for protected areas and conservation education, and benefits development agencies by providing a more realistic evaluation of the environment in development planning (Berkes 1993; Calamia 1996).

UNESCO's work on the TEK of islands has emphasized that recording and applying traditional ecological knowledge provides one approach to making more effective use of global biological wealth, particularly as a starting point for strategies of integrated conservation and sustainable development. It also recognized that there may be considerable scope for information on techniques and practices refined over generations in one part of a particular geographical ecological region, to be tested and adapted to other localities.

On some tourist islands, the commercial value of traditional knowledge and cultural property is already well recognized through its contribution to income from tourism, arts, and crafts. However, many islanders are justly concerned about the unauthorized and uncompensated use of their heritage, including the appropriation of indigenous arts and cultural expression, similar to the ongoing appropriation of indigenous biodiversity material or knowledge by industrial-world companies and researchers (so-called biopiracy). They are also concerned about the introduction of "modern" agricultural, fishing, and medicinal practices that threaten to replace traditional ways. Biopiracy has been recorded in areas used for ecotourism, and the Maldives and Pacific Island

states have been particularly vulnerable to such thefts (GRAIN and Kalpavriksh 2002). UNESCO has recognized that much of the TEK and cultural knowledge is unrecorded and unexploited, and every year part of this knowledge is lost through the loss of natural habitats and transformation of local cultures. (See also Chapter 17.)

Much of the traditional island cultures and environmental knowledge has already been lost in recent decades. For instance, in Pohnpei information on components of Micronesian life—such as planting taro, the use of plants to stun and capture fish, fermentation methods for breadfruits, and construction of outrigger canoes—has been lost as older generations have died (Lee et al. 2001). Traditional knowledge on canoe making and turtle catching are under the greatest risk of loss (Balick 2003).

23.2.5.2 Tourism

Globally, tourist arrivals increased from 25 million in 1950 to 700 million in 2003 and are expected to double by 2020 (Christ et al. 2003). Despite multiple international crises (economic recession, SARS, terrorist attacks, and the war on terrorism), international tourism has grown 4–5% in the past decade (WTO 2001). Tourism is an important contributor to or dominates the economies of many small island states. (See Table 23.1.) The Caribbean is the most tourism-dependent region in the world and accounts for about 50% of world cruise tourism berths, while the Maldives is the most tourism-dependent country.

Tourism based on the natural environment is a fast-growing component of the tourism industry. In the last decade, nature (or eco-) tourism, which can be defined as travel to unspoiled places to enjoy nature, has emerged as the fastest growing segment of the industry, with an estimated growth rate of 10–30% annually

Table 23.1. Tourist Arrivals and Tourism Expenditures as Share of GNP, 2001 (Scott 2003; WTO 2003)

Country	Tourists (thousand)	Tourism Expenditure (percent)
Anguilla	47.9	65.1[a]
Antigua and Barbuda	93.1	43.8
Aruba	691.4	42.7[a]
Bahamas	1,428.2	40.0[b]
Bahrain	2420	8.3[a]
Barbados	507.0	28.8[b]
Bermuda	274.9	23.0[a]
British Virgin Islands	296.0	41.3[c]
Cape Verde	115	4.0[a]
Cayman Islands	334	71.7[c]
Comoros	24[b]	7.0[b]
Cyprus	2697	20.2[b]
Dominica	67.9	20.2[b]
Dominican Republic	2,868.9	14.2
Fiji	348.0	12.3
French Polynesia	228.0	10.1[a]
Grenada	123.3	17.1
Guam	1,124.1	71.5
Haiti	142	1.4
Iceland	303.0[b]	4.1
Indonesia	5,153.6	2.4
Ireland	6,749.0	3.1
Jamaica	1,276.5	17.0
Japan	4,771.6	0.1
Kiribati	4.6	4.2
Madagascar	160	2.8
Maldives	461	57.3
Malta	1180	17.3[b]
Marshall Islands	5.4	3.5[b]
Mauritius	660	13.6
Montserrat	10[b]	32.4[a]
New Caledonia	100.5	0.3[b]
New Zealand	1,909.4	4.7
Papua New Guinea	54.2	3.3
Philippines	1,796.8	1.5
Puerto Rico	3551	6.5[d]
St Kitts and Nevis	74.2	22.0
St Lucia	250.1	36.9
St Vincent and Grenadines	70.6	25.6
Samoa	88.3	15.0
San Tome and Principe	7.6	18.7
Seychelles	129	20.1[b]
Singapore	7,518.6	6.1[b]
Solomon Islands	3.4	2.0[a]
Sri Lanka	336.8	12.5
Taiwan	2,562.5	1.2[a]
Tonga	32.4	4.5
Trinidad and Tobago	383	2.8
United Kingdom	22,833	1.1
Vanuatu	53.2	21.7

[a]1999 [b]2000 [c]1998 [d]1997

(Conservation International 2002). Of the various forms of nature tourism, coastal/marine tourism, including islands, is the largest component. Biodiversity plays a key role in the nature tourism development of many islands and is the major tourism attraction for islands such as Madagascar and Borneo (Christ et al. 2003). Ecotourism extends as far as the subantarctic islands, where special voyages give tourists the experience of a variety of marine and pelagic fauna, using the islands as a base.

There is a great potential in many SIDS for the further development of ecotourism, which is often a small but rapidly growing share of their market economy. Ecotourism can provide employment and generate income while helping to protect and conserve natural resources and contributing to the implementation of national biodiversity action plans (ECOSOC 2004).

Tourism has a great potential for biodiversity conservation and the promotion of the sustainable use of natural resources. In the Seychelles, for instance, tourism has been a major force and source of funding for biodiversity management and conservation, as well as ecosystem rehabilitation. In many cases, tourism is the only means by which a management infrastructure can be put in place on isolated islands to enable conservation activities. Indeed, well-informed tourists are increasingly the driving force behind the tourism industry's involvement in biodiversity management (Chafe 2004).

Rapid and uncontrolled tourism growth can be a major cause of ecosystem degradation and destruction, however, and can lead to the loss of cultural diversity. Often, such destructive development paths start with tourists discovering a destination that rapidly develops beyond its carrying capacity and eventually fails to meet tourism demands and expectations. Such developments have been spontaneous or frequently without adequate enforcement of planning laws and guidelines. Alternatively, tourism can be developed in a more careful, planned manner, with government input and a more responsible approach by developers. Smaller resorts, such as the island resorts in the Maldives (Domroes 2001) and other small tourist enterprises on many islands, have been developed successfully. Large integrated developments, such as Nusa Dua in Bali and Bintan Beach International Resort on Bintan, both in Indonesia, require large investments, depend on large tourist flows, and are more difficult to implement.

Tourism development without proper planning and management standards and guidelines poses a threat to biodiversity (Christ et al. 2003). This is compounded by the fact that environmental impacts are often not clearly visible until their cumulative effects have destroyed or severely degraded the natural resources that attract tourists in the first place, and some destinations have only recognized the costs of environmental damage after significant and often irreversible damage has been done. As a consequence, many SIDS have embarked on initiatives aimed at building a wider, more sustainable support base for the tourism industry among the local population, promoting participatory action and a sense of ownership in order to ensure the success of the industry (ECOSOC 2004). Key measures suggested include ensuring that integrated planning policies and implementation plans provide for environmental impact assessments for all tourism projects and cultural impact assessments for all large tourism operations (ECOSOC 2004).

Within the Caribbean region, fisheries are important not only as a source of food and employment for commercial and subsistence fishers but also for a growing population of recreational fishers—those fishing for pleasure and relaxation rather than for commercial gain or subsistence. Dozens of international, regional, and national fishing tournaments are held each year throughout the region. In most Caribbean countries sport fishing is promoted by tourism interests and is neither monitored nor regulated by the national fisheries administrations. In addition, watching coral reef fishes and other marine life has always been an important leisure activity of thousands of locals and tourists, and skin divers and scuba divers in the Caribbean. Over the past two decades, several countries have established marine parks and aquaria to use these resources and to promote education and conservation of their marine resource systems.

23.3 Drivers of Change in Island Systems

The stresses imposed on island systems are the result of the interplay of environmental and sociocultural factors that together have the potential to increase the impact and reduce the resilience or ability of the islands to cope with changes relative to mainland ecosystems. Like continental populations, people on islands must be able to survive the stochastic variations in their environment. However, natural hazards and anthropogenic disturbance, such as deforestation, unsustainable agricultural practices, mariculture, habitat loss, and biodiversity loss, may assume a disproportionate importance in island systems for several reasons, often depending on the latitude.

For example, some equatorial islands (such as those in the Caribbean) can have rapid rates of vegetation regeneration, as evidenced after cyclones, whereas many arid islands (such as those in the Mediterranean) do not. These differences in responses to change reflect rainfall patterns and the degree of isolation. Contrary to what happens on continents, where species immigration generally occurs rapidly, leading to recolonization, catastrophic events on some islands can have long-term effects because extinction rates are higher and rates of recolonization are much lower (Vitousek et al. 1997; Courchamp et al. 2003). Small populations are also far more prone to random nonadaptive changes in their genetic pool and consequently to chance extinction. The main drivers of island ecosystems are both natural and anthropogenically induced, and both are addressed in this section.

23.3.1 Population Issues

It is estimated that by 2025, 75% of the world's population will live within 60 kilometers of the sea, which can be considered "the global coastal zone," where more than 70% of the world's metropolises are located (UN 2002). A high proportion of this occupancy will be on islands of developing countries. (See also Chapter 19.)

Population growth in cities contributes to urbanization, and this is a serious and growing problem for some islands, particularly among those in Asia, such as Java and Luzon. Many cities on these highly populated islands cannot currently provide the basic resources for the well-being of the inhabitants. Among the Pacific islands, populations are small, but growth and rapid urbanization are also putting pressure on limited resources (Zann et al. 2000)

For many islands in the tropics, the traditional activities of the coastal population have been subsistence production, such as fishing and agriculture. With rising numbers of people moving to coastal areas and islands, conflicts over coastal resources and human values and expectations will increase in the years to come. These are further accelerated by the sociopolitical, cultural, and economic differences between the traditional inhabitants and the newly arrived populations. These differences are more evident among developing countries (where most of the tropical islands occur), which may be subject to huge and rapid internal migrations.

23.3.1.1 Outmigration

Outmigration has been a familiar process for many islands and affects the population balance beyond those related to natural birth and death rates. Outmigration may be due to the impact of outside forces or internal drivers, as when an island's economy is based on a single specialized crop that fails due to disease or market changes. For example, the Asian economic crisis of 1997–98 influenced the extent and nature of population movement between Java and the Outer Islands of Indonesia (Hugo 2000). It may in part be a response to environmental constraints at home and perceived opportunities elsewhere. For instance, migrations from many of the small island states of the South Pacific and the Caribbean have been to metropolitan countries (Connell and Conway 2000).

On many small islands dependent on fisheries, the combination of overfishing and environmental stress has led to outmigration, mainly of young adults. This can alter the size of an island's population, leading to a shift in age groups toward an older society (Hamilton et al. 2004).

However, the benefits of out-migration have also been recorded. One study on the Samoan Islands in the Pacific suggests that out-migration from rural regions generally tends to preserve the local natural environment, leading not only to a more satisfactory use of agricultural and grazing land but also to a greater retention of native species diversity (Baker and Hanna 1986). Yet people who migrate from developing rural societies to urban societies generally suffer a decline in many physical and psychological aspects of health, even though life expectancy may rise.

23.3.1.2 Role of Gender

Gender issues have not been clearly identified as affecting island societies, but this does not mean that islands have no problems related to gender inequality. Indeed, they exist and seem to be related to the same kind of issues that promote gender inequalities in continental societies. Nevertheless, the "isola effect" could increase the problems. For example, in some islands matriarchal societies appear to be driven by isolation together with other factors (such as out-migration by males). Gender inequalities are most clearly related to socioeconomic underdevelopment, a common condition of many island populations (Browne 2001; Lewis 1998).

23.3.2 Energy Issues

Although ecosystem services provide low amounts of energy nowadays (wood, biomass, and so on), energy issues are critical for islands, as unanimously recognized at the SIDS conference in Mauritius in January 2005. The availability, constraints, or scarcity of energy sources are important drivers of changes for island ecosystems and human well-being, particularly enhanced under the "isola effect." Some islands have developed around the fossil fuel industry, but in many islands the import bill for fuel alone exceeds earnings from exports (Roper 2005), lowering the capacity of improving human well-being and, consequently, increasing the pressure over natural ecosystems. Also, the potential hazard to islands linked with the operation of power plants not based on renewable sources (fossil fuels and nuclear power), where size is a factor, could outweigh the benefits of these sources to the island's people and ecosystems.

With imported petroleum being the main source of primary commercial energy (ECOSOC 2004), developing further renewable and unconventional energy sources for islands is a key issue. Islands are usually well suited to use combinations of modern renewable energy technologies and energy efficiency due to the availability of renewable energy resources and current energy consumption patterns (Roper 2005). New technologies have been developed (Cavanagh et al. 1993) that harness the energy of the sun, the wind, the earth, and the ocean, and these can be specifically targeted for usage among island systems.

The world's oceans, where islands are interspersed, represent an enormous and virtually untapped source of clean, non-polluting renewable energy (see Table 23.2). Among the facilities that the ocean offers to islanders, renewable sources of energy can be listed at the top, including the capacity of ocean tides, waves, and currents to generate energy and the extraction of power from the thermal gradient of sea water (Penny and Bharatan 1987; Sanders 1991). Technological breakthroughs, standardized plant designs, increased fossil fuel prices, market instabilities, and increased world concern over environmental issues such as climate change will increase the pace at which ocean biomass, wave, tides, the current, and ocean thermal energy conversion systems are tapped.

In a number of SIDS, small-scale solar photovoltaic power systems have been used to provide electricity in rural areas on a pilot scale, but more work on financing and institutional arrangements is needed to realize their full potential. Moreover, there is a need for technology transfer and national and regional capacity building in renewable energy and energy efficiency (ECOSOC 2004).

23.3.3 Invasive Alien Species

A major threat to oceanic island biota today is the increasing breakdown of the insularization of their habitats (Whittaker 1998). Invasive alien species are one of the primary threats to biodiversity on most islands and have caused serious ecological and economic damage and high social costs (e.g., Courchamp et al. 2003; Veitch and Clout 2002). Invasive plant and animal species often outcompete native insular species directly or indirectly for common resources and can alter the ecosystem processes of an island.

Overgrazing by introduced stock, for example, has had an adverse impact on Mediterranean islands because of the aridness of the land. The introduction of the brush-tail possum (*Trichosurus vulpecula*) to New Zealand and its offshore islands has had devastating impacts on forest systems (Atkinson 1992). Sub-Antarctic Macquarie Island, 1,500 kilometers south of Australia, has seen a host of exotic species (particularly cats, rats, rabbits, and mice) introduced either directly or indirectly by sealers (Cumpston 1968), and the impacts of these species on endemic birds and flora has been significant (Taylor 1979; Copson and Whinam 1998). Another example of the impact of invasive species is the introduction of the brown snake (*Bioga irregularis*) into the formerly snake-free island of Guam in the 1940s. This led to the loss of 10–13

species of native forest birds and several lizard species, and power outages occur frequently as the snakes contact electrical lines and generation facilities. The cost to the island's economy from the establishment of this single invasive alien species is estimated at $5 million a year (Fritts 2002).

The invasion of exotic species onto islands is a worldwide phenomenon, but it is uncertain whether islands are more susceptible to invasion than mainland sites (e.g. Sol 2000). The level of invasion depends on how and to what degree the native biotic community is disrupted and on the resilience of an island's ecosystems. In theory, the ecological impacts of invasive alien species on islands can occur in the same manner as on mainland ecosystems. However, these impacts are usually more rapid and more pronounced on islands due to their vulnerabilities (CBD 2004). Approximately 80% of documented introductions (planned or unplanned) of birds and mammals have been to islands (Ebenhard 1988). The effects of such introductions have often been so devastating on the native flora and fauna that it is claimed that invasive alien species are among the main environmental hazards to island systems (Vitousek et al. 1997) and have an ability to create an ecological homogenization of the island's ecosystem in addition to influencing other agents of global change (Mack et al. 2000).

Diseases and their impact on native flora and fauna have often been associated with recent island invasions, yet they remain understudied in insular environments. Avian malaria, introduced into Hawaii in exotic birds by settlers, spread through the endemic bird populations following the introduction of the southern house mosquito (*Culex quinquefasciatus*), which acted as vectors for the waterborne parasite (Van Riper et al. 1986). Similarly, the invasion of black rats (*Rattus rattus*) onto Christmas Island is believed to be responsible for the extinction of the bulldog rat *R. nativitatis* (Day 1981). The speed of introduction and spread is increasing. On the Galapagos Islands, the number of introduced plants has doubled since 1990 from 240 species to 483, now representing 45% of the total flora. In Hawaii, naturalized species account for about 47% of the flowering plant flora (CBD 2004).

There is a realization that invasions of alien species can sometimes be managed with adequate human intervention through good planning, adequate techniques, and sustained effort (Clout and Veitch 2002). However, the feasibility of eradications of invasive aliens on islands will depend on the size of the island, available resources, the public will to undertake control programs and ensure effective quarantine, and the secondary effects of the eradication action on non-target species and other benign alien species on the island. For example, ecological release may allow one alien species to become invasive following the removal of another invasive species, with no net benefit to native populations (Zavaleta 2002).

Three main issues have been identified that can help assess whether islands are subject to a higher risk of invasion than mainland areas: the opportunities for exotic species to reach islands, whether exotic species are more likely to establish on islands, and whether exotic species have a greater impact on island systems (D'Antonio and Dudley 1995). An analysis of those criteria suggests that some island systems are more likely to be invaded by alien species than similar mainland systems because there are fewer resources to deal with risk management (D'Antonio and Dudley 1995). However, it has also been argued that not all islands show evidence of higher invasion than mainland sites (Sol 2000).

Response measures needed to prevent and minimize the impacts of invasive alien species are generally known, though many island nations and territories lack material or human resources to prevent the introduction of or to control or eradicate alien species that threaten ecosystems, habitats, or other species (Veitch and

---

Table 23.2. Potential Energy Outcomes from Ocean Sources (POEMS 2004)

Resource	Power	Energy Density[a]
	(terawatts)	*(meters)*
Ocean currents	0.05	0.05
Ocean waves	2.7	1.5
Tides	.03	10
Thermal gradient	2.0	210
Salinity gradient	2.6	240

[a] Expressed as proportional to the length of the water column.

Clout 2002). The key to quarantine of invasive species is to use early detection mechanisms together with rapid-response mechanisms that can be managed with sufficient funds and powers to see eradication campaigns through to completion (Simberloff 2000).

23.3.4 Habitat Loss, Pollution, and Land Degradation

Island biodiversity has unique biological characteristics, since isolated islands provide ideal conditions for the development of new species with specialized traits. Habitat loss, through pollution, land clearing, and natural hazards, is clearly associated with biodiversity loss, expressed either as population declines or species extinctions.

Chemicals imported for agriculture, industry, transportation, health services, and households are a growing source of pollution among populated islands. Poorly treated sewage emptying into coastal areas is the major chronic pollutant and contributes to coastal nitrogen and phosphorus eutrophication and harmful algal blooms. On islands, the cumulative impact of household runoff from baths and sinks that eventually drains into the sea is also a major contributor to ecosystem decline but is frequently overlooked. This condition occurs because most people living within the coastal zone and the majority of the hinterland population in most island developing states are not connected to centralized sewage treatment facilities. A complementary problem is widespread bacteriological contamination of groundwater from soak-a-way septic systems. Added to this severe nutrient and bacterial pollutant load in coastal areas and groundwater is the addition of fertilizers and pesticides from industrial and agricultural activities.

Due to the short-term nature of the numerous pollution studies carried out on islands, it is difficult to assess the overall pollution condition or trend. However, a proxy indicator may be used, such as a simple sewage pollution index based on the number of people without access to safe sanitation per kilometer of coastline (equivalent to the density of a population living within 100 kilometers of the coast without access to safe sanitation divided by the length of coast).

Because of the complexity of initial attempts to use topography to define the extent of the coastal zone, this study assessed the level of direct human modification of the coastal zone by examining the population within 100 kilometers of the coast. The estimate was derived for the World Resources Institute using a spatially explicit database reflecting global human population (CIESIN et al. 2000). Figure 23.2 (in Appendix A) suggests that coastal sewage pollution is a ubiquitous problem around the world except for industrial countries in North America, Australia, and Europe. Regionally, the Philippines, Latin American, Caribbean, and African and Asian islands stand out as having significant problems.

Although the deleterious effects of pollution are generally recognized, the focus of attention is usually on the immediate aesthetic affects of very visible pollutants, such as garbage and oil spills, or on the human health effects of bathing in contaminated marine waters rather than the long-term deleterious effects of the decline in the ability of an ecosystem to sustainably provide services such as fish or coastal protection. Yet in the long term it is the chronic rather than episodic pollution that has the greatest impact on the limited ecosystem services of islands (Mohammed 2002; Burke et al. 2001).

Increased amounts of hazardous waste are often associated with limited facilities for waste disposal in island systems. As such, many inhabited and uninhabited islands face increasing problems

of coastal pollution of land origin, as well as external pollution threats, which may include hydrocarbon pollution originating from local and international shipping and offshore activities and the fast-growing threat of disposal of the toxic wastes of industrial nations in the exclusive economic zones of developing islands and at land sites from which coastal waters can become contaminated. Incidents of dangerous and illegal pollutants being discharged into streams and oceans have increased on islands, with growing urbanization and establishment of manufacturing industries, as a result of, among other factors, inappropriately sited and poorly managed garbage dumps, poorly planned development, inadequate disposal methods, and destruction of and encroachment onto coastal habitats (UNEP 2004a, 2004c). Moreover, the use of agrochemicals has become standard practice in the agricultural production systems in SIDS to respond to export requirements (CBD 2004). The fertilizers, pesticides, and herbicides required to maintain high crop yields contaminate aquifers and affect the biology of sensitive riverine and coastal ecosystems (UNEP 2004b).

Islands are also facing increased problems of coastal and beach erosion due to inappropriate forms of coastline engineering and tourism development that often use coral and beach sand as building material. The degradation of critical ecosystems like coral reefs, mangrove forests, and seagrass meadows reduces the natural defenses of the coast, increasing the potential of erosion from hurricanes and storms. Coastal problems are aggravated by the vulnerability to environmental change of many coastal habitats, such as coral reefs, seagrass beds, and mangroves. Deterioration in coral reefs, for example, is caused by sewage discharge, often aggravated by tourism, and by land runoff in the form of erosion products and chemical fertilizers and pesticides. (See Chapter 19.)

23.3.5 Economic Changes

The issues and priorities relating to economic development differ from one island group to another and reflect the nature of the island systems and the extent of use of island resources. For many small island states, constraints to economic development include a small population size, with limitations in terms of trained and skilled personnel; limited exploitable land; an often weak infrastructure (transport, energy, communications and the basic service sectors, and health and education); distance from foreign markets; a restricted and undiversified natural resource base; heavy dependence on international trade; and often amorphous exclusive economic zones with little or no protection from poaching by foreign fishing or from mineral exploration interests.

23.3.5.1 Land Ownership

Land ownership can be an important driver of island ecosystem change. For example, customary land tenure in island countries such as Samoa, Papua New Guinea, or the Solomon Islands makes it difficult to lease land for tourism, forestry, mining, or extensive farming. Within the Pacific, land is a particularly sensitive issue, especially in the tourism sector (Samoa 2003).

23.3.5.2 Access and Transportation

The geographical isolation of an island has a number of implications for its economy. The sourcing of raw materials and inputs from overseas markets can be costly when minimum volumes are required for orders. Due to this isolation from major markets, downstream processing of local products—such as agricultural produce and the drying, salting, and smoking of fish—is necessary. Increased isolation for many islands has meant more reliance on the local environment and resources.

For some remote islands, the provision of transport service by ferries and other forms of communication is expected to play a crucial role in influencing island population levels, economy, and quality of life (Cross and Nutley 1999).

23.3.5.3 Economic Diversification

Many island states have attempted to diversify their economies. One approach adopted has been the development of island eco-tourism, which currently has a small share of the global tourism market but is growing rapidly, as noted earlier. The development of offshore financial services sectors as a means to diversify island economies has also occurred in a number of islands. This is not an easy measure, however, as experienced by the Seychelles, where substantial and complex levels of legislation and reporting mechanisms have to be in place before an offshore center is deemed internationally acceptable (Seychelles 2003).

Despite their geographical isolation, some islanders have migrated to other islands or countries. Contrary to early conceptions, research has shown remittances and associated spending by returning islanders are not unproductive expenditure but can be a significant form of private transfer of capital. Returning migrants represent people endowed with capital and new skills, which can enrich the economic, social, and cultural capital stocks of island communities, offering better prospects for development than those offered solely by domestic economic opportunities (Connell and Conway 2000). Compared with purely foreign investors, capital from return migrants can be channeled into projects that rehabilitate the island systems, such as capacity building, reforestation, integrated coastal management, policies and programs to address beach erosion, sand mining, and coral reef conservation and protection.

23.3.5.4 Globalization and International Trade

Globalization presents both difficulties and opportunities and has direct and indirect impacts on the biodiversity of islands. Various forms or types of globalization have distinct consequences for island systems (Read 2004).

With increasing globalization, many SIDS are concerned with their growing vulnerability as a whole. This is the result of their size and persistent economic structural weaknesses, as small islands face enormous difficulties integrating into the global economy. Trade liberalization accompanied by the progressive removal of trade preferences (tariffs) has severe impacts. SIDS can be marginalized in a world economy, as they are unable to compete due to high costs arising from their small size and geographical isolation.

For some islands with natural and primary resources, globalization presents an opportunity to gain access to new markets, facilitate the transfer of new technologies, and increase productivity. Their development is therefore dependent on the country's capacity to participate in a world economy, especially in agriculture and tourism. For economic development in many small islands, the choice is to be part of the global economy (Prasad 2001). Their governments encourage investments through transnational firms and policies, but these are sometimes influenced by international agencies such as the International Monetary Fund and the World Bank (Prasad 2001).

Globalization has had negative impacts on many island resources, however, particularly fisheries and agriculture. For example, longline fishery operations opened to foreign investors have negative impacts on migratory species, such as tuna. Increasing fishing efforts and new technology have depleted many local fisheries (Hunt 2003). This is compounded by often-illegal access to and overexploitation of the marine resources in the exclusive economic zones of many island states.

The global agro-food complex has increasingly embraced the most isolated and peripheral small island nations in the Pacific since the 1980s. This has had a negative impact on the environment. In Tonga, for instance, foreign investment has boosted the commercial production of squash pumpkin for Japan, with dire environmental consequences through the excess use of pesticides. And on Niue, taro production for the New Zealand market has destroyed biodiversity, disturbed animal habitats, and depleted soil nutrients, while fertilizers and insecticides have had an enormous impact on the water resource of the atolls (Murray 2001).

Globalization has implications for both individual and groups of island nations. Both have to restructure their economies to tackle mounting globalization, and one way is through the creation of regional markets or partnerships (Read 2004). CARICOM is one example of a successful association of small states.

23.3.6 Short-term Disturbances and Natural Events

Natural hazards, as extreme events such as earthquakes, droughts, floods, volcanic eruptions, hurricanes, and their follow-on effects such as storm surges or tsunamis, have a major impact on natural environments and human well-being at a global scale. The consequences of those same hazards are enhanced by the "isola effect," and they can have an even more critical impact on an island's systems and the well-being of its inhabitants.

Those islands located barely above sea level, such as the Maldives, are among the most vulnerable to the effects of extreme weather conditions and other natural hazards. The limited area of islands and their isolation due to the surrounding sea further increases their vulnerability to natural hazards. The changes in temperature and rainfall projected by the IPCC, for instance, could disrupt terrestrial and marine ecosystems of most islands, especially small ones (Lal et al. 2002).

The low coastal elevations also make the populations vulnerable to vector-borne diseases associated with waterlogged conditions, such as dengue and malaria. For instance, positive correlations between the El Niño/Southern Oscillation Index and dengue fever have been reported in 10 island countries of the South Pacific (Hales et al. 1999). In addition, global changes resulting in extremes of rainfall seem likely to have exacerbated diarrhea illness in many Pacific islands (Singh et al. 2001).

The impacts of cyclones on native wildlife include, among other things, high mortality due to the cyclone itself, starvation as a result of the disappearance of food and feed for long periods after the cyclone, predation of grounded wildlife by domestic animals, hunting by humans, failure to breed, and degraded health of habitats and ecosystems (UNEP 2004a).

Many islands situated near geological subduction zones, such as the islands of Indonesia and the Philippines, are prone to disasters associated with earthquakes and volcanic eruptions. Others, such as islands in the Pacific and Caribbean, are in regions that are subject to tsunamis. Low-lying islands are especially vulnerable to climate-induced hazards from sea level rise and tropical storm surge because of their high coastal-zone-to-land ratio, thereby reducing environmental security.

In general, nature and society are adapted to local climatic conditions. While climate is often considered in terms of averages, the extremes are at least as important in determining a region's climate. For instance, it is becoming clear that a warmer atmosphere will result in a greater number of extreme heat waves. In addition, a warmer atmosphere can hold more moisture, so changes in the hydrological cycle could alter flood and drought patterns at all scales, including islands of any size.

It is already known that tropical cyclones are the major cause of storm surges that affect small islands in the Atlantic, Pacific, and Indian Oceans. The devastating effects of recent hurricanes have received extensive coverage by the media. As mean sea level rises, present extreme levels will be attained more frequently, and new higher levels will result in significant increases in the area threatened with flooding. This will be especially true in areas where the height between sea surge and populated areas is low—in other words, there is a small surge envelope, which is typical for most small islands. Under such circumstances, even incrementally small elevations in sea level would have severely negative effects on atolls and low islands (Forbes and Solomon 1997; Nicholls et al. 1999).

23.3.7 Climate Change and Sea Level Rise

Many islands are likely to be among the communities most adversely affected by climate change as a result of their small size, their economic dependence on a limited number of natural resource–based sectors (particularly agriculture, tourism, and extractive industries), and their limited human and financial capacities (IPCC 2001). Although the full impact of climate change on islands is far from certain, adverse consequences are predicted under probable scenarios for several systems (IPCC 2001). One of the most significant impacts of current climate change is sea level rise. This is predicted to lead to inundation of coastal areas and islands; shoreline erosion and the destruction of important island ecosystems such as coral reefs, wetlands, and mangroves; soil salinization; and the intrusion of saltwater into groundwater aquifers.

The IPCC Third Assessment Report indicated a sea level rise of as little as 1 millimeter a year during the twentieth century (IPCC 2001). During the last hundred years, the sea level has risen 100–150 millimeters, and a further rise in global sea level in the range of 350–1,100 millimeters between 1990 and 2100 is predicted, although local rates may vary from negative to positive values, depending on other localized effects (IPCC 2001). The "best estimate" in this range results in a 660-millimeter rise by 2100. This is mainly attributed to thermal expansion of the upper ocean layers and to melting of glaciers and small ice caps. There are, however, many uncertainties in identifying and assessing the causes of sea level trends. (See Figure 23.3.)

As most islands belong to developing countries, they are especially vulnerable to sea level rise due to their limited financial resources to respond to this and other natural hazards. Several small islands, such as the Maldives in the Indian Ocean and the Marshall Islands and Tuvalu in the Pacific, could face total inundation within this century if rates of sea level rise accelerate. This impact is also predicted for archipelagos such as the Philippines and Indonesia, where millions of inhabitants face displacement from their homes from sea level rise. Most of their populations live very close to the sea, and a rise of as little as a meter could prove devastating. Before their lands are lost underwater, some will face loss of their freshwater supply due to saltwater intrusion, contributing to an increasing shortage of the water supply. Sea level rise will also cause increased pressure on forest reserves due to loss of coastal agricultural land by salination and will lead to migration or loss of wildlife species.

Projected global temperature increases are not expected to have widespread adverse consequences on the terrestrial ecosystems in tropical SIDS (IPCC 2001). Some changes are likely to occur, however, especially alteration of species ranges, an increase

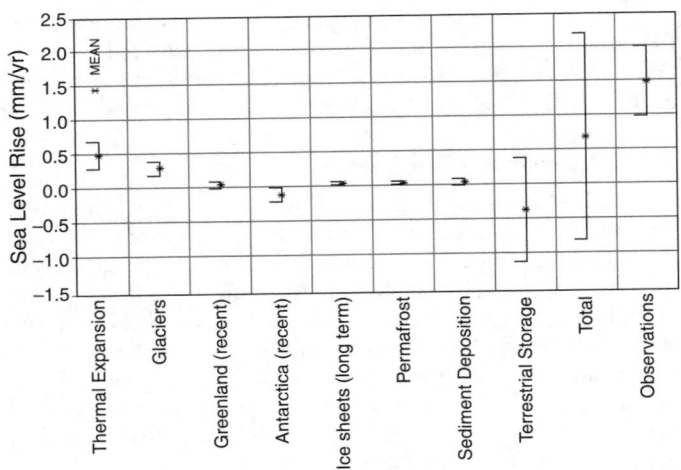

Figure 23.3. Ranges of Uncertainty for the Average Rate of Sea Level Rise and Estimated Contributions from Different Processes, 1910–90 (IPPC 2001)

in forest pest and diseases, reduction of food and water available for wildlife, and an increase in forest fire frequency, especially where precipitation remains the same or is reduced. The quantity and quality of available water supplies can affect agricultural production and human health. Similarly, changes in sea surface temperature, ocean circulation, and upwellings could affect coastal organisms such as corals, mangroves, seagrasses, and fish stocks. Tourism could also be affected through beach erosion, loss of land, and degraded reef ecosystems, as well as changes in seasonal patterns of rainfall (Nurse et al. 2001).

The SIDS account for less than 1% of global greenhouse gas emissions but are among the most vulnerable of all areas to the potential adverse effects of climate change and sea level rise (Jones 1998; Nurse et al. 1998).

Among many relevant examples, Pacific islands provide a clear case study on the issue of climate change and sea level rise. Despite having persistent trade winds and convergence zones, the climate of the Pacific Islands region continues to be dominated by interannual variability associated with El Niño/Southern Oscillation events and by extreme events such as tropical cyclones, floods, and drought. Due to the enhanced greenhouse effect, the region will likely warm by 0.6–3.5° Celsius and the climate may become a more El Niño–type, with the central and eastern equatorial Pacific warming more than the western Pacific and with a corresponding mean eastward shift of precipitation.

Future ENSO events are likely to result in anomalously wet areas becoming even wetter and unusually dry areas becoming even drier (Hay et al. 2003). Also a general increase in tropical cyclone intensity is likely with an eastward extension of their area of formation. With climate change, the Pacific will also experience a rising sea level. However, interannual variations in sea level associated with ENSO and storm surges associated with tropical storms are likely to be of greater significance than a longer-term sea level rise over decades. (See also Chapter 19.)

23.4 Human Well-being in Island Systems

Islands are not being treated here as one global ecosystem. Islands are recognized as systems each with a variety of ecosystems, and thus of ecosystem services. Combining islands of similar geographic latitudes allows for recognition of similarities in ecosystems on islands. But as there is no global "island ecosystem," common conditions as well as common trends cannot be presented at a global scale.

Another issue faced in the assessment of conditions and trends in this chapter is that only a few examples have been selected from particular islands and island groups. It was impossible to include all global islands in this assessment. However, an assessment of certain groups of islands—such as small islands off California, Mexico, and Australia as well as Pacific tropical islands (the Philippines and Indonesia) or subantarctic islands—does allow for general conclusions to be drawn. As people are a major presence in the coastal zones of islands, see also the information on coastal communities and human well-being in Chapter 19.

23.4.1 Vulnerability and Adaptation

The small size of many islands makes the people living there generally more vulnerable environmentally, economically, and socially. Their vulnerability arises from islands' limited resources, export concentration, high dependence on strategic imports, remoteness and high transportation costs, external shocks beyond their control, structural handicaps, and susceptibility to natural disasters exacerbated by climate change and sea level rise.

23.4.1.1 Adaptation to Sea Level Rise

Owing to their high vulnerability and low adaptive capacity to climate change, island communities have legitimate concerns about their future on the basis of the past and present climate model projections. Economic development, quality of life, and alleviation of poverty presently constitute the most pressing concerns of many small island states. Thus, with limited resources and low adaptive capacity, these islands face the considerable challenge of charting development paths that are sustainable and that control greenhouse gas emissions, without jeopardizing prospects for economic development and improvements in human welfare (Munasinghe 2000; Toth 2000). At the same time, islands are forced to find resources to implement strategies to adapt to increasing threats resulting from climate change, a process to which they contribute very little (Hay and Sem 1999; Sachs 2000). Consequently, the already meager resources of these island states will be placed under further pressure.

One of the most serious considerations for some small islands is whether they will have adequate potential to adapt to sea level rise within their own national boundaries (Nurse 1992; IPCC 1998). For islands where physical space is already very scarce, adaptation measures such as retreat to higher ground or a set distance separating structures from the shore would appear to have little practical utility. In extreme circumstances, sea level rise and its associated consequences could trigger abandonment and significant out-migration at great economic and social costs (Leatherman 1997; Nicholls and Mimura 1998).

23.4.1.2 Island Vulnerability

Many studies have tried to assess the ecological and economic vulnerability of island systems along different spatial scales (e.g. Sax 2001; Courchamp et al. 2003). The results of such exercises are somewhat contradictory. Many authors have suggested that the main issues are not isolation and smallness themselves, but development problems, which may be proportionately larger for island systems. According to Farrell (1991), essential problems of small island states have little to do with their smallness and may be more a matter of degree, although he admits that smallness may exacerbate the problem and its effects. An alternative view on the viability-size issue was presented by Dommen (1980), who compared a sample of small island countries with a similar set of continental countries with respect to a number of social and natural characteristics and concluded that size was an important factor.

While all developing countries face challenges in addition to the general problems of development, such as political interference, requirements for environmental friendliness, and sustainability, island nations experience specific problems arising from their limited area of land and thus have a higher exposure to interactions and exchanges of materials and energy flows than continental nations do. This vulnerability is presumed to be negatively correlated with size but is also affected by a suite of other factors: remoteness, geographical dispersion, proneness to natural disasters, the vulnerability of ecosystems to disturbance and human effects, constraints on transport and communication, isolation from markets, exposure to exogenous economic and financial shocks, a highly limited internal market, lack of land natural resources, limited freshwater supplies, heavy dependence on imports and limited commodities, depletion of nonrenewable resources, migration (particularly of skilled personnel), and a limited ability to diversify and to reap the benefits of economies of scale.

Recently, economic hegemony and new communications technology are seriously challenging these concepts of insularity. In economic terms, island-based societies like Aruba, Iceland, Bermuda, and French Polynesia are counted among the world's richest people, while those of São Tomé and Principe, Vanuatu, and the Maldives are counted amongst the poorest (Baldacchino 2004). In short, some islands with a large area but sparse human population and a limited economy that may be classified as small, while others, small in area, have gross disposable products as large as or larger than some continental economies.

There are exceptions to the typical pattern of island vulnerability as defined in terms of size, population, or GDP. For example, Singapore and Barbados have demonstrated that being physically small and lacking significant natural resources are not necessarily limitations to economic growth and prosperity. Also, some of the richest islands are heavily dependent on one or two exports such as tourism and financial services to generate considerable growth. However, all remain vulnerable to external shocks (Clayton 2004; Crowards 2004).

There have been efforts to develop a vulnerability index for small islands for a number of years, as a relatively high GNP per capita gives the impression of economic strength when in reality island economies depend on and are determined by external forces (Briguglio 1995; UWICED 2002). All SIDS have favored the development of appropriate vulnerability sub-indices as measures of the new index. Currently, the vulnerability index has yet to be accepted as an alternative to the dominant GDP per capita measure (Seychelles 2004).

Among some islands, exposure to natural hazards reflects the degree of their vulnerability to change; on others, however, this relationship is not applicable because vulnerability can be reduced through appropriate planning and preparation—as illustrated by reduced casualties from hurricanes in Cuba, which has a high level of planning and state-led provision for cyclones (Clayton 2004).

Some island states have argued for a more positive approach to the concept of vulnerability of island systems through advocating the concept of resilience—that is, increased resilience means decreased vulnerability, and vice versa (Barnett 2001). By recognizing their vulnerability, communities can build their resilience through appropriate actions and programs. Many instances of successful resilience could be emulated; they arise from a combination of factors from good governance, sound macroeconomic framework, market reform, labor productivity, social cohesion, and protection and sustainable management of the environment, including increased energy efficiency, promotion of waste management, improvement of freshwater resources management, and promotion of sustainable use of biodiversity and natural resources (Report 2004).

23.4.2 Integrated Island Systems Management

For island States, the island systems management approach, as a multidisciplinary, integrated mechanism, offers an adaptive management strategy that both addresses the issue of resource-use conflict and provides the necessary policy orientation to control the impacts of human intervention on the physical environment of islands. It was developed by the Organization of the Eastern Caribbean States and adopted by the First Ministerial Meeting on the Implementation of the Barbados Programme of Action (held in Barbados in November 1997). However, its effectiveness depends on an institutional and legal framework that coordinates the initiatives of all sectors, both public and private, to ensure the achievement of common goals through a unified approach.

The long-term development objectives of islands also need to be considered. Despite physical and natural resource limitations, important consideration will need to be given to integrated planning, social cohesion, increased attention to managing biodiversity (in particular, invasive species), and a strengthening of territorial planning if islands are to become economically, socially, and ecologically resilient and self-sufficient.

References

Algar, D.A, A.A. Burbidge , and G.J. Angus, 2002: Cat eradication on Hermite Island, Montebello Islands, Western Australia. In: *Turning the tide: the eradication of island invasive species.* C.R. Veitch and M.N. Clout (eds.), IUCN SSC Invasive Species Specialist Group, IUCN, Gland, pp. 14–18.

Aswani, S. and R.J. Hamilton, 2004: Integrating indigenous ecological knowledge and customary sea tenure with marine and social science for conservation of bumphead parrotfish (*Bolbometopon muricatum*) in the Roviana Lagoon, Solomon Islands. *Environment Conservation,* **31,** 69–83.

Atkinson, I.A.E., 1992: *Effects of possums on the vegetation of Kapiti Island and changes following possum eradication.* Department of Scientific and Industrial Research, Botany Division Report (92/52), Department of Conservation, Wellington, 68 pp.

Baker, P. and Hanna, J., 1986: Perspectives on health and behavior of Samoans. In: *The Changing Samoans: Behavior and Health in Transition.* P. Baker, J. Hanna and T. Baker (eds.), New York, Oxford University Press, pp. 419–434.

Baldacchino, G., 2004: Sustainable use practices, including tourism, in/for small islands. *INSULA,* Special Issue February 2004, pp. 15–20.

Balick, M., 2003: Traditional Knowledge: Lessons from the Past, Lessons for the Future. Conference paper presented at Biodiversity, Biotechnology and the Protection of Traditional Knowledge, Washington University, April 4–6, 30 pp. Available online at http://law.wustl.edu/centeris/Confpapers/PDF WrdDoc/Balick.doc

Barnett, J. 2001: Adapting to climate change in Pacific island countries: the problem of uncertainty. *World Development,* **29,** 977–993.

Berkes, F., 1993: *Traditional Ecological Knowledge in Perspective.* In: *Traditional Ecological Knowledge: Concepts and Cases.* J.T. Inglis (ed.), Canadian Museum of Nature/International Development Research Centre, Ottawa, pp.1–9.

Bridgewater, P. 2004: Wetland ecosystems in small islands. *INSULA,* Special Issue on Island Biodiversity, February 2004, 77–84.

Briguglio, L., 1995: Small island developing states and their economic vulnerabilities. *World Development,* 23: 1615–1632

Brookfield, H.C., 1990: An approach to islands. In: *Sustainable Development and Environmental Management of Small Islands.* W. Beller, P. d'Ayala, and P. Hein (eds.), UNESCO, Paris, pp. 23–33.

Browne, K.E. 2001: Female entrepreneurship in the Caribbean: a multisite, pilot investigation of gender and work. *Human Organisation,* **60:** 326–342.

Burke L., C. Revenga, Y. Kura, M. Spalding, K. Kassem, and D, Mc Allister. 2001: *Pilot Analysis of Global Ecosystems, Coastal Ecosystems.* World Resources Institute (WRI). Available online at: http://www.wri.org/wr2000

Calamia, M.A., 1996: Traditional ecological knowledge and geographic information systems in the use and management of Hawaii's coral reefs and fishponds. *High Plains Applied Anthropologist,* **16,** 144–164.

Case, T.J., 1975: Species numbers, density compensation, and colonizing ability of lizards on islands in the Gulf of California. *Ecology,* **56,** 3–18.

Caughley, G., 1994: Directions in conservation biology. *Journal of Animal Ecology,* **63,** 215–244.

Cavanagh, J.E., J.H. Clarke, and R. Price, 1993: Ocean Energy Systems. In: *Renewable Energy,* L. Burnham (ed.), Island Press, Washington, pp. 513–548.

CBD (Convention on Biological Diversity) 2004: *Status and trends of, and major threats to, island biodiversity.* Document no. UNEP/CBD/AHTEG-IB/1/3. Available online at http://www.biodiv.org/doc/meetings/island/tegib-01/official/tegib-01–03-en.pdf

Chafe, Z., 2004: *Consumer Demand and Operator Support for Socially and Environmentally Responsible Tourism.* CESD/TIES Working Paper No. 104, 9 pp.

Christ, C., O. Hillel, S. Matus, and J. Sweeting, 2003: *Tourism and Biodiversity: Mapping Tourism's Global Footprint.* Conservation International, Washington, D.C., 53 pp.

CIESIN (Center for International Earth Science Information Network), International Food Policy Research Institute, and World Resources Institute, 2000: Gridded Population of the World, Version 2. Palisades: NY: CIESIN, Columbia University. On-line at www.sedac.ciesin.org/plue/gpw

Clayton, A. 2004: Using economic diversification to define vulnerability. Wise Coastal Practices for Sustainable Human Development Forum. Available online at www.csiwisepractices.org/?read = 493.

Clout, M.N., and C.R. Veitch, 2002: Turning the tide of biological invasion: the potential for eradicating invasive species. In: *Turning the tide: the eradication of island invasive species.* C.R. Veitch and M.N. Clout (eds.), IUCN SSC Invasive Species Specialist Group, IUCN, Gland, pp.1–3.

Commonwealth Secretariat, 2004: Report on the International Workshop on Economic Vulnerability and Resilience of Small States, held at University of Malta, 1–3 March 2004. Organized by the Commonwealth Secretariat, the Economics Department of the University of Malta and the Islands and Small States Institute of the Foundation for International Studies, at the University of Malta. 13 pp. Available online at www.thecommonwealth.org/shared_asp_files/uploadedfiles/{6BD67478-A500–4B8C-8004-F599D23BCE57}_WorkshopReport.pdf

Connell, J., and D. Conway, 2000: Migration and remittances in island microstates: a comparative perspective on the South Pacific and the Caribbean. *International Journal of Urban and Regional Research,* **24,** 52–78.

Conservation International, 2002: Ecotourism. Cited 24 Sept 2004. Available online at http://www.conservation.org/ImageCache/CIWEB/content/publications/factsheets/ecotourism_2epdf/v1/ecotourism.pdf.

Copson, G., and J. Whinam, 1998: Response of vegetation on subantarctic Macquarie Island to reduced rabbit grazing. *Australian Journal of Botany,* **46,** 15–24.

Courchamp, F, J-L. Chapius, and M. Pascal, 2003: Mammal invaders on islands: impact, control and control impact. *Biological Reviews,* **78,** 347–383.

Cross, M. and S. Nutley, 1999: Insularity and accessibility: the small island communities of western Ireland. *Journal of Rural Studies,* **15,** 317–330.

Crowards, T., 2004: Identifying policy responses to vulnerability. Wise Coastal Practices for Sustainable Human Development Forum. Available online at http://www.csiwisepractices.org/?read = 491.

Cumpston, J.S., 1968: Macquarie Island. *ANARE Science Report Series* **A 93,** 1–380.

D'Antonio, C.M., and T.L. Dudley, 1995: Biological invasions as agents of change on islands vs. mainlands. In: *Islands: Biological Diversity and Ecosystem Function,* P.M. Vitousek, L.L. Loope, and H. Adersen (eds.), Springer, pp. 103–121.

Day, D., 1981: *The Doomsday Book of Animals.* Viking. New York, 288 pp.

Dommen, E., 1980: Some distinguishing characteristics of island states. *World Development,* **8,** 931–943.

Domroes, M., 2001: Conceptualizing state-controlled resort islands for an environment-friendly development of tourism: the Maldivian experience. *Singapore Journal of Tropical Geography,* **22,** 122–137.

Dueser, R.D., and W.C. Brown, 1980: Ecological correlates of insular rodent diversity. *Ecology,* **61(1),** 50–56.

Dulloo, M.E., S.P. Kell, and C.G. Jones, 2002: Impact and control of invasive alien species in small islands. The International Forestry Review, Special Issue on *Forestry and Small Island Developing States,* Commonwealth Forestry Association, Bicester, United Kingdom, pp. 277–285

Ebenhard, T., 1988: Introduced birds and mammals and their ecological effects. *Swedish Wildlife Research,* **13,** 1–107.

ECOSOC (Economic and Social Council), 2004: Review of progress in the implementation of the programme of action for the sustainable development of small island developing states. Report of the Secretary-General, UN. *Ecological Economics,* **35:** 135–142.

Erdelen, W.R., K. Adimihardja, H. Moesdarsono, and Sidik, 1999: Biodiversity, traditional medicine and the sustainable use of indigenous medicinal plants in Indonesia. *Indigenous Knowledge and Development Monitor,* **7(3),** 3–6.

FAO (Food and Agriculture Organization of the United Nations), 1999: Environment and Natural Resources in Small Island Developing States. Overview prepared for a Special Ministerial Conference on Agriculture in Small Island Developing States, Rome. Available online at http://www.fao.org/docrep/meeting/X1011e.htm

FAO, 2001: *Forest Resources Assessment 2000.* Main Report, FAO Forestry Study 140. Rome, Italy.

Farrell, T. 1991: *The Concept of Small States: Current Problems and Future Prospects with Special Reference to the Caribbean.* Institute of Social and Economic Studies, Cave Hill, Barbados, 30 pp.

Fisher, E., 2004: Island ecosystems conservation and sustainable use: problems and challenges. *INSULA* Special Issue on Island Biodiversity, February 2004, pp. 9–14.

Forbes, D.L., and S.M. Solomon, 1997: *Approaches to Vulnerability Assessment on Pacific Island Coasts: Examples from Southeast Viti Levu (Fiji) and Tarawa (Kiribati).* Miscellaneous Report 277, SOPAC, Suva, 21 pp.

Fritts, T.H., 2002: Economic costs of electrical system instability and power outages caused by snakes on the Island of Guam. *International Biodeterioration and Biodegradation,* **49,** 93–100.

Gibson, L.J., 1990: Species in Islands: Evidence for Change. *Geoscience Reports* (Loma Linda University, California), **12,** 1,3–6..

Gorman, M.L., 1979: *Island Ecology.* Chapman and Hall, London, 79 pp.

GRAIN and Kalpavriksh, 2002: Traditional knowledge of biodiversity in Asia-Pacific: Problems of Piracy and Protection. Available online at http://www.grain.org/briefings_files/tk-asia-2002-en.pdf

Granger, O.E., 1993: Geography of small tropical islands: implications for sustainable development in a changing world. In: *Small Islands: marine science and sustainable development,* G.A. Maul (ed.). Coastal and Marine Studies. American Geophysical Union, Washington D.C., pp. 157–187.

Hales, S., P. Weinstein, Y. Souares, and A. Woodward, 1999: El Niño and the dynamics of vectorborne disease transmission. *Environmental Health Perspectives,* **107,** 99–102.

Hamilton, L.C., C.R. Cococousis, and S.T.F. Johansen, 2004: Migration from resource depletion: the case of Faroe Islands. *Society and Natural Resources,* **17,** 443–453.

Hay, J.E., and G. Sem, 1999: *A Regional Synthesis of National Greenhouse Gas Inventories.* South Pacific Regional Environment Programme, Apia, Samoa, 29 pp.

Hay, J.E., N. Mimura, J. Campbell, S. Fifita, K. Koshy, R.F. McLean, T. Nakalevu, P. Nunn, and N. de Wet, 2003: *Climate Variability and Change and Sea-level Rise in the Pacific Islands Region: A Resource Book for Policy and Decision Makers, Educators and Other Stakeholders.* South Pacific Regional Environment Programme, Apia, Samoa, 108pp.

Hess, A.L., 1990: Overview: Sustainable Development and Environmental Management of Small Islands. In: *Sustainable Developmentand Environmental Management of Small Islands,* W.B.P. D'Ayala and P. Hein (eds.), UNESCO, Paris, pp. 3–14.

Hugo, G., 2000: The Impact of the Crisis on Internal Population Movement in Indonesia. *Bulletin of Indonesian Economic Studies,* **36,** 115–138.

Hunt, C., 2003: Economic globalization impacts on Pacific marine resources. *Marine Policy,* **27,** 79–85.

IPCC (Intergovernmental Panel on Climate Change), 1998: *The Regional Impacts of Climate Change: An Assessment of Vulnerability.* A Special Report of IPCC Working Group II. Cambridge University Press, Cambridge, 517 pp.

IPCC, 2001: *Climate Change 2001: Impacts, Adaptation, and Vulnerability.* Contribution of Working Group II to the Third Assessment Report of the Intergovernmental Panel on Climate Change, Cambridge University Press, Cambridge, 1032 pp.

Jiang, J.S., and R.S. Wang, 2003: Hydrological eco-service of rubber plantations in Hainan island and its effect on local economic development. *Journal of Environmental Sciences—China,* **15,** 701–709.

Jones, R.N., 1998: *An Analysis of the Impacts of the Kyoto Protocol on Pacific Island Countries, Part 1: Identification of Latent Sea-Level Rise Within the Climate System at 1995 and 2020.* South Pacific Regional Environment Programme, Apia, Samoa and Commonwealth Scientific and Industrial Research Organisation, Canberra, Australia, 9 pp.

Lal, M., H. Harasawa, and K. Takahashi, 2002: Future climate change and its impacts over small island states. *Climate Research,* **19,** 179–192.

Leatherman, S.P., 1997: Beach ratings: a methodological approach. *Journal of Coastal Research,* **13,** 1050–1063.

Lean, G., 2004: Disastrous weather will continue. *The Environmental Magazine.* Cited 12 September 2004. Available online at www.emagazine.com/news/archives.php?current.

Lee, R., M.J. Balick, D.L. Ling, F. Sohl, B.J. Brosi, and W. Raynor, 2001: Cultural dynamism and change—an example from the Federated States of Micronesia. *Economic Botany,* **55,** 9–13.

Lewis, N.D. 1998: Intellectual intersections: gender and health in the Pacific. *Social Science and Medicine,* **46,** 641–659.

Lillis, K.M., 1993: *Policy, Planning and Management of Education in Small States.* UNESCO, Paris.

Lugo, A., and E. Helmer, 2004: Emerging forests on abandoned land: Puerto Rico's new forests. *Forest Ecology and Management,* **190,** 145–161.

MacArthur, R.H., and E.O. Wilson, 1967: *The Theory of Island Biogeography.* Princeton University Press, Princeton, New Jersey, 224 pp.

Mack, R.N, D. Simberloff, W.M. Lonsdale, H. Evans, M. Clout, and F. Bazzaz, 2000: Biotic invasions: causes, epidemiology, global consensus and control. *Issues in Ecology* **5,** 1–25.

MacKay, K.T., 2001: Managing Fisheries for Biodiversity: Case Studies of Community Approaches to Fish Reserves among the Small Island States of the Pacific. Paper presented at Blue Millennium: Managing Global Fisheries for Biodiversity Thematic Workshop, Victoria, BC, June 2001. 45 pp. Available online at http://www.unep.org/bpsp/Fisheries/Fisheries%20Case%20Studies/MACKAY.pdf

Mackinson, S., 2001: Integrating local and scientific knowledge: an example in fisheries science. *Environmental Management,* **27,** 533–545.

Maul, G.A., 1993: Ocean science in relation to sustainable economic development and coastal area management of small islands. In: *Small Islands: Marine Science and Sustainable Development,* G.A. Maul (ed.), American Geophysical Union, Washington, D.C., pp. 1–17.

Mittermeier, R.A., N. Myers, J.B. Thomsen, G.A.B. Fonseca, and S. Olivieri, 1998: Biodiversity hotspots and major tropical wilderness areas: approaches to setting conservation priorities. *Conservation Biology,* **12,** 516–520.

Mohammed, S.M. 2002: Pollution management in Zanzibar: the need for a new approach. *Ocean and Coastal Management,* **45,** 301–311.

Munasinghe, M., 2000: Development, equity and sustainability (DES) in the context of climate change. In: *Climate Change and Its Linkages with Development, Equity and Sustainability,* M. Munasinghe, and R. Swart (eds.). Proceedings of the IPCC Expert Meeting held in Colombo, Sri Lanka, 27–29 April 1999. LIFE, Colombo, Sri Lanka; RIVM, Bilthoven, The Netherlands; and World Bank, Washington, DC, pp. 13–66.

Murray, W.E., 2001: The second wave of globalization and agrarian change in the Pacific Islands. *Journal of Rural Studies,* **17,** 135–148.

Nandwani, D., and E.J., Dasilva, 2003: Traditional knowledge and medicine in the Marshall Islands. *Agro Food Industry Hi-Tech,* **14(4),** 42–46.

Nicholls, R.J., and M. Mimura, 1998: Regional issues raised by sea-level rise and their policy implications. *Climate Change,* **11,** 5–18.

Nicholls, R.J., F.M.J., Hoozemans, and M. Marchand, 1999: Increasing flood risk and wetland losses due to global sea-level rise: regional and global analyses. *Global Environmental Change,* **9,** 69–87.

Nurse, L.A., 1992: Predicted sea-level rise in the wider Caribbean: likely consequences and response options. In: *Semi-Enclosed Seas,* P. Fabbri, and G. Fierro (eds.), Elsevier Applied Science, Essex, pp. 52–78.

Nurse, L.A., R.F. McLean, and A.G. Suarez, 1998: Small island states. In: *The Regional Impacts of Climate Change: An Assessment of Vulnerability.* A Special Report of IPCC Working Group II, R.T. Watson, M.C. Zinyowera, and R.H. Moss (eds.), Cambridge University Press, Cambridge, pp. 331–354.

Nurse, L.A., G. Sem, J.E. Hay, A.G. Suarez, P.P. Wong, L, Briguglio, and S. Ragoonaden, 2001: Small island states. In: *Climate Change 2001: Impacts, Adaptation, and Vulnerability,* J.J. McCarthy, O.F. Canziani, N.A. Leary, D.J. Dokken, and K.S. White (eds.), Cambridge University Press, Cambridge, pp. 843–875.

Penney, T. R. and D. Bharathan, 1987. Power from the sea. *Scientific American,* **256**(1), 86–92.

POEMS (Practical Ocean Energy Management Systems), 2004: Industry global view. Available online at http://www.poemsinc.org/industry.html.

Prasad, B.C. 2001: Globalization, markets, transnational corporations and environmental management: a case of small island nations in the South Pacific. *Journal of Corporate Citizenship,* **2,** 75–90.

Read, R., 2004: The implications of increasing globalization and regionalization for the economic growth of small island states. *World Development,* **32,** 365–378.

Report, 2004: *Report on the International Workshop on Economic Vulnerability and Resilience of Small States,* held at University Gozo Centre, Xewkija, Gozo, Malta, 1–3 March 2004. Organized by the Commonwealth Secretariat, the

Economics Department of the University of Malta and the Islands and Small States Institute of the Foundation for International Studies, at the University of Malta. 13 pp.

Roper, T. 2005: Energy release. *Our Planet,* **15** (1), 11.

Ronnback, P., and J.H. Primavera, 2000: Illuminating the need for ecological knowledge in economic valuation of mangroves under different management regimes—a critique. *Ecological Economics,* **35,** 135–142.

Rosabal, P., 2004: Protected areas: benefits to islanders. *INSULA* Special Issue February 2004, pp. 21–28.

Rosenzweig, M.L., 1995: *Species Diversity in Space and Time.* Cambridge University Press, Cambridge, 436 pp.

Sachs, W., 2000: Development patterns in the north and their implications for climate change. In: *Climate Change and Its Linkages with Development, Equity and Sustainability,* M. Munasinghe, and R. Swart (eds.). Proceedings of the IPCC Expert Meeting held in Colombo, Sri Lanka, 27–29 April 1999. LIFE, Colombo, Sri Lanka; RIVM, Bilthoven, The Netherlands; and World Bank, Washington, DC, pp. 163–176.

Samoa, 2003: *Barbados Programme of Action National Assessment Report : Samoa.* Available online at http://www.sidsnet.org/docshare/other/200312301548 38_Samoa_NAR_2003.pdf.

Sanders, M.M., 1991: Energy from the oceans. In: *The Energy Sourcebook,* R. Howes, and A. Fainberg (eds.), American Institute of Physics, New York.

Sax, D.F., 2001: Latitudinal gradients and geographic ranges of exotic species: implications for biogeography. *Journal of Biogeography* **28,** 139–150.

Scott, J. 2003, editor: *Travel Industry World Yearbook—The Big Picture.* Travel Industry Publishing Co., Spencertown, N.Y., 208 pp.

Seychelles, 2004: *Barbados Programme of Action + 10 Review.* Republic of Seychelles National Assessment. Available online at http://www.sidsnet.org/docshare/other/20040113105721_Seychelles_NAR_20 03.pdf.

Simberloff, D.S., 2000: Why not eradication? In : *Managing for Healthy Ecosystems,* D. J. Rapport, W.L. Lasley, D.E. Rolston, N.O. Nielsen, C.O. Qualset and A.B. Damania (eds.), CRC/Lewis Press, Boca Raton, Florida.

Singh, R.B.K., S. Hales, N. de Wet, R. Raj, M. Hearnden, and P. Weinstein, 2001: The influence of climate variation and change on diarrhoeal disease in the Pacific Islands. *Environmental Health Perspectives,* **109,** 155–159.

Sol, D., 2000: Are islands more susceptible to be invaded than continents? Birds say no. *Ecography,* **23,** 687–692.

Srinivasan, T.N., 1986: The costs and benefits of being a small, remote, island, landlocked, or ministate economy. *World Bank Research Observer,* **1,** 205–218.

Sudji, N.W., 2003: *The role of traditional villages in coastal management in Bali.* Paper presented at The East Asian Seas Congress 2003, 8–12 Dec 2003, Putrajaya, Malaysia.

Taylor, R.H., 1979: How the Macquarie Island parakeet became extinct. *New Zealand Journal of Ecology,* **2,** 42–45.

Thaman, R.R., 2002: Trees outside forests as a foundation for sustainable development in the small island developing states of the Pacific Ocean. *International Forestry Review,* **4,** 268–276.

Thomas, F.R., 2001: Remodelling marine tenure on the atolls: a case study from western Kiribati, Micronesia. *Human Ecology,* **29,** 399–423.

Toth, F.L. 2000: Development, equity and sustainability concerns in climate change decisions. In: *Climate Change and its Linkages with Development, Equity and Sustainability,* M. Munasinghe and R. Swart (eds.), IPCC 27–29 April, 1999, Colombo, Sri Lanka, pp. 263–288.

UN (United Nations), 2002: *Statistical Yearbook, 2002.* UN, Geneva.

UN, 2002: *World Population Prospects. The 2002 Revision.* United Nations Population Division, 22 pp.

UNEP (United Nations Environment Programme), 1994: *Convention On Biological Diversity.* Switzerland, UNEP/CBD, 34 pp.

UNEP, 2004a: *Atlantic and Indian Ocean Small Island Developing States Environmental Outlook (EO) 2004,* draft version distributed at the Interregional Preparatory Meeting for the 10-Year Review of Implementation of the Barbados Programme of Action for the Sustainable Development of Small Island Developing States, 26–30 January 2004.

UNEP, 2004b: *Caribbean Environmental Outlook (EO) 2004,* draft version distributed at the Interregional Preparatory Meeting for the 10-year Review of Implementation of the Barbados Programme of Action for the Sustainable Development of Small Island Developing States, 26–30 January 2004.

UNEP, 2004c: *Pacific Islands Environmental Outlook (EO) 2004,* draft version distributed at the Interregional Preparatory Meeting for the 10-year Review of Implementation of the Barbados Programme of Action for the Sustainable Development of Small Island Developing States, 26–30 January 2004.

Unit B5, 1998: Overview of traditional environmental management. *UNEP Islands Web Site, Small Island Environment Management.* Available online at http://islands.unep.ch/siemb5.htm

UWICED (University of the West Indies Centre for Environment and Development), 2002: *Final Report: Vulnerability and Small Island Developing States. For UNDP/Capacity 21,* 33 pp.

van Riper, C.S.G., M.L. van Riper, L. Goff, and M. Laird, 1986: The epizootiology and ecological significance of malaria in Hawaiian land birds. *Ecological Monographs,* **56,** 327–344.

Veitayaki, J., 2004: Building bridges: the contribution of traditional knowledge to ecosystem management and practices in Fiji. Paper presented at "Bridging scales and epistemologies: linking local knowledge and global science in multi-scale assessments', Alexandria, Egypt, 17–20 March 2004. Available online at http://www.millenniumassessment.org/en/About.Meetings.Bridging.Proceedings.aspx

Veza, J.M., 2001: Desalination in the Canary Islands: an update. *Desalination,* **133,** 259–270.

Vicente, L., 1999: Evolutionary strategies on insular environments. *Natura Croatica* 8(3), 301–323.

Vitousek, P.M., C.M. D' Antonio, L.L. Loope, M. Rejmanek, and R. Westbrooks, 1997: Introduced species: a significant component of human-caused global change. *New Zealand Journal of Ecology* **21,** 1–16.

Whittaker, R.J., 1998: *Island Biogeography: Ecology, Evolution, and Conservation,* Oxford University Press, Oxford, 304 pp.

Wilkie, M.L., C.M. Eckelmann, M. Laverdiere, and A. Mathias, 2002: Forests and forestry in small island developing states. *International Forestry Review,* **4,** 257–267.

WHO (World Health Organization), 2000: *Global Water Supply and Sanitation Assessment 2000 Report.* Geneva, WHO and United Nations Children's Fund, 124 pp.

WTO (World Tourism Organization), 2001: *Tourism 2020 Vision: Global Forecast and Profiles of Market Segment.* WTO, Madrid, 139 pp.

WTO, 2003: *Compendium of Tourism Statistics (1997–2001).* 2003 Edition, WTO, Madrid, 238 pp.

Zann, L.P., V. Vuki and C.R.C. Sheppard, 2000: The southwestern Pacific islands regions. In: *Seas at the Millennium: An Environmental Evaluation.* C.R.C. Sheppard (ed.), Pergamon, Amsterdam, v.2, pp. 705–722.

Zavaletta, E.S., 2002: It's often better to eradicate, but can we eradicate better? In: *Turning the Tide: the Eradication of Island Invasive Species.* C.R. Veitch and M.N. Clout (eds.), IUCN SSC Invasive Species Specialist Group, IUCN, Gland, pp. 393–405.

Chapter 24

Mountain Systems

Coordinating Lead Authors: Christian Körner, Masahiko Ohsawa

Lead Authors: Eva Spehn, Erling Berge, Harald Bugmann, Brian Groombridge, Lawrence Hamilton, Thomas Hofer, Jack Ives, Narpat Jodha, Bruno Messerli, Jane Pratt, Martin Price, Mel Reasoner, Alan Rodgers, Jillian Thonell, Masatoshi Yoshino

Contributing Authors: Jill Baron, Roger Barry, Jules Blais, Ray Bradley, Robert Hofstede, Valerie Kapos, Peter Leavitt, Russell Monson, Laszlo Nagy, David Schindler, Rolf Vinebrooke, Teiji Watanabe

Review Editors: Blair Fitzharris, Kedar Shrestha

---

*This appears in Appendix A at the end of this volume.

Main Messages

Half of the human population depends on mountains. Defined by elevation above sea level (minimum between 300 and 1000 meters, depending on latitude), steepness of slope (at least 2° over 25 kilometers, on the 30 arc-second grid), and excluding large plateaus, mountains occupy about one fifth of the terrestrial surface. Twenty percent (1.2 billion) of the world's human population lives in mountains or at their edges, and half of humankind depends in one way or the other on mountain resources (largely water).

Mountains are characterized by high biodiversity. Because of the compression of climatic life zones with altitude and small-scale habitat diversity caused by different topoclimates, mountain regions are commonly more diverse than lowlands and are thus of prime conservation value. They support about one quarter of terrestrial biodiversity, with nearly half of the world's biodiversity hot spots concentrated in mountains. Geographically fragmented mountains support a high ethnocultural diversity. For many societies, mountains have spiritual significance, and scenic landscapes and clean air make mountains target regions for recreation and tourism. Thirty-two percent of protected areas are in mountains (9,345 mountain protected areas covering about 1.7 million square kilometers).

Mountain ecosystems are exceptionally fragile. Mountains are subject to both natural and anthropogenic drivers of change. These range from volcanic and seismic events and flooding to global climate change and the loss of vegetation and soils because of inappropriate agricultural and forestry practices and extractive industries. Mountain biota are adapted to relatively narrow ranges of temperature (and hence altitude) and precipitation. Because of the sloping terrain and the relatively thin soils, the recovery of mountain ecosystems from disturbances is typically slow or does not occur.

Human well-being depends on mountain resources. These ecosystems are particularly important for the provision of clean water, and their ecological integrity is key to the safety of settlements and transport routes. They harbor rich biodiversity and contribute substantially to global plant and animal production. All these services depend on slope stability and erosion control provided by a healthy vegetative cover. As "water towers," mountains supply water to nearly half the human population, including some regions far from mountains, and mountain agriculture provides subsistence for about half a billion people. Key mountain resources and services include water for hydroelectricity, flood control, mineral resources, timber, and medicinal plants. Mountain populations have evolved a high diversity of cultures, including languages, and traditional agricultural knowledge commonly promotes sustainable production systems. In many mountain areas, tourism is a special form of highland-lowland interaction and forms the backbone of regional as well as national economies.

In general, both poverty and ethnic diversity are higher in mountain regions, and people are often more vulnerable than people elsewhere. Ninety percent of the global mountain population of about 1.2 billion people lives in developing countries and countries in transition—with one third of these in China and half in the Asia-Pacific region. Some 90 million mountain people—and almost everyone living above 2500 meters—live in poverty and are considered especially vulnerable to food insecurity. Land use pressure puts mountain ecosystem integrity at risk in many parts of the world. Industrial use, forest destruction, overgrazing, and inappropriate cropping practices lead to irreversible losses of soil and ecosystem function, with increased environmental risks in both mountains and adjacent lowland areas.

Mountains often represent political borders, restrict transport to narrow corridors, and are refuges for minorities and political opposition. As such they are often focal areas of armed conflicts. Further conflicts arise from the commercial exploitation of natural resources, usually by outside interests, and from ambiguity regarding traditional land use rights. Profits from extractive industries in mountains are not systematically reinvested either in the management of upland resources or the provision of benefits to mountain communities. Both poverty and remoteness are responsible for poor medical care and education systems in many mountain regions.

Strengthened highland-lowland linkages improve sustainability for both upstream and downstream populations. Lowland-highland relationships, whether formal or informal, have the potential to pay for investments in protection and sustainable use of mountain resources. When full costs are taken into account, stewardship of upland resources generally yields greater and more sustainable economic returns both to the people living in the mountain areas and to the immediate downstream economies when compared with extractive activities. In many cases, the focal point of such interactions has been based on providing a sustainable and clean supply of water, the most important and increasingly limiting mountain resource. In steep terrain, more than anywhere else, catchment quality is intimately linked to ecosystem integrity and functioning. Thus environmental conservation and sustainable land use in the world's mountains are not only a necessary condition for sustainable local livelihoods, they are also key to human well-being for nearly half the world's population who live downstream and depend on mountain resources.

24.1 Introduction and Scope of Global Mountain Systems

Since its existence, the surface of Earth has always been subject to tectonic forces that with the action of gravity and the erosive power of water have shaped landscapes into mountains, hills, lowland forelands, and old tableland. (See Figures 24.1–24.4.) Mountains are very attractive to outsiders, but the physical conditions challenge those living in these regions. Of the approximately 1.2 billion mountain people worldwide (20% of world population), only 8% inhabit places above 2,500 meters elevation. The key functions of mountains for humanity are frequently overlooked, such as the headwaters of river systems that supply nearly half of humanity with water. This chapter assesses the available knowledge on physical, biological, economic, and social conditions in the world's mountain areas and describes their likely future.

24.1.1 Definitions of Mountains and Altitude Belts

Since the transition from lowland plains to mountain terrain is usually gradual, the definition of mountains is based on convention. For the purpose of this assessment, inclusive rather than selective criteria were adopted to define the mountain system. The three major problems that needed to be resolved were latitudinal differences in climate from the equator to the poles and thus the variable altitude of different life zones (hill, montane, alpine, nival); the relative importance of elevation versus slope (high altitude plains versus steep slopes of lowland hills, for example); and, tied to both these, the definition of the lower limit for mountain terrain. For practical reasons, local climatic and topographic peculiarities could not be accommodated.

One common definition (and the one adopted by the United Nations Environment Programme World Conservation Monitoring Centre) is a lower limit of 300 meters (Kapos et al. 2000). (See Box 24.1 here and Figure 24.5 in Appendix A.) Alternatively, the lower limit has been set at 1,000 meters at the equator (the upper limit of many tropical plant species including the coconut palm), gradually decreasing to about 300 meters at the 65° northern and 55° southern latitude, reaching sea level at a short distance beyond

Figures 24.1–4. Mountains of the World. From top to bottom: Cradle Mountains, Tasmania, at 1,100 meters; Monte Rosa Glacier near Matterhorn, Switzerland, in the Central Alps at 3,000 meters; World Heritage Site Sichuan, Northwest China; Paddy field slope agriculture and deciduous montane forest (background) near Kathmandu, Nepal, at 1,200 meters.

> **BOX 24.1**
> ### Defining Mountains by Topography Only
>
> Kapos et al. (2000) used criteria based on altitude and slope in combination to represent the world's mountain environments. Topographical data from the GTOPO30 global digital elevation model (USGS EROS Data Centre 1996) were used to generate slope and local elevation range on a 30 arc-second (about 1 kilometer) grid of the world. These parameters were combined with elevation to arrive at empirically derived definitions of six elevation classes. To reduce projection distortion in the original data set, analysis was based on continental subsets in equidistant conic projection. The global mountain area thus defined is almost 40 million square kilometers, or 27% of Earth's surface. Assuming a lower mountain boundary of 1,000 meters at the equator and a linear reduction of this boundary to 300 meters at 67°N and 55°S reduced the total "mountain" land area by 5.4 million square kilometers or 3.7 % of the global land.
>
> Class 1, elevation > 4,500 meters
> Class 2, elevation 3,500–4,500 meters
> Class 3, elevation 2,500–3,500 meters
> Class 4, elevation 1,500–2,500 meters and slope ≧ 2
> Class 5, elevation 1,000–1,500 meters and slope ≧ 5 or local elevation range (7 kilometer radius) >300 meters
> Class 6, elevation 300–1,000 meters and local elevation range (7 kilometer radius) >300 meters outside 23°N—19°S
> Class 7, isolated inner basins and plateaus less than 25 square kilometers in extent that are surrounded by mountains but do not themselves meet criteria 1–6 (this seventh class was introduced in the 2002 revision of the original 2000 system)

these latitudes, where the alpine merges with the polar life zones. Ideally, however, the lower mountain limit should be defined climatically, irrespective of latitude. But this would require a world topoclimate map, which is currently not available.

The choice of convention is important because it has a large influence on the global mountain area. The UNEP-WCMC definition gives the global mountain area at about 23%, whereas under the second definition it accounts for about 19% of the global land area. For this review, flat terrain (basins or plateaus) below 2,500 meters elevation was excluded if the aerial extent of such plains exceeded 25 square kilometers. In essence, the definition used here followed that by Kapos et al. (2000).

In this global assessment, three belts were distinguished for mountain regions where precipitation regimes allow forest growth. In treeless arid or semiarid regions, analogues to these belts can be defined. (See Figure 24.6.)

- *The montane belt* (see Figure 24.7) extends from the lower mountain limit to the upper thermal limit of forest (irrespective of whether forest is present or not). This limit has a mean growing season temperature of 6.7 + 0.8°C globally, but is closer to 5.5°C near the equator and to 7.5°C at temperate latitudes. Between 40°N and 30° S, this belt covers a range of 2,000–3,000 meters of elevation. Note the difference between *mountain* and *montane*.
- *The alpine belt* (see Figure 24.8) is the treeless region between the natural climatic forest limit and the snow line. The term "alpine" has many meanings, but here it refers strictly to a temperature-driven treeless high-altitude life zone that occurs worldwide and not solely in the European Alps (the term "alp" is of pre-Indo Germanic origin). Some synonyms such

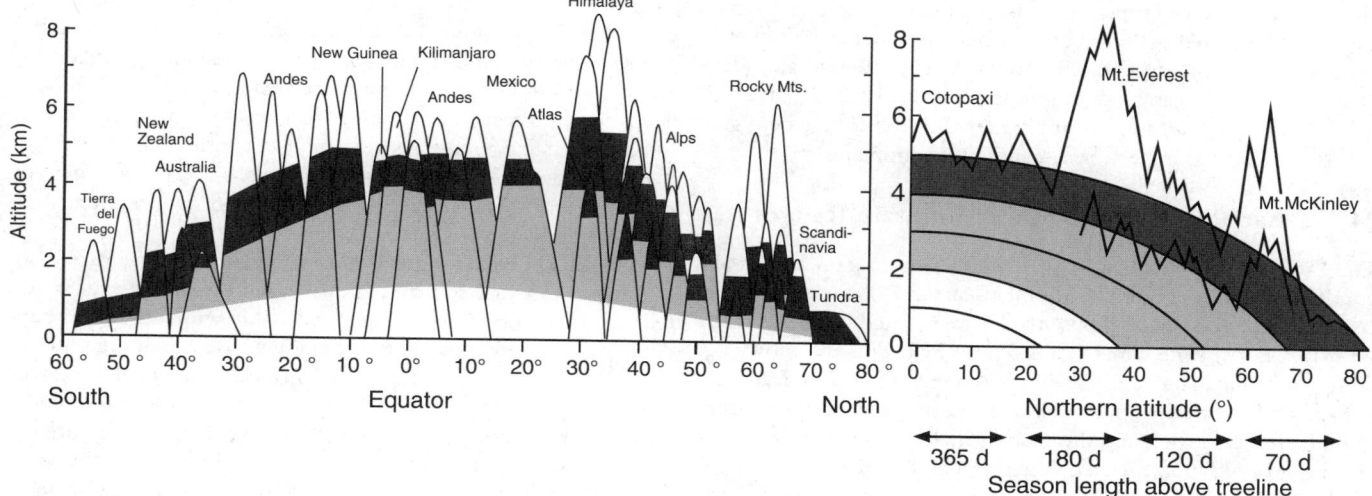

Figure 24.6. Classic Humboldt Profile of the Latitudinal Position of Altitude Belts in Mountains across the Globe and Compression of Thermal Zones on Mountains, Altitude for Latitude. Grey is montane; black is alpine; white is the nival belt. (Körner 2003)

Figure 24.7. Montane Rainforest, Kilimanjaro, at 2,600 Meters

as "andean" and "afro-alpine" are in common scientific use. Land cover is dominated by grassland or low stature shrubland. Outside subpolar regions (<60°N, <50°S), the alpine belt extends over an elevation range of 800–1,200 meters, with its lower boundary varying from about 500–4,000 meters above sea level, depending on latitude.

- *The nival belt* (see Figure 24.9) is the terrain above the snowline, which is defined as the lowest elevation where snow is commonly present all year round (though not necessarily with full cover). While the lower part of the nival belt is still rich in living organisms, usually very little plant and animal life is found beyond 1,000–2,000 meters above the tree line, although animals and flowering plants can be found up to around 6,000 meters in some parts of the world.

The critical bioclimatic reference line that permits global comparison and "calibration" is the high elevation tree line. The thermal limit for forest growth is surprisingly consistent worldwide and holds as a reference for all mountains where moisture permits tree growth. It is important to note that this may not be a visible line in many mountains, because forests have been re-

Figure 24.8. Alpine Grassland in the North Argentinean Andes at 4,100 Meters

Figure 24.9. Snow and Rock Fields in the Nival Zone in the Swiss Alps at 2,700 Meters

placed by pastures or cropland. These alternative land cover types are still categorized as "montane" when they occur below the thermal forest limit. Within the montane belt, different altitude-specific forest belts can be distinguished at lower latitudes, and the number of belts decreases toward higher latitudes. These belts are often referred to as lower-, mid-, and upper-montane.

24.1.2 Aerial Extent and Main Mountain Ranges

The Eurasian landmass has by far the largest mountain area of all continents; all of the world's mountains above 7,000 meters are in Asia, and all peaks above 8,000 meters are situated in the Greater Himalaya range. The Tibet (Qing Zang) Plateau is the most extensive inhabited land area above 2,500 meters elevation. Excluding Antarctica, South America has the second most extensive area of high elevation land, and the world's highest point outside Asia (Aconcagua, at 6,962 meters). Antarctica and Greenland also figure prominently, in part due to the extent and thickness of their icecaps.

Most of the world's mountain areas are found in the Northern Hemisphere and in temperate–sub-tropical latitudes (the Eurasian ranges and the North American cordilleras). (See Tables 24.1 and 24.2.) In addition, there are extensive mountain systems in the boreal (for example, Altai) and the subpolar (northeast Siberia, for instance) zones. In the Southern Hemisphere, the largest mountain systems are the Andes and the mountains of the Southeast Asian archipelago (such as New Guinea). Important but comparatively smaller mountain systems are also found in Africa, Australia, and New Zealand. More than a third of the montane belt is covered by forest, and perhaps as much as half could be potentially be covered by forest—that is, the cover is not limited by climate (see Table 24.3)—but has been converted to more open vegetation and agricultural lands by logging, fire, and grazing.

Slope, aspect, and altitude determine many of the fundamental characteristics of mountain environments. Gravity-driven topographic diversity adds significantly to the small-scale variation in life conditions. Geographic position such as latitude and distance from oceans affects climate and local weather patterns, making some mountains almost permanently wet, others dry, and some highly seasonal. Geological substratum adds a further dimension of geo-diversity and influences soil type and development, erosion processes, and vegetation cover.

Mountain climate shows a number of common features globally, but it can vary greatly regionally and locally. Several factors relevant to life processes change predictably with altitude and underlie the marked environmental gradients typical of high mountains. The most important common components are reduced pressure and reduced air temperature, with the associated reduction of water vapor pressure deficit. On average, temperature declines by 5.5 K per kilometer of elevation (but differs diurnally, seasonally, latitudinally, and from region to region), and air pressure (and with it, the partial pressures of oxygen and carbon dioxide) decreases by about 10% for every kilometer of elevation. Clear sky solar radiation increases with altitude, and higher maximum radiation and a greater short wave radiation (UV) are typical for higher elevations. However, clouds and fog may reverse altitudinal trends in solar radiation (Yoshino 1975; Barry 1992; Körner 2003).

Physical processes, in large part related to gravity, include erosion, landslides, mud flows, avalanches, and rockfall, and these determine life conditions in many parts of the world's mountains. At a more regional scale, volcanism and the associated sedimentation and slope processes affect biota and can have dramatic impact on people's life conditions. These physical phenomena of the mountain environment become enhanced when seismic activity comes into play, which is particularly the case in geologically

Table 24.1. Estimated Global Mountain Area by Continent Based on Topography Alone (Kapos et al. 2000)

Region	>4,500 Meters	3,500–4,500 Meters	2,500–3,500 Meters	1,500–2,500 Meters and Slope ≥ 2°	1,000–1,500 Meters and Slope ≥5° or Local Elevation Range >300 Meters	300–1,000 Meters and Local Elevation Range > 300 Meters	Total Mountain Area
				(square kilometers)			
North America	197	11,417	200,830	1,092,881	1,104,529	1,840,140	4,249,994
Central America	38	968	67,127	353,586	259,367	412,215	1,093,301
Caribbean			32	2,809	5,528	38,322	46,691
South America	154,542	583,848	374,380	454,417	465,061	970,707	3,002,955
Europe		225	497,886	145,838	345,255	1,222,104	2,211,308
Africa	73	4,859	101,058	559,559	947,066	1,348,382	2,960,997
Middle East	40,363	128,790	339,954	906,461	721,135	733,836	2,870,539
Russian Federation	31	1,122	31,360	360,503	947,368	2,961,976	4,302,360
Far East	1,409,259	741,876	627,342	895,837	683,221	1,329,942	5,687,477
Continental Southeast Asia	170,445	107,974	97,754	211,425	330,574	931,217	1,849,389
Insular Southeast Asia	22	4,366	34,376	120,405	157,970	599,756	916,895
Australia				385	18,718	158,645	177,748
Oceania			41	7,745	29,842	118,010	155,638
Antarctica	17	1,119,112	4,530,978	165,674	144,524	327,840	6,288,145
Total	**1,774,987**	**2,704,557**	**6,903,118**	**5,277,525**	**6,160,158**	**12,993,092**	**35,813,437**

Table 24.2. Global Mountain Area Based on Broad Biomes in Mountains, Using Different Classification Schemes (FAO 2001; Holdridge 1967; Ramankutty and Foley 1999; Udvardy 1975; Olson et al. 2001)

Broad Mountain Biome	Area					Share of Total				
	FAO	Holdridge	Ramankutty and Foley	Udvardy	Olson	FAO	Holdridge	Ramankutty and Foley	Udvardy	Olson
	(thousand square kilometers)					*(percent)*				
Desert	4,276	7,453	2,968	7,590	5,227	12.9	22.6	9.2	22.9	15.8
Forest and woodland	8,159[a]	15,476	14,248	13,428	15,819	24.7	46.9	44.3	40.6	47.8
Grassland, savanna, steppe	2,334	5,773	11,224	1,420	7,970	7.0	17.5	34.9	4.3	24.0
Mixed	1,834	0	0	8,470	1,213	55.4	0.0	0.0	25.6	3.7
Treeless alpine	0	4,291	3,729	2,206	2,899	0	13.0	11.6	6.7	8.8
Total	**33,104**	**32,993**	**32,168**	**33,113**	**33,128**	**100**	**100**	**100**	**100**	**100**

[a] The FAO mountain forest fraction is smaller because of a broadly defined "mixed land cover" category, which includes fragmented tree cover that is treated as forest in other statistics.

Table 24.3. Altitudinal Distribution of Land Area and Forest Cover for World's Mountains (Modified from Kapos et al. 2000)

Elevation Class	Global Land Area	Share of Global Total	Global Mountain Area	Share of Global Land Area	Mountain Forest Area
	(mill. sq. km)	*(percent)*	*(mill. sq. km)*	*(percent)*	*(thousand sq. km)*
>4,500 meters	1.8	1.0	1.8	1.2	23.3[a]
3,500–4,500 meters	2.7	2.3	2.7	1.8	141.4
2,500–3,500 meters	6.9	7.8	6.9	4.7	450.8
1,500–2,500 meters	11.9	9.5	5.3	3.6	1,551.3
1,000–1,500 meters	15.1	9.9	6.2	4.2	2,133.0
300–1,000 meters	53.3	33.6	13.0	8.8	5,179.4
0–300 meters	55.7	35.9	0.0	0.0	0
Total	**147.6**	**100.0**	**35.9**	**24.3**	**9,479.2**

[a] In large fragmented elfin wood forests by *Polylepis* (Andes) and *Juniperus* (Himalayas).

young and thus steep mountains. Economic consequences and the death toll can be dramatic, as exemplified by catastrophic events in the recent history of Rwanda, the Philippines, Nepal, India, Italy, and the United States.

24.1.3 Biota

Vegetation on lower mountain slopes may be broadly similar to that of surrounding lowlands. However, environmental gradients linked with elevation typically lead to marked zonation. In less humid regions, the availability of moisture may at first increase with elevation. In drylands, this can allow tree growth on mountains that emerge from treeless semi-desert plains. In humid regions, epiphyte-rich evergreen cloud forest may occur above more seasonal forest. With further elevation, temperature decreases to a point where tree growth cannot be sustained. There is no common altitudinal trend of precipitation. In the temperate zone it commonly increases with altitude, but in the tropics it often decreases beyond a montane maximum, often leading to semi-deserts above 4,000 meters (such as in the altiplano in the Andes or the semiarid top of Kilimanjaro).

The altitudinal temperature gradient in mountains is about 600–1,000 times higher than the corresponding latitudinal gradient. Discernible vegetation belts on mountains may commonly span an elevation range of 1,000 meters. Over such a range, the temperature change is about 5–6 K, enough to cause a full bioclimatic vegetation belt to be replaced by another (alpine by montane forest, for example). The latitudinal increase in seasonality and the annual temperature amplitude are mainly due to decreasing winter temperature, which limits the poleward extension of lower latitude species. Similarly, the colder climate of successive altitude belts restricts the growth of species from lower and warmer belts. One consequence of this is that ecosystems situated on mountain tops, with a species composition currently restricted by cold climate, are likely to disappear as a result of climate change.

Because of the compression of climatic zones along an elevation gradient, exposure effects, and large habitat diversity, species richness in mountains commonly exceeds that in the lowlands at small scales (such as hundred square meters). Within mountain regions, species richness decreases with increasing altitude, largely in proportion to the available land area (Körner 2000), but endemism often increases, due partly to topographic isolation (Gentry 1988; Peterson et al. 1993) and the often rapid formation and loss of links (corridors) in geological time.

Tree species diversity within a habitat commonly decreases with altitude; for example, in the tropical Andes there is an average decrease of nine species per 100 meters increase in altitude (Gentry 1988). Tropical mountain forests have 10 times higher species richness than temperate ones. Globally, there are some

10,000 species of flowering plants in the alpine belt alone—representing about 4% of all known species and covering about 3% of the vegetated land area (Körner 1995). Some groups of organisms (amphibia, for example, and bryophytes) may reach their highest taxonomic diversity in the montane belt. Two types of endemism can be prominent in mountain areas: palaeo-endemism (the survival of evolutionary old taxa in isolated refugia, exemplified by Tertiary relics of primitive angiosperms of the genera *Davidia, Tetracentron, Trochodendron* in Southeast Asia) and neo-endemism (more-recent speciation, for instance following the creation of new habitats due to volcanism or other major disturbances).

Thirty-two percent of all protected areas are located in mountainous regions, providing habitats for rare, relict, and endangered plants and animals (UNEP-WCMC 2002). Many species that survive in such refuges—pandas, tigers, takins, golden langurs, condors, and tapirs, for instance—are at risk from habitat fragmentation, however. Extended mountain ranges with continuous habitats provide a corridor for high altitude and cloud forest species, avoiding densely populated lowlands.

Ecological corridors that link isolated habitats are essential for many migrating species, which have extensive hunting or feeding territory requirements. Corridors can also facilitate species radiation, as shown for example in *Espeletia,* a giant rosette plant, in the Northern Cordilleras of the Andes (Cuatrecasas 1986). Connecting remote nature reserves, such corridors are effective tools to compensate for natural and human-induced fragmentation of habitats. Bhutan, for example, has nine protected areas covering 26% of the land (all in mountains), and all protected areas are linked by corridors, which cover another 9% of land area where land uses are compatible with conservation objectives (Dorji 2000).

24.1.4 Social and Economic Conditions

Twenty percent of the world's population—about 1.2 billion people—live in mountains. Most of them inhabit lower montane elevations, and almost half are concentrated in the Asia-Pacific region. Of the 8% living above 2,500 meters, almost all—about 90 million—live in poverty and are considered highly vulnerable to food insecurity. However, they have significant impact on larger populations living at lower elevations through their influence on catchments.

Low temperatures become prohibitive for people above 2,000 meters in temperate latitudes and above 3,500 meters in tropical latitudes (although there are exceptions up to 4,200 m), and human activities rarely occur above 4,500 meters. Special efforts and techniques are required to sustain agricultural production at altitudes close to the upper tree line level.

There are many historical examples of flourishing mountain economies based on mountain ecosystem services (including Berbers, Afghan and Caucasian tribes, Tibetans, Mongolians, Highland Papuas, Incas, and Aztecs), and many of these cultures still survive and in some cases even thrive. Lowland economies have generally dominated, however, because of intensive sedentary agriculture, manufacturing based on larger scales, easier transportation and trade, urbanization and associated better education, and the broader reach of common language and culture.

In most parts of the world, mountain areas are perceived as economically backward and culturally inferior. But there are some exceptions. In industrial countries, mountain areas have been rapidly transformed economically with improved access and the proliferation of recreational activities. In Africa, for instance, highland areas that grow tea and other high-value crops are more prosperous than lowlands. More often, however, mountain resources are extracted without benefit to local communities in order to support lowland economies, thereby contributing to the further marginalization of mountain people. Where extractive industries have been developed, mountain communities have often become dependent on wages for their livelihoods, and asset values and rents are usually allocated elsewhere.

With notable exceptions, particularly in areas where tourism and amenities migration (the movement of people because of a perceived high incidence of attractive or cultural resources) have created pockets of wealth, mountain communities suffer disproportionately from poverty and often lack even basic social services such as education and health care facilities. This, in part, has caused a counter movement in several mountain areas (the Andes and Himalayas) that is strongly linked to control over mountain resources (such as the movement of water in Bolivia).

Mountain communities are also insufficiently recognized as rich reservoirs of traditional knowledge and cultural and spiritual resources.

24.2 Mountain Ecosystem Services

For the purposes of this assessment, three main types of mountain ecosystem services are addressed:

- *Provisioning services*: extractive resources that primarily benefit lowland populations (water for drinking and irrigation, timber, and so on) and ecosystem production (agricultural production for local subsistence and for export; pharmaceuticals and medicinal plants; and non-timber forest products);
- *Regulating and supporting services,* such as biodiversity, watershed and hazard prevention, climate modulation, migration (transport barriers/routes), soil fertility, soil as storage reservoir for water and carbon, and so on; and
- *Cultural services*: spiritual role of mountains, biodiversity, recreation, and cultural and ethnological diversity.

Each of these mountain ecosystem services makes specific contributions to lowland and highland economies. Mountains play a key role in the water cycle, with feedback to the regional climate and by modulating the runoff regime. Tropical cloud forests are particularly significant in the latter respect. Mountain vegetation and soils play a significant role in reducing or mitigating risks from natural hazards. Mountain forests, for instance, protect from avalanches and rockfall; their waterholding capacity reduces peak stream flow; they are an important carbon pool; and they provide timber for fuelwood and non-timber products, including game and medicinal plants. Mountains are also used for grazing and subsistence farming. Mountain ecosystems are significant for global biodiversity, as noted earlier, and in addition they have intrinsic spiritual and aesthetic value (Bernbaum 1998; Daniggelis 1997).

Table 24.4 rates ecosystem services per unit of specific type of land area. This definition avoids a rating by the abundance of certain land types.

24.3 Condition and Trends of Mountain Systems

24.3.1 Atmospheric Conditions

Mountains extract moisture from the atmosphere through the orographic uplift of air masses that pass over mountain ranges. In this sense, mountains act as "water pumps" by pulling moisture from the atmosphere. Mountains also act as "water towers" by

Table 24.4. Ecosystem Services in Mountains

Mountain Type		Downslope Safety		Water		Food		Fiber		Medicinal		Cultural (Recreational, etc.)
		Safety	Dams	Fresh water	Energy	Grazing	Crop	Fuel	Timber	Wild	Cultivars	
Alpine	terrestrial	+++	+++	+++	+++	++	●	●	●	+++	●	++
	aquatic	●	+	+++	+++	●	●	●	●	●	●	+
Montane	terrestrial	+++	+	+++	+++	+++	+++	++	++	+++	+++	++
	aquatic	●	+	+++	+++	●	+	●	●	●	●	++
Hills and plateaus	terrestrial	+	+	+	+	+++	o	++	++	+	+	+++
	aquatic	●	●	+	+	●	++	●	●	●	●	+

Key: ● not relevant; + relevant; ++ important; +++ very important

storing water in mountain glaciers, permafrost, snowpacks, soil, or groundwater.

There are conflicting predictions about the rate of tropospheric warming. General circulation models predict a warming in high northern latitudes and also in the mid to upper troposphere in the tropics and sub-tropics. Many tropical and sub-tropical mountain ranges reach the levels of the troposphere where the warming is predicted, and the retreat of many of the world's glaciers is consistent with warming at higher elevations. However, reliable assessments of the status of mountain atmospheric conditions are currently limited to relatively few high-elevation meteorological stations. For example, a transect along the Cordilleras of the Americas shows that there are currently no meteorological stations positioned at elevations high enough to address the issue of potential warming in the mid-troposphere in the tropics and sub-tropics.

24.3.1.1 Trends in Atmospheric Physics (Climate)

24.3.1.1.1 Temperature trends

Temperature changes in 1951–89 between 30° and 70° N show that mean maximum temperatures increased slightly between 500 and 1,500 meters, with minor changes at higher elevations, while mean minimum temperatures rose by about 0.2 K per decade from 500 meters to above 2,500 meters (Diaz and Bradley 1997). In the tropical and sub-tropical Andes, mean annual temperature trends for 268 stations between 1° N and 23° S during 1939–98 (Vuille and Bradley 2000) showed an overall warming of about 0.1 K per decade, but the rate has tripled over the last 25 years to 0.33 K per decade. The warming trend declined with elevation, especially on the Pacific slopes of the Andes, whereas in the central Himalaya the warming trend increased with altitude (Shrestha et al. 1999). In the Swiss Alps, temperatures increased by a total of 1 K during the twentieth century, and milder winters now occur (Beniston and Rebetez 1996). Temperature effects appear to be stronger at night than during the day.

In many locations for which high-elevation monitoring data are available, the rate at which the atmosphere cools with increasing altitude (lapse rates) has shown an increase because of faster warming at lower altitudes. However, there are exceptions. For example, in the Colorado Front Range of the Rocky Mountains there has been an overall cooling at 3,750 meters but warming between 2,500 and 3,100 meters since 1952 (Pepin 2000). Generally, the increase of air temperature lapse rate on mountains at the mid-latitudes is greater in winter than in summer (Yoshino 2002).

There have also been remarkable trends in permafrost temperatures. In the Swiss Alps, for example, permafrost warmed by

about 1 K between 1880 and 1950, then stabilized, before warming accelerated between 1980 and 1994, followed by rapid cooling in 1994–96, which largely offset the previous warming (Vonder Mühll et al. 1998). Permafrost temperatures in the northern Tien Shan have risen by 0.2–0.3 K over the last 25 years (Gorbunov et al. 2000).

The position of the snow line has been similarly affected. The snowline in mountainous areas within 10° of the equator retreated by 100–150 meters between 1970 and 1986, which has been correlated with a warming of the sea surface over the eastern tropical Pacific (Diaz and Graham 1996). On the Quelccaya Ice Cap in Peru (14° S), meltwater penetration obliterated the uppermost part of an important climatic record provided by the ice core that Thompson et al. (1993) had collected only a few years earlier. Thus, some paleo records that are vital for our understanding of human-environment interactions are vanishing fast.

24.3.1.1.2 Precipitation and snowpack trends

Precipitation in mountain regions is best assessed through hydrological budgets of catchments. This is because precipitation is highly variable and strongly influenced by dominant wind direction (slope/aspect effects) and because precipitation analysis is complicated by seasonality and the occurrence of extreme events whose statistics are difficult. Records for the Alps (Gurtz et al. 2003; Beniston et al. 2003) suggest a future trend toward higher winter and lower summer precipitation and an increase in the altitude at which freezing occurs, with largest relative changes occurring in alpine catchments. Less summer precipitation in combination with higher evapotranspiration rates will lead to a reduction in soil moisture and groundwater recharge.

Trends in snowpack and snow duration reflect the interplay between temperature and precipitation. Snowpack has already diminished in montane altitudes and is likely to continue to do so, while there may even be an increase in snowpack at upper alpine/ nival elevations due to increased solid precipitations (Beniston 2003). On average, glaciers lost 6,000 to 7,000 millimeters of water from 1980 to 2000 (250–300 millimeters per year, based on the glacier area). (See Figure 24.10.) Unfortunately, few meteorological stations are situated at high altitudes that could address the extent to which the observed changes in mass balance represent increasing summer ablation of ice and snow or a decrease in the accumulation of solid precipitation.

24.3.1.2 Trends in Atmospheric Chemistry

24.3.1.2.1 Atmospheric deposition of nutrients and pollutants

Atmospheric processes control the deposition of long-distance pollutants in mountain environments (nutrient enrichment of

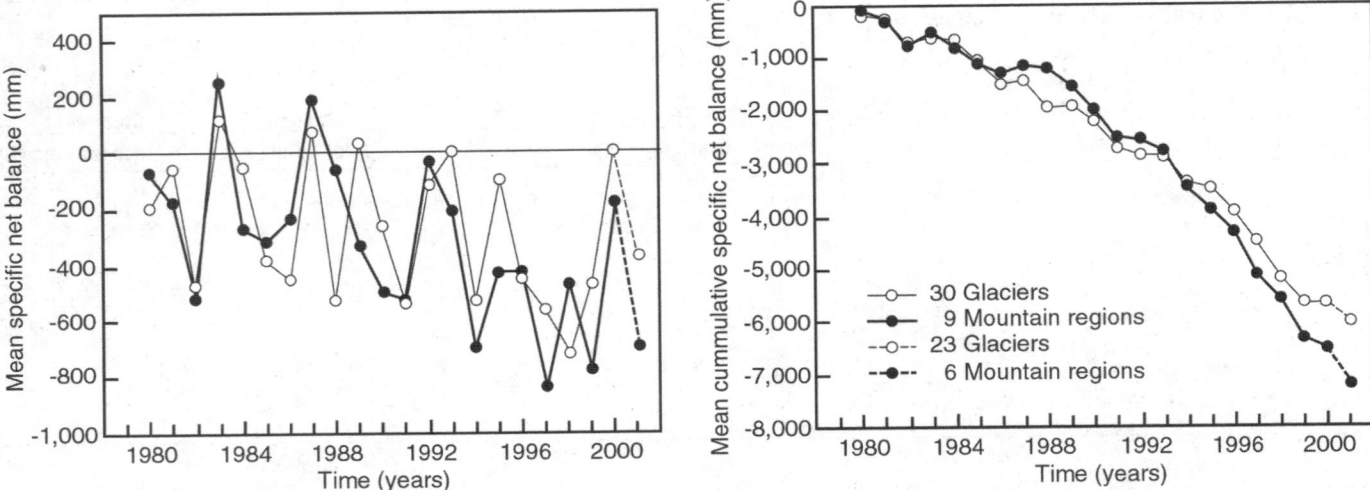

Figure 24.10. Mean Net Balance and Cumulative Mean Net Balance Continuously Measured for 1980–99 on 30 Glaciers in Nine Mountain Ranges. Data for 2000 are for 29 glaciers in eight mountain ranges and preliminary results for 2001 are for 23 glaciers in six mountain ranges. (World Glacier Monitoring Service, at http://www.geo.unizh.ch/wgms/)

mountain ecosystems and impacts on water quality). Atmospheric deposition of nitrogen and sulfur compounds, persistent organic pollutants, and metals such as mercury in mountainous areas is enhanced by proximity to anthropogenic sources and precipitation. For semi-volatile POPs, volatilization in warmer temperatures and condensation in colder temperatures results in increased deposition at high altitudes. For instance, semi-volatile organochlorine compound deposition increased 10- to 100-fold in snowpack with altitude (770–3,100 meters above sea level) in the Canadian Rockies (Blais et al. 1998).

Atmospheric deposition of acids, nutrients, organochlorines, and metals affect all components of mountain ecosystems. The loss of acid-neutralizing capacity in soils caused by the deposition of sulfur and nitrogen compounds reduces soil fertility and biodiversity and acidifies water bodies, leading to alterations in aquatic species composition. Excess nitrogen is undesirable because it induces changes in plant species composition and nutrient cycling, and it affects the ability of plants to withstand stress. Nitrogen enrichment can also increase non-native species invasions in mountainous aquatic or wetland habitats. On the other hand, nitrogen's fertilizing action can be a benefit in areas where enhanced productivity is desired, such as in commercial mountain forests. Rare or endemic species, often found in oligotrophic habitats, are likely to become suppressed by more vigorous species of wider distribution.

A preliminary examination indicates a pattern of biological accumulation of POPs in animals and foliage at high altitudes (Schindler 1999). The occurrence of the insecticide toxaphene in fish increased by 1,000-fold over a 1,500-meter elevation range in the Canadian Rockies, and similar patterns have been observed in the Alps for polychlorinated biphenyl concentrations (Grimalt et al. 2001). Organochlorines and metals are harmful to fish health and reproductive ability, and bioaccumulation transfers poisons to waterfowl, wildlife, and humans.

Atmospheric deposition in conjunction with other disturbances (such as unsuitable land use practices and floods) can create many problems, such as losses of soil nutrients or the accumulation of xenobiotic substances. For example, water from glacial catchments was recently shown to be the dominant sources of POPs in the mountain rivers of Alberta, Canada (Blais et al.

2001). At the same time, however, processes such as N enrichment may increase species invasions in mountainous aquatic or wetland habitats. Glacial runoff in summer becomes channelized on the glacier surface, funneling POPs rapidly to alpine and subalpine waters. Glacial meltwater also showed evidence of the accumulation of tritium in glacial ice during atmospheric nuclear tests in the 1960s and 1970s (Blais et al. 2001). There is a high probability that under a scenario of increasing global temperatures, glacial melt will lead to unexpectedly high concentrations of POPs in montane waters.

24.3.1.2.2 Carbon dioxide

Globally, mountain ecosystems at temperate latitudes are among the largest biotic carbon reserves because these mountains tend to be forested, accounting for, for instance, 25–50% of the contribution to the total U.S. carbon stock and up to 75% of the western U.S. carbon stock (Schimel et al. 2002). The effects of elevated CO_2 concentration on montane forests vary with nutrient availability and species. Haettenschwiler and Körner (1998) found no effect of elevated CO_2 on the growth of montane *Picea abies* (Norway spruce), and no effect was found in *Pinus uncinata* (mountain pine) after three years of in situ free air CO_2-enrichment (FACE) at the Swiss tree line. However, deciduous larch *Larix decidua* showed a continuous response, provided trees were not affected by larch bud moth. If this trend in differential growth continues over the long term, it would provide a clear example for CO_2-driven biodiversity effects (Handa et al. in press). Studies of ecosystem productivity in alpine grasslands in the Swiss Alps found no detectable increase after four years of double-ambient CO_2 concentration. However, higher CO_2 concentrations reduced forage quality, increased (compensatory) herbivory by grasshoppers during certain periods, and altered the plant community composition (Körner et al. 1997).

24.3.1.3 Consequences of Atmospheric Changes

Environmental conditions change rapidly with elevation because of the steep temperature and precipitation gradients. Thus, rapid changes in life zones occur over short vertical distances, and relatively small changes of the climate can induce large changes in the area available for a given life zone (cf. Theurillat and Guisan

2001). Because topographic effects have a greater influence than elevation effects over conditions for life above the tree line, biota do not necessarily occur at higher altitudes, but may form new mosaics, with the distribution of snow often being far more influential than temperature (Gottfried et al. 1999).

Climatic change may enhance or reduce precipitation, depending on the region. A reduction of moisture in already dry mountain regions (such as the upper *Erica* belt on Kilimanjaro) will enhance fire frequency. Some mountain forelands will receive less water, with disastrous consequences for marginal semi-arid lowlands. Enhanced activities such as El Niño may in turn expose other tropical and sub-tropical regions to excessive precipitation followed by floods and mudslides, as occurred recently in Peru, Ecuador, and Colombia. In many cases, mountains are the primary source of fresh water in these regions.

The reduction in glacier volumes is expected to have a strong impact on dry-season river flows in rivers fed largely by ice melt (Haeberli and Beniston 1998). This will very likely affect the provision of downstream water for drinking, hydropower, and irrigation. Over 65 countries use more than 75% of their available fresh water for agriculture. These include countries with large populations such as Egypt, India, and China, which rely heavily on mountain discharge (Viviroli et al. 2003). Unfortunately, mitigation efforts such as constructing dams and reservoirs are problematic in tectonically active regions where slope stability is further compromised by glacier recession and melting permafrost (Haeberli and Beniston 1998). Conversely, it is likely that some maritime mountain regions may experience increased precipitation under warmer conditions, which may lead to slope instability, mass movement, and accelerated erosion.

In the Himalayas of Nepal and Bhutan, glacier lake outburst floods are increasing in frequency due to the rapid recession of glaciers. It is anticipated that such events could reach rates of one significant glacier outburst flood each year by 2010 (Kaeaeb et al. 2005), which will impose a substantial risk to downstream communities and hydroelectric power schemes.

Winter tourism is particularly vulnerable to climatic change in areas near the lower winter snowline. For example, it is likely that a number of winter resorts in the European Alps situated below 1,500 meters above sea level will be forced to close in the near future, which in turn will increase demand on high-elevation resorts.

24.3.2 Mountain Biota

24.3.2.1 *Overview of Land Cover in Mountains*

Table 24.5 provides a summary of land cover in mountain areas, obtained by overlaying the global land cover data set for 2000 (Bartholome and Belward 2004) with the modified mountain map of Kapos et al. (2000). Based on this data set, 13.3% of the mountain area is cultivated, while the urban ("artificial") land area amounts to 0.05%, or 15,400 square kilometer, making it nearly negligible at the global scale.

Overall, it is very likely that about half of the global mountain area is under some sort of human land use (we assume that unvegetated or bare land and wetlands and other water bodies, which together amount to 14% of the global mountain area, are mostly unaffected by humans). A considerable proportion of forests, including plantation forests, woodlands, and shrubland as well as herbaceous vegetation in mountains are under various human land uses, such as silvicultural interventions and grazing. Probably half of all temperate/boreal forests and two thirds of all tropical mountain forests are under some sort of management (from selective logging and shifting cultivation to plantation forest).

Wildlife—and free-roaming domestic animals—can have a dramatic impact on land cover. For instance, much of the initial fragmentation of montane forest on Mt. Kenya was a consequence of free elephant access. By contrast, deep canyons prevent elephant access to much of the montane cloud forest on Kilimanjaro, which has remained largely intact. Through wildlife management, including hunting and the extensive use of fire, wildlife influences have been modified and have often become less important than those of domestic animals.

24.3.2.2 *Mountain Forests*

Mountain forests account for 26.5% or 9.5 x 10⁶ square kilometer (using the 300 meters low elevation threshold in the tropics) of the global closed forest area (Kapos et al. 2000). (See Table 24.6.) The upper forest limit lies at around 4,000 meters in the tropics; it gradually decreases toward the poles and ends near sea level at the polar forest limit. On a large scale, both the mountain and polar tree lines are set by temperature during the growing season (Ohsawa 1990; Körner and Paulsen 2004).

In the Southern Hemisphere, current forests often do not reach the potential climatic limit for tree growth that is predicted for the Northern Hemisphere. (In the Southern Hemisphere, introduced northern temperate trees can grow even above the indigenous species (Nothofagus)'s upper limit (cf. Wardle 1971).) So native species do not always reach their life-form limit. Since there is no land in the extreme south, deciduous and coniferous boreal forests are absent in the Southern Hemisphere, and evergreen temperate trees form the southern forest limit.

A continuous zonation of humid forests from the equator to the poleward forest limit can be observed in the mountain chains from humid, monsoon Southeast Asia to East Asia, and from the Northern to Southern Andes. In the other parts of the world, the forest area is interrupted by drylands between the equatorial and the temperate mountains.

The number of distinct elevational belts within the montane forest belt decreases toward higher latitudes (Holdrige 1967; Brown et al. 1991). Humid tropical montane forests are dominated by evergreen broad-leaved trees, from the foothills to the upper forest limit. (See Figure 24.11.) In temperate mountains, a marked altitudinal sequence of evergreen, deciduous, and coniferous forests can be found, but this varies from continent to continent. In North America and Europe, the evergreen broad leaf component is missing, while New Zealand has no conifer or deciduous belt. The presence or absence of conifers and deciduous trees in tropical versus temperate mountains suggests large differences in ecosystem functioning (seasonality, water, and nutrient relations).

24.3.2.2.1 *Natural, unmanaged forests*

Natural, unmanaged mountain forests are becoming rare. They are often isolated or fragmented but host a rich and original flora and fauna. What is commonly termed "natural" may still include some human activities, but these do not normally alter the abundance and composition of forest species. Many of these forests are relicts with varied protection status. Nearly 10% of the global mountain forests are under some sort of protection (UNEP-WCMC 2002). Recent networks of protected areas in many parts of the world aim at establishing connections using ecological corridor systems (for example, Bhutan). These forests are essential to protect fragile mountain slopes from erosion and leaching processes and as reservoirs of species for resettlement of deforested, fragmented, or newly created habitats affected by anthropogenic or natural disturbances.

Table 24.5. Land Cover in Mountains. The share of different land use types in each biogeographical zone is provided in parentheses. (Bartholome and Belward 2004, UNEP-WCMC updated by Thonell using the formation categorization by Ohsawa (1995 modified)) The criteria for the humid life zones are as follows (WI is defined as the sum of monthly mean temperature above 5° Celsius (Kira 1948), CMT is defined as the coldest monthly mean temperature (Ohsawa, unpublished)):
– Tropical life zone (latitude below 20–30° N/S): WI <15 (CMT > −1°C) = above limit of tropical forest; 15 < WI < 85 (CMT > 6°C) = tropical upper montane zone; 85 < WI <240 (CMT >12°C) = tropical lower montane zone; WI > 240 (CMT > 18°C) = tropical hill zone if elevation is above 300 m, otherwise not mountainous.
– Sub-tropical/temperate life zone (latitude above ca. 30° N/S): WI<15 = above limit of temperate forest; CMT < −7°C = temperate upper montane zone; −1°C > CMT > −7°C (WI > 15) = temperate lower montane zone; 6°C> CMT > −1°C, 12°C > CMT > −1°C and WI > 85 = temperate hill zone.
– For dry life zones, the Holdridge life zone classes (scrub, steppe, woodlands, dry tundra, and desert vegetation) have been used. (Holdridge 1967)

Biogeographical Zone	Forest and Woodland Cover	Shrub, Herbaceous Cover	Cultivated	Wetlands (mires, swamps, river basins)	Bare Areas (rock, gravel, etc.)	Water Bodies (including snow and ice)	Artificial (urban, industrial, etc.)	No Data	Total
			(square kilometers and percent of total area)						
Humid tropical hill	525,415 (56)	155,115 (15)	255,130 (27)	1,392 (0)	2,052 (0)	3,125 (0)	188 (0)	118 (0)	942,535
Humid tropical lower montane	3,003,963 (53)	1,208,793 (21)	1,421,357 (25)	13,927 (0)	13,477 (0)	28,234 (0)	2,444 (0)	3,757 (0)	5,695,952
Humid tropical upper montane	73,604 (34)	97,726 (46)	37,814 (18)	367 (0)	2,999 (1)	1,319 (1)	71 (0)	–	213,900
Humid temperate hill and lower montane	2,001,526 (51)	1,120,418 (29)	638,096 (17)	4,428 (0)	68,035 (2)	36,152 (1)	4,701 (0)	114 (0)	3,918,620
Humid temperate lower/ mid-montane	964,836 (60)	342,793 (21)	227,184 (14)	12,998 (1)	13,687 (1)	41,466 (3)	2,480 (0)	–	1,605,558
Humid temperate upper montane	3,448,322 (67)	1,080,996 (21)	267,071 (5)	70,520 (1)	178,840 (3)	135,509 (3)	642 (0)	167 (0)	5,181,900
Humid temperate alpine/nival	411,790 (25)	960,857 (58)	16,817 (1)	23,474 (1)	104,307 (6)	143,254 (9)	46 (0)	627 (0)	1,660,712
Humid tropical alpine/nival	22,027 (13)	101,128 (61)	4,070 (2)	1,744 (1)	13,486 (8)	21,648 (13)	1 (0)		164,731
Dry tropical hill	42,093 (9)	170,368 (36)	88,317 (19)	559 (0)	164,465 (35)	1,927 (0)	170 (0)	–	467,899
Dry sub-tropical hill	141,055 (7)	789,890 (38)	113,898 (5)	1,409 (0)	1,035,832 (5)	2,508 (0)	1,138 (0)	–	2,085,730
Dry warm temperate lower montane	146,996 (11)	743,592 (54)	117,200 (9)	380 (0)	353,788 (26)	5,521 (0)	650 (0)	–	1,368,127
Dry cool temperate montane	526,173 (18)	1,547,610 (53)	430,737 (15)	3,658 (0)	419,257 (14)	16,593 (1)	3,227 (0)	–	2,947,255
Dry boreal/sub-alpine	250,670 (25)	420,964 (42)	92,869 (9)	4,821 (0)	207,577 (21)	34,253 (3)	279 (0)	–	1,011,533
Dry subpolar/alpine	115,462 (36)	121,795 (38)	938 (0)	1,634 (1)	59,846 (19)	22,798 (7)	114 (0)	–	322,587
Polar/nival	458,843 (8)	2,952,626 (56)	69,345 (1)	20,717 (0)	771,042 (14)	1,305,532 (23)	52 (0)	–	5,578,157
Total area	**12,132,775 (37)**	**11,814,671 (36)**	**3,825,843 (12)**	**162,028 (0)**	**3,408,690 (10)**	**1,799,839 (5)**	**16,303 (0)**	–	**33,165,196**

24.3.2.2.2 *Semi-natural forests*

Human use has turned most natural forests into semi-natural forests, a process often completed several centuries ago in Europe but still going on in other mountain areas of the world. Thus semi-natural forests have become part of ecosystems in many parts of the world and are essential for local people to obtain both timber and non-timber forest products.

Large-scale forest statistics (e.g., FAO 2001) do not differentiate between high and low human impacts on forests, however. Sustainably managed semi-natural forests provide many ecosystem services, including tradable products such as timber and various non-timber products, and at the same time retain high biodiversity (Peterken 1981). But heavily disturbed or damaged montane forests are often invaded by fast-growing, early successional tree or shrub species. These pioneers often endanger relic species of the Tertiary (e.g., the Dove tree *Tetracentron* in western China, cf.

Tang and Ohsawa 2002). On the other hand, such species may be crucial for the conservation of disturbed mountain ecosystems through their role of covering and stabilizing steep slopes and facilitating forest regeneration (cf. Bormann and Likens 1979).

The invasion of unpalatable herbs into overgrazed pastures and of small trees/shrubs into clearings is often regarded as noxious, but, again, these species may reduce nutrient loss, enhance the resilience of the ecosystem, and facilitate restoration (Callaway et al. 2000). In some cases alien species even invade pristine forests once established in disturbed areas (e.g., exotic *Myrica faya* invades undisturbed *Metrosideros* forests in Hawaii). Shifting agriculture systems in areas of low population density can maintain ecosystem integrity similar to coppiced forest systems, provided the replenishment of soil nutrients is achieved by allowing sufficient fallow periods. Exploitative farming systems with too short a fallow period (as is the case in systems where human population density is increasing) are not sustainable.

Table 24.6. Areas of Different Forest Types Occurring in Each Mountain Class (UNEP-WCMC 2003; FAO 2003)

Elevation Class	Temperate and Boreal Evergreen Needleleaf Forests	Temperate and Boreal Deciduous Needleleaf Forests	Temperate and Boreal Deciduous Broadleaf and Mixed Forests	Tropical (and Sub-tropical) Dry Forests	Tropical (and Sub-tropical) Moist Forests	Total	Forest Area/ Mountain Area
	(thousand square kilometers)						*(percent)*
Above 3,500 meters	25.0		1.7	0.2	19.4	23.3	1.3
2,500–3,500 meters	151.8	1.2	122.9	35.3	138.8	450.8	6.5
1,500–2,500 meters	548.0	76.2	476.9	50.6	277.0	1,551.3	29.4
1,000–1,500 meters	788.7	313.9	441.1	107.3	545.7	2,133.0	34.6
300–1,000 meters	1,377.1	985.6	1,275.7	343.4	1,173.0	5,179.4	39.9
Total	**2,890.5**	**1,377.0**	**2,338.0**	**551.8**	**2,333.0**	**9,479.2**	**26.5**
			(percent)				
Share of total mountain forests	30.4	14.5	24.7	5.8	24.6	100.0	
Location	North America, Europe, Central Asia, Himalaya	Central Asia, Northeast Asia	North America, Southern Andes, Europe, Himalaya, Eastern Asia	South Africa, India	Trop. Andes, Central America, East Africa, Madagascar, Southeast Asia		

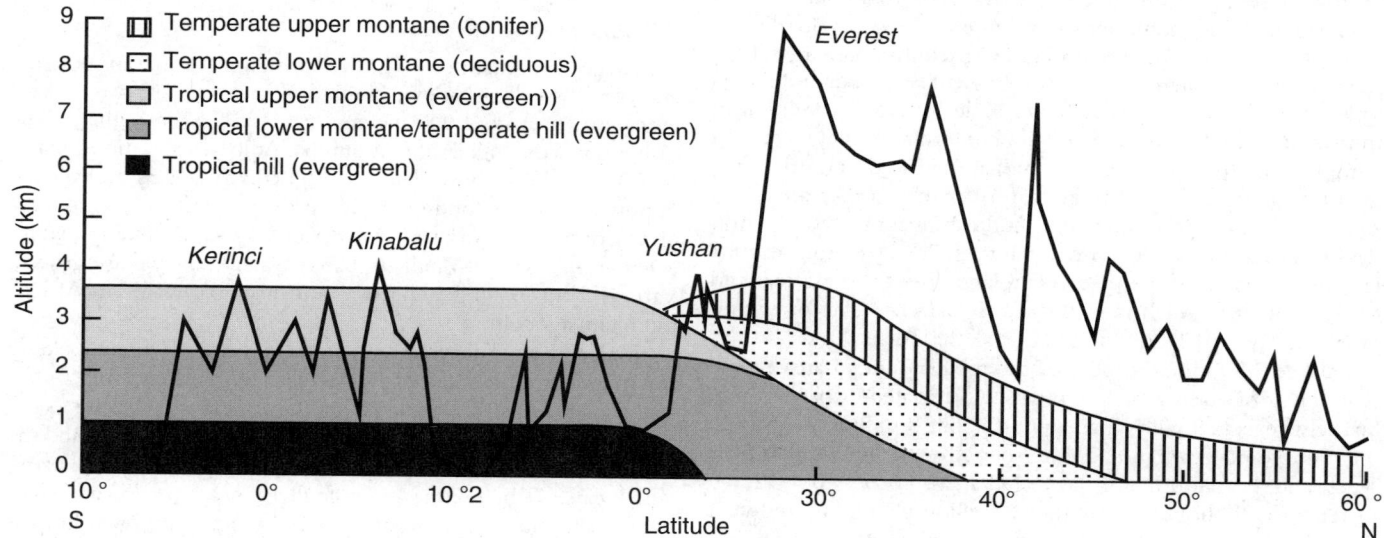

Figure 24.11. Potential Forest Life Zone Model Overlaid on Mountain Profile of Southeast to East Asia. (Ohsawa 1990, 1995) Note maximum of five sub-belts at 23°N and the latitudinal reduction of sub-belts to one at >47°N.

24.3.2.2.3 *Plantations*

Forest plantations occur mainly in temperate countries (75% of the global area of plantation forest); the rest are found in tropical countries. In sub-tropical and tropical humid mountains, the slow-growing natural hardwood forests are often replaced by fast-growing softwood species, and such plantations extract more water and reduce catchment yields, such as in South Africa, New Zealand, and some parts of the Andes (Hofstede et al. 2002; Morris 1997). In some cases, exotic plantations or ornamental trees introduce diseases and pests.

Large-scale monospecific plantations may exclude wild or domesticated herbivores and lead to a shift in their habitat selection toward the remaining natural or semi-natural forest fragments, thus causing excessive animal densities and deterioration in these more natural forests. When introduced species invade natural forests, they may suppress the regeneration of native species that protect soils from erosion and are important for biodiversity conservation. Recent trends in industrial countries suggest that productive exotic tree plantations can be converted back into less productive but lower risk systems dominated by native trees (FAO 2003). However, plantations can also have an important role in the process of restoring degraded land, as in Ethiopia (Yirdaw 2001), and act as catalysts for succession of native tree species at low population densities.

Recent changes in how forest services are valued, with a shift from simple timber production toward biodiversity, aesthetic, spiritual, and recreational aspects, may help change forest plantations to more natural forest in some cases.

24.3.2.2.4 Trends in mountain forests

Considerable changes are taking place in mountain forests as a result of overgrazing, pathogens, fire, and direct transformation of forest into other uses, such as plantations and agriculture. Development projects, such as dams, hydropower plants, roads, tourist infrastructure, and urbanization, also contribute to forest loss. In their global forest statistics, FAO (2003) distinguishes three types of new forest area: reforestation, afforestation, and natural expansion of forest. A comprehensive analysis of the change in forest area for the period 1990–2000 showed that natural forest area decreased by 6.8% in the tropics, while in temperate areas it expanded by 1.2%, mainly due to increases in forest cover in the mountainous countries of Europe (FAO 2001). (See also Chapter 21.)

24.3.2.3 Agricultural Systems

According to a recent FAO estimate, 78% of the world's mountain area is unsuitable or only marginally suitable for growing crops (Huddleston et al. 2003). Pastoralism and forestry are the predominant uses of mountain land in all regions. Nevertheless, agriculture remains important for a large number of people who will continue to depend on it as their main source of livelihood for the foreseeable future (see later sections on traditional use and vulnerability). Mountain livelihood systems are generally diverse within a variety of agricultural and nonagricultural activities. Typical mountain dwellers grow a wide range of crops and often multiple varieties of each crop. Small-scale livestock production, timber, hunting, fishing, and non-timber forest product collection complement food production. Nonagricultural activities frequently include seasonal migration of men to other areas and tourism, such as mountain guiding and nature conservation work.

In developing and transition countries, 7% of the total mountain area is currently classified as cropland; forest and grazing land cover about 25% each, and the rest is barren (33%) or within protected areas (10%) (Huddleston et al. 2003).

Above 2,500 meters, 88% of the total mountain area represents a mix of grazing land and sparsely vegetated or barren land. Sparsely vegetated high lands support about 5 million people; 29 million live off grazing land, interspersed with other land cover types; and 4 million live in protected areas. Forests above 2,500 meters provide home for another 2 million people. Mixed land use patterns—such as crop agriculture combined with exploitation of forest resources and herding of small livestock—are characteristic of some locations between 2,500 and 3,500 meters (mountain class 3) in Central and South America, in the East African and Ethiopian Highlands, and in Nepal. Although mountain people in these locations are increasingly vulnerable, their numbers are quite small. Two million people live in rural areas above 2,500 meters, which are mainly classified as cropland or mixed use. Cropland at higher elevations constitutes only 3% of the total mountain area.

Below 2,500 meters, grazing land interspersed with other land cover types accounts for 45% of the land, with a further 20% classified as mainly barren land. Around 300 million people, or two thirds of the rural mountain population below 2,500 meters, inhabit these areas and rely on livestock for income with some crop agriculture (Huddleston et al. 2003).

These findings confirm the importance of pastoralism for mountain people at all elevations. However, the loss of traditional trade routes and patterns of goods exchange, degradation of land resulting from population growth and increase in livestock numbers, and the impact of frequent, severe droughts in recent years have all greatly increased the vulnerability of this livelihood system, with mountaineering, tourism, or seasonal migration offering the only major alternatives for income generation. In the absence of men, women are forced back into crop agriculture, which is barely suited to the land.

FAO has identified 17 distinct farming systems that are significant in mountain areas at elevations above 1,000 meters. (See Table 24.7.) Together, these systems account for 67% of the total mountain area and include 82% of the total rural mountain population in developing and transition countries. The "intensive mixed highland farming system," a livestock-dependent system, is by far the most important. Stretching from northern Africa to Southeast Asia and beyond, this system has many variants that are specific to local cultures and conditions. In the Ethiopian Highlands (1,800–3,000 meters), Near East and North Africa (300–2,000 meters), Hindu Kush–Himalaya (1,500–2,500 meters), and central Andes (1,500–3,000 meters), the main features of the system usually include cereals, legumes, potatoes, fodder trees and crops, ruminant livestock, coffee, and horticultural tree crops, with the sale of wool, meat, and tree fruits constituting the main sources of income. By contrast, in western China (500–1,500 meters) rice is important, and income is derived mainly from the sale of vegetables, fruits, pigs, and poultry.

24.3.2.3.1 Mountain crops

The major mountain crops grown at high elevation are potatoes and cereals (such as barley), in some tropical and sub-tropical regions grown up to 4,000 meters (and locally even higher). One cash crop of the high Andes is quinoa. At lower elevations, maize (corn), rice, beans, peas, and sweet potato as well as cabbage are important as staple foods globally. At the lowest tropical montane elevations, crops such as taro (*Xanthosoma, Colocasia*), yams (*Dioscorea*), cassava (*Manihot*), bananas, and papayas are grown. Various forms of millet and sorghum play a key role in African mountain agriculture.

Major plantation crops grown in the lower montane belt are tea and coffee: the latter at higher altitudes in Colombia and in montane regions of Brazil, Costa Rica, Guatemala, Viet Nam, and elsewhere, while tea is grown extensively in hill areas and the highlands of Africa and Asia. The presence of traditional crops, which can also be used to produce illegal crops (marijuana, coca, poppy) are a major socioeconomic and political problem in parts of the Andes, Atlas, Afghanistan, and Central and Southeast Asia. In some parts of the high Andes, industrial potato production is threatening the Parámo flora, which is rich in species and endemism.

24.3.2.3.2 Mountain rangeland

Mountain rangelands may be natural (vegetation above the tree line) or of anthropogenic origin (below the tree line). In the Alps, for example, a 60-kilometer North-South transect across the main divide revealed that 57% of the land is covered by meadows, pastures, and other types of palatable low-stature vegetation (Körner 1989). The vegetation of these rangelands is composed of species formerly found along rivers, in naturally disturbed areas and pockets grazed by wild herbivores. These species underwent selective pressure from human land use and assembled into highly adapted plant communities, which over millennia have formed closed and extensive ground cover on nonforested land.

Table 24.7. Main Mountain Farming Systems in Developing and Transition Countries: Location, Characteristics, Area, and Rural Population. Population numbers refer to number of people living in these areas. (Huddleston et al. 2003; definition of farming systems, Dixon et al. 2001)

Mountain Farming System Categories	Farming Systems	Area	Rural Population	Elevation and Geographical Location
		(mill. sq. km.)	(million)	
Other	not specified	7.4	88.2	Below 1,000 meters—all regions
Irrigated	irrigated rice	1.0	38.6	300–2,500 meters—Madagascar and some mountain riverbanks in Africa; coastal areas of Chile, Ecuador, and Peru and of Caspian and Aral Seas; terraced hills of Mexico, and South and Southeast Asia
Maize mixed	rice-tree crop	1.4	26.1	300–2,500 meters—uplands of Kenya, Lesotho, Malawi, South Africa, Swaziland, Tanzania, Zambia, and Zimbabwe; Central Mexico, Costa Rica, El Salvador, Guatemala, Honduras, Nicaragua, and Panama
Tree crop/sparse forest	maize mixed maize-bean	3.5	37.3	300–1,500 meters—hilly areas of West African coastal countries from Côte d'Ivoire to Angola; 500–3,000 meters—highlands of Burundi, Ethiopia, Rwanda, and Uganda; Indonesia, Malaysia, Mongolia, Myanmar, Pacific Islands, northern Argentina and southern Chile, western Chile
Pastoral	tree-crop highland perennial sparse (forest)	0.3	6.4	all elevations—semiarid and arid areas in all regions; important in Central Asian CIS countries and in Hindu Kush-Himalaya highlands and plateaus
Small-scale cereal-livestock	small-scale cereal livestock	3.5	224.3	300–2,500 meters—Turkey
Highland intensive mixed	highland temperate mixed highland mixed upland intensive mixed intensive highland mixed	2.0	50.7	300–3,000 meters—Ethiopian Highlands and small pockets in Angola, Cameroon, Eritrea, Lesotho, and Nigeria; Himalayan, South Asian, Near Eastern and North African hills; Indonesia, northern Thailand, Philippines, South China, and Viet Nam; Colombia, Ecuador, and Venezuela
Highland extensive mixed	upland extensive mixed high altitude mixed	2.3	11.1	800–4,500 meters and above—Cambodia, Indonesia, Laos, Myanmar, northern Thailand, Philippines, southeastern China, Viet Nam; Bolivia, northern Chile, northwestern Argentina, and Peru
Sparse	sparse (arid) sparse (mountain)	1.0	7.3	All elevations—arid areas throughout North Africa and Near East, and in China, Kazakhstan, Mongolia, Pakistan, Turkmenistan, and Uzbekistan; Above 3,000 meters—middle and upper Himalaya slopes
Total		**22.4**	**490.0**	

Present in the upper montane and alpine belt across the globe, these ecosystems reflect traditional, sustainable land use and represent biota of high ecological, conservation, and economic value. In many places they host unique and species rich assemblages of plants and wildlife and can often support 50 species or more of higher plants in a 100-square-meter area. The best examples of such rangeland assemblages are found in Europe, including the Caucasus, and across the Himalayas into western China. These high-altitude rangeland systems and their sustained productivity depend on appropriate land use (see vulnerability section later) and represent a cultural heritage that deserves protection. As with forests, increasingly intensive use is occurring over ever increasing areas in many parts of the world, causing loss of sustainability, land transformation with associated changes in ecosystem function, and, in extreme cases, a loss of ecosystem integrity and soils.

The general trends reflect the common economic, developmental, and demographic differences between industrial and developing countries. While abandonment of high-elevation rangelands is common in the former (in France, for instance), overexploitation has reached dramatic dimensions in many developing countries (Ethiopia and Nepal, to name two), a result of unprecedented population pressure. The long-term consequences will very likely be similar to those seen today in part of the Mediterranean coastal mountains, where overutilization has often been followed by catastrophic erosion, complete loss of soils, and conversion of flora and fauna (van der Knaap and van Leeuwen 1995).

The key to sustainable management of high-elevation rangeland has proved to be strict control of grazing. Prevention of loss of soil (productivity) has high potential as a policy-making intervention. In the Kosciusko National Park and Biosphere Reserve in the Snowy Mountains of Australia, for example, phasing out inappropriate grazing helped protect vegetation in an upland catchment, increasing the value for hydroelectric generation. In the absence of urgent measures, it is very likely that developing countries risk large-scale environmental degradation from overgrazing of mountain rangeland, with stark consequences for a large fraction of their current and future populations.

Through the abandonment of traditional grazing lands, industrial countries also risk a decrease in the landscape value of their mountains and the loss of a resource for quality food production.

In 30–50 years, abandoned pastures revert back, through scrub, to montane forest as long as seed sources are present. It is nearly impossible to reverse this process, because of both high costs and the loss of traditional knowledge. Abandonment of grazing management in mountain ecosystems that have a long history of grazing can also have adverse effects on the survival of some native plant species, many of which may be endemic. Traditional use is affordable only if all benefits are accounted for.

A remarkable secondary benefit of maintaining good quality grazing land is the benefit to hydro-schemes. The extra value in hydropower often exceeds the monetary value of the agricultural yield. In the Alps, and similarly for the Caucasus, the monetary value of an intact, high-elevation, short grass pasture in terms of hydroelectric yield was estimated to be about 150 Euro per hectare, higher than that of abandoned long grass turf or forest (Körner 1989).

24.3.2.4 Alpine Biota

The area of alpine land above the natural climatic tree line accounts for about 3% of the planet's land surface. Depending on region, most arctic-alpine vegetation north of 65–70° is probably better included in the term "arctic." The alpine belt alone supports about 10,000 plant species worldwide (Körner 1995), corresponding to 4% of the total number of known species of flowering plants. Local flora of individual mountains (except for isolated volcanic peaks) throughout the world consist of 200–300 species, a surprisingly constant number (Körner 1995). The compression of climatic zones, high fragmentation and topographically diverse habitats, geographic isolation, glaciation, and varied history of species migration or evolution have led to high degrees of taxonomic richness (including endemism) in alpine biota (Nagy et al. 2003; Körner 2004).

Treeless by definition, the alpine belt is in large part composed of dwarf shrub heath and grassland, a vegetation that dominates the headwaters of most major river systems. In the tropics and sub-tropics, but also in the oceanic Southern Hemisphere temperate and subpolar zone, tall tussock grasses represent the dominant life form. In most parts of the world the wild ungulate grazers of this life zone have been replaced by domestic species. Animal trampling of fragile soils on steep terrain and fire management of tussock grasses are the major threats to the biota. In fact, the overdominance of tussocks versus more palatable low-stature grass cover has been interpreted as a consequence of excessive land use (Hofstede 1995). Inter-tussock surface erosion is a major problem associated with insufficient or weakened inter-tussock vegetation.

Since soil development is very slow at these elevations, revegetation after the loss of substrate is nearly impossible. In many parts of the world, alpine vegetation extends to below the climatic tree limit because of forest destruction. Together with forest preservation, the sustained integrity of these highland ecosystems is key to the quality and quantity of catchment discharge. Indeed, the ecosystem engineering role of species-rich alpine biota has become a focal area of internationally coordinated research (Körner and Spehn 2002; Spehn et al. 2005).

With respect to ecosystem productivity, neither atmospheric CO_2-enriched nor moderate climatic warming appear to exert significant effects on alpine systems. It seems that productivity is not limited by the availability of CO_2, and thermal conditions in the low stature ground cover are much more controlled by radiative solar heating than is the case in trees and forests, which are aerodynamically well coupled to atmospheric circulation (Körner 2003). At high latitudes, temporal and spatial patterns of alpine snow pack exert strong influences on vegetation, with patterns of precipitation, particularly during the cold season, being potentially more significant than summer temperatures. In addition to changes in precipitation and snow pack, major global threats in the alpine belt are regional nitrogen deposition and land use. Warming can open higher elevation habitats for organisms from lower altitudes, provided the summits are high enough (Grabherr et al. 1994).

24.3.2.5 Aquatic Biota

Mountain lakes, ponds, and streams provide important ecosystem services—including drinking water, fish, recreation, and aesthetic values. The productivity and diversity of algae, invertebrates, and fish decline significantly with increasing altitude because of the more extreme environmental conditions, such as low temperatures, short season, and low nutrient concentrations, and the isolation (Vinebrooke and Leavitt 1999), causing such biota to be more sensitive to any change in environmental conditions (Donald et al. 2001).Consequently, the migratory responses of mountain aquatic biota are closely tied to changes in the environment (e.g., Donald et al. 2001). Human actions, such as widespread introduction of exotic fish, led to significant biological impoverishment of montane and alpine lakes and streams during the twentieth century, which may require decades to recover from (Donald et al. 2001; Schindler and Parker 2002).

Oligotrophic aquatic ecosystems in mountains rely heavily on external inputs of nutrients (Vinebrooke and Leavitt 1998), which make them highly sensitive to human land use practices and air pollution. For example, nitrogen deposition impacts have been documented from oligotrophic mountain lakes in the western United States (Baron et al. 2000) and from streams in the Hindu-Kush Himalayas (Jenkins 2002). Cold climate aquatic biota also show pronounced accumulation of mercury and organochlorine compounds at higher elevations as a result of low temperature condensation of emissions originating from lower elevations and of release from melting glaciers in both North America (Blais et al. 2001) and Europe (Rognerud et al. 2002). However, the sensitivity of species-poor mountain aquatic ecosystems makes them excellent early indicators of environmental change.

Climate warming and drought have pronounced impacts on biota in non-glacial lakes and streams as they become clearer, warmer, less acidic, and more ephemeral because of reduced snowpack and increased mineralization rates (Sommaruga-Wögrath et al. 1997). Interactions between multiple environmental stressors, such as introduction of exotic species, air pollution, climate warming, and human land use, determine the cumulative impact of global change on aquatic mountain biota (Battarbee et al. 2002).

24.3.3 Mountain Watersheds

The relationship between vegetation, soil, and water is best expressed in the functioning of the hydrologic unit—the watershed. Watersheds integrate conditions and processes over large areas and determine the functionality of their ecosystems and the water yield for river systems, which provide essential fresh water for aquatic life (including fisheries), agriculture, hydropower generation, and industrial and domestic use for growing populations both in the mountains and in the lowlands. More than 3 billion people depend directly or indirectly on water from mountains.

24.3.3.1 The Hydrological Importance of Mountains

One of the most important services from mountain ecosystems is the provision of clean water. In 2000, the Second World Water Forum in The Hague declared that major challenges included:

- protecting the ecosystems that supply water—that is, the mountains, the water towers;
- managing the risks that have an impact on water supply and distribution—that is, the drivers of change; and
- increasing the valuation and improving the governance of water resources and watersheds.

Mountain areas typically produce about twice the discharge that could be expected from the land area they cover (Viviroli et al. 2003). Mountains account for 20–50% of the total discharge in humid areas, rising to 50–90% in semiarid and arid areas mountain watersheds (with extremes over 95% in the Nile, Colorado, Orange, Syr Darya, Amu Darya, and Rio Negro). (See also Box 24.2.) The drier the lowlands, the greater the importance of the linked and more humid mountain areas that supply them (Viviroli et al. 2003; Liniger et al. 1998). (See Figures 24.12 and 24.13.) Moreover, discharge from mountain areas greatly reduces the intra- and inter-annual variation in total discharge.

24.3.3.2 Hydrology and Forests

Natural forests are hydrologically the most effective land cover in mountain watersheds. They can reduce runoff peaks and local flooding, but this influence decreases with the increasing size of the watershed and distance from the headwaters (Hamilton with King 1983). On very shallow soil in mountains, however, runoff peaks may not be reduced. In monsoon climates, with very high amounts and intensity of precipitation, landslips and mass earth failures may occur even with full forest cover, although the incidence of landslips is greater with other types of land cover. Low-stature vegetation, such as lightly grazed grassland and well-constructed and maintained terraced cropland, is also effective in maintaining a good and balanced runoff regime.

While removal of forests may increase total water yield, such removal and subsequent land use commonly has a host of undesir-

---

BOX 24.2

Overutilization of Water from Mountain Areas Leading to Desertification

The Aral Sea Basin is an example of the importance of mountain water resources and ice and snow storage (Spreafico 1997). The basin includes parts of Afghanistan, Kazakhstan, Kyrgyzstan, Tadjikistan, Turkmenistan, and Uzbekistan and has an area of about 690,000 square kilometers with a population of 32 million.

Home to many civilizations since 6,000 BC, irrigation agriculture has been practiced for millennia (today, it covers over 8 million hectares). In the high mountains of the Tien Shan and Pamir, the annual precipitation ranges from 600 to over 2,000 millimeters, with 30% falling as snow and 60% of the total precipitation falling between December and May. The lowland deserts that cover most of the basin and are characterized by low rainfall (less than 100 millimeters a year) and high evaporation (potential evaporation as much as 1,500 millimeters a year).

In the summer, snow and glacial melt contribute to the flow of the two main rivers—the Amu Darya in the south and the Syr Darya in the north. The total annual runoff of the rivers in the basin is about 120 cubic kilometers, of which approximately 116 cubic kilometers originate in mountain areas (about 77 cubic kilometers contributed by the Amu Darya, and 39 by the Syr Darya). Thus the mountains provide more than 95% of the basin's fresh water. More than half a century of over-utilization and high evaporative losses have resulted in a massive shrinking of the Aral Sea and in large-scale desertification. (See also in Chapter 20.)

---

able consequences that offset the additional water gain, and forest removal is not usually viewed as a sustainable management option. Because of lower evapotranspiration losses, well-maintained grazing lands can yield more total water than forested land. However, sustainable grazing regimes and well-constructed and maintained terraces are rare. Moreover, erosion can become a serious problem with other land uses.

Under most conditions, continued forest cover in watersheds is essential for the maintenance of hydrological integrity, although caution is needed with plantations of exotic fast-growing species, which often use more water than native forests. Many of the world's biggest cities, including New York, Jakarta, Tokyo, Mumbai, Rio de Janeiro, Los Angeles, Barcelona, Nairobi, Melbourne, Bogota, La Paz, and Mexico City rely on protected forests in catchment areas for much of their drinking water (e.g., Velázquez 2003). In fact, 33 of the world's 105 largest cities get their drinking water directly from formally protected areas (Dudley and Stolten 2003).

Montane cloud forests capture fog or cloud water (horizontal precipitation), which can add substantial amounts of water to the hydrologic system, especially for dry climates with dependable clouds intersecting the mountains (Bruijnzeel and Proctor 1995). This process is mimicked in some dry coastal ranges (in Peru and Chile, for example) where trees or artificial nets have been established to capture this otherwise unused and unavailable water from fog.

24.3.3.3 Erosion and Sediments

Leaf litter, understory vegetation, and forest debris protect the soil from splash erosion—reducing surface, rill, and gully erosion. Moreover the shear strength provided to the soil by tree roots protects against slumping and landslips (O'Loughlin and Ziemer 1982). Soil compaction is less under forest than under other kinds of land cover. Grasslands provide an excellent cover too, but poorly controlled grazing can impair watershed quality. The quality of water delivered from a watershed may also be adversely affected by surface erosion from cropland and by the use of fertilizers and pesticides. Sediments from eroded watersheds impair water quality for many uses, affect aquatic life, and reduce reservoirs' capacity for storage, flood control (see Box 24.3), hydropower generation, and low flow augmentation.

Conservative grazing, horticulture, and cropping, using well-tested soil conservation techniques such as terracing, can result in low surface erosion rates and reduced sediment production (Whiteman 1988). Agroforestry systems, fruit orchards, and coffee plantations can reduce shallow landslip incidence by increasing root shear strength. The impact of traditional shifting agriculture depends on the length of fallow periods (full forest recovery), the size of the areas cleared, and the pattern in the landscape (Hamilton with King 1983). Vegetated riparian buffer zones are especially important in mountain watersheds as they act as sediment/nutrient traps, stream bank stabilizers, and a good habitat for many species of wildlife (Hamilton and Bruijnzeel 1997).

24.3.3.4 Dams for Hydropower Production and Irrigation

The number of reservoirs in upper watersheds and river basins, built largely for the benefit of lowland dwellers, is still increasing, although in industrial countries most of the best sites have already been exploited. Dams change the hydrology of rivers, sediment loads, riparian vegetation, patterns of stream bank erosion, migration of fish, and water temperature and have a multitude of socioeconomic impacts, often including the displacement of local inhabitants. Dams may reduce downstream flooding.

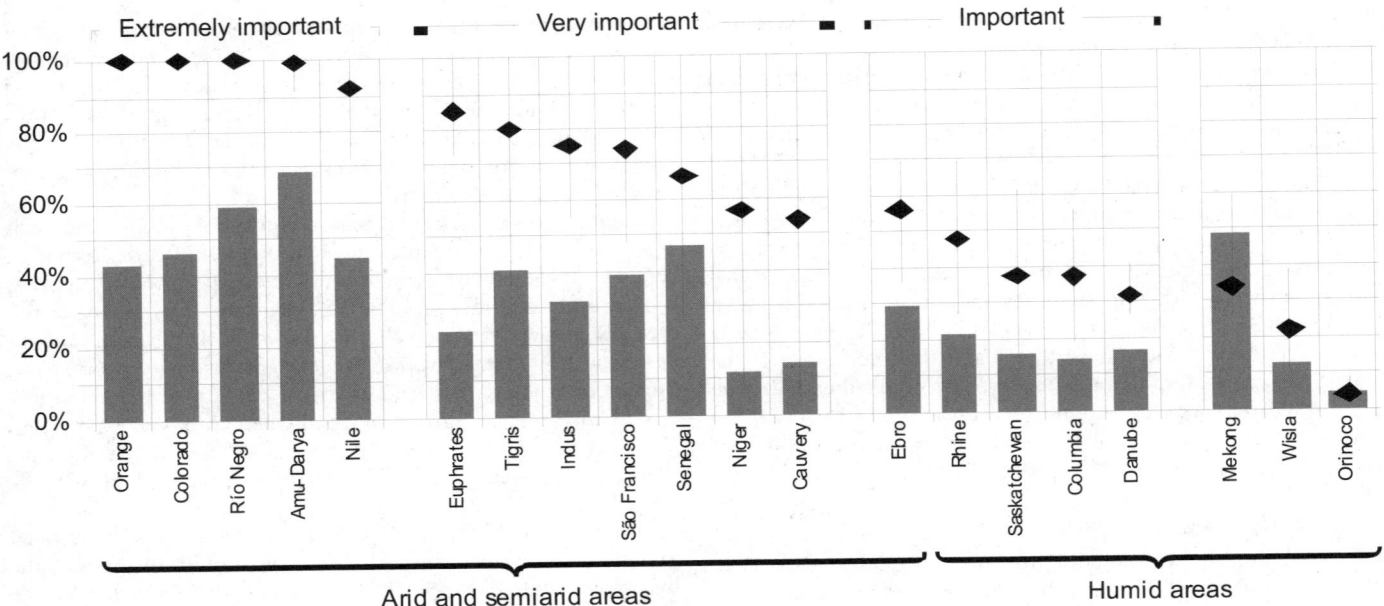

Figure 24.12. Mean Contribution (Importance) in Percent of Mountain Catchments to the Total Discharge per River System. The vertical lines denote the minimum and maximum contribution. Grey bars illustrate the areal contribution in percent of mountain area to the total area of the river system. Aridity declines from left to right. (Viviroli et al. 2003) Reprinted with the Permission of Mountain Research and Development.

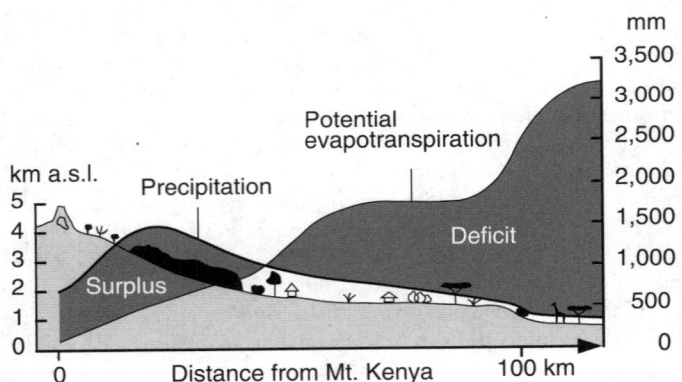

Figure 24.13. Example of the Ewaso Ng'iro River Catchment, Originating from the Slopes of Mount Kenya. Some 90% of the river flow comes from the upper montane zone (above 2,400 m) of Mount Kenya, supplying the semiarid plateau and the arid lowlands, where water resources are critically scarce during the dry season.

Often, the lifespan of a dam is shorter than calculated, and the cost performance (if one includes all indirect costs) is rather low due to siltation. Policies that provide compensation for environmental services provided by mountain communities (through tariffs on the sale of water or of hydropower, for example) have been shown to have high economic benefits. The water bodies created behind dams can also benefit tourism enterprises or support a reservoir fishery.

24.3.3.5 Watershed Management

Watersheds are fragile environments that depend on a careful balance among many different elements, including soil, water, and

> **BOX 24.3**
> **Floods, Landslides, and Reservoirs**
>
> On 19–20 July 1993, an extraordinary flood, which it was estimated would happen only once in 50–100 years, took place in eastern and central Nepal—with catastrophic effects (DHITAL et al. 1993). Several districts were affected by floods and landslides, a number of dams and roads were damaged, many people died, and many more became homeless. The high sedimentation during the flood halved the life span of the Kulekhani-Reservoir, which is located to the south of Kathmandu in the Bagmati watershed, cutting it from 50 to about 25 years.

both wild and agricultural plant and animal species. Population growth, particularly in cities, is increasing the pressure on watershed areas to provide water and food. As a result, nearly half of the world's population is affected by the degradation of watershed areas, which results from changes in water quantity and quality, sedimentation in lakes and reservoirs, loss of biodiversity, and ecological imbalances. Land abandonment following damming leaves behind complex environmental problems. In the developing world, degraded watersheds are among the most serious constraints to sustainable development.

Emigration has also played a crucial role in determining current watershed condition in developing countries. Demographic changes introduced by migrations strongly affect economic activities and trigger land use change processes that are most severe at upper watershed slopes (Li and Sun 1997).

While recognizing that watershed boundaries do not coincide with political units, there are nonetheless compelling reasons for planning and managing resources with due attention to this hydrologic unit. Easter and Hufschmidt (1985) provided this summary of the rational for a watershed approach to rural development:

- The watershed is a *functional region* defined by physical properties and boundaries.
- The watershed approach allows for evaluation of the biophysical interactions between upland and lowland activities that are linked through the hydrological cycle.
- The watershed approach is *holistic* and considers multiple facets of resource use and adopts an ecosystem view.
- Land use and natural phenomena in the highlands often result in a chain of *environmental reactions downstream* that can be readily examined in the watershed context.
- The watershed approach has a *strong economic logic.* Many of the externalities involved with alternative land management practices on an individual farm are internalized when the watershed is managed as a unit.
- The watershed provides a framework for analyzing interactions between people and the environment. Environmental impacts due to human activity within a watershed feed back to the socioeconomic system.
- The watershed approach can be *integrated with programs* of forest management, soil conservation, rural and community development, and farming systems.

As catchment limits rarely coincide with political boundaries, transboundary collaboration is normally required for sound catchment management. Effective watershed management requires accounting for interrelationships between highland and lowland areas and needs to involve local populations, as indicated in *Agenda 21* (UNCED 1992): "Promoting integrated watershed development programmes through effective participation of local people is a key to preventing further ecological imbalance. An integrated approach is needed for conserving, upgrading and using the natural resource base of land, water, plant, animal and human resources."

24.3.4 Socioeconomy in Mountains

24.3.4.1 Population

The global mountain population was estimated in 2000 to be 1.2 billion, or 20% of the total global population. (See Tables 24.8 and 24.9.) Nearly half (49%) of these people live in the Asia-Pacific region. About 30% of the mountain population occurs in each of two biogeographical zones—humid tropical lower montane and humid temperate hill and lower montane. Another 12% live in the dry cool temperate montane zone. (See Table 24.10.) Over 70% of the global mountain population lives below 1,500 meters, mainly in China. While just 8% live above 2,500 meters,

Table 24.8. Global Mountain Population Estimate and Share That Is Urban (CIESIN et al. 2004a, 2004b)

Mountain Area Class	Population	Urban
	(thousand)	(percent)
≥4,500 meters	5,405	4.6
3,500–4,500 meters	20,541	18.8
2,500–3,500 meters	63,373	27.7
1,500–2,500 meters	22,700	26.8
1,000–1,500 meters	226,292	30.3
300–1,000 meters	574,797	31.4
Total	**1,113,108**	**29.7**

Table 24.9. Mountain Population by Region and Average Mountain Population Density (CIESIN et al. 2004a)

Region	Population	Share of Total	Density
	(thousand)	(percent)	(people/sq. km.)
Asia	597,714	49	65.2
Former Soviet Union	34,851	3	6.4
Latin America	173,549	14	37.7
Northern Africa	141,113	12	52.3
OECD	119,559	10	18.3
Sub-Saharan Africa	152,613	13	43.1
World	**1,219,399**	**100**	**38.2**

this still amounts to about 90 million people, almost all considered extremely vulnerable. (See Figures 24.14–24.17 for photographs of mountain dwellers in various parts of the world.)

Most mountain people are rural, particularly in the Asia-Pacific and sub-Saharan Africa regions. Globally, 30% of mountain people are urban, and settlements in and adjacent to mountain areas are expanding. The urban proportion is particularly high in the humid tropical upper montane, dry cool temperate montane, dry sub-tropical hill, and humid temperate lower-mid montane zones (52%, 43%, 39%, and 37%, respectively). The populations of the two zones with the greatest numbers of people are more than one quarter urban.

Mountain population density generally decreases with altitude. The total population at 1,500–2,500 meters is only slightly lower than that at 1,000–1,500 meters, however, reflecting more moderate climates and healthier environments above 1,500 meters in tropical mountains. Outside of urban areas, the highest overall density in mountains occurs in the humid temperate hill and lower montane biogeographical zone (96 people per square kilometer), which is also the zone with the greatest population. High densities (57–74 people) are also found in other humid zones, both temperate and tropical. The next highest densities occur in the widespread dry cool temperate montane zone (49 people per square kilometer), while the lowest densities (under 9 people) are found in alpine and nival zones.

Population growth rates vary considerably between biogeographical zones, with an average global rate of 16% from 1990 to 2000. The highest rates are generally in dry biogeographical zones (tropical hill, sub-tropical hill, warm temperate lower montane), which have population densities somewhat lower than the average of 38 people per square kilometer. Of the two zones with the highest populations, the growth rate in the humid tropical lower montane is above the average (22%), while that in the humid temperate hill and lower montane—the most densely populated—is below it (11%). Many of the zones with high population growth rates also have infant mortality rates above the global average for mountain populations (58 deaths per 1,000). Rates are also particularly high in the dry boreal subalpine and the humid tropical lower montane (respectively, 85 and 74 deaths per 1,000).

Migration has been characteristic for many mountain peoples since the earliest historical times. The numerous wars affecting mountain regions have led to massive movements of people. While comparison of data between censuses may show gross patterns of change, demographic flows are also often recorded within statistical reporting districts and within annual cycles. Consequently, although the spatial distribution of the population may

Table 24.10. Global Mountain Population, Share That Is Urban, Population Density, Population Growth, and Infant Mortality Rate by Biogeographical Zone (CIESIN et al. 2004a, 2004b)

Biogeographical Zone	Population	Urban	Density	Population Growth Rate 1990–2000	Infant Mortality Rate
	(thousand)	(percent)	(people/sq. km.)	(percent)	(deaths/1,000)
Humid tropical hill	53,940	18.2	60.2	18.6	55.4
Humid tropical lower montane	345,082	28.5	61.4	22.0	73.5
Humid tropical upper montane	12,130	52.0	74.4	22.2	35.1
Humid temperate hill and lower montane	349,308	26.4	95.9	11.2	42.8
Humid temperate lower–mid montane	79,949	37.1	57.3	5.4	29.8
Humid temperate upper montane and pan-mixed	49,947	22.2	10.4	7.7	39.1
Humid temperate alpine nival	4,426	4.7	2.8	17.4	51.1
Humid tropical alpine nival	1,103	18.5	8.8	14.4	39.7
Dry tropical hill	14,000	17.9	33.3	24.6	73.8
Dry sub-tropical hill	45,071	38.7	22.5	31.5	58.3
Dry warm temperate lower montane	43,575	35.6	33.1	24.8	60.5
Dry cool temperate montane	139,405	42.8	49.3	18.1	54.2
Dry boreal subalpine	11,178	29.7	10.4	23.7	85.2
Dry subpolar alpine	1,317	14.2	3.7	29.8	5.8
Polar nival	10,588	13.0	2.5	19.0	48.1
Total	**1,161,019**[a]	**30.0**[a]	**38.2**	**16.3**	**57.9**

[a] Difference from Tables 24.8 and 24.9 due to rounding errors.

change considerably, the level of data aggregation often hides these movements. Trends show both increased land use pressure at high elevation as well as increasing trends of urbanization (and a decrease in rural population) (Preston 1996).

In developing countries, mountain populations are generally growing, and some of the highest global rural population densities are found in tropical mountain areas (in Central American and Ethiopian highlands, for instance). This endogenous growth, complemented by immigration, has led to conflicts over land and other resources and the introduction of inappropriate land management practices and invasive alien species. In some mountain areas, such as in Latin America, there is increasing urbanization, with the total number of people in rural areas decreasing over the last two decades.

In industrial countries, one factor that stabilizes populations in mountain regions, or that encourages growth, is tourism, which is increasingly linked to amenity migration (Moss 1994). This phenomenon is becoming evident in many mountain regions. But amenity migrants and other immigrants may not spend all, or even much, of their working time in mountain communities. Indeed, the number of mountain commuters is also growing as travel times to urban and industrial centers decrease. These trends are leading to a blurring of rural/urban populations as mountain areas, and their inhabitants, become more integrated into the mainstream economy in many areas.

24.3.4.2 Economic Conditions

Of the main economic resources of mountains, extractive resources are generally thought to have the highest economic value. National economic policies therefore generally favor extractive industries over other resources and services in mountain areas.

Economic cost-benefit analyses of extractive resources are deficient, however, because they fail to take account of indirect costs

or the distribution of benefits that largely accrue to firms and agents located far from the area of extractive activity. Indeed, the opportunities for local people to generate sustainable livelihoods are often restricted. Timber and mining in West Virginia in the United States are notable examples of this (U.S. Census Bureau and Lewis 1998, in Pratt and Shilling 2002). The Grasberg mine of PT-Freeport in Indonesia is a particularly egregious example (Kennedy et al. 1998, in Pratt and Shilling 2002).

Distortions in the distribution of benefits are evident in national accounts data. Foreign investment in Peru, for example, focuses mainly on the mining industry, with projected foreign investment of $9 billion for the period 1999–2007. This accounts for about 40–50% of exports, representing approximately 5.4% of the gross national product in 1998, and contributing 13% to government tax revenues (CONITE 1999). Little of this revenue, however, is redistributed to the mountain communities affected in the mining regions. Another example is Western New Guinea (Irian Jaya, Indonesia), a mountainous province rich in natural resources, particularly forests. On a provincial GDP basis, its per capita income is the highest in Indonesia, but its GNP per capita is the lowest. The nearly threefold difference between GDP and GNP clearly demonstrates the level of inequality between total production from the area and the proportion of goods and services that accrues to local residents.

These examples illustrate a general problem in assessing economic and social conditions in mountains: data collection is rarely undertaken on a spatial basis, and this represents a critical gap. Because mountain ecosystems and production systems are closely interrelated, geographically referenced data are essential to enable sound management. Most data linked solely to coordinates have proved to be insufficient for sound decision management. Natural entities, such as geomorphologic units as watersheds, are crucial to understanding processes in mountain regions and should form the basis for management decisions (Bocco et al. 2001).

Figures 24.14–17. Mountain People. Top left: Shifting agriculture in montane tropical forests, Papua New Guinea, at 1,200 meters. Top right: Mountain village in the Atlas Mountains, Morocco, at 1,400 meters. Bottom left: Pastoralism in the Tien Shan Mountains, South Kazakhstan, at 2,550 meters. Bottom right: A local market in Otavalo, Ecuador, at 2,000 meters.

Mountain areas tend to have greater poverty and lower levels of development than lowland areas. Starr (2004) noted that, "Poverty has long been a feature of life in many high altitude communities. But the poverty that prevails in many mountain areas today is of a peculiarly modern sort, in that it arises from a growing dependence on lowland metropolitan centers rather than from age-old self-sufficiency in a harsh environment."

Using standard economic criteria, relative poverty in many mountain areas is high—from the Appalachians in the United States to the Amerindians in the Andes and the inhabitants of the Pamirs and Caucasuses in Eurasia. Initial work by the International Livestock Research Institute in Kenya has suggested that mountain areas tend to have relatively high poverty, though the use of relatively low-resolution data does not allow strong conclusions.

The Aga Khan Rural Support Programme and the U.N. Food and Agriculture Organization have each begun to assess mountain poverty (Rasmussen and Parvez 2002; Huddleston et al. 2003). AKRSP found that the status of people in mountain areas reflects the overall level of income of the country in which they live. The higher the national income, the higher the income of mountain people. Yet most available intra-country studies indicate that mountain people do economically less well than lowland populations. There are notable exceptions, however. The mountain areas of the European Alps underwent a rapid economic transformation following improved accessibility. Similarly, traders involved in long-distance trade across some mountain barriers, such as the Himalayas, are quite rich compared with local subsistence farming communities. Noncommercial values such as better access to water, better air quality, and fewer pests such as malarial mosquitoes also add to the benefits of life in the mountains.

Traditionally, dispersion and fragmentation of mountain communities is associated with language differences (for instance, more than 800 different languages in the mountain areas of Papua New Guinea), which further inhibits cooperation on larger-scale activities that could help the improve living standards of mountain people. Mountain roads are expensive to build and maintain, and they generally serve few people per kilometer. A comparison of mountain and non-mountain areas in several countries confirms that mountain populations are more isolated, as measured by distance from roads (Huddleston et al. 2003).

Settlement fragmentation in mountain areas makes it difficult to provide basic social services, medical care, or schooling. There is evidence that malnutrition, particularly micronutrient deficiencies, is higher among mountain inhabitants than in lowland populations (Huddleston et al. 2003). Migration provides important—though often socially negative—connections between upland and lowland areas, and remittances constitute an important source of income for mountain families.

Mountain and lowland economies are interdependent today. Lowland populations depend on environmental services provided by mountain ecosystems and people, including watershed protection and recreation. Mountain populations in turn are increasingly affected by global markets, particularly where commodities are concerned—whether these be agricultural products (coffee, tea, medicinal plants), minerals, or hydropower. There is a high level of consensus that the key to achieving sustainable and acceptable standards of living in mountains has been to transfer to local people more control over mountain assets and the means to negotiate more equitable allocations of benefits. In part, this requires improving access to education and health services. This, in turn, depends on building more equitable relations with lowland political institutions and assuring a better distribution of public services. Enlightened self-interest on the part of lowland institutions and well-coordinated actions on the part of mountain people are required to achieve progress in this area.

24.3.4.2.1 Types of mountain economies

Almost all adverse environmental and social impacts of economic activities in lowlands have their mountain equivalent. What is different about mountain regions is that the constellation of adverse ecosystem and social impacts is characteristic, rather than exceptional. Anthropogenic impacts often result in permanent, or at least very long-lasting, destruction of biodiversity and productive potential. Given that every mountain range is different in the specificity, complexity, and economic potential of its ecosystem resources and services, methods that help assess the impacts of alternative management choices will need to be developed.

Forest management professionals, mining engineers, policymakers, and environmentalists often face conflicts about how to manage mountain resources. Separating economic values and environmental service values can help management choices. (See Figure 24.18.) In all but one of the cases that follow, actions needed to balance conservation and development interests are quite specific, both in nature and in terms of which actors have primary responsibility.

- *Low Export and Low In Situ Value.* The Ethiopian highlands, Tibetan plateau, and Andean altiplano exemplify the case of low export and low in situ value, where subsistence farmers have no access to markets and have poor soil and other resources. In such cases, some economic improvement is possible with appropriate technologies, when combined with some restoration of ecosystem functions. Ecosystem restoration is critical, as lowland economies simply cannot absorb the mountain populations that would be forced to migrate in the absence of a viable environment. Downstream populations benefit economically when mountain ecosystems and ecosystem services remain intact. In some cases, if transportation access is provided a region can even shift to high export value if it can find markets for its minerals, timber, or agricultural produce. Even the most minimal transportation access can help

communities gain access to local markets. In extreme cases, however, subsidies continue to be justified on humanitarian grounds.

- *Low Export and High In Situ Value.* In some of the world's most scenic ranges, such as the Cascades Park in North America and Makalu Barun National Park in Nepal, the biodiversity, watershed protection, and recreational values are clear. Low export values may result from inaccessibility to markets, making resource extraction prohibitive; but at the same time the inaccessibility enhances scenic and recreational value. Tourism is a major resource for such mountain economies, and studies in many regions have shown that protection of watersheds provides greater economic value than resource extraction (The Mountain Institute 1998). Conservation is often compatible with tourism, and generates revenue for government in addition to local employment and income. Bhutan and Rwanda, for example, have established high fees, generating substantial funds for conservation and sustainable development. And Nepal and Peru have begun using community-based tourism to improve livelihoods of local people. However, conservation and the creation of parks and protected areas are not the only choice. Management for sustainable use in agro-forestry-grazing systems has been practiced in the Alps for at least eight centuries. Such systems, however, require strong links to downstream markets.

- *High Export and Low In Situ Value.* In areas with low biodiversity, where extractive potential is high (mineral resources or hydropower, for example, and managed timber), export of resources is appropriate, such as in the arid mountains of the Peruvian Andes. Adverse environmental impacts can be addressed in environmental impact assessments, provided that mitigating measures and social safeguards are implemented. In practice, however, indigenous communities receive an equitable share of revenues only in exceptional cases, either from corporations or through government programs funded by taxes from the extractive enterprise.

- *High Export and High In Situ Value.* Areas where extractive potential and environmental benefits are both high, such as in old-growth mountain forests, are classic examples of conflict, with mining in Papua New Guinea often singled out for particular attention. In such cases, decision-makers—corporate, governmental, and civil society—encounter expensive and extended negotiations with stakeholders and are faced with difficult trade-offs in the attempt to satisfy the needs and demands of all stakeholders.

24.3.4.2.2 Economic contribution of ecosystem services

Ecosystem services of mountains, often ignored, provide greater economic benefits than extractive resource use in most cases. Intact biodiversity protects watersheds and attracts tourism, as well as furnishing rich natural resources for key industries. When measured, watershed protection values have been found to significantly outweigh extractive resource use.

While mountain people play essential roles in maintaining cultural landscapes and traditions, they may not derive significant benefits from tourism, particularly when most investments come from outside and when revenues leave the area. Governments, nonprofit organizations, and private companies have the potential to ensure benefits to local communities through investments, subsidies, low-interest loans, or training to those involved in promoting tourism and maintaining cultural landscapes and traditions.

The concept of payment for ecosystem services has received much attention in various countries as an innovative tool for the

High economic export
Low in situ value

High economic export
High in situ value

Low economic export
Low in situ value

Low economic export
High in situ value

Figure 24.18. A Typology of Mountain Economies. In situ values are ecosystem services provided by the mountain environment in its natural state (such as flood mitigation, water purification, and carbon sequestration. The top left (Lake Powell Reservoir, Utah) represents an area where water existing at high altitudes could be tapped for hydropower or irrigation downstream, but where access problems make it all but impossible for humans to reap economic benefits in situ. The top right (a montane tropical forest in Wau, Papua New Guinea) indicated the main areas of conflict, as represented symbolically by a rich montane forest with shifting cultivation supplying all local needs coming in conflict with a major cash crop, coffee, and the need to clear the forest to grow it. The high, barren landscape in the bottom left (Bolivian Altiplano at 4,200 meters; no export of water, minerals, timber, etc., and marginal local pastoralism) shows that some mountain areas have few natural resources available to inhabitants, and fewer resources that could be profitably exported. A combination of high in situ value but low economic export (bottom right, Langtang valley, Nepal Himalaya; rich forest and gardens at lower montane altitudes, but little to no export) is generally found in areas where lack of access makes export of natural resources too difficult or expensive or where legal or regulatory restrictions have been established that prohibit export of resources, as in national parks.

financing of sustainable management of land and water resources in watersheds. PES schemes consist of a payment or direct compensation by the users to the providers of the service for the maintenance or provision of an environmental service. In watersheds, this usually relates to water supply, availability, or quality. Until now, PES systems in watersheds have been applied at very different stages of the process and for various reasons, from the micro-watershed level—focusing on a very specific service—and usually managed by an NGO or local government to national programs controlled by the State.

Recent reviews of PES experiences in Latin America (FAO 2004; FAO/REDLACH 2004) show that such schemes can be sustainable in the long term if they are funded by local resources and if the service as well as the users and providers are well defined. They can contribute to conflict-solving processes by providing platforms for negotiations. Furthermore, they may transfer resources to socially and economically vulnerable sectors in upper watersheds, which offer environmental services. Important limita-

tions to the application of the PES concept include the high transaction costs during design and implementation of the scheme (biophysical studies, for instance, and assessment and system installation) and the significant uncertainties regarding the cause-effect relationships between land use and the environmental services, particularly forest-water linkages (FAO 2002b).

Modern transportation gives tourists access to almost all mountain regions. Mountain tourism is very unevenly distributed, however. For example, in the Alps 40% of communities have no tourism and only 10% have major tourist infrastructure. Generally, the former are losing population, while the latter have stable or growing populations. Mountain tourism is highly diverse, involving many often seasonal activities, and is highly competitive and sensitive to political tensions. Thus its benefits are unevenly distributed and unpredictable. Flexibility and strategic perspectives are essential, and where income from tourism is reinvested it provides livelihoods for mountain people and helps maintain environmental quality.

Finally, mountains help protect against hazards (erosion control, slope stability) and contribute to soil fertility (volcanic ashes, carbon fixation), as well as modulating climate through several mechanisms discussed earlier. The vertical upthrust of mountains captures precipitation, often creating rain shadows on one side of the range and wet areas on the other. Seasonal snowpack melting and mountain bog ecosystems are critical sources of water for agriculture as well as for drinking water.

The economic impact of these climate-modulating effects is taken for granted and implicitly factored into the costs and benefits of water use. Major changes in the ability of mountains to provide such environmental services, however, would create significant and potentially disastrous cost increases. For example, the retreat of glaciers caused by climate change is likely to cause short-term increases in water availability, leading to increased investments in irrigation and enabling the growth of downstream cities. In the medium term, however, the water supply from meltwater is likely to be markedly reduced if glaciers continue to recede, negating the value of massive infrastructure investments.

24.3.4.2.3 Extractive industries and public utilities

Mountains provide the largest share of resources for mining, forestry, water for drinking and irrigation, and hydropower, and they generate an increasing amount of wind power as well.

Water is becoming a limiting constraint to development in many parts of the world, and in some cases to life itself. The critical importance of mountain ecosystems in regulating water quality and quantity cannot be overstated. In addition, the specific economic value of hydropower depends largely on mountain water, which in turn depends on conserving mountain watersheds.

Some 6% of the world's energy and 15% of its electricity is produced from hydropower. Aside from a few hydropower projects built on swiftly flowing rivers that require no dams, most hydro projects inundate large areas for reservoirs. In mountains, the massive weight of such holdings creates a risk of induced seismic events and dam failure, sometimes resulting in catastrophic flooding. The benefits of hydropower—in terms of access to electricity and profits—accrue mostly to the lowlands. Those in mountains often suffer losses from inundation, diversion of water flows, and disruption of traditional production.

Yet mountain communities and downstream beneficiaries share a common interest in protecting upstream watersheds to assure continued productivity of both hydropower sites and upland production systems. This has proved to be one avenue to joint stewardship arrangements. Properly managed, water supply and hydropower can be a sustainable use of mountain resources. But extra efforts are usually needed to ensure that mountain people get a fair share of benefits and, in the case of hydropower projects, a connection to the power grid.

Globally, forests are probably the second most important economic resource provided by mountains, although this varies across mountain regions. Since most logging is done in pristine forests, little or nothing is paid for production costs. However, standing timber also provides valuable services—stabilization of water flow, protection of biodiversity, carbon sequestration, provision of amenities, and many non-timber products. Several studies have calculated that the economic value for such services exceeds that of the timber extracted. Harvesting primary forests is therefore like mining a resource without compensatory reinvestment. And as with hydropower and mining, many of the economic benefits do not remain in the region that provides the resource.

The loss of trees significantly alters the ecosystems of mountain areas and leads to the loss of alternative sources of income—non-timber forest products, biodiversity, tourism, and so on. Losses of ecosystem benefits are often hard to measure, however, as they are not priced or traded in markets. But some local studies have shown that such benefits often far outweigh the value of timber from logging. In Indonesia, for example, the value of alternative uses exceeds the value of logging by more than 50% (Conservation International 1999).

Communal forestry is a critical element in integrative participatory forest management in many regions. In large parts of mountainous areas in Latin America, Africa, and South and Southeast Asia, collective forest management has proved to be a successful alternative to government or commercial control. Policies that undermine these collective systems have promoted the abandonment of traditional farming systems, often with large, adverse environmental consequences.

Mineral extraction contributes a relatively small part of global GDP, but mineral revenues are often important in mountainous countries. Minerals account for 45% of Chile's exports, 49% of those from Peru, 64% of Zambian exports, and 62% of Papua New Guinea's. While some of the world's largest and most productive mines are found in low-lying areas, mines tend to cluster along mountain ranges. An exception is the Himalayan area, which is not yet cluttered with mines, where ruggedness and distance from markets—in addition to strong local opposition—makes it harder to justify mining operations. Cases in the Supreme Courts of India and Nepal against mining in the fragile Himalayan watersheds drew attention to the risks of mining in these areas (Bandyopadhyay and Shiva 1985). China is nevertheless expanding exploration in some areas on the northern side of the Himalaya.

Mines are nearly always highly destructive to the local environment and displace people living in the immediate area (Pratt 2001). More threatening still are the pollution and toxic wastes produced or accidentally released by mine operations. Toxic pollution from mines has often been recorded leaching out and contaminating large areas downstream, especially in developing countries.

There is a high level of consensus that significant benefits are obtained from better stewardship of upland resources and the appropriate reinvestment of profits from extractive industries into resource management in a way that benefits local communities. For instance, the Sierra Nevada in California produces some $2.2 billion per year in commodities and services. Water resources constitute 61% of the total, yet reinvestment in watershed management is basically zero, since water rights are not taxed as property, and commercial real estate assessments are low compared with revenues generated. In contrast, recreation and residential use provide 21%, timber 14%, and grazing 2% of total value, but reinvestment amounts to $10 million, $23 million, and $7 million (in the latter case, as subsidies) (The Mountain Institute, Investing in Mountains, 1997).

Mountain farming systems involve multiple land use activities and diversified production systems that adapt/amend the natural resources (such as through water harvesting or terracing). This has resulted in diversified and context-specific farming systems characterized by positive social system–ecosystem links. In nearly all mountain regions, non-timber forest products are an important adjunct to traditional agriculture, often providing the sole or major source of medicine for local people, as well as supplying key nutritional supplements. With globalization, the production of nuts, fruits, off-season vegetables, flowers, and cosmetic and medicinal plants has opened new economic opportunities for mountain communities. However, the value added by mountain dwellers will likely remain proportionally small unless local processing replaces the export of raw produce (Jodha 2002).

With closer integration of mountain and lowland economies, enhanced administrative interventions, and population growth, the connection of mountain communities to natural resources has declined (Berkes and Folke 2002). Increased poverty, inequity, and dependence on external support in most areas is a major consequence. Interventions to address such problems have not proved effective because they lack a mountain perspective—that is, sensitivity to mountain-specific conditions that take into account the fragility, marginality, limited accessibility, and diversity of mountain systems. Based on studies in developing countries, the International Centre for Integrated Mountain Development has identified over 50 indicators of unsustainable mountain agriculture and resource use that are emerging in these areas (Jodha 2001). These range from reduced crop productivity to a decline in the range and quality of sustenance options (excluding areas with better access and infrastructure, and where inaccessibility encourages cultivation of illicit crops). These trends are likely to accelerate as market forces gain primacy.

The rise in global demand for mountain herbs and other organic and non-timber forest products is leading to over-extraction. Disregard of customary rights, collective risk sharing arrangements, and reduced social transfers (subsidies) have increased vulnerabilities. While some mountain areas have benefited from the process, others have to face the deleterious effects of globalization (Jodha 2000). In general, agricultural production systems in mountain areas are likely to be driven and controlled by market forces rather than people's sustenance needs and preferences.

24.3.4.2.4 Relationships between drivers and changes in ecosystem condition

All the impacts just described are exacerbated by the development of modern economic infrastructures, such as highways and communications towers that interrupt conservation corridors, reduce scenic values, and contribute to erosion and pollution from traffic. Although roads reduce remoteness and inaccessibility of highland communities, inappropriate siting, construction, and maintenance often have serious adverse impacts in steep upper watersheds (Cassells 1996; Hamilton and Bruijnzeel 1997). Location in landslip- or landslide-prone environments creates both on-site and downstream sediment problems. Roads run the risk of crossing groundwater aquifers, altering hydrologic stability and interfering with water supply for both upstream and downstream watershed residents. Sealed surfaces enhance runoff, increasing the risk of flash flooding. Cut-and-fill situations are particularly erosion-prone. Soil erosion problems demand frequent and costly maintenance. To lessen hazards, the construction of underpasses and overpasses for road users and animals has proved effective. Construction or mining camps or towns for workers create environmental disturbances and also serve as a major vector for HIV/AIDS and other diseases.

24.3.4.2.5 Implications for human well-being

This assessment indicates that when full costs are taken into account, stewardship of upland resources generally yields greater and more sustainable economic returns both to the people living in the mountain areas and to the immediate downstream economies when compared with extractive activities, in cases where there is competition between the two. For mountain communities, protecting the ecosystem that they depend on requires interventions: regulatory protection has a high potential to ensure that full costs are included in project design and implementation, and promoting solid links between upstream and downstream markets creates economic opportunities that generate mutual benefits. In the long term, there is a consensus that upland and lowland ecosystems, and the economic resources and services they provide, depend on populations in both regions supporting stewardship of mountain environments. The challenge for investment and policy action has been to bridge short- and long-term interests.

The failure to protect and manage mountain resources sustainably has dire consequences that become visible only when it is already too late. Among the most important economic consequences are impacts on employment and sustainability. Conservation programs that protect environmental services have been shown to create employment for local people and, in the best cases, to strengthen cultural identity and security (for example, the UNESCO Biosphere Reserve concept). Mismanaged conservation programs can drive local people out by force or when competition for land and housing drives up prices. Those driven out tend to place burdens on nearby urban areas through unemployment and demands for public services. They are often considered "different" from local people, contributing to discrimination and ethnic violence.

Extractive industries are particularly problematic in mountains. Hydro projects frequently disrupt aquatic ecosystems but rarely offer long-term jobs for local people. Forestry provides greater local job opportunities, but mountain forests regenerate slowly or not at all, and employment disappears once the timber is gone. The loss of forest also eliminates traditional lifestyles that use hunting, fishing, and non-timber forest products to supplement farming. Mining has serious environmental consequences; but modern mining often provides scant local employment. Mining companies have shown increasing willingness to provide social services, but the benefits tend to last only until the mine is exhausted, while environmental damage is persistent.

Greater attention to conservation and to strengthening traditional production systems has the potential to generate greater economic value. Managing trade-offs among uses is often limited because most data on resource extraction and ecosystem production and services are not spatially referenced, masking the contributions of mountain regions, and leading to ecosystem degradation and economic inequities. The availability of spatially referenced data is thus a critical gap.

24.3.4.3 Cultural Issues

Many mountain communities are ethnically or culturally distinct from lowland populations, and local highland communities are often highly distinct from each other. The significance of mountains for human cultural diversity is demonstrated by the great ethnic and linguistic diversity of some mountain regions, such as the Caucasuses, the Himalaya, and the mountains of New Guinea (Association des populations des montagnes du monde 2003). Indigenous mountain populations often exhibit genotypic physiological adaptations to altitude (Beall 2002).

In mountain ranges throughout the world, traditional cultures and conservation have evolved together over the ages. Sustainable natural resource management is driven by the beliefs and behaviors of human communities, and local cultures are strengthened by their intimate connections to the natural environment that sustains them. Sacred and spiritual values are thus integral to mountain cultures; mountains are considered sacred by more than 1 billion people (Bernbaum 1998). (See also Chapter 17.)

The value of place-based mountain cultures today is in their continuing stewardship of watersheds and other mountain resources and in the wisdom they have to impart regarding the requirements for sustainability of mountain ecosystems (Pratt 1998). A major cultural element of life in the mountains is tied to animal husbandry, which also is by far the most influential form of land use (in terms of area). Quite often this involves the use of fire (Price 1999). The U.N. Commission on Sustainable Development has stated that "support is needed to recover and foster the cultural expression of mountain populations because mountain cultural diversity is a strong and valid basis for sustainable use and conservation of mountain resources" (Commission on Sustainable Development 1995). And the need to protect and support the cultural diversity of mountains has been emphasized in the declarations of numerous major international meetings (e.g., Association des populations des montagnes du monde 2003).

24.4 Drivers of Change in Mountain Systems

Environmental and economic change is a constant and familiar factor in mountains, but the magnitude and rate of change and its influence on social systems in recent times threatens to overwhelm mountain ecosystems—with serious consequences for the well-being of mountain communities as well as hundreds of millions of people downstream. Mountain systems are changing more rapidly than at any time in human history. The core issue is that more than half of humanity depends on mountains for water to drink, to grow food, to produce electricity, and to generate industry. In addition, mountain ranges represent important challenges for transportation, communications, and access.

Natural forces such as volcanic and seismic events, landslides, and flooding devastate large areas of mountainous ecosystems every year. Such changes, though vast and visible, are nevertheless dwarfed by deleterious anthropogenic changes, such as intensification of land use and overexploitation of natural resources (Messerli and Hurni 2000; FAO 2002a; Pratt and Shilling 2002).

24.4.1 Direct Drivers

Because mountains are formed by tectonic forces, it is not surprising that mountain regions are particularly susceptible to damage from earthquakes and volcanoes, which in many cases results in a significant loss of life and property. Climate change is another direct driver with special significance for mountains and serious implications for human well-being. Mountains are particularly susceptible to climate change because their biota are adapted to specific often narrow altitudinal zones, and diseases have proved to be able to move more quickly than plants and animals can adapt. Conversion of land in mountains, largely through pasture practices, the use of fire, and animal husbandry is another major driver.

Large-scale mining can massively overburden steep terrain, and often pollutes streams and damages aquatic and other wildlife. Construction of roads in mountain areas often leads to slope instability, land slides, and erosion. The large-scale building of vacation homes and resorts has had mixed impacts in mountains, while in many regions, out-migration of mountain farmers has led to reforestation and loss of alpine pastures. Elsewhere, forest is cleared for commercial timber. Such loss of vegetative cover—regardless of the cause—can have a significant adverse impact on water quality and quantity both in mountain regions and below. Threats to human well-being from these drivers range from increased risk of avalanche to loss of income from tourism.

These drivers of change are not new—there are historical accounts of human impacts on mountain environments, with drought and famine, dating back millennia (such as the felling of

mountain cedar forests in Palestine and Lebanon 5,000 years ago). Forest destruction continues, however. Major drivers of the direct anthropogenic impacts are the lack of public awareness and knowledge and indirect socioeconomic forces.

24.4.2 Indirect Drivers

Indirect drivers affecting mountains can be complex. In mountain areas from Jamaica to Nepal, mountain forests are destroyed as population increases in the lowlands, forcing poorer people into the mountains, where they cultivate marginal land for subsistence. Economic development often results in land use changes such as those just described, with consequent degradation of ecosystem services. More difficult to measure, but nonetheless important, economic development in mountains usually leads to a weakening of traditional cultures and religions that have provided the underpinnings for local sustainability.

Science and technology, on the other hand, have frequently had positive impacts in mountain areas and have the potential to provide solutions to a number of critical problems. Solar and wind energy, and especially small-scale hydropower, have brought enormous benefits in regions where they have been introduced and where extension of transmission lines is otherwise prohibitively expensive. Information technology is perhaps the single most promising technology for mountain communities, with potential to overcome access barriers that currently limit educational opportunities for tens of millions of mountain families.

The sectorally based organization of governments hampers the implementation of more holistic or ecosystem-based approaches to mountain ecosystem management (Rodgers 2002; Pratt and Shilling 2002). Overall, flawed institutional responses (lack of mandates, policies, and political will) are pervasive indirect drivers of change in mountains. The lack of structural mechanisms that can deal holistically with mountain areas has made it difficult or impossible to prevent or mitigate adverse impacts of key drivers. This can be seen in the lessons from decades of mountain water conservation initiatives that have generally failed due to sectoral fragmentation of institutional responsibilities, political interference, over-reliance on technocrats at the planning stage, too little involvement of landowners and local communities, an overemphasis on maximal instead of sustainable resource use, and a lack of knowledge on adequate farming systems (many introduced systems were originally developed for lower altitudes).

Other indirect drivers are the lack of public awareness, lack of real valuation of resources and services, and lack of knowledge transfer. Empowered and well informed local communities have proved to be key to managing changes imposed from outside; in the case of biodiversity, a functional network of protected areas is an essential starting point for genetic reservoirs and monitoring stations (Dhar 1997).

For many drivers, indicators and monitoring programs are in place, like the recent worldwide installation of monitoring sites in mountain summit regions—the Global Observation Research Initiative in Alpine Environments (Pauli et al. 2004, 2005). What is missing is the sociopolitical and economic understanding of how scientific insights can be applied and how local stakeholders can be involved on a continuing basis.

24.4.3 Property Rights to Mountains

In most mountains outside Western Europe and the United States, legal ownership of the land is retained by the state. Mountain lands are also covered with customary use rights for members of local communities (de facto ownership). Major challenges for resource governance in many parts of the world include encouraging governments to recognize customary rights and finding ways of recording and enforcing such rights. (See Box 24.4.) Over a dozen countries have enacted specific legislation related to mountains (Lynch and Maggio 2000).

Property rights to mountains are often poorly defined because mountain areas for most of history have been seen to contain few resources. Now, however, market forces are reaching into remote mountain communities, and government interest in managing ecosystems and their services has increased. Indigenous (or non-capitalist) cultures have usually developed customary law similar to property rights for specific material resources. Property rights claimed by the state have led to discrimination against indigenous peoples in some parts of the world, making their customary rights harder to defend.

The early modern states tended to pursue a policy of bundling "ground," "remainder," and specific resources (see definitions in Box 24.5) into one owner unit in their legal systems. This "dominium plenum" position on ownership and its assumed beneficial economic consequences led to the processes of enclosure and land consolidation. Applied to mountains and other areas where local people were interested only in specific resources and socio-

cultural symbols, this practice has tended to create conflicts, most notably for timber and other commercially valued resources.

Social and technological change creates new specific resources usually seen as belonging to the owner of the remainder (such as the generation of hydroelectric power), and this also leads to new regulation of ecosystem services. Such developments often conflict with customary use rights. Thus the potential for conflict is rising and is often precipitated by government interventions—for example, to protect mountain resources. Since states have made little or no effort to enforce their claims to property rights (except for timber, hydroelectric power, and mining rights), most customary uses have continued more or less uninterrupted. While the viability of local cultures depends on traditional resource use, it has often proved hard for mountain people to get recognition of their de facto and customary property rights. It should also be noted that customary systems of rights are vulnerable to the impact of market forces. For instance, the makers of local rules may respond too slowly to rapid changes in harvesting of local resources that have acquired market value.

Current trends in international law put greater emphasis on de facto rights as these are expressed in customary uses of an area (such as ILO Convention 169 concerning Indigenous and Tribal Peoples in Independent Countries). International conventions on human rights and indigenous peoples have sought to award property rights (by implication, probably in the dominium plenum tradition) to those who have, through traditional usages of an area, established use rights to specific resources.

Statistics on ownership of mountain resources do not yet exist for the world as a whole. Neither do available sub-regional figures conform to ecological boundaries or to social realities of mountain communities. Furthermore, publicly available statistics on property rights only report de jure rights. The plethora of de facto use rights is often found in separate records, if they are recorded at all.

Most land registers are based on the dominium plenum ownership concept and only register owners of the land itself (not its use). Even if approximate figures for de jure owners were collected (nations, local states, towns, collectives, or individuals), the lack of reliable information on de facto ownership of specific resources would make the presentation of de jure figures more misleading than helpful. Developments in international law, such as ILO convention 169, have tended to put emphasis on de facto possession rather on de jure claims. (See Box 24.6.) Theoretical developments in the management of complex resource systems tend to support the allocation of a high degree of autonomy to local user groups protected by property rules.

Starting with the Roman law assumption that all lands have a landlord, medieval states in Europe tried to gain control of nonarable lands. Unclaimed lands became crown lands. In many cases the early modern states (notably Sweden, Germany, and France) introduced state ownership of forestlands and strengthened the state control of the lands without owners. The result was often state ownership of mountains.

"New nations" (including the United States) have at least since 1776 routinely claimed state ownership of unimproved lands. "Improvement" (such as industrial activity or agricultural use of arable land) was needed to justify privatization. This "improvement" policy for awarding title to land has in most cases led to state property rights to mountains. Socialist states in 1917–89 routinely nationalized land. In the restitution period after 1989, many of these states either neglected to include mountain areas (outside settlements) in the process or expressly reserved these areas for state control. Likewise, many new nations created through decolonization since 1945 have nationalized land, or at

---

BOX 24.4

An Institutional Definition of Property Rights

Property rights provide legitimate allocation to particular owners, with material or immaterial objects supplying income or satisfaction to the owner. They comprise a detailed specification of rights and duties, liberties, and immunities citizens have to observe. These are defined partly by law and partly by cultural conventions, and they are different for owners and non-owners. Property rights are ultimately guaranteed by the legitimate use of power.

The dynamics and performance of economic systems are intimately linked to the kind of property rights a state is able to enforce.

BOX 24.5

A Legal Definition of Resources (Black 1990)

The technical details in the specifications of property rights are many and are important to the dynamic of the economy. They are changing through time and across space and are in general moving toward greater diversity and more detail. For management purposes, legal reasoning will divide resources into 3 types:

- *the ground* (sometimes called the soil), meaning the abstract bounded area;
- *the specific material resources* embedded in the ground, attached to the ground, or flowing over the ground (in general there are limits on how far into the ground and how far above the ground the rights reach); and
- *the remainder,* meaning the future interest in resources not yet discovered or not yet capable of being exploited.

These three types of resources are usually included in discussions of who owns what and are routinely recognized by mature legal institutions. Landlords are, at a minimum, owners of the ground and are then entitled to the ground rent. It must be emphasized that in principle there may be different owners to the ground, to every single well-specified resource, and to the remainder.

BOX 24.6

Emerging Collectively Owned Resources

Environmental legislation is at the outset independent of ownership, but it is increasingly seen to change the meaning and content of ownership by defining and taking control over two additional types of resources that can be seen as emerging from the remainder:

- *ecosystem services,* such as water control, disaster mitigation, local climate control, biodiversity, and so on, and
- *sociocultural symbols* vested in a landscape (often attached to amenity and heritage sites).

Ecosystem services are usually managed through government regulations. Sociocultural symbols are created and sustained by the local culture but now increasingly taken over by national and international bureaucracies. (See Buck 1996 and Lowenthal 1985 for more about environmental management and sociocultural symbols.)

---

least unimproved lands. Mountains and less accessible land or lands assumed to be less valuable have tended to remain in state ownership even where other lands were privatized.

24.4.4 Wars and Other Conflicts

In 1999, 23 of the 27 major armed conflicts were in mountains (FAO 2002a). Due to their usual situation of relative inaccessibility and remoteness from centers of population and government, mountains are often used by those who wish to escape the established authority. In countries where there are guerrilla movements or rebels, it is the mountains that are often their sanctuary (as in Afghanistan, Chechnya, and Colombia), and many illegal drugs are produced from crops such as coca, poppy, and marijuana grown in highland regions.

For most of the past 500 years, the main source of conflict in mountainous countries was the effort of emerging states to extend their power over mountain peoples. Starr (2004) states "Any government that thinks it can bludgeon mountain people into submission is engaging in a most destructive form of self-deception. The sense of territoriality, independence and cohesive social (often clan) relationships formed in isolated upland valleys are perfectly suited to sustain conflicts over the long haul."

Mountain ranges are often borders between nations or other political jurisdictions. Tensions along borders are common, and in some cases the location of the border has been in dispute, such as in the Peru-Ecuador conflict of 1995 in the Cordillera del Condor (Peace Parks in this area are described in the next section). Many conflicts have also arisen over natural resources in mountains, often based on issues of property rights, such as over logging on customarily used lands. The Chipko forest conservation movement in the Himalayas is a good example of such a dispute (Bandyopadhyay 1997).

24.4.5 Mountain Protected Areas

24.4.5.1 Global Network

Biological diversity, water resources, soil, and geological, cultural, and spiritual values of mountains are all maintained best in some kind of protected area situation. Protected areas in this sense are those without unbridled exploitation, where some degree of restraint is required in human use in the interest of natural or metaphysical values. IUCN–the World Conservation Union defines

protected areas as: "Areas of land and/or sea especially dedicated to the protection and maintenance of biological diversity, and of natural and associated cultural resources, and managed through legal or other effective means" (IUCN 2000). As of 2003, there are 102,000 of these sites, covering 18.8 million square kilometers, or roughly 11.5% of the world's terrestrial surface (Chape et al. 2003). They span many kinds of situations, with varying degrees of human intervention—from wilderness areas and national parks to multiple use areas and lived-in protected landscapes. The IUCN category system is shown in Table 24.11.

24.4.5.2 The Mountain PA Situation

Mountain PAs are well-represented in the global network, even though there is some ambiguity over what constitutes a "mountain" PA. The U.N. list for 2003 shows 9,345 mountain protected areas covering 1,735,828 square kilometers in the "mixed mountain systems" biome defined by Udvardy (1975); therefore about 16% of this biome is protected (Chape et al. 2003). Thorsell (1997) showed that the highest elevation Pas—based only on IUCN categories I-IV, minimum relative relief of 1,500 meters and minimum size of 10,000 hectares—had a good distribution throughout the biogeographic realms. (See Table 24.12.)

The Thorsell assessment covers "high mountains" and those with a minimum of human land use modification, though many of them do have quite intensive visitation by tourists, mountaineers, and hikers. They are most often in the ownership or under the control of some level of government.

The importance of the PAs in the remaining categories (V and VI, Table 24.11) must not be discounted. Grazing, forestry operations, and many kinds of agronomic use such as orchards, vineyards, and terraced annual crops can be conducted in non-destructive and non-resource-polluting ways. In addition, important cultural values are often maintained in mountainous protected landscapes. Agro-biodiversity, as well as much wild native biodiversity, can be conserved if sustainable land uses are in place.

Table 24.11. The Six IUCN Management Categories of Protected Areas (IUCN 2000)

Category	Name	Description
I a	Strict Nature Reserve	protected area managed mainly for science
I b	Wilderness Area	protected area managed mainly for wilderness protection
II	National Park	protected area managed mainly for ecosystem protection and recreation
III	Natural Monument	protected area managed mainly for conservation of specific natural features
IV	Habitat/Species Management Area	protected area managed mainly for conservation through management intervention
V	Protected Landscape/ Seascape	protected area managed mainly for landscape/seascape conservation and recreation
VI	Managed Resource Protected Area	protected area managed mainly for the sustainable use of natural ecosystems

Table 24.12. Mountain Parks (High Mountains)

Biogeographical Realm (Udvardy classification)	Parks	Total Area
	(number)	*(mill. hectares)*
Afrotropical	42	20.4
Antarctic	15	3.2
Australian	3	2.6
Indomalayan	42	7.2
Nearctic	96	153.8
Neotropical	103	34.5
Oceanian	8	3.6
Palearctic	164	39.1
Total	**473**	**264.5**

Water and soil resources can also be safeguarded by proper husbandry of forests and agricultural lands.

Most of this nature-friendly management will be carried out by private landowners or communities, often using traditional practices that have proved their sustainability over generations. Some are in national and state forest management areas or community forestry units. As wild areas, even in the mountains, succumb to development, much of the hope for maintaining biological and cultural diversity in mountain environments rests in proliferation of Category V and VI areas of protection. And as secular forces erode ancient cultural belief systems, mountains once protected de facto by reverence, awe, or taboo need to come under formal secular protection in all kinds of PAs. Geological heritage is often protected in the Natural Monument Category, though "geoheritage" is under-represented in the world network.

24.4.5.3 International Designations

Many Protected Areas are also designated as UNESCO Biosphere Reserves, where core zones of more strict preservation are buffered by zones of conservation use, in which sustainable land uses are promoted. Scientific research on ecosystem functioning and human-environment interactions are carried out in these reserves. As of 2003, there were 436 Biosphere Reserves (Chape et al. 2003), at least 190 of which were in mountain areas (UNEP-WCMC 2002).

Some mountain PAs are of such global significance that they have been placed on the World Heritage List of UNESCO. This designation is reserved for areas of universal value. There are 88 natural World Heritage Sites and 16 mixed (natural and cultural) ones in mountain areas (UNEP-WCMC 2002). For high mountains, Thorsell and Hamilton (2004) reported on 57 existing World Heritage Sites and identified 28 other potential sites to help fill the gaps in coverage.

24.4.5.4 Weaknesses in Protected Area System

A major weakness in the mountain PA global system is that most of the units are discrete, covering single mountains. Connectivity between these "sky-islands" is badly needed along the ranges or in biogeographic clusters. Linkages through a landscape of conservation corridors can effectively enlarge the PA, providing better protection of the full suite of biodiversity, including "umbrella" species such as large wide-ranging carnivores. Moreover, such connectivity would provide greater insurance for migration of species and genes in the face of climate change. A number of these

corridor initiatives are now in place, such as the 3,200-kilometer-long Yellowstone-to-Yukon corridor in the U.S. and Canadian Rockies and the Condor Bioreserve constellation in Ecuador.

Unfortunately, many mountain PAs were established to protect the scenic high peaks of local or national value as cultural icons or for mountaineering and tourism. Biodiversity values were not considered, and the PAs often conserve mostly rock, ice, and snow or upper montane forests and alpine meadows. Many are too small to accommodate serious natural or human disturbance or to embrace much mountain biodiversity.

The challenge is to enlarge these areas, in particular to extend them to lower elevations to achieve species, genetic, and community conservation and provide functional landscapes for wide-ranging species. Expansion and connection from summits to lowlands is also a "must" for climate change response—for instance, the corridor from the Royal Manas Tiger Reserve in the tropical lowlands of India through a series of parks and conservation areas in Bhutan up to the crest of the Himalaya in Jigme Dorji National Park, Bhutan. There are at least 36 such initiatives around the world in mountains areas.

24.4.5.5 Transborder Parks in Mountains

Since many national or sub-national borders follow mountain ranges, many mountain protected areas abut such borders and each other. There are approximately 169 complexes of internationally adjacent protected areas (Zbicz 2001). About 42 of these are in mountains. These offer good opportunities to carry out cooperative transborder planning and management to better conserve shared biodiversity and water resources and to fight fires, pests, and non-native species—none of which recognize political boundaries (Hamilton et al.1996). As indicated earlier, these offer opportunities to reduce tension and conflict between neighboring countries, as Peace Parks.

24.4.5.6 Effective Management and Monitoring of Protected Areas

The World Commission on Protected Areas of IUCN has an active Task Force on Management Effectiveness, and it has developed criteria and standards for more effective management of protected areas. Baseline data and monitoring are sorely needed as a basis for adaptive management. Far too many protected areas are "paper parks" without effective protection and little management. The IUCN World Commission on Protected Areas has recently produced a set of "Guidelines for Planning and Managing Mountain Protected Areas" (Hamilton and McMillan 2004) to help rectify this situation.

24.5 Trade-offs, Synergies, and Management Interventions in Mountain Systems

24.5.1 Highland-Lowland Interactions and Their Trade-offs

Until recently, the economic importance of mountains was generally ignored (with the exception of supplies of some minerals, timber, and water), and little attention was paid to local environmental, socioeconomic, and cultural issues. With the U.N. Conference on the Environment in 1972, changes in mountain landscapes—including deforestation, accelerating slope instability, earthquakes, landslides, and floods—began to be highlighted, but the focus was mainly on the potential destructive impacts on lowlands originating from the mountains.

These early and simplistic perceived linkages between high-lands and lowlands fell into two categories: physical processes under the influence of gravity and the exploitation of mountain resources to satisfy the needs of lowland residents.

Highlands and lowlands have widely different resources and production opportunities. This forms a natural basis for complementary economic links between them. In practice, however, the relationship has been more often characterized by inequitable power relationships, although highland communities can have significant effects on the power structure and way of life of far distant lowlands.

There are often competitive demands on mountain resources—increased resource extraction reduces the extent and value of environmental services that ecosystems can provide. Conversely, preserving ecosystem services may reduce incomes for particular interest groups. Furthermore, the relative value placed on mountain resources depends on technological developments and shifts in the world economy. In developing countries, this often creates a bias for exports, and in most mountain regions it creates a bias for extraction rather than conservation of resources.

Several factors affect the highland-lowland links:

- limited accessibility, isolation, semi-closed situation created by slope, terrain conditions, and permanent underinvestment in addressing the problem, all of which adds to the cost of logistics and other support systems to harness production opportunities and their competitiveness and equitable trade, although in some cases limited accessibility can be tied to tourist attractions (in the Everest region, for example);
- fragility—a product of slope, soil factors, and so on—which not only prevents intensification of land resources use for high productivity but obstructs infrastructure development to improve accessibility to facilitate mobility and trade at lower or competitive costs;
- marginality of production—resource limitations caused by the factors just described and socioeconomic and geopolitical marginalization of mountain habitats;
- high levels of biological, cultural, climatic, and other diversity characteristic of mountains, which creates many special economic opportunities if properly harnessed and traded; and
- major known niche resources (hydropower, timber, NTFPs, minerals, eco-tourism, and so on) with comparative advantage to highlands.

The factors just outlined also help explain the persistence of poverty in many highland areas. Lowlands invest to harness highland opportunities largely for their own benefit. This has been helped by the unequal balance between highland and lowland people as trading partners and has resulted in generally unfavorable terms of trade for the highlands. Indeed, many export flows (both traded and non-traded) from mountains are neither appropriately priced nor fully compensated (Banskota and Sharma 1999).

The Earth Summit of 1992 in Rio de Janeiro signaled a new recognition of the critical importance of highland-lowland linkages and of the need for poverty and equity to be integrated in environmental management. The underpinnings of this increased awareness stem from four concerns regarding highland-lowland linkages: water shortages, together with growing demands for hydroelectricity in various parts of the world; warfare, which is disproportionately concentrated in mountain regions (Libiszewski and Bächler 1997); catastrophic events resulting from mismanagement of mountain resources (flash floods, massive flooding, and landslides); and climate change effects, including glacier retreat and loss of snowpack (Beniston 2000; Beniston et al. 1996).

Economic disparities between highland and lowland regions are closely related, either as cause or effect, to other key changes in mountain environments (described more fully in other sections), including migration, warfare, production of illegal drugs, risks and disasters, and climate change. These are noted here briefly, but only as related to the issue of linkages.

The problems of mountain-lowland population change are exemplified by trends in the Alps, where jobs and population are concentrating into a few favorable locations (transportation corridors and nodes). At the same time, the real alpine zone, with a few notable exceptions, is losing its productive potential (Bätzing et al. 1996) due to the loss of expert knowledge to manage landscapes in a traditional way. In terms of absolute numbers, however, out-migration is more than offset by amenity migration. This is causing problems in the mountains, where new migrants have different and often inappropriate land use practices, while migrants from mountains who settle in lowlands also face problems of adjustment and assimilation (Moss 1994; Price et al. 1997).

Armed conflict, guerrilla warfare, and extreme political unrest disproportionately affect mountain regions, both in terms of total surface area and in terms of populations. There is also widespread expression of political discontent among mountain (minority) peoples. Much of this conflict is attributed—either directly or indirectly—to the growing struggle for control of water (Libiszewski and Bächler 1997). Hewitt (1997) indicates that more than 70% of the almost 8 million war deaths in mountain lands since the end of World War II have been unarmed civilians. More recently, Starr (2004) supports this overall assessment, underscoring deeper linkages to poverty and inequity in mountains.

Mountain regions worldwide are frequently the source of illegal (and legal) narcotics: marijuana production in British Columbia in Canada; opium from the sizable remnant of the Golden Triangle, including Myanmar; hashish and heroin from Afghanistan; and cocaine from the central and northern Andean countries. These are effective cash crops for cultivators, and especially for the traders in the middle, because of the high market value and low weight. Some of the most dangerous places in the world for outsiders to visit are the drug-producing areas of the northern and central Andes. The level of hostile encounters there amounts to full-scale warfare; herbicide defoliation by military aircraft adds to the scale of environmental and human loss. The effects of downstream transfer of the toxic overflow are unknown. Certainly, the movement of the products takes on the guise of a singular highland-lowland interaction.

Mountain lands include regions of exceptional risk for human activities as well as some unique dangers. Earthquakes and volcanic eruptions are central to the processes of mountain building. Hewitt (1997) has argued that mountain peoples have experienced a pronounced disproportionate share of these disasters, whether this is calculated in terms of land area or population numbers. Mountain people also are deeply implicated in responsibility for some disasters as a result of their own management practices and are the disproportionate victims of inappropriate practices introduced by outsiders. Moreover, mountain regions lack access to emergency relief compared with lowland areas.

Climatological changes in mountain ranges are likely to have much more readily apparent impacts than in the surrounding lowlands. Winter recreation, availability of water, hydroelectricity, irrigation, and the sudden release of glacier lakes as glaciers continue to thin and retreat are all potential components.

Maximizing highland-lowland complementarities is crucial for both upstream and downstream communities. Healthy mountain communities require linkages to lowland markets, and lowland populations need mountain people to serve as stewards for upland resources and watersheds. Investments that favor such positive interactions are properly treated as transfer payments, not subsidies, and have a high potential to improve sustainability.

24.5.2 Management and Interventions

Chapter 13 of *Agenda 21,* adopted by governments in 1992, draws the attention of political authorities to the special issues facing mountain regions. Government structures to deal with mountain issues were called for, but as yet there are no government departments specifically for mountains, as there are for forestry or wildlife. Land, water, forest, environment, and development policies do not generally consider the challenges facing mountain regions, and organizational divisions make it particularly difficult to deal with the integrated, systems approaches needed. Nevertheless, with significant exceptions that are described here, existing management approaches have the potential to deal with most environmental problems specific to mountains.

Where biodiversity and scenic values are high and economically valuable commercial resources are inaccessible or limited (the "high biodiversity and low extractive value" situation described earlier), conservation interventions have proved valuable. Parks and protected areas help conserve water resources while providing scenic, aesthetic, and recreational value with considerable economic returns from tourism, as well as protection of investments in water supply for downstream populations. In some cases where formal conservation mechanisms are inappropriate (for example, due to high concentrations of upland settlements), interventions have nevertheless been designed to protect watersheds. Economic incentives for stewardship have potential for effective management, with transfer payments given in exchange for maintaining environmental services.

Where biotic resources are few, as in arid regions, and where commercial resources such as mineral ores are abundant (the "low biodiversity, high extractive value" case), regulatory approaches such as environmental impact assessments have been effective. However, successful interventions have been characterized by careful attention to implementing measures that avoid or mitigate adverse impacts and by social safety nets and revenue-sharing mechanisms designed and approved with active participation of local communities.

In a few regions, natural resources are so poor or degraded, and linkages to markets are so weak or nonexistent, that adequate management options have proved elusive. In these cases, downstream inhabitants receive few environmental services, but the size and cultural distinctness of mountain communities is such that it would be difficult to absorb any massive out-migrations. Here, governments have often justified welfare payments out of humanitarian concern, and such interventions have proved successful in achieving the limited goal of alleviating at least the extremes of hardship. More rarely, environmental restoration of such degraded lands has been attempted, albeit with varying results.

Management approaches have proved generally inadequate in two areas. The most important example is in mountain regions where both biological resources and commercially valuable extractive resources are significant and important. Standard management approaches, such as regulatory protections, have proved wholly inadequate, leaving almost all stakeholders frustrated. While environmental assessments are necessary, they are insufficient to deal with the complicated trade-offs involved, which require long time frames and mechanisms that permit continuing participation of all stakeholders. Such mechanisms and processes take more time than corporations and governments are generally

comfortable with. And it has proved difficult to create "level playing fields" where local communities can negotiate on an equitable footing with national governments and private corporations. Nonetheless, providing the time and resources needed to address these highly complex situations is an urgent priority, as failure in such cases produces a disproportionate share of environmental damage in mountain regions.

A second area where interventions are lacking has to do with information for policy formulation and decision-making. In general, data are not currently collected on a spatial basis, making it difficult to "see" what needs to be done.

24.6 Mountain Systems and Human Well-being

24.6.1 Sustainability

This section is in part based on a background paper prepared for the *World Development Report 2002/2003* (Pratt and Shilling 2002), but see also the recent *Ambio* Special Report (Sonesson and Messerli 2002).

Sustainable development has been defined as "development that meets the needs of the present without compromising the ability of future generations to meet their own needs" (WCED 1987). The fragility of mountain ecosystems represents a considerable challenge to sustainable development due to the fact that the impacts of unsustainable development are more rapid, heavier, and more difficult to correct than in other ecosystems. Arriving at a comprehensive definition of sustainability in mountains, particularly one that is universally accepted, is itself a mountainous task—and not likely to be a productive effort. More useful is to identify areas that merit protection and the characteristics and attributes that contribute to the sustainable use of mountain resources for human needs, broadly defined, for the alleviation of poverty, and for a more equitable allocation of resources and power.

Human activity in mountains that is not in balance with the environment can have serious consequences, resulting, for example, in soil erosion, pollution of natural waters, disruption of water and energy balances, elimination of both animal and plant species, loss of soil productivity, increasing food deficits, malnutrition and poor standards of living. Some of these consequences can be irreversible, such as the extinction of species and the loss of soil and cultural diversity.

In looking at sustainability, it is important to recognize that there are several time spans to consider. Short-term impacts would occur over the coming 20 years, medium-term impacts over 20–50 years, and long-term impacts over a longer horizon, extending to centuries or geological time spans. Our concerns should extend over both short and long time spans: while fires, landslides, and erosion can wipe out large areas of forest and other ecosystems in a very short period of time, it takes 50–100 years for a forest to regrow in mountainous areas, if it does so at all. Roads, mines, and other constructions last 20–50 years and their impacts even longer, so decisions to undertake such activities have long-term implications.

Sustainability does not mean cessation of all change. Mountains are subject to continual natural change. They were created by massive geological forces and they are being torn down by natural forces of erosion and landslides. New species have evolved in mountains, and others have become extinct. The objective of promoting sustainability is therefore not to stop change in mountains but to manage resources in them in ways that provide livelihoods for people living there as well as the services valued in lowland areas—and to do so in ways that protect the long-term capacity of mountains to continue to provide such services.

In order to ensure sustainability in mountain areas it is necessary to reduce poverty, inequality, and marginality, to prevent deterioration of mountain natural resources and environments, and to improve the capabilities of institutions and organizations to promote conservation and sustainable mountain development. The goals are to:

- assure that people living in the mountains receive full benefit from their mountain resources so that poverty and inequity can be substantially reduced;
- preserve and enhance the long-term value of resources in mountains;
- eliminate or minimize disruptive, damaging, and polluting aspects of human interventions; and, most important,
- manage human-introduced change so that it generates benefits for current and future mountain inhabitants and for those living downstream.

Achieving environmental and human sustainability in mountains means finding ways to manage mountain resources and systems so that they can provide critical services indefinitely. While we cannot predict exactly what the future will look like or which services will be in demand, it is clear that mountains provide many essential services that will be valued for a long time and others that may increase in value (such as biological and cultural diversity, high-value forest products, and scenic beauty). Nearly all these values are tied to soil conservation—the alpha and omega of mountain integrity.

24.6.2 Vulnerability

This chapter has described how the vulnerability of mountain people has a variety of aspects and many different causes: availabil-ity of land; ownership of land; environmental constraints (climate, soils, slope, natural hazards); food insecurity; lack of access to markets, education, and health care; dependence on one single economic factor (such as only forests, livestock, or tourism); inappropriate governmental or industrial interventions; high specialization and interdependency of mountain social and land use systems; and globalization. Many elements of vulnerability are not well documented (but see Shrestha 2001 and Munir and Adhikari 2003), and there are few studies or statistics that quantify the number of mountain people vulnerable to these different elements. This discussion is based on a recent FAO study (Huddleston et al. 2003) and focuses on food insecurity, accessibility, and nutrient deficiencies.

Around 40% of mountain populations in developing and transition countries, or 271 million people, are estimated to be vulnerable to food insecurity, and of these, around half are likely to be chronically hungry. Most are rural people, with only 26 million of the vulnerable people living in mountain cities. An agriculture-based livelihoods approach has been used to locate and enumerate vulnerable people in rural mountain areas. Rural people living in areas where annual cereal production is less than 200 kilograms per capita and cattle numbers are small are considered vulnerable, as well as those living in closed forests. Work currently under way will extend the approach to cover other income sources in future vulnerability assessments. For instance, people living in protected areas can compensate through income from tourism if these monies are not channeled away to governmental agencies and operators, as is currently the case in most "trekking" destinations. However, they still remain very vulnerable in terms of food security, because tourism is unpredictable and may collapse over night.

Of the 245 million vulnerable mountain people living in rural areas, 87% live below 2,500 meters above sea level (classes 4, 5,

Table 24.13. Vulnerable Rural Mountain People in Developing and Transition Countries by Mountain Area Class.[a] Based on LandScan 2000 Global Population Database. (Huddleston et al. 2003; mountain area classes, see Box 24.1)

Region	300–1,000 Meters and Local Elevation Range > 300 meters	1,000–1,500 Meters and Slope > 5° or Local Elevation Range > 300 Meters	1,500–2,500 Meters and Slope > 2°	2,500–3,500 Meters	3,500–3500 Meters	> 4,500 Meters	Total for Mountain
	(million population)						
Asia and the Pacific	77.5	28.8	19.8	6.5	4.3	3.1	140.0
Latin America and Caribbean	9.9	5.0	8.9	4.9	4.0	0.2	32.9
Near East and North Africa	10.7	7.1	7.5	4.1	0.3	0.03	29.7
Sub-Saharan Africa	10.6	10.6	7.3	2.2	0.09	0	30.9
Countries in transition	7.7	1.9	1.0	0.4	0.2	0.02	11.2
Total vulnerable in developing and transition countries	116.4	53.5	44.6	18.1	8.8	3.4	244.7
Total rural mountain population in developing and transition countries	**241.6**	**98**	**104.8**	**31.6**	**10.1**	**4.1**	**490.3**
	(percent)						
Vulnerable in class as share of rural mountain population in class	48	55	43	57	87	82	
Vulnerable in class as share of total vulnerable	48	22	18	7	4	1	100

[a] Vulnerable rural mountain people are those living in rural areas where rain-fed cereal production is less than 20 kilograms per person per year and the bovine density index is medium to low, along with those living in closed forests of protected areas.

and 6), where they represent less than half of the mountain population at lower altitudes. (See Table 24.13.) With more than three quarters of mountain populations in developing and transition countries still classified as rural, the performance of agriculture is a crucial factor in determining the degree of their vulnerability to food insecurity. As described earlier in this chapter, pastoral systems are very important for mountain people at all elevations in developing and transition countries. At the present time, these systems are becoming increasingly vulnerable as populations grow, livestock numbers increase, the quality of pasture and browse declines, and the incidence of drought becomes more frequent and its impacts more severe.

In high mountain areas, the absolute number of vulnerable rural people is small, but they represent almost 70% of the population living above 2,500 meters, and many live in extreme poverty. The higher prevalence of vulnerability at higher elevations and the importance of these areas for the overall sustainability of mountain ecosystems warrant particular attention.

It is generally accepted that mountain people live in remote, isolated areas that are poorly served by physical infrastructure and social services. In Ethiopia, for instance, about half of the mountain population and 40% of the non-mountain population live more than 5 kilometers from roads. In Afghanistan and China, the figure for mountain people is around one third and for non-mountain people, about 20%. In Peru, however, just 20% of mountain people and 13% of non-mountain people live more than 5 kilometers from a road (Huddleston et al. 2003).

In 33 of the 40 mountainous developing countries covered by the FAO report (Huddleston et al. 2003), there has been an increase in malnutrition as the proportion of mountain people has increased, measured by the prevalence of vitamin A, iron, and iodine deficiencies (globally the most significant micronutrient deficiencies in children). There are also significant differences in the distribution of micronutrient deficiencies across regions. Vitamin A deficiency is particularly common in mountainous countries of eastern and southern Africa, where consumption of fruits and vegetables that are rich in vitamin A is low; iodine deficiency is particularly prevalent in the Himalaya, where the soils have been leached of their iodine-carrying salts; and iron deficiency is common across all regions, though with somewhat greater incidence in sub-Saharan Africa.

References

Association des populations des montagnes du mond., 2003: Quito Declaration—Charter for World Mountain People. 20 September 2002, Quito, Ecuador. Available at http://www.mtnforum.org/resources/library/dcwmp03a.htm.

Bandyopadhyay, J., 1997: Chipko—a unique movement of mountain people for sustainability. In: *Mountains of the World—A Global Priority,* B. Messerli and J.D. Ives (eds.). Parthenon Publishing Group, New York, London, pp. 127–128.

Bandyopadhyay, J. and V. Shiva, 1985: Conflict over limestone quarrying in Doon Valley, Himalaya. *Environmental Conservation,* 12(2): 1319.

Baron, J.S., H.M. Rueth, A.M. Wolfe, K.R. Nydick, E.J. Allstott, et al. 2000: Ecosystem responses to nitrogen deposition in the Colorado Front Range. *Ecosystems* 3: 352–368.

Barry, R.G., 1992: *Mountain Weather and Climate.* 2nd ed, Routledge, London.

Bartholome, E. M. and A. S. Belward, 2004 (in press): GLC2000, a new approach to global land cover mapping from earth observation data. *International Journal of Remote Sensing.*

Banskota, K. and B. Sharma, 1999: *Traded Resource Flows from Highland to Lowland: Understanding Economic Linkages.* ICIMOD, Kathmandu, 99 pp.

Battarbee, R.W., J.A. Grytnes, R. Thompson, P.G. Appleby, J. Catalan, et al. 2002: Comparing paleolimnological and instrumental evidence of climate change for remote mountain lakes over the last 200 years. *Journal of Paleolimnology,* **28:** 161–179.

Bätzing, W., M. Perlik, and M. Dekleva, 1996: Urbanization and depopulation in the Alps. *Mountain Research and Development,* **16:** 335–350.

Beall, C.M., 2002: Human dimensions of global mountain biodiversity. In: *Mountain Biodiversity. A global assessment.* C. Körner, and E.M. Spehn (eds.), Parthenon Publishing Group, London, New York, 199–210 pp.

Beniston, M., D.G. Fox, S. Adhikary, R. Andressen, A. Guisan, et al. 1996: *The impacts of climate change on mountain region.* Second assessment report of the Intergovernmental Panel on Climate Change (IPCC), Chapter 5. Cambridge University Press, pp. 191–213.

Beniston, M., 2000: *Environmental Change in Mountains and Uplands.* Oxford University Press, London, 172 pp.

Beniston, M. and M. Rebetez, 1996: Regional behaviour of minimum temperatures in Switzerland for the period 1979–1993. *Theor. Appl. Climatol.,* **53,** 231–243.

Beniston, M., F. Keller, and S. Goyette, 2003: Snow pack in the Swiss Alps under changing climatic conditions: an empirical approach for climate impact studies. *Theor. Appl. Climatol.,* **74,** 19–31.

Berkes, F. and Folke, C., 2002: Back to the Future: Ecosystem dynamics and local knowledge, In: *Panarchy: Understanding Transformations in Human and Natural Systems.* Gunderson, L.H. and Holling, C.S. (eds.), Island Press, Washington: 121–146.

Bernbaum, E., 1998: *Sacred Mountains of the World.* University of California Press, Berkeley, Los Angeles, London, 291 pp.

Black, H.C., 1990: *Black's Law Dictionary.* Sixth Edition, West Publishing, St.-Paul, Minn.

Blais, J.M., D.W. Schindler, D.C.G. Muir, D.B. Donald, and B. Rosenberg, 1998: Accumulation of persistent organochlorines in mountains of western Canada. *Nature* 395: 585–588.

Blais, J.M., D.W. Schindler, D.C.G. Muir, M. Sharp, D. Donald, et al. 2001: Glaciers are a dominant source of persistent organochlorines to a subalpine lake in Banff National Park, Canada. *Ambio* 30: 410–415.

Bocco, G., M. Mendoza and A. Velázquez, 2001: Remote sensing and GIS-based regional geomorphologic mapping – A tool for land use planning in developing countries. *Geomorphology,* **39,** 211–219.

Bormann, F.H. and Likens, G.E. 1979: *Pattern and Process in a Forested Ecosystem.* Disturbance, Development and the Steady State Based on the Hubbard Brook Ecosystem Study. Springer, Berlin, New York, 253 pp.

Brown, S., A.J.R. Gillespie, and A.E. Lugo, 1991: Biomass of tropical forests of South and Southeast Asia. *Canadian Journal of Forest Research,* **21,** 111–117.

Bruijnzeel, L. A. and J. Proctor. 1995: Hydrology and biochemistry of tropical montane cloud forests: What do we really know? In: *Tropical Montane Cloud Forests.* Hamilton, L. S., J. O. Juvik and F. N. Scatena (eds.). Springer Verlag, New York, Berlin pp. 38–78.

Buck, S., 1996: *Understanding Environmental Administration and Law.* Island Press, Washington DC.

Callaway, R.M., Kikvidze, Z., and Kikodze, D. 2000: Facilitation by unpalatable weeds may conserve plant diversity in overgrazed meadows in the Caucasus Mountains. *Oikos* **89,** 275–282.

Cassells, D. 1996: *Review of the Watershed Management Component of the Natural Resources Management Project, Colombia.* Internal Document. The World Bank, Washington, DC.

CIESIN (Center for International Earth Science Information Network), Columbia University, and Centro Internacional de Agricultura Tropical, 2004a: Gridded population of the world (GPW), version 3. CIESIN, Columbia University, Palisades, NY. Available at http://sedac.ciesin.columbia.edu/gpw.

CIESIN, Columbia University, International Food Policy Research Institute, World Bank, and Centro Internacional de Agricultura Tropical, 2004b: Global rural-urban mapping project (GRUMP), version 1 alpha: gridded population of the world (GPW), version 3, with urban reallocation. CIESIN, Columbia University, Palisades, NY. Available at http://sedac.ciesin.columbia.edu/gpw.

Chape, S., S. Blyth, L. Fish, P. Fox, and M. Spalding, 2003: *2003 United Nations List of Protected Areas.* IUCN/UNEP-WCMC, Cambridge, 27 pp.

Commission on Sustainable Development, 1995: *Economic and Social Council Official Records,* Supplement No. 12 in the CSD report on the third session, 11–28 April 1995. United Nations, New York.

Conservation International, 1999: *Valuation of Mt. Gede-Pangrango National Park, Natural Resource Management Program.* Conservation International (Indonesia), Jakarta.

CONITE (Comisión Nacional para la Inversión y Technología Extranjera), 1999: Boletín Estadístico. Lima, Perú.

Cuatrecasas, J. 1986: Speciation and radiation of the Espeletiinae in the Andes. In: *High Altitude Tropical Biogeography,* Vuilleumier, F. and Monasterio, M. (eds.), Oxford Univ. Press, New York, pp. 267–303.

Daniggelis, E., 1997: *Hidden Wealth. The Survival Strategy of Foraging Farmers in the Upper Arun Valley, Eastern Nepal.* Mandala Book Point & The Mountain Institute, 261 pp.

Dhar, U. (ed.), 1997: *Himalayan Biodiversity – An Action Plan.* The G.B. Pant Institute of Himalayan Environment and Development, Almora, India, 136 pp.

Diaz, H.F. and N.E. Graham, 1996: Recent changes of tropical freezing heights and the role of sea surface temperature. *Nature 383,* 152–155.

Diaz, H.F and R.S. Bradley, 1997: Temperature variations during the last century at high elevation sites. *Climatic Change,* **36(3–4),** 253–279.

Dixon, J., A. Gulliver, and D. Gibbon, 2001: *Farming Systems and Poverty,* Improving Farmer's Livelihoods in a Changing World. FAO and World Bank, 420 pp.

Donald, D.B., Vinebrooke R.D., R.S. Anderson, J. Syrgiannis, M.D. Graham, 2001: Recovery of zooplankton asssemblages in mountain lakes from the effects of introduced sport fish. *Canadian Journal of Fisheries and Aquatic Sciences,* **58,** 1822–1830.

Dorji, K., 2000: *Biodiversity Assessment and Conservation Planning Bhutan.* WWF Bhutan Program.

Dudley, N. and S. Stolton, 2003: *Running Pure: The Importance of Forest Protected Areas to Drinking Water.* World Bank/WWF Alliance for Forest Conservation and Sustainable Use, Washington, D.C., 114 pp.

Easter, K.W. and M.M. Hufschmidt, 1985: *Integrated Watershed Management Research for Developing Countries.* East West Center Workshop Report. Honolulu.

FAO (Food and Agriculture Organization of the United Nations), 2001: *Global Forest Resources Assessment 2000.* FAO Forestry Paper 140, Rome, 512 pp.

FAO, 2002a: *International Year of the Mountains.* Food and Agriculture Organisation of the United Nations, Rome.

FAO, 2002b: Land-water linkages in rural watersheds. *Land and Water Bulletin 9.* Food and Agriculture Organisation of the United Nations, Rome, 88 pp.

FAO, 2003: *State of the World's Forests 2003.* Food and Agriculture Organisation of the United Nations, Rome, 168 pp.

FAO, 2004: Payment schemes for environmental services in watersheds. *Land and Water Discussion Paper 3.* Food and Agriculture Organisation of the United Nations, Rome.

FAO/REDLACH, 2004: *Foro electrónico sobre Sistemas de Pago por Servicios Ambientales en Cuencas.* Informe final. Available at http://www.rlc.fao.org/foro/psa/

Gentry, A.H., 1988: Changes in plant community diversity and floristic composititon on environmental and geographical gradients. *Ann. Missouri Bot. Gard.,* **75,** 1–34.

Gorbunov, A.P., S.S. Marchenko, and E.V. Seversky, 2000: Permafrost and seasonally frozen ground response to climate changes in the northern Tien Shan. *Krisfera Zemli,* **4,** 11–17.

Gottfried, M., H. Pauli, K. Reiter, and G. Grabherr, 1999: A fine-scaled predictive model for changes in species distributions patterns of high mountain plants induced by climate warming, *Diversity and Distributions,* 5: 241–251.

Grabherr, G., M. Gottfried, and H. Pauli, 1994: Climate effects on mountain plants. *Nature* 369: 448.

Grimalt, J.O., Fernandez, P., Berdie, L., Vilanova, R.M., Catalan, et al. 2001: Selective trapping of organochlorine compounds in mountain lakes of temperate areas. *Environ. Sci. Technol.,* 35: 2690–2697.

Gurtz, J., M. Zappa, K. Jasper, H. Lang, M. Verbunt, A. Badoux and T. Vitvar, 2003: A comparative study in modelling runoff and its components in two mountainous catchments. *Hydrol. Process,* **17(2),** 297–311.

Haeberli, W. and M. Beniston, 1998: Climate change and its impacts on glaciers and permafrost in the Alps. *AMBIO,* **27(4):** 258–265.

Hamilton, L. S. (with P.A. King), 1983: *Tropical Forested Watersheds; Hydrologic and Soils Response to Major Uses or Conversions.* Westview Press, Boulder, Colorado.

Hamilton, L.S. and L.A. Bruinjzeel, 1997: Mountain watersheds integrating water, soils, gravity, vegetation, and people. In: *Mountains of the World—A Global Priority.* B. Messerli and J.D. Ives (eds.). Parthenon Publishing Group, New York, London, 337–370.

Hamilton, L.S. and L. McMillan (eds.), 2004: *Guidelines for Planning and Managing Mountain Protected Areas.* IUCN, Gland, Switzerland and Cambridge, UK

Hamilton, L.S., J.C. Mackay, G.L. Worboys, R.A. Jones, and G.B. Manson, 1996: *Transborder Protected Area Cooperation.* IUCN/Australian Alps National Parks, Canberra.

Handa, T., C. Körner, and S. Haettenschwiler, in press: Atmospheric CO₂ enrichment and defoliation: manipulating the carbon balance of alpine conifers at treeline. *Ecology.*

Hewitt, K., 1997: Risk and disasters in mountain lands. In: *Mountains of the World—A Global Priority.* B. Messerli and J.D. Ives (eds.). Parthenon Publishing Group, New York, London, pp. 371–408.

Hofstede, R.G.M., M.X. Mondragón Castillo, and C.M. Rocha Osorio, 1995: Biomass of grazed, burned and undisturbed Páramo grasslands, Colombia. I. Aboveground vegetation. *Arctic Alpine Research,* **27,** 1–12.

Hofstede, R.G.M., J.P. Groenendijk, R. Coppus, J. Fehse, and J. Sevink, 2002: Impact of pine plantations on soils and vegetation in the Ecuadorian high Andes. *Mountain Research and Development,* **22(2):** 159–167.

Huddleston, B., E. Ataman, and L. Fe d'Ostiani, 2003: *Towards a GIS-based analysis of mountain environments and populations.* Environment and Natural Resources Working Paper, No. 10. Food and Agriculture Organisation of the United Nations, Rome.

IUCN (World Conservation Union), 2000: *Guidelines for Protected Area Management Categories:* interpretation and application of the protected area management categories in Europe. 2nd edition. IUCN, EUROPARC Federation, IUCN World Commission on Protected Areas, WCMC, Grafenau: EUROPARC, 46 pp.

Jenkins, A. 2002: The sensitivity of headwater streams in the Hindu Kush Himalayas to acidification. *Water, Air, and Soil Pollution: Focus* **2,**181–189.

Jodha, N.S. 2000: Globalisation and Fragile Mountain Environments—Policy Challenges and Choices. *Mountain Research and Development,* **20(4):** 296–299.

Jodha, N.S., 2001: Sustainable agriculture in fragile resource zones: technological imperatives. In: *Life on the Edge: Sustaining Agriculture and Community Resources in Fragile Environments,* N.S. Jodha (ed.), Oxford University Press, New Delhi.

Jodha, N.S., 2002: *Rapid Globalisation and Fragile Mountains: Sustainability and Livelihood Security Implications in Himalayas.* Final narrative report of the research planning project submitted to the MacArthur Foundation, ICIMOD, Kathmandu, Nepal.

Kaeaeb, A., Reynolds, J.M. and Haeberli, W. 2005: Glacier and permafrost hazards in high mountains. In: *Global change in mountain regions: a state of knowledge overview,* Huber, U., Reasoner M. and Bugmann, H. (eds.), Springer, Dordrecht, pp. 225–234.

Kapos, V., J. Rhind, M. Edwards, M.F. Price and C. Ravilious, 2000: Developing a map of the world's mountain forests. In: *Forests in Sustainable Mountain Development: A State-of-Knowledge Report for 2000,* M.F. Price and N. Butt (eds.), CAB International, Wallingford: 4–9.

Körner, C., 1989: Der Flächenanteil unterschiedlicher Vegetationseinheiten in den Hohen Tauern: Eine quantitative Analyse grossmassstäblicher Vegetationskartierung in den Ostalpen. In: *Struktur und Funktion von Graslandökosystemen im Nationalpark Hohe Tauern,* Cernusca, A. (ed), Wagner, Innsbruck, Veröffentl Oesterr MaB-Programm, **13:** 33–47.

Körner, C., 1995: Alpine plant diversity: a global survey and functional interpretations. In: *Arctic and Alpine Biodiversity: Patterns, Causes and Ecosystem Consequences,* F.S. Chapin III, and C. Körner (eds.), Ecological Studies 113, Springer, Berlin, pp. 45–62.

Körner, C., M. Diemer, B. Schappi, P.A. Niklaus, and J. Arnone, 1997: The responses of alpine grassland to four seasons of CO₂ enrichment: a synthesis. *Acta Oecologia International Journal of Ecology,* **18:** 165–175.

Körner, C., 2000: Why are there global gradients in species richness? Mountains might hold the answer. *Trends in Ecology and Evolution.* **15:** 513–514

Körner, C. and E.M. Spehn (eds.), 2002: *Mountain Biodiversity. A Global Assessment.* Parthenon Publishing Group, London, New York.

Körner C, 2003: *Alpine Plant Life.* Springer, Berlin. 2nd edition.

Körner, C. and J. Paulsen, 2004: A world-wide study of high altitude treeline temperatures. *Journal of Biogeography,* **31,** 713–732.

Körner, C., 2004: Mountain biodiversity, its causes and function. *Ambio,* **33,** in press.

Li, X., and L. Sun, 1997: *Driving forces of arable land conversion in China.* International Institute for Applied Systems Analysis (IIASA). Available at http://www.iiasa.ac.at/cgi-bin/pubsrch?IR97076.

Libiszewski, S. and G. Bächler, 1997: Conflicts in mountain areas—a predicament for sustainable development. In: *Mountains of the World—A Global Priority.* B. Messerli and J.D. Ives (eds.). Parthenon Publishing Group, New York, London, pp. 103–130.

Liniger, H.P., R. Weingartner, M. Grosjean, C. Kull, L. MacMillan, et al., 1998: *Mountains of the World, Water Towers for the 21st Century—A Contribution to Global Freshwater Management.* Mountain Agenda, Paul Haupt, Bern, 28 pp.

Lowenthal, D., 1985: *The Past is a Foreign Country.* Cambridge University Press, Cambridge.

Lynch, O.J. and G.F. Maggio, 2000: *Mountain Laws and Peoples: Moving towards sustainable development and recognition of community-based property rights.* The

Mountain Insititue and Center for International Environmental Law. Washington, D.C.

Messerli, B. and H. Hurni, 2000: *African Mountains and Highlands: Problems and Perspectives*. African Mountains Association, Walsworth Press, Missouri.

Morris, A., 1997: Afforestation projects in highland Ecuador: patterns of success and failure. *Mountain Research and Development, 17(1):* 31–42.

Moss, L.A.G.. 1994: Beyond tourism: the amenity migrants. In: *Chaos in our Uncommon Futures,* Mannermaa, M., Inayatullah, S. and Slaughter, R. (eds.), University of Economics, Turku, 121–128.

Munir, S. and K. Adhikari, 2003: *Globalisation and Mountain Farmers: Tapping opportunities and mitigating threats.* Synthesis Report, v + 41, South Asia Watch on Trade, Economics & Environment (SAWTEE), Kathmandu, 47 pp.

Nagy, L., G. Grabherr, C. Körner, D.B.A. Thompson (eds.), 2003: *Alpine Biodiversity in Europe.* Ecological Studies 167, Springer, Berlin, Heidelberg, New York.

O'Loughlin, D. L. and R. R. Ziemer, 1982: The importance of root strength and deterioration rates upon edaphic stability in steepland forests. In: *Carbon Uptake and Allocation: A Key to Management of Subalpine Ecosystems,* R. H. Waring (ed.), Oregon State University, Corvallis, Oregon, pp. 70–78.

Olson, D.M., E. Dinerstein, E.D. Wikramanayake, N.D. Burgess, G.V.N. Powell, et al., 2001: Terrestrial ecoregions of the world: a new map of life on earth. *BioScience,* **51(11),** 933–938.

Ohsawa, M., 1990: An interpretation of latitudinal patterns of forest limits in South and East Asian mountains. *Journal of Ecology, 78:* 326–339.

Ohsawa, M., 1995: Latitudinal comparison of altitudinal changes in forest structure, leaf type, and species richness in humid monsoon Asia. *Vegetatio,* **121:** 3–10.

Pauli, H., M. Gottfried, D. Hohenwallner, K. Reiter, R. Casale, and G. Grabherr (eds.), 2004: *The GLORIA (Global Observation Research Initiative in Alpine Environments) Field Manual—Multi-Summit Approach.* European Commission, Directorate-General for Research, EUR 21213, Office for Official Publications of the European Communities, Luxembourg.

Pauli, H., M. Gottfried, D. Hohenwallner, K. Reiter, and G. Grabherr, 2005 (in press): Ecological climate impact research in high mountain environments: GLORIA (Global Observation Research Initiative in Alpine Environments)—its roots, its purpose and the long-term perspectives. In: *Global change in mountain regions: a state of knowledge overview,* Huber, U., Reasoner M. and Bugmann, H. (eds.), Springer, Dordrecht, pp. 377–386.

Pepin, N. 2000: Twentieth-century change in the climate record for the Front Range, Colorado, USA. *Arct. Antarct. Alp. Res.* **32(2),** 135–146.

Peterken, G.F. 1981: Woodland Conservation and Management. Chapman & Hall.

Peterson, A.T., V.O.A. Flores, P.L.S. Leon, B.J.E. Llorente, M.M.A. Luis, S.A.G. Navarro, C.M.G. Torres, and F.I. Vargas, 1993: Conservation priorities in Northern Middle America: Moving up in the world. *Biodiversity Letters* 1: 33–38.

Pratt, D. J. (ed.), 1998: *Sacred Mountains and Environmental Conservation, A Practitioner's Workshop.* The Mountain Institute, West Virginia.

Pratt, D. J., 2001: Corporations, Communities, and Conservation. *California Management Review,* Vol. 43, No. 3, Haas School of Business, University of California Press, Berkeley.

Pratt, D. J. and J. Shilling, 2002: *High Time for Mountains.* Background paper prepared for the *World Development Report 2002/2003.* World Bank, Washington DC

Preston, D., 1996: *Latin America Development: Geographical Perspectives.* Longman Scientific & Technical, Harlow, England, 165–187.

Price, M.F., L.A.G. Moss, and P.W. Williams, 1997: Tourism and amenity migration. In: *Mountains of the World—A Global Priority,* B. Messerli and J.D. Ives (eds.). Parthenon Publishing Group, New York, London, pp. 249–280.

Price, M., 1999: *Global Change in the Mountains.* Parthenon, New York, 217 pp.

Ramankutty, N., and J.A. Foley, 1999: Estimating historical changes in global land cover: croplands from 1700 to 1992. *Global Biogeochemical Cycles,* **13(4),** 997–1027.

Rasmussen, S.F. and S. Parvez, 2004: Sustaining mountain economies: sustainable livelihoods and poverty alleviation. In: *Key issues for Mountain Areas,* M.F Price, L. Jansky, and A.A. Iatsenia (eds.), United Nations University Press, Tokyo.

Rodgers, W.A., 2002: Development Contradictions. In: *Kilimanjaro: The Story of A Mountain.* National Geographic Press, Washington. DC, pp 230–231.

Rognerud, S., J.O. Grimalt, B.O. Rosseland, P. Fernandez, R. Hofer, et al., 2002: Mercury and organochlorine contamination in brown trout (*Salmo trutta*) and arctic charr (*Salvelinus alpinus*) from high mountain lakes in Europe and the Svalbard archipelago. *Water, Air, and Soil Pollution:* Focus 2: 209–232.

Schimel, D., Kittel T.G.F., Running S., Monson R., Turnipseed A. and D. Anderson, 2002: Carbon sequestration studied in Western US Mountains. EOS, **83:** 445–449

Schindler, D.W. 1999: From acid rain to toxic snow. *Ambio* **28(4):** 350–355.

Schindler, D.W. and B.R. Parker, 2002: Biological pollutants: aline fishes in mountain lakes. *Water, Air, and Soil Pollution:* Focus 2: 379–397.

Shrestha, A.B., C.P. Wake, P.A. Mayewski and J.E. Dibb, 1999: Maximum temperature trends in the Himalayas and its vicinity: An analysis based on temperature records from Nepal for the period 1971–1994. *Journal of Climate,* **12,** 2775–2786.

Shrestha, K.L., 2001: Global change and the Himalayan mountain ecosystem. In: *Global Change and Himalayan Mountains,* K.L. Shrestha (ed.), Proceedings of a Scoping Workshop, Kathmandu, Nepal, 2–5 October 2001, Institute for Development and Innovation, Lalitpur, Nepal.

Sommaruga-Wögrath, S., K.A. Koinig, R. Schmidt, R. Sommaruga, R. Tessadri, and R. Psenner, 1997: Temperature effects on the acidity of remote alpine lakes. *Nature* **387:**64–67.

Sonesson, M. and B. Messerli (eds.), 2002: The Abisko Agenda: Research for mountain area development. Rethinking Agenda 21, Chapter 13: Managing fragile ecosystems. A contribution to the United Nations Year of Mountains 2002. *Ambio Special Report,* **11,** 1–105.

Spreafico, M., 1997: Without mountains there is no life in the Aral basin. In: *Mountains of the World,* B. Messerli and J.D. Ives (eds.), Parthenon, London, 145.

Spehn, E.M., M. Libermann, and C. Körner (eds.), 2005 (in press): *Land use change and mountain biodiversity.* CRC Press, Boca Raton.

Starr, S.F., 2004: Conflict and peace in mountain societies. In: *Key Issues for Mountain Areas,* M.F Price, L. Jansky, and A.A. Iatsenia (eds.), United Nations University Press, Tokyo, 169–180.

Tang, C.Q. and M. Ohsawa, 2002: Tertiary relic deciduous forests on a humid subtropical mountain, Mt. Emei, Sichuan, China. *Folia Geobotanica,* **37,** 93–106.

The Mountain Institute, 1998: Community Based Mountain Tourism, Practices for Linking Conservation with Enterprise. Synthesis of a Mountain Forum Electronic Conference in Support of the Mountain Agenda. Washington, DC.

Theurillat, J.-P., and A. Guisan, 2001: Potential impact of climate change on vegetation in the European Alps: A review. *Climatic Change,* **50,** 77–109.

Thompson, L.G., E. Mosley-Thompson, M. Davis, P.N. Lin, T. Yao, M. Dyurgerov, and J. Dai, 1993: Recent warming: Ice core evidence from tropical ice cores with emphasis on central Asia. *Global Planet. Change,* **7,** 145–56.

Thorsell, J., 1997: Protection of nature in mountain regions. In: *Mountains of the World—A Global Priority.* B. Messerli and J.D. Ives (eds.). Parthenon Publishing Group, New York, London, pp 237–248.

Thorsell, J. and L. S. Hamilton, in press: A global overview of mountain protected areas on the world Heritage list. In: *Managing Mountain Protected Areas in the 21st Century,* D. Harmon and G. Worboys (eds.). Andromeda Editrice, Colledara, Italy.

Udvardy, M.D.F., 1975: A classification of the biogeographical provinces of the world. *Occasional Paper* No. 18., IUCN, Morges, Switzerland.

UNEP (United Nations Environment Programme), 2002: *Global Environment Outlook 3. Past, present and future perspectives.* GEO-3. Earthscan, London.

UNEP-WCMC (World Conservation Monitoring Centre), 2002: *Mountain Watch: environmental change & sustainable development in mountains.* Cambridge, UK.

UNEP-WCMC, 2003: Mountains derived from U.S. Geological Survey National Mapping Division, EROS Data Center (EDC), 1996: Global 30 Arc Second Elevation Data (GTOPO 30), and Forests derived from UNEP-WCMC's global forest cover dataset, an update of Iremonger, S. Ravilious, C. and Quinton, T. (eds.), 1997: *A global overview of forest conservation.* Available at: <http://www.unep-wcmc.org/habitats/mountains/statistics.htm.>

UNCED (United Nations Conference on Environment and Development), 1992: *Agenda 21. The United Nations Programme of Action from Rio.* The final text of agreements negotiated by governments 3–14 June 1992, Rio de Janeiro, Brazil. 294 pp.

van der Knaap, W.O. and J.F.N van Leeuwen, 1995: Holocene vegetation succession and degradation as responses to climatic change and human activity in the Serra da Estrela, Portugal. *Review of Palaeobotany and Palynology,* **89,** 153–211.

Velázquez, A., G. Bocco, F.J. Romero, and A.P. Vega, 2003: A landscape perspective on biodiversity conservation. *Mountain Research and Development,* **23(3),** 240–246.

Vinebrooke, R.D. and P.R. Leavitt, 1998: Direct and interactive effects of allochthonous dissolved organic matter, inorganic nutrients, and ultraviolet radiation on an alpine littoral food web. *Limnology and Oceanography,* **43,**1065–1081.

Vinebrooke, R.D. and P.R. Leavitt, 1999: Phytobenthos and phytoplankton as bioindicators of climate change in mountain lakes and ponds: an HPLC-based approach. *Journal of the North American Benthological Society,* **18,**14–32.

Viviroli, D., R. Weingartner, and B. Messerli, 2003: Assessing the hydrological significance of the world's mountains. *Mountain Research and Development,* **23(1),** 32–40.

Vonder Mühll, D.S., T. Stucki, and W. Haeberli, 1998: Borehole temperatures in alpine permafrost: a ten-year series. In: *Proceedings, The 7th International Permafrost Conference,* A.G. Lewcowitz and M. Allard (eds.), University of Laval, Quebec, pp. 1089–95.

Vuille, M. and R.S. Bradley, 2000: Mean annual temperature trends and their vertical structure in the tropical Andes. *Geophysical Research Letters,* **27(23),** 3885–3888.

Wardle, P., 1971: An explanation for alpine timberline. *New Zealand Journal of Botany,* **9,** 371–402.

WCED (World Commission on Environment and Development), 1987: *Our Common Future.* Oxford University Press, Oxford.

Whiteman, P.T.S., 1998: Mountain agronomy in Ethiopia, Nepal and Pakistan. In: *Human Impacts on Mountains.* N.J. R. Allan, G.W. Knapp and C. Stadel (eds.), Rowman and Littlefield, Totowa, New Jersey,pp. 57–82.

Yirdaw, E., 2001: Restoration of biodiversity on degraded lands in Ethiopia through the use of forest plantations. In: *World forests, markets and policies. World Forests, Society and Environment III,* M. Palo, J. Uusivuori and G. Mery (eds.), Kluwer Academic Publishers, The Netherlands, pp. 337–338.

Yoshino, M., 1975: *Climate in a small area.* Tokyo, University of Tokyo Press.

Yoshino, M., 2002: *Global warming and mountain environment.* Paper presented at the Symposium on International Year of Mountains held at the United Nations University on 1 Feb., 2002.

Zbicz, D., 2001: Global list of complexes of internationally adjoining protected areas. Appendix 1. In: *Transboundary Protected Areas for Peace and Co-operation,* Sandwith, T., C. Shine, L. Hamilton and D. Sheppard (eds.), WCPA/IUCN with Cardiff University: Gland and Cambridge, pp. 55–75.

Chapter 25

Polar Systems

Coordinating Lead Author: F. Stuart Chapin, III

Lead Authors: Matthew Berman, Terry V. Callaghan, Peter Convey, Anne-Sophie Crépin, Kjell Danell, Hugh Ducklow, Bruce Forbes, Gary Kofinas, Anthony David McGuire, Mark Nuttall, Ross Virginia, Oran Young, Sergei A. Zimov

Contributing Authors: Torben Christensen, Anna Godduhn, Eugene J. Murphy, Diana Wall, Christoph Zockler

Review Editor: Blair Fitzharris

---

*This appears in Appendix A at the end of this volume.

Main Messages

Changes in polar community composition and biodiversity are affecting human well-being (*high certainty*). Important changes include the reduction of top predators in Antarctic marine food webs, altering food resources; increased shrub dominance in Arctic wetlands, which contributes to summer warming trends and alters forage available to caribou; changes in insect abundance that alter food availability to wetland birds, energy budgets of reindeer and caribou, or productivity of forests; increased abundance of snow geese, which are degrading Arctic wetlands; overgrazing by domestic reindeer in parts of Fennoscandia, Russia, and sub-Antarctic islands; and a rapid increase in the occurrence and impact of invasive alien species, particularly in previously isolated sub-Antarctic islands.

Climate change has substantially affected ecosystem services and human well-being in polar regions (*high certainty*). Warming has been regionally variable but, on average, temperatures are warmer now than at any time in the past 400 years. Warming-induced thaw of permafrost is becoming more widespread, causing threshold changes in ecosystem services, including subsistence resources, climate feedbacks (energy and trace gas fluxes), and support for industrial infrastructure. International conventions have established mechanisms to reverse human impacts on UV-B, but international efforts to reverse human impacts on climate change have been less successful.

Regional changes in atmospheric temperatures and sea-ice extent and duration are changing the functioning of Antarctic marine ecosystems (*high certainty*). The Antarctic Peninsula, with its neighboring oceanic sectors, is one of the most rapidly warming regions on the planet (*high certainty*). It is also the area where populations of higher predators are concentrated as a result of high primary and secondary production and where the majority of exploitation of living resources has been concentrated. Changes in ecosystem structure are already occurring.

Most changes in feedback processes that occur in polar regions magnify trace-gas-induced global warming trends and reduce the capacity of polar regions to act as a cooling system for planet Earth. These climate feedbacks result from changes in the physical system (increased moisture transport to the poles, declines in the areal extent of sea ice and glaciers, and earlier snowmelt) (*high certainty*). In addition, within the Arctic, most changes in vegetation (expansion of shrubs in North America) and trace gas fluxes (release of soil carbon to the atmosphere as carbon dioxide and methane) are amplifying regional warming, although the retreat of the tree line in Russia is leading to cooling (*medium certainty*).

In the Arctic, regional warming interacts with socioeconomic change to reduce subsistence activities by indigenous and other rural people, the segments of society with the greatest cultural and economic dependence on these resources. Warming has reduced access to marine mammals (less sea ice) and made the physical and biotic environment less predictable. Industrial development has reduced the capacity of ecosystems to support subsistence activities in some locations. Other animals, such as moose (*Alces alces*) in North America, have moved northward in response to warming.

There is *high certainty* that reductions in the summer extent of sea ice will increase shipping access along northern sea routes, fostering northern development, and—together with rising sea level—will increase coastal erosion that currently threatens many coastal villages. The net effect is generally to increase the economic disparity between rural subsistence users and urban residents.

Increases in persistent organic pollutants and radionuclides in subsistence foods have increased health risks in some regions, but diet changes associated with the decline in harvest of these foods are usually a greater health risk.

Mitigation of impacts (rather than reversing changes in drivers) is the most feasible short-term strategy for protecting polar ecosystem services and human well-being because the major causes of polar change are globally distributed. Direct impacts of human activities on polar regions have been modest, and nations with Arctic lands or Antarctic obligations have the economic resources to mitigate many current and expected problems if appropriate policies are applied. Consequently, many parts of polar regions have a high potential to continue providing key ecosystem services, particularly in polar oceans and wetlands where biodiversity and resource harvest are concentrated. However, the sensitivity of polar ecosystems to disturbances associated with resource extraction makes them vulnerable to future global increases in resource demand.

25.1 Introduction

The polar systems treated in this chapter are treeless lands at high latitudes. These systems merge in the north with boreal forest (see Chapter 21) and in the south with the Southern Ocean (see Chapter 18). This chapter emphasizes the ecological processes that most directly influence human well-being within and outside polar regions. The physical processes in polar regions that influence human well-being (such as ozone effects on UV-B and changes in glaciers and sea ice) are described briefly in this chapter and more fully in the Arctic Climate Impact Assessment and in assessments of the Intergovernmental Panel on Climate Change (Anisimov et al. 2001). Because of its greater area of ice-free land and larger human population, the Arctic figures more prominently than the Antarctic in this chapter, although both are equally important when physical processes and marine ecosystems are integrated with terrestrial ecological processes.

25.1.1 The Arctic

The basic characteristics of Arctic terrestrial ecosystems were recently summarized by the Arctic Climate Impact Assessment (Callaghan et al. in press). It is a 12-million-square-kilometer treeless zone between closed boreal forests and the ice-covered Arctic Ocean. Within the Arctic, there are northward gradients of shorter snow-free seasons (from three months to one month), lower temperatures (from 10–12° Celsius to 2° Celsius in July), and less precipitation (generally from about 250 to 45 millimeters per year) (Jonasson et al. 2000). Permafrost (permanently frozen ground) is nearly continuous in most of the Arctic but becomes less continuous to the south and in maritime regions such as Scandinavia. Regional variation in Arctic climate reflects the nature of adjacent oceans. Cold waters in ocean currents flowing southwards from the Arctic depress the temperatures in Greenland and the eastern Canadian Arctic, whereas the northeasterly flowing North Atlantic Current warms the northern landmasses of Europe.

The land cover of the Arctic includes ice, barrens (which in this chapter includes polar desert and prostrate shrub tundra with less than 50% plant cover), and tundra (which in this chapter includes treeless vegetation with nearly continuous plant cover). Tundra constitutes the largest natural wetland in the world (5 million square kilometers). (See Table 25.1.) The distribution of these major Arctic land cover types is well known, although their areal extent differs substantially among authors, depending on vegeta-

Table 25.1. Polar Subtypes and Their Areas. The area of the major subtypes of Arctic and Antarctic ecosystems was estimated from the maps in Figure 25.1. The Arctic includes only areas north of 55°N and excludes forests and woodlands in that zone. Barrens are lands with less than 50% vascular plant cover, and arctic tundra has >50% plant cover. Barrens include polar desert and prostrate shrub tundra; Arctic tundra includes graminoid tundra, erect shrub tundra, and wetlands. All Arctic tundra is classified as wetlands under the Ramsar Convention. (CAVM Team 2003)

Ecosystem type	Total	Canada	United States	Greenland	Eurasia
			(mill. sq. km.)		
Arctic	10.57	3.29	1.01	2.14	4.13
Ice	2.50	0.25	0.10	1.95	0.20
Barrens	3.01	1.90	0.11	0.12	0.88
Artic tundra	5.06	1.14	0.80	0.07	3.05
Antarctic ice	12.44				

tion classification (McGuire et al. 2002; CAVM Team 2003). (See Figure 25.1 in Appendix A.)

Only 3% of the global flora and 2% of the global fauna occur in the Arctic, and their numbers decrease toward the North (Chernov 1995; Matveyeva and Chernov 2000). However, the Arctic is an important global pool of some groups, such as mosses, lichens, and springtails. The proportions of species that occur in the Arctic differ among major groups—spiders at 1.2%, for instance, insects at 0.3%, fishes at 1.8%, reptiles at <0.1%, mammals at 2.8%, and birds at 2.8%. In general, primitive groups (such as springtails, up to 12%) are better represented in the Arctic than are advanced groups such as beetles (0.1%). Exceptions to the general gradient of declining terrestrial diversity at higher latitudes include sawflies and shorebirds (Kouki 1999; CAFF 2001).

Animal species decline with increasing latitude more strongly than do vascular plants (frequently by a factor of 2.5) (Callaghan et al. in press). There are about 1,800 species of vascular plant, 4,000 species of cryptogam, 75 species of terrestrial mammal, 240 species of terrestrial bird, 2,500 species of fungus, and 3,200 species of insect (Matveyeva and Chernov 2000). Because of the low species diversity, some ecologically important species—such as the sedge *Eriophorum vaginatum,* lemmings, reindeer and caribou, and mosquitoes—have large populations with broad geographic, often circumpolar, distributions. Terrestrial food webs are often simple, with few species at a particular level in the web. Consequently, changes in abundance of one species can have many direct and indirect ecosystem consequences (Blomqvist et al. 2002). A significant attribute of Arctic biodiversity is the importance of migratory species, including most birds and marine mammals, caribou, and many key fish species such as salmon. Many of these species are important subsistence foods for local residents, and their population dynamics can be strongly affected by processes outside the Arctic.

Terrestrial net primary production and decomposition rates are low and decrease from south to north. The stocks of soil carbon are high in boreal woodlands and Arctic tundra but low in barrens. (See Table 25.2.) Because of low productivity, revegetation after human disturbance can take centuries (Forbes et al. 2001).

The Arctic has been populated throughout the Holocene. Of the 3.8 million people who live there, about 8% (300,000) are indigenous. Population density ranges from near 8 persons per square kilometer in the Murmansk Region of Russia to fewer than 0.1 person per square kilometer in the Canadian Arctic (Knapp 2000). Most people in the Arctic live in urban areas, so population densities in rural areas are extremely low (typically fewer than 0.1 person per square kilometer). The percentage of indigenous peoples is greatest in Greenland and North America, intermediate in Scandinavia, and lowest in Russia.

25.1.2 The Antarctic

The Antarctic (12.4 million square kilometers) is similar in size to the Arctic but differs in being a largely ice-covered continent surrounded by a ring of sea ice and extensive cold oceans. It has no indigenous peoples, and use of the area for harvest of marine mammals, birds, and fish began less than 200 years ago. The northern boundary of the Antarctic region is the Antarctic Polar Frontal Zone linked with the Antarctic Circumpolar Current, where the southern cold surface waters sink below warmer southern temperate waters at about 58° S (Anisimov et al. 2001). The combination of the oceanic frontal zone and the circumpolar current and westerly atmospheric circulation provides a strong barrier to the movement of both terrestrial and marine biota into or out of the region (Clarke and Crame 1989; Barnes et al. submitted). Within the Antarctic region, three zones are frequently recognized (Smith 1984; Longton 1988): the sub-Antarctic (oceanic islands close to the Polar Frontal Zone), maritime Antarctic (Scotia Arc archipelagoes and west coast of Antarctic Peninsula to about 72° S), and continental Antarctic (the remainder of the peninsula and main continental mass).

The Antarctic and Arctic experience parallel latitudinal influences on seasonal climate (day length and insolation) but otherwise have quite different environmental patterns and extremes (Convey 1996; Danks 1999), largely driven by the contrasting geography of the two regions. Several sub-Antarctic islands encircle Antarctica close to the Polar Frontal Zone. These have cold, relatively stable temperatures, with thermal variation buffered by the surrounding ocean and with high precipitation and cloud cover. Mean monthly air temperatures for most islands are positive year-round (Doran et al. 2002a; Thost and Allison in press).

The maritime Antarctic also experiences a strong oceanic influence, effectively acting as a physical barrier to the circulation of moist air from the Pacific component of the Southern Ocean. Mean monthly air temperatures in the maritime Antarctic are positive (but less than 2° Celsius) for two to four months in summer and negative for the remainder of the year, although positive air temperature may occur in any month (Walton 1984; Smith et al. 2003).

Inland, the climate of the Antarctic continent is colder than the Arctic, with average annual temperatures of −20° Celsius or lower (Hempel 1994; Doran et al. 2002b). Large parts of continental Antarctica are classified as frigid or polar deserts, with extremely low precipitation. This, combined with low humidity and strong katabatic winds, can lead to rapid ablation and extensive ice-free areas (Doran et al. 2002a; Nylen et al. 2004). The largest of these, the McMurdo Dry Valleys (about 4,800 square kilometers), contains a mosaic of perennially ice-covered lakes, ephemeral streams, and arid soils (Fountain et al. 1999). Plant and animal biomass of these valleys is low, and microbes dominate biological productivity (Doran et al. 2002b).

The Southern Ocean is covered by an expanse of sea ice that varies seasonally from 3 million to 20 million square kilometers: about the size of North America. Sea ice contains within its matrix a microbial community of algae, bacteria, and small consum-

Table 25.2. Comparisons of Carbon Pools in Arctic-Alpine Tundra with the Boreal Zone and World Total. The soil pools do not include the most recalcitrant humic fractions. (McGuire et al. 1997)

	Area	Soil	Vegetation	Soil:Veg. ratio	Total carbon		
					Soil	Veg.	Soil+Veg.
	(mill. sq. km.)	*(grams per sq. meter)*			*(trillion kilograms)*		
Arctic and Alpine tundra	10.5	9,200	550	17.0	96	5.7	102
Boreal woodlands	6.5	11,750	4,150	2.8	76	27	103
Boreal forest	12.5	11,000	9,450	1.2	138	118	256
Terrestrial Total	130.3	5,900	7,150	0.8	772	930	1,702

ers (Arrigo et al. 1997; Brierley and Thomas 2002). It also serves as a refuge for juvenile krill that browse on the microbial community (Siegel et al. 1990) and as a feeding platform for penguins and seals (Fraser and Hofmann 2003). There are marked regional interannual variations in the extent and duration of sea ice that generate changes in the functioning of the whole ecosystem (Murphy et al. 1995, 1998; Loeb et al. 1997; Fraser and Hofmann 2003). The entire Antarctic marine ecosystem, from primary producers to whales, therefore depends on the extent and duration of sea-ice cover (Quetin and Ross 2001; Smith et al. 2003; Atkinson et al. in press).

The terrestrial biodiversity of Antarctica is much lower than that in the Arctic because of its geographic isolation, the relative youth of most terrestrial habitats (formed after the retreat of Pleistocene glaciers), and the extreme environmental conditions (Convey 2001b). There are no native terrestrial vertebrates, but large populations of marine birds (penguins, petrels, gulls, terns, skuas) and seals take advantage of the absence of land-based predators, relying on terrestrial sites to breed, molt, and rest. These provide considerable nutrient input to terrestrial habitats while also imposing physical damage through trampling and manuring.

Less than 1% of the continent is seasonally ice- or snow-free, providing rock and soil habitats for life (Block 1994). Vascular plants and higher insects are poorly represented on the Antarctic continent (two native species of each, both restricted to the maritime Antarctic). Mosses, liverworts, and lichens are frequent in coastal low-altitude areas but rapidly decrease with progression into the interior or the ice-free dry valley deserts. Likewise, the more primitive or lower groups of invertebrates (such as mites, springtails, nematodes, and other soil mesofauna) assume a dominant role in food webs rarely seen elsewhere. The simplest faunal assemblages found worldwide occur in the Dry Valleys and inland continental nunataks (Freckman and Virginia 1997; Convey and McInnes in press). Many of these groups show high levels of endemism. Biodiversity on the sub-Antarctic islands is considerably higher than on the continent, though lower than at comparable latitudes in the Arctic, and rates of endemism are again extremely high (Chown et al. 1998; Bergstrom and Chown 1999).

The only permanent human residents in Antarctica are scientists and support staff who live in 37 research stations and many temporary camps. They number about 4,000 in the three- to four-month summer and 1,000 in winter (Frenot et al. in press). Their research examines processes and patterns that can only be explored in extreme conditions, such as the record of Earth's climate history preserved in ice cores, physiology and organism adaptation in extreme conditions, and the functioning of highly simplified ecosystems (Weller et al. 1987). There is a rapidly growing tourist trade, with 14,000 people visiting the Antarctic in 1999–2000 (Frenot et al. in press). Tourists predominantly visit coastal areas to view marine birds and seals and to see historic huts

and sites used by Antarctic explorers of the early twentieth century, spending most of their time on ships. Increased tourism has global causes. Economic prosperity provides individuals with the financial means to participate, while events such as the economic downturn in Russia resulted in the release of ice-strengthened research ships to support tourism.

25.2 Condition and Trends in Polar Ecosystem Services

The dramatic changes in many of the drivers that shape polar processes are having profound effects on ecosystems and the services they provide to society. This section describes the current condition and major trends in ecosystem services that are important to society, both within and beyond polar regions.

25.2.1 Climate Regulation

25.2.1.1 *Physical Feedbacks*

Polar regions play a key role in the global climate system and therefore influence human activities and well-being throughout the world. (See Chapter 13.) They act as an important cooling system for Earth by reflecting incoming radiation from ice, snow, and clouds and by radiating back to space the heat that is transported poleward by the atmosphere and oceans. The heat loss from polar regions (180 W m^{-2} annually), for example, is greater than solar input (80 W m^{-2}), with the imbalance (100 watts per square meter) coming from lower latitudes (Nakamura and Oort 1988).

The latitudinal temperature gradient is a major driving force for atmospheric and ocean circulation and therefore for heat transport from the equator to the poles. Ocean heat transport is driven both by surface winds and by the movement of cold saline surface water to depths around Antarctica and in the North Atlantic (Anisimov et al. 2001). Recent increases in Antarctic precipitation have caused a freshening of surface layers that weakens bottom-water formation (Anisimov et al. 2001). This bottom-water formation is sensitive to climate effects on sea-ice formation and decay and, in the Arctic, on freshwater discharge to the ocean. Polar ice sheets account for 68% of the fresh water on Earth, so changes in the mass balance of these ice sheets could alter sea level and the input of fresh water to zones of deepwater formation. Past variation in the strength of bottom-water formation has contributed substantially to Earth's long-term climate variation (Anisimov et al. 2001).

Until recently, the warming trend at high latitudes had little detectable effect on the mass balance of the Greenland and Antarctic ice sheets, because of measurement inaccuracy and because warming simultaneously increased inputs of snow (because the warmer atmosphere holds more water) and the melting of ice sheets (Anisimov et al. 2001). Since 1990, however, increased melting and water infiltration at the base have increased ice flows to the ocean (Krabill et al. 2000; Thomas 2004). The mass wastage in 1991–2000 was 80 cubic kilometers per year (Box et al. 2004). Mountain and subpolar glaciers have exhibited a negative mass balance of similar magnitude (90–120 cubic kilometers per year) for the past 40 years (see Figure 25.2), particularly in Alaska, Canada, high-mountain Asia, and Patagonia (Dowdeswell et al. 1997; Dyurgerov and Meier 1997; Arendt et al. 2002; Rignot et al. 2004). In contrast, the Antarctic ice sheet shows regional and temporal variation in mass balance but no overall directional trend (Rignot and Thomas 2002; Bentley 2004). Warming has been most pronounced on the Antarctic Peninsula, causing a 10,000-square-kilometer retreat of adjacent ice shelves (Anisimov et al. 2001).

There is enough ice on the Antarctic Peninsula to raise global sea level by 0.5 meters, but the time course of its melting is *speculative*. The rapid disintegration and collapse of the Larsen-B ice shelf in 2002 was unprecedented; grounded ice sheets serve as brakes on the glaciers behind them, and the glacier behind the (now absent) Larsen-B sheet has accelerated its seaward flow (Scambos et al. 2004). The total freshwater input from these ice sheets and glaciers has, however, had less effect on sea level than the thermal expansion of oceans resulting from recent climate warming has. Nonetheless, freshwater input from glacial melt has substantially increased since 1990.

The Southern Ocean has warmed faster than the global ocean at mid-depths (700–1,100 meters) has over the past 50 years (Gille 2002). In addition, the surface and shelf waters in the Ross Sea have warmed and become less saline over the past 40 years (Jacobs et al. 2002), owing to a combination of increased precipitation, a reduction of sea-ice production, and increased melting of the West Antarctic ice sheet (Jacobs et al. 2002). Ice dynamics are also sensitive to fluctuations in the Antarctic Circumpolar Current, regional warming, and ENSO-related variation in Southern Hemisphere atmospheric and oceanic conditions (Liu et al. 2002). The apparent precession around the Antarctic of anomalies in sea-ice conditions and the ocean temperatures associated with the Antarctic Circumpolar Current (termed the Antarctic Circumpolar Wave) (Murphy et al. 1995; White and Peterson 1996) are also related to ENSO and contribute to regional and interannual variation in ocean temperatures and ice conditions. Whaling records suggest a possible circumpolar retreat in Antarctic sea ice by 2.8 latitude between the mid-1950s and early 1970s, although this interpretation is debated (Anisimov et al. 2001). The sea ice then became more extensive from 1979 to 2002, but with high re-

gional variability (Liu et al. 2004) and large impacts on ecosystem processes (Smith et al. 1998).

In contrast, Arctic sea ice continues to decrease in extent by 2.9% per decade (see Figure 25.3) and has become thinner over the past 40 years (Maslanik et al. 1996; Anisimov et al. 2001). The reductions in areal extent of sea ice, glaciers, and seasonal snow cover reduce high-latitude albedo (the reflectance of incoming solar radiation) and act as a positive feedback that has *medium certainty* of amplifying the rate of high-latitude warming at both poles (Serreze et al. 2000; Mitchell et al. 2001). Increases in clouds, as sea ice retreats, could dampen this polar amplification, however (Wang and Key 2003). The reduction in summer ice cover has a *high certainty* of making commercial shipping feasible in the Northern Sea Route, likely by 2020. The reduction in sea ice over the past 40 years has reduced available habitat and hunter access to many marine mammals, which are an important subsistence and cultural resource for many coastal indigenous peoples of the Arctic (Krupnik 2002). The modest sea level rise that has occurred to date, combined with reduction in sea ice and greater storm surges, has caused considerable coastal erosion, which endangers coastal communities and increases organic carbon input to coastal oceans.

25.2.1.2 Ecosystem Feedbacks

Ecosystem processes at high latitudes influence the climate system when they cause these regions to become net sources or sinks of greenhouse gases such as carbon dioxide and methane. In the Antarctic, oceans have the strongest influence on carbon flux through both physical processes that are driven by ocean circulation and biotic processes driven by photosynthesis and respiration. Photosynthesis (carbon uptake) by marine phytoplankton converts inorganic carbon into organic matter. When algae are eaten or die, some of this carbon is respired and returns to the atmosphere and some sinks to depth as dead cells or fecal pellets of their grazers, a biological pump that sequesters carbon in the deep ocean (Ducklow et al. 2001).

Many factors interact to control the productivity of phytoplankton, as described later in this section, and therefore the carbon export to depth in the oceans around Antarctica. Spatial variability in productivity and carbon export depends on vertical mixing, mixed layer depth, krill grazing, and micronutrient (iron, for example) limitation (Arrigo et al. 1999; Prezelin et al. 2000). Temporal variation correlates with sea-ice extent and timing, with

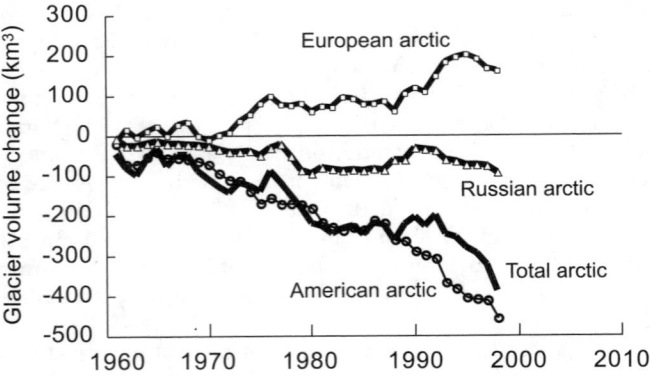

Figure 25.2. Trends in Glacier Volume for Different Arctic Regions, 1961–98 (Hinzman et al. in press)

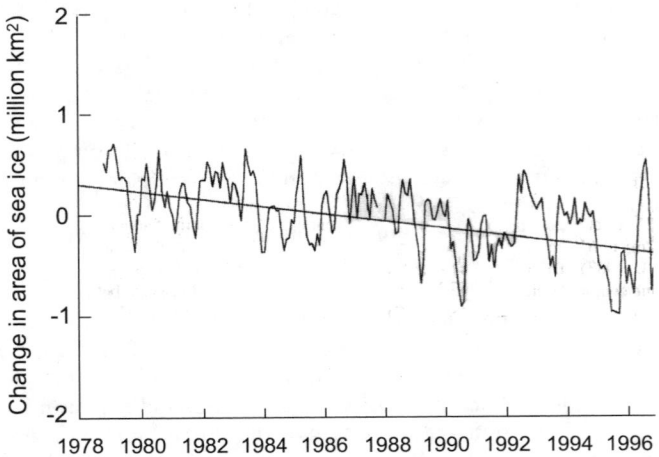

Figure 25.3. Trend in Arctic Sea Ice Extent, 1978–97 (Serreze et al. 2000)

extensive or late-melting ice favoring high productivity. Ice margins are highly productive because of high nutrient availability, whereas the productivity of the open ocean appears limited by iron availability (Boyd et al. 2000).

Warming has two counterbalancing effects on productivity and therefore on carbon export to depth. The more highly developed freshwater lens beneath melting ice enhances productivity, whereas the decreasing extent of sea ice reduces the areal extent of productive ocean. The Southern Ocean below 50°S is a net carbon sink, taking up about 0.5 petagrams of carbon annually, or about 20% of the total oceanic uptake, in 10% of its area (Takahashi et al. 2002). The intensity of uptake in the Southern Ocean highlights its role in the global carbon cycle. How it will respond to climate change remains *highly uncertain* (Sarmiento and Le Quere 1996; Sarmiento et al. 1998).

In the Arctic, terrestrial ecosystems are the major site of greenhouse gas exchange. Under current conditions, both direct measurements and global carbon models suggest that the circumpolar Arctic is neither a large source nor a large sink of carbon dioxide. Measurements suggest a modest source, whereas models suggest a small sink of 17 ± 40 grams of carbon per square meter per year (mean ± SD, spatial variance) (McGuire et al. 2000; Sitch et al. 2003; Callaghan et al. in press). Both estimates overlap zero, owing to *low certainty* and large interannual and regional variation.

Areas that have warmed and dried, such as Alaska and Eastern European Arctic, are generally a carbon source. For example, after a pattern of net carbon accumulation during most of the Holocene, Alaska became a net carbon source when regional warming began in the 1970s (Oechel et al. 1994). The strength of this source then declined as plant- and ecosystem-scale negative feedbacks increased carbon uptake by plants (Oechel et al. 2000). (See Figure 25.4.) This may reflect warming-induced nutrient release, which tends to enhance photosynthesis and net primary production (Shaver et al. 2000). Scandinavian and Siberian peatlands, which have become warmer and wetter, are a net carbon sink of 15–25 grams of carbon per square meter per year (Aurela et al. 2002; Smith et al. 2004). In Greenland, where there has been little warming (Chapman and Walsh 1993), net carbon exchange is close to zero, with sinks in wet fens balanced by carbon losses in dry heath (Christensen et al. 2000; Soegaard et al. 2000; Nordstrøm et al. 2001). Carbon fluxes in the high Arctic are extremely low—a net sink of about 1 gram of carbon per square meter per year (Lloyd 2001).

Taken together, Arctic flux measurements suggest that warming has substantially altered Arctic carbon balance but that the direction of this effect varies regionally, depending on hydrology,

with wet areas tending to gain carbon and dry areas tending to lose carbon with warming. Remote sensing and indigenous observations suggest that drying trends predominate in the North American Arctic (Hinzman et al. in press). The greatest uncertainties in estimating recent and future trends in carbon exchange relate to changes in hydrology, nutrient dynamics associated with decomposition, and disturbance effects on vegetation (Chapin et al. 2000a; Oechel et al. 2000; Callaghan et al. in press). These processes have not yet been adequately incorporated into global carbon models. In summary, the short time period of record and the incomplete inclusion of key processes in global models result in *low certainty* of long-term trends, but the balance of evidence suggests a small trend toward carbon release in the short term, with long-term trends depending on the balance between increased production and uncertain trends in respiration.

High-latitude wetlands are one of the largest natural sources of atmospheric methane, about 70 teragrams per year (Cicerone and Oremland 1988; Schlesinger 1997). Methane fluxes are highly variable, both temporally and spatially. However, methane efflux responds positively to soil moisture, summer soil temperature, and the presence of oxygen-transporting vascular plants such as wetland sedges (Christensen et al. 2003). Warming and thawing of permafrost (see Figure 25.5) increase the area of wetlands and thaw lakes, further increasing methane efflux from the Arctic (Zimov et al. 1997; Christensen et al. 2004). There is *medium certainty* that warming enhances methane release, creating a positive feedback to climate change (Christensen et al. 2003).

Recent increases in length of the snow-free season (2.6 days per decade and similar increases in other northern regions) (Keyser et al. 2000; Walther et al. 2002) and the reduced albedo (reflectance) associated with shrub expansion (Eugster et al. 2000) both tend to increase annual energy absorption. This acts as a positive feedback to high-latitude warming (Betts and Ball 1997; Chapin et al. 2000b), but the magnitude of these effects has *low certainty*. Conversion of tundra to forest creates an even larger climate feedback by replacing a snow-covered surface with a dark, more absorptive surface. (See Chapter 13.) Model simulations suggest that conversion of tundra to forest accounted for half of the high-latitude mid-Holocene warming (Foley et al. 1994). Forest expansion generally lags behind regional warming because tree establishment is slow near the climatic limit of trees (Huntley 1996; Lloyd et al. 2003b).

Russian rivers have increased their discharge to the Arctic Ocean by 7% over the past 70 years, primarily due to increases in winter discharge (Peterson et al. 2002; Yang et al. 2002). (See Figure 25.6.) Changes in permafrost distribution associated with

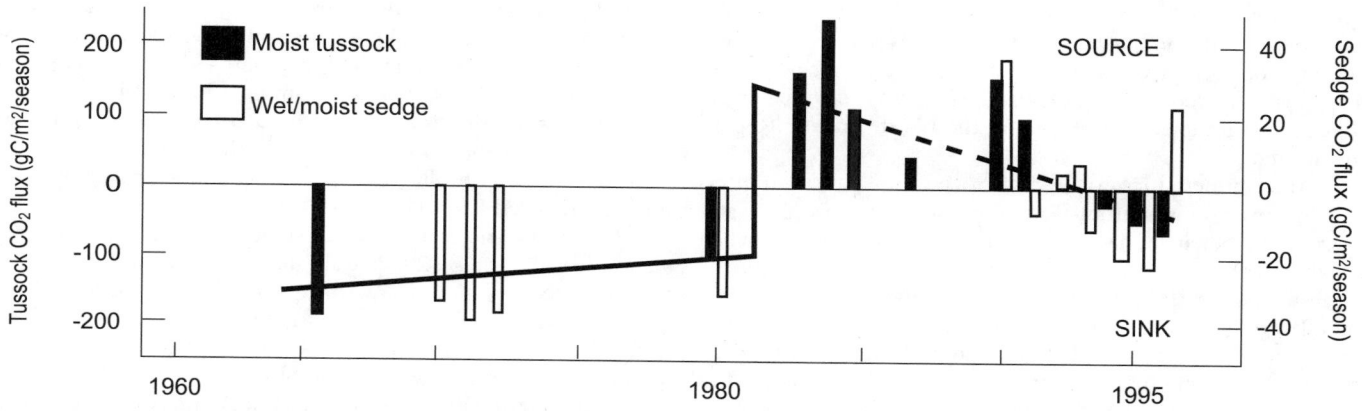

Figure 25.4. Trend in Net Annual Carbon Flux from Alaskan Arctic Tundra, 1960–95 (Oechel et al. 2000)

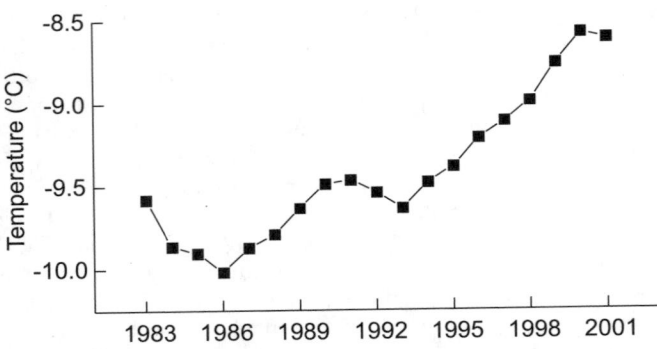

Figure 25.5. Trend in Permafrost Temperature at 20 m Depth at West Dock on the Coastal Plain of Northern Alaska, 1983–2001 (Osterkamp and Romanovsky 1999)

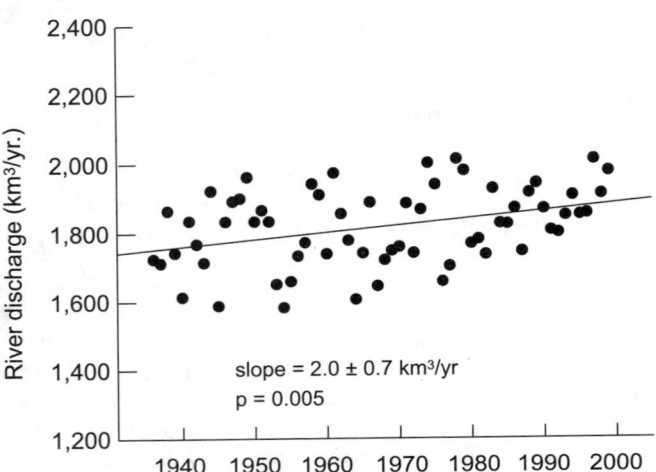

Figure 25.6. Trend in Discharge of the Six Largest Eurasian Arctic Rivers, 1935–2000 (Peterson et al. 2002)

wildfire or human-induced vegetation changes may have contributed to the increase in winter discharge (Serreze et al. 2003), although the causes are *speculative*. The discharge records for North American rivers are too short to detect long-term discharge trends. The Arctic Ocean is extremely sensitive to changes in discharge because it receives more discharge per unit ocean volume than any other ocean. Most river discharge into the Arctic Ocean eventually exits in the North Atlantic. Continued increase in the input of low-density fresh water could reduce North Atlantic bottom-water formation by the end of the twenty-first century if the discharge of Arctic rivers continues to increase at its current rate (Peterson et al. 2002). This potential change is *speculative,* but the climatic implications for Europe and the North Atlantic Region are enormous.

In the Southern Ocean, physical processes are entirely responsible for variation in bottom-water formation. Dense saline water is produced when sea ice forms beneath ice shelves and when polynyas (open water) or thin ice transfer heat rapidly from the ocean to the atmosphere. Antarctic surface waters have warmed since 1970 (Wong et al. 1999), a trend that would tend to reduce rates of bottom-water formation and thermohaline circulation. In summary, different processes at the two poles both tend to weaken bottom-water formation.

Less is known about carbon cycling in Antarctic terrestrial systems. Rates of CO_2 flux in Antarctic dry valley soils are exceed-

ingly low, and the total volume of soil in the Antarctic is tiny in comparison with that of the Arctic (Parsons et al. 2004). The soil organic matter in some dry valleys appears to be a legacy of past climates when paleo-lake production enriched soils in carbon and nutrients (Burkins et al. 2001). It is not known whether these systems are net sources or sinks of carbon because interactions between abiotic and biological controls of CO_2 flux vary with soil environment and climate (Parsons et al. 2004).

25.2.2 Fresh Water

Antarctica is the driest continent on Earth, with nearly all water locked up in ice. Dependence on fuel to melt ice for usable water limits potential for human habitation, and low water availability limits many physical and biological processes across the continent (Kennedy 1993).

In contrast to Antarctica and much of the rest of the world, inhabited portions of the Arctic generally have abundant fresh water despite low precipitation. Water is important to Arctic residents as a source of hydropower and as a transportation corridor. (See Chapter 20.) Water quality in the Arctic is generally good except in areas of industrial development.

The net transport of contaminants (persistent organic pollutants, heavy metals, and radionuclides) to polar regions, however, produces a dilute source of pollutants that often become concentrated as they move through food webs and are a potential health risk to people, as described later in this chapter. Changes in water availability in the Arctic (for example, regional drying) are important primarily through effects on ecosystem services (such as habitat for fish and water birds, or trace gas emissions).

Arctic fresh water continues to receive attention from water-deficient regions of the world as a potential freshwater supply. Russian newspapers, for example, recently reported the reactivation of 1970s plans to divert the Ob and other northbound rivers toward water-starved regions in the south.

25.2.3 Biodiversity and Species Composition

25.2.3.1 Changes in Vegetation

Polar regions have historically experienced fewer invasions of exotic plant species than most biomes because climate is a severe physiological filter (Walther et al. 2002). However, recent climate warming has facilitated invasion of new species (Robinson et al. 2003; Frenot et al. in press). On some sub-Antarctic islands exotic species account for more than 50% of vascular plant diversity, and exotic grasses may outcompete native species (Smith 1994; Chown et al. 1998; Gremmen et al. 1998; Bergstrom and Chown 1999; Gremmen and Smith 1999; Frenot et al. in press). Exotic species also occur in maritime regions of the Arctic (such as Iceland) (Wiedema 2000) and inland areas with road and rail connections (in Canada and Russia, for example) (Forbes 1995), but the frequency of invasion is known with *low certainty*. Twenty species of Arctic plants are considered globally threatened—vulnerable, endangered, or critically endangered, according to IUCN Red List criteria (CAFF 2001).

Polar plant species have also changed in their relative abundance (Walther et al. 2002; Callaghan et al. in press). On the Antarctic continent, mosses have colonized previously bare ground, and the only two native vascular plant species have expanded their ranges (Smith 1994; Convey 2001a). Repeat aerial photography demonstrates that shrubs have increased in dominance in 70% of 200 sample locations in Arctic Alaska (Sturm et al. 2001), a change confirmed by indigenous observations across much of the North American Arctic (Nickels et al. 2002; Thorpe et al. 2002). NDVI, an index of vegetation greenness, has increased by 15% since 1981 in Arctic Alaska (Jia et al. 2003) (see Figure 25.7) and to a more variable extent in the circumpolar Arctic as a whole (Myneni et al. 1997).

In Alaska, the latitudinal tree line has moved northward, converting about 2% of tundra to forest in the past 50 years (Lloyd et al. 2003a), whereas in Russia the tree line has retreated southward as a result of forest harvest and anthropogenic burning, creating about 500,000 square kilometers of wetlands superficially resembling the tundra (Callaghan et al. 2002; Vlassova 2002). Thawing of permafrost has also converted large areas of well-drained lands to wetlands (Crawford et al. 2003; Hinzman et al. in press).

Experimental manipulations of climate, nutrients, and UV-B radiation in numerous studies throughout the Arctic suggest that mosses and lichens could become less abundant when vascular plants increase their growth (Van Wijk et al. 2004). Mosses and lichens are a large component of polar plant diversity and controllers of ecosystem processes: lichens are a key winter food for caribou and reindeer, and mosses insulate the soil. Similar manipulations on the Antarctic Peninsula indicate complex responses that could include a decline in density and diversity of soil invertebrates in response to warming (Convey et al. 2002) and a decline in nematodes in response to cooling in the McMurdo Dry Valleys (Doran et al. 2002b).

25.2.3.2 Changes in Caribou and Reindeer

Many of the North American barren-ground wild caribou (*Rangifer tarandus*) herds were at historic high levels at the end of the 1980s; several herds are currently in decline (Russell et al. 2002). Interannual variation in caribou calving success of several North American herds correlates with the rate of spring vegetation growth in calving grounds, as measured by satellite-derived NDVI (Griffith et al. 2002), although regional heterogeneity in other ecological conditions also affects reproductive success (Russell et al. 2002). The Peary caribou herd that occupies polar barrens in the Canadian Arctic islands has, in contrast to other North American herds, decreased to critically low levels and is currently on Canada's endangered species list. Potential causes of the decline include the impact of climate change on extreme weather events (such as autumnal ice storms), vegetation composition, insect harassment, animal energy demands, animal behavior, and shifts in animal distribution relative to human users' access.

In Fennoscandia, reindeer are intensively managed for meat, as a cultural resource, and for recreational harvest. Here reindeer herding is the subject of political conflicts owing to the degradation of pastures, protection of predators, and indigenous people's efforts to assert their access rights to traditional herding areas (covered later in this chapter).

In the Russian North, the collapse of the former state-supported supply and marketing system has led to a decline in domesticated reindeer stock over the past 10 years from 2 million to 1 million animals (Baskin 2000). This decrease was accompanied by increases in several large wild reindeer populations of Taimyr, Yakutia, and Chukotka, leading to serious impacts on pasturelands. Wild and domesticated reindeer are typically seen as ecological antagonists, because wild reindeer lead domesticated animals away, compete for (or damage) pastures, and are a reservoir of infectious diseases. On the Yamal Peninsula, in contrast, the population of semi-domesticated reindeer increased steadily in the post-Soviet period, in part because of the cultural role of reindeer herding among the Nenets people of that region.

Oil development has contributed to trends in caribou and reindeer populations in some areas. Caribou and reindeer are sensitive to disturbance during calving (Vistnes and Nellemann 2001; Griffith et al. 2002). In Alaska, for example, concentrated calving was displaced from industrialized areas to areas of lower forage richness, with caribou returning to industrialized areas during the post-calving period (Griffith et al. 2002). The effects on population dynamics of this herd displacement during calving are debated (NRC 2003). Development conflicts associated with potential habitat loss have been resolved in some areas through "calving group protection measures" (in the Northwest Territories of Canada, for instance), whereas in other areas (such as Alaska and Russia) calving grounds hold no special policy status.

Onshore oil and gas activities also impede access to traditional hunting and herding areas and thus disrupt community activities and traditional practices (Golovnev and Osherenko 1999). Pipelines and facilities create obstacles to free movement of reindeer herds. In the intensively developed Yamal Peninsula in Western Siberia, destruction of vegetation due to construction of facilities, roads, and pipelines and to off-road vehicle traffic exceeds 2,500 square kilometers and could more than double under current development plans. The resulting concentration of reindeer herds into an ever-decreasing undeveloped area has led to overgrazing,

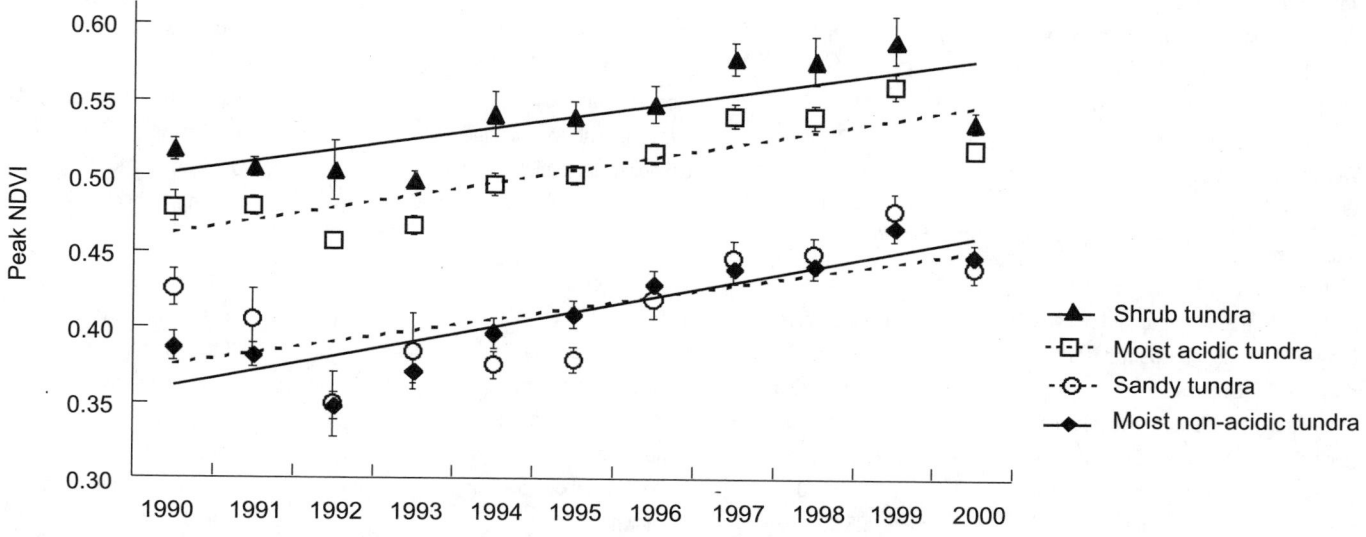

Figure 25.7. Trend in Normalized Difference Vegetation Index in Alaskan Arctic Tundra, 1990–2000 (Jia et al. 2003)

with potential long-term adverse effects on ecosystem productivity and local economies (Forbes 1999).

25.2.3.3 Changes in Other Terrestrial Mammals

Twenty-one species of Arctic mammal are considered globally threatened (CAFF 2001). However, it is difficult to assess recent population trends of polar mammals as a whole, because quantitative data are available for only a few species, mostly animals of economic importance or high conservation profile. (See Table 25.3.) Possibly in response to changes in climate or prey abundances, red fox (*Vulpes vulpes*) has expanded northward into habitats of the smaller Arctic fox (*Alopex lagopus*) in Fennoscandia and Canada, where it is thought to exploit the most productive habitats by interference competition (Hersteinsson and MacDonald 1992; Elmhagen et al. 2002). The observed expansion of shrubs favors the range expansion from adjacent biomes of some herbivore species with wide dietary flexibility (Klein 1999). Moose (*Alces alces*), for example, have expanded from the boreal forest to Arctic tundra in Alaska and eastern Canada but have declined in Siberia owing to increased hunting.

Deliberate introductions are modifying animal distributions. Musk oxen (*Ovibos moschatus*) have been introduced to new areas and to portions of their former ranges, after dramatic reductions in their numbers in the nineteenth century from many of these areas; they are expanding in population sizes and ranges in Alaska, Russia, west Greenland, Quebec, and Norway (CAFF 2001). Wood bison (*Bison bison athabascae*), the closest relative of the extinct Pleistocene bison, have also been reintroduced in several areas of Arctic Canada. Bison, like many animals, substantially disturb their habitat, so their reintroduction could lead to ecosystem changes that alter many species abundances and ecosystem processes (Zimov et al. 1995).

Fennoscandian lemmings and voles are changing in seasonal dynamics and losing the synchronous cyclic population fluctuations that characterize these species (Hanski et al. 2001), as occurred earlier in Alaska (Batzli et al. 1980). In Scandinavia, but not in Greenland, declines in lemming populations cause increased predation on water birds as an alternative prey, although changes in weather may contribute to these population changes (Summers and Underhill 1987; Soloviev et al. 1998; Yoccoz and

Table 25.3. Population Numbers and Trends of Some Key Arctic Animals. Population trends indicate moderate to high certainty that populations exhibit a directional trend. Where the trends differ regionally, the nature of these trends (increasing, decreasing, stable) is indicated. (Modified from CAFF 2001 with marine mammals from Marine Mammal Commission 2002)

Species	Total Arctic Population	Population Trend	Major Driver of Change
Birds			
Eiders (4 spp)	>3 million	declining/stable/increasing	unknown
Common murre/guillemot (*Uria aalge*)	8–11 million	declining/stable/increasing	unknown
Thick-billed murre (*U. lomvia*)	14 million	unknown	unknown
Gulls		increasing	garbage ??
Geese (15 spp)	11 million	40% of populations increasing 28% stable 18% declining 14% trend unknown	winter habitat, overharvesting
Shorebirds			
North America (21 spp)	14 million	decreasing	unknown
Eurasia (16 spp)	8 million	no clear trend; some declines	unknown
Terrestrial mammals			
Caribou/reindeer (*Rangifer tarandus*)	4.5 million	many herds increased until 1980s; Peary caribou declining	forage, weather
Muskoxen (*Ovibos moschatus*)	150,000	increasing/stable	reintroductions, weather
Brown bear (*Ursus arctos*)	170,000	declining/stable/increasing	habitat loss, poaching
Wolverine (*Gulo gulo*)	30,000	unknown/stable/increasing	unknown
Wolf (*Canis lupus*)	50,000	declining/stable/increasing	hunting, habitat loss
Marine mammals			
Right whale (*Eubalaena glacialis*)	<1,000	stable/uncertain	19th century harvest
Humpback whale (*Megaptera novaeanglie*)	thousands	increasing	19th century harvest
Gray whale (*Eschrichtius robustus*)	27,000	stable; near optimal	19th/20th century harvest
Killer whale (*Orcinus orca*)	few thousand	stable, not endangered	not widely harvested
Steller sea lion (*Eumatopius jubatus*)	30–40,000	declined 85% since 1970s	uncertain
Pacific walrus (*Odobaenus rosmarus divergens*)	230,000	trend uncertain	uncertain
Harbor seal (*Phoca vitulina richardsi*)	100,000	some populations declining	human and natural factors
Polar bear (*Ursus maritimus*)	25,000	unknown	sea ice decline
Sea otter (*Enhydra lutiris*)	100,000	decline since 1995	unknown

Ims 1999). Trends in mammalian predators (such as wolves (*Canis lupus*), bears (*Ursus* spp.), and wolverines (*Gulo gulo*)) have been regionally variable. In general, species interactions are complex and regionally variable (Blomqvist et al. 2002; Gilg 2002), making broad generalizations uncertain.

25.2.3.4 Changes in Marine Mammals

Hunting for marine mammals, especially whales, seals, and sea otters (*Enhydra lutris*), in the nineteenth century radically reduced many populations (Marine Mammal Commission 2002). The Steller sea cow (*Hydrodamalis gigas*) was hunted to extinction. The Southern right whale (*Eubalaena glacialis australis*) and Antarctic fur seal (*Arctocephalus gazella*) were almost extinct by 1830. Whaling proliferated in the nineteenth and first half of the twentieth centuries. Overexploitation led to the formation of the International Whaling Commission in 1949. There is now a moratorium on commercial whaling, although Iceland and Japan continue to harvest whales for research, and Norway, which objected to the moratorium on commercial whaling, continues commercial whaling, leading to an annual harvest of 400 ± 40 minke whales (*Balaenoptera acutorostrata*). Indigenous whaling in Greenland and North America has had no detectable effect on whale populations (Caulfield 1997).

Antarctic fur seal populations are growing as a result of decreased whale predation and are expanding their range to the south. In contrast, many whale populations have not fully recovered. The causes of recent declines in populations of southern elephant seal, Steller sea lion (*Eumatopius jubatus*), harbor seal (*Phoca vitulina richardsi*), sea otter (*E. lutris*), and other marine mammals are *speculative*, but in some cases appear to involve commercial fishing (of the southern elephant seal and sea otters, for example), reduced sea ice (for polar bears (*U. maritimus*) and walrus (*Odobaenus rosmarus divergens*)), or other changes in marine ecosystems (Estes et al. 1998).

Northern oil and gas development may also influence marine mammals. Noise from offshore oil exploration in the Beaufort Sea disturbs bowhead whales (*Balaena mysticetus*) and could deflect them from migration routes, making them less accessible to hunters. Autumn-migrating bowheads, for example, stay 20 kilometers from seismic vessels (NRC 2003). Oil spills from marine transportation or offshore oil platforms have the potential for widespread ecological damage, particularly in ice-covered Arctic waters. Spills from pipelines in temperate-zone oil basins in the headwaters of Arctic rivers such as the Ob, Pechora, and McKenzie could also contaminate Arctic waters.

25.2.3.5 Changes in Birds

There is extensive evidence of environmentally related changes in Antarctic seabird populations, although long-term exploitation of living resources may also be a contributing factor (Fraser et al. 1992). Among the most dramatic Antarctic changes have been local declines in Adélie penguins (*Pygoscelis adeliae*) combined with a southward reduction in range (Fraser et al. 1992). Chicks that fledge at a body mass of less than 2,850 grams have low survivorship. Most available penguin rookery habitat occurs in landscapes where snow deposition is enhanced during late winter and early spring storms. This causes chicks to hatch later and delays their key growth period until early February, when local krill (*Euphasia superba*) is less abundant and adults must forage greater distances for food. Adélie penguins have become locally extinct on some islands along the Antarctic Peninsula and are declining rapidly on other islands (Ainley et al. 2003). As the Adélie penguins decline,

they are being replaced by a southward shift in distributions of chinstrap and gentoo penguins (*P. antarctica* and *P. papua*).

Adélie penguin populations in the Ross Sea sector appear to oscillate with a five-year lag to sea-ice extent, which in turn is related to ENSO conditions (Wilson et al. 2001). In the longer term, the southern limit to distribution of Adélie penguin breeding colonies is determined by changes in the pattern of year-round sea-ice presence (Baroni and Orombelli 1994; Emslie et al. 2003) Extensive sea ice during winter months appears to reduce sub-adult survival. Similarly, emperor penguins (*Aptenodytes forsteri*) decline with warmer temperatures and reduced sea-ice extent (Barbraud and Weimerskirch 2001), although they hatched fewer eggs when winter sea ice was extended (Barbraud and Weimerskirch 2001). These data indicate that penguins may be quite susceptible to climatic changes. In addition, there have been serious population declines in 16 of 24 species of albatrosses, primarily as a result of incidental catches in longline fishing (Crawford and Cooper 2003).

The Arctic is a breeding ground for many migratory wetland birds that overwinter throughout the world. Several hundred million birds, including swans, geese, and ducks, migrate from southern overwintering grounds to the Arctic to breed. These birds are indirectly affected by climate changes, habitat loss, and altered food abundance throughout their range (Lindström and Agrell 1999; Zöckler and Lysenko 2000). Critical coastal stopover sites are also threatened by human activities. The increases in Arctic plant biomass, height, and density caused by recent warming and eutrophication affect wetland birds negatively (Callaghan et al. in press). Twelve Arctic bird species are globally threatened (CAFF 2001). Climate change also fosters northern migration of more southerly bird species, including the common snipe (*Gallinago gallinago*), black-tailed godwit (*Limosa limosa*), and northern lapwing (*Vanellus vanellus*) in Russia (Lebedeva 1998; Morozov 1998), as well as American robins (*Turdus migratorius*) in the polar barrens of North America (Jolly et al. 2002). Ravens (*Corvus corax*) and some gulls have become more abundant near human settlements, acting as nest predators on other birds.

The greater snow goose (*Chen caerulescens atlantica*) increased twenty-five-fold, from 28,000 birds in 1965 to 700,000 in 1998 (CAFF 2001). The overall populations of "white" geese (greater and lesser snow goose and Ross's goose) increased from about 1 million to about 8.5 million over the same time period (Batt 1997). Changes in agricultural practices, which increased food availability, are probably the most important cause of these changes. The population increase has exceeded the threshold for persistence of salt-marsh vegetation, leading to catastrophic vegetation change and salinization of many coastal wetlands (Hik et al. 1992; Srivastava and Jefferies 1995).

25.2.3.6 Changes in Fish

The most dramatic changes in fish populations have occurred in Antarctic waters, where overfishing rapidly depleted stocks of marbled notothenia (*Notothenia rossii*), mackerel icefish (*Champsocephalus gunnari*), gray notothenia (*Lepidonotothen squamifrons*), and Patagonian toothfish (*Dissostichus eleginoides*). Some stocks recovered when conservation measures were instituted by the Convention on the Conservation of Antarctic Marine Living Resources, but others remain depressed by illegal, unregulated, and unreported fishing (CCAMLR 2002). (See Chapter 18.)

Overharvesting of whales and seals may have led to population increases of krill (*E. superba*), their major food source. Subsequent reductions in krill reflect some combination of commercial fishing, recent increases in fur seal and penguin populations, changes in sea-ice duration and extent, and, perhaps, increases in UV-B associated with ozone depletion (Naganobu et al. 2000). Krill are long-lived animals whose variation in growth and reproduction are sensitive to oceanic and sea-ice conditions; they comprise the main food source for fish in Antarctic waters (Murphy et al. 1998; Quetin and Ross 2001; Fraser and Hofmann 2003). Under conditions of increased sea-ice melt, cryptomonads expand at the expense of the diatoms, which are preferred by krill, potentially resulting in krill decline (Moline et al. 2000). This tight but complex linkage of krill population dynamics to sea ice suggests that any future changes in timing, duration, or extent of sea ice will strongly affect the community composition of phytoplankton, krill, and their predators. Because advection of biological material is important in maintaining Southern Ocean ecosystems, the potential impacts of regional changes may extend well to the north of the main sea-ice-covered regions.

In the Arctic, regional warming may have contributed to recent northward range extensions of anadromous fish such as salmon (*Oncorhynchus* spp and *Salmo* spp.) (Babaluk et al. 2000; Jolly et al. 2002) and to increased abundance of salmonid parasites in Alaskan rivers. Human activities have also altered fish distributions. For example, the salmonid parasite *Gyrodactylus salari* that is native in Central Asia spread naturally to the Baltic and then further to Scandinavia with the help of humans (Johnsen and Jensen 1991; CAFF 2001). This parasite feeds on young salmonids and causes major damages in Norwegian fish farms and rivers. Another example of human impacts on fish production involves introduction of the shrimp *Mysis relicta* to high-elevation Norwegian and Swedish lakes and rivers that are regulated for hydroelectric purposes. Initially the shrimp introduction had positive effects but over time feeding by *Mysis* on zooplankton reduced this food resource, leading to a decline in fish growth (Nesler and Bergersen 1991).

25.2.3.7 Changes in Insects

The introduction of alien insects on sub-Antarctic islands is threatening some native species and vegetation communities (Ernsting et al. 1995; Hanel and Chown 1998), and introduced flora is affecting soil faunal composition (Gremmen et al. 1998). The northern limit of Arctic insect species is usually determined by climatic factors (Strathdee and Bale 1998). For example, food plants probably determine distributions in less than 3% of the macrolepidopteran (butterfly) species of Finland (Virtanen and Neuvonen 1999). Species richness decreases by 65 species for each degree of latitude northward—that is, 93 species (12% of the total) per degree of mean summer temperature.

Lepidopteran species are usually good dispersers, so climate warming will likely promote increases in their richness as species move poleward. Conversely, the distribution of northern species (11% of the Finnish species) such as sawflies may shrink in a warmer climate (Kouki et al. 1994). Among bark beetles, *Ips amitinus* and *Xylechinus pilosus* have expanded their ranges in Fennoscandia (Heliövaara and Peltonen 1999), whereas other species (such as *Tomicus minor*) are retracting southwards.

A rise in winter temperature would favor species overwintering as eggs and may increase the frequency of insect outbreaks of those species (such as *Epirrita autumnata,* a geometrid defoliator of birch) (Nilsson and Tenow 1990; Neuvonen et al. 1999; Niemelä et al. 2001). Species overwintering as pupae will likely increase the number of generations per year (Virtanen and Neuvonen 1999). Summer warming will also increase the number of generations completed; this increase has been found in experimental field manipulations to increase the overwintering population of

aphids on Svalbard by an order of magnitude (Strathdee et al. 1993).

In boreal Canada, pest-caused timber losses may be as much as 1.3–2.0 times the mean annual depletions due to fires (Volney and Fleming 2000). Global change will likely increase the frequency and intensity of outbreaks, particularly at the margins of host ranges. Changes in mosquito abundance in response to altered hydrology could strongly affect fish and waterfowl, for which they are an important food source, as well as caribou and reindeer, whose energy budgets are sensitive to insect harassment (Chernov 1985).

25.2.4 Food, Fuel, and Fiber

Documented changes in biodiversity have had negligible effects on the food supply of people in Antarctica, who bring their food, fuel, and fiber from lower latitudes. Marine harvests in Antarctic waters, however, provide food that is used globally. Fish and krill are now primary targets of human exploitation in the Southern Ocean. From 1970, when recordkeeping began, to 1998 a total of 8.7 million tons of krill and fish were harvested (CCAMLR 2000).

In contrast, indigenous peoples throughout the Arctic (and many nonindigenous residents as well) maintain strong social, cultural, and economic connections to the environment through traditional hunting, herding, fishing, trapping, and gathering of renewable resources. Local mixed economies of cash and subsistence depend strongly on household production involving harvest of local resources, food preparation and storage, distribution, consumption, and intergenerational transmission of knowledge and skills (Nuttall 1992; Caulfield 2000; Dahl 2000; Freese 2000; Nuttall et al. in press). Per capita consumption by rural Alaskans (indigenous and nonindigenous), for example, is 170 kilograms per year of wild foods (60% fish, 16% land mammals, 14% marine mammals, 10% plant products), valued at about $200 million. Urban Alaskans, in contrast, consume 22 kilograms of wild foods per capita per year. Cultivated crops are a smaller source of food except in maritime regions (such as Iceland and coastal Norway). Wood, sod, peat, and coal are used locally as fuels.

The subsistence resources used by Arctic peoples vary regionally. In the barrens, where terrestrial productivity is low, most communities are coastal and people depend primarily on fish and marine mammals (such as polar bears, seals, walrus, narwhals (*Monodon monoceros*), and beluga (*Delphinapterus leucas*), fin, and minke whales), although terrestrial mammals (such as caribou, reindeer, and musk ox), migratory birds and their eggs (such as ducks, geese, terns, and gulls), and plants and berries are seasonally important. In tundra, in contrast, people rely more heavily on fish—including salmon, Arctic char (*Salvelinus alpinus*), whitefish, and northern pike (*Esox lucius*)—migratory and sedentary birds, terrestrial mammals, and berries.

Caribou and reindeer, which include wild and domestic populations of North American barren-ground wild caribou, are arguably the most important terrestrial subsistence resource for Arctic indigenous peoples (Klein 1989; Paine 1994; Kofinas et al. 2000; Jernsletten and Klokov 2002). Many Arctic and sub-Arctic indigenous cultures co-evolved with reindeer or caribou, which provide food, shelter, clothing, tools, transportation, and other marketable goods. In North America, where indigenous subsistence hunting constitutes the primary use of caribou, there are approximately 3.2 million barren ground caribou and an estimated annual harvest of over 160,000 animals, equivalent to more than $30 million annually. In Russia, large-scale commercial hunting of wild reindeer, which began on the Taimyr in the

1970s, produced more meat than all reindeer husbandry of both Central Siberia and Yakutia but has not fulfilled the cultural role that reindeer play in reindeer-herding societies. The change from a migratory existence to permanent communities with schools, stores, and jobs alters traditional lifestyles.

25.2.5 Cultural Benefits

The numerous cultural groups in the Arctic (more than 50) reflect a diversity of historical roots and local ecological conditions provided by ecosystems. Subsistence activities, such as hunting, herding, fishing, trapping, and gathering, remain important for maintaining social relationships and cultural identity in these indigenous societies (Brody 1983; Nuttall 1992). These activities link people inextricably to their histories and their contemporary cultural settings and provide a context for thinking about sustainable livelihoods in the future (Nuttall et al. in press).

The Antarctic and Arctic provide important cultural benefits to non-Arctic residents as well. Some of these benefits are mediated by polar species that migrate to lower latitudes. In addition, non-polar residents value the near-pristine conditions of polar regions, motivating them to visit these lands (tourism) and to support legislation and lobbying efforts for their protection.

25.3 Drivers of Change in Polar Systems

The relative importance of drivers of change varies across the polar regions and depends on the stakeholders involved. For polar residents, the most important changes (in order of decreasing importance) are often climate change, industrial development, contaminants, marine fishing, and UV-B. For non-polar residents, the most important changes—again in order of importance—are often climate change, marine fishing, increased UV-B, industrial development, and introduction of exotic species. The shared concern by both polar and non-polar residents about many of the same drivers of change provides opportunities to develop agendas that enhance the well-being of residents both within and outside polar regions.

25.3.1 Climate and Hydrologic Change

Climate has warmed more dramatically in portions of the Arctic and Antarctic than in any other region on Earth, with substantial impacts on ecosystems, their services, and human well-being. The magnitude and global pattern of Arctic warming are known with *high certainty*. These warming trends are most pronounced in western North American Arctic and central Siberia (Kozhevnikov 2000; Serreze et al. 2000; Smith 2002; Convey et al. 2003). Warming has been negligible in parts of Scandinavia. (See Figure 25.8 in Appendix A.) Arctic warming results both from a general northern hemisphere warming and from a regime shift in hemispheric circulation (such as more-frequent positive phases of the North Atlantic and Arctic Oscillations) (Overland et al. 2004). Temperate air masses penetrate more frequently into the Arctic, causing increased climate variability (Overland et al. 2004) and conditions that are unfamiliar to local residents (Krupnik and Jolly 2002). Regions that previously exhibited a cooling trend (such as eastern North American Arctic and Chukotka from 1950 until 1990) are now warming.

Increases in precipitation during this time period are approximately balanced by increased evapotranspiration, suggesting only minor changes in terrestrial water balance (Serreze et al. 2000). Nonetheless, terrestrial studies suggest that changes in water balance are occurring but are regionally variable. These include increasing river runoff in Russia (Peterson et al. 2002) (as

mentioned earlier), bog expansion in western Russia (Crawford et al. 2003), and drier soils in North America (Hinzman et al. in press) (perhaps explaining recent increases in area burned (Murphy et al. 2000)). The spatial pattern and magnitude of these changes in soil moisture are known with only a *low certainty*.

The Antarctic shows complex temporal and spatial patterns of both warming and cooling. Over the past 15–20 years, 60% of continental Antarctica has been thermally stable or cooling slightly (Doran et al. 2002b; Kwok and Comiso 2002; Thompson and Solomon 2002), although this trend is debated (Turner et al. 2002). In contrast, the McMurdo Dry Valleys show general twentieth-century warming but cooling since 1985 (Bomblies et al. 2001); the sub- and maritime Antarctic islands show consistent warming (Bergstrom and Chown 1999; Quayle et al. 2002; Convey submitted); and the Antarctic Peninsula has warmed as rapidly as any place on Earth (King et al. 2003; Smith et al. 2003; Vaughan et al. 2003). Increases and decreases in precipitation have also both been reported (Turner et al. 1997; Smith 2002; Quayle et al. 2003).

Temperature and precipitation changes show teleconnections with El Niño/Southern Oscillation events in the southern Pacific Ocean (Cullather et al. 1996; Harangozo 2000). Reported Antarctic cooling may result from increased strength of the Southern Hemispheric Annular Mode, which would cause the strong westerly winds around Antarctica to spend more time in the strong-wind phase (Thompson and Solomon 2002). Such an effect could also contribute to the warming seen along the Antarctic Peninsula, as fewer cold-air outbreaks would be seen with increased advection of warm moist air from the Southern Ocean.

25.3.2 Development of Extractive Industries

Extractive industries in the Antarctic are prohibited by the Protocol on Environmental Protection (the Madrid Protocol) to the Antarctic Treaty of 1959, which set aside Antarctica for peaceful purposes and international collaboration in science. In contrast, extractive industries have been a significant driving force for ecological and socioeconomic change in the Arctic for over a century. Gold mining has contaminated streams with mercury used to amalgamate gold dust and with increased sediment loads that damage downstream aquatic ecosystems. Industrial coal and base-metal mines have caused local surface contamination. However, the greatest local effects on ecosystem services derive from smelting non-ferrous metals. Emissions from smelters on the Kola Peninsula and in Norilsk, Russia have produced local concentrations of atmospheric heavy metals among the highest in the world, resulting in areas that are entirely devoid of vegetation from the combined effects of sulfur fumigation and acid and heavy-metal deposition (Doiban et al. 1992).

As petroleum and military development spread in the latter half of the twentieth century, transportation infrastructure (roads, pipelines, airstrips, ports) contributed significantly to surface disturbance and habitat fragmentation. Between 1900 and 1950, less than 5% of the Arctic was affected by infrastructure development (Nellemann et al. 2001; Ogden in press). By 2050, some 50–80% of the Arctic is projected to be disturbed, although this level of disturbance may occur by 2020 in Fennoscandia and some areas of Russia.

Changes in world energy markets and technology have led to a rapid expansion of oil and gas development in several regions of the Arctic during the past 30 years. Most activity to date involves oil onshore along the North Slope of Alaska and in western Siberia, and offshore in the Barents and Beaufort Seas. However, the Alaskan North Slope, the McKenzie Delta of Canada, the Yamal

Peninsula of Russia, and their adjacent offshore areas hold enormous natural gas deposits that are projected to be developed during the next decade (Forbes 2004a). These developments will likely continue expansion as reductions in sea ice open new sea and river routes and reduce development and transportation costs. In addition to direct effects on vegetation and hydrology, oil and gas developments have many cumulative effects on subsistence resources and on the economies and well-being of local peoples, including increased wages to local residents, the fragmentation of habitat, and increased access by nonresidents (Walker et al. 1987; NRC 2003).Global changes in politics, corporate structure, and resource demand strongly influence the patterns and rates of resource extraction at high latitudes (Whiteman et al. 2004).

25.3.3 Contaminants

Many environmental contaminants that are produced and released to the environment at low latitudes tend to accumulate in polar regions. Persistent organic pollutants, for example, are stable, fat-soluble, carbon-based compounds that volatilize at warm temperatures and are transported poleward by wind, water, and wildlife. Old and current research stations in the Antarctic and Distant Early Warning stations in the Arctic often constitute additional local sources of contaminants (MacDonald et al. 2002). Atmospheric transport is the most rapid pathway by which persistent organic pollutants, especially volatile or semi-volatile compounds, reach the poles. Once in polar regions, POPs are deposited on particles or exchanged with water, both processes that are enhanced by low temperature. Oceanic transport occurs more slowly but is an equally or more important pathway for compounds such as hexachlorocyclohexane or toxaphene that partition strongly into water (MacDonald et al. 2002).

Fish and migratory waterfowl, which winter in more-polluted regions of the world and come to polar regions to reproduce, constitute a third pathway for polar transport. Anadromous fish also transport POPs from the ocean to high-latitude lakes and streams (Ewald et al. 1998). Animals are particularly important vectors for highly fat-soluble compounds. Marine mammals, seabirds, top carnivores, and predatory fish accumulate the largest amounts of fat-soluble contaminants because of their high trophic position in complex marine food webs (AMAP 2003). These general contaminant patterns are known with *high certainty,* but the regional variation in contaminants is not well documented. Persistant organic pollutant concentrations in Antarctic pelagic food webs (Corsolini et al. 2003) and in the air (Kallenborn et al. 1998) are much lower than those found in the Arctic, but some forms may be increasing owing to greater usage of POPs in the Southern Hemisphere (Weber and Goerke 2003).

Limited evidence suggests a current decline in polar concentrations of POPs, such as dichlorodiphenyl-trichloroethane, or DDT, the use of which has declined globally. POPs that are increasing in their global use continue to accumulate in polar regions (AMAP 2003; Chiuchiolo et al. 2004), and brominated diphenyl ethers—flame retardants whose use is not banned—occur in high concentrations in Antarctic sea ice and juvenile krill (Chiuchiolo et al. 2004). Antarctic sea ice serves as a collector and focusing mechanism that injects accumulated POPs into the plankton system at its period of maximum biological activity. The effects of climate warming on persistant organic pollutant transport to polar regions have *low certainty.* Sources of uncertainty include the dynamics of adsorption to snow and the extent to which POPs currently trapped in sea and glacial ice will be released with warming.

Heavy metals, like POPs, are persistent compounds that can be globally transported, especially by wind. They differ from POPs, however, in that they occur naturally within and outside polar regions and exhibit areas of naturally high and low concentrations. Heavy metals bind to proteins, accumulate in organs (liver, kidney, brain), and are slowly excreted in hair, feathers, nails, and claws. With the exception of mercury, heavy metals tend not to biomagnify (concentrate as they move through food webs). Global sources of mercury pose the greatest threat in polar regions because the global combustion of coal, which is its major source, is expected to continue rising throughout the next century. There are trends of increasing mercury in some Arctic species (AMAP 2003).

In East Greenland, 100% of the human population has concentrations of mercury that are unacceptable, and health advisories have recommended reduced consumption of some locally harvested resources. Heavy metal pollution inputs have declined in those portions of the Russian North where cessation of subsidies caused many extractive operations to close. Even in these areas, however, pollutants released previously remain in high concentrations in ecosystems. In the Antarctic region, burning of fossil fuels (involving NO_x emissions) at research stations might affect local systems at decadal time scales (Lyons et al. 2000).

Radionuclides such as cesium-137 and strontium-90 are stable enough to be transported globally in the atmosphere and oceans. Radionuclide concentrations increased during atmospheric testing of nuclear weapons in the 1950s and, like other forms of air pollution, drifted to polar regions. Background levels of atmospheric fallout have declined markedly since the end of atmospheric testing in 1963, but a more recent release occurred during the Chernobyl accident in 1986. Anthropogenic radionuclides are locally abundant in sediments near sites of weapons testing, storage, and nuclear-powered electricity generation facilities. Lichens, which derive most of their mineral nutrition from the air, are particularly effective in accumulating radionuclides. Caribou and reindeer, which eat lichens in winter, are an avenue by which any future nuclear contamination might affect human health (Section 25.5.4).

25.3.4 Marine Fishing

Both climate variability and commercial fishing have caused significant variations in marine mammals and fish available for commercial and subsistence harvest (Finney et al. 2002; AHDR 2004). (See also Chapter 18.) For example, over half of the Northeast Atlantic regional stocks of cod (*Gadus morhua*), haddock (*Melanogrammus aeglefinus*), whiting (*Merluccius bilinearis*), and saithe (*Pollachius virens*) are depleted below safe biological limits and are therefore threatened with collapse. In Greenland, the collapse of the cod fishery occurred when overfishing coincided with climatic deterioration (Hamilton et al. 2000). In the North Pacific, the decline in bottom fish has contributed to drops in populations of Steller sea lions, causing killer whales to shift to sea otters for food and reducing the availability of marine mammals for human harvest (Estes et al. 1998). Commercial fishing has reduced fish runs in major rivers such as the Yukon in North America, reducing their availability as a subsistence resource. As sea ice continues to decline, commercial fishing may expand northward, reducing stocks that have previously had limited human harvest. These changes have, in many cases, had dramatic socioeconomic effects, as small Arctic communities adapt to the combined effects of climate change, changing fish stocks, and emergent markets (Hamilton et al. 2000).

In Antarctica, exploitation in the nineteenth and early twentieth centuries reduced seal and whale populations almost to extinction, but the consequences have *low certainty* (May et al. 1979; Murphy et al. 1995). Stocks of krill, the major food source of whales, may have increased, but this assumes that their abundance was under top-down control by predator demand. Fishing for krill has occurred over the past 20 years, but the fishery has operated at very low levels relative to the estimated stock sizes, and there is no evidence that fishing has significantly affected local abundance and availability to predators.

Fishing in Antarctic waters, particularly by Russia, expanded rapidly in the 1960s, leading to depletion of several fin-fish stocks by the 1990s. (See Chapter 18.) The Convention on the Conservation of Antarctic Marine Living Resources limits catch sizes, but continued illegal, unregulated, and unreported high-seas fishing threatens major fish stocks and makes effective monitoring and management difficult.

25.3.5 Increased UV-B

Anthropogenic destruction of Earth's protective stratospheric ozone layer gives rise to an "ozone hole" that allows UV-B radiation to penetrate to the surface (Farman et al. 1985). The boundaries of the ozone hole that forms over Antarctica each austral spring are dynamic and can extend to southern South America, New Zealand, and southern Australia. Although the Montreal Protocol curbs emissions of the causal chlorofluorocarbons, there is still no reduction in intensity of the annual ozone hole, although recovery is predicted within about a century (Shindell et al. 1998).

The most intensively studied biological impacts of UV-B are the effects on human health (described later in the chapter) and the reduction of phytoplankton production and plankton biodiversity (Smith et al. 1992; Quartino et al. 2001; Vernet and Kozlowski 2001). On land, UV-B reduces plant height and shoot mass (Day et al. 2001; Robinson et al. 2003) and induces the formation of protective pigments (Newsham et al. 2002; Newsham 2003) that may also alter palatability to herbivores (Hessen 2002). The impacts of UV-B in the Arctic are less dramatic than in the Antarctic and potentially include decreased marine primary production, increased mortality of fish larvae of some species, increased mortality of freshwater invertebrates, reduced growth of some plant species, and changes in microbial biodiversity and biogeochemical cycling (Vernet and Kozlowski 2001; Callaghan et al. in press). The net impact of increased polar UV-B on the ecological determinants of human well-being has *low certainty*, but its effects on Antarctic marine plankton assemblages and production have *high certainty* (Weiler and Penhale 1994; Meador et al. 2002).

25.3.6 Introduction of Exotic Species

Despite Annex II of the Protocol on Environmental Protection to the Antarctic Treaty, which prohibits introduction of exotic species to Antarctica, human introductions of exotic species (including rats, slugs, reindeer, cats, temperate grasses, and other plants) are problematic on most sub-Antarctic islands, although they have not yet significantly affected the Antarctic continent (Chown et al. 1998; Vidal et al. 2003; Frenot et al. in press). The combination of increased human activity in the region and the lowering of barriers to the transfer and establishment of biota through climate change leads to predictions of further increases in rates of exotic introduction and consequential impacts on native biota and biodiversity (Frenot et al. in press).

Weedy plant species from temperate and Mediterranean ecosystems have a long history of establishment in northern boreal and Arctic regions (Forbes 1995). Their migration north of the Arctic tree line is most common along road and railway corridors with connections to the south, but numerous incidental introductions have also occurred in places such as Baffin Island, Greenland, Svalbard, and remote portions of the Russian Arctic. Many of the introductions establish first on disturbed ground. Many persist for decades or even centuries without additional anthropogenic or zoogenic influence, although an ongoing disturbance regime favors the maintenance and spread of new populations.

Introductions of exotic mammals have had substantial impacts on many polar ecosystems. Until the advent of human influence, sub-Antarctic islands and many islands in the Bering Sea had no indigenous terrestrial mammals, so seabirds nested safely on the ground or in burrows (CAFF 2001). Arctic foxes (*A. lagopus*) were deliberately introduced to many Bering Sea islands as a fur resource, and cats, sheep, rabbits, reindeer, and cattle were introduced to some sub-Antarctic islands. At the same time, rodents (rats and mice) were inadvertently introduced to many islands.

As seen elsewhere when predators or competitors were introduced to previously isolated communities, these animals generally reduced, and in some cases drove to extinction, populations of marine birds, waterfowl, and other ground-nesting birds, through either habitat alteration or direct predation (Frenot et al. in press). Exotic plants have frequently been introduced as a by-product of animal introductions, typically accidentally in forage material or as resistant stages transported within the animal's digestive system. Introduced mammals have been successfully eliminated from some islands, resulting in recovery of bird nesting success. On Svalbard in Scandinavia, the introduction of the rodent *Microtus rossiaemeridionalis* led to the introduction of a tapeworm and associated disease problems (Henttonen et al. 2001). Muskrat (*Ondatra zibethica*) was introduced in 1905 into central Europe from North America and later to Finland and several places over the Soviet Union because of its valuable pelt (Hoffmann 1958). It has spread widely beyond the areas of introduction. Sable (*Martes zibellina*) has also been introduced to the Siberian Arctic.

One of the largest crab species in the world, the king crab (*Paralithodes camtschaticus*), was transferred by the Russians from the Bering Sea to the Barents Sea beginning in about 1960. Since it was first observed in Norwegian waters in about 1980, the species has spread eastward and southward along the coast (Sundet 1998). Although this species is a valuable catch, it causes problems to fishing gear and has reduced sea urchin populations in the Barents Sea. Marine introductions have to date been of very limited impact in Antarctic seas, although this perception may derive in part from a lack of monitoring.

25.4 Trade-offs, Synergies, and Management Interventions

Given the complex web of changes in ecosystem services in polar regions and their sensitivity to changes in global and regional drivers, changes in some ecosystem services inevitably affect others, either positively or negatively. (See Figure 25.9.) Identification of these interactions, whether they are trade-offs or synergies, facilitates the design of policies that enhance a broad array of services and reduce the likelihood that policies focused on a specific service will inadvertently damage others.

25.4.1 Synergies between Climate Regulation, Subsistence, and Cultural Resources

The most pervasive synergy among ecosystem services in polar regions links climate regulation, subsistence use, and cultural re-

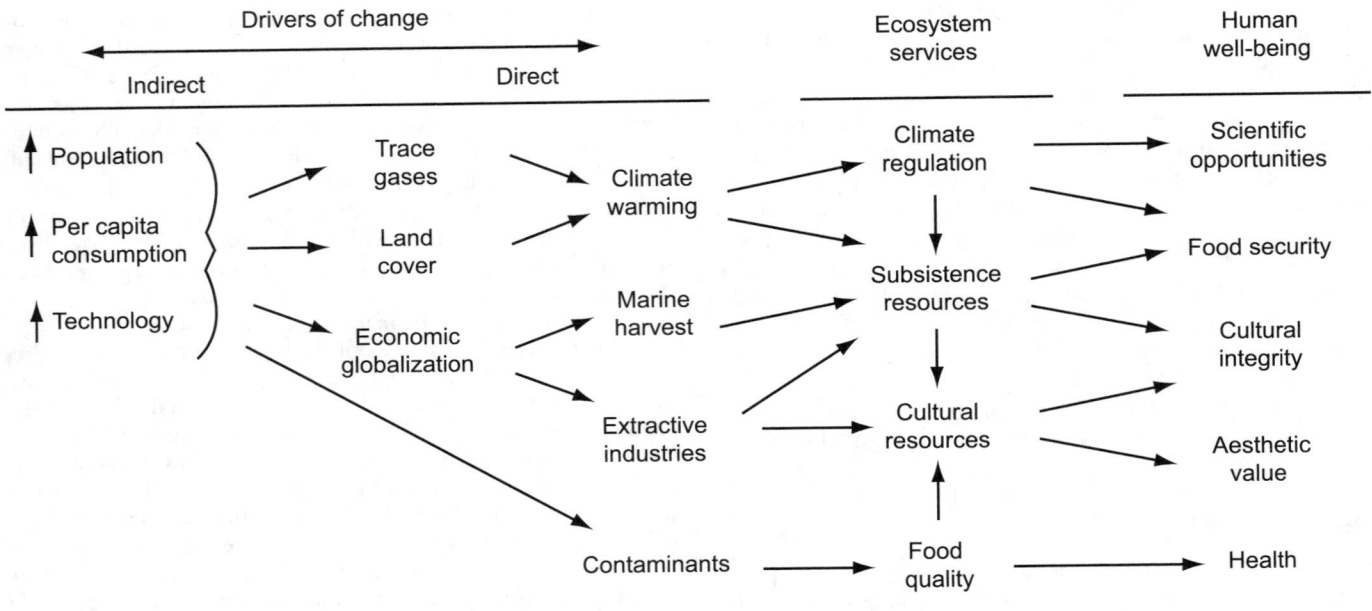

Figure 25.9. Links between Drivers of Change, Ecosystem Services, and Human Well-being in Polar Regions

sources. (See also Chapter 5.) Until recently, the magnitude of land use change in polar regions has been modest, allowing Arctic residents to maintain a substantial dependence on subsistence resources for food and clothing and for their cultural ties to the land. These same attributes make polar regions important to non-Arctic residents for existence and aesthetic values.

Fortunately, management strategies that foster the sustainability of these cultural resources have a *high certainty* of maximizing the retention of biodiversity, particularly in wetlands, and of maintaining the capacity of polar regions to serve as the planet's cooling system by maintaining current large stocks of soil carbon and preventing thaw of permafrost that would enhance methane emissions (as described earlier). Conversely, industrial development or pollution that seriously alters these ecosystems detracts from the aesthetic values that are important to both polar and non-polar residents, reduces ecosystem capacity to sustain biodiversity and subsistence resources, and increases methane and carbon emissions, a positive feedback to climate warming.

Policies that recognize these synergies sustain all these services simultaneously. For example, policies that allow local people to pursue traditional subsistence activities within protected areas contribute to biodiversity and cultural integrity while meeting the aesthetic needs of both Arctic and non-Arctic residents (Watson et al. 2003). Alternatively, policies that exclude subsistence use from protected areas create trade-offs between the subsistence needs of local residents and the aesthetic and recreational interests of non-Arctic residents. Current national regulations permit subsistence use in protected areas of the North American Arctic. In Russia, indigenous peoples are often excluded from protected areas, even though laws may permit subsistence uses (Fondahl and Poelzer 2003). Subsistence use in protected areas is generally prohibited in Fennoscandia.

25.4.2 Synergies and Trade-offs between Subsistence and Cash Economies

Residents of all Arctic nations participate in a mixed economy in which individuals and communities depend on both subsistence harvest from the land and sea and imports of food, fuel, and other

products from outside the Arctic. Among many Arctic indigenous peoples, participation in a mixed cash-subsistence economy has been a way of life for more than a hundred years, with fur trading being the original base of the cash economy. This economy has both positive and negative impacts on subsistence activities; in recent years, participation in the cash economy has increased dramatically (AHDR 2004).

On the one hand, this reduces the necessity and time available to acquire food from the land, thereby reducing the transmission of cultural traditions to younger generations. Substance abuse, television, fast foods, and other attributes of a cash economy have also detracted from the central role of subsistence in the lives of rural Arctic people. On the other hand, cash income provides access to new harvesting technologies (Chance 1987) and other material benefits. Motorboats and vehicles (such as snow machines, off-road vehicles, and in some places aircraft) give people continued access to lands for hunting, herding, fishing, and gathering.

In all Arctic nations, government policies consolidated semi-nomadic indigenous peoples into permanent settlements by the mid-twentieth century in order to provide schools and other services. If residents of these communities had not had a cash income to help them purchase motorized transport, they would have been less able to reach the large areas typically required for subsistence. However, reliance on these technologies make industrial commodities a necessity, not just a supplement to the traditional economy (Kirkvliet and Nebesky 1997). These technologies also alter the balance between hunters and the hunted, raising new challenges for conservation and game management.

The intertwining of subsistence and market economies creates winners and losers in what was, until recently, a fairly egalitarian society and introduces new challenges of balancing time and money in household production. Gender roles change in areas where women obtain wage-earning jobs more frequently than men. Continued viability of a community may require increased specialization. Some individuals who work full-time for pay, for example, have insufficient time to participate in traditional activities, whereas others who are unable to obtain jobs serve as full-time hunters to meet the overall subsistence needs of their communities (Stabler 1990; Kruse 1991; Chabot 2003). These groups are linked through a traditional system of reciprocity that redistributes market and subsistence products.

All Arctic nations recognize the potential lack of viability of market economies in northern regions when semi-nomadic indigenous peoples are consolidated in permanent settlements and have subsidized northern communities heavily (Duerden 1992). The consequence, however, is that Arctic mixed economies are now vulnerable to withdrawal of government support, as witnessed in Russia after the collapse of the Soviet Union or in cases of reduced support in Alaska (Knapp and Morehouse 1991). Products derived from locally available fish and wildlife resources often offer important sources of cash that supplement wages and transfer payments from governments.

However, subsistence economies are vulnerable to declines in global markets for these commodities, which include seal or muskrat pelts (as changes in cultural values reduce global demand for furs), salmon (as fish farming increases alternative supplies), and reindeer antler (as cultural change in Asia reduces demand) (Myers 2000). When world market prices are high, regional resource management institutions may be unable to respond to the increased incentives for unregulated or illegal harvest (such as for Kamchatka salmon or Greenland cod) (Hamilton et al. 2000) or overgrazing by reindeer (Forbes 1999). On the other hand, government policies to conserve stocks may prevent Arctic people from taking advantage of the only viable commercial activities available (as with the International Whaling Commission ban on commercial whaling).

Specialization on one or two products also increases the vulnerability to ecological change. In Greenland, for example, northward movement of warm currents in the 1930s reduced seal harvests by 60%, but local residents switched to fishing for cod, with financial assistance from the Danish government (Hamilton et al. 2000). When the cod fishery collapsed in the 1980s in response to a combination of overfishing and colder currents, those communities that had boats large enough to fish for offshore shrimp continued to prosper, whereas those communities that had only small boats and therefore no access to the offshore shrimp fishery declined in income and population (Hamilton et al. 2000)

25.4.3 Synergies and Trade-offs between Industrial Development and Cultural Resources

There is a fundamental trade-off between the bundle of services associated with environmental protection (aesthetic resources for Arctic and non-Arctic residents and the capacity of ecosystems to provide subsistence resources and to store soil carbon that contributes to climate regulation) and the services provided by industrial development that provides cash income. Most Arctic regions are ruled by nations with a non-Arctic population majority and have governments housed outside the Arctic. Therefore non-Arctic strategic and economic considerations often weigh heavily in decisions about Arctic development. Political decisions by the United States and Russia, for example, have strong effects on polar regions because of the nations' large demands for resources and strong influence on polar development (AHDR 2004).

An ongoing challenge for the governments of Arctic nations is to create options that maximize the economic benefits of industrial development, including wages and services for local residents, without seriously compromising the integrity of ecosystem services. This challenge may become more complex if warming leads to increased development and further immigration of nonindigenous people (Whiteman et al. 2004).

New technologies provide opportunities to minimize the impact of industrial development on ecosystem services of Arctic wetlands, including ice roads that minimize the areal extent of transportation infrastructure, improved pipeline designs that reduce the probability of oil spills, and lateral drilling that maximizes access to oil reservoirs from a limited number of surface installations. However, these advances are relatively ineffective in preserving the aesthetic qualities of wilderness, and some of these options are threatened by climatic warming. As winter temperatures rise, permanent roads and bridges will increasingly replace ice roads for overland access, augmenting the disturbed area for a given development. In Alaska, the number of days on which the oil industry is permitted by the state to use off-road vehicles has decreased by 50% since 1970 because of both warmer autumn conditions and increased environmental awareness by managers. (See Figure 25.10.) In Yamal in northern Russia, melting of ice-rich permafrost exacerbates deterioration of forage from trampling and overgrazing by reindeer (Forbes 1999).

Where Arctic residents have opportunities to capture some of the economic benefits from industrial development through both employment and corporate investments, benefits in the form of improved public infrastructure, educational services, and health care can be significant (as in North Slope Borough, Alaska, for instance). Trade-offs can be decreased where communities of resource users are afforded adequate authority in development planning and operation policies to ensure that community concerns are adequately addressed.

Other opportunities for synergies between conservation and development involve compromises that concentrate development in certain areas, leaving other areas protected for the subsistence, cultural, and climate-regulating services that they provide.

25.4.4 Institutional Trade-offs in Managing Ecosystems and Their Services

In the Antarctic, international conventions form the framework for decisions about resource management. The Antarctic Treaty was ratified in 1961, with additional instruments added in subsequent years. Comprehensive protection of the environment was

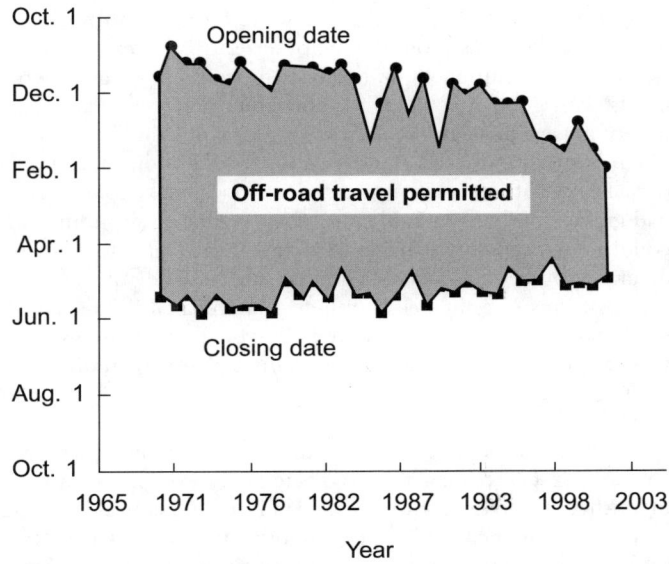

Figure 25.10. Trend in Dates Allowed for Travel by Seismic Vehicles on Alaskan Arctic Tundra, 1970–2000 (Alaska Department of Natural Resources)

achieved with the ratification of the Protocol on Environmental Protection, known as the Madrid Protocol, in 1998, which prohibited any activity relating to mineral resources, with exception for scientific research. Environmental assessment of all activities was required by the Madrid Protocol, ensuring that cumulative as well as immediate impacts are prevented. The Convention on the Conservation of Antarctic Marine Living Resources of 1980 permits fishing in the waters surrounding Antarctica, but only under the guidance of ecosystem management. This convention takes account not just of the dynamics of the exploited stocks but also their interactions in the ecosystem and the links to the physical environment.

Under these and other conventions, Antarctica now has the most stringent restrictions on use rights and the most advanced system of protected areas of any major region of the world. Nonetheless, Antarctica remains vulnerable to indirect human impacts such as pollution, stratospheric ozone destruction, and greenhouse-gas-induced climate change.

In contrast to Antarctica, institutions of the Arctic have evolved most actively at local and regional levels (AHDR 2004). Local stakeholders have asserted rights to develop their own rules and regulations governing use of natural resources and the right of their traditional institutions to be recognized by external parties. Both the outcomes of these efforts and the forms of local control differ regionally. In Alaska, Native corporations own about 12% of the land, and the federal government recognizes a rural subsistence preference regarding the use of living resources located on the 60% of the land remaining under federal jurisdiction. The U.S. federal government defends indigenous rights to use marine mammals and has established community development quotas to ensure that local groups have a stake in the region's marine fisheries.

The 1982 Canadian constitution acknowledges the validity of existing aboriginal rights, and the settlement of a number of comprehensive claims there has transferred to indigenous peoples both cash payments and title to large landholdings, along with recognized use and management rights in even larger traditional-use areas. In Greenland, there is no system of private property or well-defined use rights to land and natural resources, but the indigenous-controlled Greenland Home Rule has authority to make most decisions about the use of terrestrial and marine living resources. Devolution of authority to regional or local governments with good systems of accountability increases responsiveness of governments to the concerns of local stakeholders.

In Fennoscandia and Russia, national institutions have traditionally regulated natural resources, and local stakeholders have had greater difficulty establishing their rights. In Scandinavian countries, an ongoing struggle to secure Saami rights to land and natural resources has met with limited success. The state provides various forms of assistance to reindeer and coastal Saami. However, secure use rights have proved elusive, especially in Sweden, where the courts have generally denied claims to indigenous rights despite state recognition of these rights a century ago (Hahn 2000).

Recent federal laws in Russia theoretically allow indigenous peoples to establish long-term rights to land through the creation of both extensive Territories of Traditional Nature Use and smaller "commune lands" for pursuit of traditional activities (Kryazhkov 2000), although laws are difficult to implement in some areas and are not always enforced (Fondahl and Poelzer 2003). Recent outmigration of Russian and other nonindigenous residents may reduce pressures arising from competition over land and resource use. Efforts to devise a comprehensive system of rights and rules governing the use of land and natural resources in the region have repeatedly failed in the Duma.

A particularly significant innovation in governance in the circumpolar north—especially in the North American Arctic—has been the creation of co-management arrangements designed to forge partnerships between state governments holding formal authority to manage living resources and local user communities (such as the Porcupine caribou herd in Canada) (Osherenko 1988; Berkes et al. 1991). Co-management implies a sharing of power between resource users and state governments in various functions of resource management (monitoring, planning, enforcement, policy-making, and so on). These arrangements continue to evolve and are proving beneficial in regions where government agencies and resource users are together tracking the trends of climate change and monitoring the impacts of industrial development while exploring options for human adaptation.

In Fennoscandia, where traditional migration pathways conflict with complex patterns of private and Crown ownership, easement rights provide secure access (Hahn 2000). However, the amount of accessible, high-quality pasture available has decreased in many areas as a result of extensive forestry, tourism, hydropower, mining, and border fences and the concomitant increases in heavy trampling and grazing on the remaining lands (Forbes 2004b).

Oil, gas, and mineral developments have generally provided few long-term jobs for local residents. In North America, however, where local governments and land claims organizations provide an institutional framework for mitigation and compensation, extractive industries have provided substantial cash infusions to communities. Nevertheless, anxiety persists among Arctic residents about the cumulative effects of historical and proposed activities on resources and cultures (NRC 2003; AHDR 2004). In Russia's Yamal, where local residents have little influence over resource extraction, change has hit a crisis point, challenging the ability of local residents to adapt to the pace of oil and gas development (Krupnik 2000).

In the Arctic, the early development of international arrangements (the 1920 Svalbard Treaty guaranteeing signatories access to coal reserves, for example, or the 1973 Polar Bear Agreement committing signatories to the implementation of conservation measures) involved little input from local stakeholders. More recently, the Arctic Council and its forerunner, the Arctic Environmental Protection Strategy, have provided an important forum for the indigenous peoples of the Arctic to advance their agendas, including matters relating to the use of living resources. These international conventions then provide a basis for groups to go back to their home countries and demand the rights specified by these conventions. The Arctic Council has also devised and promoted an initiative designed to create and enlarge a Circumpolar Protected Areas Network.

25.5 Polar Systems and Human Well-being

25.5.1 Human Population Changes in the Arctic

Population in the circumpolar Arctic increased rapidly after 1960 as a result of immigration associated with expanded resource development and government activities in the north (AHDR 2004). However, change has been uneven, with much slower growth in Fennoscandia than in Russia or North America. (See Table 25.4.) Most of the population growth occurred in urban centers tied to industrial activities or public administration, so population density remains very low in rural areas across the Arctic. This pattern of growth has changed markedly since 1990. Growth has slowed in

Table 25.4. Population of Arctic Regions, 1960–2000 (Knapp 2000 for 1960–90; 2000 data from national census data and other sources as noted)

Region	1960	1970	1980	1990	2000
			(thousand)		
United States					
Arctic Alaska	81.9	96.1	112.2	151.0	158.7[a]
Canada					
Yukon Territory	14.6	18.4	23.2	27.8	30.9
Northwest Territories[b]	23.0	34.8	45.7	57.7	40.8
Nunavut[b]					26.7
Total, Arctic Canada	37.6	53.2	68.9	85.5	98.4[c]
Greenland	33.1	46.5	49.8	55.6	56.5[d]
Fennoscandia					
Norway	437.4	454.1	469.6	460.8	464.7[e]
Sweden (Norrbotten)	261.8	255.4	267.1	263.7	256.2[f]
Finland (Lappland)	205.1	197.1	194.9	200.7	187.7[g]
Total, Arctic Fennoscandia	904.3	906.6	931.6	925.2	908.6
Russia[h]					
European Arctic (Murmansk and Nenets)	613.2	838.6	1,025.0	1,201.0	934.0
Other European North	2,687.8	3,040.5	3,261.4	3,570.0	3,071.5
Asian Arctic, excluding Sakha[i]	468.2	637.5	895.6	1,298.0	951.0
Sakha Republic (Yakutia)	487.3	664.1	851.8	1,081.0	949.3
Other Asian North[j]	327.4	540.6	935.5	1,720.0	1,801.3
Total, Russian North	4,583.9	5,721.3	6,969.3	8,870.0	7,715.1
Arctic Russia (excluding Sakha)	1,081.4	1,476.1	1,920.6	2,499.0	1,885.0

[a] U.S. Census Bureau, Census 2000. Arctic Alaska includes all lands north of the Alaska Range.

[b] In 1999 Nunavut became a territory, separated from the Northwest Territories.

[c] Statistics Canada, 2001 Census, http://www.statcan.ca/.

[d] Statistics Greenland, 2001 Census, http://www.statgreen.gl/english/publ/figures/grfig-02.pdf.

[e] Statistics Norway. Arctic Norway includes provinces of Nordland, Troms and Finnmark.

[f] Population of Norrbotten: http://www.regionfakta.com/norrbotten_eng/Kapitel_09/e_a01_2500.htm.

[g] Statistics Finland: population of Lappland.

[h] Goskomstat of Russia, 2002 All-Russia Population Census, vol. 1.

[i] Yamalo-Nenets, Taimyr, Norilsk, Magadan, Chukotka, Koryak Oblasts

[j] Khanty-Mansi, Evenk, Kamchatka Oblasts

North America and Greenland, and population has declined in Arctic Fennoscandia and particularly in Russia. Although the indigenous population has grown at a rate of about 1.5% annually, its share of total population has declined. Indigenous peoples have become ethnic minorities in all Arctic regional government jurisdictions except Greenland and portions of Canada (Nunavut and the Northwest Territories). (See Table 25.5.) Nonindigenous population growth in the latter could make indigenous peoples a minority there, too, within a decade.

Drivers of population change differ significantly for indigenous and nonindigenous residents. For the nonindigenous, most of whom are of European ancestry, past cycles of population change coincided with changes in world demand for Arctic resources (Sugden 1982). In the twentieth century, the Arctic became important to international security; this led to increased military presence and to government policies that boosted industrial development (Armstrong et al. 1978; Osherenko and Young 1989). The easing of tensions after the collapse of the Soviet Union led to the scaling back and closure of military and industrial installations in most Arctic nations and to a reevaluation of regional industrial policies. This change had the greatest impact on the nonindigenous population in the Russian Arctic. Arctic areas of Russia have lost nearly 25% of their inhabitants since 1990, while the population of the Russian North as a whole has declined by 13%. The withdrawal of government support led to rapid outmigration of ethnic Russians and other nonindigenous people (Heleniak 2003). The indigenous population remained relatively stable but suffered a decline in living standards.

Net outmigration has tempered the effects on indigenous population growth of the relatively high birth rates and increasing life expectancy (AMAP 2003). Migrants seek education and jobs in northern urban centers or outside the Arctic; some, but not all, return later in life. In Alaska, for example, there may be as much as 1% per year net population outflow (Huskey et al. in press). More women than men leave small rural communities, leading to a gender imbalance among unmarried adults in many parts of the

Table 25.5. Arctic Indigenous Population, 1960–2000. Share of total population is shown in parenthesis. (Knapp 2000 for 1960–90; 2000 data from national census as noted)

Region	1960	1970	1980	1990	2000
			(thousand)		
United States	n.a.	33.9 (35%)	40.0 (36%)	50.0 (33%)	54.4[a] (34%)
Canada[b]	15.3 (41%)	21.2 (40%)	28.7 (42%)	39.1 (46%)	49.0[c] (50%)
Greenland	30.4 (92%)	38.9 (84%)	40.9 (82%)	46.1 (83%)	49.8[d] (88%)
Fennoscandia[e]	n.a.	34.0 (4%)	n.a.	50.0 (5%)	n.a.
Russia					
European Arctic[f]	n.a.	18.5 (2%)	18.2 (2%)	19.3 (2%)	19.5[g] (2%)
Asian Arctic, Excluding Sakha[h]	n.a.	66.0 (14%)	66.6 (7%)	77.3 (6%)	78.1[g] (8%)
Total Arctic	n.a.	212.5	n.a.	281.8	n.a.

[a] U.S. Census Bureau, Census 2000. Arctic Alaska includes all lands north of the Alaska Range.

[b] Approximately 9,000 Inuit living in the Nunavik region of Quebec are not included in the total. (Source: Indian and Northern Affairs Canada, Quebec Region, General Data on Indian Population available at: http://www.ainc-inac.gc.ca/qc/gui/population_e.html.) Canadian data provides for more than one definition of native origin. The definition used in this table is "aboriginal identity." Individuals that responded to "aboriginal" only are counted (no multiple responses). Figures for Nunavut count only Inuit and North American Indian ethnicity. There was no information on "aboriginal identity."

[c] Statistics Canada, 2001 Census, http://www.statcan.ca/.

[d] Statistics Greenland, 2001 Census, http://www.statgreen.gl/english/publ/figures/grfig-02.pdf.

[e] Population of the Saami in the Scandinavian North.

[f] Murmansk and Nenets Oblasts.

[g] Stepanov 2004.

[h] Yamalo-Nenets, Taimyr, Norilsk, Magadan, Chukotka, Koryak Oblasts.

Arctic (Hamilton and Seyfritt 1994). Consolidation of the population into larger settlements improves job prospects and reduces the cost of providing infrastructure and services. This trend also increases pressure on local renewable resources near population centers, as people attempt to continue hunting and herding traditions, and it weakens cultural ties to ancestral homelands. Traditions of reciprocity often continue between urban indigenous residents and their rural kin, so outmigration provides some economic benefits to rural communities, although the magnitude of these benefits is uncertain.

25.5.2 Patterns and Trends in Human Well-being

Most polar regions are governed by industrial nations with substantial economic resources. Human well-being in these regions is therefore largely the product of choices made by the people and leaders of these nations rather than a lack of national resources to fulfill human goals. The eight Arctic nations account for 40% of

global carbon emissions, so there are direct within-country links between the anthropogenic sources of climate change and the components of the global society that are currently most directly affected by this warming.

Polar regions are important to the well-being not only of the residents but also of the global population, which depends on polar regions for climate regulation and for providing extensive areas that remain wild and relatively unaffected by human activities and serve as critical areas for many culturally and otherwise important migratory species. This potential to provide for human well-being has not been fully met and is currently threatened by global human impacts on the climate system and by inadequate attention to human impacts within polar regions on ecosystems and the services they provide. The increased pressure that polar systems are experiencing implies that we are approaching critical thresholds (such as the thawing of permafrost and vegetation change), although the nature and timing of these thresholds are regionally variable and uncertain. Crossing these thresholds would likely cause a cascade of ecological changes with large effects (some negative, some positive) on human well-being. These changes could appear quickly and be irreversible (Dasgupta 2001; Chapin et al. 2004).

There is substantial variation in the well-being provided by ecosystems in different polar regions (AHDR 2004). There are few permanent residents in Antarctica, and there have been no clear trends in the mass balance of the Antarctic ice sheet on which the global population depends for climate regulation and control of sea level, as described earlier. There are indications of greater deterioration in the capacity of Arctic regions to provide a cooling system for the planet and to provide the aesthetic values of wilderness.

25.5.3 Cultural and Economic Ties to Ecosystem Services

The deterioration of cultural ties to subsistence activities among indigenous peoples is the most serious cause of decline in well-being within the Arctic. (See Figure 25.11.) There has been a gradual loss of connection to the land through change in lifestyles, loss of indigenous languages, and dominance of nonindigenous

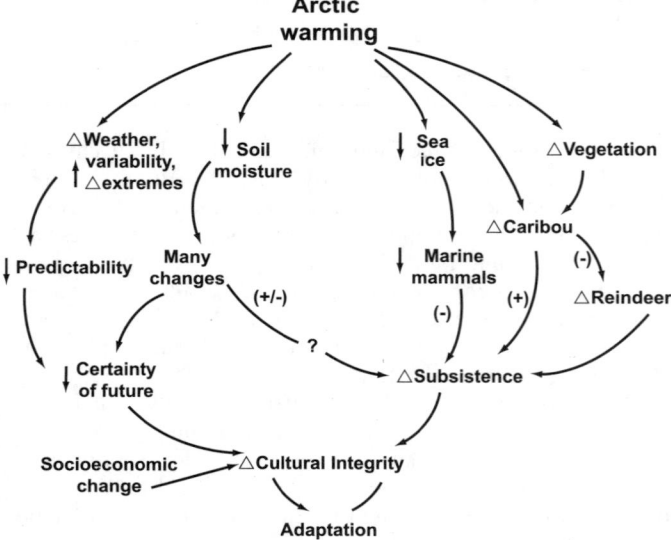

Figure 25.11. Specific Links between Arctic Warming and Well-being of Arctic Indigenous Peoples (see also Figure 25.9)

educational systems over indigenous peoples. In cultures with an important oral tradition, the death of elders and the reduced experience of a younger generation in living on the land have eroded traditional knowledge. Changes from nomadic to sedentary lifestyles (among Saami and Nenets herders, for instance), the moving of children to distant schools for education, economic infusions into communities from oil development, and migration to cities have all further weakened cultural traditions and the sense of self-sufficiency that makes these traditions meaningful.

The effects of regional climate warming on ecosystem services have contributed to the decline in cultural connections to the land. As the climate warms, traditional cues for predicting environmental variability no longer work. This strips Arctic residents of their considerable knowledge, predictive ability, and self-confidence in making a living from their resources and may ultimately leave them as "strangers in their own land" (Berkes 2002). This increases their vulnerability to both climatic and social changes (Norton 2002). Nonetheless, indigenous cultures continue to exhibit substantial resilience as a result of short-term coping mechanisms (such as prey switching in response to changing animal abundances) and long-term adaptation through application of new tools and technologies, such as global positioning system navigational aids (Sahlins 1999; Berkes and Jolly 2001).

Industrial development interacts with cultural and climatic change to create both positive and negative effects on human well-being. The greatest positive effect is increased income associated with oil development in North America and tourism in Fennoscandia and with government subsidies that are funded in part by industrial development. These wages enable people to buy snow machines, guns, boats, GPS units, and other equipment that enhances their capacity to harvest food from the land. Depending on the level of harvest, this can sustain subsistence activities or (for example, in the Russian Arctic) can lead to overharvesting. The environmental deterioration and changes in animal behavior associated with industrial development are major concerns of indigenous peoples throughout the Arctic because this threatens their subsistence lifestyles and cultural ties to the land. The jobs provided by industry and government sectors are important motivations for emigration of younger people from rural communities to urban centers.

Because of the interdependence between Arctic economies and global markets, indigenous peoples are doubly exposed: to both climate change and changes caused by the global processes affecting markets, technologies, and public policies. Having access to traditional food resources and ensuring food security will be a major challenge in an Arctic affected increasingly by climate change and global processes (Nuttall et al. in press). The low diversity of economic options in the Arctic renders it vulnerable to changes in both the local resource base and global economic trends and markets and explains the widespread northern phenomenon of boom-and-bust cycles (Chapin et al. 2004). Enhancing local economic diversification is a critical step toward increasing Arctic resilience.

The development and collapse of the Soviet economy in Russia had numerous impacts on the well-being of northern residents. Many non-indigenous residents of the Russian Arctic have left since 1990, whereas indigenous residents have largely remained, because they had no family ties outside the Arctic. Despite this reduction in demand for food and services, the economic subsidies and marketing system that were the primary economic base for Russians in the Arctic have virtually disappeared. This situation has reduced the economic incentive to maintain domestic reindeer, leading to a liquidation of some herds to support food needs. Hunting of wild mammals and birds has increased to meet food needs, straining the capacity of northern ecosystems to provide the services that northern people will require over the long term.

Resolution of the property rights of indigenous peoples in North America has been a key institutional change that strengthens their ties to the land. These changes in property rights promise enhanced resilience in the North American Arctic but remain a challenge in Eurasia (Osherenko 2001; Fondahl and Poelzer 2003).

25.5.4 Environmental Effects on Human Health

Increased UV-B radiation due to ozone depletion directly increases the risk of skin cancer and sight damage, particularly to people with fairer skin (Altmeyer et al. 1997). It also suppresses immune responses and thus increases susceptibility to disease (De Gruijl et al. 2002). Ozone depletion occurs at both poles but is most pronounced and extensive in the south. In the austral spring, when ozone depletion is strongest over Antarctica, ozone-depleted parcels of air move northward, leading to increased UV-B exposure to southern regions of Australia, New Zealand, Africa, and South America.

Bioaccumulation of contaminants in northern biota represents a potential health risk to northern peoples, as described in detail in Arctic Monitoring and Assessment Programme reports (MacDonald et al. 2002) The impacts of environmental deterioration on human health are regionally variable in the Arctic and are generally most pronounced in eastern Canada and the North Atlantic (AHDR 2004). However, studies conducted outside the Arctic have shown that even "low" concentrations of POPs and other endocrine disruptors interfere with hormone function and genetic regulation, indicating that developing organisms are easily put at risk of impaired reproductive, immune, and neurological function (Guillette and Crain 2000).

The ubiquity of POPs in the Arctic environment raises concern for the safety of future generations of Arctic peoples (AMAP 2003; Downie and Fenge 2003; Godduhn and Duffy 2003). As with POPs, the young and unborn are generally most vulnerable to subtle sublethal effects of heavy metals. In coastal Canadian and Greenland communities where marine mammals are a significant food source, concentrations of POPs in mothers' milk are high enough to be considered a health risk (Deutch and Hansen 2000; Muckle et al. 2001; AMAP 2003), and consumption of wild foods has been restricted in several cases in the eastern Canadian Arctic.

Of the small number of studies on the impact of environmental contaminants on human health in the Arctic, few have detected significant health effects. Health assessments suggest that the benefits of traditional foods, including lower risk of heart disease, generally outweigh the risks from contamination, because these foods are more nutritious than are store-bought substitutes (AMAP 2003). Regardless of the medical impacts of contaminants, the widespread perception of health risks by local residents reduces their sense of well-being.

Human populations that depend on caribou and reindeer are vulnerable to radionuclide accumulations whenever nuclear fallout occurs. Such populations suffer the physiological effects of radiation, the economic effects of lost resources, or both. For example, after the explosion at the nuclear power station at Chernobyl, cesium-137 and strontium-90 concentrations increased dramatically in lichens and reindeer in Scandinavia, where tens of thousands of reindeer had to be destroyed (MacDonald et al. 2002).

25.5.5 Aesthetic and Recreational Values

The Arctic and Antarctic include some of the largest wilderness areas on Earth. Establishment of protected areas of various types is the conventional way to protect landscapes, ecosystems, and habitats in the terrestrial environment. Treaties have provided Antarctica with a high level of protection that restricts development (as through Annex V of the Protocol on Environmental Protection to the Antarctic Treaty). Increasing tourism in Antarctica reflects the growing recognition of its aesthetic and recreational values.

Protected areas cover approximately 15% of the terrestrial Arctic. However, they are unevenly distributed across ecosystems and habitats. Over 27% of Arctic glaciers but less of the vegetated Arctic is protected (CAFF 2001). A strategic plan to establish a Circumpolar Protected Areas Network was completed in 1996 and endorsed by the eight Arctic nations (CAFF 1997). The initial burst of creating new protected areas has since come to a standstill, particularly in the Russian Arctic (WWF 2002). Owing to the rapidly changing climate, reserves cannot successfully protect species unless they have flexible boundaries or adequate interconnections that allow redistribution or migration in response to climate change (Elmqvist et al. 2003; Callaghan et al. in press). In addition to current practices of protecting rare and threatened species, conservation may become necessary for more widespread Arctic species that decline in response to climate warming.

25.5.6 Opportunities for Scientific Study

Polar regions have been key sources of new information that improve human understanding of the Earth system. Ice cores from Antarctica and Greenland have revealed detailed records of climatic and environmental changes that have occurred over the past million years (EPICA 2004). Antarctica, in particular, has been an important scientific platform for many disciplines, including space and atmospheric sciences, geomagnetism, astronomy, paleoclimatology, biogeochemistry, and human physiology. For example, subglacial lakes that have not been exposed to the atmosphere for 500,000 years may contain a unique archive of the past environment and biota (Karl et al. 1999).

Because of polar amplification of climate change, the ecological impacts of warming are evident earliest and most clearly at high latitudes, providing society with a preview of changes that may become more widespread. In a region of near-pristine wilderness, relationships between ecosystems, species, and environment are more clearly defined than in populated regions where human influences can mask these relationships. In addition, the lure of polar regions for young people wishing to experience wilderness provides a unique opportunity to involve the next generation in working toward a more sustainable future.

References

AHDR (*Arctic Human Development Report*), 2004: Stefansson Arctic Institute, Akureyri, Iceland.

Ainley, D.G., G. Ballard, S.E. Emslie, W.R. Fraser, P.R. Wilson, and E.J. Woehler, 2003: Adelie penguins and environmental change. *Science,* **300(5618),** 429.

Altmeyer, P., K. Hoffmann, and M. Stucker (eds.), 1997: *Skin Cancer and UV Radiation.* Springer-Verlag, Berlin.

AMAP (Arctic Monitoring and Assessment Programme), 2003: *AMAP Assessment 2002: Human Health in the Arctic.* Oslo, 137 pp.

Anisimov, O., B. Fitzharris, J.O. Hagen, R.L. Jefferies, H. Marchant, et al. 2001: Polar Regions (Arctic and Antarctic). In: *Climate Change 2001: Impacts, Adaptation, and Vulnerability,* J.J. McCarthy, O.F. Canziani, N.A. Leary, D.J. Dokken, and K.S. White (eds.), Cambridge University Press, Cambridge, 803–841.

Arendt, A., K. Echelmeyer, W.D. Harrison, G. Lingle, and V. Valentine, 2002: Rapid wastage of Alaska glaciers and their contribution to rising sea level. *Science,* **297,** 382–386.

Armstrong, T., G. Rogers, and G. Rowley, 1978: *The Circumpolar North.* Methuen & Co. Ltd., London.

Arrigo, K.R., D.L. Worthen, M.P. Lizotte, P. Dixon, and G. Dieckmann, 1997: Primary production in Antarctic sea ice. *Science,* **276,** 394–397.

Arrigo, K.R., D.H. Robinson, D.L. Worthen, R.B. Dunbar, G.R. DiTullio, M. van Woert, and M.P. Lizotte, 1999: Phytoplankton community structure and the drawdown of nutrients and CO_2 in the Southern Ocean. *Science,* **283,** 365–367.

Atkinson, A., V. Siegel, E. Pakhomov, and P. Rothery, In press: Long-term decline in krill stock and increase in salps within the Southern Ocean. *Nature.*

Aurela, M., T. Laurita, and J.-P. Tuovinen, 2002: Annual CO_2 balance of a subarctic fen in northern Europe: Importance of the wintertime efflux. *Journal of Geophysical Research,* **107,** 4607, doi:4610.1029/2002JD002055.

Babaluk, J.A., J.D. Reist, J.D. Johnson, and L. Johnson, 2000: First records of sockeye (*Oncorhynchus nerka*) and pink salmon (*O. gorbuscha*) from Banks Island and other records of Pacific salmon in Northwest Territories, Canada. *Arctic,* **53(2),** 161–164.

Barbraud, C. and H. Weimerskirch, 2001: Emperor penguins and climate change. *Nature,* **411,** 183–185.

Barnes, D.K.A., D.A. Hodgson, P. Convey, C. Allen, and A. Clarke, Submitted: Incursion and excursion of Antarctic biota: Past, present and future. *Oikos.*

Baroni, C. and G. Orombelli, 1994: Abandoned penguin rookeries as holocene paleoclimatic indicators in Antarctica. *Geology,* **22,** 23–26.

Baskin, L., 2000: Reindeer husbandry/hunting in Russia in the past, present, and future. *Polar Research,* **19,** 23–29.

Batt, B.D.J., 1997: *Arctic ecosystems in peril: Report of the Arctic Goose Habitat Working Group.* Arctic Goose Joint Venture Special Publication, U. S. Fish and Wildlife Service and Canadian Wildlife Service, Washington and Ottawa, 1–120 pp.

Batzli, G.O., R.G. White, S.F. MacLean, Jr., F.A. Pitelka, and B.D. Collier, 1980: The herbivore-based trophic system. In: *An arctic ecosystem: The coastal tundra at Barrow, Alaska,* J. Brown, P.C. Miller, L.L. Tieszen, and F.L. Bunnell (eds.), Dowden, Hutchinson and Ross, Stroudsburg, 335–410.

Bentley, C.R., 2004: Mass balance of the Antarctic ice sheet: Observational aspects. In: *Mass Balance of the Cryosphere: Observations and Modeling of Contemporary and Future Changes,* J.L. Bamber and A.J. Payne (eds.), Cambridge University Press, Cambridge, 459–490.

Bergstrom, D.M. and S.L. Chown, 1999: Life at the front: History, ecology and change on southern ocean islands. *Trends in Ecology and Evolution,* **14(12),** 472–477.

Berkes, F., 2002: Epilogue: Making sense of arctic environmental change? In: *The Earth Is Faster Now,* I. Krupnik and D. Jolly (eds.), Arctic Research Consortium of the United States, Fairbanks, Alaska, 335–349.

Berkes, F. and D. Jolly, 2001: Adapting to climate change: social-ecological resilience in a Canadian western Arctic community. *Conservation Ecology,* **5(2),** [online] URL: http://www.consecol.org/vol5/iss2/art18.

Berkes, F., P. George, and R. Preston, 1991: Co-management: The evolution of the theory and practice of joint administration of living resources. Research Program for Technology Assessment in the Subarctic, Hamilton, Canada.

Betts, A.K. and J.H. Ball, 1997: Albedo over the boreal forest. *Journal of Geophysical Research-Atmospheres,* **102,** 28901–28909.

Block, W., 1994: Terrestrial ecosystems: Antarctica. *Polar Biology,* **14,** 293–300.

Blomqvist, S., N. Holmgren, S.H. Akesson, A., and J. Petterson, 2002: Indirect effects of lemming cycles on sandpiper dynamics: 50 years of counts from southern Sweden. *Oecologia,* **133,** 146–158.

Bomblies, A., D.M. McKnight, and E.D. Andrews, 2001: Retrospective simulation of lake level rise in Lake Bonney based on recent 21-year record: Indication of recent climate change in the McMurdo Dry Valleys, Antarctica. *Journal of Paleolimnology,* **25,** 477–492.

Box, J., D. Bromwich, and L.E.-S. Bai, 2004: Greenland ice sheet surface mass balance 1991–2000: Application of Polar MM5 mesoscale model and *in situ* data. *Journal of Geophysical Research,* **109,** D16105, doi: 10.1029/2003JD004451.

Boyd, P.W., A.J. Watson, C.S. Law, E.R. Abraham, T. Trull, et al., 2000: A mesoscale phytoplankton bloom in the polar Southern Ocean stimulated by iron fertilization. *Nature,* **407,** 695–702.

Brierley, A. S. and D. N. Thomas 2002: Ecology of Southern Ocean pack ice. *Advances in Marine Biology,* **43,** 171–276.

Brody, H., 1983: *Maps and Dreams: Indians and the British Columbia Frontier.* Penguin, Harmondsworth.

Burkins, M.B., R.A. Virginia, and D.H. Wall, 2001: Soil organic carbon cycling in Taylor Valley, Antarctica: Quantifying soil reservoirs and soil respiration. *Global Change Biology,* **7,** 113–125.

CAFF (Conservation of Arctic Flora and Fauna International Secretariat), 1997: *Co-operative Strategy for the Conservation of Biological Diversity in the Arctic Region.* Rekjavik.

CAFF, 2001: *Arctic Flora and Fauna: Status and Conservation.* Edita, Helsinki, 272 pp.

Callaghan, T.V., B.R. Werkman, and R.M.M. Crawford, 2002: The tundra-taiga interface and its dynamics: Concepts and applications. *Ambio Special Report (Tundra-Taiga Treeline Research), 12,* 6–14.

Callaghan, T.V., L.O. Björn, Y. Chernov, F.S. Chapin, III, T. Christensen, B. Huntley, R. Ims, D. Jolly, N. Matveyeva, N. Panikov, W.C. Oechel, and G.R. Shaver, In press: Arctic tundra and polar desert ecosystems. In: *Arctic Climate Impact Assessment,* ACIA (ed.), Cambridge University Press, Cambridge.

Caulfield, R., 2000: The political economy of renewable resource harvesting in the Arctic. In: *The Arctic: Environment, People, Policy,* M. Nuttall and T.V. Callaghan (eds.), Harwood Academic Publishers, Amsterdam.

Caulfield, R.A., 1997: *Greenlanders, Whales, and Whaling : Sustainability and Self-Determination in the Arctic.* University Press of New England, Hanover.

CAVM Team, 2003: *Circumpolar Arctic Vegetation Map (Scale 1:7,500,000).* Map No. 1, Conservation of Arctic Flora and Fauna (CAFF), U.S. Fish and Wildlife Service, Anchorage, Alaska.

CCAMLR (*Convention for the Conservation of Antarctic Marine Living Resources*), 2000: www.ccamlr.org.

CCAMLR, 2002: *Statistical Bulletin.* Vol. 14, Hobart, Australia, 155 pp.

Chabot, M., 2003: Economic changes, household strategies, and social relations of contemporary Nunavik Inuit. *Polar Record, 39,* 19–34.

Chance, N.A., 1987: Subsistence research in Alaska: Premises, practices, and prospects. *Human Organization, 46,* 85–89.

Chapin, F.S., III, W. Eugster, J.P. McFadden, A.H. Lynch, and D.A. Walker, 2000a: Summer differences among arctic ecosystems in regional climate forcing. *Journal of Climate, 13,* 2002–2010.

Chapin, F.S., III, A.D. McGuire, J. Randerson, R. Pielke, Sr., D. Baldocchi, et al. 2000b: Arctic and boreal ecosystems of western North America as components of the climate system. *Global Change Biology, 6 (Suppl. 1),* 1–13.

Chapin, F.S., III, G. Peterson, F. Berkes, T.V. Callaghan, P. Angelstam, et al. 2004: Resilience and vulnerability of northern regions to social and environmental change. *Ambio, 33,* 344–349.

Chapman, W.L. and J.E. Walsh, 1993: Recent variations of sea ice and air temperature in high latitudes. *Bulletin of the American Meteorological Society, 74,* 33–47.

Chernov, Y., 1985: *The Living Tundra.* Cambridge University Press, Cambridge.

Chernov, Y.I., 1995: Diversity of the arctic terrestrial fauna. In: *Arctic and Alpine Biodiversity: Patterns, Causes and Ecosystem Consequences,* F.S. Chapin, III and C. Körner (eds.), Springer-Verlag, Berlin, 81–95.

Chiuchiolo, A.L., R.M. Dickhut, M.A. Cochran, and H.W. Ducklow, 2004: Persistent organic pollutants at the base of the Antarctic marine food web. *Environmental Science & Technology, 38,* 3551–3557.

Chown, S.L., N.M. Gremmen, and K.J. Gaston, 1998: Ecological biogeography of southern ocean islands: Species-area relationships, human impacts, and conservation. *American Naturalist, 152,* 562–575.

Christensen, T.R., T. Friborg, M. Sommerkorn, J. Kaplan, L. Illeris, H. Søgaard, C. Nordstrøm, and S. Jonasson, 2000: Trace gas exchange in a high arctic valley. 1. Variations in CO_2 and CH_4 flux between tundra vegetation types. *Global Biogeochemical Cycles, 14,* 701–714.

Christensen, T.R., T. Johansson, H.J. Akerman, M. Mastepanov, N. Malmer, T. Friborg, P. Crill, and B.H. Svensson, 2004: Thawing sub-arctic permafrost: Effects on vegetation and methane emissions. *Geophysical Research Letters, 31,* doi:10.1029/2003GL018680.

Christensen, T.R., A. Joabsson, L. Ström, N. Panikov, M. Mastepanov, et al. 2003: Factors controlling large scale variations in methane emissions from wetlands. *Geophysical Research Letters, 30.*

Cicerone, R.J. and R.S. Oremland, 1988: Biogeochemical aspects of atmospheric methane. *Global Biogeochemical Cycles, 2(4),* 299–327.

Clarke, A. and J.A. Crame, 1989: The origin of the Southern Ocean marine fauna. In: *Origins and Evolution of the Antarctic Biota,* J.A. Crame (ed.), The Geological Society, London, 253–268.

Convey, P., 1996: Overwintering strategies of terrestrial invertebrates from Antarctica—the significance of flexibility in extremely seasonal environments. *European Journal of Entomology, 93,* 489–505.

Convey, P., 2001a: Terrestrial ecosystem response to climate changes in the Antarctic. In: *"Fingerprints" of Climate Change—Adapted Behaviour and Shifting Species Ranges,* G.-R. Walther, C.A. Burga, and P.J. Edwards (eds.), Kluwer, New York, 17–42.

Convey, P., 2001b: Antarctic ecosystems. In: *Encyclopedia of Biodiversity,* S.A. Levin (ed.). 1, Academic Press, San Diego, 171–184.

Convey, P., Submitted: Terrestrial ecosystems and climate change. In: *Trends in Antarctic Terrestrial and Limnetic Ecosystems,* D. Bergstrom, P. Convey, and A. Huiskes (eds.), Kluwer, New York.

Convey, P. and S.J. McInnes, In press: Exceptional, tardigrade-dominated, ecosystems from Ellsworth Land, Antarctica. *Ecology.*

Convey, P., D. Scott, and W.R. Fraser, 2003: Biophysical and habitat changes in response to climate alteration in the Arctic and Antarctic. *Advances in Applied Biodiversity Science, 4,* 79–84.

Convey, P., P.J.A. Pugh, C. Jackson, A.W. Murray, C.T. Ruhland, F.S. Xiong, and T.A. Day, 2002: Response of antarctic terrestrial microarthropods to long-term climate manipulations. *Ecology, 83(11),* 3130–3140.

Corsolini, S., N. Ademollo, T. Romeo, S. Olmastroni, and S. Focardi, 2003: Persistent organic pollutants in some species of a Ross Sea pelagic tropic web. *Antarctic Science, 15,* 95–104.

Crawford, R.J.M. and J. Cooper, 2003. *African Journal of Marine Science, 25,* 415–426.

Crawford, R.M.M., C.E. Jeffree, and W.G. Rees, 2003: Paludification and forest retreat in northern oceanic environments. *Annals of Botany, 91,* 213–226.

Cullather, R.I., D.H. Bromwich, and M.L. van Woert, 1996: Inter-annual variations in Antarctic precipitation related to El Niño—Southern Oscillation. *Journal of Geophysical Research, 101,* 19109–19118.

Dahl, J., 2000: *Saqqaq: An Inuit Hunting Community in the Modern World.* University of Toronto Press, Toronto.

Danks, H.V., 1999: Life cycles in polar arthropods—flexible or programmed? *European Journal of Entomology, 96,* 83–102.

Dasgupta, P., 2001: *Human Well-Being and the Natural Environment.* Oxford University Press, Oxford.

Day, T.A., C.T. Ruhland, and F.S. Xiong, 2001: Influence of solar ultraviolet-B radiation on Antarctic terrestrial plants: Results from a four year field study. *Journal of Photochemistry and Photobiology B: Biology, 62,* 78–87.

De Gruijl, F.R., J. Longstreth, M. Norval, A.P. Cullen, H. Slaper, et al. 2002: Health effects from stratospheric ozone depletion and interactions with climate change. In: *Environmental Effects of Ozone Depletion and Its Interactions with Climate Change: 2002 Assessment,* UNEP, Nairobe, 25–53.

Deutch, B. and J.C. Hansen, 2000: High human plasma levels of organochlorine compounds in Greenland: Regional differences and lifestyle effects. *Danish Medical Bulletin, 47(2),* 132–137.

Doiban, V.A., M. Pretes, and A.V. Sekarev, 1992: Economic development of the Kola Region, USSR: An overview. *Polar Record, 28,* 7–16.

Doran, P.T., C.P. McKay, G.D. Clow, G.L. Dana, A.G. Fountain, T. Nylen, and W.B. Lyons, 2002a: Valley floor climate observations from the McMurdo dry valleys, Antarctica, 1986–2000. *Journal of Geophysical Research, 107,* D24, 4772, doi:10.1029/2001JD002045.

Doran, P.T., J.C. Priscu, W.B. Lyons, J.E. Walsh, A.G. Fountain, et al. 2002b: Antarctic climate cooling and terrestrial ecosystem response. *Nature, 415,* 517–520.

Dowdeswell, J.A., J.-O. Hagen, H. Björnsson, A.F. Glazovsky, W.D. Harrison, et al. 1997: The mass balance of circum-Arctic glaciers and recent climate change. *Quaternary Research, 48,* 1–14.

Downie, D.L. and T. Fenge, 2003: *Northern Lights against POPs: Combating Toxic Threats in the Arctic.* McGill-Queen's University Press, Montreal.

Ducklow, H.W., D.K. Steinberg, and K.O. Buesseler, 2001: Upper ocean carbon export and the biological pump. *Oceanography, 14,* 50–58.

Duerden, F., 1992: A critical look at sustainable development in the Canadian North. *Arctic, 45,* 219–225.

Dyurgerov, M.B. and M.F. Meier, 1997: Mass balance of mountain and subpolar glaciers: A new global assessment for 1961–1990. *Arctic and Alpine Research, 29,* 379–391.

Elmhagen, B., M. Tannerfeldt, and A. Angerbjorn, 2002: Food-niche overlap between arctic and red foxes. *Canadian Journal of Zoology, 80,* 1274–1285.

Elmqvist, T., C. Folke, M. Nyström, G. Peterson, J. Bengtsson, B. Walker, and J. Norberg, 2003: Response diversity, ecosystem change, and resilience. *Frontiers in Ecology and the Environment, 1(9),* 488–494.

Emslie, S.D., P. Ritchie, and D. Lambert, 2003: Late-Holocene penguin occupation and diet at King George Island, Antarctic Peninsula. In: *Antarctic Peninsula Climate Variability: Historical and Palaeoenvironmental Perspectives,* E. Domack, A. Burnett, A. Leventer, P. Convey, M. Kirby, and R. Bindschadler (eds.). 79, American Geophysical Union, Washington, 171–180.

EPICA, 2004: Eight glacial cycles from an Antarctic ice core. *Nature, 429,* 623–628.

Ernsting, G., W. Block, H. MacAlister, and C. Todd, 1995: The invasion of the carnivorous carabid beetle *Trechisibus antarcticus* on South Georgia (sub-Antarctic) and its effect on the endemic herbivorous beetle *Hydromedion sparsutum*. *Oecologia*, **103**, 34–42.

Estes, J.A., M.T. Tinker, T.M. Williams, and D.F. Doak, 1998: Killer whale predation on sea otters linking oceanic and nearshore ecosystems. *Science*, **282**, 473–476.

Eugster, W., W.R. Rouse, R.A. Pielke, J.P. McFadden, D.D. Baldocchi, et al. 2000: Land-atmosphere energy exchange in arctic tundra and boreal forest: available data and feedbacks to climate. *Global Change Biology*, **6 (Suppl. 1)**, 84–115.

Ewald, G., P. Larsson, H. Linge, L. Okla, and N. Szarzi, 1998: Biotransport of organic pollutants to an inland Alaska lake by migrating sockeye salmon (*Oncorhynchus nerka*). *Arctic*, **51**, 40–47.

Farman, J.C., B.G. Gardiner, and J.D. Shanklin, 1985: Large losses of total ozone in Antarctica reveal seasonal ClOx/NOx interaction. *Nature*, **315**, 207–210.

Finney, B.P., I. Gregory-Eaves, M.S.V. Douglas, and J.P. Smol, 2002: Fisheries productivity in the northeastern Pacific Ocean over the past 2,200 years. *Nature*, **416**, 729–733.

Foley, J.A., J.E. Kutzbach, M.T. Coe, and S. Levis, 1994: Feedbacks between climate and boreal forests during the Holocene epoch. *Nature*, **371**, 52–54.

Fondahl, G. and G. Poelzer, 2003: Aboriginal land rights in Russia at the beginning of the twenty-first century. *Polar Record*, **39(209)**, 111–122.

Forbes, B.C., 1995: *Effects of surface disturbance on the movement of native and exotic plants under a changing climate*. Ecosystems Research Report 10, European Commission, Brussels, 209–219 pp.

Forbes, B.C., 1999: Land use and climate change in the Yamal-Nenets region of northwest Siberia: Some ecological and socio-economic implications. *Polar Research*, **18**, 1–7.

Forbes, B.C., 2004a: Impacts of energy development in polar regions. In: *Encyclopedia of Energy*, C.J. Cleveland (ed.), Academic Press, San Diego, 93–105.

Forbes, B.C., 2004b: *RENMAN Final Report: The Challenges of Modernity for Reindeer Management*. Arctic Centre Reports 41, University of Lapland, Rovaniemi, 1–100 pp.

Forbes, B.C., J.J. Ebersole, and B. Strandberg, 2001: Anthropogenic disturbance and patch dynamics in circumpolar arctic ecosystems. *Conservation Biology*, **15**, 954–969.

Fountain, A.G., W.B. Lyons, M.B. Burkins, G.L. Dana, P.T. Doran, et al, 1999: Physical controls on the Taylor Valley ecosystem, Antarctica. *BioScience*, **49**, 961–971.

Fraser, W.R. and E.E. Hofmann, 2003: A predator's perspective on causal links between climate change, physical forcing and ecosystem response. *Marine Ecology Progress Series*, **265**, 1–15.

Fraser, W.R., W.Z. Trivelpiece, D.G. Ainley, and S.G. Trivelpiece, 1992: Increases in Antarctic penguin populations: Reduced competition with whales or a loss of sea ice due to environmental warming? *Polar Biology*, **11**, 525–531.

Freckman, D.W. and R.A. Virginia, 1997: Low-diversity Antarctic soil nematode communities: Distribution and response to disturbance. *Ecology*, **78**, 363–369.

Freese, C.H., 2000: *The consumptive use of wild species in the Arctic: Challenges and opportunities for ecological sustainability*. World Wildlife Fund Canada, Toronto, Canada, 143 pp.

Frenot, Y., S.L. Chown, J. Whinam, P.M. Selkirk, P. Convey, M. Skotnicki, and D. Bergstrom, In press: Biological invasions in the Antarctic: Extent, impacts and implications. *Biological Reviews*.

Gilg, O., 2002: The summer decline of collared lemming, *Dicrostonyx groenlandicus*, in high Arctic Greenland. *Oikos*, **99**, 499–510.

Gille, S.T., 2002: Warming of the Southern Ocean since the 1950s. *Science*, **295**, 1275–1277.

Godduhn, A. and L. Duffy, 2003: Multi-generation health risks of persistent organic pollution in the far north: Use of the precautionary approach in the Stockholm Convention. *Environmental Science and Policy*, **6**, 341–353.

Golovnev, A.V. and G. Osherenko, 1999: *Siberian Survival: The Nenets and Their Story*. Cornell University Press, Ithaca.

Gremmen, N.J., S.L. Chown, and D.J. Marshall, 1998: Impact of the introduced grass *Agrostis stolonifera* on vegetation and soil fauna communities at Marion Island, subantarctic. *Biological Conservation*, **85**, 223–231.

Gremmen, N.J.M. and V.R. Smith, 1999: New records of alien vascular plants from Marion and Prince Edward Islands, subantarctic. *Polar Biology*, **21**, 401–409.

Griffith, B., D.C. Douglas, N.E. Walsh, D.D. Young, T.R. McCabe, et al. 2002: *The Porcupine caribou herd*. Biological Science Report USGS/BRD/BSR-2002-0001, U.S. Geological Survey, Biological Resources Division.

Guillette, L. and D. Crain (eds.), 2000: *Environmental Endocrine Disruptors: An Evolutionary Perspective*. Taylor & Francis, London.

Hahn, T., 2000: *Property rights, ethics, and conflict resolution: Foundations of the Sami economy in Sweden*. Department of Economics, PhD Dissertation, Swedish University of Agricultural Sciences, Uppsala, 225 pp.

Hamilton, L. and C. Seyfritt, 1994: Coming out of the country: Community size and gender balance among Alaskan Natives. *Arctic Anthropology*, **31**, 16–25.

Hamilton, L., P. Lyster, and O. Otterstad, 2000: Social change, ecology and climate in 20th century Greenland. *Climatic Change*, **47**, 193–211.

Hanel, C. and S.L. Chown, 1998: *An Introductory Guide to the Marion and Prince Edward Island Special Nature Reserves*. Department of Environmental Affairs and Tourism, Pretoria, 1–80 pp.

Hanski, I., H. Henttonen, E. Korpimäki, L. Oksanen, and P. Turchin, 2001: Small-rodent dynamics and predation. *Ecology*, **82**, 1505–1520.

Harangozo, S.A., 2000: A search for ENSO teleconnections in the west Antarctic Peninsula climate in austral winter. *International Journal of Climatology*, **20**, 663–679.

Heleniak, T., 2003: The 2002 Census in Russia: preliminary results. *Eurasian Geography and Economics*, **45(6)**, 430–442.

Heliövaara, K. and M. Peltonen, 1999: Bark beetles in a changing environment. *Ecological Bulletins*, **47**, 48–53.

Hempel, G. (ed.), 1994: *Antarctic Science. Global Concerns*. Springer-Verlag, Berlin.

Henttonen, H., E. Fuglei, C. Gower, V. Haukisalmi, R.A. Ims, J. Niemimaa, and N.G. Yoccoz, 2001: *Echinococcus multilocularis* on Svalbard: Introduction of an intermediate host has enabled the local life-cycle. *Parasitology*, **123**, 547–552.

Hersteinsson, P. and D.W. MacDonald, 1992: Interspecific competition and the geographical distribution of red and arctic foxes *Vulpes vulpes* and *Alopex lagopus*. *Oikos*, **64**, 505–515.

Hessen, D.O., 2002: *UV Radiation and Arctic Ecosystems*. Springer-Verlag, Berlin, 321 pp.

Hik, D.S., R.L. Jefferies, and A.R.E. Sinclair, 1992: Grazing by geese, isostatic uplift and the occurrence of multiple stable states of salt marsh vegetation. *Journal of Ecology*, **80**, 395–406.

Hinzman, L.D., N. Bettez, F.S. Chapin, III, M. Dyurgerov, C.L. Fastie, et al., In press: Evidence and implications of recent climate change in terrestrial regions of the Arctic. *Climatic Change* (in press).

Hoffmann, M., 1958: *Die Bisamratte*. Akademische Verlagsgellschaft Geest & Portig, K.-G., Leipzig.

Holdgate, M.W. (ed.), 1970: *Antarctic Ecology*. Academic Press, London, 998 pp.

Huntley, B., 1996: The responses of vegetation to past and future climate changes. In: *Global Change and Arctic Terrestrial Ecosystems*, W.C. Oechel, T.V. Callaghan, T.H. Gilmanov, J. I., B. Maxwell, U. Molau, and B. Sveinbjörnsson (eds.), Springer-Verlag, New York, 290–311.

Huskey, L., M. Berman, and A. Hill, In press: Leaving home, returning home: Migration as a labor market choice for Alaska Natives. *Annals of Regional Science*.

Jacobs, S.S., C.F. Giulivi, and P.A. Mele, 2002: Freshening of the Ross Sea during the late 20th century. *Science*, **297**, 386–389.

Jernsletten, J.L. and K. Klokov, 2002: *Sustainable Reindeer Husbandry*. University of Tromso, Tromso.

Jia, G.J., H.E. Epstein, and D.A. Walker, 2003: Greening of arctic Alaska, 1981–2001. *Geophysical Research Letters*, **30**, 30: 2067, doi:10.1029/2003GL018268.

Johnsen, B.O. and A.J. Jensen, 1991: The *Gyrodactylus* story in Norway. *Aquaculture*, **98**, 289–302.

Jolly, D., F. Berkes, J. Castleden, T. Nichols, and C.o.S. Harbour, 2002: We can't predict the weather like we used to: Inuvialuit observations of climate change, Sachs Harbour, Western Canadian Arctic. In: *The Earth Is Faster Now: Indigenous Observations of Arctic Environmental Change*, I. Krupnik and D. Jolly (eds.), Arctic Research Consortium of the United States, Fairbanks, Alaska, 93–125.

Jonasson, S., T.V. Callaghan, G.R. Shaver, and L.A. Nielsen, 2000: Arctic terrestrial ecosystems and ecosystem function. In: *The Arctic: Environment, People, Policy*, T.V. Callaghan (ed.), Harwood Academic Publishers, Amsterdam, 275–314.

Kallenborn, R., M. Oeheme, D.D. Wynn-Williams, M. Schlabach, and J. Harris, 1998: Ambient air levels and atmospheric long-range transport of persistent organochlorines to Signy Island, Antarctica. *Science of the Total Environment*, **220**, 167–180.

Karl, D.M., D.F. Bird, K. Bjorkman, T. Houlihan, R. Shackelford, and L. Tupas, 1999: Micoorganisms in the accreted ice of Lake Vostok, Antarctica. *Science,* **286,** 2144–2147.

Kennedy, A.D., 1993: Water as a limiting factor in the Antarctic terrestrial environment: A biogeographical synthesis,,. *Arctic and Alpine Research,* **25,** 308–315.

Keyser, A.R., J.S. Kimball, R.R. Nemani, and S.W. Running, 2000: Simulating the effects of climatic change on the carbon balance of North American high-latitude forests. *Global Change Biology,* **6 (Suppl. 1),** 185–195.

King, J.C., J. Turner, G.J. Marshall, W.M. Connally, and T.A. Lachlan-Cope, 2003: Antarctic Peninsula climate variability and its causes as revealed by analysis of instrumental records. In: *Antarctic Peninsula Climate Variability: Historical and Palaeoenvironmental Perspectives,* E. Domack, A. Burnett, A. Leventer, P. Convey, M. Kirby, and R. Bindschadler (eds.). 79, American Geophysical Union, Washington, 17–30.

Kirkvliet, J. and W. Nebesky, 1997: Whaling and wages on Alaska's North Slope: A time allocation approach to natural resource use. *Economic Development and Cultural Change,* **46,** 651–665.

Klein, D., 1989: Subsistence hunting. In: *Wildlife Production Systems: Economic Utilisation of Wild Ungulates,* L.M. Baskin (ed.), Cambridge University Press, Cambridge, 96–111.

Klein, D.R., 1999: The roles of climate and insularity in establishment and persistence of *Rangifer tarandus* populations in the high Arctic. *Ecological Bulletins,* **47,** 96–104.

Knapp, G., 2000: The population of the circumpolar north. In: *The Arctic: Environment, People, Policy,* M. Nuttall and T.V. Callaghan (eds.), Amsterdam, Harwood Academic Publishers.

Knapp, G. and T. Morehouse, 1991: Alaska's North Slope Borough revisited. *Polar Record,* **27,** 303–312.

Kofinas, G., G. Osherenko, D. Klein, and B.C. Forbes, 2000: Research planning in the face of change: The human role in reindeer/caribou systems. *Polar Research,* **19,** 3–22.

Kouki, J., 1999: Latitudinal gradients in species richness in northern areas: Some exceptional patterns. *Ecological Bulletins,* **47,** 30–37.

Kouki, J., P. Niemelä, and M. Viitasaari, 1994: Reversed latitudinal gradient in species richness of sawflies (Hymenoptera, Symphyta). *Annales Zoologici Fennici,* **31,** 83–88.

Kozhevnikov, Y.P., 2000: Is the Arctic getting warmer or cooler? In: *Heritage of the Russian Arctic: Research, Conservation and International Cooperation,* B. Ebbinge, Y.L. Mazourov, and P.S. Tomkovich (eds.) Moscow, 348–355.

Krabill, W., W. Abdalati, E. Frederick, S. Manizade, C. Martin, et al. 2000: Greenland Ice Sheet: High elevation balance and peripheral thinning. *Science,* **289,** 428–430.

Krupnik, I., 2000: Reindeer pastoralism in modern Siberia: Research and survival in the time of crash. *Polar Research,* **19,** 49–56.

Krupnik, I., 2002: Watching ice and weather our way: Some lessons from Yupik observations of sea ice and weather on St. Lawrence Island, Alaska. In: *The Earth is Faster Now: Indigenous Observations of Arctic Environmental Change,* I. Krupnik and D. Jolly (eds.), Arctic Research Consortium of the United States, Fairbanks, 156–197.

Krupnik, I. and D. Jolly (eds.), 2002: *The Earth is Faster Now: Indigenous Observations of Arctic Environmental Change.* Arctic Research Consortium of the United States, Fairbanks, Alaska, 384 pp.

Kruse, J.A., 1991: Alaska Inupiat subsistence and wage employment patterns: Understanding individual choice. *Human Organization,* **50,** 3117–3126.

Kryazhkov, V.A., 2000: Pravo korennykh malochislennykh narodov Severa na zemlyu i drugie prirodnye resursy, Obychay i zakon. In: *Issledovaniya po Yuridicheskoy Antropologii,* N.I. Novikova and V.A. Tishkov (eds.), Strategiya, Moscow.

Kwok, R. and J.C. Comiso, 2002: Spatial patterns of variability in Antarctic surface temperature: Connections to the Southern Hemisphere Annular Mode and the Southern Oscillation. *Geophysical Research Letters,* **29(14),** 50-1–50-4.

Lebedeva, E., 1998: Waders in agricultural habitats of European Russia. *International Wader Studies,* **10,** 315–324.

Lindström, Å. and J. Agrell, 1999: Global change and possible effects on the migration and reproduction of arctic-breeding waders. *Ecological Bulletins,* **47,** 145–159.

Liu, J., J.A. Curry, and D.G. Martinson, 2004: Interpretation of recent Antarctic sea ice variability. *Geophysical Research Letters,* **31.**

Liu, J. P., X. J. Yuan, D. Rind, and D.G. Martinson, 2002: Mechanism study of the ENSO and southern high latitude climate teleconnections. *Geophysical Research Letters* **29,** 1029/2002GL015143.

Lloyd, A.H., T.S. Rupp, C.L. Fastie, and A.M. Starfield, 2003a: Patterns and dynamics of treeline advance on the Seward Peninsula, Alaska. *Journal of Geophysical Research,* **107(8161),** doi: 8110.1029/2001JD000852.

Lloyd, A.H., K. Yoshikawa, C.L. Fastie, L. Hinzman, and M. Fraver, 2003b: Effects of permafrost degradation on woody vegetation at arctic treeline on the Seward Peninsula, Alaska. *Permafrost and Periglacial Processes,* **14,** doi: 10.1002/ppp.446.

Lloyd, C.R., 2001: The measurement and modeling of the carbon dioxide exchange at a high arctic site in Svalbard. *Global Change Biology,* **7,** 405–426.

Loeb, V., V. Siegel, O. Holm-Hansen, R. Hewitt, W. Fraser, W. Trivelpiece, and S. Trivelpiece, 1997: Effects of sea-ice extent and krill or salp dominance on the Antarctic food web. *Nature* **387,** 897–900.

Longton, R.E., 1988: *Biology of Polar Bryophytes and Lichens.* Cambridge University Press, Cambridge.

Lyons, W.B., C.A. Nezat, K.A. Welch, S.T. Kottmeier, and P.T. Doran, 2000: Fossil fuel burning in Taylor Valley, southern Victoria Land, Antarctica: Estimating the role of scientific activities on carbon and nitrogen reservoirs and fluxes. *Environmental Science and Technology,* **34,** 1659–1662.

MacDonald, R., T. Harner, J. Fyfe, H. Loeng, and T. Weingartner, 2002: *AMAP Assessment 2002: The Influence of Global Change on Contaminant Pathways to, within, and from the Arctic.* Arctic Monitoring and Assessment Programme, Oslo.

Mann, M.E., R.S. Bradley, and M.K. Hughes, 1999: Northern hemisphere temperatures during the past millennium: Inferences, uncertainties and limitations. *Geophysical Research Letters,* **26,** 759–762.

Marine Mammal Commission, 2002: *Annual report to congress 2001.* Marine Mammal Commission, Bethesda Maryland, 1–253 pp.

Maslanik, J.A., M.C. Serreze, and R.G. Barry, 1996: Recent decreases in arctic summer ice cover and linkages to atmospheric circulation anomalies. *Geophysical Research Letters,* **23,** 1677–1680.

Matveyeva, N. and Y. Chernov, 2000: Biodiversity of terrestrial ecosystems. In: *The Arctic: Environment, People, Policy,* M. Nuttall and T.V. Callaghan (eds.), Harwood Academic Publishers, Amsterdam.

May, R.M., J.R. Beddington, C.W. Clark, S.J. Holt and R.M. Laws, 1979: Management of multispecies fisheries. *Science,* **205,** 267–277.

McGuire, A.D., J.S. Clein, J.M. Melillo, D.W. Kicklighter, R.A. Meier, C.J. Vorosmarty, and M.C. Serreze, 2000: Modeling carbon responses of tundra ecosystems to historical and projected climate. The sensitivity of pan-arctic carbon storage to temporal and spatial variation in climate. *Global Change Biology,* **6(Suppl. 1),** 141–159.

McGuire, A.D., J.M. Melillo, D.W. Kicklighter, Y. Pan, X. Xiao, et al. 1997: Equilibrium responses of global net primary production and carbon storage to doubled atmospheric carbon dioxide: Sensitivity to changes in vegetation nitrogen concentration. *Global Biogeochemical Cycles,* **11,** 173–189.

McGuire, A.D., C. Wirth, M. Apps, J. Beringer, J. Clein, et al., 2002: Environmental variation, vegetation distribution, carbon dynamics, and water/energy exchange in high latitudes. *Journal of Vegetation Science,* **13,** 301–314.

Meador, J., W.H. Jeffrey, J.P. Kase, J.D. Pakulski, S. Chiarello, and D.L. Mitchell, 2002: Seasonal fluctuation of DNA photodamage in marine plankton assemblages at Palmer Station, Antarctica. *Photochemistry and Photobiology,* **75(3),** 266–271.

Mitchell, J.F.B., D.J. Karoly, G.C. Hegerl, F.W. Zwiers, M.R. Allen, and J. Marengo, 2001: Detection of climate change and attribution of causes. In: *Climate Change 2001: The Scientific Basis,* C.A. Johnson (ed.), Cambridge University Press, Cambridge, 695–738.

Moline, M.A., H. Claustre, T.K. Frazer, J. Grzymski, O.M. Schofield, and M. Vernet, 2000: Changes in phytoplankton assemblages along the Antarctic Peninsula and potential implications for the Antarctic food web. In: *Antarctic Ecosystems: Models for Wider Ecological Understanding,* W. Davidson, C. Howard-Williams, and P. Broady (eds.), 263–271.

Morozov, V.V., 1998: Distribution of breeding waders in the north-east European Russian tundras. *International Wader Studies,* **10,** 186–194.

Muckle, G., P. Ayotte, E. Dewailly, S.W. Jacobson, and J.L. Jacobson, 2001: Prenatal exposure of Inuit infants from Northern Quebec to elevated levels of environmental contaminants. *Environmental Health Perspectives,* **109(12),** 1291–1299.

Murphy, E. J., A. Clarke, C. Symon, and J. Priddle, 1995: Temporal Variation in Antarctic Sea-Ice—Analysis of a Long-Term Fast-Ice Record from the South-Orkney Islands. *Deep-Sea Research Part I-Oceanographic Research Papers* **42,** 1045–1062.

Murphy, E. J., J. L. Watkins, K. Reid, P.N. Trathan, I. Everson, et al. 1998: Interannual variability of the South Georgia marine ecosystem: Biological and physical sources of variation in the abundance of krill. *Fisheries Oceanography* **7,** 381–390.

Murphy, P.J., J.P. Mudd, B.J. Stocks, E.S. Kasischke, D. Barry, M.E. Alexander, and N.H.F. French, 2000: Historical fire records in the North American boreal forest. In: *Fire, climate change and carbon cycling in the North American boreal forests*, E.S. Kasischke and B.J. Stocks (eds.), Springer-Verlag, New York, 275–288.

Myers, H., 2000: Options for appropriate development in Nunavut communities. *Inuit Studies*, **24**, 25–40.

Myneni, R.B., C.D. Keeling, C.J. Tucker, G. Asrar, and R.R. Nemani, 1997: Increased plant growth in the northern high latitudes from 1981–1991. *Nature*, **386**, 698–702.

Naganobu, M., K. Kutsuwada, Y. Sasui, S. Taguchiand, and V. Siegel, 2000: Relationships between Antarctic krill (*Esphasauia superba*) variability and westerly fluctuations and ozone depletion in the Antarctic Peninsula area. *Journal of Geophysical Research*, **104(C9)**, 20651–20665.

Nakamura, N. and A.H. Oort, 1988: Atmospheric heat budgets of the polar regions. *Journal of Geophysical Research*, **93(D8)**, 9510–9524.

Nellemann, C., L. Kullerud, J. Vistnes, B.C. Forbes, G.P. Kofinas, et al. (eds.), 2001: *GLOBIO—Global methodology for mapping human impacts on the biosphere*. United Nations Environment Programme.

Nesler, T.P. and E.P. Bergersen (eds.), 1991: *Mysis in Fisheries: Hard Lessons from Headlong Introductions*. American Fisheries Society, Bethesda, Maryland.

Neuvonen, S., P. Niemela, and T. Virtanen, 1999: Climatic change and insect outbreaks in boreal forests: The role of winter temperature. *Ecological Bulletins*, **47**, 63–67.

Newsam, K.K., 2003:: UV-B radiation arising from stratospheric ozone depletion influences the pigmentation of the Antarctic moss *Andreaea regularis*. *Oecologia*, **135**, 327–331.

Newsham, K.K., D.A. Hodgson, A.W.A. Murray, H.J. Peat, and R.I.L. Smith, 2002: Response of two Antarctic bryophytes to stratospheric ozone depletion. *Global Change Biology*, **8**, 972–983.

Nickels, S., C. Furgal, J. Castleden, P. Moss-Davies, M. Buell, et al. 2002: Putting the human face on climate change through community workshops. In: *The Earth is Faster Now: Indigenous Observations of Arctic Environmental Change*, I. Krupnik and D. Jolly (eds.), Arctic Research Consortium of the United States, Fairbanks, Alaska, 301–333.

Niemelä, P., F.S. Chapin, III, K. Danell, and J.P. Bryant, 2001: Animal-mediated responses of boreal forest to climatic change. *Climatic Change*, **48**, 427–440.

Nilsson, A. and O. Tenow, 1990: Diapause, embryo growth and supercooling capacity of *Epirrita autumnata* eggs from northern Fennoscandia. *Entomologia Experimentalis et Applicata*, **57**, 39–55.

Nordstrøm, C., H. Søgaard, T.R. Christensen, T. Friborg, and B.U. Hansen, 2001: Seasonal carbon dioxide balance and respiration of a high-arctic fen ecosystem in NE-Greenland. *Theoretical and Applied Climatology*, **70**, 149–166.

Norton, D.W., 2002: Coastal sea ice watch: Private confessions of a convert to indigenous knowledge. In: *The Earth Is Faster Now: Indigenous Observations of Arctic Environmental Change*, I. Krupnik and D. Jolly (eds.), Arctic Research Consortium of the United States, Fairbanks, Alaska, 127–155.

NRC, 2003: *Cumulative Environmental Effects of Oil and Gas Activities in Alaska's North Slope*. National Academies Press, Washington, 288 pp.

Nuttall, M., 1992: *Arctic Homeland: Kinship, Community and Development in Northwest Greenland*. University of Toronto Press, Toronto.

Nuttall, M., F. Berkes, B.C. Forbes, G.P. Kofinas, G. Wenzel, and T.K. Vlassova, In press: Hunting, herding, fishing and gathering: Indigenous peoples and renewable resource use in the Arctic. In: *Arctic Climate Impact Assessment*, ACIA (ed.), Cambridge University Press, Cambridge.

Nylen, T.H., A.G. Fountain, and P.T. Doran, 2004: Climatology of katabatic winds in the McMurdo dry valleys, southern Victoria Land, Antarctica. *Journal of Geophysical Research—Atmospheres*, **109(D3)**, D03114.

Oechel, W.C., G.L. Vourlitis, S.J. Hastings, R.C. Zulueta, L. Hinzman, and D. Kane, 2000: Acclimation of ecosystem CO_2 exchange in the Alaskan Arctic in response to decadal climate warming. *Nature*, **406**, 978–981.

Oechel, W.C., S. Cowles, N. Grulke, S.J. Hastings, W. Lawrence, et al. 1994: Transient nature of CO_2 fertilization in arctic tundra. *Nature*, **371**, 500–503.

Ogden, A.E., In press: Adaptation user's manual: Arctic terrestrial ecosystems. In: *Arctic Climate Impact Assessment*, ACIA (ed.). Cambridge University Press, Cambridge.

Osherenko, G., 1988: Can co-management save arctic wildlife? *Environment*, **30(6)**, 6–13, 29–34.

Osherenko, G., 2001: Indigenous rights in Russia: Is title to land essential for cultural survival? *Georgetown International Environmental Law Review*, **13**, 695–734.

Osherenko, G. and O.R. Young, 1989: *The Age of the Arctic: Hot Conflicts and Cold Realities*. Cambridge University Press, Cambridge.

Osterkamp, T.E. and V.E. Romanovsky, 1999: Evidence for warming and thawing of discontinuous permafrost in Alaska. *Permafrost and Periglacial Processes*, **10**, 17–37.

Overland, J.E., M. Spillane, and N.N. Soreide, 2004: Integrated analysis of physical and biological pan-Arctic change. *Climatic Change*, **63(3)**, 291–322.

Paine, R., 1994: *Herds of the Tundra: A Portrait of Saami Reindeer Pastoralism*. Smithsonian Institution Press, Washington.

Parsons, A.N., J.E. Barrett, D.H. Wall, and R.A. Virginia, 2004: Soil carbon dioxide flux in Antarctic dry valley ecosystems. *Ecosystems*, **7**, 286–295.

Peterson, B.J., R.M. Holmes, J.W. McClelland, C.J. Vörösmarty, R.B. Lammers, et al. 2002: Increasing arctic river discharge to the Arctic Ocean. *Science*, **298**, 2171–2173.

Prezelin, B.B., E.E. Hofmann, C. Mengelt, and J.M. Klinck, 2000: The linkage between upper circumpolar deep water (UCDW) and phytoplankton assemblages on the west Antarctic Peninsula Continental Shelf. *Journal of Marine Research*, **58(2)**, 165–202.

Quartino, M.L., H. Klöser, I.R. Schloss, and C. Wiencke, 2001: Biomass and associations of benthic marine macroalgae from the inner Potter Cove (King George Island, Antarctica) related to depth and substrate. *Polar Biology*, **24**, 349–355.

Quayle, W.C., L.S. Peck, H. Peat, J.C. Ellis-Evans, and P.R. Harrigan, 2002: Extreme responses to climate in Antarctic lakes. *Science*, **295**, 645.

Quayle, W.C., P. Convey, L.S. Peck, J.C. Ellis-Evans, H.G. Butler, and H.J. Peat, 2003: Ecological responses of maritime Antarctic lakes to regional climate change. In: *Antarctic Peninsula Climate Variability: Historical and Palaeoenvironmental Perspectives*, E. Domack, A. Burnett, A. Leventer, P. Convey, M. Kirby, and R. Bindschadler (eds.). 79, American Geophysical Union, Washington, 159–170.

Quetin, L.B. and R.M. Ross, 2001: Environmental variability and its impact on the reproductive cycle of Antarctic krill. *American Zoologist*, **41(1)**, 74–89.

Rignot, E. and R.H. Thomas, 2002: Mass balance of polar ice sheets. *Science*, **297(5586)**, 1502–1506.

Rignot, E., A. Rivera, and G. Casassa, 2004: Contribution of the Patagonia Icefields of South America to sea level rise. *Science*, **302**, 434–437.

Robinson, S.A., J. Wasley, and A.K. Tobin, 2003: Living on the edge—plants and global change in continental and maritime Antarctica. *Global Change Biology*, **9**, 1681–1717.

Russell, D.E., G. Kofinas, and B. Griffith, 2002: *Barren-ground caribou calving ground workshop*. Technical Report Series 39, Canadian Wildlife Service, Ottawa.

Sahlins, M., 1999: What is anthropological enlightenment? Some lessons of the twentieth century. *Annual Review of Anthropology*, **28**, 1–23.

Sarmiento, J.L. and C. Le Quere, 1996: Oceanic carbon dioxide uptake in a model of century-scale global warming. *Science*, **274**, 1346–1350.

Sarmiento, J.L., T.M.C. Hughes, R.J. Stouffer, and S. Manabe, 1998: Simulated response of the ocean carbon cycle to anthropogenic climate warming. *Nature*, **393**, 245–249.

Scambos, T.A., J.A. Bohlander, C.A. Shuman, and P. Skvarca, 2004: Glacier acceleration and thinning after ice shelf collapse in the Larsen B embayment, Antarctica. *Geophysical Research Letters*, **31**, 10.1029/2004GL020670.

Schlesinger, W.H., 1997: *Biogeochemistry: An Analysis of Global Change*. Second ed. Academic Press, San Diego, 443 pp.

Serreze, M.C., D.H. Bromwich, M.P. Clark, A.J. Etringer, T. Zhang, and R.B. Lammers, 2003: Large-scale hydro-climatology of the terrestrial arctic drainage system. *Journal of Geophysical Research*, **108**, doi:10.1029/2001 JD000919.

Serreze, M.C., J.E. Walsh, F.S. Chapin, III, T. Osterkamp, M. Dyurgerov, V. Romanovsky, W.C. Oechel, J. Morison, T. Zhang, and R.G. Barry, 2000: Observational evidence of recent change in the northern high-latitude environment. *Climatic Change*, **46**, 159–207.

Shaver, G.R., J. Canadell, F.S. Chapin, III, J. Gurevitch, J. Harte, et al.2000: Global warming and terrestrial ecosystems: A conceptual framework for analysis. *BioScience*, **50**, 871–882.

Siegel, V., B. Bergstrom, J.O. Stromberg, and P.H. Schalk, 1990: Distribution, size frequencies and maturity stages of krill, *Euphausia superba*, in relation to sea-ice in the northern Weddell Sea. *Polar Biology*, **10**, 549–557.

Sitch, S., B. Smith, I.C. Prentice, A. Arneth, A. Bondeau, et al. 2003: Evaluation of ecosystem dynamics, plant geography and terrestrial carbon cycling in the LPJ dynamic global vegetation model. *Global Change Biology*, **9**, 161–185.

Smith, L.C., G.M. MacDonald, A.A. Velichko, D.W. Beilman, O.K. Borisova, et al. 2004: Siberian peatlands a net carbon sink and global methan source since the early Holocene. *Science*, **303**, 353–356.

Smith, R.C., K.S. Baker, and S.E. Stammerjohn, 1998: Exploring sea ice indexes for polar ecosystem studies. *BioScience*, **48(2)**, 83–93.

Smith, R.C., W.R. Fraser, and S.E. Stammerjohn, 2003: Climate variability and ecological response of the marine ecosystem in the western Antarctic Peninsula (WAP) region. In: *Climate Variability and Ecosystem Response at Long-Term Ecological Research Sites,* D. Greenland, D.G. Goodin, and R.C. Smith (eds.), Oxford University Press, New York, 158–173.

Smith, R.C., B.B. Prézelin, K.S. Baker, R.R. Bidigare, N.P. Boucher, et al. 1992: Ozone depletion: Ultraviolet radiation and phytoplankton biology in Antarctic waters. *Science,* **255,** 952–959.

Smith, R.I.L., 1984: Terrestrial plant biology of the sub-Antarctic and Antarctic. In: *Antarctic Ecology,* R.M. Laws (ed.), Academic Press, London.

Smith, R.I.L., 1994: Vascular plants as bioindicators of regional warming in Antarctica. *Oecologia,* **99,** 322–328.

Smith, V.R., 2002: Climate change in the subantarctic: An illustration from Marion Island. *Climatic Change,* **52,** 345–357.

Soegaard, H., T. Friborg, B.U. Hansen, C. Nordstrøm, and T.R. Christensen, 2000: Trace gas exchange in a high arctic valley. 3. Integrating and scaling CO_2 fluxes from canopy to landscape using flux data, footprint modelling and remote sensing. *Global Biogeochemical Cycles,* **14,** 725–744.

Soloviev, M.Y., P.S. Tomkovich, and N. Davidson, 1998: An international breeding condition survey of arctic waterfowl: Progress report. *WSG Bulletin,* **87,** 43–47.

Srivastava, D.S. and R.L. Jefferies, 1995: Mosaics of vegetation and soil salinity: A consequence of goose foraging in an arctic salt marsh. *Canadian Journal of Botany,* **73,** 75–83.

Stabler, J., 1990: A utility analysis of activity patterns of Native males in the Northwest Territories. *Economic Development and Cultural Change,* **39,** 47–60.

Stepanov, V., 2004: *Russian experience in the north indigenous statistics.* Workshop of data collection and disaggregation for indigenous people, United Nations Department of Economic and Social Affairs, Division for Social Policy and Development, Secretariat of the Permanent Forum on Indigenous Issues, New York, January 2004.

Strathdee, A.T. and J.S. Bale, 1998: Life on the edge: Insect ecology in Arctic environments. *Annual Reviews of Entomology,* **43,** 85–106.

Strathdee, A.T., J.S. Bale, W.C. Block, S.J. Coulson, I.D. Hodkinson, and N.R. Webb, 1993: Effects of temperature elevation on a field population of *Acyrthosiphon svalbardicum* (Hemiptera: Aphididae) on Spitsbergen. *Oecologia,* **96,** 457–465.

Sturm, M., C. Racine, and K. Tape, 2001: Increasing shrub abundance in the Arctic. *Nature,* **411,** 546–547.

Sugden, D., 1982: *Arctic and Antarctic: A Modern Geographical Synthesis.* Basil Blackwell, Oxford.

Summers, R.W. and L.G. Underhill, 1987: Factors relating to breeding populations of Brent Geese (*Branta b. bernicla*) and waders (*Charadrii*) on the Taimyr Peninsula. *Bird Study,* **34,** 161–171.

Sundet, J.H., 1998: Kongekrabben, en nykommer fra öst. In: *Et levende hav,* S. Mortensen and P. Eide (eds.), KOM Publisher, Kristiansund, Norway.

Takahashi, T., S.C. Sutherland, C. Sweeney, A. Poisson, N. Metzl, et al. 2002: Global sea-air CO2 flux based on climatological surface ocean pCO2, and seasonal biological and temperature effects. *Deep-Sea Research II,* **49,** 1601–1622.

Thomas, R.H., 2004: Greenland: Recent mass balance observations. In: *Mass Balance of the Cryosphere: Observations and Modeling of Contemporary and Future Changes,* J.L. Bamber and A.J. Payne (eds.), Cambridge University Press, Cambridge, 393–436.

Thompson, D.W.J. and S. Solomon, 2002: Interpretation of recent Southern Hemisphere climate change. *Science,* **296,** 895–899.

Thorpe, N., N. Eyegetok, N. Hakongak, and K. Elders, 2002: Nowadays it is not the same: Inuit Quajimajatuqangit, climate and caribou in the Kitikmeot Region of Nunavut, Canada. In: *The Earth is Faster Now: Indigenous Observations of Arctic Environmental Change,* I. Krupnik and D. Jolly (eds.), Research Consortium of the United States, Fairbanks, Alaska, 201–239.

Thost, D. and I. Allison, In press: The Climate of Heard Island. In: *Heard Island: Southern Ocean,* K. Green and E.J. Woehler (eds.), Sentinel Surrey Beatty, Sydney.

Turner, J., S.R. Colwell, and S. Harangozo, 1997: Variability of precipitation over the coastal western Antarctic Peninsula from synoptic observations. *Journal of Geophysical Research,* **102,** 13999–14007.

Turner, J., J.C. King, and T.A.J. Lachlan-Cope, P. D., 2002: Recent temperature trends in the Antarctic. *Nature,* **418,** 291–292.

Van Wijk, M.T., K.E. Clemmensen, G.R. Shaver, M. Williams, T.V. Callaghan, et al., 2004: Long-term ecosystem level experiments at Toolik Lake, Alaska and at Abisko, Northern Sweden: Generalisations and differences in ecosystem and plant type responses to global change. *Global Change Biology,* **10,** 105–123.

Vaughan, D.G., G.J. Marshall, W.M. Connolley, C. Parkinson, R. Mulvaney, D.A. Hodgson, J.C. King, C.J. Pudsey, and J. Turner, 2003: Recent rapid regional climate warming on the Antarctic Peninsula. *Climatic Change,* **60,** 243–274.

Vernet, M. and W.A. Kozlowski, 2001: Ultraviolet radiation and the Antarctic coastal marine ecosystem. In: *Ecosystems and Ultraviolet Radiation,* C.S. Cockell and A.R. Blaustein (eds.), Springer-Verlag, New York, 170–194.

Vidal, E., P. Jouventin, and Y. Frenot, 2003: Contribution of alien and indigenous species to plant-community assemblages near penguin rookeries at Crozet archipelago. *Polar Biology,* **26,** 432–437.

Virtanen, T. and S. Neuvonen, 1999: Climate change and macrolepidopteran biodiversity in Finland. *Chemosphere,* **1,** 439–448.

Vistnes, I. and C. Nellemann, 2001: Avoidance of cabins, roads, and power lines by reindeer during calving. *Journal of Wildlife Management,* **65,** 915–925.

Vlassova, T.K., 2002: Human impacts on the tundra-taiga zone dynamics: The case of the Russian lesotundra. *Ambio Special Report (Tundra-Taiga Treeline Research),* **12,** 30–36.

Volney, W.J.A. and R.A. Fleming, 2000: Climate change and impacts of boreal forest insects. *Agriculture Ecosystems & Environment,* **82,** 283–294.

Walker, D.A., P.J. Webber, E.F. Binnian, K.R. Everett, N.D. Lederer, E.A. Nordstrand, and M.D. Walker, 1987: Cumulative impacts of oil fields on northern Alaskan landscapes. *Science,* **238,** 757–761.

Walther, G.-R., E. Post, P. Convey, A. Menzel, C. Parmesan, et al. 2002: Ecological responses to recent climate change. *Nature,* **416,** 389–395.

Walton, D.W.H., 1984: The terrestrial environment. In: *Antarctic Ecology,* R.M. Laws (ed.), Academic Press, London.

Wang, X.J. and J.R. Key, 2003: Recent trends in arctic surface, cloud, and radiation properties from space. *Science,* **299,** 1725–1728.

Watson, A., L. Alessa, and B. Glaspell, 2003: The relationship between traditional ecological knowledge, evolving cultures, and wilderness protection in the circumpolar north. *Conservation Ecology,* **8(1),** http://www.consecol.org/vol8/iss1/art2.

Weber, K. and H. Goerke, 2003: Persistent organic pollutants (POPs) in Antarctic fish: Levels, patterns, changes. *Chemosphere,* **53,** 667–678.

Weiler, C.S. and P.A. Penhale (eds.), 1994: *Ultraviolet Radiation in Antarctica: Measurements and Biological Effects.* Vol. 62. *Antarctic Research Series,* American Geophysical Union, Washington.

Weller, G., C.R. Bentkey, D.H. Elliot, L.J. Lanzerotti, and P.J. Webber, 1987.: Laboratory Antarctica—Research contributions to global problems. *Science,* **238,** 1361–1368.

White, W. B. and R. G. Peterson 1996: An Antarctic circumpolar wave in surface pressure, wind, temperature and sea-ice extent. *Nature* **380,** 699–702.

Whiteman, G., B.C. Forbes, J. Niemelä, and F.S. Chapin, III, 2004: Bringing feedback and resilience of high-latitude ecosystems into the corporate boardroom. *Ambio,* **33(6),** 371–376.

Wiedema, I.R. (ed.), 2000: *Introduced species in the Nordic countries.* Nord 2000, 13.

Wilson, P.R., D.G. Ainley, N. Nur, S.S. Jacobs, K.J. Barton, G. Ballard, and J.C. Comiso, 2001: Adélie penguin population change in the pacific sector of Antarctica: Relation to sea-ice extent and the Antarctic Circumpolar Current. *Marine Ecology Progress Series,* **213,** 301–309.

Wong, A.P.S., N.L. Bindoff, and J.A. Church, 1999: Large-scale freshening of intermediate waters in the Pacific and Indian Oceans. *Nature,* **400,** 440–443.

WWF, 2002: *Wanted: Protected Areas in the Arctic. Arctic Bulletin No. 1.02.* World Wildlife Fund International Arctic Programme, Oslo.

Yang, D., D. Kane, L.D. Hinzman, X. Zhang, and H. Ye, 2002: Siberian Lena River hydrologic regime and recent change. *Journal of Geophysical Research,* **107.**

Yoccoz, N.G. and R.A. Ims, 1999: Demography of small mammals in cold regions: The importance of environmental variability. *Ecological Bulletins,* **47,** 137–144.

Zimov, S.A., V.I. Chuprynin, A.P. Oreshko, F.S. Chapin, III, J.F. Reynolds, and M.C. Chapin, 1995: Steppe-tundra transition: An herbivore-driven biome shift at the end of the Pleistocene. *American Naturalist,* **146,** 765–794.

Zimov, S.A., Y.V. Voropaev, I.P. Semiletov, S.P. Davidov, S.F. Prosiannikov, et al. 1997: North Siberian lakes: a methane source fueled by Pleistocene carbon. *Science,* **277,** 800–802.

Zöckler, C. and I. Lysenko, 2000: *Water birds on the edge: First circumpolar assessment of climate change impact on Arctic Breeding Water Birds.* WCMC Biodiversity Series 11, 20 pp.

Chapter 26

Cultivated Systems

Coordinating Lead Authors: Kenneth G. Cassman, Stanley Wood

Lead Authors: Poh Sze Choo, H. David Cooper, C. Devendra, John Dixon, Joanne Gaskell, Shabaz Khan, Rattan Lal, Leslie Lipper, Jules Pretty, Jurgenna Primavera, Navin Ramankutty, Ernesto Viglizzo, Keith Wiebe

Contributing Authors: Sandra Kadungure, Nancy Kanbar, Zahia Khan, Roger Leakey, Sarah Porter, Kate Sebastian, Rebecca Tharme

Review Editors: Arsinio Balisacan, Peter Gardiner

Main Messages

Approximately 24% of Earth's terrestrial surface is occupied by cultivated systems. Cultivated areas continue to expand in some areas but are shrinking in others. As the demand for food, feed, and fiber has increased, farmers have responded by expanding the cultivated area, intensifying production (for example, higher yields per unit land-time), or both. Globally, over the past 40 years intensification of cultivated systems has been the primary source (almost 80%) of increased output. In countries with high levels of productivity and low population growth rates, the extent and distribution of land under cultivation is stabilizing or even contracting (for example, Australia, Japan, the United States, and Italy). The area in agricultural production has also stabilized and begun to contract in China. But some countries, predominantly found in sub-Saharan Africa, have had persistently low levels of productivity and continue to rely mainly on the expansion of cultivated area.

Globally, opportunities for further expansion of cultivation are reducing. Since nearly all well-suited land is currently cultivated, continued expansion draws more economically marginal land (steeper slopes, poorer soils, harsher climates, or reduced market access) into production—often with unwelcome social and environmental consequences.

Cultivated systems specialize in the provision of food, feed, and fiber, often at the expense of other ecosystem services. Cultivation has affected the provision of other services in three ways: by conversion of biologically diverse natural grasslands, wetlands, and native forests into less diverse agroecosystems; by the choice of crop species grown and the pattern of cropping in time and space; and by the manner in which crops, soil, and water resources are managed at both plot and landscape levels. For many ecosystem services, significant losses arise as a direct consequence of conversion to agriculture. Subsequent impacts are conditioned primarily by the intensity of cultivation in time and space, by the type and amount of applied inputs, including water, nitrogen, and pesticides, and by the effectiveness with which production inputs and residues are managed.

Two key "win-win" strategies have emerged to increase economic benefits to farmers while reducing negative ecosystem aspects of cultivation: first, increasing the productivity of existing cropland through intensive management of specialized cropping systems and use of improved crop, soil, and water management practices and, second, designing more diverse crop and agroforestry systems that provide improved livelihood options as well as supporting greater levels of biological diversity and other environmental services at a local level.

Because food security requires that increasing demand for food be met, difficult choices about ecosystem service trade-offs are faced when evaluating alternative cultivation strategies. For example, intensification of production to gain more output per unit land area and time runs the risk of unintended negative impacts associated with greater use of external inputs such as fuel, irrigation, fertilizer, and pesticides. Likewise, area expansion of production reduces natural habitat and biodiversity through land use conversion and decreases the other environmental services that natural ecosystems provide. Which strategy has the least overall impact on ecosystem services depends on the specific context.

This assessment strongly suggests that pursuing the necessary increases in global food output by emphasizing the development of more environmentally and ecologically sound intensification is likely to be the preferred, and in many cases the only, long-term strategy.

Improved cultivation practices can conserve biodiversity in several ways: sustaining adequate yield increases on existing cropland in order to limit expansion of cultivation, enlightened management of cultivation mosaics at the landscape scale, and increasing diversity within cropping systems. At the global level, conversion of natural habitat to agricultural uses is perhaps the single greatest threat to biodiversity. Hence, sustaining yield increases on existing farmland to meet growing human food needs will be essential for the conservation of existing biodiversity. At the local level, advances in ecological science coupled with field-based experimentation have yielded improved insights as to how farmers might configure and manage cultivated systems so as to enhance opportunities for wild biodiversity through, for example, habitat creation, wildlife corridors, refugia, and buffers around sensitive areas. More has also been learned about maintaining viable collections of wild relatives of commercially cultivated products, particularly in farming communities (in-situ conservation). But such approaches are most likely to be used where there are demonstrable benefits to farmers.

The economic benefits of pollinators, biological control of pests, soil bacteria, insects, birds, and other animals are better understood and are increasingly being articulated to farmers and the agricultural community. Successes include the rapid and extensive spread of integrated pest management in Southeast Asia and the growing acceptance of the role of sustainability-focused platforms such as eco-agriculture, agroecology, and integrated natural resource management by both subsistence and commercial farmers.

Cultivated systems have become the major global consumer of water. While rain-fed croplands might consume more or less water than the natural plant communities they replaced, irrigated areas consume significantly more. About 18% of the area of cultivated systems is irrigated, but the crop output generated by such irrigation represents about 40% of global food production. While irrigation systems divert 20–30% of the world's available water resources, chronic inefficiencies in distribution and application result in only 40–50% of that water being used in crop growth.

Growing water demand for uses other than agriculture is increasingly competing with water demand for food production in many areas, and more transparent and equitable approaches to water allocation are needed. There is significant scope to achieve substantial increases in irrigation efficiency from improvements in water delivery systems (irrigation system maintenance and design; drip irrigation) and from improvements in water application methods (improved irrigation scheduling). Water harvesting practices, including small tanks, runoff farming, and zai (dug pits that concentrate water at the plant), have also proved effective, as have structural landscape features such as shelterbelts that reduce evapotranspiration.

In addition to water quantity trade-offs, intensification of food production involving increased use of applied nutrients and agricultural chemicals can lead to water pollution that degrades downstream freshwater, estuarine, and marine ecosystems and that limits downstream water use or raises its costs. Technologies or practices that increase nutrient use efficiency and minimize the need for pesticide application can greatly reduce water pollution from intensive agriculture. Inappropriate farming practices on sloping land prone to erosion and expansion of rain-fed cropping onto sloping lands with marginal soils can result in severe erosion that also contributes to pollution of rivers, water bodies, and estuary or marine ecosystems.

Cultivation has accelerated and modified the spatial patterns of nutrient cycling. Most pressing is the disruption of the nitrogen cycle, caused primarily by the application of inorganic fertilizers, which included around 85 million tons of nitrogen in 2000. Nitrogen is the most commonly limiting plant nutrient and a major constituent of dietary protein. While some form of augmentation of naturally "fixed" N is an essential component of more productive cultivation, application of inorganic N increases emission of nitrous oxide, a potent greenhouse gas, and contributes to acid rain, soil acidification, and eutrophication and, through these changes, to biodiversity loss.

The best opportunity for limiting these negative effects is to increase the efficiency in the handling and application of fertilizers, as well as increased or rationalized use of organic sources of nitrogen (such as mulching, animal manure, and legume crops) to substitute for inorganic fertilizers and increase nitrogen use efficiency. Some landscape elements (ponds and buffer strips, for example) can also provide cost-effective means for mitigating water contamination. In some countries, notably the United States, Japan, and the Netherlands, there has been significant progress in improving N use efficiency and even in decreasing N application rates on several major cereal crops.

A clear distinction must be made, however, between the overuse or inefficient use of nitrogen in some parts of the world and the desperate need for substantial increases in the amount of nitrogen (and other nutrients) applied to crops in regions like sub-Saharan Africa where yields are low and often declining—precisely because of the cumulative depletion of soil nutrients. Phosphorus is another nutrient that must be applied to maintain crop yields on most agricultural soils, and lack of adequate phosphorus significantly limits agricultural productivity in regions where phosphorus fertilizers are not available or affordable.

The impact of cultivation on climate regulation can, as with biodiversity, best be viewed in two distinct stages. When natural ecosystems have been converted for cultivation, carbon-based greenhouse gases are generally released and carbon sequestration potential is reduced to an extent dependent upon the original land cover and the means of conversion. Thereafter, the impact of cultivation on climate regulation is intimately linked to production system choices and management practices. Frequent cultivation, irrigated rice production, livestock production, and the burning of cleared areas and crop residues now contribute about 166 million tons of carbon a year in methane and $1,600 \pm 800$ million tons in CO_2. About 70% of anthropogenic nitrous oxide gas emissions are attributable to agriculture, mostly from land conversion and nitrogen fertilizer use.

But while agriculture contributes to greenhouse gas emissions, it also represents an opportunity for mitigation. Minimum tillage and tree-based production systems are two of a growing number of practices being adopted by farmers for their direct productivity and income benefits, which also represent successful strategies for mitigating GHG emissions from cultivated systems. The cultivation of biofuels (such as corn, sorghum, and sugarcane used for ethanol production) is seen as having great potential, although these are of relatively minor significance at present. However, a growth in demand for biofuels would result in expansion of cultivated areas or displacement of traditional crops or both unless there is a concomitant acceleration in the rate of gain in crop yields to offset the grain and biomass used for biofuels and bio-based industrial products.

While better practices and new technologies have been and must continue to be developed to reduce the negative environmental impact of cultivation, such measures will only be adopted if they generate benefits for the farmer in a time frame of relevance to the socioeconomic context in which cultivation takes place. For example, adoption of improved soil and water conservation practices has been low in many developing countries where farmers are often poor and have few assets, limited access to credit, and uncertain access or rights to land and water resources. From an economic perspective, many negative ecosystem service impacts are production "externalities"—impacts whose costs, while real from a broader social perspective, are not factored into production decisions.

In richer countries, public funds are increasingly being used to provide incentives for producers to take greater account of the external negative impacts of production. These have included investments in payments to producers to help offset the additional costs of environmentally friendly practices, research and development of new technologies and practices that reduce the trade-offs between food provision and other ecosystem services, and environment-related regulation and enforcement systems for the agriculture sector. But the principle of engaging the potential beneficiaries of improved cultivation practices in some form of dialogue with producers continues to define new institutional arrangements to better manage production externalities. Examples are watershed user groups, commodity boards, organic certification systems, and trading of carbon credits.

National policies, international agreements, and market forces play a significant role in determining farmers' choices about the scale of cultivation, the selection of the cultivation system, and the level and mix of production inputs—all of which influence trade-offs among ecosystem services and external impacts on other ecosystems from cultivated systems. For example, where governments have invested in agricultural research and extension, productivity growth rates have been higher and area expansion rates often lower. Likewise, although investments in irrigation schemes and subsidized seeds, fertilizers, and pesticides have almost certainly resulted in depletion of river flows and increased salinization, eutrophication, and biodiversity loss, they have also led to greater productivity per unit of arable land, which reduces pressure for the expansion of cultivated systems into marginal areas and natural ecosystems. Agricultural subsidies in many industrial countries have encouraged overproduction while at the same time reducing the economic viability of cropping systems in poorer countries by driving down the prices of traded commodities such that unsubsidized (and sometimes domestically taxed) producers in those countries find it hard to compete.

Some, mainly richer, countries have introduced "conservation" or "set-aside" programs to encourage farmers to take environmentally sensitive land out of production. Others, such as Costa Rica, have gone further through programs that explicitly compensate farmers for delivering ecosystem services. Governments are playing an increasing role in ensuring that farmers will profit from cultivation choices that deliver the broad array of ecosystem services valued by society, including but not limited to food production.

Significant challenges will be faced at both global and regional levels to meet increasing food, feed, and fiber demand and to do so in ways that support key environmental services. At the global level, the rate of increase in cereal yields is falling below the rate of projected demand, which will likely lead to a large of expansion of cultivated area unless yields can be increased. Many more low-input systems in marginal lands may soon reach irreversibly low levels of soil quality and face increasingly erratic climatic patterns and new crop and livestock pests and diseases, such as coffee and banana wilt and avian flu. Such trends could lead to the collapse of important cash and food producing systems on a regional basis. There is also growing concern that market liberalization, coupled with the inability of farmers and governments in poorer countries to make the investments necessary to raise the productivity of their predominantly subsistence and smallholder agricultural sectors, may lead to further impoverishment of rural populations. A warmer global temperature associated with climate change is an emerging challenge to sustaining yield increases in currently favorable crop production areas and may decrease yield stability in dryland cropping systems dependent on rainfall.

26.1 Introduction

Human transformation of natural ecosystems for production of food, fiber, and fuel has occurred on a massive scale—cultivated systems now occupy 24% of Earth's terrestrial surface and are the single greatest land use by humans.[1] Although there are a wide variety of cultivated systems, this chapter focuses on those that constitute major providers of food, feed, or fiber or that have significant impacts on the provision of other ecosystem services, at regional or global scales. In this chapter, ecosystem services are divided into those that provide food, feed, fiber, and other cultivated outputs and "other services" that include, for example, biodiversity, fresh water, nutrient cycling, and cultural services.

Despite a tripling of the human population in the twentieth century, global food production capacity more than kept pace with demand. In fact, per capita food supply increased while food prices decreased in real terms. (See Chapter 8.) At the turn of the millennium, cultivated systems provided around 94% of the protein and 99% of the calories in human diets (FAOSTAT 2003). At the same time, they represented a major source of income for the estimated 2.6 billion people who depend on agriculture for their livelihoods (FAOSTAT 2004).

Despite these successes there are still many parts of the world, often the poorest, where the productive capacity of cultivated systems has stagnated or even declined in the face of increased food demand from growing populations. Local disruption of cultivation by drought, flood, pests, disease epidemics (crop, animal, or human), armed conflict, and social unrest can be catastrophic in human, economic, and environmental terms. The prospect of providing sufficient food to sustain another 2 billion people by 2020 has rightly focused attention on the very real threats to food security and income generation if the productivity of cultivated systems cannot keep pace with this demand.

But food security and concern for the more than 852 million who currently go hungry each day (FAO 2004) are only part of the challenge faced by cultivated systems. Human well-being depends not only on a sufficient and safe supply of food, feed, and fiber but also on access to clean water and air, timber, recreational opportunities, cultural and aesthetic pleasure, and so on. Cultivation often has a negative impact on provision of these services.

For example, cultivated systems tend to use more water, increase water pollution and soil erosion, store less carbon, emit more greenhouse gases, and support significantly less habitat and biodiversity than the ecosystems they replace. Hence, as the share of the world's natural ecosystems converted for cultivation has increased, the overall supply of ecosystem services other than food, feed, and fiber has fallen (Wood et al. 2000), despite growing demand for these additional services. Cultivated systems are under increasing pressure, therefore, to meet the growing need for cultivated products as well as to supply an amount and quality of other ecosystem services. Appropriately responding to this "double burden" represents a critical, long-term challenge to modern agriculture (Conway 1999; Runge et al. 2003).

This chapter assesses the global extent, distribution, and condition of cultivated systems with regard to their continued capacity to both deliver food, feed, and fiber and contribute to the broader range of ecosystems services on which human well-being depends.

26.1.1 The Emergence of Cultivation

Agriculture first emerged almost 10,000 years ago in several different regions, including Mesopotamia, eastern China, mesoAmerica, the Andes, and New Guinea (Smith 1998). The extent of agriculture and its impact on ecosystem services tracks population pressures at local, regional, and global scales. While human population and agricultural extent maintained a relatively steady rate of increase for much of human history, both increased dramatically with the rapid rates of scientific discovery, economic development, and global trade that accompanied the Industrial Revolution and as a consequence of European economic and political control (Richards 1990). The direct impact of European settlement and accompanying agricultural technologies was seen in North and South America, Southern and Eastern Africa, and Australia/New Zealand. Other parts of the world also experienced significant cropland expansion as they connected to European markets.

In 1700, most of the world's cropland was confined to the Old World. (See Figure 26.1 in Appendix A.) While indigenous peoples elsewhere modified the landscape, their impact was not as large as that of the Europeans, who used more advanced cultivation technologies that supported higher population densities. Since 1700 cropland has increased by 1,200 million hectares (466%), including major expansion in North America and the former Soviet Union, with the greatest expansion occurring in the past 150 years. Indeed, more land was converted to cropland in the 30 years after 1950 than in the 150 years between 1700 and 1850.

The rate of cropland expansion in China has been steady throughout the last three centuries. Cropland in Latin America, Africa, Australia, and South and Southeast Asia expanded very gradually between 1700 and 1850, but subsequently expanded rapidly. Since 1950, cropland area in North America has stabilized, while it has decreased in Europe and China. Cropland area increased significantly in the former Soviet Union between 1950 and 1960 but has decreased since then. In the two decades before 2000, the major areas of cropland expansion were located in Southeast Asia, parts of Asia (Bangladesh, Indus Valley, Middle East, Central Asia), in the Great Lakes region of eastern Africa, and in the Amazon Basin. The major decreases of cropland occurred in the southeastern United States, eastern China, and parts of Brazil.

Since the middle of the twentieth century there has been a major shift in emphasis away from area expansion toward intensification of agriculture, which produces greater yields per unit time and area (Ramankutty et al. 2002). This shift was made possible by widespread investment in irrigation systems, mechanization, cost-effective means of producing inorganic fertilizers (especially nitrogen), and new crop varieties that could better exploit water and nutrients. Declining availability of suitable agricultural land and growing competition for land from human settlements, industry, recreation, and conservation have also increased pressure on existing farmland.

Hence, most of the increase in food demand of the past 50 years has been met by intensification of crop, livestock, and aquaculture systems rather than expansion of production area. For example, Bruinsma (2003) states that for all developing countries over the period 1961–99, expansion of harvested land contributed only 29% to growth in crop production versus the contribution of increases in yields, which amounted to 71%. Expansion of harvested land accounts for both expansion in arable land (23%) and increases in cropping intensity (6%). Furthermore, the share of growth in crop production attributed to yield increases varies by region; sub-Saharan Africa has the smallest portion (34%) and South Asia has the largest (80%). Inclusion of industrial countries lowers the global contribution of harvested area expansion to crop production growth, as a consequence of their greater reliance on increased yields.

Today, nearly all of the world's suitable land is already under cultivation. Although Africa and Latin America contain the majority of the world's remaining stock of potentially cultivatable lands, most of this currently supports rain forest and grassland savannas that provide many other ecosystem services and are crucial habitat for fauna and flora in natural ecosystems (Bruinsma 2003). In many parts of North America, Europe, Japan, and China, where productivity has grown faster than demand, land is increasingly being withdrawn from cultivation. The decline would likely be even more rapid in the United States, the European Union, and Japan in the absence of production-related subsidies (Watkins 2003). On a global basis, there has been a steady decrease in area devoted to the major cereal crops—maize, rice, and wheat, which account for the majority of calories in human diets—amounting to 2.4 million hectares per year since 1980 (FAOSTAT 2004).

26.1.2 Typology of Cultivated Systems

While there are several global frameworks for classifying the biophysical potential of agriculture (e.g., Koeppen 1931; Papadakis 1966; FAO 1982), none fully capture the enormous diversity of cultivated systems and practices on a global basis. The most comprehensive approach to date covers the farming systems of the developing world (Dixon et al. 2001), which identifies and characterizes a total of 44 crop, livestock, mixed, forest-based, and fishery-based systems, using agroecological, management, and commodity-related criteria. However, the omission of cultivated systems of North America, Western Europe, and Oceania limits its application to a global assessment.

In the absence of a widely accepted and truly global cultivated system framework, the MA assessment makes use of a schema built on easily accessible, more highly aggregated system characteristics, based on two key dimensions of cultivated systems—the agroecological context and the enterprise/management context. (See Table 26.1.) The agroecological context is defined by (sub-)tropical and temperate conditions, reflecting broad day length, radiation, and thermal differences, and by (sub-)humid and (semi-)arid conditions, reflecting differences in rainfall and evapotranspiration regimes. The importance and distinctiveness of highland and mountain cultivated systems in the (sub-)tropics is further recognized. Cultivation enterprises and practices themselves are divided into six broad categories: four crop-based categories (irrigated, high external-input rain-fed, low external-input rain-fed, and shifting cultivation) as well as confined ("landless") livestock production and freshwater aquaculture. Combining the agroecological and enterprise/management dimensions generates a matrix into which most of the world's important cultivated systems can readily be categorized. (Extensive grazing systems are not treated here as cultivated systems but are dealt with in Chapter 22.)

26.1.2.1 Irrigated Systems

The roughly 18% (250 million hectares) of total cultivated area that is irrigated accounts for about 40% of crop production (Gleick 2002). Irrigated systems are served by water from impoundment or diversion structures, boreholes, and wells or other means of delivering water. From an investment perspective, irrigation systems range from large civil engineering works delivering water to hundreds of thousands of hectares in Pakistan and India (Barker and Molle 2004) through farm-based wells that use small pumps to tap groundwater aquifers all the way to small-scale community-based systems powered by draught animals and manual labor, such as those found in West Asia, North Africa, and the Sahel (Oweis et al. 1999). In addition to increasing and stabilizing the yields of individual crops, irrigation can extend the growing period and allow two or even three crops to be grown each year on the same piece of land, where water availability and temperature permit such intensification.

26.1.2.2 Rain-fed Systems

Rain-fed agricultural systems account for the largest share (about 82%) of the total agricultural land area and exist in all regions of the world. In Asia and the Pacific, for example, rain-fed agriculture represents about 223 million hectares, or 67% of the total arable land (Asian Development Bank 1989), and rain-fed production accounts for 16–61% of agricultural GDP in this region (excluding Pakistan as part of West Asia).

Rain-fed systems are prevalent in both high and low yield potential areas, as largely determined by the amount and distribution of precipitation in relation to crop water requirements. Lower potential lands, also referred to as marginal or less-favored lands, are discussed in Chapter 22 (and also occur in higher-altitude mountain systems, see Chapter 24). The discussion here focuses on more-favorable rain-fed areas where both high and low levels of external inputs are used to produce crops. Pressure on these systems is increasing as arable land becomes scarcer; as the productivity of existing irrigated lands declines due to a reduction in water availability or to land degradation, especially salinization; and as food demand increases.

Rain-fed systems may involve both annual and perennial crops as well as livestock. In Asia, the rain-fed humid/sub-tropical systems and arid/semiarid areas include a range of mixed crop-livestock systems, which can be categorized into lowland and upland systems. The former is more associated with crop cultivation due to higher levels of soil moisture. Rain-fed lowland rice, for example, is defined as nonirrigated, but the topography is generally flat and the soil surface is inundated for at least part of the crop cycle with sustained flooding. Rain-fed upland rice, on the other hand, is grown on well-drained fields that are never flooded. Major rice cropping systems in the rain-fed lowlands are rice-wheat, rice-pulses (including chickpea, lentil, peanut, and pigeon pea), and rice-mustard. Maize, sugarcane, and cotton are also important crops in humid lowland areas of tropical/subtropical Asia. Cropping systems that use more drought-tolerant cereal crops such as sorghum and millet are found in semiarid rain-fed lowland areas.

The uplands, by comparison, have sloping to hilly topography, and typically have less fertile soil that is easily degraded by erosion and nutrient depletion without the use of appropriate husbandry practices. (See also Chapter 24.) Although both annual crops (such as cereals, legumes, roots, and vegetables) and perennial ones (such as coconuts, oil palm, rubber, and fruit trees) are grown, agroforestry systems involving the latter are especially important. Rain-fed areas have relatively large populations of livestock, and their contribution via animal manure to crop cultivation, food security, and the livelihoods of poor people is significant (Devendra 2000). Overstocking and uncontrolled grazing of ruminants are major problems in semiarid rain-fed regions where land tenure rights are not well defined, such as in the Sahel region of sub-Saharan Africa.

26.1.2.3 Shifting Cultivation

Shifting cultivation, also called "swidden" agriculture or "slash-and-burn" agriculture, is one of the oldest forms of farming and consists of cropping on cleared plots of land, alternated with lengthy fallow periods. These systems are the dominant form of agriculture in tropical humid and subhumid upland regions and are typically associated with tropical rain forests.

Table 26.1. Global Typology of Cultivated Systems with Examples

Farming System[a]		Tropical and Sub-tropical (62%)			Temperate (38%)	
		Warm Humid/Subhumid	Warm Semiarid/ Arid	Cool/Cold (Highland/ Montane)	Humid/Subhumid	Semiarid/Arid
		(26%)	*(12%)*	*(24%)*	*(22%)*	*(16%)*
Irrigated	(18%)	rice (e.g., East, Southeast Asia) rice-wheat (e.g., Pakistan, India, Nepal)	rice (e.g., Egypt, Peru)			cotton
Rain-fed—high external input (crops, livestock, tree crops)		rice-wheat (e.g., Pakistan, India, Nepal)		tea, coffee plantations (e.g., East Africa, Sri Lanka)	maize and soybean–Argentinean pampas, U.S. corn belt	
	(82%)				small grains (wheat, barley, rapeseed, sunflower, oats) and mixed crop-livestock systems (e.g., West and North Central Europe)	
Rain-fed—low external input (crops, livestock, tree crops)		staple tropical crops in humid tropics (e.g., yam, cassava, banana in SSA)	mixed crop, livestock (e.g., Sahel, Australia)	cereals/tubers (e.g., High Andes)	mixed crop–livestock systems (e.g., Europe)	wheat–fallow systems (e.g., Central Asia, Canada, United States, Australia)
Shifting cultivation	NA	e.g., Amazon Basin, Southeast Asia				
Industrial confined livestock	NA	"landless" livestock systems, e.g., cut and carry systems, mixed low-intensity livestock/crop systems, beef feeding lots, broiler and pig houses				
Freshwater aquaculture	NA	e.g., artisanal ponds, industrial cages				

[a]High-level aggregations of the global farming systems typology developed by Dixon et al. 2001.

Notes: Agroecological characterization according to FAO Global Agroecological Zones (FAO/IIASA 2001). Total area shares shown in parentheses are for settled agriculture (e.g., excluding shifting cultivation areas). Derived from FAOSTAT 2004. Breakdown of cropland by agroecological zones from Wood et al. 2000. The MA cultivated systems do not encompass marine and coastal aquaculture (Chapter 19) and extensive grazing systems (Chapter 22).

Shifting cultivation is practiced on about 22% of all agricultural land in the tropics and is the primary source of food and income for some 40 million people (Giller and Palm 2004). While the contribution to global food security is negligible, given the low yields and general lack of infrastructure in areas where shifting cultivation predominates, this method of cultivation has a potentially large impact on regional and global ecosystem services through its effects on biodiversity, greenhouse gas emissions, and soil nutrients. (For a more comprehensive analysis of such effects in the humid tropics, see the MA Sub-Global Assessment of the Alternatives to Slash and Burn program.)

Although these systems are generally associated with soils of low fertility, they are highly sustainable and resource-conserving in areas with low population density. High population density increases the pressure on available land and resources, reducing the time available for a regenerative fallow between cropping cycles. One method used to raise productivity and reduce land degradation in areas of shifting cultivation is "alley cropping," growing tree crops in conjunction with annual crops. In the Philippines, for example, alley cropping in sloping upland rice areas with *Flemingia macrophylla* showed that over two years, average soil loss was cut down to 42 cubic meters per hectare, compared with 140 cubic meters under traditional practices, together with concurrent increases in rice yields (Labios et al. 1995).

26.1.2.4 Mixed Crop and Livestock Systems

Mixed crop-livestock farming systems, where crops and animals are integrated on the same farm, represent the backbone of smallholder agriculture throughout the developing world, supporting an estimated 678 million rural poor. In Asia, more than 95% of the total population of large and small ruminants and a sizable number of pigs and poultry are reared on small farms with mixed crop-livestock systems, which are dominant in both irrigated and rain-fed areas in humid and subhumid environments.

Mixed farming systems enable farmers to diversify agriculture, to use labor more efficiently, to have a source of cash for purchasing farm inputs, and to add value to crops or their by-products. Mixed farming systems provide the best opportunities to exploit the multipurpose role of livestock in many rural societies (Devendra 1995). A number of crop-animal interactions are important and dictate the development of mixed systems. These include animal traction for field operations, animal manure, and animal feeds from crops, as evident in sub-Saharan Africa (McIntire et al. 1992) and Asia (Devendra and Thomas 2002). These interactions have demonstrated the important contribution that animals make to increased production, income generation, and the improved sustainability of annual and perennial cropping systems.

Crop-livestock systems can be separated into those that mix animals with annual and perennial crops; of the two, the use of the latter has been more limited. Examples of integrated annual crop-animal systems include rice, maize, cattle, and sheep in West Africa; rice, wheat, cattle, sheep, and goats in India; rice, goats, duck, and fish in Indonesia; rice, buffalo, pigs, chicken, duck, and fish in the Philippines; rice, vegetables, pigs, ducks, and fish in Thailand; and vegetables, goats, pigs, ducks, and fish in Viet Nam. Examples of integrated perennial tree crop-animal systems include rubber and sheep in Indonesia; oil palm and cattle in Malaysia and Colombia; coconut, sheep, and goats in the Philippines; and coconut, fruit, cattle, and goats in Sri Lanka. In West Asia and North Africa, integration of sheep with wheat, barley, peas, and lentils is common, together with olives and tree crops.

With annual cropping systems, ruminants graze native grasses and weeds on roadside verges, on common property resources, or in stubble after the grain crop harvest. Crop residues and by-products are also fed to livestock throughout the year or seasonally, depending on availability. In the perennial tree crop systems, ruminants graze the understory of native vegetation or leguminous cover crops. Non-ruminants in these systems mainly scavenge in the villages, on crop by-products, and on kitchen waste. However, village livestock systems can evolve into more-intensive production systems depending on the availability of feeds, markets, and the development of co-operative movements. This is evident in many parts of Central America, West Africa (Nigeria), Southeast Asia (Indonesia), and South Asia (Bangladesh).

Because of the synergies between crop and livestock components, mixed crop-livestock systems have shown themselves to be both economically and environmentally robust from a smallholder perspective. It is likely that smallholder mixed farms will remain the predominant form of agricultural land use in rain-fed cropping regions in developing countries where labor is abundant.

26.1.2.5 Confined Livestock Systems

Confined livestock production systems in industrial countries are the source of most of the world's poultry and pig meat production and hence of global meat supplies. Such large-scale livestock systems are also being established in Asia to meet increasing demand for meat and dairy products. In addition, beef and mutton are produced from intensive confined feeding operations, the former mostly in North America and the transition states of Eastern Europe. The majority of sheep and goat fattening under "landless" (non-grazing) conditions occurs in the Near East and in much of Africa. Cut-and-carry, zero-grazing dairy production systems are similar to confined systems in industrial countries in that hand feeding and disposal of manure are involved. These systems involve cutting feed, crop residues, or litter and transporting them to livestock that are confined in pens on the farm.

The use of purchased cereals and oilseeds for feed in confined livestock systems allows separation of crop production and utilization of feed in livestock rations. These concentrated feeds are less perishable and easier to transport than the livestock products. Even if several kilograms of concentrates are needed to produce one kilogram of meat, it is still cheaper to establish the production system near the consumer market and to transport the feeds to the animals. A significant share of the increase in cereal imports to developing countries over recent decades has occurred to provide feed for the expanding poultry or pig industries (Delgado et al. 1999).

Animal confinement facilitates the management of nutrition, breeding, and health but increases the labor and infrastructure requirements for feeding, watering, and husbandry of the livestock.

Apart from the capital embodied in the animals, additional investment is needed in providing fencing, housing, and specialized equipment for feeding and other activities. Special equipment is also needed for animal slaughter and meat processing or for milk cooling and processing. There are economies of scale in the provision of such processing services and the associated product marketing, and possibly in the supply of inputs (feed and feed supplements) and genetic material (such as day-old chicks or semen). This has often led to either co-operative group activity or vertical integration of smallholder producers with large-scale processing and marketing organizations.

While there are good economic arguments for the concentration of large numbers of animals associated with many confined systems, there can be significant impacts on surrounding ecosystems. Problems often arise in the disposal of large amounts of manure and slaughtering by-products. While some types of manure can be recycled onto local farmland, soils can quickly become saturated with both nitrogen and phosphorus because it is too costly to transport manure, which has relatively low nutrient concentration, for long distances. Manure treatment or digestion to produce methane can help minimize pollution, but even in countries with strong regulation and enforcement systems, nutrient and bacterial leakage to water courses can occur, with consequential impacts on freshwater and aquatic systems (de Haan et al. 1997; Burton et al. 1997).

Confined systems tend to be located near markets in peri-urban areas. Distance from these centers, or from their main transport routes, has an important influence on the net prices received for farm products. Similarly, location in relation to urban centers affects access to markets for purchased inputs and the costs of such inputs (Upton 1997). Transport costs vary from one commodity to another, depending on the perishability and bulk-to-value ratio. Milk and eggs are relatively perishable and therefore are most often produced intensively in peri-urban zones. Furthermore, agricultural enterprises dependent on purchased inputs, such as concentrate feeds, are likely to be established in peri-urban zones with easy access to input markets. In contrast, ruminant meat can be produced in more-distant rural areas and transported as live animals to urban markets for slaughter.

26.1.2.6 Freshwater Aquaculture Systems

Aquaculture involves the propagation, cultivation, and marketing of aquatic plants and animals from a controlled environment and usually involves tenure and ownership, as opposed to the open-access or common property systems that occur in land agriculture. Aquaculture can be applied in coastal (mariculture), brackish, or fresh water (inland), but this chapter focuses on freshwater aquaculture. (Coastal and brackish aquaculture systems are discussed in Chapter 19.) There are four types of production systems: ponds, cages, raceways, and recirculating systems:

- Earthen ponds are most common for both small-farm and commercial production systems, and they may be specifically designed and built for aquaculture. Ponds for aquaculture (called dike or levee ponds) require an adequate amount of water of sufficient quality and clay soils that retain water. The size of a levee pond depend on its planned usage, whether as a holding, spawning, rearing, or grow-out pond.
- Cage culture uses existing water resources (lakes, ponds) but encloses the fish in a cage or basket that allows water to pass freely. Its main advantage is ease of harvesting. Small lakes, mining pits, and farm ponds may be used for cage culture. The potential for expanding cage farming is more limited in areas where freshwater bodies are already actively used.

- Rectangular raceways are mostly used in industrial countries (whereas ponds and cages are common in developing countries). Rectangular raceways are almost exclusively used for trout production and require large quantities of cheap, high-quality water. Using gravity, water passes from a spring or stream through raceways arranged in a series on slightly sloping terrain.

- Water recirculating systems are also common in industrial countries. Water is recirculated rather than passing through once; hence, less water is needed than for a pond or open raceway. Most recirculating systems are indoors, allowing growers to maintain more control on water characteristics like temperature. Clearly, this type of production requires high initial capital investment.

In addition, production systems can be distinguished by the level of production intensity or amount of inputs (labor, feed, materials, or equipment) used. Such production intensity can be extensive, where low levels of external inputs result in lower production levels, or intensive, where higher levels of inputs of technology and greater degree of management generally increase yield (FAO 2003).

Aquaculture can also be land-based or water-based. Land-based aquaculture consists mainly of ponds, rice fields, and other facilities built on dry lands. Carp and tilapia are the most commonly grown species in freshwater ponds, while shrimp and finfish are cultivated in brackish-water ponds. Water-based systems include enclosures, pens, cages, and rafts and are usually situated in sheltered coastal or inland waters. Pens and cages are made up of poles, mesh, and netting. Cages are suspended from poles or rafts that float, while pens rest on the bottom of the water body (FAO 2003).

Unlike livestock, where only a limited number of species are farmed, aquaculture production involves many species of aquatic organisms, although some predominate. In freshwater aquaculture alone, some 115 freshwater species of finfish, crustaceans, and mollusks were cultured in 2000, with finfish contributing the bulk of production. Over the period 1991–2000, carp (and other cyprinids) and tilapia (with other cichlids) ranked first and second respectively in global freshwater fish production, accounting for 76–82% and 5–6% respectively of the total (FAO 2002).

Though a number of freshwater species were cultivated, only a few freshwater species, like carp, milkfish, and tilapia, have been domesticated—that is, breeding agencies (government and private) produce fry as a source of fingerlings for aquaculture. This contrasts with the livestock and crop sectors, where selective breeding has been able to develop superior animal breeds and crop varieties suitable for intensive production. As a result, many forms of freshwater aquaculture are still very dependent on wild sources of fish spawn, seed, or young fish.

Aquaculture operations can have both positive and negative impacts on the environment. On the positive side, if aquaculture is integrated with agriculture, environmental benefits include recycling, lower net pollution, and reduced use of pesticides and fertilizers. On the other hand, some aquaculture operations can have damaging effects on water quality and quantity and aquatic biodiversity—similar to the externalities associated with confined livestock feeding operations or intensive, high-input cropping systems.

26.1.2.7 Major Cropping Systems

Among the tremendous diversity of crops and cropping systems, both in terms of agroecologies and management practices, five major cropping systems stand out in importance. These systems supply a substantial portion of the world's food, occupy a large portion of the world's cultivated lands, or both. The major systems are shifting cultivation in the forest margins of tropical Africa, Asia, and Latin America; irrigated lowland rice systems in Asia; irrigated rice-wheat systems in the Indo-Gangetic Plains of India, Pakistan, Nepal, Bangladesh, and south-central China; rain-fed wheat in north, west, and central Europe; and rain-fed maize-soybean systems in the United States, southeast Canada, Argentina, and south-central Brazil. Estimates of the scale of these systems are provided in Table 26.2.

The highly productive irrigated rice-based cropping systems are practiced in regions with fertile soils and access to supplementary ground or surface water. The wheat and maize-soybean rotations are located on deep, fertile soils in regions that typically have adequate and consistent rainfall during the growing season. Because of these natural endowments, these systems provide food to about half the human population and do so on a relatively small area. In addition to meeting local food needs, these systems account for more than 80% of all grains that enter international markets.

To sustain high yields in these systems, modern farming practices are employed, including high-yielding varieties and hybrids, substantial fertilizer inputs, and integrated pest control methods that include use of herbicides, insecticides, and fungicides when other management practices are inadequate. For example, three cereals—rice, wheat, and maize—receive 56% of all nitrogen fertilizer applied in agriculture (Cassman et al. 2003). Yield increases in these systems during the past 50 years are estimated to have avoided the need to expand cultivation by hundreds of millions of hectares globally, thus helping to maintain the ecosystem services derived from tropical and temperate forests, grasslands, and wetlands (Waggoner 1994; Evans 1998; Cassman 1999). Given the 670 million hectares of global cereal production in 2000, each 1% increase in productivity is equivalent to saving 6.7 million hectares of additional land that would be required for cereal production, keeping cropping intensity constant. (For a discussion of cropping intensity, see Bruinsma 2003:127–37.)

However, the relatively high levels of nutrients, pesticides, and water applied to these systems can deplete water resources, reduce water quality, increase greenhouse gas emissions, and accelerate the loss of terrestrial and aquatic biodiversity (Hooper et al. 2003; Mineau 2003). While nutrient losses from applied nitrogen fertilizer via denitrification release N_2O, a powerful greenhouse gas, to the atmosphere, recent studies have also demonstrated that these intensive cropping systems can sequester carbon in soil organic matter, thus reducing global emissions of CO_2 (Bronson et al. 1998; Lal et al. 2003; Paustian et al. 1997). Achieving global food security for an increasing and rapidly urbanizing population will depend on sustaining continued yield increases in these major high-potential cereal production systems.

The key challenge to sustaining cereal yield increases to meet anticipated demand while also protecting ecosystem services is to use crop and soil management practices to greatly increase the efficiency with which fertilizers, water, and other external inputs are used. For example, greater efficiency of nitrogen fertilizer use allows more grain to be produced per unit of applied nitrogen, reducing nitrogen losses and diminishing associated negative impacts on ecosystem services (Dobermann and Cassman 2002). Likewise, substantial increases in water use efficiency can be achieved by investment in improved irrigation infrastructure, better irrigation scheduling, and application equipment, such as a shift from furrow irrigation to subsurface drip irrigation (Howell 2001). Ensuring continued progress toward ecological intensification of these major cereal cropping systems requires long-term and sufficient investments in research and extension.

Table 26.2. Extent of the World's Major Cropping Systems and Population Dependent on Them as the Major Source of Cereal Supply. The population given is of the countries or regions where these cropping systems represent the predominant form of agriculture. Food production from the high potential systems, which include the two irrigated rice-based systems, rain-fed wheat, and rain-fed maize-soybean, not only accounts for a major food source for the countries and regions in which they occur but also accounts for a large majority of all traded grain that crosses international borders. (Compiled from Giller and Palm 2004; Dixon et al. 2001; Huke and Huke 1997)

Cropping System	Area	Population	Region/Countries
	(mill. hectares)	*(million)*	
Shifting cultivation	1,035	40	Forest margins in tropical Africa, Asia, Latin America
Irrigated continuous lowland rice	24[a]	1,800	Tropical/sub-tropical lowlands of Asia
Irrigated rice-wheat annual double crop	17[a]	248	Indo-Gangetic Plains of India, Pakistan, Nepal, Bangladesh, and south-central China
Rain-fed wheat	40	500	North, west, central Europe
Rain-fed maize-soybean	85	420	United States, southeast Canada, Argentina, southcentral Brazil
Total	1,227	2,968	

[a] Continuous irrigated lowland rice systems in the tropics and sub-tropical lowlands of Asia produce two and sometimes three rice crops per year. A total of 49 million hectares of harvested rice is obtained from these systems (McLean et al. 2002; Huke and Huke 1997). Similarly, the rice-wheat systems produce 17 million hectares of each crop on an annual basis, for a total harvested grain area of 34 million hectares (Huke and Huke 1997), which supports an agricultural population of 248 million (Dixon et al. 2001).

26.1.2.8 Assessing the Global Distribution and Intensity of Cultivated Systems

While the global typology is helpful in broad stratification of the major cultivated systems, the spatial extent, distribution, and condition of cultivated systems requires further analysis. For mapping purposes in this assessment, the global extent of cultivation has been defined on the basis of rain-fed and irrigated croplands only. There is no comprehensive information, even at a national scale, on the number and location of industrial livestock and freshwater aquaculture enterprises.

The global cropland map is a composite of an updated version of the 5-kilometer resolution PAGE Agroecosystems Map (Wood et al. 2000) that incorporates revisions of the underlying 1992–93 1-kilometer resolution AVHRR datasets, combined with the 10-kilometer resolution global irrigation map produced by University of Kassel and FAO (Döll and Siebert 1999).[2] (See Figure 26.2 in Appendix A.) The revisions aim to identify all occurrences of agriculture, even those that are minor cover components under the classification scheme. (This was limited by the seasonal land cover naming convention, which did not identify an agricultural component if it occupied less than 30% of the land cover class area.)

Intrinsic weaknesses in the dataset include regional variations in the reliability of the satellite data interpretation, reflecting differences in the structure of land cover and in the availability of reliable ground truthing data (Brown and Loveland 1998) (See Box 26.1.) Specific agricultural land cover types for which interpretation is considered problematic include irrigated areas, permanently cropped areas (especially tree crops in forest margins), mixed smallholder agriculture, and extensive pasture land.

For mapping purposes, the extent of cropland is defined as areas where at least 30% of a 1x1-kilometer grid cell is classified as cropland through the interpretation and ground truthing of satellite imagery. By this definition, it is likely that a significant share of the 1,035 million hectares of shifting cultivation as practiced in the humid tropics is not detected and classified as cropland because fallow periods are typically greater than five years, implying that no more than 20% of the land area is actually planted to crops at any given point in time.

Within the physical extent of cultivated systems (rain-fed and irrigated cropland), an existing agroecologically based characterization schema has been used to delimit cultivation system sub-types (Wood et al. 2000). This is a 16-class schema that provides a spatial visualization of the location and extent of most elements of the cropland typology presented in Table 26.1. (See Figure 26.3 in Appendix A.) The classification of climatic variables into tropical, sub-tropical and temperate, humid, subhumid, semiarid, and arid are made in accordance with FAO's agroecological zones approach implemented in a global spatial database (FAO/IIASA 2001). The agroecological classification also relies on slope and the presence or absence of irrigation technologies. Slope is an important attribute in terms of potential for mechanization as well as for increased surface runoff and soil erosion due to cultivation. The slope data used in this characterization are also derived from the FAO/IIASA digital agroecological database, and the irrigation data are from Döll and Siebert (1999).

Cultivated systems are extensive. Globally, they cover 36.6 million square kilometers, or approximately 27% of land area (24% if inland waters, Greenland, and Antarctica are included in the definition of land area). (See Table 26.3.) By intersecting the extent of agriculture with maps of global population, it is estimated that 74% of the world's population lives within the boundaries of cultivated systems and that cultivated systems overlap in significant ways with other systems, such as forests, mountains, and drylands. Close to half of all cultivated systems are located in dryland regions, where cultivation is strongly linked to livelihoods and resource issues—particularly poverty, land degradation, and water scarcity.

The characterization presented in Figure 26.3 identifies domains within which biophysical and environmental constraints and opportunities are broadly similar from a cultivation perspective and where common ecosystem service impacts might be faced. However, as this chapter illustrates, it is the specific type of cultivation practiced in any location and the precise ways in which cultivation is managed that ultimately determine the type and scale of impacts on ecosystem services and human well-being. Integrating additional information about the ways in which cultivated systems are managed can only be accomplished at smaller geographic scales (see, e.g., Dixon et al. (2001) for a regional ap-

BOX 26.1

Reliability of Satellite-based Global Assessments of Cultivated Systems

Obtaining global-scale information about the location and extent of cultivated systems is fraught with difficulties. Satellite-based remote sensing offers a visualization of land cover for the entire globe in a more or less uniform way, and several publicly available, coarse-resolution (1km pixel size) datasets offer opportunities for locating cropland. These include Global Land Cover 2000 (GLC2000), MODerate resolution Imaging Spectrometer (MODIS) land cover, GLCCDv2, and the cultivated systems extent used by the MA, which is a cropland-focused reinterpretation of GLCCDv2. (More information on these datasets is available at edcdaac.usgs.gov/glcc/background.asp (GLCCDv2), www-gvm.jrc.it/glc2000 (GLC2000), and edcdaac.usgs.gov/modis/mod12q1.asp (MODIS). The MA reinterpretation of GLCCDv2 is based on methods described in Wood et al. 2000.)

However, a comparison of these data sources reveals large differences in the extent and distribution of areas classified as cropland. The GLCCDv2 imagery and the MA extent represent land cover in 1992–93, while the GLC2000 and MODIS imagery are for the year 2000. Clearly, land cover change took place between these years, but the significant differences between the 1992–93 and 2000 cropland areas, as well as those between the two 2000-based assessments of cropland, cannot be explained by changes over time alone. Many of the differences result from the use of different data sources, methodologies, and classification systems. These findings raise concerns about our present ability to detect cropland reliably using globally applicable analysis of coarse-resolution data sources, and they cast extreme doubt on the possibility of assessing cropland change by comparing global data sets from different sources.

Figure A compares the share of land area that falls within the extent of cultivated area, by region, for these four different land cover data sets. The extent of cultivation is defined as any cell classified as cropland or a cropland mosaic. Cropland mosaics are areas that appear in the imagery as composites of multiple landscape types (cropland and forest, for example, or cropland and grassland). The variations across the globe are large: in sub-Saharan Africa, MODIS land cover classifies less than 3% of the total land area as cropland, while according to the classification adopted by the MA, just over 25% of land in sub-Saharan Africa falls within the extent of cultivated area.

Part of this discrepancy is definitional. For mapping purposes, the MA classifies areas as "cultivated systems" if at least 30% of the land cover grid cell appears to be cropland. MODIS land cover and GLC2000, on the other hand, use higher cutoffs of about 50% cropland. The MA considers that ecosystem services are already likely to be significantly affected by cultivation at the lower cutoff.

Most of the differences among datasets involve the "mosaic" land-cover classes. Mosaic landscapes are some of the most important ones from an ecosystem service perspective, as they are likely to be transitional areas where change is taking place or where agricultural systems exist in close proximity to natural biodiversity. They are also some of the hardest to classify. In contrast to large, intensively farmed agricultural systems found in much of the industrial world, smaller plots of agricultural land, mixed in with forest or grassland, are more difficult to distinguish. Isolated or even clusters of small cultivated plots and fields, even if they can be identified as agriculture, are easily lost due to the coarse resolution of the data.

Figure B (see Appendix A) compares the geographic location of cropland and cropland mosaic classes used by the MA to the MODIS and GLC2000 land cover classifications for Africa. Africa is among the most challenging continents for mapping cropland because of generally small field sizes and mixed cropping systems. Areas of agreement are shown in orange, and areas of near-agreement (classified as cropland in one dataset and cropland mosaic in the other) are shown in yellow. Blue areas are those classified as either cropland or cropland mosaic by the MA, but not classified as cultivated in the other data sets. Much of the additional area within the MA cultivated systems compared with other datasets can be attributed to its lower threshold for defining cultivated landscapes—that is, a grid cell is considered to be a cultivated system if at least 30% of it is cropland.

Given the difficulties in classifying cultivated areas using coarse resolution satellite data, cultivated area maps and statistics derived from such data should be interpreted with caution. Mapping changes over time presents an additional set of challenges, as it requires comparing data sets that may be mismatched in multiple ways. Much more needs to be done to improve the reliability of publicly available regional- and global-scale spatial cropland data, including improved use of higher-resolution data such as LANDSAT and SPOT imagery, the use of consistent methodologies and classification systems over time, and field validation.

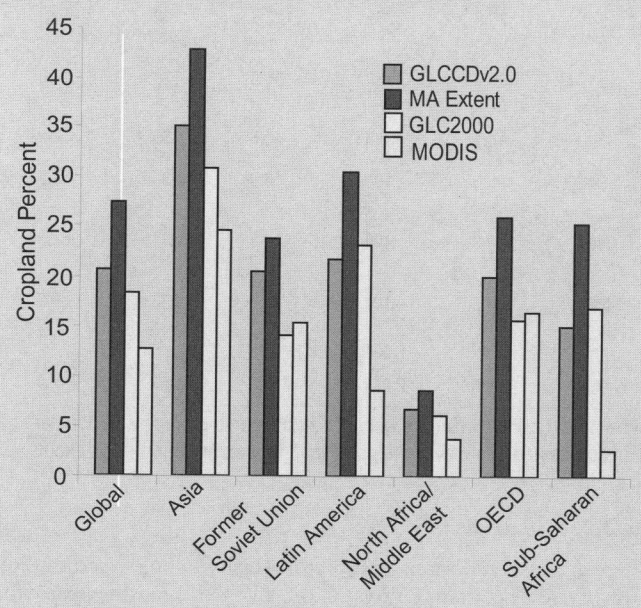

Figure A. Comparison of Cropland Estimates by Region and Data Source

proach and Wortmann and Eledu (1999) for a national application in Uganda).

26.2 Cultivated Systems and Ecosystem Services

To achieve increased production of food, feed, and fiber, cultivated systems use biodiversity and numerous supporting, regulating, and provisioning services such as pollination, nutrient cycling, soil formation, and fresh water for irrigation. While cultivated systems depend on such services, they in turn also influence the supply of a host of other services, including food, feed, and fiber; clean water; climate regulation; pollution control; flood control; viable populations of wildlife; clean air; and scenic qualities (Allen and Vandever 2003). Some types of production systems, such as

Table 26.3. Area and Population of Cultivated Systems and the Extent of Cultivation in Other MA Systems.

System	Cultivated Total		Drylands		Mountains		Coastal		Forest	
	Area	Population	Area	Population	Area	Population	Area	Population	Area	Population
	(thousand km²)	*(million)*	*(thousand km²)*	*(million)*	*(thousand km²)*	*(million)*	*(thousand km²)*	*(million)*	*(thousand km²)*	*(million)*
Temperate										
10 Irrigated and mixed irrigated	1,684	554.2	999	249.8	245	39.7	179	110.7	166	43.5
11 Rain-fed, humid and subhumid, flat	3,954	463.9	935	93.5	274	45.6	284	69.9	965	75.7
12 Rain-fed, humid and subhumid, sloping	2,380	254.5	238	21.6	680	61.2	17	4.4	1,191	83.5
13 Rain-fed, arid/dry and moist semiarid	6,041	172.4	4,832	116.6	1,104	47.3	47	3.1	1,407	27.0
Moderate cool/cool/cold tropics										
20 Irrigated and mixed irrigated	1,501	20.5	71	5.1	72	8.1	7	0.8	34	8.5
21 Rain-fed, humid and subhumid	1,098	110.8	370	16.2	743	65.4	19	2.6	417	26.6
Moderate cool/cool/cold sub-tropics										
30 Irrigated and mixed irrigated	1,428	262.7	1,042	148.1	342	31.1	73	35.5	173	39.6
31 Rain-fed, humid and subhumid	4,028	496.7	642	63.9	1,482	201.5	299	33.8	1,320	162.6
32 Rain-fed, dry and semiarid	1,684	73.9	1,407	51.9	596	27.7	48	1.5	150	6.6
Warm tropics and sub-tropics										
40 Tropics, irrigated and mixed irrigated	989	328.4	395	105.7	88	23.8	202	127.3	162	40.5
41 Sub-tropics, irrigated and mixed irrigated	1,245	509.0	714	335.3	124	36.7	133	85.3	178	43.6
42 Rain-fed, humid, flat	1,721	197.0	214	30.7	73	14.7	325	40.5	659	52.2
43 Rain-fed, subhumid, flat	2,709	168.3	646	32.5	94	66.3	100	16.8	1,237	62.5
44 Rain-fed, humid/ subhumid, sloped	2,783	192.9	293	19.3	980	66.3	100	16.8	1,237	62.5
45 Semiarid/arid, flat	4,028	262.7	3,042	199.0	460	29.3	102	12.0	983	50.2
46 Semiarid/arid, sloped	476	41.0	417	37.1	83	7.3	17	2.3	77	4.1
Total	**36,614**	**4,104.9**	**16,256**	**1,526.4**	**7,439**	**710.6**	**2,051**	**617.8**	**9,863**	**747.4**

Note: By definition, these MA systems may overlap spatially, so area totals cannot be added across columns without risk of counting areas and populations twice. Note also that the global cultivated total includes areas and populations contained in ecosystems other than those shown in the breakouts.

multitiered, tree and crop-based farming systems, can be very effective in building up soil nutrients, reducing soil erosion, enhancing water-related, climate, and flood regulation services, and even promoting biodiversity. But they often possess other features less attractive to farmers, such as high labor needs, longer establishment and payoff times, or lower food productivity.

Because cultivated systems are so extensive, pressure is growing for them to make a greater contribution to meeting human needs for services other than food, feed, and fiber. They may do this by being managed to have less impact on supporting and regulating services, by consuming fewer provisioning services, or by supplying more of all three types of services. Moreover, effects on specific services at a local level may differ from the aggregated effects of a given cultivated system at a regional or ecosystem level.

26.2.1 Biodiversity

There are several dimensions to biodiversity in cultivated systems. These systems contain cultivated or "planned" biodiversity—that is, the diversity of plants sown as crops and animals used for livestock or aquaculture. This is largely domesticated biodiversity and is supplemented by wild food sources. Together with crop wild relatives, this diversity comprises the genetic resources directly needed for food production. (See Table 26.4.)

Agricultural biodiversity is a broader term, also encompassing the "associated" biodiversity that supports agricultural production through nutrient cycling, pest control, and pollination (Wood and Lenne 1999). Sometimes biodiversity that provides broader ecosystem services such as watershed protection, as well as biodiversity in the wider agricultural landscape, is also included in this term (FAO/SCBD 1999; Cromwell et al. 2001; Convention on

Table 26.4. Biodiversity and Cultivated Systems

	Inside Cultivated Systems	Outside Cultivated Systems
Components of production	*crops, livestock, aquacultured fish*	*wild food sources*
Sources of genetic improvement	*crops and crop wild relatives*	*crop wild relatives (also ex situ collections in gene banks and breeders collections*
Biodiversity providing ecosystem services to agricultural production	"associated biodiversity" including soil biota, natural enemies of pests and pollinators, as well as alternative forage plants for pollinators; alternative prey for natural enemies	alternative forage plants for pollinators etc. in the wide landscape
	biodiversity that protects water supplies, prevents soil erosion, etc.	biodiversity that protects water supplies, prevents soil erosion, etc.
Other biodiversity	other biodiversity, including species of conservation/aesthetic interest (e.g., farmland birds)	other wild biodiversity

Key:
italics   Definition of genetic resources for food and agriculture
▨▨▨   Different definitions of "agricultural biodiversity"

Biological Diversity 2000). In addition, cultivated systems contain biodiversity beyond that used in or directly supporting production systems. Since agriculture is now so widespread, strategies for biodiversity conservation should address the maintenance of biodiversity within these largely anthropogenic systems as well as the aggregate impact of various cropping systems and management practices on biodiversity at regional levels.

The multiple dimensions of biodiversity in cultivated systems make it difficult to categorize production systems into "high" or "low" biodiversity systems, especially when spatial and temporal scales are also included. Figure 26.4 attempts to illustrate how different types of production systems relate to three biodiversity-related variables (building upon the approach of Swift and Anderson (1999)). The Figure also focuses on a single production system—tropical lowland rice—to illustrate how the levels of various dimensions of biodiversity can vary with management practice.

Thus the relationship between cultivated systems and biodiversity is manifold: biodiversity is cultivated in such systems (genetic resources for food and agriculture); biodiversity supports the functioning of cultivated systems (associated agricultural biodiversity); and cultivated systems harbor biodiversity beyond agricultural biodiversity of functional significance. In addition, cultivated systems have an impact on biodiversity outside the cultivated field in surrounding areas and through both expansion and intensification of agriculture. The following sections focus on these four issues.

26.2.1.1 Maintenance of Cultivated Species and Genetic Diversity

Diversity at species and genetic levels comprises the total variation present in a population or species in any given location. The culti-

vated species diversity of some production systems such as shifting cultivation and home gardens is high. Most major staple crops, however, are grown in monoculture. Even such systems may contain other dimensions of agricultural biodiversity: intensive rice "monocultures," for example, can support small areas of vegetable cultivation (on the dikes between paddies) as well as fish cultivation. In fact, in some rice-growing areas in South and Southeast Asia, fish may provide most of the local dietary protein. Genetic diversity can be manifest in different phenotypes and their different uses. It can be characterized by three different facets: numbers of different entities (such as the number of varieties used per crop or the number of alleles at a given locus); evenness of the distribution of these entities; and the extent of the difference between the entities (as in the case of pedigree date, for example) (UNEP/CBD 2004). Crop genetic diversity can be measured at varying scales (from countries or large agroecosystems to local communities, farms, and plots), and indicators of genetic diversity are scale-dependent.

The conservation and use of plant genetic resources for food and agriculture has been comprehensively reviewed by FAO (FAO 1998). Since 1960, there has been a fundamental shift in the pattern of intra-species diversity in farmers' fields in some regions and farming systems as a result of the Green Revolution. For major cereal crops, the germplasm planted by farmers has shifted from locally adapted and developed populations (landraces) to more widely adapted varieties produced through formal breeding systems (modern varieties) (Smale 2001, 2005; Heisey et al. 2002; Morris and López-Pereira 1999; Morris and Heisey 1998; Cabanilla et al. 1999). While there is no absolute dichotomy, traditional, landrace-based farming systems tend to contain higher levels of crop genetic diversity in situ than modernized systems. Depending upon the species (and its breeding system), traditional landrace-based farming systems also tend to include a higher number of varieties and more genetic variation within varieties.

Adoption of modern varieties among the three major cereal crops—wheat, rice, and maize—has been most rapid where land is scarce and where there is a high degree of market integration. In general, modern varieties of these crops have been adopted in "high potential" production areas, which have favorable climatic conditions, good soils, and either adequate rainfall or irrigation. They have been less successful in marginal areas, where landraces are still widely cultivated and are often the main source of crop germplasm.

Roughly 80% of the wheat area in the developing world is sown to modern semi-dwarf varieties. However, landrace varieties are grown extensively in Turkey, Iran, Afghanistan, and Ethiopia, with smaller pockets in countries of the Near Eastern and Mediterranean region (Morocco, Tunisia, Syria, Egypt, Cyprus, Spain, and Italy) (Heisey et al. 2002). Over three quarters of all rice planted in Asia is planted to improved semi-dwarf varieties, although farmers continue to grow landrace varieties in upland rain-fed areas, as well as in deep-water environments. Landrace varieties are grown in upland areas of Southeast Asia (Thailand, Laos, Viet Nam, and Cambodia), as well as in parts of West Africa (such as Mali and Sierra Leone) (Smale 2001; Cabanilla et al. 1999).

Relative to wheat and rice, maize has a much higher proportion of area planted to landraces. In Latin America, most of the maize area is planted to landraces, as is a higher proportion of the maize grown in sub-Saharan Africa (Morris and López-Pereira 1999). However, it can be expected that there has been genetic exchange between modern maize varieties or hybrids and maize landraces in some of these areas (Morris and Heisey 1998).

Diversity attributes of selected systems

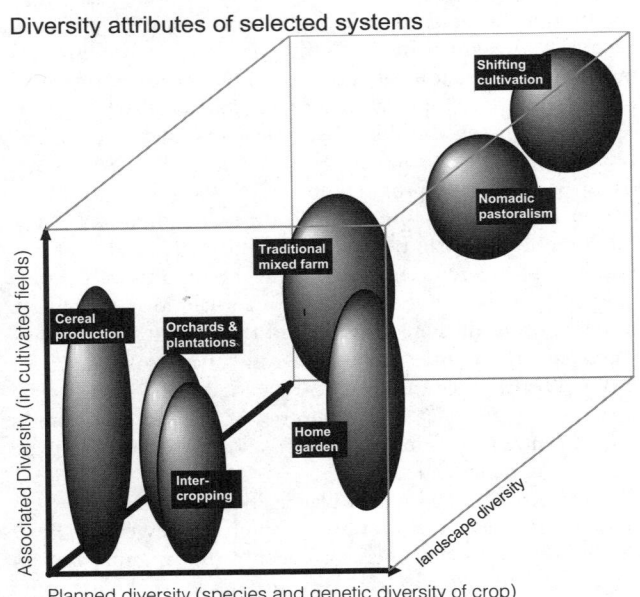

Diversity attributes of management options
in intensive tropical rice production

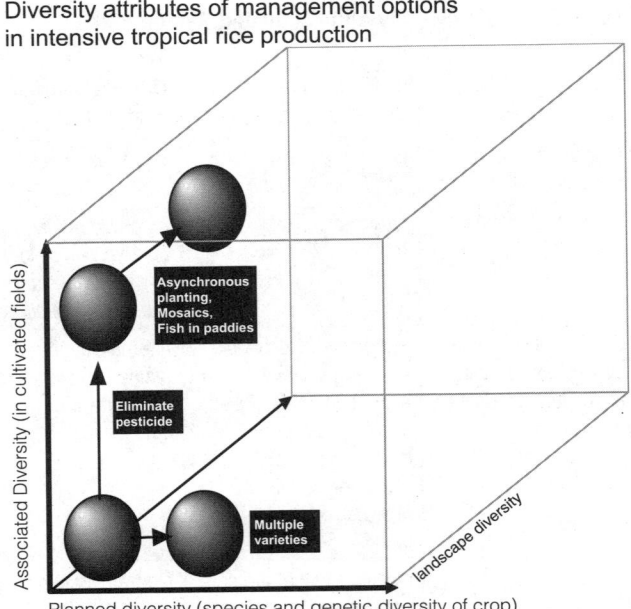

Figure 26.4. Dimensions of Diversity in Selected Production Systems

For some other major crops, such as sorghum and millet, the picture is quite different. While sorghum- and millet-growing regions of North America and parts of India have modernized systems, in the African Sahel, where these crops provide the main food source, there has been very little adoption of modern varieties and traditional practices prevail.

Farmers continue to use landraces rather than modern varieties for a range of different reasons (FAO 1998; Brush et al. 1992; Gauchan 2004; Meng 1997; Smale et al. 2001; Van Dusen 2000). First, landraces provide a wider range of end uses and have distinct culinary purposes, which may also contribute to maintaining a balanced diet. For example, in the Sahel and other dryland areas of Africa, certain varieties of sorghum produce better porridge, while others are better for boiling, and entirely different varieties are used for brewing beer. Some crop varieties have long stalks for animal feed or fencing, while others have sweet stalks that provide a refreshing snack in the field. Markets may have specific requirements, distinct from preferences at home, and some varieties have distinct cultural uses.

Second, production factors and risk management provide additional motives for continued use of landraces. Farmers frequently explain that they select different landraces to match differences in soil water regimes, even within the same field. Varieties with different maturities may be used to spread labor requirements through the season. Where weather patterns are uncertain or diseases are prevalent, planting several varieties can spread risk. Poor farmers are more likely to be faced with failures in insurance markets and thus use natural resource allocation as a means of insuring. Maintaining a diverse set of crop varieties to insure against production or market risks may be the most accessible means of insurance available to low-income households, whereas higher-income households with greater access to formal financial markets or other means of risk management may be more likely to risk a narrower portfolio of crop varieties. In addition, some traditional varieties (of millet, for instance, in West Africa) are photoperiodic: the time of maturity is set by day-length changes, allowing planting time to be varied according the

start of the rains while still ensuring that the crop is ready for harvest on time (Niangado 2001).

Finally, in some unfavorable and heterogeneous environments, appropriate modern varieties have simply not been developed or are not available. Breeding for such environments requires a decentralized approach to exploit "genotype x environment interaction" (Simmonds 1991; Ceccarelli 1994; de Vries and Toenniessen 2001). Participatory approaches to plant breeding are, however, having some success in developing suitable varieties for such areas (Ceccarelli et al. 2001; Weltzein et al. 1999; Cleveland and Soleri 2002).

Though empirical evidence is limited, both theory and observation suggest that genetic heterogeneity provides greater disease suppression when used over large areas. Some studies, including those of wheat mosaic virus (Hariri et al. 2001), fungal pathogens of sorghum (Ngugi et al. 2001), and rice blast (Zhu et al. 2000), have shown that mixed planting of resistant varieties with other varieties can reduce the disease incidence across the whole crop, while possibly extending the functional "lifespan" of the resistant genotypes. However, evolutionary interactions among crops and their pathogens mean that improvement in crop resistance to a pathogen is, in most cases, likely to be transitory. Thus, maintaining stocks of genetic diversity for plant breeding is critically important.

Cultivated systems also support a high diversity of livestock. Globally, there are 6,500 breeds of domesticated animals, including cattle, goats, sheep, buffalo, yaks, pigs, horses, chicken, turkeys, ducks, geese, pigeons, and ostriches. A third of these are under near-future threat of extinction due to their very small population size. Over the past century, it is believed that 5,000 domesticated animal and bird breeds have been lost. The situation is most serious in industrialized farming systems, with half of current breeds at risk in Europe and a third at risk in North America. While only 10–20% of current livestock breeds are at risk in Asia, Africa, and Latin America, it is likely that the risk of breed loss will increase as these countries pursue the path of economic development followed in industrial countries (FAO/UNEP 2000; Blench 2001).

26.2.1.2 Management of Associated Agricultural Biodiversity That Supports Production

The biodiversity of fauna and flora found in agroecosystems often plays an essential role in supporting crop production (Swift et al. 1996; Pimbert 1999; Cromwell et al. 2001). Earthworms and other soil fauna and microorganisms, together with plant root systems, maintain soil structure and facilitate nutrient cycling. Pests and diseases are kept in check by parasites, predators, and disease control organisms, as well as by genetic resistances in crop plants themselves. Insect pollinators also contribute to cross-fertilization of crop species that outcross.

As the examples in this section illustrate, it is not only the organisms that directly provide such services that are important, but also the associated food webs, such as alternative forage plants for pollinators (including those in small patches of wild lands within agricultural landscapes) and alternative prey for natural enemies of agricultural pests. Agroecosystems vary in the extent to which this biological support to production is replaced by external inputs. In industrial-type agricultural systems, they have been replaced to quite a significant extent by inorganic fertilizers and chemical pesticides; but in the many areas, particularly in the tropics, agricultural biodiversity provides the primary forces governing nutrient availability and pest pressure.

26.2.1.2.1 Soil biodiversity

Soil organisms contribute a wide range of essential services to the function of terrestrial ecosystems by acting as the primary driving agents of nutrient cycling and regulating the dynamics of soil organic matter formation and decomposition, soil carbon sequestration, and greenhouse gas emission. They modify soil physical structure and hydraulic properties that influence root growth and function and nutrient acquisition. In addition, many pollinators as well as natural enemies of agricultural pests spend part of their life cycle in the soil.

Soil biodiversity is responsive to the management of cultivated systems (Giller et al. 1997). Cultivation drastically affects the soil environment and hence the number and kinds of organisms present (Karg and Ryszkowski 1996; Ryszkowski et al. 2002). In general, tillage, monoculture, pesticide use, erosion, and soil contamination or pollution have negative affects on soil biodiversity. In contrast, no-till or minimal tillage, the application of organic materials such as livestock manures and compost, balanced fertilizer applications, and crop rotations generally have positive impacts on soil organism densities, diversity, and activity. Soil condition can thus be improved by farm practices and, indeed, some soils are in effect created by farmers (Brookfield 2001).

26.2.1.2.2 Pollination

Over three quarters of the major world crops rely on animal pollinators. While bees are the principal agents of pollination, flies, moths, butterflies, wasps, beetles, hummingbirds, bats, and others serve also as pollinators. Approximately 73% of the world's cultivated crops, including cashews, squash, mangos, cocoa, cranberries, and blueberries, are pollinated by bee species, 19% by flies, 6.5% by bats, 5% by wasps, 5% by beetles, 4% by birds, and 4% by butterflies and moths (Roubik 1995). Of the hundred or so crops that make up most of the world's food supply, only 15% are pollinated by domestic bees, while at least 80% are pollinated by wild bees and other wildlife. The services of wild pollinators are estimated to be worth $4.1 billion a year to U.S. agriculture alone. Wild plants and weeds provide alternative forage and nesting sites for pollinators, whose diversity is directly dependant on plant diversity, and vice versa (Kevan 1999). Forest-based pollina-

tors in Costa Rica have been shown to increase the value of coffee production from a single farm by approximately $60,000 per year by increasing yields and improving crop quality (Ricketts et al. 2004).

Many pollinating species are at risk of extinction, and pollination is now regarded as an ecosystem service in jeopardy, which requires attention in all terrestrial environments—from intensive agriculture to wilderness (Buchmann and Nabhan 1996). Pollinators are declining because of habitat fragmentation, agricultural and industrial chemicals and associated pollution, parasites and diseases, the introduction of exotic species, and declines in non-crop nectar and larval food supplies. Due to declining pollinator populations, an increasing number of farmers around the world are now paying for pollination services, importing and raising pollinators to ensure that crop seed yields are not limited by lack of pollination. Despite their tremendous importance, little is known about wild pollinator populations or the consequences of their decline (Kevan 1999; Kevan and Phillips 2001). (See Chapter 11 for further information on pollinators.)

26.2.1.2.3 Pest management

Insects, spiders, and other arthropods often act as natural enemies of crop pests. Research in the rice fields of Java has documented that other components of arthropod diversity are important in this respect (Settle et al. 1996). Without alternative food sources, populations of natural enemies would be directly dependent on the plant pest, which in turn is directly dependent on the rice plant for food. Such a linear system would be expected to give rise to seasonal oscillations in populations at the various trophic levels. In the Javanese rice fields, however, "neutral" arthropods, mostly detritivores and plankton feeders, such as midges and mosquitoes, provide an alternative source of food for the natural enemies of rice plant pests, thus stabilizing the populations of the natural enemies and providing better pest control. Furthermore, the detritivores depend on high levels of organic matter in the paddy soils, which provide the food source for an array of microorganisms (bacteria and phtytoplankton) and zooplantkton.

Further stability is provided by spatial and temporal heterogeneity at the landscape level. In Central Java, for example, the landscape is made up of a patchwork of small to intermediate sized plots of paddy rice (patches of between 10 and 100 hectares), planted continuously to rice at differing times throughout the year, with only a short fallow period and interspersed with patches or lines of trees and shrubs. There is some evidence of greater abundance of natural predators in such landscapes (as compared with more-uniform rice environments found in West Java, for instance) and that asynchronous planting of rice and the patches of uncultivated land mean that there are always alternative food supplies for natural enemies (Settle et al. 1996).

26.2.1.3 Conservation of Wild Biodiversity in Agricultural Landscapes

Besides the services required to sustain agriculture, biodiversity in agricultural ecosystems has a wider significance. Agricultural ecosystems represent substantial portions of watersheds, which are often landscapes that support recreation and tourism. They also harbor important biodiversity in their own right. Moreover, biodiversity in agricultural landscapes has powerful cultural significance, partly because of the interplay with historic landscapes associated with agriculture and partly because many people come into contact with wild biodiversity in and around farmland. In fact, in some regions elements of biodiversity now only exist in areas dominated by agriculture. Management of biodiversity in such areas is therefore an essential component of an overall approach to its conservation.

Indeed, in some parts of the world, notably Europe, biodiversity conservation has in recent years been acknowledged as one of the aims of agricultural policy. In spite of this, the negative trend of biodiversity in agricultural ecosystems, which was initiated with the intensification of agriculture in the latter part of the twentieth century, still prevails in Europe. Indicators such as the populations of farmland birds tend to show a negative trend (Birdlife International 2004). (See Figure 26.5.) Other indicators also show a loss in wildlife distribution and habitat as a consequence of intensification in agricultural production (Mankin and Warner 1999; Gall and Orians 1992). In developing countries, however, the expansion of agriculture is considered to be the greatest threat to extinction of threatened bird species, and a recent study suggests that intensification of agriculture in these areas to avoid further expansion of cropland would reduce this threat to biodiversity of bird species (Green et al. 2004).

One positive landscape-wide impact noted in sub-Saharan Africa, South Asia, and Southeast Asia is the trend of growing more trees for a wide variety of purposes. Trees can stabilize and enhance soils, can contribute to plant biodiversity in the landscape, and may provide habitat for a variety of birds, reptiles, small mammals, and insects. Some birds and small mammals can be important sources of revenue in farmlands, such as when farmers make agreements with outfitters and hunters and plan their management in an integrated way. Wildlife in cultivated systems can contribute to food security by providing an important source of animal proteins for the most marginal rural settlers. It should be noted, however, that the introduction of trees and other woody vegetation into some ecosystems, particularly remnant tracts of grassland or where area-sensitive grassland species are present, can have negative consequences to those species and become invasive woody perennials in these ecosystems (Allen 1994; Samson et al. 2004). (See Chapters 4 and 11 for more on alien invasive species.)

26.2.1.4 Impacts of Agricultural Practices on Biodiversity

Cultivated systems have large impacts on other ecosystems and on the services they supply. The most obvious impact is through

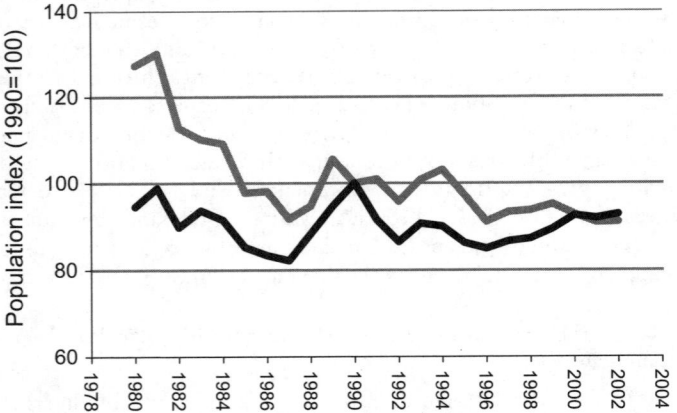

Figure 26.5. Supranational Multi-species Indicator of European Bird Populations, 1980–2002. Farmland birds (grey line: 18 countries, 23 species, ± 1.96 SE) and woodland, park, and garden birds (black line: 19 countries, 24 species, ± 1.96 SE). The index for the base year, 1990, is set to 100. Data come from the Pan-European Common Bird Monitoring Scheme, an initiative of the Royal Society for the Protection of Birds, the European Bird Census Council, and BirdLife International.

expansion of cultivated systems. Globally, agricultural land has expanded by around 130,000 square kilometers a year over the past 25 years, predominantly at the expense of natural forests and grasslands. In addition, a rapid increase in coastal aquaculture has led to the loss of many mangrove ecosystems. Though future rates of conversion are expected to be much lower in absolute terms, assuming no shortage of staple food crops, the locations of major agricultural expansion frequently coincide with remnants of those natural habitats with high biodiversity value (Myers et al. 2000).

Other externalities associated with cultivated systems include the use of water and nutrients and the pollution of ecosystems resulting from excess use of pesticides and nutrients. Irrigated agriculture is a major user of fresh water. Both the direct loss of wetland habitats from conversion and the pollution of inland waters by excess nutrients have major negative impacts on inland water biodiversity. (See Chapter 20.) Despite increases in water use efficiency, total water demand for agriculture is increasing and in many regions is projected to outstrip supply over the coming decades. (See Chapter 7.)

Agriculture is the major user of industrially fixed nitrogen, and only a fraction of this fertilizer is used and retained in food products. The excess nitrogen leads to biodiversity loss in inland water, coastal, and marine systems through eutrophication and to loss of terrestrial plant diversity through aerial deposition. (See Chapters 18, 19, and 20.) Conversely, the soils of several cultivated systems, especially in Africa South of the Sahara, are nutrient-depleted. This is especially problematic where fruits, vegetables, and other crops are exported or transferred on a large scale from rural areas to large urban centers. Significantly greater use of fertilizers will be necessary in some regions to maintain soil nutrient stocks and support increased production by these systems.

Pesticides and herbicides have direct impacts on biodiversity through the degradation of ecosystems. Birds are particularly vulnerable to losses in invertebrate populations due to the use of pesticides and herbicides (Hooper et al. 2003). Especially important are those pesticides and herbicides that are persistent organic pollutants, since they have effects on large spatial and temporal scales. Many of the more persistent chemicals are being phased out and replaced by others with much lower toxicity that are less persistent. However, the overall use of pesticides is still increasing (FAOSTAT 2004).

26.2.1.5 Concluding Comments on the Relationship between Biodiversity and Cultivated Systems

Although cultivated systems often have a negative impact on it, biodiversity remains essential for the productivity and sustainability of cultivated systems. In modern industrial agricultural systems, many components of biodiversity relevant to agriculture can be separated from the production system (such as ex situ germplasm collections and plant breeding programs). Biodiversity is still required for production, however, even if partially maintained ex situ.

In addition, some of the services provided by biodiversity can, to a certain degree, be provided by externally provided services. For example, market integration and insurance services can substitute for risk management provided by crop genetic diversity, and other forms of pest management, including the use of pesticides, can substitute for pest and disease control. But it is important to note two points. First, poor farmers often do not have the option of introducing modern methods for services provided by biodiversity because of the lack of market integration or heterogeneity of the environment or because they cannot afford the alternatives. Second, substitution of some services may not be sustainable and may have negative environmental and human health effects (for example, the reliance on toxic and persistent pesticides to control certain pests can have negative effects on the provision of services by the cultivated system and other ecosystems connected to the cultivated system).

The relative cost-effectiveness of biodiversity-based over substitute services is sensitive to many factors and may be amenable to the application of incentive measures. The levels of biodiversity in cultivated systems and the ecosystems services they provide can be manipulated directly or indirectly by management practices. For instance, practices such as integrated pest management and minimal tillage agriculture, as well as multicropping, the use of genetic diversity, and mosaic landscaping can increase biodiversity in cultivated systems. However, if such measures reduce crop yields per unit land area-time, the aggregate effects of such practices could result in the expansion of crop area at the expense of natural ecosystems, thus trading increased biodiversity within cultivated systems for a decrease in the extent of natural ecosystems and the biodiversity they contain.

26.2.2 Fresh Water

Agriculture is by far the most consumptive human use of fresh water. Water requirements for cultivation are large; it takes 500 liters, 900 liters, 1,400 liters, and 2,000 liters of transpired water to produce 1 kilogram of potatoes, wheat, maize, and rice respectively (Klohn and Appelgren 1998). Other crops, such as sugarcane and bananas, are even more water-demanding.

Cultivation both relies on and influences the provision of fresh water. Both the quantity and quality of water resources can be affected, as well as the timing and distribution of water flows in local catchments and large river basins. The impact of cultivation on freshwater quantity is much larger in irrigated than in rain-fed systems. Deforestation associated with rain-fed cultivation tends to increase the amount of water available for agriculture because of reduced transpiration losses. Impoundments for irrigation can regulate downstream flows, while seasonally bare soil and field drainage systems can accelerate runoff and reduce infiltration, resulting in more severe local flooding and decreased dry weather flows (Bruijnzeel 2001). Water quality effects have been reported for all forms of cultivation, including confined livestock systems and aquaculture, but the nature and magnitude of impact can vary substantially. Poorly managed cultivation, particularly on sloping lands, is often associated with soil erosion, high silt loading, and downstream sedimentation. Intensive, high-input production systems can result in water pollution from leaching or runoff that carries nutrients, pesticides, or animal wastes to waterways (National Research Council 2000; de Haan et al. 1997).

The negative impact of cultivation on water resources can limit options for, and increase the cost of downstream water use (for example, through reduced domestic water supply or recreational opportunities). It can also have additional negative effects on the supply of other ecosystem services (such as reducing aquatic biodiversity and increasing nutrient flows) and on the condition of other ecosystems (the integrity and productivity of inland waters and coastal systems, for instance). (See Chapters 19 and 20.)

This section briefly reviews the sources and means by which fresh water is provided to and utilized by cultivated systems, identifies some of the issues surrounding water use efficiency, and considers some of the most important environmental consequences of cultivation on water resources.

26.2.2.1 Irrigation

Irrigation involves the withdrawal of groundwater and the diversion of surface water resources to help meet the transpiration re-

quirements of crops and to lengthen the growing period when rainfall alone is insufficient to support crop growth. By mitigating moisture deficits, irrigation can significantly increase yields and total crop biomass, stabilize production and prices (by dampening the effects of rainfall variability within and across seasons), and encourage production diversity. But irrigation requires increased freshwater use and promotes more-intensive land use with regard to labor and other inputs, such as improved seeds, fertilizers, and pesticides (because such inputs are usually needed to achieve the yield increases that are possible when water limitations to crop growth are removed).

Of the 9,000–12,500 cubic kilometers of surface water estimated to be available globally for use each year (UN 1997), between 3,500 and 3,700 cubic kilometers were withdrawn in 1995 (Shiklomanov 1996). Of that total, about 70% was withdrawn for irrigation (Postel 1993). According to the World Bank (2000), the share of extracted water used for agriculture ranges from 87% in low-income countries to 74% in middle-income countries and 30% in high-income countries. By 2002, there were 276 million hectares of irrigated cropland globally—five times more than at the beginning of the twentieth century. While this irrigated area represents only 18% of all croplands, irrigated agriculture provides about 40% of the global food supply (FAOSTAT 2004; Bruinsma 2003; Wood et al. 2000).

The wide variability in freshwater endowments between regions and countries has a large influence on the potential for the development and long-term viability of irrigated agriculture. While the 206 cubic meters per capita withdrawn annually for agriculture in Africa represents 85% of total water withdrawals on that continent, the 1,029 cubic meters per capita withdrawn for agriculture in North America represents just 47% of that continent's withdrawals (World Resources Institute 2000). Compared with high-income countries, mostly located in subhumid/humid temperate and sub-tropical climates, many poor countries tend to have scarcer water supplies and relatively large agricultural demands due to the higher share of agriculture in their economies.

The simplest measure of irrigation intensity is the amount of irrigation water withdrawn (or applied) per year. This is most commonly expressed as an equivalent water depth per unit area (cubic meters of water per year divided by hectares irrigated). Using data from WRI (1998) and FAOSTAT (1999) across 118 countries, Seckler et al. (1998) calculated the mean depth of irrigation globally to be about 1 meter per year on the 276 million hectares of irrigated cropland.

Although irrigation is by far the largest global water user, the net rate of increase in irrigated area has decreased steadily in each of the past four decades and now stands at just under 1% annually (FAOSTAT 2004). Expansion in irrigated area has slowed as unexploited freshwater resources have become more limited and more expensive to develop. In addition, cereal prices have trended downwards in real terms, and environmental and social objections to the construction of large-scale impoundments have grown. There is also increasing competition for water from domestic and industrial users. Such pressures have resulted in increasing regulation of the allocation of water resources in many countries and of effluent and water quality standards (including the establishment of "minimum environmental flows" in some cases). These trends have increased public awareness of water use by agriculture and have fostered greater concern by farmers and researchers about improvements in water use efficiency in cultivated systems (Tharme 2003; Benetti et al. 2004).

Irrigation can have positive in addition to negative externalities. In some rural areas, it is the only reliable source of water for cooking and cleaning. Infiltration from rice paddy systems also contributes to groundwater reservoirs that are important sources of water in urban areas, as well as contributing to flood control and prevention of saltwater intrusion (Renault and Montginoul 2003).

Water loss also occurs with aquaculture through evaporation and pond seepage. Pond seepage may be as much as 2.5 centimeters per day, while as much as 1–3% of the fish pond volume may be lost daily (Beveridge and Phillips 1993).

26.2.2.2 Water Use Efficiency

Irrigation systems, particularly those involving surface water impoundment and conveyance, are often inefficient in terms of water loss through evaporation and leakage. Water efficiency is defined as the ratio of water used by crops to the gross quantity of water extracted for irrigation use. Global estimates of irrigation efficiency vary, but the average is around 43% (Postel 1993; Seckler et al. 1998). Seckler et al. (1998) estimate that arid agroecosystems have more efficient irrigation—for example, 54% and 58% efficiency for the two driest groups of countries, compared with 30% for the least water-constrained countries. China and India show irrigation efficiencies of around 40%, and they strongly influence the global average because of their large irrigated area. Irrigation efficiencies typically range from 25% to 45% in Asia, but up to 50–60% in Taiwan, Israel, and Japan (Seckler et al. 1998:25).

Recognizing the large potential for water efficiency improvements in agriculture, and spurred by increasing competition for water, many technologies have been developed to enhance the effectiveness of water use in both irrigated and rain-fed cultivation. Postel (1999) describes how microirrigation systems, such as drip and micro-sprinklers, often achieve efficiencies in excess of 95% compared with standard flood irrigation efficiencies of 60% or less. She cites significant water productivity gains for a wide range of crops, resulting from the shift from conventional to drip irrigation in India. For example, water use declined as much as 65% in the case of sugarcane cultivation, and water productivity increased by 255% in cotton. The reason for these increases in irrigation efficiency is that a precise water application can both reduce total water use and increase yields. Sugarcane and cotton yields increased 20% and 27% respectively, along with substantial reductions in water use. Postel (1997) indicated that as of 1991, only 0.7% of irrigated farmland worldwide was being microirrigated. While this fraction is expected to have increased since 1991, no recent, comprehensive global data are available (Gleick 2002).

Other techniques for improving water use efficiency in both irrigated and rain-fed systems have included furrow diking, land leveling, direct seeding, moisture monitoring, low-energy precision application sprinklers, low pressure sprinklers, water accounting, and stomatal control by chemical signaling (Gleick 2002; Davies et al. 2003). Complementary strategies have included the development of more drought-tolerant crop germplasm (Edmeades et al. 1999; Pantuwan et al. 2002), experimentation with policies that foster water markets or other economic or regulatory arrangements, and institutional reforms that engage farming communities more directly in improving water resource management (Postel 1997; Subramanian et al. 1997).

Water conservation methods, such as mulching, deep tillage, contour farming, and ridging, also help increase water use efficiency by ensuring that the rainwater is retained long enough to ensure infiltration into the soil root zone (Habitu and Mahoo 1999; Reij et al. 1988). These approaches can be complemented by "water harvesting" techniques involving the small-scale concentration, collection, storage, and use of rainwater runoff for both domestic and agricultural use.

Increasing effective rainfall use through improved water harvesting technologies and water conservation methods has largely been pioneered in arid and semiarid regions, and water harvesting techniques have been classified in various ways (Reij et al. 1988). Pacey and Cullis (1986) described three broad categories: external catchment systems, microcatchments, and rooftop runoff collection, the latter used almost exclusively for nonagricultural purposes. External catchment rainwater harvesting involves the collection of water from areas distant from where crops are grown (Oweis et al. 1999). Microcatchment techniques are those in which the catchment area and the cropped area are distinct but adjacent (Habitu and Mahoo 1999). Microcatchments generate higher yields per unit area than larger catchments (Bruins et al. 1986) and they are simple, inexpensive, and easily reproduced where land is available (Boers and Ben-Asher 1982). Microcatchments have been used in Asia, Africa, America, and Australia, where they are often used for medium water-demanding crops such as maize, sorghum, millet, and groundnuts (Habitu and Mahoo 1999), but evidence of large-scale adoption and impact is so far limited.

Water use efficiency can also be improved by carefully designed landscapes. Studies of processes induced by shelterbelts and woods in agricultural landscapes indicate that the structure of plant cover has an important bearing on agricultural water resources (as well as on habitat and natural biodiversity). The protective effects of trees decrease wind speeds close to Earth's surface and lower saturation vapor deficits, thus decreasing evapotranspiration from sheltered fields. Fields between shelterbelts conserve moisture (Brandle et al. 2004; Cleugh et al. 2002; Kedziora and Olejnik 2002). Shelter effects are greater under dry and warm meteorological conditions compared with wet and cool weather (Ryszkowski and Kedziora 1995).

In addition, shelterbelts have been shown to decrease surface runoff rates, protect soil against water erosion, and increase soil infiltration rates, thus improving dry-season flows (Kedziora and Olejnik 2002; Werner et al. 1997). Some studies suggest that heterogeneity of plant cover structure, including trees in agricultural landscapes, also generates meso-scale atmospheric circulation, which can increase regional or local precipitation (Pielke et al. 1991, 1998) and recycling of water in the landscape (Lawton et al. 2001; Stohlgren et al. 1998). Counterbalancing these positive effects, tree shelterbelts also compete for land, nutrients, and water with crops and also shade them, which can reduce crop yields or total crop output.

With rapid urbanization and growing competition for water resources (particularly in arid and semiarid regions), as well as budget constraints for effective treatment of growing wastewater volumes, the reuse of urban wastewater for agriculture is receiving increasing attention. Wastewater is being used as a low-cost alternative to conventional irrigation water to support vegetable production in urban and peri-urban agriculture, despite the health and environmental risks that might be associated with this practice. It is suggested that raw wastewater use in agriculture is increasing at close to the rate of urban growth in developing countries, where urban and peri-urban land is available (Scott et al. 2004).

Just how prevalent wastewater irrigation is today is a matter of conjecture, as no reliable global data exist. However, as an important step toward a global figure, Rachid-Sally et al. (2004), Cornish and Kielen (2004), and Ensik et al. (2004) present assessments of the area irrigated with wastewater at the country level, with estimates of 9,000 hectares for Viet Nam, 11,900 hectares for

Ghana, and 32,500 hectares in Pakistan. As the recycling of waste-water for irrigation grows, there are increasing concerns about the long-term human health consequences (Scott et al. 2004).

26.2.2.3 Impacts on Water Quality

Besides their effect on water quantity, cultivated systems can have negative impacts on freshwater quality through pollutants contained in the drainage water, runoff, and effluents. Where irrigation depletes rivers and aquifers that receive increased agricultural pollution, quality impacts are exacerbated because of reduced dilution capacity. Physical loading of water resources with inorganic (soil particles) and organic sediments or particulate matter, as well as chemical loading of plant nutrients, especially nitrogen, phosphorus, and pesticides, can often occur as a result of cultivation or intensive livestock and aquaculture operations (Sharpley and Halverson 1994; Owens 1994;).

Agricultural impact on water quality is also mediated through erosion brought about by poor crop cover, field drainage, and cultivation operations, particularly on sloping lands. Gleick (1993) estimates that about 22% of the annual storage capacity lost through siltation of U.S. reservoirs is due to soil erosion from cropland. Water-borne transportation of nitrates and phosphates is quite common where external nutrients are applied in excess or inefficiently and can cause eutrophication of surface waters. In some countries, such as Belgium and the Netherlands, the nitrogen input to some crops has in the past exceeded 500 kilograms per hectare (Wood et al. 2000).

Phosphorus transportation into aquatic ecosystems is the principal cause of blue-green algae blooms in reservoirs, and the anoxia in the Gulf of Mexico is one example of eutrophication attributable to nutrient enrichment (Snyder 2001). The off-site economic impact of water quality changes attributable to cultivation include damage to water-based recreational facilities, fisheries, navigation, water storage facilities, municipal and industrial water users, and water conveyance systems as well as increased flooding or inundation of low-lying urban areas and civil structures.

Salinization and waterlogging are two significant consequences of poor irrigation management and inadequate drainage (Ghassemi et al. 1995). Salinization occurs through the accumulation of salts deposited when water is evaporated from the upper layers of soils and is especially important in irrigated arid areas where evaporation rates are high. Since most crops are not tolerant of high salt levels, salinization decreases yields. This problem is particularly severe in arid and semiarid areas, such as Pakistan and Australia. Waterlogging is more common in humid environments and in irrigated areas where excessive amounts of water are applied to the land.

Ghassemi et al. (1995) estimated that around 45 million hectares, representing 20% of the world's total irrigated land, suffers from salinization or waterlogging. Losses amount to approximately 1.5 million hectares of irrigated land per year (Ghassemi 1995 quoting Dregne et al. 1991) and about $11 billion annually from reduced productivity (Postel 1999), representing about 1% of the global totals of both irrigated area and annual value of production respectively (Wood et al. 2000). Once salinization has occurred, rehabilitation for further cultivation is difficult and costly, but successes via specific vegetation strategies, using tree species, have been documented (Cacho et al. 2001; Barrett-Lenard 2002).

Freshwater aquaculture operations are strongly linked to water quality in terms of both the necessary quality of incoming water as well as the impacts of aquaculture effluents. Wells and springs are the best sources of water, but other sources are used if a number of water quality characteristics, including temperature, dissolved oxygen, ammonia, nitrites, nitrates, pH, alkalinity, and hardness, are within viable ranges. Water pollution risks arise in aquaculture when large amounts of harmful materials are added to the water body, adversely affecting its local and effluent water quality.

Fish culture operations, especially in intensive aquaculture, require fishmeal or fish feed. Feed contains nutrients such as nitrates and phosphates, and excess of these nutrients can lead to eutrophication and triggering of intense growth of aquatic plants (micro and macro). While in some aquaculture systems phytoplankton are themselves used as a food, overproduction of aquatic plants, particularly algae, causes algal blooms and can consequently lead to clogging of waterways, depletion of dissolved oxygen, and hindrance of light penetration to deeper water depths affecting photosynthetic and other metabolic functions of aquatic organisms. Thus unused feed, algal blooms, and detritus from the fish themselves impose additional pollutant loads when they discharge into external freshwater sources.

Pond and recirculation systems, as well as integrated agriculture-aquaculture systems, pose fewer risks of external pollution than the more open cage and raceway forms of aquaculture (Boyd 1985; Beveridge and Phillips 1993). In some cases, freshwater aquaculture ponds can improve water quality by acting as sinks for sediments (Stickney 1994).

To reduce the direct discharge of effluents and increase water use efficiency, wastewater from integrated agriculture-aquaculture systems has been used for irrigation. Where fresh water is available, aquaculture is a good way of using marginal land that is less suited to crop and livestock agriculture. Freshwater aquaculture ponds can be designed to contribute positively to soil and water conservation by dissipating the energy of overland flow and reducing erosion and downstream flooding.

26.2.3 Food

Trends in food provision, predominantly derived from cultivated systems, are assessed in detail in Chapter 8. This short section simply summarizes relevant key findings of that chapter.

The production of food and other products is, by design, the primary goal of cultivated systems. The global demand for food continues to be driven by population growth (albeit at a slowing rate), by the increasing real incomes of many households worldwide, and by evolving consumer preferences for more convenient, safer, and nutritious foods. Furthermore, wealthier consumers in industrial countries are increasingly willing to pay more for foods produced and marketed in ways that are perceived to be more environmentally sustainable and socially equitable.

From a food supply perspective, the scale of conversion of natural ecosystems for cultivation purposes, and the nature and extent of the trade-off between provision of food and of other ecosystem services within cultivated systems, has been shaped by the cultivation practices and technologies accessible to farmers. The decisions of most farmers about which crops to produce and how to produce them has also been influenced by a wide range of economic signals and, particularly in richer countries, by regulatory standards.

Farmers and, increasingly, scientists have accelerated the processes of domestication and adaptation of plant species and available germplasm through breeding and biotechnology to enhance food output from crops and animals across a very broad range of environmental and agronomic conditions. Use of transgenic crop varieties developed with recombinant DNA technology is increasing rapidly worldwide in both industrial and developing countries. Although this technology holds tremendous promise to increase productivity significantly and to improve end-use properties of crops for both rich and poor producers, the widespread use of transgenic crops, often referred to as genetically modified organisms, continues to generate controversy with regard to ethical, environmental, equity, and intellectual property issues.

Over the past half-century, and at a global scale, food provision has more than kept pace with growth in demand, leading to a significant, long-term decline in the real price of food and allowing an ever-growing share of a rapidly increasing world population to be fed adequately at reasonable cost. Nevertheless, there remain significant causes for concern about food provision on several fronts. First, there remains a persistent and, recently, growing population of undernourished people, estimated at 852 million for 2000/02 (FAO 2004). Second, in many of the same countries where hunger and poverty persist, population growth rates tend to be high, and expansion of food production is failing to keep pace with demand. In the face of population pressure, often compounded by limited access to resources and technologies, poor intensification practices have all too frequently degraded the productive capacity of existing cultivated areas. Depletion of soil nutrient stocks in subsistence systems has, for example, reduced the productive capacity of large areas in sub-Saharan Africa.

Third, the linear rate of increase in the yields of the three major cereals (maize, rice, and wheat) is falling below the rate of increase in demand in many of the world's major production areas (Cassman et al. 2003). Moreover, global warming from human-induced climate change may reduce crop yield potential and thus decrease the rate of yield gain (Peng et al. 2004; Lobell and Asner 2003; Rosenzeig and Parry 1996; Brown and Rosenberg 1997). Fourth, there are concerns of growing divergence, rather than convergence, between the economic, science, and technology capacities of richer and poorer nations with regard to food production. This divergence is hindering efforts to promote the emergence of profitable and sustainable smallholder agriculture in poorer countries. Finally, there is growing recognition that virtually all forms of cultivation have involved trade-offs between provision of food and provision of other ecosystem services. See Chapter 8 for further details on food provision.

26.2.4 Non-food Products

Besides producing food, cultivated systems provide other products such as fiber (cotton, flax, and jute, for instance), biofuels, medicines, pharmaceutical products, dyes, chemicals, timber, and other non-food industrial raw materials. Non-food crops account for nearly 7% of harvested crop area (Wood et al. 2000). Based on FAOSTAT 2004, the annual value of non-food crops from cultivated systems, excluding timber, is about 3.4% of total agricultural production ($50 billion, compared with $1.4 trillion for food crops).

In 2003, the reported primary production of fiber crops worldwide was about 25 million tons. Cotton is the major fiber crop and is extensively grown in China, the United States, and India, accounting for 15.6 million, 10.4 million, and 6.3 million tons, respectively, and providing 5.2 million, 3.9 million, and 2.1 million tons of cotton lint. Flax, another fiber crop, is widely grown in China and France, which produce around 500,000 and 86,000 tons respectively (FAOSTAT 2004).

In industrial countries, biofuel crops currently represent a relatively small proportion of output from cultivated systems. However, diversion of grain and crop biomass for biofuel and bio-based

industrial feedstocks could grow substantially with increasing oil prices and continued improvements in the energy efficiency of crop production and bio-fuel conversion. Another approach is through crop genetic engineering to enhance traits facilitating production of plastics and other bio-based industrial feedstocks. Current U.S. maize production systems produce a net energy surplus based on a complete life-cycle analysis, including the embodied energy content of all inputs and operations (Shapouri et al. 2003). It is likely that future gains in energy yield and in efficiency of biofuel production or conversion to feedstocks will increase competitiveness of these renewable resources, especially if fossil fuel prices rise significantly.

Improvements in crop yields and nitrogen fertilizer efficiency are the most promising avenues through which to achieve increased energy output and overall efficiency. Both these factors would also contribute to reducing the negative impact of cultivation on ecosystem services through reductions in greenhouse gas emissions, replacement of fossil fuel usage with a renewable energy source, reduction of NO_2 emissions, and a decrease in nitrogen losses via leaching, denitrification, and volatilization.

If use of grain and crop biomass for biofuel and bio-based industrial feedstocks were to expand, however, it would place additional burdens on other cultivated systems to continue to meet growing food demand and could promote additional area expansion of cultivation and, perhaps, upward pressure on food prices.

26.2.5 Nutrient Cycling and Soil Fertility

Essential nutrients are required to sustain all life and include the macronutrients such as nitrogen, phosphorus, potassium, calcium, magnesium, and sulfur, which are present in plant tissues at relatively high concentrations (0.1–2.0% on a dry weight basis), and micronutrients such as iron, zinc, and copper, which are required in very small quantities (1–50 parts per million). (See Chapter 12.) Of the essential nutrients, nitrogen and phosphorus have the greatest impact on environmental quality and ecosystem services because they can easily move from cultivated systems to other ecosystems and accumulate to potentially polluting levels.

Moreover, a large proportion of the total global load of reactive nitrogen and phosphorus cycles through agricultural systems, because these nutrients are required in large quantities to maintain crop yields. For example, nitrogen fertilizer applied to cropland represents more than 50% of the annual load of reactive nitrogen attributable to human activities (Smil 1999). Likewise, phosphorus contained in cultivated plants, livestock manure, and recycled organic matter represents 24–40% of the annual global phosphorus flux in terrestrial ecosystems (Smil 2000). While other nutrients are also important, their use in agriculture and their effects on global ecosystem services are much smaller and more localized. Hence, the discussion of nutrient cycling in this chapter will focus on nitrogen and phosphorus. (See Chapter 12 for a wider discussion on nutrient cycling and Box 26.2 for a discussion of "virtual trade" in crop nutrients.)

26.2.5.1 Nutrient Resources in Cultivated Systems

Nutrients available for uptake by crops are derived from resources and processes that are either internal or external to the cultivated system. Internal sources include the weathering of nutrients from soil minerals, which is a very slow process producing only small amounts of plant-available nutrients, and nutrients released in the decomposition of soil organic matter. All SOM is derived from the decomposition of organic materials that include crop and weed residues returned to soil, and livestock manure, mulch, and

BOX 26.2
Virtual Trade in Crop Nutrients

Nutrients are often a scarce and limiting factor in African cropping systems. Traditionally, farmers have relied on fallow or applied animal and green manure to maintain soil fertility, but pressure to expand output to meet growing food demand has led to shorter fallow periods and increased cropping intensity without corresponding increases in organic nutrient inputs. Since chemical fertilizers are too costly or not available for most farmers, cultivation has progressively led to depletion of soil nutrient stocks over much of sub-Saharan Africa. Here two dimensions of this process are assessed: the total amount of nutrients removed in the harvested component of crops and the net flux (the "virtual trade") of nutrients across national borders by examining the share of domestically produced commodities exported as well as the nutrient composition of imports.

Figure A provides a spatially disaggregated estimate of the total amount of nitrogen, phosphorus, and potassium extracted from agricultural soils each year in the harvested part of crops. Assuming crop residues are recycled, this extraction represents the lower threshold for the amount of nutrients that must be replaced to maintain soil fertility. If crop residues are also removed for fuel, fodder, or building materials, then the nutrients removed in these residues must also be replaced. For each country, the amount of each nutrient removed at harvest was assessed by applying nutrient content/concentration coefficients (amount of nutrient contained in each unit weight of harvested product) to the average annual production (1999–2001) of each of 20 regionally important crops as well as an "other" composite to represent the remaining crops.

The spatial distribution of individual crops was assessed by fusing data from sub-national production statistics; maps of cropland, irrigation, population density, and biophysical crop suitability; and other secondary information, according to the method described by You and Wood (2004). The individual estimates of harvested nutrients for each crop in each (10x10 kilometer) pixel were summed to produce a single NPK total per 100 square kilometers. Areas with larger amounts of nutrient removal are shown in darker shades. Because the amount of nutrients applied in these areas as fertilizer or organic inputs falls far short of these removal rates, failure to adequately replenish soils through the use of applied nutrients

is lowering soil fertility and reducing land productivity in many, if not most, of the areas shown.

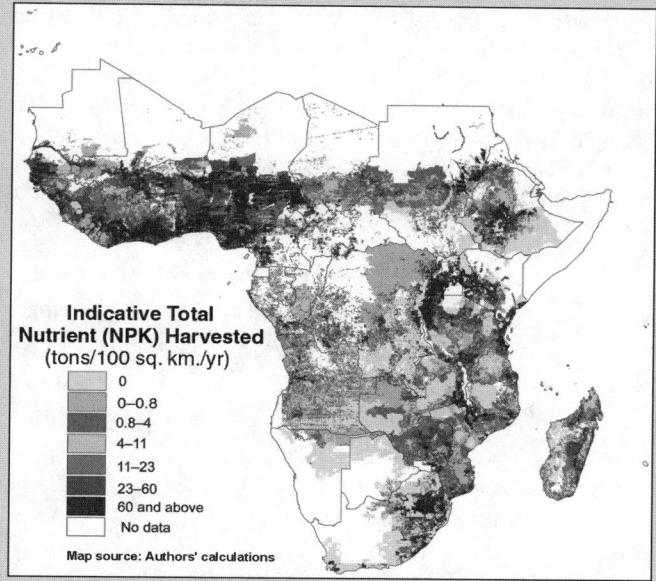

Indicative Total Nutrient (NPK) Harvested
(tons/100 sq. km./yr)

- 0
- 0–0.8
- 0.8–4
- 4–11
- 11–23
- 23–60
- 60 and above
- No data

Map source: Authors' calculations

To assess the overall flux of nutrients attributable to trade, the nutrient content coefficients were applied to the quantities of crops traded. The results for Africa (including north of the Sahara) show that a total of about 7.4 million tons of NPK are contained in the harvested crops of Africa each year (see Table), of which just over 1 million tons (14%) is contained in crop exports from the region. However, crop imports to the region contain around 2.5 million tons of NPK, about 34% of total domestic harvested removal.

This pattern of crop trade provides a net nutrient inflow into Africa of around 1.5 million tons per year. There are major geographical imbalances, however, between locations where nutrient are removed (plots and

---

compost applied to cropland. SOM contains substantial amounts of nitrogen, phosphorus, and sulfur, although these nutrients are unavailable for plant uptake when they are chemically bound in macro-modules within the organic matter. They only become available when these macro-modules are decomposed by soil microbes and microflora.

Maintaining a high SOM content and the soil microbe and microflora communities supported by SOM is therefore important for preserving soil fertility. Cropping practices that lead to a reduction in SOM result in a proportional decrease in the internal supply of nutrients and greater amounts of external inputs are needed to sustain crop yields. Conversely, cropping systems that increase SOM can reduce the need for applied nutrients. Crop residues and roots that are returned to the soil also decompose through microbial action and release nutrients that are available for crop uptake. A portion of these residues is converted to SOM, and thus the balance of organic matter input relative to SOM decomposition determines whether SOM increases, decreases, or stays the same. Burning of crop residues during fallow periods releases nutrients to the soil through the ash, although most of the nitrogen and some of the phosphorus are lost to the atmosphere in the combustion process.

Another internal source of nutrients, especially nitrogen, is biological nitrogen fixation, which is performed by symbiotic bacteria in association with forage and food legumes and by free-living nitrogen-fixing microorganisms that live in the rhizosphere of plant roots. Prior to the advent of modern farming practices and commercial fertilizers, BNF was the primary source of nitrogen in cultivated systems. In low-input cropping systems and in many natural ecosystems, however, BNF is often limited by a deficiency of phosphorus and other essential nutrients (Vitousek et al. 2002). In addition to BNF, nutrients from livestock manure are another internal source of nutrients in integrated crop-livestock farming systems.

Nutrient sources of external origin include inorganic fertilizers and livestock manure produced in confined feeding operations that are not associated with an integrated crop-livestock farming system. Secondary sources of external nutrient input include wet and dry deposition through nutrients contained, respectively, in rainfall and wind-blown dust, although the amount of nutrient addition from these sources is typically very small. Irrigation can provide substantial external inputs of nitrogen, potassium, calcium, magnesium, and sulfur in areas where groundwater or surface water used for irrigation contains relatively high concen-

BOX 26.2
continued

fields in cultivated systems) and locations where imported nutrients are used, primarily in urban areas and populated rural areas connected to the coast by a few transport corridors. Even within countries there are large movements of nutrients that exacerbate soil fertility problems. In Uganda, for example, *matooke,* the basic food staple produced by cooking the fruit of the East African highland banana, used to be prepared such that skins, stems, and other residues were recycled locally. Now, 30–50% of the country's 9 million tons per year of *matooke* enters the market system, and entire banana bunches with stems are shipped away to urban markets, primarily in Kampala. Here, though, wastes are often used as feed in confined livestock systems, especially for pigs.

	Nutrient Content of Harvested Crop Products			
	N	**P**	**K**	**Total NPK**
	(thousand tons per year)			
Production				
Eastern Africa	714	105	483	1,301
Northern Africa	738	114	383	1,235
Southern Africa	683	127	422	1,232
Western Africa	2,106	420	1,125	3,651
Harvested Product: Africa	4,241	766	2,412	7,419
Exports[a]	550	147	352	1,049
Imports[a]	1,448	439	634	2,522
Net trade flow of embodied crop nutrients[a]	+898	+292	+282	+1,427
Fertilizer consumption	2,462	953	485	3,900

[a] Derived from FAOSTAT (average 1999–2001) and nutrient content database

Comparing harvested nutrients to applied fertilizer estimates, a rough indication of the regional nutrient shortfall is apparent. Both nitrogen and potassium replenishment from fertilizers at the regional scale are significantly less than the nutrients removed in harvested crop product. This shortfall will be even greater to the extent that crop residues are not recycled and applied nutrients are not taken up by the crop (typically, nutrient uptake rates from applied nutrients are quite low). The shortfall is reduced where organic nutrients are applied but, typically, overall nutrient NPK balances in East and West Africa have been estimated at greater than −60 kilograms per hectare per year (Henao and Baanante 1999; Stoorvogel and Smaling 1990). Moreover, in African soils a large share of applied phosphorus fertilizer is fixed in the soil complex and is unavailable for plant use.

Clearly this aggregate assessment hides many important details. One is the lack of complete nutrient balances for specific crops and cropping systems. Some crops such as legumes can improve the nitrogen balance of soils through symbiotic nitrogen fixation in association with nitrogen-fixing bacteria, and high-value crops, often for export, are much more likely to receive fertilizers. Of the primary regional export crops, cotton, groundnuts, and cocoa contain the largest absolute quantities of nutrients, while of the major imports, nutrient totals are largest in wheat, soybean, and maize. Trade in oil palm and sugar accounts for a large share of nutrient flows as both import and export crops in different parts of sub-Saharan Africa. Unless the steady depletion of nutrients is reversed in the region and soil nutrient stocks are restored, it will be very difficult to sustain the rate of growth in food supply that will be required to meet food demand. In fact, Africa currently depends on the net import of more than 30 million tons of the three major cereals—rice, wheat, and maize—and the past two decades indicate an increasing trend of reliance on imported grain.

trations of these nutrients. Irrigation water also delivers sodium and chloride, important salts in the process of soil salinization.

26.2.5.2 Nutrient Balance and Maintenance of Soil Fertility

Maintenance of soil fertility is crucial for sustaining the food production capacity of cultivated systems. Harvesting of plant parts removes nutrients from the system and eventually depletes soil nutrient stocks unless nutrients are replenished through application of fertilizers or manures or, for nitrogen, by leguminous crops. Nutrient losses also occur through soil erosion and leaching of water-soluble nutrients when water percolates below the active root zone. For nitrogen, losses occur as a result of ammonia volatilization and denitrification, the latter releasing nitrous oxide, a potent greenhouse gas. The overall nutrient balance of a cultivated system is therefore determined by the difference between the inputs and outputs of each essential nutrient.

Internally generated nutrients are the primary source of nutrients in subsistence cropping systems where farmers do not have access to or cannot afford fertilizers or manure. The shifting cultivation systems practiced in remote areas in the humid and subhumid tropics are examples of subsistence systems that rely almost entirely on internal nutrient sources (Nye and Greenland 1960). Depletion of soil fertility occurs in many continuously cropped

cereal production systems practiced on soils of low inherent fertility in India, Southeast Asia, and sub-Saharan Africa that primarily produce rice, millet, and sometimes sorghum under rain-fed conditions. In these systems, yields are relatively low and highly variable because of low soil fertility and lack of adequate rainfall.

Greater nitrogen input from BNF and increased use of livestock manure are generally not feasible in these continuous cropping systems because high human population density does not allow diversion of arable land away from food crops to non-food legume cover crops or forage crops. Dual-purpose grain legumes such as cowpea and pigeonpea, which can provide an income source to farmers in addition to improving soil fertility, have provided a partial solution to this problem (Giller 2001).

On good soils with adequate rainfall or irrigation, commercial fertilizers are used to support high yields and to maintain soil fertility. From a global perspective, such systems represent the foundation of the human food supply and include the irrigated lowland rice systems of Asia, the rain-fed wheat systems of northern and central Europe, and the maize-soybean rotations in the North American prairies and comparable environments of Argentina and Brazil. Relatively high doses of nitrogen and phosphorus are applied in these systems, which can lead to substantial nutrient losses without skillful management techniques that foster high nutrient use efficiency and nutrient retention in soil. Nitrogen is the most difficult nutrient to control because it is extremely mobile and can be lost rapidly via a number of pathways (Smil 1999). Average uptake efficiency from applied fertilizer is typically only about 30–50% (Cassman et al. 2002), which means there is significant scope for increasing uptake efficiency and reducing the potential for nitrogen losses.

The past half-century has seen large increases in the application of nitrogen and phosphorus fertilizers in high-production cropping systems (Galloway and Cowling 2002; Smil 2000) (see Figure 26.6), although application rates vary markedly by region and crop. This injection of external sources of N and P to cultivated systems has expanded and accelerated global nutrient cycles and, as a result of the inefficiencies in fertilizer application and uptake and the loss of fertilizer nutrients, has played a role in reducing environmental services through decreased water quality (Di and Cameron 2002; Howarth et al. 2002; Sharpley and Halvorson 1994; Spalding and Exner 1993), in the loss of diversity in aquatic plant and animal species (Rabalais 2002), and in emissions of N_2O (Bouwman et al. 2002) and NO_x, which can cause respiratory problems in humans (Wolfe and Patz 2002).

A number of technologies have been developed to increase the efficiency with which applied nutrients are used to produce food. One challenge is to match precisely the amount of nutrients available at any given time to the immediate crop requirements, without deficiency or excess, throughout the crop growth period (Matson et al. 1997; Tilman et al. 2002; Dobermann and Cassman

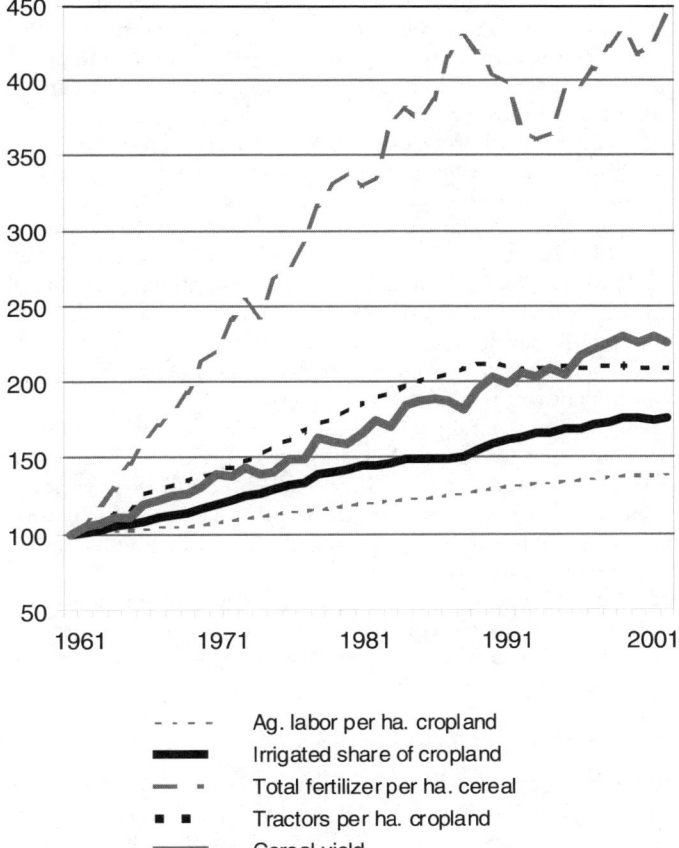

Figure 26.6. Trends in Intensification of Crop Production, 1961–2003. All variables are indexed at 100 in the base year 1961. (FAOSTAT 2004)

2002). In small-scale agriculture that is typical of high production systems in developing countries, this precision is best achieved through field-specific management because blanket recommendations cannot account for field-to-field variation in soil conditions and crop nutrient status. In large-scale agriculture that is typical of high-potential systems in industrial countries, site-specific management will be required to accommodate the substantial variation in crop and soil properties within individual large production fields.

Recent research has demonstrated in on-farm studies the potential for significant increases in nitrogen uptake efficiency using these approaches (Dobermann and Cassman 2002). Success in developing these approaches and achieving adoption by farmers, however, requires substantial long-term investments in research and extension to ensure that the improved management practices are well adapted and cost-effective to specific cropping systems and agroecological zones.

While organic nitrogen sources, such as livestock manure and legume cover crops used in "organic" agricultural production systems, can be substituted for commercial nitrogen fertilizers, these practices are not feasible in the high-potential cereal production systems of developing countries, where population density is high and arable land resources are limited. Moreover, net profit was found to decrease when organic nitrogen sources were used in place of N fertilizer; which has limited adoption of such practices in tropical lowland rice systems (Ali 1999).

In contrast, "organic" production systems that rely entirely on organic nitrogen sources are becoming more popular in Europe and North America, although they still account for less than 2% of crop production. Between 1992 and 2001, the extent of organic cropland in the United States grew by over 200%, from about 163,000 hectares to 526,000 hectares. Organic systems are feasible, and even profitable, in these countries because people can afford to pay higher prices for their food, and there is adequate land to support the crop rotations, legume cover crops, and forages that are needed to supply adequate nitrogen.

It is not clear, however, that environmental benefits would accrue from widespread adoption of organic agriculture if these systems were forced to produce as much grain as conventional systems do today, because it is just as difficult to control the fate of nitrogen from organic sources as it is from nitrogen fertilizer (Cassman et al. 2003). But use of both organic or fertilizer nitrogen need not be an "either-or" decision. In most conventional systems, farmers use organic nitrogen sources and rotate with legume crops to minimize the need for nitrogen fertilizer when it is cost-effective to do so.

Although nutrients obtained from livestock manure remain a significant source of nutrient input to cultivated land, their relative contribution has declined substantially in association with the increase in availability and use of commercial fertilizers. On a global basis, Sheldrick et al. (2003) estimate that the contribution of nutrients from livestock manure has decreased from 60% in 1961 to 25% in 1996 for nitrogen, from 50% to 38% for phosphorus, and from 75% to 57% for potassium. However, because livestock manure also contains substantial quantities of organic matter, it can help improve soil physical and chemical properties that determine soil quality. The total amount of nutrients recovered in livestock manure in 2000 was estimated to be 34 million tons of N, 9 million tons of phosphorus, and 23 million tons of potassium.

While livestock in developing countries of Africa and Latin America produce substantial quantities of nutrients in livestock manure, most of this manure originates from grazing cattle and is therefore difficult to collect and use on cultivated land. In contrast, some industrial countries, such as the Netherlands, with large livestock industries that produce cattle, pigs, and poultry in confined feeding operations produce as much nitrogen in manure as their farmers use in nitrogen fertilizer, and the phosphorus and potassium content of this manure exceeds the amount used in fertilizer. (Environmental concerns associated with nutrient losses from large-scale, confined livestock production systems are discussed in earlier in this chapter.)

In contrast to high-potential systems, environmental damage from nutrient losses is not a concern in subsistence cropping systems that are practiced on soils of low inherent fertility in large areas in the tropics. Instead, severe nutrient deficiencies and depletion of soil fertility are the major threats to ecosystem services. Deficient nutrient supply limits food production capacity and profit, which contributes to malnutrition, susceptibility to disease, and economic insecurity. Severe depletion of soil fertility results in a spiral of soil degradation that can eventually render the land unsuitable for crop production. When abandoned, such degraded land can no longer support the native plant and animal communities it previously hosted, and invasive plant species often take over (Cairns and Garrity 1999; Lumbanraja et al. 1998). Subsistence farmers who abandon such land must then cultivate additional areas, thus expanding the area at risk of degradation.

While it is possible to sustain cropping with judicious use of fertilizers (Nye and Greenland 1960; Reardon et al. 1999), access to external supplies is often limited by lack of roads, infrastructure, and markets. Likewise, it is generally not possible to maintain fertility with only organic sources of nutrients because the inherent soil fertility is too low (Sanchez 2002). Integrated use of both organic nutrient sources and fertilizers appears to be the most promising option. Success in gaining adoption of such approaches has been limited by poverty, land tenure policies, and inadequate investment in the development of basic infrastructure, markets, credit, and extension services.

26.2.6 Atmospheric and Climate Regulation

Although carbon dioxide, nitrous oxide, and methane occur naturally in the atmosphere, their recent increase is largely a result of human activity. This increase has altered the composition of Earth's atmosphere and may affect future global climate (IPCC 1996). (See also Chapter 13.) Agriculture contributes to changes in atmospheric concentrations of each of these three greenhouse gases, significantly so in the case of CH_4 and N_2O, of which it contributes 50% and 70% respectively of the total anthropogenic emissions (Bhatia et al. 2004). Frequent cultivation, irrigated rice production, livestock production, and burning of cleared areas and crop residues now release about 166 million tons of carbon per year in methane and $1,600 \pm 800$ million tons of carbon per year in CO_2. Agricultural systems emit carbon dioxide through the direct use of fossil fuels in field operations (such as tillage, harvesting, irrigation pumping, transport, and grain drying), the indirect use of embodied energy in inputs that require the combustion of fossil fuels in their production, and the decomposition of soil organic matter and crop residues. The direct effects of land use and land cover change (including conversion of forest and grasslands) also resulted in net emission of 1.7 gigatons of carbon per year in the 1980s and 1.6 gigatons annually in the 1990s (IPCC 2000). Burning of standing biomass is a pivotal component of shifting cultivation that emits nitrous oxide in addition to carbon dioxide.

Cultivated fields can be a source or a sink for carbon, depending on the specific circumstances of carbon dynamics during culti-

vation. Factors having the greatest impact on the carbon balance include crop yield levels, removal of crop residues for fuel or livestock forage, crop rotations that include a pasture phase or perennial forage legume, and tillage. During much of the past century, most cropping systems have undergone a steady net loss of soil organic matter (Lal et al. 2003; Paustian et al. 1997; Lal 2004; Lugo and Brown 1993). Global average soil organic carbon density is estimated at 102 tons of carbon per hectare of land within the extent of agriculture (Wood et al. 2000), and the total global store of soil organic carbon within the extent of agriculture is estimated at 368 gigatons, with 43% of this in temperate zones.

However, with the steady increase in crop yields, which increases crop biomass and the amount of residue returned to the soil, and with the adoption of conservation tillage and no-till cropping systems, net carbon sequestration is estimated to occur in the maize-soybean systems of North America (Paustian et al. 1997), as well as in continuous irrigated lowland rice systems where soils remain flooded for most of the year, reducing the rate of soil organic matter decomposition because of anoxic soil conditions (Bronson et al. 1998; Witt et al. 2000). Estimates of the potential to sequester carbon in cultivated systems on a global basis range from 400 million to 800 million tons per year, assuming that best management practices that foster net carbon storage are widely adopted (Paustian et al. 1997; Lal 2003), although adoption has been limited to U.S. maize-soybean and wheat systems and similar cropping systems in Argentina and Brazil.

Large quantities of agricultural crop wastes are produced from cultivated systems. Disposal systems for these wastes include burning them in the field; plowing them back into soil; composting, landfilling, and using as a biomass fuel; or selling them in supplemental feed markets. Burning crop residues releases a number of greenhouse gases, including carbon dioxide, methane, carbon monoxide, nitrous oxide, and oxides of nitrogen.

An additional impact of cultivation on greenhouse gases occurs from erosion. One ecological off-site impact of accelerated erosion is the emission of erosion-induced greenhouse gases into the atmosphere. While some of the organic carbon transported to depositional sites and aquatic ecosystems is buried and sequestered (Stallard 1998; Smith et al. 2001), a large fraction may be emitted into the atmosphere. Erosion-induced emission of CO_2 into the atmosphere may be about 1 billion tons of carbon a year (Lal 2003). Wind-borne sediments, which transport particulate matter over long distances, also adversely affect air quality.

Agriculture can also contribute to mitigation of greenhouse gases emissions by adopting practices that promote the retention of carbon in stable forms of SOM (called humus) or in standing biomass such as occurs in forest trees. These carbon sinks are promoted by the use of less aggressive tillage and by a reduction in the rate of deforestation to support an expansion of cultivated area. Further reductions could also be achieved in the more efficient use of fossil fuels in all aspects of crop and soil management, which would include greater fertilizer and irrigation efficiency as well as reduced tillage.

26.2.6.1 Methane Emissions

Atmospheric methane is second only to CO_2 as an anthropogenic source of greenhouse gases in the atmosphere, and agriculture accounts for between 44% (IPCC 1996) and 50% (Bhatia et al. 2004) of those anthropogenic emissions. The concentration of methane in the atmosphere has more than doubled over the last two centuries, with enteric fermentation in domestic livestock, manure management, rice cultivation, and field burning of agricultural crop wastes as the main causes. Several other agricultural activities, such as irrigation and tillage practices, may also contribute to methane emissions. About 80% of methane from agricultural sources is produced biologically (IPCC 1992; Yang and Chang 1999, 2001).

During digestion of feed intake, methane is produced through enteric fermentation in the rumen of cattle, buffalo, sheep, and goats, a process in which microbes that reside in the digestive system break down the feed consumed by the animal. These animals have the highest methane emissions among all animal types because they have a rumen, or large "fore-stomach," in which a significant amount of methane-producing fermentation occurs. The amount of methane produced and excreted by an individual animal also depends on the amount and type of feed it consumes and other environmental factors.

The need to increase food production in order to keep pace with population growth and changing consumer tastes has led to a large increase in animal production (FAOSTAT 1999), as noted earlier, and to problems related to disposing of increasing quantities of dung and urine. The problem is exacerbated by disassociation of crop and livestock production (Bouwman and Booij 1998; Ke 1998) such that the animal wastes cannot be directly returned to fields where the feed was grown, which recycles the nutrients for succeeding crops. Instead, livestock manures from large confined feeding operations must be transported greater distances to surrounding farmland. But the nutrient content of the manure is low relative to commercial fertilizers, which increases the cost of handling and transporting it. Moreover, care must be taken to ensure that the amount of applied manure does not lead to excessive accumulation of phosphorus in the soil, which can lead to phosphorus losses via erosion and runoff, resulting in degradation of water quality and health concerns (Burton et al. 1997).

The decomposition of organic material in animal manure in an anaerobic environment produces methane. The most important factor affecting the amount of methane produced is how the manure is managed, since certain types of storage and treatment systems promote an oxygen-free environment. In particular, liquid systems (ponds, tanks, or pits) tend to produce a significant quantity of methane. However, when manure is handled as a solid or is deposited on pastures and rangelands, it tends to decompose aerobically and produce little or no methane. Higher temperatures and moist climatic conditions also promote methane production.

Applying manure to agricultural land can lead to groundwater contamination by nitrates after nitrification of the ammonium nitrogen (NH_4–N) present and to emissions of ammonia (European Centre for Ecotoxicology and Toxicology of Chemicals 1994), methane (Chadwick and Pain 1997), and N_2O (Jarvis et al. 1994)—all of which contribute to climate change. Ammonia, after deposition on land surfaces and water bodies, and nitrification act as a secondary source of N_2O and may also decrease the capacity of soils to absorb CH_4 and act as a sink for this gas (Mosier et al. 1996).

Rice fields are large producers of methane, accounting for as much as one third of total anthropogenic methane emissions. When fields are flooded, anaerobic conditions develop in the soils, and methane is produced through anaerobic decomposition of soil organic matter mediated by soil microbes. In fact, both methane and nitrous oxide are simultaneously emitted, as irrigated rice fields offer favorable conditions for their production and emission (Cai et al. 1997; Bronson et al. 1997; Ghosh and Bhat 1998; Majumdar et al. 2000). Global methane emissions from rice fields are estimated to be 37 teragrams per year, while N_2O emissions are much lower, at 1.8–5.3 teragrams per year, although N_2O is a much more potent greenhouse gas (IPCC 1996).

26.2.6.2 Nitrous Oxide Emissions

Agriculture is the main source of nitrous oxide, a chemically active greenhouse gas, accounting for about 70% of anthropogenic emissions. Atmospheric concentration of N_2O is increasing at a rate of 0.22 ± 0.02% per year. Concern over N_2O emissions is particularly great because of its long atmospheric lifetime and high climate change potential (Bhatia et al. 2004). Although global atmospheric loading of N_2O is less than CH_4, the former is 310 times more potent as a greenhouse gas than CO_2 on a 100-year time-scale, while CH_4 is only 21 times more potent (Majumdar 2003). N_2O is produced naturally from a wide variety of biological sources in soil, water, and animal wastes and contributes to the depletion of stratospheric ozone. The release of nitrous oxide has increased in recent years due to more intensive agricultural practices, in particular land conversion and application of nitrogen fertilizer. A wide range of other agricultural and soil management practices can also affect N_2O fluxes, including irrigation, tillage practices, the burning of agricultural crop residues, and changes in land use, such as loss and reclamation of freshwater wetland areas, conversion of grasslands to pasture and cropland, and conversion of managed lands to grasslands or the fallowing of land (Mosier et al. 2004).

From the agricultural perspective, N_2O emissions from soil represent a loss of N from the soil system and a decrease in N use efficiency. Soil is considered to be one of the major sources of nitrous oxide, contributing 65% of the global emissions. Annual emissions of N due to N_2O emissions from agricultural systems amounts to 6.3 teragrams. Soil receiving chemical fertilizers and biologically fixed nitrogen contributes to nitrous oxide emissions during the processes of nitrification and denitrification, and the increasing use of fertilizers will lead to increased N_2O emissions unless N fertilizer efficiency can be increased as well.

Use of organic nitrogen sources instead of nitrogen fertilizer causes a substantial increase in methane emissions in irrigated rice systems, and it may not decrease nitrous oxide emissions (when both are applied at levels that achieve similar yields). Thus, from a purely climate change perspective, organic fertilizers should be used with caution in such systems.

In summary, agriculture may be contributing about 20% of current annual greenhouse gas–forcing potential. It is the largest source of anthropogenic CH_4 and a significant contributor to increases in atmospheric N_2O concentration. In contrast, cultivated systems play a relatively small role in total CO_2 emissions, and some systems have the potential to sequester carbon by use of improved crop and soil management practices, thus becoming a sink for carbon dioxide.

26.3 Drivers of Change in Cultivated Systems

Many factors have influenced the evolution of cultivated systems and their capacity to meet the increasing demands placed on them. (See Figure 26.7.) These factors have driven the changes that have occurred in cultivated systems and will continue to do so in the future. This section reviews the nature these drivers, their interactions and extent, and their impact on system performance.

Although the Figure is a simplification of the context and dynamics of cultivation, it illustrates three key points: that the number of drivers and interactions among them are potentially large; that important feedback mechanisms exist that influence the ability of cultivated systems to generate desired cultivated products and ecosystem services; and that individual drivers can simultaneously have positive and negative impacts (for example, a new

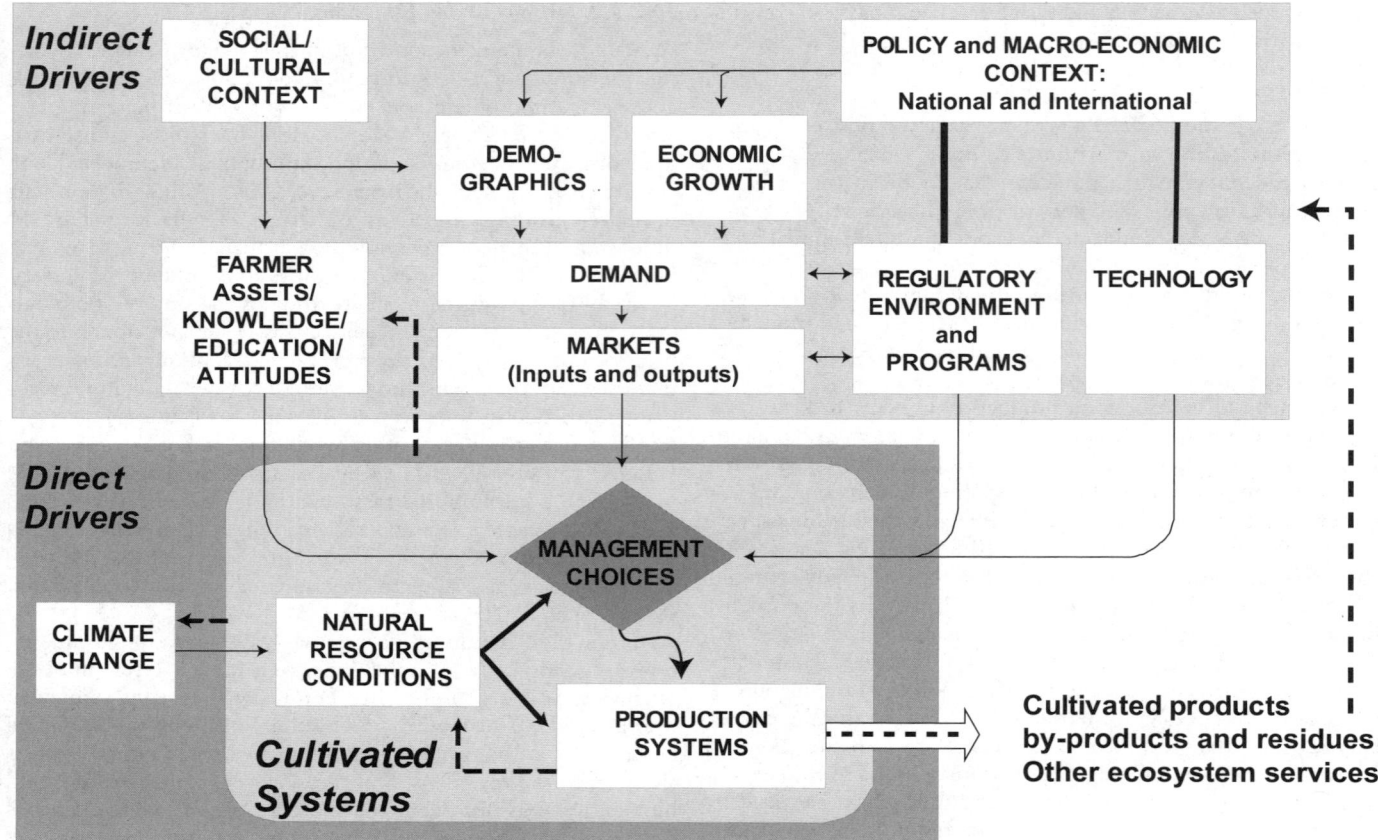

Figure 26.7. Interactions between Drivers of Cultivated Systems

technology that increases the yield of outputs might also generate a negative impact on ecosystem services).

The central role of the "manager" of the cultivated system is also highlighted in this conceptual model, whether the manager is an impoverished subsistence farmer with two or three hectares of remote hillside in East Africa or a professional agronomist in a multinational corporation that cultivates a 5,000-hectare cash crop plantation in Southeast Asia. Within any given socioeconomic and environmental context, it is the sequence of choices made by these managers about what to produce and how to produce it that drives the long-term capacity of cultivated systems to deliver products and ecosystem services. These choices are driven by farmers' incentives to take particular courses of action and by their capacity to act on those incentives. Through a better understanding of the key drivers of change, decision-makers are better placed to target policy and investment interventions for improving the economic and environmental outcomes of cultivation.

In keeping with the MA conceptual framework, drivers are grouped into two broad categories: indirect—those that influence the demand for both cultivated products and other ecosystem services, as well as the overall feasibility and attractiveness of different cultivation options—and direct—those that come into play at the actual site of cultivation.

26.3.1 Indirect Drivers

Many of the indirect drivers of change relevant to cultivated systems have already been described in Chapters 3 and 8, so this section focuses on a selective synthesis together with complementary material of more specific relevance to a cultivated systems perspective.

26.3.1.1 Demand for Cultivated Products and Other Ecosystem Services

The scale and structure of demand for cultivated products as well for other ecosystem services from agricultural landscapes has been broadly shaped by three drivers: demographic change, economic growth, and changing consumer preferences.

Over the past 50 years population growth has been the single most important global driver determining the aggregate demand for food and other cultivated products and shaping the extent and intensity of cultivation. Between 1960 and 1999, world population doubled to 6 billion, with an average growth rate of around 1.7% per year, while aggregate per capita food energy consumption grew at just over 0.5% per year. In industrial as well as developing countries, 60–70% of the total increase in calories consumed between 1961 and 2002 was accounted for by population growth (FAOSTAT 2004).

Population growth rates are declining, however, and currently stand at around 1.4% per year globally, although with major regional differences. Developing countries now account for over 95% of global population growth and hence a correspondingly greater share of the pressure to expand food output from cultivated systems. High population growth rates are negatively correlated with income levels. Hence, the population in poorer countries are typically less well nourished or even undernourished compared with populations in rich countries. Food insecurity in poor countries or regions often results from the low productivity of local cultivated systems (UN Hunger Task Force 2005). In Europe and some richer developing countries, population growth rates are stagnant or negative, so population growth is no longer a driver of food demand, and this trend will continue globally as economic development proceeds and population growth rates continue to fall (United Nations Population Fund 2004).

Economic growth is another strong stimulus of demand. As incomes rise in many developing countries, a large share of the increased income is used to purchase a greater and more diverse food supply. Compounding both population growth and increased purchasing power, social and cultural change—often linked to urbanization, increased female participation in the workplace, and increased exposure to food industry advertising and to public health and nutrition information—have changed consumer preferences with regard to the type, amount, and quality of food they demand. This includes growing preference for animal protein, (particularly chicken and pork), for fruits, vegetables, and oils, and for more processed and convenience foods and declining preference—as a share of per capita consumption—for starchy staples and cereals (FAOSTAT 2004). The rapid growth in industrial-scale, confined livestock systems and aquaculture have been direct consequences of these trends. Urbanization not only alters food preferences, it also changes the age and sex structure of rural populations and increases remittances—both factors that influence cultivation practices. (See Chapters 3 and 8.)

Many of the same demographic and socioeconomic changes have also increased the demand for a broader range of ecosystem services beyond food, such as fresh water, clean air, wildlife conservation, and recreation. Since cultivated systems now dominate many of the populated landscapes of the world, they have come under increasing pressure to play a greater role in delivering more (or consuming less) of these other services, while at the same time continuing to meet growing food needs (Wood et al. 2000).

Cutting across demographic and economic factors is the issue of poverty, which severely curtails livelihood opportunities. From a cultivation perspective, poverty limits access to production inputs such as credit and to new technologies that improve crop and soil management. Poverty is also often associated with a lack of security in terms of access to or title to land and other natural resources, in turn diminishing farmers' incentives and ability to choose production practices with long-term payoffs. Without such incentives, cultivated systems are focused on meeting short-term needs, and increasingly intensive cultivation under such conditions has often resulted in the degradation of soil and water resources that are required to maintain even low levels of productivity. This process has been called a "downward spiral" of productivity and degradation (Scherr 2000; Ehui and Pender 2005; Wiebe 2003). Ultimately such a degradation spiral can lead to abandonment of the cultivated system and migration to other locations that are likely to be of more even more marginal production potential (Barbier 1997; Chopra and Gulati 1997).

26.3.1.2 Policy, Legal, and Sociocultural Context of Cultivation

The policy, regulatory, and cultural environment have profound impacts on the incentives to produce more and higher-quality food, to engage in local, regional, and international trade, to invest in long-term productivity and enhanced cultivated system capacity, and to reduce the off-farm impacts (the externalities) of cultivation. The distinct and evolving nature of policies and institutions across and within countries influences the effectiveness of markets and hence choices about where, what, how, and how much to cultivate as well as the incentives, if any, for farmers to reduce or eliminate negative externalities caused by their cultivation practices (Uri 2001; Eicher 2000).

Agricultural, trade, and food security policies can distort incentives to produce and trade cultivated products in one way or other. These include price policies that favor either domestic rural producers (such as the U.S. Farm Bill and the EU Common Agricultural Policy) or urban consumers (such as the food price control schemes prevalent in many developing countries). The level and effectiveness of investments in education, infrastructure (roads, irrigation, rural electrification, and telecommunications), and credit have been shown to be strongly related to improvements in agricultural productivity and rural incomes (Fan et al. 2000; Fan and Hazell 2001; Zeller and Sharma 1998; Wiebe 2003). Investments in agricultural research and technology transfer have been especially strong drivers of change in cultivated systems, as described later in this section.

Resettlement policies, though now less common and certainly of lesser scale, have had significant impact on the conversion of natural ecosystems over very large areas for cultivation, with consequent, large-scale environmental consequences. The massive transmigration program from Java to the outer islands of Indonesia and the colonization policies of the Brazilian government, implemented largely during the 1960s and 1980s, are two notable examples (Fearnside 1997).

The nature and strength of land tenure and resource use laws and customs have been shown to strongly influence the willingness of farmers to engage in cultivation beyond meeting subsistence needs, as well as to invest in sustainable land management practices (Soulé et al. 2000; Meinzen-Dick et al. 2002). Similarly, the effectiveness of collective action can significantly affect the productivity and sustainability of cultivated systems. This is true not only for proper management and utilization of open access and common property resources such as pastures or woodlots, but also where the productivity of individually managed plots and fields would benefit from collective action, such as coordination of agronomic activities so as to pool labor, minimize pest and disease problems, or make the best use of available water resources (Meinzen-Dick et al. 2002).

Inheritance laws and customs are also relevant. The practice of dividing land among heirs, particularly common in Asia, has so fragmented holdings in some areas that the scale of cultivation operations limits their economic viability over the long term (Maxwell and Wiebe 1999).

In recognition of the potential environmental costs of cultivation, the growing demand for improved environmental services, and the lack of incentives for farmers to consider externalities, governments have played an increasing role in influencing crop selection and cultivation practices through both regulatory and voluntary incentive schemes. Regulatory policies have included systems of wildlife and watershed protection and conservation that have sought to exclude or restrict cultivation in areas considered to have high biodiversity, hydrological, watershed protection, or amenity value. Where cultivation pre-existed in such areas, or where land and population pressure external to such areas has been high, these policies have often created conflict with farming communities (Gillingham and Lee 2003; Maikhuri et al. 2000). This has led to the emergence of more enlightened and participatory approaches to the design and management of conservation areas in partnership with local communities (Farrington and Boyd 1997).

Other approaches have included the zoning and regulation of certain types of cultivation or cultivation practices, such as large-scale confined livestock feeding operations, that present local waste and odor problems or use of certain categories of pesticides. While such restrictions are often associated with punitive sanctions, their effectiveness has varied depending on the technical and economic validity of the regulation standards applied, the deterrent value of the sanctions, and the rigor of enforcement (Kleijn et al. 2004).

Voluntary strategies, particularly in richer countries, involve incentive payments to farmers linked to production or conservation practices that are considered to be more environmentally sound (Dobbs and Pretty 2004; Wu et al. 2004). Increasingly these policies are being aligned with the "boxes" established under the auspices of the World Trade Organization that govern permitted levels and types of domestic support to agriculture. One goal of the WTO is to "decouple" support to farmers from production and price level, as a means of reducing trade distortion (WTO 2004). The U.S. 2002 Farm Bill provides for support to farmers related to programs for resource conservation, wildlife habitats, and wetlands within cultivated systems (National Resources Conservation Service 2002), and similar programs operate in most, if not all, OECD countries. But the WTO provisions that accommodate such programs are still controversial among many developing countries: they see them as an indirect means of providing otherwise restricted or disallowed income support, which places their own farmers at a competitive disadvantage (*The Economist* 2003).

It is difficult to generalize about the net effect that national policies have had on cultivated systems. However, in those countries where policies tend to expand production to levels that would otherwise be uneconomic—such as cotton in the United States, sugar in the European Union, and rice in Japan—it is likely that more land is being kept in production and more agricultural pollution is taking place than would otherwise be the case. Another implication is that, to the extent that such subsidies distort trade, less area is allocated to the cultivation of these crops in competitor countries, such as cotton in West Africa and India, sugar in the Brazil and Australia, and rice in Viet Nam.

As subsidies and other barriers to trade are removed in these and other commodity sectors, adjustments in global patterns of production will take place. The net local and global consequences of such changes on cultivated systems and ecosystems services depends primarily on the relative yield levels (as those determine the harvested area required for a given level of production) and the specific production inputs and practices used in each location, such as the nature, management, and mix of inputs and practices for plant, soil fertility, water, and pest and weed management. These in turn depend on local markets, farmer characteristics, resource conditions, and management choices.

26.3.1.3 Markets

The existence and efficiency of markets, and the extent to which farmers are able to participate in them, provide perhaps the strongest signals shaping cultivation decisions for an ever increasing number of farmers. Even where subsistence goals dominate household production strategies, survey data indicate that households frequently engage in markets to varying degrees. Markets include those for cultivated outputs (main products and by-products), inputs (such as labor, land, seeds, fertilizers, and pesticides) and those, often at a nascent stage, for ecosystem-related services such as carbon sequestration and habitat conservation.

Essential ingredients for the development of markets include a stable monetary system, accepted procedures for establishing and enforcing contracts, viable entry costs for market participation, financially acceptable search and transaction costs, and adequate access to physical infrastructure and transportation. As described earlier, incentives for market development and participation have also been shaped by policy factors that directly affect markets

through transfers (taxes or subsidies) or other barriers related to production, consumption, or trade.

High transaction and transport costs limit market opportunities since they increase the farmgate cost of inputs and reduce the farmgate value of outputs. The geographic scope of market potential is also influenced by the bulk density and perishability of cultivated products and by the unit value of the product itself. Where markets function effectively, where products can be cultivated competitively, and where demand exists, the geographic distances between production and consumption can be very large, as seen in the global cereal markets. Other examples include the cultivation of high-value horticultural crops, of flowers and ornamentals in the cool tropical highlands of Central and northern South America for the U.S. market and in Kenya and Uganda for UK supermarkets, of out-grower schemes in Indonesia for the Dutch flower industry—all of which use air transportation—as well as the production, packaging, and shipping of fresh fruit and vegetables in California to all parts of the United States by rail and road transportation.

Thus, where it has been possible to lower marketing and transportation costs, the geographic distances between production and consumption becomes less relevant, and producers, and the cultivation systems they manage, are exposed to an increasingly broad range of market opportunities. This is often a double-edged sword—on the one hand, providing increased incentives to expand or intensify production with possible negative ecosystems consequences, while on the other hand providing greater incentives to preserve the long-term sustainability of the production base that might foster more positive outcomes for ecosystem management (Lopez 1998; Kaimowitz and Angelsen 2000).

But there remain significant obstacles for smallholders in developing countries to engage more in local, regional, and international markets. Many relate to constraints to market entry through insufficient access to credit or information about market needs and to the insufficiency and variability of the quantity and quality of farm outputs to engage in stable marketing arrangements. At the same time, there appear to be growing barriers to trade arising from stricter sanitary and phytosanitary standards being imposed by importing countries. While fears of pest, virus, and disease consequences for plant, animal, and human health are undoubtedly genuine and call for effective safeguards, these regulations have become another contentious issue under the WTO. As with subsidies and environmental payments, many developing countries regard the sanitary and phytosanitary requirements and regulations imposed by richer countries as another mechanism for imposing indirect trade barriers (Henson and Loader 1999; Athukorala and Jayasuriya 2003).

There are several market niches that link cultivated products with what are considered to be improved standards of cultivation with regard to ecosystem outcomes or ethical issues. These are products designated as, for example, organic, bird-friendly, shade-grown, fair-trade, and humane from an animal welfare perspective (Harper and Makatouni 2002; Lockie et al. 2002). The organic food movement is perhaps most developed in Europe, but it can probably be considered a global scale phenomenon, especially with richer consumers.

The term "organic" is open to many different interpretations but can include avoiding or minimizing the use of pesticides, inorganic fertilizers, antibiotics, GMOs, fossil fuels, and so on as well as promoting biodiversity at various levels. Broadly accepted standards are beginning to emerge in some markets (Guthman 1998; European Union 2000; USDA 2004), as well as widely-accepted market certification procedures, such as those of the UK's Soil Association (2004). Currently, almost 23 million hectares globally are reportedly explicitly managed according to organic principles (IFOAM 2004). Of this total, some 46% are reported in Australia/Oceania, 23% in Europe, and 21% in Latin America. While the United Kingdom and Germany have about 4% of cultivated land under organic production, the United States has less than half a percent, although these systems contribute some 3–5% of fruit and vegetable production (Greene and Kremen 2003).

26.3.1.4 Prices

The ability of farmers to respond to changes in prices of inputs and products is an important indicator of the resilience of food production systems. Producer decisions about what and how much to produce (or to harvest, in the case of wild fisheries) are strongly influenced by the relative prices of outputs (maize versus beans, for instance, or cod versus plaice), as well as of essential inputs (such as the maize/nitrogen fertilizer price ratios). Consideration of time frames and the need to maximize return on fixed assets are important determinants of the willingness and ability of producers to respond to price signals.

Output responses are quicker and stronger for short-term production cycles than for longer ones. Thus adjustments in annual cropping can be made in a short time frame, whereas decisions about changing animal herds or perennial crops that take longer to develop their economic potential are more complex. The average price of food has been on a downward trend for some 40 years, and many poor smallholders who have limited access to productivity-raising technologies and practices often face situations in which their on-farm costs per unit of product, plus the unit costs of transportation and marketing, are higher than the market price of their products. In the case of marine fisheries, prices have increased, reflecting the scarcity as more and more fisheries are fully exploited, as well as increased costs because of increased fishing effort.

Increased farmgate prices—for example, for higher-quality or better-timed products or brought about by temporary shortfall in supply—can raise producer incomes and increase incentives for more investment in the underlying production system. This could have positive or negative outcomes for ecosystem services other than food, depending on the type of investment. Furthermore, increased profits from increased productivity might be a spur to bring more land into production, including more conversion of natural ecosystems (Kaimowitz and Angelsen 2000). Ironically, falling prices can also have equally ambiguous outcomes, ranging from providing incentives to raise productivity to removing incentives to make any further investments—likely with negative ecosystem impacts.

26.3.1.5 Technology and Information

One of the most widely researched areas in the field of agriculture is the impact of technical change on the productivity of cultivated systems. Assessments at programmatic and national scales typically suggest that the contribution of technical change to overall productivity growth is in the range of 30% to 50% (Evenson and Gollin 2003; Ruttan 2002; Roe and Gopinath 2001). Technologies include better-quality inputs such as improved crop varieties with higher genetic potential and increased pest and disease resistance, improved livestock breeds and fish species, better cultivation techniques such as zero tillage, improved agronomic practices such as the timing and placement of applied nutrients and water, and better storage and other post-harvest technologies. In countries where a sufficiently large base of commercial farmers exist, the private sector plays an important role in technology development and delivery, but in virtually all countries public investment in agricultural research and extension are important and well established if not always adequately funded areas of public policy (Pardey and Beintema 2001). These investments reflect both the importance of the agricultural sector to the rural economy and the generally high levels of economic payoff to agricultural R&D investments.

During the past 50 years, crop genetic improvement and improved technologies for managing soil nutrients and pests have come predominantly from investment in research and extension conducted by public-sector institutions such as universities and national and international agricultural research centers (Pardey and Beintema 2001). In recent decades, however, investment in private-sector research has increased markedly, especially for improvement of commercial crops such as maize, soybean, and cotton that require purchase of new seeds each cropping season for achieving optimal yields.

Today, agricultural research investment in the private sector exceeds that in the public sector, with consequences on research priorities (Pardey and Beintema 2001). The private sector focuses on improving crop traits that result in greater seed sales, emphasizing relatively short-term research successes. Private-sector research has also given greater emphasis to use of modern tools of molecular genetics to develop crop varieties with traits controlled by single genes. (See Box 26.3.) Many of the major crop development constraints, such as yield potential, drought tolerance, and nitrogen use efficiency, however, are controlled by numerous genes, and progress will require greater scientific effort and longer-term investment. The private sector has few incentives at present to invest in technologies aimed at improving environmental services.

The focus of public investment is on research producing "public goods" (knowledge and technologies that can be used by all, without exclusion, such as a new soil conservation practice), as well as research that is too long-term, risky, or otherwise financially unattractive to the private sector but that would yield social benefits, such as more environmentally sustainable production practices. In the past, publicly-funded research has focused on understanding and increasing crop yield potential, achieving greater fertilizer use efficiency, protection of water and soil quality using conservation tillage systems, and reducing pesticide use through integrated pest management.

BOX 26.3

Crop Breeding and Genetics

Plant types and agricultural techniques that are better suited to farmers' needs could go a long way toward improving the productivity of cultivated systems and thus the livelihoods of farmers. Genetically modified crops provide economic gains to farmers that have shown to be large in the case of cotton and soybeans. Extending these gains to other "orphan" crops that are planted by smallholders could, in the presence of appropriate regulatory policies, have significant poverty-reducing effects. Stress-tolerant varieties have the potential to benefit producers; nutritionally enriched varieties have the potential to benefit consumers as well.

Many people are concerned about the prospects of biotechnology. Concerns center on the science itself, control over the science, access to the science, environmental effects, and human and animal health effects (FAO 2004a). Addressing these concerns separately and in a case-specific manner is essential for analyzing the costs and benefits of genetic technologies as they are applied to crops.

Studies of economic returns into public-sector agricultural research have documented substantial and consistent returns on investment as a result of higher yields and farm profits, increased labor productivity and prices, and lower prices of staple grains for consumers (Alston et al. 2000). Despite this evidence, recent trends in public funding of research and technology transfer in both industrial and developing countries show general decline at a time when constraints to sustaining yield growth while protecting environmental services are becoming more complex and scientifically challenging. If maintained, this decline will affect agricultural research outputs globally, with serious consequences for the ability of crop productivity growth to keep pace with food demand and for opportunities to improve the environmental characteristics of new technologies.

The overall efficiency of converting research investment into yield gains at the farm level is an important driver of food supply. For example, the total research investment in both public and private sector for maize genetic improvement increased 3.4-fold in inflation-adjusted dollars from the mid-1970s to the mid-1990s in the United States while the rate of gain in U.S. maize yields remained constant at about 100 kilograms per hectare per year during this period (Duvick and Cassman 1999). Therefore, the efficiency of converting investment in maize genetic improvement to greater yields at the farm level has decreased by about 70%.[3] (See Chapters 4 and 8 for further extensive treatment of the role of science and technology.)

A particular area of concern has been the relatively low adoption of many technologies designed to improve soil and water conservation or provide other improvements in ecosystem service delivery from cultivated systems. Many of these technologies—such as use of the nitrogen-fixing azolla plant to replace nitrogen fertilizer in lowland rice production, alley-cropping with leguminous trees on plot borders, use of the tree leaf mulch on subsistence cereal crops in sub-Saharan Africa, and contour bunds and vegetative field border strips to reduce erosion for hillside cropping systems—are often labor-intensive, provide benefits after several years, or provide benefits off-site. These characteristics often make them unattractive to farmers in countries where conservation efforts are not subsidized and in situations of limited assets (including labor) and insecure land tenure (Lutz et al. 1994; Antle and Diagana 2003).

Changes in cultivation systems are driven by access to various types of information: market data on prices, grades, and standards; advances in cultivation practices and technologies (both for farmers and for researchers); current weather conditions and forecasts; and information on current pest and disease threats and recommended responses. Where they are available and accessible to farmers, and farmers have the capacity to use them, all these types of information have economic value (Solow et al. 1998).

26.3.1.6 Farmer Characteristics

Ultimately, it is farmers who make decisions about the nature and management of cultivated systems, decisions that affect the delivery of both cultivated products and ecosystem services. Thus the cultural, socioeconomic, and educational background as well as the expectations, preferences, and risk attitudes of farmers and farm households all play a role in shaping cultivation decisions.

In the case of subsistence, resource-poor farmers, it is the reduction of production risks while best using family labor that is the driving force behind decision-making (Willock et al. 1999). There may also be different attitudes to risk, crop management, and even crop selection and cropping patterns on the landscape among men and women farmers. In general, women-headed households focus more on meeting food self-sufficiency needs using a diverse portfolio of products to meet a range of nutritional and domestic needs. Male-headed households have generally been shown to be less risk-averse and often focus on the production of cash crops. Farmer age and education level are also considered to be important factors in conditioning willingness to accept new ideas and technologies (Soulé et al. 2000).

Often farmers and farming communities have a very large amount of accumulated indigenous and science-based knowledge, but experience and new knowledge continue to evolve. Indeed, it is the interplay between indigenous knowledge, access to new technologies, and risk aversion that are major determinants of decisions about cultivation practices and evolution of farming systems. Considerable effort is being made to improve understanding of this process (e.g., Röling and Wagemakers 1998) and to advance it through better designed interactions between farmers, researchers, and technology support specialists (Loevinsohn et al. 2002).

26.3.2 Direct Drivers

Direct drivers are those that manifest themselves at the point of cultivation. As shown earlier in Figure 26.7, we can broadly identify three types of direct drivers:

- management choices made at plot, field, pen, or pond level about the scale of cultivation and what to cultivate and how;
- the production system itself—its specific mix of inputs including labor, production practices, and outputs in terms of cultivated products as well as other residues; and
- the natural resource base (including the local impacts of climate change) that underpins and is affected by the cultivation process.

26.3.2.1 Management Choices

To a large extent the potential productivity and ecosystem service impacts of a cultivated system are pre-determined by the crop and resource management choices farmers make, which are often extremely constrained in the case of poor farmers. Key drivers involve strategic decisions about which crops to produce and the cropping pattern in time and space, how much area to devote to cropping, and tactical decisions about specific production practices involving crop and soil management involving nutrients, water, and pest control.

In the face of growing demand for cultivated outputs, several key factors are involved in these choices from an ecosystem service perspective. First, the choice about what to produce often has direct implications on services. For example, perennial crops reduce cultivation needs and are often associated with more ground cover and less soil disturbance, which may result in less soil erosion and lower carbon emissions. The more uniform landscapes of annual crops grown in monoculture reduce biodiversity and can increase the risk of erosion on sloping land unless conservation tillage practices are used that leave adequate crop residues to protect the soil surface. Cultivation of high-value cotton and horticultural crops often uses substantial amounts of pesticides to ensure adequate product quality to meet consumer demand, and growing tobacco in some developing countries is frequently associated with high consumption of fuelwood for drying purposes.

A second strategic factor with a large impact on ecosystem services relates to choices about how much area to cultivate, and especially whether to expand production into as yet uncultivated areas—that is whether to transform natural ecosystems or semi-natural rangeland plant communities into cultivated systems. Pressures to expand are larger if land suitable for cultivation is available at low cost and if land currently under cultivation has low or declining productivity.

The third set of choices is related to production technologies and practices, which in turn are strongly linked to strategies adopted for the intensification of production. Intensification can be achieved through increased inputs and outputs (increased yields) per hectare per harvest, or by increasing the number of harvests in a given time (such as reducing fallow periods or sequential cropping within a single growing season). This is generally termed increasing the cropping intensity. Increasing cropping intensity is often the first stage in the transformation from swidden to permanently cultivated systems, and it can be one of the major consequences of irrigation in regions where rainfall is uni-modal and sufficient for only one cropping season per year.

It is the accumulation of such management decisions by many rural households over time that ultimately drives the aggregate extent and condition of cultivated systems and their impact on ecosystem services—both within the agroecosystem in question and in adjacent or even distant ecosystems that are affected by the externalities of cultivation.

Globally, 78% of the increase in crop output between 1961 and 1999 was attributable to yield increases and 22% to expansion of harvested area. Of the expansion in area harvested, roughly two thirds was accounted for by physical expansion of arable land and the remainder was due to increases in cropping intensity (Bruinsma 2003). While the pattern of yield increases outpacing harvested area increases was true for most regions, the proportions varied. For example, only 55% of total output growth was derived from yield increases in Latin American and the Caribbean compared with 80% in South Asia. In contrast, only 34% of increased output was derived from yield increases in sub-Saharan Africa and 66% from harvested area expansion. In industrial countries where the amount of cultivated land has been stable or declining, increased output was derived predominantly through increased yield and cropping intensities.

In both physical land area and proportional terms, the largest expansion of arable land took place in Latin America and the Caribbean, where expansion of the agricultural frontier accounted for about half of the increase in crop output, with cropping intensities static (Bruinsma 2003). For, example soybean production expanded by some 25 million hectares in Brazil and Argentina between 1981 and 2004, largely through expansion of arable land (Fearnside 2001; James 2004). And conversion of forest and savanna to agropastoral systems was widespread throughout South and Central America. By contrast, of the 66% of increased crop output that was due to increased harvested area in sub-Saharan Africa, about half was attributed to increase in cropping intensity and the rest to increased cultivated area.

Some of the factors that drove these trends have been described earlier: the Green Revolution in Asia; resettlement policies in Brazil; environmental conservation programs in Europe, North America, and Oceania; and so on. But a key underpinning factor in all these cases is the difference in relative endowment or scarcity of land, labor, and capital in the various regions. These endowments have, for example, shaped national technology generation strategies, such as investment in land-saving R&D in Asia and labor-saving R&D in North America (Hayami and Ruttan 1985). At the farm level, area expansion has been pursued in regions of relative land abundance, and intensification has been the preferred strategy where land or labor are scarce and capital more abundant.

Figure 26.8 illustrates the distinct levels and trends in land productivity (total value of crop and livestock outputs per unit of arable land) and labor productivity (total value of crop and live-

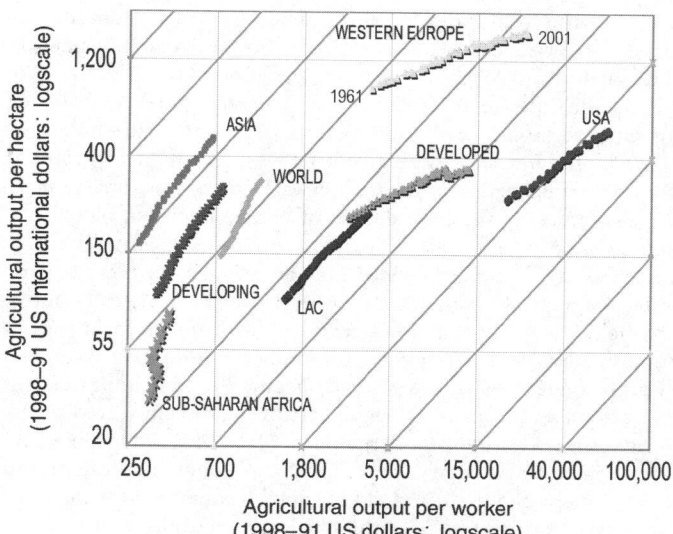

Figure 26.8. Growth in Land and Labor Productivity by Region, 1961–2001. The graph is derived from an assessment of crops and livestock only. Output values were computed for individual commodities using the FAOSTAT production time series and average world prices estimated by FAO for the period 1989–91 (FAO 1997). "Land" is the sum of arable and permanent cropland and pasture in each year. "Labor" is the population economically active in agriculture: as defined by FAO. The diagonal lines represent pathways of equal growth in land and labor productivity.

stock output per agricultural worker) from 1961 to 2001 for different regions. Western Europe, with extreme land constraints, shows high land and improving labor productivity. The United States, with high capital and limited labor, has shown high and increasing land and labor productivity. Asia, with little additional land and abundant labor, has shown high and increasing land productivity but low labor productivity. Sub-Saharan Africa remains low in both dimensions, and while some limited progress has been made in land productivity (but by 2001, only reaching levels that are still below or equal to the starting point of all other regions in 1961), virtually no gains have been made in labor productivity.

26.3.2.2 Natural Resource Conditions and Production Systems

Natural resource conditions are described briefly here because the impact of production processes, ecosystem services, and cultivated system management that influence natural resources have been covered in greater detail elsewhere in the chapter. Furthermore, many production system practices represent direct responses to changes in natural resource conditions described in previous sections (declining soil productivity, for example).

The range of feasible cultivation options open to farmers is fashioned by a number of indirect drivers: input and output markets, the regulatory environment, accessibility of usable and profitable technologies and information, the cultural context and socioeconomic condition of the farm household, and the farmer's knowledge base, goals, and attitudes to risk. But specific cultivation decisions for each site are made taking into account the set of prevailing local natural resource conditions, particularly the availability of land and water; the type and variability weather conditions; the quality of soil and water; the prevalence of pests, diseases, and weeds; and other potential natural hazards such as erosion and flooding. Key among these factors for crop-based systems are the availability of land suitable for cultivation and the depth, water-holding capacity, fertility, and workability of soils. The quality and reliability of water are important for all forms of cultivation but are most critical for irrigated systems. Rain-fed systems are subject to the usual uncertainties of weather, exacerbated by the impacts of climate change—increasing temperatures, shifting rainfall patterns, and greater variability in seasonal rainfall.

As a result of the high degree of heterogeneity of natural resource endowments and climatic variability over relatively short distances, it is difficult to make uniform management recommendations for production technologies or practices at scales above the field level. For example, recent results from on-farm tests in intensive irrigated rice systems of South and Southeast Asia confirm the benefits of taking a "field-specific" approach for nutrient management to optimize yield, fertilizer efficiency, and profit (Cassman et al. 1996; Dobermann et al. 2002). Such specificity highlights the challenge of developing and scaling up the adoption of new technologies from field to district/county and to regional scales, and the magnitude of this challenge increases in proportion to the complexity and sophistication of crop and soil management practices. In less favorable production environments where natural resource endowments are relatively poor and there is little infrastructure or market development, it is particularly difficult to support farmer adoption of new technologies. Such is the case in sub-Saharan Africa, where not only are production conditions extremely heterogenous but public and private institutions that support technology generation and transfer are often quite weak.

Some drivers related to natural resource condition are unpredictable and largely uncontrollable, such as weather variability, climate change, and the emergence of new pests and diseases, and can at best only be managed once they occur. Others include on-site conditions that both affect and are affected by production, such as soil nutrient dynamics, soil water status, erosion, and weeds. These are mostly controllable conditions if farmers have sufficient resources, access to information, and technologies that provide profitable solutions to address these constraints.

26.3.3 Summary of Drivers as Potential Points of Intervention

The drivers affecting the evolution of cultivated systems and their capacity to produce cultivated outputs and services are many and are interrelated. It is tempting to simplify the complexity of these direct and indirect drivers and focus only on the field-level issues that affect provision of food and environmental services. That would be, and indeed has been, a mistaken approach. The literature of technology adoption examined in this assessment is, despite the many significant successes, replete with examples of failed "fixes" at the farm level in broader contexts where income levels, security, property rights, equity, financial and agricultural markets, health, education systems, and so on were inadequate to provide the proper enabling environment and incentives for farmers to make the type of productive, long-term investments that are required to deliver economic and environmental benefits in a sustainable fashion.

Regardless of the productivity and profitability of cultivated systems from a farmer perspective, the central MA concerns of how best to deal with the externalities of cultivation remain a major challenge in all cultivated systems. As described earlier, many of the impacts of cultivation on ecosystem services occur away from the farm, outside the agroecosystem boundaries, which provides little or no incentive to farmers to invest in reducing them. Likewise, consumers are increasingly demanding affordable and safe food, which means continued increase in cultivated yields

and product quality at the same time as addressing concerns about negative impact on environmental quality.

Our assessment documents that a number of approaches have been used to address these concerns: the development of productive, environment-friendly, profitable ("win-win-win") technologies or practices, the regulation of farming practices on a statutory or voluntary basis (such as effluent standards and penalties versus watershed stakeholder institutions), and, more recently, payments designed to promote improved environmental outcomes. Continued experimentation with, improvements in, and integration of such strategies, involving several indirect drivers listed in Figure 26.7, will be central to progress from both a food security and environmental sustainability perspective.

26.4 Trade-offs, Synergies, and Interventions in Cultivated Systems

The preceding sections examined the condition and trends of ecosystem services and the major driving forces that shape them, highlighting how it is the response of farmers to these trends and pressures, using means that best match their opportunities and constraints, that largely determines the various outcomes of cultivation. This section summarizes some of the trade-offs that have been faced in balancing between food provision and other ecosystem services and briefly reviews a number of approaches and interventions that appear to reduce such trade-offs: integrated pest management; integrated agriculture-aquaculture; farm-scale options for mitigating carbon emissions, increasing carbon sequestration, and minimizing soil erosion; and agroforestry. Such approaches have shown results both in farmers' fields and in reducing off-site effects, but they are often very knowledge-intensive, require additional land or labor, and take time to yield benefits—all factors that can limit broader adoption unless more cost-effective interventions can be developed or non-distorting incentives can be provided.

26.4.1 Trade-offs

The world's cultivated systems embody a diverse array of biophysical constraints and production strategies. The specific quantity and mix of outputs generated by each system, including the supply of ecosystem services in both the short and the long term, is a consequence of the interaction among natural and managed processes, including the use of external inputs (chemical, physical, mechanical, or biological). The extent to which specific management interventions result in trade-offs or in synergies among system outputs (such as the impact of increased food output efficiency on water and nutrient cycling and biological diversity) is often both system- and location-specific.

Some clear trade-offs have been observed in the evolution of the world's dominant cultivation systems. For example, most flat, well-watered, fertile areas have increasingly been managed to simplify ecosystem function and to specialize in the efficiency of food production. Sustaining the high levels of food output such systems provide has generally and significantly reduced the supply of other ecosystem services from cultivated areas. High food-yielding cultivated systems have often required substantial externally derived inputs to sustain yield levels, such as additional reserves of water and nutrients, as well as the use of herbicides, insecticides, fungicides, and external energy sources.

The integration of cultivated systems into commercial food, feed, and fiber markets has usually provided the knowledge, incentives, and financial resources to maintain and often increase their already high food production capacity. However, the impact of intensive cultivation on the provision of ecosystem services both within and beyond the extent of cultivation has been equally substantial, resulting in the depletion of natural and human-made water reservoirs, water pollution, the disruption of global nutrient (particularly nitrogen) cycles, increased carbon- and nitrogen-based gas emissions, and an accelerated loss of terrestrial and aquatic biodiversity (Merrington et al. 2002). The global extent of farming and the specific trade-offs it entails imply that agriculture is the single largest threat to biodiversity and ecosystem function of any single human activity (Clay 2004).

While the evolution of the world's other dominant crop-based cultivation systems, low-input, smallholder rain-fed systems, has been markedly different, they too have increasingly been faced with significant trade-offs in the provision of ecosystem services. In general, low-input systems consume less energy and emit fewer pollutants. They also tend to accommodate higher levels of agricultural biodiversity with regard to more diverse crop mixtures and crop varieties.

Within many of these systems, increasing the provision of food would have a significant positive effect on human well-being, especially in cases where they support poor rural populations in areas with underdeveloped markets and where a lack of purchasing power prevents farmers from importing food from more-productive systems. Increases in food provision in low-input systems are likely to come from land-clearing and expansion, however, which reduces the services provided by pre-existing forest or grassland systems. (See Box 26.4.) Intensification in such low-input systems sometimes has within-system sustainability trade-offs—reduced soil fertility due to nutrient depletion when fertilizer inputs are underutilized or not available.

26.4.2 Integrated Pest Management

The goal of integrated pest management is to achieve economical protection from pest damage while minimizing hazards to crops, human health, and the environment (Kogan 1998; Bajwa and Kogan 2002). IPM takes advantage of existing ecosystem dynamics or sometimes involves the introduction of new, competing organisms to control pests.

Successful IPM practices achieve multiple goals at once, but careful monitoring and high levels of technical expertise are necessary. IPM farmers must choose from a wide array of options: cultural, biological, chemical, physical, mechanical, and genetic techniques. They must also have detailed understanding of numerous key factors:

- cropping histories (variety, seeding date, fertilization, seed treatment, tillage system); the timing and date of any pest control methods; environmental conditions before, during, and after treatment; past, present, and future plans for cropping; pesticide use history; and yield results;
- pest information, such as pest identity, growth conditions, development, reproduction and spreading modes, damage symptoms, and natural enemies; and
- field scouting, which involves systematic sampling of pest populations.

Only by understanding the ecology and economics of their cultivated systems can farmers make informed choices about appropriate levels of pesticide use (Kenmore 1996).

Following the Green Revolution, IPM scored several striking successes, notably in Indonesia, where the introduction of simple methods allowed farmers to halve the money they spent on pesticides (Orr and Ritchie 2004). Attempts to generate such successes in Africa have been mixed. In the 1970s, mealybug infestations caused crop losses of up to 80% in African cassava plants; today,

cassava mealybug damage is minimal thanks to the introduction of a parasitic wasp predator that maintains mealybug populations at a low level. This biological control method was free to farmers and environmentally benign (Herren and Neuenschwander 1991). Results were more limited elsewhere. In Malawi, for example, the major field pests of maize, beans, pigeon pea, and sweet potato were targeted using 18 different IPM strategies, but site variability, risk of reduced profitability, and overly complicated trials were all obstacles to adoption.

IPM has also had some success in industrial countries. In 1993, the U.S. government set a goal of having 75% of U.S. agriculture managed under IPM programs by 2000 (Fernandez-Cornejo and Jans 1999). While IPM has significant potential, however, that potential has yet to be fully realized. Despite extensive research into IPM programs, implementation is lagging (Sorensen 1993; Steffey 1995; Hutchins 1995). Many examples of cost-effective IPM trials exist, but in practice economic and institutional incentives are often not sufficient to encourage farmers to take on the risk of switching to integrated pest management (Sorby et al. 2003).

26.4.3 Integrated Agriculture-Aquaculture

Many freshwater species can adapt to an integrated farming system, where the wastes produced by one species are used by another species cultured in the system. IAA allows farmers to optimize resource flows and increase productivity by recycling nutrients between the various components of the system. In general, livestock manure is used as fertilizer for a crop species, the residues of which are fed to herbivorous fish. Fish excreta and other components of the pond humus are then recycled as manure for crop cultivation.

Such low-waste approaches reduce the discharge of nutrient-charged wastewaters into the environment, thus mitigating eutrophication and lowering net pollution compared with each cultivation component functioning independently. IAA systems also offer greater scope for more-efficient use of perhaps scarce water resources not only within the IAA system and but also by using IAA wastewater for irrigation. This both reuses water and delivers the residual levels of nutrients it contains directly to soil and crops. IAA systems have been developed for fish-duck farming, fish-chicken farming, fish-pig farming, rice-fish farming system in integrated areas, rice-shrimp farming, fish-vegetable farming, or fish-aquaponics farming (Lightfoot 1990).

Pig-grass-fish systems in China are used in both large-scale state-operated farms and in smaller-scale family-operated ones. Excreta from pig production is reused and treated as fertilizer for high-yielding fodder grasses, which serve as the main feed for herbivorous fish. Pig excreta are also applied directly to fish ponds, where it supports the growth of phytophankton—another source of fish feed. Wastes and residues that accumulate at the bottom of the fish ponds are harvested and recycled as manure for grass cultivation, completing the nutrient cycle. Pig-grass-fish systems are more labor-intensive than systems that use purchased feed inputs, and they also require substantial land area to grow the grass; however, their ability to simultaneously capitalize on in situ vitamins and proteins and to minimize waste makes them models of nutrient efficiency (Yang et al. 2001).

Another example of IAA systems is fish/fruit/vegetable cultivation in India, in which pond embankments are planted with fruit and vegetable crops. This provides several benefits: pond mud can be used as crop fertilizer, thus decreasing the cost of organic manures; pond water can be used to irrigate crops; fruit and vegetable residues can be used as low-cost fish feed; and plants

BOX 26.4
Aggregate Impacts of Trade-offs in Cultivated Systems: Land Use Perspective

At a regional or global scale, one measure of trade-off is the amount of land that needs to be brought into production according to different levels of food productivity. The "land-sparing" impact of modern farming practices has largely been achieved through yield increases brought about by the use of crop monocultures with improved crop varieties, fertilizer inputs, and irrigation where farmers have access to supplemental water. For example, if yields of the six major crop groups that are cultivated on 80% of the total cultivated land area had remained at yield levels farmers achieved in 1961, it would require an additional 1.4 billion hectares of land to meet global food demand in 2004. (See Figure.) This represents 34% of total land area suitable for crop cultivation and would have required conversion of large areas of uncultivated land that currently support rain forests, grassland savannas, and wetlands. In Asia alone, it would require an additional 600 million hectares, which represents 25% more land area than is suitable for cultivation on this continent. Asia would have had to be heavily dependent on food imports if crop yields had remained at 1961 yield levels.

The key ecological question is therefore whether environmental services other than food production at regional and global scales would be enhanced by focusing food production on less land under intensive management with high yields versus expanding cultivated area in lower-yielding systems that use farming practices that seek to preserve environmental services at the field and local levels. Few studies have addressed this issue using sound, ecological analytical methods. One recent study found that farming is already the greatest extinction threat to birds and evaluated the impact of land-sparing high-yield systems with "wildlife-friendly" farming practices on bird species persistence using ecological models (Green et al. 2005). The results suggest that high-yield farming may allow more bird species from a range of taxa to persist in developing countries. More such studies with other threatened fauna and flora species are needed to answer this critical question.

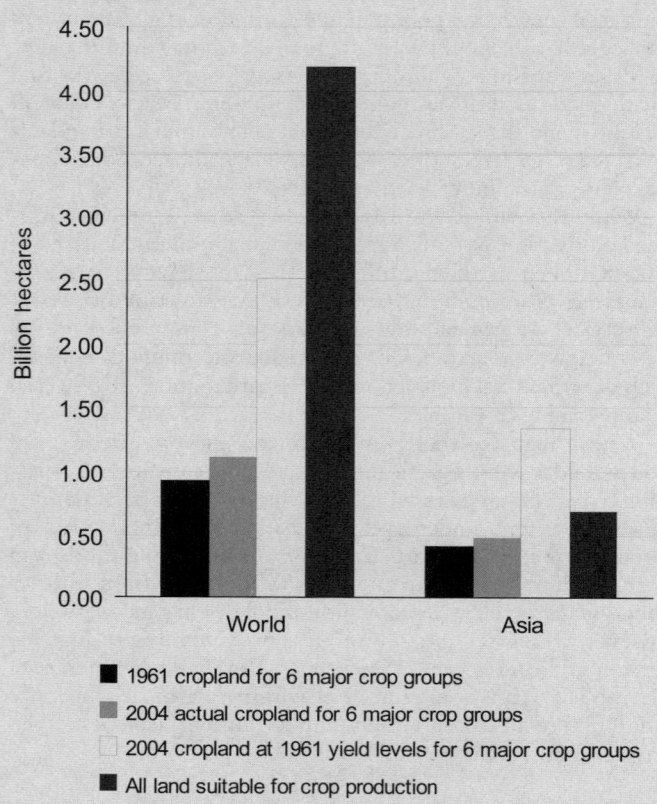

Land Used to Produce Major Crops in 1961 and 2004 and Land That Would be Needed to Produce Them in 2004 at 1961 Yield Levels. The six major crop groups included in this analysis are cereals, oil crops, pulses, root and tuber crops, sugar crops, and fiber crops. They accounted for 87% of all cropland in 1961 and 80% in 2004.

---

growing on the embankment strengthen the dikes (Tripathi and Sharma 2001).

26.4.4 Options for Mitigating Carbon Emissions and Increasing Carbon Sequestration

Several actions taken by farmers reduce overall greenhouse gas emissions. Those with the greatest potential for reducing emissions include increasing crop yields and return of crop residues; increasing the efficiency with which energy-requiring inputs (such as fertilizers and irrigation) are used; reducing or eliminating tillage operations; modifying crop rotations that include grass pastures and legumes; and increasing renewable energy production from biomass that either substitutes for consumption of fossil fuels (such as ethanol) or replaces the inefficient burning of fuelwood or crop residues and so avoids carbon emissions (Wassmann and Vlek 2004; Lal 2002; Antle et al. 2001). When considering biofuels as substitutes for fossil fuels, the greenhouse gas emissions associated with production and transport must also be taken into account to determine the net effect on greenhouse warming potential of the system.

It is notable that higher yields and input use efficiency result from farmer adoption of the best available crop and soil management technologies, and they contribute to increased profits. Reducing tillage improves yields and profits in rain-fed systems that

are often limited by drought. The viability of modified rotations and bio-energy production systems depends on a number of economic factors that are often beyond the control of farmers and typically do not favor adoption.

In addition to the actions just described, there are a wide range of mechanisms and measures for increasing carbon sinks in agriculture. (See Box 26.5.) However, there is considerable scientific uncertainty over the magnitudes and permanence of carbon sinks and emissions in cultivated systems. In addition, the economic potential for sequestration is considerably less than the technical potential, since sequestration practices are often costly (Lewandrowski et al. 2004).

26.4.5 Strategies for Minimizing Soil Erosion

Accelerated erosion has numerous adverse ecological and economic impacts to both ecosystems that are sources of erosion and those that receive sediments and sediment-borne contaminant (Lowdermilk 1953; Olson 1981; Oldeman 1998; Scherr 1999; Lal 2001). The on-site ecological impacts lead to disruption in cycles of water, carbon, nitrogen, phosphorus, sulfur, and other elements; a reduction in effective rooting depth; and a decline in soil quality. The on-site economic impacts are associated with reduction in agronomic productivity, which may be caused by reversible productivity effects due to loss of soil fertility, and re-

quality that reduce productivity cannot be easily or economically alleviated.

Estimates of the global impact of erosion on agricultural productivity vary widely because of differences in methodology. Estimates of potential yield losses in the absence of farmers' decisions are greater than estimates that account for farmers' incentives to mitigate the impacts of erosion. In the absence of farmer interventions, erosion would cost the world $523 million per annum in lost agricultural productivity (Den Biggelaar et al. 2004), or 0.3% of agricultural production per year, averaged across crops, soils, and regions. Other estimates are larger: Crosson (1995) calculated the on-farm economic costs of soil erosion on a global level at about 5% of agricultural production. Oldeman (1998) calculated the global productivity loss during the post World War II period at about 13% for cropland and 4% for pastures. Off-site damages to navigation, reservoirs, fishing, and water treatment, industrial, and municipal water facilities was estimated at $2–8 billion per year in the United States (Ribaudo 1997).

Economic analysis by Hopkins et al. (2003) finds that actual losses (when farmers respond to land degradation to maximize net returns over the long term) average 0.1% per year in the north-central United States. Global impacts of erosion are expected to similarly be less as farmers anticipate and respond to land degradation.

A number of effective soil and crop/vegetation management systems have been developed to minimize soil erosion. They include conservation tillage along with use of crop residue mulch and incorporation of cover crops in the rotation cycle on cropland; controlled grazing with appropriate stocking rates and use of improved pasture species on grazing lands; and adoption of methods of timber harvesting and logging operations that cause the least amount of soil disturbance (shear blade, tree extractors) on forestland (Lal 1998, 2001).

Planting choices have a significant impact. Frequent use of cover crops in the rotation cycle, integrated nutrient management, reduced pesticide use (through use of IPM, for instance), and use of agroforestry are important to soil and water conservation. Cover crops can limit erosion and prevent the accumulation of hazardous biogeochemical compounds, such as phenolic acids, that inhibit plant growth (Ryszowski et al. 1998). These ecological measures of minimizing risks of soil erosion may be supplemented by the installation of physical conservation structures, such as terraces and grass waterways, that reduce and direct runoff along with slope stabilization structures (Lal 1991).

Erosion control is enhanced by the adoption of management regimes that reinforce natural ecological cycles and processes in crop and rangeland systems. Soil erosion can be greatly reduced if there are minimal disruptions to water and nutrient cycles and when soil fertility and physical properties are not degraded. In cultivated systems where soils are prone to erosion, development of soil-specific farming systems and use of appropriate management practices are essential components of erosion control, as is the improvement of soil structure and enhancing biotic activity of soil fauna and flora (earthworms, termites, and so on).

26.4.6 Agroforestry

Agroforestry involves the integration of trees into farming systems in ways that create an agroecosystem succession, akin to that in natural systems (Leakey 1996). Biodiversity increases with each stage in the development of this succession (Leakey 1999). Agroforestry systems take many forms—short-term improved fallows with leguminous shrubs, medicinal, or other products in low-input tropical systems of the Amazon basin; enriched forest fal-

duction in soil organic matter and attendant water-holding capacity, versus more permanent, sometimes irreversible adverse impact on soil quality such as reduction in effective rooting depth with an accompanying decline in available water and nutrient retention capacities. While the reversible effects may be mitigated by use of additional inputs (such as fertilizers, organic amendments, and supplemental irrigation), the more permanent changes to soil

lows in Southeast Asia; intensive cash crop agroforestry systems with indigenous fruits and nuts in cocoa and coffee in West Africa; and contour strips in high-input maize/soybean systems in North America that mitigate erosion and runoff. The specific benefits of agroforestry vary by system but have included more profitable and nutritious food production, biodiversity conservation, improved soil resources, improved water quality, and carbon sequestration. Agroforestry systems have shown the ability to achieve multiple goals simultaneously, thus reducing the ecosystem service trade-offs inherent in crop production (Leakey 2001; Sanchez 1995, 2002).

Agroforestry systems have been shown to increase farmer incomes in sloped areas of Nepal (Neupane and Thapa 2001), in nutrient-poor farmlands of Africa (Sanchez 2002), and in Thailand (Wannawong et al. 1991), Cameroon (Palm et al. 2004), and Indonesia (Palm et al. 2004). Indigenous tree species are increasingly being domesticated to produce improved agroforestry tree products for local and regional food and medicinal markets. These improved species have been shown to generate household income, diversify production and the local economy, provide environmental services such as the mitigation of soil erosion, enhance carbon sequestration and biodiversity, and improve agroecosystem processes, like nutrient and water cycling. These multiple attributes of agroforestry are particularly valuable to subsistence-based livelihoods and simultaneously enhance the sustainability of crop production.

In the Philippines, the primary agroforestry practice is contour hedgerows, in which food crops are planted between hedges of woody perennials established along the contours of upland sloping farm plots. Prunings from the hedgerow trees or shrubs are placed at the up-slope base of the hedges to trap eroding soil so that, over time, natural terraces are formed (Pattanayak and Mercer 2002). Such hedgerows can improve soil conservation by 15–20% for a typical small farmer (Pattanayak and Mercer 2002). In addition to erosion control, biophysical effects of contour hedgerows on soil include maintenance or increase of organic matter and diversity, nitrogen fixation, enhancement of physical properties such as soil structure, porosity, and moisture retention, and enhanced efficiency of nutrient use (Nair 1993).

Besides agroforestry systems that combine trees with annual crops, there are those that combine trees with animals. Silvopastoral systems (defined as the integration of trees and pasture) are the most common form of agroforestry in the southern United States (Zinkhan and Mercer 1997). Silvopastoral systems are increasingly important in the developing world, especially in areas where perennial crops such as coconuts, oil palm, rubber, and fruit trees are found. In Southeast Asia, the integration of oil palm plantations with cattle and goats resulted in increased production of 3.52 tons of fresh fruit bunches per hectare, equivalent to 0.7 tons of palm oil per hectare. In Central America, most livestock farms include some silvopastoral systems that improve economic returns through diversification and the timing of cash flows (Henderson 1991).

Despite the potential benefits of agroforestry techniques, adoption has been relatively limited. Impediments fall into five categories: economic incentives, biophysical conditions, risk and uncertainty, household preferences, and resource endowments (Pattanayak et al. 2003). Additional research is required to domesticate novel tree species (and other crops) that can further enhance agroforestry systems. Identifying and domesticating such species could, for example, increase the availability and quality of traditional fruits and nuts rich in vitamins and minerals, which would improve the nutrition of smallholder farmers and their families (Leakey in press).

26.4.7 Constraints and Opportunities for Improved Interventions and Outcomes

The interventions just described span notions of high and low input or of tropical versus temperate agriculture. IPM; reduced tillage, agroforestry, and soil conservation, for example, have all been used in a range of agroecological and socioeconomic contexts globally. To these farm-scale interventions could be added the emergence of landscape-scale approaches that recognize and respond to the scale at which water and nutrient cycling and energy fluxes take place. Landscape approaches involve complementary and coordinated farm- and landscape-scale interventions as a means of improving long-term productivity and environmental sustainability (Baudry et al. 2000; Ryszkowski et al. 1999; Thenail 1996). Achieving the full potential of such approaches, however, requires continued development and integration of knowledge, strengthening of institutions, and improved feasibility and profitability for farmers. (See Box 26.6.)

Most approaches that seek to reduce food versus environment trade-offs require intensive use and integration of knowledge from the biological, agronomic, and ecological sciences together with farmer knowledge. Thus, the greater role and impact of such interventions is conditional on bridging perspectives of often productivity-focused scientific research with more ecosystem-focused perspectives—encompassing, for example the role of agroecological and eco-agriculture approaches (Conway 1999; Altieri 2002; McNeely and Scherr 2002). There is both much to learn and likely much to gain from, for example, improved understanding of the role of soil microbiology in improving water and nutrient efficiency in high-input systems (Matson et al. 1997; Woomer and Swift 1994), as well as rich possibilities of using biotechnology tools to enhance the productivity of low-input systems or orphan crops (Naylor et al. 2004).

Ultimately, decisions about the use of specific technologies and practices will depend on the opportunities and constraints of farmers, and there is evidence that here, too, more needs to be done to foster the adoption of practices that minimize trade-offs. Even where technologies have the potential to be profitable, many adoption decisions are affected by local institutions, particularly the effectiveness of local property rights systems and capacity for organizing and sustaining collective action.

Figure 26.9 plots increasingly secure property rights on the horizontal axis and increasing levels of collective action on the vertical axis. Some of the most successful agricultural technologies lie close to the origin in this figure. For example, the benefits of high yielding cereal varieties—the cornerstone of the Green Revolution—could be captured within a single agricultural season by individual farmers and hence did not require secure property rights or collective action. In tackling more complex objectives that include both yield and conservation goals, however, local institutional issues are more prominent.

Integrated pest management requires that farmers in an area work together to control pesticide use and to synchronize planting dates. The returns are relatively quick, however, so secure property rights are still not a major issue, and IPM appears in the upper left corner of the Figure. In contrast, planting of trees on farms (agroforestry) is a long-term investment that requires secure property rights. But since trees can be planted by individual farmers, agroforestry appears in the lower right-hand corner. Still other approaches, however, such as watershed conservation, require both secure property rights and effective collective action, and therefore appear in the upper right-hand quadrant. If these institutional conditions are not met, then the technology is not likely

BOX 26.6

Service Trade-offs in Cultivated Systems: A Case Study from the Argentine Pampas

The pampas agroecological zone is a vast, flat region of Argentina extending more than 50 million hectares and used predominantly for crop and cattle production (Satorre 2001). Agriculture in the pampas has a relatively short history (a little more than 100 years) comparable with that of the American Great Plains (Hall et al. 1992). Both agroecological zones were mostly native rangelands until the end of the nineteenth and the beginning of the century centuries, when lands were initially transformed for crop (cereals and oil seeds) and cattle production under rain-fed conditions. Where European tillage methods with a conventional plow were used, heavy erosion (dust bowls) occurred the first half of the last century, especially on the more fragile lands (Covas 1989).

Mixed-grain, crop-cattle production systems have now expanded to occupy most of the pampas and involve rotations of maize, wheat, and soybeans, with cattle pastures being integrated in various ways depending on local soil and climate conditions. Cattle operations vary from cow-calf to cattle finishing. The pampas suffers occasional droughts and floods that temporarily affect both crop and cattle production (Viglizzo et al. 1997).

A major challenge in sustaining the economic viability of the pampas low-input agroecosystems is to maintain soil quality that supports crop production and environmental services. Soil organic matter content is a key component of soil quality since it serves as a reservoir of nitrogen, phosphorus, and sulfur and has a large impact on soil physical properties that promote water infiltration, storage, and root function, all essential to support crop growth (Viglizzo and Roberto 1998).

Intensification of agricultural systems in the pampas during the past 50 years has involved a steady increase in farm area devoted to annual crop production and a consequent reduction in area allocated to perennial and annual pastures. Similar trends have occurred in the U.S. Corn Belt. From 1960 to 2001, grain production in the pampean provinces increased from 11.1 million to 43.5 million tons. Changes in soil organic matter and nitrogen dynamics associated with intensification provide an illustration of environmental service trade-offs. For example, leguminous pastures in a pasture-

crop rotation can promote biological nitrogen fixation such that the soil nitrogen supply fluctuates around a value determined by the length of the leguminous pasture phase. Changes in land use that reduce or eliminate leguminous pasture decrease soil organic matter and nitrogen and phosphorus supply unless there are compensating applications of fertilizer or livestock manure (Viglizzo et al. 2001). Because current levels of N fertilizer use efficiency achieved by farmers are relatively low, there is substantial risk that nitrogen losses can damage environmental services in off-farm ecosystems.

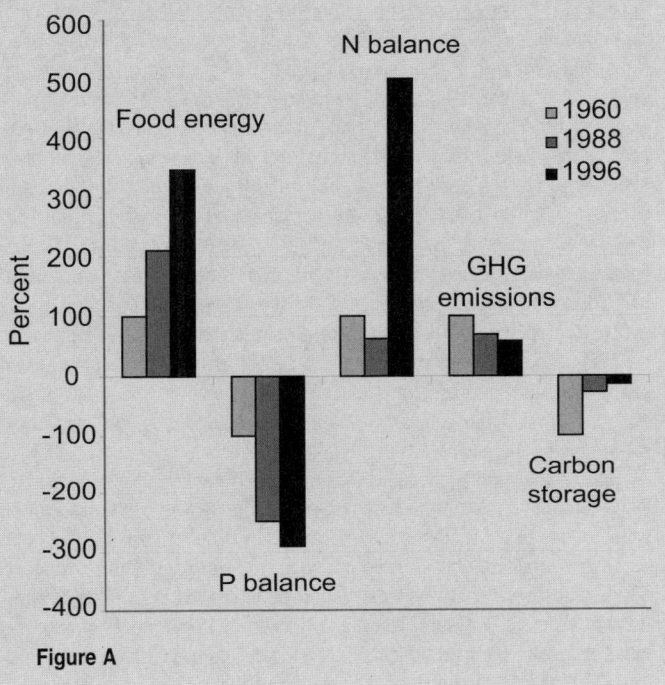

Figure A

---

to be adopted and maintained, regardless of its profitability and scientific soundness.

26.5 Cultivated Systems and Human Well-being

The ability of humans to convert natural systems to cultivated systems and to derive more food from each hectare of land has, for millennia, supported the growth of civilizations. Indeed, the first civilizations developed in the Fertile Crescent of the Middle East because local conditions were well suited to cultivation and the domestication of animals for livestock (Smith 1998; Diamond 1999). Similarly, in many parts of Asia, efficient and sustainable irrigated paddy fields have supported a number of prosperous cultures with high population densities over thousands of years. A stable food supply has always been the foundation on which human civilizations are built. Moreover, adequate nutrition is fundamental for human development and health.

For all the benefits they provide, cultivated systems can also pose risks to human well-being, most notably via direct health effects from, for example, the handling and use of pesticides and zoonotic diseases associated with certain cultivation practices, as well as through pollution of air and water. Cultural and amenity services of natural ecosystems are diminished when they are converted for cultivation, and that loss may or may not be compen-

sated for by cultural and amenity services associated with cultivated systems.

This section deals with the linkages between human well-being and cultivated systems, noting that the largest single source of human well-being derived from cultivation is through the production and consumption of affordable food, fiber, and other products. The human well-being impacts mediated through food consumption are dealt with separately and in detail in Chapter 8.

26.5.1 Economic Component of Human Well-being

Cultivated systems play a vital role in global economic well-being, especially in poorer countries. In 2000, agriculture (including forestry and fishing) represented 24% of total GDP in countries with per capita incomes less the $765 (the low-income developing countries, as defined by the World Bank) (World Bank 2003). About 2.6 billion people depend on agriculture for their livelihoods, either as actively engaged workers or as dependants (FAOSTAT 2004). In 2000, just over half (52%) of the world's population were living in rural areas and, of these, about 2.5 billion people were estimated to be living in agriculturally based households (World Bank 2003). The global agricultural labor force includes approximately 1.3 billion people, about a fourth (22%) of the world's population and half (46%) of the total labor force (Deen 2000).

BOX 26.6

continued

Analysis of 85 pampas farm systems differing in their land use patterns and level of intensification (measured in terms of energy use) reveals trade-offs between the share of land used for crop production and the provision of ecosystem services. Results show that carbon storage, greenhouse gases emissions, and annual nitrogen and phosphorus balances decrease as the cropping area, use of energy derived from fossil fuels, and the net primary productivity of systems increase. Risk of pesticide contamination and soil erosion and the human disturbance of the habitat also increase, although both risk of erosion and disturbance stabilized or decreased somewhat at the highest levels of cropping intensity.

In contrast, GHG forcing potential decreases because removal of pastures and livestock grazing is associated with a reduction in methane emissions and fire used to improve vegetation quality in pastures and grazing land. The nature of trade-offs observed in the pampas varies not only by local agroecological conditions and production systems but also over time. Figure A shows how the average level and mix of ecosystem services have changed across the pampas over time. Compared with 1960, food output has increased significantly, while phosphorus balances have worsened. Nitrogen balances were positive but declining as pasture was converted to low-input cropland, but they have surged as urea application to cropland has become increasingly necessary. GHG emissions have fallen in line with pasture and livestock decreases, while carbon stocks continue to be depleted, but at a declining rate. Given the broad adoption of no-tillage cultivation practices in recent times, carbon stocks might now be increasing (Viglizzo 2002a, 2002b).

Both agricultural production and ecosystem services have economic value that can contribute to human welfare. Hence, the costs associated with the loss of ecosystem services caused by crop intensification should be weighed against the benefits obtained from farming. The ecosystem service valuation techniques developed by Costanza et al. (1997) were used to estimate both the market and nonmarket components of ecosystem services in pampas agricultural systems. The gross margins of crop and livestock production operations during the 1990s were assessed using standard economic valuation approaches. A comparison of the dynamics of crop and livestock gross margins and ecosystem service values related to the intensity of cropping is shown in Figure B. While the gross margin of farming production increases proportionally to the intensification of cropland, there is a relatively sharp decline in the value of ecosystem services at the earlier stages of intensification, such that about half

the value of ecosystem services is lost when around 40% of the area is used for crop cultivation.

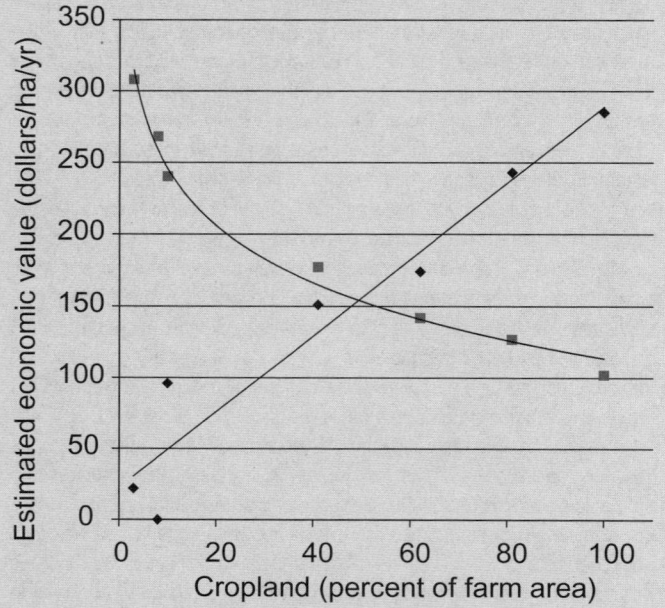

Gross margin of farming (dollars/ha/yr)
Value of agroecological services (dollars/ha/yr)

Figure B

This analysis considers only the implications of intensification within the pampas and does not examine broader geographic effects. Globally, the pampas has become a major source of grain for countries and regions where local food supply is insufficient to meet demand. Thus while reducing cropping intensity and increasing the percentage of land devoted to pastures in the pampas might improve ecosystem services locally, the loss of grain output would need to be offset by yield increases elsewhere in the pampas or by expansion of cultivated area and yields elsewhere in the world. In both cases, there would likely be negative effects on environmental services in these other locations that any comprehensive assessment of ecosystem service trade-offs would need to take into account.

In Africa, agriculture provides two thirds of all employment and half of all exports and accounts for 37% of GNP. Despite rapid urbanization and economic diversification in South Asia, agriculture continues to provide employment for over 60% of the population and generates 27% of GNP (DFID 2002). In 2000, globally, cultivated systems produced approximately $815 billion worth of food crops and $50 billion worth of non-food crops. In the same year, fisheries output was valued at $156 billion and livestock products at $576 billion. (See Chapter 8.)

Measuring the economic benefits of employment is difficult because globally comparable agricultural wage rates do not exist. One very rough proxy of gross agricultural income is the total value of agricultural production divided by the number of agricultural workers. This provides a rough estimate of the gross economic returns to labor. Globally, the average annual value of agricultural production per agricultural laborer for 1995–97 was approximately $1,027 per person (using 1989–91 average international prices). The range of estimates is quite broad—from about

$50,500 per person per year for the United States down to $411 for sub-Saharan Africa (Wood et al. 2000).

Livestock provides the main source of livelihood for 650 million farmers worldwide. Despite low productivity, livestock husbandry is one of the few means for the poor to generate income, acquire assets, and escape from poverty. Sales of livestock, animal-source food, hides, and fibers through both formal and informal markets make major contributions to household income. Evidence from in-depth field studies in Asia and Africa indicates that livestock contribute as much as 76% of household incomes in some regions, and generally a higher percentage to the incomes of poorer households (Delgado et al. 1999; Kaufmann and Fitzhugh 2004).

There is a growing consensus that poverty, hunger reduction, and increased economic growth cannot be achieved in most poor countries without more fully exploiting the productive capacity of the agricultural sector (Timmer 1989; Sarris 2001; Hazell and Haddad 2001). Agricultural growth can reduce poverty through

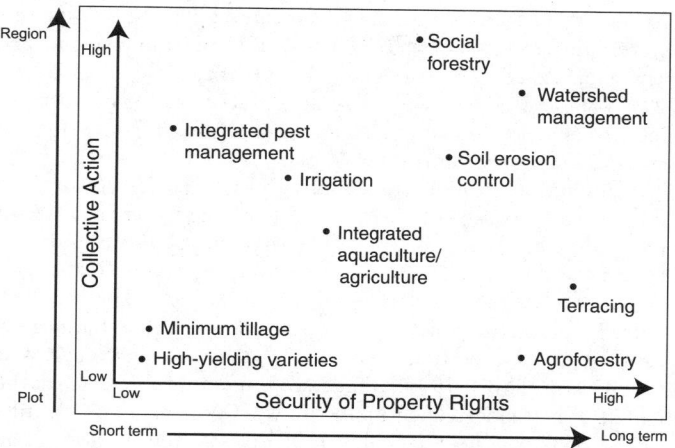

Figure 26.9. Links between Property Rights, Collective Action, and Technology Adoption in Cultivated Systems (adapted from Knox et al. 2002)

increased employment and wages and through income generated by the sale of goods produced by the poor (Datt and Ravaillon 1998). It also results in increased demand for food, services, and unskilled labor (Mellor 2000). The relationship between agricultural wages, higher yields, and poverty in the case of India is shown in Figure 26.10. Timmer (1997) has shown for 27 countries from the period 1960 to 1992 that agricultural growth reduced poverty more than growth in manufacturing did, while López and Valdéz (2000) have shown that rural growth is more effective than urban growth in reducing poverty in Peru. Growth in Peruvian agriculture was also shown to have reduced urban poverty through slower rural-to-urban migration and more affordable food prices.

Beyond the direct economic impact on employment and incomes, there are several indirect economic benefits of cultivated systems that can be even greater. These are mediated through rural growth linkages, inter-sectoral linkages including the post-harvest agribusiness sector, consumer income effects, and trade. Rural growth linkages are an important mechanism by which agricultural growth spurs growth in non-farm incomes and employ-

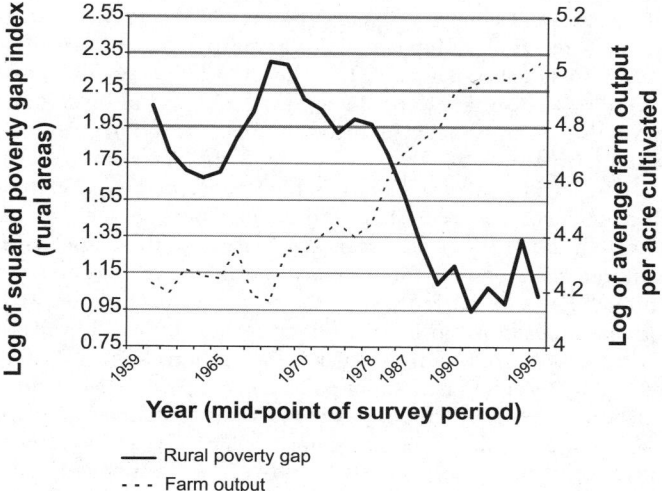

Figure 26.10. Yield Growth and Poverty Reduction in India, 1959–94 (Datt and Ravillon 1998)

ment (Hazell and Roell 1983; Mellor 1966). Growth in the use of farm inputs, processing, marketing, and transport services serve to increase rural non-farm incomes. The increasing household expenditure that results enhances consumer spending and triggers a further rise in non-farm incomes and employment. For example, a $1 increase in agricultural income resulted in increases in rural income of an additional $1.88 in Burkina Faso, $1.48 in Zambia, $0.96 in Niger, and $1.24–$1.48 in two locations in Senegal (Delgado et al. 1998).

In an analysis of Asia and the Near East, Timmer (1989) concluded that linked growth between industrial and service sectors and the rural economy could lead to increases in rural wages and more equitable income distribution. Mellor (2000) contends that the substantial lags between accelerated agricultural growth and reduction in poverty are strong evidence that agricultural growth reduces poverty more through indirect processes than direct ones. Furthermore, these linkages enhance overall economic growth. Rural growth linkages are particularly important because they benefit small labor-intensive enterprises and thus contribute to the alleviation of rural poverty.

It is increasingly recognized that estimates of agriculture's contribution to economic growth and human well-being at national levels are underestimated by economic indexes that focus only on farm-level added value. Upstream and downstream linkages with agro-industries, services, and trade are not properly accounted for. In Argentina, for example, primary agriculture is estimated to represent about 5% of total GDP, but this increases to 32% when the linkages with the food and agro-industry sectors are considered (IDB 2004). Similarly, agriculture in Brazil and the United States is estimated to contribute about 4% and 1%, respectively, to national GDP, while full accounting of food chain linkages gives estimates of 26% and 8% (IICA 2004).

The economic well-being generated from cultivated systems is not limited to those employed in the food supply sector. Food consumers have benefited enormously from the long-term decline in food prices. Cereal prices have fallen by about 40% in real terms during the past 40 years, resulting in increased disposable incomes for consumers. (See Chapter 8.) This allows increased expenditures on education, health, and better nutrition (Hazell and Roell 1983).

Economic well-being can be further improved through international agricultural and food markets. Trade in food not only broadens choices but also provides access to foods year-round that often can be grown locally only on a seasonal basis. It provides local farmers with new market opportunities, resulting in higher living standards for those able to participate. Food trade also helps provide more stable and secure access to food at competitive prices, but it cannot play such an effective role where prices are distorted or significant market barriers exist. A major limit to agriculture's role in the global economy is that agricultural trade barriers are on average 10 times higher than industrial trade barriers (Abbott 2005; Charlton and Stiglitz 2005).

Agriculture's effects on economic well-being are not all positive, and other sections of this chapter have described the negative impacts, including loss of biodiversity; soil, water, and airborne pollution; and health risks. Although it is difficult to quantify the costs of these externalities from agriculture, there can be significant financial and economic consequences that are explicitly and implicitly borne by society. Pretty et al. (2000) have estimated the external environmental and health costs of agriculture in the United Kingdom at £2.3 billion in 1996 using ecosystem service values derived from Costanza et al. (1997). These costs represent 13% of average gross farm returns and 89% of average net farm income in the country.

Improving formal measures of productivity by accounting properly for the full social costs and benefits of all inputs and outputs from cultivated systems, including environmental services and health as well as the economic multipliers from agricultural productivity, is critical for making more informed decisions about policies and investment in the agricultural sector, especially in developing countries, where sustainable agricultural development is the foundation for broader economic development.

26.5.2 Linkages between Cultivated Systems and Nutrition and Health

Cultivated systems contribute to human health and nutrition primarily by providing food either through subsistence agriculture or through commercial agriculture and food markets. The magnitude of the human well-being benefits derived from an adequate and nutritious food supply are so large that they are often taken for granted, especially in wealthy countries in which food costs represent a small proportion of disposable income. Likewise, it is very difficult to protect against environmental degradation and loss of ecological services in regions where people experience chronic food shortages. Unless chronic hunger and food insecurity are reduced, the poor will continue to exploit natural resources in the short run, thereby undermining the sustainability of natural ecosystems and consequent food security in the long run (Webb 2002).

The important linkages between food consumption, health, and nutrition are assessed in detail in Chapter 8. This section focuses on linkages between human health and factors other than food supply per se, such as the impact of production systems, production practices, and associated environmental externalities. The linkages are grouped into the health of farm workers and farm families and that of the broader population potentially affected by cultivation practices.

Agriculture is a hazardous occupation. Globally it is estimated that farm workers run at least twice the risk of dying on the job than workers in other sectors and that around 170,000 people die per year because of these work hazards (Forastieri 1999). In the United States, for example, the death rate among agricultural workers was an estimated 20.9 per 100,000 workers in 1996—more than five times the average for all industries (Reeves and Schafer 2003). Not only are mortality rates higher, but so are rates of accident and illness. In the United States, farmers and farm workers account for only 3% of the workforce but for nearly 8% of all work-related accidents; in Italy, some 10% of workers in agricultural production account for some 29% of workplace accidents, often related to the use of tractors, harvesting machinery, and power tools (Forastieri 1999). In developing countries, where farmers use smaller and less powerful equipment, there are likely to be fewer serious work-related accidents, although few data on these are available.

Apart from accidents, two other linkages between production and health are respiratory problems caused by working in barns and confined livestock systems and a variety of health problems linked to pesticide handling and use. A compilation of studies from Australia, Finland, Denmark, Sweden, Scotland, the United States, and Canada indicated very high levels of occupational respiratory problems in farm workers. In intensive dairy systems, about 20% of farm workers were reported to suffer bronchitis problems directly related to in-barn air quality, 5% from asthma problems, and just under 5% each with symptoms of organic dust toxic syndrome and "farmer's lung" disease. Health effects are comparable or larger for pig and poultry operations and arise largely through the presence of dust from the use and handling of hay, straw, and dry animal feeds (Omland 2002). Kansas farmers were found to be at an increased risk of death from prostate cancer, brain cancer, non-Hodgkin's lymphoma, and leukemia in a 1983–89 study; they were also at elevated risk of death from motor vehicle accidents, accidents resulting from falling objects, and machinery accidents (Frey 1991).

In addition to high risk of physical injury, the estimated 2 million farm workers in the United States face a greater risk of pesticide exposure than any other segment of the population (Reeves and Schafer 2003). On a global scale, it is estimated that 20,000 people die of adverse effects of pesticide exposure each year, 3 million are poisoned, and there are nearly 750,000 new cases of chronic pesticide exposure, such as cancer, each year (WHO and UNEP 1990). There are limited reliable data on the extent of pesticide-related illness anywhere due to poor identification of such illnesses, which leads to underestimations. The magnitude of health damage caused by agrochemical exposure will vary according to the type of agrochemical used, the mode of application/exposure, the individual susceptibility, and the climatic conditions—and each of these factors is related to the type of crop grown and specific pesticide use practices.

Pingali and Roger (1995) documented that pesticide use had an adverse impact on human health and subsequently on farmer productivity for rice farmers in the Philippines in 1991. Eye, pulmonary, and neurological problems are significantly associated with long-term pesticide exposure. In Northern Ecuador, heavy use of pesticides by potato farmers was the principal cause of death after traffic accidents for both men and women (Yanggen et al. 2004). In China, in 2001, decreased use of pesticide as a result of growing *Bacillus thuringiensis* cotton resulted in a lower incidence of poisonings (Hossain et al. 2004).

Farm workers and farm families are particularly susceptible to zoonoses—animal diseases that can be transmitted to humans through contact with infected animals. There are approximately 150 kinds of zoonoses, and many are transmitted by livestock. For example, avian influenza, or "bird flu," is a disease that humans may contract through direct contact with live poultry infected with the flu virus or direct contact with the feces, nasal, or eye discharges from infected birds (WHO 2004). In contrast, the risk of infection from consumption of poultry product is extremely low. Because there is no viable treatment, recent outbreaks caused by the H5N1 strain are considered to be one of the greatest potential threats to human health if human-to-human transmission of the disease becomes widespread (WHO 2004). The first documented infection of humans with an avian influenza virus occurred in Hong Kong in 1997, when the H5N1 strain caused severe respiratory disease in 18 humans, 6 of whom died. Cases have also recently been reported in Viet Nam, Thailand, and Cambodia (WHO 2004).

The incidence of zoonoses is high in developing countries because social and economic factors contribute to their spread (Langoni 2004). Poor sanitation can exacerbate these diseases in children by allowing the zoonotic agent to be disseminated through rainwater, streams, and brooks where children often play. Bovine brucellosis can be transmitted to humans and is a major zoonosis associated with livestock. Human brucellosis is characterized by fever and back/joint pain (Unger 2003). Further complications due to human brucellosis may include hepatitis (Masouridou et al. 2003). According to WHO data, the number of cases of human brucellosis worldwide was estimated to be about 500,000 (WHO 2005). The advent of HIV/AIDS has increased the prevalence of many zoonoses in humans because HIV can increase susceptibility to zoonotic agents by depressing the human immune system (Langoni 2004).

Contamination of surface and groundwater by pesticides and fertilizers is also reported to affect public health (Ongley 1996). Excessive waterborne nitrogen has been linked to respiratory ailments, cardiac disease, and several cancers (Townsend et al. 2003). Nitrate levels have grown in some countries to the point where more than 10% of the population is exposed to nitrate levels in drinking water that are above the 10 milligrams per liter guideline (WHO 1993). Although WHO finds no significant links between nitrates and nitrites and human cancers, the drinking water guideline is established to prevent methemoglobinemia in infants (blue baby syndrome) (WHO 1993). Water polluted by waste and run-off from grazing areas and stockyards can also cause disease. The most common diseases associated with contaminated waters are cholera, typhoid, ascariasis, amebiasis, giardiasis, and enteroinvasive *Escherichia coli*. Four million children die every year as a result of diarrhea caused by waterborne infection, although the share attributable to agriculture is unknown (Ongley 1996).

Irrigation systems provide sources of water that can improve sanitation, and thus human health, but they can also serve as a breeding ground for disease vectors. Increases in malaria have been linked to reservoir construction (De Plaen 1997; Reiff 1987). Schistosomiasis (bilharziasis), a parasitic disease that spends part of its lifecycle in a snail species and that affects more than 200 million people in 70 tropical and sub-tropical countries, has also been demonstrated to increase significantly following reservoir construction for irrigation and hydroelectric power production (DFID 1997). The two groups at greatest risk of schistosomiasis infection are farm workers involved in the production of rice, sugarcane, and vegetables and children who bathe in infested water.

Water contamination is not restricted to the developing world. The total cost of drinking water contamination from agriculture has been estimated at £120 million in the United Kingdom due to pesticides and £16 million due to nitrate from fertilizers (Pretty et al. 2000).

Recently, HIV/AIDS has added another dimension to the relationship between agriculture and human health. Gillespie and Haddad (2002) suggest that improved nutrition for agricultural workers with HIV/AIDS is important for improving their quality of life. However, ill health as a result of HIV/AIDS also affects agricultural production through reduced stamina and strength of sick farm workers and the diversion of household resources and time to care for the sick and for funerals. Subsequent decreased labor productivity can in turn affect human well-being since households may resort to growing less nutritious or less lucrative crops because they are less labor-intensive.

26.5.3 Equity and Distributional Aspects of Cultivation

At the scale of the farm and community, linkages with equity and distribution are conditioned by the existing distribution of assets and the limited access of poor people and vulnerable groups to cultivation-related resources and opportunities such as land, credit, extension, and markets. At the scale of the country and region, there are often biases in political and economic power against rural areas and against specific marginalized groups. At the international scale, there are imbalances among richer and poorer countries with regard to their ability to promote competitive agriculture through publicly-funded domestic farm support, influence on trading patterns, and the strength of public and private systems delivering improved production technologies and practices.

In theory, agricultural growth should eventually lead to more equitable distribution of both income and resources (Kuznets

1955). However, empirical evidence of agriculture's effects on promoting equity is ambiguous (von Braun 2003) or marginal at best (Deininger et al. 2004; Tsur and Dinar 1995; Bautista et al. 1998). For example, a number of studies have shown that administrative land reform was not effective in transferring land to the poor in Colombia and Ethiopia (Castagnini et al. 2004; Adenew et al. 2003). In Viet Nam, despite the rapid growth of agricultural wages in the 1990s, wage inequality fell modestly (Gallup 2002). In China, however, long-term government investments across multiple sectors, including agricultural research and development, irrigation, rural education, and infrastructure (including roads, electricity, and telecommunications) contributed not only to agricultural growth but also to the reduction of rural poverty and regional inequality (Fan et al. 2002).

The marginalization of vulnerable groups such as women and children is also a constraint to more equitable sharing of benefits from farming. Women are especially vulnerable to existing inequities in terms of wages, access to and control of production technologies, gender segregation in labor markets, and access to property and entitlement in their own right (Quisumbing and Meinzen-Dick 2001).The central role of improving gender equity in African agriculture, where women are productive farmers and key food producers, is now widely recognized (Kabutha 1999). And in Cambodia, 90% of children worked in agriculture or agriculturally related activities during 1996 (ILO 1997). Lack of education propagates vulnerability and promotes widening inequality. While widespread use of child labor in agriculture is an economic necessity in many countries as families are too poor to pay for schooling, adult males often migrate to urban areas seeking employment. In addition, sickness and care-giving—especially related to HIV/AIDS—reduces the pool of family labor in sub-Saharan Africa.

Persistent barriers to agricultural trade across international boundaries, such as export subsidies and import restrictions, limit more equitable agricultural income distribution among countries by, for instance, limiting developing-country access to EU and U.S. markets, as described earlier. But the impact of trade liberalization on the distribution of income within developing countries varies according to country-specific policy conditions and socio-economic structure. In Latin America, for example, analysis suggests that trade liberalization has had positive effects on income equality in nine countries and negative effects in five countries (von Braun 2003).

There are growing concerns about inequalities with regard to the capacity to generate and gain access to new scientific information and technology (von Braun 2003). An increasing share of agricultural R&D globally is being funded by the private sector at the same time that the science needed to make key advances becomes more complex, costly, and, particularly for biotechnology, increasingly proprietary in nature. The fear is that large bioscience companies have few incentives to focus on the crops, constraints, and technologies most appropriate for poor farmers in tropical areas but have proprietary rights over processes and components of technologies that need to be used (Pardey and Beintema 2001).

These trends, compounded by the long-term underinvestment in agricultural R&D by most developing countries despite the economic importance of agriculture, are widening already significant gaps in scientific capacity compared with industrial countries. Increasingly it is only the larger developing countries, such as Brazil, China, India and South Africa, who can muster the investments in R&D and human capacity needed to keep their farmers competitive.

As trade liberalization proceeds, increased reliance is placed on knowledge, science and technology, and technology transfer to keep farmers in business. In addition, emerging trends in global food retail and agro-processing markets, increasing demand for food safety, and shifts in diets and preferences toward processed foods are raising concerns about the long-term future of smallholder farmers in developing countries (Lipton 2005). In part, these concerns arise from the disproportionately negative impact of structural adjustment programs on smallholders during the 1980s and 1990s brought about by the wholesale withdrawal of public-sector services (disappearing market, input services, and credit).

With regard to the specific case of genetically modified crops, recent studies have documented substantial economic benefits from the most widely adopted transgenic crops—*Bt* cotton and herbicide-resistant soybean. Non-GMO cotton varieties are highly susceptible to yield loss from bollworm and boll weevil insect pests that require as many as five or six applications of highly toxic pesticides to avoid severe yield loss. In contrast, the *Bt* cotton varieties have been transformed to contain a bacterial gene that produces a protein that is toxic to these insect pests when they feed on the plant's tissues.

Reduced cost of insecticide applications and higher yields contribute to substantial increases in profit for the farmer and lower prices for cotton, which benefit the consumer. As a result, the economic benefits from use of insect-resistant *Bt* cotton varieties were found to be evenly distributed among farmers, private-sector seed companies, and consumers in both industrial and developing countries (Huang et al. 2002; Falck-Zepeda et al. 2000). Similar studies of the distribution of economic benefits from herbicide-resistant GMO soybean have also documented balanced distribution among farmers, seed companies, and consumers (Qaim and Traxler 2005).

There are also linkages between the impact of cultivation on ecosystems services and equity. The poor and the vulnerable are likely to suffer most from environmental externalities of production, such as downstream water depletion and pollution and loss of habitat and biodiversity—particularly since the landless rely more on wild sources of food (Grimble et al. 2002).

26.5.4 Cultural Aspects of Cultivation

Cultivated systems and human culture are inextricably linked. Religious and ethical values, cultural backgrounds, and philosophical convictions are important factors linked to the sustainability of cultivated systems, rural development, and food security. Cultural practices and traditions are often integrated into cultivation norms and practices, into land inheritance, ownership, and access, and into access to other productive resources. Cultural factors and preferences can have a large influence on the demand and value of various food products in the marketplace. Likewise, traditional food taboos and food distribution along age and gender lines can have a substantial impact on nutrition by affecting the types of food that are available or culturally acceptable. (Chapter 8 contains more discussion on food and culture.)

Farmers' close ties to the land and their intimate relationship with it is an intangible aspect of farming that can outweigh maximizing short-term economic gain. In northern New Mexico, for example, small livestock operations are a critical aspect of families' and communities' way of life, maintaining cultural heritage and traditional values as well as passing those values on to future generations. Keeping land in the family and upholding traditional values are regarded more highly than material possessions or monetary gain (Raish and McSweeny 2003). Despite the commercialization of agriculture, agricultural societies still exist that value the cultural aspects of farming and, as a result, have created home-

steads, communities, collective action mechanisms, and alternative technologies that allow continuation of traditional peasant agriculture (Schwarz and Schwarz 1999). It is not just farm households who value the existence of agriculture—in richer countries, the policy of publicly funded payments to farmers for "environmental stewardship" is broadly supported, reflecting findings of formal studies of the public's willingness to pay for maintained agricultural landscapes (Drake 1992; Olsson and Ronningen 1999).

Changing cultural attitudes toward agriculture can be traced back to the Industrial Revolution, when technological innovations that control the environment began replacing the spiritual relationship of farmers to the land. Ultimately, the evolution of agriculture in industrial countries has profoundly changed culture, eliminating the need for millions of farmers and farm workers and displacing entire communities (Bailey 1999).

The farming of animals involves a culture of its own, and ancient myths surrounding animals have deeply influenced animal production (Fraser 2001). Nurturing animals is an integral part of the ecology and economy of many farming systems; it was and is regarded as a moral responsibility in many religious and cultural traditions, with different species serving important and complementary functions. Animals are also important for moral education, because children often learn responsibility by caring for animals. These values are now embodied in broader social concerns for animal welfare, particularly with regard to industrial livestock systems.

Just one example of the strong sense of rural community spirit can be seen in *gotong royong,* Indonesia's traditional spirit of mutual help. The underlying philosophy is that people cannot live a solitary existence; they need each other, particularly family members and relatives. The practice of *gotong royong* originates from the traditional peasant subculture, which is characterized by subsistence farming, family-oriented grouping, and strong social interdependence. Even in commercial smallholder areas of Indonesia, *gotong royong* is still practiced widely in farming operations—including during land preparation, pest management, water management, weeding, and harvesting.

An example of the strong spiritual connection to cultivation is the growing of rice in Asia. For the Balinese, rice is much more than just a staple food; it is an integral part of the Balinese culture. The rituals associated with the cycle of planting, maintaining, irrigating, and harvesting continue to enrich the cultural life of Bali after thousands of years and despite the strong external cultural influences associated with tourism. Before planting, throughout the growing season, and at harvest, ceremonies are held and offerings are presented to Dewi Sri, the goddess of rice. In the middle of most rice fields, even far from villages, shrines are well maintained with flowers, fruit, and other offerings for Dewi Sri.

Gender-related cultural norms and practices also play an important role in the functioning of cultivated systems. For example, in the rural Philippines land is preferentially given to sons because rice farming requires intensive male labor (Estudillo et al. 2001). In many patrilineal African communities, the cultural custom of *lévirat* dictates that if a woman becomes a widow, she has to remarry one of her husband's brothers, which allows the woman continued access to land and food security; otherwise she would have to leave the family on the death of her husband (Estudillo et al. 2001).

Cultivated systems have a long history. Since as early as 10,000 years ago, crops have been carefully and deliberately managed by people, who in turn have reaped the benefits of increased food production. It has been argued that domesticated seed varieties and agricultural technologies were some of the most important factors in shaping the evolutionary course of civilizations (Diamond 1999). Domesticated, nutritious crops are capable of supporting larger populations, which in turn promotes innovation and technological advances. The social, cultural, economic, and political patterns and institutions that underlie both traditional rural societies and modern nation-states are in many ways products of humans' evolving ability to manage plants and animals for the production of food and other services (Diamond 1999).

Notes

1. If inland waters, Greenland, and Antarctica are excluded from this analysis, the coverage rises to approximately 27%. According to the MA definition, an area is considered cultivated if at least 30% of the underlying 1x1-kilometer land cover grid cell has been classified as cropland. This definition seeks to identify landscapes where a significant degree of ecosystem transformation has already taken place. The MA definitions of ecosystems allow for overlapping geographical extents of terrestrial systems.

2. The AVHRR-derived Global Land Cover Characterization Database V1.2 was produced by the EROS Data Center of the U.S. Geological Survey (Loveland et al. 2000) with revisions for Latin America (USGS EDC 1999). This dataset identifies approximately 200 seasonal land cover regions per continent (for example, 167 for South America and 205 for North America) based on the interpretation of a series of satellite images captured every 10 days from April 1992 to March 1993.

3. This conversion efficiency is embedded in the IFPRI IMPACT model, which was used to provide the food supply projections for the MA scenarios. (See MA *Scenarios*, Chapter 6). The conversion efficiency used in the IMPACT model was estimated by evaluating the impact of research investment on genetic improvement of major crops over the past 30 years. The recent evidence cited, however, suggests that conversion efficiencies have decreased markedly as average crop yields have increased.

References

Abbott, P., 2005: Agriculture and the WTO: creating a trading system for development. *Agricultural Economics, 32*(2), 211–219.

Adenew, B., 2003: *Market and Nonmarket Transfers of Land in Ethiopia: Implications for Efficiency, Equity, and Nonfarm Development.* Working Paper 2992. World Bank. Washington DC.

Ali, M., 1999: Evaluation of green manure technology in tropical lowland rice systems. *Field Crops Research, 61,* 61–78.

Allen, A.W., 1994: Agroforestry and wildlife: opportunities and alternatives. In: *Agroforestry and Sustainable Systems: Symposium Proceedings.* W.J. Rietveld (ed.), Tech. Coord. U.S. Department of Agriculture, Forest Service.

Allen, A. and W.M.W. Vandever, 2003: *A National Survey of Conservation Reserve Program (CRP) Participants on Environmental Effects, Wildlife Issues, and Vegetation Management on Program Lands.* Ed. USGS/BRD/BSR.-0001, 51 pp. Denver, CO, U.S. Government Printing Office. Biological Science Report.

Alston, J.M., C. Chan Kang, M.C. Marra, P.G. Pardey and T.J. Wyatt, 2000: *A Meta-Analysis of Rates of Return to Agricultural R&D: Ex Pede Herculem?* Research Report 113, International Food Policy Research Institute, Washington, DC. 148 pp.

Altieri, M., 2002: Agroecology: the science of natural resource management for poor farmers in marginal environments. *Agriculture, Ecosystems and Environment (in press), 1971,* 1–24.

Antle, J.M., and B. Diagana, 2003: Creating incentives for the adoption of sustainable agricultural practices in developing countries: the role of soil carbon sequestration. *American Journal of Agricultural Economics, 85*(5), 1178–1184.

Antle, J.M., S.M. Capalbo, S. Mooney, E.T. Elliott, and K.H. Paustian, 2001: Economic analysis of soil carbon sequestration: An integrated assessment approach. *Journal of agricultural and resource economics, 26*(2), 344–367.

Asian Development Bank, 1989: *Rainfed Agriculture in Asia and the Pacific.* Manila, Philippines: 644 pp.

Athukorala, P. and S. Jayasuriya, 2003: Food safety issues, trade and WTO rules: a developing country perspective. *The World Economy, 26*(9), 1395.

Bajwa, W.I. and M. Kogan, 2002: *Compendium of IPM Definitions (CID—What is IPM and how is it defined in the Worldwide Literature?* IPPC Publication No. 998, Integrated Plant Protection Center, Oregon State University, Corvallis, OR.

Barbier, E.B., 1997: The economic determinants of land degradation in developing countries. *Philosophical Transactions: Biological Sciences 352*(1356), 891–899.

Barker, R. and F. Molle, 2004: *Evolution of Irrigation in South and Southeast Asia.* Research Report No. 5. Comprehensive Assessment Secretariat. Colombo. Sri Lanka. 55 pp.

Barrett-Lenard, E.G., 2002: Restoration of saline land through revegetation. *Agricultural Water Management, 53* (1), 213–226.

Baudry, J., F. Burel, C. Thenail, and D. Le Coeur, 2000: A holistic landscape ecological study of the interactions between farming activities and ecological patterns in Brittany, France. *Landscape and Urban Planning,* 50(1–3), 119–128.

Bautista, R.M., H. Lofgren, and M. Thomas, 1998: *Does Trade Liberalization Enhance Income Growth and Equity in Zimbabwe? The Role of Complementary Policies.* IFPRI. Discussion Paper No. 32 International Food Policy Research Institute, Washington DC, Available at http://www.ifpri.org/divs/tmd/dp/tmdp32.htm.

Benetti, A.D., A.E. Lanna, and M.S. Cobalchini, 2004: Current practices for establishing environmental flows in Brazil. *River Research and Applications,* **20(4):** 427–444.

Beveridge, M.C.M. and M.J. Phillips, 1993: Environmental impact of tropical inland aquaculture. In: *Environment and Aquaculture in Developing Countries,* R.S.V. Pullin, H. Rosenthal, and J.L. Maclean (eds.), ICLARM Conference Proceedings 31, pp. 213–236.

Bhatia, A., H. Pathak, and P.K. Aggarwal, 2004: Inventory of methane and nitrous oxide emissions from agricultural soils of India and their global warming potential. *Current Science,* **87(3),** 317–324.

Birdlife International, 2004: *The State of the World's Birds 2004: Indicators for Our Changing World.* Birdlife International, Cambridge, UK, 73 pp.

Blench, R., 2001: Why conserve livestock biodiversity? In: *Living Off Biodiversity,* I. Koziell and J. Saunders (eds.), International Institute for Environment and Development, London.

Boers, T.M. and J. Ben-Asher, 1982: A review of rainwater harvesting. *Agricultural Water Management, 5,* 145–158.

Bouwman, A.F. and H. Booij, 1998: Global use and trade of feedstuffs and consequences for the nitrogen cycle. *Nutrient Cycling in Agroecosystems,* **52(2–3),** 261–267.

Bouwman, A.F., L.J.M. Boumans, and N.H. Batjes, 2002: Emissions of N2O and NO from fertilized fields: summary of available measurement data. *Global Biogeochemical Cycles,* **16(4),** Art. No. 1058.

Boyd, C.E., 1985: Chemical budget for channel catfish ponds. *Transaction of the American Fisheries Society,* **114:** 291–298.

Brandle, J.R., L. Hodges, X.H. Zhou, M.R. Rao, and L.E. Buck, 2004: Windbreaks in North American agricultural systems. In: *New Vistas in Agroforestry.* P.K.B. Nair (ed.), Kluwer, Dordrecht: 65–78.

Bronson, K.F., K.G. Cassman, R. Wassmann, D.C. Olk, M. van Noordwijk, and D.P. Garrity, 1998: Soil carbon dynamics in different cropping systems in principal ecoregions of Asia. pp. 35–57. In R. Lal, J.M. Kimble, R.F.Follett, and B.A. Stewart (eds.). *Management of Carbon Sequestration in Soil.* CRC Press. Boca Raton, FL.

Brookfield, H., 2001: *Exploring Agrodiversity.* Columbia University Press, New York, 608 pp.

Brown, J. and T. Loveland, 1998: USGS/EROS Data Center (EDC). Personal Communication.

Brown, R.A. and N.J. Rosenberg, 1997: Sensitivity of crop yield and water use to change in a range of climatic factors and CO2 concentrations: A simulation study applying IPIC to the central USA. *Agricultural and Forest Meteorology,* **83(3–4),** 171–203.

Bruijnzeel, L.A., 2001: Forest hydrology. In: J. Evans (ed.), *The Forests Handbook 1,* Blackwell Science, Oxford, pp. 301–343.

Bruins, H.J., M. Evenari, and U. Nessler, 1986: Rainwater-harvesting agriculture for food production in arid zones: the challenge of the African famine. *Applied Geography* **6(1):** 13–32.

Bruinsma, J. (ed.), 2003: *World Agriculture: Towards 2015/30—An FAO Perspective.* London, UK., Earthscan Publications Ltd., 432 pp.

Brush, S.B., J.E. Taylor, and M.R. Bellon, 1992: Technology adaptation and biological diversity in Andean potato agriculture. *Journal of Development Economics* 39,365–87.

Buchmann, S.E. and G.P. Nabhan, 1996: *The Forgotten Pollinators.* Island Press, Washington, DC, 292 pp.

Burton, C.H., J. Beck, P.F. Bloxham, P.J.L. Derikx, and J. Martinez (eds.), 1997: *Manure Management: Treatment Strategies for Sustainable Agriculture.* Silsoe Research Institute, Wrest Park, Silsoe, Bedford, UK.

Cabanilla, V.L., M. Hossain, and G.S. Khush, 1999: Diffusion of breeding materials and genetic composition of improved rice in South and Southeast Asia 1965–98. Paper presented at the IAEG Germplasm Research Impact Study Workshop in Nashville, Tennessee, 7–8 August 1999.

Cacho, O., R. Greiner, and L. Fulloon, 2001: An economic analysis of farm forestry as a means of controlling dryland salinity. *Australian Journal of Agricultural and Resource Economics,* **45(2):** 233–256.

Cai, Z.C., G.X. Xing, X.Y. Yan, H. Xu, H. Tsuruta, K. Yagi, and K. Minami, 1997: Methane and nitrous oxide emissions from rice paddy fields as affected by nitrogen fertilizers and water management. *Plant and Soil,* **196(1),** 7–14.

Cairns, M. and D.P. Garrity, 1999: Improving shifting cultivation in Southeast Asia by building on indigenous fallow management strategies. *Agroforestry Systems,* **47(1–3),** 37–48.

Cassman, K.G., 1999: Ecological intensification of cereal production systems: Yield potential, soil quality, and precision agriculture. *Proceedings of the National Academy of Sciences,* **96,** 5962–5959.

Cassman, K.G., A. Dobermann, D.T. Walters, and H. Yang, 2003: Meeting cereal demand while protecting natural resources and improving environmental quality. *Annual Review of Environmental Resources,* **28,** 315–358.

Cassman, K.G., A. Dobermann, and D.T. Walters, 2002: Agroecosystems, nitrogen-use efficiency, and nitrogen management. *Ambio,* **31,** 131–140.

Cassman, K.G., G.C. Gines, M.A. Dizon, M.I. Samson, and J.M. Alcantara, 1996: Nitrogen-use efficiency in tropical lowland rice systems: Contributions from indigenous and applied nitrogen. *Field Crops Research,* **47,** 1–12.

Castagnini, R. et al., 2004: *Comparing Land Reform and Land Markets in Colombia: Impacts on Equity and Efficiency.* Working Paper 3258. World Bank Washington DC. Available at http://econ.worldbank.org/working_papers/34462/.

Ceccarelli, S., 1994: Specific adaptation and breeding for marginal conditions. *Euphytica,* **77,** 205–219.

Ceccarelli, S., S. Grando, A. Amri, F.A. Assad, A. Benbelkacem, et al., 2001: Decentralized and participatory plant breeding for marginal environments. In H.D. Cooper, T. Hodgkin, and C. Spillane (eds.), *Broadening the Genetic Base of Crop Production.* CABI/IPGRI/FAO, 480 pp.

Chadwick, D.R. and B.F. Pain, 1997: Methane fluxes following slurry applications to grassland soils: Laboratory experiments. *Agriculture, Ecosystems and Environment,* **63(1),** 51–60.

Charlton, A.H. and J.E. Stiglitz, 2005: A development-friendly prioritisation of the Doha Round proposals. *The World Economy,* **28(3),** 293.

Chopra, K. and S.C. Gulati, 1997: Environmental degradation and population movements: the role of property rights, *Environmental & Resource Economics,* **9(4),** 383–408.

Clay, J., 2004: *World Agriculture and the Environment.* Island Press, Washington, DC, 570 pp.

Cleugh, H., R. Prinsley, R.P. Bird, S.J. Brooks, P.S. Carberry, et al., 2002: The Australian National Windbreaks Program: overview and summary of results. *Australian Journal of Experimental Agriculture,* **42(6),** 649–664.

Cleveland, D.A. and D. Soleri, 2002: *Farmers, Scientists and Plant Breeding—Integrating Knowledge and Practice.* CABI publishing, Wallingford UK. 368 pp.

Convention on Biological Diversity, 2000: Programme of work on agricultural biodiversity. Annex to Decision V/5 of the Conference of the Parties of the Convention on Biological Diversity.

Conway, G., 1999: *The Doubly Green Revolution: Food for all in the 21st Century.* Cornell University Press, Ithaca, NY, 360 pp.

Cornish, G.A. and N.C. Kielen, 2004: Wastewater irrigation- hazard or lifetime? Empirical results from Nairobi, Kenya and Kumasi, Ghana. In: *Wastewater Use in Irrigated Agriculture.* C.A. Scott, N.I. Faruqui, and L. Raschid-Sally (eds). CAB International, Wallingford, UK, pp. 69–79.

Costanza, R., R. D'Arge, R. de Groot, S. Farber, M.Grasso, B. Hannon, et al., 1997: The value of the world's ecosystem services and natural capita. *Nature,* **387,** 253–260.

Covas, G., 1989: Evolución del manejo de suelos en la región pampeana semi-árida . *Actas de las Primeras Jornadas de Ssuelos en Zonas Aridas y Semiaridas.* INTA, pp. 1–11. Santa Rosa, La Pampa.

Cromwell, E., D. Cooper, and P. Mulvany, 2001: Agricultural biodiversity and livelihoods: issues and entry points for development agencies. In I. Koziell and J. Saunders (eds.), *Living Off Biodiversity: Exploring Livelihoods and Biodiversity Issues in the Natural Resources Management,* International Institute for Environment and Development, London.

Crosson, P., 1995: *Soil Erosion and Its On-farm Productivity Consequences. What Do We Know?* Discussion Paper 95–29, Resources for The Future, Washington, DC.

Datt, G. and M. Ravallion, 1998: *Farm Productivity and Rural Poverty in India.* FCND Discussion Paper No. 42. International Food Policy Research Institute, Washington, DC.

Davies, W.J., S. Wilkinson, and B. Loveys, 2003: Stomatal control by chemical signaling and the exploitation of this mechanism to increase water use efficiency in agriculture. *New Phytologist,* **153(3),** 449–460.

Deen, T., 2000: *Development: Agriculture Workers Too Poor to Buy Food.* UN IPS, New York.

de Haan, C., H. Steinfeld, and H. Blackburn, 1997: *Livestock and the Environment, Finding a Balance.* European Commission Directorate-General for Development, 115 pp.

Deininger, K., R. Castagnini, and M. A. Gonzalez, 2004: Comparing land reform and land markets in Colombia: Impacts on equity and efficiency. *Policy Research Working Paper Series No 3258,* **World Bank,** Washington DC.

Delgado C., M. Rosegrant, H. Steinfeld, S. Ehui, and C. Courbois, 1999: *Livestock to 2020: The Next Food Revolution.* Food, Agriculture and the Environment. Discussion Paper 28. 2020 Vision. International Food Policy Research Institute. Washington, DC, 72 pp.

Delgado, L.C., J. Hopkins, and V.A. Kelly, 1998: *Agricultural Growth Linkages in Sub-Saharan Africa.* International Food Policy Research Institute Research Report No. 107. Washington, DC, International Food Policy Research Institute.

den Biggelaar, C., R. Lal, K.D. Wiebe, H. Eswaran, V. Breneman, and P. Reich, 2004: The global impact of soil erosion on productivity. *Advances in Agronomy,* **81,** 49–95.

De Plaen, R., 1997: *Hidden Links: Irrigation, Malaria and Gender.* International Development Research Council, Ottawa.

Devendra, C., 2000: Animal production and rainfed agriculture in Asia: potential opportunities for productivity enhancement. *Outlook on Agriculture,* 29161–29175.

Devendra, C., 1995: Mixed farming and intensification of animal production in Asia. In *Proceedings of FAO/ILRI Roundtable on Livestock Development in Low Income Countries.* Addis Ababa, Ethiopia: International Livestock Research Institute.

Devendra, C. and D. Thomas, 2002: Crop- animal interactions in mixed farming systems in Asia. *Agricultural Systems,* 7127–7140.

de Vries, J. and G. Toenniessen, 2001: *Securing the Harvest: Biodiversity, Breeding, and Seed Systems for African Crops.* CAB International, Wallingford, UK.

DFID (Department for International Development), 2002: *Better Livelihoods for Poor People: The Role of Agriculture.* London, 32 pp.

DFID, 1997: *Schistosomiasis Host Snail Control in Irrigation Night Storage Reservoirs.* Report no. 83. HR Wallingford Group Ltd. Wallingford.

Di, H.J. and K.C. Cameron, 2002: Nitrate leaching in temperate agroecosystems: sources, factors and mitigating strategies. *Nutrient Cycling in Agroecosystems,* **64(3),** 237–256.

Diamond, J., 1999: *Guns, Germs and Steel. The Fates of Human Societies.* W. W. Norton and Company, New York, 480 pp.

Dixon, J., A. Gulliver, and D. Gibbon, 2001: *Farming Systems and Poverty.* M. Hall (ed.), FAO and World Bank, Rome, 412 pp.

Dobbs, T. L. and J.N. Pretty, 2004: Agri-environmental stewardship schemes and "multifunctionality." *Review of Agricultural Economics,* **26(2),** 220–237.

Dobermann, A. and K.G. Cassman, 2002: Plant nutrient management for enhanced productivity in intensive grain production systems of the United States and Asia. *Plant Soil* **247:** 153–175.

Dobermann A., C. Witt, D. Dawe, G.C. Gines, R. Nagarajan, et al., 2002: Site-specific nutrient management for intensive rice cropping systems in Asia. *Field Crops Research* **74,** 37–66.

Döll, P. and S. Siebert, 1999: *A Digital Global Map of Irrigated Areas.* Report No. A9901, Centre for Environmental Systems Research, Germany, University of Kassel.

Drake, L., 1992: The non-market value of the Swedish agricultural landscape. *European Review of Agricultural Economics,* **19,** 351–364.

Dregne, H., M. Kassas, and B. Razanov, 1991: A new assessment of the world status of desertification. *Desertification Control Bulletin,* **20,** 6–18.

Duvick, D.N. and K.G. Cassman, 1999: Post-green-revolution trends in yield potential of temperate maize in the north-central United States. *Crop Science,* **39,** 1622–1630.

Edmeades, G.O., J. Bolanos, S.C. Chapman, H.R. Lafitte, and M. Banziger, 1999: Selection improves drought tolerance in tropical maize populations. *Crop Science,* **39(5),** 1306–1315.

Ehui, S. and J. Pender, 2005: Resource degradation, low agricultural productivity, and poverty in sub-Saharan Africa: pathways out of the spiral. In: D. Colman and N. Vink (eds.), *Reshaping Agriculture's Contributions to Society: Proceedings of the Twenty-Fifth International Conference of Agricultural Economists,* pp. 225–242.

Eicher, K., 2000: Institutions and the African Farmer. Third Distinguished Economist Lecture. International Maize and Wheat Improvement Center. *Issues in Agriculture* **14.** Consultative Group on International Agricultural Research, Washington, DC.

Ensik, J.H.J., T. Mahmoud, W. van der Hoek, L. Rashid-Sally, and R.P. Amerasingh, 2004: A nation-wide assessment of wastewater use in Pakistan: an obscure activity or a vitally important one? *Water Policy* **6**, 1–10.

Estudillo, J.P. et al., 2001: Gender differences in land inheritance, schooling and lifetime income: evidence from the rural Philippines. *Journal of Development Studies.* Available at http://static.highbeam.com/j/journalofdevelopment studies/april012001/genderdif ferencesinlandinheritanceschoolingandlife/.

European Centre for Ecotoxicology and Toxicology of Chemicals, 2004: *Ammonia emissions to air in western Europe.* Technical Report 62. Brussels.

European Union, 2000: *Organic Farming: Guide to Community Rules.* Directorate General for Agriculture. Belgium. ISBN 92–894–0363–2.

Evans, L.T., 1998: *Feeding the Ten Billion: Plants and Population Growth.* Cambridge University Press, Cambridge, UK.

Evenson, R.E. and D. Gollin, 2003: Assessing the impact of the Green Revolution, 1960 to 2000. *Science,* **300(5620),** 758–762.

Falk-Zepeda, J., G. Traxler, and R. Nelson, 2000: Surplus distribution from the introduction of a biotechnology innovation. *American Journal of Agricultural Economics,* **82,** 360–369.

Fan, H. and P. Hazell, 2001: Returns to public investments in the less-favored areas of India and China. *American Journal of Agricultural Economics,* **83(5),** 1217–1222.

Fan, S., L. Zhang, and X. Zhang, 2002: *Growth Inequality and Poverty in China.* Research Report No. 125. International Food Policy Research Institute, Washington DC.

Fan, S.G., P. Hazell, and S. Thorat, 2000: Government spending, growth and poverty in rural India. *American Journal of Agricultural Economics,* **82(4),** 1038–1051.

FAO (Food and Agriculture Organization), 2004: *The State of Food Insecurity in the World. Monitoring Progress towards the World Food Summit and the Millennium Development Goals.* Rome, 43 pp.

FAO, 2004a: *The State of Food and Agriculture 2003–2004: Biotechnology, Meeting the Needs of the Poor?* Rome.

FAO, 2003: *Review of the State of World Aquaculture.* Fisheries Circular No. 886, Rev 2. Fishery Resource Division. Rome.

FAO, 2002: *The State of World Fisheries and Aquaculture.* Rome, 150 pp.

FAO, 2000: *Report of the Conference on Aquaculture in the Third Millennium.* FAO Fisheries Report No. 661. Rome.

FAO, 1998: *The State of the World's Plant Genetic Resources for Food and Agriculture.* Rome, 208 pp.

FAO, 1997: Computer Printout of FAOSTAT's International Commodity Prices 1989–91. Personal communication via Technical Advisory Committee, CGIAR. Rome.

FAO, 1982: *Report of the Agro-Ecological Zones Project—Methodology and Results for Africa.* Rome.

FAO/IIASA (International Institute for Applied Systems Analysis), 2001: Global Agro-ecological Zoning [CD-ROM]. FAO Land and Water Digital Media Series #11. Global Agroecological Zoning.

FAO/SCBD (Secretariat of the Convention on Biological Diversity), 1999: W. Aarnink, S. Bunning, L. Collette, and P. Mulvany (eds.), *Sustaining Agricultural Biodiversity and Agro-ecosystem Functions: Opportunities, Incentives and Approaches for the Conservation and Sustainable Use of Agricultural Biodiversity in Agro-ecosystems and Production Systems.* FAO, Rome.

FAOSTAT, 2004: Food and Agriculture Organization of the United Nations, Statistical Databases, Online at http://faostat.fao.org.

FAOSTAT, 2003: Food and Agriculture Organization of the United Nations, Statistical Databases, Online at http://faostat.fao.org.

FAO/UNEP (United Nations Environment Programme), 2000: World watch list for domestic animal biodiversity. Available at http://www.fao.org/dad-is.

Farrington, J. and C. Boyd, 1997: Scaling-up the participatory management of common pool resources. *Development Policy Review,* **15,** 371–391.

Fearnside, P.M., 2001: Soybean cultivation as a threat to the environment in Brazil. *Environmental Conservation,* **28(1),** 23–38.

Fearnside, P.M., 1997: Transmigration in Indonesia: lessons from its environmental and social impacts. *Environmental Management* **21(4)** 553–570.

Fernandez-Cornejo, J. and S. Jans, 1999: *Pest Management in U.S. Agriculture.* Economic Research Service. U.S. Department of Agriculture. Agricultural Handbook No. 717. Washington, DC.

Forastieri, V., 1999: *Safework: The ILO Program on Occupational Safety and Health in Agriculture.* ILO. Geneva.

Fraser, D., 2001: The culture and agriculture of animal production. *Advances in Pork Production,* **12,** 17.

Frey, R.S., 1991: Accidental deaths among Kansas farmers. *Sociology and Social Research,* **75(4),** 210–212.

Gall, G.A.E. and G.H. Orians, 1992: Agriculture and biological conservation. *Agriculture, Ecosystems and Environment,* **42,** 1–8.

Galloway, J.N. and E.B. Cowling, 2002: Reactive nitrogen and the world: 200 years of change. *Ambio,* **31(2),** 64–71.

Gallup, J.L., 2002: *The Wage Labor Market and Inequality in Vietnam in the 1990s.* Working Paper 2896, World Bank, Washington, DC.

Gauchan, D., 2004: *Conserving Crop Genetic Resources On-farm: The Case of Rice in Nepal.* PhD Thesis. University of Birmingham, UK.

Ghassemi, F., A.J. Jakeman, and H.A. Nix, 1995: *Salinisation of Land and Water Resources: Human Causes, Extent, Management, and Case Studies.* Wallingford, CAB International.

Ghosh, B.C. and R. Bhat, 1998: Environmental hazards of nitrogen loading in wetland rice fields. *Environmental Pollution,* **102,** 123–126, Suppl. 1.

Giller, K.E., 2001: *Nitrogen Fixation in Tropical Cropping Systems,* 2nd ed. CAB International, Wallingford, 423 pp.

Giller, K. and C. Palm, 2004: Cropping systems: slash-and-burn cropping systems of the tropics. *Encyclopedia of Plant and Crop Science,* UK, CPL Scientific Publishing Services, Ltd., pp. 363–366.

Giller, K.E., M.H. Beare, P. Lavelle, A-M.N. Izac, and M.J. Swift, 1997: Agricultural intensification, soil biodiversity and agroecosystem function. *Applied Soil Ecology,* **6(1):** 3–16.

Gillespie, S. and L. Haddad, 2002: Food security as a response to AIDS. In: *AIDS and Food Security,* International Food Policy Research Institute, Washington, DC, pp. 10–16.

Gillingham, S. and P.C. Lee, 2003: People and protected areas: a study of local perceptions of wildlife crop-damage conflict in an area bordering the Selous Game Reserve, Tanzania. *Oryx,* **37,** 316–325.

Gleick, P., 2002: *The World's Water: The Biennial Report on Freshwater Resources 2002–2003.* Island Press, Washington, DC, 334 pp.

Gleick, P.H. (ed.), 1993: *Water in Crisis: A Guide to the World's Fresh Water Resources.* Oxford University Press, New York.

Green, R.E., S.J. Cornell, J.P.W. Scharlemann, and A. Balmford, 2005: Farming and the fate of wild nature. *Science* **307,** 550–555.

Greene, C. and A. Kremen, 2003: *U.S. Organic Farming in 2000–2001: Adoption of Certified Systems,* Agriculture Information Bulletin 780, Economic Research Service, US Department of Agriculture, Washington, DC, 55 pp.

Grimble, R. et al., 2002: *Poor People and the Environment: Issues and Linkages.* Policy Series 16, Natural Resources Institute, University of Greenwich, Chatham, UK.

Guthman, J., 1998: Regulating meaning, appropriating nature: the codification of California organic agriculture. *Antipode,* **30(2),** 135–154.

Habitu, N. and H. Mahoo, 1999: Rainwater harvesting technologies for agricultural production: a case for Dodoma, Tanzania. In *Conservation Tillage with Animal Traction,* ed. P.G. Kambutho, and T.E. Simalenga. Harare, Zimbabwe: A resource book of Animal Traction Network for Eastern and Southern Africa.

Hall, A.J., C.M. Rebella, C.M. Ghersa, and J. Ph. Culot, 1992: Field-crop systems of the Pampas. In: *Field Crop Ecosystems,* Series: Ecosystems of the World (C.J. Pearson Ed.). Elsevier Science Publishers B.V., Amsterdam.

Hariri, D., M. Fouchard, and H. Prud'homme, 2001: Incidence of soil-borne wheat mosaic virus in mixtures of susceptible and resistant wheat cultivars. *European Journal of Plant Pathology,* **107,** 625–631.

Harper, G.C. and A. Makatouni, 2002: Consumer perception of organic food production and farm animal welfare. *British Food Journal.* **104(3–5),** 287–299.

Hayami, Y. and V.W. Ruttan, 1985: *Agricultural Development.* 2nd ed. Johns Hopkins University Press, Baltimore, MD, 506 pp.

Hazell, P. and L. Haddad, 2001: *Agricultural Research and Poverty Reduction.* Food, Agriculture, and the Environment Discussion Paper No. 34. International Food Policy Research Institute, Washington, DC.

Hazell, P.B.S. and A. Roell, 1983: *Rural Growth Linkages: Household Expenditure Patterns in Malaysia and Nigeria.* Research Report No. 41, International Food Policy Research Institute, Washington DC.

Heisey, P.W., M.A. Lantican, and H.J. Dubin, 2002: *Impacts of International Wheat Breeding Research in Developing Countries, 1966–97.* International Maize and Wheat Improvement Center, Mexico.

Henao, J. and C. Baanante, 1999: *Estimating Rates of Nutrient Depletion in Soils of Agriculture Lands in Africa.* International Fertilizer Development Center, Muscle Shoals, AL.

Henderson, D., 1991: Opportunities for agroforestry in the Mid-South. In: Henderson D.R. (ed) *Proceedings: Mid-South Conference on Agroforestry Practices and Policies.* Winrock International Institute for Agriculture Development, West Memphis, AR, pp 85–100.

Henson, S. and R. Loader, 1999: Impact of sanitary and phytosanitary standards on developing countries and the role of the SPS Agreement. *Agribusiness,* **15(3).**

Herren, H.R. and P. Neuenschwander, 1991: Biological control of cassava pests in Africa. *Annual Review of Entomology,* **36,** 257–283.

Hooper, M., P. Mineau, M.E. Zaccagnini, and B.Woodbridge, 2003: Pesticides and international migratory bird conservation, in *Handbook of Ecotoxicology,* 2nd ed., D.J.Hoffman (ed.), Lewis Publishers, Boca Raton, FL, pp. 737–754.

Hopkins, J.W., G.D. Schnitkey, M.J. Miranda, B.G. Sohngen, and L.G. Tweeten, 2003: Farmers' Incentives to Conserve Soil Resources: A Dynamic Model Applied to the North-Central United States. In: *Land Quality, Agricultural Productivity, and Food Security.* Keith Wiebe (ed.). Edward Elgar Publishing Ltd., Northampton, MA, 461 pp.

Hossain, F., C.E. Pray, Y.M. Lu, J.K. Huang, and C.H. Fan, 2004: Genetically modified cotton and farmers' health in China. *International Journal of Occupational and Environmental Health,* **10,** 296–303.

Howarth, R.W., A. Sharpley, and D. Walker, 2002: Sources of nutrient pollution to coastal waters in the United States: Implications for achieving coastal water quality goals. *Estuaries,* **25(4B),** 656–676.

Howell, T.A., 2001: Enhancing water use efficiency in irrigated agriculture. *Agronomy Journal,* **93,** 281–289.

Huang, J., S. Rozelle, C. Pray, and Q. Wang, 2002: Plant biotechnology in China. *Science,* **295,** 674–677.

Huke, R.E. and E.H. Huke, 1997: *Rice Area by Type of Culture: South, Southeast and East Asia. A Revised and Updated Database.* International Rice Research Institute, Manila.

Hutchins, S.H., 1995: Free enterprise: the only sustainable solution to IPM implementation. *Journal of Agricultural Entomology,* **12,** 211–217.

IFOAM (International Federation of Organic Agriculture Movements), 2004: *The World of Organic Agriculture: Statistics and Emerging Trends–2004.* 6th ed.. M. Yussefi and H. Willer (eds.).

IICA (Inter-American Institute for Cooperation and Agriculture), 2004: *More Than Food on the Table: Agriculture's True Contribution to the Economy.* Interagency Group on Rural Development, San Jose, Costa Rica.

ILO (International Labour Organization), 2000: *Safety and Health in Agriculture,* Report VI (1) to 88th Session, 30 May–15 June 2000.

ILO, 1997: *World of Work,* No 22.

Ingco, M.D. and J.D. Nash, 2004: *Agriculture and the WTO: Creating a Trading System for Development.* World Bank and Oxford University Press, Washington DC.

IPCC (Intergovernmental Panel on Climate Change), 2000: *Land Use, Land Use Change, and Forestry.* R.T. Watson, L.R. Noble, B. Bolin, et al. (eds.), Cambridge University Press, Cambridge, UK.

IPCC, 1996: *Climate Change 1995: Impacts, Adaptations and Mitigation of Climate Change: Scientific-Technical Analyses.* Cambridge University Press, Cambridge, UK.

IPCC, 1992: *Greenhouse Gasses: Sources and Sinks,* Supplementary Report. Cambridge University Press, Cambridge, UK.

James, C., 2004: *Preview: Global Status of Commercialized Biotech/GM Crops: 2004.* ISAAA Briefs. No. 32. International Service for the Acquisition of Agri-biotech Applications, Ithaca, NY.

Jarvis, S.C., D.J. Hatch, B.F. Pain, and J.V. Klarenbeek, 1994: Denitrification and the evolution of nitrous-oxide after the application of cattle slurry to a peat soil. *Plant and Soil,* **166(2),** 231–241.

Kabutha, C., 1999: *The Importance of Gender in Agricultural Policies, Resource Access and Human Nutrition.* A Paper Presented to a Seminar on "Agricultural Policy, Resource Access and Nutritional Outcomes," Addis Ababa, November 3–5.

Kaimowitz, D., and A. Angelsen, 2000: Economic models of tropical deforestation: review. *Environment and Development Economics,* **6,** 147–153.

Karg, J. and L. Ryszkowski, 1996: Animals in arable land. In: *Dynamics of an Agricultural Landscape.* L. Ryszkowski, N. French, and A. Kedziora (eds.), Panstwowe Wydawnictwo Rolnicze i Lesne. Poznan: 138–172.

Kaufmann, R. and H. Fitzhugh, 2004: Importance of livestock for the world's poor. In: Colin Scanes and John Miranowski (eds.). *Perspectives in World Food and Agriculture 2004.* Iowa State Press.

Ke, B., 1998: Area-wide integration of crop and livestock: case study Beijing. In: Ho, Y. and Y. Chan (ed), *Proceeding of the Regional Workshop on Area-wide integration of Crop-Livestock activities.* FAO, Bangkok.

Kedziora, A. and J. Olejnik, 2002: Water balance in agricultural landscape and options for its management by change in plant cover structure of landscape. In: *Landscape Ecology in Agroecosystem Management.* L. Ryszkowski (ed). CRC Press, Boca Raton, FL, pp. 57–110.

Kenmore, P.E., 1996: Integrated pest management in rice. In: *Biotechnology and Integrated Pest Management.* Perseley, G.J. (ed.) Wallingford, UK, CAB International, 475 pp.

Kevan, P.G., 1999: Pollinators as bioindicators of the state of the environment: species, activity and diversity. *Agriculture, Ecosystems and Environment* **74,** 373–393.

Kevan, P.G. and T.P. Phillips, 2001: The economic impacts of pollinator declines: An approach to assessing the consequences. *Conservation Ecology,* **5(1),** Art. No. 8.

Kleijn, D., F. Berendse, R. Smit, N. Gilissen, J. Smit, B. Brak, and R. Groeneveld, 2004: Ecological effectiveness of agri-environment schemes in different agricultural landscapes in the Netherlands. *Conservation Biology,* **18(3),** 775–786.

Klohn, W.E. and B.G. Appelgren, 1998: Challenges in the field of water resource management in agriculture. In: *Sustainable Management of Water in Agriculture: Issues and Policies.* OECD Proceedings. Paris, OECD, p. 33.

Knox, A., R. Meinzen-Dick, and P. Hazell, 2002: Property rights, collective action and technologies for natural resource management: a conceptual framework. In R. Meinzen-Dick et al. (eds), *Innovation in Natural Resources: The Role of Property Rights and Collective Action in Developing Countries.* Johns Hopkins University Press for International Food Policy Research Institute, Baltimore, MD.

Kogan, M., 1998: Integrated pest management: Historical perspectives and contemporary development. *Annual Review Entomology,* **43,** 243–70.

Köppen, W., 1931: Klimakarte der Erde. *Grundriss der Klimakunde,* 2nd Ed., Berlin and Leipzig.

Kuznets, S., 1955: Economic growth and income inequality. *American Economic Review,* **45,** 1–28.

Labios, R.V., J.G. Motesur, and R.O. Retales, 1995: Alley cropping in sloping upland rice areas in the Philippines. In *Proceedings of Fragile Lives in Fragile Ecosystems.* International Rice Research Institute, Los Banos, Philippines, pp. 451–457.

Lal, R., 2004: Carbon sequestration in dryland ecosystems. *Environmental Management,* **33,** 528–44.

Lal, R., 2003: Soil erosion and the global carbon budget. *Environment International,* 29437–29450.

Lal, R., 2002: Soil carbon dynamics in cropland and rangeland. *Environmental Pollution,* **116(3),** 353–362.

Lal, R., 2001: Soil degradation by erosion. *Land Degradation and Development,* **12,** 519–539.

Lal, R., 1998: Mulching effects on runoff soil erosion, and crop response on alfisols in western Nigeria. *Journal of Sustainable Agriculture,* **11(2–3),** 135–154.

Lal, R., 1991: Tillage and agricultural sustainability. *Soil and Tillage Research,* **20(2–4),** 133–146.

Lal, R., R.F. Follett, and J.M. Kimble, 2003: Achieving soil carbon sequestration in the United States: A challenge to the policy makers. *Soil Science,* **12,** 827–45.

Langoni, H., 2004: Zoonoses and human beings. *J. Venom. Anim. Toxins incl. Trop. Dis.* **10(2),** 111.

Lawton, R.D., U.S. Nair, R.A. Pielke, and R.M. Welch, 2001: Climatic impact of tropical lowland deforestation on nearly montane cloud forests. *Science,* **294,** 584–587.

Leakey, R.R.B., in press: Domestication and marketing of novel crops for Ecoagriculture. In: *State of the Art of Ecoagriculture. Proceedings of International Ecoagriculture Conference and Practitioners Fair.* 27 September–1 October 2004, Nairobi, Kenya.

Leakey, R.R.B., 2001: Win:Win landuse strategies for Africa: 2. Capturing economic and environmental benefits with multistrata agroforests, *International Forestry Review,* **3,** 11–18.

Leakey, R.R.B., 1999: Agroforestry for biodiversity in farming systems, *Biodiversity in Agroecosystems,* W.W. Collins and C.O. Qualset (eds.), CRC Press, New York, pp. 127–145.

Leakey, R.R.B., 1996: Definition of agroforestry revisited, *Agroforestry Today,* **8(1),** 5–7.

Lewandrowski, J., M. Peters, C. Jones, R. House, M. Sperow, M. Eve, and K. Paustian, 2004: *Economics of Sequestering Carbon in the U.S. Agricultural Sector,* Technical Bulletin 1909, Economic Research Service, U.S. Department of Agriculture, Washington, DC.

Lightfoot, C., 1990: Integration of aquaculture and agriculture, a route to sustainable farming systems. *Naga, ICLARM Quarterly,* **13(1),** 9–12.

Lipton, M., 2005. *The family farm in a globalizing world: The role of crop science in alleviating poverty.* 2020 Discussion Paper No. 40. International Food Policy Research Institute. Washington, DC.

Lobell, D. and G. Asner, 2003: Climate and management contributions to recent trends in US agricultural yields. *Science,* **299,** 1032.

Lockie, S., K. Lyons, G. Lawrence, and K. Mummery, 2002: Eating 'Green': Motivations behind organic food consumption in Australia. *Sociologia Ruralis,* **42(1),** 23–40.

Loevinsohn, M., J.A. Berdegué, and I. Guijt, 2002: Deepening the basis of rural resource management: learning processes and decision support. *Agricultural Systems,* **73,** 3–22.

Lopez, R., 1998: Agricultural intensification, common property resources and the farm-household. *Environmental and Resource Economics,* **11(3–4),** 443–458.

Lopez, R. and A. Valdez (eds.), 2000: *Rural Poverty in Latin America: Analytics, New Empirical Evidence and Policy.* World Bank, Washington, DC.

Lowdermilk, W.C., 1953: *Conquest of the Land through 7000 Years,* USDA Soil Conservation Service, Washington, DC, 99.

Lugo, A.E. and S. Brown, 1993: Management of tropical soils as sinks or sources of atmospheric carbon. *Plant and Soil,* 14927–14941.

Lumbanraja, J., T. Syam, H. Nishide, A.K. Mahi, M. Utomo, and M. Kimura, 1998: Deterioration of soil fertility by land use changes in South Sumatra, Indonesia from 1970–1990. *Hydrological Processes,* **12(13–14),** 2003–2013.

Lutz, E., S. Pagiola, and C. Reiche, 1994: The costs and benefits of soil conservation: the farmers' viewpoint. *The World Bank Research Observer,* 9(2), 273–295.

Maclean, J.L., D.C. Dawe, B. Hardy, and G.P. Hettel, 2002 : Rice Almanac. 3rd ed. CABI Publishing, Wallingford, Oxon.

Maikhuri, R.K., S. Nautiyal, K.S. Rao, K. Chandrasekhar, R. Gavali, and K.G. Saxena, 2000: Analysis and resolution of protected area–people conflicts in Nanda Devi Biosphere Reserve, India. *Environmental Conservation,* **27(1),** 43–53.

Majumdar, D., 2003: Methane and nitrous oxide emission from irrigated rice fields: Proposed mitigation strategies. *Current Science,* **84(10),** 1317–1326.

Majumdar, D., S. Kumar, H. Pathak, M.C. Jain, and U. Kumar, 2000: Reducing nitrous oxide emission from an irrigated rice field of North India with nitrification inhibitors. *Agriculture, Ecosystems and Environment,* **81(3),** 163–169.

Mankin, P.C. and R.E. Warner, 1999: Responses of eastern cottontails to intensive row-crop farming. *Journal of Mammalogy* 80(3), 940–949.

Masouridou, S., P. Andriopoulos, S. Fokas, M. Kalkani, M. Dinosopoulou, G. Asimakopoulos, and M. Tsironi, 2003: *Presentation of Acute Brucellosis: A Review of 144 Cases.* Brucellosis 2003 International Research Conference. September 15–17, University of Navarra, Pamplona, Spain.

Matson, P.A., W.J. Parton, A.G. Power, and M.J. Swift, 1997: Agricultural intensification and ecosystem properties. *Science,* **277,** 504–509.

Maxwell, D. and K.D. Wiebe, 1999: Land tenure and food security: exploring dynamic linkages. *Development and Change,* **30(4),** 825–849.

McIntire, J., D. Bourzat, and P. Pingali, 1992: *Crop-Livestock Interactions in Sub-Saharan Africa.* World Bank, Washington, DC.

McNeely, J. and S.J. Scherr, 2002: *Ecoagriculture: Strategies to Feed the World and Save Wild Biodiversity.* Island Press, Washington, DC, 323 pp.

Meinzen-Dick, R., A. Knox, F. Place, and B. Swallow, 2002: *Innovation in Natural Resources Management: The Role of Property Rights and Collective Action.* Johns Hopkins University Press, Baltimore, MD, 336 pp.

Mellor, J.W., 2000: *Agricultural Growth, Rural Employment, And Poverty Reduction—Non-Tradables, Public Expenditure, And Balanced Growth.* Paper Prepared for the World Bank Rural Week 2000, Poverty or Prosperity: Rural People in a Globalized Economy.

Mellor, J.W., 1966: *The Economics of Agricultural Development.* Cornell University Press, Ithaca, NY.

Meng, E.C., 1997: *Land Allocation Decisions and In Situ Conservation of Crop Genetic Resources: The Case of Wheat Landraces in Turkey.* Ph.D. Dissertation, University of California, Davis, CA.

Merrington, G., L. Winder, R. Parkinson, and M. Redman, 2002: *Agricultural Pollution: Environmental Problems and Practical Solutions.* Spon Press, London and New York, 243 pp.

Mineau, P., 2003: Avian species. In: *Encyclopedia of Agrochemicals.* J. R. Plimmer, D. W. Gammon, and N. N. Ragsdale (eds.) John Wiley / Sons, Inc., 1970 pp.

Morris, M.L. and P.W. Heisey, 1998: Achieving desirable levels of crop diversity in farmers' fields: factors affecting production and use of commercial seed. In *Farmers, Gene Banks and Crop Breeding: Economic Analyses of Diversity in Wheat, Maize, and Rice.* M. Smale (ed.). Kluwer Academic Press and International Maize and Wheat Improvement Center, Boston. pp. 217–238.

Morris, M.L. and M.A. López-Pereira, 1999: *Impacts of Maize Breeding Research in Latin America, 1966–67.* International Maize and Wheat Improvement Center, Mexico.

Mosier, A.R., J.K. Syers, and J.R. Freney, 2004: *Agriculture and the Nitrogen Cycle: Assessing the Impacts of Fertilizer Use on Food Production and the Environment.* Scientific Committee on Problems of the Environment, SCOPE 65. Island Press, Washington DC.

Mosier, A.R., J.M. Duxbury, J.R. Freney, O. Heinemever, and K. Minami, 1996: Nitrous oxide emissions from agricultural fields: assessment, measurement and mitigation. *Plant Soil,* **181,** 95–108.

Myers, N.A., R.A. Mittermeier, C.G. Mittermeier, G.A.B. da Fonseca, and J. Kent, 2000: Biodiversity hotspots for conservation priorities. *Nature,* **403,** 853–858.

Nair, P., 1993: *An Introduction to Agroforestry.* Kluwer, Boston, MA, 520 pp.

National Research Council, 2000: *Clean Coastal Waters: Understanding and Reducing the Effects of Nutrient Pollution.* National Academy Press, Washington, DC, 428 pp.

Naylor, R.L., W. Falcon, R. Goodman, M. Jahn, T. Sengooba, H. Tefera, and R.J. Nelson, 2004: Biotechnology in the developing world: a case for increased investment in orphan crops. *Food Policy,* **29,** 13–44.

Neupane, R.P. and G.B. Thapa, 2001: Impact of agroforestry intervention on farm income under the subsistence farming system of the middle hills, Nepal. *Agroforestry Systems,* **53,** 31–37.

Ngugi, H.K., S.B. King, J. Holt, and A.M. Julian, 2001: Simultaneous temporal progress of sorghum anthracnose and leaf blight in crop mixtures with disparate patterns. *Phytopathology,* **91,** 720–729.

Niangado, O., 2001: The state of millet diversity and its use in West Africa. In H.D. Cooper, T. Hodgkin, and C. Spillane (eds.). *Broadening the Genetic Base of Crop Production.* CABI/IPGRI/FAO, 480 pp.

NRCS (National Resource Conservation Service), 2002: 2002 Farm Bill Conservation Provisions. [online] United States Department of Agriculture. Cited January 2005. Available at http://www.nrcs.usda.gov/programs/farmbill/2002/.

Nye, P.H. and D.J. Greenland, 1960: *The Soil under Shifting Cultivation.* Wallingford, CAB International, 156 pp.

Oldeman, L.R., 1998: *Soil Degradation: A Threat to Global Food Security,* ISRIC, Wageningen, Netherlands.

Olson, W.G.J., 1981: Archaeology–lessons on future soil use. *Journal of Soil and Water Conservation,* **36(5),** 261–264.

Olsson, G.A. and K. Ronninge, 1999: *Environmental Values in Norwegian Agricultural Landscapes.* Report prepared for the Ministry of Agriculture, Norwegian University of Science and Technology, Trondheim, 49 pp.

Omland, O., 2002: Exposure and respiratory health in farming in temperate zones–a review of the literature. *Annals of Agricultural and Environmental Medicine,* **9(2),** 119–136.

Ongley, E.D., 1996: *Control of Water Pollution from Agriculture,* Irrigation and Drainage Paper 55, Food and Agriculture Organization of the United Nations, Rome.

Orr, A. and M. Ritchie, 2004: Learning from failure: smallholder farming systems and IPM in Malawi. *Agricultural Systems,* **79,** 31–54.

Oweis, T., A. Hachum, and J. Kijne, 1999: *Water Harvesting and Supplementary Irrigation for Improved Water Use Efficiency in Dry Areas.* SWIM Paper 7. International Water Management Institute, Colombo, Sri Lanka.

Owens, L.B., 1994: Impacts of soil N management on the quality of surface and sub-surface water. In: *Soil Processes and Water Quality,* R. Lal and B.A. Stewart (eds.), Lewis Publishers, Boca Raton, FL, pp. 137–162.

Pacey, A. and A. Cullis, 1986: *Rainwater Harvesting: The Collection of Rainfall and Runoff in Rural Areas.* Intermediate Technology Publications, London.

Palm, C.A., T. Tomich, M. van Noordwijk, S. Vosti, J. Gockowski, J. Alegre, and L. Verchot, 2004: Mitigating GHG emissions in the humid tropics: case studies from the Alternatives to slash and Burn Program (ASB). *Environment, Development and Sustainability,* **6,** 145–162.

Pantuwan, G. S. Fukai, M. Cooper, S. Rajatasereekul, and J.C. O'Toole, 2002: Yield response of rice (*Oryza sativa* L.) genotypes to different types of drought under rainfed lowlands. *Field Crops Research,* **73(2–3),** 153–200.

Papadakis, J., 1966: Climates of the world and their agricultural potentialities. Published by the author, Buenos Aires.

Pardey, P.G. and N.M. Beintema, 2001: *Slow Magic: Agricultural R&D a Century after Mendel.* Food Policy Report, International Food Policy Research Institute, Washington DC, 30 pp.

Pattanayak, S.K. and D.E. Mercer, 2002: Indexing soil conservation: farmer perceptions of agroforestry benefits. *Journal of Sustainable Forestry,* **15(2),** 63–85.

Pattanayak, S.K., D.E. Mercer, E. Sills, and J.C. Yang, 2003: Taking stock of agroforestry adoption studies. *Agroforestry Systems,* **57(3),** 137–150.

Paustian, K., E. Levine, W.M. Post, and I.M. Ryzhova, 1997: The use of models to integrate information and understanding of soil C at the regional scale. *Geoderma,* 79227–79260.

Peng, S., J. Huang, J.E. Sheehy, R. Laza, R.M. Visperas, et al., 2004: Rice yields decline with higher night temperature from global warming. *Proceedings of the National Academy of Sciences,* **101,** 9971–9975.

Pielke, R.A., R. Avissar, M. Raupach, A.J. Dolman, X. Zeng, and A.A. Denning, 1998: Interactions between the atmosphere and terrestrial ecosystems: influence on weather and climate. *Global Change Biology,* **4,** 461–475.

Pielke, R.A., G.A. Daki, J.S. Snook., T.J. Lee, and T.G.F. Kittel, 1991: Non-linear influence of mesoscale land use on weather and climate. *Journal of Climate,* **4,** 1053–1069.

Pimbert, M., 1999: Sustaining the multiple functions of agricultural biodiversity. Background paper for the FAO/Netherlands Conference on the multifunctional character of agriculture and land. Food and Agriculture Organization of the United Nations.

Pingali, P.L. and Roger, P.A. (eds.), 1995: *Impact of Pesticides on Farmer Health and the Rice Environment.* International Rice Research Institute, Philippines.

Postel, S., 1999: *Pillar of Sand: Can the Irrigation Miracle Last?* W.W. Norton & Company. New York. 313 pp.

Postel, S., 1997: *Last Oasis: Facing Water Scarcity.* W.W. Norton & Company, New York, 239 pp.

Postel, S., 1993: Water and agriculture. In: P.H. Gleick, (ed.), *Water in Crisis: A Guide to the World's Fresh Water Resources.* Oxford University Press, Oxford, pp. 56–66.

Pretty, J.N., A.S. Ball, L. Xiaoyun, and N.H. Ravindranath, 2002: The role of sustainable agriculture and renewable resource management in reducing greenhouse gas emissions and increasing sinks in China and India. *Phil. Trans. Roy. Soc. Lond.* A **360,** 1741–1761.

Pretty, J.N., C. Brett, D. Gee, R.E. Hine, C.F. Mason, et al., 2000: An assessment of the total external costs of UK agriculture. *Agricultural Systems,* **62(2),** 113–136.

Qaim, M. and G. Traxler, 2005: Roundup Ready soybeans in Argentina: farm level and aggregate effects. *Agricultural Economics,* **32(1),** 73–86.

Quisumbing, A.R. and R. S. Meinzen-Dick, 2001: *Empowering Women to Achieve Food Security,* 2020 Focus 6, Brief 1 of 12, International Food Policy Research Institute, Washington, DC.

Rabalais, N.N., 2002: Nitrogen in aquatic ecosystems. *Ambio,* **31(2),** 102–112.

Raish, C. and A.M. McSweeney, 2003: *Economic, Social, and Cultural Aspects of Livestock Ranching on the Española and Canjilon Ranger Districts of the Santa Fe and Carson National Forests: A Pilot Study.* U.S. Department of Agriculture, Washington, DC.

Ramankutty, N., J.A. Foley, and N.J. Olejnickzak, 2002: People on the land: changes in population and global croplands during the 20th century. *Ambio,* 31251–31257.

Raschid-Sally, L., D. Doan Tuan, and S. Abayawardana, 2004: National assessments on wastewater use in agriculture and an emerging typology: the Vietnam case study. In: *Wastewater Use in Irrigated Agriculture.* C.A. Scott, N.I. Faruqui, and L. Raschid-Sally (eds.). CAB International, Wallingford, UK, pp. 81–90.

Reardon, T., C. Barrett, V. Kelly, and K.Savadogo, 1999: Policy reforms and sustainable agricultural intensification in Africa. *Development Policy Review,* **17** (4), 375.

Reeves, M. and Schafer, K.S., 2003: Greater risks, fewer rights: US farmworkers and pesticides. *International Journal of Occupational and Environmental Health,* **9(1),** 30–39.

Reiff, F.M., 1987: Health aspects of waste-water reuse for irrigation of crops. In: *Proceedings of the Interregional Seminar on Non-conventional Water Use in Developing Countries, 22–28 April 1985,* Series No. 22, United Nations, New York, pp. 245–259.

Reij, C., P. Mulder, and L. Begemann, 1988: *Water Harvesting for Agriculture.* Technical Paper No. 91. World Bank, Washington, DC.

Renault, D. and M. Montginoul, 2003: Positive externalities and water service management in rice-based irrigation systems of the humid tropics. *Agricultural Water Management,* **59(3),** 171–189.

Ribaudo, M., 1997: Water quality. In: *Agricultural Resources and Environmental Indicators, 1996–1997.* M. Anderson and R. Magleby (eds.), U.S. Department of Agriculture, Economic Research Service, Natural Resources and Environment Division, Washington, DC.

Richards, J.F., 1990: Land Transformation. In: *The Earth as Transformed by Human Action.* Ed. B.L. Turner, Cambridge University Press, New York, 729 pp.

Rickets, T.H., G.C. Daily, P.R. Ehrlich, and C.D. Michener, 2004: Economic value of tropical forest to coffee production. *Proceedings of the National Academy of Sciences,* **101(34),** 12579–12582.

Roe, T. and M. Gopinath, 1998: The "miracle" of U.S. agriculture. *Minnesota Agricultural Economist,* **69,** 1–4.

Röling, N.G. and M.A.E. Wagemakers (eds.), 1998: *Facilitating Sustainable Agriculture: Participatory Learning and Adaptive Management in Times of Environmental Uncertainty.* Cambridge University Press, Cambridge, UK, 344 pp.

Rosenzweig, C. and M. Parry, 1996: Potential impact of climate change on world food supply. *Nature,* **367,** 133–138.

Roubik, D.W. (ed.), 1995: *Pollination of Cultivated Plants in the Tropics.* FAO Agricultural Services Bulletin No. 118. FAO, Rome. 196 pp.

Runge, C.F., B. Senaur, P. Pardey, and M.W. Rosegrant, 2003: *Ending Hunger in Our Lifetime: Food Security and Globalization.* Johns Hopkins University Press, Baltimore, MD, 204 pp.

Ruttan, V., 2002: *Technology, Growth and Development: An Induced Innovation Perspective.* Oxford University Press, New York, 672 pp.

Ryszkowski, L., 2002: Integrity and sustainability of natural and man-made ecosystems. In: *Just ecological integrity.* P. Miller and L. Westra (eds), Rowman and Littlefield Publishers, New York.

Ryszkowski, L. and A. Kedziora, 1995: Modification of the effects of global climate change by plant cover structure in an agricultural landscape. *Geographia Polonica* **65:**5–34.

Ryszkowski, L., A. Bartoszewicz, and A. Kedziora, 1999: Management of matter fluxes by biogeochemical barriers at the agricultural landscape level. *Landscape Ecology,* **14,** 479–492.

Ryszkowski, L. Szajdak, and J. Karg, 1998: Effects of continuous cropping of rye on soil biota and biochemistry. *Critical Reviews in Plant Sciences,* **17,** 225–244.

Samson, F.B., F.L. Knopf, and W.R. Ostile, 2004: Great Plains ecosystems: past, present and future. *Wildlife Society Bulletin,* 326–15.

Sanchez, P.A., 2002: Soil fertility and hunger in Africa. *Science,* **295(5562),** 2019–2020.

Sanchez, P., 1995: Science in agroforestry. *Agroforestry Systems,* **30(1 and 2),** 5–55.

Sarris, A.H., 2001: The role of agriculture in economic development and poverty reduction: an empirical and conceptual foundation. Paper prepared for the Rural Development Department of the World Bank, Washington, DC.

Satorre, E., 2001: Production systems in the Argentine pampas and their ecological impact, In: *The Impact of Global Change and Information on the Rural Environment.* O.T. Solbrig, F. Di Castri, R. and Paarlberg (eds.), Harvard University Press, Cambridge, MA.

Scherr, S.J., 2000: A downward spiral? Research evidence on the relationship between poverty and natural resource degradation. *Food Policy,* **25,** 479–498.

Scherr, S.J., 1999: Soil degradation: A threat to developing country food security by 2020, 27; International Food Policy Research Institute, Washington, DC.

Schwarz, W. and D. Schwarz, 1999: *Ideas about food in the global culture of living lightly.* University of Essex. Available at http://www2.essex.ac.uk/ces/ConfsVisitsEvsGrps/LocalFoodSystems/localfoodw&ds. htm.

Scott, C.A., N.I. Faruqui, and L. Raschid-Sally (eds.), 2004: *Wastewater Use in Irrigated Agriculture: Confronting the Livelihood and Environmental Realities.* CAB International, Wallingford, UK, 206 pp.

Seckler, D., U. Amarasinghe, D. Molden, R. de Silva, and R. Barker, 1998: *World Water Demand and Supply, 1990 to 2025: Scenarios and Issues.* Research Report No 19. International Water Management Institute, Colombo, Sri Lanka.

Settle, W.H. et al., 1996: Managing tropical rice pests through conservation of generalist natural enemies and alternative prey. *Ecology,* **77(7),** 1975–1988.

Shapouri, H., J.A. Duffield, and M. Wang, 2003: The energy balance of corn ethanol revisited. *American Society of Agricultural Engineers,* **46(4),** 959–968.

Sharpley, A.N. and A.D. Halvorson, 1994: The management of soil phosphorus availability and its impact on surface water quality. In: *Soil Processes and Water Quality,* R. Lal and B.A. Stewart (eds.), Lewis Publishers, Boca Raton, FL, pp. 7–90.

Sheldrick, W., J.K. Syers, and J. Lingard, 2003: Contribution of livestock excreta to nutrient balances. *Nutrient Cycling in Agroecosystems,* **66,** 119–131.

Shiklomanov, I.A., 1996: *Assessment of Water Resources and Water Availability in the World.* Report for the Comprehensive Global Freshwater Assessment of the United Nations. State Hydrological Institute, St. Petersburg, Russia (February draft).

Simmonds, N., 1991: Selection for local adaptation in a breeding programme. *Theoretical and Applied Genetics,* **90,** 142–145.

Smale, M. (ed.), 2005: *Valuing Crop Biodiversity: On-farm Genetic Resources and Economic Change.* CABI International, Wallingford, UK.

Smale, M., 2001: Economic incentives for conserving crop genetic diversity on farms: issues and evidence. In: G.H. Peters and P. Pingali (eds.), *Tomorrow's Agriculture: Incentives, Institutions, Infrastructure and Innovations.* Proceedings of the Twenty-fourth International Conference of Agricultural Economists. Ashgate, VT, pp. 287–305.

Smale, M., M.R. Bellon, and J.A. Aguirre Gómez, 2001: Maize diversity, variety attributes, and farmers' choices in southeastern Guanajuato, Mexico. *Economic Development and Cultural Change,* 50201–50225.

Smil, V., 2000: Phosphorus in the environment: natural flows and human interferences. *Annual Review of Energy and the Environment,* **25,** 23–88.

Smil, V., 1999: Nitrogen in crop production: an account of global flows. *Global Biogeochemical Cycles,* **13(2),** 647–662.

Smith, B.D., 1998: *The Emergence of Agriculture.* New York, Scientific American Library, 231 pp.

Smith, S.V., W.H. Renwick, R.W. Buddemeier, and C.J. Crossland, 2001: Budgets of soil erosion and deposition for sediments and sedimentary organic carbon across the conterminous United States. *Global Biogeochemical Cycles,* **15(3),** 697–707.

Snyder, S.C., 2001: Hypoxia, fertilizer, and the Gulf of Mexico. *Science,* **292,** 1485–1486.

Soil Association, 2004: Soil Association Certification Limited. [online]. Cited January 2005. Available at UK. http://www.soilassociation.org/sacert. Accessed January 2005.

Solow, A.R., R.F. Adams, K.J. Bryant, D.M. Legler, J.J. O'Brien, B.A. McCarl, W. Nayda, and R. Weiher, 1998: The value of improved ENSO prediction to U.S. agriculture. *Climatic Change,* **39,** 47–60.

Sorby, K., G. Fleischer, and E. Pehu, 2003: *Integrated Pest Management in Development: Review of Trends and Implementation Strategies.* Agriculture and Rural Development Working Paper 5. World Bank, Washington, DC.

Sorensen, A.A., 1993: *Regional Producer Workshops: Constraints to the Adoption of Integrated Pest Management.* National Foundation for IPM Management Education, Austin, TX.

Soulé, M.J., A. Tegene, and K.D. Wiebe, 2000: Land tenure and the adoption of conservation practices. *American Journal of Agricultural Economics,* **82,** 4, 993.

Spalding, R.F. and M.E. Exner, 1993: Occurrence of nitrate in groundwater–A Review. *Journal of Environmental Quality,* **22(3),** 392–402.

Stallard, R.F., 1998: Terrestrial sedimentation and the carbon cycle: coupling weathering and erosion to carbon burial. *Global Biogeochemical Cycles,* **12(2),** 231–257.

Steffey, K.L., 1995: IPM today: are we fulfilling expectations? *Journal of Agricultural Entomology,* **12,** 183–190.

Stickney, R.R., 1994: *Principles of Aquaculture.* John Wiley, New York, 502 pp.

Stohlgren, T.J., T.N. Chase, R.A. Pilke, G.F. Kittels, and J.S. Baron, 1998: Evidence that local land use practices influence regional climate vegetation and stream flow patterns in adjacent natural areas. *Global Change Biology,* **4,** 495–504.

Stoorvogel, J.J. and E.M.A. Smaling, 1990: *Assessment of Soil Nutrient Depletion in Sub-Saharan Africa: 1983–2000.* Report 28. Winand Staring Centre, Wageningen, Netherlands.

Subramanian, A., N.V. Jagannathan, and R. Meinzen-Dick, 1997: *User Organizations for Sustainable Water Services.* Technical Paper 354. World Bank, Washington, DC.

Swift, M.J. and J.M. Anderson, 1999: Biodiversity and ecosystem function in agricultural systems. In: *Biodiversity and Ecosystem Function.* E.D. Schulze and H.A. Mooney (eds.), Berlin, Pringer-Verlag.

Swift, M.J. et al., 1996: Biodiversity and agroecosystem function. In: H.A. Mooney, J.H. Cushman, E. Medina, O.E. Sala, and E-D. Schulze (eds.), *Functional Roles of Biodiversity: A Global Perspective.* SCOPE/UNEP, Wiley, Chichester.

Tharme, R.E., 2003: A global perspective on environmental flow assessment: emerging trends in the development and application of environmental flow methodologies for rivers. *River and Research Applications,* **19(5–6),** 397–441.

The Economist, 2003: The Cancun challenge. September 4.

Thenail, C.B.J., 1996: Consequences on landscape pattern of within farm mechanisms and of land use changes. In: *Land Use Change in Europe and Its Ecological Consequences.* European Centre for Nature Conservation, Tilburg, pp. 242–258.

Tilman, D., K.G. Cassman, P.A. Matson, R. Naylor, and S. Polasky, 2002: Agricultural sustainability and intensive production practices. *Nature,* **418,** 671–677.

Timmer, P., 1997: *How Well Do the Poor Connect to the Growth Process?* CAER Discussion Paper 17, Harvard Institute for International Development, Cambridge, MA.

Timmer, P.C., 1989: *The Role of Agriculture in Employment Generation and Income Distribution in Asia and the Near East.* Harvard Institute for International Development, Cambridge, MA.

Townsend, A.R., R.W. Howarth, F.A. Bazzaz, M.S. Booth, C.C. Clevland, et al., 2003: Human health effects of a changing global nitrogen cycle. *Frontiers in Ecology and the Environment,* **1(5),** 240–246.

Tripathi, S.D. and B.K. Sharma, 2001: Integrated fish-horticulture farming in India. In: *Integrated Agriculture-Aquaculture.* FAO Fisheries Technical Paper 407, 149 pp.

Tsur, Y. and A. Dinar, 1995: *Efficiency and Equity Considerations in Pricing and Allocating Irrigation Water.* World Bank Working Paper 1460. World Bank, Washington DC.

UN (United Nations), 1997: *United Nations Comprehensive Assessment of the Freshwater Resources of the World.* Stockholm, Commission for Sustainable Development, Stockholm Environment Institute, section 37.

UN Hunger Task Force, 2005: *Halving Hunger: It Can Be Done.* Earthscan, London, 224 pp.

UNEP/CBD (United Nations Environment Program/Convention on Biological Diversity), 2004. *Indicators for Assessing Progress towards the 2010 Target: Trends in Genetic Diversity of Domesticated Animals, Cultivated Plants, and Fish Species of Major Socio-economic Importance.* Subsidiary Body on Scientific, Technical and Technological Advice. Montreal. Canada.

Unger, F., 2003: *Risk Associated with Bovine Brucellosis in Selected Study Herds and Market Places in Four Countries of West Africa.* Animal Health Working Paper 2. ITC: Banjul.

United Nations Population Fund, 2004: *State of World Population 2004: The Cairo Consensus at Ten – Population, Reproductive Health and the Global Effort to End Poverty.* New York.

Upton, M., 1997: Intensification or extensification: which has the lowest environmental burden? *World Animal Review,* **88(1),** 21–29.

Uri, N.D., 2001: Note on soil erosion and its environmental consequences in the United States. *Water, Air, & Soil Pollution,* **129(1–4),** 181–197.

USDA (United States Department of Agriculture), 2004: http://www.ams .usda.gov/nop/NOP/standards.htm. Accessed January 2005.

Van Dusen, M.E., 2000: *In Situ Conservation of Crop Genetic Resources in the Mexican Milpa System.* Ph.D. Dissertation, University of California, Davis, CA.

Vitousek, P.M., K. Cassman, C. Cleveland, T. Crews, C.B. Field, et al., 2002: Towards an ecological understanding of biological nitrogen fixation. *Biogeochemistry,* **57(1),** 1–45.

Viglizzo, E.F. and Z.E. Roberto, 1998: On trade-offs in low-input agroecosystems. *Agricultural Systems,* **56,** 253–264.

Viglizzo, E.F., Z.E. Pordomingo, M.G. Castro, and F.A. Lértora, 2002a: *The Environmental Sustainability of Agriculture in the Argentine Pampas.* Ediciones INTA, Buenos Aires, 84 pp.

Viglizzo, E.F., Z.E. Pordomingo, M.G. Castro, F.A. Lértora, and O.T. Solbrig, 2002b: *Environmental Sustainability of Argentine Agriculture: Patterns, Gradients and Tendencies 1960–2000.* The David Rockefeller Center for Latin American Studies, Harvard University. Working Papers on Latin America, Paper No. 01/02–2, Cambridge, MA.

Viglizzo, E.F., F.A. Lértora, A.J. Pordomingo, J. Bernardos, Z.E. Roberto, and H. Del Valle, 2001: Ecological lessons and applications from one century of low external-input farming in the pampas of Argentina, *Agriculture, Ecosystems & Environment,* **81,** 65–81.

Viglizzo, E.F., Z.E. Roberto, F. Lértora, E. López Gay, and J. Bernardos, 1997: Climate and land-use change in field-crop ecosystems of Argentina. *Agriculture, Ecosystems and Environment,* **66,** 61–70.

von Braun, J., 2003: *Agricultural Economics and Distributional Effects.* Presidential address at the 25th Conference of the International Association of Agricultural Economists, August 16–22, 2003, Durban, South Africa. Available at www.ifpri.org/pubs/speeches/200308vonbraun.htm.

Waggoner, P.E., 1994. *How Much Land Can Ten Billion People Spare for Nature?* Task Force Rep. 121. Council for Agricultural Science and Technology, Ames, IA.

Wannawong, S., G.H. Belt, and C.W. McKetta, 1991: Benefit-cost analysis of selected agroforestry systems in northeastern Thailand. *Agroforestry Systems,* **16(1),** 83–94.

Wassmann, R. and P.L.G Vlek, 2004: Mitigating greenhouse gas emissions from tropical agriculture: scope and research priorities. Environment, Development and Sustainability, **6(1–20),** 1–9.

Watkins, K., 2003: *Northern Agricultural Policies and World Poverty: Will the Doha Development Round Make a Difference?* Paper presented at the Annual World Bank Conference of Development Economics, 15–16 May, Washington, DC.

Webb, P., 2002: *Cultivated Capital: Agriculture, Ecosystems and Sustainable Development.* Discussion Paper No. 15. Tufts University, Boston.

Weltzein E., M. Smith, L.S. Meitzner, and L. Sperling, 1999: Technical and institutional issues in participatory plant breeding from the perspective of formal plant breeding – a global analysis of issues, results and current experience. CGIAR Systemwide programme on participatory research and gender analysis for technology development and institutional innovation. Working document number 3, Cali, Colombia.

Werner, A., F. Eulenstein, U. Schindler, L. Müller, L. Ryszkowski, and A. Kedziora, 1997: Grundwasserneubildung und Landnutzung. *Zeitschrift für Kulturtechnik und Landentwicklung,* **38,** 106–113.

WHO, 2004: Avian influenza: frequently asked questions. World Health Organization, Geneva.

WHO, 2005: Brucellosis. Fact Sheet No. 173. World Health Organization, Geneva.

WHO, 1993: *Guidelines for Drinking-Water Quality, Volume 1: Recommendations.* 2nd ed., World Health Organization, Geneva.

WHO and UNEP (World Health Organization and United Nations Environment Programme), 1990: *Public Health Impact of Pesticides Used in Agriculture.* Geneva, 85–89.

Wiebe, K., 2003: *Linking Land Quality, Agricultural Productivity, and Food Security.* Agricultural Economic Report 823, Economic Research Service, U.S. Dept. of Agriculture, Washington, DC, 63 pp.

Willock, J., I.J. Deary, M.M. McGregor, A. Sutherland, G. Edwards-Jones, O. Morgan, B. Dent, R. Grieve, G. Gibson, and E. Austin, 1999: Farmers' attitudes, objectives, behaviors, and personality traits: the Edinburgh study of decision making on farms. Journal of Vocational Behavior, **54(1),** 5–36.

Witt, C., K.G. Cassman, D.C. Olk, U. Biker, S.P. Liboon, M.I. Samson, and J.C.G. Ottow, 2000: Crop rotation and residue management effects on carbon sequestration, nitrogen cycling, and productivity of irrigated rice systems. *Plant Soil,* 263–278.

Wolfe, A.H. and J.A. Patz, 2002: Reactive nitrogen and human health: Acute and long-term implications. *Ambio,* **31(2),** 120–125.

Wood, D. and J.M. Lenne (eds.), 1999: *Agrobiodiversity: Characterization, Utilization and Management.* New York, CABI, Wallingford, 490 pp.

Wood, S., K. Sebastian, and S.J. Scherr, 2000: *Pilot Analysis of Global Ecosystems (Agroecosystems),* International Food Policy Research Institute and World Resources Institute, Washington, DC, 110 pp.

Woomer, P.L. and M.J. Swift, 1994: *The Biological Management of Tropical Soil Fertility.* John Wiley and Sons, New York, 252 pp.

World Bank. 2003: *World Development Indicators* [CD-ROM] World Bank, Washington, DC.

World Bank, 2000: *World Development Indicators* [CD-ROM] World Bank, Washington, DC.

Wortmann, C.S. and C.A. Eledu, 1999: Uganda's agroecological zones: a guide for planners and policy makers. Kampala, Uganda, Centro Internacional de Agricultura Tropical, 56 pp.

WRI (World Resources Institute), 1998: *World Resources 1998–99: A Guide to the Global Environment.* World Resources Institute, Washington, DC.

WTO (World Trade Organization), 2004: *Agriculture.* Available at http://www.wto.org/english/tratop_e/agric_e/agric_e.htm.

Wu, J., R.M. Adams, C.L. King, and K. Tanaka, 2004: From microlevel decisions to landscape changes: an assessment of agricultural conservation policies. *American Journal of Agricultural Economics.* **86(1),** 26–41.

Yang, S.S. and H.L. Chang, 2001: Methane emission from paddy fields in Taiwan. *Biology and Fertility of Soils,* **33(2),** 157–165.

Yang, S.S. and H.L. Chang, 1999: Diurnal variation of methane emission from paddy fields at different growth stages of rice cultivation in Taiwan. *Agriculture Ecosystems and Environment,* **76(2–3),** 75–84.

Yang, H.Z., Y.X. Fang, and Z.L. Chen, 2001: Integrated grass-fish farming systems in China. In: *Integrated agriculture-aquaculture.* FAO Fisheries Technical Paper 407, Rome, 149 pp.

Yanggen, D., D. Cole, C. Crissman, and S. Sherwood, 2004: Pesticide use in commercial potato production: reflections on research and intervention efforts towards greater ecosystem health in northern Ecuador. *EcoHealth,* **1(2),** SU72-SU83.

You, L. and S. Wood, 2004: *Assessing the Spatial Distribution of Crop Production Using a Cross-entropy Method.* Discussion Paper No 126. Environment and Production Technology Division, International Food Policy Research Institute, Washington, DC.

Zeller, M., and M. Sharma, 1998: The role of rural financial services for improving household food security. Concept, evidence, and policy implications. In: M. Schulz and U. Kracht (eds.). *Overcoming World Hunger: Challenges, Opportunities, and Strategies.* Münster, Westfalen: Lit Verlag.

Zhu, Y.Y., H.R. Chen, J.H. Fan, Y.Y. Wang, Y. Li, et al., 2000: Genetic diversity and disease control in rice. *Nature,* **406,** 718–722.

Zinkhan, F.C. and D.E. Mercer, 1997: An assessment of agroforestry systems in the southern USA. *Agroforestry Systems,* **35,** 303–321.

Chapter 27
Urban Systems

Coordinating Lead Authors: Gordon McGranahan, Peter Marcotullio

Lead Authors: Xuemei Bai, Deborah Balk, Tania Braga, Ian Douglas, Thomas Elmqvist, William Rees, David Satterthwaite, Jacob Songsore, Hania Zlotnik

Review Editors: Jerry Eades, Exequiel Ezcurra

*This appears in Appendix A at the end of this volume.

Main Messages

Urbanization and urban growth continue to be major demographic trends. The world's urban population increased from about 200 million (~15% of world population) in 1900 to 2.9 billion (~50% of world population) in 2000, and the number of cities with populations in excess of 1 million increased from 17 in 1900 to 388 in 2000. As people are increasingly living in cities, and as cities act as both human ecosystem habitats and drivers of ecosystem change, it will become increasingly important to foster urban systems that contribute to human well-being and reduce ecosystem service burdens at all scales.

Urbanization is not in itself inherently bad for ecosystems. Many ecosystems in and around urban areas are more biodiverse than rural monocultures are, and they can also provide food, water services, comfort, amenities, cultural values, and so on, particularly if they are well managed. Moreover, urban areas currently only account for about 2.8% of the total land area of Earth, despite containing about half the world's population.

Urban demographic and economic growth has been increasing pressures on ecosystems globally, but affluent rural and suburban living often places even more pressure on ecosystems. Dense urban settlement is considered to be less environmentally burdensome than urban and suburban sprawl are. At the same time, urban centers facilitate human access to and management of ecosystem services through, for example, the scale and proximity economies of piped water systems.

Urban development trends do pose serious problems with respect to ecosystem services and human well-being. Ecosystem processes that provide services to urban residents tend to be neglected as a result of the continued lack of understanding and appreciation of the complex processes involved, many of which take place at some distance from the urban consumers; the difficulties that private enterprises encounter in owning, trading, and negotiating over ecosystem services (and burdens), which rarely conform to property boundaries; the difficulties that public agencies encounter in managing and regulating ecosystem services, which also tend to cross administrative and sectoral boundaries; and the fact that the people most adversely affected by the loss of ecosystem services tend to be the least influential economically and politically (such as the urban poor, future generations, and residents living far from where the decisions are being made).

The problems documented in this chapter include severe environmental health problems within urban settlements resulting from inadequate access to ecosystem services (such as clean water), the degradation of ecosystems adjoining urban areas resulting from urban expansion and demands, and pressures on distant ecosystems resulting from urban production, consumption, and trade. In affluent countries, the historical trend has been for the negative impacts of urban settlements on ecosystem services and human well-being to become more delayed and dispersed. Urban developments in other parts of the world have been taking place in a different technological and economic context, with different environmental consequences, but this trend is still globally significant.

Interrelated problems involving local water, sanitation, waste, and pests still contribute a large share of the urban burden of disease, especially in low-income countries. This typically reflects a combination of degraded or increasingly scarce ecological services generated within the urban area, minimal infrastructure (such as water pipes) tapping more-distant ecosystem services, and differential access to the ecological and derived services available within the urban area.

Problems relating to the degradation of ecosystems adjoining urban settlements are undermining their capacity to supply ecosystems services, espe-

cially in and around large, industrializing, and motorizing cities (often of middle income). This typically reflects a combination of the geographic displacement of intraurban environmental scarcities and "wastes" and increasing levels of activities using or degrading ecosystems.

The consumption and production activities that are driving long-term, global ecosystem change are concentrated in urban centers, and especially those located in upper- and middle-income countries. This typically reflects a combination of environmental displacement and increasing rates of direct and indirect consumption of energy, materials, and ecosystem services.

Although there are many examples of successful responses to urban environmental challenges, concerted responses (such as sanitary reform in many cities in the nineteenth century) have been motivated by serious crises rather than by precautionary planning and have often succeeded by displacing rather than eliminating the environmental burdens. As an increasing number of urban environmental burdens, particularly from affluent urban centers, are likely to fall on future generations and are already of global scale (such as urban emissions of greenhouse gases), past successes based on displacing the burdens spatially are of declining relevance.

When urban systems are managed more equitably and the loss of ecosystem services is purposefully addressed, the benefits to human well-being can be substantial. It is *well established* that motivated governmental and nongovernmental agencies can implement radical changes in local environmental management that reduce local burdens and benefit vulnerable groups. The experience of urban sanitary reform provides strong evidence for this. At the same time, the historical record also illustrates a tendency to displace the burdens spatially and temporally.

The regulation of urban air and water pollution, in part motivated by popular pressures, has also resulted in better air and water quality in and around cities in some parts of the world, although again there is a tendency to displace the burdens spatially and temporally.

There is comparatively little evidence of significant steps to reduce the global ecosystem burdens of cities, although there are many examples of measures that, if replicated on a large scale, would result in appreciable reductions in those burdens.

27.1 Introduction

Urban systems are centered in urban areas; in terms of ecosystem services, urban areas are primarily sites of consumption. This contrasts with the other systems assessed in this report (such as cultivated systems, drylands, and coastal systems), which primarily generate and supply ecosystem services. Urban systems exist at several scales and can be identified with individual urban settlements or networks of such settlements.

Urban settlements are agglomerations of people and their activities; although urban areas may contain a wide variety of species, it is the humans that make them urban. About half the people in the world live in areas defined as urban (see Box 27.1 for definitions), up from less than 15% at the start of the twentieth century. Combined with population growth, this has meant an almost fifteenfold increase in the world's urban population, from 200 million in 1900 to 2.9 billion a hundred years later. Over the same period, the rural population more than doubled, increasing from 1.4 billion to 3.2 billion (United Nations 2002).

The share of Earth's land area that is urban is also growing, but it remains only about 2.8% (this figure is based on the urban-rural mapping described later). Urbanization has, however, in-

BOX 27.1

Defining Urban Areas and Populations

In line with other MA systems (such as cultivated, dryland, and mountain systems), urban systems are associated with particular spatial locations, in this case urban areas. Urban areas are in turn associated with urban settlements and populations. This Box focuses on how urban areas and populations are distinguished from rural areas and populations. There are no hard and fast rules on this; although conceptual clarity is important, it must also be recognized that the dividing line between urban and rural is inevitably somewhat arbitrary. For example, many people move regularly between locations classified as urban and rural.

It is generally agreed that urban agglomerations (cities and other urban centers) tend to have larger populations than rural agglomerations (villages) do, are more likely to be the site of large facilities (such as hospitals) and higher-level administrative functions (national or local government offices, for example), and create comparatively densely settled areas, with a higher share of built-up area. Furthermore, urban residents are less likely to work in agriculture and more likely to work in industry or services. It is also agreed that there are more and less urban lifestyles and cultures. There is no international agreement, however, on the defining characteristics of urban, nor are there any scientifically accepted criteria by which to identify urban areas and populations. Moreover, many urban researchers believe that the distinction between rural and urban is becoming less relevant (Cohen 2003) and that the boundary definitions are inevitably somewhat arbitrary.

Cut-off points for identifying urban areas or populations vary within the different criteria. Minimum population density criteria commonly range between 400 and 1,000 persons per square kilometer; minimum size criteria typically range between 1,000 and 5,000 residents; and maximum agricultural employment is usually in the vicinity of 50–75%. In each case, however, cut-off points outside these ranges can easily be found.

According to a recent report on world urbanization prospects (United Nations 2002), 109 of the 228 countries covered use an administrative criterion to distinguish urban from rural localities, and 89 of these use it as the sole criterion. Population size or density was used as a criterion in 96 countries, and as the sole criterion in 46. The administrative and population-based criteria are themselves different in different countries: for example, the lower limit above which a settlement was considered urban ranged between 200 and 50,000 persons.

There are also differences in the manner in which localities are identified and settlement populations calculated. Some countries report city populations on the basis of the boundaries of the city proper; others in terms of a metropolitan area; and still others in terms of a (usually larger) urban agglomeration. These differences have persisted for many years. In the 2001 revision of the U.N. report, about half the countries in the world used estimates based on "city proper," with most of the remaining claiming to be applying the concept of urban agglomeration.

Although country-specific definitions will remain central to defining and assessing urban centers for many years to come, the basis for a more uniform definition is emerging from work using remote sensing and geographical information systems. This chapter relies on two different delineations of urban to examine urban conditions and trends: one based on the country-specific definitions used by the United Nations, and the other based on a preliminary urban-rural split developed as part of a broader mapping and indicator exercise being undertaken by the Center for International Earth Science Information Network (Balk et al. 2004).

The country-specific definitions provide the basis for the statistics on historic urban populations and short-term projections, as well as a number of the descriptive statistics on urban population, such as the share with access to improved water sources. The geospatial estimates are used for the map in Figure 27.1 and for the Tables situating urban populations in relation to coastal zones, dryland, mountains, and other MA system categories. Although the geospatial estimates remain provisional and are unlikely to be adopted by national governments in the foreseeable future, they have a number of potential advantages, including better international comparability and local verifiability, as well as the ability to portray, for example, how sets of urban centers themselves are concentrated spatially.

Many economic and social characteristics once considered quintessentially urban are increasingly found among residents of what must demographically be classified as rural areas. Alternatively, many people living in large cities do not have access to what is sometimes considered defining "urban" infrastructure, such as piped water and sewerage. Such phenomena are important to recognize, but for the purposes of this chapter people are identified as urban or rural depending on their primary residence rather than their economic or sociocultural characteristics. Thus, for example, a Kansas farmer living in a rural area but with "a university degree, hooked up to the Internet and a fax machine, with a barn full of expensive machinery, who keeps strict accounts and sells his grain on the Chicago Mart" (Friedmann 2002) would be identified as a rural person with the accoutrements of an urban lifestyle rather than an urban person living in a rural location.

---

volved profound changes in human ecology, and urban land area is a very poor indicator of the ecological significance of urban systems.

As illustrated in Figure 27.1 (in Appendix A) (see also Table 27.6, later in this chapter), urban settlements are themselves concentrated regionally. Africa and Oceania have the lowest shares of area in urban systems, whereas Asia has the largest. Within regions, there is also a high degree of variation. The global map also indicates that urban areas tend to be on or near the coasts.

Human systems within urban boundaries are not functionally complete ecosystems. Urban areas have been described as the human equivalent of the livestock feedlot: a spatially limited area characterized by a large population of humans living at a high density and supported by biophysical processes mostly occurring somewhere else (Rees 2003). Large urban agglomerations, or cities, are intense nodes of energy and material transformation and consumption. However, the biologically productive part of the human ecosystem, which sustains both the human and the industrial metabolism of the city, is located primarily in rural areas, as well as in oceans and other uninhabited locations. Urban dwellers still rely on rural residents to transform and tap these rural ecosystems, and much human activity in rural areas responds to urban demands. Historically, the development of urban centers has been tightly bound up with changes in the surrounding ecosystems (Cronon 1992). Increasingly, urban systems are also linked to more distant ecosystems scattered across the globe.

Although it is the concentration of humans that makes an area urban, urban areas are also home to many nonhuman species. Many studies of urban ecosystems have focused on these species and their relations with each other and with nonliving components of the urban environment. (See Box 27.2.) These may include, for example, forest, wetland, or grassland ecosystems that exist in and around cities and towns. These ecosystems provide services to humans, such as recreation in urban parks and fresh

BOX 27.2
Chapter Topic: Ecosystems in Urban Systems, Urban Systems as Ecosystems, or Ecosystems and Urban Systems?

There is a significant literature on ecosystems within urban areas and an emerging literature on the ecology of urban areas, treating urban systems as ecosystems (Hejný et al. 1990; Platt et al. 1994; Pickett et al. 1997; Brennan 1999; Grimm et al. 2000; Pickett et al. 2001). Whereas the former is consistent with traditional ecosystems analysis, which has treated human activity as disturbing rather than constituting ecosystem dynamics, the latter is more consistent with the MA, which clearly situates humans within ecosystems (MA 2003).

This chapter does not restrict itself to examining the conditions and trends of ecosystem remnants in and around urban centers, nor does it rely on treating urban systems as ecosystems or treat all urban services as ecosystems services. Related to this, although it is accepted that urban systems have social as well as biophysical dimensions, ecosystems are understood to be biophysical systems, and the value of ecosystem ser-

vices is assumed to be distinct from the value intentionally added through the application of human labor. Thus the urban conditions and trends of primary concern in this chapter are biophysical; social conditions and trends that do not have clear relation to these biophysical conditions and trends are not examined in any detail, even when they have important implications for human well-being.

As the result, this chapter only covers some of the ways in which urban development affects human well-being. There is no mention of urban violence, for example, and urban inequality is only considered to the extent that it affects access to or pressures on ecosystem services. Similarly, there is no systematic treatment of the benefits that derive from urban manufacturing or commercial services, except to the extent that they rely upon ecosystem services, as defined in the glossary of this report.

water from nearby watersheds, some of which are difficult to appropriate from ecosystems that are farther away.

Understanding these ecosystems in urban areas can not only help with ecosystem management within urban areas, it can also help us understand how urban systems function more broadly (Berkowitz et al. 2003). From this perspective, not only are urban systems characterized by a varied landscape, consisting of a range of ecosystems and habitats, but the shift from rural to urban can be conceived of in terms of a series of gradients rather than a single threshold or boundary. Typical rural-urban gradients include not only increasing human population density and increasing shares of impermeable land cover, but also decreasing population density for many nonhuman species (McDonnell and Pickett 1991, 1997; Blair 1996; Natuhara and Imai 1996; Rolando et al. 1997; Denys and Schmidt 1998; Luck and Wu 2002) as well as changes in species diversity. The dynamics of ecosystem change in and around urban centers are also influenced by a number of features characteristic of how urban landscapes change, such as a high rate of introduction of alien species, high habitat diversity and fragmentation, and a high rate of (human-induced) habitat disturbance (Rebelee 1994; Niemela 1999).

Humans themselves are host to many microorganisms; changing urban settlement patterns influence the relations between humans and these microorganisms, some of which cause human diseases. Just as early shifts from hunting and gathering changed the infectious disease profile of early agriculturalists, changing patterns of human movement and settlement are still influencing infectious as well as noninfectious diseases (Anderson and May 1991; McNeill 1993). (See also Chapters 13 and 14.)

Shifts in the drivers of change in urban systems at very different scales often combine to create new challenges for humans. Urban development and trade, for example, enabled the epidemics that devastated Europe in the Middle Ages and introduced people in large parts of the world to infectious diseases they had never encountered and to which they were particularly susceptible (McNeill 1989). The affluent cities of the nineteenth century spearheaded industrialization and economic "modernization," but on average the people living in these cities lived shorter lives than did their rural contemporaries (Bairoch 1988; Woods 2003). The emergence and spread of HIV/AIDS toward the end of the twentieth century has itself been an urban-centered phenomenon, with large urban populations one of the conditions resulting in its emergence as a significant human disease and with a higher urban

than rural prevalence in sub-Saharan Africa, where there are indications that it is beginning to affect urbanization levels (Dyson 2003). Despite all these challenges and setbacks and the significant number of countries where life expectancies have declined in recent years (McMichael et al. 2004), on average people living in urban centers today live longer and healthier lives than ever before.

Urban areas are more spatially scattered than most other systems assessed in the MA. On the other hand, urban centers are closely interlinked. Indeed, geographers have long viewed urban systems not as individual settlements but as networks of urban centers connected by flows of capital, people, information, and commodities regionally (Armstrong and McGee 1985).

27.1.1 Classifying Urban Settlements: Size, Economic Status, and Location

Of the many ways of classifying urban settlements, three of the most common are population size, economic condition, and location. All three are used in this chapter, with the particular emphases as described in this section.

27.1.1.1 Urban Population Size

Table 27.1 summarizes the distribution of the world's population by the population class of urban settlement in the year 2000, according to both the 2001 revision of *World Urbanization Prospects* from the Population Division of the United Nations (United Nations 2002) and the urban-rural mapping undertaken by the Center for International Earth Science Information Network in support of the MA (Balk et al. 2004). As indicated, and contrary to some of the more inflated rhetoric about "exploding cities," more than half of the world's urban population lives in settlements of less than half a million inhabitants, and well under 10% live in cities of more than 10 million. Because of the higher densities of the latter, the share of urban land area accounted for by these large cities is considerably less than this. As described in later sections, however, even these figures reflect a continuing demographic shift toward urban living, and toward large cities in particular, that is having a profound effect on both the socioeconomic organization of the human world and the biophysical organization of the world as a whole.

Although settlement size is clearly related to the ability of an urban center to play certain roles, the size distributions of urban settlements show little evidence of clustering around certain sizes.

Table 27.1. Distribution of World Population by Size Class of Settlement, 2000 (CIESIN et al. 2004a, 2004b; United Nations 2002)

Size Class of Urban Settlement	UN	GRUMP[a]			UN	GRUMP[a]
	Total Population	Settlements	Population Density		Share of Total Population	
	(million)	*(number)*	*(persons per sq. km.)*		*(percent)*	
Urban area	2,862	2,828	24,176	770	47.3	46.7
10 million or more	225	426	23	2,192	3.7	7.0
5 to 10 million	169	265	39	1,571	2.8	4.4
1 to 5 million	675	729	353	1,223	11.1	12.0
500,000 to 1million	290	280	395	821	4.8	4.6
under 500,000	1,503	1,128	23,366		24.8	18.6
100,000 to 500,000		568	2,792	706		9.4
50,000 to 100,000		223	3,199	517		3.7
20,000 to 50,000		229	7,297	419		3.8
5,000 to 20,000		108	10,078	183		1.8
Rural area	3,195	3,224		25	52.7	53.3
Total	**6,057**	**6,052**		**46**		

[a] Global Rural-Urban Mapping Project.

Instead, in many countries there is an approximately log-linear relationship between a ranking of the sizes of urban settlements and their actual populations. This has come to be known as the rank-size distribution. When the slope is exactly negative one, this implies that the largest city is twice the size of the second largest, three times the size of the third largest, and so on. The claim that the slope does tend to be exactly one is sometimes referred to as Zipf's Law and appears to hold for some countries, such as the United States. More generally, although the slopes may vary over time and between countries, given comparable definitions and criteria for including urban centers, rank size distributions do generally conform to a log-linear relation (Brakman et al. 2001).

27.1.1.2 Urban Economic Conditions

Economic conditions are not quite such obvious features of an urban center as its population size and are less easy to define and measure.

Table 27.2 summarizes the distribution of urban population based on the World Bank's classification of low-, lower middle-, upper middle-, and high-income countries. In principle, national income accounts can be adapted to urban centers and used to calculate the urban equivalent of GNP and GDP per capita. Because the required statistics are often not available at the appropriate level, however, it is more common to refer to the per capita income of the country where the urban centers are located. This will tend to be lower than the average income of the urban population, because in most countries average incomes are higher in urban centers than in rural areas, and in some countries the disparities are very significant (Eastwood and Lipton 2000).

As indicated, urbanization levels are higher in higher-income countries, although even in low-income countries almost one third of the population lives in urban areas, and about 60% of the world's urban population lives in low- or lower middle-income countries, implying a national income of less than $3,000 per capita in 2001.

Table 27.2. Population in Urban Areas and Percentage of Total Population That Is Urban, 2000. Economies are divided according to 2001 GNI per capita, calculated using the World Bank Atlas method. The groups are low-income, $745 or less; lower-middle-income, $746–2,975; upper-middle-income, $2,976–9,205; and high-income, $9,206 or more. (Based on figures from www.world bank.org/data/countryclass/countryclass.html in May 2003; data on percentage urban based on United Nations 2002)

Economy Group	Urban Population	Share Urban
	(million)	*(percent)*
Low-income countries	718	31.2
Lower-middle-income countries	949	44.6
Upper-middle-income countries	365	77.0
Upper-income countries	731	79.2
Undesignated	99	41.7
World	**2,862**	**47.2**

27.1.1.3 Urban Location

In international statistics, urban location is conventionally summarized in terms of countries or regions. This typically means that coastal-zone settlements are combined with inland settlements, mountain with lowland settlements, and so on. Following a brief overview of the regional distribution, this section uses a newly constructed urban-rural database (Balk et al. 2004) to examine the distribution of urban populations in relation to the principal nonurban systems used in the MA. The bases for the system boundaries are summarized in Chapter 1. These systems are not mutually exclusive.

Table 27.3 provides a summary of urban population by continent, based on the regions used by the United Nations Population Division. The two lowest-income continents (Africa and Asia)

Table 27.3. Population in Urban Areas, and Share Urban, for World and Major Areas, 2000, Comparison of U.N. and GRUMP Statistics (CIESIN et al. 2004a, 2004b; United Nations 2002)

Region	UN		GRUMP[a]	
	Urban Population	Share Urban	Urban Population	Share Urban
	(million)	(percent)	(million)	(percent)
Africa	295	37.2	304	38.4
Asia	1,376	37.5	1,378	37.5
Latin America and Caribbean	391	75.4	352	67.9
Northern America	243	77.4	256	81.5
Europe	534	73.4	514	70.9
Oceania	23	74.1	22	70.8
World	**2,862**	**47.2**	**2,828**	**46.7**

[a] Global Rural-Urban Mapping Project.

are also the least urbanized. Whereas Latin America has conventionally been combined with Asia and Africa in discussions of "developing countries," its level of urbanization is comparable to that of Europe and North America. Furthermore, the overall and urban population densities in Latin America are comparable (26 and 656 persons per square kilometer, respectively) to those found in Europe (32 and 588) and North America (17 and 289). In contrast, sparsely populated Africa (27 persons per square kilometer overall) has urban areas as dense as those found in densely populated Asia (120 overall): both over 1,250 persons per square kilometer. Rural densities tend to be fairly consistent by continent with the exception of Asia, where population densities are more than four times greater than that of any other continent.

The coastal system is disproportionately more urban than other systems assessed in the MA. (See Table 27.4.) Population densities in both urban and rural areas are especially high in coastal areas because of the services available from coastal systems and the access to transportation. The population density of urban areas in the coastal zones is about 45% greater than the average density of urban areas globally.

The coastal zone is also the system with the greatest share of urban land area (10.2% globally), as indicated in the far-right columns in Table 27.4. Cultivated agricultural systems and inland water zones also have urban land areas that are higher than average. It is noteworthy that in addition these systems are more

densely populated than average. Coastal, cultivated, and inland water zones tend to support the world's largest cities, as shown in Table 27.5. Conversely, mountain, forest, and dryland systems tend to support smaller settlements than the other systems. However, as Table 27.4 also shows, coastal and inland water systems are less populated than cultivated zones and thus do not contain such large shares of urban dwellers.

Many of the differences between the MA systems globally are also evident within individual continents. (See Table 27.6.) In every continent except North America, for example, the highest shares of urban population and land are in the coastal zones; even in North America the figures for the coastal zone are well above the continent average. The differences are more accentuated in some continents than others, however. Thus, whereas populations of the coastal zones of Europe and North America are only slightly more urbanized than their continental averages (84% and 90% in the coastal zones, compared with 71% and 81% on average), in Asia and even more in Africa the differences are far more striking (56% and 72% in the coastal zones, compared with 37% and 38% on average). The net result, when combined with other factors, is that the two continents with the lowest shares of population living in urban areas have the greatest number of coastal urban dwellers per square kilometer of coastal zone.

Total urban population distribution also tends to reflect a region's underlying system characteristics. For example, more than

Table 27.4. Population Estimates, Densities, and Land Areas for MA Systems, by Urban and Rural (CIESIN et al. 2004a, 2004b)

System	Population				Population Density			Land Area			
	Total	Urban	Rural	Share Urban	Overall	Urban	Rural	Total	Urban	Rural	Share Urban
	(million)			(percent)	(persons per sq. km.)			(square kilometers)			(percent)
Coastal zone	1,147	744	403	64.9	175	1,119	69	6,538,097	664,816	5,873,281	10.2
Cultivated	4,233	1,914	2,309	45.3	119	793	70	35,475,983	2,412,618	33,063,350	6.8
Dryland	2,149	963	1,185	44.8	36	749	20	59,990,129	1,286,421	58,703,698	2.1
Forest	1,126	401	725	35.6	27	478	18	42,092,529	839,094	41,253,435	2.0
Inland Water	1,505	780	726	51.8	51	826	25	29,439,286	943,518	28,495,767	3.2
Mountain	1,154	349	805	30.3	36	636	26	32,083,873	548,559	31,535,242	1.7
World	**6,052**	**2,828**	**3,224**	**46.7**	**46**	**770**	**25**	**130,669,507**	**3,673,155**	**126,996,316**	**2.8**

Note: Population numbers for each ecosystem will not add to total as systems are not mutually exclusive. Island systems are excluded.

Table 27.5. Population and Share of Various Population Sizes in Urban Areas within Selected MA Systems, 2000 (CIESIN et al. 2004a, 2004b)

System	Urban Population	Urban Population by Settlement Size							
		5,000–20,000	20,000–50,000	50,000–100,000	100,000–500,000	500,000–1 million	1 million–5 million	5 million–10 million	10 million or more
		(thousand)							
Coastal zone	744,000	13,000 (1.7%)	28,000 (3.7%)	33,000 (4.4%)	112,000 (15.0%)	69,000 (9.2%)	196,000 (26.4%)	119,000 (16.0%)	175,000 (23.5%)
Cultivated	1,914,000	75,000 (3.9%)	175,000 (9.1%)	166,000 (8.6%)	411,000 (21.5%)	183,000 (9.5%)	484,000 (25.3%)	172,000 (9.0%)	249,000 (13.0%)
Dryland	963,000	39,000 (4.1%)	84,000 (8.7%)	88,000 (9.2%)	224,000 (23.3%)	111,000 (11.5%)	260,000 (27.0%)	71,000 (7.4%)	85,000 (8.9%)
Forest	401,000	22,000 (5.5%)	43,000 (10.7%)	37,000 (9.2%)	83,000 (20.8%)	41,000 (10.1%)	98,000 (24.3%)	26,000 (6.4%)	52,000 (12.9%)
Inland water	780,000	24,000 (3.1%)	49,000 (6.2%)	48,000 (6.1%)	151,000 (19.4%)	81,000 (10.4%)	193,000 (24.8%)	79,000 (10.2%)	154,000 (19.8%)
Mountain	349,000	21,000 (6.1%)	47,000 (13.3%)	35,000 (10.1%)	77,000 (22.1%)	34,000 (9.7%)	85,000 (24.4%)	25,000 (7.2%)	24,000 (7.0%)

Note: Urban population figures have been rounded to nearest million, therefore total population does not equal the sum of populations in all settlement sizes. Percent columns do not sum to 100. Island systems are excluded.

half of Africa's urban population lives in dryland or cultivated systems because these systems predominate in Africa, even though the total population of these systems is only about 20% urban. Similar patterns are observed globally. Further, predominating ecosystems—drylands in Africa, for example—may be home to Africa's largest cities, even though they tend to be less urban overall than other systems, simply by virtue of the size of the system (and the constraints imposed by political borders).

Although it is beyond the scope of this chapter to examine the differential impact of cities across the ecosystems they inhabit, which depends heavily on local conditions, monitoring urban locations in relation to ecosystems is a potentially important contribution to policy debate and decision-making. More attention needs to be paid to preventing or restricting urban growth where this threatens ecosystem services, such as in watersheds or ecologically fragile areas. It may also be possible to identify locations where cities can benefit more from ecosystem services. It should be kept in mind, however, that urban growth in a liberal market economy occurs where investors decide to locate job-creating enterprises rather than where planners decide that growth ought to occur. Political processes can influence urban development, but urban location is not itself a policy decision.

27.1.2 Urbanization Trends

During the twentieth century, the world's urban population increased almost fifteenfold, rising from less than 15% to close to half the total population. In most middle- and high-income nations, the majority of the population to live in and work in urban areas. Many aspects of urban change during the twentieth century were unprecedented, including the size of each region's urban population, the number of nations having predominantly urban populations and economies, and the size and number of very large cities. For Europe, North America, and parts of Latin America, the most rapid urban change was mostly in the first half of the

century; for most of the rest of the world, it was in the second half. During the past 50 years, most nations in Africa, Asia, and Latin America experienced rapid urban change, including cities whose population grew more than tenfold, and a growing share of the world's urban population and its largest cities have been in Africa, Asia, and Latin America. (See Table 27.7.) By 2000, most of the world's largest cities were, once again, found in Asia, not in Europe and North America. (See Table 27.8.)

Although there have been numerous cities with over 1 million inhabitants during the past 2,000 years, until recently they were rare (at most only one or two within the world at any one time), and they still existed within predominantly rural societies, except for a few city-states. Only in the late nineteenth century did London emerge as the first city with several million inhabitants; the megacities with 10 million or more inhabitants only emerged in the second half of the twentieth century. By 2001, there were 17 of these (United Nations 2002). Only with the industrial revolution did the increasing concentration of population (and production) in urban areas become commonplace.

Changes in urbanization levels were underpinned by large economic, social, political, and demographic changes. The main driver of urbanization (understood as an increase in the proportion of a population living in urban areas) is economic growth; in general, the most urbanized nations are those with the highest per capita incomes, and the nations with the largest increases in their levels of urbanization are those with the largest economic growth. Decolonization and the development of independent nation-states had large influences on urbanization levels for all of Africa and much of Asia, in part as controls on the rights of the inhabitants to live in or move to urban centers were dismantled, and in part because the building of the institutions for independent governments increased urban employment. In some countries, such as Australia, colonizing populations concentrated in urban settlements from the start, resulting in somewhat different patterns

Table 27.6. Urban Population and Land Percentages and Densities by Selected MA Systems and Continent (CIESIN et al. 2004a, 2004b)

System	Africa	Asia	Latin America	Oceania	Europe	North America	World
Share of population that resides in urban areas							
				(percent)			
Coastal	71.5	55.7	82.1	89.2	83.7	90.4	64.9
Cultivated	40.5	36.6	68.8	71.1	71.6	97.5	45.3
Dryland	43.5	37.7	67.0	54.2	67.4	88.2	44.8
Forest	22.7	23.2	58.9	47.0	56.2	69.3	35.6
Inland water	51.2	41.3	74.6	80.5	79.1	85.3	51.8
Mountain	21.7	23.0	57.8	12.4	47.9	66.1	30.3
Overall	38.4	37.5	67.9	70.8	70.9	81.5	46.7
Urban land as share of total land							
				(percent)			
Coastal	5.4	13.0	8.8	3.3	11.6	11.6	10.2
Cultivated	1.8	6.9	4.6	1.9	9.7	13.0	6.8
Dryland	0.6	3.0	2.7	0.1	5.0	4.1	2.1
Forest	0.5	2.6	1.2	1.0	1.9	4.2	2.0
Inland water	1.2	5.0	2.8	1.0	3.2	3.8	3.2
Mountain	1.1	1.6	2.7	0.4	1.7	1.8	1.7
Overall	0.8	3.5	2.6	0.6	3.9	4.7	2.8
Urban population density							
				(persons per square kilometer)			
Coastal	2,123	1,934	789	610	640	497	1,119
Cultivated	1,279	1,352	548	300	630	258	793
Dryland	1,200	1,034	541	159	522	265	749
Forest	997	956	685	300	387	206	478
Inland water	1,647	1,536	655	451	604	302	826
Mountain	810	879	746	191	387	154	636
Overall	1,278	1,272	656	427	588	289	770
Average population density							
				(persons per square kilometer)			
Coastal	160	451	83	23	89	64	175
Cultivated	56	255	36	8	85	34	119
Dryland	18	82	21	0	39	12	36
Forest	23	105	14	6	13	12	27
Inland water	37	185	25	6	25	13	51
Mountain	42	60	34	6	14	4	36
Overall	27	120	26	4	32	17	46
Share of urban dwellers in cities over 1 million							
				(percent)			
Coastal	56.1	69.6	54.9	67.7	50.9	79.0	65.3
Cultivated	49.8	47.5	40.6	37.0	44.0	55.5	47.0
Dryland	50.3	41.6	39.4	27.2	38.4	59.6	43.3
Forest	25.9	39.9	53.7	39.9	36.7	54.8	42.9
Inland water	54.6	56.7	45.9	56.0	46.1	61.3	54.4
Mountain	19.8	34.1	59.8	0.7	23.3	46.5	38.5
Overall	45.9	50.6	49.3	57.4	44.5	61.5	49.8

Table 27.7. Distribution of World's Urban Population by Region, 1950–2010 (Satterthwaite 2002, with statistics from United Nations 2002)

Region	1950	1970	1990	2000	Projection for 2010
Urban population			(million)		
Africa	32	82	197	295	426
Asia	244	501	1,023	1,376	1,784
Europe	287	424	521	534	536
Latin America and the Caribbean	70	164	313	391	273
Northern America	110	171	213	243	273
Oceania	8	14	19	23	26
World	751	1,357	2,286	2,862	3,514
Share of population living in urban areas			(percent)		
Africa	14.7	23.1	31.8	37.2	42.7
Asia	17.4	23.4	32.2	37.5	43.0
Europe	52.4	64.6	72.1	73.4	75.1
Latin America and the Caribbean	41.9	57.6	71.1	75.4	79.0
Northern America	63.9	73.8	75.4	77.4	79.8
Oceania	61.6	71.2	70.8	74.1	75.7
World	29.8	36.8	43.5	47.2	51.5
Share of world's urban population			(percent)		
Africa	4.3	6.1	8.6	10.3	12.1
Asia	32.5	37.0	44.8	48.1	50.8
Europe	38.3	31.3	22.8	18.7	15.3
Latin America and the Caribbean	9.3	12.1	13.7	13.7	13.4
Northern America	14.6	12.6	9.3	8.5	7.8
Oceania	1.0	1.0	0.8	0.8	0.8

of urban growth. In general, rapid demographic growth influenced growth rates for urban populations but had little influence on urbanization levels.

Although the general trend worldwide is toward increasingly urbanized societies, the aggregate statistics in Tables 27.7 and 27.8 obscure the great diversity in urban trends between nations (and how these change over time) and within nations, especially the large-population nations. Nations may have been urbanizing more slowly than anticipated in the past two decades because of poor economic performance, but this may not register in the official estimates of countries that have not had recent population censuses. Many of the world's largest cities have had significant decelerations in their population growth rates and have much smaller populations in 2000 than had been anticipated. In 1978, for example, the United Nations Population Division projected Mexico City's population in 2000 to be 31 million and that of São Paulo to be 26 million, whereas the population of both these cities in 2000 was estimated at 18 million (Satterthwaite 2002).

Most of the world's largest cities either have key global roles (the command-and-control centers for global or regional economies) or are centers linking large national economies with the global economy. The exceptions tend to be national capitals or former national capitals in large-population nations (such as Cairo, Lagos, and Delhi), and there are also locations that have major roles in the global economy without very large cities, as shown by Zurich and Silicon Valley. However, the world's large cities will increasingly be those that are successful in concentrating enterprises able to compete in the global economy, and the low-

and middle-income nations that urbanize most will be those with more successful economies (Satterthwaite 2002).

The low-income nations in Africa, Asia, and Latin America that do not have successful economies are unlikely to urbanize much unless civil conflict or famine drives people to urban centers. However, recent trends in Africa may appear to contradict this assessment. In the most recent U.N. Population Division figures for urban trends, it appears that most of sub-Saharan Africa continued urbanizing rapidly during the 1990s, despite very poor economic performance. However, a lack of reliable census data means that most urban population statistics for the region from 1990 onwards are based on projections from older census data. In many sub-Saharan African nations, there is only one census available between 1959 and the present; for many more, there are only two. In the most recent U.N. Population Division compendium of urban statistics (United Nations 2004), very few sub-Saharan African nations had new census data from the past 10 years. Thus it is likely that many sub-Saharan African nations urbanized much more slowly during the 1990s than the U.N. figures (based on projections) suggest.

Although most urbanization will take place in the nations with growing economies, this is likely to be less concentrated in very large cities than in the past, or at least less concentrated in what are today the world's largest cities. In successful economies with good transport and communications systems and increasingly competent local authorities outside the larger cities, new investment is often targeted outside the largest cities, and most large cities are also becoming more dispersed (McGee and Robinson

Table 27.8. Distribution of World's Largest Cities by Region over Time. The statistics for 2000 in this Table are an aggregation of national statistics, many of which draw on national censuses held in 1999, 2000, or 2001, but some are based on estimates or projections from statistics drawn from censuses held around 1990. There is also a group of countries (mostly in Africa) for which there are no census data since the 1970s or early 1980s, so all figures for their urban populations are based on estimates and projections. (Satterthwaite 2002; data for 1950 and 2000 from United Nations 2002; data for 1800 and 1900 from IIED database, drawn from various sources, including Chandler et al. 1974, Chandler 1987, and Showers 1979)

Region	1800	1900	1950	2000
Number of "million cities"				
Africa	0	0	2	35
Asia	1	4	31	195
Europe	1	9	29	61
Latin America and the Caribbean	0	0	7	50
Northern America	0	4	14	41
Oceania	0	0	2	6
World	2	17	85	388
Regional distribution of the world's largest 100 cities				
Africa	4	2	3	8
Asia	65	22	36	45
Europe	28	53	35	15
Latin America and the Caribbean	3	5	8	17
Northern America	0	16	16	13
Oceania	0	2	2	2
Average size of the world's 100 largest cities	187,000	725,000	2.1 million	6.2 million

1995). Within wealthier nations, urbanization has become obscured as increasing numbers of rural dwellers do not work in typical "rural" occupations such as forestry and farming, including those who commute to urban areas, those who are retired and live supported by pensions, and those able to work in rural areas because of advanced telecommunications systems (such as industrial or service enterprises located in greenfield sites in what are officially classified as "rural" areas or those who telecommute) (Pahl 1965).

27.1.3 Urban Systems and Ecosystem Services

The net flow of ecosystem services is invariably into rather than out of urban systems. These flows have increased even more rapidly than has urban population growth in recent centuries, and the average distance of these flows has increased substantially as well.

In the fourteenth century, Ibn Khaldûn could advise the planners of his day to locate new towns in well-protected locations with wholesome air, ample freshwater resources, and easy access to pastures for livestock, arable fields for grain, and forests for fuel (Khaldûn 1981). Few modern urban populations can still rely on local ecosystem services to meet their fuel, food, or water needs. The scale of the relationship between urban centers and ecosystem services has expanded, and while the global linkages may be especially evident for the more affluent cities, this expansion has been experienced by virtually all urban areas.

In light of these changing spatial relations, it is useful to distinguish among the linkages between urban systems and ecosystem services that exist within urban areas, between urban centers adjoining nonurban ecosystems, and between urban centers and distant ecosystems. Moreover, to appreciate the importance of relations between urban systems and ecosystem services, it is important to consider the negative as well as the positive effects that

urban systems can have on ecosystem services. Even if urban systems are not major producers of ecosystem services, urban activities can alter the supply of ecosystem services at every scale, from within to far beyond the bounds of the urban area itself.

Within urban areas, the primary issue from the perspective of human well-being is whether the urban settlements provide a healthy and satisfying living environment for residents. Urban development can easily threaten the quality of the air, the quality and availability of water, the waste processing and recycling systems, and many other qualities of the ambient environment that contribute to human well-being. Certain groups (such as low-income residents) are particularly vulnerable, and certain services (such as those not easily traded or brought in from outside—the recreational services provided through urban parks, for instance) are of concern to all urban dwellers. Moreover, even for easily traded products, local ecosystems can be important, especially for households that lack the monetary income to purchase imports. Agriculture practiced within urban boundaries, for example, contributes significantly to food security in urban Sub-Saharan Africa (Bakker et al. 2000).

The urban area and its surrounding region is a better scale at which to understand the relations between urban development and local ecosystems. People in urban areas have historically been heavily dependent on adjoining systems for food, clean water, waste disposal, and a range of other services. The intensity of interaction between an urban system and its surroundings tends to fall off with increasing distance. Interaction also tends to be more intense along certain corridors (such as rivers and roads) and within environmentally bounded areas, such as watersheds. Because most urban centers are growing in population and extent, the peri-urban areas where the systems adjoining urban systems are located are also undergoing a twofold transformation, with arable land coming under increasingly intense cultivation and both arable and nonarable land being increasingly built over to provide space for commercial, industrial, and residential establishments and for roads and parking facilities. The more populated an urban area is, the greater its influence is likely to be on surrounding areas, although other characteristics, such as industrial production levels and per capita incomes, can also be important (Hardoy et al. 2001). (Peri-urban is used in this chapter to refer to land around the edges of an urban area, either just within or beyond urban boundaries, where land use patterns are often in the process of changing from more rural (agriculture) to more urban (buildings).)

In order to capture all the ecosystem service flows into urban areas, it is also necessary to consider the global scale, as many of the ecosystem services contributing to urban well-being do not depend upon the condition of local environments and ecosystems. Many products and amenities used in urban areas, including food, are traded extensively, and their availability depends primarily on the purchasing power of local residents. By importing goods, urban consumers are effectively drawing on ecosystem services from other parts of the planet. The institutions and practices controlling the ecosystems of origin remain outside the political reach of urban consumers (although emerging exceptions include certification systems (Bass et al. 2001)). Affluent consumers (and producers) are also increasingly likely to be contributing to pollutants whose impacts are themselves spread out across increasingly large areas (Wackernagel and Rees 1996; McGranahan et al. 2001).

Table 27.9 presents problems relating to urban systems and ecosystem services that are potentially of critical importance to human well-being at each of these three scales. Superficially, these problems can seem to represent a temporal sequence from past problems (such as bad urban water and sanitation) toward modern and even future problems (such as climate change). However, there are many urban centers today that still have bad urban water and sanitation, and the more obvious description of the current state of affairs is that all sets of problems coexist, but with different severity in different parts of the world. Similarly, although super-

ficially there can seem to be a shift from issues involving the provisioning of private goods (such as water for home consumption) to those involving the provision of public goods (such as global climate stability), at their own scales all these issues involve externalities and public goods.

All these problems also involve issues of human well-being and social justice, with distinct spatial dimensions. Unhealthy and unpleasant living conditions involve the most vulnerable groups living in urban areas and the risks they face when local ecosystem services are lacking and alternatives are inaccessible. At the second scale, when urban development harms ecosystems in the surrounding region, there are more extensive issues of spatial injustice, although most of those affected are likely to be of the same nationality. The third scale involves burdens that urban activities impose on distant people and future generations by reducing their access to ecosystem services, either because these services are diverted to urban uses or because the ecosystems themselves are degraded: this raises international issues of spatial justice as well as issues of temporal justice.

As described in Box 27.3, the patterns described in Table 27.9 can also be represented in terms of stylized urban environmental transitions, displaying a historical tendency for more economically successful urban settlements to create more extensive and delayed environmental burdens. There is also a great deal of variation between urban centers of similar economic status, and no reason to assume that this stylized transition represents current or future urban development patterns. Currently, however, it is rare to find a very poor urban community that does not face serious environmental health hazards or a very affluent urban community that does not impose a large ecological footprint (as described later in this chapter).

27.2 Condition and Trends of Urban Systems and Ecosystems

This description of condition and trends builds on the spatial classifications presented earlier. The first section examines the condi-

Table 27.9. Priority Problems in Urban Systems and Ecosystem Services at Three Different Spatial Scales

Problem and Characteristics	Intra-Urban (Urban Systems as Human Habitats)	Urban-Region (Urban Systems and Their Biospheres)	Urban-Globe (Urban Systems and Global Ecosystems)
Priority problem identified	unhealthy and unpleasant living environments	deteriorating relations with adjoining ecosystems	excessive "ecological footprints"
Urban areas most closely associated with problems	low-income cities and neighborhoods	large, middle-income, industrial cities	affluent cities and suburbs
Indirect driving forces	demographic change, inequality; trade and development that ignores ecology of infectious diseases and urban ecosystem services	industrialization, motorization; trade and development that ignores impacts on adjoining ecosystems	material affluence, waste generation; trade and development that ignores global ecosystem impacts
Direct driving forces	inadequate household access to safe water, sanitation, clean fuels, land for housing	ambient air pollution, groundwater degradation, river pollution, resource plundering, land use pressures	greenhouse gas emissions, import of resource and waste intensive goods (linear vs. circular flows)
Negative impacts associated with problem	spread of infectious diseases, loss of human welfare and dignity	loss of natural ecosystem services, "modern" diseases, declining agroecosystem productivity	global climate change, loss of biodiversity, depletion of globally scarce natural resources
Temporal characterization of key processes	rapid	varied	slow
Example of historically relevant response	sanitary reform	pollution controls	sustainable cities?

BOX 27.3

The Relationship between Economic Conditions and Urban Environmental Burdens

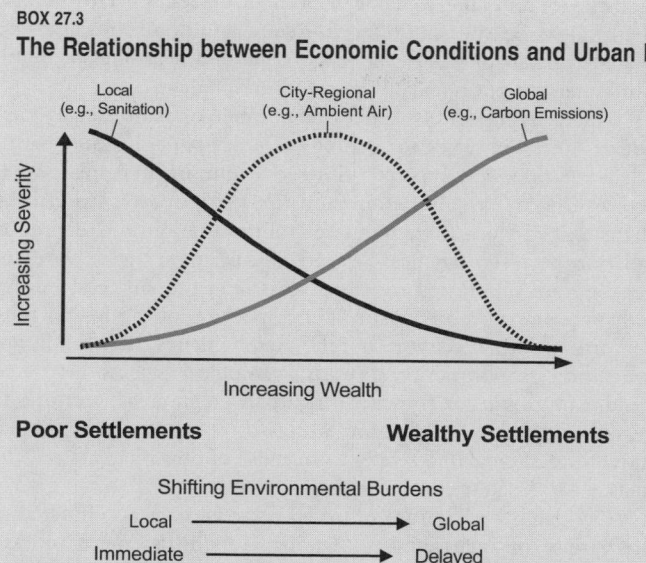

Shifting Environmental Burdens

Local ⟶ Global

Immediate ⟶ Delayed

Threaten Health Directly ⟶ Threaten Life Support Systems

(McGranahan et al. 2001)

This stylized diagram portrays local environmental health burdens declining with increasing urban affluence, global burdens increasing, and city-regional burdens first increasing and then declining (McGranahan et al. 2001). As with many ecological systems, the "big" processes are "slow" whereas the "small" processes are "fast" (MA 2003, pp. 114–17), with the result that the burdens associated with low-income settlements affect poor people in the present, whereas those associated with affluent settlements affect a more extensive public after a considerable delay.

A number of similar diagrams have been generated to describe urban "transitions". Smith and Lee (1993) presented an urban environmental risk transition in terms of two curves—one representing declining traditional risks (such as fecally contaminated water and food, or indoor air pollution due to smoky cooking fuels) and the other increasing modern risks (such as exposure to industrial pollution). Holdren and Smith (2000) more recently adapted the Figure reproduced here to present a risk transition that incorporates the risks arising from global threats such as climate change.

Bai and Imura (2000) and Bai (2003) presented similar curves to provide a conceptual illustration of a staged evolutionary trajectory of environmental problems in cities as they become more affluent and as their environmental burdens shift from poverty-related to production-related and finally to consumption-related challenges. Marcotullio (2004) and Marcotullio and Lee (2003) used the Figure as a basis for exploring the differences between western and rapidly developing Asian urban environmental transition histories, hypothesizing that transitions in rapidly developing countries are occurring sooner (at lower incomes), faster (increasing more rapidly over time), and more simultaneously (with greater overlaps among sets of environmental burdens) than those experienced by what are now industrial-world cities.

tion and dynamics of ecosystems in and around urban settlements. The second focuses on the condition and trends in the suitability of urban areas as habitats for humans. The third section examines how urban systems relate to adjoining ecosystems and their ecosystem services. The fourth and last section focuses on the global pressures that urban systems exert on ecosystems.

27.2.1 Condition and Dynamics of Ecosystems within Urban and Peri-urban Areas

Ecosystems in urban and peri-urban areas often display distinctive characteristics and dynamics that are neither of human design nor a hold-over from some rural past. Historically, urban planners have been inclined to focus on the purposefully designed and built components of urban systems and to ignore all but the most glaring changes in local ecosystems. Their environmental critics, on the other hand, have been inclined to focus on the loss of rural ecosystems and services. The combined effect has been the neglect of new opportunities that urban development creates for nonhuman species and ecosystems (not all of which benefit humans, of course).

27.2.1.1 Nonhuman Species and Ecosystems in Urban Areas

Although a concentration of humans is a defining feature of them, urban areas typically contain numerous opportunities for the persistence of native nonhuman species as well as the invasion or introduction of exotic species. Similarly, although a high share of built-over land is often used to identify areas as urban, urban land is far from impervious and often includes a range of different land use types, including gardens, grassland, wooded land, and agricultural land.

Urban and peri-urban ecosystems are heavily influenced by environmental change driven by people, but they also reflect the ability of plants, animals, and microorganisms to survive and exploit these changes. Urban construction and production typically conflict with wildlife and habitat conservation (Thompson 2003). They frequently result in the loss of critical wetland habitats, forest environments, and coastal sites and threaten many aspects of biodiversity. Urban development does not eliminate natural processes, however. It creates different opportunities and allows for new combinations of species through both introductions of exotic species to parks and gardens and the migration of invasive species. Many new opportunities are provided by the habitat shifts created by modification of biogeochemical cycles and the adjustment of micro- and local climates to make human life more comfortable, and some species are more able than others to exploit urban opportunities (see Rydell 1992 and Shapiro 2002 for examples relating to bats and butterflies).

Urban structures themselves provide opportunities for organisms—from the rats in the sewers to the birds nesting under the eaves of buildings. Spillages in factories, retail outlets, transport depots, and homes can create abundant food sources. Important mineral nutrients, such as calcium and magnesium, are common in building materials and find their way into urban soils (Bradshaw 2003). Parks, gardens, and zoos provide sites for a variety of plants and animals, including introduced and exotic species. Vacant and derelict sites are also colonized, and urban areas often include disused sites at various stages of succession. (See Table 27.10.)

The thermal properties of built-over land surfaces result in more solar energy being stored and converted to sensible heat (the heat energy stored in a substance as a result of an increase in its temperature), and the removal of shrubs and trees reduces the natural cooling effects of shading and evapotranspiration. The average ambient temperature in urban systems is generally 2–3 degrees higher than in nonurban systems, which can cause discomfort and even health risks in locations that are already warm

Table 27.10. Essential Steps in the Process of Natural Succession in Urban Areas (Bradshaw 2003)

Ecosystem Attribute	Processes Involved
Colonization by species	immigration of plant species
	establishment of those plant species adapted to the local condition
Growth and accumulation of resources	surface stabilization and accumulation of fine mineral materials
	accumulation of nutrients, particularly nitrogen
Development of the physical environment	accumulation of organic matter
	immigration of soil flora and fauna causing changes in soil structure and function
Development of recycling process	development of soil microflora and fauna
	possible difficulties in urban areas
Occurrence of replacement process	negative interactions between species by competition
	positive interaction by facilitation
Full development of the ecosystem	further growth
	new immigration, including aliens
Arrested succession	effect of external factors
	reduction of development
Final diversification	the city as a mosaic of environments
	high biodiversity as a result

or can lead to greater use of air-conditioning and attendant energy consumption. It can also exacerbate urban air pollution, alter rainfall patterns in and around urban centers, and change the composition of urban wildlife (see, e.g., Nowak et al. 2002). On the other hand, the heat island effect can attract warmth-seeking species to urban areas, and for people, too, more warmth is not always a disadvantage.

Within built-up areas, complex mosaics of land use emerge. The inner areas of many older North American cities, for example, have abandoned areas that may be totally neglected or derelict but that provide opportunities to create urban gardens that can be beneficial to local residents' well-being and health (Dinno 2000). In addition to abandoned spaces, there are numerous areas of vegetation that are planted or managed to some degree, ranging from roadside verges and canal and railway corridors to formal gardens, parks, urban woodlands, ponds, and lakes.

There is a trend to develop areas of more natural vegetation in cities, establishing urban nature reserves, such as the 1-hectare Camley Street Reserve adjacent to Kings Cross Station in London and the 1,215-hectare Rock Creek Park in Washington, D.C., which is 86% wooded. Preserved natural areas can become symbolic for cities, such as the 164-hectare Bukit Timah Reserve in Singapore and the 11-hectare Bukit Nanas Reserve in Kuala Lumpur. Other areas are totally ephemeral, being invaded by vegetation while awaiting development. The value of these ephemeral patches of vegetation depends on how long they are left

undisturbed and on the character of their substrates. Many derelict chemical works sites can offer unusually acid or alkaline sites that provide niches for plants associated with acid peat bogs or highly saline marshes not normally found in urban areas.

Collins et al. (2000) suggest a number of reasons why new human-imposed scales for ecological processes are found within urban areas. First, compared with ecosystems in rural areas, urban ecosystems are highly patchy and the spatial patch structure is characterized by a high point-to-point variation and degree of isolation between patches. Second, disturbances such as fire and flooding are suppressed in urban areas, and human-induced disturbances are more prevalent. Third, because of the higher temperatures in urban systems, in temperate climates there are longer vegetation growth periods. Fourth, ecological successions are altered, suppressed, or truncated in urban green areas, and the diversity and structure of communities of plants and animals may show fundamental differences from those of nonurban areas (Niemela 1999; Pickett et al. 2001).

The patchwork nature of urban ecosystems is accentuated by the variety of agencies, landowners, individuals, and businesses responsible for parcels of urban land, ranging from municipal parks and gardens departments, to public hospitals and educational institutions, to private individuals and corporations. Their differing goals and practices create diversity among these managed spaces. The urban environment is thus full of ecological discontinuities. Many species overcome these, simply surmounting obstacles—for example, the way urban foxes dash across main roads to get at other food sources. For some, however, migration corridors are important, and many planning strategies incorporate green corridors along streams or public utility easements. Preliminary results of investigations in Birmingham, England, suggest that the River Cole "wildlife corridor" does not enhance the number of wetland specialist species but it may act to increase and stabilize the number of habitat generalist species (Small 2000).

Compared with relatively simple temperate forest ecosystems, temperate industrialized agroecosystems, or tropical plantations, urban areas tend to be high in species richness as a result of the high habitat diversity of urban areas (Rebele 1994). However, some of the species richness is due to introduced species and is not always conducive to high levels of biodiversity at larger scales. An imported species that initially increases the species diversity within an urban area can, in certain circumstances, become an invasive species that reduces biodiversity in the surrounding areas. For example, on March 6, 1890, 40 pairs of the European Starling (*Sturnus vulgaris*) were released in New York's Central Park. Within a few years the starlings had spread from coast to coast, and they are now one of the most common birds in the United States, competing with native species (Kieran 1995; Mittelback and Crewdson 1997).

27.2.1.2 Contrasting Urban and Peri-urban Areas

In several studies, species diversity along an urban-rural gradient has tentatively been found to be hump-shaped in distribution, with the highest diversity in areas between rural areas and the urban core (Blair 1996; Blair and Launer 1997). The generality of this pattern across taxa has not been thoroughly investigated, nor have the mechanisms leading to such a pattern. It is, however, consistent with a more general observation that peri-urban areas are more varied and changing than are central urban or more distant rural locations.

In the heart of built-up areas, there is often a large share of fixed or long-term land uses. Although one building may replace another, comparatively few green spaces are built on, few new roads are created, and few new plants are introduced. It is in the peripheries and suburbs that the most rapid land use changes typically occur, usually with a loss of gardens and other open spaces and with increases in paved and roofed impermeable areas, as apartments or compound housing units replace single family dwellings and as retail land office buildings get larger. Sometimes new green spaces are created in such redevelopment projects, although these are often landscaped and maintained.

At the edge of the built-up area, large tracts of land are affected by transient uses. These peri-urban areas (except for many protected areas, which often include river valley and transportation corridors) undergo a change from rural to increasingly urban uses. In many North American and Australian peri-urban areas, the transition typically begins with the building of isolated houses for comparatively affluent townspeople. In much of Africa, Asia, and South America, on the other hand, migrants and a mix of long-standing residents frequently occupy peri-urban areas. Extremely poor people build temporary dwellings on any land from which they are not immediately evicted, provided there are employment opportunities. They may also cultivate food for themselves, hoping to sell the surplus at the roadside or in urban markets. (See Box 27.4.) Such land cover changes introduce not only plants but also a variety of waste and other materials to the local environment, which is likely to alter both the character and the dynamics of local ecosystems.

Alternatively, land-market economics can lead to agricultural land around the city being taken out of production while the owner waits for the price for urban uses to rise. In urbanizing rice-growing areas of Asia, this process can result in a patchwork of developed former rice fields, abandoned rice fields, and rice fields that are still being cultivated. Settlements and even regions may combine characteristically urban and rural features (McGee 1991).

With tight greenbelt planning regulations around many European cities (Hall 2002), the pace of peri-urban land cover change is often not as obvious as elsewhere in the world, such as in the United States, where it has been estimated that urban area has doubled since 1960 (Heimlich and Anderson 2001). In many rapidly growing cities in low- and middle-income countries, areas that were totally rural 10 years ago may be part of suburbia today. The peri-urban transition zone migrates out from the city almost relentlessly unless tight regulations are enforced or transport costs are high, as they were when many of the more compact cities developed (Bairoch 1988; Newman and Kenworthy 1999).

Rapid urban development also creates peri-urban demands for earth resources, especially aggregates and brick-making clays, and for land for disposal of wastes. Frequently, industries with high levels of pollution or hazardous wastes are also located in peri-urban zones, so that there is a high risk of contamination from industrial chemicals and toxic substances. Thus peri-urban zones may accommodate potentially conflicting activities. The juxtaposition of emissions of chemicals, disposal of waste, and peri-urban agriculture can lead to many health hazards. Vermin from waste dumps can be a threat to crops, domestic animals, and human beings.

Where urban settlements are themselves combining into large conurbations, the distinction between urban and peri-urban areas often ceases to be meaningful. In a multicentered agglomeration, for example, an area near the middle of the agglomeration may be peripheral to several sub-centers and retain at least some features characteristic of peri-urban areas. Green areas often remain between the original major towns, particularly where there is a history of greenbelt designation. At the same time, the inner areas of the city show successive waves of building, demolition, and rebuilding as needs change and as industry and business activity grows and declines.

27.2.1.3 Ecosystem Services in Urban and Peri-urban Areas

Whereas urban development is driven by deliberate human activity, most of the ecosystem changes that occur in and around cities are unintentional. These changes affect the supply of ecosystem services, including the regulation and ecology of human diseases.

By the time a given area is urbanized, most pre-existing ecosystems are likely to be severely disrupted, if not entirely transformed. Even the ecosystems associated with lands that remain comparatively undisturbed are likely to be altered by the habitat fragmentation and pollution that typically accompany urbanization (Bradshaw 2003). New opportunities for native species may arise, but in a different context and with potentially important implications for local ecosystem dynamics. Newly introduced species find opportunities beyond their area of introduction and compete with native species.

In urban areas and their margins, ecosystems can provide an especially wide range of services. The most widely recognized services are associated with green spaces and are recreational and cultural. Parks have become a central part of the identity of many urban centers, and greenbelts are an increasingly accepted means of providing outdoor recreation facilities for urban dwellers. In the MA sub-global assessment for Stockholm, 10 potential urban ecosystem services in National City Park have so far been examined, including not only recreational and cultural values but also air filtration, regulation of microclimate, noise reduction, surface-water drainage, nutrient retention, genetic library, pollination, seed dispersal, and insect pest regulation (Bolund and Hunhammar 1999). The sustainable supply of these ecosystem services depends not only on the presence of the parkland but also on the resilience of the ecosystems that provide them. This resilience could be undermined by insufficient conservation of parkland and by increased fragmentation. Alternatively, many of these services are also provided to at least some degree by non-park land, even if a park is one of the urban sites where the scope for managing and enhancing the value of these services is greatest (Elmqvist et al. 2004).

Just as not all ecosystem services arising from urban ecosystems are from green spaces, not all green spaces are ecologically beneficial. Urban and suburban lawns, for example, provide recreational services to their homeowners, aesthetic value to the neighborhood, and a number of other ecosystem services. However, large quantities of water, fertilizer, and pesticides are applied to maintain the aesthetics of the green lawns, especially in affluent countries, with numerous adverse consequences (Robbins et al. 2001). Indeed, fertilizers and pesticides are applied more intensively to lawns in the United States than to arable lands in large parts of the world (Robbins et al. 2001).

The lack of some ecosystem services in urban systems makes them more valuable. For example, the at-field value of urban agricultural produce is greater than that produced elsewhere because it does not need to be transported so far to reach the consumer. Some ecosystem services in urban areas can become so degraded through overuse that changes that would otherwise increase service delivery are of no avail. For example, once groundwater is no longer used for drinking purposes due to low quality, a further loss in the capacity of local ecosystems to filter and clean the water is less directly relevant. The result is the importation of the service from other ecosystems (in this case, through piped or bottled water), often at higher overall cost. Moreover, high population densities and the fact that some of these services provide spatially

---

BOX 27.4

Urban Agriculture, Vulnerability, and Recycling

For many of today's urban dwellers, urban agriculture provides an important source of food and supplementary income, especially in times of economic crisis. Although urban agriculture is associated with environmental health risks, it also has many environmental advantages and can help to provide a range of ecosystem services within urban areas.

There are no reliable estimates of the land used, the labor applied, or the outputs produced by urban agriculture. This is not surprising. The extent of urban agriculture is particularly sensitive to where urban boundaries are drawn, because a large share is located on the margins of urban areas. More centrally located agriculture is often spatially scattered and involves a large number of small plots (or even pots and pools) and animals (domestic fowl, for instance) that are difficult to identify, let alone monitor. In many cities, agriculture is officially banned, further complicating any attempt to collect reliable statistics.

There is, however, a growing body of research detailing the importance of urban agriculture in particular locations and for particular groups (United Nations Development Programme 1996; ETC—Urban Agriculture Programme 2001). This research suggests that urban agriculture can provide a number of major benefits: income and food security for producers; employment for under- or unemployed residents; lower prices for urban consumers; environmental improvements such as reduced runoff; and avoided costs of wastewater treatment and solid waste disposal. Urban farming takes place not only in peri-urban fields but also on rooftops, in backyards, in community vegetable and fruit gardens, and on unused or public spaces. It produces high-value products like fruit, vegetables, and fish, staples such as cassava, maize, and beans, and supplementary products such as berries, nuts, herbs, and spices. Urban agricultural enterprises range from highly commercialized operations to small informal and occasional enterprises. These latter operations are typically managed by long-term urban residents, by the moderately poor, and often by women.

From the perspective of current human well-being, the most significant contribution of urban agriculture probably does not lie in the share of overall agriculture production for which it accounts, but in the food security and supplementary income it can provide to cash-strapped urban residents, and to women in particular. In response to the economic deprivations of recent decades, urban agriculture in sub-Saharan Africa provided an important safety net for those who could find the land (Maxwell 1999).

Urban agriculture reportedly grew rapidly in many African cities (Howorth et al. 2001; Page 2002; Bryld 2003), largely as an informal activity involving either on-plot cultivation in more densely settled areas or off-plot cultivation on urban peripheries and marginal lands (Rogerson 1995). The urban farmers are often not from the poorest groups (Flynn 2001), and

indeed in some cities the very poor find it difficult to gain access to land. Regardless, urban agriculture has helped many urban Africans weather the continuing crisis, and where the data have been examined there is at least some evidence that urban agriculture is contributing to food security (Maxwell et al. 1998). Urban agriculture has often played a similar role in other parts of the world. For example, it emerged in Cuba in response to the decline of Soviet aid and trade and the persistence of the American trade embargo (Altieri et al. 1999; Moskow 1999).

Increasingly, proponents of urban agriculture also emphasize its ecological benefits (Smit and Nasr 1992). One of the ecological disadvantages of urban development is that it tends to replace circular flows with linear ones: ecological cycles are disrupted; materials previously returned to the soil as nutrients become urban waste; substances that are hazardous at high concentrations accumulate. Urban agriculture provides the opportunity to recycle the nutrients in urban organic waste (Eaton and Hilhorst 2003) and can be combined with "ecological sanitation" to improve public health (Esrey 2002). The cultivation of plants in urban areas can also provide other ecosystem services of particular value to urban dwellers, such as cooling and pollution reduction.

Urban agriculture does pose various problems. It can create environmental health problems including food contamination, water pollution, and the increased prevalence of disease vectors such as malarial mosquitoes (Birley and Lock 1998). When combined with ecological sanitation, there is the potential for improving public health if the systems are well managed, but there are also severe risks if management is poor. Moreover, theft can be an especially serious problem for urban agriculture. The fact that it remains illegal in a great many cities where it is practiced constrains its potential (Allen 1999). On the other hand, when agriculture is formally allowed, it often has difficulty competing against alternative urban land uses (Midmore and Jansen 2003); in some contexts, making urban agriculture legal could make it even harder for the poorest residents to gain access to the land.

In affluent cities, urban agriculture now tends to be associated with the production of high-value products supplied fresh to discerning local consumers. Because arable land is scarce and costly, farm sizes tend to be small and yields tend to be high. In the United States, for example, farming in metropolitan areas accounts for less than one fifth of the cropland but for about one third of farms and one third of the value of agricultural production (Heimlich and Anderson 2001). Even in the United States, however, urban agriculture is sometimes promoted as a means of improving food security for some of the more deprived urban communities (Allen 1999).

---

delimited public benefits (that is, one person's use does not detract from use by others) also help make ecosystem services more valuable in urban areas.

Some ecosystem services in urban locations are especially important to vulnerable groups. As described in Box 27.4, urban agriculture can enhance food security and supplement the livelihoods of the urban poor. In many urban areas, although the more affluent residents no longer use it, local groundwater continues to be used by those living in poverty despite being heavily contaminated. In Jakarta, for example, bottled water has become popular for those who can afford it, and piped water is available at a price in many parts of the city, but a large share of low-income households still rely on shallow groundwater and are very aware of differences in quality even among different wells in the neighborhood (McGranahan et al. 1998). Local ecosystem filtration clearly

affects the quality of water, even if it often cannot make it truly potable. More generally, those who cannot afford to purchase alternatives are more dependent on local ecosystem services. This does not mean that protecting ecosystems and ecosystem services in urban areas will necessarily benefit vulnerable groups: vulnerable people are sometimes evicted from their homes in the name of environmental protection. It does mean, however, that there is the potential in many cities for policies that would both enhance ecosystem services and benefit vulnerable groups.

Among planners and decision-makers, there has historically been a strong tendency to neglect ecosystem services and other relations between ecosystems and human well-being, at least until a local or international crisis has forced such concerns onto the policy agenda. In many urban settlements, the quality of the urban groundwater and of the ambient air, for example, has been allowed to deteriorate considerably before any action has been taken (Melosi 2000; Tarr 1996). The policy debates are often very poorly informed, partly because of underlying ignorance about the processes involved, and partly because they are driven by vested interests rather than by sincere attempts to understand the nature of the problem (see, e.g., Davis 2002).

27.2.2 Urban Systems as Habitats for Humans

As long as people continue to live in urban areas, it is important for their well-being that the urban air be healthy to breathe, that there be sufficient water of adequate quality to meet domestic needs, that the urban landscape be pleasing to the eye, that the urban climate be comfortable, and generally that the urban environment be healthy and pleasant for people to live in. With urbanization, the ability of local ecosystems to provide these services tends to decline, even as the number of people per unit of area, and hence the need for these services, increases.

Many of the services once provided by local ecosystems are now provided by some combination of more-distant ecosystem services (such as water diverted to the city through constructed waterways) and manufactured services (such as water treatment plants). Even if the health benefits of the economic growth that has accompanied urbanization in most countries has outweighed the local loss of ecosystem services, these losses have been extremely important historically and remain important to this day, as described later in this section.

Urban development can bring major investments in public health infrastructure and measures to reduce exposure to environmental hazards. Without such investments and measures, urban areas would still be far less healthy than rural areas. With them, however, urban habitats are on average healthier (Montgomery et al. 2003, Chapter 7). Yet the benefits from urban investments in public health infrastructure are very unevenly distributed (Hardoy et al. 2001).

Although urban living is often associated in people's minds with industrial and motor vehicle pollution, the role of cities in facilitating the spread of infectious diseases has probably been more important to human health, and it remains important today. Humans are exceptional among animals in the high proportion of their deaths due to disease (see, e.g., McKeown 1988). Changes in human densities and travel patterns are implicated in the emergence of many of the most devastating infectious diseases.

Without sufficiently large urban settlements, a number of diseases, including measles and smallpox, could not be maintained in human populations (the measles virus, for example, can only persist in one person for a couple of weeks, and so at least 26 times a year it must move to a person who has not been previously infected) (Mascie-Taylor 1993). Trade and urban conditions helped spread the vectors and eventually the plagues that beset Europe during its early urbanization, and in the more extreme cases killed upward of 25% of the population (McNeill 1989). Urban settlements are still important to the spread of epidemics and pandemics, including, for example, HIV/AIDS (Dyson 2003; see also Chapter 14).

The water, sanitation, and hygiene problems described in the next section provide an example of urban conditions that historically created some of the most serious health problems and that remain significant to this day. In a recent World Health Organization ranking of leading risk factors in terms of attributable disease burdens, unsafe water, sanitation, and hygiene ranked sixth (WHO 2002). Water and sanitation deficiencies tend to be a particular risk in economically deprived areas, both rural and urban.

Urban exposure to chemical pollution is also important to human health and well-being. Urban development often leads to unintentional threats to health and well-being by increasing waste generation or by bringing people into closer contact with waste products, some of which contain harmful chemicals (as well as pathogens). These waste products may be in the urban air, water, or land, in public or private spaces, and in relatively more or less frequented locations. Exposure to health-threatening ambient urban air pollution is highest in large industrialized and motorized cities (McGranahan and Murray 2003), whereas health-threatening indoor air pollution is particularly severe in homes where smoky fuels are used without adequate ventilation (Saxena and Smith 2003). Neither surface nor groundwater is potable in most urban settlements, with chemical water pollution a particularly serious problem in industrial centers. Solid waste can contribute to urban air pollution (through burning, for instance) or water pollution (such as through leaching) or can result in direct exposure.

The ambient air pollution problems described in more detail later provide an example of an environmental health problem often considered quintessentially urban. The burden of disease attributable to urban air pollution is estimated to be less than one quarter of the burden from water and sanitation problems (WHO 2002). The burden of illness due to indoor air pollution, by contrast, is nearly the same as that due to water and sanitation problems (WHO 2002).

The later sections on air pollution and on water, sanitation, and health attempt to provide simplified accounts of how urban conditions affect ecosystem services (such as clean air and water) and hence human health. It is also important to recognize, however, that complex environmental interactions and enormous inter- and intraurban variation in environmental health conditions are themselves characteristic of the challenges encountered in urban habitats.

First, in the evolution of an epidemic or of an individual's health, there are discontinuities and thresholds in relations between environmental conditions and health outcomes. Thus, for example, declining sanitary conditions may initially increase the burden of endemic diarrheal diseases, and then, after crossing some threshold, allow a cholera epidemic to break out. An urban settlement's role in epidemics also depends on its size, with larger cities acting as reservoirs of disease and providing a source for outbreaks in smaller settlements below the threshold necessary to maintain the infection in the human population (Cliff et al. 1998). Alternatively, whereas the conditions in the United States and Europe were such that relatively small shifts in the ecology of malaria could lead to its disappearance, in the parts of sub-Saharan Africa where malaria is holoendemic (that is, in an equilibrium where the disease is endemic at a high level among children and adults show less evidence of the disease), the disease can persist in the face of far larger shifts (Bradley 1991).

Second, in service-deprived low-income neighborhoods the conventional boundaries between environmental health problems do not apply (McGranahan et al. 2001). As the result of home industries, occupational health hazards are often encountered in people's homes. When fecal material is not separated off and contained or flushed away, it can contaminate water supplies, become mixed with the solid waste, and attract flies and other pests. Where water is not piped into people's homes, it is more easily contaminated with fecal material and less likely to provide for good hygiene within the home, and there is a risk that water storage containers will become a breeding site for vectors of diseases such as dengue and dengue hemorrhagic fever. When solid waste is not contained and carried away, there is a significant likelihood that it will create unsightly, malodorous, and inconvenient accumulations of refuse and become a breeding ground for pests. And it will also cause air pollution when it is burned or flooding when it is washed into the drains. Combined with crowded housing, smoky fuels, the use of pesticides, and food storage problems, these multiple hazards often create extremely unhealthy living environments, especially for infants and children who have not yet built up resistance to infectious diseases (see, e.g., Cairncross and Feachem 1993; McGranahan et al. 2001; McGranahan and Murray 2003).

Third, many of the environmental conditions that facilitate the transmission of infectious diseases in deprived urban areas lie in the public domain, such as those associated with poor sanitation and solid waste removal, and create local public health risks that private actions cannot address effectively (Pickering et al. 1987; Bateman et al. 1993; McGranahan et al. 2001). Others, on the other hand, involve transmission within households. The relative importance of public and domestic routes of transmission varies, depending on the disease (Cairncross et al. 1995). Much the same applies to exposure to chemical pollutants; it is noteworthy that whereas indoor air pollution was identified as one of major risks to health in the most recent burden of disease estimates (WHO 2002), it was not even included in previous estimates (Murray and Lopez 1996).

The very nature of these interconnections makes the resulting hazards difficult to address, either through the privately negotiated trades that have historically underpinned the success of market economies or through public agency. Those who are most affected tend to have very little income or assets with which to trade and comparatively little political power with which to influence government agencies or political processes. In any case, pathogens, pests, and toxins respect neither the boundaries of private property nor those of organized communities, administrative areas, or ministerial responsibilities. These difficulties are important factors when considering both the history of urban environmental health and the contemporary situation with regard to urban water, sanitation, and air pollution.

27.2.2.1 Urban Water, Sanitation, and Hygiene and Human Health

In most parts of the world, there have been enormous improvements in urban water and sanitation since the mid-nineteenth century, when urban water and sanitation problems first gained international prominence. Nevertheless, according to the most recent global burden of disease assessment, unsafe water, sanitation, and hygiene still account for almost 6% of the burden of disease in "high-mortality developing regions," exceeding all but two other risk factors (Ezzati et al. 2002). Although the "urban penalty"—the increase in mortality rates associated with living in urban areas (see, e.g., Dobson 1997; Woods 2003)—that helped to motivate reform in the nineteenth century is no longer evident, eliminating unhealthy conditions in African, Asian, and Latin American urban areas remains a major challenge.

Urban poverty, particularly when combined with rapid urban population growth, is still closely associated with unsafe water and sanitation. Reducing the share of the population without adequate water and sanitation services is still central to the development goals and targets that have been adopted internationally, including most notably the Millennium Development Goals. (See MA *Policy Responses*, Chapter 20.)

In Table 27.11, the left-hand columns summarize the water and sanitation statistics that were used in developing exposure estimates for the burden of disease just mentioned and also used in starting to monitor progress toward the water and sanitation targets associated with the MDGs. These statistics might seem to suggest that only a small minority of urban dwellers lack provision of clean water and sanitation. Even in Africa, 85% of the urban population had "improved" provision for water and 84% had "improved" provision for sanitation by 2000. Problems are probably much more serious in rural areas, where most of the 1.1 billion people without access to improved drinking water and most of the 2.4 billion people without access to improved sanitation live (WHO and UNICEF 2000).

Unfortunately, these statistics are based on a definition of "improved" provision for water and sanitation that includes conditions where the risk of human contamination from fecal-oral pathogens remains high (Prüss et al. 2002). The Global Assessment from which the statistics are taken acknowledges that, because of the lack of internationally comparable data, it was not able to calculate the proportion of people with "adequate" provision or with "safe" water (WHO and UNICEF 2000).

For water supply, access to "improved" supplies was defined as being able to obtain at least 20 liters of water per person per day from a household connection, public standpipe, borehole, protected dug well, protected spring, or rainwater collector within 1 kilometer of the user's dwelling (WHO and UNICEF 2000). In many low-income urban settings, however, standpipes or other publicly available water sources available within a kilometer may be shared with hundreds and occasionally thousands of people, and there are often serious deficiencies in the quality of the water and the regularity of the supply (Hardoy et al. 2001; UN-Habitat 2003a).

For sanitation, "improved" provision was defined as access to a private or shared toilet with connection to a public sewer or a septic tank or access to a private or shared pour-flush latrine, simple pit latrine, or ventilated improved pit latrine (WHO and UNICEF 2000). In many urban settings, however, dozens of households share each latrine, making access difficult and maintenance inadequate, sometimes causing people and especially children to avoid using the latrines (UN-Habitat 2003a).

Moreover, detailed case studies often indicate levels of provision that are difficult to reconcile with the national estimates used in calculating the figures on "improved" provision in Table 27.11, even accepting the definitions of "improved" supply. For instance, the national estimates for Bangladesh show that 99% of its urban population had access to "improved" water supplies in 2000 (WHO and UNICEF 2000), whereas detailed studies in its two largest cities (Dhaka and Chittagong) show large sections of their populations having to rely on poor-quality water that was difficult to obtain (UN-Habitat 2003a). Similarly, the national estimates for Tanzania and Kenya show that virtually all their urban populations had "improved sanitation," but detailed studies in their major cities and smaller urban centers showed otherwise, especially in the large informal areas within urban settlements where a high proportion of the population of Dar es Salaam and Nairobi live (UN-Habitat 2003a). The numbers in the right-hand columns of Table 27.11 are very crude estimates, but they suggest a far higher level of water and sanitation deprivation (UN-Habitat 2003a).

Such statistics, even if they are rigorously defined and measured, can misleadingly imply that the underlying problem is a lack of infrastructure. In effect, health risks arising from the local ecology of waterborne or water-related diseases are ascribed to the absence of the presumed solution: more extensive piped water and sanitation systems (or other "improved" technologies). As part of this more general tendency to oversimplify, in policy discussions it is often presumed that "waterborne" diseases, which

Table 27.11. Different Estimates of Number of Urban Dwellers Lacking Provision for Water and Sanitation, 2000 (WHO and UNICEF 2000; UN-Habitat 2003b)

Region	Number (and Share) of Urban Dwellers without "Improved" Provision for:		Indicative Estimates for the Number (and Share) of Urban Dwellers without "Adequate" Provision for:	
	Water	Sanitation	Water	Sanitation
Africa	44 million (15%)	46 million (16%)	100–150 million (c. 35–50%)	150–80 million (c. 50–60%)
Asia	98 million (7%)	297 million (22%)	500–700 million (c. 35–50%)	600–800 million (c. 45–60%)
Latin America and the Caribbean	29 million (7%)	51 million (13%)	80–120 million (c. 20–30%)	100–150 million (c. 25–40%)

include most diarrheal diseases, are contracted by people drinking water contaminated with fecal material.

In fact, although waterborne diseases can be spread via drinking water, they can also spread through person–to–person contact, and often by other means (Cairncross and Feachem 1993). Many waterborne diseases can be transmitted mechanically by insects, and there is some evidence that the presence of flies can make a large difference to their prevalence (Cohen et al 1991; Levine and Levine 1991; Crosskey and Lane 1993). Contaminated food is quite possibly an even greater problem than contaminated water. Insufficient water for washing is probably more important to health than poor-quality drinking water. Better sanitation facilities are unlikely to achieve their potential health improvements unless they are accompanied by changes in hygiene behavior; in some circumstances, changes in behavior are the most significant factor in reducing the prevalence of fecal-oral diseases (Curtis et al. 2000; Curtis and Cairncross 2003).

27.2.2.2 Urban Air Pollution and Human Health

Serious exposure to air pollution began with the advent of burning fuels for cooking and heat within unventilated abodes. Air pollution was a major nuisance for many and a serious concern for some in the industrializing cities of the nineteenth century (Mosley 2001), but concerted efforts to address ambient air pollution only began in the twentieth century. In particular, it was the urban air pollution episodes between the late 1940s and mid-1960s in Donora (in the state of Pennsylvania), London, Osaka, and New York City, among other locations, when many died or were hospitalized, that prompted public concern and responses including clean air legislation, regulations, and other actions.

Table 27.12 provides the sources of indoor and outdoor air pollution associated with some of the principal pollutants. The distribution, magnitude, and trends of many of these pollutants within ecosystems are addressed in Chapter 13. This section focuses on their generation and health impacts within urban systems.

Recent estimates of the global burden of disease suggest that approximately 5% of trachea, bronchus, and lung cancer, 2% of cardiorespiratory mortality, and about 1% of respiratory infections are attributed to urban outdoor air pollution (WHO 2002, and see also Ezzati et al. 2002). This amounts to about 800,000 deaths (1.4% of the total) and about 0.8% of the total global burden of disease. This burden falls predominantly on low- and middle-income countries, with 42% occurring in parts of the WHO Western Pacific Region and 19% occurring in parts of the WHO Southeast Asian Region.

Although these figures suggest that outdoor urban air pollution is an important health concern, the burden of indoor air pollution is estimated to be considerably higher (Smith and Akbar 2003; WHO 2002). Indoor air pollution concentrations tend to be highest in low-income settings, and more specifically where smoky fuels are used in homes with poor ventilation (Saxena and Smith 2003). Nearly half the world cooks with biofuels, including more than 75% of those living in India, China, and nearby countries, and 50–75% of those living in parts of South America and Africa (WHO 2002). Exposure to pollutants from burning these fuels is particularly intense for women and young children, who spend much of their time indoors, and is in aggregate substantially greater than exposure to outdoor air pollution in cities with severe air pollution problems (Smith and Akbar 2003). Although ambient air pollution is usually worse in urban centers, overall exposure to air pollution (both indoor and ambient) is higher in rural areas because most of these biofuel users are rural (Saxena and Smith 2003).

There is also considerable variation in exposure to air pollution between and within urban centers, depending on geographical factors as well as the types of activities undertaken in and around the urban centers and the fuels used to power them. Ambient air pollution has reached excessively high levels in many large cities in Asia, Africa, and Latin America (Krzyzanowski and Schwela 1999), where concentrations of ambient air pollution often rival and exceed those experienced in industrial countries in the first half of the twentieth century. Pollution from industries and power plants can account for a large share of urban emissions, and it also tends to be the target of initial pollution control measures. Vehicular pollution is also a chief contributor to overall local and regional ambient air pollution (NO_x, O_3, CO, volatile organic compounds, and suspended particles).

In general, low- and middle-income countries account for only 10% of the world's vehicles, including 20% of the buses (Elsom 1996). Growth rates for vehicle ownership, however, are two to three times higher in these countries than in high-income countries. For example, during the 1980s Pakistan experienced an annual average vehicle growth rate of 9%, Brazil's was 11%, China 14%, Kenya 26%, and both the Republic of Korea and Thailand 30%, compared with 2–3% growth in the United Kingdom and the United States (Elsom 1996). In 1990, there were 700,000 private cars in China and 5 million other motor vehicles. By 2001, this had risen to more than 5 million private cars and some 13 million other vehicles. For the next 20 years, East and Southeast Asia are expected to have the fastest-growing car markets in the world (Walsh 2003). The number of motor vehicles

Table 27.12. Sources of Outdoor and Indoor Emissions and Principal Pollutants (Murray and McGranahan 2003)

Sources	Principal Pollutants
Predominantly outdoor	
Fuel combustion, smelters	sulfur dioxide and particles
Photochemical reactions	ozone
Trees, grass, weeds, plants	pollens
Automobiles	lead, manganese
Industrial emissions	lead, cadmium
Petrochemical solvents, vaporization of unburned fuels	volatile organic compounds, polycyclic aromatic hydrocarbons
Both indoor and outdoor	
Fuel burning	nitrogen oxides and carbon monoxide
Fuel burning, metabolic activity	carbon oxides
Environmental tobacco smoke, re-suspension, condensation of vapors and combustion products	particles
Biologic activity, combustion, evaporation	water vapor
Volatilization, fuel burning, paint, metabolic action, pesticides, insecticides, fungicides	volatile organic compounds
Fungi, moulds	spores
Predominantly indoor	
Soil, building construction materials, water	radon
Insulation, furnishing, environmental tobacco smoke	formaldehyde
Fire-retardant, insulation	asbestos
Cleaning products, metabolic activity	ammonia
Environmental tobacco smoke	polycyclic aromatic hydrocarbons, arsenic, nicotine, acrolein
Adhesives, solvents, cooking, cosmetics	volatile organic compounds
Fungicides, paints, spills, or breakages of mercury-containing products	mercury
Consumer products, house dust	aerosols
House dust, animal dander	allergens
Infections	viable organisms

worldwide is expected to increase from around 660 million in 1990 to 1 billion by 2030 (Faiz et al. 1990).

Besides absolute numbers, the quality and fuel efficiency of motor vehicles also affects emissions and hence ambient air quality. High emissions per vehicle are associated with outdated technologies, older vehicles, poorly surfaced or badly maintained roads, weaker environmental legislation or weak enforcement of the regulations, poor vehicle maintenance (as vehicle emission inspections are less rigorous or nonexistent), and the dominance of low-quality fuels (such as diesel with high sulfur content) (Elsom 1996). These circumstances tend to be more common in low- and middle-income cities than in high-income cities.

Leaded fuels are also more common in low-income cities and account for most atmospheric lead in countries where they are still in use.

27.2.3 Urban Systems Interrelating with Surrounding Regions

Partly because of the demands that urban systems place on ecosystems in the surrounding region, cities and towns are often presented as environmentally damaging. This is misleading, particularly if human well-being is a central concern. If urban activities and residents moved to rural areas, the demands placed on ecosystems would be more dispersed, but not reduced. Yet even if, from an ecosystems perspective, urbanization is preferable to most rural alternatives involving similar economic production levels, urban pressures are increasing rapidly as the result of population growth, economic growth, and urbanization. Moreover, for adjoining ecosystems, the concentration of people and activities in urban areas can be a particular burden. Urban centers in the vicinity of fragile ecosystems are especially problematic. Cities associated with highly polluting industries typically have a greater impact on nearby ecosystems than those dominated by service industries. Poorly managed urban development can be especially destructive to nearby ecosystems.

27.2.3.1 Urban Systems and Rural Lands

In peri-urban areas, the influence of urban development is visible and often involves the conversion of land to urban uses, as described earlier, but the less direct urban influence on somewhat more distant rural lands can be just as great and extends from demand-driven land use changes to the effects of urban remittances on rural development patterns.

In an ecological history of Chicago and the "Great West," Cronon (1992) describes how innovations in grain markets were linked to the loss of species diversity in the grasslands, how developments in meat handling and marketing affected animal stocks and living conditions on the farms, and how the urban lumber industry led to the decline of the White Pine forests on which it depended. When such changes occur, it is not just the increasing size of urban demands that influences the surrounding ecosystems, but the changing qualities of urban demands, including, for example, the tendency of many urban markets to demand standardized produce, thereby favoring monoculture and reducing biological diversity.

Although most contemporary cities do not have as great an influence on their hinterlands as Chicago once did, urban development remains a major influence on agricultural systems. A recent study of peri-urban agriculture in Hanoi documents a process that is likely to be present in the peripheries of most developing urban centers: the shift by farmers to producing higher-value goods in response to consumer demand in the urban areas (van den Berg et al. 2003). Such goods include vegetables, milk, and other perishable commodities (including from fish farming, shrimp farming, and flower production). Here, as in and around many other cities, agriculture is also disrupted by land speculation or the conversion of land to urban uses (including farmers who sub-divide and sell their land for housing, sometimes illegally). However, cities often provide surprising new opportunities for farmers: for instance, the demand for "turf" (sod) and ornamental plants for middle-class gardens in Mexico City and the demand by international tourists for authentic "pre-Columbian" food produce new opportunities for farmers around Mexico City (Losada et al. 1998; Losada et al. 2000).

While increasing urban demand for agricultural produce can be expected to lead to a larger expanse of agricultural land, urban demands for marketable wood products are often assumed to reduce tree cover. In the 1980s and 1990s, for example, urban demands for fuelwood and charcoal were often presented as leading to "rings of deforestation" around African and Asian cities where charcoal is a major cooking fuel (Cleaver and Schreiber 1994). However, as described in Chapter 21, such outcomes depend on the institutions guiding the resource use. Increasing urban demand can contribute to institutional conflicts over forest use. On the other hand, it can also motivate efforts to protect and plant forests. Although the slow growth of most woody products may inhibit private investment, trees can be planted, and in some circumstances an increasing demand for wood will result in an expansion of forest area.

From the perspective of human well-being, many of the more destructive relations between urban and adjoining systems involve interrelations that are neither valued within the market economy nor given priority by government agencies. Urban water demands often conflict with agricultural demands, and in many circumstances the institutions for reconciling such conflicts are neither equitable nor efficient (Baumann et al. 1998). Urban water pollution can damage downstream agriculture; conversely, the use of agricultural fertilizers and pesticides pollutes urban water sources. Cultivated systems can also lead to erosion, siltation, and more flooding in downstream urban areas, as well as damage to water storage facilities and water conveyance services. (See Chapter 26.) Urban air pollution contributes to acid precipitation, affecting forests and croplands with low buffering capacity; forest fires contribute to urban pollution concentrations as well as to urban fire risks.

Although some of these negative relationships have little to do with urban settlement patterns per se, others are directly related to the spatial concentration of urban consumption and production. Urban centers rely on adjacent ecosystem services to break down their biodegradable wastes, but when the capacities of these local ecosystems are overwhelmed, people living in downstream settlements are put at risk. These same biodegradable wastes may represent the loss of nutrients from agricultural and forestry systems. Urban consumption and production can also result in the accumulation of nondegradable and sometimes toxic substances (such as heavy metals) at waste sites, where they may leach into the groundwater or result in human exposure through some other means. Even relatively small urban centers face such problems, but they are magnified in large cities, and particularly in large industrial cities.

Again, it is important to distinguish between the often negative impacts that urban development has on ecosystems services and the often positive comparisons that can be made between well-managed urban development and alternative, less urban, development options. It has been suggested, for example, that urban development in drylands should lead to a reduced risk of desertification when compared with agricultural development (Portnov and Safriel 2004). In some circumstances, urban development can also provide the justification for expensive investments in water infrastructure, providing the basis for other developments.

27.2.3.2 Urban Development and Regional Water Systems

Historically, urban centers have often been founded near water sources and waterways, both to provide for urban water demands and to take advantage of water transport. As described earlier, the coastal zone is not only the most urbanized of all of the major systems identified for the MA, it is also the most densely populated with rural dwellers. A disproportionate number of urban centers, including especially large urban centers (over 500,000 people), are located at or near river mouths, which are also ecologically critical sites, particularly for some migratory aquatic and bird species. (See Chapter 20.)

Water is also a resource with a strong regional dimension. Freshwater resources from surrounding regions are still the major source of water for urban consumption, unlike many other resources that can more easily be imported great distances. Intraregional water flows provide critical connections between urban systems and the surrounding regions; as indicated earlier, unintentional changes to these water flows can create serious problems. Even groundwater aquifers can have a regional dimension. The flow of water represents the largest material flow in and out of urban areas, and it has been estimated that water represents about 90% of all material entering megacities (Decker et al. 2000).

In assessing the water relations between urban centers and their surrounding regions, it is important to consider:

- Urban→Upstream: how measures designed to meet urban demands for water and hydropower have changed the upstream water flows, affecting, for example, the availability of water for urban and nonurban users upstream.

- Upstream→Urban: how upstream water and land use changes not specifically designed to change urban water conditions have affected the qualities and quantities of water available to or flowing through urban centers.

- Urban→Downstream: how urban water and land use changes have affected the qualities and quantities of water available downstream (including coastal waters).

There are some changes that are not captured by these three categories. Thus, for example, dams and water diversions created to serve urban demands affect not just upstream users, but also those downstream, in all the river basins affected. While competition for good-quality water is often central to these relations, changes in urban water regimes can also influence flood risks, biodiversity, wetland and delta ecosystems, fisheries, migratory aquatic species, and a range of other less obvious water-related issues. Moreover, groundwater depletion often leads to land subsidence, which can have severe consequences in urban areas.

During the nineteenth and twentieth centuries, rapidly growing and economically successful urban centers relied on bringing in water from increasingly distant sources (Tarr 1996). Conflicts between urban and nonurban users have been common. Urban water use requires a higher-quality and more stable supply than that in most rural uses (for irrigation, for example), and the social, economic, and political importance of cities often ensures that their demands are given priority. The manner in which the water demands of Los Angeles were allowed to dominate over those of Owens Valley provides a well-documented example (Kahrl 1982; Reisner 2001).

When water is diverted from agriculture to urban areas, agricultural productivity can be severely affected. For example, in the Hai river basin in China, most of the freshwater resources captured by large reservoirs are directed to meet the increasing water demands from Beijing and Tianjin, and access to water has become the limiting factor in the region's agricultural productivity (Bai and Imura 2001). On the other hand, with decreases in water availability, farmers have traded off grain production for other more economically productive uses for their land and their time, not all of which are so dependent on water, and despite declining water resources, incomes have been increasing (Nickum 2002).

In most parts of the world, the spatial range of urban water withdrawals is expanding. In countries with capital-intensive water infrastructure, some of the regional water systems have become so closely integrated that it is no longer meaningful to link urban centers with spatially delimited supply networks: as with electricity systems, they are simply "attached to the network" (Baumann et al. 1998). Even where there is less water infrastructure, cities are reaching further upstream for more and fresher water resources, sometimes even from other river basins. In Africa, where inadequate infrastructure is often cited as a major problem, in the early 1970s many urban centers still used groundwater supplies as their primary water sources, but by the 1990s the primary water sources were more likely to be rivers, and increasingly these river sources were more than 25 kilometers away (Showers 2002).

Investment in tapping water supplies that are further away is often undertaken when less costly alternatives have not been explored. Moreover, when cities and surrounding rural areas compete for water resources, ecological water requirements (the water needed to maintain ecosystem function and local hydrological cycles) are often neglected. In many situations, demand-side management is an inexpensive means of freeing up water supplies and could be used to avoid tapping distant water supplies or undermining ecological functions (Baumann et al. 1998). Alternatively, economic analysis of other measures to improve water supplies in New York City found that in many cases it would be cheaper for the city's residents to pay upstream individuals and enterprises operating in the city's upper watershed to adopt less damaging practices than to invest in more water supply and treatment facilities. Investing in water filtration in New York is estimated to cost approximately $6 billion for design and construction and $300 million in annual operating expenses (NRC 2000; Pires 2004).

Urban centers themselves can cause a wide range of problems for people and ecosystems downstream, including those in other urban locations. Urban areas usually have a high percentage of paved areas, which concentrates rainwater rather than dissipating it. This tends to intensify flooding and can cause flash floods. Changes in the water flows can also affect downstream fish stocks, recreational opportunities, and biodiversity. (See Chapter 20.) Sewers convey human waste out of urban locations, often releasing it untreated in local waterways or coastal waters. Human waste not only poses a health risk for people who might come to ingest the contaminated water, it causes eutrophication and damages aquatic ecosystems downstream. (See Chapter 12.) Chemical water pollution is also a major problem, particularly around large industrial centers.

Coastal zones are among the worst affected by urban development, and they combine many of the most critical land and water issues. As indicated earlier, the share of land in coastal zones that is urban is particularly high, and land conversion and habitat losses of coastal wetlands, dune systems, and coral reefs are often irreversible. (See Chapter 19.) Urban areas at river mouths often constitute bottlenecks for aquatic migratory species. Other important situations related to urban areas in coastal systems are the development of ports out of natural harbors, the dredging of shipping channels, and the development of industrial centers in the coastal fringe. Port development also creates the risks of species invasion, with large ships in harbors acting as vectors for species introduced via ballast-water transfer and hull fouling.

Although high levels of urbanization are not in and of themselves a problem, urban development undertaken with little regard for its ecological implications can be extremely destructive. Dispersed rural settlements can bring about more vegetation fragmentation than population concentration in urban areas, with a strong negative impact on the health of inland water systems. The concentration of population in urban areas makes it easier to treat

wastewater and avoid pollution, as point pollution sources are more likely to be controlled or eliminated. There are also many opportunities in urban areas for reusing wastewater and for engaging in demand-side management for conservation and for improving well-being. There are indications that water-management systems are slowly changing, with more attention being given to improving water use efficiency and productivity and less of a tendency to assume that water shortages must be met by more water infrastructure (Gleick 2003). On the other hand, concentrating settlement concentrates the burdens, and where urban development is poorly managed, concentration will make local disturbances even worse.

27.2.4 Urban Systems Creating Global Ecosystem Pressures

If the environmental shortcomings of the affluent city in the nineteenth century were unsanitary slums and environmental health problems within the city, and those of the affluent city in the twentieth century were urban pollution and environmental degradation in and around the city, then the major environmental burden of the affluent city in the twenty-first century is likely to be the global burden it imposes, often on ecosystems far removed from the city itself.

The importance of global trade and of global environmental burdens has grown considerably over the past two centuries, and especially in the last few decades. (See Chapter 3.) Urban development has been an integral part of this process; all urban centers are engaged to some degree in the production and consumption of internationally traded goods and in contributing to globally burdensome wastes such as greenhouse gases, persistent organic pollutants, and ozone-depleting substances. In general, however, global ecosystem pressures derive from the consumption and wastage undertaken to support the lifestyles of the world's more affluent residents, most of whom live in the urban centers of high-income countries. As much as two thirds of total consumption and pollution can be traced to cities in rich countries alone (Rees 1997).

Increasing long-distance trade spreads the ecological burden of consumption, but it also increases the likelihood that consumers will neglect the costs of ecological pressures and damage. If a population depends on local ecosystems and degrades these through excessive growth or overexploitation, the negative consequences (declining productivity) are more likely to inhibit further growth. Trade serves to short-circuit such negative feedback and may even lead to positive feedback. Urbanites who live mainly on imported goods lose their incentive to conserve remaining local or regional stocks of natural capital (biophysical resources). Thus, a city interested in promoting economic growth may sacrifice prime cropland on the urban fringe to "highest and best (economic) use," permanently destroying the land's agricultural potential. Second, people living on imports are less likely to be aware of the negative ecological or social consequences of unsustainable production processes in the distant regions that are supplying them. The most successful traders are those who seek out and find the least-cost supplies, whether the low costs are based on a real comparative advantage or on the fact that the loss of ecosystem services is not being costed into the supply chain.

Urban systems are also vulnerable to global environmental shifts, including climate change and its consequences. In addition to the direct effects of warming on the habitability of urban centers, many cities are vulnerable to flooding from sea level rise or to damage from tropical storms. Nevertheless, most assessments of the global pressures of urban development focus first and foremost on impacts outside urban boundaries, recognizing that in the long run these impacts too will affect the well-being of urban residents.

27.2.4.1 *Ecological Footprint Analysis*

Ecological footprint (or eco-footprint) analysis is a quantitative tool that estimates the load imposed on the ecosphere by any specified human population in terms of the land and water (eco-system) area dedicated to supporting that population (Rees 1992; Rees and Wackernagel 1994; Wackernagel and Rees 1996). It does not capture the dynamics of either ecosystems or markets, nor does it provide the basis for assessing whether any given ecological burden involves economic externalities or can be justified in terms of human well-being. Summing up ecological footprints is inevitably complicated by the diversity of services that any given ecological area can provide. Ecological footprint estimates are revealing, however, not only in demonstrating how much more extensive urban footprints are than the urban areas themselves, but also in allowing the ecological pressures of different urban centers or different population groups to be compared, at least roughly.

An eco-footprint analysis begins with the quantification of the material and energy resources required to support the consumption demands of the study population at its present material standard of living. The method is based on the fact that many of these resource and waste flows can then be converted into a corresponding productive land and water area. Thus, the ecological footprint of a specified population can be formally defined as "the area of land and water ecosystems required, on a continuous basis, to produce the resources that the population consumes, and to assimilate the wastes that the population produces, wherever on Earth the relevant land/water is located" (Rees 2001). A complete eco-footprint analysis would therefore include the ecosystem area that the population effectively appropriates to supply its needs through all forms of economic activity, including trade, plus the area it needs to provide its share of certain (usually free) land- and water-based services of nature, such as the carbon sink function.

The size of the eco-footprint depends on four factors: population, the average material standard of living, the productivity of the land/water base (whether local or "imported" in trade goods), and the efficiency of resource harvesting, processing, and use. Regardless of the relative importance of these factors and how they interact, every population has an ecological footprint. (For full details of the methodology, see Wackernagel et al. 1999; Rees 2001, 2003; World Wide Fund for Nature 2002; Monfreda et al. 2004.)

Eco-footprint analysis reveals that the residents of the more urban high-income countries impose a vastly larger load on Earth than do the residents of low-income countries. The citizens of high-income countries such as the United States and Canada have average ecological footprints of 8–10 hectares, or up to 20 times larger than the eco-footprints of the citizens of the world's poorest countries such as Bangladesh or Mozambique (see, e.g., World Wide Fund for Nature 2002).

Because consumption, production, and trade data are generally compiled at the national level by domestic statistical offices and international agencies, it is easiest to estimate national eco-footprints. Data specific to lower-level political entities such as states, provinces, or cities are generally much harder to come by. Nevertheless, a rough estimate of the eco-footprint of any given city can be made by multiplying the city's human population by the national per capita ecological footprint, and methods do exist

for developing estimates at sub-national levels (see method applied in Wackernagel 1998).

Various researchers have estimated urban eco-footprints by using different assumptions and levels of detail. Despite methodological differences, such studies invariably show that the eco-footprints of typical modern cities are two to three orders of magnitude larger than those of the geographic or political areas they occupy (see, e.g., Rees 2003).

A city may represent as little as 0.1% of the area of the host ecosystems that sustain it. Such fractions emphasize that even in a stable world, no city or urban region as presently configured could be sustainable on its own. Moreover, the combined requirements of urban systems are increasingly unsustainable in the long run; in a politically unstable world, dependence on extensive and often distant ecosystems raises issues of shorter-term sustainability.

27.2.4.2 *The Urban Sustainability Multiplier*

Although cities, particularly high-income cities, have large eco-footprints, they also provide many opportunities to lighten the human load on Earth's ecosystems. To begin with, cities are concentrations of buildings and associated infrastructure, and the "built environment" is a key consumer of materials and energy, with considerable scope for savings. To a first approximation, the construction, operation, and maintenance of the built environment accounts for 40% of the materials used by the world economy and for about one third of energy consumption. Studies indicate that buildings in the United States account for between 15% and 45% of the total environmental burden in each of eight major categories of impact used for life-cycle assessment (an integrated "cradle to grave" approach to assessing the environmental performance of products and services) (Levin et al. 1995; Levin 1997). However, given equivalent levels of consumption, increased human density is associated with lower eco-footprints.

Many other attributes of urban life provide leverage in dealing with the energy and material dimensions of sustainability. Together these factors contribute to what might be called the "urban sustainability multiplier" and include the following (Rees 2003):

- high population densities, which reduce the per capita demand for occupied land;
- lower costs per capita of providing piped treated water, sewer systems, waste collection, and most other forms of infrastructure and public amenities;
- a high proportion of condominiums, apartment buildings, and other multiple-family dwelling units, which reduces per capita consumption of building materials and service infrastructure;
- increasing interest in eco-neighborhoods and forms of cooperative housing, which reduces demand for appliances and personal automobiles;
- easy access to the necessities for life and to urban amenities by walking, cycling, and public transit. This further reduces the demand for private automobiles, thereby lowering fossil energy consumption and air pollution (some residents even adopt an auto-free lifestyle);
- a high density and diversity of communication infrastructure, reducing the need for energy-intensive travel to face-to-face meetings;
- greater possibilities and a greater range of options for material recycling, reuse, remanufacturing and a concentration of the specialized skills and enterprises needed to make these things happen;
- economies of scale and agglomeration economies that make electrical co-generation possible and facilitate the use of waste

process-heat from industry or power plants for local (neighborhood) water and space heating, thus reducing demand for energy; and

- the opportunity to implement the principles of low-throughput "industrial ecology" (that is, the creation of closed-circuit industrial parks in which the waste energy or materials of some firms are the essential feed-stocks for others).

Walker and Rees (1997) provide a graphic illustration of the economies associated with housing type and attendant urban form. They show that the increased density and consequent energy and material savings associated with high-rise apartments, compared with single-family houses, reduce the part of the per capita urban ecological footprint associated with housing type and related transportation needs by about 40%. Such gains are independent of building materials used. Similarly, Kenworthy and Laube (1996) detail how personal energy consumption associated with transportation needs is dramatically inversely related to urban density. The sprawling cities of Australia and the United States feature vastly less energy-efficient transportation systems than can be found in wealthy, compact Asian cities. European cities generally fall somewhere in between.

27.3 Important Processes Driving Change

This section explores some of the important drivers of urban system change and their impact on urban systems and, at least indirectly, ecosystem services. Drivers can have either direct or indirect impacts on urban development. (See Chapter 1.) Among indirect drivers being considered in this assessment, those associated with globalization, technological change, political shifts (including institutional and legal framework changes), and demographic shifts are of particular importance for urban systems. Direct drivers for urban centers include, among other things, changes in land use (the expansion of cities and urban areas) and user rights and structures.

Contemporary urban development around the world also reflects the fossil fuel economy; energy use and availability are primary urban drivers. Without petroleum-based fuels and the transportation systems they underpin, existing urban systems would be inconceivable, not just because they would be unsustainable but because they would be dysfunctional (Droege 2004). More generally, energy underpins economic growth, globalization flows, and technological advances, all of which operate through urban centers. Globally, urban activities, including intra- and interurban transport, consume approximately 75 percent of the world's fossil fuel production (Droege 2004). Urban activities in high-income countries account for a disproportionate share of this consumption, but even urban centers in very low-income countries have levels of energy consumption that are higher than the historical norm.

Globalization and other drivers are experienced differently in different parts of the world. The urbanization and feminization of poverty (United Nations Centre for Human Settlements 2001; Chant 2003), the aging of populations (Lutz et al. 2004), and the adoption of telecommunication technologies are all examples of trends that are common to many different types of cities but that take different forms (see, e.g., United Nations Centre for Human Settlements 1996, 2001).

The following sections examine drivers and urban development in low-, middle- and high-income cities. This division has been selected because of its political significance as much as its empirical relevance.

27.3.1 Low-income Cities and Middle-income Cities with Economic Difficulties

Not all nations and their cities have benefited economically from globalization. Many nations have not benefited from contemporary trade and foreign direct investments flows, which underpin globalization, and therefore their urban centers are not considered "world cities" (see, e.g., Friedmann 1986, 1995). These include, for example, cities in the lowest-income economies of Africa, Asia, and Latin America (Lo 1994; Gilbert 1996; Rakodi 1997). Macroeconomic conditions and national debt burdens have played an important role in the recent development of some of these cities, and the predominance of agricultural and mineral trade and the lack of manufacturing have made them more dependent on rural than on urban economic activity and more susceptible to changes in commodity market prices. When global primary commodity prices fell in the early 1980s and interest rates increased, many of these countries experienced recessions combined with high inflation and increased debt. Foreign and domestic investment typically slowed, housing finance became less available, and infrastructure deteriorated. Without foreign investment, the application of new technologies slowed, particularly digital information and communication technologies, which further exacerbated the "digital divide" (United Nations Centre for Human Settlements 2001).

Regulatory frameworks, including policy documents, laws, traditions, regulations, standards, and procedures, influence urban development, although often not in a straightforward manner. The lack of provision of adequate shelter and the generation of slums themselves are often due in part to inappropriate regulatory frameworks (Payne and Majale 2004). The lack of appropriate institutional structures helps drive ecological and environmental trends within and around cities in this category. Ability to finance infrastructure and provide public services depends heavily on intragovernmental institutional arrangements, because capital markets rarely provide an adequate source of finance in low-income countries and because users, or potential users, in these cities are typically unable to pay the large costs associated with construction (United Nations Centre for Human Settlements 1996). Lack of access to land, because of its high cost or inadequate property rights and land tenure arrangements, has facilitated the development and expansion of slum and squatter areas within urban settlements (Hardoy et al. 2001).

These direct and indirect pressures have combined to affect cities in this category in a number of ways. One important outcome has been a reshaping of urban structure through the expansion of urban slums and squatter (sub-) settlements. The UN Habitat suggests, for example, that squatter settlements house 40–50% of the people living in Calcutta; 50–60% of those living in Bombay, Delhi, Lagos, and Lusaka; and 60% or more of those living in Dar es Salaam, Kinshasa, Addis Ababa, Cairo, Casablanca, and Luanda (see "Global Trends" from the Global Urban Observatory, at www.unchs.org/habrdd/global.html). Informal areas within urban settlements are not aberrations; rather, they are the dominant means through which cities in this category are growing. Moreover, many of these are expanding in ecologically sensitive areas, such as on steep hillsides, riverbanks, and wetlands (Hardoy et al. 2001).

The results of these pressures on human ecology have been just as stark as changes in urban form. The increase in size of cities without adequate infrastructure has put pressure on the basic ecosystem services necessary for healthy life. Inadequate and unsafe piped water supply, a lack of proper sewerage and stormwater drainage, lack of provision for garbage collection and disposal, and indoor air pollution that results from burning biomass have all affected human health and well-being (Bartone et al. 1994; McGranahan et al. 1996). The shift of population to urban areas, the lack of access to safe drinking water, the lack of the simplest latrines, the spread of preventable diseases and health risks, inadequate secure and healthy shelter, and hunger have combined to move considerable poverty and poverty-related problems from rural to urban settings (United Nations Centre for Human Settlements 2001).

Flooding and a general proneness to other natural disasters affect these communities more than more-wealthy settlements. Natural disasters act as both a driver and a consequence of current development patterns. Without adequate governance, regulation, and public spending, increasing numbers of the urban population are locating on floodplains, within swamps, and in other ecologically sensitive areas. This development pattern is making parts of a large number of cities vulnerable to natural disasters. Among the total reported deaths from disasters between 1993 and 2002, more than 53% were in countries at low levels on the Human Development Index, compared with just 4% of deaths in countries with high levels and 42% in countries with medium levels of human development (International Federation of Red Cross and Red Crescent Societies 2001).

Although most of the ecological challenges remain localized and associated with health and safety issues, activities within cities in this category have generated city-wide ecological impacts. Natural resources in and directly around these cities are being depleted as urban populations search for space to live, biofuels for cooking, and water for daily needs (World Commission on Environment and Development 1987; Hardoy et al. 2001). The pressures inherent in consumption patterns of citizens in cities of this category are far less than those created by more affluent urban residents, but the context can mean that these pressures have serious consequences both locally and throughout the metropolitan region.

In some cities in this category there are promising signs of economic growth. For example, since the 1980s many if not most Latin American countries have experienced positive growth, notwithstanding fluctuations (Tulchin 1994). Recently, some of the cities in Asia that were previously disconnected from the world economic system of trade and foreign direct investment have become connected to it, and cities such as Phnom Penh and Hanoi are on the brink of rapid economic growth. This growth, however, does not ensure either improved well-being for all or decreased ecosystem impact. Indeed, citizens in rapidly developing cities may be experiencing intensive local ecological deterioration while increasing their burdens on hinterlands and ecosystems at larger scales and farther away.

27.3.2 Rapidly Growing Middle-income Cities

Cities in this category are undergoing rapid development and are sometimes included as "second-tier" world cities (Taylor 2004). They are the locations of the world's new industrial production systems, and many of them are in Asia. The most obvious impacts of indirect drivers are increasingly rapid economic development. For example, during the mid-1990s Jakarta's economy was growing at 8.2% annually (Soegijoko and Kusbiantoro 2001) and Shanghai's at 14.2% annually (Ning 2001). This growth has been facilitated by the transnational connections between these cities and international investment capital. Many of the cities in this category are linked to the global economy and have experienced the world city formation process—the process by which the world's capital accumulates in cities through global trade and

investments (Friedmann and Wolff, 1982), as they become manufacturing production centers (Lo and Yeung 1996; Lo and Marcotullio 2001).

Rapid economic growth has been driven, in some cases, by the emergence of clusters of private-sector organizations applying new technologies to production processes in and around cities (Hall 1995). For example, new industrial regions have sprung up in and around some cities within this category: Guangzhou, Kuala Lumpur, Seoul, Singapore, and Taipei, which are among the leaders of the high-technology and computer industries.

Rapid growth has been associated with rapid urbanization, rapid urban growth rates, and large urban sizes. The trend, particularly within the rapidly developing world, has been toward very large urban agglomerations, or megacities and megaurban regions (Fuchs et al. 1994; McGee and Robinson 1995; Gilbert 1996).

Although cities in this category have been able to compete successfully within the global economy, they have paid the price in terms of disruption to local ecosystems (Lo and Marcotullio 2001). The growth of the largest cities has slowed over the past few decades, but at the same time their internal ecosystem service conditions have worsened, suggesting that rapid demographic growth is not the primary source increasing urban environmental burdens (Brennan 1999). For example, the larger, slower-growing Asian megacities are among the world's most severely environmentally distressed (Asian Development Bank 1997).

In contrast to the experiences of cities in industrial countries, where the emergence of ecological challenges appeared over longer periods of technological and socioeconomic change and in a more sequential order, rapidly developing cities may experience a new mix of environmental challenges at lower levels of income that increasingly appear concurrently (Marcotullio 2004). Within a single city there are often pressing challenges associated with basic sewage and sanitation, industrial water and air pollution, greenhouse gas emissions, and green space.

In Bangkok, for example, transport accounts for 70% of urban energy consumption. The level of the city's GDP is three times the national average, and 70–80% of the city's population has been described as "middle class" (Plumb 1999), with the associated increased consumption levels that wealth brings. At the same time, Bangkok, like many of these cities in this category, has substantial numbers of people without water supply and sanitation services. Further, Bangkok is a major source of pollution, both industrial and residential, for the Chao Phraya River and a growing source of consumption, with related wider ecological impacts. Hence, Bangkok is experiencing several different sets of burdens simultaneously (Marcotullio 2003).

Given the forecasts for urbanization and industrialization in the Asian region, the overlapping burdens experienced within cities in this category are not likely to disappear in the medium term. The same may be said for rapidly developing cities in other parts of the world. From the policy perspective, this phenomenon has produced important questions concerning how to manage this mix of environmental problems in order to ensure the well-being of the urban population, the ecological integrity of urban region, and the long-term viability of economic growth.

27.3.3 High-income Cities

Cities that are at the top of the global networked hierarchy are also rapidly growing as producers of business services for global capital and finance (Sassen 1991; Honjo 1998). During the recent past, those in North America have been undergoing restructuring, as some have moved from industrial to service-dominated economies and others have grown as rapidly as cities in the developing world. Other than the influence of globalization flows, particularly important drivers of urban system change in this category include demographic shifts, technological advances, and institutional and policy changes.

Important demographic shifts include the reduced birth rate, the increase in the older population, and the decrease in household size. Most of the OECD nations have stopped growing in size and are experiencing increases in the proportion of aged populations (Lutz et al. 2004). This has substantially affected urban population structures. For instance, while the proportion of the elderly is greater in the nonmetropolitan regions of the United States than in the metropolitan regions, more than 74% of the older population resides in metropolitan areas, and the number and share of the elderly are increasing in both (Glasgow 2000). In 1990, the U.S. population over 65 years of age living in metropolitan areas reached 23.1 million. This has social and economic consequences: increasing numbers and proportions of the elderly translate into both proportionately fewer middle-aged care providers and proportionately fewer contributors of social security funding.

Accompanying the low or even negative overall natural increase in urban populations in the United States and Europe, average household size has shrunk to fewer than three, and the number of single-person households has increased to 25% in Europe and 20% in the United States. In Europe, the increased demand for new households is expected to account for 12.5 million new dwelling units in 2000–05 and for 11.5 million units in 2005–10. Despite the stabilization of the size of the urban population, the increasing number of households is still driving the demands for construction materials, space for building, energy, transport, and natural resources. Satisfying these demands can have an adverse impact on biodiversity (Keilman 2003; Liu et al. 2003).

Another important driver for urban system change is technological advance. Advances in transportation and communications have allowed the decentralization of industry and the loss of manufacturing jobs from urban centers. At the same time, the increasing application of information and communication technologies has facilitated both an expansion of markets and control over national and international economic space (globalization) and the rise of the service sector dominated by advanced business services (Sassen 1991). Moreover, within the OECD, new industrial regions based upon these technologies have sprung up around the older industrial areas (such as the Western Crescent around London, the southwest sector of Paris, the south of France, the Munich region, Silicon Valley, and Los Angeles) and some older cities (such as Munich and Tokyo) have transformed themselves into high-tech centers. Contrary to former predictions that these technologies would make cities unnecessary, dense human concentrations within the new global economy are increasingly important (see, e.g., Sassen 1991; Taylor 2004).

Environment-related policies within cities of this category have played an important role in providing cleaner and more habitable internal urban environments and have therefore become important ecosystem drivers. User charges, including utility charges for water and wastewater and pollution charges for solid waste, affect the use of these services. (See Chapter 7.) Air quality is protected in many urban areas by user charges such as road tolls (used as a traffic regulator) and pollution charges used to control and reduce emissions. Land use is regulated through user charges such as betterment charges and access fees to parks and beaches, which are widely used, in part to protect landscapes. Zoning and transferable development policies have been extensively used to conciliate environmental and development aims, including conservation of greenbelts, wetlands, cultural heritage sites, and coastal areas, and for the preservation of open space, green space, and farmland in urban hinterlands (World Resources Institute 1996).

On the other hand, decreasing household sizes, increasing vehicle ownership and usage, highway development, air-conditioning, mall construction, and new building technologies combine to facilitate land use change at urban fringes (Rusk 2001) and hence directly affect hinterland ecosystems. U.S. cities are consuming land at rates faster than that of population growth. For example, the Washington, D.C., area lost 85,000 hectares of farmland, forest, wetlands, and other open spaces during the 1980s, and California continues to lose wetlands at a rate of almost 2,000 hectares per year. Environmentally related impacts of sprawl are increasingly evident, as urbanization in these economies is associated with the degradation of water resources and water quality, changes in hydrology, increased inputs of water pollution and nutrients, and increased acidity and higher water temperatures of lakes, ponds, and streams (US Environmental Protection Agency 2001). There are also growing concerns about the health implications of this urban sprawl (Frumkin et al. 2004).

Moreover, increased transportation has also had significant global ecosystem impacts through greenhouse gas emissions. In 1997, the U.S. transportation sector accounted for 32% of national carbon emissions from fossil fuels, or 473.1 million tons of carbon. From 1984 to 1997, carbon emissions from transportation increased by 25% (rising from 379 million tons). In addition, vehicle use contributes to emissions of two other greenhouse gases, methane and nitrous oxide. Total emissions of these gases for the United States in 1997 were 213 tons of methane and 205 tons of nitrous oxide (US Environmental Protection Agency 2001).

27.4 Responding to the Environmental and Ecological Burdens of Urban Systems

Many of the reasons why ecosystem services and environmental problems in general have tended to be neglected in urban systems are similar to the reasons they have been neglected elsewhere.

- *Ecosystem services are provided through complex and poorly understood processes, taking place mostly beyond urban boundaries.* It is often difficult to understand how ecosystem changes affect human well-being, and hence which ecosystems are of particular value and which changes will result in the greatest losses (MA 2003). In urban areas, people are also likely to be even less aware than rural dwellers of how dependent they are on ecosystem services and how their actions affect distant ecosystems.

- *Ecosystems services (and environmental burdens that affect ecosystems) are difficult for private agencies to own and trade.* Only a small share of ecosystem services accrue to the owner of property where the ecosystem is located, and it is rarely feasible for the owner to charge beneficiaries for the services that they receive. It is equally difficult for those affected by ecosystem change to negotiate with those who are causing the changes to take place. Thus property owners rarely have the incentive to take account of how they are altering the availability of ecosystem services. Attempts have been made to develop markets for ecological or environmental services, but these remain rudimentary.

- *Ecosystem services are difficult for public agencies to manage or regulate.* Ecosystems located on private land are difficult for government agencies to regulate in the best of circumstances. Even where public land is involved, the benefits of ecosystem services typically cross administrative and sectoral boundaries, and often no agency has the responsibility and capacity to care for the relevant ecosystems.

- *The groups most vulnerable physically and socially tend to be the least influential economically and politically.* The urban dwellers most dependent on local environmental services and conditions are the urban poor in low-income countries. Alternatively, among those most likely to be affected by global ecological degradation and resource depletion are future generations, who do not even have a political or economic presence.

There have been some notable successes despite these difficulties. This applies at all scales, from the intraurban environmental health issues so common in low-income areas to the global pressures associated with affluent urban lifestyles. The enormous variation in and among urban centers makes it difficult to generalize about the relevance of these successes. Moreover, even if they reflect the potential for addressing urban environmental and ecological challenges, the relevance of past successes to future responses is limited.

The concerted responses to sanitary threats that emerged in a number of cities in the nineteenth century are testimony to the potential for changing urban environmental management when the need arises. They included elements addressing each of the four challenges just noted and combined them in what came to be known as the sanitary movement (Melosi 2000).

Enormous progress was made in the study of urban public health issues: the nineteenth century saw eventual ascendance of the bacterial theory of disease over alternatives such as miasma theory, which held the diseases were contracted from the vapors emitted by, for example, urban filth (Rosen 1993). Municipal and national governments were put under pressure to organize sanitary improvements, and although there was successful resistance to demands to control air pollution (Mosley 2001), water and sewerage improvements received widespread support (Melosi 2000). There was public debate over the appropriate organization form for local water and sanitation networks (Jacobson 2000), and in many cities private enterprises helped undertake major public infrastructure projects. The increasing efforts to address water and sanitation were linked to changes that had given more political influence to the urban residents at risk (Szreter 1998, 2002).

Despite these successes, the problems that motivated the nineteenth-century sanitary reforms remain some of the major challenges for twenty-first-century urban areas. It has been estimated that more than 900 million people live in urban slums, characterized as having inadequate housing and basic services (UN-Habitat 2003b, 2003c). As indicated, a poorly documented but appreciable share of urban and rural dwellers do not have access to adequate water and sanitation, and this remains one of the major causes of preventable illness and death globally. The ecology of disease in low-income urban settlements remains poorly understood, debate over the appropriate roles of the public and private sectors continues, and the political influence of slum dwellers is still not sufficient to secure needed improvements. Moreover, conventional infrastructure projects, involving water-borne sewage systems and long-distance water conveyance, can have large environmental as well as economic costs.

The urban environmental problems that helped motivate the environmental movement in industrial countries in the twentieth century also remain a major challenge. There is a considerable body of work on sustainable cities (Haughton and Hunter 1994; Satterthwaite 1999; Beatley 2004), and the agenda for action agreed on at the United Nations Conference on Environment and Development in 1992 (*Agenda 21*) explicitly attempted to ground global aspirations in local initiatives (*Local Agenda 21*s).

Moreover, numerous technologies have been developed that, if adopted on a large scale, would radically reduce the ecological footprint of urban settlements. For example, industrial parks that promote the use of wastes from one production process as inputs into another are emerging in Canada, Denmark, the United States, and many countries in Asia, following the model set by Kalundborg in Denmark (Cohen-Rosenthal and Musnikow 2003). Rapid bus transit systems in Latin American cities such as Bogotá, Quito, and Curitiba provide successful alternative examples to the previous solutions of heavy and light rail (Fjellstrom 2003). Fuel-efficient cars and cleaner fuels are becoming increasingly available, creating a potentially significant influence on reducing local and global burdens of urban transport systems. Indeed, some have argued that the technological bases for reducing environmental pressures manyfold, without sacrificing human well-being, have been available for some time now (Weizsäcker et al. 1997).

There is no evidence, however, that the tendency toward increasing ecological footprints per capita has been reversed, even in upper-income countries where footprints are the largest and least sustainable. In part this is due to the large use of materials to produce the tools necessary for high-technology products. For example, manufacturing one desktop computer and a 17-inch cathode ray tube monitor requires at least 240 kilograms of fossil fuels, 22 kilograms of chemicals, and 1,500 liters of water. In terms of mass, the total amount of materials used is about equal to that of a midsize car (Williams 2003). The large energy and material inputs for these products, without recycling, does not suggest a dematerialization of the "knowledge society" in the near future.

In summary, although past trends demonstrate considerable potential for addressing urban environmental and ecological burdens, they do not indicate whether this potential will be realized. Historically, concerted responses have been a reaction to crises rather than the result of forward thinking. Sanitary reform, for example, emerged at a time when epidemics flourished as the result of unsanitary urban conditions, and it has proved far more difficult to maintain in the face of less dramatic diseases. It is possible to interpret the history of some of the world's affluent cities as a series of victories over ecological and environmental challenges: first a sanitary revolution (nineteenth century), then a pollution revolution (twentieth century), and now an anticipated sustainability revolution. It is also possible, however, to interpret this same sequence as a process of displacement that has left most contemporary cities in a very difficult situation.

References

Allen, P., 1999: Reweaving the food security safety net: Mediating entitlement and entrepreneurship. *Agriculture and Human Values,* 16(2), 117–129.

Altieri, M.A., N. Companioni, K. Canizares, C. Murphy, P. Rosset, M. Bourque, and C.I. Nicholls, 1999: The greening of the "barrios": urban agriculture for food security in Cuba. *Agriculture and Human Values,* 16(2), 131–140.

Anderson, R.M. and R.M. May, 1991: *Infectious Diseases of Humans: Dynamics and Control.* Oxford University Press, Oxford, 757 pp.

Armstrong, W.R. and T.G. McGee, 1985: *Theatres of Accumulation. Studies in Asian and Latin American Urbanization.* Methuen, London.

Asian Development Bank, 1997: *Emerging Asia: Challenges and Changes.* Asian Development Bank and Oxford University Press, Hong Kong.

Bai, X., 2003: The process and mechanism of urban environmental change: an evolutionary view. *International Journal of Environment and Pollution,* 19(5), 528–541.

Bai, X. and H. Imura, 2000: A Comparative Study of Urban Environment in East Asia: Stage Model of Urban Environmental Evolution. *International Review for Environmental Strategies,* 1(1), 135–158.

Bai, X. and H. Imura, 2001: Towards sustainable urban water resource management: a case study in Tianjin, China. *Sustainable Development,* 9, 24–35.

Bairoch, P., 1988: *Cities and Economic Development: From the Dawn of History to the Present.* Mansell Publishing, London, 574 pp.

Bakker, N., M. Dubelling, S. Gündel, U. Sabel-Koschella, and H. de Zeeuw (eds.), 2000: *Growing Cities, Growing Food: Urban Agriculture on the Policy Agenda.* Deutsche Stiftung für Entwicklung (DSE), Feldafing.

Balk, D., F. Pozzi, G. Yetman, U. Deichmann, and A. Nelson, 2004: *The Distribution of People and the Dimension of Place: Methodologies to Improve the Global Estimation of Urban Extents.* CIESIN, New York.

Bartone, C., J. Bernstein, J. Leitmann, and J. Eigen, 1994: *Toward Environmental Strategies for Cities: Policy Consideration for Urban Environmental Management in Developing Countries.* UNDP/UNCHS/World Bank Urban Management Programme 18, World Bank, Washington, DC.

Bateman, O.M., S. Smith, and P. Roark, 1993: *A Comparison of the Health Effects of Water Supply and Sanitation in Urban and Rural Areas of Five African Countries.* WASH Field Report 398, Water and Sanitation for Health Project, WASH Operations Center, Arlington, VA, 1993.

Baumann, D.D., J.J. Boland, and W.M. Hanemann (eds.), 1998: *Urban Water Demand Management and Planning.* McGraw-Hill, New York, 350 pp.

Beatley, T. (ed.), 2004: *The Sustainable Urban Development Reader.* Routledge, London, 348 pp.

Berkowitz, A.R., C.H. Nilon, and K.S. Hollweg (eds.), 2003: *Understanding Urban Ecosystems: A New Frontier for Science and Education.* Springer-Verlag, New York, 523 pp.

Birley, M.H. and K. Lock, 1998: Health and peri-urban natural resource production. *Environment and Urbanization,* **10(1),** 89–106.

Blair, R.B., 1996: Land use and avian species diversity along an urban gradient. *Ecological Applications,* **6(2),** 506–519.

Blair, R.B. and A.E. Launer, 1997: Butterfly diversity and human land use: species assemblages along and urban gradient. *Biological Conservation,* **80,** 113–125.

Bolund, P. and S. Hunhammar, 1999: Ecosystem services in urban areas. *Ecological Economics,* **29,** 293–301.

Bradley, D.J., 1991: Malaria. In: *Disease and Mortality in Sub-Saharan Africa,* R.G. Feachem and D.T. Jamison (eds.), Oxford University Press for the World Bank, Oxford, 190–202.

Bradshaw, A.D., 2003: Natural Ecosystems in Cities: A Model for Cities as Ecosystems. In: *Understanding Urban Ecosystems,* A.R. Berkowitz, C.H. Nilon, and K.S. Hollweg (eds.), Springer-Verlag, New York, 77–94.

Brakman, S., H. Garretsen, and C. van Marrewijk, 2001: *An Introduction to Geographical Economics: Trade, Location and Growth.* Cambridge University Press, Cambridge, 350 pp.

Brennan, E., 1999: *Population, Urbanization, Environment, and Security: A Summary of the Issues.* Occasional Paper Series, Comparative Urban Studies 22, Woodrow Wilson International Center for Scholars, Washington DC.

Brenner, N., 1999: Globalization as reterritorialisation: the re-scaling of urban governance in the European Union. *Urban Studies,* **36(3),** 431–451.

Bryld, E., 2003: Potentials, problems, and policy implications for urban agriculture in developing countries. *Agriculture and Human Values,* **20(1),** 79–86.

Cairncross, S. and R.G. Feachem, 1993: *Environmental Health Engineering in the Tropics: An Introductory Text.* 2nd ed. John Wiley & Sons, Chichester, 306 pp.

Cairncross, S., U. Blumenthal, P. Kolsky, L. Moraes, and A. Tayeh, 1995: The public and domestic domains in the transmission of disease. *Tropical Medicine and International Health,* **39,** 173–176.

CIESIN (Center for International Earth Science Information Network), Columbia University; IFPRI (International Food Policy Research Institute); World Bank; and CIAT (Centro Internacional de Agricultura Tropical) 2004a. Global Rural-Urban Mapping Project (GRUMP): Gridded Population of the World, version 3, with Urban Reallocation (GPW-UR) [online] CIESIN, Columbia University, Palisades, NY. Available at http://beta.sedac.ciesin.columbia.edu/gpw.

CIESIN, Columbia University; IFPRI; World Bank; and CIAT 2004b. Global Rural-Urban Mapping Project (GRUMP): Urban Extents. [online] CIESIN, Columbia University, Palisades, NY. Available at http://beta.sedac.ciesin.columbia.edu/gpw.

Chant, S., 2003: *Female household headship and feminization of poverty: facts, fictions and forward strategies.* Gender Institute, London School of Economics, London.

Cleaver, K. and G. Schreiber, 1994: *Reversing the Spiral: The Population, Agriculture and Environment Nexus in Sub-Saharan Africa.,* World Bank, Washington DC.

Cliff, A.D., P. Haggett, and M. Smallman-Raynor, 1998: *Deciphering Global Epidemics: Analytical Approaches to the Disease Records of World Cities, 1888–1912.* Cambridge University Press, Cambridge, 469 pp.

Cohen, B., 2003: Urban Growth in Developing Countries: A Review of Current Trends and a Caution Regarding Existing Forecasts. *World Development,* **32(1),** 23–51.

Cohen, D., M. Green, C. Block, R. Slepon, R. Ambar, S.S. Wasserman, and M.M. Levine, 1991: Reduction of transmission of shigellosis by control of houseflies (Musca domestica). *The Lancet,* **337,** 993–997.

Cohen-Rosenthal, E. and J. Musnikow (eds.), 2003: *Eco-industrial strategies : unleashing synergy between economic development and the environment.* Greenleaf, Sheffield, 384 pp.

Collins, J.P., A. Kinzig, N.B. Grimm, W.F. Fagan, D. Hope, J.G. Wu, and E.T. Borer, 2000: A new urban ecology. *American Scientist,* **88(5),** 416–425.

Cronon, W., 1992: *Nature's Metropolis: Chicago and the Great West.* W.W. Norton, New York, 530 pp.

Crosskey, R.W. and R.P. Lane, 1993: House-flies, blow-flies and their alies (calyptrate Diptera). In: *Medical Insects and Arachnids,* R.P. Lane and R.W. Crosskey (eds.), Chapman & Hall, London, 403–428.

Curtis, V. and S. Cairncross, 2003: Effect of washing hands with soap on diarrhea risk in the community: a systematic review. *Lancet Infectious Diseases,* **3(5),** 275–281.

Curtis, V., S. Cairncross, and R. Yonli, 2000: Review: Domestic hygiene and diarrhea—pinpointing the problem. *Tropical Medicine & International Health,* **5(1),** 22–32.

Davis, D., 2002: *When Smoke Ran Like Water, Tales of Environmental Deception and the Battle Against Pollution,* New York: Basic Books.

Decker, E.H., S. Elliott, F.A. Smith, D.R. Blake, and F.S. Rowland, 2000: Energy and material flow through the urban ecosystem. *Annual Review of Energy and the Environment,* **25,** 685–740.

Denys, C. and H. Schmidt, 1998: Insect communities on experimental mugwort (Artemisia vulgaris L.) plots along an urban gradient. *Oecologia,* **113(2),** 269–277.

Dinno, A., 2000: *Health and community based urban residential restoration: an investigation in the utility of the traditional epidemiological approach.* Urban Resources Initiative Working Paper, Yale University, School of Forestry and Environmental Studies, New Haven, May.

Droege, P., 2004: Renewable energy and the city. *Encyclopedia of Energy, Volume 5,* 301–311.

Dyson, T., 2003: HIV/AIDS and Urbanization. *Population and Development Review,* **29(3),** 427–442.

Eastwood, R. and M. Lipton, 2000: *Rural-Urban Dimensions of Inequality Change.* Working Paper, World Institute for Development Economics Research (WIDER), Helsinki, 60 pp.

Eaton, D. and T. Hilhorst, 2003: Opportunities for managing solid waste flows in the peri-urban interface of Bamako and Ouagadougou. *Environment and Urbanization,* **15(1),** 53–63.

Elmqvist, T., J. Colding, S. Barthel, S. Borgström, A. Duit, et al. 2004: The dynamics of social-ecological systems in urban landscapes: Stockholm and the National Urban Park, Sweden. *Annals of New York Academy of Sciences,* **1023,** 308–322.

Elsom, D., 1996: *Smog Alert, Managing Urban Air Quality.* Earthscan, London.

Esrey, S.A., 2002: Philosophical, ecological and technical challenges for expanding ecological sanitation into urban areas. *Water Science and Technology,* **45(8),** 225–228.

ETC—Urban Agriculture Programme, 2001: *Annotated Bibliography on Urban Agriculture.* ETC Netherlands, Leusden, The Netherlands, Updated March 2003, 804 pp.

Ezzati, M., A.D. Lopez, A. Rodgers, S. Vander Hoorn, and C.J.L. Murray, 2002: Selected major risk factors and global and regional burden of disease. *Lancet,* **360(9343),** 1347–1360.

Faiz, A., K. Sinha, M. Walsh, and A. Varma, 1990: *Automotive Air Pollution: Issues and Options for Developing Countries.* World Bank, Washington DC.

Fjellstrom, K., 2003: Sustainable transport for cities in ASEAN. Paper presented at the *Workshop on Environmentally Sustainable Cities in ASEAN,* 2–4 December. ASEAN, Singapore.

Flynn, K.C., 2001: Urban agriculture in Mwanza, Tanzania. *Africa,* **71(4),** 666–691.

Friedmann, J., 1986: "The world city hypothesis," *Development and Change,* 17(1): 69–83

Friedmann, J., 1995: Where we stand: a decade of world city research, in P. L. Knox and P. J. Taylor (eds). *World Cities in a World-System,* Cambridge, Cambridge University Press, pp. 21–47.

Friedmann, J., 2002: *The Prospect of Cities.* University of Minnesota Press, Minneapolis, 194 pp.

Friedmann, J. and G. Wolff, 1982: World city formation: an agenda for research and action, *International Journal of Urban and Regional Research,* 6(3): 309–344.

Frumkin, H., L. Frank, and R. Jackson, 2004: *Urban Sprawl and Public Health: Designing, Planning and Building for Healthy Communities.* Island Press, Washington D.C., 338 pp.

Fuchs, R.J., E. Brennan, J. Chamie, F.-c. Lo, and J.I. Uitto (eds.), 1994: *Mega-City Growth and the Future.* United Nations University Press, Tokyo, 428 pp.

Gilbert, A. (ed.), 1996: *The Mega-City in Latin America.* United Nations University, Tokyo.

Glasgow, N., 2000: Rural/urban patterns of aging and caregiving in the United States. *Journal of Family Issues,* **21(5),** 611–631.

Gleick, P.H., 2003: Water Use. *Annual Review of Environment and Resources,* **28,** 275–314.

Grimm, N.B., J.M. Grove, S.T.A. Pickett, and C.L. Redman, 2000: Integrated approaches to tong-term studies of urban ecological systems. *BioScience,* **50(7),** 571–584.

Hall, P., 1995: Towards a general urban theory. In: *Cities in Competition, Productive and Sustainable Cities for the 21st Century,* J. Brotchie, M. Batty, E. Blakely, P. Hall, and P. Newton (eds.), Longman Australia, Melbourne, 3–31.

Hall, P., 2002: *Urban and Regional Planning.* 4th ed. Routledge, London, 237 pp.

Hardoy, J.E., D. Mitlin, and D. Satterthwaite, 2001: *Environmental Problems in an Urbanizing World.* Earthscan, London.

Haughton, G. and C. Hunter, 1994: *Sustainable Cities.* Jessica Kingley Publishers, London, 357 pp.

Heimlich, R.E. and W.D. Anderson, 2001: *Development at the Urban Fringe and Beyond: Impacts on Agriculture and Rural Land.* 803, Economic Research Service, U.S. Department of Agriculture, Washington D.C., 80 pp.

Hejný, S., H. Sukopp, and I. Kowarik (eds.), 1990: *Urban Ecology: Plants and Plant Communities in Urban Environments.* SPB Academic, The Hague, 282 pp.

Holdren, J.P. and K.R. Smith, 2000: Energy, the Environment and Health. In: *World Energy Assessment: Energy and the Challenge of Sustainability,* J. Goldemberg (ed.), UNDP, New York, 62–110.

Honjo, M., 1998: The growth of Tokyo as a world city. In: *Globalization and the World of Large Cities,* F.-c. Lo and Y.-M. Yeung (eds.), UNU Press, Tokyo, 109–131.

Howorth, C., I. Convery, and P. O'Keefe, 2001: Gardening to reduce hazard: Urban agriculture in Tanzania. *Land Degradation & Development,* **12(3),** 285–291.

International Federation of Red Cross and Red Crescent Societies, 2001: *World Disasters Report.* Kumarian Press, Bloomfield, CT.

Jacobson, C.D., 2000: *Ties that Bind: Economic and Political Dilemmas of Urban Utility Networks, 1800–1990.* University of Pittsburgh Press, Pittsburgh, 282 pp.

Kahrl, W., L., 1982: *Water and Power: The Conflict over Los Angeles' Water Supply in the Owens Valley.* University of California Press, Berkeley.

Keilman, N., 2003: Biodiversity: The threat of small households. *Nature,* **421(6922),** 489–490.

Kenworthy, J.R. and F.B. Laube, 1996: Automobile dependence in cities: An international comparison of urban transport and land use patterns with implications for sustainability. *Environmental Impact Assessment Review: Special Issue: Managing Urban Sustainability,* **16(4–6),** 279–308.

Khaldûn, I., 1981: *The Muqaddimah: An Introduction to History.* Bollingen Series, Princeton, New Jersey, 465 pp.

Kieran, J., 1995: *A natural history of New York city : a personal report after fifty years.* 2nd ed. Fordham University Press, New York, 428 pp.

Krzyzanowski, M. and D. Schwela, 1999: Patterns of air pollution in developing countries. In: *Air Pollution and Health,* S.T. Holgate, H.S. Koren, J.M. Samet, and R.L. Maynard (eds.), Academic Press, London, 105–113.

Levin, H., 1997: Systematic Evaluation and Assessment of Building Environmental Performance (SEABEP). Paper presented at the *Buildings and Environment,* 9–12 June, 1997. Hal Levin and Associates, Santa Cruz, Ca, Paris.

Levin, H., A. Boerstra, and S. Ray, 1995: Scoping US Buildings Inventory Flows and Environmental Impacts in Life Cycle Assessment. Paper presented at the *World Congress of the Society for Environmental Toxicology and Chemistry (SETAC),* November, Vancouver, B.C.

Levine, O.S. and M.M. Levine, 1991: Houseflies (musca domestica) as mechanical vectors of shigellosis. *Reviews of Infectious Diseases,* **13 (July–August),** 688–696.

Liu, J.G., G.C. Daily, P.R. Ehrlich, and G.W. Luck, 2003: Effects of household dynamics on resource consumption and biodiversity. *Nature,* **421(6922),** 530–533.

Lo, F.-c., 1994: The impacts of current global adjustment and shifting techno-economic paradigm on the world city system. In: *Mega-City Growth and The Future,* R.J. Fuchs, E. Brennan, J. Chamie, F.-c. Lo, and J.I. Uitto (eds.), United Nations University Press, Tokyo, 103–130.

Lo, F.-c. and Y.-m. Yeung (eds.), 1996: *Emerging World Cities in Pacific Asia.* UNU Press, Tokyo.

Lo, F.-c. and P.J. Marcotullio (eds.), 2001: *Globalization and the Sustainability of Cities in the Asia Pacific Region.* United Nations University Press, Tokyo.

Losada, H., R. Bennett, R. Soriano, J. Vieyra, and J. Cortes, 2000: Urban agriculture in Mexico City: Functions provided by the use of space for dairy based livelihoods. *Cities,* **17(6),** 419–431.

Losada, H., H. Martinez, J. Vieyra, R. Pealing, R. Zavala, and J. Cortes, 1998: Urban agriculture in the metropolitan zone of Mexico City: changes over time in urban, suburban and peri-urban areas. *Environment and Urbanization,* **10(2),** 37–54.

Luck, M. and J.G. Wu, 2002: A gradient analysis of urban landscape pattern: a case study from the Phoenix metropolitan region, Arizona, USA. *Landscape Ecology,* **17(4),** 327–339.

Lutz, W., W.C. Sanderson, and S. Scherbov (eds.), 2004: *The End of World Population Growth in the 21st Century, New Challenges for Human Capital Formation & Sustainable Development.* Earthscan, London.

MA (Millennium Ecosystem Assessment), 2003: *Ecosystems and Human Wellbeing: A Framework for Assessment.* Island Press, Washington, DC, 245 pp.

Marcotullio, P.J., 2003: Globalisation, urban form and environmental conditions in Asia- Pacific cities. *Urban Studies,* **40(2),** 219–247.

Marcotullio, P.J., 2004: Why the Asian urbanization experience should make us think differently about planning approaches. In: *Towards Sustainable Cities, East Asian, North American and European Perspectives on Managing Urban Regions,* A. Sorenson, P.J. Marcotullio, and J. Grant (eds.). 38–58, Ashgate Publishers, LTD, Aldershot, UK.

Marcotullio, P.J. and Y.-s.F. Lee, 2003: Urban environmental transitions and urban transportation systems: a comparison of the North American and Asian experiences. *International Development Planning Review.*

Mascie-Taylor, C.G.N. (ed.), 1993: *The Anthropology of Disease. Biosocial Society Series 5,* Oxford University Press, Oxford, 169 pp.

Maxwell, D., 1999: The political economy of urban food security in Sub-Saharan Africa. *World Development,* **27(11),** 1939–1953.

Maxwell, D., C. Levin, and J. Csete, 1998: Does urban agriculture help prevent malnutrition? Evidence from Kampala. *Food Policy,* **23(5),** 411–424.

McDonnell, M.J. and S.T.A. Pickett, 1991: Comparative analysis of ecosystems along gradients of urbanization: Opportunities and limitations. In: *Comparative Analyses of Ecosystems, Patterns, Mechanisms and Theories,* J.J. Cole, G.M. Lovett, and S.E.G. Findlay (eds.), Springer Verlag, New York, 351–255.

McDonnell, M.J. and S.T.A. Pickett, 1997: Ecosystem processes along an urban-to-rural gradient. *Urban Ecosystems,* **1,** 21–36.

McGee, T.G., 1991: The emergence of Desakota regions in Asia: expanding a hypothesis. In: *The Extended Metropolis: Settlement Transition in Asia,* S. Ginsburg Norton, T.G. McGee, and B. Koppel (eds.), University of Hawaii Press, Honolulu.

McGee, T.G. and I. Robinson (eds.), 1995: *The mega-urban regions of southeast Asia.* UBC Press, Vancouver, B.C., 384 pp.

McGranahan, G. and F. Murray (eds.), 2003: *Air Pollution and Health in Rapidly Developing Countries.* Earthscan, London, 227 pp.

McGranahan, G., J. Songsore, and M. Kjellén, 1996: Sustainability, poverty and urban environmental transitions. In: *Sustainability, the Environment and Urbanization,* C. Pugh (ed.), Earthscan, London, 103–133.

McGranahan, G., J. Leitmann, and C. Surjadi, 1998: Green grass and brown roots: Understanding environmental problems in disadvantaged neighbourhoods. *Journal of Environmental Planning and Management,* **41(4),** 505–518.

McGranahan, G., P. Jacobi, J. Songsore, C. Surjadi, and M. Kjellén, 2001: *The Citizens at Risk: From Urban Sanitation to Sustainable Cities.* Earthscan, London, 200 pp.

McMichael, A.J., M. McKee, V. Shkolnikov, and T. Valkonen, 2004: Mortality trends and setbacks: global convergence or divergence, *The Lancet,* **363,** 1155–59.

McNeill, W.H., 1989: *Plagues and Peoples.* Doubleday, New York, 340 pp.

McNeill, W.H., 1993: Patterns of disease emergence in history. In: *Emerging Viruses,* S.S. Morse (ed.), Oxford University Press, New York, 29–36.

Melosi, M.V., 2000: *The Sanitary City: Urban Infrastructure in America from Colonial Times to the Present.* The Johns Hopkins University Press, Baltimore, 578 pp.

Midmore, D.J. and H.G.P. Jansen, 2003: Supplying vegetables to Asian cities: is there a case for peri-urban production? *Food Policy,* **28(1),** 13–27.

Mittelback, M. and M. Crewdson, 1997: *Wild New York: A Guide to the Wildlife, Wild Places and Natural Phenomena of New York City.* Crown Publishers, New York, 131 pp.

Monfreda, C., M. Wackernagel, and D. Deumling, 2004: Establishing national natural capital accounts based on detailed ecological footprint and biological capacity accounts. *Land Use Policy,* **21,** 231–236.

Moskow, A., 1999: Havana's self-provision gardens. *Environment and Urbanization,* **11(2),** 127–133.

Montgomery, M.R., R. Stren, B. Cohen, and H. Reed (eds.), 2003: *Cities Transformed: Demographic Change and Its Implications in the Developing World.* The National Academies Press, Washington, D.C., 529 pp.

Mosley, S., 2001: *The Chimney of the World: A History of Smoke Pollution in Victorian and Edwardian Manchester.* White Horse, Cambridge, 271 pp.

Murray, C.J.L. and A.D. Lopez, 1996: *Global Health Statistics: a compendium of incidence, prevalence and mortality estimates for over 200 conditions.* Global Burden of Disease and Injury Series. Volume II, Harvard School of Public Health on behalf of the World Health Organization and the World Bank, Cambridge, MA, 1996.

NRC (National Research Council), 2000: *Watershed Management for Potable Water Supply: Assessing New York City's Approach.* National Academy Press, Washington DC.

Natuhara, Y. and C. Imai, 1996: Spatial structure of avifauna along urban-rural gradients. *Ecological Research,* **11(1),** 1–9.

Newman, P. and J. Kenworthy, 1999: *Sustainability and Cities: Overcoming Automobile Dependence.* Island Press, Washington, 442 pp.

Nickum, J., 2002: Water and sustainability in Asian megalopolises: the case of Beijing. In: *1.18 Human Resource System Challenge VII: Human Settlement Development, in Encyclopedia of Life Support Systems (EOLSS),* S. Sassen (ed.), Developed under the auspices of the UNESCO, EOLSS Publishers, [http://www.eolss.net]. Oxford, UK.

Niemela, J., 1999: Ecology and urban planning. *Biodiversity and Conservation,* **8(1),** 119–131.

Ning, Y.m–, 2001: Globalization and the sustainable development of Shanghai. In: *Globalization and the Sustainability of Cities in the Asia Pacific Region,* F.-c. Lo and P.J. Marcotullio (eds.), UNU Press, Tokyo, 271–310.

Nowak, D.J., D.E. Crane, J.C. Stevens, and M. Ibarra, 2002: *Brooklyn's Urban Forest.* General Technical Report NE-290, U.S. Department of Agriculture, Forest Service, Northeastern Research Station, Newtown Square, PA, 107 pp.

Page, B., 2002: Urban agriculture in Cameroon: an anti-politics machine in the making? *Geoforum,* **33(1),** 41–54.

Pahl, R., 1965: *Urbs in Rure.* Weidenfeld and Nicolson, London.

Payne, G. and M. Majale, 2004: *The Urban Housing Manual: Making Regulatory Frameworks Work for the Poor.* Earthscan, London.

Pickering, H., R.J. Hayes, A.M. Tomkins, D. Carson, and D.T. Dunn, 1987: Alternative measures of diarrhoeal morbidity and their association with social and environmental factors in urban children in The Gambia. *Transactions of the Royal Society of Tropical Medicine and Hygiene,* **81,** 853–859.

Pickett, S.T.A., W.R. Burch Jr, S.E. Dalton, T.W. Foresman, J.M. Grove, and R. Rowntree, 1997: A conceptual framework for the study of human ecosystems in urban areas. *Urban Ecosystems,* **1(4),** 185–199.

Pickett, S.T.A., M.L. Cadenasso, J.M. Grove, C.H. Nilon, R.V. Pouyat, W.C. Zipperer, and R. Costanza, 2001: Urban ecological systems: Linking terrestrial ecological, physical, and socioeconomic components of metropolitan areas. *Annual Review of Ecology and Systematics,* **32,** 127–157.

Pires, M., 2004: Watershed protection for a world city: the case of New York. *Land Use Policy,* **21,** 161–175.

Platt, R.H., R.A. Rowntree, and P.C. Muick (eds.), 1994: *The Ecological City: Preserving and Restoring Urban Biodiversity.* University of Massachusetts Press, Amherst, 291 pp.

Plumb, C., 1999: Bangkok. In: *Cities in the Pacific Rim, Planning Systems and Property Markets,* J. Berry and S. McGreal (eds.). 129–156, E & F Spon, London.

Portnov, B.A. and U.N. Safriel, 2004: Combating desertification in the Negev; dryland agriculture vs. dryland urbanization. *Journal of Arid Environments,* **56,** 659–680.

Prüss, A., D. Kay, L. Fewtrell, and J. Bartram, 2002: Estimating the burden of disease from water, sanitation and hygiene at a global level. *Environmental Health Perspectives,* **110(5),** 537–542.

Rakodi, C., 1997: Global forces, urban change and urban management in Africa. In: *The Urban Challenge in Africa, Growth and Management of its Large Cities,* C. Rakodi (ed.), UNU Press, Tokyo, 17–73.

Rebele, F., 1994: Urban ecology and special features of urban ecosystems. *Global Ecology and Biogeography Letters,* **4(6),** 173–187.

Rees, W.E., 1992: Ecological footprints and appropriated carrying capacity: what urban economics leaves out. *Environment and Urbanization,* **4(2),** 121–130.

Rees, W.E., 1997: Is "sustainable city" an oxymoron? *Local Environment,* **2(3),** 303–310.

Rees, W.E., 2001: Ecological Footprint, Concept of. In: *Encyclopedia of Biodiversity,* S.A. Levin (ed.). 2, Academic Press, San Diego, 229–244.

Rees, W.E., 2003: Understanding urban ecosystems: an ecological economics perspective. In: *Understanding Urban Ecosystems: A New Frontier for Science and Education,* A.R. Berkowitz, C.H. Nilon, and K.S. Hollweg (eds.), Springer-Verlag, New York, 115–136.

Rees, W.E. and M. Wackernagel, 1994: Ecological footprints and appropriated carrying capacity: Measuring the natural capital requirements of the human economy. In: *Investing in Natural Capital: The Ecological Economics Approach to Sustainability,* A.-M. Jannson, M. Hammer, C. Folke, and R. Contanza (eds.), Island Press, Washington, DC, 362–390.

Reisner, M., 2001: *Cadillac Desert: The American West and its Disappearing Water.* Revised ed. Pimlico, London, 582 pp.

Robbins, P., A. Polderman, and T. Birkenholtz, 2001: Lawns and toxins—An ecology of the city. *Cities,* **18(6),** 369–380.

Rogerson, C.M., 1995: Globalization or informalization?: African urban economies in the 1990s. In: *Managing Urban Growth in Africa,* C. Rakodi (ed.), 1–39.

Rolando, A., G. Maffei, C. Pulcher, and A. Giuso, 1997: Avian community structure along an urbanization gradient. *Italian Journal of Zoology,* **64(4),** 341–349.

Rosen, G., 1993: *A History of Public Health.* Expanded ed. Johns Hopkins University Press, Baltimore, 535 pp.

Rusk, D., 2001: Foreword. In: *Planning for a New Century, The Regional Agenda,* J. Barnett (ed.), Island Press, Washington DC, ix–xiii.

Rydell, J., 1992: Exploitation of insects around streetlamps by bats in Sweden. *Functional Ecology,* **6(6),** 744–750.

Sassen, S., 1991: *The Global City: New York, London, Tokyo.* Princeton University Press, Princeton.

Satterthwaite, D. (ed.), 1999: *The Earthscan Reader in Sustainable Cities.* Earthscan, London, 472 pp.

Satterthwaite, D., 2002: *Coping with rapid urban growth.* Royal Institution of Chartered Surveyors (RICS), London.

Saxena, S. and K.R. Smith, 2003: Indoor air pollution. In: *Air Pollution and Health in Rapidly Developing Countries,* G. McGranahan and F. Murray (eds.), Earthscan, London, 227.

Shapiro, A.M., 2002: The Californian urban butterfly fauna is dependent on alien plants. *Diversity and Distributions,* **8(1),** 31–40.

Showers, K.B., 2002: Water scarcity and urban Africa: an overview of urban-rural water linkages. *World Development,* **30(4),** 621–648.

Small, E., 2000: Understanding the ecology of urban green space: Decreasing habitat quality and increasing isolation? Cited August 28 2003. Available at http://urgent.nerc.ac.uk/Meetings/2000/2000Proc/ecology/small.htm.

Smit, J. and J. Nasr, 1992: Urban agriculture for sustainable cities: using wastes and idle lands and water bodies as resources. *Environment and Urbanization,* **4(2),** 141–151.

Smith, K. and S. Akbar, 2003: Health-Damaging Air Pollution: A matter of scale. In: *Health and Air Pollution in Rapidly Developing Countries,* G. McGranahan and F. Murray (eds.), Earthscan, London.

Smith, K.R. and Y.-S.F. Lee, 1993: Urbanization and the Environmental Risk Transition. In: *Third World Cities: Problems, Policies and Prospects,* J.D. Kasarda and A.M. Parnell (eds.), Sage Publications, Newbury Park, CA, 161–179.

Soegijoko, B. and B.S. Kusbiantoro, 2001: Globalization and the sustainability of Jabotabek, Indonesia. In: *Globalization and the Sustainability of Cities in the Asia Pacific Region,* F.-c. Lo and P.J. Marcotullio (eds.), UNU Press, Tokyo, 311–363.

Szreter, S., 1998: Health and welfare during industrialisation. *Economic History Review,* **51(2),** 432–433.

Szreter, S., 2002: The state of social capital: Bringing back in power, politics, and history. *Theory and Society,* **31(5),** 573–621.

Taylor, P.J., 2004: *World City Network, A Global Urban Analysis.* Routledge, London.

Thompson, S., 2003: Environmental impacts of construction on habitats—future priorities. *International Journal of Environmental Studies,* **60(3),** 277–286.

Tulchin, J.S., 1994: *Global Forces and the Future of the Latin American City.* 4, Woodrow Wilson International Center for Scholars, Washington DC.

UN-Habitat, 2003a: *Water and Sanitation in the World's Cities: Local Action for Global Goals.* Earthscan, London, 304 pp.

UN-Habitat, 2003b: *Guide to Monitoring Target 11: Improving the lives of 100 million slum dwellers.* United Nations Human Settlements Programme, Nairobi, 15 pp.

UN-Habitat, 2003c: *The Challenge of Slums: Global Report on Human Settlements 2003.* United Nations Human Settlements Programme, Nairobi, 310 pp.

United Nations, 2002: *World Urbanization Prospects: The 2001 Revision.* E/ESA/WP.191, United Nations, New York.

United Nations, 2004: *World Urbanization Prospects: The 2003 Revision.* E/ESA/WP.191, United Nations, New York.

United Nations Centre for Human Settlements, 1996: *An Urbanizing World: Global Report on Human Settlements, 1996.* Oxford University Press for United Nations Centre for Human Settlements (HABITAT), Oxford, 559 pp.

United Nations Centre for Human Settlements, 2001: *Cities in a Globalizing World, Global Report on Human Settlement in 2001.* Earthscan, London.

UNEP (United Nations Development Programme), 1996: *Urban Agriculture: Food, Jobs and Sustainable Cities.* Vol. One, *Publication Series for Habitat II,* United Nations Development Programme, New York, 302 pp.

U.S. Environmental Protection Agency, 2001: *Our Built and Natural Environments, A Technical Review of the Interactions between Land Use, Transportation and Environmental Quality.* EPA 231-R-01–002, EPA, Washington DC, January.

van den Berg, L.M., M.S. van Wijk, and P. Van Hoi, 2003: The transformation of agriculture and rural life downstream of Hanoi. *Environment and Urbanization,* **15(1),** 35–52.

Wackernagel, M., 1998: The Ecological Footprint of Santiago de Chile. *Local Environment,* **3(1),** 7–25.

Wackernagel, M. and W. Rees, 1996: *Our Ecological Footprint.* New Society Publishers, Gabriola Island, Canada.

Wackernagel, M., L. Onisto, P. Bello, A.S. Linares, I.S.L. Falfán, J.M. Garcia, A.I.S. Guerrero, and M.G.S. Guerrero, 1999: National natural capital accounting with the ecological footprint concept. *Ecological Economics,* **29,** 375–390.

Walker, L. and W.E. Rees, 1997: Urban density and ecological footprints: An analysis of Canadian households. In: *Eco-city Dimensions: Healthy Communities, Healthy Planet,* M. Roseland (ed.), New Society, Gabriola Island, B.C.

Walsh, M.P., 2003: Vehicle emission and health in development countries. In: *Air Pollution & Health in Rapidly Developing Countries,* G. McGranahan and F. Murray (eds.), Earthscan, London, 146–175.

Weizsäcker Ernst, U.v., B. Lovins Amory, and L.H. Lovins, 1997: *Factor four : doubling wealth—halving resource use : the new report to the Club of Rome.* Earthscan, London, 322 pp.

WHO (World Health Organization), 2002: *The World Health Report 2002: Reducing Risks, Promoting Healthy Life.* Geneva.

WHO and UNICEF (United Nations Children's Fund), 2000: *Global Water Supply and Sanitation Assessment 2000 Report.* Geneva and New York, 79 pp.

Williams, E., 2003: Environmental Impacts in the Production of Personal Computers. In: *Computers and the Environment: Understanding and Managing their Impacts,* R. Kuehr and E. Williams (eds.), Kluwer Academic, Dordrecht, 41–72.

Woods, R., 2003: Urban-rural mortality differentials: An unresolved debate. *Population and Development Review,* **29(1),** 29–46.

World Commission on Environment and Development, 1987: *Our Common Future.* Oxford University Press, Oxford, 400 pp.

WRI (World Resources Institute), 1996: *World Resources 1996–97; The Urban Environment.* Oxford University Press, New York, 365 pp.

World Wide Fund for Nature, 2002: *Living Planet Report 2002.* World Wide Fund for Nature International, Gland, Switzerland.

Chapter 28

Synthesis: Condition and Trends in Systems and Services, Trade-offs for Human Well-being, and Implications for the Future

Coordinating Lead Authors: Anthony C. Janetos, Roger Kasperson
Lead Authors: Tundi Agardy, Jackie Alder, Neville Ash, Ruth DeFries, Gerald Nelson
Review Editors: Brian Huntley, Richard Norgaard

---

*This appears in Appendix A at the end of this volume.

28.1 Introduction

This chapter draws on previous chapters to present a portrait of the major changes in the condition of ecosystems, the services they provide, the drivers of change, and the prospects for sustainability. It is organized around major questions and issues rather than around services and systems per se. The time frame for considering these issues is generally the last several decades, although in some instances this has been expanded where data allow or if needed to illustrate some particularly important points.

The chapter examines some of the implications of the observed trends and trade-offs for the continued capacity of ecosystems to provide services for human well-being but does not consider different trajectories into the future or possible policy responses, as these are considered in the *Scenarios* and *Policy Responses* volumes of the MA. Finally, the chapter outlines some of the major needs for continued research and data, focusing primarily on the areas that most severely hampered the ability of the MA Condition and Trends Working Group to reach conclusions.

28.2 How Have Ecosystem Services Contributed to Recent Improvements in Human Well-being?

Substantial improvements in human well-being in many parts of the globe have been apparent over the last half of the twentieth century. (See Chapter 5.) World population has more than doubled over the last 50 years (see Table 28.1), and consumption of many ecosystem services has grown even more. (See Chapters

Table 28.1. Global Population and Life Expectancy, 1950, 1975, and 2000 (United Nations 2003)

Population size

	1950	1975	2000
	(million)		
Industrial countries	813	1,047	1,194
Developing countries	1,706	3,021	4,877
Africa	221	408	796
Asia	1,398	2,398	3,680
Europe	547	676	728
Latin America and Caribbean	167	322	520
North America	172	243	316
Oceania	13	22	31
World	**2,519**	**4,068**	**6,071**

Life expectancy, both sexes combined

	1950–55	1970–75	1995–2000
	(years)		
Industrial countries	66.1	71.4	74.8
Developing countries	41.0	54.7	62.5
Africa	37.8	46.2	50.0
Asia	41.4	56.3	65.7
Europe	65.6	71.0	73.2
Latin America and Caribbean	51.4	60.9	69.4
North America	68.8	71.6	76.4
Oceania	60.3	65.8	73.2
World	**46.5**	**58.0**	**64.6**

7–17.) Life expectancy has increased in most areas of the world and infant mortality rates have declined almost everywhere, with the exceptions of the former Soviet Union and sub-Saharan Africa. Famines have become less common, and hunger has declined in absolute terms, although it remains a significant problem in some specific regions. (See Chapters 6 and 8.) Arrayed against this aggregate pattern of gains is the growing disparity of wealth, both between and within countries, and the failure of Africa in particular to share in the decades of major economic gains and the growth in social and institutional capabilities. (See Chapters 5 and 6.)

Numerous factors have contributed to these overall global improvements in human well-being, including major gains in manufactured and social capital, increased efficiencies associated with research and technology development, and the emergence of more-effective national and international institutions. (See Chapter 3.) But the human capacity to exploit ecosystem services has played a central role, and people have been extraordinarily successful in most of the world in using those services to meet a wide range of needs, such as food, clean air and water, shelter and protection from natural hazards, and cultural fulfillment. As human populations and individual well-being have increased, so has the consumption of ecosystem services, leading to increasing demands on ecosystems to provide for people.

The use of ecosystem services has also changed in its nature, in large part due to research and technology development, allowing for more-efficient use and production of services such as clean water and food and for partial substitutes to be developed for other services, such as for some fibers and for some cultural services. However, the parallel increased efficiency of use of many ecosystem services has been offset by increases in the absolute amounts of consumption of services, giving rise to serious concerns about the sustainability of their supply. (See Chapters 7–17.) In addition, the increasing consumption of some ecosystem services is clearly resulting in trade-offs in other services and thereby in the availability of certain benefits to people around the world. Such trade-offs in services, and thereby human well-being, are occurring across temporal and spatial scales; some areas and some people have gained, but at the same time some areas and some people are losing out.

Social and institutional issues are important in the effective use of ecosystem services. The enormous expansion of human capital has been crucial in securing both greater efficiency in and simply greater use of natural resources. Social institutions play an important role in facilitating growth, avoiding overexploitation, and managing environmental impacts. A decision-making environment with transparency, extensive flows of information, accountability, and minimization of corruption is a key to efficient uses of ecosystem services. The weight of evidence, however, suggests that current international disparities in income and wealth and in gender equality seriously constrain the development of institutional capabilities to support major future gains in human well-being. Well-functioning legal and judicial systems also contribute to sustainable development and the protection of environmental assets.

28.3 Which Areas Have Seen the Biggest Changes in Ecosystems over the Last Several Decades?

Changes in land cover, driven by the way people use land, are perhaps the most important single change in terrestrial ecosystems, affecting the supply of services. (See Chapters 21–27.) Although there has been a rapid expansion in the availability of data and information on ecosystems, there has not yet been a systematic examination, using global and regional observations, of the status and trends in land cover. This section summarizes a synthesis of the available information on rapid land cover change for about 1980 to 2000. Coastal and marine systems have also undergone dramatic changes during this period—for example, through the loss of mangrove forests and degradation of coral reefs in coastal areas (see Chapter 19), the declines in the abundance of marine fishes, and degradation of the sea bed in many areas (see Chapter 18). Of all the broad ecosystem types, inland waters are thought to be the most altered by human actions, particularly through the decline in water quality (see Chapters 7, 15, and 20) and the loss and fragmentation of wetlands (see Chapter 20).

The types of land use change included in the terrestrial analysis presented here are forest cover changes (deforestation and forest expansion), land degradation in drylands, and expansion of urban settlements. (See Box 28.1.) Information is also presented on recent changes in cropland extent.

Deforestation has been the most frequently measured process of land cover change on a regional scale; more datasets are available for the tropics than the boreal zones. During the 1990s, deforestation and forest degradation appeared to be more frequent in the tropics than elsewhere. (See Figure 28.1 in Appendix A and Chapter 21.) In particular, the Amazon Basin and Southeast Asia had large concentrations of deforestation "hotspots."

It is possible that areas of deforestation or degradation in the boreal or temperate regions (such as Canada or Siberia) are underrepresented. For example, forest degradation in Siberia, mostly related to logging activities, has been rapidly increasing in recent years. Moreover, the frequency of major fires, which are a natural disturbance factor in boreal forests, has increased globally since 1960. (See Chapter 16.)

There also remains uncertainty even within regions for which data are available. National statistical data indicating no net loss of forest can mask more detailed trends, such as a combination of deforestation in some regions and expansion of plantation forests in others. Some European countries, the United States, and Canada all experienced an overall increase in forest cover at the national level. (See Chapter 21.)

Most of the known areas of degraded dryland are found in Asia (see Figure 28.2 in Appendix A and Chapter 22), although not all the drylands of the world have been well studied. Major gaps occur in knowledge of dryland areas around the Mediterranean basin, in eastern Africa, in parts of South America (North of Argentina, Paraguay, Bolivia, Peru, and Ecuador) and in the United States. If all drylands were equally well studied, the global distribution of the most degraded land might turn out to be different, but the patterns currently observed in Asia would likely remain the same. The available data do not support the claim that the African Sahel is currently a desertification "hotspot." (See Chapter 22.)

The most populated areas of the world are located in the Gangetic plain of northern India, on the plain and north plateau of China, and on the island of Java in Indonesia. The most populated cities, with more than 750,000 inhabitants, are mainly located on the eastern coast of the United States, in Western Europe, and in India and East Asia. The most rapidly growing cities are located around the tropical belt, however, rather than in temperate regions. (See Figure 28.3 in Appendix A and Chapter 27.)

Data availability to determine cropland expansion and abandonment varies regionally. However, it is likely that the largest and most rapid areas of cropland increase are found in Southeast Asia, followed by Bangladesh, the Indus Valley, parts of the Middle East and Central Asia, the region of the Great Lakes of eastern Africa, and the southern border of the Amazon Basin in Latin America. (See Chapter 26.) North America accounts for most of the main areas of decrease in cropland (in the southeastern United States), followed by Asia (eastern part of China) and South America (parts of Brazil and Argentina). Areas of decrease in cropland extent are located in most other continents, although Africa is the only continent where no decrease in cropland was identified. Data quality issues are extremely important in evaluating these trends, and a graphic presentation of these changes, together with the associated data and an assessment of the data, is available at www.MAweb.org.

BOX 28.1
Constructing the Land Cover Change Maps

Deforestation and forest degradation. The map of the main areas of deforestation and forest degradation is based on three types of data sources: expert opinion gathered through formal procedures, remote sensing-based products, and national statistics. To avoid the coarse scale of national statistics, priority was given to the remote sensing and expert opinion data. The final map identifies, for each "forested" grid cell, how many input datasets covered the area and how many times it was considered as a main area of deforestation or forest degradation by these different datasets. A second color code represents the reliability (estimated in terms of convergence of evidence) of the information—that is, the frequency of detection as a hotspot relative to the number of data sets covering the area (see legend for color code). The information based on (sub)-national statistics provides average annual rates of deforestation and should be considered as secondary to the other sources because it is not at a fine resolution. When that rate is higher than 3% per year, the area is considered as rapid change.

Desertification. The map of the main areas of degraded land is constrained by lack of reliable data compared with the maps on deforestation and cropland extent. Most available data are quite heterogeneous in terms of monitoring methods or indicators used.

Changes in urban extent. While urban areas are defined as any region with population density greater than a threshold, the impact of urbanization on land cover is better measured by the change in built-up area. The final map shows the spatial distribution of the population density in 1995 and identifies the most populated and most rapidly changing cities of more than 750,000 inhabitants.

28.4 What are the Main Drivers of Change That Affect Ecosystems, Ecosystem Services, and Associated Human Well-being?

The human and environmental forces that drive changes in ecosystems, and thereby changes in ecosystem services and human well-being, are highly variable from place to place. As such, generic explanations or statements of causality are difficult to create. Driving forces are almost always multiple and interactive, so that a one-to-one linkage between particular driving forces and changes in ecosystems and ecosystem services is usually not possible. (See Chapters 3, 4, and 11.) Similarly, the linkage between particular changes in ecosystem services and various indicators of human well-being is often not well understood. (See Chapters 5 and 7–17.) In both cases, the causal linkage is almost always highly mediated by other factors, thereby complicating statements of causality or attempts to establish the proportionality of various contributors to changes. Analyses of driving forces generally distinguish between direct factors (the immediate causal agents of change) and the more indirect underlying factors that themselves cause change in the direct drivers. (See Chapter 3.)

Drivers of biodiversity loss are reasonably well understood, at least qualitatively. It is *well established,* for example, that habitat conversion, degradation, and fragmentation on land (usually for agricultural expansion) and in the oceans (mostly associated with fishing activities) have been the most important direct drivers of biodiversity loss globally in the recent past. Although habitat change will continue to be an important threat to biodiversity, the impacts of climate change, invasive non-native species, pollution, and nutrient overload are all increasingly important. (See Chapter 4.)

Some other linkages between drivers and impacts are also well known. In the case of impacts on human health, for example, it is known that burning wood, charcoal, and dung in poorly designed stoves has serious adverse effects on indoor air quality and is thought to be responsible for some 1.6 million deaths per year, almost all in developing countries. (See Chapters 9 and 27.) There is a complex set of driving forces affecting the increasing risk of infectious diseases in humans, including logging, dam and road building, agricultural expansions, urban sprawl, and pollution of coastal areas. (See Chapter 14.) With regard to adverse impacts on inland waters, land use change, nutrient overload, and pollution have been the key human driving forces. (See Chapter 20.)

Population growth, economic development, and increasing consumption and production are all important indirect drivers of change in ecosystems and ecosystem services. However, increasing attention is being given to the role of government policies relating to, for example, investments in rural roads, irrigation, credit systems, and agricultural research and extension, which have often served to expand food production. (See Chapter 8.) Policies that restrict trade, capital, and labor flows have conditioned access to international markets and have structured the international food system and global patterns of food production and consumption. Small-scale food producers in many poorer countries have been particularly affected by many such policies, as have general patterns of nutrient cycling. (See Chapter 12.)

28.5 Is There Evidence That Changes Made to Ecosystems in Order to Increase Provisioning Services Have Altered Regulating, Cultural, or Supporting Services?

The growth in human well-being at the global scale over the last several decades has come in large part because of increases in provisioning services from several major ecosystems. These changes have been particularly significant in cultivated systems, where the largest changes in the recent past have been as a result of intensification rather than the large-scale conversion of uncultivated land to agriculture (see Chapters 8 and 26), and in coastal and marine systems, where changes have occurred due to increased fish harvest and the increasing addition of nutrients in coastal systems (see Chapters 8, 12, 18, and 19). Over the last 50 years forests have also been changed dramatically, with the largest changes in tropical and sub-tropical forests, where there has been substantial clearing and transformation of previously forested land for agricultural and timber production. (See Chapters 9 and 21.)

Changes in forests in the temperate and boreal zone, while regionally important, have generally been small in the recent past compared with changes in the sub-tropics and tropics, and they have often involved increases in forest cover. (See Chapter 21.)

This section explores the major changes in ecosystem provisioning services over the last several decades and the trade-offs that have come as a result of the increased focus of human management on provisioning services. Over the past 50 years, there have also been substantial changes in some of the regulating, cultural, and supporting services that ecosystems provide, and those that are considered to be the most important are presented here. In some cases, these have been the direct result of managing ecosystems primarily for their provisioning services. In other cases, they are the direct result of transformation of ecosystems and habitat to other uses entirely. The major trade-offs in other services for the enhancement of provisioning services have come from the influence of ecosystems on atmospheric composition and climate feedbacks, nutrient cycling, the spread of disease, and biodiversity itself.

28.5.1 Food

Globally, ecosystems have met the rising demand for food over the last 50 years. The availability of basic food items such as cereals has increased faster than population growth, and the price of staple food items for many people is lower. (See Chapter 8.) There are significant regional differences in the accessibility to food, however.

Marine capture fisheries are the exception to the general increase in food availability. Globally, fish catches have declined over the last 10–15 years, and prices for fish from capture fisheries have risen. (See Chapters 8 and 18.) Although the cultivation of some of the targeted marine species has the potential to offset this decline at cheaper prices than for capture fisheries, the current high dependence on wild capture fisheries suggests that this potential will not be realized in the next few years.

The last 50 years has witnessed major successes in global agriculture, largely the result of improved crop varieties, synthetic fertilizer, irrigation, and other agricultural technologies, although expansion of land area under cultivation has played a role in many developing countries, particularly in sub-Saharan Africa. (See Chapter 26.) Food production from croplands has outpaced population growth over the last several decades when viewed in the global aggregate, with increases in food output per capita most rapid in East Asia. Yet in the world's poorest regions, especially in sub-Saharan Africa, yields have not benefited from advances seen elsewhere, and food insecurity persists. In sub-Saharan Africa, per capita food output has declined. (See Chapter 8.)

The intensification of agriculture underlies the enormous increase in the flow of food provisioning services over the last several decades, including of crops, livestock, and aquaculture. The trade-offs associated with the increases in this service are many, most prominently:

- major impacts on nutrient cycling, as rapid growth in the application of excess synthetic fertilizer contributes nitrogen and phosphorus to inland waters and coastal systems (see Chapters 12, 19, and 20), and
- loss of biodiversity in cultivated systems with monocultures associated with intensive agriculture (see Chapters 4 and 26).

The increase in land area under cultivation associated with agricultural extensification has involved a different set of trade-offs, most notably:

- habitat fragmentation and loss of biodiversity as land (mainly forest) is cleared for cropland (see Chapters 4 and 21), and

- impacts on atmospheric composition, particularly the greenhouse gas carbon dioxide, and climate regulation as biomass is cleared for cropland (see Chapters 13, 21, and 26).

Over the long term, declines in supporting and regulating ecosystem services, such as soil fertility, water cycling, and genetic resources, potentially undermine the ability of food production to keep pace with population growth in the absence of new, major technological advancements in agriculture. Those relying on subsistence agriculture are among the poorest and the most directly vulnerable to reductions in these ecosystem services, as their lack of economic resources limits access to alternative food sources. (See Chapter 7.)

28.5.2 Water

Over the last 50 years, people's access to water has also improved globally, although—like food—there are regional differences in quantity and quality of supply. (See Chapter 7.) The regulating and provisioning of water and its associated benefits (such as for food production through irrigation, or energy production through hydropower) has been a key factor in improving human well-being. The provisioning services of cultivated and dryland systems in particular have been possible primarily through the delivery and regulation of water through irrigation and flood control. The capacity of ecosystems to provide clean and reliable sources of water, however, is in decline in many parts of the world. (See Chapters 15, 16, 20, 21, and 24.)

Human well-being has improved through managing water use, controlling floods, providing transportation, irrigation, generating hydroelectricity and pollution control. The trade-offs for these improvements have included habitat fragmentation and loss, biodiversity loss, increases in certain human health risks, and declines in sediment supplies to the coastal zone. Levels of organic pollution (such as human and animal wastes and excess fertilizers) and inorganic pollution (such as pesticides, heavy metals, and PCBs) are increasing, much of which is from cultivated and urban systems. (See Chapters 15, 26, and 27.) Many of these pollutants are ultimately deposited in freshwater, coastal, and marine systems, affecting aquatic habitats, fish stocks, and the health of local human populations. (See Chapters 15, 19, and 20.)

The deterioration of the quantity or quality of fresh water is especially acute in cultivated systems, dryland systems, urban systems, and wetlands. (See Chapters 20, 22, 26, and 27.) The per capita availability of water has declined significantly since 1960 and the trend is expected to continue, albeit more slowly, until at least 2010. (See Chapter 7.) Economic development, including food provisioning and other water uses, will be most affected in those areas experiencing or at risk of water scarcity, particularly drylands. (See Chapter 22.)

28.5.3 Fish

Total fish supply has increased over the last 50 years but the cost to the sustainability of fish stocks and to the quality of many coastal and marine environments has been high. (See Chapters 18 and 19.) Technological changes followed by economic subsidies have fueled the expansion of fisheries into every ocean. Many fishing fleets are continuing to fish further offshore and deeper to sustain catches and to meet the growing demand for fish products, and this has led to a number of targeted stocks in all oceans having collapsed due to overfishing. (See Chapter 18.) More recently, the demand for selected marine products throughout the year has fueled the growth of inland, coastal, and offshore aquaculture, with consequential impacts to ecosystems. For example, the development of shrimp aquaculture has accounted for a significant loss of coastal habitat, in particular mangroves, in many tropical countries. (See Chapters 8 and 19.)

Globally, per capita consumption of fish is increasing, especially in Asia and the Americas, while in Europe and Africa consumption growth is moderate. Aquaculture production is increasing in both freshwater and marine systems at a rate of approximately 1 million tons per year , and it now supplies almost one third of all fish consumed, thereby sustaining increasing per capita consumption. (See Chapter 19.) However, much of the increase in marine aquaculture is in high-value species such as shrimp and salmon and therefore is not necessarily meeting the needs of poor consumers. Capture fisheries also provide employment and subsistence opportunities for many of the world's poor, many of whom are without access to property or property rights.

Provisioning of fish for food directly and indirectly (via fishmeal, animal feed, and fertilizers) has in many places resulted in degradation of coastal and marine systems and of other ecosystem services. (See Chapters 18 and 19.) Overfishing of many fish stocks in shallow coastal shelf systems has changed highly diverse, complex, and robust coastal ecosystems into systems of reduced diversity and resilience. (See Chapter 19.) Due to fishing pressures, for example, the Gulf of Thailand has changed since 1970 from a system with a high diversity of fish, including top predators, to one dominated by small, short-lived species that support a low-value fishery from which catches are mainly used for feed in the high-value invertebrate aquaculture industry. Such a reduced-diversity system may be more sensitive to external impacts and has a lower capacity to deliver ecosystem services. (See Chapter 11.)

Many coral reefs have shifted to algal-dominated systems where recovery is highly unlikely due to a combination of overfishing, disease, and climate change. (See Chapter 19.) The impact of destructive fishing practices such as bottom trawling and bombing exacerbates the problem of overfishing and restoration of depleted stocks. (See Chapter 18.) Although our understanding of the impact of overfishing in deep water and pelagic systems on regulating and supporting services is limited, the exploited species in some deepwater systems such as seamounts that were fished in the 1970s have not been recorded since then (see Chapter 18), suggesting that such systems are unable to recover over the short to medium term.

Overfishing has also affected the cultural services provided by marine and coastal systems. Many communities whose culture is based on a long history of fishing are in decline, with many fishers and their families migrating to urban areas. Those who choose to remain find that their social and economic conditions often decline. (See Chapter 18.)

28.5.4 Atmospheric Composition and Climate Feedbacks

Ecosystems influence air quality and climate, both through natural processes that maintain the status quo and through management-induced changes that can be either detrimental or beneficial to human well-being. (See Chapter 13.) Ecosystems play an important role in cleansing the atmosphere of pollutants. Changes in ecosystems as a result of human activity are one of the main drivers of change in climate and air quality, along with fossil fuel burning and industrial emissions, and concentrations of key greenhouse gases and other atmospheric constituents will continue to change in the future as a result of human impacts on ecosystems.

The most important changes in ecosystems leading to changes in atmospheric composition and climate feedbacks have been land clearing for agriculture, ranching, and urbanization—mostly through deforestation and biomass burning, although draining of wetlands has also been important, as has the increase in rice cultivation and livestock production. (See Chapters 9, 13, and 21.) These activities have resulted in an increase in concentrations of many trace gases in the atmosphere that subsequently change the chemistry of the troposphere and reduce the atmosphere's own capacity to remove pollutants (atmospheric cleansing capacity). These trace gases also act as pollutants themselves, act as fertilizing agents, change ozone concentration in the troposphere, and affect global climate (through impacts on radiative forcing, greenhouse gases, and cloud formation). Finally, ecosystem changes such as decreasing forest cover alter the physical properties of Earth's surface that in turn influence climate and hydrology, although these effects have likely been relatively smaller than the direct effects on atmospheric composition. (See Chapter 13.)

28.5.5 Nutrient Cycling

Nutrient cycling has been affected significantly in the last few decades, mainly from large-scale changes in agriculture and its inputs. (See Chapters 12 and 26.) As such, most of the trade-offs with other services can be tracked by focusing on areas where agriculture has changed substantially.

Changes in the extent and management of cultivated lands are nothing new. Before the Industrial Revolution, both because human populations were smaller and because there was a need to maintain forests and woodlands for fuel, the expansion of agricultural land was relatively small. Between 1950 and 1980, however, more land was converted to cropland than in the 150 years between 1700 and 1850 (see Chapter 26), due to both the needs of growing human populations and the availability of coal as an energy source. Cropland expansion has been estimated at about 1,200 million hectares over roughly the last 300 years.

With the exception of many tropical regions, which have experienced recent expansion of croplands, much of the potentially productive agricultural land appears already to have come under cultivation, thereby limiting potential for further expansion. (See Chapter 26.) Instead, there have been major increases in the intensification of the management of cultivated lands, including the increased addition of nutrients. This is reflected in part in the available data for the application of nitrogen-based fertilizer and in the observation that cropland yields have largely continued to rise on a per-hectare basis even as the rate of cropland expansion has dropped in much of the world. (See Chapters 8 and 26.)

The changes in the nitrogen cycle have been dramatic. Fertilizers spread on agricultural land, enhancement of N-fixation by planted legumes, biomass burning, fossil fuel combustion, land clearance, and wetland drainage have contributed to a doubling of natural inputs of nitrogen to ecosystems. (See Chapter 12.) The increase of biologically available nitrogen due to these sources, coupled with the leaching of phosphorus fertilizers and wastes, has led to substantially increased eutrophication in a number of aquatic systems. (See Chapters 19 and 20.) The increase in number, extent, and severity of periodically anoxic zones in estuarine systems around the world, for example (see Chapter 19), is a direct consequence of this trade-off.

Impairment of the nutrient cycling service is due to a disruption of a number of regulatory mechanisms that operate at different spatial and temporal scales. (See Chapter 12.) In soils, for example, synchrony between nutrient supply and demand by plants is a complex system whose stability requires a minimum biodiversity of plants and soil organisms. The intermediate storage of nutrients in soil aggregates is often decreased by a severe deple-

tion of the abundance and diversity of the plants and soil organisms that create this structure. Another effect of human activities on nutrient cycling has been the impairment or removal of the buffers, such as riparian forests and wetlands, that naturally ensure a close cycling of nutrients and thereby reduce losses to other compartments of the biosphere. (See Chapter 12.)

28.5.6 Spread of Disease

Historically, many diseases have emerged from altered ecosystems or domesticated animals (such as tuberculosis, measles, plague, and HIV), while other disease agents are in the process of adapting to human-dominated systems. Newly resurgent diseases deserve special attention if they have recently increased in incidence or emerged in a new geographical location, if they are of major public health importance and economic impact, or if they are difficult to control (such as antibiotic-resistant strains). Among these, malaria, leishmaniasis, dengue, and schistosomiasis are of major concern for the tropics; West Nile virus and Lyme disease in North America and Europe; and Japanese encephalitis in Asia. (See Chapter 14.) Also, food- and water-borne diseases stemming from intensive livestock or fish production are a growing concern. (See Chapter 8.) Approximately 75% of all emerging diseases are zoonotic (coming from animals), thus stressing the importance of further investigation of the role of biodiversity and ecological dynamics that are now recognized as central to disease prevention.

The most important drivers of ecosystem change that have affected infectious disease risk include tropical deforestation, road building, expansion of irrigation and dam building, local and regional weather anomalies, intensification of animal production systems, urban sprawl, poor sanitation, and pollution of coastal zones. (See Chapter 14.) The groups most vulnerable to disease risks from these changes include poor populations with little shelter or sanitation, which increases exposure to these risks, who have few financial resources to respond adequately. Activities such as international trade and travel have also led to an increase in infectious diseases. West Nile virus and monkeypox in North America and SARS globally are prime examples.

The magnitude and direction of altered disease incidence depends both on the type of land use change and the size of the human population exposed. Migration to a newly accessible forest or shoreline of a dam, for example, results in higher potential for disease epidemics. And the type of land use change, whether from mining, irrigation, dam construction, deforestation, or other causes, will promote specific diseases depending on geographic location. Major changes in habitat can both increase or decrease the risk of a particular infectious disease. (See Chapter 14.)

While many systems, such as forests, drylands, or cultivated systems, contain a distinct set of infectious diseases, several major diseases (including malaria and dengue) are more ubiquitous occurring across many ecosystems. For example, malaria is transmitted by 26 differing species of Anopheline mosquitoes that are each dominant in varying habitats and geographic locations. While *Anopheles gambiae* and *An. funestus* are the primary malaria vectors in Africa, *An. darlingi* is the primary carrier in South America, and *An. dirus* is in parts of Southeast Asia. Each species responds differently to a specified land use change, and it is therefore difficult to generalize ecosystem change effects across many regions.

On the other hand, some diseases such as yellow fever can be transferred across ecosystems. The natural zoonotic cycle of yellow fever occurs between mosquitoes and monkeys high in forest canopies, but the disease can move into savanna, agricultural, and even urban areas facilitated by human economic activities such as logging or forest clearing for crops and livestock.

Large preserved natural systems, due to their physical and biological characteristics, are relatively unreceptive to the introduction of many invasive human and animal pathogens that are brought in through human migration and settlement. For example, schistosomiasis, Kala-azar, and cholera have been introduced in the Amazon region but have not been able to become established in the natural forest system. (See Chapter 14.)

Higher levels of biodiversity can reduce the risk of some vector-borne diseases via a "dilutional effect" or by maintaining natural predators. (See Chapters 11 and 14.) The former effect has been documented in the case of Lyme disease in North America and is likely true for many diseases where the capability of an animal host to be infected and carry a disease agent varies greatly across the intermediate animal hosts on which an insect vector must feed. Schistosomiasis in Lake Malawi provides a good example of the relationship of disease emergence to natural predators, as it rose rapidly following overfishing of fish predators on the snail intermediate host for the parasite that causes the disease.

28.5.7 Biodiversity

It is *well established* that losses in biodiversity are occurring globally at all levels, from ecosystems through species, populations, and genes. (See Chapter 4.) The current documented rate of species extinction is two orders of magnitude higher than the average rate of species extinction from the fossil record, and there is a continuing trend for conversion of naturally occurring, species-rich ecosystems into more intensively managed habitats with reduced biodiversity. Losses at the population level are variable but substantial in certain systems, such as marine, freshwater, and agricultural systems, and more common in large and long-lived species. Figure 28.4 (in Appendix A) presents the aggregate global trends in populations of well-studied species. The extent of loss of genetic diversity is less well understood and is mainly inferred from declines in higher levels of biodiversity organization, although recorded losses in agricultural genetic diversity are widespread. (See Chapters 4 and 26.)

There are some instances where biodiversity is increasing, in terms of extent of habitat and species composition, such as in temperate forest areas in the northern hemisphere. (See Chapters 4 and 21.) However, available data show aggregate global declines in both the distribution and diversity of biomass. Although widespread, losses of biodiversity are currently particularly prevalent in areas of high species richness, such as tropical forests and coral reef systems. (See Chapters 4, 19, and 21.) Inland waters are also likely to be experiencing high levels of biodiversity loss in most parts of the world. (See Chapter 20.) Habitat conversion (generally for agricultural expansion), degradation, and fragmentation continue to be the most important direct drivers of biodiversity loss globally, although there is an increasing impact of invasive non-native species, of nutrient pollution, and of climate change in many systems. Island systems in particular have historically been affected by the introduction of exotic species, with widespread negative impacts on native island biodiversity.

The impacts of current trends of biodiversity loss on human well-being are multifaceted. While people benefit directly from components of biodiversity in the form of provisioning services and are therefore affected directly by declines in availability of those elements of biodiversity that are providing those goods, the more fundamental role of biodiversity is in the functioning of ecosystems and thereby in the capacity of ecosystems to provide the full range of ecosystem services. (See Chapters 4 and 11.) Both the amount of living material (biomass) and its diversity and distribution play important roles in determining the capacity of systems to provide services now and into the future. Evidence suggests that decreases in the amount of live biomass have directly affected the capacity of some ecosystems to provide services, such as the capacity of tropical forests to regulate local and regional climate (see Chapter 13) and to protect from natural hazards (see Chapter 17) or the capacity of marine, coastal, and inland water systems to provide food (see Chapters 8, 18, 19, and 20). And these services show declining trends.

Current understanding of the consequences of losses in the diversity of biomass is poor and on the whole is limited to a selection of species that play particularly obvious roles in ecosystem functions, such as pollinators for the provision of food services. (See Chapter 11.) The consequences of losses in most rare or restricted-range species are likely to be subtle. Indeed, some provisioning services, such as timber and food, appear most efficiently produced from less diverse systems, such as forest plantations and agricultural landscapes, as evidenced by the wide coverage of cultivated systems. (See Chapter 26.)

However, all ecosystem services are ultimately reliant on ecosystem functions and on the interactions between elements of biodiversity. (See Chapter 1.) It is also likely that more diverse systems are more resilient, resistant, and adaptable to changes in drivers. (See Chapter 11.) Among the most important factors identified is the degree of functional redundancy found between species within an ecosystem. For example, in many ecosystems there are several species that fix nitrogen. If the loss of any one of them is compensated for by the growth of others and there is no overall loss in nitrogen fixation, then the impacts of the loss of the species are reduced, in terms of the system's capacity to fix nitrogen.

There may, of course, be other consequences of the loss of these species. The possibility of significant losses of function increases as more species, and variability within species, are lost and as redundancy is reduced—that is, there is an asymptotic relationship between biodiversity and ecosystem functioning. Greater functional redundancy represents greater insurance that an ecosystem will continue to provide both higher and more predictable levels of services. (See Chapter 11.)

Although there have been significant benefits derived from the commercial exploration of biodiversity across a range of industrial sectors, and particularly from pharmaceutical bioprospecting, the consequences of losses of genetic and species diversity on bioprospecting potential remains largely speculative, especially in marine and coastal areas. (See Chapter 10.) Unlike some other services where there are minimum thresholds, however, overall declines in the amount of biodiversity proportionally reduce the resource base from which commercial exploration is possible.

There have been significant impacts of declining biodiversity on cultural services. Globally, many cultural values associated with the conservation of components of biodiversity, particularly relating to the amount and diversity of natural systems and of species and populations, continue to be affected directly through the current trends in declining biodiversity. (See Chapter 17.)

28.6 Is There Evidence That the Capacity of Ecosystems to Provide Services Is Reaching Critical Levels?

In addition to documenting the actual trade-offs that have been made while managing ecosystems for different services, the extent to which those trade-offs have resulted in a reduction in the capacity of ecosystems to provide services is also important to consider. If the underlying capacity of ecosystems to provide a range of services has been reduced, then there are obvious implications for the future not only of the ecosystems and services in question, but also for human well-being.

Of the services and systems examined in this report, it is clear that at a global level there are two issues where the capacity to continue to provide services has most clearly declined. One is marine and coastal capture fisheries, as described earlier in this chapter and in Chapter 18. It is now *well established* that the capacity of the oceans to provide fish for food has declined substantially and in some regions showing no sign of recovery. The other is the loss of biodiversity, in large part because the rates of loss (of species diversity) are so much more rapid than the creation of new diversity through evolutionary processes. (See earlier section and also Chapter 4.) The implications of this loss are less immediately clear than those of the decline of marine fisheries, but over the long run they are likely to be considerably more important. In addition, some systems have eroded their capacity to provide services on a regional basis, such as inland waters (Chapter 20), forests (Chapter 21), and drylands (Chapter 22). The implications of this regional reduction in capacity are explored more fully in the next section.

Understanding how the decline in the capacity of ecosystems to provide services has occurred is as important as documenting where the capacity has declined. A good deal of ecosystem change and corresponding decline in ecosystem services is gradual and occurs over long time frames. Such chronic loss of ecosystem services certainly affects human well-being but over decadal or intergenerational time frames. However, some ecosystem changes are nonlinear or abrupt and sometimes irreversible. The reasons for such nonlinearities include:

- intrinsic features of the ecology of certain ecosystems (that is, ecological thresholds),
- the magnitude and nature of the impact causing change (such as changes occurring in response to technological advances), and
- the features of the drivers of change (such as social and cultural "ratchets" that allow change in only one direction).

While each of these is described here separately, it should be noted that they are often present in tandem, particularly when large-scale ecosystem changes occur.

Significant changes in ecosystem structure and function can occur when certain triggers result in changes in the dominant species. An excellent example of this is provided by coral reef ecosystems that undergo rather sudden shifts from coral-dominated to algal-dominated reefs. The trigger for such phase shifts, which are for all intents and purposes irreversible, is usually multifaceted and includes increased nutrient input leading to eutrophic conditions and the removal of herbivorous fishes that maintain the balance between corals and algae. (See Chapter 19.) Once the thresholds for the two ecological processes of nutrient loading and herbivory (one an upper threshold and one a lower threshold) are passed, the regime shift occurs within months, and the resulting ecosystem, though stable, is less productive and less diverse than before the transformation. Human well-being is affected not only by reductions in food supply and income from reef-related industries (such as diving and snorkeling, aquarium fish collecting), but also by increased costs accruing from the decreased ability of reefs to protect shorelines, as algal reefs are more prone to being broken up in storm events, leading to shoreline erosion and seawater breaches of land.

Nonlinear changes in ecosystem structure and function can also occur at the regional level, affecting, for example, regional climate. (See Chapter 13.) The vegetation in a region influences climate through albedo, transpiration, and the aerodynamic prop-erties of the surface. In the Sahel region of North Africa, vegetation cover is almost completely controlled by rainfall. When vegetation is present, rainfall is quickly recycled, generally increasing precipitation and in turn leading to a denser vegetation canopy. Model results suggest that land degradation leads to a substantial reduction in water recycling and may have contributed to the observed trend in rainfall reduction in the region over the last 30 years. In tropical regions, deforestation generally leads to decreased rainfall. Since forest existence crucially depends on rainfall, the relationship between tropical forests and precipitation forms a positive feedback, which, under certain conditions, theoretically leads to the existence of two steady states: rainforest and savanna. Some models suggest only one stable climate-vegetation state in the Amazon.

Introduced invasive species can also act as a trigger for dramatic changes in ecosystem structure, function, and delivery of services. (See Chapters 3, 4, and 11.) In marine systems, species are commonly brought into new areas through ballast water discharges from ships, and they sometimes establish in new areas through outcompeting native species for food and space. One example is the rapid and irreversible change in the Black Sea, where the carnivorous ctenophore *Mnemiopsis leidyi* caused the loss of 26 major fisheries species and has been implicated (along with other factors) in subsequent growth of the anoxic "dead zone." (See Chapter 19.)

Certain kinds of human activity can lead directly to largely irreversible changes—the most obvious being habitat loss from conversion to urban environments, tourist resorts, ports and harbors, reservoirs, and agricultural lands. Habitat loss results in loss of not only the ecosystem services provided by the affected habitat, but often also the services provided by associated habitats. For instance, development of a port in an estuary may prevent people from obtaining food from the estuary and at the same time affect nearby fisheries whose target species depend on the estuary for nursery habitat. Even minor losses in species or habitat extent may reduce the capacity of ecosystems for adjustment to changing environments, with consequences for ecosystem function, services, and human-well being. (See Chapter 11.)

Finally, technological advances and other changes in drivers can have nonlinear impacts. For instance, societies moving from subsistence harvesting to harvesting with improved technology can cause very sudden and large changes in the rate of resource exploitation. These jumps in exploitation rates often pass the threshold for sustainability and result in crashes of harvested populations, such as fish stocks. The changes in drivers are also, in effect, irreversible, since a return to previous, low-tech methods is unlikely.

Another driver with a potentially disproportionate effect on environmental degradation and loss in services is human migration. This outcome can occur when in-migrants originate from culturally or ethnically different groups than local residents and have neither the same vested interests in environmental protection nor societal self-regulatory mechanisms, as do local communities. Such effects of migration are often not captured in assessments of population impact on environment, since migration, and especially internal migration, is often difficult to monitor and rarely shows up in population census data.

28.7 Are There Parts of the World in Which Recent Declines or Stagnation in Human Well-being Can Be Attributed to Changes in Ecosystem Services?

This assessment shows that global environmental change is highly variable among the world's regions. Some trends and their impacts become apparent only when the scale of analysis shifts from the global to regional and local scales. This section details some of the major findings emerging from the assessment at these lower scales and illustrates the needs to look at ecosystem changes at multiple scales.

Population growth rates are now declining nearly everywhere in the world, but with substantial regional differences. Substantial population growth is still expected in sub-Saharan Africa, South Asia, and the Middle East over the next few decades. The share of the global population represented by North America and Europe is expected to decline from 17% to 10%, while Africa's share is expected to increase from 13% to 23%. Urban populations are growing three times as fast as the population as a whole, creating ecological and socioeconomic problems in cities and surrounding hinterlands. Most of this growth will occur in developing countries. The most critical environmental effects, meanwhile, are local, such as unsafe water supply and indoor air pollution, where residents are also drawing down ecosystem services at ever-greater distances.

The trend toward an increasing proportion of urban dwellers can also most easily be seen on continental and regional scales. Over the past 50 years, there have been relatively modest increases in the proportion of the population living in urban areas in Europe (from 52% to 73%) and North America (64% to 77%). But both the absolute numbers and the percentages have increased dramatically in the developing regions of Africa (from 15% to 37%), Asia (17% to 38%), and Latin America and the Caribbean (42% to 75%). By the year 2000, Asia had 195 cities of more than a million inhabitants (up from 31 in 1990) out of a global total of 388 such cities. Asia also had 45 of the 100 largest cities in the world, while Europe had 15, North America 13, and Latin America 17. (See Chapter 27.)

Global agriculture has registered many successes for the provisioning of food from the world's ecosystems, but again a regional perspective reveals problems among generally favorable aggregate trends. Farmers in some of the world's poorest countries and other resource-poor regions have not shared in the yield increases over the past several decades. Per capita food output has actually declined in sub-Saharan Africa.

Meanwhile, policy distortions and market failures are important problems contributing to the highly variable pattern of food provisioning and food security. OECD countries provide extensive subsidies to their agricultural sectors and also protect them through tariffs, quotas, and export subsidies. The worldwide consequences of these institutional arrangements have been lower prices for internationally traded commodities, higher tax bills, and the overuse of agricultural inputs, such as fertilizers, with associated consequences on ecosystems. (See Chapters 12, 19, and 20.)

Despite widespread increases in the use of fertilizers in other parts of the world, sub-Saharan Africa stands out as a particular problem area where declining soil fertility is a principal constraint for sustaining food production and where soil nutrient stocks are being used unsustainably and fertilizer application remains low. (See Chapter 26.) Meanwhile, rising populations, declining use of fallow, the cultivation of fragile lands, and limited conservation investments are all apparent in many developing regions.

Although there is substantial reason to believe that the world in the coming decades can produce sufficient food to feed its growing population, important regional issues exist in the global pattern of cultivated land. Many regions have experienced declines in the area of cultivated lands over the last several decades, as described earlier. Due partly to increased competition for alternative land uses and to reduced availability of suitable land (particularly outside the tropics), there is declining potential for further

expansion of agricultural lands; future increasing agricultural productivity in many parts of the world is likely to come largely from the intensification of the management of existing cultivated lands. (See Chapters 8 and 26.)

Access to sufficient clean fresh water is a problem at a global scale, but also one with a strong regional component. The essential functions of clean water are several: maintaining human and environmental health; supporting essential economic production, such as agriculture, energy, and industry; diluting and transporting wastes; and contributing to religious and cultural activities. (See Chapters 7, 15, 17, 20, and 26.) Despite their critical importance, these freshwater services are under threat throughout the world, and the world's poor people, concentrated in developing countries, are at particular risk. In the mid-1990s, some 80 countries with 40% of the world's population were already suffering from serious water shortages. North Africa and the Middle East, in particular, face great pressure and demand on already overstressed water resources. Azerbaijan, Egypt, and Libya, for example, were already using 55%, 110%, and 770%, respectively, of their sustainable water supplies in the early 1990s.

Water pollution is exacerbating local water scarcity in many parts of the world and is expected to accelerate in coming decades. In developing countries, 90% of sewage continues to be discharged into rivers, lakes, and coastal areas. In Africa, where such problems are particularly intense, major pollution sources include fecal contamination and toxic pollution from cities, industrial centers, and mining sites. (See Chapters 7 and 15.) It is clear, meanwhile, that environmental water is often the loser worldwide, for water needs also to be left in rivers and lakes to maintain the health of ecosystems and fisheries, which are already under heavy pressure from population and economic growth. (See Chapter 20.)

Differential vulnerability to environmental change and poverty also reveal strong regional and local variability. (See Chapter 6.) In the face of general global food availability and the progress in reducing the incidence of famine, many developing countries are experiencing declines in agricultural production and food security, especially among small-scale farmers, isolated rural populations, and those living on marginal lands. The global trends in natural disasters reveal that Asia is disproportionately affected, with more than 70% of all lives lost from natural disasters occurring there. (See Chapters 6 and 16.) In China alone, floods affect more than 100 million people on average annually.

Sea level rise from climate change poses risks for low-lying coastal areas throughout the world (see Chapters 19 and 23), but particularly vulnerable are the small islands of the Pacific, coral reefs, and the deltas of such areas as Egypt and Bangladesh. And the Intergovernmental Panel on Climate Change has concluded that the adverse impacts of ongoing and future climate change will occur largely in the developing countries already beset by poor sanitation, water stresses, financial pressures for needed development programs, and poor health and inadequate medical services.

28.8 What Are the Most Critical Gaps in Knowledge and the Most Crucial Research Needs?

There are many limitations on the scientific community's ability to provide a comprehensive judgment about the conditions and trends in ecosystems and the services they provide. Although it is certain that human well-being is affected by changes in the provision of ecosystem services, the details of that relationship remain difficult to untangle except in the simplest cases, such as the ability of ecosystems to provide food to increasing numbers of people. This section categorizes the main types of uncertainties that limit the ability to synthesize the results presented in the individual chapters on systems and services. Two major needs are focused on: data and information, and processes and understanding.

28.8.1 Data and Information

The most basic limitation is that there are many important features of today's world for which no information is available, much less the high quality, well-documented, and comparable information that is necessary to understand crucial problems. For example, we have relatively little replicable data on forest extent that can be tracked over time. Methods of measuring forest extent vary from country to country, and in spite of large efforts by international agencies to harmonize the information, the ability to document changes in this is surprisingly poor for much of the world.

A similar situation exists for cropland, where methodological issues and significant data gaps cloud the picture of cropland conversion and the use of cropland over time in most regions. The global distribution of wetlands remains unknown, as does the actual current distributions of many important plant and animal species, much less their changes over time.

All these gaps in information result in significant constraints on documenting the trade-offs between provisioning and non-provisioning services. While there is high certainty in some cases relating to large-scale trends such as climate change, species losses, and land degradation, the weakness in documentation and information on regional trends remains a serious handicap. Interestingly, local data and information can in turn be of extremely high quality, in part because the scales of measurement are more amenable to traditional sampling technologies and methods. However, the ability to generalize from local information to regional and global information is limited.

28.8.2 Processes and Understanding

The ability to provide a clear picture of the trade-offs among ecosystem services, and therefore information that is relevant to the continued management of ecosystems for human well-being, is also constrained by limits in understanding of the relevant processes and underlying relationships. For example, while it has become much clearer over the past decade that biodiversity is important for ecosystem functioning, there is limited understanding of the ways in which biodiversity regulates ecosystem functioning at local and regional scales, and there is intrinsic difficulty in predicting unexpected, accelerated, and some times irreversible changes triggered by alterations of local and regional biodiversity. The response of ecosystems to changes in the availability of important nutrients, including carbon, especially through increasing atmospheric pathways is not broadly understood and cannot be deduced strictly through model simulations.

One of the most critical needs for further information is an improved understanding of the factors governing the capacity of ecosystems to provide services. Documenting threshold changes in ecosystems and understanding the structural and dynamic characteristics of systems that lead to threshold and irreversible changes is clearly important in this respect and is currently not well understood. Equally important is the development of both conceptual and quantitative models that can begin to give both scientific and policy communities advance warning of when the capacity of systems is beginning to be eroded or thresholds are likely to be reached, so that action may be taken before significant adverse trade-offs have occurred.

Table 28.2 on page 838 synthesizes some of the main recommendations for data and information on particular ecosystems and services that are the most important ones for building the capacity to forecast changes in ecosystem services. The table identifies a selection of the critical issues to be resolved by further research, why those issues are important, and what processes or questions are important to understand in terms of those issues. Finally, the minimum data and information that will be required to address the issues raised are identified.

References

Lepers, E., E.F. Lambin, A.C. Janetos, R. DeFries, F. Achard, N. Ramankutty, and R.J. Scholes, 2005: A synthesis of rapid land-cover change information for the 1981–2000 period. *BioScience,* **55 (2),** 19–26.

Loh, J. and M. Wackernagel (eds.) 2004: *The Living Planet Report 2004.* Gland, Switzerland, and Cambridge, UK: World Wide Fund for Nature and United Nations Environment Programme World Conservation Monitoring Centre.

United Nations, 2003: World Population Prospects: The 2002 Revision Population Database. New York.

Table 28.2. Critical Issues Requiring Research and Data Collection to Forecast Significant Impacts of Ecosystem Change on Human Well-being. This list is intended to highlight the most critical issues and does not identify many other important research and data needs.

Which ecosystems and services most critically require research and data collection to forecast significant impacts on human well-being?	Why is the issue critical?	What do we need to know?	What types of data are required?
Drylands in all semiarid areas	direct dependence of poor on ecosystem services and coping capacity during times of stress; intensifying use for rangeland and cropland	Where is productivity of drylands declining from intensifying use? What are the feedbacks from intensifying rangeland and cropland on soil fertility, climate regulation, and access to water? Where and how is human well-being changing in response to changes in productivity?	indicators of land productivity; indicators of human well-being attributable to changes in land productivity
Coastal and marine systems in all parts of the world	recipient of nutrients from food production systems; importance for fisheries	Where is nutrient input and overharvesting affecting fisheries? Where are declining fisheries affecting access to protein?	nutrient inputs; trends in fish populations
Tropical forests	most significant repository of biodiversity, with high cultural value and potential implications for ecosystem function; rapid conversions expected to continue; non-timber forest products can be major contributor to household food, medicine, traditional livelihoods	Where is habitat loss causing declines in biodiversity? What are the effects of habitat loss on species and populations over what time scales?	indicators of habitat extent and quality, indicators of biodiversity
Boreal forest	acceleration of natural disturbances (e.g., pests and fire)	What drivers are most important for accelerating disturbance regimes? What are anticipated responses to climatic changes?	information on interaction of disturbance regimes and global change
Inland waters	water is basic societal and biological need; human health and economic development issue; direct provision of food	What are contemporary and historical patterns of water infrastructure, use, and supply? What is the access of humans to surface and groundwater sources? Sustainability of fisheries?	basic surface hydrography and well-log data from around the world; extent of fisheries; water quality data
Polar systems	most sensitive to climate change; potentially significant feedbacks to climate regulation; importance of traditional livelihoods	How are polar systems responding to climate change? How does the climate regulation function of polar systems change in response to climate change?	species composition; changes in hydrology and trace gas fluxes' rate and extent of change in permafrost and peat
Food and cultivated systems	cultivated systems occupy a large global area; food provision critical for human well-being; large changes in ecosystem states and services due to provision of food	What are the trade-offs in other services inherent in different management practices? What is the overall sustainability of current fisheries management?	contemporary simultaneous data on outputs and other services from different cultivation and grazing systems
Emergence of disease related to ecosystem change	potentially large implications for human health	Which types of ecosystem changes trigger the emergence of disease? Which types of diseases and which ecosystems are potentially the most significant?	occurrences of diseases related to ecosystem change

Appendix A
Color Maps and Figures

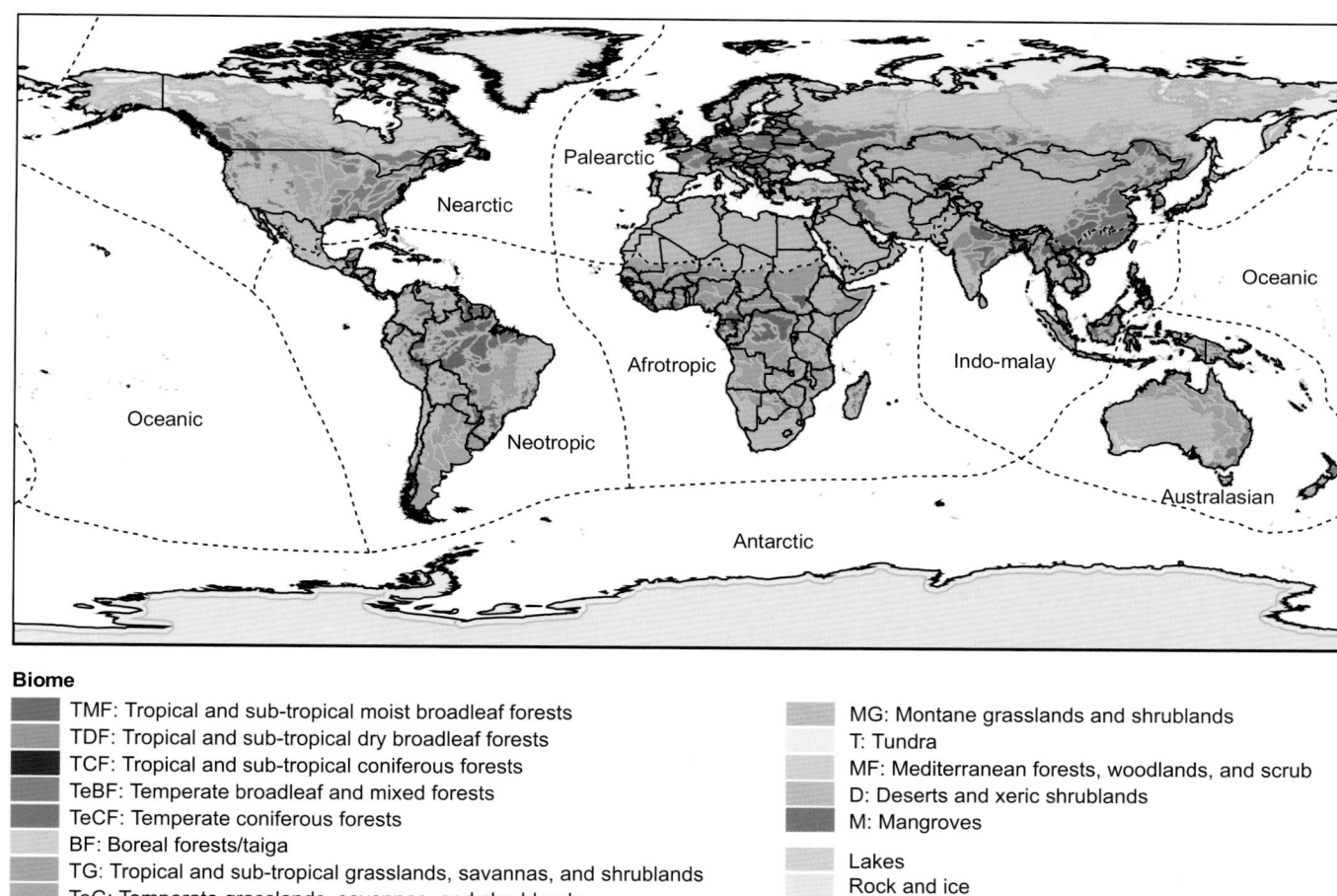

Biome

- TMF: Tropical and sub-tropical moist broadleaf forests
- TDF: Tropical and sub-tropical dry broadleaf forests
- TCF: Tropical and sub-tropical coniferous forests
- TeBF: Temperate broadleaf and mixed forests
- TeCF: Temperate coniferous forests
- BF: Boreal forests/taiga
- TG: Tropical and sub-tropical grasslands, savannas, and shrublands
- TeG: Temperate grasslands, savannas, and shrublands
- FG: Flooded grasslands and savannas

- MG: Montane grasslands and shrublands
- T: Tundra
- MF: Mediterranean forests, woodlands, and scrub
- D: Deserts and xeric shrublands
- M: Mangroves

- Lakes
- Rock and ice

- Biogeographic realm
- Country
- Ecoregions

Figure 4.3. The 14 WWF Biomes and Eight Biogeographic Realms of the World. Biomes are coded in colors and listed with abbreviations that will be used in following figures and tables (e.g., TMF). Biogeographic realms are named in the figure. Ecoregions are nested within both biomes and realms.

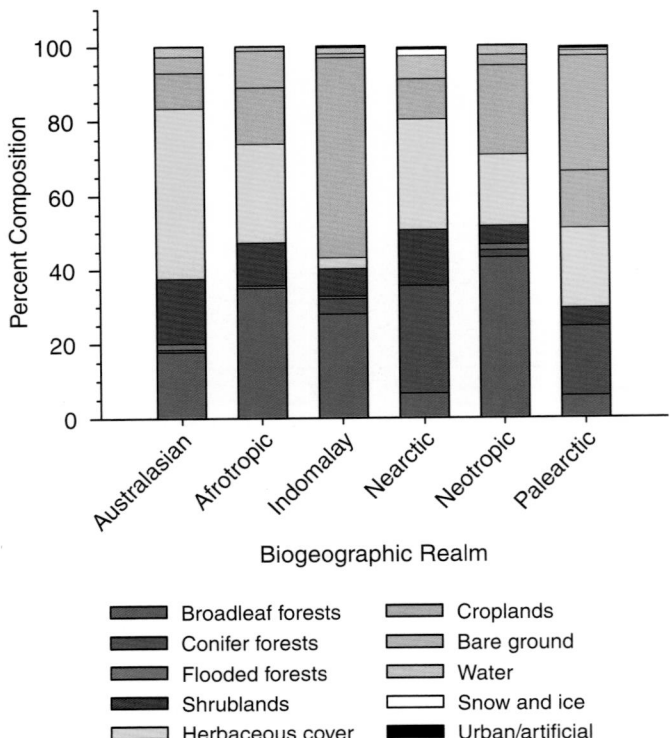

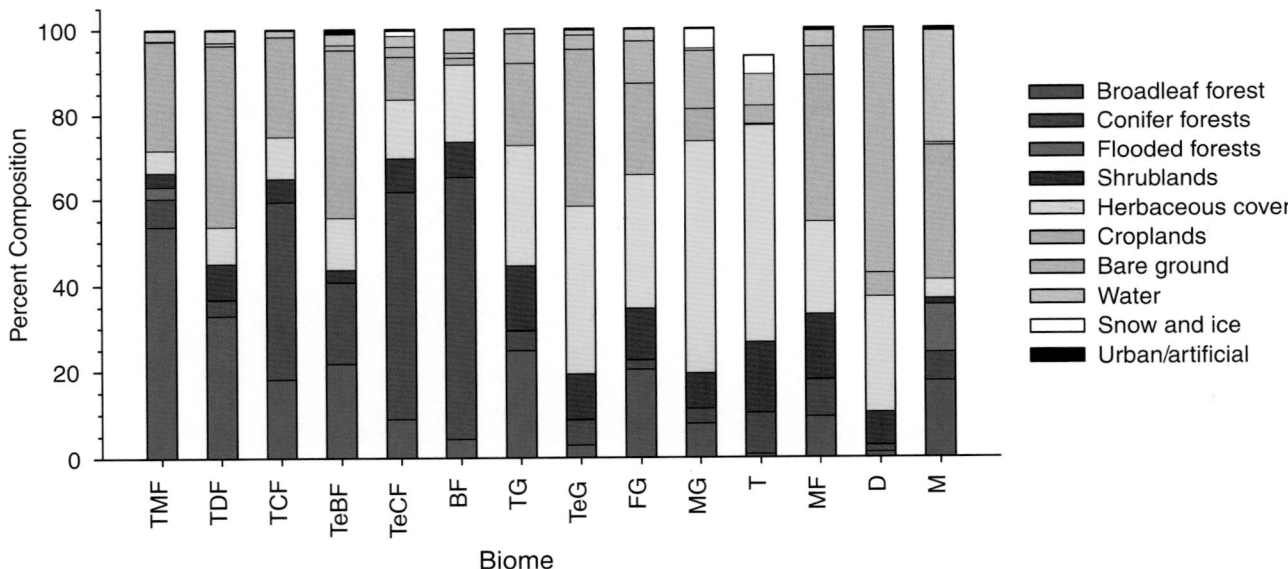

Figure 4.4. Land Cover Composition of Six of the Eight Terrestrial Biogeographic Realms. Oceania and Antarctica are omitted because land cover data were not available.

Figure 4.6. Land Cover Composition of 14 Terrestrial Biomes. Biome codes as in Figure 4.3. Tundra bar does not reach 100% because 7% of this biome was unclassified by the land cover dataset.

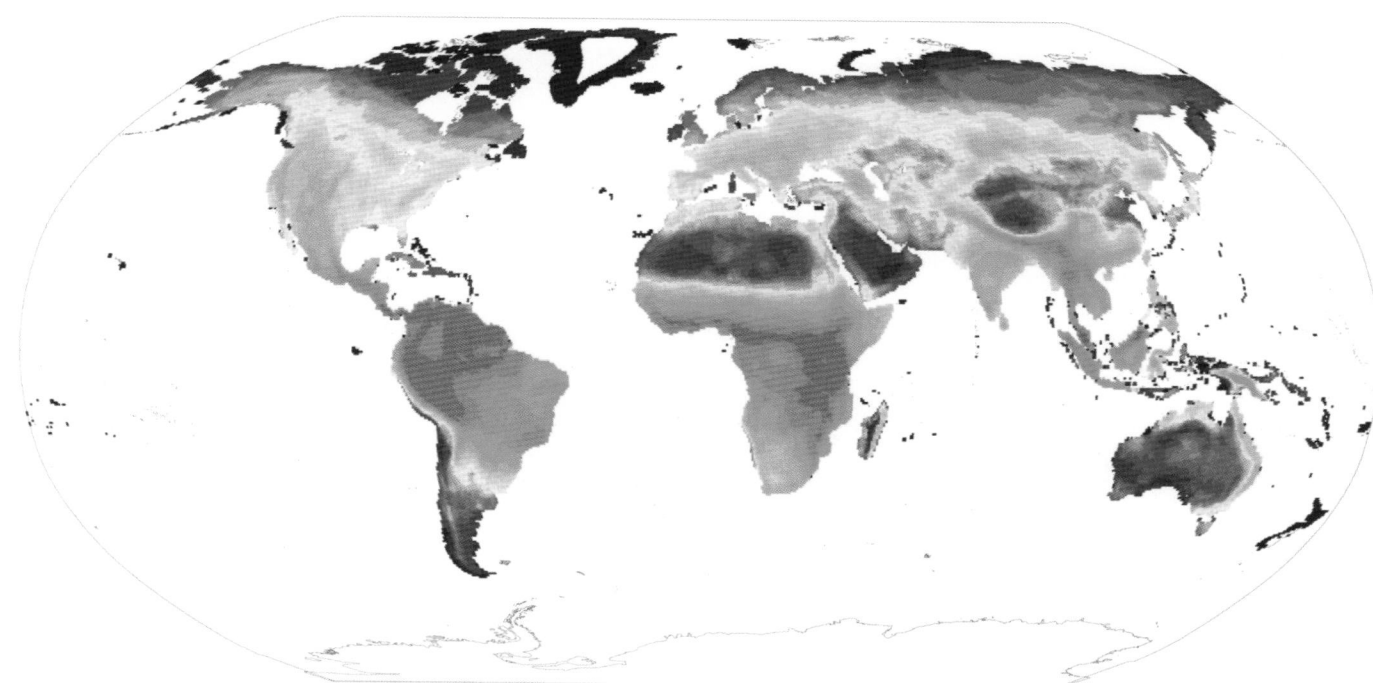

Figure 4.10. Global Species Richness of Terrestrial Mammals per Half-degree Cell. N = 4,734. Dark orange colors correspond to higher richness, dark blue colors correspond to lower richness. Maximum richness equals 258 for mammals. Color scale based on 20 equal-area classes. (Baillie et al. 2004)

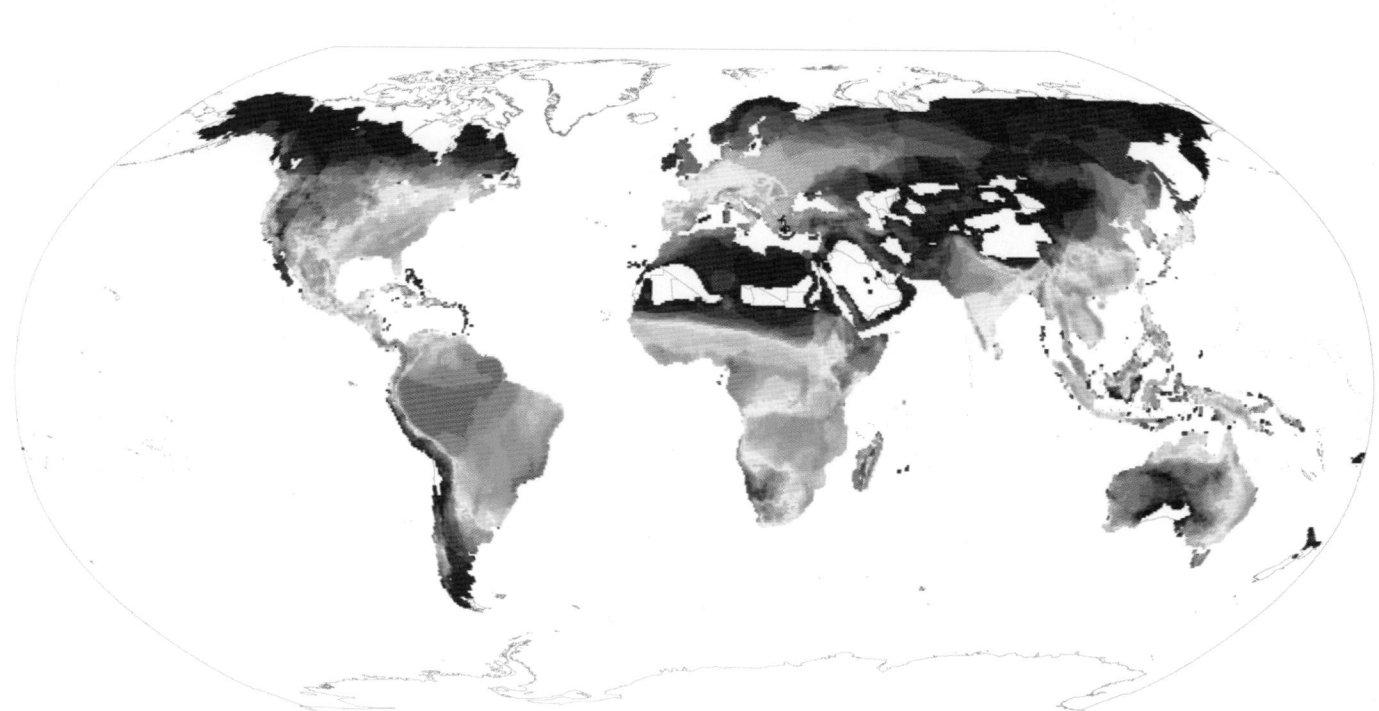

Figure 4.11 Global Species Richness of Amphibians per Half-degree Cell. N = 5,743. Dark orange colors correspond to higher richness, dark blue colors correspond to lower richness. Maximum richness equals 142 for amphibians. Color scale based on 20 equal-area classes. (Baillie et al. 2004).

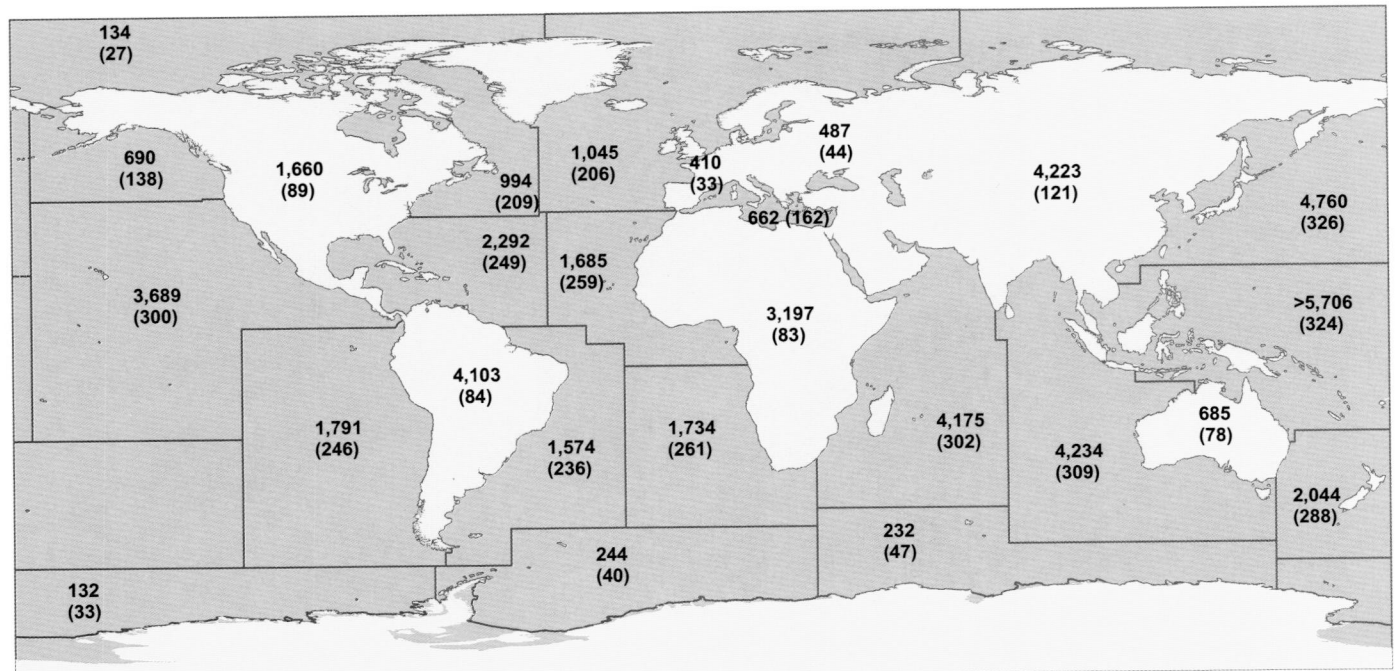

Figure 4.12. Global Richness of Finfish Species (and Finfish Families in Parentheses) across FAO Areas (data source Froese and Pauly 2003)

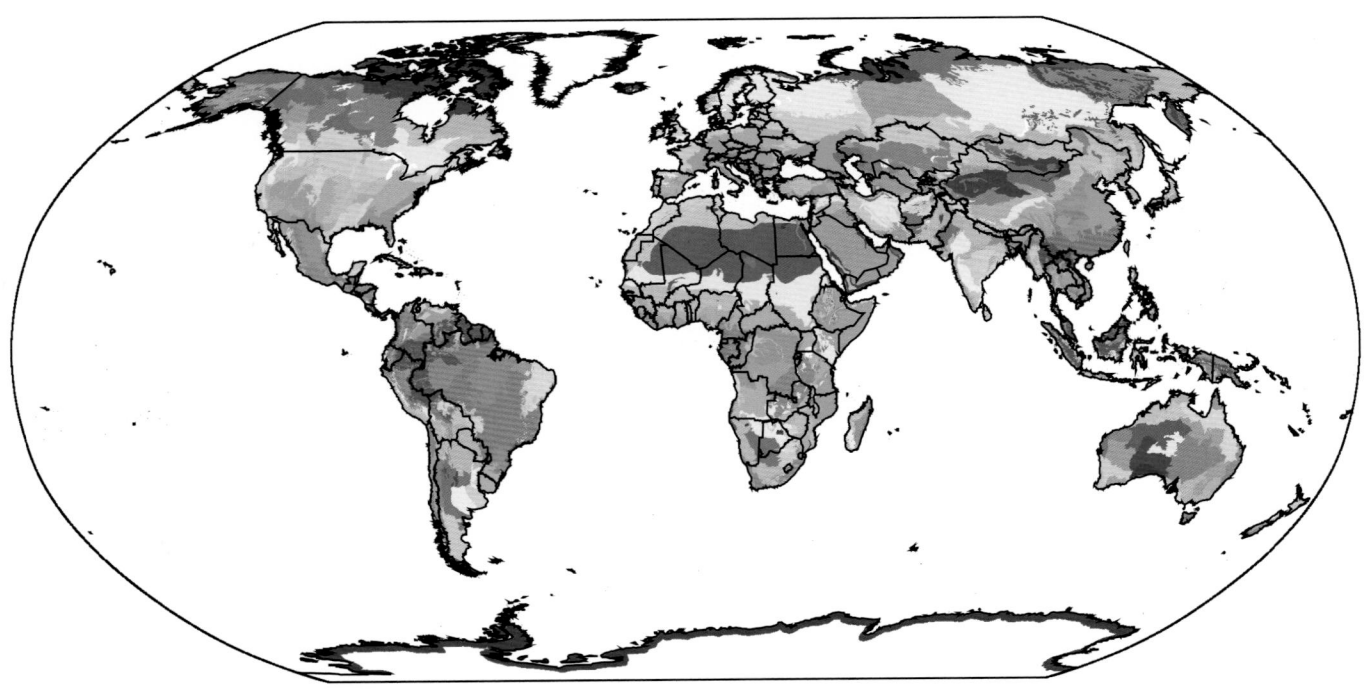

Figure 4.13. Global Species Richness of Vascular Plants Modeled and Mapped across Ecoregions. Dark orange colors correspond to higher richness, dark blue colors correspond to lower richness. Maximum richness equals 10,000 for plants. Color scale based on 20 equal-area classes. (Kier et al. 2002; Olson et al. 2001)

Anthropogenic Drivers	Scales of Ecological Organization		
	Genes	**Populations/Species**	**Biomes**
Habitat change	↑ 4	↑ 3	↑ 1
Fragmentation/Dam construction	↑ 2	↑ 2	? 2
Invasive alien species	? 4	↑ 4	↑ 4
Exploitation	? 4	↑ 2	↑ 2
Inputs (fertilizer, acid rain, pollution)	? 2	↑ 2	↑ 2
Disease	? 2	↑ 3	? 3
Climate change	? 5	↑ 5	↑ 5

Figure 4.16. Major Anthropogenic Variables Acting as Drivers of Change on Different Scales of Ecological Organization or Biodiversity Levels. Color=degree of driver impact on ecological scale (red=maximum impact followed by orange, then yellow); ↑=upward trend of driver impact on ecological scale; 1 to 5=degree of impact reversibility (5=least reversible); Shading=degree of certainty based on expert knowledge (dark shading=least certain); ?=information on trends unknown. Impact indices were based on a year 2010 timeframe.

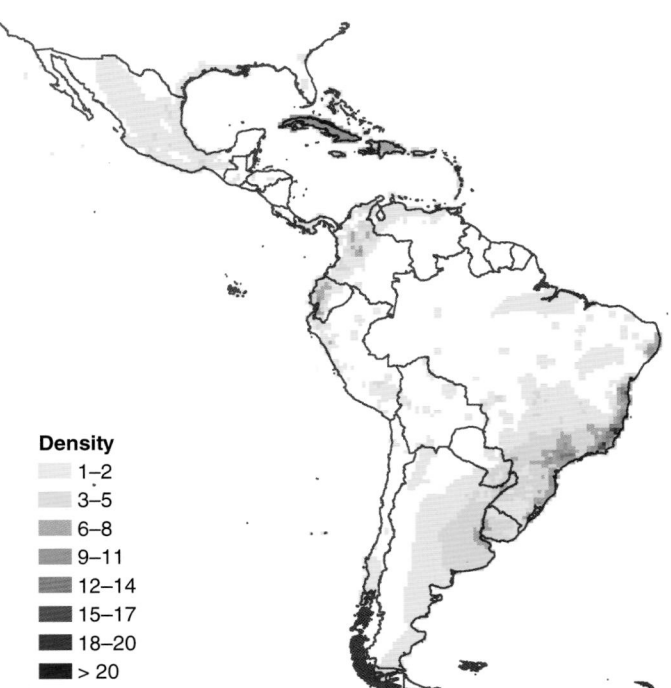

Density
- 1–2
- 3–5
- 6–8
- 9–11
- 12–14
- 15–17
- 18–20
- > 20

Figure 4.18. Density Map on Extent to Which the Ranges of Threatened Bird Species Have Contracted in Central and South America. The color scale indicates the number of threatened bird species that used to occur in a pixel, but now no longer do so. (BirdLife International 2004a; unpublished data from BirdLife's World Bird Database)

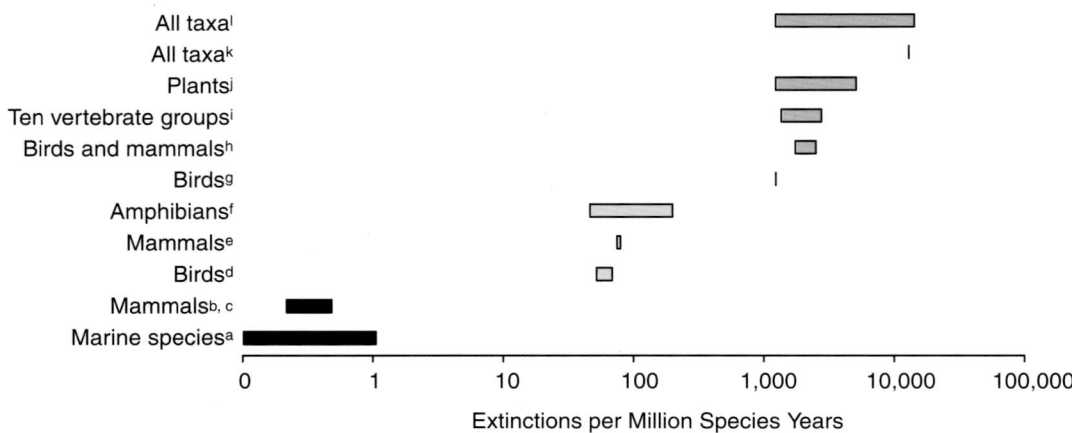

Figure 4.22. Background and Contemporary Extinction Rates. Background extinction rates are in black, extinction rates based on observed extinctions over the past 100 years are in yellow and estimated contemporary extinction rates using a number of different approaches are in orange. Based on background extinction rates from the fossil record: [a]May (1995), [b]Alroy (1998) (lower estimate of 0.21), [c]Foote (1997) (higher estimate of 0.46). Observed extinctions over the past 100 years: [d], [e], [f]Baillie et al. (2004). Projections based on threatened species: [g]Pimm and Brooks (1997), [h]Smith et al. (1993) (also uses recently extinct species), [i]Mace (1994). Plant extinctions using species-area curve with assumptions about habitat loss from agricultural/urban expansion and from climate change: [j]MA *Scenarios*, Chapter 10. Increased energy consumption: [k]Ehrlich (1994). Species-area relationship from deforestation rates: four studies in [l]Reid (1992).

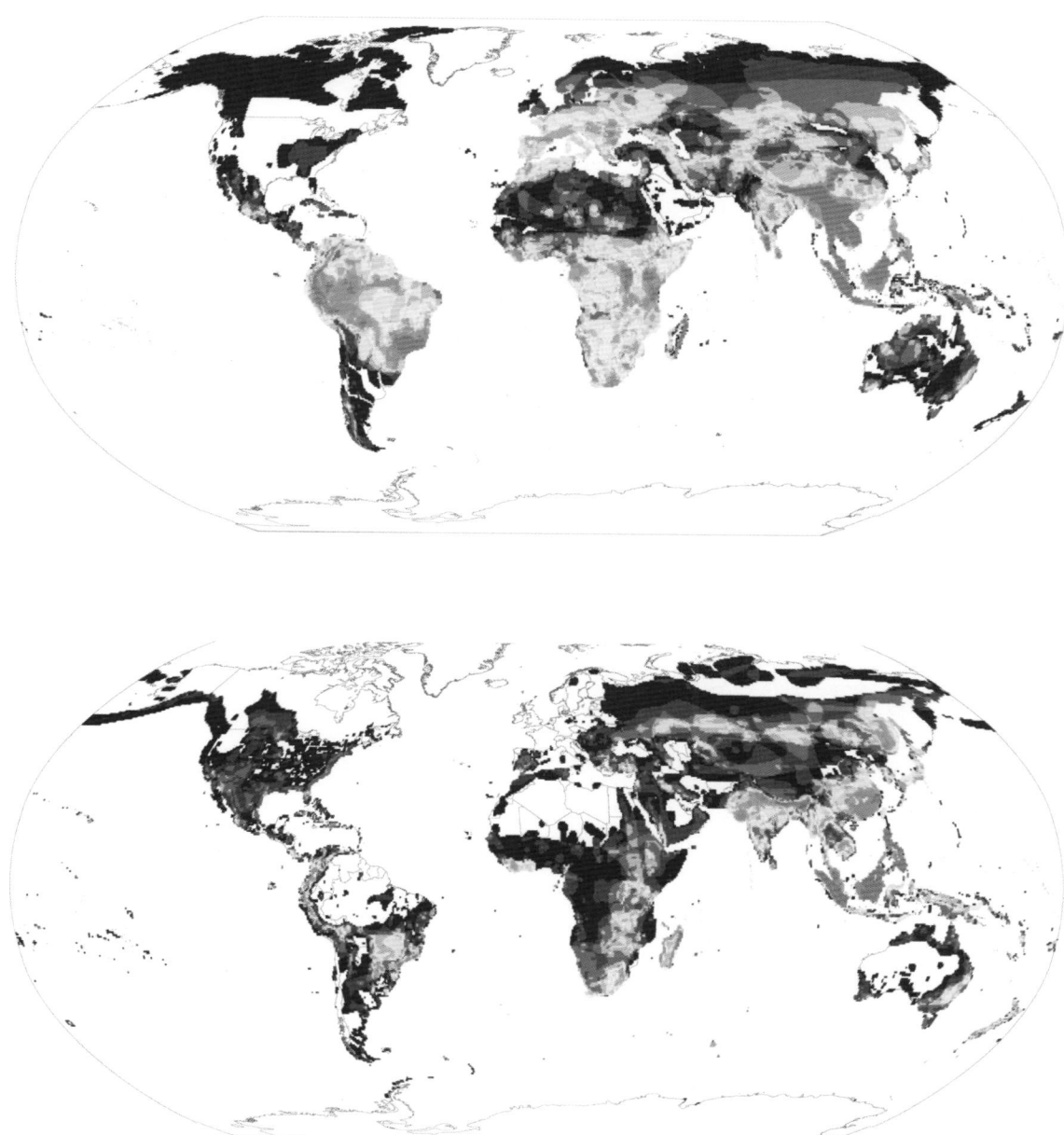

Figure 4.25. Density Distribution Map of Globally Threatened Mammal and Bird Species Mapped at a Resolution of 1/4 Degree Grid Cell. N = 1,063 mammals and 1,213 birds. Dark orange colors correspond to higher richness, dark blue colors to lower richness. Maximum richness equals 25 for mammals and 25 for birds. Color scale based on 10 equal-area classes. (Baillie et al. 2004)

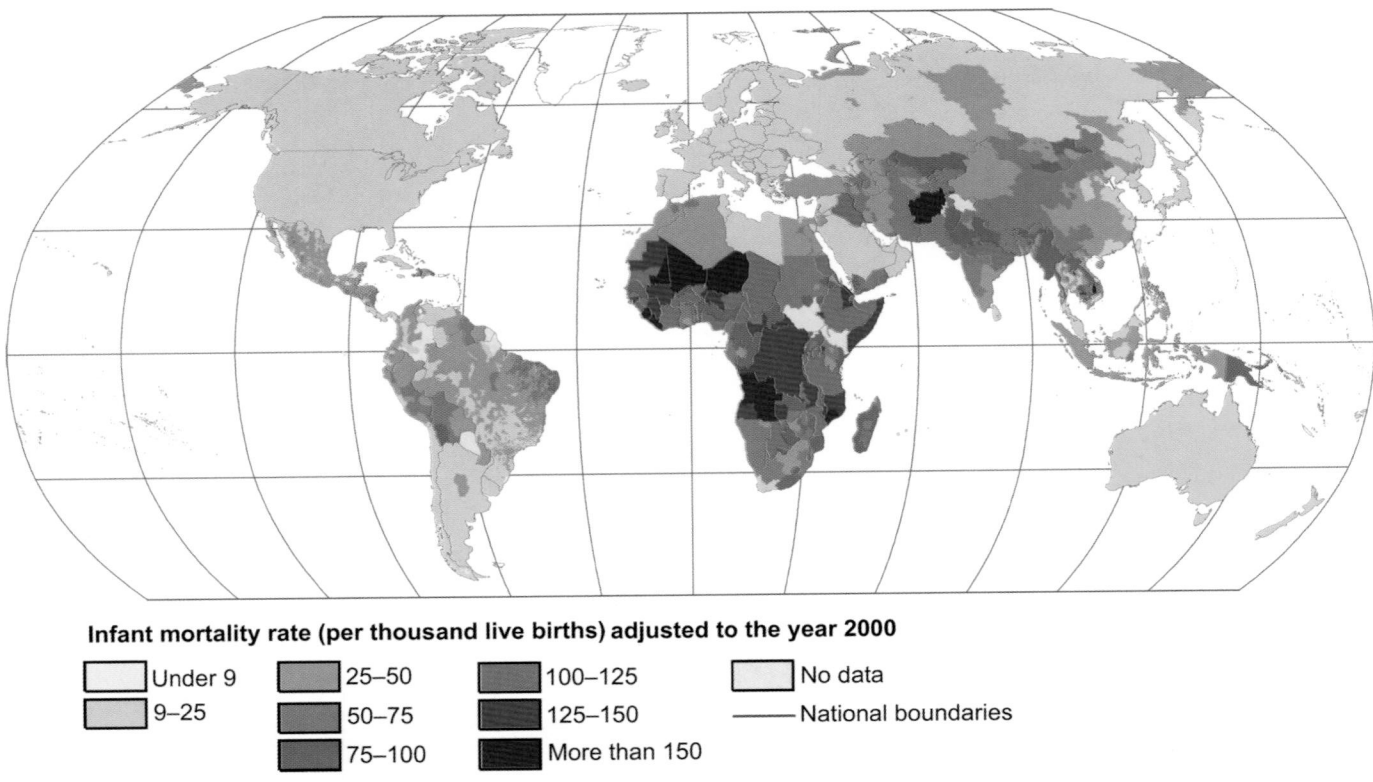

Infant mortality rate (per thousand live births) adjusted to the year 2000

Under 9   25–50   100–125   No data

9–25   50–75   125–150   National boundaries

75–100   More than 150

Figure 5.5. Global Distribution of Infant Mortality Rate (Robinson Projection; UNICEF, DHS, NSOs, NHDRs)

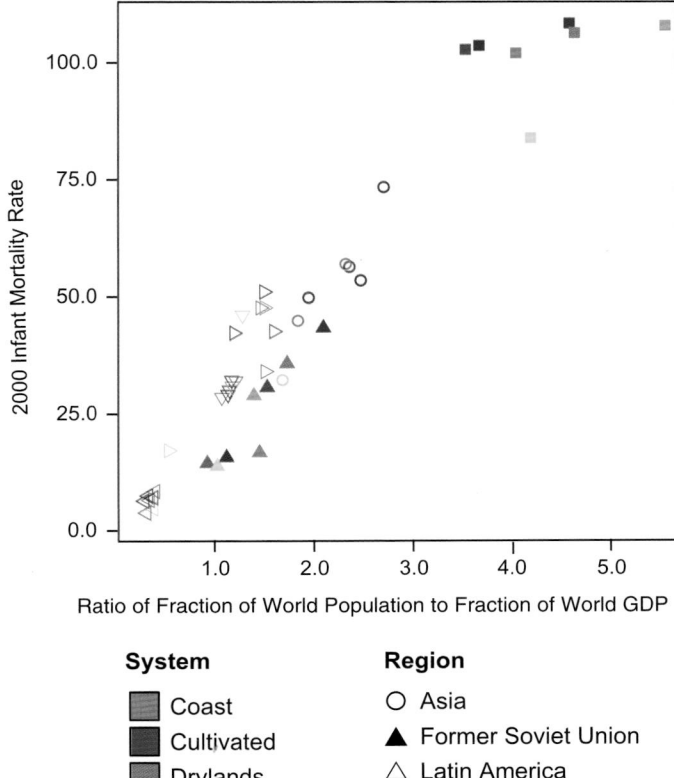

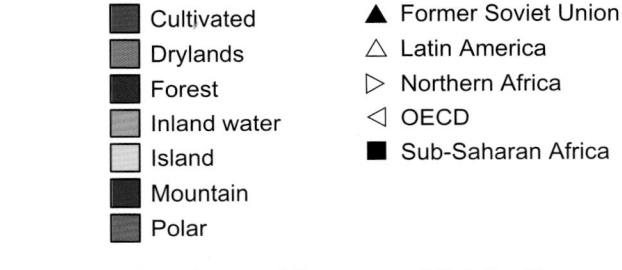

Figure 5.8. MA Regions and Systems and Relative Measures of Well-being

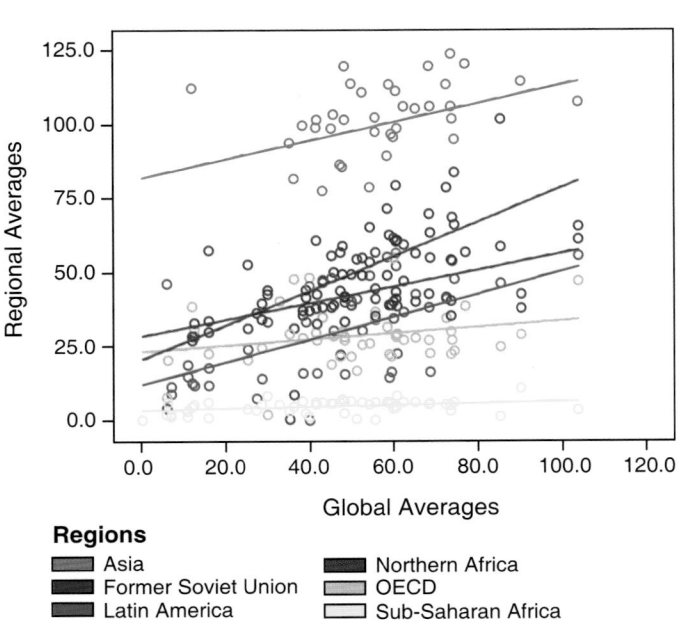

Figure 5.9. Infant Mortality Rate in MA Subsystems, Regional Averages Compared with Global Averages

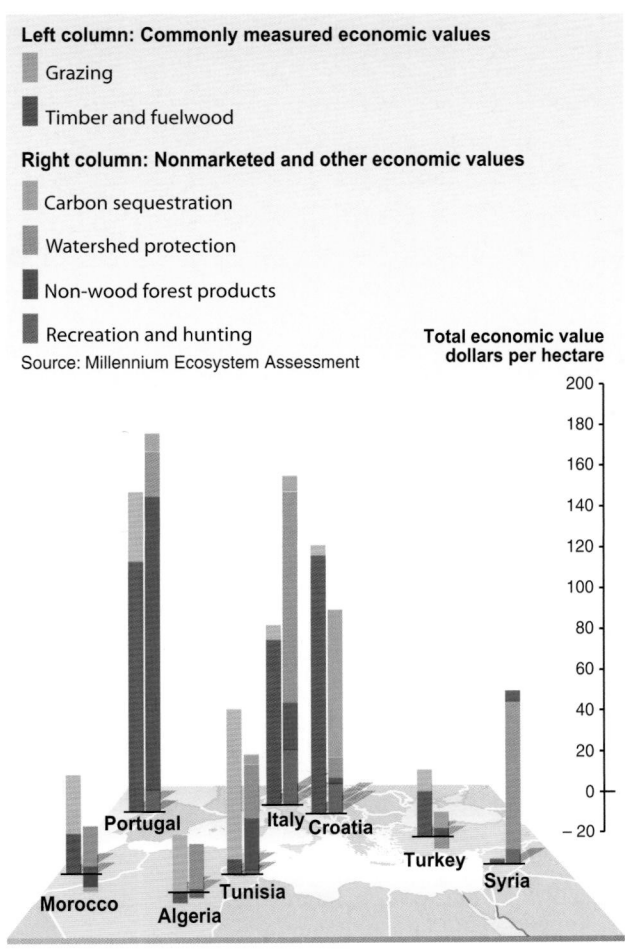

Box 5.2 Figure A

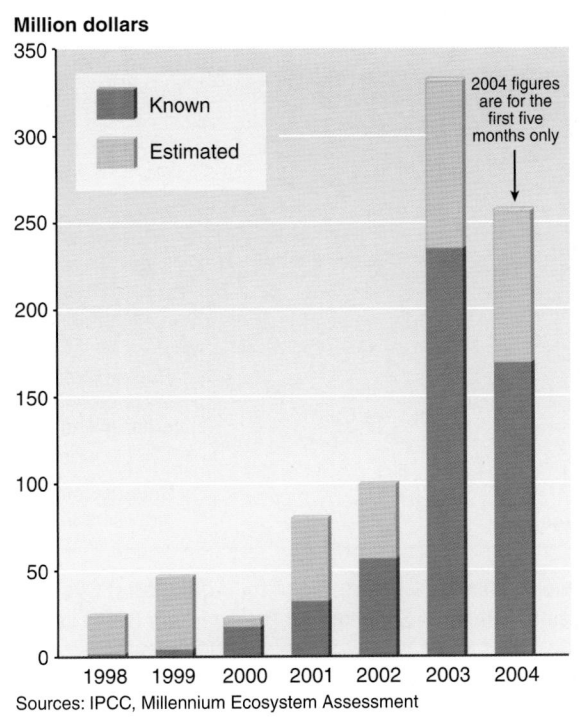

Box 5.2 Figure B

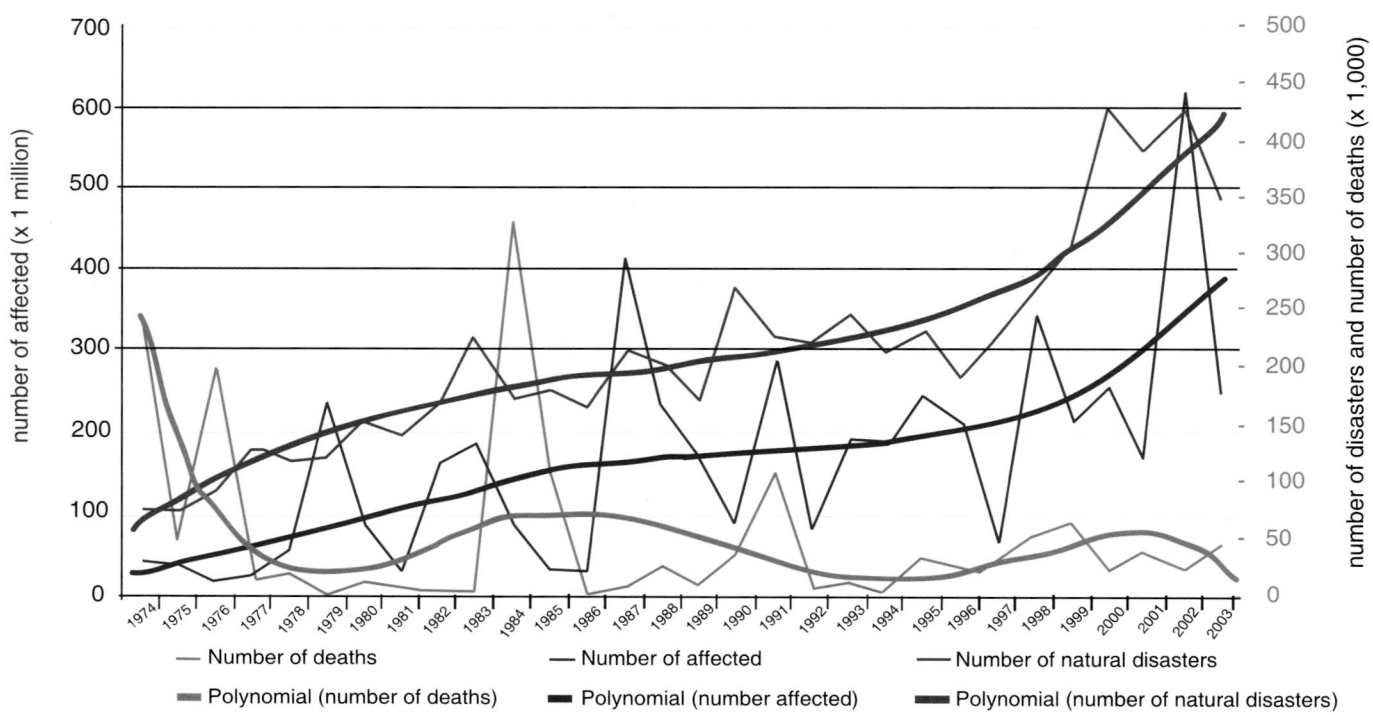

Figure 6.3. Polynomial Trends in the Numbers of Natural Disasters, Persons Killed, and Persons Affected, 1974–2003 (CRED 2003)

Global warming ° C	Reasons for Concern				
	I: Global food production	II: National agricultural economies and market trade	III: Effects on natural resource base	IV: Food security among vulnerable livelihoods	V: Impacts of large-scale droughts and floods
6	Increased potential for shortfalls	Large increases in trade and dependence on imports	Widespread increase in desertification	Increased variability and costs in some regions threaten food security	Potential for large-scale, prolonged events to trigger migration and economic collapse
5					
4	Increased risks in periods of adverse weather	Risks to economies with existing stresses (water shortages, high temperatures)			
3			Increased competition for water, depleted surface water	Regional risks are significant for many livelihoods	Prolonged events create serious economic and societal crises
2	Little threat to global food supply	Some risk to small economies, e.g., small island states			
1			Locally significant water conflicts; widespread soil degradation	Some livelihoods already in crisis	Prolonged events have significant costs at present
0 (present)		Underdevelopment prevalent in LDCs			

Figure 6.6. Climate Change Risks for Agricultural Systems. The five "Reasons for Concern" follow the IPPC's template from the Third Assessment Report's Summary for Policymakers (IPPC 2001). (Downing 2002)

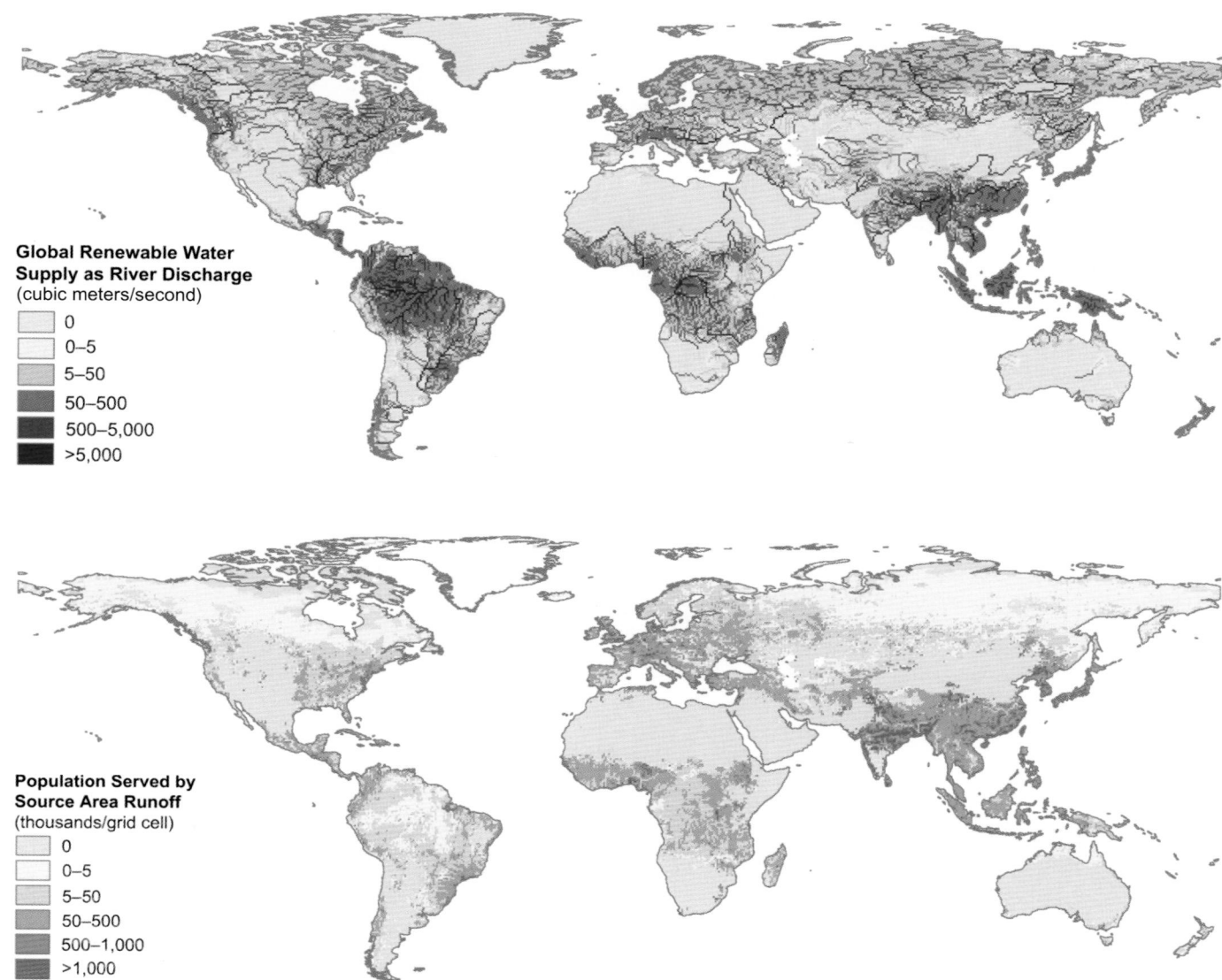

Figure 7.1. Global Renewable Water Supply as River Discharge and Populations Dependent on Accessible Runoff at Point of Origin.
River flows, or total blue water (B_t) is that water passing through 50 km x 50 km grid cells. The top map depicts the global renewable water supply. The bottom map depicts total renewable blue water that is accessible to humans (B_a). Due to their remoteness, some high runoff-generating regions (e.g., Amazonia) fail to support significant populations and are effectively inaccessible. Populations served by nonrenewable groundwater or desalinization are not shown. Table 7.2 gives aggregated regional summaries of the geographic distributions shown here. (Dividing by 31.7 converts values in the top map into units of cubic kilometers per year.)

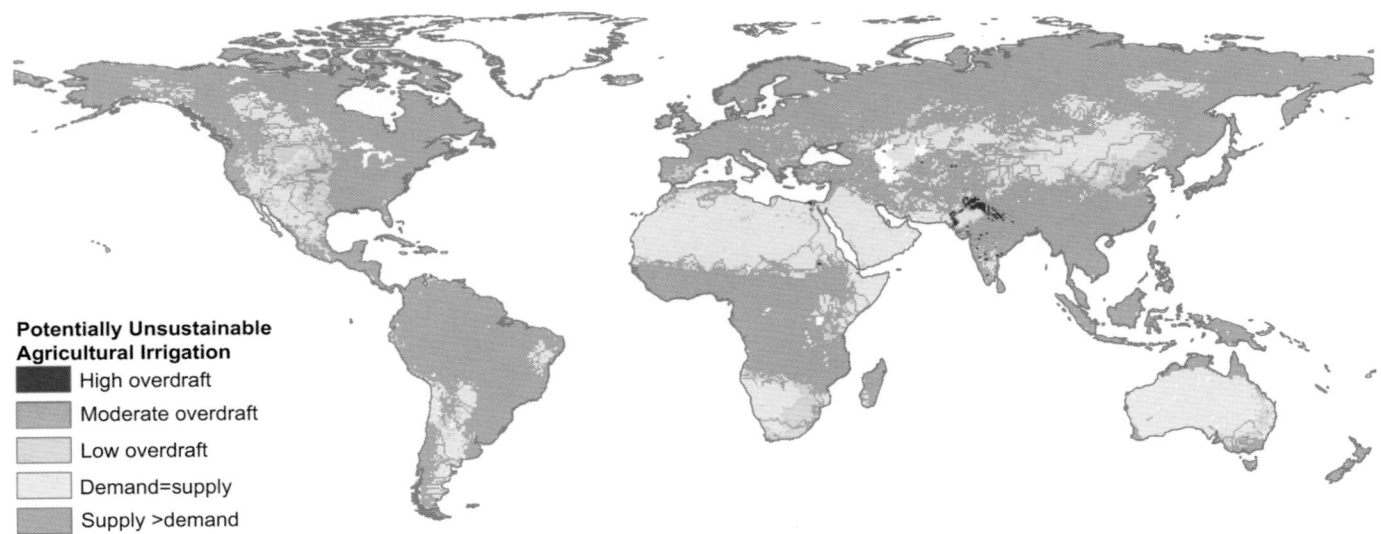

Figure 7.3. Contemporary Geography of Non-sustainable Withdrawals for Irrigation. The following divisions based on calculated consumptive use by crops were used: High overdraft: <1 km³/yr; Moderate: 0.1–1 km³/yr; Low: 0–0.1 km³/yr. All estimates made on ca. 50 km x 50 km resolution grids. The map indicates where there is insufficient fresh water to fully satisfy irrigated crop demands. The imbalance in long-term water budgets necessitates diversion of surface water or the tapping of groundwater resources. The areas shown with moderate-to-high levels of non-sustainable use occur over each continent and are known to be areas of aquifer mining and/or major water transfer schemes.

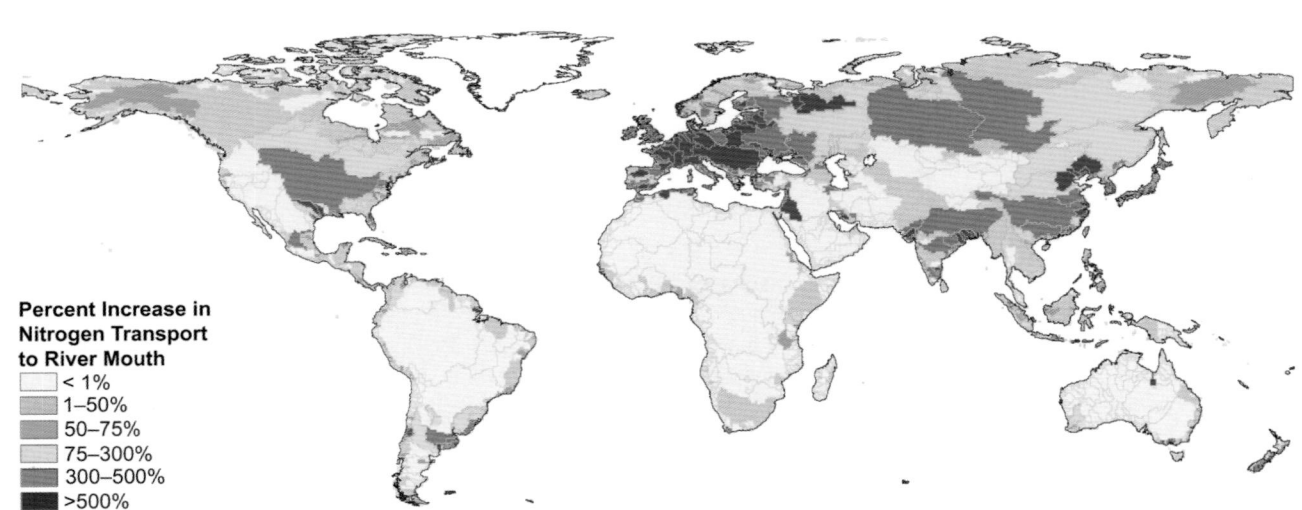

Figure 7.5. Contrast between Mid-1990s and Pre-disturbance Transports of Total Nitrogen through Inland Aquatic Systems Resulting from Anthropogenic Acceleration of This Nutrient Cycle. While peculiarities of individual pollutants, rivers, and governance define the specific character of water pollution, the general patterns observed for nitrogen are representative of anthropogenic changes to the transport of waterborne constituents through inland waterways. Elevated contemporary loading to one part of the system (e.g., to croplands) often reverberate through other parts of the system (e.g., coastal zones), exceeding the capacity of natural systems to assimilate additional constituents. (Green et al. 2004)

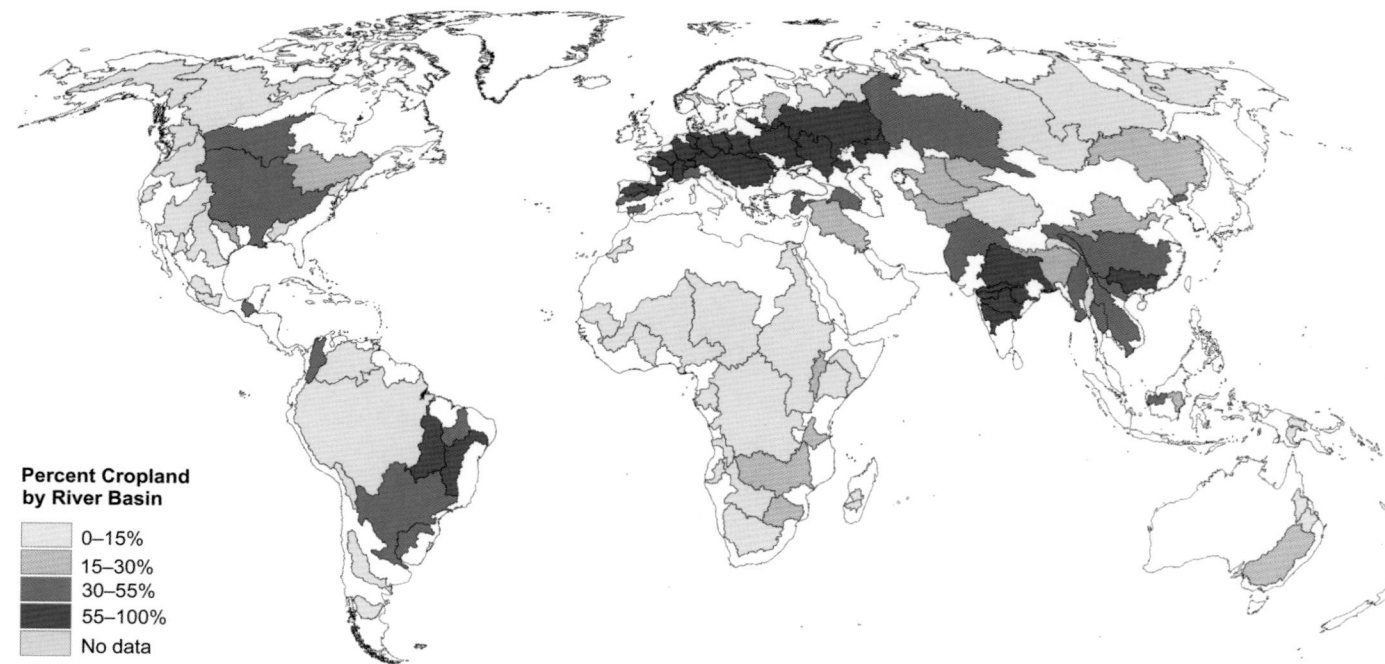

Figure 7.10. Percentage of Cropland Area by River Basin. Cropland areas exclude those with more balanced mosaics of cropland and natural vegetation. (Revenga et al. 2000)

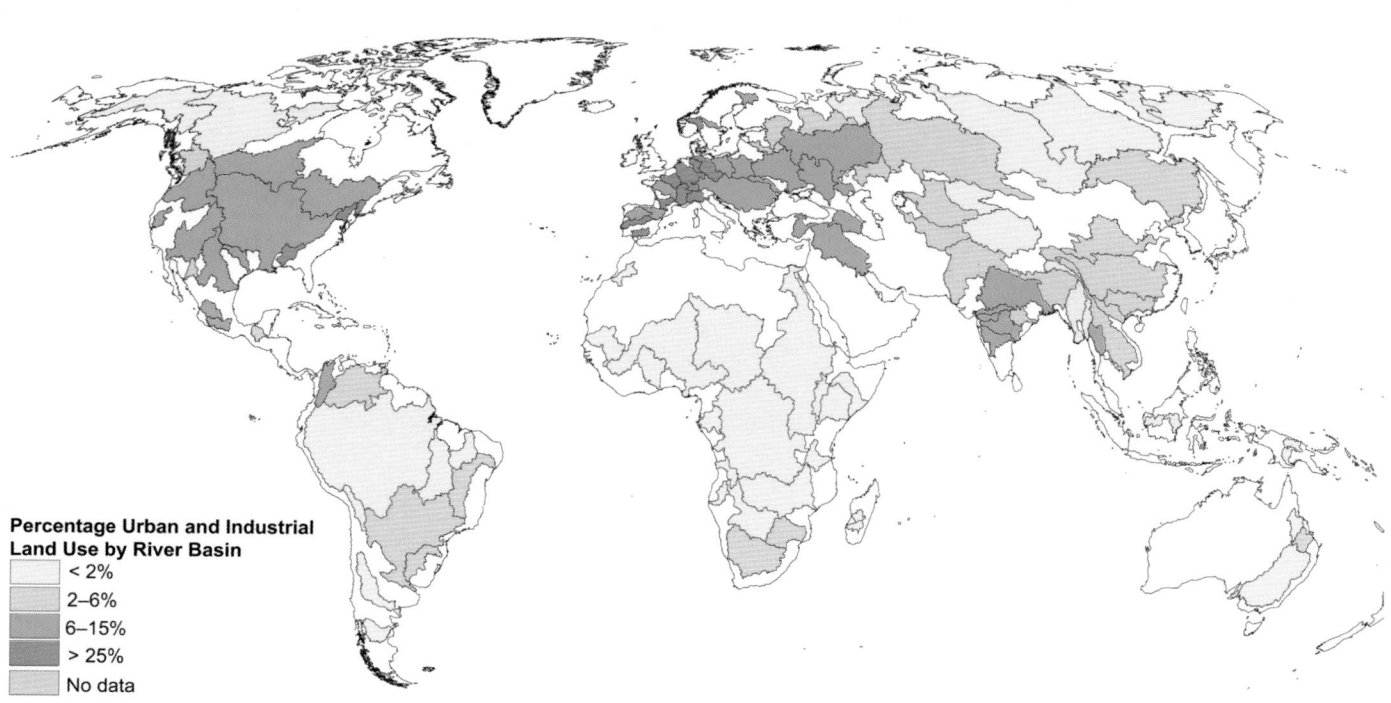

Figure 7.11. Percentage Urban and Industrial Land Use by River Basin (Revenga et al. 2000)

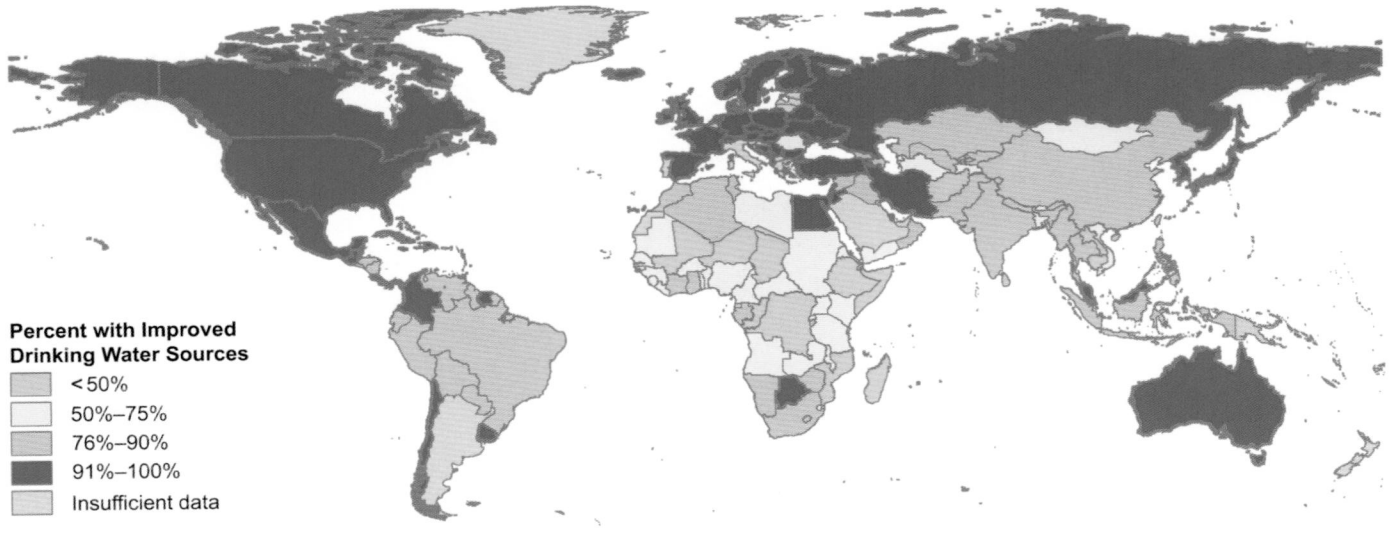

Figure 7.13. Proportion of Population with Improved Drinking Water Supply, 2002 (WHO/UNICEF 2004)

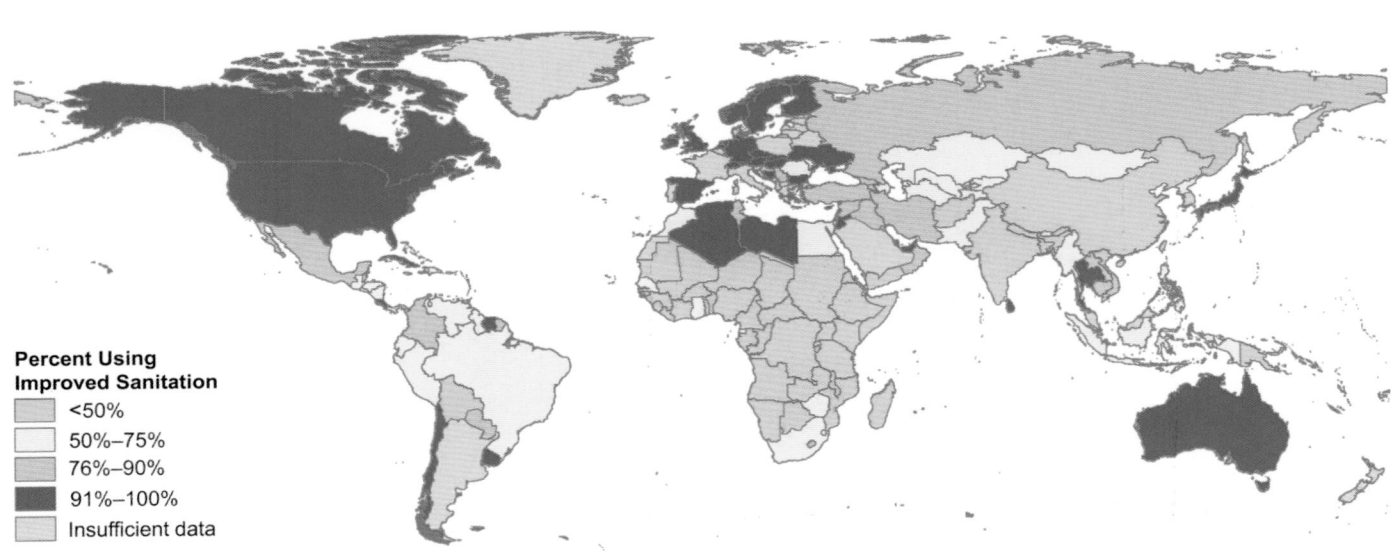

Figure 7.14. Proportion of Population with Improved Sanitation Coverage, 2002 (WHO/UNICEF 2004)

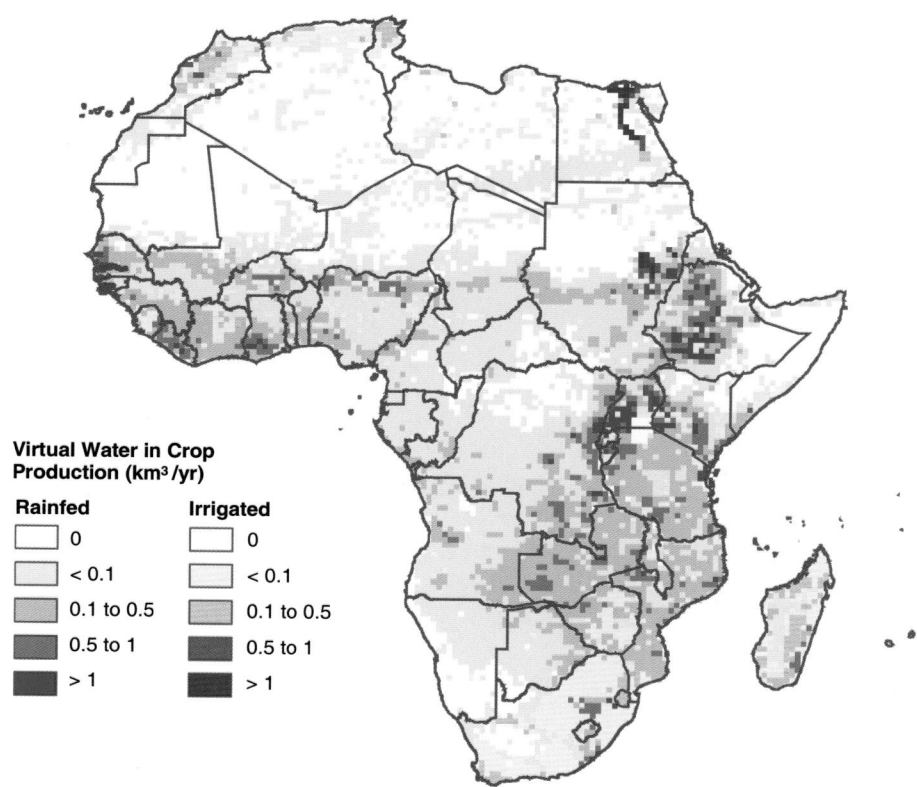

Virtual Water in Crop
Production (km³/yr)

Rainfed	Irrigated
0	0
< 0.1	< 0.1
0.1 to 0.5	0.1 to 0.5
0.5 to 1	0.5 to 1
> 1	> 1

Virtual Water for Africa (km³/yr)

	Crops [1]	Meat [2]	Total
Production	1326	289	1615
Percent of AET [3]	9%	2%	11%
Imports	404	21	425
Exports	50.5	0.3	50.8
VW Balance	1680	309	1989

[1] VW in crops = AET over rainfed cropland +PET over irrigated cropland.
[2] VW in meat = VW in feed/fodder + 30% AET over grazing land.
[3] AET = actual evapotranspiration; percent relative to continental total.

Box 7.4 Figure A

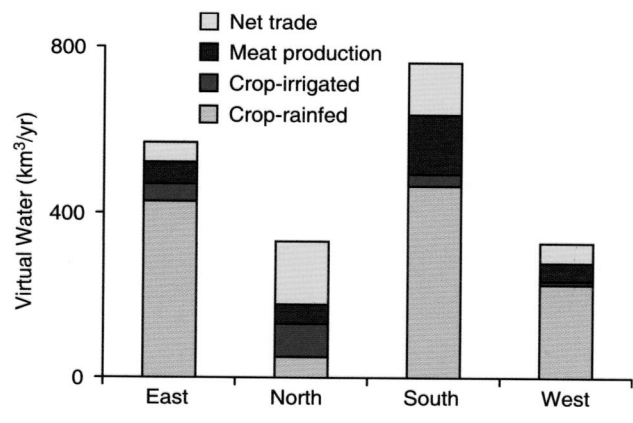

Box 7.4 Figure B

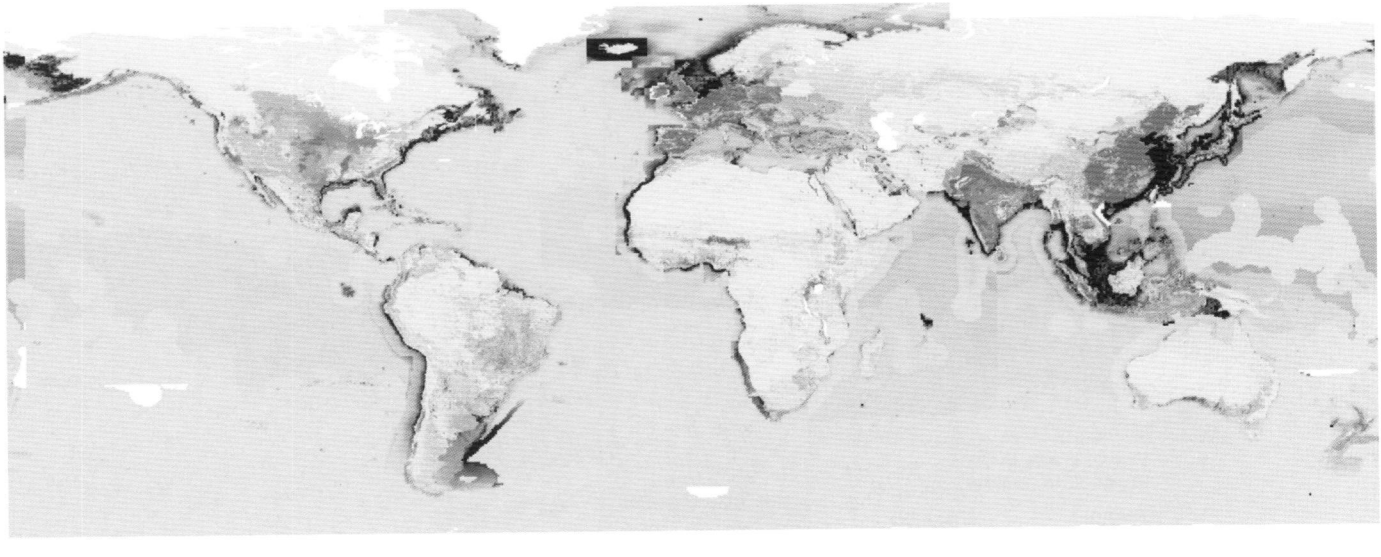

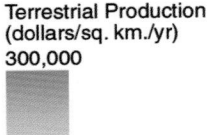

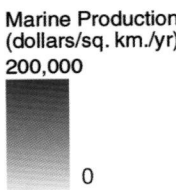

Figure 8.2 Spatial Distribution of Value of Food Production for Crops, Livestock, and Fisheries, 2000. The map shows the approximate value of production in year 2000 using FAOSTAT (2003) production data for all food crops and livestock products weighted by a set of 1989–91 global average commodity prices denominated in International US dollars. These prices are used by FAO to compute its Production Indices. The image was constructed from a composite rainfed-irrigated cropland surface using the global 1992–3 cropland map of Ramankutty and Foley (1998) intersected by the global irrigation map of Doell and Siebert (1999). Crop production was allocated by country in proportion to the share of each 100 km² occupied by rainfed and irrigated agriculture assuming irrigated agriculture is, on average, twice as productive as rainfed agriculture. Livestock production was allocated across a global pasture dataset (Foley et al. 2003) by country, assuming production was distributed into each pixel in proportion to its area of pasture/rangeland.

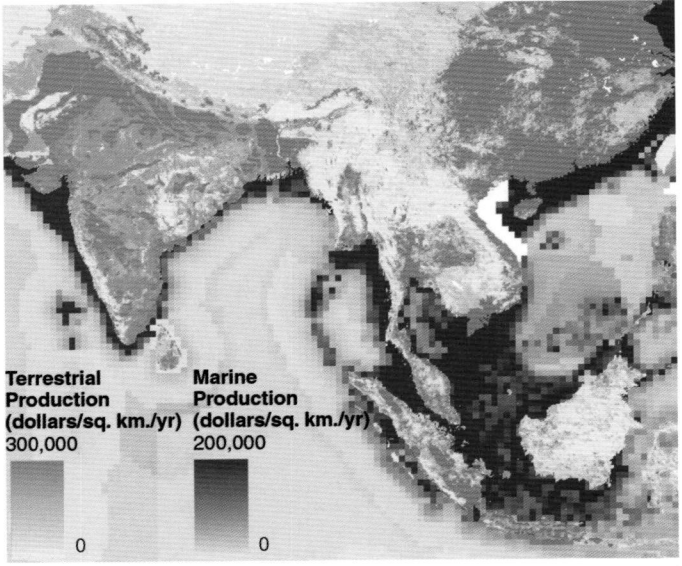

Figure 8.3 Spatial Distribution of Food Production in Parts of Asia for Crops, Livestock, and Fisheries, 2000. This map shows a detail of Figure 8.1. Notice the high value of food production—both marine and terrestrial—in coastal areas.

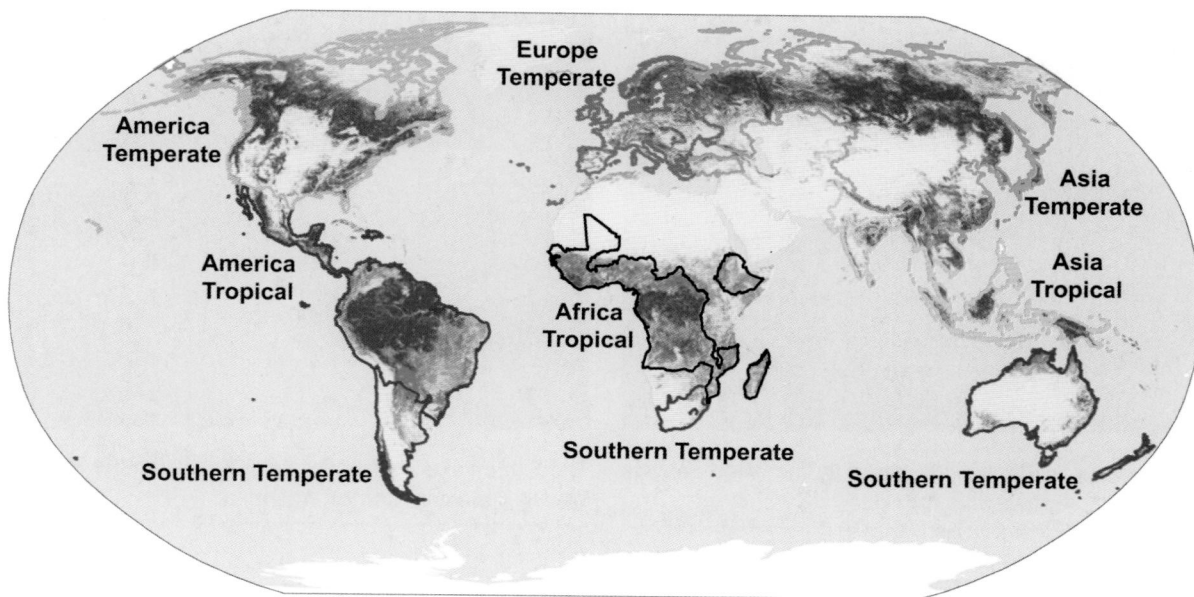

Figure 9.1 Forest Regions Used in Wood Products Analyses. Regions are based on closed forest cover, continents, climate, and national boundaries. (FAO 2001a; cartography, P. Gonzalez)

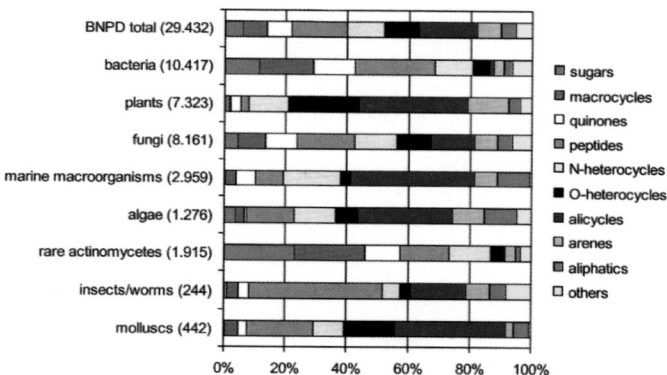

Figure 10.1 Summary of Different Kinds of Natural Product Structures Produced by Different Organisms (Bioactive Natural Product Database, Szenzor Management Consulting Company, Budapest, cited in Henkel et al. 1999)

Box 10.3 Figure Lake in Australia Covered by the Weed Salvinia
(Photos from CSIRO, Australia)

The Same Lake Six Months Later Following the Release of a Weevil Biological Control Agent

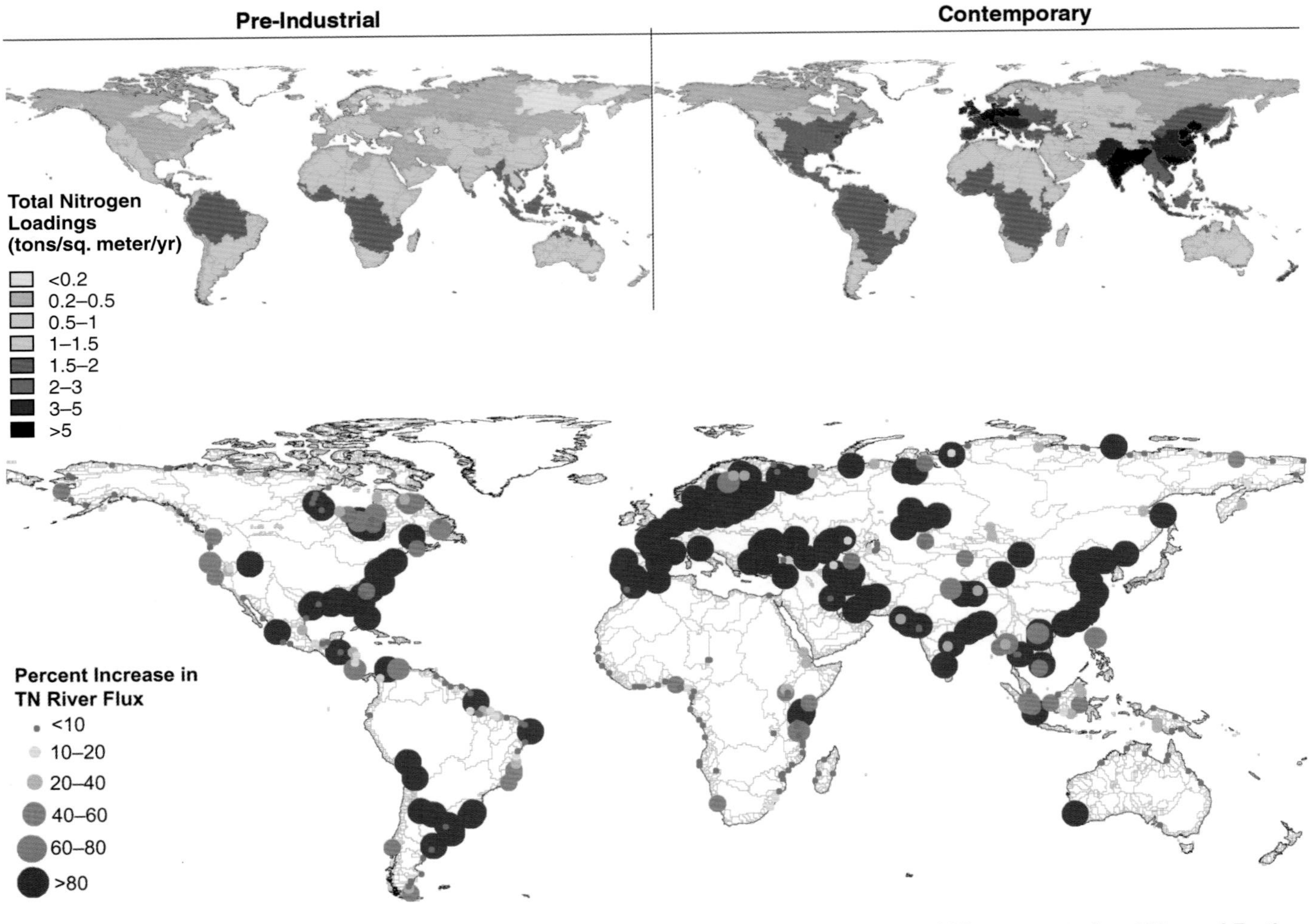

Figure 12.3 Contrast between Contemporary and Preindustrial Loadings of Easily Transported Nitrogen onto Land Mass of Earth and Geography of Relative Increases in Riverborne Nitrogen Fluxes Resulting from Anthropogenic Acceleration of Cycle. Contemporary time is from the mid-1990s. While the peculiarities of individual pollutants, rivers, and governance define the specific character of water pollution, the general patterns observed for nitrogen are representative of anthropogenic changes to the transport of waterborne constituents. Elevated contemporary loadings to one part of the system (e.g., to croplands) often reverberate to other parts of the system (e.g., coastal zones), exceeding the capacity of natural systems to assimilate additional constituents. (Green et al. 2004)

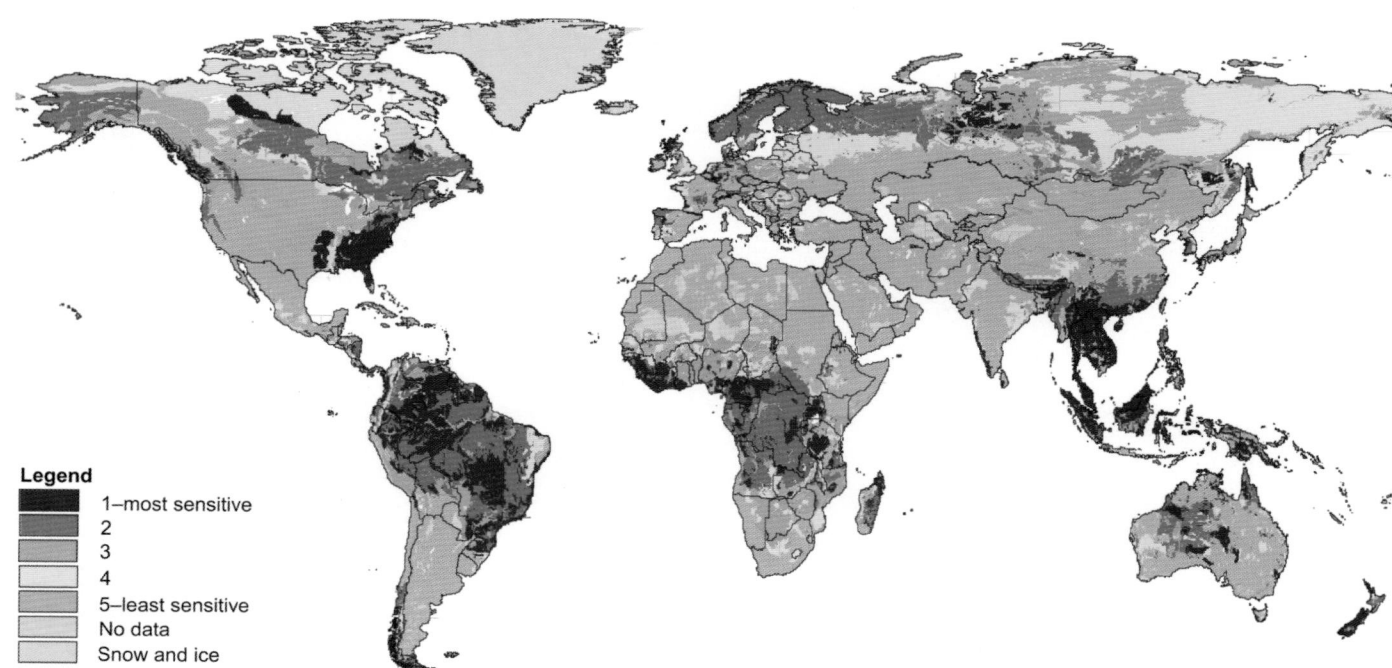

Figure 12.6 Global Map of Soil Sensitivity to Acidic Inputs from Atmospheric Sulfur and Nitrogen Deposition. This map shows the ability of the soil to buffer acid deposition. Problems of acidification are most likely to arise where high projected rates of deposition coincide with high sensitivity—for instance, in Southeast Asia. (Kuylenstierna et al. 2001)

Modeled Map of Carbon Dioxide Sources and Sinks, Excluding Fossil Fuels

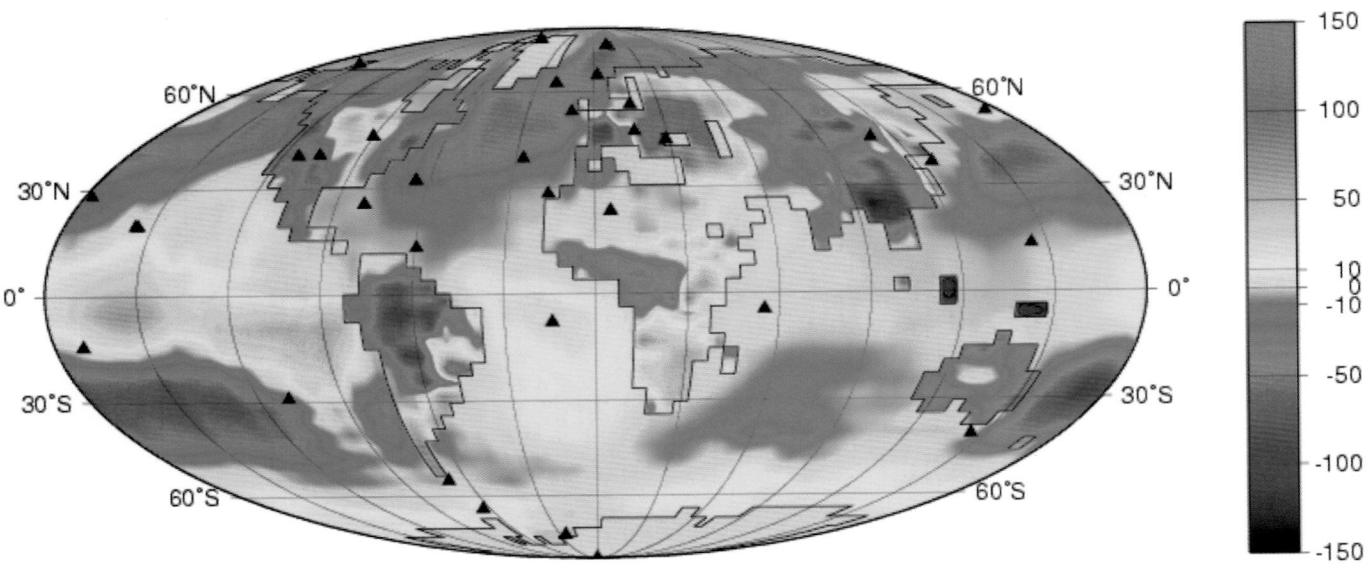

Modeled Map of Carbon Dioxide Sources and Sinks, Including Fossil Fuels

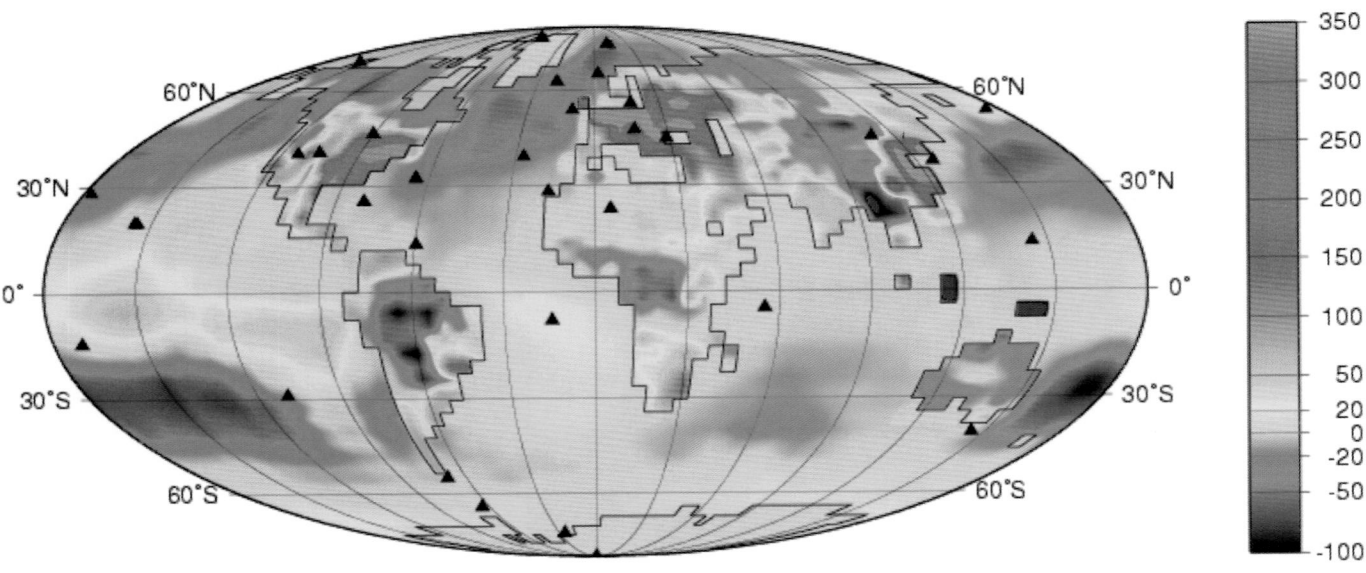

Figure 13.5 Maps of Carbon Dioxide Fluxes Estimated from Atmospheric Measurements, July 1995 to June 2000 (in gC/m²/year) (Rödenbeck et al. 2001). The spatial allocation of sources and sinks of CO_2 is derived from measurements of atmospheric concentrations from a network of sites over the globe using a technique known as inverse modeling. This technique gives the sum of all fluxes. Positive numbers denote a source into the atmosphere; negative numbers denote a sink from the atmosphere. The magnitude and spatial allocation of fluxes is very sensitive to the number of measuring sites and the time period of the analysis. The top figure is the total flux excluding fossil fuel emissions to highlight the terrestrial vegetation fluxes. The bottom figure includes fossil fuel emissions; therefore land areas appear to be sources or smaller sinks.

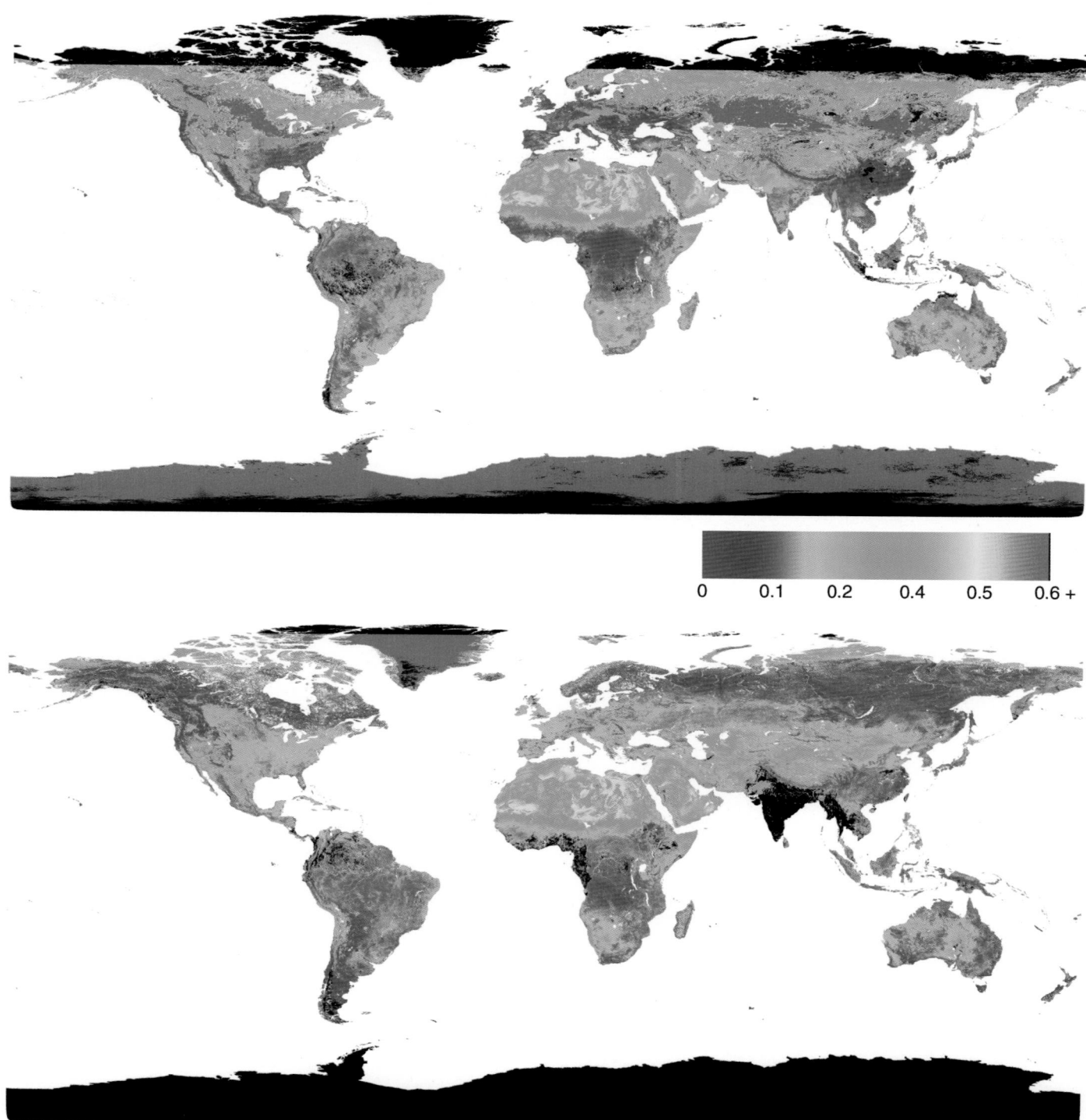

Figure 13.8 Map of Land Surface Albedo Captured by the MODIS Satellite Instrument (Schaaf et al. 2002; Lucht et al. 2000). Albedo is the fraction of solar radiation reflected back into the atmosphere from Earth's surface. Higher albedo means that more energy leaves the planetary boundary layer (net cooling of the atmosphere). Regions where there were no data available, e.g., due to clouds, are indicated by black. The top figure is of data sampled in January 2001. In the northern areas during the winter season, snow albedo is very high (up to 0.8, red). The boreal forest belt can be clearly seen in blue and green since trees mask snow, reduce albedo, and warm the surface air during the snow season. The bottom figure is of data sampled in June 2001. In comparison with January 2001, the northern land areas have a much lower albedo due to the absence of snow. In this map, the area with the highest albedo (up to 0.5, green and yellow) is the Sahara desert. High albedo in this region suppresses rainfall during the summer rain season.

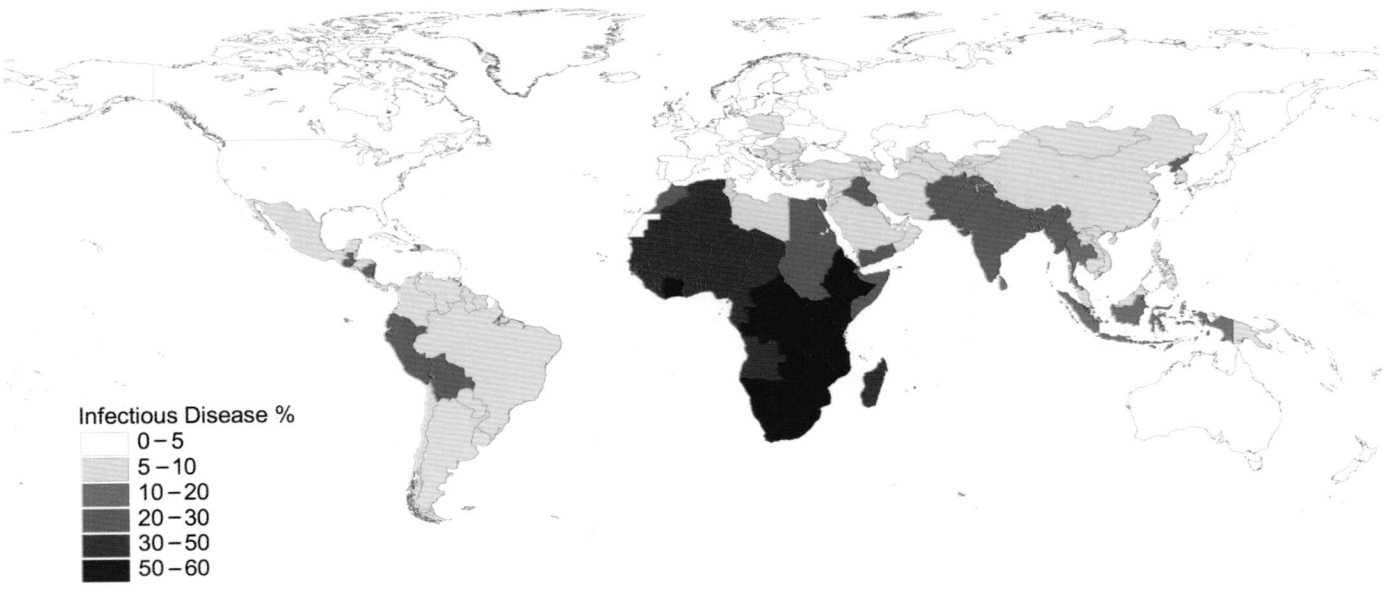

Figure 14.1 Current Map of Infectious and Parasitic Diseases

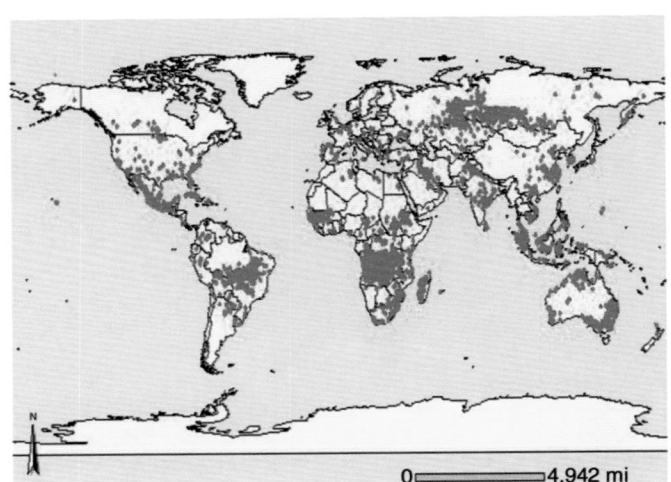

Figure 16.2 MODIS Fire Pixels Detected May 20–22, 2004 (Image courtesy of MODIS Rapid Response Project at NASA/GSFC)

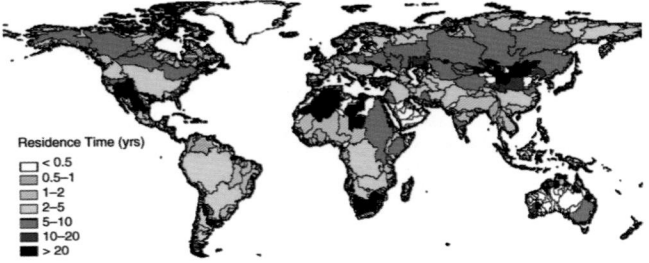

Figure 16.3 Residence Time in Lakes, Reservoirs, and Soils, by Basin (Green et al. 2004)

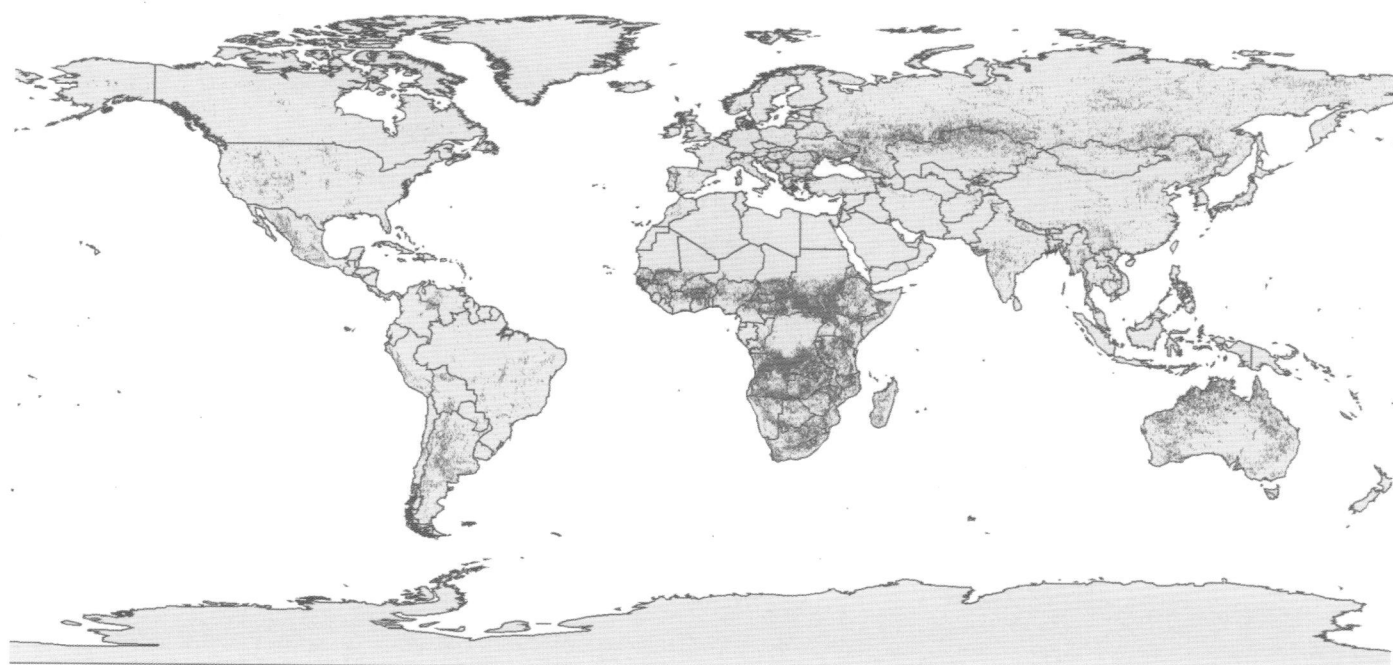

Figure 16.6 Global Patterns of Burned Area in 2000, Based on the GBA2000 Product (Grégoire et al. 2003)

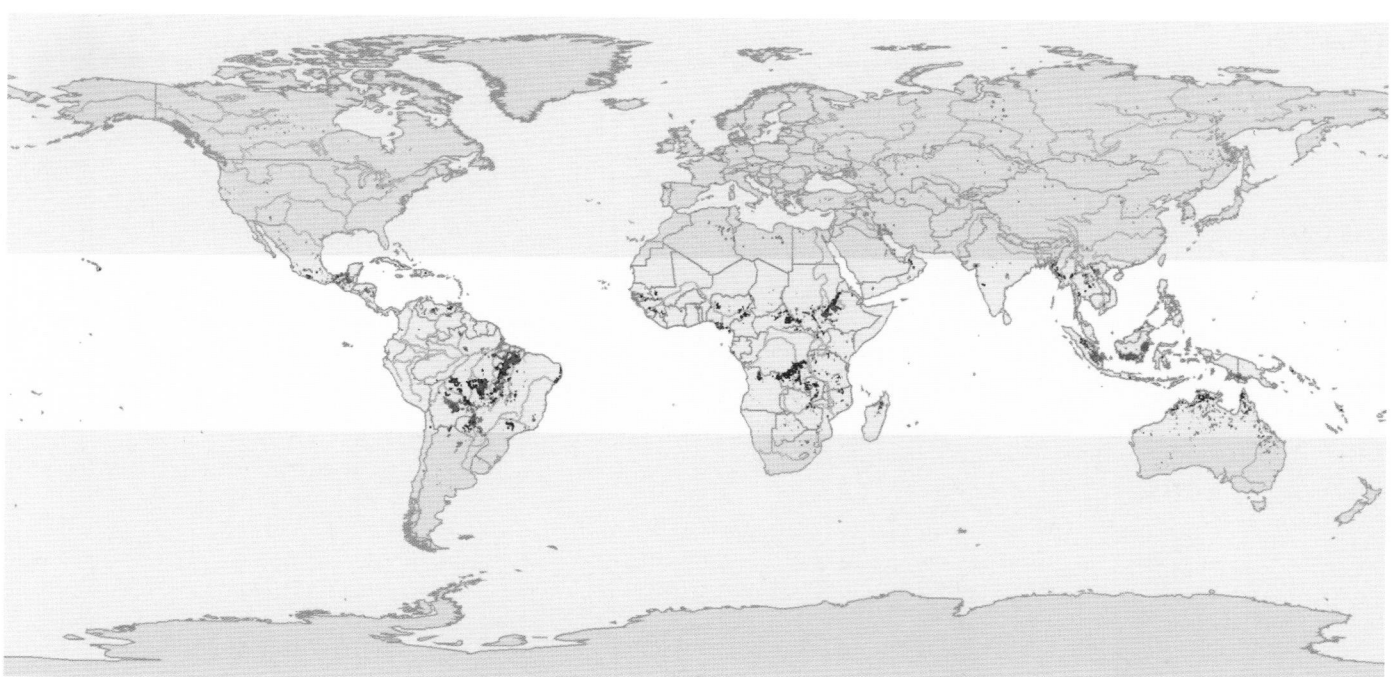

Figure 16.7 Map of Most Frequent and Exceptional Fire Events in the Tropics, 1997–2000 (Lepers 2003)

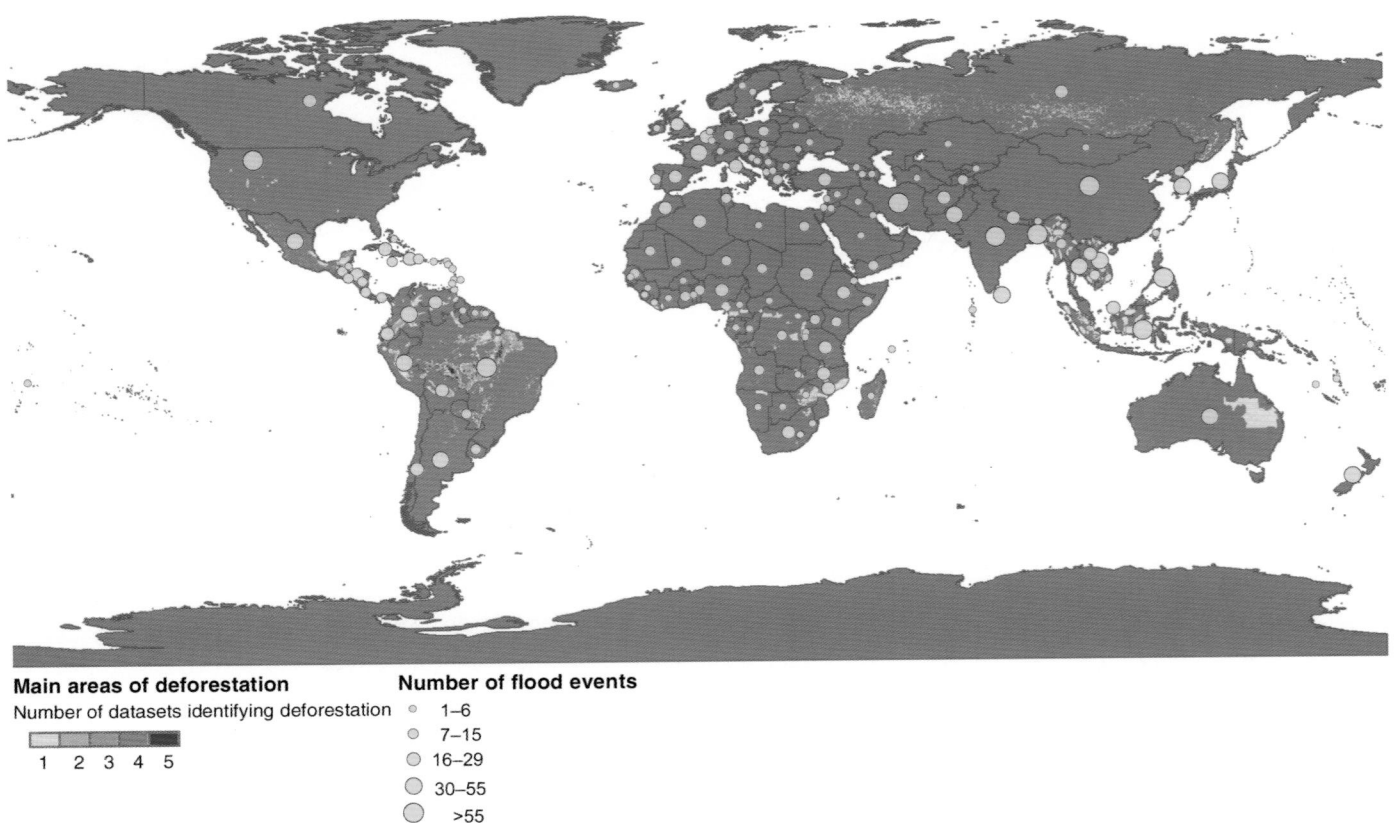

Main areas of deforestation
Number of datasets identifying deforestation

1 2 3 4 5

Number of flood events

○   1–6
○   7–15
◯   16–29
◯   30–55
◯   >55

Figure 16.10 Main Areas of Deforestation and Forest Degradation and Number of Floods, by Country, 1980–2000 (Lepers 2003; OFDA/CRED)

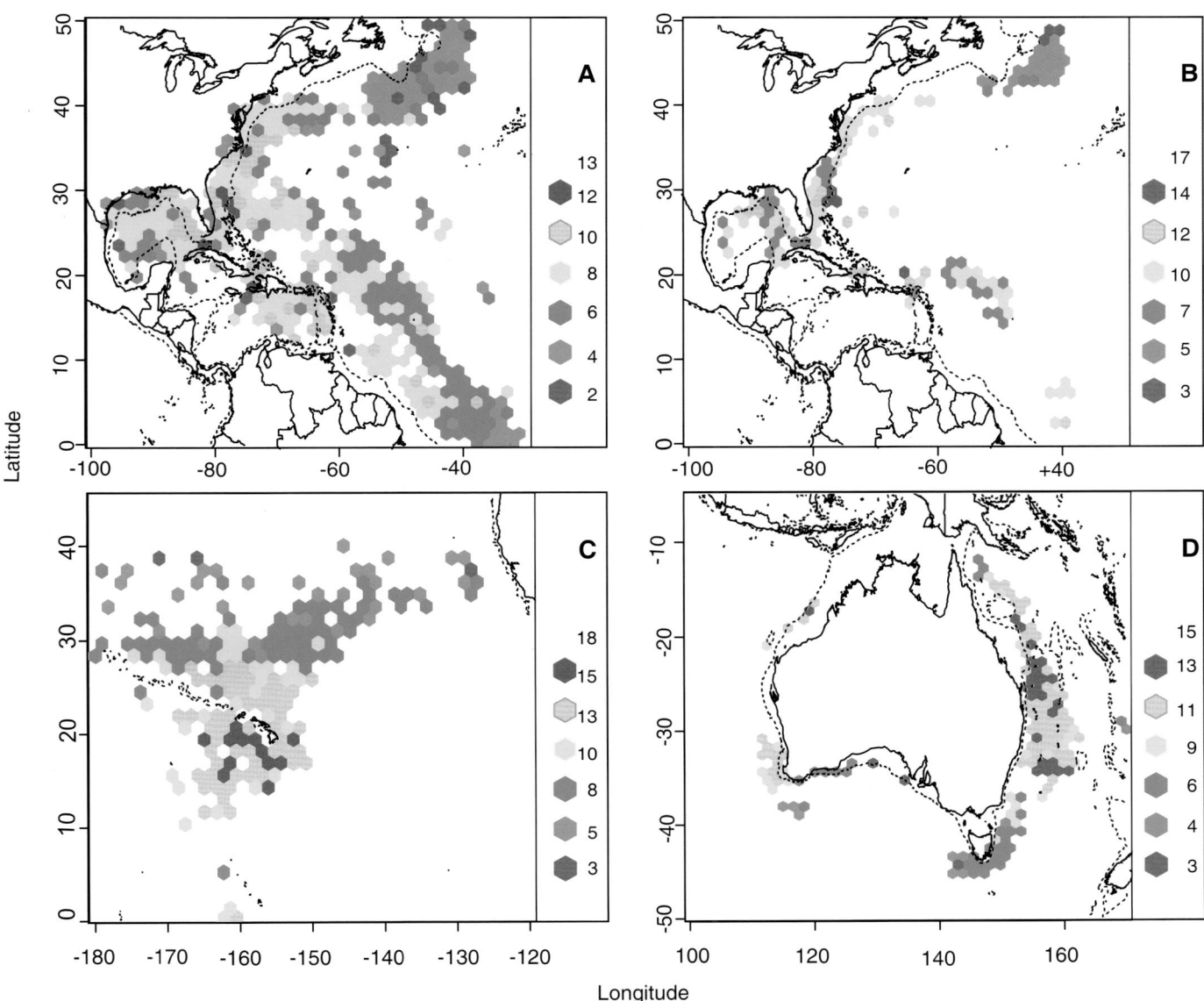

Figure 18.7 Predator Diversity in the Ocean. Predicted from the Northwest Atlantic Longline Logbook (A), Observer Data (B), Hawaiian Observer Data (C), and Australian Observer Data (D). Codes indicate level of species diversity. Dotted line represents 1,000-m isobaths, identifying the outer margins of continental slopes. (Worm et al. 2003)

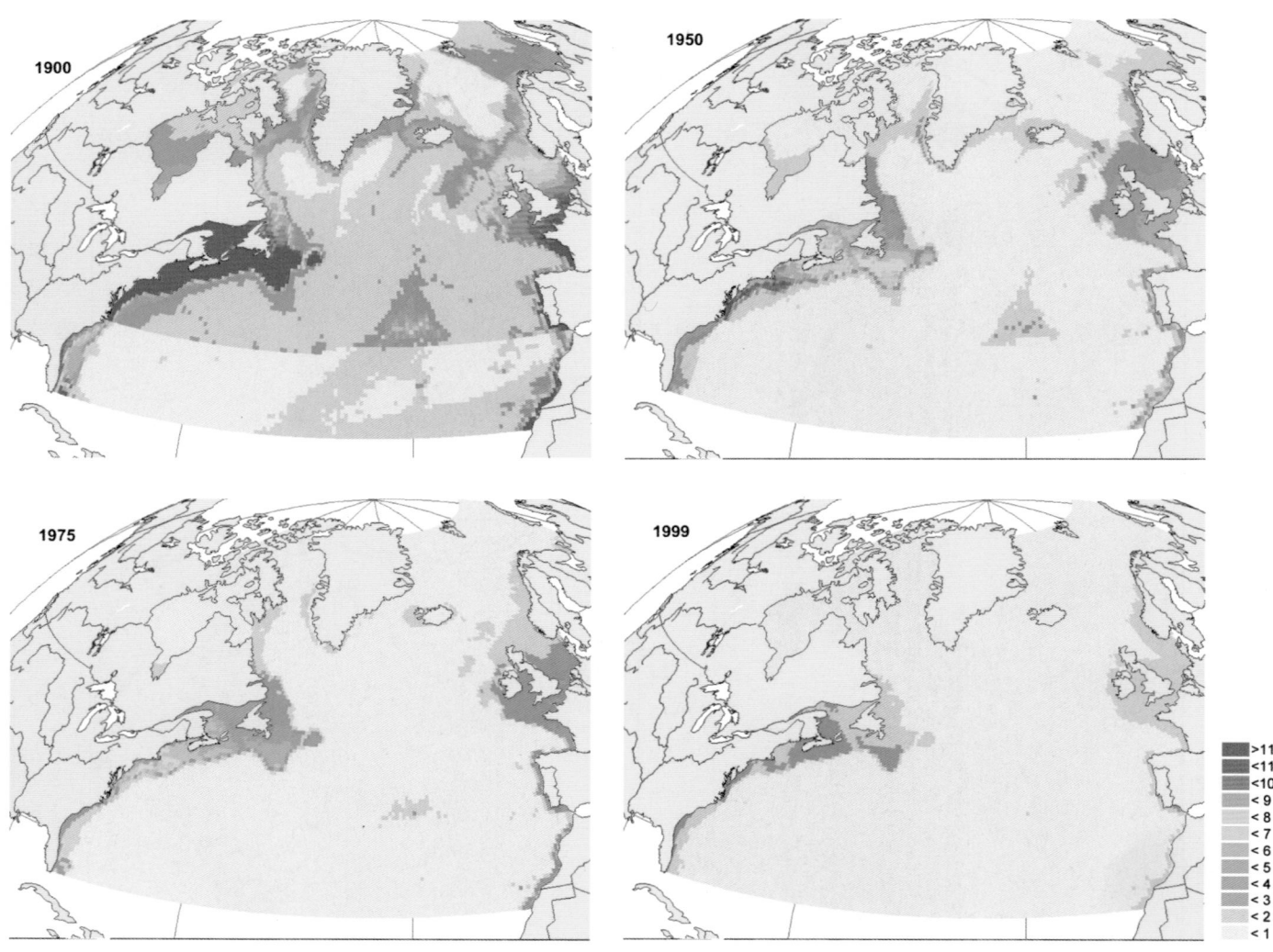

Figure 18.8 Changes in Marine Biomass in North Atlantic in 1900, 1950, 1975, and 1999 (in tons per square kilometer) (Christensen et al. 2003)

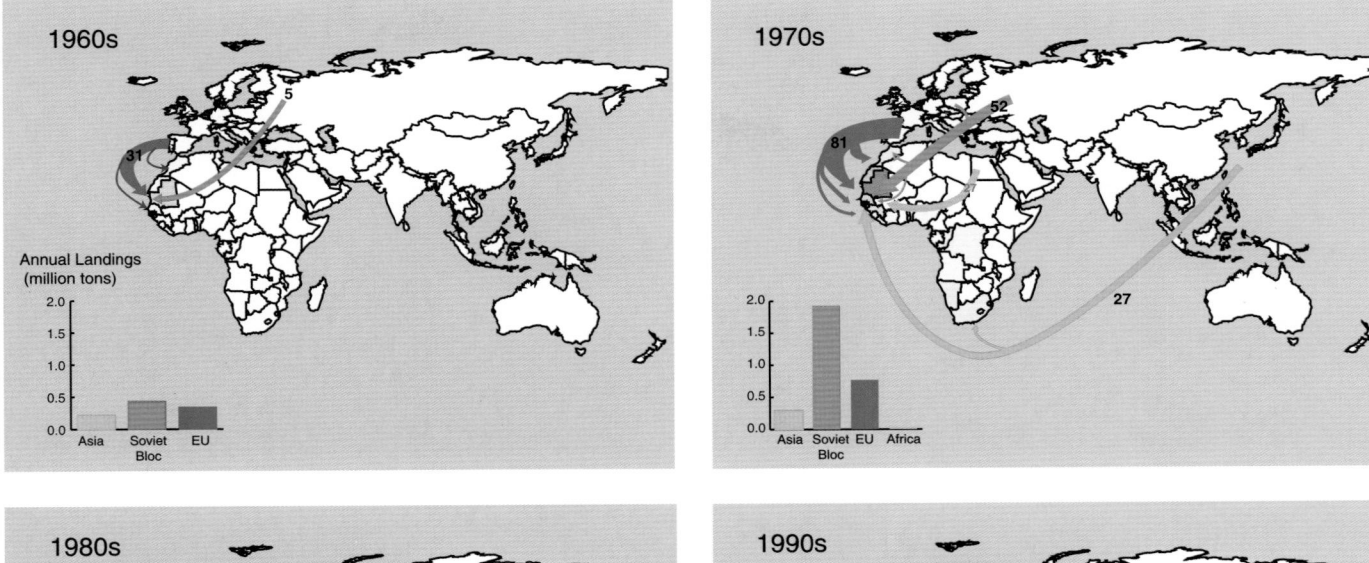

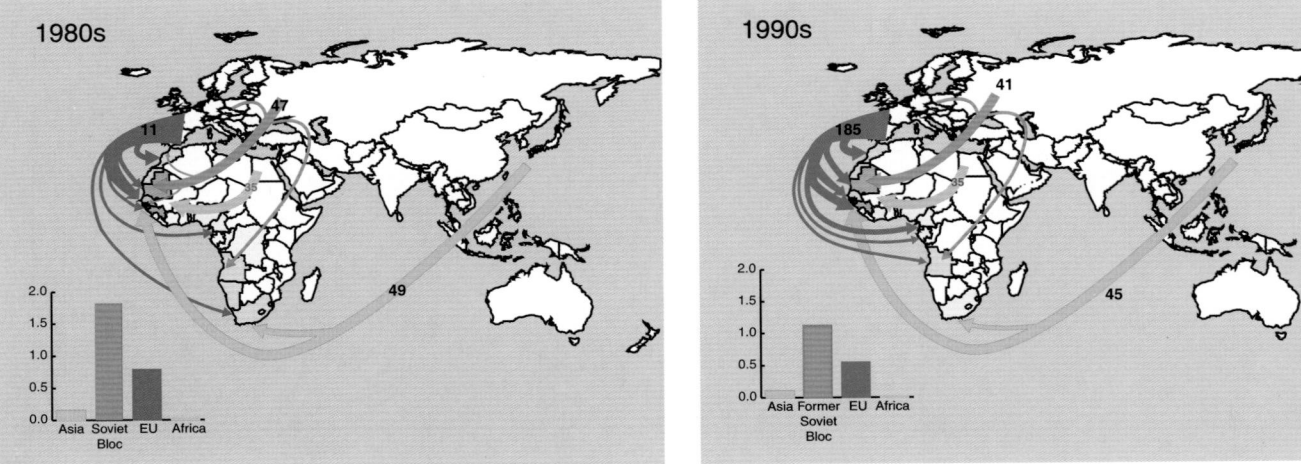

Figure 18.17 Changes in Distant Water Fleet Access as Number of Agreement Years for 1960s, 1970s, 1980s, and 1990s (Alder and Sumaila 2004)

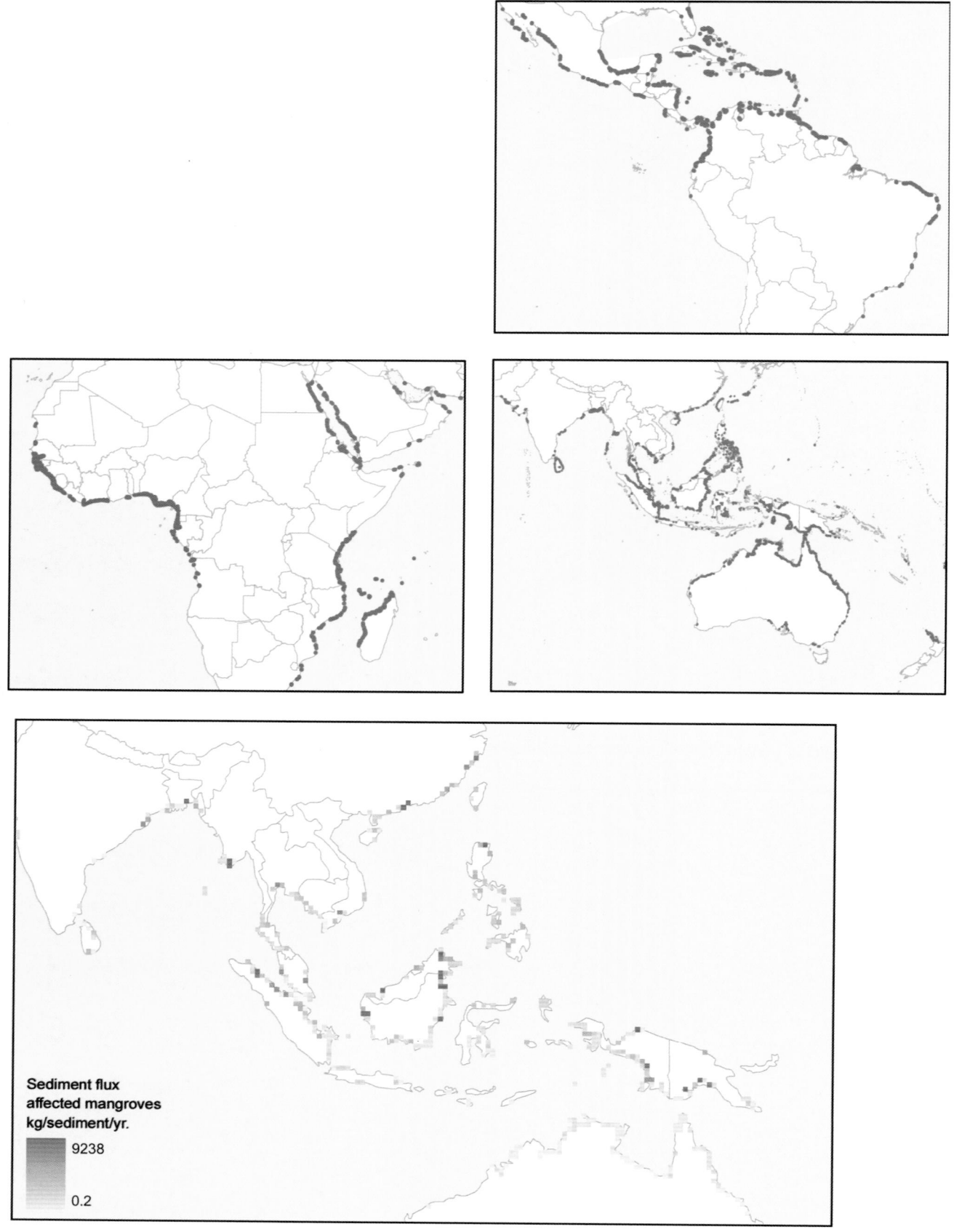

Figure 19.5 Global Distribution of Mangrove Forests, and Levels of Sediment Loading on Mangroves in the Asia–Pacific Region
(UNEP-WCMC 2003a; Syvitski et al. 2005)

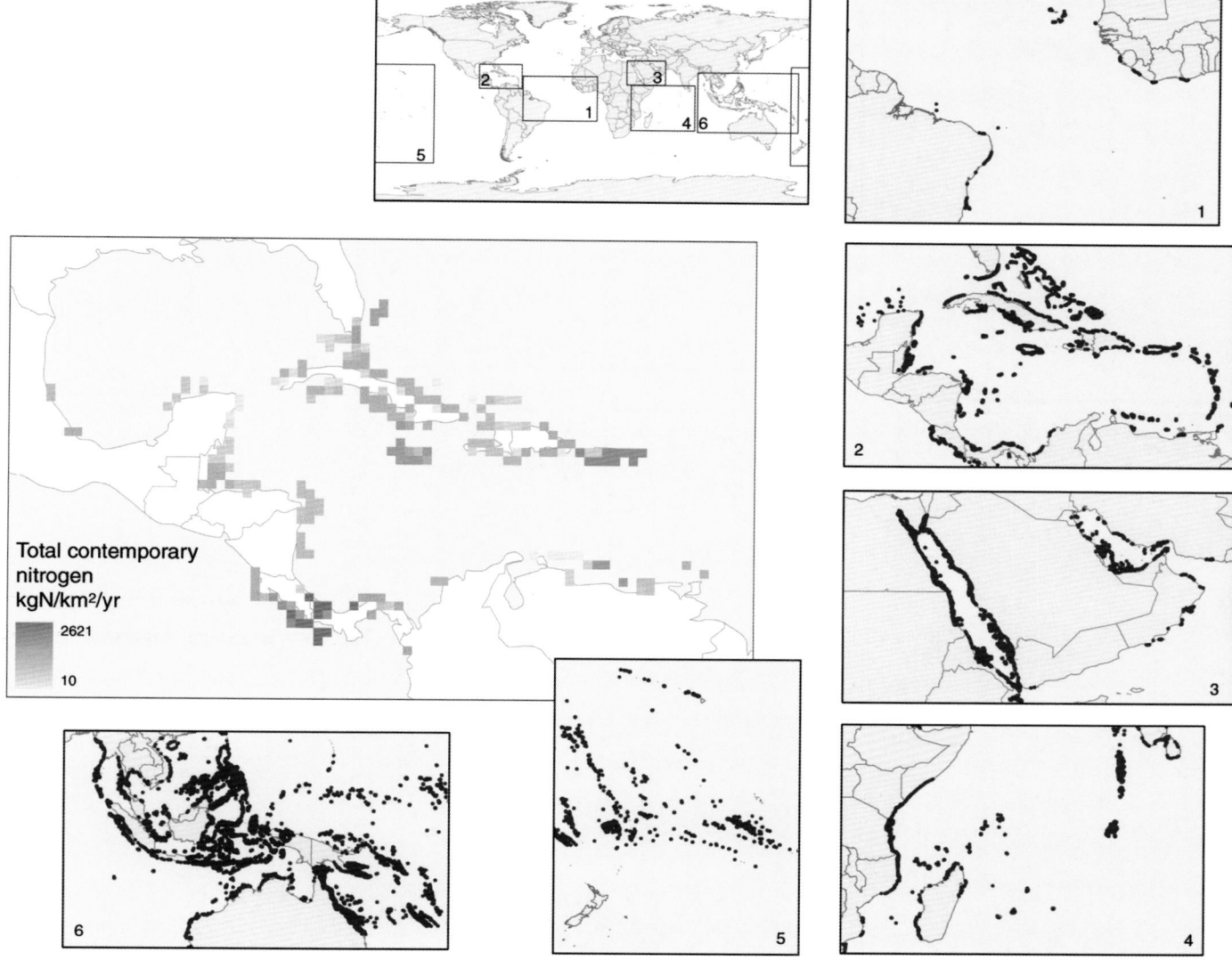

Figure 19.6 Global Distribution of Major Coral Reefs and Levels of Nitrogen on Caribbean Coral Reefs (UNEP-WCMC 2003d; Syvitski et al. 2005)

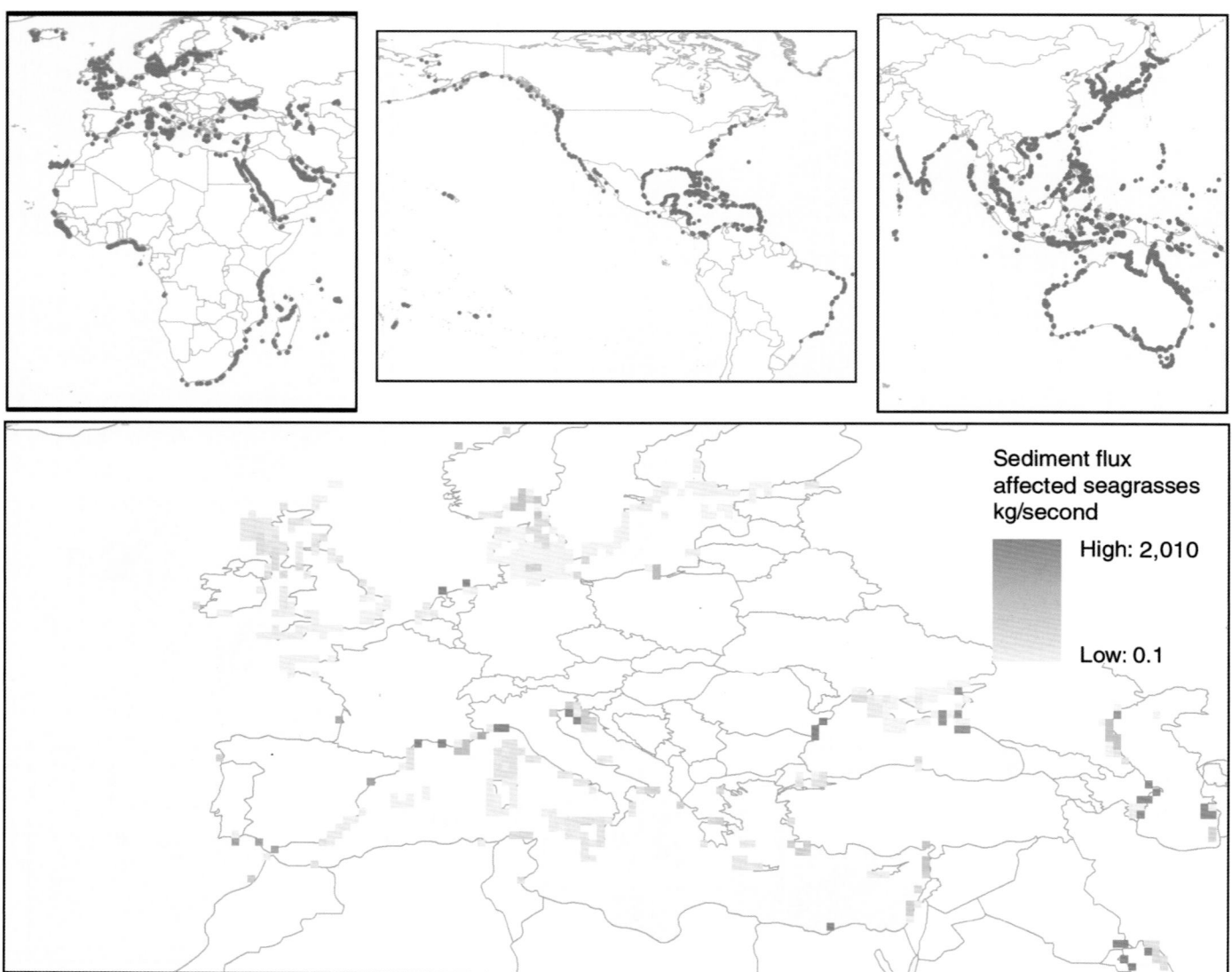

Figure 19.7 Global Distribution of Seagrasses, and Levels of Sediment Loading on European Seagrass Areas (UNEP-WCMC 2003c; Syvitski et al. 2005)

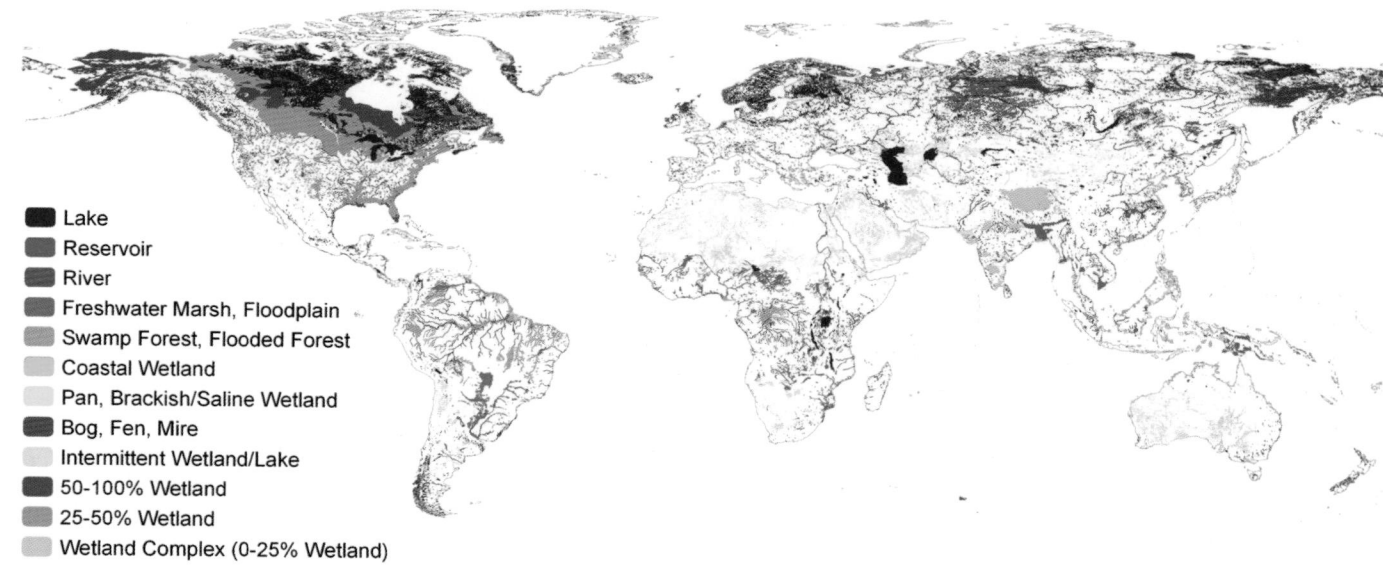

Figure 20.1 Distribution of Inland Water Systems Described as Large Lakes, Reservoirs, and Wetlands (Adapted from Lehner and Döll 2004 and LakeNet)

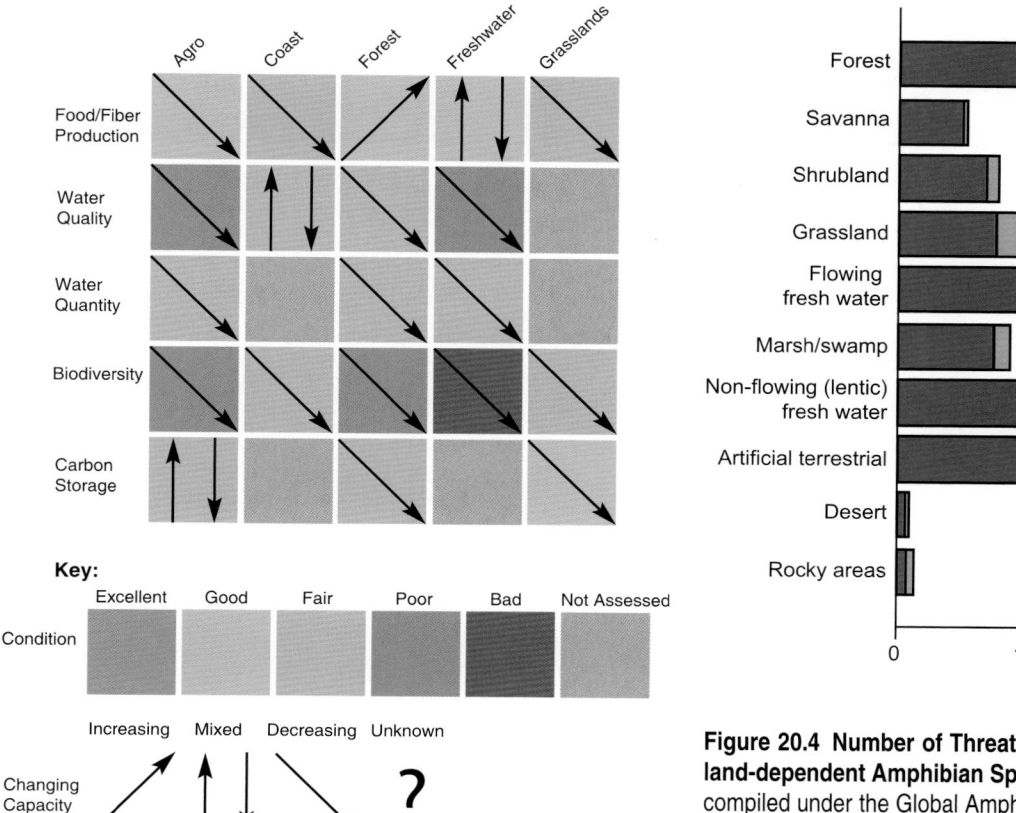

Figure 20.2 Summary Analysis of Capacity of a Range of Ecosystems to Produce Services (WRI et al. 2000)

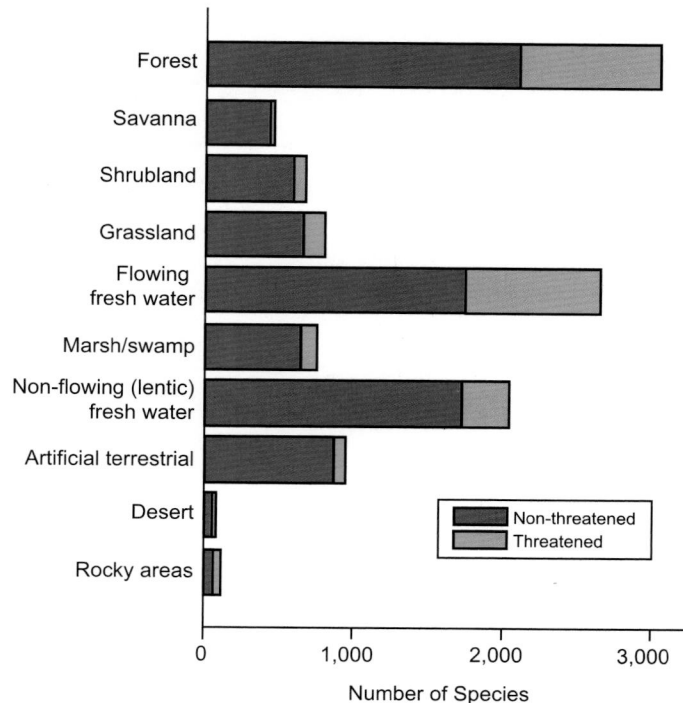

Figure 20.4 Number of Threatened versus Non-threatened Wetland-dependent Amphibian Species by Major Habitat Type I (Data compiled under the Global Amphibian Assessment; IUCN et al. 2004)

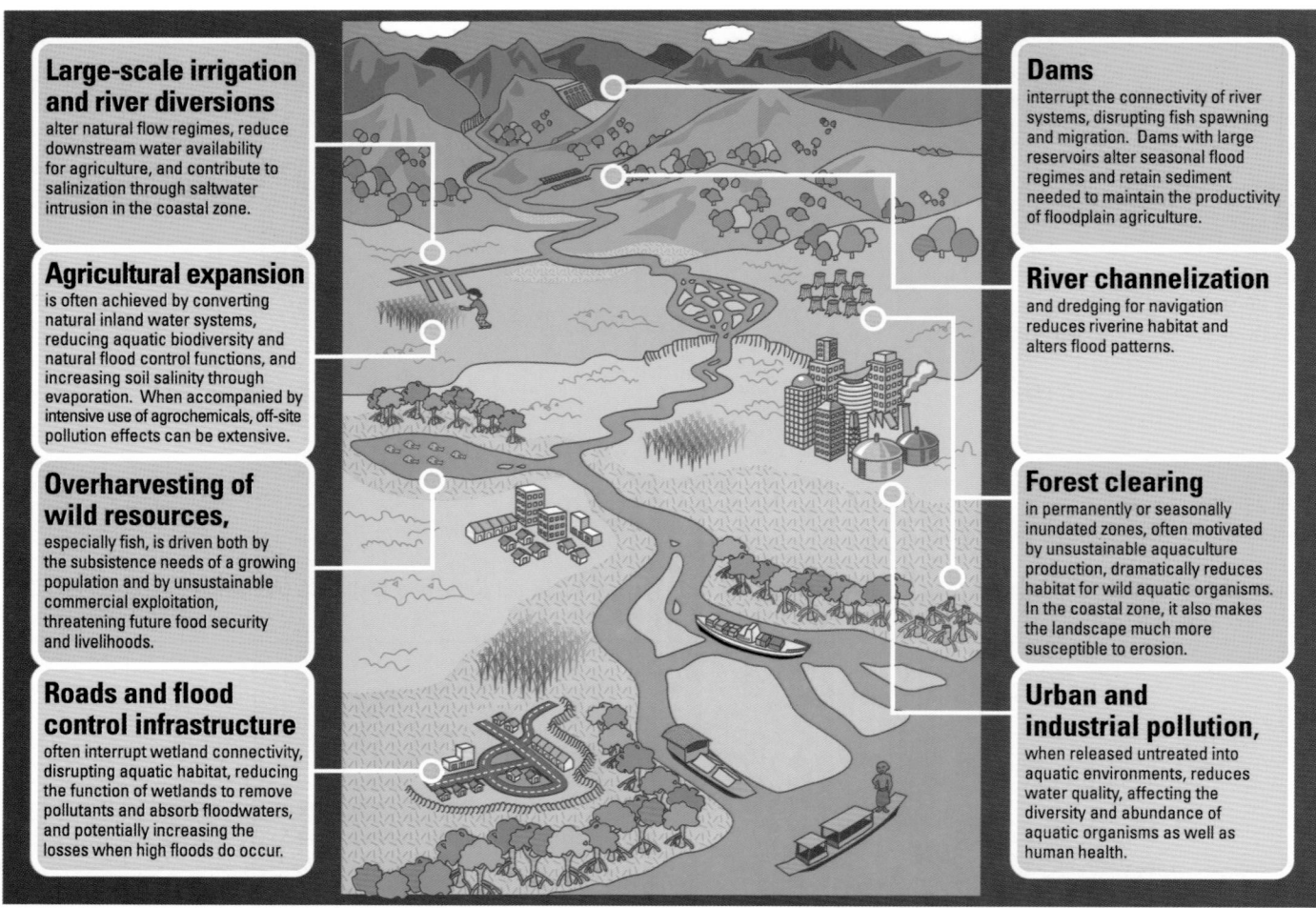

Large-scale irrigation and river diversions alter natural flow regimes, reduce downstream water availability for agriculture, and contribute to salinization through saltwater intrusion in the coastal zone.

Agricultural expansion is often achieved by converting natural inland water systems, reducing aquatic biodiversity and natural flood control functions, and increasing soil salinity through evaporation. When accompanied by intensive use of agrochemicals, off-site pollution effects can be extensive.

Overharvesting of wild resources, especially fish, is driven both by the subsistence needs of a growing population and by unsustainable commercial exploitation, threatening future food security and livelihoods.

Roads and flood control infrastructure often interrupt wetland connectivity, disrupting aquatic habitat, reducing the function of wetlands to remove pollutants and absorb floodwaters, and potentially increasing the losses when high floods do occur.

Dams interrupt the connectivity of river systems, disrupting fish spawning and migration. Dams with large reservoirs alter seasonal flood regimes and retain sediment needed to maintain the productivity of floodplain agriculture.

River channelization and dredging for navigation reduces riverine habitat and alters flood patterns.

Forest clearing in permanently or seasonally inundated zones, often motivated by unsustainable aquaculture production, dramatically reduces habitat for wild aquatic organisms. In the coastal zone, it also makes the landscape much more susceptible to erosion.

Urban and industrial pollution, when released untreated into aquatic environments, reduces water quality, affecting the diversity and abundance of aquatic organisms as well as human health.

Figure 20.7 Pictorial Presentation of the Direct Drivers of Change in Inland Waters (Ratner et al. 2004)

1973

1986

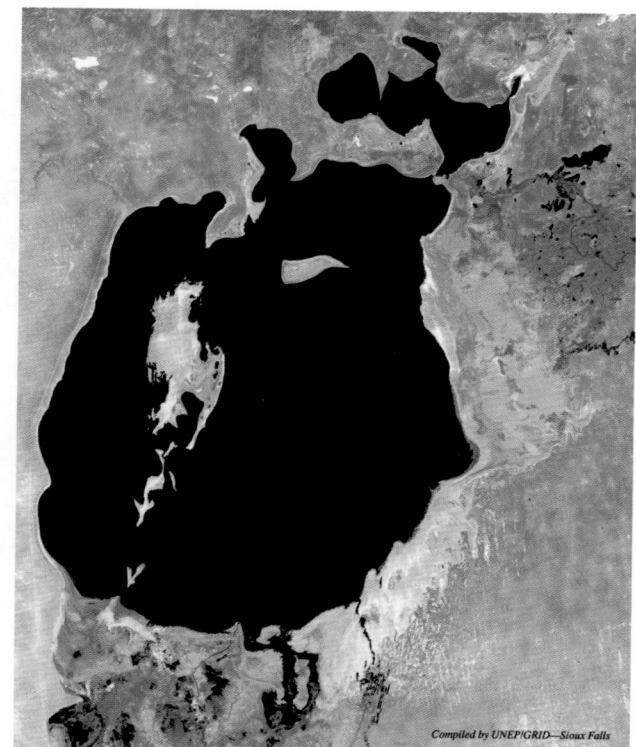

1999

2001

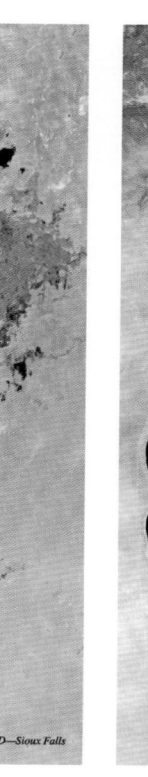

Figure 20.8 Changes in the Aral Sea, 1960–2001 (UNEP 2002)

1963   **1973**   **1999**

December 2002   **February 2003**   **May 2003**

August 2003

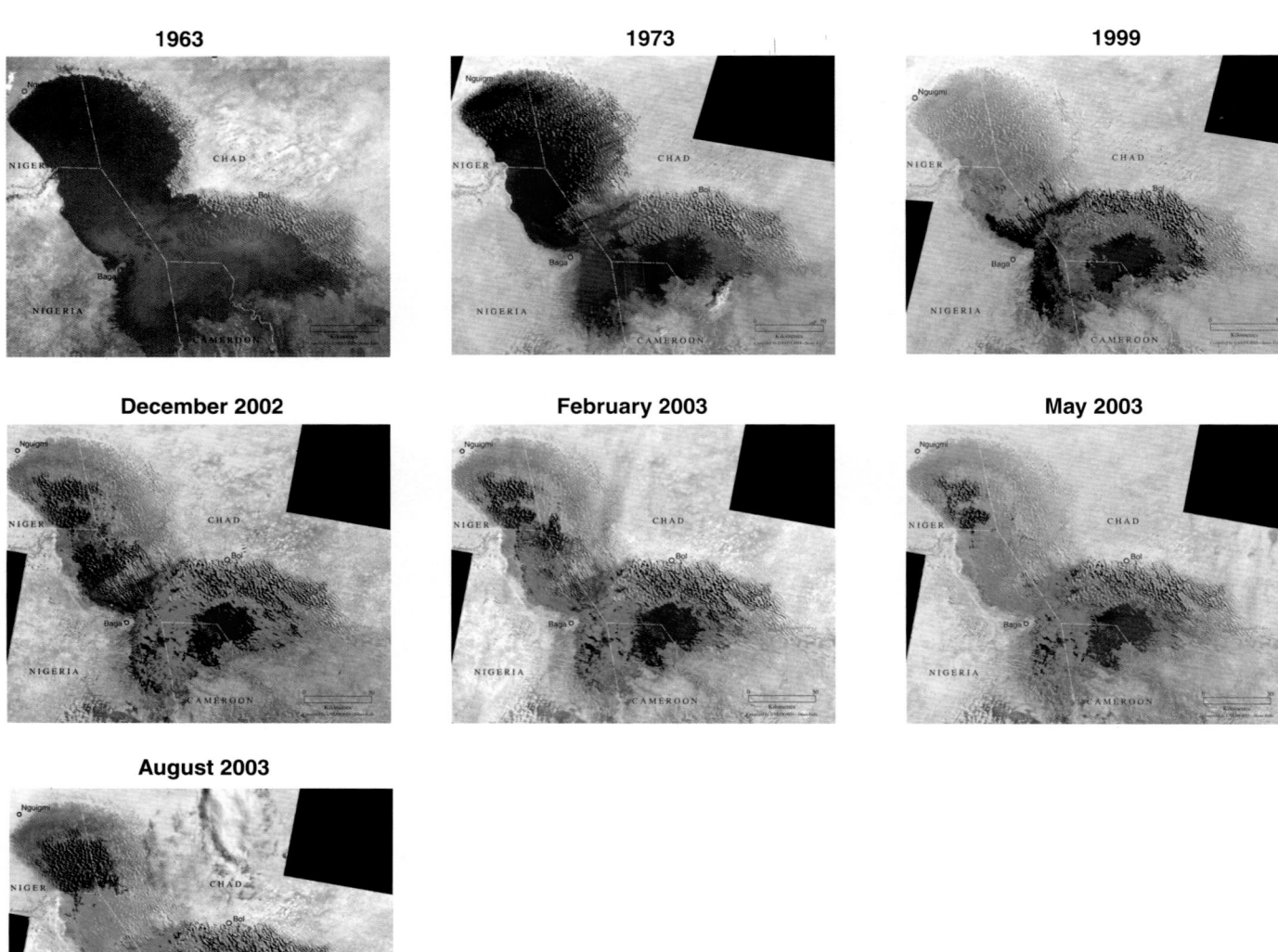

Figure 20.9 Changes in Area of Water in Lake Chad, 1963–2001 (UNEP 2002)

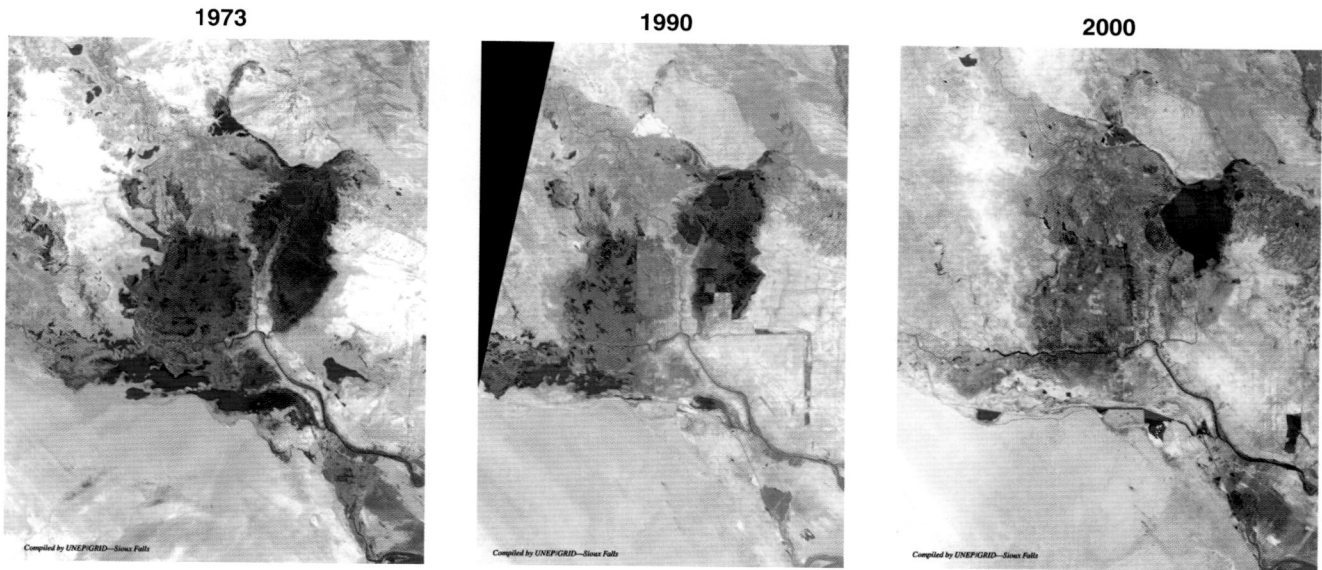

Figure 20.10 Changes in the Mesopotamian Marshes, 1973–2000 (Partow 2001)

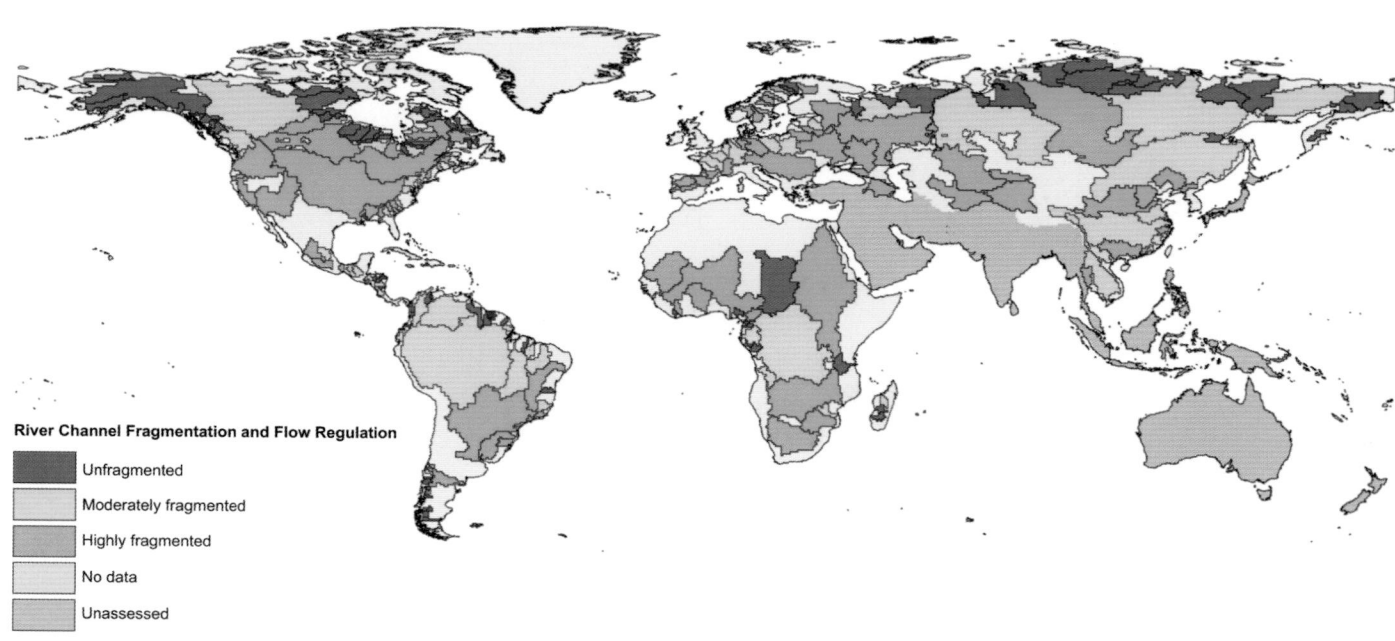

Figure 20.11 Fragmentation and Flow Regulation of Global Rivers (Revenga et al. 2000)

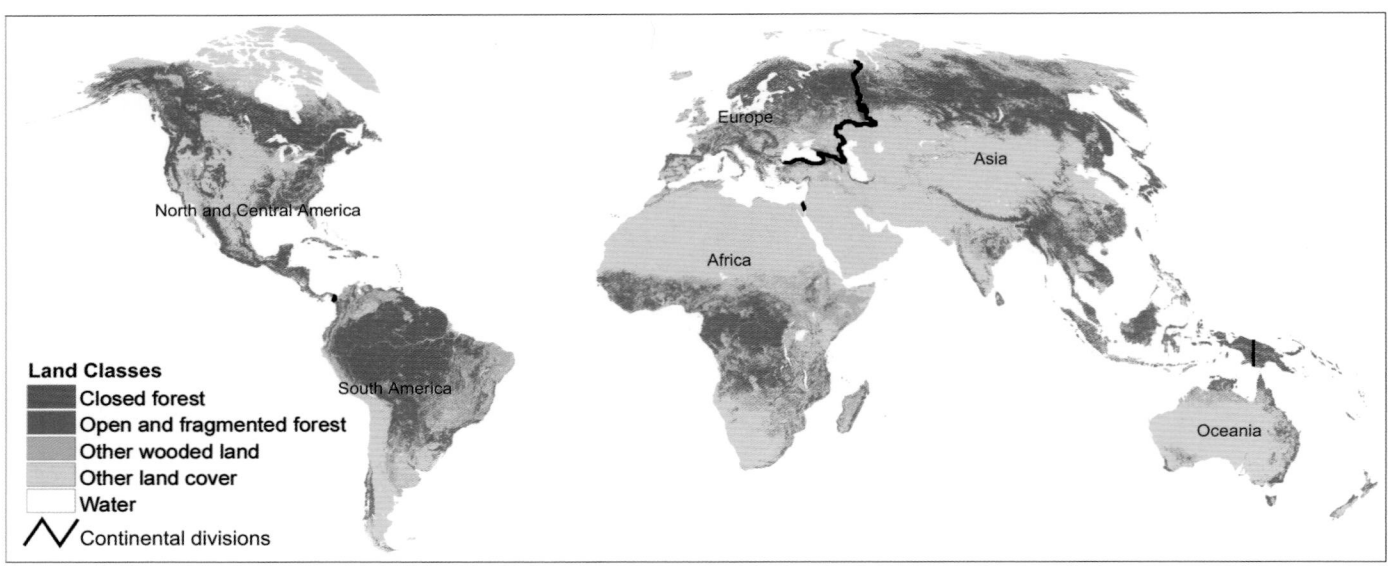

Figure 21.1 Global Forest and Woodland Cover by Aggregated Category and Continent. Open forests and fragmented forests have a canopy closure from 10–40%, and closed forests have a canopy closure of less than 40%. (FRA 2000 datasets)

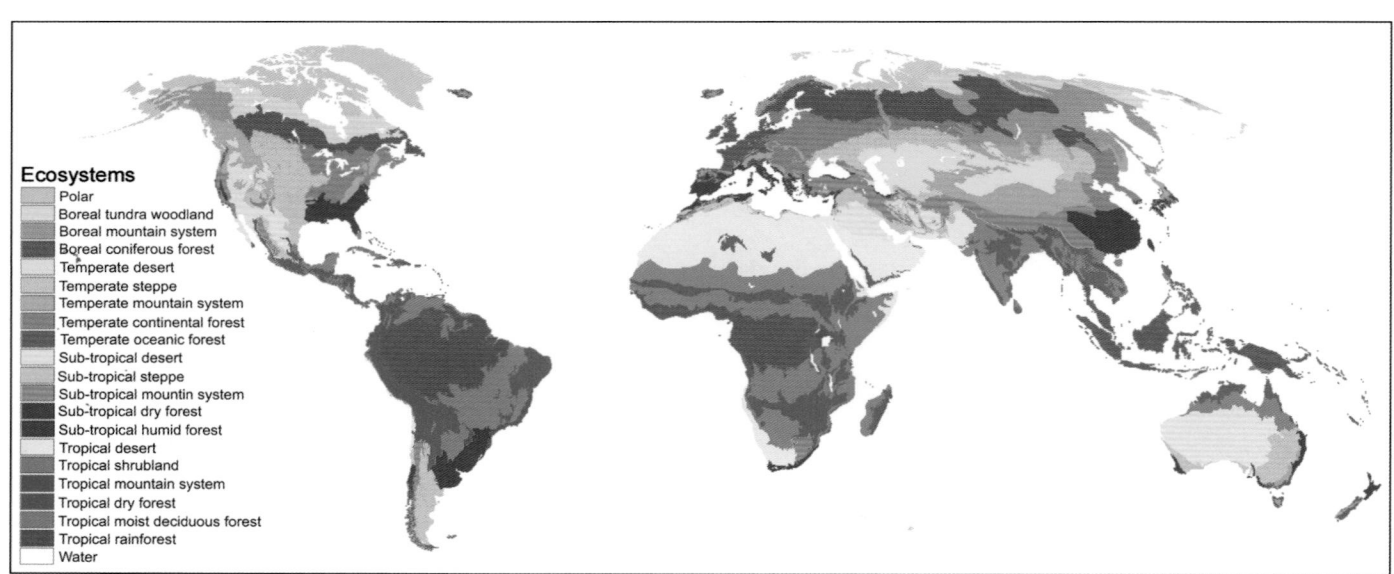

Figure 21.2 Distribution of Global Forests by Ecological Zone (FRA 2000 datasets)

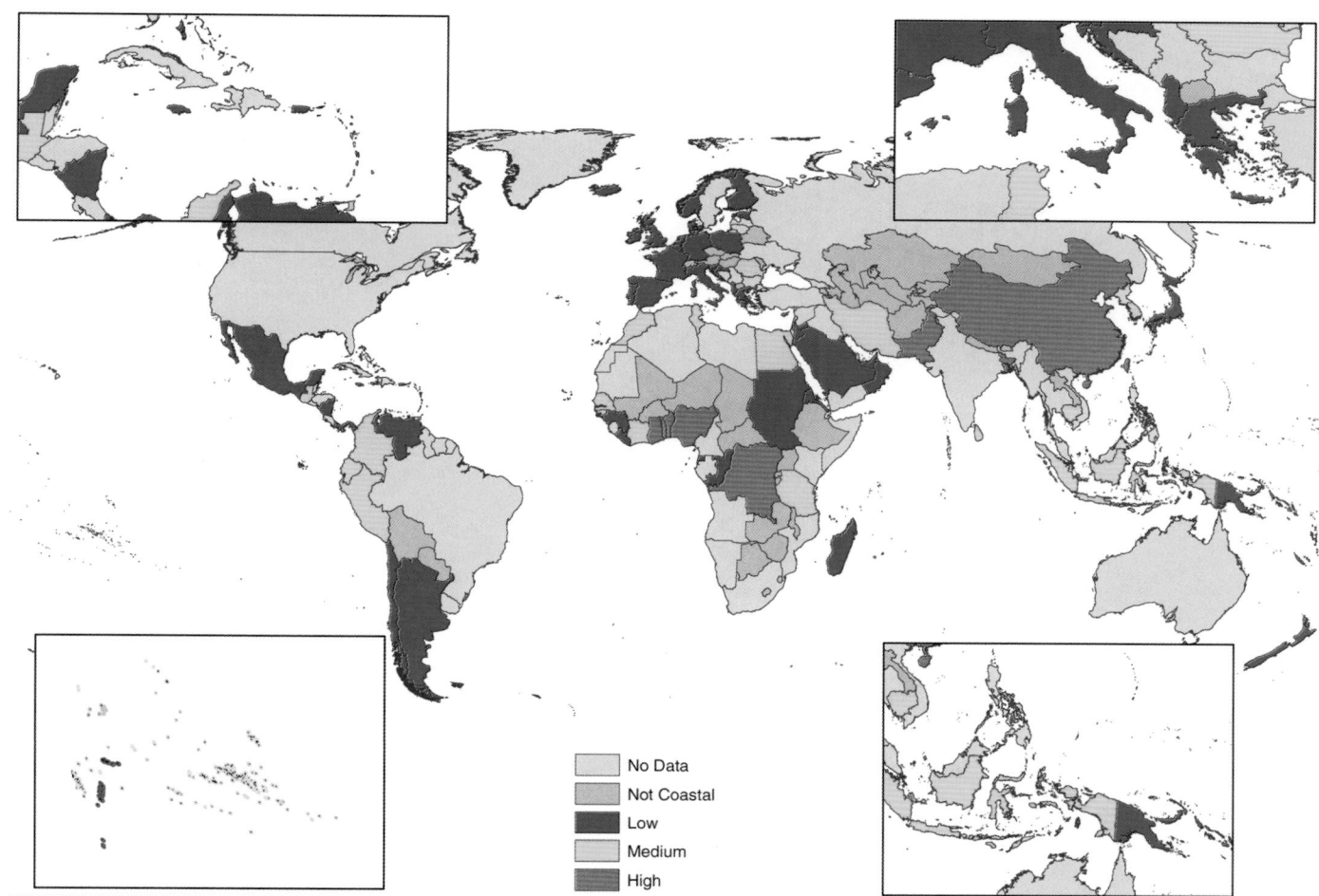

Figure 23.1 Fishing Pressure in Coastal Areas Based on the Number of People Actively Fishing per Kilometer of Coastline

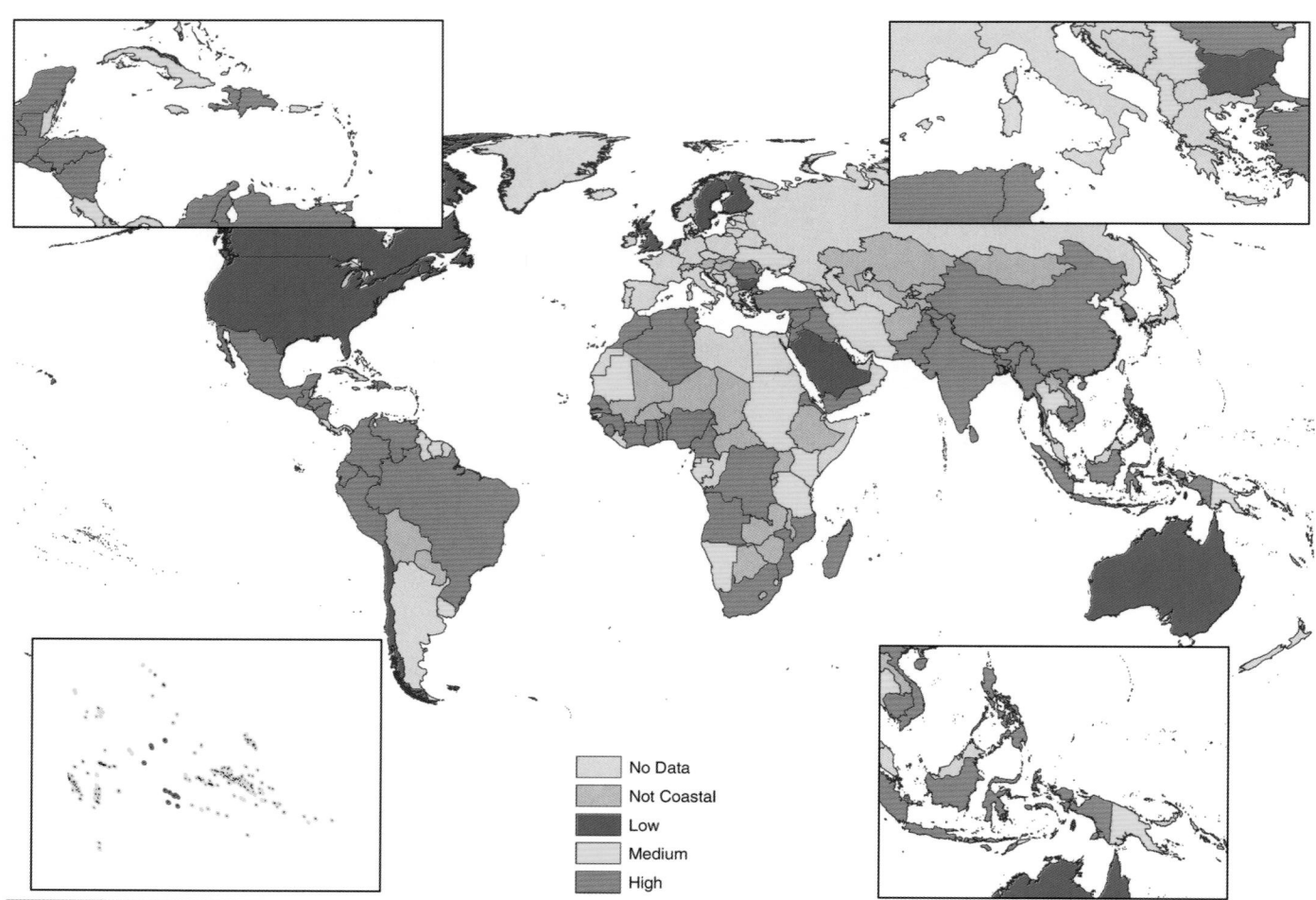

Figure 23.2 Sewage Pollution Index for Coastal Areas

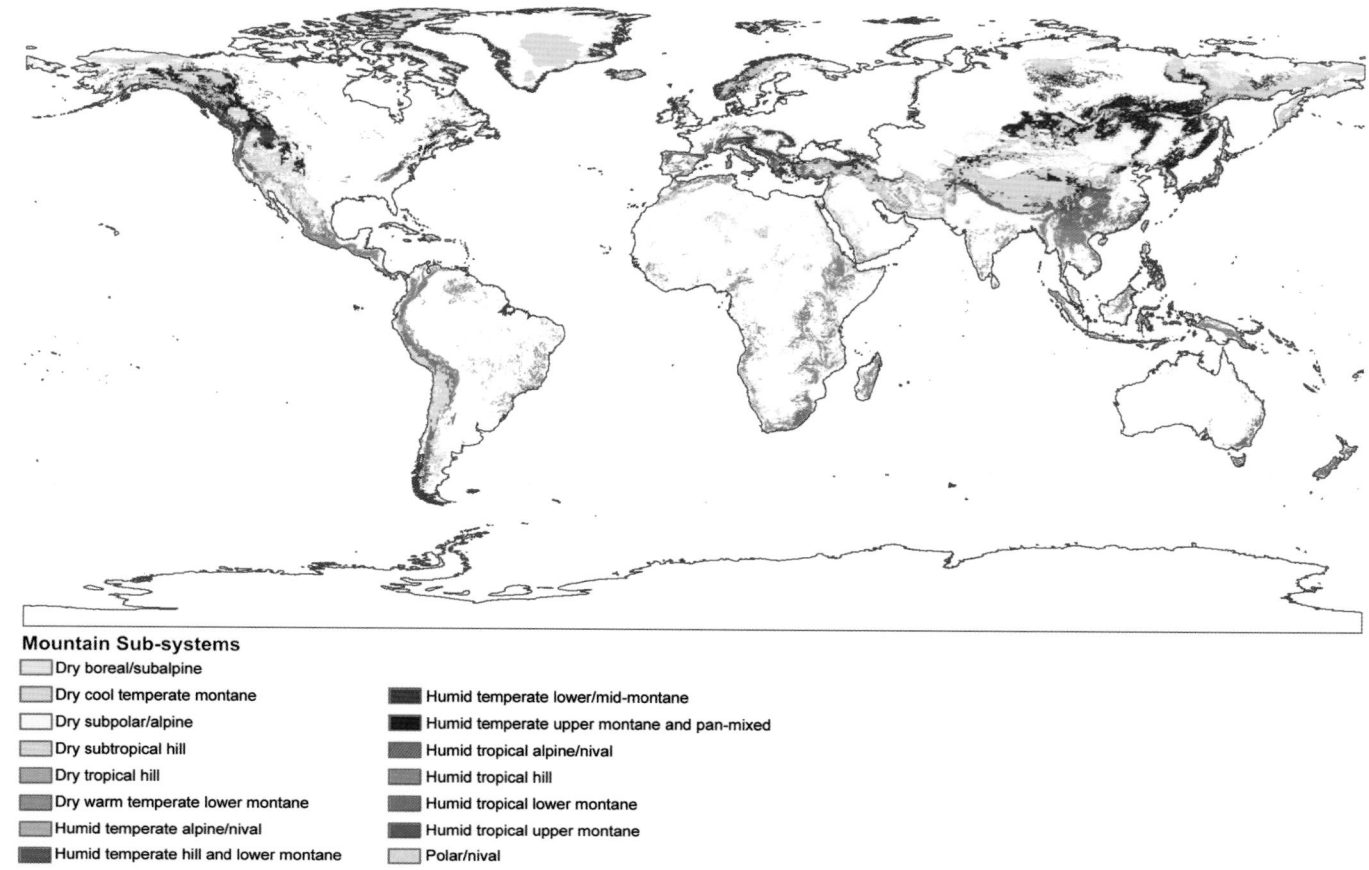

Mountain Sub-systems

Dry boreal/subalpine

Dry cool temperate montane

Dry subpolar/alpine

Dry subtropical hill

Dry tropical hill

Dry warm temperate lower montane

Humid temperate alpine/nival

Humid temperate hill and lower montane

Humid temperate lower/mid-montane

Humid temperate upper montane and pan-mixed

Humid tropical alpine/nival

Humid tropical hill

Humid tropical lower montane

Humid tropical upper montane

Polar/nival

Figure 24.5 Mountains of the World Based on Topography Alone (Kapos et al. 2000) Copyright UNEP-WCMC, Cambridge, UK

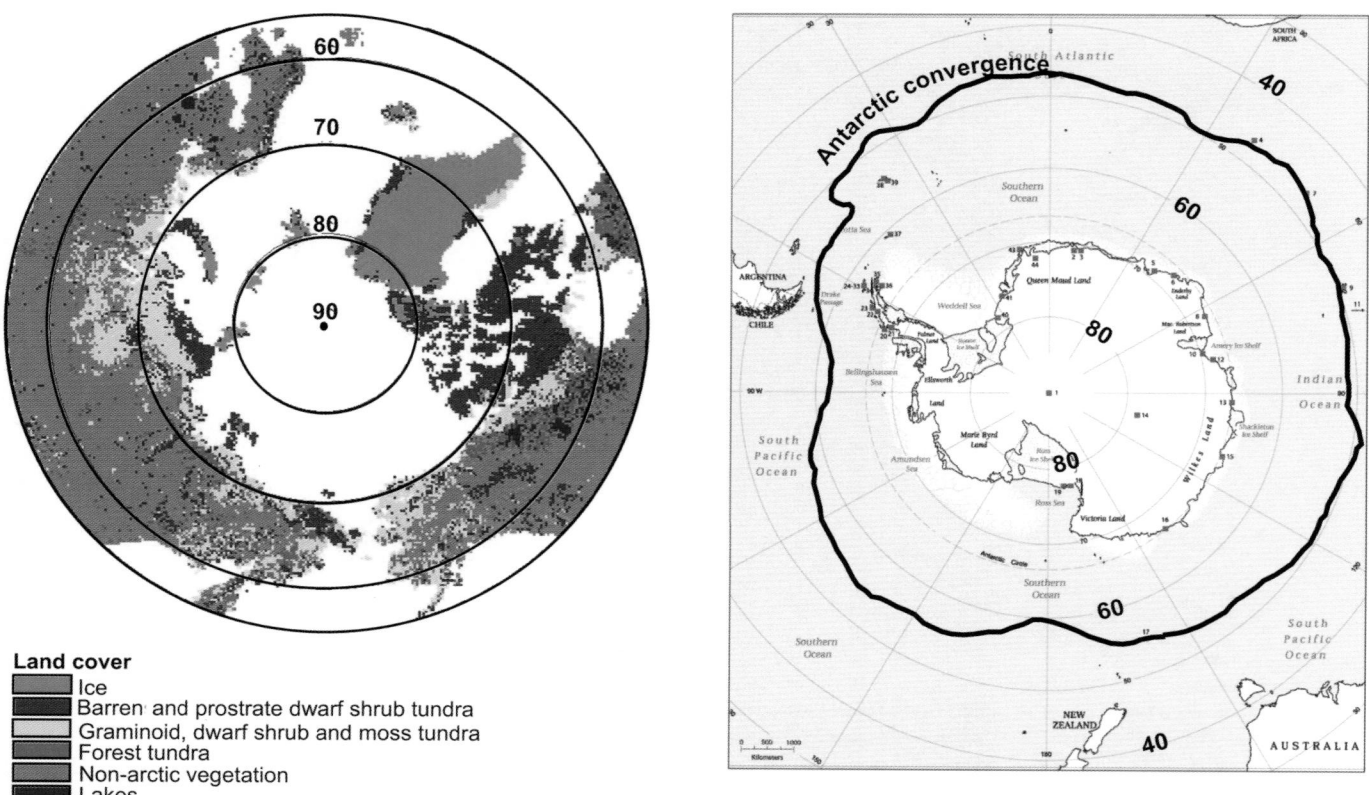

Figure 25.1 Major Subtypes of Arctic and Antarctic Terrestrial Ecosystems (Arctic modified from McGuire et al. 2002; Antarctic modified from Holdgate 1970)

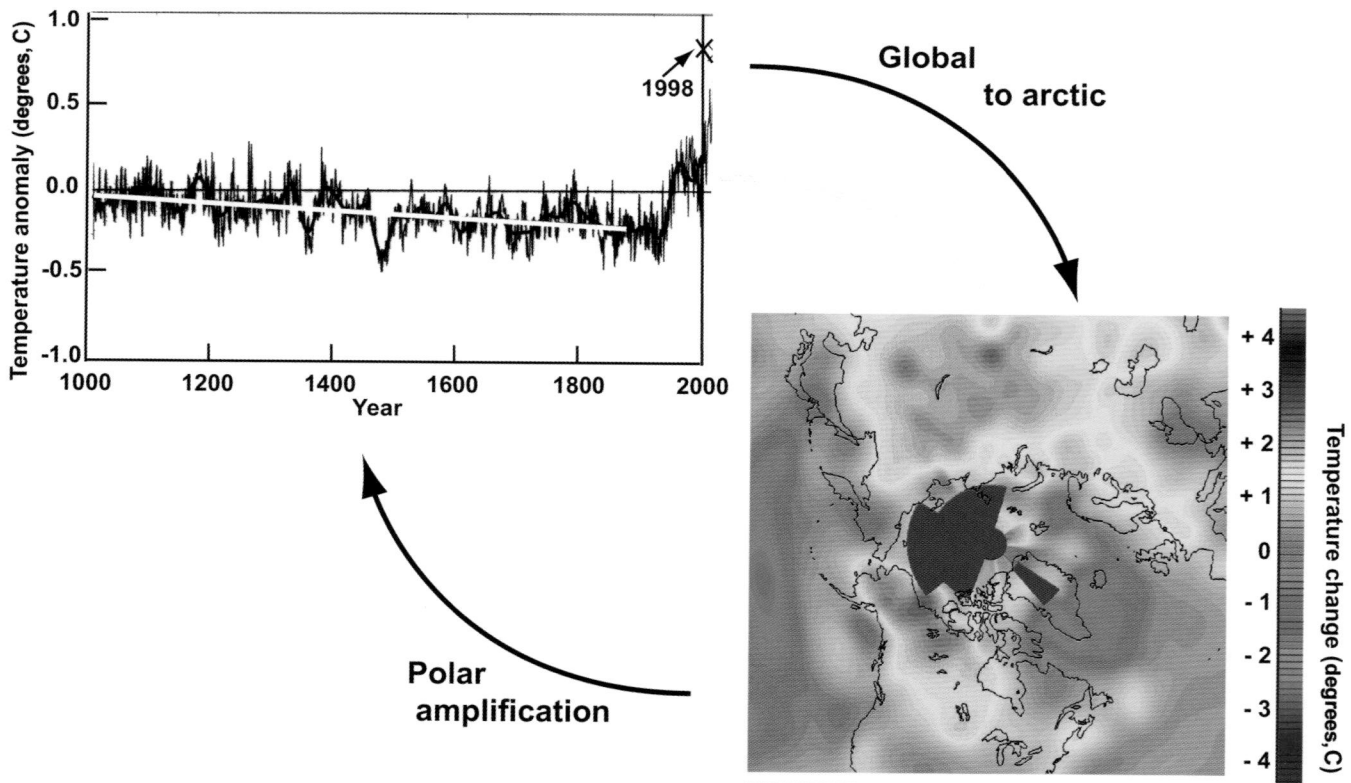

Figure 25.8 Interaction of Global and Northern Hemisphere Temperature Trends (Hinzman et al. in press; Mann et al. 1999)

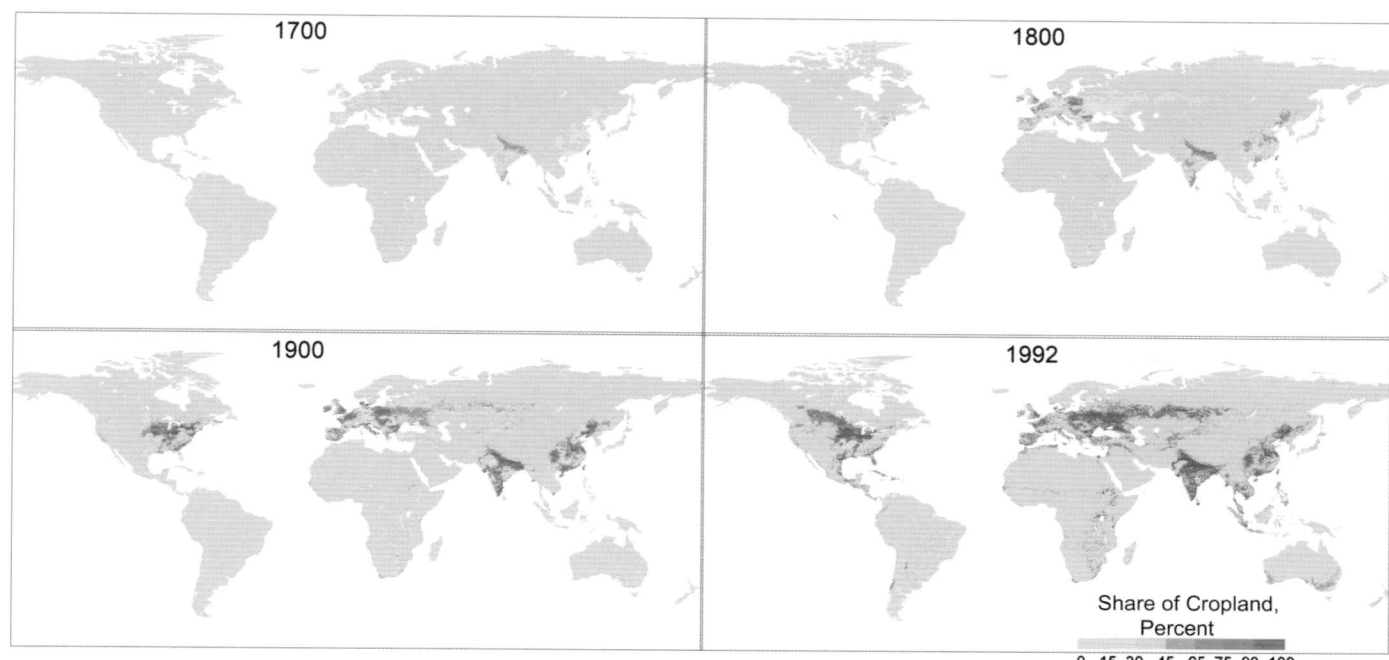

Figure 26.1 Evolution of Cultivated Systems from Pre-Industrial to Contemporary Times (Ramankutty et al. 2002)

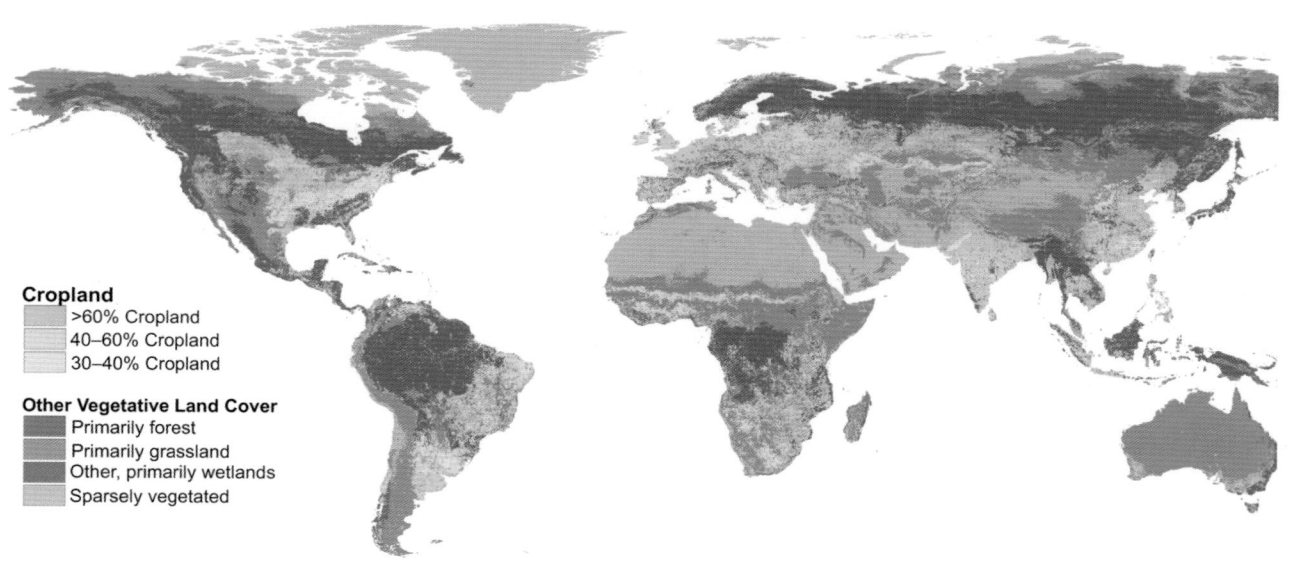

Figure 26.2 Contemporary Global Extent of Cultivated Systems (Wood et al. 2000)

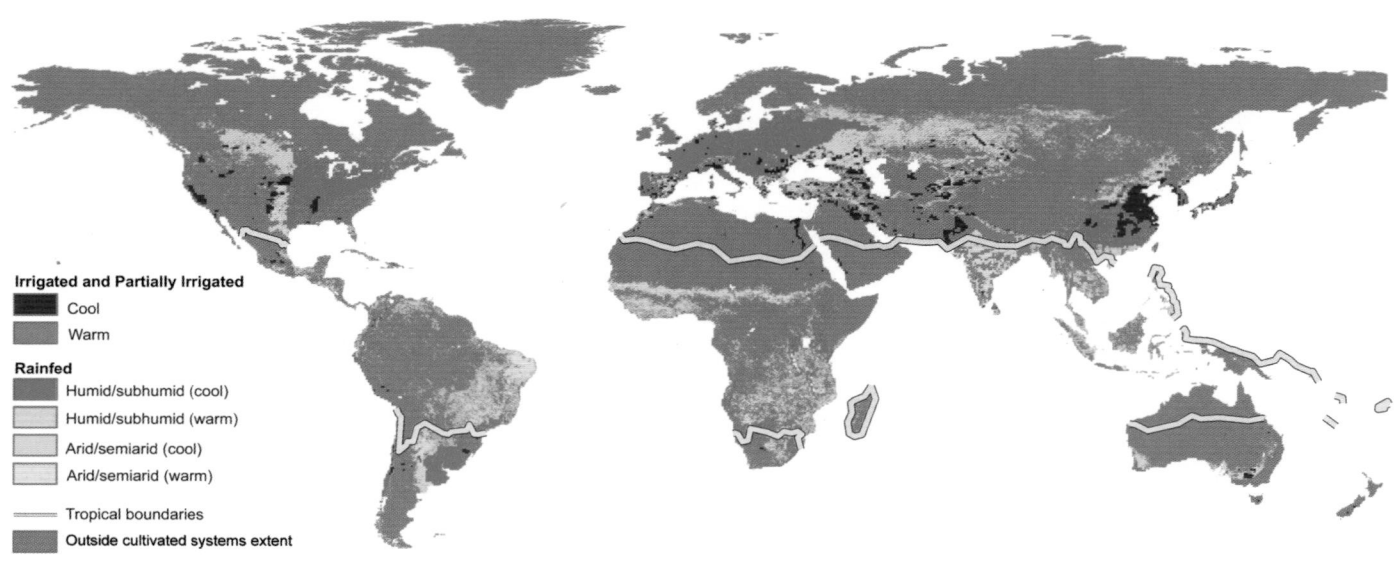

Figure 26.3 Agroecological Characterization of Cultivated Systems. This map classifies cultivated systems according to their major agroecological characteristics. Farm management practices have not been mapped consistently at a global scale, but important agroecological sub-divisions determined by climate, rainfall, irrigation, and slope are broadly indicative of the type of cultivation opportunities and constraints. This typology also gives some indication of potential productivity and of cultivation externalities—e.g., irrigation suggests higher productivity and a more intensive use of freshwater resources; cultivation in semiarid and sloping areas may have lower productivity and higher potential for soil erosion (Wood et al. 2000). The map is a composite of the 1 km. resolution global irrigation map produced by Kassel University and FAO (Doell and Siebert 2002), climate data from the Global Agroecological Zones project (FAO/IIASA 2001) and the PAGE agricultural extent (Wood et al. 2000).

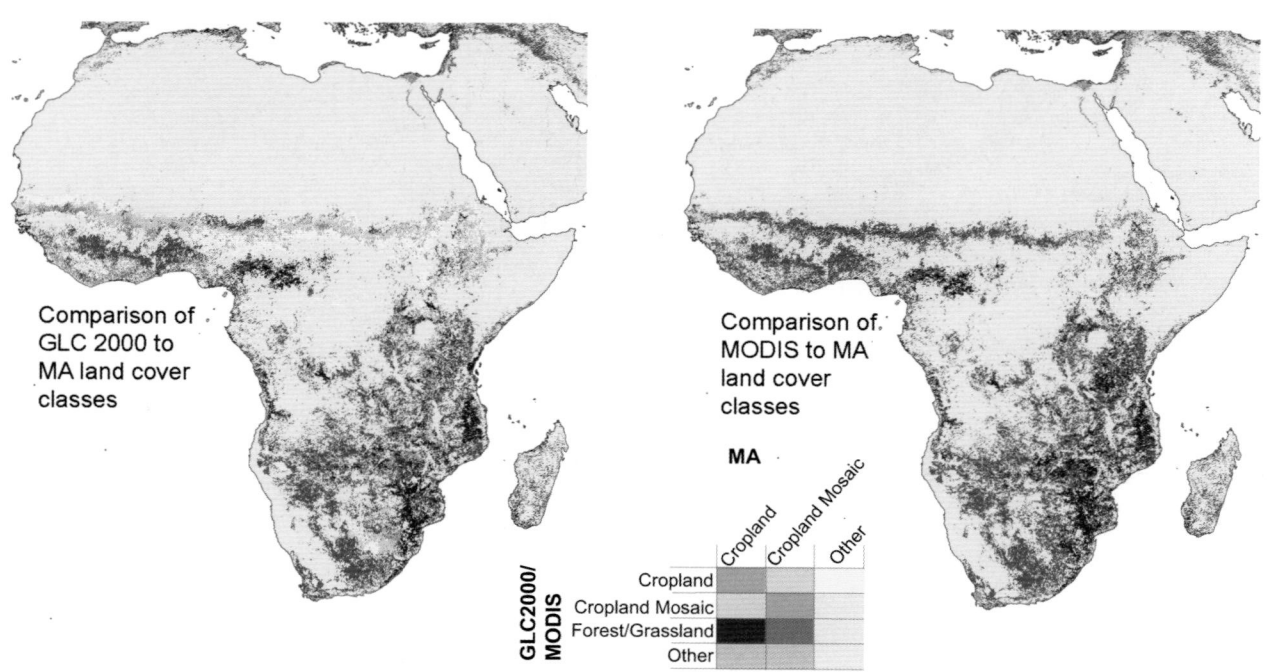

Box 26.1 Figure B. Comparison of Cropland and Cropland Mosaic Areas by Data Source

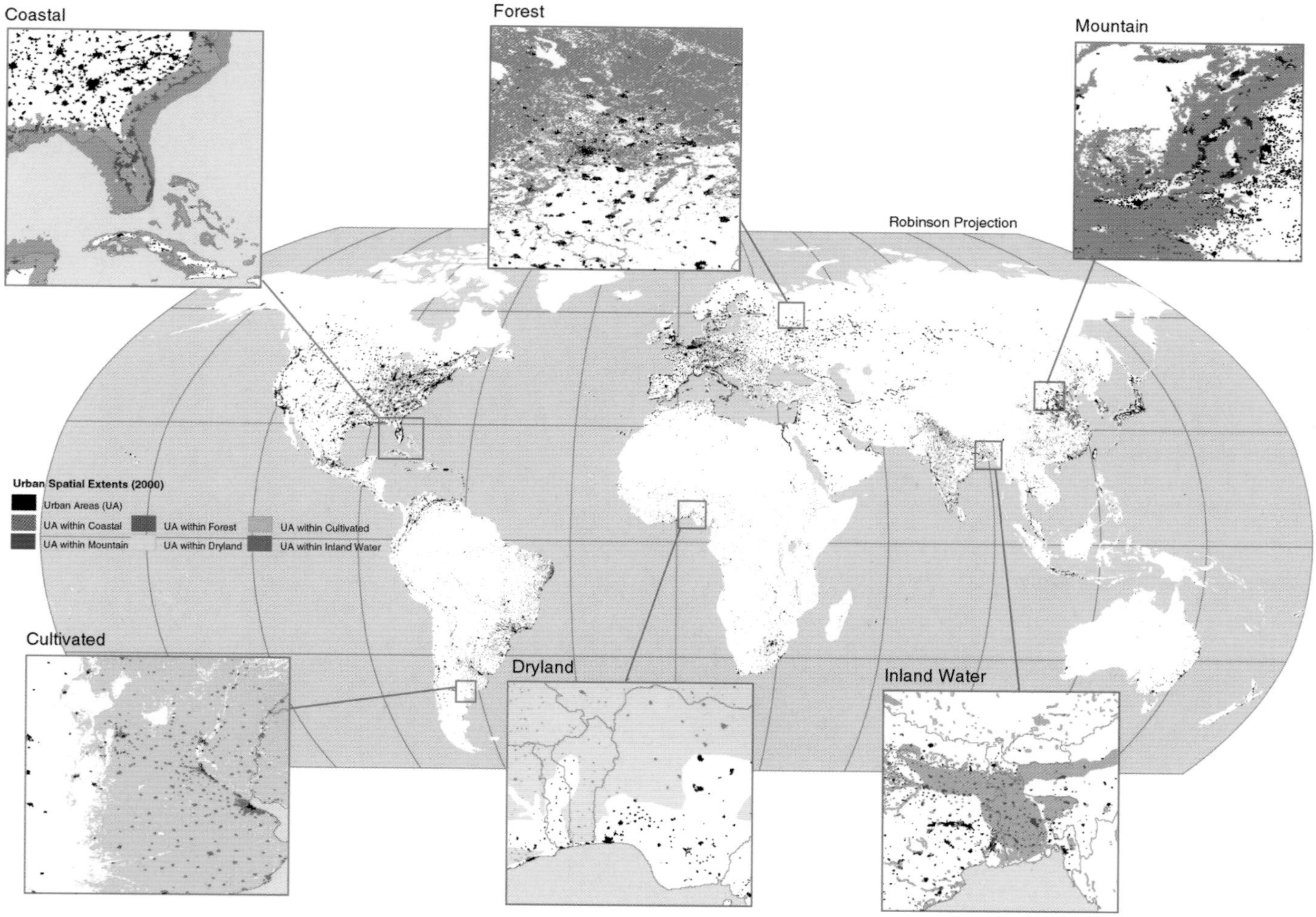

Coastal

Forest

Mountain

Robinson Projection

Urban Spatial Extents (2000)

■ Urban Areas (UA)

▨ UA within Coastal   UA within Forest   UA within Cultivated

▨ UA within Mountain   UA within Dryland   UA within Inland Water

Cultivated

Dryland

Inland Water

Figure 27.1 Urban Areas across the Globe (Copyright 2004: The Trustees of Columbia University in the City of New York. Center for International Earth Science Information Network, Columbia University; International Food Policy Research Institute, World Bank; and Centro Internacional de Agricultura Tropical, 2004. Global Rural-Urban Mapping Project: Urban Mask version 1. Palisades, NY: CIESIN, Columbia University. Available at http://sedac.ciesin.columbia.edu/gpw.)

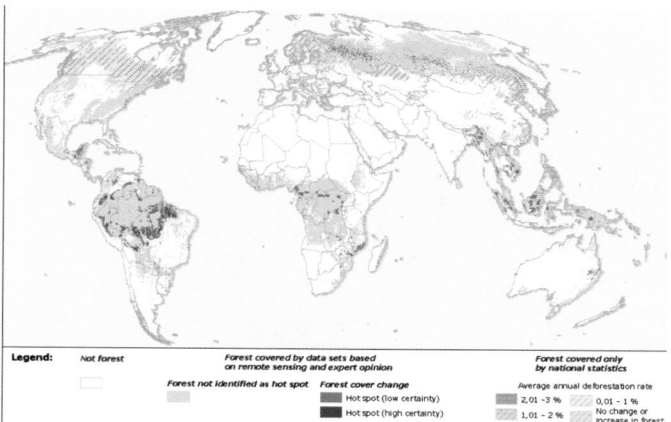

Figure 28.1 Areas of Rapid Land Cover Change Involving Deforestation and Forest Degradation (Lepers et al. 2005)

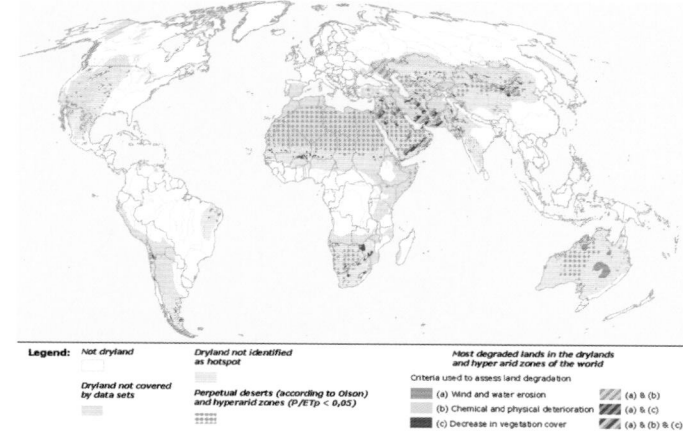

Figure 28.2 Areas of Rapid Land Cover Change Involving Desertification and Land Degradation (Lepers et al. 2005)

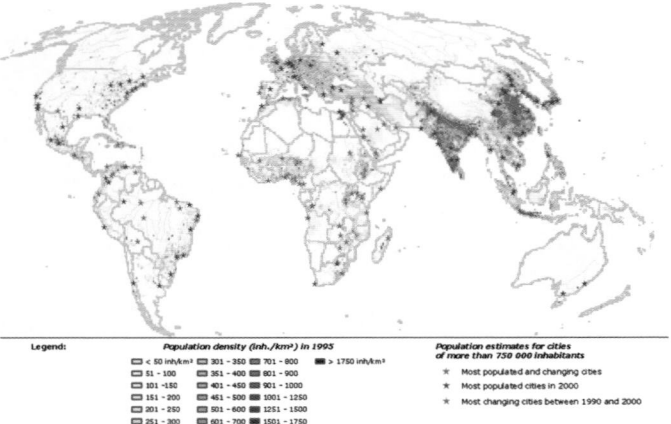

Figure 28.3 Areas of Rapid Land Cover Change Involving Changes in Urban Extent (Lepers et al. 2005)

Population Index = 100 in 1970

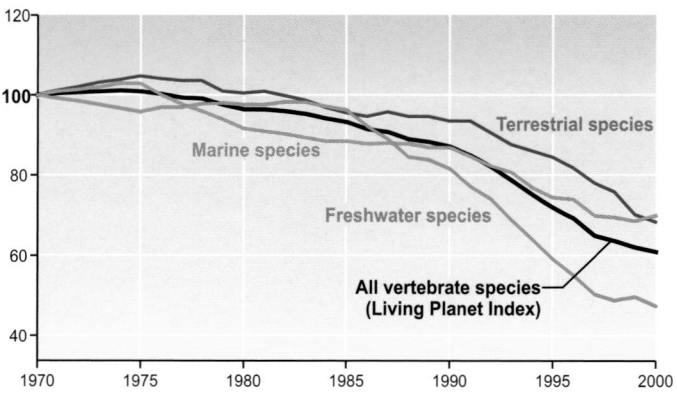

Figure 28.4 The Living Planet Index, 1970–2000. The Living Planet Index is an indicator of the state of the world's biodiversity: it measures trends in populations of vertebrate species living in terrestrial, freshwater, and marine ecosystems.

Authors

Argentina
Daniel Cáceres, Universidad Nacional de Córdoba
Sandra Díaz, Universidad Nacional de Córdoba—CONICET
Thomas Kitzberger, Universidad Nacional del Comahue
Ernesto Viglizzo, National Institute for Agricultural Technology

Australia
Andrew Beattie, Macquarie University
Steve Blaber, Commonwealth Scientific and Industrial Research Organisation
Nesa Boopalan Eliezer, Independent Scholar
C. Max Finlayson, Environmental Research Institute of the Supervising Scientist
Alex Hyatt, Commonwealth Scientific and Industrial Research Organisation
Shahbaz Khan, Commonwealth Scientific and Industrial Research Organisation
Roger Leakey, James Cook University
Kerrie Wilson, University of Queensland

Austria
Ian McCallum, International Institute for Applied Systems Analysis
Sten Nilsson, International Institute for Applied Systems Analysis
Anatoly Shvidenko, International Institute for Applied Systems Analysis

Botswana
Hillary Masundire, University of Botswana

Brazil
Tania Braga, Universidade Candido Mendes
Ulisses Confalonieri, Fundação Oswaldo Cruz
Elaine Elisabetsky, Universidade Federal do Rio Grande do Sul
Gustavo Fonseca, Federal University of Minas Gerais
Mauro Galetti, University of Sao Paulo
Paulo Lana, Universidade Federal do Paraná
Flavio Luizao, INPA-Ecologia
Eduardo Marone, Centro de Estudos do Mar
Maria Assuncao Silva Dias, Universidade de Sao Paulo
Pedro Vasconcelos, Instituto Evandro Chagas
Christina Whiteman, Universidade Federal Rural da Amazônia

Burkina Faso
Laurent Yameogo, World Health Organisation

Cambodia
Sarah Porter, IUCN—The World Conservation Union

Canada
W.L. Adamowicz, University of Alberta
Zafar Adeel, United Nations University

Jacqueline Alder, University of British Columbia
Sabrina Barker, Environment Canada
Jules Blais, University of Ottawa
Villy Christensen, University of British Columbia
David Cooper, Secretariat of the Convention on Biological Diversity
Paola Deda, Secretariat of the Convention on Biological Diversity
Katia Freire, University of British Columbia
Jack Ives, Independent Scholar
Tim Johns, McGill University
Joseph Kalemani Mulongoy, Secretariat of the Convention on Biological Diversity
Kristin Kaschner, University of British Columbia
Caroline King, United Nations University
Adrian Kitchingman, University of British Columbia
Peter Leavitt, University of Regina
Randy Milton, Nova Scotia Department of Natural Resources
Parastu Mirabzadeh, Independent Scholar
Maria-Lourdes Paloma, University of British Columbia
Daniel Pauly, University of British Colombia
William Rees, University of British Columbia
Dave Schindler, University of Alberta
Rashid Sumaila, University of British Colombia
Ian Thompson, Canadian Forest Service
Peter Tyedmers, Dalhousie University
Marjo Vierros, Secretariat of the Convention on Biological Diversity
Rolf Vinebrooke, University of Alberta
Colette Wabnitz, University of British Columbia
David Waltner-Toews, University of Guelph
Reg Watson, University of British Columbia
Boris Worm, Dalhousie University

Chad
Mahamat Hassane Idriss, Department of the Environment

China
Chen Ma, Peking University
Yvonne Sadovy, The University of Hong Kong
Tianxiang Yue, Chinese Academy of Sciences

Colombia
Juan Restrepo, EAFIT University
Maria Rivera, Ministry of the Environment

Costa Rica
Edgar Gutierrez-Espeleta, Universidad de Costa Rica

Denmark
Thomas Theis Nielsen, University of Copenhagen

Ecuador
Fander Falconi, Facultad Latinoamericana de Ciencias Sociales
Robert Hofstede, International Potato Centre

Egypt
Mohamed Tawfic Ahmed, Suez Canal University
Patrick Dugan, WorldFish Center

Estonia
Martin Zobel, Tartu University

Finland
Bruce Forbes, University of Lapland
Pekka Kauppi, University of Helsinki

France
Anne-Marie Izac, Agricultural Research Centre for International Development
Patrick Lavelle, University of Paris, and Institut de Recherche pour le Développement
Jacques Lemoalle, Institut de Recherche pour le Développement
Christian Lévêque, Institut de Recherche pour le Développement
Michel Meybeck, University of Paris
Mechtild Rässler, UNESCO World Heritage Centre
Paul Treguer, University of Brest
Luís Vicente, Museum National d'Histoire Naturelle

Germany
Almut Arneth, Max-Planck Institute for Biogeochemistry
Wilhelm Barthlott, Nees-Institute for Biodiversity of Plants
Victor Brovkin, Potsdam Institute for Climate Impact Research
Wolfgang Cramer, Potsdam Institute for Climate Impact Research
Laurens Ganzeveld, Max-Planck Institute for Biogeochemistry
Johann Goldammer, Max Planck Institute for Biogeochemistry
Joanna House, Max Planck Institute for Biogeochemistry
Wolfgang Junk, Max-Planck-Institut fuer Limnologie
Karl-Hermann Kock, Federal Research Centre for Fisheries
Luis Jose Mata, Center for Development Research, University of Bonn
Colin Prentice, Max Planck Institute for Biogeochemistry
Ulf Riebesell, Institute for Marine Research
Klaus Riede, Alexander Koenig
Ina Tegen, Max Planck Institute for Biogeochemistry
Mark Winslow, Desertification, Drought, Poverty and Agriculture (DDPA) Consortium

Ghana
Samuel Benin, International Livestock Research Institute
Jacob Songsore, University of Ghana

Greece
Kostas Stergiou, Aristotle University of Thessaloniki

India
Kavita Iyenga, Asian Development Bank
P. Ramakrishnan, Jawaharlal Nehru University

Indonesia
Roland Offrell, APRIL Pulp & Paper

Israel
Esther Lachman, Hebrew University of Jerusalem
Boris Portnov, University of Haifa
Uriel Safriel, Hebrew University of Jerusalem

Italy
Steffen Fritz, EU Joint Research Council
Barbara Gemmill, Food and Agriculture Organization of the United Nations

Richard Grainger, Food and Agriculture Organization of the United Nations
Thomas Hofer, Food and Agriculture Organization of the United Nations
John Latham, Food and Agriculture Organization of the United Nations
Leslie Lipper, Food and Agriculture Organization of the United Nations
Joachim Otte, Food and Agriculture Organization of the United Nations
Stefania Vannuccini, Food and Agriculture Organization of the United Nations
Ulf Wijkstrom, Food and Agriculture Organization of the United Nations

Japan
Xuemei Bai, Institute for Global Environmental Strategies
Peter Marcotullio, United Nations University Institute of Advanced Studies
Masahiko Ohsawa, University of Tokyo
Taikan Oki, University of Tokyo
Nophea Kim Phat, University of Hyogo
Teiji Watanabe, Hokkaido University
Masatoshi Yoshino, United Nations University

Kazakhstan
Elena Kreuzberg-Mukhina, The Regional Environmental Centre for Central Asia (CAREC)

Kenya
Mohamed Bakarr, World Agroforestry Centre

Kyrgyzstan
Vladislav N. Talskikh, Central Asia Transboundary Biodiversity Project,

Laos
Alvin Lopez, IUCN—The World Conservation Union

Malaysia
Kaw Bing Chua, International Medical University
Rebecca D'Cruz, Aonyx Environmental
C. Devendra, International Livestock Research Institute (ILRI)
Apsara Ram Gopal, Kshetra Academy
Teck Kheng Chua, Sarawak Biodiversity Centre
Thayanithi Kulenthran, Greenfields Consulting
Poh Sze Choo, WorldFish Center

Mali
Barry Shapiro, International Crops Research Institute for the Semi-Arid Tropics

Mexico
Patricia Balvanera Levy, Universidad Nacional Autonoma de Mexico
Rodolfo Dirzo, Universidad Nacional Autonoma de Mexico
John Dixon, International Maize and Wheat Improvement Centre
José Garcia Calderón, UAM-Iztapalapa
Guadalupe Lanza Espino, Universidad Nacional Autonoma de Mexico
Manuel Maass, Universidad Nacional Autonoma de Mexico
Eduardo Mestre Rodriguez, Red Latinoamericana de Organizaciones de Cuenca

Morocco
Said Hajib, Department of Water, Forests and Land

Namibia
Julianne Ziedler, Natuye—Institute for the Environment

Nepal
Narpat Singh Jodha, International Centre for Integrated Mountain
Development

Netherlands
Rob Alkemade, National Institute for Public Health and the
Environment
Bas Amelung, University of Maastricht
Joost Brouwer, Brouwer Environmental and Agricultural Consultancy
Rudolf de Groot, Wageningen University
Ward Hagermeijer, Wetlands International
Henk Ketelaars, Water Storage Company Brabantse Biesbosch
David Niemeijer, Wageningen University
Marcel Silvius, Wetlands International
Agnes van den Berg, Wageningen University
Dirk Wascher, Alterra—Green World Research

New Zealand
Roberta Farrell, University of Waikato
Gamini Wijesuriya, Department of Conservation

Nigeria
Oladele Osibanjo, University of Ibadan

Norway
Erling Berge, Norwegian University of Science and Technology
Siri Eriksen, University of Oslo

Pakistan
Abid Hussainy, Aga Khan University Institute for Educational
Development
Syed Saifullah, University of Karachi

Panama
Scott Muller, Conservation and Development Strategy Associates

Philippines
Jurgenne Primavera, Asian Fisheries Development Centre
Juan Pulhin, University of the Philippines
Agnes Rola, University of the Philippines

Portugal
Henrique Miguel Pereira, Universidade de Lisboa

Russian Federation
N.V. Aladin, Zoological Institute Russian Academy of Sciences
I.S. Plotnikov, Zoological Institute Russian Academy of Sciences
Sergei Zimov, North East Science Station

Senegal
Racine Kane, United Nations Development Programme

Singapore
Poh Poh Wong, National University of Singapore

Slovenia
Gordana Beltram, Ministry of the Environment and Spatial Planning

South Africa
Emma Archer, University of the Witwatersrand
Allan Batchelor, B&M Environmental Services (Pty) Ltd

William Bond, University of Cape Town
George Branch, University of Cape Town
Connal Eardley, Agricultural Research Council
Rashid Hassan, University of Pretoria
Clifford Mutero, International Water Management Institute
Belinda Reyers, Council for Scientific and Industrial Research
Bob Scholes, Council for Scientific and Industrial Research
Mary Scholes, University of the Witwatersrand
Berndt van Rensburg, University of Pretoria
Coleen Vogel, University of Witwatersrand
Maureen Wolfson, South African National Biodiversity Institute
Gina Ziervogel, Dept of Enviro & Geographical Science

Spain
Juan Puigdefabregas, Higher Council for Scientific Research
Montserrat Vila, Autonomous University of Barcelona

Sri Lanka
Felix Amerasinghe, International Water Management Institute
Rebecca Tharme, International Water Management Institute

Sweden
Terry Callaghan, Abisko Scientific Research Station
Torben Christensen, Lunds University
Anne-Sophie Crépin, The Beijer International Institute of Ecological
Economics
Kjell Danell, Independent Scholar
Thomas Elmqvist, Stockholm University
Carl Folke, Stockholm University
Guoyi Han, Stockholm Environment Institute
Christer Nilsson, Umea University
Reidar Persson, Swedish University of Agricultural Sciences
Cathy Reidy, Umea University

Switzerland
Robert Bos, World Health Organization
Harald Bugmann, Swiss Federal Institute of Technology Zurich
Nick Davidson, Ramsar Convention on Wetlands
Fortunat Joos, University of Bern
Christian Körner, University of Basel
Bruno Messerli, University of Bern
David Pitt, Independent Scholar
Mel Reasoner, Independent Scholar
Eva Spehn, Global Mountain Biodiversity Assessment, Basel

Tanzania
Mwele Malecela-Lazaro, National Medical Research Institute
Alan Rodgers, United Nations Development Programme

Thailand
Miguel Fortes, United Nations Educational, Scientific and Cultural
Organisation

Trinidad and Tobago
John Agard, University of the West Indies
Sherry Heileman, Independent Scholar

Turkey
H. Resit Akcakaya, Applied Biomathmatics

Ukraine
Petro Lakyda, Ukrainian National Agrarian University
Iskandar Mirabdullaev, Independent Scholar
Raisa Toryannikova, Independent Scholar

United Kingdom

Neville J. Ash, UNEP-World Conservation Monitoring Centre
Jonathan Baillie, Institute of Zoology, Zoological Society of London
Andrew Balmford, Cambridge University
Richard Betts, Hadley Centre for Climate Prediction and Research
Nadia Bystriakova, University of Cambridge
John Chilton, British Geological Survey
Zoe Cokeliss, Institute of Zoology, Zoological Society of London
Peter Convey, British Antarctic Survey
Sarah Cornell, University of Bristol
Peter Daszak, Consortium for Conservation Medicine
Ian Douglas, University of Manchester
Tom Downing, Stockholm Environment Institute, University of York
Rhys Green, Cambridge University
Brian Groombridge, UNEP-World Conservation Monitoring Centre
Craig Hilton-Taylor, IUCN—The World Conservation Union
Joanna House, University of Bristol
Sandra Kadungure, University of Leeds
Valerie Kapos, UNEP-World Conservation Monitoring Centre
Georgina Mace, Institute of Zoology, Zoological Society of London
Gordon McGranahan, International Institute for Environment and Development
Douglas McNab, Overseas Development Institute
David Molyneux, Liverpool School of Tropical Medicine
Dorian Moro, University of Wales
Musonda Mumba, University College London
Eugene Murphy, British Antarctic Survey
Laszlo Nagy, McConnell Associates
Mark Nuttall, University of Aberdeen
Simon Potts, University of Reading
Ghillean Prance, Independent Scholar
Jules Pretty, University of Essex
Martin Price, Perth College
Andy Purvis, Imperial College London
David Satterthwaite, International Institute for Environment and Development
Marko Scholze, University of Bristol
Alison Stattersfield, Birdlife International
Kerry ten Kate, Insight Investment
Jillian Thonell, UNEP-World Conservation Monitoring Centre
Jinliang Wang, Institute of Zoology, Zoological Society of London
Andrew Wilby, University of Reading
Christoph Zockler, UNEP-World Conservation Monitoring Centre

United States

Robin Abell, World Wildlife Fund US
Tundi Agardy, Sound Seas
Tom Allnutt, World Wildlife Fund US
Agustin Arcenas, World Bank
Asmeret Asefaw Berhe, University of California, Berkeley
Suresh Chandra Babu, International Food Policy Research Institute
Andrew Bakun, Rosenstiel School of Marine and Atmospheric Science
Deborah Balk, Center for International Earth Science Information Network
Charles Barber, U.S. Agency for International Development
David Read Barker, LakeNet
Jill Baron, Colorado State University
Roger Barry, University of Colorado
Matt Berman, University of Alaska
Charles Birkeland, University of Hawaii
Dee Boersma, University of Washington
Raymond Bradley, University of Massachusetts
Thomas Brooks, Conservation International
Sandra Brown, Winrock International

Manoel Cardoso, University of New Hampshire
Ed Carpenter, San Francisco State University
Kenneth Cassman, University of Nebraska
Christopher Caudill, University of Idaho
Janice Chanson, IUCN–World Conservation Union and Conservation International
Terry Chapin, University of Alaska
Louis Codispoti, University of Maryland
Robert Costanza, University of Vermont
Neil Cox, IUCN–World Conservation Union and Conservation International
Sara Curran, Princeton University
Gretchen Daily, Stanford University
Paul Dayton, Scripps Institution of Oceanography
Ruth DeFries, University of Maryland
Andrew Dobson, Princeton University
Ellen Douglas, University of New Hampshire
Kirstin Dow, University of South Carolina
Hugh Ducklow, College of William and Mary
Richard Dugdale, San Francisco State University
Laurie Duker, LakeNet
Kristie Ebi, Exponent Health Group
Simeon Ehui, World Bank
Joseph Fargione, University of New Mexico
Jon Foley, University of Wisconsin
Gustavo Fonseca, Conservation International
Joanne Gaskell, International Food Policy Research Institute
Anna Goddhun, University of Alaska
Patrick Gonzalez, The Nature Conservancy
Terry Griswold, Utah State University
Kirk Hamilton, World Bank
Lawrence Hamilton, IUCN World Commission on Protected Areas
Drew Harvell, Cornell University
Kenneth Hinga, U.S. Department of Agriculture
Mike Hoffmann, Conservation International
Elisabeth Holland, National Center for Atmospheric Research
George Hurtt, Univeristy of New Hampshire
Lloyd Irland, The Irland Group and Yale University
Thomas Jaenisch, Johns Hopkins University
Anthony Janetos, The H. John Heinz III Center for Science, Economics, and the Environment
Nancy Kanbar, International Food Policy Research Institute
Michael Kaplowitz, Michigan State University
Roger Kasperson, George Perkins Marsh Institute, Clark University
Elizabeth Kennedy, Conservation International
Zahia Khan, International Food Policy Research Institute
Gary Kofinas, University of Alaska Fairbanks
Karen Kohfeld, City University of New York
Claire Kremen, Princeton University
Rattan Lal, Ohio State University
John Lamoreux, World Wildlife Fund US
William Laurance, Smithsonian Tropical Research Institute
Marc Levy, Columbia University
Noelle Lewis, Independent Scholar
Colby Loucks, World Wildlife Fund US
Pamela Mace, National Marine Fisheries Service
William Masters, Columbia University
Emily Matthews, World Resources Institute
Anthony McGuire, University of Alaska
Charles Mitchell, University of North Carolina
Russell Monson, University of Colorado
Shahid Naeem, Columbia University
Roz Naylor, Stanford University
Gerald Nelson, University of Illinois Urbana-Champaign

Scott Nixon, University of Rhode Island
Stefano Pagiola, World Bank
Jonathan Patz, University of Wisconsin
William F. Perrin, Southwest Fisheries Science Center
Michael Pilson, University of Rhode Island
Alison Power, Cornell University
Daniel Prager, World Resources Institute
Jane Pratt, EcoLogica Inc.
Steve Prince, University of Maryland
Navin Ramankutty, University of Wisconsin
Walter V. Reid, Millennium Ecosystem Assessment
Inbal Reshef, University of Maryland
Carmen Revenga, The Nature Conservancy
Valerie Rhoe, International Food Policy Research Institute
Taylor H. Ricketts, World Wildlife Fund US
Ana Rodrigues, Conservation International
Joshua Rosenthal, Fogarty International Center
Neil Sampson, The Sampson Group, Inc.
Crystal Schaaf, Boston University
Kate Sebastian, International Food Policy Research Institute
Wes Sechrest, IUCN–World Conservation Union and
University of Virginia
Roger Sedjo, Resources for the Future
Debdatta Sengupta, Penn State University
R. David Simpson, U.S. Environmental Protection Agency
Melanie Steinkamp, U.S. Fish and Wildlife Service
Simon Stuart, IUCN–World Conservation Union and
Conservation International

Madeleine Thomson, International Research Institute for Climate
Prediction
David Tilman, University of Minnesota
Juan Valdes, University of Arizona
Ross Virginia, Dartmouth College
Charles Vörösmarty, University of New Hampshire
Diana H. Wall, Colorado State University
Bess Ward, Princeton University
Konard Wessels, University of Maryland
Robin White, U.S. Geological Survey
Keith Wiebe, U.S. Dept of Agriculture
Matthew Wilson, University of Vermont
Stanley Wood, International Food Policy Research Institute
Ulrike Wood-Sichra, International Food Policy Research Institute
Cai Ximing, University of Illinois
Oran Young, University of California
Hania Zlotnik, United Nations Population Division

Uzbekistan
Vyacheslav Aparin, Complex Geological and Ecological Expedition
Elena Bykova, Institute of Zoology, Uzbekistan Academy of Science
Nikolay E. Gorelkin, GEF-Project, Transboundary Water-Monitoring
in Aral Sea Region
Alex Kreuzberg, Institute of Zoology, Uzbekistan Academy of Science

Venezuela
Lelys Bravo de Guenni, Universidad Simón Bolívar
Yasmin Rubio-Palis, Universidad de Carabobo

Appendix C
Abbreviations and Acronyms

AI	aridity index
AKRSP	Aga Khan Rural Support Programme
AMF	arbuscular mycorrhizal fungi
ASB	alternatives to slash-and-burn
ASOMPH	Asian Symposium on Medicinal Plants, Spices and Other Natural Products
AVHRR	advanced very high resolution radiometer
BCA	benefit-cost analysis
BGP	Biogeochemical Province
BII	Biodiversity Intactness Index
BMI	body mass index
BNF	biological nitrogen fixation
BOOT	build-own-operate-transfer
BRT	Bus Rapid Transit (Brazil)
BSE	bovine spongiform encephalopathy
Bt	*Bacillus thuringiensis*
C&I	criteria and indicators
CAFO	concentrated animal feeding operations
CAP	Common Agricultural Policy (of the European Union)
CAREC	Central Asia Regional Environment Centre
CBA	cost-benefit analysis
CBD	Convention on Biological Diversity
CBO	community-based organization
CCAMLR	Commission for the Conservation of Antarctic Marine Living Resources
CCN	cloud condensation nuclei
CCS	CO_2 capture and storage
CDM	Clean Development Mechanism
CEA	cost-effectiveness analysis
CENICAFÉ	Centro Nacional de Investigaciones de Café (Colombia)
CFCs	chlorofluorocarbons
CGIAR	Consultative Group on International Agricultural Research

CIFOR	Center for International Forestry Research
CITES	Convention on International Trade in Endangered Species of Wild Fauna and Flora
CMS	Convention on the Conservation of Migratory Species of Wild Animals (Bonn Convention)
CONICET	Consejo de Investigaciones Científicas y Técnicas (Argentina)
COP	Conference of the Parties (of treaties)
CPF	Collaborative Partnership on Forests
CSIR	Council for Scientific and Industrial Research (South Africa)
CV	contingent valuation
CVM	contingent valuation method
DAF	decision analytical framework
DALY	disability-adjusted life year
DDT	dichloro diphenyl trichloroethane
DES	dietary energy supply
DHF	dengue hemorrhagic fever
DHS	demographic and health surveys
DMS	dimethyl sulfide
DPSEEA	driving forces-pressure-state-exposure-effect-action
DPSIR	driver-pressure-state-impact-response
DSF	dust storm frequency
DU	Dobson Units
EEA	European Environment Agency
EEZ	exclusive economic zone
EGS	ecosystem global scenario
EHI	environmental health indicator
EIA	environmental impact assessment
EID	emerging infectious disease
EKC	Environmental Kuznets Curve
EMF	ectomycorrhizal fungi

E/MSY	extinctions per million species per year	**HWB**	human well-being
ENSO	El Niño/Southern Oscillation	**IAA**	integrated agriculture-aquaculture
EPA	Environmental Protection Agency (United States)	**IAM**	integrated assessment model
EPI	environmental policy integration	**IBI**	Index of Biotic Integrity
EU	European Union	**ICBG**	International Cooperative Biodiversity Groups
EU ETS	European Union Emissions Trading System	**ICDP**	integrated conservation and development project
FAO	Food and Agriculture Organization (United Nations)	**ICJ**	International Court of Justice
FAPRI	Food and Agriculture Policy Research Institute	**ICRAF**	International Center for Research in Agroforestry
FLEGT	Forest Law Enforcement, Governance, and Trade	**ICRW**	International Convention for the Regulation of Whaling
FRA	Forest Resources Assessment	**ICSU**	International Council for Science
FSC	Forest Stewardship Council	**ICZM**	integrated coastal zone management
GATS	General Agreement on Trade and Services	**IDRC**	International Development Research Centre (Canada)
GATT	General Agreement on Tariffs and Trade	**IEA**	International Energy Agency
GCM	general circulation model	**IEG**	international environmental governance
GDI	Gender-related Development Index	**IEK**	indigenous ecological knowledge
GDP	gross domestic product	**IFPRI**	International Food Policy Research Institute
GEF	Global Environment Facility	**IGBP**	International Geosphere-Biosphere Program
GEO	*Global Environment Outlook*	**IIASA**	International Institute for Applied Systems Analysis
GHG	greenhouse gases	**IK**	indigenous knowledge
GIS	geographic information system	**ILO**	International Labour Organization
GIWA	Global International Waters Assessment	**IMF**	International Monetary Fund
GLASOD	Global Assessment of Soil Degradation	**IMPACT**	International Model for Policy Analysis of Agricultural Commodities and Trade
GLC	Global Land Cover	**IMR**	infant mortality rate
GLOF	Glacier Lake Outburst Flood	**INESI**	International Network of Sustainability Initiatives (hypothetical, in *Scenarios*)
GM	genetic modification		
GMO	genetically modified organism	**INTA**	Instituto Nacional de Tecnología Agropecuaria (Argentina)
GNI	gross national income		
GNP	gross national product	**IPAT**	impact of population, affluence, technology
GPS	Global Positioning System	**IPCC**	Intergovernmental Panel on Climate Change
GRoWI	*Global Review of Wetland Resources and Priorities for Wetland Inventory*	**IPM**	integrated pest management
		IPR	intellectual property rights
GSG	Global Scenarios Group	**IRBM**	integrated river basin management
GSPC	Global Strategy for Plant Conservation	**ISEH**	International Society for Ecosystem Health
GtC-eq	gigatons of carbon equivalent	**ISO**	International Organization for Standardization
GWP	global warming potential	**ITPGR**	International Treaty on Plant Genetic Resources for Food and Agriculture
HDI	Human Development Index		
HIA	health impact assessment	**ITQs**	individual transferable quotas
HIPC	heavily indebted poor countries	**ITTO**	International Tropical Timber Organization
HPI	Human Poverty Index	**IUCN**	World Conservation Union
HPS	hantavirus pulmonary syndrome	**IUU**	illegal, unregulated, and unreported (fishing)
		IVM	integrated vector management

IWMI	International Water Management Institute
IWRM	integrated water resources management
JDSD	Johannesburg Declaration on Sustainable Development
JI	joint implementation
JMP	Joint Monitoring Program
LAC	Latin America and the Caribbean
LAI	leaf area index
LARD	livelihood approaches to rural development
LDC	least developed country
LEK	local ecological knowledge
LME	large marine ecosystems
LPI	Living Planet Index
LSMS	Living Standards Measurement Study
LULUCF	land use, land use change, and forestry
MA	Millennium Ecosystem Assessment
MAI	mean annual increments
MBI	market-based instruments
MCA	multicriteria analysis
MDG	Millennium Development Goal
MEA	multilateral environmental agreement
MENA	Middle East and North Africa
MER	market exchange rate
MHC	major histocompatibility complex
MICS	multiple indicator cluster surveys
MIT	Massachusetts Institute of Technology
MPA	marine protected area
MSVPA	multispecies virtual population analysis
NAP	National Action Program (of desertification convention)
NBP	net biome productivity
NCD	noncommunicable disease
NCS	National Conservation Strategy
NCSD	national council for sustainable development
NDVI	normalized difference vegetation index
NE	effective size of a population
NEAP	national environmental action plan
NEP	new ecological paradigm; also net ecosystem productivity
NEPAD	New Partnership for Africa's Development
NFAP	National Forestry Action Plan

NFP	national forest programs
NGO	nongovernmental organization
NIH	National Institutes of Health (United States)
NMHC	non-methane hydrocarbons
NOAA	National Oceanographic and Atmospheric Administration (United States)
NPP	net primary productivity
NSSD	national strategies for sustainable development
NUE	nitrogen use efficiency
NWFP	non-wood forest product
ODA	official development assistance
OECD	Organisation for Economic Co-operation and Development
OSB	oriented strand board
OWL	other wooded land
PA	protected area
PAH	polycyclic aromatic hydrocarbons
PCBs	polychlorinated biphenyls
PEM	protein energy malnutrition
PES	payment for environmental (or ecosystem) services
PFT	plant functional type
PNG	Papua New Guinea
POPs	persistent organic pollutants
PPA	participatory poverty assessment
ppb	parts per billion
PPI	potential Pareto improvement
ppm	parts per million
ppmv	parts per million by volume
PPP	purchasing power parity; also public-private partnership
ppt	parts per thousand
PQLI	Physical Quality of Life Index
PRA	participatory rural appraisal
PRSP	Poverty Reduction Strategy Paper
PSE	producer support estimate
PVA	population viability analysis
RANWA	Research and Action in Natural Wealth Administration
RBO	river basin organization
RIDES	Recursos e Investigación para el Desarrollo Sustentable (Chile)
RIL	reduced impact logging
RLI	Red List Index
RO	reverse osmosis

RRA	rapid rural appraisal		**TSU**	Technical Support Unit
RUE	rain use efficiency		**TW**	terawatt
SADC	Southern African Development Community		**UMD**	University of Maryland
SADCC	Southern African Development Coordination Conference		**UNCCD**	United Nations Convention to Combat Desertification
SAfMA	Southern African Millennium Ecosystem Assessment		**UNCED**	United Nations Conference on Environment and Development
SAP	structural adjustment program		**UNCLOS**	United Nations Convention on the Law of the Sea
SAR	species-area relationship		**UNDP**	United Nations Development Programme
SARS	severe acute respiratory syndrome		**UNECE**	United Nations Economic Commission for Europe
SBSTTA	Subsidiary Body on Scientific, Technical and Technological Advice (of CBD)		**UNEP**	United Nations Environment Programme
SEA	strategic environmental assessment		**UNESCO**	United Nations Educational, Scientific and Cultural Organization
SEME	simple empirical models for eutrophication		**UNFCCC**	United Nations Framework Convention on Climate Change
SES	social-ecological system		**UNIDO**	United Nations Industrial Development Organization
SFM	sustainable forest management		**UNRO**	United Nations Regional Organization (hypothetical body, in *Scenarios*)
SIDS	small island developing states			
SMS	safe minimum standard		**UNSO**	UNDP's Office to Combat Desertification and Drought
SOM	soil organic matter		**USAID**	U.S. Agency for International Development
SRES	Special Report on Emissions Scenarios (of the IPCC)		**USDA**	U.S. Department of Agriculture
SSC	Species Survival Commission (of IUCN)		**VOC**	volatile organic compound
SWAP	sector-wide approach		**VW**	virtual water
TAC	total allowable catch		**WBCSD**	World Business Council for Sustainable Development
TBT	tributyltin		**WCD**	World Commission on Dams
TC	travel cost		**WCED**	World Commission on Environment and Development
TCM	travel cost method		**WCMC**	World Conservation Monitoring Centre (of UNEP)
TDR	tradable development rights		**WFP**	World Food Programme
TDS	total dissolved solids		**WHO**	World Health Organization
TEIA	transboundary environmental impact assessment		**WIPO**	World Intellectual Property Organization
TEK	traditional ecological knowledge		**WISP**	weighted index of social progress
TEM	terrestrial ecosystem model		**WMO**	World Meteorological Organization
TESEO	Treaty Enforcement Services Using Earth Observation		**WPI**	Water Poverty Index
TEV	total economic value		**WRF**	white rot fungi
TFAP	Tropical Forests Action Plan		**WSSD**	World Summit on Sustainable Development
TFP	total factor productivity		**wta**	withdrawals-to-availability ratio (of water)
TFR	total fertility rate		**WTA**	willingness to accept compensation
Tg	teragram (10^{12} grams)		**WTO**	World Trade Organization
TK	traditional knowledge		**WTP**	willingness to pay
TMDL	total maximum daily load		**WWAP**	World Water Assessment Programme
TOF	trees outside of forests		**WWF**	World Wide Fund for Nature
TRIPS	Trade-Related Aspects of Intellectual Property Rights		**WWV**	World Water Vision

Appendix D
Glossary

Abatement cost: See *Marginal abatement cost.*

Abundance: The total number of individuals of a taxon or taxa in an area, population, or community. Relative abundance refers to the total number of individuals of one taxon compared with the total number of individuals of all other taxa in an area, volume, or community.

Active adaptive management: See *Adaptive management.*

Adaptation: Adjustment in natural or human systems to a new or changing environment. Various types of adaptation can be distinguished, including anticipatory and reactive adaptation, private and public adaptation, and autonomous and planned adaptation.

Adaptive capacity: The general ability of institutions, systems, and individuals to adjust to potential damage, to take advantage of opportunities, or to cope with the consequences.

Adaptive management: A systematic process for continually improving management policies and practices by learning from the outcomes of previously employed policies and practices. In active adaptive management, management is treated as a deliberate experiment for purposes of learning.

Afforestation: Planting of forests on land that has historically not contained forests. (Compare *Reforestation.*)

Agrobiodiversity: The diversity of plants, insects, and soil biota found in cultivated systems.

Agroforestry systems: Mixed systems of crops and trees providing wood, non-wood forest products, food, fuel, fodder, and shelter.

Albedo: A measure of the degree to which a surface or object reflects solar radiation.

Alien species: Species introduced outside its normal distribution.

Alien invasive species: See *Invasive alien species.*

Aquaculture: Breeding and rearing of fish, shellfish, or plants in ponds, enclosures, or other forms of confinement in fresh or marine waters for the direct harvest of the product.

Benefits transfer approach: Economic valuation approach in which estimates obtained (by whatever method) in one context are used to estimate values in a different context.

Binding constraints: Political, social, economic, institutional, or ecological factors that rule out a particular response.

Biodiversity (a contraction of biological diversity): The variability among living organisms from all sources, including terrestrial, marine, and other aquatic ecosystems and the ecological complexes of which they are part. Biodiversity includes diversity within species, between species, and between ecosystems.

Biodiversity regulation: The regulation of ecosystem processes and services by the different components of biodiversity.

Biogeographic realm: A large spatial region, within which ecosystems share a broadly similar biota. Eight terrestrial biogeographic realms are typically recognized, corresponding roughly to continents (e.g., Afrotropical realm).

Biological diversity: See *Biodiversity.*

Biomass: The mass of tissues in living organisms in a population, ecosystem, or spatial unit.

Biome: The largest unit of ecological classification that is convenient to recognize below the entire globe. Terrestrial biomes are typically based on dominant vegetation structure (e.g., forest, grassland). Ecosystems within a biome function in a broadly similar way, although they may have very different species composition. For example, all forests share certain properties regarding nutrient cycling, disturbance, and biomass that are different from the properties of grasslands. Marine biomes are typically based on biogeochemical properties. The WWF biome classification is used in the MA.

Bioprospecting: The exploration of biodiversity for genetic and biochemical resources of social or commercial value.

Biotechnology: Any technological application that uses biological systems, living organisms, or derivatives thereof to make or modify products or processes for specific use.

Biotic homogenization: Process by which the differences between biotic communities in different areas are on average reduced.

Blueprint approaches: Approaches that are designed to be applicable in a wider set of circumstances and that are not context-specific or sensitive to local conditions.

Boundary organizations: Public or private organizations that synthesize and translate scientific research and explore its policy implications to help bridge the gap between science and decision-making.

Bridging organizations: Organizations that facilitate, and offer an arena for, stakeholder collaboration, trust-building, and conflict resolution.

Capability: The combinations of doings and beings from which people can choose to lead the kind of life they value. Basic capability is the capability to meet a basic need.

Capacity building: A process of strengthening or developing human resources, institutions, organizations, or networks. Also referred to as capacity development or capacity enhancement.

Capital value (of an ecosystem): The present value of the stream of ecosystem services that an ecosystem will generate under a particular management or institutional regime.

Capture fisheries: See *Fishery.*

Carbon sequestration: The process of increasing the carbon content of a reservoir other than the atmosphere.

Cascading interaction: See *Trophic cascade.*

Catch: The number or weight of all fish caught by fishing operations, whether the fish are landed or not.

Coastal system: Systems containing terrestrial areas dominated by ocean influences of tides and marine aerosols, plus nearshore marine areas. The inland extent of coastal ecosystems is the line where land-based influences dominate, up to a maximum of 100 kilometers from the coastline or 100-meter elevation (whichever is closer to the sea), and the outward extent is the 50-meter-depth contour. See also *System.*

Collaborative (or joint) forest management: Community-based management of forests, where resource tenure by local communities is secured.

Common pool resource: A valued natural or human-made resource or facility in which one person's use subtracts from another's use and where it is often necessary but difficult to exclude potential users from the resource. (Compare *Common property resource.*)

Common property management system: The institutions (i.e., sets of rules) that define and regulate the use rights for common pool resources. Not the same as an open access system.

Common property resource: A good or service shared by a well-defined community. (Compare *Common pool resource.*)

Community (ecological): An assemblage of species occurring in the same space or time, often linked by biotic interactions such as competition or predation.

Community (human, local): A collection of human beings who have something in common. A local community is a fairly small group of people who share a common place of residence and a set of institutions based on this fact, but the word 'community' is also used to refer to larger collections of people who have something else in common (e.g., national community, donor community).

Condition of an ecosystem: The capacity of an ecosystem to yield services, relative to its potential capacity.

Condition of an ecosystem service: The capacity of an ecosystem service to yield benefits to people, relative to its potential capacity.

Constituents of well-being: The experiential aspects of well-being, such as health, happiness, and freedom to be and do, and, more broadly, basic liberties.

Consumptive use: The reduction in the quantity or quality of a good available for other users due to consumption.

Contingent valuation: Economic valuation technique based on a survey of how much respondents would be willing to pay for specified benefits.

Core dataset: Data sets designated to have wide potential application throughout the Millennium Ecosystem Assessment process. They include land use, land cover, climate, and population data sets.

Cost-benefit analysis: A technique designed to determine the feasibility of a project or plan by quantifying its costs and benefits.

Cost-effectiveness analysis: Analysis to identify the least cost option that meets a particular goal.

Critically endangered species: Species that face an extremely high risk of extinction in the wild. See also *Threatened species.*

Cross-scale feedback: A process in which effects of some action are transmitted from a smaller spatial extent to a larger one, or vice versa. For example, a global policy may constrain the flexibility of a local region to use certain response options to environmental change, or a local agricultural pest outbreak may affect regional food supply.

Cultivar (a contraction of cultivated variety): A variety of a plant developed from a natural species and maintained under cultivation.

Cultivated system: Areas of landscape or seascape actively managed for the production of food, feed, fiber, or biofuels.

Cultural landscape: See *Landscape.*

Cultural services: The nonmaterial benefits people obtain from ecosystems through spiritual enrichment, cognitive development, reflection, recreation, and aesthetic experience, including, e.g., knowledge systems, social relations, and aesthetic values.

Decision analytical framework: A coherent set of concepts and procedures aimed at synthesizing available information to help policy-makers assess consequences of various decision options. DAFs organize the relevant information in a suitable framework, apply decision criteria (both based on some paradigms or theories), and thus identify options that are better than others under the assumptions characterizing the analytical framework and the application at hand.

Decision-maker: A person whose decisions, and the actions that follow from them, can influence a condition, process, or issue under consideration.

Decomposition: The ecological process carried out primarily by microbes that leads to a transformation of dead organic matter into inorganic matter.

Deforestation: Conversion of forest to non-forest.

Degradation of an ecosystem service: For *provisioning services,* decreased production of the service through changes in area over which the services is provided, or decreased production per unit area. For *regulating* and *supporting services,* a reduction in the benefits obtained from the service, either through a change in the service or through human pressures on the service exceeding its limits. For *cultural services,* a change in the ecosystem features that decreases the cultural benefits provided by the ecosystem.

Degradation of ecosystems: A persistent reduction in the capacity to provide ecosystem services.

Desertification: land degradation in drylands resulting from various factors, including climatic variations and human activities.

Determinants of well-being: Inputs into the production of well-being, such as food, clothing, potable water, and access to knowledge and information.

Direct use value (of ecosystems): The benefits derived from the services provided by an ecosystem that are used directly by an economic agent. These include consumptive uses (e.g., harvesting goods) and nonconsumptive uses (e.g., enjoyment of scenic beauty). Agents are often physically present in an ecosystem to receive direct use value. (Compare *Indirect use value.*)

Disability-adjusted life years: The sum of years of life lost due to premature death and illness, taking into account the age of death compared with natural life expectancy and the number of years of life lived with a disability. The measure of number of years lived with the disability considers the duration of the disease, weighted by a measure of the severity of the disease.

Diversity: The variety and relative abundance of different entities in a sample.

Driver: Any natural or human-induced factor that directly or indirectly causes a change in an ecosystem.

Driver, direct: A driver that unequivocally influences ecosystem processes and can therefore be identified and measured to differing degrees of accuracy. (Compare *Driver, indirect.*)

Driver, endogenous: A driver whose magnitude can be influenced by the decision-maker. Whether a driver is exogenous or endogenous depends on the organizational scale. Some drivers (e.g., prices) are exogenous to a decision-maker at one level (a farmer) but endogenous at other levels (the nation-state). (Compare *Driver, exogenous.*)

Driver, exogenous: A driver that cannot be altered by the decision-maker. (Compare *Driver, endogenous.*)

Driver, indirect: A driver that operates by altering the level or rate of change of one or more direct drivers. (Compare *Driver, direct.*)

Drylands: See *Dryland system.*

Dryland system: Areas characterized by lack of water, which constrains the two major interlinked services of the system: primary production and nutrient cycling. Four dryland subtypes are widely recognized: dry sub-humid, semiarid, arid, and hyperarid, showing an increasing level of aridity or moisture deficit. See also *System.*

Ecological character: See *Ecosystem properties.*

Ecological degradation: See *Degradation of ecosystems.*

Ecological footprint: An index of the area of productive land and aquatic ecosystems required to produce the resources used and to assimilate the wastes produced by a defined population at a specified material standard of living, wherever on Earth that land may be located.

Ecological security: A condition of ecological safety that ensures access to a sustainable flow of provisioning, regulating, and cultural services needed by local communities to meet their basic capabilities.

Ecological surprises: unexpected—and often disproportionately large—consequence of changes in the abiotic (e.g., climate, disturbance) or biotic (e.g., invasions, pathogens) environment.

Ecosystem: A dynamic complex of plant, animal, and microorganism communities and their non-living environment interacting as a functional unit.

Ecosystem approach: A strategy for the integrated management of land, water, and living resources that promotes conservation and sustainable use. An ecosystem approach is based on the application of appropriate scientific methods focused on levels of biological organization, which encompass the essential structure, processes, functions, and interactions among organisms and their environment. It recognizes that humans, with their cultural diversity, are an integral component of many ecosystems.

Ecosystem assessment: A social process through which the findings of science concerning the causes of ecosystem change, their consequences for human well-being, and management and policy options are brought to bear on the needs of decision-makers.

Ecosystem boundary: The spatial delimitation of an ecosystem, typically based on discontinuities in the distribution of organisms, the biophysical environment (soil types, drainage basins, depth in a water body), and spatial interactions (home ranges, migration patterns, fluxes of matter).

Ecosystem change: Any variation in the state, outputs, or structure of an ecosystem.

Ecosystem function: See *Ecosystem process.*

Ecosystem interactions: Exchanges of materials, energy, and information within and among ecosystems.

Ecosystem management: An approach to maintaining or restoring the composition, structure, function, and delivery of services of natural and modified ecosystems for the goal of achieving sustainability. It is based on an adaptive, collaboratively developed vision of desired future conditions that integrates ecological, socioeconomic, and institutional perspectives, applied within a geographic framework, and defined primarily by natural ecological boundaries.

Ecosystem process: An intrinsic ecosystem characteristic whereby an ecosystem maintains its integrity. Ecosystem processes include decomposition, production, nutrient cycling, and fluxes of nutrients and energy.

Ecosystem properties: The size, biodiversity, stability, degree of organization, internal exchanges of materials, energy, and information among different pools, and other properties that characterize an ecosystem. Includes ecosystem functions and processes.

Ecosystem resilience: See *Resilience.*

Ecosystem resistance: See *Resistance.*

Ecosystem robustness: See *Ecosystem stability.*

Ecosystem services: The benefits people obtain from ecosystems. These include *provisioning services* such as food and water; *regulating services* such as flood and disease control; *cultural services* such as spiritual, recreational, and cultural benefits; and *supporting services* such as nutrient cycling that maintain the conditions for life on Earth. The concept "ecosystem goods and services" is synonymous with ecosystem services.

Ecosystem stability (or ecosystem robustness): A description of the dynamic properties of an ecosystem. An ecosystem is considered stable or robust if it returns to its original state after a perturbation, exhibits low temporal variability, or does not change dramatically in the face of a perturbation.

Elasticity: A measure of responsiveness of one variable to a change in another, usually defined in terms of percentage change. For example, own-price elasticity of demand is the percentage change in the quantity demanded of a good for a 1% change in the price of that good. Other common elasticity measures include supply and income elasticity.

Emergent disease: Diseases that have recently increased in incidence, impact, or geographic range; that are caused by pathogens that have recently evolved; that are newly discovered; or that have recently changed their clinical presentation.

Emergent property: A phenomenon that is not evident in the constituent parts of a system but that appears when they interact in the system as a whole.

Enabling conditions: Critical preconditions for success of responses, including political, institutional, social, economic, and ecological factors.

Endangered species: Species that face a very high risk of extinction in the wild. See also *Threatened species.*

Endemic (in ecology): A species or higher taxonomic unit found only within a specific area.

Endemic (in health): The constant presence of a disease or infectious agent within a given geographic area or population group; may also refer to the usual prevalence of a given disease within such area or group.

Endemism: The fraction of species that is endemic relative to the total number of species found in a specific area.

Epistemology: The theory of knowledge, or a "way of knowing."

Equity: Fairness of rights, distribution, and access. Depending on context, this can refer to resources, services, or power.

Eutrophication: The increase in additions of nutrients to freshwater or marine systems, which leads to increases in plant growth and often to undesirable changes in ecosystem structure and function.

Evapotranspiration: See *Transpiration.*

Existence value: The value that individuals place on knowing that a resource exists, even if they never use that resource (also sometimes known as conservation value or passive use value).

Exotic species: See *Alien species.*

Externality: A consequence of an action that affects someone other than the agent undertaking that action and for which the agent is neither compensated nor penalized through the markets. Externalities can be positive or negative.

Feedback: See *Negative feedback, Positive feedback,* and *Cross-scale feedback.*

Fishery: A particular kind of fishing activity, e.g., a trawl fishery, or a particular species targeted, e.g., a cod fishery or salmon fishery.

Fish stock: See *Stock.*

Fixed nitrogen: See *Reactive nitrogen.*

Flyway: Areas of the world used by migratory birds in moving between breeding and wintering grounds.

Forest systems: Systems in which trees are the predominant life forms. Statistics reported in this assessment are based on areas that are dominated by trees (perennial woody plants taller than five meters at maturity), where the tree crown cover exceeds 10%, and where the area is more than 0.5 hectares. "Open forests" have a canopy cover between 10% and 40%, and "closed forests" a canopy cover of more than 40%. "Fragmented forests" refer to mosaics of forest patches and non-forest land. See also *System.*

Freedom: The range of options a person has in deciding the kind of life to lead.

Functional diversity: The value, range, and relative abundance of traits present in the organisms in an ecological community.

Functional redundancy (= functional compensation): A characteristic of ecosystems in which more than one species in the system can carry out a particular process. Redundancy may be total or partial—that is, a species may not be able to completely replace the other species or it may compensate only some of the processes in which the other species are involved.

Functional types (= functional groups = guilds): Groups of organisms that respond to the environment or affect ecosystem processes in a similar way. Examples of plant functional types include nitrogen-fixer versus non-fixer, stress-tolerant versus ruderal versus competitor, resprouter versus seeder, deciduous versus evergreen. Examples of animal functional types include granivorous versus fleshy-fruit eater, nocturnal versus diurnal predator, browser versus grazer.

Geographic information system: A computerized system organizing data sets through a geographical referencing of all data included in its collections.

Globalization: The increasing integration of economies and societies around the world, particularly through trade and financial flows, and the transfer of culture and technology.

Global scale: The geographical realm encompassing all of Earth.

Governance: The process of regulating human behavior in accordance with shared objectives. The term includes both governmental and nongovernmental mechanisms.

Health, human: A state of complete physical, mental, and social well-being and not merely the absence of disease or infirmity. The health of a whole community or population is reflected in measurements of disease incidence and prevalence, age-specific death rates, and life expectancy.

High seas: The area outside of national jurisdiction, i.e., beyond each nation's Exclusive Economic Zone or other territorial waters.

Human well-being: See *Well-being.*

Income poverty: See *Poverty.*

Indicator: Information based on measured data used to represent a particular attribute, characteristic, or property of a system.

Indigenous knowledge (or local knowledge): The knowledge that is unique to a given culture or society.

Indirect interaction: Those interactions among species in which a species, through direct interaction with another species or modification of resources, alters the abundance of a third species with which it is not directly interacting. Indirect interactions can be trophic or nontrophic in nature.

Indirect use value: The benefits derived from the goods and services provided by an ecosystem that are used indirectly by an economic agent. For example, an agent at some distance from an ecosystem may derive benefits from drinking water that has been purified as it passed through the ecosystem. (Compare *Direct use value.*)

Infant mortality rate: Number of deaths of infants aged 0–12 months divided by the number of live births.

Inland water systems: Permanent water bodies other than salt-water systems on the coast, seas and oceans. Includes rivers, lakes, reservoirs wetlands and inland saline lakes and marshes. See also *System.*

Institutions: The rules that guide how people within societies live, work, and interact with each other. Formal institutions are written or codified rules. Examples of formal institutions would be the constitution, the judiciary laws, the organized market, and property rights. Informal institutions are rules governed by social and behavioral norms of the society, family, or community. Also referred to as organizations.

Integrated coastal zone management: Approaches that integrate economic, social, and ecological perspectives for the management of coastal resources and areas.

Integrated conservation and development projects: Initiatives that aim to link biodiversity conservation and development.

Integrated pest management: Any practices that attempt to capitalize on natural processes that reduce pest abundance. Sometimes used to refer to monitoring programs where farmers apply pesticides to improve economic efficiency (reducing application rates and improving profitability).

Integrated responses: Responses that address degradation of ecosystem services across a number of systems simultaneously or that also explicitly include objectives to enhance human well-being.

Integrated river basin management: Integration of water planning and management with environmental, social, and economic development concerns, with an explicit objective of improving human welfare.

Interventions: See *Responses.*

Intrinsic value: The value of someone or something in and for itself, irrespective of its utility for people.

Invasibility: Intrinsic susceptibility of an ecosystem to be invaded by an alien species.

Invasive alien species: An alien species whose establishment and spread modifies ecosystems, habitats, or species.

Irreversibility: The quality of being impossible or difficult to return to, or to restore to, a former condition. See also *Option value, Precautionary principle, Resilience,* and *Threshold.*

Island systems: Lands isolated by surrounding water, with a high proportion of coast to hinterland. The degree of isolation from the mainland in both natural and social aspects is accounted by the *isola effect.* See also *System.*

Isola effect: Environmental issues that are unique to island systems. This uniqueness takes into account the physical seclusion of islands as isolated pieces of land exposed to marine or climatic disturbances with a more limited access to space, products, and services when compared with most continental areas, but also includes subjective issues such as the perceptions and attitudes of islanders themselves.

Keystone species: A species whose impact on the community is disproportionately large relative to its abundance. Effects can be produced by consumption (trophic interactions), competition, mutualism, dispersal, pollination, disease, or habitat modification (nontrophic interactions).

Land cover: The physical coverage of land, usually expressed in terms of vegetation cover or lack of it. Related to, but not synonymous with, *land use.*

Landscape: An area of land that contains a mosaic of ecosystems, including human-dominated ecosystems. The term cultural landscape is often used when referring to landscapes containing significant human populations or in which there has been significant human influence on the land.

Landscape unit: A portion of relatively homogenous land cover within the local-to-regional landscape.

Land use: The human use of a piece of land for a certain purpose (such as irrigated agriculture or recreation). Influenced by, but not synonymous with, *land cover.*

Length of growing period: The total number of days in a year during which rainfall exceeds one half of potential evapotranspiration. For boreal and temperate zone, growing season is usually defined as a number of days with the average daily temperature that exceeds a definite threshold, such as 10° Celsius.

Local knowledge: See *Indigenous knowledge.*

Mainstreaming: Incorporating a specific concern, e.g. sustainable use of ecosystems, into policies and actions.

Malnutrition: A state of bad nourishment. Malnutrition refers both to undernutrition and overnutrition, as well as to conditions arising from dietary imbalances leading to diet-related noncommunicable diseases.

Marginal abatement cost: The cost of abating an incremental unit of, for instance, a pollutant.

Marine system: Marine waters from the low-water mark to the high seas that support marine capture fisheries, as well as deepwater (>50 meters) habitats. Four sub-divisions (marine biomes) are recognized: the coastal boundary zone; trade-winds; westerlies; and polar.

Market-based instruments: Mechanisms that create a market for ecosystem services in order to improving the efficiency in the way the service is used. The term is used for mechanisms that create new markets, but also for responses such as taxes, subsidies, or regulations that affect existing markets.

Market failure: The inability of a market to capture the correct values of ecosystem services.

Mitigation: An anthropogenic intervention to reduce negative or unsustainable uses of ecosystems or to enhance sustainable practices.

Mountain system: High-altitude (greater than 2,500 meters) areas and steep mid-altitude (1,000 meters at the equator, decreasing to sea level where alpine life zones meet polar life zones at high latitudes) areas, excluding large plateaus.

Negative feedback: Feedback that has a net effect of dampening perturbation.

Net primary productivity: See *Production, biological.*

Non-linearity: A relationship or process in which a small change in the value of a driver (i.e., an independent variable) produces a disproportionate change in the outcome (i.e., the dependent variable). Relationships where there is a sudden discontinuity or change in rate are sometimes referred to as abrupt and often form the basis of thresholds. In loose terms, they may lead to unexpected outcomes or "surprises."

Nutrient cycling: The processes by which elements are extracted from their mineral, aquatic, or atmospheric sources or recycled from their organic forms, converting them to the ionic form in which biotic uptake occurs and ultimately returning them to the atmosphere, water, or soil.

Nutrients: The approximately 20 chemical elements known to be essential for the growth of living organisms, including nitrogen, sulfur, phosphorus, and carbon.

Open access resource: A good or service over which no property rights are recognized.

Opportunity cost: The benefits forgone by undertaking one activity instead of another.

Option value: The value of preserving the option to use services in the future either by oneself (option value) or by others or heirs (bequest value). Quasi-option value represents the value of avoiding irreversible decisions until new information reveals whether certain ecosystem services have values society is not currently aware of.

Organic farming: Crop and livestock production systems that do not make use of synthetic fertilizers, pesticides, or herbicides. May also include restrictions on the use of transgenic crops (genetically modified organisms).

Pastoralism, pastoral system: The use of domestic animals as a primary means for obtaining resources from habitats.

Perturbation: An imposed movement of a system away from its current state.

Polar system: Treeless lands at high latitudes. Includes Arctic and Antarctic areas, where the polar system merges with the northern boreal forest and the Southern Ocean respectively. See also *System*.

Policy failure: A situation in which government policies create inefficiencies in the use of goods and services.

Policy-maker: A person with power to influence or determine policies and practices at an international, national, regional, or local level.

Pollination: A process in the sexual phase of reproduction in some plants caused by the transportation of pollen. In the context of ecosystem services, pollination generally refers to animal-assisted pollination, such as that done by bees, rather than wind pollination.

Population, biological: A group of individuals of the same species, occupying a defined area, and usually isolated to some degree from other similar groups. Populations can be relatively reproductively isolated and adapted to local environments.

Population, human: A collection of living people in a given area. (Compare *Community (human, local)*.)

Positive feedback: Feedback that has a net effect of amplifying perturbation.

Poverty: The pronounced deprivation of well-being. Income poverty refers to a particular formulation expressed solely in terms of per capita or household income.

Precautionary principle: The management concept stating that in cases "where there are threats of serious or irreversible damage, lack of full scientific certainty shall not be used as a reason for postponing cost-effective measures to prevent environmental degradation," as defined in the Rio Declaration.

Prediction (or forecast): The result of an attempt to produce a most likely description or estimate of the actual evolution of a variable or system in the future. See also *Projection* and *Scenario*.

Primary production: See *Production, biological*.

Private costs and benefits: Costs and benefits directly felt by individual economic agents or groups as seen from their perspective. (Externalities imposed on others are ignored.) Costs and benefits are valued at the prices actually paid or received by the group, even if these prices are highly distorted. Sometimes termed "financial" costs and benefits. (Compare *Social costs and benefits*.)

Probability distribution: A distribution that shows all the values that a random variable can take and the likelihood that each will occur.

Production, biological: Rate of biomass produced by an ecosystem, generally expressed as biomass produced per unit of time per unit of surface or volume. Net primary productivity is defined as the energy fixed by plants minus their respiration.

Production, economic: Output of a system.

Productivity, biological: See *Production, biological*.

Productivity, economic: Capacity of a system to produce high levels of output or responsiveness of the output of a system to inputs.

Projection: A potential future evolution of a quantity or set of quantities, often computed with the aid of a model. Projections are distinguished from "predictions" in order to emphasize that projections involve assumptions concerning, for example, future socioeconomic and technological developments that may or may not be realized; they are therefore subject to substantial uncertainty.

Property rights: The right to specific uses, perhaps including exchange in a market, of ecosystems and their services.

Provisioning services: The products obtained from ecosystems, including, for example, genetic resources, food and fiber, and fresh water.

Public good: A good or service in which the benefit received by any one party does not diminish the availability of the benefits to others, and where access to the good cannot be restricted.

Reactive nitrogen (or fixed nitrogen): The forms of nitrogen that are generally available to organisms, such as ammonia, nitrate, and organic nitrogen. Nitrogen gas (or dinitrogen), which is the major component of the atmosphere, is inert to most organisms.

Realm: Used to describe the three major types of ecosystems on earth: terrestrial, freshwater, and marine. Differs fundamentally from *biogeographic realm*.

Reforestation: Planting of forests on lands that have previously contained forest but have since been converted to some other use. (Compare *Afforestation*.)

Regime shift: A rapid reorganization of an ecosystem from one relatively stable state to another.

Regulating services: The benefits obtained from the regulation of ecosystem processes, including, for example, the regulation of climate, water, and some human diseases.

Relative abundance: See *Abundance*.

Reporting unit: The spatial or temporal unit at which assessment or analysis findings are reported. In an assessment, these units are chosen to maximize policy relevance or relevance to the public and thus may differ from those upon which the analyses were conducted (e.g., analyses conducted on mapped ecosystems can be reported on administrative units). See also *System*.

Resilience: The level of disturbance that an ecosystem can undergo without crossing a threshold to a situation with different structure or outputs. Resilience depends on ecological dynamics as well as the organizational and institutional capacity to understand, manage, and respond to these dynamics.

Resistance: The capacity of an ecosystem to withstand the impacts of drivers without displacement from its present state.

Responses: Human actions, including policies, strategies, and interventions, to address specific issues, needs, opportunities, or problems. In the context of ecosystem management, responses may be of legal, technical, institutional, economic, and behavioral nature and may operate at various spatial and time scales.

Riparian: Something related to, living on, or located at the banks of a watercourse, usually a river or stream.

Safe minimum standard: A decision analytical framework in which the benefits of ecosystem services are assumed to be incalculable and should be preserved unless the costs of doing so rise to an intolerable level, thus shifting the burden of proof to those who would convert them.

Salinization: The buildup of salts in soils.

Scale: The measurable dimensions of phenomena or observations. Expressed in physical units, such as meters, years, population size, or quantities moved or exchanged. In observation, scale determines the relative fineness and coarseness of different detail and the selectivity among patterns these data may form.

Scenario: A plausible and often simplified description of how the future may develop, based on a coherent and internally consistent set of assumptions about key driving forces (e.g., rate of technology change, prices) and relationships. Scenarios are neither predictions nor projections and sometimes may be based on a "narrative storyline." Scenarios may include projections but are often based on additional information from other sources.

Security: Access to resources, safety, and the ability to live in a predictable and controllable environment.

Service: See *Ecosystem services*.

Social costs and benefits: Costs and benefits as seen from the perspective of society as a whole. These differ from private costs and benefits in being more inclusive (all costs and benefits borne by some member of society are taken into account) and in being valued at social opportunity cost rather than market prices, where these differ. Sometimes termed "economic" costs and benefits. (Compare *Private costs and benefits*.)

Social incentives: Measures that lower transaction costs by facilitating trust-building and learning as well as rewarding collaboration and conflict resolution. Social incentives are often provided by bridging organizations.

Socioecological system: An ecosystem, the management of this ecosystem by actors and organizations, and the rules, social norms, and conventions underlying this management. (Compare *System*.)

Soft law: Non-legally binding instruments, such as guidelines, standards, criteria, codes of practice, resolutions, and principles or declarations, that states establish to implement national laws.

Soil fertility: The potential of the soil to supply nutrient elements in the quantity, form, and proportion required to support optimum plant growth. See also *Nutrients*.

Speciation: The formation of new species.

Species: An interbreeding group of organisms that is reproductively isolated from all other organisms, although there are many partial exceptions to this rule in particular taxa. Operationally, the term *species* is a generally agreed fundamental taxonomic unit, based on morphological or genetic similarity, that once described and accepted is associated with a unique scientific name.

Species diversity: Biodiversity at the species level, often combining aspects of species richness, their relative abundance, and their dissimilarity.

Species richness: The number of species within a given sample, community, or area.

Statistical variation: Variability in data due to error in measurement, error in sampling, or variation in the measured quantity itself.

Stock (in fisheries): The population or biomass of a fishery resource. Such stocks are usually identified by their location. They can be, but are not always, genetically discrete from other stocks.

Stoichiometry, ecological: The relatively constant proportions of the different nutrients in plant or animal biomass that set constraints on production. Nutrients only available in lower proportions are likely to limit growth.

Storyline: A narrative description of a scenario, which highlights its main features and the relationships between the scenario's driving forces and its main features.

Strategies: See *Responses.*

Streamflow: The quantity of water flowing in a watercourse.

Subsidiarity, principle of: The notion of devolving decision-making authority to the lowest appropriate level.

Subsidy: Transfer of resources to an entity, which either reduces the operating costs or increases the revenues of such entity for the purpose of achieving some objective.

Subsistence: An activity in which the output is mostly for the use of the individual person doing it, or their family, and which is a significant component of their livelihood.

Subspecies: A population that is distinct from, and partially reproductively isolated from, other populations of a species but that has not yet diverged sufficiently that interbreeding is impossible.

Supporting services: Ecosystem services that are necessary for the production of all other ecosystem services. Some examples include biomass production, production of atmospheric oxygen, soil formation and retention, nutrient cycling, water cycling, and provisioning of habitat.

Sustainability: A characteristic or state whereby the needs of the present and local population can be met without compromising the ability of future generations or populations in other locations to meet their needs.

Sustainable use (of an ecosystem): Human use of an ecosystem so that it may yield a continuous benefit to present generations while maintaining its potential to meet the needs and aspirations of future generations.

Symbiosis: Close and usually obligatory relationship between two organisms of different species, not necessarily to their mutual benefit.

Synergy: When the combined effect of several forces operating is greater than the sum of the separate effects of the forces.

System: In the Millennium Ecosystem Assessment, reporting units that are ecosystem-based but at a level of aggregation far higher than that usually applied to ecosystems. Thus the system includes many component ecosystems, some of which may not strongly interact with each other, that may be spatially separate, or that may be of a different type to the ecosystems that constitute the majority, or matrix, of the system overall. The system includes the social and economic systems that have an impact on and are affected by the ecosystems included within it. For example, the Condition and Trend Working Group refers to "forest systems," "cultivated systems," "mountain systems," and so on. Systems thus defined are not mutually exclusive, and are permitted to overlap spatially or conceptually. For instance, the "cultivated system" may include areas of "dryland system" and vice versa.

Taxon (pl. taxa): The named classification unit to which individuals or sets of species are assigned. Higher taxa are those above the species

level. For example, the common mouse, *Mus musculus,* belongs to the Genus *Mus,* the Family Muridae, and the Class Mammalia.

Taxonomy: A system of nested categories (*taxa*) reflecting evolutionary relationships or morphological similarity.

Tenure: See *Property rights,* although also sometimes used more specifically in reference to the temporal dimensions and security of property rights.

Threatened species: Species that face a high (*vulnerable species*), very high (*endangered species*), or extremely high (*critically endangered species*) risk of extinction in the wild.

Threshold: A point or level at which new properties emerge in an ecological, economic, or other system, invalidating predictions based on mathematical relationships that apply at lower levels. For example, species diversity of a landscape may decline steadily with increasing habitat degradation to a certain point, then fall sharply after a critical threshold of degradation is reached. Human behavior, especially at group levels, sometimes exhibits threshold effects. Thresholds at which irreversible changes occur are especially of concern to decision-makers. (Compare *Non-linearity.*)

Time series data: A set of data that expresses a particular variable measured over time.

Total economic value framework: A widely used framework to disaggregate the components of utilitarian value, including *direct use value, indirect use value, option value,* quasi-option value, and *existence value.*

Total factor productivity: A measure of the aggregate increase in efficiency of use of inputs. TFP is the ratio of the quantity of output divided by an index of the amount of inputs used. A common input index uses as weights the share of the input in the total cost of production.

Total fertility rate: The number of children a woman would give birth to if through her lifetime she experienced the set of age-specific fertility rates currently observed. Since age-specific rates generally change over time, TFR does not in general give the actual number of births a woman alive today can be expected to have. Rather, it is a synthetic index meant to measure age-specific birth rates in a given year.

Trade-off: Management choices that intentionally or otherwise change the type, magnitude, and relative mix of services provided by ecosystems.

Traditional ecological knowledge: The cumulative body of knowledge, practices, and beliefs evolved by adaptive processes and handed down through generations. TEK may or may not be indigenous or local, but it is distinguished by the way in which it is acquired and used, through the social process of learning and sharing knowledge. (Compare *Indigenous knowledge.*)

Traditional knowledge: See *Traditional ecological knowledge.*

Traditional use: Exploitation of natural resources by indigenous users or by nonindigenous residents using traditional methods. Local use refers to exploitation by local residents.

Transpiration: The process by which water is drawn through plants and returned to the air as water vapor. Evapotranspiration is combined loss of water to the atmosphere via the processes of evaporation and transpiration.

Travel cost methods: Economic valuation techniques that use observed costs to travel to a destination to derive demand functions for that destination.

Trend: A pattern of change over time, over and above short-term fluctuations.

Trophic cascade: A chain reaction of top-down interactions across multiple tropic levels. These occur when changes in the presence or absence (or shifts in abundance) of a top predator alter the production at several lower trophic levels. Such positive indirect effects of top predators on lower tropic levels are mediated by the consumption of mid-level consumers (generally herbivores).

Trophic level: The average level of an organism within a food web, with plants having a trophic level of 1, herbivores 2, first-order carnivores 3, and so on.

Umbrella species: Species that have either large habitat needs or other requirements whose conservation results in many other species being conserved at the ecosystem or landscape level.

Uncertainty: An expression of the degree to which a future condition (e.g., of an ecosystem) is unknown. Uncertainty can result from lack of information or from disagreement about what is known or even knowable. It may have many types of sources, from quantifiable errors in the data to ambiguously defined terminology or uncertain projections of human behavior. Uncertainty can therefore be represented by quantitative measures (e.g., a range of values calculated by various models) or by qualitative statements (e.g., reflecting the judgment of a team of experts).

Urbanization: An increase in the proportion of the population living in urban areas.

Urban systems: Built environments with a high human population density. Operationally defined as human settlements with a minimum population density commonly in the range of 400 to 1,000 persons per square kilometer, minimum size of typically between 1,000 and 5,000 people, and maximum agricultural employment usually in the vicinity of 50–75%. See also *System*.

Utility: In economics, the measure of the degree of satisfaction or happiness of a person.

Valuation: The process of expressing a value for a particular good or service in a certain context (e.g., of decision-making) usually in terms of something that can be counted, often money, but also through methods and measures from other disciplines (sociology, ecology, and so on). See also *Value*.

Value: The contribution of an action or object to user-specified goals, objectives, or conditions. (Compare *Valuation*.)

Value systems: Norms and precepts that guide human judgment and action.

Voluntary measures: Measures that are adopted by firms or other actors in the absence of government mandates.

Vulnerability: Exposure to contingencies and stress, and the difficulty in coping with them. Three major dimensions of vulnerability are involved: exposure to stresses, perturbations, and shocks; the sensitivity of people, places, ecosystems, and species to the stress or perturbation, including their capacity to anticipate and cope with the stress; and the resilience of the exposed people, places, ecosystems, and species in terms of their capacity to absorb shocks and perturbations while maintaining function.

Vulnerable species: Species that face a high risk of extinction in the wild. See also *Threatened species*.

Water scarcity: A water supply that limits food production, human health, and economic development. Severe scarcity is taken to be equivalent to 1,000 cubic meters per year per person or greater than 40% use relative to supply.

Watershed (also catchment basin): The land area that drains into a particular watercourse or body of water. Sometimes used to describe the dividing line of high ground between two catchment basins.

Water stress: See *Water scarcity*.

Well-being: A context- and situation-dependent state, comprising basic material for a good life, freedom and choice, health and bodily well-being, good social relations, security, peace of mind, and spiritual experience.

Wetlands: Areas of marsh, fen, peatland, or water, whether natural or artificial, permanent or temporary, with water that is static or flowing, fresh, brackish or salt, including areas of marine water the depth of which at low tide does not exceed six meters. May incorporate riparian and coastal zones adjacent to the wetlands and islands or bodies of marine water deeper than six meters at low tide laying within the wetlands.

Wise use (of an ecosystem): Sustainable utilization for the benefit of humankind in a way compatible with the maintenance of the natural properties of the ecosystem

Index

Italic page numbers refer to Figures, Tables, and Boxes. Bold page numbers refer to the Summary.